P9-DGJ-370

McGraw-Hill Encyclopedia of ENERGY

McGraw-Hill Encyclopedia of

Sybil P. Parker
EDITOR IN CHIEF

energy

Second Edition

McGraw-Hill Book Company

New York St. Louis San Francisco

Auckland Bogotá Guatemala Hamburg Johannesburg
Lisbon London Madrid Mexico Montreal
New Delhi Panama Paris San Juan São Paulo
Singapore Sydney Tokyo Toronto

Printed in the United States of America

1234567890 DODO 86543210

Library of Congress Cataloging in Publication Data:

Main entry under title:

McGraw-Hill encyclopedia of energy.

Includes bibliographies and index.
1. Power resources—Dictionaries. 2. Power (Mechanics)—Dictionaries. I. Parker, Sybil P.
II. Title: Encyclopedia of energy.
TJ163.2.M3 1980 333.79'03'21 80-18078
ISBN 0-07-045268-7

Contents

Editorial Staff

Consulting Editors

Preface

THE ENERGY OUTLOOK FOR THE 1980S is less than optimistic. There are worldwide shortages of fossil fuels; oil surpluses no longer exist. Under these circumstances, the United States faces continued dependence on oil imports at costs that are straining the economy. During this decade, energy alternatives must be evaluated and their implications assessed to determine the future course for energy systems.

Energy can be obtained from a variety of sources to provide the light, heat, and power needs of an industrial society. The principal fossil fuels—petroleum, coal, natural gas, and lignite—are nonrenewable and reserves are, therefore, finite. Nonfuel energy sources include radioactive elements, wastes, water, wind, geothermal deposits, biomass, and solar heat. These nonfuel sources contribute very little energy, but will become of greater importance in the future since they are renewable.

A number of these alternative energy sources have the potential for development to the level of commercialization; some have already been tapped on a modest scale. Nuclear power plants are operational, but the future of nuclear energy is uncertain since it is clouded by questions of environmental impact. Solar energy technology has reached the marketplace and meets technical, economic, and environmental criteria; there are, however, political issues to be resolved. Research and development are ongoing in the areas of geothermal energy and synthetic fuels; while a few commercial plants have been constructed, there are serious limitations to large-scale operations. A number of plants utilizing wastes as fuel have been built. Tidal power has become a commercial reality in France, which developed one site to a capacity of 240 megawatts-electric.

The energy crisis cannot be abated in the next 10 years because conversion to renewable energy sources is a slow process. Improvements can be realized by immediate adoption of conservation measures, such as reduced consumption, energy storage and recycling, and improved efficiency of current systems. These approaches are practicable and could ease the strain during the energy transition period of the 1980s.

In this completely revised and updated second edition of the *Encyclopedia of Energy*, leading international authorities from government, industry, and the academic community address the technological, environmental, economic, social, and political aspects of energy systems. In the first section of the book—Energy Perspectives—the major issues are discussed in six feature articles: Risk of energy production, Energy conservation, Exploring energy choices, Outlook for fuel reserves, Protecting the environment, and Energy consumption. The second section—Energy Technology—details, in 300 alphabetically arranged entries, the technologies for energy discovery, development, and distribution. Topics include nuclear power, solar energy, synthetic fuels, electric power generation, tidal power, cogeneration, waste heat management, pollution, engines, turbines, and superconductivity. The text is supplemented by hundreds of illustrations. All information is readily accessible through the detailed index and cross-references. An appendix provides SI conversions and furnishes outlines of Federal energy agencies and legislation.

It is hoped that this book will be useful to scientists, engineers, students, teachers, librarians, and the general public needing current and accurate information on the state of energy in the world.

Sybil P. Parker
EDITOR IN CHIEF

Energy Perspectives

Energy Conservation

Eric Hirst

Energy conservation, probably the least understood element in debates over United States energy policy, refers to practices and measures that increase the efficiency with which energy is used in all sectors of the economy. Increased efficiency implies an improvement in overall productivity—doing more with less.

It is important to distinguish between energy conservation and energy curtailment. Curtailments (for example, Sunday closings of gas stations or mandatory reductions in winter temperature settings in commercial buildings) involve immediate short-term mandatory reductions in energy use and occur primarily because of policy failures to deal with long-term problems. Thus, curtailments may reduce economic output and well-being; energy conservation generally increases economic productivity and improves well-being.

Some conservation practices (for example, lower indoor temperature settings in winter), when compared with historical patterns, imply doing with less. However, when considered in the context of present and future higher fuel prices and reduced fuel availability, these measures are more attractive than the alternatives.

This article first discusses historical trends in energy consumption and energy efficiency, with a focus on the 1970s. Next, several specific energy conservation opportunities in the industrial, transportation, residential, and commercial sectors of the United States are presented. Some of the barriers to widespread and prompt adoption of these cost-effective conservation opportunities are considered, as well as different government policies that can be (and are being) employed to overcome these barriers. Finally, alternative projections of United States energy use between now and the end of the century are examined.

The purpose of this discussion is to make two major points. First, the wide array of existing cost-effective energy conservation measures represents an enormous—and largely untapped—energy resource. Adoption of these measures can substantially reduce growth in energy use, save money for consumers, and have only slight life-style effects on consumers. In addition, conservation resources can reduce the adverse environmental effects of energy production and conversion and provide additional time to develop new energy resources.

Second, the best means to obtain these benefits is unclear. Decision-makers lack sufficient information to decide on the proper roles for private enterprise and governments. The best mix of government activities is not known. These activities deliver information on energy use choices; regulate efficiency of equipment, buildings, and motor vehicles; provide financial incentives to purchase energy-efficient systems; and fund research projects on new technologies for energy production.

HISTORY OF ENERGY USE

Figure 1 shows trends in United States energy consumption by type of fuel from 1948 through 1978. Between 1948 and 1973 energy use grew at an average annual rate of 3.3%. Since then growth has been erratic (and negative for 1974 and 1975) with an average growth rate of 0.9% per year. Table 1 shows growth rates in United States energy use, by sector and by fuel, for the 1950–1973 and 1973–1978 periods. In 1978 total energy use was 78 quads (1 quad equals 10^{15} Btu or 1.055×10^{18} J).

The industrial sector dominates energy use (Fig. 2). This sector, which includes manufacturing, agriculture, and mining, uses 37% of the nation's total energy budget. More than half of all industrial energy use is consumed in three industries—chemicals, primary metals, and petroleum refining. Transportation, both personal and freight, accounts for another 26%. The commercial sector, that part of the economy which delivers services (such as schools, hospitals, offices, and retail trade), accounts for 16%. Finally, space heating, water heating, and operation of appliances in homes use the remaining 21%. Households directly consume about one-third of the nation's energy budget in their homes and automobiles. The other two-thirds is consumed indirectly through purchases of goods and services.

Energy is used for a variety of purposes in each sector. Major uses in the industrial sector are direct heat, process steam, and mechanical drive. Automobiles use half of all the transportation fuel. Space heating accounts for half of all energy use in residential and commercial buildings. Thus, a few major end uses account for a sizable portion of United States energy consumption.

To a large extent, the abrupt change after 1973 in energy growth trends shown in Fig. 1 can be explained by changes in fuel prices. Figures 3 and 4 show trends in "real" fuel prices during the past 3 decades. (Real prices are adjusted for the effects of inflation.) Figure 3 shows prices of basic energy resources as they leave the ground. The trends are

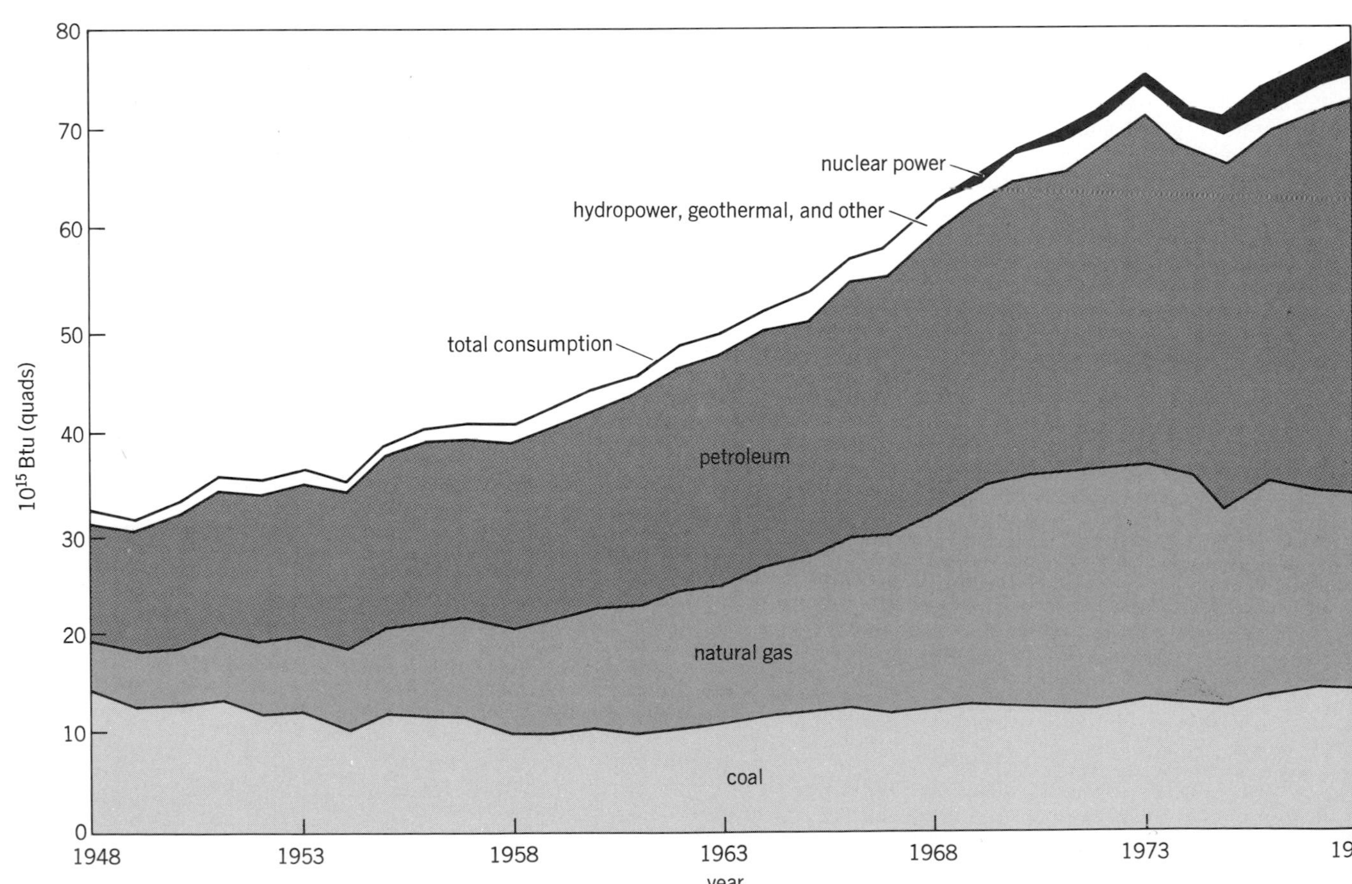

Fig. 1. United States energy consumption according to fuel type for the period extending from 1948 to 1978. *(From U.S. Department of Energy, Annual Report to Congress 1978, 1979)*

clear: between 1950 and the early 1970s, fuel prices were generally declining; since then, however, fuel prices have been increasing. The effect of the 1973–1974 Arab oil embargo on crude oil prices is particularly dramatic.

Figure 4 shows trends in prices of fuels sold to residential customers. Note the dramatic increases in oil prices between 1973 and 1974, and again between 1978 and 1979. Between 1973 and mid-1979, retail prices of all fuels increased: 85% for oil, 45% for natural gas, and 15% for electricity. Gasoline prices (not shown in these figures) increased by 40% during this period.

Figure 5 shows overall trends in United States energy efficiency, as measured by the ratio of total energy consumption to gross national product (GNP), from 1950 to 1978. Energy efficiency improved during the 1950s and early 1960s, declined during the late 1960s, and increased steadily during the 1970s.

Improvements in overall energy efficiency between 1973 and 1978 cut 1978 energy use by about 10%. The biggest contribution to this 5-year improvement in efficiency came from the industrial sector. Although industrial output increased by 12% during this period, industrial energy consumption actually fell by 6% (Table 1). This was due to improvements within individual industries and a shift in industrial output away from energy-intensive products.

Energy use in the transportation sector increased 9% during the 1973–1978 period. Overall efficiency improved by 9%, primarily because of improvements in automobile fuel economy. Based on Federal government tests, new car fuel economy increased from 14 mi/gal (6.0 km/liter) in 1973 to almost 20 mi/gal (8.5 km/liter) in 1978. (Actual road tests show a smaller gain in fuel economy.)

Household energy use increased 14% between 1973 and 1978, while the number of households grew by 11%. Thus, household energy use grew very slowly at 0.5% per year during this period, compared with growth of almost 2% per year between 1950 and 1973. In a similar fashion, energy use in commercial buildings per square foot of floor space grew more slowly between 1973 and 1978 (0.5% per year) than it had between 1950 and 1973 (0.8% per year).

ENERGY CONSERVATION MEASURES

There are two basic ways to improve energy efficiency. The first involves changes in the way that existing systems are operated. Such changes include higher thermostat settings on room air conditioners, more frequent tune-ups of automobile engines, repair of leaks in steam lines and steam traps in industrial plants, and reductions in lighting levels in industrial and commercial buildings. These changes are characterized by their low (or zero) capital cost, the speed with which they can be implemented, and the fact that all require operational (human) changes.

The second type of conservation measure involves improvement in the technical efficiency of energy-using systems—space heating and cooling equipment, appliances, building structures, transportation equipment, and industrial process equipment. Examples include additional insulation in attics and walls, replacement of electric resistance heating systems with heat pumps, use of sodium vapor lamps rather than fluorescent lights in industrial and commercial buildings, use of lighter materials (such as plastics and aluminum) to reduce automobile weight, improved aerodynamic designs for new airplanes, waste heat recovery equipment in industrial plants, and central power plants that generate both electricity and usable thermal energy. These measures are characterized by minimal behavioral change, some (occasionally high) capital costs; and delays in implementation until existing equipment wears out.

Conservation can also include long-term efforts to restructure society in a more energy-efficient fashion. This might include substitution of some travel with electronic communication, and changes in land use patterns to reduce energy use for space conditioning (such as high-density dwellings and total energy systems) and for transportation.

Operational changes. Consider temperature settings in a typical single-family house as an example of operational changes to save energy. Reducing the winter temperature from 72 to 70°F (22.2 to 21.1°C) would cut annual heating fuel bills

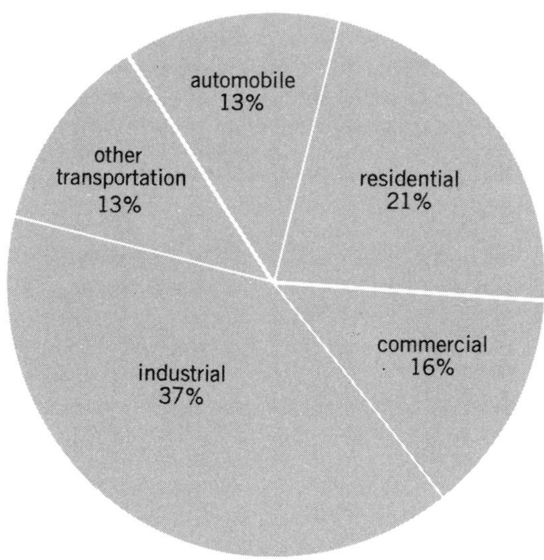

Fig. 2. Distribution of United States energy consumption by major economic sector. (*From E. Hirst, Understanding Residential/Commercial Energy Conservation: The Need for Data, Oak Ridge National Laboratory, October 1979*)

Table 1. Historical trends in energy consumption growth rates*

Area	Average growth rates, %/yr	
	1950–1973	1973–1978
Overall	+3.5	+0.9
By sector		
Residential/commercial	+4.1	+2.6
Industrial	+3.1	−1.2
Transportation	+3.3	+1.7
By fuel		
Coal	+0.1	+1.2
Gas	+5.8	−2.5
Oil	+4.2	+1.6
Nuclear power	−	+26.8
Hydro power	+3.3	+0.9
Electricity consumption	+7.7	+3.1

*From U.S. Department of Energy, *National Energy Plan II*, May 1979.

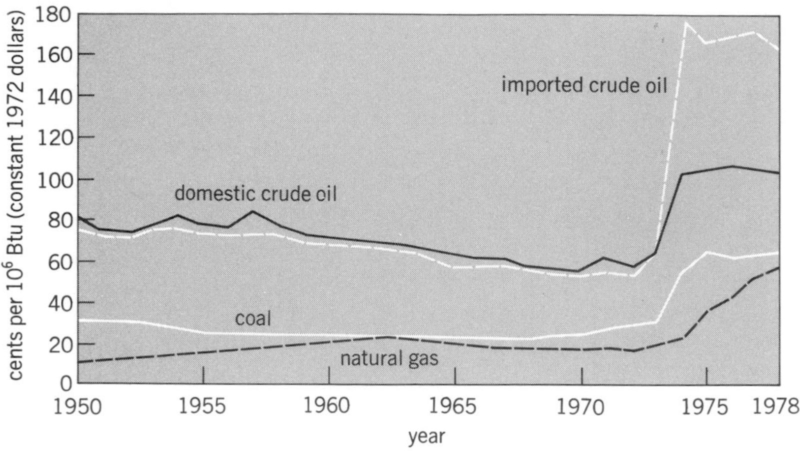

Fig. 3. Primary fuel prices in the United States, 1950–1978. Fuel prices are expressed in terms of 1972 dollars to adjust for the effects of inflation. *(From U.S. Department of Energy, National Energy Plan II, 1979)*

by 10% in a typical home in Kansas City (Fig. 6). Lowering the temperature to 68°F (20.0°C) would increase the saving to 20%. Lowering the nighttime (10 P.M. to 6 A.M.) temperature from 68 to 60°F (20.0°C to 15.6°C) would cut energy use in this home by an additional 10%, bringing the total energy saving (relative to 72°F or 22.2°C) to a remarkably high 30%. No investments are required to implement these temperature reductions, except perhaps a clock thermostat and additional sweaters for the family. *See* COMFORT HEATING.

As Fig. 6 shows, the energy benefits of winter temperature reductions vary from region to region depending on the severity of the winter. In northern climates, the percent reduction in heating fuel is relatively small; however, the absolute value of the energy and dollar savings will be larger than in mild southern climates.

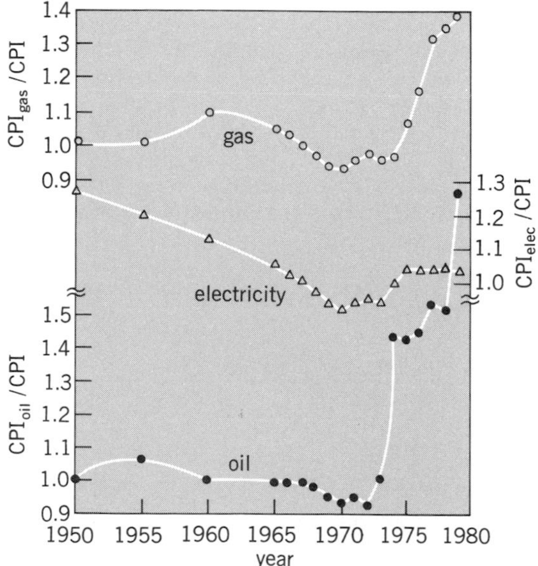

Fig. 4. Residential fuel prices in the United States, 1950–1979. Fuel prices are adjusted by the Consumer Price Index to correct for the effects of inflation. *(From E. Hirst, Understanding Residential/Commercial Energy Conservation: The Need for Data, Oak Ridge National Laboratory, October 1979)*

The same kinds of energy and dollar savings are possible by setting home temperatures up in the summer to reduce air conditioner use. As a rule of thumb, a 1°F (0.6°C) increase in temperature setting will reduce air conditioner use by 5%.

Similar savings in heating and cooling energy use are possible in commercial buildings. In fact, because these buildings are generally unoccupied at night and on weekends, the potential savings are even larger. As an example, consider a typical office building, also located in Kansas City (Fig. 7). Lowering thermostat settings from 6 P.M. to 6 A.M. on weekdays and all weekend cuts annual energy use by 37%. Reductions in lighting levels and in ventilation rates plus adjustments to hot and cold air temperatures increase the potential saving to 58%. In other words, energy use can be cut in half with essentially no capital costs.

Airline fuel efficiency has improved sharply during the past few years (motivated in part by rapidly rising fuel prices and in part by deregulation of commercial aviation). The average load factor (percentage of all seats occupied) increased from 52% in 1973 to 61% in 1978, a 17% increase in energy efficiency. In addition, airlines cut fuel use per aircraft hour by 3% during this 5-year period by better control of airplanes on the ground and by optimizing flight paths with respect to fuel use. These and other operational changes led to a 20% reduction in airline fuel use.

Improvements in technical efficiency. The second class of conservation measures includes improvements in the technical efficiency of equipment, appliances, structures, and transportation vehicles. This involves both modifications to systems already in use (retrofit) and improvements to new systems.

Consider retrofit measures first, as applied to the Kansas City home discussed above. Assume that the home was built in the late 1960s and meets the Federal standards in effect then. This house would have 3 in. (76 mm) of insulation in the attic, none in the walls or floor, and no storm windows or doors. Adding 3 in. (76 mm) of insulation to the floor, 6 in. (152 mm) to the attic, and installing storm windows would cut annual heating bills by about one-fourth. It would take 10 years to repay this investment in a gas-heated home and less than 4 years in an electrically heated home. (Of course, as fuel prices rise, the payback periods will be shorter.)

Now consider a new house, again in Kansas City. It is easy to install wall insulation, and so the exterior wall cavities are filled with 3 1/2 in. (89 mm) of insulation. The house is oriented on the building site so that most of its windows are on the long south face with a broad overhang. This ensures that the Sun's energy will enter the house in winter, providing free energy—often called passive solar energy. In the summer, the overhang blocks the Sun's rays, ensuring that the Sun imposes no additional load on the air conditioner. Insulation is put in the attic and floor and storm windows are installed, as for the retrofit house. Finally, the heating system is properly sized (traditionally, heating systems have had twice the capacity needed).

These changes reduce the heating load by over

40% and the air conditioning load by 15% relative to the original 1960s design (Fig. 8). These reductions mean that smaller heating and cooling equipment can be installed in the new house, and this provides a dollar saving that partially offsets the higher cost of the improved structure. Thus, the new house costs only a few hundred dollars more than the original. For a gas-heated home this investment is repaid in 4 years; for an electrically heated home the payback period is 2 years. *See* AIR CONDITIONING; CENTRAL HEATING AND COOLING.

Similar options exist for heating and cooling equipment and also for major household appliances such as water heaters and refrigerators (Table 2).

Similar opportunities exist with commercial buildings and their systems. Figure 9 shows the relationship between annual energy use and capital cost for the Kansas City office building considered earlier. The dashed curve shows the relationship between energy use and capital cost for systems installed in new buildings; the solid curve refers to modifications of systems in existing buildings. Clearly, it is less expensive to save energy when designing a new building than when modifying an existing one.

Some of the design changes save remarkable amounts of energy with very short payback periods. For example, a combination of measures can cut energy use by more than 40%. The payback period is 1 year in a new office building and 2 years in an existing building. These results also suggest that reductions in energy use beyond 50% become progressively more expensive (that is, the curve becomes very flat, indicating long payback periods at today's fuel prices).

The relationships between energy use and initial cost shown in Figs. 8 and 9 are applicable to a variety of systems, including automobiles, appliances, heating equipment, and structures. These relationships all show that energy savings increase with increasing investment, but at an ever-declining rate—a manifestation of the law of diminishing marginal returns. This implies that there is generally an optimum level of energy conservation. Beyond that point, the cost to save energy is greater than the value of the energy saved.

Consider waste heat recovery as an example of a cost-effective conservation measure in the industrial sector. An aluminum company had a melt furnace that exhausted 650°F (343°C) flue gases 24 hours a day. Originally, the flue gases were exhausted to the atmosphere with no recovery of waste heat. The company installed a cross-flow heat exchanger in the exhaust stream. The energy is used to heat air to 180°F (82°C), which is then used for space heating in the factory. The annual saving in fuel bills is sufficient to repay the investment in 5 years. *See* WASTE HEAT MANAGEMENT.

These conservation measures are all available today. They do not require further research and development before they are ready for commercial applications. The array of present-day techniques for cutting energy use is impressive. Perhaps even more impressive are the economic benefits that generally accompany implementation of these measures.

In addition to the technologies presently available, many more are waiting in the wings. Both private industry and the Federal government (Department of Energy) are conducting research programs to develop new technologies that provide end use services (heat, transportation, and so on) with smaller energy inputs. Some examples include electrodeless fluorescent lamps, gas-fired heat pumps, and improved control systems.

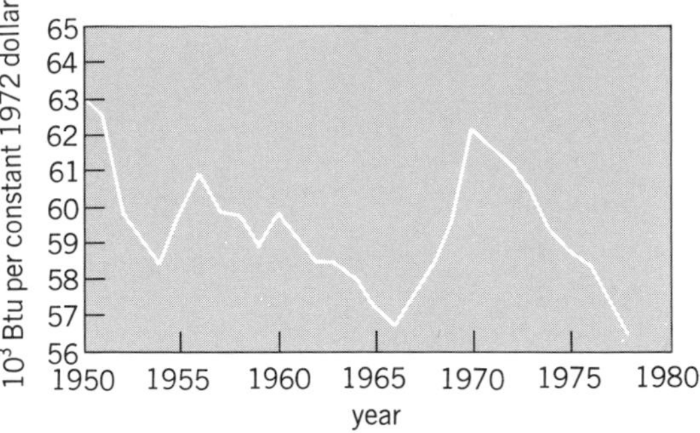

Fig. 5. Overall trends in United States energy efficiency as measured by the ratio of total energy consumption to total gross national product, 1950–1978. *(From U.S. Department of Energy, National Energy Plan II, 1979)*

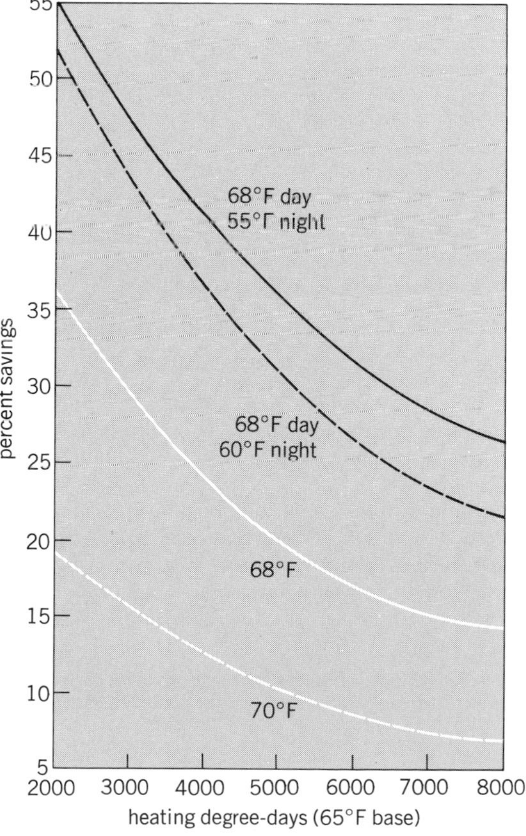

Fig. 6. Percent savings in residential space heating fuel use as a function of temperature settings and heating degree-days. The reference condition is 72°F (22.2°C), 24 hours a day for the entire space heating season. Nighttime setbacks refer to the 10 P.M.–6 A.M. period. Heating degree-day values for several cities are: Atlanta 3000, Boston 5600, Dallas 2400, Denver 6400, Kansas City 4700, Minneapolis 8400, Seattle 4400. *(From D. Pilati, The Energy Conservation Potential of Winter Thermostat Reductions and Night Setbacks, Oak Ridge National Laboratory, February 1975)*

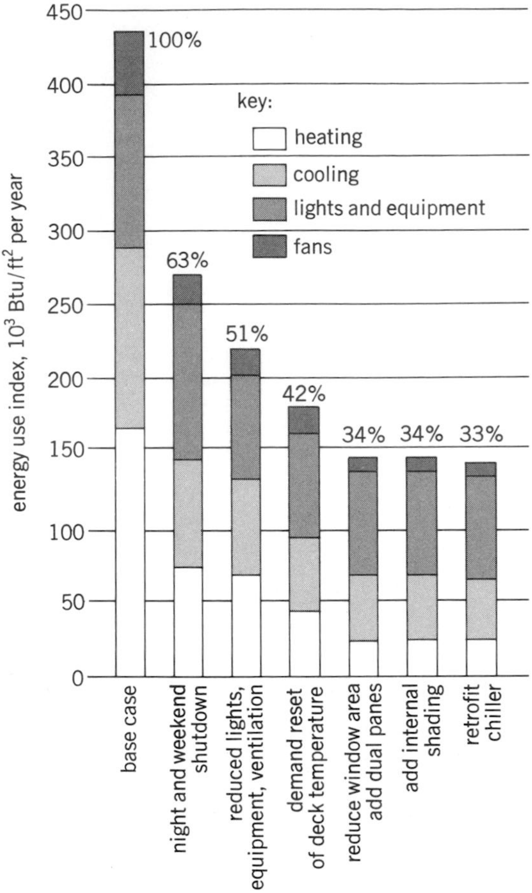

Fig. 7. Annual energy use in a typical office building in Kansas City. Energy savings are shown for several operational and retrofit changes. *(From W. Johnson and F. Pierce, Energy and Cost Analysis of Commercial Building Shell Characteristics and Operating Schedules, Oak Ridge National Laboratory, December 1979)*

BARRIERS TO CONSERVATION

The variety and cost effectiveness of the available conservation measures lead one to ask whether they are being generally adopted. If not, one must ask why.

Although data on recent trends in energy use and conservation actions are hard to interpret, it appears that improvements in energy efficiency are occurring slowly, particularly in the buildings sectors. Several plausible reasons exist for the slow pace of increasing energy efficiency.

History. First, there is a long history of low and declining fuel prices (Figs. 3 and 4). Indeed, until the early 1970s, the energy cost of operating systems was a small—and declining—fraction of the total cost (purchase, maintenance, and operation). Therefore, little attention was devoted to energy efficiency. Now with fuel prices suddenly much higher—and likely to rise further—changes in thinking are needed with respect to the design, purchase, and operation of systems. However, it will take time to overcome inertia and tradition and to adjust to the higher fuel prices.

Information. The need for information and the ability to effectively process that information are also important. Adoption of cost-effective conservation measures can take place only when decision-makers (homeowners, tenants, businesses, and governments) have accurate, site-specific information concerning the costs and benefits of their options. Such information is not now readily available, particularly to building owners and operators. However, many programs are underway to remedy that deficiency. Several states have energy "hotlines" that provide telephone answers to frequently asked questions. In addition, many electric and gas utilities offer residential and commercial customers on-site audits of their buildings.

Life-cycle costing. While information is a necessary prerequisite to adoption of wise conservation measures, it is not sufficient. Decision-makers also need to overcome the hurdle associated with the higher initial costs of most technical efficiency improvements. Offsetting this higher first cost is the subsequent reduction in operating energy costs. The balancing of initial versus operating costs can be effectively handled with an approach called life-cycle costing. The life-cycle cost of a particular system incorporates both the initial cost and the lifetime operating (including energy) cost. In principle, the system with the lowest life-cycle cost should be chosen.

Traditionally, purchase decisions have been based primarily on initial cost. As more and more decision-makers recognize the upward trend in fuel prices, the notion of life-cycle costing is likely to become the foundation of energy-related purchase decisions.

Buildings. Another obstacle to speedy adoption of conservation measures is the fragmented nature of the building sectors. The design, construction, financing, and operation of residential and commercial buildings involve a multitude of participants: homeowners, renters, commercial firms, developers, architects, engineers, builders, contractors, labor unions, financial institutions, insurance firms, and government agencies. Each participant has a different perspective and is involved with different issues related to building energy use.

Almost every community in the United States has its own building code, originally established to ensure compliance with fire, health, and safety requirements. These codes could spur energy efficiency of new buildings by incorporating requirements on the thermal performance of structures and the efficiency of heating systems. However, the multiplicity of building codes and differ-

Table 2. Potential energy and economic effects of equipment design changes with present-day technologies relative to typical 1975 designs*

Equipment	Reduction in annual energy use, %	Payback period, years
Space heating		
Gas furnace	25	6
Oil furnace	20	6
Heat pump	40	2–9
Water heating		
Electric	15	2
Gas	25	3
Refrigerator	50	2
Room air conditioner	35	6

*From E. Hirst and D. O'Neal, Contributions of improved technologies to reduced residential energy growth, *Energy and Buildings*, vol. 2, no. 3, 1980.

ences among codes and how they are administered complicate and reduce this potential.

When the owner of a building is also the sole occupant of that building, there is little ambiguity about who pays the costs and who receives the benefits of adopting conservation measures. However, for a building that is tenant-occupied, this is a problem. If the tenant pays for utilities, then the owner has little incentive to invest in energy-efficient systems; the costs would accrue to the owner but the benefits would accrue to the tenant. On the other hand, the tenant has little incentive to invest in conservation. Although the tenant will enjoy lower fuel bills while remaining in that building, the investment will not be recovered when the tenant leaves—another case where the costs and benefits accrue to different parties. The fact that 35% of the nation's housing stock is occupied by renters suggests the significance of this unresolved issue.

Transportation. Automobile production and purchase represent a different kind of problem. Automobile production is dominated by two major manufacturers in Detroit and a few foreign carmakers. Because economic power is concentrated in the hands of a few firms, the incentive for any one of them to innovate and produce fuel-efficient vehicles is small. Perhaps because of the imbalance in economic power between producers and purchasers of new automobiles, the U.S. Congress mandated strict fuel economy standards for new cars.

Encouraging people to shift part of their travel from automobiles to mass transit involves a somewhat paradoxical problem. The level of service offered by most urban transit systems in the United States is very poor compared with the services offered by the private car (in terms of speed, comfort, and frequency of service). This is both a consequence and a cause of low ridership. If transit ridership were higher (as it was before World War II), then service could be greatly improved (such as more frequent service and additional routes). However, transit commissions cannot afford to increase service until the ridership is present to help pay these costs.

Energy/GNP ratio. Another obstacle to conservation is the belief that energy use and economic growth are tied together by an "iron link." Although energy use and GNP show significant variation (Fig. 5), many still believe there is a direct, even inevitable, correlation between economic growth and energy use. Belief that the energy/GNP ratio is constant undoubtedly leads to considerable reluctance to adopt energy conservation measures.

A related obstacle is the belief that conservation means doing without. This is a difficult barrier to overcome because future patterns of energy use will probably involve some behavioral changes (for example, lower winter temperatures in buildings and use of smaller cars) that are less comfortable than historical behaviors. However, these conservation practices are desirable and necessary in a world with scarce and expensive energy resources.

Uncertainty. A final obstacle to energy conservation is uncertainty over the future. Decisions about the future are always complicated by uncertainty, and this is particularly true for energy decisions because of changes in fuel prices and availabilities. Decision-makers are likely to underinvest when faced with great uncertainty over future fuel prices and related government policies.

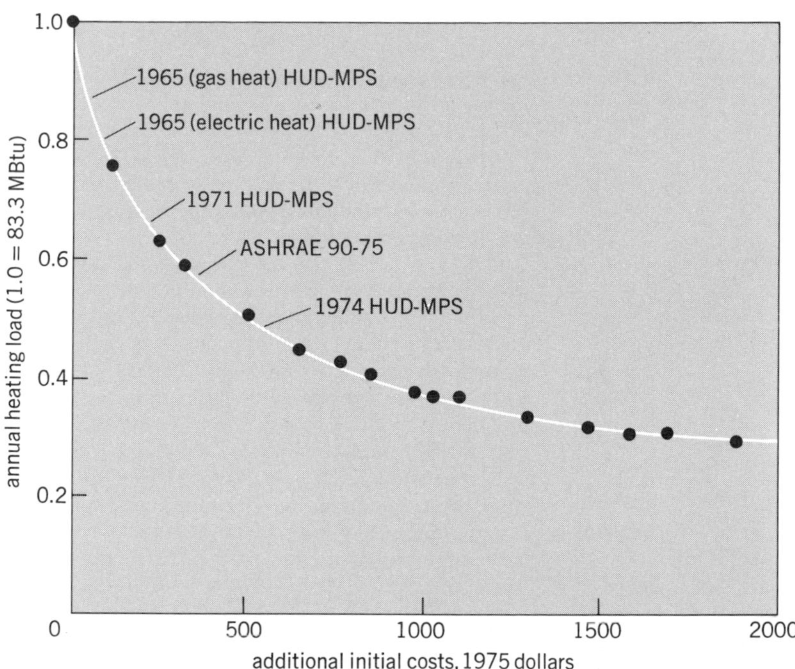

Fig. 8. Annual space heating load for a new Kansas City single-family house as a function of the initial cost to improve thermal performance. Each point represents a different combination of storm windows, storm doors, attic insulation, wall insulation, and floor insulation. HUD-MPS refers to standards promulgated by the U.S. Department of Housing and Urban Development, their Minimum Property Standards. ASHRAE 90-75 refers to the standard developed in 1975 by the American Society of Heating, Refrigerating, and Air Conditioning Engineers. *(From P. Hutchins, Jr. and E. Hirst, Analysis of single-family dwelling thermal performance, Resources and Energy, vol. 2, no. 1, 1980)*

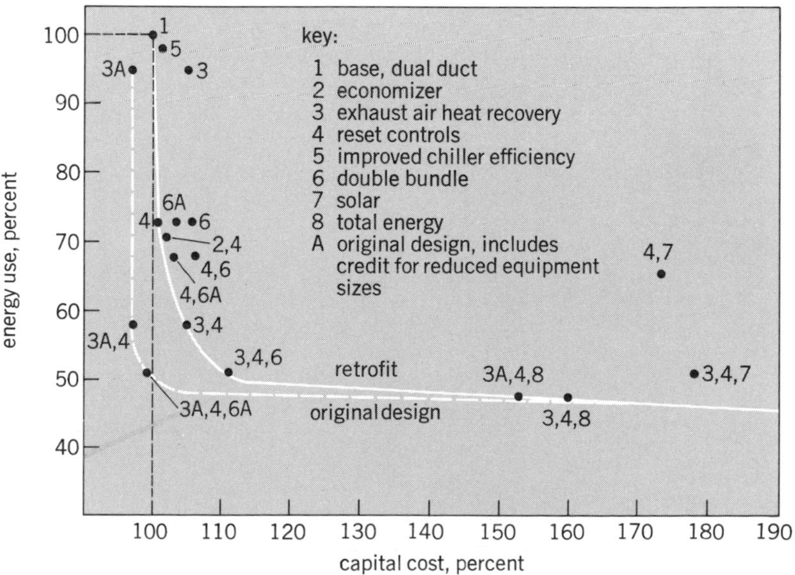

Fig. 9. Energy use for heating, ventilation, and air conditioning (HVAC) in a typical office building in Kansas City. The building is heated with natural gas and air-conditioned with electricity. Base energy use is 326,000 Btu/ft² (3.70×10^9 J/m²). *(From R. Lyman, Jr., Energy and Cost Analysis of Commercial HVAC Equipment: Offices and Hospitals, University of Tennessee, March 1979)*

GOVERNMENT POLICIES

Governments at all levels (Federal, state, and local) conduct programs to help overcome barriers to conservation. This section deals primarily with Federal policies related to information, fuel prices, financial incentives, regulation of energy efficiency in new equipment and structures, and research and development.

Prices. Perhaps the most important policy that the Federal government can employ to encourage greater energy efficiency concerns pricing of fuels. Historical policies that continue today serve to keep fuel prices below replacement costs. For example, the average residential price of natural gas in Minnesota was about \$3 per 10^6 Btu (\$2.85 per 10^9 J) in late 1979. However, the price of gas imported from Canada was more than 50% higher than the average price. Because households see the average price and not the marginal price, they have a tendency to underinvest in conservation measures.

The 1978 National Energy Act began the process of deregulating natural gas prices and modifying electric utility rate structures. The President's plan to decontrol oil prices by 1981 will also help to align fuel prices with fuel costs. However, even with these actions, fuel prices will still not reflect their full social costs. The foreign policy and national security implications of continued dependence on OPEC oil and the adverse environmental effects of energy production and conversion are not yet part of the direct price that consumers pay for fuels.

Fig. 10. A prototype electric heat-pump water heater. *(From R. Dunning, F. Amthor, and E. Doyle, Research and Development of a Heat Pump Water Heater, Energy Utilization Systems, Inc., August 1978)*

Although changes in fuel pricing are essential to improve operation of the free enterprise system with respect to energy consumption and investment decisions, other policies are also needed. Table 3 lists some of the major nonfuel price programs now underway because of recent Federal legislation. The list is limited to programs affecting energy use in residential and commercial buildings, primarily because more government attention is devoted to the buildings sectors than to the transportation and industrial sectors.

Information. Considerable government attention is devoted to dissemination of information related to energy conservation options (examples are the Energy Extension Service and the State Energy Conservation Plan). Workshops, seminars, telephone hotlines, movies, filmstrips, and publications are used to inform consumers about their energy-related choices.

Federal legislation requires that all new cars have labels showing their fuel economy rating. Beginning in 1980, most residential appliances will also have energy efficiency labels. To the extent that conservation actions are inhibited by lack of information, these labels should influence conservation investments.

Efficiency standards. The Federal government is implementing minimum efficiency standards for various energy-using systems. Standards are now in place (as part of the 1975 Energy Policy and Conservation Act) for new-car fuel economy. The standards require that each manufacturer reach minimum corporate averages of 20 mi/gal (8.5 km/liter) in 1980 and 27.5 mi/gal (11.7 km/liter) in 1985, compared to 14 mi/gal (6.0 km/liter) in 1973. Other standards deal with thermal performance of new buildings (both residential and nonresidential) and energy efficiency of new residential appliances.

Financial incentives. The 1978 Energy Tax Act (part of the National Energy Act) created a 15% Federal income tax credit for home energy conservation expenditures. The maximum credit is \$300 on an expenditure of \$2000. The Federal government also has a weatherization assistance program, which provides energy audits and retrofit services for homes occupied by low-income households.

The 1978 National Energy Conservation Policy Act created a grants program to conduct audits and retrofit institutional buildings such as schools and hospitals, and buildings owned by local governments and public care institutions. The authorization for this program included \$900,000,000 of Federal funds to be matched 50:50 by state and local contributions.

Research. In addition to these programs aimed at improving energy efficiency in the short-term, the Federal government sponsors a variety of research programs to produce long-term improvements in energy-using systems. These include basic research on combustion and materials to improve industrial processes; and applied research on efficient residential appliances and heating equipment, alternative engines for automobiles, and improved insulating materials for new and existing buildings.

Not all research is devoted to development of new technologies. Considerable effort is aimed at developing improved systems (for example, better use of daylight for lighting purposes within buildings, and passive solar designs for new homes) and at better understanding of how consumers make energy-related decisions.

A particularly interesting example of applied research is the heat-pump hot water heater (Fig. 10). The heat-pump unit is similar to a room air conditioner in size, operation, and function. As with a room unit, energy is extracted from cool, ambient air and "pumped" into a medium at a higher temperature. However, instead of rejecting heat to air outside the house, energy is transferred to water inside the hot water tank. The heat-pump compressor and evaporator are mounted directly above the conventional hot water tank. Thus, the hot water heat pump takes up the same amount of floor space as does a conventional water heater. *See* HEAT PUMP.

The Department of Energy is sponsoring laboratory tests with these systems, extensive field tests in homes throughout the country, and economic analyses to see how cost-effective such systems are. Preliminary results suggest that these systems will use only half as much energy as conventional electric water heaters; the extra cost of the heat pump unit will be repaid by lower electricity bills within a couple of years.

ENERGY USE TO THE YEAR 2000

How will higher fuel prices, new technologies, and government conservation programs affect future energy use? Although there are no facts about the future, it is instructive to examine alternative scenarios to see how different assumptions might affect future growth.

The Energy Information Administration (EIA) of the U.S. Department of Energy produces forecasts of energy use in its annual report to Congress. The 1979 report shows a range in estimated energy use of 115–140 quads in 2000. Growth rates in energy use between 1978 and 2000 range from 1.8 to 2.7% per year. Even the high projection has a growth rate substantially lower than the rate from 1948–1973 of 3.3% per year. Were energy consumption to resume its preembargo growth rate, it would reach 170 quads in 2000.

The EIA projections differ in their assumptions concerning future levels of energy supplies and demands. All assume that the world price for crude oil will reach about $30 per barrel ($190 per cubic meter) in 2000 (expressed in terms of 1978 dollars). Considering the 1979 Iranian supply interruption and subsequent increases in both OPEC and spot-market prices, $30 seems low.

In 1978 a committee working on energy demand for the National Research Council (NRC) published four scenarios of future United States energy use. They differ in terms of future fuel prices and levels of government conservation programs and private conservation actions. The lowest forecast (Table 4) shows a decline in energy use between now and the year 2000, reaching a level of about 60 quads. The highest projection is 117 quads, which is 4% lower than the mid-range EIA forecast.

The NRC forecasts are lower than the EIA forecasts because of assumptions on both future fuel prices and the effectiveness and vigor of government and private conservation activities.

The wide range in forecasts of energy use for the year 2000–60 to 140 quads—suggests the large influence that society might exert over its energy future. Future trends in energy use are not predetermined.

The discussion of alternative futures has so far

Table 3. Federal energy conservation programs dealing with residential and commerical buildings

Program name	Federal legislation	Purpose of program
State Energy Conservation Plan	1975 Energy Policy and Conservation Act	Grants to states for five required conservation measures and a variety of state conservation programs
Supplemental State Energy Conservation Plan	1976 Energy Conservation and Production Act	Grants to states for three additional mandatory conservation measures and other state conservation programs
Energy Extension Service	1977 National Energy Extension Service Act	Initial grants to 10 pilot states, with national program now underway; grants to provide conservation information and services to small energy users
Weatherization Assistance Program	1976 Energy Conservation and Production Act	Grants to states to retrofit homes occupied by low-income families
Institutional Buildings Grants Program	1978 National Energy Conservation Policy Act	Grants to states to audit and retrofit schools, hospitals, local government buildings, and public care institutions
Residential Energy Conservation Tax Credit	1978 Energy Tax Act	Federal income tax credit of 15% of first $2000 spent on residential energy conservation measures
Buildings Energy Performance Standards	1976 Energy Conservation and Production Act	Thermal performance standards for construction of all new buildings, to be implemented in 1981
Appliance Efficiency Performance Standards	1978 National Energy Conservation Policy Act	Minimum efficiency standards for 13 classes of residential appliances, to be implemented in 1981
Residential Conservation Service	1978 National Energy Conservation Policy Act	Mandatory program for large gas and electric utilities (also involves small utilities and fuel oil dealers) to offer retrofit services to residential customers
Federal Energy Management Program	Executive Orders 1978 National Energy Conservation Policy Act	Improvements in efficiency of new and existing buildings owned and leased by Federal government

Table 4. Alternative projections of United States energy demand from the Committee on Nuclear and Alternative Energy Systems*

Energy price ratio 1975/2000	Energy conservation policy	Primary energy consumption in 2000, quads†
3	Very aggressive; aimed at reduced demand, and requiring some life-style changes	62
3	Aggressive; aimed at maximum efficiency plus minor life-style changes	73
1.5	Slowly incorporates more measures to increase energy efficiency	87
1	Unchanged from present policies	117

*From U.S. energy demand: Some low energy futures, *Science*, vol. 200, no. 14, Apr. 14, 1978.
†Total United States energy consumption was 71 quads in 1975 and 78 quads in 1978. 1 quad = 10^{15} Btu = 1.055×10^{18} J.

been quite general. To provide a more specific example of future energy use, consider the residential sector. A continuation of fuel price increases as projected by EIA, with no government conservation programs, is expected to lead to a level of residential energy use of 23 quads in 2000, compared to 17 quads in 1978. The growth rate between 1978 and 2000 is 1.5% per year, considerably lower than the historical rate of 3.9% per year between 1950 and 1973.

If the Federal residential conservation programs listed in Table 3 are vigorously and effectively implemented and if existing research projects lead to commercialization of new technologies, energy use will increase even more slowly. The projection for 2000 is less than 20 quads, with a growth rate of 0.8% per year between 1978 and 2000.

Figure 11 shows how the energy savings—due to government regulations, financial incentives, and information programs, and to development of new technologies—increase over time. Clearly, research benefits differ substantially from those attributable to government conservation programs. Energy savings from government programs increase rapidly through the early 1980s and then only slightly from 1985 to 2000. Research benefits, on the other hand, grow slowly at first, but then in the 1990s grow much more rapidly. In the year 2000 the residential sector energy saving due to research (1.7 quads) is equal to the saving from conservation programs; in 1985 the saving is only a tenth of that due to Federal programs.

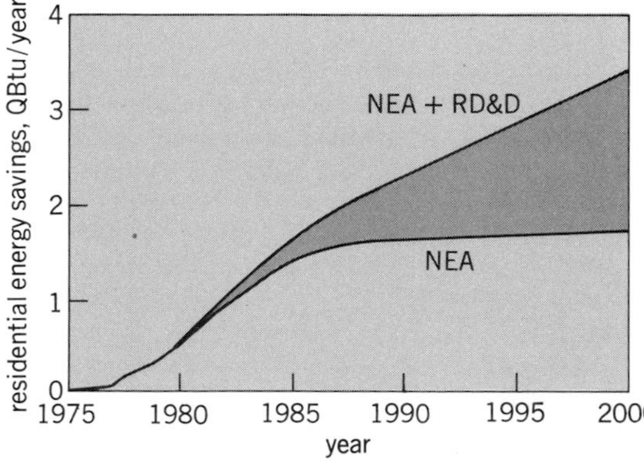

Fig. 11. Estimated energy savings in the residential sector due to implementation of Federal conservation programs and development of new technologies. *(From E. Hirst, Effects of the National Energy Act on energy use and economics in residential and commercial buildings, Energy Systems and Policy, vol. 3, no. 2, 1979)*

The energy savings shown in Fig. 11 also yield large economic benefits to households. Between 1978 and 2000, fuel bills are reduced because of these programs by $41,000,000,000. On the other hand, the higher cost of more efficient equipment and structures amounts to $22,000,000,000. Thus, the net benefit to the nation's households of these conservation efforts is $19,000,000,000. The nation saves energy with these actions and consumers also save money.

CONCLUSIONS

Many options exist to sharply reduce energy use in all sectors of the United States economy. These options are of two kinds: technical improvements to equipment, structures, and transportation vehicles; and operational changes in the ways that existing systems are used. The first generally involves an increase in the capital cost of the system. Fortunately, the reduction in annual fuel bills because of higher efficiency will generally repay this investment within a few years.

The operational changes associated with improved energy efficiency are generally minor. They involve small changes in temperature settings for space heating and air conditioning systems in homes and nonresidential buildings, greater attention to operation and maintenance practices in industrial operations, and more human ingenuity in general. These small changes often yield surprisingly large energy and dollar savings.

There is considerable uncertainty over the degree to which the market system will respond to changes in fuel prices. There is also uncertainty over the effectiveness of present and proposed government conservation programs. Nevertheless, a few conclusions emerge concerning future trends in United States energy use. It is virtually certain that growth in energy use between now and the end of the century will be at a much slower rate than between World War II and the Arab oil embargo. Historically, United States energy use grew at a rate above 3% per year; between now and the year 2000 growth will probably be less than 2.5% per year. It is even possible—assuming vigorous public and private actions to overcome existing barriers to conservation—that energy use in 2000 could be at roughly today's level.

All this suggests that energy conservation can play a major role in resolving the "energy crisis." Conservation saves energy and money, reduces the amount of pollution generated, reduces vulnerability to unstable foreign sources of fuel, and buys time during which to develop alternative sources of energy that are environmentally benign, abundant, and publicly acceptable. No small accomplishment! [ERIC HIRST]

Bibliography: Demand and Conservation Panel of the Committee on Nuclear and Alternative Energy Systems, U.S. energy demand: Some low energy futures, *Science*, 200(14):142–151, Apr. 14, 1978; Ford Foundation, *Energy: The Next Twenty Years*, 1979; L. Schipper and J. Darmstadter, The logic of energy conservation, *Technol. Rev.*, pp. 41–50, January 1978; R. Stobaugh and D. Yergin, *Energy Future*: *Report of the Energy Project at the Harvard Business School*, 1979.

Exploring Energy Choices

Todd Doscher

The first and rather sudden intrusion of the matter of energy into the minds of most Americans occurred in 1973, when the Arab oil-producing states placed an embargo on shipments of crude oil to the United States. Well before the embargo, however, a number of studies had predicted the present crisis.

In *U.S. Energy Outlook, 1971–1985*, prepared by the National Petroleum Council, the finding was made that despite the newly discovered North Slope and anticipated future discoveries, "in order to meet growing demands for petroleum liquids, imports would have to increase more than fourfold by 1985, reaching a rate of 14.8 million barrels per day." It is unfortunate that such conclusions, presented in the literature throughout the 1960s, had been overshadowed by the much greater attention given to the presumed effect of government regulations and incentives for profitability on the discovery and development of new supplies of petroleum.

SUPPLIES AND UTILIZATION

There are many ways of characterizing the deep-seated nature of the energy issue for the United States and the rest of the world, but first the role of energy supplies and their utilization must be addressed. The energy consumption in the United States increased steadily after World War II. Large new supplies of electric power from nuclear generators and geothermal plants were added to the supply of coal. The use of coal, except during the Great Depression of the 1930s, has remained almost constant since 1920 and coal was not surpassed as the most important fuel in the United States until 1951. Between 1920 and 1978, the per capita consumption of energy in the United States increased approximately fourfold and the population approximately doubled, increasing total consumption by a factor of 8. Although the life-style of Americans was considerably improved as a result of this energy usage, the United States did turn out a third of the world's goods by using a third of the world's energy supply. The energy was not wasted.

Nevertheless, the usage rate was enormous. Total annual consumption was 75 quads (one quad = 10^{15} Btu). On a per capita basis, each person in the United States annually consumed the energy of a 25-megaton bomb. On the average, each person consumed 50 lb (23 kg) of oil and gas each and every day of the year.

The sudden halt in the growth of the supplies of crude oil and natural gas and some slight curtailment in these supplies have created temporary discomforts in the United States in recent years: waiting in line for gasoline, somewhat greater reliance on mass transportation, somewhat lower discretionary spending. In less developed countries, however, the impact of the increased cost of nitrogenous fertilizer, most of which is produced from natural gas, on the production of food has had a far more pronounced effect on starvation and longevity. High-protein diets on a crowded Earth require the expenditure of more energy than is developed by the food intake; for beef, the ratio is about 1:10.

It has been customary for economists to compare energy resources according to cost. The price in the marketplace is in fact a measure of the difficulty of extraction, transportation, and utilization. However, the question of absolute availability and absolute cost is more fundamental than comparative cost. The cost of a given unit of energy in terms of human labor had never been as low in the long span of human existence as in the 1970s. The operating cost of allowing a well in Saudi Arabia to flow a barrel of oil and tanking it halfway around the Earth to the east coast of the United States is trivial; even after allowing for the capital cost of the well and the tanker, the unit cost of energy reached a minimum value in the 1970s. This low cost has fueled economic developments and implicitly has been at the core of assumptions of great expectations for the nations of the world.

The increasing price which oil now demands is not at this time a reflection of any significant increase in the cost of current production, but is a result of the realization by energy producers that supplies are becoming more and more limited and that any future replacement of current supplies is virtually impossible at current production costs.

There is thus a very real concern that these increased real costs in the near future will lead to a decreased energy consumption and ultimately to a slowing or stopping of economic development itself. As the 1970s drew to a close, there were significant threats to the economic stability of the United States and the Western World which were unequivocally related to the increased cost and curtailment of energy supplies. The effect of changes in the energy supply on the economy of a nation where the growth of services rather than the growth of industrial production has been the basis for recent overall growth will, however, be restrained. The adverse effects will be far more significant and more quickly felt in the developing countries, where industrial production is the only key to new economic growth.

Therefore the availability and cost of energy and the ultimate limitations on its use pose grave questions for the well-being of humankind in the long and short term. The energy crisis is a crisis which far transcends any local and temporary implications. The whole energy question poses a challenge of the most fundamental sort to world social structure and to society's ability to make a new technological and scientific response.

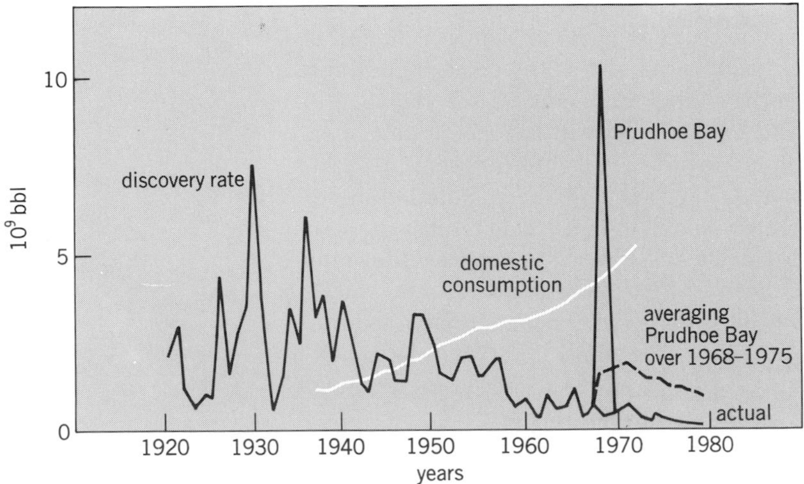

Fig. 1. Crude oil discovery rate and domestic demand.

UNITED STATES CRUDE OIL SUPPLY

During the 1920s and 1930s it was easy to become convinced that the supplies of petroleum fluids in the United States were limitless. For 2 decades, the drill found far more oil than was consumed (Fig. 1), and the excess was counted as proved reserves.The glut of oil was so great that at times the selling price was driven to absurdly low values. Crude oil had no intrinsic value to the entrepreneur who invested in the drilling and completion of the well other than its sale at whatever price would recoup the investment. Government intervention, perhaps for the first time on so grand a scale with respect to the sale of a commodity, prevented utter waste and collapse of the petroleum industry in the early 1930s. The state of Texas delegated to its railroad commission complete and total responsibility for deciding just how much oil could be produced from any well drilled within the state, and how closely wells could be drilled. The railroad commission was given this role since most oil during that decade was shipped by railroad, and the controlling mechanism was obvious.

Discovery and exploitation trends. For many years, it was not appreciated that the ease with which prolific reservoirs were discovered in Texas in the 1930s was due to the unique physical evolution of the sedimentary basins beneath the state. Meanwhile, discovery trends were extrapolated from this experience to predict a most sanguine future for the discovery of oil in the rest of the United States. After World War II, when the nation could again devote resources to the search for new oil, the results were very different (Fig. 1). Even the discovery of the greatest reservoir in the Western Hemisphere, the Sadlerochit on the North Slope, was not able to divert the monotonic average decrease in finding rate. That this could not be due to a lack of interest in drilling exploration wells is shown in Fig. 2. Exploration drilling was actually increasing during the 1950s and 1960s, when discoveries were falling to lower levels than had been experienced ever since reliable data was available.

The immediate significance of the lowered exploration success in the 1950s and 1960s was some-

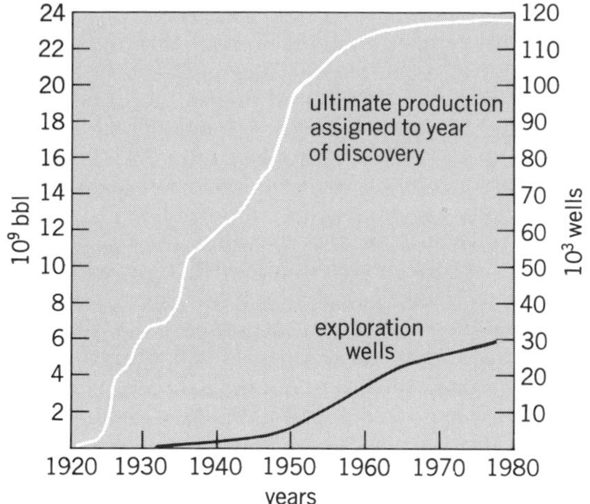

Fig. 2. The history of exploration drilling in the West Texas–southeast New Mexico area, the single most important producing province in the United States.

how not recognized. Despite the fact that demand was outstripping the rate of new discoveries, the reserves (the amount of crude that was known to be economically recoverable as a result of completed drilling activities) were being maintained. This anomalous situation resulted from the implementation of secondary recovery operations (mostly waterflooding) and increasing imports of crude oil (Fig. 3).

Despite the significant level of imported crude, the production rate in America did reach a peak in 1973. This had been anticipated by prior studies, which were based on the self-evident premise that the supplies of crude oil, like those of any resource in the Earth's crust, are limited. Once this is conceded, it readily follows that continued exploitation of that resource will lead to its exhaustion. Exhaustion will be a very gradual process because, as the availability of readily accessible accumulations of the resource are depleted, attention will shift to smaller deposits, deeper deposits, and accumulations located in more remote and hostile areas (beneath the oceans and in the arctic latitudes). The cost of new production will become higher, demand will fall off, and production rates will gradually decrease. This cycle can be fitted to a mathematical growth curve as a function of time. Plotting the annual production rate as a function of time must result in a bell-shaped curve, with the area under the curve being the amount of recoverable resource.

Future supplies. The proved reserves in the United States at the close of 1978 were estimated to be slightly less than 29×10^9 barrels (bbl) or 4.6×10^9 m^3, and cumulative production was slightly above 115×10^9 bbl (18.3×10^9 m^3). If domestic production were to be continued at 3×10^9 bbl (4.8×10^8 m^3) a year, the proved reserves would last for somewhat less than 10 years. A very important question, then, is how much additional crude oil may be discovered. During the 1970s, various sources estimated the discoverable reserve to range as high as 200×10^9 additional barrels (32×10^9 m^3). However, extrapolation according to the methodology presented above indicates a potential of only 30 to 50×10^9 additional barrels (4.8 to 7.9×10^9 m^3) to be discovered beneath the United States and its surrounding continental shelves.

It is anticipated that only 32% of the approximately 450×10^9 bbl (72×10^9 m^3) that have been discovered can be recovered by conventional primary and secondary (waterflood) operation. Much attention has therefore been given to attempts to develop technology that will recover additional quantities of crude oil from already discovered and producing reservoirs.

The primary reasons for the limited recovery is the interplay of the parameters that characterize the reservoir fluids and reservoir lithology and control the flow of oil through the porous medium. The engineer is unable to exercise any intimate control over these factors and is limited to attempting to control the displacement and ultimate production of oil by remote means: the location and density of drainage points (producing wells), the rate of pressure drawdown or fluid withdrawal, and the location and density of injection wells as well as the nature (physical state and composition) of the fluids injected.

The several technologies that have been proposed for increasing recovery are chemical flooding, solvent flooding, and thermal flooding. In chemical flooding, a system of surfactants in water is injected to reduce the capillary forces that have resulted in the isolation and immobilization of residual crude oil. In many respects, it is like laundering garmets soiled with oily residues. In solvent flooding, like dry cleaning, the oil is to be dissolved by the injected solvent. The only economically feasible solvent at this time is carbon dioxide at supercritical pressures. Finally, the injection of steam or propagation of a combustion wave is intended to raise the temperature of the oil and thus render it more mobile.

It has been difficult to conduct laboratory experiments that effectively recreate reservoir conditions encountered in tertiary recovery operations using surfactants and carbon dioxide. Laboratory results have proved to be too optimistic.

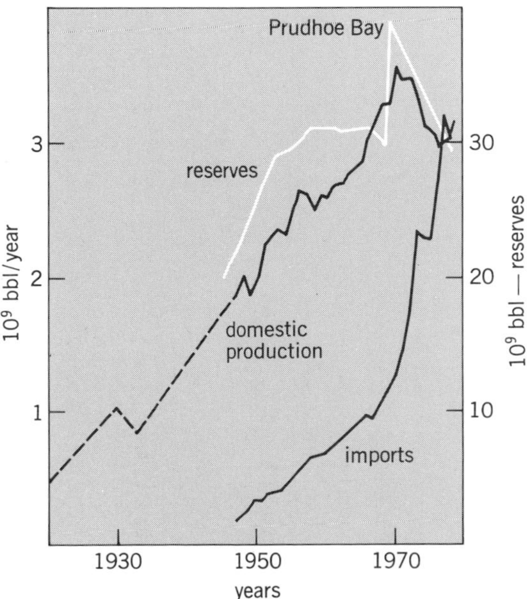

Fig. 3. Crude oil production, imports, and reserves.

To date, only technology of relatively limited applicability, the injection of steam into reservoirs containing very viscous crude oils, has proved effective in field operations. Although early estimates, based on laboratory research, of the amount of oil that might be recovered by enhanced recovery technology have been as high as 100×10^9 bbl (16×10^9 m^3), recent field pilot experience suggests that a recovery of 20×10^9 additional barrels (3.2×10^9 m^3) will represent a major technical achievement. Also affecting future supplies may be the positive or negative revisions in reserve estimates that may occur as reservoirs reach levels of depletion with which there has been no extensive experience. *See* PETROLEUM ENHANCED RECOVERY.

The projected domestic production has been estimated by using the preceding observations on historical and current oil production, and estimating potential success of tertiary operations and positive revisions (Fig. 4). The role that future discoveries must play in order to maintain current levels of production is obvious; it also appears to be unachievable.

In order to maintain production at roughly current levels by the discovery of new resources, four or five reservoirs each equivalent to Prudhoe Bay would have to be discovered in the coming decade. Considering the fact that only one such giant has been found in the Western Hemisphere in the 150 years of exploration for petroleum, the disappointing experience in such frontier areas as the Gulf of Alaska and the Baltimore Canyon, and the density of drilling in the contiguous 48 states, the probability of making such finds is remote.

WORLD CRUDE OIL SUPPLIES

The world picture is more uncertain than that of the United States because of the more limited data available for study. However, the most reliable data indicate that crude oil reserves in the world slightly more than doubled between 1960 and 1972, from 290 to 633×10^9 bbl (46.1 to 100.6×10^9 m^3) and since 1972 the reserves, although fluctuating, rose only to 650×10^9 bbl (103.3×10^9 m^3). This leveling out of current reserves, together with an increasing rate of production, put the reserves/production ratio for the entire world at about 30 at the end of 1978.

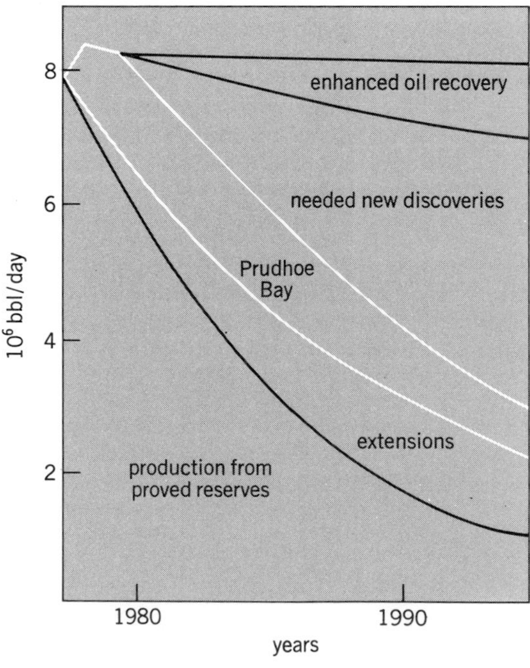

Fig. 4. Projected domestic crude oil production.

The enormity of the situation for the United States and the rest of the world is appreciated when it is realized that the cycle of discovery and exploitation of petroleum resources in the Mideast, in Africa, and in South America will duplicate that of the United States. In 1972 R. L. Jodry, using the methodology introduced by M. K. Hubbert, concluded that the peak in production will occur in the 1990s and that 80% of the total recoverable resource will be exhausted by the year 2023.

It has already been reported that production has started to decline in the Soviet Union, the greatest oil producer in the world in the 1970s. In midsummer 1979 the Secretary of Energy of the United States claimed that production from the OPEC states is "very close to its maximum production right now at about 31.5 million barrels [5.0×10^6 m^3] per day." Further, he predicted that oil prices of $40 a barrel, in constant dollars, could be anticipated by 1990 and that the supply situation in the United States would become much worse if OPEC countries hold down exports. Already in December 1979 some spot market prices for crude oil exceeded $40 per barrel ($250 per cubic meter) and a number of OPEC producers have already indicated that their production will be limited. From the point of view of optimizing profitability, the policies of the OPEC nations are in harmony with the best business practices of their customers.

Some relief for the diminishing world supply of crude oil may come from the exploitation of vast deposits of "heavy oil," that is, tar sands, which are known to occur in Canada (the Athabasca bituminous sands) and in Venezuela (the Orinoco Tar Sand Belt). Although estimates of the oil in place in these combined accumulations exceed 10^{12} bbl (1.6×10^{11} m^3), feasible economic technology for producing a significant fraction of these resources is still to be developed. It is not certain at this time what probability of success can be assigned to such development work. *See* PETROLEUM RESERVES.

ALTERNATIVES TO OIL

As the possibilities of substituting other energy sources for oil are addressed, it becomes obvious that there are many obstacles to what appears to some to be easy solutions. Certainly there are adequate coal supplies in the United States; in terms of Btu, they could satisfy the nation's needs for a century or more. Eventually, it is believed, coal will of necessity again be viewed as the prime energy source for the United States. However, the substitution of coal for petroleum fluids entails a significant change in living habits, life-styles, and centers of population. A profound but necessary change in values for American society must precede and accompany the transition from petroleum fuels to coal. Arguments for the massive substitution of the so-called renewable resources, such as solar energy and biomass energy, for petroleum fluids fail to recognize some of the basic facts about the use of energy and the cost of capital in developing such sources. Moreover, until the development of a breeder reactor is shown to be technologically and economically feasible, the supplies of fuels for conventional reactors within the United States appear to be as limited, in energy utilization, as petroleum fluids.

Finally, it is appropriate to deal with the matter of conservation of energy, which would reduce the overall dependence of the nation on energy supplies. There is little question that conservation, wise and deliberate, would ameliorate the long-term increase in per capita needs for energy. However, the short-term outlook for conservation is not a substantial one. *See the feature article* ENERGY CONSERVATION.

Natural gas. At the end of 1976, 506×10^{12} ft^3 (1.43×10^{13} m^3) of natural gas had been produced in the United States, and reserves stood at 216×10^{12} ft^3 (6.1×10^{12} m^3). An analysis for the U.S. Department of Energy concluded that inferred reserves of 84×10^{12} ft^3 (2.4×10^{12} m^3) would be added by the year 2000, and 105×10^{12} ft^3 (3.0×10^{12} m^3) would be added from new discoveries and inferred reserves based on new discoveries. The resulting prediction for the production of natural gas in the United States is shown in Fig. 5. The anticipated production of some 8×10^{12} ft^3 (2.3×10^{11} m^3) a year from conventional sources may, however, be supplemented by an almost equal amount of gas produced from nonconventional sources (Fig. 5). Thus, there is the promise that domestic natural gas supplies may extend further into the 21st century than will the supplies of domestic crude oil.

There are large quantities of natural gas in the world which are only beginning to be exploited at this time. Of the $2000-2500 \times 10^{12}$ ft^3 reserve, over 65% is in the Soviet Union and Iran. Large-scale distribution systems for the Soviet reserves in Siberia are now bringing some of the gas into eastern Europe, and Iran exports about 10^{12} ft^3 (2.8×10^{10} m^3) a year to the Soviet Union. Plans for expanding this latter trade have recently been abandoned.

The export of relatively small quantities of liquefied natural gas (LNG) has been underway for the past years. The gas, which is produced from fields relatively close to deep-water ports, is transported in refrigerated tankers so that pressure containment is not required. It is anticipated that LNG imports into the United States may approach 2×10^{12} ft^3 (5.7×10^{11} m^3) a year by 1990, a valuable addition to domestic supplies but insufficient to supply more than a marginal demand. *See* NATURAL GAS.

Coal. Coal is more widespread than petroleum, and the resource is considerably larger. In 1977 the recoverable reserve in the United States was estimated at 283×10^9 tons (257×10^9 metric tons), which at the then annual rate of production of 660×10^6 tons (599×10^6 metric tons) would last for some 430 years. Future production rates of coal will depend upon the extent to which the nation elects to use it, both for combustion and as a source material for the manufacture of synthetic liquid and gaseous fuels. Most predictions of the future use of coal (or any other energy source) are

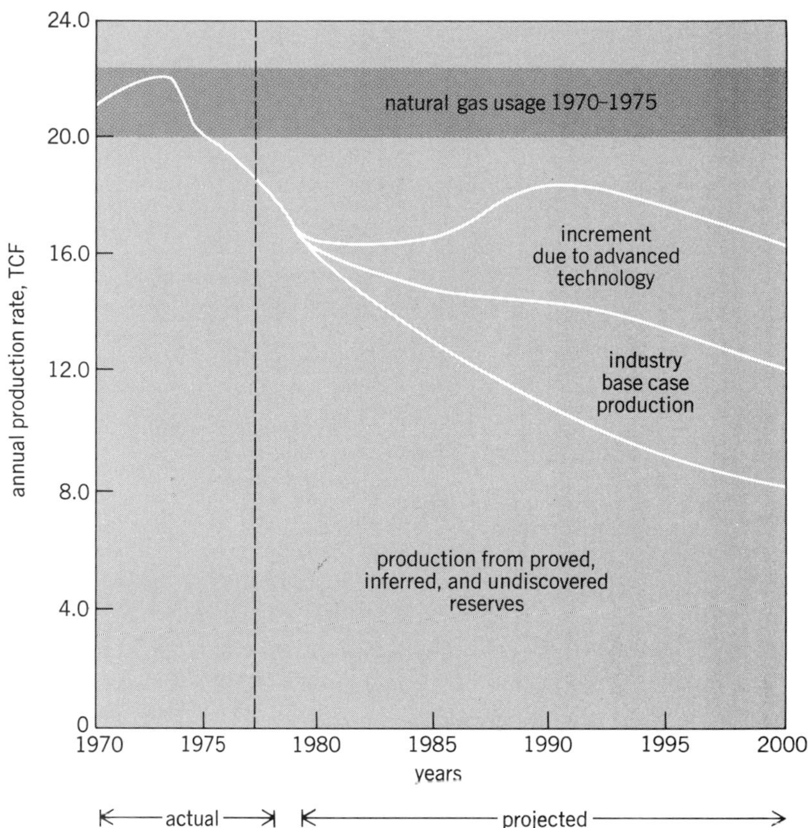

Fig. 5. The potential of unconventional gas sources under advanced technology, at gas prices of \$3.00/Mcf. Mcf = 10^3 ft^3 = 28.3 m^3. Tcf = 10^{12} ft^3 = 2.83×10^{10} m^3.

derived from scenarios in which assumptions are made as to the growth of the energy demand, the growth of competitive energy sources, and the anticipated price of the most flexible and overall most desirable source, namely petroleum and natural gas. Such scenarios fail to convey the significance and the ultimate importance of coal as a prime source of energy for the United States.

Consumption. Figure 6 shows the energy supply and demand in the United States for 1975, a year of considerable prosperity. The consumption of energy is shown in quads. It should also be noted that the figure under "electric" is the measure of the total energy input into power (including the energy losses, which are substantial – about 65% for conventional power plants). It will be quickly observed that only 55% of the oil is used for transportation and 45% is used for electric power generation and residential/commercial and industrial activities. The 45% represents 14.8 quads (15.6×10^{18} J). Seventy-five percent of this could easily be replaced by coal. If, together with this shift in usage, transportation requirements were reduced by, say, 50% by mandating the use of more efficient (Btu per passenger mile) automobiles and electrically driven mass transportation systems, the use of crude oil could be reduced to 12 quads (12.7×10^{18} J) per year, or only 2×10^9 bbl (3.2×10^8 m^3) a year (5.5×10^6 bbl or 8.7×10^5 m^3 a day versus the 20×10^6 bbl or 3.2×10^6 m^3 being consumed today). This could take place with no basic change in the life-styles of 1975.

Coal consumption would have to increase from

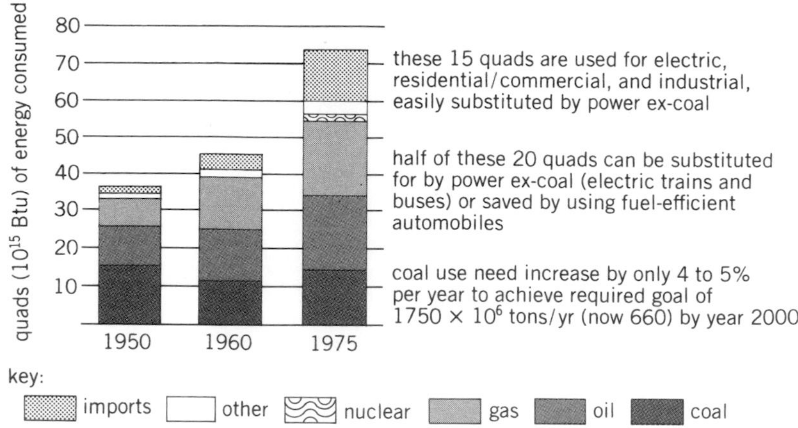

Fig. 6. Energy supply and demand in the United States.

15.3 quads (16.2 × 10^{18} J) to some 40 quads (42 × 10^{18} J) per year after allowing for the energy losses in the conversion of the necessary part of this increase in electric power generation. To raise the consumption of coal to 40 quads (42 × 10^{18} J) from 15.3 quads (16.1 × 10^{18} J) in 20 years requires an increase in coal production of only 5% per year, reaching 1750 × 10^6 tons (1588 × 10^6 metric tons) a year from the present 660 × 10^6 tons (599 × 10^6 metric tons). The remaining reserve, if coal were to continue to be used at that rate, would be still adequate for 150 years. This would presumably give the nation sufficient time for the basic readjustment to essentially new sources, such as the breeder reactor or solar energy. Although some significant changes in life-style will accompany the transition to a coal-fueled society, coal can support a very high level of economic and social well-being in the United States.

Conversion. Coal can be converted to liquids or methane, both of which are more desirable and flexible fuels than a solid fuel. Large quantities of hydrogen must be added to the coal in order to effect such conversion, and a third of the energy in the coal is consumed in these hydrogenation processes. Some of this energy loss would be offset by the relatively lower cost of transporting the liquid and gaseous products, since the rail transportation of solid coal is far more expensive. Some of the offset would be lost by using large coal slurry pipelines or by generating electricity at logistically placed large central stations and distributing the electricity through efficient grid systems.

In addition to the high energy losses in coal conversion, the capital costs for the high-pressure, high-temperature conversion plants are high; hence coal liquids, although technologically feasible, will be a high-cost energy supply. *See* COAL GASIFICATION; COAL LIQUEFACTION.

Disadvantages. The potential of coal reserves in the United States to satisfy the energy needs of the nation is apparent, but mention of its use raises for many the evils associated with its past use: land destruction, labor unrest, coal barons, spectacular accidents involving the loss of many lives, and black lung diseases. That the past history of coal exploitation has been characterized by such associations is true, but it is self-defeating to assume that the future use of coal cannot proceed with a significantly improved social and environmental impact.

Fatalities in coal mines had already decreased from 514 in 1952 to 100 in 1977. Enforcement of maximum dust standards in underground mines is expected to lead to a steady reduction in black lung diseases. Spot-checks of mines have revealed that since 1973, 90% of underground mines have been in compliance with Federal standards, but monitoring must be implemented on a continuous basis to ensure round-the-clock compliance. *See* UNDERGROUND MINING.

There are many statistical studies which have attempted to estimate the effect on human health and the environment of compounds that are emitted when coal is burned: sulfur oxides, sulfates, carbon dioxide, fine particulates, and many more. Short-term tests on cultures and tests on animals can employ high levels of pollutants and thus can measure easily identifiable acute effects. On the other hand, human epidemiological studies suffer from problems associated with heterogeneous populations, poorly measured pollutant exposures, and the coexistence of other factors with pollution and occupation. Statistical studies usually do not take into account the effect of psychological, behavioral, and organizational factors, and the implementation of Federal standards by both workers and management.

Because of the overstatement of the significance of such statistical studies, it has become increasingly difficult to mount the appropriate development and commercialization programs which are ultimately the only way in which the exact nature of the problems can be defined and solved. Already there have been developments, such as the fluidized-bed combustion of coal admixed with carbonate rocks; this effectively removes sulfur as calcium sulfate before the sulfur oxides go up the stack. The increased use of western coals will contribute to an overall greater level of safe operating conditions because of their low sulfur content and because they will be strip-mined. Reclamation of the land, since it is not glacial sediment, will be both technologically feasible and affordable. *See* FLUIDIZED-BED COMBUSTION.

The fortuitous availability of huge deposits of coal in the United States and the potential that they have for underpinning a continued high level of social and economic welfare require a mature approach to definition and solution of the problems associated with coal utilization. *See* COAL.

Oil shale. Oil shales are fine-grained sedimentary rocks (calcareous marlstone) containing significant quantities of organic material, kerogen. Upon being heated to temperatures in the neighborhood of 900°F (480°C), the shale will yield a viscous liquid hydrocarbon. The latter can be transformed into a synthetic crude oil by the addition of a small percentage of hydrogen. The oil shales in the United States, principally in the Piceance Basin of western Colorado, have been estimated to contain the equivalent of 0.5 to 1.0 × 10^{12} bbl (0.8 to 1.6 × 10^{11} m^3) of crude oil.

Recovery and conversion. The technological problems of converting this resource to petroleum are very great. A ton (0.9 metric ton) of high-grade material will yield a half barrel (0.08 m^3) of petroleum; thus, to produce a mere 100,000 bbl

(1.59×10^4 m^3) of crude oil a day, 150,000 yd^3 (115,000 m^3) of oil shale would have to be processed. That amount of material would cover a football field to a thickness of 30 ft (9 m), the height of a 3-story building, each and every day. For perspective, consider that the United States currently consumes approximately 20×10^6 bbl (3.2×10^6 m^3) of crude oil a day.

Room-and-pillar mining of the oil shale will probably not achieve a level of production commensurate with the needs of the nation because of the great problems to be encountered in the logistics of moving such large quantities of material through tunnels, followed by the perhaps greater problems of disposing of the spent shale after retorting. Open-pit mining is perhaps the only way to successfully mine and process the material on the surface. Considering that at the center of the Piceance Basin the oil shale reaches thicknesses of more than 1000 ft (300 m), and it is overlain by an equal thickness of overburden, it is obvious that massive mining projects will have to be organized in order to win this resource. At this time, government leasing policy, which restricts leases to 5100 acres (2064 ha), thwarts consideration of the massive earth-moving projects that are required for efficient recovery of the oil shale.

After mining, the oil shale must be crushed and then retorted. Numerous schemes for retorting the oil shale have been pilot-tested. Some schemes involve the passage of air through the crushed oil shale. A combustion front develops the required heat for converting the kerogen, and the viscous oil is entrained in the flow of the combustion gases. In another approach, heat is transferred to the oil shale by circulating ceramic balls which crush the ore; they are heated by combustion of the off-gases produced in the retorting and subsequent upgrading of the shale oil.

The development of true in-place recovery technology for oil shales is restricted because of the absence of any significant permeability, which in turn prevents the injection of air or steam for retorting. However, development work is under way on a modified in-place recovery technique. Horizontal shafts are cut below and above the column to be exploited, and about 20% of the column is mined out vertically to permit expansion upon subsequent retorting. Using proprietary explosive technology, the column [exceeding 100 ft (30 m) in height and 100 ft^2 (9.3 m^2) in cross section in the proposed commercial operation] is rubbled. Air is then introduced at the top of the column, a combustion front is initiated much as in a surface retort, and the produced oil is carried downward with the combustion gases. The oil is collected at the base of the column.

Just as in the case of surface retorts, the principal problem in carrying out the conversion is the need for uniform flow of air through the retort. Channeling results in nonuniform heating, some sintering, and very low recovery efficiency. Pilot operations to date have revealed a steady drop in recovery to less than 30% as the columns have been increased in size to approach the size required for a commercially viable operation. At best, the overall recovery of the resource by a modified in-place process is limited by the need to leave pillars of material between adjacent retorts.

The most positive approach to the recovery of oil shale is by open-pit mining. A large portion of the basin would be treated as a single unit for exploitation. There are certainly specific problems that would have to be overcome, such as the dewatering of the rather large quantity of brackish water in the basin, the stockpiling and retention of the spent shale, land reclamation, and the reestablishment of drainage systems. These problems are already well defined and can be solved by appropriate engineering expertise. It is only the ultimate cost of the synthetic crude which is in doubt. Estimated costs for producing the oil shale have always been somewhat more than the then current price of crude oil. With each increase in the cost of crude oil, the estimated price of oil from oil shale has been increased. This was certainly to be expected, since an increase in the level of energy costs would cause corresponding increases in the cost of capital goods and labor. However, with a foreseeable future characterized by diminished supplies where the price of energy is not determined competitively, oil shale costs will probably become internally competitive. *See* OPEN-PIT MINING.

Environmental and health hazards. A more difficult problem for solution is the dangers to environment and health that have been attributed to oil shale development. As in the case of coal, the predictions of such threats are based on statistical studies and the identification of trace materials in effluents and emissions. A far more positive program of pilot development is required to establish the real risks to United States society in an oil shale program. The greatest threat to the society may be in not taking advantage of this resource, which can provide centuries of liquid fuels, in turn providing social and economic benefits. *See* OIL SHALE.

Nuclear power. The popular concept of nuclear power is that it is essentially an infinite resource. The actual picture is considerably more complicated. The facts are that, if the present forms of nuclear power reactor are to be relied upon, and if the produced power is to be competitive with power produced from coal, the current reserves of uranium in the United States would not be sufficient to sustain much greater growth of the nuclear power industry.

Uranium. Current reactor technology requires fuel which has been enriched by the very rare ^{235}U isotope. The enrichment of available ores is a costly process, and as the uranium content of the ore decreases, the ultimate cost of the fuel concentrate increases. To produce competitive power, there is a minimum concentration of uranium which must be present in the ore for it to be considered as a reserve. During 1975 the concentration of uranium in ores used in the United States was 1500 ppm or higher. The assured reserves amounted to 400,000 tons (360,000 metric tons). Omitting the initial fill, a 1000-MWe reactor requires 175 tons (159 metric tons) per year; therefore, there was sufficient reserve to operate the then installed 36-MWe capacity for 75 years.

Although nuclear capacity was expected to grow phenomenally during the late 1970s and early 1980s, hundreds of thousands of megawatts were canceled, and growth has been only about 25%.

The OECD Nuclear Energy Agency and the International Atomic Energy Agency completed a study in 1979 which reported the reasonably assured reserves of uranium in all of North America to be 830,000 tons (750,000 metric tons), of which about two-thirds were in the United States. Thus, had the capacity of the American industry reached a level of 100,000 MWe, the reserve would be sufficient for only 30 years. Although exploration for new uranium reserves continues, most of the cheap, high-grade ores within the country may have been located. It is difficult to ascertain how much of the canceled nuclear power capacity is due to this factor and how much is due to the more commonly accepted factors of licensing delays and implementation of safety regulations.

There are indeed much larger resources of uranium available at costs of more than $50 per kilogram (see the table), and it has been customary for many to assume that these will become competitive as the cost of fossil fuels increases. This is true as long as the swing fuel is priced arbitrarily, as is anticipated for crude oil, but it is not true if the swing fuel is domestic coal, which will be priced according to the actual costs of operations and capital. The larger reserves of relatively cheaper coal will be capable of producing cheaper power. *See* URANIUM.

*Breeder reactor.*The true potential of nuclear energy as a source for power can only be realized by the development of acceptable breeder reactors. The breeder does not need ^{235}U enrichment, since it is capable of breeding its own fuel from virtually all the uranium in an ore. The value of any ore for delivering power is increased 150-fold when the ore is used in the breeder reactor; hence, reserves are similarly increased. The development of such reactors in the United States has been delayed for numerous reasons: some technical, some associated with the testing of various safety aspects of the design, and some associated with still more basic problems of safety and security.

The breeder has been widely criticized for being unsafe and vulnerable to theft of fissile material, and for the inadequacy of procedures for handling the high-level long-term radioactive wastes. Although the same criticism has been leveled against the ^{235}U reactors, it is perhaps more justified in the case of the breeder. The plutonium fissile material is a notorious pathogen, and it is inherently more difficult to provide adequate safeguards for the breeder than for the ^{235}U reactors. This issue is, of course, at the heart of much of the design effort in the development of breeder reactors, and in addition it forms the core of the debate about their widespread construction. In all these cases, the safety problem arises not from the potential of the reactor to become a nuclear bomb, but from the possible occurrence of incidents such as ordinary chemical explosions which might release portions of the radioactive core into the atmosphere. *See* NUCLEAR MATERIALS SAFEGUARDS.

Environmental questions. The two other environmental questions surrounding the nuclear option have not received the attention which has been given to safety. One concerns the possibility of employing the fissile material used in a breeder reactor for making bombs. This problem takes on many dimensions. For example, it is possible to develop a nuclear weapon from the plutonium materials produced in a commercial reactor. This was emphasized by the explosion of a nuclear device by India in 1975. Theft of fissile or highly radioactive reactor materials is also a major hazard.

The second problem concerns the disposal of the long-term high-level radioactive wastes produced in any reactor. If nuclear reactors are used on a massive scale for hundreds of years, the human race will accumulate significant amounts of radioactive materials which will remain radioactive for a time comparable to that which has elapsed since the appearance of the human race. In late 1979 only one state in America had elected not to close down its nuclear dump. *See* RADIOACTIVE WASTE MANAGEMENT.

Solar power. The energy incident upon the Earth from the Sun provides a constantly renewable source of energy. It is in addition relatively benign from a health and environmental standpoint. It has therefore attracted much attention, particularly from those who have elected to find the coal and nuclear options unacceptable.

Energy potential. If all the solar energy falling on the United States each day could be converted to electricity, each square foot of land would receive on average the equivalent of 160 kWh a year. A 140-ft² (13-m²) surface area would supply the 22,000 kWh a year used by a modern, all-electric house equipped with air conditioning. An area of only 6500 mi² (16,800 km²), less than 0.5% of the arable land in the United States, could supply an amount of energy equal to that used in the entire

Estimated world uranium resources, in 10³ metric tons

Continent	RAR*	EAR†	Speculative resources
Africa	572	200	1300– 4000
Asia and Far East‡	37	24	200– 1000
Australia and Oceania	296	49	2000– 3000
Europe	88	300	300– 1300
South and Central America	60	14	700– 1900
North America	830	1711	2100– 3600
Totals	1883	2248	6600–14,800
Eastern Europe, Soviet Union, and People's Republic of China			3300– 7300

*RAR = Reasonably assured resources, less than $80/kg.
†EAR = Estimated additional resources, $80–$130/kg.
‡Excluding People's Republic of China and Asian Soviet Union.

United States (but not usable for all the functions to which current energy supplies are put). Solar cells were used to power the Skylab.

Technological problems. The preceding description of the potential of the Sun's radiant energy has undoubtedly been the basis for the high expectations for solar power which many people have. However, there are numerous technological problems that are not conveyed by such an abstract statement of solar energy's possibilities.

The average incidence of sunshine is far from even across the United States. The land in New Mexico probably gets more than twice the average sunshine of the land in northern Minnesota. And even before sunset, there is no usable sunshine.

There are two ways to trap the Sun's energy: letting it fall on a flat plate, which is colored black to absorb the radiant energy, and then having the black plate heat air or water circulating underneath it; or allowing it to impinge on photovoltaic cells that convert the radiant energy to electrical energy.

Flat plate collectors, the systems for heating, are indeed basically simple affairs; but to secure a modest degree of efficiency they must become rather complex, and require sophisticated design and fabrication. First of all, the black surface must be covered with a protective, radiation-transparent coating or sheet of glass or plastic. The space between the transparent cover and the black surface may be filled with air, but air transmits the heat back from the hot surface and the efficiency is decreased. Therefore an optimum system is one in which the space between the black surface and the cover is evacuated. If the black surface is to transmit its heat to circulating air, then fins should extend down into the air stream; if the heat is to be transmitted to water, pipes must be embedded in the black surface. The back side of the water pipes or air ducts must be insulated to prevent loss of heat, and where freezing temperatures are encountered during the evening it is necessary to use antifreeze liquids rather than water in the circulating system.

The temperature of the water or air cannot be allowed to exceed 175°F (79°C), otherwise the efficiency of the collector decreases; therefore, the heated water or air can be used only for relatively low-level heating and cannot be used to operate absorption-cycle air-conditioning systems or for high-load heating. Since most heating is required during the colder evening hours, the heat captured during the day must be stored. This can be done by using hot water heaters of large capacity, tanks of molten salts, or beds of rock and gravel. The size of the last must be carefully chosen so as to permit circulation of fluids but not be so large that heat transfer is limited by conduction within the rock.

High-rise apartment houses and other multiple dwelling units may not have sufficient external roof area per resident to permit the installation of an adequate solar heating system; hence, the use of solar heating appears to be restricted to single-family units, or specially constructed new buildings.

For the development of electricity the matter is still more profound, since the conversion efficiencies are theoretically limited to about 25%, and a conversion efficiency of 10% has not yet been practically realized. One of the problems is that the efficiency of photovoltaic conversion decreases as temperature is increased, and therefore the collectors must be cooled by circulating air or water. Again, there is the need for adequate storage capacity, since most electric consumption occurs toward sunset and in the early evening hours. This requires the installation within the solar house of a large number of electric batteries which are superior in performance to most storage batteries used in automobiles. The batteries would occupy the volume of a small utility room.

The electricity generated by the solar collectors is direct current; in modern houses, it is not useful as such for anything but powering incandescent lights. Therefore, alternators having the capacity to convert the direct current at peak loads must be built into the system.

Subsequent operating costs will not include the cost of any energy purchase (although maintenance may be more than trivial), with the exception that an alternate utility power system will still be required. This requirement exists because again there are extended periods of cloudiness and shortened hours of daylight, in many areas occurring at the same period when the need for power is greatest. Since a significant cost of utility power is the amortization and maintenance of the generators and the distributing system, a lowered sales volume will entail a higher unit cost. Just how much of the savings due to the use of solar collectors will be offset by the higher unit cost of the utility power is difficult to estimate on a generalized basis.

It is very likely that solar energy will supply the hot water heat for numerous upper-income households in the Sun Belt within the coming decades. The more widespread utilization of solar power for space heating and electricity will probably be limited to homes and commercial buildings where significant discretionary capital or subsidies are available.

Raised collectors have been designed which use focused mirrors; large banks of these may reflect the sunlight onto the elevated boiler of a central power station. From a strictly economic standpoint, such units cannot compete with coal-fired or nuclear power stations in the foreseeable future. *See* SOLAR ENERGY; SOLAR HEATING AND COOLING.

Other sources. Of the other possibilities which are often discussed as sources of energy, one is nuclear fusion, in which energy is obtained by the fusion of two light nuclei to form a heavier nucleus. Energy produced from a controlled fusion reaction is thought to have an advantage over ordinary nuclear power because it seems to generate smaller amounts of nuclear wastes. Also, the hydrogen in the ocean might conceivably be used in nuclear fusion, and of course such a supply would be inexhaustible. Fusion has not yet been shown to be a feasible method of energy production; in the best of circumstances, it is a very long-term alternative which could become important only well into the 21st century. *See* NUCLEAR FUSION.

Energy from geothermal heat is another form which can be important locally. Since the interior of the Earth is very hot, there is, relatively speaking, an inexhaustible supply of energy in the heat of the Earth itself. Of course, the technical and economic problems of using this heat are enormous, except in certain locations where volcanic intrusions have heated the rock very close to the surface. Geothermal energy thus does not appear likely to be an important source of energy nationally. *See* GEOTHERMAL POWER.

Other schemes have also been discussed for producing additional energy, such as the use of tides and the expansion of hydroelectric power generation, the installation of large wind turbines in the high-wind belt of the nation, the decomposition and combustion of biomasses, and the fermentation of grains. The eventual impact of the total energy production is not expected to become significant because of either limited regional availability or noncompetitive costs. *See* ENERGY SOURCES; TIDAL POWER; WATERPOWER; WIND POWER.

Conclusion. Any consideration of alternative energy sources would not be complete without mention of the potential of conservation as a means for reducing the overall need for energy to maintain the economic status of society. Any real reduction in energy needs without significant alteration of society requires the investment of significant amounts of capital. Although theoretically possible and desirable, it is doubtful that the required amounts of capital can be developed for any but the truly great tasks, such as the creation of mass transportation systems. Even for such goals, the assignment of the required capital is not to be achieved easily. Other significant conservation schemes will be synonymous with reductions in the standard of living. One further exception to this rule is the enforcement of energy savings systems where new construction is involved, and the adaptation of industrial plants and power generators to co-generation schemes wherein energy is recovered from otherwise wasted heat. *See the feature article* ENERGY CONSUMPTION.

[TODD DOSCHER]

Bibliography: T. M. Doscher, Domestic oil reserves: A look at the past, a guess at the future, *American Petroleum Institute Annual Meeting Papers*, 1976; Federal Energy Administration, *National Energy Outlook*, 1976; Lewin & Associates and U.S. Department of Energy, *Symposium on Enhanced Oil Recovery in the Year 2000*, 1979; Office of Technological Assessment, U.S. Congress, *The Direct Use of Coal*, 1979; R. Stobaugh and D. Yergin (eds.), *Energy Future: Report of the Energy Project at the Harvard Business School*, 1979.

Risk of Energy Production

Herbert Inhaber

Certainly one of the earliest stories of the risk associated with energy production is that of Prometheus, brother of the Greek god Atlas. For stealing fire from Heaven and giving it to humans, he was chained to a mountain rock by Zeus, the father of the gods, and an eagle fed daily upon his liver.

For many centuries in the course of history, the gathering of wood, which was the chief energy source of the human race, resulted in accidents, but other dangers to humans such as disease loomed so large that little attention was paid to the risk associated with energy production. But gradually, both the general public and scientists began to consider the risk of different energy systems.

Indeed, risk analysis was established on a firm basis only after the beginning of civilian uses of nuclear power, in the late 1950s and early 1960s. It was clear that this newest of energy forms could pose substantial health risks if it were mishandled, so researchers in the nuclear industry began comparing its risk to that of other energy systems.

Perhaps some of the best-known early analyses were those of C. Starr and colleagues in California in the early 1970s. In one paper, Starr considered the relative risk of oil-fired electricity generation and nuclear reactors; he found that the health risk of the latter was substantially lower than that of the former, when both regular operating risks and potential accidents affecting the public were taken into account. This paper and others on the subject laid the groundwork for many of the analyses which followed.

Risk and public policy. There are many reasons why the risk of energy systems is studied. One of the most important is public policy. At present, energy systems are chosen on the basis of many criteria, such as cost, availability of resources, and esthetics. Risk or potential risk seems to play a small part. In the case of nuclear reactors, this statement should be amended, since considerable effort has gone into reducing the potential risk to public health from these systems. Whether these efforts are adequate is a matter of debate, but unquestionably the effort has been made. However, there has been comparatively little discussion by decision makers on choosing energy systems even partly on the basis of the occupational or public risk.

This omission is remarkable, since most industrialized societies spend considerable money on medical research and pass stringent laws dealing with public and occupational health. It is likely that, as more attention is paid to all aspects of energy in future years, more consideration will be given to risk.

Perhaps one reason why risk analysis has not had much effect on decision makers is that it often involves the comparison of one energy system with another. Proponents of the various energy systems have been willing to have these systems ranked on the basis of cost or other economic criteria, but not on the basis of damage to human health. Comparisons are always somewhat invidious, and it remains to be seen whether this attitude will change.

Transscientific subject. Risk analysis is, almost by definition, a transscientific subject in the sense defined by A. Weinberg; that is, it cuts across the boundaries of many disciplines. For example, to conduct a risk analysis of a given energy system, one might need knowledge of (1) occupational health statistics; (2) the biological effects of pollutants; (3) the detailed engineering of generating plants; (4) reliability history of these plants; and (5) the way in which materials needed for the system are transported, and their associated public and occupational risk. Because of the complexity of the data and the judgments required, some risk analyses vary in their conclusions. Such variations probably arise in all investigations involving transscientific subjects, for they cannot be reduced to precise laboratory experiments. A more complete discussion of this problem is given below.

Risk acceptability. This article deals only with risk analysis, that is, the objective measures of damage to human health. Society may perceive these risks in a nonobjective way. To take one prominent example, studies which are discussed below indicate that the overall occupational and public risk in using coal as an energy source is relatively high. Yet governments in industrialized countries have stated that they will rely on this energy source for a considerable period in the future. In effect, the risk attributable to coal has been deemed acceptable by society.

Exactly why a mental ranking of risk involved in an energy system does not always accord with a more objective ranking is partly a psychological question, and one which deserves more consideration than can be given here. However, two brief points can be made.

First, to some extent an individual's perception of risk is related to what can happen to that individual personally. To use coal as an example again, very few people have ever worked in a coal mine, so most tend not to think about its occupational risk. As another example, somewhat removed from the field of energy, automobile accidents are one of the major risks to human health. Yet those jurisdictions which have introduced the compulsory use of seat belts, a recognized method of saving lives, have often encountered resistance from the public. As far as some people are concerned, accidents happen only to someone else.

Second, for nuclear power energy systems, there is apparently a dread factor which seems to have a strong effect on perceptions of risk. This effect, noticed by P. Slovik while polling the public on the risks of modern society, plays a role in how some people judge the acceptability of risk. There has probably been more controversy about the risk of nuclear power than about any other form of energy, and this is often reflected in the conflicting opinions about its acceptability.

Ground rules for risk analysis. Before the possible ground rules for risk analysis are discussed, the concept of risk, as used here, should be defined more clearly. Risk is occasionally confused with consequences. An event in an energy system, such as an accident at an oil refinery or a hydroelectric dam failure, can have health consequences such as death, injury, or disease. The event in question may have a probability of happening once in x years. The probability multiplied by the number of deaths, injuries, or illnesses (the consequences) can be taken as a measure of the risk, or risk index.

Although no set of ground rules can be complete, the following points have at least been considered in recent risk analyses. Some of them may appear obvious or trivial, but they are nonetheless useful in organizing information:

1. Final outputs should be the same. Most energy risk studies have assumed that electricity would be the final product, but in principle the technique can be applied to other forms of energy, such as liquid fuels or mechanical motion. For example, the risk due to gasoline production can be compared with that of methanol, or that of producing heat by different methods can be contrasted. On the other hand, the risk from gas-fired electricity should not be compared with the risk from heating homes with gas unless appropriate efficiency factors are considered.

2. The risk should be expressed as a quantity of damage to health per unit of energy, such as joules, kilowatt-hours, or megawatt-years. Any unit can be used as long as consistency is maintained.

3. The risk should be annualized, or normalized over the lifetime of the energy system. For example, consider a coal-fired electricity station. A certain number of deaths, accidents, and illnesses are attributable to building the turbines for the station, excavating the coal mines to supply it, and laying the track for the railroad which will haul the fuel. If the expected lifetime of the system is 30 years, the number of deaths, accidents, and illnesses resulting from capital construction must be divided by 30 to produce an annualized value. To this value is added the annual risk of mining coal, transporting it, operating the station, and so forth. In a simple formula, the annual risk per unit energy might be expressed as $(C/n) + A$, where C is the total risk attributable to capital construction, n is the lifetime in years of this construction, and A is the annual operating risk, both occupational and to the public. A more sophisticated model would take account of the different lifetimes of the capital components. In effect, this procedure is a method of properly allocating risk, sometimes given the name risk accounting.

4. The same time period should be used, as much as possible, for each energy system being compared. Again, this may appear obvious, but the

adoption of this rule has strong implications for risk analysis. To take a few examples, the accident rate in coal mining has decreased substantially during the last half-century in most countries. The choice of date used in an analysis can alter the calculated risk of any system using this fuel. On the other hand, the occupational death rate has remained constant or even risen for many industries in recent years.

As another example, consider nonconventional energy systems like solar, wind, and ocean thermal. (In this article, nonconventional systems are not now in widespread use; conventional systems are all others.) Their performance data (which, as will be shown below, play a part in risk calculation) are known for present conditions. Some investigators expect that this performance can be improved in the future. If this is the case, the calculated risk of nonconventional energy systems will decrease. Although either present or predicted conditions can be used in risk calculations, the choice should be made clear.

5. Environmental effects should be separated from health effects. It is true that many environmental effects are closely related to health effects. For example, air pollution from burning fossil fuels falls into both categories. However, some environmental effects of producing energy, such as forms of water pollution, do not always have an immediate health effect. Separating environmental from health effects is not meant to downplay the importance of environmental considerations. Indeed, environmental analyses, in conjunction with risk analyses, have been carried out for various energy systems. However, it is useful from the viewpoint of simplicity to separate the two effects. *See the feature article* PROTECTING THE ENVIRONMENT.

6. Public and occupational risk should be distinguished from one another, although they can later be combined to indicate the total risk. Public risk is sometimes thought of as involuntary risk, and occupational as voluntary. The origin of occupational risk is clear, but public risk can arise from a number of sources, for example, the risks of air pollution from burning fossil fuels, in transporting materials and fuels, and of accidents in certain energy systems, such as hydroelectric dam failures, releases of radioactivity from nuclear reactors, and natural gas explosions. In the view of some people, public risk is the only aspect deserving of discussion since, in a sense, occupational risk always exists in society. On the other hand, it can be contended that the loss to a nation is the same whether there is a death of a worker or of a member of the general public. These are philosophical arguments that cannot be resolved in a simple way. In any case, showing the two types of risk separately allows appropriate judgments to be made.

7. Data should be confined, if possible, to one country or continent. Data used in risk calculations are often quite variable even within one country. Extending them across international borders can make the drawing of conclusions more difficult. As an example, a recent study noted that coal mining death rates in Norway were considerably higher than those in the United States, which in turn were higher than those in Poland. International comparisons of risk can still be made; however, transnational data cannot be mixed indiscriminately.

8. Long-term effects should be handled in the same way for each system being compared. By long-term is meant those effects which can alter health over centuries or more. Some examples might be the release of carbon dioxide from burning fossil fuels, which may alter the Earth's climate; the emission of radioactive radon gas from uranium mine tailings; and the storage or disposal of wastes from nuclear reactors. It is often difficult, if not impossible, to calculate the risk of these long-term effects because of their inherent uncertainty. However, they should be considered in risk analyses wherever possible.

9. The absolute value of risk should be indicated. Some commentators have suggested that the supplementary risk of an energy system is more significant than the absolute value. The supplementary risk is that incurred occupationally beyond that in normal employment. Using this reasoning, if an oil worker had an employment risk higher than the average worker, the difference would be attributed to producing energy from oil. Most risk analysts have not used this viewpoint, although E. E. Pochin in the United Kingdom has given it consideration. It can be shown that using the concept of supplementary risk in a relative discussion merely subtracts a constant quantity from each energy system, and leaves the rankings unchanged.

Comprehensiveness of risk analysis. One object of risk analysis is to be comprehensive, ensuring that no significant contribution to risk is overlooked. In principle, risk analysis should take into account all sources, from mining raw materials to disposal of any wastes. But it can never be proved that a given analysis is comprehensive.

A diagram of some typical sources of risk is shown in Fig. 1. The relative importance of each source depends on the energy system. All energy systems require raw materials for their construction. Mining and processing these materials pro-

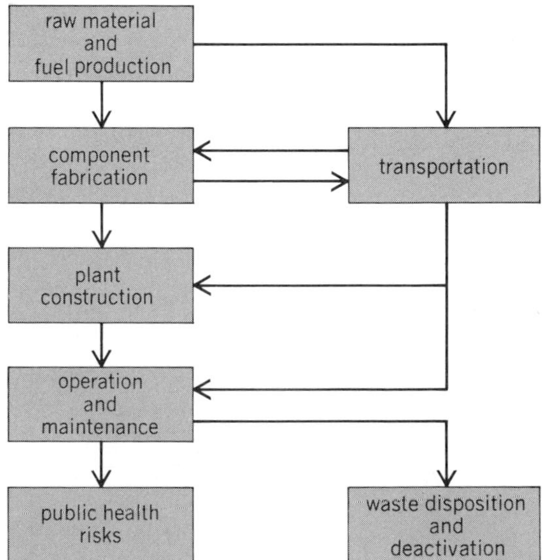

Fig. 1. Some typical sources of risk in energy production.

duce risk. Most conventional systems require fuels, although some, such as hydroelectric systems, do not. Generally speaking, nonconventional systems do not need fuel, but some, such as methanol systems, do require it.

Fabrication of components of the system can be risk-intensive, depending on the industry. This aspect of risk, along with construction and installation, is generally part of capital needs, and is annualized (as discussed earlier). Operation and maintenance risk can be substantial for some energy systems. All of the materials required for each of the risk sources mentioned so far must be transported to their destination. Transportation risk can be both occupational and public, the latter occurring at railroad crossings and on the highways.

Many energy systems pose a public health risk in addition to the public transportation risk. This risk can derive from a number of factors. Fossil fuels produce air pollution, and the public health risk due to coal and oil is primarily from this cause. In systems such as hydroelectric, natural gas, or nuclear power, there is a small but finite chance of accidents affecting the public. However, other public health risks also exist. If steel is used in construction, coal (or coke) is generally employed in the steelmaking process. The use of coal in this process produces emissions which can affect public health. This indirect risk should be attributed to the appropriate energy systems. For some nonconventional systems, such as solar space heating or wind power, which apparently require large amounts of steel per unit energy output, this risk can be substantial.

Finally, waste disposal is a risk in some systems. Considerable attention has been paid to the potential risk from nuclear waste disposal, but during the last few decades, tragic accidents involving wastes in the coal fuel cycle have occurred. *See* RADIOACTIVE WASTE MANAGEMENT.

Determining the boundaries of risk is thus a problem in this type of study. The sources shown in Fig. 1 correspond approximately to the categories used by a number of analysts. Although energy-related problems like resource depletion and the proliferation of nuclear weapons are important, they are generally not included in the quantitative aspects of risk studies, and are deserving of more detailed study.

Typical risk analyses. Tables 1 and 2 show typical risk analyses. Both tables are based on extensive compilations. Some general points can be made: (1) Both tables refer to electricity as an end product. Although coal, oil, and natural gas can be used to provide heat for homes, there has been comparatively little risk analysis of the nonelectrical uses of these resources. (2) Because of space limitations, only deaths are shown. Similar tables can be constructed for injuries due to accidents and for disease. Deaths can be assigned an arbitrary number of days lost (usually about 6000), and the days lost because of accidents and disease combined with deaths in this way. The relative ranking of energy systems in terms of this combined risk is generally similar to that produced by evaluating only deaths. (3) In the case of Table 1, data were available only for a generating station with a maximum power of 1000 MW. No system produces at a constant maximum rate, so a typical load factor (the ratio of average power to maximum power) of 0.7 was assumed. (4) Natural gas seems to have the lowest overall risk; nuclear power has the next-lowest risk. Oil and coal have substantially higher risk, with oil slightly lower. The two fossil fuels have higher occupational and public risk than either nuclear or natural gas does. (5) The ratio of public to occupational risk varies substantially for the four energy systems. For natural gas, the ratio seems to be zero. More detailed analysis has indicated that there is a small public risk, from the transmission of gas. However, the ratio is still small. This point is discussed in greater detail below. For coal and oil, the ratio is high, primarily because of the health risk of air pollution. (6) The terminology varies somewhat from one analysis to another. The term Harvesting in Table 2 is approxi-

Table 1. Risk analysis #1: Deaths associated with 1000 MW-years net electrical energy*

	Coal	Oil	Natural gas	Nuclear
Occupational				
Extraction				
Accident	0.64–1.4	0.09–0.30	0.03–0.30	0.07–0.29
Disease	0–5	–	–	0.003–0.14
Transport				
Accident	0.08–0.57	0.04–0.14	0.03	0.003
Processing				
Accident	0.03–0.06	0.06–1.4	0.009–0.014	0.004–0.29
Disease	–	–	–	0.019–0.47
Conversion				
Accident	0.01–0.04	0.014–0.053	0.014–0.053	0.014
Disease	–	–	–	0.034
Subtotal				
Accident	0.77–2.1	0.20–1.9	0.08–0.40	0.09–0.60
Disease	0–5	–	–	0.06–0.64
Public				
Transport	0.79–1.9	–	–	–
Processing	1.4–14	–	–	–
Conversion	0.1–143	1.4–140	–	0.014–0.23
Subtotal	2.3–159	1.4–140	–	0.014–0.23
Total	2.9–166	1.6–142	0.08–0.40	0.16–1.4

*From C. L. Comar and L. A. Sagan, Health effects of energy production and conversion, *Annu. Rev. Energy*, vol. 1, pp. 581–600, 1976. Numbers may not add exactly due to rounding.

Table 2. Risk analysis #2: Deaths associated with 1000 MW-years net electrical energy*

	Coal	Oil	Nuclear
Harvesting			
Accident	0.8–2.2 (–)	0.003 (–)	0.1–0.27 (†)
Disease	0–0.06 (–)	– (–)	0.003–0.006 (†)
Upgrading			
Accident	0.02–0.04 (–)	0.04–0.06 (–)	0.007 (–)
Disease	– (–)	– (–)	0.002 (†)
Transport			
Accident	1.6–5.0 (0.7)	0.06–0.07 (–)	0.003–0.012 (0.01)
Disease	– (–)	– (–)	0–0.004 (–)
Conversion			
Accident	0.01–0.09 (–)	0.01–0.05 (–)	0.013–0.017 (–)
Disease	– (0.2–36)	– (0.04–37)	0.03–0.11 (0.0003–0.013)
Waste			
Accident	† (–)	– (–)	0.0002 (–)
Disease	– (0–13)	– (–)	0.07 (0.0008–0.0019)
Subtotal			
Accident	2.4–7.3 (0.7)	0.11–0.18 (–)	0.12–0.31 (0.01)
Disease	0–0.06 (0.2–49)	– (0.04–37)	0.11–0.19 (0.001–0.015)
Total, public and occupational	3.3–57	0.15–37	0.24–0.53

*Occupational value is given first; public value is given in parentheses.
†Unknown.
SOURCE: K. R. Smith, J. Weyant, and J. P. Holdren, *Evaluation of Conventional Power Sources*, Energy and Resources Program, University of California, Berkeley, July 1975.

mately equivalent to Extraction in Table 1, and Upgrading is similar to Processing. In spite of these small differences in labels, the major steps in proceeding from ore or fuel in the ground to the final product are fairly clear. (7) There is a variation in the estimated risk, both in the entries in each table and between the two tables themselves. These variations result from a number of factors: First, there are different models of energy systems to which the estimates apply. For example, a nuclear reactor can be built in a number of ways. Second, the data may refer to somewhat different times. Third, the pollution-control equipment may vary. This is important in estimating the public risk of fossil fuels like coal or oil. And, finally, the type of required ore or fuel may vary. For example, the risk per ton of coal from strip mining is much less than that of underground mining. For these

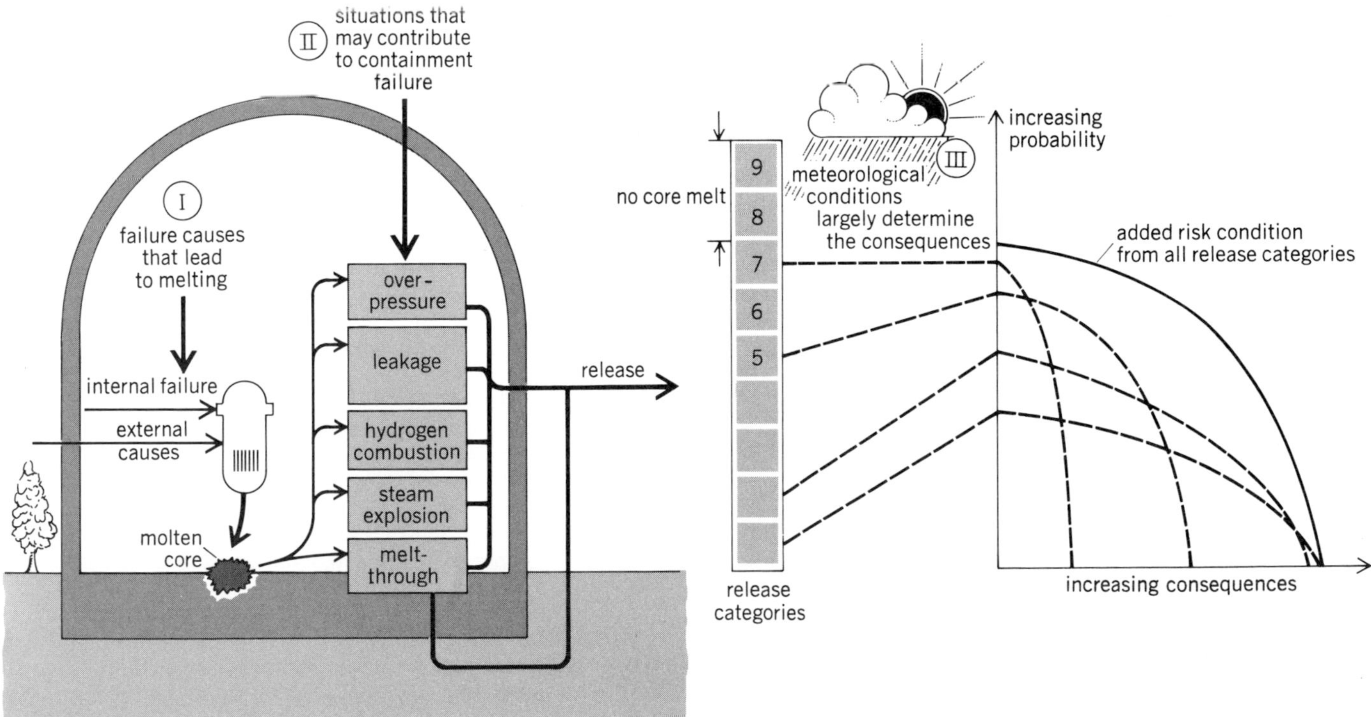

Fig. 2. Basis for calculating public risk from nuclear power, involving operating characteristics of reactors, barriers and measures for preventing radioactivity releases, and meteorological conditions. The relationship between the probability and consequences to health of reactor accidents is shown schematically. (*From Nuclear Power and Safety, Norwegian Nuclear Power Commission, 1978*)

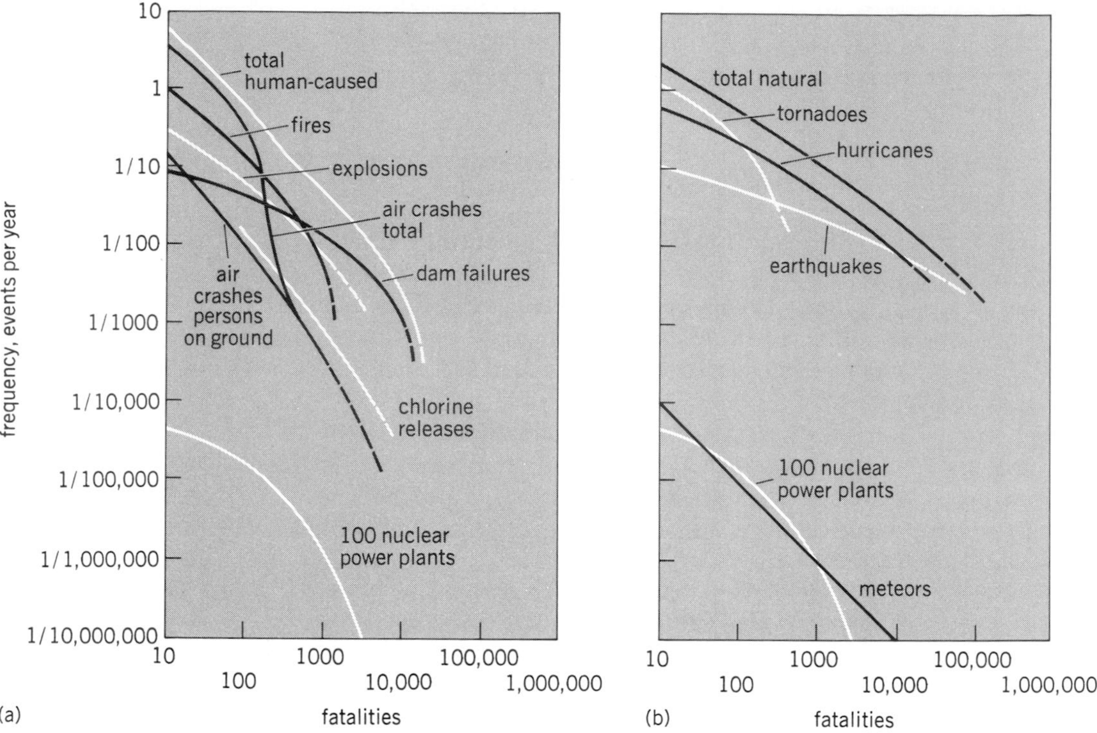

Fig. 3. Comparisons of risk from human-caused and natural events with the risk from 100 nuclear power plants in the United States. (*a*) Estimated frequency of fatalities due to human-caused events. (*b*) Estimated frequency of fatalities due to natural events. *(From Nuclear Power and Safety, Norwegian Nuclear Power Commission, 1978)*

and other reasons, there is an inevitable variation among risk analyses. However, the variations diminish in apparent magnitude when less attention is paid to the absolute values and the systems are ranked with respect to one another. *See* COAL; NATURAL GAS; PETROLEUM; STRIP MINING; UNDERGROUND MINING.

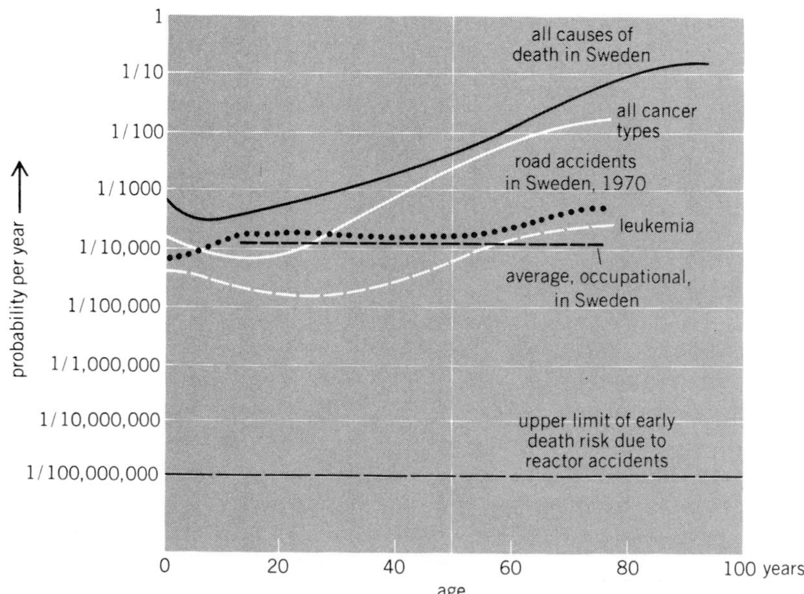

Fig. 4. Individual death risk in Sweden due to various causes. The risk is compared with the estimated early death risk due to accident in 580-MW reactor. *(From Nuclear Power and Safety, Norwegian Nuclear Power Commission, 1978)*

Nuclear public risk. There has probably been more controversy about the entries for nuclear public risk in Tables 1 and 2 than about all the other entries combined. Much of the discussion has centered on the Reactor Safety Study directed by N. C. Rasmussen (U.S. Nuclear Regulatory Commission, 1975). A predecessor, WASH-740 (U.S. Atomic Energy Commission, 1957), calculated the risk of the worst case, that is, a severe accident at a reactor. However, the average risk from nuclear power cannot be calculated on the basis of worst cases alone, although they must be included in the calculations, along with the less severe accidents.

In contrast to WASH-740, the Rasmussen report showed a spectrum of accidents, from small ones (in terms of risk) that were hypothesized to occur relatively frequently, to large ones that were expected to take place only rarely. The theoretical calculations were based on operating characteristics of reactors, barriers to prevent release of radioactivity, and meteorological effects on potential releases, as shown in Fig. 2.

The graph of probability versus consequences (or risk) shown in Fig. 2 is only schematic. Figure 3 shows the calculated values, compared with other risks of society. The frequency of events causing a given number of fatalities is plotted as a function of these fatalities. For example, it is estimated that, with 100 reactors in operation, there is about 1 chance in 1,000,000 that an accident will cause more than 1000 fatalities. The results are shown on a logarithmic scale, which diminishes differences between different types of risks. Although the results are shown as single lines, there is considerable uncertainty in much of the data. This is particularly the case for nuclear power, for which the values are theoretical, in contrast to the empirical values for the other risks. Thus, it would be appropriate to replace some of the lines with broad bands.

The calculated risk from nuclear power is shown in relation to other societal hazards in Fig. 4, as a function of an individual's age. Again, the graph is crude, and not much emphasis can be placed on the exact values. Although the nonnuclear sources of death are not directly related to energy production, they give some idea of the magnitudes involved. As an example, there is about 1 chance in 10,000 of dying of leukemia at age 55 in Sweden. Again, the results are plotted on a logarithmic scale, which diminishes differences. There are delayed deaths from reactor accidents in addition to those shown in Fig. 4, but the total probability from this source is expected to be smaller than those indicated at the top of the graph. Thus, if the calculations on which Fig. 4 is based are correct, the risk at all ages from reactor accidents is fairly small.

The key question is how reliable are the Rasmussen risk estimates? As with any theoretical risk values, this question cannot be answered with certainty. Although there has been less experience with nuclear power than with other conventional systems like coal or oil, this limited experience can be used to put bounds on the risk up to the present, if not to predict the future. It has been suggested that the risk record of nuclear power, even including the accident at the Three Mile Island (Pennsylvania) reactor in the United States in early 1979, indicates that the Rasmussen estimates do not substantially underestimate the risk. Only time can verify or disprove these contentions.

If the Rasmussen estimates as approximated in Table 1 are reasonably correct, an interesting point about nuclear power can be made: The occupational risk is much greater than the potential catastrophic risk to the public. This is contrary to the perception of some people. As mentioned above, there is sometimes a divergence between calculated risk and the way it is understood. *See* NUCLEAR POWER; NUCLEAR REACTOR.

Risk of other energy systems. In general, the methodology and assumptions outlined above can be extended to any energy system, even those about which less is known, such as solar, wind, geothermal, and fusion. The sources of risk as outlined in Fig. 1 should be applicable.

Acquisition and construction risks. The risk attributable to gathering and processing materials may be calculated as shown in Fig. 5. For each ton of steel or other material which goes into an energy system, there is an associated calculable risk of death, accident, or disease, as shown in Table 3. To find the risk for each material,

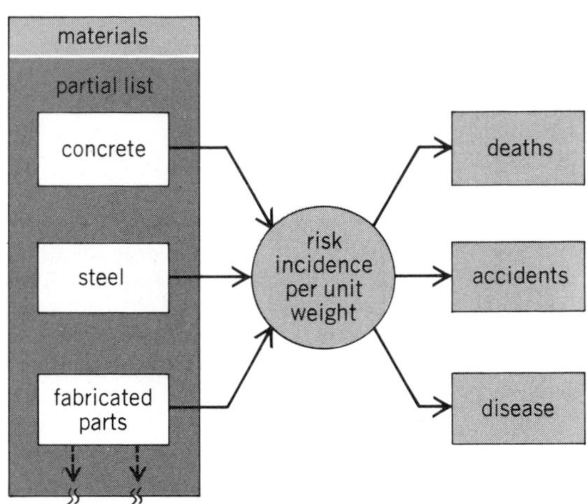

Fig. 5. Diagram of calculations of risk from acquisition of materials.

the estimated weight of concrete, steel, fabricated parts, and so forth, is multiplied by the death or accident rate for that material. For example, it takes about 10 worker-hours to produce a metric ton of steel in the United States. Suppose a given energy system required 100 metric tons of steel per unit energy output. This would correspond to $10 \times 100 = 1000$ worker-hours, or 0.5 worker-years of labor, assuming 2000 worker-hours per worker-year. From Table 3, 100 worker-years produces, on average, 0.007 deaths. Therefore 0.5 worker-year produces $0.007 \times 0.5/100 = 0.000035 = 3.5 \times 10^{-5}$ deaths.

The number of accidents and illnesses for each material is calculated in the same way. The total risk is found by adding the risk for each of the required materials. If the death rate per unit weight and the weight per unit energy output of materials 1, 2, 3, . . ., are respectively D_1 and W_1; D_2 and W_2; D_3 and W_3; . . .; the total risk is $D_1W_1 + D_2W_2 + D_3W_3 + \cdots$. As can be seen, no abstruse mathematics or computer programs are needed—merely multiplication and addition.

Account must also be taken of raw and intermediate materials necessary for the final product. For example, coal, iron ore, and other substances are required to produce steel. The risk attributed to coal, iron ore, bauxite, and so forth, should be attributed to the energy system in which the finished product is used. For example, to produce 1 ton of steel one might need 1.5 tons of coal, 1.7 tons of iron ore, and other materials.

Similar reasoning can be applied to the construction of energy facilities. A schematic diagram of the calculations is shown in Fig. 6. Each of the labor trades which are used in the construction of an energy system has an associated calculable

Table 3. Partial list of occupational health and safety statistics, per 100 worker-years

Industry	Worker-days lost, accidents	Worker-days lost, illness	Deaths
Steel	80	2.0	0.007
Aluminum	82	2.1	0.009
Hard coal	200	5.2	0.120
Cement	67	1.8	0.007

risk. The number of worker-years required for each trade is multiplied by the risk per unit time. The total construction risk is the sum of the risks associated with each trade.

The calculations in Tables 1 and 2 do not include these risks, but only those attributable to obtaining fuel and operating the system. When these additional components are included, the total risk from coal and oil changes only slightly, since it is already large. The risk from nuclear power and natural gas increases by a larger fraction, but the relative ranking of these two systems with respect to the risk from coal and oil remains the same.

The calculations suggested in Figs. 5 and 6 are of interest in terms of nonconventional energy systems, in which there often is no fuel in the ordinary sense. In addition, nonconventional systems have risk attributable to transportation of materials and operation and maintenance, as do conventional systems. As mentioned above, a further contribution to the risk of all systems that use steel is the air pollution from the emissions produced from the coal (or coke) used to smelt this steel.

Special properties of nonconventional systems. A major difference in terms of risk between conventional and nonconventional systems lies in the variability of the latter. When the Sun does not shine and the wind does not blow, the consumer still wants energy. This implies that either an energy storage system or a backup supply made up of conventional sources, or both, may be required to ensure consistency of operation. The risk attributable to any storage and backup can be included in the risk analysis of a complete nonconventional system. R. Caputo (1977) assumed a coal-fired backup, but it is possible to assume a weighted backup, made up of the present-day proportions of electricity generation. It is clear that a number of models can be used in calculating risk.

The energy in sunlight or wind is much more diluted than that in a lump of coal, a liter of oil, or a uranium fuel rod. As a result, a relatively large collecting apparatus, per unit energy output, is required for some nonconventional systems. In turn, the large amount of apparatus requires considerable material and fabrication labor. Caputo found that much greater quantities of materials were needed in building systems such as solar thermal electric (the tower-of-power concept) and solar photovoltaic than in a coal or nuclear power system. This determination has implications for some nonconventional systems since, as indicated above, the greater the material requirements, the greater the risk, all other factors being equal. *See* SOLAR CELL.

Tables 1 and 2 show risk per unit energy. Some nonconventional energy collectors, such as roof panels or small windmills, are relatively minuscule in size compared to commercial coal- or oil-fired generating plants. As a result, one might expect the risk per panel or windmill, even when all the materials and labor are accounted for, to be small—and it probably is. However, the energy production for each of these units is not large. If the associated risk is divided by the energy output in the usual way, the results may be in conflict with intuitive results. *See* SOLAR ENERGY; SOLAR HEATING AND COOLING; WIND POWER.

Analyses of nonconventional systems. Comparatively little risk analysis has been done for nonconventional systems, in part because less is known about them than about conventional technologies. In addition, the existing data tend to be variable, and thus any results of risk calculations of nonconventional systems should be viewed with caution. Caputo considered two nonconventional but centralized systems: solar thermal electric and solar photovoltaic. He found that their risk per unit energy, when appropriate backup and storage were included, was higher than that shown for natural gas-fired electric and nuclear power in Tables 1 and 2. On the other hand, the risk of these two solar systems was lower than that for coal- or oil-fired electricity, primarily because of the air-pollution effects of the latter two systems.

Inhaber (1978) found similar results. In addition to the four systems shown in Tables 1 and 2 and the two considered by Caputo, he evaluated solar space heating, wind power, methanol, ocean thermal, and one conventional system, hydroelectricity. Inhaber found the following ranking of total deaths (occupational plus public) per unit energy for 11 energy systems considered, which are listed in the order of descending risk: coal, oil, methanol, wind, solar space heating, solar photovoltaic, solar thermal electric, hydroelectricity, ocean thermal, nuclear, and natural gas. The relative ranking of coal, oil, nuclear power, and natural gas is similar to that shown in Tables 1 and 2. In between is a group of seven which has risk between that of coal and oil on the one hand and natural gas and nuclear on the other. Most systems in the middle group have about 10 deaths per 1000 MW-years net energy. Hydroelectricity and ocean thermal are somewhat lower. Since there is uncertainty in the results for nonconventional energy systems due to variability in data sources, as well as overlap between them, the numerical values are not of great significance. Although the results in terms of nonconventional energy systems may prove to be

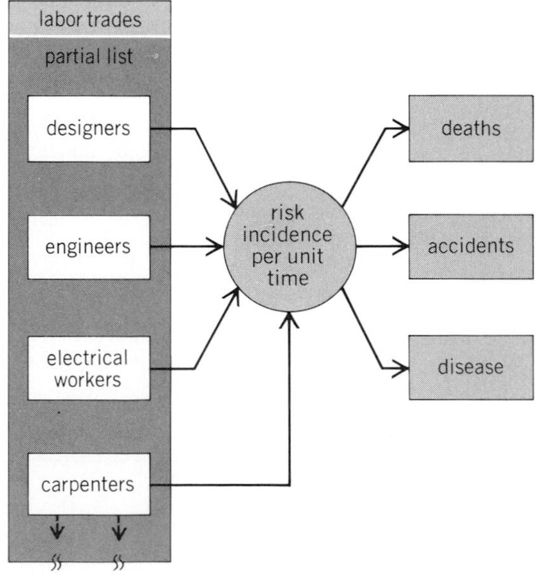

Fig. 6. Diagram of calculations of risk from construction.

surprising to some, they can be explained in terms of the large amounts of materials and construction labor that these systems apparently require, as well as the backup and storage facilities that a few of them need. Since less information is available on nonconventional systems than on their conventional counterparts, further analysis of these results is needed. *See* HYDROELECTRIC POWER; OCEAN THERMAL ENERGY CONVERSION.

There are many energy systems for which risk analyses are skimpy or have not been undertaken at all. They are geothermal, fusion, solar space satellites, coal liquefaction, coal gasification, and many others. Comparatively little is known about how certain proposed systems, such as solar satellite power (Fig. 7), would operate, and thus analysis of such systems is difficult. However, the same type of calculations that have been applied to more conventional systems could, in principle, be applied to them. As previously mentioned, risk to health is only one factor to consider when present or future energy systems are evaluated. Yet this analysis should be performed to determine what energy systems are doing *to* the population as well as *for* the population. *See* COAL GASIFICATION; COAL LIQUEFACTION; GEOTHERMAL POWER; NUCLEAR FUSION.

Fig. 7. Design concept for solar satellite power. *(From D. M. Considine, ed., Energy Technology Handbook, McGraw-Hill, 1977)*

Summary. The field of risk analysis is fairly new, but comparisons of energy systems can already be made. Fossil fuels like coal or oil tend to have a relatively high risk to human health per unit energy, as contrasted with natural gas or nuclear power. Less is known about the risk of new or nonconventional energy systems. However, recent studies have indicated that, if the entire energy or fuel cycle is considered, the risk of these systems is not negligible. As in any other human activity, nothing is for free. [HERBERT INHABER]

Bibliography: C. L. Comar and L. A. Sagan, Health effects of energy production and conversion, *Annu. Rev. Energy*, 1:581–599, 1976; Environmental Protection Agency, *Accidents and Unscheduled Events Associated with Non-Nuclear Energy Resources and Technology*, Rep. EPA-600/7-77-016, February 1977; H. Inhaber, *Risk of Energy Production*, Atomic Energy Control Board, Ottawa, Rep. AECB 1119, 1978; H. Inhaber, Risk with energy from conventional and non-conventional sources, *Science*, 203:718–723, Feb. 23, 1979; Norges Offentlige Utredninger, *Nuclear Power and Safety*, State Printing Office, Oslo, Norway, Rep. Nou 1978:35C, 1978; W. Ramsey, *Unpaid Costs of Electrical Energy: Health and Environmental Impacts of Coal and Nuclear Power*, 1979; K. R. Smith, J. Weyant, and J. P. Holdren, *Evaluation of Conventional Power Plants*, Energy and Resources Program, University of California at Berkeley, Rep. ERG 75-5, July 1975; C. Starr et al., *Public Health Risks of Thermal Power Plants*, University of California at Los Angeles, Rep. UCLA-ENG-7242, May 1972; U.S. Atomic Energy Commission, *Comparative Risk-Cost-Benefit Study of Alternative Sources of Electrical Energy*, Rep. WASH-1224, December 1974; U.S. Nuclear Regulatory Commission, *An Assessment of Accident Risks in U.S. Commerical Nuclear Power Plants*, Rep. WASH-1400, October 1975.

Energy Consumption

John C. Fisher

People have adapted energy to a wide range of personal and industrial uses. The most significant personal uses are for cooking, comfort heating and cooling, illumination, transportation, hot water, refrigeration, and communication. These uses extend far beyond the bare essentials for life, and they provide increasingly for comfort and convenience. The most significant industrial uses are for heat and power.

Nonindustrialized societies still are heavily dependent on the traditional energy sources —local solar energy that is made available through the agencies of food, work-animal feed, nonmineral fuels (wood, dung, and agricultural wastes), wind power, and direct waterpower. Energy consumption per person is very small, only a few times the food energy required to sustain life.

In contrast, industrialized societies use large quantities of fossil fuel (coal, oil, and natural gas) and electricity, and consumption of energy per person is as much as a hundred times the energy contained in food. Figure 1 illustrates the tremendous per capita consumption of fossil fuels and hydropower in the industrialized nations compared with the rest of the world. These two forms of energy provide a twelvefold increase in energy for the industrialized regions, compared with a twofold increase for the nonindustrialized regions. When one speaks of energy in an industrialized society, one ordinarily refers only to energy for heat, light, power, and communication, leaving aside the energy content of food. In keeping with this custom, food energy will not be further considered in this article.

Fairly accurate records exist for the overall energy consumption of the United States, particularly in recent decades, since it is known how much coal, oil, natural gas, hydropower, nuclear power, and other forms of energy are consumed each year. But the records are incomplete with respect to energy consumption for most specific purposes or end uses. These are good records for some, for example, energy in the form of gasoline

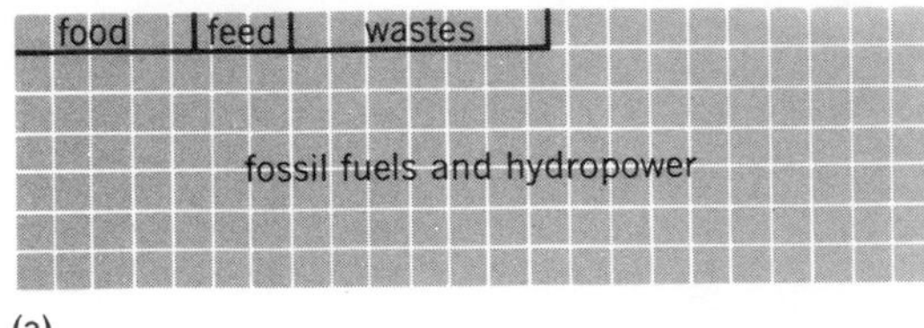

(a)

food | feed | wastes | fossil and hydropower

(b)

Fig. 1. Per capita consumption of energy in 1970 for (*a*) industrialized regions (30% of world population) and (*b*) nonindustrial regions (70% of world population). Each square represents 1,000,000 Btu per person per year. (*From J. C. Fisher, Energy Crises in Perspective, 1974*)

for automobiles. Suppliers know how much energy in the form of electricity is delivered to each home, but the proportions that are used for cooking, heating, light, refrigeration, television, and other purposes can only be estimated.

Table 1 shows the approximate pattern of energy consumption in the United States during the mid-1970s. Energy can be transformed to electricity before it is used, as for lighting and for powering machine tools in industry. Wherever this is done, the table shows the energy content of the fuel required to make the electricity. There is no doubt that the major features of the nation's energy consumption pattern are correctly portrayed in the table, but individual percentage entries are probably not accurate to better than one percentage point. Wherever there is a dash in the table, the energy consumption for that segment of the economy is estimated to be less than ½% of the nation's total consumption.

THE FLOW OF ENERGY

A number of different sources have provided significant energy inputs to the United States at one time or another. In approximate order of their historical development, they are:

Solar energy: Conversion via fuel wood, work-animal feed, wind power, waterpower.

Fossil fuel: Combustion of coal, petroleum, natural gas.

Nuclear fuel: Fission of uranium.

Other sources of potential significance for large-scale energy production include:

Solar energy: Conversion via new technologies.

Fossil fuel: Combustion of hydrocarbons from oil shale, tar sand.

Nuclear fuel: Fission of thorium.

Sources of energy are judged to be potentially significant where the available quantities are large and where technological and economic considerations show that costs are competitive or close to competitive. Other potential sources such as tidal power, geothermal power, fusion power, and trash combustion are likely to be of less significance for large-scale energy production because of limited availability or because of economic or technological barriers, although they may have limited applications at special locations or in special situations.

Some energy sources are more abundant than others. Solar energy is dilute, but large in magnitude and unlimited in time. Fossil fuels are concentrated and inexpensive to recover, but can become exhausted after several centuries. Nuclear fuels are practically inexhaustible, particularly if breeder reactors are able to utilize the common isotopes of uranium and thorium. Broadly speaking, for the industrialized societies of the world, the years of significance for fuel wood, work-animal feed, and wind power have passed; and the years of significance for nuclear fuels are just beginning (Table 2).

SHIFTS IN EMPHASIS

The 1970 emphasis on fossil fuels in the United States represents a strong shift from the 1850 emphasis on wood (for heat) and animal feed (for farm work and transportation) (Table 3). The years from 1850 to 1970 saw five major substitutions of the new energy forms for the old, as shown in Fig. 2. Fuel wood, used primarily for heating, was largely replaced by coal between 1850 and 1910. Since 1910, coal has been progressively replaced by fluid hydrocarbons (gas and oil). Work-animal feed, used primarily for motive power in transportation and on farms, was partially replaced by railroad coal in the late 1800s and early 1900s. Then, as the country adopted automobiles and tractors and as railroads converted to oil, both animal feed and

Table 1. Consumption pattern of energy for significant end uses, as an approximate percentage of total consumption, in the United States during the mid-1970s*

End use	Segment of the Economy: Industrial	Residential and personal	Commercial and public	Total
Transportation	1	16	9	26
Comfort heat	2	11	7	20
Process steam	16	—	—	16
Direct heat	11	—	—	11
Electric drive	9	—	—	9
Lighting	1	1	3	5
Hot water	—	3	1	4
Air conditioning	—	1	2	3
Refrigeration	—	1	1	2
Cooking	—	1	—	1
Electrochemistry	1	—	—	1
Other (mostly electric)	—	1	1	2
Total	41	35	24	100

*Based on a study by the Stanford Research Institute updated by a task force of the National Academy of Engineering, and on data obtained by the U.S. Bureau of the Census.

Table 2. Sources of energy for the United States in 1970

Source	Percent
Fossil fuel	
Coal	18
Petroleum products	49
Natural gas	25
Solar energy	
Hydroelectricity	4
Nuclear energy	
Uranium	4
	100

Table 3. Sources of energy for the United States in 1850

Source	Percent
Solar energy	
Fuel wood	64
Work-animal feed	22
Wind and water	7
Fossil fuel	
Coal	7
	100

railroad coal were largely replaced by gasoline and other distillate motor fuels in the years 1920–1950. Direct wind power and waterpower were replaced by hydroelectricity in the years 1890–1940.

A final substitution, not shown in Fig. 2, is a steady increase in the proportion of energy converted to electricity prior to its ultimate consumption. The percentage of primary energy input converted to electricity grew steadily from about 9% in 1925 to over 30% in 1978 (Fig. 3). It is important to keep in mind that most of the growth in electric energy consumption has resulted from this substitution of electric energy for other forms of energy. Waterpower used to be harnessed directly to factory machinery by waterwheels, pulleys, and belts, but now it is harnessed indirectly through electricity. Hydroelectric power has substituted for direct waterpower. Fuels used to be burned on the users' premises for illumination, stationary work, and heat, but increasingly fuels are burned off the users' premises in electric power plants.

Energy flows through the United States economy from the sources shown in Fig. 2 to the end uses in Table 1. The relative proportions of the major energy flows are illustrated in Fig. 4. The nation's energy sources are converted for useful applications by means of various energy conversion facilities. These include furnaces, heaters, and stoves for generating heat, internal combustion engines for generating power, and steam engines and other heat engines for generating power in electric power plants. In the process of conversion there is a flow of unavailable energy in the form of low-temperature heat which is lost up stacks and chimneys and is also lost in the conversion of high-temperature heat to mechanical power. *See* ENERGY FLOW.

QUANTIFICATION OF ENERGY

When considering the quantitative aspects of energy consumption, miners deal in tons of coal, oil suppliers in barrels of oil, gas suppliers in cubic feet of gas, and electric utility people in kilowatt-hours of hydroelectricity. Some uniform standard of measurement is required for comparing the quantities of energy from these various sources.

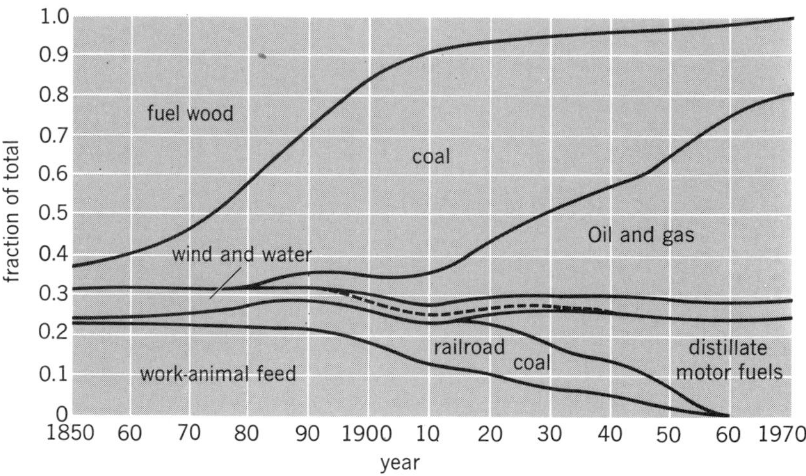

Fig. 2. Segmentation of fuel input to the United States, 1850 to 1970. (*From J. C. Fisher, Energy Crises in Perspective, 1974*)

Sources that customarily are used for the production of heat can be quantified by the amounts of heat they are capable of generating. More specifically, the numerical energy values for fossil fuels, fuel wood, and animal feed are the amounts of heat they would generate during combustion. The values for nuclear fuels are the amounts of heat generated by nuclear fission in electric power plants.

Hydroelectricity presents a special problem. Its energy content can be measured either by the amount of heat it would generate in an electric heater or by the larger amount of heat that would be required to generate the same amount of electricity in a fuel-burning power plant. Except for Figs. 4 and 14, the second of these measures—the fuel equivalent of hydroelectricity—is used in this article because it more accurately reflects hydroelectricity's economic significance.

Heat and other forms of energy can be measured in terms of British thermal units (Btu). One Btu is the amount of energy it takes to warm up 1 lb of water (approximately 1 pint of water) 1°F. In the metric system, energy is measured by joules, with a joule equal to 1 watt-second of energy. Since there are 3412 Btu in a kilowatt-hour, 1 Btu represents 1055 joules. Approximate energy contents for different energy sources, measured in Btu, are shown in Table 4.

The United States consumes so much energy

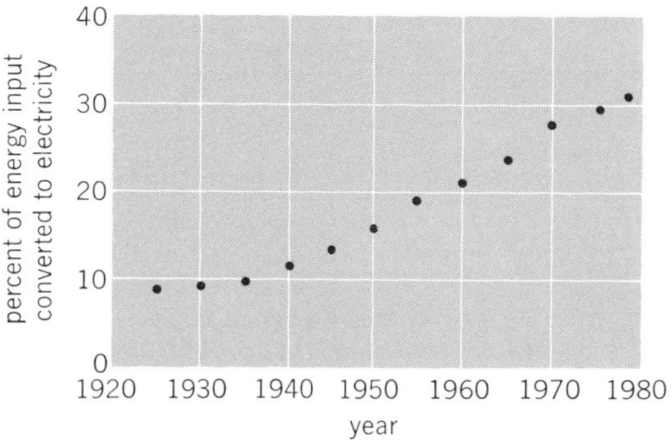

Fig. 3. Conversion of primary energy into electric energy in the United States, 1925 to 1978 (*Updated from J. C. Fisher, Energy Crises in Perspective, 1974*)

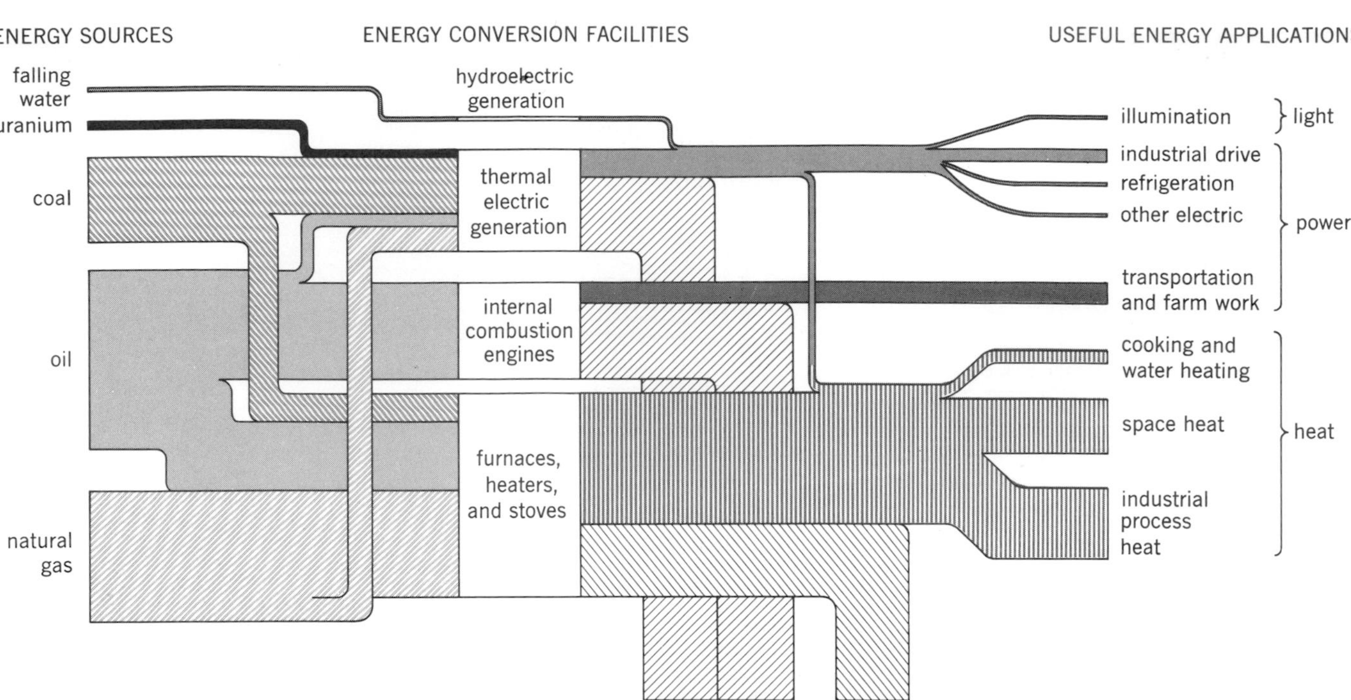

Fig. 4. Flow of energy through the United States economy in the mid-1970s, from major energy sources (left) through conversion facilities (middle) to useful applications (right), and unavailable energy at the bottom. Width of a channel is proportional to the amount of energy.

Table 4. Approximate energy contents for selected energy sources

Source	Approximate energy content
Coal	11,300 Btu per pound
Petroleum products	5,530,000 Btu per barrel
Natural gas	1,020 Btu per cubic foot
Hydroelectricity	3,412 Btu per kilowatt-hour
Fuel equivalent	10,500 Btu per kilowatt-hour

Table 5. Energy consumption in the United States in 1978

Source	Conventional quantity	Energy content, quad (=10^{15} Btu)
Fossil fuel		
Coal	624×10^6 tons	14.1
Petroleum products	6.84×10^9 bbl	37.8
Natural gas	19.4×10^{12} ft^3	19.8
Solar energy		
Hydroelectricity	302×10^9 kWhr	3.2
Nuclear energy		
Uranium	276×10^9 kWhr	3.0
Other		0.1
		78.0

Table 6. Energy consumption in the world in 1970

Source	Energy content: Quad (=10^{15} Btu)	%
Fossil fuel		
Coal	65	30
Petroleum	77	36
Natural gas	38	18
Solar energy		
Hydroelectricity	13	6
Traditional		
Wood, waste, feed	22	10
	215	100

(Table 5) that the annual amount in Btu is a very large number. To bring such large numbers down to size, it is more convenient to measure energy in quads (1 quad = 1 quadrillion Btu = 10^{15} Btu). Overall in 1978, the energy input to the United States amounted to 78.0 quads. World energy consumption in 1970 amounted to about 215 quads, including 66.8 quads consumed that year in the United States (Table 6).

WORK AND HEAT

As mentioned, the major uses for energy are for the production of work or heat, and it is important

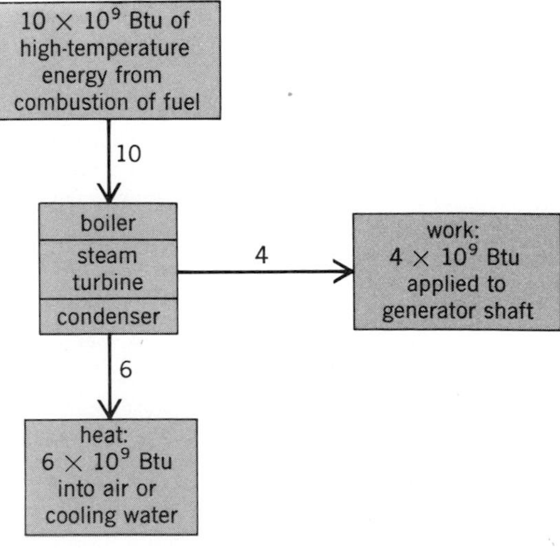

Fig. 5. Flow of energy through a modern boiler-turbine-condenser heat engine. (*From J. C. Fisher, Energy Crises in Perspective, 1974*)

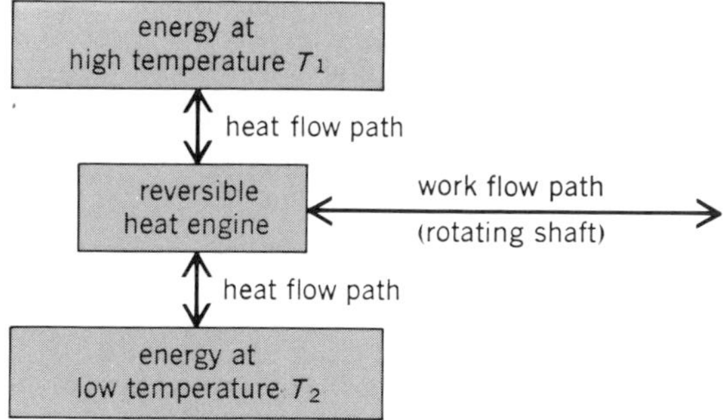

Fig. 6. Idealized reversible heat engine. *(From J. C. Fisher, Energy Crises in Perspective, 1974)*

to understand the relationships between them. The flow of energy is called work when it exerts a force. The flow of energy is called heat when it does not exert a force (for example, the flow of energy from a hot oven to a cold potato).

Because work and heat are alternate modes for the flow of energy, they can be quantified by the amount of energy that flows via each mode. Consider as an example the operation of a modern steam turbine used to drive an electric generator, shown in Fig. 5. The fuel burned in the boiler generates 10×10^9 Btu of energy each hour of operation. A portion of this energy, amounting to 4×10^9 Btu per hour, flows through the rotating turbine shaft in the form of work, where it is used to turn the shaft of an electric generator. The rest of the energy, amounting to 6×10^9 Btu per hour, is discharged into the air or into cooling water. Thus, only 40% of the energy in the fuel was actually used for the purpose for which it was intended. The engine therefore has an efficiency of 40%. *See* HEAT; WORK.

Reversible heat engines. Many engines, including jet engines, automobile engines, and steam turbines, receive energy at high temperature, transform some to work, and discharge the rest of it as heat at a lower temperature. Much study has been devoted to the potential efficiency of these engines, and the concept of a "reversible heat engine" has emerged as an idealization against which the lesser performance of real engines can be measured. Imagine a reservoir of energy at a high temperature T_1 and a second reservoir of energy at a low temperature T_2. Imagine that a reversible heat engine, with a rotatable shaft along which work can flow (Fig. 6), is in contact with both reservoirs and is able to exchange heat with both.

When utilized to generate work, a reversible heat engine draws heat from the high-temperature reservoir, delivers work along the rotating shaft, and discharges heat to the low-temperature reservoir. It is the idealization of a steam turbine power plant. Figure 7 shows what happens to a quantity Q_1 of heat that flows into the engine from the hotter reservoir: part of it, Q_2, goes to the colder reservoir and part of it, $Q_1 - Q_2$, comes out as work.

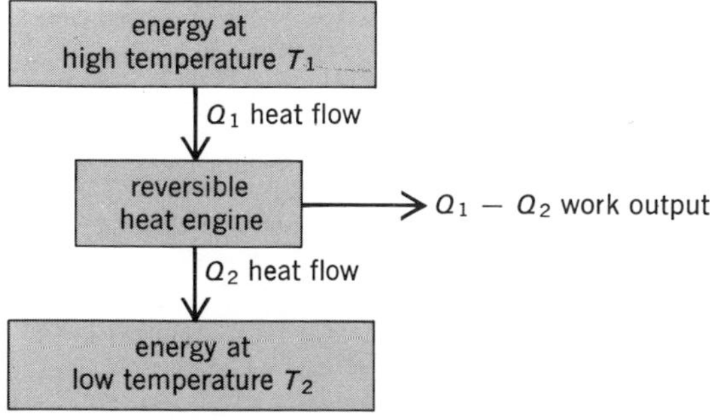

Fig. 7. Reversible heat engine used to perform work. *(From J. C. Fisher, Energy Crises in Perspective, 1974)*

Now suppose that the shaft is twisted in the opposite direction by means of some outside agency, so that work flows into the engine instead of out. The engine then draws heat from the low-temperature reservoir and delivers heat to the high-temperature reservoir. It operates as a heat pump, the idealization of an air conditioner or refrigerator. Figure 8 shows the flows of work and heat associated with the delivery of a quantity Q_1 of energy to the hotter reservoir: part of it, Q_2, comes from the colder reservoir and part, $Q_1 - Q_2$, comes in through the shaft as work. This picture is the same as the previous one except that everything is flowing in the opposite direction. This is the meaning of reversibility.

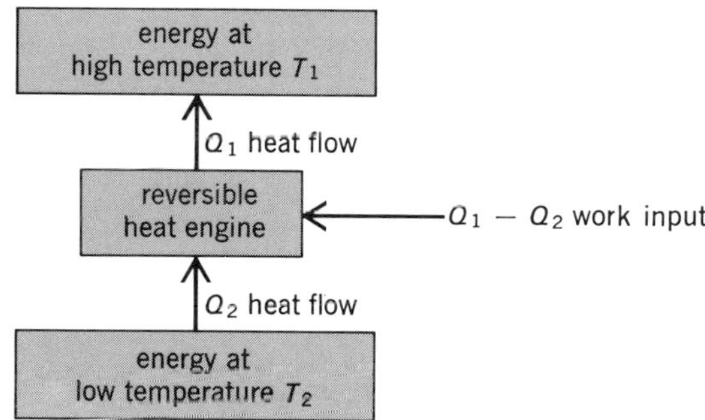

Fig. 8. Reversible heat engine used to pump heat. *(From J. C. Fisher, Energy Crises in Perspective, 1974)*

Heat engine efficiency. The efficiency of a reversible heat engine depends only on the absolute (Kelvin) temperatures T_1 and T_2, measured from the absolute zero of temperature at −273°C, as shown in Eq. (1).

$$\text{Theoretical efficiency} = \frac{T_1 - T_2}{T_1} \quad (1)$$

No practical heat engine can actually achieve this theoretical maximum efficiency. For a steam turbine where T_1 is the temperature of the steam in the boiler (540°C = 813 K) and T_2 is the temperature in the condenser where cold steam is converted back to water (50°C = 323 K), the theoretical maximum efficiency is about 60%. Modern engineering practice has achieved a respectable 40%, but there is still room for improvement.

Heat pump performance. In addition to their role in generating mechanical power, heat engines are used to pump energy from a cooler place to a warmer place. When the purpose is to cool an area, the heat engine is an air conditioner or a refrigerator. When the purpose is to warm an area, the heat engine is simply a heat pump.

The performance of a heat pump is evaluated by the coefficient of performance (COP), that is, the ratio of the energy delivered to the warmer reservoir (the desired result) to the work required to

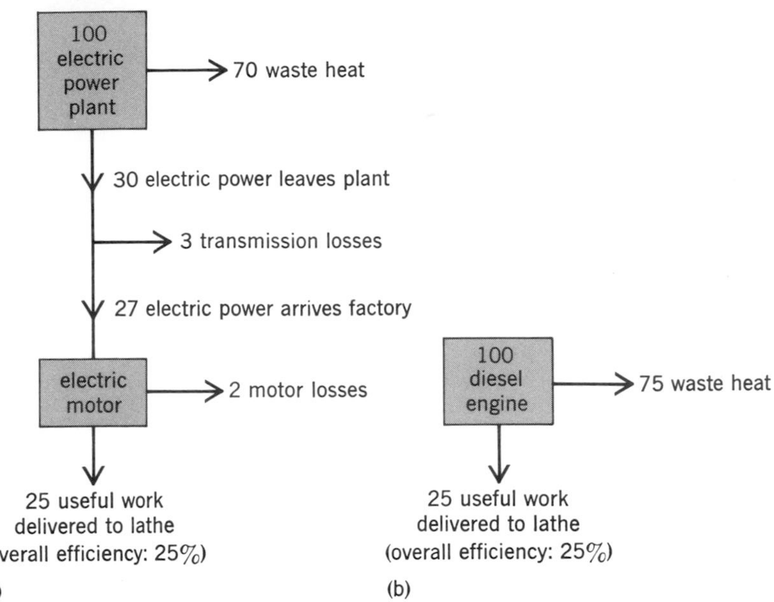

Fig. 9. Overall efficiencies of (*a*) electric power and (*b*) diesel engine power (illustrative). (*From J. C. Fisher, Energy Crises in Perspective, 1974*)

operate the pump (the necessary input). When a reversible heat engine is used as a heat pump, its COP is simply the reciprocal of its efficiency as an engine. Hence, for a reversible heat pump, COP is determined by Eq. (2). This relationship shows the

$$\mathrm{COP} = \frac{T_1}{T_1 - T_2} \qquad (2)$$

theoretical maximum amount of energy that can be pumped into a reservoir at temperature T_1 per unit work input. *See* HEAT PUMP.

As an example, consider the possibility of heating a house by means of a heat pump. Suppose that the outside air temperature (T_2) is −7°C (266 K) and the inside temperature (T_1) is 21°C (294 K). Then, using Eq. (2), the maximum theoretical ratio of heat (delivered into the house) to work (required for doing it) would be 10.5. The work put into running the pump would be multiplied more than tenfold in delivering heat to the house. For each 10.5 units of heat delivered indoors, 1.0 unit would come from the work expended in driving the heat pump and 9.5 units would come from the air outdoors. However, this is the theoretical limit. So far, a commercial heat pump working between 7 and 21°C is able to deliver only about three times as much energy in the form of heat as the energy content of the electricity that drives it.

ROLE OF ELECTRICITY

Electricity is not a primary source of energy, but rather the most highly refined form of energy. There is no alternative to electricity for some purely electrical and electronic end uses of energy. But for most other end uses of energy, consumers have a choice of burning fuel on their own premises or of utilizing electricity generated in an electric power plant, and over the years consumers have opted increasingly for electricity.

Stationary work. The flow of electric energy is equivalent to the flow of work. Work can be transmitted from one place to another by a long rotating shaft, or by a belt stretched over pulleys, or by electricity along conducting wires. Once generated, electricity can be utilized with very little additional waste.

Consider two alternatives for delivering work to a lathe in a factory: (1) burning oil in a power plant to make electricity, then transmitting the electricity to a factory where it turns an electric motor that turns a lathe; and (2) burning oil in a diesel engine that turns the lathe directly. The two alternatives are compared in Fig. 9. The comparison uses an older power plant with only 30% efficiency, which is characteristic of the average around the country. It uses a small diesel engine with only 25% efficiency, also characteristic of the average around the country. As far as generating unavailable heat is concerned, there is an approximate standoff. The power plant is more efficient than the diesel engine, but there are additional losses in transmission and in the electric motor that tend to even

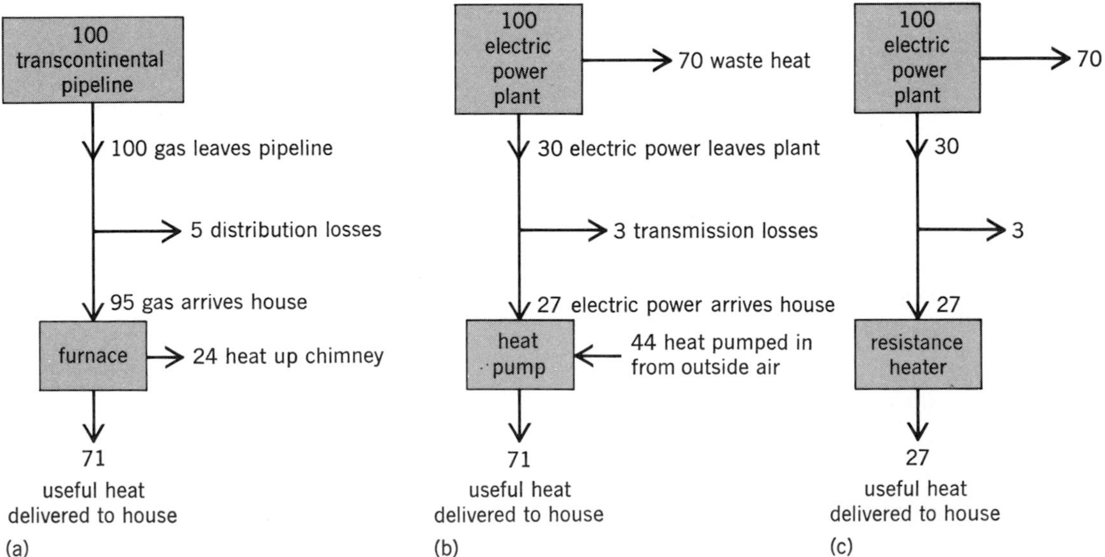

Fig. 10. Overall performance of (*a*) direct combustion of natural gas, (*b*) electric heat pump, and (c) electric resistance heating (illustrative). (*From J. C. Fisher, Energy Crises in Perspective, 1974*)

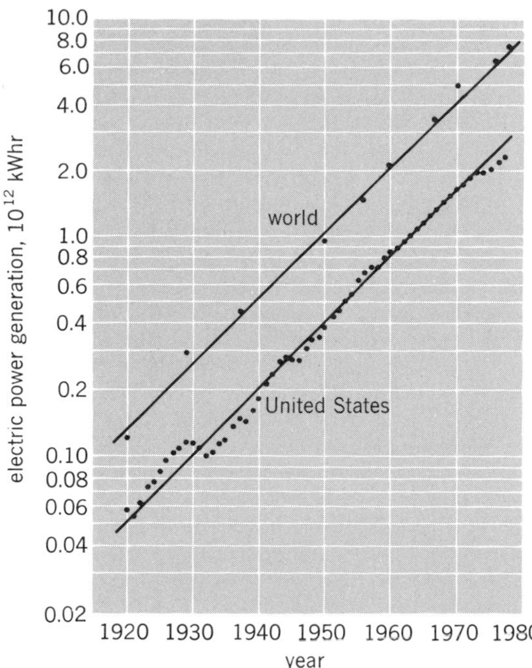

Fig. 11. Generation of electric power over the past half century, for selected years worldwide and annually for the United States. (*Updated from J. C. Fisher, Energy Crises in Perspective, 1974*)

Table 7. Approximate electrification of the United States in 1968

End uses	Percent electrified	Percent of United States energy	Percent of United States electricity
Electrified			
Electric drive			
Lighting			
Air conditioning	nearly		
Refrigeration	100%	22.0	84.4
Electrochemistry			
Other			
Heating	8% average		
Clothes drying	70%	0.4	1.0
Cooking	40%	1.4	2.1
Hot water	38%	4.2	6.2
Direct heat	6%	11.0	2.6
Comfort heat	5%	20.8	3.3
Process steam	—	14.6	—
		52.4	15.2
Transportation			
All forms	½%	25.6	0.4

things up. Hence electrification of industrial drive does not increase or decrease the amount of unavailable or waste heat associated with powering industry, but merely shifts its location from factories to power plants.

Overall cost. From the standpoint of overall cost, the comparison favors electricity. Fuel is cheaper for the power plant, and per unit of work delivered to a lathe, the cost of a power plant plus transmission system plus electric motor is less than the cost of a diesel engine. This is largely because the power plant is more fully utilized. Whereas a diesel engine might run 40 hr a week at an average 30% of its maximum rating for an overall utilization factor of about 7%, the power plant, by providing electricity to a number of users whose demands occur at different and partly overlapping times, might have an overall utilization factor of 65%, thereby achieving a much better utilization of the invested capital. Operating and maintenance costs are less for the power plant for much the same reason. Factory layout can be rearranged much more easily and cheaply when electricity provides the power. And, of increasing importance as the country turns its concern to pollution abatement, large power plants generally are able to burn fuel more thoroughly than many small engines operating independently, substantially lessening pollution.

Illumination. Illumination is another instance of substantial benefits and savings through electrification. Compared with the alternative of direct combustion, electricity gives more light at less cost with less bother, less pollution, and greater safety. Fuel is also conserved, for a gallon of oil burned in a power plant gives more light from an electric lamp than a gallon burned directly in an oil lamp.

Heating. Electric resistance heating is of particular value for producing very high temperatures, including the 5000°C plasma in a mercury-vapor lamp and even the 100,000,000°C plasmas being studied in thermonuclear fusion research. Resistance heating stays competitive down to lower temperatures such as 1540°C at which iron melts, even though such temperatures can be reached by combustion, because combustion heating tends to become less efficient as the temperature increases through loss of hot combustion products up the chimney.

At low-to-moderate temperatures, however, combustion heating tends to be more efficient because the products of combustion can give up a larger proportion of their energy as they cool down before going up the chimney. For end uses such as space heating, hot-water supply, cooking, and much industrial heating, direct combustion is inexpensive and efficient. Electricity has found only limited application to these end uses, and it is instructive to compare the two methods of electric heating—resistance heating and the electrically driven heat pump—with direct combustion to see why electricity has made so little headway (Fig. 10).

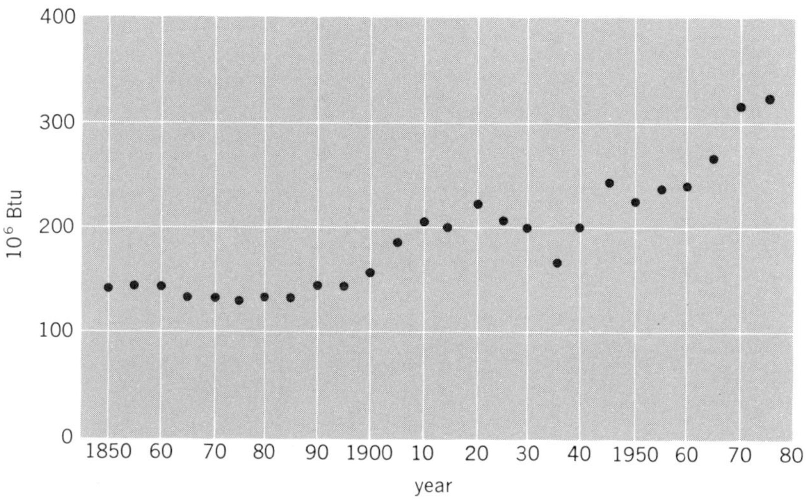

Fig. 12. Per capita consumption of energy in the United States, 1850 to 1975. 1 Btu = 1055 J. (*Updated from J. C. Fisher, Energy Crises in Perspective, 1974*)

From the standpoint of fuel conservation and overall cost, electric resistance heating is not attractive. More fuel is required, and the cost is greater. Yet where only small amounts of heat are desired and where convenience is an important consideration, resistance heating is frequently chosen. Heat pumps and direct combustion stand more or less at a draw when fuel conservation is considered, but heat pumps are more expensive than furnaces. In centrally air-conditioned buildings this does not matter as much, because a reversible heat pump can both cool and heat, but cost nevertheless remains a factor in favor of direct combustion.

Electrification patterns. In viewing the pattern of electrification of the United States, it is instructive to classify end uses according to the degree of electrification, as is done in Table 7. Three groupings emerge: completely electrified uses, heating uses, and transportation.

First consider the electrified end uses. These uses include those that are possible only through the agency of electricity, and those in which electricity has clear advantages over direct combustion.

Heating applications have not been electrified to so great an extent, and overall the penetration of electricity is small. When convenience is particularly important and energy consumption is modest, as in clothes drying, cooking, and hot-water supplying, electricity has won moderate acceptance. But when consumption is greater and cost looms larger relative to convenience, direct combustion of fuel on the premises is ordinarily the rule. For electric heating to make solid inroads against direct combustion, technological progress along two lines is essential: the efficiency of electric generation from fossil fuels must be increased, and the performance of heat pumps must be improved.

Transportation resists electrification because it is difficult to devise an arrangement of wires to supply electricity to a moving motor. If the vehicle runs on rails or some other well-defined track, electric wires can be run parallel to the track and electrification is possible. But the trend in the United States has been in the opposite direction for half a century, with vehicles of all sorts—automobiles, trucks, aircraft, ships—increasingly free to go their own ways. Electric storage batteries carried on a vehicle offer one possibility for electrification of transportation, but aside from low-speed, short-haul uses such as forklift trucks and golf carts, this method has not made much headway. For highway transportation, the development of the internal combustion engine pulled ahead of battery development in the 1890s, when electric automobiles started losing ground.

Largely through the electrification of stationary engines and of illumination, the past century has seen a steady worldwide growth in the consumption of electricity. When measured in terms of kilowatt-hours of electric energy as is the practice of the electric utility industry, United States and world consumption have been growing at an average annual rate of about 7% for half a century (Fig. 11). This growth rate corresponds to a doubling of electric energy consumption every decade. However, since the proportion of energy turned to electricity in the United States is already over 30% and cannot increase beyond 100%—it may indeed level off well short of 100%—the growth rate can be expected to slow down in future decades. *See* ELECTRIC POWER GENERATION.

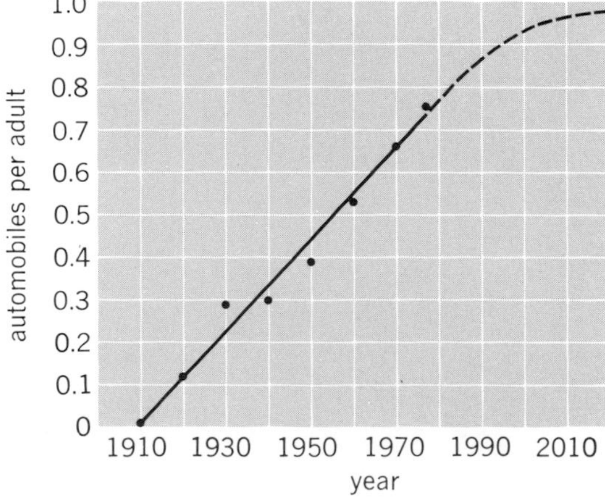

Fig. 13. Average number of automobiles per adult (age 18 and over) in the United States, 1910 to 1977 and projected to 2020.

TRENDS IN ENERGY CONSUMPTION

A hundred years ago it took about the same amount of energy to heat a house as it takes today. It took about half as much energy to feed the family horse as it now takes to power the family car. It took about the same amount of energy to cook a meal. People use more energy today, partly because they drive more and partly because they work in offices and factories instead of in open fields, but they still only use about $2\frac{1}{2}$ times as much per person.

Figure 12 shows how energy consumption per person grew between 1850 and 1975. It has been growing very rapidly in recent years. If this growth were to continue, the supplying of the required coal, oil, gas, and uranium would create a strain. The supply problem would not be so serious if energy consumption per person were to level off.

Personal automobile driving is likely to level off by the time every adult has a car to drive. Figure 13 shows how the average number of cars per adult has increased from practically nothing in 1910 to about 0.75 car per adult in the 1970s.

Job-related energy consumption has gone up as more factories and offices have been built. The fraction of the population employed in factories and offices amounted to only about 10% in 1850, but it rose to about 30% a hundred years later. Since 1960 it has risen to about 36% as more and more women have taken jobs outside the home. This trend has a natural limit at about 45% of the population, when everybody of working age will have a job in an office or factory. Growth of the nonfarm labor force will slow down to match overall population growth, and growth of job-related energy consumption will tend to do the same.

As affluence increases, partly through more jobs per family, more energy tends to be consumed in the home, mostly for hot water and for comfort heating and cooling. When these basic energy needs are met, energy consumption in the home rises more slowly with increased affluence.

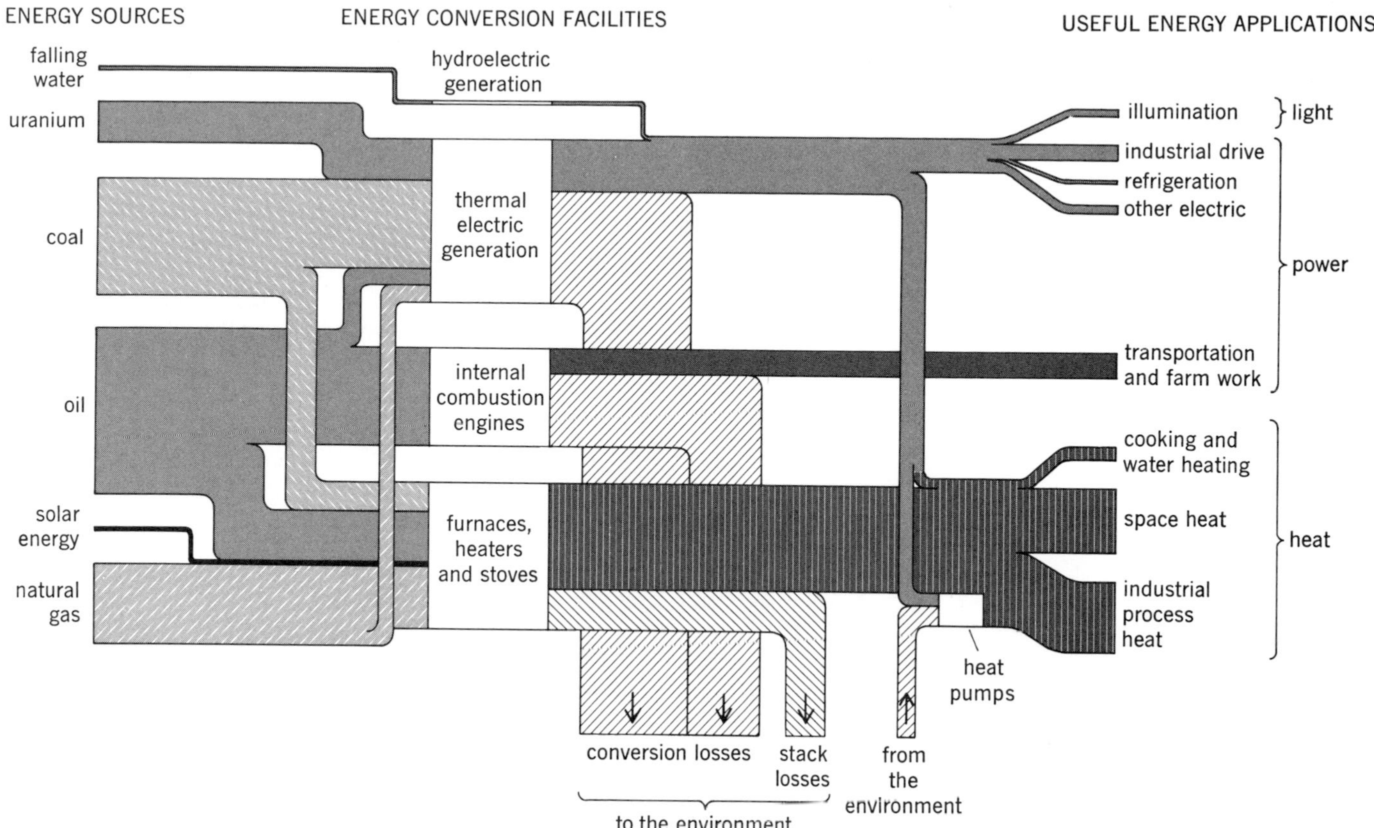

Fig. 14. Possible flow of energy through the United States economy in the 1990s.

It is anticipated that overall per capita energy consumption may level off. The conservation movement is a welcome expression of people's desire to limit and control energy consumption; indeed, the vitality of the movement may be a symptom, as much as a cause, of the growing achievement of sufficient energy for personal use in society.

The pattern of energy consumption may change in still other ways over the next several decades. Progressive electrification of energy usage is likely to continue. This is the best way to make use of nuclear energy, and improved technology can be expected to increase the efficiencies of electric power generation and application so that electricity will be chosen more often over direct fuel combustion. There is a potential for limited use of solar power, primarily for supplying hot water and for comfort heating.

A possible future pattern for the flow of energy through the United States economy is shown in Fig. 14. Compared with the present as shown in Fig. 4, uranium and coal may provide more of the energy. The efficiency of conversion facilities may improve. Heat pumps, by drawing heat from the air, may augment the effectiveness of electrical heat. More efficient use of energy, as projected in Fig. 14, combined with a leveling off of per capita energy consumption and slower population growth, will tend to moderate the nation's overall energy consumption.

[JOHN C. FISHER]

Bibliography: J. C. Fisher, *Energy Crises in Perspective*, 1974; Stanford Research Institute, for the Office of Science and Technology, Executive Office of the President, Washington, D.C., *Patterns of Energy Consumption in the United States*, January 1972; Task Force on Energy of the National Academy of Engineering, *U.S. Energy Prospects: An Engineering Viewpoint*, 1974; U.S. Department of Energy, Energy Information Administration, *Annual Report to Congress 1978*, vol. 2: *Data*.

Outlook for Fuel Reserves

M. King Hubbert
David H. Root

The significance of energy in human affairs can best be appreciated when it is realized that energy is involved in everything that happens on the Earth—everything that moves. The Earth is essentially a closed material system composed of the naturally occurring 92 chemical elements, all but a minute fraction of which are nonradioactive and hence obey the rules of conservation of matter and nontransmutability of the elements of classical chemistry. Into and out of the Earth's surface environment there occurs a continuous influx, degradation, and efflux of energy in consequence of which the mobile materials of the Earth's surface undergo either continuous or intermittent circulation. In addition, there are certain large chemical, thermal, and nuclear stores of energy within minable or drillable depths beneath the Earth's surface.

EARTH'S ENERGY SYSTEM

This total energy system of the Earth's surface is depicted graphically in Fig. 1. The horizontal bar near the bottom of the chart represents the surface of the Earth, below which are the energy stores of the fossil fuels and of geothermal, gravitational, and nuclear energy. The upper part of the chart is an energy flow diagram. The main energy influxes into the Earth's surface environment are three: the solar radiation intercepted by the Earth's diametral plane; tidal energy derived from the combined potential and kinetic energy of the Earth-Moon-Sun system; and terrestrial (especially geothermal) energy from inside the Earth. The magnitudes of these three inputs are: solar, $174{,}000 \times 10^{12}$ thermal watts; geothermal, 32×10^{12} thermal watts; and tidal, 3×10^{12} thermal watts. Thus, it is seen that the rate of energy influx from the Sun is roughly 5000 times the sum of the other two.

Of the solar power influx, about 30%, the albedo, is reflected and scattered into outer space as short-wavelength visible radiation. The remaining solar-energy flux of approximately $120{,}000 \times 10^{12}$ thermal watts, and the tidal and geothermal sources, are effective in terrestrial processes. With one small ex-

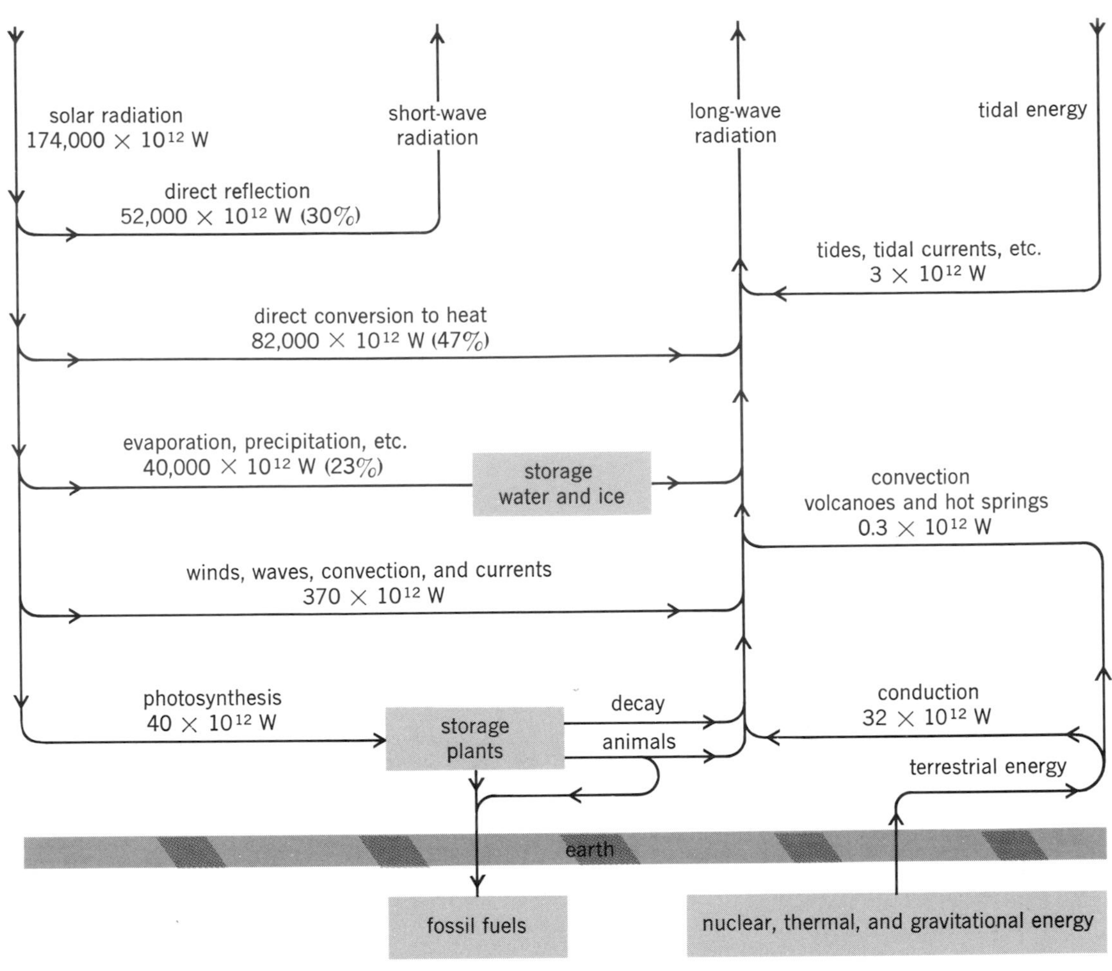

Fig. 1. Energy flow-sheet for the Earth. (*From M. K. Hubbert, U.S. energy resources: A review as of 1972, pt. 1, in A National Fuels and Energy Policy Study, U.S. 93d Congress, 2d Session, Senate Committee on Interior and Insular Affairs, ser. no. 93-40 (92-75), 1974*)

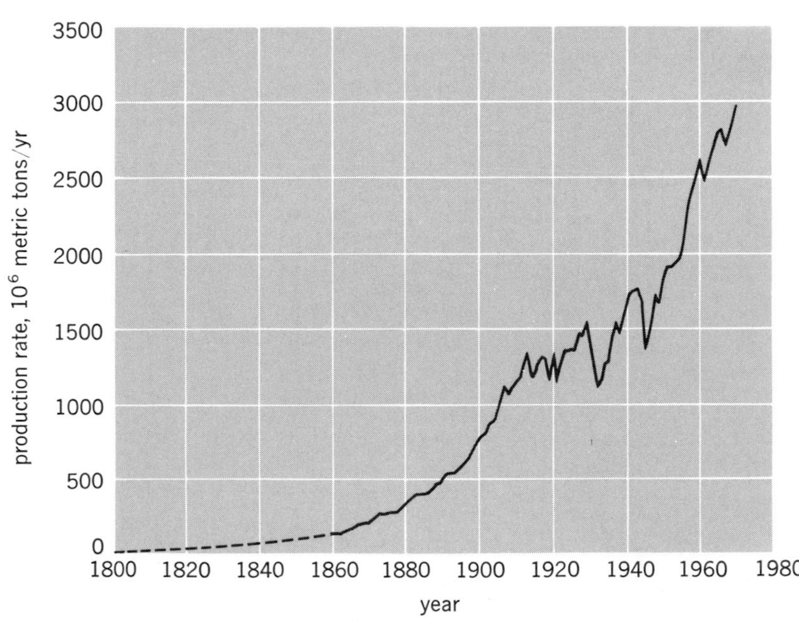

Fig. 2. World production of coal and lignite. Annual statistics are difficult to assemble for years prior to 1860 and have been estimated based on 2% average growth rate during the preceding 8 centuries. (*From M. K. Hubbert, U.S. energy resources: A review as of 1972, pt. 1, in A National Fuels and Energy Policy Study, U.S. 93d Congress, 2d Session, Senate Committee on Interior and Insular Affairs, ser. no. 93–40 (92–75), 1974*)

ception, the energy from all of these sources undergoes a series of transformations and degradations until it becomes heat at the lowest environmental temperature, after which it leaves the Earth as low-temperature thermal radiation.

The greater part of this energy flux serves to warm the atmosphere, the oceans, and the ground, and to produce atmospheric, oceanic, and hydrologic circulations. Of particular significance, however, is the 40×10^{12} W of solar power which is captured by the green leaves of plants and which by the process of photosynthesis drives the reaction whereby the inorganic compounds H_2O and CO_2 are synthesized into carbohydrates in which the solar energy becomes stored chemically. This then becomes the basic energy source for the physiological requirements of the entire plant and animal kingdoms, including the human species. *See* SOLAR ENERGY.

Nearly all the plant and animal material decays by oxidation and returns to its original constituents, H_2O and CO_2, at the same average rate as it is formed, and the stored energy is released as heat. The small exception pertains to the minute quantities of biologic materials which become deposited in peat bogs or other oxygen-deficient environments where complete oxidation is impossible and the energy of the material is preserved.

This process has been occurring during the last several hundred million years of geologic time, and the accumulated organic debris, after burial under great thicknesses of sedimentary sands, muds, and limes, has been transformed into the Earth's present supply of fossil fuels. *See* FOSSIL FUEL.

FOSSIL FUELS

The basic energy for the physiological requirements of the human species—its food supply—is obtained from the photosynthetic channel. However, during the last 2,000,000 years or so, the ancestors of the present human species have been progressively tampering with the Earth's energy system. Initially this consisted in the use of tools and weapons, and clothing and housing, whereby ever-larger fractions of the energy of the photosynthetic channel could be converted to human uses. Later, the ancient Egyptians, Greeks, and Romans began using the channel of wind power, and the Romans that of water power. This made possible a continuous increase in the human population, both in areal density and in geographical extent, but only a slight increase in the energy use per capita. *See the feature article* ENERGY CONSUMPTION.

A large increase in the energy per capita was not possible until exploitation of the large, concentrated quantities of energy stored in the fossil fuels was begun. The exploitation of coal as a continuing enterprise began in northeast England near Newcastle-upon-Tyne about 900 years ago; and the production of petroleum, the second major fossil fuel, was begun in Rumania in 1857 and in the United States in 1859.

World production. A graph of the rate of world production of coal is shown in Fig. 2. Scattered statistics exist to show that the cumulative production by 1860 was about 7×10^9 metric tons. Cumulative coal production through 1976 amounted to 157×10^9 metric tons. Of this, the amount of coal produced since 1943 exceeds somewhat all of the coal produced during the preceding 9 centuries.

During the period from 1860 to World War I, annual coal production increased steadily at an average growth rate of 4.2% per year, with a doubling period of 16.5 years. From 1914 to 1945 the growth rate was only about 0.8% per year. Since 1945 it has been at an intermediate rate of about 3% per year.

World production of crude oil from 1880 through 1978 is shown in Fig. 3. Since 1900 the annual production has grown at a uniform rate of almost 7% per year, with a doubling period of about 10 years. At such a growth rate the cumulative production also doubles in about 10 years. In fact, more than half of the crude oil produced prior to 1979 was produced in the 11 years 1968–1978; that is, 208×10^9 bbl out of a total of 405×10^9 bbl. Since 1973 the growth rate in production has slowed appreciably, averaging only 1% per year.

In terms of their energy contents as measured by the heats of combustion, the contribution of crude oil as compared with that of coal was barely significant until about 1900. Subsequently, the energy contribution of crude oil increased more rapidly than that of coal, and became greater than that of coal by 1967. Were the additional energy contributions of natural gas and natural-gas-liquids to be added to that of crude oil, the energy of

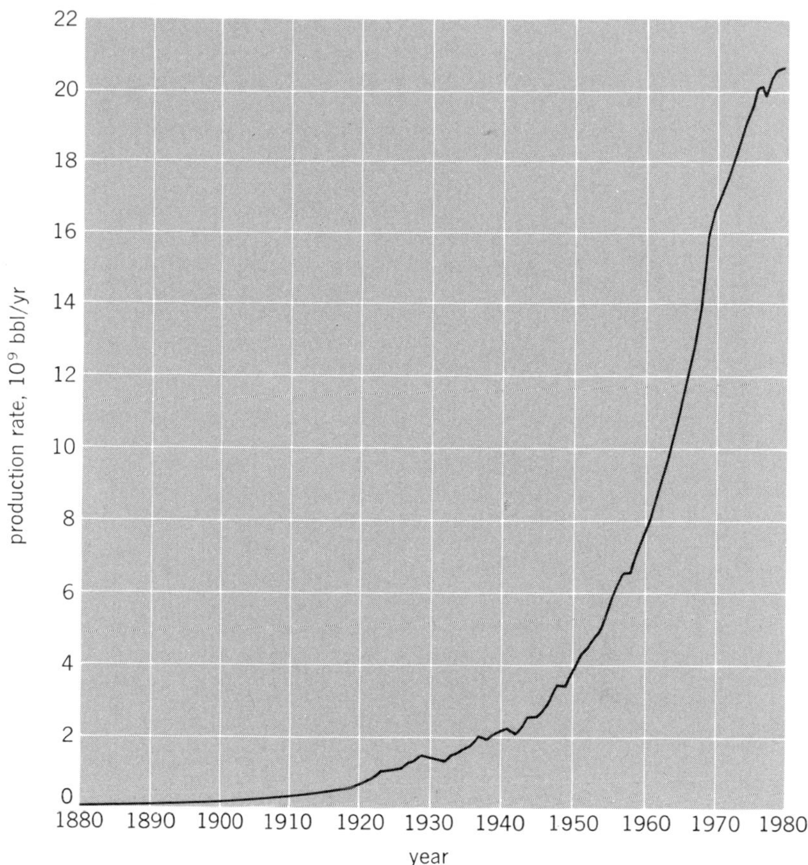

Fig. 3. World crude oil production. (*From M. K. Hubbert, U.S. energy resources: A review as of 1972, pt. 1, in A National Fuels and Energy Policy Study, U.S. 93d Congress, 2d Session, Senate Committee on Interior and Insular Affairs, ser. no. 93–40 (92–75), 1974*)

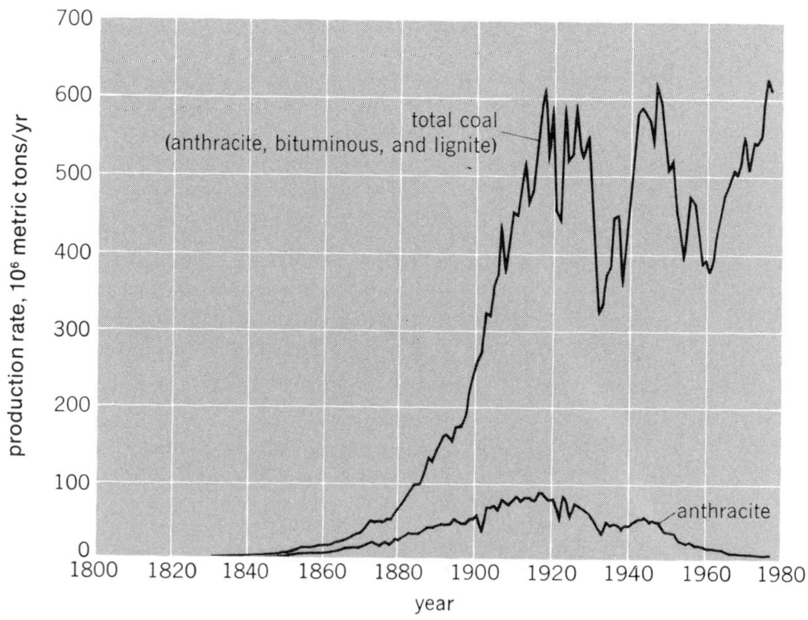

Fig. 4. United States production of coal and lignite. (*From M. K. Hubbert, U.S. energy resources: A review as of 1972, pt. 1, in A National Fuels and Energy Policy Study, U.S. 93d Congress, 2d Session, Senate Committee on Interior and Insular Affairs, ser. no. 93–40 (92–75), 1974*)

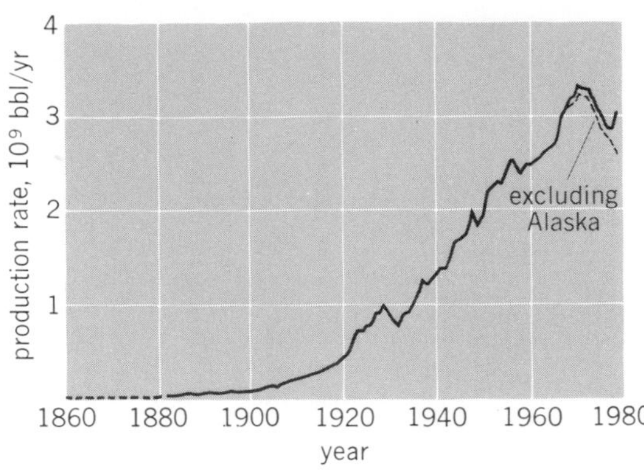

Fig. 5. United States crude oil production; figures are estimated between 1860 and 1880. (*From M. K. Hubbert, U.S. energy resources: A review as of 1972, pt. 1, in A National Fuels and Energy Policy Study, U.S. 93d Congress, 2d Session, Senate Committee on Interior and Insular Affairs, ser. no. 93-40 (92-75), 1974*)

petroleum fluids, by 1976, would represent about two-thirds and coal about one-third of the total rate of energy production from the fossil fuels.

United States production. Coal production and crude oil production in the United States are shown in Figs. 4 and 5. Coal mining in the United States began about 1820 and increased exponentially until about 1910 at a mean rate of 6.7% per year, with a doubling period of 10.4 years. After World War I United States coal production fluctuated about a constant rate of 500×10^6 metric tons per year until the late 1960s, when it once again began to grow and was 629×10^6 metric tons in 1977. See COAL MINING.

Figure 5 shows the growth in the annual production of crude oil in the United States since 1860. From 1880 to 1929 the production rate increased at a steady rate of 8.3% per year, with a doubling period of 8.4 years. After 1929 there was a drop in production during the Depression, and then a gradual slowing down in the growth rate until the peak in the production rate was reached in 1970. After that annual production has declined each succeeding year in the contiguous 48 states, although in 1977 and 1978 the decline there was more than offset by new production in Alaska.

In the United States, as in the world, the rate of energy production from crude oil, natural gas, and natural-gas liquids has increased much faster than that of coal. In 1972, of the total energy consumed in the United States, the contribution from coal reached a low of 17.3%. In 1978 coal contributed 18.6% of United States energy consumption; nuclear and water power contributed 6.7%. The remaining 74.7% was contributed by oil and natural gas.

In the light of the rates of growth in world coal and oil production, the question unavoidably arises: About how long can the rates of production of these fuels continue to increase, and over what period of time can the fossil fuels serve as major sources of industrial energy?

Various methods of analysis have been developed which, when applied individually or conjointly, are capable of giving reasonably reliable answers to these questions. These methods are: (1) the complete cycle analysis; (2) analysis of cumulative statistical data of production, reserves, and recovery; and (3) discoveries per foot of exploratory drilling. These methods are not applicable to all types of fossil fuels.

COMPLETE CYCLE ANALYSIS

One of these methods is based upon the properties of the complete cycle of production of any exhaustible and nonrenewable resource. The fossil fuels are ideal examples of such a resource. The Earth's present deposits of both coal and oil are of finite magnitudes and have required hundreds of millions of years for their geological accumulation, whereas the time required for their depletion is measurable in centuries, or at most, millennia. Although the same natural processes by which the fossil fuels were initially accumulated are still operative, the rates are so slow that no significant additions to the world's coal and oil resources can occur within the next few thousand years. Hence, the exploitation of the fossil fuels amounts to the progressive depletion of an initial stockpile.

Because of this, for any given region or for the entire world, the curve of the rate of production of coal or oil as a function of time must, during the complete cycle of production and depletion, have the following properties: The curve must begin initially at zero, and then rise until it reaches one or more maxima. Then, as the resource approaches exhaustion, the curve must gradually decline back to zero. A simplified illustration of such a complete cycle is shown in Fig. 6.

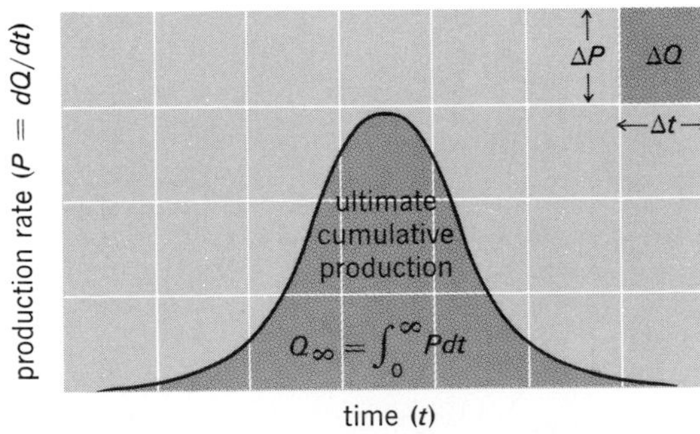

Fig. 6. Mathematical properties of arithmetical graph of production rate P versus time t for the complete production cycle of an exhaustible resource. (*After M. K. Hubbert, Nuclear Energy and the Fossil Fuels: Drilling and Production Practice, American Petroleum Institute, 1956; M. K. Hubbert, U.S. energy resources: A review as of 1972, pt. 1, in A National Fuels and Energy Policy Study, U.S. 93d Congress, 2d Session, Senate Committee on Interior and Insular Affairs, ser. no. 93-40 (92-75), 1974*)

A fundamental property of such a curve may be seen as follows. At some time t, let a small amount of time Δt (say one year) be taken on the time axis, and upon Δt as a base let a narrow vertical band be erected to the production-rate curve. The altitude of this band at time t will be

$$P = \Delta Q / \Delta t \quad (1)$$

where ΔQ is the quantity produced during Δt. Then the area of this band will be the product of its base by its altitude, or

$$\Delta A = P \Delta t \quad (2)$$

But from Eq. (1), $\Delta Q = P \Delta t$. Therefore, the area ΔA is a graphical measure of the quantity ΔQ produced during the time interval Δt. Hence, the area between the curve and the time axis to any given time is a measure of the cumulative production to that time. Likewise, the total area beneath the curve for the complete production cycle will be a measure of the total quantity Q_∞ produced during the entire cycle of production.

A graphical scale relating an area ΔA to the resource quantity ΔQ is given by the grid square in the upper-right-hand corner of Fig. 6. The quantity ΔQ represented by this area is

$$\Delta Q = \Delta t \times \Delta P \qquad (3)$$

This signifies that if a constant production rate ΔP were to be sustained for a period Δt, the quantity produced would be ΔQ.

If for a complete cycle of production the area beneath the rate of production should be n grid squares, or rectangles, then the ultimate cumulative production would be

$$Q_\infty = n\Delta Q \qquad (4)$$

Conversely, if from geological or other information the magnitude of the ultimate quantity Q_∞ to be produced in a given region can be estimated, the number of grid squares beneath the complete-cycle curve would be $n = Q_\infty/\Delta Q$, and the curve must be drawn subject to this constraint.

World coal estimates. This principle can be applied to the world production of coal. Because coal occurs in stratified seams which often crop out on the surface and may be continuous underground for tens of kilometers, reasonably good estimates of the quantities of coal in given regions can be made by surface geological mapping and a small number of deep drill holes. Such studies of coal resources have been made during the present century in all the countries of the world, and the results of these studies were compiled by P. Averitt in 1969.

Figure 7 is a graphical representation of Averitt's estimates of the recoverable coal (assuming a 50% extraction) initially present in major geographical regions of the world. The areas of the columns are proportional to the quantities of recoverable coal initially present. It will be noted that the total for the world is given as 7.64×10^{12} metric tons, and for the United States as 1.486×10^{12}, or 19% of the world total.

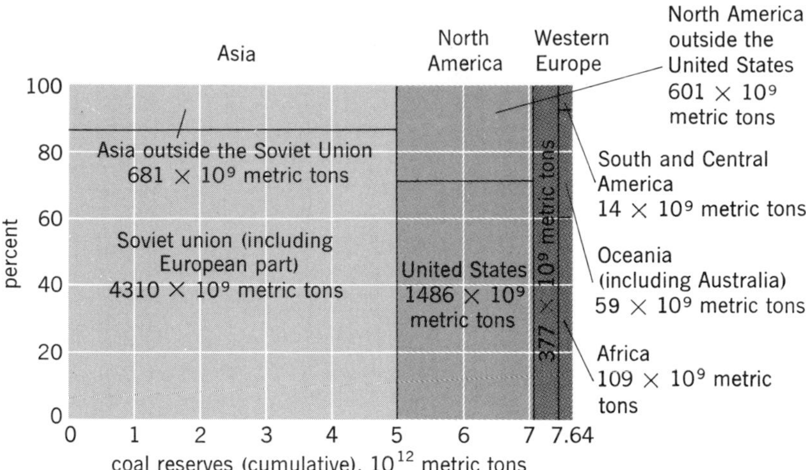

Fig. 7. Estimates of initial world resources of recoverable coal in beds 12 or more inches thick occurring at depths of 6000 ft or less. (*From P. Averitt, Coal Resources of the United States, Jan. 1, 1967, U.S. Geological Surv. Bull. 1275, 1969; M. K. Hubbert, U.S. energy resources: A review of 1972, pt. 1, in A National Fuels and Energy Policy Study, U.S. 93d Congress, 2d Session, Senate Committee on Interior and Insular Affairs, ser. no. 93-40 (92-75), 1974*)

These estimates, however, may be unrealistically high in terms of coal mining because they include seams as thin as 12 in. (0.3 m) and occurring at depths to 4000 ft and in some cases to 6000 ft (1200 and 1800 m). In view of this fact, Averitt compiled a separate estimate in 1972 of the amount of coal in the United States occurring at depths of 1000 ft (305 m) or less and in seams of not less than 28 in. (0.71 m) thick for anthracite and bituminous coal, and not less than 5 ft (1.5 m) thick for subbituminous coal and lignite. The initial amount of recoverable coal in these categories was reduced to 390×10^9 metric tons as compared with the earlier figure of 1.486×10^{12} metric tons—a reduction of 74%. Assuming that the same reduction

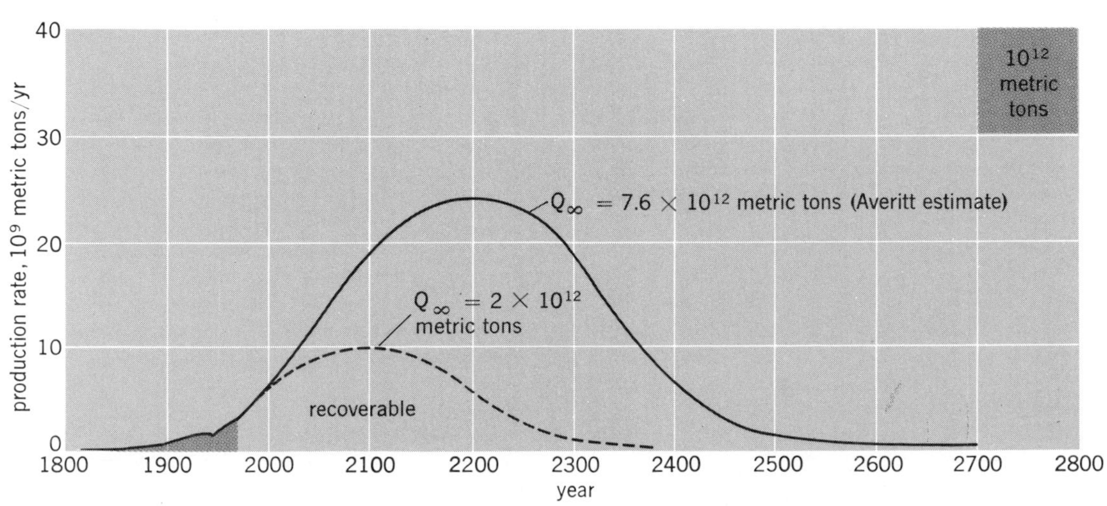

Fig. 8. Two complete cycles of world coal production based upon Averitt higher and lower estimates of initial resources of recoverable coal. (*From M. K. Hubbert, U.S. energy resources: A review as of 1972, pt. 1, in A National Fuels and Energy Policy Study, U.S. 93d Congress, 2d Session, Senate Committee on Interior and Insular Affairs, ser. no. 93-40 (92-75), 1974*)

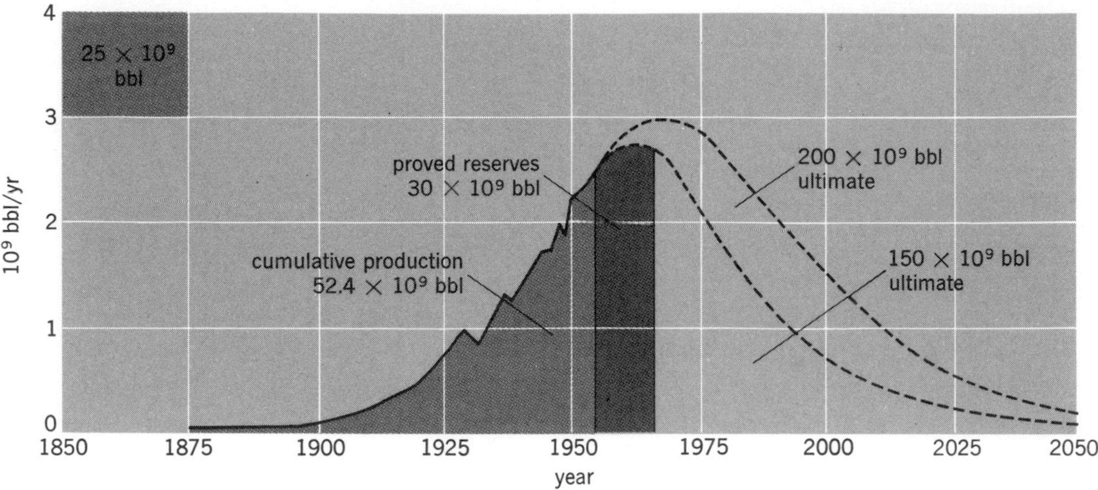

Fig. 9. Hubbert prediction of 1956 of future production of crude oil in the conterminous United States and adjacent continental shelves. (*After M. K. Hubbert, Nuclear Energy and the Fossil Fuels: Drilling and Production Practice, American Petroleum Institute, 1956*; *M. K. Hubbert, U.S. energy resources: A review as of 1972, pt. 1, in A National Fuels and Energy Policy Study, U.S. 93d Congress, 2d Session, Senate Committee on Interior and Insular Affairs, ser. no. 93-40 (92-75), 1974*)

ratio would also be valid for the world, a reduced figure of 2.0×10^{12} metric tons is obtained for world coal of the specified minimum thickness occurring at depths of 1000 ft or less.

Using Averitt's high and low figures of 7.64×10^{12} and 2.0×10^{12} metric tons for Q_∞, two complete-cycle curves for world coal production can be drawn. These are given in Fig. 8. In this figure, for one grid square, $\Delta Q = 10^{10}$ metric ton/yr $\times\ 10^2$ yr $= 10^{12}$ metric tons. Therefore, for $Q_\infty = 7.6 \times 10^{12}$ metric tons, the area beneath the complete-cycle curve will be $(7.6 \times 10^{12})/10^{12} = 7.6$ squares. For the smaller value of 2.0×10^{12} metric tons for Q_∞, the number of squares will be but 2. The curves of Fig. 8 are constructed accordingly. Obviously, the shapes are not unique, but for a fixed number of grid squares, the larger the peak rate of production the shorter the time span for the complete cycle.

In the complete cycle of coal production, long periods of time—possibly a thousand years—will be required to produce the first and last 10 percentiles of Q_∞. A much briefer time, however, will be required to produce the middle 80%. According to Fig. 8, for any likely magnitude of the maximum production rate, the date of the peak of production will probably occur within the next 100 to 200 years, and the time span for the middle 80% will

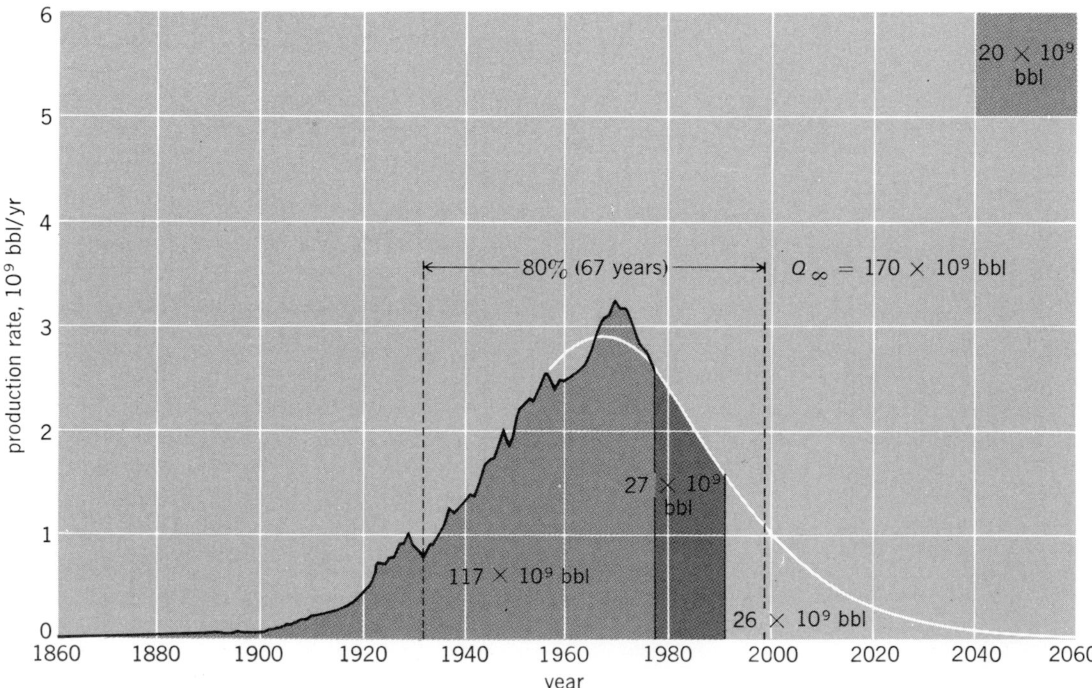

Fig. 10. Complete cycle of crude oil production in conterminous United States as of 1971. (*From M. K. Hubbert, U.S. energy resources: A review as of 1972, pt. 1, in A National Fuels and Energy Policy Study, U.S. 93d Congress, 2d Session, Senate Committee on Interior and Insular Affairs, ser. no. 93-40 (92-75), 1974*)

Table 1. Estimates of ultimate amounts of energy contents of crude oil, natural-gas liquids, and natural gas to be produced in the United States and bordering continental shelves

	Conterminous United States	Alaska	Total United States
Crude oil, 10^9 bbl	170	43	213
Natural-gas liquids, 10^9 bbl	34	5	39
Total hydrocarbon liquids, 10^9 bbl	204	48	252
Natural gas, 10^{12} ft^3	1050	134	1184
Energy contents			
Energy of liquids, 10^{18} thermal joules	1208	284	1492
Energy of natural gas, 10^{18} thermal joules	1143	146	1289
Total energy, 10^{18} thermal joules	2351	430	2781

SOURCE: M. K. Hubbert, in *A National Fuels and Energy Policy Study*, U.S. 93d Congress, 2d Session, Senate Committee on Interior and Insular Affairs, ser. no. 93-40 (92-75), 1974.

probably be not more than about 3 centuries.

Complete-cycle curves for United States coal production are not shown graphically, but based upon Averitt's high and low estimates for Q_∞, the time scales for the U.S. production are about the same as those for the world.

Oil and gas estimates. The problem of estimating the ultimate amounts of crude oil and natural gas to be produced in a given region is much more difficult than for estimates of coal, because accumulations of oil or gas occur in porous sedimentary rocks in limited regions of underground space with horizontal dimensions from 100 m to more than 100 km, and at depths ranging from about 100 m to 7.5 km. However, as exploration and drilling proceed in a given region, the eventual decline of the discoveries per unit of exploratory effort affords a basis for estimates of the ultimate amounts of oil or gas that a given region is likely to produce.

In the United States, by 1956, the cumulative production since the initial oil discovery in 1859 amounted to 52.4×10^9 bbl (1 bbl = 0.159 m^3) and the production rate was continuously increasing. Nevertheless, the cumulative experience in petroleum exploration led to a consensus among petroleum geologists and engineers that the value of Q_∞ for crude oil to be produced in the conterminous United States and adjacent continental shelves would probably be within the range of 150–200 $\times 10^9$ bbl. Figure 9 shows two complete cycles for United States crude oil production based upon these two figures made in 1956 by M. K. Hubbert. Since each grid square in this figure represents 25 $\times$ 10^9 bbl, for the lower figure of 150 $\times$ 10^9 bbl for Q_∞ there could be but six squares beneath the curve, two of which had already been used by cumulative production, leaving but four more for the future. To satisfy these con-

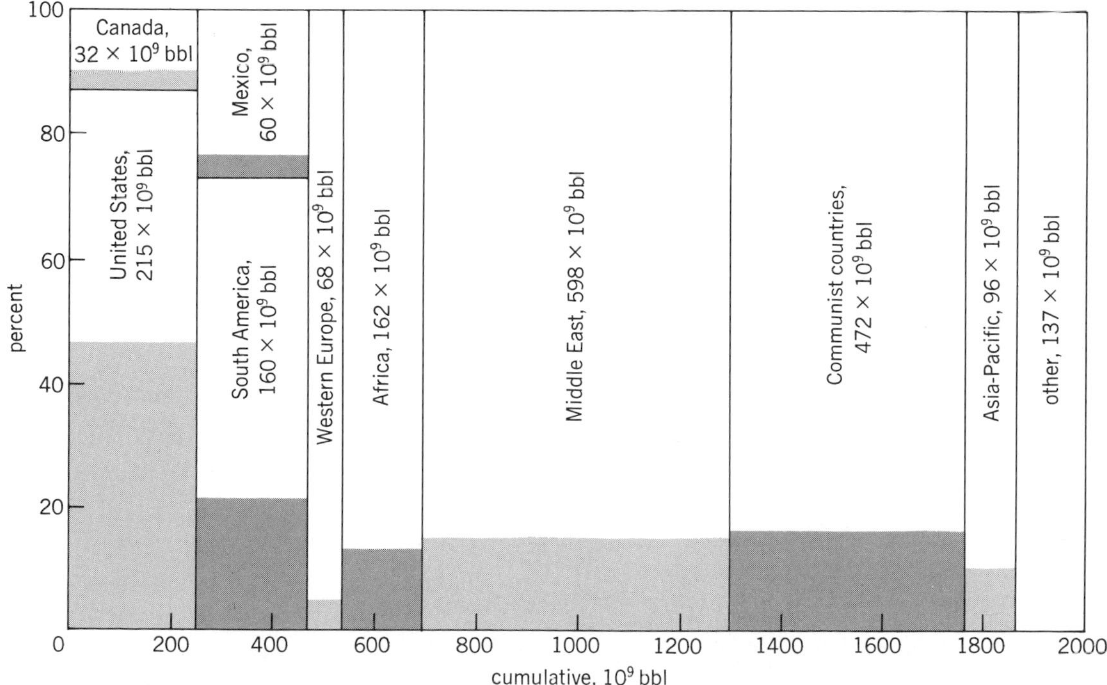

Fig. 11. Graphical representation of Jodry estimate of world ultimately recoverable crude oil. The shaded areas at the foot of each column or sector represent quantities consumed already. (*From M. K. Hubbert, U.S. energy resources: A review as of 1972, pt. 1, in A National Fuels and Energy Policy Study, U.S. 93d Congress, 2d Session, Senate Committee on Interior and Insular Affairs, ser. no. 93-40 (92-75), 1974*)

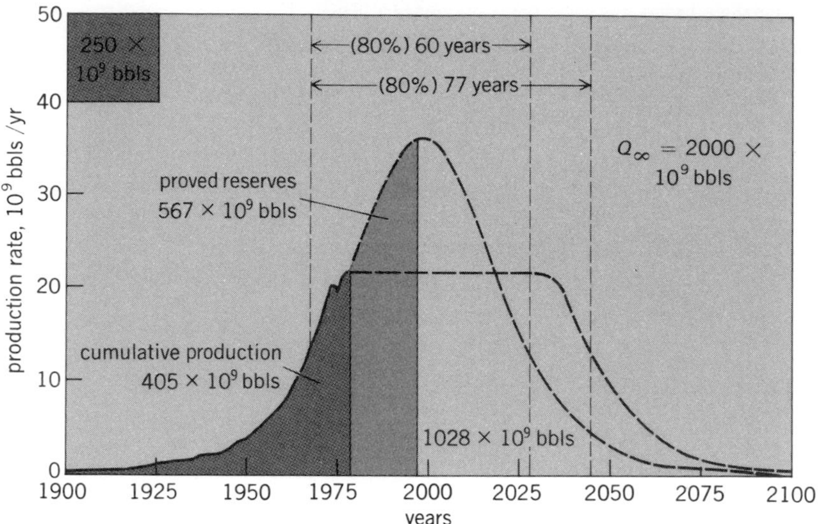

Fig. 12. Estimate as of 1972 of complete cycle of world crude oil production. *(From M. K. Hubbert, U.S. energy resources: A review of 1972, pt. 1, in A National Fuels and Energy Policy Study, U.S. 93d Congress, 2d Session, Senate Committee on Interior and Insular Affairs, ser. no. 93-40 (92-75), 1974)*

ditions, the peak in the production rate would have to occur within about 10 years, or at about 1966. For the higher figure of 200 × 10^9 barrels, two more squares would be added, but the date of peak production would be retarded by only about 5 years, or to 1971. Hence, from prevailing estimates of 1956 for Q_∞, it was possible to predict that the peak in the rate of United States crude oil production would probably occur within the period 1966–1971.

A difficulty in the application of the complete-cycle method to petroleum estimates is that it requires an independent estimate of Q_∞. Even so, for any reasonable estimates for Q_∞, the time scales obtained by this method are comparatively insensitive to error.

Crude oil production cycle. The complete cycle analysis has also been applied to the production of crude oil and ultimate amounts of crude oil, natural gas, and natural-gas liquids. The complete cycle of crude oil production for the conterminous United States, based upon 170 × 10^9 bbl for Q_∞, is shown in Fig. 10. Of particular significance is the time required to produce the middle 80% of Q_∞. This is estimated to be the 67-year period from about 1932 to 1999.

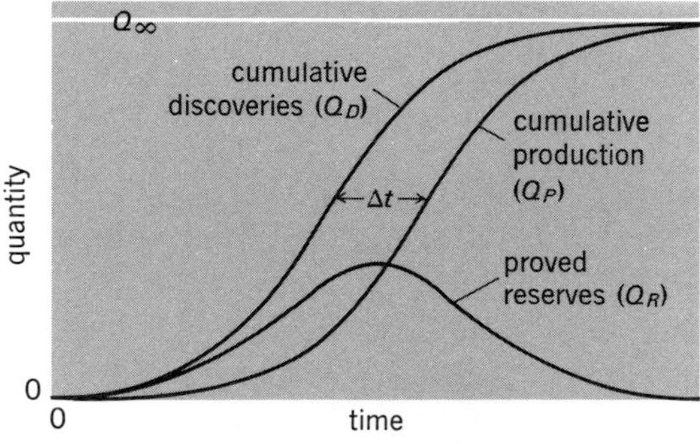

Fig. 13. Variation with time of proved reserves Q_R, cumulative production Q_P, and cumulative proved discoveries, Q_D, during a complete cycle of petroleum production. *(After M. K. Hubbert, Energy resources: A report to the Committee on Natural Resources, Nat. Acad. Sci.–Nat. Res. Counc. Publ. 1000-D, 1962; M. K. Hubbert, U.S. energy resources: A review of 1972, pt. 1, in A National Fuels and Energy Policy Study, U.S. 93d Congress, 2d Session, Senate Committee on Interior and Insular Affairs, ser. no. 93-40 (92-75), 1974).*

Estimates of the ultimate cumulative production of natural gas in the United States have been made on the basis both of the ratio of gas discoveries to crude oil discoveries and the prior estimate of Q_∞ for oil, and of gas discoveries per foot of exploratory drilling. These methods give a range from about 1×10^{15} to 1.1×10^{15} ft³ (1 ft³ = 0.02832 m³) for Q_∞ for natural gas for the conterminous United States. The peak in natural-gas proved reserves was reached in 1967, and the peak in the production rate occurred in 1973.

Estimates of natural-gas liquids are obtained from the prior estimates for natural gas and the gas-to-liquids ratio. For the conterminous United States, the estimated value for Q_∞ for natural-gas liquids is about 34 × 10^9 bbl. of which 18.4 × 10^9 bbl has been produced by the end of 1978.

For Alaska, which is still in its early stages of petroleum exploration, only rough estimates can be given at present of the ultimate quantities of petroleum fluids that may be produced. Including both land and offshore areas, such rough estimates are the following: crude oil, 43 × 10^9 bbl; natural gas, 134 × 10^{12} ft; natural-gas liquids, 5 × 10^9 bbl.

Table 1 gives a summary of the approximate magnitude of the ultimate quantities of crude oil, natural gas, and natural-gas liquids, and their energy contents, to be produced in the United States and the bordering continental shelves.

Ultimate world crude oil production. For the ultimate world production of crude oil, 12 estimates by geologists, petroleum analysts, and international oil companies were published between 1965 and 1978. Ten of these 12 estimates indicate that ultimate world production of crude oil will probably fall between 1.5 × 10^{12} and 2.5 × 10^{12} bbl. Figure 11 shows the estimated geographical distribution of the world's oil. The areas of the separate columns are proportional to the estimated ultimate oil production, totaling for the world 2 × 10^{12} bbl.

North America, with an estimated 307 × 10^9 bbl, is especially significant. This represents only 15.4% of the world total, of which about two-thirds is in the United States. Yet the United States, with only about 10% of the world's oil initially, has been until 1974 the world's largest producer as well as the world's largest consumer of oil. It is, accordingly, not surprising that the United States has already consumed half of its oil and is the farthest toward ultimate depletion of its oil of any of the major oil-producing countries.

World crude oil production cycle. Figure 12 shows two alternative complete cycles of production based upon a round figure of 2 × 10^{12} bbl for Q_∞. One cycle in Fig. 12 assumes an orderly evolution of the industry, with production rising to a peak and then declining. The other assumes that production is stabilized at the 1978 level for several decades before declining. The areas under the two curves are the same. In the case of an orderly evolution, the peak production of 36 × 10^9 bbl/yr occurs in 1997, and the middle 80% of the world's crude oil is consumed in the 60-year period from 1967 to 2027.

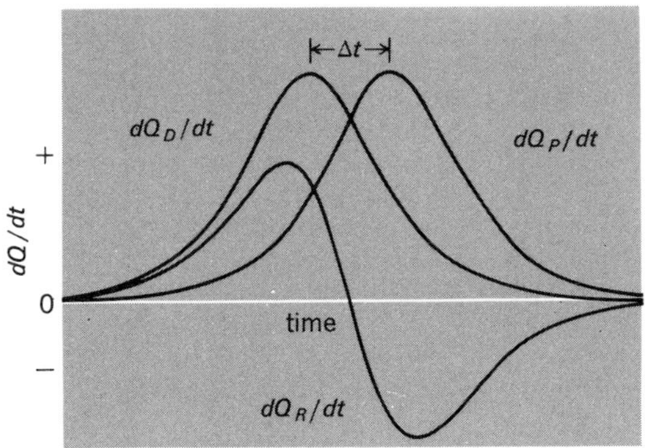

Fig. 14. Variation of rates of production, of proved discovery, and of rate of increase of proved reserves of crude oil or natural gas during a complete production cycle. (*After M. K. Hubbert, Energy resources: A report to the Committee on Natural Resources, Nat. Acad. Sci.–Nat. Res. Counc. Publ. 1000-D, 1962; M. K. Hubbert, U.S. energy resources: A review of 1972, pt. 1, in A National Fuels and Energy Policy Study, U.S. 93d Congress, 2d Session, Senate Committee on Interior and Insular Affairs, ser. no. 93-40 (92-75), 1974*)

CUMULATIVE STATISTICAL DATA

A different approach to petroleum estimation is based upon the use of cumulative statistical data on drilling, discovery, proved reserves, and production as a means of determining how far advanced the petroleum industry may be in its complete cycle. Figure 13 shows the evolution during the complete cycle of three statistical quantities, cumulative production Q_P, proved reserves Q_R, and cumulative proved discoveries Q_D. In the United States, statistical data on annual crude oil production are available from 1860 to the present. The sum of annual productions to any given year gives the cumulative production. Statistics on the proved reserves at the close of each year have been issued by a nationwide committee of petroleum engineers of the American Petroleum Institute since 1936, and approximate estimates are available annually since 1900. Proved reserves at any given time represent the amount of oil that almost certainly is present in fields already discovered, and producible by equipment already installed. It is, therefore, a working inventory; it is the difference between cumulative additions to reserves and withdrawals by means of production.

Cumulative proved discoveries is a derived quantity defined by

$$Q_D = Q_P + Q_R \qquad (5)$$

that is, all the oil that can be regarded as having been proved to be discovered by a given time is the oil produced to that time plus proved reserves.

The approximate nature of the variation with time of the three quantities, Q_D, Q_P, and Q_R, during a complete cycle is shown in Fig. 13. This is based upon the assumption that the complete cycle is one of a single maximum in the rate of production. Here, the Q_D and Q_P curves are logistic-type growth curves beginning at zero and ending asymptotically to the value of Q_∞. The Q_R curve begins at zero, reaches a maximum at about midrange, and then declines to zero at the end of the cycle. Also, in the midrange there is a time delay of Δt years between the discovery curve and the production curve.

The rates of discovery, of production, and of the

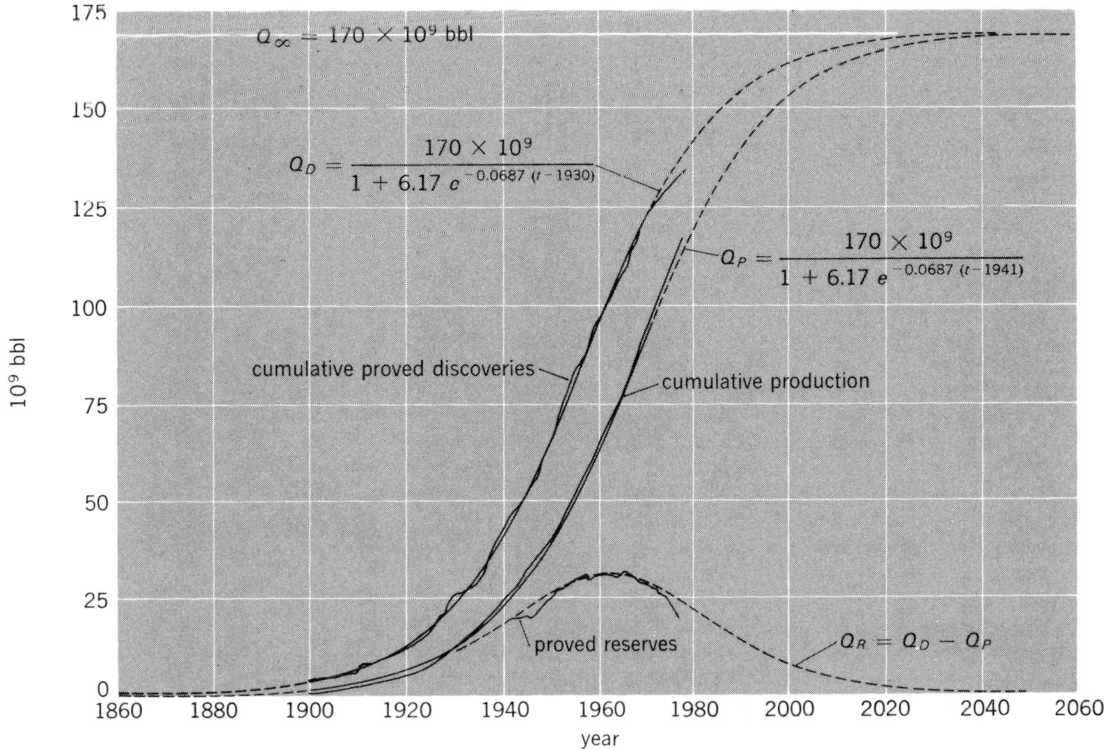

Fig. 15. Logistic equations and curves of cumulative production, cumulative discoveries, and proved reserves for crude oil from the conterminous United States, 1900–1971. (*M. K. Hubbert, U.S. energy resources: A review of 1972, pt. 1, in A National Fuels and Energy Policy Study, U.S. 93d Congress, 2d Session, Senate Committee on Interior and Insular Affairs, ser. no. 93-40 (92-75), 1974*)

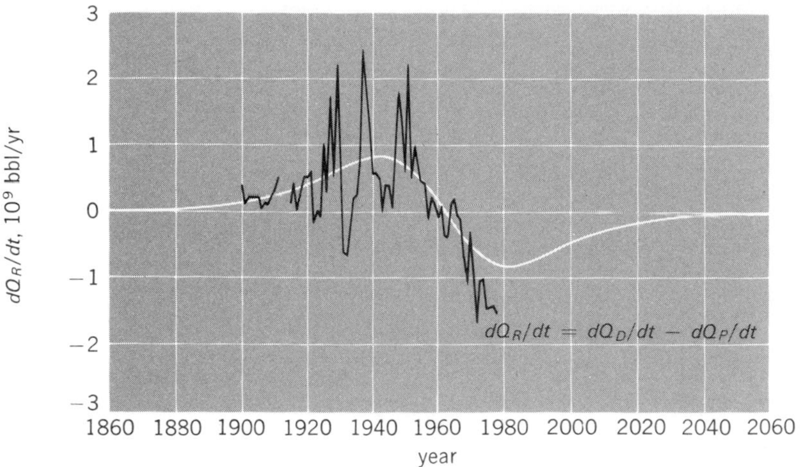

Fig. 16. Comparison of annual increases of proved reserves of conterminous United States, 1900–1971, with theoretical curve derived from logistic equations. (*From M. K. Hubbert, U.S. energy resources: A review of 1972, pt. 1, in A National Fuels and Energy Policy Study, U.S. 93d Congress, 2d Session, Senate Committee on Interior and Insular Affairs, ser. no. 93-40 (92-75), 1974*)

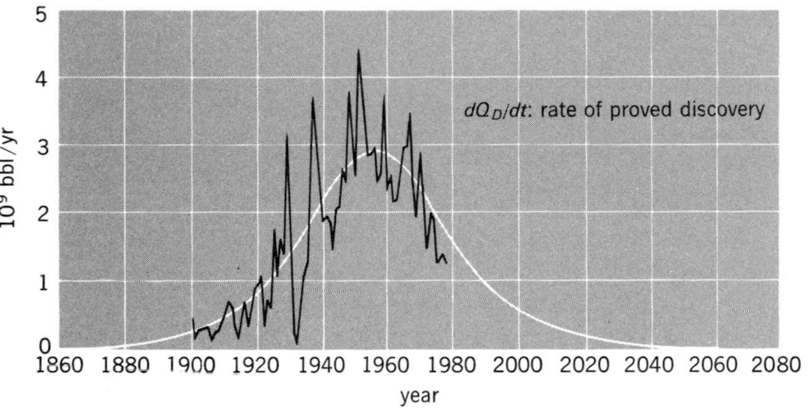

Fig. 17. Comparison of annual proved discoveries of crude oil in the conterminous United States, 1900–1971, with corresponding theoretical curve derived from logistic equation. (*From M. K. Hubbert, U.S. energy resources: A review of 1972, pt. 1, in A National Fuels and Energy Policy Study, U.S. 93d Congress, 2d Session, Senate Committee on Interior and Insular Affairs, ser. no. 93-40 (92-75), 1974*)

increase of proved reserves are equal to the slopes of the respective curves in Fig. 13. Mathematically, from Eq. (5), these are

$$dQ_D/dt = dQ_P/dt + dQ_R/dt \tag{6}$$

Graphs of these curves are shown in Fig. 14. It is to be noted that the peak in the rate of production occurs approximately Δt years later than the peak in the rate of discovery. The curve of rate of increase of proved reserves, dQ_R/dt, has a positive loop while reserves are increasing, crosses the zero line when reserves are at their maximum, and has a negative loop while reserves are decreasing. At the time when reserves are at their maximum, the rate of increase of proved reserves is

$$dQ_R/dt = 0 \tag{7}$$

At that time, by Eq. (6), the rates of discovery and production are

$$dQ_D/dt = dQ_P/dt \tag{8}$$

and these two curves cross one another, the rate of production still increasing but the rate of discovery already declining. The date of this event is about halfway between the discovery peak and the production peak.

In the earlier stages of the cycle, these two sets of curves are not very informative, but they become increasingly so from about the time of the peak in the rate of discovery onward.

These two sets of curves were constructed in 1962 by Hubbert, using the petroleum industry data to the end of 1961. Ten years later, the corresponding curves were constructed using cumulative data to the end of 1971 and reported by Hubbert in 1974. The results of these two separate sets of analyses are given in Table 2.

The actual data, as of 1978, for the Q_D, Q_P, and Q_R curves are shown in Fig. 15. The derivative, or rate, curves for the rate of increase of proved reserves and the rate of discovery, respectively, are given in Figs. 16 and 17. The theoretical curves in Figs. 15, 16, and 17 were calculated from data through 1971, and it is apparent that for the last 6 years actual discoveries were below the projections. More recent calculations indicate that the ultimate production in the conterminous United States will be less than the 170×10^9 bbl of crude oil by 8×10^9 to 10×10^9 bbl.

DISCOVERIES PER FOOT OF EXPLORATORY DRILLING

A different kind of analysis has been used to estimate ultimate production of oil or gas. This consists in determining the quantity of oil discovered per foot of exploratory drilling, dQ/dh, as a function of cumulative depth h of exploratory drilling. The area beneath this curve also is a measure of cumulative discoveries. Figure 18 shows the average numbers of barrels of crude oil discovered in the United States for each 10^8 ft of exploratory drilling from 1860 to 1972. The area of each column in the figure represents the quantity of oil discovered during each 10^8-ft interval of drilling.

As is seen from the figure, the discovery rate during the first four drilling units averaged about 225 bbl/ft. This was followed by a drastic decline to a final figure of only 30 bbl/ft by the last, or seventeenth, unit of drilling. The cumulative discoveries, defined as the sum of cumulative production plus proved reserves plus an estimated additional amount of oil in fields already discovered, by the

Table 2. Crude oil estimates as of 1962 and 1972 for conterminous United States, based upon analyses of Q_D, Q_P, and Q_R

Entity estimated	Date of estimation: 1962	Date of estimation: 1972	Observed
Date of maximum discovery rate	1957	1957	1957
Time lag Δt between discovery and production	10.5 years	11.0 years	
Date of peak of proved reserves	1962	1962	1962
Date of maximum production rate	1968	1968	1970
Ultimate cumulative production Q_∞	170×10^9 bbl	170×10^9 bbl	

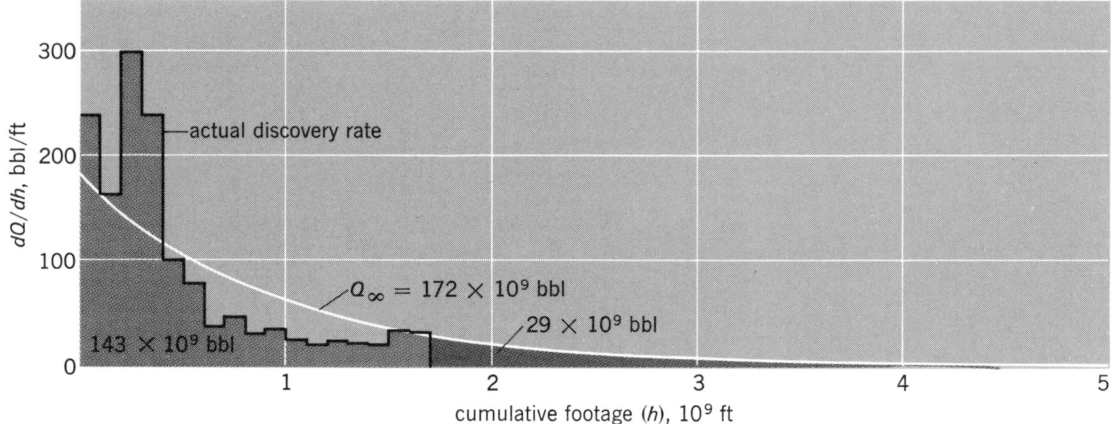

Fig. 18. Estimation of ultimate crude oil production of the conterminous United States by means of the curve ploratory drilling. (*From M. K. Hubbert, U.S. energy resources: A review of 1972, pt. 1, in A National Fuels and Energy Policy Study, U.S. 93d Congress, 2d Session, Senate Committee on Interior and Insular Affairs, ser. no. 93-40(92-75), 1974*)

17×10^8 ft of drilling, amounted to 143×10^9 bbl. The rate of decline in discoveries per unit of drilling shown in Fig. 18 is roughly a negative exponential. The best fit for such a curve, as shown in the figure, is one that equalizes the excesses and deficiencies, and passes through the last point of 30.4 bbl/ft. Extrapolation of this decline curve for unlimited future drilling gives an additional 29×10^9 bbl as the estimated future discoveries. Adding this to the 143×10^9 bbl already discovered gives a sum of 172×10^9 bbl for Q_∞. This is practically identical with the figure of 170×10^9 bbl obtained by the previous, quite different method of analysis. *See* PETROLEUM.

OTHER FOSSIL FUELS

Besides coal and lignite and the petroleum fluids (crude oil, natural gas, and natural-gas liquids) the principal remaining fossil fuels are tar or heavy-oil sands, oil shales, and minor quantities of the solid hydrocarbon gilsonite.

Tar sands. The world's largest known despoits of tar or heavy-oil sands are the Athabasca sands of northern Alberta, Canada. These consist essentially of heavy crude oil, filling the pore spaces of coarse-grained quartz sands, which is too viscous to flow into wells. These tar sands occur at various depths ranging from surface outcrops along the Athabasca River in northeastern Alberta to depths

Table 3. Oil content of known shale oil resources of world land areas

Continent	Recoverable under 1965 conditions, 10^9 bbl (10 to 100 gal/ton)	Marginal and submarginal, 10^9 bbl		
		25 to 100 (gal/ton)	10 to 25 (gal/ton)	5 to 10 (gal/ton)
Africa	10	90	*	*
Asia	20	70	14	†
Australia and New Zealand	*	*	1	†
Europe	30	40	6	†
North America	80	520	1600	2200
South America	50	*	750	†
Total	190	720	2400‡	2200

SOURCE: D. C. Duncan and V. E. Swanson, *Organic-rich shale of the United States and World Land Areas*, U.S. Geol. Surv. Circ. 523, 1965; M. K. Hubbert, in *A National Fuels and Energy Policy Study*, U.S. 93d Congress, 2d Session, Senate Committee on Interior and Insular Affairs, ser. no. 93–40 (92–75), 1974.
*Small. †No estimate. ‡Rounded.

Table 4. Reserves of recoverable oil from the Green River Formation, in Colorado, Utah, and Wyoming

Location	Shale oil reserves (at 60% recovery), 10^9 bbl			
	Class 1	Class 2	Class 3	Total
Piceance Basin, Colorado	20	50	100	170
Uinta Basin, Colorado and Utah	—	7	9	16
Green River Basin, Wyoming	—	—	2	2
Total	20	57	111	188

SOURCE: National Petroleum Council, Oil Shale Task Group, 1972; M. K. Hubbert, in *A National Fuels and Energy Policy Study*, U.S. 93d Congress, 2d Session, Senate Committee on Interior and Insular Affairs, sec. no. 93–40 (92–75), 1974.

Table 5. Magnitudes and energy contents of the world's initial recoverable fossil fuels

Fuel and quantity	Energy content: 10^{21} thermal joules	Energy content: 10^{15} thermal kilowatt-hours	Percent: Maximum	Percent: Minimum
Coal plus lignite				
Maximum 7.6 × 10^{12} metric tons	188.0	52.2	85.44	–
Minimum 2 × 10^{12} metric tons	49.6	13.8	–	60.78
Crude oil, 2 × 10^{12} bbl	12.2	3.4	5.56	14.95
Tar sand oil, 450 × 10^9 bbl	2.3	.6	.98	2.82
Shale oil, 300 × 10^9 bbl	1.8	.5	.82	2.21
Natural-gas liquids, 400 × 10^9 bbl	1.7	.5	.82	2.08
Natural gas, 12.8 × 10^{15} ft³	14.0	3.9	6.38	17.16
Total: Maximum	220.0	61.1	100.00	–
Minimum	81.6	22.7	–	100.00

SOURCE: M. K. Hubbert, in *A National Fuels and Energy Policy Study*, U.S. 93d Congress, 2d Session, Senate Committee on Interior and Insular Affairs, sec. no. 93–40 (92–75), 1974; and Canadian Department of Energy, Mines, and Resources, *Oil and Natural Gas Resources of Canada 1976*, Rep. EP 77–1, 1977.

up to 2000 ft (600 m) in deposits farther west. Estimates of recoverable quantities of these oils given by T. F. Scott in 1974 are the following: oil-in-place, 625 × 10^9 bbl; oil extractable, 148 × 10^9 bbl (by mining, 38 × 10^9, by at-site methods, 110 × 10^9). These raw oils must be coverted into synthetic crude oils by preliminary refining. According to Scott, only 70 bbl of synthetic crude oil are obtainable from 100 bbl of raw oil. Hence, the figure of 148 × 10^9 bbl of raw oil is equivalent to about 104 × 10^9 bbl of crude oil.

Oil is being produced from tar sands by only two companies, Great Canadian Oil Sands, Ltd., and Syncrude Canada, Ltd., both of which are mining the Athabasca deposit. Their combined production capacity is 170,000 bbl/day. There are heavy oil deposits in the Cold Lake region of Alberta, estimated by the Alberta Energy Resources Conservation Board to contain 165 × 10^9 bbl, of which one-third might be recoverable. There are plans to begin small-scale production.

According to reports in 1973 and 1974, a second major deposit of heavy oil underlies a region roughly 85 km wide by 600 km long extending east and west, north of and parallel to the Orinoco River in eastern Venezuela. Within this region there are four areas in which the principal quantities of this oil occur. In the westernmost area the thickness of the deposit is about 82 m; in the other three the thicknesses are about 100 m. These deposits are estimated to contain about 700 × 10^9 bbl of oil-in-place, of which about 10%, or 70 × 10^9 bbl, may be recoverable. After extensive drilling, estimates in trade publications of the oil-in-place have increased to as much as 1.8 to 2 × 10^{12} bbl and possibly even to 3 × 10^{12} bbl. Assuming synthetic crude oil equivalent to 10% of the oil-in-place can be recovered, then the oil from this source should amount to as much as 200 × 10^9 bbl. *See* OIL SAND.

Oil shale. The "oil" in oil shales differs from ordinary petroleum oils in that it occurs in a solid form, kerogen, rather than as a liquid. When a few chips of an oil shale are heated in a test tube, a dense vapor is distilled off which condenses on the walls of the tube as an amber-colored liquid. This is raw shale oil. Like tar sand oil, this too must be refined into a synthetic crude oil before it can be sent to a conventional oil refinery.

Table 3 gives a summary of the world's known shale oil deposits as compiled by D. C. Duncan and V. E. Swanson in 1965. The oil contents of the deposits range from 5 to 100 gal of oil per ton of rock. The estimated total amount of oil within these grades is given as 5.3 × 10^{12} bbl. Of this,

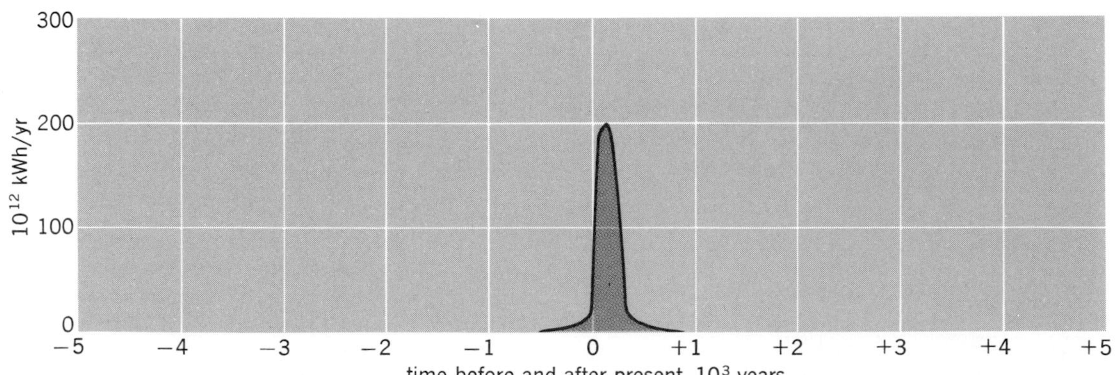

Fig. 19. The epoch of fossil-fuel exploitation as it appears on a time scale of human history ranging from 5000 years ago to 5000 years into the future. (*From M. K. Hubbert, U.S. energy resources: A review of 1972, pt. 1, in A National Fuels and Energy Policy Study, U.S. 93d Congress, 2d Session, Senate Committee on Interior and Insular Affairs, ser. no. 93-40 (92-75), 1974*)

however, Duncan and Swanson consider only about 190×10^9 bbl as being recoverable under 1965 conditions.

The largest known oil shale deposits in the world are those of the Green River shales of Eocene age occurring in three localities: southwestern Wyoming, western Colorado, and northeastern Utah. The approximate quantities of oil in these basins in classes 1 to 3 of decreasing favorability, as given by the Oil Shale Task Group of the National Petroleum Council in 1972, are shown in Table 4.

Class 1 comprises beds at least 30 ft (9 m) thick having an average oil content of 35 gal/ton. Class 2 comprises beds at least 30 ft thick having an average oil content of at least 30 gal/ton. Class 3 comprises shales comparable to those of class 2, only less well defined. The Oil Shale Task Group considered only the shales of the class 1 group, occurring in western Colorado, and containing an estimated 20×10^9 bbl of oil as being suitable for exploitation at present.

In appraising this, it should be borne in mind that 20×10^9 bbl of oil is only about a 4-year's supply for the United States at present rates of consumption. In 1973 several Federal leases of these oil shales of 5000 acres (20.235 km^2) each to different groups of oil companies were announced. These were to produce about 50,000 bpd each by the early 1980s. As of October 1979, only two pilot projects were in operation, and no commercial plants.

To appraise the significance of shale oil as a means of meeting the domestic oil requirements of the United States, account needs to be taken of the fact that by the mid-1970s the rate of consumption of petroleum liquids in the United States was about 19×10^6 bpd. To produce even 1×10^6 bbl of shale oil per day would require 20 plants with capacities of 50,000 bpd each, and even this rate of production would be barely significant with respect to domestic requirements.

According to a 1975 report on a proposed project, it will require 10,000 metric tons of rock to produce 6000 bbl of oil. This is a ratio of 1.67 metric tons, or 0.73 m^3, of rock to be mined per barrel of oil produced. At a production rate of 1×10^6 bpd, a volume of 730,000 m^3 of rock would have to be mined each day. Upon retorting to extract the oil, this shale would expand to about 1×10^6 m^3 of cinders produced per day, or about one-third of a cubic kilometer of cinders produced per year. For the estimated 20×10^9 bbl of shale oil obtainable from the Piceance Basin of western Colorado, the total volume of these wastes would amount to about 20 km^3, one-quarter of which would comprise highly alkaline calcium and magnesium oxides. Through these wastes would occur a flow of groundwater, leaching the alkaline salts and discharging into the drainage system of the Colorado River.

The two pilot projects currently operating are attempting to retort the shale in place. Some mining is still required, but it is hoped that the amount of spent shale to be disposed of at the surface can be reduced by 80%. Even with in-place retorting, it is questionable whether the small amount of oil obtainable from the Green River oil shales can compensate for the environmental damage which must accrue. *See* OIL SHALE.

SUMMARY OF WORLD FOSSIL FUELS

The approximate magnitudes of the world's recoverable fossil fuels and their energy contents are summarized in Table 5. It will be noted that coal is the largest energy source of any of the fossil fuels, representing about 85% of the total if Averitt's higher estimate is used, but only about 60% if the reduced estimate is used.

FOSSIL FUELS IN HUMAN HISTORY

The role of the fossil fuels in the longer span of human history can best be appreciated if one considers the period extending from 5000 years in the past to 5000 years in the future. On such a time scale the epoch of the fossil fuels is shown graphically in Fig. 19. This appears as a spike with a middle-80% width of about 3 centuries. It is thus seen that the epoch of the exploitation of the fossil fuels is but an ephemeral event in the totality of human history. It is a unique event, nonetheless, in geological history. Moreover, it is responsible for the world's present technological civilization and has exercised the most profound influence ever experienced by the human species during its entire biological existence.

[M. KING HUBBERT; DAVID H. ROOT]

Bibliography: P. Averitt, *Coal Resources of the United States, Jan. 1, 1967*, USGS Bull. 1275, 1969; Canadian Department of Energy, Mines, and Resources, *Oil and Natural Gas Resources of Canada 1976*, Rep. EP77-1, 1977; D. C. Duncan and V. E. Swanson, *Organic-Rich Shale of the United States and World Land Areas*, USGS Circ. 523, 1965; B. Grossling, *Window on Oil*, Financial Times, Ltd., 1976; T. A. Hendricks, *Resources of Oil, Gas, and Natural-Gas Liquids in the United States and the World*, USGS Circ. 522, 1965; M. K. Hubbert, *Energy Resources: A Report to the Committee on Natural Resources*, Nat. Acad. Sci.–Nat. Res. Counc. Publ. 1000-D, 1962, reprinted as U.S. Dep. Commer. Rep. PB 222401, 1973; M. K. Hubbert, Energy resources, in *Resources and Man*, National Academy of Sciences–National Research Council, 1969; M. K. Hubbert, Measurement of energy resources, *J. Dynamic Syst. Measure. Control (Transact. ASME)*, 101:16–30, March 1979; M. K. Hubbert, *Nuclear Energy and the Fossil Fuels: Drilling and Production Practice*, American Petroleum Institute, 1956; M. K. Hubbert, U.S. energy resources: A review as of 1972, pt. 1, in *A National Fuels and Energy Policy Study*, U.S. 93d Congress, 2d Session, Senate Committee on Interior and Insular Affairs, ser. no. 93-40 (92-75), 1974; H. D. Klemme, World oil and gas reserves from giant fields and petroleum basins (provinces), in *The Future Supply of Nature-Made Petroleum and Gas*, 1977; Large-scale action nearer for Orinoco, *Oil Gas J.*, 71(33):44–45, 1973; J. D. Moody and H. H. Emmerich, Giant oil fields of the world, in *Proceedings of the 24th International Geological Congress*, Montreal, sec. 5: Mineral Fuels, 1972; J. D. Moody and R. W. Esser, An estimate of the world's recoverable crude oil resource, in *Proceedings of the 9th World Petroleum Congress*, vol. 13: *Exploration and Transportation*; R. Nehring, *Giant Oil Fields and World Oil Resources*, Rand Co. Rep. no. R-2284-CIA, 1978; New data indicate Orinoco belt exceeding expectations,

Oil Gas J., 72(45):134, 1974; P. Odell, Estimating world oil discoveries up to 1999: The question of method, *Petrol. Times*, 79(2001):26–29, 1975; Rio Blanco oil-shale plan due by year-end, *Oil Gas J.*, 73(30):48, 1975; T. F. Scott, Athabasca oil sands to A.D. 2000, *Can. Min. Met. Bull.*, pp. 98–102, October 1974; R. T. Tippee, Tar sands, heavy-oil push building rapidly in Canada, *Oil Gas J.*, 76(5):87–91, 1978; H. R. Warman, Future problems in petroleum exploration, *Petrol. Rev.*, 25 (291):96–101, 1971.

Protecting the Environment

Congressman Mike McCormack

Few realize how much the production of fuels and energy can affect the quality of the environment. Beginning with the removal of coal, oil, natural gas, and uranium from the ground by mining or drilling and pumping, there are environmental effects associated with these activities and with the subsequent transportation, processing, conversion, transmission, and waste disposal activities. Uncontrolled production of fossil and nuclear fuels can rip up the land, release dangerous or objectionable pollutants into air and water, produce mountains of waste, and generally degrade the quality of the environment near the mines or wells. Waste heat from the generation of electricity or other industrial activities, which is usually dissipated into bodies of water, can change water life, depending upon the amount of heat and the volume and dynamics of the receiving water body. Impoundments of water to generate electricity also can affect the environment, as can the use of geothermal steam. If solar energy becomes widely used, it too will affect the environment. Depending upon the size of energy activities, the environmental effects can range from negligible or unobjectionable to effects so pronounced that the environment becomes dangerous to the health of workers or the public.

The purpose of this article is to briefly identify the environmental effects of supplying fuels and energy, to indicate some measures for keeping these effects within socially acceptable limits, and to identify major legislative approaches to this end.

PRODUCTION OF FUELS

Aside from the small amount (4%) of energy supplied by hydroelectric plants, and some negligible inputs from animals, windmills, and solar units, the energy used in the United States comes from the burning of fossil fuels (coal, oil, and natural gas) or the fissioning of uranium. Of the energy used in the United States in 1978, 48.5% came from oil, 25.5% from natural gas, 17.9% from coal, and 8.1% from nuclear, hydro, and other. Over the next few decades, the energy is expected to come increasingly from coal and uranium, with the amount from oil and gas decreasing.

Coal. The mining, cleaning, and transportation of coal all have some environmental effects, the major effects being those from mining. Coal is mined in two different ways which have substantially different environmental effects: by underground mining and by surface mining. *See* COAL MINING.

Underground mining. Underground mining can cause unexpected cave-ins and subsidence of the surface above the mined areas. In extreme cases, this subsidence can cause buildings to collapse, disturb water supplies, prevent use of land for agriculture, and, if sudden, cause earth tremors. Approximately a third of the total area mined for coal already has experienced some subsidence. Experiments with backfilled mined areas have yet to produce an acceptable solution. However, the use of long-wall underground mining techniques promises controlled subsidence and stabilization of the surface.

Coal mining is also a source of water pollution. Water pumped from mines, or water that leaks from abandoned mines, may contain sulfuric acid and other water pollutants. The water pollutants from deep mines may also include trace quantities of elements and chemicals not found in the local environment. These pollutants can be toxic to plant and animal life in water and on land. The piles of waste that accumulate at deep mines also can be a source of water pollution from the leaching of pollutants into runoff water, or of air pollution if they catch fire and smolder.

Surface mining. Surface mining can horribly scar the environment unless carefully carried out with positive reclamation measures. In area or strip mining, which is the method of surface mining used in flat areas, the overburden of soil removed in one cut is usually dumped into the empty space left by a previous cut so that reclamation can be a continuing process. In addition to disruption of the surface, a major environmental impact from strip mining is the runoff of rainwater carrying silt which can pollute and clog streams and prematurely fill reservoirs. In contour mining, which is used in hilly regions such as Appalachia, the overburden is dumped downslope, destroying vegetation and property below, clogging streams, causing mud slides, and aggravating problems of runoff. Contour mining often is done in wilderness areas which are difficult to restore.

With the growing demand for coal, surface mining can be expected to affect large land areas. While reclamation is fairly straightforward in the eastern United States, where water is available to initiate quick growth of new plants, the scarcity of water in the arid western states is likely to make restoration much more difficult, expensive, and time-consuming.

Other effects. Coal mines can also be a local source of dust, particularly when compressed air is used to clean the coal or when the coal is heated to drive off moisture. Fine coal particles contained in wash water from coal preparation can adversely affect water life unless impounded or treated.

The concept of large coal refineries near mines promises economic benefits to mining areas. But such "complexes" are likely to be accompanied by objectionable emissions into local air and water unless they include pollution control systems.

Finally, the transportation of coal has some environmental impacts, but these are small in comparison with those from mining and processing. The effect of long and frequent coal trains upon road and highway traffic can become a serious local nuisance.

Oil. Historically, most oil wells have been drilled on land, but an increasing number are now being drilled at sea. In the future, the outer continental shelf off both coasts is expected to supply an increasing share of the United States oil supply. Drilling for oil on land can have local environmental effects from the release of drilling fluid, or "mud," which may contain chemical agents that are water pollutants. Uncontrolled release of brine often found in oil deposits can cause water pollution on land, while blowouts and fires can cause local contamination. These environmental impacts are multiplied for wells drilled into the seabed. There, any oil released from drilling accidents or from pipelines or tanks can contaminate distant shorelines. The environment at the interface between land and sea is particularly susceptible to adverse impacts from oil spills, as are the fish, birds, and plant life of the coastal zone. *See* OIL AND GAS WELL DRILLING.

Oil refineries, unless properly designed and operated, can also be the source of objectionable air pollution from emission of hydrocarbon vapors, other chemicals, and combustion products from fuel burned to supply process heat. Oil refineries also can be sources of water pollution from oil leaks and chemical wastes. Some refining processes produce a heavy sludge which constitutes a solid-waste disposal problem.

The pumping out of oil fields along coastal areas can produce subsidence of the surface and increase the risk of flooding that land at high tides and during storms.

Past controversy over the anticipated environmental effects of building and operating the Trans-Alaska Pipeline highlights both the direct effects of such operations in remote regions and the indirect effects upon the environment caused by the sudden building of communities and increases in the local population.

Natural gas. Natural gas is the cleanest and most versatile of fossil fuels for stationary use. However, the production of natural gas entails many of the environmental impacts associated with production of oil, for oil and gas are frequently found together and produced from the same wells. Transportation of natural gas has only minor environmental effects except for land taken for rights of way, and for local impacts from leaks which may injure local vegetation or cause explosions and fires. Pipeline explosions, however, can present a serious risk to public safety in populated areas. The growing transportation of liquefied natural gas also can pose potentially catastrophic risks to the public health and safety from accidental release and possible fire or explosion. *See* LIQUEFIED NATURAL GAS (LNG); NATURAL GAS.

Uranium. Uranium ores are mined by both underground and surface techniques. A unique environmental effect of mining and milling of uranium derives from the presence of radium in the ores and in the tailings from mills. This radium emits a radioactive gas, radon, which with its decay products can collect in underground mines and present an occupational health hazard. Overexposure is linked to lung cancer. The waste piles (tailings) from uranium mills, unless stabilized, can also become a source of radioactive contamination. Land areas may be contaminated by wind erosion of the tailings piles, and local streams may become polluted by runoff from the piles containing radioactive materials. Uranium mills also can produce liquid wastes which, unless processed or impounded, can introduce radioactive or toxic chemicals into streams.

The subsequent chemical and metallurgical processing of uranium compounds and the enrichment of uranium for nuclear fuel have environmental effects comparable to industrial operations in the chemical industries, for which control and collection technologies exist. These operations also generate some mildly radioactive wastes which are collected, packaged in steel drums, and sent to licensed disposal sites for burial. *See* URANIUM; URANIUM METALLURGY.

Other heat and fuel sources. Oil shales and tar sands are mentioned often as potential sources of liquid fuels. However, aside from still unfavorable economics, the production of liquid fuels from oil shales and tar sands would involve substantial environmental effects. The mining and processing of oil shale is complicated by the expansion of the volume of wastes during processing, so that the original mined areas cannot accommodate the return of all the solid wastes. Mining of oil shales and tar sands could also affect the environment through water runoff from open workings and waste piles. Also, unless controlled, oil shale refineries would be a source of dust and air pollutants. Restoration of mined areas in arid climates is likely to be difficult. Additionally, processing of oil shale requires large amounts of water, ranging from 1.4 to 4.5 barrels of water per barrel of oil produced. Finally, the establishment of large facilities to make oil from oil shale or tar sands would bring with them supporting communities and their impacts upon the environment. It should be noted that most of the environmental effects associated with the production of liquid fuels from oil shale are likely to be considerably reduced if in-place technologies can be commercially developed. *See* OIL SAND; OIL SHALE.

During 1979, the Carter administration proposed a major effort to develop and demonstrate ways to make synthetic fuels from coal and from oil shales. Legislation proposed to carry out this program included an office endowed with extraordinary authority to cut through red tape that might delay contemplated demonstrations. This authority, however, did not extend to setting aside the environmental protection requirements of Federal laws. *See* SYNTHETIC FUEL.

Geothermal energy, another potential energy source for some locations, will have some adverse environmental effects, including the possible release of foul-smelling gases, brines to contaminate water, waste heat, and subsidence of the surface. The reinjection of geothermal waters into the ground after use could cause contamination of underground waters and perhaps affect local geological stability. *See* GEOTHERMAL POWER.

PRODUCTION OF ENERGY

Heat energy from the fuel sources discussed above can be used directly as process heat by industry, to heat and cool buildings, or for conversion into electrical and mechanical energy. Common to all of these uses are the environmental effects from the burning of fossil fuels or the fissioning of uranium. *See* ENERGY FLOW.

Burning coal. Of all the fuels, coal is the dirtiest to burn, and the environmental effects of its uncontrolled burning have caused public concern, opposition, and regulation in both historical and recent times. The particulates create smoke and haze, soil buildings and materials, and increase cleaning costs. Fine particulates may be a health hazard and are suspected of accelerating the attack of corrosive gases upon buildings and structural materials. Emission of particulates can be controlled by collecting fly ash with electrostatic precipitators and filters. However, small particulates from some kinds of coal are difficult to trap.

Sulfur emission. Probably the most noticeable gas from the combustion of coal is sulfur dioxide, which can, in some cases, cause corrosion and damage to plant life, and is also suspected of causing or aggravating respiratory illness in human beings. It is the source of acid rain that is harming trees, plants, and lakes in Europe and the United States. The simplest way to keep sulfur dioxide emissions within acceptable limits is to burn coal of a low sulfur content. However, there is an insufficient supply of low-sulfur coal, so other control technologies must be used.

Constant emission limitation, or permanent control, involves imposing a fixed limit on the rate of sulfur dioxide emissions from a furnace or boiler. Constant emission limitation may be accomplished by burning low-sulfur coal to the extent it is available, or by installing stack-gas cleaning systems, or "scrubbers," to remove much of the sulfur dioxide before release to the atmosphere. Commercially, scrubber technology is relatively new, and there is still some controversy over how well it works.

An alternative to scrubbers is the supplementary control system (SCS), in which the emissions from a furnace are discharged through a stack high enough that the concentration at ground level usually is within acceptable limits. The electrical utility industry regards the supplementary control system as less expensive, less energy-consuming, and more reliable than scrubbers. The Environmental Protection Agency claims that sulfur dioxide in the air may be transformed into sulfates that may be injurious to the public health. Using SCS does not reduce the total atmospheric burden of sulfur, so EPA is strongly opposed to supplementary control systems, advocating permanent controls instead.

Another alternative is to trap sulfur during combustion. The fluidized-bed combustion process holds promise for doing so, but the engineering and economic practicability of large units has yet to be demonstrated.

Nitrogen oxides. Oxides of nitrogen are corrosive gases that are one cause of photochemical smog and are suspected of having unfavorable health effects. Most of these oxides (97%) are produced by natural decay processes; some (3%) are caused by the burning of fuels. Production of nitrogen oxides can be controlled by regulation of the combustion process and keeping the maximum temperature below 2800°F (1538°C), which is the threshold for formation of nitrogen oxide. Unfortunately, the drive to increase conversion efficiency of power plants requires going to higher temperatures. At present there are no well-established means of removing nitrogen oxide from combustion products.

Carbon dioxide. Another potentially troublesome product from burning any fossil fuel is carbon dioxide. Some scientists speculate that the increasing carbon dioxide content of the atmosphere will increase its capacity to absorb solar energy, leading to a gradual warming of the troposphere (the greenhouse effect). If the concentration of carbon dioxide were to become high enough, some scientists warn, the temperature increase could melt the polar icecaps, with catastrophic flooding of coastal areas throughout the world. However, other scientists point to the increase of human-produced dust in the atmosphere, which tends to reflect some of the solar energy, and thereby to decrease air temperatures. What the ultimate balance between these effects will be is not now known. There is no practical way to limit the emission of carbon dioxide from burning fuels. However, the carbon dioxide released by burning fuels during the last 50 years has not reversed the long-term cooling trend which is now being experienced.

Other emissions. Because many impurities are present in coal, a variety of trace metals and elements may occur in uncontrolled exhausts from furnaces and boilers. Those elements most likely to appear are mercury, lead, zinc, and beryllium. Incomplete combustion can also cause the presence of hydrocarbons in the exhaust. Some of these, such as benzo-alpha-pyrene, are known to cause cancer in experimental animals.

The ash from burning coal, including fly ash collected by emission control systems, forms a solid waste. Although some of this material is used to make cement or in road building or construction, most of it is piled on land or dumped at sea. Some very fine particles of ash are not removed by scrubbers, and there is concern that they can become trapped in the lung and impair health.

Burning liquid fuels. In burning, gasoline, jet fuels, fuel oils, and residual oils all produce some sulfur dioxide, nitrogen oxide, particulates, and carbon dioxide. In addition, burning of gasoline in internal combustion engines usually produces carbon monoxide. Such engines also discharge small amounts of incompletely burned hydrocarbons. For stationary use, emissions of sulfur dioxide can be controlled through the use of low-sulfur fuel or control systems, while production of nitrogen oxides can be limited by control of combustion process and temperature. Systems also exist for control of carbon monoxide emissions.

Burning natural gas. The environmental effects from burning natural gas are mainly limited to production of carbon dioxide and nitrogen oxides, whose effects were mentioned above. Despite its desirable minimal environmental impacts, the growing shortage of natural gas is likely to cause it increasingly to be reserved for special uses. Within a decade or so, supplies of synthetic natural gas made from coal are likely to appear. The environmental effects of burning this fuel should be indistinguishable from the burning of natural gas, although there will be some adverse effects resulting from the conversion of coal to synthetic gas. *See* COAL GASIFICATION; SUBSTITUTE NATURAL GAS (SNG).

Burning solid wastes. Urban and rural solid wastes may also be used as fuel to produce process heat or steam. The environmental effects from burning such wastes approximate those from burning coal. In addition, the collection and holding of large volumes of solid wastes might create a public nuisance or public health hazard from decaying organic matter and from insects and rodents attracted by it.

Fissioning of uranium. In addition to heat, the fission process produces large amounts of intensely radioactive materials, which in routine operations are virtually all confined within the nuclear fuel. Small amounts do escape from the fuel and are collected, packaged, and sent to special burial grounds for disposal. Some small quantities of radioactive materials, mainly gases such as tritium, are released to the environment in concentrations that are low in comparison with emission limits set by the Nuclear Regulatory Commission (NRC). The environmental effects of small routine emissions from a nuclear power plant are believed to be so small as to be indistinguishable from effects of natural radioactivity, although some observers have expressed concern that these small quantities might be concentrated in some animal or plant life form and thus become a health hazard to human beings or to other creatures in the environment. *See* NUCLEAR FISSION.

Accidental release. The radioactive fission products produced in nuclear power plants could have substantial environmental effects if a large amount were to be released in an accident or by some other mechanism. Various protective mechanisms are employed to provide multiple barriers against the release of hazardous quantities of radioactive materials. To date, there has been no such release in the United States or elsewhere in the world. However, in March 1979, an accident at the Three Mile Island nuclear power plant near Harrisburg, PA, caused fears of such a release before the plant was brought under control.

Reprocessing. After most of the usable energy has been extracted from the nuclear fuel, the spent fuel must be removed from the power plant. At that time it is highly radioactive because of the fission products it contains. The spent fuel is stored in pools of water at the nuclear power plants until, in principle, its radioactivity has decayed enough for the fuel to be transported to a reprocessing plant. However, in April 1977, President Carter, in a statement of nuclear policy, announced that the United States would forego reprocessing to set an example for other nations. His intent was to dissuade them from reprocessing, particularly nations that do not have nuclear weapons, because the plutonium recovered from spent fuel might be used to make nuclear weapons. Thereafter his administration tried to make acceptable arrangements for the temporary storage of spent fuel away from reactors (AFR storage) and for ultimate permanent long-term disposal of spent fuel, probably into stable layers of salt or rock deep beneath the surface of the Earth.

To complicate the situation, the Federal government has on hand inventories of highly radioactive liquid wastes from its reprocessing plants that produce plutonium for United States nuclear weapons. Some of this material has leaked from storage tanks, but with no appreciable damaging effects to the environment. Nonetheless, sooner or later something will have to be done about the final disposition of these wastes. What is done with them could show how to handle comparable wastes from commercial plants if and when reprocessing is again permitted.

Commercial reprocessing plants are likely to have the same environmental effects as any large industrial establishment, and they will present some risk of release of intensely radioactive materials to the environment in a major accident. There may also be some releases of radioactivity from normal operations that are small in comparison with background radioactivity. *See* NUCLEAR FUELS REPROCESSING.

On the whole, the risks of accidental release of radioactive materials from nuclear power plants and associated facilities are regarded by the government as small and worth the benefits of the electricity supplied. This assessment, however, is disputed by some antinuclear advocates who claim that the potential effects of accidents are so severe that even a minute risk is unacceptable. *See* NUCLEAR POWER; NUCLEAR REACTOR.

CONVERSION OF HEAT INTO MECHANICAL ENERGY

Heat from burning can be converted into mechanical energy in internal combustion engines, gas turbines, and steam turbines. Heat from the fissioning of uranium can also be converted into mechanical energy through steam turbines. The environmental effects of these machines largely take the form of air pollution and noise, with some effects also caused by waste heat. Automobile engines are a major source of nitrogen oxide, carbon monoxide, and unburned hydrocarbons in the air. Their emissions are a major component of the photochemical smogs that plague some urban areas. Emission control systems and modification of engines to reduce emissions are two alternatives. A more radical alternative is to use electric vehicles which have virtually zero environmental effect in the vicinity of the vehicle, although supplying the electricity needed to power this mode of transportation would have some environmental impacts from the expanded use of power plants.

CONVERSION OF HEAT INTO ELECTRICITY

The principal impacts on the environment from the large-scale generation of electricity from heat energy result from the burning of fossil fuels or fission of nuclear fuels as described, and from the release of waste heat produced in these processes. For a conventional steam electric power plant, about two-thirds of the heat produced by burning fuel or fissioning uranium is released to the environment, and only about one-third is converted to electricity. The waste heat is removed by pumping through the power plant enormous amounts of water which emerges warm to the touch. The heat can then be dissipated by returning the cooling water to a river, bay, lake or reservoir. *See* ELECTRIC POWER GENERATION.

The effects of waste heat rejection systems upon bodies of water are twofold: the increase in temperature can affect plant and animal life in the waters, driving some species away and attracting others; and the powerful pumps that supply the cooling water can suck in marine life which may be killed or injured at the pump, or may succumb to or be injured by exposure to high temperatures in the plant's heat exchangers. Cooling systems also can be a source of toxic wastes from chemicals used to clean marine growth from the insides of pipes and other surfaces.

Alternative ways to dissipate waste heat into the environment are to discharge it directly into the air through evaporative or dry cooling towers, or into artificial ponds that are isolated from other bodies of water. However, these alternatives can produce undesirable effects of their own. Cooling towers have already become the huge and unsightly symbols of electric power plants. Evaporating waters of cooling towers or ponds can produce local mist or icing in cold weather. Those cooling towers that use salt water produce a salt-laden mist than can corrode buildings and structural materials and contaminate agricultural land downwind. Some cooling towers employ large, noisy fans.

Dry cooling towers can avoid some of the problems. They have yet to be used on a large enough scale to observe their overall environmental effects. Some scientists theorize that the plumes of dry, heated air above such towers could cause local atmospheric instability in some weather conditions and perhaps trigger tornadoes. *See* COOLING TOWER; WASTE HEAT MANAGEMENT.

Another environmental impact of electric power plants is commitment of the land required. A large coal-fired power plant with controlled emission systems would require some 940 acres (3.8 km^2) for the plant plus about another 17,000 acres (68.8 km^2) for associated transmission lines, depending on the distance from its load center.

The diesel engines and gas turbines that are sometimes used to drive electric generators, because they are small in comparison with central power plants, produce comparatively small environmental effects.

OTHER ENERGY SOURCES

Looking ahead, other sources which may supply useful amounts of energy by the year 2000 include hydroelectric power; tidal power; and solar energy, which can be obtained directly from the Sun, or indirectly from fuel crop systems, by tapping wind power, and ocean heat and currents. Some additional hydroelectric plants may be built as well as some tidal plants.

Hydroelectricity. Water power has long been valued as a clean and inexpensive source of electricity. However, hydroelectric plants have environmental effects which can be objectionable to some parts of society. Dams can interfere with waterlife movement up and down a stream, increase the nitrogen content of the water with adverse effects upon fish, permanently flood productive, scenic, or historical lands, and alter water flow downstream. At deep reservoirs, water released from bottom layers may be very cold and devoid of oxygen. In addition, the enormous weight of dams and impounded waters could cause geological distortions in some locations. Of course, not all environmental effects of hydroelectric plants are considered detrimental. Reservoirs can provide recreational opportunities and expanded fishing opportunities. Flood control and irrigation water supplies from dams are usually also counted as positive benefits. *See* DAM; HYDROELECTRIC POWER.

Tidal power. The harnessing of tidal energy to generate electricity is a possibility for a few places such as Passamoquoddy Bay in Maine. However, the dams for such ventures and the system of locks would interfere with the normal flow of waters and would change the kinds and populations of water animals and plant life present. Scenic values could also be affected. The world's only operating large-scale tidal power plant is located on the estuary of the Rance River in France. While technically a success, it is not economical in comparison with conventional power sources because of its high capital costs. *See* TIDAL POWER.

Solar power. Although solar energy can already supply heat for water and space in individual buildings, there remains an open and controversial question whether it will prove a practical source for larger amounts of energy for industry or to generate electricity.

Solar energy, because it is so diffuse, requires large collecting surfaces to produce useful amounts of fuel or energy. A solar system involves large land areas. The land requires for a solar electric plant of 1000 MW output range from 10 mi^2 (26 km^2) for a thermal conversion system, to 30 mi^2 for photovoltaic conversion and up to 500 mi^2 for a fuel crop system. The land most likely to be used for central power stations is in the semidesert Southwest. Building the installations and access roads would disturb these lands. Once in place, the structures would shade the earth beneath them and probably change the kind of wildlife living there. In some places there is already local regulation to govern the shading effect of new buildings upon solar energy rights of neighbors. *See* SOLAR ENERGY.

Fusion power. Over the past few years significant scientific and engineering accomplishments have occurred which make it possible to conceive of practical power production using magnetic fusion before the end of the 20th century. Since fusion energy uses deuterium found in ocean water as its fuel source, this energy technology can supply the energy required by all humanity essentially forever. The demonstration of this technology will be the second most important event in the history of humanity—second only to the controlled use of fire. *See* NUCLEAR FUSION.

Wind power. Many of the schemes proposed for large-scale use of wind power involve construction of chains of tall towers. Their principal known environmental effects would be esthetic. Such tall structures would also increase hazards to air navigation at low altitudes and to bird flights, and might affect radar transmission and other means of communication. Wind machines would have little effect upon air and water quality, but might require land for their structures and for rights-of-way of transmission lines. *See* WIND POWER.

LEGISLATION FOR ENVIRONMENTAL PROTECTION

In the United States much of the present control of objectionable effects of increasing supplies of fuels and energy is based upon legislation by the Federal and state governments. Today the entrepreneur who arranges for the financing of energy facilities and the engineers who design them face limitations set by society, limitations which often can be as demanding as those of nature, materials, or technology.

Historically, in the Federal system of government, the control over health and environmental effects of producing and using fuels and energy resided in the States. This began to change in the mid-1940s with a growing presence of the Federal government in such regulation. The Atomic Energy Act of 1946 preempted to the Federal government the control of radioactive emissions from the use of nuclear power except for mining and refining of uranium. This plenary authority was continued by the Atomic Energy Act of 1954, and has been confirmed by the Supreme Court.

As for production and use of other fuels and sources of energy, since the 1950s there has been a stream of Federal legislation to strengthen the States and to give the Federal government more voice in protecting the quality of air, water, and earth. Overarching this still-evolving Federal legislation is the National Environmental Policy Act of 1969 (NEPA), which has provided the public and special-interest groups with unprecedented amounts of information about possible environmental effects of activities that require some kind of Federal license or approval.

Because Federal legislation to protect the environment is still evolving, the following discussion is limited to early landmarks. For the status of current Federal controls and regulations affecting the supply and use of fuels and energy, the latest regulations as they appear in the Code of Federal Regulations should be consulted.

National environmental quality. The National Environmental Policy Act of 1969 (NEPA) is notable in three respects: it declared a governmental mandate and responsibility for environmental quality; it created the Council on Environmental Quality; and it required all Federal agencies to prepare an environmental impact statement on proposed actions which could significantly affect the environment. These concepts, particularly that of the environmental impact statement, have since been incorporated into similar legislation in many states.

NEPA requires Federal officials to examine many factors in the preparation of an environmental impact statement, among them: (1) the environmental impact of the proposed actions; (2) any adverse environment impacts that cannot be avoided, should the proposal be implemented; (3) alternatives to the proposed action; (4) the relationship of local short-term uses of the environment and the maintenance and enhancement of long-term productivity; (5) any irreversible and irretrievable commitments of resources which would be involved in the proposed action.

Preparation of environmental impact statements is now a recognized component of proceedings for many major construction projects at the Federal, state, and local government level, and has significantly contributed to protection of the environment. These benefits have, of course, been obtained in exchange for some costs—direct ones in the preparation of such statements, and indirect, lost-opportunity costs during the additional time required to bring energy facilities into service.

Air quality. Federal legislation to protect air quality began in 1955 when Congress established the principle that state and local governments are responsible for air-pollution control, with the Federal government providing leadership, information, and support. The Clean Air Act Amendments of 1970 provided that while the state governments retain their fundamental responsibility for air-pollution control, the Federal government has a stronger voice. The Environmental Protection Agency now has the responsibility for establishing national standards. EPA in 1971 established standards for major air pollutants, including sulfur dioxide, particulates, carbon monoxide, hydrocarbons, nitrogen oxides, and photochemical oxidants. These standards form the basis for state control. In essence, the legislation requires polluters to install control technology or to take other steps which will permit air quality standards to be achieved.

The 1970 amendments also are the source of the controversial requirement for the substantial reduction of emissions of pollutants from motor vehicles.

Another legislative act affecting both the environment and the supply of fuels and energy is the Energy Supply and Environmental Coordination Act of 1971. This act prohibited power plants or other major fuel-burning installations from burning oil if conversion to coal was practicable, if coal was available, and if reliability of power was not impaired. However, a conversion cannot be ordered until it has been determined that the plant can comply with EPA air-pollution requirements and an environmental impact statement has been prepared. *See* AIR-POLLUTION CONTROL.

Water quality. The Water Quality Act of 1965 required the state governments to establish water-pollution control standards for all interstate waters, and placed most of the enforcement responsibility with the states. However, the state governments moved slowly. Congress, therefore, in 1972 revised the act with legislation which defined the respective responsibilities of the Federal and state governments and set 1985 as the date for attainment of a national goal of eliminating all discharges of pollutants into the waters. In this respect, waste heat is defined as a water pollutant. All point sources of effluents, which include power plants, are to be limited to levels achievable through use of "best practicable" technology, and to achieve by 1985, compliance by "best available" technology.

Strip mining control. The Surface Mining Control and Reclamation Act, passed by Congress in 1975, established strict controls over strip mining practices and requires reclamation of strip-mined land. The regulations promulgated by the Office of Surface Mining (the new agency created by the legislation) have been strongly opposed by the mining industry, and an amendment to significantly alter the law was passed by the Senate in 1979.

Nuclear power control. The Atomic Energy Act of 1954, as amended, and the Energy Reorganization Act of 1974 are the primary basis for regulation of the construction and operation of nuclear power plants. Together, these acts provide the framework of Federal authority and organization to regulate the safety and environmental impacts of nuclear power plants and associated facilities. The licensing process, which is administered by the Nuclear Regulatory Commission, includes the issuance of a construction permit and an operating license. One major part of the proceedings for both permits is the formulation, review, and consideration of environmental impact statements. Generally, the NRC analysis of a proposed nuclear power plant and the NRC regulations are intended to keep emissions of radioactive materials within acceptable limits and to control the manner in which waste heat is dissipated.

Coastal zone management. Oil spills and discharge of waste heat are of particular concern in coastal zones. In addition to controls established by other legislation, the Coastal Zone Management Act of 1972 requires that applicants for a Federal license or permit needed for an energy operation must certify that the proposed activity will be conducted in a manner consistent with the goals of the coastal zone management program. Another requirement forbids a Federal agency to issue a license or permit until the state involved has concurred with the applicant's certification.

CONCLUSION

The environmental effects inherent in the various forms of energy usage range from inconsequential to substantial. While it is obviously advantageous to control and reduce these effects to the maximum extent practicable, it must at the same time be recognized that environmental control measures are not without impacts of their own in terms of costs, jobs, resources, or other matters. Control could conceivably be carried to the unacceptable extreme where society is denied the use of the fuel or energy source in question. On the other hand, uncontrolled production and use of fuels and energy without any concern for impacts on the environment is equally unacceptable. The problem, then, is to balance society's need for energy against the need for a livable environment, at the same time giving appropriate attention to important economic, technical, and social factors. *See the feature articles* EXPLORING ENERGY CHOICES; THE RISK OF ENERGY PRODUCTION; *see also* AIR POLLUTION; WATER POLLUTION.

[MIKE McCORMACK]

Bibliography: A. P. Carter (ed.), *Energy and the Environment: A Structural Analysis*, 1976; J. Holdren and P. Herrera, *Energy: A Crisis in Power*, 1971; W. Ramsay, *Unpaid Costs of Electrical Energy: Health and Environmental Impacts from Coal and Nuclear Power*, 1978; R. S. Rouse and R. O. Smith, *Energy: Resource, Slave, Pollutant—A Physical Science Text*, 1975; E. H. Thorndike, *Energy and Environment: A Primer for Scientists and Engineers*, 1976; A. J. Van Tassel (ed.), *The Environmental Price of Energy*, 1975.

Energy Technology

A-Z

Air conditioning

The maintenance of certain aspects of the environment within a defined space to facilitate the intended function of that space. Environmental conditions generally encompassed by the term air conditioning include air temperature and motion, radiant heat energy level, moisture level, and concentration of various pollutants, including dust, germs, and gases. Because these environmental factors are associated with air itself, and because air temperature and motion are the factors most readily sensed, simultaneous control of all these factors is called air conditioning, although space conditioning is more descriptive of the activity.

Comfort air conditioning refers to control of spaces inhabited by people to promote their comfort, health, or productivity. Spaces in which air is conditioned for comfort include residences, offices, institutions, sports arenas, hotels, and factory work areas. Process air conditioning systems are designed to facilitate the functioning of a production, manufacturing, or operational activity. For example, heat-producing electronic equipment in an airplane cockpit must be kept cool to function properly, while the occupants of the cockpit are maintained at comfortable conditions. The environment around a multicolor printing press must have constant relative humidity to avoid paper expansion or shrinkage for accurate registration, while press heat and ink mists must be conducted away for the health of pressmen. Maintenance of conditions within surgical suites of hospitals and in "clean" or "white" rooms of manufacturing plants, where an almost germ- or dust-free atmosphere must be maintained, has become a specialized subdivision of process air conditioning.

Calculation of loads. Engineering of an air-conditioning system starts with selection of design conditions; air temperature and relative humidity are principal factors. Next, loads on the system are calculated. Finally, equipment is selected and sized to perform the indicated functions and to carry the estimated loads.

Design conditions are selected on the bases discussed above. Each space is analyzed separately. A cooling load will exist when the sum of heat released within the space and transmitted to the space is greater than the loss of heat from the space. A heating load occurs when the heat that is generated within the space is less than the loss of heat from it. Similar considerations apply to moisture.

Heat generated within the space consists of body heat, approximately 250 Btu/hr/person (73 W/person), heat from all electrical appliances and lights, 3.41 Btu/hr/watt (1 W of heat power per watt of electric power), and heat from other sources such as gas cooking stoves and industrial ovens. Heat is transmitted through all parts of the space envelope, which includes walls, floor, ceiling, and windows. Whether heat enters or leaves the space depends upon whether the outside surfaces are warmer or cooler than the inside surfaces. The rate at which heat is conducted through the space envelope is a function of the temperature difference across the envelope and the thermal conductance of the envelope. Conductances, which depend on materials of construction and their thicknesses along the path of heat transmission, are a large factor in walls and ceilings exposed to the outdoors in cold winters and hot summers. In these cases insulation is added to decrease the overall conductance of the envelope.

Solar heat loads are an especially important part of load calculation because they represent a large percentage of heat gain through walls and roofs, but are very difficult to estimate because solar irradiation is constantly changing. Intensity of radiation varies with the seasons (it rises to 457 Btu/hr/ft² or 1442 W/m² in midwinter and drops to 428 Btu/hr/ft² or 1347 W/m² in midsummer). Intensity of solar irradiation also varies with surface orientation. For example, the half-day total for a horizontal surface at 40 degrees north latitude on January 21 is 353 Btu/ft² (4.01 MJ/m²) and on June 21 it is 1121 Btu/ft² (12.73 MJ/m²), whereas for a south wall on the same dates comparable data are 815 and 311 Btu/ft² (9.26 and 3.53 MJ/m²), a sharp decrease in summer. Intensity also varies with time of day and cloud cover and other atmospheric phenomena.

The way in which solar radiation affects the space load depends also upon whether the rays are transmitted instantly through glass or impinge on opaque walls. If through glass, the effect begins immediately but does not reach maximum intensity until the interior irradiated surfaces have warmed sufficiently to reradiate into the space, warming the air. In the case of irradiated walls and roofs, the effect is as if the outside air temperature were higher than it is. This apparent temperature is called the sol-air temperature, of which tables are available.

In calculating all these heating effects, the object is proper sizing and intelligent selection of equipment; hence, a design value is sought which will accommodate maximums. However, when dealing with climatic data, which are statistical, historical summaries, record maximums are rarely used. For instance, if in a particular locality the recorded maximum outside temperature was 100°F (37.8°C), but 95°F (35°C) was exceeded only four times in the past 20 years, 95°F may be chosen as the design summer outdoor temperature for calculation of heat transfer through walls. In practice, engineers use tables of design winter and summer outdoor temperatures which list winter temperatures exceeded more than 99% and 97.5% of the time during the coldest winter months, and summer temperatures not exceeded 1%, 2.5%, and 5% of the warmest months. The designer will select that value which represents the conservatism required for the particular type of occupancy. If the space contains vital functions where impairment due to occasional departures from design space conditions cannot be tolerated, the more severe design outdoor conditions will be selected.

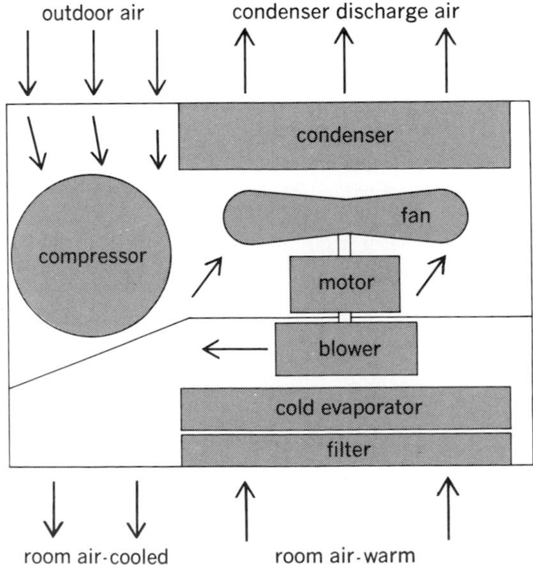

Fig. 1. Schematic of room air conditioner. (*American Society of Heating, Refrigerating and Air-Conditioning Engineers, Inc., Guide and Data Book, 1967*)

In the case of solar load through glass, but even more so in the case of heat transfer through walls and roof, because outside climate conditions are so variable, there may be a considerable thermal lag. It may take hours before the effect of extreme high or low temperatures on the outside of a thick masonry wall is felt on the interior surfaces and space. In some cases the effect is never felt on the inside, but in all cases the lag exists, exerting a leveling effect on the peaks and valleys of heating and cooling demand; hence, it tends to reduce maximums and can be taken advantage of in reducing design loads.

Humidity as a load on an air conditioning system is treated by the engineer in terms of its latent heat, that is, the heat required to condense or evaporate the moisture, approximately 1000 Btu/lb (2.3 MJ/kg) of moisture. People at rest or at light work generate about 200 Btu/hr (59 W). Steaming from kitchen activities and moisture generated as a product of combustion of gas flames, or from all drying processes, must be calculated. As with heat, moisture travels through the space envelope, and its rate of transfer is calculated as a function of the difference in vapor pressure across the space envelope and the permeability of the envelope construction. To decrease permeability where vapor pressure differential is large, vapor barriers (relatively impermeable membranes) are incorporated in the envelope construction.

Another load-reducing factor to be calculated is the diversity among the various spaces within a building or building complex served by a single system. Spaces with east-facing walls experience maximum solar loads when west-facing walls have no solar load. In cold weather, rooms facing south may experience a net heat gain due to a preponderant solar load while north-facing rooms require heat. An interior space, separated from adjoining spaces by partitions, floor, and ceiling across which there is no temperature gradient, experiences only a net heat gain, typically from people and lights. Given a system that can transfer this heat to other spaces requiring heat, the net heating load may be zero, even on cold winter days.

Air conditioning systems. A complete air conditioning system is capable of adding and removing heat and moisture and of filtering dust and odorants from the space or spaces it serves. Systems that heat, humidify, and filter only, for control of comfort in winter, are called winter air conditioning systems; those that cool, dehumidify, and filter only are called summer air conditioning systems, provided they are fitted with proper controls to maintain design levels of temperature, relative humidity, and air purity.

Design conditions may be maintained by multiple independent subsystems tied together by a single control system. Such arrangements, called split systems, might consist, for example, of hot-water baseboard heating convectors around a perimeter wall to offset window and wall heat losses when required, plus a central cold-air distribution system to pick up heat and moisture gains as required and to provide filtration for dust and odor.

Air conditioning systems are either unitary or built-up. The window or through-the-wall air conditioner (Fig. 1) is an example of a unitary summer air conditioning system; the entire system is housed in a single package which contains heat removal, dehumidification, and filtration capabilities. When an electric heater is built into it with suitable controls, it functions as a year-round air conditioning system. Unitary air conditioners are manufactured in capacities as high as 100 tons (1 ton of air conditioning equals 12,000 Btu/hr or 3.517 kW) and are designed to be mounted conveniently on roofs, on the ground, or other convenient location, where they can be connected by ductwork to the conditioned space.

Built-up or field-erected systems are composed of factory-built subassemblies interconnected by means such as piping, wiring, and ducting during final assembly on the building site. Their capacities range up to thousands of tons of refrigeration and millions of Btu per hr of heating. Most large buildings are so conditioned.

Another important and somewhat parallel distinction can be made between incremental and central systems. An incremental system serves a single space; each space to be conditioned has its own, self-contained heating-cooling-dehumidifying-filtering unit. Central systems serve many or all of the conditioned spaces in a building. They range from small, unitary packaged systems to serve single-family residences to large, built-up or field-erected systems serving large buildings.

When many buildings, each with its own air conditioning system which is complete except for a refrigeration and a heating source, are tied to a central plant that distributes chilled water and hot water or steam, the interconnection is referred to as a district heating and cooling system. This system is especially useful for campuses, medical complexes, and office complexes under a single management.

Conditioning of spaces. Air temperature in a space can be controlled by radiant panels in floor, walls, or ceiling to emit or absorb energy, depending on panel temperature. Such is the radiant panel system. However, to control humidity and air purity, and in most systems for controlling air temperature, a portion of the air in the space is withdrawn, processed, and returned to the space to mix with the remaining air. In the language of the engineer, a portion of the room air is returned (to an air-handling unit) and, after being conditioned, is supplied to the space. A portion of the return air is spilled (exhausted to the outdoors) while an equal quantity (of outdoor air) is brought into the system and mixed with the remaining return air before entering the air handler.

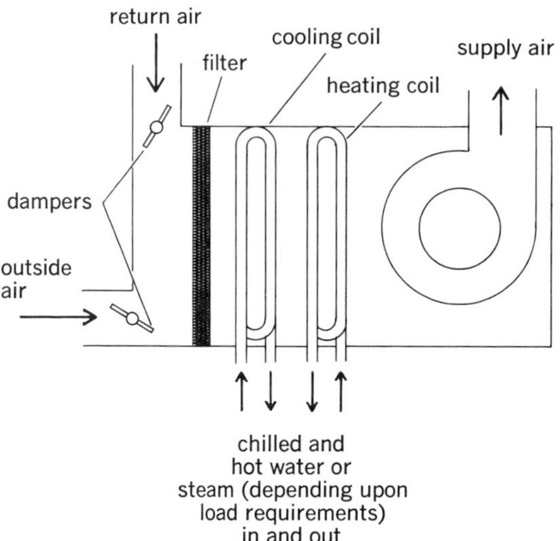

Fig. 2. Schematic of central air-handling unit.

Typically, the air-handling unit contains a filter, a cooling coil, a heating coil, and a fan in a suitable casing (Fig. 2). The filter removes dust from both return and outside air. The cooling coil, either containing recirculating chilled water or boiling refrigerant, lowers air temperature sufficiently to dehumidify it to the required degree. The heating coil, in winter, serves a straightforward heating function, but when the cooling coil is functioning, it serves to raise the temperature of the dehumidified air (to reheat it) to the exact temperature required to perform its cooling function. The air handler may perform its function, in microcosm, in room units in each space, as part of a self-contained, unitary air conditioner, or it may be a huge unit handling return air from an entire building. *See* AIR COOLING.

There are three principal types of central air-conditioning systems: all-air, all-water, and air-water. In the all-air system all return air is processed in a central air-handling apparatus. In one type of all-air system, called dual-duct, warm air and chilled air are supplied to a blending or mixing unit in each space. In a single-duct all-air system air is supplied at a temperature for the space requiring the coldest air, then reheated by steam or electric or hot-water coils in each space.

In the all-water system the principal thermal load is carried by chilled and hot water generated in a central facility and piped to coils in each space; room air then passes over the coils. A small, central air system supplements the all-water system to provide dehumidification and air filtration. The radiant panel system, previously described, may also be in the form of an all-water system.

In an air-water system both treated air and hot or chilled water are supplied to units in each space. In winter hot water is supplied, accompanied by cooled, dehumidified air. In summer chilled water is supplied with warmer (but dehumidified) air. One supply reheats the other.

All-air systems preceded the others. Primary motivation for all-water and air-water systems is their capacity for carrying large quantities of heat energy in small pipes, rather than in larger air ducts. To accomplish the same purpose, big-building all-air systems use high velocities and pressures, requiring much smaller ducts. *See* SOLAR HEATING AND COOLING. [RICHARD L. KORAL]

Air cooling

Lowering of air temperature for comfort, process control, or food preservation. Air and water vapor occur together in the atmosphere. The mixture is commonly cooled by direct convective heat transfer of its internal energy (sensible heat) to a surface or medium at lower temperature. In the most compact arrangement, transfer is through a finned (extended surface) coil, metallic and thin, inside of which is circulating either chilled water, antifreeze solution, brine, or boiling refrigerant. The fluid acts as the heat receiver. Heat transfer can also be directly to a wetted surface, such as water droplets in an air washer or a wet pad in an evaporative cooler. *See* AIR CONDITIONING; HEAT TRANSFER.

Evaporative cooling. For evaporative cooling, nonsaturated air is mixed with water. Some of the sensible heat transfers from the air to the evaporating water. The heat then returns to the airstream as latent heat of water vapor. The exchange is thermally isolated (adiabatic) and continues until the air is saturated and air and water temperatures are equal. With suitable apparatus, air temperature approaches within a few degrees of the theoretical limit, the wet-bulb temperature. Evaporative cooling is frequently carried out by blowing relatively dry air through a wet mat (Fig. 1). The technique is employed for air cooling of machines where higher humidities can be tolerated; for cooling of industrial areas where high humidities are required (textile mills); and for comfort cooling in hot dry climates, where partial saturation results in cool air at relatively low humidity.

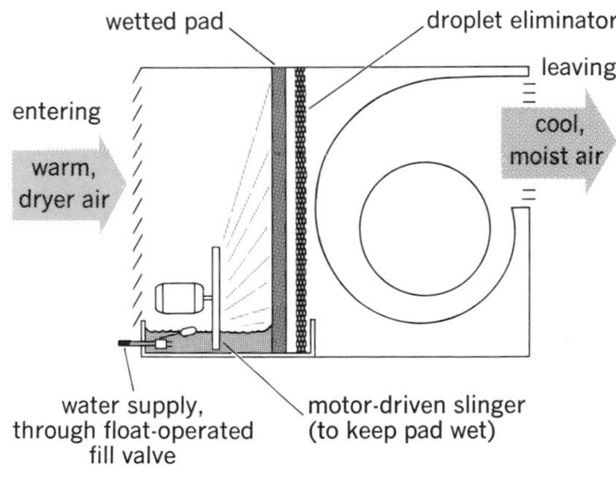

Fig. 1. Schematic view of simple evaporative air cooler.

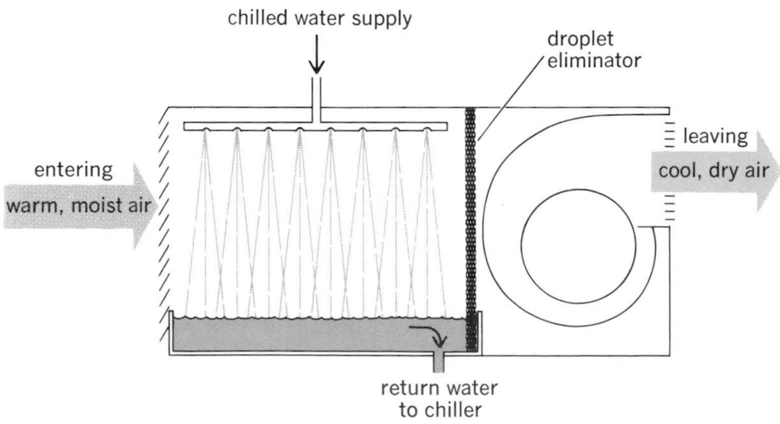

Fig. 2. Schematic of air washer.

Air washer. In the evaporative cooler the air is constantly changed and the water is recirculated, except for that portion which has evaporated and which must be made up. Water temperature remains at the adiabatic saturation (wet-bulb) temperature. If water temperature is controlled, as by refrigeration, the leaving air temperature can be controlled within wide limits. Entering warm, moist air can be cooled below its dew point so that, although it leaves close to saturation, it leaves with less moisture per unit volume of air than when it entered. An apparatus to accomplish this is called an air washer (Fig. 2). It is used in many industrial and comfort air conditioning systems, and performs the added functions of cleansing the airstream of dust and of gases that dissolve in water, and in winter, through the addition of heat to the water, of warming and humidifying the air.

Air-cooling coils. The most important form of air cooling is by finned coils, inside of which circulates a cold fluid or cold, boiling refrigerant (Fig. 3). The latter is called a direct-expansion (DX) coil. In most applications the finned surfaces become wet as condensation occurs simultaneously with sensible cooling. Usually, the required amount of dehumidification determines the temperature at which the surface is maintained and, where this results in air that is colder than required, the air is reheated to the proper temperature. Droplets of condensate are entrained in the airstream, removed by a suitable filter (eliminator), collected in a drain pan, and wasted.

In the majority of cases, where chilled water or boiling halocarbon refrigerants are used, aluminum fins on copper coils are employed. Chief advantages of finned coils for air cooling are (1) complete separation of cooling fluid from airstream, (2) high velocity of airstream limited only by the need to separate condensate that is entrained in the airstream, (3) adaptability of coil configuration to requirements of different apparatus, and (4) compact heat-exchange surface.

Defrosting. Wherever air cooling and dehumidification occur simultaneously through finned coils, the coil surface must be maintained above 32°F (0°C) to prevent accumulation of ice on the coil. For this reason, about 35°F (1.7°C) is the lowest-temperature air that can be provided by coils (or air washers) without ice accumulation. In cold rooms, where air temperature is maintained below 32°F (0°C), provision is made to deice the cooling boils. Ice buildup is sensed automatically; the flow of cold refrigerant to the coil is stopped and replaced, briefly, by a hot fluid which melts the accumulated frost. In direct-expansion coils, defrosting is easily accomplished by bypassing hot refrigerant gas from the compressor directly to the coil until defrosting is complete.

Cooling coil sizing. Transfer of heat from warm air to cold fluid through coils encounters three resistances: air film, metal tube wall, and inside fluid film. Overall conductance of the coil, U, is shown in the equation below, where K_o is film conductance

$$\frac{1}{U}=\frac{1}{K_o}+r_m+\frac{R}{K_i}$$

of the outside (air-side) surface in Btu per (hr)(sq ft)(F); r_m is metal resistance in (hr)(sq ft)(F) per Btu, where area is that of the outside surface; K_i is film conductance of the inside surface (water, steam, brine, or refrigerant side) in Btu per (hr)(sq ft)(F); U is overall conductance of transfer surface in Btu per (hr)(sq ft)(F), where area again refers to the outside surface; and R is the ratio of outside surface to inside surface.

Values of K_o are a function of air velocity and typically range from about 4 Btu per (hr)(sq ft)(F) [23 W/(m²)(°C)] at 100 feet per minute (fpm) (0.51 m/sec) to 12 (68) at 600 fpm (3.05 m/sec). If condensation takes place, latent heat released by the condensate is in addition to the sensible heat transfer. Then total (sensible plus latent) K_o increases by the ratio of total to sensible heat to be transferred, provided the coil is well drained.

Values of r_m range from 0.005 to 0.030 (hr)(sq ft)(F) per Btu, depending somewhat on type of metal but primarily on metal thickness.

Typical values for K_i range from 250 to 500 Btu per (hr)(sq ft)(F) [1.4 to 2.8 kW/(m²)(°C)] for boiling refrigerant. In 40°F (4.4°C) chilled water, values range from 230 Btu per (hr)(sq ft)(F) [1.3 kW/(m³)(°C)] when water velocity is 1 foot per second (fps) (0.3 m/sec) to 1250 (7.1) when water velocity is 8 fps (2.4 m/sec).

Use of well water. Well water is available for air cooling in much of the world. Temperature of water from wells 30 to 60 ft (9 to 18 m) deep is approximately the average year-round air temperature in the locality of the well, although in some regions overuse of available supplies for cooling purposes and recharge of ground aquifers with higher-temperature water has raised well water temperature several degrees above the local normal. When well water is not cold enough to dehumidify air to the required extent, an economical procedure is to use it for sensible cooling only, and to pass the cool, moist air through an auxiliary process to dehumidify it. Usually, well water below 50°F (10°C) will dehumidify air sufficiently for comfort cooling. Well water at these temperatures is generally available in the northern third of the United States, except the Pacific Coast areas.

Ice as heat sink. For installations that operate only occasionally, such as some churches and meeting halls, water recirculated and cooled over ice offers an economical means for space cooling (Fig. 4). Cold water is pumped from an ice bunker through an extended-surface coil. In the coil the water absorbs heat from the air, which is blown across the coil. The warmed water then returns to the bunker, where its temperature is again reduced by the latent heat of fusion (144 Btu/lb) to 32°F. Although initial cost of such an installation is low, operating costs are usually high.

Refrigeration heat sink. Where electric power is readily available, the cooling function of the ice, as described above, is performed by a mechanical refrigerator. If the building complex includes a steam plant, a steam-jet vacuum pump can be used to cause the water to evaporate, thereby lowering its temperature by the latent heat of evaporation (about 1060 Btu/lb or 2440 kW/kg, depending on temperature and pressure). High-pressure steam, in passing through a primary ejector, aspirates water vapor from the evaporator, thereby maintaining the required low pressure that causes the water to evaporate and thus to cool itself (Fig. 5).

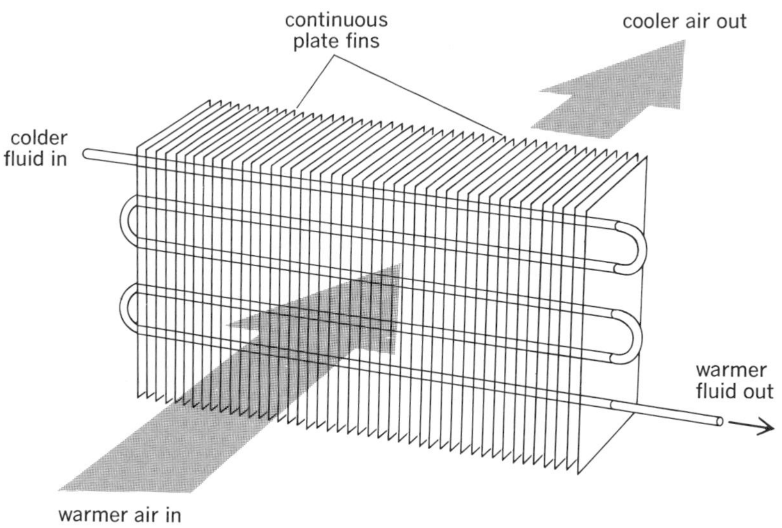

Fig. 3. Typical extended-surface air-cooling coil.

Where electric power is costly compared to low-temperature heat, such as by gas, absorption refrigeration may be used. Two fluids are used: an

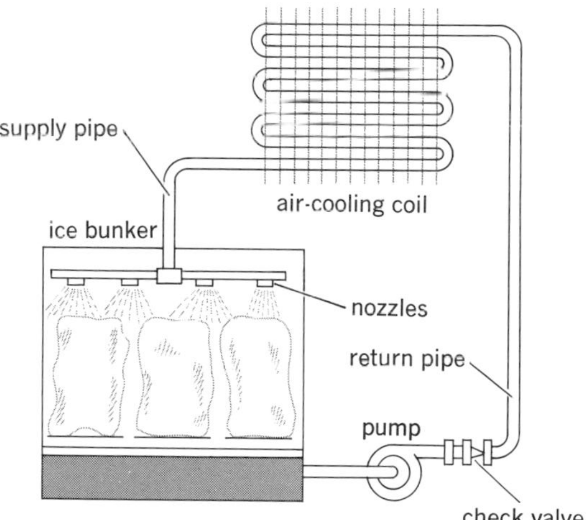

Fig. 4. Air cooling by circulating ice-cooled water.

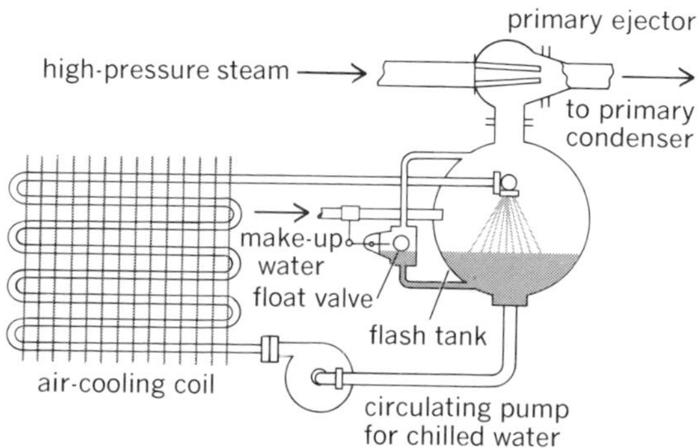

Fig. 5. Air cooling by circulating water that is cooled, in turn, by evaporation in flash tank.

absorbent and a refrigerant. The absorbent is chosen for its affinity for the refrigerant when in vapor form, for example, water is used as the absorber with ammonia as the refrigerant. Concentrated ammonia water is pumped to a high pressure and then heated to release the ammonia. The high-pressure ammonia then passes through a condenser, an expansion valve, and an evaporator, as in a mechanical system, and is reabsorbed by the water. The cycle cools air circulated over the evaporator. [RICHARD L. KORAL]

Bibliography: Air Conditioning and Refrigeration Institute, *Refrigeration and Air Conditioning*, 1979; American Society of Heating and Air Conditioning Engineers, *Guide and Data Book*, annual; S. Elonka and Q. W. Minich, *Standard Refrigeration and Air Conditioning Questions and Answers*, 2d ed., 1973.

Air pollution

Alteration of the atmosphere by the introduction of natural and artificial particulate contaminants. Most artificial impurities are injected into the atmosphere at or near the Earth's surface. The atmosphere cleanses itself of these quickly, for the most part. This occurs because in the troposphere, that part of the atmosphere nearest to the Earth, temperature decreases rapidly with increasing altitude (Fig. 1), resulting in rapid vertical mixing: the rainfall sometimes associated with these conditions also assists in removing the impurities.

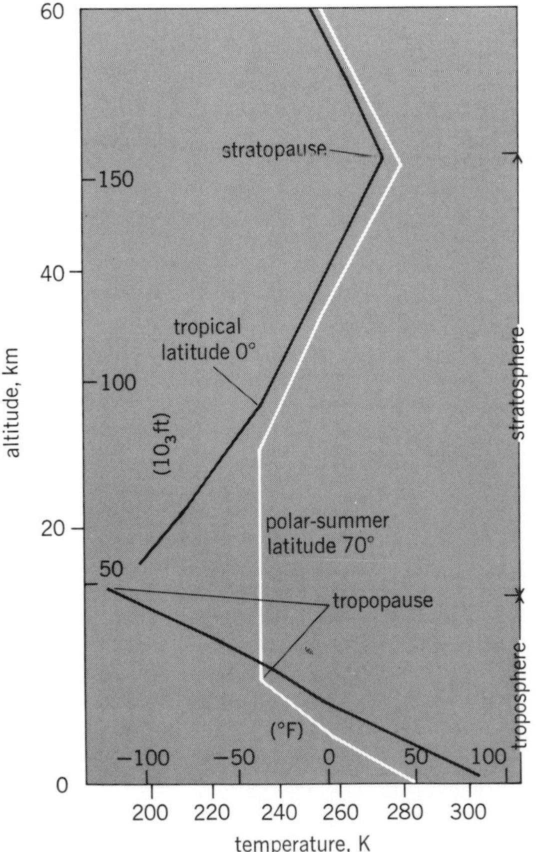

Fig. 1. The atmosphere's temperature-altitude profile. *(Adapted from R. E. Newell, Radioactive Contamination of the Upper Atmosphere, in Progress in Nuclear Energy, ser. 12: Health Physics, vol. 2, p. 538, Pergamon Press, 1969)*

Exceptions, such as the occasional temperature inversion layer over the Los Angeles Basin, may have notably unpleasant results.

In the stratosphere, that is, above the altitude of the temperature minimum (tropopause), either temperature is constant or it increases with altitude, a condition that characterizes the entire stratosphere as a permanent inversion layer. As a result, vertical mixing in the stratosphere (and hence, self-cleansing) occurs much more slowly than that in the troposphere. Contaminants introduced at a particular altitude remain near that altitude for periods as long as several years. Herein lies the source of concern: the turbulent troposphere cleanses itself quickly, but the relatively stagnant stratosphere does not. Nonetheless, the pollutants injected into the troposphere and stratosphere have impact on humans and the habitable environment.

All airborne particulate matter, liquid and solid, and contaminant gases exist in the atmosphere in variable amounts. Typical natural contaminants are salt particles from the oceans or dust and gases from active volcanoes; typical artificial contaminants are waste smokes and gases formed by industrial, municipal, household, and automotive processes, and aircraft and rocket combustion processes. Another postulated important source of artificial contaminants is certain fluorocarbon compounds (gases) used widely as refrigerants and as propellants for aerosol products. Pollens, spores, rusts, and smuts are natural aerosols augmented by humans' land-use practices.

Sources and types. Sources may be characterized in a number of ways. A frequent classification is in terms of stationary and moving sources. Examples of stationary sources are power plants, incinerators, industrial operations, and space heating. Examples of moving sources are motor vehicles, ships, aircraft, and rockets. Another classification describes sources as point (a single stack), line (a line of stacks), or area (a city).

Different types of pollution are conveniently specified in various ways: gaseous, such as carbon monoxide, or particulate, such as smoke, pesticides, and aerosol sprays; inorganic, such as hydrogen fluoride, or organic, such as mercaptans; oxidizing substances, such as ozone, or reducing substances, such as oxides of sulfur and oxides of nitrogen; radioactive substances, such as iodine-131, or inert substances, such as pollen or fly ash; or thermal pollution, such as the heat produced by nuclear power plants.

Air contaminants are produced in many ways and come from many sources. It is difficult to identify all the various producers. For example, it is estimated that in the United States 60% of the air pollution comes from motor vehicles and 14% from plants generating electricity. Industry produces about 17% and space heating and incineration the remaining 9%. Other sources, such as pesticides and earth-moving and agricultural practices, lead to vastly increased atmospheric burdens of fine soil particles, and of pollens, pores, rusts, and smuts; the latter are referred to as aeroallergens because many of them induce allergic responses in sensitive persons.

The annual emission over the United States of many contaminants is very great (Fig. 2). As mentioned, motor vehicles contribute about 60%

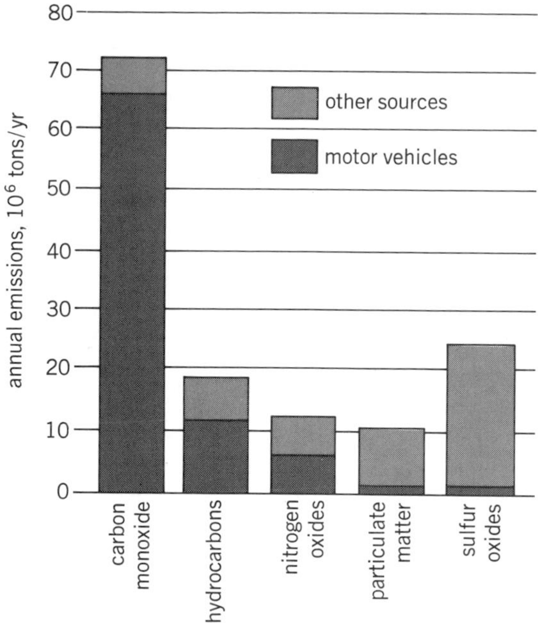

Fig. 2. Motor vehicles' contribution to five atmospheric contaminants in the United States.

of total pollution: nearly all the carbon monoxide, two-thirds of the hydrocarbons, one-half of the nitrogen oxides, and much smaller fractions in other categories.

Pollution in the stratosphere. Sources of contaminants in the stratosphere are effluents from high-altitude aircraft such as the supersonic transport (SST), powerful nuclear explosions, and volcanic eruptions. There are also natural and artificial sources of gases which diffuse from the troposphere into the stratosphere.

Table 1 lists the natural burden of gases and particles injected into the stratosphere by high-flying aircraft, assuming the consumption of 2×10^{11} kg of fuel during a period of one year. It is to be noted that the percentage increase over the natural burden is substantial for NO, NO_2, HNO_3, and SO_2. Since the concentrations are substantial, they may adversely affect humans' environment.

Other manufactured pollutants which diffuse from the troposphere into the stratosphere are the halogenated hydrocarbons, specifically dichloromethane (CF_2Cl_2) and trichlorofluoromethane ($CFCl_3$) gases which are used as propellants in many of the so-called aerosol spray cans for deodorants, pesticides, and such, and as refrigerants. These gases have an average residence time (residence time is the time required for a substance to reduce its concentration by $1/e$, approximately 1/3) of 1000 years or more. $CFCl_3$ is one of a family of halogenated hydrocarbons (also known as fluorocarbons) which are widely used. Of the total amount of these compounds produced worldwide, almost all are ultimately released to the atmosphere.

The nitrous oxide (N_2O) and the halogenated hydrocarbons reach the upper regions of the stratosphere, where they are photodissociated by the Sun's radiation to produce nitric oxide (NO) and chlorine (Cl). The NO and Cl react with ozone, as in reactions (1)–(6). Ozone is destroyed by

$$N_2O + O \rightarrow 2NO \quad (1)$$

$$NO + O_3 \rightarrow NO_2 + O_2 \quad (2)$$

$$NO_2 + O \rightarrow NO + O_2 \quad (3)$$

$$CF_2Cl_3 + h\nu \text{ (solar energy)} \rightarrow CFCl_2 + Cl \quad (4)$$

$$Cl + O_3 \rightarrow ClO + O_2 \quad (5)$$

$$ClO + O \rightarrow Cl + O_2 \quad (6)$$

NO and Cl respectively, whereas NO and Cl are conserved. Stratospheric ozone is valuable as a filter for solar ultraviolet (uv) radiation. A decrease of its concentration results in an increase in the amount of uv impinging on the surface of the Earth. As an illustration based on theoretical considerations, an injection 2×10^9 kg/yr of $NO_x (NO_x = NO + NO_2)$ at 17 km (see Fig. 1) will result in a reduction of the total amount of ozone by about 3%. This represents an increase of vertically incident uv on the Earth of 6%. An increase of uv could adversely affect both humans and plants.

SO_2-aerosol-climate relations. Aircraft engine effluents contain SO_2, as shown in Table 1, and could add a considerable amount of aerosol particles at the end of this century for the predicted aircraft fleet sizes. Table 1 indicates aircraft effluents could increase the natural background by

Table 1. The natural stratospheric background of several atmospheric gases from 13 to 24 km compared to engine emissions

Gas	Mass mixing ratio	Natural burden, kg		Increase in mass due to aircraft emission, %‡
		IDA*	Penndorf†	
CO_2	480 E-6	500 E-12§	480 E-12	0.1
H_2	2.7 E-6	2 E-12	2.7 E-12	9
CH_4	0.55 E-6	1 E-12	0.55 E-12	0.02
CO	0.05–0.1 E-6	30 E-9	50–100 E-9	0.6–1.2
NO	0.5 E-9	1 E-9	0.52 E-9	100
NO_2	1.6 E-9	3 E-9	1.8 E-9	100
HNO_3	4 E-9	<10 E-9	3.6 E-9	85
SO_2	1 E-9	4 E-9 (?)	1.4 E-9	10–40
Aerosol ($\alpha > 0.1\ \mu m$)	2 E-9	0.3 E-9	2 E-9	10

*Estimation by R. Oliver, Institute for Defense Analyses, 1974.

†Estimation by R. Penndorf, *CIAP Atmospheric Monitoring and Experiments, The Program and Results*, DOT-TST-75-106, pp. 4–7, 1975.

‡One-year fuel consumption by stratospheric aircraft of 2×10^{11} kg.

§Read 500 E-12 as 5×10^{12}.

10–40%—seemingly large, yet small compared to volcanic injections, for which estimates range up to 10,000%.

Why are particles so important? They scatter and absorb (in specific wavelength regions) solar radiation, and thereby influence the radiative budget of the Earth-atmospheric system, and finally perhaps the climate on the ground. The particles formed from aircraft effluents may increase the optical thickness of the layer, the upwelling and downwelling infrared radiation, and the albedo of the Earth. The average global albedo of the Earth-atmospheric system has been measured as 28% with probably some short-term variation of unknown but small magnitude. It has been calculated that for an additional mass of 0.1 μg/m³ particles over a 10-km layer from 15 to 25 km (equivalent to 5.1×10^8 kg for the whole Earth, or about 20% of the "natural" background concentration), the albedo increases by about 0.05% (from 28 to 28.05%) at low latitudes all year and at high latitudes in summer, but by about 0.1–0.15% from September to February in high latitudes. If the added mass is larger than 0.1 μg/m³, the albedo increases proportionally to the cited numbers. For the optical thickness of the stratosphere, a value of 0.02 is generally assumed. While the present subsonic flights increase this value by a very small amount (10^{-4}), a large fleet of high-flying aircraft (Table 1) could increase it by about 10%.

The chemistry of the natural stratospheric aerosols is dominated by sulfate, presumably of volcanic origin. These naturally occurring aerosols are concentrated in thin layers at altitudes between 15 and 25 km. R. Cadle has described the chemical composition of stratospheric particles at an altitude of 20 km during the period 1969–1973 as consisting of 48%, by mass, of sulfate; 24% of stony elements (such as silicon, aluminum, calcium, and magnesium); and 20% of chlorine; other constituents make up the remainder. The amount of sulfate introduced into the stratosphere as a result of a large fleet of SSTs operating at about 18–20 km may, in a worst-case estimate, equal the total amount occurring naturally. The influence of dynamic motions of the stratosphere on the distribution of aerosols is indicated by Fig. 2, which illustrates high correlation of the aerosol and ozone-rich layers in the 15-km region. Moreover, the water vapor mixing ratio also increases in layers at about 15–20 km. Since there is no known chemical link between the production of aerosols and that of water vapor and ozone, the observation illustrated by Fig. 3 may be a dynamic rather than a chemical effect, the implication being that the dynamic effects are far more important in determining the relative profiles of these constituents than any chemical effect at this altitude.

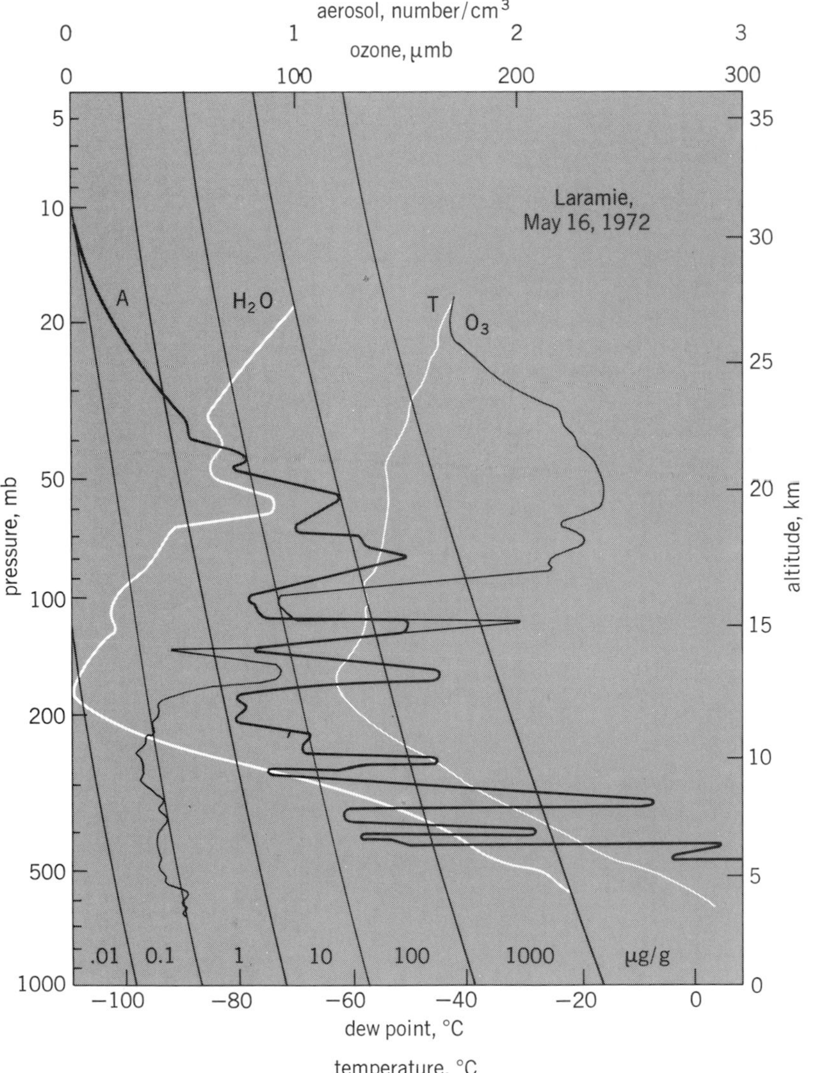

Fig. 3. A simultaneous measurement of the vertical distribution of ozone (O_3), water vapor (H_2O), and temperature (T). The aerosol (A) sounding was made about 5 hr before the ozone–water vapor soundings. The smooth curves are lines of constant water vapor mixing ratio. 1 mb = 10^2 Pa; 1 μmb = 10^{-4} Pa. *(From T. J. Pepin, J. M. Rosen, and D. H. Hoffman, The University of Wyoming Global Monitoring Program, in Proceedings of the AIAA/AMS International Conference of the Environmental Impact of Aerospace Operations in the High Atmosphere, AIAA Pap. no. 73-521, 1973)*

The perturbation of the lower stratosphere by the engine effluents of a large fleet of vehicles may strongly increase its optical thickness to visual-band solar radiation, which has a natural value of about 0.02. An increase in optical thickness results in a reduction of solar radiation, the principal source of atmospheric heating, by about the same fraction. This effect can be likened to a reduction in solar constant by about the same amount at the subsolar point, and about double that value when the solar zenith angle is 60° or greater, as it may be at high latitudes. The sensitivity of the troposphere to changes in the solar constant has been studied by M. MacCracken. In the modeling for which the computed precipitation is illustrated by Fig. 4, a 10% reduction of solar constant leads to a reduction of average temperatures, from 3°C at the Equator to 10°C average from latitude 40° to the pole. The winds are substantially weakened, and total precipitation reduced, mainly in the summer. Snowfall increases, and the winter snow line moves lower in latitude by about 5°.

An increase in the solar constant of 13% results in an annual mean temperature increase of 2–5°C at all latitudes. Although the overall precipitation increases, as illustrated in Fig. 4, the relative humidity decreases slightly, and thus cloudiness decreases. The total snowfall decreases, and the snow line moves higher in latitude

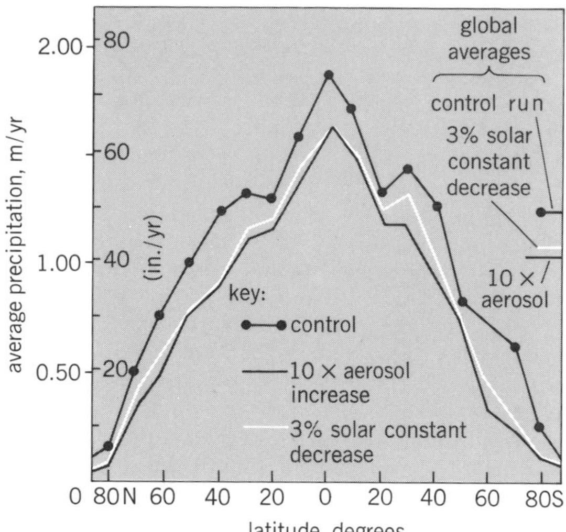

Fig. 4. Latitudinal distribution of total precipitation. *(From M. C. MacCracken, in Report of Findings: Effects of Stratospheric Pollution by Aircraft, DOT-TST-75-50, 1974)*

by 10°. Extensive melting of the polar ice caps also begins. The effects of doubling of the stratospheric optical thickness, for example, are an order of magnitude smaller than the changes of precipitation and snowfall depicted in Fig. 4.

Sinks. A sink is defined as a process by which gases or particles are removed from a given volume of atmosphere. It could be chemical, homogeneous or heterogenous (gas-solid reactions), or dynamical, such as dispersion (transport), diffusion, gravity, or precipitation.

In the stratosphere, three important contaminants are NO, Cl, and SO_2. The sink for NO_2 is a chemical reaction by which NO_2 is combined by a complex method with (OH)—a derivative of water (H_2O) present in the stratosphere in parts per million (ppm)—to form nitric acid (HNO_3).

The SO_2 reacts chemically with oxygen (O), water, and its derivatives (OH, HO_2) to form H_2SO_4. Then, water molecules are absorbed by H_2SO_4 to form $nH_2O \cdot (H_2SO_4)$ cluster or aerosol (where n, number, is equal to about 2). Aggregates of these polymolecules, growing larger with each successive collision, eventually become aerosols; when the diameter is greater than 0.1 μm, they act as scatterers of the visual band of sunlight.

The chlorine (Cl) chemically reacts with oxygen and water to form Cl, ClO, and HCl. In the stratosphere, these reactions involving small concentrations have not been measured adequately.

In the troposphere, the predominant sinks are dispersion, transport, and precipitation. The chemical reactions in the troposphere are less active in general than in the stratosphere.

Dispersion. Dispersion of pollution is dependent on atmospheric conditions. Winds transport and diffuse contaminants; rain may wash them to the surface; and under cloudless skies, solar radiation may induce important photochemical reactions.

Wind direction, speed, and turbulence influence atmospheric pollution. Wind direction determines the area into which the pollution is carried. Dilution of contaminants from a source is directly proportional, other factors being constant, to wind speed, which also determines the intensity of mechanical turbulence produced as the wind flows over and around surface objects, such as trees and buildings.

Eddy diffusion by wind turbulence is the primary mixing agency in the troposphere; molecular diffusion is negligible in comparison. In addition to mechanical turbulence, there is thermal turbulence which occurs in an unstable layer of air. Thermal turbulence and associated intense mixing develop in an unsaturated layer in which the temperature decreases with height at a rate greater than 1°C/100 m, the dry adiabatic rate of cooling. When the temperature decreases at a lower rate, the air is stable, and turbulence and mixing, now primarily mechanical, are less intense. If the temperature increases with height—which is the normal behavior of temperature in the stratosphere, creating a condition known as an inversion—the air is stable, and horizontal turbulence and mixing are still appreciable, but vertical turbulence and mixing are almost completely suppressed.

Inversion. Precipitation, fog, and solar radiation exert secondary meteorological influences. Falling raindrops may collect particles with radii greater than 1 μm or may entrain gases and smaller particles and carry them to the ground. Gas reactions with aerosols also occur; neutralizing cations in fog droplets or traces of ammonia (NH_3) in the air act as catalysts to accelerate reaction rates leading to rapid oxidation of sulfur dioxide (SO_2) in fog droplets. For highly polluted city air, it is estimated that in the presence of NH_3, the oxidation of the SO_2 to ammonium sulfate, $(NH_4)_2SO_4$, is completed in 1 hr for fog droplets 10 μm in radius. Photochemical oxidation of hydrocarbons in sunlight is frequent. Most hydrocarbons do not have appropriate absorption bands for a direct photochemical reaction; nitrogen dioxide (NO_2), when present, acts as an oxidation catalyst by absorbing solar radiation strongly and subsequently transferring the light energy to the hydrocarbon and thereby oxidizing it.

Natural ventilation in the atmosphere is best when the winds are strong and turbulent so that mixing is good, and when the volume in which mixing occurs is large so that dilution of pollution is rapid. As cities have grown in size, air pollution has become more widespread. It has become necessary to think of whole urban complexes as large area sources of pollution. The rate of natural ventilation of an urban area is dependent on two quantities: the wind speed and the mixing volume over the city. Active mixing upward is often limited by a stable layer, perhaps even a very stable inversion layer, aloft. The upward extent of this region of active mixing, known as the mixing height, determines the magnitude of the mixing volume of the city.

The number of air changes per unit time in this mixing volume specifies the rate of natural ventilation of the urban area. The problems of air pollution become highly complex, however, because the mixing height is rarely constant for long. Some of the factors causing it to vary are described below.

At night when the sky is clear and the wind light, Earth's surface loses heat by long-wave radi-

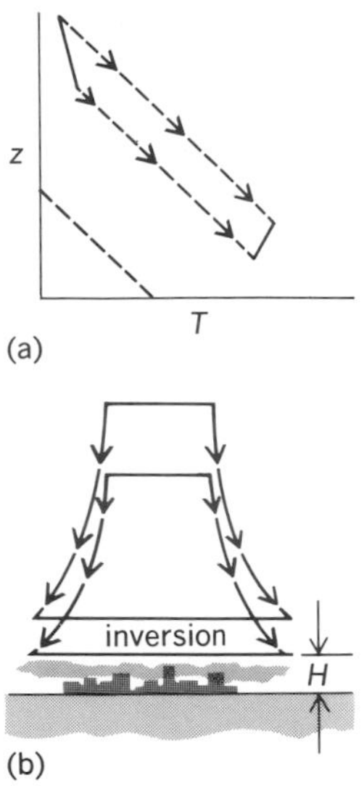

AIR POLLUTION

Fig. 5. Subsidence inversion. (*a*) Solid lines show temperature *T* and height *z* before and after dry adiabatic descent of air; dashed lines represent dry adiabatic rate of heating. (*b*) Inversion limits mixing height *H* over city.

ation to space. As a result, the ground cools and a surface radiation inversion is formed. The inversion inhibits mixing, so that pollution accumulates. Solar heating of the ground causes a reversal of the lapse rate, which may exceed the dry adiabatic rate of cooling and enhances active mixing in the unstable layer.

The mixing may bring pollution from aloft, causing a temporary peak in the surface concentrations, a process known as an inversion breakup fumigation. By midafternoon, the height of mixing is a maximum for the day, and surface concentrations tend to be low as the natural ventilation improves. In the evening, the lapse rate becomes stable, and accumulation of contaminants may begin again.

Subsidence inversion. The accumulation of pollution for longer periods of time is especially likely to occur if a persistent inversion aloft exists. Such an inversion aloft is the subsidence inversion formed by the sinking and vertical convergence of air in an anticyclone, illustrated in Fig. 5. A layer of air at high levels descends, diverging horizontally and hence converging in the vertical, and warms at the dry adiabatic rate of heating of 1°C/100 m. Figure 5*a* shows how a low-level inversion may result from this process, while Fig. 5*b* depicts how the mixing height *H* is limited in vertical extent by the subsidence inversion aloft, so that pollution accumulates within and just above the city. It is the presence of such a subsidence inversion aloft associated with the Pacific subtropical anticyclone which is the primary cause of Los Angeles and other California smogs; these are made even worse by local mountain and valley sides which prevent horizontal dispersion.

Fog. The worst pollution occurs when, in addition to subsidence inversions accompanying slowly moving or stationary anticyclones, fog also develops. All the major air pollution disasters, such as those listed in a later section, took place when fog persisted during protracted stagnant anticyclonic conditions. The reasons for the adverse influence of fog are shown in Fig. 6. When there is no fog (Fig. 6*a*), solar radiation heats the ground, which in turn causes a lapse rate equal to, or greater than, the dry adiababic rate of cooling, with good mixing and hence a substantial mixing height *H*. On the other hand, with a fog layer (Fig. 6*b*), up to 70% of the solar radiation incident at the top of the fog is reflected to space, with relatively little left to heat the fog and ground below. With the cloudless skies characteristic of anticyclonic weather, there is a continuous loss of heat to outer space from the upper surface of the fog bank, which acts radiatively as an elevated ground surface. More heat is lost to space than is gained from the Sun, and an inversion develops above the fog and persists night and day until the anticyclone dissipates or moves away. If the air is polluted, the fog particles may become acids and salts in solution; the saturation vapor pressure over such particles may decrease to 90 or 95% of the pure water value, so the smog becomes even more persistent than if it remained as a pure water fog. Disastrous concentrations of contaminants may accumulate during prolonged foggy conditions of this kind.

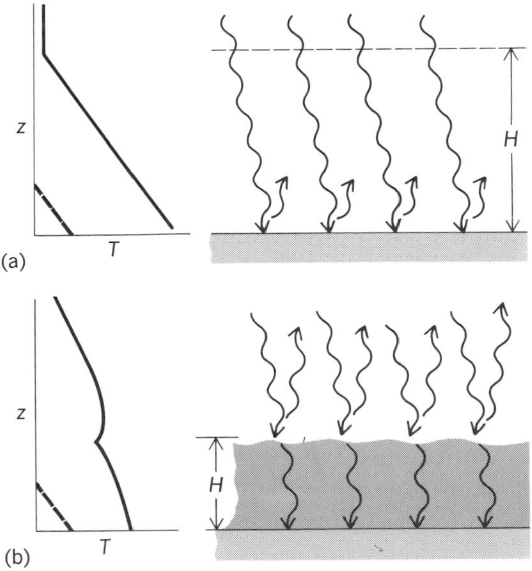

Fig. 6. The influence of fog in reducing mixing height *H*. (*a*) Without fog. (*b*) With fog. The dashed lines represent the dry adiabatic rate of cooling. Arrows represent solar and reflected radiation.

Warm fronts. Another significant inversion aloft is associated with a slowly moving warm frontal surface. Consider two cities, one lying to the southwest and the other lying to the northeast of a warm front extending from the southeast to the northwest (and moving in a northeast direction), as illustrated in Fig. 7. City B lies in the cool air, with the warm frontal above it. In the cool air ahead of the warm front, the pollution from City B is trapped below the warm frontal inversion and may travel for many miles with large surface concentrations. On the other hand, the prevailing southwest winds in the warm sector will carry pollution from City A up and above the warm frontal inversion, which effectively prevents its diffusion downward to the surface. This situation brings out an important point: an inversion layer may be advantageous, not disadvantageous, if it inhibits diffusion down to the ground.

Stack dispersion. Dispersion from an elevated point source, such as a stack, is conveniently expressed by Eq. (7), where χ is ground level concentration of contaminant in mass per unit volume;

$$\chi = \frac{Q}{\pi \sigma_y \sigma_z \bar{u}} \exp\left[-\frac{1}{2}\left(\frac{y^2}{\sigma_y^2} + \frac{h^2}{\sigma_z^2}\right)\right] \quad (7)$$

Q is the source strength in mass per unit time; $\bar{u}$ is mean wind speed; y is the horizontal direction perpendicular to the mean wind; σ_y and σ_z are diffusion coefficients expressed in length units in the y and z directions, respectively, the z direction being vertical; and h is the height of the source above ground.

This diffusion equation should be used only under the simplest conditions, for example, in flat uniform terrain and well away from hills, slopes, valleys, and shorelines. Table 2 lists various meteorological categories, and Table 3 gives values of the diffusion coefficients appropriate for each category. It should be noted that the values to be used depend on distance from the source at which concentrations are to be calculated. A variety of other forms is available for more complex conditions of terrain and meteorology.

Natural cleansing processes. Pollution is removed from the atmosphere in such ways as washout, rain-out, gravitational settling, and turbulent impaction. Washout is the process by which contaminants are washed out of the atmosphere by raindrops as they fall through the contaminants; in rain-out the contaminants unite with cloud droplets, which may later grow into precipitation.

Gravitational settling is significant mainly for large particles, those having a diameter greater than 20 μm. Agglomeration of finer particles may result in larger ones which settle out by gravitation. Fine particles may also impact on surfaces by centrifugal action in very small turbulent eddies. Gases may be converted to particulates, as by photochemical action of sunlight in Los Angeles, Denver, and Mexico City. These particulates may then be removed by settling or impaction.

The rate of natural cleansing may be slower than the rate of injection of pollutants into the atmosphere, in which case pollution may increase on a global scale. There is evidence that the concentration of atmospheric carbon dioxide has been increasing slowly since the beginning of the century because of combustion of fossil fuels. The tropospheric burden of very small particles and of Freon gas may also be increasing.

Effects of stratospheric pollution. Pollution of the stratosphere with nitrogen oxide causes reduction of stratospheric ozone. Ozone reduction in the stratosphere has been linked to biological effects such as skin cancer in two steps: (1) reduced ozone in the stratosphere causes an increase in uv radiation reaching the Earth's surface, and (2) increased uv radiation enhances the normal biological effects of natural uv radiation.

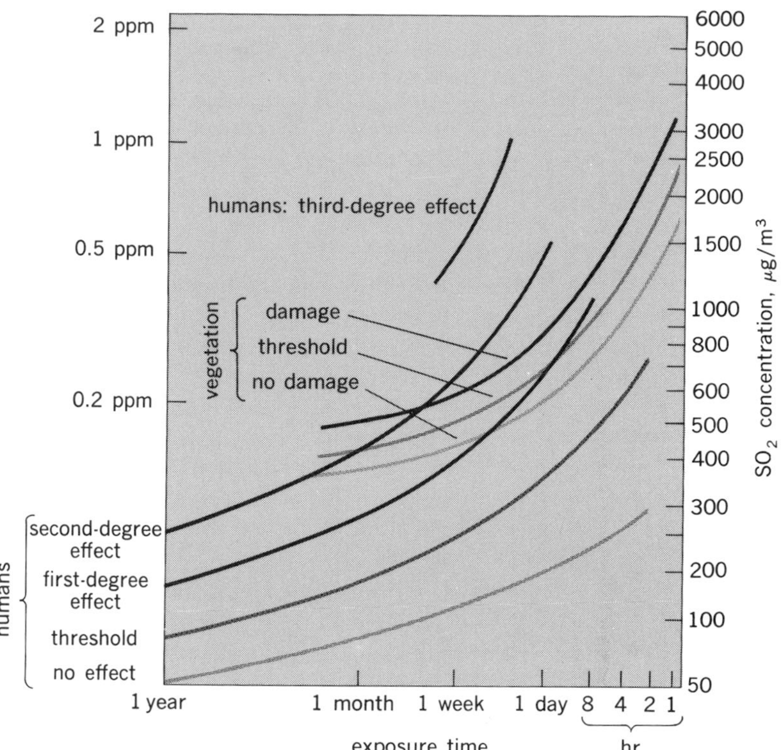

Fig. 7. Effects of sulfur dioxide on humans and vegetation. (*L. J. Brasser et al.*)

Biological damage. In step 1, the relation between the reduction of stratospheric ozone and the increase in solar uv flux effective in causing sunburn and, presumably, also skin cancer is readily calculable. The factors of interest here are

Table 2. Meteorological categories*

Surface wind speed, m/s	Daytime insolation: Strong	Moderate	Slight	Thin overcast or ≥ 4/8 cloudiness†	≤ 3/8 cloudiness
<2	A	A-B	B		
2	A-B	B	C	E	F
4	B	B-C	C	D	E
6	C	C-D	D	D	D
<6	C	D	D	D	D

*A = Extremely unstable conditions. D = Neutral conditions (applicable to heavy overcast, day or night).
B = Moderately unstable conditions. E = Slightly stable conditions.
C = Slightly unstable conditions. F = Moderately stable conditions.

†The degree of cloudiness is defined as that fraction of the sky above the local apparent horizon which is covered by clouds.

Table 3. Values of diffusion coefficients

Distance from source, m	Diffusion coefficient, m², in various meteorological categories: A	B	C	D	E	F
10^2, σ_y	22	16	12	8	6	4
10^3	210	150	105	75	52	36
10^4	1700	1300	900	600	420	360
10^5	11,000	8500	6300	4100	2800	2000
10^2, σ_z	14	11	7.6	4.8	3.6	2.2
10^3	500	120	70	32	24	14
10^4	—	—	420	140	90	46
10^5	—	—	2100	440	170	92

the uv wave band of 290–320 nanometers (nm), the middle latitudes (where the summer sun is nearly overhead and where the worst cases of skin cancer occur), and small decreases in ozone. For these factors of interest, the percent increase in solar ultraviolet flux is about twice the percent decrease in ozone.

Step 2, from uv radiation to the enhancement of skin cancer incidence, involves the assumption, supported by some scientific evidence but not proven by experiments on humans, that skin cancer in humans is induced by exposure to uv radiation of the same wavelength (290–320 nm) that causes erythema (sunburn), and that the relative effectiveness of the various wavelengths for carcinogenesis is the same as that for sunburn. Estimates of biological damage to humans from exposure to uv radiation are based on inferences from the statistics of epidemiological surveys of humans and of laboratory experiments with animals. Nonmelanomic skin cancer, which is almost never fatal if given proper care, occurs primarily on sun-exposed areas of the skin, especially the face and hands. It is relatively common—about 250 cases per 100,000 persons occur in fair-skinned Caucasians in the United States. The incidence of nonmelanomic skin cancer is correlated with latitude—and, therefore, with sunlight, including uv flux, since average sunlight varies with latitude.

Climatic changes. Pollution of the stratosphere may involve also a climate chain of cause-and-effect relations by which aircraft engine effluents, notably sulfur dioxide (SO_2), and to a lesser degree water vapor (H_2O) and nitrogen oxides (NO_x), affect climatic variables such as temperature, wind, and rainfall.

If enough particles larger than 0.1 μm in diameter were added to the stratosphere, they could alter the radiative heat transfer of the Earth-Sun system, and thereby influence climate. Particles of this size are produced by several constituents of engine emissions, in particular those of SO_2. When considering the large numbers of aircraft postulated for future operations in the stratosphere, the amount of particles developed from the SO_2 engine emissions are potentially serious, unless the fuels used have a sulfur content smaller than that of today's fuels.

The sequence in the climate cause-and-effect chain is proposed as follows. Stratospheric SO_2, after first being oxidized, interacts with the abundant water vapor exhaust from jet engines to produce solid sulfuric acid particles that build up to sizes greater than 1 μm. These particles disperse within the stratosphere, principally within the hemisphere in which they are injected, where they remain for periods as long as 3 years, depending on their altitude.

The effects of SO_2 emissions are summarized in Table 4, where the stratosphere's opacity to sunlight is represented by optical thickness. The natural optical thickness of the stratosphere is about 0.02, which means that sunlight and heat are reduced by about 2% while passing through the stratosphere.

The present subsonic fleet of about 1700 aircraft operates at times in the low stratosphere, burning a total of about 2×10^{10} kg (2×10^7 tons) of fuel per year and producing an estimated increase in the stratosphere's optical thickness of about 0.0001 or 0.5% of the natural value.

As a matter of history, the variability of temperature due to natural causes is a substantial fraction of 1°C, even over a few decades. The year-to-year variation is several tenths of a degree. In the 1890–1940 period, a warming of ½°C occurred. During the 1940–1960 period, a general cooling of ⅕°C took place.

Effects of tropospheric pollution. A number of the many effects of tropospheric pollution are described briefly below.

Humans. The effects of many pollutants on human health under most ordinary circumstances of living, rural or urban, are difficult to specify with confidence. In the United States, a number of animal experiments under controlled conditions and epidemiological studies have been made, but the results are difficult to interpret in terms of human health. A study on the lower East Side of New York City indicated that, in children less than 8 years old, the occurrence of respiratory symptoms was associated with the levels of particulate matter and of carbon monoxide in the atmosphere. With heavy smokers, however, eye irritation and headache were directly related to increasing concentrations of carbon monoxide.

Air pollution is suspected as a causative agent in the occurrence of chronic bronchitis, emphysema, and lung cancer, but the evidence is not clear cut. On the other hand, from mid-August to late September the potent aeroallergen, ragweed pollen, is a substantial cause of allergic rhinitis and bronchial asthma for over 10,000,000 persons living east of the Rockies. In Los Angeles County the effects of smog on health are becoming better understood. For example, people living in less smoggy areas of

Table 4. Estimated increase in stratospheric optical thickness per 100 aircraft

Subsonic aircraft type*	Fuel burned, kg/yr†	Altitude, km (10^3 ft)	Maximum SO_2 EI‡ without controls, g/kg fuel	Percent change in stratospheric optical thickness in Northern Hemisphere: Without controls	With future EI controls achieving only 1/20 of emission of present-day aircraft
707/DC-8	1×10^9	11 (36)	1	0.023	0.0012
DC-10/L-1011	1.5×10^9	11 (36)	1	0.032	0.0016
747	2×10^9	11 (36)	1	0.044	0.0022
747-SP	2×10^9	13.5 (44)	1	0.10	0.0050

*The present subsonic fleet consists of 1217 707/DC-8s, 232 DC-10/L-1011s, and 232 747s flying at a mean altitude of 11 km (36,000 ft), and is estimated to cause an increase in stratospheric optical thickness of 0.5%.

†Subsonics are assumed to operate at high altitude, 5.4 hr per day, 365 days per year.

‡EI is emission index, which is defined as grams of pollutants per kilogram of fuel used.

the county survive heart attacks more readily than others: In 1958 in high-pollution areas the mortality rate per 100 hospital admissions for heart attacks averaged 27.3 in comparison with 19.1 for low-smog areas. Other studies show a small but significant relation between motor vehicle accidents and oxidant levels. Medical authorities in Los Angeles are becoming concerned about the long-term influences of various pollutants, including photochemical smog, despite the lack of comprehensive knowledge of the nature of such effects.

In Great Britain there is a similar lack of precise knowledge. There are indications that emphysema, bronchitis, and other respiratory diseases are not caused primarily by increased atmospheric pollution, but because more people are living longer. Despite a substantial reduction in the concentrations of atmospheric particulates since the Clean Air Act was passed in 1956, the respiratory disease rate continues to rise. These facts do not prove that air pollution is not a factor, but that its influence may be synergistic and therefore difficult to identify precisely. For example, it is known that the combined effect of sulfur oxides and particulates is substantially greater than the sum of the two separate effects, and many other such synergistic effects doubtless occur.

In the Netherlands special efforts have been made to relate SO_2, both concentration values and exposure times, to effects on humans and on vegetation. The results of these studies, based on investigations in England, the United States, West Germany, Italy, and the Netherlands, are illustrated in Fig. 7. Influences on humans are shown in the lower family of curves: A first-degree effect is a small increase in functional disturbances, symptoms, illnesses, diseases, and deaths; a second-degree effect is a more prevalent or more pronounced effect of the same kind; and a third-degree effect is a substantial increase in the number of deaths. It should be emphasized that the exposures were, in general, to SO_2 in dusty and sooty atmospheres.

Under extreme circumstances when stagnant atmospheric conditions with persistent low wind and fog exist, major disasters involving many deaths occurred, as in and around London in 1873, 1880, 1891, 1948, 1952, 1956, and 1962. Similar diasters occurred in the Meuse Valley of Belgium in 1930 and at Donora, Pa., in 1948.

Atmospheric pollution has a substantial influence on the social aspects of human life and activity. For example, the distribution of urban populations is being increasingly affected by such pollution, and recreational patterns are similarly influenced. The atmospheric burden of pollution is thus becoming more and more important as a determinant in social decision making.

Animals. Studies of the response of laboratory animals to specified concentrations of pollutants have been conducted for many years, but the interpretation of the results in terms of corresponding human response is most difficult. Assessment of the effects of certain contaminants on livestock is relatively straightforward, however. Thus contamination of forage by airborne fluorides and arsenicals from certain industrial operations has led to the loss of large numbers of cattle in the areas adjacent to such chemical industries.

Plants. Damage to vegetation by air pollution is of many kinds. Sulfur dioxide may damage such field crops as alfalfa, and trees such as pines, especially during the growing season; some general relations are presented in Fig. 7. Both hydrogen fluoride and nitrogen dioxide in high concentrations have been shown to be harmful to citrus trees and ornamental plants which are of economic importance in central Florida. Ozone and ethylene are other contaminants which cause damage to certain kinds of vegetation.

Materials. Corrosion of materials by atmospheric pollution is a major problem. Damage occurs to ferrous metals; to nonferrous metals, such as aluminum, copper, silver, nickel, and zinc; to building materials; and to paint, leather, paper, textiles, dyes, rubber, and ceramics.

Weather. Tropospheric pollution may affect weather in a number of ways. Heavy precipitation at Laporte, Ind., is attributed to a substantial source of air pollution there, and similar but less pronounced effects have been observed elsewhere. Industrial smoke reduces visibility and also ultraviolet radiation from the Sun, and polluted fogs are more dense and more persistent than natural fogs occurring under similar conditions. Possible major effects of air pollution on Earth's climate have been mentioned earlier.

Controls. Four main methods of air-pollution control are indicated below.

Prevention. This method was originally applied mainly to reduce pollution from combustion processes. Improved equipment design and smokeless fuels have contributed to the reduction of pollution from both industrial and motor vehicle sources.

Collection. Collection of contaminants at the source has been one of the important methods of control. Many types of collectors have been employed successfully, such as settling chambers, cyclone units employing centrifugal action, bag filters, liquid scrubbers, gas-solid adsorbers, ultrasonic agglomerators, and electrostatic precipitators. The optimum choice for a given industrial process depends on many factors. A major problem is disposal of the collected materials. Sometimes they can be used in by-product manufacture on a profitable or a break-even basis.

Containment. This method is useful for pollutants whose noxious characteristics may decrease with time, such as radioactive contaminants from nuclear power plants. For contaminants with a short half-life, containment may allow the radioactivity to decay to a level which permits their release to the atmosphere. Containment, with destruction or conversion of the offending substances, often malodorous or toxic, is used in certain chemical, oil refining, and metallurgical processes and in liquid scrubbing.

Dispersion. Atmospheric dispersion as a control method has a number of advantages, especially for industrial processes which can be varied to take advantage of the periods when dispersion conditions are so good that contaminants may be distributed very widely in such small concentrations that they inconvenience no one. Some coal-burning electrical power stations are building high stacks, up to 1000 ft (300 m), to lift the SO_2-bearing stack gases well above the ground. Some plants store low-sulfur anthracite coal for use when atmospheric dispersion is poor. *See* AIR-POLLUTION CONTROL.

Laws. Many laws designed to limit air pollution have been enacted. Major steps forward were taken by the Netherlands in 1952, by Great Britian in 1956, by Germany in 1959 and 1962, by France in 1961, by Norway in 1962, by the United States in 1963 and 1967, and by Belgium in 1964.

Efforts to control air pollution by legal means commenced many years ago in Great Britain. In 1906 the Alkali Act consolidated and extended previous similar acts, the first of which was passed in 1863. This calls for the annual registration of scheduled industrial processes, and requires that the escape of contaminants to the atmosphere from scheduled processes must be prevented by the "best practicable means." The Alkali Act functions by interpretation and not by statutory requirement, the Alkali Inspector being the sole judge of the "best practicable means." The Clean Air Act of 1956 provided more effective ways of limiting air pollution by domestic smoke, industrial particulates, gases and fumes from the processes registrable under the Alkali Act, and smoke from diesel engines. This legislative program has had considerable success in alleviating air-pollution problems in Great Britain.

In the United States, air-pollution control had been considered to be a matter of local concern only. By 1963 only one-third of the states had air-pollution control programs, most of which were relatively ineffective. Only in California were local programs, at the city and county level, supported adequately. The Clean Air Act of 1963 brought the Federal government into a regulatory position of increased scope by granting the Secretary of Health, Education, and Welfare specific abatement powers under certain circumstances. It also established a Federal program of financial assistance to local control agencies and recommended more vigorous action to combat pollution by motor vehicle exhausts and by smoke from incinerators.

The Air Quality Act of 1967 brought the Federal government into a more substantial regulatory role. One of its important effects has been to change the emphasis in legislation from standards based on emissions from sources, such as stacks, to standards based on concentrations of contaminants in the ambient air which result from such emissions. The Air Quality Act of 1967 consists of three main portions, listed below.

Title I: Air-Pollution Prevention and Control. The first section of the Air Quality Act amends the Clean Air Act to encourage cooperative activities by states and local governments for the prevention and control of air pollution and the enactment of uniform state and local laws; to establish new and more effective programs of research, investigation, training, and related activities; to give special emphasis to research related to fuel and vehicles; to make grants to agencies to support their programs; to provide strong financial support for interstate air-quality agencies and commissions; to define atmospheric areas and to assist in establishing air quality control regions, criteria, and control techniques; to provide for abatement of pollution of the air in any state or states which endangers the health or welfare of any persons and to establish the necessary procedures; to establish the President's Air Quality Advisory Board and Advisory Committees; and to provide for control of pollution from Federal facilities.

Title II: National Emission Standards Act. This section is concerned mainly with pollution from motor vehicles which accounts for some 60% of the total for the United States. The act covers such matters related to motor vehicle emissions as the following: establishment of effective emission standards and of procedures to ensure compliance by means of prohibitions, injunction procedures, penalties, and programs of certification of new motor vehicles or motor vehicle engines and registration of fuel additives. The act also calls for a comprehensive report on the need for, and the effect of, national emission standards for stationary sources.

Title III. The final section is general and covers matters such as comprehensive economic cost studies, definitions, reports, and appropriations.

There is no doubt that this far-reaching legislative program, stimulating new approaches at the local, state, and Federal levels, will play a major role in controlling air pollution within the United States. The other industrial nations of the world are preparing to meet their growing air-pollution problems by initiatives appropriate to their own particular cirumstances.

Other acts. The Clean Air Act of 1970 set specific deadlines for the reduction of certain hazardous automobile emissions. That year the Environmental Protection Agency (EPA) assumed control over air-pollution programs formerly administered by the Department of Health, Education, and Welfare.

In April 1973 EPA granted a 1-year extension of the strict auto emission standards. In 1975 the Energy Supply and Environmental Coordination Act relaxed emission requirements for another year.

The Clean Air Act Amendments of 1977 set new deadlines for compliance with emission limits for both industrial and automobile emissions.

[A. J. GROBECKER; S. C. CORONITI; E. WENDELL HEWSON]

Bibliography: L. J. Brasser et al., *Sulphur Dioxide: To What Level Is It Acceptable?*, Research Institute of Public Health Engineering, Delft, Netherlands, Rep. no. G300, 1967; J. H. Chang and H. Johnston, *Proceedings of the 3d CIAP Conference*, DOT-TSC-OST-74-15, pp. 323–329, 1974; R. E. Dickinson, in *Proceedings of the AIAA/AMS International Conference on the Environmental Impact of Aerospace Operations in the High Atmosphere*, AIAA Pap. no. 73-527, 1973; Federal Task Force on Inadvertent Modification of the Stratosphere, *IMOS Report*, prepared for the Federal Council for Science and Technology, 1975; J. Friend, R. Liefer, and M. Tichon, *Atmos. Sci.*, 30:465–479, 1973; D. Garvin and R. F. Hampson, *Proceedings of the AIAA/AMS International Conference on the Environmental Impact of Aerospace Operations in the High Atmosphere*, AIAA Pap. no. 73-500, 1973; A. J. Grobecker, S. C. Coroniti, and R. H. Cannon, *Report of Findings: The Effects of Stratospheric Pollution by Aircraft*, U.S. Department of Transportation, DOT-TST-75-50, 1974; P. A. Leighton, *Photochemistry of Air Pollution*, 1961; M. C. MacCracken, *Tests of Ice Age Theories Using a Zonal Atmospheric Model*, UCRL-72803,

Lawrence Livermore Laboratory, 1970; A. R. Meetham, *Atmospheric Pollution*, 1964; M. J. Molina and F. S. Rowland, *Geophys. Rev.*, pp. 810–812, 1974; National Academy of Sciences, *Environmental Impact of Stratospheric Flight*, pp. 128–129, 1975; P. A. O'Connor (ed.), *Congress and the Nation*, vols. 3 and 4, 1973, 1977; R. Scorer, *Air Pollution*, 1968; A. R. Smith, *Air Pollution*, 1966; A. C. Stern (ed.), *Air Pollution*, 1968; R. S. Stolarski and R. J. Cicerone, *Can. J. Chem.*, 52:1610–1615, 1974; S. C. Wofsy and M. B. McElroy, *Can. J. Chem.*, 52:1582–1591, 1974.

Air-pollution control

Air pollution, according to the definition developed by the Engineers Joint Council, means the presence in the outdoor atmosphere of one or more contaminants, such as dust, fumes, gas, mist, odor, smoke, or vapor, in quantities, of characteristics, and of duration such as to be injurious to human, plant, or animal life or to property, or to interfere unreasonably with the comfortable enjoyment of life and property. The sources of airborne wastes are many. They may be roughly divided into natural, industrial, transportation, agricultural activity, commercial and domestic heat and power, municipal activities, and fallout.

Sources of pollution. Natural sources include the pollen from weeds, water droplet or spray evaporation residues, wind storm dusts, meteoritic dusts, and surface detritus. Industrial sources include ventilation products from local exhaust systems, process waste discharges, and heat, power, and waste disposal by combustion processes. Transportation sources include motor vehicles, rail-mounted vehicles, airplanes, and vessels. Agricultural activity sources include insecticidal and pesticidal dusting and spraying, and burning of vegetation. Commercial heat and domestic heat and power sources include gas-, oil-, and coal-fired furnaces used to produce heat or power for individual dwellings, multiple dwellings, commercial establishments, utilities, and industry. Municipal activity sources include refuse disposal, liquid waste disposal, road and street plant operations, and fuel-fired combustion operations. Fallout is a term applied to radioactive pollutants in mass atmosphere resulting from thermonuclear explosion.

The sources are so varied that pollution of the atmosphere is a matter of degree. Pollution from natural sources is in effect a base line of pollution. The major problems of pollution are associated with community activity as opposed to rural activity, because community air is generally more grossly polluted and may contain harmful and dangerous substances affecting property, plant life, and, on occasion, health. Environment is made less desirable by the polluting influence, and there is ample reason to conserve the air resource in many ways parallel to the need for conservation of the water resource. In actuality, the engineer is concerned with engineering management of the air resource, a broader concept than the control of air pollution.

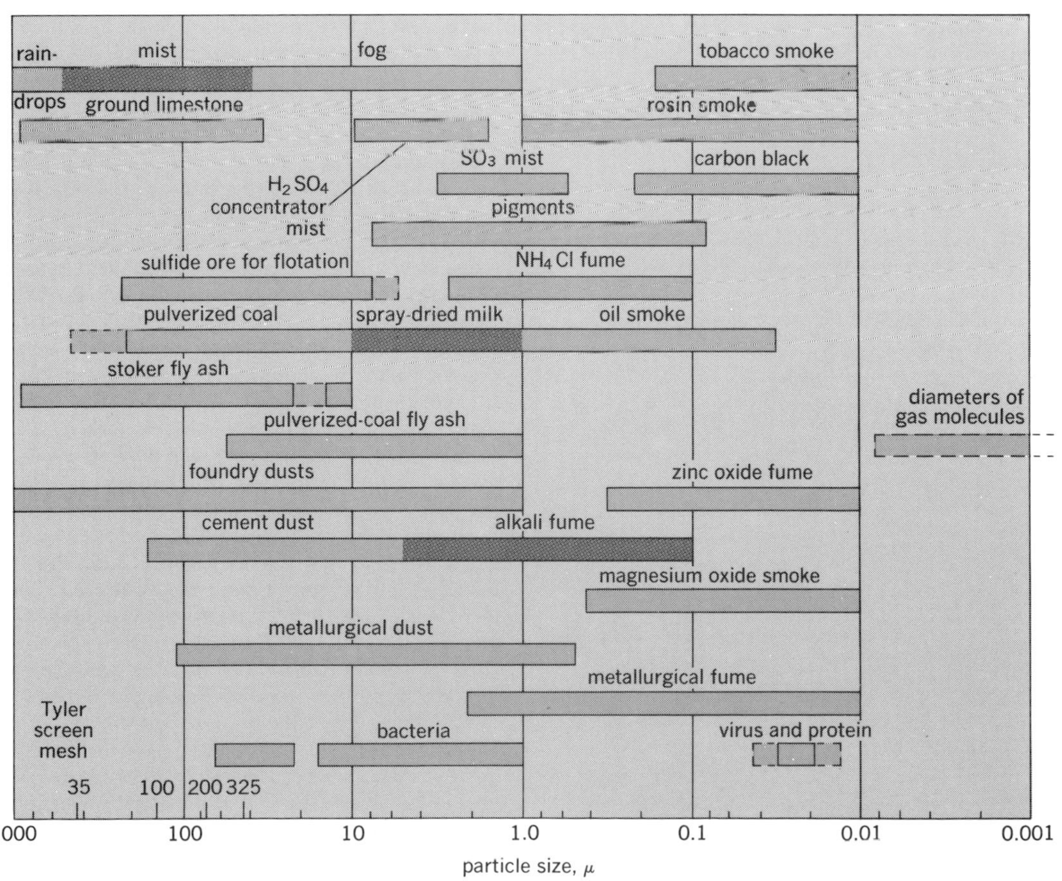

Fig. 1. Particle size ranges applicable to aerosols, dusts, and fumes. (*From W. L. Faith, Air Pollution Control, copyright © 1959 by John Wiley & Sons, Inc.; reprinted with permission*)

Control. Air-pollution control suggests in its simplest form a background of knowledge concerning ideal atmospheres and criteria of clean air, the existence of specific standards setting limits on the allowable degree of pollution, means of precise measurement of pollutants, and practical means of treating polluting sources to maintain the desired degree of air cleanliness. There are many areas in the above listing that are under research at the present time. University research foundations, Federal, state, and municipal air-pollution control agencies, and all of the professional engineering societies are actively engaged in the development of criteria, standards, design factors, and equipment for the control of air pollution.

Reduced visibility has been a focal point of air pollution for over 700 years. The burning of soft coal in England combined with the fog of the atmosphere forms a particularly opaque mixture which may at times reduce visibility to zero. The word smog has been coined for this mixture.

Microscopic water droplets condense about nucleating substances in the air to form aerosols. An aerosol is a liquid or solid submicron particle dispersed in a gaseous medium. An atmosphere having an aerosol concentration of about 1 mg/m^3 has been estimated to limit visibility to 1600 ft (488 m). The mass would contain perhaps 16,000 particles/cm^3. Restriction in visibility is the result of light scattering by these particles. Chemical condensation of reaction products in the air may also nucleate and grow to size that will bring about light scattering. Sulfur dioxide is also a nucleating substance as it oxidizes and hydrolyzes to form sulfuric acid mist.

Elimination of sources of pollution has been one of the favored means of controlling pollution. There are many means of accomplishing the reduction of pollution, but complete elimination is not always practicable. Sulfur dioxide release can be reduced by choosing low sulfur-bearing fuel. An industrial process with a gaseous effluent can be changed to eliminate the gaseous waste. Gases and particulates can be removed from a gas stream by air-cleaning equipment.

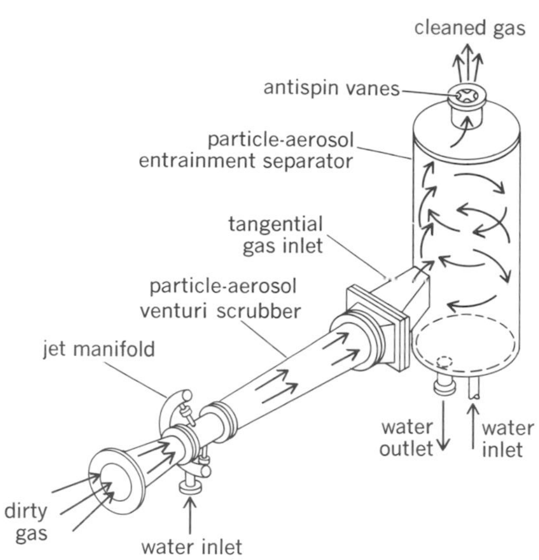

Fig. 3. Typical venturi scrubber. (*From W. L. Faith, Air Pollution Control, copyright © 1959 by John Wiley & Sons, Inc.; reprinted with permission*)

Air-cleaning devices. Air-cleaning devices to remove particulates are selected to remove particles and aerosols on the basis of their size (Fig. 1). Screens will remove coarse solids. Settling chambers are containers which by expanded cross section reduce velocity below 10 feet per second (fps; 3m/s) and thereby allow particles to settle. Particles down to 10 μm in size may be recovered with such chambers. Cyclone separators operate by injecting a gas stream tangentially at the top of a cylindrical chamber. A high-velocity spiral motion is created. Particles are centrifuged out of the gas stream, hit the side wall, and fall to a conical bottom out of the airflow, which turns up through the core or vortex beginning at the bottom and flows to the top through a pipe inserted into the core and extending into the body of the cyclone. Particles from 10 to 200 μm are removed with 50–90% efficiency. Filters are made of cloth, fiber, or glass. Air velocities are low and efficiency is about 50% for dry fiber filters. Efficiency is increased by using a low volatile oil viscous coating. Cloth filters are usually tubular and a number of bags are enclosed in a large chamber. Particles are trapped as air passes through the cloth from inside to outside. Dust is knocked down by shaking and falls to a hopper. Bag filters remove 99% of particles above 10-μm size. Wet collectors, or scrubbers, operate by passing and contacting the gas with a liquid. Water is sprayed, atomized, or distributed over a geometric shape. Deflectors may be added to provide an impinging surface. Scrubbers are efficient on 1- to 5-μm size particles (Figs. 2 and 3). Electrostatic precipitators operate by charging or ionizing particles as the gas flow passes through the unit (Fig. 4). Opposite-pole high-voltage plates, or electrodes, are provided to trap particles. Pre-

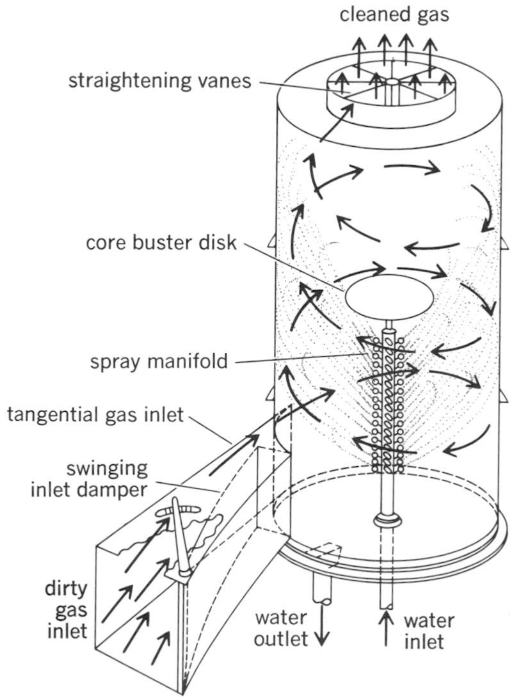

Fig. 2. Typical cyclonic spray scrubber. (*From W. L. Faith, Air Pollution Control, copyright © 1959 by John Wiley & Sons, Inc.; reprinted with permission*)

cipitators operate at 80–99% efficiency of ionizable aerosols down to 0.1-μm size.

Scrubbers may also remove water-soluble gases. Chemicals may be added to the liquid to provide improved absorption. Filters packed with activated charcoal are used to adsorb gases.

Packed towers, plate towers, and spray towers are also used to absorb gaseous pollutants from a gas stream. These devices provide for mixing a gas stream under treatment with water or a chemical solution, so that gases are taken into solution and possibly converted chemically as well.

Atmospheric dilution. This provides another means of reducing air pollution. Meteorology of a region, local topography, and building configuration are critical factors in determining suitability of atmosphere as a dispersal, diffusion, and dilution medium. Basic meteorological conditions of atmosphere that must be considered include wind speed and direction, gustiness of wind, and vertical temperature distribution. Humidity is also important under certain circumstances.

In general, diffusion theories predict that the ground concentration of a gas or a fine particle effluent with very low subsidence velocity is inversely proportional to the mean wind speed. Vertical temperature distribution is an important factor, determining the distance from stack of known height at which maximum ground concentration occurs. Temperature of the stack gas has the effect of increasing stack height, as does stack gas velocity. Gas does not normally come to the ground under inversion conditions, but may accumulate aloft under calm or near calm conditions and be brought down to the surface as the Sun heats the ground in the early morning. Effect of building configuration is shown in Fig. 5. The turbulence introduced by buildings and topography is so complex that it is difficult to make theoretical calculations of effect. Model studies in wind tunnels have been used successfully to make predictions based on measurements of gas concentration and visible pattern of smoke (Fig. 6).

Nonventilating conditions may be present over an area for several days as a result of certain meteorological phenomena. During such periods the pollution emitted from various sources, such as fuel-fired combustion and automobile exhaust, continues to increase in concentration until ventilation sufficient to dilute the accumulated gases and particulates takes place. Figure 7 illustrates the record of sulfur dioxide–concentration measurement in the atmosphere over New York City during one such period of poor ventilation lasting for several days.

Incineration. The need for municipalities to find a means of disposing of refuse when land values are high and little land is available for sanitary landfill has resulted in increased use of incineration for refuse disposal. Incineration introduces problems of air pollution that are quite different from those of fuel-fired combustion. The material is not homogeneous, and has a wide variation in fuel value ranging from 600 to 6500 Btu/lb (1.4 to 15.1 MJ/kg) of refuse as fired. Volatiles are driven off by destructive distillation and ignite from heat of the combustion chamber. Gases pass through a series of oxidation changes in which time-temperature relationship is important. The gases must be heated above 1200°F (650°C) to destroy odors. End products of refuse combustion pass out of the stack at 800°F (430°C) or less after passing through expansion chambers, fly ash collectors, wet scrubbers, and in some instances electrostatic precipitators. The end products include carbon dioxide; carbon monoxide; water; oxides of nitrogen; aldehydes; unoxidized or unburned hydrocarbons;

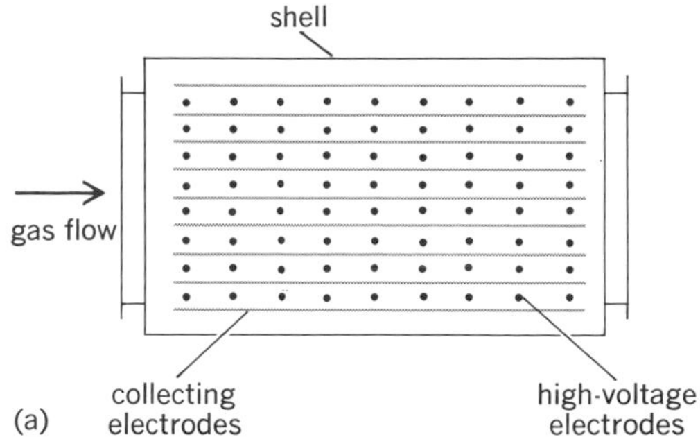

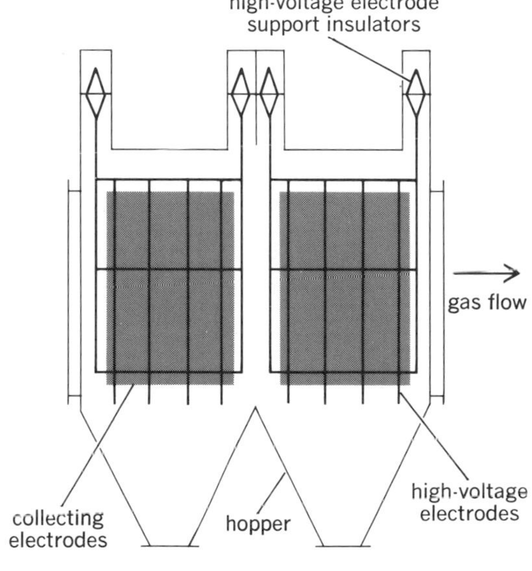

Fig. 4. Diagram of horizontal-flow electrostatic precipitator. (*a*) Plan. (*b*) Elevation. (*From W. L. Faith, Air Pollution Control, copyright © 1959 by John Wiley & Sons, Inc.; reprinted with permission*)

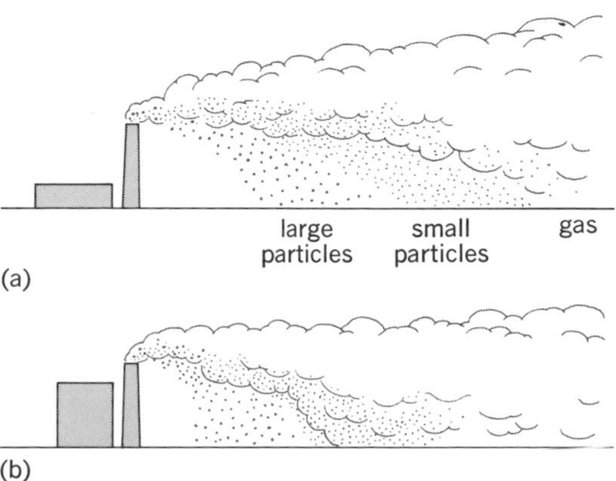

Fig. 5. Effect of building configuration on dispersal of gas plume. (*a*) Favorable configuration. (*b*) Unfavorable configuration. (*Research Division, New York University School of Engineering and Science*)

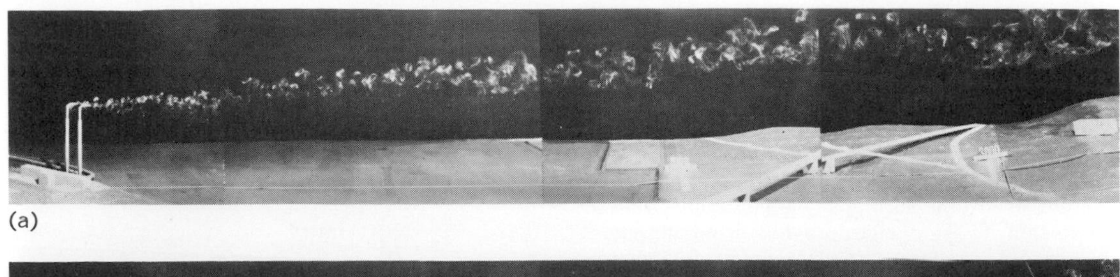

(a)

(b)

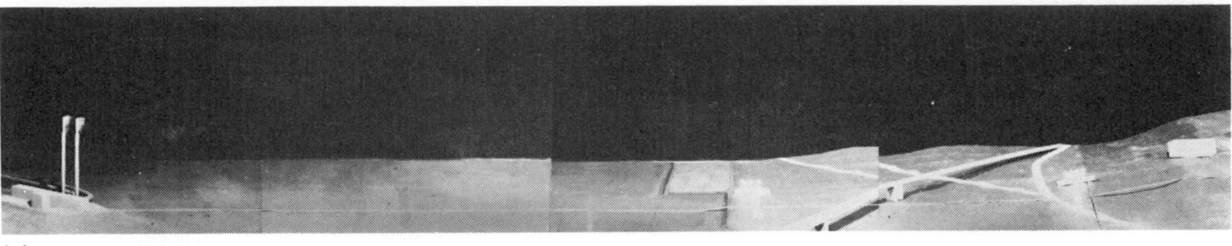

(c)

Fig. 6. Photographs of wind tunnel demonstration of dispersal patterns of smoke at three specified wind speeds: (*a*) 20 mph or 8.9 m/s. (*b*) 25 mph or 11.2 m/s. (*c*) 30 mph or 13.4 m/s.

particulate matter comprising unburned carbon, mineral oxides, and unburned refuse; and unused or excess air. Particulates are reduced in quantity. Normally only micrometer-size and submicrometer particles should escape with the flue gases. Care in operation is required to hold down particulate loading. Dust emissions in stacks may be in the range of 2–3 lb/ton/hr (1–15 kg/metric ton/hr) of refuse charged at a well-operated unit equipped with air-cleaning devices.

Incinerator design. There are several types of incinerator design promoted by manufacturers of incinerator equipment. Kiln shape may be round, rectangular, or rotary. The hearth may be horizontal fixed with grates, traveling with grates, multiple, step movement, or barrel-type rotary (Fig. 8). Drying hearths are provided on some types. Feed into the incinerator may be continuous, stoker, gravity, or batch.

It is necessary to know or estimate water content, percentage combustible material and inert material, Btu content, and weight of refuse to complete a rational design of incinerators. Heat balance can be calculated from several estimates based on averages. Available heat from the refuse must be balanced against the heat losses due to radiation, as well as from moisture, excess air, flue gas, and ash. Each type design has recommended sizings suggested by the manufacturer. There is fair agreement on the need for over 100% excess air. An allowance of 20,000 Btu/ft^3 (745 MJ/m^3) has been suggested for approximating chamber volume, and an allowance of 300,000 Btu/ft^2 (3.4 GJ/m^2) for grate area. Incinerator loading rates of 40–70 lb/(ft^2 grate area) (hr) [195–341 kg/(m^2 grate area) (hr)] have been used. Small incinerators for apartment houses and institutions are loaded at much lower rates. The Incinerator Institute of America in its standards has suggested loading rates for household or domestic-type refuse from 20 lb/(ft^2) (hr) [98 kg/(m^2) (hr)] in 100 lb/hr [45 kg/hr] burning units up to 30 lb/(ft^2) (hr) [146 kg/(m^2) (hr)] in 1000 lb/hr [454 kg/hr] units.

The Building Research Advisory Board (National Academy of Sciences, National Research Council) suggests that apartment house single-chamber incinerators should be sized on the basis of 0.375 ft^3 (10.62 liters) capacity per person, 0.075 ft^2 (69.7 cm^2) grate area per person, and heat release rate of not more than 18,000 Btu/ft^3 (670 MJ/m^3) of capacity, where the burning period is 10 hr or less.

Air-monitoring instruments. Air-sampling methods may be classified generally as those for sampling particulates or gases or both concurrently. The samples may be analyzed for specific pollutants or for general pollution levels. Sampling devices have been constructed with many variants. Generally, however, they follow reasonably well-defined principles which include gravity and suction-type collection, with passage through thermal and electrostatic precipitators; impingers and impactors; cyclones; absorption and adsorption trains; scrubbing apparatus; filters of various materials, such as paper, glass, plastic, membrane, and wool; glass plates; and impregnated papers. Combination instruments that measure wind direction and velocity and direct air samples into multiple sample units, each of which represents a wind sector, are used for general sampling and locating of emission sources. Samples may be taken as single samples, or as a composite over a predetermined time period, or as a continuing monitoring operation. Some instruments are designed to extract a sample from the air, analyze it automatically, and record the result on a chart.

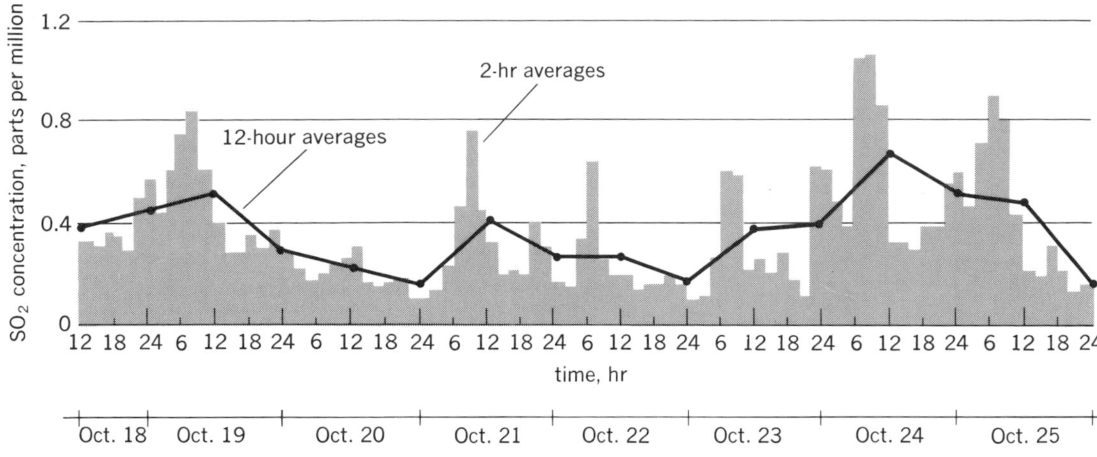

Fig. 7. Air-pollution episode in New York City, Oct. 18–25, 1963. The SO_2 values were at Christodora Station, 189 ft (57.6 m) above ground. (*Research Division, New York University School of Engineering and Science*)

Others take a sample which must be examined in a laboratory.

Many instruments have been developed during the 1950s and 1960s that are mechanized, automatic, and recording, so that they can be used with a minimum of attendance and manipulation. Such instruments require careful initial calibration with standard test substances that are to be measured, and a continuing field check with recalibration at frequent intervals to maintain accuracy.

Particulate samples may be analyzed for weight, size range, effect on visibility, chemical character, shape, and other specific information required. Gas samples may be analyzed to determine the presence of a specific gas or of a group of gases of the same chemical family; that is, nitrogen oxide may be determined or the concentration of all oxides of nitrogen may be established by analysis, or total hydrocarbons may be measured, or by more specific analysis the fractions of several specific hydrocarbons making up the total may be found.

Several types of units with air pumps drawing air through paper tapes mounted on a spool have been developed. The tape is moved automatically so that successive samples on fresh paper are taken at timed intervals.

High-volume samplers are used at many sampling network stations in the United States. The electron microscope has been employed for the examination of aerosols and fine particles. Spectrographic instruments are used for analyzing hydrocarbons and oxides of nitrogen and carbon. Automatically operated units take a sample and then pump chemicals into it at the appropriate time to produce a succession of chemical reactions; they are used to obtain a continuous record of concentration fluctuation of gaseous pollutants, such as sulfur dioxide and others where a wet-chemical analytical method is appropriate. Other instruments use the principle of conductance for measurement of a gas dissolved in a liquid medium. The sample is passed into the liquid medium and a change in electrical energy is measured and recorded. Such instruments measure the effect of any substance that is ionizable in the medium. Orsat analyses are made on flue gases. Photoelectric cells are used to control alarm systems connected to stacks.

Analytical instruments that use several principles of measurement have made it possible to take more data at less cost and manpower. Systems are being developed whereby data from a number of monitoring stations can be transmitted to a central point, transferred to computer program operations, and become statistical information concerning air-pollution concentration. Other systems under development provide for measurement of certain index pollutants while in motion by utilizing automatic instruments mounted in vehicles.

Methods of sampling and methods of analysis have yet to achieve widespread agreement or standardization. The American Society for Testing and Materials (ASTM) has published some 17 standard methods of tests applicable to atmospheric analysis. ASTM has also published definitions of terms relating to air sampling and analysis. Numerous industrial associations and professional organizations are in the process of bringing together in published form the multitude of sampling and testing methods in use.

Air-quality control. This is predicated on standards or guides that take form in official regulations and laws according to three control approaches: restriction of all sources of pollution so that pollution levels in community air are not in excess of certain levels chosen as a safe standard of air quality; limitation on the amount of a specific pollutant that may be present in the exhaust gas from a duct or a stack; or limitation on the amount of impurity in raw materials whose residues reach the community air. These approaches are frequently com-

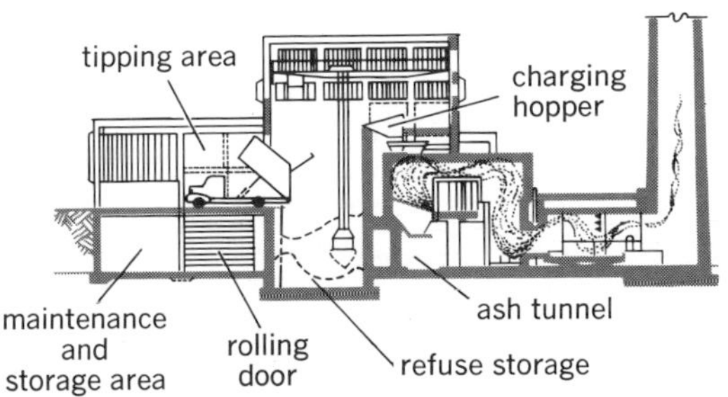

Fig. 8. Diagram of incinerator with rectangular grate. (*American Society of Civil Engineers*)

bined in an attempt to achieve maximum control. The guides or standards may consist of one or more of the following: ambient air-quality standards, emission standards, and material-quality standards.

Official control agencies at local, state, and Federal levels are now in the process of establishing ambient air-quality levels for such pollutants as sulfur dioxide and particulates. Many municipal control agencies have specified the limits of pollutants such as sulfur dioxide, particulates, and solvents that may be emitted from a single source. Many control agencies at municipal and state levels have adopted standards limiting the amount of sulfur in fuel. The U.S. Department of Health, Education, and Welfare publishes from time to time a digest of state air-pollution control laws. All major cities of the United States have adopted laws and regulations based on one or more of the quality control approaches mentioned.

Standards of ambient air quality may vary. In the state of New York, for example, there is a recognized difference in the kind and quantity of pollutants that may be emitted in rural areas, as opposed to highly urbanized areas. Within any region, land use may vary as to its industrial, commercial, residential, or rural components. Subregions based on the predominant land use and on air-quality objectives to be obtained may be established, each having its own air-quality guides. *See* AIR POLLUTION.

[WILLIAM T. INGRAM]

Bibliography: Air Quality Committee of the Manufacturing Chemists Association, *Source Materials for Air Pollution Control Laws*, 1968; American Industrial Hygiene Association, Air Pollution Committee, *Air Pollution Manual*, pt. 1: Evaluation, 1960, pt. 2: Control equipment, 1968; American Society for Testing and Materials Standards, *ASTM Standards on Methods of Atmospheric Sampling and Analysis*, pt. 23, 1967; M. R. Bethea, *Air Pollution Control*, 1978; R. J. Bibbero and I. G. Young, *Systems Approach to Air Pollution Control*, 1974; W. T. Ingram et al., Adaption of Technicon Auto-Analyzer for continuous measurement while in motion, *Technicon Symposia, 1967*, vol. 1: *Automation in Analytical Chemistry*, 1968; W. T. Ingram, C. Simon, and J. McCarroll, *Air Research Monitoring Station System*, J. Sanit. Eng. Div., Proc. Amer. Soc. Chem. Eng. 93, no. SA2, 1967; Interbranch Chemical Advisory Committee, U.S. Department of Health, Education, and Welfare, *Selected Methods for the Measurement of Air Pollutants*, Environmental Health Series, 1965; W. C. McCrone et al., *The Particle Atlas*, 6 vols., 1973–1978; A. C. Stern (ed.), *Air Pollution*, vols. 2 and 5, 3d ed., 1977; C. D. Yaffe et al. (eds.), *Air Sampling Instruments for Evaluation of Atmospheric Contaminants*, 2d ed., 1962.

Aircraft engine

A component of an aircraft that develops either shaft horsepower or thrust and incorporates design features most advantageous for aircraft propulsion. An engine developing shaft horsepower requires an additional means to convert this power to useful thrust for aircraft, such as a propeller, a fan, or a helicopter rotor. It is common practice in this case to designate the unit developing shaft horsepower as the aircraft engine, and the combination of engine and propeller, for example, as an aircraft power plant. In case thrust is developed directly as in a turbojet engine, the terms engine and power plant are used interchangeably.

The characteristics primarily emphasized in an aircraft engine are high takeoff thrust and low specific weight, low specific fuel consumption, and low drag of the installed power plant at the aircraft speeds and altitudes desired. Reliability and durability are essential, as is emphasis on high output and light weight, so that a premium is placed on quality materials and fuels, as well as on design and manufacturing skills and practices. *See* AIRCRAFT FUEL; PROPULSION.

Air-breathing types of aircraft engines use oxygen from the atmosphere to combine chemically with fuel carried in the vehicle, providing the energy for propulsion, in contrast to rocket types in which both the fuel and oxidizer are carried in the aircraft. Air-breathing engines suffer decreased power or thrust output with altitude increase, due to decreasing air density, unless supercharged. *See* INTERNAL COMBUSTION ENGINE; RECIPROCATING AIRCRAFT ENGINE.

[RONALD HAZEN]

Aircraft fuel

The source of energy required for the propulsion of airborne vehicles. This energy is released in the form of heat and expanding gases that are products of a combustion reaction that occurs when fuel combines with oxygen from ambient air. The exhaust gases are water vapor formed from hydrogen in the fuel, carbon dioxide formed from carbon in the fuel, traces of carbon monoxide and nitrogen oxides, and heated but uncombusted components of the intake air. Aircraft fuel is burned with ambient air and is thereby distinct from rocket propellants, which carry both fuel and oxidant. An important criterion for aircraft fuel is that its energy density, or heat of combustion per unit of weight, be high. This allows reasonable expenditures of fuel during takeoff, efficient performance in flight, and long range or flight duration.

There are two general types of aircraft fuels in conventional use: gasolines for reciprocating (piston) engines, and kerosinelike fuels (called jet fuels) for turbine engines. Because of anticipated limitations in the supply of these crude-oil-derived fuels, alternative fuels are being considered for future aircraft.

Piston engine fuels. Piston engine fuels, or aviation gasolines, are special blends of gasoline stocks and additives that produce a high-performance fuel that is graded by its antiknock quality. The gasoline blending stocks are virgin (that is, uncracked) naphtha, alkylate, and catalytically cracked gasoline. Naphthas are mixtures of hydrocarbons distilled directly from crude oil; alkylates are branched paraffin compounds synthesized by refining processes; and catalytically cracked gasolines contain ring compounds called aromatics (such as benzene). In general, the chemical composition of aviation gasoline can be approximated as $C_xH_{1.9x}$, where the number of carbon atoms x is between 4 and approximately 10. Tetraethyllead (Tel) is a common additive used in concentrations of up to 4 ml/gal (1.057 ml Tel/liter) of fuel to increase the antiknock quality of the fuel. *See* GASOLINE; PETROLEUM PROCESSING.

Antiknock quality. In reciprocating aircraft engines the fuel-to-air ratio can be varied from lean (for maximum economy) to rich (for maximum power). Fuel combustion is more likely to detonate—or knock—under fuel-lean conditions. Knocking, if allowed to occur extensively, can harm an engine. Aviation gasolines, like automotive gasolines, are rated according to their antiknock quality as compared with a reference fuel (isooctane or isooctane plus specified amounts of Tel). *See* OCTANE NUMBER.

Aviation gasolines are rated by the (minimum observed) octane number for both lean and rich conditions. The American Society for Testing and Materials (ASTM) and American National Standards Institute (ANSI) have specified standards and testing procedures for determining lean and rich knock values. For example, a fuel rated as 115/145 has a minimum octane rating of 115 when tested under fuel-lean conditions and 145 under fuel-rich conditions. Often, aviation gasolines are graded and designated by their lean octane number of 80 and a rich octane number of 87; Grade 100 and Grade 100LL have a lean octane number of 100 and a rich octane rating at least that of isooctane plus 1.28 ml Tel/gal (0.338 ml Tel/liter). Grade 80 may contain up to 0.5 ml Tel/gal (0.132 ml Tel/liter) and is dyed red. Grade 100 may contain up to 4 ml Tel/gal (1.057 ml Tel/liter) and is dyed yellow. Grade 100LL may contain up to 2 ml Tel/gal (0.528 ml Tel/liter) and is dyed blue.

Volatility. Aviation gasolines must be sufficiently volatile to evaporate quickly and blend with air in the engine manifolds and must be distributed evenly among all cylinders. They cannot be too volatile, however, or the fuel will boil in the tanks or lines. Gasolines boiling over a range of about 100°F (38°C) to 325°F (163°C) meet these requirements, and all grades of aviation gasolines have identical ANSI/ASTM distillation specifications. The tendency of a gasoline to boil is characterized by its Reid vapor pressure (RVP), which is approximately the absolute pressure that the gasoline will exert at 100°F (37.778°C). The RVP is between 5.5 and 7.0 psi (38 and 48 kPa) for all grades of aviation gasoline. Aviation gasolines must also have low freezing points to be stable in storage; the ANSI/ASTM specification for this is −72°F (−58°C).

Heat of combustion. The heat of combustion of all grades of aviation gasoline is about 18,700 Btu/lb (43.5 MJ/kg). This is the net or low heating value at which all combustion products are gaseous. A gallon of aviation gasoline weighs about 6.1 lb, and thus has an energy content of about 114,000 Btu (1 liter weighs about 0.73 kg and has an energy content of about 31.8 MJ).

Turbine engine fuels. Turbine engine fuels are distillate hydrocarbon fuels, like kerosines, used to operate turbojet, turbofan, and turboshaft engines. While all piston engine fuels have the same volatility but differ in combustion characteristics, jet fuels differ primarily in volatility; differences in their combustion qualities are minor. The volatility characteristics of several grades of jet fuel are shown in Table 1. For fuels in which the RVP is too low for accurate measurement, the flash point is given. This is the temperature to which a fuel must be heated to generate sufficient vapor to form a flammable mixture in air. The characteristics listed for Jet A and Jet B are the 1978 ANSI/ASTM standard specifications. *See* KEROSINE.

Fuels JP-1 and JP-3 are no longer used, and JP-2 never achieved specification status. JP-1 was the kerosine first used by the military and is substantially the same as Jet A or Jet A-1, which is now the most widely used commercial jet fuel. JP-3 proved to be too volatile for high-altitude usage. JP-4 and Jet B are military and commercial fuels, respectively, with nearly the same specifications. They are sufficiently volatile that explosive mixtures are present at most ground storage conditions and many flight conditions. JP-4 is used by the Air Force in subsonic aircraft, but Jet B has seen little commercial use in the United States. JP-5, the least volatile of the turbine fuels, is the Navy service fuel. JP-6 is the Air Force fuel for supersonic aircraft. *See* JET FUEL.

Composition. Production of distillate turbine fuel uses up to about 5% of crude oil input to a refinery. This percentage could be increased at added incremental costs and with a concurrent reduction in the output of motor gasoline and diesel fuel. Turbine fuel contains aromatic hydrocarbons; limits are placed on this content owing to concerns about smoke and coke formation. For military jet fuels the limit on aromatics is 25% by volume, and for commercial fuel the limit is 20% (except by mutual agreement between supplier and purchaser, in which case the content may not exceed 25% for Jet A or 22% for Jet A-1 or Jet B). Smoke can be an atmospheric pollutant, but its formation does not represent an appreciable loss in combustion efficiency. Coke is a carbonaceous deposit that adheres to the internal parts of the combustor and can reduce engine life.

High-temperature stability. An increasingly important requirement is to provide a fuel that is stable at relatively high temperatures. In subsonic jets the fuel is used to cool the engine lubricant, and the temperature of the fuel can be raised by about 200°F (93°C). In supersonic jets the fuel is used as a heat sink for the engine lubricant, for cabin air conditioning, and for cooling the

Table 1. Volatility characteristics of jet fuels

Jet fuel grade	Distillation range, °F (°C)	RVP, psia (kPa, absolute)	Flash point, °F (°C)
JP-1	325–450 (163–230)	—	120 (49)
JP-3	100–500 (38–260)	6 (41)	—
JP-4	150–500 (65–260)	2½ (17)	—
JP-5	350–500 (177–260)	—	150 (65)
JP-6	300–500 (149–260)	—	100 (38)
Jet A	—	—	100 (38)
Jet B	—	—	100

hydraulic systems. For very-high-speed flight, the fuel may be used to cool additional engine components and critical air frame areas, such as the leading edges of wings. Therefore, depending on flight speeds and aircraft design, turbine fuels can be heated from 300°F (150°C) to 500°F (260°C) before they are burned. When they are heated to this degree, small amounts of solids may form, and foul the heat exchangers and clog the filters and fuel injectors. There are specifications to indicate the temperature at which solids are first formed and the amount of solids formed with time. In a specification test, fuel is preheated and passed through a heated filter for 5 hr. No significant amount of solids may form in the preheater, and the pressure drop across the filter must stay within limits. For JP-5 and Jet A, A-1, and B, the test temperatures are 300°F (148.9°C) for the preheater and 400°F (204.4°C) for the filter. For JP-6, these temperatures are 425°F (218.3°C) and 525°F (273.9°C), respectively.

Freezing point. Turbine fuels must have low freezing points: −40°F (−40°C) for Jet A and −58°F (−50°C) for Jet A-1 and Jet B. There is also a limit on sulfur content: 0.3% by weight.

Heat of combustion. The heat of combustion of all jet fuels is about 18,400 Btu/lb (42.8 MJ/kg). This is the net low heating value. A gallon of turbine fuel weighs about 6.7 lb and thus has an energy content of about 123,000 Btu (1 liter weighs about 0.80 kg and has an energy content of about 34.4 MJ).

Alternative fuels. Alternative fuels for aircraft are under active consideration by the National Aeronautics and Space Administration (NASA) and several aircraft manufacturers. One of the driving incentives is the relationship between aircraft life cycles and estimates of future world production of crude oil. Industry experience has shown that a new aircraft design and development effort beginning in 1985, for example, will require about 10 years from inception to the beginning of production. The operational service life of the aircraft could approach 30 years, extending from 1995 until about 2025. Many forecasts of world crude production indicate a peak or plateau in the oil production rate before 2010; hence, the availability of crude-based fuels for aircraft is not certain. As a result, alternative fuels made from coal, oil shale, or solar or nuclear energy plus a suitable raw material are being considered.

Liquid hydrogen and liquid methane. Two candidate alternative fuels are cryogenic liquid hydrogen and liquid methane. Table 2 compares some pertinent properties of these fuels with those of the conventional Jet A fuel. As indicated in the table, liquid hydrogen has an energy density (heat of combustion) 2.8 times that of Jet A. Its volume is relatively large, and for the same energy content, the volume of liquid hydrogen would be four times that of Jet A. This, however, is of less importance than the high energy density, which is the predominant benefit of liquid hydrogen. Methane, with properties intermediate between those of Jet A and liquid hydrogen, could be more attractive than hydrogen on the basis of cost. Engineering studies currently in progress show that methane from coal will cost less to produce, to liquefy, and probably, to store and deliver than hydrogen would. Whether the energy density benefit of hydrogen more than compensates in terms of increased aircraft range and payload remains to be determined.

Hydrogen can be produced from water through electrolysis, with commercially available electrolyzers. For such production to make sense in terms of energy utilization, the electric power for operating the electrolyzers should be generated with a nonpetroleum energy source like falling water or perhaps nuclear heat. An experimental process at the laboratory development stage is thermochemical water splitting, in which nuclear or solar heat is used to operate a closed-loop sequence of chemical reactions that result in the splitting of water into hydrogen and oxygen.

Both hydrogen and methane can be made from coal. Detailed process studies have been made for various coal feeds, operating conditions, and plant capital and operating costs. Examples of commercially available coal-to-hydrogen processes are the Koppers-Totzek suspension gasification system and the Texaco partial-oxidation process, both with modified process trains to produce a pure, liquefiable grade of hydrogen. Examples of newer-technology, pilot plant–status processes that might be used to produce fuel-grade hydrogen are the Institute of Gas Technology's continuous fluidized-bed steam-iron process, which also produces by-product electric power, and Bituminous Coal Research Inc.'s BI-GAS slagging gasifier modified with a hydrogen process train. Methane, the principal constituent of natural and substitute natural gas (SNG), can be made from coal with the commercially available Lurgi plants. Newer-technology coal-to-methane processes with pilot-plant status are the synthane (U.S. Bureau of Mines) process, the HYGAS (Institute of Gas Technology) process, and the CO_2 acceptor (Consolidation Coal Company) process. *See* COAL GASIFICATION; METHANE; SUBSTITUTE NATURAL GAS (SNG).

Liquid hydrocarbons from oil shale or coal. Another option is to produce a synthetic crude or mixture of liquid hydrocarbons from oil shale or coal. When followed by refining steps, these liquefaction processes might be used to produce a synthetic turbine fuel similar to Jet A, as well as other fuels and chemicals.

A Fischer-Tropsch type of synthesis has been commercialized by the South African Coal, Oil,

Table 2. Comparison of cryogenic fuels and Jet A fuel

Fuel	Chemical composition	Heat of combustion, Btu/lb (MJ/kg)	Boiling point, °F (°C)	Density, lb/ft³ (kg/liter)	Specific heat capacity, Btu/lb-°F (kJ/kg-°C)
Jet A	$CH_{1.9}$	18,400 (42.8)	572 (300)*	51 (0.82)	0.47 (1.97)
Liquid hydrogen	H_2	51,600 (120.0)	−423 (−253)	4.4† (0.070)	2.3 (9.6)
Liquid methane	CH_4	21,500 (50.0)	−258 (−161)	26* (0.42)	0.84 (3.52)

*Final maximum boi'ing point. †At normal boiling point.

and Gas Corporation (SASOL). In this process, coal is gasified to a mixture of carbon monoxide and hydrogen called synthesis gas, which is catalytically converted via the synthol process to liquid hydrocarbons. Portions of this liquid product could be hydrogenated to change the relatively unstable olefins to paraffins; then isomerization would likely be necessary to create enough branched-chain isomers to meet turbine fuel specifications. *See* COAL LIQUEFACTION; FISCHER-TROPSCH PROCESS.

In addition to its coal reserves, the United States has extensive deposits of oil shale. The richest shales are in Colorado, where assays of extensive deposits exceed 35 gal of kerogen per ton of shale (146 liters per metric ton). These shales can be processed in place by retorting or by various processes, including retorting after mining. The kerogen in the shale is a waxy organic substance that can be thermally decomposed by pyrolysis at temperatures above 800°F (462°C) to yield a liquid oil product called crude shale oil. Synthetic crude oil, or syncrude, is the upgraded oil product that results from hydrogenation of crude shale oil. Extensive refining steps would be needed to convert significant fractions of the shale syncrude to Jet A fuel. However, blending of syncrude with conventional crudes and blending during refining could reduce the overall cost and improve the yield of aircraft turbine fuel. Although it is difficult to upgrade and refine, shale oil may become the most practical source of alternative liquid hydrocarbon fuels in the near future. The total United States resource of just those reserves with an assay above 35 gal per ton (146 liters per metric ton) existing over a continuous 30-ft (9.1-m) interval is over 3×10^{10} bbl (4.8×10^9 m^3). *See* AIRCRAFT PROPULSION; JET PROPULSION; OIL SHALE; PROPULSION.

[JON B. PANGBORN]

Bibliography: American Society for Testing and Materials, *Standard Specifications for Aviation Gasolines*, ANSI/ASTM D910-76, 1976; American Society for Testing and Materials, *Standard Specifications for Aviation Turbine Fuels*, ANSI/ASTM D1655-78, 1978; M. Newman and J. Grey, *Utilization of Alternative Fuels for Transportation*, AIAA Aerospace Assessment Series, vol. 2, 1979; Southwest Research Institute, *Identification of Probable Automobile Fuels Composition: 1985–2000*, U.S. Depart. Energy Rep. HCP/W3684-01/1, pt. A: Synthetic Hydrocarbon Fuels, 1978.

Aircraft propulsion

Flying machines obtain their propulsion by the rearward acceleration of matter. This is an application of Newton's third law: For every action there is an equal and opposite reaction.

Aircraft requirements. In propeller-driven aircraft, the propulsive medium is the ambient air which is accelerated to the rear by the action of the propeller. The acceleration of the air that passes through the engine provides only a secondary contribution to the thrust.

In the case of turbojet and ramjet engines, the ambient air is again the propulsive medium, but the thrust is obtained by the acceleration of the air as it passes through the engine. After being compressed and heated in the engine, this air is ejected rearward from the engine at a greater velocity than it had when it entered. *See* JET PROPULSION.

Rockets carry their own propulsive medium. The propellants are burned at high pressure in a combustion chamber and are ejected rearward to produce thrust. In every case, the thrust provided is equal to the mass of propulsive medium per second multiplied by the increase in its velocity produced by the propulsive device. This is substantially Newton's second law.

The common types of aircraft propulsion systems are:

Reciprocating or piston engine
Compound engine, a combination of the reciprocating engine and the exhaust gas turbine
Turboprop engine
Turbofan engine
Turbojet engine
Ramjet engine
Turboramjet engine
Scramjet engine
Pulsejet engine
Nuclear engine
Rocket engine

Other propulsion systems have been proposed and studied, but have played a lesser role in the development of the airplane.

The airplane lift-drag ratio L/D is a primary factor that determines the thrust required from the propulsion system to fly a given airplane:

$$\frac{L}{D} = \frac{\text{airplane lift}}{\text{airplane drag}}$$

To sustain flight, the airplane lift must be equal to airplane gross weight, and the engine thrust must be equal to the airplane drag. Hence, if the L/D of an airplane type is known, the required engine thrust F can be computed from the airplane gross weight W_g by Eq. (1).

$$F = (D/L)W_g \tag{1}$$

The airplane lift-drag ratio depends upon the flight speed, the state of the art of aircraft design, the capability of the designer, and in some measure upon the application of the airplane, which may be reflected in configuration compromises that affect the drag. The higher the lift-drag ratio,

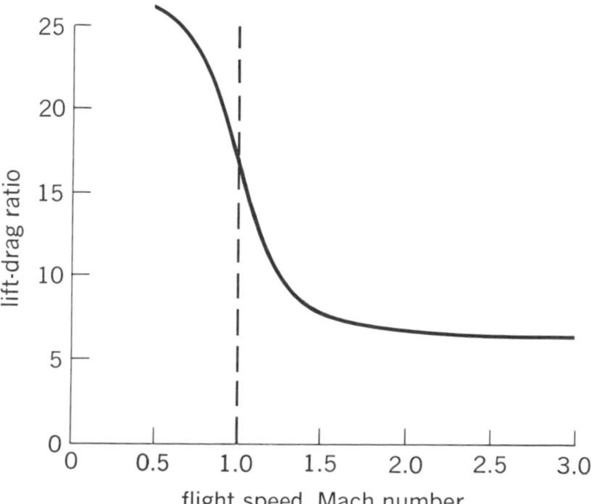

Fig. 1. Lift-drag ratios of aircraft. Curve represents envelope of a variety of designs of various wing sweep angles. (*Taken from 27th Wright Brothers Lecture by G. S. Schairer, J. Aircraft, vol. 1, no. 2, 1964*)

the more efficient is the airplane. Figure 1 represents a typical curve of lift-drag ratios against flight Mach number for aircraft of good design for the specific Mach number.

A sharp reduction in L/D occurs with increase in flight Mach number in the vicinity of a Mach number of unity, and this is reflected in a sharp increase in the thrust required for flight. Flight Mach number is the ratio of the airplane speed to the speed of sound in the ambient atmosphere. At standard sea-level conditions, the speed of sound is 773 mph (346 m/s).

Aircraft performance. In selecting a propulsion system for a given application, one is concerned with the following ratios of engine characteristics: (1) engine weight to engine thrust, designated specific engine weight, lb/lb of thrust (or kg/N); (2) engine frontal area to engine thrust, designated specific frontal area, ft²/lb of thrust (or m²/N); (3) fuel-flow rate to engine thrust, designated thrust specific fuel consumption, SFC, lb/hr/lb of thrust (or kg/hr/N).

Low values of these ratios are desired: For example, a reduction in specific frontal area is reflected in a reduction in airplane drag, and hence in an increase in L/D.

The gross weight W_g of an airplane can be broken into component weights as shown in Eq. (2),

$$W_g = W_{st} + W_e + W_f + W_p \quad (2)$$

where W_{st} is structural weight of empty airplane less engines, and is usually 25–30% of gross weight; W_e is weight of engine or propulsion system; W_f is weight of fuel; and W_p is weight of payload.

For a given gross weight, a reduction in engine specific weight allows a corresponding increase in payload or fuel load. An increase in fuel load increases flight range.

An increase in flight range for a given fuel load can be achieved by reducing the engine specific fuel consumption, since less fuel is burned per hour per pound of thrust.

Aircraft speed requirement. The competition among nations and among commercial airlines has created a continuing demand for increased flight speed. The reduction in aircraft L/D that accompanies an increase in speed (Fig. 1) requires an increase in engine thrust for an airplane of a given gross weight, as shown by Eq. (1). For a given engine specific weight, an increase in required engine thrust results in an increase in engine weight and hence a reduction in fuel load and payload that an airplane of a given gross weight can carry, as shown by Eq. (2). If the engine weight becomes so large that no fuel can be carried by the airplane, the airplane has zero flight range regardless of the efficiency of the engine. At some speed before this point is reached, it becomes advantageous to shift to an engine type that has a lower specific engine weight even at the cost of an increased specific fuel consumption.

At low subsonic flight speeds, the piston-type reciprocating engine, because of its low specific fuel consumption, provides the best airplane performance in terms of payload and flight range. As flight speed increases, specific weight of reciprocating engines increases because of falling propeller efficiency. This effect, coupled with reduction in L/D which accompanies increase in flight speed, results in the weight of reciprocating engines becoming excessive at a flight speed of about 400 mph (179 m/s). At about this speed it is advantageous to shift to the lighter-weight turboprops even if the efficiency of the latter is poorer. (With the lighter engines, the increase in fuel weight that can now be carried more than compensates for the higher specific fuel consumption, with the result that the flight range is increased.) At about 550 mph (245 m/s) it is advantageous to shift from the turboprop to the lighter but less efficient turbojet. Intermediate between the turboprop and turbojet in the spectrum of flight speeds is the turbofan. It has been displacing the turboprop in its regime of application, in spite of a higher specific fuel consumption, because it generates less noise and vibration and thus enhances passenger comfort. The increasing cost of fuel is currently reversing this trend. At high subsonic speeds, the turbofan, in spite of its greater complexity, has largely displaced the turbojet because its greater thrust in takeoff reduces takeoff run distance and hazard, and because it has a lower specific fuel consumption. *See* RECIPROCATING AIRCRAFT ENGINE; TURBOFAN; TURBOJET; TURBOPROP.

By burning additional fuel in the exhaust pipe of a turbojet or turbofan engine (called afterburning), the thrust is increased and the engine weight per pound of thrust is decreased at the cost of an increase in specific fuel consumption. Afterburning is advantageous for operation at supersonic flight speeds because of the sharp reduction in L/D in this regime (Fig. 1). Turbofans with afterburning are competitive with turbojets with afterburning up to flight Mach numbers of about 2.5, and the latter are effective to flight Mach numbers somewhat higher than 3. Between Mach numbers of approximately 3 to 8, the ramjet takes over, and above Mach 8 the scramjet engine predominates (Fig. 2). *See* RAMJET.

The rocket engine is useful in aircraft for brief bursts of thrust either for assisting takeoff or for accelerating the aircraft quickly to very high flight speeds.

For a given angle of attack, the lift of an airplane wing per square foot of surface area is proportional to the product of the atmospheric density and the

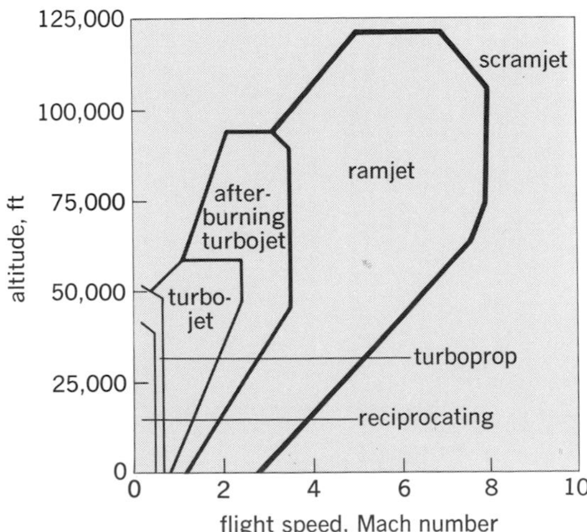

Fig. 2. Flight regime for aircraft propulsion systems; 1 ft =0.3048 m.

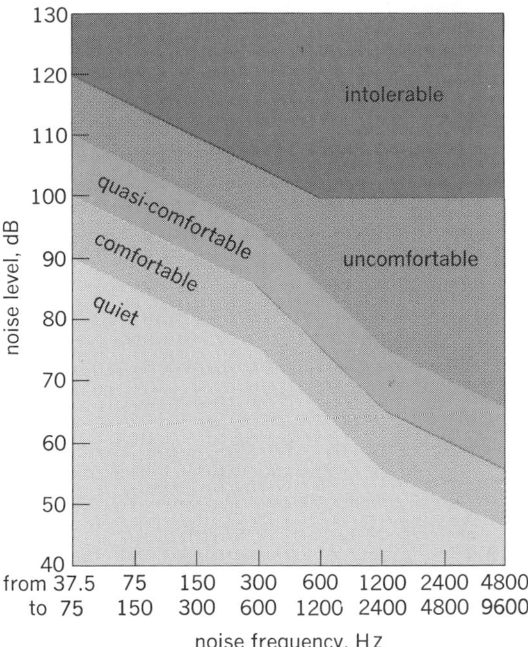

Fig. 3. Noise-level criteria for aircraft. *(From S. Lippert and M. M. Miller, A method for evaluating aircraft acoustical comfort, J. Aviat. Med., 23:54–66, February, 1952)*

square of the flight speed. As aircraft fly faster, they can go to higher altitudes where the atmospheric density is lower. The takeoff condition dictates the amount of wing area required for an airplane. It is therefore desirable to design the higher-speed aircraft for higher flight altitudes in order to make the lift, when flying at angle of attack for maximum L/D, compatible with the required takeoff lift.

These considerations relating to the shift in engine type with design flight speed are exemplified by the current trend in engines for commercial and military aircraft summarized in the table.

Engine installation. In single-engine aircraft, the engine is mounted in the fuselage. In multiengine aircraft, two, three, four, or six engines are usually employed. These engines are contained within streamlined housings called nacelles and are mounted on the wings. The engine installation can be incorporated directly within the wing structure or suspended from the wings by streamlined supports. In a few cases, especially seaplanes, the engines are mounted above the wings. In some two-engine installations, both engines are mounted in the fuselage. In the three-engine configurations two engines are located under the wing or at either side of the rear section of the fuselage, and the third engine is located at the rear of the airplane, usually above the fuselage at the base of the vertical fin.

Propulsion system noise. Propulsion system noise is a serious problem (Fig. 3). In the case of the reciprocating piston engine, the noise mainly originates from the engine exhaust and from vortices shed from the propeller. In the case of the jet engine, the noise is generated mainly from vortices formed at the boundaries of the exhaust jet where the air is sheared by the high-velocity jet.

Radiated sound power varies with the eighth exponential power of the exhaust-jet velocity. The intensity of the noise from a given engine depends on the distance and angular position relative to the engine. Exhaust noise in the order of 125 decibels (dB) is produced by turbojet engines (Fig. 4). A noise reduction in the order of 10 dB has been effected on turbojet engines by discharging the exhaust from a multiplicity of small nozzles rather than from a single large nozzle and by means of lobe-type nozzles in combination with ejectors. By this means the exhaust jet is dissipated in a shorter distance from the engine. This noise reduction is obtained at some loss in nozzle efficiency. In the turbofan engine the noise of the primary jet exhaust stream is much lower than that of the equivalent turbojet engine because of the high extraction of energy from the primary gas to drive the fan. However, the fan contributes considerable noise because of vortices shed by the fan blades. The noise from the fan can be decreased by reducing the fan blade tip speeds at the penalty of requiring additional fan stages to provide the desired fan pressure ratio. The fan noise can be further reduced by applying a sound-absorbing liner to the inside surface of the fan duct. The personnel within the airplane can be shielded from the noise by using soundproofing materials inside the fuselage. The noise of engines is, however, still a serious problem around airports.

Trend in aircraft types

Designation	Speed, statute mph*	Engine type
Passenger transport		
DC-6	307	Reciprocating
DC-7C	346	Reciprocating compound
Electra	405	Turboprop
DC-8	566	Turboprop or turbofan
DC-10	610	Turbofan
747	601	Turbofan
L-1011	600	Turbofan
Military cargo transport		
C-130	386	Turboprop
C-5A	553	Turbofan
Fighter		
F-4	Over Mach 2	Turbojet with A/B
F-15	Over Mach 2.5	Turbofan with A/B
Bomber		
B-52	650	Turbojet
B-58	1380	Turbojet with A/B

*1 mph = 0.447 m/s.

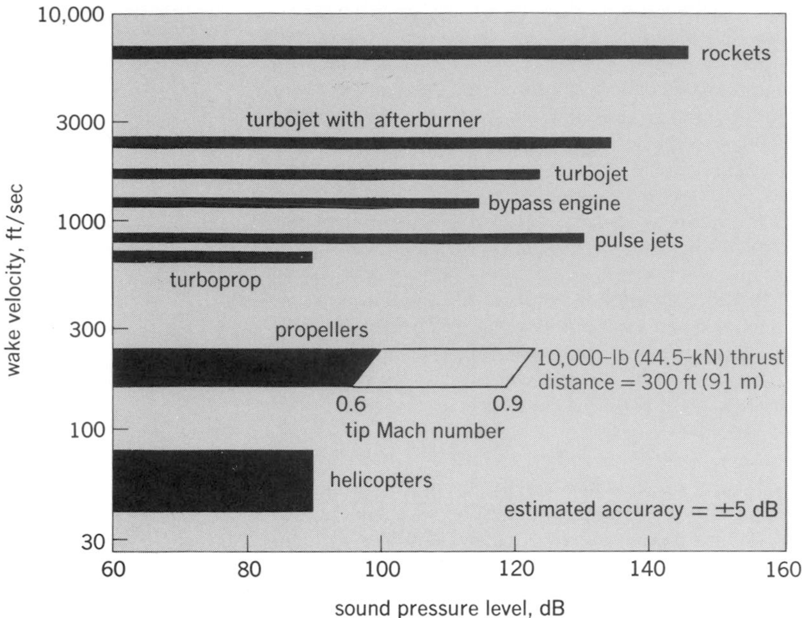

Fig. 4. Sound-pressure levels at equal distances along angle of maximum sound radiation as function of wake or slipstream velocity for aircraft noise sources of equal thrust; 1 ft/s = 0.3048 m/s. *(From C. M. Harris, Handbook of Noise Control, McGraw-Hill, 1957)*

Fuels. The fuels currently used for engines other than rockets are derived from petroleum. The specifications for these fuels call for a sufficiently low viscosity at −76°F (−60°C) to ensure that the fuel remains fluid and that it can be pumped to the engine at the lowest atmospheric temperatures normally encountered. A specification is also placed on the vapor pressure in order to limit the amount of fuel lost on hot days or at high altitudes. The trend toward higher-altitude flight has caused adjustment in the specification of turbojet engine fuels toward lower volatility.

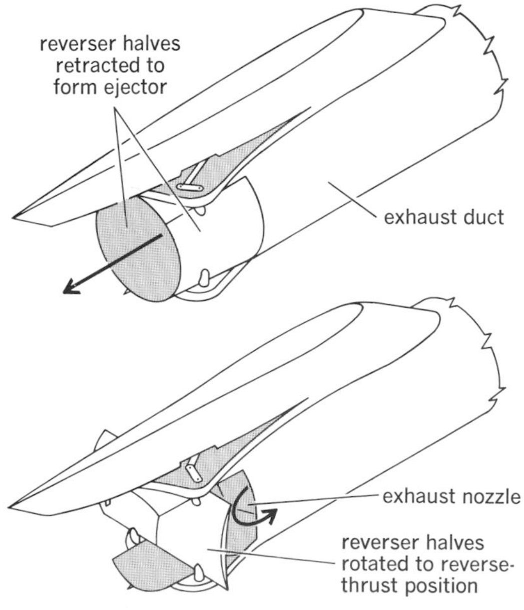

Fig. 5. Principle of target-type thrust reversers which are installed on a turbojet engine.

In reciprocating engines, an important fuel requirement is its knock rating. A poor fuel causes sharp explosions in the engine cylinder, accompanied by a pinging noise. These explosions, called detonation or knock, cause structural deterioration of the engine components, such as the pistons and cylinder walls, exposed to the exploding gas. *See* AIRCRAFT FUEL; OCTANE NUMBER.

Some fuel additives increase the knock ratings appreciably. Tetraethyllead is a very effective antiknock additive and is used extensively. Some difficulty has resulted from fouling of spark plugs by lead compounds formed in the combustion of the tetraethyllead. This problem has been countered by the addition of scavenging agents such as bromides to the fuel.

Knock is not a problem in the turboprop, turbojet, and ramjet engines, so a wider cut of the petroleum output is acceptable. Carbon deposition in the combustion chamber is a problem in the turbine engines. Consequently, restrictions are placed on the proportion of constituents in the fuel that have high carbon-to-hydrogen ratios and very high boiling temperatures.

Restrictions are also placed on impurities in the fuels, such as gum, sulfur, and bromine, which cause corrosion or clogging of the fuel system. Vanadium, present as an impurity in petroleum from some sources, burns to vanadium pentoxide which deposits as a corrosive slag on turbine blades. Water in the fuel can freeze in cold weather and clog fuel systems with ice.

Studies are in progress on fuels with higher energy content per pound than the currently used hydrocarbon fuels. These experimental fuels include methane and hydrogen for turbine engines, and boron-based fuels and hydrogen for the ramjet and scramjet engines. These higher-energy fuels promise substantial increases in aircraft range over current fuels.

Aircraft fire. The aircraft fuel, the lubricating oil, and some hydraulic fluids are combustible. If they should leak from their systems, they are fire hazards. Inerting systems, which can spray chemicals such as carbon dioxide into engine nacelles for smothering fires, are installed on military and commercial aircraft. Considerable progress has been made in research on suppression of crash fires. Fatalities can be avoided in many crashes if no fire results.

Thrust reversal. To stop an aircraft after it lands on a slippery runway where wheel brakes are ineffective, thrust is reversed on propeller-driven aircraft by changing the pitch of the blades to blow the air forward. In turbojet aircraft, thrust reversal is obtained by devices that can turn the exhaust jet to discharge in a forward direction (Fig. 5). Commercial aircraft like the Douglas DC-8 are equipped with turbojets having thrust-reversal devices. A retarding thrust equal to about 40% of the normal engine thrust is currently obtained by these devices. Although higher reverse thrusts have been obtained on experimental thrust reversers, 40% is currently considered adequate for stopping aircraft within the confines of existing airfields. *See* PROPULSION.

[BENJAMIN PINKEL]

Bibliography: H. C. Barnett and R. R. Hibbard, *Properties of Aircraft Fuels*, Nat. Adv. Comm.

Aeronaut. Tech. Note no. 3276, 1956; A. Ferri, Review of scramjet propulsion technology, *J. Aircraft*, vol. 5, January-February 1968; W. R. Hawthorne and W. T. Olson, *Design and Performance of Gas Turbine Power Plants*, 1960; O. E. Lancaster, *Jet Propulsion Engines*, 1959; D. L. Nored, Propulsion, *Astronaut. Aeronaut.*, vol. 16, July/August 1978; R. E. Pendley and A. H. Marsh, Turbofan-engine noise suppression, *J. Aircraft*, vol. 5, May-June 1968; C. D. Perkins and R. E. Hage, *Airplane Performance, Stability and Control*, 1949; B. Pinkel (ed.), *Performance and Ranges of Applications of Various Types of Aircraft-Propulsion Systems*, Nat. Adv. Comm. Aeronaut. Tech. Note no. 1349, 1947; I. I. Pinkel et al., *Origin and Prevention of Crash Fires in Turbojet Aircraft*, Nat. Adv. Comm. Aeronaut. Tech. Note no. 3973, 1957; I. I. Pinkel, G. M. Preston, and G. J. Pesman, *Mechanism of Start and Development of Aircraft Crash Fires (Reciprocating Engines)*, Nat. Adv. Comm. Aeronaut. Rep. no. 1133, 1954; F. P. Povinelli, J. M. Klineberg, and J. J. Kramer, Improving aircraft energy efficiency, *Astronaut. Aeronaut.*, vol. 14, February 1976; E. C. Simpson and R. J. Hill, The answer to the energy deficiency question, *Astronaut. Aeronaut.*, vol. 16, January 1978; J. W. R. Taylor (ed.), *Jane's All the World's Aircraft*, revised periodically; R. J. Weber, NASA propulsion research for supersonic cruise aircraft, *Astronaut. Aeronaut.*, vol. 14, May 1976; M. J. Zucrow, *Principles of Jet Propulsion and Gas Turbines*, 1948.

Alkylation (petroleum)

As applied to the petroleum refining industry, the reaction of isoparaffins with olefins to produce higher-boiling isoparaffinic compounds. Specifically, C_3, C_4, and C_5 olefins are reacted with isobutane to form alkylates which are important components of fuels for internal combustion engines and jet aircraft engines. The desirable properties of alkylate include high octane number, excellent stability, high heat of combustion, and low vapor pressure.

The classic petroleum alkylation reaction is shown in reactions (1)–(3). A proton is added to the double bond of an isobutylene molecule to form a carbonium ion as in reaction (1). This ion combines with another olefin molecule to form a new, eight-carbon atom carbonium ion as in reaction (2). The second carbonium ion abstracts a hydride ion, H^-, from an isobutane molecule to form 2,2,4-trimethyl pentane (isooctane) and another carbonium ion, as in reaction (3), which continues the reaction.

$$CH_3-C(CH_3)=CH_2 + H^+ \rightarrow CH_3-C^+(CH_3)-CH_3 \quad (1)$$

$$CH_3-C^+(CH_3)-CH_3 + CH_3-C(CH_3)=CH_2 \rightarrow CH_3-C(CH_3)_2-CH_2-C^+(CH_3)-CH_3 \quad (2)$$

$$CH_3-C(CH_3)_2-CH_2-C^+(CH_3)-CH_3 + CH_3-CH(CH_3)-CH_3 \rightarrow CH_3-C(CH_3)_2-CH_2-CH(CH_3)-CH_3 + CH_3-C^+(CH_3)-CH_3 \quad (3)$$

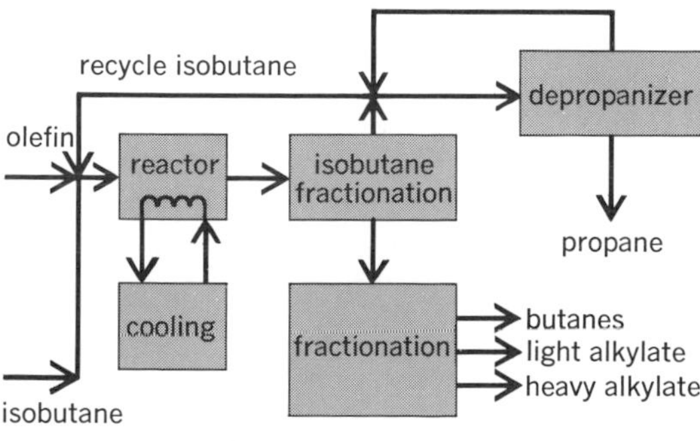

Alkylation, a petroleum fuels process.

Processes. The proton for the above reaction is furnished by an acid catalyst, either H_2SO_4 or HF. Although thermal alkylation is possible at temperatures of 900–975°F/(482–524°C) and pressures of 3000–8000 psig/(18–55 MPa gage), use of the catalyst allows the process to operate under much less severe conditions of temperature and pressure, and the process has therefore found commercial acceptance. Operating temperatures are 100°F (38°C) or below, and the pressure is only high enough to maintain liquid phases.

The HF-catalyzed alkylation process has become increasingly popular, largely because of simpler regeneration techniques that can reduce acid consumption below 0.05 lb/bbl (0.14 kg/m^3 of alkylate produced, and the fact that refrigeration facilities are not needed. The percentage of all alkylation plants using HF for catalyst increased to 41% of the total plants in existence in 1977.

The illustration schematically shows the alkylation process in a typical plant, including the reaction, isoparaffin separation and recycle, and product fractionation. Cooling is employed because the reaction is highly exothermic.

Conditions. Operating conditions are very important for yielding the largest possible volume of high-octane products. The reaction is favored by low temperatures, high acid strength, and low velocity in the reactor. Typical conditions are given in the table.

The catalysts employed have a considerable polymerization activity for olefins. Polymerization produces high-molecular-weight products that lead to loss of catalyst activity and high catalyst consumption. Polymerization is controlled by maintaining a high ratio of isobutane to olefin through recycle of isobutane. Ratios up to 12:1 have a pronounced effect on product yield and octane rating.

Feed quality is also important. The components in the olefin feedstock significantly affect the product properties. Alkylation of pure 1-butene will produce alkylate with a 93 research octane num-

Alkylation process conditions

Catalyst	H_2SO_4	HF
Temperature	2–16°C (36–61°F)	16–52°C (61–126°F)
iC_4/olefin ratio	3–12	3–12
Olefin contact time, min	20–30	8–20
Catalyst acidity, wt %	88–95	80–95

ber, whereas processing pure 2-butene will yield alkylate with a 98 research octane number.

Propylene, ethylene, and amylene in the feedstock convert to alkylate components with a lower octane number than that of alkylates from the butylenes, although the total volume of product is larger. Catalysts are subject to deactivation or destruction by several extraneous components in the feed. Compounds such as dienes, acetylenes, hydrogen sulfide, mercaptans, and excess water adversely affect the alkylation process.

Developments. The effort to change the fuel used in automobiles in the United States to gasoline without the tetraethyllead antiknock additive may change the application of the alkylation process. Although the unleaded octane number of alkylate is higher than that of many other gasoline blending stocks, the refining industry has depended upon the high susceptibility of alkylates to the lead additive to raise the octane number of the final gasoline product.

Successful energy savings in the process have been concentrated at the isobutane fractionation step. A less pure theoretical separation here results in significant operating cost reductions with little influence on product quality. *See* PETROLEUM PROCESSING. [ROBERT S. NEWSHAM]

Bibliography: L. F. Hatch and S. Matar, From Hydrocarbons to Petrochemicals, pt. 4, *Hydrocarb. Process.*, August 1977; W. L. Nelson, *Petroleum Refinery Engineering*, 4th ed. 1958; NPRA '76 Views Processes, *Hydrocarb. Process.*, March 1977.

Alternating current

Electric current that reverses direction periodically, usually many times per second. Electrical energy is ordinarily generated by a public or a private utility organization and provided to a customer, whether industrial or domestic, as alternating current.

One complete period, with current flow first in one direction and then in the other, is called a cycle, and 60 cycles per second (60 hertz, or Hz) is the customary frequency of alternation in the United States and in all of North America. In Europe and in many other parts of the world, 50 Hz is the standard frequency. On aircraft a higher frequency, often 400 Hz, is used to make possible lighter electrical machines.

When the term alternating current is used as an adjective, it is commonly abbreviated ac, as in ac motor (and direct current is abbreviated dc).

Advantages. The voltage of an alternating current can be changed by a transformer. This simple, inexpensive, static device permits generation of electric power at moderate voltage, efficient transmission for many miles at high voltage, and distribution and consumption at a conveniently low voltage. With direct (unidirectional) current it is not possible to use a transformer to change voltage. On a few power lines, electric energy is transmitted for great distances as direct current, but the electric energy is generated as alternating current, transformed to a high voltage, then rectified to direct current and transmitted, then changed back to alternating current by an inverter, to be transformed down to a lower voltage for distribution and use.

In addition to permitting efficient transmission of energy, alternating current provides advantages in the design of generators and motors, and for some purposes gives better operating characteristics. Certain devices involving chokes and transformers could be operated only with difficulty, if at all, on direct current. Also, the operation of large switches (called circuit breakers) is facilitated because the instantaneous value of alternating current automatically becomes zero twice in each cycle and an opening circuit breaker need not interrupt the current but only prevent current from starting again after its instant of zero value.

Sinusoidal form. Alternating current is shown diagrammatically in Fig. 1. Time is measured horizontally (beginning at any arbitrary moment) and the current at each instant is measured vertically. In this diagram it is assumed that the current is alternating sinusoidally; that is, the current i is described by Eq. (1), where I_m is the maximum in-

$$i = I_m \sin 2\pi f t \qquad (1)$$

stantaneous current, f is the frequency in cycles per second (hertz), and t is the time in seconds.

A sinusoidal form of current, or voltage, is usually approximated on practical power systems because the sinusoidal form results in less expensive construction and greater efficiency of operation of electric generators, transformers, motors, and other machines.

Measurement. Quantities commonly measured by ac meters and instruments are energy, power, voltage, and current. Other quantities less commonly measured are reactive volt-amperes, power factor, frequency, and demand (of energy during a given interval such as 15 min).

Energy is measured on a watt-hour meter. There is usually such a meter where an electric line enters a customer's premises. The meter may be single-phase (usual in residences) or three-phase (customary in industrial installations), and it displays on a register of dials the energy that has passed, to date, to the system beyond the meter. The customer frequently pays for energy consumed according to the reading of such a meter. *See* WATT-HOUR METER.

Power is measured on a wattmeter. Since power is the rate of consumption of energy, the reading of the wattmeter is proportional to the rate of in-

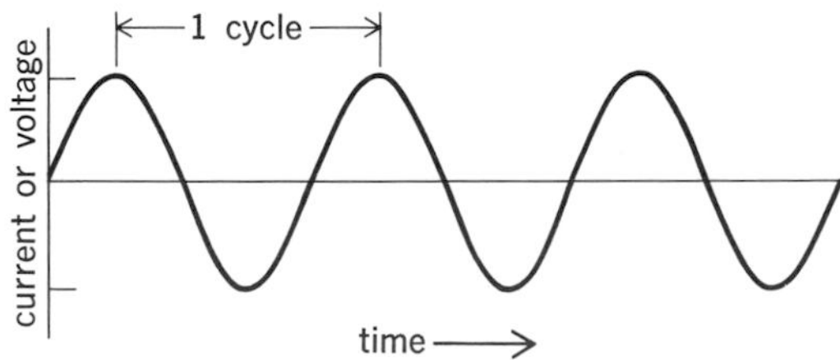

Fig. 1. Diagram of sinusoidal alternating current.

crease of the reading of a watt-hour meter. The same relation is expressed by saying that the reading of the watt-hour meter, which measures energy, is the integral (through time) of the reading of the wattmeter, which measured power. A wattmeter usually measures power in a single-phase circuit, although three-phase wattmeters are sometimes used. *See* WATTMETER.

Current is measured by an ammeter. Current is one component of power, the others being voltage and power factor, as in Eq. (5). With unidirectional (direct) current, the amount of current is the rate of flow of electricity; it is proportional to the number of electrons passing a specified cross section of a wire per second. This is likewise the definition of current at each instant of an alternating-current cycle, as current varies from a maximum in one direction to zero and then to a maximum in the other direction (Fig. 1.) An oscilloscope will indicate instantaneous current, but instantaneous current is not often useful. A dc (d'Arsonval-type) ammeter will measure average current, but this is useless in an ac circuit, for the average of sinusoidal current is zero. A useful measure of alternating current is found in the ability of the current to do work, and the amount of current is correspondingly defined as the square root of the average of the square of instantaneous current, the average being taken over an integer number of cycles. This value is known as the root-mean-square (rms) or effective current. It is measured in amperes. It is a useful measure for current of any frequency. The rms value of direct current is identical with its dc value. The rms value of sinusoidally alternating current is $I_m/\sqrt{2}$, where I_m is the maximum instantaneous current. See Fig. 1 and Eq. (1).

Voltage is measured by a voltmeter. Voltage is the electrical pressure. It is measured between one point and another in an electric circuit, often between the two wires of the circuit. As with current, instantaneous voltage in an ac circuit reverses each half cycle and the average of sinusoidal voltage is zero. Therefore the root-mean-square (rms) or effective value of voltage is used in ac systems. The rms value of sinusoidally alternating voltage is $V_m/\sqrt{2}$, where V_m is the maximum instaneous voltage. This rms voltage, together with rms current and the circuit power factor, is used to compute electrical power, as in Eqs. (4) and (5).

The ordinary voltmeter is connected by wires to the two points between which voltage is to be measured, and voltage is proportional to the current that results through a very high electrical resistance within the voltmeter itself. The voltmeter, actuated by this current, is calibrated in volts.

Phase difference. Phase difference is a measure of the fraction of a cycle by which one sinusoidally alternating quantity leads or lags another. Figure 2 shows a voltage v which is described in Eq. (2) and a current i which is described in Eq. (3).

$$v = V_m \sin 2\pi ft \qquad (2)$$

$$i = I_m \sin (2\pi ft - \varphi) \qquad (3)$$

The angle φ is called the phase difference between the voltage and the current; this current is said to lag (behind this voltage) by the angle φ. It would be equally correct to say that the voltage leads the current by the phase angle φ. Phase difference can be expressed as a fraction of a cycle or in degrees of angle, or as in Eq. (3), in radians of angle, with corresponding minor changes in the equations.

If there is no phase difference, and $\varphi = 0$, voltage and current are in phase. If the phase difference is a quarter cycle, and $\varphi = \pm 90$ degrees, the quantities are in quadrature.

Power factor. Power factor is defined in terms of the phase angle. If the rms value of sinusoidal current from a power source to a load is I and the rms value of sinusoidal voltage between the two wires connecting the power source to the load is V, the average power P passing from the source to the load is shown as Eq. (4). The cosine of the phase

$$P = VI \cos \varphi \qquad (4)$$

angle, cos φ, is called the power factor. Thus the rms voltage, the rms current, and the power factor are the components of power.

The foregoing definition of power factor has meaning only if voltage and current are sinusoidal. Whether they are sinusoidal or not, average power, rms voltage, and rms current can be measured, and a value for power factor is implicit in Eq. (5).

$$P = VI(\text{power factor}) \qquad (5)$$

This gives a definition of power factor when V and I are not sinusoidal, but such a value for power factor has limited use.

If voltage and current are in phase (and of the same waveform), power factor equals 1. If voltage and current are out of phase, power factor is less than 1. If voltage and current are sinusoidal and in quadrature, power factor equals zero.

The phase angle and power factor of voltage and current in a circuit that supplies a load are determined by the load. Thus a load of pure resistance, as an electric heater, has unity power factor. An inductive load, such as an induction motor, has a power factor less than 1 and the current lags behind the applied voltage. A capacitive load, such as a bank of capacitors, also has a power factor less than 1, but the current leads the voltage, and the phase angle φ is a negative angle.

If a load that draws lagging current (such as an induction motor) and a load that draws leading current (such as a bank of capacitors) are both connected to a source of electric power, the power factor of the two loads together can be higher than that of either one alone, and the current to the combined loads may have a smaller phase angle from the applied voltage than would currents to either of the two loads individually. Although power to the combined loads is equal to the arithmetic sum of power to the two individual loads, the total current will be less than the arithmetic sum of the

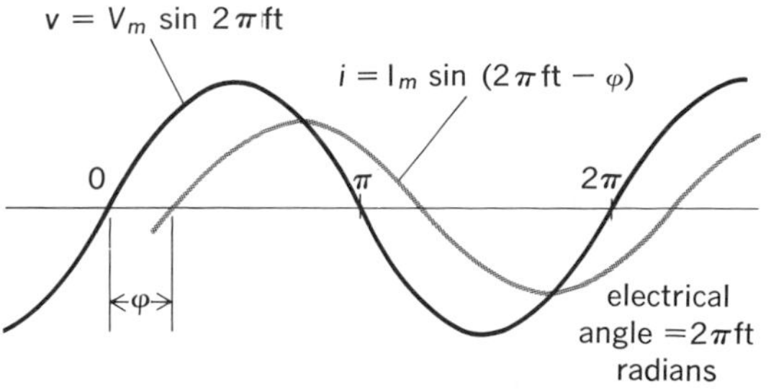

Fig. 2. The phase angle φ.

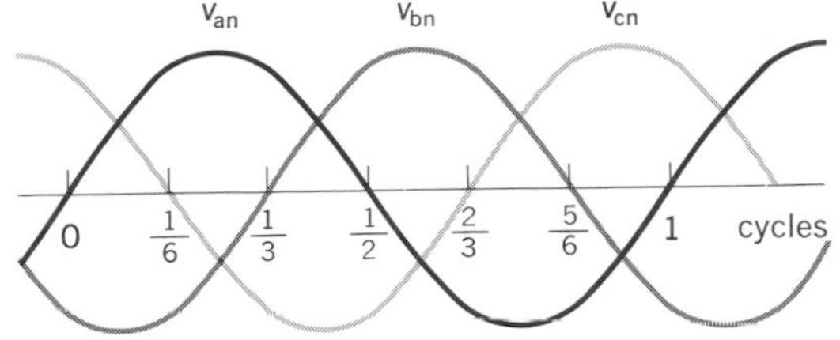

Fig. 3. Voltages of a balanced three-phase system.

two individual currents (and may, indeed, actually be less than either of the two individual currents alone). It is often practical to reduce the total incoming current by installing a bank of capacitors near an inductive load, and thus to reduce power lost in the incoming distribution lines and transformers, thereby improving efficiency.

Three-phase system. Three-phase systems are commonly used for generation, transmission, and distribution of electric power. A customer may be supplied with three-phase power, particularly if he uses a large amount of power or if he wishes to use three-phase loads. Small domestic customers are usually supplied with single-phase power.

A three-phase system is essentially the same as three ordinary single-phase systems (as in Fig. 2, for instance) with the three voltages of the three single-phase systems out of phase with each other by one-third of a cycle (120 degrees) as shown in Fig. 3. The three voltages may be written as Eqs. (6), (7), and (8), where $V_{an(\max)}$ is the maximum

$$v_{an} = V_{an(\max)} \sin 2\pi ft \tag{6}$$

$$v_{bn} = V_{bn(\max)} \sin 2\pi (ft - 1/3) \tag{7}$$

$$v_{cn} = V_{cn(\max)} \sin 2\pi (ft - 2/3) \tag{8}$$

value of voltage in phase *an*, and so on. The three-phase system is balanced if relation (9) holds,

$$V_{an(\max)} = V_{bn(\max)} = V_{cn(\max)} \tag{9}$$

and if the three phase angles are equal, 1/3 cycle each as shown.

If a three-phase system were actually three separate single-phase systems, there would be two wires between the generator and the load of each system, requiring a total of six wires. In fact, however, a single wire can be common to all three systems, so that it is only necessary to have three wires for a three-phase system (*a*, *b*, and *c* of Fig. 4) plus a fourth wire *n* to serve as a common return or neutral conductor. On some systems the earth is used as the common or neutral conductor.

Each phase of a three-phase system carries current and conveys power and energy. If the three loads on the three phases of the three-phase system are equal and the voltages are balanced, then the currents are balanced also. Figure 2 can then apply to any one of the three phases. It will be recognized that the three currents in a balanced system are equal in rms (or maximum) value and that they are separated one from the other by phase angles of 1/3 cycle and 2/3 cycle. Thus the currents (in a balanced system) are themselves symmetrical, and Fig. 3 could be applied to line currents i_a, i_b, and i_c as well as to the three voltages indicated in the figure. Note, however, that the three currents will not necessarily be in phase with their respective voltages; the corresponding voltages and currents will be in phase with each other only if the load is pure resistance and the phase angle between voltage and current is zero; otherwise some such relation as that of Fig. 2 will apply to each phase.

It is significant that, if the three currents of a three-phase system are balanced, the sum of the three currents is zero at every instant. Thus if the three curves of Fig. 3 are taken to be the currents of a balanced system, it may be seen that the sum of the three curves at every instant is zero. This means that if the three currents are accurately balanced, current in the common conductor (n of Fig. 4) is always zero, and that conductor could theoretically be omitted entirely. In practice, the three currents are not usually exactly balanced, and either of two situations obtains. Either the common neutral wire *n* is used, in which case it carries little current (and may be of high resistance compared to the other three line wires), or else the common neutral wire *n* is not used, only three line wires being installed, and the three phase currents are thereby forced to add to zero even though this requirement results in some inbalance of phase voltages at the load.

It is also significant that the total instantaneous power from generator to load is constant (does not vary with time) in a balanced, sinusoidal, three-phase system. Power in a single-phase system that has current in phase with voltage is maximum when voltage and current are maximum and it is instantaneously zero when voltage and current are zero; if the current of the single-phase system is not in phase with the voltage, the power will reverse its direction of flow during part of each half cycle. But in a balanced three-phase system, regardless of phase angle, the flow of power is unvarying from instant to instant. This results in smoother operation and less vibration of motors and other ac devices.

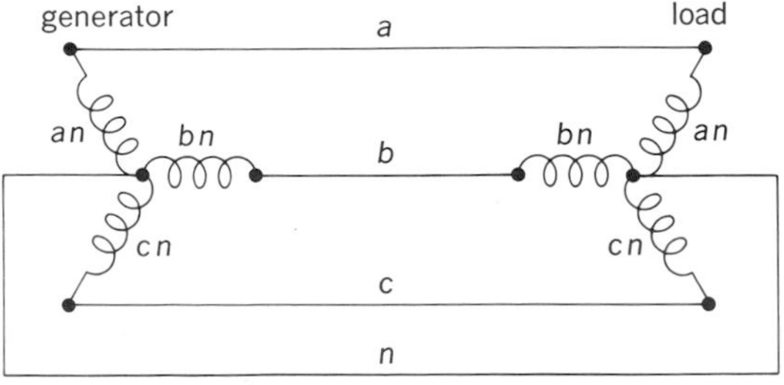

Fig. 4. Connections of a simple three-phase system.

Three-phase systems are almost universally used for large amounts of power. In addition to providing smooth flow of power, three-phase motors and generators are more economical than single-phase machines. Polyphase systems with two, four, or other numbers of phases are possible, but they are little used except when a large number of phases, such as 12, is desired for economical operation of a rectifier.

Ideal circuit. An ideal power circuit should provide the customer with electric energy always available at unchanging voltage of constant waveform and frequency, the amount of current being determined by the customer's load. High efficiency is greatly desired. *See* CIRCUIT.

[H. H. SKILLING]

Alternating-current generator

A machine which converts mechanical power into alternating-current electric power. The most common type, sometimes called an alternator, is the synchronous generator, so named because its operating speed is proportional to system frequency. Another type is the induction generator, the speed of which varies somewhat with load for constant output frequency.

Synchronous generators. These generators usually have the field winding mounted on the rotor and a stationary armature winding mounted on the stator. In small ratings where high reliability is imperative, the magnetic field may be created by permanent magnets. Small alternators are also being built with stationary field windings and rotating armatures with their leads brought out through collector rings or fed directly to rotating rectifiers, which supply direct current to the rotating field of a much larger alternator, as in the brushless excitation system. Still another type, the inductor alternator described below, has both its field and armature windings in the stator. Although synchronous generators may be single-phase, they are usually two- or three-phase; most are three-phase. Single-phase generators are rare because they are larger than polyphase machines of the same kilovolt-ampere ratings, because they have a pulsating torque and are noisy, and because single-phase power is not well suited to self-starting ac motors other than those of fractional horsepower sizes. For theory of synchronous machines *see* SYNCHRONOUS MOTOR.

In the usual type with rotating field windings, a pair of field poles must pass a given point on the armature in 1 cycle. Hence the number of poles required are determined from the frequency f in hertz (Hz) and speed by Eq. (1). The speed se-

$$\text{Number of poles} = 120f/\text{rpm} \qquad (1)$$

lected is that best suited to the prime mover. Steam turbines operate most economically at high speed, hence two- or four-pole generators are used for this service, running at 3600 or 1800 rpm, respectively, for an output frequency of 60 Hz. For hydraulic turbine or engine drive, slow-speed machines having many poles are customary.

High-speed synchronous generators. Generators of this type have cylindrical rotors of solid alloy-steel forgings, with radial slots machined along their length to receive the field windings, as shown in Fig. 1. The field coils are of bare, hard-drawn, strip copper installed turn by turn, within an insulating channel in each slot. Mica plate, epoxy-bonded woven-glass laminate, or similar insulation is commonly used between turns.

The slot portion of the windings is supported against centrifugal force by nonmagnetic wedges, while the coil ends are retained by nonmagnetic metal rings lined with insulation. Since most high-speed generators over 15,000 kVA are now either hydrogen-cooled or water-cooled, the insulated field leads are brought out to the collector rings through an axial hole in the shaft, with the aid of gastight radial studs. Fans or blowers are mounted on the rotor in most cases, employing hydrogen as a cooling medium.

The stator of a large, high-speed, synchronous generator uses a steel yoke, which also serves as an enclosure for the ventilating medium. A cylindrical core of laminated electrical sheet steel is stacked on dovetail bars within the yoke and tightly clamped to minimize magnetic noise. The insulated armature coils of large machines are usually of one turn, lap wound in rectangular slots around the inner periphery of the core. Coil ends are securely lashed to supporting brackets, and both ends of each phase are brought out through terminal bushings. Hydrogen seals are mounted in the end covers, and the entire stator is designed to resist explosion pressures.

Cooling methods. Air-cooled and hydrogen-cooled generators under 100 MVA are most commonly indirectly cooled, that is, the cooling medium contacts metal surfaces and exterior surfaces of the coil insulation. Heat generated in the windings must flow through the major insulation, which is a poor heat conductor. In 1948 S. Beckwith designed a 60-MW 3600-rpm rotor in which hydrogen was forced by a powerful blower to flow at high velocity through ducts within the conductors.

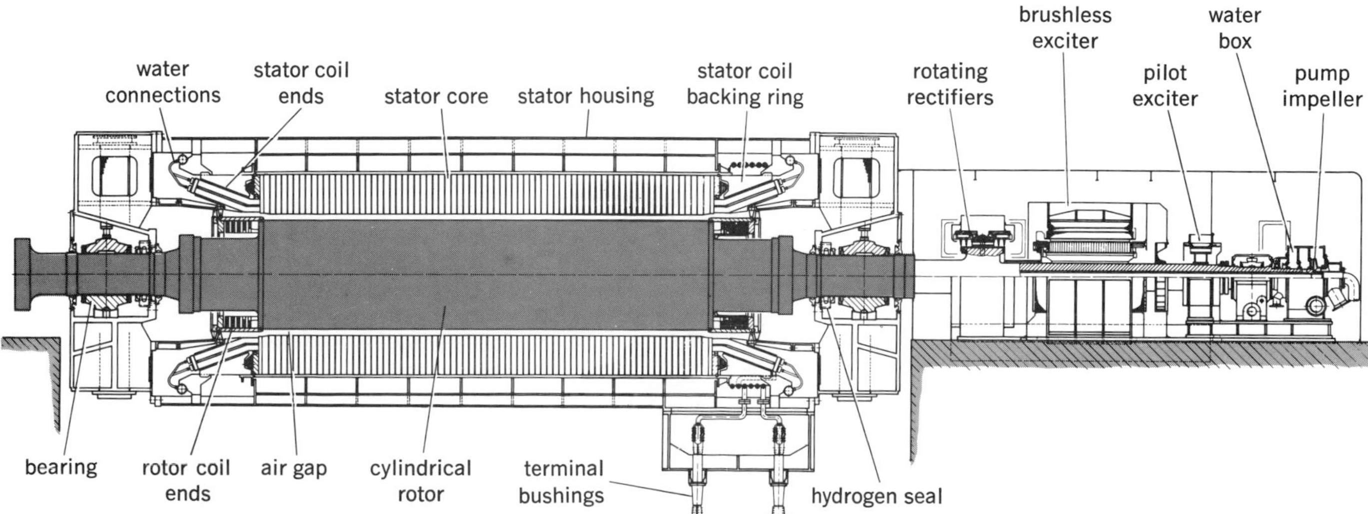

Fig. 1. An 1800-MVA four-pole (operating at 1800 rpm) synchronous generator; it has a water-cooled stator and rotor windings, and brushless excitation equipment. (*Allis-Chalmers Power Systems*)

A tremendous increase in output was made possible by this highly effective cooling method, which is now called direct cooling. When applied also to the stator windings and augmented with higher hydrogen pressures, much greater gains in generator output were achieved.

In parallel with this development, direct water cooling of stator windings was also adopted in the 1950s, and later applied to rotors as well. Figure 1 illustrates an 1800-MVA four-pole generator with water-cooled stator and rotor windings, together with its brushless excitation equipment.

Slow-speed synchronous generators. The field poles of these generators are usually the salient-pole type. When driven by reciprocating engines, they sometimes require flywheels to minimize the pulsations transmitted to the power system. In hydroelectric generators the rotors are sometimes required to withstand high overspeed because of the time delay in closing the gates.

Generated voltage. The voltage generated per phase in a synchronous generator can be derived from Faraday's law. If it is assumed the flux distribution over each pole is sinusoidal, a sinusoidal voltage is induced in each coil side. However, the coil voltages must be added vectorially because of their time-phase displacement. The effective root-mean-square voltage per phase can be shown to be that in Eq. (2), where Φ_m is peak flux per

$$E = 2.22(a/c)\Phi_m f k_d k_p \qquad (2)$$

pole in webers, a is total conductors per phase, f is frequency in Hz, c is number of parallel circuits, k_d is distribution factor, and k_p is pitch factor.

If the flux distribution is not a perfect sine wave, any irregularites appear as odd harmonics in the generated voltage. Although seldom harmful to the generator, harmonics sometimes interfere with telephone communication. Perhaps the most troublesome harmonics result from pulsation in airgap reluctance as the poles move across the slots and teeth. These are known as slot harmonics. They occur in pairs at frequencies equal to (the slots per pair of poles ± 1) times the fundamental. Slot harmonics can be minimized by careful design of pole contour, fractional-slot windings, and skewed stator slots. Interference effects are often reduced with external resonant shunts or wave traps.

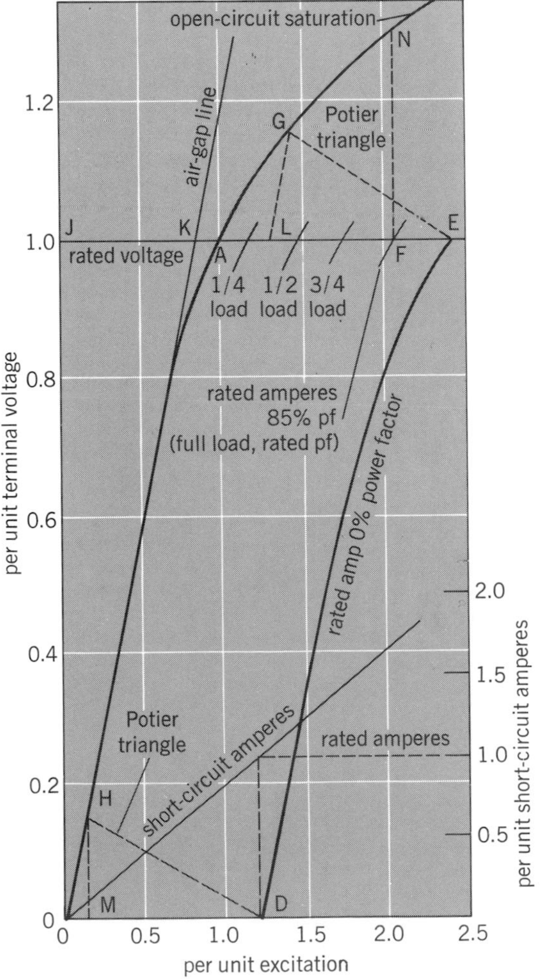

Fig. 2. Characteristic curves of a synchronous generator.

Characteristics. The characteristic curves of a synchronous generator are shown in Fig. 2. The open-circuit and short-circuit curves are readily found from tests at no-load. The saturation curve at rated amperes and 0% power factor is obtained by electrical connection to another synchronous machine. The field currents of both machines are adjusted at each successive test voltage point to circulate rated amperes between them, over-excited on the unit being tested and underexcited on the other machine. The curves show how field current varies with load and voltage, and indicate some of the constants. The Potier reactance drop, used in determining saturation, is the height MH of the Potier triangle OHD.

The triangle is found from the 0% power factor saturation curve DE as follows: (*a*) Lay off length OD to left of point E to find L. (*b*) Draw line LG parallel to the airgap line to find G, its intersection with the open circuit saturation curve. (*c*) Complete the triangle by drawing line GE. (*d*) Make OH equal to LG, thus establishing the Potier reactance drop MH.

An approximate value of the Potier reactance may be calculated from Eq. (3), where x_l = leakage reactance and x'_d = transient reactance.

$$x_p = x_l + 0.63(x'_d - x_l) \qquad (3)$$

Unsaturated per unit synchronous reactance is OD/JK, and the short circuit ratio, whose reciprocal is often used in steady-state stability studies, is JA/OD. The voltage regulation is given by FN/OJ for 85% power-factor lagging load.

Inductor alternator. A synchronous generator in which the field winding is fixed in magnetic position relative to the armature conductors is known as an inductor alternator. There are two types: the homopolar, in which the dc field coil is concentric with the shaft, and the heteropolar, in which the dc windings are distributed. In both types the ac windings are distributed, and generate their induced voltage from the pulsation in the flux caused by the change in position of the salient poles on the rotor. Inductor alternators are used for high-frequency power and, in conjunction with static rectifiers, as a maintenance-free power source for ac excitation systems.

Induction generators. These nonsynchronous ac generators are driven above synchronous speed by external sources of mechanical power. The construction of these machines is identical to that of induction motors. They can operate either as motors or generators, depending on whether the speed is below or above synchronous speed. In some frequency-changer sets, induction generators may operate as a motor part of the time. This is accomplished by coupling them to a two-speed synchronous machine capable of operating at system frequency as a motor at the higher speed and as a generator at the lower speed. Induction generators are not common in large sizes because of their poor power factor. They can deliver only leading currents. Moreover, they require a power supply from which to obtain their magnetizing current. This normally requires that they be operated in parallel with a synchronous source. Capacitors may be used to minimize the current taken from the source. Under such conditions the frequency of the induction generator is determined by the frequency of the synchronous source and the output of the generator is determined by the mechanical input to its shaft. Induction generators do not require any dc excitation and their control can be very simple. For this reason they are well suited to small, unattended hydroelectric units, where they are easily operated by remote control. *See* INDUCTION MOTOR. [LEON T. ROSENBERG]

Bibliography: D. G. Fink and H. W. Beatty (eds.), *Standard Handbook for Electrical Engineers*, 11th ed., 1978; A. E. Fitzgerald et al., *Electrical Machinery*, 3d ed., 1971; D. G. Gehmlich and S. B. Hammond, *Electromechanical Systems*, 1967; E. Levi and M. Panzer, *Electromechanical Power Conversion*, 1974; J. Rosenblatt and M. H. Friedman, *Direct and Alternating Current Machinery*, 1963; G. J. Thaler and M. L. Wilcox, *Electric Machines: Dynamics and Steady State*, 1966.

Alternating-current motor

An electric rotating machine which converts alternating-current (ac) electric energy to mechanical energy; one of two general classifications of electric motor. Because ac power is widely available, ac motors are commonly used. They are made in sizes from a few watts to thousands of horsepower (hp) (Fig. 1). *See* DIRECT-CURRENT MOTOR.

CLASSIFICATIONS OF AC MOTORS

Each type of ac motor has special properties. These motors are generally classified by application, construction, principle of operation, or operating characteristics.

Induction motor. The most common type of ac motor is the induction motor. Current is induced in a rotor as its conductors cut lines of magnetic flux created by currents in a stator. Three-phase induction motors are simple, reliable motors with

ALTERNATING-CURRENT MOTOR

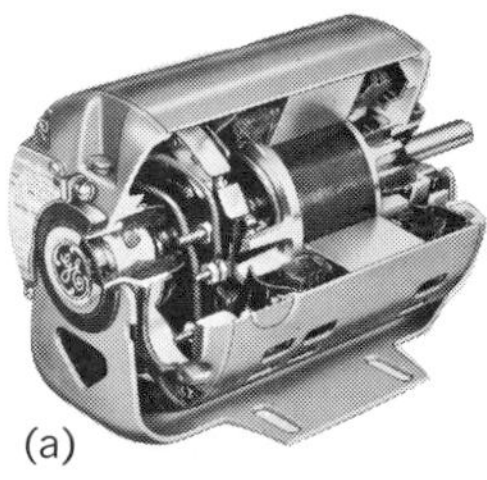

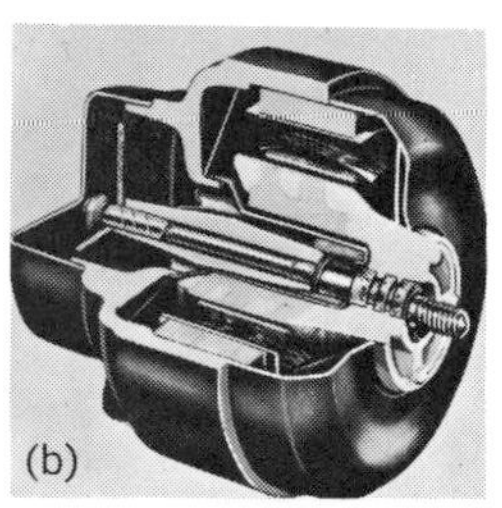

Fig. 1. Typical ac motors. (*a*) One-half-hp split-phase induction motor, about 6 in. (15 cm) in diameter. (*b*) One-hundredth-hp motor, about 1 in. (2.5 cm) in diameter.

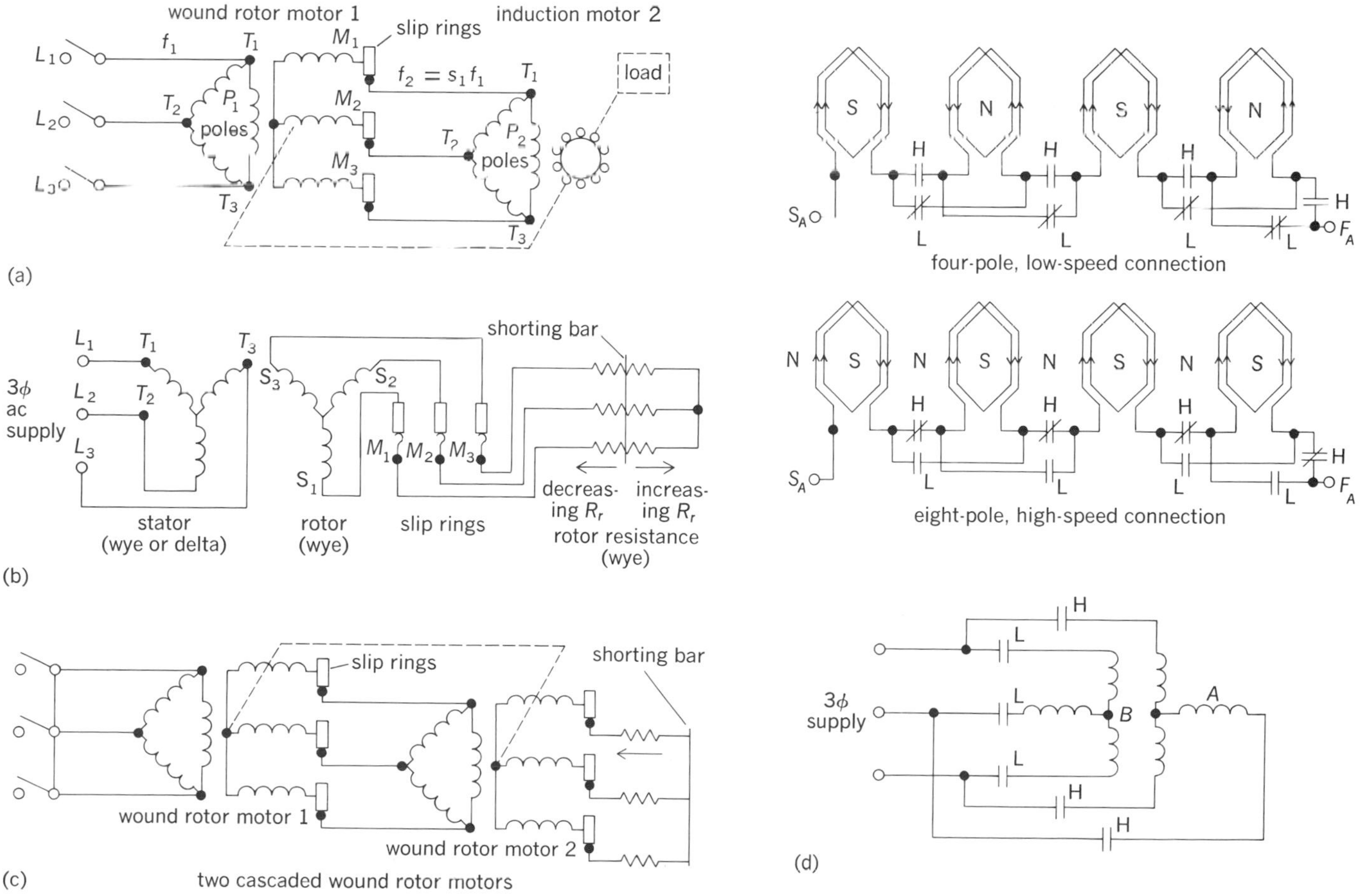

Fig. 2. Schematics of methods of speed control for polyphase ac motors. (*a*) Frequency control using cascade-wound rotor and squirrel-cage motors. (*b*) Slip control using wound-rotor induction motor external rotor resistance. (*c*) Foreign voltage slip control using secondary concatenation. (*d*) Phase *A* of a three-phase winding showing consequent pole connections for multispeed induction motors.

a fairly constant speed over the rated load range. They are self-starting and are widely used in industry. Single-phase induction motors require special means for starting, but are widely used in fractional and small integral horsepower sizes, especially in homes. *See* INDUCTION MOTOR.

Synchronous motor. Where constant speed is essential, a synchronous motor is used. It runs at a fixed speed in synchronism with the frequency of the power supply. Large synchronous motors used in industry employ dc fields on their rotors and three-phase armature (or stator) windings. Efficiency and power factor of these motors are high. The reluctance motor and the Permasyn motor, either single- or three-phase, come in fractional and lower integral hp sizes. The reluctance motor has low efficiency and power factor, but is simple and inexpensive. The Permasyn has permanent magnets embedded in the squirrel-cage rotor to provide the equivalent of a dc field. The hysteresis motor is used only in small sizes where its quiet operation is especially desired, such as in phonographs. *See* SYNCHRONOUS MOTOR.

AC series motor. For operation from either ac or dc power, the series motor has its field winding connected in series with the armature winding through a commutator and brush arrangement, as in a dc series motor. Field and armature iron are laminated. This universal motor has high starting torque. Speed can be controlled by adjusting the applied voltage. They are used in sizes up to 2000 hp in electric railways, and in small sizes for domestic appliances.

Repulsion motor. This is also a commutator motor, but the brushes and commutator are short-circuited and not connected in series with the stator. Armature current is set up by induction from the stator rather than by conduction, as in the series motor. The rotor of a repulsion motor differs from the squirrel-cage rotor of the single-phase induction motor; it is similar to a dc armature with commutator and brushes. Torque is developed by action of induced armature current on stator flux. The repulsion motor has high starting torque. Its speed can be controlled by changing the applied voltage or by shifting the brushes or both. Its no-load speed is above synchronism, but is lower than the no-load speeds of ac series and universal motors. In normal operation the brushes are located 15–25 electrical degrees off the stator position. Rotation is reversed by shifting the brushes to the opposite side of the stator axis. Although repulsion motors have been used on single-phase electric railways, by far the widest application in the past has been as a starting arrangement for single-phase induction motors. When used for this purpose, a centrifugal device short-circuits the commutator bars at about 70% of full speed. Combinations of series and repulsion connections have been used for controlling speed, commutation, and characteristics; occasionally a squirrel-cage winding is added on the armature. For starting duty, the repulsion motor draws low starting current, but despite this advantage the repulsion motor has been superseded by the simpler, cheaper capacitor motor. *See* REPULSION MOTOR.

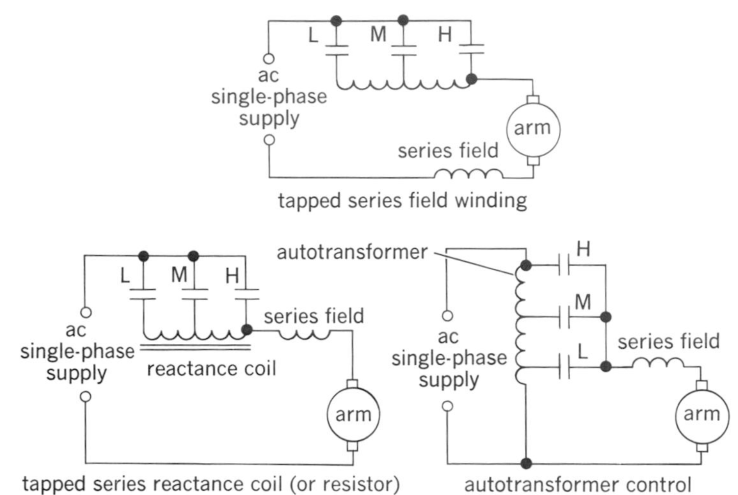

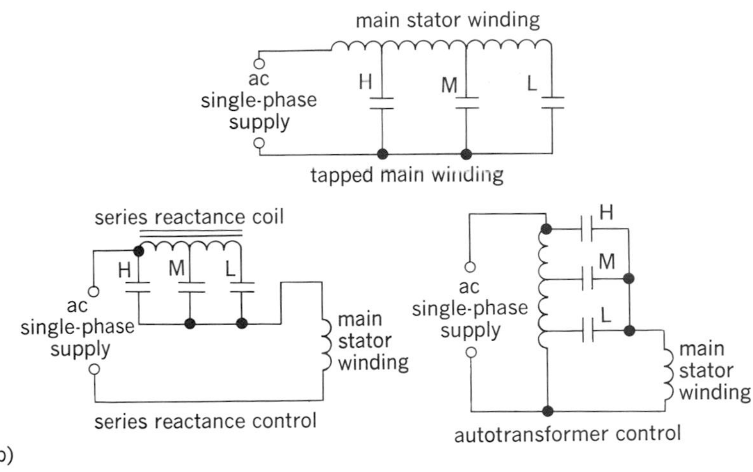

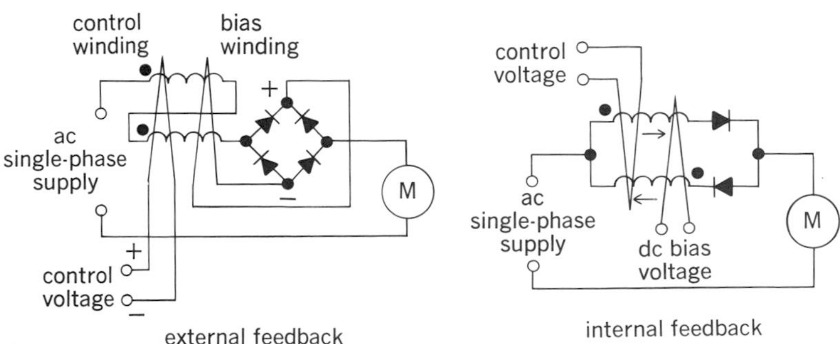

Fig. 3. Schematics for various methods of speed control for single-phase ac motors operating from fixed supplies. (*a*) Controls for ac series motors or universal motors. (*b*) Slip control for shaded-pole, reluctance, and split-phase motors. (*c*) Magnetic amplifier for voltage control of single-phase motors.

Standard ranges of hp, speeds, and slips at 60 Hz*

No. of phases	Type of motor	Power output, hp	Synchronous speed, rpm	Slip, %
Single	Capacitor	1–10	1800–1200	0–5
	Split-phase	0–0.5	1800–1200	0–5
	Shaded-pole	0–0.25	1800–900	11–14
	Repulsion-start	0–25	1800–1200	0–5
Three	Induction	0–100,000	3600–450	0–5
Three	Synchronous	20–30,000	3600–80	0

*National Electrical Manufacturers Association. †1 hp = 0.75 kW.

The table gives a comparison of available sizes and characteristics of some common ac motors.

For general principles *see* ELECTRIC ROTATING MACHINERY; MOTOR. [ALBERT F. PUCHSTEIN]

SPEED CONTROL OF AC MOTORS

The polyphase synchronous motor is a constant-speed (synchronous), variable-torque, doubly excited machine. The stator armature is excited with polyphase ac of a given frequency, and the rotor field is excited with dc. The rotor speed of the synchronous motor is a direct function of the number of stator and rotor field poles and the frequency of the ac applied to the stator. Since the number of rotor poles of the polyphase synchronous motor is not easily modified, the change of frequency method is the only way to control synchronous speed of the motor.

The polyphase induction motor is also a doubly excited machine, whose stator armature is excited with polyphase ac of line frequency and whose rotor is excited by induced ac of variable frequency, depending on rotor slip. The speed of the polyphase induction motor is an asynchronous speed varied by one or more of the following methods: (1) by changing the applied frequency to the stator (same method as for a synchronous motor, already discussed); (2) by controlling the rotor slip by means of rheostatic rotor resistance control (used for wound-rotor induction motors); (3) by changing the number of poles of both stator and rotor; and

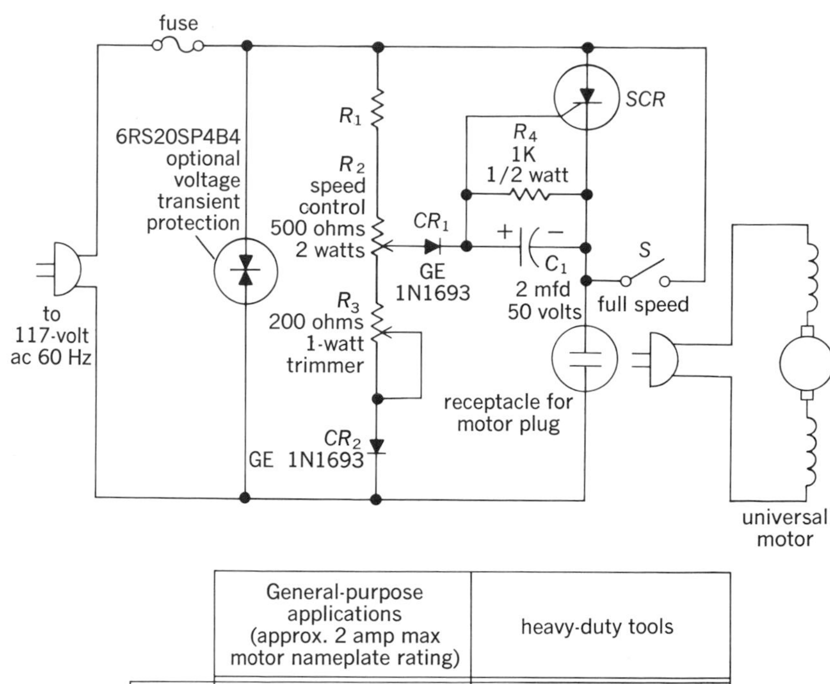

	General-purpose applications (approx. 2 amp max motor nameplate rating)	heavy-duty tools
SCR	GE C15B	GE 2N1846(C36B)
R_1	4000 ohms, 2 watts	1000 ohms, 5 watts

Fig. 4. Plug-in speed control with feedback using a silicon-controlled rectifier (SCR) for voltage control of single-phase motors.

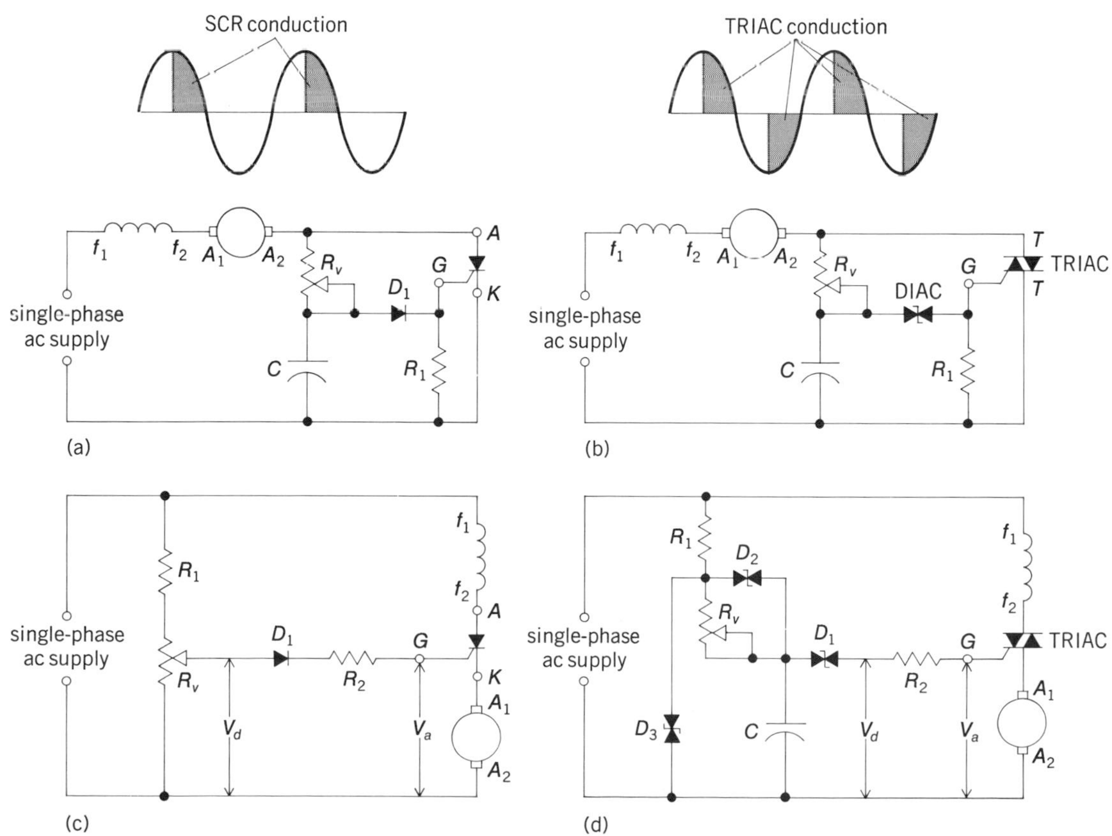

Fig. 5. Single-phase, universal or dc motor control using half- and full-wave (ac) controls, with and without voltage feedback. (*a*) Half-wave circuit. (*b*) Full-wave circuit. (*c*) Half-wave circuit with voltage feedback. (*d*) Full-wave circuit with voltage feedback. (*From I. L. Kosow, Control of Electric Machines, 1973*)

(4) by "foreign voltage" control, obtained by conductively or inductively introducing applied voltages of the proper frequency to the rotor. Figure 2 illustrates with schematics these methods of speed control.

It should be noted that a number of electromechanical or purely mechanical methods also serve to provide speed control of ac motors. One is the Rossman drive, in which an induction motor stator is mounted on trunnion bearings and driven with an auxiliary motor, providing the desired change in slip between stator and rotor. Polyphase induction and synchronous motors having essentially constant speed characteristics at rated voltage are also assembled in packaged drive units employing gears, cylindrical and conical pulleys, and even hydraulic pumps to produce a variable speed output.

In addition to reversal of rotation, some of these units employ magnetic slip clutches and solenoids to control the various mechanical and hydraulic arrangements through which a relatively smooth control of speed is achieved. A discussion of such electromechanical and mechanical speed-control methods is beyond the scope of this article.

The principal method of speed control used for fractional hp single-phase induction-type, shaded-pole, reluctance, series, and universal motors is the method of primary line-voltage control. It involves a reduction in line voltage applied to the stator winding (of the induction type) or to the armature of series and universal motors. In the former this produces a reduction of torque and increase in rotor slip. In the latter it is simply a means of controlling speed by armature voltage control or field flux control or both. The reduction in line voltage is usually accomplished by any one of five methods shown in Figs. 3 and 4; these are autotransformer control, series reactance control, tapped main-winding control, saturable reactor (or magnetic amplifier) control, and silicon-controlled rectifier feedback control.

Electronic speed control techniques. The development of the thyristor, or silicon-controlled rectifier (SCR), has created unlimited possibilities for control of virtually all types of motors (single-phase, dc, and polyphase). SCRs in sizes up to 1600 amperes (rms) with voltage ratings up to 1600 volts are available.

SCR control of series motors. Fractional horsepower universal, ac and dc series motors may be speed-controlled from a single-phase 110- or 220-volt supply using the half-wave circuit involving a diode D_1 and SCR (Fig. 5*a*). The trigger point of the SCR is adjusted via potentiometer R_v, which phase-shifts the gate turn-on voltage of the SCR whenever its anode A is positive with respect to cathode K. During the negative half-cycle of input voltage, SCR conduction is off and the gating pulse is blocked by diode D_1. When the positive half-cycle is initiated once again, capacitor C charges to provide the required gating pulse at the time

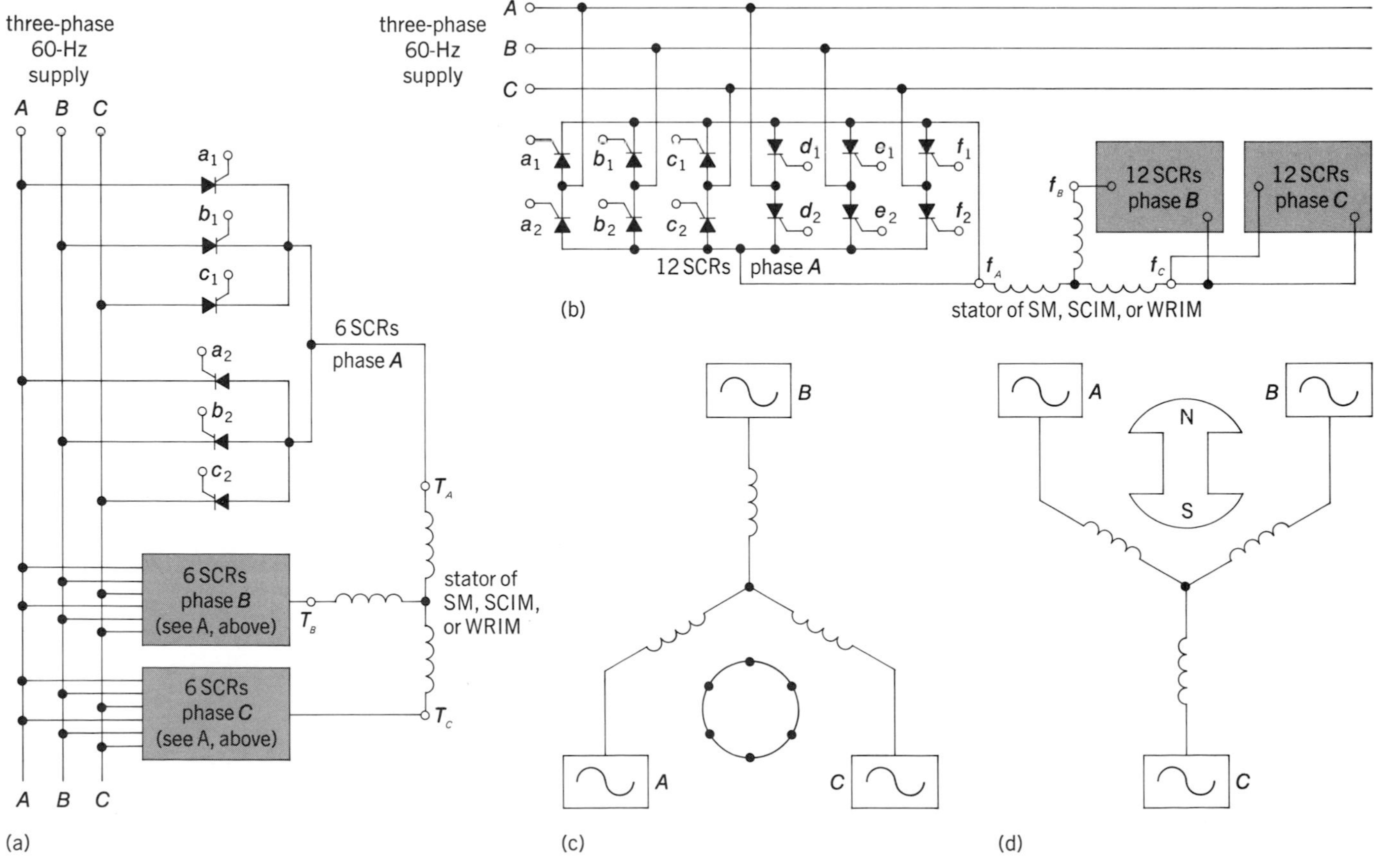

Fig. 6. Half-wave and full-wave cycloconverters with graphical symbols for simplified representation. (*a*) Half-wave cycloconverter (exclusive of triggering of SCR gates). (*b*) Full-wave cycloconverter (exclusive of SCR gating). (*c*) Symbol for SCIM driven by a cycloconverter. (*d*) Symbol for SM driven by a cycloconverter. (*From I. L. Kosow, Control of Electric Machines, 1973*)

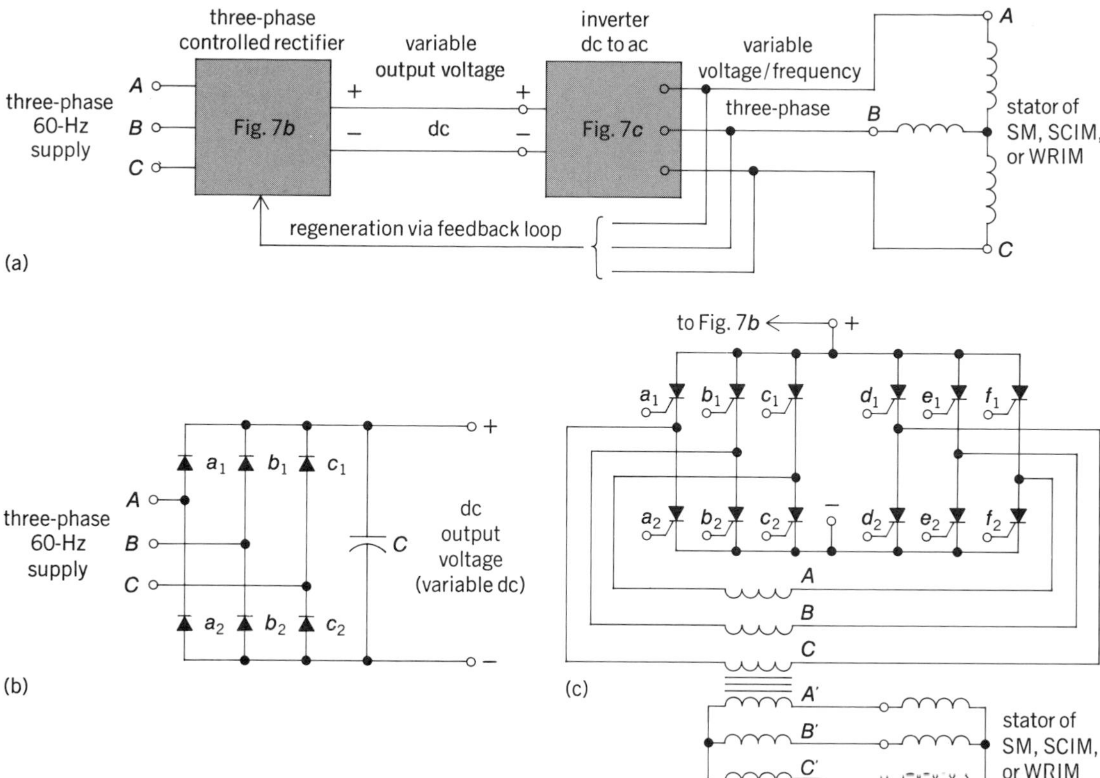

Fig. 7. Rectifier-inverter circuitry for variable voltage/frequency control of three-phase motors. (*a*) Block diagram of rectifier-inverter. (*b*) Three-phase full-wave controlled rectifier. (*c*) Basic inverter dc to three-phase ac for phase-shift voltage control. (*From I. L. Kosow, Control of Electric Machines, 1973*)

preset by the time constant R_vC.

Replacing diode D_1 by a DIAC and the SCR by a TRIAC converts the circuit from half-wave to full-wave operation and control (Fig. 5*b*). Compared with Fig. 5*a*, this circuit provides almost twice the torque and improved speed regulation. Neither circuit, however, is capable of automatic maintenance of desired speed because of the inherently poor speed regulation of series-type ac, universal, or dc motors, with application or variation of motor loading.

Automatic speed regulation is achieved by using the half-wave feedback circuit shown in Fig. 5*c*. This circuit requires, however, that both field leads (f_1, f_2) and both armature leads (A_1, A_2) are separately brought out for connection as shown in Fig. 5*c*. The desired speed is set by potentiometer R_v, in terms of reference voltage V_d across diode D_1. The actual motor speed is sensed in terms of the armature voltage V_a. Diode D_1 conducts only when the reference voltage V_d (desired speed) exceeds the voltage V_a to restore the motor speed to its desired setting. Thus, when the motor load is increased and the speed drops, the SCR is gated earlier in the cycle because diode D_1 conducts whenever V_d exceeds V_a.

As with the previous half-wave circuit, the feedback circuit may be converted to full-wave operation by replacing diode D_1 with a DIAC and the SCR with a TRIAC (Fig. 5*d*). This last circuit also is shown with additional minor modifications to improve regulation of speed at lower speeds, namely addition of DIACs D_2, D_3 and capacitor C. The advent of high-power TRIACs and SCRs has extended use of the circuit shown in Fig. 5*d* to larger, integral horsepower ac series motors as well.

When dc motors are operated by using electronic techniques shown in Fig. 5, it is customary to derate motors proportionately because of heating effect created by ac components in both half- and full-wave waveforms. Note that there is no reversal of direction in either the ac or dc series motor under full-wave operation because both the armature and field currents have been reversed.

SCR control of polyphase ac motors. There are fundamentally only three types of polyphase ac motors, namely synchronous motors (SMs), squirrel-cage induction motors (SCIMs), and wound-rotor induction motors (WRIMs). All three employ identical stator constructions. As a result, larger polyphase motors (up to 10,000 hp) are presently being controlled by SCR packages which employ some so-called universal method of speed control. In these methods, both the frequency and stator voltages are varied (in the same proportion) to maintain constant polyphase stator flux densities and saturation, thus eliminating the possibility of overheating.

Two major classes of solid-state adjustable voltage/frequency drives have emerged, namely the cycloconverter (an ac/ac package) and the rectifier-inverter (an ac-dc-ac package). Both packages convert three-phase fixed-frequency ac to three-phase variable voltage/variable frequency.

The half-wave cycloconverter shown in Fig. 6*a* is capable of supplying from zero to 20 Hz to an ac polyphase motor. It uses 6 SCRs per phase but is incapable of either phase-sequence reversal or frequencies above 20 Hz. The full-wave converter, shown in Fig. 6*b*, uses twice the number of SCRs (12 per phase) and possesses advantages of wider frequency variation (from +30 Hz down to 0 and up to −30 Hz), potentialities for dynamic braking, and capability of power regeneration.

The symbol for a polyphase SCIM driven by a cycloconverter is shown in Fig. 6*c*, and an SM in Fig. 6*d*. Note that the symbol implies frequency/voltage control of identical stators, and the nature of the motor is determined solely by its rotor.

The second class of solid-state drive package is the rectifier-inverter, which converts fixed-frequency polyphase ac to dc (variable voltage). The dc is then inverted by a three-phase inverter (using a minimum of 12 SCRs, commercially) to produce variable-voltage, variable-frequency three-phase ac for application to the motor stator. The block diagram of a solid-state rectifier-inverter package is shown in Fig. 7*a*. The three-phase 60-Hz supply is first rectified to produce variable dc (Fig. 7*b*) by appropriate phase shift of SCR gates. The variable dc is then applied to the dc bus of Fig. 7*c*. Inversion is accomplished by appropriate phase shift of gates of the 12 SCRs to produce phase and line voltages displaced, respectively, by 120°. This three-phase output voltage is applied to a transformer whose secondary is applied to the stator of the motor (Fig. 7*c*).

The rectifier-inverter is capable of power regeneration as noted by the feedback loop shown in Fig. 7*a*. There is no clear-cut choice between cycloconverters and rectifier-inverter packages, as of this writing.

Brush-shifting motors. By use of a regulating winding, a commutator, and a brush-shifting device, speed of a polyphase induction motor can be controlled similarly to that of a dc shunt motor. Such motors are used for knitting and spinning machines, paper mills, and other industrial services that require controlled variable-speed drive. The primary winding on the rotor is supplied from the line through slip rings. The stator windings are the secondary windings (S_1, S_2, S_3), and the third winding, also in the rotor, is an adjusting winding provided with a commutator (Fig. 8). Voltages collected from the commutator are fed into the secondary circuit. Brushes 1, 2, and 3 are mounted 120 electrical degrees apart on a movable yoke. Brushes 4, 5, and 6 are similarly mounted on a separate movable yoke. Each set of brushes can be moved as a group. Thus both the spacing between sets of brushes and the angular position of the brushes are adjustable. Brush spacing determines the magnitude of the voltage applied to the secondary. When brush sets are so adjusted that pairs of brushes are in contact with the same commutator segment, the secondary is short-circuited and no voltage is supplied. Under these conditions the motor behaves as an ordinary induction motor. The speed can be reduced by separating the brushes so that secondary current produces a negative torque. The machine can be operated above synchronism by interchanging the position of the brushes, so the voltage collected is in a direction to produce a positive torque. The motor can be reversed by reversing two of the leads supplying the primary.

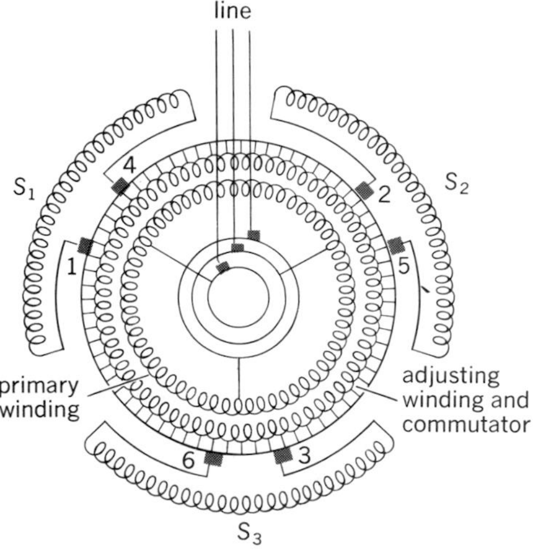

Fig. 8. Schematic of adjustable-speed brush-shifting motor. (*From A. E. Fitzgerald and C. Kingsley, Jr., Electric Machinery, McGraw-Hill, 1952*)

[IRVING L. KOSOW]

Bibliography: D. G. Fink and H. W. Beatty (eds.), *Standard Handbook for Electrical Engineers*, 11th ed., 1978; A. E. Fitzgerald et al, *Electric Machinery*, 3d ed, 1971; I. L. Kosow, *Electric Machinery and Transformers*, 1972; I. L. Kosow, *Control of Electric Machines*, 1973.

Atomic energy

The energy released in the rearrangement of the particles making up the nucleus of an atom, popularly referred to as atomic energy, but preferably called nuclear energy. Atomic properly refers to phenomena involving the orbital electrons of the atom, but not involving any transformation of the nucleus. *See* NUCLEAR POWER. [JAMES A. LANE]

Atomic nucleus

The central region of an atom. Atoms are composed of negatively charged electrons, positively charged protons, and electrically neutral neutrons. The protons and neutrons (collectively known as nucleons) are located in a small central region known as the nucleus. The electrons move in orbits which are large in comparison with the dimensions of the nucleus itself. Protons and neutrons possess approximately equal masses, each roughly 1840 times that of an electron. The number of nucleons in a nucleus is given by the mass number A and the number of protons by the atomic number Z. Nuclear radii r are given approximately by $r = 1.4 \times 10^{-13} A^{1/3}$ cm.

[HENRY E. DUCKWORTH]

Automotive engine

An integral major component of and source of power for automotive vehicles. Several types of engines are available for passenger and commercial vehicles, but most vehicles are driven by reciprocating internal combustion gasoline engines operating upon the four-cycle principle worked out by Nikolaus Otto in 1876. Many large commercial vehicles use diesel engines, with the fuel ignited by the heat caused by high air compression, for more economical operation and longer life, rather than gasoline engines with ignition by spark plugs. Diesel engines are also available in some imported passenger cars. *See* DIESEL ENGINE; OTTO CYCLE.

Domestic passenger car engines evolved from an assortment of electrics and steamers, designed with two or four cycles. Present four-, six-, and eight-cylinder engines are four-cycle types of in-line, V-block, slant, or horizontally opposed "pancake" configurations. In 1900, electrics and steamers far outnumbered gasoline engines, but within 2 decades they faded from the scene. Formerly popular in-line eights and V-12 and V-16 powerplants were discontinued when higher compression ratios were developed. Redesign of six- and eight-cylinder engines provided equivalent horsepower, with incidental reduction in weight.

As vehicles grew in size and type, and greater road speed was demanded, engines tended to increase in size and weight and had to improve in performance. In an attempt to reduce engine weight, aluminum cylinder blocks were tried, but found faulty. Present cylinder blocks and heads are gray iron castings, as are numerous other engine parts. Weight reduction has been achieved by decreasing cylinder wall sections and substituting stampings for castings, and plastics for metal parts.

Higher engine compression ratios produced greater combustion temperatures, and ignition knocks which could be eliminated only with higher-octane fuel, made possible by introduction of gasoline additives. Such developments created a new family of high-performance engines. This developmental trend was modified by the Clean Air Act of 1970, which mandated specific emission controls and use of nonleaded fuel, and encouraged manufacturers to get more miles per gallon.

Engine compartments have limited space available and this has been encroached upon by increasing "hang-on" hardware and by front suspension units and similar space-robbing components. The current tendency toward compact and subcompact cars accentuates this problem, which will undoubtedly require revolutionary changes. Rotary and other types of engines have been available for more than 2 decades, with approximately 40% fewer parts and 50% less bulk.

In the 1970s a new engine vocabulary came into being, as well as a plethora of sophisticated devices: catalytic converters, lean-burn engines, electronic spark advance (ESA), computer controlled timing (CCT), stratified charge (SC), electronic fuel injection (EFI), electronic fuel metering (EFM), and a myriad of complicated systems still in development stages. *See* INTERNAL COMBUSTION ENGINE.

[PAUL A. OTT, JR.]

Battery

A device which transforms chemical energy into electric energy. The term is usually applied to a group of two or more electric cells connected together electrically. In common usage the term battery is also applied to a single cell, such as a flashlight battery.

Types. There are in general two types of batteries, primary batteries and secondary storage or accumulator batteries. Primary types, although sometimes consisting of the same plate-active materials as secondary types, are constructed so that only one continuous or intermittent discharge can be obtained. Secondary types are constructed so that they may be recharged, following a partial or complete discharge, by the flow of direct current through them in a direction opposite to the current flow or discharge. By recharging after discharge, a higher state of oxidation is created at the positive plate or electrode and a lower state at the negative plate, returning the plates to approximately their original charged condition.

Primary and secondary cells may be constructed from several materials. For the more important of these types *see* PRIMARY BATTERY; STORAGE BATTERY.

Applications. Primary cells or batteries are used as a source of dc power where the following requirements are important.

1. Electrical charging equipment or power is not readily available.

2. Convenience is of major importance, as in the case of the hand or pocket flashlight.

3. Stand-by power is desirable without cell deterioration during periods of nonuse for days or years. Reserve-electrolyte designs may be necessary, as in torpedo, guided-missile, and some emergency light and power batteries. *See* RESERVE BATTERY.

4. The cost of a discharge is not of primary importance.

Secondary cells or batteries are used as a source of dc power where the following requirements are important.

1. The battery is the primary source of power and numerous discharge-recharge cycles are required, as in industrial hand or rider trucks, electric street trucks, mine or switching locomotives, and submarines.

2. The battery is used to supply large, short-time (or relatively small, longer-time), repetitive power requirements, as in automotive and airplane batteries.

3. Stand-by power is required and the battery is continuously connected to a voltage controlled dc circuit. With proper voltage the battery is said to "float" (drawing from the dc circuit only sufficient current to compensate automatically for the battery's own internal self-discharge). Telephone exchange, central-station circuit breaker, and emergency light and power batteries are in this category.

4. Long periods of low-current-rate discharge followed subsequently by recharge are required, as in buoy service.

5. The very large capacitance is beneficial to the circuit, as in telephone exchanges.

Size. Both primary and secondary cells are manufactured in many sizes and designs, from the small electric wristwatch battery and the small penlight battery to the large submarine battery, where a single cell has weighed 1 ton. In all applications the cell must be constructed for its particular service, so that the best performance may be obtained consistent with cost, weight, space, and operational requirements. Automotive and aircraft batteries generally use thin positive and negative plates with thin separation to conserve space and weight and to provide high rates of current discharge at low temperatures. Stand-by batteries use thick plates and thick separators to provide long life. Notable size and weight reductions have been made through use of new plastic materials, active materials, and methods of construction.

Ratings. Since the power that can be obtained from a cell varies with its temperature and the rate of current discharge, the power-output rating is very important. Common secondary-battery practice is to rate cells in terms of ampere-hours (discharge rate in amperes times hours of discharge) and to specify the hourly rate of discharge. A popular automotive battery capable of giving 2.5 amp for 20 hr is rated at 50 amp-hr at the 20-hr rate. This same battery may provide an engine-cranking current at 150 amp for only 8 min at 80°F (27°C) or for 4 min at 0°F (−18°C), giving a service of only 20 and 10 amp-hr, respectively. By multiplying ampere-hours by average voltage during discharge, watt-hour rating is obtained. Ratings must be made to a specified final voltage, which is either at the point of rapid voltage drop or at minimum usable voltage. The rating of primary batteries is generally stated as the number of hours of discharge which can be obtained when discharging through a specified fixed resistance to a specified final voltage.

Life. Life of cells varies from the single discharge obtainable from primary types to 10,000 or more discharge-charge cycles obtainable from some secondary cells operating at very high rates for very short times. Automotive batteries may generally be expected to give approximately 300 cycles, or to last 2 years. Industrial-truck sizes may be expected to give 1500 to 3000 cycles in 5–10 years. Stand-by sizes may be expected to float across the dc bus 8–30 years. Generally the most costly, largest, heaviest cells are the longest-lived.

To obtain life from batteries, certain precautions are necessary. The stated shelf life and temperature of wet primary cells must not be exceeded. For dry reserve-electrolyte primary cells and secondary cells of the dry construction with charged plates, the cell or battery container must be protected against moisture, and storage must be within prescribed temperature limits. Wet, charged secondary batteries require periodic charging and water addition, depending upon the kind of construction.

Reliability. Batteries are probably the most reliable source of power known. In fact, most critical electric circuits are protected in some manner by battery power. There are no moving parts and, with good quality control in component materials and construction, one can be assured of power, particularly since adequate checks to indicate the condition of the cells usually exist. To ensure reliability, manufacturer's stipulations on storage and maintenance must be followed.

For other sources of electric energy known as batteries or cells *see* FUEL CELL; NUCLEAR BATTERY; SOLAR CELL. [HAROLD C. RIGGS]

Blowout coil

A coil that produces a magnetic field in an electrical switching device for the purpose of lengthening and extinguishing an electric arc formed as the contacts of the switching device part to interrupt the current. The magnetic field produced by the coil is approximately perpendicular to the arc. The interaction between the arc current and the magnetic field produces a force driving the arc in the direction perpendicular to both the magnetic flux and the arc current (Fig. 1).

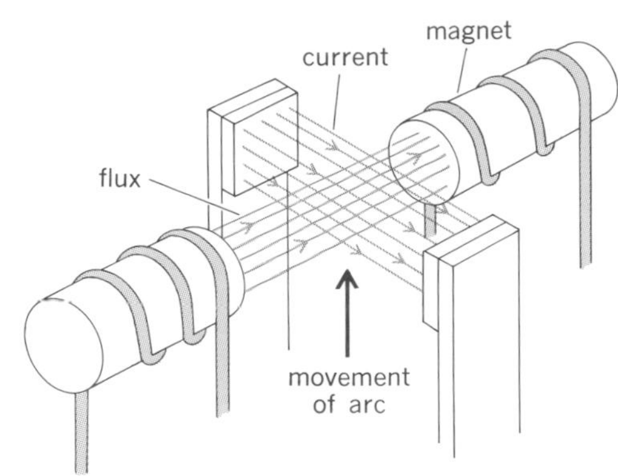

Fig. 1. Relation of directions of current, magnetic flux, and movement of arc.

In alternating-current circuits the arc usually extinguishes at a natural current zero (when the current passes through zero before reversing its direction). If, within a short time around current zero, sufficient energy can be removed from the arc, conduction will cease. The function of the blowout coil is to move the arc into a region inside a circuit interrupter, such as arc chutes in circuit breakers where the energy removal process takes place. *See* CIRCUIT BREAKER.

In direct-current circuits there are no natural current zeros. When a switching device opens, an arc strikes. As long as the voltage across the open contacts is sufficient to sustain the arc, current will flow. To interrupt this dc arc, the arc must be

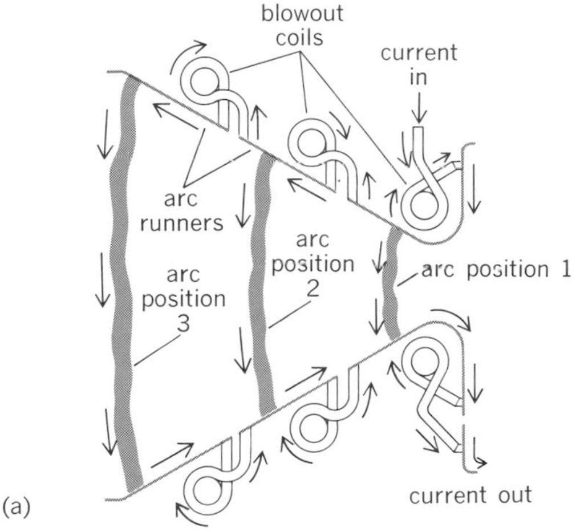

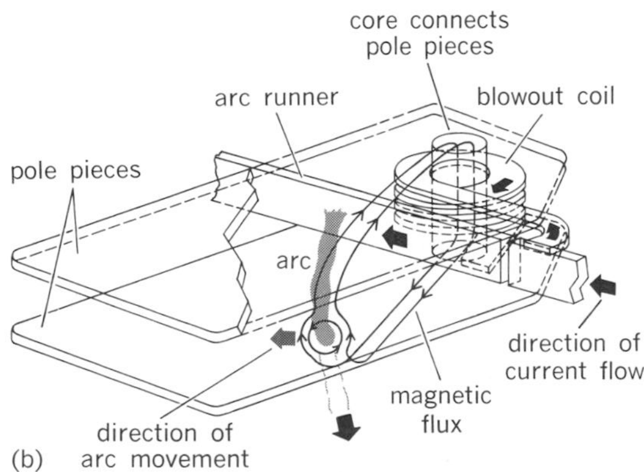

Fig. 2. Arc suppressor: (*a*) cross section and (*b*) cutaway schematic. As the arc progresses along the runners, it draws current through the blowout coils whose magnetic fields then drive the arc further out until it extinguishes.

converted to a form that, to continue, requires more than the available voltage. Thus, in a dc switching device, the function of the blowout coil is to lengthen the arc, which increases the arc voltage, and also to move the arc into regions, inside the interrupter, where the arc voltage can be further increased.

The blowout coil is either connected in series with the contacts or is inserted as the arc moves along an arc runner (Fig. 2). [THOMAS H. LEE]

Boiler

A pressurized system in which water is vaporized to steam, the desired end product, by heat transferred from a source of higher temperature, usually the products of combustion from burning fuels. Steam thus generated may be used directly as a heating medium, or as the working fluid in a prime mover to convert thermal energy to mechanical work, which in turn may be converted to electrical energy. Although other fluids are sometimes used for these purposes, water is by far the most common because of its economy and suitable thermodynamic characteristics.

The physical sizes of boilers range from small portable or shop-assembled units to installations comparable to a multistory 200-ft-high (60-m) building equipped, typically, with a furnace which can burn coal at a rate of 6 tons/min (90 kg/sec). In terms of steam production capacities, commercial boilers range from a few hundred pounds of steam per hour to more than 6,000,000 lb/hr (750 kg/sec). Pressures range from 0.5 psi (3.4 kPa) for domestic space heating units to 5000 psi (34 MPa) for some power boilers. The latter type will deliver steam superheated to 1100°±F (593°C) and reheated to similar values at intermediate pressures. Large units are field-assembled at the installation site but small units (frequently referred to as package boilers) are shop-assembled to minimize the overall boiler price.

Boilers operate at positive pressures and offer the hazardous potential of explosions. Pressure parts must be strong enough to withstand the generated steam pressure and must be maintained at acceptable temperatures, by transfer of heat to the fluid, to prevent loss of strength from overheating or destructive oxidation of the construction materials. The question of safety for design, construction, operation, and maintenance comes under the police power of the state and is supplemented by the requirements of the insurance underwriters. The ASME Boiler Construction Code is the prevalent document setting basic standards in most jurisdictions.

Being in the class of durable goods, boilers that receive proper care in operation and maintenance function satisfactorily for several decades. Thus the types of boilers found in service at any time represent a wide span in the stages of development in boiler technology.

The earliest boilers, used at the beginning of the industrial era, were simple vats or cylindrical vessels made of iron or copper plates riveted together and supported over a furnace fired by wood or coal. Connections were made for offtake of steam and for the replenishment of water. Evolution in design for higher pressures and capacities led to the use of steel and to the employment of tubular members in the construction to increase economically the amount of heat-transferring surface per ton of metal. The earliest improvement was the passage of hot gases through tubes submerged in the water space of the vessel, and later, arrangements of multiple water-containing tubes which were exposed on their outer surface to contact with hot gases. *See* FIRE-TUBE BOILER; WATER-TUBE BOILER.

The overall functioning of steam-generating equipment is governed by thermodynamic properties of the working fluid. By the simple addition of heat to water in a closed vessel, vapor is formed which has greater specific volume than the liquid, and can develop increase of pressure to the critical value of 3208 psia (22.12 MPa absolute pressure). If the generated steam is discharged at a controlled rate, commensurate with the rate of heat addition, the pressure in the vessel can be maintained at any desired value, and thus be held within the limits of safety of the construction.

Addition of heat to steam, after its generation, is accompanied by increase of temperature above the saturation value. The higher heat content, or enthalpy, of superheated steam permits it to develop a higher percentage of useful work by expansion through the prime mover, with a resultant gain in efficiency of the power-generating cycle.

If the steam-generating system is maintained at pressures above the critical, by means of a high-pressure feedwater pump, water is converted to a vapor phase of high density equal to that of the water, without the formation of bubbles. Further heat addition causes superheating, with corresponding increase in temperature and enthalpy. The most advanced developments in steam-generating equipment have led to units operating above critical pressure, for example, 3600–5000 psi (25–34 MPa).

Superheated steam temperature has advanced from 500°±F (260°C) to the present practical limits of 1050–1100°F (566–593°C). Progress in boiler design and performance has been governed by the continuing development of improved materials for superheater construction having adequate strength and resistance to oxidation for service at elevated temperatures. For the high temperature ranges, complex alloy steels are used in some parts of the assembly.

Steam boilers are built in a wide variety of types and sizes utilizing the widest assortment of heat sources and fuels. *See* NUCLEAR POWER; STEAM-GENERATING UNIT. [THEODORE BAUMEISTER]

Bibliography: American Society of Mechanical Engineers, *Boiler Construction Code*, 1968; Babcock and Wilcox Co., *Steam: Its Generation and Use*, 1955; T. Baumeister, *Standard Handbook for Mechanical Engineers*, 7th ed., 1967; Combustion Engineering, Inc., *Combustion Engineering*, revised periodically.

Boiler air heater

A component of a steam-generating unit that absorbs heat from the products of combustion after they have passed through the steam-generating and superheating sections. Heat recovered from the gas is recycled to the furnace by the combustion air and is absorbed in the steam-generating unit, with a resultant gain in overall thermal efficiency. Use of preheated combustion air also accelerates ignition and promotes rapid burning of

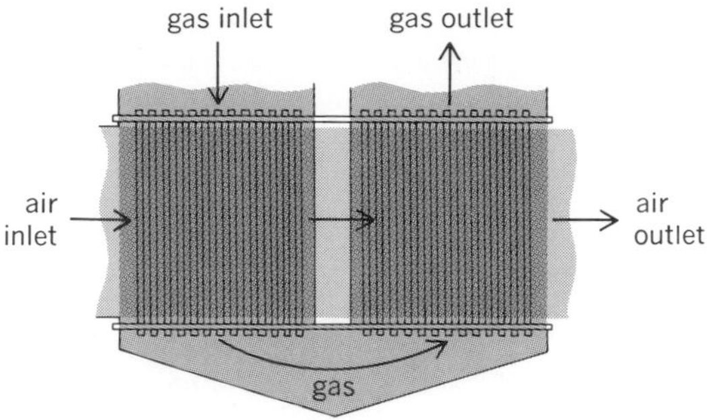

Fig. 1. Tubular air heater, two-gas single-air pass.

the fuel. Air heaters often are used in conjunction with economizers, because the temperature of the inlet air is less than that of the feedwater to the economizer, and in this way it is possible to reduce further the temperature of flue gas before it is discharged to the stack. *See* BOILER ECONOMIZER.

Air heaters are usually classed as recuperative or regenerative types. Both types depend upon convection transfer of heat from the gas stream to a metal or other solid surface and upon convection transfer of heat from this surface to the air. In the recuperative type, exemplified by tubular- or plate-type heaters, the metal parts are stationary and form a separating boundary between the heating and cooling fluids, and heat passes by conduction through the metal wall (Fig. 1). In rotary regenerative air heaters (Fig. 2) heat-transferring members are moved alternately through the gas and air streams, thus undergoing repetitive heating and cooling cycles; heat is transferred to or from the thermal storage capacity of the members. Other forms of regenerative-type air heaters, which seldom are used with steam-generating units, have stationary elements, and the alternate flow of gas and air is controlled by dampers, as in the refractory stoves of blast furnaces; or they may employ, as in the pebble heaters used in the petroleum industry for high-temperature heat exchange, a flow of solid particles which are alternately heated and cooled.

In convection heat-transfer equipment, higher heat-transfer rates and better utilization of the heat-absorbing surface are obtained with a counterflow of gases through small flow channels. The rotary regenerative air heater readily lends itself to the application of these two principles and offers high performance in small space. However, leakage of air into the gas stream necessitates frequent maintenance of seals between the moving and stationary members, and fly ash often is transported into the combustion air system. These problems are not experienced with recuperative air heaters of the tubular type. See STEAM-GENERATING UNIT. [GEORGE W. KESSLER]

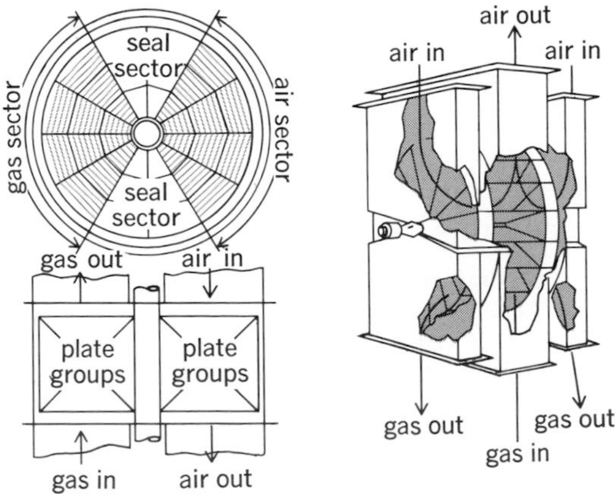

Fig. 2. Two types of rotary regenerative air heaters.

Boiler economizer

A component of a steam-generating unit that absorbs heat from the products of combustion after they have passed through the steam-generating and superheating sections. The name, accepted through common usage, is indicative of savings in the fuel required to generate steam.

An economizer is a forced-flow, once-through, convection heat-transfer device to which feedwater is supplied at a pressure above that in the steam-generating section and at a rate corresponding to the steam output of the unit. The economizer is in effect a feedwater heater, receiving water from the boiler feed pump and delivering it at a higher temperature to the steam generator or boiler. Economizers are used instead of additional steam-generating surface because the feedwater, and consequently the heat-receiving surface, is at a temperature below that corresponding to the saturated steam temperature; therefore, the economizer further lowers the flue gas temperature for additional heat recovery. *See* THERMODYNAMIC CYCLE.

Generally, steel tubes, or steel tubes fitted with externally extended surface, are used for the heat-absorbing section of the economizer; usually, the economizer is coordinated with the steam-generating section and placed within the setting of the unit.

The size of an economizer is governed by economic considerations involving the cost of fuel, the comparative cost and thermal performance of alternate steam-generating or air-heater surface, the feedwater temperature, and the desired exit gas temperature. In many situations it is economical to utilize an economizer and an air heater together. [G. W. KESSLER]

Brayton cycle

A thermodynamic cycle (also variously called the Joule or complete expansion diesel cycle) consisting of two constant-pressure (isobaric) processes interspersed with two reversible adiabatic (isentropic) processes (Fig. 1). The ideal cycle performance, predicated on the use of perfect gases, is given by relationships (1) and (2). Thermal efficiency η_T, the work done per unit of heat added, is given by Eq. (3). In these relationships

$$V_3/V_2 = V_4/V_1 = T_3/T_2 = T_4/T_1 \quad (1)$$

$$\frac{T_2}{T_1} = \frac{T_3}{T_4} = \left(\frac{V_1}{V_2}\right)^{k-1} = \left(\frac{V_4}{V_3}\right)^{k-1} = \left(\frac{p_2}{p_1}\right)^{\frac{k-1}{k}} \quad (2)$$

$$\eta_T = [1-(T_1/T_2)] = \left[1-\left(\frac{1}{r^{k-1}}\right)\right] \quad (3)$$

V is the volume in cubic feet, p is the pressure in pounds per square foot, T is the absolute temperature in degrees Rankine, k is the c_p/c_v, or ratio of specific heats at constant pressure and constant volume, and r is the compression ratio, V_1/V_2.

The thermal efficiency for a given gas, air, is solely a function of the ratio of compression (Fig. 2). This is also the case with the Otto cycle. For the diesel cycle with incomplete expansion, the thermal efficiency is lower, as shown for comparison in Fig. 2. The overriding importance of high compression ratio for intrinsic high efficiency is clearly demonstrated by these data.

A reciprocating engine, operating on the cycle of Fig. 1, was patented in 1872 by G. B. Brayton and was the first successful gas engine built in the United States. The Brayton cycle, with its high inherent thermal efficiency, requires the maximum volume of gas flow for a given power output. The Otto and diesel cycles require much lower gas flow rates, but have the disadvantage of higher peak pressures and temperatures. These conflicting elements led to many designs, all attempting to achieve practical compromises. With a piston and cylinder mechanism the Brayton cycle, calling for the maximum displacement per horsepower, led to proposals such as compound engines and variable-stroke mechanisms. They suffered overall disadvantages because of the low mean effective pressures. The positive displacement engine consequently preempted the field for the Otto and diesel cycles.

With the subsequent development of fluid acceleration devices for the compression and expansion of gases, the Brayton cycle found mechanisms which could economically handle the large volumes of working fluid. This is perfected today in the gas turbine power plant. The mechanism (Fig. 3) basically is a steady-flow device with a centrifugal or axial compressor, a combustion chamber where heat is added, and an expander-turbine element. Each of the phases of the cycle is accomplished with steady flow in its own mechanism rather than intermittently, as with the piston and cylinder mechanism of the usual Otto and diesel cycle engines. Practical gas-turbine engines have various recognized advantages and disadvantages which are evaluated by comparison with alternative engines available in the competitive market place. *See* GAS TURBINE; INTERNAL COMBUSTION ENGINE.

The net power output P_{net}, or salable power, of the gas-turbine plant (Fig. 3) can be expressed as shown by Eq. (4), where W_e is the ideal power out-

$$P_{net} = W_e \times \text{eff}_e - \frac{W_c}{\text{eff}_c} \qquad (4)$$

put of the expander (area b34a, Fig. 1), W_c the ideal power input to the compressor (area a12b, Fig. 1), eff_e the efficiency of expander, and eff_c the efficiency of compressor. This net power output for the ideal case, where both efficiencies are 1.0, is represented by net area (shaded) of the p-V cycle diagram of Fig. 1. The larger the volume increase from point 2 to point 3, the greater will be the net power output for a given size compressor. This volume increase is accomplished by utilizing the maximum possible temperature at point 3 of the cycle.

The difference in the two terms on the right-hand side of Eq. (4) is thus basically increased by the use of maximum temperatures at the inlet to the expander. These high temperatures introduce metallurgical and heat-transfer problems which must be properly solved.

The efficiency terms of Eq. (4) are of vital practical significance. If the efficiencies of the real compressor and of the real expander are low, it is entirely possible to vitiate the difference in the ideal powers W_e and W_c, so that there will be no useful output of the plant. In present practice this means that for adaptations of the Brayton cycle to acceptable and reliable gas-turbine plants, the

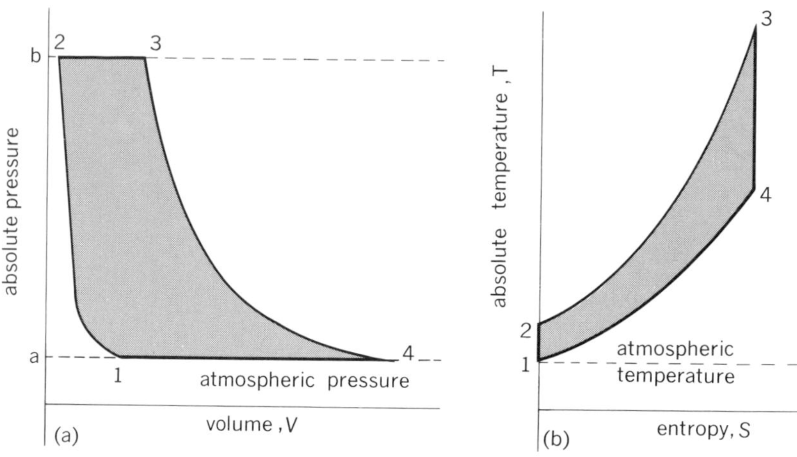

Fig. 1. Brayton cycle, air-card standard: *(a)* *p-V* diagram; *(b)* *T-S* diagram. Phases: 1–2, compression; 2–3, heat addition; 3–4, expansion; and 4–1, heat abstraction.

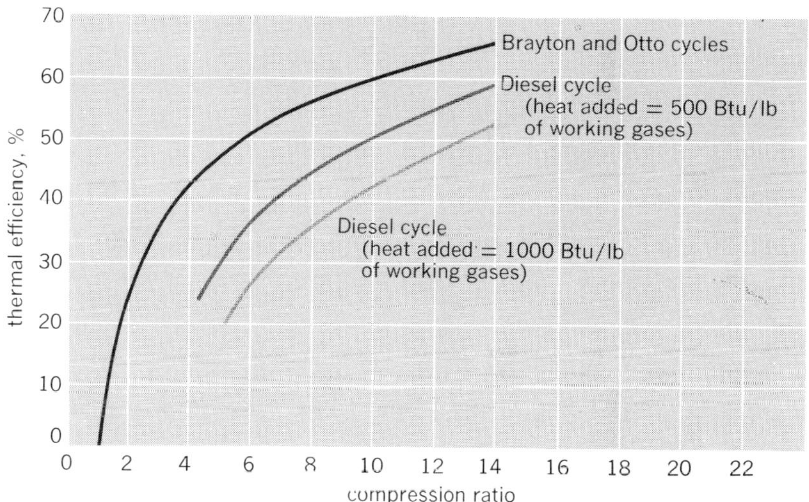

Fig. 2. Thermal efficiency of Brayton, Otto, diesel ideal cycles; air-card standard. 1 Btu/lb = 2326 J/kg.

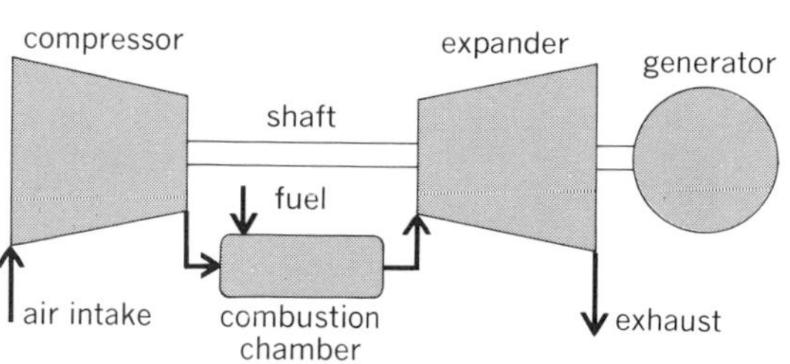

Fig. 3. Simple, open-cycle gas-turbine plant.

engineering design must provide for high temperatures at the expander inlet and utilize high built-in efficiencies of the compressor and expander elements. No amount of cycle alteration, regeneration, or reheat can offset this intrinsic requirement for mechanisms which will safely operate at temperatures of 1500°±F (815°C). *See* CARNOT CYCLE; DIESEL CYCLE; OTTO CYCLE; THERMODYNAMIC CYCLE.

[THEODORE BAUMEISTER]

Bibliography: T. Baumeister (ed.), *Standard Handbook for Mechanical Engineers*, 8th ed., 1978; J. B. Jones and G. A. Hawkins, *Engineering Thermodynamics*, 1960; J. H. Keenan, *Thermodynamics*, 1970.

British thermal unit (Btu)

A unit quantity of energy. Usually the expression is used in connection with energy as heat, but it may be used when referring to energy as work or energy in any other form. Before 1929 the 60° Btu was defined as that quantity of energy as heat required to increase the temperature of 1 lb of water from 59.5°F to 60.5°F, the water being held under a constant pressure of 1 atm. Because of the difficulty of experimentally determining the exact value of the Btu, and especially its relation to the foot-pound, the Btu is now defined in terms of electrical units. At the International Steam Table Conference in London in 1929, 1 Btu was defined as 251.996 IT (International Steam Table) cal, or 778.26 ft-lb. By definition 1 IT cal = 1/860 watt-hour: hence a Btu is equivalent to approximately 1/3 watt-hour. *See* ELECTRICAL MEASUREMENTS; HEAT.

[HAROLD C. WEBER]

Bus-bar

An electric conductor that serves as a common connection between load circuits and the source of electric power of one polarity in direct-current systems or of one phase in alternating-current systems. Two or more bus-bars which serve all polarities or phases are collectively called a bus (see illustration).

Bus-bars must be designed to carry the continuous current without overheating. The highest continuous current can be in the order of hundreds of thousands of amperes. Bus-bars must also be designed to withstand the mechanical forces caused by short-circuit currents. Special metallic shields are designed to enclose the bus, thus minimizing the effect of short-circuit forces. Such a bus is known as an isolated phase bus.

[THOMAS H. LEE]

Three-phase isolated phase bus.

Carnot cycle

A hypothetical thermodynamic cycle originated by Sadi Carnot and used as a standard of comparison for actual cycles. The Carnot cycle shows that, even under ideal conditions, a heat engine cannot convert all the heat energy supplied to it into mechanical energy; some of the heat energy must be rejected.

In a Carnot cycle, an engine accepts heat energy from a high-temperature source, or hot body, converts part of the received energy into mechanical (or electrical) work, and rejects the remainder to a low-temperature sink, or cold body. The greater the temperature difference between the source and sink, the greater the efficiency of the heat engine.

The Carnot cycle (Fig. 1) consists first of an isentropic compression, then an isothermal heat addition, followed by an isothermal expansion, and concludes with an isentropic heat rejection process. In short, the processes are compression, addition of heat, expansion, and rejection of heat, all in a qualified and definite manner.

Processes. The air-standard engine, in which air alone constitutes the working medium, illustrates the Carnot cycle. A cylinder of air has perfectly insulated walls and a frictionless piston. The top of the cylinder, called the cylinder head, can either be covered with a thermal insulator, or, if the insulation is removed, can serve as a heat transfer surface for heating or cooling the cylinder contents.

Initially, the piston is somewhere between the top and the bottom of the engine's stroke, and the air is at some corresponding intermediate pressure but at low temperature. Insulation covers the cylinder head. By employing mechanical work from the surroundings, the system undergoes a reversible adiabatic, or an isentropic, compression. With no heat transfer, this compression process raises both the pressure and the temperature of the air, and is shown as the path *a-b* on Fig. 1

After the isentropic compression carries the piston to the top of its stroke, the piston is ready to reverse its direction and start down. The second process is one of constant-temperature heat addition. The insulation is removed from the cylinder head, and a heat source, or hot body, applied that is so large that any heat flow from it will not affect its temperature. The hot body is at a temperature just barely higher than that of the gas it is to heat. The temperature gradient is so small it is considered reversible; that is, if the temperature changed slightly the heat might flow in the other direction, from the gas into the hot body. In the heat addition process, enough heat flows from the hot body into the gas to maintain the temperature of the gas while it slowly expands and does useful work on the surroundings. All the heat is added to the working substance at this constant top temperature of the cycle. This second process is shown as *b-c* on Fig. 1.

Part way down the cylinder, the piston is stopped; the hot body is removed from the cylinder head, and an insulating cover is put in its place.

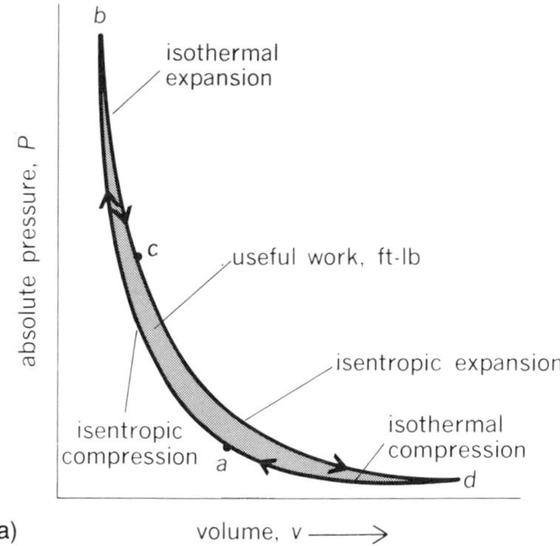

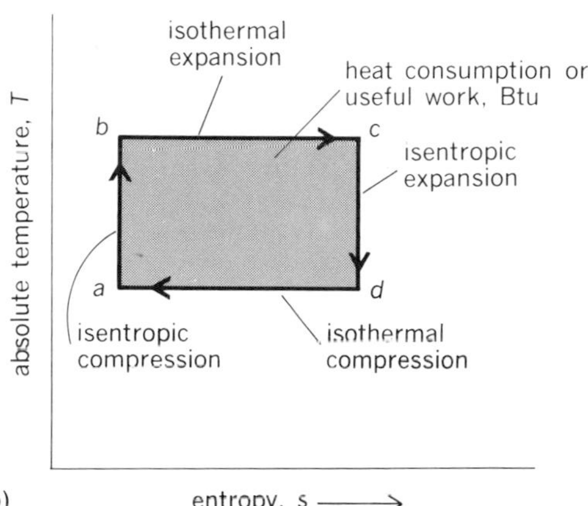

Fig. 1. Carnot cycle for air: (a) *P-v* diagram; (b) *T-s* diagram.

Then the third process of the cycle begins; it is a frictionless expansion, devoid of heat transfer, and carries the piston to the bottom of its travel. This isentropic expansion reduces both the pressure and the temperature to the bottom values of the cycle. For comparable piston motion, this isentropic expansion drops the pressure to a greater extent than the isothermal process would do. The path *c-d* on Fig. 1 represents this third process, and is steeper on the *P-v* plane than process *b-c*.

The last process is the return of the piston to the same position in the cylinder as at the start. This last process is an isothermal compression and simultaneous rejection of heat to a cold body which has replaced the insulation on the cylinder head. Again, the cold body is so large that heat flow to it does not change its temperature, and its temperature is only infinitesimally lower than that of the gas in the system. Thus, the heat rejected during the cycle flows from the system at a constant low-temperature level. The path *d-a* on Fig. 1 shows this last process.

The net effect of the cycle is that heat is added at a constant high temperature, somewhat less heat is rejected at a constant low temperature, and the algebraic sum of these heat quantities is equal to the work done by the cycle.

Figure 1 shows a Carnot cycle when air is used as the working substance. The *P-v* diagram for this cycle changes somewhat when a vapor or a liquid is used, or when a phase change occurs during the cycle, but the *T-s* diagram always remains a rectangle regardless of phase changes or of working substances employed.

It is significant that this cycle is always a rectangle on the *T-s* plane, independent of substances used, for Carnot was thus able to show that neither pressure, volume, nor any other factor except temperature could affect the thermal efficiency of his cycle. Raising the hot-body temperature raises the upper boundary of the rectangular figure, increases the area, and thereby increases the work done and the efficiency, because this area represents the net work output of the cycle. Similarly, lowering the cold-body temperature increases the area, the work done, and the efficiency. In practice, nature establishes the temperature of the coldest body available, such as the temperature of ambient air or river water, and the bottom line of the rectangle cannot circumvent this natural limit.

The thermal efficiency of the Carnot cycle is solely a function of the temperature at which heat is added (phase *b-c*) and the temperature at which heat is rejected (phase *d-a*) (Fig. 1). The rectangular area of the *T-s* diagram represents the work done in the cycle so that thermal efficiency, which is the ratio of work done to the heat added, equals $(T_{hot} - T_{cold})/T_{hot}$. For the case of atmospheric temperature for the heat sink ($T_{cold} = 500°R$), the thermal efficiency, as a function of the temperature of the heat source, T_{hot}, is shown in Fig. 2.

Carnot cycle with steam. If steam is used in a Carnot cycle, it can be handled by the following flow arrangement. Let saturated dry steam at 500°F (260°C) flow to the throttle of a perfect turbine where it expands isentropically down to a pressure corresponding to a saturation temperature of the cold body. The exhausted steam from the turbine, which is no longer dry, but contains several percent moisture, is led to a heat exchanger called a condenser. In this device there is a constant-pressure, constant-temperature, heat-rejection process during which more of the steam with a particular predetermined amount of condensed liquid is then handled by an ideal compression device. The isentropically compressed mixture may emerge from the compressor as completely saturated liquid at the saturation pressure corresponding to the hot-body temperature. The cycle

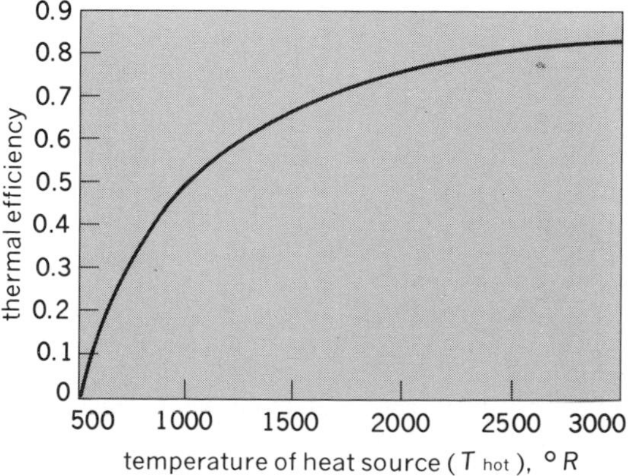

Fig. 2. Thermal efficiency of the Carnot cycle with heat-rejection temperature T_{cold} equal to 500°R.

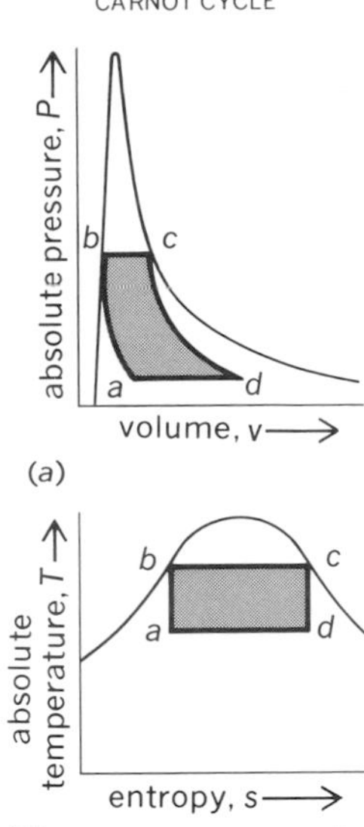

Fig. 3. Carnot cycle with steam: (a) P-v diagram; (b) T-s diagram.

is closed by the hot body's evaporating the liquid to dry saturated vapor ready to flow to the turbine.

The cycle is depicted in Fig. 3 by *c-d* as the isentropic expansion in the turbine; *d-a* is the constant-temperature condensation and heat rejection to the cold body; *a-b* is the isentropic compression; and *b-c* is the constant-temperature boiling by heat transferred from the hot body. *See* STEAM ENGINE.

Carnot cycle with radiant energy. Because a Carnot cycle can be carried out with any arbitrary system, it has been analyzed when the working substance was considered to be a batch of radiant energy in an evacuated cylinder. If the system boundaries are perfectly reflecting thermal insulators, the enclosed radiant energy will be reflected and re-reflected with no loss of radiant energy and no change of wavelength.

The electromagnetic theory of radiation asserts that the radiant energy applies pressure P to the cylinder walls. This radiation pressure is equal to $u/3$, where u represents the radiant energy density, or the amount of radiant energy per unit volume.

The piston moves so that the cylinder boundaries expand, and additional radiant energy is supplied to the system so that the temperature remains constant. Then, cutting off any further supply of radiant energy, the system undergoes a further infinitesimal expansion with its associated pressure drop and temperature drop. The third process is an isothermal compression at the low-temperature level, accompanied by some rejection of radiant energy. One last process closes the cycle with a reversible compression that raises the temperature to the original level. The assumption is made throughout this analysis that energy density is a function of temperature alone. Thus, because this last compression process increases the energy density, it raises the temperature.

A record of all energy quantities in this radiant-energy Carnot cycle indicates that energy density u is proportional to the fourth power of the absolute temperature. Consequently the total rate of emission of radiant energy from the surface of a blackbody is also proportional to the fourth power of its absolute temperature, thereby using a Carnot cycle to provide a relationship by theoretical analysis, the same relationship which had previously been determined experimentally and labeled as Stefan's law.

Conversion of heat to electricity. Several physical phenomena convert heat energy directly into electrical energy. The extent of this direct conversion of heat to electricity is limited by the temperature levels between which the process operates. The ideal efficiency of such direct-conversion thermoelectric cycles equals the efficiency of a Carnot cycle that operates between the same temperature limits of heat source and heat sink. However, the conversion efficiency obtained in practice is only a small fraction of the ideal efficiency at the present stage of development.

The most widely known physical arrangement for direct thermoelectric generation is the thermocouple, which produces an electromotive force, or voltage, when one junction of two dissimilar conductors is heated while the other junction is kept cool. Thermocouples made of metals are inefficient converters of heat energy to electric energy, because metals that have good electrical conductivity unfortunately have equally good thermal conductivity, which permits heat loss by conduction from the hot to the cold junction. In contrast, thermocouples made of semiconductors offer the prospect of operation at high temperatures and with high temperature gradients, because semiconductors are good electrical conductors but poor heat conductors. Semiconductor thermocouples may have as large a junction potential as the metal couples do.

Thermionic emission is another phenomenon that permits the partial conversion of heat energy directly into electrical energy. Externally applied heat imparts kinetic energy to electrons, liberating them from a surface. The density of the emission current is a function of the absolute temperature and work function of the emitter material.

For many years thermionic emission received little attention as a source of power because of its very low conversion efficiency. However, interest has been stimulated by development of a contact potential thermionic emission cell. In this device, current flows between the surfaces of two materials which have different work functions. These materials are held at different temperatures, and the gap between the electrode surfaces is filled with gas at low pressure.

Such direct-conversion techniques show promise of becoming small-scale, if unconventional power sources. Although these techniques ideally can convert heat to electrical energy with the efficiency of a Carnot cycle operating between the same temperature levels, laboratory devices do not surpass about 8% efficiency, which is just a fraction of the ideal performance.

Reversed Carnot cycle. A Carnot cycle consists entirely of reversible processes; thus it can theoretically operate to withdraw heat from a cold body and to discharge that heat to a hot body. To do so, the cycle requires work input from its surroundings. The heat equivalent of this work input is also discharged to the hot body.

Just as the Carnot cycle provides the highest efficiency for a power cycle operating between two fixed temperatures, so does the reversed Carnot cycle provide the best coefficient of performance for a device pumping heat from a low temperature to a higher one. This coefficient of performance is defined as the ratio of the quantity of heat pumped to the amount of work required, or it equals $T_{hot}/(T_{hot} - T_{cold})$ for a warming machine, and $T_{cold}/(T_{hot} - T_{cold})$ for a cooling machine, where all temperatures are in degrees absolute. This is one of the few engineering indices with numerical values greater than unity. *See* HEAT PUMP; REFRIGERATION CYCLE.

Practical limitations. Good as the ideal Carnot cycle may be, there are serious difficulties that emerge when one wishes to make an actual Carnot engine. The method of heat transfer through the cylinder head either limits the operation of the engine to very low speeds, or requires an engine with a huge bore and small stroke. Moreover, the material of the cylinder head is subjected to the full top temperature of the cycle, imposing a metallurgical upper limit on the cycle's temperature.

A practical solution to the heat transfer difficulties which beset the Carnot cycle is to burn a fuel in the air inside the engine cylinder. The result is an internal combustion engine that consumes and replaces its working substance while undergoing a periodic sequence of processes.

The working substance in such an internal combustion engine can attain very high temperatures, far above the melting point of the metal of the cylinder walls, because succeeding lower-temperature processes will keep the metal parts adequately cool. Thus, as the contents change temperature rapidly between wide extremes, the metal walls hover near a median temperature. The fuel-air mixture can be ignited by a spark, or by the rise in temperature produced by the compression stroke. *See* DIESEL CYCLE; DIESEL ENGINE; INTERNAL COMBUSTION ENGINE; OTTO CYCLE.

Even so, the necessarily high peak pressures and temperatures limit the practical thermal efficiency that an actual engine can achieve. The same theoretical efficiency can be obtained from a cycle consisting of two isobaric processes interspersed by two isentropic processes. The isobaric process requires that the cycle handle large volumes of low-pressure gas, which can best be done in a rotating turbine. *See* BRAYTON CYCLE; GAS TURBINE.

Although the Carnot cycle is independent of the working substance, and hence is applicable to a vapor cycle, the difficulty of efficiently compressing a vapor-liquid mixture renders the cycle impractical. In a steam power plant the sequence of states assumed by the working substance is (1) condensate and feedwater are compressed and pumped into the boiler, (2) heat is added to the water first at constantly increasing temperature and then at constant pressure and temperature, (3) the steam expands in the engine, and (4) the cycle is completed by a constant-temperature heat-rejection process which condenses the exhausted steam.

By comparison to the Carnot cycle, in which heat is added only at the highest temperature, this steam cycle with heat added over a range of temperatures is necessarily less efficient than is theoretically possible. An analysis of engine operations in terms of thermodynamic cycles indicates what efficiencies can be expected and how the operations should be modified to increase engine performance, such as high compression ratios for reciprocating internal combustion engines and high steam temperatures for steam engines. *See* POWER PLANT; THERMODYNAMIC CYCLE; THERMODYNAMIC PRINCIPLES. [THEODORE BAUMEISTER]

Bibliography: T. Baumeister, *Standard Handbook for Mechanical Engineers*, 8th ed., 1978; J. B. Jones and G. A. Hawkins, *Engineering Thermodynamics*, 1960; H. C. Weber and H. P. Meissner, *Thermodynamics for Chemical Engineers*, 1975; M. W. Zemansky, *Heat and Thermodynamics*, 5th ed., 1968.

Catalytic reforming

A process used for upgrading gasoline by improving its antiknock characteristics (increasing the octane number). It is also widely used for the production of aromatic hydrocarbons for the petrochemical industry.

Process. The process utilizes a supported platinum catalyst and involves a number of different reactions. These reactions convert, or reform, the low-octane-number feed components, such as the paraffins, to components with increased octane number. Reactions (1)–(5) show typical examples.

Reaction (1) illustrates isomerization (chain branching) of paraffins. Hydrocracking, shown in reaction (2), is usually held to a minimum because it entails formation of some butane and propane, which boil below gasoline and hence represent a yield loss. The most important reactions in catalytic reforming are those leading to the formation of aromatic hydrocarbons because these are high-octane components as well as valuable petrochemical intermediates. Aromatics are formed by rear-

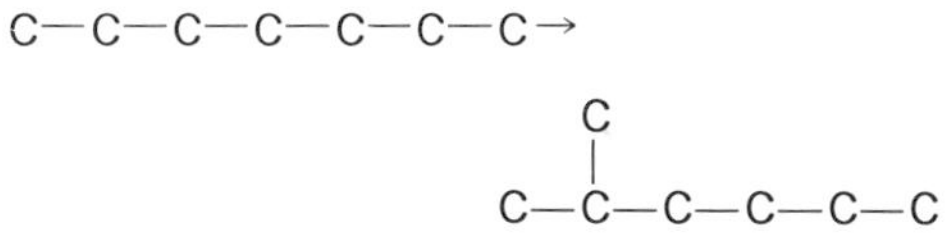

C—C—C—C—C—C—C→

C (branch on second carbon)
C—C—C—C—C—C (1)

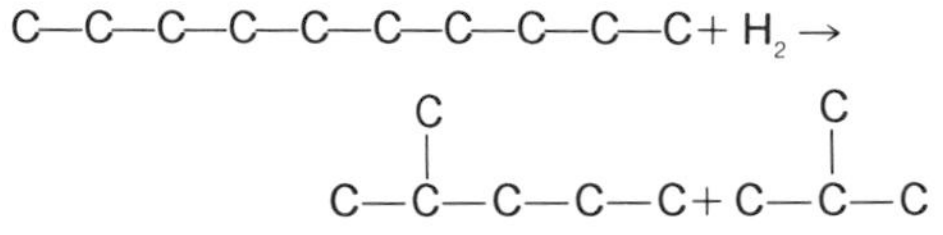

C—C—C—C—C—C—C—C—C—C + H_2 →

C (branch) C (branch)
C—C—C—C—C + C—C—C (2)

[methylcyclopentane ring: C, C, C, C, C with CH_3] → [benzene ring: six C with alternating double bonds] + $3H_2$ 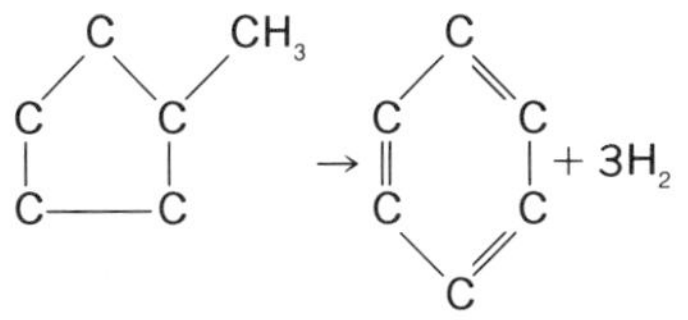 (3)

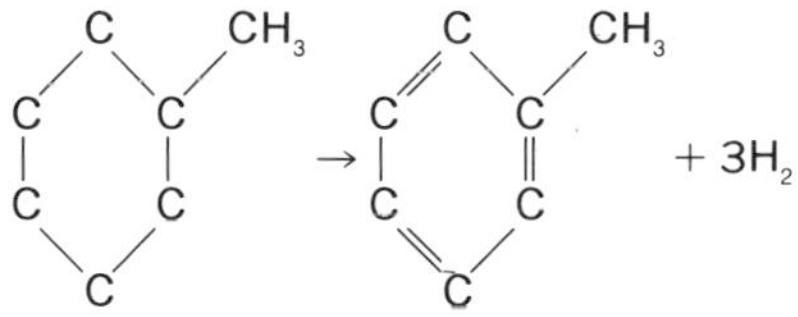

[methylcyclohexane ring: six C with CH_3] → [toluene ring: six C with alternating double bonds and CH_3] + $3H_2$ (4)

C (branch on second carbon)
C—C—C—C—C—C—C—C→

[ring: six C with CH_3, CH_3, H_3C substituents] →

[xylene ring: six C with alternating double bonds, CH_3 and H_3C] + $3H_2$ + CH_4 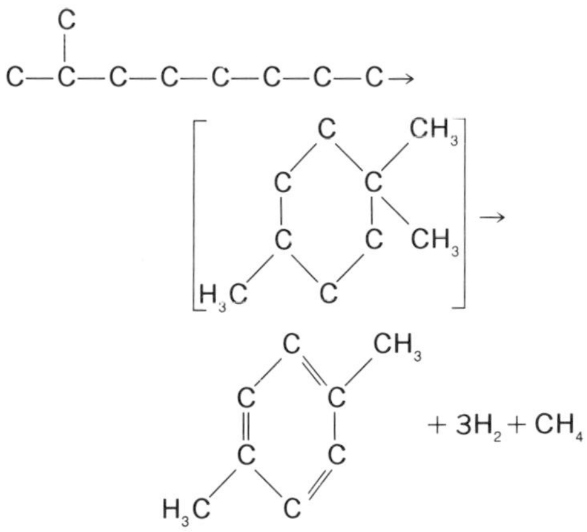 (5)

rangement and dehydrogenation of five-membered ring cycloparaffins, reaction (3); by dehydrogenation of six-membered ring cycloparaffins, reaction (4); and by cyclization (dehydrocyclization) of paraffins, reaction (5).

The product from catalytic reforming consists essentially of aromatics and branched paraffins. The higher-boiling fractions are richer in aromatics; the paraffins, particularly in severe reforming, are concentrated in the lower-boiling fractions.

The catalysts commonly employed consist essentially of a platinum dehydrogenation component on an acidic support. The dehydrogenation activity and acid activity of such catalysts are carefully balanced to achieve the highest possible yield at a given product octane number or aromatics concentration. A number of bimetallic catalysts have been introduced in which one or more additional elements, such as rhenium, are used to modify and stabilize the platinum component. This results in greatly improved catalyst stability and, in some cases, improvements in activity and liquid-product yield. Such catalysts have found wide acceptance in the refining industry.

Catalytic reforming units usually have a fractionation section, where the fresh feed is distilled to remove overhead pentane and lower-boiling hydrocarbons and to reject material boiling above the gasoline range (> 400°F or 204°C). Most modern installations also include a pretreatment section in which sulfur compounds and other impurities are removed by reaction with hydrogen over a hydrotreating catalyst, such as cobalt-molybdena-alumina. This pretreating step also removes from the feed arsenic compounds which otherwise poison the platinum catalyst.

The pretreated C_6 to 400°F (204°C) fraction is mixed with hydrogen and preheated to the desired temperature, and the reforming is carried out in three or four reactors in series. Intermediate reheating (between stages) is necessary since the overall reaction is endothermic. The reactions are carried out at temperatures of 800–1050°F (427–566°C) and pressures of 100–500 psig (700–3400 kPa).

The effluent from the reactors goes through heat exchangers to a separator where the liquid product is separated from the hydrogen and other light gases. It is then sent to a stabilizer to produce a finished gasoline as the bottoms product while removing, as overhead, the propane and butane produced during the reaction. Part of the separator gas (consisting mostly of hydrogen) is withdrawn from the system; the rest is recycled.

This recycle of a hydrogen-rich gas is an important feature of catalytic reforming because it acts to suppress those side reactions which tend to form carbonaceous deposits on the catalyst. The usual hydrogen recycle (3–6 moles per mole of hydrocarbon) is sufficient to maintain an active catalyst surface for 6–24 months in normal operation. When a catalyst becomes fouled, the carbonaceous deposits are burned off and the catalyst is then returned to service.

Modern requirements for higher-octane gasoline and higher aromatics yields demand higher-severity reforming, which leads to increased fouling rates despite the use of modern bimetallic catalysts. Many reformers are therefore designed with swing reactors which permit regeneration of the catalyst while the unit remains on-stream. In the most recent version, the catalyst is withdrawn continuously from an operating unit, treated in a separate but integral regenerator, and returned to the reaction section. A large percent of the more recently installed units utilize this continuous catalyst regeneration technology.

The use of platinum-containing catalysts for the reforming of gasoline has grown rapidly since introduction in 1949. In 1979 the catalytic reforming capacity in the Western world was 8,100,000 barrels per day (bpd) or 1,290,000 m³ per day. The United States capacity was about 3,800,000 bpd (600,000 m³/day), representing nearly 40% of total domestic gasoline production.

Aromatics production. Although catalytic reforming is primarily used to upgrade gasoline, the products have also become an exceedingly important source of aromatic hydrocarbons — in fact, the single most important source. These aromatics are used as intermediates in the manufacture of plastics, explosives, detergents, phenols, and other chemicals. The processing schemes are the same, except that somewhat lower pressures are used. The charge stock is a much narrower boiling cut, in the range 200–300°F (93–149°C). The main products include benzene, toluene, and xylenes. The aromatics are concentrated by extraction of the product with a solvent, or by absorption techniques.

Liquid petroleum gas (LPG). LPG consists of propane and butanes and is usually derived from natural gas. In locations where there is no natural gas and LPG is more important than gasoline, or in refineries that require additional isobutane for alkylation, naphtha can be converted to LPG (primarily isobutane) by catalytic reforming. The catalyst is modified in the direction of higher acidity, thus promoting the hydrocracking reactions. Under suitable conditions it is possible to convert 40% of the naphtha to LPG, the by-product being high-octane gasoline. *See* CRACKING; HYDROCRACKING; LIQUEFIED PETROLEUM GAS (LPG); PETROLEUM PROCESSING.

[ERNEST L. POLLITZER; VLADIMIR HAENSEL]

Bibliography: G. D. Hobson (ed.), *Modern Petroleum Technology*, 4th ed., 1973; Oil and Gas Journal, *Handbook on Catalytic Reforming*, 1966; *Refining Petroleum for Chemicals*, Advances in Chemistry Series 97, pp. 2–37, 1970.

Central heating and cooling

The use of a single heating or cooling plant to serve a group of buildings, facilities, or even a complete community through a system of distribution pipework that feeds each structure or facility. Central heating plants are basically of two types: steam or hot-water. The latter type uses high-temperature hot water under pressure and has become the more usual because of its considerable advantages. Steam systems are only used today where there is a specific requirement for high-pressure steam. Central cooling plants utilize a central refrigeration plant with a chilled water distribution system serving the air-conditioning systems in each building or facility.

Benefits. Advantages of a central heating or cooling plant over individual ones for each building or facility in a group include reduced labor cost, lower energy cost, less space requirement, and simpler maintenance. Though a central plant may require a 24-hr shift of operators, the total number of employees can be substantially less than that required to operate and maintain a number of individual plants.

Firing efficiencies of 85–93%, dependent upon such factors as fuel, boiler, and plant design, are usual with large central heating plants. Corresponding efficiencies for small individual heating boiler plants average 50–70%. Fuel in bulk quantity has a lower unit cost, and single handling for one large plant as distinct from multiple handling for many small plants saves appreciably in labor and transportation. Maintenance costs on a single central plant are considerably lower than for the aggregate of small plants of equal total capacity.

The disadvantages of a central heating plant concern mainly the maintenance of the distribution system where steam is used. Corrosion of the condensate water return lines shortens their life, and the steam drainage traps need particular attention. These disadvantages do not occur with high-temperature hot-water installations.

Central cooling plants, using conventional, electrically driven refrigeration compressors, have the advantage of utilizing bulk electric supply, at voltages as high as 13.5 kV, at wholesale rates. Additionally, their flexible load factor, resulting from load divergency in the various buildings served, results in major operating economies.

Design. Winter heat-load requirements are calculated by the addition of the following for each individual building or facility: (1) winter heat losses, (2) domestic water-heating requirements, and (3) industrial or other special heat requirements. To the sum of these for all buildings or facilities must be added the system distribution heat losses.

The summer load on the central cooling plant is calculated by the addition for each individual building or facility of the summer air-conditioning requirements. To the sum of these for all buildings or facilities must be added the system summer distribution heat losses.

To the individual winter and summer totals a diversity factor of 60–80% is applied because not all loads peak simultaneously. Winter heat loss due to weather conditions must be taken at its maximum. Water-heating and industrial requirements vary throughout the daily cycle. The summer air conditioning load must also be taken at its maximum. System distribution losses must be calculated both for the winter and the summer loads. The individual characteristics of the system must be considered in the diversity factor used.

Individual boiler sizing should allow the best arrangement to meet load variations as between individual 24-hr peaks for winter loads. Standby capacity is essential. Fuel selection, usually oil, coal, or gas, depends on local conditions and costs, taking into account labor and firing efficiencies to be expected with each fuel. Individual refrigeration machine sizing must allow for cooling load variation over the 24-hr cycle and the turn-down range of the machine itself. Considerable flexibility in machine choice thus results.

Distribution pipework sizing follows normal practice using suitable pressure drops with allowance for load variations and diversity factors as indicated.

Economics. The economics of each system must be individually computed because of the many factors involved. There is no average cost unitary basis applicable. The following major factors affect each plant's economics study: (1) system type, that is, steam, low pressure or high pressure; medium- or high-temperature hot water; chilled water temperature used; and so forth; (2) cost and type of energy used, that is, coal, gas, oil, and electricity; (3) type and occupancy of facilities served; (4) labor costs and conditions; (5) terrain; (6) climatic conditions; (7) plant first cost; (8) system life and maintenance costs. *See* DISTRICT HEATING.

Boiler plant. Both high-pressure (125 psi or 860 kPa saturated) and low-pressure (15 psi or 103 kPa saturated) steam plants are used, although the former is the more common. Both types follow conventional design. Feedwater and firing auxiliaries are of conventional type. Chemical treatment of feedwater is necessary.

Either conventional hot-water heating boilers or high-temperature hot-water boilers, depending on size, are used. Design and components for high-temperature hot-water plants are more complex and include nitrogen pressurization and gland-cooled circulating pumps. However, standard manufactured equipment is available for both conventional and high-temperature plants. Hot-water circulation, due to losses in the distribution pipework, should be at maximum temperature-pressure limitation. Circulation through the distribution system is by centrifugal pump with standby equipment being furnished. High-temperature hot-water installations may operate in 400–500°F (200–260°C) range.

Fuel handling, firing, and control arrangements also follow conventional design. Capacity of fuel storage should be on a minimal 3–4 week basis.

Central refrigeration plant. The central refrigeration plant may be electrically driven, or it may be of the steam-turbine-driven centrifugal type or steam absorption type. Steam as the prime energy source can be used where the plant is installed adjacent to a central boiler plant. Otherwise, medium-tension electricity, bought at bulk rates, forms an economical energy source. Gas turbines may also power centrifugal refrigeration compressors.

All equipment, including cooling towers and pumps, is of conventional design.

Distribution systems. Both overhead and underground pipework are used for distribution, although the latter is more usual except for industrial plants. Overhead mains must be strongly supported, insulated, and weatherproofed. Underground heating mains must be insulated and carefully waterproofed, particularly in damp areas. They must be structurally adequate. Steam distribution requires proper drainage of mains and often pumped return of condensate when gravity flow is not practical. In municipal distribution systems in larger cities where steam is used, it is frequently considered uneconomical to return the condensate to the central boiler plant due to pumping costs.

Hot-water distribution systems have the advantage that they are not affected by grade variations, that is, they can be run both uphill and downhill. The circulating pump pressures must be calculated accordingly.

A number of prefabricated insulated piping systems are commercially available. Each requires special handling for its installation and jointing. Hydrocarbon fill may also be used for heating mains insulation. Proper depth of burial must be arranged to give adequate protection against surface loads. Arrangements for expansion by loops or sliding fittings in access manholes are essential, and frequently cathodic protection is necessary. Where high-temperature hot water is used, it may be necessary to provide heat exchangers at each building or facility to furnish secondary heat at more moderate temperatures for uses such as heating systems and hot water. Where low-pressure steam is required, such exchangers may furnish this on the secondary side, if the primary hot water is at sufficiently high temperature.

For chilled water distribution systems, asbestos-cement plastic-lined piping can be used. The plastic lining has a low flow friction coefficient. If the piping is buried at depths of 4 ft (1.2 m) or greater, no insulation is required, provided soil is not waterlogged.

Heat sales. Various methods of heat sale are in use where central plants service public communities or facilities. With steam distribution steam meters to each individual building or facility served are usual, and a utility type of sliding scale rate per pound of steam sold is charged.

For high-temperature and chilled water distribution a combination meter measuring both water flow and temperature differential between supply and return mains may be used. This measures directly Btu per hour furnished. If a constant temperature differential is maintained between supply and return water mains, metering may be by flow only, although this is not very accurate. *See* AIR CONDITIONING; BOILER; COMFORT HEATING; STEAM HEATING; WARM-AIR HEATING SYSTEM.

[JOHN K. M. PRYKE]

Bibliography: *ASHRAE Systems Handbook*, 1976; P. L. Geiringer, *High Temperature Water Heating*, 1963.

Cetane number

A number that indicates the ability of a diesel engine fuel to ignite quickly after being injected into the cylinder. In high-speed diesel engines, a fuel with a long ignition delay tends to produce rough operation. *See* COMBUSTION CHAMBER; DIESEL FUEL.

To determine cetane number of a fuel sample, a specially designed diesel engine is operated under specified conditions with the given fuel. The fuel is injected into the engine cylinder each cycle at 13° before top center. The compression ratio is adjusted until ignition takes place at top center (13° delay). Without changing the compression ratio, the engine is next operated on blends of cetane (*n*-hexadecane), a short-delay fuel, and heptamethylnonane, which has a long delay. When a blend is found that also has a 13° delay under these conditions, the cetane number of the fuel sample may be calculated from the quantity of cetane required in the blend. [AUGUSTUS R. ROGOWSKI]

Bibliography: American Society for Testing and Materials, *ASTM Manual for Rating Diesel Fuels by the Cetane Method*, 1963.

Chain reaction

A succession of generation after generation of acts of division (called fission) of certain heavy nuclei. The fission process releases about 200 MeV (3.2×10^{-4} erg $= 3.2 \times 10^{-11}$J) in the form of energetic particles including two or three neutrons. Some of the neutrons from one generation are captured by fissile species (^{233}U, ^{235}U, ^{239}Pu) to cause the fissions of the next generations. The process is employed in nuclear reactors and nuclear explosive devices. *See* NUCLEAR FISSION.

The ratio of the number of fissions in one generation to the number in the previous generation is the multiplication factor k. The value of k can range from less than 1 to less than 2, and depends upon the type and amount of fissile material, the rate of neutron absorption in nonfissile material, the rate at which neutrons leak out of the system, and the average energy of the neutrons in the system. When $k = 1$, the fission rate remains constant and the system is said to be critical. When $k > 1$, the system is supercritical and the fission rate increases. *see* REACTOR PHYSICS.

A typical water-cooled power reactor contains an array of uranium rods (about 3% ^{235}U) surrounded by water. The uranium in the form of UO_2 is sealed into zirconium alloy tubes. The water removes the heat and also slows down (moderates) the neutrons by elastic collision with hydrogen nuclei. The slow neutrons have a much higher probability of causing fission in ^{235}U than faster (more energetic) neutrons do. In a fast reactor, no light nuclei are present in the system and the average neutron velocity is much higher. In such systems it is possible to use the excess neutrons to convert ^{238}U to ^{239}U. Then ^{239}U undergoes radioactive decay into ^{239}Pu, which is a fissile material capable of sustaining the chain reaction. If more than one ^{239}Pu atom is provided for each ^{235}U consumed, the system is said to breed (that is, make more fissile fuel than it consumes). In the breeder reactor, the isotope ^{238}U (which makes up 99.3% of natural uranium) becomes the fuel. This increases the energy yield from uranium deposits by more than a factor of 60 over a typical water-moderated reactor, which mostly uses the isotope ^{235}U as fuel.

A majority of the power reactors in the world today use water as both the moderator and the coolant. However, a limited number of reactors use heavy water instead of light water. The advantage of this system is that it is possible to use natural uranium as a fuel so that no uraniun enrichment is needed. Some other power reactors are gas-cooled by either helium or carbon dioxide and are moderated with graphite. *See* NUCLEAR REACTOR. [NORMAN C. RASMUSSEN]

Bibliography: R. D. Evans, *The Atomic Nucleus*, 1955; A. J. Henry, *Nuclear-Reactor Analysis*, 1975; J. R. Lamarsh, *Introduction to Nuclear Engineering*, 1975.

Chemical energy

In most chemical reactions, heat is either taken in or given out. By the law of conservation of energy, the increase or decrease in heat energy must be accompanied by a corresponding decrease or increase in some other form of energy. This other form is the chemical energy of the compounds involved in the reaction. The rearrangement of the atoms in the reacting compounds to produce new compounds causes a change in chemical energy. This change in chemical energy is equal numerically and of opposite sign to the heat change accompanying the reaction.

Most of the world's available power comes from the combustion of coal or of petroleum hydrocarbons. The chemical energy released as heat when a specified weight, often 1 g, of a fuel is burned is called the calorific value of the fuel.

Depending on whether the pressure or volume of the system is kept constant, differing quantities of heat are liberated in a chemical reaction. The heat of reaction at constant pressure q_p is equal to minus the change in the chemical energy at constant pressure ΔH, called the change in the enthalpy; the heat of reaction at constant volume q_v is equal to minus the change in the chemical energy at constant volume ΔE, called the change in the internal energy. *See* ENTHALPY; INTERNAL ENERGY.

It is not possible to measure an absolute value for the chemical energy of a compound; only changes in chemical energy can be measured. It is therefore necessary to make some arbitrary assumption as a starting point. One such assumption would be to take the chemical energies of the free atoms as zero and measure the chemical energies of all elements and compounds relative to this standard. If this were done, all chemical energies would then be negative quantities. Heat is given out when all elements and compounds are formed from their atoms. Since chemical energy can be regarded as a form of potential energy, it is interesting that the formation of chemical bonds is always accompanied by a decrease of potential energy. The reason, in qualitative terms at least, lies in the quantum theory. When chemical bonds are formed between atoms, for example, in the formation of a chlorine molecule from two chlorine atoms, electrons are shared between the two atoms, and this electron-sharing produces a lowering of the potential energies of the shared electrons. This lowering of potential energy, in turn, causes a lowering of the potential energy of the molecule relative to the free atoms.

Although it would be more fundamental to take the chemical energies of the separated atoms as zero, in practice it is more convenient to take a more arbitrary starting point and to assume that the chemical energies of the elements are zero. For precision, the standard states of the elements at 25°C and 1 atm (101.325 kPa) pressure are chosen. Thus, in the case of carbon the chemical energy of graphite, not diamond, is said to be zero at 25°C and 1 atm. Diamond then has a definite enthalpy value. *See* ENERGY SOURCES.

[THOMAS C. WADDINGTON]

Chemical fuel

The principal fuels used in internal combustion engines (automobiles, diesel, and turbojet) and in the furnaces of stationary power plants are organic fossil fuels. These fuels, and others derived from them by various refining and separation processes, are found in the earth in the solid (coal), liquid (petroleum), and gas (natural gas) phases.

Special fuels to improve the performance of combustion engines are obtained by synthetic chemical procedures. These special fuels serve to increase the fuel specific impulse of the engine (specific impulse is the force produced by the engine multiplied by the time over which it is produced, divided by the mass of the fuel) or to increase the heat of combustion available to the engine per unit mass or per unit volume of the fuel. A special fuel which possesses a very high heat of combustion per unit mass is liquid hydrogen. It has been used along with liquid oxygen in rocket engines. Because of its low liquid density, liquid hydrogen is not too useful in systems requiring high heats of combustion per unit volume of fuel ("volume-limited" systems). In combination with liquid fluorine, liquid hydrogen produces extremely large specific impulses, and rocket engines using this combination are under development. *See* AIRCRAFT FUEL.

Liquid fuels and their associated oxidizers

Fuel	Oxidizer
Ammonia	Liquid oxygen
95% Ethyl alcohol	Liquid oxygen
Methyl alcohol	87% Hydrogen peroxide
Aniline	Red fuming nitric acid
Furfural alcohol	Red fuming nitric acid

A special fuel which produces high flame temperatures of the order of 5000°C is gaseous cyanogen, C_2N_2. This is used with gaseous oxygen as the oxidizer. The liquid fuel hydrazine, N_2H_4, and other hydrazine-based fuels, with the liquid oxidizer nitrogen tetroxide, N_2O_4, are used in many space-oriented rocket engines. The boron hydrides, such as diborane, B_2H_6, and pentaborane, B_5H_9, are high-energy fuels which are being used in advanced rocket engines.

For air-breathing propulsion engines (turbojets and ramjets), hydrocarbon fuels are most often used. For some applications, metal alkyl fuels which are pyrophoric (that is, ignite spontaneously in the presence of air), and even liquid hydrogen, are being used.

A partial list of additional currently used liquid fuels and their associated oxidizers is shown in the table.

Fuels which liberate heat in the absence of an oxidizer while decomposing either spontaneously or because of the presence of a catalyst are called monopropellants and have been used in rocket engines. Examples of these monopropellants are hydrogen peroxide, H_2O_2, and nitro-methane, CH_3NO_2.

Liquid fuels and oxidizers are used in most large-thrust (large propulsive force) rocket engines. When thrust is not a consideration, solid-propellant fuels and oxidizers are frequently employed because of the lack of moving parts such as valves and pumps, and the consequent simplicity of this type of rocket engine. Solid fuels fall into two broad classes, double-base and composites. Double-base fuels are compounded of nitroglycerin (glycerol trinitrate) and nitrocellulose, with no separate oxidizer required. The nitroglycerin plasticizes and swells the nitrocellulose, leading to a propellant of relatively high strength and low elongation. The double-base propellant is generally formed in a mold into the desired shape (called a grain) required for the rocket case. Composite propellants are made of a fuel and an oxidizer. The latter could be an inorganic perchlorate such as ammonium perchlorate, NH_4ClO_4, or potassium perchlorate, $KClO_4$, or a nitrate such as ammonium nitrate, NH_4NO_3, potassium nitrate, KNO_3, or sodium nitrate, $NaNO_3$. Fuels for composite propellants are generally the asphalt-oil-type, thermosetting plastics (phenol formaldehyde and phenol-furfural resins have been used) or several types of synthetic rubber and gumlike substances. Recently, metal particles such as boron, aluminum, and beryllium have been added to solid propellants to increase their heats of combustion and to eliminate certain types of combustion instability. *See* LIQUID FUEL.

[WALLACE CHINITZ]

Bibliography: S. Penner, *Chemistry Problems in*

Jet Propulsion, 1957; H. W. Ritchey and J. M. McDermott, Solid propellant rocket technology, F. I. Ordway, III (ed.), *Advances in Space Science and Technology*, 1963; G. P. Sutton, *Rocket Propulsion Elements*, 3d ed., 1963.

Circuit

A general term referring to a system or part of a system of conducting parts and their interconnections through which an electric current is intended to flow. A circuit is made up of active and passive elements or parts and their interconnecting conducting paths. The active elements are the sources of electric energy for the circuit; they may be batteries, direct-current generators, or alternating-current generators. The passive elements are resistors, inductors, and capacitors. The electric circuit is described by a circuit diagram or map showing the active and passive elements and their connecting conducting paths.

CIRCUIT

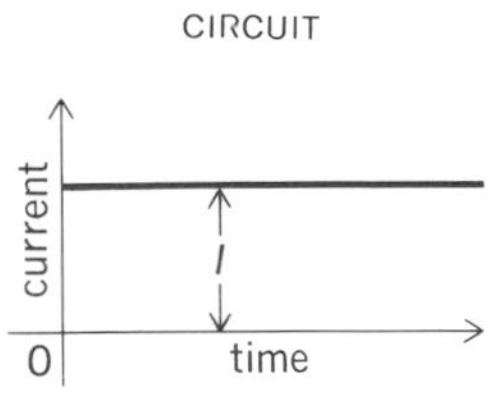

Fig. 1. Direct current.

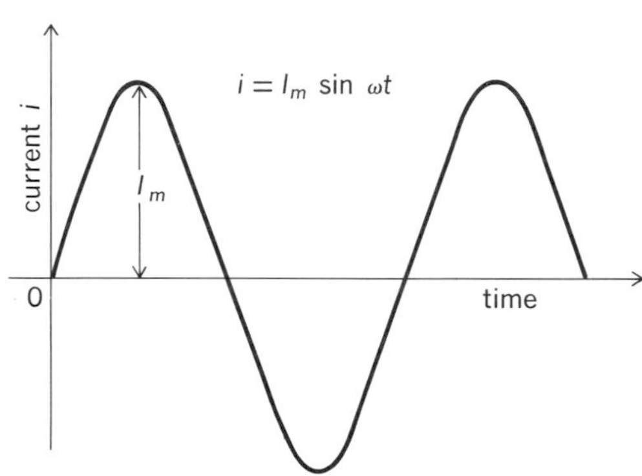

Fig. 2. Alternating current.

Devices with an individual physical identity such as amplifiers, transistors, loudspeakers, and generators, are often represented by equivalent circuits for purposes of analysis. These equivalent circuits are made up of the basic passive and active elements listed above.

Electric circuits are used to transmit power as in high-voltage power lines and transformers or in low-voltage distribution circuits in factories and homes; to convert energy from or to its electrical form as in motors, generators, microphones, loudspeakers, and lamps; to communicate information as in telephone, telegraph, radio, and television systems; to process and store data and make logical decisions as in computers; and to form systems for automatic control of equipment.

CIRCUIT

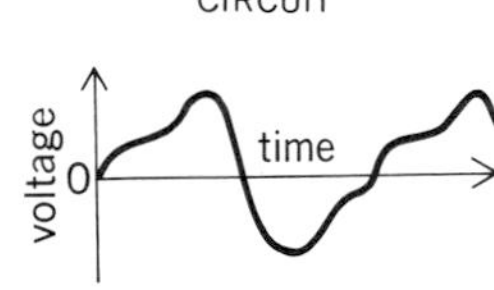

Fig. 3. Nonsinusoidal voltage wave.

Electric circuit theory. This includes the study of all aspects of electric circuits, including analysis, design, and application. In electric circuit theory the fundamental quantities are the potential differences (voltages) in volts between various points, the electric currents in amperes flowing in the several paths, and the parameters in ohms or mhos which describe the passive elements. Other important circuit quantities such as power, energy, and time constants may be calculated from the fundamental variables.

CIRCUIT

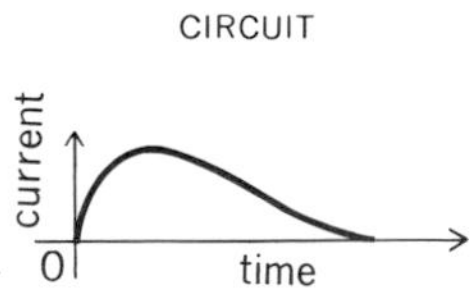

Fig. 4. Transient electric current.

Electric circuit theory is an extensive subject and is often divided into special topics. Division into topics may be made on the basis of how the voltages and currents in the circuit vary with time; examples are direct-current, alternating-current, nonsinusoidal, digital, and transient circuit theory. Another method of classifying circuits is by the arrangement or configuration of the electric current paths; examples are series circuits, parallel circuits, series-parallel circuits, networks, coupled circuits, open circuits, and short circuits. Circuit theory can also be divided into special topics according to the physical devices forming the circuit, or the application and use of the circuit. Examples are power, communication, electronic, solid-state, integrated, computer, and control circuits.

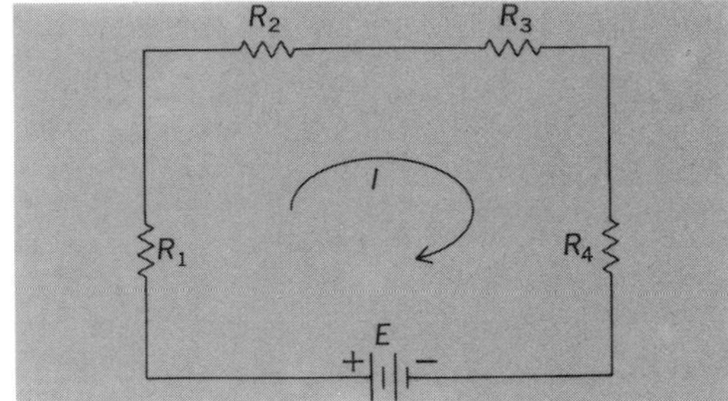

Fig. 5. Series circuit.

Direct-current circuits. In dc circuits the voltages and currents are constant in magnitude and do not vary with time (Fig. 1). Sources of direct current are batteries, dc generators, and rectifiers. Resistors are the principal passive element.

Magnetic circuits. Magnetic circuits are similar to electric circuits in their analysis and are often included in the general topic of circuit theory. Magnetic circuits are used in electromagnets, relays, magnetic brakes and clutches, computer memory devices, and many other devices. For a detailed treatment *see* MAGNETIC CIRCUITS.

Alternating-current circuits. In ac circuits the voltage and current periodically reverse direction with time. The time for one complete variation is known as the period. The number of periods in 1 sec is the frequency in cycles per second. A cycle per second has recently been named a hertz (in honor of Heinrich Rudolf Hertz's work on electromagnetic waves).

Most often the term ac circuit refers to sinusoidal variations. For example, the alternating current in Fig. 2 may be expressed by $i = I_m \sin \omega t$.

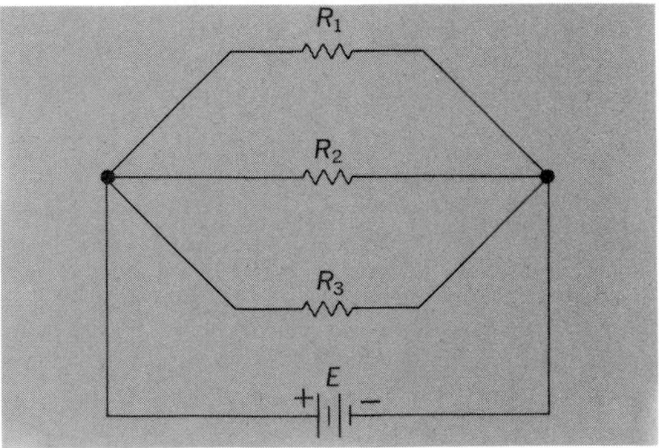

Fig. 6. Parallel circuit.

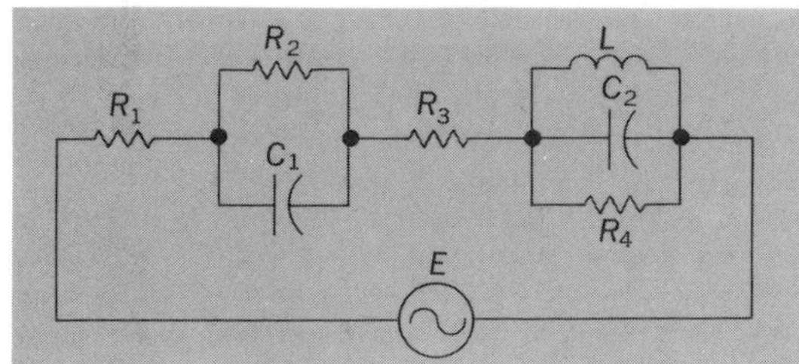

Fig. 7. Series-parallel circuit.

Sinusoidal sources are ac generators and various types of electronic and solid-state oscillators; passive circuit elements include inductors and capacitors as well as resistors. The analysis of ac circuits requires a study of the phase relations between voltages and currents as well as their magnitudes. Complex numbers are often used for this purpose.

Nonsinusoidal waveforms. These voltage and current variations vary with time but not sinusoidally (Fig. 3). Such nonsinusoidal variations are usually caused by nonlinear devices, such as saturated magnetic circuits, electron tubes, and transistors. Circuits with nonsinusoidal waveforms are analyzed by breaking the waveform into a series of sinusoidal waves of different frequencies known as a Fourier series. Each frequency component is analyzed by ac circuit techniques. Results are combined by the principle of superposition to give the total response.

Electric transients. Transient voltage and current variations last for a short length of time and do not repeat continuously (Fig. 4). Transients occur when a change is made in the circuit, such as opening or closing a switch, or when a change is made in one of the sources or elements.

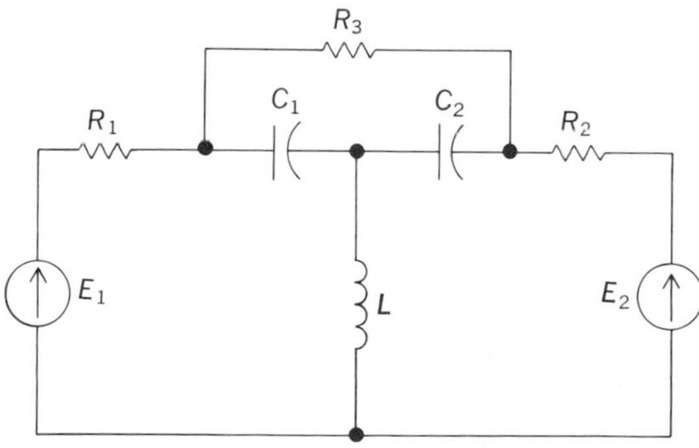

Fig. 8. A three-mesh electric network.

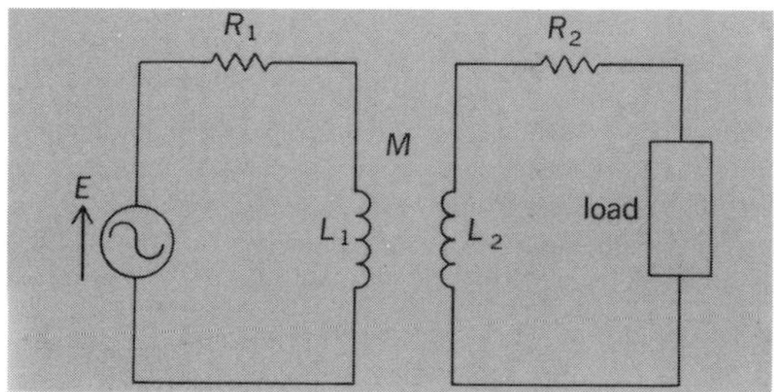

Fig. 9. Inductively coupled circuit.

Series circuits. In a series circuit all the components or elements are connected in an end-to-end arrangement and carry the same current, as shown in Fig. 5.

Parallel circuits. Parallel circuits are connected so that each component of the circuit has the same potential difference (voltage) across its terminals, as shown in Fig. 6.

Series-parallel circuits. In a series-parallel circuit some of the components or elements are connected in parallel, and one or more of these parallel combinations are in series with other components of the circuit, as shown in Fig. 7.

Electric network. This is another term for electric circuit, but it is often reserved for the electric circuit that is more complicated than a simple series or parallel combination. A three-mesh electric network is shown in Fig. 8.

Coupled circuits. A circuit is said to be coupled if two or more parts are related to each other through some common element. The coupling may be by means of a conducting path of resistors or capacitors or by a common magnetic linkage (inductive coupling), as shown in Fig. 9.

Open circuit. An open circuit is a condition in an electric circuit in which there is no path for current flow between two points that are normally connected.

Short circuit. This term applies to the existence of a zero-impedance path between two points of an electric circuit.

Integrated circuit. The integrated circuit is a recent development in which the entire circuit is contained in a single piece of semiconductor material. Sometimes the term is also applied to circuits made up of deposited thin films on an insulating substrate.

[CLARENCE F. GOODHEART]

Bibliography: D. Bell, *Fundamentals of Electric Circuits*, 1978; E. Brenner and M. Javid, *Analysis of Electric Circuits*, 2d ed., 1967; A. E. Fitzgerald and D. E. Higginbotham, *Electrical and Electronic Engineering Fundamentals*, 1964; W. Hayt, Jr., and J. E. Kemmerly, *Engineering Circuit Analysis*, 3d ed., 1978; R. E. Scott, *Elements of Linear Circuits*, 1965; R. Smith, *Circuits, Devices, and Systems*, 3d ed., 1976.

Circuit breaker

A device to open or close an electric power circuit either during normal power system operation or during abnormal conditions. A circuit breaker serves in the course of normal system operation to energize or deenergize loads. During abnormal conditions, when excessive current develops, a circuit breaker opens to protect equipment and surroundings from possible damage due to excess current. These abnormal currents are usually the result of short circuits created by lightning, accidents, deterioration of equipment, or sustained overloads.

Formerly, all circuit breakers were electromechanical devices. In these breakers a mechanism operates one or more pairs of contacts to make or break the circuit. The mechanism is powered either electromagnetically, pneumatically, or hydraulically. The contacts are located in a part termed the interrupter. When the contacts are parted, opening the metallic conductive circuit, an

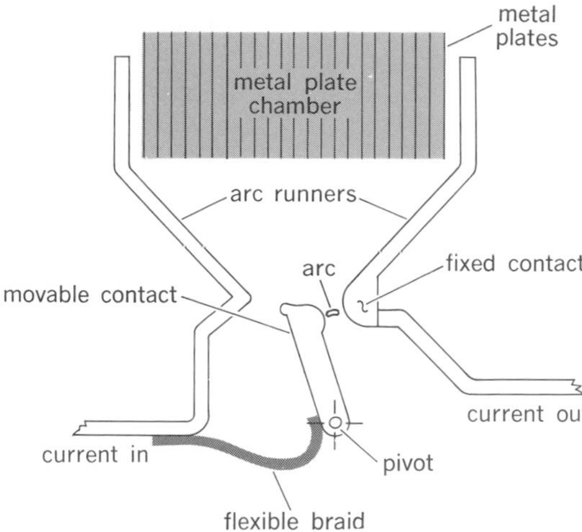

Fig. 1. Cross section of interrupter for a typical medium-voltage circuit breaker.

electric arc is created between the contacts. This arc is a high-temperature ionized gas with an electrical conductivity comparable to graphite. Thus the current continues to flow through the arc. The function of the interrupter is to extinguish the arc, completing circuit-breaking action.

Current interruption. In alternating-current circuits, arcs are usually extinguished at a natural current zero, when the ac voltage applied across the arcing contacts reverses polarity. Within a short period around a natural current zero, the power input to the arc, equal to the product of the instantaneous current and voltage, is quite low. There is an opportunity to remove more energy from the arc than is applied to it, thus allowing the gas to cool and change from a conductor into an insulator.

In direct-current circuits, absence of natural current zero necessitates the interrupter to convert the initial arc into one that could only be maintained by an arc voltage higher than the system voltage, thus forcing the current to zero. To accomplish this, the interrupter must also be able to remove energy from the arc at a rapid rate.

Different interrupting mediums are used for the purpose of extinguishing an arc. In low- and medium-voltage circuits (110–15,000 volts) the arc is driven by the magnetic field produced by the arc current into an arc chute (Fig. 1). In the arc chute the arc is either split into several small arcs between metal plates or driven tightly against a solid insulating material. In the former case the splitting of the arc increases the total arc voltage, thus increasing the rate of energy dissipation. In the latter case the insulating material is heated to boiling temperatures, and the evaporated material flows through the arc, carrying a great deal of energy with it.

For outdoor applications at distribution and subtransmission voltage (10 kV and above), oil breakers are widely used. In the United States, bulk oil breakers are used (Fig. 2), and in Europe and many other countries, "low-oil-content" breakers are quite popular. The principles of operation of both kinds of oil breakers are basically the same. Only the amount of oil used and the detailed engineering design differ. In oil breakers, the arc is drawn in oil. The intense heat of the arc decomposes the oil, generating high pressure that produces a fluid flow through the arc to carry energy away. At transmission voltages below 345 kV, oil breakers used to be popular. They are increasingly losing ground to gas-blast circuit breakers such as air-blast breakers (Figs. 3 and 4) and SF_6 circuit breakers (Fig. 5). Even though a 765-kV low-oil-content breaker design has been announced, it has seen little practical application.

In air-blast circuit breakers, air is compressed to high pressures (50 atm approximately; 1 atm = 1.01×10^5 Pa). When the contacts part, a blast valve is opened to discharge the high-pressure air to ambient, thus creating a very-high-velocity flow near the arc to dissipate the energy. In SF_6 circuit breakers, the same principle is employed, with SF_6 as the medium instead of air. In the "puffer" SF_6 breaker, the motion of the contacts compresses the gas and forces it to flow through an orifice into the neighborhood of the arc. Both types of SF_6 breakers have been developed for ehv (extra high voltage) transmission systems.

For ehv systems, it was discovered that closing of a circuit breaker may cause a switching surge which may be excessive for the insulation of the system. The basic principle is easy to understand. If a breaker is closed at the peak of the voltage wave to energize a single-phase transmission line which is open-circuited at the far end, reflection can cause the transient voltage on the line to reach twice the peak of the system voltage. If there are

Fig. 2. Bulk oil circuit breaker for 138-kV application.

Fig. 3. Air blast circuit breaker rated for 500 kV.

Fig. 4. Air-blast circuit breaker, 800-kV.

Fig. 5. SF_6 circuit breakers, 500-kV 3-kA.

trapped charges on the line, and if it happens that at the moment of breaker-closing the system voltage is at its peak, equal in magnitude to but opposite to the polarity of the voltage due to the trapped charges left on the line, the switching surge on the line can reach a theoretical maximum of three times the system voltage peak.

One way to reduce the switching surge is to insert a resistor in series with the line (Fig. 6) for a short time. When switch A in Fig. 6 is closed, the voltage is divided between the resistance and the surge impedance Z of the line by the simple relationship $V_L = V[Z/(R+Z)]$, where V is voltage impressed, V_L is voltage across the transmission line, R is resistance of the resistor, and Z is surge impedance of the line. Only V_L will travel down the line and be reflected at the other end. The magnitude of the switching surge is thus considerably reduced.

Vacuum and solid-state breakers. Two other types of circuit breakers have been developed. The vacuum breaker, another electromechanical device, uses the rapid dielectric recovery and high dielectric strength of vacuum. A pair of contacts is hermetically sealed in a vacuum envelope (Fig. 7). Actuating motion is transmitted through bellows to the movable contact. When the contacts are parted, an arc is produced and supported by metallic vapor boiled from the electrodes. Vapor particles expand into the vacuum and condense on solid surfaces. At a natural current zero the vapor particles disappear, and the arc is extinguished. Vacuum breakers of up to 242 kV have been built (Fig. 8).

The other type of breaker uses a thyristor, a semiconductor device which in the off state prevents current from flowing but which can be turned on with a small electric current through a third electrode, the gate. At the natural current zero, conduction ceases, as it does in arc interrupters. This type of breaker does not require a mechanism. Semiconductor breakers have been built to carry continuous currents up to 10,000 A.

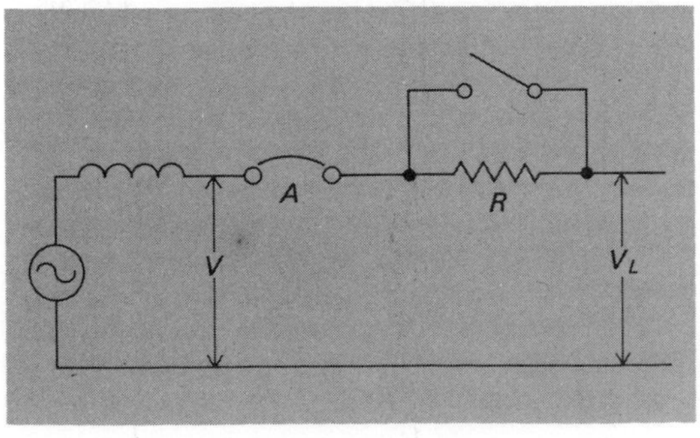

Fig. 6. Resistor insertion to reduce switching surge.

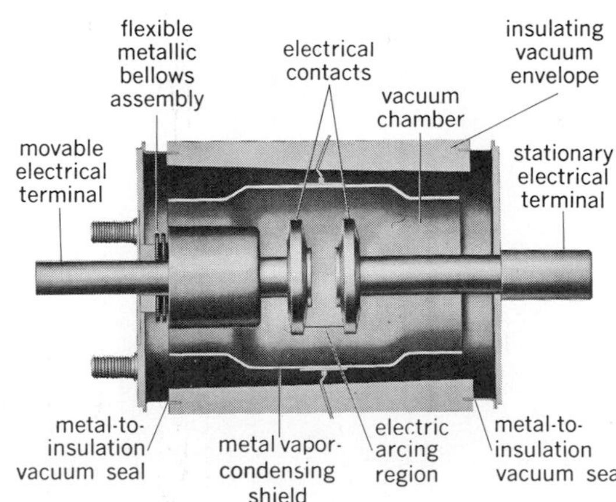

Fig. 7. Cutaway view of vacuum interrupter.

Fig. 8. Vacuum circuit breaker, 242-kV 40-kA.

Fig. 10. Semiconductor switching device for thermonuclear fusion research.

Semiconductor circuit breakers can be made to operate in microseconds if the commutation principle is applied. Figure 9 illustrates the commutation principle for an hvdc (high-voltage direct-current) circuit, but it can easily be extended to ac circuits. During normal operation, the circuit breaker (CB in the diagram) would be closed and the load would be supplied with the current from the hvdc source. In this diagram, inductances L_1 and L_2 represent the circuit inductance on either side of the breaker. Suppose that a fault occurs which applies a short circuit between points A and B. The current will commence to increase, its rate being determined by L_1 and L_2. When the increased current is detected, the contacts of the circuit breakers are opened, drawing an arc, and the switch (S in the diagram) is closed, causing the precharged capacitors C to discharge through the circuit breaker. The current I_2 so produced is traveling in such a direction as to oppose I_1 and drive it to zero, thereby giving the circuit breaker an opportunity to interrupt.

Thyristors can be used for both circuit breakers and switches. The closing and opening operations are, of course, not mechanical but are controlled by the gates. Such a circuit breaker has been built for very special switching applications, such as thermonuclear fusion research (Fig. 10).

[THOMAS H. LEE]

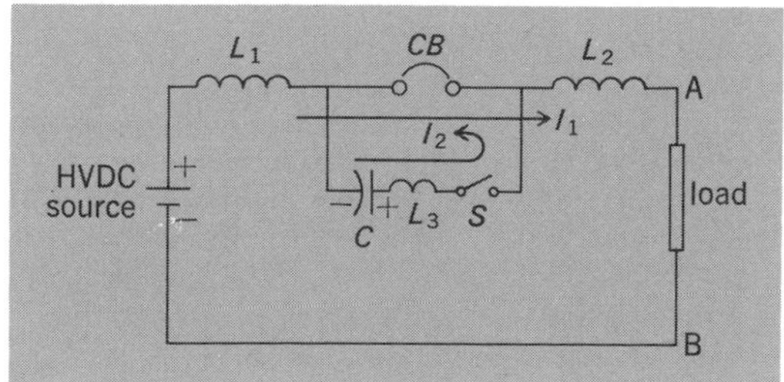

Fig. 9. Commutation principles of circuit interruption.

Bibliography: Thomas H. Lee, *Physics and Engineering of High Power Switching Devices*, 1975.

Coal

A brown to black combustible rock that had its origin in the accumulation and physical and chemical alteration of vegetation. Original accumulations of vegetation, usually in a swamp or moist environment which reduced decay, resulted in formation of peat, the precursor of coal. Peat is converted to coal after burial and subsequent geological processes, including increased pressure and temperature, which progressively compressed and indurated and otherwise altered the material through a series of coal varieties to an extreme of graphite or graphitelike material. In American terminology, the rank varieties of coal constituting the coalification series consist of lignitic (including brown coal); subbituminous; bituminous (including high-, medium-, and low-volatile); and anthracitic coal (including semianthracite, anthracite, and metaanthracite or graphitic coal). *See* LIGNITE.

Formation. Although coal may originate from isolated fragments of a variety of vegetation, most coal represents the coalification of woody plants accumulated in peat beds. Peat accumulations are principally of two kinds: autochthonous deposits represent accumulations of vegetation in the area in which the plants grew, such as those in the Great Dismal Swamp of Virginia and the Okefenokee Swamp of Georgia and Florida; and allochthonous deposits accumulated elsewhere than at the place of growth by the drifting action of stream, lake, or sea currents, such as the Red River "rafts." Generally, autochthonous coals overlie an underclay or other seat rock which contains traces of plant roots (commonly called *Stigmaria* in coals of Paleozoic age).

Transformation of peat. Biochemical activity modifies the character of the unsubmerged, lightly submerged, or lightly buried peat. This process consists in part of general oxidation, but mainly of attack by aerobic bacteria and fungi, which can live only where oxygen is available, and of anaerobic bacteria where water or thin sediments cover the peat. Fires set by lightning or other causes may consume part of the peat from time to time, leaving in places a residue of charcoal which may eventually be incorporated into the coal bed in the form of fusain, known to miners as mineral charcoal or mother-of-coal.

The theory of forest fire origin of fusain is not universally accepted to account for all types of fusain. For some occurrences, many believe the presence of certain combustible components in fusain, such as resins, indicates that chemical causes operating under special conditions bring about the formation of fusain. This high-carbon component is found in all ranks of coal, with relatively little difference in composition. Because of its porosity, fusain is commonly mineralized into a hard and heavy substance: unmineralized fusain is soft and light. Fusain occurs in a wide range of sizes, from barely visible dimensions to aggregates forming fairly continuous thin sheets or lenses several feet across and up to several inches thick. Similar material is observed under the microscope to be of very fine sizes.

The material composing the peat which is finally transformed into coal varies with differences in source material (kinds of plant materials), in conditions of accumulation and diagenesis, in the length of time involved prior to burial, and in the depth and time of burial. Figure 1 shows the relative resistance of the principal peat-forming plant substances to microbic decomposition. Based upon the peat-forming components, which are more or less segregated into bands, several common types or varieties of coal can be recognized.

Banded coal. Banded coal may be either bright or dull. Banded bituminous coal is a humic type of coal produced by thin lenses of highly lustrous coalified wood or bark called vitrain, if of megascopically distinguishable thickness (1/2 mm or 1/50 in.), and layers of more or less striated bright or dull coal—clarain or durain (attrital coal)—consisting of comminuted plant debris. Bright attrital coal, also called clarain, contains a predominance of fine vitrainlike laminae or lenses; dull attrital coal, also called durain, has little of the fine vitrainlike material. Dullness may also result from a predominance of mineral matter or from a relatively high content of more highly volatile constituents such as spores, cuticles, resins, and waxes. Likewise, the presence of microscopically opaque matter or finely disseminated fusain contributes to dullness in banded coal.

Nonbanded coal. Nonbanded coal is a sapropelic type of coal, produced by the subaqueous deposition of spores, algal remains, or reworked peat material. Such coals are also called canneloid in reference of cannel coals, which from microscopic studies are determined to be composed mainly of spores from plants. Boghead coals are similar in appearance and origin but consist mainly of algal material. Canneloid coals are distinguished by their greasy luster and blocky, conchoidal fracture.

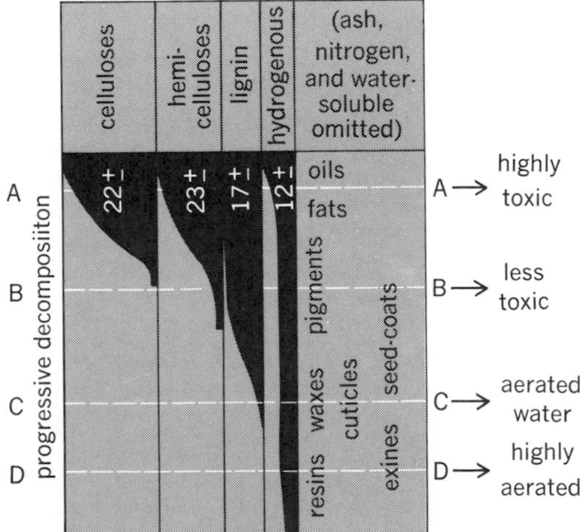

Fig. 1. Approximate proportions of principal peat-forming components in the dry-ingredient plant debris plotted according to relative resistance to microbic decomposition and order of disappearance. (*After D. White*)

Coalification. This term refers to the process of coal metamorphism brought about by increasing weight of overlying sediments, by tectonic movements, by an increase in temperatures resulting from depth of burial, or from close approach to, or contact with, igneous intrusions or extrusions. Increase in pressure affects principally the physical properties of the coal, that is, hardness, strength, optical anisotropy, and porosity. Increase in temperature acts chiefly to modify the chemical composition by increasing the carbon content and decreasing the content of oxygen and hydrogen (increasing the amount of fixed carbon and decreasing the volatile matter) and increasing the calorific value to a maximum in those coals having 15–30% volatile matter (dry, mineral-matter-free). Rapid metamorphism or coalification of coal, effected by close approach to or contact with igneous intrusions or extrusions, may result in the formation of natural coke. Natural coke is found in some coal fields but is relatively rare. *See* COKE.

Analyses. The classification and description of coal depend largely upon information supplied by chemical analysis by systematically standardized procedures and the results of a number of empirical tests. Chemical analyses are of two principal kinds: the elementary or ultimate analysis and the proximate or commercial type of analysis. The ultimate analysis is limited to the percent of hydrogen, carbon, oxygen, nitrogen, and sulfur, exclusive of the mineral matter or ash. The proximate analysis, which is empirical in character, is the prevailing form of analysis in North America. In this analysis, values are reported for volatile matter, fixed carbon, moisture, and ash, all of which add up to 100%. Calorific and sulfur values are also determined, the latter sometimes in terms of forms of sulfur such as pyritic, organic, and sulfate sulfur.

It is common practice to use qualifying terms denoting the basis of reported analysis, such as "as-received," "moisture-free" (mf), "moisture-and-ash-free" (maf), and dry-mineral-matter-free (dmmf) so-called "pure coal," in presenting the results of proximate and ultimate analysis. Usually values are determined initially on an "mf" or "dry" basis, the other forms of analysis being calculated with the use of determined moisture and

Fig. 2. Low-temperature ashing equipment that removes 99% or more organic matter at temperatures below 150°C, leaving a residue of essentially unaltered minerals from the coal.

ash values. The maf value sometimes has been used improperly on the mistaken assumption that it represents the heat value or composition of the pure coal material, as though the formation of ash from the original mineral matter in the coal were without calorific or other effects. To make a rapid and convenient allowance for the errors inherent in the maf values, various procedures have been devised to arrive at a mineral-matter-free (mf) basis of comparison and classification. Formulas in common use to determine analysis values corrected for the influence of mineral matter on an mmf basis are indicated below. The calculations are

Moist mmf Btu/lb =

$$\frac{100(\text{Btu/lb} - 50 \text{ sulfur } \%)}{100 - (1.08 \text{ ash } \% + 0.55 \text{ sulfur } \%)} \quad (1)$$

Dry mmf fixed carbon =

$$\frac{100 (\text{fixed carbon } \% - 0.15 \text{ sulfur } \%)}{100 - (\text{moisture } \% + 1.08 \text{ ash } \% + 0.55 \text{ sulfur } \%)} \quad (2)$$

all based upon the as-received values, and both moist and dry mmf values are used in the standard classification of coal by rank (Table 1). This correction is applicable to most coals, but unusual mineralization in some coals requires further correction.

Mineral matter. Mineral matter is the inorganic material present in coal that is the source of ash when coal is burned. It occurs predominantly as discrete mineral phases, but also as chemical elements combined with organic matter and as dissolved elements in the pore water. Mined coals commonly contain between 5 and 15% mineral matter, although some have greater and lesser amounts. Early studies attempted to estimate original mineral content from chemical analysis of ash.

Minerals in coal are now more directly studied by optical microscopy and by scanning electron microscopy in order to determine the morphology and distribution of the minerals in the coal. X-ray diffraction methods are employed to identify and quantify the minerals present. Minerals are commonly separated for studies by use of a low-temperature (<150°C) ashing device that generates an activated oxygen atmosphere in which organic matter is removed from the coal, leaving a mineral residue suitable for mineralogical and chemical studies (Fig. 2).

Mineral phases. Pyrite, quartz, calcite, and various types of clay minerals are the major components of mineral matter in coals. These and other minerals in coals are grouped according to their origin as either detrital, syngenetic, or epigenetic, or as products of weathering. Detrital minerals are those which were washed into the peat swamp by rivers or blown in by winds. These are generally very fine grains of quartz, clays, and lesser amounts of other minerals such as mica, feldspar, zircon, rutile, tourmaline, and apatite. Syngenetic minerals formed in place from colloidal, chemical, and biochemical processes as the organic matter accumulated in the swamp. Pyrite is the most common type, and in some coals certain clay minerals are syngenetic in origin. Epigenetic minerals form in coals at some later stage of coalification. These minerals are commonly calcite, dolomite, ankerite, siderite, pyrite, marcasite, sphalerite, barite, certain clay minerals (especially kaolinite), various chlorides (mainly NaCl), and others. These minerals commonly precipitated from solutions as they percolated through the peat during the late peat stage or along joints and pores in the coal. These minerals are sometimes coarsely crystalline, and they also occur in nodules of various sizes. The most common mineral products of weathering associated with coal are gypsum and iron sulfates which form from pyrite, marcasite, and calcite and other carbonates.

Elements. Inorganic elements complexly bound within the organic matter in coals, the so-called inherent mineral matter, was in large measure incorporated in the tissues of the living plants that formed the coal. These elements are mainly calcium, magnesium, sodium, sulfur, and iron, but include many others in trace amounts.

Effect on fusion temperature. The character of the nonorganic matter in the coal has a considerable effect upon the fusion temperature of the ash, the determination of which is one of the common subsidiary tests in coal analysis procedure. In general, for combustion uses of coal, ash that fuses at high temperatures is preferred to low-fusion ash, the temperature being below 2000°F (1093°C) for low-fusion ash and above 2400°F (1316°C) for high-fusion ash; the range is about 1600-2800°F (871–1538°C).

The properties of the minerals that are present may be critical factors in causing formation of deposits of ash, and also resulting in corrosion of boiler tubes in large steam boilers. High alkali or alkali chloride content of coal has been associated with these problems in boilers. Mineral contributions to gaseous wastes of coal combustion have been of environmental concern for many years, especially sulfur. Some trace elements in coals also may be potentially deleterious to the environment.

Catalysts in processing. The behavior of mineral matter during liquefaction or gasification of coals may be important. Some compounds such as those of titanium, occurring in coals in trace amounts, are known to poison catalysts in some processes. Constituents of mineral matter may act as catalysts in some processes.

Information source. Mineral constituents in coal may provide clues to environment of deposition (marine versus nonmarine) and also information on

Table 1. Classification of coals by rank[a]

Class	Group	Fixed carbon limits, % (dry, mineral-matter-free basis)		Volatile matter limits, % (dry, mineral-matter-free basis)		Calorific value limits, Btu/lb[f] (moist,[b] mineral-matter-free basis)		Agglomerating character
		Equal or greater than	Less than	Greater than	Equal or less than	Equal or greater than	Less than	
I. Anthracitic	Metaanthracite	98	—		2	—	—	Nonagglomerating
	Anthracite	92	98	2	8	—	—	Nonagglomerating
	Semianthracite[c]	86	92	8	14	—	—	Nonagglomerating
II. Bituminous	Low-volatile bituminous coal	78	86	14	22	—	—	Commonly agglomerating[e]
	Medium-volatile bituminous coal	69	78	22	31	—	—	Commonly agglomerating[e]
	High-volatile A bituminous coal	—	69	31	—	14,000[d]	—	Commonly agglomerating[e]
	High-volatile B bituminous coal	—	—	—	—	13,000[d]	14,000	Commonly agglomerating[e]
	High-volatile C bituminous coal	—	—	—	—	11,500	13,000	Commonly agglomerating[e]
						10,500	11,500	Agglomerating
III. Subbituminous	Subbituminous A coal	—	—	—	—	10,500	11,500	Nonagglomerating
	Subbituminous B coal	—	—	—	—	9,500	10,500	Nonagglomerating
	Subbituminous C coal	—	—	—	—	8,300	9,500	Nonagglomerating
IV. Lignitic	Lignite A	—	—	—	—	6,300	8,300	Nonagglomerating
	Lignite B	—	—	—	—	—	6,300	Nonagglomerating

[a]This classification does not include a few coals, principally nonbanded varieties, which have unusual physical and chemical properties and which come within the limits of fixed carbon or calorific value of the high-volatile bituminous and subbituminous ranks. All these coals either contain less than 48% dry, mineral-matter-free fixed carbon or have more than 15,500 moist, mineral-matter-free Btu/lb.

[b]Moist refers to coal containing its natural inherent moisture but not including visible water on the surface of the coal.

[c]If agglomerating, classify in low-volatile group of the bituminous class.

[d]Coals having 69% or more fixed carbon on the dry, mineral-matter-free basis shall be classified according to fixed carbon, regardless of calorific value.

[e]It is recognized that there may be nonagglomerating varieties in these groups of the bituminous class, and there are notable exceptions in high volatile C bituminous group.

[f]1 Btu/lb = 2326 J/kg.

SOURCE: From *Annual Book of ASTM Standards*, pt. 26, 1979.

geologic history of coal development. Knowledge of elemental associations in minerals in coal may provide valuable information for coal beneficiation.

Rank. The term rank refers to the stage of coalification reached in the course of metamorphism. Classification of coal by rank is fundamental in coal characterization and generally is based upon the chemical composition as stated in the previous section on coalification, and upon proximate values based upon mmf coal obtained from standard face channel samples (Table 1). In the case of lower-rank bituminous (high-volatile C) and subbituminous coals, rank also is based upon agglomerating characteristics and, in the case of higher- and lower-rank lignite, upon calorific value separation at 6300 Btu/lb (14.65 MJ/kg). The higher-rank coals are classified with respect to fixed carbon on the dry basis (mmf), and the low-rank coals according to calorific value on the moist basis (mmf). In Europe, coking properties are a more common basis of classification than in North America.

The difference between the banded and the nonbanded or canneloid types of coal of the same rank, which may occur in the same general region, is mainly in the higher hydrogen and slightly higher calorific value of the latter type.

The calorific values of coal are expressed in British thermal units (Btu) per pound in some English-speaking countries, and in calories per gram generally elsewhere. Heat values, determined by a calorimeter, vary for coal material from about 6300 Btu/1b (14.65 MJ/kg) for California lignite to about 16,300 Btu/lb (37.91 MJ/kg) for some maf cannel coal. Proximate and ultimate analyses of representative United States coals by rank are given in Table 2, the analyses being on the as-received basis. *See* BRITISH THERMAL UNIT (BTU)

Lithologic characteristics. The lithologic characteristics of coal concern the structural aspects of

Table 2. Proximate and ultimate analyses of samples of each rank of common banded coal in the United States*

Rank	State	Proximate analysis, %				Ultimate analysis, %					Heating value, Btu/lb†
		Moisture	Volatile matter	Fixed carbon	Ash	S	H	C	N	O	
Anthracite	Pa.	4.4	4.8	81.8	9.0	0.6	3.4	79.8	1.0	6.2	13,130
Semianthracite	Ark.	2.8	11.9	75.2	10.1	2.2	3.7	78.3	1.7	4.0	13,360
Bituminous coal											
Low-volatile	Md.	2.3	19.6	65.8	12.3	3.1	4.5	74.5	1.4	4.2	13,220
Medium-volatile	Ala.	3.1	23.4	63.6	9.9	0.8	4.9	76.7	1.5	6.2	13,530
High-volatile A	Ky.	3.2	36.8	56.4	3.6	0.6	5.6	79.4	1.6	9.2	14,090
High-volatile B	Ohio	5.9	43.8	46.5	3.8	3.0	5.7	72.2	1.3	14.0	13,150
High-volatile C	Ill.	14.8	33.3	39.9	12.0	2.5	5.8	58.8	1.0	19.9	10,550
Subbituminous coal											
Rank A	Wash.	13.9	34.2	41.0	10.9	0.6	6.2	57.5	1.4	23.4	10,330
Rank B	Wyo.	22.2	32.2	40.3	4.3	0.5	6.9	53.9	1.0	33.4	9,610
Rank C	Colo.	25.8	31.1	38.4	4.7	0.3	6.3	50.0	0.6	38.1	8,580
Lignite	N. Dak.	36.8	27.8	30.2	5.2	0.4	6.9	41.2	0.7	45.6	6,960

*From *Technology of Lignitic Coals*, U.S. Bur. Mines Inform. Circ. no. 769, 1954. Sources of information omitted.

†1 Btu/lb = 2326 J/kg.

the coal bed and texture as determined by the megascopic and microscopic physical constitution of the coal itself.

Bed structure. Structurally the coal bed (or seam) is a geological stratum characterized by the same irregularities in thickness, uniformity, and continuity as other strata of sedimentary origin. Thickness varies greatly. German brown coals occasionally approach 300 ft (100 m). A drill hole in the Lake De Smet area in Wyoming penetrated 223 ft (68 m) of lignite or subbituminous coal essentially in one bed. Major surface mines in Wyoming are operating in coal with thicknesses as much as 50–100 ft (15–30 m). Most coal beds mined in the eastern part of the United States are between 3 and 8 ft (0.9 and 2.4 m) thick.

Coal beds may consist of essentially uniform continuous strata or, like other sedimentary deposits, may be made up of distinctly different bands or benches of varying thickness. The benches may be separated by thin layers of clay, shale, fusain, pyrite, or other mineral matter, commonly called partings by the miner. Like other sedimentary strata, coal beds may be structurally disturbed by folding and faulting so that the originally approximate horizontality of position is lost to the extent that beds may become vertically oriented or even overturned, as in the anthracite fields of the eastern United States, and in other places in the world where similar high-rank coals are found.

Coal texture. The texture of the coal itself is determined by the character, grain, and distribution of its megascopic and microscopic components. In general, banded coals composed of relatively coarse, highly lustrous vitrain lenses (1/4 in. or 6 mm, or more, in thickness) are considered coarsely textured. As the thickness of the vitrain bands progressively lessens, the texture becomes fine-banded and then microbanded, with bright laminae composed of fine vitrainlike material. Coals displaying homogeneous texture are regarded as nonbanded or canneloid coals. In general, similar textural units are observed in bituminous and anthracite coals.

The names employed for the lithotypes, that is, vitrain, clarain, and durain, are somewhat less suitable for lignitic coals, because in these ranks of coal the bands of material that appear as vitrain in bituminous coals and anthracite commonly have the unmistakable appearance of tree trunks and pieces of wood or bark.

Microscopic texture is determined by the physical composition of the lithotypes as noted above, or of the anthraxylon and attritus, in terms of the microscopic constituents. The microscopy and petrology of coal are concerned very largely with these constituents. This textural aspect of coal is of importance in accurate coal description and classification, and for an understanding of the behavior of coal in its preparation and utilization.

Botanical and petrologic entities. The fundamental physical constituents of coal have been investigated from two points of view. In North America the point of view established by Reinhardt Thiessen and followed principally by the U.S. Bureau of Mines is microscopic, regarding coal as an aggregate of original botanical entities identifiable only by microscopic means. This ap-

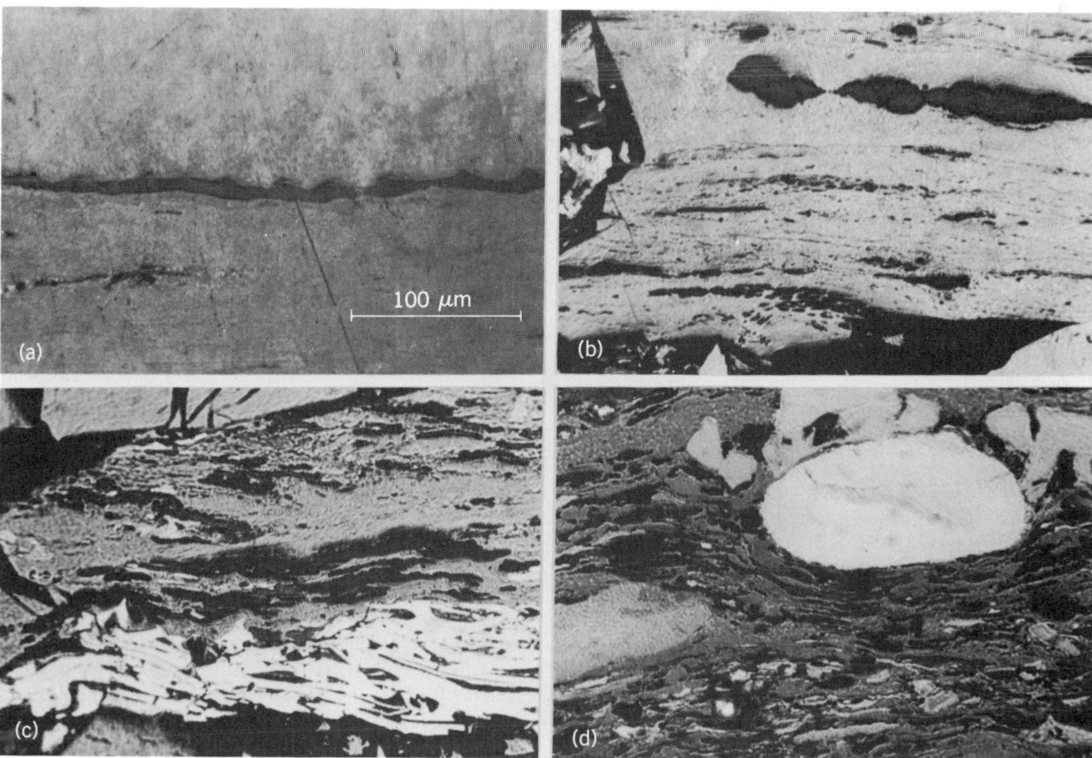

Fig. 3. Macerals in polished surfaces of bituminous coals in reflected light, all enlarged to the same magnification with oil immersion lens. *(a)* Thin, "sawtoothed" cutinite (dark) enclosed in vitrinite. *(b)* Elongate globs of resinite (dark) and other smaller exinite macerals (dark) enclosed in vitrinite. The dark area around the edges of the picture is epoxy mounting medium. *(c)* Sporinite (dark gray laminated lenses) in vitrinite (gray) with fusinite (white) on lower side of photo. *(d)* Large macrinite (white), smaller semifusinite (light gray), and sporinite and other exinites (dark gray) interlayered in vitrinite (gray).

Table 3. Maceral reflectance classes and reactivity during carbonization*†

Reactives			Inerts		
Group macerals	Macerals	Reflectance class	Group macerals	Macerals	Reflectance class
Vitrinite		V0 to V21	Inert vitrinite		V22 to V80
	Collinite	C0 to C21		Inert resinite	R22 to R80
	Telinite	T0 to T21	Inertinite		I18 to I80
				Fusinite	F40 to F80
				Micrinite	M18 to M80
				Semifusinite‡	SF22 to SF80
Exinite		E0 to E15		Sclerotinite	Sc22 to Sc80
	Sporinite	St0 to St15			
	Cutinite	Ct0 to Ct15			
			Group minerals	Minerals	
	Alginite	At0 to At15	Sulfides	Pyrite, etc.	
	Resinite	R0 to R15	Carbonates	Calcite, etc.	
			Silicates	Illite, etc.	
Fusible inertinite	Semifusinite‡	SF0 to SF21			
	Micrinite	M0 to M18			

*From J. A. Harrison, H. W. Jackman, and J. A. Simon, *Predicting Coke Stability from Petrographic Analysis of Illinois Coals*, Ill. State Geol. Surv. Circ. no. 366, 1964.

†Nomenclature as defined in Glossary of International Committee for Coal Petrology and based primarily on Stopes-Heerlen system of classification. Range of reflectance values of macerals based on values of N. Schapiro and R. J. Gray, 1960.

‡Estimated values; reactive group is about one-third and inert group about two-thirds of semifusinite total.

proach has commonly been referred to as coal microscopy, although the use of the phrase coal petrography has become more frequent in recent years. The other point of view, that assumed by Marie C. Stopes of England and accepted in much of the world, is based upon the concept of the four megascopic ingredients now called lithotypes: vitrain, clarain, durain, and fusain. Since these concern coal materials as rock substances, the concept provides a basis for the petrologic study of coal.

Microscopy. The microscopic study of coal, as developed in North America, is primarily concerned with the botanical entities or phyterals of coal, with fusain regarded as a coal substance of unique character, and with appropriate consideration being given to mineral mater. Thiessen established three categories of microscopic components—anthraxylon, attritus, and fusain—and recognized and described most of the botanical constituents of coal. This established a system of nomenclature, description, and classification which has been followed in most U.S. Bureau of Mines publications concerned with coal microscopy. Anthraxylon consists of coal occurring in bands in which wood or bark structure is microscopically evident. All vitrain of the European classification, if more than 14 micrometers thick, has been regarded as anthraxylon; below this threshold it has been classified as attritus. The attritus consists of finely textured coalified plant entities or phyterals not classified as anthraxylon (less than 14 μm) or fusian (less than 40 μm). Attritus, therefore, may contain very fine shreds of anthraxylonlike material, fine particles of fusian, disintegrated or macerated humic material or "humic degradation matter" (HDM), and such constituents as resins, waxes, cuticles, spore and pollen exines, algae, opaque matter, fungal bodies such as sclerotia, and fine mineral matter of various kinds with clay minerals usually predominating. The subdivision of attritus into translucent and opaque attritus on the basis of its content of opaque matter is the only subdivision made of the attritus with respect to variations in its heterogeneous constitution. In North America, banded bituminous coals with 30% or more of opaque matter have been classified as splint coals; those with 20–30% opaque matter have been classified as semisplint coals. Generally, no equivalent classification is recognized in other parts of the world. These distinctions in regard to opaque attritus were applied because splint coals are generally not amenable to hydrogenation and commonly not to carbonization.

Coal microscopy technique depends almost entirely on the use of thin, translucent sections of coal, a technique which is not adapted for use with the higher-ranking bituminous and anthracitic coals. Maceration has also been used as a means of breaking down the coal and isolating the more resistant constituents such as fossil spores that are the basis of the science of palynology.

Petrology and petrography. The use of the word petrology as applied to coal assumes that coal can correctly be regarded as a rock substance; its description is therefore consistently regarded as petrographic, that is, as a field of petrography.

Maceral concept. The initial contribution of Stopes to the field of coal microscopy included the adoption of a petrographic concept of coal as a rock substance composed of banded "ingredients" now called lithotypes—vitrain, clarain, durain, and fusian—and also mineral matter. Stopes also introduced the maceral concept, macerals being the individual components of the lithotypes, comparable to minerals composing most rocks (Table 3).

Of primary importance is that macerals are recognized by their optical properties and morphology. They differ from minerals in that macerals possess no fixed chemical composition or crystal structure such as those possessed by minerals. Each maceral becomes progressively modified chemically and physically as the rank of the coal advances. Hence it has become the practice in coal petrology to indicate the rank position of individual macerals solely by reliance upon measurement of some physical attribute, power of reflectance now being the most favored because of the relative simplicity of its application.

Modern petrographic methods of analysis of coals involve the determinations of the maceral composition (in terms of volume percent) and the reflectance of the vitrinite or other macerals. Procedures of these analyses have been standardized by the American Society for Testing and Materials (ASTM). Photomicrographs typical of many of the macerals which are common in bituminous coal and anthracitic coals in the United States are shown in Fig. 3.

Nomenclature. There has been wide acceptance of thc Stopes-Heerlen nomenclature of coal petrography. But the recognition that this nomenclature makes no provision for even the major subdivisions of coals by rank—anthracite, bituminous coal, and lignite—has led to at least one important series of proposals for modification of the terminology by W. Spackman. The Spackman system accepts the general validity of the three maceral groups of the Stopes-Heerlen system—vitrinite, exinite, and inertinite—but prefers to designate these as suites and to use the term liptinite rather than exinite, as the former term includes both resinite and exinite. The suite subdivision, at least with respect to the vitrinite suite, on thc basis of rank is subdivided into anthrinoid, vitrinoid, and xylinoid groups for anthracite, bituminous coal, and lignite respectively. Within the three maceral groups are various maceral types identified not by name but by reflectivity.

Applied coal petrography. A wide range of coal petrography applications has been developed to evaluate coals and solve various problems encountered in utilization of different types of available ores.

Coke. The usefulness of a bituminous coal for the production of metallurgical coke is primarily determined by its rank, that is, by its position in the lignite to anthracite series. Methods for predicting the production of satisfactory metallurgical coke from coal are based on the rank of the coal determined by reflectance of certain macerals, particularly vitrinite. The usefulness of this method with respect to the coals of the United States has been thoroughly investigated and developed. These investigations of many coals involved determination of reflectance, particularly of vitrinite; maceral analysis; chemical analysis; and determination of stability or strength of laboratory cokes produced from the coal.

In the early years of activity in this field of investigation, emphasis was mainly on the determination of the reflectance of vitrinite because of its common occurrence and conspicuous reflectance. However, as investigations progressed, it was found that the presence and abundance of other macerals, and even of minerals, also affected coke strength even though some did not actually produce coke.

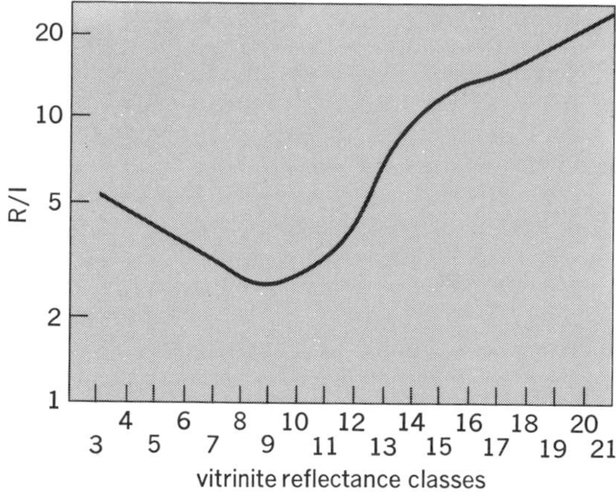

Fig. 4. Optimum ratio of the reactives to inerts (*R/I*) for each vitrinite reflectance class. (*Modification by J. A. Harrison, H. W. Jackman, and J. A. Simon, Predicting Coke Stability from Petrographic Analysis of Illinois Coals, Ill. State Geol. Surv. Circ. no. 366, 1964, of a figure by N. Schapiro, R. S. Gray, and G. R. Eusner, 1961*)

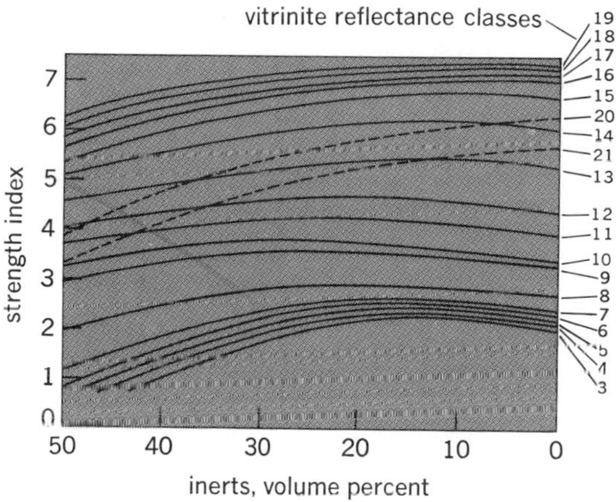

Fig. 5. Strength index of vitrinite reflectance classes depending on the amounts of inerts present. (*Modification by J. A. Harrison, H. W. Jackman, and J. A. Simon, Predicting Coke Stability from Petrographic Analysis of Illinois Coals, Ill. State Geol. Surv. Circ. no. 366, 1964, of curves by N. Schapiro, R. S. Gray, and G. R. Eusner, 1961*)

Many investigations concerned with the reflectance of the coal since 1958 consisted largely in the accumulation of petrographic and experimental data which provided the basis for graphs used in predicting the suitability of a particular coal or coal blends for the production of metallurgical coke suitable for use in the steel industry.

For the preparation of the graphs shown in Figs. 4, 5, and 6, it was necessary to accumulate a large volume of laboratory data on reflectance, maceral analyses, chemical analyses, and the stability of the coke made from many bituminous coals. These are representative of similar types of curves which have been developed in many coal petrography laboratories, particularly by major steel companies.

Of the three graphs that have come into use for predicting the suitability of particular coals or one of fundamental importance. The experimentally determined position of the optimum coke for each reflectance class of vitrinite and the best ratio of reactives to inerts (R/I) is determined by the position of the curve. When a coal or blend of coals contains reactives of several reflectance classes, as is usually the case, the inert index N is derived from Eq. (3), where Q=total percent by volume of

$$N=\frac{Q}{P_1/M_1+P_2/M_2+...+P_{21}/M_{21}} \qquad (3)$$

inerts in coal blend from analysis; P_1, P_2, etc. = percent of reactives (exinite, resinite, and one-third of total semifusinite) from analysis; and M_1, M_2, etc. = ratio of reactives to inerts to produce optimum coke. Inerts consist of fusinite, two-thirds of semifusinite, micrinite, and ash. (For the application of the inert index see Fig. 6.)

Figure 5 consists of a set of curves, based upon experimental data, designed to show how the volume percent of the inerts (not the inert index) affects the strength of the coke (sometimes called the coking coefficient) made from coals of various reflectance classes. The position, spacing, and curvature of the curves for vitrinite reflectance classes are based upon experimental determinations with reference to an arbitrary scale of strength index (0–10) and to the volume of inerts (0–50% in steps of 5%). The strength index K_T (coking coefficient) of a coal or blend of coals composed of reactives of various reflectance classes can be determined from Eq. (4) in conjunction with

$$K_T=\frac{(K_1 \cdot P_1)+(K_2 \cdot P_2)+...+(K_{21} \cdot P_{21})}{P_T} \qquad (4)$$

Fig. 5. Here P_T= total percentage of reactives; K_1, K_2, etc. = strength index of reactives in the reflectance classes present in the coal sample, obtained from the family of curves in Fig. 5; and P_1, P_2, etc. =percentage of reactives in reflectance classes present in the coal sample. The values K_1, K_2, etc. are obtained from Fig. 5 on the basis of the volume of inerts reported in the analytical data. A vertical line projected from the abscissa at this position will successively intersect the curves representing progressively higher reflectance classes present in the sample. By projecting lines horizontally from the points of intersection to the ordinate, the strength index of the successive reflectance classes can be read, assuming the same percentage of inerts. The analytical data supply values for P_1, P_2, etc. The products indicated by the calculation K_1, P_1, etc. are the results obtained by multiplying the strength index of a particular reflectance class by the percentage of such class as provided by the petrographic analysis. Summation of the resulting values for the various reflectance classes provides a figure representing the total strength index (K_T) of the reactives. This value divided by the total amount of reactives (P_T) provides the calculated strength index.

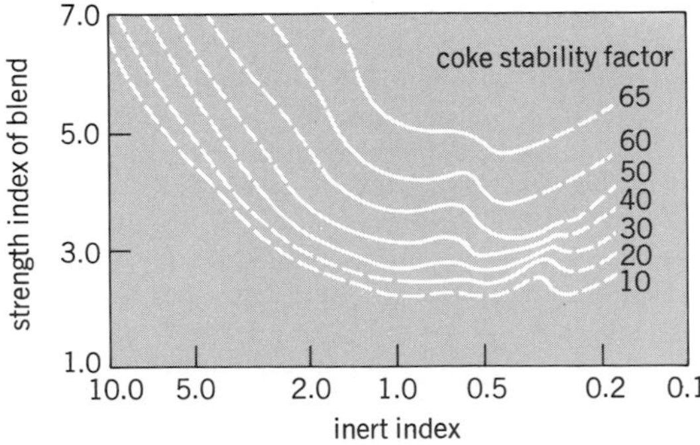

Fig. 6. Curves showing the relation between the strength index, inert index, and stability factor. (*Modification by J. A. Harrison, H. W. Jackman, and J. A. Simon, Predicting Coke Stability from Petrographic Analysis of Illinois Coals, Ill. State Geol. Surv. Circ. no. 366, 1964, of a figure by N. Schapiro, R. S. Gray, and G. R. Eusner, 1961*)

Many tests in a number of laboratories provide the basis for graphs similar to Fig. 6. The purpose of such graphs is to predict the stability factor of coke made from various coals and from blends of coals. The curves are based on the results obtained in laboratory practice in terms of stability factors varying between 10 and 65%, strength indexes between 2 and 10, and inert indexes between 0.2 and 10.0. In order to apply this chart, the coal or coals must be subjected to the various petrographic tests and analyses already cited whereby the strength and inert indexes are obtained. By the use of these data and Fig. 6, a close approach to the actual stability of the coke can be forecast by using relatively small samples of coal in the laboratory.

Formed coke. Petrographic analyses are also being used to evaluate the properties of formed coke, a formulated material which may become a substitute for conventional coke. Formed cokes are carbonized briquettes made at elevated temperatures from finely crushed coal mixed with certain additives under various conditions. Formed coke offers possibilities for the future use of lower-rank coals for the production of iron. Petrography can distinguish the presence or absence of significant factors, composition, and conditions within the briquetting process that will aid in optimizing conditions for production of high-quality and consistent briquettes.

Coal for liquefaction. Petrographic analyses are useful for the evaluation of coals for conversion to liquid and gaseous hydrocarbons. Bituminous coals and others of low rank that are enriched in exinite and vitrinite macerals are thought to be the desirable characteristics of coals for liquefaction. These types of coals yield liquid hydrocarbons when mixed with a little oil and heated under pressure. In this reaction, the exinite, most of the vitrinite, and some of the inertinite macerals dissolve in the oil. The resulting liquid is filtered to remove mineral and unreacted maceral particles and is refined to gasoline and other useful products. Petrographic characteristics of the minerals and macerals in the coal aids in the design and operation of the system and also provides information basic to the mechanism of the conversion to liquids.

Coal grindability. Coal petrography is useful for solving certain problems encountered in the mining, crushing, cleaning, and sizing of coal. Abundant fusain bands in the coal bed that are not mineralized constitute planes along which breakage occurs, and this leads to abundant fusinite in the fine particle sizes, thus giving the coal a dusty character. Noteworthy too are applications of specific recommendations of crushing and sizing, based on petrographic data, to maximize the liberation of pyrite (FeS_2 particles) from the coal. Additionally the grindability of a coal is an important index useful to design preparation plants, and the reflectance of vitrinite is a useful guide to this property.

Coaly particles in sedimentary rocks. A major application of the methods of coal petrography concerns the geologic evaluation of coaly particles that occur in shales and other sedimentary rocks. This evaluation is mainly focused on the determination of the relative degree of metamorphism of the organic matter in these rocks as an index of maximum paleotemperature. Knowledge of maximum paleotemperature can assist the petroleum geologist in identifying favorable areas for exploration for oil and gas based on microscopic analyses of dispersed coaly materials. Excessive paleotemperatures provide information on areas or depths where oil would not be anticipated but where gas may be present. The mean reflectance of vitrinite and the fluorescence spectra of the exinites provide the most useful indexes of the maturity of the organic matter, or "rank" of the rock. This application is based on the known behavior of organic matter in laboratory heating tests and other indirect measures.

Utilization. The uses for coal have varied widely with rank, but three general areas of use may be distinguished. The major use is for combustion to generate electricity and heat (electric utilities, industrial, domestic, and railroads). A significant percentage of the world's coal with suitable properties is used in the manufacture of metallurgical-grade coke. Coal was formerly used extensively to manufacture gas, and new research appears to be leading to major developments for conversion of coal to liquid and gaseous fuels to supplement supplies of oil and natural gas. Coal was an important source of a wide range of chemicals before being largely supplanted by oil and gas, but this possible renewed future use is similar to processes of conversion to liquid and gaseous fuels.

During World War II, Germany produced a significant amount of liquid fuels by coal conversion. Commercial-scale facilities are now in operation in South Africa and East Germany. In the United States several facilities are in various stages of planning for technical and economic demonstration for the production of both gaseous and liquid fuels from a broad range of ranks of coal. A number of pilot-scale plants are in operation which produce liquid and gaseous fuels from coal, and one is producing a so-called solvent-refined coal. *See* COAL GASIFICATION; COAL LIQUEFACTION.

Occurrence. Coal is found on every continent, in the islands of Oceania, and in the West Indies. Most coal resources of the world are contained in strata from the Carboniferous (Pennsylvanian-Mississippian of the United States) and younger strata, although minor occurrences have been reported in Devonian and older strata. Most of the coals in the major fields of the eastern United States are Pennsylvanian in age, but most of the coals in the western states are of Cretaceous and Tertiary age. Approximately 95% of the known coal resources and 85% of recoverable reserves of the world lie in the Northern Hemisphere.

The amount of coal in the various countries varies greatly. In 1976, 11 countries accounted for about 91% of world production of coal, including lignite (Soviet Union, United States, China, East Germany, Poland, West Germany, United Kingdom, Czechoslovakia, India, Australia, and South Africa). A total of 53 countries reported production of coal in that year, which was estimated to total about 3.2×10^9 metric tons.

Estimates of total coal resources of the world are very large, but vary widely depending on assumptions on which the estimates are based. Such estimates made in the past decade range from about 10 to more than 16×10^{12} tons. Recoverable coal reserves of the world based on current mining practice have been estimated to total over 800×10^9 tons. It is estimated that the countries cited above have over 93% of estimated recoverable reserves. *See the feature article* EXPLORING ENERGY CHOICES. [JACK A. SIMON]

Bibliography: American Society for Testing and Materials, *Annual Book of ASTM Standards*, pt. 26, 1979; I. I. Ammosov et al., Calculation of coking charges on the basis of petrographic characteristics of coals, *Koks i Khimya*, no. 12, p. 9, 1957; N. H. Bostick, Microscopic measurements of the level of catagenesis of solid organic matter in sedimentary rocks to aid exploration for petroleum and to determine former burial temperatures: A review, *Aspects of Diagenesis*, Soc. Econ. Paleontol. Mineralog. Spec. Pub. no. 26, pp. 17–43, 1979; F. Chayes, *Petrographic Model Analysis: An Elemental Statistical Appraisal*, 1956; A. C. Fieldner and W. A. Selvig, *Methods of Analyzing Coal and Coke*, U.S. Bur. Mines Bull. no. 492, 1951; P. H. Given et al., Dependence of coal liquefaction behavior on coal characteristics: 2. Role of petrographic composition, *Fuel*, 54(1):34–41, 1975; J. A. Harrison, H. W. Jackman, and J. A. Simon, *Predicting Coke Stability from Petrographic Analysis of Illinois Coals*, Ill. State Geol. Surv. Circ. no. 366, 1964; International Committee for Coal Petrology, *International Handbook of Coal Petrography*, 2d ed. 1963, suppl. 1971, 2d suppl. 1975; M.-Th. Mackowsky, The application of coal petrography in technical processes, *Stach's Textbook of Coal Petrology*, 2d ed., Gebruder Borntraeger, Stuttgart, 1975; M.-Th. Mackowsky, Prediction methods in coal and coke microscopy, *J. Microsc.*, 109 (pt. 1), 119–137, 1977; B. C. Parks and H. J. O'Donnell, *Petrography of American Coals*, U.S. Bur. Mines Bull. no. 550, 1956; N. Schapiro and R. J. Gray, Petrographic classification applicable to coals of all ranks, *Proceedings of the Illinois Mining Institute*, 1960; N. Schapiro, R. J. Gray, and G. R. Eusner, Recent developments in coal petrography, *Blast Furnace, Coke Oven and Raw Materials Conference: Proceedings of the American Institute of Mining Engineering*, vol. 20 pp. 89–112, 1961; U.S. Bureau of Mines, *Minerals Yearbook*, vol. 1, 1976; D. W. van Krevelen and J. Schuyer, *Coal Science*, 1957; World Energy Conference, *Survey of Energy Resources*, 1976, R. and R. Clock Ltd., Edinburgh, 1976.

Coal gasification

The conversion of coal, coke, or char to gaseous products by reaction with air, hydrogen, oxygen, steam, carbon dioxide, or a mixture of these. Products consist of carbon dioxide, carbon monoxide, hydrogen, methane, and some other chemicals in a ratio dependent upon the particular reactants employed and the temperatures and pressures within the reactors, as well as upon the type of treatment which the gases from the gasifier undergo subsequent to their leaving the gasifier. Strictly speaking, reaction of coal, coke, or char with air or oxygen to produce heat plus carbon dioxide might be called gasification. However, that process is more properly classified as combustion, and thus is not included in this coal gasification summary.

Industrial uses. In most coal gasification processes which are available for use or under development, the reactions are endothermic. Air or oxygen is typically supplied to the gasifier to provide the necessary heat input. From the industrial user's viewpoint, the net result is to produce either low-Btu or intermediate-Btu gas. Low-Btu gas typically has a heating value on the order of 4.5 kilojoules/m³ (150 Btu/ft³). Because of the absence of the nitrogen diluent, intermediate-Btu gas has twice the heating value, or about 9 kJ/m³. Also, because the inert nitrogen is not present, intermediate-Btu gas can be upgraded to produce substitute natural gas (SNG) with a heating value of about 30 kJ/m³ (1000 Btu/ft³).

Each of these gas types has attracted industrial interest. The electric industry is principally investigating the production of clean, low-Btu gas from coal with the object of burning it in a combined-cycle power generation system. Such a combined-cycle system would use a gas turbine plus a waste heat boiler and steam turbine to drive separate electric generators, thereby providing overall station efficiencies approaching 45%. The natural gas industry views intermediate-Btu gas production with strong interest since this same gasification product can also be upgraded to synthetic natural gas by using relatively conventional and proved gas-processing steps. Heavy industry is studying the feasibility of using both low- and intermediate-Btu gas for many applications, thereby freeing critical volumes of natural gas and also reducing imported oil consumption. Parts of the chemical industry are examining intermediate-Btu gas as a source of hydrogen and carbon monoxide to offset potentially reduced supplies of these synthetic chemical building blocks, which are currently obtained primarily from steam reforming of natural gas, petroleum, natural gas liquids, or petroleum derivitives. Various projects are also being studied which would involve coproduction of SNG and intermediate-Btu gas for industrial or chemical use, electric power, coal-derived liquids, and various other products.

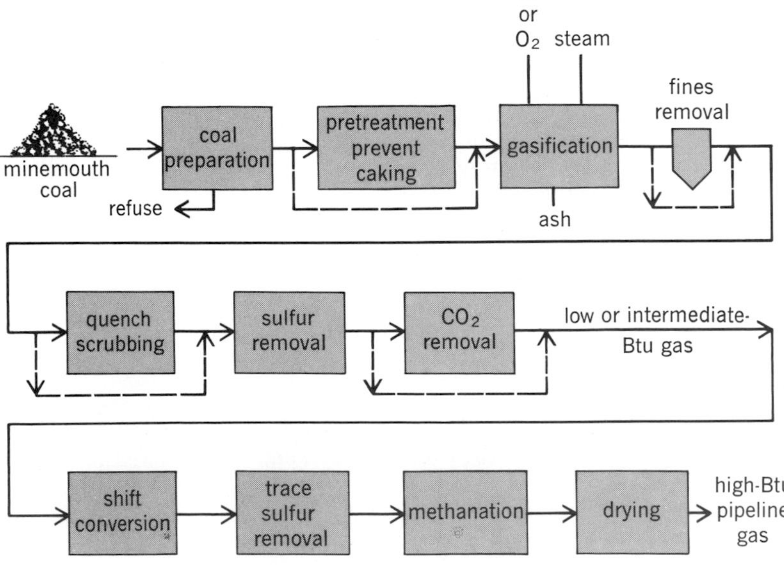

Fig. 1. Schematic representation of the processing steps in coal gasification.

Gasification processes. Some of the basic technology for many of the coal gasification processes under active consideration today is quite old and well known. Much of it derived directly from the large body of manufactured gas technology. In 1965 Bituminous Coal Research, Inc., completed a study for the Office of Coal Research (OCR) of the U.S. Department of the Interior and published a report containing the results of a literature search on 65 different coal gasification processes. In spite of the large number of processes described and identified at that time, there continues to be a variety of new processes under development. In nearly all of the processes the chemistry of the high-temperature gasification is the same (Fig. 1). The basic reactions are:

Coal reactions

$$\text{Coal} \xrightarrow{\text{Heat}} \text{gases } (CO, CO_2, CH_4, H_2) + \text{liquids} + \text{char} \quad (1)$$

$$\text{Coal} + H_2 \xrightarrow{\text{Catalyst}} \text{liquids} + (\text{char}) \quad (2)$$

$$\text{Coal} + H_2 \text{ (from a hydrogen donor)} \rightarrow \text{liquids} + (\text{char}) \quad (3)$$

$$\text{Coal} + H_2 \xrightarrow[\text{destruction}]{\text{Noncatalytic}} CH_4 + \text{char} \quad (4)$$

Char reactions

$$C\text{ (char)} + 2H_2 \rightarrow CH_4 \text{ exothermic} \quad (5)$$

$$C\text{ (char)} + H_2O \rightarrow CO + H_2 \text{ endothermic} \quad (6)$$

$$C\text{ (char)} + CO_2 \rightarrow 2CO \text{ endothermic} \quad (7)$$

$$C\text{ (char)} + O_2 \rightarrow CO_2 \text{ exothermic} \quad (8)$$

Gaseous reactions

$$CO + H_2O \xrightarrow{\text{Catalyst}} H_2 + CO_2 \text{ exothermic} \quad (9)$$

$$CO + 3H_2 \xrightarrow{\text{Catalyst}} CH_4 + H_2O \text{ exothermic} \quad (10)$$

$$CO_2 + 4H_2 \xrightarrow{\text{Catalyst}} CH_4 + 2H_2O \text{ exothermic} \quad (11)$$

$$xCO + yH_2 \xrightarrow{\text{Fe}} \text{hydrocarbon gases and/or liquids} + zCO_2 \text{ exothermic} \quad (12)$$

Thermodynamics. From a thermodynamic standpoint in coal gasification, at least one simplifying assumption is customarily made; namely, coal can be considered as carbon in the form of coke. This assumption is made because coal is a chemically ill-defined material that does not fit into the regime of rigorous thermodynamic deduction. Errors associated with this assumption are not likely to be very large. Thus the individual high-Btu gas-producing step, reaction (5) is considered to be of great interest. Not only is this reaction exothermic, but it also has a large negative standard free energy change, indicating that it is a spontaneous reaction. Unfortunately, the rate of this reaction at ordinary temperatures and in the absence of a catalyst is nearly zero. In order to force the reaction to proceed at a fast rate, the temperature must be raised considerably. Some processes are designed to operate at as high as 1100°C. Another important consideration in regard to reaction (5) is that very large quantities of energy must be used to obtain the hydrogen for this reaction. The most widely available source of hydrogen is water, large quantities of which are necessary for a gasification plant. In fact, a plant producing 7,000,000 m^3 (250,000,000 ft^3) of pipeline-quality gas per day requires about 11,300 m^3 (3,000,000 gal) of water per day to supply hydrogen. Substantially greater volumes of water are required for associated process operations such as cooling tower makeup for low-level heat rejection.

Reaction (13) is highly endothermic and also

$$H_2O \rightarrow H_2 + \tfrac{1}{2}O_2 \qquad (13)$$

requires large quantities of energy. As a result, nearly all of the gasification processes under investigation today do not decompose water directly according to reaction (13) and then follow with a reaction of hydrogen with coal or char. Rather, the water is decomposed according to reaction (6). Products of this reaction are then treated as shown in reactions (9), (10), and (11). It is currently necessary to use a multistep process as outlined in Fig. 1, although some processes are under development which rearrange these blocks in order to enhance methane formation in the gasifier.

Gasification step. The various gasification processes which have already been developed or which are in the development stage vary with respect to the gasification step. In fact, to a large degree, the mechanical and engineering variations of this step, particularly the features designed for supplying the heat for the endothermic reaction $C + H_2O$ and for handling solids in the gasifier, characterize the processes. Heat can be furnished by one of several methods: by partial combustion of coal or char with air or oxygen, by an inert (pelletized ash, ceramic spheres, and so on) heat carrier, by heat released from the reaction of a metal oxide with CO_2, by high-temperature waste heat from a nuclear reactor, or by heat produced when an electric current passes through the coal or char bed between electrodes immersed in the bed. Processes have been investigated for gasification at atmospheric and elevated pressures by using molten-media, moving-bed, fluidized-bed, or entrained suspended, dilute-phase operations in the gasifier.

Research and development. The cost of gas made by coal gasification is sufficiently high to justify continuing research and development to try to decrease capital and operating costs of gasification plants. In the United States, calculated costs for a plant capable of producing 7,000,000 m^3 (250,000,000 ft^3) per day of high-Btu gas are in the range of $1,500,000,000. Most development work on processes, or on parts of processes, is directed toward lowering the plant costs or increasing the efficiency of certain process steps. One of the major programs for developing improved coal gasification is funded jointly by the Gas Research Institute (GRI) and DOE.

In addition to the GRI-DOE program, other research and development is underway. Slagging gasifier studies have been carried out on a Lurgi unit at Westfield, Scotland, and on a COGAS unit being developed by the Coal Utilisation Research Laboratory (formerly BCURA) at Leatherhead, England. Texaco and Shell-Koppers have pilot plants under commissioning in Europe. This area is being emphasized because it is believed that slagging gasifiers will produce gas at a faster rate than nonslagging gasifiers can.

A discussion of some of the coal gasification processes, as well as some of their pertinent features, follows.

COGAS process. In the COGAS process, low-pressure coal pyrolysis is combined with char gasification to produce liquid fuel and synthetic pipeline gas. Synthesis gas is produced from the char, with air used instead of oxygen to provide the heat for the steam-carbon reaction through a separately fired combustor which heats a circulating stream of gasifier char. The operation may be carried out at the relatively low pressure of 450 kPa absolute (50 psig).

Commercially available process technology is reportedly used for hydrogen generation, gas purification and dehydration, sulfur and oil recovery, and oil hydrotreating. The COGAS technology is being considered by DOE for the construction of an 1800-metric-ton-per-day SNG demonstration plant to be built in Illinois.

Consolidated synthetic gas (CSG) process. A basic feature of this process, also known as the CO_2 acceptor process, is the use of lime or calcined dolomite to react exothermally with the CO_2 liberated in gasification to supply a portion of the heat needed for the carbon-steam reaction. The removal of CO_2 by this method enhances the water gas shift reaction and methane formation, both of which are exothermic. The combination of these effects allows the gasification system to operate without an oxygen supply. The spent "acceptor" (lime or dolomite) is withdrawn from the gasification vessel and calcined separately, with residual char and air supplying the necessary heat. The process also embodies the contacting of the incoming raw coal with the hydrogen-rich synthesis gas to form a portion of the methane directly.

The operating conditions of the CO_2 acceptor process are limited to about 870°C (1600°F) and 2000 kPa (300 psia) because of melt-formation problems in the dolomite system. Since the rate of the uncatalyzed steam-carbon reaction for gasification of bituminous coals is prohibitively slow—870°C or lower—the process is not competitive for the gasification of these types of coals.

Exxon catalytic process. The Exxon catalytic process differs significantly from other SNG gasification methods in that a potassium catalyst solution is sprayed on the feed coal in order to prevent agglomeration and catalyze reactions (6) and (10) in a fluid-bed gasifier with dry ash removal. The catalyst enables the gasifier to operate at the attractively mild process conditions of 700°C (1300°F) and 3550 kilopascals (500 psig). The resulting gasification reactions are almost thermally neutral with the minor amount of required energy being supplied by preheating the fluidizing gas, which is steam plus recycled H_2 and CO. This eliminates the need for oxygen injection or other indirect heat transfer to supply heat to the gasifier. The gasifier off-gas is treated by conventional acid gas removal processing and the methane product cryogenically separated, with H_2 and CO recycled to the gasifier. The catalyst-char-ash material which is removed from the gasifier is digested at 150°C (300°F) and washed countercurrently with water to recover about 90% of the potassium catalyst for reuse.

The process has been studied in a fluid-bed test

gasifier, and Exxon is constructing a fully integrated 1-metric-ton-per-day process development unit close to its refinery in Baytown, TX. Construction of a 90-metric-ton-per-day large pilot plant is under consideration by DOE.

HYGAS. Pretreated coal is slurried with a light oil (produced as a by-product of the process), and the slurry is pumped to 3550–10,450 kPa (500–1500 psig) and injected into a fluidized bed at the top of the vertical multistage unit, where the oil is driven off and recovered for reuse.

The oil-free coal then passes downward through two hydrogasification stages which provide countercurrent treatment (coal passes down, gases produced pass upward and are drawn off).

In the first hydrogasification stage, dried coal is flash-heated to the reaction temperature 620–705°C (1200–1300°F) by dilute-phase contact with hot reaction gas and recycled hot char. Volatile matter in the coal and active carbon are converted to methane in a few seconds. Active carbon, which is present during the first moments of gasification, gasifies at a rate in excess of 10 times that of the carbon in the less reactive char.

The solids then pass down to the second HYGAS stage and enter a dense-phase, fluidized-bed reactor at temperatures of 925–980°C (1700–1800°F), where formation of methane from partially depleted coal char continues simultaneously with a steam-carbon reaction that produces hydrogen and carbon monoxide.

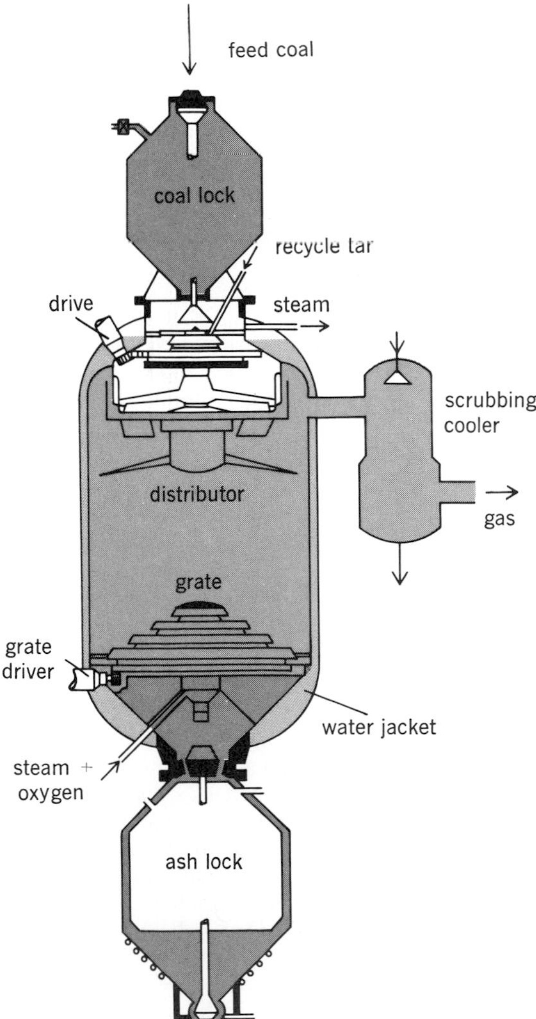

Fig. 2. Diagram of a Lurgi pressure gasifier.

Hot gases produced in the lower hydrogasification stage rise, passing through the first-stage hydrogasification reactor and into the fluidized drying bed, where much of their heat is used to dry the feed coal. After leaving the hydrogasifier, the raw gas is shifted to the proper CO/H_2 ratio in preparation for methanation; the oils, carbon dioxide, unreacted steam, sulfur compounds, and other impurities are removed; and the purified gas is catalytically methanated. Sulfur is recovered in elemental form.

The partially depleted coal char is used to produce a hydrogen-rich stream required in the process by steam-oxygen gasification of the char. Ash is withdrawn from the system as high-mineral-matter char.

Lurgi. The dry-bottom Lurgi gasifier employs a bed of crushed coal traveling downward through the gasifier, and operates at pressures up to 3200 kPa (450 psig). Steam and oxygen are admitted through a revolving steam-cooled grate which also removes the ash produced at the bottom of the gasifier. The gases are made to pass upward through the coal bed, carbonizing and drying the coal. Steam is used to prevent the ash from clinkering and the grate from overheating, and a hydrogen-rich gas is produced. Because of the pressure, some of the coal is hydrogenated into methane (in addition to that distilled from the coal), releasing heat which in turn is given up to the coal, minimizing the oxygen requirements.

Most gas produced by coal gasification comes from Lurgi reactors of the general type shown in Fig. 2. The internal design and operating conditions of these reactors may be altered for coals of different caking characteristics.

The dry-bottom Lurgi is considered particularly attractive for western coals and lignite. A number of gasification projects utilizing dry-bottom Lurgi technology have been proposed or are in various stages of consideration.

A number of moving-bed gasifiers which are similar to the dry-bottom Lurgi—but of atmospheric pressure—such as Wellman-Galusha and Woodall-Duckham, have been used worldwide and have been utilized or considered for on-site industrial gasification in the United States. Dry-bottom Lurgis are used at the world's largest gasification facility, a multibillion-dollar SASOL coal-to-oil plant near Johannesburg, South Africa.

An extension of this technology is the BGC-Lurgi slagging gasifier, which is under development at Westfield, Scotland. In this process the grate assembly and dry ash lock shown in Fig. 2 are replaced by a series of tuyeres firing into the lower side of the gasifier, a slag tap hole at the bottom of the gasifier, and a water-filled quench tank and slag lockhopper system. Steam and oxygen are injected through the tuyeres into a 1260°C (2300°F) 2800 kPa (400-psig) reaction zone at the bottom of the moving bed. Ash in the form of molten slag is removed through the tap hole and granulated by water in the quench vessel. This process is considered to be far more attractive than the dry-bottom Lurgi for eastern coals, which generally have lower ash fusion temperatures and reactivities. The

BGC-Lurgi slagging gasification technology is being considered by DOE for an SNG demonstration plant to be built in Noble County, OH.

Rocketdyne hydrogasification process. The Rocketdyne process uses rocket engine injector technology to rapidly mix unpretreated coal in dense-phase transport with 1100°C (2000°F) hydrogen at up to 10,400 kPa (1500 psig) pressure for use in a short-residence (about 1500-ms) entrained flow reactor tube. Conventional heat exchange technology is used to heat the inlet hydrogen to 815°C (1500°F), followed by partial combustion with injected oxygen to reach required reactor inlet temperatures. Depending on reactor temperatures, pressures, and residence times, the reactor product consists of various mixtures of unconsumed excess hydrogen, light hydrocarbon gases which have included up to 90% methane, vapors of light and heavy hydrocarbon liquids, and a dry char.

The process is being developed under DOE sponsorship at 0.22- and 0.9-metric-ton-per-hour experimental reactors which are operated at Rockwell International's laboratory in Canoga Park, CA. Designs are also under way for a 90-metric-ton-per-day experimental reactor system. In commercial concepts, hydrogen to operate the Rocketdyne process is projected to be supplied from a separate oxygen-based char-coal gasification system. Commercial concepts have generally been based on the coproduction of SNG and benzene-like liquids.

Texaco process. In the Texaco process, a coal-water slurry is reacted with oxygen in an entrained gasifier with ash removal as a granulated slag. The pilot test gasifier at Montebello, CA, is a downward-fired, two-zone gasifier. The top zone is a reaction chamber, and the bottom zone a water-filled quench vessel. The molten slag and any ungasified carbon soot by-product are removed by the water. Saturated product gas exits the quench chamber at about 925°C (1700°F). The granulated slag is removed periodically by a lockhopper and the soot by an aqueous slurry which can be concentrated and recycled to the gasifier. The clarified quench water stream reportedly contains no tars or phenols.

The Texaco coal gasification process is the result of a development program based on the Texaco synthesis gas generation process, which has been licensed to numerous worldwide plants for feedstocks ranging from natural gas to asphalts. Several different-sized generators for these feedstocks are available which can be operated at over 8400 kPa (1200 psig) and produce more than 3,000,000 m^3 (110,000,000 ft^3) per day of hydrogen and carbon monoxide in a single generator. The Electric Power Research Institute (EPRI) has used Texaco's Montebello test unit to demonstrate gasification of coal liquefaction residues. A Texaco coal gasification pilot plant of 136-metric-tons-per-day capacity is being demonstrated in West Germany.

Westinghouse process. The Westinghouse gasification process consists of a fluid-bed devolatilizer operating at 870°C (1600°F) and 1650 kPa (225 psig). The devolatilizer is fluidized by hot reducing gases from a gasifier in which ash is removed as an agglomerate. Unpretreated coal enters the devolatilizer, where it is rapidly mixed with char from the fluidized bed in a draft tube in order to prevent agglomeration as the coal is heated through its plastic stage. Char is continuously withdrawn from the devolatilizer to feed the gasifier-agglomerator. In the bottom portion of this reactor, a portion of the char is combusted with air at a temperature of 1065°C (1950°F), at which ash particles agglomerate and defluidize. The spherical ash agglomerates are continuously removed from the gasifier after being cooled with steam, which is used to gasify the remainder of the char in the gasifier-agglomerator and to moderate the combustion zone temperature.

The Westinghouse process is being developed under the auspices of DOE at a 13.6-metric-ton-per-day pilot unit at Waltz Mill, PA. The process has been initially developed for air gasification to supply fuel for electric production from combustion turbines or combined cycle systems. However, tests have been run using oxygen in both the two-stage integrated configuration of coupled devolatilizer and gasifier-agglomerator vessels and in an advanced single-vessel configuration.

Other processes. A number of other gasification processes are either in use worldwide or under development. (These gasification systems are characterized parenthetically in this section by their reaction method/pressure/type of ash removal.) Examples of systems in worldwide use include Koppers-Totzek (entrained/atmospheric/combination fritted slag and char-ash) and Winkler (fluid-bed/atmospheric/char-ash). Examples of developmental processes which are of interest either because of their industrial support or because of their technological features include: Atomics International (molten salt bath/pressurized/aqueous slurry ash), Combustion Engineering (entrained-flow air-fired/pressurized/fritted slag), Otto Szarberg (molten slag bath/pressurized/fritted slag), and Shell-Koppers (entrained/pressurized/fritted slag).

Underground gasification. At the request of the Bureau of Mines, a detailed survey of underground gasification technology was prepared in 1971. This survey of worldwide activities revealed no large-scale active program of either basic or applied research. However, there has been a renewal of interest in the United States on the part of industrial concerns, and there are indications of one or two exploratory experimental programs on a modest scale. Underground coal gasification has been practiced on commercial scale in the Soviet Union's steeply dipping coal beds.

The Bureau of Mines has begun work in this field again, with particular emphasis on the utilization of new technology developed since the mid-1950s. This new technology includes an understanding of the nature and direction of subsurface fracture systems and the means to calculate underground fluid movement. Directional drilling techniques have been advanced; and chemical explosive fracturing, a new method of preparing underground formations, has been introduced by the Bureau of Mines. DOE has conducted field tests using linked vertical wells near Hanna, WY. Underground coal gasification is generally considered most applicable to low-rank nonswelling coals found in the western United States. Additionally, field experiments have established the technical possibility of in-place recovery of crude oil and shale oil, and modern methods of surface-processing of coal to high-Btu gases have been demonstrated.

At the same time, an active program on fracture characterization and gas flow in coal beds is under way at the Morgantown Energy Research Center.

[RICHARD H. MC CLELLAND]

Bibliography: American Gas Association – GRI-DOE-IGU, *Proceedings of the 7th, 8th, 9th and 10th Synthetic Gas—Pipeline Gas Symposiums*, 1975–1978; Coal Technology *Conference Papers*, Vols. 1 and 2, 1978; DOE—Morgantown Energy Research Center, *An Engineering Assessment of Entrainment Gasification*, Rep. MERC/RI-78/2, 1978; Dravo-DOE, *Handbook of Gasifiers and Gas Treatment Systems*, 1976, Rep. FE-1772-11, 1976.

Coal liquefaction

The conversion of most types of coal, anthracite excepted, by direct or indirect technologies primarily to produce synthetic low-sulfur, low-ash liquid or solid fuels to replace some of the fuel coal used by electric utilities. Coal liquefaction shows promise for making syncrude, a material suitable for use as a refinery feedstock and for petrochemical production.

Although the term liquefaction implies liquid products, several processes also produce gaseous and solid products. The advantage of coal liquefaction is that a wide range of liquid products, especially heavy fuels for electric utilities, distillate fuel oil, and gasoline, can be produced by varying the operating conditions of different processes. The atomic hydrogen:carbon (H:C) ratios of the products are equal or greater than the ratio in the feed coal, depending on the extent of hydrogenation used. Coal liquefaction requires less chemical transformation than gasification, and has an approximate 78% energy-conversion efficiency versus 60–70% for gasification.

Direct liquefaction. The Bergius single-stage process for liquefaction and hydrogenation of coal was developed in Germany in the 1920s, and was commercially proved in World War II by producing a maximum of 110,000 barrels per day (bpd; 17,500 m^3 per day) of transportation fuels for the German war effort. There are three direct-liquefaction second-generation progeny of the Bergius process in the United States, (none of which is predicted to attain commercial status before 1990): Gulf Solvent Refined Coal (SRC); H-Coal; and Exxon Donor Solvent (EDS). A lesser-known one-step approach is the molten zinc chloride process of Shell Development and Conoco Coal Development. In all the Bergius-derived processes, coal particles are first slurried with recycled liquid product prior to the initial pyrolytic breakup of the organic coal matrix. The slurry then is supplied with either gaseous or donor-furnished hydrogen (H_2), thereby producing heavy, highly aromatic liquids containing unacceptable levels of sulfur and nitrogen. Respectively, these processes use tubular-flow, continuous-flow slurry-bed, and continuous-tubular-flow, trickle-bed-flow reactors. The typical catalyst for SRC can be coal mineral matter, fly ash, or pyrite; the H-Coal and EDS catalysts, $CoO-MoO_3/Al_2O_3$, are expensive and difficult to recover. Typical liquefaction conditions are 450°C and 120–200 atm (12-20 MPa). The SRC scheme requires no added catalyst; the H-Coal process is noted for rapid catalyst aging; and EDS uses a separate reactor for selective hydroprocessing with relatively slow catalyst deterioration.

SRC products. The first SRC product, SRC-I, a solid melting near 350°F (177°C) and containing about 0.2% ash and 0.8% sulfur, performed successfully in a 3000-ton (2700-metric ton) test in a utility boiler. With the increased use of H_2 in the process, SRC-II can be produced. This is a product reported to be pumpable near 100°F (38°C) and containing much smaller amounts of ash and sulfur than found in SRC-I. The development of a process variation termed two-stage liquefaction (TSL) has been proposed that would use SRC-I as a building block for petrochemicals. Full upgrading would occur in an expanded bed of catalyst operating isothermally at 750–850°F (400–455°C) and with gaseous H_2 addition. The products contain less sulfur and nitrogen than those produced at comparable costs by one-stage liquefaction processes. Another variation would fractionate SRC-I into light and heavy oils and gas. The solid bottoms would be passed through a furnace and a drum for producing green coke, which later would be calcined to make anode-grade material.

H-Coal. H-Coal liquids have been made only by bench-scale systems and process development units. In terms of feedstock flexibility, the H-Coal process is rated higher than the EDS process, and much higher than SRC-II. A pilot plant of 250–600 tons per day (tpd) of coal capacity (225–550 metric tons per day) in Catlettsberg, KY, is designed to produce 750–1800 bpd (119–286 m^3 per day) of liquids. At severe operating conditions a large amount of gasoline and distillate will be produced; at low severity the primary product will be a replacement for No. 6 fuel oil.

EDS. This process, to be tested in a 250-tpd (225 metric tons per day) coal capacity pilot plant in Baytown, TX, differs from both SRC-II and H-Coal in that it uses a catalyst indirectly. In practice, a catalyst, a solvent, and about half the process H_2 are mixed in a pressure vessel. The mixture is slurried with coal that enters a process reactor together with more H_2. The solvent mixture gives up its H_2 in this process reactor, whereby 2–3 barrels (bbl; 0.3–0.5 m^3) of liquids are produced from 1 ton (0.9 metric ton) of coal. The pilot plant, provided with commercial components so far as practical. is expected to acelerate proof of the technical feasibility and to provide useful engineering data.

Molten zinc chloride. This is a process for liquefying coal to gasoline in a single stage. A 1-tpd (0.9 metric ton per day) unit has started operation to confirm cost estimates based on a continuous bench-scale unit.

Indirect liquefaction. The chief processes for indirect liquefaction of coal are the Sasol processes and the Mobil-M process.

Sasol-I. The only commercial coal liquefaction plant in the world (and the only one using the indirect liquefaction route) is the 9000-bpd (1430 m^3 per day) plant of the South African Coal, Oil & Gas Company (Sasol) in the Republic of South Africa. The plant, called Sasol-I, started operations in 1955, based in part on German techniques used during World War II to produce aviation and other fuels.

Sasol-I had some major problems when it first began operating. The plant had two Fischer-Tropsch reactor systems in parallel. One, a fixed-bed system developed by Lurgi in West Germany, produced higher-boiling hydrocarbons, including a range of solid waxes. This system has worked well since it was put into operation. *See* FISCHER-TROPSCH PROCESS.

The difficulties were with the other reactor system—the so-called Synthol fluid-bed reactors intended to produce lighter hydrocarbons concentrated in the gasoline and diesel oil ranges. They were designed to perform the Fischer-Tropsch synthesis. When first started, they did not work. Much of the problem was with the catalyst, which was based on certain iron ores that varied in composition. Initially the composition of the product also varied, and only uneconomically short production runs were attained. With the development of an entirely new catalyst, also based on iron, and further research, the initial two Synthol reactors were put into reliable commercial production. A third reactor was added later. These Synthol reactors are now considered completely reliable.

There are a number of steps in the Sasol process. Crude syngas is produced in Lurgi, high-pressure, steam-oxygen gasifiers operating at 350–450 psi (35–45 M Pa) to produce a gas mixture consisting mainly of carbon monoxide (CO) and H_2. Tar and oil are then removed by water-quenching of the hot, crude syngas. The crude gas is purified by a low-temperature Rectisol process to remove the last traces of tar and oil, 98% of the carbon dioxide (CO_2), and all of the hydrogen sulfide (H_2S), organic sulfur, ammonia, phenolics, hydrocarbons, and entrained solids. The purified gases are then subjected to the water-gas shift reaction to produce mixtures, averaging about 84% H_2 plus CO, and 13.5% methane (CH_4), that are contacted in the fixed-bed or fluid-bed reactors with highly pyrophoric iron catalysts whose activity is enhanced by a small amount of certain metals. By varying the H_2/CO ratios and using the temperature and pressure best suited to the fixed-bed reactor, the mix of its products can be made to include straight-chain, high-boiling hydrocarbons, some medium-boiling and diesel oils, liquefied petroleum gas (LPG), organic acids, and oxygenated alcohols. The products of the Synthol fluid-bed reactors are mainly C_1-C_4 hydrocarbons, and gasoline containing a substantial content of oxygenated products and aromatics.

Sasol synfuels are generally considered the most expensive high-tonnage hydrocarbons in the world, and the Sasol process can make economic headway only in an area in which there is a large differential between the cost of indigenously produced coal and the price of imported oil.

Sasol-II and Sasol-III. Sasol-II, at Secunda, east of Johannesburg, went into operation in the early part of 1980; a site has been prepared for Sasol-III. Both Sasol-II and Sasol-III are modified and enlarged versions of Sasol-I. Each is designed to have the equivalent output of a 60,000-bpd (9500 m³ per day) refinery. Daily output of transportation fuels—gasoline, diesel oil, and jet fuel—from each plant is to be about 40,000 bbl (6400 m³). The annual output of other products from Sasol-II will include 400 × 10⁶ lb (180 × 10⁶ kg) of ethylene, about 190 × 10⁶ lb (85 × 10⁶ kg) of other chemicals, nearly 400 × 10⁶ lb (180 × 10⁶ kg) of tar products, 110,000 tons (100,000 metric tons) of nitrogen [as ammonia (NH_3)], and 100,000 tons (90,000 metric tons) of sulfur. In Sasol-III the production of nonfuel products will be minimized. Each new plant will have a total of seven Synthol reactors.

Of the 14×10^6 ton (12.7×10^6 metric ton) annual coal consumption, 9.2×10^6 tons (8.3×10^6 metric tons) will be gasified, and 4.8×10^6 tons (4.4×10^6 metric tons)—too small for use in the gasification stage—will generate steam and one-third of the plant's electric power needs. The low-quality subbituminous coal for the Sasol-II and -III plants contains 10% water and 25% ash. Coal reserves are sufficient to supply both plants for at least 60 years.

Sasol-II has thirty-six 4-m-diameter Lurgi pressurized gasifiers in which coal, oxygen (O_2), and steam will react to produce a raw gas comprising 57% H_2 plus CO, 9.4% CH_4, 32.0% CO_2, 0.7% H_2S, 0.3% nitrogen (N_2), and 0.5% hydrocarbons. The methane reforming stage demands 3700 tpd (3400 metric tons per day) O_2, and 9500 tons (8600 metric tons) will be used in the Lurgi gasifiers.

The Sasol-II plus Sasol-III complex is one of the largest industrial projects ever undertaken. Sasol-II and -III are about a decade ahead of any other way for making oil from coal. Comparably sized plants using theoretically more efficient, direct-liquefaction processes cannot be operating until the end of the 1980s at the earliest. Primarily because of this big head start, Sasol-type plants capable of making a maximum of 58,000 bpd (9200 m³ per day) of transportation fuels could have a future role in the United States.

Mobil-M. Another proposed indirect-liquefaction process is the Mobil-M process. In brief, CO and H_2 would be produced from coal in gasifiers operated at 500 psi (50 MPa) and 3000°F (1650°C). The process would provide more complete carbon conversion than other types, while producing practically no tars or hydrocarbon emissions. The purified syngas can be converted catalytically to crude methanol containing 17% water. The methanol solution is then passed over another catalyst, where it is converted almost quantitatively to gasoline and water. The selectivity of the catalyst is obtained by precisely varying the size of the catalyst pores.

Notable improvements claimed for the Mobil-M process over the Sasol process are that the process is much more selective in making gasoline, and its product has a high octane number (research octane number = 93–96). This high octane number is attained because the Mobil process gasoline has a high aromatics content, typically 40%. *See* OCTANE NUMBER.

Production constraints. All direct-liquefaction projects, regardless of the specific process, have encountered combinations of inherent production constraints: costly alloys are indicated in many construction areas; the coal feed must be dry and essentially less than 200-mesh particle size; exchangers, feeders, slurry pumps, filters, valves, piping, and pressure-letdown devices are particularly subject to erosion, corrosion, or both; chemical and physical changes occur in the product because of aging or various thermal histories; and product yields are quire sensitive to process variables such as temperature, pressure, slurry composition, and loss of catalyst activity.

The separation of unreacted coal and ash from the hydrogenated slurry is a major problem common to all liquefaction schemes. Hydroclones remove only about two-thirds of the solids, centrifuges are too expensive, and magnetic separation results are not good. Filtration is less effective than vacuum distillation. Experimental proprietary solvent-precipitation and gravity settling techniques were installed in an SRC pilot plant in 1979.

Prospects and issues. The establishment of a commercially competitive coal liquefaction industry in the near future is unlikely. Studies indicate that, for liquefaction to be attractive in the United States, the world price of oil must increase two or three times in real terms.

Based on acquisitions beginning in 1964, 11 oil companies own 25% of all coal in the United States. Further, 26 oil companies in 1978 accounted for 143×10^6 short tons (130×10^6 metric tons) of the 656×10^6 ton (595×10^6 metric ton) production in the United States. Because of its current ownership of 55.3×10^9 tons (50×10^9 metric tons) of recoverable reserves, the oil industry can exert much pressure on future coal industry developments. Most of the major American oil companies have significant coal expansion programs in progress. However, if the petroleum industry were divested of its coal holdings, its willingness to commit itself to an aggressive plan for coal liquids commercialization would be reduced.

Fuel suppliers (primarily oil companies) show more interest in markets for gasoline substitutes than markets for electric utility fuels. The capital costs per unit of capacity of gasoline from coal liquefaction plants will be higher, and it is likely that the production cost for coal liquids will be greater than for petroleum products (even those refined from offshore or shale oil) until the 1990s.

An often overlooked aspect of coal liquefaction is that 50–60% of the raw coal liquids produced are basically utility-grade fuel oils. Although somewhat different in composition from No. 5 fuel oil or residual oil, these synthetics should have a potential use for direct substitution under existing electric utility boilers. With minimal additional refining, they can be made compatible for use in combustion turbines. Thus, the primary and major fuels produced by coal liquefaction can be "natural" utility fuels.

Electric utilities are likely to be the first customers for a substantial portion of coal liquids production, and they are the key to the early commercialization of liquefaction processes. It has not been decided whether utilities should own or partially finance the construction of demonstration and pioneer plants, or whether they should guarantee a floor for coal liquids prices. It has also not been determined if the utilities can show that coal liquids are truly economical for midrange duty and thereby obtain regulatory commission approval to pay a premium price for them.

Reports in 1979 by Resources for the Future and the Harvard Business School showed that the proposed synfuels crash program involves: serious technical weaknesses; the inefficient use of funds; hazards to the environment; and a widespread distrust of the synfuel options as promises are not met. The researchers warned against opting for what seems the lowest-cost technology on the basis of only laboratory and pilot plant tests.

Hydrogen for liquefaction processes. The technologies for producing H_2 (electrolysis of water, iron-steam reformation, or reforming natural gas) are not economical or practical means to supply the enormous volumes of H_2 needed for commercial-size liquefaction processes. To increase the 0.7–0.9 ratio of atomic H:C in coal to the approximate 1.6 ratio of crude petroleum or its refined products requires 15,000–20,000 SCF (standard cubic feet) per ton (47–62 m^3 per metric ton, under standard conditions) of coal feed. Plant designers consider the major future source of H_2 will come from the energy-intensive catalytic oxidation, in the presence of steam and O_2, of carbon sources such as coal, unreacted coal plus ash, or vacuum residuums in pressure gasifiers.

The gasification step would preferably be done in a suspension-type bed at 3000°F (1650°C) and 13.6 atm (1.38 MPa) pressure to produce a syngas containing principally H_2 and CO. Following removal of H_2S and CO_2, most of the purified syngas, following shift conversion and CO_2 removal, is added to the coal-solvent slurry fed to the hydrogenation reactor, while the remainder provides H_2 for the hydrogenation of light distillates. Proved gasification processes, used for many years in commercial plants all outside the United States, are the Lurgi or Winkler processes, or the Koppers-Totzek scheme that is preferred because of its higher yield of H_2. The energy conversion efficiency of these gasification processes is 50% or less, and the cost of the energy equivalent of the H_2 produced is about two to three times the cost per unit energy of the carbonaceous feed to the gasifier.

Modifications of Lurgi coal gasification practices so as to produce H_2 only is predicted by 1990, when the economics should be favorable. Meanwhile, heavy emphasis is directed to the development and commercial demonstration of this concept. Three gasification schemes that may be in active competition in the early 1990s are: the Lurgi dry-bottom system that is now commercial; the British Gas/Lurgi slagging gasifier that is expected to be in the demonstration stage; and finally one or more pressurized, entrained-bed gasifiers will probably have been proved commercially.

All of the Lurgi variations pass purified raw syngas through a shift converter to reduce the CO content to 8%, and to convert sulfur to H_2S, which is removed together with most of the CO_2 in a Rectisol system. To produce 100×10^6 mm scf per day (2.83×10^6 m^3 per day) H_2 in the modified Lurgi plants would require 3150 tpd (2850 metric tons per day) of subbituminous coal having a heat value of 9000 Btu/lb (20.9×10^6 J/kg) as received or 13,000 Btu/lb (31.4×10^6 J/kg) on a dry and ash-free basis. The energy conversion efficiency of such a modified Lurgi plant is claimed to exceed 55%.

Exxon and the Danish firm Haldor Topsoe in 1976 developed a new shift catalyst for converting syngas to H_2. The low-temperature catalyst is promoted by sulfur, rather than poisoned as are most catalysts. The new catalyst is expected to reduce the production cost of H_2 from coal-derived syngas.

Costs. The production costs of several processes for making coal-derived liquid fuels were issued in a 1979 report by the Engineering Societies' Commission on Energy (ESCOE). The report concluded that the costs to produce 10^6 Btu (10^9 J) of energy are competitive with 1 bbl of imported oil, which yields 5.8×10^6 mm Btu (the equivalent of 38.5×10^9 J/m^3). However, companies involved in the design of coal liquefaction plants in the United States have estimated the production cost of liquid synfuels as being considerably higher (25–50%) than the cost of imported oil. Some optimists contend that as petroleum and gas supplies diminish and their market values increase, coal-derived fuels could become price-competitive in the 1980–1990 period. In real life, theoretical economic forecasts are impractical; cost estimates and forecasts are at best only approximate even when based on unbiased data accrued by the operation of a commercial-size plant during a several-year period, in which period inflation and government pressures are not major factors.

Cost estimating for process and plant construction is practically useless when based on laboratory or pilot plant data, both subject to various changes. There is a wide cost breach in the interim between the inception of a process and its embodiment as a commercial-size plant. The realization of this situation probably prompted the National Coal Association in 1979 to suggest that the government make firm, long-term commitments to encourage the private sector to build commercial-scale plants for the liquefaction of coal. Such commitments would include tax incentives, completion guarantees, loans, loan guarantees, rolled-in pricing, an all-events tariff, and direct government funding or cost sharing.

Modifications. Several modifications of second-generation direct-liquefaction processes have been developed.

1. A process (U.S. Patent 4,155,832) has been developed for the contacting of coal with an organic solvent containing both H_2 and a nickel-Ziegler catalyst in solution. The solution when mixed with gaseous H_2 forms a complex that reacts with coal more efficiently than does molecular H_2.

2. The Brookhaven National Laboratory demonstrated that the flash pyrolysis of coal, at 60% effective carbon conversion and 76–100% effective energy efficiency, principally yielded C_6H_6, aromatic oils, CH_4, and C_2H_6.

3. Work by the Atlantic Richfield Company indicated that fluid catalytic cracking (FCC) may potentially be the most important of the three main cracking processes used to upgrade syncrudes from coal.

4. The Kerr-McGee Corporation developed a process, intended for licensing, to separate mineral matter and unreacted coal from coal liquids. The process, termed critical solvent deashing (CSD), was successfully used to produce 130 metric tons of 0.09% ash product from vacuum still bottoms. The process employs solvents at or near their critical pressures.

5. C-E Lummus developed an antisolvent deashing process that is completely free of the mechanical complexities of other solid-liquid separation techniques, and that offers potential operational and economic advantages over other filtration techniques. The process was commissioned at Gulf Corporation's Fort Lewis, WA, SRC pilot plant in 1979.

6. The Great Britain National Coal Board's (NCB) Gas Extraction of Coal process contacts coal with compressed solvent gas maintained above its critical temperature. When brought to the supercritical gas-phase state, toluene, *ortho*-xylene, dodecane, *para*-cresol, and so on, easily penetrate the pore structure of the coal and boost by as much as 10,000 times the volatility of those liquid coal components that are formed when coal is heated above 750°F (400°C) in the absence of air. The volatilized coal extract is recovered by depressuring the supercritical gas phase, thereby causing the coal extract to condense from it. The solvent gas is recycled to treat another batch of coal.

In investigations by NCB's Coal Research Establishment (CRE), 5–10 kg/hr of high-volatile bituminous coal have been continuously treated to yield, up to half of the coal weight, a low-molecular-weight liquid together with 1–2% by weight of water and gas typical of pyrolysis (thermal depolymerization) at temperatures of 350–450°C and atmospheric pressure. For comparison of liquid yields, less than 10% by weight of the coal is recovered as liquid during normal pyrolysis, without solvent and at ambient pressure. The residue remaining after extraction of the coal liquids is a porous solid, with size distribution similar to that of the original coal, and containing all the mineral matter associated with the coal. The low-molecular-weight material so recovered from coal is referred to as coal extract. At ambient temperature, it has the appearance of a glassy solid with a softening temperature of about 90°C. The material has a generally open-chain polynuclear aromatic structure linked by ether and methylene groups. It appears that the extract is in a structural state substantially unaltered from that in which it existed within the coal framework. Laboratory-scale studies have demonstrated that conventional catalytic hydrogenation techniques can be employed to convert the coal extract to a distillable oil. Fractions of this oil can be used as chemical feedstocks or refined to premium liquid fuels.

Several factors affect the choice of the gas solvent so that a high solvent performance is attained when operating at a low, reduced temperature. Therefore, the chosen gases should have critical temperatures in the 300–400°C range. Aromatic (ring) compounds are more effective than aliphatic (straight-chain) compounds, presumably because aromatic compounds are compatible with the aromatics present in coal liquids. Other factors affecting the choice of gas are stability at the reaction temperature and availability. Typical solvents employed include common compounds derived from coal tar that are liquids at ambient temperature.

Developing third-generation processes. Research on the direct liquefaction of coal or lignite has continued to be very active both in the United States and Europe. In Europe, development of a modification of the I. G. Farben liquefaction technology (designated as BOTTROP) has been undertaken, and construction of a 200-tpd (181 metric ton per day) pilot plant has been started near Essen, Germany. The IG process features the use of a low-cost, disposable solid catalyst and a pressure of 300 atm (4410 psi or 30.4 MPa) instead of the 700 atm (10,300 psi or 70.9 MPa) used in German plants during World War II. *See* COAL; COAL GASIFICATION; SYNTHETIC FUEL.

[FRANKLIN D. COOPER]

Bibliography: L. J. Carter, Synfuels crash program viewed as risky. *Science*, 205(4410):978–979, 1979; M. Heylin, South Africa commits to oil-from-coal process, *Chem Eng. News*, 57(38):13–16, 1979; R. R. Maddocks, J. Gibson, and D. F. Williams, Supercritical extraction of coal, *Chem. Eng. Prog.*, 75(6):49–55, 1979; T. J. Pollaert, Hydrogen from coal, *Chem. Eng. Prog.*, 74(8):95–98, 1978; R. G. Schweiger, Burning tomorrow's fuels, *Power*, 123(2):S1–S24, 1979; W. Worthy, Synfuels: Uncertain and costly fuel option, *Chem, Eng. News*, 57(35):20–26, 28, 1979.

Coal mining

The technical and mechanical job or removing coal from the earth and preparing it for market. Coal, the most abundant and traditionally the most economical source of power in the world, is found in varying amounts throughout the globe. It lies in veins of various thickness and richness beneath the crust of the Earth. The product of fossilized plant material mixed with various mineral matter, coal rests in giant subterranean sandwiches, shallow or deep, flat or pitched. At the present time, there are coal mines in the United States operating at the 1500-ft level in Virginia and shafts under development at the 1800-ft level in Alabama. Obtaining coal in sufficient quantities at a competitive expense and making it the proper grade for the market demand is, in its simplest terms, the science of coal mining. Mammoth mining machines require technically qualified manpower. The pick-and-shovel miner has all but disappeared from the American scene, along with the stereotyped characterization of him as an oppressed laborer living in a company town. Today's miner is a skilled, well-paid technician who handles complex, costly, and highly efficient machines.

Prospecting and planning. Of major importance in coal property development is the accumulation of seam information from borehole drilling. Chemical analysis of the cores will provide details concerning moisture, volatile material, fixed carbon, ash, sulfur, Btu per pound, and fusion temperature of the ash. Washability data will give the percentage of float (coal) and sink (foreign material) for each size coal at each specific gravity used (generally from 1.35 to 1.60), together with the amounts of ash and sulfur and the Btu value on a dry basis for each increment of specific gravity. *See* BORING AND DRILLING (MINERAL).

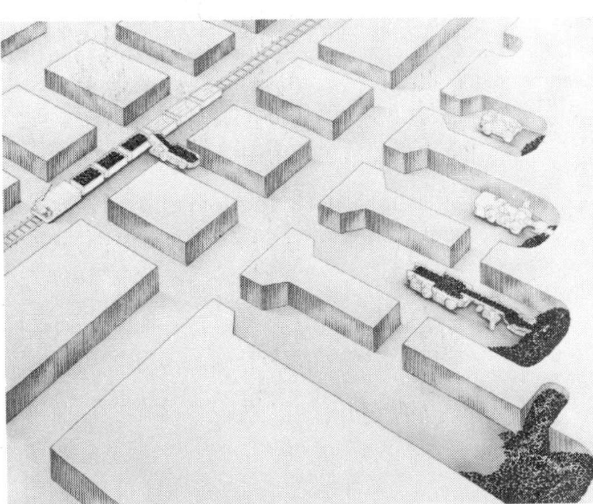

Fig. 1. Conventional coal mining system, for underground mining, shuttle car and train haulage. (*Joy Manufacturing Co.*)

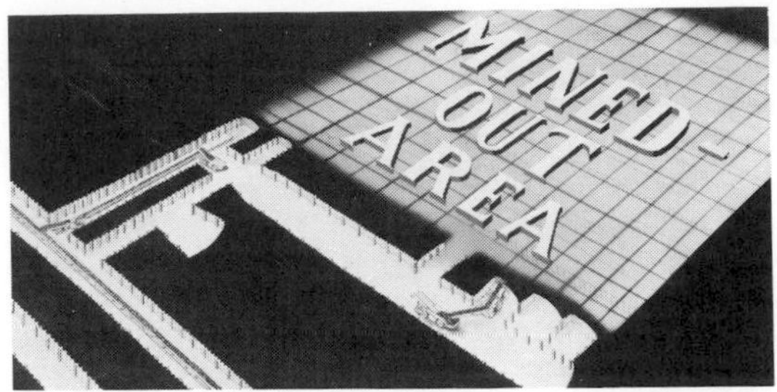

Fig. 2. Coal mining system with continuous miner using shuttle car and belt haulage. (*Joy Manufacturing Co.*)

After preliminary analyses and correlations, the engineering details are planned. Production requirements and the extent of coal reserves generally determine the method of mining and the type and capacity of equipment to be used. Amortization of the investment and an adequate return on the investment are also influential factors. The cost of opening, developing, and operating the mine is estimated. A reasonable evaluation can then be based on the projected costs and markets available.

Development is commonly planned for the life of the property. This involves projecting and estimating the working sequence of the various parts of the area.

The type of mining equipment to be used; transportation to be required; the water drainage; ventilation and roof control (underground); overburden analysis (surface) are the other items that must be projected and estimated. The interrelations of these factors with costs and quality of the coal are a necessary part of the planning.

Coal is extracted from the earth by three basic methods: (1) underground, (2) strip, and (3) auger. Approximately two-thirds of America's coal comes from underground mining. The remaining one-third is mined from the surface, either by strip or auger mining.

Underground mining. The systems of underground coal mining generally in use are room-and-pillar, longwall, and, in a few places in Europe, hydraulic. In the room-and-pillar system, tunnels are carved into the seam, leaving pillars of coal for support. In some mines these pillars are removed in subsequent mining, allowing the overlying strata to collapse; in others the pillars are not recovered. In longwall mining, widely spaced tunnels are driven, leaving large blocks of coal. Later, these blocks are completely extracted, allowing the roof material to collapse behind the coal face as it is removed. In hydraulic mining, as now practiced, a stream of water is directed against the coal face with sufficient pressure to dislodge the coal. The water also acts as a transporting medium.

The room-and-pillar system of mining has been the most widely practiced method in the United States, whereas the longwall system has been used for many years in Europe. However, the demand for metallurgical coal has necessitated mining deeper into the earth. With the advent of mechanical plows and shearers, it has become economically feasible to use the longwall system in the United States.

Underground mining equipment. The type of equipment used at the mining face is governed by a complex of factors. Outstanding are the relative difficulties of supporting the immediate roof, the height of the seam, grades of coal, maintenance required on machinery, and productivity expected of manpower using different types of machines. Today, mechanical equipment falls into two classes, the so-called conventional and the continuous miner machine (Figs. 1 and 2).

In the conventional method several machines are used in a cycle of operation: undercutting or top-cutting the seam with cutting machines; blasting down the face; loading coal with a mechanical loader; and transporting it from the face by shuttle car or conveyor. The continuous miner (Fig. 3) has been used in many seams. It bores or rips to dislodge the coal from the face without blasting, then loads the coal and puts it into the transportation system.

The longwall mining machine (Fig. 4) employs a plow or shearer which is pulled back and forth across a working face several hundred feet long. The loosened coal is dropped into a conveyor. Self-advancing hydraulic jacks support the roof and follow the machine as it slices into the coal on a wide front. With longwall mining in overburdens of 1000 ft or more, it is possible to mine extremely gassy seams. Because of the large tonnage being extracted from a single face, large volumes of air are employed to dilute the liberated gases. The massive jacks used in longwalling control the weight of the roof and also create the falls necessary to remove the pressure at the coal face, thereby providing protection from roof falls.

The power source for operating mining machinery can be liquid fuels, seldom used in coal mining, or electricity, either alternating or direct current. Power distribution for a mining property has to be planned for a whole mine in the same manner as the actual mining operation.

Prospecting information and actual experience determine the amount and type of roof control necessary for the protection of workmen. The old type of roof support is so-called timber; this may be simply wooden posts, or may consist of beams of steel or wood supported by posts or fastened to the walls. Currently, the use of roof bolts has replaced a great amount of timber support. Holes are drilled into the roof, and steel rods, held by expansion devices, are screwed in until the rod head is tight against the surface. The size, spacing, and length of the bolts are governed by the roof conditions to be controlled. Roof bolting combines many roof materials into one large beam instead of letting the various strata act separately. For permanent support at some mines the roof is gunited, particularly when the roof materials are subject to deterioration by weathering if exposed to the air currents.

Mined coal is transported from the working face to the main transportation system by shuttle cars (underground trucks) (Fig. 5) or by conveyor. Shuttle cars are used predominantly, although many properties with steep grades or thin coal find it more economical to employ conveyors. In hydraulic mining in Europe, the water used to break down the coal also conveys the coal away from the face.

Fig. 3. Continuous miner. (*Lee-Norse Co.*)

The main transportation to the surface or to shafts or slopes is generally provided by mine cars which are pulled along tracks by locomotives. Although not as common, there are many instances where coal is transported on conveyor belts. Main-line belt transportation eliminates a great deal of grading required for track haulage but requires essentially the same amount of maintenance and roof protection.

The largest cars that seam height and width of working areas will permit are used in track systems. Modern mine cars hold 6–20 tons of material. Most have solid bodies that are emptied by tipping or rotary dumping. Motive power is generally provided by electric locomotives of 20-tons weight.

When the coal does not have a surface outcrop, it must be elevated by a slope or shaft. Almost all slopes are on inclinations that permit the use of belts to transport the coal, which is dumped at the bottom of the slope, to the surface. These belts are installed with sufficient capacity to carry the maximum production of the mine on a daily or shift basis. When the coal is brought to a shaft, it is dumped into skip buckets and hoisted to a dump at the surface. Modern installation in these shafts are completely automatic as to loading and dumping.

An underground coal mine consists of a great number of spaces and openings in which people must work safely. In many cases, however, explosive gas is emitted. Hence, artificial ventilation over the entire mine is required to maintain a normal atmosphere and to dilute and carry away such gases. Numerous shafts and fans provide the necessary volume of air.

To prevent water from entering and to eliminate it from the mine are the aims of mine drainage. Removing excess water from mine properties may be difficult and expensive; in extreme cases over 30 tons of water have to be removed for each ton of coal mined. *See* MINING MACHINERY; UNDERGROUND MINING.

Strip mining. Where the coal seam lies close to the surface, it is more economical to remove the overburden of earth and rock that covers the coal

COAL MINING

Fig. 4. Longwall mining equipment. Plow works back and forth across working face. (*Mining Progress, Inc.*)

Fig. 5. Shuttle car hauling from working face area in a coal mine to the main transportation system. (*National Mine Service Co.*)

seam. For this job, power shovels, draglines, or wheel excavators are used. As a farmer plows his field in furrows, a shovel excavates a "furrow" and casts the overburden parallel to the cut. Draglines sit on the bank above the coal seam and remove the overburden from the seam. Bucket-wheel machines excavate the material from above the coal seam. It then flows in a continuous stream via a transfer to the conveyor system, which in turn transports it to the discharge point. When the seam has been exposed, the coal is loaded by smaller power shovels into trucks and then hauled to preparation plants or loading bins. *See* STRIP MINING.

COAL MINING

Fig. 6. Sketch of 200-yd^3 (168-m^3) dragline. (*Bucyrus-Erie Co.*)

Strip mining is an efficient way to mine coal. Seventy-foot-thick seams of coal in Wyoming are now being strip-mined; seams such as these will furnish the United States with a substitute for natural gas. Gigantic machines have been developed to efficiently mine coal by the strip-mining method. The biggest shovel in 1968 weighed 27,000,000 lb (12.2×10^6 kg), towering 220 ft (67 m) above the coal seams in which it worked, moving 180 yd^3 (138 m^3) of earth every 50 sec. A dragline is being erected which will take as much as 220 yd^3 (168 m^3) of earth per bite (Fig. 6); it will employ a 310-ft (94-m) boom. Off-the-highway haulage trucks in the 100-ton (90-metric-ton) category are common; units capable of carrying 240 tons (218 metric tons) are available. Auxiliary equipment which is used to complement the excavating machines has increased in capacity in a corresponding manner.

Augering. For coal seams which continue under rising land too thick for economical strip mining and where underground mining cannot burrow further to the surface because of the shallow and more treacherous roof conditions, a relatively new procedure of mining, the auger method, was developed in 1951. The auger miner (Fig. 7) twists huge drills like carpenter's bits into a hillside coal seam, drawing out the coal to a conveyor which loads it into trucks. Section by section, the drills bore into the hillside. Augering in general will recover 40–60% of the coal seam. The development of the dual-headed auger and the multiheaded auger, with progressive increase in size until the giants of today have been manufactured with head diameters up to 96 in. (2.44 m), has made it possible to penetrate to a depth of 300 ft (90 m) in a level seam. Auger recovery gives additional tonnage at minimal cost for equipment and labor; moreover, it permits recovery of coal that might not be recovered by other methods.

Cleaning, grading, and shipping. After the coal is mined, it may be loaded directly into transportation facilities for the market if the foreign material in it is not excessive. However, coal from most seams requires preparation to provide a desired and uniform quality.

A number of washing devices are used for cleaning coal, all of which operate on the basic principle of the difference of specific gravity between coal and foreign material. The coal is floated in a vessel and the foreign material, being heavier, drops to the bottom. Washed coal in sizes less than 3/8 in. (9.5 mm) carries sufficient water from the washing circuit so that it must be dewatered to be acceptable on the market. Where a market can accept a moisture content of 7–8%, mechanical devices can meet the requirements. If a lower moisture is required, the product is thermally dried of surface moisture to the required extent after being mechanically dewatered. Once the coal is ready for market, it is loaded for shipment to the customer

Fig. 7. Single-head auger. (*Salem Tool Co.*)

by railroad, water transportation, or in some cases belt conveyor systems.

The 108-mi (174-km) coal slurry pipe line between Cadiz and Cleveland, both in Ohio, pioneered a new concept in coal transportation. A 273-mi (439-km) coal slurry pipe line is also in operation between Arizona and Nevada. This innovation in coal transportation has proved to be a more economical method than building railroads.

One innovation in the industry has been the adoption of the unit train concept for coal shipments. A unit train, in its true sense, is a complete train of cars (usually privately owned) with assigned locomotives. It operates only on a regularly scheduled cycle movement between a single origin and a single destination each trip.

Future trends. In view of improvements in the past, the following future developments can be expected: (1) Conversion of coal into oil. Present coal reserves in the United States are known to be 3.2×10^{12} tons (2.9×10^{12} metric tons). With 1 ton of coal converting into approximately 4.5 bbl (1 metric ton into 0.79 m^3) of oil, present reserves will yield more than 14×10^{12} bbl (2.2×10^{12} m^3), which is more oil than presently known to exist in the world. (2) Conversion of coal into gas to be placed in the underground storage fields and delivered into the major pipelines which intersect the coal fields. (3) Further research into the possibilities of using coal tars and shales as a source for synthetic oils. *See* COAL GASIFICATION.

[JAMES D. REILLY]

Bibliography: I. A. Given, *Mechanical Loading of Coal Underground*, 1943; E. S. Moore, *Coal*, 2d ed., 1940; National Coal Association, *Yearbook*, 1968; P. Pfleider, *Surface Mining*, 1968.

Cogeneration

Cogeneration, an old technology with a new name, began in 1977–1978 to challenge the underlying concepts of appropriate power generation. Defined as the simultaneous generation of electric energy and useful low-grade heat from the same source, cogeneration is reemerging as a viable alternative to central generation of electric power. This 60-year-old technology of combined heat and electric power generation was renamed on Apr. 19, 1977, when President Jimmy Carter, in unveiling the National Energy Plan, introduced cogeneration to the electric power generation lexicon.

This article explores several features of cogeneration, including on-site and mass-produced plants, district heating, energy efficiencies, past economic deterrents, other remaining obstacles, and the challenge to the underlying concepts of electric generation.

To cogenerate, or recapture and use the low-grade heat remaining after generating electric power, two approaches are possible: moving electrical generation to the user's facility, and moving the waste heat to the user.

Cogeneration at user's site. This first approach has been commonly practiced with large industries, but has declined as electric rates fell. In 1950, 15% of United States electricity was cogenerated in large on-site industrial plants, but the percentage fell to 4% by 1970. These plants typically have 15–150 MW of capacity and make high-pressure steam, creating electricity via a back-pressure turbine which drives an electric alternator, and then feeds the 1.03–1.38-megapascal (150–200-psig) exhaust steam to plant processes—using the industrial plant as a condenser.

Mass-produced plants. Since 1960, roughly 600 on-site cogenerating plants using diesels or combustion engines have been constructed in the 200-kW to 6-MW range, using prime movers or engines that were mass-produced for mobile power markets. To these prime movers—diesels, natural gas engines, and gas turbines—are attached alternators to generate electricity, heat recovery devices, and suitable controls (Fig. 1).

Any two of the diesel generator sets in Fig. 1 can meet peak facility electrical load using 11,000–11,600 kilojoules per kilowatt-hour (10,500–11,000 Btu/kWh) generated. A control panel monitors key parameters and automatically matches input fuel to facility electric loads, holding frequency constant. The internal logic automatically shuts down engines which are unneeded or malfunctioning, starts up engines, and electrically parallels generators as loads rise.

Cooling water circulates through the engine block, gaining roughly 2530 kJ (2400 Btu) of power per kilowatt-hour of electricity produced. Heated block water then passes through exhaust recovery boilers, capturing heat from the 480–530°C exhaust gases.

The heated engine coolant passes through heat exchangers, and warms facility-process or space-heating water. Any remaining heat is removed in a cooling tower before the coolant is pumped back to the engine circuit. A variation in plumbing produces up to 1.03-MPa (150-psig) steam from the exhaust gases or low-pressure steam from the engine and exhaust.

When the prime mover is a turbine, higher-input fuel per kilowatt-hour is required, but all waste heat is in exhaust gases, allowing recapture of high temperature hot water or steam up to 200°C.

Cogeneration by district heating. District heating plants that centrally generate electricity and simultaneously supply steam or hot water via a network of underground pipes represent the other approach to cogeneration. Such plants have operated in the United States since 1877, typically in dense center-city areas. About 250 utilities operate district heating plants in the United States today, over 100 cogeneration plants feed district heating systems in Germany, and over 30% of all power generated in the Soviet Union comes from district heating cogeneration plants. The older systems supplying 1.03-MPa (150-psig) steam with no condensate return span a maximum radius of 3.5 mi (5.7 km). The efficiency of these older systems is hampered by the relatively low thermal efficiency of electric power generation, escaping steam vapor, losses to surrounding earth, and loss of condensate. The present United States systems nearly all suffer from age and are generally considered unprofitable. The inherent energy inefficiencies and the cost of maintaining older systems have recently combined to make the cost of central steam greater than the cost of noncogenerated steam from on-site boilers. This form of cogeneration has thus been losing market penetration and declining in both absolute and relative terms.

Newer district heating systems use high-temper-

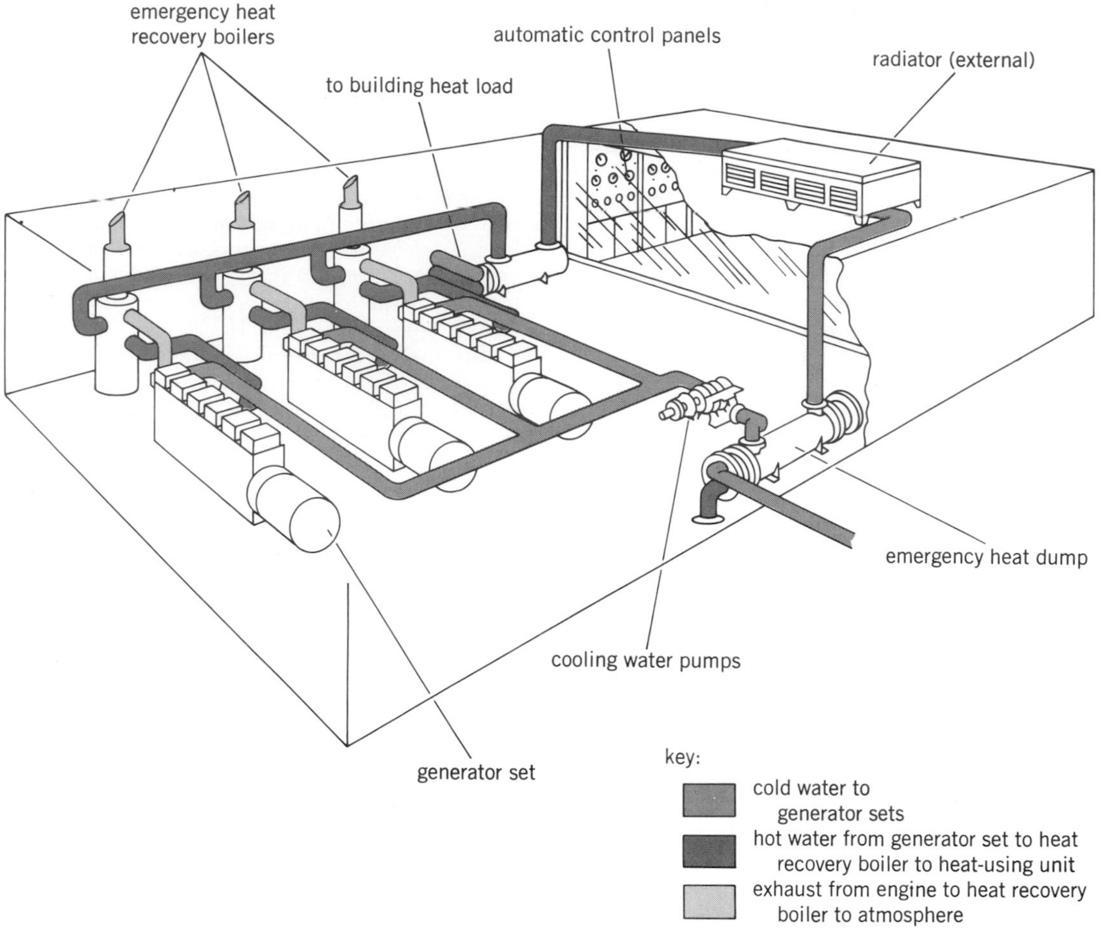

Fig. 1. Typical three-diesel-engine cogeneration plant.

ature water, return cooled water to the generating stations, and can operate at 65–70% overall efficiency. Their use is economically attractive, for example, on campuses where single owners control all land. However, given high pipe costs and many legal and physical barriers to installing new systems in builtup areas, district heating as a form of cogeneration faces difficult expansion prospect.

Fuel efficiencies of generating technology. Current central generation of electricity without heat recapture delivers on average only 29% of the input heat energy to the end user. By contrast, cogeneration has theoretical efficiencies of up to 75–80% and commonly delivers 60–65% of input energy. Some currently utilized energy converters and their efficiencies are given in the table. In each case, the waste energy can be utilized for low-temperature-process heat, space heating, or cooling with absorption chillers. However, low-grade by-product energy does not transport over long distances efficiently and requires large expenditures in underground insulated piping if the heat user is any distance from the generation site.

Past economic deterrents. Oil-fired cogeneration, replacing conventional oil-fired systems of central generation and on-site boilers, saves on average 111–127 m³ (700–800 barrels) of oil per MM (million) kWh, and can save 160 m³ (1000 bar-

Relative fuel efficiencies of various generating technologies

Type of energy converter	Gross kJ/kWh (Btu/kWh)	Recoverable kJ/kWh (Btu/kWh)	Net electrical heat rate, kJ/kWh* (Btu/kWh)	Net electrical heat rate as a percent of central stations
Central fossil boiler, nondistrict heating	12,500 (11,800)	–	12,500 (11,800)	100
Central fossil boiler, district heating	13,200–17,950 (12,500–17,000)	6300–9000 (5950–8500)	4800–5950 (4550–5650)	39–48
Low-speed diesel (6–30 MW each)	8650–9080 (8200–8600)	1270–1310 (1200–1240)	6950–7350 (6600–6950)	56–59
High-speed diesel (0.2–1-MW capacity)	11,100–12,100 (10,500–11,500)	4650–4850 (4400–4600)	4900–5650 (4650–5350)	39–45
Combustion turbines (0.8–2.5 MW)	18,500–21,100 (17,500–20,000)	8500–11,100 (8000–10,500)	6300–7150 (5950–6750)	50–58

*Assumes a 75% boiler efficiency in producing equivalent recoverable heat with direct-heat fossil-fired boiler.

rels) of oil per MM kWh. Nonetheless, cogeneration declined in use between 1950 and 1970. Several reasons explain why larger-scale industrial cogeneration has declined and why cogeneration in smaller sizes is only now emerging.

Between 1950 and 1969, the average cost per kilowatt of central generating capacity fell in real terms, due to economies of scale and a stable design climate, while real energy prices fell by 42%. Since the late 1960s, new central generating stations have encountered new environmental requirements, inability to further increase plant size, litigation delays, and requirements for highly specific designs causing quintupling of their installed cost per kilowatt. Figure 2 shows the change in price of central generating capacity between 1967 and 1978. On-site cogeneration plants' average cost per kilowatt were about $275 in 1967, or 2.2 times the cost of new central capacity. By 1978, on-site plants at $600 per kilowatt of capacity were costing under 50% of the cost of new central capacity.

Equally significant to cogeneration's economic viability has been the trend toward increasing the value of low-grade heat. Between 1950 and 1972, real energy prices fell from an index of 100 to 58, lowering the value of recovered heat every year. In the past 5 years, the cost of heat energy has increased 5–18 times. In 1970, interstate pipeline gas sold for 10¢ per MM kJ (11¢ per MM Btu), and diesel oil at 10¢ per gallon ($26 per cubic meter) cost 68¢ per MM kJ (72¢ per MM Btu). By 1978, pipeline gas, when available, cost $1.89 per MM kJ ($2.00 per MM Btu), an increase of 1800%. Diesel oil has risen 500%. The impact has been to increase the recovered heat value of cogeneration.

Finally, prior to the mid 1970s, smaller cogeneration plants were largely uneconomical since they required daytime or even full-time operators, costing $40,000 to $175,000 per year. This effectively eliminated all but larger facilities from on-site cogeneration prior to 1975–1976.

Unattended cogeneration plants in the 200-kW to 6-MW range have been made possible by improvements in control technology and general increases in product reliability. One major manufacturer of diesel engines now offers unattended, automatically operated congeneration plants with guarantees of fuel efficiency and reliability. These new options remove a major hurdle for on-site cogeneration at smaller plants.

Other obstacles to cogeneration. Several problems which bar widespread moves to cogeneration at this time include shortage of experienced well-financed vendors of complete cogeneration systems, current central utility overcapacity, environmental concerns, and national fuel use policies.

System development. Cogeneration systems combine several distinctly different pieces of equipment, typically of diverse manufacture, which then must operate together, such as engines or boilers, generators, control equipment, heat exchangers, pumps, sensors. Many past attempts to cogenerate experienced extensive debugging and unplanned system startup problems. Vendors of complete cogeneration systems who have credible backing are only beginning to emerge, and their absence has slowed commericial development of cogeneration.

Central utility overcapacity. A second major problem is the current central utility overcapacity. Since 1973, the growth of peak electrical demand has been sharply curtailed. Central utilities have been attempting to slow completion of new capacity under construction, but in each year since 1973, have added more new capacity then required to cover that year's growth in peak loads. National generation capacity was 25% over peak demand at the end of 1973, and rose to 33% over peak in 1978. Reserve capacity of 20% is considered adequate for central system reliability, with the excess capacity thought to be unproductive use of capital. Given this large and growing overcapacity, managements have focused on delaying construction of planned capacity and have not been in the market for new small capacity, regardless of its energy efficiency. *See* ELECTRICAL UTILITY INDUSTRY.

Rate structures. Existing rate structures provide a third block to widespread moves to cogeneration. Present utility standby rates are typically significantly higher than the cost of amortizing on-site redundant capacity. This forces most potential cogenerators to isolate themselves from the utility grid in order to achieve acceptable economic returns. The National Energy Act of 1978 mandates state public utility commissions to review standby rates by 1981, ensuring that the rates are equitable and based on statistical probability of outage.

Environmental impact. A fourth concern is impact on environment of any large-scale move to cogeneration. In moving generation to the site of heat use, cogeneration also moves some extra emissions to the site, often a builtup area. Although cogenerators burn 40–60% less fuel than conventional energy conversion systems, environ mental impacts are in dispute. Total emission of SO_2, hydrocarbons, and particulates are all lower from cogeneration than from equivalent conventional energy conversion, CO production appears to be roughly equal, and oxides of nitrogen can be 30–150% higher from cogeneration. Cogeneration reduces both CO_2 and thermal emissions proportional to the oil saved. Street-level effects are theoretically predicted by dispersion equations, but only to accuracies of a factor of 3 in urban settings. Ultimate judgment of on-site cogeneration's environmental effect awaits empirical testing.

Challenge to underlying concepts. Cogeneration provides a case study in the problems any technology faces in displacing older, less efficient approaches. Existing regulations were not drawn with cogeneration in mind and do not recognize its inherent fuel efficiency. No automatic emission offsets are allowed under the Federal Clean Air Act Amendments of 1977, creating the paradox that an existing facility can be prevented from cogenerating even though to do so will lower most local emissions. Users with long histories of reliable service from central utilities are reluctant to self-generate. State and local governments often use central utilities as tax collectors, and cogeneration typically escapes some local taxes, while paying more Federal income taxes on users' savings.

The substitution of efficient on-site cogeneration capacity for less efficient but existing central generating stations is perhaps the most difficult policy conundrum. The central utilities enjoy a near monopoly of power generation in their franchise areas, and while on-site generation is often com-

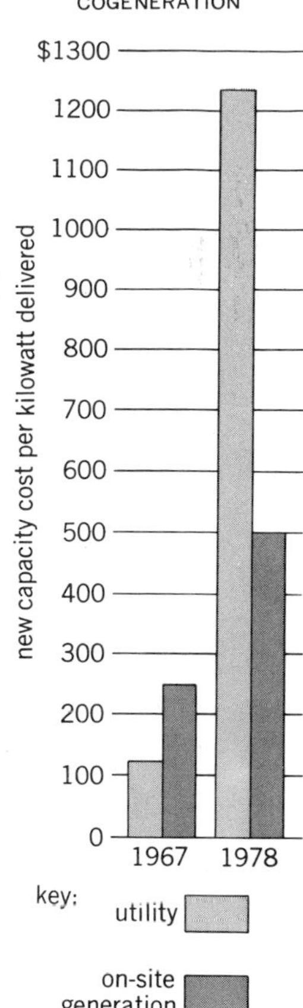

Fig. 2. Relative costs for electric generating capacity in 1967 and 1978.

petitive, it may or may not include utility involvement. To the extent on-site, privately owned cogeneration replaces central-utility-owned stations, those stations become economically less viable. Regulated utilities, unlike competitive business, do not have a history of removing operating capacity, however inefficient, from their rate base. New cogeneration, either utility-owned or privately owned, that displaces present and planned load will confront the regulatory bodies and the utility managements with a challenging period of innovative thinking. *See* ELECTRIC POWER GENERATION.

[THOMAS R. CASTEN]

Bibliography: *Cogeneration: Its Benefits to New England—Final Report of the Governor's Commission on Cogeneration*, October 1978; International District Heating Association (Pittsburgh), *District Heating Mag.*, all issues; U.S. Department of Energy, Office of Conservation and Solar Applications, *The Potential for Cogeneration Development in Six Major Industries by 1985*, Resource Planning Associates, Inc., Cambridge, MA, December 1977.

Coke

A coherent, cellular, carbonaceous residue remaining from the dry (destructive) distillation of a coking coal. It contains carbon as its principal constituent, together with mineral matter and residual volatile matter. The residue obtained from the carbonization of a noncoking coal, such as subbituminous coal, lignite, or anthracite, is normally called a char. Coke is produced chiefly in chemical-recovery coke ovens (see figure), but a small amount is also produced in beehive or other types of nonrecovery ovens.

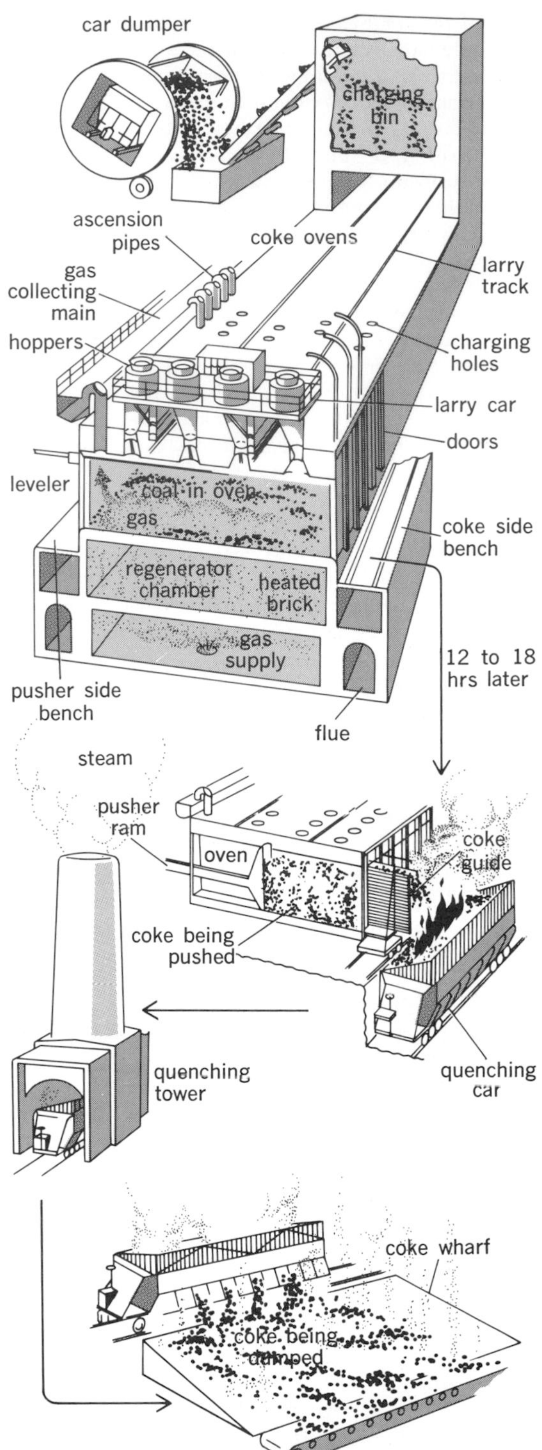

Steps in the production of coke from coal. After the coal has been in the oven for 12–18 hr, the doors are removed, and a ram mounted on the same machine that operates the leveling bar shoves the coke into a quenching car for cooling. (*American Iron and Steel Institute*)

Uses and types. Coke is used predominantly as a fuel reductant in the blast furnace, in which it also serves to support the burden. As the fuel, it supplies the heat as well as the gases required for the reduction of the iron ore. It also finds use in other reduction processes, the foundry cupola, and househeating. About 91% of the coke made is used in the blast furnace, 4% in the foundry, 1% for water gas, 1% for househeating, and 3% for other industries, such as calcium carbide, nonferrous metals, and phosphates. Approximately 1300 lb of coke is consumed per ton of pig iron produced in the modern blast furnace, and about 200 lb of coke is required to melt a ton of pig iron in the cupola.

Coke is classified not only by the oven in which it is made, chemical-recovery or beehive, but also by the temperature at which it is made. High-temperature coke, used mainly for metallurgical purposes, is produced at temperatures of 900–1150°C. Medium-temperature coke is produced at 750–900°C, and low-temperature coke or char is made at 500–750°C. The latter cokes are used chiefly for househeating, particularly in England. The production of these is rather small as compared to high-temperature coke and usually requires special equipment. Coke is also classified according to its intended use, such as blast-furnace coke, foundry coke, water-gas coke, and domestic coke.

Production. The United States has approximately 15,000 chemical-recovery ovens distributed among 75 plants capable of producing a total of about 85,000,000 tons (77×10^6 metric tons) of coke per year. Beehive-oven coke production fluctuates between 2,000,000 and 8,000,000 tons ($2-7 \times 10^6$ metric tons) per year, depending principally upon the demands of the blast furnaces.

Coke is formed when coal is heated in the absence of air. During the heating in the range of 350–500°C, the coal softens and then fuses into a solid mass. The coal is partially devolatilized in this temperature range, and further heating at temperatures up to 1000–1100°C reduces the volatile matter to less than 1%. The degree of softening attained during heating determines to a large extent the character of the coke produced.

In order to produce coke having desired properties, two or more coals are blended before charging into the coke oven. Although there are some exceptions, high-volatile coals of 32–38% volatile-matter content are generally blended with low-volatile coals of 15–20% volatile matter in blends containing 20–40% low-volatile coal. In some cases, charges consist of blends of high-, medium-, and low-volatile coal. In this way, the desirable properties of each of the coals, whether it be in impurity content or in its contribution to the character of the coke, are utilized. The low-volatile coal is usually added in order to improve the physical properties of the coke, especially its strength and yield. In localities where low-volatile coals are lacking, the high-volatile coal or blends of high-volatile coals are used without the benefit of the low-volatile coal. In some cases, particularly in the manufacture of foundry coke, a small percentage of so-called inert, such as fine anthracite, coke fines, or petroleum coke, is added.

In addition to the types of coals blended, the carbonizing conditions in the coke oven influence the characteristics of the coke produced. Oven temperature is the most important of these and has a significant effect on the size and the strength of the coke. In general, for a given coal, the size and shatter strength of the coke increase with decrease in carbonization temperature. This principle is utilized in the manufacture of foundry coke, where large coke of high shatter strength is required.

Properties. The important properties of coke that are of concern in metallurgical operations are its chemical composition, such as moisture, volatile-matter, ash, and sulfur contents, and its physical character, such as size, strength, and density. For the blast furnace and the foundry cupola, coke of low moisture, volatile-matter, ash, and sulfur content is desired. The moisture and the volatile-matter contents are a function of manner of oven operation and quenching, whereas ash and sulfur contents depend upon the composition of the coal charged. Blast-furnace coke used in the United States normally contains less than 1% volatile matter, 85–90% fixed carbon, 7–12% ash, and 0.5–1.5% sulfur. If the coke is intended for use in the production of Bessemer or acid open-hearth iron, the phosphorus content becomes important and should contain less than 0.01%.

The requirements for foundry coke in analysis are somewhat more exacting than for the blast furnace. The coke should have more than 92% fixed carbon, less than 8% ash content, and less than 0.60% sulfur.

Blast-furnace coke should be uniform in size, about $2\frac{1}{2}$–5 in. (6–13 cm), but in order to utilize more of the coke produced in the plant, the practice has been, in some cases, to charge separately into the furnace the smaller sizes after they were closely screened into sizes such as $2\frac{1}{2} \times 1\frac{3}{4}$ in. (64 × 44 mm), $1\frac{3}{4} \times 1$ in. (44 × 25 mm), down to about $\frac{3}{4}$ in. (19 mm) in size. There is now a trend, when pelletized ore is used, to crush and screen the coke to a more closely sized material, such as $2\frac{1}{2} \times \frac{3}{4}$ in. (64 × 19 mm).

The blast-furnace coke should also be uniformly strong so that it will support the column of layers of iron ore, coke, and stone above it in the furnace without degradation. Standard test methods, such as the tumbler test and the shatter test, have been developed by the American Society for Testing and Materials for measuring strength of coke. Good blast-furnace cokes have tumbler test stability factors in the range of 45 to 65 and shatter indices (2-in. or 5-cm sieve) in the range of 70 to 80%. These are not absolute requirements since good blast-furnace performance is obtained in some plants with weaker coke—but with the ore prepared so as to overcome the weakness.

Foundrymen prefer coke that is large and strong. The coke for this purpose ranges in size from 3 to 10 in. (8 to 30 cm) and larger. The size used usually depends upon the size of the cupola and, in general, the maximum size of the coke is about one-twelfth of the diameter of the cupola. High shatter strength is demanded by the cupola operator; shatter indices of at least 97% on 2-in. (5-cm) and 80% on 3-in. (8-cm) sieves have been specified.

Other important requirements of foundry coke are high reactivity toward oxygen so that carbon dioxide is produced with high heat evolution, and low reactivity toward carbon dioxide so that the amount of carbon monoxide formed is minimized.

The requirements for coke intended for house-heating vary in different localities. Usually the coke is of a narrow size range, and the size used depends upon the size and type of appliance. In general, a high ash-softening temperature is desirable (more than 2500°F or 1370°C).

Coke intended for water-gas generators should be over 2 in. (5 cm) in size, and should have an ash content below 10% and a moderately high ash-softening temperature. *See* COAL; FOSSIL FUEL.

[MICHAEL PERCH]

Bibliography: H. H. Lowry (ed.), *Chemistry of Coal Utilization*, suppl. vol., 1963; P. J. Wilson and J. H. Wells, *Coal, Coke and Coal Chemicals*. 1950.

Coking (petroleum)

A process for thermally converting the heavy residual bottoms of crude oil entirely to lower-boiling petroleum products and by-product petroleum coke. The heavy residual bottoms cannot be cata-

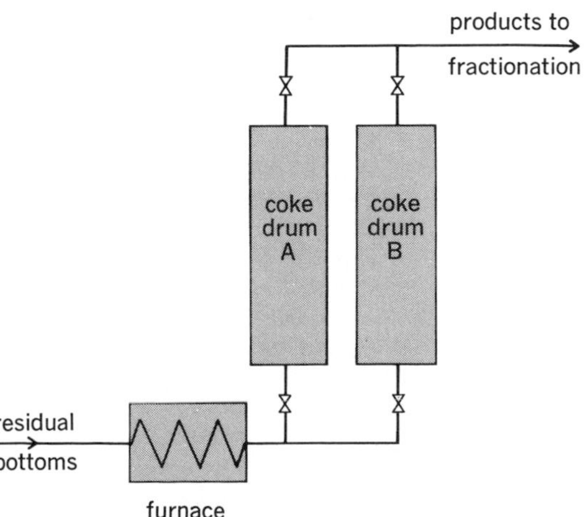

Fig. 1. Delayed coking, in simplified flow plan.

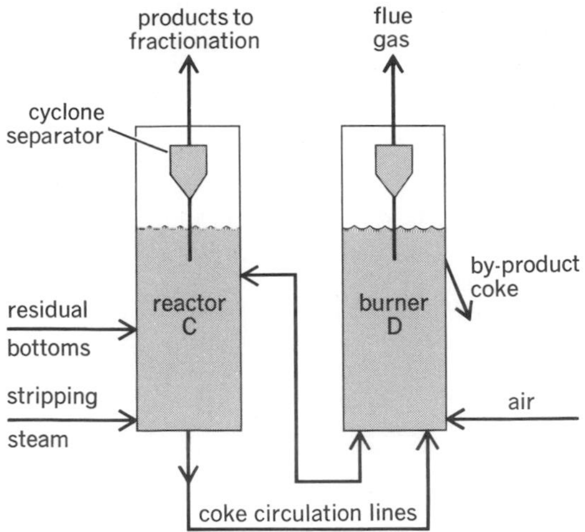

Fig. 2. Fluid coking, in simplified flow plan.

lytically cracked, principally because the metals present are potent catalyst poisons. Traditionally, the residual bottoms have been blended with lighter stocks and marketed as low-quality heavy fuel oils. However, in the United States the demands for gasoline, heating oil, and other products have increased substantially over the years. To meet this shift in demand, refiners have resorted to deeper vacuum flashing, vis-breaking, solvent deasphalting, and coking. The major products from coking are fuel gas, gasoline, gas oil, and petroleum coke. Generally speaking, the gasoline is of poor quality and must be further treated before use in modern premium fuels. The gas oil is commonly used as catalytic-cracking feed stock. *See* FUEL GAS; GASOLINE.

The two coking processes in commercial use are delayed coking and fluid coking. In delayed coking (Fig. 1), residual bottoms are heated to temperatures of about 900°F (480°C) in a conventional furnace. The hot oil is charged directly to the bottom of a large coke drum A, where the coking reaction takes place. Coke is deposited in the drum while the lighter products are removed overhead to fractionation. Approximately every 24 hr, hot oil is diverted to the second coke drum B, and coke is removed from drum A by hydraulic drilling. In fluid coking (Fig. 2), residual bottoms are charged directly to the reactor C containing a bed of fluidized coke particles at about 950°F (510°C). The hot coke particles supply the heat of reaction and also a surface for coke deposition. Converted light products are removed overhead through cyclone separators to fractionation. The coke particles are continuously circulated between the reactor C and the burner D. In the burner D, air is used to burn a portion of the coke produced, thereby maintaining the system in heat balance. Normally, more coke is produced in the reactor than is required for heat balance. The excess coke is continuously withdrawn from the burner as by-product petroleum coke.

Petroleum coke is used principally as fuel or, after calcining, for carbon electrodes. The crude from which the coke is produced governs its chemical composition (high-sulfur crudes yield high-sulfur cokes). However, all petroleum cokes are characterized by low total ash contents. *See* CRACKING; PETROLEUM PROCESSING.

[J. F. MOSER, JR.]

Combustion

The burning of any substance, whether it be gaseous, liquid, or solid. In combustion, a fuel is oxidized, evolving heat and often light. The oxidizer need not be oxygen per se. The oxygen may be a part of a chemical compound, such as nitric acid (HNO_3) or ammonium perchlorate (NH_4ClO_4), and become available to burn the fuel during a complex series of chemical steps. The oxidizer may even be a non-oxygen-containing material. Fluorine is such a substance. It combines with the fuel hydrogen, liberating light and heat. In the strictest sense, a single chemical substance can undergo combustion by decomposition, with emission of heat and light. Acetylene, ozone, and hydrogen peroxide are examples. The products of their decomposition are carbon and hydrogen for acetylene, oxygen for ozone, and water and oxygen for hydrogen peroxide.

Solids and liquids. The combustion of solids such as coal and wood occurs in stages. First, volatile matter is driven out of the solid by thermal decomposition of the fuel and burns in the air. At usual combustion temperatures, the burning of the hot, solid residue is controlled by the rate at which oxygen of the air diffuses to its surface. If the residue is cooled by radiation of heat, combustion ceases.

The first product of combustion at the surface of char, or coke, is carbon monoxide. This gas burns to carbon dioxide in the air surrounding the solid, unless it is chilled by some surface. Carbon monoxide is a poison and it is particularly dangerous because it is odorless. Its release from poorly designed, or malfunctioning, open heaters constitutes a serious hazard to human health.

Liquid fuels do not burn as liquids but as vapors above the liquid surface. The heat evolved evaporates more liquid, and the vapor combines with the oxygen of the air.

Spontaneous combustion. This occurs when certain materials are stored in bulk. The oxidizing action of microorganisms often produces the initial heat.

As the temperature increases, the air trapped in the material takes over the oxidation process, liberating more heat. Because the heat cannot be dissipated to the surroundings, the temperature of the material rises still more and the rate of oxidation increases. Eventually the material reaches an ignition point and bursts into flame. Coal is subject to spontaneous combustion and is generally stored in shallow piles to dissipate the heat of oxidation.

Gases. At ordinary temperatures, molecular collisions do not usually cause combustion. At elevated temperatures the collisions of the thermally agitated molecules are more frequent. More important as a cause of chemical reaction is the greater energy involved in the collisions. Moreover, it has been reasonably well established that there is very little combustion attributable to direct reaction between the molecules. Instead, a high-energy collision dissociates a molecule into atoms, or free radicals. These molecular fragments react with greater ease, and the combustion process proceeds generally by a chain reaction involving these fragments. An illustration will make this clear. The combustion of hydrogen and oxygen to form water does not occur in a single step, Eq. (1). In this

$$2H_2 + O_2 = 2H_2O \tag{1}$$

seemingly simple case, some fourteen reactions have been identified. A hydrogen atom is first formed by collision; it then reacts with oxygen molecules, Eq. (2), forming an OH radical. The latter

$$H + O_2 = OH + O \tag{2}$$

in turn reacts with a hydrogen molecule, Eq. (3),

$$OH + H_2 = H_2O + H \tag{3}$$

forming water and regenerating the H atom which repeats the process. This sequence of reactions constitutes a chain reaction. Sometimes the O atom reacts with a hydrogen molecule to form an OH radical and another H atom, Eq. (4). Thus

$$O + H_2 = OH + H \tag{4}$$

a single H atom can form a new H atom in addition to regenerating itself. This process constitutes a branched-chain reaction. Atoms and radicals recombine with each other to form a neutral molecule, either in the gas space or at a surface after being adsorbed. Thus, chain reactions may be suppressed by proximity of surfaces; and the number and length of the chains may be controlled by regulating the temperature, the composition of the mixtures, and other conditions.

Under certain conditions, where the rate of chain branching equals or exceeds the rate at which chains are terminated, the combustion process speeds up to explosive proportions; because of the rapidity of molecular events, a large number of chains are formed in a short time so that essentially all of the gas undergoes reaction at the same time; that is, an explosion results. The branched-chain type of explosion is similar in principle to atomic explosions of the fission type, where more than one neutron is generated by the reaction occurring between a neutron and a uranium nucleus.

Another cause of explosion in gaseous combustion arises when the rate at which heat is liberated in the reaction is greater than the rate at which the heat dissipates to the surroundings. The temperature increases, accelerates the reaction rate, liberates more heat, and so on, until the entire gas mixture reacts in a very short time. This type of explosion is known as a thermal explosion. There are cases intermediate between branched-chain and thermal explosions which depend upon the type and proportion of gases mixed, the temperature, and the density.

In slow combustion, intermediate products can be isolated. Aldehydes, acids, and peroxides are formed in the slow combustion of hydrocarbons, and hydrogen peroxide in the slow combustion of hydrogen and oxygen. At the relatively low temperature of combustion of paraffin hydrocarbons (propane, butane, ethers) a bluish glow is seen. This light, which results from activated formaldehyde formed in the process, is called a cool flame.

In the gaseous combustion and explosive reactions described above, the processes proceed simultaneously throughout the vessel. The gas mixture in a vessel may also be consumed by a combustion wave which, when initiated locally by a spark or a small flame, travels as a narrow intense reaction zone through the explosive mixture. The gasoline engine operates on this principle. Such combustion waves travel with moderate velocity, ranging from 1 ft/s (0.3 m/s) in hydrocarbons and air to 20–30 ft/s (6–9 m/s) in hydrogen and air. The introduction of turbulence or agitation accelerates the combustion wave. The accelerating wave sends out compression or shock waves which are reflected back and forth in the vessel. Under certain conditions these waves coalesce and change from a slow combustion wave to a high-velocity detonation wave. In hydrogen and oxygen mixtures, the speed is almost 2 mi/s (3.2 km/s). The pressure created by detonation can be very high and dangerous.

Combustion mixtures can be made to react at lower temperatures by employing a catalyst. The molecules are adsorbed on the catalyst, where they may be dissociated into atoms or radicals, and thus brought to reaction condition. An example is the catalytic combination of hydrogen and oxygen at ordinary temperatures on the surface of platinum. The platinum glows as a result of the heat liberated in the surface combustion.

[BERNARD LEWIS]

Bibliography: N. A. Chigier (ed.), *Progress in Energy and Combustion Science*, vol. 1, 1977, and vol. 2, 1978; R. M. Fristom and A. A. Wistenberg, *Flame Structure*, 1965; B. Lewis and G. von Elbe, *Combustion, Flames and Explosions of Gases*, 2d ed., 1961. B. Lewis, R. N. Pease, and H. S. Taylor (eds.), *Combustion Processes*, vol. 2, 1956; F. A. Williams, *Combustion Theory*, 1965.

Combustion of light metals

The combustion of light metals in air or oxygen is a spectacular heat-release process, with a very hot and intensely luminous flame and copious "smoke" of finely divided metal oxide. These characteristics of burning metals make them useful in a broad range of applications: as additives to rocket fuels, as the light source in photographic flashbulbs, as constituents of commercial explosives, as well as in fireworks and military flares. Aluminum and magnesium are most commonly used in these ways, but the combustion of other metals such as boron, beryllium, titanium, and zirconium is also of interest. The light metals are very reactive and ignite easily when finely divided, shredded into thin foil, or milled into a powder. Large pieces of metal are very difficult to ignite, and they very rarely burn.

However, accidental fires involving chips and shavings such as those produced in machining of magnesium have occurred. Such fires must be put out with specially formulated extinguishing powders rather than water or even wet sand, because the burning metals are so reactive they can burn in steam. Even though burning metals release much energy, they cannot be exploited as an energy source because they are not found in pure form in nature and must be extracted by processes which absorb even more energy from other sources.

Fuel characteristics. Light metals generally have very large heats of combustion. Figure 1 shows a comparison of the light metals as fuels with a number of other substances on the basis of the energy released per unit weight of the fuel plus the necessary oxygen, when the substances burn to produce condensed oxides. This basis of comparison is particularly relevant to rockets, which must carry both fuel and oxygen on board. The figure shows clearly that lithium (Li), beryllium (Be), boron (B), magnesium (Mg), silicon (Si), calcium (Ca), and titanium (Ti) are all more energetic than gasoline, and zirconium (Zr) is only slightly less energetic. *See* METAL-BASE FUEL.

Flame temperature. The temperature of a flame depends on the heat of combustion, which can be thought of as the energy available to heat up the products of combustion from ambient temperature to flame temperature. Since light metals have very large heats of combustion, their flame temperatures are very high. For example, aluminum burning in oxygen at atmospheric pressure can produce a flame temperature as high as 3800 K (over 3500°C or 6300°F). Magnesium can burn at over 3300 K and zirconium at well over 4000 K. For comparison, the highest flame temperature for acetylene-oxygen is 3430 K, and for propane in air, it is only 2265 K.

Light-metal oxides are refractory substances with very high melting points. Instead of boiling, they generally decompose at extremely high temperatures (depending on the pressure) and absorb much energy in the process. This property of the oxides generally limits the flame temperature in the combustion of metals to the decomposition temperature of the oxide formed. The heat of combustion of the metal is sufficient to heat all of the oxide to the decomposition temperature and then decompose only some portion of it. The remainder remains in a condensed state. For example, when aluminum burns in oxygen the products of combustion which exist in the flame are droplets of molten alumina (Al_2O_3) and various gaseous species such as Al, O, Al_2O, and AlO. Of course, as they cool down to ambient temperature, these fragments recombine, and outside the flame all of the oxide appears as a white smoke, actually an aerosol of alumina.

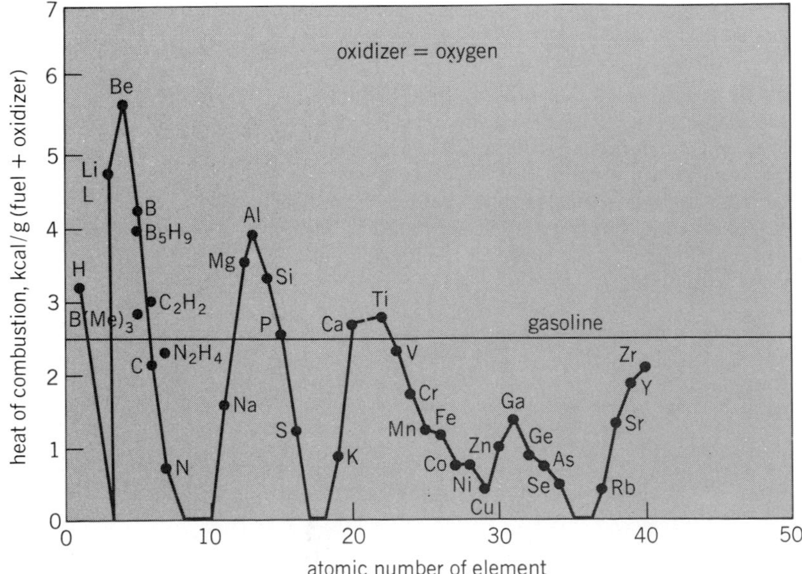

Fig. 1. Comparison of the light metals and other fuels on the basis of the heat of combustion per unit weight of the fuel and the required oxygen.

Light source. The particles of condensed metal oxide which exist in a flame act in the same way as the mantle in a gas lantern or the particles of burning soot in the luminous flame of a coal oil lamp. They all emit radiation which is much more like that from a solid surface than from a thin layer of hot gas. However, because of their much higher temperatures, the metal-oxygen flames are much more intense sources of light than even the most luminous hydrocarbon-air flames. The intensity of thermal radiation increases in proportion to the fourth power of the source temperature. The fraction of this radiation which is visible light also increases with source temperature. For example, a source at 2200 K can emit as much as 130 watts per square centimeter of surface, of which about 1% is visible light. A source at 3800 K can emit up to 1100 W/cm^2 of which 14% is visible. Therefore, as a source of light the aluminum-oxygen flame can be more than 100 times brighter than a hydrocarbon-air flame of the same size. The much hotter zirconium-oxygen flames are even brighter.

Ignition. When a metal is heated in oxygen, a layer of oxide grows on its surface. An oxide such as magnesium oxide (MgO), which has a smaller volume than the original metal, is porous and presents no barrier to the continuing diffusion of oxygen or metal vapor. By contrast, aluminum oxide (Al_2O_3) has a greater volume than the metal which was oxidized to form it. It forms a protective layer on the remaining metal and inhibits further oxidation. As a result of these differences in the properties of the oxide layer, magnesium and aluminum ignite in very different ways.

Ignition occurs when the rate of heat release in the oxidizing reaction first exceeds the rate of heat loss from the reaction zone. Magnesium continues to oxidize by the same mechanism from room temperature right up to ignition. The ignition temperature depends on the size of the metal particles, their concentration, and the amount of oxygen in the oxidizing atmosphere. For example, magnesium particles 35 μm in diameter ignite at 960 K in air at very low concentrations and at 920 K when their concentration is 160 mg/liter. Particles of 7-μm size at the same concentration ignite at 800 K. For comparison, the melting point of Mg is 923 K.

In the case of aluminum, ignition occurs when the protective layer of alumina melts at 2320 K, regardless of particle size and concentration. In this process, hot molten aluminum (melting point 932 K) is suddenly exposed to oxygen and ignites instantly.

Vapor phase. Metals generally burn on the surface of the original particle. However, the more volatile metals can also burn in the vapor phase. Magnesium is a very volatile metal. Its boiling point at 1 atm (101,325 Pa) is 1381 K, far less than the decomposition temperature of magnesium oxide. Aluminum is less volatile, but its boiling point of 2740 K is still much less than the decomposition temperature of its oxide. As a result, it is possible for both aluminum and magnesium to burn by a vapor-phase diffusion mechanism, in

Fig. 2. Magnesium ribbon burning in the vapor phase in air at 0.08 atm (8 kPa). The ribbon broke on igniting, and the two halves, now hanging down, are burning in separate but similar flames.

which heat transfer from the flame evaporates the metal and the metal vapor diffuses to react with oxygen in a flame which surrounds the metal particle. Such a flame is shown in Fig. 2 for the case of magnesium burning in air at the low pressure of 0.08 atm. The resemblance between this flame and a candle flame is not a coincidence. Except for the details of chemistry, the processes are similar.

Energy relationship. Estimates show that on the average 54 kcal (1 kcal = 4184 joules) of energy are required to produce 1 g of pure aluminum from the ore in its natural state. Figure 1 shows that the heat release on combustion of Al to $Al_2O_3(c)$ is 4 kcal per gram of $Al + O_2$, or about 15 kcal per gram of Al. As a result, a net debit of at least 40 kcal must be drawn from other sources of energy for each gram of aluminum burned. Thus, aluminum may be burned because its flame has characteristics which are useful in particular applications, but it is not a source of energy. The same conclusion holds for the other metals.

[T. A. BRZUSTOWSKI]

Bibliography: M. F. Elliott-Jones, Aluminum, in J. G. Myers et al. (eds.), *Energy Consumption in Manufacturing*, chap. 31, 1974; *International Symposium on Combustion*, published annually; *12th International Symposium on Combustion*, Combustion Institute, Pittsburgh, pp. 39–81, 1969, *13th Symposium*, pp. 833–868, 1971, *14th Symposium*, pp. 1389–1412, 1973, *15th Symposium*, pp. 479–514, 1975; H. G. Wolfhard and I. Glassman (eds.), *Heterogeneous Combustion*, pp. 3–279, 1964.

Combustion turbine

A machine for generating mechanical power in rotary motion from the energy of high-pressure, high-temperature combustion gases. The combustion turbine is a versatile prime mover. In addition to the tens of thousands of units that serve daily in propjet and jet engines for military and commercial aircraft, and in the propulsion systems of many ships, boats, trains, buses, and trucks, combustion turbines also are a basic building block of industry. Thousands of combustion turbines, ranging in size from 25 to over 100,000 hp (1 hp = 746 W), drive pumps, compressors, and electric generators in base-load and peaking electrical service.

Turbine cycles and performance. The simplest combustion turbine arrangement consists of a compressor, combustion chamber, and turbine (Fig. 1). The combustion turbine is designed to convert the heat energy of fuel into mechanical energy. Air enters the system and is compressed before entering the combustor, where it is heated at constant pressure. The compressor and combustion chamber produce a high-energy working fluid that can be expanded in the turbine to develop mechanical energy, just as steam is expanded in the more familiar steam turbine. *See* STEAM TURBINE.

Volume of the working fluid is smallest at the compressor outlet, although its temperature is higher than at the compressor inlet. A high excess-air ratio in the combustor keeps gas temperatures at levels which are compatible with turbine materials. At the turbine outlet, exhaust gas is at atmospheric pressure and gas volume is at maximum. The main elements of the combustion turbine must be proportioned to handle the flow of working fluid with minimum pressure loss through the system.

Work output. For a simple-cycle plant, the fuel represents net energy input. Work output (net useful output of the cycle) plus exhaust energy equals the fuel energy input. Compressor-shaft input work simply circulates within the cycle. In a sense, this is the catalyst that makes the combustion turbine cycle workable. The turbine must always be capable of generating the mechanical energy needed to drive the compressor and pressurize the air. Whatever it generates above this is available to drive an external load. Of the total shaft work developed by the turbine, roughly two-thirds goes to drive the compressor and one-third to drive the load. This may seem like a high proportion of mechanical work to make the cycle operate, but the same relation holds for any heat engine using hot gases directly to generate useful mechanical work output.

Efficiency. Thermal efficiency (the ratio of work output to heat input) of the cycle depends on its pressure ratio—that is, the ratio of the absolute compressor discharge pressure to the absolute compressor inlet pressure. Figure 2 shows how thermal efficiency rises with increasing pressure

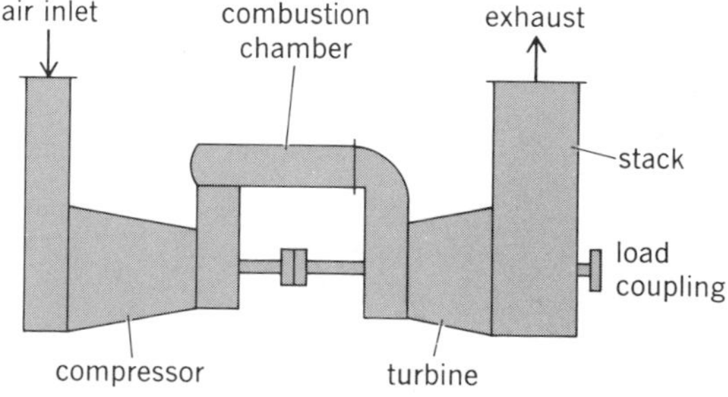

Fig. 1. Diagram of the simplest type of combustion turbine.

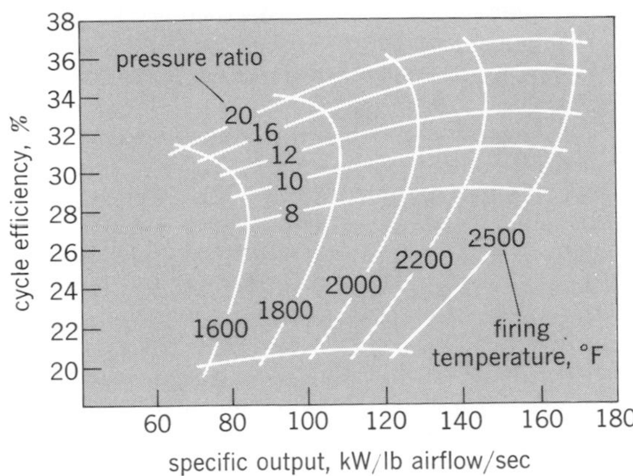

Fig. 2. Thermal efficiency of a typical aircraft-derivative combustion turbine.

ratio and firing temperature. For a typical turbine with a pressure ratio of 8 : 1 and a firing temperature of 1600°F (870°C) the efficiency is approximately 27.2%. By increasing the firing temperature of this machine to 2500 °F (1370°C), a small efficiency gain is achieved (about 2%). But when the pressure ratio is doubled to 16 at the 2500°F temperature (conditions representative of today's most advanced machines), efficiency jumps to about 35%.

Regenerators. Regenerative cycles make use of the exhaust heat discharged from a simple-cycle combustion turbine by means of a heat exchanger called a regenerator (Fig. 3). The regenerator conserves energy by heating high-pressure compressor discharge air with low-temperature turbine exhaust gases. Savings can be considerable. A regenerator designed with an 80% air-side effectiveness, for example, reduces fuel consumption approximately 30% over a comparable simple-cycle unit. This corresponds to an improvement of about 25% in cycle efficiency for some large combustion turbines.

Combined cycles. Combined cycles are also becoming popular for increasing the efficiency of gas turbines (Fig. 4). They are employed where steam is required in addition to the mechanical energy of the combustion turbine. The exhaust gas from a combustion turbine flows to a fired or unfired heat exchanger called a waste-heat boiler. Heat is captured by finned tubes which transfer the heat from

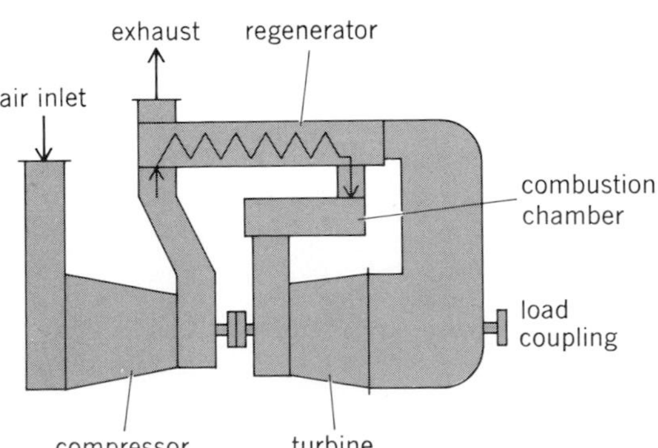

Fig. 3. Diagram of a simple-cycle combustion turbine modified by the addition of a regenerator.

the high-temperature exhaust gas to water flowing through the tubes, thus producing steam. The steam is used to propel a conventional steam turbine to produce electricity.

Working parts. The three major components common to all gas turbines are compressors, combustors, and turbines.

Compressors. Efficient compression of a large-volume flow of air is the key to a successful gas turbine cycle. This has been achieved in two types of compressors, the axial-flow and the centrifugal. The axial-flow type is most common in industrial applications. In it, blades on the compressor rotor have airfoil shapes similar to those on aircraft propellers. They "bite" into the airflow, speed it up, and push it into the succeeding stationary-blade passages. These are shaped to form diffusers that slow up the incoming air to make it compress and pressurize itself by "catching up with the air ahead of it" in smooth fashion. In passing through both moving- and stationary-blade passages, the direction of airflow changes slightly. As a whole, the air follows a helical path through the compressor passages.

Blades of the axial compressor are twisted to accommodate the vortex-flow pattern of the air (higher circumferential velocity nearer the inner edge of the annular passage) and the different speeds of blade sections. Blade heights shrink as air flows through the compressor because of diminishing specific volume of the air at higher pressures. Blade shapes must be precise and clean to achieve high-compression efficiencies; surface fouling retards performance substantially.

Centrifugal compressors take in air at the center of the bladed impeller. The centrifugal force of the blades moves the air out radially at high speed into the stationary diffuser, where the slowing action pressurizes the air. Rotor blades and diffuser passages have varying shapes depending on the pressure characteristic desired with changing rotor speed.

Both types of compressors have pumping limits. When airflow drops below this limit at any rotor speed, the flow becomes unstable because of secondary circulation created in the blade passages. This causes the flow as a whole to pulse and may damage the blading. Axials are more liable to surge at low flows, so they are often bled at an intermediate stage during starting to induce a higher airflow and prevent damage. The air bled from an intermediate stage is wasted to the atmosphere.

Combustors. Combustors burn fuel in the pressurized air after it leaves the compressor. The lack of suitable high-temperature materials limits the exit temperatures to between 1900 and 2200°F (1040 and 1205°C) for base loading of the turbine. Since fuels burn with a flame at about 3000°F (1650°C), careful design is needed to ensure complete combustion. Only about 30% of the air takes part in oxidizing the fuel, with the remainder acting as a film coolant to preserve the combustor basket and as a diluent to cool the gas going to the turbine nozzles and buckets.

Combustors must be built to work with a low pressure drop of the working air, from compressor outlet to turbine inlet. To achieve this, adequate passage flow areas and a minimum number of changes in flow direction are required. These

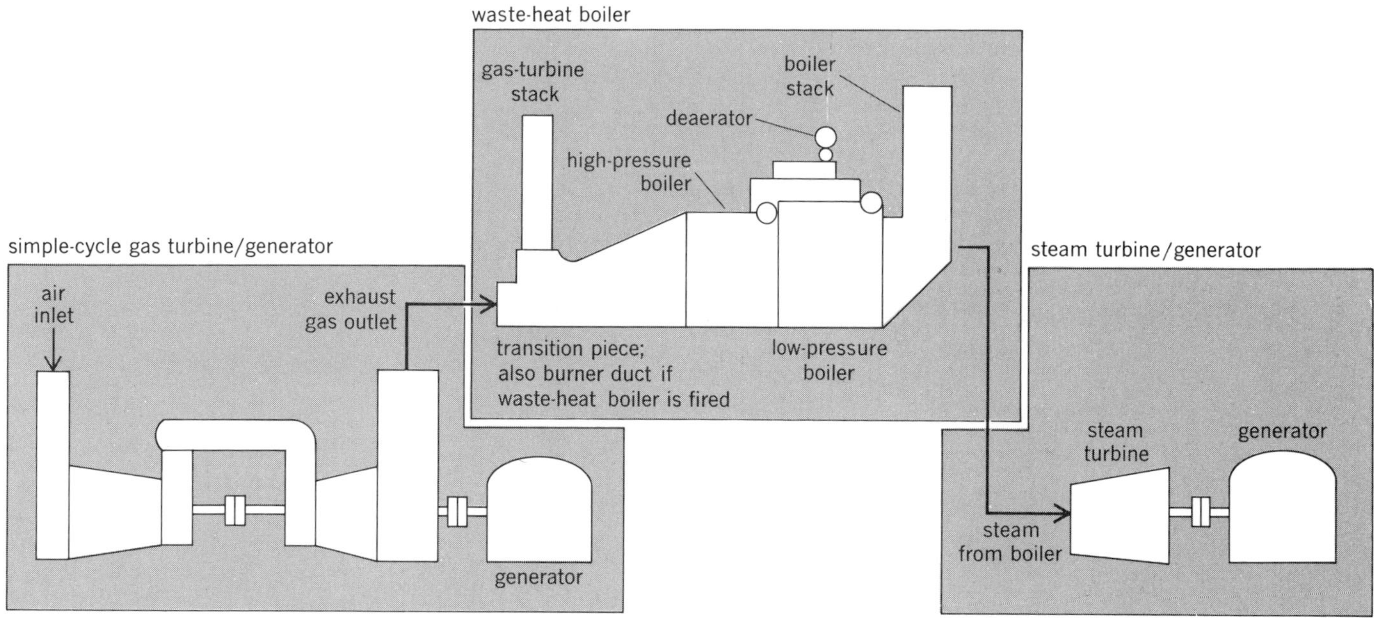

Fig. 4. Operation of a combined-cycle gas turbine.

needs are met by arrangements varying from large single combustors to paralleled multiple combustors.

Ensuring continuous flame action depends largely on fuel nozzle design and proper circulation of combustion air to and through the burning zone. For turbine protection, the combustion gas must be mixed thoroughly with the remaining excess air to avoid large temperature differentials across the flow passages and consequent destructive thermal stresses.

Combustors have igniters to start combustion turbines from a cold condition. Once stable ignition is achieved, the igniters are removed from service.

Fuels for combustors range from natural gas to distillate and residual oils. Coal is still the subject of some experimental investigation since it presents serious problems with respect to disposal of the ash and the effect of ash on blading. Gasification or liquefaction of coal, however, probably will be the means by which these difficulties are overcome. *See* COAL GASIFICATION; COAL LIQUEFACTION.

Fuel must meet certain basic quality standards to ensure troublefree operation. Problems that can be caused by impurities in fuels include high-temperature corrosion, ash deposition on critical turbine components such as blades, and fuel system fouling. High-temperature corrosion is caused by some trace metallic elements found in fuel. They form compounds during combustion that melt on hot-gas-path components and dissolve the protective oxide coatings, leaving blade surfaces in a condition conducive to corrosion. Sodium, potassium, vanadium, and lead, in conjunction with sulfur, are the biggest offenders. Impurities may also form ash deposits which gradually decrease the cross-sectional area of the gas path. The rate of ash deposition depends on ash chemistry

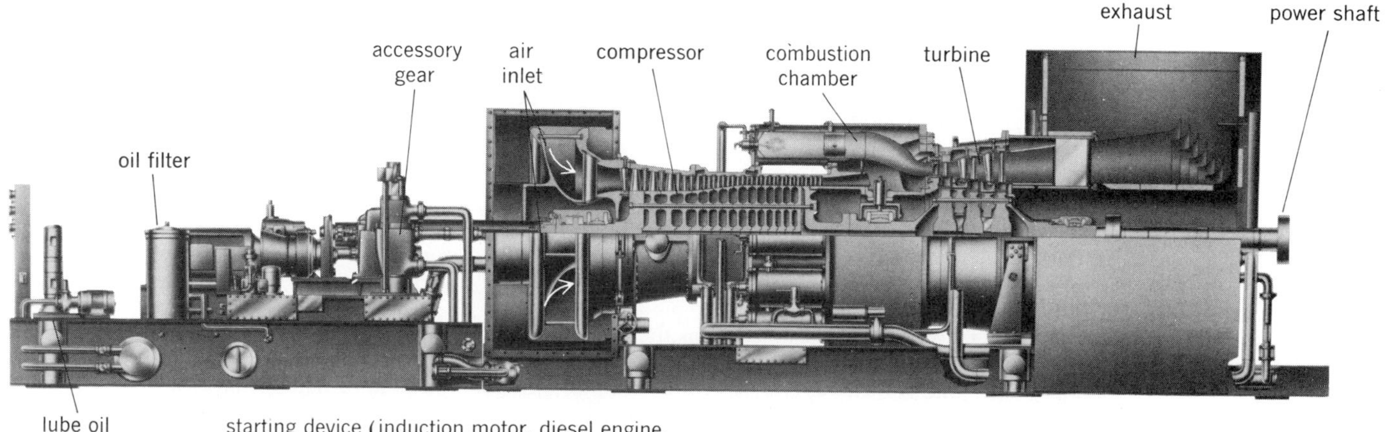

Fig. 5. Components of a simple-cycle single-shaft gas turbine, which is designed primarily for electric generation and produces more than 50,000 kW. (*General Electric Co.*)

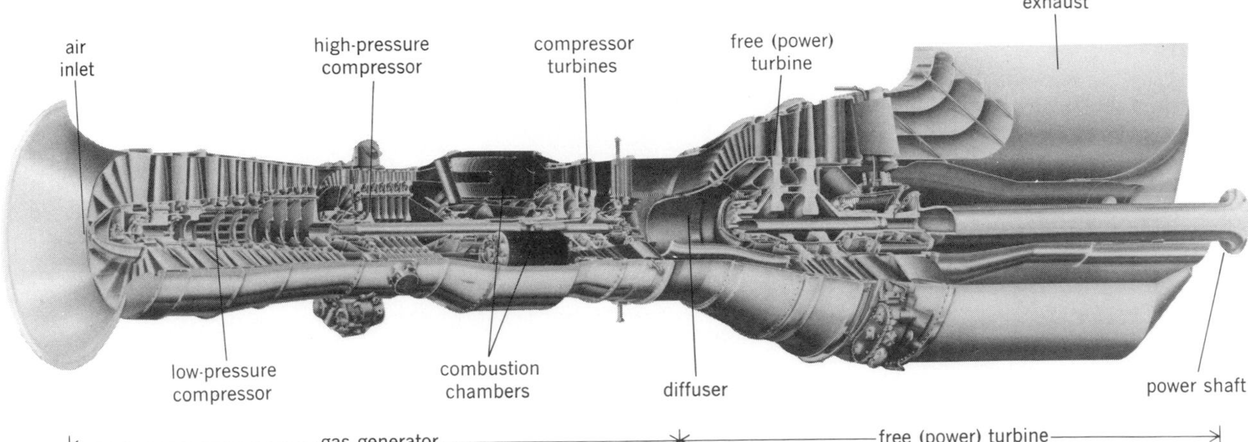

Fig. 6. Components of a two-shaft aircraft-derivative gas turbine, which is used primarily for generating electricity in central stations. (*Turbo Power & Marine Systems Inc.*)

and other operating conditions, such as the quantity of additives used, particularly the vanadium-corrosion inhibitors. Fouling occurs in fuel systems when inorganic particulates, such as solid oxides, silicates, sulfides, and related compounds, are not adequately removed prior to reaching the turbine. These materials clog controls, fuel pumps, flow diverters, and nozzles.

Turbine. The turbine portion of the combustion turbine uses impulse and reaction principles, but because it works with lower pressure drops than the steam turbine, it has fewer stages and less change in blade height from inlet to exhaust.

For best performance, the combustion turbine must work with high inlet-gas temperatures, and this poses severe materials design problems. Advanced alloys and coatings and claddings are commonly used to protect blades for turbine inlet temperatures below 2200°F (1040°C). It is expected that improvements in metallurgy will permit inlet-gas temperatures to increase to 2400°F (1315°C). Beyond that, ceramic materials or sophisticated cooling techniques probably will be required.

Advanced cooling techniques include film, water, liquid-metal, and transpiration cooling. Transpiration cooling is thought by some to show a great deal of promise for units at temperatures of 2500°F (1370°C) and above. In this technique, the cooling air effuses through pores in the airfoil into the boundary layer between the blade and the hot gases. Having cooled the airfoil structure, this air forms a relatively cool film that insulates the blade airfoil surface from the hot gas stream. Clean combustion gases and cooling air are essential to prevent air passages in the airfoil surface from becoming plugged.

Industrial designs. Combustion turbines used by industry are of two basic designs: the industrial and the aircraft-derivative. The so-called industrial combustion turbine was developed from the bedplate up as a heavy-duty machine for industrial applications (Fig. 5). The aircraft-derivative combustion turbine is an adaptation of the aircraft jet engine to economical stationary service (Fig. 6). One distinguishing characteristic of aircraft-derivative engines is that they are of two-shaft design. That is, the external load—pump, compressor, or electric generator—is driven by turbine stages on a shaft separate from that carrying the machine's compressor and its turbine drive. This design offers advantages when the external load has varying power requirements. Industrial combustion turbines for electric generation service often are of single-shaft design—the compressor and the turbine driving the compressor, as well as the external load, are on one shaft; two-shaft units often drive pumps and compressors.

Environmental considerations. An outstanding feature of the combustion turbine is that it has little adverse impact on the environment. For example, the use of clean fuels precludes the formation of significant amounts of sulfur oxides and particulates. Oxides of nitrogen are formed, but these can be brought within the strictest limits imposed by regulatory bodies, with tight control of flame temperature in the combustor. Water cooling is one method used when clean air codes are extremely stringent, but redesigned combustors probably will eliminate the need for it in the future. Water pollution is not a problem since the combustion turbine requires no cooling water for operation. Noise is of some concern, but problems can be solved easily with soundproofed buildings, inlet silencers, and exhaust silencers. *See* ELECTRIC POWER GENERATION; TURBINE.

[ROBERT G. SCHWIEGER]

Bibliography: American Gas Association, *Gas Turbine Manual*, 1965; G. M. Dusinberre and J. C. Lester, *Gas Turbine Power*, 1958; J. W. Javetski, The changing world of gas turbines, *Power*, Special Report, September 1978; J. W. Sawyer, *Gas Turbine Engineering Handbook*, 1966; B. G. A. Skrotzki, Gas turbines, *Power*, Special Report, December 1963; H. A. Sorensen, *Gas Turbines*, 1961; *Today's Technology; Gas Turbines*, a compilation of articles from *Power*, 1975.

Comfort heating

The maintenance of the temperature in a closed volume, such as a home, office, or factory, at a comfortable level during periods of low outside temperature. Two principal factors determine the amount of heat required to maintain a comfortable inside temperature: the difference between inside and outside temperatures and the ease with which heat can flow out through the enclosure.

Heating load. The first step in planning a heating system is to estimate the heating requirements. This involves calculating heat loss from the space, which in turn depends upon the difference between outside and inside space temperatures and upon the heat transfer coefficients of the surrounding structural members.

Outside and inside design temperatures are first selected. Ideally, a heating system should maintain the desired inside temperature under the most severe weather conditions. Economically, however, the lowest outside temperature on record for a locality is seldom used. The design temperature selected depends upon the heat capacity of the structure, amount of insulation, wind exposure, proportion of heat loss due to infiltration or ventilation, nature and time of occupancy or use of the space, difference between daily maximum and minimum temperatures, and other factors. Usually the outside design temperature used is the median of extreme temperatures.

The selected inside design temperature depends upon the use and occupancy of the space. Generally it is between 66 and 75°F (19 and 24°C).

The total heat loss from a space consists of losses through windows and doors, walls or partitions, ceiling or roof, and floor, plus air leakage or ventilation. All items but the last are calculated from $H_l = UA\,(t_i - t_o)$, where heat loss H_l is in British thermal units per hour (or in watts), U is overall coefficient of heat transmission from inside to outside air in Btu/(hr)(ft²)(°F) (or J/s · m² · °C), A is inside surface area in square feet (or square meters), t_i is inside design temperature, and t_o is outside design temperature in °F (or°C).

Values for U can be calculated from heat transfer coefficients of air films and heat conductivities for building materials or obtained directly for various materials and types of construction from heating guides and handbooks.

The heating engineer should work with the architect and building engineer on the economics of the completed structure. Consideration should be given to the use of double glass or storm sash in areas where outside design temperature is 10°F (−12°C) or lower. Heat loss through windows and doors can be more than halved and comfort considerably improved with double glazing. Insulation in exposed walls, ceilings, and around the edges of the ground slab can usually reduce local heat loss by 50–75%. Table 1 compares two typical dwellings. The 43% reduction in heat loss of the insulated house produces a worthwhile decrease in the cost of the heating plant and its operation. Building the house tight reduces the normally large heat loss due to infiltration of outside air. High heating-energy costs may now warrant 4 in. (10 cm) of insulation in the walls and 8 in. (20 cm) or more in the ceiling.

Humidification. In localities where outdoor temperatures are often below 36°F (2°C), it is advisable to provide means for adding moisture in heated spaces to improve comfort. The colder the outside air is, the less moisture it can hold. When it is heated to room temperature, the relative humidity in the space becomes low enough to dry out nasal membranes, furniture, and other hygroscopic materials. This results in discomfort as well as deterioration of physical products.

Various types of humidifiers are available. The most satisfactory type provides for the evaporation of the water to take place on a mold-resistant treated material which can be easily washed to get rid of the resultant deposits. When a higher relative humidity is maintained in a room, a lower dry-bulb temperature or thermostat setting will provide an equal sensation of warmth. This does not mean, however, that there is a saving in heating fuel, because heat from some source is required to evaporate the moisture.

Some humidifiers operate whenever the furnace fan runs, and usually are fed water through a float-controlled valve. With radiation heating, a unitary humidifier located in the room and controlled by a humidistat can be used.

Insulation and vapor barrier. Good insulating material has air cells or several reflective surfaces. A good vapor barrier should be used with or in addition to insulation, or serious trouble may result. Outdoor air or any air at subfreezing temperatures is comparatively dry, and the colder it is the drier it can be. Air inside a space in which moisture has been added from cooking, washing, drying, or humidifying has a much higher vapor pressure than cold outdoor air. Therefore, moisture in vapor form passes from the high vapor pressure space to the

Table 1. Effectiveness of double glass and insulation*

		Heat loss, Btu/hr‡	
Heat-loss members	Area, ft²†	With single-glass weather-stripped windows and doors	With double-glass windows, storm doors, and 2-in. (5.1-cm) wall insulation
Windows and doors	439	39,600	15,800
Walls	1,952	32,800	14,100
Ceiling	900	5,800	5,800
Infiltration		20,800	20,800
Total heat loss		99,000	56,500
Duct loss in basement and walls (20% of total loss)		19,800	11,300
Total required furnace output		118,800	67,800

*Data are for two-story house with basement in St. Louis, Mo. Walls are frame with brick veneer and 25/32-in. (2.0-cm) insulation plus gypsum lath and plaster. Attic floor has 3-in. (7.6-cm) fibrous insulation or its equivalent. Infiltration of outside air is taken as a 1-hr air change in the 14,400 ft³ (408 m³) of heating space. Outside design temperature is −5°F (−21°C); inside temperature is selected as 75°F (24°C). †1 ft² = 0.0929 m². ‡1 Btu/hr = 0.293 W.

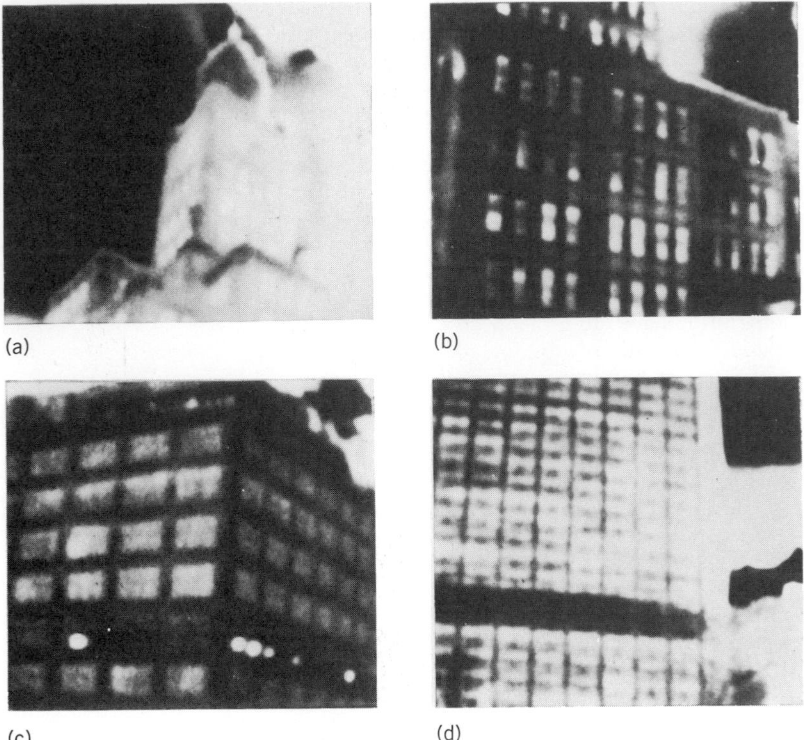

Thermograms of building structures: (*a–c*) masonry buildings; (*d*) glass-faced building. Black indicates negligible heat loss; gray, partial loss; and white, excessive loss. *(Courtesy of A. P. Pontello)*

lower pressure space and will readily pass through most building materials. When this moisture reaches a subfreezing temperature in the structure, it may condense and freeze. When the structure is later warmed, this moisture will thaw and soak the building material, which may be harmful. For example, in a house that has 4 in. (10 cm) or more of mineral wool insulation in the attic floor, moisture can penetrate up through the second floor ceiling and freeze in the attic when the temperature there is below freezing. When a warm day comes, the ice will melt and can ruin the second floor ceiling. Ventilating the attic helps because the dry outdoor air readily absorbs the moisture before it condenses on the surfaces. Installing a vapor barrier in insulated outside walls is recommended, preferably on the room side of the insulation. Good vapor barriers include asphalt-impregnated paper, metal foil, and some plastic-coated papers. The joints should be sealed to be most effective.

Table 2. Infiltration loss with 15-mph outside wind*

Building item	Infiltration, $ft^3/(ft)(hr)$*
Double-hung unlocked wood sash windows of average tightness, non-weather-stripped including wood frame leakage	39
Same window, weather-stripped	24
Same window poorly fitted, non-weather-stripped	111
Same window poorly fitted, weather-stripped	34
Double-hung metal windows unlocked, non-weather-stripped	74
Same window, weather-stripped	32
Residential metal casement, 1/64-in. (0.4 mm) crack	33
Residential metal casement, 1/32-in. (0.8 mm) crack	52

*15 mph = 6.7 m/s. 1 $ft^3/(ft)(hr) = 25.8\ cm^3/(m)(s)$.

Thermography. Remote heat-sensing techniques evolved from space technology developments related to weather satellites can be used to detect comparative heat energy losses from roofs, walls, windows, and so on. A method called thermography is defined as the conversion of a temperature pattern detected on a surface by contrast into an image called a thermogram (see illustration). Thermovision is defined as the technique of utilizing the infrared radiation from a surface, which varies with the surface temperatures, to produce a thermal picture or thermogram. A camera can scan the area in question and focus the radiation on a sensitive detector which in turn converts it to an electronic signal. The signal can be amplified and displayed on a cathode-ray tube as a thermogram.

Normally the relative temperature gradients will vary from white through gray to black. Temperatures from −22° to 3540°F (−30° to 2000°C) can be measured. Color cathode-ray tubes may be used to display color-coded thermograms showing 10 different isotherms. Permanent records are possible using a photographic camera or magnetic tape.

Infrared thermography is used to point out where energy can be saved, and comparative insulation installations and practices can be evaluated. Thermograms of roofs are also used to indicate areas of wet insulation caused by leaks in the roof.

Infiltration. In Table 2, the loss due to infiltration is large. It is the most difficult item to estimate accurately and depends upon how well the house is built. If a masonry or brick-veneer house is not well caulked or if the windows are not tightly fitted and weather-stripped, this loss can be quite large. Sometimes, infiltration is estimated more accurately by measuring the length of crack around windows and doors. Illustrative quantities of air leakage for various types of window construction are shown in Table 2. The figures given are in cubic feet of air per foot of crack per hour.

Design. Before a heating system can be designed, it is necessary to estimate the heating load for each room so that the proper amount of radiation or the proper size of supply air outlets can be selected and the connecting pipe or duct work designed.

Heat is released into the space by electric lights and equipment, by machines, and by people. Credit to these in reducing the size of the heating system can be given only to the extent that the equipment is in use continuously or if forced ventilation, which may be a big heat load factor, is not used when these items are not giving off heat, as in a factory. When these internal heat gain items are large, it may be advisable to estimate the heat requirements at different times during a design day under different load conditions to maintain inside temperatures at the desired level.

Cost of operation. Design and selection of a heating system should include operating costs. The quantity of fuel required for an average heating season may be calculated from

$$F = \frac{Q \times 24 \times \mathrm{DD}}{(t_i - t_o) \times \mathrm{Eff} \times H}$$

where F = annual fuel quantity, same units as H
Q = total heat loss, Btu/hr (or J/s)
t_i = inside design temperature, °F (or °C)
t_o = outside design temperature, °F (or °C)
Eff = efficiency of total heating system (not just the furnace) as a decimal
H = heating value of fuel
DD = degree-days for the locality for 65°F (19.3°C) base, which is the sum of 65 (19.3) minus each day's mean temperature in °F (or °C) for all the days of the year.

If a gas furnace is used for the insulated house of Table 1, the annual fuel consumption would be

$$F = \frac{56{,}500 \times 24 \times 4699}{[75 - (-5)] \times 0.80 \times 1050}$$
$$= 94{,}800 \text{ ft}^3 \text{ (2684 m}^3\text{)}$$

For a 5°F (3°C), 6- to 8-hr night setback, this consumption would be reduced by about 5%.

[GAYLE B. PRIESTER]

Bibliography: American Society of Heating, Refrigerating, and Air Conditioning Engineers, *Handbook of Fundamentals*, 1977; A. P. Pontello, Thermography: Bringing energy waste to light, *Heat./Piping/Air Condit.*, 50(3):55–61, 1978.

Conductor (electricity)

Metal wires, cables, rods, tubes, and bus-bars used for the purpose of carrying electric current. Although any metal assembly or structure can conduct electricity, the term conductor usually refers to the component parts of the current-carrying circuit or system.

Types of conductor. The most common forms of conductors are wires, cables, and bus-bars.

Wires. Wires employed as electrical conductors are slender rods or filaments of metal, usually soft and flexible. They may be bare or covered by some form of flexible insulating material. They are usually circular in cross section; for special purposes they may be drawn in square, rectangular, ribbon, or other shapes. Conductors may be solid or stranded, that is, built up by a helical lay or assembly of smaller solid conductors (Fig. 1).

Cables. Insulated stranded conductors in the larger sizes are called cables. Small, flexible, insulated cables are called cords. Assemblies of two or more insulated wires or cables within a common jacket or sheath are called multiconductor cables.

Bus-bars. Bus-bars are rigid, solid conductors and are made in various shapes, including rectangular, rods, tubes, and hollow squares. Bus-bars may be applied as single conductors, one bus-bar per phase, or as multiple conductors, two or more bus-bars per phase. The individual conductors of a multiple-conductor installation are identical.

Sizes. Most round conductors less than ½ in. (1 in. = 2.54 cm) in diameter are sized according to the American wire gage (AWG)—also known as the Brown & Sharpe gage. AWG sizes are based on a simple mathematical law in which intermediate wire sizes between no. 36 (0.0050-in. diameter) and no. 0000 (0.4600-in. diameter) are formed in geometrical progression. There are 38 sizes between these two diameters. An increase of three gage sizes (for example, from no. 10 to no. 7) doubles the cross-sectional area, and an increase of six gage sizes doubles the diameter of the wire.

Sizes of conductors greater than no. 0000 are usually measured in terms of cross-sectional area. Circular mil (cmil) is usually used to define cross-sectional area and is a unit of area equal to the area of a circle 1 mil (0.001 in.) in diameter.

Wire lengths are usually expressed in units of feet or miles in the United States. Bus-bar sizes are usually defined by their physical dimensions—height and width in inches or fractions of an inch, and length in feet.

Materials. Most wires, cables, and bus-bars are made from either copper or aluminum. Copper, of all the metals except silver, offers the least resistance to the flow of electric current. Both copper and aluminum may be bent and formed readily and have good flexibility in small sizes and in stranded constructions. Typical conductors are shown in Fig. 1.

Aluminum, because of its higher resistance, has less current-carrying capacity than copper for a given cross-sectional area. However, its low cost and light weight (only 30% that of the same volume of copper) permit wide use of aluminum for bus-bars, transmission lines, and large insulated-cable installations.

Metallic sodium conductors were used in 1965 on a trial basis for underground distribution insulated for both primary and secondary voltages. Sodium cable offered light weight and low cost for equivalent current-carrying rating compared with other conductor metals. Because of marketing problems and a few safety problems—the metal is reactive with water—the use of this cable was abandoned temporarily.

For overhead transmission lines where superior strength is required, special conductor constructions are used. Typical of these are aluminum conductors, steel reinforced (ACSR), a composite construction of electrical-grade aluminum strands surrounding a stranded steel core. Other constructions include stranded, high-strength aluminum alloy and a composite construction of aluminum strands around a stranded high-strength aluminum alloy core (ACAR).

For extra-high-voltage (EHV) transmission lines, conductor size is often established by corona performance rather than current-carrying capacity. Thus special "expanded" constructions are used to provide a large circumference without excessive weight. Typical constructions use helical lays of widely spaced aluminum strands around a stranded steel core. The space between the expanding strands is filled with paper twine, and outer layers of conventional aluminum strands are applied. In another construction the outer conductor stranding is applied directly over lays of widely separated helical expanding strands, without filler, leaving substantial voids between the stranded steel core and the closely spaced outer conductor layers. Diameters of 1.6 to 2.5 in. are typical. For lower reactance, conductors are "bundled," spaced 6–18 in. apart, and paralleled in groups of two, four, or more per phase. Figure 2 shows views of typical aluminum-conductor—steel-reinforced and expanded constructions.

Bare conductors. Bare wires and cables are used almost exclusively in outdoor power transmission and distribution lines. Conductors are supported on or from insulators, usually porcelain, of various designs and constructions, depending

CONDUCTOR (ELECTRICITY)

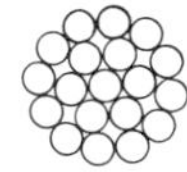

19-strand

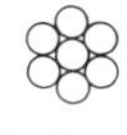

7-strand

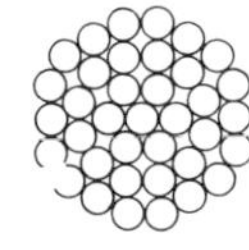

37-strand

Fig. 1. End views of stranded round conductor.

CONDUCTOR (ELECTRICITY)

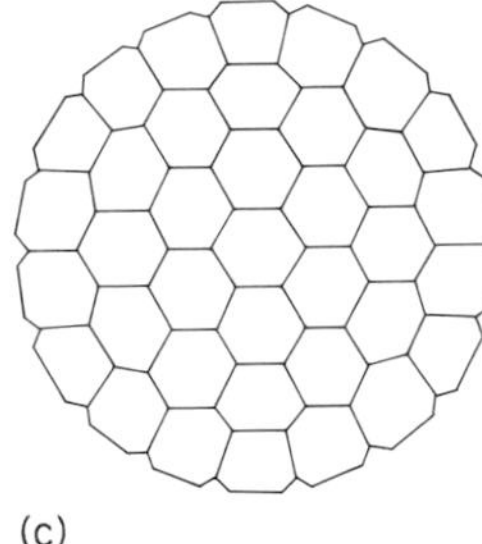

(a)

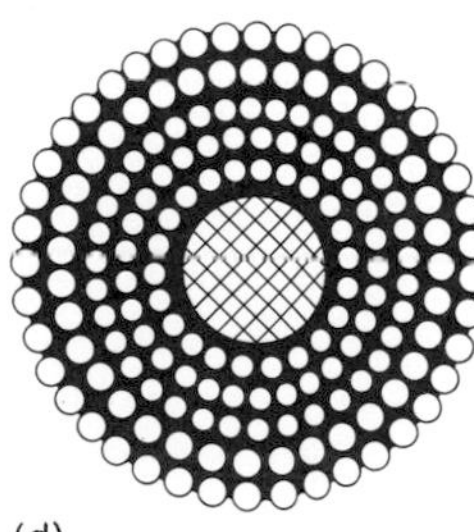

(b)

(c)

(d)

Fig. 2. Aluminum conductors. (*a*) ACSR. (*b*) ACAR. (*c*) Compact concentric stranded conductor. (*d*) Expanded-core concentric stranded conductor.

upon the voltage of the line and the mechanical considerations involved. Voltages as high as 765 kV are in use, and research has been undertaken into the use of ehv transmission lines, with voltages as high as 1500 kV.

Bare bus-bars are used extensively in outdoor substation construction, in switchboards, and for feeders and connections to electrolytic and electroplating processes. Where dangerously high voltages are carried, the use of bare bus-bars is usually restricted to areas which are accessible only to authorized personnel. Bare bus-bars are supported on insulators of a design which is suitable for the voltage.

Insulated conductors. Insulated electric conductors are provided with a continuous covering of flexible insulating material. A great variety of insulating materials and constructions has been developed to serve particular needs and applications. The selection of an appropriate insulation depends upon the voltage of the circuit, the operating temperature, the handling and abrasion likely to be encountered in installation and operation, environmental considerations such as exposure to moisture, oils, or chemicals, and applicable codes and standards.

Magnet wires, used in the windings of motors, solenoids, transformers, and other electromagnetic devices, have relatively thin insulations, usually of enamel or cotton or both. Magnet wire is manufactured for use at temperatures ranging from 105 to 200°C.

Conductors in buildings. Building wires and cables are used in electrical systems in buildings to transmit electric power from the point of electric service (where the system is connected to the utility lines) to the various outlets, fixtures, and utilization devices. Building wires are designed for 600-volt operation but are commonly used at utilization voltages substantially below that value, typically 120, 240, or 480 volts. Insulations commonly used include thermoplastic, natural rubber, synthetic rubber, and rubberlike compounds. Rubber insulations are usually covered with an additional jacket, such as fibrous braid or polyvinyl chloride, to resist abrasion. Building wires are grouped by type in several application classifications in the National Electrical Code.

Classification is by a letter which usually designates the kind of insulation and, often, its application characteristics. For example, Type R indicates rubber or rubberlike insulation. TW indicates a thermoplastic, moisture-resistant insulation suitable for use in dry or wet locations; THW indicates a thermoplastic insulation with moisture and heat resistance.

Other insulations in commercial use include silicone, fluorinated ethylene, propylene, varnished cambric, asbestos, polyethylene, and combinations of these.

Building wires and cables are also available in duplex and multiple-conductor assemblies; the individual insulated conductors are covered by a common jacket. For installation in wet locations, wires and cables are often provided with a lead sheath (Fig. 3).

For residential wiring, the common constructions used are nonmetallic-sheath cables, twin-and multiconductor assemblies in a tough abrasion-resistant jacket; and armored cable with

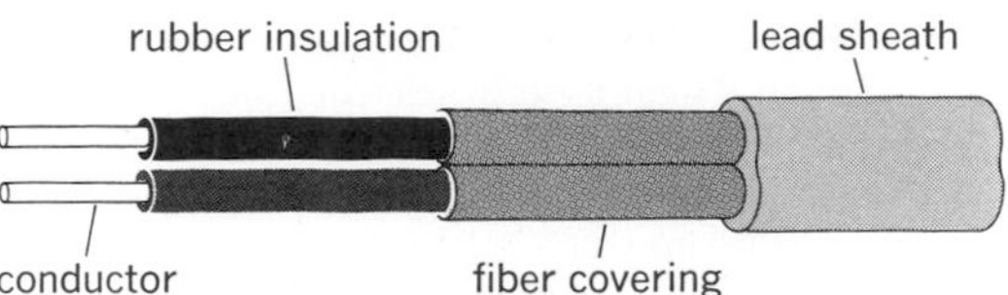

Fig. 3. Rubber-insulated, fiber-covered, lead-sheathed cable, useful for installation in wet places.

twin-and multiconductor assemblies encased in a helical, flexible steel armor as in Fig. 4.

Power cables. Power cables are a class of electric conductors used by utility systems for the distribution of electricity. They are usually installed in underground ducts and conduits. Power cables are also used in the electric power systems of industrial plants and large buildings.

Power-cable insulations in common use include rubber, paper, varnished cambric, asbestos, and thermoplastic. Cables insulated with rubber (Fig. 5), polyethylene, and varnished cambric (Fig. 6) are used up through 69 kV, and impregnated paper to 138 kV. The type and thickness of the insulation for various voltages and applications are specified by the Insulated Power Cable Engineer Association (IPCEA).

Spaced aerial cable. Spaced aerial cable systems are employed for pole-line distribution at, typically, 5 to 15 kV, three-phase. Insulated conductors are suspended from a messenger, which may also serve as a neutral conductor, with ceramic or plastic insulating spacers, usually of diamond configuration. For 15 kV, a typical system may have conductors with 10/64-in. polyethylene insulation, 9-in. spacing between conductors, and 20 ft (6 m) between spacers.

High-voltage cable. High-voltage cable constructions and standards for installation and application are described by IPCEA. Insulations include (1) paper, solid type; (2) paper, low-pressure, gas-filled; (3) paper, low-pressure, oil-filled; (4) pipe cable, fully impregnated, oil-pressure; (5) pipe cable, gas-filled, gas-pressure; (6) rubber or plastic with neoprene or plastic jacket; (7) varnished cloth; (8) AVA and AVL (asbestos-varnished cloth).

Underground transmission cables are in service at voltages through 345 kV, and trial installations have been tested at 500 kV. Research has also been undertaken to develop cryoresistive and superconducting cables for transmitting power at high density. Preliminary designs have been prepared for a three-phase, 345-kV, 3660-MVA cable.

Enclosed bus-bar assemblies, or busways, are extensively used for service conductors and feeders in the electrical distribution systems of industrial plants and commercial buildings. They consist of prefabricated assemblies in standard lengths of bus-bars which are rigidly supported by

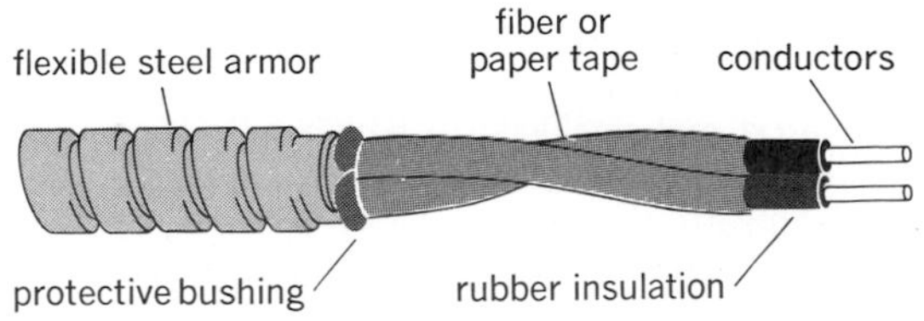

Fig. 4. Two-conductor armored cable.

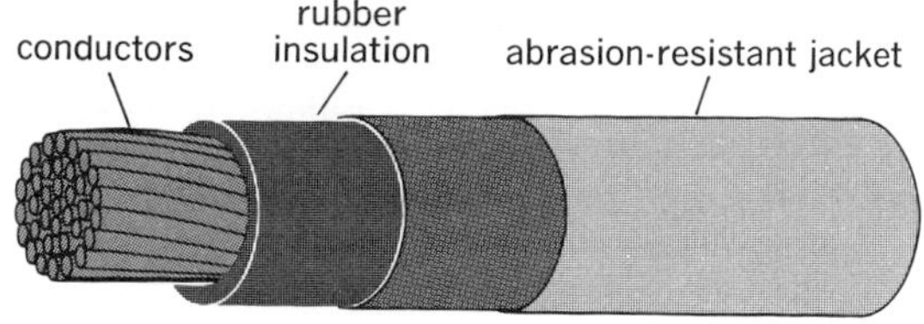

Fig. 5. Rubber-insulated power cable.

solid insulation and enclosed in a sheet-metal housing.

Busways are made in two general types, feeder and plug-in. Feeder busways have no provision for taps or connections between the ends of the assembly. Low-reactance feeder busways are so constructed that conductors of different phases are in close proximity to minimize inductive reactance. Plug-in busways have provisions at intervals along the length of the assembly for the insertion of bus plugs.

Voltage drop in conductors. In electric circuits the resistance and (in ac circuits) the reactance of the circuit conductors result in a reduction in the voltage available at the load (except for capacitive loads). Since the line and load resistances are in series, the source voltage is divided proportionally. The difference between the source voltage and the voltage at the load is called voltage drop.

Electrical utilization devices of all kinds are designed to operate at a particular voltage or within a narrow range of voltages around a design center. The performance and efficiency of these devices are adversely affected if they are operated at a significantly lower voltage. Incandescent-lamp light output is lowered; fluorescent-lamp light output is lowered and starting becomes slow and erratic; the starting and pull-out torque of motors is seriously reduced.

Voltage drop in electric circuits caused by line resistance also represents a loss in power which appears as heat in the conductors. In excessive cases, the heat may rapidly age or destroy the insulation. Power loss also appears as a component of total energy use and cost.

Thus, conductors of electric power systems must be large enough to keep voltage drop at an acceptable value, or power-factor corrective devices—such as capacitors or synchronous condensers—must be installed. A typical maximum for a building wiring system for light and power is 3% voltage drop from the utility connection to any outlet under full-load condition.

[H. WAYNE BEATY]

Bibliography: Aluminum Association, *Aluminum Electrical Conductor Handbook*, September 1971; American National Standards Institute, *National Electrical Safety Code*, ANSI C2, 1977; Electric Power Research Institute, *Research Progress Report TD-3*, September 1975; D. Fink and H. W. Beaty (eds.), *Standard Handbook for Electrical Engineers*, 11th ed., 1978; J. McPartland, *McGraw-Hill's National Electric Code Handbook*, 16th ed., 1979.

Conservation of energy

The principle of conservation of energy states that energy cannot be created or destroyed, although it can be changed from one form to another. Thus in any isolated or closed system, the sum of all forms of energy remains constant. The energy of the system may be interconverted among many different forms—mechanical, electrical, magnetic, thermal, chemical, nuclear, and so on—and as time progresses, it tends to become less and less available; but within the limits of small experimental uncertainty, no change in total amount of energy has been observed in any situation in which it has been possible to ensure that energy has not entered or left the system in the form of work or heat. For a system that is both gaining and losing energy in the form of work and heat, as is true of any machine in operation, the energy principle asserts that the net gain of energy is equal to the total change of the system's internal energy. *See* THERMODYNAMIC PRINCIPLES.

Application to life processes. The energy principle as applied to life processes has also been studied. For instance, the quantity of heat obtained by burning food equivalent to the daily food intake of an animal is found to be equal to the daily amount of energy released by the animal in the forms of heat, work done, and energy in the waste products. (It is assumed that the animal is not gaining or losing weight.) Studies with similar results have also been made of photosynthesis, the process upon which the existence of practically all plant and animal life ultimately depends.

Conservation of mechanical energy. There are many other ways in which the principle of conservation of energy may be stated, depending on the intended application. Examples are the various methods of stating the first law of thermodynamics, the work-kinetic energy theorem, and the assertion that perpetual motion of the first kind is impossible. Of particular interest is the special form of the principle known as the principle of conservation of mechanical energy (kinetic E_k plus potential E_p) of any system of bodies connected together in any way is conserved, provided that the system is free of all frictional forces, including internal friction that could arise during collisions of the bodies of the system. Although frictional or other nonconservative forces are always present in any actual situation, their effects in many cases are so small that the principle of conservation of mechanical energy is a very useful approximation. Thus for a missile or satellite traveling high in space, the dissipative effects arising from such sources as the residual air and meteoric dust are so exceedingly small that the loss of mechanical energy E_k+E_p of the body as it proceeds along its trajectory may, for many purposes, be disregarded. *See* ENERGY.

Mechanical equivalent of heat. The mechanical energy principle is very old, being directly derivable as a theorem from Newton's law of motion. Also very old are the notions that the disappearance of mechanical energy in actual situations is always accompanied by the production of heat and that heat itself is to be ascribed to the random motions of the particles of which matter is composed. But a really clear conception of heat as a form of energy came only near the middle of the 19th century, when J. P. Joule and others demonstrated the equivalence of heat and work by showing experimentally that for every definite amount of work done against friction there always appears a definite quantity of heat. The experiments usu-

CONDUCTOR (ELECTRICITY)

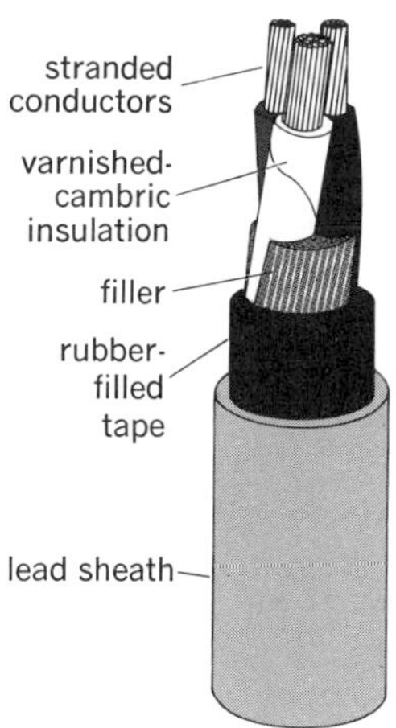

Fig. 6. Varnished cambric-insulated, lead-sheathed power cable, for voltages up to 28 kV.

ally were so arranged that the heat generated was absorbed by a given quantity of water, and it was observed that a given expenditure of mechanical energy always produced the same rise of temperature in the water. The resulting numerical relation between quantities of mechanical energy and heat is called the Joule equivalent, or mechanical equivalent of heat. The present accepted value is one 15° calorie $= 4.1855 \pm 0.0004$ joules.

Conservation of mass-energy. In view of the principle of equivalence of mass and energy in the restricted theory of relativity, the classical principle of conservation of energy must be regarded as a special case of the principle of conservation of mass-energy. However, this more general principle need be invoked only when dealing with certain nuclear phenomena or when speeds comparable with the speed of light (3×10^{10} cm/sec) are involved.

If the mass-energy relation, $E = mc^2$, where c is the speed of light, is considered as providing an equivalence between energy E and mass m in the same sense as the Joule equivalent provides an equivalence between mechanical energy and heat, there results the relation, 1 kg $= 9 \times 10^{16}$ joules.

Laws of motion. The law of conservation of energy has been established by thousands of meticulous measurements of gains and losses of all known forms of energy. It is now known that the total energy of a properly isolated system remains constant. Some parts or particles of the system may gain energy but others must lose just as much. The actual behavior of all the particles, and thus of the whole system, obeys certain laws of motion. These laws of motion must therefore be such that the energy of the total system is not changed by collisions or other interactions of its parts. It is a remarkable fact that one can test for this property of the laws of motion by a simple mathematical manipulation that is the same for all known laws: classical, relativistic, and quantum mechanical.

The mathematical test is as follows. Replace the variable t, which stands for time, by $t + a$, where a is a constant. If the equations of motion are not changed by such a substitution, it can be proved that the energy of any system governed by these equations is conserved. For example, if the only expression containing time is $t_2 - t_1$, changing t_2 to $t_2 + a$ and t to $t_1 + a$ leaves the expression unchanged. Such expressions are said to be invariant under time displacement. When daylight-saving time goes into effect, every t is changed to $t + 1$ hr. It is unnecessary to make this substitution in any known laws of nature, for they are all invariant under time displacement.

Without such invariance laws of nature would change with the passage of time, and repeating an experiment would have no clear-cut meaning. In fact, science, as it is known today, would not exist.

[DUANE E. ROLLER/LEO NEDELSKY]

Bibliography: K. R. Atkins, *Physics*, 3d ed., 1976; G. P. Harnwell and G. J. F. Legge, *Physics: Matter, Energy and the Universe*, 1967; G. Laundry et al., *Physics: An Energy Introduction*, 1979; E. P. Wigner, Symmetry and conservation laws, *Phys. Today*, 17(3):34–40, March 1964.

Contact condenser

A device in which a vapor is brought into direct contact with a cooling liquid and condensed by giving up its latent heat to the liquid. In almost all cases the cooling liquid is water, and the condensing vapor is steam. Contact condensers are classified as jet, barometric, and ejector condensers, as illustrated. In all three types the steam and cooling water are mixed in a condensing chamber and withdrawn together. Noncondensable gases are removed separately from the jet condenser, entrained in the cooling water of the ejector condenser, and removed either separately or en-

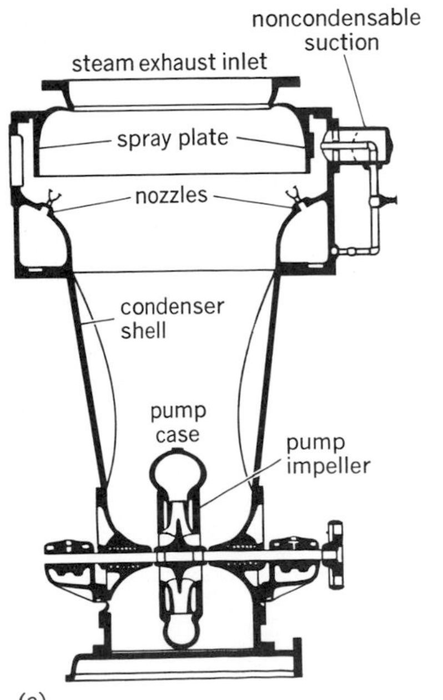

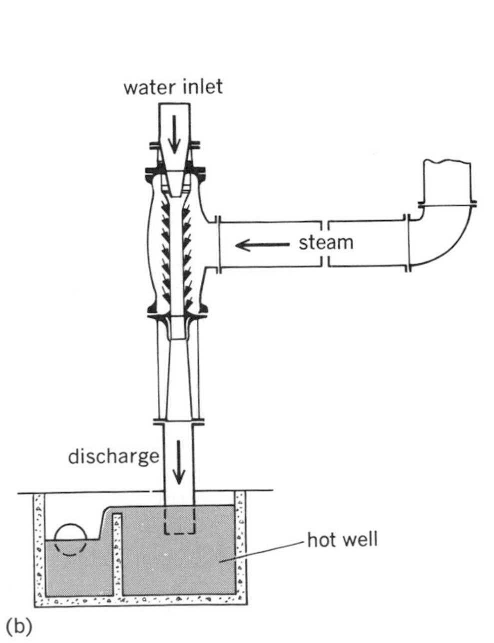

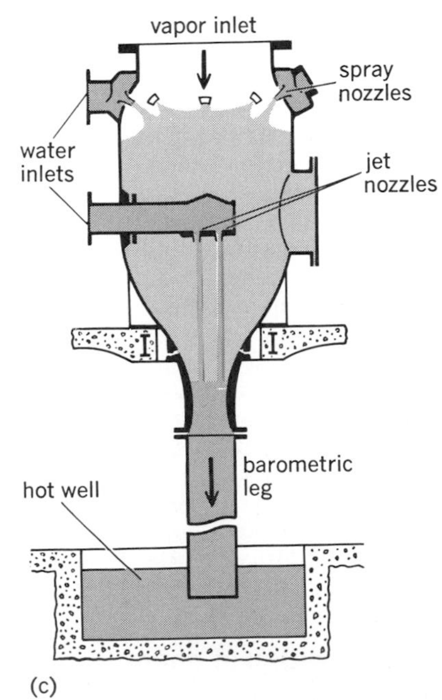

Three basic types of contact condenser. *(a)* Low-level jet condenser *(C. H. Wheeler Manufacturing Co.)*. *(b)* Single-jet ejector condenser; *(c)* multijet barometric condenser *(Schutte and Koerting Co.)*

trained in the barometric condenser. The jet condenser requires a pump to remove the mixture of condensate and cooling water and a vacuum breaker to avoid accidental flooding. The barometric condenser is self-draining. The ejector condenser converts the energy of high-velocity injection water to pressure in order to discharge the water, condensate, and noncondensables at atmospheric pressure. *See* VAPOR CONDENSER.

[JOSEPH F. SEBALD]

Cooling tower

A tower- or buildinglike device in which atmospheric air (the heat receiver) circulates in direct or indirect contact with warmer water (the heat source) and the water is thereby cooled. A cooling tower may serve as the heat sink in a conventional thermodynamic process, such as refrigeration or steam power generation, or it may be used in any process in which water is used as the vehicle for heat removal, and when it is convenient or desirable to make final heat rejection to atmospheric air. Water, acting as the heat-transfer fluid, gives up heat to atmospheric air, and thus cooled, is recirculated through the system, affording economical operation of the process.

Basic types. Two basic types of cooling towers are commonly used. One transfers the heat from warmer water to cooler air mainly by an evaporation heat-transfer process and is known as the evaporative or wet cooling tower. The other transfers the heat from warmer water to cooler air by a sensible heat-transfer process and is known as the nonevaporative or dry cooling tower. These two basic types are sometimes combined, with the two cooling processes generally used in parallel or separately, and are then known as wet-dry cooling towers.

Cooling process. With the evaporative process, the warmer water is brought into direct contact with the cooler air. When the air enters the cooling tower, its moisture content is generally less than saturation; it emerges at a higher temperature and with a moisture content at or approaching saturation. Evaporative cooling takes place even when the incoming air is saturated, because as the air temperature is increased in the process of absorbing sensible heat from the water, there is also an increase in its capacity for holding water, and evaporation continues. The evaporative process accounts for about 65–75% of the total heat transferred; the remainder is transferred by the sensible heat-transfer process.

The wet-bulb temperature of the incoming air is the theoretical limit of cooling. Cooling the water to within 5–20°F (1°F = 0.56°C) above wet-bulb temperature represents good practice. The amount of water evaporated is relatively small. Approximately 1000 Btu (1 Btu = 1055 J) is required to vaporize 1 lb (0.45 kg) of water at cooling tower operating temperatures. This represents a loss in water of approximately 0.75% of the water circulated for each 10°F cooling, taking into account the normal proportions of cooling by the combined evaporative and sensible heat-transfer processes. Drift losses may be as low as 0.01–0.05% of the water flow to the tower (recent performance of 0.001% has been achieved) and must be added to the loss of water by evaporation and losses from blowdown to account for the water lost

Fig. 1. Counterflow natural-draft cooling tower at Trojan Power Plant in Ranier, OR. (*Research-Cottrel*)

Fig. 2. Cross-flow mechanical-draft cooling tower. (*Marley Co.*)

from the system. Blowdown quantity is a function of makeup water quality, but it may be determined by regulations concerning its disposal. Its quality is usually expressed in terms of the allowable concentration of dissolved solids in the circulating cooling water and may vary from about two to six concentrations with respect to the dissolved solids content of the cooling water makeup.

Fig. 3. Mechanical-draft cooling towers. (*a*) Conventional rectangular cross-flow evaporative induced-draft type; (*b*) circular cross-flow evaporative induced-draft type. (*Marley Co.*)

With the nonevaporative process, the warmer water is separated from the cooler air by means of thin metal walls, usually tubes of circular cross section, but sometimes of elliptical cross section. Because of the low heat-transfer rates from a surface to air at atmospheric pressure, the air side of the tube is made with an extended surface in the form of fins of various geometries. The heat-transfer surface is usually arranged with two or more passes on the water side and a single pass, cross flow, on the air side. Sensible heat transfer through the tube walls and from the extended surface is responsible for all of the heat given up by the water and absorbed by the cooling air. The water temperature is reduced, and the air temperature increased. The nonevaporative cooling tower may also be used as an air-cooled vapor condenser and is commonly employed as such for condensing steam. The steam is condensed within the tubes at a substantially constant temperature, giving up its latent heat of vaporization to the cooling air, which in turn is increasing in temperature. The theoretical limit of cooling is the temperature of the incoming air. Good practice is to design nonevaporative cooling towers to cool the warm circulating water to within 25 to 35°F of the entering air temperature or to condense steam at a similar temperature difference with respect to the incoming air. Makeup to the system is to compensate for leakage only, and there is no blowdown requirement or drift loss.

With the combined evaporative-nonevaporative process, the heat-absorbing capacity of the system is divided between the two types of cooling towers, which are selected in some predetermined proportion and usually arranged so that adjustments can be made to suit operating conditions within definite limits. The two systems, evaporative and nonevaporative, are combined in a unit with the water flow arranged in a series relationship passing through the dry tower component first and the wet tower second. The airflow through the towers is in a parallel-flow relationship, with the discharge air from the two sections mixing before being expelled from the system. Since the evaporative process is employed as one portion of the cooling system, drift, makeup, and blowdown are characteristics of the combined evaporative-nonevaporative cooling tower system, generally to a lesser degree than in the conventional evaporative cooling towers.

Of the three general types of cooling towers, the evaporative tower as a heat sink has the greatest thermal efficiency but consumes the most water and has the largest visible vapor plume. When mechanical-draft cooling tower modules are arranged in a row, ground fogging can occur. This can be eliminated by using natural-draft towers, and can be significantly reduced with modularized mechanical-draft towers when they are arranged in circular fashion.

The nonevaporative cooling tower is the least efficient type, but it can operate with practically no consumption of water and can be located almost anywhere. It has no vapor plume.

The combined evaporative-nonevaporative cooling tower has a thermal efficiency somewhere between that of the evaporative and nonevaporative cooling towers. Most are of the mechanical-draft type, and the vapor plume is mitigated by mixing the dry warm air leaving the nonevaporative section of the tower with the warm saturated air leaving the evaporative section of the tower. This retards the cooling of the plume to atmospheric temperature; visible vapor is reduced and may be entirely eliminated. This tower has the advantage of flexibility in operation; it can accommodate variations in available makeup water or be adjusted to atmospheric conditions so that vapor plume formation and ground fogging can be reduced.

Evaporative cooling towers. Evaporative cooling towers are classified according to the means employed for producing air circulation through them: atmospheric, natural draft, and mechanical draft.

Atmospheric cooling. Some towers depend upon natural wind currents blowing through them in a substantially horizontal direction for their air supply. Louvers on all sides prevent water from being blown out of these atmospheric cooling towers, and allow air to enter and leave independently of wind direction. Generally, these towers are located broadside to prevailing winds for maximum sustained airflow.

Thermal performance varies greatly because it

is a function of wind direction and velocity as well as wet- and dry-bulb temperatures. The normal loading of atmospheric towers is 1–2 gal/min (0.06–0.13 liter/s) of cooling water per square foot of cross section. They require considerable unobstructed surrounding ground space in addition to their cross-sectional area to operate properly. Because they need more area per unit of cooling than other types of towers, they are usually limited to small sizes.

Natural draft. Other cooling towers depend for their air supply upon the natural convection of air flowing upward and in contact with the water to be cooled. Essentially, natural-draft cooling towers are chimneylike structures with a heat-transfer section installed in their lower portion, directly above an annular air inlet in a counterflow relationship with the cooling air (Fig. 1), or with the heat-transfer section circumscribing the base of the tower in a cross-flow relationship with the cooling airflow (Fig. 2). Sensible heat absorbed by the air in passing over the water to be cooled increases the air temperature and its vapor content and thereby reduces its density so that the air is forced upward and out of the tower by the surrounding heavier atmosphere. The flow of air through the tower varies according to the difference in specific weights of the ambient air and the air leaving the heat-transfer surfaces. Since the difference in specific weights generally increases in cold weather, the airflow through the cooling tower also increases, and the relative performance improves in reference to equivalent constant-airflow towers.

Normal loading of a natural-draft tower is 2–4 gal/(min)(ft^2) [1.4–2.7 liters/(s)(m^2)] of ground-level cross section. The natural-drafting cooling tower does not require as much unobstructed surrounding space as the atmospheric cooling tower does, and is generally suited for both medium and large installations. The natural-draft cooling tower was first commonly used in Europe. Subsequently, a number of large installations were built in the United States, with single units 385 ft (117 m) in diameter by 492 ft (150 m) high capable of absorbing the heat rejected from an 1100-MWe light-water-reactor steam electric power plant.

Mechanical draft. In cooling towers that depend upon fans for their air supply, the fans may be arranged to produce either a forced or an induced draft. Induced-draft designs are more commonly used than forced-draft designs because of lower initial cost, improved air-water contact, and less air recirculation (Fig. 3). With controlled airflow, the capacity of the mechanical-draft tower can be adjusted for economic operation in relation to heat load and in consideration of ambient conditions.

Normal loading of a mechanical-draft cooling tower is 2–6 gal/(min)(ft^2) [1.4–4.1 liters/(s)(m^2)] of cross section. The mechanical-draft tower requires less unobstructed surrounding space to obtain adequate air supply than the atmospheric cooling tower needs; however, it requires more surrounding space than natural-draft towers do. This type of tower is suitable for both large and small installations.

Nonevaporative cooling towers. Nonevaporative cooling towers are classified as air-cooled condensers and as air-cooled heat exchangers, and are further classified by the means used for producing air circulation through them. *See* HEAT EXCHANGER; VAPOR CONDENSER.

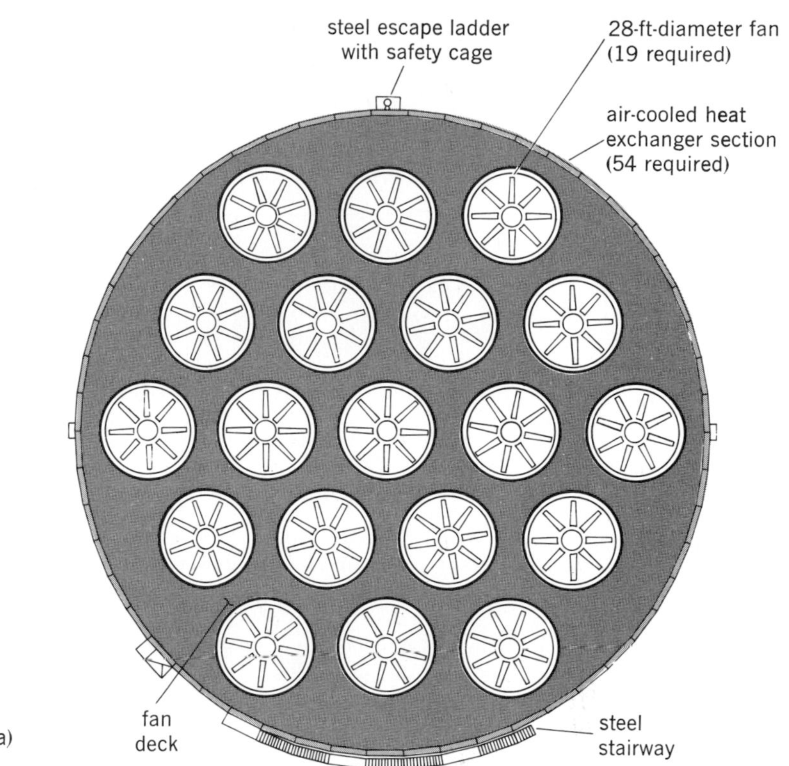

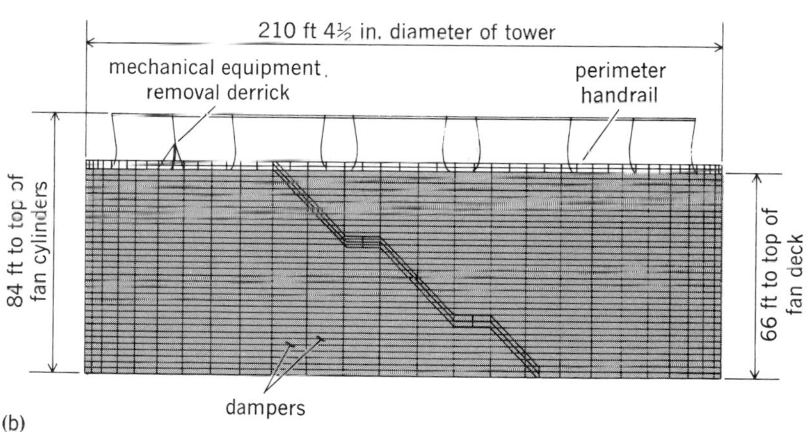

Fig. 4. Multifan circular nonevaporative cooling tower. *(a)* Plan. *(b)* Elevation. 1 ft = 0.3048 m. 210 ft 4½ in. = 64.12 m. *(Marley Co.)*

Air-cooled condensers and heat exchangers. Two basic types of nonevaporative cooling towers are in general use for power plant or process cooling. One type uses an air-cooled steam surface condenser as the means for transferring the heat rejected from the cycle to atmospheric cooling air. The other uses an air-cooled heat exchanger for this purpose. Heat is transferred from the air-cooled condenser, or from the air-cooled heat exchanger to the cooling air, by convection as sensible heat.

Nonevaporative cooling towers have been used for cooling small steam electric power plants since the 1930s. They have been used for process cooling since 1940; a complete refinery was cooled by the process in 1958. Until recently, the nonevapor-

Fig. 5. Wet-dry cooling tower at Atlantic Richfield Company. (*Marley Co.*)

ative cooling tower was used almost exclusively with large steam electric power plants in Europe. Interest in this type of tower is increasing in the United States, and a 330-MWe plant in Wyoming using nonevaporative cooling towers was completed in the late 1970s.

The primary advantage of nonevaporative cooling towers is that of flexibility of plant siting. There is seldom a direct economic advantage associated with the use of nonevaporative cooling tower systems in the normal context of power plant economics. They are the least efficient of the cooling systems used as heat sinks.

Cooling airflow. Each of the two basic nonevaporative cooling tower systems may be further classified with respect to type of cooling airflow. Both types of towers, the direct-condensing type and the heat-exchanger type, can be built as natural-draft or as mechanical-draft tower systems.

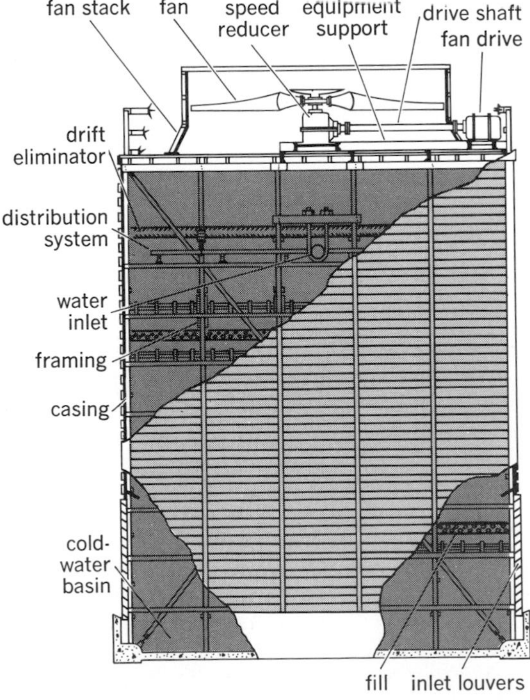

Fig. 6. Induced-draft counterflow cooling tower, showing the component parts.

Design. The heat-transfer sections are constructed as tube bundles, with finned tubes arranged in banks two to five rows deep. The tubes are in a parallel relationship with each other and are spaced at a pitch slightly greater than their outside fin diameter, either in an in-line or a staggered pattern. For each section, two headers are used, with the tube ends secured in each. The headers may be of pipe or of a box-shaped cross section and are usually made of steel. The bundles are secured in an open metal frame.

The assembled tube bundles may be arranged in a V shape, with either horizontal or inclined tubes, or in an A shape, with the same tube arrangement. A similar arrangement may be used with vertical tube bundles. Generally, inclined tubes are shorter than horizontal ones. The inclined-tube arrangement is best suited to condensing vapor, the horizontal-tube arrangement best for heat-exchanger design. The A-shaped bundles are usually used with forced-draft airflow, the V-shaped bundles with induced-draft airflow. With natural-draft nonevaporative cooling towers there is no distinction made as to A- or V-shaped tube bundle arrangement; the bundles are arranged in a deck above the open circumference at the bottom of the tower, in a manner similar in principle to that used in the counterflow natural-draft evaporative cooling tower (Fig. 1). A natural-draft cooling tower with a vertical arrangement of tube bundles has been built, but this type of structure is not generally used.

Modularization of the bundle sections has become general practice with larger units; a recent installation of a circular module–type unit is shown in Fig. 4.

Combined evaporative-nonevaporative towers. The combination evaporative-nonevaporative cooling tower is arranged so that the water to be cooled first passes through a nonevaporative cooling section which is much the same as the tube bundle sections used with nonevaporative cooling towers of the heat-exchanger type. The hot water first passes through these heat exchangers, which are mounted directly above the evaporative cooling tower sections; the water leaving the nonevaporative section flows by force of gravity over the evaporative section.

The cooling air is divided into two parallel flow streams, one passing through the nonevaporative section, the other through the evaporative section to a common plenum chamber upstream of the induced-draft fans. There the two airflow streams are combined and discharged upward to the atmosphere by the fans. A typical cooling tower of this type is shown in Fig. 5.

In most applications of wet-dry cooling towers attempts are made to balance vapor plume suppression and the esthetics of a low silhouette in comparison with that of the natural-draft evaporative cooling tower. In some instances, these cooling towers are used in order to take advantage of the lower heat-sink temperature attainable with evaporative cooling when an adequate water supply is available, and to allow the plant to continue to operate, at reduced efficiency, when water for cooling is in short supply.

Tower components. The more important components of evaporative cooling towers are the

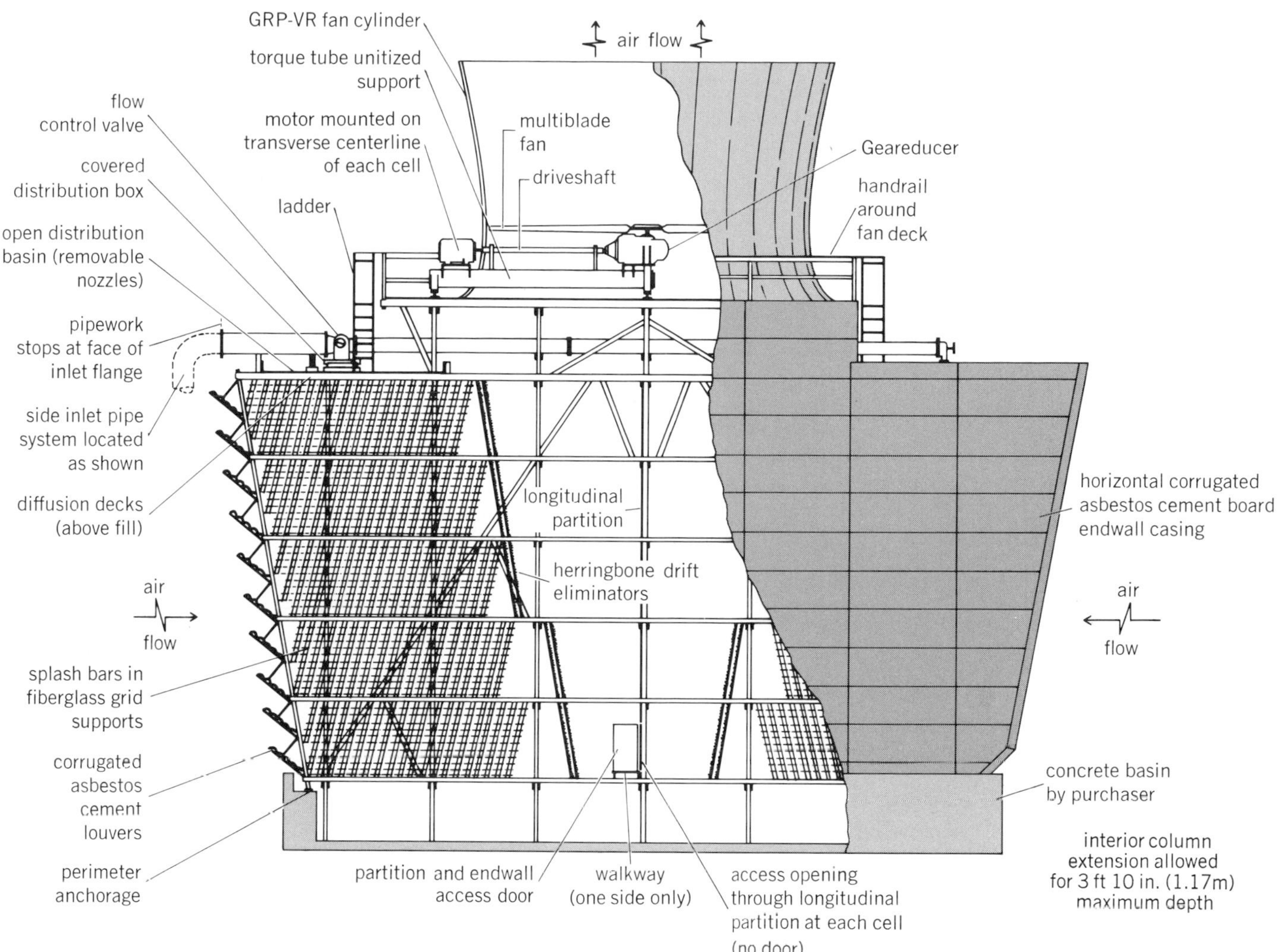

Fig. 7. Transverse cross section of a cross-flow evaporative cooling tower. (*Marley Co.*)

supporting structure, casing, cold-water basin, distribution system, drift eliminators, filling and louvers, discharge stack, and fans (mechanical-draft towers only). The counterflow type is shown in Fig. 6; the cross-flow type is shown in Fig. 7.

Treated wood, especially Douglas fir, is the common material for atmospheric and small mechanical-draft cooling towers. It is used for structural framing, casing, louvers, and drift eliminators. Wood is commonly used as filler material for small towers. Plastics and asbestos cement are replacing wood to some degree in small towers and almost completely in large towers where fireproof materials are generally required. Framing and casings of reinforced concrete, and louvers of metal, usually aluminum, are generally used for large power plant installations. Fasteners for securing small parts are usually made of bronze, copper-nickel, stainless steel, and galvanized steel. Distribution system may be in the form of piping, in galvanized steel, or fiber glass–reinforced plastics; and they may be equipped with spray nozzles of noncorrosive material, in the form of troughs and weirs, or made of wood, plastic, or reinforced concrete. Structural framing may also be made of galvanized or plastic-coated structural steel shapes. Natural-draft cooling towers, especially in large sizes, are made of reinforced concrete. The cold-water basins for ground-mounted towers are usually made of concrete; wood or steel is usually used for roof-mounted towers. Fan blades are made of corrosion-resistant material such as monel, stainless steel, or aluminum; but most commonly fiber glass–reinforced plastic is used for fan blades.

The heat-transfer tubes used with nonevaporative cooling towers are of an extended-surface type usually with circumferential fins on the air side (outside). The tubes are usually circular in cross section, although elliptical tubes are sometimes used. Commonly used tube materials are galvanized carbon steel, ferritic stainless steel, and various copper alloys. They are usually made with wrapped aluminum fins, but steel fins are commonly used with carbon steel tubes and galvanized. Most designs employ a ratio of outside to inside surface of 20:25. Outside-diameter sizes range from $^3/_4$ to $1^1/_2$ in. (19 to 38 mm).

Tube bundles are made with tube banks two to five rows deep. The tubes are in parallel relationship with each other and secured in headers, either of steel pipe or of weld-fabricated steel box

Fig. 8. Nonevaporative cooling tower, Utrillas, Spain. (*GEA*)

headers. The tube bundle assemblies are mounted in steel frames which are supported by structural steel framework. The bundle assemblies may be arranged in a V pattern, requiring fans of the induced-draft type, or in an A pattern, requiring fans of the forced-draft type, Figure 8. Fans and louvers are similar to those described for evaporative cooling towers.

Nonevaporative cooling towers may also be used with natural airflow. In this case, the tube bundles are usually mounted on a deck within the tower and just above the top of the circumferential supporting structure for the tower. In this application, the tower has no cold-water basin, but otherwise it is identical with the natural-draft tower used for evaporative cooling with respect to materials of construction and design.

Performance. The performance of an evaporative cooling tower may be described by the generally accepted equation of F. Merkel, as shown below, where a = water-air contact area, ft^2/ft^3; h =

$$\frac{KaV}{L} = \int_{T_2}^{T_1} \frac{dT}{h'' - h}$$

enthalpy of entering air, Btu/lb; h'' = enthalpy of leaving air, Btu/lb; K = diffusion coefficient, lb/(ft^2)(hr); L = water flow rate, lb/(hr)(ft^2); T = water temperature, °F; T_1 = inlet water temperature, °F; T_2 = outlet water temperature, °F; and V = effective volume of tower, ft^3/ft^2 of ground area. The Merkel equation is usually integrated graphically or by Simpson's rule.

The performance of a nonevaporative cooling tower may be described by the generally accepted equation of Fourier for steady-state unidirectional heat transfer, using the classical summation of resistances formula with correction of the logarithmic temperature difference for cross-counterflow design in order to calculate the overall heat transfer. It is usual practice to reference the overall heat-transfer coefficient to the outside (finned) tube surface.

Evaluation of cooling tower performance is based on cooling of a specified quantity of water through a given range and to a specified temperature approach to the wet-bulb or dry-bulb temperature for which the tower is designed. Because exact design conditions are rarely experienced in operation, estimated performance curves are frequently prepared for a specific installation, and provide a means for comparing the measured performance with design conditions.

[JOSEPH F. SEBALD]

Bibliography: J. D. Guerney and I. A. Cotter, *Cooling Towers in Refrigeration*, International Ideas, Philadelphia, 1966; D. Q. Kern and A. D. Kraus, *Extended Surface Heat Transfer*, 1972; R. D. Landon and J. R. Houx, Jr., Plume abatement and water conservation with the wet-dry cooling towers, *Proc. Amer. Power Conf.*, Chicago, 35:726–742, 1973; Marley Company, *Cooling Tower Fundamentals and Application Principles*, 1969; J. I. Reisman and J. C. Ovard, Cooling towers and the environment: An overview, *Proc. Amer. Power Conf.*, Chicago, 35:713–725, 1973; J. F. Sebald, *Site and Design Temperature Related Economies of Nuclear Power Plants with Evaporative and Nonevaporative Cooling Tower Systems*, Energy Research and Development Administration, Division of Reactor Research and Development, C00-2392-1, January 1976; E. C. Smith and M. W. Larinoff, Power plant siting, performance and economies with dry cooling tower systems, *Proc. Amer. Power Conf.*, Chicago, 32:544–572, 1970; W. Stanford and G. B. Hill, *Cooling Towers: Principles and Practice*, 1970.

Cracking

A process used in the petroleum industry to reduce the molecular weight of hydrocarbons by breaking molecular bonds. Cracking is carried out by thermal, catalytic, or hydrocracking methods. Increasing demand for gasoline and other middle distillates relative to demand for heavier fractions makes cracking processes important in balancing the supply of petroleum products.

Thermal cracking depends on a free-radical mechanism to cause scission of hydrocarbon carbon-carbon bonds and a reduction in molecular size, with the formation of olefins, paraffins, and some aromatics. Side reactions such as radical saturation and polymerization are controlled by regulating reaction conditions. In catalytic cracking, carbonium ions are formed on a catalyst surface, where bond scissions, isomerizations, hydrogen exchange, and so on, yield lower olefins, isoparafins, isoolefins, and aromatics. Hydrocracking, a relative newcomer to the industry, is based on catalytic formation of hydrogen radicals to break carbon-carbon bonds and saturate olefinic bonds. Hydrocracking converts intermediate- and high-boiling distillates to middle distillates, high in paraffins and low in cyclics and olefins. Hydrocracking also causes hydrodealkylation of alkyl-aryl components in heavy reformate to produce benzene and naphthalene. *See* HYDROCRACKING.

Thermal cracking. This is a process in which carbon-to-carbon bonds are severed by the action of heat alone. It consists essentially in the heating of any fraction of petroleum to a temperature at which substantial thermal decomposition takes place through a thermal free-radical mechanism followed by cooling, condensation, and physical separation of the reaction products.

There are a number of refinery processes based primarily upon the thermal cracking reaction. They differ primarily in the intensity of the thermal conditions and the feedstock handled.

Visbreaking is a mild thermal cracking operation (850–950°F; 454–510°C) where only 20–25% of the residuum feed is converted to mid-distillate and lighter material. It is practiced to reduce the volume of heavy residuum which must be blended with low-grade fuel oils.

Thermal gas-oil or naphtha cracking is a more severe thermal operation (950–1100°F; 510–593°C) where 45% or more of the feed is converted to lower molecular weight. Attempts to crack residua under these conditions would coke the furnace tubes.

Steam cracking is an extremely severe thermal cracking operation (1100 to 1400°F; 593–760°C) in which steam is used as a diluent to achieve a very low hydrocarbon partial pressure. Primary products desired are olefins such as ethylene and butadiene.

Fluid coking is a thermal operation where the residuum is converted fully to gas-oil products boiling lower than 950°F (510°C) and coke. The thermal conversion is carried out on the surface of a fluidized bed of coke particles.

Delayed coking is a thermal cracking operation wherein a residuum is heated and sent to a coke drum, where the liquid has an infinite residence time to convert to lower-molecular-weight hydrocarbons which distill out of the coke drum, and to coke which remains in the drum and must be periodically removed.

In fluid coking and delayed coking, there is total conversion of the very heavy high-boiling end of the residuum feed.

Although there are many variations of visbreaking and thermal cracking, most commonly a feedstock that boils at higher temperatures than gasoline is pumped at inlet pressures of 75–1000 psig (1 psi = 6895 Pa) through steel tubes so placed in a furnace as to allow gradual heating of the coil to temperatures in the range 850–1100°F (454–593°C). The flow rate is controlled to provide sufficient time for the required cracking to lighter products; the time may be extended by subsequently passing the hot products through a reaction chamber that is maintained at a high temperature. To achieve optimum process efficiency, part of the overhead product ordinarily is returned to the cracking unit for further cracking (Fig. 1).

Crude oils differ in their compositions, both in molecular weight and molecular type of hydrocarbon. Since refiners must make products in harmony with market demand, they often need to alter the molecular structure of the hydrocarbons. The cracking of heavy distillates and residual oils increases the yields of gasoline and the light intermediate distillates used as diesel fuels and domestic heating oils, as well as providing low-molecular-weight olefins needed for the manufacture of chemicals and polymers.

Beginning in 1912, thermal cracking proved for many years to be eminently suitable for this purpose. During the period 1920–1940, more efficient automobile engines of higher compression ratios were developed. These engines required higher-octane-number gasolines, and thermal cracking operations in the United States were expanded to meet this need. Advantageously, thermal cracking reactions produce olefins and aromatics, leading to gasolines generally of higher octane number than those obtained by simple distillation of the same crude oils. The general nature of the hydrocarbon products and the basic mechanism of thermal cracking is well described by the free-radical theory of the pyrolysis of hydrocarbons.

In the early 1930s, the petrochemical industry

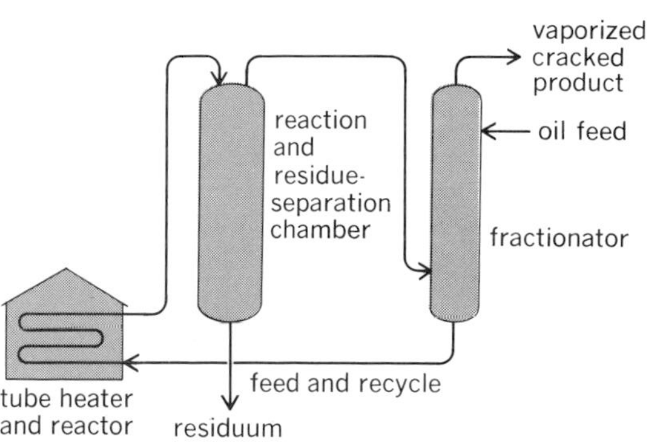

Fig. 1. Thermal cracking unit.

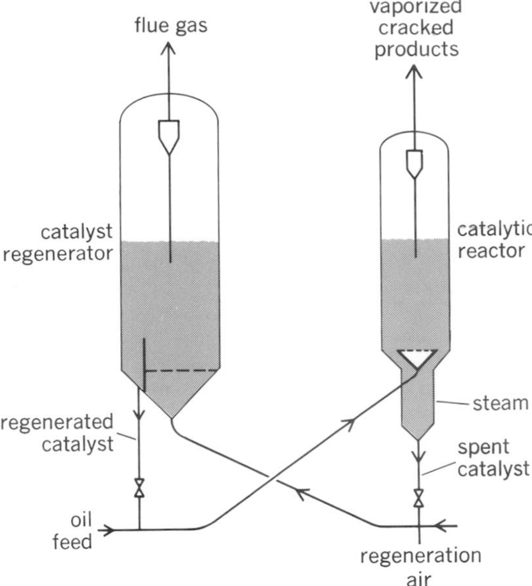

Fig. 2. Fluid catalytic cracking unit.

began its growth. Olefinic gases from thermal cracking operations, especially propylene and butylenes, were used as the chief raw materials for the production of aliphatic chemicals. Simultaneously, practical catalytic processes were invented for polymerizing propylene and butylenes to gasoline components, and for dimerizing and hydrogenating isobutylene to isooctane (2,2,4-trimethylpentane), the prototype 100-octane fuel. Just prior to World War II, the alkylation of light olefins with isoparaffins to produce unusually high-octane gasoline components was discovered and extensively applied for military aviation use. These developments resulted in intense engineering efforts to bring thermal cracking operations to maximum efficiency, as exemplified by a number of processes made available to the industry.

Since World War II, thermal cracking has been largely supplanted by catalytic cracking, both for the manufacture of high-octane gasoline and as a source of light olefins. It is, however, still used widely for the mild cracking of heavy residues to reduce their viscosities and for final cracking of gas oils derived from catalytic cracking.

Since carbon-to-hydrogen bonds are also severed in the course of thermal cracking, two hydrogen atoms can be removed from adjacent carbon atoms in a saturated hydrocarbon, producing molecular hydrogen and an olefin. This reaction prevails in ethane and propane cracking to yield ethylene and propylene, respectively. Methane cracking is a unique case wherein molecular hydrogen is obtained as a primary product and carbon as a coproduct. These processes generally operate at low pressures and high temperatures and in some cases utilize regenerative heating chambers lined with firebrick, or equipment through which preheated refractory pebbles continuously flow. Such conditions also favor the production of aromatics and diolefins from normally liquid feedstocks and are applied commercially to a limited extent despite relatively low yields of the desired products.

Catalytic cracking. This is the major process used throughout most of the world oil industry for the production of high-octane quality gasoline by the conversion of intermediate- and high-boiling petroleum distillates to lower-molecular-weight products. Oil heated to within the lower range of thermal cracking temperatures (850–1025°F; 454–551°C) reacts in the presence of an acidic inorganic catalyst under low pressures (10–35 psig). Gasoline of much higher octane number is obtained than from thermal cracking, a principal reason for the widespread adoption of catalytic cracking. All nonvolatile carbonaceous materials are deposited on the catalyst as coke and are burned off during catalyst regeneration.

In contrast to thermal cracking, residual oils are not generally processed, because excessive amounts of coke are deposited on the catalyst, and inorganic components of these oils contaminate the catalyst. Feedstocks usually are restricted to distillates boiling above gasoline.

Catalytic cracking, as conceived by E. J. Houdry in France, reached commercial status in 1936 after extensive engineering development by American oil companies. In its first form, the process used a series of fixed beds of catalyst in large steel cases. Each of these alternated between oil cracking and catalyst-coke burning at intervals of about 10 min and had suitable provision for heat and temperature control.

Successful operation led to major engineering improvements, and the goals of much improved efficiency, enlarged capacity, and ease of operation were achieved by two different systems. One employs a moving bed of small pellets or beads of catalyst traveling continuously through the oil-cracking vessel and subsequently through a regeneration kiln. The beads are lifted mechanically or by air to the top of the structure to flow down through the vessels again. This process has two commercially engineered embodiments, the Thermofor (or TCC) and the Houdriflow processes, which are similar in general process arrangements.

These moving-bed processes were limited in size

Representative yield structures for three different processing objectives in catalytic cracking

Process objective	Light gases	Gasoline	Light cycle oil
Feed	Light gas oil	Gas oil	Gas oil
Reactor temperature, °F	990	950–990	890–900
Light gases, wt %	4.5	2.8	1.6
Propane/propylene, vol %	*15*	10.0	7.5
Butane/butylene, vol %	*22*	16.4	11.2
Gasoline, vol %	46	*69.5*	32.6
Catalytic diesel oil, vol %	18	10	*43.6*
Bottom, vol %	5	5	5

and are now technically obsolete. They are being replaced by another type of unit, a fluid solids unit, as dictated by economic considerations, and no moving-bed catalytic cracking units are being constructed. In the fluid solids unit, a finely divided powdered catalyst is transported between oil-cracking and air-regeneration vessels in a fluidized state by gaseous streams in a continuous cycle. This system employs the principle of balanced hydrostatic heads of fluidized catalyst between the two vessels. Catalyst is moved by injecting heated oil vapors into the transport line from the regenerator to the reactor, and by injecting air into the transport line the reactor stripper to the regenerator. Large amounts of catalyst can be moved rapidly; cracking units of total oil intake as great as 180,000 bbl/day (28,800 m^3/day) are in operation (Fig. 2).

In both the moving-bed and fluidized systems, the circulating catalyst provides the cracking heat. Coke deposited on the catalyst during cracking is burned at controlled air rates during regeneration; heat of combustion is converted largely to sensible heat of the catalyst, which supplies the endothermic heat of cracking in the reaction vessel.

Gasoline of 90–95 research octane number without tetraethyllead is rather uniformly produced by catalytic cracking of fractions from a wide variety of crude oils, compared with 65–80 research octane number via thermal cracking, the latter figures varying with crude oil source.

Although the primary objective of catalytic cracking is the production of maximum yields of gasoline concordant with efficient operation of the process, large amounts of normally gaseous hydrocarbons are produced at the same time. The gaseous hydrocarbons include propylene and butylenes, which are in great demand for chemical manufacture. Isobutane and isopentane are also produced in large quantities and are valuable for the alkylation of olefins, as well as for directly blending into gasoline as high-octane components.

The other chief product is the material boiling above gasoline, designated as catalytically cracked gas oil. It contains hydrocarbons relatively resistant to further cracking, particularly polycyclic aromatics. The lighter portion may be used directly or blended with straight-run and thermally cracked distillates of the same boiling range for use as diesel and heating oils. Part of the heavier portion is recycled with fresh feedstock to obtain additional conversion to lighter products. The remainder is withdrawn for blending with residual oils so as to reduce the viscosity of heavy fuel, otherwise it is subjected to a final step of thermal cracking.

Thus, the catalytic cracking process is used in refineries to shift the production of products to match swings in market demand. It can process a wide variety of feeds to different product compositions. For example, light gases, gasoline, or diesel oil can be emphasized by varying process conditions, feedstocks, and boiling range of products as shown in the table.

To account for the difference between the product compositions obtained by catalytic and thermal cracking, the mechanism of cracking over acidic catalysts has been investigated intensively. In thermal cracking, free radicals are reaction intermediates, and the products are determined by their specific decomposition patterns. In contrast, catalytic cracking takes place through ionic intermediates, designated as carbonium ions (positively charged free radicals) generated at the catalyst surface. Although there is a certain parallelism between the modes of cracking of free radicals and carbonium ions, the latter undergo rapid intramolecular rearrangement reactions prior to cracking. This leads to more highly branched hydrocarbon structures than those from thermal cracking, and to important differences in the molecular weight distribution of the cracked products. Furthermore, the cracked products undergo much more extensive secondary reactions in the presence of the catalyst.

The catalytic cracking mechanism also favors the production of aromatics in the gasoline boiling range; these reach quite high concentrations in the higher-boiling portion. This characteristic, together with the copious production of branched aliphatic hydrocarbons especially in the lower-boiling portion, is largely responsible for the high octane rating of catalytically cracked gasoline.

Cracking catalysts must have two essential properties: (1) a chemical composition capable of maintaining a high degree of acidity, preferably as readily available hydrogen ions (protons); and (2) a physical structure of high porosity (high surface area). Mechanical durability is also necessary for industrial use.

Cracking catalysts are essentially silica-alumina compositions. A dramatic improvement in catalytic unit performance occurred with the switch from acid-treated clays (montmorillonite or kaolinite) to synthetic silica-aluminas. After 1960, a new group of aluminosilicates, molecular sieve zeolites, were introduced into the catalyst formation. These crystalline materials (Fig. 3) have cracking activity 50 to 100 times the previous amorphous catalyst. They permit cracking to greater conversion levels, producing more gasoline, less coke, and less gas.

As the catalyst particles pass through the reactor regenerator system every 3 to 15 min they are gradually deactivated, through loss of surface area by the effect of heat and steam and contamination through the effects of heat and steam, and the particles are contaminated by the trace metallic components on the feedstocks, mainly nickel, vanadium, and copper. They also undergo mechanical attrition, and fines are lost in the reactor and regenerator gases. To compensate, fresh catalyst is added.

The zeolite catalysts resulted in considerable change in the process itself. The catalysts are more resistant to thermal degradation, and regenerator temperatures can be safely raised to the 1350°F (732°C) level. The carbon on regenerated catalyst is reduced to the 0.05 wt % level resulting in improved gasoline yields. In addition, all the carbon monoxide produced at lower generation temperatures (10% concentration in the regenerator flue gas) can be combusted in the regenerator, making for a more efficient recovery of combustion heat and reduced atmospheric pollution. The effluent CO concentration can be reduced to less than 0.05 vol %.

The high-activity zeolite catalyst has permitted units to be designed with all riser cracking (Fig. 4), wherein all the cracking reaction takes place in a

CRACKING

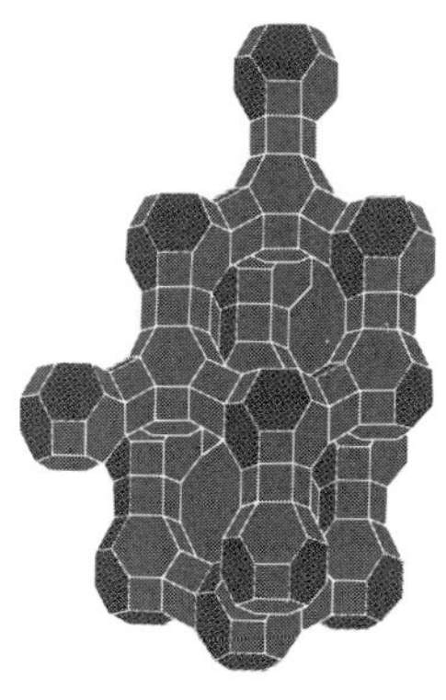

Fig. 3. Model of zeolite type Y.

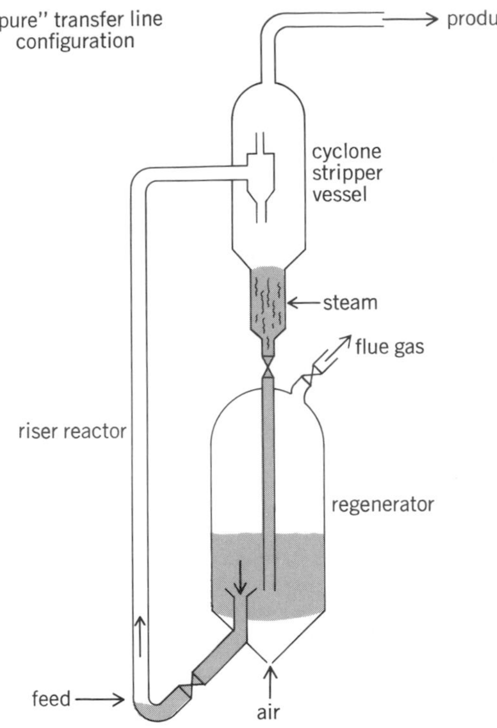

Fig. 4. Riser catalytic cracker.

relatively dilute (less than 2 lb/ft³) catalyst suspension in a 2- or 5-sec residence time. No dense-bed (10–15 lb/ft³) cracking exists.

Many old units are being converted to riser crackers, and virtually all new units feature riser cracking. Some riser crackers are also provided with small dense beds to achieve an optimum yield pattern.

With added emphasis on protection of the environment, complex facilities are provided to remove pollutants from the effluent regeneration gases. Third-stage cyclone collectors, electrostatic precipitators, and scrubbers are used commercially to meet regulations.

To conserve energy in the process, flue gas expanders are used in the flue gas system to provide more than enough energy to compress the regeneration air.

With the coming emphasis on conservation of petroleum resources, catalytic cracking is assuming a more important role. The heavier products from a refinery will probably be replaced by coal-, shale-, or tar-sand-derived products. Thus, there will be a need for greater conversion of petroleum to gasoline. Emphasis will be on catalytic cracking since it is the cheapest major conversion process available. *See* ALKYLATION (PETROLEUM); CATALYTIC REFORMING; PETROLEUM PROCESSING.

[BERNARD S. GREENSFELDER/ EDWARD LUCKENBACK]

Bibliography: F. H. Blanding, Reaction rates in catalytic cracking of petroleum, *I and EC*, 45: 1185–1197, June 1953; G. D. Hobson and W. Pohl *Modern Petroleum Technology*, 4th ed., 1973; K. A. Kobe and J. J. McKetta, Jr., *Advances in Petroleum Chemistry and Refining*, vols. 5 and 6, 1962; W. L. Nelson, *Petroleum Refinery Engineering*, pp. 759–818, 1958; M. Sittig, Catalytic cracking techniques in review, *Petrol. Refinery*, 37:263–316, 1952; J. W. Ward and S. A. Quader (eds.), *Hydrocracking and Hydrotreating*, 1975.

Critical mass

That amount of fissile material (U^{233}, U^{235}, or Pu^{239}) which permits a self-sustaining chain reaction. The critical mass ranges from about 950 g of U^{235} or a smaller amount of Pu^{239} for dissolved compounds through 16 kg for a solid metallic sphere of U^{235}, and up to hundreds of tons for some power reactors. It is increased by the presence of such neutron absorptive materials as admixed U^{238}, aluminum pipes for flow of coolant, and boron or cadmium control rods. It is reduced by a moderator, such as graphite or heavy water, which slows down the neutrons, inhibits their escape, and indirectly increases their chance to produce fission. *See* CHAIN REACTION; NUCLEAR FISSION.

[JOHN A. WHEELER]

Cyclone furnace

A water-cooled horizontal cylinder in which fuel is fired and heat is released at extremely high rates. When firing coal, the crushed coal, approximately 95% of which is sized at 1/4 in. (6 mm) or less, is introduced tangentially into the burner at the front end of the cyclone (see illustration). About 15% of the combustion air is used as primary and tertiary air to impart a whirling motion to the particles of coal. The whirling, or centrifugal, action on the fuel is further increased by the tangential admission of high-velocity secondary air into the cyclone.

The combustible is burned from the fuel in the cyclone furnace, and the resulting high-temperature gases melt the ash into liquid slag, a thin layer of which adheres to the walls of the cyclone. The incoming fuel particles, except those fines burned in suspension, are thrown to the walls by centrifugal force and caught in the running slag. The secondary air sweeps past the coal particles embedded in the slag surface at high speed. Thus, the air required to burn the coal is quickly supplied and the products of combustion are rapidly removed.

The products of combustion are discharged through a water-cooled reentrant throat at the rear of the cyclone into the boiler furnace. The part of the molten slag that does not adhere to the cyclone walls also flows toward the rear and is discharged through a taphole into the boiler furnace.

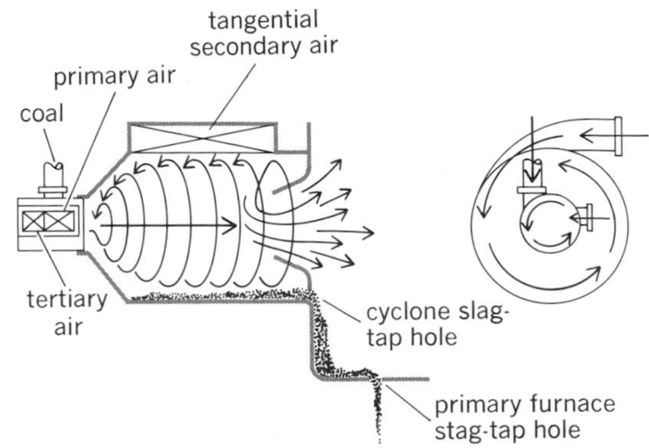

Schematic diagram of cyclone furnace. (*From T. Baumeister, ed., Standard Handbook for Mechanical Engineers, 7th ed., McGraw-Hill, 1967*)

Essentially, the fundamental difference between cyclone furnaces and pulverized coal–fired furnaces is the manner in which combustion takes place. In pulverized coal–fired furnaces, particles of coal move along with the gas stream; consequently, relatively large furnaces are required to complete the combustion of the suspended fuel. With cyclonic firing, the coal is held in the cyclone and the air is passed over the fuel. Thus, large quantities of fuel can be fired and combustion completed in a relatively small volume, and the boiler furnace is used to cool the products of combustion.

Gas and oil can also be burned in cyclone furnaces at ratings and with performances equaling those of coal firing. When oil is the fuel, it is injected adjacent to the secondary air ports and directed downward into the secondary airstream. The oil is picked up and sufficiently atomized by the high-velocity air. Gas is fired similarly through flat, open-ended ports located in the secondary air entrance. *See* STEAM BOILER; STEAM-GENERATING FURNACE; STEAM-GENERATING UNIT.

[G. W. KESSLER]

Dam

A structure that bars or detains the flow of water in an open channel or watercourse. Dams are constructed for several principal purposes. Diversion dams divert water from a stream; navigation dams raise the level of a stream to increase the depth for navigation purposes (Fig. 1); power dams raise the level of a stream to create or concentrate hydrostatic head for power purposes; and storage dams store water for municipal and industrial use, irrigation, flood control, river regulation, recreation, or power production. A dam serving two or more purposes is called a multiple-purpose dam. Dams are commonly classified by the material from which they are constructed, such as masonry, concrete, earth, rock, timber, and steel. Most dams now are built either of concrete or of earth and rock.

Concrete dams. Concrete dams may be typed as gravity, arch, or buttress type. Gravity dams depend on weight for stability against overturning and for resistance to sliding on their foundations (Figs. 2 and 3). An arch dam may have a near-vertical face or, more usually, one that curves concave downstream (Figs. 4 and 5). The dam acts as an arch to transmit most of the horizontal thrust from the water pressure against the upstream face of the dam to the abutments of the dam. The buttress type of concrete dam includes the slab-and-buttress, or Ambursen, type; round- or diamond-head buttress type; multiple-arch type; and multiple-dome type. Buttress dams depend on the weight of the structure and of the water on the dam to resist overturning and sliding.

Forces acting on concrete dams. Principal forces acting on a concrete dam are (1) vertical forces from weight of the structure and vertical component of water pressure against the upstream and downstream faces of the dam, (2) uplift pressures under the base of the structure, (3) horizontal forces from the horizontal component of the water pressure against the upstream and downstream faces of the dam, (4) forces from earthquake accelerations in regions subject to earthquakes, (5) temperature stresses, (6) pressures from silt

Fig. 1. John Day Lock and Dam, looking upstream across the Columbia River at the Washington shore. In the foreground the navigation lock may be seen, beyond it the spillway dam, and then the powerhouse. The John Day multiple-purpose project boasts the highest single-lift navigation lock in the United States. (*U.S. Army Corps of Engineers*)

deposits and earth fills against the structure, and (7) ice pressures.

The uplift pressure under the base of a dam varies with the effectiveness of the foundation drainage system and with the perviousness of the foundation.

Earthquake loads are usually selected after consideration of the accelerations which may be expected at the site as indicated by the geology, proximity to major faults, and the earthquake history of the region. Conventionally, earthquake forces have been treated as static forces representing the effects of the acceleration of the dam itself and the hydrodynamic force produced against the dam by water in the reservoir. Such horizontal forces often are assumed to equal 0.05–0.10 the force of gravity, with a somewhat smaller vertical force. Dynamic analysis procedures have been developed which determine the structure's response to combined effects of the contemplated ground motion and the structure's dynamic properties.

Stresses resulting from temperature changes must be considered in analyzing arch dams. These stresses are usually disregarded in the design of concrete gravity dams, but must be controlled to acceptable limits by concreting and curing methods, discussed below.

Pressure from silt deposited in the reservoir against the dam is considered only after sedimentation studies indicate that it may be a significant factor. Backfill pressures are important where a concrete gravity dam ties into an embankment.

Ice pressure, applied at the maximum elevation at which the ice will occur in project operations, is considered when conditions indicate that it would be significant. The pressure, commonly assumed to be 10,000–20,000 lb per linear foot (150–300 kN/m), results from the thermal expansion of the ice sheet and varies with the rate and magnitude of temperature rise and thickness of the ice.

Fig. 2. Green Peter Dam, a concrete gravity type on the Middle Santiam River, Willamette River Basin, OR. Gate-controlled overflow-type spillway is constructed through crest of dam; powerhouse is at downstream toe of dam. (*U.S. Army Corps of Engineers*)

Stability and allowable stresses. Stability of a concrete gravity dam is evaluated by analyzing the available resistance to overturning and sliding. To satisfy the former, the resultant of forces is required to fall within the middle third of the base under normal load conditions. Sliding stability is assured by requiring available shear and friction resistance to be greater by a designated safety factor than the forces tending to produce sliding. The strengths used in computing resistance to sliding are based on investigation and tests of the foundation. Bearing strength of the foundation for a gravity dam is a controlling factor only for weak foundations or for high dams. Because an arch dam depends on the competency of the abutments, the rock bearing strength must be sufficient to provide an adequate safety factor for the compressive stresses, and the resistance to sliding along any weak surface must be great enough to provide an adequate safety factor.

Concrete stresses control the design of arch dams, but ordinarily not gravity dams. Stresses adopted for concrete arch dams are conservative. A safety factor of 4 on concrete compressive strength is commonly used for normal load conditions.

Concrete temperature control. Volume changes accompanying temperature changes in a concrete dam tend to cause the development of tensile stresses. A major factor in development of temperature changes within a concrete mass is the heat developed by chemical changes in the concrete after placement. Uncontrolled temperature changes can cause cracking which may endanger the stability of a dam, cause leakage, and reduce durability. Temperatures are controlled by using cementing materials having low heat of hydration, and by artificial cooling by precooling the concrete mix or circulating cold water through pipes embedded in the concrete or both.

Concrete dams are constructed in blocks, with the joints between the blocks serving as contraction joints (Fig. 6). In arch dams the contraction joints are filled with cement grout after maximum shrinkage has occurred to assure continuous bearing surfaces normal to the compressional forces set up in the arch when the water load is applied to the dam.

Quality control. During construction, continuing testing and inspection are performed to ensure that the concrete will be of required quality. Tests are also made on materials used in manufacture of the concrete, and concrete batching, mixing, transporting, placing, curing, and protection are continuously inspected.

Earth dams. Earth dams have been used for water storage since early civilizations. Improvements in earth-materials techniques, particularly the development of modern earth-handling equipment, have brought about a wider use of this type of dam, and today as in primitive times the earth embankment is the most common dam (Figs. 7 and 8). Earth dams may be built of rock, gravel, sand, silt, or clay in various combinations.

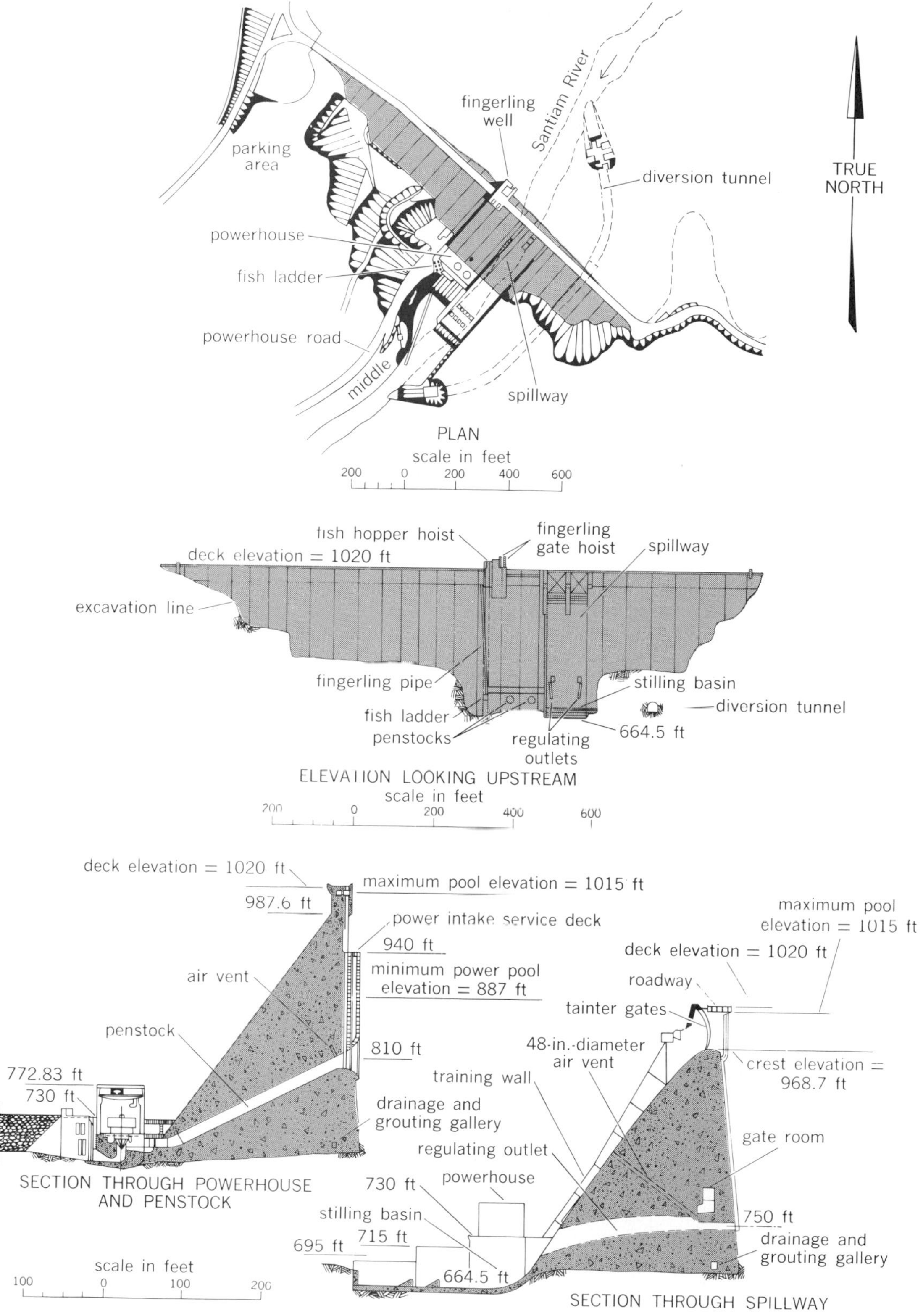

Fig. 3. Plan and sections of Green Peter Dam. 1 ft = 0.3048 m. 48 in. = 122 cm. (*U.S. Army Corps of Engineers*)

Most earth dams are constructed with an inner impervious core, with upstream and downstream zones of more pervious materials, sometimes including rock zones. Earth dams limit the flow of water through the dam by use of fine-grained soils. Where possible, these soils are formed into a relatively impervious core. When there is a sand or gravel foundation, the core may be connected to bedrock by a cutoff trench backfilled with compacted soil. If such cutoffs are not economically feasible because of the great depth of pervious foundation soils, then the central impervious core is connected to a long horizontal upstream impervious blanket that increases the length of the seepage path. The impervious core is often encased in pervious zones of sand, gravel, or rock fill for stability. When there is a large difference in the particle sizes of the core and pervious zones, transition zones are required to prevent the core material from being transported into the pervious zones by seeping water. In some cases where pervious soils are scarce, the entire dam may be a homogeneous fill of relatively impervious soil. Downstream pervious drainage blankets are provided to collect seepage passing through, under, and around the abutments of the dam.

Materials can be obtained from required excavations for the dam and appurtenances or from borrow areas. Rock fill is generally used when large quantities of rock are available from required excavation or when soil borrow is scarce.

Earth-fill embankment is placed in layers and compacted by sheepsfoot rollers or heavy pneumatic-tire rollers. Moisture content of silt and clay soils is carefully controlled to facilitate optimum compaction. Sand and gravel fills are compacted in slightly thicker layers by pneumatic-tire rollers, vibrating steel drum rollers, or placement equipment. The placement moisture content of pervious fills is less critical than for silts and clays. Rock fill usually is placed in layers 1–3 ft (0.3–0.9 m) deep and is compacted by placement equipment and vibrating steel drum rollers.

Spillways. A spillway releases water in excess of storage capacity so that the dam and its foundation are protected against erosion and possible failure. All dams must have a spillway, except small ones where the runoff can be safely stored in the reservoir without danger of overtopping the dam. Ample spillway capacity is of particular importance for large earth dams, which would be destroyed or severely damaged by being overtopped. Failure of a large dam could result in severe hazards to life and property downstream.

Types. Spillways are of two general types: the overflow type, constructed as an integral part of the dam; or the channel type, located as an independent structure discharging through an open chute or tunnel. Either type may be equipped with gates to control the discharge. Various control structures have been used for channel spillways, including the simple overflow weir, side-channel overflow weir, and drop or morning-glory inlet where the water flows over a circular weir crest and drops directly into a tunnel.

Unless the discharge end of a spillway is remote from the toe of the dam or erosion-resistant bedrock exists at shallow depths, some form of energy dissipator must be provided to protect the toe of the dam and the foundation from spillway discharges. For an overflow spillway the energy dissipator may be a stilling basin, a sloping apron downstream from the dam, or a submerged bucket. When a channel spillway terminates near the dam, it usually has a stilling basin. A flip bucket is used for both overflow and channel spillways when the flow can be deflected far enough downstream, usually onto rock, to prevent erosion at the toe of the dam or end of the spillway.

Gates. Several types of gates may be used to regulate and control the discharge of spillways (Fig. 9). Tainter gates are comparatively low in cost and require only a small amount of power for operation, being hydraulically balanced and of low friction. Drum gates, which are operated by reservoir pressure, are costly but afford a wide, unobstructed opening for passage of drift and ice over the gates. Vertical-lift gates of the fixed-wheel or roller type are sometimes used for spillway regulation, but are more difficult to operate than the others. Floating ring gates control the discharge of morning-glory spillways. Like the drum gate, this type offers a minimum of interference to the passage of ice or drift over the gate and requires no external power for operation.

Reservoir outlet works. These are used to regulate the release of water from the reservoir; they consist essentially of an intake and an outlet connected by a water passage, and are usually provided with gates. Outlet works usually have trashracks at the intake end to prevent clogging by debris. Bulkheads or stop logs are commonly provided to close the intakes so that the passages may be unwatered for inspection and maintenance. A stilling basin or other type of energy dissipator is usually provided at the outlet end.

Locations. Outlets may be sluices through concrete dams with control valves located in chambers in the dam or on the downstream end of the sluices, tunnels through the abutments of the dam, or cut-and-cover conduits extending along the foundation through an earth-fill dam. In the last case, the control valves are usually located within the dam or at the upstream end of the conduit, and special precautions must be taken to prevent leakage of water along the outside of the conduit.

Outlet control gates. Various gates and valves are used for regulating the release of water from reservoirs, including high-pressure slide gates, tractor gates (roller or wheel), and radial or tainter gates (Fig. 10); also needle valves of various kinds,

Fig. 4. East Canyon Dam, a thin-arch concrete structure on the East Canyon River, UT. Note uncontrolled overflow-type spillway through crest of dam at right center of photograph. (*U.S. Bureau of Reclamation*)

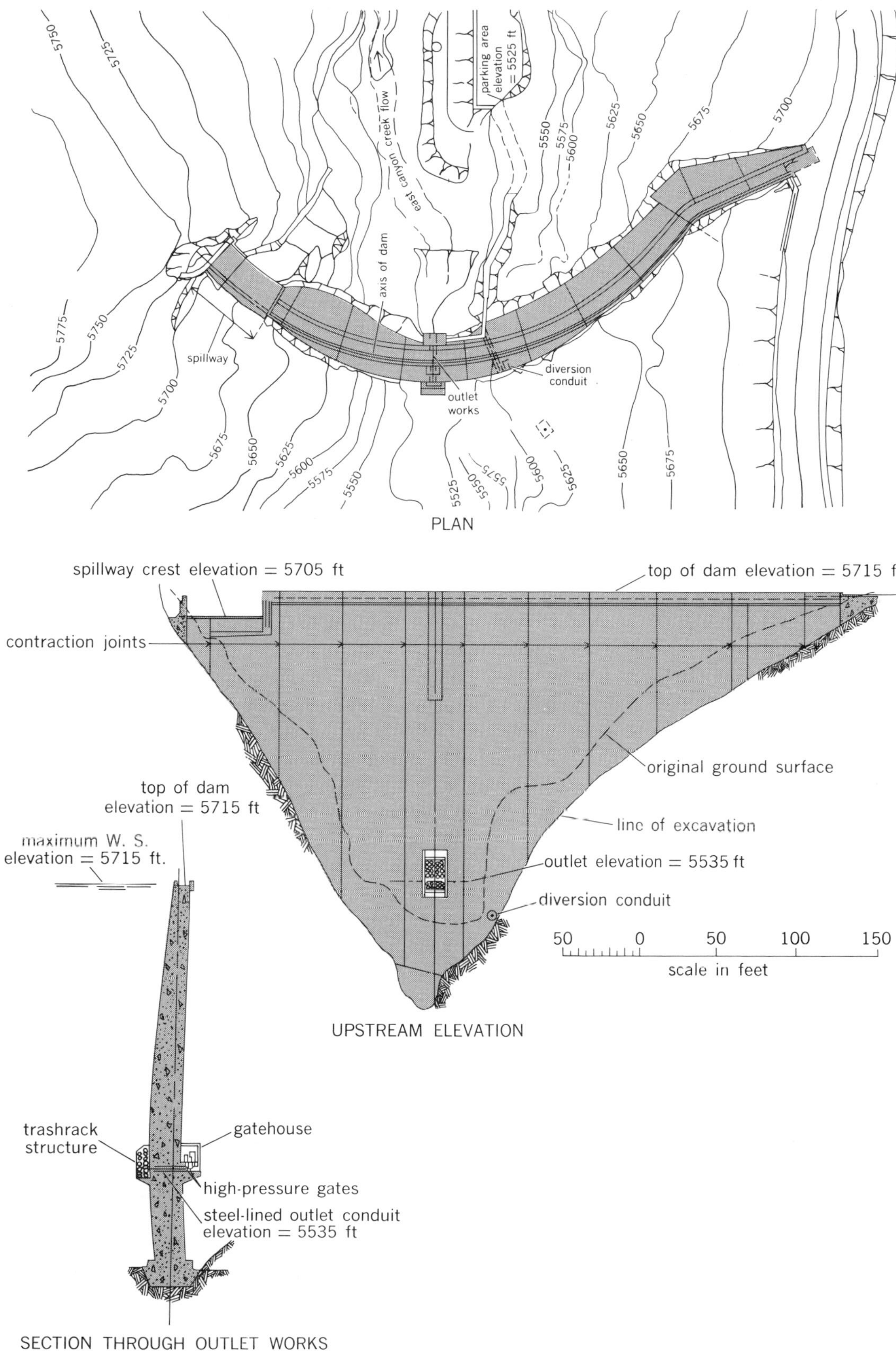

Fig. 5. Plan and sections of East Canyon Dam. 1 ft = 0.3048 m. (*U.S. Bureau of Reclamation*)

butterfly valves, fixed cone dispersion valves, and cylinder or sleeve valves. They must be capable of operating, without excessive vibration and cavitation, at any opening and at any head up to the maximum to which they may be subjected. They also must be capable of opening and closing under the maximum operating head. Emergency gates generally are used upstream of the operating gates,

Fig. 6. Block method of construction on a typical concrete gravity dam. *(U. S. Army Corps of Engineers)*

where stored water is valuable, so that closure can be made if the service gate should fail to function.

The slide gate, which consists of a movable leaf that slides on a stationary seat, is the most commonly used control gate. The high-pressure slide gate is of rugged design, having corrosion-resisting metal seats on both the movable rectangular leaf and the fixed frame. This gate has been used for regulating discharges under heads of over 600 ft.

Provision of low-level outlet. The usual storage reservoir has low-level outlets near the elevation of the stream bed to enable release of all the stored water. Some power and multiple-purpose dams have relatively high-level dead storage pools and do not require low-level outlets for ordinary operation. In such a dam, provision of a capability for emptying the reservoir in case of an emergency must be weighed against the additional cost.

Penstocks. A penstock is a pipe that conveys water from a forebay, reservoir, or other source to

Fig. 7. Aerial view of North Fork Dam, a combination earth and rock embankment on the North Fork of Pound River, VA. Channel-type spillway (left center) has simple overflow weir. *(U.S. Army Corps of Engineers)*

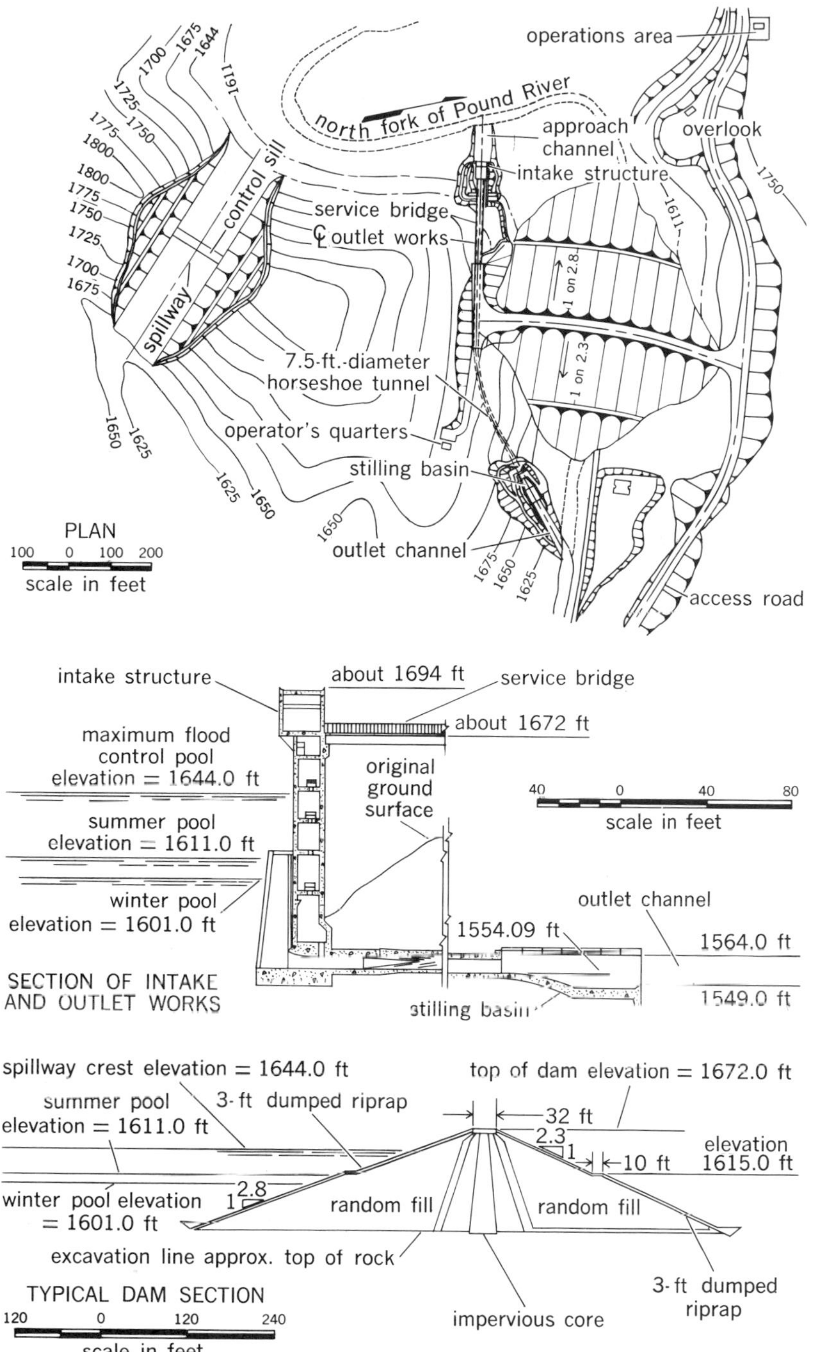

Fig. 8. Plan and sections of North Fork of Pound Dam. 1 ft = 0.3048 m. *(U.S. Army Corps of Engineers)*

a turbine in a hydroelectric plant. It is usually made of steel, but reinforced concrete and wood-stave pipe have also been used. Pressure rise and speed regulation must be considered in the design of a penstock.

Pressure rise, or water hammer, is the pressure change that occurs when the rate of flow in a pipe or conduit is changed rapidly. The intensity of this pressure change is proportional to the rate at which the velocity of the flow is accelerated or decelerated. Accurate determination of the pressure changes that occur in a penstock involves consideration of all operating conditions. For example, one important consideration is the pressure rise that occurs in a penstock when the turbine wicket gates are closed subsequent to the loss of load.

Selection of dam site. This depends upon such factors as hydrologic, topographic, and geologic conditions; storage capacity of reservoir; accessibility; cost of lands and necessary relocations of prior occupants or uses; and proximity of sources of suitable construction materials. For a storage dam the objective is to select the site where the desired amount of storage can be most economically developed. Power dams must be located to develop the desired head and storage. For a diversion dam the site must be considered in conjunc-

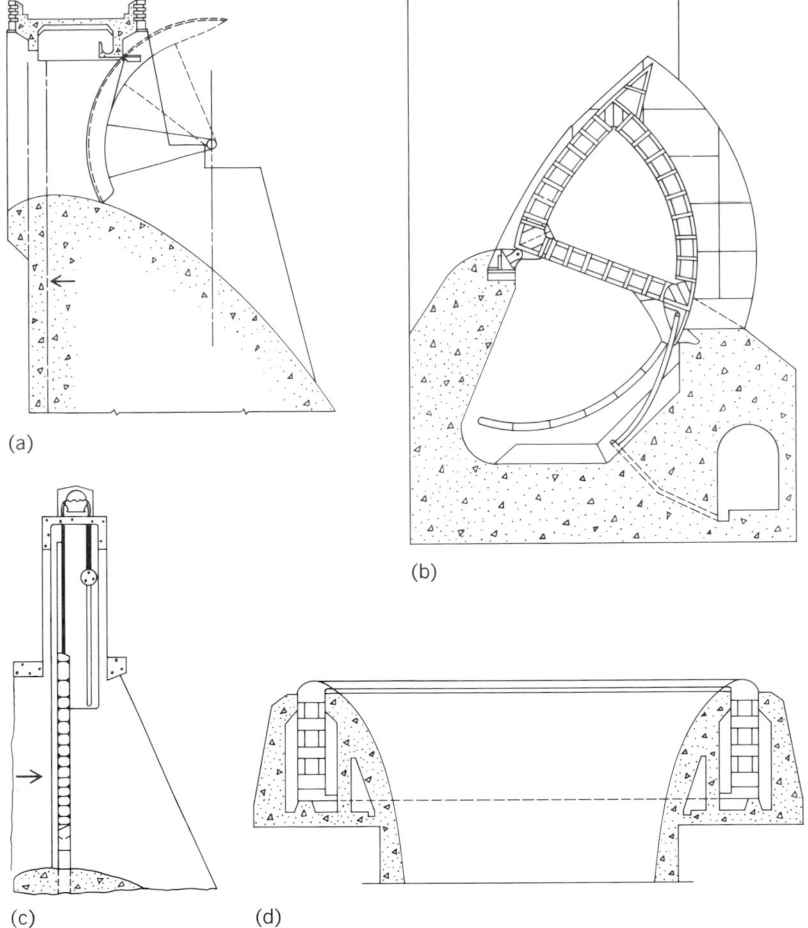

Fig. 9. Spillway gates. *(a)* Tainter gate. *(b)* Drum gate. *(c)* Vertical lift gate. *(d)* Ring gate. *(U.S. Army Corps of Engineers and U.S. Bureau of Reclamation)*

tion with the location and elevation of the outlet canal or conduit. Site selection for navigation dams involves special factors such as desired navigable depth and channel width, slope of river channel, natural river flow, amount of bank protection, amount of channel dredging, approach and exit conditions for tows, and locations of other dams in the system.

Unless topographic and geologic conditions for a proposed storage, power, or diversion dam site are satisfactory, hydrological features may need to be subordinated. Important topographic characteristics include width of the floodplain, shape and height of valley walls, existence of nearby saddles for spillways, and adequacy of reservoir rim to retain impounded water. Controlling geologic conditions include the depth, classification, and engineering properties of soils and bedrock at the dam site, and the occurrence of sinks, faults, and major landslides at the site or in the reservoir area. The elevation of the groundwater table is also significant because it will influence the construction operations and suitability of borrow materials. The beneficial effect of reservoir water on groundwater recharge may become an important consideration, as well as the adverse effects on existing or potential mineral resources and developments that would be destroyed or require relocation at the site or within the reservoir.

Selection of type of dam. This is made on the basis of the estimated costs of various types. The most important factors are topography, foundation conditions, and the accessibility of construction materials. In general, a hard-rock foundation is suitable for any type of dam, provided the rock has no unfavorable jointing, there is no danger of movement in existing faults, and foundation underseepage can be controlled at reasonable cost. Rock foundations of high quality are essential for arch dams because the abutments receive the full thrust of the water pressure against the face of the dam. Rock foundations are necessary for all medium and high concrete dams. An earth dam may be built on almost any kind of foundation if properly designed and constructed.

The chance of an embankment dam being most economical is improved if large spillway and outlet capacities are required and topography and foundation are favorable. In a wide valley a combination of an earth embankment dam and a concrete dam section containing the spillway and outlets often is economical. Availability of suitable construction materials frequently determines the most economical type of dam. A concrete dam requires adequate quantities of suitable concrete aggregate and reasonable availability of cement, while an earth dam requires sufficient quantities of both pervious and impervious earth materials. If quantities of earth materials are limited and enough rock is available, a rock-fill dam with an impervious earth core may be the most economical.

Determination of dam height. The dam must be high enough to (1) store water to the normal full-pool elevation required to meet intended functions of the project, (2) provide for the temporary storage needed to route the spillway design flood through the dam, and (3) provide sufficient freeboard height above the maximum surcharge elevation to assure an acceptable degree of safety against possible overtopping from waves and runup.

Physical characteristics of the dam and reservoir site or existing developments within the reservoir area may impose upper limits in selecting the normal full-pool level. In other circumstances economic considerations govern.

With the normal full-pool elevation established, flood flows of unusual magnitude may be passed by providing spillways and outlets large enough to discharge the probable maximum flood or other spillway design flood without raising the reservoir above the normal full elevation or, if it is more economical, by raising the height of the dam and obtaining additional lands to permit the reservoir to temporarily attain surcharge elevations above the normal pool level during extreme floods. Use of temporary surcharge storage capacity also serves to reduce the peak rates of spillway discharge.

Freeboard height is the distance between the maximum reservoir level and the top of the dam. Usually 3 ft (0.9 m) or more of freeboard is provided to avoid overtopping the dam by wind-generated waves. Additional freeboard may be provided for possible effects of surges induced by earthquakes, landslides, or other unpredictable events.

Diversion of stream. During construction the dam site must be unwatered so that the foundation may be prepared properly and materials in the structure may be moved easily into position. The stream may be diverted around the site through tunnels, passed through or around the construction area by flumes, passed through openings in the dam, or passed over low sections of a partially completed concrete dam. Diversion may be conducted in one or more stages, with a different method used for each stage. Initial diversion is conducted during a period of low flow to avoid the necessity for passing large flows.

Foundation treatment. The foundation of a dam must support the structure under all operating conditions. For concrete dams, following removal of unsatisfactory materials to a sound foundation surface, imperfections such as adversely oriented rock joints, open bedding planes, localized soft seams, and faults lying on or beneath the foundation surface receive special treatment. Necessary foundation treatment prior to dam construction may include "dental excavation" of surface weaknesses, or shafting and mining to remove deeper localized weaknesses, followed by backfilling with concrete or grout. Such work is sometimes supplemented by pattern grouting of foundation zones after construction of the dam. Foundation features such as rock joints, bedding planes, or faults that do not require preconstruction treatment are made relatively water-tight by curtain grouting from a line of deep grout holes located near the upstream heel and extending the full length of the dam. Although a grout curtain controls seepage at depth, the effectiveness of the grouting or its permanency cannot be relied upon alone to reduce hydrostatic pressures acting on the base of the dam. As a result, drain holes are drilled into the foundation just downstream of the grout curtain to intercept seepage passing through it and to reduce hydrostatic pressure. Occasionally chemical solutions such as acrylamide, sodium silicate, chromelignin, and polyester and epoxy resins are used for consolidating soils or rocks with fine openings.

The foundation of an earth dam must safely support the weight of the dam, limit seepage of stored water, and prevent transportation of dam or foundation material away or into open joints or seams in the rock by seepage. Earth-dam foundation treatment may include removal of excessively weak surface soils to prevent both potential sliding and excessive settlement of the dam, excavation of a cutoff trench to rock, and grouting of joints and seams in the bottom and downstream side of the cutoff trench. The cutoff trenches and grouting extend up the abutments, which are first stripped of weak surface materials.

When weak soils in the foundation of an earth dam cannot be removed economically, the slopes of the embankment must be flattened to reduce shear stress in the foundation to a value less than the soil strength. Relief wells are installed in pervious foundations to control seepage uplift pressures and to reduce the danger of piping when the depth of the pervious material is such as to preclude an economical cutoff.

Instrumentation. Instruments are installed at dams to observe structural behavior and physical conditions during construction and after filling, to check safety, and to provide information for design improvement.

In concrete dams instruments are used to measure stresses either directly or to measure strains from which stresses may be computed. Plumb lines are used to measure bending, and clinometers to measure tilting. Contraction joint openings are measured by joint meters spanning between two adjacent blocks of a dam. Temperatures are measured either by embedded electrical resistance thermometers or by adapting strain, stress, and joint measuring instruments. Water pressure on the base of a concrete dam at the contact with the foundation rock is measured by uplift pressure cells. Interior pressures in a concrete dam are measured by embedded pressure cells. Measurements are also made to determine horizontal and vertical movements; strong-motion accelerometers are being installed on and near dams in earthquake regions to record seismic data.

Instruments installed in earth-dam embankments and foundations are piezometers to determine pore water pressure in the soil or bedrock during construction and seepage after reservoir

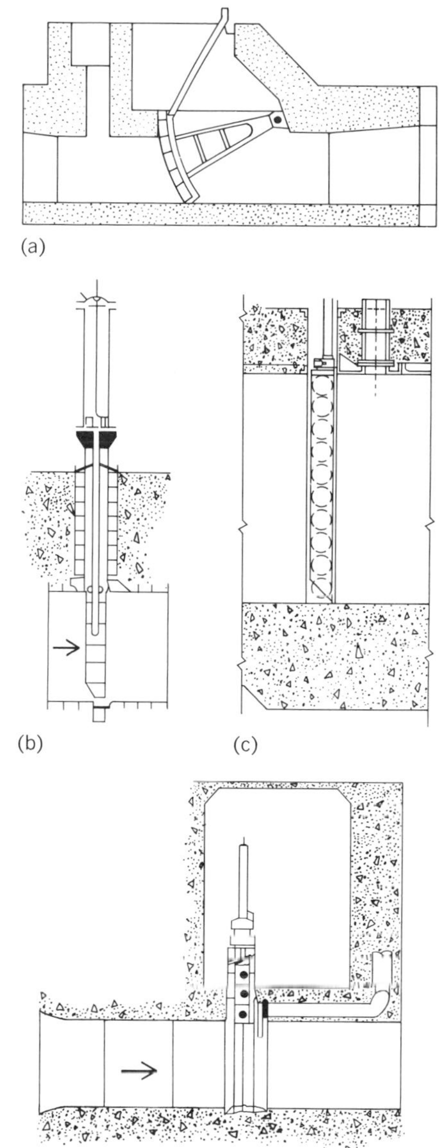

Fig. 10. Outlet gates. *(a)* Tainter gate. *(b)* High-pressure slide gate. *(c)* Tractor gate. *(d)* Jet flow gate. *(U.S. Army Corps of Engineers and U.S. Bureau of Reclamation)*

impoundments; settlement gages to determine settlements of the foundation of the dam under dead load; vertical and horizontal markers to determine movements, especially during construction; and inclinometers to determine horizontal movements along a vertical line.

Inspection of dams. Because failure of a dam may result in loss of life or property damage in the downstream area, it is essential that dams be inspected systematically both during construction and after completion. The design of dams should be reviewed to assure competency of the structure and its site, and inspections should be made during construction to ensure that the requirements of the design and specifications are incorporated.

After completion and filling, inspections may vary from cursory surveillance during day-to-day operation of the project to regularly scheduled comprehensive inspections. The objective of such inspections is to detect symptoms of possible distress in the dam at the earliest time. These symptoms include significant sloughs or slides in embankments; evidence of piping or boils near embankments; abnormal changes in flow from drains; unusual increases in seepage quantities; unexpected changes in pore water pressures or uplift pressures; unusual movement or cracking of embankments or abutments; significant cracking of concrete structures; appearance of sinkholes or localized subsidence near foundations; excessive deflection, displacement, erosion, or vibration of concrete structures; erratic movement or excessive deflection or vibration of outlet or spillway gates or valves; or any other unusual conditions in the structure or surrounding terrain.

Detection of symptoms of distress should be followed by an investigation of the causes, probable effects, and remedial measures required. Inspection of a dam and reservoir is particularly important following significant seismic events. Systematic monitoring of the instrumentation installed in dams is essential to the inspection program. [JACK R. THOMPSON]

Bibliography: American Concrete Institute, *Symposium on Mass Concrete*, Spec. Publ. SP-6, 1963; W. P. Creager, J. D. Justin, and J. Hinds, *Engineering for Dams*, 1945; C. V. Davis, *Handbook of Applied Hydraulics*, 3d ed., 1968; A. R. Golze (ed.), *Handbook of Dam Engineering*, 1977; J. L. Sherard et al., *Earth and Earth-Rock Dams*, 1963; G. B. Sowers and G. F. Sowers, *Introductory Soil Mechanics and Foundations*, 3d ed., 1970; U.S. Bureau of Reclamation, *Design of Small Dams*, 2d ed., 1973; U.S. Bureau of Reclamation, *Trial Load Method of Analyzing Arch Dams*, Boulder Canyon Proj. Final Rep., pt. 5, Bull. no. 1, 1938; H. M. Westergaard, Water pressures on dams during earthquakes, *Trans. ASCE*, 98:418–472, 1933.

Dehumidifier

Equipment designed to reduce the amount of water vapor in the atmosphere.

The atmosphere is a mechanical mixture of dry air and water vapor, the amount of water vapor being limited by air temperature. Water vapor is measured in either grains per pound of dry air or pounds per pound of dry air (7000 gr = 1 lb, 1 gr = 64.80 mg, 1 lb = 453.6 g).

There are three methods by which water vapor may be removed: (1) the use of sorbent materials, (2) cooling to the required dew point, and (3) compression with aftercooling.

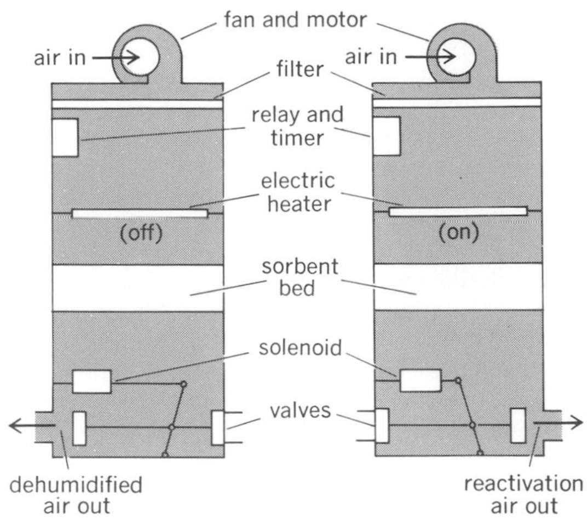

Fig. 1. Single-bed solid-sorbent dehumidifier. Dehumidifying cycle on left and reactivation cycle on right.

Sorbent type. Sorbents are materials which are hygroscopic to water vapor; they are available in both solid and liquid forms. Solid sorbents include silica gels, activated alumina, and aluminum bauxite. Liquid sorbents include halogen salts such as lithium chloride, lithium bromide, and calcium chloride, and organic liquids such as ethylene, diethylene, and triethylene glycols and glycol derivatives.

Solid sorbents may be used in static or dynamic dehumidifiers. Bags of solid sorbent materials within packages of machine tools, electronic equipment, and other valuable materials subject to moisture damage constitute static dehumidifiers. An indicator chemical may be included to show by a change in color when the sorbent is saturated. The sorbent then requires reactivation by heating at 300–350°F (150–180°C) for 1–2 hr before reuse.

A dynamic dehumidifier for solid sorbent consists of a main circulating fan, one or more beds of sorbent material, reactivation air fan, heater, mechanism to change from dehumidifying to reactivation, and aftercooler.

A single-bed dehumidifier (Fig. 1) operates on an

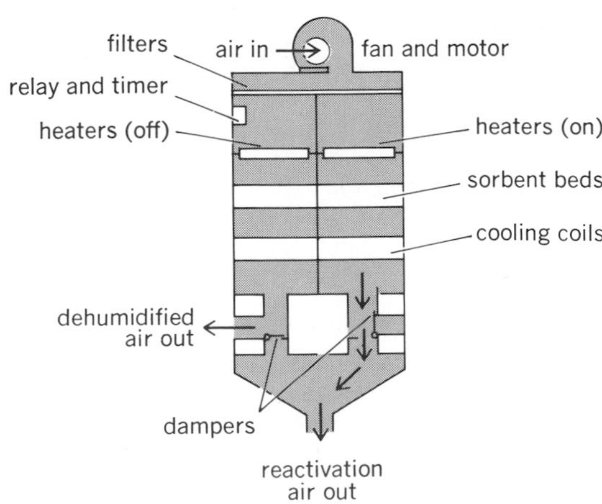

Fig. 2. Dual-bed solid-sorbent dehumidifier. Air is being dehumidified through the left bed at the same time that the right bed is being reactivated.

intermittent cycle of dehumidifying for 2–3 hr and then swtiches to the reactivation cycle for 15–45 min. No dehumidification is obtained during the reactivation cycle. A single-bed unit is used for small areas where moisture is a problem. The moist reactivation air is discharged to the outside.

The dual-bed machine is larger in capacity than the single-bed unit and has the advantage of providing a continuous supply of dehumidified air (Fig. 2). While one bed is dehumidifying, the other bed is reactivating. After a predetermined time interval, the air cycle is switched to pass the air through the reactivated bed for dehumidification and to reactivate the saturated bed.

The dew point of the effluent air of a fixed-bed machine is lowest at the start of a cycle immediately after the reactivated bed has been placed in service. The dew point gradually rises as the bed absorbs the water vapor and eventually would be the same as the entering dew point when the vapor pressure of the sorbent reached the vapor pressure of the air and could no longer absorb moisture from the air. The cutoff point at which the absorbing bed is changed over to reactivation is fixed by the maximum allowable effluent dew point.

A multibed unit with short operating cycles will reduce the range of effluent dew point to within a few degrees. A unit with rotating cylindrical bed maintains a reasonably constant effluent dew point.

The liquid-sorbent dehumidifier consists of a main circulating fan, sorbent-air contactor, sorbent pump, and reactivator including contactor, fan, heater, and cooler (Fig. 3). This unit will control the effluent dew point at a constant level because dehumidification and reactivation are continuous operations with a small part of the sorbent constantly bled off from the main circulating system and reactivated to the concentration required for the desired effluent dew point.

Cooling type. A system employing the use of cooling for dehumidifying consists of a circulating fan and cooling coil. The cooling coil may use cold water obtained from wells or a refrigeration plant, or may be a direct-expansion refrigeration coil. In place of a coil, a spray washer may be used in which the air passes through two or more banks of sprays of cold water or brine, depending upon the dew-point temperature required.

When coils are used, the leaving dew point is seldom below 35°F (2°C) because of possible build-up of ice on the coil. When it is necessary to use coils for temperatures below 35°F, as in cold-storage rooms, either two coils are used so one can be defrosted while the other is in operation, or only one coil is used and dehumidifying is stopped during the defrost period.

A brine-spray dehumidifier or brine-sprayed coil can produce dew-point temperatures below 35°F without frosting if properly operated and maintained.

Compression type. Dehumidifying by compression and aftercooling is used when the reduction of water vapor in a compressed-air system is required. This is particularly important, for example, if the air is used for automatic control instruments or cleaning of delicate machined parts.

If air is compressed and the heat of compression removed to bring the temperature of the air back to the temperature entering the compressor, condensation will take place and the remaining water

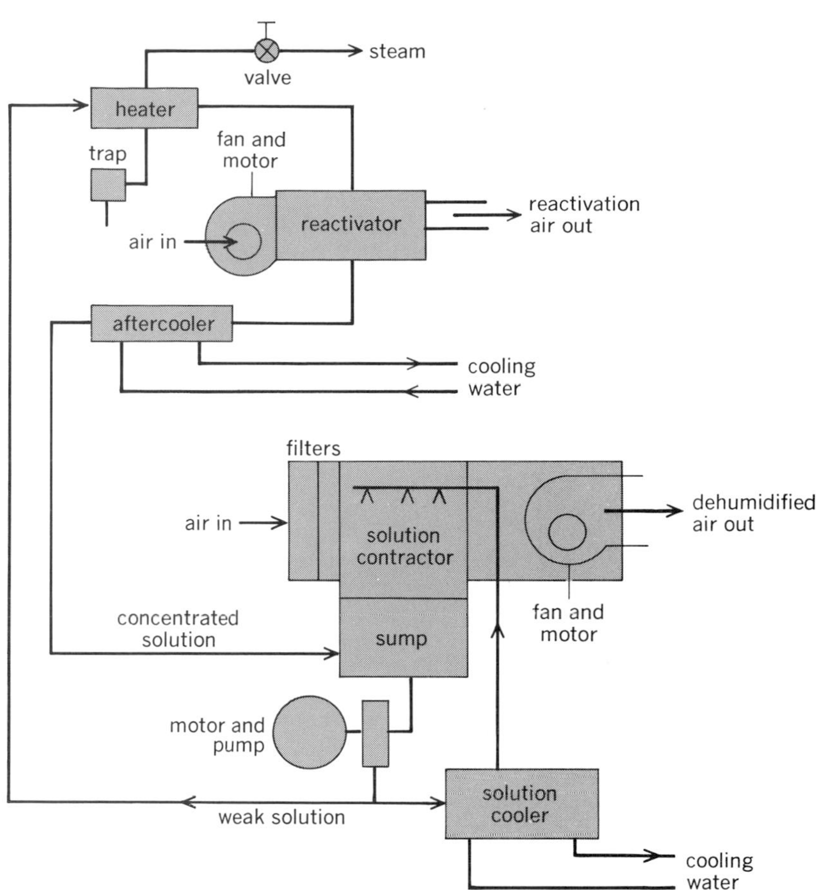

Fig. 3. Liquid-sorbent dehumidifier with continuous dehumidifying and reactivation.

vapor content will be directly proportional to the absolute pressure ratio of the compressed air (Fig. 4).

For example, if saturated air at 70°F or 21°C (111 gr/lb or 15.9 g/kg of dry air) is compressed from atmospheric pressure (14.7 pounds per square inch absolute, psia, or 101.3 kPa absolute) to 88 psia or 607 kPa absolute (6:1 compression ratio) and cooled to 70°F, the remaining water vapor in the compressed air will be 111/6 = 18.5 gr/lb (2.65 g/kg) of dry air. If the air is expanded back to atmospheric pressure and 70°F, the dew point will be 24°F (−4°C).

The power required for compression systems is

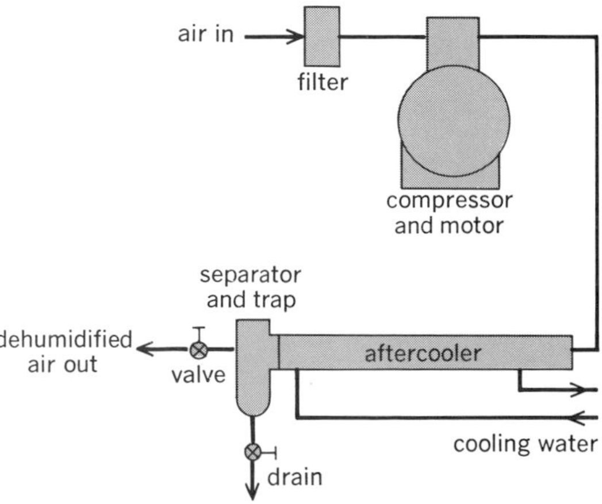

Fig. 4. Dehumidifying by compression and aftercooling.

so high compared to power requirements for dehumidifying by either the sorbent or refrigeration method that the compression system is not an economical one if dehumidifying is the only end result required.

[JOHN EVERETTS, JR.]

Bibliography: American Society of Heating, Refrigerating, and Air-Conditioning Engineers, *Handbook and Product Directory*: *Equipment*, 1979; ASHRAE, *Heating, Ventilating, Air-Conditioning Guide*, annual; ASHRAE, Symposium on dehumidification: Journal section, *Heat. Piping Air Cond.*, 29(4):152–162, 1957; J. Everetts, Jr., Dehumidification methods and applications, *Heat. Piping Air Cond.*, 18(12):121–124, 1964.

Destructive distillation

The primary chemical processing of materials such as wood, coal, oil shale, and some residual oils from refining of petroleum. It consists in heating material in an inert atmosphere at a temperature high enough for chemical decomposition. The principal products are (1) gases containing carbon monoxide, hydrogen, hydrogen sulfide, and ammonia, (2) oils, and (3) water solutions of organic acids, alcohols, and ammonium salts.

Crude shale oil obtained by destructive distillation of carboniferous shales, is being produced on a commercial scale in Scotland, Latvia, and Sweden and on a pilot plant scale in the United States. Crude shale oil may be subjected to a destructive, or coking, distillation to reduce its viscosity and increase its hydrogen content. Subsequent catalytic hydrogenation (cobalt molybdatealumina catalyst, about 400°C, 100–1500 psi pressure) lowers the nitrogen and sulfur contents so that the oil can then be refined by normal petroleum refinery operations. Residual oils from petroleum refinery operations are subjected to coking-distillation to reduce the carbon content. The coke is used for the manufacture of electrode carbon and the oil is returned to the feed for normal petroleum refining. *See* COKING (PETROLEUM); OIL SHALE.

[H. H. STORCH/H. W. WAINWRIGHT]

Dewaxing of petroleum

The process of separating hydrocarbons which solidify readily (waxes) from petroleum fractions. Removal of wax is usually necessary to produce lubricating oil which will remain fluid down to the lowest temperature of use. It is therefore an important step in the manufacture of lubricating oils. The wax removed may be purified further to produce commercial paraffin or microcrystalline waxes.

Most commercial dewaxing processes utilize solvent dilution, chilling to crystallize the wax, and filtration. The MEK process (methyl ethyl ketone-toluene solvent) is most widely used. Wax crystals are formed by chilling through the walls of scraped surface chillers, and wax is separated from the resultant wax-oil-solvent slurry by using fully enclosed rotary vacuum filters. In a relatively new process modification, most of the chilling is accomplished by multistage injection of very cold solvent into the waxy oil with vigorous agitation, resulting in more uniform and compact wax crystals which filter faster.

In the propane process, part of the propane diluent is allowed to evaporate by reducing pressure, so as to chill the slurry to the desired filtration temperature, and rotary pressure filters are employed.

Other solvents in commercial use for dewaxing include MEK-MIBK (methyl isobutyl ketone), acetone-benzene, dichloethane-methylene dichloride, and propylene-acetone.

Older dewaxing processes are centrifugal dewaxing, applicable only to heavy residual stocks, utilizing naphtha dilution, indirect chilling, and centrifugal separation; and cold pressing, applicable only to low-viscosity light lube fractions, in which the crystallized wax is separated from the chilled, undiluted oil in plate-and-frame-type pressure filters.

Complex dewaxing requires no refrigeration, but depends upon the formation of a solid urea-*n*-paraffin complex which is separated by filtration and then decomposed. This process is used, to a limited extent, to make low-viscosity lubricants which must remain fluid at very low temperatures (refrigeration, transformer, and hydraulic oils). Similar use is anticipated for the catalytic dewaxing process, which is based on selective hydrocracking of the normal paraffins; it uses a molecular sieve-based catalyst in which the active hydrocracking sites are accessible only to the paraffin molecules. *See* PETROLEUM; PETROLEUM PROCESSING.

[STEPHEN F. PERRY]

Bibliography: R. N. Bennett, G. J. Elkes, and G. J. Wanless, New process produces low pour oils, *Oil Gas J.*, Jan. 6, 1975; J. F. Eagen et. al., Successful development of two new lubricating oil dewaxing processes, Paper at World Petroleum Congress, Japan, 1975; G. D. Hobson and W. Pohl, *Modern Petroleum Technology*, 1973; Hydrocarbon processing, *1974 Refining Process Handbook*, September 1974.

Diesel cycle

An internal combustion engine cycle in which the heat of compression ignites the fuel. Compression-ignition engines, or diesel engines, are thermodynamically similar to spark-ignition engines. The sequence of processes for both types is intake, compression, addition of heat, expansion, and exhaust. Ignition and power control in the compression-ignition engine are, however, very different from those that are found in the spark-ignition engine.

Usually, a full unthrottled charge of air is drawn in during the intake stroke of a diesel engine. A compression ratio between 12 and 20 is used, in contrast to a ratio of 4 to 10 for the Otto spark-ignition engine. This high compression ratio of the diesel raises the temperature of the air during the compression stroke. Just before top center on the compression stroke, fuel is sprayed into the combustion chamber. The high temperature of the air ignites the fuel, which burns almost as soon as it is introduced, adding heat. The combustion products expand to produce power, and exhaust to complete the cyle.

Performance of a diesel engine is anticipated by analyzing the action of an air-standard diesel cycle. An insulated cylinder equipped with a frictionless piston contains a unit air mass. The metal cylinder head is alternately insulated and then uncovered for heat transfer.

Air is compressed until the piston reaches the top of the stroke. Then the air receives heat through the cylinder head and expands at constant pressure along path *a-b* as shown in Fig. 1, moving the piston part way down through the cylinder. Then the cylinder head is insulated, and the air completes its expansion along path *b-c* at constant entropy. The cylinder head is uncovered, and with the piston at the bottom of its stroke, a constant-volume heat rejection takes place on path *c-d*. The insulation is replaced, and the cycle is completed with an isentropic compression on path *d-a*.

An increase in compression ratio $r=v_d/v_a$ increases efficiency η, the increase becoming less at higher compression ratios. Another characteristic of the diesel cycle is the ratio of volumes at the end and at the start of the constant-pressure heat-addition process. This cutoff ratio $r_c=v_b/v_a$ measures the interval during which fuel is injected. For an engine to develop greater power output, the cutoff ratio is increased and heat continues to be added further into the expansion stroke. The air-standard cycle shows that, with less travel remaining during which to expend the additional heat energy as mechanical energy, the efficiency of the engine is reduced. Conversely, efficiency increases as the cutoff ratio decreases, so that a diesel engine is most efficient at light loads. Specifically, the equation below may be written, where $k=c_p/c_v$ the ratio

$$\eta=1-\frac{1}{r^{k-1}}\left[\frac{r_c^{\,k}-1}{k(r_c-1)}\right]$$

of specific heat of the working substance at constant pressure to its specific heat at constant volume (Fig. 2). In the limiting case when cutoff ratio r_c approaches unity, diesel cycle efficiency approaches Otto cycle efficiency for cycles of the same compression ratio.

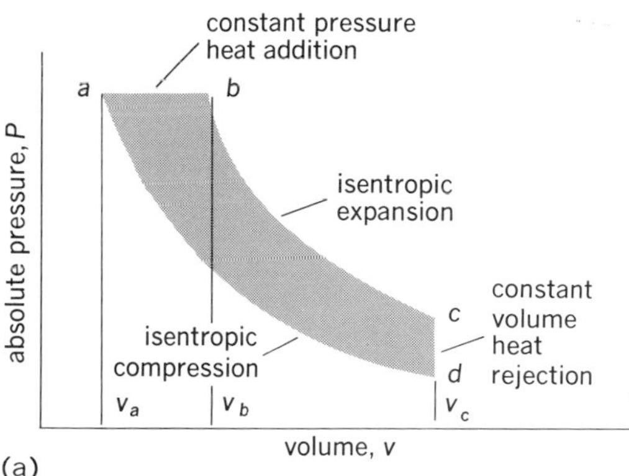

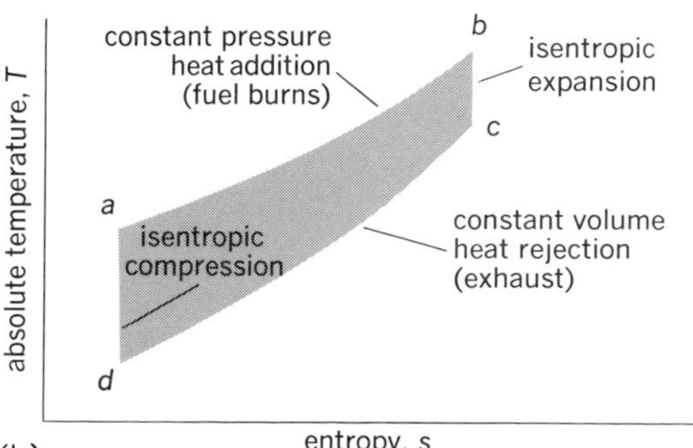

Fig. 1. Ideal diesel cycle, with (*a*) pressure-volume and (*b*) temperature-entropy bases.

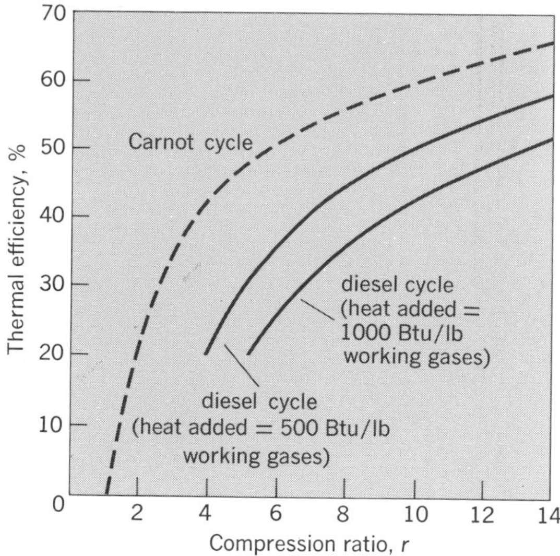

Fig. 2. Thermal efficiency of an ideal diesel cycle. 1 Btu/lb = 2326 J/kg.

In an actual engine with a given compression ratio, the Otto engine has the higher efficiency. However, fuel requirements limit the Otto engine to a compression ratio of about 10, whereas a diesel engine can operate at a compression ratio of about 15 and consequently at a higher efficiency.

In addition, heat can be added earlier in the cycle by injecting fuel during the latter part of the compression process *d-a*. This mode of operation is the dual-combustion or semidiesel cycle. With most of the heat added near peak compression, semi-diesel efficiency approaches Otto cycle efficiency at a given compression ratio. *See* OTTO CYCLE. [THEODORE BAUMEISTER]

Bibliography: T. Baumeister (ed.), *Standard Handbook for Mechanical Engineers*, 8th ed., 1978; L. Lichty, *Combustion Engine Processes*, 7th ed., 1967; D. S. Williams (ed.), *The Modern Diesel: Development and Design*, 14th ed., 1977.

Diesel engine

An internal combustion engine operating on a thermodynamic cycle in which the ratio of compression ($R_v = 15\pm$) of the air charge is sufficiently high to ignite the fuel subsequently injected into the combustion chamber. The engine differs essentially from the more prevalent mixture engine in which an explosive mixture of air and gas or air and the vapor of a volatile liquid fuel is made externally to the engine cylinder, compressed to a point some 200°F (111°C) below the ignition temperature, and ignited at will as by an electric spark. The diesel engine utilizes a wider variety of fuels with a higher thermal efficiency and consequent economic advantage under many service applications. The true diesel engine, as projected by R. Diesel and as represented in most low-speed engines, such as about 300 rpm, uses a fuel-injection system where the injection rate is delayed and controlled to maintain constant pressure during combustion. Adaptation of the injection principle to higher engine speeds, such as 1000–2000 rpm, has necessitated departure from the constant pressure specification because the time available for fuel

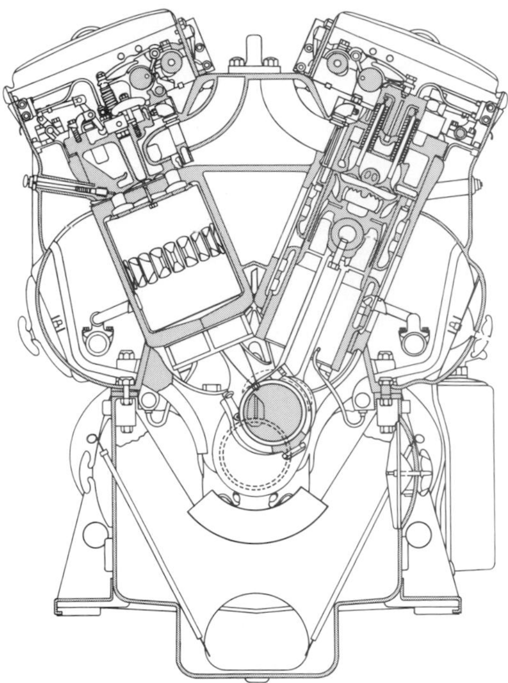

Fig. 1. Section through a locomotive diesel engine. (*General Motors, Electromotive Division*)

injection is so short (milliseconds). Combustion proceeds with little regard to the constant-pressure specification. High peak pressures may be developed. Yet nonvolative (distillate) fuels are burned to advantage in these engines which cannot be rigorously identified as true diesels but which properly should be called commercial diesels. In ordinary parlance all such engines are classified as diesels. *See* DIESEL CYCLE; OTTO CYCLE.

Identifying alternative features of diesel engine types include: (1) two-cycle or four-cycle operation; (2) horizontal or vertical piston movement; (3) single or multiple cylinder; (4) large (5000 hp or 3.7 MW) or small (50 hp or 37 kW); (5) cylinders in line, opposed, V, or radial; (6) single acting or double acting; (7) high (1000–2000 rpm), low (100–300 rpm), or medium speed; (8) constant speed or variable speed; (9) reversible or nonreversible; (10) air injection or solid injection; (11) supercharged or unsupercharged; and (12) single or multiple fuel. Section drawings of two representative engines are given in Figs. 1 and 2, and selected performance data are given in Table 1.

Maximum diesel engine sizes (5000 kW) are less than steam turbines (1,000,000 kW) and hydraulic turbines (300,000 kW). They give high instrinsic and actual thermal efficiency (20–40%); a sample comparative heat balance is shown in Table 2; variation in performance with load is shown in Fig. 3. Control of engine output is by regulation of the fuel supplied but without variation of the air supply (100±% excess air at full load). Supercharging (10–15 psi or 69–103 kPa) increases cylinder weight charge and consequently power output for a given cylinder size and engine speed. With two-cycle constructions, scavenging air (approximately 5 psi) is delivered by crankcase compression, front end compression, or separate rotary, reciprocating, or centrifugal blowers. The engine cylinder may be without valves and with complete control of admission of scavenging air and release of spent gases in a two-port construction, the piston covering and uncovering the ports; or the cylinder may have a single port (for admission or release) uncovered by the main piston at the outer end of its stroke and conventional cam-operated valve in the cylinder head. The objective is to replace spent gases with fresh air by guided flow and high turbulence. The four-cycle engine, with its complement of admission and exhaust valves on each cylinder, is most effective in scavenging. But the sacrifice of one power stroke out of every two is a frequent deterrent to its selection. Valves are exclusively of the poppet type with the burden of tightness and cooling dominant in the exhaust valve designs. Cylinder heads become complicated structures because of valve porting, jacketing, and spray-valve locations and the accommodation of these to effective combustion, heat transfer, and internal bursting pressures.

Distillate fuel (40° API, 19,000–19,500 Btu/lb or 44–45 MJ/kg, 135,000–140,000 Btu/gal or 38–39 MJ/liter) prevails with locomotive, truck, bus, and automotive applications. Lower-speed engines (stationary and motorship service) burn heavier fuels (for example, 20° API, 18,500–19,000 Btu/lb or 43–44 MJ/kg, 145,000–150,000 Btu/gal or 40–42 MJ/liter). Alternative fuels are burned in dual-fuel and gas diesel engines for stationary service. The main fuel is typically natural gas (90–95%) with oil (5–10%) used to control burning and to stabilize ignition. In the more prevalent liquid-

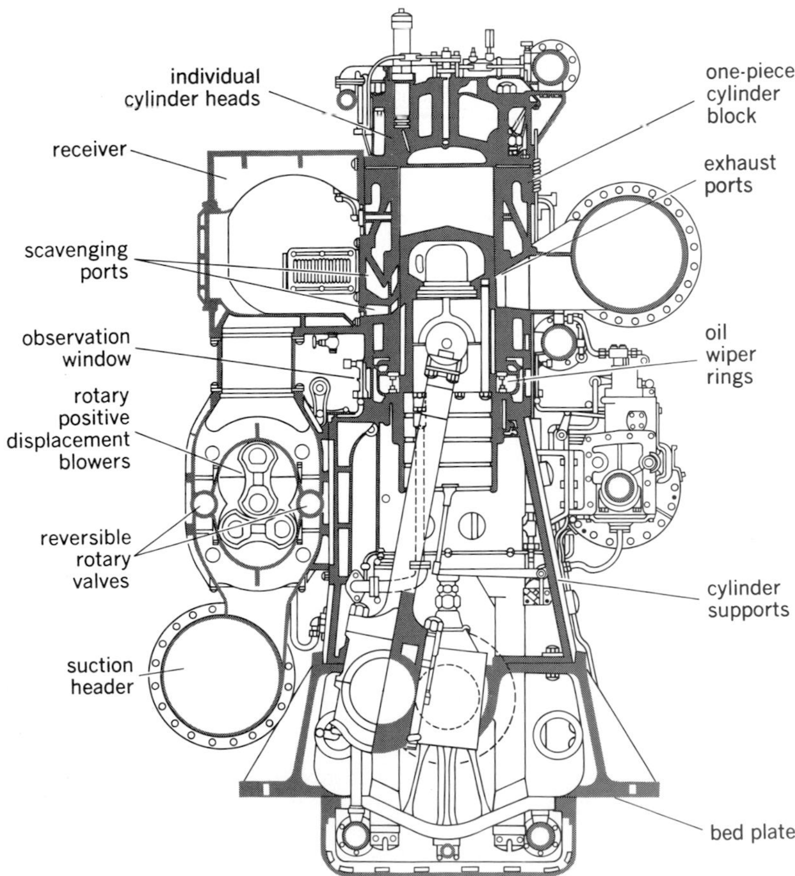

Fig. 2. Section through a Busch-Sulzer two-cycle diesel engine.

Table 1. Performance of selected diesel engine plants*

Type of plant	Shaft horse-power (shp)	Ratio of compres-sion, R_v	Brake mean effective pressure, psi	Piston speed, ft/min	Weight, lb/in.³ displace-ment	Weight, lb/shp	Overall thermal efficiency, %
Air injection engine	300–5000	12–15	50–75	600–1000	3–8	25–200	30–35
Solid injection, compression ignition							
Automotive	20–300	12–15	75–100	800–1800	2.5–4	7–25	25–30
Railroad	200–2500	12–15	60–90	800–1800	2.5–4	10–40	30–35
Stationary							
Unsupercharged	50–2500	12–15	70–80	600–1500	2.5–5	10–100	30–35
Supercharged	60–4000	10–13	110–125	600–1500	2.5–5	7.5–75	32–40
Dual fuel, stationary							
Unsupercharged	50–2500	12–15	80–90	600–1500	2.5–5	10–100	30–35
Supercharged	60–4000	10–13	120–135	600–1500	2.5–5	7.5–75	32–40

*1 hp = 0.7457 kW; 1 psi = 6.895 kPa; 1 ft/min = 5.08 mm/s; 1 lb/in³ = 27.7 g/cm³; 1 lb/hp = 0.6083 kg/kW.

Table 2. Approximate allocation of losses in internal combustion engine plants

Type of loss	Mixture engines, %	Injection (diesel) engines, %
Output	20	33
Exhaust losses	40	33
Cooling system losses	40	33
Other	<1	1
Total (input)	100	100

fuel-injection system, the technical problems are numerous and embrace such elements as pumps, spray nozzles, and combustion chambers for the delivery, atomization, and burning of the fuel in the hot compressed air. There must be accurate timing (measured in milliseconds) for the entire process to give clean, complete combustion without undue excess air. Combustion characteristics of fuels are defined by rigorous specifications and include such factors as viscosity, flash point, pour point, ash, sulfur, basic sediment, water, Conradsen carbon number, cetane number, and diesel index.

Small size (<200 hp or 150 kW) engines are conveniently started by an electric motor and storage battery. Larger engines use compressed air (about 200 psi or 150 kW) introduced through valves in the cylinder head. Starting, with engine-driven generator sets, may be accomplished by motoring the generator.

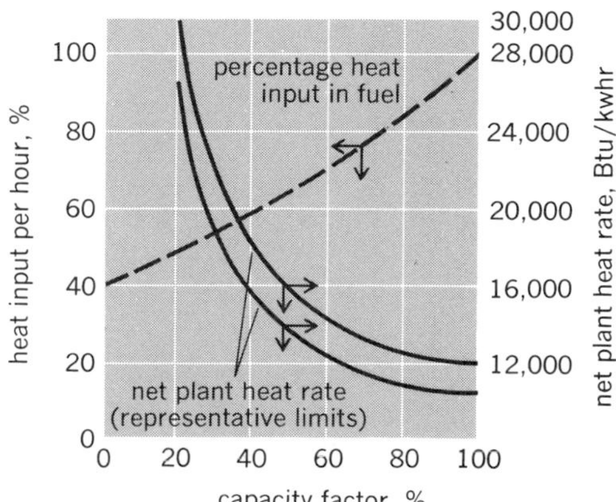

Fig. 3. Heat rate and heat input curves for selected diesel-engine generating plants.

Cooling systems use water at 120–180°F (50–80°C) with radiators, cooling towers, and cooling ponds employed for conservation and reclamation. Lubrication costs can become prohibitive with inadequate engine maintenance. Foundations must be designed to handle stress loadings and to reduce vibration. Exhaust systems should be equipped with wavetrap silencers or mufflers. Filters on air and fuel supply are good insurance for engine reliability. *See* INTERNAL COMBUSTION ENGINE. [THEODORE BAUMEISTER]

Bibliography: T. Baumeister (ed.), *Standard Handbook for Mechanical Engineers*, 8th ed., 1978; Diesel Engine Manufacturers Association, *Standard Practices for Stationary Diesel and Gas Engines*, 1958; L. C. Lichty, *Combustion Engine Processes*, 7th ed., 1967; D. S. Williams (ed.), *The Modern Diesel: Development and Design*, 14th ed., 1977.

Diesel fuel

A broad class of petroleum products which includes distillate or residual materials (or blends of these two) from the refining of crude oil. Diesel fuel generally has a distillation range between 190° and 380°C and a specific gravity 15°C/15°C range between 0.760 and 0.935. In addition to these two primary criteria, other properties which are used to define diesel fuel are: viscosity (1.4 to 26.5 mm²/s), sulfur content (usually < 1.0% by weight), cetane number (30 to 60), ash content (usually <0.10% by weight), water and sediment content (usually <0.5% by volume), flash point (usually >55°C), and cloud point (that temperature at which wax crystals begin to precipitate as the oil is cooled under prescribed conditions) or pour point (the lowest temperature at which the oil will flow when cooled under prescribed conditions). The properties of diesel fuel greatly overlap those of kerosine, jet fuels, and burner fuel oils; thus all these products are generally referred to as intermediate distillates. *See* FUEL OIL; JET FUEL; KEROSINE; PETROLEUM PROCESSING; PETROLEUM PRODUCTS.

The table contains the limiting requirements of the three grades of diesel fuel which are most commonly used in the United States. Although diesel engines usually operate on a wide range of fuels, they are designed to operate most cleanly and efficiently on one of the fuels described in this

Detailed requirements for diesel fuel oils

Grade of diesel fuel oil	Flash point, °F (°C)	Cloud point, °F (°C)	Water and sediment, vol %	Carbon residue on, 10% residuum, %	Ash, weight, %	Distillation temperatures, °F(°C) 90% point		Viscosity at 100°F (37.8°C) Kinematic, cSt (or SUS)		Sulfur,[d] wt %	Copper strip corrosion	Cetane number[e]
	Min	Max	Max	Max	Max	Min	Max	Min	Max	Max	Max	Min
No. 1-D, a volatile distillate fuel oil for engines in service requiring frequent speed and load changes.	100 or legal (37.8)	[b]	0.05	0.15	0.01	—	550 (287.8)	1.4	2.5 (34.4)	0.50 or legal	No. 3	40[f]
No. 2-D, a distillate fuel oil of lower volatility for engines in industrial and heavy mobile service	125 or legal (51.7)	[b]	0.05	0.35	0.01	540[c] (282.2)	640 (338)	2.0[c] (32.6)	4.3 (40.1)	0.50 or legal	No. 3	40[f]
No. 4-D, a fuel oil for low- and medium-speed engines	130 or legal (54.4)	[b]	0.50	—	0.10	—	—	5.8 (45)	26.4 (125)	2.0	—	30[f]

SOURCE: American Society for Testing and Materials. *Standard Specifications for Diesel Fuel Oils*, ASTM D-975.

[a]To meet special operating conditions, modifications of individual limiting requirements may be agreed upon between purchaser, seller, and manufacturer.

[b]It is unrealistic to specify low-temperature properties that will ensure satisfactory operation on a broad basis. Satisfactory operation should be achieved in most cases if the cloud point (or wax appearance point) is specified at 10°F above the tenth percentile minimum ambient temperature for the area in which the fuel will be used. Some ability properties should be agreed on between the fuel supplier and purchaser for the intended use and expected ambient temperatures.

[c]When cloud point less than 10°F (−12.2°C) is specified, the minimum viscosity shall be 1.8 cSt. and the 90% point shall be waived.

[d]In countries ouside the United States, other sulfur limits may apply.

[e]Where cetane number by Method D 613, is not available, ASTM Method D 976, Calculated Cetane Index of Distillate Fuels, may be used as an approximation. Where there is disagreement, Method D 613 shall be the referee method.

[f]Low-atmospheric temperatures as well as engine operation at high altitudes may require use of fuels with higher cetane ratings.

table; using a fuel which departs from that recommended by the engine manufacturer will have a negative effect on engine life, performance, exhaust emissions, noise, and so on. For example, increasing sulfur content causes an increase in corrosive wear of piston rings, cylinder walls, and bearings. Decreasing cetane number causes hard starting, rough operation, increased exhaust emissions, and increased noise. Increasing ash or wax content, or water and sediment, causes fuel system wear and fuel-handling problems. Some heavier (higher-boiling-range) fuels, such as Navy distillate fuel, MIL-F-24397 (ships), and marine diesel fuel oil, MIL-F-16884), are used in low- and medium-speed diesel engines, but they are generally not recommended for high-speed engines.

The diesel engine, with compression ratios ranging from 12:1 up to 22.1, ignites its fuel by injecting it into the charge air, which is heated to 500–550°C by the heat of compression; thus diesel engines are referred to as compression ignition engines. This method of ignition requires that the fuel ignite spontaneously and quickly (within 1 to 2 ms in a high-speed engine). The time lag between the initiation of injection and the initiation of combustion is called ignition delay. Two major factors characterize ignition delay: a mechanical factor which is influenced by such things as compression ratio, motion of the charge air during injection, and ability of the injector to atomize the fuel; and a chemical factor which is influenced by such things as the fuel's autoignition temperature, specific heat, density, thermal conductivity, surface tension, and coefficient of friction. *See* DIESEL ENGINE.

The fuel's effect on the chemical portion of ignition delay is expressed by a quantity called the cetane number. Cetane (hexadecane), which has a high-ignition quality (short chemical ignition delay), has arbitrarily been assigned a cetane number of 100, whereas heptamethylnonane has been assigned a cetane number of 0. The cetane number of a diesel fuel is determined by comparing it to a blend of cetane and heptamethylnonane which has the same ignition quality. This comparison is made with the use of an ASTM-CFR engine. The cetane number is the percentage by volume of cetane in the blend which has an ignition quality equal to the test fuel.

The cetane number of the paraffinic hydrocarbon compounds is generally high, whereas the cetane number of aromatic and naphthenic hydrocarbon compounds is low. Diesel fuel normally contains 60–80% by volume of paraffinic compounds and 20–40% by volume of aromatic and naphthenic compounds. When the paraffinic content is toward the low end of this range, the natural cetane number of the fuel is low. A low cetane number can be counteracted to some degree by the addition of a cetane improver additive, such as amyl nitrate or hexyl nitrate, which increases the cetane number of the fuel, in much the way antiknock additives increase the octane number of gasoline.

It has been found that cetane number can be approximated from the physical properties of the fuel. The best correlation is with the specific gravity and mid-boiling point.

[ROBERT TEASLEY, JR.]

Bibliography: American Society for Testing and Materials, *Annual Book of ASTM Standards*, pt. 23, 1975; American Society for Testing and Materials, *Diesel Fuel Oils*, STP 413, 1967; British Petroleum Co., Ltd., *Medium and High Speed Diesel Engines*, 1970.

Direct current

Electric current which flows in one direction only through a circuit or equipment. The associated direct voltages, in contrast to alternating voltages, are of unchanging polarity. Direct current corresponds to a drift or displacement of electric charge

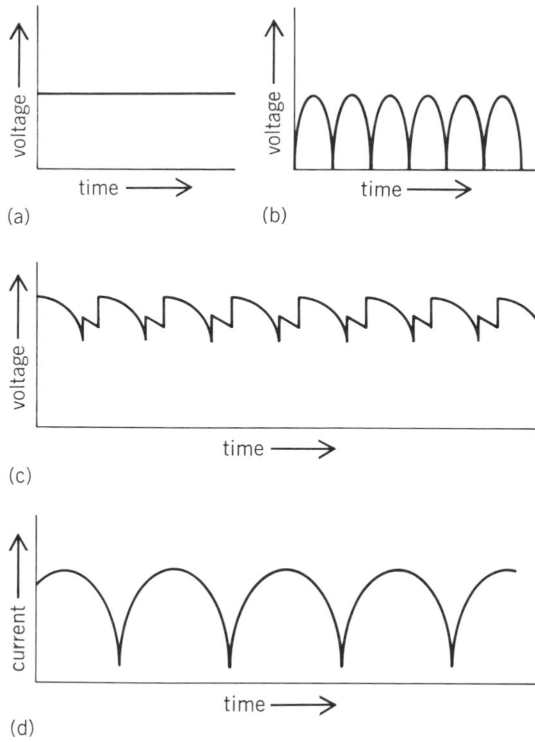

Typical direct currents and voltages. (*a*) Output from a battery; (*b*) full-wave rectified voltage; (*c*) output of the rectifier station of a high-voltage dc transmission link (*from E. W. Kimbark, Direct Current Transmission, vol. 1, Wiley-Interscience, 1971*). (*d*) Current in a rectifier-supplied dc motor (*from A. E. Fitzgerald, C. Kingsley, and A. Kusko, Electric Machinery, 3d ed., McGraw-Hill, 1971*).

in one unvarying direction around the closed loop or loops of an electric circuit. Direct currents and voltages may be of constant magnitude or may vary with time.

Batteries and rotating generators produce direct voltages of nominally constant magnitude (illustration *a*). Direct voltages of time-varying magnitude are produced by rectifiers, which convert alternating voltage to pulsating direct voltage (illustration *b* and *c*). *See* BATTERY; GENERATOR.

Direct current is used extensively to power adjustable-speed motor drives in industry and in transportation (illustration *d*). Very large amounts of power are used in electrochemical processes for the refining and plating of metals and for the production of numerous basic chemicals.

Direct current ordinarily is not widely distributed for general use by electric utility customers. Instead, direct-current (dc) power is obtained at the site where it is needed by the rectification of commercially available alternating current (ac) power to dc power. Solid-state rectifiers ordinarily are employed to supply dc equipment from ac supply lines. Rectifier dc supplies range in size from tiny devices in household electronic equipment to high-voltage dc transmission links of at least hundreds of megawatts capacity. *See* ELECTRIC POWER SYSTEMS.

Many high-voltage dc transmission systems have been constructed throughout the world since 1954. Very large amounts of power, generated as ac and ultimately used as ac, are transmitted as dc power. Rectifiers supply the sending end of the dc link; inverters then supply the receiving-end ac power system from the link. High-voltage dc transmission often is more economical than ac transmission when extremely long distances are involved. *See* ALTERNATING CURRENT; TRANSMISSION LINES. [D. D. ROBB]

Direct-current generator

A rotating electric machine which delivers a unidirectional voltage and current. An armature winding mounted on the rotor supplies the electric power output. One or more field windings mounted on the stator establish the magnetic flux in the air gap. A voltage is induced in the armature coils as a result of the relative motion between the coils and the air gap flux. Faraday's law states that the voltage induced is determined by the time rate of change of flux linkages with the winding. Since these induced voltages are alternating, a means of rectification is necessary to deliver direct current at the generator terminals. Rectification is accomplished by a commutator mounted on the rotor shaft. *See* ELECTRIC ROTATING MACHINERY; GENERATOR.

Carbon brushes, insulated from the machine frame and secured in brush holders, transfer the armature current from the rotating commutator to the external circuit. Brushes are held against the commutator under a pressure of $2-2\frac{1}{2}$ psi (14–17 kPa). Armature current passes from the brush to brush holder through a flexible copper lead. In multipolar machines all positive brush studs are connected together, as are all negative studs, to form the positive and negative generator terminals. In most dc generators the number of brush studs is the same as the number of main poles. In modern machines brushes are located in the neutral position where the voltage induced in a short-circuited coil by the main pole flux is zero. The brushes continuously pick up a fixed, instantaneous value of the voltage generated in the armature winding.

The generated voltage is dependent upon speed n in revolutions per minute, number of poles p, flux per pole Φ in webers, number of armature conductors z, and the number of armature paths a. The equation for the average voltage generated is

$$E_g = \frac{np\Phi z}{60a} \quad \text{volts}$$

The field windings of dc generators require a direct current to produce a magnetomotive force

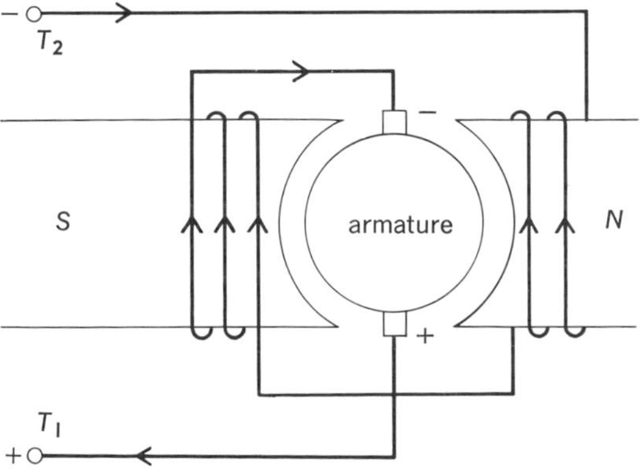

Fig. 1. Series generator.

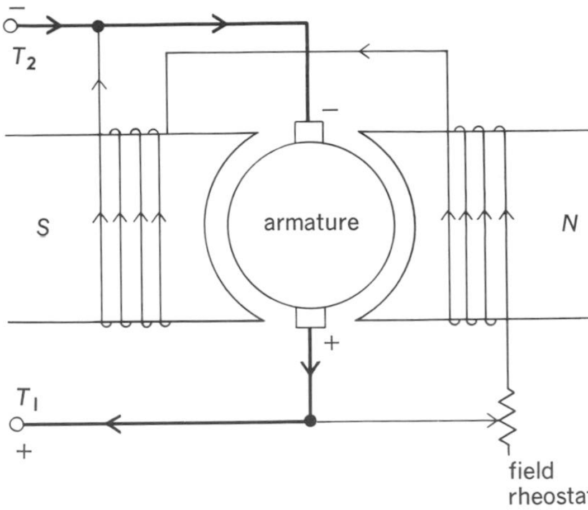

Fig. 2. Shunt generator.

(mmf) and establish a magnetic flux path across the air gap and through the armature. Generators are classified as series, shunt, compound, or separately excited, according to the manner of supplying the field excitation current. In the separately excited generator, the field winding is connected to an independent external source. Using the armature as a source of supply for the field current, dc generators are also capable of self-excitation. Residual magnetism in the field poles is necessary for self-excitation. Series, shunt, and compound-wound generators are self-excited, and each produces different voltage characteristics.

When operated under load, the terminal voltage changes with change of load because of armature resistance drop, change in field current, and armature reaction. Interpoles and compensating, or pole-face, windings are employed in modern generators in order to improve commutation and to compensate for armature reaction.

Series generator. The armature winding and field winding of this generator are connected in series, as shown in Fig. 1. Terminals T_1 and T_2 are connected to the external load. The field mmf aids the residual magnetism in the poles, permitting the generator to build up voltage. The field winding is wound on the pole core with a comparatively few turns of wire of large cross section capable of carrying rated load current. The magnetic flux and consequently the generated emf and terminal voltage increase with increasing load current. Figure 4 shows the external characteristic or variation of terminal voltage with load current at constant speed. Series generators are suitable for special purposes only, such as a booster in a constant voltage system, and are therefore seldom employed.

Shunt generator. The field winding of a shunt generator is connected in parallel with the armature winding, as shown in Fig. 2. The armature supplies both load current I_t and field current I_f. The field current is 1–5% of the rated armature current I_a, the higher value applying to small machines. The field winding resistance is fairly high since the field consists of many turns of small-cross-section wire. For voltage buildup the total field-circuit resistance must be below a critical value; above this value the generator voltage cannot build up. The no-load voltage to which the generator builds up is varied by means of a rheostat in the field circuit. The external voltage characteristic (Fig. 4) shows a reduction of voltage with increases in load current, but voltage regulation is fairly good in large generators. The output voltage may be kept constant for varying load current conditions by manual or automatic control of the rheostat in the field circuit. A shunt generator will not maintain a large current in a short circuit in the external circuit, since the field current at short circuit is zero.

The shunt generator is suitable for fairly constant voltage applications, such as an exciter for ac generator fields, battery charging, and for electrolytic work requiring low-voltage and high-current capacity. Prior to the use of the alternating-current generator and solid-stage rectifying devices in automobiles, a shunt generator, in conjunction with automatic regulating devices, was used to charge the battery and supply power to the electrical system. Shunt-wound generators are well adapted to stable operation in parallel.

Compound generator. This generator has both a series field winding and a shunt field winding. Both windings are on the main poles with the series winding on the outside. The shunt winding furnishes the major part of the mmf. The series winding produces a variable mmf, dependent upon the load current, and offers a means of compensating for voltage drop. Figure 3 shows a cumulative-compound connection with series and shunt fields aiding. A diverter resistance across the series field is used to adjust the series field mmf and vary the degree of compounding. By proper adjustment a nearly flat output voltage characteristic is possible. Cumulative-compound generators are overcompounded, flat-compounded, or undercompounded, as shown by the external characteristics in Fig. 4. The shunt winding is connected across the armature (short-shunt connection) or across the output terminals (long-shunt connection). Figure 3 shows the long-shunt connection.

Voltage is controllable over a limited range by a rheostat in the shunt field circuit. Compound generators are used for applications requiring constant voltage, such as lighting and motor loads. Generators used for this service are rated at 125 or 250 volts and are flat or overcompounded to give a

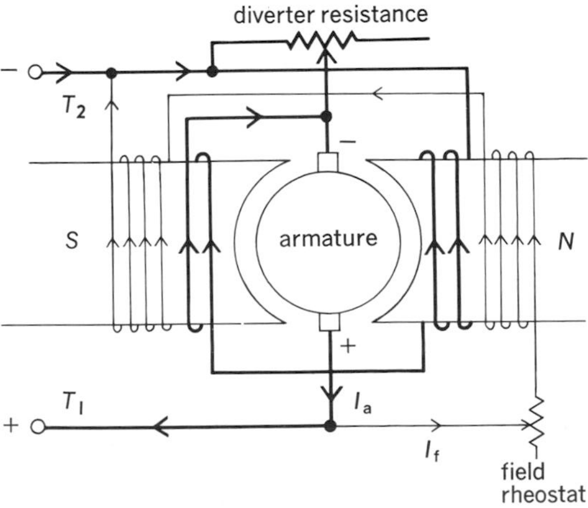

Fig. 3. Cumulative compound generator. The long-shunt connection is seen in this example.

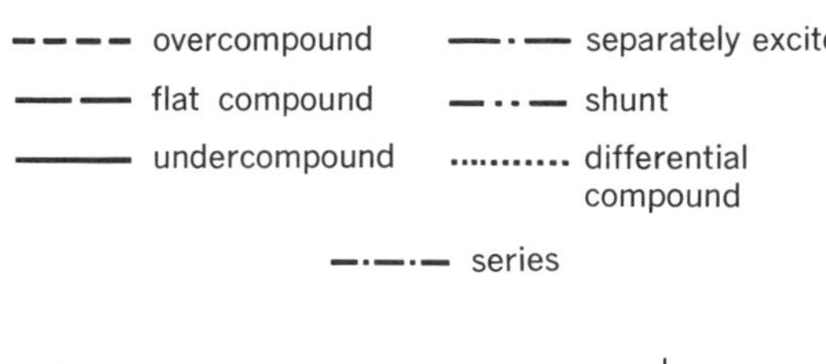

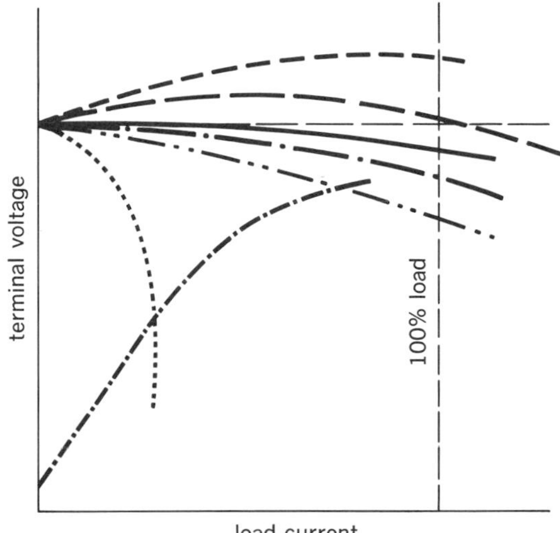

Fig. 4. External characteristics of dc generators.

regulation of about 2%. An important application is in steel mills which have a large dc motor load. Cumulative-compound generators are capable of stable operation in parallel if the series fields are connected in parallel by an equalizer bus.

In the differentially compounded generator the series field is connected to oppose the shunt field mmf. Increasing load current causes a large voltage drop due to the demagnetizing effect of the series field. The differentially compounded generator has only a few applications, such as arc-welding generators and special generators for electrically operated shovels.

Separately excited generator. The field winding of this type of generator is connected to an independent dc source. The field winding is similar to that in the shunt generator. Separately excited generators are among the most common of dc generators, for they permit stable operation over a very wide range of output voltages. The slightly drooping voltage characteristic (Fig. 4) may be corrected by rheostatic control in the field circuit. Applications are found in special regulating sets, such as the Ward Leonard system, and in laboratory and commercial test sets.

Special types. Besides the common dc generators discussed in this article, there are a number of special types, including the homopolar, third-brush, diverter-pole, and Rosenberg generators. *See* DIRECT-CURRENT MOTOR.

Commutator ripple. The voltage at the brushes of dc generators is not absolutely constant. A slight high-frequency variation exists, which is superimposed upon the average voltage output. This is called commutator ripple and is caused by the cyclic change in the number of commutator bars contacting the brushes as the machine rotates. The ripple decreases as the number of commutator bars is increased and is usually ignored. In servomechanisms employing a dc tachometer for velocity feedback, the ripple frequency is kept as high as possible.

[ROBERT T. WEIL, JR.]

Bibliography: D. G. Fink and H. W. Beaty (eds.), *Standard Handbook for Electrical Engineers*, 11th ed., 1978; A. E. Fitzgerald and C. Kingsley, *Electric Machinery*, 3d ed., 1971; A. S. Langsdorf, *Principles of Direct-Current Machines*, 5th ed., 1940; M. Liwschitz-Garik and R. T. Weil, Jr., *D-C and A-C Machines*, 1952; M. Liwschitz-Garik and C. C. Whipple, *Direct-Current Machines*, 2d ed., 1956.

Direct-current motor

An electric rotating machine energized by direct current and used to convert electric energy to mechanical energy. It is characterized by its relative ease of speed control and, in the case of the series-connected motor, by an ability to produce large torque under load without taking excessive current. Output is given in horsepower, the unit of mechanical power. Normal full-load values of voltage, current, and speed are generally given.

Direct-current motors are manufactured in several horsepower-rating classifications: (1) subfractional, approximately 1–35 millihorsepower (mhp) (0.75–26 W); (2) fractional, 1/40–1 horsepower (hp) (19–750 W); and (3) integral, 1/2 to several hundred horespower (370 W to several hundred kilowatts). The standard line voltages applied to dc motors are 6, 12, 27, 32, 115, 230, and 550 volts. Occasionally they reach higher values.

Normal full-load speeds are 850, 1140, 1725, and 3450 rpm. Variable-speed motors may have limiting rpm values stated.

Protection of the motor is afforded by several types of enclosures, such as splash-proof, drip-

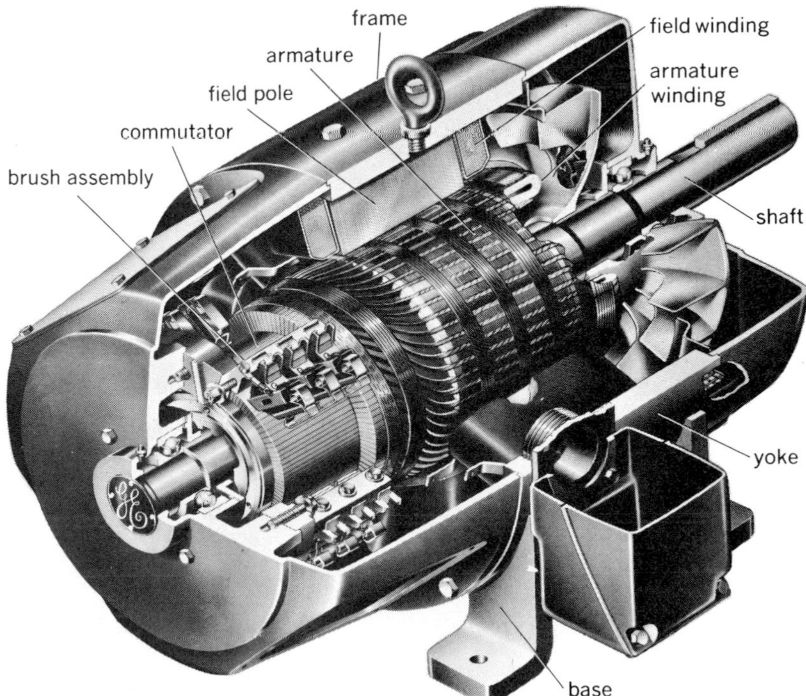

Fig. 1. Cutaway view of typical dc motor. (*General Electric*)

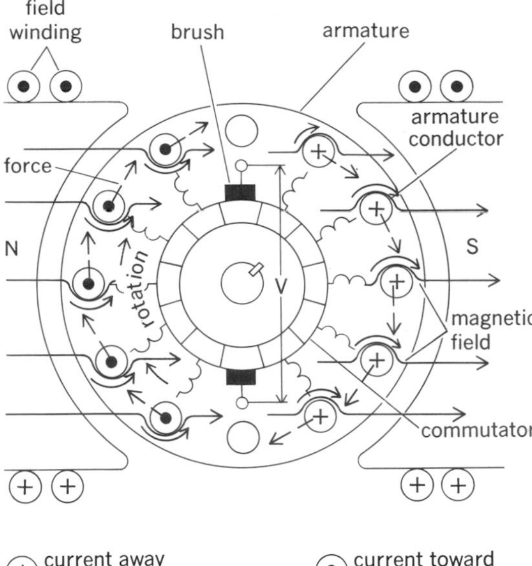

Fig. 2. Rotation in a dc motor.

proof, dust-explosion-proof, dust-ignition-proof, and immersion-proof enclosures. Some motors are totally enclosed.

The principal parts of a dc motor are the frame, the armature, the field poles and windings, and the commutator and brush assemblies (Fig. 1). The frame consists of a steel yoke of open cylindrical shape mounted on a base. Salient field poles of sheet-steel laminations are fastened to the inside of the yoke. Field windings placed on the field poles are interconnected to form the complete field winding circuit. The armature consists of a cylindrical core of sheet-steel disks punched with peripheral slots, air ducts, and shaft hole. These punchings are aligned on a steel shaft on which is also mounted the commutator. The commutator, made of hard-drawn copper segments, is insulated from the shaft. Segments are insulated from each other by mica. Stationary carbon brushes in brush holders make contact with commutator segments. Copper conductors placed in the insulated armature slots are interconnected to form a reentrant lap or wave style of winding.

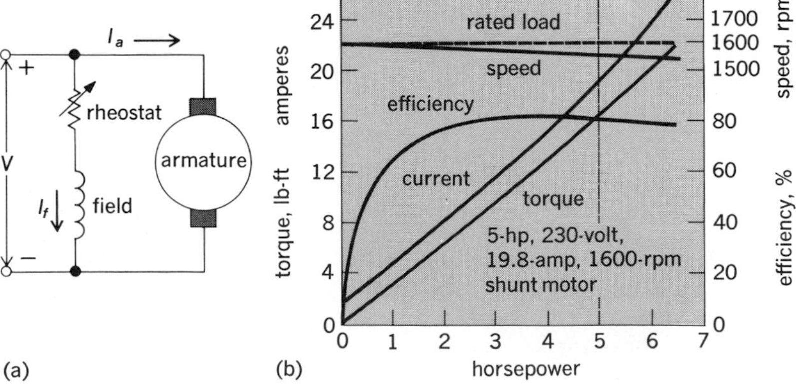

Fig. 3. Shunt motor. (a) Connections. (b) Typical operating characteristics. 1 hp = 0.7457 kW. 1 lb-ft = 1.356 N-m.

PRINCIPLES

Rotation of a dc motor is produced by an electromagnetic force exerted upon current-carrying conductors in a magnetic field. For basic principles of motor action *see* MOTOR, ELECTRIC.

In Fig. 2, forces act on conductors on the left path of the armature to produce clockwise rotation. Those conductors on the right path, whose current direction is reversed, also will have forces to produce clockwise rotation. The action of the commutator allows the current direction to be reversed as a conductor passes a brush.

The net force from all conductors acting over an average radial length to the shaft center produces a torque T given by Eq. (1), where K_t is a conver-

$$T = K_t \Phi I_a \tag{1}$$

sion and machine constant, Φ is net flux per pole, and I_a is the total armature current.

The voltage E, induced as a counter electromotive force (emf) by generator action in the parallel paths of the armature, plus the voltage drop I_aR_a through the armature due to armature current I_a and armature resistance R_a, must be overcome by the total impressed voltage V from the line. Voltage relations can be expressed by Eq. (2).

$$V = E + I_aR_a \tag{2}$$

The counter emf and motor speed n are related by Eq. (3), where K is a conversion and machine constant.

$$n = \frac{E}{K\Phi} \tag{3}$$

Mechanical power output can be expressed by Eq. (4), where n is the motor speed in rpm and T is the torque developed in pound-feet.

$$\text{HP} = \frac{2\pi nT}{33{,}000} \quad \text{horsepower} \tag{4}$$

By use of these four equations, the steady-state operation of the dc motor may be determined.

TYPES

Direct-current motors may be categorized as shunt, series, compound, or separately excited.

Shunt motor. The field circuit and the armature circuit of a dc shunt motor are connected in parallel (Fig. 3*a*). The field windings consist of many turns of fine wire. The entire field resistance, including a series-connected field rheostat, is relatively large. The field current and pole flux are essentially constant and independent of the armature requirements. The torque is therefore essentially proportional to the armature current.

In operation an increased motor torque will be produced by a nearly equal increase in armature current, Eq. (1), since K_t and Φ are constant. Increased I_a produces an increase in the small voltage I_aR_a, Eq. (2). Since V is constant, E must decrease by the same small amount resulting in a small decrease in speed n, Eq. (3). The speed-load curve is practically flat, resulting in the term "constant speed" for the shunt motor. Typical characteristics are shown in Fig. 3*b*.

Typical applications are for load conditions of fairly constant speed, such as machine tools, blowers, centrifugal pumps, fans, conveyors, wood- and metalworking machines, steel, paper, and cement mills, and coal or coke plant drives.

Series motor. The field circuit and the armature circuit of a dc series motor are connected in series (Fig. 4*a*). The field winding has relatively few turns per pole. The wire must be large enough to carry the armature current. The flux Φ of a series motor is nearly proportional to the armature current I_a which produces it. Therefore, the torque, Eq. (1), of a series motor is proportional to the square of the armature current, neglecting the effects of core saturation and armature reaction. An increase in torque may be produced by a relatively small increase in armature current.

In operation the increased armature current, which produces increased torque, also produces increased flux. Therefore, speed must decrease to produce the required counter emf to satisfy Eqs. (1) and (3). This produces a variable speed characteristic. At light loads the flux is weak because of the small value of armature current, and the speed may be excessive. For this reason series motors are generally connected permanently to their loads through gearing.

The characteristics of the series motor are shown in Fig. 4*b*. Typical applications of this motor are to loads requiring high starting torques and variable speeds, for example, cranes, hoists, gates, bridges, car dumpers, traction drives, and automobile starters.

Compound motor. A compound motor has two separate field windings. One, generally the predominant field, is connected in parallel with the armature circuit; the other is connected in series with the armature circuit (Fig. 5).

The field windings may be connected in long or short shunt without radically changing the operation of the motor. They may also be cumulative or differential in compounding action. With both field windings, this motor combines the effects of the shunt and series types to an extent dependent upon the degree of compounding. In Fig. 6 its typical speed characteristics are compared with those of the shunt and series types. Applications of this motor are to loads requiring high starting torques and somewhat variable speeds, such as pulsating loads, shears, bending rolls, plunger pumps, conveyors, elevators, and crushers. *See* DIRECT-CURRENT GENERATOR.

Separately excited motor. The field winding of this motor is energized from a source different from that of the armature winding. The field winding may be of either the shunt or series type, and adjustment of the applied voltage sources produces a wide range of speed and torque characteristics. Small dc motors may have permanent-magnet fields with armature excitation only. Such motors are used with fans, blowers, rapid-transfer switches, electromechanical activators, and programming devices.

STARTING AND SPEED CONTROL

Except in small dc motors, it is necessary to limit armature current when the motor is started. Therefore a load resistance must be in the armature circuit until the motor reaches full speed.

Starting. Direct-current motors are usually started with a rheostat in series with the armature circuit. This motor-starting resistor is of the proper rating in watts and ohms to withstand starting currents (Fig. 7).

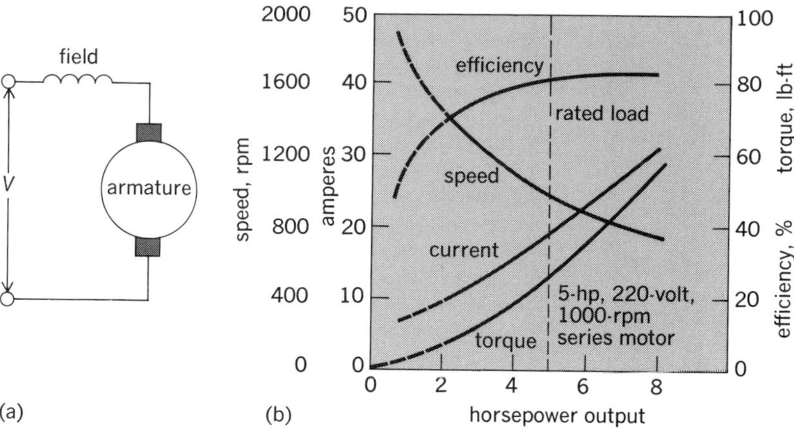

Fig. 4. Series motor. (*a*) Connection. (*b*) Typical operating characteristics. 1 hp = 0.7457 kW. 1 lb-ft = 1356 N-m.

When a dc motor is started, the field winding is fully excited. Since there is no rotation of the armature, no counter emf is generated. Therefore, the armature current would be dangerously high unless an additional starting resistance were placed in the armature circuit, Eq. (2). This rheostat is manually or automatically cut out of the circuit as the motor approaches full speed. Small motors which have low armature inertia reach full speed rapidly and do not require starting resistors. Separately excited motors may be started by control of the voltages applied to the armature.

Speed control. Speed of a dc motor may be controlled by changing the flux or counter emf of the motor, Eq. (3). Adjustment of the armature voltage V will affect the counter emf E, Eq. (2), by approximately the same amount. The speed n is affected by the change in counter emf according to Eq. (3). Insertion of a resistor in the armature circuit would also affect the speed but is seldom used because of the large power losses in the resistor. Speed control by adjustment of the applied armature voltage is used extensively where separate, adjustable voltage sources are available.

A change in flux Φ will also affect speed n, Eq. (3). Flux may be changed by a variable resistor in series with the shunt field of a shunt or compound motor. This field rheostat should have a total resistance comparable to that of the shunt field and be of sufficient capacity to withstand the relatively small shunt-field current.

Ward Leonard. In this system the armature voltage of a separately excited dc motor is controlled by a motor-generator set. A typical circuit (Fig. 8) shows a prime mover M_1, often a three-phase induction motor, mechanically coupled to a dc generator G and to an exciter generator E. The latter provides field excitation for the dc machines. Control of the generator field rheostat R_1 affects the output voltage of the generator G. This voltage may be smoothly varied from a low value to a value above normal. When this voltage is applied to the armature of the motor M_2, the speed of this motor will be variable over a wide range. Additional speed control of motor M_2 may be gained by adjustment of rheostat R_2.

DIRECT-CURRENT MOTOR

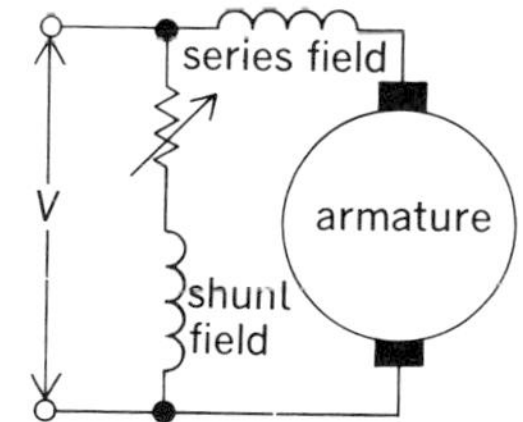

Fig. 5. Connection of a compound motor.

DIRECT-CURRENT MOTOR

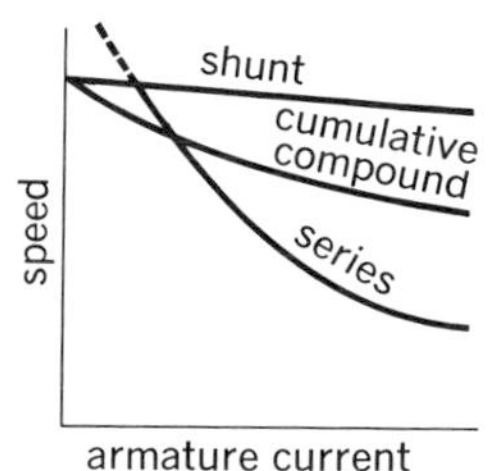

Fig. 6. Comparative speed-current curves of dc motors.

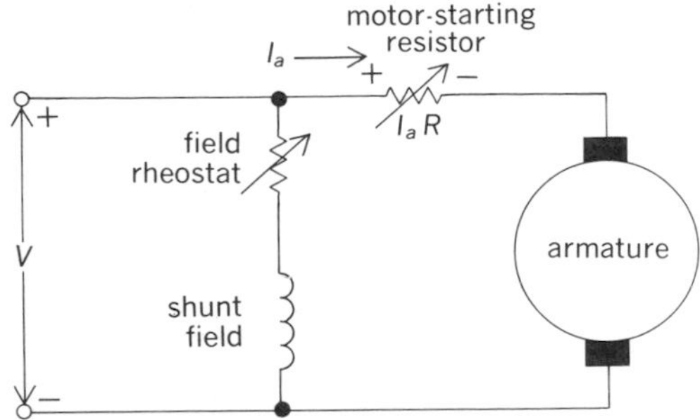

Fig. 7. Connection for starting a dc shunt motor.

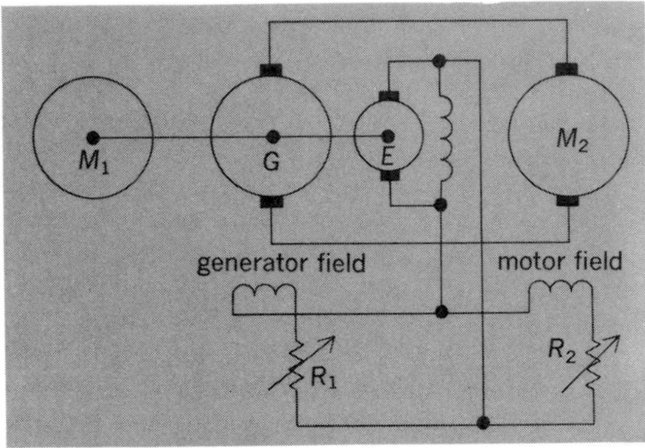

Fig. 8. Ward Leonard speed control system.

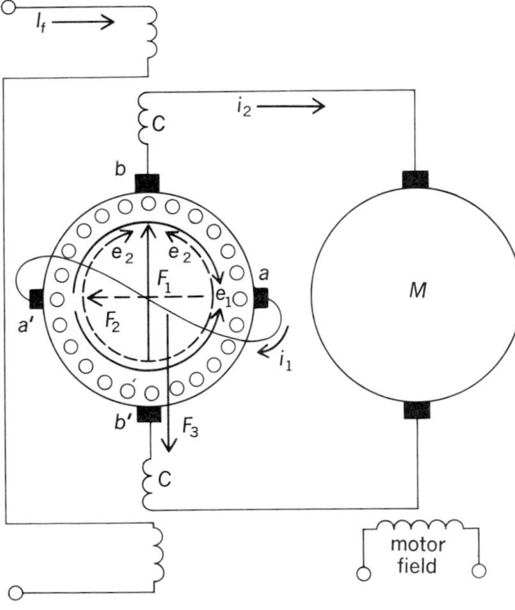

Fig. 9. Operation of the Amplidyne.

DIRECT-CURRENT MOTOR

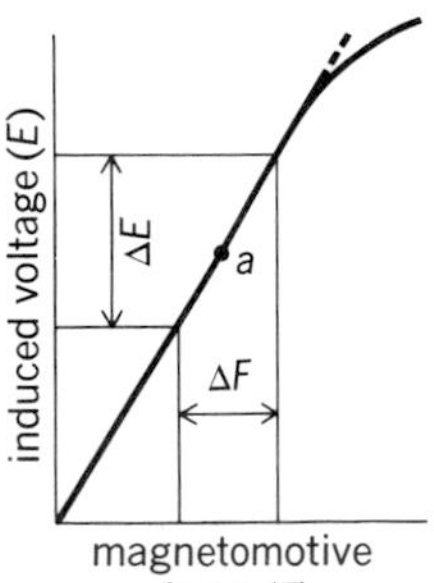

Fig. 10. Regulex magnetization curve.

The disadvantages of this system are the added equipment and maintenance costs it entails. However, the wide range and fineness of control in a low-current circuit make it applicable to high-speed passenger elevators, large hoists, power shovels, steel-mill rolls, drives in paper or textile mills, and the propulsion of small ships.

Amplidyne. The dynamoelectric amplifier (Amplidyne) is a rotating, two-stage, power amplifier in which a small change in field power in a dc generator results in a large change in output armature power. A large motor connected to the output of the generator may be controlled in speed by adjustment of the relatively small field power of the Amplidyne.

In Fig. 9 the control field current I_f produces an mmf F_1. The resultant flux and the short circuit of brushes *aa′* cause the induced voltage e_1 to force a large current i_1 through armature circuit. Because of the magnetic core design, current i_1 produces mmf F_2 and its resultant flux which induces voltage e_2 between brushes *bb′*. Motor *M* connected across brushes *bb′* will draw a current i_2 which produces an mmf F_3 tending to weaken the original mmf F_1. However, compensating windings *C* energized by i_2 will produce an mmf to oppose F_3 and restore the value of F_1.

This dynamic amplifier may produce amplifications of 10,000 to 1 or higher. It is applied to a variety of servomotors to control starting, acceleration, and deceleration. Other typical applications include voltage regulation of large ac generators, dc voltage control in cold-strip mills, speed control of paper mills, positioning control of gun turrets, machine-tool drives, and power-factor control of synchronous generators.

Regulex. The regulating exciter (Regulex) is a dc generator acting as a power amplifier. By proper design of the machine magnetic core, an extensive linear portion of the voltage buildup curve is obtained (Fig. 10). A small change in mmf *F* will produce a large change in induced voltage *E* resulting in a degree of amplification. Critical-value adjustment of the field rheostat *R* (Fig. 11) will cause the generator to operate on this linear portion. A reference field F_2 and an opposing field F_3 combine with field F_1 to establish a point of operation, such as point *a* in Fig. 10.

Departure from this balance because of a variation in the control field F_3 will produce the large change in voltage *E*. The output of this device may be used to drive a dc motor *M*, which has its speed translated into voltage by means of a small pilot generator coupled to the motor shaft. By proper feedback of this voltage to control field F_3, the motor speed may be maintained at a constant value.

Rototrol. The rotating control (Rototrol) is a dc generator acting as a power amplifier. It is similar to the Regulex, but the self-excited field is a series type in contrast to the shunt-type field of the Regulex (Fig. 12).

Solid-state control for motor speed. Solid-state devices such as diodes and thyristors, including silicon-controlled rectifiers (SCRs), may be used in a number of circuit applications to control the speed of dc motors.

Pulse control. One such circuit supplies a number of unidirectional voltage pulses to the motor whose speed is adjusted by variation of pulse frequency or pulse width. In the circuit of Fig. 13, the thyristor acts as an ON-OFF switch. The relative on-to-off time determines the width of the voltage pulses. The start of thyristor conduction is determined by a gating voltage, but a negative voltage must be applied across the thyristor by the commutating circuit in order to stop its conduction. Controlling the width of the voltage pulses or the number of pulses per second (several hundred) determines the average value of the voltage applied to the motor and hence its speed. During the

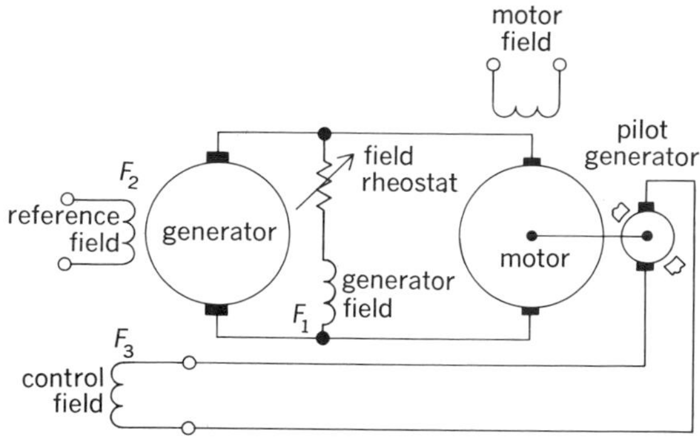

Fig. 11. Typical Regulex circuit.

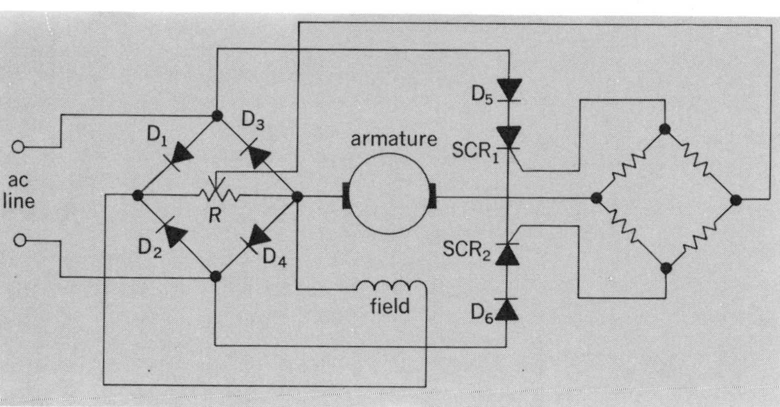

Fig. 14. SCR motor speed control.

thyristor OFF period, diode D allows the energy stored in the motor field coils to be discharged through the motor armature. Current is continuous, increasing during the ON period and decreasing during the OFF period when the current is the result of the energy coming from the coils, but the average voltage is independent of the current.

Pulse control is applied to several types of electrically operated vehicles and affords an efficient and smooth control of speed.

Full-wave rectifier control. Thyristors (SCRs) are used in full-wave rectifier circuitry to control the motor speed, as indicated in Fig. 14.

Diodes D_1, D_2, D_3, and D_4 form a bridge rectifier to supply a full-wave rectified dc voltage to the field of the motor. A second bridge rectifier circuit also using diodes D_3 and D_4 as well as SCR_1 and SCR_2 provides adjustable current to the armature.

Control of the firing of the SCRs is accomplished by amplitude changes in the SCR gate voltages due to adjustment of potentiometer *R*. If the gate voltages exceed the back emf generated in the armature of the motor for a given load condition, the SCRs will fire. The average value of armature current and hence the speed is thus under control.

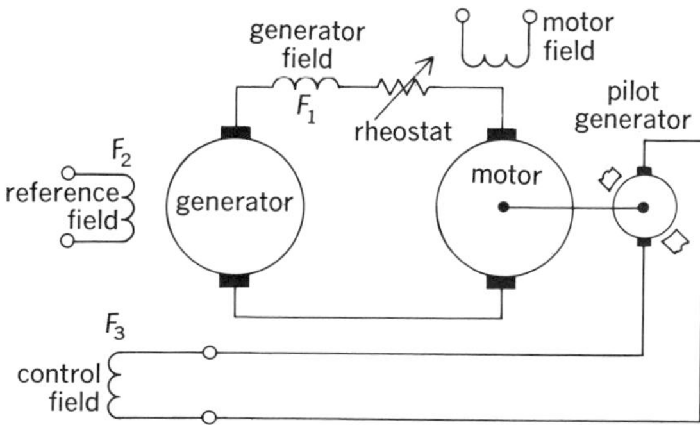

Fig. 12. Typical Rototrol circuit.

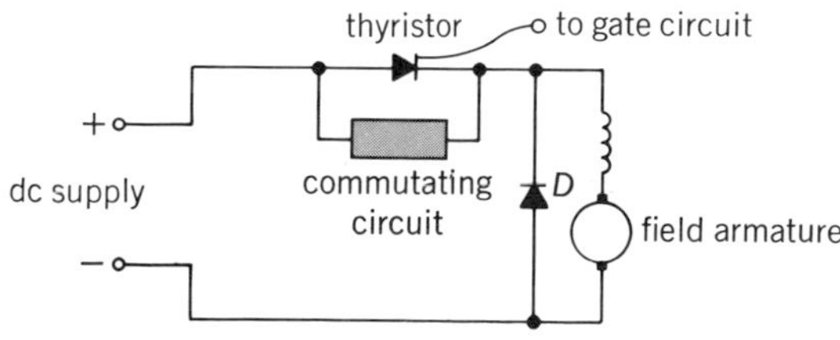

Fig. 13. Elementary Thymatrol circuit.

Damage to the SCRs due to large values of reverse currents is prevented by diodes D_5 and D_6, which block the flow of such reverse currents.

[L. F. CLEVELAND]

Bibliography: A. E. Fitzgerald, D. E. Higginbotham, and A. Grabel, *Basic Electrical Engineering*, 4th ed., 1975; A. E. Fitzgerald, C. Kingsley, and A. Kusko, *Electric Machinery*, 3d ed., 1971; B. C. Kuo and J. Tal (eds.), *Incremental Motion Control: DC Motors and Controls*, vol. 1, 1978; J. Rosenblatt and M. H. Friedman, *Direct and Alternating Current Machinery*, 1963; A. Kusko, *Solid-State DC Motor Drives*, 1969.

Direct-current transmission

The conduction of electric power by means of electric currents which flow in one direction only. *See* DIRECT CURRENT.

Early use. The first commercial applications of electric power used direct current (dc). They included groups of series-connected arc lamps and groups of parallel-connected carbon-filament incandescent lamps energized by electricity from dc generators, usually driven by steam engines. Another early application was street railways.

Sometimes storage batteries were used with the low-voltage dc systems to provide against emergencies such as failures of a generator or of the conductors (mains) from the generator to the load. The chief limitation of these early electric power systems was their low voltage, which limited the economic distance between generator and loads. One improvement was the adoption of the three-wire circuit consisting of both a positive conductor and a negative conductor instead of one or the other. The load could be divided fairly equally between these two conductors, giving a much lower current in the neutral conductor, which was grounded. This system effectively doubled the voltage of the distribution system without creating the hazard of a higher voltage to ground at the incandescent lamps accessible to the customers.

Replacement of dc by ac. Still more improvement in efficiency was required by increased loads and by the demands for service in suburban and rural areas having lower load density than the city. This problem was solved by the invention of the transformer and the use of alternating current (ac). The transformer is both efficient and simple. Its use permits any number of voltage levels. For example, the generators can have one voltage, the transmission lines another, higher voltage, the local distribution system a lower but moderately high voltage, and the loads the same low voltage that they had when dc was used. The ac generators have no commutators, and this fact makes it possible to build more powerful generators operating at higher speed. They are driven by steam turbines instead of reciprocating steam engines. *See* ALTERNATING CURRENT; TRANSFORMER.

Another notable improvement in electric power systems was the use of polyphase systems, especially of three-phase systems. A three-phase generator has a higher rated power than does a single-phase generator of equal size. The polyphase induction motor, especially that with a squirrel-cage rotor, is simple, rugged, and self-starting, and most of these motors can be started safely by applying their full rated voltages. *See* ALTERNATING-CURRENT GENERATOR; ALTERNATING-CURRENT MOTOR.

Polyphase high-voltage transmission lines make it practical to generate electric power from water power at remote sites and to transmit this power efficiently over long distances to the sites (usually in or near large towns) where there is a demand for electric power.

Low-voltage dc networks persisted in the central districts of large cities into the 1970s, but most of these have now been converted to ac networks.

Advantages. After the almost complete replacement of dc by ac, however, interest in dc power transmission began increasing in 1954. One reason is that the cost of a dc line is about two-thirds that of the corresponding ac line. This is because the usual dc bipolar line has two conductors, compared with three conductors of an ac single-circuit three-phase line. In neither case are the overhead ground wires for intercepting lightning strokes counted. In emergencies the bipolar line can be operated for considerable time by use of one pole with return path through the ground.

The dc line requires additional apparatus at its terminals, not required for an ac line, which increases the cost. These are the converters for changing ac to dc at the sending end and dc to ac at the receiving end. Considering the combined costs of the line and of the terminal equipment, the dc link is thus more expensive than ac for a short line but less expensive for a long line. The costs are equal at a "break-even" distance of approximately 500 mi (800 km), but this distance varies because of the variability of the costs. The costs of converters are expected to decrease in the future because of the increasing number of suppliers and the experience that they will gain. This would decrease the break-even distance. However, the choice between ac and dc usually depends on other factors.

Transmission over submarine cables. Direct-current lines are used to perform functions for which ac lines are unsuited, such as transmitting power across a wide body of water, using submarine cables. Cables have a much larger shunt capacitance per unit of length than overhead lines have. When an ac voltage is applied between the two conductors of a cable, an alternating "charging" current flows even if the distant end is not connected to a load. When the length of the cable exceeds some distance—about 30 mi (50 km)—the thermal limit is reached, even though no power is transmitted. This was a significant factor in the choice of dc for connecting the island of Gotland to the mainland of Sweden, Denmark to Sweden, England to France, and the South Island of New Zealand to the North Island.

System interconnections. Direct-current lines are also used to provide interconnections of two ac systems of different nominal frequencies, for example, 50 Hz and 60 Hz, as was done in Japan, or interconnections between two ac systems having the same nominal frequency but different frequency controls. An ac tie of the latter type might require a much higher power rating than the greatest power which was desired to be interchanged. In some cases, the two converters are installed in the same station, so that really there is no dc line. Such stations are called frequency-changer stations if there are two different nominal frequencies, and asynchronous ties if the nominal frequencies are equal, as in New Brunswick, Canada, or Steagle, Nebraska.

Corona. Overhead dc lines differ from ac in the effect of corona in producing radio interference. An important difference between ac and dc corona is the effect of rainfall, which greatly increases the amount of radio noise on an ac line but makes a negligible change on a dc line. In addition, even with equal readings on a noise meter, on an ac line the modulation at the power frequency increases the annoyance to a human observer by an amount equivalent to 3 dB.

Converters. A converter consists of transformers and of valves which conduct current in only one direction, called the forward direction, but not in the opposite direction, called the inverse direction. The valves used in dc transmission are controlled valves; each valve has a control electrode which can prevent conduction from starting even though the voltage across the valve is in the proper direction of conduction. However, once the valve begins to conduct, the control electrode cannot stop conduction or control the value of the instantaneous current.

The early dc transmission schemes (1954 to 1974) used mercury-arc valves in steel tanks with mercury-pool cathodes. Most of these valves are still in use. However, all schemes commissioned since the late 1970s use thyristors, which are solid-state devices. These require less maintenance and less expensive auxiliary devices.

The most advantageous method of interconnecting the several valves in a converter is the three-phase two-way circuit, also known as the Graetz

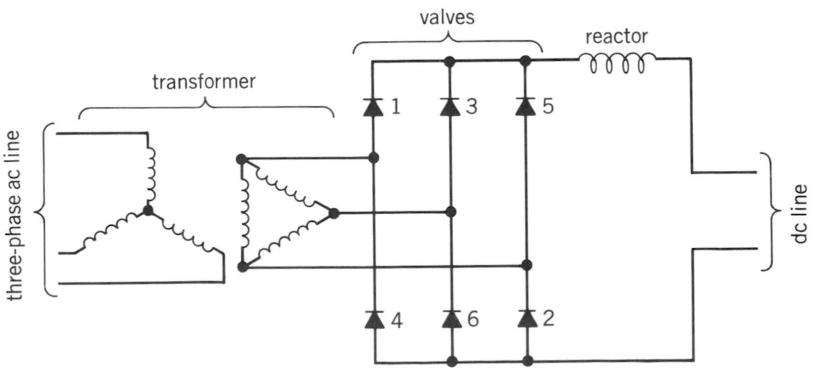

Three-phase two-way or Graetz converter circuit.

circuit, having six valves (see illustration). They are ignited in the sequence indicated by the valve numbers.

The direction of power flow is usually reversed on a dc line by reversing the polarity of the direct voltage, since the current cannot be reversed except by switching.

Fault clearance. Most faults on dc lines, as well as on ac lines, are temporary and are cleared by deenergizing the line for a short while to extinguish the fault arc. Circuit breakers are not used for this purpose on dc lines. Instead, the valves are controlled so as to tend to reverse the current in the faulted conductor. Since the valves can conduct only in the normal direction, the fault current is brought to zero. After allowance of a short time for deionization of the arc path, the line is reenergized. *See* ELECTRIC POWER SYSTEMS; TRANSMISSION LINES. [EDWARD W. KIMBARK]

Bibliography: C. Adamson and N. G. Hingorani, *High Voltage Direct Current Power Transmission*, 1960; B. J. Cory (ed.), *High Voltage Direct Current Convertors*, 1965; E. W. Kimbark, *Direct Current Transmission*, vol. 1, 1971; E. Uhlmann, *Power Transmission by Direct Current* (transl. by R. Clark), 1975.

Distillate fuel

A broad term for any one of the wide variety of fuels obtained from fractions boiling above gasoline in the distillation of petroleum. The most important distillate fuels are kerosine, furnace oils, and diesel fuels. Formerly, heavy naphtha-kerosine distillates of low octane number (known as engine distillates or tractor fuels) were used as fuels for low-compression engines of farm tractors, farm lighting units, and small boats. Such power units have been largely replaced by high compression gasoline engines or small diesels, requiring different fuels. *See* DIESEL FUEL; KEROSINE; PETROLEUM PRODUCTS. [MOTT SOUDERS]

District heating

The supply of heat, either in the form of steam or hot water, from a central source to a group of buildings. As an industry, district heating began in the early 1900s with distribution of exhaust steam, previously wasted to atmosphere, from power plants located close to business and industrial districts. Advent of condensing-type electrical generating plants and development of long-distance electrical transmission led to concentration of electrical generation in large plants remote from business districts. Most district heating systems in the United States rely on separate steam generation facilities close to load centers. In some cities, notably New York, high-pressure district steam (over 120 psi or 830 kPa) is used extensively to feed turbines that drive pumps and refrigerant compressors. Although some district heating plants serve detached residences, the cost of underground piping and the small quantities of heating service required make this service generally unfeasible.

District heating, apart from utility-supplied systems in downtown areas of cities, is accepted as efficient practice in many prominent colleges, universities, and government complexes. Utility systems charge their customers by metering either the steam supply or the condensed steam returned from the heating surfaces.

District heating is growing faster in Europe than in the United States because of economic considerations; because fuel costs are higher in Europe, more efficient combustion associated with central facilities is that much more attractive. Further, labor costs are lower there, and this affects the high capital costs of plant and distribution systems. A substantial part of the heat distribution is by hot water from plants combining power and heat generation. Heat from turbine exhaust or direct boiler steam is used to generate high-temperature (270–375°F or 132–191°C) water, often in cascade heat convertors, where live steam is mixed with return hot water. Hot water distribution systems have the advantage of high thermal mass (hence, storage for peak periods) and freedom from problems of condensate return. On the other hand, pumping costs are high and two mains are required, supply and return. (The cost of condensate return piping and pumping is so high that many steam distributors in the United States make no provision for condensate return at all.)

Reykjavik, Iceland, is completely heated from a central hot water system supplied from hot springs. Average water temperature is 170°F (77°C).

The most extensive European developments in district heating have been in the Soviet Union, mainly in the form of combination steam-electric plants serving new towns. The Soviets have developed a single-pipe high-temperature water district heating system. At each building served, superheated water from the main is diluted with return water from the building system. Excess water is used in kitchens, laundries, bathrooms, and swimming pools.

An important extension of the principle of district heating is the development of district heating and cooling plants. Usually owned and operated by the natural gas utilities, central plants distribute chilled water for air conditioning in parallel conduits with hot water or steam for heating. The system serving Hartford, Conn., was the pioneer installation. Because the costs of these four-pipe distribution systems are extremely high, development is slow and restricted to small areas of high density usage within a few blocks of plant. *See* COMFORT HEATING. [RICHARD L. KORAL]

Bibliography: International District Heating Association, *District Heating Handbook*, 4th ed., 1970.

Dry cell

A primary cell in which the electrolyte is absorbed in a porous medium, or is otherwise restrained from flowing. Common practice limits the term dry cell to the Leclanché cell, which is the major commercial type. Other dry cells, not discussed in this article, include the mercury cell, the alkaline-zinc-manganese dioxide cell, and the air-depolarized cell. These cells all use aqueous electrolytes immobilized in absorbent materials or gels. By order of the Federal Trade Commission, "leakproof" must not be printed on any dry cell. *See* PRIMARY BATTERY.

Construction. The Laclanché cell is made in a variety of sizes in either round or flat shapes. American National Standard Specifications for Dry Cells and Batteries. C.18.1–1972, lists the smallest cylindrical cell as the 0 cell, 0.328 cm^3.

0.9 g, to the largest, the no. 6 cell, 483 cm^3, 990 g. Flat-type cells, which are avilable as multiple-cell batteries only, range in size from the F 12 cell, 2 cm^2 in cross section and 0.71 cm thick, to the F 100, which is 27.35 cm^2 by 1.04 cm. The energy-volume ratio of flat-cell batteries is about twice that of batteries made with round cells because of (1) the absence of an expansion chamber and carbon rod and (2) because of their rectangular form, which eliminates waste space in assembled batteries and voids between cells.

The negative electrode is usually solid zinc, either in cup form (cylindrical cell) or in sheet form (flat cell). The carbon positive electrode is embedded in a black mixture of manganese dioxide and carbon black. The carbon is either a rod located in the center of a cylindrical cell or a coating on the back of the flat zinc electrode in the flat cell. The separator between the black mix and the zinc electrode consists of a paper barrier coated with cereal or methyl cellulose. In the older construction the separator was composed of a gelatinous paste which also held the electrolyte. Not only does the paper barrier give about 30% more volume for the depolarizer mass, but it also decreases the internal resistance of the cell and reduces the number of manufacturing operations.

The electrolyte is a solution of ammonium chloride and zinc chloride in water. The cell enclosure consists of a top seal in the cylindrical cell or thin plastic wrappings for the flat cell.

Zinc electrode. Zinc cans (cup form of electrode) are made by drawing or impact extrusion. A typical alloy for extruded cans contains 1.0% lead, 0.05% cadmium, and remainder zinc. The zinc used is of 99.99% purity. The alloying elements improve the mechanical properties and decrease wasteful corrosion. The cell capacity is less than that theoretically possible from the weight of zinc. This is because zinc is used as the container, the limiting factor being the manganese dioxide. For example, the no. 6 can weighs 110 g, giving a theoretical capacity of 90 ampere-hours. The actual capacity is in the range of 40–50 A-hr, depending on the depolarizer composition and rate of discharge.

Black mix. The black mix is composed of manganese dioxide mixed with carbon black. The manganese dioxide is usually obtained from natural ore (mainly from Gabon, Greece, and Mexico), but may be a synthetic product prepared by chemical precipitation or by electrolytic methods. Mixtures of the natural and synthetic oxides are also used. The carbon black is usually acetylene black made by the thermal decomposition of acetylene. Graphite is used to a lesser extent. Manganese dioxide has a theoretical capacity of about 0.3 A-hr/g. The practical capacity is somewhat less. The carbon black is used in varying proportions, depending on design factors. It serves the double purpose of increasing the conductivity of the manganese dioxide and absorbing the electrolyte. For cells that require a high capacity at low current drains (for example, transistor use), the ratio of manganese dioxide to carbon may be 10:1; and, at the other extreme, for cells requiring a very high flash current (for example, photoflash), the ratio may be as low as 1:1. In addition to these components, the black mix also contains electrolyte amounting to about 25% of the total weight.

Carbon rod. The carbon rod used in a cylindrical cell serves as the conductor of electricity for the positive electrode; it also serves as a vent to allow gas to escape. Carbon rods are usually made from petroleum coke which is calcined, ground, and mixed with pitch. The "green" rods are baked to form a hard carbon, having low electrical resistance. They may be partially waterproofed by impregnation with oil or paraffin wax to prevent capillary creepage of electrolyte out of the cell.

Flat cells are usually made with duplex electrodes. The zinc is coated on one side with a carbonaceous coating, which serves to conduct electricity between the zinc and the black mix of the adjacent cell.

Separator. The modern method of separating the two electrodes is to use kraft paper which has been coated on the side adjacent to the zinc with a film of cereal or methyl cellulose containing mercurous (or mercuric) chloride. The latter corresponds to a mercury concentration of 0.155 mg/cm^2 of paper area.

In the older method a paste was made from a mixture of electrolyte with corn starch and wheat flour. The paste was added to the cell in liquid form and gelatinized by heating the cell in a water bath. By increasing the zinc chloride concentration in the electrolyte, the gelatinization of the paste could take place at ambient temperatures. Mercuric chloride was added to the paste, but it was not possible to control the intensity or uniformity of amalgamation to the same degree as with the paper-lined cells. In a typical assembly line the cereal-coated paper is formed into a cylinder on a mandrel and inserted into the zinc can with a bottom washer. The calculated amount of wet depolarizer mix is injected into the lined can, followed by insertion of the carbon rod. Simultaneously with insertion of the carbon rod, the black mix is compressed. This gives a solid mass and forces sufficient electrolyte out of the mix to completely wet the cereal-coated paper barrier. A cardboard washer is placed on the black mix and an air space (expansion chamber) is enclosed over this washer by a further washer over which is poured a layer of bitumen. This seal is necessary to minimize moisture loss. Flat cells are made by placing treated paper containing the paste between the black-mix cake and the zinc of each cell.

Electrolyte. The electrolyte is made by dissolving ammonium chloride and zinc chloride in water. For paste cells a small amount of mercuric chloride is usually added. This component, however, converts to zinc chloride as soon as the zinc and electrolyte come into contact. Mercury then plates out on the zinc. The composition of the electrolyte depends on the cell designs.

During discharge the composition of the electrolyte changes. In one test in which a D-size cell was discharged through a 4-ohm resistance, the pH of the paste layer next to the zinc changed from 5.7 to 3.8 (more acid) while the pH of the innermost portion of the mix went from 5.8 to 10.1 (more alkaline).

Ordinary dry-cell electrolyte has a resistivity of 2.42 ohm-cm at 20°C. For low-temperature operation special electrolytes have been developed. An electrolyte of 12% zinc chloride, 15% lithium chloride, 8% ammonium chloride, and 65% water

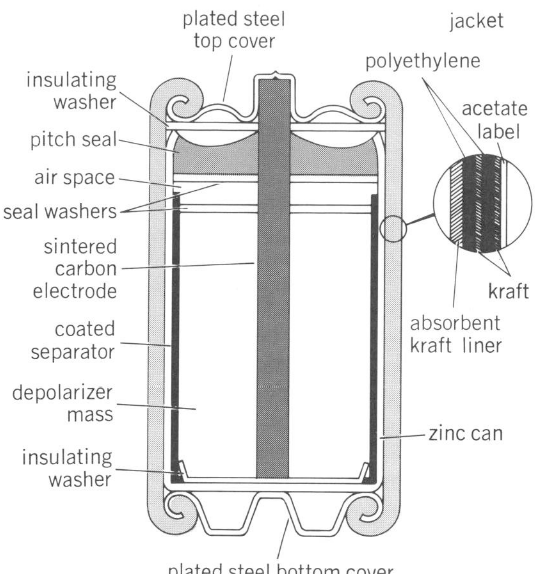

Modern Leclanché dry cell. *(Bright Star Industries Inc.)*

is fluid at −40°C. Other electrolytes for low-temperature operation use a mixture of calcium chloride, zinc chloride, and ammonium chloride solutions.

Cell enclosure. Whereas the cylindrical cell was originally wrapped in a paper jacket, modern methods are more sophisticated in order to resist leakage which could damage valuable equipment, for example, cameras, tape recorders, and record players. One method uses an absorbent board wrap and an outer jacket of sheet steel. The cells are finally sealed in these containers with tinplate top and bottom closures. The other method (see illustration) uses a jacket consisting of a laminate of absorbent paper, polyethylene, and kraft paper covered with a cellulose acetate–coated label. As with the steel-jacketed cells, tinplate covers are placed top and bottom to make contact with the carbon rod and zinc can, respectively.

Flat cells use thin plastic wrappings around the edges of each cell. This confines the electrolyte to individual cells and avoids internal discharge in a stack. The wrappings are sufficiently gas-permeable to prevent the building up of pressure in the cell. After the requisite number of cells are stacked, the stack is bound together by tapes, and dipped in molten wax for further moistureproofing.

Cell chemistry. At the anode (zinc) the zinc oxidizes to zinc ion and simultaneously liberates electrons to the external circuit, at a rate proportional to the current. For each ampere which flows, 1.2 g of zinc per hour is converted to zinc ion.

At the cathode (manganese dioxide), the electrons from the external circuit reduce the manganese dioxide to three different substances, depending on circumstances which have not yet been thoroughly explained. Studies have shown, however, that the total ampere-hour output of the cell can be accounted for by analyzing the cathode mix for the following substances: soluble manganese (Mn^{++}), each gram of which accounts for nearly 1 A-hr of discharge; insoluble manganite (MnOOH), each gram of which accounts for about 0.3 A-hr of discharge; insoluble hetaerolite ($ZnO \cdot Mn_2O_3$), each gram of which accounts for about 0.22 A-hr of discharge.

The electrochemical reduction of the manganese dioxide (MnO_2) has been reported to occur as the reaction to form soluble manganese, shown as reaction (1). This occurs only when the cell delivers current.

$$MnO_2 + 4H^+ + 2e^- \rightarrow Mn^{++} + 2H_2O \qquad (1)$$

Two secondary reactions, (2) and (3), can then occur.

$$MnO_2 + Mn^{++} + 2OH^- \rightarrow 2MnOOH \qquad (2)$$

$$MnO_2 + Mn^{++} + 4OH^- + Zn^{++} \rightarrow ZnO \cdot Mn_2O_3 + 2H_2O \qquad (3)$$

Reaction (3) can occur only if zinc is in solution in the cathode mix.

Operating characteristics. The service capacity of dry cells is not a fixed number of ampere-hours, but varies with current drain, operating schedule, cutoff voltage, operating temperature, and storage conditions prior to use. Most cells are tailor-made for their rated end use. For example, D-size cells can be rated as general-purpose, industrial flashlight, transistor, electronic flash, and photoflash. No. 6 cells are specially formulated for bell ringing, telephone, protective alarm, ignition, and general purpose. Minimum outputs of D cells are specified by the General Service Administration are as follows:

(1) General-purpose flashlight – $2\frac{1}{4}$ ohms for 5-min periods at 24-hr intervals until the closed-circuit voltage drops to 0.65 V; 400 min. (2) Light industrial flashlight – 4 ohms for 4-min periods beginning at hourly intervals for 8 consecutive hours each day until the closed circuit voltage drops to 0.9 V, 950 min. (3) Heavy industrial flashlight – 4 ohms for 4-min periods beginning at 15 min intervals for 8 consecutive hours each day until the closed-circuit voltage drops to 0.9 V; 800 min. (4) Photoflash bulb test – 0.15 ohm for 1 s each minute for 1 hr at 24-hr intervals for 5 consecutive days each week until the closed-circuit voltage falls below 0.5 V; 800 s. (5) Electronic photoflash – 1.0 ohm for 15 s each minute for 1 hr at 24-hr intervals for 5 consecutive days each week until the closed-circuit voltage falls below 0.75 V; 275 15-s discharges. (6) Transistor test – $83\frac{1}{3}$ ohms during a continuous period of 4 hr daily until the closed-circuit voltage falls below 0.9 V; 200 hr.

Temperature effect. The higher the temperature during discharge, the greater is the energy output. Conversely, the lower the temperature, the lower is the output. At −23°C the battery is virtually inoperative. However, shelf life is influenced in the reverse direction by environmental temperatures.

Better low-temperature output can be obtained with special electrolytes and cell structures giving a high ratio of electrode area to mix thickness, and special types of manganese dioxide.

Shelf life. This is the period of time that a battery can be stored before it drops to 90% of its capacity when tested fresh at 21°C and 50% relative humidity.

Deterioration in a dry cell occurs in a number of ways: (1) Zinc can oxidize by reaction with the electrolyte; this reaction produces hydrogen.

(2) Manganese dioxide can be reduced by carbon and by the organic materials used in the cells; this can produce carbon dioxide. (3) Water can be evaporated from the electrolyte; this increases the cell resistance and alters the composition of the electrolyte unfavorably.

In general, shelf life decreases as the cell size becomes smaller: with well-constructed cells a shelf life of 3 years with a no. 6 telephone cell and 10 months with a penlight cell. Other sizes can be prorated. Flat cells have a shorter shelf life than cylindrical cells: 6–9 months for all sizes. American National Standard Specifications quote 30 hr as the minimum initial output for a 9-V transistor radio battery (six flat cells), and 28 hr after 6 months' delay. High temperatures reduce shelf life, and at 32°C the shelf life of a battery is about one-third that of one stored at 21°C. Low-temperature storage increases the shelf life of batteries considerably. Batteries stored at 4°C (sealed polyethylene bags should be used to prevent condensation with subsequent corrosion of the terminals and metal jacket or degradation of the paper jacket) have their shelf life increased two or three times. Tests have been conducted by military agencies in many countries showing that, when batteries are frozen, they suffer no deterioration for about 10 years. They must, however, be allowed to reach room temperature before use.

Recharging. It is possible to recharge dry-cell batteries for five or six cycles provided the following precautions are taken: The battery should be subject only to shallow discharges between cycles, used immediately after charging, charged over a 10- to 15-hr period at constant current, and not overcharged.

Zinc chloride cell. The zinc chloride cell is a variation of the standard Leclanché carbon zinc cell, differing mainly in the quantity of zinc chloride and ammonium chloride in the electrolyte. In fact, whereas the standard carbon zinc cell uses mostly ammonium chloride with a small percentage of zinc chloride, the zinc chloride cell uses primarily zinc chloride with little or no ammonium chloride.

Most physical and electrical characteristics of the zinc chloride cell are essentially the same as those of the standard carbon zinc cell, for example, watt-hours per kilogram, watt-hours per cubic centimeter, emf (open circuit voltage), voltage under load, and flash current. However, the zinc chloride cell does have significant advantages in the following areas: (1) better low-temperature performance; for example, at −18°C typically 45% of the 21°C capacity is available; (2) better continuous and high drain capacity due to more efficient depolarization; and (3) better leakage resistance because water is consumed along with the active materials, making the cell practically dry at the end of discharge. The major disadvantage of the zinc chloride cell is the need for an improved seal because of its high sensitivity to moisture loss.

Modifications. In addition to those systems discussed, the following combinations have been developed: (1) The magnesium + magnesium perchlorate or bromide + manganese dioxide modification is more expensive than the Leclanché type. It exhibits the delayed voltage at the commencement of discharge characteristic of magnesium cells and has excellent storage properties. It has a higher initial voltage than the standard dry cell. (2) Another example of cell modification is magnesium + magnesium perchlorate or bromide + metadinitrobenzene. The organic depolarizer used can be produced at a cost that gives the same number of watt-hours per dollar as electrolytic manganese dioxide. This factor becomes important as supplies of battery-grade natural ore diminish. (3) The cell modification of aluminum + aluminum and chromic chlorides and ammonium chromate + manganese dioxide is attractive on account of the lower density and electrochemical equivalent of aluminum compared with zinc.

[JACK DAVIS; KENNETH FRANZESE]

Bibliography: American National Standard Specifications for Dry Cells, C.18.1, 1972; Nicholas Branz, *Modern Primarbatterien*, 1951; D. H. Collins (ed.), *Power Sources*, biennial; *Eveready Battery Applications and Engineering Data*, 1968; A. Fleischer (ed.), *Proceedings of the Power Sources Conference*, annual; G. W. Vinal, *Primary Batteries*, 1950.

Dynamo

An electric machine for the conversion of electrical energy into mechanical energy or, conversely, mechanical energy into electrical energy. It is called a generator if it converts mechanical into electrical energy, and it is called a motor if it converts electrical into mechanical energy. *See* ELECTRIC ROTATING MACHINERY; GENERATOR; MOTOR.

[ARTHUR R. ECKELS]

Efficiency

The ratio, expressed as a percentage, of the power output to the power input. When only mechanical efficiency is concerned, the difference, or loss, between the input and output power is due to friction and is dissipated in the form of heat.

[RICHARD M. PHELAN]

Electric charge

A basic property of elementary particles of matter. One does not define charge but takes it as a basic experimental quantity and defines other quantities in terms of it. The early Greek philosophers were aware that rubbing amber with fur produced properties in each that were not possessed before the rubbing. For example, the amber attracted the fur after rubbing, but not before. These new properties were later said to be due to "charge." The amber was assigned a negative charge and the fur was assigned a positive charge.

According to modern atomic theory, the nucleus of an atom has a positive charge because of its protons, and in the normal atom there are enough extranuclear electrons to balance the nuclear charge so that the normal atom as a whole is neutral. Generally, when the word charge is used in electricity, it means the unbalanced charge (excess or deficiency of electrons), so that physically there are enough "nonnormal" atoms to account for the positive charge on a "positively charged body" or enough unneutralized electrons to account for the negative charge on a "negatively charged body."

The rubbing process mentioned "rubs" electrons off the fur onto the amber, thus giving the amber a surplus of electrons, and it leaves the fur with a deficiency of electrons.

In line with the previously mentioned usage, the total charge q on a body is the total unbalanced charge possessed by the body. For example, if a sphere has a negative charge of 1×10^{-10} coulomb, it has 6.24×10^8 electrons more than are needed to neutralize its atoms. The coulomb is used as the unit of charge in the meter-kilogram-second (mks) system of units. *See* ELECTRICAL UNITS AND STANDARDS.

The surface charge density σ on a body is the charge per unit surface area of the charged body. Generally, the charge on the surface is not uniformly distributed, so a small area ΔA which has a magnitude of charge Δq on it must be considered. Then σ at a point on the surface is defined by the equation below.

$$\sigma = \lim_{\Delta A \to 0} \frac{\Delta q}{\Delta A}$$

The subject of electrostatics concerns itself with properties of charges at rest, while circuit analysis, electromagnetism, and most of electronics concern themselves with the properties of charges in motion. [RALPH P. WINCH]

Electric current

The net transfer of electric charge per unit time. It is usually measured in amperes. The passage of electric current involves a transfer of energy. Except in the case of superconductivity, a current always produces heat in the medium through which it passes.

On the other hand, a stream of electrons or ions in a vacuum, which also may be regarded as an electric current, produces no local heating. Measurable currents range in magnitude from the nearly instantaneous 10^5 or so amperes in lightning strokes to values of the order of 10^{-16} A, which occur in research applications.

All matter may be classified as conducting, semiconducting, or insulating, depending upon the ease with which electric current is transmitted through it. Most metals, electrolytic solutions, and highly ionized gases are conductors. Transition elements, such as silicon and germanium, are semiconductors, while most other substances are insulators.

Electric current may be direct or alternating. Direct current (dc) is necessarily unidirectional but may be either steady or varying in magnitude. By convention it is assumed to flow in the direction of motion of positive charges, opposite to the actual flow of electrons. Alternating current (ac) periodically reverses in direction.

Conduction current. This is defined as the transfer of charge by the actual motion of charged particles in a medium. In metals the current is carried by free electrons which migrate through the spaces between the atoms under the influence of an applied electric field. Although the propagation of energy is a very rapid process, the drift rate of the individual electrons in metals is only of the order of a few centimeters per second. In a superconducting metal or alloy the free electrons continue to flow in the absence of an electric field after once having been started. In electrolytic solutions and ionized gases the current is carried by both positive and negative ions. In semiconductors the carriers are the limited number of electrons which are free to move, and the "holes" which act as positive charges.

Displacement current. When alternating current traverses a condenser, there is no physical flow of charge through the dielectric (insulating material), but the effect on the rest of the circuit is as if there were a continuous flow. Energy can pass through the condenser by means of the so-called displacement current. James Clerk Maxwell introduced the concept of displacement current in order to make complete his theory of electromagnetic waves. *See* ALTERNATING CURRENT; DIRECT CURRENT.

[JOHN W. STEWART]

Electric distribution systems

That part of an electric power system that supplies electric energy to the individual user or consumer. The distribution system includes the primary circuits and the distribution substations that supply them: the distribution transformers; the secondary circuits, including the services to the consumer; and appropriate protective and control devices. The four general classes of individual users are residential, industrial, commercial, and rural.

Systems. The three-phase, alternating-current (ac) system (Fig. 1) is practically universal, although a small amount of two-phase and direct-current systems from early days are still in operation. Three-phase transmission and substransmission lines require three wires, termed phase conductors. Most of the three-phase distribution systems consist of three phase conductors and a common or neutral conductor, making a total of four wires. Single-phase branches (consisting of two wires) supplied from the three-phase mains are used for single-phase utilization in residences, small stores, and farms. Loads are connected in parallel to common supply circuits. *See* ELECTRIC POWER SYSTEMS.

Substation. The distribution substation is an assemblage of equipment for the purpose of switching, changing, and regulating the voltage from subtransmission to primary distribution. More important substations are designed so that the failure of a piece of equipment in the substation or one of the subtransmission lines to the substation will not cause an interruption of power to the load.

Primary voltages. The primary system leaving the substation is most frequently in the 11,000–15,000-V range. A particular voltage used is 12,470-V line-to-line and 7200-V line-to-neutral (conventionally written 12,470Y/7200 V). Some utilities use a lower voltage, such as 4160Y/2400 V. The use of voltages above the 15-kV class is increasing. Several percent of primary distribution circuits are in the 25- and 35-kV classes; all are four-wire systems. Single-phase loads are connected line-to-neutral on the four-wire systems.

Secondary voltages. Secondary voltages are derived from distribution transformers connected to the primary system and they usually correspond

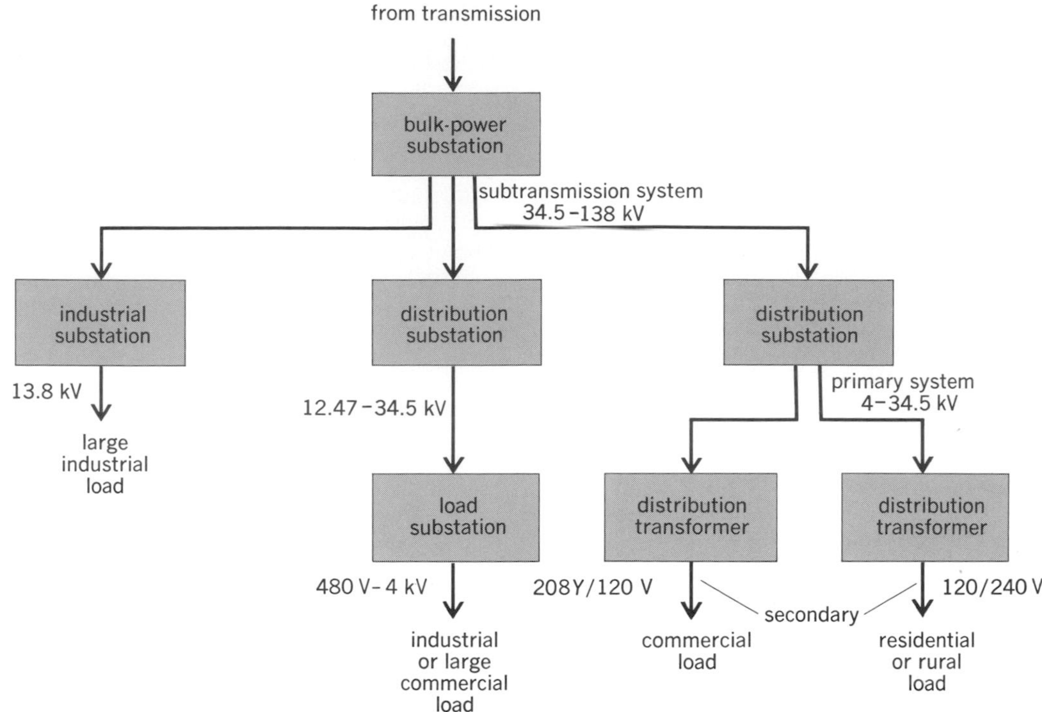

Fig. 1. Typical three-phase power system from bulk-power source to consumer's switch.

to utilization voltages. Residential and most rural loads are supplied by 120/240-V single-phase three-wire systems (Fig. 2). Commercial and small industrial needs are supplied by either 208Y/120-V or 480Y/277-V three-phase four-wire systems.

The secondary voltage is usually used to supply multiple street lights, in addition to service to consumers. Photoelectric controls are employed to turn the street lamps on and off.

Good voltage. Good voltage means that the average voltage level is correct, that variations do not exceed prescribed limits, and that sudden momentary changes in level do not cause objectionable light flicker. Utilization voltage varies with changing load on the system, but a voltage variation of less than 5% at the consumer's meter is common. To achieve this result, distribution systems are designed for a plus and minus voltage spread from the nominal voltages shown above. This is accomplished by proper wire size for the circuits, application of capacitors, both permanently connected and switched, and use of voltage regulators. Voltage regulators may be used at the substation or at a point along the circuit.

Good continuity. Service continuity is the providing of uninterrupted electric power to the consumer; therefore, good continuity is doing this a high percentage of the time. This is accomplished for large industrial and commercial loads by use of some form of duplicate power supply. Downtown commercial areas are supplied from three-phase 208Y/120-V grid networks (Fig. 3). These networks are fed from a number of primary feeders, stepping down through network transformers and automatic-reclosing secondary circuit breakers (protectors) to the secondary grid formed by cables under the streets. The system is arranged so that the failure of a primary feeder will not cause a loss of load on the secondary. Commercial buildings and shopping centers are often served by spot networks. All of the transformers and protectors are at the same location. Residential and rural loads are usually supplied by a radial system. Good continuity for them is obtained by sectionalizing the system with fuses, circuit breakers, and manual switches to reduce the extent of an outage due to a failure.

Elements of distribution systems. In distribution systems there are a number of elemental parts or subsystems, which are discussed below.

Primary feeders. Power is carried by primary feeders from distribution substations to the load areas where the consumers are located.

Distribution transformers. The distribution transformer located near the consumers changes the voltage from the primary distribution voltage to the secondary distribution voltage.

Secondary mains. This element of electric distribution is a low-voltage system which connects the

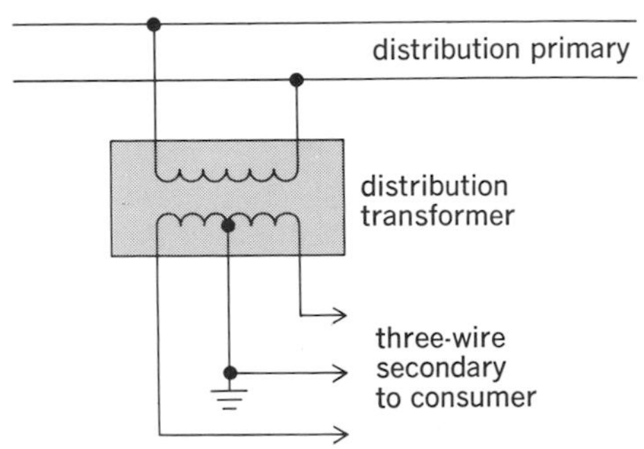

Fig. 2. Single-phase three-wire secondary circuit.

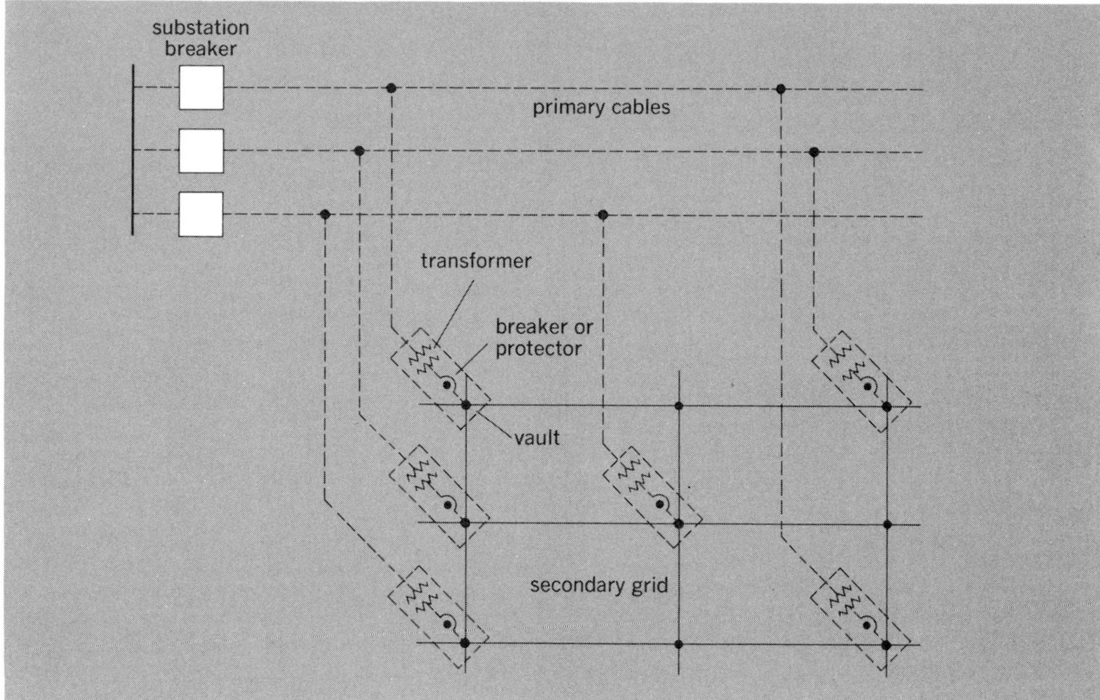

Fig. 3. Typical three-phase 208Y/120-volt secondary grid network system.

secondary winding of distribution transformers to the consumers' services.

Overhead construction. The majority of distribution systems already built in residential, industrial, and rural areas has been overhead construction. This construction utilizes poles treated with pentachlorophenol or creosote, distribution transformers mounted near the top of the pole, and bare primary and secondary conductors strung from pole to pole. Aluminum conductors have been used in place of copper because of their lower cost.

For esthetic reasons, underground distribution systems are being installed in most new residential developments. *See* TRANSMISSION LINES.

Underground construction. Commercial areas and certain main thoroughfares usually have underground construction. Construction for these areas is usually a system of concrete or fiber ducts and vaults. Impregnated paper protected by lead sheaths or synthetic compounds is generally used as conductor insulation.

In residential areas the cable is usually insulated with polyethylene or chemically crosslinked polyethylene, which is a thermosetting material. Primary cables supplying distribution transformers in the development area are likely to be single-phase, consisting of an insulated phase conductor with a concentric bare neutral conductor taken from a 12,470-V three-phase primary feeder. The cables are often directly buried in the earth. The distribution transformers used are either the submersible type installed in a hole with a liner or pad-mounted and located on the surface of the ground. *See* ELECTRIC POWER SUBSTATION.

[HAROLD E. CAMPBELL]

Bibliography: Edison Electric Institute, *Underground Systems Reference Book*, EEI Publ. no. 55-15, 1967; B. G. A. Skortizki (ed.), *Electric Transmission and Distribution*, vol. 3, 1954; U.S. Standards Institute, *Definition of Electrical Terms: Transmission and Distribution (Group 35)*, USASI C42.35, 1957; U.S. Standards Institute, *Preferred Voltage Ratings for A-C Systems and Equipment: Guide for (EEI R-6-1949) (IEC38)*, USASI C84.1, 1954; E. Vennard, *The Electric Power Business*, 2d ed., 1970; B. M. Weedy, *Electric Power Systems*, 3d ed., 1979.

Electric energy measurement

The measurement of the integral, with respect to time, of the power in an electric circuit. The absolute unit of measurement of electric energy is the joule, or the charge in coulombs times the potential difference in volts. The joule, however, is too small (1 watt-second) for use in commercial practice, and the more commonly used unit is the watt-hour (3.6×10^3 joules).

The most common measurement application is in the utility field, where it is estimated that over 2×10^{12} kWh were measured and sold during 1978 to industry and residential consumers in the United States.

Electric energy is one of the most accurately measured commodities sold to the general public. Many methods of measurement, with different degrees of accuracy, are possible. The choice depends on the requirements and complexities of the measurement problems. Basically, measurements of electric energy may be classified into two categories, direct-current power and alternating-current power. The fundamental concepts of measurement are, however, the same for both.

Methods of measurement. There are two types of methods of measuring electric energy: electric instruments and timing means, and electricity meters.

Electric instruments and timing means. These make use of conventional procedures for measuring electric power and time. The required accuracy of measurement dictates the type and quality of the measuring devices used (for example, portable instruments, laboratory instruments, potentiometers, stopwatches, chronographs, and electronic timers). Typical methods are listed below. *See* ELECTRIC POWER MEASUREMENT.

1. Measurement of energy on a direct-current circuit by reading the line voltage and load current at regular intervals over a measured period of time. The frequency of reading selected (such as one per second, one per 10 seconds) depends upon the steadiness of the load being measured, the time duration of the test, and the accuracy of measurement desired. The first dc electric energy meter was an electroplating cell in series with the load. It deposited a mass of metal on an electrode exactly proportional to the total charge transported to the load. The electrode was weighed periodically to determine the energy used. Errors were introduced if line voltage was not constant. It was replaced by more convenient instruments. *See* CURRENT MEASUREMENT; VOLTAGE MEASUREMENT.

In electric energy measurements, the losses in the instruments must be considered. Unless negligible from a practical standpoint, they should be deducted from the total energy measured. If the voltmeter losses are included in the total energy measured, then watt-hours $= (VI - V^2R)t/3600$, where V is the average line voltage (volts), I is the average line current (amperes), R is the voltmeter resistance (ohms), and t is the time (seconds).

2. Measurement of energy on a direct-current circuit by controlling the voltage and current at constant predetermined values for a predetermined time interval. This method is common for controlling the energy used for a scientific experiment or for determining the accuracy of a watt-hour meter. For best accuracy, potentiometers and electronic timers are desirable.

3. Measurement of energy on an alternating-current circuit by reading the watts input to the load at regular intervals over a measured period of time. This method is similar to the first, except that the power input is measured by a wattmeter.

4. Measurement of energy on an alternating-current circuit by controlling the voltage, current, and watts input to the load at constant predetermined values. This method is similar to the second, except that the power input is measured by a wattmeter. A common application of this method is to determine the standard of measurement of electric energy, the watt-hour.

5. Measurement of energy by recording the watts input to the load on a linear chart progressed uniformly with time. This method makes use of a conventional power record produced by a recording wattmeter. The area under the load record over a period of time is the energy measurement.

Electricity meters. These are the most common devices for measuring the vast quantities of electric energy used by industry and the general public. The same fundamentals of measurement apply as for electric power measurement, but in addition the electricity meter provides the time-integrating means necessary for electric energy measurement.

A single meter is sometimes used to measure the energy consumed in two or more circuits. However, multistator meters are generally required for this purpose. Totalization is also accomplished with fair accuracy, if the power is from the same voltage source, by paralleling secondaries of instrument current transformers of the same ratio at the meter. Errors can result through unbalanced loading or use of transformers with dissimilar characteristics.

Watt-hour meters are generally connected to measure the losses of their respective current circuits. These losses are extremely small compared to the total energy being measured and are present only under load conditions.

Other errors result from the ratio and phase angle errors in instrument transformers. With modern transformers these errors can generally be neglected for commercial metering. If considered of sufficient importance, they can usually be compensated for in adjusting the calibration of the watt-hour meter. For particularly accurate measurements of energy over short periods of time, portable standard watt-hour meters may be used. Errors may also arise due to integral-cycle control. This is a well-established method of power control in which a silicon-controlled rectifier circuit acts to turn the power on for several cycles and off for a different number of cycles. The main source of error in the induction meter is the difference between the mechanical time-constants for two on-off intervals, making the meter read high.

Watt-hour meters used for the billing of residential, commercial, and industrial loads are highly developed devices. Over the last decade many significant improvements have been made, including improvements in bearings, insulating materials, mechanical construction, and new sealing techniques which exclude dust and other foreign material. As a result of the higher degree of accuracy and dependability achieved by modern meters, the utility industries have adopted statistical sampling methods for testing of in-service accuracy as sanctioned by ANSI C12-1965. *See* WATT-HOUR METER.

Automatic remote reading. Various aspects of the energy crisis have spurred active development of automatic meter-reading systems with the functional capability of providing meter data for proposed new rate structures, for example, time-of-day pricing, and initiating control of residential loads, such as electric hot-water heaters.

Automatic meter-reading systems under development generally consist of a utility-operated, minicomputer-controlled reading center, which initiates and transmits commands over a communication system to a terminal at each residential meter. The terminal carries out the command, sending the meter reading back to the reading center, or activating the control of a residential load.

Several communication media are being proposed and tested by system developers, including radio, CATV, use of the existing subscriber phone lines, and communication over the electric distribution system itself.

The rapid advances of technology, coupled with the increasing needs for improved management of energy usage, indicate that automatic meter reading and control systems may start replacing conventional manual meter reading within the next few years.

Quantities other than watt-hours. Included in the field of electric energy measurement are demand, var hours, and volt-ampere hours.

Demand. The American National Standards Institute defines the demand for an installation or system as "the load which is drawn from the source of supply at the receiving terminals, averaged over a suitable and specified interval of time. Demand is expressed in kilowatts, kilovolt-amperes, amperes, kilovars and other suitable units" (ANSI C12-1965).

This measurement provides the user with information as to the loading pattern or the maximum loading of equipments rather than the average loading recorded by the watt-hour meter. It is used by the utilities as a rate structure tool.

Var hour. ANSI defines the var hour (reactive volt-ampere hour) as the "unit for expressing the integral of reactive power in vars over an interval of time expressed in hours" (ANSI C12-1965).

This measurement is generally made by using reactors or phase-shifting transformers to supply to conventional meters a voltage equal to, but in quadrature with, the line voltage.

Volt-ampere hour. This is the unit for expressing the integral of apparent power in volt-amperes over an interval of time expressed in hours. Measurement of this unit is more complicated than for active or reactive energy and requires greater compromises in power-factor range, accuracy, or both. Typical methods include: (1) Conventional watt-hour meters with reactors or phase-shifting transformers tapped to provide an in-phase line voltage and current relationship applied to the meter at the mean of the expected range of power-factor variation. (2) A combination of a watt hour and a var hour meter mechanically acting on a rotatable sphere to add vectorially watt-hours and var-hours to obtain volt-ampere hours, volt-ampere demand, or both.

Measurement of volt-ampere hours is sometimes preferred over var-hours because it is a more direct measurement and possibly gives a more accurate picture of the average system power factor. This would not necessarily be true, however, where simultaneous active and reactive demand are measured and recorded. *See* ELECTRICAL MEASUREMENTS. [WILLIAM H. HARTWIG]

Bibliography: T. S. Banghart and R. E. Riebs, Practical aspects of large-scale automatic meter reading using existing telephone lines, *Proceedings of the American Power Conference*, vol. 36, pp. 945–951, 1974; W. C. Downing, Watthour meter accuracy on SCR controlled resistance loads, *IEEE Trans. Power App. Syst.*, 93(4):1083–1089, 1974; A. E. Emanuel, B. M. Hynds, and F. J. Levitsky, Watthour meter accuracy on integral-cycle-controlled resistance loads, *IEEE Trans. Power App. Syst.*, 98(5):1583–1590, 1979. S. G. Hardy, *New Developments in Automatic Meter Reading*, paper presented to the EEI/AEIC Meter and Service Committee, Atlanta, Apr. 9, 1973; P. B. Robinson, Progress in automatic meter reading, *Proceedings of the American Power Conference*, vol. 36, pp. 959–964, 1974; Utilities Act of 1975, House of Representatives Bill HR2650, Feb. 4, 1975.

Electric field

A condition in space in the vicinity of an electrically charged body such that the forces due to the charge are detectable. An electric field (or electrostatic field) exists in a region if an electric charge at rest in the region experiences a force of electrical origin. Since an electric charge experiences a force if it is in the vicinity of a charged body, there is an electric field surrounding any charged body.

Field strength. The electric field intensity (or field strength) **E** at a point in an electric field has a magnitude given by the quotient obtained when the force acting on a test charge q' placed at that point is divided by the magnitude of the test charge q'. Thus, it is force per unit charge. A test charge q' is one whose magnitude is small enough so it does not alter the field in which it is placed. The direction of **E** at the point is the direction of the force **F** on a positive test charge placed at the point. Thus, **E** is a vector point function, since it has a definite magnitude and direction at every point in the field, and its defining equation is Eq. (1).

$$\mathbf{E} = \frac{\mathbf{F}}{q'} \qquad (1)$$

Principle of superposition. As applied to electric fields, this principle states that the total **E** at a point P due to the combined influence of a distribution of point charges is the vector sum of the electric field intensities that the individual point charges would produce at P if each acted alone. Thus, using the rationalized mks system of units, Eq. (2) holds, where $\epsilon^0 \cong 8.85 \times 10^{-12}$

$$\mathbf{E} = \frac{1}{4\pi\epsilon_0} \sum_{i=1}^{n} \frac{q_i}{r_i^2} \quad \text{vector sum} \qquad (2)$$

coulomb²/newton-m² is the permittivity of empty space, q_i is the ith charge (in coulombs) in the distribution, and r_i is the distance in meters from q_i to P. The units of **E** in the mks system are newtons/coulomb, which are the same as volts/meter. A common method of solving for **E** in a particular known distribution of charges is to evaluate the vector sum in Eq. (2). In many cases, however, Gauss' theorem affords a more powerful and convenient method.

Electric displacement. Electric flux density or electric displacement **D** in a dielectric (insulating) material is related to **E** by either of the equivalent equations shown as Eqs. (3), where **P** is the polarization of the medium, and ϵ is the permittivity of the dielectric which is related to ϵ_0 by the equation $\epsilon = k\epsilon_0$, k being the relative dielectric constant of the dielectric. In empty space, $\mathbf{D} = \epsilon_0\mathbf{E}$. The units of **D** are coulombs/meter².

$$\mathbf{D} = \epsilon_0\mathbf{E} + \mathbf{P} \qquad \mathbf{D} = \epsilon\mathbf{E} \qquad (3)$$

In addition to electrostatic fields produced by separations of electric charges, an electric field is also produced by a changing magnetic field. The relationship between the **E** produced and the rate of change of magnetic flux density $d\mathbf{B}/dt$ which produces it is given by Faraday's law of induced electromotive forces (emfs) in Eq. (4), where ds is a vector element of path length directed along the

$$\oint \mathbf{E} \cdot ds = -\int_A \frac{d\mathbf{B}}{dt} \cdot d\mathbf{A} \qquad (4)$$

path of integration in the general sense of **E**. Thus $\oint \mathbf{E} \cdot ds$ is the emf induced in this closed path of integration. The area of the surface bounded by the path of integration is A and the direction of $d\mathbf{A}$, an infinitesimal vector element of this area, is the direction of the thumb of the right hand when the fingers encircle the path of integration in the general sense of **E**. The right side of Eq. (4) is seen to be the negative of the time rate of change of the magnetic flux linking the path of integration chosen for the left side.

In an electrostatic field, $\oint \mathbf{E} \cdot ds$ is always zero. *See* ELECTRIC CHARGE; ELECTROMAGNETIC INDUCTION. [RALPH P. WINCH]

Bibliography: R. P. Winch, *Electricity and Magnetism*, 2d ed., 1963.

Electric heating

Methods of converting electric energy to heat energy by resisting the free flow of electric current. Electric heating has no upper limit to the obtainable temperature except for the materials used. It has the advantage of electrical temperature control and provides uniform heating.

Any comparison of costs between electric heating and other methods should consider total costs and not just the cost of obtaining equal heat energies from electricity and fuels. The cost per unit of manufactured product is the true measure. This involves labor, quality, time, cleanliness, safety, and maintenance costs. Electric heat can usually excel in all of these.

Types of electric heaters. There are four major methods of electric heating.

1. Resistance heaters produce heat by passing an electric current through a resistance. Resistance heaters have an inherent efficiency of 100% in converting electric energy into heat. A high proportion of this heat can be transferred to the work material by conduction and by radiation.

2. Dielectric heaters use currents of high frequency which generate heat by dielectric hysteresis within the body of a nominally nonconducting material. The power factor varies but, due to inherent losses, can never approach unity.

3. Induction heaters produce heat by means of a periodically varying electromagnetic field within the body of a nominally conducting material. The power factor can never approach unity, because of inherent reactances.

4. Electric-arc heating is really a form of resistance heating in which a bridge of vapor and gas carries an electric current between electrodes. The arc has a property of resistance. Both electrodes may be of carbon, or one may be the conducting work material in a furnace.

Thermal problems. Electric heating differs radically from other methods that have a constant temperature. The heat-energy input of those methods depends on the temperature difference, which decreases as the work temperature rises. The heat input decreases as the work temperature increases.

Electric heating has a constant heat-energy input (for a given voltage), and the surface temperature of the heater rises to compensate for work temperature rise. Therefore, some control function is necessary to prevent overheating. To avoid damage, a suitable heat density per unit of heating surface must be maintained. With fluids, heat is distributed by natural thermal or forced convection currents. This helps heat transfer by wiping away surface films, and greater heat density can be tolerated. Heat capacity and mass play large parts in satisfactory heating. In large masses, such as tanks of oil or water, the temperature changes slowly and seldom presents a serious control problem. With small masses, uneven heat zones, or variable heating time, the control of the input heat becomes more critical. Heat control is required when heating such materials as gas vapor and paper. Moisture content also affects the heat requirements.

The heater rating can be found from the heat required for the process. This must include the heat required to heat the material and its container, any heat of fusion or vaporization in the process, and the heat losses. The heat required for the material and its container can be found from the weight, the specific heat, and the temperature rise of the material and container. The total heat required for the process is the sum of all the pertinent factors and can be put in units of kilowatt-hours. The number of kilowatt-hours divided by the hours allowed for the process will then give the required rating of the heater in kilowatts. The heater rating therefore depends on the time allowed for the process. A longer time will permit use of a heater of lower rating and will thus result in reduced costs of equipment and operation.

General design features. All electrical parts must be well protected from contact by operators, work materials, and moisture. Terminals must be enclosed within suitable boxes, away from the high heat zone, to protect the power supply cables. Repairs and replacements should be possible without tearing off heat insulations.

Resistance heaters are often enclosed in pipes or tubes suitable for immersion or for exposure to difficult external conditions. Indirect heating is done by circulating a heat transfer medium, such as special oil or Dowtherm (liquid or vapor), through jacketed vessels. This permits closer control of heating-surface temperature than is possible with direct heating.

Some conducting materials can be heated by passing electric current through them, as is done in the reduction of aluminum. Some conducting liquids can be heated by passing an electric current between immersed electrodes. Heat is produced by the electrical resistance of the liquid.

The supply of necessary electric power for large heating installations necessitates consultation with the utility company. The demand, the power factor of the load, and the load factor all affect the power rates. Large direct-current or single-phase alternating-current loads should be avoided. Polyphase power at 440–550 volts permits lower current and reduced costs. [LEE P. HYNES]

Bibliography: E. R. Ambrose, *Heat Pumps and Electric Heating*, 1966; Edison Electric Institute, *Power Sales Manual: Induction and High Frequency Resistance Heating*, 1961; D. G. Fink and H. W. Beaty, *Standard Handbook for Electrical Engineers*, 11th ed., 1978; L. P. Hynes, Industrial electric resistance heating, *AIEE Trans.*, 67:1359–

1361, 1948; L. P. Hynes, Some unusual designs of electric resistance heating, *AIEE Trans.*, pap. no. 59-77.

Electric power generation

The production of bulk electric power for industrial, residential, and rural use. Although limited amounts of electricity can be generated by many means, including chemical reaction (as in batteries) and engine-driven generators (as in automobiles and airplanes), electric power generation generally implies large-scale production of electric power in stationary plants designed for that purpose. The generating units in these plants convert energy from falling water, coal, natural gas, oil, and nuclear fuels to electric energy. Most electric generators are driven either by hydraulic turbines, for conversion of falling water energy; or by steam or gas turbines, for conversion of fuel energy. Limited use is being made of geothermal energy, and developmental work is progressing in the use of solar energy in its various forms. Electric power generating plants are normally interconnected by a transmission and distribution system to serve the electric loads in a given area or region.

An electric load is the power requirement of any device or equipment that converts electric energy into light, heat, or mechanical energy, or otherwise consumes electric energy as in aluminum reduction, or the power requirements of electronic and control devices. The total load on any power system is seldom constant; rather, it varies widely with hourly, weekly, monthly, or annual changes in the requirements of the area served. The minimum system load for a given period is termed the base load or the unity load-factor component. Maximum loads, resulting usually from temporary conditions,

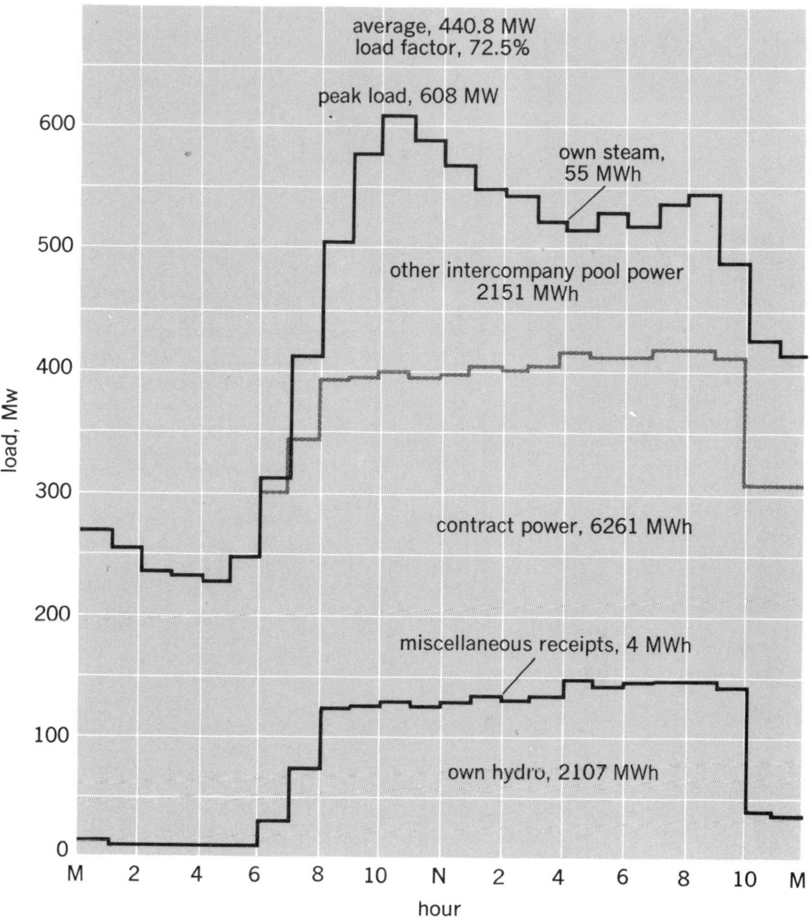

Fig. 1. Load graph indicates net system load of a metropolitan utility for typical 24-hr period (midnight to midnight), totaling 10,578 MWh. Such graphs are made to forecast probable variations in power required.

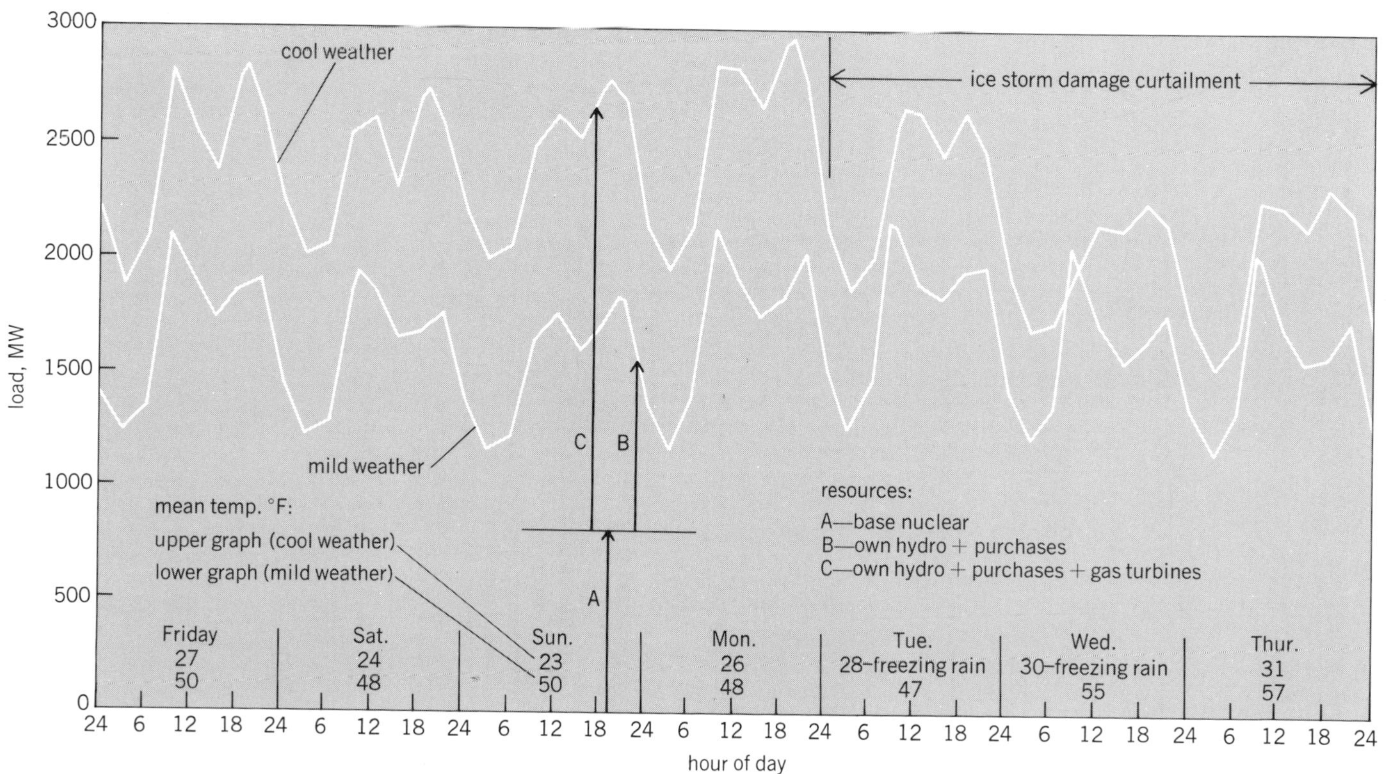

Fig. 2. Examples of northwestern electric utility weekly load curves showing same-year weather influence.

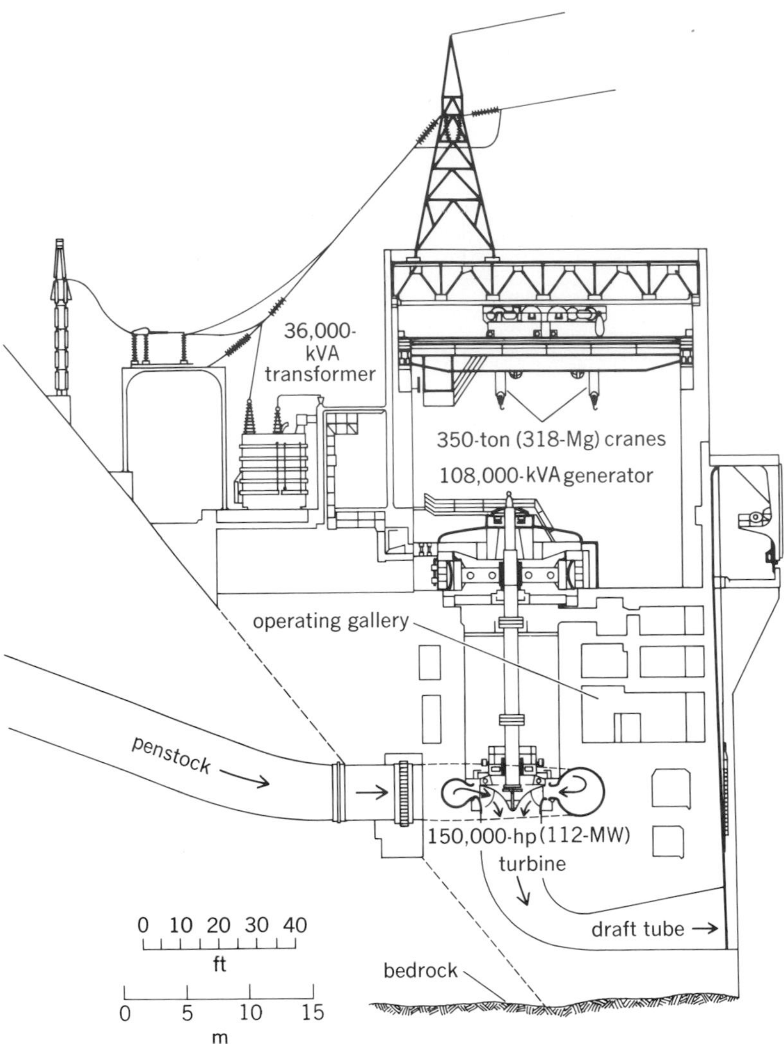

Fig. 3. Typical layout and apparatus arrangement in a hydroelectric generating plant. Cross section is made through the powerhouse and dam at a main generator position in the original Grand Coulee plant.

are called peak loads. Electric energy cannot feasibly be stored in large quantities; therefore the operation of the generating plants must be closely coordinated with fluctuations in the load.

Actual variations in the load with time are recorded, and from these data load graphs are made to forecast the probable variations of load in the future. A study of hourly load graphs (Figs. 1 and 2) indicates the generation that may be required at a given hour of the day, week, or month, or under unusual weather conditions. A study of annual load graphs and forecasts indicates the rate at which new generating stations must be built. Load graphs and forecasts are an inseparable part of utility operation and are the basis for decisions that profoundly affect the financial requirements and overall development of a utility.

Generating plants. Often termed generating stations, these plants contain apparatus that converts some form of energy to electric energy in bulk. Three significant types of generating plants are hydroelectric, fossil-fuel-electric, and nuclear-electric.

Hydroelectric plant. This type of generating plant utilizes the potential energy released by the weight of water falling through a vertical distance called head. Ignoring losses, the power, in horsepower (hp) and kilowatts (kW), obtainable from falling water is shown in the equations below (metric quantities in brackets).

$$\text{hp} = \frac{\begin{pmatrix}\text{quantity of water} \\ \text{in ft}^3\text{/s [m}^3\text{/s]}\end{pmatrix}\begin{pmatrix}\text{vertical head} \\ \text{in ft [m]}\end{pmatrix}}{8.8\ [0.077]}$$

$$\text{kW} = 0.746\ \text{hp}$$

A plant consists basically of a dam to store the water in a forebay and create part or all of the head, a penstock to deliver the falling water to the turbine, a hydraulic turbine to convert the hydraul-

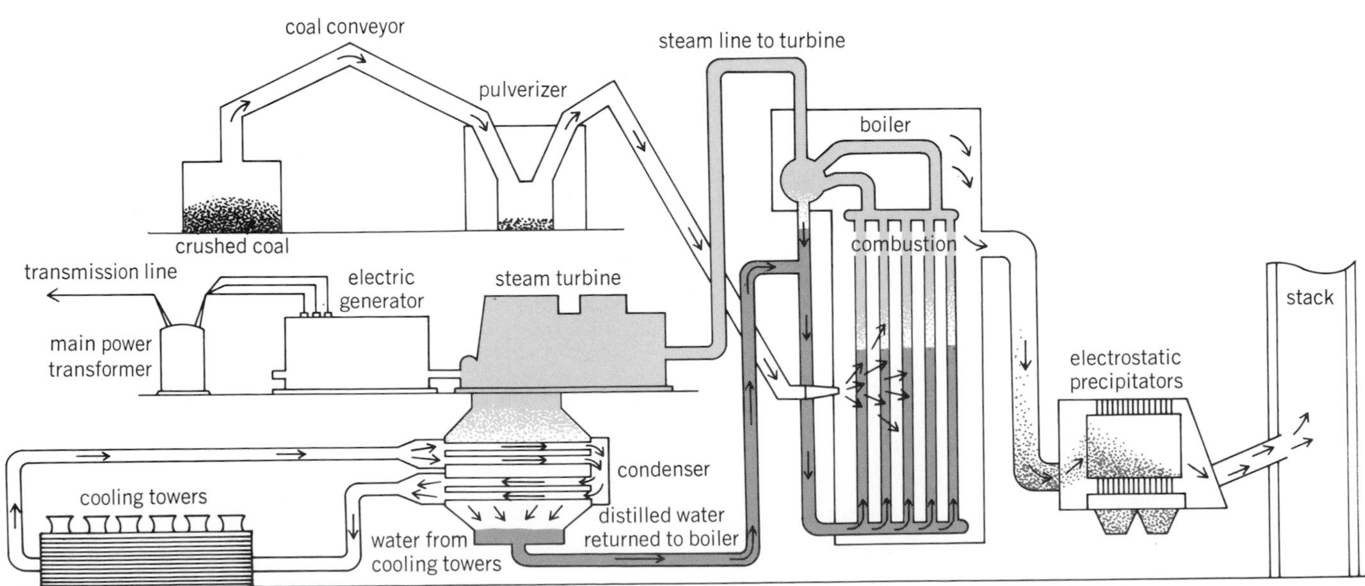

Fig. 4. Schematic of typical coal-fired steam electric power plant. (*Pacific Power & Light Co.*)

ic energy released to mechanical energy, an alternating-current generator (alternator) to convert the mechanical energy to electric energy, and all accessory equipment necessary to control the power flow, voltage, and frequency, and to afford the protection required (Fig. 3).

Pumped storage hydroelectric plants are being used increasingly. Under suitable geographical and geological conditions, electric energy can, in effect, be stored by pumping water from a low to a higher elevation and subsequently releasing this water to the lower elevation through hydraulic turbines. These turbines and their associated generators are reversible. The generators, operating in reverse direction as motors, drive their turbines as pumps to elevate the water. When this water is released through the turbines, electric power is produced by the generators. A relatively high overall cycle efficiency can be attained, usually of the order of 65–75%.

Since system peak loads are usually of relatively short duration (Figs. 1 and 2), the high output available for a short time from pumped storage can be used to supply this peak. During off-peak hours, that is, 10 P.M. to 7 A.M., the surplus generating capacity of the most economical system energy resources can be used to return the water by pumping to the elevated storage space for use on the next peak. This type of operation assists in maintaining a high capacity factor on prime generation with resulting best economy.

Pumped storage plants can be brought up to load much faster than large steam plants and, hence, contribute to system reliability by providing an immediately available reserve against the unscheduled loss of other generation.

Fossil-fuel-electric plant. This type utilizes the energy of combustion from coal, oil, or natural gas. A typical large plant (Fig. 4) consists of fuel processing and handling facilities, a combustion furnace and boiler to produce and superheat the steam, a steam turbine, an alternator, and the accessory equipment required for plant protection and for control of voltage, frequency, and power flow.

A steam plant can frequently be built near a convenient load center, provided an adequate supply of cooling water and fuel is available, and is usually readily adaptable to either base loading or intermediate or peak loading. Environmental constraints require careful control of stack emissions with respect to sulfur oxides, and particulates. Cooling towers or ponds are often required for waste heat dissipation. Gas turbine plants do not require condenser cooling water (unless combined with a steam cycle), have a relatively low unit capital cost and relatively high unit fuel cost, and are widely used for peaking service. Progress is being made in the development of magnetohydrodynamic (MHD) "topping" generators to be used in conjuction with normal steam turbines to improve the overall thermal conversion efficiency.

Nuclear electric plant. In this type of plant one or more of the nuclear fuels are utilized in a suitable type of nuclear reactor, which takes the place of the combustion furnace in the typical steam electric plant. The heat exchangers and boilers (if not combined in the reactor), the turbines, and alternating-current generators, complete with controls, accessories, and auxiliaries, make up the atomic electric plant. Large-scale fission reaction plants have been developed to the point where they are economically competitive in much of the United States, and many millions of kilowatts of capacity are under construction and more on order. The current and projected future growth of the nuclear power industry in the United States is shown graphically in Figs. 5 and 6. Although in 1978 only approximately 9% of the total generating capacity was nuclear, by 1989 it is predicted to become over 21%. However, in 1978 about 14% of the net electric energy production was nuclear and is forecast to become more than 22% by 1988. The coal-fired share in 1978 was approximately 45% and is forecast to become about 50% in 1988. Conventional hydro generation as a source of prime electric energy had reached near-saturation by 1978, and growth in fossil-fuel generation shows a declining trend. However, nuclear generation ex-

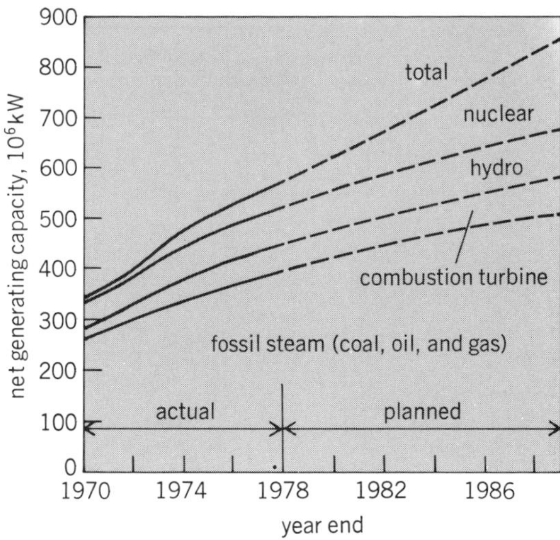

Fig. 5. Net generating capacity of the United States electric power industry. Hydro includes pumped storage. Combustion turbine includes internal combustion.

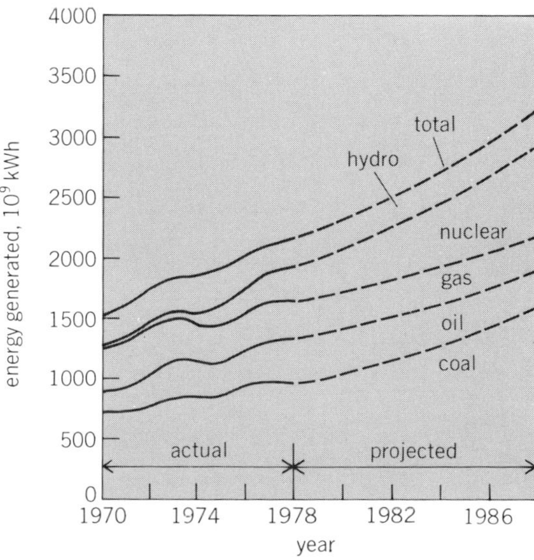

Fig. 6. Net energy production of the United States electric power industry.

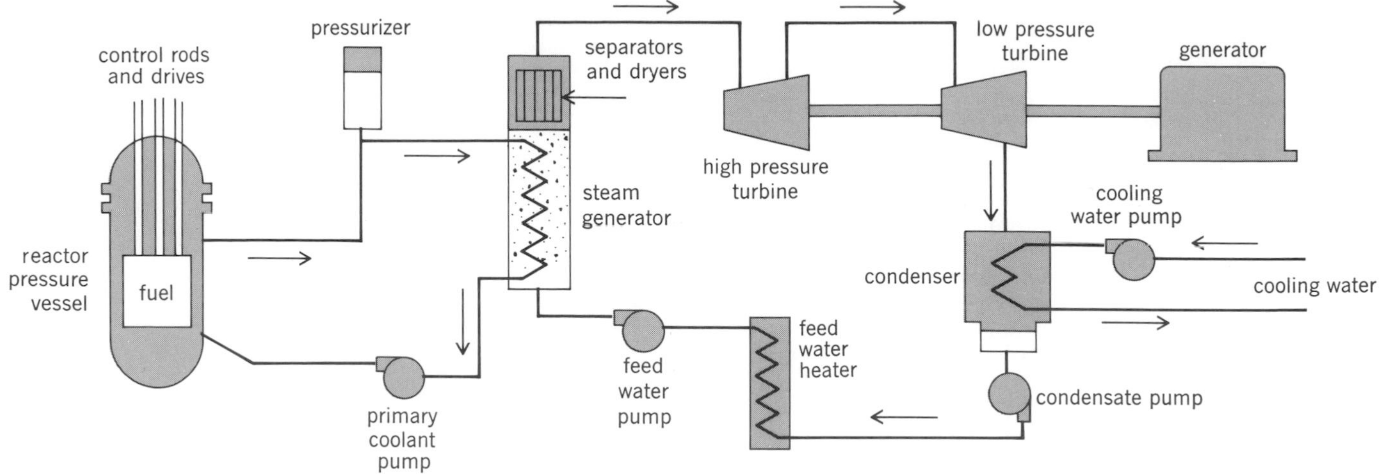

Fig. 7. Schematic of a pressurized-water reactor plant. *(From the Nuclear Industry, USAEC, WASH 1174–73, 1973)*

hibits a doubling time of about 51/2 years during a considerable period after 1978. Recent operating events may slow this growth somewhat.

Three types of nuclear plants are in general use and a fourth is being developed actively. The pressurized and boiling light-water types, shown schematically in Figs. 7 and 8, have been highly developed in the United States. The pressurized heavy-water type has been well developed in Canada. The gas-cooled type, one version of which is shown schematically in Fig. 9, has been highly developed and extensively used in Great Britain. The liquid-metal (sodium) cooled type is under intensive development and will be featured prominently in the fast or breeder reactor field. It is shown schematically in Fig. 10. As in the high-temperature gas-cooled reactor system, steam temperatures and pressures in the liquid-metal systems can be essentially the same as those recorded in the plants powered by fossil fuel and can give similar thermal efficiencies.

Several other types of nuclear fission reactor systems are receiving attention, and some may be expected to become commercially competitive. Fusion reaction nuclear plants are in the early research and development stage with possible commercialization early in the next century. Direct conversion from nuclear reaction energy to electric energy on a commercial scale for power utility service is a future possibility but is not economically feasible at present.

Generating unit sizes. The size or capacity of electric utility generating units varies widely, depending upon type of unit, duty required, that is base-, intermediate-, or peak-load service, and system size and degree of interconnection with neighboring systems. Base-load, nuclear or coal-fired units may be as large as 1200 MW each, or more. Intermediate-duty generators, usually coal-, oil-, or gas-fueled steam units, are typically of 200 to 600 MW capacity each. Peaking units, combustion turbines or hydro, range from several tens of mega-

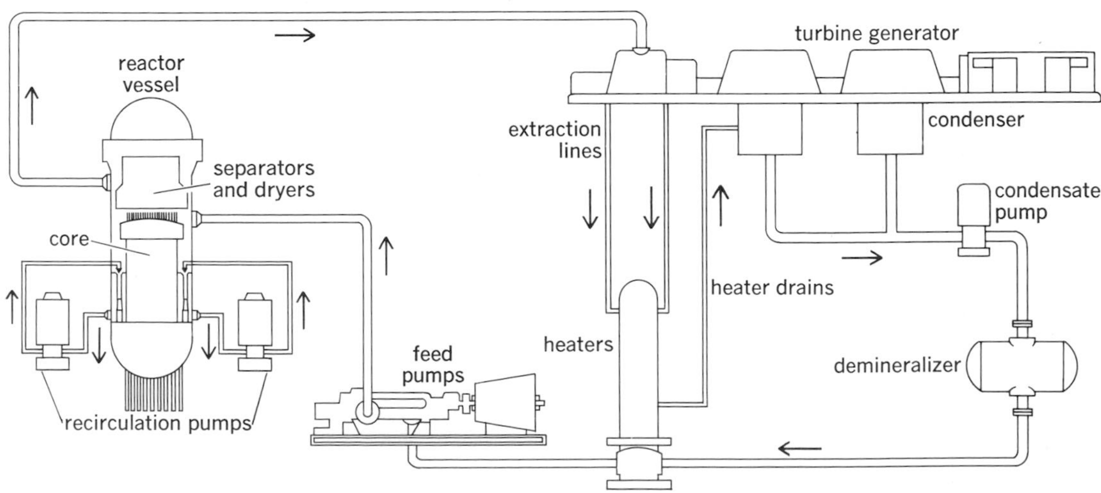

Fig. 8. Single-cycle boiling-water reactor system flow diagram. *(General Electric Co.)*

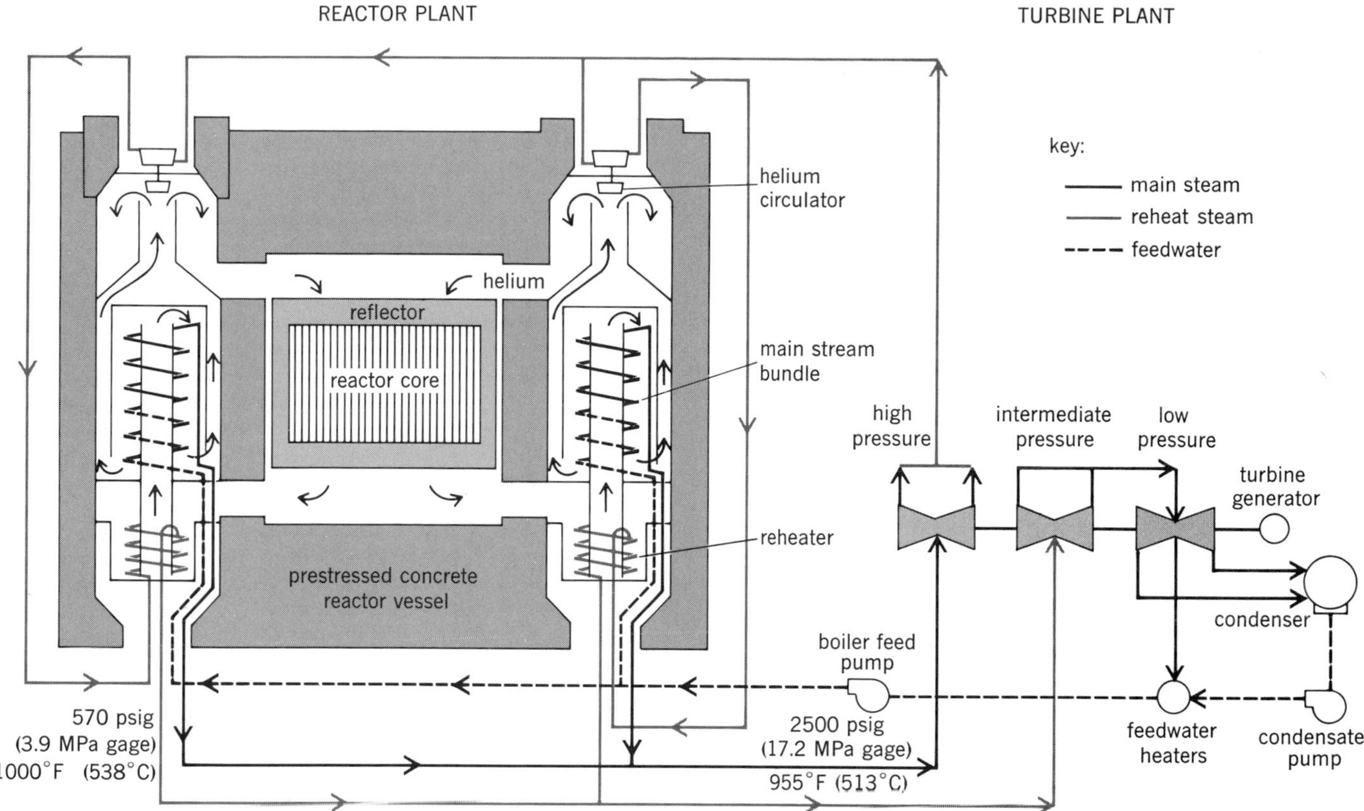

Fig. 9. Schematic diagram for a high-temperature gas-cooled reactor power plant. (*General Atomic Corp.*)

watts for the former to hundreds of megawatts for the latter. Hydro units, which are in both base-load and intermediate service, range in size up to 700 MW.

The total installed generating capacity of a system is typically 20 to 30% greater than the annual predicted peak load in order to provide reserves for maintenance and contingencies.

Power-plant circuits. Both main and accessory circuits in power plants can be classified as follows:

1. Main power circuits to carry the power from the generators to the step-up transformers and on to the station high-voltage terminals.
2. Auxiliary power circuits to provide power to the motors used to drive the necessary auxiliaries.
3. Control circuits for the circuit breakers and other equipment operated from the control room of the plant.
4. Lighting circuits for the illumination of the plant and to provide power for portable equipment required in the upkeep and maintenance of the plant. Sometimes special circuits are installed to supply the portable power equipment.
5. Excitation circuits, which are so installed that they will receive good physical and electrical protection because reliable excitation is necessary for the operation of the plant.
6. Instrument and relay circuits to provide values of voltage, current, kilowatts, reactive kilovolt-amperes, temperatures, and pressures, and to serve the protective relays.
7. Communication circuits for both plant and system communications. Telephone, radio, transmission-line carrier, and microwave radio may be involved.

It is important that reliable power service be provided for the plant itself, and for this reason station service is usually supplied from two or more sources. To ensure adequate reliability, auxiliary power supplies are frequently provided for start-up, shut-down, and communication services.

Generator protection. Necessary devices are installed to prevent or minimize other damage in cases of equipment failure. Differential-current and ground relays detect failure of insulation, which may be due to deterioration or accidental overvoltage. Overcurrent relays detect overload currents that may lead to excessive heating; overvoltage relays prevent insulation damage. Loss-of-excitation relays may be used to warn operators of low excitation or to prevent pulling out of synchronism. Bearing and winding overheating may be detected by relays actuated by resistance devices or thermocouples. Overspeed and lubrication failure may also be detected.

Not all of these devices are used on small units or in every plant. The generator is immediately deenergized for electrical failure and shut down for any over-limit condition, all usually automatically.

Voltage regulation. This term is defined as change in voltage for specific change in load (usually from full load to no load) expressed as percentage of normal rated voltage. The voltage of an electric generator varies with the load and power factor; consequently, some form of regulating equipment is required to maintain a reasonably constant and predetermined potential at the distribution stations or load centers. Since the inherent regulation of most alternating-current generators is rather poor (that is, high percentagewise), it is necessary to provide automatic voltage control. The rotating or magnetic amplifiers and voltage-sensitive circuits of the automatic regulators, together with the exciters, are all specially designed to respond quickly to changes in the alternator voltage and to make the necessary changes in the main exciter output, thus providing the required adjustments in voltage. A properly designed automatic regulator acts rapidly, so that it is possible to maintain desired voltage with a rapidly fluctuating load without causing more than a momentary change in voltage even when heavy loads are thrown on or off.

Electronic voltage control has been adapted to some generator and synchronous condenser installations. Its main advantages are its speed of operation and its sensitivity to small voltage variations. As the reliability and ruggedness of electronic components are improved, this form of voltage regulator will become more common.

Generation control. Computer-assisted (or on-line controlled) load and frequency control and economic dispatch systems of generation supervision are being widely adopted, particularly for the larger new plants. Strong system interconnections greatly improve bulk power supply reliability but require special automatic controls to ensure adequate generation and transmission stability. Among the refinements found necessary in large, long-distance interconnections are special feedback controls applied to generator high-speed excitation and voltage regulator systems.

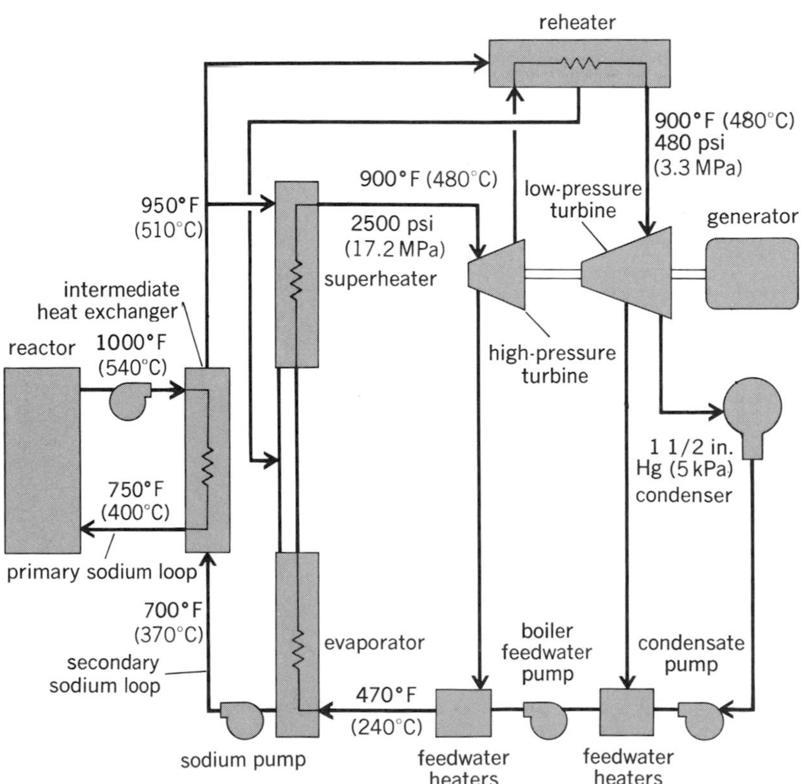

Fig. 10. Diagram for liquid-metal-cooled fast breeder reactor power plant. *(North American Rockwell Corp.)*

Synchronization of generators. Synchronization of a generator to a power system is the act of matching, over an appreciable period of time, the instantaneous voltage of an alternating-current generator (incoming source) to the instantaneous voltage of a power system of one or more other generators (running source), then connecting them together. In order to accomplish this ideally the following conditions must be met:

1. The effective voltage of the incoming generator must be substantially the same as that of the system.
2. In relation to each other the generator voltage and the system voltage should be essentially 180° out of phase; however, in relation to the bus to which they are connected, their voltages should be in phase.
3. The frequency of the incoming machine must be near that of the running system.
4. The voltage wave shapes should be similar.
5. The phase sequence of the incoming polyphase machine must be the same as that of the system.

Synchronizing of ac generators can be done manually or automatically. In manual synchronizing an operator controls the incoming generator while observing synchronizing lamps or meters and a synchroscope, or both. Voltage (potential) transformers may be used to provide voltages at lamp and instrument ratings. Lamps properly connected between the two sources are continuously dark when voltage, phase, and frequency are properly matched. Wave shape and phase sequence are determined by machine design and rotation or terminal sequence. Large units generally are provided with voltmeters and frequency meters for matching these quantities, and a synchroscope connected to both sources to indicate phase relationship. Lamps may also be included. The standard synchroscope needle revolves counterclockwise when the incoming machine is slow and clockwise when fast. The needle points straight up when the two sources are in phase. The operator closes the connecting switch or circuit breaker as the synchroscope needle slowly approaches the in-phase position.

Automatic synchronizing provides for automatically closing the breaker to connect the incoming machine to the system, after the operator has properly adjusted voltage (field current), frequency (speed), and phasing (by lamps or synchroscope). A fully automatic synchronizer will initiate speed changes as required and may also balance voltages as required, then close the breaker at the proper time, all without attention of the operator. Automatic synchronizers can be used in unattended stations or in automatic control systems where units may be started, synchronized, and loaded on a single operator command. *See* ELECTRIC POWER SYSTEMS.

[EUGENE C. STARR]

Bibliography: H. C. Barnes et al., Alternator-rectifier exciter for Cardinal Plant 724-MVA generation, *IEEE Trans. Power App. Syst.*, PAS-87 (4): 1189, 1968; J. G. Brown, *Hydro-Electric Engineering Practice*, 1958; P. H. Cootner, *Water Demand for Steam Electric Generation*, 1966; K. Fenton, *Thermal Efficiency and Power Production*, 1966; S. Glasstone and A. Sesonske, *Nuclear Reactor*

Engineering, 1967; R. F. Grundy (ed.), *Magnetohydrodynamic Energy for Electric Power Generation*, Noyes Data Corporation, 1978; G. E. Kholodovskii, *Principles of Power Generation*, 1965; A. H. Lovell, *Generating Stations: Economic Elements of Electrical Design*, 4th ed., 1951; F. T. Morse, *Power Plant Engineering: The Theory and Practice of Stationary Electric Generating Plants*, 3d ed., 1953; F. R. Schleif et al., Excitation control to improve powerline stability, *IEEE Trans. Power App. Sys.*, PAS-87(6):1426–1432, 1968; B. G. A. Skrotzki, *Electric Generation: Hydro, Diesel, and Gas-Turbine Stations*, 1956; P. Sporn, *Energy: Its Production, Conversion and Use in the Service of Man*, 1963; *Standard Handbook for Electrical Engineers*, 10th ed., sect. 6, 8, 9, and 10, 1968; C. D. Swift, *Steam Power Plants: Starting, Testing and Operation*, 1959; H. G. Tak, *Economic Choice Between Hydroelectric and Thermal Power Developments*, 1966.

Electric power measurement

The measurement of the time rate at which work is done or energy is dissipated in an electric system. The work done in moving an electric charge is proportional to the charge and the voltage drop through which it moves. Charge per unit time defines electric current; electric power p is therefore defined as the product of the current i in a curcuit and the voltage e across its terminals at a given instant. Expressed symbolically, $p = ei$.

A second important definition of power follows directly from Ohm's law. $p = i^2R$, where R is the resistance of the circuit.

The practical unit of electric power is the watt. The watt represents a rate of expending energy, and thus it is related to all other units of power; for example, in mechanics 1 watt = 1 joule/s and 746 watts = 1 horsepower. Commonly used small units are the milliwatt (0.001 watt) and the microwatt (0.000001 watt). Large units are the kilowatt (1000 watts) and the megawatt (1,000,000 watts).

Power measurements must cover the frequency spectrum from direct current through the conventional power frequencies, the audio and the lower radio frequencies, to the highest frequencies (up to 25,000 gigahertz). In general, different techniques are required in each frequency range, and this article is divided into sections dealing with these frequency ranges. *See* ELECTRICAL MEASUREMENTS.

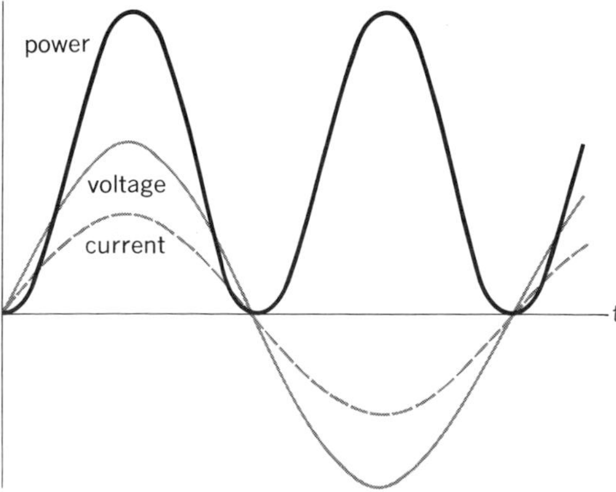

Fig. 1. Curves of instantaneous current, voltage, and power in an ac circuit; current and voltage in phase.

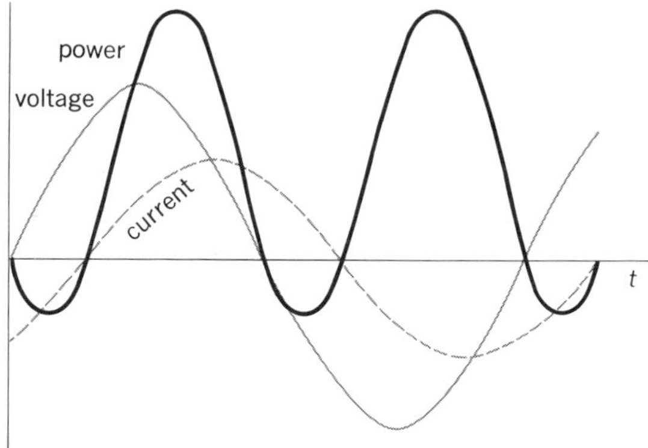

Fig. 2. Curves of instantaneous current, voltage, and power in an ac circuit; current and voltage out of phase.

DC AND AC POWER FREQUENCIES

In the measurement of power in a dc circuit not subject to rapid fluctuations, there is usually no difficulty in making simultaneous observations of the true values of voltage and current using common types of dc voltmeters and ammeters. The product of these observations then gives a sufficiently accurate measure of power in the given circuit, except that, if great accuracy is required, allowance must be made for the power used by the instruments themselves.

If in a circuit the voltage e, or current i, or both are subject to rapid variations, instantaneous values of power are difficult to measure and are usually of no interest. The important value is the average value, which is expressed mathematically in Eq. (1), where T is the period or time interval

$$P = \frac{1}{T}\int_0^T ei\, dt \qquad (1)$$

and t is time. This relation holds true for any waveform of current and voltage. In circuits with rapidly varying direct currents, pulsating rectified current, or, in general, alternating currents, the continuous averaging over short periods of time and the automatic multiplication of current and voltage values is accomplished by the wattmeter. *See* WATTMETER.

In ac circuits with steady effective values of voltage and current, the voltmeter-ammeter method may be used as in the dc case, except that, of course, ac meters are used, and a phase meter is also required to measure phase angle unless current and voltage are in phase. Because ac ammeters and voltmeters actually measure root-mean-square, or effective, values, these lead directly to values of average power.

Sinusoidal ac waves. Figure 1 illustrates the case of a sinusoidal voltage and current in a circuit containing only a resistive load. Here the current wave is entirely symmetrical with the voltage wave, and the power curve formed from the product of the voltage and current at each instant appears as a double-frequency wave on the positive side of the zero axis.

In Fig. 2, an inductance is assumed in the meas-

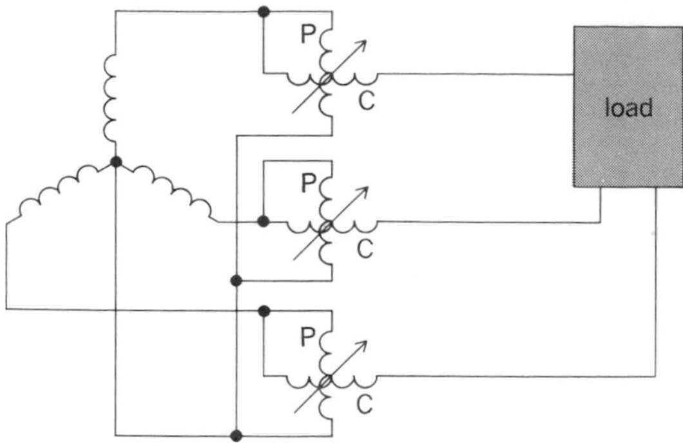

Fig. 3. Three wattmeters in three-phase, three-wire circuit. C and P refer to current and potential coils.

ured circuit. The current wave lags behind the voltage wave in what is called the negative out-of-phase, or quadrature, condition. A pure capacitance, however, would produce a positive out-of-phase, or quadrature, condition in which the current wave leads the voltage wave. In general, circuits contain elements of resistance, inductance, and capacitance in varying amounts and they must, therefore, assume some intermediate condition of phase angle between voltage and current. In this case the power wave, as shown in Fig. 2, dips below the zero line and becomes negative, indicating that during that part of the cycle power feeds back into the circuit. Measurements based on readings of conventional ac ammeters and voltmeters do not account for these negative excursions, and therefore, in general, the product of steady effective voltage and current readings in an ac circuit differs from, and is greater than, the reading of a wattmeter in such a circuit.

These relationships in the general ac circuit may be expressed as Eqs. (2) and (3), where I_m and E_m

$$i = I_m \sin 2\pi ft - \sqrt{2}\, I \sin 2\pi ft \qquad (2)$$

$$e = E_m \sin (2\pi ft + \phi) = \sqrt{2}\, E \sin (2\pi ft + \phi) \qquad (3)$$

are maximum values of current and voltage, f is frequency in hertz, and ϕ is the phase angle by which the current leads (+) or lags behind (−) the voltage in the circuit.

But by definition Eq. (1) holds. Substituting and carrying out the indicated operations, Eq. (4) is obtained where E and I are effective values of voltage and current.

$$P = EI \cos \phi \qquad (4)$$

The expression for P in Eq. (4) is the real or active power in the circuit and is distinguished from the simple product EI, which is called the apparent or virtual power, by the factor cos ϕ, which is

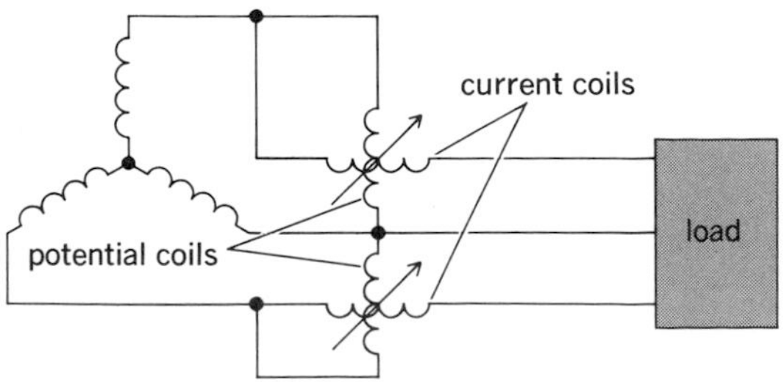

Fig. 4. Wattmeters in three-phase, three-wire circuit.

called the power factor. It is obvious from the previous formula that Eqs. (5) hold.

$$\cos \phi = P/EI \qquad (5a)$$

or

$$\text{Power factor} = \frac{\text{real power}}{\text{apparent power}} \qquad (5b)$$

Negative, or reactive, power due to inductance and capacitance in a circuit, is given by the relation $EI \sin \phi$.

The units for these quantities are, for real power, watts; for apparent power, volt-amperes; and, for reactive power, reactive volt-amperes or vars.

Polyphase power measurement. Summation of power in the separate phases of a polyphase circuit is accomplished by combinations of single-phase wattmeters, or wattmeter elements, disposed according to the general rule, called Blondel's theorem, as follows: If energy is supplied to any system of conductors through N wires, the total power in the system is given by the algebraic sum of N wattmeters, so arranged that each of the N wires contains one current coil, the corresponding potential coil being connected between that wire and some point on the system which is common to all the potential circuits. If this common point is on one of the N wires and coincides with the point of attachment of the potential lead to that wire, the measurement may be effected by the use of $N-1$ wattmeters.

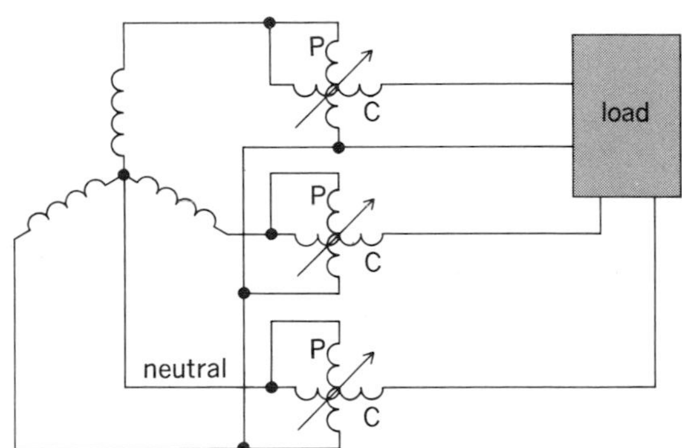

Fig. 5. Wattmeters in three-phase, four-wire circuit. C and P refer to current and potential coils.

Considering measurement in three-phase circuits as an example of polyphase practice in wide application in the power industry, the two most common systems are the three-phase, three-wire system in which the source may be Y-connected or delta-connected to three load wires; or the three-phase, four-wire system which also has three load wires and, in addition, a fourth wire, or neutral, which may or may not carry current to the load. If the neutral does not carry current, the circuit may be treated as a three-wire system.

Before applying the rule to commonly used circuits, an exception may be noted in the case of a balanced circuit in which the effective values of the currents and voltages and the phase relationships between them remain constant. In other words, the loads on the separate phases are equal. In this special case, power may be measured by a single wattmeter connected in one phase and the reading multiplied by three.

To measure power in either three-wire or four-wire systems, a wattmeter may be connected in each of the power-receiving circuits, as in Fig. 3. The sum of the three readings gives the total power.

Alternatively, in a three-wire system total power may be measured by two wattmeters, each having its current coil connected in one of the line conductors and its potential circuit connected between the line conductor in which its current coil is connected and the third line conductor (Fig. 4). The algebraic sum of the readings of the two wattmeters indicates the total power in the three power-receiving circuits.

In a four-wire system, three wattmeters may also be effectively used by connecting the current coils in each of two of the line conductors and in the neutral conductor, as in Fig. 5. The potential coils are connected between each of the line conductors and the neutral conductor in which the respective current coils are connected and the third line conductor.

In the last three cases the methods are correct for any value of balanced or unbalanced load and for any value of power factor.

A variety of other circuit connections is available for polyphase power measurement for various special conditions of use.

AUDIO AND RADIO FREQUENCIES

At frequencies above those used in the ordinary power-distribution systems, dynamometer-type wattmeters become inaccurate. For measurements of transmitted power at audio and the lower radio frequencies, no generally satisfactory substitute has been developed and measurements are therefore confined to determinations of power dissipated in a load, or available from a source, and are deduced from measurements of impedance and current or voltage.

Power-output meters. Power-output meters combine resistive loads, which can be adjusted to various known values, and voltmeters calibrated to indicate the power dissipated therein. They are used at audio frequencies to determine the maximum power output that can be obtained from a source, or to measure output power in studies of harmonic distortion, intermodulation, overload, frequency characteristic, and so forth.

Figure 6 shows an illustrative schematic. A voltmeter V is fed from an attenuator having a constant input resistance R and a voltage ratio k. The voltage V at the meter corresponds to a voltage kV at the attenuator input, and the meter scale is calibrated in power input to the attenuator, $P=(kV)^2/R$. By adjusting the attenuator one can obtain convenient scale multiplying factors k^2. The tapped transformer T provides different turns ratios n_1/n_2 to adjust the effective input resistance of the instrument over a range of values.

By adjusting this ratio the user can obtain a maximum meter reading when the corresponding input impedance is approximately equal to the output impedance of the source. If the output impedance is purely resistive, this condition occurs when the impedances are exactly equal, or matched, and the power indicated is the maximum power output that can be obtained from the source.

Standard-signal generators. At radio frequencies the principle of impedance matching is frequently used to determine the power input to a re-

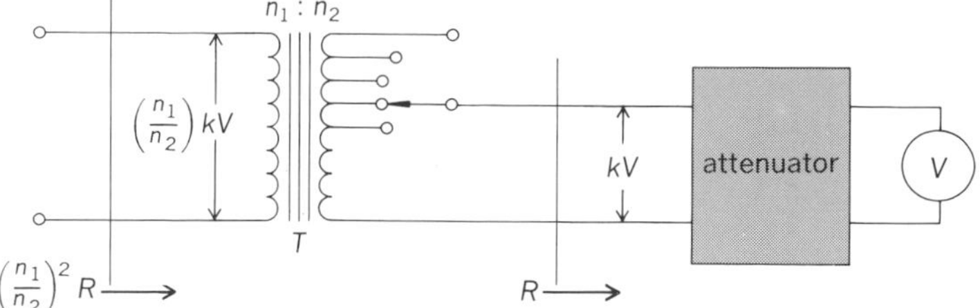

Fig. 6. Schematic representation of audio-frequency power-output meter.

ceiver, with a standard-signal generator serving as source. The circuit of Fig. 7 shows the arrangement.

A standard-signal generator produces a known voltage E "behind" an output resistance R. Impedances at radio frequencies frequently have significant reactive components; thus the impedance-matching network serves the dual function of tuning out the reactive component X_x of the receiver input impedance Z_x and multiplying the resistive component R_x by a factor that depends upon the values of the component inductance and capacitances. When the matching network is so adjusted that the receiver output is a maximum, the reactive component is nullified and the input resistance of the matching network equals the output resistance R of the standard-signal generator. The power input is therefore $P=E^2/4R$.

Calorimetric method. The power dissipated in a resistive load can be determined from the heat developed therein, and several ingenious schemes for measuring this heat have been derived.

A widely used technique depends upon dissipating the power in a resistive element of relatively high temperature coefficient of resistivity. The power is deduced from the change in resistance by one of several methods. Measurements that are of this general nature are known as bolometric methods.

Another method depends upon using as the load an incandescent lamp that will emit visible light at power inputs of the order of those to be measured. The temperature of the filament can be determined by pyrometric techniques, or the light can be measured with a photocell. The indicating system can then be calibrated in terms of power input at direct current or low-frequency alternating current, where accurate moving-coil wattmeters can be used. Such calibrations will continue to hold good as the frequency of the source is raised until dielectric losses in the glass and eddy-current losses in the metal parts other than the filament become significant compared with the heating of the filament proper, or until the wavelength be-

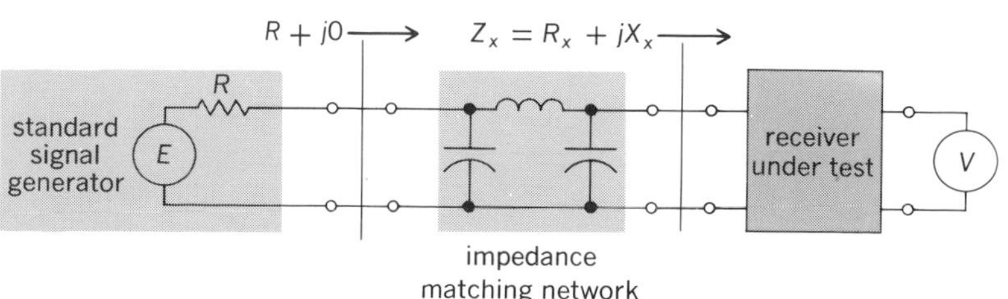

Fig. 7. Schematic showing standard-signal generator and impedance matching set up to produce known power input to receiver.

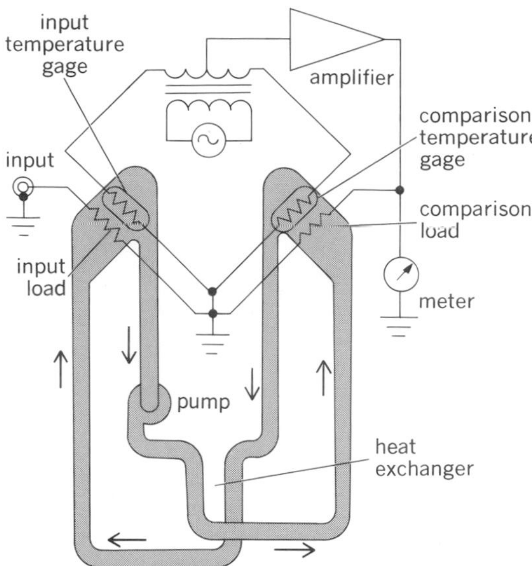

Fig. 8. Commercial self-balancing calorimeter.

comes so short that the current flow in the filament becomes nonuniform along its length. Reasonable accuracy can be attained at frequencies as high as a few hundred megahertz.

A variant of this method uses a diode in which a pure-metal filament is heated by the power to be measured. When not space-charge-limited, the dc plate current is a sensitive indicator of filament temperature. The temperature rise in a heater element can also be measured by thermocouples.

The classical calorimetric method depends upon measuring the temperature rise in a liquid coolant circulated through an enclosure containing the load and so insulated that heat losses other than to the coolant are negligible.

Measurements of rate of temperature rise can be used for absolute determination of the power input in terms of the specific heat and volume of the liquid, or the liquid can be used as a transfer mechanism to compare the radio-frequency power input with a known dc or low-frequency power developed elsewhere in the system.

Figure 8 shows a commercial instrument, which uses a self-balancing bridge to provide a comparison temperature rise that can be accurately calibrated in terms of a low-frequency source.

MICROWAVE FREQUENCIES

Measurement of dissipated power at microwave frequencies can be carried out by methods similar to those used at lower frequencies. Because voltage and impedance become increasingly difficult to define precisely as the frequency is increased, however, calorimetric methods generally, and bolometric methods specifically, are almost universally used.

In contrast to the lower radio frequencies, convenient methods of measuring transmitted power are available at microwave frequencies. These use the properties of waves propagated along transmission lines. A commonly used device for this purpose is the directional coupler, a simple example of which is shown in Fig. 9. This comprises an auxiliary line, coupled at two points spaced one-fourth wavelength apart to the main line down which power flows. For the sake of simplicity no coupling means is shown between the main and auxiliary lines other than holes in the outer conductors; in practice, either capacitive probes or small coupling loops may be used to transfer energy from one line to the other. For the flow of power illustrated, a wave entering the right-hand hole and propagating to the left in the auxiliary line arrives at the left-hand hole 180° out of phase with the wave entering the left-hand hole. If there are no losses in the system, and the couplings through the two holes are equal, the two signals cancel, and no signal reaches the left-hand termination. On the other hand, a wave entering the left-hand hole and propagating to the right in the auxiliary line arrives at the right-hand hole in phase with the wave entering the right-hand hole, and the two signals add and propagate to the right in the auxiliary line. Terminating the auxiliary line in a resistance equal to the characteristic impedance Z_0 prevents any of this power from reflecting in the auxiliary line and reaching the left-hand termination. Power traveling to the right in the main line therefore produces a signal voltage only at the right-hand auxiliary-line termination.

Conversely, a signal that is propagating to the left in the main line reaches only the left-hand termination. If the two terminations are used as bolometers, for instance, they can then measure the individual powers traveling each way in the main line and the net power reaching the load, which is equal to the difference between them. *See* TRANSMISSION LINES.

This measurement can also be made with a standing-wave detector. The voltage maximum is equal to the sum of the voltages of the incident and reflected waves, $V_I + V_R$, whereas the voltage minimum is equal to the difference, $V_I - V_R$. The power transmitted is the difference in power given by expression (6), which is equal to $V_{\max}V_{\min}/Z_0$. The

$$\frac{(V_I)^2}{Z_0} - \frac{(V_R)^2}{Z_0} \tag{6}$$

relation between the probe voltage and the line voltage can be established by measuring power in matched loads. In contrast to methods using directional couplers, this has the advantage of requiring substantially no power; however, it requires manipulation to obtain a measurement.

[WILLIAM H. HARTWIG]

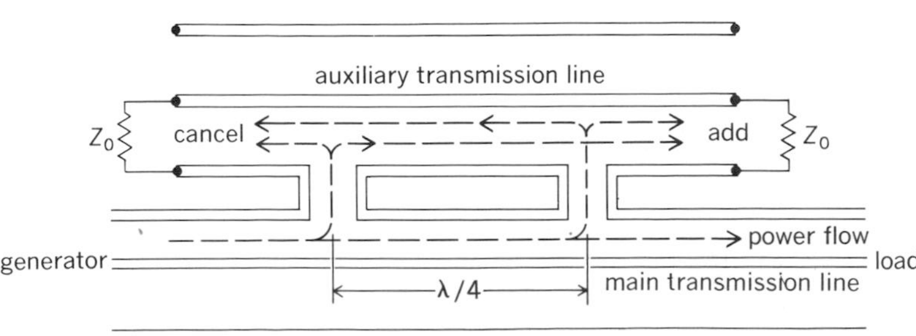

Fig. 9. Cross-sectional view of simplified two-hole directional coupler.

Bibliography: H. Buckingham and E. M. Price, *Principles of Electrical Measurements*, 1957; E. W. Golding, *Electrical Measurements and Measuring*

Instruments, 1955; F. K. Harris, *Electrical Measurements*, 1952, reprinted 1975; P. Kantrowitz et al., *Electronic Measurements*, 1979.

Electric power substation

An assembly of equipment in an electric power system through which electrical energy is passed for transmission, distribution, interconnection, transformation, conversion, or switching. *See* ELECTRIC POWER SYSTEMS.

Specifically, substations are used for some or all of the following purposes: connection of generators, transmission or distribution lines, and loads to each other; transformation of power from one voltage level to another; interconnection of alternate sources of power; switching for alternate connections and isolation of failed or overloaded lines and equipment; controlling system voltage and power flow; reactive power compensation; suppression of overvoltage; and detection of faults, monitoring, recording of information, power measurements, and remote communications. Minor distribution or transmission equipment installation is not referred to as a substation.

Classification. Substations are referred to by the main duty they perform. Broadly speaking, they are classified as: transmission substations (Fig. 1), which are associated with high voltage levels; and distribution substations (Fig. 2), associated with low voltage levels. *See* ELECTRIC DISTRIBUTION SYSTEMS.

Substations are also referred to in a variety of other ways, as follows.

Transformer substations. These are substations whose equipment includes transformers.

Switching substations. These are substations whose equipment is mainly for various connections and interconnections, and does not include transformers.

Customer substations. These are usually distribution substations located on the premises of a larger customer, such as a shopping center, large office or commercial building, or industrial plant.

Converter stations. The main function of converter stations is the conversion of power from ac to dc and vice versa. Converter stations are complex substations required for high-voltage direct-current (HVDC) transmission or interconnection of two ac systems which, for a variety of reasons, cannot be connected by an ac connection. The main equipment includes converter valves usually located inside a large hall, transformers, filters, reactors, and capacitors. Figure 3 shows a ±250-kV 500-MW HVDC converter station together with a 245-kV ac substation. The building in the center of the converter station contains all the thyristor valves. On each side of the building against the wall are converter transformers connected to the valves inside. Further away from the building are filters and capacitors. Going away from the building is the dc line, with a typical dc line tower in the uppermost part of the figure. *See* DIRECT-CURRENT TRANSMISSION.

Fig. 1. A typical transmission substation (345 kV).

Air-insulated substations. Most substations are installed as air-insulated substations, implying that the bus-bars and equipment terminations are generally open to the air, and utilize insulation properties of ambient air for insulation to ground. Modern substations in urban areas are esthetically designed with low profiles and often within walls, or even indoors.

Metal-clad substations. These are also air-insulated, but for low voltage levels; they are housed in

Fig. 2. View of a residential outdoor distribution substation with double-ended underground supply to transformers which are positioned at either end of the metal-clad switchgear.

Fig. 3. View of ±250-kV 500-MW HVDC converter station (upper part) and 245-kV ac substation (lower part). The large building of the converter station contains all the thyristor valves. (*General Electric*)

metal cabinets and may be indoors or outdoors (Fig. 2).

Gas-insulated substations. Acquiring a substation site in an urban area is very difficult. Land in many cases is either unavailable or very expensive. Therefore, there has been a definite trend toward increasing use of gas-insulated substations, which occupy only 5–20% of the space occupied by the air-insulated substations. In gas-insulated substations, all live equipment and bus-bars are housed in grounded metal enclosures, which are sealed and filled with sulfur hexafluoride (SF_6) gas, which has excellent insulation properties. These substations have the appearance of a large-scale plumbing (Fig. 4). The disk-shaped enclosures in Fig. 4 are circuit breakers.

Mobile substations. For emergency replacement or maintenance of substation transformers, mobile substations are used by some utilities. They may range in size from 100 to 25,000 kVA, and from 2400 to 220,000 V. Most units are designed for travel on public roads, but larger substations are built for travel on railroad tracks. These substations are generally complete with a transformer, circuit breaker, disconnect switches, lightning arresters, and protective relaying.

Substation switching arrangement. An appropriate switching arrangement for "connections" of generators, transformers, lines, and other major equipment is basic to any substation design. There are seven switching arrangements commonly used (Fig. 5). Each breaker is usually accompanied by two disconnect switches, one on each side, for maintenance purposes. Selecting the switching arrangement involves considerations of cost, reliability, maintenance, and flexibility for expansion.

Single bus. This involves one common bus for all connections and one breaker per connection. This is the least costly arrangement, but also the least desirable for considerations of reliability and maintenance.

Double bus, single breaker. This involves two common buses, and each connection of the line equipment and so forth can be made to either bus through one or the other breaker. However, there is only one breaker per connection, and when a breaker is out for maintenance, the connected equipment or line is also removed from operation.

Double bus, double breaker. This arrangement has two common buses and two breakers per connection. It offers the most reliability, ease of maintenance, and flexibility, but is also the most expensive.

Main and transfer bus. This is like the single-bus arrangement; however, an additional transfer bus is provided so that a breaker can be taken out of service by transferring its connection to the transfer bus which is connected to the main bus through a breaker between the two buses.

Ring bus. This may consist of four, six, or more breakers connected in a closed loop, with the same number of connection points. In this case, a breaker can be taken out of service without also taking out any connection.

Fig. 4. A 345-kV ac gas-insulated substation. (*Gould-BBC*).

Breaker-and-a-half. This arrangement involves two buses, between which three breaker bays are installed. Each three-breaker bay provides two circuit connection points—thus the name breaker-and-a-half.

Breaker-and-a-third. In this arrangement, there are four breakers and three connections per bay.

Substation equipment. A substation includes a variety of equipment. The principal items are listed and briefly described below.

Transformers. These involve magnetic core and windings to transfer power from one side to the other at different voltages. Substation transformers range from small sizes of 1 MVA to large sizes of 2000 MVA. Most of the transformers and all those above a few MVA size are insulated and cooled by oil, and adequate precautions have to be taken for fire hazard. These precautions include adequate distances from other equipment, fire walls, fire extinguishing means, and pits and drains for containing leaked oil. *See* TRANSFORMER.

Circuit breakers. These are required for circuit interruption with the capability of interrupting the highest fault currents, usually 20–50 times the normal current, and withstanding high voltage surges that appear after interruption. Switches with only normal load-interrupting capability are referred to as load break switches. *See* CIRCUIT BREAKER.

Disconnect switches. These have isolation and connection capability without current interruption capability.

Bus-bars. These are connecting bars or conductors between equipment. Flexible conductor buses are stretched from insulator to insulator, whereas more common solid buses (hollow aluminum alloy tubes) are installed on insulators in air or in gas-enclosed cylindrical pipes. *See* BUS-BAR.

Shunt reactors. These are often required for compensation of the line capacitance where long lines are involved.

Shunt capacitors. These are required for compensation of inductive components of the load current.

Current and potential transformers. These are for measuring currents and voltages and providing proportionately low-level currents and voltages at ground potential for control and protection.

Control and protection. This includes (a) a variety of protective relays which can rapidly detect faults anywhere in the substation equipment and lines, determine what part of the system is faulty, and give appropriate commands for opening of circuit breakers; (b) control equipment for voltage and current control and proper selection of the system configuration; (c) fault-recording equipment; (d) metering equipment; (e) communication equipment; and (f) auxiliary power supplies. *See* ELECTRIC PROTECTIVE DEVICES; RELAY.

Many of the control and protection devices are solid-state electronic types, and there is a trend toward digital techniques using microprocessors. Most of the substations are fully automated locally with a provision for manual override. The minimum manual interface required, along with essential information on status, is transferred via communications channels to the dispatcher in the central office.

Other items which may be installed in a substation include: phase shifters, current-limiting reactors, dynamic brakes, wave traps, series capacitors, controlled reactive compensation, fuses, ac to dc or dc to ac converters, filters, and cooling facilities.

Substation grounding and shielding. Good substation grounding is very important for effective relaying and insulation of equipment; however, the

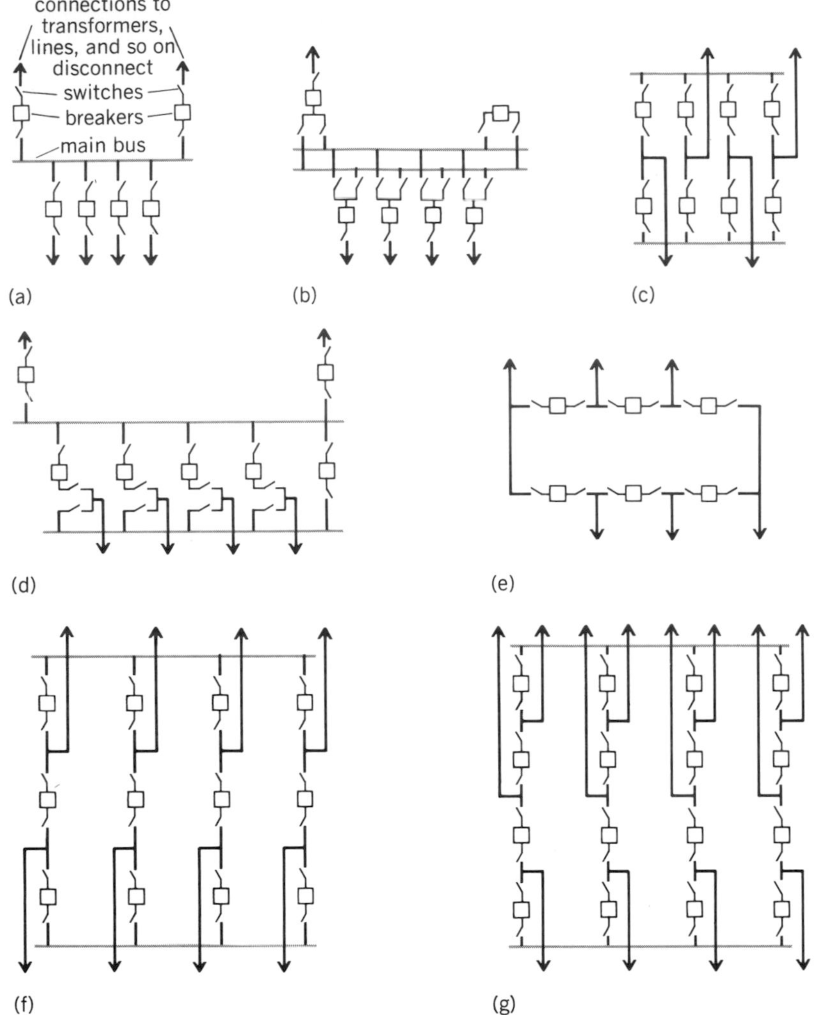

Fig. 5. One-line diagrams of substation switching arrangements. *(a)* Single bus. *(b)* Double bus, single breaker, *(c)* Double bus, double breaker. *(d)* Main and transfer bus. *(e)* Ring bus. *(f)* Breaker-and-a-half. *(g)* Breaker-and-a-third.

safety of the personnel is the governing criterion in the design of substation grounding. It usually consists of a bare wire grid, laid in the ground; and all equipment grounding points, tanks, support structures, fences, shielding wires and poles, and so forth, are securely connected to it. The grounding resistance is reduced to be low enough that a fault from high voltage to ground does not create such high potential gradients on the ground, and from the structures to ground, to present a safety hazard. Good overhead shielding is also essential for outdoor substations, so as to virtually eliminate the possibility of lightning directly striking the equipment. Shielding is provided by overhead ground wires stretched across the substation or tall grounded poles. *See* GROUNDING; LIGHTNING AND SURGE PROTECTION. [NARAIN G. HINGORANI]

Bibliography: C. Adamson and N. G. Hingorani, *High Voltage Direct Current Power Transmission*, 1960; C. H. Flurscheim, *Power Circuit Breaker Theory and Design*, 1975; E. W. Kimbark, *Direct Current Transmission*, 1971; S. A. Stigant and A. C. Franklin, *THE J&P Transformer Book*, 10th ed., 1974.

Electric power systems

A complex assemblage of equipment and circuits for generating, transmitting, transforming, and distributing electrical energy. In the United States, electrical energy is generated to serve more than 87,625,000 customers. The investment represented by these facilities has grown rapidly over the years until in 1978 it was close to $270,000,000,000, with about 75% spent on the power systems of investor-owned utilities and the remainder spread among systems built by governmental agencies, municipal electric departments, and rural electric cooperatives. Principal elements of a typical power system are shown in Fig. 1.

Generation. Electricity in the large quantities required to supply electric power systems is produced in generating stations, commonly called power plants. Such generating stations, however, should be considered as conversion facilities in which the heat energy of fuel (coal, oil, gas, or uranium) or the hydraulic energy of falling water is converted to electricity. *See* ELECTRIC POWER GENERATION.

Steam stations. About 89% of the electric power used in the United States is obtained from generators driven by steam turbines. The largest such unit in service in 1978 was rated 1300 MW, equivalent to about 1,730,000 hp. But 650-, 800- and 950-MW units are commonplace for new fossil-fuel–fired stations, and 1150–1300-MW units are the most commonly installed units in nuclear stations.

Coal was the fuel for nearly 51% of the steam turbine generation in 1978, and its share should increase somewhat because of the projected long-term shortage of natural gas and both the sharp rise in the cost of fuel oil and governmental policy of restricting firing of oil. Natural gas, used extensively in the southern part of the United States, fueled about 16% of the steam turbine generation, and heavy fuel oil, 19%, largely in power plants able to take delivery from ocean-going tankers or river barges. The remaining 14% was generated from the radioactive energy of slightly enriched uranium, which, for many power systems, produces electricity at a lower total cost than either coal or fuel oil at current delivered prices. As a consequence, some 46% of the generating capability additions planned as of Jan. 1, 1979, according to an *Electrical World* survey, will be nuclear and largely in the 900- to 1300-MW-per-unit range. As these nuclear units go into commercial operation, the contribution of uranium to the electrical energy supply will rise, probably to more than 25% of the total fuel generated output by the mid-1980s.

Nuclear steam systems used by United States utilities are mostly of the water-cooled-and-moderated type, in which the heat of a controlled nuclear reaction is used to convert water into steam to drive a conventional turbine generator; such units are presently limited to about 1300 MW by the thermal limit placed on nuclear reactors by the Nuclear Regulatory Commission of the U.S. Department of Energy.

Hydroelectric plants. Waterpower during 1978 supplied about 10.3% of the electric power consumed in the United States. But this share can only decline in the years ahead because very few

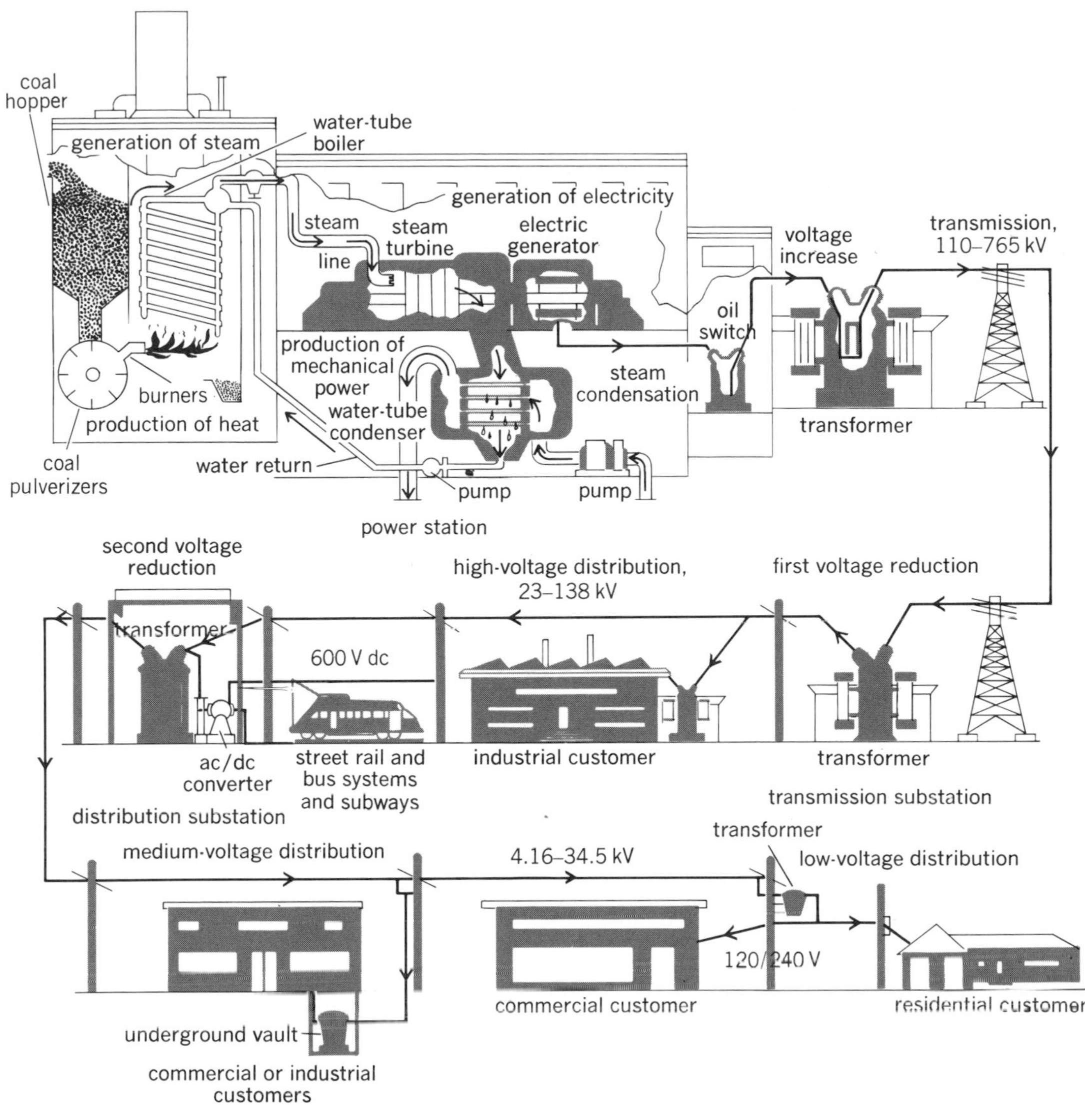

Fig. 1. Major steps in the generation, transmission, and distribution of electricity.

sites remain undeveloped where sufficient water drops far enough in a reasonable distance to drive reasonably sized hydraulic turbines. Consequently, the generating capability of hydro plants, 10.4% of the utility industry's total as of Dec. 31, 1978, is slated to fall off to 8.9% by the mid-1980s because of its very small share of the planned additions. Much of this additional hydro capability will be used at existing plants to increase their effectiveness in supplying peak power demands, and as a quickly available source of emergency power.

Some hydro plants totaling 10,640 MW as of Dec. 31, 1978, actually draw power from other generating facilities during light system-load periods to pump water from a river or lake into an artificial reservoir at a higher elevation from which it can be drawn through a hydraulic station when the power system needs additional generation. These pumped-storage installations consume about 50% more energy than they return to the power system and, accordingly, cannot be considered energy sources. Their use is justified, however, by their ability to convert surplus power that is available during low-demand periods into prime power to serve system needs during peak-demand intervals—a need that otherwise would require building more generating stations for operation during the relatively few hours of high system demand. Installations now planned should double the existing capacity by the late 1980s to 22,684 MW.

Combustion turbine plants. Gas-turbine-driven generators, now commonly called combustion turbines because of the growing use of light oil as fuel, have gained wide acceptance as an economical source of additional power for heavy-load periods. In addition, they offer the fastest erection time and the lowest investment cost per kilowatt of installed capability. Offsetting these advantages, however, is their relatively less efficient consumption of more costly fuel. Typical unit ratings in the United States have climbed rapidly in recent years until some units operating in 1979 are rated at 80 MW. Some turbine installations involve a group of smaller units totaling, in one case, 260 MW. Combustion turbine units, even in the larger ratings, offer extremely flexible opera-

Table 1. Power capability of typical three-phase open-wire transmission lines

Line-to-line voltage, kV	Capability, MVA
115 ac	60
138 ac	90
230 ac	250
345 ac	600
500 ac	1200
765 ac	2500
800 dc*	1500

*Bipolar line with grounded neutral.

tion and can be started and run up to full load in as little as 10 min. Thus they are extremely useful as emergency power sources, as well as for operating during the few hours of daily load peaks. Combustion turbines totaled 8.5% of the total installed capability of United States utility systems at the close of 1978 and supplied less than 3% of the total energy generated.

In the years ahead, however, combustion turbines are slated for an additional role. Several installations in the 1970s have used their exhaust gases to heat boilers that generate steam to drive steam turbine generators. Such combined-cycle units offer fuel economy comparable to that of modern steam plants and at considerably less cost per kilowatt. In addition, because only part of the plant uses steam, the requirement for cooling water is considerably reduced. A number of additional combined-cycle installations have been planned, but wide acceptance is inhibited by the doubtful availability of light fuel oil for them. This barrier should be resolved, in time, by the successful development of systems for fueling them with gas derived from coal. *See* COAL GASIFICATION.

Internal combustion plants. Internal combustion engines of the diesel type drive generators in many small power plants. In addition, they offer the ability to start quickly for operation during peak loads or emergencies. However, their small size, commonly about 2 MW per unit although a few approach 10 MW, has limited their use. Such installations account for about 1% of the total power-system generating capability in the United States, and make an even smaller contribution to total electric energy consumed.

Three-phase output. Because of their simplicity and efficient use of conductors, three-phase 60-Hz alternating-current systems are used almost exclusively in the United States. Consequently, power-system generators are wound for three-phase output at a voltage usually limited by design features to a range from about 11 kV for small units to 30 kV for large ones. The output of modern generating stations is usually stepped up by transformers to the voltage level of transmission circuits used to deliver power to distant load areas.

Transmission. The transmission system carries electric power efficiently and in large amounts from generating stations to consumption areas. Such transmission is also used to interconnect adjacent power systems for mutual assistance in case of emergency and to gain for the interconnected power systems the economies possible in regional operation. Interconnections have expanded to the point where most of the generation east of the Rocky Mountains, except for a large part of Texas, regularly operates in parallel, and over 90% of all generation in the United States, exclusive of Alaska and Hawaii, and in Canada can be linked.

Transmission circuits are designed to operate up to 765 kV, depending on the amount of power to be carried and the distance to be traveled. The permissible power loading of a circuit depends on many factors, such as the thermal limit of the conductors and their clearances to ground, the voltage drop between the sending and receiving end and the degree to which system service reliability depends on it, and how much the circuit is needed to hold various generating stations in synchronism. A widely accepted approximation to the voltage appropriate for a transmission circuit is that the permissible load-carrying ability varies as the square of the voltage. Typical ratings are listed in Table 1.

Transmission as a distinct function began about 1886 with a 17-mi (27-km) 2-kV line in Italy. Transmission began at about the same time in the United States, and by 1891 a 10-kV line was operating (Fig. 2). In 1896 an 11-kV three-phase line brought electrical energy generated at Niagara Falls to Buffalo, 20 mi (32 km) away. Subsequent lines were built at successively higher levels until 1936, when the Los Angeles Department of Water and Power energized two lines at 287 kV to transmit 240 MW the 266 mi (428 km) from Hoover Dam on the Colorado River to Los Angeles. A third line was completed in 1940.

For nearly 2 decades these three 287-kV lines were the only extra-high-voltage (EHV). lines in

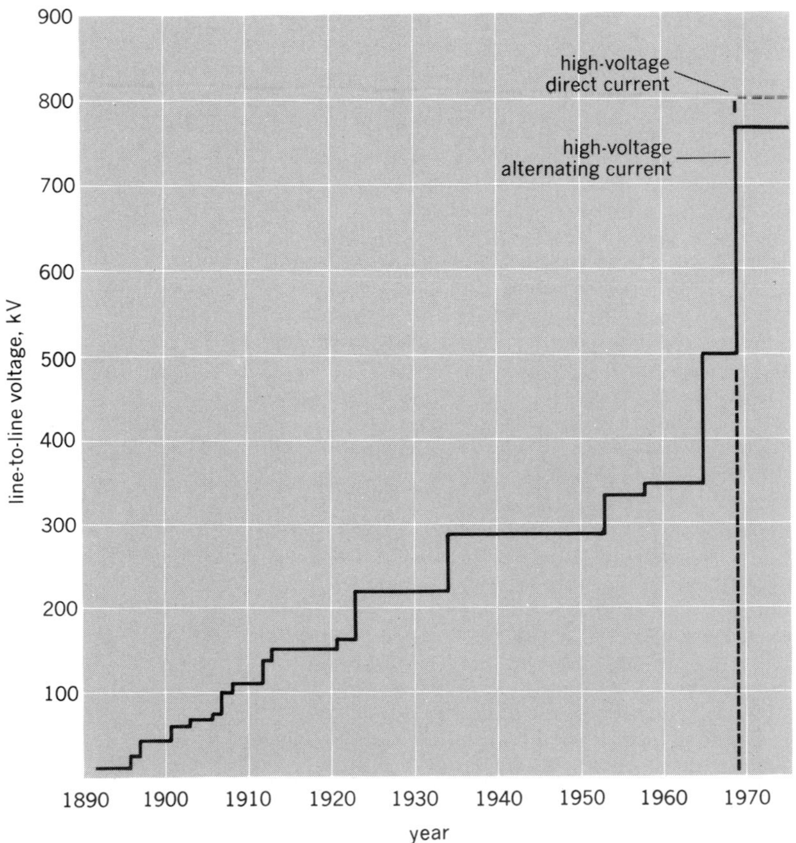

Fig. 2. Growth of ac transmission voltages from 1890.

North America, if not in the entire world. But in 1946 the American Electric Power (AEP) System inaugurated, with participating manufacturers, a test program up to 500 kV. From this came the basic design for a 345-kV system, the first link of which went into commercial operation in 1953 as part of a system overlay that finally extended from Roanoke, VA, to the outskirts of Chicago. By the late 1960s the 345-kV level had been adopted by many utilities interconnected with the AEP System, as well as others in Illinois, Wisconsin, Minnesota, Kansas, Oklahoma, Texas, New Mexico, Arizona and across New York State into New England.

The development of 500-kV circuits began in 1964, even as the 345-kV level was gaining wide acceptance. One reason for this was that many utilities that had already adopted 230 kV could gain only about 140% in capability by switching to 345 kV, but the jump to 500 kV gave them nearly 400% more capability per circuit. The first line energized at this new level was by Virginia Electric & Power Company to carry the output of a new mine mouth station in West Virginia to its service area. A second line completed the same year provided transmission for a 1500-MW seasonal interchange between the Tennessee Valley Authority and a group of utilities in the Arkansas-Louisiana area. Lines at this voltage level now extend from New Jersey to Texas and from New Mexico via California to British Columbia.

The next and latest step-up occurred in 1969 when the AEP System, after another cooperative test program, completed the first line of an extensive 765-kV system to overlay its earlier 345-kV system. The first installation in this voltage class, however, was by the Quebec Hydro-Electric Commission to carry the output at 735 kV (the expected international standard) from a vast hydro project to its load center at Montreal, some 375 mi (604 km) away.

Transmission engineers are anticipating even higher voltages of 1100 to 1500 kV, but they are fully aware that this objective may prove too costly in space requirements and funds to gain wide acceptance. Experience already gained at 500 kV and 765 kV verifies that the prime requirement no longer is insulating the lines to withstand lightning discharges, but insulating them to tolerate voltage surges caused by the operation of circuit breakers. Audible noise levels, especially in rain or humid conditions, are high, requiring wide buffer zones. Environmental challenges have been brought on the basis of possible negative biological effects of the electrostatic field produced under EHV lines, though research to date has not shown any such effects.

Experience has indicated that, within about 10 years after the introduction of a new voltage level for overhead lines, it becomes necessary to begin connecting underground cable. This has already occurred for 345 kV; the first overhead line was completed in 1953, and by 1967 about 100 mi (160 km) of pipe-type cable had been installed to take power received at this voltage level into metropolitan areas. The first 500-kV cable in the United States was placed in service in 1976 to take power generated at the enormous Grand Coulee hydro plant to a major switchyard several thousand feet (1 ft = 0.3 m) away. And 765-kV cable, after extensive testing above the 500-kV level at the Waltz Mill Cable Test Center operated by Westinghouse Electric Corporation for the Electric Power Research Institute (EPRI), will be available when required.

In anticipation of the need for transmission circuits of higher load capability, an extensive research program is in progress, spread among several large and elaborately equipped research centers. Among these are: Project UHV for ultra-high-voltage overhead lines operated by the General Electric Company near Pittsfield, MA, for EPRI; the Frank B. Black Research Center built and operated by the Ohio Brass Company near Mansfield; the above-mentioned Waltz Mill Cable Test Center; and an 1100 kV, 1-mi (1.6 km) test line built by the Bonneville Power Authority. All include equipment for testing full-scale or cable components at well over 1000 kV. In addition, many utilities and specialty manufacturers have test facilities related to their fields of operation.

A relatively new approach to high-voltage long-distance transmission is high-voltage direct current (HVDC), which offers the advantages of less costly lines, lower transmission losses, and insensitivity to many system problems that restrict alternating-current systems. Its greatest disadvantage is the need for costly equipment for converting the sending-end power to direct current, and for converting the receiving-end direct-current power to alternating current for distribution to consumers. Starting in the late 1950s with a 65-mi (105-km) 100-kV system in Sweden, HVDC has been applied successfully in a series of special cases around the world, each one for a higher voltage and greater power capability. The first such installation in the United States was put into service in 1970. It operates at 800 kV line to line, and is designed to carry a power interchange of 1440 MW over a 1354-km overhead tie line between the winter-peaking Northwest Pacific coastal region and the summer-peaking southern California area. These HVDC lines perform functions other than just power transfer, however. The Pacific Intertie is used to stabilize the parallel alternating-current transmission and lines, permitting an increase in their capability; and back-to-back converters with no tie line between them are used to tie together two systems in Nebraska that otherwise could not be synchronized. The first urban installation of this technology was energized in 1979 in New York City.

In addition to these high-capability circuits, every large utility has many miles of lower-voltage transmission, usually operating at 110 to 345 kV, to carry bulk power to numerous cities, towns, and large industrial plants. These circuits often include extensive lengths of underground cable where they pass through densely populated areas. Their design, construction, and operation are based upon research done some years ago, augmented by extensive experience.

Interconnections. As systems grow and the number and size of generating units increase, and as transmission networks expand, higher levels of bulk-power-system reliability are attained through properly coordinated interconnections among separate systems. This practice began more than 50

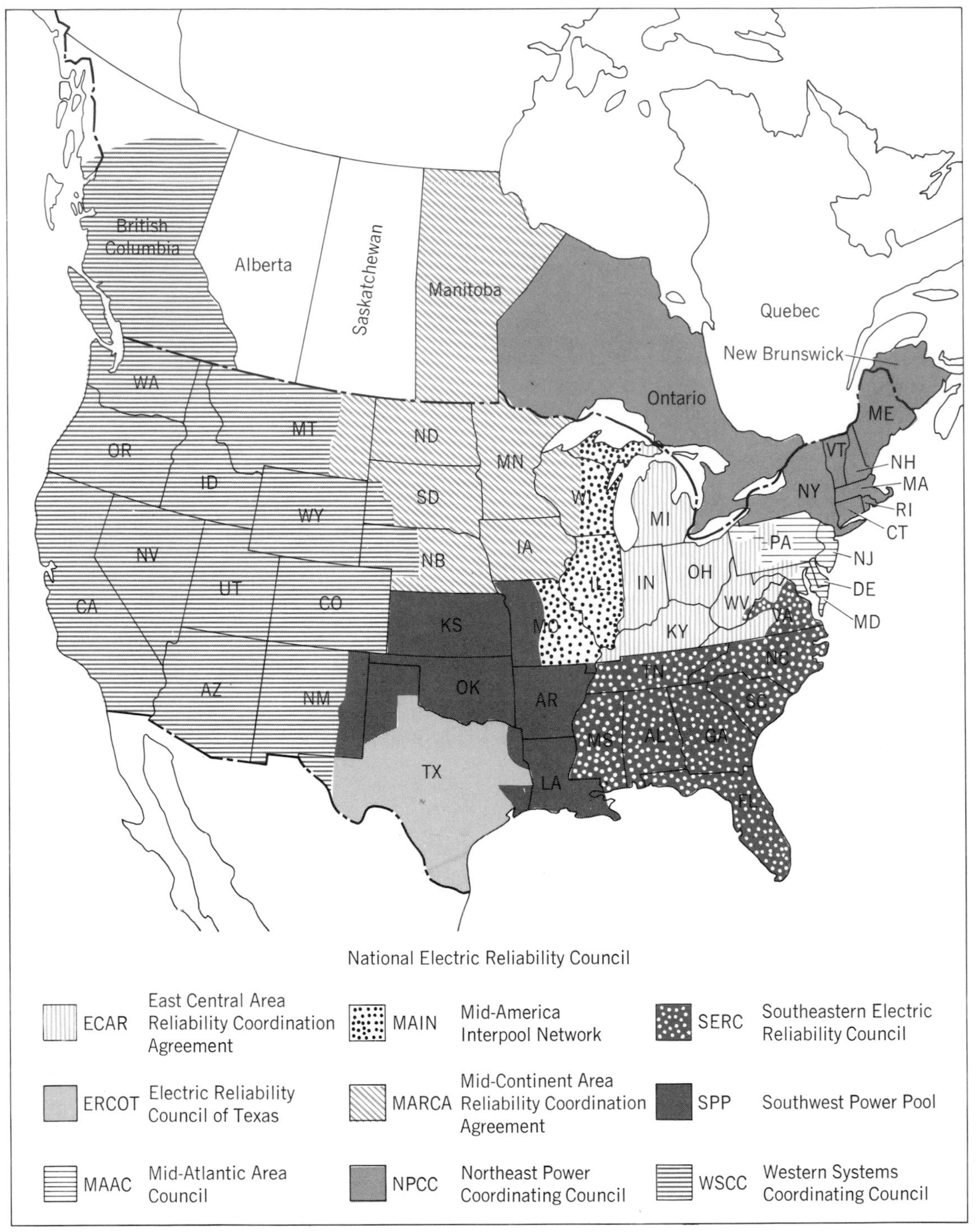

Fig. 3. Areas served by the nine regional reliability councils, coordinated by the National Electric Reliability Council, that guide the coordination and operation of generation and transmission facilities.

years ago with such voluntary pools as the Connecticut Valley Power Exchange and the Pennsylvania-New Jersey-Maryland Interconnection. Most of the electric utilities in the contiguous United States and a large part of Canada now operate as members of power pools, and these pools (except one in Texas) in turn are interconnected into one gigantic power grid known as the North American Power Systems Interconnection. The operation of this interconnection, in turn, is coordinated by the North American Power Systems Interconnection Committee (NAPSIC). Each individual utility in such pools operates independently, but has contractual arrangements with other members in respect to generation additions and scheduling of operation. Their participation in a power pool affords a higher level of service reliability and important economic advantages.

Regional and national coordination. The Northeast blackout of Nov. 9, 1965, stemmed from the unexpected trip-out of a key transmission circuit carrying emergency power into Canada and cascaded throughout the northeastern states to cut off electric service to some 30,000,000 people. It spurred the utilities into a chain reaction affecting the planning, construction, operation, and control procedures for their interconnected systems. They soon organized regional coordination councils, eventually nine in number, to cover the entire contiguous United States and four Canadian provinces (Fig. 3). Their stated objective was "to further augment reliability of the parties' planning and operation of their generation and transmission facilities."

Then, in 1968, the National Electric Reliability Council (NERC) was established to serve as an effective body for the collection, unification, and dissemination of various reliability criteria for use by individual utilities in meeting their planning and operating responsibilities. NAPSIC was reorganized shortly afterward to function as an advisory group to NERC.

Increased interconnection capability among power systems reduces the required generation reserve of each of the individual systems. In most utilities the loss-of-load probability (LOLP) is used to measure the reliability of electric service, and it is based on the application of probability theory to unit-outage statistics and load forecasts. A common LOLP criterion is 1 day in 10 years when load may exceed generating capability. The LOLP decreases (that is, reliability increases) with increased interconnection between two areas until a saturation level is reached which depends upon the amount of reserve, unit sizes, and annual load shape in each area. Any increase in interconnection capability beyond that saturation level will not cause a corresponding improvement in the level of system reliability.

Traditionally, systems were planned to withstand all reasonably probable contingencies, and operators seldom had to worry about the possible effect of unscheduled outages. Operators' normal security functions were to maintain adequate generation on-line and to ensure that such system variables as line flows and station voltages remained within the limits specified by planners. However, stronger interconnections, larger generating units, and rapid system growth spread the transient effects of sudden disturbances and increased the responsibilities of operators for system security.

System security is concerned with service continuity at standard frequency and voltage levels. The system is said to be insecure if a contingency would result in overloading some system components, in abnormal voltage levels at some stations, in change of system frequency, or in system instability, even if there is adequate capability as indicated by some reliability index.

Energy control centers. The majority of large electric systems have computerized energy centers whose functions are to control operation of the system so as to optimize economy and security. In some large interconnected networks of individual utility companies, such as those in the Pennsylvania–New Jersey–Maryland power pool, regional control centers have been established to perform the same function for a group of companies. New generation centers now use digital computers, or hybrid analog-digital units, rather than the analog

Fig. 4. Matrix mimic board and CRT displays. *(Courtesy of Ferranti-Packard Inc. and Cleveland Electric Illuminating Co.)*

real time, this means calculation of the cost penalty factors for line losses, machine efficiencies, and fuel costs, thereby minimizing cost of power. Modern systems also include predictive programs that produce probabilistic forecasts of hourly system load for several days into the future, using historical data on weather-load correlations. This permits scheduling startup and shutdown of available generating units to optimize operating economy and to establish necessary maintenance schedules.

System security and analysis programs permit the system operators to simulate problems on the system as it actually exists currently, thereby preparing them for appropriate actions should such emergencies occur. The most advanced centers today use techniques such as state estimation, which processes system data to calculate the probabilities of emergencies in the near-term future, thus permitting the operators to take preventive action before the event occurs.

Interface between the operators and computers are through data loggers, cathode-ray tubes (CRTs), plotters, or active mimic boards (Fig. 4). Dynamic mimic boards, in which the displays are driven by the computer output, have generally displaced the older static representation systems which used supervisory lights to display conditions. The CRTs display the system in up to seven colors, and the operator can interact with them by using a cursor or light pen either to analog the effect of changes to the system or actually to operate the equipment.

machines previously used, and redundant systems are required to ensure security.

Control can be be broken into two phases: operations and security. In the operational area, a typical center provides a completely updated analysis of power flows in the system at 10-min intervals. In

Substations. Power delivered by transmission circuits must be stepped down in facilities called substations to voltages more suitable for use in industrial and residential areas. On transmission systems, these facilities are often called bulk-power substations; at or near factories or mines, they are termed industrial substations; and where they supply residential and commercial areas, distribution substations.

Basic equipment in a substation includes circuit breakers, switches, transformers, lightning arresters and other protective devices, instrumentation, control devices, and other apparatus related to specific functions in the power system.

Distribution. That part of the electric power system that takes power from a bulk-power substation to customers' switches, commonly about 35% of the total plant investment, is called distribution. This category includes distribution substations, subtransmission circuits that feed them, primary circuits that extend from distribution substations to every street and alley, distribution transformers, secondary lines, house service drops or loops, metering equipment, street and highway lighting, and a wide variety of associated devices.

Primary distribution circuits usually operate at 4160 to 34,500 V line to line, and supply large commercial institutional and some industrial customers directly. The lines may be overhead open wire on poles, spacer or aerial cable, or underground cable. Legislation in more than a dozen states now requires that all new services to developments of five or more residences be put underground. Most existing lines are overhead, however, and will remain so for the indefinite future.

At conveniently located distribution transformers in residential and commercial areas, the line voltage is stepped down to supply low-voltage secondary lines, from which service drops extend to supply the customers' premises. Most such service is at 120/240 V, but other common voltages are 120/208 V, 125/216, and, for larger commercial and industrial buildings, 240/480, 265/460, or 277/480 V. These are classified as utilization voltages.

Electric utility industry. In the United States, which has the third highest per-capita use of electricity in the world and more electric power capability than any other nation, the electric capability

Table 2. Electric utility industry statistics, 1971–1978*

Category	1971	1972	1973	1974	1975	1976	1977	1978
Generating capacity installed at year's end, 10^3 kW†	367,396	399,606	438,492	474,574	504,393	531,162	557,174	574,366
Electric energy output, 10^6 kWh	1,617,500	1,754,900	1,878,500	1,884,000	1,919,912	2,037,674	2,124,026	2,232,774
Energy sales, 10^6 kWh								
Total	1,466,440	1,577,714	1,703,203	1,700,769	1,733,024	1,849,625	1,950,791	2,020,610
Residential	179,080	511,423	554,171	554,960	591,108	613,072	652,345	680,874
Small light and power	333,752	361,859	396,903	392,716	421,088	440,625	469,227	481,561
Large light and power	592,699	639,467	687,235	689,435	656,440	725,169	757,168	783,682
Other	60,909	64,965	64,894	63,558	67,270	70,758	72,051	74,492
Customers at year's end, $\times 10^3$	74,265	76,150	78,461	80,102	81,845	83,613	85,590	87,629
Revenue, $\$10^6$	24,725	27,921	31,663	39,127	46,853	53,463	62,610	69,865
Average residential use, kWh/year	7,380	7,691	8,079	7,907	8,176	8,360	8,693	8,873
Average residential rate, cents/kWh	2.19	2.29	2.38	2.83	3.21	3.45	3.78	4.02
Coal (and equivalent) burned, lb/kWh‡	0.918	0.911	0.918	0.946	0.952	0.949	0.968	0.984
Capital expenditures, $\$10^6$	15,130	16,651	18,723	20,556	20,155	25,189	27,711	30,250

*From *Elec. World*, p. 61, Sept. 15, 1978, and p. 51, Mar. 15, 1979; and Edison Electric Institute.
†Fuel and hydro. ‡1 lb = 454 g.

systems as measured by some criteria are the largest industry (Table 2). Total plant investment, as of Dec. 31, 1978, was about $270,000,000,000. The electric utility industry spends approximately $30,000,000,000 a year for new plants to supply the growing load, and collects nearly $70,000,000,000 per year from more than 87,000,000 customers. This industry comprised about 3115 public and investor-owned systems producing electricity in about 3100 generating plants with a combined operating capability of 574,365 MW at the close of 1978.

The 3115 systems included 301 investor-owned companies operating 78.3% of the generation and serving 77.5% of the ultimate customers. The remaining 22.5% were served by 1769 municipal systems, about 928 rural electric cooperatives, 58 public power districts, 7 irrigation districts, 40 United States government systems, 9 state-owned authorities, 1 county authority, and 2 mutual systems.

The industry's annual output reached 2,232,-774,000,000 kWh in 1978, and its sales to ultimate customers totaled 2,020,610,000,000 kWh at an average of 3.46 cents/kWh. Of this, about 39% was consumed by industrial and other large power customers, the remainder going mostly to residential (34%) and commercial customers. Residential usage has climbed steadily for many years, reaching a record 8873 kWh/average customer in 1978.

[WILLIAM C. HAYES]

Bibliography: A. S. Brookes et al., *Proceedings of International Conference on Large High Voltage Electric Systems*, Paris, Aug. 21–29, 1974; J. F. Dopazo, State estimator screens incoming data, *Elec. World*, 185(4):56–57, Feb. 15, 1976; *EHV Transmission Line Reference Book*, Edison Electric Institute, 1968; The electric century, 1874–1974, *Elec. World*, 181(11):43–431, June 1, 1974; Electric Research Council, *Electric Transmission Structures*, Edison Electric Institute, 1968; *Electrical World Directory of Electric Utilities, 1977–1978*, 1977; *Electrostatic and Electromagnetic Effects of Ultra-High-Voltage Transmission Lines*, Electric Power Research Institute, 1978; L. W. Eury, Look one step ahead to avoid crises, *Elec. World*, 187(10):50–51, May 15, 1977; G. F. Friedlander, 15th Steam Station Design Survey, *Elec. World*, 190(10):73–88, Nov. 15, 1978; G. F. Friedlander, 20th Steam Station Cost Survey, *Elec. World*, 188(10):43–58, Nov. 15, 1977; 1979 Annual Statistical Report, *Elec. World*, 191(6):51–82, Mar. 15, 1979; W. P. Rades, Convert to digital control system, *Elec. World*, 186(1):50–51, July 1, 1976; N. D. Reppon et al., *Proceedings of the Department of Energy's System Engineering for Power: States and Prospects Conference*, Henniker, NH, Aug. 17–22, 1975; W. D. Stevenson, Jr., *Elements of Power System Analysis*, 1975; R. L. Sullivan, *Power System Planning*, 1977; *Transmission Line Reference Book, HVDC to ±600 kV*, Electric Power Research Institute & Bonneville Power Administration, 1977; 29th Annual Electric Utility Industry Forecast, *Elec. World*, 190(6):62–76, Sept. 15, 1978.

Electric power systems engineering

Electricity plays a key role in future energy strategy because of its versatility with respect to input energy form. Electricity can be produced with coal, nuclear, hydro, geothermal, fission, fusion, biomass, wind, or solar energy as well as oil or gas. Electrical supply also offers the opportunity of total environmental enhancement compared to other energy use patterns. *See the feature article* ENERGY CONSUMPTION.

In meeting the need of an increasing electrical penetration, the electrical utility industry faces critical driving forces of increased capital costs, financial and environmental restraints, increasing fuel costs, and regulatory delays. The totality of these driving forces leads to the need for more comprehensive understanding and analysis of electric utility systems. Developments in systems analysis and synthesis as well as in related digital, analog, and hybrid computer techniques provide important tools aiding the systems engineer in meeting these more exacting challenges.

This article discusses tools that aid long-range electric utility conceptual planning leading to design studies and equipment specification. Also discussed are concepts in systems engineering applied to improving system operation. The specific computer programs used to illustrate the discussion were developed by the General Electric Company; however, the general concepts and techniques covered here are used throughout the electrical utility industry. *See* ELECTRICAL UTILITY INDUSTRY.

SYSTEM PLANNING

The tools to be discussed are directed toward the bulk power system consisting of generating sources, loads which are supplied from substations, interconnection points with neighbors, and a transmission system which interconnects these elements.

In the area of planning, considerations facing the system planner include: the amount and type of generating capacity to be added; the optimum size of generating units; the generation types or combinations thereof to be used (nuclear, gas turbine, conventional steam, pumped hydro, solar, wind, and so forth); the environmental impact of various generation alternatives; the location of new generation; the size of the interconnections with neighboring systems; the transmission lines to be constructed and their most economic voltage levels; the impact of major facility additions upon the financial structure of the utility; and the effect of utility requirements on targets of performance for new technologies.

Simulation programs. The various system simulation programs used in long-range system planning are shown in Fig. 1. The load shape modeling program tabulates the daily peak loads throughout the year for use in the generation expansion program. The load shape program also determines daily load shapes for weekdays and weekends throughout the year for use in the production costing program.

The generation expansion program, based upon generation and load data, undertakes a probabilistic simulation to determine the size and installation dates for future units and interconnections to achieve adequate reliability levels throughout the expansion.

Generation production costs are computed by the production costing program. This program simulates the operation of future systems to pro-

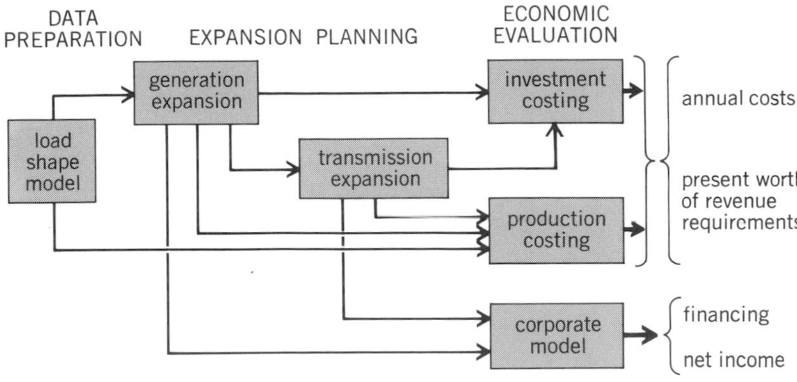

Fig. 1. System planning programs.

vide a knowledge of fuel costs and start-up and maintenance costs. Generation investment costs are treated in the investment cost program, which determines annual costs of investments. Transmission expansion programs have been developed to synthesize the transmission system required to serve future loads and connect future generation into the network. Transmission investment costs are treated in the investment cost program. The programs described determine the annual costs and the present worth of future revenue requirements, typically for a 20-year future period. However, engineering economic analyses of this type are frequently insufficient to fully judge alternative plans. The translation of various technical options into measures meaningful to chief executives is accomplished by a corporate model, which predicts the effects of technological decisions on the utility's financial statements.

Generation expansion program. The generation expansion program determines the timing and sizing of generation additions and interconnections from an analysis of system reserves through the use of probability mathematics to determine the ability of generation to meet daily peak loads. Capacity and load models measure the service reliability expressed as a loss of load probability, which states the probability of insufficient capacity to meet the load. This modeling may be accomplished on either a single-area or multiarea basis.

In the single-area program, generation is assumed to be located on a single bus. Subsequent programs plan an integrated transmission system to meet this assumption. Principal input data are the unit size, forced outage rate, maintenance requirements, peak load characteristics, and load-forecasting uncertainity.

The output of this single-area program includes the system maintenance schedule, the system loss of load probability, the load-carrying capability of the system for a given risk level, and the effective capability of the added units. This program may be used to automatically add units to meet a given index of reliability. Multiarea modeling is used to examine in detail the interconnection requirements between areas. *See* ELECTRIC POWER GENERATION.

Production costing program. The production costing program simulates the minimum-cost daily economic operation of a power system to determine the expense incurred for fuel, start-up, and operation and maintenance labor. Capacity factors are also determined. As the initial step in this program, a load model is prepared, with the peaks for each month of the year represented as a fraction of the annual peak for each of the 12 months. Next, annual maintenance schedules are developed. Units are taken out for maintenance to levelize and minimize risk throughout the year. The next step involves modeling the daily load cycles. Each month is represented in terms of a Sunday, peak weekday, average weekday, and Saturday shape. The procedure next involves unit commitment. In operating the electric utility system, the total generating capacity placed in service exceeds the load by an amount designated as spinning reserve. This reserve is required to safeguard the operation of the system, considering various unexpected forced outages that may occur. Based upon spinning reserve rules, the minimum-cost thermal commitment is determined. The actual thermal commitment will exceed the minimum because of operating constraints such as minimum downtimes and uptimes. Once the unit commitment is given, the units are dispatched according to equal incremental costs in order to achieve minimum overall costs.

Production costing output may be yearly or monthly, and includes information for a given unit and the overall system as well. Unit information includes unit identity, rating, maintenance period, number of starts, number of hours on the line, capacity factor, average operating heat rate, energy produced, start-up costs, operation and maintenance costs, and total costs. System output includes the energy produced, fuel cost, total cost, and energy sold or purchased.

Multiarea production models have also been developed to study interconnection requirements based upon economy flows between areas.

Application to wind and solar generation. In the use of the generation expansion, production cost, and investment costing programs, various alternatives are formulated by the planner and the effect of various input assumptions determined. Application studies have involved the assessment of wind and solar photovoltaic generating sources. Plant performance models were developed that calculated the hourly output of such plants based upon hourly weather data and the plant characteristics. The generation expansion program measured the expected number of hours per year of capacity deficiency. This calculation was performed hourly to recognize the hourly performance of solar and wind plants. Capacity and energy benefits were evaluated in order to determine power plant break-even values.

Figure 2 presents the wind and photovoltaic capacity factors for various locations. The capacity factor is determined by calculating the ratio of the kilowatt-hours generated to the kilowatt-hours that would be generated if the plant operated at full load for all of the hours of the year. The results shown are for the best candidate wind and photovoltaic plant designs. The plant capacity factors vary from 20 to 50% and are very site-specific.

Effective capability for 5% penetration is shown in Fig. 3 for the same sites as Fig. 2. The percent penetration is the percent of total system generat-

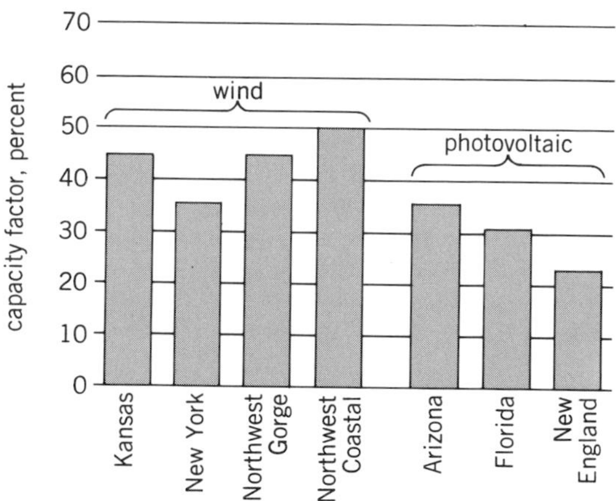

Fig. 2. Wind and photovoltaic capacity factors based on weather data for a typical year.

ing capacity represented by the designated type of generating unit. The percent effective capability represents the amount of capacity displacement that corresponds to a perfect conventional unit. For example, at Northwest Coastal, with 20% effective capability, the addition of 100 MW of wind generation would be equivalent to adding a perfect conventional unit of 20 MW. Of particular importance is the point that capacity credits are present for these intermittent forms of generation. Dedicated storage for photovoltaic and wind is often advocated. Dedicated storage can increase the effective capability, but dedicated storage is not economically attractive.

Figure 4 presents illustrative results showing effective capability in percent of rated capacity plotted as a function of penetration. The effective capability and therefore the capacity credit decrease rapidly with penetration. This function is very site-specific and is dependent upon the load, wind, and solar characteristics.

The final results of such studies are power plant break-even values, such as those shown in Fig. 5. The bottom part of each column is the energy or production cost value; the top part, the capacity value. The energy value results from the savings in fuel costs. The capacity value results from the investment savings realized by the reduction in installed generation of the other types of units in the system. These data show that photovoltaic power will require a technological breakthrough to compete as an energy source. Wind generation could compete in some locations. However, long blade life has not been demonstrated. The studies indicate that photovoltaic and wind power will play a modest role before the year 2000. *See* SOLAR CELL; SOLAR ENERGY; WIND POWER.

Optimized generation planning program. The generation expansion, generation production cost, and generation investment cost programs have been combined into one overall program designated as the optimized generation planning program (OGP; Fig. 6). Based on economics, the OGP automatically seeks an optimum mix of various power generation types for a given reliability, and equipment and fuel costs assumptions. New generation unit data, giving performance and costs, together with data on the existing system, are inputs to the OGP. Output from each run of the program is an optimum expansion pattern listing unit sizes, installation dates, annual cost details, and the total present worth of the expansion. Time and effort expended in finding an optimum is a fraction of that required with earlier, separate step-by-step cut-and-try programs and procedures. Studies can be made quickly and easily, showing cost sensitivity to changes in input parameters such as load shape and growth, unit size, unit types, installed unit costs, fuel costs, reliability index, forced outage rate, and environmental constraints.

The OGP provides environmental outputs on a system-wide basis, including waste heat rejection, emission discharges (sulfur dioxide, nitrogen oxides, carbon monoxide, and particulates), and cooling water consumption. Scrubbers, cooling towers, and precipitators are modeled with allowance for a total system-wide evaluation of these abatement devices. Environmental system dispatching schemes to minimize environmental effects can be modeled and assessed.

The OGP synthesizes a future electric generation system and provides economic results utilizing the present worth of future revenue requirements. To analyze the financial aspects of an electric utility system, a financial simulation program (FSP) has been computer-linked to the OGP, as is illustrated in Fig. 6. The FSP performs a financial analysis of the electric utility and presents the results in terms of balance sheets, income statements, and cash reports. The FSP is a simplified corporate model, developed to produce a realistic utility financial simulation but with less detailed input data requirements than a comprehensive corporate model. When the FSP is used with the OGP, much of the input data is developed by the OGP and transferred to the FSP automatically.

Applications of the combined OGP-FSP include optimum generation mix, parametric sensitivity tests, long-range fuel requirements, load management, evaluation of new technologies, unit slippages, unit size, and the financial impact on nonoptimum additions.

Transmission planning. The next major area of electric utility planning is that of transmission

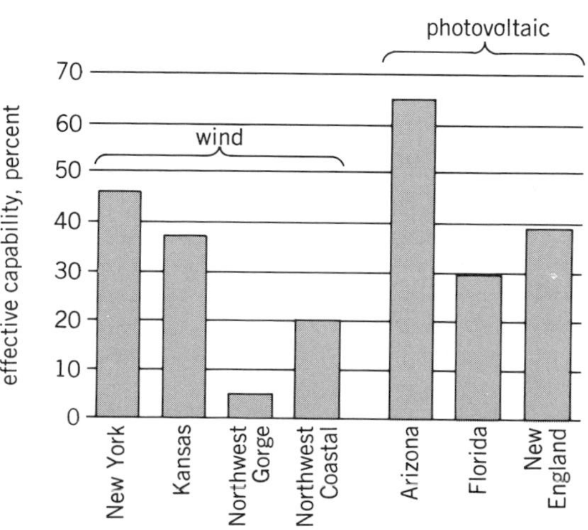

Fig. 3. Wind and photovoltaic effective capability for 5% penetration, based on weather data for a typical year.

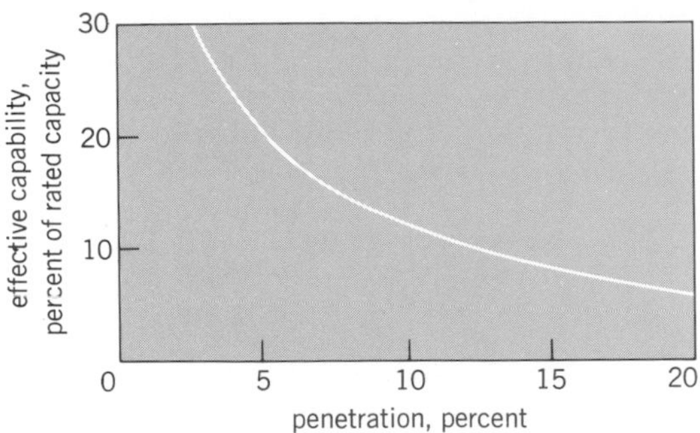

Fig. 4. Effective capability versus penetration for wind and photovoltaic power.

planning. The transmission planning program is a key tool directed toward determining: when new lines will be required; the location of lines in the network; and the voltage level to be used.

The direct-synthesis procedure illustrated in Fig. 7 starts with the initial system, and a projection of loads and generation for the horizon year (say, 20 years hence). A horizon plan is synthesized, with recognition of normal as well as line and unit outage conditions. The program then moves through time, year by year, to specify the annual additions. As appropriate, elements from the horizon-year design are used to meet the yearly requirements. The total plan will require elements in addition to the horizon elements to recognize intermediate-year generation load unbalances.

Two essential blocks make up the synthesis procedure: a flow estimation program and rules for circuit selection. Flow estimation is a linear approximation to a power flow designed to answer the following questions about the transmission network: (1) Does the network have capacity shortages? (2) If so, what is the most economic route to utilize, thereby eliminating the shortages? (3) What are the shortage magnitudes? Circuit additions are made according to programmed rules that recognize both normal and emergency operating conditions.

The results are presented as a computer-drawn map as well as a list of circuits added each year, including the terminals involved, voltage level,

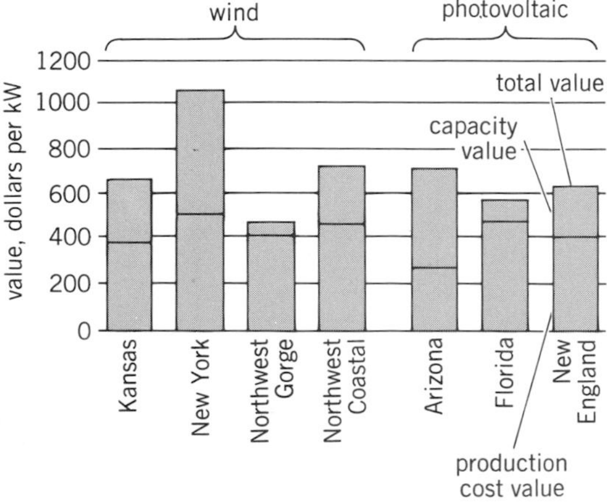

Fig. 5. Wind and photovoltaic power plant breakeven values, based on weather data for a typical year (in 1978 dollars per kilowatt).

length, and reason for the addition. Potential loading information for new rights-of-way is also given.

Illustrative studies include evaluation of the effect of generation siting and transmission voltage levels on transmission system costs. One study utilized the transmission expansion program to determine the transmission feasibility and requirements of large-scale energy parks as compared to conventional dispersed generation. Also, this program has been used to identify transmission requirements of small solar and wind generation plants dispersed on an electric utility system.

Power flow and stability programs. In general, after a transmission plan is selected based upon the transmission expansion program, a thorough check of system performance is made with a large-capacity digital ac power flow program and a companion stability program. The load flow program predicts the steady-state ac behavior of the network. Systems planners through use of power flow programs confirm the adequacy of the proposed transmission network, considering line loadings, bus voltages, and reactive power supply under normal and outage conditions.

Of course, the adequacy of the system must also be examined upon the occurrence of faults or the loss of major blocks of generation. Digital transient stability programs evaluate system performance after the disturbance to ascertain both transient stability and dynamic stability. Stability studies start with the steady-state solution calculated by the power flow program for the conditions existing immediately before the system disturbance.

Power flow and stability programs define line and transformer loadings, generator excitation requirements, series and shunt compensation needs, fault clearing and reclosing time requirements, and the performance of interconnected ac/dc systems. These programs were used to demonstrate the superior stability characteristics of the Square Butte Project, which is the first solid-state high-voltage direct-current (HVDC) transmission system in the United States. Square Butte generates electric power at the lignite coal fields in Center, ND, and delivers power by dc over a distance of 456 mi (734 km) to Duluth, MN. The project is rated ± 250 kV dc, 1000 A, 500 MW. The improvement in stability is shown in Fig. 8. Generator rotor angle is shown as a function of time after initiation of a disturbance. The generator recovers after the disturbance with the combined HVDC and ac system and is less stable with the all-ac system. *See* DIRECT-CURRENT TRANSMISSION.

Research studies affecting stability calculations include the determination of load characteristics, the development of system equivalents, and the development of the capability of modeling multiterminal dc systems. Future HVDC applications, particularly in the western part of the United States, will utilize multiterminal concepts.

System behavior under severe conditions. Following a major disturbance, a power system may experience large variations in frequency and voltage and heavy loadings on the transmission system. These abnormal conditions impose stresses on the system and its components which may result in cascading, thereby leading to the formation of electrical islands, loss of generation, and underfrequency load shedding.

The next planning step involves analyzing the

system behavior under extremely severe conditions where system split-up and possible shutdown are imminent. Research efforts examining a large number of actual occurrences have identified the predominant processes involved in the long-term dynamics of power systems. Quantitative models of these processes have been developed leading to a long-term dynamics computer simulation program. This program has been used to evaluate measures for preventing breakups. Also, simulations of actual disturbances have been used to test and validate the program.

Figure 9 illustrates the elements of the power system represented in the program. The prime mover models include fossil-fired steam, nuclear, hydro, and gas turbine units. The generator and excitation system models determine the voltage and reactive power output and represent the controls in both automatic and manual regulation models. The block that is designated "system connection relations" solves the transmission system power flow with acceleration constraints to determine the system frequency, acceleration, line real and reactive power flows, and bus voltages. Dispatch center activities are modeled, thereby providing signals to change the generation of the controlled prime movers.

This program is a powerful tool for system planners to assess the effect of long-term disturbances.

SYSTEM DESIGN

Once the transmission system configuration has been selected, the major transmission elements require careful study. The evaluation of switching overvoltages is an important initial step.

Transient network analyzer. The transient network analyzer (TNA) is a special analog computer used to predict the transient and harmonic overvoltages which can occur from switching operations. The TNA has been used to design the vast majority of the extra-high-voltage systems in the United States and Canada. It is a three-phase analog of a transmission system, with each element of the actual system represented by a model counterpart. All components are interconnected at a common connection panel to form the desired system.

TNA studies provide guidance on such design considerations as line insulation levels, lightning arrester duties, apparatus insulation levels, reactor ratings, and locations. They also help define improved system operating procedures.

Parallel to improvements to the analog TNA, developmental work has been devoted to extending the digital simulation of such transients. The digital program has been found particularly helpful in studying the behavior of circuit breakers and other equipment where the analysis of rapidly changing voltages and currents is required. Two outstanding advantages of analysis by digital means are the ease in changing equipment parameters in the study and the ability to model low-loss transmission lines and transformers.

Improved transmission-line models extend the range of the TNA up to 1500-kV systems. Totally electronic shunt reactor models represent linear or nonlinear reactors and are continuously adjustable over a wide range of inductances, saturation levels, saturated inductances, and quality factors. Surge arrester models have been developed to electronically represent present silicon carbide arresters with current-limiting gaps as well as arresters using zinc oxide valve elements. The digital data acquisition system provides a fast means of obtaining statistical distributions of switching surge magnitudes. A statistical presentation of switching surge overvoltages is of primary importance in the probabilistic designs of extra-high-voltage transmission lines, and will be essential when evaluating transmission-line designs at the ultra-high-voltage level. *See* ELECTRIC PROTECTIVE DEVICES; LIGHTNING AND SURGE PROTECTION.

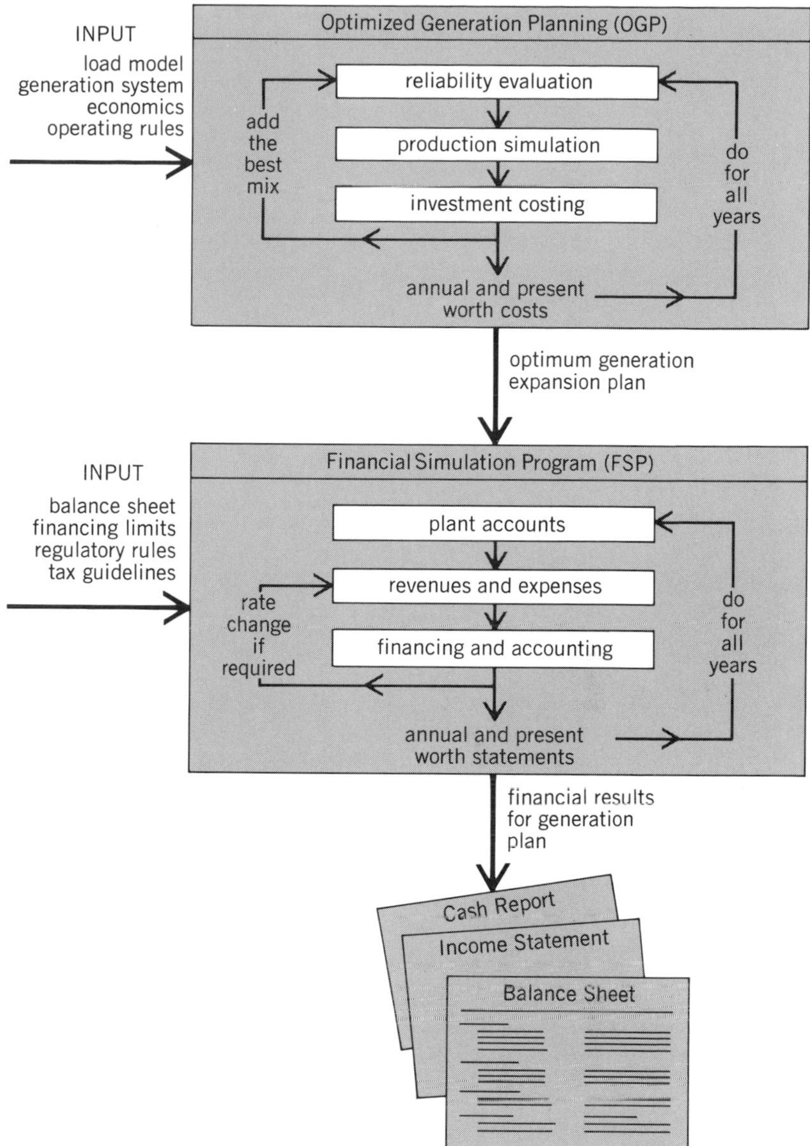

Fig. 6. Overall structure of the linked optimized generation planning program and financial simulation program.

Transmission-line design. With respect to line design, a series of digital programs examine statistical strength/stress criteria of line insulation, including weather influence. Transmission line design programs also examine environmental aspects of line design, including corona phenomena, radio/television noise, audible noise, and ground-level electric fields. These programs enable optimum designs of new lines, or changes needed to uprate

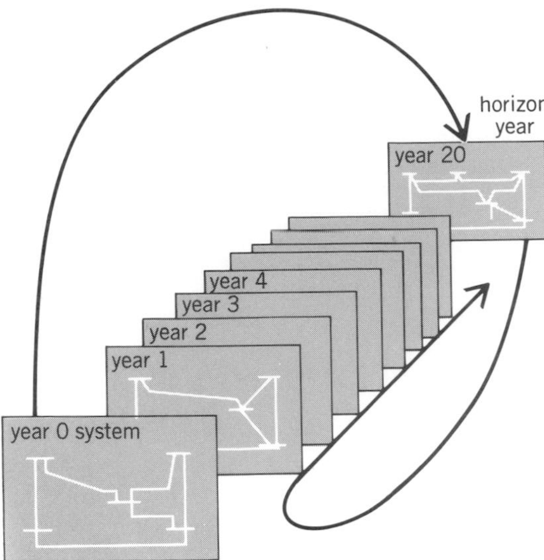

Fig. 7. Transmission expansion program, using direct-synthesis procedure.

existing lines. The analysis of transmission-line performance with such digital programs is considerably enhanced by the use of research results from the Project UHV research facility (Pittsfield, MA). These data cover the area of transient and steady-state insulation, corona performance, and environmental concerns such as ground-level electric fields. Figure 10 illustrates some of the facilities of Project UHV being used to study the behavior of high-voltage dc transmission. *See* TRANSMISSION LINES.

HVDC simulator. The HVDC simulator plays a key role during the design stages of a dc system. It may be interconnected with the TNA. The HVDC simulator is a model HVDC system complete with converter control, dc and ac lines, ac and dc harmonic filters, HVDC circuit breakers, and independent nonsynchronous generation sources. In addition, it has a switching and sequencing system to control disturbances and a measurement system to record system response. This simulator was initially constructed for use in the design stages of the West Coast HVDC Intertie. Since that time, in modified and improved form, it has been extensively used to study a wide range of HVDC system application problems. Improvements have included adding General Electric HVDC system control, additional converter bridges with circuitry facilitating short-pulse firing of silicon controlled rectifiers, low-loss filter and line components, and instrumentation advances. The development of an electronic machine model representing the direct and quadrature axis machine equations and coupled through an amplifier to the model ac/dc system has further improved system representation capability.

The HVDC simulator is used for such studies as insulation coordination, surge arrester design requirements, ac and dc filter performance, diode/thyristor systems, parallel ac/dc transmission systems, HVDC control techniques, recovery of ac/dc systems from faults, multiterminal dc systems, and HVDC breaker application problems.

Hybrid computer. Another useful tool is the hybrid facility. This configuration consists of two general-purpose analog computer consoles and an associated digital computer for monitoring and controlling the analog simulation.

This analytical tool is extensively used for control and dynamical studies, and provides fast turnaround time. The hybrid computer is very cost-effective for the analysis of system dynamics, particularly when high-frequency components are of interest. Thus, the hybrid computer is used extensively for studies of systems involving converters and other thyristor-controlled components in which current and voltage waveshapes are important. Examples include studies of excitation systems, static reactive power systems, and HVDC controls.

Machine and network stability. A program, entitled MANSTAB (machine and network stability), makes it possible to study in one simulation many complex interactions over an extended frequency range which occur between turbine generators and the transmission system. As shown in Fig. 11, the MANSTAB program models balanced transmission networks by differential equations and represents the torsional dynamics of the turbine generator as well. In addition to time domain results, MANSTAB generates root locus and frequency response characteristics. A fundamental aspect of this approach is a technique of linearizing a system about its operating point and constructing the matrix equations of the linearized system directly from the nonlinear time-simulation model. The state matrix equations are then used to obtain eigenvalues and eigenvectors of the system, and frequency response of selected transfer functions. This program has been particularly helpful in the study of subsynchronous resonance.

Subsynchronous resonance. Subsynchronous resonance between the transmission system electrical natural frequency and a shaft torsional natural frequency may occur in systems with series-compensated lines, and has resulted in cracked turbine generator shafts. This phenomenon can be analyzed with the MANSTAB program by representing the turbine generator as a mechanical spring-mass system connected through a transformer to a transmission line which has series compensation. Because of the presence of series compacitors, the electrical system has a natural

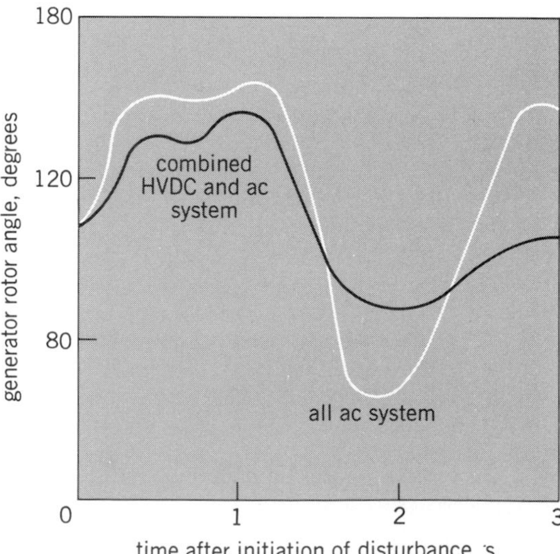

Fig. 8. Stability results for Square Butte Project.

frequency generally in the range of 15–40 Hz. The turbine generator also has torsional natural frequencies in this range. Electrical transients normally decay away in about half a second, but while they are doing so they produce mechanical torques in the shaft. When these torques are on, or close to, one of the natural shaft torsional frequencies, the stresses produced in the shaft can be enormous. Furthermore, the torsional oscillation of the shaft, in response to these torques, sets up voltages in the generator which act to reinforce the electrical disturbances. As a consequence, torsional instability may result.

Analysis of this phenomenon with the MANSTAB program led to the development of a new concept of protection utilizing a series-blocking power filter between the generator and the power transmission system. This subsynchronous filter blocks the electrical frequencies that correspond to the mechanical torsional resonances. The first subsynchronous filter was installed in 1976 at the Navajo Project (Page, AZ), and field tests have confirmed the validity of the digital simulation.

MANTRAP. For problems requiring representation of the network as separate phases, the program MANTRAP (machine and network transient program) may be applied to provide a time simulation. Examples of problems to be studied by MANTRAP include unbalanced faults, multiple faults, reclosing, resynchronizing, generator load rejection, and generator out-of-step operation.

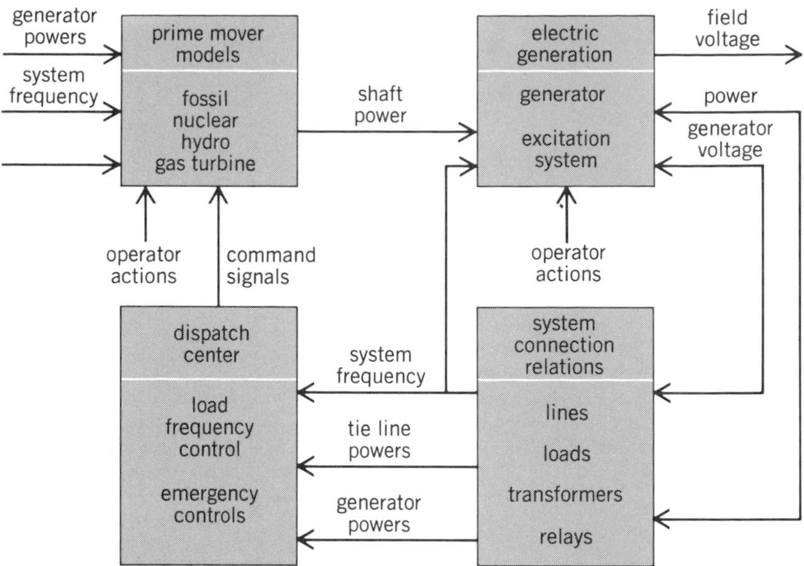

Fig. 9. Elements of long-term dynamics computer simulation program.

SYSTEM OPERATION

Analytical techniques and automation technology are applied to improve the system operation of electric power system facilities. Day-to-day operation and control entails utilizing these installed facilities to meet customer load requirements at maximum reliability and minimum costs within constraints such as equipment limitations, maintenance needs, and interconnection agreements.

Applying the system planning techniques described previously has led to the interconnection of individual power systems into large grids, thereby obtaining economies in capital and operating expenses as well as improved reliability. Fully exploiting these benefits presents ever more complex problems to the power system operator.

It is helpful to think of the system operation problems in terms of operations planning, operations control, and operations accounting and review as shown in Fig. 12. These problems involve economic and security aspects as well as environmental factors. Operations planning involves the problems of looking ahead to the next hour, day, week, or month. Examples are daily load forecasting and unit commitment scheduling. Operations control is concerned with second-by-second real-time actions such as frequency control. Operations accounting and review is concerned with after-the-fact evaluations such as interconnection billing and postdisturbance analysis.

Automation technology. There has been a rapid evolution of automation technology in assisting the operator in solving these problems. The operator's first tool was the telephone involving manual instructions. Then came remote supervisory systems.

Following this, analog dispatch computers were applied which allocated and controlled generation to minimize total system fuel costs and held system frequency and interchange at desired values. These functions are called economic dispatch and load frequency control. With the development of process-control digital computers, digital computers were next used to drive analog circuits to pro-

Fig. 10. Project UHV dc test set built for the Department of Energy.

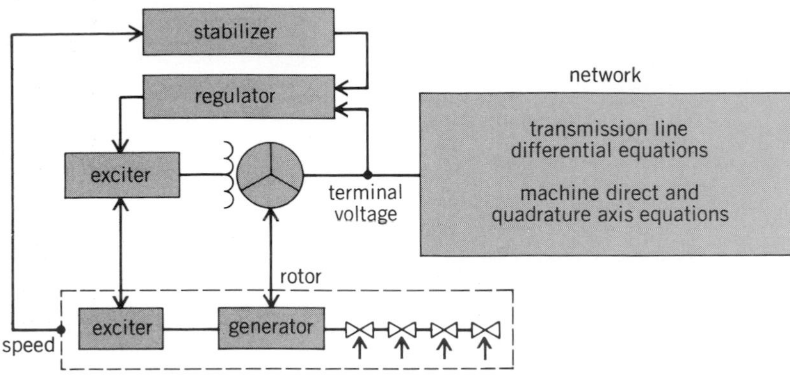

Fig. 11. MANSTAB program.

vide the operations control functions of load frequency control and economic dispatch. This same computer was also used in a shared-time manner to undertake problems of operations planning, and accounting and review as well. As digital process-control computers increased in capability, the analog regulating functions were taken over by the digital computer. Next the digital computer assumed control of the supervisory systems as well.

Early justification of computers in the dispatch center was primarily based on improvements in fuel economy, whereas emphasis now primarily involves security of system operation. However, with rapidly escalating costs of fuel, the benefits of economic dispatching are greatly increasing.

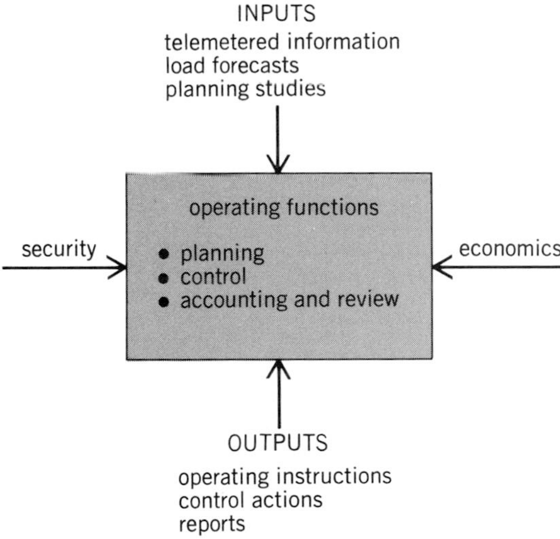

Fig. 12. Operating functions.

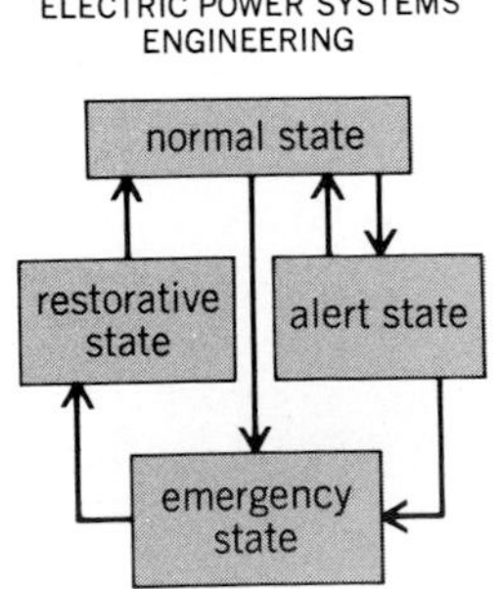

Fig. 13. System states.

System security. In thinking about power system security, it is helpful to consider the system to be in one of the four states shown in Fig. 13. In the normal state, all customer demands are met, no apparatus or lines are overloaded, and there are no impending emergencies. The objective in this state is to continue to meet customer demands, to operate the apparatus and lines within limits, to operate the system at minimum costs within the constraints, and to minimize the effects of possible future contingent events.

The alert state is similar to the normal state in that all customer demands are met and no apparatus or lines are overloaded. However, a potential emergency has been detected by noting that a line loading has reached a limit set below its rating or that an assumed contingency would result in a transmission overload. The objective is to impose constraints and return to the normal state in a minimum time.

There will, however, be situations in which the system will reach the emergency state where customer demands are not met or apparatus is overloaded. In this case, the objective is to prevent the spread of the emergency. In the restorative state, the emergency has been stabilized; yet customer demands may not be served or apparatus may be overloaded. The objective here is to return to normal in minimum time.

The normal state is the only secure state since it implies a low risk of failure. Steps leading to improved system security have the goals of minimizing the probability of leaving the normal state, and minimizing the time required to return to the normal state once the system is in some other state. It is apparent that the first goal requires that steps be taken while the system is in the normal state.

A step-by-step computer application approach has been developed involving: (1) status monitor and display; (2) contingency evaluation; (3) corrective strategy; and (4) automatic control. Significant improvement in system security is obtained through increased monitoring of information and more meaningful display of system information. In the second step, the computer is called upon to predict the effect of contingencies and planned outages and to alert the operator to potential problems in the system. Here the computer is used primarily in a predictive mode rather than in the mode of reacting after a disturbance. As a third step, the computer formulates corrective strategies. At this stage, the operator calls for the execution of the strategy he or she chooses. In the fourth step, the computer uses the communication network to execute automatically the computer-formulated strategies.

Hierarchy of control centers. A hierarchy of action centers as shown in Fig. 14 exists within a typical power company. Overall responsibility for the bulk generation-transmission system resides with a company dispatching center. This center directs the second-by-second control of generation to maintain tie-line schedules and frequency, and is responsible for minimizing the cost of power generation within constraints imposed by security.

Operation of the substation and transmission facilities is accomplished within transmission divisions. Unstaffed substations are monitored and controlled at a division dispatching office by means of supervisory control equipment. Staffed stations may or may not have telemetry equipment installed to enable automatic monitoring by the division.

Generating plants, with the exception of remote hydro or gas-turbine installations, are staffed with operators and maintenance personnel. The operation of large thermal plants involves the local control of many plant variables, and the high degree of automation achieved has significantly contributed to their reliability and economy.

Viewed from the company dispatch center,

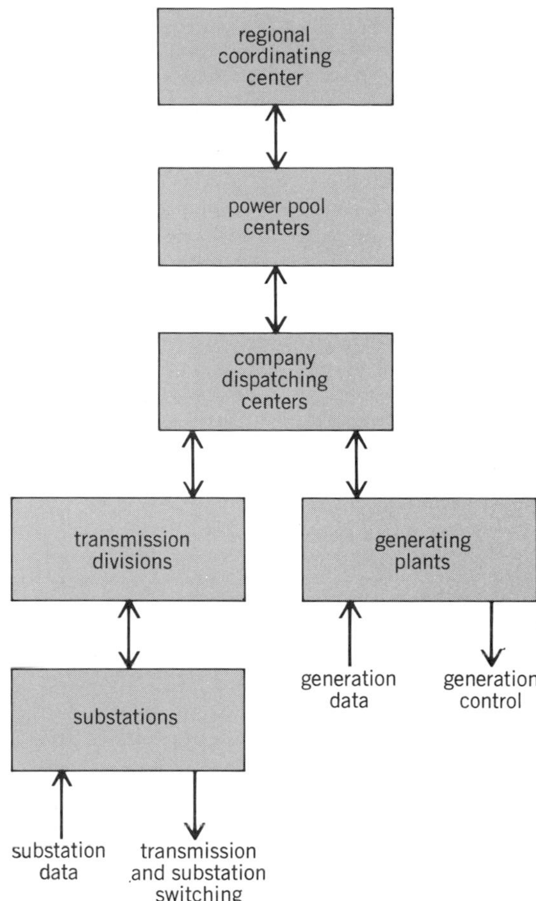

Fig. 14. Hierarchy of action centers.

these plants provide a point at which generation is controlled as well as a source of data on plant status.

As interconnections among utilities become stronger, power pools have been formed in which the facilities of individual member companies are scheduled and operated in a manner which minimizes the total cost of the pool and improves its reliability. The pool is responsible for improving overall economics, and has direct control over total generation for the purpose of maintaining net pool tie flow and frequency on schedule. For this reason, the general responsibilities and facilities of a power pool center are much like those of a company dispatch center, the difference being one of degree.

More recently, the electric utility industry has organized regional coordination centers which have responsibilities over very large geographic areas. At present, these centers are largely concerned with studies of future operating conditions, coordinating emergency procedures, and analyzing disturbances.

The four-step plan previously described is being implemented with data collection, communication devices, and computers organized in a multilevel structure such as the hierarchy of Fig. 14.

Local levels are associated with local control and the collection of primary data. Higher levels receive data from levels below, process it, and act as a data source for levels above. Reliability is enhanced by placing direct control at the lowest feasible level. Such a multilevel computer configuration is designed to supplement automatic control already in place such as normal protective relays, load shedding relays, and excitation controls. Computer technology is becoming of increasing importance in improving system operations ranging from microcomputers at the local primary levels to large-scale scientific computers at the company, pool, and regional levels. *See* ELECTRIC POWER SYSTEMS. [L. K. KIRCHMAYER]

Electric protective devices

A particular type of equipment applied to electric power systems to detect abnormal conditions and to initiate appropriate action to correct the abnormal condition. From time to time, disturbances in the normal operation of electric power systems occur. These may be caused by natural phenomena, such as lightning, wind, or snow; by accidental means traceable to reckless drivers, inadvertent acts by plant maintenance personnel, or other acts of human beings; or by conditions produced in the system itself, such as switching surges, load swings, or equipment failure. Protective devices must therefore be installed on a power system to ensure continuity of electric service, to prevent injury to personnel, and to limit damage to equipment when abnormal situations develop.

Protective devices, like any type of insurance, are applied commensurately with the degree of protection desired. For this reason, application of protective devices varies widely.

This article first introduces the concept of zone protection, followed by a discussion of protective relays as a basic device used in protective systems. Eight common abnormal conditions for which protection is desired are then discussed. For other areas of protection *see* GROUNDING; LIGHTNING AND SURGE PROTECTION.

Zone protection. For the purpose of applying protection, power systems are divided into areas, or zones. Each component of a power system falls into one of five different zone classifications: generator, transformer, bus, transmission line, or motor. Protective devices are applied to each zone to detect abnormal system conditions within that zone and to initiate action for removal of that zone from the rest of the system. Figure 1 illustrates the principle of zone protection on a simple power system. Note that complete protection is afforded by overlapping zones. Removal of only the malfunctioning part of the system ensures maximum electric service continuity.

Protective relays. These are used to sense changes in the voltages and currents on a power system. Sufficiently large variations from normal

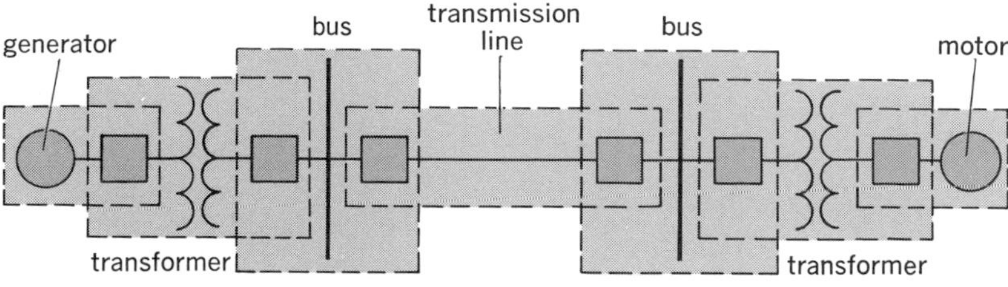

Fig. 1. Zones of protection on simple power system.

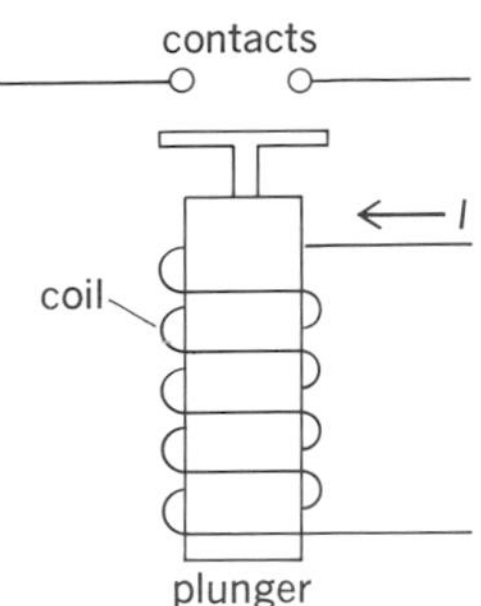

Fig. 2. Plunger relay.

in these quantities cause the relay to operate. Operation of the relay results in opening of circuit breakers to isolate that portion of the power system experiencing an abnormal voltage or current condition. A fault in one part of the system affects all other parts of the system. Therefore relays throughout the power system must be coordinated to ensure the best quality of service to the loads, and to isolate equipment near the fault to prevent excessive damage or personal hazard.

Electromechanical relays. These relays are built to respond to voltage, current, or a combination of voltage and current. Operation of the relay either opens or closes a contact. Two basic principles are used in the construction of electromechanical relays. The simplest type of relay operates on the electromagnetic attraction principle. This relay is composed of a coil, plunger, and set of contacts, as shown in Fig. 2. When current (I) flows in the coil, a force is produced that causes the plunger to move and close the relay contacts. These relays are characterized by their fast operating time.

The electromagnetic induction principle is also used as a basic building block in construction of induction relays. This type of relay responds to alternating current only, whereas the relays discussed above respond to either direct or alternating current. Briefly, an induction relay consists of an electromagnetic circuit, a disk or other form of rotor made of a nonmagnetic current-carrying material, and contacts. A schematic of an induction-type relay is shown in Fig. 3. The main coil is connected to an external source. When current flows in the main coil, transformer action induces current in the secondary circuit connected to the upper poles. Fluxes produced by the currents flowing in the upper pole circuit induce eddy currents in the rotor disk. Interaction between rotor eddy currents and the flux from the lower pole produces torque on the rotor, causing it to move and thus closing the contacts. This will be recognized as the split-phase motor principle, where two out-of-phase fluxes produce torque in a rotor. The upper pole may be supplied from another source to permit comparison of two quantities. A spring relay automatically resets the disk after the relay has operated. By use of the principles of electromagnetic attraction and electromagnetic induction, protective relays can be built to respond to all abnormal conditions that may occur. *See* RELAY.

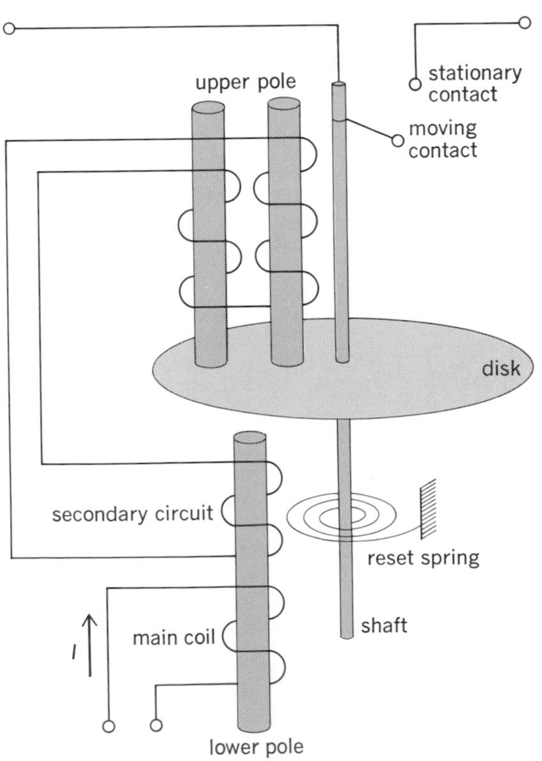

Fig. 3. Induction relay.

Solid-state relays. These relays, also known as static relays, essentially perform the same function as electromechanical relays. The main difference is that solid-state electronic circuits are used to detect abnormal currents and voltages or a combination of these two quantities. Silicon-controlled rectifiers may be used in place of the contacts, reducing the operating time of the relay. The main advantages of solid-state relays are faster operating times, greater flexibility, longer life, lower maintenance, improved accuracy, and a relative immunity to environment.

Overcurrent protection. This must be provided on all systems to prevent abnormally high currents from overheating and causing mechanical stress on equipment. Overcurrent in a power system usually indicates that current is being diverted from its normal path by a short circuit. In low-voltage, distribution-type circuits, such as those found in homes, adequate overcurrent protection can be provided by fuses that melt when current exceeds a predetermined value.

Small thermal-type circuit breakers also provide overcurrent protection for this class of circuit. As the size of circuits and systems increases, the problems associated with interruption of large fault currents dictate the use of power circuit breakers. Normally these breakers are not equipped with elements to sense fault conditions, and therefore overcurrent relays are applied to measure the current continuously. When the current has reached a predetermined value, the relay contacts close. This actuates the trip circuit of a particular breaker, causing it to open and thus isolating the fault. *See* CIRCUIT BREAKER.

Either induction or plunger relays can be used to detect overcurrent conditions. As the current in either type of relay increases, the resultant force also increases. When sufficient force is available, the relay contacts close. Relays have a well-defined time-current characteristic; that is, a longer time is required to close the contacts on a relay measuring a slight overcurrent. A shorter time is required to close the contacts on a relay measuring a heavy overcurrent.

Overvoltage protection. This is usually applied on generators that are subject to overspeed when the load is lost. A voltage which is higher than normal places a severe stress on insulation. If the insulation should fail, a current path to ground would result and above-normal current would then produce additional damage to the equipment. Overvoltage relays are installed to detect this condition at locations where overvoltage conditions would be harmful. Either induction or plunger relays can be set to trip appropriate circuit breakers at a predetermined value of voltage.

Undervoltage protection. This must be provided on circuits supplying power to motor loads. Low-voltage conditions cause motors to draw excessive currents, which can damage the motors. If a low-voltage condition develops while the motor is running, the relay senses this condition and removes the motor from service.

Undervoltage relays can also be used effectively prior to starting large induction or synchronous motors. These types of motors will not reach their rated speeds if started under a low-voltage condition. Relays can be applied to measure terminal voltage, and if it is below a predetermined value, the relay prevents starting of the motor.

Underfrequency protection. This may be necessary in industrial plants where the load is supplied by a combination of local generators and a tie to an outside power company. If the power-company tie is disconnected, local generators become overloaded and the frequency drops. Underfrequency relays detect this condition and act to disconnect part of the load, thereby preventing damage to the generators. Underfrequency protection is also used to disconnect certain selected loads automatically or to sectionalize a transmission system when system frequency drops below a predetermined value. Induction-type relays are used for underfrequency protection.

Reverse-current protection. This is provided when a change in the normal direction of current indicates an abnormal condition in the system. Plunger relays applied to dc circuits sense a change in direction of current by a reversal of the direction of force on the plunger if the plunger is polarized. In an ac circuit, reverse current implies a phase shift of the current of nearly 180° from normal. This is actually a change in direction of power flow and can be directed by ac directional relays.

A common application of reverse-current protection is shown in Fig. 4. In this example, a utility supplies power to an industrial plant having some

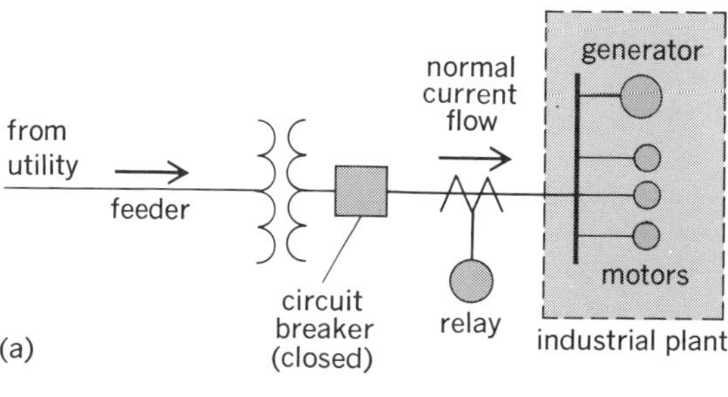

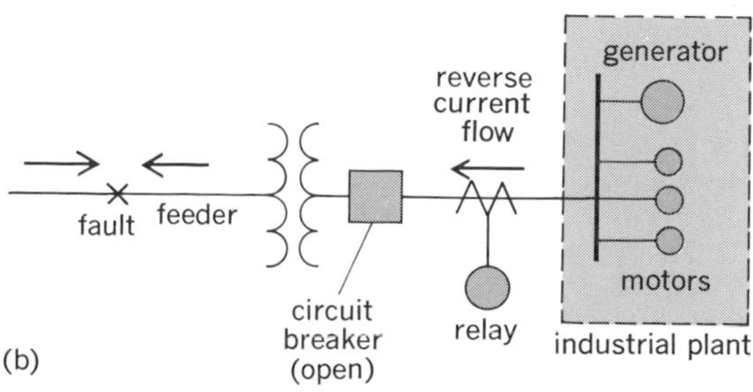

Fig. 4. Reverse-current protection. (*a*) Normal load conditions. (*b*) Internal fault condition; relay trips circuit breaker under reverse-current condition.

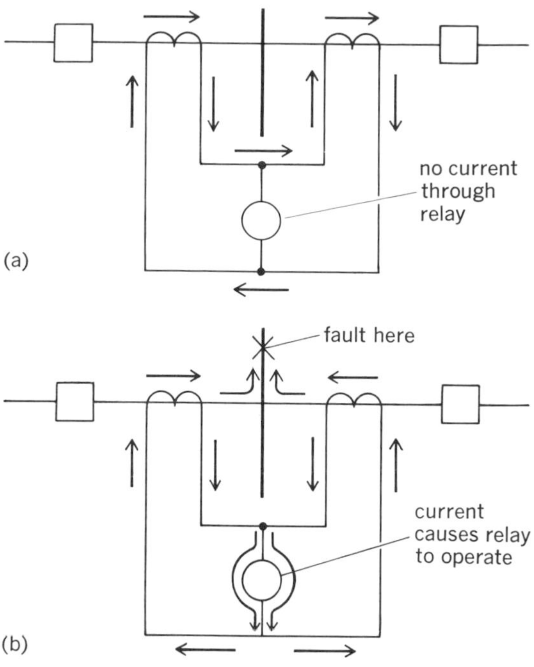

Fig. 5. Simple differential scheme for unbalanced current protection. (*a*) Normal load or external fault condition. (*b*) Bus fault condition.

generation of its own. Under normal conditions, current flows from the utility to the plant (Fig. 4*a*). In the event of a fault occurring on the utility feeder (Fig. 4*b*) the current reverses direction and flows from the plant to the fault location. The relay operates and trips the circuit breaker, isolating the plant from the utility, thus preventing an excessive burden on the plant generator.

Unbalanced-current protection. This type of protection employs the differential relay scheme used extensively in protecting generators, transformers, buses, and large motors. The principle involved is illustrated in Fig. 5. For normal load conditions, currents flow as shown in Fig. 5*a*. Since the current flowing into the bus is equal to the current flowing out of the bus, no current flows through the relay. When a fault occurs on the bus, current flowing into the bus no longer equals the current out of the bus. The difference of these two currents flows through the relay, causing it to operate as indicated in Fig. 5*b*. If the current enters or leaves the bus on more than one line, current transformers must be arranged to compare the sum of the currents entering with the sum of the currents leaving the bus, including the loads. This arrangement is highly sensitive to low-current, internal faults.

Reverse-phase-rotation protection. Where direction of rotation is important, electric motors must be protected against phase reversal. A reverse-phase-rotation relay is applied to sense the phase rotation. This relay is a miniature three-phase motor with the same desired direction of rotation as the motor it is protecting. If the direction of rotation is correct, the relay will let the motor start. If incorrect, the sensing relay will prevent the motor starter from operating.

Thermal protection. Motors and generators are particularly subject to overheating due to overloading and mechanical friction. Excessive temperatures lead to deterioration of insulation and increased losses within the machine. Temperature-sensitive elements, located inside the machine, form part of a bridge circuit used to supply current to a relay. When a predetermined temperature is reached, the relay operates, initiating opening of a circuit breaker or sounding of an alarm.

[JAMES M. CLAYTON]

Bibliography: C. R. Mason, *The Art and Science of Protective Relaying*, 1956; H. Ungrad, *Electronics in Protective Systems*, vol. 51, 1963; Westinghouse Electric Corp., *Applied Protective Relaying*, 1968; Westinghouse Electric Corp., *Electrical Transmission and Distribution Reference Book*, 4th ed., 1950.

Electric rotating machinery

Any form of apparatus, having a rotating member, which generates, converts, transforms, or modifies electric power. The most common forms are motors, generators, synchronous condensers, synchronous converters, rotating amplifiers, phase modifiers, and combinations of these in one machine. The capacity, or rating, is usually indicated on a nameplate and denotes the maximum continuous duty which can be sustained without overheating or other injury. Motors are rated in horsepower. They are built in sizes from a small fraction of a horsepower to more than 230,000 hp. Generators are rated in kilowatts or kilovolt-amperes (kva). The maximum output of alternating-current generators being built exceeds 1,250,000 kVA. Other types of electric rotating machines fall within these limits.

Construction. Most rotating machines consist of a stationary member, called the stator, and a rotating member, called the rotor. The rotor may be supported in bearings at both ends, or it may be supported at one or both ends by the shaft of another machine. *See* GENERATOR; MOTOR.

The illustration shows a typical rotating machine having a bracket bearing at one end and an arrangement for coupling to a turbine shaft at the other. Although small machines sometimes employ antifriction bearings, larger units are built with sleeve bearings generally lined with babbitt. Vertical shaft machines use thrust bearings to support the rotating member. Lubrication in slow- or medium-speed units is often supplied from an oil reservoir contained within the bearing housing. Where bearing losses are high, water-cooling coils may be immersed in the oil to prevent overheating. High-speed machines are often lubricated from a pressurized oiling system, which also supplies the shaft seals in hydrogen-cooled units.

To function properly, rotating machines must have a magnetic circuit, usually involving both rotor and stator, and one or more insulated electrical circuits which interlink the magnetic circuit. To afford a low-reluctance magnetic path, the rotor and stator are separated only by a small clearance, called the air gap.

The windings are insulated electrically with materials such as enamel, cotton, varnished cambric, mica, asbestos, dacron, and glass fabric. The most common impregnants are shellac, asphaltum base varnish, and epoxy, polyester, or phenolic resins. External partially conducting varnish is sometimes applied to high-voltage coils for corona shielding.

Electrical-mechanical energy conversion. The force F in newtons produced on a conductor located at right angles to a magnetic field is $F = BIL$ newtons, where B is the flux density in webers per square meter in the vicinity of the conductor, I is the conductor current in amperes, and L is the length of the conductor, in meters, exposed to the flux. In a motor the magnetic field created by one member exerts a force on the current-carrying conductors of the other, producing a mechanical torque which drives the load. In a generator the changing magnetic field induces voltage in the armature windings when the rotor is driven by a source of mechanical power. Little power is required at no-load, but as the load current builds up, the prime mover must supply the torque to overcome the forces in the equation between the field and conductors.

Ventilation. Rotating machinery must be ventilated to avoid overheating from internal losses. The principal cooling medium, usually air or hydrogen,

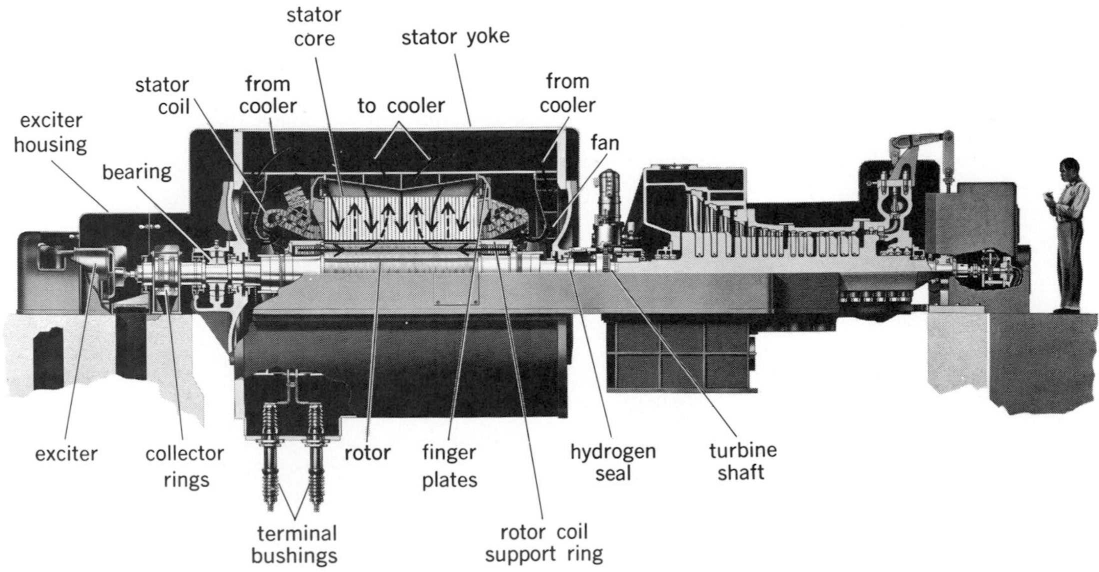

Cross section of a typical electric rotating machine. (*Allis-Chalmers*)

is circulated by fans or blowers mounted on the rotor or separately driven. The illustration shows axial-flow fans at each end of the rotor, with arrows indicating the path taken by the gas. With conventional cooling, the cooling medium is blown over exposed surfaces of the insulated windings and core. In conductor cooling, the cooling medium flows in ducts situated within the major insulation wall.

In large machines the superior effectiveness of conductor cooling is essential. In addition to hydrogen at pressures up to several atmospheres for cooling the rotor and the stator core iron, hydrogen at far greater pressure or a liquid such as oil or water is circulated through the stator conductors in the largest ratings. Generators having liquid-cooled rotor conductors are also being built. *See* ALTERNATING-CURRENT GENERATOR.

Losses. In all rotating machines, losses occur. Among them are I^2R losses, called copper losses, in the windings, connections, and brushes; stray load losses in windings, solid metal structures, and frame; core loss in the magnetic material and structural parts; windage and friction loss; and exciter and rheostat losses.

I^2R losses (in watts) in each path of the windings are equal to the square of the effective current in amperes times the resistance in ohms. Brush I^2R loss is the product of the potential drop in volts times the current in amperes. Stray load losses are caused mainly by eddy currents, which are due to variable magnetic fields (produced by the load current) within the conductors, pole surface, structural members, end shields, frame, and so forth.

Windage and friction losses are the result of circulation and turbulence of the cooling medium and friction of bearings, seals, and brushes. Windage loss is relatively large in air-cooled high-speed machines. In hydrogen the loss is only 7–15% of that in air within the operating range of purity. Bearing and seal friction losses are generally absorbed by the lubricating oil. To avoid excessive friction or overheating of bearings, an inlet oil temperature of 100–120°F (38–49°C) is often recommended for large machines, with discharge at a temperature of about 150°F (66°C).

[LEON T. ROSENBERG]

Bibliography: D. G. Fink and H. W. Beaty (eds.), *Standard Handbook for Electrical Engineers*, 11th ed., 1978; E. Levi and M. Panzer, *Electromechanical Power Conversion*, 1974; J. Rosenblatt and M. H. Friedman, *Direct and Alternating Current Machinery*, 1963; S. Seely, *Electromechanical Energy Conversion*, 2d ed., 1973.

Electrical conduction

The passage of electric charge. This can occur by a variety of processes.

In metals the electric current is carried by free electrons. These are not bound to any particular atom and can wander throughout the metal. In general, the conductivity of metals is higher than that of other materials. At very low temperatures certain metals become superconductors, possessing infinite conductivity. The free electrons are able to move through the crystal lattice without any resistance whatsoever.

In semiconductors (germanium, silicon, and so on) there are a limited number of free electrons and also "holes," which act as positive charges, available to carry current. The conductivity of semiconductors is much smaller than that of metals and, in contrast to most metals, increases with rising temperature.

Aqueous solutions of ionic crystals readily conduct electricity by means of the positive and negative ions present, for example, Na^+ and Cl^- in an ordinary solution of sodium chloride. Solid ionic crystals are themselves fair conductors. These crystals have sufficient lattice vacancies so that a few of the ions are able to migrate through the crystal under the influence of an applied electric field.

A strong electric field ionizes gas molecules, and thereby permits a flow of current through the gas in which the ions are the charge carriers. If sufficient ions are formed, there may be a spark.

Electric current can flow across a vacuum, for example, in a vacuum tube. The charge carriers are electrons emitted by the filament. The effective conductivity is low because of low available current densities at the normal temperatures of electron-emitting filaments.

[JOHN W. STEWART]

Electrical measurements

The measurement of any one of the many quantities by which the behavior of electricity is characterized. The knowledge of the quantitative behavior of electricity is essential to scientific and technical progress. Electrical measurements play a major role in industry, communications, and even in such unrelated fields as medicine.

Many electrical measurements can be made with direct-indicating instruments merely by connecting the instrument properly in the circuit. Thus a volt-meter provides a pointer which moves over a scale calibrated in volts, and an ammeter in the same way presents a reading of current in amperes. Other direct-reading instruments are wattmeters, frequency meters, power-factor or phase-angle meters, and ohmmeters. Many electrical quantities are measured both as instantaneous values and as values integrated over time. Some electrical measurements must be made with various specialized devices or systems requiring adjustment or balancing to obtain the measured value. Typical of these are potentiometers and bridges which are found in many standard and specialized forms.

Because of differences in instruments and techniques, it is convenient to divide measurements into direct-current (dc) and alternating-current (ac) classes.

DC measurements. In dc circuits the measurement of voltage and current often suffices to define the operation of the circuit. The product of the two represents power. In the commercial sale of dc electricity the measurement of energy must be made with a dc watt-hour meter. Occasional use is made of a dc ampere-hour meter in battery-charging installations.

To measure high values of current, shunts are used to bypass all but a small fraction of the current around the measuring instrument. A newer technique employs a form of saturable reactor energized by alternating current to measure large direct currents. *See* ELECTRIC ENERGY MEASUREMENT; ELECTRIC POWER MEASUREMENT.

AC measurements. Alternating-current circuits involve more variables and hence more measurements than dc circuits. The most common measurements are voltage, current, and power; the last requires a wattmeter, as ac power cannot always be calculated directly from voltage and current. Also measured are frequency and power factor (or phase angle) and sometimes waveform or harmonic content. Energy is measured by means of the ac induction watt-hour meter. In general, ac instruments differ in principle and design from dc instruments, although many ac instruments may be used to measure dc quantities. Direct-current instruments do not respond to ac quantities, but some may be adapted by the addition of rectifiers to convert alternating current to direct. The thermocouple is another form of convertor by which a dc instrument may be made to read ac quantities.

If alternating voltages and currents above the normal ranges of self-contained instruments are to be measured, instrument transformers may be used to extend the ranges of those instruments. In the study of ac waveform a qualitative evaluation may be made with an oscillograph or a cathode-ray oscilloscope. Quantitative measurement of harmonic content requires the use of a harmonic analyzer.

Accuracy of measurements. Accuracy denotes the degree of compliance of the instrument reading with the true value of the measured quantity. It is common to describe the instrument's accuracy by stating the maximum allowable error. Thus an instrument with a maximum error of 2% is often described as having an accuracy of 2%. For many applications a small panel or miniature electric instrument with a maximum error of 2–5% of full-scale calibration will suffice. More refined instruments are available with maximum errors of 1, 0.5, or 0.25%. When measurements of higher accuracy than this are required, measurement systems, such as potentiometers and bridges, must be used. Direct current and voltage can readily be measured in this way to an accuracy of 0.01%. Alternating-current measurements can be readily made to an accuracy of 0.1% or better.

Laboratory measurements. In a laboratory, emphasis is normally placed on accuracy and on completeness of facilities to deal with all types of measurements. There is relatively little limitation on the size and complexity of equipment used. If standardizing service is a function of the laboratory, the equipment must include standard cells and precision standard resistors and also suitable potentiometers, bridges, shunts, and volt boxes (voltage-dividing resistors to extend the range of voltage measurements). With these, the calibration of dc instruments can be performed with high accuracy. Extension of calibration service to ac instruments requires transfer standards (instruments having negligible difference in performance when operated on alternating current or direct current).

Field measurements. This term is used to designate all measurements made outside a laboratory as in generating stations and substations, service shops, factory testing areas, ships, and aircraft. For these uses equipment is chosen to perform only specialized services. Accuracy well below that of laboratory measurements is usually permissible. Convenience, compactness, and often portability are prime considerations in choosing equipment. Electric instruments for this kind of service are often of the panel type, sometimes in miniature sizes. Multipurpose and multirange instruments, like the volt-ohm-milliammeter, are handy for service measurements where 2–5% error is permissible.

For field measurements of alternating current and voltage at power frequencies, the hook-on voltammeter provides readings within 2% maximum error. As a voltmeter it is connected to the circuit with spring clips; as an ammeter it operates on the current-transformer principle. The core, which is circular and extends outside the instrument case, has a hinged link that may be opened to slip around a conductor carrying current and then closed again. Thus there is no need to break the circuit under measurement; the conductor itself becomes a one-turn primary winding in the transformer measurement system. The measurement of power can also be made with a hook-on wattmeter.

On power systems, field measurements may be desired continuously over a period of time, sometimes at unattended locations. For this purpose recording instruments may be used or the reading may be telemetered to a manned station. Any instrument that is made in indicating form can be made in recording form, but the greater power necessary to drive a marking device over a chart may call for some kind of amplification.

The operation of a power system also calls for the recording of disturbances due to lightning strokes, insulation flashover, short circuits, and other transient phenomena. Recording oscillographs used for this purpose can be triggered by any condition that deviates from normal operation, thus making a record of the disturbance.

Frequency considerations. The measurement of voltage and current is commonly made over a frequency range from a few hertz (Hz) up through 2000 megahertz (MHz). The frequency at which measurements are to be made dictates the type of equipment needed, the precautions to be taken, and the degree of accuracy which may reasonably be expected. Alternating-current instruments of the moving-iron, fixed-coil type are intended primarily for 60 Hz applications but may be used with only moderate errors up to several hundred hertz. Electrodynamic (moving-coil, fixed-coil) instruments, which are generally of greater accuracy, may also be used in this frequency range. Errors in such instruments result mainly from reactance effects, which may be minimized by special design to permit operation to several kilohertz (kHz). Rectifier-type instruments possess only small frequency errors up to several kilohertz in relation to their overall accuracy rating, which is of the order of 2–5% error. Electronic voltmeters, which are of the same general accuracy, are especially suited for use over a wide range of frequencies.

Circuit loading. All electric instruments draw some power from the circuits to which they are connected, and ac instruments generally take more power than dc instruments. This circuit loading may alter appreciably the quantity being measured. For instance, an ac voltmeter rated 150 volts and having a resistance of 3000 ohms is perfectly suitable to measure voltage on a 120-volt house lighting circuit. However, if the same voltmeter is connected to the terminals of a small power amplifier with a maximum output of 200 milliamperes (mA), the voltmeter will load the source and seriously reduce its voltage. To avoid this error the measurement should be made with a rectifier voltmeter (about 150,000 ohms resistance) or an electronic voltmeter (above 0.5 megohm resistance).

In the measurement of current, consideration must be given to the voltage drop in the ammeter. If it is appreciable in relation to the source voltage, the current is not the same as that if the ammeter were not connected in the circuit. However, the magnitude of this error can usually be evaluated, and minimized by proper choice of instruments.

[ISAAC F. KINNARD/EDWARD C. STEVENSON]

Time dependence. When voltage and current are variable functions of time, the measurement of frequency, wavelength, and waveform is of importance, in addition to phase-angle measurements. At frequencies where the physical dimensions of the electric circuit are small compared to the wavelength of the voltage and current, the frequency is said to be low and various forms of frequency meters are employed, depending upon the range involved.

At higher frequencies it often becomes necessary to measure frequency and wavelength independently since their product is not always a constant under this condition.

The waveform of the electrical quantity being measured is of importance. Many indicating instruments are calibrated to give correct readings only for sine-wave inputs. If the waveform is nonsinusoidal, it is necessary to consider the principles of operation of the particular instrument to interpret the meter indication correctly. For example, an electronic voltmeter, which measures peak values but is calibrated in root-mean-square (rms) values based on sinusoidal wave shape, would not give correct rms values for a nonsinusoidal wave.

Measurement of parameters. The parameters of any electric circuit are the resistances, inductances, and capacitances along and between the conducting branches of the circuit, including any ground plane that may be near or surrounding the circuit. The measurement of these parameters may be classified according to the apparent disposition of the parameters, which is a function of frequency. For any given circuit there is some frequency below which the circuit can be treated as having lumped parameters or circuit elements; above this frequency the parameters must be considered as being distributed throughout the circuit.

Lumped parameters. The measurement of lumped parameters may be subdivided according to the measurement accuracy desired. If errors of several percent are permissible, direct-reading instruments, which indicate the value of the parameter directly on a calibrated meter scale, are available. Inductance and capacitance measurements are made at some convenient frequency. Of this class of instruments the ohmmeter, used for measuring resistance at zero frequency (direct current), is the only one in common use.

For greater accuracies bridge measurements are preferred. Direct-current measurements are made with the Wheatstone bridge with maximum errors on the order of 0.01%. Resistances of less than a few ohms can be satisfactorily measured only with a bridge, regardless of the degree of accuracy desired.

Most circuit designers prefer to measure inductance and capacitance in the particular operating frequency range under consideration, and the ac bridge is commonly used. Bridge measurements provide numerous advantages over other methods, including high accuracy and the ability to compare the unknown to a known standard. Bridges are designed to operate in various frequency ranges from direct current to several hundred kilohertz, from several kilohertz to several megahertz, from 1 or 2 to several hundred megahertz, and from several hundred to several thousand megahertz. At the higher frequencies the application of the bridge becomes more complicated, and considerable caution and planning are necessary if reliable results are to be obtained.

A unique instrument known as a Q meter is available for measuring inductance or capacitance and effective resistance at radio frequencies.

[ROBERT L. RAMEY]

Distributed parameters. Electrical systems can be completely described by their associated electric and magnetic fields, the properties of the materials involved, the physical dimensions, and the velocity of light. When the dimensions are small compared with the wavelength, however, it is more convenient to treat them as circuits composed of lumped parameters.

At low frequencies lumped inductance and capacitance can be used, although they are rigorously derived only for nonvarying currents and voltages, respectively. At high frequencies the finite propagation velocity of electromagnetic waves cannot be neglected, and the derivations break down. If only one system dimension is comparable with the wavelength of the electrical disturbance, restricted conditions permit a rigorous definition of distributed inductance and capacitance. These distributed parameters combine with the resistance of a pair of conductors and the conductance between them to define the behavior of a transmission line for plane-wave propagation and to relate the voltage between conductors at any point on the line to the voltage at any other point. *See* TRANSMISSION LINES.

The concept of distributed parameters is also useful at low frequencies when it must be recognized that a circuit component, nominally representable by a single parameter, is actually modified by the presence of residual parameters. Thus a coil has not only inductance, but capacitance and resistance as well. This capacitance is definable by low-frequency analysis, since the dimensions are small, but it cannot be localized and represented as a unique lumped parameter because the winding is not an equipotential surface.

For example, a coil mounted over a ground plane has one terminal grounded. The voltage between winding and ground increases from zero at the grounded terminal to maximum at the other. Capacitance near the grounded terminal is therefore less effective than capacitance near the other. The resultant effective terminal capacitance is, in consequence, only one-third of the total capacitance for uniformly distributed capacitance. For other conditions of grounding, the ratio of effective capacitance to total distributed capacitance will again be different.

Values of distributed parameters are inferred from the behavior of the system that they define. For transmission lines, measurements may involve observing the voltage distribution along the line under different terminal conditions. For circuit elements, impedance may be measured at different frequencies or under different conditions of adjustment of some known lumped parameter.

[DONALD B. SINCLAIR]

Bibliography: R. F. Field and D. B. Sinclair, A method for determining the residual inductance and resistance of a variable air condenser at radio frequencies, *Proc. IRE*, 24(2), 1936; F. K. Harris, *Electrical Measurements*, 1975; P. Kantrowitz et al., *Electronic Measurements*, 1979; T. Laverghetta, *Microwave Measurements and Techniques*, 1979; F. A. Laws, *Electrical Measurements*, 1938; G. R. Partridge, *Principles of Electronic Instruments*, 1958.

Electrical resistance

That property of an electrically conductive material that causes a portion of the energy of an electric current flowing in a circuit to be converted into heat. In 1774 A. Henley showed that current flowing in a wire produced heat, but it was not until 1840 that J. P. Joule determined that the rate of conversion of electrical energy into heat in a conductor, that is, the rate of power dissipation, could be expressed by relation (1).

$$H/t \propto I^2R \qquad (1)$$

The day-to-day determination of resistance by measuring the rate of heat dissipation is not practical. However, this rate of energy conversion is also VI, where V is the voltage drop across the element in question and I the current through the element, as in Eq. (2), from which the more conventional

$$H/t \propto I^2R = VI \qquad (2)$$

relationship that is implied by Ohm's law, Eq. (3),

$$R = V/I \qquad (3)$$

is apparent. [CHARLES E. APPLEGATE]

Electrical units and standards

The standard in terms of which electrical quantities are evaluated, the quantities so adopted being known as units. The ohm, for example, is a unit of electrical resistance. The electrical units in practical use today, and also in extensive theoretical use, were designated by the Eleventh General Conference of Weights and Measures in 1960 as members of the International System of Units (Système International d'Unités, abbreviated SI in all languages). This action by the General Conference was the culmination of an effort initiated by A. Giorgi at the beginning of this century to bring the practical electrical units into a coherent system with appropriate mechanical units of the metric system.

ELECTRICAL UNITS

To accomplish the above objective, the base units for mechanical quantities were arbitrarily selected: the meter for the unit of length, the kilogram for the unit of mass, and the second for the unit of time. Units for other mechanical quantities are derived from these units in accordance with physical laws and concepts such as the unit of speed, the meter per second, and the unit for acceleration, the meter per second per second.

This system was originally called the mks system to distinguish it from the cgs system (based on the centimeter, gram, and second).

Meter-kilogram-second system. Acting under authority given it by the Eighth General Conference of Weights and Measures, the International Committee of Weights and Measures in 1937 proceeded to define a unit for force (now called the newton, N) and units for energy and power in mechanical terms. The theoretical magnitudes of these units are given below.

Unit of force. The force which gives to a mass of 1 kilogram an acceleration of 1 meter per second per second.

Joule (J). The work done when the point of application of the mks unit of force is displaced a distance of 1 meter in the direction of the force.

Watt (W). The power which gives rise to the production of energy at the rate of 1 joule per second.

The Committee then proceeded to define electric and magnetic units in terms of these mechanical units. The revised units were to replace the definitions which had been in effect for many years such as the "mercury ohm" and the "silver ampere." The revised definitions of electrical and magnetic units which have been accepted since 1948 were given by the Committee as follows.

Ampere (A). The constant current which, if maintained in two straight parallel conductors of infinite length, of negligible circular sections, and placed 1 meter apart in a vacuum, would produce between these conductors a force equal to 2×10^{-7} mks unit of force per meter of length.

Volt (V). The difference of electric potential between two points of a conducting wire carrying a constant current of 1 ampere, when the power dissipated between these points is equal to 1 watt.

Ohm (Ω). The electric resistance between two points of a conductor when a constant difference of potential of 1 volt, applied between these two points, produces in the conductor a current of 1 ampere, the conductor not being the seat of any electromotive force.

Coulomb (C). The quantity of electricity transported in 1 second by a current of 1 ampere.

Farad (F). The capacitance of a capacitor between the plates of which there appears a difference of potential of 1 volt when it is charged by a quantity of electricity equal to 1 coulomb.

Henry (H). The inductance of a closed circuit in which an electromotive force of 1 volt is produced when the electric current in the circuit varies uniformly at a rate of 1 ampere per second.

Weber (Wb). The magnetic flux which, linking a circuit of 1 turn, produces in it an electromotive force of 1 volt as it is reduced to zero at a uniform rate in 1 second.

The revised definitions were intended solely to fix the magnitudes of the units and not the methods to be followed for their practical realization. This realization is effected in accord with the well-known laws of electromagnetism. For example, the definition of the ampere represents only a particular case of the general formula expressing the forces which are developed between conductors carrying electric currents, chosen for the simplicity of its verbal expression. It serves to fix the constants in the general formula which has to be used for the realization of the unit.

A special name was added to the list by the Eleventh General Conference of Weights and Measures in 1960, the tesla (T) for the unit of magnetic flux density (one weber per square meter).

Centimeter-gram-second systems. Two systems of electric and magnetic units have been in use in scientific circles for a long time but both are rapidly giving way to the International System. They are the electrostatic system of units (esu) and the electromagnetic system of units (emu).

The electrostatic system defines a unit charge as that charge which exerts 1 cgs unit of force (1 dyne, which is equivalent to 10^{-5} newton) on another unit charge when separated from it by a distance of 1 centimeter in a vacuum. All other units of the system are derived from this definition by assigning unit coefficients in equations relating electric and magnetic quantities to each other. The units so derived are often referred to in terms of the SI units with the prefix "stat," for example, statvolts, statohms, and statamperes.

The electromagnetic system defines a unit magnetic pole (a highly fictitious concept) as that pole which exerts 1 cgs unit of force on another unit pole when separated from it by a distance of 1 centimeter in a vacuum. All other units of the system are derived from this definition in accord with the principles set forth above for the electrostatic system. Units so derived are often referred to in terms of the SI units with the prefix "ab," for example, abvolts, abohms, and abamperes. Special names are given to some magnetic units of the emu system such as the maxwell and the gauss, which correspond, respectively, to the weber and the tesla of SI, although differing from them in magnitude.

The magnitudes of corresponding units of the electrostatic and electromagnetic systems differ from each other by a factor theoretically equal to the speed of light, c, (3×10^{10} centimeters per second, approximately) or its square. Thus, 1 abampere is equal to 3×10^{10} statamperes; 1 statvolt is equal to 3×10^{10} abvolts; and hence 1 statohm is equal to 9×10^{20} abohms.

The esu system is found convenient for handling purely electrostatic problems. A combination of the two systems in which electrostatic quantities are expressed in esu and magnetic and electromagnetic quantities in emu, with appropriate use of the conversion constant c between the two systems, is called the Gaussian system. All of these systems are rapidly giving way to use of SI units in treatment of electric, magnetic, and electromagnetic phenomena.

ELECTRICAL STANDARDS

Electrical standards are the physical embodiments by means of which the electrical units are realized and maintained. The ampere is unique in this system, since an arbitrary constant, other than unity is employed in its definition to bring the entire system into agreement with the mechanical units while still adhering substantially to the old value for the unit.

Not all standards for electrical quantities are maintained in the national standards laboratories, such as the National Bureau of Standards; the only standards maintained are those for the volt, the ohm, the farad, and sometimes the henry, since very stable standards for these quantities can be produced. The other electrical quantities are determined from suitable combinations of these standards.

Determination of the ampere. Since the ampere is defined in terms of the force between two current-carrying wires, the conventional means of determining the ampere is by some kind of current balance in which the force between the current-carrying elements is compared with the force of gravity on a known mass. One form of current balance consisting of two fixed coils and a movable coil supported by one arm of an equal arm balance is shown in Fig. 1. An electric current, supplied by a battery and controlled by a rheostat, is sent through the fixed and movable coils. The current flows through the fixed coils in opposite directions so that the magnetic fields produced by them are in opposition in the region of the movable coil. Here the magnetic field is directed horizontally and radially with respect to the axis of the coil system. The direction of the current in the movable coil is controlled by a reversing switch. When the current flowing through the movable coil is in one direction, the force exerted on it by the currents in the other coils is downward; but when the current is reversed, the force is upward. The force of gravity on the mass of the movable coil is balanced by a tare weight on the other arm of the balance.

The currents and the balancing weights are adjusted so that when the force due to the current is upward on the coil, the tare weight just balances the coil; but when the current is reversed, the additional weight must be added to achieve balance.

When balance is achieved, the current through

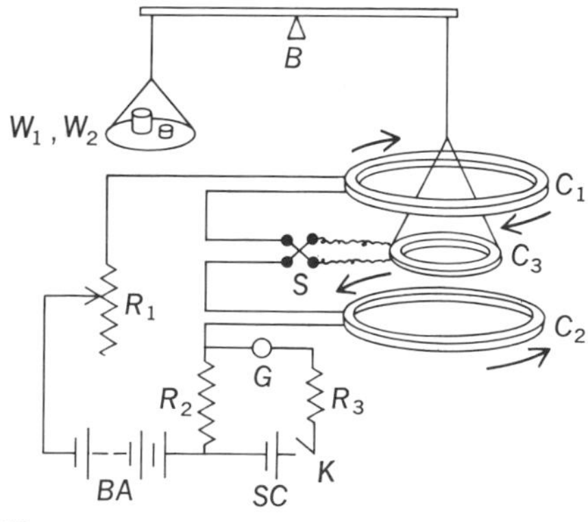

Key:

B = weight beam
W_1 = tare weight
W_2 = balancing weight
C_1, C_2 = fixed coils
C_3 = moving coil
R_1 = adjusting resistor
R_2 = standard resistor
R = protective resistor
BA = battery
S = reversing switch
SC = standard cell
G = galvanometer
K = key

Fig. 1. Current balance, with schematic.

the coil is given by Eq. (1), in which C is the con-

$$2\,Ci^2 = mg \tag{1}$$

stant computed from the dimensions of the coil assembly, i is the current in amperes, m is the mass of the small weight, and g is the acceleration of gravity at the place where the experiment is performed.

Current circulating in the coil system is passed through a standard resistor having a resistance of 1 ohm, approximately, but its value must be accurately known.

A standard cell, that is, an electrochemical cell which produces a constant emf, with a protective resistor and a galvanometer and key in series with it, is placed across the known resistor. If no deflection of the galvanometer is observed when the key is closed in this circuit, then the emf of the standard cell is given by Eq. (2), where r is the re-

$$ir = v \tag{2}$$

sistance of the known resistor. Thus the experiment for the determination of the ampere is actually an experiment for determining the emf of a standard cell.

Several such standard cells are calibrated by means of the current balance, and they in turn preserve the unit of voltage when the current balance is not in use. A current of 1 ampere may thus be established at anytime by sending a current of such strength through a 1-ohm resistor that it gives rise to exactly 1-volt drop across it.

In modern versions of the current balance, both the fixed coils and the movable coils are wound in a single layer on cylinders of marble, pyrex, or fused silica, in which accurate grooves have been lapped to maintain the wires in fixed positions so that the dimensions of the coils can be accurately measured.

Another form of the current balance, the Pellat balance, is called an electrodynamometer (Fig. 2). It consists of a long solenoid within which another cylindrical coil is balanced on knife edges so that it is free to rotate. The axis of the inner coil is at right angles to the axis of the solenoid. When the electric current is sent through both coils, a torque is exerted on the movable coil which is balanced by a weight suspended at the end of an arm extending out horizontally from the rotatable coil. When balance is attained, the current through the coil system is given by Eq. (3), where C is the computed

$$Ci^2 = mgl \tag{3}$$

constant of the coil system, and l the length of the lever arm on which the weight is supported. When the current in the rotatable coil is reversed, the torque is reversed and the balancing weight is moved to an arm on the opposite side of the coil. The electrical circuit used with the Pellat balance is essentially identical with that used with the other type. Experiments at the National Bureau of Standards with the two forms of current balance gave agreement to within about 1 part in a million.

Similar experiments in the national standardizing laboratories of other countries have served to establish the value for the unit of voltage for those countries, but there were known differences in values assigned to the standards in various countries, as demonstrated by international comparisons. Increased accuracy with which such experiments have been conducted since World War II has permitted assignment of values to the electrical standards in much closer conformity with the theoretically defined values of the units. By international agreement, decision was made that electrical standards in use throughout the world should be referred to a uniform basis after Jan. 1, 1969, bringing them into closer agreement with the units they embody. The new volt at the National Bureau of Standards is 8.4 parts in a million smaller than the old volt. Thus, a standard cell which was assigned a value of, say, 1.0183000 volts before the changeover would now be assigned a value of 1.0183086 volts.

Determination of the ohm. Several methods have been employed for the determination of the ohm. The impedance of ohms of an electric circuit containing only resistive elements for direct or alternating current is given by Eq. (4), where r is the

$$z = r \tag{4}$$

resistance in ohms. But for a circuit containing only inductive or capacitive elements the impedance for alternating current is, respectively, given by Eqs. (5) and (6), where L is the inductance in

$$z = \omega L \tag{5}$$
$$z = 1/\omega C \tag{6}$$

henries, C is the capacitance in farads, and ω is the angular frequency of the alternating current in radians per second.

The inductance of an arrangement of current-carrying elements or the capacitance of an arrangement of charge-bearing elements can be cal-

Fig. 2. Electrodynamometer, the current balance used in absolute determination of the ampere at the National Bureau of Standards.

culated readily from electromagnetic principles and the geometry of the arrangement in units of inductance (henries) or in units of capacitance (farads). The impedance in ohms of either of these arrangements may be calculated for given alternating current frequencies from Eqs. (5) or (6). If one of these calculated inductors or capacitors is placed in one arm of an alternating-current bridge, its impedance can be compared with that of a resistor in another arm of the bridge, thus establishing the value of the resistor in ohms.

For many years the values of resistors were obtained from the values of computable inductors in most cases. The discovery of the Thompson-Lampard theorem in electrostatics in 1956 led to a great improvement in the art. In comprehending this theorem, a new form of capacitor should be visualized, consisting of a long metal tube divided into four segments by longitudinal cuts coplanar with the axis of the tube. If C_1 is the capacitance per unit length between one opposite pair of segments when the other pair is grounded, and C_2 is the capacitance per unit length between the other pair of opposite segments when the first pair is grounded, then Eq. (7) holds, where ϵ_0 is the electric constant.

$$e^{-\Pi C_1/\epsilon_0} + e^{-\Pi C_2/\epsilon_0} = 1 \qquad (7)$$

If C_1 and C_2 are nearly equal, the tube length is reduced as shown in Eq. (8), where $\Delta C = C_1 - C_2$.

$$\overline{C} = \frac{C_1 + C_2}{2} = \epsilon_0 \frac{\ln 2}{\Pi}\left(1 + 0.087\frac{\Delta C^2}{\overline{C}^2}\right) \qquad (8)$$

The tube need not be of cylindrical form. It is possible to replicate this arrangment with carefully machined parts such as cylindrical gage blocks insulated from each other, each block corresponding to one segment of the tube. With this arrangement C_1 and C_2 can be made nearly equal and ΔC becomes vanishingly small.

Since the length of cylindrical gage blocks may be measured with very great accuracy, the capacitance of this type of capacitor can be calculated from the length measurement with correspondingly great accuracy. The greatest uncertainty in calculation of the capacitance arises from uncertainty of the knowledge of the speed of light because the constant ϵ_0 is implicitly defined in the International System by Eq. (9), where c is the speed of light

$$c^2 = 1/\epsilon_0\mu_0 \qquad (9)$$

and μ_0 has the arbitrarily assigned numerical value $4\Pi \times 10^{-7}$.

The impedance of a practical-size capacitor of this type is very great, about 10^8 ohms. However, its impedance can be compared with that of a 1-ohm resistor in successive steps so that nearly the full accuracy of the calculation can be realized.

[ALVIN G. MC NISH]

Bibliography: F. L. Hermech and R. F. Dziuba, *Precision Measurement and Calibration: Electrical*, Nat. Bur. Stand. Spec. Publ. no. 300, vol. 3, 1968.

Electrical utility industry

Three events during 1979 portend major changes in the electrical utility industry in the United States. A major accident at the Three Mile Island nuclear plant of Metropolitan Edison Company near Harrisburg, PA, was the first occurrence in a commercial nuclear station to affect the public; President Jimmy Carter requested that Congress make mandatory a 50% reduction by 1990 in the use of oil burned by utilities; and, for a second year, annual peak growth was below 3%, touching only 0.5% for 1979.

Three Mile Island accident. The Three Mile Island accident occurred when a combination of equipment malfunctions and operator misjudgments resulted in a partial meltdown of the unit's nuclear fuel and a subsequent major release of radioactive gases into the atmosphere. Though Federal authorities insist that the exposure of the public to radiation was within permissible limits, this has become a major area of contention. The interior of the containment vessel was so severely contaminated by the presence of 500,000 gal (1.9×10^6 liters) of highly radioactive water that entry might not be possible for a year or more afterward.

The effects on the industry are manifold. All reactors of this design are now subject to shutdown for modifications designed to prevent recurrence of this kind of accident. National legislation imposing a moratorium on nuclear construction in any state that does not have an evacuation plan for nuclear disasters has been passed. And, most importantly, the financial costs attributed directly or indirectly to the accident must heavily influence other utilities' policy on nuclear capacity. Replacement power that the Three Mile Island station owners have to buy from neighboring utilities costs as much as $500,000 a day, and the $1,200,000,000 installation has been removed from the rate base, pushing the utility to the brink of bankruptcy. *See* NUCLEAR POWER.

Mandate for reduced oil use. The proposed mandate for a 50% reduction in use of oil could have an even more profound effect than the nuclear accident, however. Utilities now depend on oil for about 19% of their total electrical generation. This cannot be displaced by nuclear units because such units ordered now could not be in service until after 1992. A lead time of 12–14 years is now normal. The industry has little confidence that enough coal could be mined and shipped to displace oil as fuel; and in any event, environmental constraints would make it impossible to burn that much coal and meet air-quality standards. The policy, therefore, poses a major question whose answer is not readily apparent at this time.

Growth rate. The low growth rate, compared to consensus forecasts of 4.5–5.0% or higher, may presage a new plateau of growth for the future. Many utilities feel that the summer of 1979 did not display the pattern of hot periods that cause high demands. It was, however, a hot summer. Should this 2-year succession of low peak growth represent a new plateau, utility finances would be greatly affected, since construction programs could be substantially lower.

Ownership. Ownership of electric utility facilities in the United States is pluralistic, being shared by private investors, customer-owned cooperatives, and public bodies on city, district, state, and Federal levels. Investor-owned companies constitute by far the major portion of the industry. They serve 67,923,000 customers, representing a 77.5% share of all electric customers, and own 78.3% of the installed generating capacity. Cooperatives serve 9.7% of the total, but only own about 2% of the generating capacity. Public bodies at all levels serve 12.8% of the total electric customers, and own 19.7% of the installed generating capacity. Of this amount, Federal agencies hold 9.5%, and all others 10.2%.

The small amount of generating capacity owned by the cooperatives reflects the fact that most such organizations are distribution companies which buy their power either from investor-owned utilities at wholesale rates, from special generation and transmission cooperatives, or from publicly owned utilities.

There is a growing tendency for cooperatives and, to some extent, municipal utilities to purchase shares in large generating units built by investor-owned utilities. This arrangement permits small utilities to share in the economies of scale of very large units and in the lower costs of nuclear units. For the investor-owned builder, the arrangement eases the financial drain, since cooperatives and public entities have access to lower cost financing and avoid antitrust requirements. Typical is the joint ownership of the Black Fox nuclear plant now under construction in Oklahoma. Public Service Company of Oklahoma owns a 700-MW share, Associated Electric Cooperative owns 250 MW, and Western Farmers Electric Cooperative owns the remaining 200 MW.

United States electric power industry statistics for 1979*

Parameter	Amount	Increase compared with 1978, %
Generating capability, $\times 10^3$ kW		
Conventional hydro	65,207	8.7
Pumped-storage hydro	12,795	12.0
Fossil-fueled steam	408,655	2.9
Nuclear steam	61,053	16.1
Combustion turbine and internal combustion	54,839	1.2
Total	602,549	4.9
Energy production, $\times 10^6$ kWh	2,301,000	4.3
Energy sales, $\times 10^6$ kWh		
Residential	702,500	3.4
Commercial	493,300	2.6
Industrial	813,800	4.0
Miscellaneous	78,400	5.2
Total	2,100,000	3.5
Revenues, total; $\times 10^6$ dollars	77,238	10.6
Capital expenditures, total; $\times 10^6$ dollars	32,866	8.9
Customers, $\times 10^3$		
Residential	79,997	2.9
Total	89,670	2.3
Residential usage, kWh	8,904	0.3
Residential bill, ¢/kWh (average)	4.26	5.9

From 30th annual electrical industry forecast, *Elec. World*, 192(6)69–84, Sept. 15, 1979, and extrapolations from monthly data of the Edison Electric Institute—The Association of Electric Companies.

Capacity additions. Utilities had a total generating capability of 574,365 MW at the end of 1978, having added 23,935 MW during that year (see table). By the end of 1979, industry capability had increased to 602,549 MW. The annual 1979 summer peak demand for the entire United States was 416,400 MW which, with the substantial capacity additions in 1979, gave a new reserve margin of 34.3%. This is a slight increase from the 33.7% recorded in 1978. The normal target is taken by most utilities to be 25%, though the norm can vary regionally from 13 to 28%.

A major concern in 1979 was the timing of the Environmental Protection Agency's imposition of fines against utilities which have coal-fired plants that are not in compliance with the latest air quality standards for point sources of particulates and of sulfur dioxides. These noncompliance fines will be levied against some 70 plants, of an aggregate capacity of about 25 GW, representing about 10% of all coal-fired plants in the United States. The magnitude of the proposed fines would render continued operation of these plants financially unfeasible in many cases. Should utilities elect to shut any major portion of their plants down rather than pay the fines, conditions of severe power inadequacy could result in several regions of the country. The application of these fines was protested by the utility industry, but unsuccessfully, and the expectation was that they would probably be imposed in early 1980.

The 28,184 MW of capacity added during 1979 consisted of 11,708 MW of fossil-fuel capacity, 8486 MW of nuclear units, 5196 MW of conventional hydroelectric, 2155 MW of pumped-storage hydroelectric, 527 MW of combustion turbines, and only 112 MW of diesels.

Added capacity by type of ownership was 14,839 MW by investor-owned utilities, 2644 MW by cooperatives, 7711 MW by Federal agencies, and 2990 MW by public bodies.

The total plant types as of the end of 1979 were 67.9% of fossil-fueled (of this capacity, 59% was coal, 27% was oil-fired, and 14% was gas-fired), 10% nuclear, 10.8% conventional hydroelectric, 2.1% pumped-storage hydroelectric, 8.2% combustion turbines, and 1% internal combustion engines such as diesels (see illustration).

Fossil-fueled capacity. Fossil-fuel units constitute 41.6% of the total new capacity added in 1979. A total of 32 individual units went into service, of which 24 were coal-fired, 3 were oil-fired, and 1 was fueled by gas; 2 small (16 MW each) units that burn waste came into service, and 2 geothermal plants with a total of 190 MW were started up.

The $10,679,336,000 expended in 1979 for fossil-fired construction, though up from the $9,705,817,000 expended in 1978, was an increase of only 1.6%.

Nuclear power. Utilities added 7 more nuclear units in 1979 to bring the total of reactors operating in the United States to 79. The total capacity of the plants brought into service during 1979 was 8486 MW, raising the total operating to 61,043 MW. Of the units added, 2 were boiling-water reactors (BWR) and 5 were pressurized-water reactors (PWR). There are now 46 PWRs and 28 BWRs operating in the United States: the remaining 5 units use other technologies. Nuclear units planned or in construction have a total capacity of 186,998 MW which, if current plans hold, will bring nuclear capacity to about 22% of all installed capacity by 1995. During 1978 nuclear units generated a total of 2.8×10^{11}kWh, accounting for 13% of that year's total.

Combustion turbines. Combustion turbines have served admirably as peak-load units because of their quick-start capabilities, that is, the ability to go from cold to full load in 2–3 min, and their low initial capital cost of less than \$200/kW. Utilities keep an average of about 9% of their peak demand in gas turbine capacity, using them for several hundred hours a year to meet annual peak loads, or to go on line quickly to supply load in emergency situations. However, because of the uncertainty of the future supply of the distillate oil or gas that these machines burn, and the uncertainty of national policies concerning the permissible use of petroleum fuels and the permissible levels of the nitrogen oxides that these machines produce in their exhaust gases, this percentage will undoubtedly decline substantially.

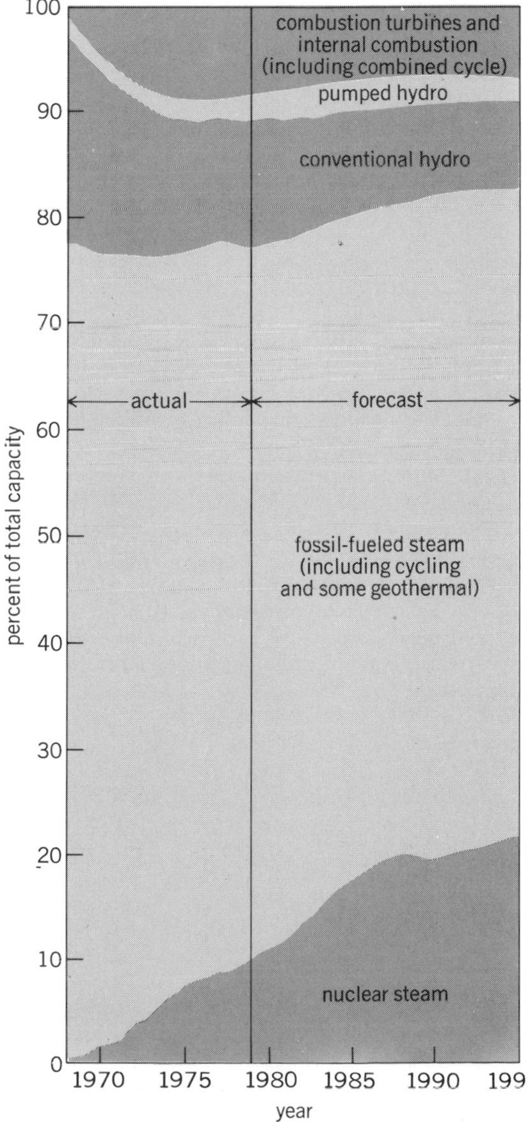

Probable mix of net generating capacity. *(From 30th annual electric industry forecast, Elec. World, 192(6):69–84, Sept. 15, 1979; used with permission of the Edison Electric Institute—The Association of Electric Companies)*

Units installed in 1979 ranged from 20 to 79 MW. Utilities brought 2172 and 527 MW into service in 1978 and 1979, respectively. The 527 MW in 1979 was made up of 12 individual units. The total installed combustion turbine capacity for the entire industry is now 48,683 MW, or 8.5% of total capacity. Of the total, however, 1314 MW represent the combustion turbine portion of combined cycle installations, in which the 900–1000° (482–538°C) exhaust gas of the turbines is used to generate steam for a conventional steam turbine. Utilities spent \$263,000,000 on combustion turbine construction in 1979.

Hydroelectric installations. Installation of conventional hydroelectric capacity continued, with 5196 MW coming on line in 1979, and a total of about 17,207 MW additional capacity planned for the future. Hydroelectric capacity, amounting to 60,011 MW, provided 10.4% of total industry capacity. However, as sites become increasingly more difficult to find and develop, this percentage is expected to decline to 9% by 1985. During 1979 the industry spent \$602,695,000 on hydroelectric projects, of which \$341,330,000 was by Federal agencies and \$218,810,000 by investor-owned utilities. The future of hydroelectric power will depend on economical development of low-head, or run-of-river, turbines, since environmentally acceptable high dam sites are almost nonexistent. During 1979 the smallest unit brought on line was rated 2 MW. The largest, at Grand Coulee Dam in Washington, was rated at 700 MW.

Pumped storage. Pumped storage represents one of the few methods by which electrical energy can be stored. In the mechanical analog, water is pumped to an elevated reservoir during off-peak periods and is released through hydraulic turbines during the subsequent peak period, recovering 65% of the original fuel energy. Although the 12,795 MW installed at the end of 1979 was only 2% of the industry's total capacity, another 12,000 MW are planned for the future, and utilities continue to spend about \$321,000,000 annually on such projects. The world's largest pumped-storage installation is Ludington Station on Lake Michigan, rated at 970 MW.

Rate of growth. Electric demand rose appreciably less than had been anticipated for 1979, inching up only 0.5%. This was the second year of low growth—although 1978 grew somewhat more at 2.3%—and may signal a basic change in growth pattern. It is clear that the price of electricity has had a substantial suppressant effect on peak demand. It is still not entirely clear, however, whether this is a permanent long-range change or reflects the short-term psychology of current sensitivity to the energy situation, emphasized by the gasoline shortage.

The low growth rate of peak demand will have the effect of further shaving expectation for long-range growth, which had been projected in 1978 at 5.0%. The annualized compound growth rate of the last 5 years since the OPEC oil embargo of late 1973, was 3.5%. These 5 years were marked by the embargo itself, the radical upward movement of the price of electricity (the first real period of rise in history), introduction of double-digit inflation, and the longest and deepest recession since the 1930s. Longer-term growth in the future, in the face of such strength during such adversity, should be considerably higher, or about 4.7% over the next decade. The pattern will be one of higher early growth, tapering off as energy-efficient appliances, equipment, and processes phase in and as conservation cuts use in existing installations.

The pattern of peak growth generally follows that of growth in total usage. Though it is not possible to separate out the residential, commercial, and industrial components of peak demand, it appears that in 1979 industrial growth helped sustain what little growth there was, with residential and commerical sectors lagging. This is the opposite pattern from 1978. Industry is operating at close to full capacity because of the lack of capital expansion over the last few years, and may have found little opportunity to reduce peak demand as a result. Residential and commercial load, on the other hand, is considerably more flexible and has been able to effect conservation.

Regionally, only the Pacific Northwest and the Arizona–New Mexico areas attained substantial growth. The Pacific Northwest, rebounding from a low-growth year in 1978, rose 4.8%, while the Arizona–New Mexico area rose 6%. The entire northern tier of industrialized states had negative growth in 1979, however, and the southeastern and south-central regions, normally regions of above-average growth, barely attained the peak levels of 1978. In the latter regions, the weather did not reach forecast basis—that is, there were lower temperatures, and high temperatures were sustained for shorter periods than had been forecast. Rainy weather throughout the Gulf Coast states reduced both air-conditioning and irrigation loads. But the fact remains that a new low peak growth in peacetime was achieved by the total electric utility industry.

Usage. Sales of electricity rose at a higher rate than peak demand, reaching 3.5% for 1979. Total usage for 1979 was 2.1×10^{12} kWh. Energy sales for the first half of the year were pushed higher than in 1978 by cold weather, recovery from the coal strike, and a sharp increase in industrial sales. Total industrial sales for the year were 8.138×10^{11} kWh, compared with 7.821×10^{11} kWh in 1978, for an increase of 4%.

Residential and industrial growth were much stronger than that of the commercial sector. Residential usage rose to 7.025×10^{11} kWh for 1979, and commercial use to 4.933×10^{11} kWh—increases of 3.4% and 2.6%, respectively, over 1978. Electric heating continued to contribute strongly to total industry sales, accounting for 1.38×10^{11} kWh. In the residential sector it accounted for over 20% of sales. Residential revenues also rose from $29,800,000,000 in 1978 to $29,900,000,000 in 1979, a 0.3% rise in constant 1979 dollars. The 1979 sales reflect an average annual use per residential customer of 8904 kWh and an average annual bill of $379.31.

In the industrial sector generation by industrial plants contributed 6.98×10^{10} kWh, down from the 7.23×10^{10} kWh of 1978. This cut industrial generation's share of the total industrial kilowatt-hour use about 0.5%, to 7.9%. The long-term trend of industrial generation as percent of total usage is consistently downward, however, and by 1989 is expected to decline to only 5.2% of the total.

Fuels. The 1973 oil embargo brought tremendous Federal pressures on utilities to convert from petroleum fuels to coal, and the cutoff of Iranian oil greatly increased this pressure. President Carter called for a 50% cut in use of oil as boiler fuel by 1990. As a result, all new planned stations have been designed to fire coal. The use of coal in 1978 rose from 4.770×10^8 tons (4.327×10^8 metric tons) in 1977 to 4.824×10^8 tons (4.376×10^8 metric tons) in 1978, a rise of only 1.1%. Oil also rose from 6.242×10^8 bbl (9.925×10^7 m^3) in 1977 to 6.333×10^8 bbl (10.069×10^7 m^3) in 1978, a rise of 1.5%. Gas used rose from 3.1912×10^{12} ft^3 (9.037×10^{10} m^3) in 1977 to 3.224×10^{12} ft^3 (9.129×10^{10} m^3) in 1978.

Energy generated by each of these major fuels and their percentage share of total generation were as follows for 1978: coal, 9.802×10^{13} kWh (50.7%); oil, 3.621×10^{11} kWh (18.8%); gas, 3.089×10^{11} kWh (16.0%); and nuclear, 2.800×10^{11} kWh (14.5%). Converting these quantities to coal equivalent gives an equivalent total fuel consumption of 9.519×10^8 tons (8.635×10^8 metric tons) of coal for the entire industry in 1978. During that year, it took only 0.984 lb (446 g) of coal to produce 1 kWh.

Transmission. Utilities spent $3,590,000,000 on transmission construction in 1979. This included $754,000,000 for overhead lines below 345 kV, $835,000,000 for overhead lines at 345 kV and above, $44,500,000 for underground construction, and $896,000 for substations. These costs bought 5121 km of overhead lines at 345 kV and above, and 13,216 km at lower voltages. Only 217 mi (349 km) of underground cables were installed, primarily because of the 8:1 ratio of underground to overhead costs. Utilities also installed a total of 81.13 GVA of substation capacity during the year. Maintaining existing lines cost utilities $420,000,000 in 1979.

Distribution. Distribution facilities required the expenditure of $4,833,000,000 in 1979. Of this, $1,258,000,000 was spent to build 36,342 km of three-phase equivalent overhead primary lines ranging from 5 to 69 kV, with the majority at 15 kV. Of all overhead lines constructed in 1979, 3% was rated at 4 kV, 80% at 15 kV, 10% at 25 kV, and 7% at 34.5 kV. Expenditures for underground primary distribution lines amounted to $646,670,000 in 1979. In underground construction of the 14,138 km of three-phase equivalent lines built, 4.6% was rated at 5 kV, 76.3% at 15 kV, 12.2% at 25 kV, and 6.9% at 34.5 kV. Utilities energized 20,738 MVA of substation distribution capacity in 1979 at a total cost of $588,352,000. Maintenance costs for distribution were $1,555,000,000 in 1979.

Capital expenditures. Utilities increased their capital expenditures in 1979 to $34,527,000,000, up 3.9% in constant 1979 dollars from 1978. Of this total, $24,441,000,000 was for generation, $3,635,000,000 for transmission, $4,833,000,000 for distribution, and $1,618,000,000 for miscellaneous uses, such as headquarters buildings, services, and vehicles, which cannot be directly posted to the other categories. Total assets for the investor-owned segment of the industry rose from $171,093,000,000 in 1977 to $221,000,000,000 in 1978, the last year for which figures are available.

See ELECTRIC POWER GENERATION; ELECTRIC POWER SYSTEMS; ENERGY SOURCES; TRANSMISSION LINES.

[WILLIAM C. HAYES]

Bibliography: Edison Electric Institute, *Statistical Yearbook of Electric Utility Industry*, 1979; 1979 annual statistical report, *Elec. World*, 191(6): 51–82, Mar. 15, 1979; 30th annual electric industry forecast, *Elec. World*, 192(6):69–84, Sept. 15, 1979; 20th steam station cost survey, *Elec. World*, 188(10):43–58, Nov. 15, 1977.

Electricity

Electricity comprises those physical phenomena involving electric charges and their effects when at rest and when in motion. Electricity is manifested as a force of attraction, independent of gravitational and short-range nuclear attraction, when two oppositely charged bodies are brought close to one another. It is now known that the elementary (nondivisible) electric charges are possessed by electrons and protons. The charge of the electron is equal in magnitude to that of the proton, but is electrically opposite. The electron's charge is arbitrarily termed negative, and that of the proton, positive. Magnetism, those physical phenomena involving magnetic fields and their effects upon materials, manifests itself in the presence of moving electric charge. For this reason, magnetism was originally considered to be a part of electricity. *See* ELECTRIC CHARGE.

Historical development. The earliest observations of electric effects were made on naturally occurring substances. Magnetism was observed in the attraction of metallic iron by the iron ore magnetite. The natural resin amber was found to become electrified when rubbed (triboelectrification) and to attract lightweight objects. Both of these phenomena were known to Thales of Miletus (640–546 B.C.). Jerome Cardan in 1551 first clearly distinguished the difference between the attractive properties of amber and magnetite, thus presaging the division of electric and magnetic effects. He also envisioned electricity as a type of fluid, a viewpoint that was developed more extensively in the late 18th and early 19th centuries. In 1600 W. Gilbert observed variations in the amounts of electrification of various substances. He divided substances into two classes, according to whether they did or did not electrify by rubbing. The division actually is into poor and good conductors, respectively. A two-fluid theory was first proposed by C. F. duFay in 1733. A one-fluid theory of electricity was propounded in 1747 by Benjamin Franklin, who called an excess of the fluid positive electrification, and a deficiency of fluid negative electrification. This theory fell into disrepute, but the choice of positive and negative remains. Although fluid theories of electricity were superseded at the end of the 19th century, the concept of electricity as a substance persists.

The quantitative development of electricity began late in the 18th century. J. B. Priestley in 1767 and C. A. Coulomb in 1785 discovered independently the inverse-square law for stationary charges. This law serves as a foundation for electrostatics.

In 1800 A. Volta constructed and experimented with the voltaic pile, the predecessor of modern batteries. It provided the first continuous source of electricity. In 1820 H. C. Oersted demonstrated magnetic effects arising from electric currents. The production of induced electric currents by changing magnetic fields was demonstrated by M. Faraday in 1831. In 1851 he also proposed giving physical reality to the concept of lines of force. This was the first step in the direction of shifting the emphasis away from the charges and onto the associated fields. *See* ELECTROMAGNETIC INDUCTION; ELECTROMAGNETISM.

In 1865 J. C. Maxwell presented his mathematical theory of the electromagnetic field. This theory proposed a continuous electric fluid. It remains valid today in the large realm of electromagnetic phenomena where atomic effects can be neglected. Its most radical prediction, the propagation of electromagnetic radiation, was convincingly demonstrated by H. Hertz in 1887. Thus Maxwell's theory not only synthesized a unified theory of electricity and magnetism, but also showed optics to be a branch of electromagnetism.

The developments of theories about electricity subsequent to Maxwell have all been concerned with the microscopic realm. Faraday's experiments on electrolysis in 1833 had indicated a natural unit of electric charge, thus pointing toward a discrete rather than continuous charge. Thus, the groundwork for exceptions to Maxwell's theory of electromagnetism was laid even before the theory was developed. H. A. Lorentz began the attempt to reconcile these viewpoints with his electron theory in 1895. He postulated discrete charges, called electrons. The interactions between the electrons were to be determined by the fields as given by Maxwell's equations. The existence of electrons, negatively charged particles, was demonstrated by J. J. Thomson in 1897 using a Crookes tube. The existence of positively charged particles (protons) was shown shortly afterward (1898) by W. Wien, who observed the deflection of canal rays. Since that time, many particles have been found having charges numerically equal to that of the electron. The question of the fundamental nature of these particles remains unsolved, but the concept of a single elementary charge unit is apparently valid. Of these many particles only two, the electron and the proton, exist in a stable condition on Earth.

A second departure from classical Maxwell theory was brought on by M. Planck's studies of the electromagnetic radiation emitted by "black" bodies. These studies led Planck to postulate that electromagnetic radiation was emitted in discrete amounts, called quanta. This quantum hypothesis ultimately led to the formulation of modern quantum mechanics. The most satisfactory fusion of electromagnetic theory and quantum mechanics was achieved in 1948 with the work of J. Schwinger and R. Feynman in quantum electrodynamics, which suppressed the particle aspect and emphasized the field.

Sources. The sources of electricity in modern technology depend strongly on the application for which they are intended.

The principal use of static electricity today is in the production of high electric fields. Such fields are used in industry for testing the ability of components such as insulators and condensers to with-

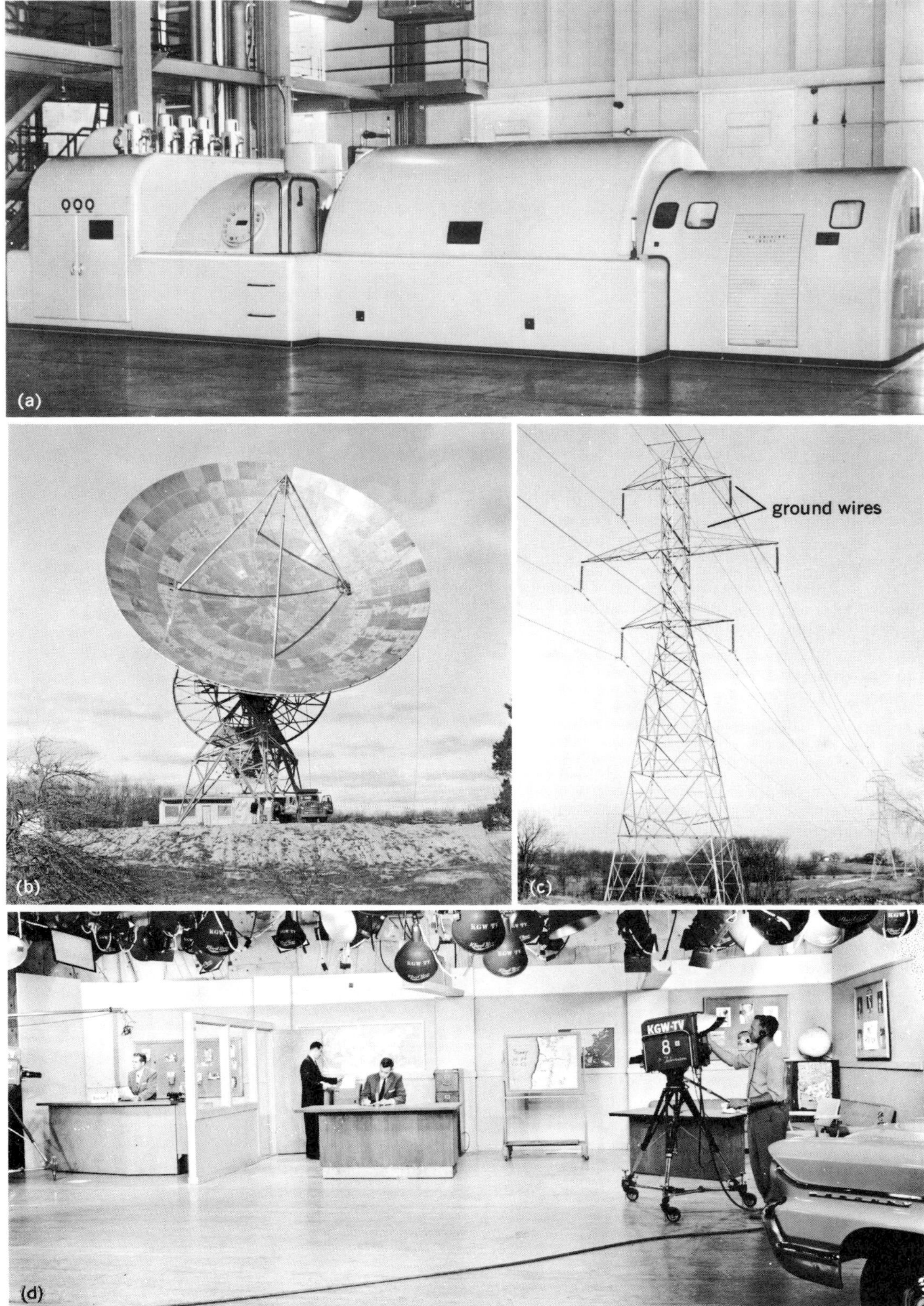

Some examples of large-scale electrical equipment. (*a*) Turbogenerator (*Allis-Chalmers*). (*b*) Radio telescope (*U.S. Office of Naval Research and University of Michigan*). (*c*) Transmission line (*Indiana and Michigan Electric Co.*). (*d*) General-purpose television studio equipment (*RCA Corporation*)

stand high voltages, and as accelerating fields for charged-particle accelerators. The principal source of such fields today is the Van de Graaff generator.

The major use of electricity today arises in devices using electric currents alternating at low or zero frequency. The use of alternating current, introduced by S. Z. de Ferranti in 1885–1890, allows power transmission over long distances at very high voltages with a resulting low percentage power loss followed by highly efficient conversion to lower voltages for the consumer through the use of transformers. Large amounts of zero-frequency current, that is, direct current, are used in the electrodeposition of metals, both in plating and in metal production, for example, in the reduction of aluminum ore. To avoid power transmission difficulties, such facilities are frequently located near sources of abundant power. *See* ALTERNATING CURRENT; DIRECT CURRENT; ELECTRIC CURRENT.

The principal sources of low-frequency electricity are rotary generators whose operation is based on the Faraday induction principle. The force to drive such generators derives from the flow of water or the expansion of gases, as in steam and internal combustion engines. The primary heat has been derived principally from fossil fuels. Economic considerations, particularly the cost of natural gas and oil and the need to conserve these for petrochemical purposes, are leading to increased reliance on nuclear reactors as the heat source. In addition, the use of coal is reemerging, and intensive efforts are being made for the discovery and development of geothermal sources. Other sources, such as fusion, solar, and oceanic, appear several decades away from significant application.

A more direct method of using fission or fusion reactors is the direct conversion of the energy released in the nuclear process into electricity. This has been achieved on a laboratory scale in the case of fission reactors. *See* GENERATOR; NUCLEAR FUSION; NUCLEAR REACTOR.

Many high-frequency devices, such as communications equipment, television, and radar, involve the consumption of only moderate amounts of power, generally derived from low-frequency sources (see illustration). If the power requirements are moderate and portability is needed, the use of ordinary chemical batteries is possible. Ion-permeable membrane batteries are a later development in this line. Fuel cells, particularly hydrogen-oxygen systems, are being developed. They have already found extensive application in Earth satellite and other space systems. The successful use of thermoelectric generators based on the Seebeck effect in semiconductors has been reported in the Soviet Union and in the United States. In a particularly compact low-power device constructed in the United States, the heat needed for the operation of such a generator has been supplied by the energy release in the radioactive decay of suitably encapsulated isotopes produced in fission reactors. *See* BATTERY; THERMOELECTRICITY.

The Bell solar battery, also a semiconductor device, has been used to provide charging current for storage batteries in telephone service and in communications equipment in artificial satellites. *See* SOLAR CELL.

There are a number of other effects which might also serve to convert various forms of energy into electrical energy, but they do not seem generally practicable.

The changing magnetic flux required for the Faraday induction may be produced by an oscillating (rather than rotary) mechanism or by varying the temperature of a magnetic circuit whose components are made of substance with a highly temperature-dependent permeability. It has been proposed to extract the energy of the fission (or possibly the fusion) reaction directly by inducing currents in external circuits by the changing magnetic field of bursts of ions from the reaction.

Direct conversion of mechanical energy into electrical energy is possible by utilizing the phenomena of piezoelectricity and magnetostriction. These have some application in acoustics and stress measurements. Pyroelectricity is a thermodynamic corollary of piezoelectricity.

Some other sources of electricity are those in which charged particles are released with some energy and collected in some manner. Charged particles are suitably released in radioactive decay, in the photoelectric effect, and in thermionic emission, among other ways. The photovoltaic effect may also be in this group.

The differences of work functions of various materials can be used for energy conversion. The contact potential difference may be used to convert heat directly to electricity or to provide improved collection for currents arising from some other source such as radioactivity.

Other possible sources of electricity arise from the existence of electrokinetic potentials in flowing fluids and of phase-transition potentials such as occur in the Workman-Reynolds effect. The possibilities of combining several effects also exist as exemplified in thermogalvanic potentials. It also appears that organic materials (as distinguished from the inorganic materials for which most of the work already described was done) merit investigation. A primitive type of organic solar battery has been developed. *See* CIRCUIT; ELECTRIC POWER MEASUREMENT; ELECTRICAL CONDUCTION; ELECTRICAL UNITS AND STANDARDS.

[WALTER ARON]

Bibliography: P. H. Abelson (ed.), *Energy: Use, Conservation and Supply*, American Association for the Advancement of Science, 1974; B. I. Bleaney and B. Bleaney, *Electricity and Magnetism*, 3d ed., 1976; R. P. Feynman et al., *Feynman Lectures on Physics*, vol. 2, 1964; E. M. Pugh and E. W. Pugh, *Principles of Electricity and Magnetism*, 2d ed., 1970.

Electrodynamics

The study of the relations between electrical, magnetic, and mechanical phenomena. This includes considerations of the magnetic fields produced by currents, the electromotive forces induced by changing magnetic fields, the forces on currents in magnetic fields, the propagation of electromagnetic waves, and the behavior of charged particles in electric fields and magnetic fields.

Classical electrodynamics deals with fields and charged particles in the manner first systematically described by J. C. Maxwell, whereas quantum electrodynamics applies the principles of quantum mechanics to electrical and magnetic phenomena. Relativistic electrodynamics is concerned with the behavior of charged particles and fields when the velocities of the particles approach that of light. Cosmic electrodynamics is concerned with electromagnetic phenomena occurring on celestial bodies and in space. *See* ELECTROMAGNETISM.

[JOHN W. STEWART]

Electromagnetic induction

The production of an electromotive force either by motion of a conductor through a magnetic field in such a manner as to cut across the magnetic flux or by a change in the magnetic flux that threads a conductor.

Motional electromotive force. A charge moving perpendicular to a magnetic field experiences a force that is perpendicular to both the direction of the field and the direction of motion of the charge. In any metallic conductor, there are free electrons, electrons that have been temporarily detached from their parent atoms.

If a conducting bar (Fig. 1) moves through a magnetic field, each free electron experiences a force due to its motion through the field. If the direction of the motion is such that a component of the force on the electrons is parallel to the conductor, the electrons will move along the conductor. The electrons will move until the forces due to the motion of the conductor through the magnetic field are balanced by electrostatic forces that arise because electrons collect at one end of the conductor, leaving a deficit of electrons at the other. There is thus an electric field along the rod, and hence a potential difference between the ends of the rod while the motion continues. As soon as the motion stops, the electrostatic forces will cause the electrons to return to their normal distribution.

From the definition of magnetic induction (flux density) B, the force on a charge q due to the motion of the charge through a magnetic field is given by Eq. (1), where the force F is at right angles to a

$$F = Bqv \sin\theta \tag{1}$$

plane determined by the direction of the field, and the component $v \sin\theta$ of the velocity is perpendicular to the field. When B is in webers/m^2, q is in coulombs, and v is in meters/sec, the force is in newtons. *See* MAGNETIC INDUCTION.

The electric field intensity E due to this force is given in magnitude and direction by the force per unit positive charge. The electric field intensity is equal to the negative of the potential gradient along the rod. In motional electromotive force (emf), the charge being considered is negative. Thus, Eqs. (2) hold. Here l is the length of the con-

$$E = \frac{F}{-q} = -Bv\sin\theta = -\frac{\mathscr{E}}{l} \qquad \mathscr{E} = Blv\sin\theta \tag{2}$$

ductor in a direction perpendicular to the field, and $v \sin\theta$ is the component of the velocity that is perpendicular to the field. If B is in webers/m^2, l is in meters, and v is in meters/sec, the emf $\mathscr{E}$ is in volts.

This emf exists in the conductor as it moves through the field whether or not there is a closed circuit. A current would not be set up unless there were a closed circuit, and then only if the rest of the circuit does not move through the field in exactly the same manner as the rod. For example, if the rod slides along stationary tracks that are connected together, there will be a current in the closed circuit. However, if the two ends of the rod were connected by a wire that moved through the field with the rod, there would be an emf induced in the wire that would be equal to that in the rod and opposite in sense in the circuit. Therefore, the net emf in the circuit would be zero, and there would be no current.

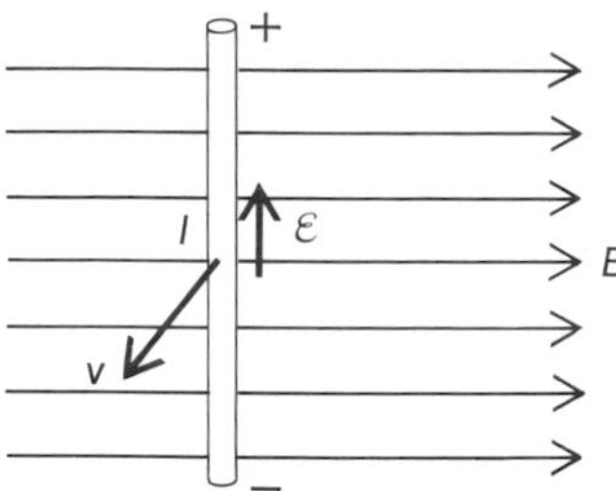

Fig. 1. Flux density B, motion v, and induced emf$_e$ when a conductor of length l moves in a uniform field. *(From R. L. Weber, M. W. White, and K. V. Manning, Physics for Science and Engineering, McGraw-Hill, 1957)*

The emf due to change of flux. When a coil is in a magnetic field, there will be a flux Φ threading the coil whose magnitude will depend upon the area of the coil and its orientation in the field. The flux is given by $\Phi = BA\cos\theta$, where A is the area of the coil and θ is the angle between the normal to the plane of the coil and the magnetic field. Whenever there is a change in the flux threading the coil, there will be an induced emf in the coil while the change is taking place. The change in flux may be caused by a change in the magnetic induction of the field or by a motion of the coil. The magnitude of the induced emf, Eq. (3), depends upon the

$$\mathscr{E} = -N\frac{d\Phi}{dt} \tag{3}$$

number of turns of the coil N and upon the rate of change of flux. The negative sign in Eq. (3) refers to the direction of the emf in the coil; that is, it is always in such a direction as to oppose the change that causes it, as required by Lenz's law. If the change is an increase in flux, the emf would be in a direction to oppose the increase by causing a flux in a direction opposite to that of the increasing flux; if the flux is decreasing, the emf is in such a direction as to oppose the decrease, that is, to produce a flux that is in the same direction as the decreasing flux.

Consider the case of a flat coil of area A rotating with uniform angular velocity ω about an axis perpendicular to a uniform magnetic field of flux density B. For any position of the coil, the flux threading the coil is $\Phi = BA\cos\theta = BA\cos\omega t$, where the zero of time is taken when θ is zero and the normal to the plane of the coil is parallel to the field. Then the emf induced as the coil rotates is given by Eq. (4).

$$\begin{aligned}\mathscr{E} &= -N\frac{d\Phi}{dt} \\ &= -NBA\frac{d(\cos\theta)}{dt} = NBA\omega\sin\omega t\end{aligned} \tag{4}$$

The induced emf is sinusoidal, varying from zero when the plane of the coil is perpendicular to the field to a maximum value when the plane of the coil is parallel to the field.

Self-induction. If the flux threading a coil is produced by a current in the coil, any change in that current will cause a change in flux, and thus there will be an induced emf while the current is changing. This process is called self-induction. The emf of self-induction is proportional to the rate of change of current. The ratio of the emf of induction to the rate of change of current in the coil is called the self-inductance of the coil.

Mutual induction. The process by which an emf is induced in one circuit by a change of current in a neighboring circuit is called mutual induction. Flux produced by a current in a circuit A (Fig. 2) threads or links circuit B. When there is a change of current in circuit A, there is a change in the flux linking coil B, and an emf is induced in circuit B while the change is taking place. Transformers operate on the principle of mutual induction. *See* TRANSFORMER.

The mutual inductance of two circuits is defined as the ratio of the emf induced in one circuit B to the rate of change of current in the other curcuit A.

Coupling coefficient. This refers to the fraction of the flux of one circuit that threads the second circuit. If two coils A and B having turns N_A and N_B, respectively, are so related that all the flux of either threads both coils, the respective self-inductances are given by Eqs. (5) and the mutual inductance of the pair is given by Eq. (6). Then Eqs. (7) hold.

$$L_A=\frac{N_A\Phi_A}{I_A} \qquad L_B=\frac{N_B\Phi_B}{I_B} \tag{5}$$

$$M=\frac{N_A\Phi_B}{I_B}=\frac{N_B\Phi_A}{I_A} \tag{6}$$

$$\begin{aligned} M^2&=\frac{N_AN_B\Phi_A\Phi_B}{I_AI_B}=\frac{N_A\Phi_A}{I_A}\frac{N_B\Phi_B}{I_B}=L_AL_B \\ M&=\sqrt{L_AL_B} \end{aligned} \tag{7}$$

In general, not all the flux from one circuit threads the second. The fraction of the flux from circuit A that threads circuit B depends upon the distance between the two circuits, their orientation with respect to each other, and the presence of a ferromagnetic material in the neighborhood, either as a core or as a shield. It follows that for the general case that Eq. (8) holds.

$$M\leqq\sqrt{L_AL_B} \tag{8}$$

The ratio of the mutual inductance of the pair to the square root of the product of the individual self-inductances is called the coefficient of coupling K, given by Eq. (9). The coupling coefficient has a

$$K=\frac{M}{\sqrt{L_AL_B}} \tag{9}$$

maximum value of unity if all the flux threads both circuits, zero if none of the flux from one circuit threads the other. For all other conditions, K has a value between 0 and 1.

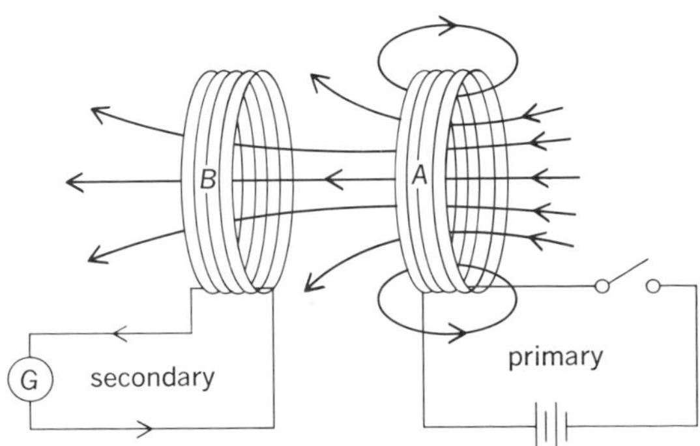

Fig. 2. Mutual induction. An emf is induced in the secondary when the current changes in the primary. (*R. L. Weber, M. W. White, K. V. Manning, Physics for Science and Engineering, McGraw-Hill, 1957*)

Applications. The phenomenon of electromagnetic induction has a great many important applications in modern technology. For example, *see* GENERATOR; MOTOR.

[KENNETH V. MANNING]

Bibliography: S. S. Attwood, *Electric and Magnetic Fields*, 3d ed., 1966; B. I. Bleaney and B. Bleaney, *Electricity and Magnetism*, 3d ed., 1976; E. M. Pugh and E. W. Pugh, *Principles of Electricity and Magnetism*, 2d ed., 1970; E. M. Purcell, *Electricity and Magnetism*, vol. 2, 1965.

Electromagnetism

The branch of science dealing with the observations and laws relating electricity to magnetism. Electromagnetism is based upon the fundamental observations that a moving electric charge produces a magnetic field and that a charge moving in a magnetic field will experience a force.

The magnetic field produced by a current is related to the current, the shape of the conductor, and the magnetic properties of the medium around it by Ampère's law.

The magnetic field at any point is described in terms of the force that it exerts upon a moving charge at that point. The electrical and magnetic units are defined in terms of the ampere, which in turn is defined from the force of one current upon another.

The association of electricity and magnetism is also shown by electromagnetic induction, in which a changing magnetic field sets up an electric field within a conductor and causes the charges to move in the conductor. *See* ELECTRICITY; ELECTROMAGNETIC INDUCTION; MAGNETIC CIRCUITS; MAGNETIC FIELD; MAGNETIC INDUCTION; MAGNETOMOTIVE FORCE.

[KENNETH V. MANNING]

Electromotive force (emf)

The electromotive force, represented by the symbol $\mathcal{E}$, around a closed path in an electric field is the work per unit charge required to carry a small positive charge around the path. It may also be defined as the line integral of the electric intensity around a closed path in the field. The abbreviation emf is preferred to the full expression since emf, also called electromotance, is not really a force. The term emf is applied to sources of electric energy such as batteries, generators, and inductors in which current is changing.

The magnitude of the emf of a source is defined as the electrical energy converted inside the source to some other form of energy (exclusive of electrical energy converted irreversibly into heat), or the amount of some other form of energy converted in the source into electrical energy, when a unit charge flows around the circuit containing the source. In an electric circuit, except for the case where an electric current is flowing through resistance and thus electrical energy is changed irreversibly into heat energy, electrical energy is converted into another form of energy only when current flows against an emf. On the other hand, some other form of energy is converted into electrical energy only when current flows in the same sense as an emf.

For a discussion of motional emf *see* ELECTROMAGNETIC INDUCTION. For information on back, or counter, emf *see* DIRECT-CURRENT MOTOR. *See also* MAGNETOMOTIVE FORCE. [RALPH P. WINCH]

Bibliography: E. M. Purcell, *Electricity and Magnetism*, 1965; F. W. Sears, *Electricity and Magnetism*, 1951.

Energy

The capacity for doing work. Energy is possessed, for example, by a body that is in motion, for stopping it provides work; by a compressed or stretched spring, for it is capable of doing work in returning to its ordinary configuration; by gunpowder or a bomb, because of the work it can do in exploding; by a charged electrical capacitor, for it can do work while being discharged. Energy, like work, is a scalar quantity. Its units are the same as those of work and include the foot-pound, foot-poundal, erg, joule, and kilowatt-hour. *See* WORK.

Because a system may possess an enormous store of energy that is not available for doing work, energy is better defined as that property of a system which diminishes when the system does work on any other system by an amount equal to the work so done. Although energy may be exchanged among various bodies or may undergo transformation from one form to another, it has the tremendously important property that it cannot be created or destroyed. *See* CONSERVATION OF ENERGY.

Energy occurs in several well-defined forms, as internal energy, kinetic energy, and potential energy.

Internal energy. This is the energy present within a body or system, such as a fuel, steam, or compressed gas, by virtue of the motions of, and forces between, the molecules and atoms of the body or system. Internal energy is sometimes erroneously referred to as the "heat energy" of a body. It is a property of any given state of a system, is evidenced by certain other properties of the system, notably temperature, and is to be distinguished from any kinetic or potential energy possessed by the system as a whole in its relation to other systems. According to the first law of thermodynamics, the gain of internal energy in any given process is equal to the difference between the heat gained by the system and the work done by the system on other systems external to it. *See* THERMODYNAMIC PRINCIPLES.

Kinetic energy. This is the term applied to the capacity for doing work that matter possesses because it is in motion. As everyday experience shows, the more massive a body is and the higher its speed, the more work it will do upon striking and being slowed down by an obstacle, and hence the larger is its kinetic energy. Specifically, for a body of mass m moving with a speed v, the kinetic energy E_k is given by Eq. (1).

$$E_k = \tfrac{1}{2}mv^2 \qquad (1)$$

Thus if a car of mass 60 slugs (about 1900 lb) is moving with a speed of 90 ft/sec (about 60 mi/hr), its kinetic energy relative to the ground is 243,000 slug-ft^2/sec^2, or 243,000 ft-lb. Now the change in the kinetic energy of a body during a given displacement is equal to the work done by the net, or resultant, force applied to the body during this displacement, a statement that is usually referred to as the work–kinetic energy theorem. Thus, to bring the car in the example to a stop in, say, 100 ft would require an average retarding force of 2430 lb, or about 1.2 tons of force. It will be noted that the value of the kinetic energy is always dependent on the body's speed relative to some chosen reference body. Thus the kinetic energy of a car relative to a man sitting in it is zero; only if the car changes speed relative to the man can it do work on him because of its motion.

To derive Eq. (1), suppose that a body of mass m and moving with speed v is brought to rest within a distance s by applying to it a constant force of magnitude f. The work done is fs and, by the work–kinetic energy theorem, is equal to the change in the body's kinetic energy E_k. By Newton's second law of motion, $f=ma$; and since the acceleration a is assumed here to be constant, $s=v^2/2a$. Therefore $E_k=fs=ma\cdot v^2/2a=\tfrac{1}{2}mv^2$. This expression for the kinetic energy is valid for all speeds except those comparable to the speed of light.

The kinetic energy of rotation of a body that is turning about a fixed axis with angular velocity of magnitude ω (radians/sec) is given by $E_k=\tfrac{1}{2}I\omega^2$, where I is the body's moment of inertia with respect to the axis in question.

Potential energy. This form of energy, as contrasted with kinetic energy, is the capacity to do work that a body or system has by virtue of its position or configuration. Thus elastic potential energy is possessed by a coiled spring that is compressed or stretched (Fig. 1); indeed, frictional forces in a spring are often so small that more than 99% of the work of deformation is recovered when the spring is released. Gravitational potential energy is possessed by a body that has been raised above the Earth's surface, for the body can do work in falling to the ground; it can strike and drive a nail or a pile, or can compress a spring. Electrical, magnetic, chemical, and nuclear systems may also possess potential energy.

In general, the potential energy of one configuration of a system relative to another configuration of it may be defined as equal to the work done against the conservative forces of the system when

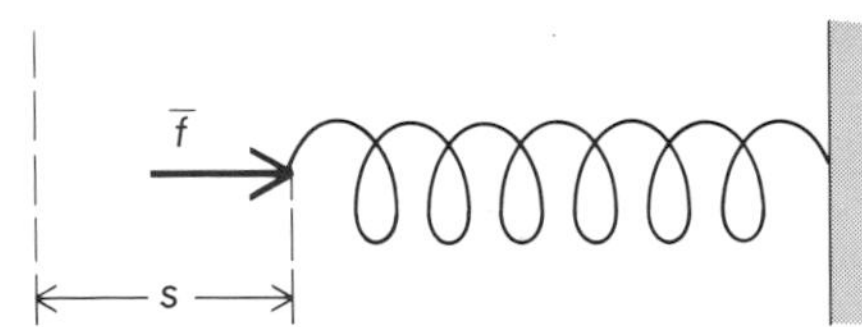

Fig. 1. A compressed spring possesses potential energy. Here s is the distance that the spring has been compressed from its normal length, and $\bar{f}$ is the average force of compression. It can be shown from Hooke's law that $f=\tfrac{1}{2}ks$, where k is a constant known as the stiffness coefficient of the spring. It follows that the elastic potential energy of the compressed spring is $\tfrac{1}{2}ks^2$.

its parts change from the one configuration to the other. Conservative forces are forces such as those of gravity or the force exerted by a spring, where the work is recoverable, that is, the net work done in a round trip is zero. As this definition implies, the potential energy of a system when in a particular configuration must always be computed with respect to some other arbitrarily selected configuration or position of the system; moreover, its value is a function only of the initial and final positions, and not of the paths followed by the parts in changing position.

Gravitational potential energy. Suppose that a body of mass m is at a height h above the ground and that h is small in comparison with the Earth's radius so that the body's weight mg does not change appreciably with the height. The gravitational potential energy E_p of the system body-Earth will then be mgh, for this is the work done against the weight mg in lifting the body to the height h and, in the absence of air resistance, is the work the body can do in returning to the ground. For example, if a 1-kg object is 1000 m above sea level, E_p with respect to sea level is $mgh = 1.0\ \text{kg} \times 9.8\ (\text{m/sec}^2) \times 1000\ \text{m} = 9800$ joules. But relative to a land surface of altitude, say, 500 m above sea level, E_p is only half as much, or 4900 joules, this being the work the body could do in dropping to this reference level rather than to the sea (Fig. 2).

If a body is at a great distance from the Earth, as when a missile is fired to a very high altitude, account must be taken of the change in the body's weight with distance r from the center of the Earth as given by the Newtonian law of gravitation, $f_g = Gmm'/r^2$, where f_g is the gravitational force of attraction, G is the gravitational constant, and m and m' are the masses of the body and Earth, respectively. If the distance between the body and the Earth's center is increased from r_1 to r_2 (Fig. 3), the work W done against gravitational attraction, and therefore the increase ΔE_p in the gravitational potential energy of the system body-Earth, is shown by Eq. (2). Equation (2) holds for any two bodies

$$W = \Delta E_p = \int_{r_1}^{r_2} f_g\, dr = E_{p2} - E_{p1} = Gmm' \left(\frac{1}{r_1} - \frac{1}{r_2}\right) \qquad (2)$$

that can be treated as particles and therefore is applicable to the Earth as one of the attracting bodies to the extent that the Earth can be regarded as spherical and made up of concentric shells, each of which is homogeneous as regards density. In the theory of gravitation the results are often simplified when the reference distance r_1 between the two attracting bodies is infinitely great, that is, the gravitational potential energy of a body is assumed to be zero when it is far removed from all other bodies. Then, from Eq. (2), $E_{p2} = -Gmm'/r_2$, and since r_2 may be any distance, the subscripts can be dropped and the equation rewritten as $E_p = -Gmm'/r$. The negative sign means simply that E_p at any finite distance of separation r is smaller than it is at infinity.

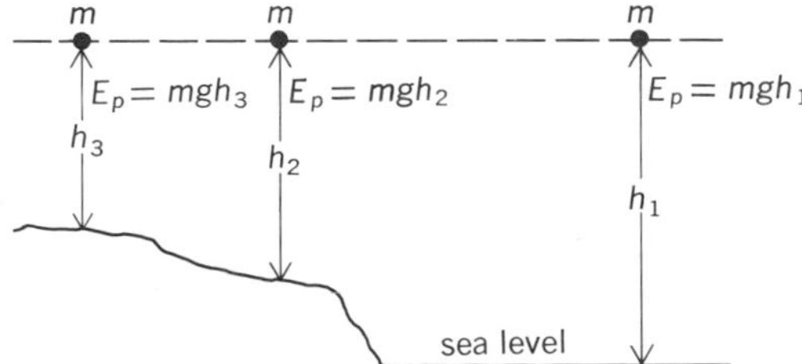

Fig. 2. The gravitational potential energy E_p of a body of weight mg for different reference levels.

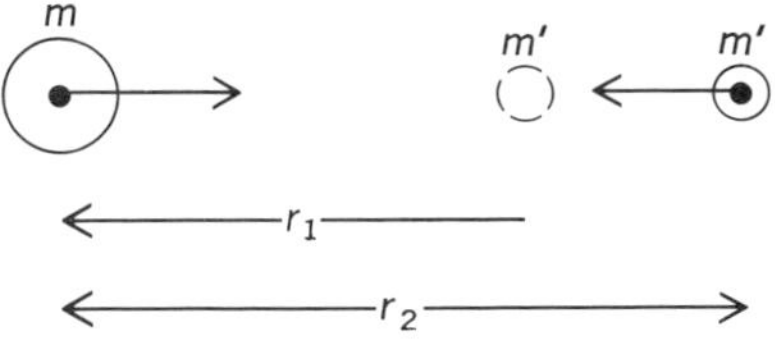

Fig. 3. Finding E_p for two particles of masses m and m'.

Electrical potential energy. The electrical potential energy of two particles in vacuum having charges q and q' can similarly be shown, with the help of Coulomb's electrostatic law, to be $E_p = \pm k_0 qq'/r$, where k_0 is a constant; the algebraic sign is negative or positive, according as the charges have unlike or like signs. *See* ENERGY SOURCES.

[DUANE E. ROLLER/LEO NEDELSKY]

Bibliography: G. P. Harnwell and G. J. F. Legge, *Physics: Matter, Energy and the Universe*, 1967; E. J. Hoffman, *The Concept of Energy: An Inquiry into Origins and Applications*, 1977; F. A. Kaempffer, *The Elements of Physics: A New Approach*, 1967; E. M. Rogers, *Physics for the Inquiring Mind*, 1960.

Energy conversion

The process of changing energy from one form to another. There are many conversion processes that appear as routine phenomena in nature, such as the evaporation of water by solar energy or the storage of solar energy in fossil fuels. In the world of technology the term is more generally applied to man-made operations in which the energy is made more usable, for instance, the burning of coal in power plants to convert chemical energy into electricity, the burning of gasoline in automobile engines to convert chemical energy into propulsive energy of a moving vehicle or the burning of a propellant for ion rockets and plasma jets.

There are well-established principles in science which define the conditions and limits under which energy conversions can be effected, for example, the law of the conservation of energy, the second law of thermodynamics, the Bernoulli principle, and the Gibbs free-energy relation. Recognizable forms of energy which allow varying degrees of conversion include chemical, atomic, electrical, mechanical, light, potential, pressure, kinetic, and heat energy. In some conversion operations the transformation of energy from one form to another, more desirable form may approach 100% efficiency, whereas with others even a "perfect" device or system may have a theoretical limiting efficiency far below 100%.

The conventional electric generator, where solid metallic conductors are rotated in a magnetic field, actually converts 95–99% of the mechanical energy input to the rotor shaft into electric energy at the generator terminals. On the other hand, an automobile engine might operate at its best point with only 20% efficiency, and even if it could be made perfect, might not exceed 60% for the ideal thermal cycle. Wherever there is a cycle which involves heat phases, the all-pervading limitation of the Carnot criterion precludes 100% conversion efficiency, and for customary temperature conditions the ideal thermal efficiency frequently cannot exceed 50 or 60%. *See* CARNOT CYCLE.

In the prevalent method of producing electric energy in steam power plants, there are many energy-conversion steps between the raw energy of fuel and the electricity delivered from the plant, for example, chemical energy of fuel to heat energy of combustion; heat energy so released to heat energy of steam; heat energy of steam to kinetic energy of steam jets; jet energy to kinetic energy of rotor; and mechanical energy of rotor to electric energy at generator terminals. This is a typical, elaborate, and burdensome series of conversion processes. Many efforts have been made over the years to eliminate some or many of these steps for objectives such as improved efficiency, reduced weight, less bulk, lower maintenance, greater reliability, longer life, and lower costs. For a discussion of major technological energy converters *see* POWER PLANT. *See also* ELECTRIC POWER GENERATION.

Efforts to eliminate some of these steps have been stimulated by needs of astronautics and of satellite and missile technology and need for new and superseding devices for conventional stationary and transportation services. Space and missile systems require more compact, efficient, self-contained power systems which can utilize energy sources such as solar and nuclear. With conventional services the emphasis is on reducing weight, space, and atmospheric contamination, on improving efficiency, and on lowering costs. The predominant objective of energy conversion systems is to take raw energy from sources such as fossil fuels, nuclear fuels, solar energy, wind, waves, tides, and terrestrial heat and convert it into electric energy. The scientific categories which are recognized within this specification are electromagnetism, electrochemistry (fuel cells), thermoelectricity, thermionics, magnetohydrodynamics, electrostatics, piezoelectricity, photoelectricity, magnetostriction, ferroelectricity, atmospheric electricity, terrestrial currents, and contact potential. *See* ENERGY SOURCES.

The electromagnetism principle today dominates the field. Electric batteries are an accepted form of electrochemical device of small capacity, for example, 1 kW in automobile service. Other categories are in various stages of development. Extensive efforts and funds are being given to some fields with attractive prospects of practical adaptation. *See* BATTERY; ELECTRIC ROTATING MACHINERY; FUEL CELL; MAGNETOHYDRODYNAMIC POWER GENERATOR; SOLAR THERMIONIC POWER GENERATOR; THERMOELECTRIC POWER GENERATOR.

[THEODORE BAUMEISTER]

Bibliography: R. C. Bailey, *Energy Conversion Engineering*, 1978; T. Baumeister (ed.), *Standard Handbook for Mechanical Engineers*, 8th ed., 1978; S. S. L. Chang, *Energy Conversion*, 1963; W. Mitchell, Jr. (ed.), *Fuel Cells*, 1963.

Energy flow

Energy flows through the United States economy from coal, oil, natural gas, falling water, uranium, and other primary sources to provide heat, light, and power. Some primary energy is used directly as fuel: most of the heat is generated by burning coal, oil, or natural gas in stoves and furnaces, and most of the power for automobiles, trucks, airplanes, ships, and trains is generated by burning oil products in internal combustion engines. Other primary energy is converted to electric power before it is used; most lighting has been electrified, as has most of the power for machine tools, refrigeration plants, elevators, appliances, television, and other stationary machines and equipment. And some energy is lost in the processes of conversion and use. *See* ENERGY CONVERSION; ENERGY SOURCES.

Transforming energy for use. Energy is used for three main purposes—heat, light, and mechanical power. Heat and light are easy to generate by means of fire, and people have done so for thousands of years, using wood and other organic materials as fuels. Mechanical power is more difficult to generate, but people have done so for thousands of years, using waterpower, windmills, sails, work animals, and their own muscles.

The ability to generate mechanical power from heat is relatively recent. On a practical scale, the transformation of heat to mechanical power began with steam engines in England in the 1770s, marking the beginning of the industrial revolution. Over a period of 2 centuries, steam engines and other heat engines replaced wind power and work animals throughout the societies that are now called the developed nations. This gradual replacement has been completed only recently. In the early 1900s the United States still obtained much of its mechanical power from work animals and windmills. Complete reliance on heat engines and waterpower came only since the 1950s. *See* ENGINE; WATERPOWER.

The transformation of mechanical power into electricity, which can be transmitted long distances over wires and converted to light or to mechanical power again at a distant location, is the most recent development in energy conversion technology. It began on a practical scale in the 1880s, and now nearly all illumination and all mechanical power except for farm work and transportation have been electrified. *See* ELECTRIC POWER GENERATION.

Heat is still obtained from combustion, but fossil fuels—coal, oil, and natural gas—are burned instead of organic materials. Some mechanical power is still obtained from falling water, but now in the form of electric power. Mechanical power for farm work and transportation comes from internal combustion engines burning oil products, and most other mechanical power comes in the form of electric power from steam engines whose boilers are fueled by coal, natural gas, or oil. Nuclear fuels have only just begun to be utilized as heat sources for stream engine boilers. *See* COMBUSTION; INTERNAL COMBUSTION ENGINE; NUCLEAR FUELS

U.S. energy flow chart. The relative proportions of the major energy flows for the United States in the mid-1970s are shown in the chart. At the left

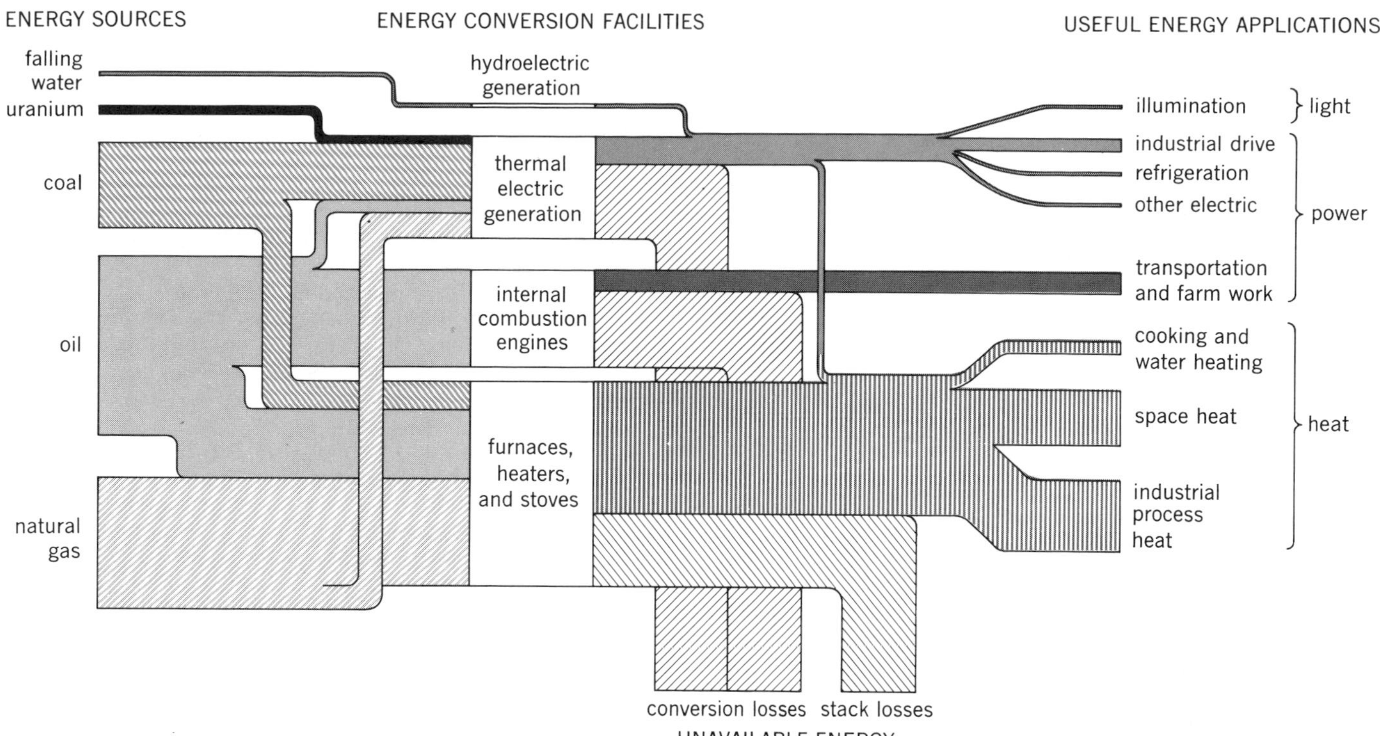

Flow of energy through the United States economy in the mid-1970s, from major energy sources, through conversion facilities, to useful applications, with some resultant unavailable energy. The width of each channel is proportional to the amount of energy that flows along it each year.

are the nation's energy sources—fossil fuels and uranium as sources of high-temperature heat, and falling water as a source of mechanical power. In the middle are the nation's energy conversion facilities. On the right are the flows of energy into major useful applications: light, power, and heat. At the bottom of the figure are the flows of unavailable energy: low-temperature heat lost up stacks and chimneys and low-temperature heat lost in the conversion of high-temperature heat to mechanical power. *See* WASTE HEAT MANAGEMENT.

Most of the power from heat engines and from falling water is used for light and power applications. A small amount is used for electric heating applications, primarily for high-temperature industrial heat that cannot be achieved economically by direct combustion, for residential cooking and water heating where convenience and cleanliness are of importance, and for a limited amount of space heating where convenience, cleanliness, and low installation cost of heating equipment are important.

Efficiency considerations. The laws of nature limit the efficiency with which heat engines can convert heat into mechanical power or work. High-temperature heat input is converted partly to mechanical work output (the desired end-product) and partly to low-temperature heat output (the unavailable residue). The higher the temperature of the input heat and the lower the temperature of the output heat, the larger the proportion of work output that a heat engine can achieve. Because electric power plants are able to reject their unavailable heat at so low a temperature—only slightly warmer than the outdoor temperature—they are more efficient than internal combustion engines in converting input heat into output power. As a result, most stationary power applications have been electrified. *See* EFFICIENCY; POWER PLANT.

Illumination has been electrified, partly for greater convenience, cleanliness, and safety, but also for greater overall efficiency and lower cost. More light can be obtained from a gallon of oil burned in a power plant to make electricity that operates an electric lamp than from a gallon of oil burned directly in an oil lamp.

In future years nuclear energy sources are likely to grow in significance. Improvements in the efficiency of heat engines may increase the proportion of useful power relative to unavailable energy, and electricity may be more widely used for heating applications. Overall energy use may grow as the population and the general level of prosperity increase. *See the feature articles* ENERGY CONSUMPTION; EXPLORING ENERGY CHOICES.

[JOHN C. FISHER]

Bibliography: John C. Fisher, *Energy Crises in Perspective*, 1974; National Academy of Engineering, *U.S. Energy Prospects: An Engineering Viewpoint*, 1974.

Energy sources

Sources from which energy can be obtained to provide heat, light, and power. Sources of energy have evolved from human and animal power to fossil fuels, uranium, water power, wind, and the Sun. The industrial age has been based on the substitution of fossil fuels for human and animal power. Future generations will have to increase use of solar energy and nuclear power as the finite reserves of fossil fuels are exhausted.

Table 1. United States consumption of energy during 1972–1978

Year	Total consumption, quads*	Consumption per capita, 10^6 Btu†
1972	71.625	343.0
1973	74.605	354.6
1974	72.348	341.4
1975	70.706	331.1
1976	74.163	344.1
1977	76.562	353.1
1978	78.014	357.0

*1 quad = 10^{15} Btu = 1.05506 × 10^{18} J.
†1 Btu = 1.05506 × 10^3 J.

Table 2. United States consumption of fuels and energy in 1978

Fuel	Quads*	Percent
Refined petroleum products	37.79	48.44
Natural gas	19.82	25.40
Coal	14.09	18.06
Nuclear power	2.98	3.81
Hydro power	3.15	4.03
Geothermal and other	0.07	0.09
Net imports of coke	0.13	0.17
TOTAL	78.01	100.00

*1 quad = 10^{15} Btu = 1.05506 × 10^{18} J.

The principal fossil fuels are coal, lignite, peat, petroleum, and natural gas—all of which were formed in finite amounts millions of years ago. Other potential sources of fossil fuels include oil shale and tar sands. Nonfuel sources of energy include wastes, water, wind, geothermal deposits, biomass, and solar heat. At the present time the nonfuel sources contribute very little energy, but as fossil fuels become depleted, nonfuel sources and fission and fusion sources will become of greater importance since they are renewable. Nuclear power based on the fission of uranium, thorium, and plutonium, and the fusion power based on the forcing together of the nuclei of two light atoms such as deuterium, tritium, or helium-3, could become principal sources of energy in the 21st century.

As the world has become more industrialized, the consumption of fuels to produce power and energy has increased at a rapid rate. World energy demand amounted to 132 quads (132 × 10^{15} Btu or 139 × 10^{18} J; 1 quad = 10^{15} Btu = 1.055 × 10^{18} J) in 1961 and increased to 238 quads (251 × 10^{18} J) in 1972, for an average annual growth of 5.5%. During the same period of time, population and real gross national product increased at rates of 1.9 and 5.1%, respectively. In 1973 the Arab embargo and the subsequent quadrupling of the price of oil threw the world into economic turmoil, and as a result, industrial growth slowed drastically and in some countries there was no growth for several years.

Table 3. United States consumption by sector in 1978

Sector	Quads*	Percent
Residential and commercial	15.10	19.35
Industrial	18.62	23.87
Transportation	20.54	26.33
Electric utilities	23.75	30.45
TOTAL	78.01	100.00

*1 quad = 10^{15} Btu = 1.05506 × 10^{18} J.

An example of the effect of much higher prices for fuels and energy on consumption can be seen in Table 1. The absolute consumption of energy in the United States declined for 3 years after 1973, but increased at an average rate of 3.45% per year in the 3 years after 1975.

The United States consumed a total of 78 quads (82 × 10^{18} J) in 1978, or about 30% of the world's total energy consumption (Table 2).

Electric power generation in the United States in 1978 required 30.5% of the total fuels and energy consumed in that year. The consumption of fuels and energy by sector are shown in Table 3.

The predictions for total United States energy demand by 1990 range from 96 to 100 quads (101 to 106 × 10^{18} J), depending upon the average annual percent gain in energy requirements.

In 1978 coal continued to be the dominant source of fuel for electric power production, providing 44% of the total. In 1973 coal supplied 46% of the total. Fuels consumed in the United States by the electric utility industry in these years are given in Table 4. The disposition of the electricity generated is given in Table 5. *See* ELECTRIC POWER GENERATION.

Petroleum. The production of crude petroleum and natural gas liquids in the United States in 1978 was 3.8 × 10^9 barrels (1 bbl has approximate mass of 0.136 metric ton). Crude petroleum and natural gas liquids production in the United States peaked at 4 × 10^9 bbl in 1970 and had been declining ever since. The consumption of petroleum products totaled 5.4 × 10^9 bbl in 1974, which was accomplished by importing 2.2 × 10^9 bbl of crude oil and refined products. However, by 1978, consumption had reached 6.9 × 10^9 bbl, and it was necessary to import 3 × 10^9 bbl of crude oil and refined products in order to supply the demand. Oil imports accounted for over 40% of total oil consumption in 1974 and 44% in 1978, and imports of oil will continue to grow as domestic oil production declines and demand for oil continues to increase. The economic effects of oil imports in the United States are shown in Table 6.

The United States proved reserves of crude oil and natural gas liquids had declined to a level of 33 × 10^9 bbl at the end of 1978, a life rate of less than 9 years at current levels of production.

The decline in petroleum production in the United States is the result of less and less drilling after 1960 and lower finding rates. This decline in drilling activity was caused by government actions which made it uneconomical to drill for oil and gas in the United States. These actions included wellhead ceiling prices on oil and gas, repeal of depletion allowance for the oil and gas industry, and a generally unfavorable climate for the oil industry in the United States. As a result, the oil companies sent their drilling rigs to foreign countries, particularly in the Middle East and Africa, where the cost of drilling was less and the expectations and probability of finding oil in large quantities were greater. Once it was established that very large reserves of oil were present in these areas of the world and that they could be produced for 10 to 15 cents per barrel rather than several dollars and higher in the United States, the die was cast, and the future of world oil production predominantly in the Middle East became a reality.

World production of crude oil was over 22 × 10^9

Table 4. United States electric utility supply in 1973 and 1978

Supply production	1973 10⁹ kWh*	1973 Percent	1978 10⁹ kWh*	1978 Percent
Coal	848	45.57	976	44.24
Petroleum	313	16.82	365	16.54
Natural gas	341	18.32	305	13.83
Nuclear power	83	4.46	276	12.51
Hydro power	272	14.62	281	12.74
Geothermal and other	4	.21	3	.14
TOTAL PRODUCTION	1861	100.00	2206	100.00
Imports of electricity	17		20	
Transmission and other losses	(162)		(224)	
TOTAL SUPPLY	1716		2002	

*1 kWh = 3.6×10^6 J.

bbl in 1978. While the United States produced only 3.2×10^9 bbl, or 15% of the total world production, it consumed 6.9×10^9 bbl, or 31% of total world consumption. World proved recoverable reserves of crude oil as of Jan. 1, 1979, amounted to 85×10^9 metric tons, or 628×10^9 bbl. The United States proved reserves of crude oil as of Jan. 1, 1979, were 27.8×10^9 bbl. *See* PETROLEUM.

Natural gas. At the end of 1978 the United States proved natural gas reserves still remaining amounted to 237×10^{12} ft³ (6.7×10^{12} m³). The annual production of natural gas in the United States peaked in 1973 at about 23×10^{12} ft³ (6.5×10^{11} m³). Since 1973, natural gas production in the United States has declined at the rate of over 5% per year. By June 1975, production had declined to a level of about 20×10^{12} ft³ (5.6×10^{11} m³) per year. Since that time, the production of natural gas in the United States has stabilized at about 20×10^{12} ft³ (5.6×10^{11} m³) per year, and it appears that the partial decontrol of wellhead prices for natural gas has had a beneficial effect on its production. In other words, a free market price and full decontrol of all new natural gas production should keep the production at the level of about 20×10^{12} ft³ (5.6×10^{11} m³) per year.

In 1975 industry experienced its worst shortage of natural gas. The shortage first appeared in 1970 when curtailments amounted to only 100×10^9 ft³ (2.8×10^9 m³), or less than 1% of demand. In 1974, curtailments totaled 2×10^{12} ft³ (5.6×10^{10} m³), or 10% of demand. For 1975 and 1976, the shortage amounted to 2.9×10^{12} ft³ (8.2×10^{10} m³), or 15% of demand.

The economic impact of the natural gas shortage was severe during the 1975–1977 period, and many industrial plants which could not burn other fuels were shut down for many days.

The American Gas Association forecast of May 1978 showed that a number of supplemental sources of gas could be made available over the next 20 years. The sources, as forecast by the AGA, are shown in Table 7. *See* NATURAL GAS.

Coal. Coal production in the United States ranged between 500 and 600×10^6 short tons (450 and 540×10^6 metric tons) per year between 1965 and 1975. By 1978, coal production was down to 654×10^6 tons (593×10^6 metric tons), from 691×10^6 tons (627×10^6 metric tons) in 1977. The forecast for 1979 was for production of coal in the United States to reach 740×10^6 tons (671×10^6 metric tons). The projection for 1983 is 812 to 940×10^6 (737 to 853×10^6 metric tons), with forecasts of future United States coal production by 1985 ranging from 850×10^6 to 10^9 short tons (770 to 910×10^6 metric tons) per year. In order to attain this future level of production, it will be necessary to develop hundreds of new deep mines in the east and hundreds of new surface mines in the east and west. Failure to reach the 1985 goal of 1.2×10^9 tons (1.09×10^9 metric tons) of coal production will result in an increasing dependence upon imported oil, with more and more coming from the Middle East. *See* COAL.

The U.S. Bureau of Mines estimated that as of Jan. 1, 1974, the demonstrated coal reserve base in the United States was 434×10^9 tons (394×10^9 metric tons) of coal, of which 297×10^9 tons (269×10^9 metric tons) is minable by surface methods. The world's total proved and currently recoverable reserves of coal in place amount to 2.376×10^{12} short tons (2.155×10^{12} metric tons) of coal, and on this basis, the United States contains 18% of the world's proved and currently recoverable reserves.

The domestic consumption of coal in the United States in 1978 totaled 619×10^6 short tons (562×10^6 metric tons). The consumption by sectors for the years 1975–1979 and National Coal Association projections for 1983 are shown in Table 8.

Nuclear energy. In 1975 nuclear energy in the United States supplied 7% of the electric power, (nearly 2% of the total energy) in the nation. By 1978, nuclear power provided 12.5% of the electric power production and 3.8% of the total United States energy supply. Seventy-one nuclear reactors with a total capacity of 50,721 MWe were in existence in the United States as of June 1979. At

Table 5. United States electric utility disposition in 1973 and 1978

Disposition	1973 10⁹ kWh*	1973 Percent	1978 10⁹ kWh*	1978 Percent
Exports of electricity	3		3	
Sales of electricity by sector:				
Residential	579	33.8	669	33.5
Commercial	388	22.7	457	22.8
Industrial	686	40.0	799	40.0
Other sales	59	3.5	74	3.7
TOTAL SALES	1713	100.0	1999	100.0
TOTAL DISPOSITION	1716		2002	

*1 kWh = 3.6×10^6 J.

Table 6. Salient characteristics of United States oil imports during 1970–1979*

Year	Crude oil and refined petroleum, 10^6 bpd†	Net energy as percent of total energy	Cost of imports, 10^9 dollars	Cost of oil imports as percent of total exports
1970	3.0	10.1	2.8	4.3
1971	3.8	12.4	3.3	4.8
1972	4.7	15.6	4.3	5.6
1973	6.3	17.0	7.6	6.9
1974	6.1	16.7	24.3	16.6
1975	6.0	18.6	24.8	15.9
1976	7.3	19.5	31.8	18.6
1977	8.7	23.5	41.5	22.7
1978	8.2	21.7	39.1	17.9
1979	8.5	24.7	60.0	24.6

*From Institute of Gas Technology, *Energy Topics*, Oct. 1, 1979.
†1 bpd = 0.159 m^3/day.

one time it was projected that by 1985 at least 200,000 MWe of nuclear plant capacity would be completed and operating and would be supplying 20% of the total United States electric power production. Nuclear power originally was expected to supply over 11% of the total United States energy by 1985. However, due to the accident at Three Mile Island near Harrisburg, PA, in March 1979, there has been a substantial slowdown in nuclear plant construction in the nation and in future plans for nuclear plants by electric utilities.

The present United States nuclear program is based principally on light-water reactors. The cutback in orders for power plants in 1975 and in subsequent years by electric utilities resulted in a number of both light-water reactors and high-temperature gas-cooled reactors being delayed or canceled.

The importance of nuclear power to the United States and to the world at large can be seen by comparing of costs of electric power in the United States by the principal sources of fuel: coal, oil, gas, and nuclear fuel.

The cost of generating electricity from nuclear plants in the United States remained stable in 1978 for the third year in a row, while those of coal- and oil-produced electricity rose moderately; the data apply to utilities that have both nuclear capacity and fossil-fired plants.

United States nuclear plants supplied more than 275 × 10^9 kWh, or about 12.5% of the country's electricity in 1978. This offset consumption in 1978 of the equivalent of 135 × 10^6 tons (122 × 10^6 metric tons) of coal, or 2.9 × 10^{12} ft^3 (8.2 × 10^{10} m^3) of natural gas, or 470 × 10^6 bbl (75 × 10^6 m^3) of oil.

In addition to costs, reliability of nuclear plants has been compared with large coal-fired stations. The net capacity factor (weighted averages for base-loaded plants) for nuclear units in 1978 was 68%, compared to 66.2% in 1977 and 61.6% in 1976. For coal units, the 1978 net capacity factor was 55.1%, compared to 57.1% in 1977 and 58.7% in 1976.

Reserves of uranium are deemed to be inadequate to support light-water reactor (LWR) nuclear power plants in the numbers planned for the next 25 years since these reactors make use of only a small percentage of the energy which is potentially available in nuclear fuels. Maximum use of the energy potential of nuclear fuels would require development of the breeder reactor, which produces more fuel than it consumes. Two different types of breeder reactors are being considered: the thermal breeder, which operates on a thorium-uranium fuel cycle and would employ either water or molten salt as a coolant; and the fast breeder, which operates best with a uranium-plutonium fuel cycle and would use gas or liquid metal as a coolant. *See* NUCLEAR POWER; NUCLEAR REACTOR.

Fusion power. Fusion power is expected to be a major source of energy in the future, but commercialization is not expected until the early part of the 21st century. The fusion process is based on the principle that when the nuclei of two light atoms are forced together to form one or two nuclei with a smaller mass, energy is released. The principal fuels in a fusion reactor are the gases deuterium, tritium, or helium-3. Deuterium, a heavy isotope of hydrogen, is found in ordinary sea water. It can be extracted relatively cheaply as deuterium oxide, or heavy water. When used in a fusion reactor, the amount of deuterium in sea water will provide an inexhaustible supply of energy for the world. *See* NUCLEAR FUSION.

Oil shale. Oil shale deposits have been found in many areas of the United States, but the only deposits of sufficient potential oil content considered as near-term potential resources are those of the Green River Formation in Colorado, Wyoming, and Utah. The U.S. Geological Survey estimates that about 80 × 10^9 bbl of oil are recoverable under present economic conditions. When marginal and submarginal reserves of all shale are included, it is estimated that reserves of 600 × 10^9 bbl of oil equivalent are present in economically recoverable oil shale deposits in the United States. It is projected that production of synthetic crude oil (syncrude) from United States oil shale will total 148,000 barrels per day by 1990 and 374,000 bpd by 2000. *See* OIL SHALE.

Tar sands. Tar sands of the world represent the largest known supply of liquid hydrocarbons. Extensive resources are located throughout the world, but primarily in the Western Hemisphere. The total world reserves in 1971 were estimated to contain 2 × 10^{12} bbl.

The best-known deposit is the Athabasca tar sands in northeastern Alberta, Canada. The oil in place is estimated to amount to 95 × 10^9 metric tons (over 700 × 10^9 bbl). Commercial operations have been under way for a number of years at a production level of 50,000 bpd and by 1979 had

Table 7. Potential supplemental sources of gas for United States, in 10^{12} ft^3 per year*

	Actual	1980	1985	1990	1995	2000
Canadian imports	1.0	1.4	1.4	1.1	1.0	0.8
SNG	0.3	0.5	0.9	0.9	0.9	0.9
LNG imports	0.01	0.6	1.6	2.4	3.0	3.0
Mexican imports	—	0.4	0.7	1.0	1.0	1.0
Alaskan gas						
Southern	—	—	0.1	0.2	0.3	0.6
North Slope	—	—	0.7	1.4	2.2	3.0
Coal gasification	—	—	0.2	1.2	2.4	4.0
New technologies	—	—	0.1	0.5	1.0	1.5
TOTAL	1.31	2.9	5.7	8.7	11.8	14.8
48 states with decontrol	20.0	19.6	20.0	20.1	20.0	20.0
TOTAL SUPPLY	21.31	22.5	25.7	28.8	31.8	34.8

*10^{12} ft^3 = 2.832 × 10^{10} m^3.

reached a level of 150,000 bpd. Several other commercial plants were announced in 1979. It is projected that by 1995 production of syncrude from Canadian tar sands will total 570,000 bpd and will reach 850,000 bpd by 2000.

The principal United States reserves are located in Utah. It is estimated that the resources in place amount to 18 to 28 $\times 10^9$ bbl in five major formations. *See* OIL SAND.

Solar energy. Solar energy has always been a potential source of limitless, clean energy. However, commercial development of solar energy has been slow because of storage requirements and high capital cost requirements. Past availability of low-cost fossil fuels resulted in solar energy being considered uneconomical. With the sudden massive increase in world oil and gas prices, solar energy economics, particularly for low-level heating and cooling, are reaching the stage where commercial applications are economical. Nevertheless, only a small percentage of the total United States energy supply will come from solar energy even by the year 2000.

The U.S. Department of Energy predicts that by 1985 solar energy could provide between 0.3 and 0.6 quads (0.3 and 0.6 $\times 10^8$ J) of space heating. This would still amount to only one-fourth to one-half of 1% of the energy required by the United States in 1985. It is well to keep the use of commercial solar energy in perspective and recognize that the conventional fossil fuels—coal, oil, and gas—are still going to be providing the country with most of its energy even in the period 1985–2000. *See* SOLAR ENERGY.

Geothermal energy. The heat content of the Earth is immense and would appear to many to be an inexhaustible and plentiful supply for much of the energy needs. However, for a variety of reasons, it appears highly unlikely that the amount of geothermal energy will ever supply more than a few percent of the United States energy requirements.

The only area in the nation where geothermal energy has been commercialized is in California, yet there is considerable doubt that geothermal energy will be able to supply more than a few percent of the state's electrical generating requirements.

It is important, nevertheless, to develop all available economic geothermal resources since they represent an important source of renewable energy. *See* GEOTHERMAL POWER.

Synthetic fuels. Fuels which do not exist in nature are known as synthetic fuels. They are synthesized or manufactured from varieties of fossil fuels which cannot be used conveniently in their original forms. Substitute natural gas (SNG) is manufactured from coal, peat, or oil shale. Synthetic liquid fuels can be produced from coal, oil shale, or tar sands. Both gaseous and liquid fuels can be synthesized from renewable resources, collectively called biomass. These carbon sources are trees, grasses, algae, plants, and organic waste. Production of synthetic fuels, particularly from renewable resources, increases the scope of available energy sources.

It should always be kept in mind that the capital investment required for synthetic fuels is many times more than the investment required to drill for oil and gas and obtain production by conventional means. Before embarking on a very capital-intensive program such as synthetic fuels, attention should be paid to the possibility of a shortage of capital which could result from the demands of the program. Full decontrol of prices for all forms of petroleum, including crude oil and refined products, and full decontrol of natural gas prices must be instituted before embarking on the very expensive program of synthetic fuel production. *See* SYNTHETIC FUEL.

Table 8. United States consumption of coal by sectors, in 10^6 short tons*

	Actual 1975	1977	1978	Forecast 1979	Projection range 1983
Electric utilities	404	476	479	525	615–685
Coking coal	83	78	71	76	70–78
Industrial and retail	69	67	68	70	85–125
TOTAL DOMESTIC USE	556	621	618	671	770–888
Exports to:					
Canada		17	15	17	16–21
Overseas		37	25	37	30–43
Subtotal exports	65	54	40	54	46–64
Stock changes	27	—	—	—	— —
TOTAL CONSUMPTION	648	675	658	725	816–952
Production:					
East	547	527	469	525	552–606
West	101	164	185	215	260–334
TOTAL PRODUCTION	648	691	654	740	812–940
TOTAL IMPORTS		4	11	9	11–28

*1 short ton = 0.907 metric ton.

Other energy sources. Several other sources of energy are worth considering.

Waste. Refuse in the form of residential, commercial, industrial, and agricultural wastes has been underutilized in the past. However, as the cost of conventional fossil fuels increases, the use of refuse as an alternative fuel becomes economically attractive. A number of plants utilizing wastes as fuel have been built to produce steam and electric energy, and more plants are being considered.

Total available combustible solid wastes expected in the United States annually amount to 800 $\times 10^9$ lb, or 400 $\times 10^6$ tons (360 $\times 10^6$ metric tons). At 5000 Btu per pound (1.163 $\times 10^7$ J/kg) heating value, the potential fuel available is 4 quads (4.2 $\times 10^{18}$ J) per year. Consumption of energy in the United States in 1979 was about 79 $\times 10^{15}$ Btu (83 $\times 10^{18}$ J), so refuse represents about 5% of the total United States energy consumption.

Wind. Wind power's energy potential is very large. However, wind power is not expected to add significantly to the United States energy capability by 1990 or 2000. *See* WIND POWER.

Tides. Tidal power, while potentially large, has not become a commercial reality except in a few areas of the world such as France, which has developed one site to 240 MWe of capacity. *See* TIDAL POWER.

Biomass. Biomass conversion to energy is technically feasible and is potentially economical. Use of biomass on a significant scale would not require a large research or demonstration effort or excessively large capital and worker-power investments. Production of alcohol from grains and other biomass material became economic in 1979 and sparked the sale of gasohol in the United States. Even though alcohol for this use will increase in the future, usage will still remain relatively small and insignificant. *See* GASOHOL.

Energy management. Energy has become a major item of cost in every industrial process, commercial establishment, and home. Energy management now includes not only the procurement of fuels on the most economical basis, but the conservation of energy by every conceivable means. Whether this is done by squeezing out every Btu through heat exchangers, or by room-temperature processes instead of high-temperature processes, or by greater insulation to retain heat which has been generated, each has a role to play in requiring less energy to produce the same amount of goods and materials.

Conservation of energy is as important as the finding of new sources of energy. It will be necessary to give tax incentives to encourage industry to install the very expensive equipment required to achieve the maximum utilization of fuels and energy. The alternative is higher and higher levels of imported oil which will eventually become too expensive to use. *See the feature articles* ENERGY CONSERVATION; ENERGY CONSUMPTION.

[GERARD C. GAMBS]

Bibliography: Atomic Industrial Forum, *Status Report: Energy Resources and Technology*, March 1975; Council on Environmental Quality, *The Good News about Energy*, 1979; Federal Energy Administration, *Project Independence Report*, November 1974; Institute of Gas Technology, *International Gas Technology*, September 1979; National Electric Reliability Council, *Forecast of Fuel Requirements*, July 1979; National Petroleum Council, *U.S. Energy Outlook*, December 1972; *9th World Energy Conference*, September 1974; Power Magazine, *The 1975 Energy Management Guidebook*, August 1975; U.S. Bureau of Mines, *Mineral Industry Surveys*, 1979; U.S. Bureau of Mines, *Minerals in the U.S. Economy*, July 1975; U.S. Department of Energy, *Energy Data Report: Annual Energy Balance 1978*, April 1979.

Energy storage

The general method and specific techniques for storing energy derived from some primary source in a form convenient for use at a later time when a specific energy demand is to be met, often in a different location.

In the past, energy storage on a large scale had been limited to storage of fuels. For example, large amounts of natural gas are routinely stored under pressure in underground reservoirs during the summer and used to meet increased demands for heating fuel in the colder seasons. Petroleum and its products are stored at several points in the energy system, from the strategic petroleum reserve to the fuel tanks of automobiles. Since gasoline is a highly concentrated and readily portable form of energy, this method of energy storage makes the automobile independent of the supply system for appreciable distances and times. On a smaller scale, electric energy is stored in batteries that power automobile starters and a great variety of portable appliances. *See* OIL AND GAS STORAGE.

In the future, energy storage in many forms is expected to play an increasingly important role in shifting patterns of energy consumption away from scarce to more abundant and renewable primary resources. For example, automobiles are likely to store transportation energy in the form of coal-derived synthetic fuels or as electricity in batteries charged with electric power from coal or nuclear power plants. Solar energy can already be made more usable by accumulating it during the day as warm water that can be stored for later use. An example of growing importance is the storage of electric energy generated at night by coal or nuclear power plants to meet peak electric loads during daytime periods. This is achieved by pumping water from a lower to a higher reservoir at night and reversing this process during the day, with the pump then being used as a turbine and the motor as a generator. As shown in Fig. 1, this example can be used to illustrate the conversion and storage functions of an energy storage system.

Broader application and use of new methods of energy storage could reduce oil consumption in the major energy use sectors collectively by perhaps 2.5×10^6 bbl (4×10^5 m^3) per day in the year 2000, as much as $5-8 \times 10^6$ bbl ($8-13 \times 10^5$ m^3) per day if all likely applications of storage can be fully developed. Probably the largest practical potential for oil displacement is in the heating and cooling of buildings through storage of heat or coolness generated on site with solar energy or off-peak electricity (singly or in combination). In principle, the potential for oil displacement is largest in transportation, since highway vehicles consume around 9×10^6 bbl (1.43×10^6 m^3) of fuel per day, more than 50% of the United States' oil use. In practice, extensive displacement of conventional automobiles will be very difficult because of the very large energy storage capacity of fossil fuels compared to the most attractive alternative, electric storage batteries: a 20-gal (76-liter) tank of gasoline will give the average United States car a range of 200–300 mi (300–500 km), at minimal extra cost for the tank itself. On the other hand, a lead-acid

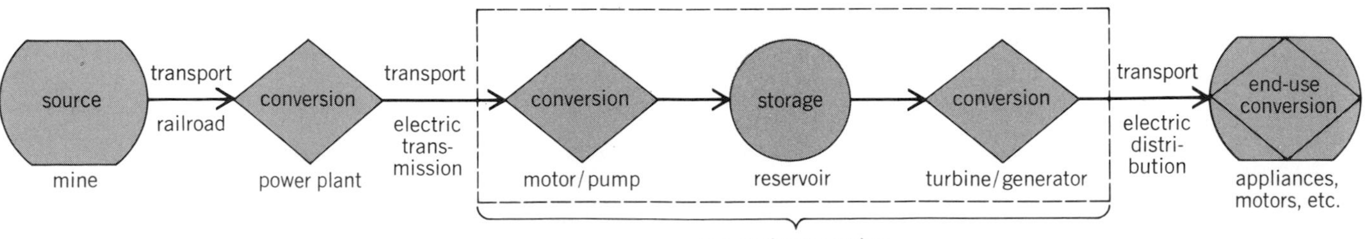

Fig. 1. Principal components of an energy storage system with specific reference to an electric power system.

battery providing a range of just 30–60 mi (50–100 km) weighs nearly a ton (0.9 metric ton) and costs between $1200 and $1500. The contrast between the two systems is illustrated in Fig. 2, which compares their specific ranges. Yet, because of their independence of oil and their more efficient as well as cleaner use of energy, electric vehicles powered by batteries are a possible alternative to conventional automobiles, especially for urban driving.

Energy storage in electric power systems has excellent potential for reducing the use of oil and gas for power generation, especially if pumped hydro storage can be supplemented by more broadly applicable techniques such as underground pumped hydro, compressed-air storage, and batteries. These techniques also could provide utilities with special operating advantages not found in conventional generating plants. Storage technologies that could find broad application in one or more of the major energy-use sectors are discussed below.

Pumped hydro storage. Until the late 1970s the only bulk energy storage method used by electric utilities, pumped hydro goes back to 1929 in the United States; the largest such facility in the world—with a capacity of 15×10^6 kWh—is the Ludington pumped storage plant on Lake Michigan. The construction of such facilities is increasingly limited by a variety of geographic, geologic, and environmental constraints, all of which may be considerably reduced if the power generation equipment and discharge reservoir are placed deep underground (Fig. 3). The size of the surface reservoir required for such an underground pumped hydroelectric system will be several times smaller than that used in the conventional system because of the greater pressure available to run the generators. The difference in height between reservoirs in the underground system may be several thousand feet (1 ft = 0.3 m), compared to several hundred feet common in the conventional pumped hydroelectric system. *See* PUMPED STORAGE.

Compressed-air storage. Off-peak electric energy can also be converted into mechanical energy by pumping air into a suitable cavern where it is stored at pressures up to 80 atm (8 MPa). Compared to pumped hydro storage, compressed air offers several advantages, including a wider choice of suitable geologic formations, lower volume requirements, and a smaller minimum capacity to be economically attractive. The world's first commercial compressed-air energy storage system (Fig. 4) is in operation in Huntorf, West Germany. In this installation, expanding air must be heated by the combustion of natural gas before it can be used to drive the generator and turbines. In a more efficient concept, the heat evolved during air compression would be stored in a bed of pebbles and reused to heat the expanding air.

Batteries. The development of advanced batteries with characteristics superior to those of the familiar lead-acid battery could result in use of battery energy storage on a large scale. For example, batteries lasting 2000 or more cycles could be used in installations of several-hundred-thousand–kilowatt-hour capacity in various locations on the electric power grid, as an almost universally applicable method of utility energy storage. Batteries combining these characteristics with energy

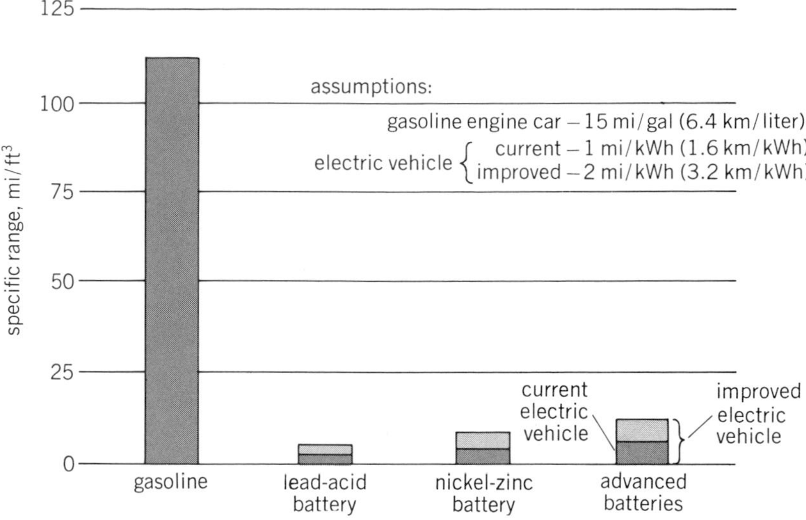

Fig. 2. Specific range of vehicles using energy stored in various forms. 1 mi/ft³ = 57 km/m³.

densities (storage capacity per unit weight and volume) well above those of lead-acid batteries could provide electric vehicles with greater range, thus removing a major barrier to their broader use.

The development of several new battery types has been undertaken to achieve systems with superior characteristics. Candidates include nickel-zinc and nickel-iron systems that promise modest improvements over lead-acid. The zinc-chlorine, sodium-sulfur, and lithium–iron sulfide systems could yield substantial improvements over lead-acid; of these more advanced systems the zinc-chlorine battery (Fig. 5) is farthest along in development. The battery operates near ambient temperature with an aqueous solution of zinc chloride as the electrolyte. On charge, zinc is deposited on a graphite substrate and chlorine is released as a slightly soluble gas at the opposite electrode, composed of porous graphite. Chlorine is stored in a separate vessel by freezing it out of the aqueous solution as a solid compound of water (ice) and chlorine at about 9°C; in this form, chlorine is relatively safe to handle. On discharge, zinc redissolves and chlorine is reduced at the porous graphite cathode, reforming the original zinc chloride aqueous solution. The active materials and other components offer the promise of a low-cost battery; however, the complexity of the system (which uses pumps, valves, and other auxiliaries) suggests that operating reliability could be a possible problem.

The sodium-sulfur battery is another battery under intensive development in the United States and Europe. Sodium-sulfur operates at temperatures above 300°C, where its electrodes exist as liquids separated by a unique solid ceramic electrolyte. Small experimental cells have already been tested through 1000 charging cycles while showing no appreciable degradation. A disadvantage of the sodium-sulfur battery, however, is that it must operate at a temperature of 300–350°C. The operating costs and safety characteristics of large batteries consisting of thousands of single cells in electric series and parallel connection are also potential drawbacks.

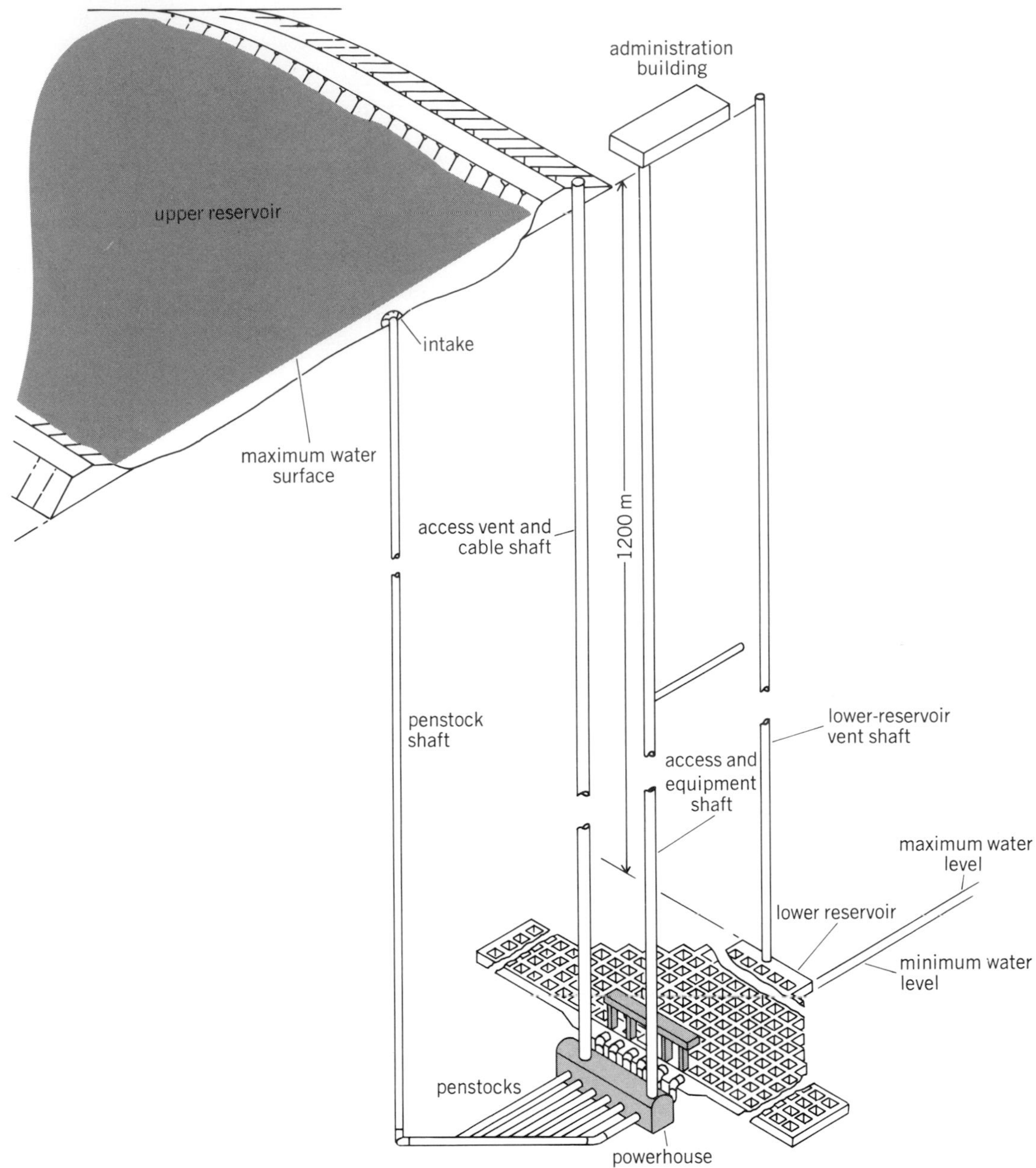

Fig. 3. Concept of an underground pumped hydroelectric storage system.

Two important national programs have been aimed at bringing advanced batteries into commercial use. The Electric and Hybrid Research, Development and Demonstration Act of 1976 provided funds for a 5-year program to demonstrate viable electric vehicles. This act provided major funding for the development and evaluation of batteries such as nickel-zinc and nickel-iron that could become available in the mid-1980s to increase the range of electric vehicles. To assist ongoing efforts to develop and commercially introduce advanced batteries for utility energy storage service on the multimegawatt level, the establishment of a major new facility, the Battery Energy Storage Test (BEST) Facility, has been undertaken to test a zinc-chlorine battery, and other advanced batteries when they become available, in the service area of Public Service Electric and Gas Company of Newark, NJ.

In parallel with this facility is the Storage Battery Electric Energy Demonstration (SBEED) project, designed to demonstrate a 10-MW, 30-MWh lead-acid battery energy storage system connected to the electrical system of the Wolverine Power Cooperative in northern Michigan. This project should go far in establishing the basic characteristics and advantages of battery energy storage for utilities. *See* BATTERY; STORAGE BATTERY.

Thermal storage. Ceramic brick "storage heaters" that store off-peak electricity in the form of heat have gained wide acceptance for heating buildings in Europe, and the barriers to their increased use in the United States are more institutional and economic than technological.

Testing has been undertaken on prototype "coolness" storage systems, which use electric refrigeration to chill water or produce ice at night. Experimental installations indicate that daytime

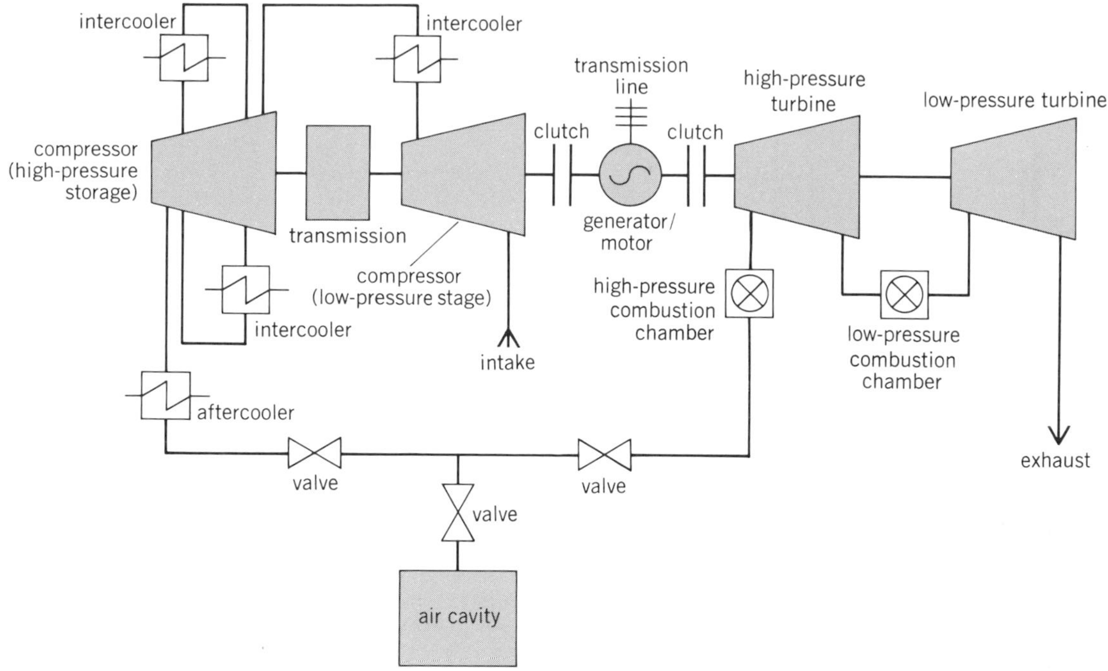

Fig. 4. Schematic layout of compressed-air storage plant at Huntorf, West Germany.

electric power demand for air conditioning could be reduced up to 75% by using such systems, but these systems are still relatively bulky and expensive.

Solar hot-water storage is technically simple and commercially available. However, the use of solar energy for space heating requires relatively large storage systems, with water or rock beds as storage media, and difficulties can arise in integrating this storage with existing buildings while keeping costs within acceptable limits. Innovative designs that may help overcome these problems include new types of heat transfer equipment, storage of heat in the walls and ceilings of buildings (passive solar heating), and the use of low-cost, roof-mounted water bags. *See* SOLAR ENERGY; SOLAR HEATING AND COOLING.

Chemical reaction systems. Heat or electricity may be stored by using these energy forms to force certain chemical reactions to occur. Such reactions are chosen so that they can be reversed readily with release of energy; in some cases the products can be transported from the point of generation to that of consumption, adding flexibility to the ways the stored energy can be used. For example, reactions which produce hydrogen could become attractive since hydrogen could be stored for extensive periods of time and then conveniently used in either combustion devices or in fuel cells.

Another possibility is to apply heat to a mixture of methane and water, converting them to hydrogen and carbon monoxide, which can be stored and transported to the end-use site, where a catalyst permits the reverse reaction to occur spontaneously with the release of heat. While not yet in practical use, chemical reaction systems are under extensive investigation as economic and flexible ways of storing energy. *See* FUEL CELL; HYDROGEN-FUELED TECHNOLOGY; PHOTOELECTROLYSIS.

Superconducting magnets. Electrical energy can be stored directly in the form of large direct currents used to create fields surrounding the superconducting windings of electromagnets. In principle such devices appear attractive because their storage efficiency is high, plant life could be long, and utilities would have few difficulties establishing the necessary conversion equipment. However, the need for maintaining the system at temperatures approaching absolute zero and, particularly, the need to physically restrain the coils of the magnet when energized require auxiliary equipment (insulation, vacuum vessels, and structural supports) which will represent a large cost. Development of such systems is still in the early stages, and even the discovery of new superconducting materials with high critical currents and higher cryogenic temperatures may not be sufficient to reduce costs to acceptable levels since this would not significantly affect the structural containment requirements. *See* SUPERCONDUCTIVITY.

Flywheels. Storage of kinetic energy in rotating mechanical systems such as flywheels is attractive where very rapid absorption and release of the stored energy is critical. However, research indicates that even advanced designs and materials are likely to be too expensive for utility energy storage on a significant scale, and applications will probably remain limited to systems where high power capacity and short charging cycles are the prime consideration. Such applications do exist in pulse power supplies and in electric transportation for recovery of braking energy.

Combined systems. The rising importance of energy storage comes from its potential for shifting demand from scarce to plentiful primary energy sources. As such, the most successful storage devices are likely to be those that are adopted as components of larger systems designed specifically for resource conservation. In electric power genera-

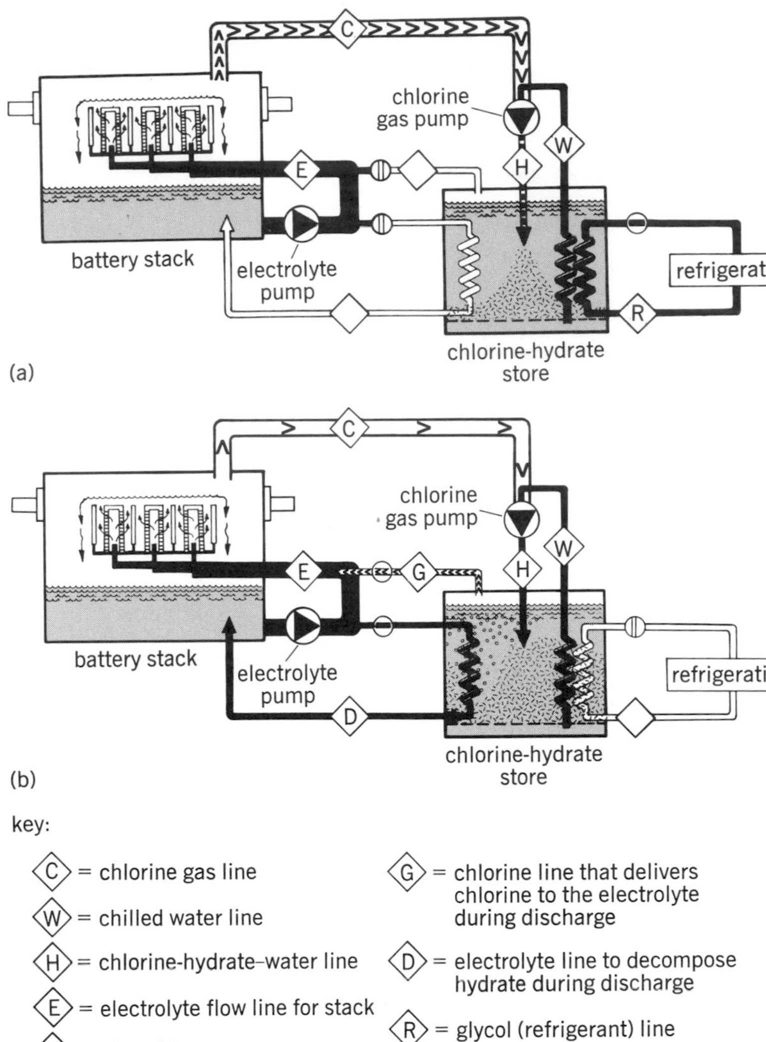

Fig. 5. Schematic layout of zinc-chloride battery, showing two main compartments in the design, the battery stack and the chlorine hydrate store. (*a*) Charge cycle. (*b*) Discharge cycle. (*Gulf and Western Company*)

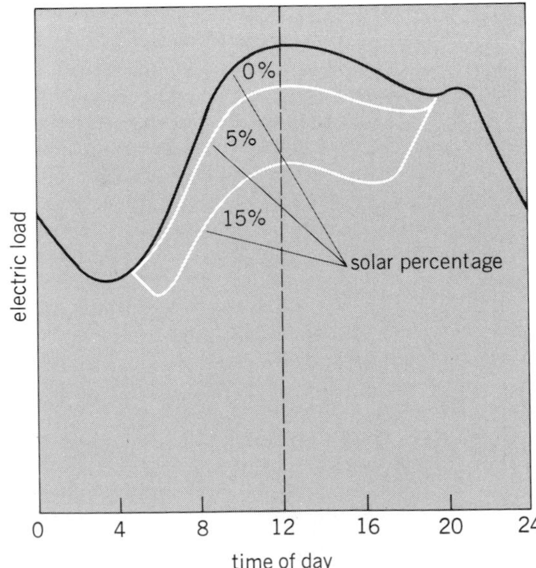

Fig. 6. Impact of solar-electric generation on utility load curve (schematic).

tion, for example, system-wide storage may be combined with solar-electric systems to flatten load curves and reduce oil consumption (Fig. 6). If such complex systems are eventually to be developed and commercially introduced, financial incentives and more coherent national energy policy planning must be adopted. Thus, while advanced energy storage systems could have a major impact on United States energy consumption patterns before the end of the century with continued, successful research, nontechnical factors such as regulatory strategies and pricing policies are likely to become decisive to their large-scale use. *See the feature articles* ENERGY CONSERVATION; EXPLORING ENERGY CHOICES.

[FRITZ KALHAMMER; THOMAS R. SCHNEIDER]

Bibliography: J. R. Birk, K. Klunder, and J. C. Smith, Superbatteries: A progress report, *IEEE Spectrum*, 16(3):49–55, March 1979; J. R. Bolton, Solar fuels, *Science*, 202(4369):705–711, Nov. 17, 1978; E. J. Cairns and H. Shimotake, High temperature batteries, *Science*, 164(3886):1347–1355, June 20, 1969; R. A. Hein, Superconductivity: Large-scale applications, *Science*, 185(4147):211–222, July 19, 1974; F. R. Kalhammer, Energy-storage systems, *Sci. Amer.*, 241(6):56–65, December 1979; F. R. Kalhammer and T. R. Schneider, Energy storage, *Annu. Rev. Energy*, 1:311–343, 1976; G. D. Whitehouse, M. E. Council, and J. D. Martinez, Peaking power with air, *Power Eng.*, 72(1):50–52, January 1968.

Engine

A machine designed for the conversion of energy into useful mechanical motion. The principal characteristic of an engine is its capacity to deliver appreciable mechanical power, as contrasted to a mechanism such as a clock or analog computer whose significant output is motion. By usage an engine is usually a machine that burns or otherwise consumes a fuel, as differentiated from an electric machine that produces mechanical power without altering the composition of matter. Similarly, a spring-driven mechanism is said to be powered by a spring motor; a flywheel acts as an inertia motor. According to this definition a hydraulic turbine is not an engine, although it competes with the engine as a prime source of mechanical power. *See* HYDRAULIC TURBINE; MOTOR; WATERPOWER.

Applications. A fuel-burning engine may be stationary, as a donkey engine used to lift cargo between wharf and ship, or it may be mobile, like the engine in an aircraft or automobile. Such an engine may be used for both fixed service and mobile operation, although accessory modifications that adapt the engine to its particular purpose are preferable. For example, the fan that draws air through the radiator of a water-cooled fixed engine is large and fitted in a baffle, whereas the fan of a similar but mobile engine can be small and unbaffled because considerable air is driven through the radiator by means of ram action as the engine propels itself along. *See* AIRCRAFT ENGINE; AUTOMOTIVE ENGINE.

Some types of engine can be designed for economic efficiencies in fixed service but not in mobile operation. Thus, the steam engine is widely used in central electric generator stations but is obsolete in mobile service. This is chiefly because, in a large ground installation, the furnace and boiler can be fitted with means for using most of the available heat. The engine proper can be a reciprocating (piston) or a rotating (turbine) type. Because shaft rotation is by far the most used form of mechanical motion, the turbine is the more common form of modern steam engine. In the operation of railroads the steam engine has given place to diesel and gasoline internal combustion engines and to electric motors. *See* POWER PLANT; STEAM ENGINE; STEAM-GENERATING UNIT; STEAM TURBINE.

Types. Traditionally, engines are classed as external or internal combustion. External combustion engines consume their fuel or other energy source in a separate furnace or reactor. *See* NUCLEAR REACTOR.

Strictly, the furnace or reactor releases chemical or nuclear energy into thermal energy, and the engine proper converts the heat into mechanical work. The principal means for the conversion of heat to work is a gas or vapor, termed the working fluid. By extension, an engine which derives its heat energy from the Sun by solar radiation to working fluid in a boiler can be considered an external combustion type. To avoid loss of or contamination from nuclear fuel, the reactor and boiler are separated from (and may also be shielded from) the engine.

The working fluid takes on energy in the form of heat in the boiler and gives up energy in the engine, the engine proper being a thermodynamic device. The device may be a turbine for stationary power generation on a nozzle for long-range vehicular propulsion. *See* NUCLEAR POWER.

In an engine used for propulsion, the rearward velocity with which the working fluid is ejected and, thus, the forward acceleration imparted to the vehicle depend on the temperature of the fluid. For practical purposes, temperature is limited by the engine materials that serve to contain the chemical combustion or nuclear reaction. To achieve higher exhaust velocity, the working fluid may be contained by nonmaterial means such as electric and magnetic fields, in which case the fluid must be electrically conductive.

The engine proper is then a magnetohydrodynamic device receiving electric energy from a separate fuel-consuming source such as a gas turbine and electric generator or a nuclear reactor and electric generator, or possibly from direct electric conversion of nuclear or solar radiation. *See* THERMOELECTRICITY.

A further basis of classification concerns the working fluid. If the working fluid is recirculated, the engine operates on a closed cycle. If the working fluid is discharged after one pass through boiler and engine, the engine operates on an open cycle. Closed-cycle operation assures the purity of the working fluid and avoids the discharge of harmful wastes. The open cycle is simpler. Thus the commonest types of engine use atmospheric air in open cycles as the principal constituent of the working fluids and as oxidizer for the fuels.

If open-cycle operation is used, the next modification is to heat the working fluid directly by burning fuel in the fluid; the engine becomes its own furnace. Because this internal combustion type engine uses the products of combustion as part of the working fluid, the fuel must be capable of combustion under the operating conditions in the engine and must produce a noncorrosive and nonerosive working fluid. Such engines are the common reciprocating gasoline and diesel units. *See* DIESEL ENGINE; INTERNAL COMBUSTION ENGINE; STIRLING ENGINE.

At low speeds the combustion process is carried out intermittently in a cylinder to drive a reciprocating piston. At high speed, however, friction between piston and cylinder walls and between other moving parts dissipates an appreciable portion of the developed power. Thus, where high power is developed at high speed, performance is improved by continuous combustion to drive a turbine wheel. *See* BRAYTON CYCLE; CARNOT CYCLE; DIESEL CYCLE; GAS TURBINE; OTTO CYCLE.

Engine shaft rotation may be used in the same way as in a reciprocating engine. However, for high-velocity vehicular propulsion, the energy of the working fluid may be converted into thrust more directly by expulsion through a nozzle. Once the vehicle is in motion, the turbine can be omitted. Alternatively, instead of drawing atmospheric oxygen into the combustion chamber, the engine may draw both oxidizer and fuel from storage tanks which are contained within the vehicle, or the combustion chamber may contain the full supply of fuel and oxidizer. *See* RAMJET; TURBOJET; TURBOPROP.

Despite all the variation in structure, mode of operation, and working fluid—whether of moving parts, moving fields, or only moving working fluid—these machines are basically means for converting heat energy to mechanical energy. *See* THERMODYNAMIC PROCESSES.

[FRANK H. ROCKETT]

Bibliography: D. H. Marter, *Engines*, 1962.

Fire-tube boiler

A steam boiler in which hot gaseous products of combustion pass through tubes surrounded by boiler water. The water and steam in fire-tube boilers are contained within a large-diameter drum or shell, and such units often are referred to as shell-type boilers. Heat from the products of combustion is transferred to the boiler water by tubes or flues of relatively small diameter (approximately 3–4 in.) through which the hot gases flow. The tubes are connected to tube sheets at each end of the cylindrical shell and serve as structural reinforcements to support the flat tube sheets against the force of the internal water and steam pressure. Braces or tension rods also are used in those areas of the tube sheets not penetrated by the tubes.

Fire-tube boilers may be designed for vertical,

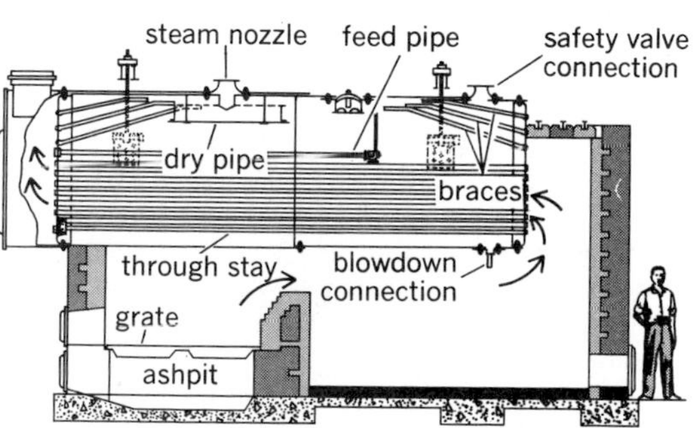

Fig. 1. Horizontal-return-tube boiler.

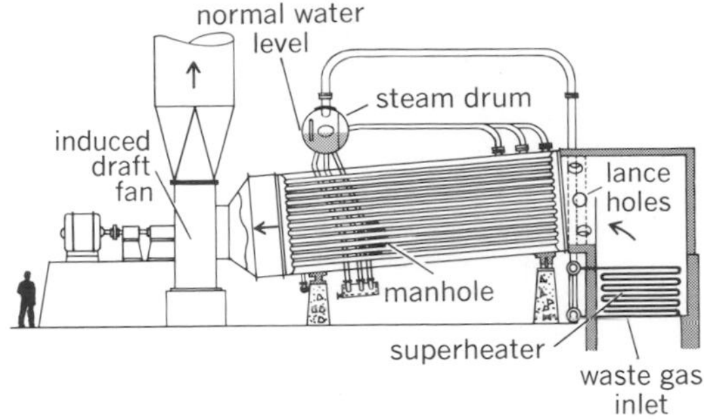

Fig. 2. Fire-tube waste-heat boiler.

inclined, or horizontal positions. One of the most generally used types is the horizontal-return-tube (HRT) boiler (Fig. 1). In the HRT boiler, part of the heat from the combustion gases is transferred directly to the lower portion of the shell. The gases then make a return pass through the horizontal tubes or flues before being passed into the stack.

In the Scotch marine boiler, one or more large flues (approximately 18–24 in. in diameter) are used for furnaces or combustion chambers within the shell, and the hot gases are returned through groups of small-diameter tubes or flues. The flues that form the combustion chamber are corrugated to prevent collapse when subjected to the water and steam pressure. These boilers may be oil-fired, or solid fuel can be burned on grates set within the furnace chambers. Scotch marine boilers have, with few exceptions, been superseded by water-tube marine boilers.

Gas-tube boilers are sometimes used for the absorption of waste heat from process gases or the exhaust from internal combustion engines, particularly in those cases in which their installation provides a simple and economical means for the recovery of low-grade heat. The boiler shell may be augmented by an external steam-separating drum and downcomer and riser connections to provide for proper circulation (Fig. 2).

Shell-type boilers are restricted in size to 14 ft in diameter and an operating pressure of 300 psi by the stresses in the large-diameter shells and the necessity to design the flat-tube sheets for practicable thicknesses. They are best suited for low-pressure heating service and for portable use because the integrated structure can be transported easily and the generous reserve in water capacity requires minimum attendance. However, there is the risk of catastrophic damage if overheating and rupture of the shell or tube sheet occur as a result of low water levels or the formation of an insulating layer of internal scale produced by water impurities. *See* STEAM-GENERATING UNIT.

[GEORGE W. KESSLER]

Fischer-Tropsch process

The synthesis of hydrocarbons and, to a lesser extent, of aliphatic oxygenated compounds by the catalytic hydrogenation of carbon monoxide. The synthesis was discovered in 1923 by F. Fischer and H. Tropsch at the Kaiser Wilhelm Institute for Coal Research in Mülheim, Germany. The reaction is highly exothermic, and the reactor must be designed for adequate heat removal to control the temperature and avoid catalyst deterioration and carbon formation. The sulfur content of the synthesis gas must be extremely low to avoid poisoning the catalyst. The first commercial plant was built in Germany in 1935 with cobalt catalyst, and at the start of World War II there were six plants in Germany producing more than 4,000,000 bbl/year (6.4×10^5 m^3) of primary products. Iron catalysts later replaced the cobalt.

Following World War II, considerable research was conducted in the United States on the iron catalysts. One commercial plant was erected at Brownsville, Tex., in 1948, which used a fluidized bed of mill scale promoted by potash. Because synthetic oil was not competitive with petroleum, the plant was shut down within a few years. A Fischer-Tropsch plant constructed in South Africa (SASOL) about the same time has continued to produce gasoline, waxes, and oxygenated aliphatics. The SASOL plant gasifies inexpensive coal in Lurgi generators at elevated pressure. After purification, the gas is sent to entrained and fixed-bed synthesis reactors containing iron catalysts. *See* COAL GASIFICATION.

Research at the U.S. Bureau of Mines Pittsburgh Coal Research Center has resulted in the development of active and low-cost iron catalysts of steel lathe turnings and flame-sprayed magnetite. With these catalysts, improved reactors have been tested by use of gas recycle to control the operating temperature while maintaining a low pressure drop.

Methanation. Since about 1960, interest has grown in the United States in catalytic methanation to produce high-Btu gas from coal. Synthesis gas containing three volumes of hydrogen to one volume of carbon monoxide is reacted principally to methane and water, as shown in Eq. (1). Nickel catalysts, first discovered by P. Sabatier and J. B. Senderens in 1902, are still the principal catalysts for methanation. Active precipitated catalysts were developed at the British Fuel Research Station, and Raney nickel catalysts were tested in fluid and fixed-bed reactors by the Bureau of Mines and the Institute of Gas Technology at the Illinois Institute of Technology. The Bureau of Mines has developed a technique for applying a thin layer of Raney nickel on plates and tubes by flame-spraying the powder. This has led to development of efficient gas recycle and tube-wall reactors. Commercial production of high-Btu gas from coal in the United States is expected to start about 1980 to supplement natural gas supplies.

Reactions. Typical Fischer-Tropsch reactions for the synthesis of paraffins, olefins, and alcohols are the pairs of reactions labeled (1), (2), and (3), respectively.

$$\begin{aligned} &(2n+1)H_2 + nCO \xrightarrow{\text{Co catalysts}} C_nH_{2n+2} + nH_2O \\ &(n+1)H_2 + 2nCO \xrightarrow{\text{Fe catalysts}} C_nH_{2n+2} + nCO_2 \end{aligned} \tag{1}$$

$$\begin{aligned} &2nH_2 + nCO \xrightarrow{\text{Co catalysts}} C_nH_{2n} + nH_2O \\ &nH_2 + 2nCO \xrightarrow{\text{Fe catalysts}} C_nH_{2n} + nCO_2 \end{aligned} \tag{2}$$

$$\begin{aligned} &2nH_2 + nCO \xrightarrow{\text{Co catalysts}} C_nH_{2n+1}OH + (n-1)H_2O \\ &(n+1)H_2 + (2n-1)CO \xrightarrow{\text{Fe catalysts}} \\ &\qquad C_nH_{2n+1}OH + (n-1)CO_2 \end{aligned} \tag{3}$$

The primary reaction on both the cobalt and the iron catalysts yields steam, which reacts further on iron catalysts with carbon monoxide to give hydrogen and carbon dioxide. On cobalt catalysts, at synthesis temperatures (about 200°C, or 392°F), the reaction $H_2O + CO = CO_2 + H_2$ is much slower than on iron catalysts at synthesis temperatures (250–320°C, or 482–617°F). All these synthesis reactions are exothermic, yielding 37–51 kcal/mole (155–213 kJ/mole) of carbon in the products or 4700–6100 Btu/lb (10.9–14.2 MJ/kg) of product.

The hydrocarbons formed in the presence of iron catalysts contain more olefins than those formed in cobalt catalyst systems. The products from both catalysts are largely straight-chain aliphatics; branching is about 10% for C_4, 19% for C_5, 21% for C_6, and 34% for C_7. Aromatics appear in small amounts in the C_7 and in larger amounts in the higher boiling fractions. Operating conditions and special catalysts required, such as nitrides and carbonitrides of iron, have been ascertained for the production of higher proportions of alcohols. [JOSEPH H. FIELD]

Bibliography: H. A. Dirksen and H. R. Linden, *Inst. Gas Technol. Res. Bull.* no. 31, July, 1963; J. H. Field et al., *Ind. Eng. Chem. Prod. Res. Develop.*, 3:150–153, June, 1964; J. C. Hoogendoorn and J. M. Salomon, *Brit. Chem. Eng.*, pp. 238–243, May, 1957; L. A. Moignard and F. J. Dent, *Gas Times*, pp. 40–41, May 11, 1946; H. H. Storch, N. Golumbic, and R. B. Anderson, *The Fischer-Tropsch and Related Syntheses*, 1951.

Fluidized-bed combustion

A method of burning fuel in which the fuel is continually fed into a bed of reactive or inert material while a flow of air passes up through the bed, causing it to act like a turbulent fluid. Fluidized beds have long been used for the combustion of low-quality, difficult fuels and have become a rapidly developing technology for the clean burning of coal.

Fluidization process. A fluidized-bed combustor is a furnace chamber whose floor is slotted, perforated, or fitted with nozzles. Air is forced through the floor and upward through the chamber. The chamber is partially filled with particles of either reactive or inert material, which will fluidize at an appropriate air flow rate. When fluidization takes place, the bed of material expands (bulk density decreases) and exhibits the properties of a liquid. As air velocity increases, the particles mix more violently, and the surface of the bed takes on the appearance of a boiling liquid. If air velocity were increased further, the bed material would be blown away.

Once the bed is fluidized, its temperature can be increased with ignitors until a combustible material can be injected to burn within the bed. Proper selection of air velocity, operating temperature, and bed material will cause the bed to act as a chemical reactor. In the case of coal combustion the bed is generally limestone, which reacts with and absorbs the sulfur in the coal, reducing sulfur dioxide emissions.

Fluidized-bed combustion can proceed at low temperatures (1400–1500°F, or 760–840°C), so that nitrogen oxide formation is inhibited. In the limestone/sulfur-absorbing designs the waste product is a relatively innocuous solid, although the large volume produced does have environmental implications. Both Federal and privately funded programs are exploring techniques for utilizing the waste product, for example, determining its potential value as an agricultural supplement or bulk aggregate. Low combustion temperatures minimize furnace corrosion problems. At the same time, intimate contact between the hot, turbulent bed particles and the heat-transfer surfaces results in high heat-transfer coefficients. The result is that less surface area is needed, and overall heat-transfer tube requirements and costs are lower. Intensive development of fuel injection systems has been undertaken, particularly for pressurized operation, where the movement of solids through the pressure barrier poses special problems. These efforts hold the promise of early solutions.

Although frequently identified as a sulfur-reduction process, fluidized-bed combustion has application to low-sulfur fuels as well. In firing a low-sulfur lignite fuel, for example, the bed material would simply be coal ash.

Applications. There are three broad areas of application: incineration, gasification, and steam generation.

Coal gasification. The original application of fluidized combustion was to the gasification of coal. A gasifier burns coal in a fluidized bed with

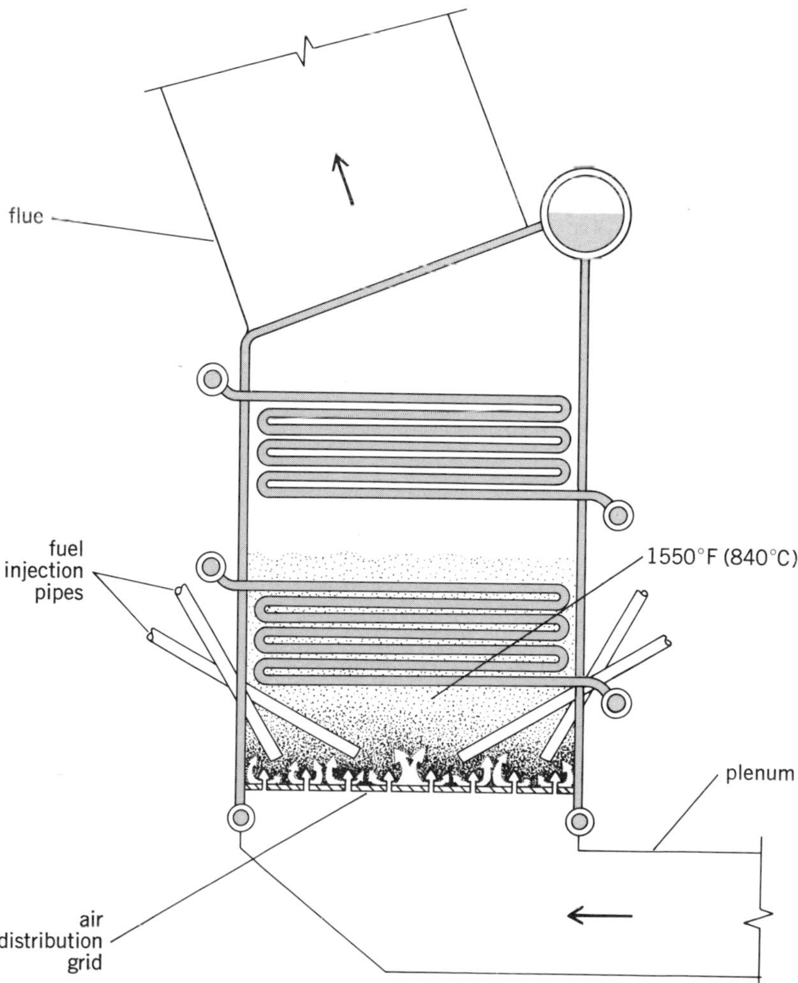

Fig. 1. Fluidized-bed steam generator.

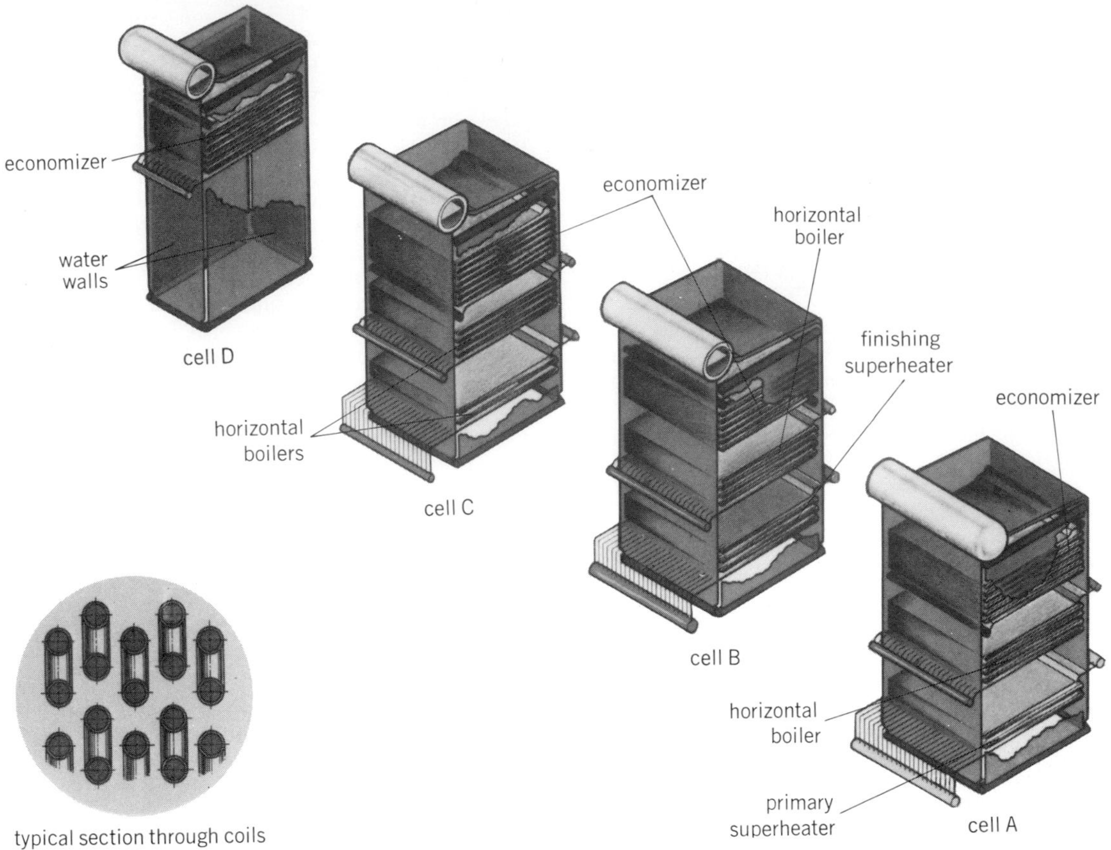

Fig. 2. Multicell fluidized-bed boiler.

less air than that required for complete combustion. This results in a high concentration of combustible gases in the exhaust. This principle is being applied in a number of pilot plants which are for the most part directed at producing low-sulfur-content gases of medium heating value (150–300 Btu per standard cubic foot, or 5590–11,170 kJ/m³). *See* COAL GASIFICATION.

Incineration. The use of this technology for efficient incineration of waste products, such as food processing wastes, kraft process liquors, coke breeze, and lumber wastes, has increased rapidly. The process has the advantages of controllability, lower capital cost, more compact design, and reduced emissions.

Steam generation. Immersing heat-transfer tubes within the hot fluidized bed converts the furnace from a simple chemical reactor to a steam boiler (Fig. 1). The use of a number of relatively small cells or beds permits the construction of large-capacity boilers (Fig. 2). Cells may be arranged horizontally, vertically, or both ways.

The largest fluidized-bed boiler in operation is at Rivesville, WV. This unit generates 300,000 lb/hr (136,200 kg/hr) of high-pressure superheated steam for power generation. The boiler is a horizontally arranged multicell unit. Since initial start-up of this boiler in March 1977, tests performed while burning high-sulfur fuels resulted in both NO_x and SO_2 emissions below Environmental Protection Agency (EPA) limits.

Designs of steam generators for larger plants, supplying 200, 600, and 800 MWe (megawatts of electric power), have also been undertaken; the first such unit, planned by the Tennessee Valley Authority, should be operational in 1985. The steam generator furnace pressure designs are generally for operation at or near atmospheric pressure.

Application of the technology to small boilers should permit the burning of coal instead of gas and oil in industrial-sized boilers. A 100,000 lb/hr (45,000 kg/hr) high-sulfur-coal–burning boiler was put into operation at Georgetown University in Washington, DC. In Renfrew, Scotland, a conventional 40,000 lb/hr (18,160 kg/hr) stoker-fired boiler was converted to fluidized combustion by the substitution of an air distribution floor for the stoker.

In the so-called pressurized fluidized-bed combustor, operating the combustion chamber at elevated pressures permits the use of the heated exhaust gases to drive gas turbines. Tests have been undertaken on a 13-MWe pilot plant using this combined cycle. Designs for larger plants have also been undertaken (Fig. 3). *See* GAS TURBINE; STEAM-GENERATING UNIT.

Other applications. Fluidized-bed combustion techniques have had a long history of application in the chemical industries, especially in catalytic petroleum cracking, and in extractive metallurgy, for example, in ore roasters and calciners. Development of a fluidized-bed combustor for the spent graphite fuel rods from high-temperature gas-cooled nuclear reactors has been undertaken.

Competitiveness of process. Since nearly any combustible product, including such difficult fuels

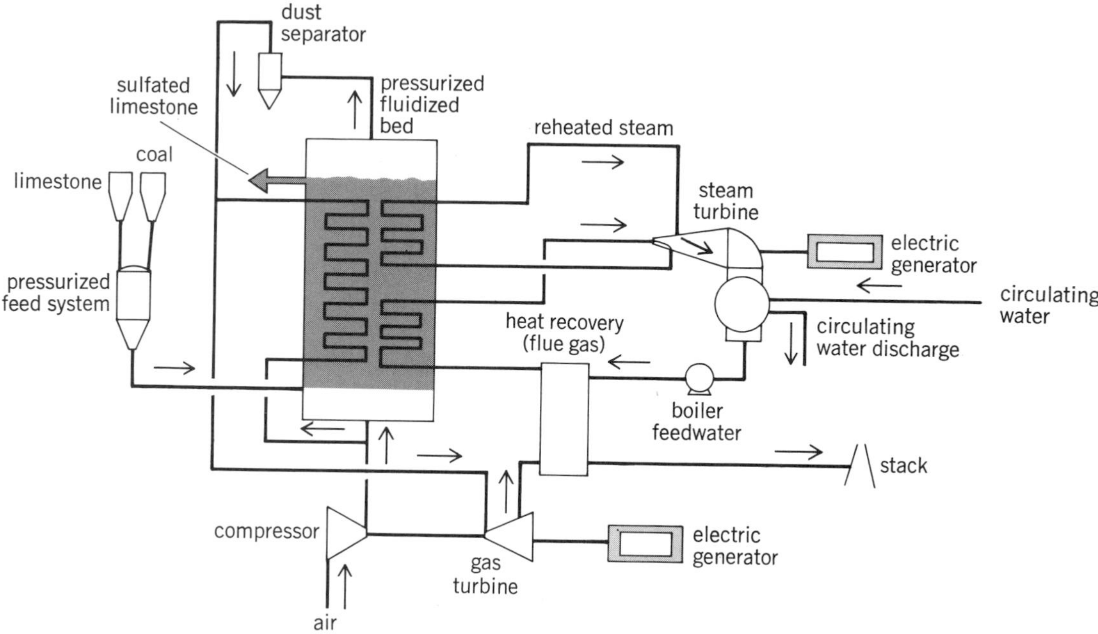

Fig. 3. Flow diagram of 600-MW combined-cycle system.

as oil shales, anthracite wastes, and residual oils, can be effectively burned, the potential of fluidized-bed combustion as a major energy conversion process is great. Economically and environmentally, its competitiveness with flue gas desulfurization, synthetic fuel preparation, and similar alternatives has been demonstrated. *See* COMBUSTION.

[MICHAEL POPE]

Forest resources

Forest resources consist of two separate but closely related parts: the forest land and the trees (timber) on that land. In the United States, forests cover one-third of the total land area of the 50 states, in total, about 740 × 10^6 acres (10^6 acres = 4046 km^2). The fact that 1 of every 3 acres of the United States is tree-covered makes this land and its condition a matter of importance to every citizen. Recognizing this importance, Congress has charged the U.S. Forest Service with the responsibility of making periodic appraisals of the national timber situation, a charge that was substantially increased in scope by the Forest and Rangeland Renewable Resources Planning Act of 1974 and the National Forest Management Act of 1976. Most of the data used here are taken from the review draft of the 1978 edition of *Forest Statistics of the U.S., 1977.* A new assessment of the timber situation was scheduled for publication in 1980.

Some 488 × 10^6 acres, or about two-thirds of the total forest area in the United States, is classified as commercial forest land, that is, forest land capable of producing at least 20 ft^3 (10 ft^3 = 0.28 m^3) of industrial wood per acre per year, and not reserved for uses incompatible with timber production. The 252 × 10^6 acres of noncommercial forest includes about 24 × 10^6 acres which meets the growth criteria for commercial timberland but has been set aside for parks, wilderness areas, and such. The remaining 228 × 10^6 acres is incapable of producing a sustained crop of industrial wood, but is valuable for watershed production, grazing, and recreation use. Table 1 gives details of the distribution of forest land in the four main regions of the United States.

Most of the noncommercial forests are in public ownership, including approximately 19.5 × 10^6 productive forest acres legally withdrawn for such

Table 1. Distribution of forest land in the United States*

Type of forest land	Area, 10^6 acres†				
	Total‡	Pacific Coast§	North	South	Rocky Mountains
Commercial forest	487.7	70.8	170.8	188.4	57.8
Noncommercial forest	—	—	—	—	—
Productive reserved	19.5	4.1	5.1	1.9	8.4
Productive deferred	4.6	1.2	.2	.1	3.2
Other forest	228.3	138.4	4.8	16.7	68.4
Total noncommercial	252.4	143.7	10.1	18.7	80.0
Total forest land	740.1	214.5	180.8	207.1	137.7

*From *U.S. Forest Service, Forest Statistics of the U.S., 1977,* USDA, review draft, 1978.
†10^6 acres = 4046 km^2.
‡Totals may not agree in last figure due to rounding.
§Pacific Coast includes both Alaska and Hawaii.

uses as national parks, state parks, and national forest wilderness areas. Another 4.6 × 10^6 acres, classed as "productive deferred," is under study for possible inclusion in the wilderness system. Of the remaining noncommercial forest, about 108 × 10^6 acres are in Alaska (part of the Pacific Coast region). Most discussions of forest resources, however, concentrate upon the commercial areas, and these lands are the primary concern of the following discussion.

The South alone has about 38.6% of the total commercial area; 35% is in the North, and the remaining 26.4% in the West. Within this general pattern are very large differences. In Maine, for example, 81% of the land surface is covered with commercial forests, while North Dakota is at the opposite extreme, with only 1% of its area similarly utilized.

For many years, changes in United States agriculture led to abandonment of marginal farms, which rapidly reverted to forest. This "new forest" was more than enough to offset those areas lost to highways, pipelines, urban development, and such. In fact, between 1943 and 1963 the total commercial forest area increased by about 31.6 × 10^6 acres. Between 1963 and 1970, however, the total area of commercial forest land declined about 8.4 × 10^6 acres, mostly in the South and in the Rocky Mountains, and by 1977 a further reduction of 12.2 × 10^6 acres had occurred. Some of the early reduction in the West was the result of shifting public forest land to a reserved or deferred class to meet demands for recreation uses. A later reclassification, however, restored 3.2 × 10^6 acres in the West to commercial status. In all regions, substantial areas have been taken over by suburban development, highways, reserves, and other nontimber uses, while in the South (where much of the early decline occurred) timberland has been cleared for crop production and for pasture.

Forest types. There are literally hundreds of tree species used for commercial purposes in the United States. The most general distinction made is that between softwoods (the conifers, or cone-bearing trees, such as pine, fir, and spruce) and hardwoods (the broad-leafed trees, such as maple, birch, oak, hickory, and aspen). Viewed nationally, 51% of United States commercial forest land is occupied by eastern hardwood types. Softwoods of various kinds make up 42%, western hardwoods only 3%; and 4% of the area is unstocked. Oak-hickory stands cover the largest area, accounting for 23% of all commercial timberland in 1977. The oak-pine type (14% of the eastern hardwood area) is mostly in the South and is primarily the residual resulting from the cutting of merchantable pine from mixed pine-hardwood forests. During the last few decades many of these stands have been converted to pine by the killing or cutting of hardwoods and, often, by the planting of pine, although little change, percentage-wise, has occurred since 1970.

Of the eastern hardwood forests, 44% are oak-hickory types, containing a large number of species but characterized by the presence of one or more species of oak or hickory. Other important eastern types are the maple-birch-beech (found throughout the New England, Middle Atlantic, and Lake states regions), the oak-gum-cypress forests (primarily in the Mississippi Delta and other southern river bottoms), and the aspen-birch type of the Lake states (relatively short-lived species that followed logging and fires). The bottomland hardwoods (oak-gum-cypress type) were reduced about 20% between 1962 and 1970, primarily by the clearing of forests for agriculture. For many years these forests have supplied much of the quality hardwoods in the United States.

Softwood types dominate the western forests, altogether occupying 83% of the region's commercial forest area. Douglas fir and ponderosa pine, the principal types, together constitute 45% of the region's commercial timberland. The western softwood types are the principal sources of lumber and plywood in the United States. Nearly all the commercial forest area of coastal Alaska is of the hemlock–Sitka spruce type. Hardwoods, mostly in Washington and Oregon, occupy only 12% of the West's commercial forest area, but have increased substantially since 1962 as the Douglas fir forests have been cut.

Growth. Growth on these areas has been more than enough to match the harvest since the mid-1960s. This is commendable, but the fact that growth and drain are in approximate balance provides no assurance that all is well. Much of the growth is still on low-quality hardwoods in the East, while about two-thirds of United States demand for the raw materials of the forest is met by the softwood production of the West and South. Potentially the most productive forest land is that of the Pacific Coast states, where it is estimated that 25 × 10^6 acres are capable of growing more than 120 ft^3/acre/year, and that another 22 × 10^6 acres could produce more than 85 ft^3 acre/year. The next most productive area is the South, with only 12 × 10^6 acres of highest-quality forest (more than 120 ft^3 acre/year) but some 46 × 10^6 acres of good-quality land (85–120 ft^3/acre/year), and 101 × 10^6 acres producing 50–85 ft^3/acre per year. The North is considerably less well off as far as forest growth rates are concerned.

Growth per acre on forests of the United States has been rising steadily—in all regions and on all ownerships. The increase has been particularly dramatic over the past quarter-century: since 1952, average per-acre growth has increased from 28 to 45 ft.[3] This 61% rise is due at least in part to efforts made by American forestry on both public and private lands. But there is still far to go: net growth per acre is far below that possible under intensive management, and only three-fifths of what could be achieved from fully stocked natural stands.

Timber inventory. Only soil productivity exceeds stocking (the number of trees per acre) and the age of the trees in importance as factors determining timber production.

Stocking. The commercial forests contained a truly vast amount of sound wood—800,803 × 10^6 ft^3 (22,676 × $10^6 m^3$) at the beginning of 1977. Only 8.4% of this consisted of trees that were dead, diseased, or in such poor condition they were not commercially useful. Another 2.6% consisted of trees which were dead but still usable for timber. Some 63% of the total volume was in trees sufficiently large to yield at least one sawlog: such trees are called sawtimber (Fig. 1). Another 25%

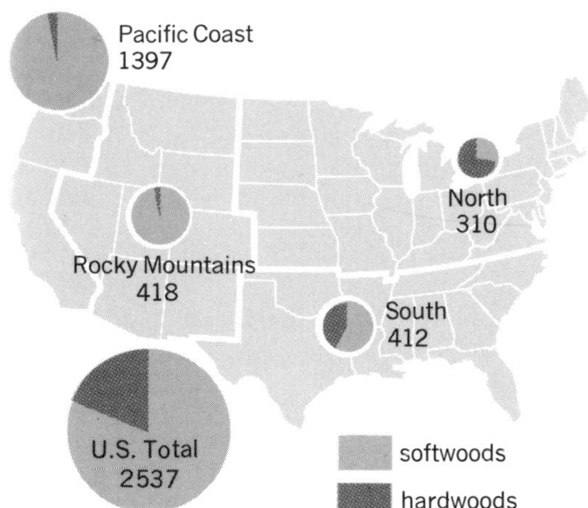

Fig. 1. Map indicating distribution of sawtimber in billions of board feet for all species as of Jan. 1, 1963. (*U.S. Department of Agriculture*)

was in pole timber—trees ranging from 5 in. (13 cm) in diameter (breast high) up to sawtimber size. Softwoods accounted for 64% of the total growing stock (pole timber plus sawtimber) and for 73% of the portion in sawtimber sizes, with 76% of the latter in the western states. This distribution highlights an important facet of timber distribution: the western states have only 26% of the commercial forest land and 23% of the total growing stock of the nation. But these western forests have over 75% of the important softwood sawtimber, and 69% of the nation's softwood growing stock (Table 2). The East, on the other hand, has 91% of the total hardwood growing stock (about equally divided between the North and South) and 89% of the hardwood sawtimber volume.

Age. Age is the other factor that must be considered before the significance of growth (and removals) can be understood. The virgin softwood stands of the West are growing relatively slowly, but as these old stands are replaced with young, vigorous trees, growth on western forests will begin to balance the cut—although almost surely at a reduced level. In 1976, net growth of western softwood growing stock averaged 90% of softwood removals, but removals of sawtimber exceeded growth by fully 45% (nearly 10^{10} bd ft or 2.36×10^7 m^3). Harvest of western sawtimber cannot be maintained at present levels unless forest management is greatly intensified. In the East, however, net growth of softwood growing stock exceeded removals by 52%, while softwood sawtimber growth was 33% above removals. Net growth of all eastern hardwood growing stock was fully 215% above removals, and growth in sawtimber sizes was 159% of removals. Here it should be noted that hardwood harvesting is concentrated in high-valued species such as walnut, cherry, gum, yellow birch, and maple, and in trees of large diameter, while much of the growth is in trees of less preferred species, and of small size. Annual growth and removal data are summarized in Table 2.

Ownership. The future condition of forest land in the United States is dependent to a very large extent on the decisions of the people who own these areas. The key factors in understanding the forest situation, therefore, are the forest land ownership pattern and the attitude these owners have toward forest management and hence toward the future of the forest lands they hold.

About 72% of the commercial forest is held in private ownership (351.1×10^6 acres). The remaining 28% is held by various public owners, with about 18% in national forests, 2% in other Federal ownership, 5% in state holdings, 2% under county and municipal control, and slightly over 1% under native-American sovereignty. Among private owners, three major classes can be distinguished: industrial, farm, and a very heterogeneous group labeled miscellaneous. About 14% of the commercial forest land is in the hands of industry, the

Table 2. Summary of net annual growth and removals of growing stock and sawtimber, by species group and region*

	Growing stock, 10^9 ft^3†			Sawtimber, 10^9 bd ft‡		
Section and species group	Net growth	Removals	Growth-removal ratio	Net growth	Removals	Growth-removal ratio
North						
Softwoods	1.6	0.7	2.3	4.0	2.2	1.8
Hardwoods	3.9	1.9	2.1	8.7	5.7	1.5
South						
Softwoods	6.3	4.5	1.4	24.2	19.0	1.3
Hardwoods	4.9	2.2	2.2	13.6	8.3	1.6
Rocky Mountains						
Softwoods	1.6	0.8	1.9	6.4	4.8	1.3
Hardwoods	0.1	—	18.9	0.4	0.1	3.7
Pacific Coast						
Softwoods	2.9	4.2	0.7	14.6	25.7	0.6
Hardwoods	0.5	0.1	4.3	1.7	0.4	3.9
United States						
Softwoods	12.4	10.2	1.2	49.2	51.7	0.9
Hardwoods	9.5	4.3	2.2	24.4	14.5	1.7

*From U.S. Forest Service, *An Assessment of the Forest and Range Land Situation in the United States*, USDA, review draft, 1978.

†10^9 ft^3 = 28,316,846 m^3.

‡10^9 bd ft = 2,359,737 m^3, as nominal recovered lumber.

remainder being divided between farmers (24%) and miscellaneous private owners (34%). Table 3 gives details about forest ownership, as well as the distribution of privately owned forests, in the four major regions of the United States.

Private owners. Some of the most productive forest land of the United States is in private industrial holdings. Pulp and paper companies lead the forest-based industries, with much of their forest land concentrated in the southern states, where nearly 36×10^6 acres is industrially owned. As forest industries become integrated, it is increasingly difficult to make distinctions between pulp and paper, lumber, and plywood companies, and certainly the distinctions are much less meaningful than in former years.

The miscellaneous private category, which holds 34% of the privately owned forest land, consists of a tremendous variety of individuals and groups, ranging from housewives to mining companies. Relatively few of these owners are holding this land for commercial timber production. Some owners, such as railroad companies or oil corporations, may indeed be interested in producing timber while holding the subsurface mineral rights, but miscellaneous private owners are usually interested in other, nontimber, objectives.

Most farm forests have been cut over several times; and, because farmers are primarily interested in the production of other kinds of crops, they tend to be less concerned with the condition of their woodlands than are other owners. The condition of farm-owned woodlands has been a source of much disappointment, discussion, and considerable action on the part of public conservation agencies. Whether this situation is of crucial significance to the production of forest products in the United States is a matter of dispute. In past years some professional foresters have argued that these farm woodlands should somehow be made to contribute their full share to the timber supply of the nation. Others, of a more economic persuasion, have argued that, as long as farmers have opportunities for investing time and money in ways that will yield a greater return, they should not be expected to worry about timber production. Farm owners are often unfamiliar with forest practices, usually lack the capital required for long-term investments, and in many cases are simply not interested in growing trees.

A recent study, however, indicated that when compared with other forest land of the same quality and within the same region, farm woodlot management levels were only slightly, if at all, below that given other ownership categories.

Public owners. Public agencies of several kinds hold large forest areas, the most important being the national forests. These contain 89×10^6 acres of commercial forest land which is managed and administered by the U.S. Forest Service, a bureau of the U.S. Department of Agriculture. The Bureau of Land Management oversees about 5.8×10^6 acres. Various other agencies, especially those of the armed services, administer 4.85×10^6 acres under Federal supervision. State ownerships total 23.6×10^6 acres, and counties and municipalities control another 7.22×10^6 acres. The remainder, somewhat over 6×10^6 acres, is under native-American sovereignty. Most of the public holdings are managed under multiple-use principles. As on most forest land, wood has always been the principal product of the national forests. With the passage of the 1974 Planning Act and the 1976 National Forest Management Act, the nontimber uses were given increased official attention, and the debate between those favoring timber production and those primarily concerned with range, water, wildlife, and recreation has been much intensified.

Public holdings now contain 54% of the softwood growing stock but only 17% of the hardwoods. Sixty four percent of all softwood sawtimber is publicly owned, and most is in national forests. The high concentration of sawtimber in these areas makes many wood-based industries in the United States highly dependent upon government owned raw material supplies.

World forest resources. Since much of the world's forested area has yet to be surveyed, data concerning volume, species, or even area undoubtedly contain substantial errors. Recent advances in remote sensing, particularly the information being relayed from satellites, give promise of vast

Table 3. Area of commercial timberland in the United States by type of ownership and section, Jan. 1, 1977*

	Total United States		Region			
	Area, 10^3 acres†	Proportion %	North, 10^3 acres†	South, 10^3 acres†	Rocky Mountains, 10^3 acres†	Pacific Coast, 10^3 acres†
National forests	89,007	18.3	10,121	10.955	36,436	31,496
Bureau of Land Management	5,799	1.2	15	3	1,667	4,113
Other	4,849	1.0	1,079	3,343	75	352
All Federal	99,655	20.4	11,215	14,300	38,179	35,961
Native-American	6,089	1.3	1,028	185	2,711	2,165
State	23,642	4.9	13,129	2,519	2,203	5,791
County and municipal	7,216	1.5	5,945	738	75	458
All public	136,602	28.0	31,318	17,742	43,167	44,374
Forest industry	67,976	13.9	17,777	35,754	2,096	12,349
Farm	116,785	23.9	45,384	57,217	8,311	5,872
Miscellanous private	166,364	34.1	76,290	77,720	4,191	8,163
All private	351,124	72.0	139,451	170,691	14,598	26,384
All ownerships	487,726	100.0	170,769	188,433	57,765	70,758

*From U.S. Forest Service, *Forest Statistics of the U.S., 1977*, USDA, review draft, 1978.
†10^3 acres = 405 ha.

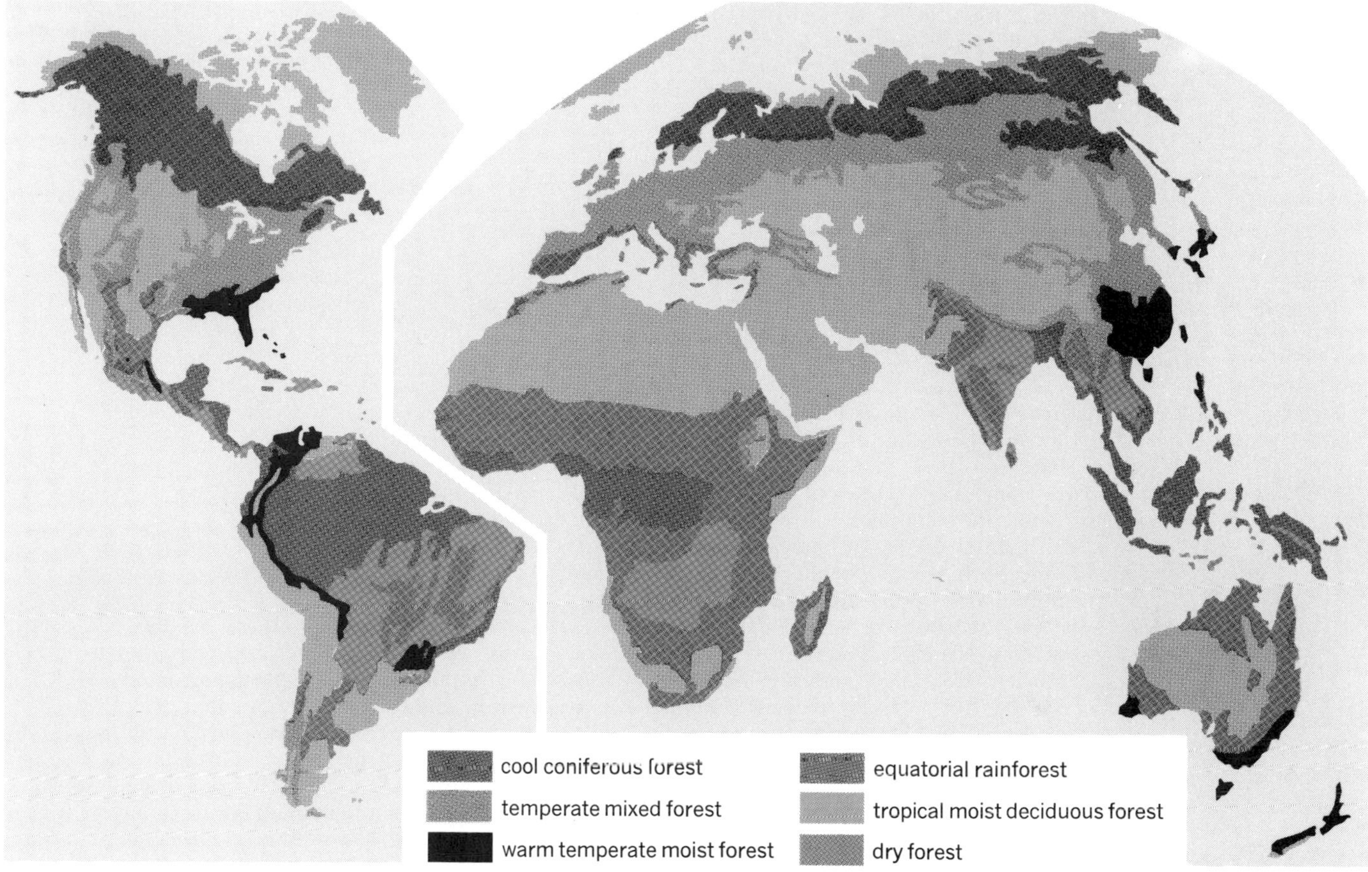

Fig. 2. The world's forests. *(From Economic Atlas of the World, 3d ed., Oxford University Press, 1965)*

improvement in knowledge of how the world's forests are faring. Currently, about 7.5×10^9 acres have a 20% or more tree crown cover—roughly one third of the world's land surface and about the same average percentage as that for the United States as a whole. Latin America and the tropical regions of Africa and Southeast Asia have most of the hardwood forests, while the softwood areas are concentrated in North America and the Soviet Union.

Latin America has about 43% of the world's hardwood growing stock, and this is nearly half again as much as the hardwood forested area of Africa—the continent next richest in hardwood. Together, Latin America and Africa account for 72% of the world's hardwoods, with Southeast Asia adding another 15%. North America has only 5% of the total. In the important softwood category, however, the Soviet Union and North America come in first and second, and together account for 83% of the world's softwoods. The Soviet Union has by far the largest share—about 2½ times as much as North America. Because of low productivity, great distance from markets, and rugged terrain, much of the softwood in the Soviet Union will be difficult to bring into commercial use.

Forests reflect differences in soil, climate, situation, and past land use and merely listing the forest types of the world would take several pages. Yet there is value in distinguishing very generalized forest types by broad locational patterns (Fig. 2).

Coniferous and temperate mixed forests. In Europe, the Soviet Union, North America, and Japan, the forests are predominantly coniferous, a fact that has been of great significance in shaping the pattern and nature of wood use in the industrialized part of the world. Closely associated with the industrialized nations are the temperate mixed forests, which normally contain a high proportion of conifers along with a few broad-leafed species.

Most of the temperate mixed forests are in use, for these heavily populated areas have well-developed transportation systems. Growth rates in both the coniferous and temperate mixed forests of the North Temperate Zone are similar to those already given for the United States, which is an excellent example of the general region.

Tropical rain forests. These forests are made up exclusively of broad-leafed species and include the bulk of the volume of the world's broad-leafed woods. They are concentrated in and around the Amazon Basin in South America, in western and west-central Africa, and in Southeast Asia. Generally characterized by sparse population and only slight industrial development, these forests have been little used. The fact that they typically contain many species within a small area (as many as a hundred species per acre) has also served to limit their use, even though tropical rain forests include some of the most valuable of all woods, such as mahogany, cedar, and greenheart (South America); okoume, obeche, lima, and African mahogany (Africa); and rosewood and teak (Asia).

The rate of growth in the tropics can be very high, and someday it may be possible to obtain a major part of the world's wood fiber needs from the more than 2×10^9 acres of these forests. Today, however, so little is known about many of the species, and so few areas are under any form of systematic management, that little reliance can be placed on these vast areas to satisfy the future wood needs of humanity.

Savanna. Most of the other forests in the tropics and subtropics are dry, open woodlands, or savannas. These forests contain low volumes per acre, mostly in small sizes, and only a few of the species are of commercial value. Much of Africa (excluding western and west-central Africa) is of this low-yielding type.

Management. In recent years intensive afforestation has received considerable attention. The use of measures such as careful soil preparation prior to planting, application of fertilizer, and even irrigation to adapt the environment to high-yielding species can result in an enormous increase in returns from forests. For example, yields of 400–500 ft^3/acre are common with eucalyptus in both South America and Africa, or with poplars in southern Europe. Only slightly smaller yields are obtained from the fast-growing pines under similarly intensive management. The high-yield potential of cultivated forests makes them much more important than their area might indicate. According to reports, it is the intention of most nations to have these forests make an even greater contribution in the future.

While fast-growing plantations could add a new component to the world's industrial wood supply, a major threat to virtually all forests in developing nations stems from the pressing need for fuelwood. Expanding populations in Africa, Latin America, and Southeast Asia are destroying forests at a truly alarming rate as increasing numbers seek the means to cook their food and warm their families. Rising petroleum prices have increased this pressure, and there is a very real danger that in some areas the deforestation could lead to ecological changes that would be difficult or impossible to reverse. Forest removal from hill regions and drought-prone areas has far-reaching consequences, since soil and wind erosion and drastic changes in stream flow are almost inevitable. These, in turn, threaten food supplies for major segments of the world's population.

[G. ROBINSON GREGORY]

Bibliography: S. L. Pringle, Tropical moist forests in world demand, supply and trade, *Unasylva*, vol. 28, no. 112–113, 1976, *An Assessment of the Forest and Range Land Situation in the United States*, USDA, review draft, 1979. *U.S. Forest Service*, *Forest Statistics of the U.S.*, *1977*, USDA, review draft, 1978.

Fossil fuel

A material whose combustion is used to supply heat for any purpose. Fossil fuels are carbon-containing materials that are burned with air or with oxygen derived from the air.

Classification. Fuels are characterized (1) by their physical form at normal temperatures, whether they are solids, liquids, or gases; (2) by their heating value, that is, by the amount of heat given off when a unit weight or volume of the material is burned under standard conditions; and (3) by their combustion characteristics, which cover such points as ease of ignition, rate of combustion, flame temperature, and flame luminosity.

Fuels may be used to supply heat directly, as in a furnace. They may also be used to supply heat to perform some other function, such as to produce steam in a boiler, the steam then being used to produce power in a turbine or engine. The heat may be used to raise the temperature and pressure of the products of the combustion to provide a driving force, as in the internal combustion engine or gas turbine. *See* ENGINE; TURBINE.

Solid fuels today comprise primarily the various ranks of coal (anthracite, bituminous, subbituminous, and lignite) and the coke or char derived from them. Wood, charcoal, peat, and various other plant products, while important under special circumstances, represent only a small fraction of total fuel consumed. The important factors governing choice of coal for a given use are cost per unit of heating value, size range, moisture content, ash content, and amount of smoke evolved. *See* COAL.

Liquid fuels are derived almost entirely from petroleum. Historically, coal oil (a kerosine derived from coal tar) was used extensively for heat and light, but it is now largely supplanted by petroleum products. In general, the liquid fuels suitable for internal combustion engines command a premium, and the fuels designed for external combustion are derived from those portions of the petroleum that have the least potential for gasoline and diesel fuel manufacture. Factors in choice of liquid fuels are heating value, fluidity, boiling range, impurities such as sulfur, and freedom from water and sediment. For fuels for internal combustion engines *see* DIESEL FUEL; FUEL OIL; GASOLINE; LIQUID FUEL; OIL SAND; OIL SHALE; PETROLEUM PRODUCTS.

Of the gaseous fuels, the most important are those derived from natural gas or petroleum, chiefly methane. Availability of these has increased sharply since World War II, and they have largely supplanted manufactured fuel gas for general distribution by public utilities. Manufactured gas was usually coke-oven gas or carbureted water gas made from coke, steam, and petroleum oil. During the era of gas illumination, the carbureting step yielded a gas which burned with a luminous flame. Other gaseous fuels are those that are by-products of some manufacturing process and therefore available at a cost that makes them competitive with natural gas. These include coke-oven gas, blast-furnace gas, and producer gas. *See* FUEL GAS; NATURAL GAS.

Heating value. The heating value of gaseous fuels varies widely with composition. For this reason it is generally important to have an accurate and continuous determination of the heating value, and much effort has been devoted to the development of an apparatus for this purpose. Carefully metered streams of gas and combustion air are burned under constant temperature conditions in a submerged chamber. The temperature rise of a metered stream of cooling air is used as a measure of the heating value and is usually recorded continuously on a chart. This value in the United States is usually expressed as British thermal units (Btu) per cubic foot of gas (see table). *See* BRITISH THERMAL UNIT (BTU).

Heating values of solid and liquid fuels are determined by oxidation of a weighed sample in a system whose heat capacity is known and whose

Approximate heating values of representative fuels

Fuel	Heating value*
Anthracite coal	13,500 Btu/lb
Bituminous coal (low-volatile)	14,500 Btu/lb
Lignite	7,200 Btu/lb
Fuel oil	19,000 Btu/lb
Natural gas	1,100 Btu/ft^3
Producer gas	129 Btu/ft^3
Butane	3,200 Btu/ft^3

*1 Btu/lb = 2326 J/kg. 1 Btu/ft^3 = 37.26 kJ/m^3.

temperature rise can be measured. The oxidant may be gaseous oxygen or an oxidizing material such as sodium peroxide.

Heating values are reported as gross (or higher) heating value when the water formed in combustion is condensed, and as net (or lower) heating value when the water from combustion leaves the system as a vapor and the heat of condensation is not included. *See* COMBUSTION; ENERGY SOURCES; HEATING VALUE.

[HOWARD R. BATCHELDER]

Bibliography: W. Francis, *Fuels and Fuel Technology*, vols. 1 and 2, 1965; D. A. Williams and G. Jones, *Liquid Fuels*, 1963.

Fuel cell

An electric cell that converts the chemical energy of a fuel directly into electric energy in a continuous process. The efficiency of this conversion can be made much greater than that obtainable by thermal-power conversion. In the latter the chemical reaction is made to produce heat by combustion. The heat is then transformed partially into mechanical energy by a heat engine, which drives a generator to produce electric energy. Further loss is involved if the direct current generated is converted into alternating current.

Although, in principle, the nature of the reactants is not limited, the fuel-cell reaction almost always involves the combination of hydrogen with oxygen, as shown by reaction (1). At 25°C and 1

$$H_2(g) + {}^1\!/_2O_2(g) \rightarrow H_2O(l) \quad (1)$$

atm pressure, that is, standard temperature and pressure (STP), the reaction takes place with a free energy change ΔG of −56.69 kcal/mole, that is, 237,000 J/mole water.

If the reaction is harnessed in a galvanic cell working at 100% efficiency, a cell voltage of 1.23 V results. In actual service such cells have shown steady-state potentials in the range 0.9–1.1 V, with reported coulombic efficiencies of the order 73–90%.

Fuel cells are of 200–500 W capacity and 50–100 mA/cm^2 current density. Larger prototypes have been produced, some as large as 15 kW capacity, while a system under study is expected to provide 100 kW.

In the present stage of development, it is difficult to make a classification of the fuel-cell types. The most popular and successful type remains the classical H_2-O_2 fuel cell of the direct or indirect type. In the direct type, hydrogen and oxygen are used as such, the fuel being produced in independent installations. The indirect type employs a hydrogen-generating unit which can use as raw material a wide variety of fuel. The reaction taking place at the anode is as in reaction (2), and at the cathode as in reaction (3).

$$2H_2 + 4OH^- \rightarrow 4H_2O + 4e^- \quad (2)$$

$$O_2 + 2H_2O + 4e^- \rightarrow 4OH^- \quad (3)$$

Because of the low solubility of H_2 and O_2 in electrolytes, the reactions take place at the interface electrode-electrolyte, requiring a large area of contact. This is obtained with porous materials called upon to fulfill the following main duties: the materials must provide contact between electrolyte and gas over a large area, catalyze the reaction, maintain the electrolyte in a very thin layer on the surface of the electrode, and act as leads for the transmission of electrons.

The porosity is obtained by the Raney technique or by sintering. When flooding of the pores is feared, the electrode is made with double porosity, fine at the electrolyte side and coarse at the gas side. The catalytic effect is obtained with noble metals, mainly platinum, silver, nickel, cobalt, and palladium.

The thickness of the electrolyte layer, on which depends the internal resistance of the cell, is controlled by pore size, wetting properties, and pressure of the fuel gas. When pressure is used, care must be taken not to increase it to the extent that gas is allowed to bubble through the electrolyte because of the danger of forming an explosive hydrogen-oxygen mixture.

The fuel cells may work with acid or alkaline electrolytes.

The acid electrolytes require costly corrosion-resistant construction materials but are not sensitive to CO and CO_2 in the fuel, which may lead to the buildup of carbonates. Some models using phosphoric acid proved quite successful. The alkaline electrolytes are more practical, and they are found in most fuel cells produced industrially at present.

Some fuel cells are designed to work with molten carbonates as electrolyte, at temperatures as high as 800°C. These cells are attractive because they can use reformed hydrocarbon fuels, require a small investment, and can be made as large units. They are insensitive to carbon oxides but, at least in the present situation, are affected by important shortcomings. Mainly, these shortcomings are excessive size, rapid corrosion of metallic parts, and long periods of heating required before useful service.

The cells may use any alkali metal carbonate or eutectic mixtures of the same. The reactions taking place are those known for the systems involving hydrogen and oxygen or carbon monoxide and oxygen. In the case of carbon monoxide, the reaction step at the anode is as in reaction (4), and at the cathode as in reaction (5).

$$CO + CO_3^{--} \rightarrow 2CO_2 + 2e^- \quad (4)$$

$$CO_2 + {}^1\!/_2O_2 + 2e^- \rightarrow CO_3^{--} \quad (5)$$

The total cell reaction is given by Eq. (6), with a

$$2CO + O_2 \rightarrow 2CO_2 \quad (6)$$

free energy change ΔG of −61.45 kcal/mole of CO. At 100% efficiency, this gives a theoretical cell voltage of 1.34 V at STP.

Principal fuel-cell reactions. The principal overall reactions which have been employed in

Table 1. Theoretical cell potentials at various temperatures

Reaction	Cell potential, V					
	25°C	100°C	250°C	500°C	750°C	1000°C
$C+O_2 \rightarrow CO_2$	1.02	1.02	1.02	1.02	1.02	1.01
$2C+O_2 \rightarrow 2CO$	0.71	0.75	0.82	0.93	1.04	1.15
$2CO+O_2 \rightarrow 2CO_2$	1.33	1.30	1.23	1.11	1.00	0.88
$2H_2+O_2 \rightarrow 2H_2O$	1.23	1.18	1.12	1.05	0.97	0.90

fuel-cell work are summarized in Tables 1 and 2.

The direct anodic use of carbon has been practically abandoned in modern fuel-cell work. Carbon potentials seem entirely due either to carbon monoxide, CO, or to hydrogen, H_2, formed at high temperature by direct reaction between the carbon and the electrolyte. For example, in the Jacques cell, which consists of carbon electrodes and iron (air) electrodes in molten sodium hydroxide, H_2 is liberated at the carbon by reaction with the electrolyte. It is this H_2 which is responsible for the observed potential.

Modern fuel cells use gaseous fuels, either H_2 or CO or mixtures of these gases. The oxidizer is normally oxygen or air. Hydrocarbons have not been made to function anodically. Where potentials have been measured, they are attributed to decomposition of the hydrocarbon to liberate H_2. For example, methane decomposes at high temperatures, as shown by reaction (7).

$$2CH_4 \rightarrow 2C + 4H_2 \quad (7)$$

For technical reasons, it is simpler to use the carbon or hydrocarbon fuel in a chemical reactor to produce the active gases, H_2 and CO, than to attempt to operate a cell under the conditions best suited for the chemical reaction. Typical chemical production of the active gases might be as in reaction (8).

$$2C + O_2 \rightarrow 2CO \quad (8)$$

Theoretically, CO has a requirement of 1.15 lb/kWh (0.52 kg/kWh) when reacted anodically against an oxygen cathode. To produce 1.15 lb (0.52 kg) CO, 0.493 lb (0.22 kg) carbon is needed. Hence a perfect process, starting with 0.493 lb (0.22 kg) carbon, would yield 1 kWh. A mixture of H_2 and CO can also be produced by reacting carbon with steam, as shown by reaction (9).

$$C + H_2O \rightarrow CO + H_2 \quad (9)$$

The complete engineering design of the chemical reactor in conjunction with the fuel cell has been extensively studied. The main difficulties in the past were due to the large concentration of CO in the reformed fuel, which poisons the Pt catalyst often used in the fuel cell proper. At present, various new catalysts have been developed, such as Pt-Rh, Pt-Ir, and Pt-Ru, well capable of processing H_2-CO fuel mixtures.

Table 2. Theoretical material consumption

Reaction	Temperature, °C	Consumption, kg/kWh		
		Anode	Cathode	Total
$C+O_2 \rightarrow CO_2$	750	0.156	0.417	0.573
$2C+O_2 \rightarrow 2CO$	750	0.215	0.287	0.502
$2CO+O_2 \rightarrow 2CO_2$	750	0.523	0.298	0.821
$2H_2+O_2 \rightarrow 2H_2O$	100	0.032	0.255	0.287
	750	0.039	0.309	0.348

The present fuel generators are capable of covering most of the needs of established fuel cells from almost pure hydrogen to 1:1 mixtures of H_2 and CO. One project, sponsored by the Office of Coal Research, U.S. Department of the Interior, has as its object a 100-kW coal reactor—solid electrolyte fuel-cell system.

Hydrogen-oxygen fuel cell. Work with H_2 has established that it can operate efficiently at moderate temperatures, with polarization decreasing as the temperature is increased. This permits the use of aqueous solutions. One has been reported to have, at 25°C, the following characteristics:

Current density, mA/cm²	0	1.1	10.8	54.0
Cell voltage	1.12	1.01	0.95	0.70

From the point of view of the working principle, three well-developed systems should be mentioned: those of General Electric, Allis-Chalmers, and Bacon. Some of their general characteristics are listed in Table 3.

In the same category should be mentioned the hydrazine-air fuel cell under investigation by Monsanto Research Corporation and Allis-Chalmers, together with some research centers of the U.S. Army.

This cell is based on the reaction shown in reaction (10). It is a medium-size system intended

$$N_2H_4 + O_2 \rightarrow N_2 + 2H_2O \quad (10)$$

to produce 60–300 W for the Monsanto project and up to 3 kW for the Allis-Chalmers project. The unit cell voltage is 0.6–0.7 V, and its greatest advantage is that it uses a condensed fuel, convenient in some applications. A different concept is used in the alkaline metal—oxygen fuel cells which were developed by the M. W. Kellogg Company. In an actual unit the electrochemical process involves oxidation of Na with oxygen from air, with NaOH as electrolyte. The reaction at the anode is shown in reaction (11) and at the cathode in reaction (12), with the cell reaction given as reaction (13).

$$4Na \rightarrow 4Na^+ + 4e^- \quad (11)$$

$$O_2 + 4e^- + 2H_2O \rightarrow 4OH^- \quad (12)$$

$$4Na + O_2 + 2H_2O \rightarrow 4NaOH \quad (13)$$

Because Na as such is too reactive, the cell uses a sodium amalgam which is quite stable in concentrated NaOH. The sodium amalgam–oxygen fuel cell possesses some remarkable features. It provides almost 1.5 V at steady state and very high current densities of the order of 200 mA/cm², is insensitive to water quality, and requires a small gas consumption.

Ion-exchange membrane types. An interesting fuel cell in the laboratory stage is the one which uses inorganic ion-exchange membranes. This type of cell offers two potential advantages: it tolerates higher-temperature operation and has a higher ionic conductivity because of the higher density of ion-exchange sites.

An advanced laboratory fuel-cell model developed by Armour Research Foundation uses zirconyl phosphate as an ion-exchange electrolyte, hydrogen and oxygen as fuels, and platinum black

Table 3. Characteristics of three fuel-cell systems

Type	Principle	Temperature, °C		Application
General Electric	Ion-exchange membrane	25–35	100–1000	Gemini
Allis-Chalmers	Porous Ni electrodes and porous electrolyte vehicle	90–100	2000	Space flight
Bacon (Pratt and Whitney)	Porous Ni electrodes	200–220	500–1500	Apollo

as catalyst. The fuel cell is capable of delivering almost 1 V, but the current density reached so far is too small. The inorganic ion-exchange membrane requires water for its operation.

In the cells using solid electrolytes of the ionic conductive type, the need for water is eliminated. An electrolyte studied by Westinghouse Electric is a mixture of zirconium oxide and calcium oxide (0.85:0.15) and is expected to work at about 1000°C. Another promising electrolyte is the mixture zirconium oxide and yttrium oxide (0.9:0.1).

Closed-cycle types. The last fuel-cell type to be considered is the closed-cycle type, in the frame of which the reactants are recovered by an auxiliary process. Some systems, such as those developed by Electro-Optical Systems Division of Xerox, are simply made of two converse cycles operating one at a time. The regeneration consists in producing hydrogen and oxygen by electrolysis when electrical energy is available and shifting to cell performance on stored fuel when the production of electrical energy becomes necessary. In the regenerative system developed by the United Aircraft, the electrolysis current is provided by a solar-energy converter made of regular silicon solar cells.

Another possibility is that of radiochemical regeneration. A model under development by Union Carbide is that proposed by J. A. Ghormley, using ferrous sulfate irradiated by gamma radiation. The electrical efficiency reported to the gamma radiation absorbed had been evaluated at 3%. The unit cell voltage is about 0.6 V, and the module containing six units should deliver 5 W over a period of 2 years.

Problem areas. In the development of fuel cells, there are some general difficulties which must be solved before they become mature industrial propositions. The main problems are as follows.

Catalyst. Its importance increases while the temperature decreases. With few exceptions the catalyst is a very expensive constituent of the cell, and not only will its price increase when industrial production requires increased amounts, but it may even become unavailable at any price. This suggests that future models will tend to work at higher temperatures and pressures (like Bacon's fuel cell) at which cheaper catalysts (Ni, NiO, and so on) can be used.

Capital cost. Since it requires a large per-kilowatt investment, the fuel cell is not commercially attractive. However, by various standard improvements, its capital cost should be reduced so that it will become competitive with other electrical energy sources.

Heat transfer. In most fuel cells the reaction product is water in gaseous or liquid form. At 100% efficiency there are 421 g of water per kilowatt-hour to dispose of. In addition, about 30% of the heat of reaction, that is, about 260 kcal/kW, must be disposed of in order to maintain the working temperature at a level at which the electrolyte is not decomposed and the material does not become too susceptible to corrosion.

High-temperature fuel cells. The use of carbon monoxide has been limited to high-temperature cells. One carbon monoxide cell using air as the cathodic material operated at about 700°C to yield 0.75–0.85 V at 0.034 A/cm^2. It has been demonstrated that molten-salt electrolyte cells can operate at 550–800°C on inexpensive hydrocarbons if steam is admitted to prevent carbon deposition. *See* DRY CELL.

[J. DAVIS; L. ROZEANU]

Bibliography: D. H. Archer and R. L. Zahradnik, The design of a 100-kilowatt, coal-burning fuel cell system, *Chem. Eng. Progr.*, 63:55, 1967; N. P. Chopey, What you should know about fuel-cells, *Chem. Eng.*, no. 125, May 25, 1964; E. Findl and M. Klein, Electrolytic regenerative hydrogen-oxygen fuel-cell battery, *Proceedings of the 20th Annual Power Sources Conferences*, 1966; M. I. Gillibrand and G. B. Lomax, Factors affecting the life of fuel cells, *Proceedings of the 20th Annual Power Sources Conferences*, 1966; D. W. McKee and A. G. Scarpellino, Electrocatalysts for hydrogen/carbon monoxide fuel cell anodes, *Electrochem. Technol.*, 6:101, 1968; E. Yeager and W. Mitchell, Jr. (eds.), *Fuel Cells*, 1963.

Fuel gas

A fuel in the gaseous state whose potential heat energy can be readily transmitted and distributed through pipes from the point of origin directly to the place of consumption. The development and use of fuel gases is closely associated with the progress of civilization. As shown in the illustration, such gases have become especially prominent in the industrial development period since 1900. Natural gas provides 30% of the energy needs of the United States, with LP (liquefied petroleum) fuel gases providing nearly another 2%.

The types of fuel gases are natural gas, LP gas, refinery gas, coke oven gas, and blast-furnace gas. The last two are used in steel mill complexes. Typical analyses of several fuel gases are presented in the table. Since these analyses are based on dry gases, the heating value of gases saturated with water vapor would be slightly lower than the values shown. *See* HEATING VALUE.

Most fuel gases are composed in whole or in part of the combustibles hydrogen, carbon monoxide, methane, ethane, propane, butane, and oil vapors and, sometimes, of mixtures containing the inerts nitrogen, carbon dioxide, and water vapor.

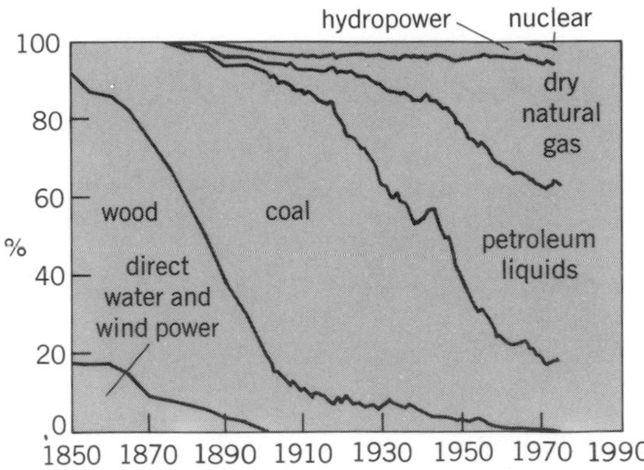

Contribution of various sources to primary energy consumption in the United States, 1850–1974.

Natural gas. The generic term "natural gas" applies to gases commonly associated with petroliferous geologic formations. As ordinarily found, these gases are combustible, but nonflammable components such as carbon dioxide, nitrogen, and helium are often present. Natural gas is generally high in methane. Some of the higher paraffins may be found in small quantities.

The olefin hydrocarbons, carbon monoxide, and hydrogen are not present in American natural gases. The term "dry natural gas" indicates less than 0.1 gal (1 gal, U.S. = 0.003785 m^3) of gasoline vapor occurs per 1000 ft^3 (1 ft^3 = 0.028 m^3); "wet natural gas" indicates more than 0.1 gal/1000 ft^3. "Sweet" and "sour" are terms that indicate the absence or presence of hydrogen sulfide.

There is no single composition which might be termed typical natural gas. Methane and ethane constitute the bulk of the combustible components; and CO_2 and nitrogen, the inerts. The net heating value of natural gas served by a utility company is often 1000–1100 Btu/ft^3 (1 Btu = 1055 joules).

Natural gas is an ideal fuel for heating because of its cleanliness, ease of transporation, high heat content, and the high flame temperature. *See* LIQUEFIED NATURAL GAS (LNG); NATURAL GAS.

LP gas. This term is applied to certain specific hydrocarbons, such as propane, butane, and pentane, which are gaseous under normal atmospheric conditions but can be liquefied under moderate pressure at normal temperatures. *See* LIQUEFIED PETROLEUM GAS (LPG).

Oil gas. This term encompasses a group of gases derived from oils by exposure of such oils to elevated temperatures. Refinery oil gases are those obtained as by-products during the thermal processing of the oil in the refinery. They are used primarily for heating equipment in the refinery. Gas made by thermal cracking of oil was formerly very important as an urban fuel gas, but has been almost totally displaced by natural gas, and the equipment for its production dismantled. A typical oil gas consists of saturated and unsaturated hydrocarbons and has a heating value of 1300–2000 Btu/ft^3. Methane, ethane, propane, butane, ethylene, and propylene are the main constituents.

Coal gases. Until about 1940, gas produced from coal was an important part of the energy mix in the United States. These gases were rapidly replaced by natural gas in the distribution systems of gas utilities serving the residential, commercial, and industrial markets. During 1971 it became clear that the supply of natural gas would probably be insufficient to meet the ever-increasing demand. Therefore, as of the late 1970s, it is expected that gas will once again be produced from coal. In this period, the older processes will be replaced by modern methods capable of very high production rates at improved efficiency and reduced cost.

Coal gasification can be accomplished in a large number of ways, including pyrolysis or partial oxidation with air or oxygen and steam. Various processes operate as fixed beds or fluidized beds or with the coal entrained. The pressure may vary from near atmospheric to 1000 psig (1 psi = 6895 Pa) or more. All operate at high temperatures (1200–3000°F; 650–1645°C). The direct products of gasification vary in heating value from 120 to 150 Btu/ft^3 (low-Btu gas), to 300 Btu/ft^3 (medium-Btu gas or synthesis gas), to as high as 600 Btu/ft^3. Low-Btu gas can be used as fuel for industrial processes or for production of electrical power by electric utilities. Medium-Btu gas, which consists principally of carbon monoxide and hydrogen, can be used directly as a fuel, or it can be upgraded by catalytic methanation to essentially pure methane, which is, for all practical purposes, identical to natural gas. Such a product gas is commonly called pipeline-quality gas or substitute natural gas (SNG). *See* COAL GASIFICATION.

Low-Btu gas. Producer gas is made by oxygen-deficient combustion of coal or coke, in which process a mixture of air and steam is blown upward through a thick hot bed of coal or coke. The gas is high in nitrogen introduced in the air. Its heating value is low, its specific gravity is high, and the percentage of inerts is high. Producer gas, which contains 23–27% carbon monoxide, is used as it

Typical gas analyses*

Type	Analysis, % vol								Specific Gravity	Btu/ft^3	
	CO_2	O_2	CO	H_2	CH_4	C_2H_6	C_3H_8 and C_4H_{10}	N_2		Gross	Net
Dry natural gas	0.2				99.2			0.6	0.56	1007	906
Propane (LP)						2.6	97.3	0.1	1.55	2558	2358
Refinery oil gas		0.2	1.2	6.1	4.4	72.5	15.0	0.6	1.00	1650	1524
Coke oven gas	2.0	0.3	5.5	51.9	32.3		3.2	4.8	0.40	569	509
Blast-furnace gas	11.5		27.5	1.0				60.0	1.02	92	92
Producer gas	8.0	0.1	23.2	17.7	1.0			50.0	0.86	143	133

*From L. Shnidman (ed.), *Gaseous Fuels*, 2d ed., American Gas Association, 1954.

comes from the generators after some preliminary purification. It was once the cheapest form of industrial gas, and could again become important; it may be possible to produce it in modernized equipment on a large scale.

Blast-furnace gas is a by-product from the manufacture of pig iron. Like producer gas, it is derived from the partial combustion of coke. Some of the combustibles in the gas are used to reduce the iron ore; thus the final gas contains about 27% carbon monoxide and more than 70% of inert gases (CO_2 and N_2), giving it the lowest heating value, less than 100 Btu/ft^3, of any usable fuel gas. It is used for the operation of gas engines, heating by-product coke ovens, steel plant heating, steam raising, and crude heating. *See* FOSSIL FUEL.

Medium-Btu gases. Coke oven gas is the only important gas that has an intermediate heating value. The gas is made by destructive distillation of a packed bed of coal out of contact with air. The process results in the formation of coke, which is used in the blast furnace. The gas is utilized totally within the steel-making complex.

The combination of the 1974 oil embargo, the energy crisis, and restrictions on air pollution resulted in the proposal of a large number of modern processes. Nearly all these processes use a mixture of steam and oxygen to combust coal. Less oxygen than is required for complete combustion is used; therefore, the products are primarily carbon monoxide and hydrogen. Depending on the end use of the gas, moderate amounts of methane and liquid products are produced.

SNG. Gas which is to be used within the modern natural gas transmission and distribution system must be essentially methane. The medium-Btu gas processes can be used as precursors to produce substitute natural gas (SNG), as noted above. Other new processes which operate at very high pressure (1000 psig) are being developd to produce very large quantities of gas (250,000,000 ft^3/day of 1000 Btu/ft^3 gas) to supplement the declining supplies of natural gas. Some of these processes use the partial combustion of oxygen as a direct source of heat, and others use air combustion in a variety of indirect modes. *See* SUBSTITUTE NATURAL GAS (SNG).

Nonfossil sources. In view of the immediate shortage of natural gas and the longer-range (A.D. 2030–2060) forecast that all fossil fuels will be consumed, there are several projects which may ensure a perpetual supply of fuel gas. These include conversion of waste materials to gas by pyrolysis or by anaerobic digestion. Large efforts have been initiated to convert solar energy to fuel gas. This would be accomplished by growing plants on land or sea and converting the harvest to gas in processes similar to those used on wastes. Finally, there has been a large effort to develop a "hydrogen economy." Hydrogen would be produced by thermochemical processes using heat from various sources such as atomic energy reactors.

[JACK HUEBLER]

Bibliography: *5th Synthetic Pipeline Gas Symposium*, American Gas Association cat. no. L51173, October 1973; Hammond et al., *Energy and the Future*, American Association for the Advancement of Science, 1973; W. J. Mead et al., *Transporting Natural Gas from the Arctic*, 1977.

Fuel oil

Any of the petroleum products which are less volatile than gasoline and are burned in furnaces, boilers, or other types of heaters. The two primary classes of fuel oils are distillate and residual. Distillate fuel oils are composed entirely of material which has been vaporized in a refinery distillation tower. Consequently, they are clean, free of sediment, relatively low in viscosity, and free of inorganic ash. Residual fuel oils contain fractions which cannot be vaporized by heating. These fractions are black and viscous and include any inorganic ash components which are in the crude. In some cases, whole crude is used as a residual fuel.

Uses. Distillate fuel oils are used primarily in applications where ease of handling and cleanliness of combustion are more important than fuel price. The most important use is for home heating. They are used in about 35% of United States homes with central heating systems, the remainder being heated with gas or electricity. Distillate fuel oils are also used in certain industrial applications where low sulfur or freedom from ash is important. Certain types of ceramics manufacture are examples. Increasing amounts of distillate fuel oils have been burned in gas turbines used for electricity generation. *See* ELECTRIC POWER GENERATION; GAS TURBINE.

Residual fuel oils are used where fuel cost is an important enough economic factor to justify additional investment to overcome the handling problems they pose. They are particularly attractive where large volumes of fuel are used, as in electric power generation, industrial steam generation, process heating, and steamship operation.

Combustion. Fuel oils are burned efficiently by atomizing them into fine droplets, about 50–100 μm in diameter, and injecting the droplet spray into a combustion chamber with a stream of combustion air. Small home heating units use a pressure atomizing nozzle. Larger units use pressure, steam, or air atomizing nozzles, or else a spinning cup atomizer. Older units employ natural draft to induce combustion air, but new units usually use forced draft, which gives better control of air-to-fuel ratio and increased turbulence in the combustion chamber. This permits more efficient operation and clean, smoke-free combustion. *See* OIL FURNACE.

Tests and specifications. Fuel oils are blended to meet certain tests and specifications which ensure that they can be handled safely and easily, that they will burn properly, and that they will meet air-pollution regulations. These tests include (1) flash point, which determines that the fuel can be stored and handled without danger of explosion; (2) pour point, which determines that it will not solidify under normal handling conditions; (3) carbon residue, which determines that it will not coke during handling or form undue amounts of carbon particulates during combustion; (4) sediment, which determines that it will not clog pumps and nozzles; (5) sulfur content, which determines how much sulfur dioxide will be emitted during combustion; (6) ash content; and (7) viscosity or resistance to flow.

Viscosity. Viscosity is an important factor in determining the grade of a fuel oil. Distillate fuels and light grades of residuals have a low viscosity so that they can be handled and atomized without heating. Heavier grades have high viscosity and must be heated in order to be pumped and atomized. The grade classification of the fuel oil determines what viscosity range it is blended to, and therefore the temperature level required in the preheat system.

Ash content. Fuel ash content may be important in residual fuels, even though ash content is negligible compared with coal, because certain ash components may cause slagging and corrosion problems. Vanadium and sodium are frequently present in oil ash. During combustion, they are converted to materials which tend to accumulate on boiler and superheater tubes. Under certain conditions, they may cause corrosion of these tubes.

Sulfur content. Sulfur content of fuel oil is carefully controlled in order to meet air-pollution regulations. This has required a drastic reduction in sulfur content of residual fuel oils in many areas since 1969. To achieve the reduction, extensive processing and changes in blending procedure have been required, which have caused major increases in the price of fuel oil.

In spite of the higher price, fuel oil demand has increased rapidly since 1970 because alternate fuels, coal and gas, either cannot meet the regulations or are not available in sufficient quantities to meet the demand. *See* COAL; NATURAL GAS; OIL ANALYSIS; PETROLEUM PRODUCTS.

A detailed list of specifications for fuel oils is given in the *Annual Book of ASTM Standards*, published by the American Society for Testing and Materials, Philadelphia. [C. W. SIEGMUND]

Furnace construction

A furnace is an apparatus in which heat is liberated and transferred directly or indirectly to a solid or fluid mass for the purpose of effecting a physical or chemical change. The source of heat is the energy released in the oxidation of fossil fuel (commonly known as combustion) or the flow of electric current through adjacent semiconductors or through the mass to be heated. In recent years, scientific and engineering effort has been made to utilize nuclear and solar energy for heating purposes. Therefore, according to the source of heat and method of its application, there are four categories of furnaces: combustion, electric, nuclear, and solar, in the order of their present commercial or industrial importance.

Furnaces employing combustion vary widely in construction, depending upon the application of the heat released, whether direct or indirect. Direct heat transfer is used, in regenerative refractory-type heaters, in flow systems in which reactants are injected into the combustion gases. Indirect heat transfer is employed in heaters in which the mass to be heated is kept separate from the combustion gases and made to flow in tubes which absorb and transmit the heat to the fluid to be heated.

Furnaces developed for indirect heat transfer can be divided into two classes. One class of heaters is used solely for general utility purposes, such as all types of boilers; and the second class is applied in the petroleum and chemical industries as an essential unit operation in refining or processing plants. The second category of furnaces will be discussed in this article.

Directly fired furnaces are employed in oil refineries and chemical process units whenever the temperature level to which a fluid must be heated is above that attainable with utility steam. Furnaces as a heat-transfer apparatus generally cost more per thermal unit of heat transferred than conventional tubular heat exchangers. According to the kind of service, furnaces in process units should be divided into two classes:

1. Those which perform solely a heating duty, that is, raising the temperature of a fluid and effecting essentially no change of state or of chemical composition. Furnaces which perform this one duty may be termed conventional heaters.

2. Those which handle a fluid undergoing a change during heating. The physical or chemical changes constitute an essential performance requirement. Typical applications of this order are associated with such processes as distillation or preheating of temperature-sensitive materials, pyrolysis of hydrocarbons or organic chemicals, and catalytic steam-gas reforming for the production of synthesis gas.

The conventional tubular heater came into industrial use, especially in oil refineries, about 1925 and was termed a tube still to distinguish it from the shell still, a horizontal cylindrical vessel mounted on top of a firebox. The design of these tube stills or furnaces, which is essentially the same as originally conceived by analogy with developments in steam-generating equipment, consists of pipes connected by 180° return bends forming a continuous coil and arranged in a refractory furnace setting, partly in the combustion chamber. Heat is absorbed mainly by radiation and partly in a confined flue-gas passage from the combustion chamber in which heat is absorbed mainly by convection. The fluid flow through the coil is generally countercurrent to the flow of combustion gases, first through convection tubes and then through the radiant-tube section; thus a reasonable thermal efficiency can readily be attained by providing convection tube surface to an economically justifiable extent. *See* PETROLEUM PROCESSING; STEAM-GENERATING FURNACE.

The furnaces provided in modern process units vary considerably in outer shape, arrangement of tubes, and type and location of burners, depending mainly upon the desire of the designer or manufacturer to be identified with a specific model. Some distinctive furnace designs are shown diagrammatically in Fig. 1.

Design. The furnace design for a given performance or thermal efficiency is usually evolved by the following procedure: (1) determination of the composition of the combustion products and the amount of the liberated heat which must be utilized to meet the postulated thermal efficiency; (2) allocation of heat to be absorbed by the heating elements located in the combustion or radiant chamber and in the convection section; (3) determination of the heat-transfer rate and heating surface area in the radiant section; and (4) determination of the heat-transfer rate and tube surface area in the convection section or sections.

Source of heat. The combustion products vary in composition according to the type of fossil fuel burned and the excess air used in the oxidation

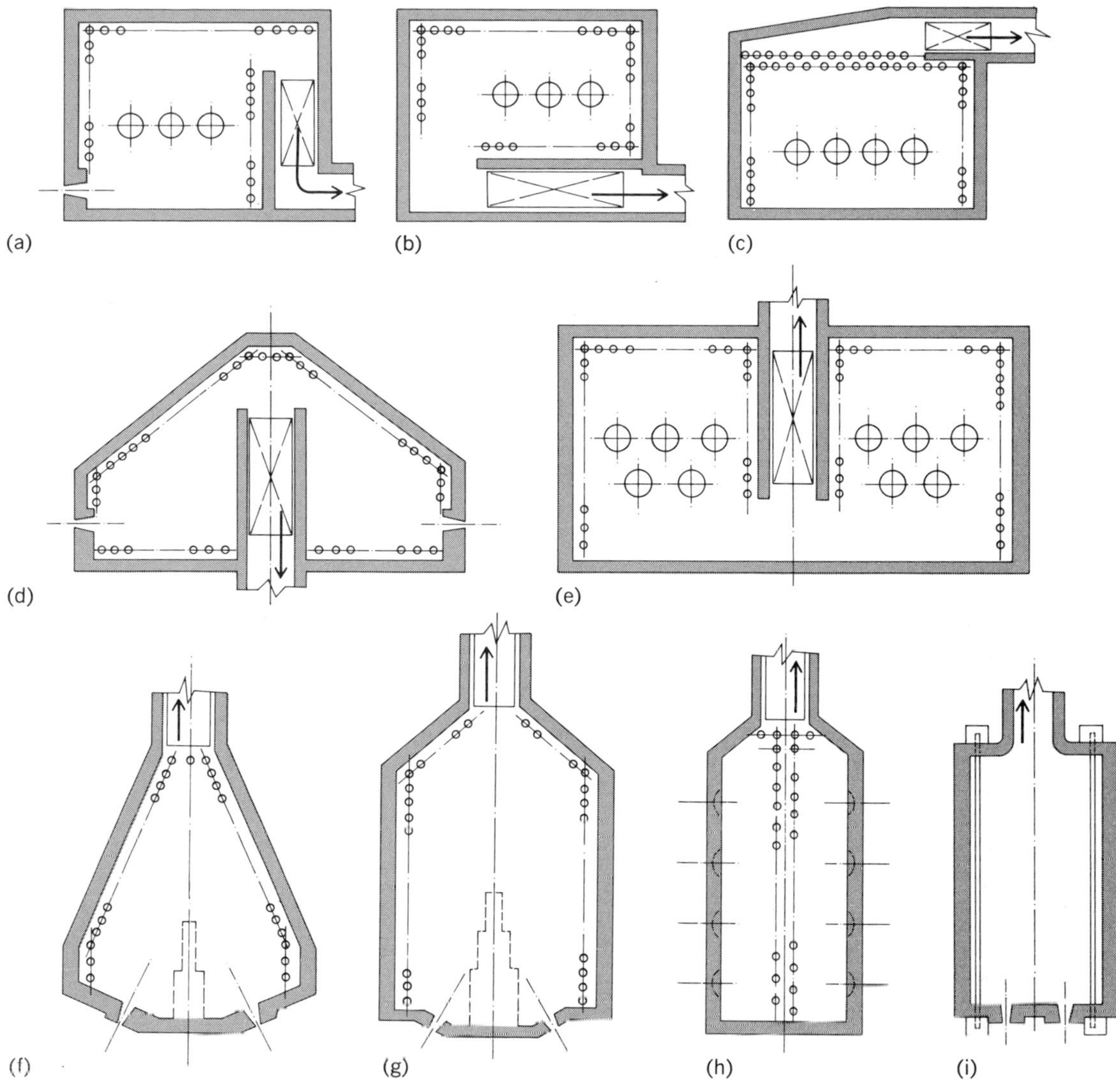

Fig. 1. Typical process furnace designs and characteristic features. (*a*) Radiant, updraft, vertical-convection bank. (*b*) Radiant, downdraft, horizontal-convection bank. (*c*) Radiant, updraft, with high convection effect in radiant roof tubes. (*d*) Large refractory heater, dual radiant sections, sloping roofs. (*e*) Large conventional refinery heater, dual radiant sections. (*f*) Updraft furnace, slanting walls, two parallel coils. (*g*) Updraft furnace, vertical walls, two parallel coils. (*h*) Updraft furnace, tubes in center, burners in wall, and with good heat-intensity distribution. (*i*) Updraft circular, all-radiant furnace, with the burners in the floor.

process. The carbon and hydrocarbon content of the fuel governs the combustion-gas composition, and many formulas relating it to the elemental, oxidizable constituents of the fuel have appeared in the literature. Complete combustion is a prerequisite for high thermal efficiency; to ensure this, air is used in excess of the minimum or stoichiometric requirements, depending upon the type of fuel and the combustion equipment or system (Table 1).

The type of fuel, its heating value, and the excess air applied in combustion determine the theoretical flame temperature which would prevail if the oxidation were instantaneous. However, combustion is a rate process requiring time, and the actual flame temperature is considerably lower because of the radiation from the combustion zone. The theoretical flame temperature can readily be determined from the heating value of the fuel and the amount of combustion products evolved, including the excess air. Accurate enthalpy data on the combustion gases are required and are presented in the literature. *See* COMBUSTION; FOSSIL FUEL.

In order to meet a required thermal efficiency, the heat losses must be appraised. There is a certain loss through the furnace setting which varies with the surface area of the enclosure, the heating capacity, and the atmospheric environment. The wall construction of most types of furnaces, that is, application of insulating material, is such that the specific heat loss, Btu/(ft^2)(hr) of outer wall area, is practically the same for all operating temperature levels; hence the heat loss will vary principally only with the furnace capacity. Table 2 shows the average values that will be encountered in industrial furnace constructions. The setting heat loss ΔH_{sl} can be calculated from the relation $\Delta H_{sl} = \lambda H_{fl}$, where λ is the fraction of heat released, and H_{fl} is the enthalpy of the combustion gas at the theoretical flame temperature.

The flue-gas effluent temperature from the

Table 1. Fuel-air mixtures

Type of fuel	Type of burner	Normal range of excess air, %
Natural or refinery gas	Air and fuel gas premixed	5–15
Natural or refinery gas	Air induced by natural draft	10–20
Distillate fuel oils	Steam atomization air by natural draft	15–25
Distillate fuel oils	Mechanical atomization air by natural draft	15–30
Residual fuel oils	Steam atomization air by natural draft	20–35

furnace can then be determined in accordance with the postulated thermal efficiency η, as shown in the equation below, where H_{st} is the enthalpy of

$$H_{st} = H_{fl} \times (1-\eta) - \Delta H_{sl}$$

the combustion gas at the convection-bank exit temperature.

Heat load. The distribution of heat load between the radiant and convection sections of conventional heaters is associated with the manufacturer's design approach and service requirements. There are all-radiant heaters, and furnaces with tubes arranged mostly for heat absorption by convection. The amount and disposition of heating surface in the combustion or radiant chamber has a distinct influence upon the radiant-heat absorption and the supplementary heat to be recovered by convection for the required thermal efficiency. Therefore, the allocation of heating duties to the radiant and convection sections is more or less an empirical matter. Conventional process heaters are generally designed for approximately 75% thermal efficiency in the United States, where fuel costs are relatively low. Most box-type heaters are designed for moderate radiant-heat-transfer rates; approximately 65–75% of the heat load is carried by the radiant section and 25–35% by the convection bank. In some cases, high radiant-heat intensities are used as a matter of design principle, and in these furnaces, the heat absorbed in the radiant section may be only 50% of the total duty.

Radiant section. The present procedure for determining the radiant-heat-transfer rate and heating surface rests on a long series of scientific investigations and evaluation of operating data.

The combustion gases radiate to the solid-body environment, tubes, and refractories. Carbon dioxide and water vapor are the principal constituents with emissive power. Their emissivity depends upon the partial pressure and the thickness of the gas layer, and at 1900°F (1038°C) has a value of approximately 25% of the black body radiation intensity. The terms luminous- and nonluminous-flame radiation used in scientific treatises are somewhat misleading; the mere luminosity or light has no significance as far as heat radiation is concerned, because practically all emissivity and certainly carbon dioxide, CO_2, and water, H_2O, radiation is in the infrared wavelength region, even at flame temperatures encountered in process heaters. The maximum gas radiation for an infinite value of the product of the partial pressure and thickness of the gas layer, frequently termed black gas, amounts to only 21–33% of the complete blackbody emissivity at temperatures of 2500 and 1100°F (1371 and 593°C), respectively. The mean radiating temperature in the combustion chamber lies within this temperature range.

The consideration of heat transmission by radiation alone, from the combustion gas and refractory walls by reradiation, has not resulted in the derivation of a universally applicable formula for the heat-transfer rate in the radiant section. This is because heat is also transmitted to the tubes by convection which is in the order of 10% to as high as 35% of the heat absorbed in the radiant section, depending upon (1) the shape of the combustion chamber, (2) the arrangement of tubes, and (3) the type and location of burners. The convection effect is a linear function of the temperature difference (not the difference of temperatures to the fourth power) and can be varied for the same furnace design by the amount of excess air employed.

Convection heating surface. The rate of heat transfer solely by convection, with flow of gas at a right angle to tubes, increases with the gas velocity and the temperature level, and is greater for small than for large tubes.

There are radiation effects in the convection-tube bank augmenting the heat absorption, especially when the flue-gas temperatures or the temperature differences between tube wall and flue gases are large. These include (1) radiation to the front two rows of tubes, which can be appreciable if they are exposed to the combustion chamber; (2) radiation from the hot gases surrounding the tubes; and (3) the radiation from the refractory enclosure to the tube bank as a whole.

Process furnaces. The design of process furnaces, although following generally the procedure outlined earlier for conventional heaters, requires careful consideration of the transitory state of the fluid being heated.

Vacuum distillation of a temperature-sensitive fluid or the stripping of solvent from the raffinate and extract of a selective solvent-separation process are typical cases in which temperature limitations are encountered in the heating operation. Partial vaporization of the fluid must take place in the tubes under relatively mild heat-transfer rates and tube-wall temperatures, not greatly in excess of the bulk-flowing temperature. Aside from the knowledge and postulation of the heat-intensity pattern and distribution on the combustion-gas side, the proper design of the furnace coil involves exacting pressure-drop calculation for mixed vapor-liquid flow in conjunction with phase-equilibrium determination of the amount and composition of vapor and liquid at any given point in the coil. The pressure drop for mixed vapor-liquid flow

Table 2. Heat loss from setting

Heat liberated, Btu/hr*	Loss as a fraction of heat released, λ $\Delta H_{sl} = \lambda H_{fl}$
15,000,000	0.030
30,000,000	0.0275
60,000,000	0.0225
75,000,000+	0.020

*Btu/hr = 0.293 J/s.

is reasonably well established, and vapor-liquid equilibrium relationships are known for many binary and ternary systems. The determination of temperature, pressure, and transition state of the fluid throughout the coil involves stepwise trial-and-error calculations which can best be executed by a modern digital computer for various coil dimensions and heat-intensity patterns. The coil design which satisfies all technical criteria and also cost aspects can then be selected from the computer evaluations.

Gas reforming is another industrially important heating process involving a chemical change of the fluid. Methane or natural gas and steam are decomposed in the presence of excess steam in tubes filled with a catalyst. The principal products are hydrogen, H_2, carbon monoxide, CO, and CO_2, and maximum conversion of hydrocarbons is attempted. The chemical transformation proceeds at approximately 1400–1450°F (760–790°C) and is highly endothermic, having a heat of reaction of 87,500 Btu/(mole of methane converted) [92.3 MJ/mole]. The tubes are located in the center of the radiant or combustion chamber and are heated from two sides to effect an even circumferential heat-intensity distribution. The average rate of heat transfer is in the order of 18,000 Btu/(ft^2)(hr) [57 kJ/(m^2)(sec)].

Pyrolysis of hydrocarbons, the thermal conversion of ethane, propane, butane, and heavier hydrocarbons to produce olefins and diolefins for the petrochemical industry, is another example of a process furnace. The modern pyrolysis furnace is an outgrowth of the oil-cracking heater, the mainstay of motor-fuel production from petroleum before the advent of catalytic cracking. The design of a pyrolysis furnace must take thermochemical equilibria and reaction rates into consideration in order to realize a high conversion and a product distribution most favorable with respect to yield of olefins. The conversion of higher-molecular-weight hydrocarbons which have a high rate of decomposition involves the problem of attaining the coil outlet temperature dictated by equilibrium relationships. A shorter coil-residence time and correspondingly higher heat-transfer rates are means to this effect, but still the basic principle of moderating the heat intensity in the pyrolysis coil zone, in which the reactant is already in a high state of conversion, must be observed. Figure 2 gives a typical temperature-pressure and heat-flux pattern for a pyrolysis coil cracking a naphtha boiling in the range 190–375°F for maximum ethylene yield. The design of such a pyrolysis coil can be executed with greatest precision or refinement by the use of a modern digital computer, permitting the rapid evaluation of all design variables. *See* CRACKING.

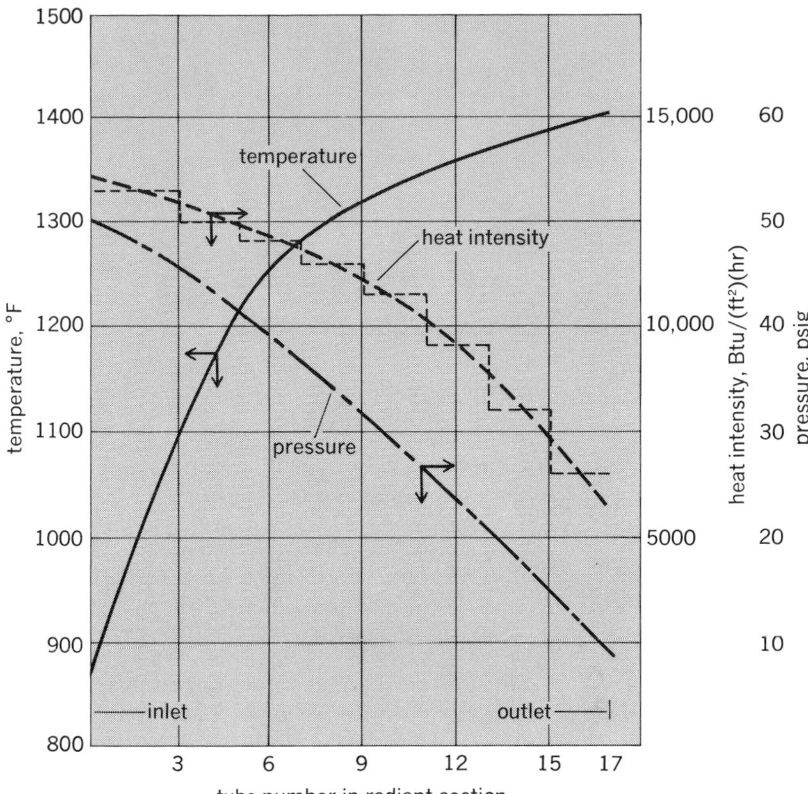

Fig. 2. Typical temperature and heat-flux pattern for naphtha pyrolysis coil. Temperature (°C)=5/9[temperature (°F)− 32]. 1 Btu/(ft^2)(hr)= 3.15 J/(m^2)(s). 1 psi=6.895 kPa.

Table 3. Heating-coil materials

Tube-wall temperature range, °F (°C)	Tube material*
875 and under (468 and under)	Carbon steel, A-161
875–1150 (468–621)	Low-chrome steel, A-200 (1.25–2.25% Cr, 0.5% Mo)
1150–1550 (621–843)	18/8 stainless steel, A-271, type 304
1550–1800 (843–982)	25/20 stainless steel, A-271, type 310

*American Society for Testing and Materials specifications.

Mechanical construction. The design of the furnace setting and application of refractories and insulating materials generally follow standard practice prevailing in boiler-furnace construction. Fully suspended wall construction and the use of light-weight firebrick with bulk densities of 30–65 lb/ft^3 or 480–1040 kg/m^3 (as compared with 125–150 lb/ft^3 or 2000–2400 kg/m^3 for fireclay bricks) are generally preferred for process heaters to reduce the heat storage in the setting, start-up, and shutdown time. The burners are carefully selected and must offer great flexibility with respect to firing rate and excess air. The use of a multiplicity of small-capacity burners is rather common for specific process heaters such as gas-reforming and pyrolysis furnaces.

The design of the heating coil for high-temperature service involves the application of alloy steels. Tube materials are usually selected according to the schedule in Table 3. Tubes of carbon and low-chrome steel are usually connected by forged-welding return bends, and those of stainless steel by cast 25/12 chrome-nickel alloy steel bends, also weldable to the tubes.

The flue-gas exhaust is normally carried to the atmosphere by a stack of sufficient height to induce a draft equivalent to the pressure drop in the flue-gas passage. For highest heat economy a process furnace can readily be equipped with an air preheater or waste-heat boiler usually requiring air and flue-gas exhaust blowers. Preheated combustion air results in higher flame and mean effective radiation temperatures which aid in attaining high radiant-heat-transfer rates required in certain pyrolysis operations. These furnace accessories are provided in the same manner as in boiler design practice. *See* CHIMNEY; GAS FURNACE; HEAT EXCHANGER; OIL FURNACE.

[H. C. SCHUTT/H. L. BEGGS]

Bibliography: D. Q. Kern, *Process Heat Transfer*, 1950; L. Lichty, *Combustion Engineering Processes*, 1967; W. H. McAdams, *Heat Transmission*, 3d ed., 1954; W. L. Nelson, *Petroleum Refinery Engineering*, 4th ed., 1958.

Gas field and gas well

Petroleum gas, one form of naturally occurring hydrocarbons of petroleum, is produced from wells that penetrate subterranean petroleum reservoirs of several kinds. Oil and gas production are commonly intimately related, and about one-third of gross gas production is reported as derived from wells classed as oil wells. If gas is produced without oil, production is generally simplified, in part at least because the gas flows naturally without lifting, and also because of fewer complications in reservoir problems. As for all petroleum hydrocarbons, the term field designates an area underlain with little interruption by one or more reservoirs of commercially valuable gas. *See* NATURAL GAS; OIL AND GAS FIELD EXPLOITATION; OIL AND GAS WELL DRILLING; PETROLEUM; PETROLEUM GEOLOGY; PETROLEUM RESERVOIR ENGINEERING.

[CHARLES V. CRITTENDEN]

Gas furnace

An enclosure in which a gaseous fuel is burned. Domestic heating systems may have gas furnaces. Some industrial power plants are fired with gases that remain as a by-product of other plant processes. Utility power stations may use gas as an alternate fuel to oil or coal, depending on relative cost and availability. Some heating processes are carried out in gas-fired furnaces. *See* STEAM-GENERATING FURNACE.

Among the gaseous fuels are natural gas, producer gas from coal, blast furnace gas, and liquefied petroleum gases such as propane and butane. Crude industrial heating gases carry impurities that corrode or clog pipes and burners. Solid or liquid suspensoids are removed by cyclones or electrostatic precipitators; gaseous impurities are removed chemically. The cleaned gas may be mixed with air in the furnace, in the burners, or in a blower before going to the burners. The gas and air may be supplied at moderate pressure, or one or both at high pressure. The high-pressure component may serve to induce the other component into the furnace. The burner may be a single center-fire type or a multispud type with numerous small gas parts, depending on how the heat is to be concentrated or distributed in the furnace. Crude uncleaned gases are fired through burners with large ports; in such cases, the burners are removable for cleaning. *See* FUEL GAS.

[RALPH M. HARDGROVE]

Gas turbine

A heat engine that converts some of the energy of fuel into work by using gas as the working medium and that commonly delivers its mechanical output through a rotating shaft.

Cycle. In the usual gas turbine, the sequence of thermodynamic processes consists basically of compression, addition of heat in a combustor, and expansion through a turbine. The flow of gas during these thermodynamic changes is continuous in the basic, simple, open-cycle arrangement (Fig. 1*a*).

This basic open cycle can be modified through the addition of heat exchangers and multiple components for reasons of efficiency, power output, and operating characteristics (Fig. 1*b–d*). A regenerator recovers exhaust heat and returns it to the cycle by heating the air after compression and before it enters the combustor. An intercooler reduces the work of compression by removing some of the heat of compression. A second stage of heating can be added between sections of the turbine, called an afterburner in aircraft turbines. In various combinations, the auxiliary features provide means for meeting a wide range of operating needs.

Types. The various gas-turbine cycle arrangements can be operated as open, closed, or semiclosed types.

Open cycle. In the open-cycle gas turbine, there is no recirculation of working medium within the structural confines of the power plant, the inlet and exhaust being open to the atmosphere (Fig. 1). This cycle offers the advantage of a simple control and sealing system. It also can be designed for high power-to-weight ratios (aircraft units) and for operation without cooling water. Most gas turbine plants are of this type.

Closed cycle. In the closed-cycle gas turbine, essentially all the working medium (except for seal leakage, bleed loss, and any addition or extraction of working medium for control purposes) is continuously recycled (Fig. 2*a*). Heat from a source such as fossil fuel (or, possibly, nuclear reaction) is transferred through the walls of a closed heater to the cycle. The closed cycle can be charged with gases other than air such as helium, carbon dioxide, or nitrogen. This is a particular advantage with a nuclear heat source. Other advantages of the closed cycle are (1) clean working fluid; (2) control of the pressure and composition of the working fluid; (3) high absolute pressure and density of the working fluid; and (4) constant efficiency over wide load range. A precooler is required to reduce the temperature of the working fluid before recompression. The higher densities of the working fluid increase the horsepower capacity of a plant of given volume. Changing the absolute pressure level at the compressor inlet changes the weight of working fluid circulated without changing the compression ratio or the temperatures, which results in relatively constant efficiency over a wide load range. The major disadvantage of the closed-cycle gas-turbine plant is the cost and size of the required high-temperature heater.

Semiclosed cycle. In the semiclosed cycle gas turbine, a portion of the working fluid is recirculated (Fig. 2*b*). This type requires a precooler for the recirculated gas, and a charging compressor to provide the necessary air for combustion. The semiclosed cycle can operate at high densities. The major disadvantages of this cycle are the corrosion and fouling which occur with the recirculation of the products of combustion, particularly when the fuels used have high sulfur or ash content.

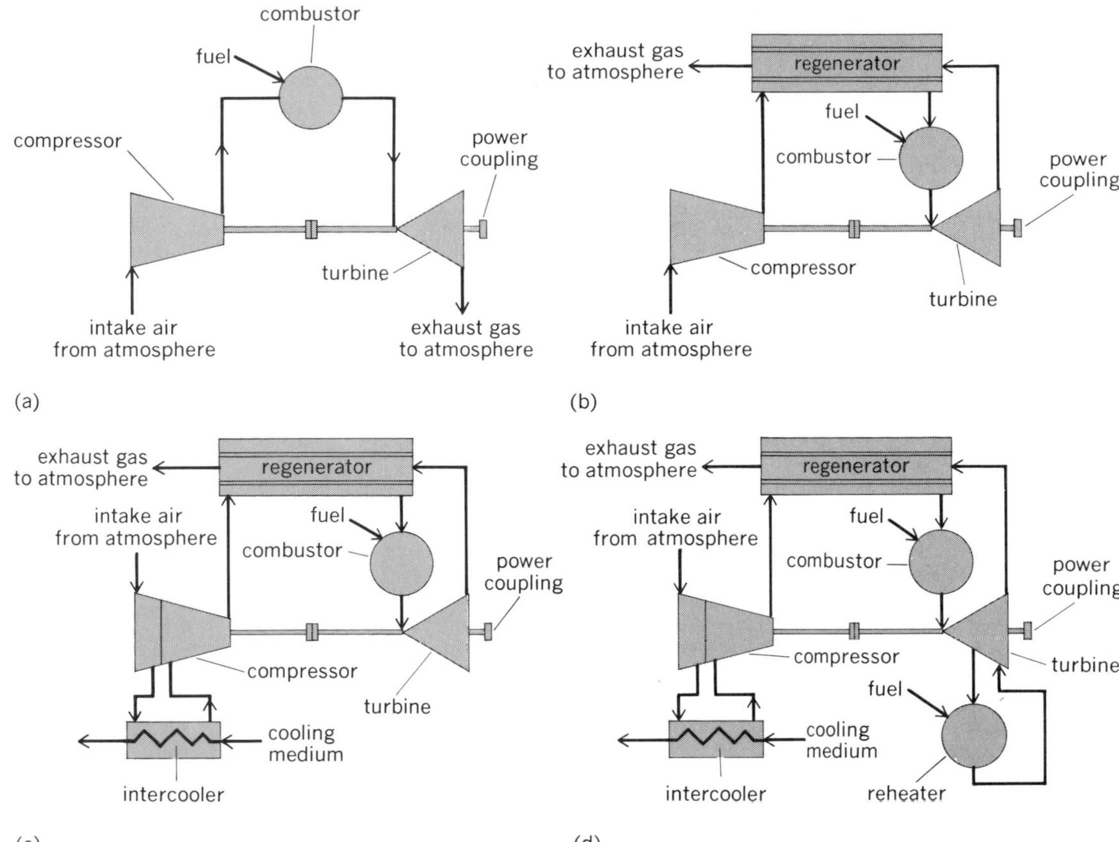

Fig. 1. Series-flow, single-shaft, gas-turbine power plant. (*a*) Simple open cycle. (*b*) Open cycle with regenerator. (*c*) Open cycle with intercooler and regenerator. (*d*) Open cycle with intercooler, regenerator, and reheater.

Generalized performance. The overall performance of a given gas-turbine power-plant cycle depends basically on component efficiencies, pressure and leakage losses, pressure ratio (ratio of the highest to the lowest pressure), and temperature level.

For the simple cycle, the influence of pressure ratio on performance is illustrated in Fig. 3*a*. The curves are based on ambient inlet conditions of 80°F and 1000 ft altitude, 85% compressor efficiency, 90% turbine efficiency, 95% combustion efficiency, and 5% combustor pressure loss.

The effect of pressure ratio on plant efficiency for the four basic gas-turbine cycle arrangements illustrated in Fig. 1 is shown in Fig. 3*b*; the effect of inlet temperature on cycle efficiency for the four basic cycles is shown in Fig. 3*c*. The curves are drawn for the optimum pressure ratio for each cycle arrangement at the various turbine inlet temperatures.

Components. To achieve such overall performance, each process is carried out in the engine by a specialized component (Fig. 4). Air for the combustion chamber is forced into the engine by a compressor. In an aircraft, the intake may advance into the air fast enough to ram air into the engine. Fuel is mixed with the compressed air and burned in combustors. The heat energy thus released is converted by the turbine proper into rotary energy. Because of the high initial temperature of the combustion products, excess air is used to cool the combustion products to the allowable turbine inlet design temperature. To improve efficiency, heat exchangers can be added on the gas turbine exhaust to recover heat energy and to return it to the working medium after compression and prior to its combustion.

Compressors. Two basic types of compressors are used in gas turbines: axial and centrifugal. In a few special cases a combination type known as a mixed wheel, which is partially centrifugal and partially axial, has been used. The axial-flow compressor is the most widely used because of its ability to handle large volumes of air at high efficiency. For small gas turbines in the range of 500 hp and less, the centrifugal replaces the axial because it has comparable efficiency when handling reduced volume flow, and is smaller and more compact.

Axial and centrifugal compressors both have a stall or pumping limit where flow reverses. This limit is usually encountered on starting with stationary power units, and at high altitude and high speed with aircraft units. This stall limit is shifted out of the operating range of the gas turbine by the use of compressor bleed, variable compressor vanes, or water injection. All three methods are applied to reduce the aerodynamic loading on the stalled compressor stages.

Ram effect. Aircraft gas turbines moving at high speeds obtain a pressure rise from the ram effect in addition to the compressor pressure rise. The ram effect is the recovery of part of the air velocity, due to the forward motion of the plane, and conversion of this velocity energy to pressure.

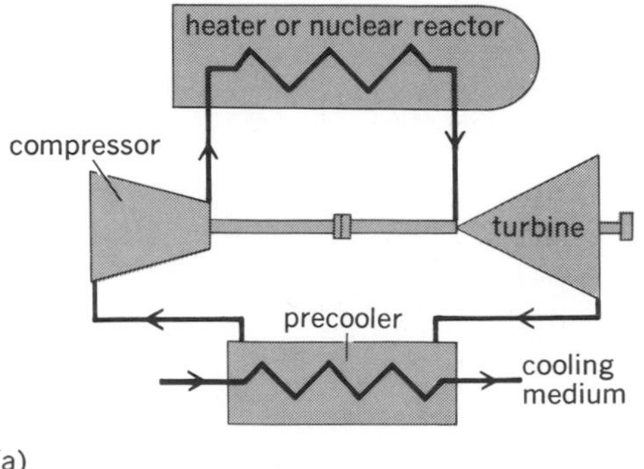

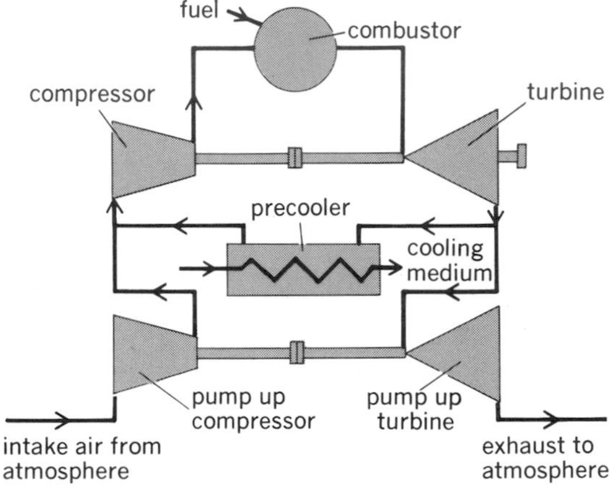

Fig. 2. Gas-turbine power plant. (*a*) Closed cycle with precooler, series flow, single shaft. (*b*) Semiclosed cycle with precooler, series flow, single shaft.

Combustors. Combustors, sometimes referred to as combustion chambers, for gas turbines take a wide variety of shapes and forms. All contain fuel nozzles to introduce and meter the fuel to the gas stream and to atomize or break up the fuel stream for efficient combustion (Fig. 5). All are designed for a recirculating flow condition in the region of the nozzle to develop a self-sustaining flame front in which the gas speed is lower than the flame propagation speed. In addition to being designed to burn the fuel efficiently they also uniformly mix excess air with the products of combustion to maintain a uniform turbine-inlet temperature.

The combustor must bring the gas to a controlled, uniform temperature with a minimum of impurities and a minimum loss of pressure. In the open-type gas turbine, the large excess of air must be controlled to avoid chilling the flame before complete combustion has taken place. Extremely high rates of combustion are common, the heat release being 1,000,000–5,000,000 Btu/ft^3/hr/atm (100–500 J/m^3/s/Pa), or 5–20 times that of high-output, steam-boiler furnaces. Refractory linings are unsuitable because they cannot withstand the vibrations and velocities. Metal liners and baffles cooled by the incoming air are often used.

Turbine wheels. Two types of gas turbine wheels are used: radial-inflow and axial-flow. Small gas turbines use the radial-flow wheel. For large volume flows, axial turbine wheels are used almost exclusively. Although some of the turbines used in the small gas turbine plants are of the simple impulse type, most high performance turbines are neither pure impulse nor pure reaction. The high-performance turbines are normally designed for varying amounts of reaction and impulse to give optimum performance. This usually results in a blade which is largely impulse at the hub diameter and almost pure reaction at the tip. *See* TURBINE.

Turbine cooling. Gas turbines all employ cooling to various extents and use a liquid or gas coolant to reduce the temperature of the metal parts. The cooling systems vary from the simplest form where only first-stage disk cooling is involved to the more complex systems where the complete turbine (rotor, stator, and blading) is cooled. At high temperatures of operation the turbine material is designed to have a predetermined life, which can vary from a few minutes for missile applications to 100,000 hr or more for industrial units.

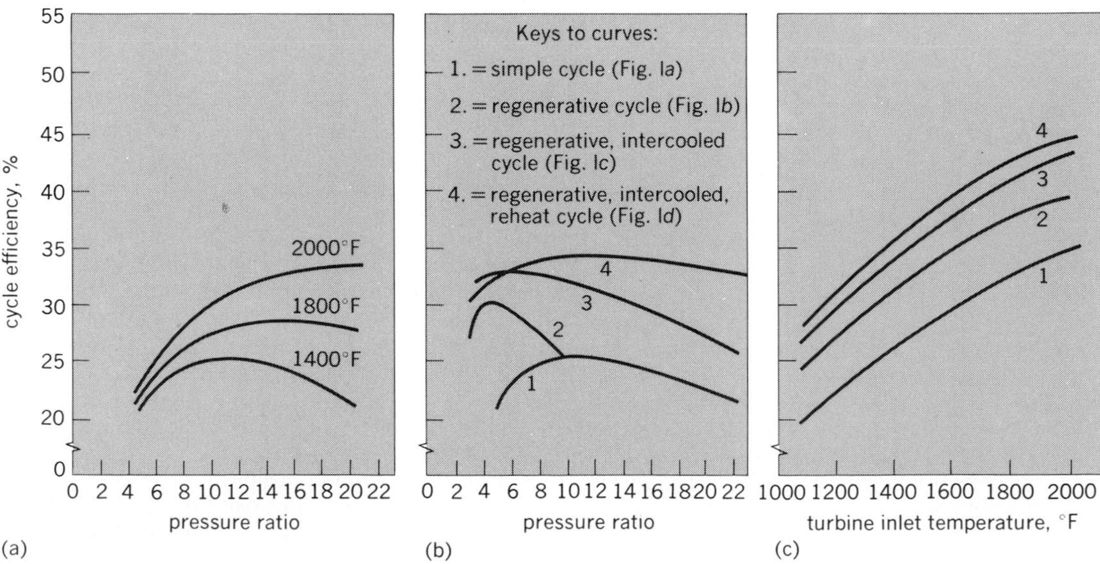

Fig. 3. Effects of pressure ratio on thermal efficiency. (*a*) Simple gas-turbine cycle at various turbine-inlet temperatures. (*b*) Various gas-turbine cycles at constant turbine-inlet temperature. (*c*) Effect of turbine-inlet temperature on thermal efficiency for various gas-turbine cycles at optimum pressure ratios.

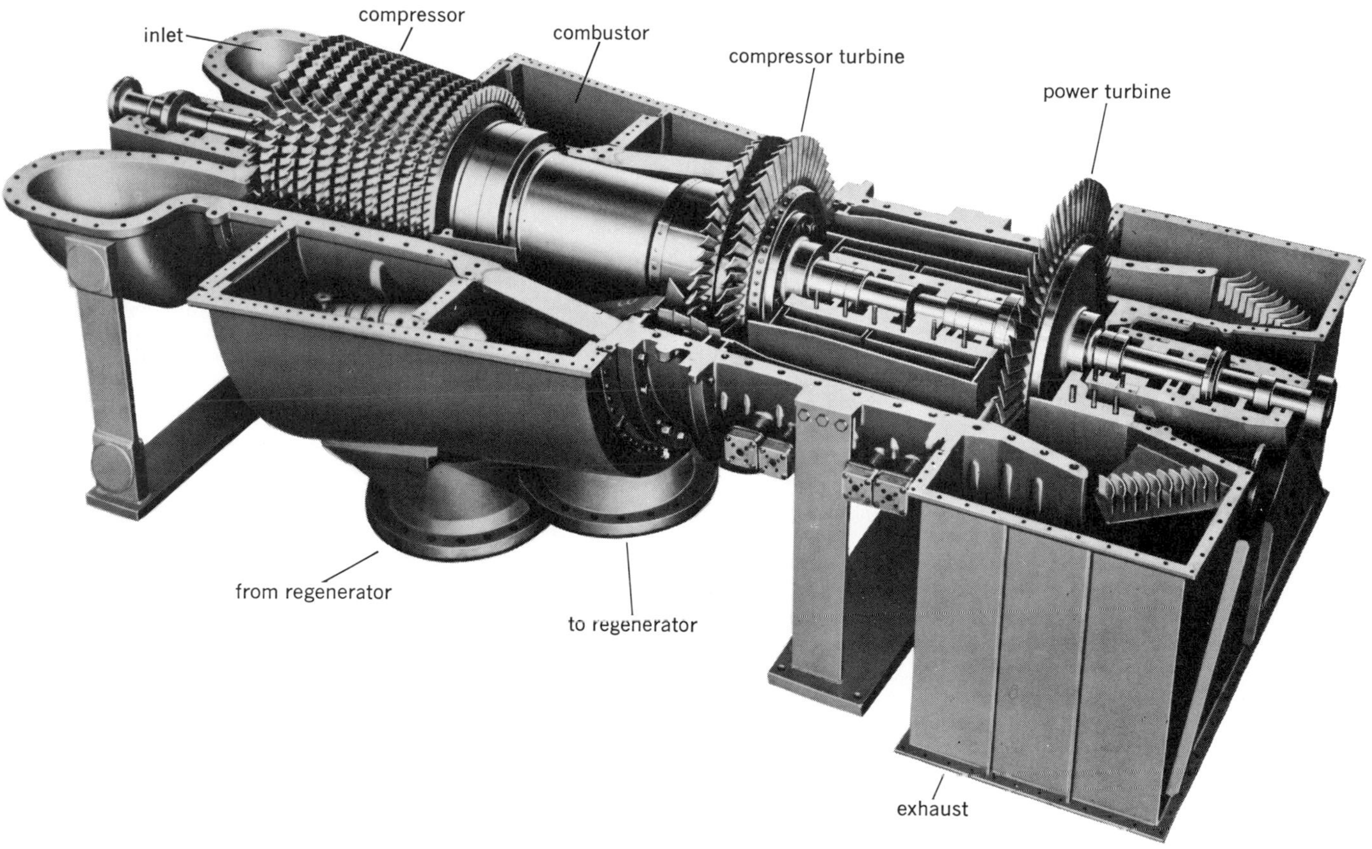

Fig. 4. Dual-shaft open-cycle 5000-hp (3.7-MW) gas turbine with regenerator.

Heat exchangers. Two basic types of heat exchangers are used in gas turbines: gas-to-gas and gas-to-liquid. An example of the gas-to-gas type is the regenerator, which transfers heat from the turbine exhaust to the air leaving the compressor. The regenerator must withstand rapid large temperature changes and must have low pressure drop. Regenerators of both the shell-and-tube and the extended-surface type are used. Rotary regenerators having high performance and reduced weight are under development. Intercoolers, which are used between stages of compression, are air-to-liquid units. They reduce the work of compression and the final compressor discharge temperature. When used with a regenerator, they increase both the capacity and efficiency of a gas turbine power plant of a given size.

Controls and fuel regulation. The primary function of the subsystem that supplies and controls the fuel is to provide clean fuel, free of vapor, at a rate appropriate to engine operating conditions. These conditions may vary rapidly and over a wide range. As a consequence, fuel controls for gas turbines are, in effect, special-purpose computers employing mechanical, hydraulic, or electronic means, frequently all three in combination.

In aircraft the pilot controls the turbine engine through a throttle lever, which usually establishes engine speed for each position of the lever. To achieve this relation between throttle position and engine speed under the design envelope range of flight conditions, the fuel control senses numerous engine conditions in addition to engine speed in revolutions per minute. Among these conditions are compressor inlet pressure and ambient air temperature. From these inputs, together with the position of the pilot's throttle, the fuel control continuously adjusts the rate of fuel flow to the engine. When the pilot moves the throttle to change engine power, the control senses the rate of engine acceleration or decleration and provides for the required rate of change in fuel rate.

The fuel control, as well as maintaining the engine at the speed called for by the throttle, should also prevent overspeeding, excessive temperature, or loss of the fire. Under some conditions, the rate at which a turbine engine can accelerate may be limited by the rate at which heat can be released by fuel combustion: Excessive internal engine

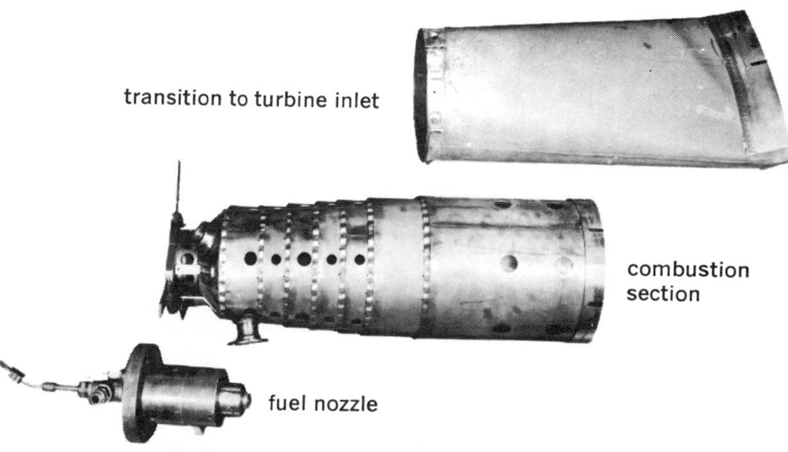

Fig. 5. Gas-turbine combustor components.

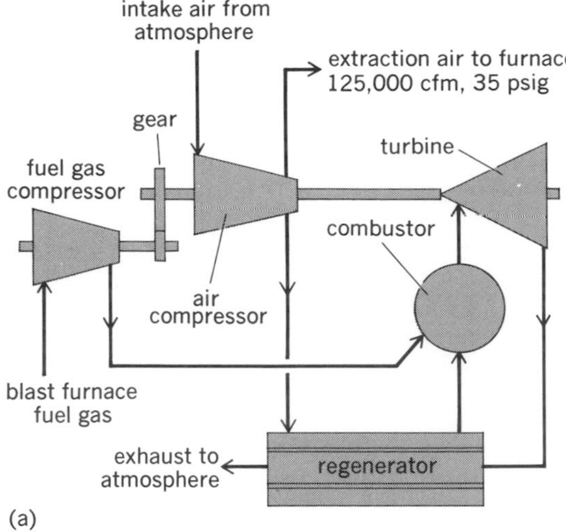

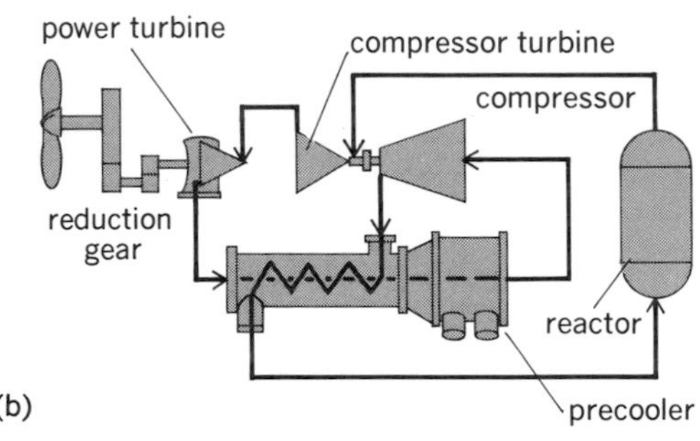

Fig. 6. Cycle arrangements (*a*) for a 125,000-cfm (ft³/min) [59 m³/sec] extraction unit and (*b*) for a closed-cycle gas turbine utilizing a gas-cooled reactor.

temperatures could damage such engine parts as the turbine wheels. Under other conditions, the rate at which an engine can decelerate may be limited by the necessity of maintaining sufficient fuel flow to hold the flames; too low a fuel rate for the prevailing gas velocity in the combustors could blow out the flame.

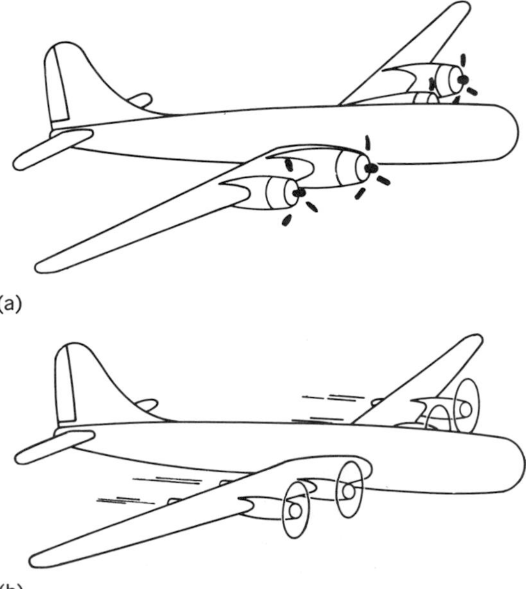

Fig. 7. Types of aircraft engine. (*a*) Reciprocating engine has greater frontal area for a given horsepower than (*b*) internal-combustion gas-turbine engine.

Arrangements. Gas turbines can be constructed for either single-shaft or multiple-shaft, and can be arranged to supply power, high-pressure air, or hot-exhaust gases, singly or in combination.

A single-shaft unit, consisting of a compressor, combustor, and turbine, is a compact, lightweight power plant. It is capable of rapid starting and loading and has no standby losses. It can be arranged to use little or no cooling water, which makes it particularly attractive for powering transportion equipment and as a mobile standby and emergency power plant. This type of plant can compete efficiently with small steam plants. Its simplicity involves a minimum of station operating personnel; some of these plants are arranged for completely remote operation.

To improve efficiency. the energy in the exhaust gases can be used either in a waste-heat boiler, or in combination with other processes. For example, a unit arranged to supply process air at 35 psig (240 kPa gage pressure) for operation of blast furnaces can use blast-furnace gas for its fuel (Fig. 6*a*).

Gas turbines might also be arranged to use a nuclear heat source. For aircraft application, the cycle is open because of weight and space considerations. The combustor is replaced with a nuclear reactor (Fig. 2*a*). For shipboard and stationary units, the cycle is closed so that fission products can be contained within the power plant in the event of a nuclear accident or fuel element failure. A closed-cycle, gas-turbine power plant for marine service could operate directly with a high-temperature, gas-cooled reactor (Fig. 6*b*).

Fuels. In the open-cycle plant, products of combustion come in direct contact with the turbine blading and heat-exchanger surfaces. This requires a fuel in which the products of combustion are relatively free of corrosive ash and of residual solids that could erode or deposit on the engine surfaces. Natural gas, refinery gas, blast-furnace gas, and distillate oil have proved to be ideal fuels for open-cycle gas turbines. Combustion chamber and fuel nozzle maintenance are negligible with these types of fuel. Residual fuel oil treated to avoid hot ash corrosion and deposition is also satisfactory although it requires frequent cleaning of fuel nozzles and higher combustor maintenance. Vanadium pentoxide and sodium sulfate are the principle ashes that have been found to cause corrosion and deposition in the 1250–1600°F (680–870°C) temperature ranges of modern-day gas turbines. The treatment of residual oil to make it suitable consists of washing with water to remove sodium and introducing additives to raise the fusion temperature of the ash.

In closed-cycle turbines, gaseous fuel, distillate oils, and coal are satisfactory. Residual fuels also must be treated to avoid corrosion in the heater parts. Ash erosion is not a major problem because of the lower velocity of the combustion gases over the heater surfaces.

Fuels for semiclosed-cycle turbines must be selected to minimize both erosion and corrosion. Because the products of combustion are circulated through the entire cycle, the fuel must be selected to avoid corrosion at all pressures and temperatures of the cycle. These requirements limit the use of the semiclosed plant to fuels of low ash and sulfur content.

Applications. Gas-turbine power plants have been successfully applied in the following industries, where their characteristics have proved superior to competitive power plants.

Aviation. The gas turbine finds its most important application in the field of aviation. In the propulsion of aircraft, the gas-turbine power plant as a turboprop or turbojet engine has replaced the reciprocating engine in large, high-speed airplanes. This change is due primarily to its high power-to-weight ratio and its ability to be built in large horsepower sizes with high ratio of thrust per frontal area (Fig. 7). It makes use of the ram effect on the compressor inlet to give almost constant thrust at all aircraft speeds.

Gas pipeline transmission. Gas turbines have been installed along piplines to drive centrifugal compressors. Turbine sizes range from 1800–14,000 hp (1350–10,500 kW). Gas fuel is normally used, though units have been arranged for dual fuel firing using either natural gas or distillate oil.

Petroleum. In the petroleum industry the gas turbine generates power, drives air and gas compressors in refineries, supplies extraction air or exhaust gases for process, and in the oil fields drives gas compressors to maintain well pressure.

Steel. In the steel industry gas turbines drive air compressors to furnish extraction air for processes such as blast-furnace operation, as mentioned above, and drive electric generators for various plant services. The primary fuel in these applications is blast-furnace gas with a number of units arranged for dual fuel firing.

Marine. Prototype gas turbines are being tested for marine use in ratings up to 6000 ship horsepower (shp). These units operate with residual oil as fuel processed through fuel treatment equipment. A regenerative-cycle gas turbine with a separate power turbine provides the wide speed range required by such applications as marine service.

Marine installations of gas turbines range from small units used for propulsion of short-range, high-speed craft, emergency generator drive, minesweeper boat propulsion and generator drive, deicing, smoke generation, fire-pump drive, and pneumatic power applications, to large boost engines used in combination with steam turbines. *See* MARINE ENGINE.

Electric utilities. Gas turbines perform a variety of functions in electric-power generation. Peaking service uses simple single-shaft units in sizes up to 25,000 kW. Gas turbines have been installed as hydro-standby. Aircraft propulsion elements can be applied as gas generators, driving electric generator units of up to 300,000 kW. Rail-mounted mobile power plants provide 5000–6000 kW for emergency service. The compactness and simplicity of the gas turbine make it possible to house these units in a single cab. Units from 5000 to 30,000 kW carry base load. These units have been simple cycle machines in low fuel cost areas; units with regenerative, intercooled cycles are used in high fuel cost areas.

Combined steam and gas plants. Efficiency of a power plant is improved by combining a gas turbine with a steam turbine. Various combinations are possible; gas-turbine exhaust can heat feed water for the steam turbine, fuel-fired boilers can use gas turbine exhaust as combustion air (Fig. 8*a*), or in supercharged boiler plants the high-pressure boiler can serve as the pressurized combustor for the gas turbine (Fig. 8*b*). The gain in efficiency by using turbine exhaust (Fig. 8*a*) depends on efficiency of the overall steam plant to which the combined cycle is compared. For the most efficient plant using the best steam conditions and largest number of feedwater heaters, a gain of 2% is possible, while for the less efficient plant, a gain of 5% is possible.

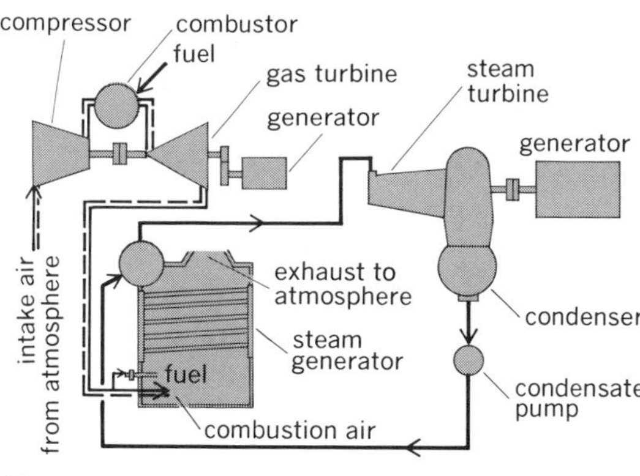

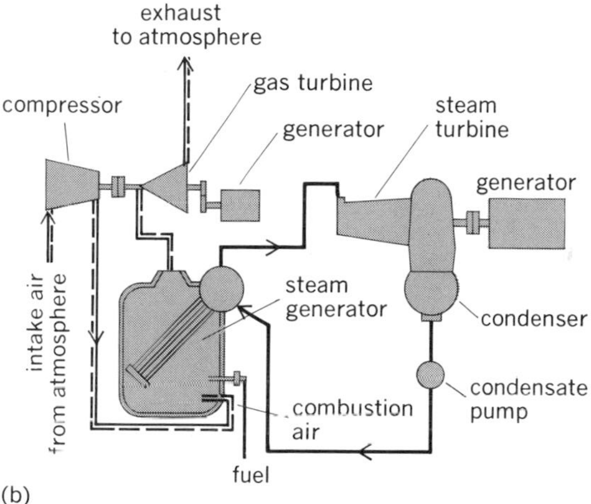

Fig. 8. Cycle arrangements (*a*) for combined steam- and gas-turbine cycle with gas-turbine, exhaust fired steam generator and (*b*) for combined steam- and gas-turbine cycle utilizing a pressurized-steam generator.

Where the steam generator is pressurized with air from the gas turbine compressor and the hot gases from the boiler are then expanded through the gas turbine (Fig. 8*b*), a gain in overall efficiency of 4–8% is realized, depending on the plant used for comparison. These improvements can be made to all steam plant cycles, including the most efficient in operation, which embloy the superpressure cycle. [THOMAS J. PUTZ/JAY A. BOLT]

Bibliography: T. Baumeister (ed.), *Standard Handbook for Mechanical Engineers*, 8th ed., 1978; H. Cohen et al., *Gas Turbine Theory*, 2d ed., 1979; D. G. Fink and H. W. Beaty (eds.), *Standard Handbook for Electrical Engineers*, 11th ed., 1978; C. W. Grennan (ed.), *Gas Turbine Pumps*, 1972; W. R. Hawthorne and W. T. Olson (ed.), *Designs and Performance of Gas Turbine Power Plants*, 1960; A. W. Judge, *Small Gas Turbines and Free Piston Engines*, 1960; D. G. Shepherd, *Introduction to the Gas Turbine*, 2d ed., 1960; C. W. Smith, *Aircraft Gas Turbines*, 1956; H. A. Sorensen, *Gas Turbines*, 1951.

Gasohol

A solution of one part ethanol (grain or beverage alcohol) in nine parts of gasoline. In most automobiles it can be used instead of gasoline with only minor differences in performance, although some automobiles might require significant part replacement due to damage to pump diaphragms and engine gaskets, and may require carburetor adjustments. It could be expected that, if gasohol were to become a standard fuel, the automobile fleet would become progressively altered to accommodate the new fuel with no problems.

Since ethanol releases 20% less energy than gasoline hydrocarbons, a given amount of gasohol will liberate 4% less energy than the same amount of gasoline, and consequently the mileage can be expected to be 4% less. Some of this loss might be offset by fine-tuning the engine to a leaner air-fuel ratio in order to gain a higher thermal efficiency. Also, since the octane rating of gasohol is some three to four octane numbers higher than gasoline, if gasohol were to become a standard fuel, auto makers could redesign engines to take advantage of the potentially higher efficiency of the higher octane rating. *See* OCTANE NUMBER.

Gasohol must be prepared with anhydrous ethanol, since as little as 0.3% water in the mixture will cause phase separation of the water. This would, of course, cause significant problems both with regard to corrosion and engine stalling. Methanol (wood alcohol) could also be used to dilute gasoline, but only 1 part of methanol could be mixed with 19 parts of gasoline because otherwise a slight contamination of the fuel with water would lead to separation of the methanol from the gasoline. Ethanol itself could also be substituted for gasoline.

Production. Since the primary goal of substituting alcohol for gasoline, in part or in whole, is to effect a savings in gasoline consumption, it is important to consider the sources of ethanol and the fuel required to produce it. First, the feedstock (principally corn and to a far lesser extent spoiled grain and certain food-processing wastes) is treated to produce a sugar solution. The sugar is then fermented by inoculation with yeast and bacteria to produce alcohol and carbon dioxide. The ethanol is recovered from the fermented batch by distillation. Because of the physicochemical interaction of ethanol and water, the distilled alcohol contains 4.4% water. Therefore, the azeotrope (95.6% alcohol) is redistilled after adding a chemical that prevents the association of ethanol and water. Anhydrous alcohol is then recovered. If the feedstock was grain, the material remaining after distillation of the alcohol, stillage, may be used as a protein additive for animal feed.

Although there is a physical possibility of producing as much as $5-10 \times 10^9$ gal ($1.9-3.8 \times 10^7$ m^3) of ethanol (3.5 to 7% the energy equivalent of gasoline) annually in the United States, it is unlikely that more than $1-2 \times 10^9$ gal ($3.8-7.6 \times 10^6$ m^3) could be produced from good cropland without profoundly aggravating the inflation in the food and feed markets. In the 1990s, or even earlier, the dedication of cropland to the production of such quantities of fuel ethanol would seriously interfere with the production of the anticipated required amount of food.

Although ethanol can be produced from cellulosic feedstocks such as crop residues and wood, which would not have a direct impact on food prices, these processes are not competitive at this time. The capital requirements for such plants are several times higher, and the yields of alcohol per unit of feedstock are lower. Development work on novel processes to increase the efficiency of converting these feedstocks is continuing. However, the justification for this work usually does not include attention to the fact that intensive farming of the feedstocks will require increased production of expensive fertilizers and increased use of herbicides to assure rapid and strong growth. Herbicide use in tree and other farming activities is already under serious question due to its possible effect on human and animal health.

Costs and energy balance. It must be determined, however, whether or not energy, particularly in the form of gasoline and natural gas, is saved by the substitution of ethanol for gasoline as an engine fuel. The evidence indicates strongly that it requires as much energy in the form of gasoline and natural gas to produce ethanol as can be secured from the produced ethanol.

For each gallon (3.8 liters) of ethanol derived from corn, the processes of farming, which include seeding, fertilizing, tilling, harvesting, and drying, are estimated to consume 0.29 gal (1.1 liters) of gasoline. Stand-alone ethanol refineries will consume an additional 0.4–0.6 gal (1.5–2.3 liters) of gasoline for processing (mostly for distillation). Depending upon the proximity of the refinery, transportation of the grain to the refinery may require an additional use of gasoline or diesel fuel. Thus, the equivalent of no less than 0.69–0.89 gal (2.6–3.4 liters) of gasoline (this may be in part diesel fuel or natural gas) are required to produce a gallon of ethanol. On the other hand, the Btu (energy) content of a gallon (or liter) of ethanol is only 0.8 of that of a gallon of gasoline; hence at the very best the production of a gallon of ethanol saves about a tenth of a gallon of gasoline, and at the worst it requires the additional use of a small amount of gasoline. When capital and operating costs for the production of ethanol are included, it is apparent that the ethanol is more costly than gasoline or other fluid hydrocarbon equivalents.

Most of the $10-20 \times 10^9$ gal ($3.8-7.6 \times 10^7$ m^3) of ethanol fuel produced in the United States originates at one major refinery, which is part of a food-processing complex. By an arbitrary assignment of energy consumption within the complex, the nonbeverage-grade ethanol is available at a competitive price with gasoline. However, the extension of such processing is unlikely because of market saturation with the primary product, corn oil. Certainly, coal and other energy sources should be substituted in the refining of ethanol. The net energy balance would be worsened because of the lower thermal efficiencies in the stills, but a significant reduction in gasoline consumption could be achieved.

A final consideration must be given to the much

heralded potential of on-farm production of ethanol for farmers' own use. Automated package units for fermentation and distillation are possible, but the cost of safe and foolproof units would be high. For such systems to be economical, farm labor expertise would of course have to be heightened and the labor valued very poorly. The support given to gasohol production by farming groups is probably based on the hope that increased demand for corn would raise the price and profitability of such crops.

Conclusion. It may be concluded that gasohol does not hold any significant promise as a substitute motor fuel in the United States. In an overall plan for optimum energy utilization it may play a small component role, probably for the most part as an octane booster. In other countries there may be a greater opportunity for the utilization of crops as a source of essential supplies of engine fuels. *See* LIQUID FUEL; SYNTHETIC FUEL.

[TODD M. DOSCHER]

Bibliography: Gasohol in Brazil, *Chem. Eng. Prog.*, April 1979; Office of Technological Assessment, U.S. Congress, *Gasohol: A Technical Memorandum*, September 1979.

Gasoline

A mixture of hydrocarbons used to fuel the spark-ignited internal combustion engine. This mixture consists of more than 100 different hydrocarbons. The lightest, that is, lowest-boiling, is isobutane, while the heaviest hydrocarbons consist of a variety of substituted naphthalenes. Figure 1 shows a drawing of a gas chromatographic trace of a typical gasoline; in practice, there are three to four times as many peaks, shoulders, and bumps. The area of each peak relates directly to the amount of that particular compound present in the mixture. The subsequent discussion of gasoline emphasizes its end use and considers its various properties from the point of view of how they affect its performance in an engine.

All the hydrocarbons in gasoline are the products of a limited number of refining processes designed to increase the yield or quality of gasoline components (naphthas). These processes are listed in Table 1 in order of decreasing volume output in United States refining; in addition, Table 1 briefly describes the primary chemical transformation and octane quality resulting from the process. *See* PETROLEUM PROCESSING.

With these various components at hand, the refiner chooses a combination which will meet the

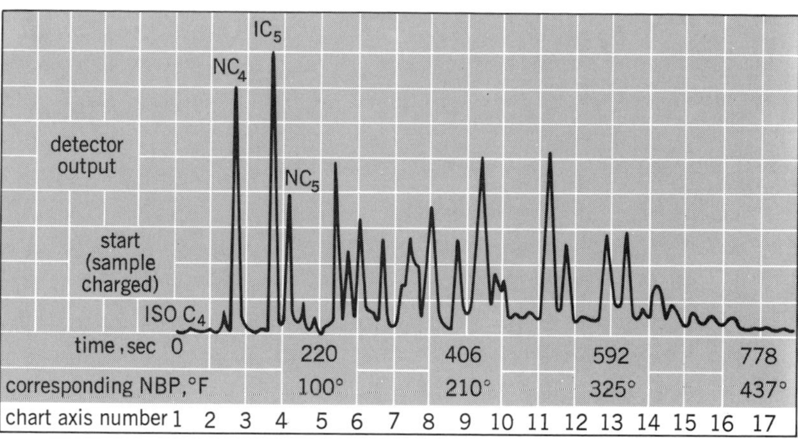

Fig. 1. Chromatogram of a typical gasoline.

quality requirements of the particular market being supplied. The American Society for Testing and Materials (ASTM) specification on motor gasoline, D-439, provides general guidelines for gasoline quality. The qualities imparted to current motor gasoline can be grouped in three categories: volatility, octane, and additive derived.

Volatility. The volatility of a modern gasoline is controlled to provide a balance between a lack of vehicle performance due to insufficient vaporization and poor performance due to an excess of vapor. Figure 2 shows the summer-to-winter variation between gasoline distillation curves and identifies the critical upper and lower limits along the whole curve. In order for the spark-ignited internal combustion engine to function, there must be a gaseous mixture of fuel vapor and air that sustains combustion when ignited by the spark discharge. That is, in cold weather there must be enough low-boiling hydrocarbons to form a flammable mixture in the cold cylinder. *See* INTERNAL COMBUSTION ENGINE.

Shortly after the engine is running, it is expected to produce power and propel the vehicle. As power is required, greater quantities of vapor (up through the middle of the boiling range) are required. However, once the engine is fully warmed up, this same gasoline which was volatile enough to start the cold engine must not generate so much vapor as to create another set of operating problems.

A vehicle's carburetor meters liquid fuel as a function of the engine's air demand. For best operation, considering performance, fuel economy, and

Table 1. Common gasoline components

Naphtha	Chemical change	Major characteristic
Catalytic cracked	Heavy fractions are cracked to naphtha	Olefins and aromatics Good octane
Catalytic reformed	Paraffins are cyclized and dehydrogenated	Aromatics Very good octane
Straight run	Distillation from crude	Paraffins Low octane
Alkylate	Isobutane is added to a light olefin	Isoparaffin Good octane
Hydrocracked	Heavy fractions are cracked with hydrogen	Isoparaffins Good octane
Catalytic isomerized	Straight chains are branched	Isoparaffins Good octane

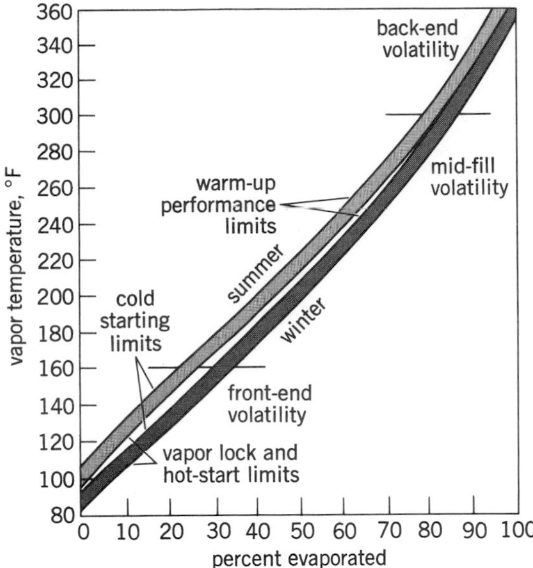

Fig. 2. A gasoline distillation versus performance.

exhaust emissions, the ratio of air to fuel (A/F ratio) must be kept within narrow limits. If the gasoline in use creates excessive vapors in the warmed-up fuel system (these are not metered by the carburetor), the engine will operate with an overly rich fuel-to-air mixture. The result of a mild form can be poor fuel economy, high exhaust emissions, and roughness at idle. In the extreme, the engine will cease to operate (vapor lock).

Satisfactory vehicle performance as the engine warms to full operating temperature is enhanced by good midrange volatility, which can be characterized as the volume percent of a gasoline which has distilled at 212°F (100°C). This feature is particularly important to later-model vehicles with quick-acting chokes and generally lean carburetion. In the extreme, a high volume percent distilled at 212°F can limit the amount of aromatics that can be used in gasoline. Only benzene, the lightest of the aromatics (use limited because of toxicity), boils below 212°F.

The last portion of a gasoline to vaporize consists primarily of substituted aromatics. These compounds improve the total fuel energy content and thus its fuel economy. They do this because they are denser and have higher heat contents than the total fuel. Again, there is a trade-off; these heavier aromatics contribute a disproportionately large share to the buildup of combustion chamber deposits. Many states have adopted a legal maximum final boiling point of 437°F (225°C).

The volatility measurements referred to so far are all based on the Engler distillation, ASTM D-86. This is a simple distillation from a specified flask at an imprecisely defined rate of heat input. That is, the Engler distillation is a poorly defined analytical technique. There is increasing interest in a gas chromatographic technique for a more precise definition of the vaporization characteristics of complex mixtures of hydrocarbons. It is likely that the simple, undefined distillation of gasoline will be replaced by a carefully controlled gas chromatographic technique, which will more precisely define the vaporization of complex mixtures of hydrocarbons.

Octane. A gasoline's octane number is an indication of the fuel's ability to prevent the occurrence of spark knock in an engine. Spark knock is an audible sound resulting from the explosion of a portion of the air-fuel charge prior to the arrival of the flame front; that is, the timed spark ignites a flame which propagates smoothly through the charge, except when spark knock occurs. Spark knock takes place when chain-branching reactions lead to an increasingly rapid buildup of hydroperoxide radicals (HO_2) in the region ahead of the advancing flame front. Tetraethyllead is an effective antiknock agent, because its decomposition gives rise to a nonstoichiometric lead oxide which acts as a radical trap.

The octane quality of a gasoline can be measured by two laboratory procedures which employ a single cylinder, CFR (Coordinating Fuels Research) knock rating engine (Table 2). The numerical octane scale used is based on blends of *n*-heptane and isooctane (2,2,4-trimethylpentane). Isooctane is assigned a value of 100, while the *n*-heptane value is designated 0; blends of the two standards are used to match the knock intensity of the test fuel.

For commercial gasolines, research octane numbers (RON) are higher than motor octane numbers (MON). According to the Energy Resources Development Agency (ERDA), premium gasoline averages 98.9 RON and 91.4 MON. Unfortunately, the average vehicle on the road does not operate under the narrowly specified conditions employed in the knock testing engines. Thus it is not surprising that neither RON nor MON alone is adequate to define antiknock quality for cars on the road; both are required.

Research octane number generally provides an indication of low-speed performance at full throttle, while motor octane number relates to part throttle (cruising and moderate acceleration) operation. A combination of both methods is required to define octane quality for vehicles equipped with automatic transmissions. Such a combination of laboratory octane numbers is said to define the road octane requirement of a vehicle, as shown in the equation below. The road octane number of a

$$\text{Road ON} = a\,(\text{RON}) + b\,(\text{MON}) + k$$

gasoline is the octane number of the primary reference fuel (*n*-heptane/isooctane) which produces the same knock intensity as the test fuel in a vehicle.

When refinery components are blended together, the octane number of the blend may be greater than, equal to, or less than that calculated from the volumetric average of the octane numbers of the individual blend components. Blending deviations of several octane numbers between experimentally observed and calculated antiknock rating can occur when blending refinery fuel components. Numerous calculational procedures have been developed over the years for predicting both research and motor octane numbers of multicomponent motor fuel blends. Some are based upon only the research and motor octane numbers of the components; others additionally consider composition data, such as olefins and aromatic contents, or detailed gas chromatographic analyses. Octane blending is nonlinear by nature, and blend deviations from linearity can be positive or negative, or

Table 2. Conditions of laboratory knock methods

Condition	Research method	Motor method
Engine speed, rpm	600	900
Mixture temperature, °F	100 (approx.)	300
Spark timing, °BTC	13	Varied with compression ratio

both, over the range of blend compositions. These deviations are attributed to the effect of hydrocarbon type and concentration and their interactions on the chain-branching reactions occurring ahead of the flame front. *See* OCTANE NUMBER.

Additive derived properties. A host of properties exhibited by current gasolines results from the use of additives. Gasolines are colored with oil-soluble dyes to differentiate between the various grades marketed. Gasolines are made less susceptible to oxidation; this allows the gasoline to be stored for many weeks without excessive "gum" formation. In the absence of an oxidation inhibitor, the chemically reactive hydrocarbons and the trace quantities of oxygenated compounds undergo free-radical-catalyzed reactions which produce minor amounts (0.005–0.05%) of partially oxidized high-molecular-weight hydrocarbons. The "gums," now only partially soluble in gasoline, can deposit on critical parts of the carburetor and cause various parts to stick and thus fail to operate properly. Oxidation inhbitors, materials which are preferentially oxidized, are used to inhibit gum formation. The inhbitors when oxidized form stable compounds which remain gasoline-soluble. Thus the reactive hydrocarbons do not have the chance to oxidize and eventually form gasoline-insoluble gums.

Despite efforts to maintain a dry gasoline distribution system, small amounts of water (0.005–0.1%) are normally present in gasoline as a result of processing and condensation. Accordingly, rusting is an ever-present problem of the gasoline distribution system and ultimately in the end-use vehicle. Aside from shortening the useful life of the various containers involved, the corrosion products become finely dispersed and are transported by the fuel. This particulate matter can lodge in small passages in a carburetor and cause engine malfunction.

Numerous organic compounds with an affinity for polar surfaces are used as rust inhibitors. These materials used in concentrations of a few parts per million reduce corrosion throughout the system and thus reduce the particulate content of the fuel. This means longer life for the vehicle's fuel filter and less likelihood that critical carburetor passages will become blocked.

In addition to deposits that can result from the autoxidation of relatively unstable fuel components, a vehicle's carburetor can become deposited with a mixture of airborne dirt, engine compartment fumes, and in heavy traffic the exhaust products of other vehicles. These deposits build up in the throttle area of the carburetor. As these deposits accumulate, they change the metering characteristics of the carburetor. Generally, they restrict airflow and thus cause the engine to run overly fuel-rich. The positive crankcase ventilation (PCV) systems required on all vehicles sold in the United States since 1963 further aggravate induction system deposits by discharging the crankcase blowby in the intake manifold. In the absence of a carburetor cleaner or detergent, these deposits result in rough idle and poorer fuel economy and higher exhaust emissions. Carburetor detergents encompass a variety of chemical compound types, all of which can be generally described as having hydrocarbon tails for solubility and nonhydrocarbon heads containing polar oxygen and nitrogen functions which associate with the deposits. The combination of polar and nonpolar groups in a single molecule account for their ability to orient at surfaces. As these molecules concentrate at the surface, they loosen the deposits which are swept away to be burned in the engine.

Gasoline marketed in cold weather must have high volatility in order to provide adequate starting and drivability prior to the engine's becoming warmed up. During the first 10–15 minutes of operation on such a volatile fuel, ice can form in and around the carburetor throat. As the gasoline evaporates, it removes heat from the incoming airstream. With a volatile fuel the incoming air temperature can drop from 40°F (4.5°C) to 20°F (−6.7°C). Under these conditions most of the water vapor being carried by the 40°F (4.5°C) air condenses in the carburetor throat to liquid water at a temperature significantly below its freezing point of 32°F (0°C). As the supercooled water freezes, the ice buildup in carburetor or throat begins to restrict airflow to the engine. This becomes very apparent at idle when the engine may no longer operate.

Many of the compounds which function as rust inhibitors and carburetor detergents also help in this latter instance. The surfactant modifies ice-crystal growth, fostering the growth of small crystals which are more readily sloughed off the throttle plates and throats into the airstream.

Most of the desirable features of current gasolines require the deliberate application of technology during the refining and marketing of gasoline. The particular balance of properties arrived at will not be the same in all instances. Technical capability and economic considerations both play a part in determining which properties are imparted and emphasized.

[HUGH F. SHANNON]

Bibliography: American Society for Testing and Materials, *Book of Standards*, pts. 23 and 47, 1975; Energy Resources Development Agency, Office of Public Affairs, Technical Information Center, *Motor Gasolines*, summer 1975.

Generator

Any machine by which mechanical power is transformed into electric power. Generators fall into two main groups, alternating-current (ac) and direct-current (dc). They may be further classified by their source of mechanical power, called the prime mover. Generators are usually driven by steam turbines, hydraulic turbines, engines, gas turbines, or motors. Small generators are sometimes powered from windmills or through gears, belts, friction, or direct drive from parts of vehicles or other machines.

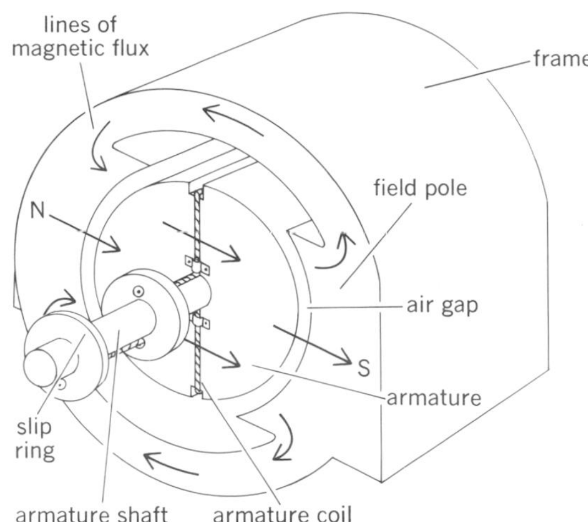

Fig. 1. Elementary generator.

Theory of operation. The theory of operation of most electric generators is based upon Faraday's law. When the number of webers of magnetic flux linking a coil of wire is caused to change, an electromotive force proportional to the product of the number of turns times the rate of change of flux is generated in the coil. The instantaneous induced voltage is given in Eq. (1), where n is the

$$e = -\,n\,(d\phi/dt)\ \text{volt} \tag{1}$$

number of turns, ϕ is the flux in webers, and t is time in seconds. The minus sign indicates that the induced voltage opposes the effect which produced it. Voltage is induced in the windings of a generator by mechanically driving one member relative to the other, thereby causing the magnetic flux linking one set of coils, called the armature windings, to vary, pulsate, or alternate. The magnetic flux may originate from a permanent magnet, a dc field winding, or an ac source. *See* MAGNETIC INDUCTION.

Figure 1 shows an elementary generator having a stationary field and a single, rotating, armature coil. It is apparent that the magnetic flux threading the coil reverses direction twice per revolution, thereby generating one cycle of voltage in the armature coil for each revolution. If this variation in flux is expressed as a function of time, the voltage generated in the coil would be given by Eq. (1). For example, let the flux vary as given by Eq. (2),

$$\phi = \Phi_m \cos 2\pi f t \tag{2}$$

where Φ_m is the maximum flux, f is the frequency, and t is time. By differentiating Eq. (2) with respect to time, and substituting in Eq. (1), the instantaneous voltage in the coil is found to be Eq. (3). This

$$e = 2\pi f n \Phi_m \sin 2\pi f t \tag{3}$$

ac voltage can be taken from the armature by brushes on the slip rings shown. If the coil terminals were brought to a two-segment commutator instead of slip rings, a pulsating dc voltage would appear at the brushes. *See* ALTERNATING-CURRENT GENERATOR; DIRECT-CURRENT GENERATOR.

Construction. In practice, permanent-magnet fields are used only in small generators. Large generators, except induction generators, are equipped with dc field windings. The field coils are wound on the stators of most dc generators to permit mounting the armature coils and commutator on the rotor. On ac generators, the field coils are normally located on the rotors. Field coils require only low voltage and power, and only two lead wires. They are more easily insulated and supported against rotational forces and are better suited to sliding contacts than the relatively high-voltage armature windings, which often have six leads brought out.

Any part of the magnetic circuit not subject to changing flux may be of solid steel. This includes the field poles of dc machines and portions or all of the rotating field structure of some ac generators. In machines with small air gaps the poles are frequently of laminated steel, even though their flux may be substantially constant. Laminations help minimize pole-face losses arising from tooth-frequency pulsations. The armature core is almost

Fig. 2. Turbogenerator unit with direct-connected exciter, rated 22,000 kW. (*Allis-Chalmers*)

always composed of thin sheets of high-grade electrical steel in order to reduce core loss. *See* CORE LOSS.

The windings are insulated from the magnetic structure, and are either embedded in slots distributed around the periphery or mounted to encircle the field poles. The terminals from the stator windings and from the brush holders are usually brought to a convenient terminal block for external wiring connections.

Turbogenerators. Generators driven by steam or gas turbines are sometimes called turbogenerators. Although in small sizes these may be gear-driven, and some may be dc generators, the term turbogenerator generally means an ac generator driven directly from the shaft of a steam turbine. A typical turbogenerator unit is shown in Fig. 2.

In order to achieve maximum efficiency, the steam turbine must operate at high speed. Consequently, direct-connected turbogenerators are seldom built to operate below 1500 rpm. To minimize windage loss and to keep rotational stresses down to a safe level, turbogenerator rotors are usually long and slender, in some cases five to six times the diameter in length of active iron. Long rotors operate above their first critical speed, and in some cases near or above their second, thereby introducing mechanical problems in balancing and resonance. To shorten the length of turbogenerators, conductor cooling has proved effective. *See* ELECTRIC POWER GENERATION; ELECTRIC ROTATING MACHINERY.

[LEON T. ROSENBERG]

Bibliography: S. L. Dixon, *Fluid Mechanics, Thermodynamics of Turbomachinery*, 2d ed., 1974; D. G. Fink and H. W. Beaty, *Standard Handbook for Electrical Engineers*, 11th ed., 1978; M. G. Say, *Alternating Current Machines*, 4th ed., 1976.

Geothermal power

Thermal or electrical power produced from the thermal energy contained in the Earth (geothermal energy). Use of geothermal energy is based thermodynamically on the temperature difference between a mass of subsurface rock and water and a mass of water or air at the Earth's surface. This temperature difference allows production of thermal energy that can be either used directly or converted to mechanical or electrical energy.

CHARACTERISTICS AND USE

Temperatures in the Earth in general increase with increasing depth, to 200–1000°C at the base of the Earth's crust and to perhaps 3500–4500°C at the center of the Earth. Average conductive geothermal gradients to 10 km (the depth of the deepest wells drilled to date) are shown in Fig. 1 for representative heat-flow provinces of the United States. The heat that produces these gradients comes from two sources: flow of heat from the deep crust and mantle; and thermal energy generated in the upper crust by radioactive decay of isotopes of uranium, thorium, and potassium. The gradients of Fig. 1 represent regions of different conductive heat flow from the mantle or deep crust. Some granitic rocks in the upper crust, however, have abnormally high contents of U and Th and thus produce anomalously great amounts of thermal energy and enhanced flow of heat toward the Earth's surface. Consequently, thermal gradients at shallow levels above these granitic plutons can be somewhat greater than shown on Fig. 1.

The thermal gradients of Fig. 1 are calculated under the assumption that heat moves toward the Earth's surface only by thermal conduction through solid rock. However, thermal energy is also transmitted toward the Earth's surface by movement of molten rock (magma) and by circulation of water through interconnected pores and fractures. These processes are superimposed on the regional conduction-dominated gradients of Fig. 1 and give rise to very high temperatures near the Earth's surface. Areas characterized by such high temperatures are the primary targets for geothermal exploration and development.

Natural geothermal reservoirs. Commercial exploration and development of geothermal energy to date have focused on natural geothermal reservoirs—volumes of rock at high temperature (up to 350°C) and with both high porosity (pore space, usually filled with water) and high permeability (ability to transmit fluid). The thermal energy is tapped by drilling wells into the reservoirs. The thermal energy in the rock is transferred by conduction to the fluid, which subsequently flows to the well and then to the Earth's surface.

Natural geothermal reservoirs, however, make up only a small fraction of the upper 10 km of the Earth's crust. The remainder is rock of relatively low permeability whose thermal energy cannot be produced without fracturing the rock artificially by means of explosives or hydrofracturing. Experiments involving artificial fracturing of hot rock have been performed, and extraction of energy by circulation of water through a network of these artificial fractures may someday prove economically feasible.

There are several types of natural geothermal reservoirs. All the reservoirs developed to date for electrical energy are termed hydrothermal convection systems and are characterized by circulation

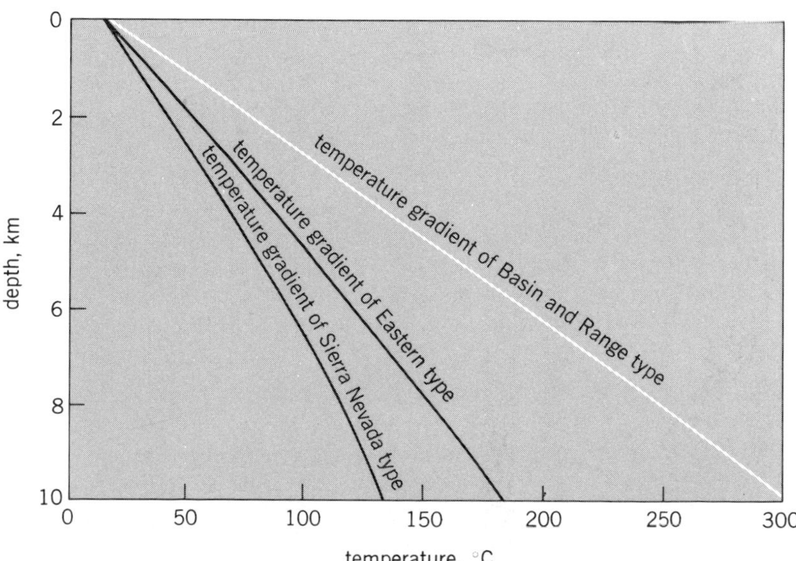

Fig. 1. Calculated average conductive temperature gradients to a depth of 10 km in representative heat-flow provinces of the United States. (*Adapted from D. E. White and D. L. Williams, eds., Assessment of Geothermal Resources of the United States—1975, USGS Circ. 726, 1975*)

of meteoric (surface) water to depth. The driving force of the convection systems is gravity, effective because of the density difference between cold, downward-moving, recharge water and heated, upward-moving, thermal water. A hydrothermal convection system can be driven either by an underlying young igneous intrusion or by merely deep circulation of water along faults and fractures. Depending on the physical state of the pore fluid, there are two kinds of hydrothermal convection systems: liquid-dominated, in which all the pores and fractures are filled with liquid water that exists at temperatures well above boiling at atmospheric pressure, owing to the pressure of overlying water; and vapor-dominated, in which the larger pores and fractures are filled with steam. Liquid-dominated reservoirs produce either water or a mixture of water and steam, whereas vapor-dominated reservoirs produce only steam, in most cases superheated.

Natural geothermal reservoirs also occur as regional aquifers, such as the Dogger Limestone of the Paris Basin in France and the sandstones of the Pannonian series of central Hungary. In some rapidly subsiding young sedimentary basins such as the northern Gulf of Mexico Basin, porous reservoir sandstones are compartmentalized by growth faults into individual reservoirs that can have fluid pressures exceeding that of a column of water and approaching that of the overlying rock. The pore water is prevented from escaping by the impermeable shale that surrounds the compartmented sandstone. The energy in these geopressured reservoirs consists not only of thermal energy, but also of an equal amount of energy from methane dissolved in the waters plus a small amount of mechanical energy due to the high fluid pressures. *See* AQUIFER; GROUNDWATER.

Use of geothermal energy. Although geothermal energy is present everywhere beneath the Earth's surface, its use is possible only when certain conditions are met: (1) The energy must be accessible to drilling, usually at depths of less than 3 km but possibly at depths of 6–7 km in particularly favorable environments (such as in the northern Gulf of Mexico Basin of the United States). (2) Pending demonstration of the technology and economics for fracturing and producing energy from rock of low permeability, the reservoir porosity and permeability must be sufficiently high to allow production of large quantities of thermal water. (3) Since a major cost in geothermal development is drilling and since costs per meter increase with increasing depth, the shallower the concentration of geothermal energy the better. (4) Geothermal fluids can be transported economically by pipeline on the Earth's surface only a few tens of kilometers, and thus any generating or direct-use facility must be located at or near the geothermal anomaly.

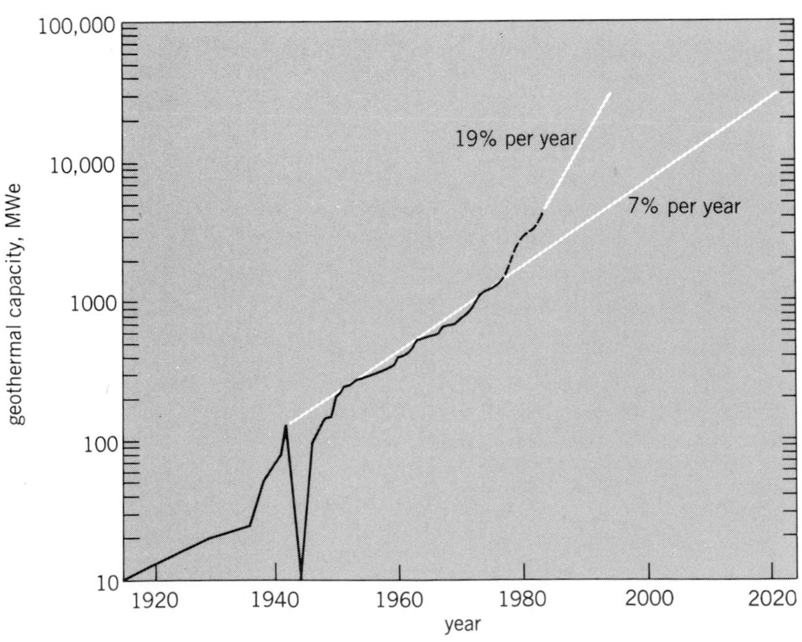

Fig. 2. Graph showing growth of geothermal electrical generating capacity of the world. *(From L. J. P. Muffler, ed., Assessment of Geothermal Resources of the United States–1978, USGS Circ. 790, 1979)*

Electric power generation. The most conspicuous use of geothermal energy is the generation of electricity. Hot water from a liquid-dominated reservoir is flashed partly to steam at the Earth's surface, and this steam is used to drive a conventional turbine-generator set. In the relatively rare vapor-dominated reservoirs, superheated steam produced by wells can be piped directly to the turbine without need for separation of water. Electricity is most readily produced from reservoirs of 180°C or greater, but reservoirs of 150°C or even lower show promise for electrical generation, either by using steam directly or by transferring its heat to a working fluid of low boiling point such as isobutane or Freon. Installed geothermal electrical capacity in mid-1979 is shown in the table, and the increase in worldwide geothermal electrical capacity with time is shown in Fig. 2. The importance of geothermal electricity to a small country is illustrated by El Salvador, where in 1977 the electricity generated from the Ahuachapán geothermal field represented 32% of the total electricity generated in that country. *See* ELECTRIC POWER GENERATION.

Direct use. Equally important worldwide is the direct use of geothermal energy, often at reservoir temperatures less than 100°C. Geothermal energy is used directly in a number of ways: to heat buildings (individual houses, apartment complexes, and even whole communities); to cool buildings (using lithium bromide absorption units); to heat greenhouses and soil; and to provide hot or warm water for domestic use, for product processing (for example, the production of paper), for the culture of shellfish and fish, for swimming pools, and for therapeutic (healing) purposes.

Major localities where geothermal energy is directly used include Iceland (30% of net energy consumption, primarily as domestic heating), the Paris Basin of France (where 60–70°C water is used in district heating systems for the communities of Melun, Creil, and Villeneuve la Garenne), and the Pannonian Basin of Hungary.

Prospects. In any analysis of the possible contribution of geothermal energy to human energy needs, one must keep in mind that the geothermal resource (that is, the potentially usable geothermal energy) is only a fraction of the thermal energy in a subsurface volume of rock and water. For favorable hydrothermal convection systems, this frac-

Installed geothermal electrical generating capacity of the world

Country	Megawatts-electrical	
	Early 1979	Projected by 1984
Italy	420.6	482
New Zealand	202.6	302
Mexico	75	235
Japan	166	216
Iceland	64	64
Soviet Union	5	28
Chile		30
China	1	?
El Salvador	60	95
Nicaragua		35
Philippines	3	548
Taiwan	0.6	5.6
Turkey	0.5	15.5
Indonesia		30
United States	608	2008
Total	1606.3	4094

tion can be 25% or greater, but for systems of restricted permeability the fraction is likely to be far smaller. Only this recoverable energy can be meaningfully compared with the thermal energy equivalent of barrels of recoverable oil, cubic meters of recoverable gas, tons of minable coal, or kilograms of minable uranium.

Use of geothermal energy is likely to increase greatly in many countries as other sources of energy become less abundant and more expensive and as geothermal reservoirs become better defined through systematic exploration and resource assessment. In the United States, the U.S. Geological Survey has estimated that the geothermal resources in identified and undiscovered hydrothermal convection systems are 2400×10^{18} joules, equivalent to the energy in 430×10^9 barrels (68×10^9 m^3) of oil or 47 years of oil consumption at the projected 1980 rate. Geopressured-geothermal resources (both thermal energy and energy from dissolved methane) of the northern Gulf of Mexico Basin are estimated to be between 430 and 4400×10^{18} joules, equivalent to 74 to 780×10^9 barrels (11.8 to 124×10^9 m^3) of oil or 8 to 85 years of oil consumption at the projected 1980 rate. Any energy that might be developed in the future from rock of low permeability or from magma would be in addition to this amount. Clearly, geothermal energy can play an important role in the energy economy of the United States as well as in the economies of many other countries throughout the world.

[L. J. PATRICK MUFFLER]

PRODUCTION AND POLLUTION PROBLEMS

The chief problems in producing geothermal power involve mineral deposition, changes in hydrological conditions, and corrosion of equipment. Pollution problems arise in handling geothermal effluents, both water and steam.

Mineral deposition. In some water-dominated fields there may be mineral deposition from boiling geothermal fluid. Silica deposition in wells caused problems in the Salton Sea, California, field; more commonly, calcium carbonate scale formation in wells or in the country rock may limit field developments, for example, in Turkey and the Philippines. Fields with hot waters high in total carbonate are now regarded with suspicion for simple development. In the disposal of hot wastewaters at the surface, silica deposition in flumes and waterways can be troublesome.

Hydrological changes. Extensive production from wells changes the local hydrological conditions. Decreasing aquifer pressures may cause boiling of water in the rocks (leading to changes in well fluid characteristics), encroachment of cool water from the outskirts of the field, or changes in water chemistry through lowered temperatures and gas concentrations. After an extensive withdrawal of hot water from rocks of low strength, localized ground subsidence may occur (up to several meters) and the original natural thermal activity may diminish in intensity. Some changes occur in all fields, and a good understanding of the geology and hydrology of a system is needed so that the well withdrawal rate can be matched to the well's long-term capacity to supply fluid.

Corrosion. Geothermal waters cause an accelerated corrosion of most metal alloys, but this is not a serious utilization problem except, very rarely, in areas where wells tap high-temperature acidic waters (for example, in active volcanic zones). The usual deep geothermal water is of near-neutral pH. The principal metal corrosion effects to be avoided are sulfide and chloride stress corrosion of certain stainless and high-strength steels and the rapid corrosion of copper-based alloys. Hydrogen sulfide, or its oxidation products, also causes a more rapid degradation than normal of building materials, such as concrete, plastics, and paints.

Pollution. A high noise level can arise from unsilenced discharging wells (up to 120 decibels adjusted), and well discharges may spray saline and silica-containing fluids on vegetation and buildings. Good engineering practice can reduce these effects to acceptable levels.

Because of the lower efficiency of geothermal power stations, they emit more water vapor per unit capacity than fossil-fuel stations. Steam from wellhead silencers and power station cooling towers may cause an increasing tendency for local fog and winter ice formation. Geothermal effluent waters liberated into waterways may cause a thermal pollution problem unless diluted by at least 100:1.

Geothermal power stations may have four major effluent streams. Large volumes of hot saline effluent water are produced in liquid-dominated fields. Impure water vapor rises from the station cooling towers, which also produce a condensate stream containing varying concentrations of ammonia, sulfide, carbonate, and boron. Waste gases flow from the gas extraction pump vent.

Pollutants in geothermal steam. Geothermal steam supplies differ widely in gas content (often 0.1–5%). The gas is predominantly carbon dioxide, hydrogen sulfide, methane, and ammonia. Venting of hydrogen sulfide gas may cause local objections if it is not adequately dispersed, and a major geothermal station near communities with a low tolerance to odor may require a sulfur recovery unit (such as the Stretford process unit). Sulfide dispersal effects on trees and plants appear to be small. The low radon concentrations in steam (3–200 nanocuries/kg or 0.1–7.4 kilobecquerels/kg), when dispersed, are unlikely to be of health significance. The mercury in geothermal stream (often 1–10 μg/kg) is finally released into the atmosphere, but the concentrations created are unlikely to be hazardous. *See* AIR POLLUTION.

Geothermal waters. The compositions of geothermal waters vary widely. Those in recent volcanic areas are commonly dilute (<0.5%) saline solutions, but waters in sedimentary basins or active volcanic areas range upward to concentrated brines. In comparison with surface waters, most geothermal waters contain exceptional concentrations of boron, fluoride, ammonia, silica, hydrogen sulfide, and arsenic. In the common dilute geothermal waters, the concentrations of heavy metals such as iron, manganese, lead, zinc, cadmium, and thallium seldom exceed the levels permissible in drinking waters. However, the concentrated brines may contain appreciable levels of heavy metals (parts per million or greater).

Because of their composition, effluent geothermal waters or condensates may adversely affect potable or irrigation water supplies and aquatic life. Ammonia can increase weed growth in waterways and promote eutrophication, while the entry of boron to irrigation waters may affect sensitive plants such as citrus. Small quantities of metal sulfide precipitates from waters, containing arsenic, antimony, and mercury, can accumulate in stream sediments and cause fish to derive undesirably high (over 0.5 ppm) mercury concentrations. *See* WATER POLLUTION.

Reinjection. The problem of surface disposal may be avoided by reinjection of wastewaters or condensates back into the countryside through disposal wells. Steam condensate reinjection has few problems and is practiced in Italy and the United States. The much larger volumes of separated waste hot water (about 50 metric tons per megawatt-electric) from water-dominated fields present a more difficult reinjection situation. Silica and carbonate deposition may cause blockages in rock fissures if appropriate temperature, chemical, and hydrological regimes are not met at the disposal depth. In some cases, chemical processing of brines may be necessary before reinjection. Selective reinjection of water into the thermal system may help to retain aquifer pressures and to extract further heat from the rock. A successful water reinjection system has operated for several years at Ahuachapan, El Salvador. [A. J. ELLIS]

Bibliography: H. C. H. Armstead, *Geothermal Energy*, 1978; H. C. H. Armstead (ed.), *Geothermal Energy: Review of Research and Development*, UNESCO, 1973; R. C. Axtmann, *Science*, 187: 795–803, 1975; A. J. Ellis and W. A. J. Mahon, *Chemistry and Geothermal Systems*, 1977; L. J. P. Muffler (ed.), *Assessment of Geothermal Resources of the United States—1978*, USGS Circ. 790, 1979; *Proceedings of the First United Nations Symposium on the Development and Utilization of Geothermal Resources, Pisa, Italy, Sept. 1970*, spec. issue no. 2 of *Geothermics*, 2 vols., 1973; *Proceedings of the Second United Nations Symposium on the Development and Use of Geothermal Resources, San Francisco, May 1975*, U.S. Government Printing Office, 3 vols., 1976; E. F. Wahl, *Geothermal Energy Utilization*, 1977; D. E. White and D. L. Williams (eds.), *Assessment of Geothermal Resources of the United States—1975*, USGS Circ. 726, 1975.

Graybody

An energy radiator which has a blackbody energy distribution, reduced by constant factor, throughout the radiation spectrum or within a certain wavelength interval. The designation "gray" has no relation to the visual appearance of a body but only to its similarity in energy distribution to a blackbody. Most metals, for example, have a constant emissivity within the visible region of the spectrum and thus are graybodies in that region. The graybody concept allows the calculation of the total radiation intensity of certain substances by multiplying the total radiated energy (as given by the Stefan-Boltzmann law) by the emissivity. The concept is also quite useful in determining the true temperatures of bodies by measuring the color temperature.

[HEINZ G. SELL; PETER J. WALSH]

Grounding

Intentional electrical connections to a reference conducting plane, which may be earth (hence the term ground), but which more generally consists of a specific array of interconnected electrical conductors, referred to as the grounding conductor. The symbol which denotes a connection to the grounding conductor is three parallel horizontal lines, each of the lower two being shorter than the one above it (Fig. 1). The electric system of an airplane or ship observes specific grounding practices with prescribed points of grounding, but no connection to earth is involved. A connection to such a reference grounding conductor which is independent of earth is denoted by use of the symbol shown in Fig. 2.

The subject of grounding may be conveniently divided into two categories: system grounding and equipment grounding. System grounding relates to a grounding connection from the electric power system conductors for the purpose of securing superior performance qualities in the electrical

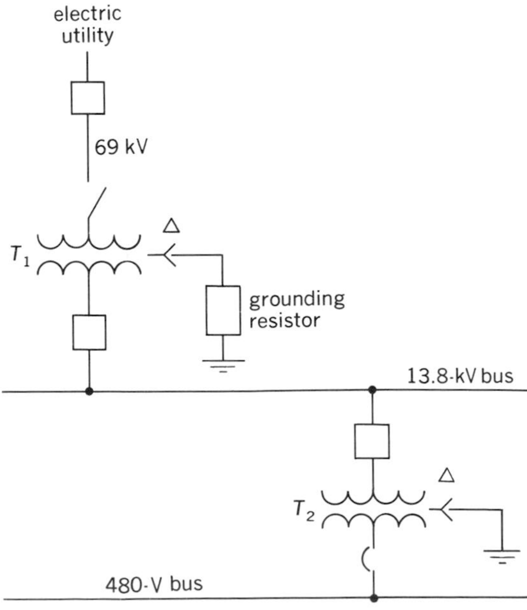

Fig. 1. Each conductively isolated portion of a distribution system requires its ground.

system. Equipment grounding relates to a grounding connection from the various electric-machine frames, equipment housings, metal raceways containing energized electrical conductors, and closely adjacent conducting structures judged to be vulnerable to contact by an energized conductor. The purpose of such equipment grounding is to avoid environmental hazards such as electric shock to area occupants, fire ignition hazard to the building or contents, and sparking or arcing between building interior metallic members which may be in loose contact with one another. The design of outdoor open-type installations presents special problems.

System grounding. Appropriately applied, a grounding system can (1) avoid excess voltage stress on electrical insulation within the system, leading to longer apparatus life and less frequent breakdown; (2) improve substantially the operating quality of the overcurrent protection system; and (3) greatly diminish the magnitude of arc fault heat-energy release at an insulation breakdown point, lessening arc burning damage and fire ignition possibility.

Sectionalization. Each voltage transformation point employing an insulating transformer interrupts the continuity of the system grounding circuit. Figure 1 illustrates the point. System grounding connections made on the 69-kV electric service companies' lines extend their influence only to the 69-kV winding of transformer T_1. The grounding connections established at the 13.8-kV winding of transformer T_1 extends its influence only to the 13.8-kV winding of transformer T_2. The grounding connections established at the 480-V secondary winding of transformer T_2 apply to the 480-V conductor system only. Two distinct advantages result. First, the system grounding arrangement of each voltage-level electric system is independent of all others. Second, the type and pattern of system grounding to be used with any individual voltage-level electric system can be tailored to optimize the performance of that particular electric-system section. *See* ELECTRIC POWER SYSTEMS.

It is preferable to locate the grounding connection at the source-point electrical neutral of the particular voltage-level system, and mandatory to do so at the service entrance point if the point of origin is outside the local building.

Common patterns. The great majority of system grounding patterns fall into one of the varieties that are shown in Fig. 3. The most used varieties of system grounding impedance are illustrated in Fig. 4.

The use of solid grounding exclusively for grounding patterns of Fig. 3*a*, *b*, and *d* are influenced by two considerations: (1) Overcurrent protection is present in only the phase conductor of single-phase, one-side grounded circuits. (2) The National Electrical Code (NEC) requires any electric system that can be solidly grounded in a manner which will avoid a phase-to-ground potential in excess of 150 volts to be so grounded.

Solid grounding is also used almost exclusively in the case of operating voltages of 69 kV and higher, as in Fig. 3*c*, to achieve the most rigid control of overvoltage stress. Such control allows reduced-rating lightning arresters, which in turn permits the successful use of reduced insulation level on station apparatus.

The high, unrestricted magnitude of short-circuit current created by a line-to-ground (L-G) fault on a solidly grounded system can pose severe design problems with costly solutions. The desire to artificially reduce the magnitude of L-G fault current is the chief reason for the use of other grounding impedances.

A low-reactance (inductive) grounding connection impedance can be used to effect a moderate reduction in the L-G short-circuit current, particularly for the purpose of avoiding excess short-circuit current flow in a phase winding of a rotating machine or of accomplishing a desired distribution of neutral unbalanced load current among source machines. Use of reactance to achieve the reduction in L-G fault current to below about 40% of the three-phase value enters the high-reactance region, which is subject to the generation of damaging transient overvoltages as a result of a ground fault condition, unless appropriate resistive damping circuits are added. Only one special case of high reactance grounding is free of overvoltage

Fig. 2. Symbol to denote connection to a reference ground that is independent of earth.

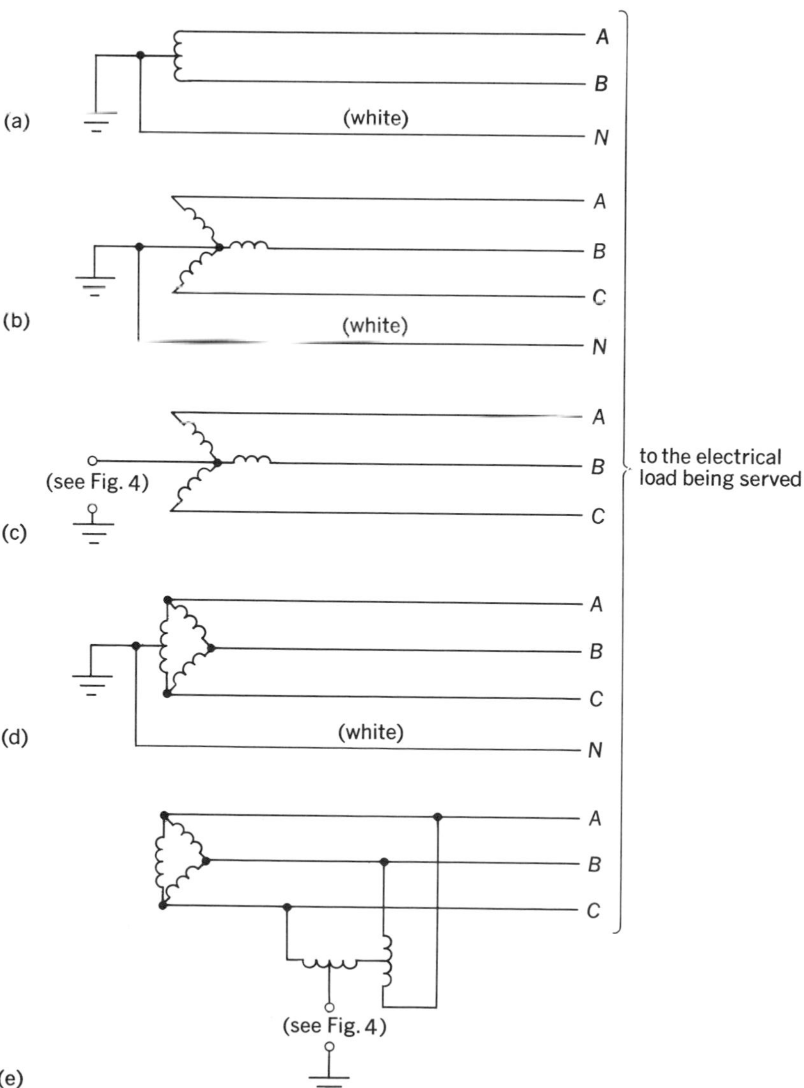

Fig. 3. Commonly used patterns of grounding. (*a*) Single-phase three-wire 240/120-V service. (*b*) Three-phase four-wire 208/120- or 480/277-V service. (*c*) Three-phase three-wire pattern. (*d*) Three-phase four-wire Δ 240-V line-to-line pattern. (*e*) Three-phase Δ with derived neutral.

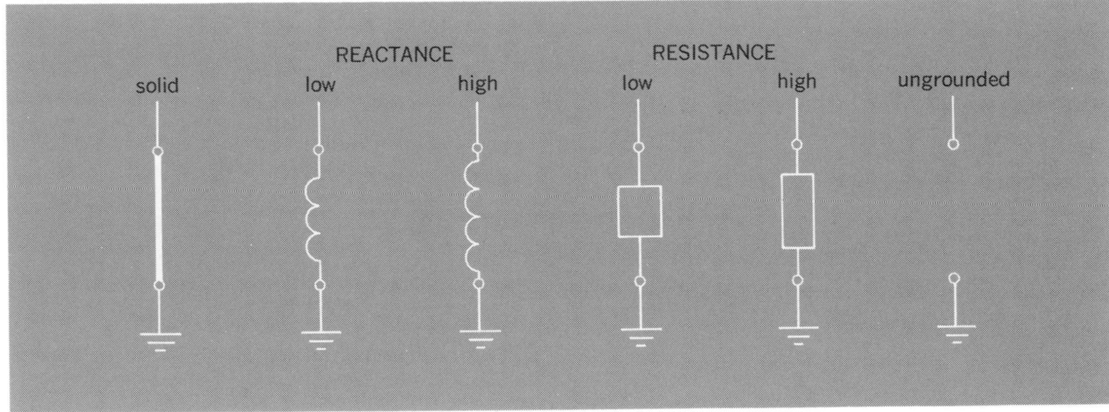

Fig. 4. Varieties of system grounding impedances in common use.

trouble: the ground-fault neutralizer case, in which the reactance magnitude is carefully matched with the electric-system L-G capacitance to create a natural frequency of oscillation almost exactly equal to power system operating frequency. It is critical, however, because a small deviation from the resonant value will destroy the overvoltage immunity.

Resistance grounding. By substituting a resistive grounding impedance (a totally dissipative impedance), much greater reductions in the L-G fault current can be intentionally created without danger of transitory overvoltages.

The low-resistance region is characterized by an established level of available ground-fault current well below the three-phase fault value yet ample to properly operate protective devices responsive to ground-fault current flow. Typical current values in use range from a few thousand amperes downward. Present-day protective practices allow the current value to be set at 400 A for general purpose medium-voltage electric systems widely used in industry. The far more critical electric-shock hazards incident to electric power supply to portable excavating machinery have led to the selection of a much lower level of available ground-fault current, typically in the 25–50-A region.

Most electrical breakdowns occur line-to-ground and many remain so throughout the interval of detection and isolation. A summary of the operating advantages achieved by intentional reduction of available ground-fault current is given below and illustrated in Fig. 5. (1) Low heat-energy release at the fault location because of the low current magnitude. (2) No noticeable dip in the system line-to-line voltages, which means no disturbance to the operation of all healthy load circuits. (3) Diminished interrupting duty on the circuit interrupter (low current and high power factor), which contributes to infrequent maintenance requirements. (4) Diminished duty imposed on the equipment grounding conductor network, which allows superior performance achievement at less cost.

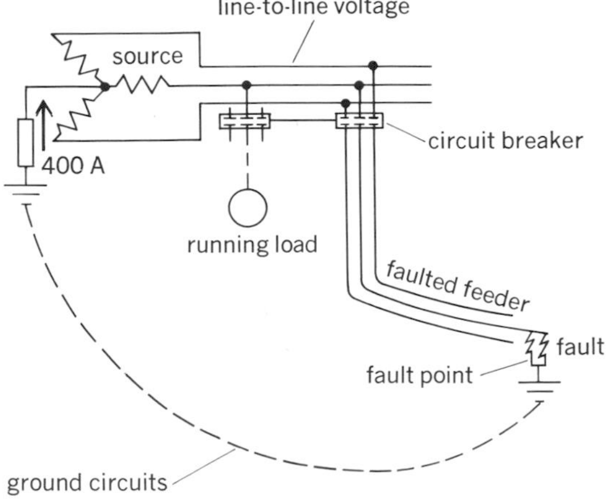

Fig. 5. Resistance grounding. (*General Electric Co.*)

High-resistance grounding relates to a mode of operation in which the fault location and subsequent corrective action are undertaken manually by skilled maintenance personnel. It is used principally in electric systems that serve critical continuous-process machines.

The available ground-fault current is reduced to the same order as the electric system charging current to ground (generally less than 5 A). This resistive component of fault current is sufficient to arrest the generation of transient overvoltages by L-G fault disturbances and provides a positive signal for identifying the presence of an L-G fault somewhere on the system. Successful results depend upon the presence of a skilled maintenance crew who can respond quickly to the ground-fault alarm, promptly locate the faulty circuit element, and take effective action to remove this faulty circuit from the system before a second insulation failure is induced.

Ungrounded system. Although a system may have no intentional grounding connection, and hence is named an ungrounded system, it is in fact unavoidably capacitively grounded. The layer of insulation surrounding every energized conductor metal constitutes the dielectric film of a minute distributed capacitor between power conductors and ground. In the aggregate this capacitance can amount to a substantial fraction of a microfarad for the complete metallically connected system. Surge voltage suppression filters typically include line-to-ground-connected capacitors which add to the distributed capacitors inherent in the system proper. Unless modified by the stabilizing qualities of grounding connections previously described, L-G fault disturbances can create dangerous overvoltage transients (impressed on system insulation) in about the same ways possible had the grounding connection impedance of Fig. 4 been a capacitor. The equivalent circuit of one phase conductor of an ungrounded three-phase power supply (relative to ground) takes the form illustrated in Fig. 6.

Equipment grounding. At each electrical equipment, grounding serves to establish a near-zero potential reference plane (even during L-G fault conditions). This reference extends to the outer reaches of the particular voltage-level electric system to which a solid grounding connection can be made from the metal frames of served electric machines, the metal housings that contain switching equipment or other electric-system apparatus, and the metal enclosures containing energized power conductors; these enclosures may be metal cable sheaths or metal raceways.

The purposes of this interconnected mesh conducting network, drawn as the heavy lines in Fig. 7, are listed below.

1. To avoid electric shock hazard to any occupant of the area who may be making bodily contact with a metallic structure containing energized conductors, one of which has made an electric fault connection to the mentioned structure.
2. To provide an adequately low impedance to the return path of L-G fault current so as not to interfere with the operation of system overcurrent protectors.
3. To provide ample conductivity (cross-sectional area) to carry the possible magnitude of ground-fault current for the duration controlled by the overcurrent protectors in the electric system.
4. To avoid, by installation, with appropriate geometric spacing (relative to phase conductors) dangerous amounts of ground-fault current diverted into paralleling conductive paths.

The fast-growing use of low-signal-level input high-speed electronic computing and data-processing systems places added emphasis on the need to minimize the transmission of stray electrical noise from the electric power circuits to the surrounding space in which these critical equipments are located. The presence of fast-acting solid-state switching devices among the electric-system switching components can aggravate the problem by intensifying the amount of high-frequency disturbance present with the electric-system voltage carried by the power conductors.

The complete metal enclosure of the electric power system, as shown in Fig. 8, contributes immensely to the elimination of electrical noise. The NEC requires that all such metal enclosures be interconnected to form an adequate continuous electrical circuit, and also that they be grounded (with some exceptions). The effect of this construction is to enclose the entire electric power system conductor array within a continuous shell of grounded metal that functions as a Faraday shield to confine the electrostatic and electromagnetic fields associated with the power conductors to the space within the metallic shell. The contribution of electrical noise external to the enclosures is reduced to almost zero.

To avoid a by-pass circuit by which power-conductor noise voltages might be conductively transmitted to the outside of the metallic enclosure, a careful check of the integrity of the insulation of the grounded power conductor (white wire) throughout the building interior is warranted. The NEC prescribes that the power-system grounded conductor be connected to the grounding conductor at the point of service entrance to the building, and at no other point within the building beyond the service equipment, with some exceptions. Only if a supply feeder extends from one building into another is it permissible to reground the white wire, and then only at the point of entrance to the second building.

Protection considerations. The importance of limited magnitude L-G fault current in easing the problem of electric shock exposure control will be evident from the discussion below. Section 230-95 of the NEC (1978) contains a mandatory requirement for installation of automatic ground-fault-responsive tripping of the power supply at the service equipment for all solidly grounded wye-connected electric services of more than 150 V L-G but not those exceeding 600 V line to line (L-L), for each service disconnecting means rated 1000 A or more. In the interests of assuring superior electric shock protection, section 210-8 of the NEC mandates the installation of ultrasensitive ground-responsive tripping features (type-GFCI personal protectors) on certain 120-V one-sided grounded

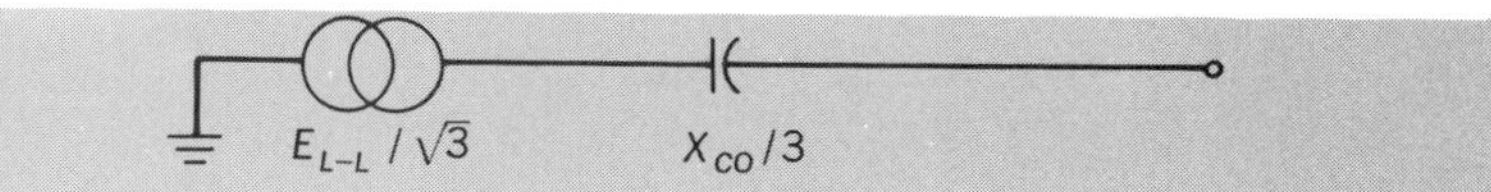

Fig. 6. Equivalent supply circuit to one phase conductor (relative to ground) of an "ungrounded" electric power system. $E_{L\text{-}L}$ = line-to-line voltage; $X_{co}/3$ = capacitive reactance coupling to ground.

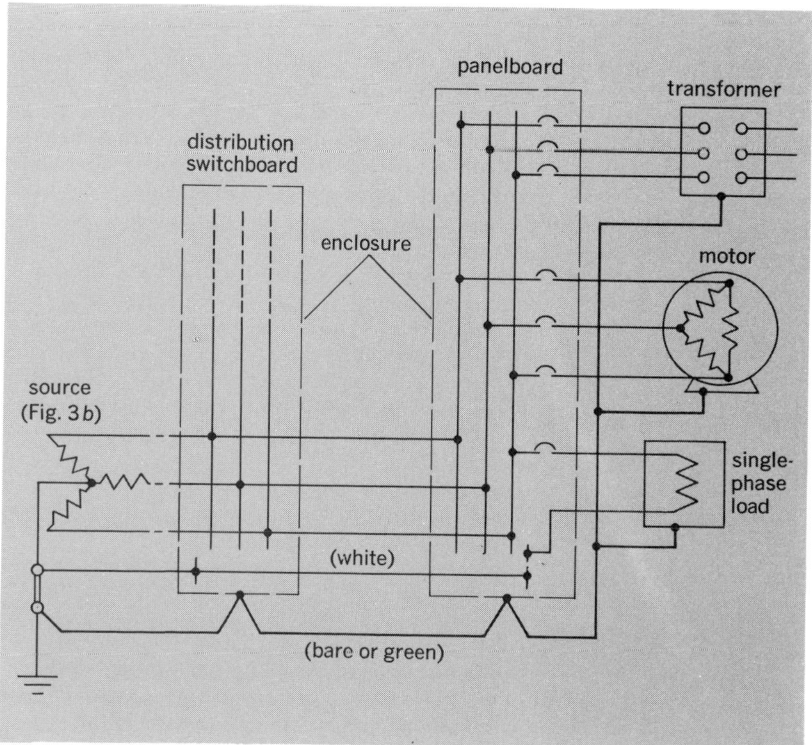

Fig. 7. Heavy line shows typical equipment grounding conductor.

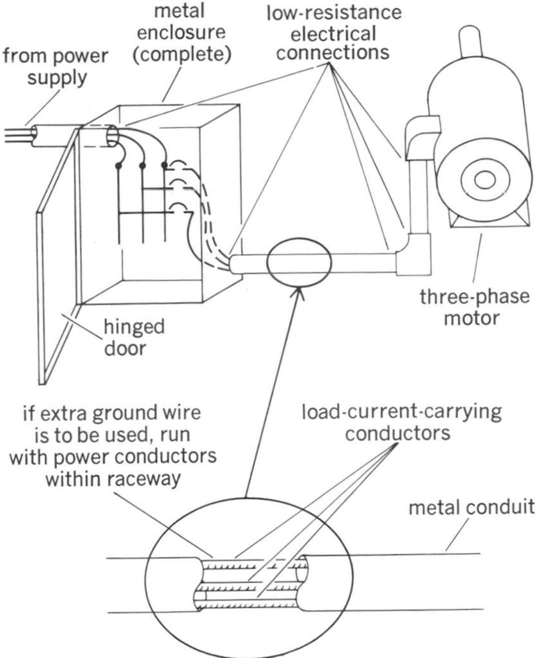

Fig. 8. Complete enclosure of power conductors within a continuous grounded steel shell confines electric and magnetic effects of power currents.

circuits. Protection is accomplished by deenergizing the supply power to the receptacle if even a minute amount of the circuit current (a few milliamperes) becomes diverted to a return path other than the grounded power conductor of the circuit.

Grounding of outdoor stations. Installations in which earth is used as a reference ground plane present special problems. To design an earth "floor surface" for an outdoor open-type substation which will be free of dangerous electric shock voltage exposure to persons around the station is a difficult task.

Personal electric shock danger. A beginning step in dealing with the shock hazard is to establish the magnitude of electric shock exposure which can be accepted by a person without ill effect. The evaluation of dangers from electric shock is aided by the simple "person model" (Fig. 9). The organs which can be vitally disoriented are the heart and lungs. Lung muscles can be propelled to a tight closed spasm, shutting off the respiratory action. However, the lungs resume normal action once the severe shock voltage is removed. On the other hand, a single short-time excess-value electric shock incident may throw the heart muscles into a state of fibrillation from which they are unable to automatically recover normal action, resulting in death within a few minutes unless antifibrillation is accomplished.

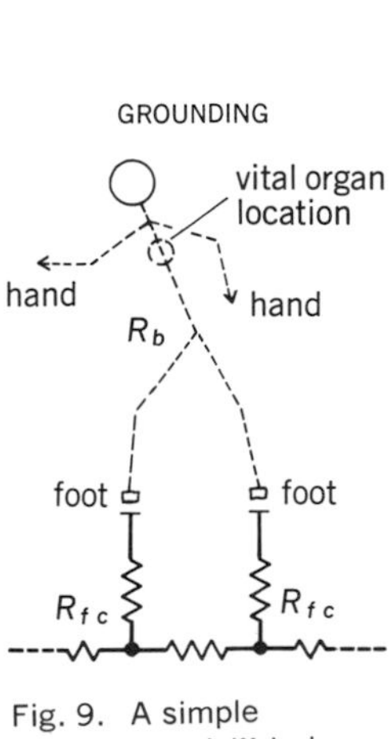

Fig. 9. A simple "person model" to be used in the evaluation of electric shock voltage-exposure intensity.

Electric shock is communicated to the body via electrical contact to the body extremities (Fig. 9), hand to hand, hand to foot, or foot to foot, each of which creates nearly the same shock intensity at the central chest area. The hand-to-foot case applies to a person standing on the substation floor and touching nearly conducting parts of the station with the extended arms; this applied shock hazard is named the E_{touch} exposure. The foot-to-foot case applies to a person walking across the substation floor during which activity the person's two feet, with considerable separation distance between, contact the substation surface; this shock hazard is named the E_{step} exposure. The effect of a surface layer of higher-resistance material over the active area of the station may be represented by an added resistance element R_{fc} (Fig. 9).

The basic limits of tolerable electric shock exposure are expressed by Eq. (1), where i is the current magnitude through the body and t the elapsed time in seconds, or, in similar form, by Eq. (2).

$$i = \frac{0.116 \text{ ampere}}{\sqrt{t}} \quad (1)$$

$$i^2 t = 0.0134 \text{ ampere}^2 \text{ seconds} \quad (2)$$

The sources of potential electric shock hazard are commonly expressed in terms of volts (rather than amperes), but the limiting shock exposure tolerance can be converted to terms of voltage and time if the body resistance R_b (Fig. 9) is known. The value of R_b may go up into the megohm region in the absence of excess moisture or perspiration, but drops drastically when these substances are present on the skin. Assuming the lower limit of the body resistance to be 1000 ohms, the maximum allowable impressed body voltage V_s is given by Eq. (3).

$$V_s = i(R_b) = i(1000) \text{ volts} = \frac{116 \text{ volts}}{\sqrt{t}} \quad (3)$$

Use of ground rods. The voltage exposure values of both E_{touch} and E_{step} depend heavily on the magnitude of voltage gradients along the surface of the outdoor station floor. An outdoor station might receive incoming power at 69 kV and an L-G circuit fault at this terminal would result in the injection of 3000 A into the grounding conductor at the station. During this current flow the station grounding conductor system could rise above mean earth potential by as much as 2500 V or more. The resulting potential distribution pattern across the station floor must be evaluated and tailored to achieve the limiting values of E_{touch} and E_{step}.

As a first step, one may consider driving a standard ground rod (8 ft or 2.4 m long, 3/4 in. or 19 mm in diameter) in the center of the substation floor area. In a typical soil makeup this might be found to establish a 25-ohm connection to earth. It takes only 100 A injected into such a connection to produce a voltage drop of 2500 V, about half of which takes place within a radius of about 2½ ft (0.8 m). As a voltage probe is moved away from the rod on the station surface, the potential drops rapidly in a conical pyramidal fashion.

The earth-connection properties of a single driven ground rod are thus controlled, almost totally, by a small cylinder of earth, about 10 ft (3 m) in diameter, immediately surrounding it. The resistance value is usually substantial (on the order of 25 ohms) and is reduced only slightly by a second ground rod within the influence zone of the first. A low-value-resistance connection to earth therefore requires a multiplicity of distributed ground rods, spaced to be nearly independent of the influence fields of each other.

To meet the electric shock safety requirements in open-type outdoor stations usually requires an

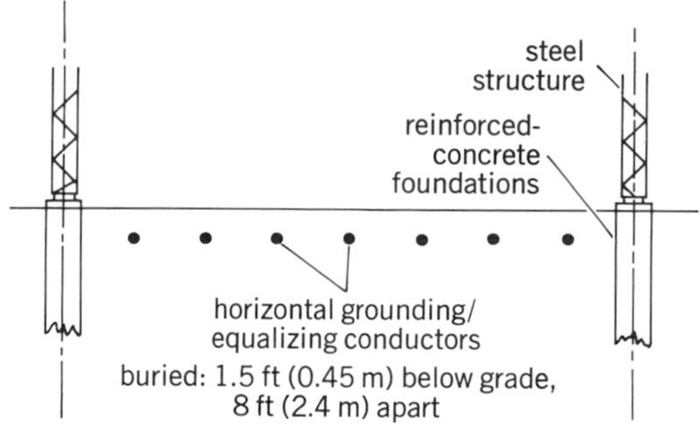

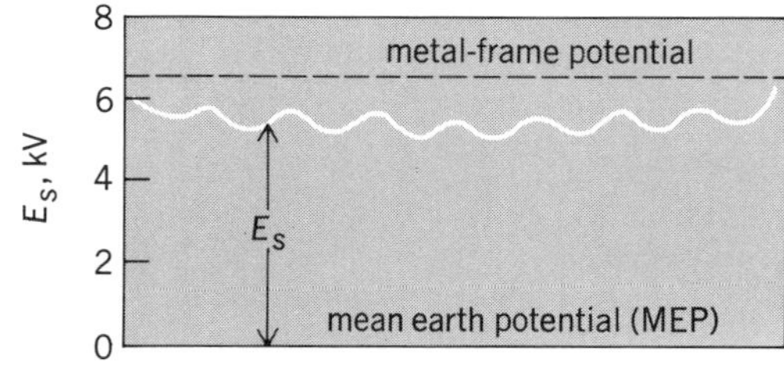

Fig. 10. Surface-potential control in an earth-floor electrical station obtained by a mesh grid of buried grounding/equalizing conductors. (*a*) Physical pattern of conductors in one bay. (*b*) Station surface potential E_s above mean earth potential during a local L-G fault.

elaborate array of metallic grounding conductors buried in the surface soil of the station area. The array is commonly composed of parallel horizontal conductors with perhaps 8 ft (2.4 m) horizontal spacing at a depth of some 1 to 2 ft (0.3 to 0.6 m) below the surface (Fig. 10*a*). The geometry of this conductor system usually matches that of the station structure and makes use of the below-grade concrete reinforcing steel involved in the construction of the station. The ground-surface voltage, within one specific bay, relative to the station grounding conductor potential and above the mean earth potential, resembles that shown in Fig. 10*b*. An artificially controlled level of available ground-fault current at 400 A is common in industrial electric power system design, in contrast to a value of 3000 A, not uncommon elsewhere, representing a 7-to-1 advantage.

Hazards near station boundary. There is the possibility that unacceptable levels of electric shock voltage exposure E_{step} and E_{touch} may be found in some places external to the outdoor station area. To ensure the absence of electric shock danger to persons who may frequent these areas adjoining the substation, it is important to check out suspect spots and institute corrective measures as necessary.

[R. H. KAUFMANN]

Bibliography: L. E. Crawford and M. S. Griffith, A closer look at "the facts of life" in ground mat design, *IEEE Trans., Ind. Appl.*, IA-15:241, 1979; C. F. Dalziel and W. R. Lee, Reevaluation of lethal electric currents, *IEEE Trans. Ind. Gen. Appl.*, IGA-4:467–476, 1968; Institute of Electrical and Electronics Engineers, *IEEE Guide for Safety in AC Substation Grounding*, IEEE Stand. no. 80, 1976; Institute of Electrical and Electronics Engineers, *IEEE Recommended Practice for Electric Power Systems in Commercial Buildings*, IEEE Stand. no. 241, 1974; Institute of Electrical and Electronics Engineers, *IEEE Recommended Practice for Grounding of Industrial and Commercial Power Systems*, IEEE Stand. no. 142, 1972; R. H. Kaufman, Some fundamentals of equipment grounding, *IEEE Trans. Appl. Ind.*, vol. 73, pt. 2, pp. 227–231, 1954; National Fire Protection Association, *National Electrical Code*, NFPA no. 70, 1978; F. J. Shields, The problem of arcing faults in low-voltage power distribution systems, *IEEE Trans. Ind. Gen. Appl.*, IGA-3(1):15, 1967; R. B. West, Grounding for emergency and standby power systems, *IEEE Trans. Ind. Appl.*, IA-15(2):124, 1979.

Health physics

The science and profession that deals with problems of protection from the hazards of radiation or prevention of damage from exposure to this radiation while making it possible for humans to make full use of the various forms of energy. Initially health physics dealt only with ionizing radiations (α, β, γ, neutrons, mesons, and so forth), but it has been extended to include nonionizing radiations (ultraviolet, visible, infrared, radio-frequency, microwave, long-wave, and sonic, ultrasonic, and infrasonic radiations). Health physics may be considered a border field of physics, biology, chemistry, mathematics, medicine, engineering, and industrial hygiene.

It is concerned with radiation protection problems involving research, engineering, education, and applied activities. It involves research on the effects of ionizing radiation on matter with a goal of developing a coherent theory of radiation damage. It deals with methods of measuring and assessing radiation dose, devices for reducing or preventing radiation exposure, the effects of ionizing radiation on humans and their environment, radioactive waste disposal, the establishment of maximum permissible exposure levels, and radiation risks associated with the nuclear energy industry, medical or hospital physics, high-voltage accelerator physics, and practical applications of radionuclides to science and industry. Many health physicists have become involved in the measurement and evaluation of exposure from nonionizing radiations and the setting of standards and appropriate exposure limits to reduce the hazards associated with these radiations.

Health physics began in 1942 along with the nuclear energy and reactor programs at the University of Chicago. It is estimated that by 1979 there were more than 5000 practicing health physicists in the United States, and 16,000 throughout the world. In 1956 the Health Physics Society was organized in the United States and Canada and now has more than 4000 members. It is one of 26 organizations affiliated with the International Radiation Protection Association, which has about 12,000 members in 70 countries.

Health physicists are employed in industry, national laboratories, state and Federal agencies, hospitals, universities, military organizations, and space programs, and in private practice. *See* MONITORING OF IONIZING RADIATION.

[KARL Z. MORGAN]

Bibliography: J. D. Abbott et al., *Protection against Radiation*, 1961; H. Cember, *Introduction to Health Physics*, 1969; J. S. Handloser, *Health Physics Instrumentation*, 1959; K. Z. Morgan and J. E. Turner, *Principles of Radiation Protection*, 1967; 2d International Conference on Peaceful Uses of Atomic Energy, Geneva, 1958, *Progress in Nuclear Energy*, ser. 12: *Health Physics*, vol. 1, 1959.

Heat

For the purposes of thermodynamics, it is convenient to define all energy while in transit, but unassociated with matter, as either heat or work. Heat is that form of energy in transit due to a temperature difference between the source from which the energy is coming and the sink toward which the energy is going. The energy is not called heat before it starts to flow or after it has ceased to flow. A hot object does contain energy, but calling this energy heat as it resides in the hot object can lead to widespread confusion. *See* ENERGY.

Heat flow is a result of a potential difference between the source and sink which is called temperature. Work is energy in transit as a result of a difference in any other potential such as height. Work may be thought of as that which can be completely used for lifting weights. Heat differs from work, the other type of energy in transit, in that its conversion to work is limited by the fundamental second law of thermodynamics, or Carnot efficiency. This natural law is that the fraction of the heat Q convertible to work is determined by the relation $dW = Q(dI/T)$ for processes where the source and sink are differentially different in temperature, or by the relation $dW = dQ(T_1 - T_2)/T_1$ where the source (at T_1) and the sink (at T_2) differ by a finite temperature interval. *See* WORK.

For the above relations to be valid, temperature must be expressed on a thermodynamic temperature scale. Conversely, any temperature scale for which the above relations are valid, irrespective of the substance or material under investigation, is a thermodynamic temperature scale. The perfect gas law defines a scale in which the temperature is proportional to the thermodynamic temperature. In order to make the two scales be identical, the triple point of water (temperature and pressure at which ice, water, and vapor are in equilibrium) is defined to be at 273.16 kelvins on both the ideal-gas and the thermodynamic scales. *See* THERMODYNAMIC PRINCIPLES.

[HAROLD C. WEBER; WILLIAM A. STEELE]

Bibliography: C. O. Bennett and J. E. Myers, *Momentum, Heat, and Mass Transfer*, 2d ed., 1974; P. W. Bridgman, *The Nature of Thermodynamics*, 1941.

Heat balance

A particular form of an energy balance. The heat balance is generally useful in science but of special importance in process engineering. It is basic to the design and analysis of operating equipment since it provides a relationship between the energy terms in a process. As a simple example, the heat loss from a pipe carrying a hot fluid is difficult to measure directly but is easily calculated by a heat balance if the fluid properties at the ends of the pipe are known.

The conservation of energy requires that for any system the accumulation of energy within must equal the difference between energy entering and leaving. Energy terms are conveniently classified as: the heat Q transferred into the system across the boundary, the work W put into the system, and the energy of the material streams flowing to and from the system. The flowing streams carry internal energy U, potential energy Z by virtue of their height, kinetic energy $mu^2/2$, where u is velocity and m is the mass of the fluid, and a flow work term pV which arises from forcing a volume of material V into or out of the system under the restraint of the pressure p.

Frequently, the kinetic energy, potential energy, and work terms are negligible or cancel out so that the energy balance simplifies to Eq. (1), where

$$U_1 + p_1V_1 + Q = U_2 + p_2V_2 + \Delta E \qquad (1)$$

subscripts 1 and 2 refer to input and output streams, respectively, and ΔE represents accumulation of energy in the system. This simplified energy balance is generally referred to as a heat balance.

The term $U + pV$ occurs in balance equations frequently and is given the name enthalpy, designated by H, so that the heat balance may be written as Eq. (2).

$$Q = H_2 - H_1 + \Delta E \qquad (2)$$

The balance in this form is termed an enthalpy balance and is especially convenient since values of H are tabulated for many of the common fluids and are easily calculated for other materials. For a steady-state flow process such as usually occurs in the continuous operation of boilers, reactors, and so forth, ΔE is zero and the heat exchange is simply equal to the enthalpy difference of the streams.

[WILLIAM F. JAEP]

Heat capacity

The quantity of heat required to raise a unit mass of homogeneous material one unit in temperature along a specified path, provided that during the process no phase or chemical changes occur is known as the heat capacity of the material in question. The unit mass may be 1 g, 1 lb, or 1 gram-

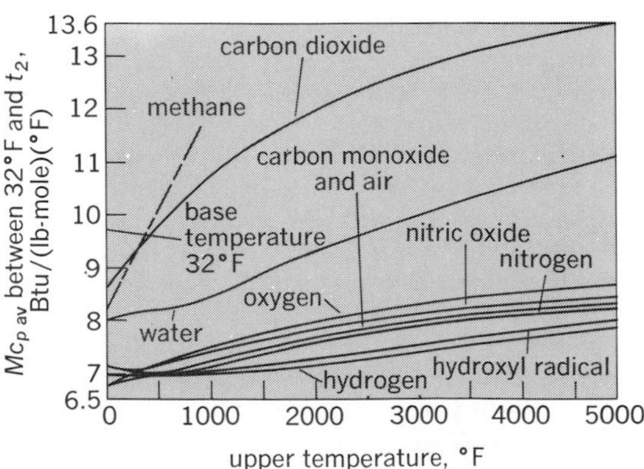

Fig. 1. Variation of average molar heat capacity with temperature for several common gases. (*Based on data of H. C. Hottel, MIT*)

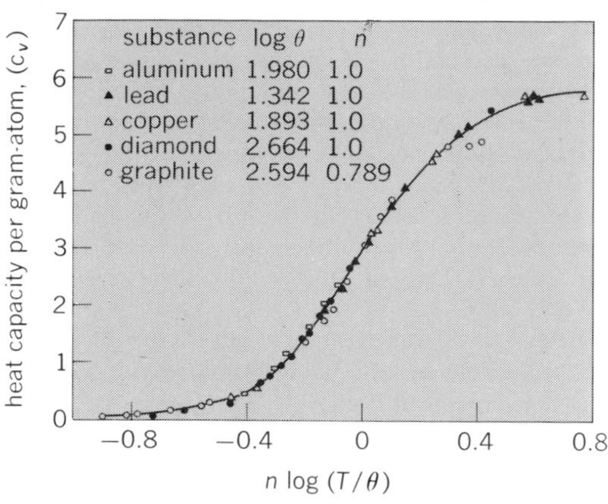

Fig. 2. Atomic heat capacity as a function of temperature (°K). (*Based on data of G. N. Lewis and G. E. Gibson, J. Amer. Chem. Soc., 39:2554–2581, 1917*)

molecular weight (1 mole). Moreover, the path is so restricted that the only work effects are those necessarily done on the surroundings to cause the change to conform to the specified path. The path, except as noted later, is at either constant pressure or constant volume. This definition conforms to an average heat capacity for the chosen unit change in temperature.

Instantaneous heat capacity at a particular temperature is defined as the rate of heat addition relative to the temperature change at the temperature in question; that is, on a plot of heat addition Q as a function of temperature T, instantaneous heat capacity is given by the slope of the curve at the temperature in question. Units of heat capacity are energy units per unit mass of material per unit change in temperature.

In accordance with the first law of thermodynamics, heat capacity at constant pressure C_p is equal to the rate of change of enthalpy with temperature at constant pressure $(\partial H/\partial T)_p$. Heat capacity at constant volume C_v is the rate of change of internal energy with temperature at constant volume $(\partial U/\partial T)_v$. Moreover, for any material, the first law yields the relation in Eq. (1).

$$C_p - C_v = \left[P + \left(\frac{\partial U}{\partial V}\right)\right]_T \left(\frac{\partial U}{\partial T}\right)_P \quad (1)$$

Gases. For one mole of a perfect gas, the preceding relation becomes $MC_p - MC_v = R$, where M is the molecular weight of the gas under discussion and R is the perfect gas law constant. For gases the ratio C_p/C_v is usually designated by the symbol K.

For monatomic gases at moderate pressures, MC_v is about 3, K is about 1.67, and the heat capacity changes but little with temperature. For diatomic gases, MC_v is approximately 5 at 20°C and moderate pressures. Change of heat capacity with temperature is usually small. The value of K is between 1.40 and 1.42. For triatomic gases at moderate pressures, MC_v varies from 6 to 7 and changes rapidly with temperature. The value of K varies but is always smaller than that for the less complex molecules at the same conditions of pressure and temperature.

For gases with more than three atoms per molecule, no generalizations are reliable. However, as molecular complexity increases, heat capacity increases, the influence of temperature on heat capacity increases, and K decreases. Figure 1 shows average MC_p for several common gases. Up to pressures of a few atmospheres, the effect of pressure on heat capacity of gases is small and is usually neglected.

Heat capacities of some elements

Element	Heat capacity
All heavy elements	6.4
Boron	2.7
Carbon	1.8
Fluorine	5.0
Hydrogen	2.3
Oxygen	4.0
Phosphorus and sulfur	5.4
Silicon	3.5

Solids. For solids, the atomic heat capacity (heat capacity when the unit mass under discussion is 1 at.wt) may be closely approximated by an equation of type (2), where $n=1$ for elements of simple crys-

$$C_v = J\left(\frac{T}{\theta}\right)^n \quad (2)$$

talline form, but has a smaller value for those of more complex structures; θ is characteristic of each element; J is a function that is the same for all substances; T is absolute temperature. Figure 2 compares measured with calculated values.

For all solid elements at room temperature, C_v is about 6.4 calories per gram atom per degree Celsius. This approximation may be used when no experimental data are available, but errors may be considerable, particularly for elements with atomic weights less than 39. Kopp's law states that for solids the molal heat capacity of a compound at room temperature and pressure approximately equals the sum of heat capacities of the elements in the compound. Errors are considerable but may be reduced by judicious choice of atomic heat capacities for the lighter elements. Recommended values for some of these are given in the table of constants for Kopp's law. Use of Kopp's law is justified only when no experimental data are available.

Figures 3 and 4 give instantaneous heat capacities for some industrially important solids.

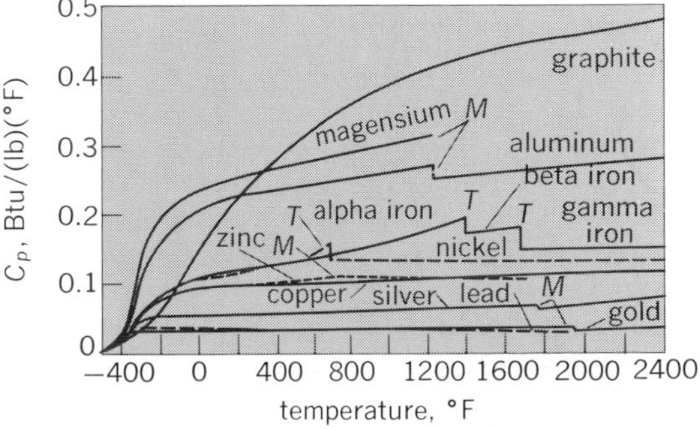

Fig. 3. Change in heat capacity of some industrially important solids with temperature. M= melting point; T= transition temperature. (*Based on data of K. K. Kelly, U.S. Bur. Mines Bull., no. 371, 1934*)

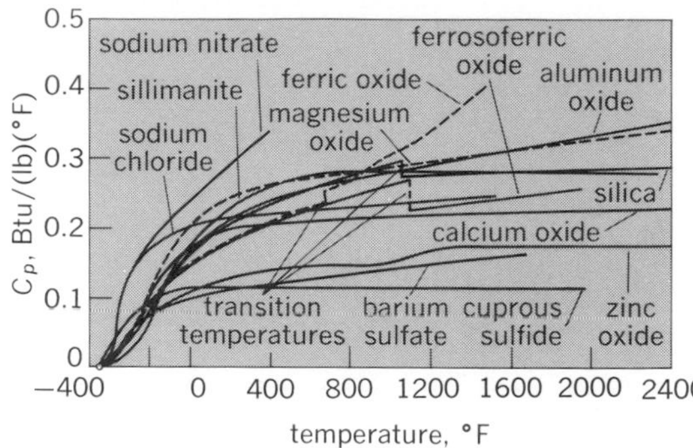

Fig. 4. Change in heat capacity of compounds with temperature. (*Based on data of K. K. Kelly, U.S. Bur. Mines Bull., no. 371, 1934*)

Liquids. For liquids and solutions no useful generally applicable approximations are available. For aqueous solutions of inorganic salts the approximate heat capacity of the solution may be estimated by assuming the dissolved salt to have negligible heat capacity. Thus, in a 20% by weight solution of any salt in water 0.8 would be the estimated heat capacity.

Effect of pressure on heat capacities at any temperature may be calculated by the relations in Eqs. (3*a*) and (3*b*).

$$\left(\frac{\partial MC_p}{\partial P}\right)_T = -T\left(\frac{\partial^2 V}{\partial T^2}\right)_P \qquad (3a)$$

$$\left(\frac{\partial MC_v}{\partial V}\right)_T = T\left(\frac{\partial^2 P}{\partial T^2}\right)_V \qquad (3b)$$

Constant temperature. Not so familiar as C_p and C_v are the heat necessary to cause unit change in pressure in a unit mass of material at constant temperature and the heat required to cause unit change in volume at constant temperature. These are designated $\partial Q_P/\partial P$ and $\partial Q_T/\partial V$. Similarly, $\partial Q_V/\partial P$ and $\partial Q_P/\partial V$ may be called heat capacities. *See* THERMODYNAMIC PRINCIPLES.

[HAROLD C. WEBER]

Bibliography: W. H. Brown, *Thermodynamics and Heat Engines*, 1964; K. E. Bett et al., *Thermodynamics for Chemical Engineers*, 1975.

Heat exchanger

A device used to transfer heat from a fluid flowing on one side of a barrier to another fluid (or fluids) flowing on the other side of the barrier.

When used to accomplish simultaneous heat transfer and mass transfer, heat exchangers become special equipment types, often known by other names. When fired directly by a combustion process, they become furnaces, boilers, heaters, tube-still heaters, and engines. If there is a change in phase in one of the flowing fluids—condensation of steam to water, for example—the equipment may be called a chiller, evaporator, sublimator, distillation-column reboiler, still, condenser, or cooler-condenser.

Heat exchangers may be so designed that chemical reactions or energy-generation processes can be carried out within them. The exchanger then becomes an integral part of the reaction system and may be known, for example, as a nuclear reactor, catalytic reactor, or polymerizer.

Heat exchangers are normally used only for the transfer and useful elimination or recovery of heat without an accompanying phase change. The fluids on either side of the barrier are usually liquids, but they may also be gases such as steam, air, or hydrocarbon vapors; or they may be liquid metals such as sodium or mercury. Fused salts are also used as heat-exchanger fluids in some applications.

With the development and commercial adoption of large, air-cooled heat exchangers, the simplest example of a heat exhanger would now be a tube within which a hot fluid flows and outside of which air is made to flow for the purpose of cooling. By similar reasoning, it might be argued that any container of a fluid immersed in any fluid could serve as a heat exchanger if the flow paths were properly connected, or that any container of a fluid exposed to air becomes a heat exchanger when a temperature differential exists. However, engineers will insist that the true heat exchanger serve some useful purpose, that the heat recovery be meaningful or profitable.

Most often the barrier between the fluids is a metal wall such as that of a tube or pipe. However, it can be fabricated from flat metal plate or from graphite, plastic, or other corrosion-resistant materials of construction. If, as is often the case, the barrier wall is that of a seamless or welded tube, several tubes may be tied together into a tube bundle (see diagram) through which one of the fluids flows distributed within the tubes. The other fluid (or fluids) is directed in its flow in the space outside the tubes through various arrangements of passes. This fluid is contained by the heat-exchanger shell. Discharge from the tube bundle is to the head (heads) and channel of the exchanger. Separation of tube-side and shell-side fluids is accomplished by using a tube sheet (tube sheets).

Applications. Heat exchangers find wide application in the chemical process industries, including petroleum refining and petrochemical processing; in the food industry, for example, for pasteurization of milk and canning of processed foods; in the generation of steam for production of power and electricity; in nuclear reaction systems; in aircraft and space vehicles; and in the field of cryogenics for low-temperature separation of gases. Heat exchangers are the workhorses of the field of heating, ventilating, air-conditioning, and refrigeration.

Classifications. The exchanger type described in general terms above and illustrated by the diagram is the well-known shell-and-tube heat exchanger. Shell-and-tube exchangers are the most numerous, but constitute only one of many types. Exchangers in use range from the simple pipe within a pipe—with a few square feet of heat-transfer surface—up to the complex-surface exchangers that provide thousands of square feet of heat-transfer area.

In between these extremes is a broad field of shell-and-tube exchangers often specifically named by distinguishing design features; for example, U tube, fin tube, fixed tube sheet, floating head, lantern-ring packed floating head, socket-and-gland packed floating head, split-ring internal floating head, pull-through floating head, nonremovable bundle with floating head or U-tube construction, and bayonet type.

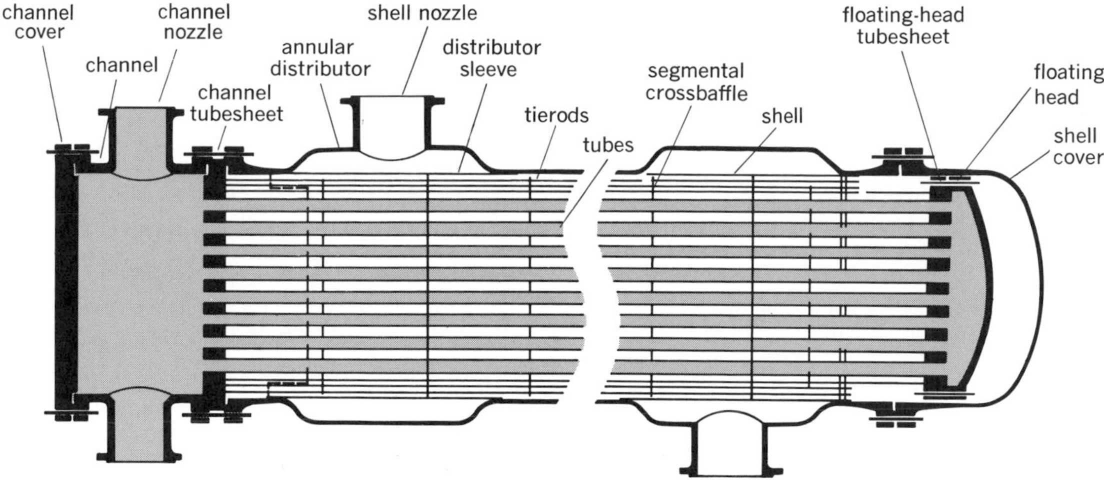

Schematic diagram of heat exchanger.

Also, varying pass arrangements and baffle-and-shell alignments add to the multiplicity of available designs. Either the shell-side or tube-side fluids, or both, may be designed to pass through the exchanger several times in concurrent, countercurrent, or cross flow to the other fluids.

The concentric pipe within a pipe (double pipe) serves as a simple but efficient heat exchanger. One fluid flows inside the smaller-diameter pipe, and the other flows, either concurrently or countercurrently, in the annular space between the two pipes, with the wall of the larger-diameter pipe serving as the shell of the exchanger.

To solve new processing problems and to find more economical ways of solving old ones, new types of heat exchangers are being developed. There has been much emphasis on cramming more heat-transfer surface into less and less volume. Extended-surface exchangers, such as those built with fin tubes, are finding wide application.

Water shortage has added a new dimension to heat-exchanger design. In its four newest refineries (all outside the United States) a large oil company has not used a single water-cooler exchanger. All are air-cooled. On the other hand, in the United States, two new plants on opposite banks of the same river disagreed on the merits of air-cooling. One chose water-cooling and the other air-cooling.

Plate-type heat exchangers, long used in the milk industry for pasteurization and skimming, are moving into the chemical and petroleum industries. Coiled tubular exchangers and coiled-plate heat exchangers are winning new assignments. Spiral exchangers offer short cylindrical shells with flat heads, carrying inlets and outlets leading to internal spiral passages. These passages may be made with spiral plates or with spiral banks of tubes. Exchangers with mechanically scraped surfaces are finding favor for use with very viscous and pastelike materials.

A somewhat unusual type of plate heat exchanger is one in which sheets of 16-gage metal, seam- and spot-welded together, are embossed to form transverse internal channels which carry the heat-transfer medium. This type is often used for immersion heating in electroplating and pickling.

Materials of construction. Every metal seems to be a possible candidate as a material of construction in fabrication of heat exchangers. Most often, carbon steels and alloy steels are used because of the strength they offer, especially when the exchanger is to be operated as a pressure vessel. Because of excellent heat conductance, brass and copper find wide use in exchanger manufacture.

Corrosion plays a key role in the selection of exchanger construction materials. Often, a high-priced material will be selected to contain a corrosive tube-side fluid, with a cheaper material being used on the less corrosive shell side.

For special corrosion problems, exchangers are built from graphite, ceramics, glass, bimetallic tubes, tantalum, aluminum, nickel, bronze, silver, and gold.

Problems of use. Each of the fluids and the barrier walls between them offers a resistance to heat transfer. However, another major resistance that must be considered in design is the formation of dirt and scale deposited on either side of the barrier wall. This resistance may become so great that the exchanger will have to be removed from service periodically for cleaning.

Chemical and mechanical methods may be used to remove the dirt and scale. For mechanical cleaning, the exchanger is removed from service and opened up. Perhaps the entire tube bundle is pulled from the exchanger shell if the plant layout has provided space for this to be done. If the deposit is on the inside of straight tubes, cleaning may be accomplished merely by forcing a long worm or wire brush through each tube.

More labor is required to remove deposits on the shell side. After removal of the tube bundle, special cleaning methods such as sandblasting may be necessary.

Much engineering effort has gone into the design of heat exchangers to allow for fouling. However, D. Q. Kern has suggested that methods are available to design heat exchangers that, by accommodating a certain amount of dirt in a thermal design, will allow heat exchangers to run forever without shutdown for cleaning. Commercial units designed in this fashion are operating today.

Another operating problem is allowance for differential thermal expansion of metallic parts. Most operating difficulties arise during the startup or shutdown of equipment. Therefore, M. S. Peters suggests the following general rules:

1. Startup. Always introduce the cooler fluid first. Add the hotter fluid slowly until the unit is up to operating conditions. Be sure the entire unit is fillcd with fluid and there are no pockets or trapped inert gases. Use a bleed valve to remove trapped gases.

2. Shutdown. Shut off the hot fluid first, but do not allow the unit to cool too rapidly. Drain any materials which might freeze of solidify as the exchanger cools.

3. Steam condensate. Always drain any steam condensate from heat exchangers when starting up or shutting down. This reduces the possibility of water hammer caused by steam forcing the trapped water through the lines at high velocities.

Standardization. Users have requested that heat exchangers be made available at lower prices through the standardization of designs. Organizations active in this work are the Tubular Exchangers Manufacturers' Association, American Petroleum Institute, American Standards Association, and the American Institute of Chemical Engineers. *See* COOLING TOWER; HEAT TRANSFER.

[RAYMOND F. FREMED]

Bibliography: W. E. Glausser and J. A. Cortright, How to specify heat exchangers, *Chem. Eng.*, 62(12):203–206, 1955; W. M. Kays and A. L. London, *Compact Heat Exchangers*, 2d ed., 1964; D. Q. Kern, Speculative process design, *Chem. Eng.*, 66(20):127–142, 1959; K. Kornwell, *The Flow of Heat*, 1977; M. S. Peters, *Elementary Chemical Engineering*, 1954; D. J. Portman and E. Ryznar, *An Investigation of Heat Exchange*, 1971; J. C. Smith, Trends in heat exchangers, *Chem. Eng.*, 61(6):232–238, 1953.

Heat insulation

Materials whose principal purpose is to retard the flow of heat. Thermal- or heat-insulation materials may be divided into two classes, bulk insulations and reflective insulations. The class and the material within a class to be used for a given application depend upon such factors as temperature of operation, ambient conditions, mechanical strength requirements, and economics.

Examples of bulk insulation include mineral wool, vegetable fibers and organic papers, foamed plastics, calcium silicates with asbestos, expanded vermiculite, expanded perlite, cellular glass, silica aerogel, and diatomite and insulating firebrick. They retard the flow of heat, breaking up the heat-flow path by the interposition of many air spaces and in most cases by their opacity to radiant heat.

Reflective insulations are usually aluminum foil or sheets, although occasionally a coated steel sheet, an aluminumized paper, or even gold or silver surfaces are used. Refractory metals, such as tantalum, may be used at higher temperatures. Their effectiveness is due to their low emissivity (high reflectivity) of heat radiation.

Thermal insulations are regularly used at temperatures ranging from a few degrees above absolute zero, as in the storage of liquid hydrogen and helium, to above 3000°F in high-temperature furnaces. Temperatures of 4000–5000°F are encountered in the hotter portions of missiles, rockets, and aerospace vehicles. To withstand these temperatures during exposures lasting seconds or minutes, insulation systems are designed that employ radiative, ablative, or absorptive methods of heat dissipation.

Heat flow. The distinguishing property of bulk thermal insulation is low thermal conductivity. Under conditions of steady-state heat flow the empirical equation that describes the heat flow through a material is Eq. (1), where q = time rate of

$$\frac{q}{A} = -k\frac{(\theta_2 - \theta_1)}{l} \tag{1}$$

heat flow, A = area, θ_1 = temperature of warmer side, θ_2 = temperature of colder side, l = thickness or length of heat-flow path, and k = thermal conductivity, representative values being listed in the table. For a given thickness of material exposed to a given temperature difference, the rate of heat flow per unit area is directly proportional to the thermal conductivity of the material.

In the unsteady state, or transient heat flow, the density and specific heat of a material have a strong influence upon the rate of heat flow. In such cases, thermal diffusivity $\alpha = k/\rho\, C_p$ is the important property. Here ρ = density and C_p = specific heat at constant pressure. In the simple case of one-dimensional heat flow through a homogeneous material, the governing equation is Eq. (2), where

$$\frac{d\theta}{dt} = \alpha \left.\frac{d^2\theta}{dx^2}\right|_0^l \tag{2}$$

t = time and x is measured along the heat-flow path from 0 to l.

Thermal conductivities of selected solids*

Material	Density, lb/ft³	°F	Conductivity, k† Btu/(hr)(ft²)(°F/in.)
Asbestos cement board	120	75	4
Cotton fiber	0.8–2.0	75	0.26
Mineral wool, fibrous rock, slag, or glass	1.5–4.0	75	0.27
Insulating board, wood, or cane fiber	15	75	0.35
Foamed plastics	1.6	75	0.29
Glass			3.6–7.32
Hardwoods, typical	45	75	1.10
Softwoods, typical	32	75	0.80
Cellular glass	9	75	0.40
Fine sand (4% moisture content)	100	40	4.5
Silty clay loam (20% moisture content)	100	40	9.5
Gypsum or plaster board	50	75	1.1

*1 lb/ft³ = 16.0 kg/m³. 75°F = 24°C, 40°F = 4°C. 1 Btu/(hr)(ft²)(°F/in.) = 0.1442 W/(m²)(°C/m).

†Typical; suitable for engineering calculations.

SOURCE: From American Society of Heating, Refrigerating, and Air Conditioning Engineers, *Heating, Ventilating and Air Conditioning Guide*, 1959.

Thermal conductivity. In general, thermal conductivity is not a constant for the material but varies with temperature. Generally, for metals and other crystalline materials, conductivity decreases with increasing temperature; for glasses and other amorphous materials, conductivity increases with temperature. Bulk insulation materials in general behave like amorphous materials and have a positive temperature coefficient of conductivity.

Thermal conductivity of bulk insulation depends upon the nature of the gas in the pores. The conductivities of two insulations, identical except for the gases filling the pore spaces, will differ by an amount approximately proportional to the difference in the conductivities of the two gases.

Increasing the pressure of the gas in the pores of a bulk insulation has little effect on the conductivity even with pressures of several atmospheres. Decreasing the pressure has little effect until the mean free path of the gas is in the order of magnitude of the dimensions of the pores. Below this pressure the conductivity decreases rapidly until it reaches a value determined by radiation and solid conduction. A few materials have such fine pores that at atmospheric pressure their dimensions are smaller than the mean free path of air. Such insulations may have conductivities less than the conductivity of still air. *See* HEAT TRANSFER.

[HARRY F. REMDE]

Bibliography: H. S. Carlslaw and J. C. Jaeger, *Conduction of Heat in Solids*, 2d ed., 1959; P. E. Glaser et al., *Investigation of Materials for Vacuum Insulators up to 4000F*, ASD TR-62-88, 1962; W. H. MacAdams, *Heat Transmission*, 3d ed., 1954; E. M. Sparrow and R. D. Cess, *Radiation Heat Transfer*, 1967; H. M. Strong, F. P. Bundy, and H. P. Bovenkerk, Flat panel vacuum thermal insulation, *J. Appl. Phys.*, 31:39, 1960; J. D. Vershoor and P. Greebler, Heat transfer by gas conduction and radiation in fibrous insulations, *Trans. ASME*, 74(6):961–968, 1952; G. B. Wilkes, *Heat Insulation*, 1950.

Heat pump

The thermodynamic counterpart of the heat engine. A heat pump raises the temperature level of heat by means of work input. In its usual form a compressor takes refrigerant vapor from a low-pressure, low-temperature evaporator and delivers it at high pressure and temperature to a condenser (Fig. 1). The pump cycle is identical with the customary vapor-compression refrigeration system. *See* REFRIGERATION CYCLE.

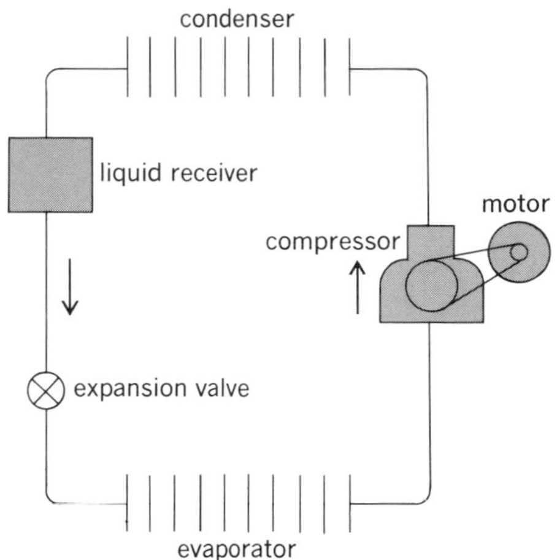

Fig. 1. Basic flow diagram of heat pump with motor-driven compressor. For summer cooling, condenser is outdoors and evaporator indoors; for winter heating, condenser is indoors and evaporator outdoors.

Application to comfort control. For air-conditioning in the comfort heating and cooling of space, a heat pump uses the same equipment to cool the conditioned space in summer and to heat it in winter, maintaining a generally comfortable temperature at all times (Fig. 2). *See* AIR CONDITIONING; COMFORT HEATING.

This dual purpose is accomplished, in effect, by placing the low-temperature evaporator in the conditioned space during the summer and the high-temperature condenser in the same space during the winter (Fig. 3). Thus, if 70°F is to be maintained in the conditioned space regardless of the season, this would be the theoretical temperature of the evaporating coil in summer and of the condensing coil in winter. The actual temperatures on the refrigerant side of these coils would need to be below 70°F in summer and above 70°F in winter to permit the necessary transfer of heat through the coil surfaces.

If the average outside temperatures are 100°F in summer and 40°F in winter, the heat pump serves to raise or lower the temperature 30° and to deliver the heat or cold as required. The ultimate ideal cycle for estimating performance is the same Carnot cycle as that for heat engines. The coefficient of performance COP_c as cooling machine is given in Eq. (1), and the coefficient COP_w as a warming machine is given in Eq. (2), where T is temperature in

$$COP_c = \frac{\text{refrigeration}}{\text{work}} = \frac{T_c}{T_h - T_c} \quad (1)$$

$$COP_w = \frac{\text{heat delivered}}{\text{work}} = \frac{T_h}{T_h - T_c} \quad (2)$$

degrees absolute and the subscripts c and h refer to the cold and hot temperatures, respectively.

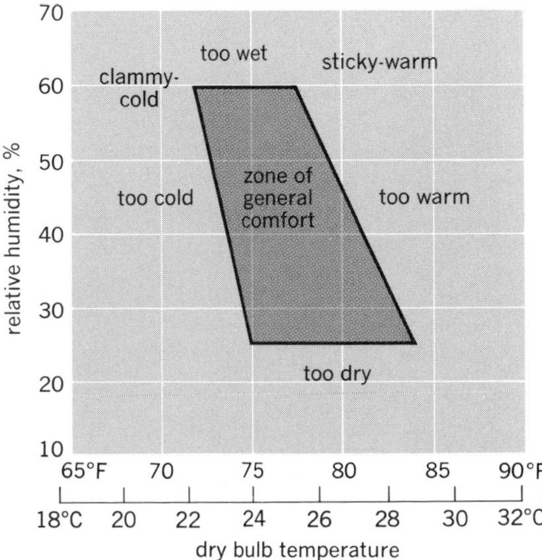

Fig. 2. Indoor climatic conditions acceptable to most people when doing desk work; continuous air motion with 5–8 air changes per hour.

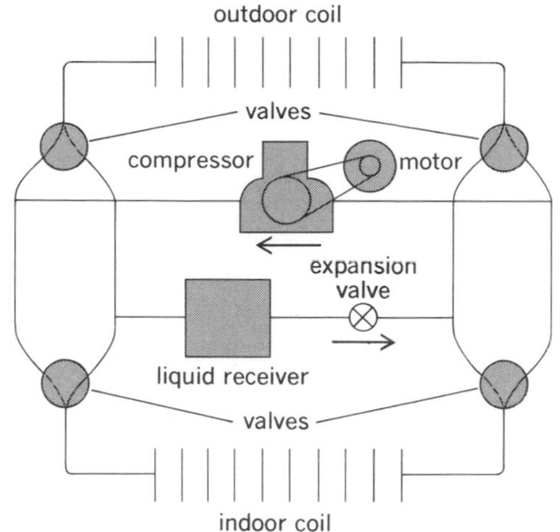

Fig. 3. Air-to-air heat pump installation; fixed air circuit with valves in the summer positions (the broken lines show the winter positions).

For the data cited, the theoretical coefficients of performance are as in Eqs. (3). The significance of

$$\begin{aligned} COP_c &= \frac{460+70}{(460+100)-(460+70)} = 17.7 \\ COP_w &= \frac{460+70}{(460+70)-(460+40)} = 17.7 \end{aligned} \quad (3)$$

these coefficients is that ideally for 1 kilowatt-hour (kwhr) of electric energy input to the compressor there will be delivered $3413 \times 17.7 = 60{,}000$ Btu/hr as refrigeration or heating effect as required. This is a great improvement over the alternative use of resistance heating, typically, where 1 kwhr of electric energy would deliver only 3413 Btu. The heat pump uses the second law of thermodynamics to give a much more substantial return for each kilowatt-hour of electric energy input, since the electric energy serves to move heat which is already present to a desired location. *See* THERMODYNAMIC PRINCIPLES.

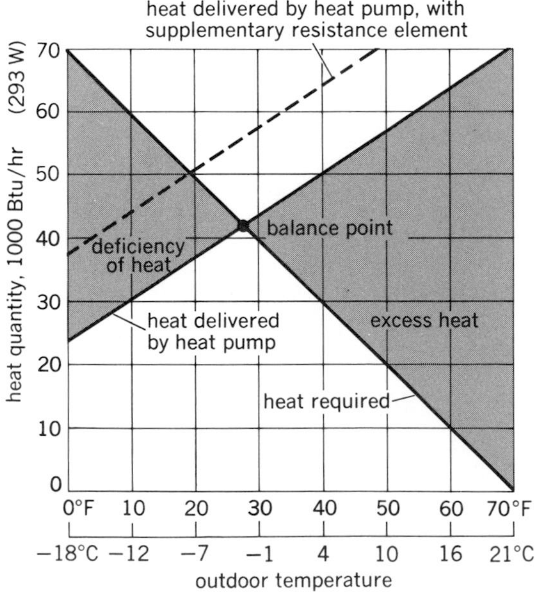

Fig. 4. Performance of air-to-air, self-contained, domestic heat pump on the heating cycle.

Effect of seasonal loads. With the prevalent acceptance of summer comfort cooling of space, it is entirely possible and practical to use the same compressor equipment and coils for winter heating and for summer cooling and to dispense with the need for direct-fired apparatus using oil, gas, coal, or wood fuel.

For an economical installation, equipment must be of correct size for both the summer cooling and winter heating loads. Climatic conditions have a significant influence and can lead to imbalance on sizing. If the heating and cooling loads are equal, the equipment can be selected with minimum investment. However, generally the loads are not balanced; in the temperate zone the heating load is usually greater than the cooling load. This necessitates (1) a large, high horsepower compressor fitted to the heating demand, (2) a supplementary heating system (electrical resistance or fuel), or (3) a heat-storage system.

If well water or the ground serves as the heat source, the imbalance is less severe than when atmospheric air is the source. However, the uncertain heat transfer rates with ground coil, the impurities, quality, quantity, and disposal of water, and the corrosion problems mitigate the use of these sources.

Atmospheric air as a heat source is preferable, particularly with smaller domestic units. A self-contained, packaged unit of this type offers maximum dependability and minimal total investment. The performance of such a unit for the heating cycle is illustrated in Fig. 4. Curves of heat required and heat available show the limitations on capacity. The heat delivered by the pump is less than the heat required at low temperatures, so that there is a deficiency of heat when the outside temperature goes below, in this case, about 28°F (−2°C). The intersection of the two curves is the balance point. There is an area of excess heat to the right and of deficiency of heat to the left of the balance point. Many devices and methods are offered to correct this situation, such as storage systems, supplementary heaters, and compressors operating alternatively in series or in parallel.

In temperate regions heat-pump installations achieve coefficients of performance in the order of 3 on heating loads when all requirements for power, including auxiliary pumps, fans, resistance heaters, defrosters, and controls are taken into account. Automatic defrosting systems, when air is the heat source, are essential for best performance, with the defrost cycle occurring twice a day.

Heat pumps are uneconomical if used for the sole purpose of comfort heating. The direct firing of fuels is generally more attractive from an overall financial viewpoint. The investment in heat-pump equipment is higher than that for the conventional heating system. Unless the price of electric energy is sufficiently low or the price of fuels very high, the heat pump cannot be justified solely as a heating device. However, if there is also need for comfort cooling of the same space in summer, the heat pump, to do both the cooling and heating, becomes attractive. The widespread use of air conditioning will probably lead to an increase in heat pumps.

The heat pump is also used for a wide assortment of industrial and process applications such as low-temperature heating, evaporation, concentration, and distillation. [THEODORE BAUMEISTER]

Solar energy–assisted system. The use of a heat pump for space heating with ambient air as the source is attractive in the temperate regions where ambient temperatures do not go significantly below the freezing point of water for extended periods of time. However, in the colder regions where ambient temperatures remain about (or below) 0°F (−18°C) for extended periods, the heat pump with outside air as a source presents problems. With a decrease in source temperature, both the capacity of the heat pump and the coefficient of performance (COP) fall. Figure 5 illustrates these characteristics for a commercially available heat pump unit. At an outside air temperature of 0°F (−18°C), the capacity of the heat pump unit is only about 43% of the capacity at 40°F (4°C). The COP also falls from 2.92 at 40°F to 1.8 at 0°F. Thus, when the heating load increases at decreasing ambient temperatures, the capacity of the heat pump decreases and the heat pump requires greater electrical energy to run it. Under these conditions there is also the problem of ice buildup on the evaporator coils, necessitating frequent defrosting. For the heat pump to be more attractive in colder regions, the source temperature for the heat pump should be increased. One possible method for achieving this is solar energy.

Solar collectors. For space heating utilizing solar energy, flat-plate collectors are generally used. A typical flat-plate collector contains a metallic plate painted black, with one or two glass covers and the sides and bottom of the collector well insulated. Solar energy is transmitted through the glass, and a significant part of that reaching the metallic plate—about 90%—is absorbed by the plate, increasing its temperature. The energy absorbed by the plate is, in turn, transferred to a working fluid—usually air or water. If air is the working fluid, it passes over the metallic plate; if water, it flows through tubes attached to the metallic plate.

Because of the heat transfer from the collector to the surroundings, not all the radiant energy reaching the collector is transferred to the working fluid. The efficiency of a collector, defined as the ratio of the heat transfer to the working fluid to the radiant energy reaching the collector surface, decreases with an increase in the absorber temperature. The higher the temperature at which the working fluid is operated, the higher is the temperature of the absorber surface and the lower the efficiency. The efficiency of a typical flat-plate collector with one glass cover and air as the working fluid is shown in Fig. 6, where the abscissa represents the average temperature of the working air. An ambient temperature of 20°F (−7°C) and a solar insolation of 200 Btu/hr ft² (630 W/m²) have been assumed. If warm air is to be directly used for space heating, its temperature should be about 110°F (43°C). At this temperature, the useful energy collected is 80 Btu/hr ft² (250 W/m²), corresponding to an efficiency of 40%. But when the temperature of the working air is 60°F (15.5°C), the useful energy collected (other conditions being the same as before) increases to 166 Btu/hr ft² (523 W/m²), corresponding to an efficiency of 83%. To make use of this increased solar energy at 60°F for space heating, a heat pump is needed. *See* SOLAR HEATING AND COOLING.

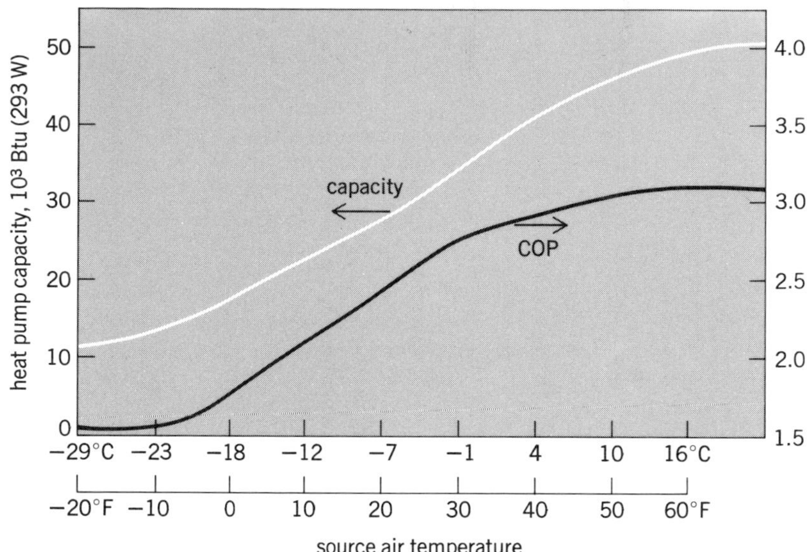

Fig. 5. Heat pump characteristics.

Advantages. The solar energy—assisted heat pump system (SAHPS) offers several advantages. Most buildings constructed during and since the 1970s are equipped with central air-conditioning systems, and with some modifications the air-conditioning system can serve as the heat pump system. As solar energy can be utilized either directly or through the heat pump at high values of COP during sunny days, electric power consumption during such days is reduced, thus contributing to load leveling at the central generating stations. Since the collectors operate at high efficiencies at low temperatures, the solar energy contribution from a given collector area can be increased. As the absorber surface temperature is lower, only one glazing may be sufficient in the majority of cases.

Working fluid and size. A heat pump system utilizing solar energy for space heating is schematically illustrated in Fig. 7. Air is the working fluid for the solar collectors, and the system uses an air-to-air heat pump. Energy storage is necessary as, during some periods, solar energy may not be available at all, such as during the night, or may be

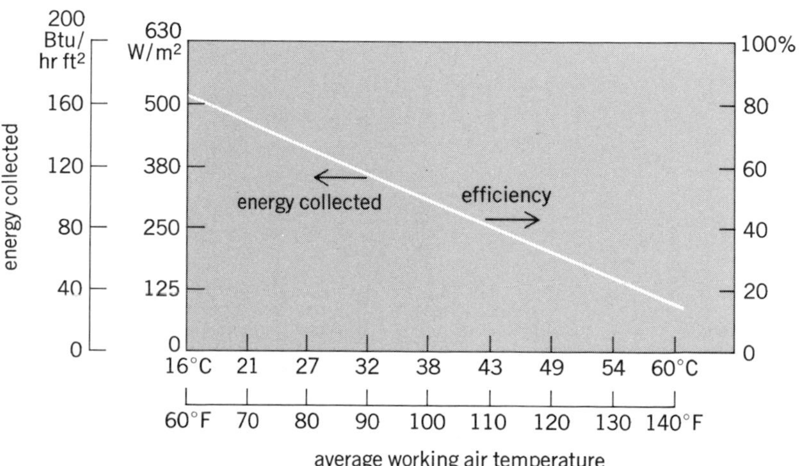

Fig. 6. Solar collector characteristics. An ambient temperature of 20°F (−7°C) and a solar insolation of 200 Btu/hr ft² (630 W/m²) have been assumed.

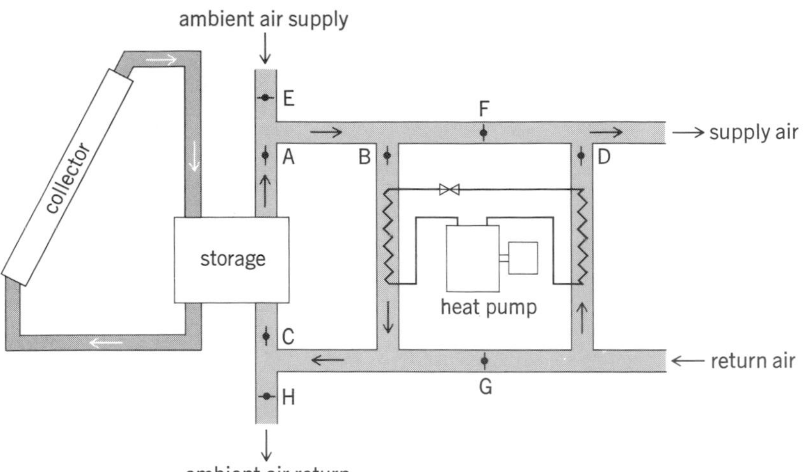

	open	close	mode
1	A,B,C,D	E,F,G,H	temperature of stored energy too low for direct use but higher than ambient air
2	A,C,F,G	E,B,D,H	stored solar energy temperature sufficiently high; direct utilization of solar energy (bypass heat pump)
3	B,D,E,H	A,C,F,G	stored solar energy temperature lower than ambient air temperature; heat pump source is ambient air

Fig. 7. Schematic arrangement of solar-assisted heat pump system.

available in very limited quantities, such as on cloudy days. The system can operate in three different modes as indicated in Fig. 7. In mode 1, the solar energy stored is not at a sufficiently high temperature for direct use but at a temperature higher than ambient temperature—conditions likely to be met on cold, sunny days. The heat pump uses the stored solar energy in this mode. In mode 2, the stored solar energy is at a sufficiently high temperature for direct utilization—probable conditions during sunny days in the fall and spring. The heat pump is then bypassed. In mode 3—during extended periods of cloudy days or very cold weather—the stored solar energy is depleted and outside air is the source for heat pumps.

Sizing the heat pump to meet the entire heating needs at the lowest expected ambient temperature, on the basis of heat pump output–source temperature characteristics, may lead to a heat pump which is significantly oversized for periods when ambient temperatures are higher. To avoid such oversizing, heat pump capacity may be determined on the basis of a source temperature higher than the lowest expected ambient temperatures, and the system may be equipped with an auxiliary heating source which will supply the required amount of energy to raise the temperature of the ambient air to the designated temperature during extremely cold days.

Besides air-to-air heat pumps, water-to-air or water-to-water heat pumps can also be used. Because of the better heat transfer characteristics when water is used (with an antifreeze additive to prevent freezing) as the working fluid, a higher efficiency for the collector and a higher COP for the heat pump can be realized.

Thermal energy storage. As indicated earlier, some form of energy storage is essential for any system utilizing solar energy, since solar energy is available only intermittently. The amount of energy storage to be provided in a given system depends on collector area and weather conditions. Storage plays a significant role in determining the performance of the system. For systems with air as the working fluid in the collector, the commonly used thermal energy storage (TES) material is crushed rocks, which are inexpensive and readily available. With water as the working fluid, water itself is probably the best TES material.

TES can be either on the low-temperature (evaporator) side or high-temperature (condenser) side of the heat pump. High-temperature storage requires that the heat pump operate during periods of sunshine even when there is no need for space heating. This is avoided if thermal storage is on the low-temperature side, thus reducing the operation of the electrically driven heat pump during periods of peak electric demand.

Use. Some interest was shown in solar-assisted heat pump systems in the 1950s but declined as a result of alternate sources of inexpensive energy. Since the mid-1970s, however, there has been a renewal of interest in solar energy utilization for space heating, and several heat pump systems utilizing solar energy are in use in New Mexico, New York, Pennsylvania, Vermont, and other states.

Performance studies. Several performance studies—both analytical and experimental—have been made. Performance tests were conducted over two partial winter heating seasons during 1974–1975 and 1975–1976 on an office building in Alburquerque, NM, that was equipped with a SAHPS and completed in 1956, and was subsequently enlarged to a floor area of 5000 ft² (465 m²). All building heating requirements were met by the SAHPS with negligible use of auxiliary heat. The average COP for the heat pump alone was 4.2, but was somewhat lower if all the power to run the auxiliary equipment was included.

A computational study of a SAHPS for a two-story elementary school building in Boston, MA, with 75,000 ft² (6970 m²) of floor area, a collector area of 7500 ft² (697 m²), and a water storage of 22,500 gal (85 m³) indicated that about 75% of the heating requirements could be met by the SAHPS. Retrofitting of a block of 19 town houses at the Heron Gate project, Ottawa, Canada, is designed to supply about 40% of the space heating requirements from the SAHPS. Collectors operate in the range of 70–100°F (20–40°C). A water-to-water heat pump is employed, and thermal energy is provided by a 7000-gal (26-m³) underground storage tank on the low-temperature side, with a collector area of 2300 ft² (214 m²).

Although there are several studies relating to SAHPS, because of the many variables—weather, available solar energy in the specific location, collector area, thermal storage, and so forth—it is not yet possible to easily predict the performance of a particular system; performance predictions can be made only after a suitable simulation employing a computer.

Solar-assisted heat pumps were not competitive at 1977 costs, but with subsequent increases in prices of home heating oil, these systems have become economical under certain conditions in certain areas of the United States. With improvements in heat pump performance—particularly their reliability and ability to operate at relatively high source temperatures—such systems are likely to become attractive in many areas.

[N. V. SURYANARAYANA]

Domestic use. Developments in the energy field have brought about an increase in the use of heat pumps. The greatly increased cost of fossil fuels, as well as the limitations on gas usage in many areas, has resulted in cost justification of the slightly more expensive heat pump over a much larger geographical area of the United States. In addition, new types of heat pump systems and innovations in design have come about as a direct result of the energy picture.

Product applications. A number of factors have changed the application of heating-cooling products in the residential market. For example, the average size of homes has decreased since the mid-1970s. Associated with this, the heat loss per square foot has been decreased even faster. Some of the energy-conserving methods that have been responsible are the increased use of insulation in walls, floors, and ceilings; use of smaller and fewer windows; placement of the larger walls toward the southern exposure; and reduced infiltration due to tighter construction. *See the feature article* ENERGY CONSERVATION.

The heating-cooling industry has nationally experienced the impact of smaller homes in the capacity mix of unitary product shipments. Until the mid-1970s a nominal 3-ton (10.6-kW) cooling system was representative of a manufacturer's normal mix of products. This dropped to 2½ tons (8.8 kW) by 1978, and is expected to continue to drop and possibly level out at 2 tons (7 kW) by the mid-1980s.

Probably the most significant breakthrough in heat pump applications has come from the design and marketing of "add-on" heat pumps. These involve a split-system set of components, very similar to a standard air-source split-system heat pump, and are capable of being added to any existing forced-air furnace. The system consists of an outdoor section, indoor coil, and a control box. The indoor coil is mounted on the discharge-air side of the existing furnace, and the control box is the interface between the heat pump and furnace. Various designs have been marketed in the add-on category. The most efficient allows for heat pump operation below the balance point, since a large number of annual heating hours fall below this temperature; higher-efficiency heat pumps give a utilization efficiency higher than the existing furnace in this range. The advantages of this type of system include the following: (1) When mated with a fossil fuel furnace, the seasonal utilization efficiency of the two different fuels is maximized, greatly increasing the net efficiency of the raw resources (minimum of 35%). Thus, nonrenewable energy sources, such as fuel oil and natural gas, can be preserved for longer periods. (2) Application in combination with an existing electric forced-air furnace gives essentially the same utilization efficiency as a conventional heat pump. This permits improvement of the seasonal COP from 1.0 to approximately 2.0, for a 50% energy savings.

New product design considerations. Previous heat pump designs were simply air conditioners which were modified to become reversible refrigeration systems. Today, the heat pump designer must evaluate the cost effectiveness of each component in the system to justify the higher first cost resulting from improved efficiency. Due to the many hours of operating during the cooling cycle and relatively few in the heating cycle in the southern United States, a unit designed for this location would emphasize efficient operation during the cooling season. The reverse would be true for the northern market. A heat pump designed for universal use throughout the United States would compromise to provide equal improvements for both operating cycles, resulting in nearly equal annual operating cost savings regardless of the geographical location. The same high-efficiency unit would provide greater energy savings in the South during the summer and in the North during the winter. Improved heat pump efficiencies are achieved by using larger heat transfer surface; more efficient compressors, fans, and fan motors; and better thermal insulation. Care must be exercised to limit the amount of heat transfer surface so that system reliability will not be significantly affected. Larger surface area also requires more refrigerant charge, and compressor life is inversely proportional to the system charge.

Good product planning will anticipate energy costs in future years to justify higher first cost for equipment in a reasonable payback period. The typical time-span from original conception through development to production is from 2½ to 3 years. Since heat pump sales have now penetrated colder climates, there is a greater need for system reliability under more severe operating conditions. In addition, the high energy cost for standby electric heat during system outages and the increasing labor cost for service have demanded more emphasis on reliable operation.

The system reliability of heat pump designs has been improved in a variety of ways. Compressors have been designed specifically for heat pump duty, where stress conditions and operating hours are more demanding than for cooling-only units. Compressor improvements include heavy-duty connecting rods and wrist pins to withstand high compression ratios encountered in heating operation, nonfoaming lubricants, better motor-winding insulation, and motor overload protectors that are more sensitive to unsafe operating conditions.

Heat pumps have been designed with solid-state controls which can monitor and regulate unit operation more completely than previously was possible with electromechanical controls. Also included are such functions as short-cycling protection; excessive temperature and pressure protection; avoidance of nuisance defrosts due to wind gusts; overriding capability of defrost termination for gusty winds; protection against repeated defrost cycles during snow drifts; and system lockout and visual indication for abnormal operation.

Design testing of heat pumps has become much more extensive so as to better simulate field conditions. Test facilities can operate over a wider range of temperature, humidity, and precipitation conditions. Feedback from malfunctioning field installations and teardown of failed systems have given engineers insights into further improvements.

Until the mid-1970s most heat pumps were built with the time-temperature method of defrost initiation, because of its low cost and simplicity. At outdoor temperatures below about 42°F (6°C), a time clock would initiate a defrost cycle approximately every 60 min of compressor operation, whether the coil surface was frosted or not. Since this wastes energy and penalizes the operating cost, there has been an increase in the use of the true demand-type method of defrost. Here, a defrost cycle is initiated only when there is significant frost buildup on the coil surface. The sensor generally measures an increase in air resistance across the coil due to frost buildup. Field tests have indicated that a time-temperature defrost system results in 10–15 times as many defrost cycles per year as a well-designed demand system. [JERRY E. DEMPSEY]

Bibliography: E. R. Ambrose, *Heat Pumps and Electric Heating*, 1966; T. Baumeister (ed.), *Standard Handbook for Mechanical Engineers*, 8th ed., 1978; E. E. Drucker et al., *Commercial Building Unitary Heat Pump System with Solar Heating*, Rep. NSF/RANN/SE/GI-43895/FR/75/3 to ERDA, October 1975; T. L. Freeman et al., Performance of combined solar heat pump systems, *Solar Energy*, 22:125–135, 1979; S. F. Gilman, M. W. Wildin, and E. R. McLaughlin, Field study of a solar energy assisted heat pump heating system, *Solar Cooling and Heating: A National Forum*, Miami Beach, ERDA Doc. C00-2704-4, Dec. 14, 1976; F. C. McQuiston and J. D. Parker, *Heating, Ventilating and Air Conditioning: Analysis and Design*, 1977; S. Walters (ed.), Ottawa town houses heated by solar energy, *Mech. Eng.*, 101(11):64–65, November 1979.

Heat transfer

Heat, a form of kinetic energy, is transferred in three ways: conduction, convection, and radiation. Heat can be transferred only if a temperature difference exists, and then only in the direction of decreasing temperature. Beyond this, the mechanisms and laws governing each of these ways are quite different. This article gives introductory information on the three types of heat transfer (also called thermal transfer) and on important industrial devices called heat exchangers.

Conduction. Heat conduction involves the transfer of heat from one molecule to an adjacent one as an inelastic impact in the case of fluids, as oscillations in solid nonconductors of electricity, and as motions of electrons in conducting solids such as metals. Heat flows by conduction from the soldering iron to the work, through the brick wall of a furnace, through the wall of a house, or through the wall of a cooking utensil. Conduction is the only mechanism for the transfer of heat through an opaque solid. Some heat may be transferred through transparent solids, such as glass, quartz, and certain plastics, by radiation. In fluids, the conduction is supplemented by convection, and if the fluid is transparent, by radiation.

The conductivities of materials vary widely, being greatest for metals, less for nonmetals, still less for liquids, and least for gases. Any material which has a low conductivity may be considered to be an insulator. Solids which have a large conductivity may be used as insulators if they are distributed in the form of granules or powder, as fibers, or as a foam. This increases the length of path for heat flow and at the same time reduces the effective cross-sectional area, both of which decrease the heat flow. Mineral wool, glass fiber, diatomaceous earth, glass foam, Styrofoam, corkboard, Celotex, and magnesia are all examples.

Convection. Heat convection involves the transfer of heat by the mixing of molecules of a fluid with the body of the fluid after they have either gained or lost heat by intimate contact with a hot or cold surface. The transfer of heat at the hot or cold surface is by conduction. For this reason, heat transfer by convection cannot occur without conduction. The motion of the fluid to bring about mixing may be entirely due to differences in density resulting from temperature differences, as in natural convection, or it may be brought about by mechanical means, as in forced convection.

Most of the heat supplied to a room from a steam or hot-water radiator is transferred by convection. In fact, the heat from the fire in the furnace heating the hot water or steam is transferred to the boiler wall by convection, and the hot water or steam transfers heat from the boiler to the radiator by convection. Iced tea is cooled and soup heated by convection.

Radiation. Solid material, regardless of temperature, emits radiations in all directions. These radiations may be, to varying degrees, absorbed, reflected, or transmitted. The net energy transferred by radiation is equal to the difference between the radiations emitted and those absorbed.

The radiations from solids form a continuous spectrum of considerable width, increasing in intensity from a minimum at a short wavelength through a maximum and then decreasing to a minimum at a long wavelength. As the temperature of the object is increased, the entire emitted spectrum decreases in wavelength. As the temperature of an iron bar, for example, is raised to about 1000°F, the radiations become visible as a dark red glow. As the temperature is increased further, the intensity of the radiation increases and the color becomes more blue. This process is quite apparent in the filament of a light bulb. When the bulb is operated at less than normal voltage, the light appears quite red. As the voltage is increased, the filament temperature increases and the light progressively appears more blue.

Liquids and gases only partially absorb or emit these radiations, and do so in a selective fashion. Many liquids, especially organic liquids, have selective absorption bands in the infrared and ultraviolet regions.

Transfer of energy by radiation is unique in that no conducting substance is necessary, as with conduction and convection. It is this unique property that makes possible the transfer of large amounts of energy from the Sun to the Earth, or the transfer of heat from a radiant heater in the home. It is the ready transfer of heat by radiation from a California orange grove to outer space on a clear night that sometimes results in a frost. The presence of a shield of clouds will tend to prevent this loss of heat and often prevent the frost. By means of heat lamps and gold-plated reflectors, heat may be transferred deep into the layer of enamel on a car body, with resultant hardening of the enamel from the inside out. It is also the transfer over great distance of quantities of radiant energy that makes the atomic bomb so destructive.

Design considerations. By utilizing a knowledge of the principles governing the three methods of heat transfer and by a proper selection and fabrication of materials, the designer atttempts to obtain the heat flow required for his purposes. This may involve the flow of large amounts of heat to some point in a process or the reduction in flow in others. It is possible to employ all three methods of heat transfer in one process. In fact, all three methods operate in processes that are commonplace. In summer, the roof on a house becomes quite hot because of radiation from the Sun, even though the wind is carrying some of the heat away by convection. Conduction carries the heat through the roof where it is distributed to the attic by convection. The prudent householder attempts to reduce the heat that enters the rooms beneath by reducing the heat that is absorbed in the roof by painting the roof white. He may apply insulation to the underside of the roof to reduce the flow of heat through the roof. Further, heated air in the attic may be vented through louvers in the roof.

Heat transferred by convection may be transferred as heat of the convecting fluid or, if a phase change is involved, as latent heat of vaporization, solidification, sublimation, or crystallization. The human body can be cooled to less than ambient temperature by evaporation of sweat from the skin. Dry ice absorbs heat by sublimating the carbon dioxide. Heat extracted from the products of combustion in the boiler flows through the gas film and the metal tube wall and converts the water inside the tube to steam, all without greatly changing the temperature of the water.

Heat exchangers. In industry it is generally desired to extract heat from one fluid stream and add it to another. Devices used for this purpose have passages for each of the two streams separated by a heat-exchange surface in the form of plates or tubes and are known as heat exchangers. Needless to say, the automobile radiator, the hot-water heater, the steam or hot-water radiator in a house, the steam boiler, the condenser and evaporator on either the household refrigerator or air conditioner, and even the ordinary cooking utensils in everyday use are all heat exchangers. In power plants, oil refineries, and chemical plants, two commonly used heat exchangers are the tube-and-shell and the double-pipe exchangers. The first consists of a bundle of tubes inside a cylindrical shell. One fluid flows inside the tubes and the other between the tubes and the shell. The double-pipe type consists of one tube inside another, one fluid flowing inside the inner tube and the other flowing in the annular space between tubes. In both cases, the tube walls serve as the heat-exchange surface. Heat exchangers consisting of spaced flat plates with the hot and cold fluids flowing between alternate plates are also in use. Each of these exchangers essentially depends upon convection heat flow through a film on each side of the heat-exchange surface and conduction through the surface. Countless special modifications, often also utilizing radiation for heat transfer, are in use in industry.

In these exchangers, the fluid streams may flow parallel concurrently or in mixed flow. In most cases, the temperatures of the various streams remain essentially constant at a given point, and the process is said to be a steady-state process. As the streams move through the exchangers, unless there is a phase change, the fluids are continuously changing in temperature, and the temperature gradient from one stream to the other may be continuously varying. To determine the amount of surface needed for a given process, the designer must evaluate the effective temperature gradient for the particular condition and exchanger.

With extremely high temperatures, or with gas streams carrying suspended solids, the use of conventional heat exchangers becomes impractical. Under these conditions, the transfer of heat from one stream to another becomes more economical by the alternate heating and cooling of refractory solids or by checkerwork as in the blast-furnace hot stove, in the glass-furnace regenerator, or in the Royster stove. At lower temperatures, metal packing is frequently employed, as in the Ljungstrom preheaters or in regenerators for liquid-air production. In petroleum refining and in the metallurgical industry, exchangers are being employed in which one or more of the streams are fluidized beds of solids, the large area of the solids tending to produce very high rates of heat exchange. In some of these devices and also in nuclear power reactors, large quantities of heat are being generated in the exchangers. Here one of the principal problems involves the rapid removal of this heat before the temperature rises to the point where the equipment is damaged or destroyed.

Often the heating or cooling of a body is desired. In this case, the body representing the second stream does not remain at constant temperature, the heat being transferred representing a change in the heat content of the body. Such a process is known as an unsteady-state process. The heating or cooling of food and canned products in utensils, refrigerators, and sterilizers; the heating of steel billets in metallurgical furnaces; the burning of brick in a kiln; and the calcination of gypsum are examples of this type of process. *See* HEAT; HEAT EXCHANGER.

[RALPH H. LUEBBERS]

Bibliography: S. Banerjee and J. T. Rogers (eds.), *Heat Transfer Nineteen Seventy-Eight: Proceedings*, 8 vols., 1979; G. M. Dusinberre, *Numerical Analysis of Heat Flow*, 1949; E. R. G. Eckert, *Introduction to Heat and Mass Transfer*, 1963; J. P. Holman, *Heat Transfer*, 3d ed., 1978; M. Jakob, *Heat Transfer*, vol. 1, 1949; D. Q. Kern, *Process Heat Transfer*, 1950.

Heating value

The energy per unit mass of material (fuel) that can be released from it as heat under specified conditions, where heat is simply energy in transit due to a temperature difference. In a sense, the heating value of a fuel can be thought of as energy which can potentially be released as heat if certain things are done to the fuel. For example, a typical bituminous coal has a heating value of about 7500 kcal/kg (13,500 Btu/lb). This means that 7500 kcal (13,500 Btu) of energy per kg (lb) of the coal can be released as heat. For a fossil fuel, such as coal, it is assumed that the energy is released upon complete combustion of the fuel at atmospheric pressure. On the other hand, 1 kg (2.2 lb) of nuclear fuel may have a heating value of 2.0×10^{10} kcal (7.92×10^{10} Btu), but this energy is released as heat upon fission, rather than combustion. *See* COMBUSTION; ENERGY; NUCLEAR FISSION.

The units used in specifying heating value are defined as follows. One kilocalorie (kcal) is the quantity of heat necessary to raise the temperature of 1 kilogram of water 1 centigrade degree. One British thermal unit (Btu) is the quantity of heat necessary to raise the temperature of 1 pound (mass) of water 1 Fahrenheit degree. (It is the normal convention that the 1 C° is from 14.5 to 15.5°C, and the 1 F° from 59.5 to 60.5°F.) By direct conversion, 1 Btu/lb is equal to 0.555 kcal/kg. *See* BRITISH THERMAL UNIT (BTU).

Fossil fuels. The heating value of fossil fuels generally refers to the heat released upon complete combustion with oxygen at atmospheric pressure. This is called the heat of combustion at atmospheric pressure. It would be possible to react the fossil fuel with another material, such as chlorine, and this would release a different amount of heat. Strictly speaking, it would not be wrong to call this heat the heating value of the fossil fuel; however, convention dictates that the heating value of fossil fuels refers to combustion with oxygen. The reason for this convention is that fossil fuels such as coal are generally burned with oxygen in air, as in a home or industrial furnace at atmospheric pressure. *See* FOSSIL FUEL.

Ideally, measurement of the heat of combustion of fossil fuels at atmospheric pressure would require a device that could maintain constant atmospheric pressure of the gases released when the fuel is combusted. Because of the difficulty of doing this, the heat of combustion is measured at constant volume instead of constant atmospheric pressure, and then the heat of combustion at atmospheric pressure (which is the practical value wanted) is calculated from the measurement using thermodynamics. *See* HEAT.

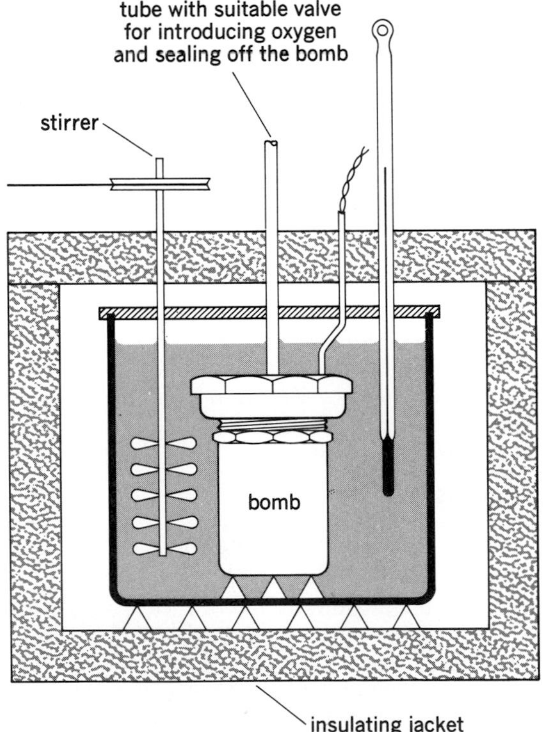

Diagram of a bomb calorimeter. *(From G. Shortley and D. Williams, Principles of College Physics, 2d ed., p. 305, Prentice-Hall, Inc., 1967)*

The heat of combustion at constant volume is measured using a bomb calorimeter as shown in the illustration. In this device, a measured quantity of the fossil fuel is placed in the bomb, which is a massive steel cylinder capable of withstanding the high pressures of the expanding gases released upon combustion. Enough oxygen is introduced to the bomb to ensure complete combustion of the fuel, and the combustion is started by heating the fuel with an electric current running through a fine wire in the bomb. The heat of combustion at constant volume (the volume of the bomb is constant) is calculated from the final temperature rise, after thermal equilibrium is reached. The result of this measurement is called the gross or higher heating value of the fossil fuel. In general, reported heating values of fossil fuels in the literature are the gross or higher heating values. Two corrections must be applied to obtain the net or lower heating value, which refers to the heat of combustion at atmospheric pressure. The first correction accounts for the latent heat of vaporization of the water in the combustion products. In the bomb calorimeter, water as steam is formed as one of the gaseous products. When the bomb reaches thermal equilibrium, the steam condenses and releases its heat of vaporization. This must be subtracted to obtain the net heating value of the fossil fuel. The second correction accounts for the fact that the measured value is at constant volume, while the desired value is for constant (atmospheric) pressure which occurs under practical use.

The heating value of gaseous fossil fuels is more conveniently reported in energy per standard cubic meter (SCM) or foot (SCF), where SCM (SCF) is 1 cubic meter (foot) of the gas at standard pressure and temperature. For example, natural gas, which is largely methane, has a heating value of 8900 kcal/SCM (1000 Btu/SCF), and synthesis gas, which is largely carbon monoxide and hydrogen, may have a heating value of 3000 kcal/SCM (337 Btu/SCF). The actual heating value of synthesis gas depends upon the percentage of carbon monoxide and hydrogen.

Table 1 gives some typical values of the heat of

Table 1. Typical values* of heat of combustion of fossil fuels

Substance	kcal/kg	Btu/lb	kcal/SCM	Btu/SCF
Solid fuels				
Anthracite	7,900	14,220		
Bituminous coal	7,500	13,500		
Lignite	4,100	7,380		
Charcoal	7,000	12,600		
Liquid fuels				
Gasoline	11,100	19,980		
Kerosine	11,400	20,520		
No. 6 fuel oil	10,500	18,900		
Benzol	10,000	18,000		
Gaseous fuels				
Methane			9,016	1,013
Hydrogen			2,892	325

*Higher heating value.

Table 2. Range of some characteristics of coal classes

Coal class	Heating value* kcal/kg	Heating value* Btu/lb	Fixed carbon content, wt. %	Moisture content, wt. %
Anthracite	Up to 8900	Up to 16,020	86–98	1–3
Bituminous	6100–8300	10,980–14,940	50–86	3–12
Subbituminous	4400–6700	7920–12,060	40–60	20–30
Lignite	3100–4400	5580–7920	<40	Up to 40

*Higher heating value.

combustion of some fossil solid, liquid, and gaseous fuels. As usual in reporting heating values, these are at constant volume and are the higher heating values.

The heating value of a fossil fuel such as coal may vary widely because of the wide variance in the proportion of fixed carbon it contains. This variance is illustrated for coal in Table 2. (Fixed carbon has a large heating value, while moisture has none.) Crude oil does not have as large a variance, because its composition does not vary much. Its heating value is about 36,000 kcal/gal (143,000 Btu/gal). Organic material such as coal can be converted in various ways (for example, by reacting it with steam and oxygen) to fossil fuel gases with heating values ranging from 1150 to 8900 kcal/SCM (129–1000 Btu/SCF). The variance in the heating values is due to the variance in the composition of the gases.

Other ways of obtaining heat. Fossil fuels are presently the greatest practical source of heat, but nuclear fission is being used more and more. In nuclear fission, the nucleus of an unstable heavy element such as uranium is broken up into lighter nuclei by absorbing a neutron. The energy released as heat comes from the conversion of mass into energy (that is, the difference between the mass of the reactants and that of the products becomes energy). One kg (2.2 lb) of fissionable material has about 2.0×10^{10} kcal (7.92×10^{10} Btu), which is the same as the energy obtained from burning 2930 metric tons of 7500 kcal/kg (13,500 Btu/lb) coal. In general, nuclear heat is used to generate steam, which can then be used to drive a turbogenerator to produce electricity.

There are other sources of energy which may be used for heat, but these have not been classically called fuels. Hot geothermal fluids may be used for home heating, as in Iceland, or to generate electrical power. Similarly, solar radiation brings an average of 3800 kcal/m^2 day (1400 Btu/ft^2 day) onto the surface of the United States. This can be used for direct heating by means of solar collectors or for electrical power by means of solar cells or focused solar collectors and a steam cycle. Any electrical power from geothermal energy, solar energy, nuclear energy, or wind energy can be used for resistance heating. There is also the energy available in the tides, which is a form of gravitational energy. This, too, could eventually be converted to heat. So far, no one has specified heating values for these sources, but they are real sources of heat and will become more and more important as fossil fuel resources are used up. *See the feature article* EXPLORING ENERGY CHOICES; *See also* GEOTHERMAL POWER; NUCLEAR POWER; SOLAR RADIATION; TIDAL POWER; WIND POWER.

[ROBERT C. BINNING; RICHARD S. HOCKETT]

Bibliography: J. P. Holman, *Thermodynamics*, 2d ed., 1974; H. H. Lowry (ed.), *Chemistry of Coal Utilization*, Suppl. Vol., 1963; R. H. Perry and C. H. Chilton (eds.), *Chemical Engineers' Handbook*, 5th ed., 1973; G. Shortley and D. Williams, *Elements of Physics*, 2 vols., 5th ed., 1971.

Horsepower

The unit of power in the British engineering system of units. One horsepower (hp) equals 550 ft-lb/sec, or 746 watts. *See* POWER.

The horsepower is a unit of convenient magnitude for measuring the power generated by machinery. As an example of the size of the horsepower, consider a 200-lb man walking up a stairway, the top of which is 10 ft higher than the bottom. In walking up the stairway, the man does 2000 ft-lb of work. If he climbs the stairs in 5 sec, his rate of work is 400 ft-lb/sec, or about 3/4 hp.

[PAUL W. SCHMIDT]

Hot-water heating system

A heating system for a building in which the heat-conveying medium is hot water. Heat transfer in British thermal units (Btu) equals pounds of water circulated times drop in temperature of water. For other liquids, the equation should be modified by the specific heats. The system may also be modified to provide cooling.

A hot water heating system consists essentially of water-heating or -cooling means and of heat-emitting means such as radiators, convectors, baseboard radiators, or panel coils. A piping system connects the heat source to the various heat-emitting units and includes a method of establishing circulation of the water or other medium and an expansion tank to hold the excess volume of water as it is heated and expands. Radiators and convectors have such different response characteristics that they should not be used in the same system.

Types. In a one-pipe system (Fig. 1), radiation units are bypassed around a one-pipe loop. This type of system should only be used in small installations.

In a two-pipe system (Fig. 2), radiation units are connected to separate flow and return mains, which may run in parallel or preferably on a reverse return loop, with no limit on the size of the system.

In either type of system, circulation may be provided by gravity or pump. In gravity circulation each radiating unit establishes a feeble gravity cir-

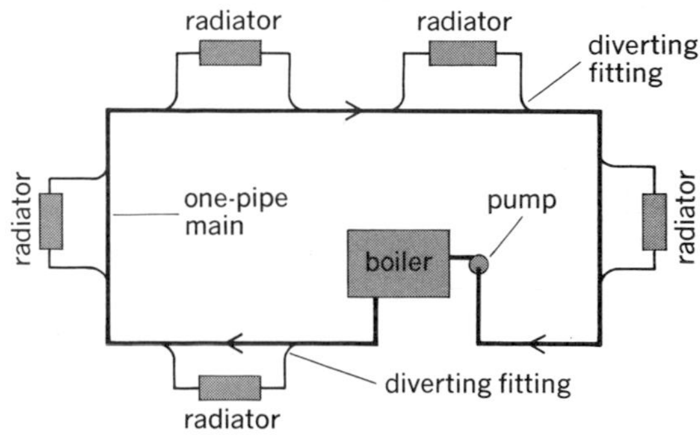

Fig. 1. One-pipe hot-water heating system.

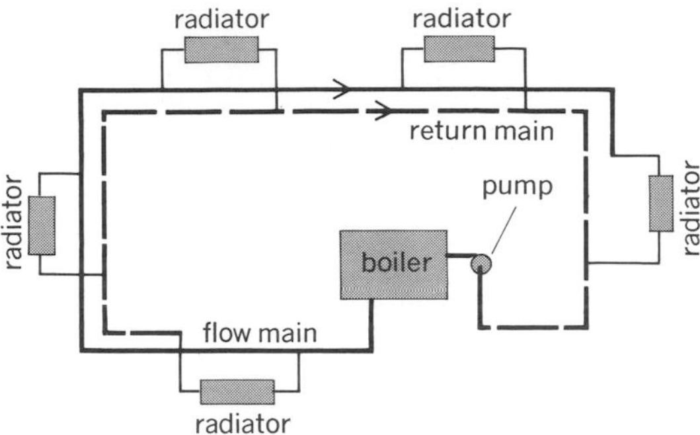

Fig. 2. Two-pipe reverse return system.

culation; hence such a system is slow to start, is unpredictable, and is not suitable for convectors, baseboard radiation, or panel coils because circulating head cannot be established, and circulation cannot be supplied to units below the mains. The pipes must be large in size. For these reasons gravity systems are no longer being used.

In forced circulation a pump is used as the source of motivation. Circulation is positive and units may be above or below the heat source. Smaller pipes are used.

Operation. For perfect operation it is imperative that the friction head from the heat source through each unit of radiation and back to the heat source be the same. To achieve this condition usually requires careful balancing after installation and during operation.

Expansion tanks may be open or closed. Open tanks are vented to the atmosphere and are used where the water temperature does not exceed 220°F (at sea level). They provide the safest operation, practically free from explosion hazards. Closed tanks, used for higher water temperatures, must be provided with safety devices to avoid possible explosions.

One outstanding advantage of hot-water systems is the ability to vary the water temperature according to requirements imposed by outdoor weather conditions, with consequent savings in fuel. Radiation units may be above or below water heaters, and piping may run in any direction as long as air is eliminated. The system is practically indestructible. Flue-gas temperatures are low, resulting in fuel savings. The absence of myriads of special steam fittings, which are costly to purchase and to maintain, is also an important advantage. Hot water is admirably adapted to extensive central heating where high temperatures and high pressures are used and also to low-temperature panel-heating and -cooling systems.

Circulating hot-water pumps must be carefully specified and selected. On medium-size installations, it is recommended that two identical pumps be used, each capable of handling the entire load. The pumps operate alternately but never together in parallel. On large installations three or more pumps may be used in parallel, provided they are identical and produced by the same manufacturer. The casings and runners must be cast from the same molds, and the metals and other features that affect their temperature characteristics must be identical. All machined finishes must be identical, and the pumps must be thoroughly shop-tested to ensure that they operate with identical characteristics.

When the system is in operation, the pump can be disconnected from the boiler by throttling down the valve at the boiler return inlet; it should not be closed completely. This procedure permits the water in all boilers to be at the same temperature so that when a boiler is thrown back into service, the flue gases do not impinge on any cold surfaces, thus producing soot and smoke to further contaminate the outdoor atmosphere. *See* COMFORT HEATING; OIL BURNER.

[ERWIN L. WEBER]

Bibliography: American Society of Heating and Ventilating Engineers, *Heating, Ventilating, Air Conditioning Guide*, vol. 37, 1959; Building Research Advisory Board–Federal Construction Council, *High-Temperature Water for Heating and Light Process Loads*, 1960; F. E. Giesecke, *Hot-water Heating and Radiant Heating and Radiant Cooling*, 1947.

Hydraulic turbine

A machine which converts the energy of an elevated water supply into mechanical energy of a rotating shaft. Most old-style waterwheels utilized the weight effect of the water directly, but all modern hydraulic turbines are a form of fluid dynamic machinery of the jet and vane type operating on the impulse or reaction principle and thus involving the conversion of pressure energy to kinetic energy. The shaft drives an electric generator, and speed must be of an acceptable synchronous value. *See* GENERATOR.

The impulse or Pelton unit has all available energy converted to the kinetic form in a few stationary nozzles and subsequent absorption by reversing buckets mounted on the rim of a wheel (Fig. 1). Reaction units of the Francis or the Kaplan types run full of water, submerged, with a draft tube and a continuous column of water from head race to tail race (Figs. 2 and 3). There is some fluid acceleration in a continuous ring of stationary nozzles with full peripheral admission to the moving nozzles of the runner in which there is further acceleration. The draft tube produces a negative pressure in the runner with the propeller or Kaplan units acting as suction runners; the Francis inward-flow units act as pressure runners. Mixed-flow units produce intermediate degrees of both rotor pressure drop and fluid acceleration. *See* IMPULSE TURBINE.

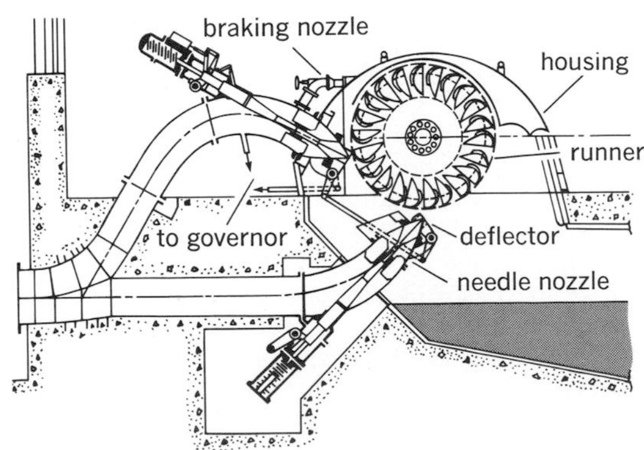

Fig. 1. Cross section of an impulse (Pelton) type of hydraulic-turbine installation.

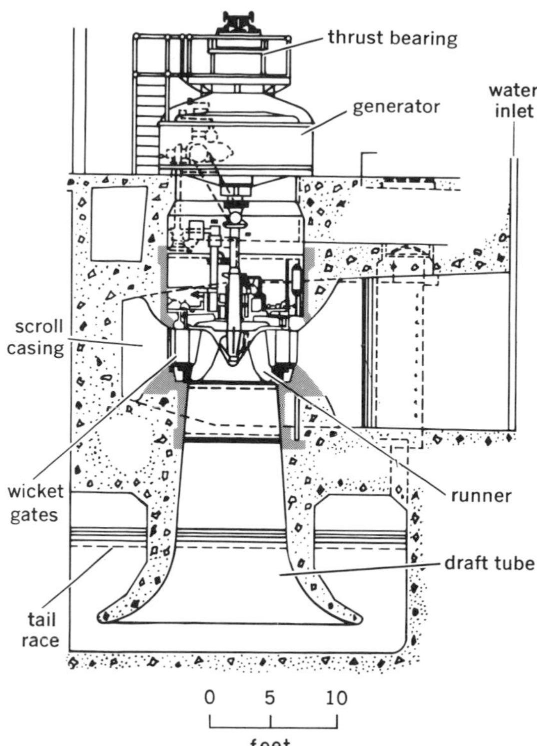

Fig. 2. Cross section of a reaction (Francis) type of hydraulic-turbine installation. 1 ft = 0.3 m.

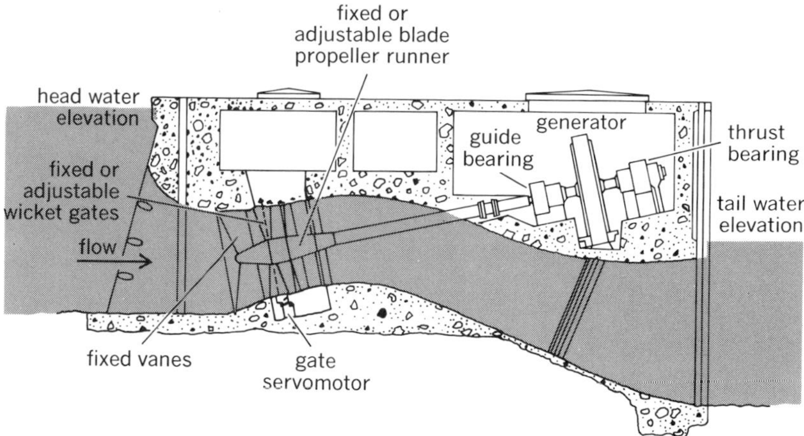

Fig. 4. Axial-flow tube-type hydraulic-turbine installation.

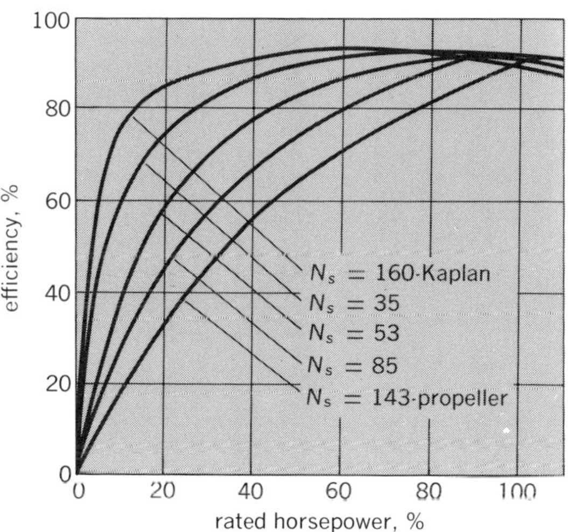

Fig. 5. The efficiency characteristics of some selected hydraulic-turbine types.

For many years reaction turbines have generally used vertical shafts for better accommodation of the draft tube whereas Pelton units have favored the horizontal shaft since they cannot use a draft tube. Vertical-shaft Pelton units have found increasing acceptance in large sizes because of multiple jets (for example, 4–6) on a single wheel; these provide reduced runner windage and friction losses and, consequently, higher efficiency. Axial-flow (Fig. 4) and diagonal-flow reaction turbines offer improved hydraulic performance and economic powerhouse structures for large-capacity low-head units. Kaplan units employ adjustable propeller blades as well as adjustable stationary nozzles in the gate ring for higher sustained efficiency (Fig. 5). Pelton units are preferred for high-head service (1000± ft or 300± m), Francis runners for medium heads (200± ft or 60± m), and propeller or Kaplan units for low heads (50± ft or 15± m).

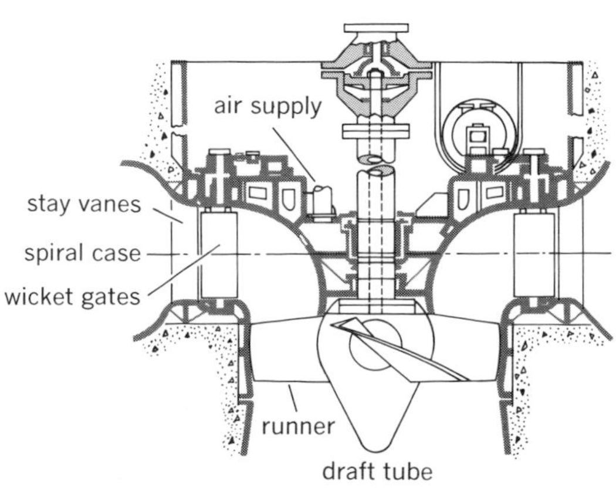

Fig. 3. Cross section of a propeller (Kaplan) type of hydraulic turbine installation.

Hydraulic-turbine performance is rigorously defined by characteristic curves, such as the efficiency characteristic (Fig. 5). The proper selection of unit type and size is a technical and economic problem. The data of Fig. 6 are significant because they show synoptically the relationship between unit type and site head as the result of accumulated experience on some satisfactory turbine installations. Specific speed N_s is a criterion or coefficient which is uniquely applicable to a given turbine type and relates head, power, and speed, which are the basic performance data in the selection of any hydraulic turbine. Specific speed is defined in the equation shown, where rpm is revolutions per minute, shp is shaft horsepower, and head is head on unit in feet. Specific speed is usually identified for a unit at the point of maximum efficiency. Cavitation must also be carefully scrutinized in any practical selection.

$$N_s = \frac{\text{rpm} \times (\text{shp})^{0.5}}{(\text{head})^{1.25}}$$

The draft tube (Fig. 2) is a closed conduit which

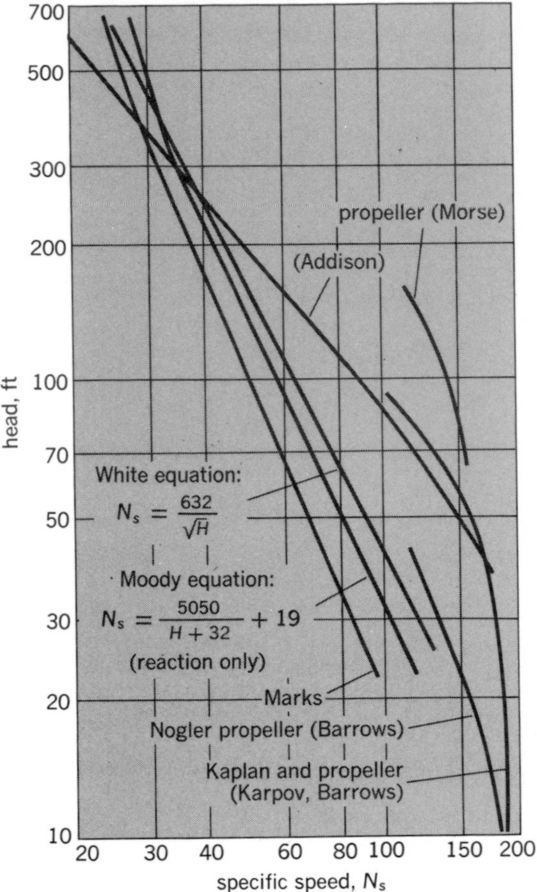

Fig. 6. Hydraulic-turbine experience curves, showing specific speed versus head. 1 ft = 0.3 m.

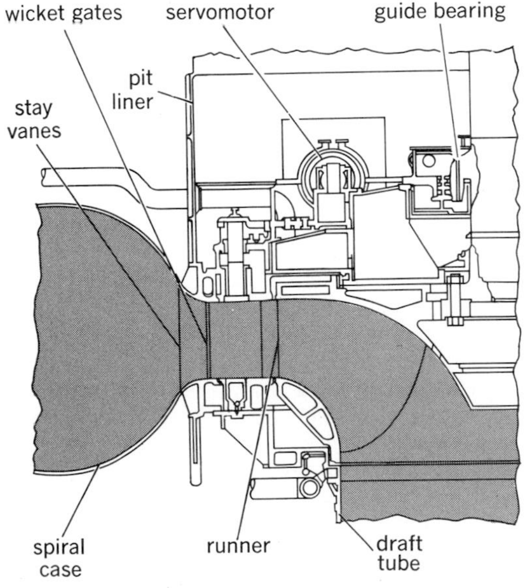

Fig. 7. Half-section of a reversible pump-turbine for pumped-storage application.

(1) permits the runner to be set safely above tail water level, yet to utilize the full head on the site, and (2) is limited by the atmospheric water column to a height substantially less than 30 ft. When made flaring in cross section, the draft tube will serve to recover velocity head and to utilize the full site head.

Efficiency of hydraulic turbine installations is always high, more than 85% after all allowances for hydraulic, shock, bearing, friction, generator, and mechanical losses. Material selection is not only a problem of machine design and stress loading from running speeds and hydraulic surges, but is also a matter of fabrication, maintenance, and resistance to erosion, corrosion, and cavitation pitting.

Governing problems are severe, primarily because of the large masses of water which are involved, their positive and negative acceleration without interruption of the fluid column continuity, and the consequent shock and water-hammer hazards.

Pumped-storage hydro plants have employed various types of equipment to pump water to an elevated storage reservoir during off-peak periods and to generate power during on-peak periods where the water flows from the elevated reservoir through hydraulic turbines. Although separate, single-purpose, motor-pump and turbine-generator sets give the best hydraulic performance, the economic burden of investment has led to the development of reversible pump-turbine units, as shown in Fig. 7. Components of the conventional turbine are retained, but the modified pump runner gives optimum performance when operating as a turbine. Compromises in hydraulic performance, with some sacrifice in efficiency, are more than offset by the investment savings with the dual-purpose machines. *See* WATERPOWER.

[THEODORE BAUMEISTER]

Bibliography: H. K. Barrows, *Water Power Engineering*, 3d ed., 1943; T. Baumeister (ed.), *Standard Handbook for Mechanical Engineers*, 8th ed., 1978; D. G. Fink and H. W. Beaty, *Standard Handbook for Electrical Engineers*, 11th ed., 1978.

Hydrocracking

A catalytic, high-pressure process flexible enough to produce either of the two major light fuels—high octane gasoline or aviation jet fuel. It proceeds by two main reactions: adding hydrogen to molecules too massive and complex for gasoline and then cracking them to the required fuels. The process is carried out by passing oil feed together with hydrogen at high pressure (1000–2500 psig or 6.9–17.2 MPa gage pressure) and moderate temperatures (500–750°F or 260–400°C) into contact with a bifunctional catalyst, comprising an acidic solid and a hydrogenating metal component. Gasoline of high octane number is produced, both directly and through a subsequent step such as catalytic reforming; jet fuels may also be manufactured simply by changing conditions with the same catalysts. The process is characterized by a long catalyst life (2–4 years), though a slow decline in activity occurs, caused by the deposition of carbonaceous material on the catalyst. Regeneration at intervals by burning off these deposits restores the activity, but eventually the catalyst porosity is destroyed and it must be replaced.

Generally, the process is used as an adjunct to catalytic cracking. Oils, which are difficult to convert in the catalytic process because they are highly aromatic and cause rapid catalyst decline, can be easily handled in hydrocracking, because of the low cracking temperature and the high hydrogen pressure, which decreases catalyst fouling. Usually, these oils boil at 400–1000°F (200–540°C), but it is possible to process even higher-boiling feeds if very high hydrogen pressures are used. However, the most important components in any feed are the nitrogen-containing compounds, as these are severe poisons for hydrocracking catalysts and must be removed to a very low level.

Hydrocracking was carried out on a practical scale in Germany and England starting in the 1930s. In this early work, a common hydrocracking catalyst was tungsten disulfide on acid-treated clay; thus, both hydrogenation and acidic components were present. Generally, a light oil from coal or coking products was vaporized and passed over the catalyst at high pressure. After separation of gasoline from the products, the unconverted material was returned to the reactor with a fresh portion of feed. Because this catalyst was not very active, the process had to be carried out at very high pressures and temperatures (4000 psig or 28 MPa gage pressure; 750°F or 400°C). It was costly and the products were not of high quality.

Research in the United States concentrated on the development of much more active catalysts, a different mode of operation, and the use of heavier oil feeds. As a result, the reaction is carried out in two separate, consecutive stages; in each, oil and hydrogen at high pressure flow downward over fixed beds of catalyst pellets placed in large vertical cylindrical vessels.

First stage. In the first, or pretreating, stage the main purpose is conversion of nitrogen compounds in the feed to hydrocarbons and to ammonia by hydrogenation and mild hydrocracking. Typical conditions are 650–740°F (340–390°C), 150–2500 psig (1.0–17.2 MPa gage pressure), and a catalyst contact time of 0.5–1.5 hr; up to 1.5 wt % hydrogen is absorbed, partly by conversion of the nitrogen compounds, but chiefly by aromatic compounds which are hydrogenated. It is most important to reduce the nitrogen content of the product oil to less than 0.001 wt % (10 parts per million). This stage is usually carried out with a bifunctional catalyst containing hydrogenation promotors, for example, nickel and tungsten or molybdenum sulfides, on an acidic support, such as silica-alumina. The metal sulfides hydrogenate aromatics and nitrogen compounds and prevent deposition of carbonaceous deposits; the acidic support accelerates nitrogen removal as ammonia by breaking carbon-nitrogen bonds. The catalyst is generally used as 1/8 × 1/8 in. (3.2 × 3.2 mm) or 1/16 × 1/8 in. (1.6 × 3.2 mm) pellets, formed by extrusion.

Second stage. Most of the hydrocracking is accomplished in the second stage, which resembles the first but uses a different catalyst. Ammonia and some gasoline are usually removed from the first-stage product, and then the remaining oil, which is low in nitrogen compounds, is passed over the second-stage catalyst. Again, typical conditions are 600–700°F (320–370°C), 1500–2500 psig (10.3–17.2 MPa gage pressure) hydrogen pressure, and 0.5–1.5 hr contact time; 1–1.5 wt % hydrogen may be absorbed. Conversion to gasoline or jet fuel is seldom complete in one contact with the catalyst, so the lighter oils are removed by distillation of the products and the heavier, high-boiling product combined with fresh feed and recycled over the catalyst until it is completely converted.

The catalyst for the second stage is also a bifunctional catalyst containing hydrogenating and acidic components. Metals such as nickel, molybdenum, tungsten, or palladium are used in various combinations, dispersed on solid acidic supports such as synthetic amorphous or crystalline silica-aluminas, such as zeolites. These supports contain strongly acidic sites and sometimes are enhanced by the incorporation of a small amount of fluorine. A long period (for example, 1 year) between regenerations is desirable; this is achieved by keeping a low nitrogen content in the feed and avoiding high temperatures, which lead to excess cracking with consequent deposition of coke on the catalyst. When activity of the catalyst does decrease, it can be restored by carefully controlled burning of the coke.

The catalyst is the key to the success of the hydrocracking process as now practiced, particularly the second-stage catalyst. Its two functions must be most carefully balanced for the product desired; that is, too much hydrogenation gives a poor gasoline but a good jet fuel. The oil feeds are composed of paraffins, other saturates, and aromatics—all complex molecules boiling well above the required gasoline or jet-fuel product. The catalyst starts the breakdown of these components by forming from them carbonium ions, that is, positively charged molecular fragments, via the protons (H^+) in the acidic function. These ions are so reactive that they change their internal molecular structure spontaneously and break down to smaller fragments having excellent gasoline qualities. The hydrogenating function aids in maintaining and controlling the ion reactions and protects the acid function by hydrogenating coke precursors off the catalyst surface, thus maintaining catalyst activity. Any olefins formed in the carbonium ion decomposition are also hydrogenated.

Products. The products from hydrocracking are composed of either saturated or aromatics compounds; no olefins are found. In making gasoline, the lower paraffins formed have high octane numbers; for example, the 5- and 6-carbon number fractions have leaded research octane numbers of 99–100. The remaining gasoline has excellent properties as a feed to catalytic reforming, producing a highly aromatic gasoline which, with added lead, easily attains 100 octane number. Both gasolines are suitable for premium-grade motor gasoline. Another attractive feature of hydrocracking is the low yield of gaseous components, such as methane, ethane, and propane, which are less desirable than gasoline. When making jet fuel, more hydrogenation activity of the catalysts is used, since jet fuel contains more saturates than gasoline. *See* GASOLINE.

The hydrocracking process is being applied in other areas, notably, to produce lubricating oils and to convert very asphaltic and high-boiling residues to lower-boiling fuels. Its use will certainly increase greatly in the future, since it accomplishes two needed functions in the petroleum-fuel economy: Large, unwieldly molecules are cracked, and the needed hydrogen is added to produce useful, high-quality fuels. *See* CRACKING.

[CHARLES P. BREWER]

Bibliography: *Advan. Petrol. Chem. Refining*, 8:168–191, 1964; W. F. Bland and R. L. Davidson (eds.), *Petroleum Processing Handbook*, sec. 3, pp. 16–25, 1967; *Hydrocarbon Process.*, 47(9): 139–144, 1968; Hydroprocesses, *Kirk-Othmer Encyclopedia of Chemical Technology*, 2d ed., vol. 11, 1966; J. W. Ward and S. A. Quader (eds.), *Hydrocracking and Hydrotreating*, 1975.

Hydroelectric generator

An electric rotating machine that transforms mechanical power from a hydraulic turbine or water wheel into electric power. Hydroelectric generators may have horizontal or vertical shafts, depending upon the turbine. The most common type in large ratings is the vertical-shaft, synchronous generator. The hydroelectric generator may be arranged to operate as a motor during periods of low power demand. In such a case the turbine serves as a pump, raising water to a high elevation for reuse to generate power at a period of peak load. Such a unit is termed a pump turbine. *See* ELECTRIC POWER GENERATION; GENERATOR; HYDRAULIC TURBINE; TURBINE; WATERPOWER.

[LEON T. ROSENBERG]

Hydroelectric power

Electric power generated through the use of flowing water. Water stored by dams is released to turn modern waterwheels called turbines, coupled to generators which supply electricity.

Place in energy economy. Hydroelectric power developments in the United States, including pumped-storage, had a generating capability of more than 63×10^6 kW at the end of 1979. This accounts for more than 13% of the United States power industry's total capability.

Worldwide, about one-fifth of all energy utilized is in the form of electric power, and about one-fifth of this electric power is generated from hydroelectric power plants. Thus, while hydro power produced only about 1/25 of the energy utilized by humans, it is still an important power source with particular characteristics which assure its future development. *See* ELECTRIC POWER GENERATION.

Hydro power is the oldest known mechanical power source. It is usually the lowest cost form of bulk energy. It is in most cases the most efficient form of power. It is the least polluting form of power. It is the most responsive—easy to start and stop—of any electric generating source. And finally, it is the most compatible power where large river projects are being built for nonpower purposes such as flood control, irrigation, navigation, or water supply. Hydro power, in the form of pumped-storage, is to date the only practical means to store excess electric power for use at a later time.

Theory of hydro power. Hydro power is the opposite of a ship's propeller or a pump. A propeller or a pump uses power to move water; a hydro turbine uses moving water to produce mechanical power. This mechanical power is almost invariably used to rotate an electrical generator, which produces electricity. Physically, hydro turbines and propellers or pumps resemble each other closely, comprising flow-controlling blades mounted on a rotating shaft connected to a power-using device. In some cases, hydro turbines actually are designed for reverse-direction rotation use as pumps. The largest and highest lift pumps in the world are such "reversible pump-turbines." These machines, rated at millions of horsepower, can literally pump a river to the top of a mountain. *See* HYDRAULIC TURBINE.

Hydro power is a form of nondepleting, self-replenishing energy—actually a form of solar energy. In this view, the water-covered seven-eighths of the Earth's surface receives a continuous inflow of solar radiant energy. Part of this inflow is radiated back out into space, and most of the rest is utilized to evaporate (and thus desalt!) sea water, which becomes clouds. These clouds pass over land masses, which receive rain from the clouds. The topography, and drainage pattern, of the continents concentrates this "solar energy in the form of rain" into narrow, powerful rivers. Hydro power plants placed strategically on these rivers can utilize a highly concentrated—and therefore both powerful and economical—form of solar energy. *See* SOLAR ENERGY.

History of waterpower. Wind and waterpower are the most ancient form of nonanimal mechanical (and motive) power used by humans. Sail and ocean currents are prehistoric means of navigation, and the use of windmills and waterwheels is roughly coincident with the birth of civilization. Early civilizations such as the Chinese, Greek, Persian, and Roman were masters of wind or water mills, or both.

Where rivers were available, waterpower was more useful because of its inherent lower cost, greater reliability, and greater force or power. As the technological and industrial revolutions developed, wind power was progressively displaced by steam, internal combustion, or electrical power. But waterpower has kept pace with the advances of technology and is more used today than ever before. Thus, waterpower is both the most ancient and one of the most modern forms of power used for the good of humankind. *See* WATERPOWER; WIND POWER.

Modern waterpower had its beginnings in the early 19th century with the development of the hydraulic reaction turbine, and the modernizing of the impulse wheel. A radial outward-flow turbine was installed at Pont sur l'Ognon in France in 1827 after 4 years of experimentation by M. Fourneynon. James B. Francis in 1847 significantly improved an inward-flow turbine patented in 1836, and is considered by many as the originator of the American-type reaction turbine.

In 1853 Jearum Atkins of the United States began studying the scientific theory behind the ancient flutter wheel, which in plan resembled a wagon wheel and moved when a water jet from a spout hit the wheel's flat, vertical vanes near their ends. Practical development of this wheel is credited to Lester A. Pelton, who commenced his radical improvements in 1882.

The first hydroelectric central generating station in the United States was built at Appleton, WI, in 1882. It furnished enough power to light 250 electric lights.

Before 1897 nearly 300 hydroelectric plants

World's largest hydroelectric plants ranked in order of planned rated capacity

Rank	Name	River or basin	Country	Rated capacity (1978), MW	Rated capacity (planned), MW
1	Itaipu*	Paraná	Brazil/ Paraguay		12,600
2	Grand Coulee	Columbia	United States	3,463	9,780
3	Guri (final stage)	Caroni	Venezuela	2,553	8,853
4	Tucurui*	Tocantins	Brazil		6,480
5	Sayano -Shushenskaya*	Yenisei	Soviet Union		6,400
6	Corpus*		Argentina/ Paraguay		6,000
7	Krasnoyarsk	Yenisei	Soviet Union	6,000	
8	La Grande 2*	La Grande	Canada		5,328
9	Churchill Falls		Canada	5,225	
10	Bratsk	Angara	Soviet Union	4,600	
11	Ust-Ilimsk	Angara	Soviet Union	720	4,500
12	Yacyreta-Apipe*	Paraná	Paraguay/ Brazil		4,050
13	Cobora-Bassa	Zambezi	Mozambique	2,000	4,000
14	Chief Joseph		United States	1,500	3,669
15	Oak Creek*		United States		3,600
16	Rogunsky*	Vakhsh	Soviet Union		3,600
17	Inga		Zaire	1,360	3,500
18	Paulo Alfonso		Brazil	1,524	3,409
19	Ilha Solteira	Paraná	Brazil	3,200	
20	Brumley Gap*		United States		3,000
21	Powell Mountain*		United States		3,000
22	John Day		United States	2,160	2,700
23	Nurek	Vakhsh	Soviet Union	900	2,700
24	Revelstoke*	Illecillewaet	Canada		2,700

*Initial operation planned for 1980–1985.

were in operation in the world. By 1900 the first pumped-storage plant had been constructed in Europe. Since the turn of the century hydroelectric power has developed and advanced with the electric power industry. Giant plants—Hoover, Grand Coulee, Chief Joseph, Churchill Falls (Canada), Krasnoyarsk (Soviet Union), Guri (Venezuela), Aswan (Egypt)—have been built and are being expanded. Developing countries have found hydroelectric power the most economical means to bring industrial advancement to their people. The table gives the world's largest operating or planned hydroelectric plants, ranked in order of planned rated capacity.

Typical hydroelectric plant. A typical hydroelectric project includes a water-diverting structure, such as a dam or canal, a conduit to transport water to the turbines, turbines and governors, generators, control and switching apparatus, a powerhouse to protect the equipment, transformers, and transmission lines to carry the generated power to distribution centers (see Fig. 1). In most cases a forebay or a vertical shaft is provided to store the surge of water which occurs when the plant is shut down or its power output is reduced, thus reducing water hammer. Water hammer in a hydro plant is similar to (but on a grander scale than) that common in home plumbing systems not provided with surge pipes near water faucets. If the power station is some distance from the water source, the forebay also serves as the initial supply of water for startup, or when increasing power output, while water in the canal or pipeline accelerates. An intake structure at the head of the water conduit houses trash racks to keep debris from the turbines, and also gates or valves to stop the flow of water for conduit inspection and maintenance. *See* DAM.

The water conduit may be very short, as when the powerhouse is an integral part of a concrete dam. Many conduits are longer, ranging from a few feet to several miles. Those made of steel are called penstocks. Generally, the steel-lined portions of water passages through a concrete dam or of tunnels are also referred to as penstocks.

Hydraulic turbines are classed as either impulse or reaction turbines. The impulse turbine is represented today by the Pelton-type waterwheel. In an impulse turbine the water is discharged through one or more nozzles as one or more jets in free air which act upon the runner (a series of buckets around the circumference of a hub). A housing prevents splash and guides the discharge, which falls freely into the tail water. *See* IMPULSE TURBINE; REACTION TURBINE.

In reaction turbines the entire flow from head water to tail water occurs in a closed conduit system. Whereas with a Pelton wheel all available head is converted into kinetic energy, with a reaction turbine only part of the available head is converted to kinetic energy at the entrance to the runner. A substantial part of the available head remains as pressure head which varies throughout the passage through the turbine.

Reaction turbines are represented by two types: the Francis turbine and the vertical-flow propeller turbine. Propeller turbines are subdivided into fixed-blade and adjustable-blade (Kaplan) types.

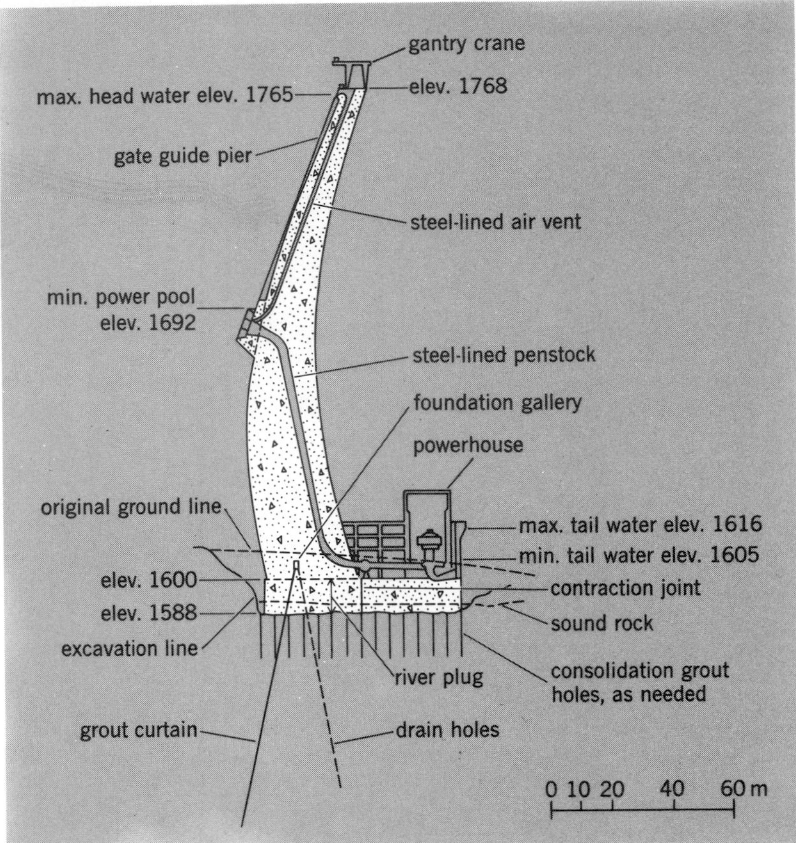

Fig. 1. Section of Karadj River arch dam and powerhouse, in Iran. Three Francis-type turbines by Harland of Scotland, each rated 49,700 hp (37,000 kW), 482 ft (147 m) net head. Three generators by Hitachi of Japan. Total plant capacity, 113,000 kW. *(From C. V. Davis and K. E. Sorensen, Handbook of Applied Hydraulics, 3d ed., McGraw-Hill, 1969)*

Diagonal-flow turbines have also been developed. In the Francis turbine (also vertical-flow), flow passes from the penstock to a spiral case and then inwardly around the inside circumference through a series of pivoting gates (wicket gates) to the runner, which resembles the blade arrangement in a jet airplane engine. The curved runner blades react to the impulse of the rotating flow of water, causing the runner to rotate. The blades also deflect the water to the tailrace. In the propeller turbine, flow also passes through wicket gates but then is deflected downward onto the propeller blades and through the blades to the draft tube (see Fig. 2). The draft tube is a diverging discharge passage connecting the runner with the tailrace. It is shaped to decelerate the flow of water to the tailrace, and at the same time accelerate the momentum of the turbine runner by converting the kinetic energy remaining in the discharge into suction head, thereby increasing the pressure difference on the runner. Draft tubes are not used with impulse wheels, so that the head from nozzle to tail water is not utilized. Water turbines are highly efficient devices, realizing average efficiencies of more than 90% and peak efficiencies of 97%.

The generator is located above the turbine except in plants having horizontal-flow turbines. Generator and turbine are connected by a common shaft. Thus when the turbine turns, the generator rotor turns. The rotor, passing through a magnetic field, generates electricity. This electricity is carried by cables to the control and switching apparatus, and thence to transformers from which it is transmitted over power lines to the distribution centers of the power system. *See* HYDROELECTRIC GENERATOR; TRANSMISSION LINES.

Hydro power classifications. Hydroelectric developments are classified as multipurpose or single-purpose, as base-load or peak-load, as run-of-river or storage, as high-head or low-head, as indoor or outdoor, as aboveground or underground, and as conventional hydro or pumped-storage or tidal. Any project may fit several of these classifications.

A multipurpose project is developed for a number of uses, including flood control, irrigation, navigation, power and silt removal. Grand Coulee Project on the Columbia River, the largest hydroelectric plant in the United States, with new construction bringing it to an installed capacity of 6×10^6 kW, was built for flood control, irrigation, and power. The Hoover Project, with a capacity of 1.34×10^6 kW, is used for silt removal as well as the uses cited for Grand Coulee. Guri Project on the Rio Caroni in Venezuela, which will exceed Grand Coulee in size (9×10^6 kW), is a single-purpose hydroelectric project built for power only.

A base-load plant supplies power almost continuously to satisfy the minimum constant power demand in a system. Several power plants may be required to furnish a system's base load. A peak-load plant generates power only during periods of peak demand, that is, when the power needed in a system exceeds that generated by the base-load plants. Fossil fuel and nuclear power projects are difficult to shut down and to start up and, therefore, are not efficient peaking plants. *See* POWER PLANT.

A run-of-river project utilizes the flow passing the plant for hydro power but does not alter the amount or intensity of flow to areas downstream. A storage project retains water so river flow downstream may be regulated. This assures farmers or other water users downstream they will have an adequate water supply during periods of low flow or drought.

Description of a project as high-head or low-head refers to the differences in water levels just upstream (head water) and downstream (tail water) of the plant. High-head plants have heads over 100 ft (1 ft = 0.3 m), while low-head plants have heads of 100 ft or less. The amount of head has a direct effect on the type of hydraulic turbine adopted. Francis turbines of the reaction type have been applied for heads up to 2205 ft (Rosshag power station, Austria), impulse turbines of the Pelton type up to 5800 ft (Reisseck plant, Austria), and propeller turbines up to 289 ft (Kaplan turbine at Nembia plant, Italy). Generally, for any one turbine type, the higher the head, the smaller the turbine needed to furnish the same power.

Designation of a plant as indoor or outdoor depends upon whether the powerhouse superstructure encloses the powerhouse overhead crane and generator housing. If the crane is outdoors but the generator housing is indoors, the project is referred to as semi-outdoor.

Aboveground plants are those where the powerhouse is at ground surface. Certain project sites

Fig. 2. Model of turbine and generator for Priest Rapids power station, Columbia River, Washington. Kaplan-type turbine rated 114,000 hp (85,000 kW); 78 ft (23.4 m) net head, 85.7 rpm, runner diameter 284 in. (7.2 m). Ten units by English Electric Company, Ltd. Total plant capacity, 790,000 kW. (*From C. V. Davis and K. E. Sorensen, Handbook of Applied Hydraulics, 3d ed., McGraw-Hill, 1969*)

make it less expensive to excavate a cavern to house the generating equipment than to build a surface structure. Sometimes a site, such as a narrow gorge, lacks space for the powerhouse aboveground, and belowground may be a good alternative, foundation conditions permitting. Generally, underground powerhouses are constructed in sound rock rather than in poor rock or soft ground.

Conventional hydroelectric power plants include all the hydro projects which fit the description of a typical hydro project given above. Pumped-storage projects, though almost as old as conventional hydro in Europe, have had in the United States the greatest development since 1960. They are designed to provide peaking power. To make power available from a pumped-storage plant during periods of high demand, water is pumped to an upper reservoir during off-peak hours, utilizing unneeded power generated by base-load plants, and then released as required to provide the energy necessary to turn the turbines. In Europe most pumped-storage plants have separate pumps and turbines. In the United States, however, reversible pump-turbines are used. Examples of such plants are the Kinzua or Seneca projects on the Allegheny River in western Pennsylvania, Muddy Run on the lower Susquehanna River in south-central Pennsylvania, and Ludington on the eastern shore of Lake Michigan. *See* PUMPED STORAGE.

A tidal power station harnesses the energy available from water-level fluctuations. Two tidal power developments have been constructed: La Rance in the Bay of St. Malo on the north coast of Brittany, and Kislogubskaya in the Soviet Union. At La Rance, the amplitude of the tides in spring reach 44 ft, but average head is 27 ft. Considerable distance from areas needing power seems to be the deterrent to construction at most potential sites. *See* TIDAL POWER.

Future of hydroelectric power. Waterpower enjoys a great future for development. It is almost always the lowest-cost and least-polluting form of power. Advanced, industrialized countries have already utilized the "cream" of their waterpower potential, but modern economics of thermal and nuclear power are making the remaining waterpower potential projects—which were often marginal economically—now look much more economical and rewarding. And the largest rivers and therefore the greatest potential hydro projects are in the underdeveloped and developing countries such as Brazil, Mozambique, Zaïre, Venezuela, Argentina, and Paraguay. The largest potential water power resources of any country are possessed by the People's Republic of China, and these resources are now one of the main development targets of that country. By the year 2000 such construction will raise world hydroelectric generation from 1.3×10^9 MW-hr to 3.4×10^9 MW-hr, with a projected ultimate potential of 5×10^9 MW-hr. This work will require decades, and absence of war, to complete. But the technology for development is in hand, economic need for such projects is evident, and initial projects are underway. *See the feature article* EXPLORING ENERGY CHOICES.

[RICHARD D. HARZA; H. CLARK DEAN]

Bibliography: A. M. Angelini, The utilization of hydraulic resources still available in the world, 9th World Energy Conference, Detroit, September 1974, quoted in *Water Power Dam Constr.*, p. 47, February 1975; Corps tallies 512,000-MW hydro resource, *Eng. News-Rec.*, 203 (13):14, Sept. 27, 1979; J. Cotillon, La Rance: Six years of operating a tidal power plant in France, *Water Power (London)*, 26(10):314–322, October 1974; C. V. Davis and K. E. Sorensen, *Handbook of Applied Hydraulics*, 3d ed., 1969; Federal Power Commission, *Hydroelectric Power Resources of the United States, Developed and Undeveloped*, FPC P-42, Jan. 1, 1972; International Commission on Large Dams, *The World's Highest Dams, Largest Earth and Rock Dams, Greatest Man-made Lakes, Largest Hydroelectric Plants,* compiled by T. W. Mermel, Scientific Affairs, Bureau of Reclamation, Department of the Interior, May 1973; T. W. Mermel, Major dams of the world, *Water Power Dam Constr.*, 30(8):43–54, August 1978; 1975 Statistical Report, *Elec. World*, 191(6):66–71, Mar. 15, 1979.

Hydrogen-fueled technology

Hydrogen may play a prominent role as a synthetic chemical fuel as fossil fuels become depleted. Future attractive energy sources (solar energy, nuclear fusion, and so on) are generally characterized by their immobility and large scale, so that their broad utility will be dependent on energy carriers, including electricity and synthetic chemical fuels such as hydrogen. The advantages and disadvantages of hydrogen as a synthetic fuel are explored below, and the current state of knowledge in its application to various energy-use sectors in society, especially transportation, is reviewed. The economics of hydrogen production and use are discussed, and a perspective on the place of hydrogen in an overall self-consistent, steady-state future energy metabolism is given. *See* SYNTHETIC FUEL.

Properties of hydrogen. The physical and chemical properties of hydrogen are recalled in the table, where particular attention is given to hydrogen as a cryogenic liquid. The advantages of hydrogen as a chemical fuel are that: (1) it has the greatest specific energy (energy per unit mass) of any chemical fuel (Fig. 1); (2) its combustion results in only water as a product, with CO, CO_2, and unburned hydrocarbons not present in principle; even oxides of nitrogen are reduced greatly, relative to those accompanying hydrocarbon combustion; and (3) it may be used in every application where hydrocarbons are now used, often with greater combustion efficiency, and lends itself to low-temperature catalytic combustion not practical with hydrocarbons. In addition, hydrogen can be burned with pure oxygen to provide a turbine power system with exceptional thermal efficiency and can be oxidized in fuel cells to produce electricity. Nonenergy uses of hydrogen include desulfurization of coal and petroleum and clean metal ore reduction. Currently, the largest industrial use of hydrogen is in ammonia synthesis (for fertilizer) and for hydrogenation of oils.

The chief disadvantage of hydrogen is its extremely low boiling point (20 K), requiring handling only in special cryogenic containers or dewar vessels as a liquid, or its storage as a metal hydride. Even as a liquid, its energy density per unit volume is about three times poorer than liquid hydrocarbons, due to its very low density. A second disadvantage often cited is the fire and explosion hazard, and indeed, the tragedy of the 1937 burning of

Properties of liquid hydrogen

Property	Value
Boiling point	20.4 K
Liquid density	0.0708 g/cm³
Latent heat of vaporization	108 cal/g or 450 J/g
Energy release upon combustion	29,000 cal/g or 2050 cal/cm³ or 1.21×10^5 J/g
Flame temperature	2483 K
Autoignition temperature	858 K

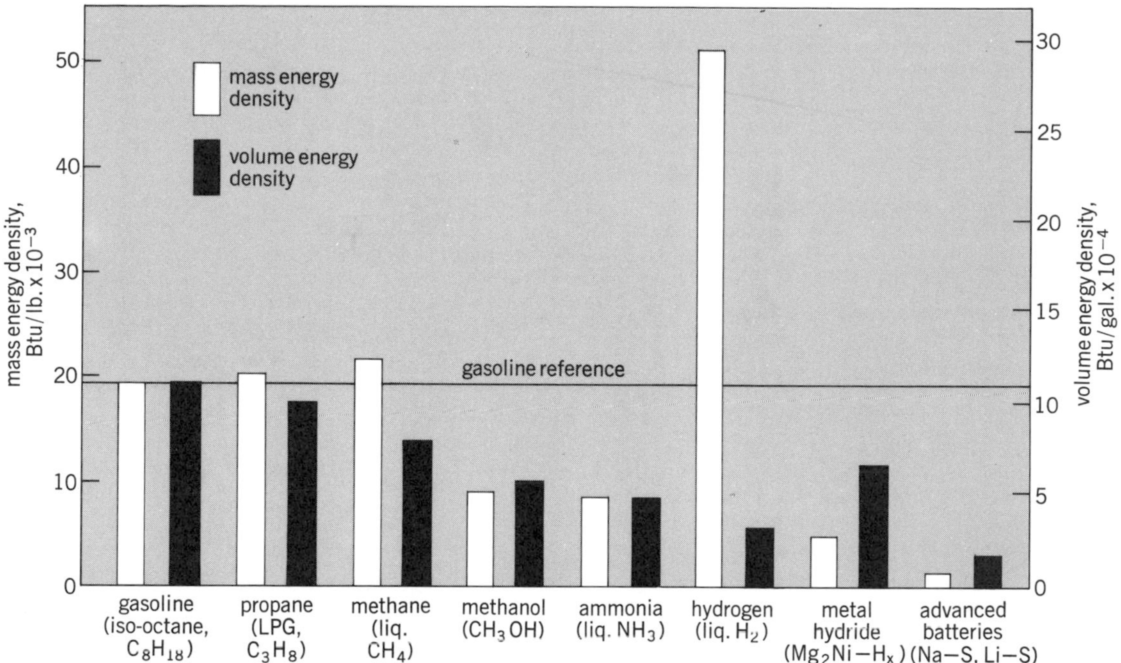

Fig. 1. A graph comparing the energy density characteristics of a number of fuels. ("Advanced batteries" includes adjustment for different energy conversion efficiencies.) 1 Btu/lb = 2326 J/kg; 1 Btu/gal = 279 J/liter.

the German zeppelin *Hindenburg* still haunts the public memory. While the safety problems with hydrogen are different than with liquid hydrocarbons, they are not a priori more severe, and the safety record of liquid hydrogen as a rocket fuel in the United States space program has been excellent.

Hydrogen storage. Hydrogen may be stored as a compressed gas, as a liquid, or as a hydride. Compressed gas storage is the least attractive, except for very small systems, in view of the very low energy density of even highly compressed hydrogen gas and the great weight of the required pressure vessel.

Liquid hydrogen has been used extensively for over a decade, primarily in the space program. Double-walled vessels with vacuum (dewar vessels) or foam insulation are widely used in sizes from small laboratory vessels, through transport truck and rail freight cars of 50,000 to 100,000 liters, to large storage dewars of up to 3,000,000 liters. The loss from small dewars is 1–2% per day through evaporation (due to thermal leakage). Hydrogen liquefaction is currently carried out with at best about 40% of the Carnot (ideal) efficiency, and since the energy difference between hydrogen gas at ambient temperature and atmospheric pressure and liquid hydrogen is about 12% of the chemical energy content of the gas, the energy required for liquefaction is about 30% of the available chemical energy. In principle, part of this energy could be beneficially recovered in applications of liquid hydrogen as a fuel. *See* CARNOT CYCLE.

Hydrogen may also be combined with metals to form loosely bound hydrides, which may then be dissociated at elevated temperature. Such hydrides can store hydrogen at a volume density as great as or greater than liquid hydrogen, however at a weight characteristic of the metal. Attractive hydrides are based on Mg-Ni and Mg-Cu alloys. The dissociation temperatures at 1 atm (100,000 N/m^2) are about 250°C. This dissociation heat might be obtained from the exhaust of a hydrogen-fueled engine. Metal hydrides have practical problems, such as the required dissociation heat, and there is much less engineering experience with them than with liquid hydrogen storage. Nevertheless, they provide an alternative storage worthy of serious research. *See* ENERGY STORAGE.

Aircraft fuel. The most appealing application of hydrogen is as a replacement fuel for aircraft, in view of its excellent mass energy density. Several

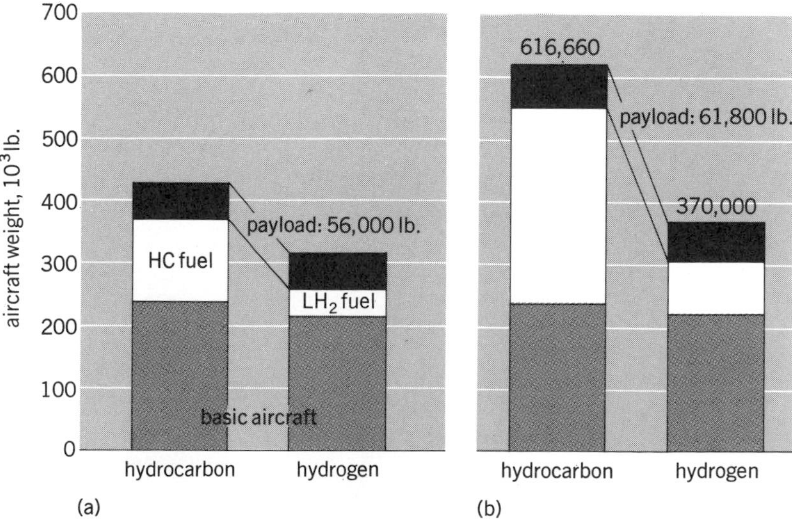

Fig. 2. Comparative aircraft weight breakdowns for subsonic and supersonic transport designs, based on Lockheed California Company estimates. (10^3 lb = 454 kg.) (*a*) Subsonic transport (modified present design). (*b*) Supersonic transport (new design).

aircraft companies have considered airframe designs for subsonic commercial jet aircraft and concluded that the same payload and range could be achieved with a much lighter fuel load, hence a lighter overall aircraft, requiring shorter runways and resulting in overall fuel economy. The craft would be more bulky, requiring either attached wing-pod tanks or an expanded fuselage to accommodate the more bulky liquid hydrogen tanks. The relative sophistication of airport crews and the relative security of airports simplify the cryogenic handling and safety problems of hydrogen. Experiments have been made on hydrogen-fueled jet engines, both in static tests and in flight. Even greater advantages accrue in the use of hydrogen fuel in supersonic and hypersonic aircraft, as the cyrogenic liquid may then be used to cool the leading-edge airfoil surfaces which are heated by the air friction (see Fig. 2). An early start on design and development of a hydrogen-fueled airframe will be necessary to develop a commercial aircraft which will operate through the first quarter of the next century when global petroleum production is expected to be declining. *See* AIRCRAFT FUEL.

Hydrogen-burning automobiles. The private automobile represents the most difficult application of hydrogen fuel because of the storage problem. There appear at this time two options, either cryogenic or metal hydride storage. In either case, the storage is bulky and, based on present technology, expensive. However, the performance of the hydrogen-burning automobile (Fig. 3) is excellent. The common, noxious pollutants that have come to be identified with the urban environment are absent altogether or (in the case of oxides of nitrogen) greatly reduced. Hydrogen burns well in a lean mixture (in a fuel-air ratio as low as half the stoichiometric value) and provides efficient power at reduced throttle and idling speeds. Because of this and because the freedom from pollution permits high compression ratios and more efficient exhaust systems, experience with hydrogen-powered automobiles points to a fuel energy efficiency as much as 50% greater than with gasoline for a typical driving cycle. Although gas carburetion has been most widely used in hydrogen experiments (as with methane combustion), fuel injection has also been used successfully.

The application of hydrogen fuel to fleet vehicles and to the surface transportation vehicles now powered by diesel engines (trucks, buses, construction and earthmoving vehicles, and railroad locomotives) is in every case simpler technologically and economically than the application to private automobiles. However, private autos are the single largest transportation consumer of energy, and their conversion to a clean fuel would have the greatest impact on air quality of any other single improvement in the energy use patterns of the United States. *See* AIR POLLUTION; INTERNAL COMBUSTION ENGINE.

Fig. 3. A Chevrolet modified to burn hydrogen by Billings Energy Research, Inc., of Provo, UT. Hydrogen is stored onboard in a 50-liter liquid dewar.

Energy storage. One interesting application of hydrogen fuel technology is in electric power generation for energy storage. The demand for electric power in a typical metropolitan area fluctuates with diurnal, weekly, and seasonal variations by over a factor of 2. Hence, the installed generating capacity must be about twice what would be needed for the minimum demand, or about 30% greater than the average. Electric power companies often make use of special, small gas-turbine-driven generators which can be started quickly to handle power demand peaks, and the main steam-generating boilers can be throttled back for power demand minima. Pumped hydroelectric storage is also used in those areas where the terrain is appropriate. The problem becomes more severe when nuclear fission power plants supply most of an electricity grid's power. Fission reactors are sensitive to thermal cycling and are less able to tolerate a "throttling back" than conventional boilers. *See* NUCLEAR POWER; PUMPED STORAGE.

Hydrogen could play a role if surplus power were used to electrolyze water and the resulting hydrogen and oxygen were stored. During demand peaks, these gases would be burned to power turbines to generate the required electric power. Water electrolysis is currently done on a large scale with commercially available units with an energy efficiency of 70–85%. The storage of hydrogen and oxygen might be in pressurized tanks but more probably underground in salt domes, depleted oil or gas wells, or limestone caverns. For power plants near or on large bodies of water, storage under water has been proposed wherein large plastic domes would be anchored to the bottom. By using hydrogen and oxygen rather than hydrogen and air, a combustion temperature of over 3500°C could be achieved; the actual temperature would be regulated by adding water to the combustion chamber. With an initial temperature of 2000°C, a hydrogen-oxygen-driven turbine could generate electricity with an energy efficiency of 60–70% (as compared with a conventional steam boiler-turbine generator efficiency of 40%). Hence, the overall efficiency of the water electrolysis and reburning of hydrogen and oxygen to generate electricity might exceed 50%.

Hydrogen may be combined with pure oxygen or with air in a fuel cell to provide energy with 50–60% efficiency, and indeed it is the fuel of choice for fuel cells. Although they are energy sources of low power density, and hence less attractive for transportation, fuel cells are very promising for electric power peaking systems with hydrogen energy storage. *See* FUEL CELL.

Hydrogen production. Hydrogen might be produced in one of several ways. Currently, the industrial production of about 20,000,000 tons (20×10^9 kg) of hydrogen per year is predominantly by steam reforming of natural gas or petroleum. Where hydrogen is required for its unique properties as in ammonia synthesis or as a rocket fuel, this method is efficient and economical. However, this is not a defensible means of producing hydrogen as a fuel to replace hydrocarbons; the hydrocarbons might better and more economically be used as fuels directly.

An exception to this statement concerns the substitution of hydrogen for petroleum products in regions where climatic conditions exacerbate pollution problems; here the superficial economic liabilities of hydrogen fuels may be more than offset by less obvious but equally real health benefits. One estimate suggests that each gallon (1 gal = 3.8 liters) of gasoline burned may lead to $1.00 in added health costs in some circumstances.

The use of hydrogen fuel is primarily considered as a replacement for hydrocarbon fluid fuels following their depletion or exhaustion. Hydrogen can be readily produced from coal, however, and would be one of several alternative fuels possible as a result of coal gasification or liquefaction. Steam reacts with coal at about 950°C and high pressure to produce CO, CO_2, and H_2. Subsequent reactions result in only easily separated CO_2 and nearly pure hydrogen. See COAL GASIFICATION; COAL LIQUEFACTION.

Where electricity is the immediate product of a thermal cycle (the primary energy source may be nuclear fission, including the breeder reactor, or solar or geothermal energy), hydrogen may be produced by electrolysis with an energy efficiency of at least 70%. The energy efficiency of electrolysis may be increased, for example, by operating at elevated temperature and pressure, albeit with increased first costs. Nuclear fusion may evolve as a major energy source in the 21st century, and the fast neutron energy could then also supply heat for a thermal cycle. An alternative has been proposed whereby the direct radiolytic dissociation of steam or water would be produced by the fast fusion neutrons. It has been noted that recombination of the dissociated hydrogen and oxygen presents a serious problem, and a more complex series of radiolytic chemical reactions have been proposed. The result might be a complex thermochemical-radiolytic cycle engineered to produce hydrogen with a greater overall efficiency than a thermochemical cycle alone. See NUCLEAR FUSION.

The heat developed by a nuclear reactor, solar furnace, or such may also be used directly to drive a series of chemical reactions which result in the separation of water into hydrogen and oxygen. For example, one such sequence, proposed by G. Beni and C. Marchetti, is shown in the reactions:

$$CaBr_2 + 2H_2O \rightarrow 2HBr + Ca(OH)_2 (730°C)$$
$$2HBr + Hg \rightarrow H_2 + HgBr_2 (250°C)$$
$$Ca(OH)_2 + HgBr_2 \rightarrow HgO + CaBr_2 + 2H_2O (200°C)$$
$$HgO \rightarrow Hg + \tfrac{1}{2}O_2 (600°C)$$

The overall energy efficiency of such processes might be 40–60%, significantly better than the overall efficiency of hydrogen production via electric power and electrolysis. A large number of other cycles have been proposed and have been the subject of serious study in many laboratories.

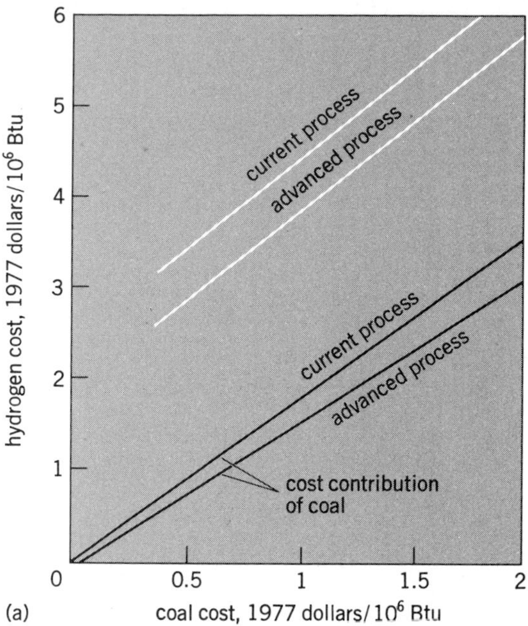

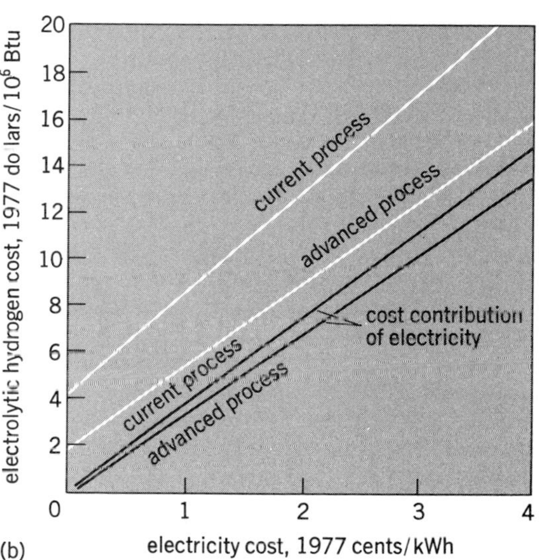

Fig. 4. Dependence of hydrogen production costs on costs of primary energy sources. *(a)* Hydrogen production cost versus coal cost. *(b)* Hydrogen production cost versus electricity cost. 10^6 Btu = 1.055×10^9 J. *(From J. J. Donnelly, Jr., et al., Hydrogen-powered vs. battery-powered automobiles, Int. J. Hydrogen Energy, 4:411–443, 1979)*

Hydrogen-fuel economy. Inevitably, a driving force of the hydrogen economy, or any other future energy scenario, will be the relative costs of different energy options. The cost estimates for hydrogen production vary, of course, with the costs of the primary energy sources, and these have risen sharply since the early 1970s. A useful representation is given in Fig. 4, where the cost of hydrogen in terms of costs of coal or electricity is plotted. Hydrogen from natural gas or petroleum is currently cheaper than the options shown but is not included for the reasons given above. Hydrogen from coal, the plausible large-scale interim source until solar and fusion energy sources mature, is next in attractiveness. Hydrogen from water electrolysis is naturally closely tied to the cost of elec-

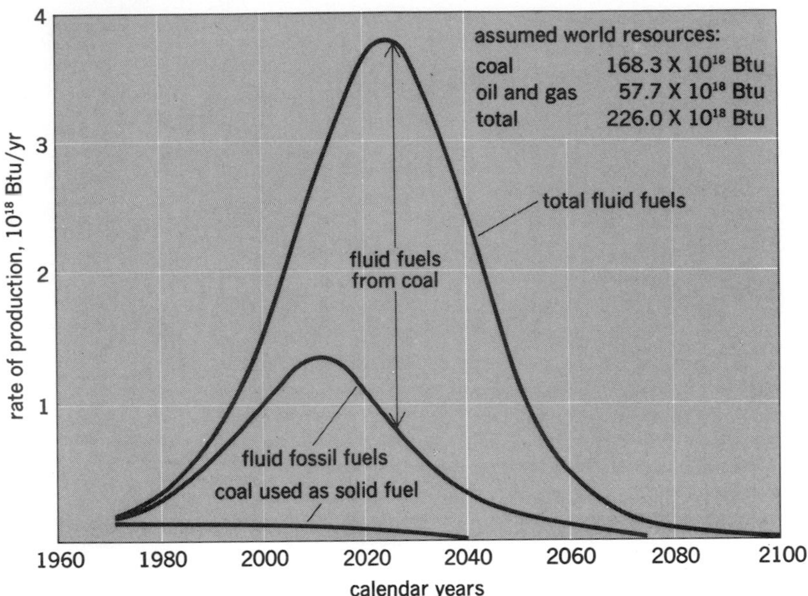

Fig. 5. Projection of rate of production of fluid fossil fuels with and without coal conversion to fluid fuels, 10^{18} Btu = 1,055 × 10^{21} J. (*From M. A. Elliot and N. C. Turner, Estimating the future rate of production of the world's fossil fuels, Texas Eastern Transmission Corporation*)

tric energy. Cost estimates for direct thermal dissociation are more speculative, as no pilot plant even has been built. The cost of liquefaction is noted as a separate item.

Distribution of hydrogen would probably be by gas pipeline, with local liquefaction near the marketing or supply area. Gas pipeline distribution is about 15% more expensive than the corresponding distribution of natural gas, and much cheaper than the corresponding energy distribution costs of electricity. This observation has led to the suggestion that some energy users might be supplied only with pipeline gas hydrogen and might use it to generate electricity locally in a fuel cell. It is then a matter of economic balance whether this option is preferable to parallel distribution of hydrogen and centrally generated electricity.

It has been suggested that a total community energy system could be derived from pipeline hydrogen gas, with space heat provided by direct combustion and electricity from fuel cells (or perhaps via a thermal cycle in conjunction with space heat, that is, cogeneration). The transportation sector would then also use hydrogen as liquid or hydride in vehicles and aircraft. Consideration of such a scenario illustrates the feasibility of hydrogen as an energy storage and carrier for virtually every niche in the energy metabolism of a society. *See* COGENERATION.

The driving motivation for the hydrogen fuel economy is the search for a cyclic, indefinitely viable energy metabolism pattern to replace the current reliance on fossil fuels as these approach exhaustion and their recovery ceases to be economically attractive.

While coal is widely heralded as an economically attractive long-range replacement for fossil hydrocarbons (with synthetic hydrocarbon fluid fuels manufactured from coal), it is also limited, albeit at a more future date than petroleum (Fig. 5). Even here, however, caution is necessary in view of the possible buildup of CO_2 in the atmosphere to a level that would severely modify the global climate. Data collected since 1950 show that atmospheric CO_2 has been rising at a rate of 0.3–0.4% per year; between 1958 and 1979 it rose by 6.4%. This corresponds to about half of the carbon burned as fuel remaining in the atmosphere as CO_2. Extrapolation to the end of the century indicates that atmospheric CO_2 will increase another 10–15%, and that by about 2030 it may be double the present level. This could lead to a global warming of about 3°C, resulting in major and deleterious large-scale climatic changes. It therefore seems most prudent to endeavor to understand totally the consequence of carbon release on the global climate, as well as in the meantime seriously to pursue alternative energy systems not reliant on carbon.

An energy pattern for the 21st century wherein nuclear fusion or solar power provides primary energy, and electricity and synthetic chemical fuels serve as energy carriers, seems logical and perhaps inevitable. Hydrogen appears in many respects to be the most attractive synthetic chemical fuel. *See the feature article* EXPLORING ENERGY CHOICES.

[LAWRENCE W. JONES]

Bibliography: G. D. Brewer, A plan for active development of LH_2 for use in aircraft, *Int. J. Hydrogen Energy*, 4:169–177, 1979; D. P. Gregory, *Sci. Amer.*, 228:1, 13, 1973; J. Hord (ed.), *Selected Topics on Hydrogen Fuel*, NBSIR 75-803, Cryogenics Division, Institute of Basic Standards, National Bureau of Standards, January 1975; L. W. Jones, *Environ. Plan. Pollut. Control*, 1:12, 1973; L. W. Jones, *Science*, 174:367, 1971; J. Michel (ed.), *Hydrogen and Other Synthetic Fuels*, Rep. TID 26136, U.S. Government Printing Office, September 1972; T. Nejat Veziroglu (ed.), *Hydrogen Energy*, pts. A and B, 1975.

Hysteresis motor

A synchronous motor without salient poles and without dc excitation which makes use of the hysteresis and eddy-current losses induced in its hardened-steel rotor to produce rotor torque. The stator and stator windings are similar to those of an induction motor and may be polyphase, shaded-pole, or capacitor type. The rotor is usually made up of a number of hardened steel rings on a nonmagnetic arbor. The hysteresis motor develops constant torque up to synchronous speed. The motor can, therefore, synchronize any load it can accelerate. These motors are built in small sizes, for instance, for electric clocks. *See* INDUCTION MOTOR; SYNCHRONOUS MOTOR.

[LOYAL V. BEWLEY]

Impulse turbine

A prime mover in which fluid (water, steam, or hot gas) under pressure enters a stationary nozzle where its pressure (potential) energy is converted to velocity (kinetic) energy. The accelerated fluid then impinges on the blades of a rotor, imparting its energy to the blades to produce rotation and overcome the connected rotor resistance. The impulse principle is basic to many turbines.

The impulse principle can be distinguished from the reaction principle by considering the flow of water from a hole near the bottom of a bucket (Fig.

IMPULSE TURBINE

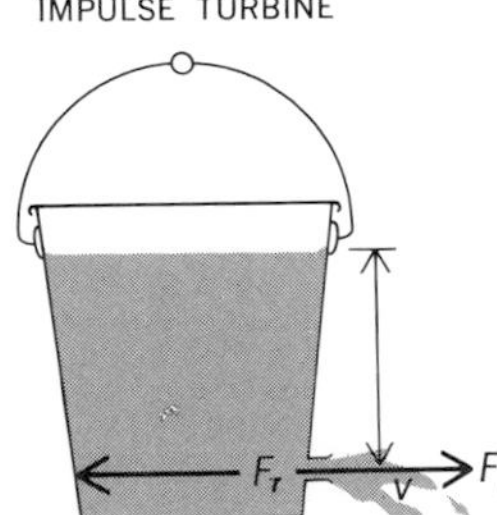

Fig. 1. Water escaping through a nozzle near the base of a bucket, free to swing, illustrating the impulse F_i and reaction F_r forces of the jet issuing with a velocity v.

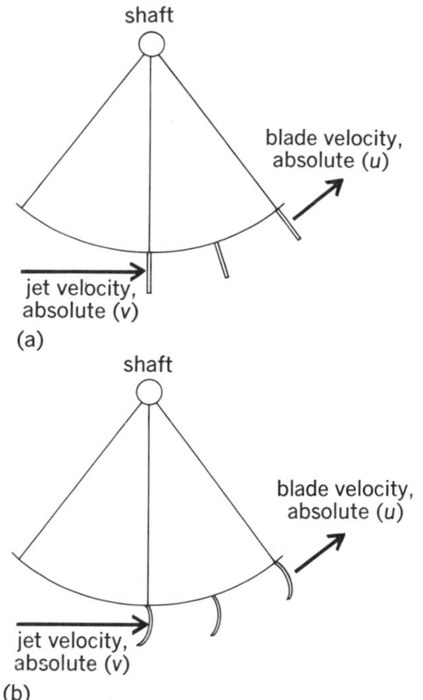

Fig. 2. Jet impinging on a series of (*a*) flat blades and (*b*) curved blades on periphery of a wheel.

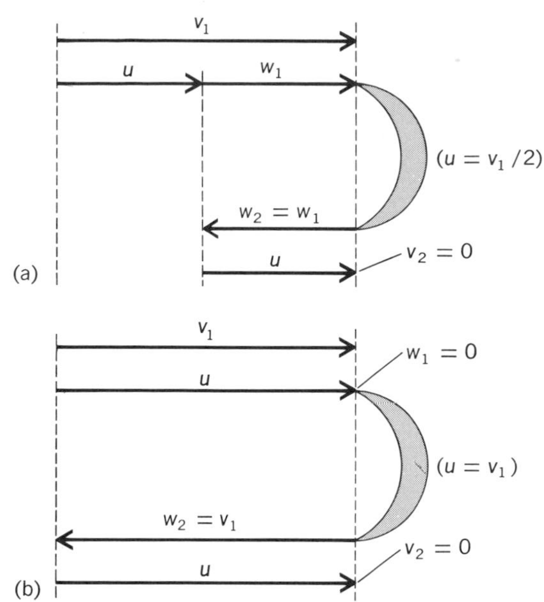

Fig. 4. Velocity vector diagrams for idealized axial fluid flow on (*a*) impulse-turbine blading and (*b*) reaction-turbine blading; 180° jet reversal. In *b* the fluid is accelerated to *w* across the moving blades (nozzles). In both cases the leaving absolute fluid velocity *v* is zero, giving maximum efficiency.

1). The hole is a nozzle that serves to convert potential energy to the kinetic form ΔE; the impulse force F_i in the issuing jet is given by the expression $\Delta E = F_i\, v/2$, where v is the velocity of the jet.

If the jet is allowed to impinge on a series of vanes mounted on the periphery of a wheel, the impulse force can overcome the resistance connected to the shaft (Fig. 2). The efficiency of the device is ideally dependent upon the vane curvature and the absolute vane velocity. For flat blades (Fig. 2*a*), the efficiency cannot exceed 50%. With curved blades and complete reversal of the jet (Fig. 2*b*), the efficiency will be 100% when the vane velocity u is one-half the jet velocity v.

If the bucket in Fig. 1 is suspended and free to move, there will be, by Newton's third law of motion, a reaction force F_r equal and opposite to the impulse force F_i. For maximum efficiency (100%) the swinging bucket will have to move with an absolute tip velocity u equal to the jet velocity, v. This is the reaction principle and is demonstrated by the Barker's mill (Fig. 3).

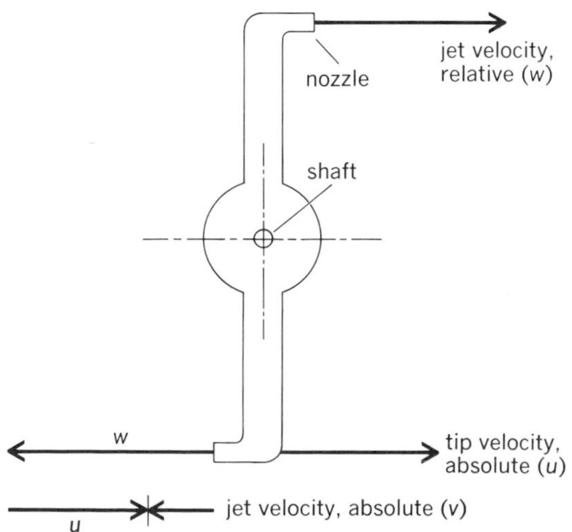

Fig. 3. Barker's mill, a reaction-type jet device, illustrating the application of moving nozzles and the resultant absolute and relative velocities.

The basic difference between the reaction principle and the impulse-turbine principle is determined by the presence or absence of moving nozzles. A nozzle is a throat section device in which there is a drop in pressure with consequent acceleration of the emerging fluid. An impulse turbine has stationary nozzles only. A reaction turbine must have moving nozzles but may have stationary nozzles also so that the fluid can reach the moving nozzle. This is the usual condition for any practical reaction turbine.

The idealized vector diagrams of Fig. 4 demonstrate distinguishing features for the construction which uses a row of blades mounted on the periphery of a wheel and for which flow is axially through the blade passages from one side of the wheel to the other. The theoretical condition further presupposes complete (180°) reversal of the jet and no friction losses. In Fig. 4, v is the absolute velocity of the fluid, u is the absolute velocity of the moving blade, and w is the relative velocity of the fluid with respect to the moving blade. Subscripts 1 and 2 apply to entrance and exit conditions, respectively. The vectors of the illustration demonstrate that maximum efficiency (zero residual absolute fluid velocity v_2) obtains (1) when the blade speed is half the jet speed for impulse turbines and (2) when the blade speed is equal to the jet speed for reaction turbines.

Figure 5 shows the situation more practically as complete reversal of the jet is not realistic. The vector diagrams of Fig. 5 show speed ratios of 0.49, 0.88, and 1.22 for reasonable degrees of jet reversal and no friction losses. By varying the angle of

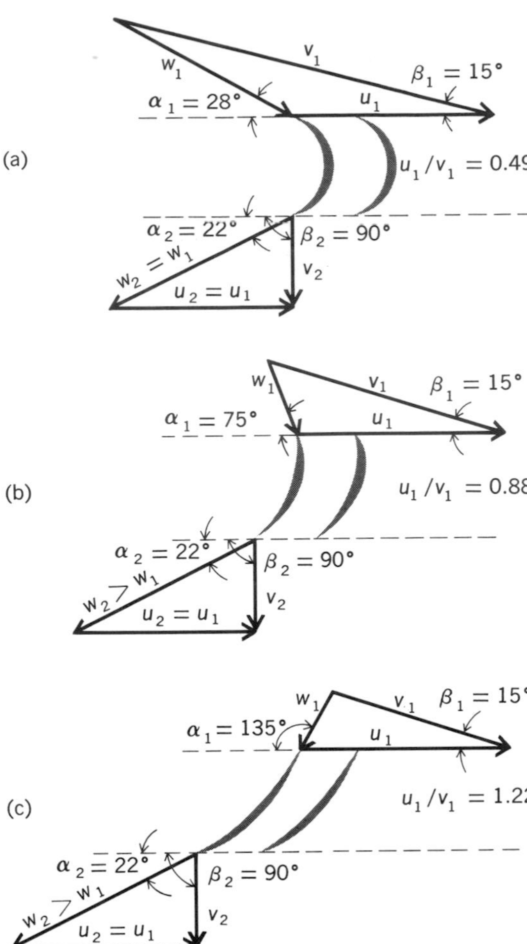

Fig. 5. Vector diagrams for idealized (frictionless) axial fluid flow on (a) impulse-turbine blading and (b, c) reaction-turbine blading; jet reversal less than 180°. Exit vector triangles are identical with consequent equal efficiencies. Resultant speed ratios (u_1/v_1) are 0.49, 0.88, and 1.22, respectively.

entrance α_1 to the moving blades, a wide range of speed ratios is available to the designer. In each case shown in Fig. 5, the stationary nozzle entrance angle β_1 is fixed at 15°; the exit vector triangle from the moving blades is identical in all cases. The data in Fig. 5c demonstrate that it is entirely possible to have speed ratios greater than unity with the reaction principle. This condition is utilized in many designs of hydraulic turbines, for example, propeller and Kaplan units.

The choice of details is at the discretion of the turbine designer. He can determine the extent to which the basic principles of impulse and reaction should be applied for a practical, reliable, economic unit in the fields of hydraulic turbines, gas turbines, and steam turbines. *See* GAS TURBINE; HYDRAULIC TURBINE; REACTION TURBINE; STEAM TURBINE. [THEODORE BAUMEISTER]

Bibliography: T. Baumeister (ed.), *Standard Handbook for Mechanical Engineers*, 8th ed., 1978; G. T. Csanady, *Theory of Turbomachines*, 1964; D. G. Shepherd, *Principles of Turbomachinery*, 1956; W. G. Stelz (ed.), *Turbomachinery Developments in Steam and Gas Turbines*, 1977; V. L. Streeter, *Handbook of Fluid Dynamics*, 1961.

Induction motor

An alternating-current motor in which the currents in the secondary winding (usually the rotor) are created solely by induction. These currents result from voltages induced in the secondary by the magnetic field of the primary winding (usually the stator). An induction motor operates slightly below synchronous speed and is sometimes called an asynchronous (meaning not synchronous) motor.

Induction motors are the most commonly used electric motors because of their simple construction, efficiency, good speed regulation, and low cost. Polyphase induction motors come in all sizes and find wide use where polyphase power is available. Single-phase induction motors are found mainly in fractional-horsepower sizes, and those up to 25 hp are used where only single-phase power is available.

POLYPHASE INDUCTION MOTORS

There are two principal types of polyphase induction motors: squirrel-cage and wound-rotor machines. The differences in these machines is in the construction of the rotor. The stator construction is the same and is also identical to the stator of a synchronous motor. Both squirrel-cage and wound-rotor machines can be designed for two- or three-phase current.

Stator. The stator of a polyphase induction motor produces a rotating magnetic field when supplied with balanced, polyphase voltages (equal in magnitude and 90 electrical degrees apart for two-phase motors, 120 electrical degrees apart for three-phase motors). These voltages are supplied to phase windings, which are identical in all respects. The currents resulting from these voltages produce a magnetomotive force (mmf) of constant magnitude which rotates at synchronous speed. The speed is proportional to the frequency of the supply voltage and inversely proportional to the number of poles constructed on the stator.

Figure 1 is a simplified diagram of a three-phase, two-pole, Y-connected stator supplied with currents I_1, I_2, and I_3. Each stator winding produces a pulsating mmf which varies sinusoidally with time. The resultant mmf of the three windings (Fig. 1c) is constant in magnitude and rotates at synchronous speed. Figure 1b shows the direction of the mmf in the stator for times t_1, t_2, and t_3 shown in Fig. 1a and shows how the resultant mmf rotates. The synchronous speed N_s is shown by Eq. (1), where f is the frequency in hertz and p is the

$$N_s = \frac{120f}{p} \text{ rpm} \tag{1}$$

number of stator poles. For any given frequency of operation, the synchronous speed is determined by the number of poles. For 60-Hz frequency, a two-pole motor has a synchronous speed of 3600 rpm; a four-pole motor, 1800 rpm; and so on.

Squirrel-cage rotor. Figure 2 shows the bars, end rings, and cooling fins of a squirrel-cage rotor. The bars are skewed or angled to prevent cogging (operating below synchronous speed) and to reduce noise. The end rings provide paths for currents that result from the voltages induced in the rotor bars by the stator flux. The number of poles

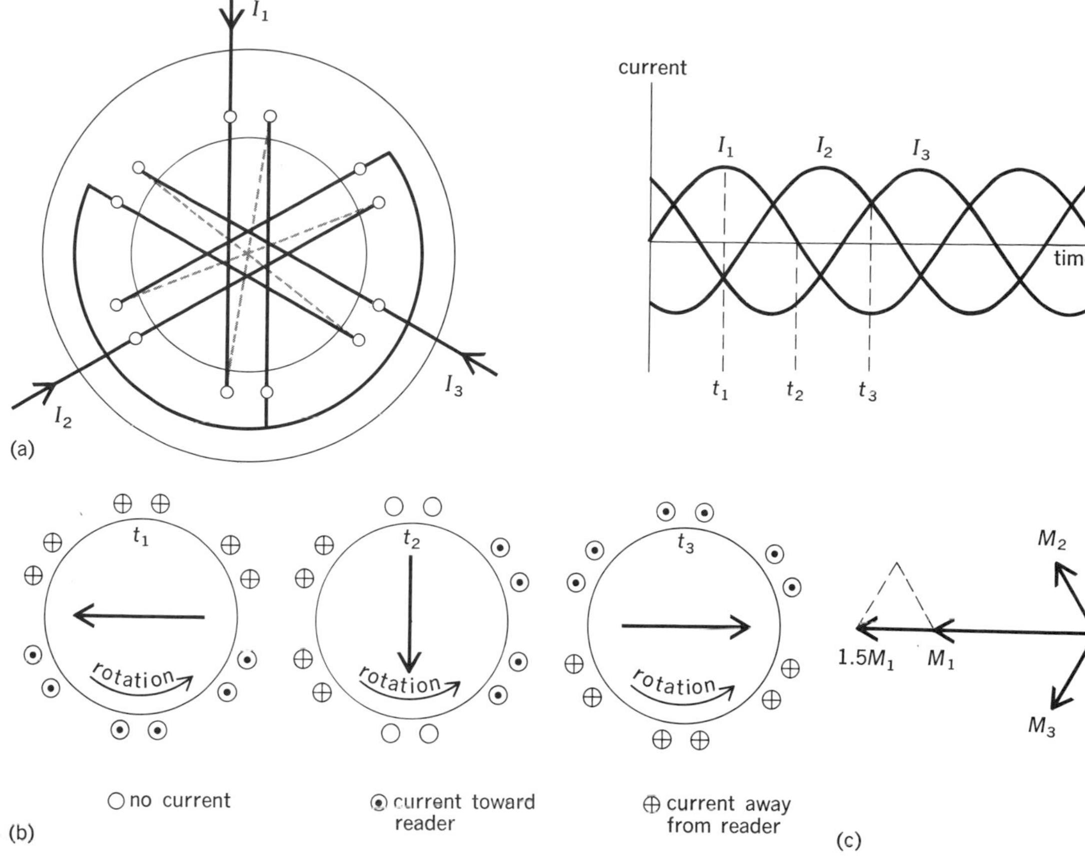

Fig. 1. Three-phase, two-pole, Y-connected stator of induction motor supplied with currents I_1, I_2, and I_3. (*a*) Stator windings and currents. (*b*) Rotating field. (*c*) Magnetomotive forces produced by stator winding.

on a squirrel-cage rotor is always equal to the number of poles created by the stator winding.

Figure 3 shows how the two motor elements interact. A counterclockwise rotation of the stator flux causes voltages to be induced in the top bars of the rotor in an outward direction and in the bottom bars in an inward direction. Currents will flow in these bars in the same direction. These currents interact with the stator flux and produce a force on the rotor bars in the direction of the rotation of the stator flux.

When not driving a load, the rotor approaches synchronous speed N_s. At this speed there is no motion of the flux with respect to the rotor conductors. As a result, there is no voltage induced in the rotor and no rotor current flows. As load is applied, the rotor speed decreases slightly, causing an increase in rotor voltage and rotor current and a consequent increase in torque developed by the rotor. The reduction in speed is therefore sufficient to develop a torque equal and opposite to that of the load. Light loads require only slight reductions in speed; heavy loads require greater reduction. The difference between the synchronous speed N_s and the operating speed N is the slip speed. Slip s is conveniently expressed as a percentage of synchronous speed, as in Eq. (2).

$$s=\frac{N_s-N}{N_s}\times 100\% \qquad (2)$$

When the rotor is stationary, a large voltage is induced in the rotor. The frequency of this rotor voltage is the same as that of the supply voltage. The frequency f_2 of rotor voltage at any speed is shown by Eq. (3), where f_1 is the frequency of the

$$f_2=f_1 s \qquad (3)$$

supply voltage and s is the slip expressed as a decimal. The voltage e_2 induced in the rotor at any speed is shown by Eq. (4), where e_{2s} is the rotor

$$e_2=(e_{2s})s \qquad (4)$$

voltage at standstill. The reactance x_2 of the rotor is a function of its standstill reactance x_{2s} and slip, as shown by Eq. (5). The impedance of the rotor at

$$x_2=(x_{2s})s \qquad (5)$$

any speed is determined by the reactance x_2 and the rotor resistance r_2. The rotor current i_2 is shown by Eq. (6). In the equation, for small

$$\begin{aligned} i_2 &= \frac{e_2}{\sqrt{r_2^2+x_2^2}} \\ &= \frac{(e_{2s})s}{\sqrt{r_2^2+(x_{2s})^2s^2}} = \frac{e_{2s}}{\sqrt{\left(\frac{r_2}{s}\right)^2+(x_{2s})^2}} \end{aligned} \qquad (6)$$

values of slip, the rotor current is small and possesses a high power factor. When slip becomes large, the r_2/s term becomes small, current increases, and the current lags the voltage by a large phase angle. Standstill (or starting) current is large and lags the voltage by 50–70°. Only in-phase, or

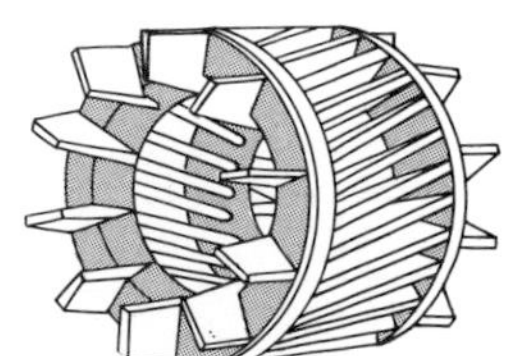

Fig. 2. Bars, end rings, and cooling fins of a squirrel-cage rotor.

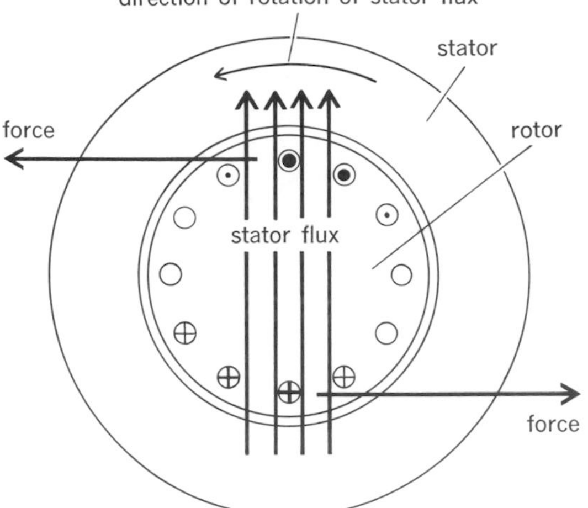

Fig. 3. Forces on the rotor winding.

unity powerfactor, rotor currents are in space phase with the air-gap flux and can therefore produce torque. The current i_2 contains both a unity power-factor component i_p and a reactive component i_r. The maximum value of i_p and therefore maximum torque are obtained when slip is of the correct value to make r_2/s equal to x_{2s}. If the value of r_2 is changed, the slip at which maximum torque is developed must also change. If r_2 is doubled and s is doubled, the current i_2 is not changed and the torque is unchanged.

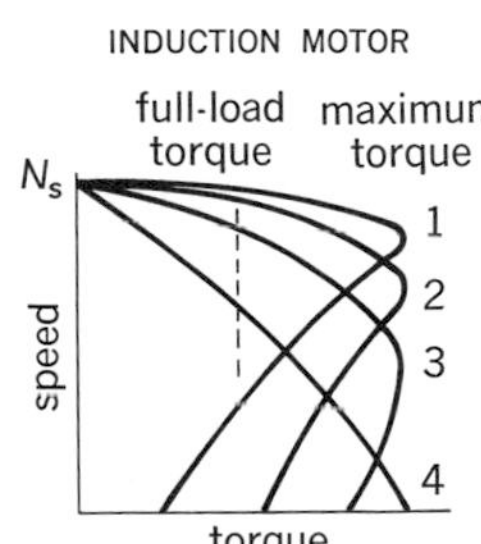

Fig. 4. Speed-torque characteristic of polyphase induction motor.

This feature provides a means of changing the speed-torque characteristics of the motor. In Fig. 4, curve 1 shows a typical characteristic curve of an induction motor. If the resistance of the rotor bars were doubled without making any other changes in the motor, it would develop the characteristic of curve 2, which shows twice the slip of curve 1 for any given torque. Further increases in the rotor resistance could result in curve 3. When r_2 is made equal to x_{2s}, maximum torque will be developed at standstill, as in curve 4. These curves show that higher resistance rotors give higher starting torque. However, since the motor's normal operating range is on the upper portion of the curve, the curves also show that a higher-resistance rotor results in more variation in speed from no load to full load (or poorer speed regulation) than the low-resistance rotor. Higher-resistance rotors also reduce motor efficiency. Except for their characteristic low starting torque, low-resistance rotors would be desirable for most applications.

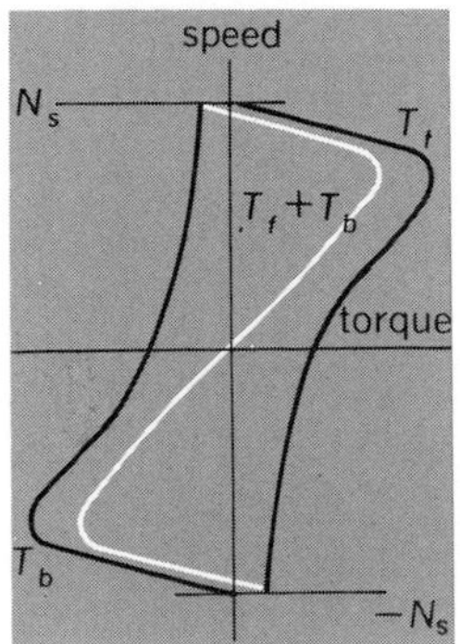

Fig. 5. Torques produced in the single-phase induction motor.

Wound rotor. A wound-rotor induction motor can provide both high starting torque and good speed regulation. This is accomplished by adding external resistance to the rotor circuit during starting and removing the resistance after speed is attained.

The wound rotor has a polyphase winding similar to the stator winding and must be wound for the same number of poles. Voltages are induced in these windings just as they are in the squirrel-cage rotor bars. The windings are connected to slip rings so that connections may be made to external impedances, usually resistors, to limit starting currents, improve power factor, or control speed.

A rheostat is used to bring a wound-rotor motor up to speed. The rheostat limits the starting current drawn from the supply to a value less than that required by a squirrel-cage motor. The resistance is gradually reduced to bring the motor up to speed. By leaving various portions of the starting resistances in the circuit, some degree of speed control can be obtained, as in Fig. 4. However, this method of speed control is inherently inefficient and converts the motor into a variable-speed motor, rather than an essentially constant-speed motor. For other means of controlling speed of polyphase induction motors and for other types of ac motors *see* ALTERNATING-CURRENT MOTOR.

SINGLE-PHASE INDUCTION MOTORS

Single-phase induction motors display poorer operating characteristics than polyphase machines, but are used where polyphase voltages are not available. They are most common in small sizes (1/2 hp or less) in domestic and industrial applications. Their particular disadvantages are low power factor, low efficiency, and the need for special starting devices.

The rotor of a single-phase induction motor is of the squirrel-cage type. The stator has a main winding which produces a pulsating field. At standstill, the pulsating field cannot produce rotor currents that will act on the air-gap flux to produce rotor torque. However, once the rotor is turning, it produces a cross flux at right angles in both space and time with the main field and thereby produces a rotating field comparable to that produced by the stator of a two-phase motor.

An explanation of this is based on the concept that a pulsating field is the equivalent of two oppositely rotating fields of one-half the magnitude of the resultant pulsating field. In Fig. 7, ϕ_m is the maximum value of the stator flux ϕ, which is shown only by its two components ϕ_f and ϕ_b, which represent the two oppositely rotating fields of constant equal magnitudes of $\phi_m/2$. Each component ϕ_f and ϕ_b produces a torque T_f and T_b on the rotor. Figure 5 shows that the sum of these torques is zero when speed is zero. However, if started, the sum of the torques is not zero and rotation will be maintained by the resultant torque.

This machine has good performance at high speed. However, to make this motor useful, it must have some way of producing a starting torque. The method by which this starting torque is obtained designates the type of the single-phase induction motor.

Split-phase motor. This motor has two stator windings, the customary main winding and a starting winding located 90 electrical degrees from the main winding, as in Fig. 6*a*. The starting winding has fewer turns of smaller wire, to give a higher resistance-to-reactance ratio, than the main winding. Therefore their currents I_m (main winding) and I_s (starting winding) are out of time phase, as in Fig. 6*c*, when the windings are supplied by a common voltage V. These currents produce an elliptical field (equivalent to a uniform rotating field superimposed on a pulsating field) which causes a

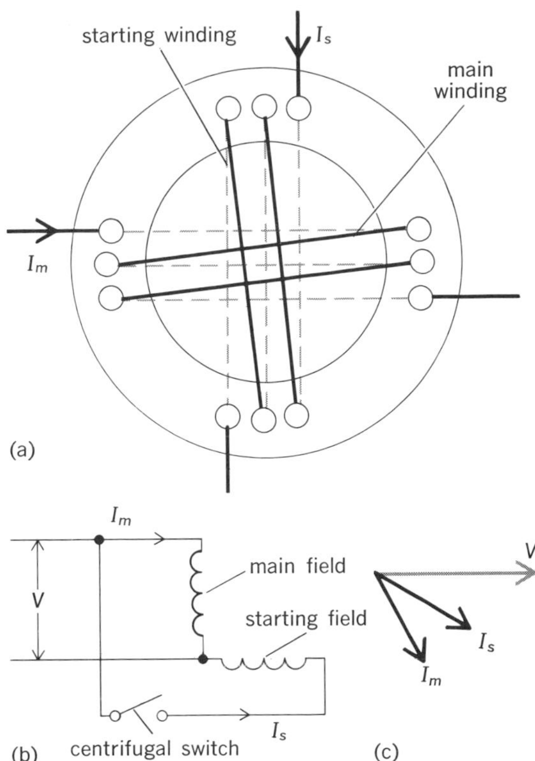

Fig. 6. Split-phase motor. (*a*) Windings. (*b*) Winding connections. (*c*) Vector diagram.

unidirectional torque at standstill. This torque will start the motor. When sufficient speed has been attained, the circuit of the starting winding can be opened by a centrifugal switch and the motor will operate with a characteristic illustrated by the broken line curve of Fig. 5.

Capacitor motor. The stator windings of this motor are similar to the split-phase motor. However, the starting winding is connected to the supply through a capacitor. This results in a starting winding current which leads the applied voltage. The motor then has winding currents at standstill which are nearly 90° apart in time, as well as 90° apart in space. High starting torque and high power factor are therefore obtained. The starting winding circuit can be opened by a centrifugal switch when the motor comes up to speed.

In some motors two capacitors are used. When the motor is first connected to the voltage supply, the two capacitors are used in parallel in the starting circuit. At higher speed one capacitor is removed by a centrifugal switch, leaving the other in series with the starting winding. This motor has high starting torque and good power factor.

Shaded-pole motor. This motor is used extensively where large power and large starting torque are not required, as in fans. A squirrel-cage rotor is used with a salient-pole stator excited by the ac supply. Each salient pole is slotted so that a portion of the pole face can be encircled by a short-circuited winding, or shading coil.

The main winding produces a field between the poles as in Fig. 8. The shading coils act to delay the flux passing through them, so that it lags the flux in the unshaded portions. This gives a sweeping magnetic action across the pole face, and consequently across the rotor bars opposite the pole face, and results in a torque on the rotor. This torque is much smaller than the torque of a split-phase motor, but it is adequate for many operations. A typical characteristic of the motor is shown in Fig. 8*b*.

For another type of single-phase alternating-current motor *see* REPULSION MOTOR. For synchronous motors built for single-phase *see* HYSTERESIS MOTOR; RELUCTANCE MOTOR.

Linear motor. Figure 9 illustrates the arrangements of the elements of the polyphase squirrel-cage induction motor. The squirrel cage (secondary) is embedded in the rotor in a manner to provide a close magnetic coupling with the stator winding (primary). This arrangement provides a small air gap between the stator and the rotor. If the squirrel cage is replaced by a conducting sheet as in Fig. 9*b*, motor action can be obtained. This machine, though inferior to that of Fig. 9*a*, will function as a motor. If the stator windings and iron are unrolled (rectangular laminations instead of circular laminations), the arrangement of the elements will take a form shown in Fig. 9*c*, and the field produced by polyphase excitation of the primary winding will travel in a linear direction instead of a circular direction. This field will produce a force on the conducting sheet that is in the plane of the sheet and at right angles with the stator conductors. A reversal of the phase rotation of the primary voltages will reverse the direction of motion of the air-gap flux and thereby reverse the force on the secondary sheet. No load on the motor corresponds to the condition when the secondary sheet is moving at the same speed as the field produced

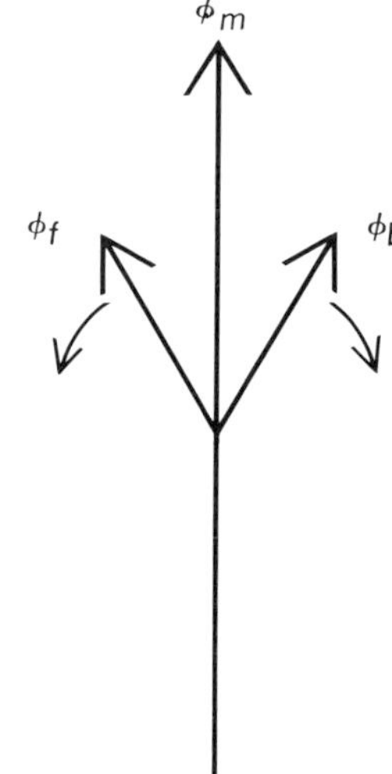

Fig. 7. Fluxes associated with the single-phase induction motor.

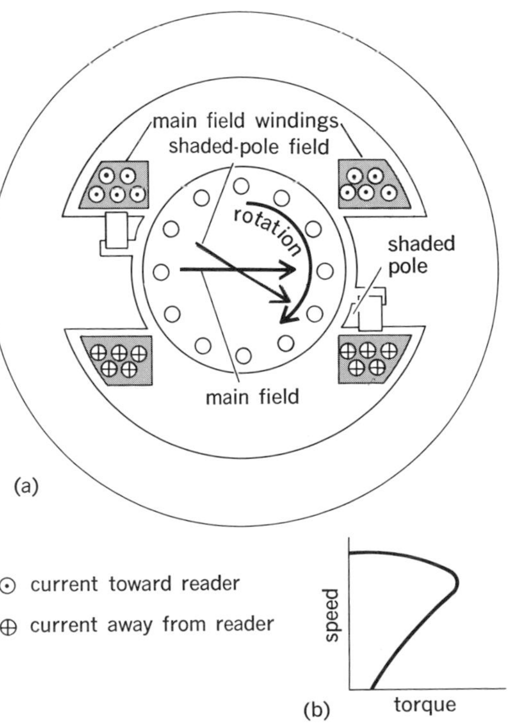

Fig. 8. Shaded-pole motor. (*a*) Cross-sectional view. (*b*) Typical characteristic.

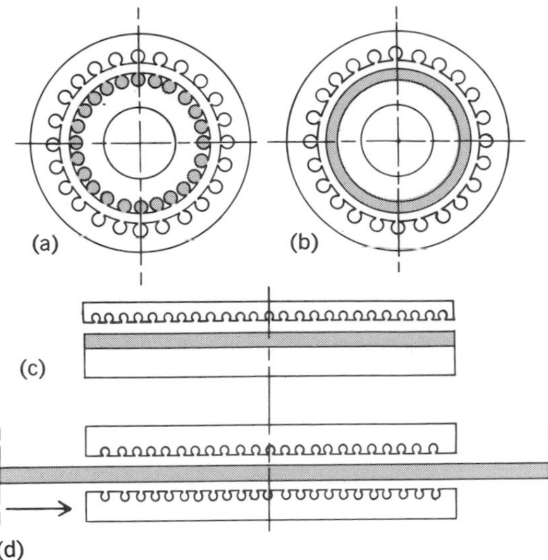

Fig. 9. Evolution of linear induction motor. (*a*) Polyphase squirrel-cage induction motor. (*b*) Conduction sheet motor. (*c*) One-sided long-secondary linear motor with short-flux return yoke. (*d*) Double-sided long-secondary linear motor.

by the primary. For the arrangement of Fig. 9*c*, there is a magnetic attraction between the iron of the stator and the iron of the secondary sheet. For some applications, this can be a serious disadvantage of the one-sided motor of Fig. 9*c*. This disadvantage can be eliminated by use of the double-sided arrangement of Fig. 9*d*. In this arrangement the primary iron of the upper and lower sides is held together rigidly, and the forces that are normal to the plane of the sheet do not act on the sheet.

In conventional transportation systems, traction effort is dependent on the contact of the wheels with the ground. In some cases, locomotives must be provided with heavy weights to keep the wheels from sliding when under heavy loads. This disadvantage can be eliminated by use of the linear motor. Through use of the linear motor with air cushions, friction loss can be reduced and skidding can be eliminated. Figure 10 illustrates the application of the double-sided linear motor for high-speed transportation.

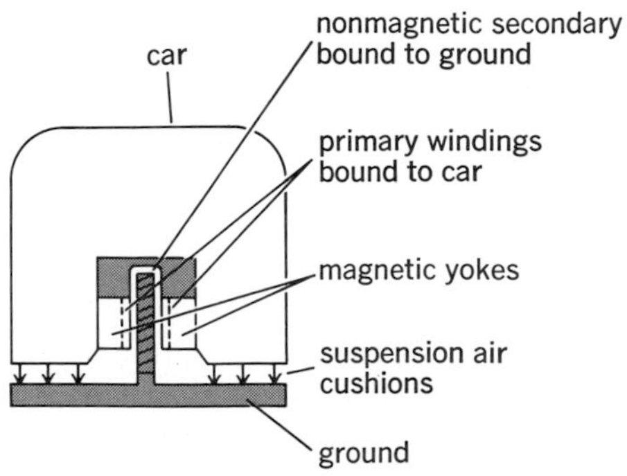

Fig. 10. Drawing of linear induction-motor configuration. Magnetic interaction supplies the forces between the car and the ground. (*Le Moteur Lineaire and Société de l'Aerotrain*)

Conveying systems that are operated in limited space have been driven with linear motors. Some of these, ranging from 1/2 to 1 mi (805–1609 m) in length, have worked successfully.

Because of the large effective air gap of the linear motor, its magnetizing current is larger than that of the conventional motor. Its efficiency is somewhat lower, and its cost is high. The linear motor is largely in the experimental stage.

[ALBERT G. CONRAD]

Bibliography: P. D. Agarwal and T. C. Wang, Evaluation of fixed and moving primary linear induction motor systems, *Proc. IEEE*, pp. 631–637, May 1973; P. L. Alger, *Induction Machines: Behavior and Uses*, 1970; J. H. Dannon, R. N. Day, and G. P. Kalman, A Linear-induction motor propulsion system for high-speed ground vehicles, *Proc. IEEE*, pp. 621–630, May 1973; E. R. Laithwaite, Linear electric machine: A personal view, *Proc. IEEE*, pp. 220–290, February 1975; M. G. Say, *Alternating Current Machines*, 4th ed., 1976; David Schieber, Principles of operation of linear induction devices, *Proc. IEEE*, pp. 647–656, May 1973.

Internal combustion engine

A prime mover, the fuel for which is burned within the engine, as contrasted to a steam engine, for example, in which fuel is burned in a separate furnace. *See* ENGINE.

The most numerous of internal combustion engines are the gasoline piston engines used in passenger automobiles, outboard engines for motor boats, small units for lawn mowers, and other such equipment, as well as diesel engines used in trucks, tractors, earth-moving, and similar equipment. This article describes the gasoline piston and diesel types of engines. For other types of internal combustion engines *see* GAS TURBINE; TURBINE PROPULSION.

The aircraft piston engine is fundamentally the same as the piston engine which is used in automobiles, but it is engineered for light weight and is usually air cooled. *See* RECIPROCATING AIRCRAFT ENGINE.

ENGINE TYPES

Characteristic features common to all commercially successful internal combustion engines include (1) the compression of air, (2) the raising of air temperature by the combustion of fuel in this air at its elevated pressure, (3) the extraction of work from the heated air by expansion to the initial pressure, and (4) exhaust. William Barnett first drew attention to the theoretical advantages of combustion under compression in 1838. In 1862 Beau de Rochas published a treatise that emphasized the value of combustion under pressure and a high ratio of expansion for fuel economy; he proposed the four-stroke engine cycle as a means of accomplishing these conditions in a piston engine (Fig. 1). The engine requires two revolutions of the crankshaft to complete one combustion cycle. The first engine to use this cycle successfully was built in 1876 by N. A. Otto. *See* OTTO CYCLE.

Two years later Sir Dougald Clerk developed the two-stroke engine cycle by which a similar combustion cycle required only one revolution of the crankshaft. In this cycle, exhaust ports in the cyl-

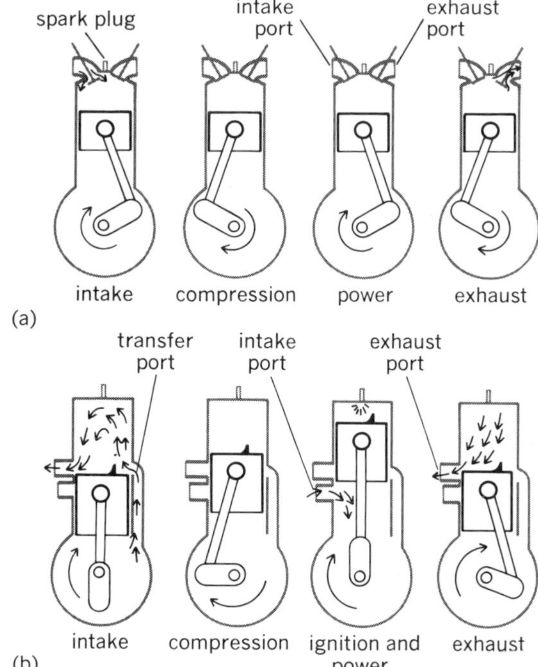

Fig. 1. (*a*) The four strokes of a modern four-stroke engine cycle. For intake stroke the intake valve (left) has opened and the piston is moving downward, drawing air and gasoline vapor into the cylinder. In compression stroke the intake valve has closed and the piston is moving upward, compressing the mixture. On power stroke the ignition system produces a spark that ignites the mixture. As it burns, high pressure is created, which pushes the piston downward. For exhaust stroke the exhaust valve (right) has opened and the piston is moving upward, forcing the burned gases from the cylinder. (*b*) The same action is accomplished without separate valves and in a single rotation of the crankshaft by a three-port two-cycle engine. (*From M. L. Smith and K. W. Stinson, Fuels and Combustion, McGraw-Hill, 1952*)

inder were uncovered by the piston as it approached the end of its power stroke. A second cylinder then pumped a charge of air to the working cylinder through a check valve when the pump pressure exceeded that in the working cylinder.

In 1891 Joseph Day simplified the two-stroke engine cycle by using the crankcase to pump the required air. The compression stroke of the working piston draws the fresh combustible charge through a check valve into the crankcase, and the next power stroke of the piston compresses this charge. The piston uncovers the exhaust ports near the end of the power stroke and slightly later uncovers intake ports opposite them to admit the compressed charge from the crankcase. A baffle is usually provided on the piston head of small engines to deflect the charge up one side of the cylinder to scavenge the remaining burned gases down the other side and out the exhaust ports with as little mixing as possible.

Engines using this two-stroke cycle today have been further simplified by use of a third cylinder port which dispenses with the crankcase check valve used by Day. Such engines are in wide use for small units where fuel economy is not as important as mechanical simplicity and light weight. They do not need mechanically operated valves and develop one combustion cycle per crankshaft revolution. Nevertheless they do not develop twice the power of four-stroke cycle engines with the same size working cylinders at the same number of revolutions per minute (rpm). The principal reasons for this are (1) the reduction in effective cylinder volume due to the piston movement required to cover the exhaust ports, (2) the appreciable mixing of burned (exhaust) gases with the combustible mixture, and (3) the loss of some combustible mixture through the exhaust ports with the exhaust gases.

Otto's engine, like almost all internal combustion engines developed at that period, burned coal gas mixed in combustible proportions with air prior to being drawn into the cylinder. The engine load was generally controlled by throttling the quantity of charge taken into the cylinder. Ignition was accomplished by a device such as an external flame or an electric spark so that the timing was controllable. These are essential features of what has become known as the Otto or spark-ignition combustion cycle.

Ideal and actual combustion. In the classical presentation of the four-stroke cycle, combustion is idealized as instantaneous and at constant volume. This simplifies thermodynamic analysis, but fortunately combustion takes time, for it is doubtful that an engine could run if the whole charge burned or detonated instantly.

Detonation of a small part of the charge in the cylinder, after most of the charge has burned progressively, causes the knock which limits the compression ratio of an engine with a given fuel. *See* COMBUSTION.

The gas pressure of an Otto combustion cycle using the four-stroke engine cycle varies with the piston position as shown by the typical indicator card in Fig. 2*a*. This is a conventional pressure-volume (*PV*) card for an 8.7:1 compression ratio. For simplicity in calculations of engine power, the average net pressure during the working stroke, called the mean effective pressure (mep), is frequently used. It may be obtained from the average net height of the card, which is found by measurement of the area with a planimeter and by division

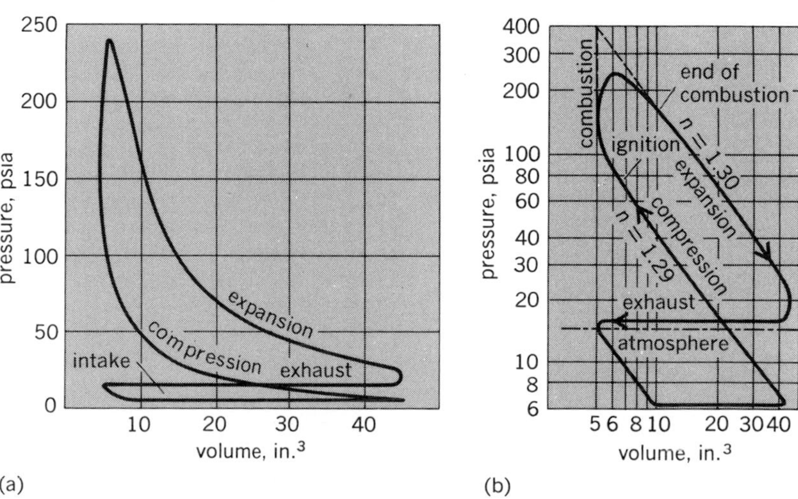

Fig. 2. Typical pressure-volume indicator card (*a*) plotted on rectangular coordinates and (*b*) plotted on logarithmic coordinates.

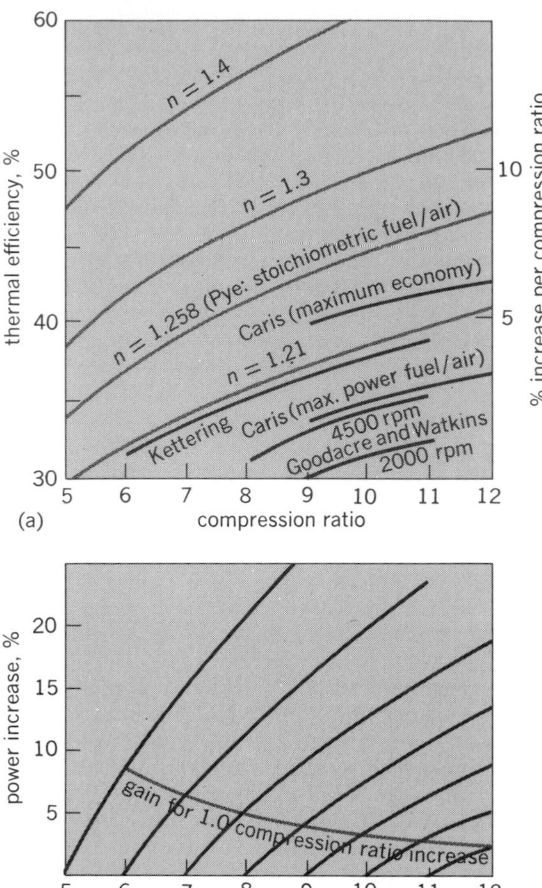

Fig. 3. (*a*) Effect of compression ratio on thermal efficiency as calculated with different values of n and compared with published experimental data. (*b*) Increase in power from raising compression ratio as calculated with $n = 1.3$. Percentage values are but little altered by calculating with different values of n.

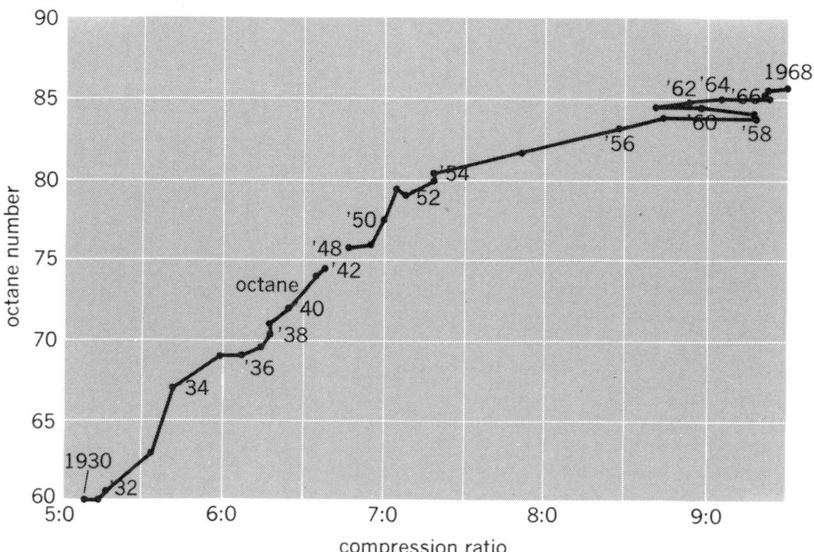

Fig. 4. Year-to-year relation between average compression ratio of cars and average octane number (ASTM research method) of regular-grade summer gasolines sold in the United States. (*Automotive Industries, March, 1954, 1969*)

of this area by its length. Similar pressure-volume data may be plotted on logarithmic coordinates as in Fig. 2*b*, which develops expansion and compression relations as approximately straight lines, the slopes of which show the values of exponent n to use in equations for pressure-volume relationships.

The rounding of the plots at peak pressure, with the peak developing after the piston has started its power stroke, even with the spark occurring before the piston reaches the end of the compression stroke, is due to the time required for combustion. The actual time required is more or less under control of the engine designer, because he can alter the design to vary the violence of the turbulence of the charge in the compression space prior to and during combustion. The greater the turbulence the faster the combustion and the lower the antiknock or octane value required of the fuel, or the higher the compression ratio that may be used with a given fuel without knocking. On the other hand, a designer is limited as to the amount he can raise the turbulence by the increased rate of pressure rise, which increases engine roughness. Roughness must not exceed a level acceptable for automobile or other service.

Compression ratio. According to classical thermodynamic theory, thermal efficiency η of the Otto combustion cycle is given by Eq. (1), where

$$\eta = 1 - \frac{1}{r^{n-1}} \qquad (1)$$

the compression ratio r_c and expansion ratio r_e are the same ($r_c = r_e = r$). When theory assumes atmospheric air in the cylinder for extreme simplicity, exponent n is 1.4. Efficiencies calculated on this basis are almost twice as high as measured efficiencies. Logarithmic diagrams from experimental data show that n is about 1.3. Even with this value, efficiencies achieved in practice are less than given by Eq. (1). This is not surprising considering the differences found in practice and assumed in theory, such as instantaneous combustion and 100% volumetric efficiency.

Attempts to adjust classical theory to practice by use of variable specific heats and consideration of dissociation of the burning gases at high temperatures have shown that this exponent should vary with the fuel-air mixture ratio, and to some extent with the compression ratio. G. A. Goodenough and J. B. Baker have shown that, for an 8:1 compression ratio, the exponent should vary from about 1.28 for a stoichiometric (chemically correct) mixture to about 1.31 for a lean mixture. Similar calculations by D. R. Pye showed that, at a compression ratio of 5:1, n should be 1.258 for the stoichiometric mixture, increasing with excess air (lean mixture) to about 1.3 for a 20% lean mixture and to 1.4 if extrapolated to 100% air. Actual practice gives thermal efficiencies still lower than these, which might well be expected because of the assumed instantaneous changes in cyclic pressure (during combustion and exhaust) and the disregard of heat losses to the cylinder walls. These theoretical relations between compression ratio and thermal efficiency, as well as some experimental results, are shown in Fig. 3*a*. The data published by C. F.

Kettering and D. F. Caris are about 85% and 82%, respectively, of the theoretical relations for the corresponding fuel-air mixtures.

Figure 3*b* gives the theoretical percentage gain in indicated thermal efficiency or power from raising the compression of an engine from a given value. They were plotted from Eq. (1) with $n = 1.3$, but would differ only slightly if obtained from any of the curves shown in Fig. 3*a*. The dotted line crossing these curves shows the diminishing gain obtainable by raising the compression ratio one unit at the higher compression ratios.

Experimental data indicate that a change in compression ratio does not appreciably change the mechanical efficiency or the volumetric efficiency of the engine. Therefore, any increase in thermal efficiency resulting from an increase in compression ratio will be revealed by a corresponding increase in torque or mep; this is frequently of more practical importance to the engine designer than the actual efficiency increase, which becomes an added bonus.

Compression ratio and octane rating. For years compression ratios of automobile engines have been as high as designers considered possible without danger of too much customer annoyance from detonation or knock with the gasoline on the market at the time (Fig. 4). Engine designers continue to raise the compression ratios of their engines as suitable gasolines come on the market.

Little theoretical study has been given to the effect of engine load on indicated thermal efficiency. Experimental evidence reveals that it varies little, if at all, with load, provided that the fuel-air-ratio remains constant and that the ignition time is suitably advanced at reduced loads to compensate for the slower rate of burning which results from dilution of the combustible charge with the larger percentages of burned gases remaining in the combustion space and the reduced turbulence at lower speeds.

Ignition timing. Designers obtain high thermal efficiency from high compression ratios at part loads, where engines normally run at automobile cruising speeds, with optimum spark advance, but avoid knock on available gasolines at wide-open throttle by use of a reduced or compromise spark advance. The tendency of an engine to knock at wide-open throttle is reduced appreciably when the spark timing is reduced 5–10° from optimum, as is shown in Fig. 5. Advancing or retarding the spark timing from optimum results in an increasing loss in mep for any normal engine as shown by the solid curve. The octane requirement falls rapidly as the spark timing is retarded, the actual rate depending on the nature of the gasoline as well as on the design of the combustion chamber. The broken-line curves A and B show the effects on a given gasoline of the use of moderate- and high-turbulence combustion chambers, respectively, with the same compression ratio. Because the mep curve is relatively flat near optimum spark advance, retarding the spark for a 1–2% loss is considered normally acceptable in practice because of the appreciable reduction in octane requirement.

In Fig. 6 similar data are shown by curve A for another engine with changes in mep plotted on a percentage basis against octane requirement as

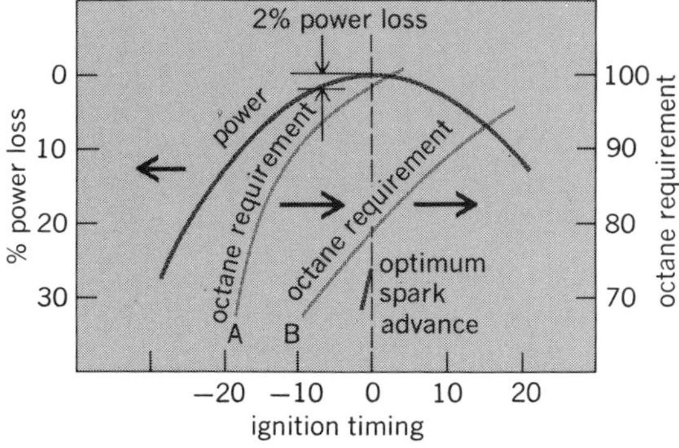

Fig. 5. Effects of advancing or retarding ignition timing from optimum on engine power and resulting octane requirement of fuel in an experimental engine with a combustion chamber having typical turbulence (A) and a highly turbulent design (B) with the same compression ratio. Retarding the spark 7° for a 2% power loss reduced octane requirement from 98 to 93 for design A.

the spark timing was charged. Point a indicates optimum spark timing, where 85 octane was required of the gasoline to avoid knock. By raising the compression ratio, the power and octane requirement were also raised as shown by the broken-line curve B. Although optimum spark required 95 octane (point b), retarding the spark timing and thus reducing the octane requirement to 86 (point c) developed slightly more power than with the original compression ratio at its optimum spark advance. The gain may be negligible at wide-open throttle, but at lower loads where knock does not develop the spark timing may be advanced to optimum (point b), where appreciably more power may be developed by the same amount of fuel.

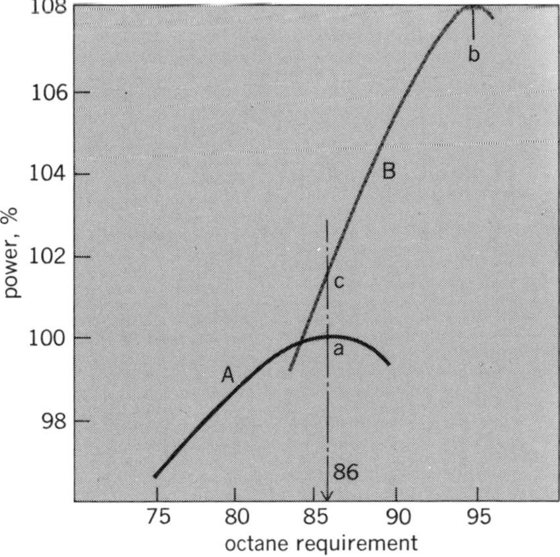

Fig. 6. Effect of raising compression ratio of an experimental engine on the power output and octane requirement at wide-open throttle. While an 86-octane fuel was required for optimum spark advance (maximum power) with the original compression ratio, the same gasoline would be knock-free at the higher compression ratio by suitably retarding the ignition timing.

In addition to the advantages of the higher compression ratio at cruising loads with optimum spark advance, the compromise spark at full load may be advanced toward optimum as higher-octane fuels become available, and a corresponding increase in full-throttle mep enjoyed. Such compromise spark timings have had much to do with the adoption of compression ratios of 10:1 to 13:1.

Fuel-air ratio. A similar line of reasoning shows that a fuel-air mixture richer than that which develops maximum knock-free mep will permit use of higher compression ratios. However, the benefits derived from compromise or superrich mixtures vary so much with mixture temperature and the sensitivity of the octane value of the particular fuel to temperature that it is not generally practical to make much general use of this method. Nevertheless it has been the practice with piston-type aircraft engines to use fuel-air mixture ratios of 0.11 or even higher during takeoff, instead of about 0.08, which normally develops maximum mep in the absence of knock.

Compression-ignition engines. About 20 years after Otto first ran his engine, Rudolf Diesel successfully demonstrated an entirely different method of igniting fuel. Air is compressed to a pressure high enough for the adiabatic temperature to reach or exceed the ignition temperature of the fuel. Because this temperature is in the order of 1000°F (540°C), compression ratios of 12:1 to 20:1 are used commercially with compression pressures generally over 600 psi (4.2 MPa). This type of engine cycle requires the fuel to be injected after compression at a time and rate suitable to control the rate of combustion.

Conditions for high efficiency. The classical presentation of the diesel engine cycle assumes combustion at constant pressure. Like the Otto cycle, thermal efficiency increases with compression ratio, but in addition it varies with the amount of heat added (at the constant pressure) up to the cutoff point where the pressure begins to drop from adiabatic expansion. *See* DIESEL CYCLE; DIESEL ENGINE.

Practical attainments. Diesel engines were highly developed in Germany prior to World War I, and made an impressive performance in submarines. Large experimental single-cylinder engines were built in several European countries with cylinder diameters up to 1 m. As an example, the two-stroke Sulzer S100 single-acting engine with a bore of 1 m and a stroke of 1.1 m developed 2050 gross horsepower (1530 kW) at 150 rpm. Multiple cylinder engines developing 15,000 hp (11.2 MW) are in marine service. Small diesel engines are in wide use also.

Fuel injection. In early diesel engines, air injection of the fuel was used to develop extremely fine atomization and good distribution of the spray. However, the need for injection air at pressures in the order of 1500 psi (10 MPa) necessitated the use of expensive and bulky multistage air compressors and intercoolers.

A simpler fuel-injection method was introduced by James McKechnie in 1910. He atomized the fuel as it entered the cylinder by use of high fuel pressure and suitable spray nozzles. After considerable development it became possible to atomize the fuel sufficiently to minimize the smoky exhaust which had been characteristic of the early solid-injection engines. By 1930 solid or airless injection had become the generally accepted method of injecting fuel in diesel engines.

Contrast between diesel and Otto engines. There are many characteristics of the diesel engine which are in direct contrast to those of the Otto engine. The higher the compression ratio of a diesel engine, the less the difficulties with ignition time lag. Too great an ignition lag results in a sudden and undesired pressure rise which causes an audible knock. In contrast to an Otto engine, knock in a diesel engine can be reduced by use of a fuel of higher cetane number, which is equivalent to a lower octane number. *See* OCTANE NUMBER.

The larger the cylinder diameter of a diesel engine, the simpler the development of good combustion. In contrast, the smaller the cylinder diameter of the Otto engine, the less the limitation from detonation of the fuel.

High intake-air temperature and density materially aid combustion in a diesel engine, especially of fuels having low volatility and high viscosity. Some engines have not performed properly on heavy fuel until provided with a super charger. The added compression of the supercharger raised the temperature and, what is more important, the density of the combustion air. For an Otto engine, an increase in either the air temperature or air density increases the tendency of the engine to knock and therefore reduces the allowable compression ratio.

Diesel engines develop increasingly higher indicated thermal efficiency at reduced loads because of leaner fuel-air ratios and earlier cutoff. Such mixture ratios may be leaner than will ignite in an Otto engine. Furthermore, the reduetion of load in an Otto engine requires throttling of the engine, which develops increasing pumping losses in the intake system.

TRENDS IN AUTOMOBILE ENGINES

Cylinder diameters of average American automobile engines prior to 1910 were over $4\frac{1}{4}$ in (108 mm). By 1917 they had been reduced to only a little over $3\frac{1}{4}$ in. (83 mm), where they stabilized until after 1945. Since then the increased demand for more power, with the number of cylinders limited to eight for practical mechanical reasons, the diameters have been increased from year to year until they averaged 3.98 in. (101 mm) in 1969, with a maximum of 4.36 in. (111 mm).

Stroke-bore ratio. Experimental engines differing only in stroke-bore ratio show that this ratio has no appreciable effect on fuel economy and friction at corresponding piston speeds. Practical advantages which result from the short stroke include (1) the greater rigidity of crankshaft from the shorter crank cheeks, with crankpins sometimes overlapping main bearings, and (2) the narrower as well as lighter cylinder block which is possible. On the other hand, the higher rates of crankshaft rotation for an equivalent piston speed necessitate greater valve forces and require stronger valve springs. Also the smaller depth of the compression space for a given compression ratio increases the surface-to-volume ratio and the proportion of heat lost by radiation during combustion. Nevertheless, stroke-bore ratios have been decreasing for more than 25 years and in 1969 reached 0.9 for the average automobile in the United States.

Cylinder number and arrangement. Engine power may be raised by increasing the number of cylinders as well as the power per cylinder. The minimum number of cylinders has generally been four for four-cycle automobile engines, because this is the smallest number that provides a reasonable balance for the reciprocating pistons. Many early cars had four-cylinder engines. After 1912 six-cylinder in-line engines became popular. They have superior balance of reciprocating forces and more even torque impulses. By 1940 the eight-cylinder 90° V engine had risen in popularity until it about equaled the six-in-line. After 1954, the V-8 dominated the field for American automobile engines. There are several important reasons for this besides the increased power. For example, the V-8 offers appreciably more rigid construction with less bearing deflection at high speeds, provides more uniform distribution of fuel to all cylinders from centrally located downdraft carburetors, and has a short, low engine that fits within the hood demanded by style trends. With the introduction of the smaller "compact" cars in 1959, where the power and cost of eight-cylinder V-type engines were not required, six-cylinder designs increased. By 1969 about 38% of all engine designs were of the six-cylinder in-line type. The evolution of cylinder arrangements included for a short period the V-12 and even a V-16 cylinder design, but experience showed that in their day there was too much practical difficulty in providing good manifold distribution of fuel, especially when starting cold, and too much difficulty in keeping all spark plugs firing.

Compression ratio. The considerable increase in power of the average automobile engine over the years shown in Fig. 7 together with the compression ratios which have had much to do with the increased mep. Such ratios approach practical limits imposed by phenomena other than detonation, such as preignition, rumble, and other evidences of undesirable combustion.

The modern trend toward high compression ratios, with their small compression volumes, has dictated the universal use of overhead valves in all American engine designs. High compression ratios also tend to restrict cylinder diameters because the longer flame travel increases the tendency to knock (Fig. 8).

Improved breathing and exhaust. Added power output has been brought about by reducing the pressure drop in the intake system at high speeds and by reducing the back pressure of the exhaust systems (Fig. 9). These results were accomplished by larger valve areas and valve ports, by larger venturi areas, and by more streamlined manifolds. Larger valve areas were achieved by higher lift of the valves and by larger valves. Larger venturi areas in the carburetors were achieved by use of one or more two-stage carburetors; in these, sufficient air velocity was developed to meter the fuel on one venturi at low power; the second venturi was opened for high power. Better streamlining of the manifold passages between carburetors and valves, especially at the cylinder ports, and the use of dual exhausts and mufflers with reduced back pressure have also improved engine breathing and exhaust.

Valve timing. The times of opening and closing of the valves of an engine in relation to the piston position are usually selected to develop maximum power over a desired speed range at wide-open throttle. For convenience the timing of these

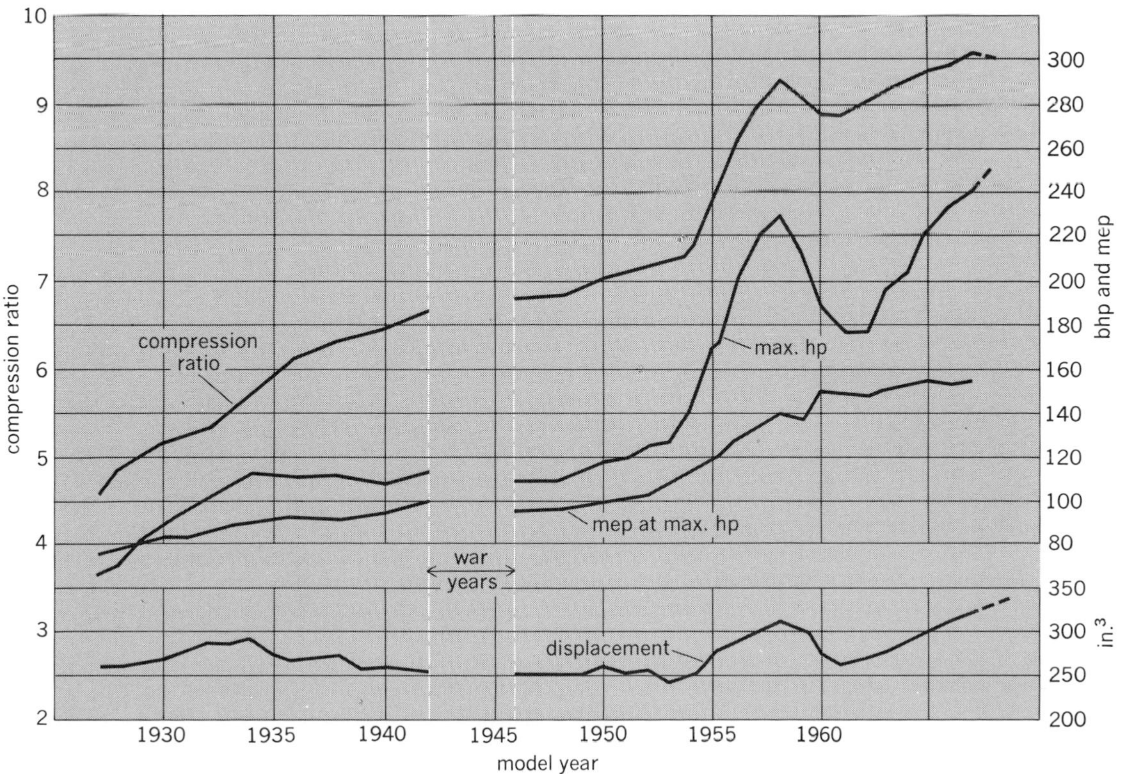

Fig. 7. Trend toward considerable increases in compression ratio, mean effective pressure (mep), and power of average United States automobile engines weighted for production volume. (1 in³ = 16.4 cm³; 1 hp = 0.7457 kW; 1 psi = 6.895 kPa). *(Ethyl Corp. data)*

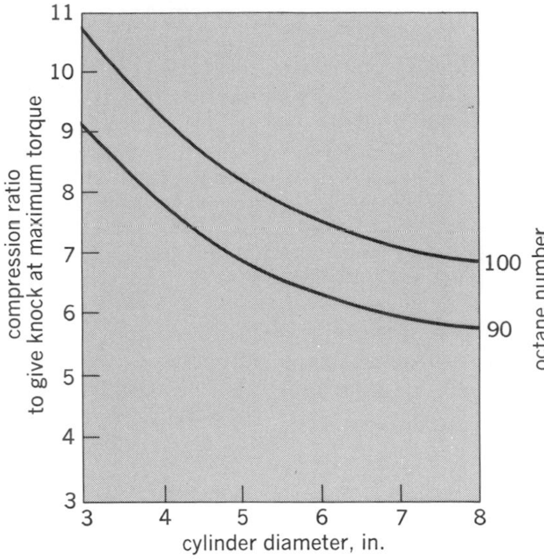

Fig. 8. Relation between cylinder diameter and limiting compression ratio for engines of similar design, one using 90- and the other 100-octane gasoline. 1 in. = 25.4 mm. (*From L. L. Brower, unpublished SAE paper, 243, 1950*)

events is expressed as the number of degrees of crankshaft rotation before or after the piston reaches the end of one of its strokes. Because of the time required for the flow of the burned gas through the exhaust valve at the end of the power or expansion stroke of a piston, it is customary to start opening the valve considerably before the end of the stroke. If the valve should be opened when the piston is nearer the lower end of its stroke, power would be lost at high engine speeds because the piston on its return (exhaust) stroke would have to move against gas pressure remaining in the cylinder. On the other hand, if the valve were

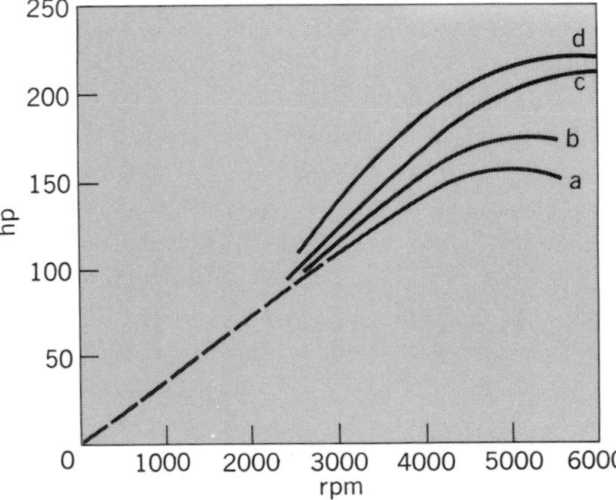

Fig. 9. Increases in peak power of a six-cylinder engine (210 in.^{3} displacement) by successive reductions in resistance to airflow through it. Curve a, power developed by original engine with two $1\frac{3}{4}$ in. carburetors; curve b, higher valve-lift valves and dual exhaust systems; curve c, larger valves, smoother valve ports, and two 2-in. carburetors; curve d, three double-barrel carburetors. 1 hp = 0.7457 kW; 1 in. = 25.4 mm; 1 in.^{3} = 16.4 cm^3. (*Adapted from W. M. Heynes, The Jaguar engine, Inst. Mech. Eng., Automot. Div. Trans., 1952–1953*)

opened before necessary, the burned gas would be released while it is still at sufficient pressure to increase the work done on the piston. Thus for any engine there is a time for opening the exhaust valve which will develop the maximum power at some particular speed. Moreover, the power loss at other speeds does not increase rapidly. It is obvious that, when an engine is throttled at part load, there will be less gas to discharge through the exhaust valve and there will be less need for it to be opened as early as at wide-open throttle.

The timing of intake valve events is normally selected to trap the largest possible quantity of air or combustible mixture in the cylinder when the valve closes at some desired engine speed and at wide-open throttle. The intermittent nature of the flow through the valve subjects it to alternate accelerations and retardation which require time. During the suction stroke, the mass of air moving through the pipe leading to the intake valve is given velocity energy which may be converted to a little pressure at the valve when the air mass still in the pipe is stopped by its closure. Advantage of this phenomenon may be obtained at some engine speed to increase the air mass which enters the cylinder. The engine speed at which the maximum volumetric efficiency is developed varies with the relative valve area, closure time, and other factors, including the diameter and particularly the length of this pipe (Fig. 10). These curves reveal the characteristic falling off at high speeds from the inevitable throttling action as air flows at increased velocities through any restriction such as a valve or intake pipe or particularly the venturi of a carburetor.

The curves shown in Fig. 10 are smoothed averages drawn through data obtained from experiments, but if they had been drawn through data taken at many more speeds, they would have revealed a wavy nature (Fig. 11) due to resonant oscillations in the air column entering the cylinder; these oscillations are somewhat similar to those which develop in an organ pipe.

Volumetric efficiency has a direct effect on the mep developed in a cylinder, on the torque, and on the power that may be realized at a given speed. Since power is a product of speed and torque, the peak power of an engine occurs at a higher speed than for maximum torque, where the rate of torque loss with any further increase in speed will exceed the rate of speed increase. This may be seen in Fig. 12, showing the torque and power curves for a typical six-cylinder engine. This engine developed its maximum power at a speed about twice that for maximum torque. The average 1969 V-8 engine developed its maximum power at a speed about 60% higher than its maximum torque.

It has been shown that the engine speeds at which maximum torque and power are desired in practice require closure of the intake valve to be delayed until the piston has traveled almost half the length of the compression stroke. At engine speeds below those where maximum torque is developed by this valve timing, some of the combustible charge which has been drawn into the cylinder on the suction stroke will be driven back through the valve before it closes. This reduces the effective compression ratio at wide-open throttle—thus the increasing tendency of an engine to develop a fuel knock as the speed and the resulting gas turbulence are reduced.

Another result of this blowback into the intake pipe is the possible reversal of the flow of some of the fuel-air mixture through the carburetor, which will draw fuel each time it passes through the metering system, thereby producing a much richer mixture than if it had passed through the same carburetor only once. The increased fuel supplied to the air reaching the intake valve because of this reversed air flow may be over 100% greater for a single cylinder engine at wide-open throttle than if there were no reversal of flow. Throttling the engine reduces and almost eliminates the blowback through the metering system and the ratio of fuel to air approaches that which would be expected from air flow in one direction only. Manifolding more than one cylinder to the metering system of one carburetor also reduces the blowback through a carburetor because it averages the blowback from individual cylinders. When six cylinders are supplied by a single carburetor, there may be practically no blowback through the carburetor or enrichment of the combustible mixture by this phenomenon. However, when three carburetors are installed on the same six-cylinder engine, with two cylinders supplied by each, the enrichment may be 80–90%. When four cylinders are supplied by a single metering system, as with most V-type eight-cylinder engines, the enrichment may be 10–50% and varies with many factors besides the closing time of the intake valve, such as the exhaust valve events and exhaust back pressure, so that carburetor settings and compensation must usually be made on the engine which it is to supply, and preferably as installed in the car, if optimum power and economy are to be realized. Unfortunately, such settings can not be predicted by simple calculations of fuel flow through an orifice into an airstream flowing at constant velocity.

Intake manifolds. Intake manifolds for multicylinder engines should meet several requirements for the satisfactory performance of spark-ignition engines. They should (1) distribute fuel equally to all cylinders at temperatures where unvaporized fuel is present, as when starting a cold engine or during the warm-up period; (2) supply sufficient heat to vaporize the liquid fuel from the carburetor as soon after starting as possible; (3) distribute the vaporized fuel-air mixture evenly to all cylinders during normal operation and at low speeds; (4) offer minimum restriction to the mixture flow at high power; and (5) provide equal ram or dynamic boost to volumetric efficiency of all cylinders at some desired part of the engine speed range. This requires that each branch from the carburetor to the valve port should be equal in length, as may be inferred from Fig. 10. Accordingly, no cylinder port should be siamesed with another at the end of a leg of the manifold.

For the warming-up period with liquid fuel present, rectangular sections are desirable to impede spiraling of liquid fuel along the walls, and right-angle bends should be sharp, at least at their inner corner, so as to throw the liquid flowing along the inner wall back into the air stream, and there should be an equal number in each branch.

Manifold heat. Intake manifolds of most American automobile engines are heated to the temperature required to vaporize the fuel from the carburetor (120–140°F or 49–60°C) by exhaust gas passing through a suitable passage in the manifold casting, particularly at the first T beyond the carburetor where the liquid fuel impinges before turning to side branches.

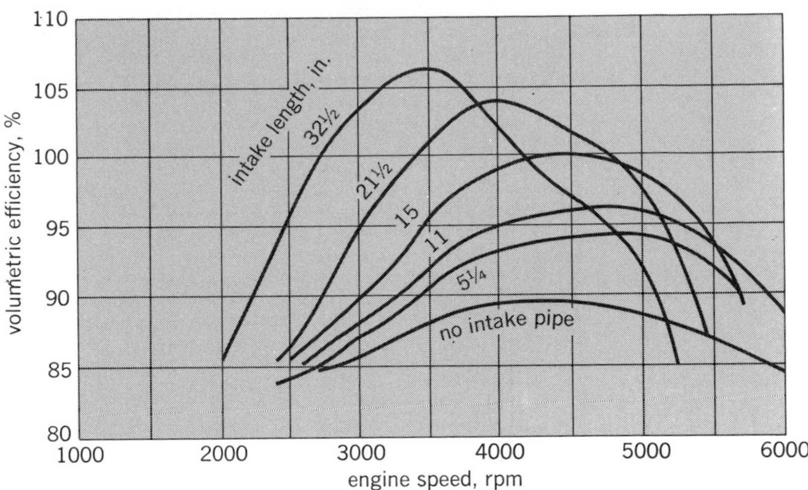

Fig. 10. Effect of intake-pipe length and engine speed on volumetric efficiency of one cylinder of a six-cylinder engine. 1 in. = 25.4 mm. (*From E. W. Downing, Proc. Automot. Div., Inst. Mech. Eng., no. 6, 0. 170, 1957–1958*)

To speed the warm-up process, thermostatically operated valves are generally placed in the engine exhaust system so as to force most of the exhaust gases through the intake manifold heater passages when the engine is cold. After the intake manifold has reached the desired temperature, such valves are intended to open and permit only the necessary small portion of exhaust gases to continue passing through the heater. This is an important feature, for too much heat causes a loss of engine power and aggravates the tendency for the engine toward knock and vapor lock.

On some engines, the intake manifolds are heated by water jackets taking hot water from the engine cooling system. This gives uniform heating over a wide range of operating conditions without danger of the overheating that might result from exhaust gas heat if the thermostatic exhaust valve should fail to open. It has the disadvantage, however, of requiring more time to reach normal manifold temperature, even though the water supply from the cylinder heads is short-circuited through the manifold jacket by a suitable water thermostat during warm-up.

One of the advantages of the V-8 engine is the excellent intake manifold design permitted by the centrally located carburetor with but small differences in the lengths of the passage between the carburetor and each cylinder, and an equal number of right-angle bends in each, as in a typical intake manifold using a dual carburetor (Fig. 13). With the usual firing order shown in Fig. 20, the firing intervals for each of the lower branches (shown dotted) are evenly spaced 360 crankshaft degrees apart, but for each of the upper branches two cylinders fire 180° and then 540° apart.

Icing. Because gasoline has considerable latent heat of vaporization, it lowers the air temperature

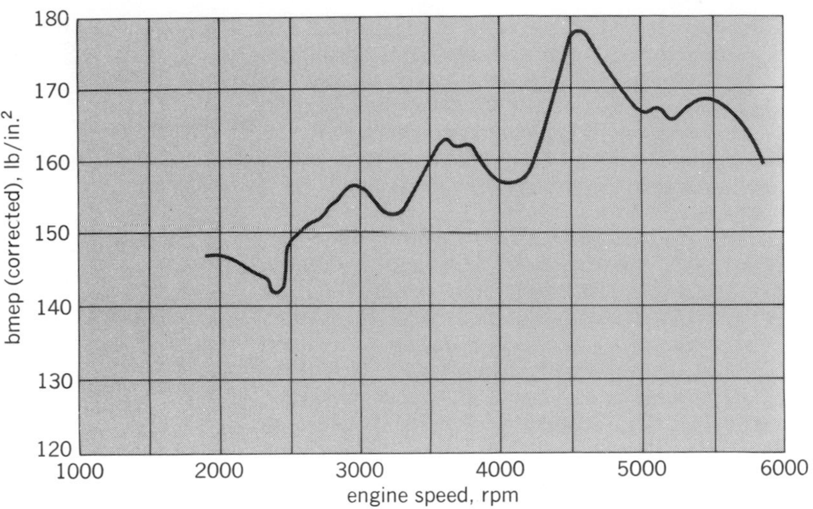

Fig. 11. Evidence of resonant oscillations in the intake pipe to a single cylinder shown by readings taken at small speed increments. 1 lb/in.² = 6.895 kPa. (*From E. W. Downing, Proc. Automot. Div., Inst. Mech. Eng.*, no. 6, p. 170, 1957–1958)

as it evaporates. This is true even at the low temperatures, where only a small part of it is vaporized. It is therefore possible for moisture which may be carried by the air to freeze under certain conditions. Ice is most likely to form when the atmosphere is almost saturated with moisture at temperatures slightly above freezing and up to about 40°F (4°C). When ice forms around or near the throttle, it can seriously interfere with the operation of an engine. For this reason small passages have been provided on some engines for jacket water, or exhaust gas from the heating supply for the intake manifold, to warm at least the flange of the carburetor. Here, again, too much heat would produce vapor lock and this would interfere with normal fuel metering. This is one of the reasons for designing some carburetors with separate casting for the throttle bodies which are heated by the manifold through only a thin gasket, while a thick gasket acting as a heat barrier is inserted between it and the float chamber containing the fuel metering systems.

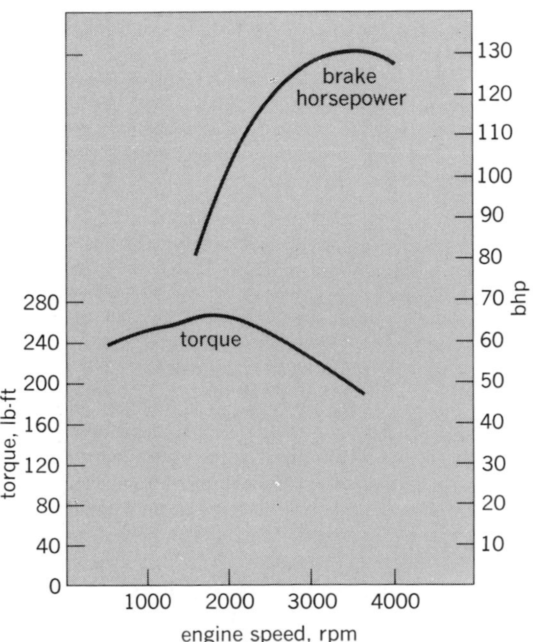

Fig. 12. Typical relation between engine speeds for maximum torque and maximum horsepower. 1 lb-ft = 1.356 N · m; 1 hp = 0.7457 kW.

FUEL CONSUMPTION AND SUPERCHARGING

Fuel consumption at loads throughout the operating range of an engine provide insight into such characteristics as friction loss within the engine. Volumetric efficiency of an engine can be increased by use of supercharging.

Part-load fuel economy. When the fuel consumption of a spark-ignition engine is plotted against brake horsepower, straight lines may generally be drawn through the test points at given speeds, as shown in Fig. 14, provided that the tests are run with optimum spark advances and at constant fuel-air ratios. Such lines are similar to the Willans lines long used for the steam consumption of steam engines.

For practical purposes the lines at various speeds may be considered parallel over a wide range of speeds. The assumption that the negative power indicated by extrapolating these lines to zero fuel consumption reveals the power absorbed by internal friction of the engine would be justified only when the thermal efficiency remains constant over the load range. On these coordinates, lines radiating from the origin represent constant ratios of fuel consumed to power developed and therefore constant specific fuel consumption (sfc). Several such lines are indicated in Fig. 14, from which the sfc at various loads may be read directly where they cross the performance lines at the various speeds.

Similar plots of even greater utility may be drawn on an indicated horsepower basis, as has been done in Fig. 15 for the same data. For many engines, a single performance line may be drawn through all test points at a given fuel-air ratio over a considerable range of speeds. When extrapolated, the performance line passes through the origin as it does for the engine shown in Fig. 15; the indicated sfc and thermal efficiency of the engine remain constant over the load range covered. Frequently a performance line for an engine passes a little to the left of the origin because of conditions causing a decrease in thermal efficiency as the load is reduced, such as insufficient turbulence or too low a manifold velocity. For a more complete picture of the fuel consumption performance of an engine, similar plots may be made on an mep basis. When this is done and both fuel consumption and horsepower are divided by the engine factor which converts horsepower to mep, the slope and nature of the fuel performance line remain unchanged. The fuel consumption scale then becomes equivalent to the product of mep and sfc. Such plots may be on the basis of either indicated mep (imep) or brake mep (bmep). Figure 16 shows the same data as Figs. 14 and 15 plotted on an imep basis for two different fuel-air ratios.

The fuel consumption performance of diesel engines at part loads may be shown on similar bases, but the plots should not be expected to be straight because the effective fuel-air ratio varies with load. This is illustrated in Fig. 17. It is charac-

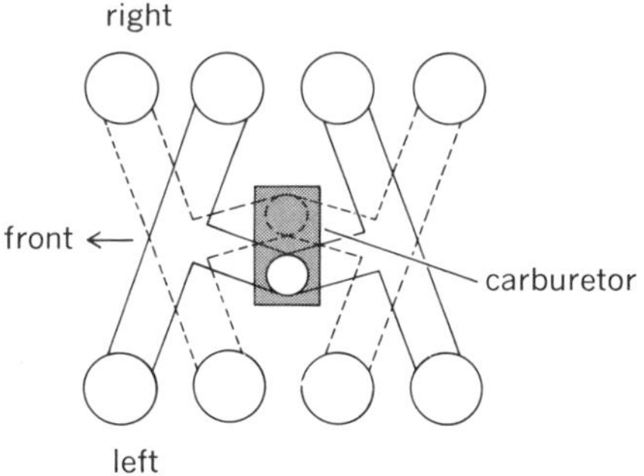

Fig. 13. Schematic of typical intake manifold with dual carburetor on a V-8 engine.

teristic of most diesel engines that the curvature of the plot generally flattens out at low loads so that it becomes tempting to extrapolate it to zero fuel consumption, and to consider the negative power intercept as friction. Such an intercept would represent friction only if the efficiency did not change with load, as for the engine characteristics shown in Fig. 14. If the thermal efficiency of a diesel engine improves as the load is reduced, as it would in theory for the classical diesel cycle, the zero fuel intercept for a curve such as shown in Fig. 17 would be to the right of the negative power representing engine friction. Although these fuel consumption performance plots are of considerable utility for recording such data for an engine, the fact that they are curved requires at least three or even more points to fix their location on the plot.

Supercharging spark-ignition engines. Volumetric efficiency and thus the mep of a four-stroke spark-ignition engine may be increased over a part of or the whole speed range by supplying air to the engine intake at higher than atmospheric pressure. This is usually accomplished by a centrifugal or rotary pump. The indicated power of an engine increases directly with the absolute pressure in the intake manifold. Because fuel consumption increases at the same rate, the indicated sfc is generally not altered appreciably by supercharging.

The three principal reasons for supercharging four-cycle spark-ignition engines are (1) to lessen the tapering off of mep at higher engine speed; (2) to prevent loss of power due to diminished atmospheric density, as when an airplane (with piston engines) climbs to high altitudes; and (3) to develop more torque at all speeds.

In a normal engine characteristic, torque rises as speed increases but falls off at higher speeds because of the throttling effects of such parts of the fuel intake system as valves and carburetors. If a supercharger is installed so as to maintain the volumetric efficiency at the higher speeds without increasing it in the middle-speed range, peak horsepower can be increased.

The rapid fall of atmospheric pressure at increased altitudes causes a corresponding decrease in the power of unsupercharged piston-type aircraft engines. For example, at 20,000 ft (6 km) the air density, and thus the absolute manifold pressure and indicated torque of an aircraft engine, would be only about half as great as at sea level. The useful power developed would be still less because of the friction and other mechanical power losses which are not affected appreciably by volumetric efficiency. By the use of superchargers, which are usually of the centrifugal type, sea-level air density may be maintained in the intake manifold up to considerable altitudes. Some aircraft engines drive these superchargers through gearing which may be changed in flight, from about 6.5 to 8.5 times engine speed. The speed change avoids oversupercharging at medium altitudes with corresponding power loss. Supercharged aircraft engines must be throttled at sea level in order to avoid any damage from detonation or excessive overheating caused by the high mep which would otherwise be developed.

Normally an engine is designed with the highest compression ratio allowable without knock from the fuel expected to be used. This is desirable for the highest attainable mep and fuel economy from an atmospheric air supply. Any increase in the volumetric efficiency of such an engine would cause it to knock unless a fuel of higher octane

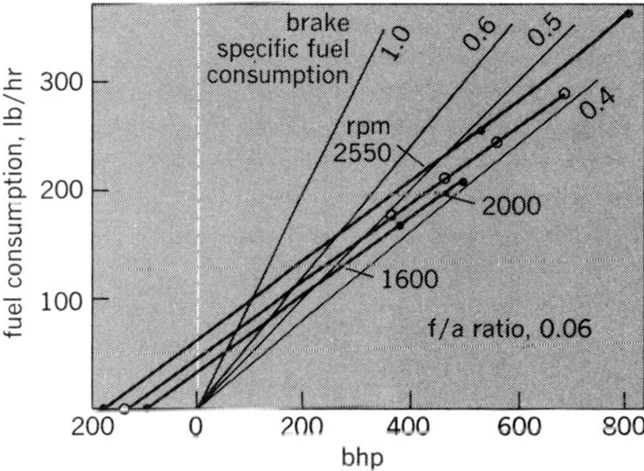

Fig. 14. Fuel consumption of an engine at part loads, plotted as typical Willans lines against brake horsepower. Data were taken with optimum spark advance and with fuel-air ratio adjusted at each test point. 1 hp = 0.7457 kW; 1 lb/hr = 0.454 kg/hr.

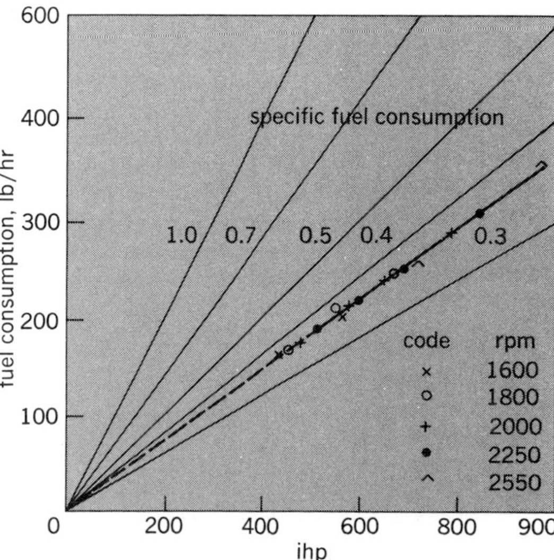

Fig. 15. The fuel consumption data of Fig. 14 plotted against the indicated horsepower, by which speed differences are neutralized. 1 hp = 0.7457 kW; 1 lb/hr = 0.454 kg/hr.

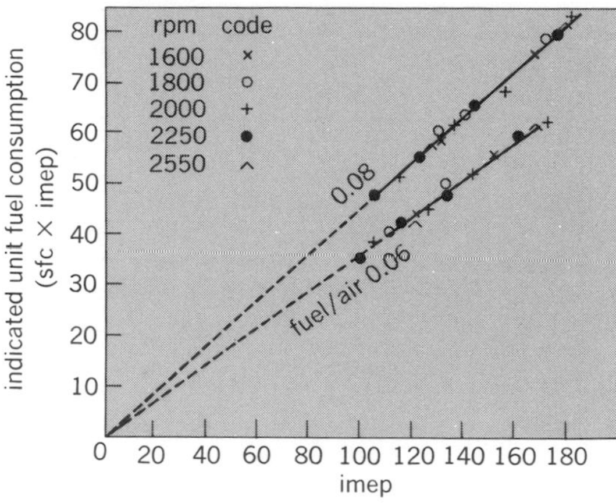

Fig. 16. The fuel consumption data of Fig. 15 plotted on the basis of imep by dividing the fuel scale and the power scale by the same factor (k= ihp/imep). Line for 0.08 fuel-air ratio has been added to show effect of fuel-air ratio on slope of such plots.

number were used or the compression ratio were lowered. When the compression ratio is lowered, the knock-limited mep may be raised appreciably by supercharging but at the expense of lowered thermal efficiency. There are engine uses where power is more important than fuel economy, and supercharging becomes a solution. The principle involved is illustrated in Fig. 18 for a given engine. With no supercharge this engine, when using 93-octane fuel, developed an imep of 180 psi at the border line of knock at 8:1 compression ratio. If the compression ratio were lowered to 7:1, the mep could be raised by supercharging along the 7:1 curve to 275 imep before it would be knock-limited by the same fuel. With a 5:1 compression ratio it could be raised to 435 imep. Thus the imep could be raised until the cylinder became thermally limited by the temperatures of critical parts, particularly of the piston head.

Supercharged diesel engines. Combustion in a four-stroke diesel engine is materially improved by supercharging. In fact, fuels which would smoke badly and misfire at low loads will burn otherwise satisfactorily with supercharging.

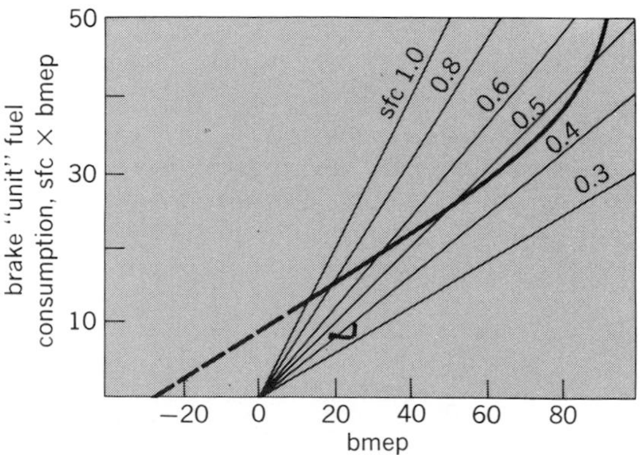

Fig. 17. Fuel consumption of a diesel engine at part loads showing the curvature, typical of such engines on these coordinates, caused by changing effective fuel-air ratios as the loads are increased.

The imep rises directly with the supercharge pressure, until it is limited by the rate of heat flow from the metal parts surrounding the combustion chamber, and the resulting temperatures. A practical application of this limitation was made on a locomotive built by British Railways where the powers, and thus the heats developed, were held reasonably constant over a considerable speed range by driving the supercharger at constant speed by its own engine. In this way the supercharge pressure varied inversely with the speed of the main engine. The corresponding torque rise at reduced speed dispensed with much gear-shifting which would have been required during acceleration with a conventional engine.

When superchargers of either the centrifugal or positive-displacement type are driven mechanically by the engine, the power required becomes an additional loss to the engine output. Experience shows that there is a degree of supercharge for any engine which develops maximum efficiency; too high a supercharge absorbs more power in the supercharger than is gained by the engine, especially at low loads. Another means of driving the supercharger which is becoming quite general is by an exhaust turbine, which recovers some of the energy that would otherwise be wasted in the engine exhaust. This may be accomplished with so small an increase of back pressure that little power is lost by the engine. This type of drive results in an appreciable increase in efficiency at loads high enough to develop the necessary exhaust pressure.

Supercharging a two-cycle diesel engine requires some means of restricting or throttling the exhaust in order to build up cylinder pressure at the start of the compression stroke, and is used on a few large engines. Most medium and large two-cycle diesel engines are usually equipped with blowers to scavenge the cylinders after the working stroke and to supply the air required for the subsequent cycles. These blowers, in contrast to superchargers, do not build up appreciable pressure in the cylinder at the start of compression. If the capacity of such a blower is greater than the engine displacement, it will scavenge the cylinder of practically all exhaust products, even to the extent of blowing some air out through the exhaust ports. Such blowers, like superchargers, may be driven by the engine or by exhaust turbines.

Engine balance. Rotating masses such as crank pins and the lower half of a connecting rod may be counterbalanced by weights attached to the crankshaft. The vibration which would result from the reciprocating forces of the pistons and their associated masses is usually minimized or eliminated by the arrangement of cylinders in a multicylinder engine so that the reciprocating forces in one cylinder are neutralized by those in another. Where these forces are in different planes, a corresponding pair of cylinders is required to counteract the resulting rocking couple.

If piston motion were truly harmonic, which would require a connecting rod of infinite length, the reciprocating inertia force at each end of the stroke would be as in Eq. (2), where W is the total

$$F = 0.000456WN^2s \qquad (2)$$

weight of the reciprocating parts in one cylinder, N is the rpm, and s is the stroke in inches. Both F

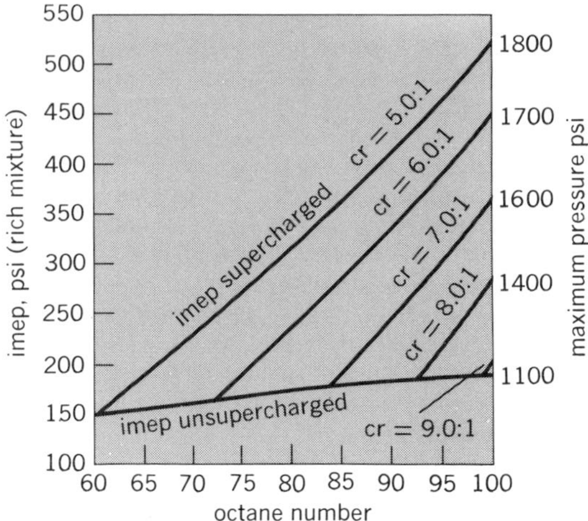

Fig. 18. Graph showing the relationship between compression ratio and knock-limited imep for given octane numbers, obtained by supercharging a laboratory engine. 1 psi = 6.895 kPa. *(From H. R. Ricardo, The High-Speed Internal Combustion Engine, 4th ed., Blackie, 1953)*

and W are in pounds. But the piston motion is not simple harmonic because the connecting rod is not infinite in length, and the piston travels more than half its stroke when the crankpin turns 90° from firing dead center. This distortion of the true harmonic motion is due to the so-called angularity a of the connecting rod, shown by Eq. (3), where r is the

$$a=\frac{r}{l}=\frac{s}{2l} \tag{3}$$

crank radius, s the stroke, and l the connecting rod length, all in inches.

Reciprocating inertia forces act in line with the cylinder axis and may be considered as combinations of a primary force—the true harmonic force from Eq. (2)—oscillating at the same frequency as the crankshaft rpm and a secondary force oscillating at twice this frequency having a value of Fa, which is added to the primary at firing dead center and subtracted from it at inner dead center. In reality there is an infinite but rapidly diminishing series of even harmonics at 4, 6, 8, . . . , times crankshaft speed, but above the second harmonic they are so small that they may generally be neglected. Thus, for a connecting rod with the average angularity of the 1969 automobile engines, where $a = 0.291$, the inertia force caused by a piston at firing dead center is about 1.29 times the pure harmonic force, and at inner dead center it is about 0.71 times as large.

Where two pistons act on one crankpin, with the cylinders in 90° V arrangements, as in Fig. 19*a*, the resultant primary force is radial and of constant magnitude and rotates around the crankshaft with the crankpin. Therefore, it may be compensated for by an addition to the weight required to counterbalance the centrifugal force of the revolving crankpin and its associated masses. The resultant of the secondary force of the two pistons is 1.41 times as large as for one cylinder, and reciprocates in a horizontal plane through the crankshaft at twice crankshaft speed.

In engines with opposed cylinders, if the two connecting rods operate on the same crankpin, as in Fig. 19*b*, the primary forces are added and are twice as great as for one piston, but the secondary forces cancel. If the pistons operate on two crankpins 180° apart, as in Fig. 19*c*, all reciprocating forces are balanced. However, as they will be in different planes, a rocking couple will develop unless compensated by an opposing couple from another pair of cylinders. Double-opposed piston pairs, operating in a single cylinder on two crankshafts as in Fig. 19*d* (with a cross shaft to maintain synchronism), are in perfect balance for primary and secondary reciprocating forces as well as for rotating masses and torque reactions.

In the conventional four-cylinder in-line engines with crankpins in the same plane, the primary reciprocating forces of the two inner pistons (2 and 3) cancel those of the two outer pistons (1 and 4), but the secondary forces from all pistons are added. They are thus equivalent to the force resulting from a weight about $4a$ times the weight of one piston and its share of the connecting rod, oscillating parallel to the piston movement, having the same stroke, but moving at twice the frequency. A large a for this type of engine is advantageous. Where the four cylinders are arranged alternately on each side of a similar crankshaft, and in the same plane, both primary and secondary forces are in balance.

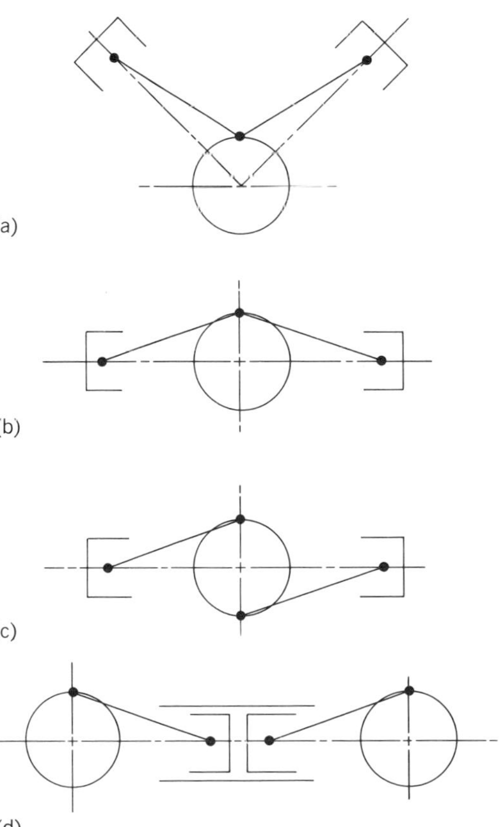

Fig. 19. Arrangements of two cylinders. (*a*) A 90° V formation with connecting rods operating on the same crankpin. (*b*) Opposed cylinders with connecting rods operating on the same crankpin. (*c*) Opposed cylinders with pistons operating on crankpins 180° apart. (*d*) Double-opposed pistons in the same cylinder but with pistons operating on separate crankshafts.

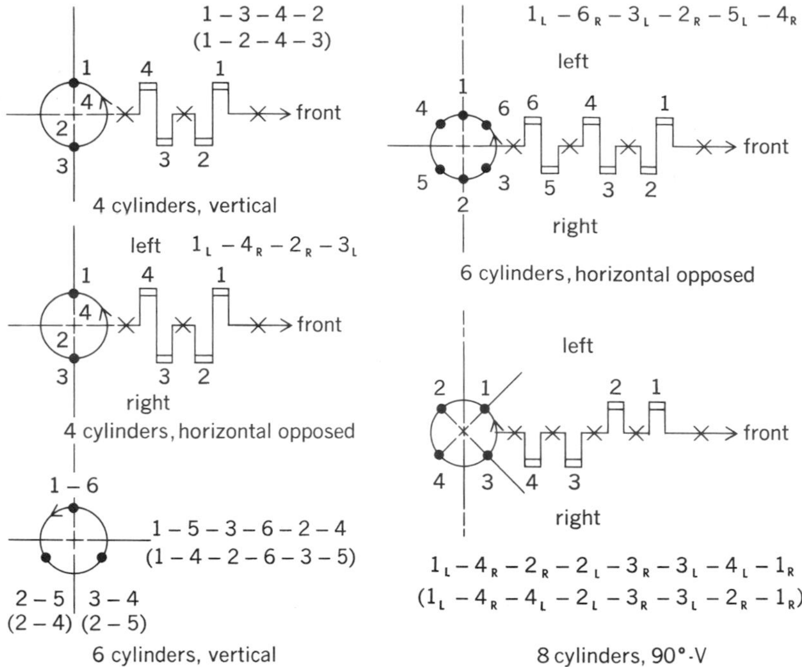

Fig. 20. Typical cylinder arrangements and firing orders.

Six cylinders in line also balance both primary and secondary forces.

Early eight-cylinder 90° engines with four crankpins in the same plane, like those of the early four-cylinder engine, had unbalanced horizontal secondary forces acting through the crankshafts, which were four times as large as those from one pair of cylinders.

In 1927 Cadillac introduced a crank arrangement for its V-8 engines, with the crankpins in two planes 90° apart. Staggering the 1 and 2 crankpins 90° from each other equalizes secondary forces, but the forces are in different planes. The couple thus introduced is cancelled by an opposite couple from the pistons operating on crankpins 3 and 4. This arrangement of crankpins is now universally used on V-8 engines.

Torsion dampers. In addition to vibrational forces from rotating and reciprocating masses, vibration may develop in an engine from torsional resonance of the crankshaft at various critical speeds. The longer the shaft for given bearing diameters, the lower the speeds at which these vibrations may develop. Such vibrations are dampened on most six- and eight-cylinder engines by a vibration damper which is similar to a small flywheel on the crankshaft at the end opposite to the main flywheel but coupled to the shaft only through rubber so arranged as to reduce the torsional resonances. Vibration dampers of this type are usually combined with the pulley that is driving the cooling fan and generator of the engine.

Even though the majority of American automobile engines are now dynamically balanced, it has been general practice for several years to mount them in the chassis frame on rubber blocks. This reduces the transmission of the small-amplitude high-frequency vibrations in torque reaction as well as small unbalance of reciprocating parts in individual cylinders, so that a low noise level is developed in the car from the operation of the engine.

Firing order. Cylinder arrangements are generally selected for even firing intervals and torque impulses, as well as for balance. As a result, the cylinder arrangements and firing orders shown in Fig. 20 may be found in automobile use. It is generally customary to identify cylinder banks as left or right as seen from the driver's seat, and to number the crankpins from front to rear. Manufacturers do not agree on methods of numbering individual cylinders of V-type engines. However, the arrangements and firing orders shown are in general use with the addition in parentheses of alternate arrangements only occasionally used.

Typical American automobile engine. American four-, six-, and eight-cylinder engines have many features in common. They generally have overhead valves located in a removable cylinder head and cylinder blocks cast integral with the upper crankcase. The valves are operated from a chain-driven camshaft in the crankcase through push rods and rockers which have arm lengths giving a valve lift about 50–75% greater than the cam lift. The valve action is usually silenced by use of hydraulic valve lifters.

Many designs provide means for rotating at least the exhaust valve to improve valve life. Cylinder barrels are completely surrounded by the jacket cooling water, and most designs extend it the full length of the bore. Main bearings support the crankshaft between each crankpin. Most designs lock the wrist pin to the connecting rod; the others permit it to "float" in both the piston and connecting rod. The use of three piston rings above the wrist pin has become general practice, two being narrow compression rings and the lowest an oil scraper. Compression rings are

Selected average dimensions of 1979 engines*

Category	Four-cylinder	Six-cylinder	Six-cylinder V	Eight-cylinder V
Cylinder bore, mm	92	91	96	101
Cylinder stroke, mm	80	99	86	86
Stroke-bore ratio	0.87	0.92	0.90	0.85
Displacement, liters	2.1	3.9	3.7	5.6
Net brake power, kW/rpm	60/4733	77/3509	95/3825	115/3722
Net torque, N · m/rpm	145/2667	252/1655	279/2127	358/2104
Connecting rod length, mm	140	159	140	153
Piston pin diameter, mm	23	23	24	24
Main bearing diameter, mm	58	65	63	66
Crankpin diameter, mm	49	54	55	54

*Adapted from *Automotive Industries*, April 1979.

Fig. 21. Typical V-8 overhead valve automobile engine with air filter. (*Pontiac Div., General Motors Corp.*)

generally of cast iron about 0.078 in. (1.98 mm) wide and provided with a coating such as chromium or tin to prevent scuffing the surface during the wearing-in period of a new engine and when it is started cold with but little oil on the cylinder wall. Oil-scraper rings are about 3/16 in. (5 mm) wide, are provided with a nonscuffiing surface, and have drain holes through the piston for the return of the excess oil scraped from the cylinder wall.

The highest power engines are of the eight-cylinder V-type. A typical engine of this type is shown in cross section in Fig. 21. The right and left banks of cylinders are staggered to enable connecting rods of opposing cylinders to be located side by side on the same crankpin. The V arrangement provides a short and very ridged structure, which is important for high engine speeds because of the minimized deflection of the main bearings. It also makes possible efficient intake-manifold designs and almost symmetrical and equal-length branches to each cylinder port from a centrally located downdraft carburetor. The short length and low height of these engines are also important features for car styling. Some of the principal dimensions and other statistics of these engines have been averaged in the table. [NEIL MacCOULL]

Bibliography: T. Baumeister (ed.), *Standard Handbook for Mechanical Engineers*, 8th ed., 1978; W. H. Crouse, *Automotive Engines*, 5th ed., 1975; W. H. Crouse, *Automotive Mechanics*, 7th ed., 1975; L. C. Lichty, *Combustion Engine Processes*, 7th ed., 1967; E. F. Obert, *Internal Combustion Engines*, 3d ed., 1968; H. R. Ricardo, *The High-Speed Internal Combustion Engine*, 1931, 4th ed., 1953; C. F. Taylor, *The Internal Combustion Engine in Theory and Practice*, vol. 1, 2d ed., vol. 2, 1977.

Isotope (stable) separation

The physical separation of different stable isotopes of an element from one another. Many chemical elements always occur in nature as a mixture of several isotopes. The isotopes of any given element have identical chemical properties, but there are slight differences in their physical properties because of the differences in mass of the individual isotopes. Thus it is possible to separate physically the isotopes of an element to produce material of isotopic composition different from that which occurs in nature. Although these separation processes are all quite difficult and expensive to carry out, they are not inherently different from the usual operations employed in the chemical process industries. *See* ISOTOPE.

The separation of isotopes is particularly important in the nuclear energy field because individual isotopes may have completely different nuclear properties. For example, uranium-235 is used as a fuel for nuclear chain reactors, heavy water (deuterium oxide) is used as a neutron moderator in nuclear chain reactors, and deuterium gas is a possible fuel for thermonuclear reactors. Separated isotopes are also used widely for research concerned with the structure and properties of the nucleus.

The process which is best suited for separating the isotopes of a given element depends upon the mass of the element and the desired quantity of separated material. Research quantities of separated isotopes are best prepared by electromagnetic separation in a mass spectrometer. For example, gram quantities of many separated isotopes have been prepared at Oak Ridge National Laboratory using the large electromagnetic separators which were built during World War II. The electromagnetic process has the advantage that a fairly complete separation of two isotopes can be obtained in one operation.

When moderate quantities of a separated isotope are desired, thermal diffusion may be used. Although thermal diffusion requires a large energy input, this is more than offset by the simplicity of the equipment, absence of moving parts, and high separation obtained in a small volume.

In the large-scale separation of stable isotopes, the best processes are those which have the highest thermodynamic efficiencies. Reversible processes involving distillation and chemical exchange are best for separating the light isotopes such as deuterium. For heavy isotopes such as those of uranium, however, no appreciable separation is obtained by the reversible processes and some type of irreversible process such as gaseous diffusion must be used. Although reversible processes have in general higher efficiencies than irreversible ones, the absolute efficiency of any isotope separation process is very small and the cost is very high in comparison to the usual operations that are employed in the chemical process industries.

Gaseous diffusion. This process has turned out to be the most economical for the separation of the isotopes of uranium. It is based on the fact that in a mixture of two gases of different molecular weights, molecules of the lighter gas will on the average be traveling at higher velocities than those of the heavier gas. If there is a porous barrier with holes just large enough to permit passage of the individual molecules but without permitting bulk flow of the gas as a whole, the probability of a gas molecule passing through the barrier will be directly proportional to its velocity. From kinetic theory it can be shown that the velocity of a gas molecule is inversely proportional to the square root of its molecular weight, so that the efficiency of gaseous diffusion will depend on the ratio of the square roots of the molecular weights of the two gases present.

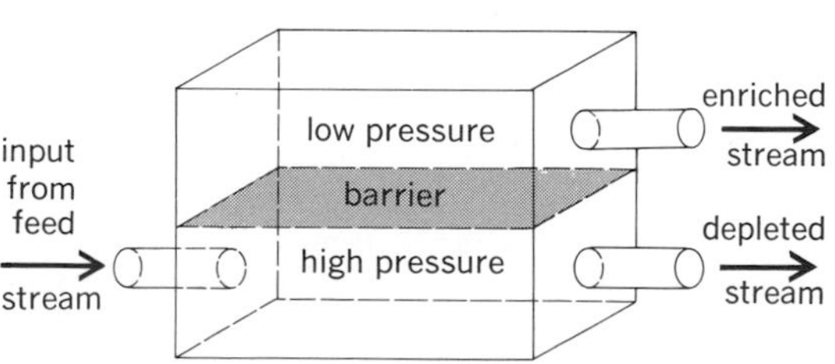

Fig. 1. Gaseous diffusion stage. *(From H. Etherington, ed, Nuclear Engineering Handbook, McGraw-Hill, 1958)*

The only uranium compound which is a gas at a reasonable temperature and pressure is uranium hexafluoride, UF_6. The two isotopes to be separated are $U^{235}F_6$ and $U^{238}F_6$, and the efficiency of separation depends on the quantity in the equation below. Since this number is close to unity, the

$$\sqrt{U^{238}F_6/U^{235}F_6} = 1.0043$$

separation is very small in any one step of the process.

The separation of the isotopes of uranium in the United States is carried out in the three plants operated for the Atomic Energy Commission which are located at Oak Ridge, Tenn.; Paducah, Ky.; and Portsmouth, Ohio. In each of these installations natural uranium containing 0.71% U^{235} and the balance U^{238} in the form of UF_6 gas is separated into an enriched uranium product containing more than 90% U^{235}, and a waste containing about 0.3% U^{235}. It is also possible to use "depleted" uranium (uranium recovered from plutonium production reactors has lower U^{235} content than the natural) as feed to a gaseous diffusion plant. Britain has a gaseous diffusion plant at Capenhurst, and the Soviet Union has facilities at an undisclosed location.

The success of the gaseous diffusion process is dependent on the performance of the single diffusion stage. In each stage, UF_6 gas is compressed, passed through a cooler to remove the heat of compression, and then admitted to the vessel containing the porous barrier (Fig. 1). About half the gas entering the vessel diffuses through the barrier and passes to the next higher stage. This diffused gas contains slightly more of the U^{235} isotope. The undiffused gas is slightly depleted in the U^{235} isotope, and passes to the next lower stage.

Several thousand individual stages are required to bring about the necessary overall change in composition. The combination of stages is known as a cascade, and the cascade which brings about the separation with the least work is known as an ideal cascade. The size of the stages varies tremendously; those feeding the natural uranium into the cascade are the largest and the final product stages are the smallest.

Nickel-clad piping and process equipment are used to handle the UF_6 gas. Thousands of pumps, coolers, and control instruments are required. The electric power requirements of the gaseous diffusion process are large; for many years approximately 10% of the total electric power output of the United States was required to operate the three diffusion plants.

The gaseous diffusion process was originally developed as a means of producing highly en-

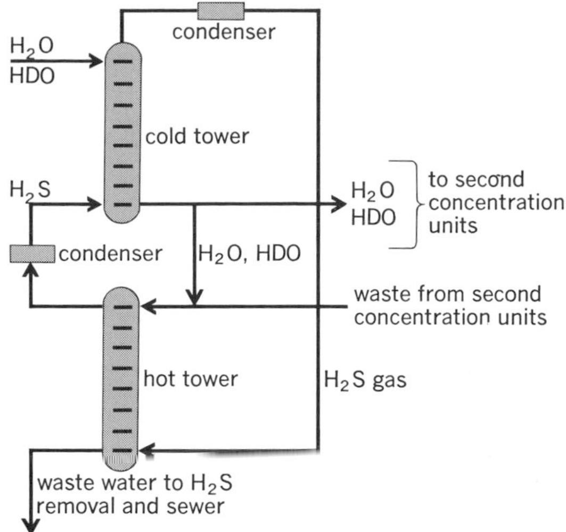

Fig. 2. Dual-temperature H_2S–HDO exchange. The end product is a mixture of H_2O, HDO, and D_2O, with the relative abundance of D_2O increasing in the later stages. *(From R. Stephenson, Introduction to Nuclear Engineering, 2d ed., McGraw-Hill, 1958)*

riched uranium for atomic bombs. At present, the gaseous diffusion plants are being modified to produce partially enriched uranium to fuel nuclear power reactors. There has also been a substantial decrease in the output from the three plants.

Chemical exchange. The chemical exchange process has proved to be the most efficient for separating isotopes of the lighter elements. This process is based on the fact that if equilibrium is established between, for example, a gas and a liquid phase, the composition of the isotopes will be different in the two phases. Thus, if hydrogen gas is brought into equilibrium with water, it is found that the ratio of heavy hydrogen (deuterium) to light hydrogen is several times greater in the water than in the hydrogen gas. By repeating the process in a suitable cascade, it is possible to effect a substantial separation of the isotopes with a relatively small number of stages.

The chief use of chemical exchange is in the large-scale production of heavy water. A dual-temperature exchange reaction between water, which contains HDO and D_2O molecules as well as H_2O molecules, and hydrogen sulfide (H_2S) gas for the primary separation of heavy water is used at the Savannah River plant. The separation is carried out in a series of hot towers operating at the boiling point and cold towers operating at room temperature (Fig. 2). In each tower liquid water passes downward countercurrent to the rising H_2S gas. The relative distribution of heavy and light water is affected by temperature, and the success of the process is determined by the difference between the concentrations in the hot and the cold towers.

Chemical exchange has also been used for the large-scale separation of other isotopes. For example, the isotopes of boron have been separated by fractional distillation of the boron trifluoride–dimethyl ether complex.

Distillation. The separation of isotopes by distillation is much less efficient than separation by other methods. Distillation was used during World War II to produce heavy water, but the cost was high and the plants are no longer in existence. Fractional distillation has been used at Savannah River to concentrate the product from the dual-temperature process (12–16% D_2O) up to 95–98% D_2O.

Electrolysis. Electrolysis of water is the oldest large-scale method of producing heavy water. Under favorable conditions, the ratio of hydrogen to deuterium in the gas leaving a cell in which water is electrolyzed is eight times the ratio of these isotopes in the liquid. In spite of this high degree of separation, electrolysis can be used only where electricity is very cheap, as in Norway, because of the large power consumption per pound of D_2O produced. Electrolysis is used in the United States only as a finishing step to concentrate to final-product specifications.

Electromagnetic process. Electromagnetic separation was the method which was first used to prove the existence of isotopes. The mass spectrometer and mass spectrograph are still widely used by physicists as a research tool. In the electromagnetic process, vapors of the material to be analyzed are ionized, accelerated in an electric field, and enter a magnetic field which causes the ions to be bent in a circular path. Since the light ions have less momentum than the heavy ions, they will be bent through a circle of smaller radius, and the two isotopes can be separated by placing collectors at the proper location. *See* MASS SPECTROSCOPE.

During World War II, a large electromagnetic separation plant was built on Oak Ridge to separate the isotopes of uranium. The large mass spectrometers used there were referred to by the code name Calutron, a contraction of California University cyclotron (Fig. 3). The first kilogram quantities of U^{235} were produced in 1944. With the completion of the gaseous-diffusion plant at Oak Ridge, the electromagnetic process was found to be uneconomical and was abandoned in 1946. However, some of the equipment is still being used to produce gram quantities of separated isotopes for research purposes.

Thermal diffusion. The separation of isotopes by thermal diffusion is based on the fact that when a temperature gradient is established in a mixture of uniform composition, one component will concentrate near the hot region and the other near the

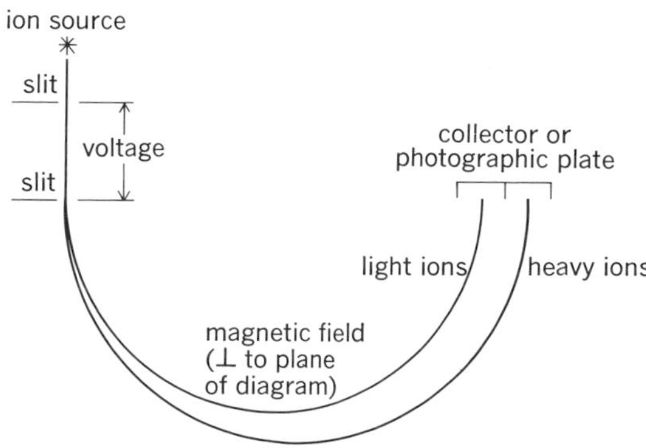

Fig. 3. Diagrammatic representation of Calutron mass spectrometer. *(From R. Stephenson, Introduction to Nuclear Engineering, 2d ed., McGraw-Hill, 1958)*

cold region. Thermal diffusion is carried out in the annular space between two vertical concentric pipes, the inner one heated and the outer one cooled. Because of thermal convection of the fluid, there is a countercurrent flow which greatly increases the separation obtained in simple thermal diffusion and makes possible substantial separations in a reasonable column height.

Thermal diffusion has been used to separate small quantities of isotopes for research purposes. In 1944 a plant was built at Oak Ridge to separate the isotopes of uranium by thermal diffusion. However, the steam consumption was very large, and the plant was dismantled when the gaseous-diffusion facilities were completed.

Centrifugation. The use of a centrifuge to separate isotopes has one major advantage, namely that the separation depends only on the difference in masses of the two isotopes, and not on the ratio of their masses. Thus it is no more difficult to separate the isotopes of uranium than those of the light elements. A disadvantage of the centrifuge method is that a very high speed of rotation is required to obtain any substantial separation of isotopes in a single unit.

A centrifuge pilot plant was built during World War II, but further work on the process was discontinued because of engineering problems involved in the operation of high-speed rotors, the low capacity of the individual machines, and the large power input which is required to overcome friction.

There has been a renewal of interest in centrifugation, particularly in Europe where several new approaches are being studied. There is some evidence that this will eventually lead to a simple and cheap way to produce enriched U^{235}. Politically, this has been a very sensitive subject since, if successful, it would enable even the smallest nations to manufacture nuclear weapons for possible military use.

Nozzle process. Isotopes can be separated by allowing a gaseous compound to exhaust through a properly shaped nozzle. Preliminary calculations indicate that this relatively new method may be competitive with gaseous diffusion for separating the isotopes of uranium. *See* NUCLEAR FUEL.

Laser methods. Security-classified for many years, laser methods for separating isotopes are under intensive study throughout the world. Most of the effort is directed toward the separation of U^{235}. For example, by the use of a tunable dye laser, U^{235} atoms (or possibly molecules) can be raised to an excited state while the accompanying U^{238} atoms remain unaffected. The U^{235} can then be separated out by conventional electromagnetic means. Laser methods have the advantage of a high rate of separation in one step, but they also have the potential problems of other beam processes such as the obsolete electromagnetic process.

[RICHARD M. STEPHENSON]

Bibliography: M. Benedict and T. H. Pigford, *Nuclear Chemical Engineering*, 1957; K. P. Cohen, *The Theory of Isotope Separation as Applied to the Large-scale Production of* U^{235}, 1951; H. Etherington, *Nuclear Engineering Handbook*, 1958; R. Stephenson, *Introduction to Nuclear Engineering*, 2d ed., 1958.

Jet fuel

Fuel blended from the light distillates fractionated from crude petroleum. There are two general types, a wide-cut heavy naphtha-kerosine blend used by the U.S. Air Force as JP-4 (or commercially as Jet B) and a kerosine used by the world's airlines as Jet A (or Jet A-1) or by the U.S. Navy as JP-5.

Since 1970, commercial kerosine of 38°C flash point has grown from a small-volume household-heating fuel into a major product of commerce rivaling gasoline in importance as a source of transportation energy. During the 1960–1975 period, JP-4 diminished in relative importance, and commercial use of Jet B also declined to a small percent. *See* KEROSINE.

All jet fuels must meet the stringent performance requirements of aircraft turbine engines and fuel systems, which demand extreme cleanliness and freedom from oxidation deposits in high-temperature zones. Combustors require fuels that atomize and ignite at low temperatures, burn with adequate heat release and controlled radiation, and produce neither smoke nor attack of hot turbine parts. The operation of the aircraft in long-duration flights at high altitude imposes a special requirement of good low-temperature flow behavior; this need establishes Jet A-1 which has a freezing point of −50°C (wax) as an international flight fuel; Jet A which has a freezing point of −40°C (wax) can serve shorter domestic routes. *See* AIRCRAFT ENGINE.

Fuels pumped through long multiproduct pipelines or delivered by tanker are usually clay-filtered to ensure freedom from surfactants. Many stages of filters operate to ensure clean, dry product as the fuel moves into airport tanks, hydrant systems, and finally aircraft. Because high-speed filtration can generate static charges, fuels may contain an electrical conductivity additive to ensure rapid dissipation of charge.

[W. G. DUKEK]

Jet propulsion

Propulsion of a body by means of force resulting from discharge of a fluid jet. This fluid jet issues from a nozzle and produces a reaction (Newton's third law) to the force exerted against the working fluid in giving it momentum in the jet stream. Turbojets, ramjets, and rockets are the most widely used jet-propulsion engines.

Jet nozzles. In each of these propulsion engines a jet nozzle converts potential energy of the working fluid into kinetic energy. Hot high-pressure gas escapes through the nozzle, expanding in volume as it drops in pressure and temperature, thus gaining rearward velocity and momentum. This process is governed by the laws of conservation of mass, energy, and momentum and by the pressure-volume-temperature relationships of the gas-state equation.

For propulsion systems in which the pressure of the working fluid is not more than approximately twice the absolute ambient pressure, a converging nozzle is used (Fig. 1*a*). The mass flow from this nozzle in terms of conditions at sections 1 and 2 is given in Eq. (1), and the velocity of the jet leaving the nozzle is given in Eq. (2).

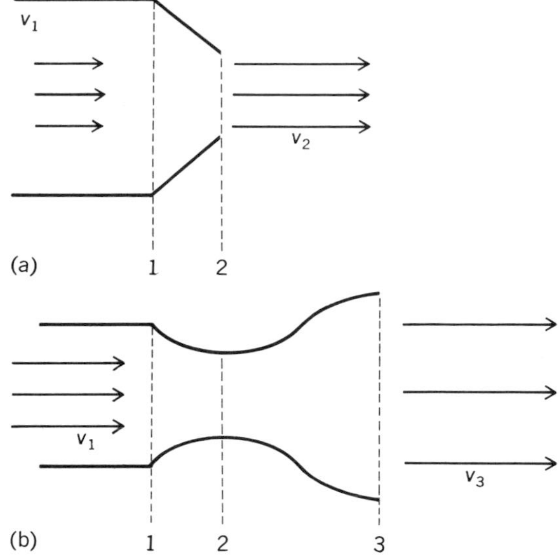

Fig. 1. Propulsion nozzles. (a) Low-pressure converging nozzle. (b) High-pressure converging-diverging nozzle.

$$m = A_2\rho_1 \sqrt{2gJC_pT_{1t}} \sqrt{\left(\frac{p_2}{p_1}\right)^{2/\gamma} - \left(\frac{p_2}{p_1}\right)^{(\gamma+1)/\gamma}} \tag{1}$$

$$v_2 = \sqrt{2gJC_pT_{1t}} \sqrt{1 - \left(\frac{p_2}{p_1}\right)^{(\gamma-1)/\gamma}} \tag{2}$$

Here m = mass flow, slug/sec
A = cross-sectional area, ft^2
ρ = density, $slug/ft^3$
J = work equivalent of heat, ft-lb/Btu
T_t = total temperature, °R
C_p = specific heat at constant pressure, Btu/(°F)(lb)
γ = ratio of specific heats
p = static pressure, $lb/in.^2$
v = velocity, ft/sec

Maximum flow P occurs when Eq. (3*a*) holds. This is the critical pressure at which flow velocity in the nozzle throat is equal to local sound velocity. For air and most combustion gas mixtures, Eq. (3*b*) holds.

$$p_2 = p_1\left(\frac{2}{\gamma+1}\right)^{\gamma/(\gamma-1)} = p_c \tag{3a}$$

$$p_c \cong 0.5p_1 \tag{3b}$$

For propulsion systems in which the working fluid pressure is high compared to ambient pressure, a converging-diverging nozzle is used (Fig. 1*b*). In this nozzle the working fluid continues to expand from the critical throat pressure to ambient pressure at section 3 with a further increase in velocity beyond the sonic throat velocity, as in Eq. (4).

$$v_3 = \sqrt{2gJC_pT_{1t}} \sqrt{1 - \left(\frac{p_3}{p_1}\right)^{(\gamma-1)/\gamma}} \tag{4}$$

Turbojet. The turbojet is an air-breathing propulsion engine used in most military fighters, bombers, and transports and in modern commercial airline transports. Thrust ratings range from a few hundred pounds to more than 50,000 lb (220 kN) for engines for the experimental stage. The engine operates best at high subsonic or supersonic flight speeds, where the high-velocity jet achieves good propulsion efficiency. Specialized versions of turbojets are used for subsonic flight and others for supersonic flight. The turbofan, or bypass engine, operates with lower exhaust-gas velocity and provides improved efficiency for subsonic flight. In some turbojets more heat is added to the exhaust stream in an afterburner to increase propulsion power output for efficient supersonic flight. *See* PROPULSION; TURBOFAN.

The turbojet is a heat engine. Air enters the inlet diffuser and is compressed adiabatically there and in the rotating compressor (Figs. 2 and 3). Heat is added by burning fuel at constant pressure in the combustor. The hot gas expands in the turbine, where energy is extracted to drive the rotating compressor. Further expansion through the jet nozzle converts the remaining available energy of the gas stream into high velocity, producing thrust for propulsion power. Additional propulsion power can be realized by heat added in an afterburner. *See* BRAYTON CYCLE; TURBOJET.

During high-speed flight the inlet diffuser decreases the relative velocity of the entering air, increasing its pressure in a process that is the reverse of nozzle expansion (Fig. 3). For supersonic speeds, a converging-diverging passage is required. The diffuser inlet and throat areas may be varied mechanically to match the airflow requirements of varying flight speed. *See* SUPERSONIC DIFFUSER.

The axial-flow compressor has alternate rows of rotating and stationary blades which compress the air further. The individual action of the blades is like that of an airplane wing in deflecting the air passing over it. The rotating blades add kinetic energy to the airstream, and the stationary blades convert some of this kinetic energy into a pressure rise. Normally, each rotating and stationary blade row produces a pressure rise as the air moves

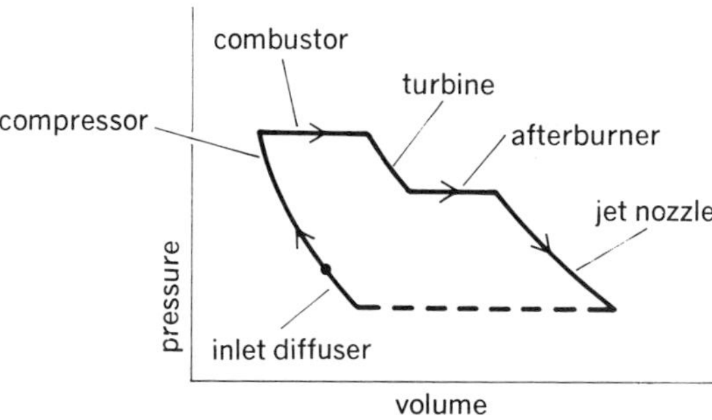

Fig. 2. Turbojet pressure-volume diagram.

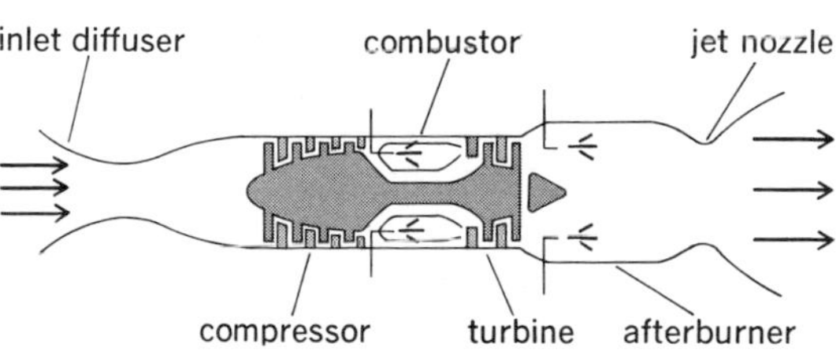

Fig. 3. Simplified schematic diagram of turbojet.

through the tapered annular passage between the rotor drum and stator casing.

Most of the total air compression is provided by the rotating compressor during subsonic flight. As the flight speed increases, the inlet diffuser performs an increasing amount of the total compression. At high supersonic speeds, the inlet diffuser does most of the compression work and the rotating compressor becomes less important in producing efficient engine performance.

In the combustor, jet fuel (a kerosinelike petroleum fraction) is sprayed, vaporized, and burned. Air from the compressor discharge is fed into the combustion space through various shaped openings in sheet-metal combustion liners. These openings serve to mix the air and fuel vapor for efficient combustion and to cool the liner which protects the outer combustion casing from the heat of the flame.

The stationary turbine nozzle vales expand the hot gas to a high-velocity stream which is directed toward the rotating blades in a near-tangential direction. The gas is turned in the blades and imparts rotational energy to the turbine wheel by this momentum change. Energy absorbed by the turbine wheel is used to provide power to drive the compressor through a connecting shaft (Fig. 4). The number of turbine stages is determined by the amount of power required to drive the compressor. The trend in modern jet engines is toward higher pressure ratio, higher work compressors, and turbines for improved economy in fuel consumption. The implied increase in number of stages has been offset by increasing the work potential of individual stages.

In the afterburner, following the turbine, fuel is injected into the hot gas stream through spray bars. V-section flame-stabilizing channels are mounted downstream from the fuel spray bars. These channels produce eddies in the gas stream to promote stable burning and prevent flame blowout at high altitude. A louvered sheet-metal liner protects the afterburner casing from overheating. In some turbojets the extra propulsion power from the afterburner is not required; the hot gas from the turbine flows directly into the exhaust-jet nozzle. *See* TURBORAMJET.

Table 1. Characteristics of medium-sized turbojet

Characteristics	With afterburner	Without afterburner
Weight, lb (kg)	3,600 (1630)	2,900 (1320)
Length, in. (m)	200 (5.08)	110 (2.79)
Diameter, in. (m)	38 (97)	37 (94)
Compressor stages	17	17
Turbine stages	3	3
Takeoff thrust, lb (kN)	15,500 (68.9)	11,000 (48.9)
Takeoff SFC*	2.0 (02)	0.8 (0.08)
Best cruise thrust, lb (kN)	2,600 (11.6)	2,500 (11.1)
Best cruise SFC (Mach 0.9 at 35,000 ft or 10.7 km altitude)	1.0 (0.1)	0.9 (0.09)
Military thrust, lb (kN)	6,000 (26.7)	
Military SFC (Mach 2.0 at 55,000 ft or 16.8 km altitude)	2.3 (0.23)	

*Specific fuel consumption, pounds of fuel per pound of thrust per hour (or kilograms of fuel per newtons of thrust per hour).

For subsonic flight applications, a converging jet nozzle of fixed dimensions is used. For supersonic flight the converging-diverging type jet nozzle is required for good performance. The dimensions of the throat and exit may be varied to match the flow and expansion requirements of varying flight speeds and altitudes.

Accessories to the engine are driven by an arrangement of shafts and gears taking power from the compressor-turbine shaft. These accessories include fuel and lube-oil pumps, tachometer, and in some installations an electric starter-generator.

An engine control system senses air pressure and temperature from the inlet diffuser, rotor speed, and throttle setting. It computes from these, by mechanical or electric analog means, the required fuel rates and meters fuel flow to the combustor and afterburner. The system senses gas temperature (entering the turbine and in the afterburner), from which it operates to limit fuel rates so as to prevent overheating or overspeeding. Engine characteristics representative of a modern turbojet are given in Table 1.

Ramjet. The ramjet is the simplest air-breathing propulsion system and is used principally in guid-

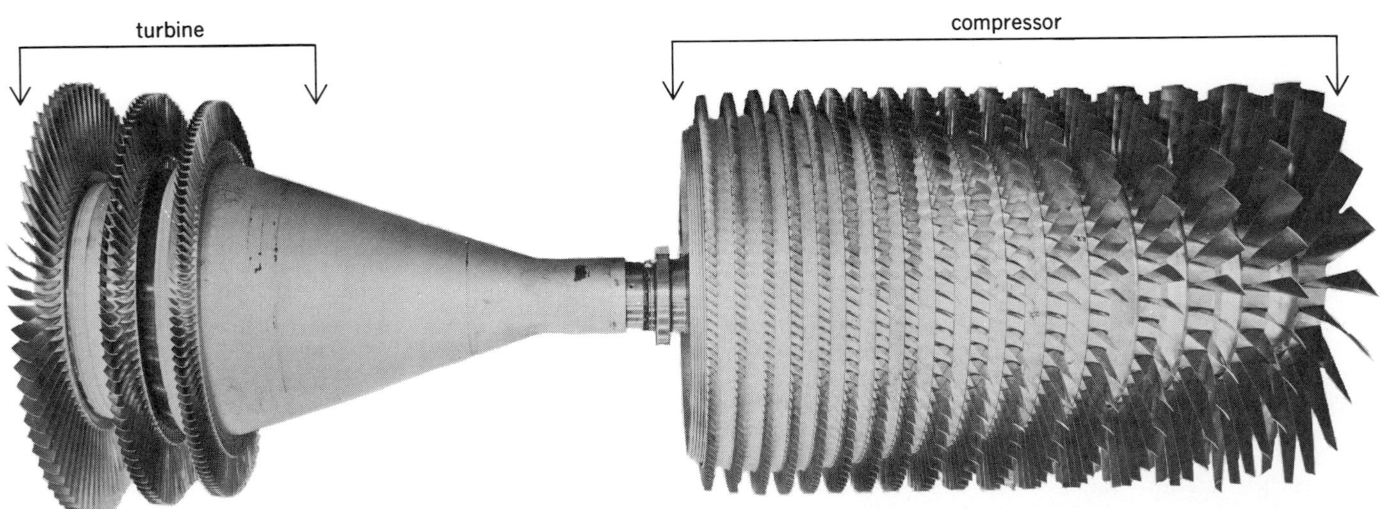

Fig. 4. Compressor and turbine rotors assembled.

Fig. 5. Cutaway model of ramjet. (*Marquardt Aircraft Co.*)

ed missiles (Fig. 5). It travels at high velocity, compressing air in its inlet diffuser, burning fuel in the air, and discharging it through a jet nozzle. There are no rotating compressor and turbine elements in the ramjet. The ram air-pressure ratios achieved by its forward velocity are high enough for efficient operation on the Brayton cycle; therefore, no additional air compression is required from a rotating compressor. However, the ramjet must be accelerated up to operating speeds by other means. In most missile applications this is accomplished by a rocket-powered booster stage. In other applications the acceleration of the ramjet can be accomplished by turbojet power.

Characteristics of a modern ramjet are a weight of 1000 lb (454 kg), a length of 150 in. (3.81 m), and a diameter of 36 in. (0.76 m); traveling at a speed of Mach 4 at 75,000 ft (22.9 km), it would have a thrust of 2000 lb (8.9 kN) and a specific fuel consumption of 2.0.

The ramjet is usually designed to cruise within a fairly narrow speed range. The combustor is similar to a turbojet afterburner with fuel injectors, V-channel flame holders, and a louvered liner for heat insulation. Controls and accessories, housed in the bullet nose of the inlet, are powered by a very small high-speed turbine drive by ram air bled from the inlet. *See* PULSE JET; RAMJET.

Liquid-propellant rocket engine. The liquid-propellant rocket engine is a propulsion system used to power missiles and space vehicles. It is used primarily to accelerate a load to high velocity, and to do this it delivers a large thrust for a relatively short time.

Liquid fuel (usually a hydrocarbon resembling kerosine) and oxidizer (usually liquid oxygen) are pumped from tanks to a fuel injector, which sprays them into a combustion chamber, in which they burn. The hot combustion gases escape through a converging-diverging jet nozzle, leaving at very high velocity (Fig. 6). The reaction forces developed on nozzle wall and injector head are transmitted to the body of the vehicle. Rocket vehicles are sometimes steered by mounting the entire engine on gimbals.

Fuel and oxidizer pumps have high-speed centrifugal impellers driven through speed-reduction gears by a small, single-stage gas turbine. High-pressure gas to drive the turbine usually is obtained by burning fuel and oxidizer bled from the pump discharge to a small combustion chamber

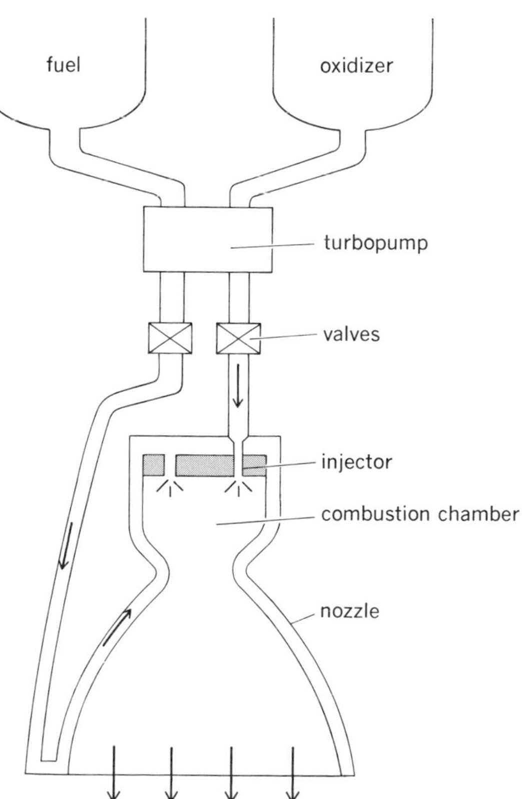

Fig. 6. Liquid-propellant rocket engine. Arrows represent direction of flows. (*General Electric Co.*)

feeding the turbine. Starting occurs by burning the propellants from gas-pressurized tanks. In some rocket engines the turbine working fluid is decomposed hydrogen peroxide. *See* PROPELLANT.

The effectiveness of a rocket propellant combination is indicated by its theoretical specific impulse. Fuels and oxidizers of low molecular weight are preferred because, for a given energy release and mass flow, the lightweight molecules achieve greatest velocity and consequently highest specific impulse (I_{sp}) (Table 2).

Typical characteristics of a large, liquid-propellant rocket engine follow.

Propellant	*Liquid oxygen – JP-1*
Chamber pressure, psia (MPa, absolute)	900 (6.2)
Thrust, lb (kN)	200,000 (890)
Height, in. (m)	120 (3.05)
Nozzle exit diameter, in. (m)	60 (1.52)
Overall I_{sp}, sec (at 100,000 ft altitude)	284
Burn time, sec	200
Weight (not including fuel or fuel tanks), lb (kg)	2000 (907)

Solid-propellant rocket engine. The solid-propellant rocket engine is a propulsion system used in missiles and space vehicles. It can deliver large thrust and is instantly ready for operation.

This engine consists of a casing filled with a mixture of propellant chemicals in solid form. These burn, generating hot high-pressure gases which escape through nozzles as high-velocity jets, creating a forward-thrust reaction. To steer the vehicle, some nozzles are constructed so that they can be tilted or swiveled during firing.

A large engine, such as the first stage of a 1500-mi (2900-km) ballistic missile, might have the following general characteristics.

Propellant	*Potassium perchlorate – polyurethene and additives*
Length, in. (m)	220 (5.6)
Diameter, in. (m)	50 (1.27)
Gross weight, lb (kg)	20,000 (9100)
Average thrust, lb (kN)	80,000 (360)
Burn time, sec	60
Overall I_{sp}, sec	230

Propellant combinations are selected to contain as much fuel and oxidizer and as little inert substance as possible. As with liquid propellants, combinations with low molecular weights are preferred because they result in greater specific impulse for a given energy release. The propellant mixture must burn at a relatively slow, uniform rate; certain inhibitors are added to control this.

There are now in use a variety of propellants of two general classes. Double-base types contain nitrocellulose and nitroglycerin plus additives for stability and for control of combustion rates. Composite types contain ammonium or potassium perchlorate granules embedded in a rubberlike hydrocarbon compound.

Table 2. Theoretical values for several fuel-oxidizer combinations (1000 psi combustion chamber pressure)‡

Combinations	I_{sp}, sec
Liquid oxygen – JP-1*	286
Liquid oxygen – liquid hydrogen	388
Hydrogen peroxide – JP-1	266
N_2O_4 – UDMH†	274

*JP-1 is a kerosinelike hydrocarbon.

†Nitrogen tetroxide – unsymmetrical dimethylhydrazine.

‡1000 psi = 6.895 MPa.

[J. W. BLANTON]

Bibliography: N. E. Borden, Jr., *Jet Engine Fundamentals*, 1967; J. V. Casmassa and R. D. Bent, *Jet Aircraft Power Systems*, 3d ed., 1965; J. W. Hesse and N. V. S. Mumford, Jr., *Jet Propulsion for Aerospace Applications*, 2d ed., 1964; C. W. Smith, *Aircraft Gas Turbines*, 1956; G. P. Sutton, *Rocket Propulsion Elements*, 2d ed., 1956; P. H. Wilkinson, *Aircraft Engines of the World*, rev. ed., 1967; M. J. Zucrow, *Aircraft and Missile Propulsion*, vol. 2, 1958.

Kerosine

A refined petroleum fraction used as a fuel for heating and cooking, jet engines, lamps, and weed burning, and as a base for insecticides. Kerosine, known also as lamp oil, is recovered from crude oil by distillation. It boils in the approximate range of 350–550°F (180–290°C). Most marketed grades, however, have narrower boiling ranges. The specific gravity is about 0.8. Determined by the Abel tester, the flash point is not below 73°F (23°C), but usually a higher flash point is specified. Down to a temperature of −25°F (−32°C), kerosine remains in the liquid phase. Components are mainly paraffinic and naphthenic hydrocarbons in the C_{10}–C_{14} range. A low content of aromatics is desirable except when kerosine is used as tractor fuel.

Specifications are established for specific grades of kerosine by government agencies and by refiners. Since these specifications are developed from performance observations, they are adhered to rigidly to assure satisfactory operation. For use in lamps, for example, a highly paraffinic oil is desired because aromatics and naphthenes give a smoky flame; and for satisfactory wick feeding, a viscosity no greater than 2 centipoises is required in this application. Furthermore, the nonvolatile components must be kept low. In order to avoid atmospheric pollution, sulfur content must be low; a minimum flash point of 100°F (38°C) is desirable to reduce explosion hazards.

Today, kerosine represents only a little over 4% of the total petroleum-products production in the United States, whereas it was the major product in the 1800s. Tractor fuel now represents an insignificant percentage of the total production. Use of kerosine as a jet fuel continues to increase as more jet planes are put into operation. Kerosine also is the principal hydrocarbon fuel for rockets.

The price of kerosine has followed the price of crude oil, and this cost continues to be the dominant price factor; however, changing use patterns and specifications may exert an additional upward force on the price of kerosine. *See* DISTILLATE FUEL; JET FUEL; PETROLEUM PRODUCTS.

[HAROLD C. RIES]

Laser fusion

A process (now primarily theoretical) in which a laser would be used to heat a plasma to a sufficiently high temperature for a minimum critical length of time to cause an efficient thermonuclear fusion reaction to occur throughout the plasma. The energy given off in the resulting thermonuclear microexplosion could then be converted to useful energy as is done presently by nuclear fission reactors. The achievement of thermonuclear fusion by a technique known as magnetic confinement has been under extensive investigation as a potential energy source since about 1950, but has not reached the conditions whereby an efficient thermonuclear reaction could be obtained. With the development of high-power pulsed lasers, the concept of laser fusion has indicated enough promise to be considered by many as a possible alternative to the magnetic confinement process. *See* NUCLEAR REACTOR.

Investigations of laster-induced fusion have taken two directions, both requiring plasma temperatures of the order of 10^8°C. The most extensively studied technique involves the concepts of implosion and compression of the plasma to increase its density and temperature and of inertial confinement to provide enough time to allow the reaction to occur throughout the entire mass of the plasma. The other approach is to use the laser as an auxiliary heating source to heat a very long, magnetically confined plasma. This process involves some of the ideas from both the laser implosion technique and the magnetic confinement technique, along with some new concepts and problems of its own.

Deuterium, proposed as a fuel from which the plasma is to be created, is a low cost, relatively clean fuel that can be obtained from a virtually inexhaustible supply in the oceans. The higher energy yield and lower contamination from the fusion reaction, as compared to fission reactions, makes the difficult program involved in attempting to initiate such a reaction a worthwhile task.

Fission. Present-day nuclear power plants use a fission reaction as their source of energy. This reaction occurs spontaneously in a special isotope of uranium in which the heavy uranium nucleus breaks apart into two nuclei of lighter elements. During this process a large amount of extra energy is produced that results in the emission of radiation and the impartation of high velocities to the newly formed particles. When the rapidly moving particles are absorbed within the reaction chamber, heat is produced that can be used to generate useful power. The high velocities of the particles help speed up reactions of surrounding atoms, and if the mass density is high enough a thermonuclear explosion could result. It is thus essential to control the reaction rate by diluting the uranium mass in the reaction region. *See* NUCLEAR FISSION.

Fusion. In the fusion reaction, rather than causing a heavy element to divide, two light elements are fused together to form a heavier element, again giving off energy in the form of radiation and of kinetic energy of the newly formed particles. The fusion reaction occurring with the fastest rate involves the fusion of a deuterium ion (hydrogen with one extra neutron) and a tritium ion (hydrogen with two extra neutrons) resulting in a helium atom and a neutron. The 10^8°C temperature required to initiate this reaction is essential in order to provide high enough impact velocities when the deuterium and tritium ions collide so as to overcome the strong repulsive forces caused by the positive electrical charges of both ions. When the nuclei fuse together to form a helium atom, most of the extra energy is given off to the emitted neutron. This neutron can transfer its energy to neighboring deuterium and tritium ions, causing the reaction to spread if the density is high enough and the confinement time is long enough. The Lawson criterion is a useful guide to determine the density-time product which must be satisfied in addition to the minimum temperature requirement. If n is the plasma density and τ is the confinement time of the plasma, $n\tau = 10^{14}$ cm^{-3}s is the Lawson criterion for the deuterium-tritium reaction. The energy released in a deuterium-tritium fusion reaction is approximately seven times that obtained in a uranium fission reaction.

Fusion reactions commonly occur in stars where the necessary temperatures are already present and the confinement time occurs as a result of gravitational pressure. These reactions were first produced by humans in the development of thermonuclear weapons. The ideas soon evolved as to how such a reaction could be controlled on a much smaller scale and the energy harnessed for peaceful uses. Since 1950 the extensive development and testing of magnetic confinement techniques, in which a plasma is created and magnetically confined within a "magnetic bottle" for times of the order of seconds, has not yet achieved the requirements for a sustained thermonuclear reaction. Thus the high-power laser has become available at an opportune time to allow new methods to be explored in achieving the goal of useful, efficient electrical power generation using thermonuclear energy. *See* NUCLEAR FUSION.

Fusion power by laser implosion. The laser implosion technique, which is the most extensively studied laser technique, involves the creation of a miniature explosion uniformly over the surface of a very small pellet (50–100 μm diameter) consisting of a mixture of frozen deuterium and tritium. Some newer pellet designs incorporate hollow spheres or layered regions of other materials to improve the burning characteristics. In the explosion process, electrons in the surface region absorb energy from the laser pulse. This energy is then rapidly transferred by collisions to the heavier surrounding ions which then move outward with explosive force at very high velocities. The opposite, momentum-conserving reaction is an inward-moving shock wave that compresses and heats the remaining major portion of the pellet mass. Compressions of 10,000 times liquid density corresponding to a decrease in the pellet diameter to approximately 1/20 of its original value should yield densities of the order of 10^{26} cm^{-3} in the compressed region. From the Lawson criterion the minimum confinement time for such a density would have to be of the order of 10^{-11} to 10^{-12} s. Such a confinement time would inherently occur during the compression because the particle velocities of 10^8 cm/s would result in very little movement for such short time periods. This type of confinement is referred to as inertial confinement. When the high-temperature requirement and the Lawson criterion are simultaneously satisfied, the nuclear reaction or burn will occur throughout the entire pellet mass, releasing the maximum possible energy to the chamber walls. Although such a laser-induced reaction has never been successfully demonstrated in the laboratory, detailed mathematical models have been used to simulate the reaction with computers. Enough details are known from previous studies of nuclear reactions to place a high degree of confidence in these models.

One of the most crucial requirements indicated by the theoretical models is that of having the proper amount of laser energy arriving at the target at each instant of time. If too much energy arrives at the target too soon, the outer surface can be heated sufficiently rapidly that only part of the pellet will be "burned." If too little energy is available, the ignition temperature will not be reached at all. It is thus essential to "tailor" the laser pulse shape to create the appropriate burn sequence. This places severe requirements upon the laser system. Extremely high energies, of the order of 5000 J, will probably be needed to achieve a gain of 1 (which means that as much energy is obtained from the fusion reaction as was put into the reaction by the laser). To achieve a gain of 75, calculations indicate that the laser must provide a 300,000-J pulse lasting for 1 nanosecond, with an efficiency of 10%. From such a laser system a 1000-MW power plant might be possible if 100 pellets are exploded every second.

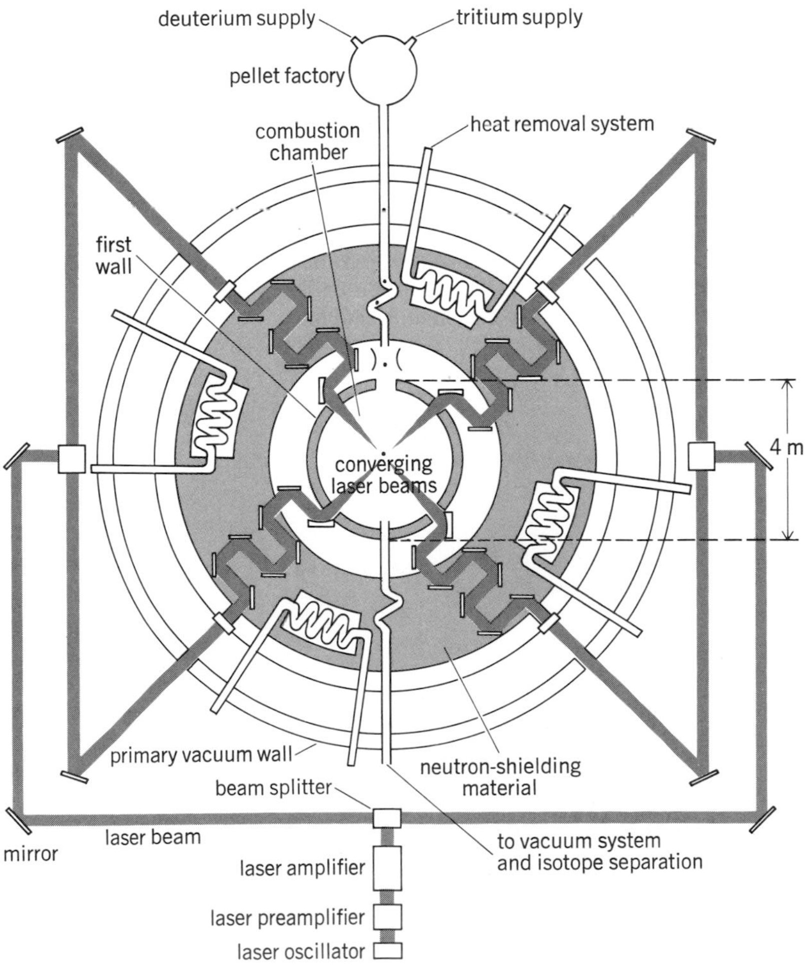

Fig. 1. Diagram of a laser fusion power plant. *(From J. L. Emmett, J. Nuckolls, and L. Wood, Fusion power by laser implosion, Sci. Amer., 230(6):24–37, 1974)*

Laser technology. There are no lasers in existence that can provide the above-mentioned (3×10^5 J) energy or the high efficiency (10%) for short (1-ns) pulses. However, lasers approaching that energy are in the design stages at some laboratories, and breakthroughs in obtaining high efficiency have been made in a new class of ultraviolet lasers. The general design approach toward achieving high energies has been to build a single-laser oscillator that produces the proper pulse shape and length at low energy. The output of this oscillator is then divided into many separate pulses (all still of the same shape and length) and sent through a large number of separate, identical amplifiers, producing many identical high-energy laser beams. The beams are individually focused from various directions to produce a uniform irradiation of the spherical fusion target pellet.

The neodymium-doped glass lasers operating at a wavelength of 1.06 μm (in the near infrared) have achieved energies of up to 1.5×10^4 J but with efficiencies of only up to 0.1%. Large lasers of this type have been built in the United States, Soviet Union, France, England, China, Japan, and Poland to test many of the ideas relating to pellet implosion. The carbon dioxide laser operating at 10.6 μm (middle infrared) has a potential short-pulse efficiency of somewhere between 5 and 10%. A 10^4 J version of this laser operates at a United States laboratory, while lower-energy versions are used in Italy and Japan. A West German laboratory has undertaken development of an atomic iodine laser, operating at 1.31 μm, for fusion experiments.

In addition, several new kinds of molecular lasers, operating in the near ultraviolet, may prove to be useful lasers for fusion if the high-energy capabilities and efficiency of smaller laboratory-size lasers can be shown to scale to larger dimensions. In particular, laboratory-size models of the krypton-fluoride laser, operating at 248 nm, have yielded energies up to 200 J and efficiencies up to 10%. Problems with this laser, for fusion applications, include its inherently long pulse length (10 to 100 times the desired length) and the possibility of damage to optical components, such as lenses and mirrors, from the high-intensity ultraviolet radiation. The problem of long pulse lengths may be overcome by using either a technique known as backward-wave Raman pulse compression (thus converting the 248-nm radiation to 268-nm) or a technique called pulse stacking. The pulse compression involves using the well-studied effect of stimulated Raman scattering, whereas the pulse stacking is a new technique in which various 1-ns oscillator pulses are delayed by different amounts of time and then passed through the amplifier to extract different segments of the 100-ns amplifier gain. The pulses are then delayed again relative to each other so that they all arrive at the target at the same time. Also, a revived interest in developing solid-state lasers, having potentially high energy-storage capabilities, may lead to useful fusion lasers in materials such as vanadium-doped magnesium fluoride.

The most desirable laser wavelength for laser fusion is not yet known, but it is thought to be somewhere between 0.3 and 0.8 μm (ranging from near ultraviolet to near infrared). Problems of optical breakdown, plasma instabilities, and production of suprathermal electrons tend to favor shorter wavelengths, whereas two-photon absorption and optical damage of materials tend to favor longer wavelengths.

A group in the Soviet Union demonstrated the first production of high-energy neutrons from a laser-irradiated target in 1968, and reported indications of implosion and compression in 1972. Later a United States laboratory demonstrated compression by a factor of 100 using a deuterium-tritium pellet irradiated from opposite sides by two neodymium-glass lasers, each having an energy of 50 J. Other United States laboratories have since confirmed the compression and neutron production results with their own tests, but the neutron yields are still many orders of magnitude below that required to reach the break-even point. Experimental studies are directed toward understanding the physics involved in the surface region of the pellet when the laser pulse arrives and the laser energy is absorbed, so that theoretical models can be improved upon.

Power plant. A simplified drawing of a possible design for a laser fusion power plant is shown in Fig. 1. Although the figure shows only four beams converging upon the spherical pellet, tens or even hundreds of beams may be used in an operating power plant, each beam being amplified separately and then focused on the target from a different direction. The converging laser beams would arrive at the central target area via zigzag mirrored paths through a neutron shielding wall. The frozen deuterium-tritium pellets would be dropped from a pellet factory guided electrostatically on a zigzag path through the wall, and timed to arrive at the target in the central region of the chamber simultaneously with the laser pulse. There they are symmetrically irradiated and imploded by the converging laser beams. Compression and heating of the pellet to 10^8°C will cause a thermonuclear fusion reaction to occur throughout the pellet mass, releasing large amounts of energy. Energy would be extracted by a circulating liquid, such as liquid lithium, which could then be used for making steam to drive a turbogenerator. If 100 microexplosions could be produced every second, a 1000-MW power plant would be possible.

Fusion power by laser heating and magnetic confinement. The development of the high-power, long-pulse, CO_2 laser made possible the serious consideration of another technique for initiating laser-induced fusion. This technique involves the use of a long-wavelength laser as an auxiliary source for heating a magnetically confined plasma. Because the plasma would consist primarily of deuterium and tritium ions, the magnetic field would trap these charged particles for long periods of time, and thus the entire laser heating process could be performed much more slowly than in the pellet implosion scheme, thereby making lower densities possible. Much of the technology for laser heating with magnetic confinement can be obtained from existing knowledge available from the extensive studies of magnetic confinement devices. Such devices have been steadily improved upon for many years, but the plasmas produced have never quite reached the appropriate temperature and density for the necessary time

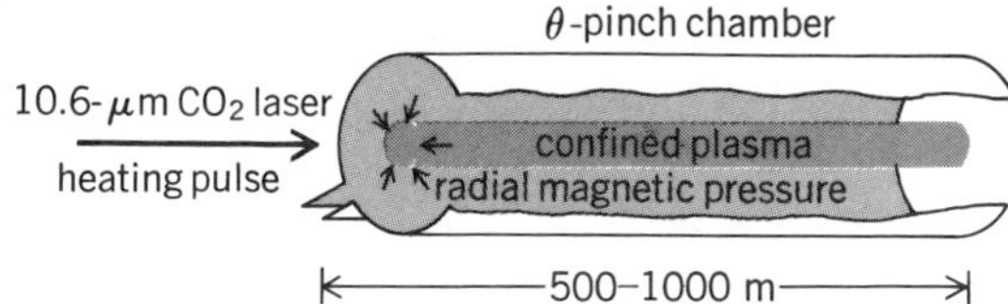

Fig. 2. Diagram of a laser-augmented pinch showing a long-wavelength laser pulse arriving from the left.

(Lawson criterion) to obtain significant energy yields from the fusion reaction. The devices under consideration for laser heating are those that would use a pulsed magnetic field for confinement, which are referred to as laser-augmented pinches, and those that use steady fields, which are termed laser-heated solenoids. In either case, the laser would be used to provide the necessary increase in heating that has not so far been possible with magnetic confinement alone.

A possible scheme is shown in Fig. 2, where the pulsed CO_2 laser arrives from the left and enters the already formed plasma within the θ-pinch. Plasma temperatures of the order of 10^8°C and densities of $10^{17}-10^{18}/\text{cm}^3$ must be maintained for times ranging from 1 to 0.1 ms to satisfy the Lawson criterion. In this density range the absorption coefficient for 10.6 μm radiation would be relatively small and therefore a plasma length of 500–1000 m would be required in order to use the CO_2 laser energy efficiently. At 10^8°C a thermonuclear burn will occur, releasing high-energy neutrons that will heat the chamber wall. This heat energy can be used to produce steam to drive a turbogenerator. Also, for these densities, confinement could be achieved with available magnetic field technology. Plasma leaks from the ends of the cylinder would be minimized by the long plasma lengths and also possibly by incorporating magnetic mirrors at the ends of the cylinder.

One of the problems in this laser-heating scheme is that of keeping the laser beam from being refracted out of the magnetically confined region while it is being absorbed over the long heating path length. It was suggested that a lower plasma density on the cylinder axis might cause the beam to be guided and therefore trapped in the central region due to the refractive effects of the plasma. Energy losses would therefore be minimized as the laser beam was being efficiently absorbed over the entire plasma length. Experiments showing that such guiding occurs have been performed in United States and Canadian laboratories.

Once the ignition temperature is reached in this long plasma, the energy from the high-energy neutrons could be collected in the chamber walls in a similar fashion to that of the laser implosion scheme. The heat would be extracted to produce steam which could then drive a turbogenerator. *See the feature article* EXPLORING ENERGY CHOICES. [WILLIAM T. SILFVAST]

Bibliography: J. L. Emmett, J. Nuckolls, and L. Wood, Fusion power by laser implosion, *Sci. Amer.*, 230(6):24–37, 1974; M. S. Feld, A. Javan, and N. A. Kurnit (eds.), *Fundamental and Applied Laser Physics*, 1973; W. C. Gough and B. J. Eastlund, The prospects of fusion power, *Sci. Amer.*, 224(2):50–64, 1971; S. Jacobs, M. Sargent III, and M. O. Scully (eds.), *High Energy Lasers and Their*

Applications, 1974; W. F. Krupke, E. V. George, and R. A. Haas, Advanced lasers for fusion, in M. Stitch (ed.), *Laser Handbook*, vol. 3, 1979; Thermonuclear neutrons from laser implosion, *Phys. Today*, 27(8):17–19, 1974; C. Yamanaka (ed.), *Laser Interaction with Matter*, Japan Society for the Promotion of Science, 1973.

Lawson criterion

A necessary but not sufficient condition for the achievement of a net release of energy from nuclear fusion reactions in a fusion reactor. As originally formulated by J. D. Lawson, this condition simply stated that a minimum requirement for net energy release is that fusion fuel charge must combust for at least a long enough time for the recovered fusion energy release to equal the sum of energy invested in heating that charge to fusion temperatures, plus other energy losses that occur during the combustion period.

The result is usually stated in the form of a minimum value of $n\tau$ that must be achieved for energy break-even, where n is the fusion fuel particle density and τ is the confinement time. Lawson considered bremsstrahlung (x-ray) energy losses in his original definition. For many fusion reactor cases, this loss is small enough to be neglected compared to the heating energy. With this simplifying assumption, the basic equation from which the Lawson criterion is derived is obtained by balancing fusion energy release against heat input to the fuel plasma. Assuming hydrogenic isotopes, deuterium and tritium at densities n_D and n_T respectively, with accompanying electrons at density n_e, all at a maxwellian temperature T, one obtains Eq. (1),

$$n_D n_T \langle\sigma v\rangle Q \tau \eta_r \geq \left[\frac{3}{2}kT\left(n_D + n_T + n_e\right)\right]\frac{1}{\eta_h} \qquad (1)$$

where the recovered fusion energy release is set equal to or greater than the energy input to heat fuel. Here $\langle\sigma v\rangle$ is the reaction cross section as averaged over the velocity distribution of the ions, Q is the fusion energy release, η_r is the efficiency of recovery of the fusion energy, η_h is the heating efficiency, and k is the Boltzmann constant.

For a fixed mixture of deuterium and tritrium ions, Eq. (1) can be rearranged in the general form of Eq. (2). For a 50-50 mixture of deuterium and tritium the minimum value of $T/\langle\sigma v\rangle$ occurs at about 25 keV ion kinetic temperature (mean ion energies of about 38 keV). Depending on the assumed efficiencies of the heating and recovery processes, the lower limit values of $n\tau$ range typically between about 10^{14} and 10^{15} cm^{-3}. These values therefore serve as a handy index of progress toward fusion, although their achievement does not alone guarantee success. Under special circumstances (unequal ion temperatures, unequal deuterium and tritium densities, and nonmaxwellian ion distributions), lower $n\tau$ values may be adequate for nominal break-even.

$$n\tau \geq F(n_r, n_h, Q)\left[\frac{T}{\langle\sigma v\rangle}\right] \qquad (2)$$

The discussion up to this point has been oriented mainly to situations in which the fusion reactor may be thought of as a driven system, that is, one in which a continuous input of energy from outside the reaction chamber is required to maintain the reaction. Provided the efficiencies of the external heating and energy recovery systems are high, a driven reactor generally would require the lowest $n\tau$ values to produce net power. An important alternative operating made for a reactor would be an ignition mode, that is, one in which, once the initial heating of the fuel charge is accomplished, energy directly deposited in the plasma by charged reaction products will thereafter sustain the reaction. For example, in the D-T reaction, approximately 20% of the total energy release is imparted to the alpha particle; in a magnetic confinement system, much of the kinetic energy carried by this charged nucleus may be directly deposited in the plasma, thereby heating it. Thus if the confinement time is adequate, the reaction may become self-sustaining without a further input of energy from external sources. Ignition, however, would generally require $n\tau$ products with a higher range of values, and is thus expected to be more difficult to achieve than the driven type of reaction. However, in all cases the Lawson criterion is to be thought of as only rule of thumb for measuring fusion progress; detailed evaluation of all energy disruptive and energy recovery processes is required in order properly to evaluate any specific system. *See* NUCLEAR FUSION.

[RICHARD F. POST]

Lightning and surge protection

Means of protecting electrical systems, buildings, and other property from lightning and other high-voltage surges.

The destructive effects of natural lightning are well known. Studies of lightning and means of either preventing its striking an object or passing the stroke harmlessly to ground have been going on since the days when Franklin first established that lightning is electrical in nature. From these studies, two conclusions emerge: (1) Lightning will not strike an object if it is placed in a grounded metal cage. (2) Lightning tends to strike, in general, the highest objects on the horizon.

One practical approximation of the grounded metal cage is the well-known lightning rod or mast (Fig. 1). The effectiveness of this device is evaluated on the cone-of-protection principle. The protected area is the space enclosed by a cone having the mast top as the apex of the cone and tapering out to the base. Laboratory tests and field experi-

Fig. 1. Lightning rods on a house.

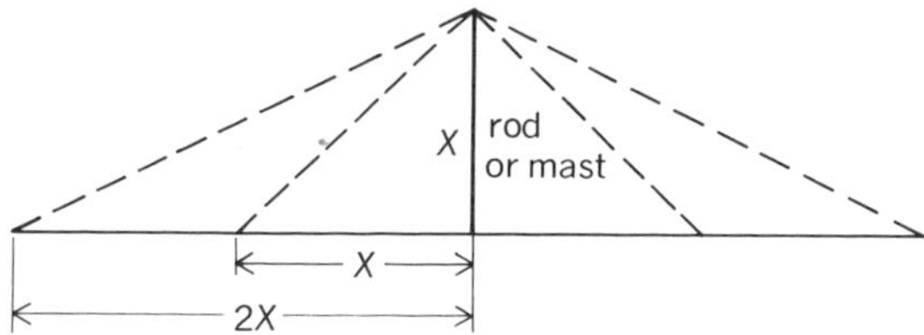

Fig. 2. Cone of protection of a lightning rod or mast.

ence have shown that if the radius of the base of the cone is equal to the height of the mast, equipment inside this cone will rarely be struck. A radius equal to twice the height of the mast gives a cone of shielding within which an object will be struck occasionally. The cone-of-protection principle is illustrated in Fig. 2.

A building which stands alone, like the Empire State in New York City, is struck many times by lightning during a season. It is protected with a mast and the strokes are passed harmlessly to ground. It is interesting to note, however, that lightning has been observed to strike part way down the side of this building (Fig. 3). This shows that lightning does not always strike the highest object but rather chooses the path having the lowest electrical breakdown.

The probability that an object will be struck by lightning is considerably less if it is located in a valley. Therefore, electric transmission lines which must cross mountain ranges often will be routed through the gaps to avoid the direct exposure of the ridges. *See* ELECTRIC DISTRIBUTION SYSTEMS; ELECTRIC POWER SYSTEMS.

Overhead lines of electric power companies are vulnerable to lightning. Lightning appears on these lines as a transient voltage, which, if of sufficient magnitude, will either flash over or puncture the weakest point in the system insulation.

Many of the troubles that cause service interruptions on electrical systems are the results of flashovers of insulation wherein no permanent damage is done at the point of fault, and service can be restored as soon as the cause of the trouble has disappeared. A puncture or failure of the insulation, on the other hand, requires repair work, and damaged apparatus must be removed from service.

There are a number of protective devices to limit or prevent lightning damage to electric power systems and equipment. The word protective is used to connote either one or two functions: the prevention of trouble, or its elimination after it occurs. Various protective means have been devised either to prevent lightning from entering the system or to dissipate it harmlessly if it does.

Overhead ground wires and lightning rods. These devices are used to prevent lightning from striking the electrical system.

The grounded-metal-cage principle is approached by overhead ground wires, preferably two, installed over the transmission phase conductors and grounded at each tower. The ground wires must be properly located with respect to the phase conductors to provide a cone of protection and have adequate clearance from them, both at the towers and throughout the span. If the resistance to ground of the tower is high, the passing of high lightning currents through it may sufficiently elevate the tower in potential from the transmission line conductors so that electrical flashover can occur. Since the magnitude of the lightning current may be defined in terms of probability, the expected frequency of line flashover may be predicted from expected storm and lightning stroke frequency, the current magnitudes versus probability, and the tower footing resistance. Where the expected flashover rate is too high, means to lower the tower footing resistance are employed such as driven ground rods or buried wires connected to the base of the tower (Fig. 4).

The ground wires are often brought in over the terminal substations. For additional shielding of the substations, lightning rods or masts are installed. Several rods are usually used to obtain the desired protection. *See* ELECTRIC POWER SUBSTATION.

Lightning arresters. These are protective devices for reducing the transient system overvoltages to levels compatible with the terminal-apparatus insulation. They are connected in parallel with the apparatus to be protected. One end of the arrester is grounded and also connected to the case of the equipment being protected; the other end is connected to the electric conductor (Fig. 5).

Lightning arresters provide a relatively low discharge path to ground for the transient overvoltages, and a relatively high resistance to the power system follow current, so that their operation does not cause a system short circuit.

In selecting an arrester to protect a transformer, for example, the voltage levels that can be maintained by the arrester both on lightning surges and surges resulting from system switching must be coordinated with the withstand strength of the transformers to these surges. Arrester gap sparkover voltages, discharge voltages on various magnitudes of discharge currents, the inductance of connections to the arrester, voltage wave shape, and other factors must all be evaluated.

Rod gaps. These also are devices for limiting the magnitude of the transient overvoltages. They

Fig. 3. Multiple lightning stroke A to Empire State and nearby buildings, Aug. 23, 1936. Single, continuing stroke B to Empire State Building, Aug. 24, 1936.

Fig. 4. Double-circuit suspension tower on the Olive–Gooding Grove 345-kV transmission line, with two circuits strung and with two ground wires. (*Indiana and Michigan Electric Co., American Electric Power Co.*)

usually formed of two 1/2-in.^{2} (3.2-cm^2) rods, one of which is grounded and the other connected to the line conductor but may also have the shape of rings or horns. They have no inherent arc-quenching ability, and once conducting, they continue to arc until the system voltage is removed, resulting in a system outage.

Fig. 5. Installation of three single-pole station lightning arresters rated 276 kV for lightning and surge protection of large 345-kV power transformer. (*General Electric Co.*)

These devices are applied on the principle that, if an occasional flashover is to occur in a station, it is best to predetermine the point of flashover so that it will be away from any apparatus that might otherwise be damaged by the short-circuit current and the associated heat.

The flashover characteristics of rod gaps are such that they turn up (increase of breakdown voltages with decreasing time of wavefront) much faster on steep-wavefront surges than the withstand-voltage characteristics of apparatus, with the result that if a gap is set to give a reasonable margin of protection on slow-wavefront surges, there may be little or no protection for steep-wavefront surges. In addition, the gap characteristics may be adversely affected by weather conditions and may result in undesired flashovers.

Immediate reclosure. This is a practice for restoring service after the trouble occurs by immediately reclosing automatically the line power circuit breakers after they have been tripped by a short circuit. The protective devices involved are the power circuit breaker and the fault-detecting and reclosing relays.

This practice is successful because the majority of the short circuits on overhead lines are the result of flashovers of insulators and there is no permanent damage at the point of fault. The fault may be either line-to-ground or between phases. Reclosing relays are available to reenergize the line several times with adjustable time intervals between reclosures.

If the relays go through the full sequence of reclosing and the fault has not cleared, they lock out. If the fault has cleared after a reclosure, the relays return to normal.

Permanent faults must always be removed from a system and the accepted electric protective devices are power circuit breakers and suitable protective relays. *See* ELECTRIC PROTECTIVE DEVICES.

[GLENN D. BREUER]

Bibliography: E. Beck, *Lightning Protection for Electric Systems*, 1954; Institute of Electrical and Electronics Engineers, *Revised Report on Standards for Value and Expulsion Type Lightning Arresters*, IEEE Stand. no. 28B, 1956; W. W. Lewis, *The Protection of Transmission Systems Against Lightning*, rev. ed., 1965.

Lignite

Soft and porous carbonaceous material intermediate between peat and subbituminous coal, with a heat value less than 8300 Btu/lb (19,300 kJ/kg) on a moist, mineral-matter-free basis. In North America, lignitic coals are classified as lignite A or lignite B, depending upon whether they have calorific value of more or less than 6300 Btu/lb (14,650 kJ/kg). Elsewhere, lignites are commonly called brown coals, with two major classes being recognized, namely, hard and soft brown coals. The soft brown coals are described as either earthy, resembling peat, or as fragmentary; the hard brown coals are regarded as dull (matte) or bright (glance). The soft brown coals correspond in a general way with lignites of class B, and the hard brown coals with lignites of class A, but the two classifications should not be regarded as strictly interchangeable.

Brown coals generally have a brown color compared with the usual black color of bituminous coal.

Botanical and petrographic studies of lignites have been most vigorously pursued in Europe, where major production has occurred in these coals for many years. The International Committee of Coal Petrology has standardized much of the nomenclature of constituents of lignite. Fluorescence microscopy, particularly applicable to the lignitic coals, has greatly expanded capabilities for description of lignitic coals. Accelerating interest in development of large deposits of lignite in the United States and Canada is increasing the level of studies similar to the levels extensively applied to higher ranks of coal.

Resources and reserves. Generally, less detailed information is available on total world resources and reserves of lignite than for higher-rank coals, but total resources of lignite are large. Europe, excluding the Soviet Union, accounts for about 72% of world production, with East Germany, West Germany, and Czechoslovakia accounting for well over two-thirds of this production. The Soviet Union accounts for about 18% of world production. North America (principally the United States), Asia, and Oceania, each with about 3% of production, account for the remainder of production.

The lignite of North America is mainly of lignite class A. Such resources have been reported in 17 states, but North Dakota and Montana, constituting the principal area of occurrence in the northern Great Plains (and the adjacent area in Canada), have been estimated to have about 97% of total United States lignite resources. Total United States resources of lignite have been estimated to be about 478×10^9 short tons (434×10^9 metric tons), although total coal in the ground now considered economically minable (reserve base) has been estimated at about 8% of resources: 33.6×10^9 short tons (30.5×10^9 metric tons).

Uses. The main uses of lignite have been for generation of electrical energy and for domestic and industrial heating. If shipped any great distance, lignite commonly requires drying. Plans are well advanced to construct a plant near mines in North Dakota to convert lignite to pipeline-quality gas through a consortium of five pipeline companies serving about one-third of the nation's gas consumers. Successful commercialization could see greatly expanded use of vast deposits of lignite in North Dakota and adjacent areas. *See* COAL; COAL GASIFICATION. [JACK A. SIMON]

Bibliography: International Coal Trade, U.S. Bureau of Mines, *ICT*, vol. 47, no. 5, May 1978; *International Handbook of Coal Petrology*, supplements, 1971, 1975.

Liquefied natural gas (LNG)

A product of natural gas which consists primarily of methane. Its properties are those of liquid methane, slightly modified by minor constituents. One property which differentiates liquefied natural gas (LNG) from liquefied petroleum gas (LPG), which is principally propane and butane, is the low critical temperature, about −100°F (−73°C). This means that natural gas cannot be liquefied at ordinary temperatures simply by increasing the pressure, as is the case with LPG; instead, natural gas must be cooled to cryogenic temperatures to be liquefied and must be well insulated to be held in the liquid state.

Peak shaving system. The earliest commercial use for LNG was for storage to meet winter home-heating peak demands. In the construction of long-distance gas pipelines normal practice is to size the line somewhat larger than the average yearly demand rate. Capacity can be increased by 15–20% by operating all compression equipment at the maximum possible flow rate. However, during the coldest days of winter the demand for heating fuel is extremely high and some temporary additional supply of gas is required. Several different techniques are available to supply this temporary peak demand. One reasonable approach is to collect gas during the summer months when demand is low and to store it until winter when it can be withdrawn to meet demands. Since 600 ft^3 of natural gas condenses to less than 1 ft^3 of liquid (600 m^3 to less than 1 m^3), this form of storage is relatively convenient.

Such a peak shaving system was constructed in Cleveland, Ohio, in the early 1940s and operated successfully for several years until one of the storage tanks developed a leak and the escaping liquid caught fire, producing a very destructive conflagration. It was almost a generation before this approach to peak shaving was used again. In the many LNG peak shaving systems that have been built since then, careful attention has been given to separation of storage tanks and other equipment, so that even if there were complete destruction by fire of one storage tank, it would not be likely to result in damage in any other area of the plant.

The large insulated storage tanks constitute the main cost element in LNG peak shaving. Conventional double-walled metal or prestressed concrete tanks with perlite powder insulation between the walls have proved to be satisfactory for units of all sizes. In the development of LNG technology, a number of novel storage schemes have been attempted, but all have weaknesses that result in unsafe or uneconomical operations. The most radical of these attempts utilized frozen earth to form the tank lining for an excavated reservoir. Thermal insulation would theoretically be provided by the earth itself. With large enough tanks, the ratio of surface area to volume declines sufficiently so that the heat-conduction rate of natural soil results in a tolerably low evaporation rate of the liquefied gas. Several attempts to exploit this scheme have failed because the intrusion of an underground stream into the frozen earth area created an unacceptable increase in heat input and liquid evaporation rate. A second weakness of this scheme is that the soil around the tank becomes saturated with hydrocarbon vapors, greatly increasing the potential for an uncontrollable conflagration should a fire start in the tank area.

Systems relying on reservoirs lined either with special concrete or with plastic films or foams have all failed. In these experiments the unequal contraction of the liner during cool-down to the cryogenic temperature of the LNG has caused cracks to open in the liner. Subsequent loss of LNG into the insulation space or into the soil around the tank has made the projects both unsafe and uneconomical.

A tragic accident occurred at an installation on Staten Island, NY, in 1973. Workers were attempting to repair the plastic film liner in the tank. The tank had been emptied and held out of service for a year before the repair work was started. Somehow, a fire originated in the equipment being used inside the tank. The fire spread, and eventually the roof collapsed; all of the men working in the tank were killed. The circumstances were unlike those of the earlier accident in Cleveland, for no LNG was in the tank or anywhere in the plant site at the time. No LNG facility has had a fire or other serious problem since the Cleveland disaster. With more than 100 LNG storage facilities operating around the world, this is an exemplary safety record.

Liquefaction equipment represents a relatively small portion of total project cost and has not produced much variation in design. One novel approach has been used successfully by the San Diego Gas and Electric Co. The peak shaving plant is located adjacent to a large electrical generating station, which is fueled by natural gas supplied at transmission-line pressure of about 600–800 psi. The gas is first processed by the liquefier through heat exchangers and expansion turbines. A fraction of the gas is liquefied, and the remainder returns through the heat exchangers to pass on to the electric generating station at low pressure. In this way the liquefaction plant uses the otherwise wasted pressure of the pipeline. The maximum amount of LNG which can be produced in any given location is limited to at most 20% of the quantity of gas which drops from pipeline pressure to low pressure.

Ocean transport. In Algeria, Libya, Alaska, and other oil-producing areas there is not much demand for the natural gas that is produced along with the oil. One economical way of transporting the surplus gas to markets in industrial centers is to liquefy the gas and ship it in specially designed insulated tankers.

The original transoceanic LNG system transported liquefied gas from Algeria to England and northern France. Since 1965 about 150,000,000 ft^3 (4.2×10^6 m^3) of gas has been liquefied daily for shipment aboard three specially insulated tank ships. Larger systems have been built to supply Italy and Spain from Libya, Japan from Alaska and Borneo, and southern France from Algeria. Smaller quantities have been supplied to cities in the eastern United States from Libya and Algeria. Liquefaction capacity is expanding in these source areas and in other areas rich in gas resources.

Other uses. As peak shaving and LNG tanker systems have developed, attention has been directed to the potential use of LNG as a fuel in different types of vehicles. LNG has a number of attractive characteristics as an engine fuel. It has an antiknock value well over 100, without any additives. It is extremely clean-burning, resulting in low maintenance costs and a minimum of air pollution from engine exhaust. In gas turbines it provides a large heat sink and burns with a relatively nonluminous flame, both of which assist in engine cooling. Its specific energy per pound of fuel is 15% higher than for gasoline or kerosine. In rocket engines it provides the highest specific impulse of any hydrocarbon fuel.

The primary limitation on the development of LNG as an engine fuel is the general shortage of natural gas in the United States, Europe, and Japan. LNG imported into these areas is intended to supplement the available supplies of natural gas for home heating and cooking. Studies on the use of LNG engine fuel for a variety of transport vehicles have shown its advantages. However, manufacturers are reluctant to undertake engine and vehicle redesign efforts since adequate supplies of LNG cannot be assured. *See* LIQUEFIED PETROLEUM GAS (LPG).

[ARTHUR W. FRANCIS]

Bibliography: C. H. Gatton, *Liquefied Natural Gas Technology and Economics*, 1967; National Fire Protection Association, *Liquefied Natural Gas: Production, Storage, and Handling*, 1972; J. W. White and A. E. S. Neumann (eds.), *Proceedings of 1st International Conference on Liquefied Natural Gas*, Chicago, 1968.

Liquefied petroleum gas (LPG)

A product of petroleum gases, principally propane and butane, which must be stored under pressure to keep it in a liquid state. At atmospheric pressure and above freezing temperature, these substances would be gases. Large quantities of propane and butane are now available from the gas and petroleum industries. These are often employed as fuel for tractors, trucks, and buses and mainly as a domestic fuel in remote areas. Because of the low boiling point (−44 to 0°C) and high vapor pressure of these gases, their handling as liquids in pressure cylinders is necessary. Owing to demand from industry for butane derivations, LPG sold as fuel is made up largely of propane. On a gallonage basis, production of LPG in the United States exceeds that of kerosine and approaches that of diesel fuel.

Operating figures for gasoline, diesel, and LPG fuels show that LPG compares favorably in cost per mile. LPG has a high octane rating, making it useful in engines having compression ratios above 10:1.

Another factor of importance in internal combustion engines is that LPG leaves little or no engine deposit in the cylinders when it burns. Also, since it enters the engine as a vapor, it cannot wash down the cylinder walls, remove lubricant, and increase cylinder-wall, piston, and piston-ring wear. Nor does it cause crankcase dilution. All these factors reduce engine wear, increase engine life, and keep maintenance costs low. However, allowances must be made for the extra cost of LPG-handling equipment, including relatively heavy pressurized storage tanks, and special equipment to fill fuel tanks on the vehicles. *See* GAS TURBINE; INTERNAL COMBUSTION ENGINE; PETROLEUM PRODUCTS. [MOTT SOUDERS]

Liquid fuel

Any of the liquids burned to produce usable energy in the form of heat or light. Ease of ignition, clean burning, and adaptability to transportation and storage have all been favorable features of the hot flame and the bright light from liquid fuels. Animal oils, vegetable oils, and petroleum have been used historically, but of all the liquid fuels used, petroleum has become the dominant basis for extensive industrial development and private convenience. This may be the century history will designate as "the age of petroleum." In the United States, beginning with production of petroleum from wells in Pennsylvania (1859), the extensive production of the 20th century has allowed escape from dependence upon animal oils. A naturally occurring hydrocarbon, petroleum ascended quickly to its principal role as an energy source as a result of the development of internal combustion engines and the emergence of machines of transportation. *See* INTERNAL COMBUSTION ENGINE.

Characteristics of petroleum. Petroleum is chemically a very complex mixture of carbon and hydrogen compounds. Minor impurities of oxygen, nitrogen, and sulfur vary in different crude oils. The term "crude oil" is in common usage to distinguish between the natural oil derived from rocks and the refined lubricants available from the neighborhood service station. All crude oil is lighter than water, immiscible with water, and soluble in ether, naptha, or benzine. Variations from solid black gilsonite to viscous black and brown asphalts, tar, and pitch, to light green and yellow crude oils are to be found in different localities where oil is produced in the United States. Terms such as "hydrocarbon" and "bitumen" are used interchangeably with the word "petroleum." However, petroleum (from the Latin *petra* meaning rock or stone and *oleum* meaning oil) is the most common usage. Petroleum occurs at the surface as springs and seepages, and in subsurface rocks where it is in the openings between the grains and in cracks in the rock. *See* PETROLEUM.

Origin of petroleum. Although it is generally agreed that petroleum is derived from organic matter mixed with sediments that later form the sedimentary rocks of the Earth's crust, there is no common agreement as to whether that organic matter was derived from animals or plants. Whether the organic matter was accumulated exclusively in marine sedimentary environments or in brackish-water continental-marginal sediments is also in contention. Conversion of organic matter into petroleum is associated with the heat, pressure, and fluids involved in converting sediments into sedimentary rock. Solid organic matter is dispersed through many sedimentary rocks, but the intermediate steps of conversion to liquid petroleum have not been observed.

Petroleum geology. Wide variation between crude oils makes a common origin doubtful, but there are recognizable requisites for the accumulation of usable quantities of petroleum in the crustal rocks of the Earth. Permeable rocks must exist which allow the passage of fluids through openings between the particles of the rock or through cracks and cavities that exist. For the accumulation of oil pools, rock porosity and permeability, then, are vital to the mechanism of accumulation. The rock layer or rock body in which an accumulation has occurred is referred to as reservoir rock. Equally important in the entrapment of petroleum is an impervious capping which impedes or stops the upward movement of fluids in the rock mass.

Petroleum is driven into an entrapped pool by the movement of water and gas in the subsurface. Oil migration and entrapment is most common within layers of sedimentary rocks and, hence, the principal production of the world is from sandstones, limestones, conglomerates, and other common sedimentary layers. Folded and faulted sedimentary layers provide potential sites for traps of oil. Subtle changes of porosity and permeability within sedimentary rock layers will also impede fluid movements and result in entrapment. These so-called stratigraphic traps are some of the most difficult sites to discover from surface investigations, yet many pools, still to be discovered, will be of this type.

Alternates to petroleum. There is disturbing evidence that over half of the oil that is to be discovered in the United States has already been produced. Since 1965, an increasing proportion of society's needs were met by importation of crude oil. Over a third of the crude oil used in the United States is imported. Along with the natural petroleum upon which the culture has become dependent, society must now accept the costly technology and availability of synthetic liquid fuels. Oils distilled from the kerogens of oil shale are an alternative to meet part of the demand for energy from liquid fuels. Crude oils and refined products from conversion of coal represent an energy resource within reach of modern-day technology. Liquefied natural gas, primarily methane, together with liquefied petroleum gas, propane and butane, must also be made available as a supplement for domestic oil supplies. The technology of producing liquefied natural gas, although not entirely efficient, is being perfected under the support of massive Federal and industrial research grants. As domestic production of petroleum gradually diminishes and prices escalate, it is expected that the substitution of these synthetic liquid fuels will become necessary. *See* COAL GASIFICATION; COAL LIQUEFACTION; LIQUEFIED NATURAL GAS (LNG); LIQUEFIED PETROLEUM GAS (LPG); OIL SHALE; SYNTHETIC FUEL. [ORLO E. CHILDS]

Bibliography: M. King Hubbert, *The Environmental and Ecological Forum 1970–1971*, 1972; K. Landes, *Petroleum Geology*, 2d ed., 1975; A. I. Levorsen, *Geology of Petroleum*, 1954; F. Park, Jr., *Earthbound*, 1975.

Lithium primary cell

A primary cell whose anode is composed of lithium. The lithium cell is a development which has a number of advantages over other primary cell systems. Lithium is an attractive anode because of its reactivity, light weight, and high voltage (between 1.6 and 3.6 V, depending on the other electrode).

The advantages include high energy density, flat discharge characteristics, excellent service over a wide temperature range (as low as −40°C), and good shelf life (up to 5 years without refrigeration).

Nonaqueous solvents are used as the electrolyte because of the solubility of lithium in aqueous solutions. Organic solvents, such as acetonitrile and propylene carbonate, and inorganic solvents, such as thionyl chloride, are typical. A compatible solute is added to provide the necessary electrolyte conductivity. A number of different materials—sulfur dioxide, carbon monofluoride, vanadium pentoxide, manganese dioxide, copper sulfide, and so forth—are used as the active cathode materials.

The table lists the important cell systems. The three systems thought to be the most important are the types that use sulfur dioxide (SO_2), thionyl chloride ($SOCl_2$), and iodine (I_2).

$Li\text{-}SO_2$ cell. In the $Li\text{-}SO_2$ system the SO_2 is used for the cathode; acetonitrile (CH_3CN) and lithium bromide (LiBr) are used for the electrolyte; lithium foil is used for the anode; and polypropylene is used as the separator. The cell reactions are given by reaction (1).

$$\text{Anode:} \quad 2Li \rightarrow 2Li^+ + 2e^- \tag{1a}$$

$$\text{Cathode:} \quad 2SO_2 + 2e^- \rightarrow S_2O_4^{-2} \tag{1b}$$

$$\text{Overall:} \quad 2Li + 2SO_2 \rightarrow \underset{\text{Lithium dithionite}}{Li_2S_2O_4} \tag{1c}$$

The good shelf life of this cell is attributed to the protective film formed by the initial reaction of lithium and sulfur dioxide, which prevents further reaction or loss of capacity during storage. Claimed energy densities have been reported to be as high as .5 Wh/cm³ and 330 Wh/kg.

This cell is initially pressurized from 2 to 4 atmospheres (200 to 400 kilopascals) and is capable of very high currents (50 amperes for D size). The combination of these two facts leads to a serious safety problem because excessive heat generated by continuous-high-rate or short-circuit discharge can cause extremely high internal pressures (over 30 atm or 3 MPa). Therefore, the use of a vent to limit the pressure buildup to preset levels is required. In additition, when multiple-cell arrangements are needed, the use of diodes and fuses is recommended to prevent accidental cell charging, short circuits, and damage from cell reversals.

$Li\text{-}SOCl_2$ cell. In the $Li\text{-}SOCl_2$ system the $SOCl_2$ is used as both the cathode and the electrolyte; lithium foil is used as the anode; and lithium aluminum chloride ($LiAlCl_4$) is used as the solvent. The cell reaction is given by reaction (2). The

$$4Li + 2SOCl_2 \rightarrow 4LiCl + SO_2 + S \tag{2}$$

cathodic reaction forms sulfur monoxide (SO), an unstable biradical which dimerizes and decomposes, undergoing exothermic reaction (3).

$$SOCl_2 + 2e^- \rightarrow SO + 2Cl \tag{3a}$$

$$2SO \rightarrow (SO)_2 \tag{3b}$$

$$(SO)_2 \rightarrow S + SO_2 \tag{3c}$$

Claimed energy densities have been reported to be as high as .9 Wh/cm³ and 420 Wh/kg.

This cell is not initially pressurized, but has a protective film similar to the $Li\text{-}SO_2$ cell which assures excellent shelf life. Some variations in construction have significantly reduced the danger of explosion, but nevertheless this cell (like the $Li\text{-}SO_2$) should never be deliberately charged, forced open, or disposed of in fire.

One other disadvantage found in lithium cells is a delay in voltage brought about by the same film which aids shelf life. The delay can be only seconds ($Li\text{-}SO_2$) or as much as 10 or more minutes ($Li\text{-}SOCl_2$).

$Li\text{-}I_2$ cell. In the $Li\text{-}I_2$ system the I_2 is used as the cathode, LiI as the solid-state electrolyte, lithium for the anode, and poly-2-vinylpyridine (P2VP) as the separator. The P2VP and LiI are actually bound together in a charge transfer complex and as such simultaneously serve the functions of cathode, depolarizer, and separator. The cell reactions are given by reaction (4). Claimed energy densities

$$\text{Anode:} \quad 2Li \rightarrow 2Li^+ + 2e^- \tag{4a}$$

$$\text{Cathode:} \quad 2Li^+ + 2e^- + P2VP \cdot nI_2 \rightarrow P2VP \cdot (n-1)I_2 + 2LiI \tag{4b}$$

$$\text{Overall:} \quad 2Li + P2VP \cdot nI_2 \rightarrow P2VP \cdot (n-1)I_2 + 2LiI \tag{4c}$$

have been reported to be as high as .8 Wh/cm³ and 230 Wh/kg.

This cell is designed primarily for cardiac pacemakers and works with a typical current drain of 30 microamperes, with a self discharge said to be approximately 10% in 10 years. *See* BATTERY; PRIMARY BATTERY; SOLID-STATE BATTERY.

[JACK DAVIS; KENNETH FRANZESE]

Bibliography: P. Bro, Heat generation in Li/$SOCl_2$ cells, *Power Sources 7: Proceedings of the 11th International Power Sources Symposium*, pp. 571–582, 1979; R. W. Graham (ed.), *Primary Batteries: Recent Advances*, Chem. Technol. Rev. no. 105: Energy Technol. Rev. no. 25, pp. 126–187, 1978; C. C. Liang and C. F. Holmes, The lithium/iodine pacemaker battery, *Progress in Batteries and Solar Cells*, vol. 2, pp. 50–53, 1979; L. F. Martin (ed.), *Dry Cell Batteries: Chemistry and Design*, pp. 86–91, 1973; R. T. Mead, C. F. Holmes, and W. Greatbatch, Design evolution of the lithium iodine pacemaker battery, *Proceedings of the Symposium on Battery Design and Optimization: Electrochemical Society Proceedings*, vol. 79–1, pp. 327–333, 1979; H. Taylor, Stability of Li/SO_2 cell, *Proceedings of the 12th Intersociety Energy Conversion Engineering Conference*, pp. 288–295, 1977; H. Taylor, S. Simena, and E. Ralto, Performance characteristics of small Li/$SOCl_2$ cells, *Proceedings of 12th Intersociety Energy Conversion Engineering Conference*, pp. 296–301, 1977.

Lithium primary cell systems

Type	Nominal voltage	Nominal energy density, Wh/cm³
$Li\text{-}SO_2$	2.9	.5
Li-CuO	1.6	.6
Li-(CF)x*	2.8	.6
Li-CuS	2.0	.5
$Li\text{-}Ag_2CrO_4$	3.3	.6
$Li\text{-}I_2$	2.8	.25–.8
$Li\text{-}PbI_2PbS$	1.9	.5
$Li\text{-}SOCl_2$	3.5	.9
$Li\text{-}V_2O_5$	3.4	.7
$Li\text{-}MoO_3$	2.9	.6
$Li\text{-}MnO_2$	2.8–3.4	.6

*Variable chemical composition.

Machine

A combination of rigid or resistant bodies having definite motions and capable of performing useful work. The term mechanism is closely related but applies only to the physical arrangement that provides for the definite motions of the parts of a machine. For example, a wristwatch is a mechanism, but it does no useful work and thus is not a machine.

Machines vary widely in appearance, function, and complexity from the simple hand-operated paper punch to the ocean liner, which is itself composed of many simple and complex machines. No matter how complicated in appearance, every machine may be broken down into smaller and smaller assemblies, until an analysis of the operation becomes dependent upon an understanding of a few basic concepts, most of which come from elementary physics. [RICHARD M. PHELAN]

Bibliography: R. M. Phelan, *Fundamentals of Mechanical Design*, 3d ed., 1970; J. E. Shigley, *Mechanical Engineering Design*, 1972.

Magnetic circuits

Closed paths of magnetic flux; also a design method using such paths to compute the magnetic field of a core geometry that is often encountered, for instance, high-permeability flux-path segments (and their associated air gaps), each segment having reasonably definite length and area. Examples of magnetic circuits are transformer cores, relay frames, and iron parts of electrical machinery. The magnetic-field equations for these devices look so similar to dc circuit equations that they are called magnetic circuits.

Reluctance. If NI ampere-turns link one closed core, then Eqs. (1) hold, where l is length, A is

$$\Phi \times \frac{l}{\mu A} = NI$$

$$\text{(Flux)} \times \text{(reluctance)} = \text{magnetomotive force (mmf)} \quad (1)$$

$$\Phi \times R = \text{mmf}$$

cross section, and μ is permeability large enough so that the flux Φ is nearly completely confined to the core (Fig. 1).

If the flux path is a sequence of dissimilar segments, its total reluctance is the sum of segment reluctances as given by Eq. (2).

$$R = \sum \frac{l_n}{\mu_n A_n} \quad (2)$$

If the (constant) flux divides among several parallel flux paths, it does so in inverse proportion to their reluctances.

These series and parallel reluctances have the same algebra as do series and parallel dc resistances; hence the fundamental equation is often called Ohm's law for magnetic circuits, even though mmf is not a force, Φ is not a flow, and the equation is not Ohm's but Hopkinson's.

Specific use of magnetic-circuit reluctance is nearly always qualitative; it is handy for explanation and discussion. How much flux change occurs when reluctance is altered is usually determined by a nonlinear calculation which does not compute a numerical value for reluctance itself. Accordingly, there is no common name for a reluctance unit.

Quantitative calculations. Three principles are applied:

1. Magnetic-flux lines are endless: Φ is the same at every point in the flux bundle, even though the lines are specially crowded at some places.
2. The relation between flux density B and field vector H at any point is a property of the matter at that point. From (experimental) charts for the material, either B or H can be found from the other (Fig. 2).
3. Ampère's line integral $\oint H \cdot dl = \Sigma I_{\text{linked}}$ is taken along the path followed by the flux bundle being analyzed.

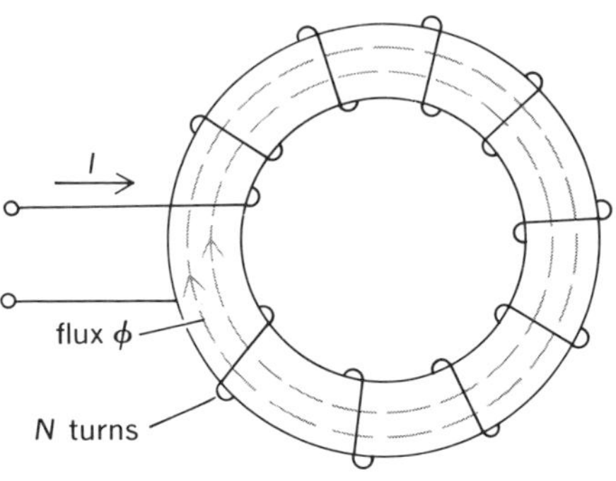

Fig. 1. Diagram of a toroidal magnetic circuit. (*A. E. Fitzgerald, D. E. Higginbotham, and A. Grabel, Basic Electrical Engineering, McGraw-Hill, 1967*)

Given total flux Φ, flux-density in the nth segment is $B_n = \Phi/A_n$. Each H_n is then found from its B_n by chart, and the Ampère integral is evaluated as shown in Eq. (3).

$$\oint H \cdot dl = \Sigma H_n l_n = NI \quad (3)$$

Units. Numerical evaluation of the Ampère integral requires consistent units. Modern practice uses the ones listed in the table.

Many published curves still use the older cgs units: Φ in maxwells, B in gauss, and H in oersteds, with $\mu_{\text{air}} = 1$. Conversion to one of the systems in the table is recommended; it avoids the absolute amperes and factors of 4π that occur in the cgs equations. This conversion, given below, is straightforward because the flux lines keep their identity.

One weber is 10^5 kilolines (10^8 lines or 10^8 maxwells).

One weber/m^2 is 64.5 kilolines/in.^{2} (10^4 lines/cm^2 or 10^4 gauss).

One amp-turn/m is 2.51×10^{-2} amp-turns/in. and corresponds to $4\pi \times 10^{-7}$ weber/m^2 in air (or to $4\pi \times 10^{-3}$ gauss in air or $4\pi \times 10^{-3}$ oersteds).

Special situations. Some rules of thumb allow for leakage flux and for reduction of flux density when the lines spread in air gaps. Magnetic circuits that include permanent magnets or superconductors or plasmas need care in their analysis. For example, when a slice of permanent-magnet material in inserted as one of the flux-path segments, it supplies a magnetomotive force that requires calculation. Instead of puzzling over what H is in a permanent magnet, it is better to replace with a slice of normal material (that is, of the same dimensions and same incremental permeability) plus ampere-turns around its edge to make the same magnetic moment as the original magnet slice.

In each piece of superconducting material, two things happen: (1) Unless the superconductor is very thin, all the flux lines detour around the out-

Commonly employed units for magnetic circuits

Quantity	mks units	Engineering units
Flux density B	Webers/m^2	Kilolines/in.^{2}
Flux Φ	Webers	Kilolines
Field vector H	Amp-turns/m	Amp-turns/in.
$\mu_{\text{air}} = B/H$ in air	$4\pi \times 10^{-7} = \frac{1}{7.95 \times 10^5}$	$3.18 \times 10^{-3} = \frac{1}{313}$
Length	Meters	Inches

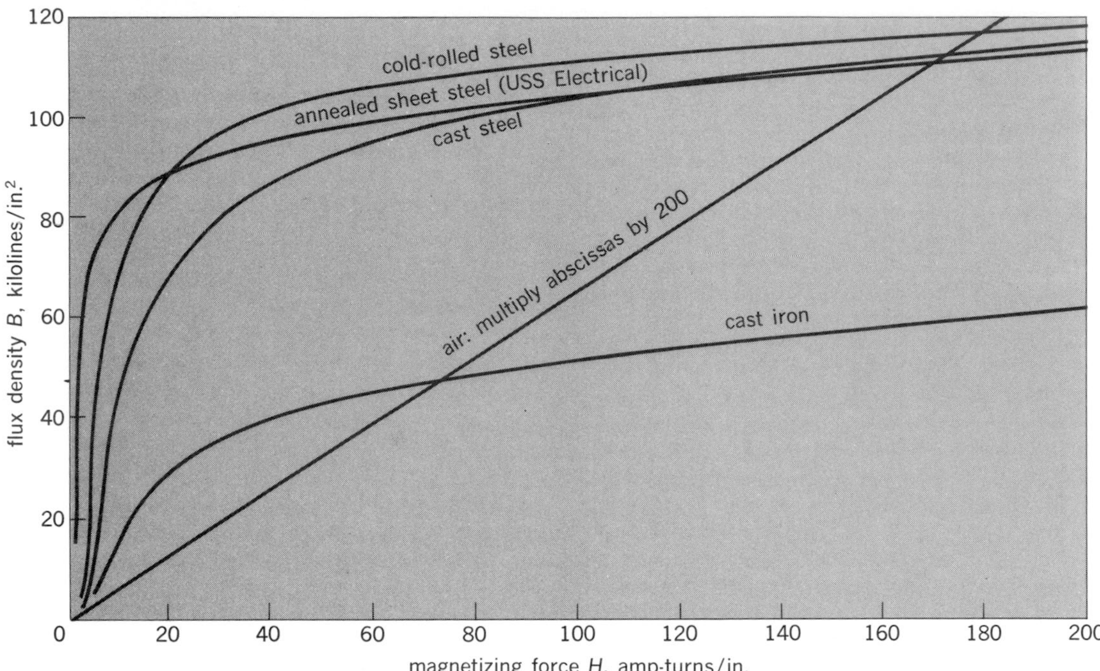

Fig. 2. Graph of the normal magnetization curves for some magnetic materials. (*A. E. Fitzgerald, D. E. Higginbotham, and A. Grabel, Basic Electrical Engineering, McGraw-Hill, 1967*)

side of it, and (2) any attempt to change Φ near a superconductor starts nondecaying eddy currents that hold Φ to (almost) its initial value. More accurate (quantum) descriptions of these phenomena have been published. The flux rejection of a superconductor can be used to channel the flux when this is desired; several devices employ very thin superconductors into which some flux can be driven in order to modify the onset of superconductivity.

The added ampere-turns from a few ions in the flux path can be found by eddy-current methods if time variation is not too violent. However, when there are enough of these ions to be called a plasma they interact by collision and the resulting currents and magnetomotive force are not easily found.

[MARK G. FOSTER]

Bibliography: A. E. Fitzgerald, D. E. Higginbotham, and A. Grabel, *Basic Electrical Engineering*, 4th ed., 1975; J. Hopkinson and E. Hopkinson, Dynamo-electric machinery, *Phil. Trans. Roy. Soc.*, pt. 1, 1886; R. J. Smith, *Circuits, Devices and Systems*, 3d ed., 1976; R. Stein and W. T. Hunt, *Electrical Power System Components: Transformers and Rotating Machines*, 1979; M. Tinkham, *Introduction to Superconductivity*, 1975.

Magnetic field

A condition existing in the vicinity of a magnetic body (or a current-carrying medium) whereby the magnetic forces due to the body (or current) are detectable. In a magnetic field, there is a force on a moving charge in addition to the electrostatic (Coulomb) forces between charges, or there is a force on a magnetic pole. The description of the field may be in terms of magnetic induction (flux density) or it may be in terms of magnetic field strength.

The magnetic induction B of the field is defined from the force F on a moving charge q or current element of length l carrying a current I by Eq. (1).

$$B=\frac{F}{qv\sin\theta}=\frac{F}{Il\sin\theta} \tag{1}$$

The direction of the magnetic induction is that direction in which the force on the moving charge is zero. The factor $v\sin\theta$ is then the component of the velocity of the charge in a direction perpendicular to B. The mks unit of B is the newton/ampere-meter or weber/meter^2.

The magnetic induction at a given point due to a current may be found from Ampère's law, Eq. (2), which may also be expressed as Eq. (2*b*). Equation (2*a*) gives the contribution of a current element Idl

$$dB=\frac{\mu_0}{4\pi}\frac{Idl\sin\theta}{r^2} \tag{2a}$$

$$B=\frac{\mu_0}{4\pi}\int\frac{Idl\sin\theta}{r^2} \quad \text{(vector sum)} \tag{2b}$$

to the magnetic induction, where μ_0 is the permeability of empty space, r the distance of the point to the current element, and θ the angle between the current element and the line joining the element to the point.

The direction of B, from Ampère's law, is perpendicular to the plane determined by a line tangent to dl and the line joining the current element to the point at which B is determined.

A second magnetic vector quantity, the magnetic field strength or intensity H, may be defined in part from Ampère's law. The part of the field strength that is due to currents is given by the vector sum from Eq. (3). Thus, H is computed in the

$$H=\frac{1}{4\pi}\int\frac{Idl\sin\theta}{r^2} \tag{3}$$

same manner as the flux density B, but without the factor μ_0. The direction is specified in exactly the same manner as B. Thus, H depends only on the current present, and not upon the properties of the surrounding medium. This definition of H is only partial because it does not include contributions by magnetic poles if they are present in the neighborhood.

The mks unit of magnetic field strength appears from the defining equation, when current is in amperes and dl and r are in meters, as ampere per meter. Because many of the equations that are derived from the defining equation involve the number of turns N of a coil times the current, the ampere-turn per meter (amp-turn/m) is also used as an equivalent unit.

Magnetic poles. A body can be magnetized by bringing it into a magnetic field due to currents or magnets. Except in the case of a ring magnetized along its circumference, the field associated with a magnetized body extends to the region surrounding the body. The external effect usually appears in limited regions of the body called poles. A magnetized bar of iron has two poles, one at either end; and from the fact that the bar will set itself in an approximate north-south direction in the Earth's field, it appears that there are two kinds of poles. The pole that is at the north end of the bar is called a north-seeking pole; that at the south end is called a south-seeking pole. The two poles at the ends are merely indications of the continuous magnetization within the body. An indication of the validity of this statement is the fact that when a bar magnet is broken into two parts, two new poles appear at the break, and the orientation of the poles in each fragment is the same as it was in the original magnet.

It is observed that magnetic poles exert forces on each other and upon moving charges in the region near the poles. There is a field near the pole, and the pole may be considered as the cause of that field.

If the poles of a magnetized body are small enough that they may be considered point poles, the force that one pole exerts upon another is found to be proportional to the product of the pole strengths m and m' and inversely proportional to the square of the distance r between them. This statement is called Coulomb's law of magnetostatics and is written as Eq. (4). The proportionality

$$F=k'\frac{mm'}{r^2} \tag{4}$$

factor k' depends upon the units used and upon the medium between the poles. For empty space in the mks system, k' is assigned the value $\frac{1}{4}\pi\mu_0$. The unit of pole strength associated with this choice is the weber.

The magnetic field strength or magnetic intensity H may be expressed as the force per unit north-seeking magnetic pole as in Eq. (5). Then,

$$H=\frac{F}{m'} \tag{5}$$

from Coulomb's law, the contribution of a point pole of strength m to the magnetic field strength near the pole is given by Eq. (6). The direction of H

$$H=\frac{F}{m'}=\frac{1}{4\pi\mu_0}\frac{m}{r^2} \tag{6}$$

is away from north-seeking poles and toward south-seeking poles. The contribution of several poles is the vector sum of the contributions of the individual poles as shown in Eq. (7). If the H due to

$$H=\frac{1}{4\pi\mu_0}\sum\frac{m}{r^2} \quad \text{(vector sum)} \tag{7}$$

poles is to be in the same units as the H due to currents, the unit of pole strength must be chosen properly. If H is to be in amperes per meter when μ_0 is in webers per ampere-meter and r is in meters, then m must be in webers.

If the poles are distributed over surfaces or throughout volumes, the summation becomes an integral as shown in Eq. (8).

$$H=\frac{1}{4\pi\mu_0}\int\frac{dm}{r^2} \quad \text{(vector sum)} \tag{8}$$

The general expression for the field strength due to both currents and poles is given by Eq. (9), where the integrals represent vector sums.

$$H=\frac{1}{4\pi}\int\frac{Idl\sin\theta}{r^2}+\frac{1}{4\pi\mu_0}\int\frac{dm}{r^2} \tag{9}$$

If a toroidal coil has an iron core, the iron is magnetized by the current of the coil. The magnetic field strength within the core is given entirely by the first term of the equation for H, since there are no poles. For a long straight solenoid with an air core (illustration *a*), H is entirely due to the current and is found by integration to be $H_I=NI/l$, where N is the number of turns of the coil, and l is the length of the solenoid. When an iron core is inserted into the solenoid (illustration *b*), the iron becomes magnetized, and poles appear at each end. The contribution of the current to H remains the same as before, but the second term now contributes to H components that are opposite in direction to H_I and that vary along the bar because of the variation in distances from the poles and because of the variation of the permeability of the iron. The effect of the poles on H is greatest at the ends near the poles. The magnetization of the iron is not uniform. The effect of the poles is essentially demagnetizing; it is large for short magnets and negligible at the center of a long magnet.

Lines of force. As in the case of magnetic induction B, which can be represented by lines called magnetic flux, the vector quantity H may be represented by lines called lines of force. The number of lines of force per unit area of a surface perpendicular to H is made equal to the value of H. The direc-

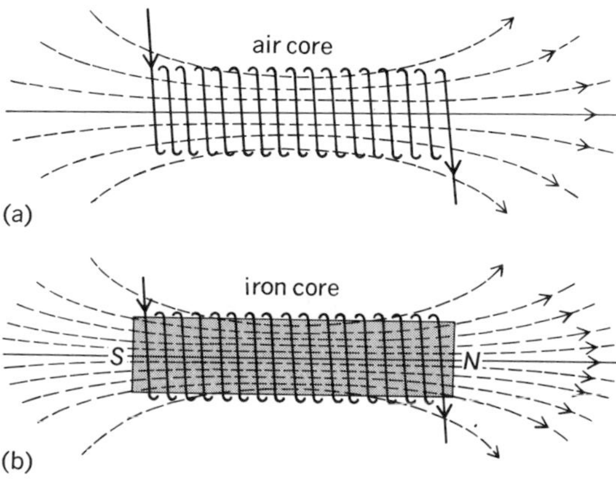

Flux in a solenoid (*a*) containing only air (essentially vacuum) and (*b*) containing an iron core.

tion of the lines of force is the direction of the field. The lines of force are closed curves, as are the lines of induction.

Energy of a magnetic field. Consider a Rowland ring (a ring-shaped sample of magnetic material) surrounded by a coil of N turns in which there is a current I. The field strength H within the ring is given from Ampère's law by Eq. (10), where l is the mean circumference of the ring.

$$H=\frac{NI}{l} \tag{10}$$

In building up the current in the coil, energy must be supplied that becomes energy of the magnetic field. This energy is given by Eq. (11), where

$$W=\tfrac{1}{2}LI^2 \tag{11}$$

L is the self-inductance of the coil.

But L is defined by Eq. (12), where Φ is the flux,

$$L=\frac{N\Phi}{I} \tag{12}$$

and W by Eq. (13), where V is the volume of the

$$W=\frac{1}{2}\frac{N\Phi}{I}I^2=\tfrac{1}{2}N\Phi I=\tfrac{1}{2}NBAI$$

$$=\frac{1}{2}\frac{NI}{l}BV=\tfrac{1}{2}HBV \tag{13}$$

core, and A is the mean cross-sectional area of the ring. The energy per unit volume of the field is then given by Eq. (14). If μ is the permeability of

$$\frac{W}{V}=\tfrac{1}{2}HB \tag{14}$$

the core, then $B=\mu H$ and the energy density of the field may be written as shown in Eq. (15).

$$\frac{W}{V}=\tfrac{1}{2}HB=\tfrac{1}{2}\mu H^2=\frac{1}{2}\frac{B^2}{\mu} \tag{15}$$

Magnetic potential. When a magnetic pole is in a magnetic field, there will be a force F acting on it. If it is moved a distance ds in the field, work dW done against the field is given by Eq. (16), where θ

$$dW=-F\cos\theta\, ds=-Hm\cos\theta\, ds \tag{16}$$

is the angle between the positive sense of H and the positive direction of s. The magnetic potential difference V may be defined as the work done per unit pole in taking the pole from one point to the other as in Eq. (17*a*) or (17*b*).

$$dV=\frac{dW}{m}=-H\cos\theta\, ds \tag{17a}$$

$$V=-\int_{s_1}^{s_2} H\cos\theta\, ds \tag{17b}$$

The resulting equation for the magnetic potential does not include the concept of the magnetic pole, and can be used as a defining equation for the magnetic potential. The integral represents the sum along a path of the products of ds and the component of H in the direction of ds. Such an integral is called a line integral and is represented by the symbol $\oint$. The defining equation for magnetic potential may be written as Eq. (18). From this

$$V=-\oint H\cos\theta\, ds \tag{18}$$

equation the relationship between magnetic potential and H may be written as Eq. (19). Thus the

$$H\cos\theta=-\frac{dV}{ds} \tag{19}$$

component of field strength in any direction is the negative of the magnetic potential gradient in that direction. This statement is similar to the relation between electric field intensity and electric potential gradient.

The rise in magnetic potential around any closed path in a magnetic field may be deduced from consideration of the field about a long straight conductor.

For this special case, the lines of force are concentric circles about the conductor and at a distance a from the conductor and Eq. (20) holds. For

$$H=\frac{I}{2\pi a} \tag{20}$$

a path that follows a circular line of force in a sense opposite to that of H, $\theta=180°$ and $\cos\theta=-1$. Then the rise in magnetic potential around the closed path is given by Eq. (21). This result is inde-

$$V=-\oint H\cos\theta\, ds=\int_0^{2\pi a}\frac{I}{2\pi a}\, ds=I \tag{21}$$

pendent of the radius of the circle followed and, in fact, is independent of the shape of the path, since any path may be resolved into components that are along circles and along radii. The contributions of the radial parts are zero, since they are perpendicular to H.

The result deduced from this special case may be generalized for a closed loop of any shape. The rise in magnetic potential around the path is equal to the line integral of $H\cos\theta$ around the path, and this in turn is equal to the current through the surface bounded by the path as shown in Eq. (22). If

$$\oint H\cos\theta\, ds=I \tag{22}$$

the path taken includes no current, the integral is zero. If there are N equal currents inside the path, the integral becomes NI.

By analogy to an electric circuit, in which the line integral of the electric intensity around the circuit is the electromotive force of the circuit, the rise in magnetic potential around the closed path, that is, the line integral of $H\cos\theta$ around the path, is called the magnetomotive force (mmf). Thus, the mmf of a coil of N turns in which there is a current I is NI. *See* MAGNETOMOTIVE FORCE.

[KENNETH V. MANNING]

Bibliography: S. S. Attwood, *Electric and Magnetic Fields*, 3d ed., 1966; *Berkeley Physics Course*, vol. 2: *Electricity and Magnetism*, 1970; B. I. Bleaney and B. Bleaney, *Electricity and Magnetism*, 1976; E. M. Pugh and E. W. Pugh, *Principles of Electricity and Magnetism*, 2d ed., 1970; R. Resnick and D. Halliday, *Physics*, 1977; F. W. Sears, *University Physics*, pt. 2, 1976.

Magnetic induction

A vector quantity that is used as a quantitative measure of a magnetic field. It is defined in terms of the force on a charge moving in the field by Eq. (1), where B is the magnitude of the magnetic induction and F is the force on the charge q, which is moving with speed v in a direction making an angle θ with the direction of the field. The direction of the vector quantity B is the direction in which the force on the moving charge is zero.

$$B=\frac{F}{qv \sin \theta} \qquad (1)$$

The magnetic induction may also be expressed in terms of the force F on a current element of length l and current I, as in Eq. (2). The meter-kilogram-second (mks) unit of magnetic induction is derived from this equation by expressing the force in newtons, the current in amperes, and the length in meters. The unit of B is thus the newton/ampere-meter.

$$B=\frac{F}{Il \sin \theta} \qquad (2)$$

Magnetic flux density is the magnetic flux per unit area through a surface perpendicular to the magnetic induction. Magnetic flux density and magnetic induction are equivalent terms. *See* MAGNETIC FIELD.

The magnetic induction may be represented by lines that are drawn so that at every point in the field the tangent to the line is in the direction of the magnetic induction. To represent the magnetic induction qualitatively, as many such lines may be drawn as are necessary to portray the field. If, however, the lines are to represent the magnetic induction quantitatively, an arbitrary choice must be made for the number of lines to represent a given condition. One such choice is that in which the number of lines per square meter of a surface perpendicular to B is set equal to the value of B. These lines are called magnetic flux. One line of induction as here selected is called a flux of 1 weber. The corresponding unit of flux density is then the weber per square meter. From the manner of defining flux used here, it follows that the weber per square meter is equivalent to the newton/ampere-meter.

Another unit of flux density is defined by using centimeter-gram-second (cgs) units in both the defining equation for magnetic induction and in the area in which there is unit flux. This cgs unit of flux density is called the gauss. The relationship between the gauss and the weber is given by 1 weber/$m^2 = 10^4$ gauss. [KENNETH V. MANNING]

Bibliography: D. S. Parasuis, *Magnetism*, 1961; E. R. Peck, *Electricity and Magnetism*, 1953; E. M. Purcell, *Electricity and Magnetism*, vol. 2, 1965; F. W. Sears, *Principles of Physics*, vol. 2, 1951.

Magnetism

Magnetism comprises those physical phenomena involving magnetic fields and their effects upon materials. Magnetic fields may be set up on a macroscopic scale by electric currents or by magnets. On an atomic scale, individual atoms cause magnetic fields when their electrons have a net magnetic moment as a result of their angular momentum. A magnetic moment arises whenever a charged particle has an angular momentum. It is the cooperative effect of the atomic magnetic moments which causes the macroscopic magnetic field of a permanent magnet.

History. The name magnet was used by the Greeks for a stone which was capable of attracting other pieces of the same material and iron as well. This original magnet (lodestone) is the naturally occurring magnetic iron oxide, magnetite. Some of the properties of magnetite were discovered earlier than 600 B.C., although it is only in the 20th century that physicists have begun to understand why substances behave magnetically. Magnetism is one of the earliest known physical phenomena of solid materials.

William Gilbert (1540–1603) was the first to apply scientific methods to a systematic exploration of magnetic phenomena. His greatest contribution was the discovery that the Earth is itself a magnet. He distinguished clearly between electricity and magnetism. Quantitative studies of magnetism began in the early 18th century. Charles Coulomb (1736–1806) established the inverse-square law of force between magnetic poles and also between electric charges. S. D. Poisson (1781–1840) set up the basis of magnetostatics through application of potential theory to the problem of the forces between magnetized bodies. In the 19th century James Clerk Maxwell (1831–1879) established the relationship between magnetism and electricity by developing the science of electromagnetism. *See* ELECTRICITY; ELECTROMAGNETISM.

Magnetic field. A magnetic field is said to occupy a region when the magnetic effect of an electric current or of a magnet upon a small test magnet which is brought in the vicinity is detectable. Because of the magnetic effect, a torque will be exerted on the test magnet until it becomes oriented in a particular direction. The magnitude of this torque is a measure of the strength of the magnetic field, and the preferred direction of orientation is the direction of the field. In electromagnetic units, the magnitude of the torque is given by Eq. (1),

$$L=-|\boldsymbol{\mu}||\mathbf{H}| \sin \theta \qquad (1)$$

where $\boldsymbol{\mu}$ is the magnetic moment of the test magnet, $\mathbf{H}$ is the magnetic field strength, and θ is the angle between the direction of $\boldsymbol{\mu}$ and the direction of $\mathbf{H}$.

When a magnetic material is placed in a magnetic field $\mathbf{H}$, it becomes magnetized; that is, it becomes itself a magnet. The intensity of this induced magnetism is called the magnetization $\mathbf{M}$. More precisely, $\mathbf{M}$ is the magnetic moment per unit volume of the material. A vector field, the magnetic induction $\mathbf{B}$, is often defined to describe the magnetic forces anywhere in space. In electromagnetic units, the definition of $\mathbf{B}$ is given by Eq. (2). *See* MAGNETIC FIELD; MAGNETIC INDUCTION.

$$\mathbf{B}=\mathbf{H}+4\pi\mathbf{M} \qquad (2)$$

Classification of substances. Materials can be grouped according to their magnetic behavior, although there is some overlap among groups.

1. Diamagnetic substances have a negative magnetic susceptibility; they are magnetized in a direction opposite to that of an applied magnetic field.

2. Paramagnetic substances have a positive magnetic susceptibility such that their magnetization is parallel and usually proportional to the applied magnetic field.

3. A ferromagnetic substance is one with net atomic moments which, within a certain temperature range, tend to line up in such a way that there

can exist a net magnetization in the absence of an applied field. This category includes all ferromagnets and ferrimagnets and some helimagnets. At sufficiently high temperatures all these substances become paramagnetic.

4. An antiferromagnetic substance is one with net atomic moments which, within a certain temperature range, tend to line up in such a way that there can exist no net magnetization in the absence of an applied field. This category includes all antiferromagnets and some helimagnets.

[ELIHU ABRAHAMS; FREDERIC KEFFER]

Bibliography: L. F. Bates, *Modern Magnetism*, 4th ed., 1961; A. H. Morrish, *The Physical Principles of Magnetism*, 1965.

Magnetohydrodynamic power generator

A system for the generation of electrical power through the interaction of a flowing, electrically conducting fluid with a magnetic field. As in a conventional electrical generator, the Faraday principle of motional induction is employed, but solid conductors are replaced by an electrically conducting fluid. The interactions between this conducting fluid and the electromagnetic field system through which electrical power is delivered to a circuit are determined by the magnetohydrodynamic (MHD) equations, while the properties of electrically conducting gases or plasmas are established from the appropriate relationships of plasma physics. Major emphasis has been placed on MHD systems utilizing an ionized gas, but an electrically conducting liquid or a two-phase flow can also be employed. *See* ELECTROMAGNETIC INDUCTION.

Improvement of the overall thermal efficiency of central station power plants has been the continuing objective of power engineers. Conventional plants based on steam turbine technology are limited to about 40% efficiency, imposed by a combination of working-fluid properties and limits on the operating temperatures of materials. Application of the MHD interaction to electrical power generation removes the restrictions imposed by the blade structure of turbines and enables the working-fluid temperature to be increased substantially. This enables the working fluid to be rendered electrically conducting and yields a conversion process based on a body force of electromagnetic origin. *See* ELECTRIC POWER GENERATION; STEAM TURBINE.

Principle. Electrical conductivity in an MHD generator can be achieved in various ways. At the heat-source operating temperatures of MHD systems (1000–3000 K), the working fluids usually considered are gases derived from fossil fuel combustion, noble gases, and alkali metal vapors. For combustion gases, a seed material such as potassium carbonate is added in small amounts, typically about 1% of the total mass flow. The seed material is thermally ionized and yields the electron number density required for adequate electrical conductivity above about 2500 K. With monatomic gases, operation at temperatures down to about 1500 K is possible through the use of cesium as a seed material. In plasmas of this type, the electron temperature can be elevated above that of the gas (nonequilibrium ionization) to provide adequate electrical conductivity at lower temperatures than with thermal ionization. In so-called liquid metal MHD, electrical conductivity is obtained by injecting a liquid metal into a vapor or liquid stream to obtain a continuous liquid phase.

When combined with a steam turbine system to serve as the high-temperature or topping stage of a binary cycle, an MHD generator has the potential for increasing the overall plant thermal efficiency to around 50%, and values higher than 60% have been predicted for advanced systems. This follows from the increased efficiency made available by the higher source temperature. Thus, the MHD generator is a heat engine or electromagnetic turbine which converts thermal energy to a direct electrical output via the intermediate step of the kinetic energy of the flowing working fluid. *See* ENGINE; THERMODYNAMIC PROCESSES.

The MHD generator itself consists of a channel or duct in which a plasma flows at about the speed of sound through a magnetic field. The power output per unit volume W_e is given by $W_e = \sigma v^2 B^2 k(1-k)$, where σ is the electrical conductivity of the gas, v the velocity of the working fluid, B the magnetic flux density, and k the electrical loading factor (terminal voltage/induced emf). Typical values are $\sigma = 10-20$ mhos/m; $B = 5-6$ tesla; $v = 600-1000$ m/s; and $k = 0.7-0.8$. These values yield W_e in the range 25–150 MW/m^3. The high magnetic field strengths required are to be provided by a superconducting magnet, and the development of suitable electrodes which will conduct current into and out of the gas is one of the major development problems.

Types. When this generator is embedded in an overall electrical power generation system, a number of alternatives are possible, depending on the heat source and working fluid selected. The temperature range required by MHD can be achieved through the combustion of fossil fuels with oxygen or compressed preheated air. The association of MHD with nuclear heat sources has also been considered, but in this case limitations on the temperature of nuclear fission heat sources with solid fuel elements has thus far precluded any practical scheme being developed where a plasma serves as the working fluid. The possibility of coupling MHD to a fusion reactor has been explored, and it is possible that 21st-century central station power systems will comprise a fusion reactor and an MHD energy conversion system. *See* NUCLEAR FUSION; NUCLEAR REACTOR.

Development efforts on MHD are focused on fossil-fired systems, the fossil fuel selected being determined by national energy considerations. In the United States, coal is the obvious candidate, whereas in the Soviet Union, where ample reserves of natural gas are available for electric power generation, this is the preferred fuel. In Japan, major emphasis is on the use of petroleum-based fuels.

MHD power systems are classified into open- or closed-cycle systems, depending respectively on whether the working fluid is utilized on a once-through basis or recirculated via a compressor. For fossil fuels, the open-cycle system offers the inherent advantage of interposing no solid heat exchange surface between the combustor and the MHD generator, thus avoiding any limitation being placed on the cycle by the temperature attainable over a long period of operation by construction materials in the heat exchanger. Closed-cycle systems were originally proposed for nuclear heat sources, and the working fluid can be either a seeded noble gas or a liquid metal – vapor mixture.

Features. The greatest development effort in MHD power generation has been applied to fossil-fired open-cycle systems, but sufficient progress has been made in closed-cycle systems to establish their potential and to identify the engineering problems which must be solved before they can be considered practical. The rest of this article discusses fossil-fired open-cycle systems.

In addition to offering increased power plant efficiencies, MHD power generation also has important potential environmental advantages. These are of special significance when coal is the primary fuel, for it appears that MHD systems can utilize coal directly without the cost and loss of efficiency resulting from the processing of coal into a clean fuel required by competing systems. The use of a seed material to obtain electrical conductivity in the working fluid also places the requirement on the MHD system that a high level of recovery be attained to avoid adverse environmental impact and also to ensure acceptable plant economics. The seed recovery system required by an MHD plant also serves to recover all particulate material in the plant effluent. A further consequence of the use of seed material is its demonstrated ability to remove sulfur from coal combustion products. This occurs because the seed material is completely dissociated in the combustor, and the recombination phenomena downstream of the MHD generator favor formation of potassium sulfate in the presence of sulfur. Accordingly, seed material acts as a built-in vehicle for removal of sulfur. Laboratory experiments have shown that the sulfur dioxide emissions can be reduced to levels below those experienced with natural gas-fired plants. A further important consideration is the reduction of the emission of oxides of nitrogen through control of combustion and the design of component operating conditions. Laboratory scale work has demonstrated that these emissions can be controlled to the most exacting standards prescribed by the Environmental Protection Administration (EPA). *See* AIR POLLUTION.

While not a property of the MHD system in itself, the potential of MHD to operate at higher thermal plant efficiencies has the consequence of substantial deduction in thermal waste discharge, following the relationship that the heat rejected per unit of electricity generated is given by $(1-\eta)/\eta$, where η is the plant efficiency. As the technology of MHD is developed along with that of advanced gas turbines, there also exists the possibility that MHD systems can dispense entirely with the need for large amounts of cooling water for steam condensation through the coupling of MHD generators with closed-cycle gas turbines. *See* WATER POLLUTION.

State of development. A number of test facilities and experimental installations have developed engineering information on the MHD process and demonstrated operation of the generator and other components of the complete system. The most technically advanced installation is located on the northern outskirts of Moscow. Known as the U-25 installation, it is of the open-cycle type and has delivered its rated power of 20.5 MW to the Moscow grid using natural gas as the fuel. Operation with direct coal firing has been successfully demonstrated in the United States, and a test generator has been operated in Japan with a 5-tesla superconducting magnet. [WILLIAM D. JACKSON]

Bibliography: International MHD Liaison Group, Nuclear Energy Agency, OECD, Paris, *1976 MHD Status Report*, 1976; W. D. Jackson and P. S. Zygielbaum, Open cycle MHD power generation: Status and engineering development approach, *Proc. Amer. Power Conf.*, 37:1058–1071, 1975; J. Raeder (ed.), *MHD Power Generation: Selected Problems of Combustion MHD Generation*, 1975; R. J. Rosa, *Magnetohydrodynamic Energy Conversion*, 1968.

Magnetomotive force

The magnetomotive force (mmf) around a magnetic circuit is the work per unit magnetic pole required to carry the pole once around the circuit. It is the analog of electromotive force. *See* MAGNETIC CIRCUITS.

It is expressed mathematically in Eq. (1), where

$$\text{mmf} = \oint H \cos \theta \, ds \tag{1}$$

$H \cos \theta$ is the component of magnetic field strength in the direction of a length of path ds. The line integral is taken around any closed path in the field.

The magnetomotive force is the rise in magnetic potential around the path. For a discussion of magnetic potential *see* MAGNETIC FIELD.

For a path that encloses a current I, Eq. (2)

$$\oint H \cos \theta \, ds = I \tag{2}$$

holds, and for a path that encloses N equal currents, for example, a path that loops through a coil of N turns, Eq. (3) is valid. If no current is enclosed by the path, the line integral is zero.

$$\oint H \cos \theta = NI \tag{3}$$

The meter-kilogram-second (mks) unit of magnetomotive force is the ampere-turn.

[KENNETH V. MANNING]

Marine engine

An engine that propels a waterborne vessel. Even in small craft the marine engine must have the following characteristics: reliability, light weight, compactness, fuel economy, low maintenance, long life, relative simplicity for operating personnel, ability to reverse, and ability to operate steadily at low or cruising speed. The relative importance of these characteristics varies with the service performed by the vessel, but reliability is of prime importance.

Steam engines. Steam, used to drive the earliest powered vessels, is still a common type of propulsion for large ships. The diesel engine has gained wide acceptance in foreign merchant ships, but in the United States the majority of seagoing vessels use steam propulsion.

Reciprocating steam engines. Early engines commonly used steam flowing in series through as many as four cylinders whose pistons had the same stroke but were of increasing diameters. This system provided for an expansion or increase in steam volume which accompanied the decrease modern, multicylinder, uniflow marine steam engine, with complete expansion in each cylinder, shows better steam economy. Because it has the same diameter for all cylinders (two to six in number), it is preferable from a manufacturing viewpoint. Equal power is developed by each cylinder; units of four cylinders or more have good torque and balance characteristics. A steam rate of 10 lb/hp/hr (6.1 kg/kW/hr) with 275 psi (1.90 MPa) at 240°F (116°C) superheat is attained. Uniflow engines as large as 5000 hp (3.7 MW) have been used on shipboard. Normally, steam engines are double-acting; that is, steam acts on each side of the pistons. With superheated steam, piston-cylinder lubrication must be provided. Pure feedwater is required by modern, high-capacity boilers; therefore, an effective oil filter is installed where the condensate must be returned to the boiler.

Steam turbines. The marine steam turbine has the advantages of direct rotary motion, little or no rubbing contact of pressure-confining surfaces, and ability to use effectively both highly superheated steam and steam at low pressure, that is, at a high vacuum where specific volumes of over 400 ft^3/lb are reached.

For good efficiency of steam turbines, high rotative speeds are required. This requirement led to the introduction of the reduction geared turbine and turboelectric drive. These systems give efficient turbine speeds and efficient propeller rpm. With geared turbines, for example, turbine rotor speeds range from 3000 to 10,000 rpm, while propeller rpm is reduced to the 80 to 400 range.

Steam is generally supplied to the turbine at 850 psi (5.9 MPa) and 950°F (510°C) by a pair of oil-fired marine water-tube boilers, and the exhaust from the turbine is usually at 1.5 in. Hg (5.1 kPa) absolute. Forced draft fans and other auxiliaries are usually motor-driven, except for the main feed pumps usually driven by an auxiliary turbine. Electric power is provided by a separate turbogenerator.

In low-powered geared turbines, steam completes its expansion in one rotor and casing. Such a design has been used in geared turbines of up to 8000 shaft horsepower (shp) or 6.0 MW of shaft power. However, series flow through two or even three casings is preferable in most steam turbines. This arrangement provides more flexibility in turbine design, allowing for different and optimum revolutions for high- and low-pressure rotors. Also, in a seagoing vessel in case of casualty to one turbine or its high-speed pinion, the vessel usually can make port with the remaining turbine in operation.

A steam turbine is made up of fixed blades, usually called nozzles, and rotating blades. A stage is generally one stationary row and one moving row. Impulse staging has all the steam pressure drop taking place in the fixed blades. The moving row then absorbs the kinetic energy produced. Reaction staging results when some of the pressure drop occurs in the moving blades, the degree of reaction depending on the design.

Modern marine practice favors impulse staging in the high-pressure end of the turbine because of the reduced parasitic losses with this type, but as the steam progresses toward the low-pressure end where the volume is much greater, the reaction stage is more efficient. The usual arrangment has a cross compound system with a high-pressure unit of 7 to 10 stages and a low-pressure unit of 6 to 8 stages, each driving a pinion of a reduction gear. *See* IMPULSE TURBINE; REACTION TURBINE.

A turbine is capable of operating in only one direction. In order to provide reverse power, a second turbine is installed on the shaft of the low-pressure, ahead turbine. The astern turbine is usually not more than three moving rows of blades, but it may be only two. It produces about 40% of the normal ahead horsepower. Since this unit is turning backward in normal ahead operation, it is located in the low-pressure end of the low-pressure turbine. The steam at this point has a very low density, and hence the astern turbine has a low windage loss.

The propeller is reversed by closing the steam valve to the ahead turbine and opening the valve to the astern turbine.

Gas turbines. The gas turbine is a relative newcomer to the marine field. It generally consists of an axial compressor discharging compressed air to a combustion chamber where fuel is burned, adding heat. The products of combustion at high temperature and pressure then pass through a gas turbine that drives the compressor and load. Generally, the term "gas turbine" is applied to the entire plant. If lower pressure ratios (final pressure leaving the compressor divided by initial pressure entering) are used, a large amount of heat is available in the exhaust gas which may be recovered by heating the compressed air before it enters the combustion chamber. This is done in a regenerator. With higher pressure ratios the expansion through the turbine is so great that the exhaust gas temperature is insufficient to heat the compressed air.

Two distinct types of gas turbines are appearing in the marine field: the aircraft-derived type and the industrial type. The aircraft-derived type uses a jet engine as a gas generator, which discharges to a gas turbine driving the load. This type of plant offers simplicity and light weight but must burn high-quality fuel. The industrial gas turbine is a more rugged machine designed for long life and is capable of using low grades of fuel, properly washed. This plant usually uses a regenerator. The gas turbine offers simplicity, ease of control, and efficiency, but requires special fuel or special treatment of the fuel. Large amounts of air and exhaust gas are used, and as a result uptakes and air supply are a special problem. The aircraft-derived gas turbine seems destined to drive a large number of naval combatant ships. Selection of this type for a new class of naval destroyers has been announced. The industrial type will be used for merchant vessels where its greater weight will be of little disadvantage. *See* GAS TURBINE.

Internal combustion engines. Both diesel and gasoline internal combustion engines are used in marine applications. Many moderate- and low-power marine installations use automotive or locomotive engines designed for variable load and intermittent service. High-power marine propulsion units normally are called on to operate continuously under load. Therefore, the brake horsepower (bhp) rating of units selected for marine service should be conservative.

The gasoline engine is the most common power plant for pleasure craft. It is inexpensive to buy and maintain. Because of its widespread use in automobiles, most parts are readily available. In most areas gasoline costs slightly more than diesel fuel, but the cost differential is usually insufficient to make up the difference between the cost of gasoline and diesel engines. Gasoline presents an explosion and fire hazard, which is its major disadvantage. *See* INTERNAL COMBUSTION ENGINE.

Direct-drive diesels. For typical commercial freight vessels, direct-drive diesels provide economical service. For good propeller efficiency, the propeller rpm should be under 120±. Such a top limit on engine revolutions results in a large, heavy, bulky, slow-rpm engine. However, the direct-drive diesels have a lower fuel oil consumption than do higher-rpm units, and with suitable fuel treatment will operate on the better grades of the cheaper fuel oil burned in boilers. *See* DIESEL ENGINE.

Slow-speed, direct-drive diesels are favored by many European owners and shipbuilders. Turbocharged, two-cycle, single-acting diesel engines of 50,000 bhp (37 MW of brake power) are now available; such engines weigh more than 100 lb/bhp (61 kg/kW of brake power). For high horsepower the total machinery weight for diesels is more than the weight of geared turbine machinery, including boilers and auxiliaries.

Moderate-speed diesels. Diesel engines of 250–500 rpm are available in two- and four-cycle, single-acting types, generally with trunk pistons (Fig. 1). In some marine applications they are connected directly to the propeller and thus fitted only with reverse gear. However, they are also employed with geared diesel and diesel-electric drive. The weight of such engines runs about 35–70 lb/bhp (21–43 kg/kW of brake power).

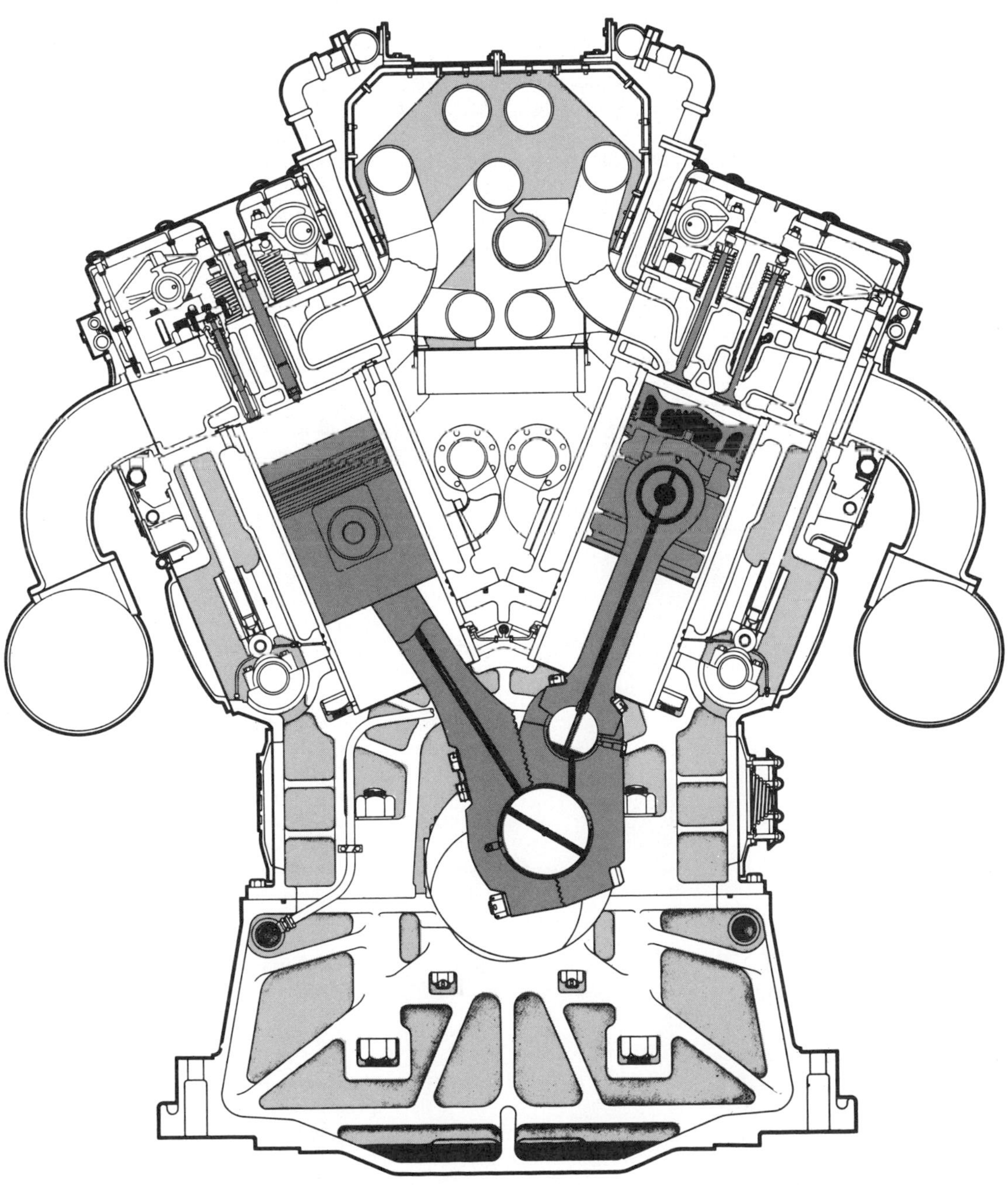

Fig. 1. Cross section of four-cycle V-type diesel.

High-speed diesels. Many high-speed diesel engines of 600 rpm and more (some types originally developed for truck and locomotive service) are available for marine propulsion. Opposed piston types have been developed; other manufacturers favor a V type to reduce weight. Such engines are of two- and four-cycle types and usually weigh 10–40 lb/bhp (6–24 kg/kW of brake power). Because of less efficient scavenging, lower powers, and other factors, their fuel and lubricating oil rates are higher than for large, low-speed diesels.

Except for direct drive in moderate or fairly high-speed craft, marine applications of diesel engines are fitted with either mechanical reduction gearing or with diesel-electric drive to provide good propeller efficiency. Because their pistons, valves, and other components are small, standardized, and carried in stock, repairs are readily made, with the result that engines of this type are popular for nonoceangoing services.

Oil consumption and starting. Lubricating oil consumption of diesel engines is high because of the cylinder-piston lubrication that must be provided and the contamination of the crankcase oil with residues blown by the piston rings. In large engines this contamination is avoided by using piston rod–crosshead construction so that the crankshaft, connecting rods, and crossheads operate in a closed casing separated from the working cylinder. These engines are started and maneuvered by pressure from one or more reservoirs filled with air at about 250 psi (1.72 MPa). To make it feasible to start and readily reverse, two-cycle, single-acting marine engines should have at least four cylinders; four-cycle engines should have five or more cylinders.

Nuclear power. Very successful installations of nuclear power have been made in submarines and a few surface ships. Operation of the first nuclear merchant ship, the *Savannah*, was successful technically but not commercially. (The operating crew required special training, and it would be difficult to replace with only one commercial nuclear ship in service.)

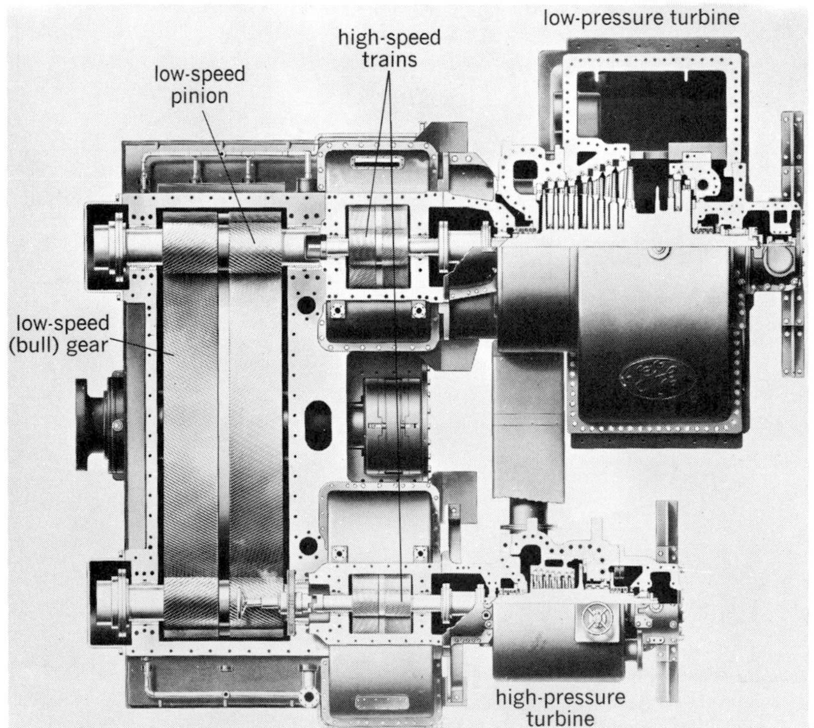

Fig. 2. Double reduction articulated gear design for use with modern high-rpm marine steam turbines. (*General Electric Co.*)

Mechanical reduction gears. Reduction gearing for diesel and gasoline engines allows the use of a relatively high engine speed and lower, more efficient propeller speed. Speed reduction ratios of 1.8:1 to 4:1 are common, preferably with helical teeth to give better wear and quieter performance. A reverse gear device often is incorporated in low-power gears for astern operation. Other methods for providing reverse rotation capability may utilize a direct reversing engine or a controllable-pitch propeller.

One, two, three, or four engines may drive the same gear through individual pinions. The use of a friction, electromagnetic, pneumatic, or hydraulic coupling serves to disconnect any engine. By reversing one or more engines, ready maneuvering, including astern operation, is provided for by the use of the respective coupling.

The high rpm (3000–9000) of modern marine steam turbines and the low revolutions of an effective propeller (as low as 80 rpm) require the use of two-stage gearing. Gear teeth are of involute form, with the pinion teeth of harder material than the gear. The gear trains are of the double helical type to avoid heavy axial thrust (Fig. 2). Double reduction gears are constructed with flexible couplings between the high-speed train and the low-speed elements.

Mechanical reduction gears are carefully constructed to close tolerances. They have forced lubrication in sprays ahead of the meshing teeth, to the bearings, and to the flexible couplings. Tests have shown that bearings represent at least half the power loss of the entire gear set.

Turboelectric drive. This type of drive, comprising one or more steam turbine generators and ac propulsion motors, is also used for ship propulsion. It was installed in many United States tankers during World War II because of available manufacturing facilities. The synchronous motors are provided with an induction winding for starting and reversing. Relatively large changes in propeller revolutions are made by alteration of the turbogenerator speed.

Motor, generator, exciter, and cooling equipment losses result in several percent lower efficiency than with geared steam turbines. Weights and costs are generally 25–30% higher than for the comparable turbine gear arrangement. Electric drive is not employed unless it offers significant operational or design advantages. These include flexibility of control and the independence of the location of the turbo generator relative to the propeller shaft or propulsion motor.

Diesel-electric drive. This type of drive, composed of one or more dc diesel generator sets and often a double-armature propulsion motor, is used in tugs, dredges, Coast Guard cutters, and icebreakers, where maneuvering and a wide range in propeller speed are necessary. For slow-speed operating during maneuvering, the engine speed is often reduced and the generator field excitation is altered to provide wide variation in the motor output and propeller speed.

Control arrangements. Bridge or pilothouse control of the ship propulsion unit, without action by the engineer on watch, is used with diesel-electric drive and for small, low-powered, direct-drive, and mechanically geared diesel installations. This is the customary arrangement for tugs and dredges.

Modern practice for large diesel and steamships is to provide pilothouse control of the main engines and no engine-room watch except day workers and when entering and leaving port. Propulsion plant monitoring is usually provided in a central control space, with chart recorders or a data storage system. All levels of automation are now being used, from simple manual surveillance of all systems in a fully manned engine space to a completely unmanned system recording data and monitoring trends to pinpoint possible trouble.

Governors. Above the operating rpm, ship propeller torque increases faster than engine or turbine torque, and thus ship propeller drive is inherently stable. Because of the ship's pitching, the propeller may lift partially out of the water and the engine may tend to race. To allow for this situation, or for propeller shafting failure, American Bureau of Shipping regulations require that a governor be fitted to limit overspeed to 15% above the rated speed.

A common type of governor uses oil pressure developed by small pumps incorporated with the main turbine rotors to activate the governor; low lubricating oil pressure will also shut off the steam supply.

With turboelectric and diesel-electric drive, there is no mechanical connection between the generator set and the propulsion motor and propeller. The operating governor holds generator speed at the set value by throttling turbogenerator steam or the amount of fuel injected in the diesel engine cylinders.

[JENS T. HOLM]

Bibliography: Babcock and Wilcox Co., *Steam*, 39th ed., 2d rev. printing, 1978; T. Baumeister (ed.), *Standard Handbook for Mechanical Engineers*, 8th ed., 1978; W. J. Fox and S. C. McBirnie, *Marine Steam Engines and Turbines*, 1970; R. F. Latham, *Introduction to Marine Engineering*, 1958; C. C. Pounder, *Marine Diesel Engines*, 1972; P. de W. Smith, *Modern Marine Electricity and Electronics*, 1966; D. Wright, *Marine Engines and Boating Mechanics*, 1977.

Marine resources

The potential for energy from the sea's motion and processes has long been apparent, and designs for wave-power and tidal-power devices can be traced back for hundreds of years. Wind was enlisted, of course, by the first sailor. The period from mid-1800 to mid-1900 was particularly fertile for ocean-energy technology and included discovery of the power potential from ocean-thermal gradients. World War II and the period of cheap oil and gas following it pushed many of these ideas aside. With the 1973 Oil Producing and Exporting Countries (OPEC) embargo and recognition of the limited reserves of fossil fuels, efforts into the use of alternate, renewable energy resources, including the oceanic ones, have redoubled.

Categorizing resources. The categorizing of ocean energy is somewhat arbitrary. The following are not truly marine: Offshore oil and gas are in principle not different from onshore oil and gas but require greater rigor in exploration and production; about 15% of the world's oil is produced offshore, and extraction capabilities are advancing. Coal deposits, known as extensions of land deposits, are mined under the sea floor in Japan and England. Geothermal resources are known to exist offshore; they are presently not being used, and their prospects are only dimly perceived. Biomass energy in the form of methane from giant kelp is under active investigation. Another extensive energy resource is potentially available in the fissionable and fusible elements contained in sea water; they may provide power for 10^5 and 10^9 years, respectively.

In a strict sense, ocean energy is expressed in the processes of the ocean, such as in the currents, tides, waves, thermal gradients, and the only recently recognized salinity gradients. Solar energy drives them, except the tides, which are fueled by the fossil kinetic energy of the Earth-Moon system. Estimates of the intensities of the processes are given in Fig. 1. Figure 2 shows the size of the resources.

Currents. It is evident that currents constitute the smallest and weakest resource. Low-pressure turbines of 170-m diameter have been proposed for the center of the Gulf Stream, but it is doubtful that ocean currents can be profitably harnessed. A few straits possess fast tidal currents—the Seymour Narrows in British Columbia and the Apolima Straits in Western Samoa, for instance—and offer better prospects.

Tides. Feasible tidal power (see Fig. 2) is limited to a few sites with high tides. Tides now provide the only source of commercial ocean power in the Rance River tidal plant in Brittany, France. Since 1968 it has produced moderate amounts of power in its bank of twenty-four 10-MW turbines, and is still being fine-tuned for greater efficiency. The plant is successful technically and, increasingly, economically, and the experience gained there is being considered in renewed studies for tidal-power development in Australia, England, Canada, India, the United States, and the Soviet Union. The

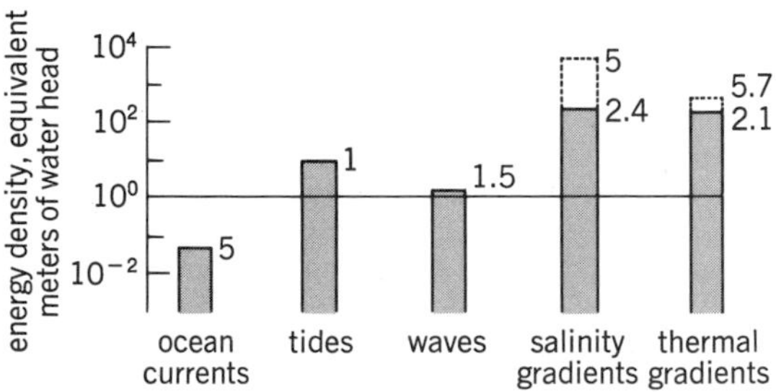

Fig. 1. Intensity or concentration of energy expressed as equivalent head of water. "Ocean currents" shows the driving head of major currents. For tides, the average head of favorable sites is given. For waves, the head represents a spatial and temporal average. The salinity-gradients head is for fresh water versus sea water, the dotted extension for fresh water versus brine (concentrated solution). The thermal-gradients head is for 12°C, and that for 20°C is dotted; both include the Carnot efficiency. *(From G. L. Wick and W. R. Schmitt, Prospects for renewable energy from the sea, Mar. Technol. Soc. J., 11(546):16–21, 1977)*

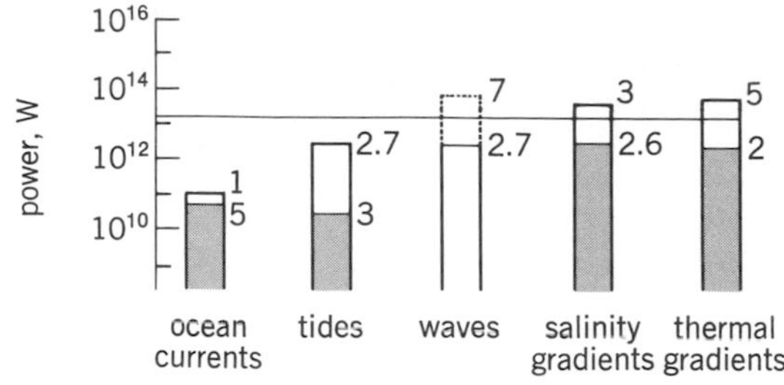

Fig. 2. On the ocean-currents bar, the shading represents the power contained in concentrated currents such as the Gulf Stream. Estimated feasible tidal power is shaded. The dotted extension on "waves" indicates that wind waves are regenerated as they are cropped. "Salinity gradients" includes all gradients in the ocean; the large ones at river mouths are shown by shading. Not shown is the undoubtedly large power if salt deposits are worked against fresh or sea water. On "thermal gradients," the shading indicates the unavoidable Carnot-cycle efficiency. The line at 1.5×10^{13} W is a projected global electricity consumption for the year 2000. *(From G. L. Wick and W. R. Schmitt, Prospects for renewable energy from the sea, Mar. Technol. Soc. J., 11(546):16–21, 1977)*

Soviets have an experimental 400-kW plant near Murmansk, and the Chinese have a number of very small plants. *See* TIDE.

Waves. Considerable efforts are being directed toward power from waves in England, Japan, Sweden, and the United States. Advanced lab and model testing is taking place in the British Isles, which are subjected to some of the most powerful waves of the Atlantic; the British hope to supply as much as 30% of their electricity needs from this source. As shown in Fig. 2, wave power is a unique resource in that it could expand under use, because in the open sea the waves could be rebuilt by the winds that cause them.

Salinity gradients. Salinity-gradient energy is present between aqueous solutions of different salinities. Vast amounts of this power are being dissipated in the estuaries of large rivers. Some conversion processes involve semipermeable and ion-selective membranes; they will require considerable development, however. One conversion process does not require membranes and is akin to an open-cycle thermal-gradient process; advances there will thus help. At the moment, salinity-gradient energy appears to be the most expensive of the marine options. Its high intensity and potential, especially when brines and salt deposits are included (see Figs. 1 and 2), which yield energy by the same principle, should, however, spur a strong level of research.

Thermal gradients. Thermal-gradient power, which would use the reservoirs of warm surface water and cold deep water, was first worked on by the French. Georges Claude spent much time and his own money on it in the 1920s, with generally bad luck. Now the United States government has made it the cornerstone of its ocean-energy program because, unlike tidal and wave power, it has the capacity for supplying base-load power. It also has a high intensity and potential. Floating tethered units of 100 megawatts electric are foreseen for locations in the Gulf of Mexico and the Hawaiian Islands. The design as much as possible employs conventional and proved technology, but the continuous operation of a complex system in a corrosive, befouling, and hazardous environment poses a formidable challenge.

Problems. There are some problems common to most marine alternatives. The principal one is transferring the power to shore. Indirect power harvest, such as by tanked or piped hydrogen or by energy-intensive products, from distant units or arrays is often suggested. But the detailed evaluation in the United States' thermal-gradient program points to electricity, the highest-value product, to justify the investment and to build up capacity and experience. Other problems pertain to corrosion in saline water, befouling by marine organisms, and deployment in often violent seas.

Environmental impact. Environmental effects of ocean-energy conversion are generally negligible since the individual sources are large compared with initial demand. Only under extensive conversion would some impacts possibly be felt, such as downstream climatic changes from major ocean-current use, bay and estuarine ecosystem upsets from tidal-basin use, possible diminishing of dissolved oxygen from extensive use of waves, and short-circuiting of the ocean's internal heat transfer from thermal-gradient use. Conflicts would also arise with present uses of the sea through hazards to navigation and fisheries operations. Juridical issues will further complicate wide ocean-energy development. Some inadvertent effects may be beneficial, however. Upwelling of nutrients by thermal-gradient plants and the sheltering of organisms by ocean structures could enhance the productivity of fish and invertebrate stocks and their harvest, for instance.

[WALTER R. SCHMITT]

Bibliography: Committee on Science and Technology, U.S. House of Representatives, *Energy from the Ocean*, 1978; G. L. Wick and W. R. Schmitt, Prospects for renewable energy from the sea, *Mar. Technol. Soc. J.*, vol. 11, no. 546, 1977.

Mass defect

The difference between the mass of an atom and the sum of the masses of its individual components in the free (unbound) state. The mass of an atom is always less than the total mass of its constituent particles; this means, according to Albert Einstein's well-known formula, that an energy of $E = mc^2$ has been released in the process of combination, where m is the difference between the total mass of the constituent particles and the mass of the atom, and c is the velocity of light.

The mass defect, when expressed in energy units, is called the binding energy, a term which is perhaps more commonly used. [W. W. WATSON]

Mercury battery

A primary dry-cell battery consisting of a zinc anode, a cathode of mercuric oxide (HgO) mixed with graphite, and an electrolyte of potassium hydroxide (KOH) saturated with zinc oxide (ZnO). With carefully purified materials and balanced amounts of ZnO and HgO, the cell has very low self-discharge and makes efficient use of the active materials.

In some cells which require long-term continuous drains, for example, in hearing aid use, MnO_2

is added to the HgO. Here the open circuit voltage is slightly above 1.4 volts, compared with 1.35 volts for cells using 100% HgO as depolarizer.

Within the steel can, the active materials are separated by a porous material which prevents migration of conducting particles from the mercuric oxide pellet. Dense dialysis paper and porous polyvinyl chloride have been used for this purpose. The electrolyte is completely absorbed in the active materials, separator, and absorbent materials. The steel can serves as the contact to the HgO. A metal top with a concentric neoprene grommet closes off the top of the can and serves as the contact to the zinc.

The electrochemical system may be written as below. This does not involve the electrolyte. The

Anode: $Zn + 2OH^- \rightarrow ZnO + H_2O + 2e^-$
Cathode: $HgO + H_2O + 2e^- \rightarrow Hg + 2OH^-$
Overall: $Zn + HgO \rightarrow ZnO + Hg$

cell potential, therefore, does not change appreciably with different concentrations of alkali.

The cutaway view shown in Fig. 1 is of the flat pellet structure. Two other types are manufactured: the cylindrical structure, which also uses pressed amalgamated zinc powder but is made in sizes corresponding to the N, AA, and D Leclanché cells; and the wound anode flat structure, with large diameter and high surface area, giving superior performance at low temperatures.

When current flows, the ZnO formed in the cell reaction quickly saturates the small amount of electrolyte and then precipitates out. This maintains a constant composition of the electrolyte. Offsetting this is the transport of water away from the anode by the solvated potassium ions. The equilibrium under steady current flow results from complex exchanges through the separator, making cell-voltage characteristics under load dependent on initial electrolyte composition.

The mercury cell has a theoretical output of 0.247 amp-hr/g of HgO. In practice, the cathode pellet contains about 95% HgO and 5% graphite, having a theoretical output of 0.234 amp-hr/g. The anode is 90% zinc and 10% mercury. This has a theoretical output of 0.738 amp-hr/g.

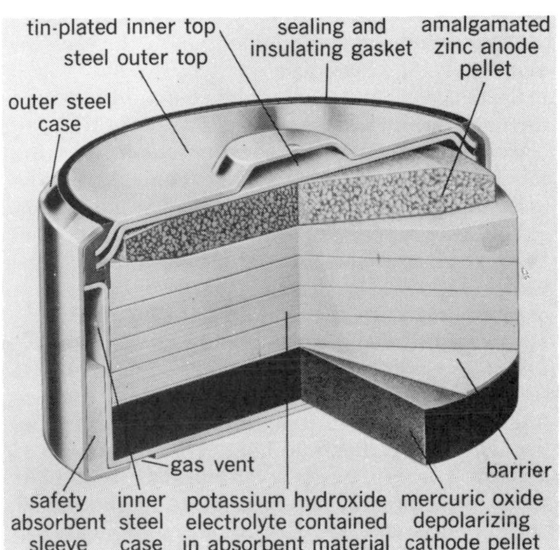

Fig. 1. Cutaway view of mercury cell.

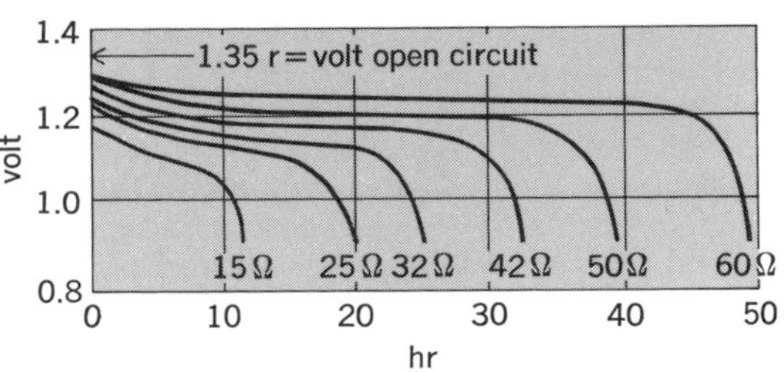

Fig. 2. Voltage-discharge characteristics of mercury cells under continuous load conditions at 70°F (21°C). At 1.25, equivalent current drains for resistances are: 15 ohms, 83 milliamperes; 25 ohms, 50 mA; 32 ohms, 40 mA; 42 ohms, 30 mA; 50 ohms, 25 mA; 60 ohms, 20 mA.

As built, the cells have slight excess of cathodic capacity. A discharged cell will then have no zinc left to react with the electrolyte and evolve hydrogen. Thus a cell with 12.5 g of cathodic material has 3.6 g of zinc amalgam, compared with 3.98 needed for exact balance.

The electrolyte used is about 1 ml/amp-hr. One composition is 100 g KOH, 100 ml H_2O, 16 g ZnO. The actual cell capacity is considerably less than the theoretical. The overall cell output is approximately 66 amp-hr/kg and 0.3 amp-hr/cm³.

The ampere-hour capacity of mercury cells is relatively unchanged with variation of discharge schedule and to some extent with variation of discharge current. The cells have a relatively flat discharge characteristic, as shown in Fig. 2. The outstanding features of the mercury cell include flat discharge curve, small variation in capacity with intermittent or continuous discharge, shelf life of several years, and good high-temperature characteristics. *See* DRY CELL; PRIMARY BATTERY; RESERVE BATTERY.

[JACK DAVIS; KENNETH FRANZESE]

Bibliography: H. Balters and G. Schneider, Mercury oxide-zinc cell, in R. W. Graham (ed.), *Primary Batteries, Recent Advances*, Chem. Technol. Rev. 105, Energy Technol. Rev. 25, 1978; R. R. Clune, Recent developments in the mercury cell, *Proc. Power Sources Conf.*, 14:117, 1960; R. R. Clune and D. Naylor, Alkaline zinc-mercuric oxide cells and batteries, *Proc. Int. Symp. Batteries*, 1958; L. F. Martin (ed.), *Dry Cell Batteries, Chemistry and Design*, pp. 86–91, 1973; D. Naylor, Wound anodes for mercury cells, *Proc. Int. Symp. Batteries*, 1960; S. Ruben, Balanced alkaline dry cells, *Trans. Electrochem. Soc.*, 92:183, 1947; G. W. Vinal, *Primary Batteries*, 1950.

Metal-base fuel

A fuel containing a metal of high heat of combustion as a principal constituent. High propellant performance in either a rocket or an air-breathing engine is obtained when the heat of combustion of the fuel is high. Chemically, high heats of combustion are attained by the oxidation of the low-atomic-weight metals in the upper left-hand corner of the periodic table. The generally preferred candidates are lithium, beryllium, boron, carbon, magnesium, and aluminum.

Performance parameters. For the rocket where the total propellant containing both fuel and oxidizer is carried on board, the performance parameter of interest is the heat of combustion per unit mass of combustion product. In air-breathing engines, however, where only the fuel is carried on board and the oxidizer in the form of air is attained free, the figure of merit is the heat of combustion per unit mass of fuel. There is still a third figure of merit for the air-breathing engine in which most of the flight involves a cruise operation in which the thrust just balances the drag. In this case, it is desirable to minimize the cross-sectional area (consequently volume) and maximize the heat of combustion so that a given total energy output can be realized in a minimum-volume vehicle to present low drag. The figure of merit, in this case, is the heat of combustion per unit volume of fuel. *See* PROPULSION.

The table shows a listing of the heats of combustion of various candidate metals along with three nonmetallized high-performance fuels for comparison. The three different heats of combustion described above are figures of performance merit, depending upon the vehicle mission and type of propulsion system employed. In general terms, for an accelerating rocket, the important figure of merit is Btu/lb pdts (pound of products); for the accelerating air breather, it is Btu/lb fuel; and for the cruising air breather it is Btu/in.^{3} fuel. *See* PROPELLANT.

The metal with the highest heat release for all applications is beryllium, whereas boron shows essentially equivalent performance for air-breathing applications. For rocket applications where Btu/lb pdts is important, lithium, boron, and aluminum are close to beryllium. As discussed later, however, because of inefficiencies in the combustion heat release as well as thermodynamic limitations, not all of these candidates are equivalent.

Method of use. The metallized additive can be used in either a liquid or solid propellant. When the pure metal is added to liquid fuels, an emulsifying or gelling agent is employed which maintains the particles in uniform suspension. Such gelling agents are added in small concentrations and so do not negate the beneficial effect of the fuel. Particle sizes in such suspensions are generally in the range 0.1–50 μm, and stable suspensions have been made which do not separate for long periods of time. When a metallic compound is employed, the compounding group is chosen so that the additive is soluble in the fuel. Thus, if gasoline is the fuel base, the suitable soluble boron compound is ethyldecaborane and, for aluminum, triethylaluminum may be employed. In both cases, the metallic additive contains a hydrocarbon constituent providing solubility in the hydrocarbon base fuel.

Heats of combustion of the lightweight metals compared with those of some other fuels

Fuel	Specific gravity	Product	Heat of combustion* Btu/lb pdts	Btu/lb fuel	Btu/in.^{3} fuel
Lithium	0.53	Li_2O	8,570	18,500	360
Beryllium	1.85	BeO	10,450	29,100	1,950
Boron	2.35	B_2O_3	7,820	25,200	2,140
Carbon	2.00	CO_2	3,840	14,100	1,020
Magnesium	1.74	MgO	6,430	10,700	670
Aluminum	2.70	Al_2O_3	7,060	13,400	1,310
Gasoline	0.75	CO_2, H_2O	4,600	20,400	550
Hydrazine	1.00	N_2, H_2O	4,180	9,360	340
Hydrogen	0.07	H_2O	6,830	61,500	155

*1 Btu/lb = 2326 J/kg; 1 Btu/in.^{3} = 17.29 kJ/cm^3.

When used in composite solid rocket propellants, the metal powder is usually mixed with the oxidizer and unpolymerized fuel, and the propellant is then processed in the usual way. If a compound of the metal additive is employed, it is convenient to dissolve it in the fuel, and then to process the fuel containing the metallic compound and oxidizer by normal procedures. Homogeneous solid propellants also employ metallizing constituents, both as the pure metal and as a compound.

Compounds of interest. Early compounding of metallized propellants employed the free metal itself. As a result of extensive research in metalloorganic compounds, however, several classes of metallic compounds have been employed. Two major reasons for the use of such compounds are that the solubility of the metal in the fuel can be realized, resulting in a homogeneous propellant, and performance higher than that for the pure metal can be obtained in some cases.

The major classes of metallic compounds of interest as high-performance propellants include the hydrides, amides, and hydrocarbons. Additional classes include mixtures either of two metals or of two chemical groups, such as an amine hydrocarbon. Some examples of the hydrides include LiH, BeH_2, B_2H_6, B_5H_9, $B_{10}H_{14}$, MgH_2, and AlH_3. These compounds form the more exotic class of metallics because they are capable of yielding the highest performances, exceeding those of the pure metals. The presence of hydrogen provides the low-molecular-weight gases needed for high thermodynamic performance. Examples of the propellant amides include $LiNH_2$ and $B(NH_2)_2BNHCH_3$. The largest class is the hydrocarbons, which includes LiC_2H_5, $Be(CH_3)_2$, $B(CH_3)_3$, and $Al(C_2H_5)_3$. The major advantage of this class is the potential solubility in conventional hydrocarbon fuels.

Operational limitations. Most metallic fuels are costly and many, because of particle-size requirements or synthesis in a specific compound, are in limited supply. The combustion gases all produce smoky exhausts which may be objectionable in use. Engine development problems are also increased because of the appearance of smoke and deposits in the engine. These engine deposits can sometimes occur on the injector face, nozzle, and turbine blades of turbojet engines. Careful adjustment of operating conditions is required in such cases to minimize these difficulties with resultant added development costs.

Combustion losses. In rocket engines, where the flame temperatures are high, vaporization and dissociation are particularly pronounced. In the tabular listing of total heats of combustion, the heat value is given as measured in a laboratory calorimeter, the end product being the condensed oxide. At high flame temperatures, however, the oxides vaporize, producing a lower effective heat of combustion. In the case of boron oxide, B_2O_3, for example, the normal boiling point is about 2500 K (4000°F), whereas the flame temperature in most high-performing rocket engines exceeds 3000 K. These effects are illustrated in the graph, which compares the total heat of combustion Q of gasoline-boron mixtures when the B_2O_3 is solid or liquid to the heat when it is a vapor. At still higher flame temperatures, in addition to vaporization, the oxides undergo endothermic dissociation such as shown in Eqs. (1), (2), and (3).

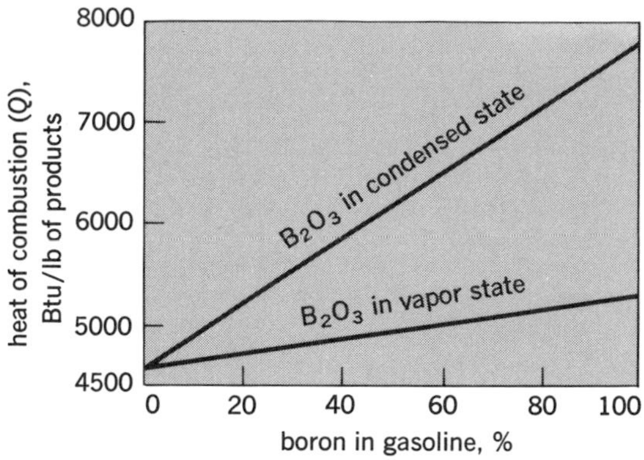

Total heat of combustion of gasoline-boron mixtures.

$$B_2O_3(g) = B_2O_2(g) + 1/2\,O_2 - 108\text{ kcal} \quad (1)$$

$$B_2O_2\,(g) = 2BO(g) - 126\text{ kcal} \quad (2)$$

$$Al_2O_3(l) = Al_2O(g) + O_2 - 330\text{ kcal} \quad (3)$$

The net result of these reactions is a further reduction in flame temperature to an extent that sensitively depends on the temperature itself. At temperatures below about 2500 K, as in air-breathing engines operating below Mach 3, vaporization and dissociation for the oxides of beryllium, aluminum, boron, and magnesium occur hardly at all, and these limitations do not apply.

In actual engines, a certain fraction of the useful energy which is lost by vaporization and dissociation is recovered by condensation and recombination in nozzles. If thermodynamic equilibrium prevails and if the nozzle area ratio is infinite, all of this energy loss is recovered. In actual cases, however, because of the finite area ratio and finite reaction and condensation rates, only a part of the energy is recovered for propulsion.

Two-phase flow losses. A second type of propulsive efficiency loss occurs as a result of the appearance of a condensed phase. Because the acceleration of the gases in the nozzle results from pressure-volume work, a reduction in gas content reduces the thermodynamic efficiency. The combustion gas behaves as if the molecular weight were increased. A second consideration is involved with the momentum exchange between condensed phase and gas. If the particles are large and lag appreciably behind the gas in velocity, impulse is lost. This effect can be appreciable when the percentage of condensed phase is high.

The net result of the thermodynamic limitations is to partially reduce in some cases the improvement in performance expected on the basis of heat of combustion alone. As additives, metals are most effective when mixed with the lower-performing propellants because the lower temperatures minimize the vaporization and dissociation losses. For this reason, the greatest use of metal additives has been in solid propellants, which develop 10–20% lower specific impulses than do liquid propellants. The result of the use of such additives has been to close the gap in performance between the solid and liquid propellants. The second attractive application is in air-breathing engines, where flame temperatures are relatively low.

[DAVID ALTMAN]

Bibliography: C. duP. Donaldson, J. V. Charyk, and M. Summerfield (eds.), *High Speed Aerodynamics and Jet Propulsion*, vol. 2, 1956; M. Gilbert, L. Davis, and D. Altman, Velocity lag of particles in linearly accelerated combustion gases, *Jet Propul.*, 25(1):26–30, 1955; F. E. Marble, The role of approximate analytical results in the study of two-phase flow in nozzles, *Proceedings of the A. F. Rocket Propulsion Laboratory Two-Phase Flow Conference*, 1967; W. R. Maxwell, W. Dickinson, and C. F. Caldin, Adiabatic expansion of a gas stream containing solid particles, *Aircraft Eng.*, 18:350–351, 1946.

Methane

A member of the alkane or paraffin series of hydrocarbons with the formula shown below.

```
    H
    |
H—C—H
    |
    H
```

Methane is called marsh gas because it forms by anaerobic bacterial decomposition of vegetable matter in swampy land. Coal miners know it as firedamp because mixtures with air are combustible. It is a major constituent of natural gas (50–90%) and of coal gas. It forms in large amounts in the activated-sludge process of sewage disposal. As a liquid it freezes at −182.6°C and boils at −161.6°C.

In addition to its use as a fuel, methane is important as a source of organic chemicals and of hydrogen. Its reaction with steam at high temperatures in the presence of catalysts yields carbon monoxide and hydrogen (synthesis gas), which can be catalytically converted to liquid alkanes (Fischer-Tropsch process) or to methanol and other alcohols. The catalytic reaction of the synthesis gas with olefins yields higher alcohols (the Oxo process), and reaction with steam produces additional hydrogen and carbon dioxide.

The incomplete combustion of methane with air produces finely divided carbon, called carbon black, hundreds of millions of pounds of which are used annually as a reinforcing and filling agent in compounding rubber and as a pigment in printing ink.

Chlorination of methane yields methyl chloride, methylene chloride, chloroform, and carbon tetrachloride. *See* FISCHER-TROPSCH PROCESS.

[LOUIS SCHMERLING]

Mining

The taking of minerals from the earth, including production from surface waters and from wells. Usually the oil and gas industries are regarded as separate from the mining industry. The term mining industry commonly includes such functions as exploration, mineral separation, hydrometallurgy, electrolytic reduction, and smelting and refining, even though these are not actually mining operations.

The use of mineral materials, and thus mining, dates back to the earliest stages of man's history, as shown by artifacts of stone and pottery and gold ornaments. The products of mining are not only basic to communal living as construction, mechanical, and raw materials, but salt is necessary to life itself, and the fertilizer minerals are required to feed a populous world.

Methods. Mining is broadly divided into three basic methods: opencast, underground, and fluid mining. Opencast, or surface mining, is done either from pits or gouged-out slopes or by strip mining, which involves extraction from a series of successive parallel trenches. Dredging is a type of strip mining, with digging done from barges. Hydraulic mining uses jets of water to excavate material.

Underground mining involves extraction from beneath the surface, from depths as great as 10,000 ft (3.05 km), by any of several methods.

Fluid mining is extraction from natural brines, lakes, oceans, or underground waters; from solutions made by dissolving underground materials and pumping to the surface; from underground oil or gas pools; by melting underground material with hot water and pumping to the surface; or by driving material from well to well by gas drive, water drive, or combustion. Most fluid mining is done by wells. A recent type of well mining, still experimental, is to wash insoluble material loose by underground jets and pump the slurry to the surface. *See* OPEN-PIT MINING; STRIP MINING; UNDERGROUND MINING.

The activities of the mining industry begin with exploration which, since mankind can no longer depend on accidental discoveries or surficially exposed deposits, has become a complicated, expensive, and highly technical task. After suitable deposits have been found and their worth proved, development, or preparation for mining, is necessary. For opencast mining this involves stripping off overburden; and for underground mining the sinking of shafts, driving of adits and various other underground openings, and providing for drainage and ventilation. For mining by wells drilling must be done. For all these cases equipment must be provided for such purposes as blasthole drilling, blasting, loading, transporting, hoisting, power transmission, pumping, ventilation, storage, or casing and connecting wells. Mines may ship their crude products directly to reduction plants, refiners, or consumers, but commonly concentrating mills are provided to separate useful from useless (gangue) minerals.

Economics. Mining is done by hand in places where labor is cheap, but in the more industrialized countries it is a highly mechanized operation. Some surface mines use the largest and most expensive machines ever developed—unless a large ship can be called a machine.

There are many small- and medium-sized mines but also a growing number of large ones. The trend to larger mines and particularly to large opencast operations is due to the great demand for mineral products, depletion of high-grade reserves, technological progress, and the need for economies of scale in mining low-grade deposits. The largest open-pit coal mine moves 300,000 tons (270,000 metric tons) of material per day and the largest metal mine, 300,000 tons. The largest underground coal mine produces 17,000 tons (15,000 metric tons) of coal per day, and the largest metal mine 50,000 tons (45,000 metric tons) of ore.

The quality of deposits which can be mined economically depends on the market value of the contained valuable minerals, on the costs of mining and treatment, and on location. Alluvial gold gravels may run as low as 1/350 oz of gold/yd^3 (0.106 g/m^3) and gold from lodes as low as 1/10 oz/ton (3.125 g/metric ton). Uranium ore containing 1.5 lb of uranium oxide per ton (0.75 kg per metric ton) is mined underground, and copper-molybdenum ore as low grade as 0.35% copper and 0.05% molybdenum is taken from open pits. On the other hand, iron ore is rarely mined below 25% iron, aluminum ore below 30% aluminum, and coal less than 90% pure. Crude petroleum is usually over 95% pure.

Use of natural resources. A unique feature of mining is the circumstance that mineral deposits undergoing extraction are "wasting assets," meaning that they are not renewable as are other natural resources. This depletability of mineral deposits not only requires that mining companies must periodically find new deposits and constantly improve their technology in order to stay in business, but calls for conservational, industrial, and political policies to serve the public interest. Depletion means that the supplies of any particular mineral, except those derived from oceanic brine, must be drawn from ever-lower-grade sources. Consciousness of depletion causes many countries to be possessive about their mineral resources and jealous of their exploitation by foreigners. Depletion also accounts for some controversial attitudes toward conservation. Some observers would reduce the scale of domestic production and increase imports in order to extend the lives of domestic deposits. Their opponents argue that encouragement of mining through tariff protection, subsidies, or import quotas is desirable, on the grounds that only a dynamic industry in being can develop the means of mining low-grade deposits or meet the needs of a national emergency. They point out that protection encourages the extraction of marginal resources that would otherwise be condemned through abandonment.

Despite its essential nature, mining today is being constrained as to where it can operate by wilderness lovers and by rapidly expanding urbanization. Concern over pollution, some of which is caused by mine water, mining wastes, and smelter effluents, has grown rapidly. More and more objection is developing to defacement of landscapes by surface mining, and many states now require restoration of the surface to a cultivable or forested condition after strip mining. People in residential areas usually resist the development of industries near their homes, particularly if such industries employ blasting, produce smoke or fumes, or cause traffic by large vehicles.

In countries with Anglo-Saxon traditions, title to minerals on private lands is vested in the private owners, but in many countries minerals are state property. Where minerals are privately owned, mining is commonly done by purchase of title or under lease and royalty contracts. State-owned minerals are mined under claims acquired through discovery and denouncement, leases, or concessions. For centuries governments have found it desirable to encourage prospecting and mining to lure the adventurous into these challenging, useful arts. [EVAN JUST]

MINING POWER

Power is applied to mining in six ways: electricity, diesel power, compressed air, hydraulic power, steam power, and hydroelectricity.

Electricity. Both alternating and direct electric current are used in modern mining. Direct current (dc) is used for locomotive haulage and for the major portion of underground coal operations because of the high torque developed in dc series motors under heavy loads and starting. Alternating current (ac) is used less extensively although some mines employ both in a dual system.

In the United States, 6000-, 4160-, and 2300-volt three-phase alternating current is generally received from service companies at the surface substation. In a well-planned mine, all the power is metered at one point and the power factor is adjusted between 90 and 95% by the use of synchronous motors and capacitors. Neoprene-covered cables, type SHD, 15,000 volts, are recommended to take the power underground through boreholes or power shafts. Each of three insulated conductors is covered with copper shielding braiding to eliminate static stresses, and a ground conductor for each power conductor is placed in the interstices of the cable.

Converting ac to dc underground may be accomplished by several methods. A 250- or 500-volt substation can be provided by converters, motor-generator sets, mercury-arc rectifiers including the glass-bulb type, and dry rectifiers made of selenium, germanium, or silicon (Fig. 1). Portable conversion units offer increased convenience and efficiency.

Trolley and feeder lines supply power throughout the main roadways. These distribution lines should not be extended more than 3500–4000 ft (1070–1220 m) from the power source to avoid low voltage at the end of the line. If local laws permit, 500-volt systems are sometimes used to minimize line voltage drop. This supplies twice the load for the same size and length of cable, or conversely, the same load can be supplied at twice the distance. Trailing cables, fastened to the trolley nipping stations or power centers, supply power to machines in sections of the mine where explosive gases may be present.

Sectionalization is a method of distributing mine power so that power cables can be isolated for reasons of faults or repairs without shutting off the main supply to several working sections. For main distribution, circuit breakers, disconnect switches, and various overcurrent protective devices are essential for properly installed sectionalization. At face areas, safety circuit centers and associated intrinsically safe circuitry make it possible to connect or disconnect short cables without danger of causing incendiary arcing which could ignite gas.

Alternating-current power for mining is increasing in popularity because its equipment is less costly than dc and its maintenance is simpler. Alternating-current motors, for example, cost one-third to one-half as much as the equivalent dc and are more compact. High voltage is transformed to 440 or 220 volts at underground substations. Portable units are also utilized. Sectionalization is applicable to ac distribution.

Alternating-current power is commonly used in strip mining. High voltage of 33,000 volts is stepped down to 7300, 6600, 4160, or 2300 volts. Permanent substations equipped with lightning arresters, circuit breakers, ground protective equipment, and other protective devices, and semiportable substations are employed for distribution. Power is distributed by pole lines or cable systems, or a combination of the two.

Fig. 1. A 300-kW Westinghouse ignitron rectifier installed at a mine.

Diesel power. This type is rapidly gaining favor in metal and noncoal mining because of its flexibility. Some states require that underground diesel-powered equipment be approved by the U.S. Bureau of Mines. Details of these standards include explosion-proof housings, mine ventilation necessary to dilute exhaust gases, control of hot exhaust gases and surface temperatures, concentrations of toxic constituents in exhaust gases, and recommendations for the selection, handling, and storage of diesel fuel oil.

Compressed air. As a mine facility, pneumatic power is utilized in a variety of applications, mostly in the metal and noncoal fields. It is used to power drills and hoists; pneumatic tools, such as grinders, drills, riveters, chippers, pneumatic diggers, and spades; air-driven sump pumps; direct-acting and air-lift pumps; pile drivers for shaft sheathing; air pistons for unloading cars; drill-steel sharpeners; air motors; compressed-air locomotives; shank and detachable-bit grinders; mine ventilation; and in supplying air for blowing converters; starting diesel engines; and coal preparation. Compressed air is used in coal mining for blasting. This method works with 9000–10,000 pounds per square inch (psi) [62–69 MPa] and the air is released from a tube which is inserted in a hole in the face of the coal when a metal diaphragm ruptures. The force is released through slots in the tube and breaks up the coal which had been previously undercut.

Hydraulic power. For mining, hydraulic applications are rather limited in usage and may employ either water or oil as a fluid. Oil types are used in connection with small tools, lifts, and in the intricate operation of continuous mining machines and other equipment. An available waterfall may be directed to power equipment such as air compressors. A unique use of hydraulic power, called jet mining, uses air pressure to force water through 1/4-in. (6-mm) nozzles under 2000 psi (14 MPa) pressure. This jet action has been developed to cut a material called gilsonite, a solid hydrocarbon, by use of water at a rate of 300 gal/min (1.9×10^{-2} m^3/s).

Steam power. Although formerly used to drive compressors, hoists, generators, and other equipment, steam is now rarely used in mining but has definite, although limited, application in some coal mines in which there is an abundance of waste fuel or unmarketable coal. Parts of certain coal seams contain impure bands that must be rejected, or that are difficult to clean, but they can be burned under boilers with special firing equipment.

Hydroelectricity. As applied to mining, hydroelectricity is used mostly in the electrometallurgical fields, where vast amounts of power are required at a reasonable cost. Some hydroelectricity has been used for normal, electrical, mining power, but mostly it is used for processes beyond the mining operation such as beneficiation, smelting, and refining. In certain cases these are done near the mine mouth in isolated areas in order to reduce bulk before shipment. [ROBERT S. JAMES]

MINE VENTILATION

The purpose of mine ventilation is to provide comfortable, safe, and healthful atmospheric conditions at places where men work or travel.

Airflow fundamentals. The following points summarize airflow principles for mine ventilation; (1) Airflow is induced by a pressure difference between intake and exhaust; (2) the pressure created must be sufficient to overcome the system resistance and may be either negative or positive; and (3) air flows from the point of higher to lower pressure. Also, mine airflow is considered turbulent and follows the square law relationship between volume and pressure; that is, a doubled volume requires four times the pressure.

The principles of fans may be summarized as follows: (1) Air quantity varies directly as fan speed, and is independent of air density; (2) pressures induced vary directly as the square of the fan speed, and directly as the air density; (3) the fan power input varies directly as the air density and the cube of the fan speed; (4) the fan mechanical efficiency is independent of fan speed and density.

The amount of air movement induced will be dependent upon the fan characteristic and mine resistance as shown by Fig. 2. The pressure H required to pass a quantity of air Q through a mine or segment is expressed by the formula $H = RQ^2$, where R, the mine resistance factor, may be calculated from known pressure losses, or from the common ventilation formula $R = KlO/5.2A^3$ in which K is the frictional coefficient, l is the length of the airway, O is the perimeter of the airway, and A is the cross-sectional area of the airway. For simplicity of calculation, frictional coefficients should be expressed without decimals, and the quantity Q of air in cubic feet per minute (cfm = 0.472×10^{-3} m^3/s) should be divided by 100,000 before using. The table gives reasonable frictional coefficients; the table's use may be exemplified by an application to the following case: What pressure is necessary to induce 60,000 cfm through a single airway 6 ft high, 12 ft wide, and 2500 ft long in sedimentary rock? The airway is straight, has average irregularities, and is moderately obstructed. K from table is 70; l is 2500; perimeter is 36; area is 72 ft². Q is 60,000 cfm ÷ 100,000 = 0.60.

The pressure H is given by Eq. (1).

$$H = RQ^2 = \frac{KlOQ^2}{5.2A^3} = \frac{70 \times 2500 \times 36 \times (0.6)^2}{5.2 \times (72)^3} = 1.17 \text{ in. water pressure} \qquad (1)$$

Parallel air flow can be determined by the square-law relationship. For example, the pressure H required to pass 60,000 cfm through 2500 ft of single entry is 1.17 in. of water. For two identical entries the pressure is given by Eqs. (2).

$$H_1 \left(\frac{1}{2}\right)^2 = H_2$$
$$H_2 = \frac{H_1}{4} = \frac{1.17}{4} = 0.292 \text{ in. water} \qquad (2)$$

When resistance factors are known or entries are not identical, the formula is Eq. (3). For example, what is the combined resistance factor of two entries 1000 ft long? One entry is substantially larger, $R_1 = 1.50$, $R_2 = 4.0$. Also, what pressure is required to pass 60,000 cfm through 2500 ft of the combined entries, assuming average conditions throughout? Computation is as follows:

$$\frac{1}{\sqrt{R}} = \frac{1}{\sqrt{R_1}} + \frac{1}{\sqrt{R_2}} + \frac{1}{\sqrt{R_3}} + \cdots + \frac{1}{\sqrt{R_n}} \qquad (3)$$

$$\frac{1}{\sqrt{R}} = \frac{1}{\sqrt{1.5}} + \frac{1}{\sqrt{4.0}} = \frac{1}{1.22} + \frac{1}{2.00}$$
$$= 0.82 + 0.50 = 1.32$$
$$\frac{1}{\sqrt{R}} = 1.32 \qquad R = 0.58 \text{ per 1000 ft entry}$$
$$R \text{ for 2500 ft} = 2.5 \times 0.58 = 1.45$$
$$\text{Pressure} = 1.45 \times (0.60)^2 = 0.52 \text{ in. water}$$

Air quantity requirements. Unless covered by state laws, the common criteria for adequate ventilation are absence of smoke and dust with moderate air temperatures in metal mines and the absence of methane, smoke, and dust in coal mines. Natural conditions of gas, rock temperatures, dust, and operating practices determine requirements. Good quality air is not deficient of oxygen and is free of harmful amounts of physiological or explosive contaminants.

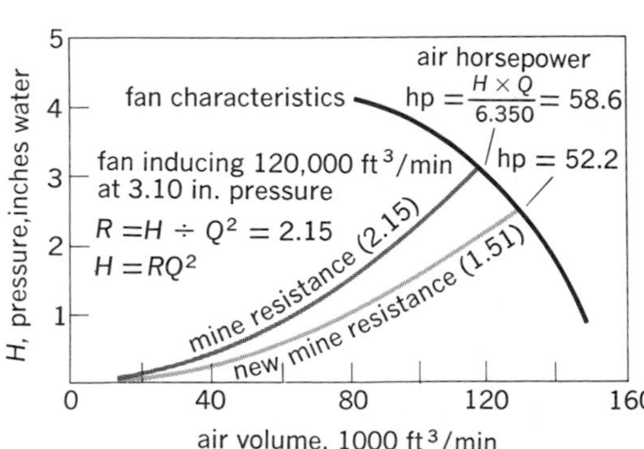

Fig. 2. Fan characteristics versus mine resistance. 1000 ft³/min = 0.472 m³/s; 1 hp = 745.7 W; pressure of 1 in. water = 249 Pa.

Reasonable frictional coefficients

Type of airway	Irregularities	Straight: Clean	Straight: Slightly obstructed	Straight: Moderately obstructed	Sinuous or curved, Moderate: Clean	Sinuous or curved, Moderate: Slightly obstructed	Sinuous or curved, Moderate: Moderately obstructed	Sinuous or curved, High degree: Clean	Sinuous or curved, High degree: Slightly obstructed	Sinuous or curved, High degree: Moderately obstructed
Smooth-lined	Minimum	10	15	25	25	30	40	35	40	50
	Average	15	20	30	30	35	45	40	45	55
	Maximum	20	25	35	35	40	50	45	50	60
Sedimentary rock or coal	Minimum	30	35	45	45	50	60	55	60	70
	Average	55	60	70	70	75	85	80	85	95
	Maximum	70	75	85	85	95	100	95	100	110
Timbered (5-ft centers)*	Minimum	80	85	95	95	100	110	105	110	120
	Average	95	100	110	110	115	125	120	125	135
	Maximum	105	110	120	120	125	135	130	135	145
Igneous rock	Minimum	90	95	105	105	110	120	115	120	130
	Average	145	150	160	160	165	175	170	175	195
	Maximum	195	200	210	210	215	225	220	225	235

*5 ft = 1.524 m.

Mine gases. Important contaminants of mine air are carbon dioxide, hydrogen sulfide, methane, carbon monoxide, and sulfur dioxide.

Carbon dioxide, CO_2, specific gravity 1.529, is produced by oxidation and combustion of organic compounds and is occluded in the rock strata of certain mines. It is heavy, colorless, and odorless and is usually found in low, poorly ventilated areas.

Hydrogen sulfide, H_2S, specific gravity 1.191, is the product of decomposition of sulfur compounds. It is colorless, has the odor of rotten eggs, and may be found near areas of stagnant water in poorly ventilated areas.

Methane, CH_4, specific gravity 0.554, is a natural constituent of all coals. It may be occluded in carbonaceous shales and sandstones and may infiltrate into metal mines at contacts with carbonaceous rocks. It is colorless, odorless, and may be found in high, poorly ventilated cavities.

Carbon monoxide, CO, specific gravity 0.967, is not a normal constituent of mine air, but is produced in mines by the incomplete combustion of carbonaceous matter, mine fires, or from gas or dust explosions. It is colorless and odorless.

Sulfur dioxide, SO_2, specific gravity 2.264, is not common, but may be found in sulfur mines and in mines with rich sulfide ores as the result of fires. It is a water-soluble and colorless gas with a suffocating odor.

Blackdamp is a common term applied to oxygen-deficient atmospheres; it is not a specific gas mixture but may contain any of many gases produced by oxidation and processes that use oxygen and liberate carbon dioxide.

Small quantities of air contaminants must be determined by laboratory analysis of air samples. On-the-spot safety determinations for carbon dioxide and oxygen deficiency may be made by flame safety lamp or small portable absorption instruments. Methane can be detected by flame safety lamp and commercial testers; carbon monoxide and hydrogen sulfide by special hand-held colorimetric indicators.

Dust and dust hazards. Dust is defined as the solid particulate matter thrown into suspension by mining operations. The size of particles may range upward from less than 1 micron (0.001 mm) diameter (as shown in Fig. 3, the Frank chart); particles larger than 10 μ can usually be seen by the naked eye. The dust hazard may be physiological or explosive or both physiological and explosive as is coal dust. The common physiological diseases are mostly various pneumoconioses. Preventive measures are to suppress dust at the source with water sprays, foam, fog, wetting agents, or dust collectors. Additional protection may be provided through suitable respirators. Accumulations of explosive dust should be removed and inert material, such as rock dust, applied to surface areas. The dust hazard can be determined by systematic sampling of airborne dust at critical points. The impinger, a common instrument used, draws a known volume of air through a liquid or filter to remove the dust. The dust concentrations of the sample can then be determined by count using a microscope or microprojector. *See* DUST AND MIST COLLECTION.

Air temperature and humidity. Temperature rise in mine workings is from (1) heat conducted from surrounding strata because of the thermal gradient (depth per °F or °C rise in temperature); (2) adiabatic compression of descending air columns (approximately 5.5°F/1000 ft or 10°C/km); and (3) heat from oxidation of minerals.

The factors that influence humidity are: rise in dry-bulb temperature; volume changes caused by pressure and temperature changes; and moisture picked up from shafts and roadways.

The air temperature approximates the temperature of adjacent walls; seasonal temperature changes are noticeable only short distances underground. Workers' efficiency (Fig. 4) is dependent upon temperature, humidity, and motion velocity of air. A solution to excessive air temperatures is air conditioning by passing the air currents through heat exchangers filled with chilled air.

Natural ventilation. Natural ventilation is induced by differences of total weights of air columns for the same vertical distance. Natural draft may operate with or against the mechanical draft (Fig. 5) or may be the only source of pressure. Natural draft pressures at standard density can be estimated as 0.03 in. water-gage for each 10°F average temperature difference per 100 ft increment vertical elevation (or 4.2 mm water-gage for each 1°C per meter). For accurate determinations the average air densities of influencing air columns must be calculated.

The formula for determining air density is given by Eq. (4), where d is the density in lb/ft³, T is the

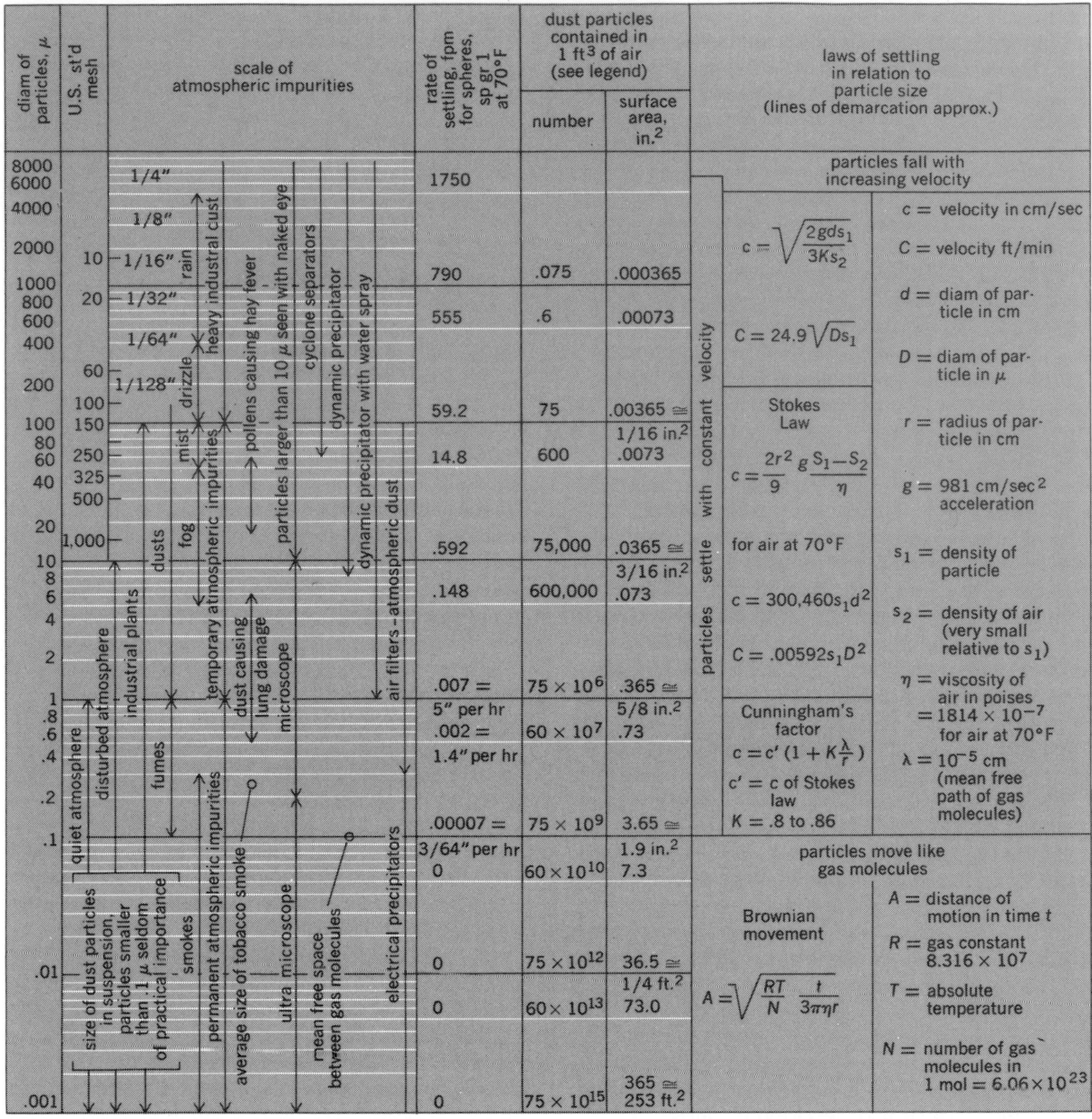

Fig. 3. Size and characteristics of airborne solids (compiled by W. G. Frank). It is assumed that particles are of uniform spherical shape having specific gravity 1 and that dust concentration is 0.6 gr/1000 ft³ (2.288 g/1000 m³) of air, average for metropolitan districts. 1″ = 25.4 mm; 70°F = 21°C; 1 ft³ = 2.83 × 10⁻³ m³; 1 in.² = 6.45 cm²; 1 ft/min = 5.08 mm/s; 1 poise = 0.01 Pa·s.

$$d=\frac{1.327}{460+T}(B-0.378\mathrm{VP}) \qquad (4)$$

dry-bulb temperature, B is the barometric pressure, in inches of mercury, and VP is the vapor pressure of water at the dew point, in inches of mercury.

With most calculations, the vapor pressure influence can be ignored. The simplified formula then is Eq. (5).

$$d=\frac{1.327}{460+T}(B) \qquad (5)$$

Auxiliary ventilation. This term applies to booster fans and auxiliary fans. A booster fan is placed underground and operated in series with the main fan to increase ventilating pressure of one or more splits of the ventilating current. The booster fan in effect reduces the mine resistance, thereby increasing the air quantity circulated. An auxiliary fan is a small fan installed in the air current to divert, through air tubing or ducts, a part of the ventilating current to ventilate some particular place or places. In metal mines, they are used to ventilate developing drifts, raises, crosscuts, and stopes. In coal mines, they are used to conduct air to working faces. [DONALD S. KINGERY]

Ventilation system evaluation. Accurate analysis and evaluation of a ventilation system, for the purpose of initiating improvements or projecting the system, require pressure and quantity measurements from which actual resistances can be determined. Proper utilization of such data permit efficient and economic changes in the present system and the accurate determination of requirements for the projected system. Although friction coefficients from the table enable engineers to design, with good results, ventilation systems for new mines, the continuously changing conditions following mining make it almost impossible to apply these values in evaluating older systems.

The aneroid barometer measures absolute static

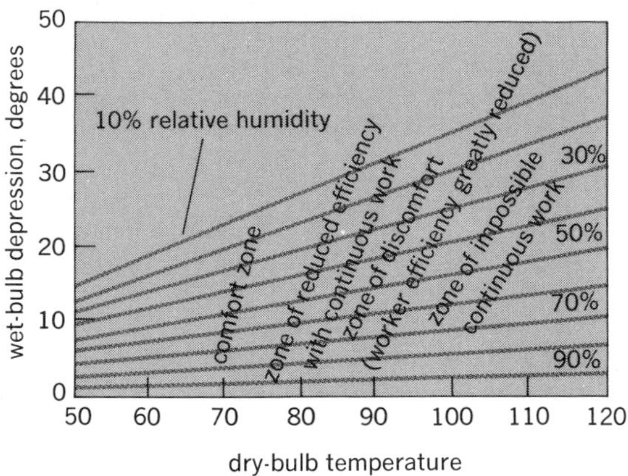

Fig. 4. Graph showing the influence of humidity and temperature on worker efficiency.

pressure and is the instrument usually used for mine-pressure surveys. The instrument is rugged and portable but relatively few have sufficient sensitivity and precision for pressure surveys. A variety of graduated scales are provided, the most common being "inches of mercury" and "feet of air." Instruments with the latter scale are termed altimeters. Regardless of scale on the instrument, mine pressure and pressure differentials are generally converted to inches of water for analysis of pressure data.

The surface absolute static pressure (barometer reading) is primarily a function of elevation and air density. Underground, within the ventilation system, absolute static pressure is a function of the same conditions plus the static pressure resulting from pressure generated by the fan. By compensating for elevation differences and atmospheric changes, the ventilating pressure at any point and between points resulting from fan operation can be calculated. By supplementing the pressure determinations with water-gage reading across stoppings and regulators, pressure losses for any part of the system can be readily determined.

Analysis of the pressure data by means of a pressure gradient will pinpoint high-resistance areas. The pressure gradient is constructed by plotting distance of air travel against pressure drop. Portions of the gradient with a steep slope indicate high-resistance airways.

For a complete analysis of any system, pressure data must be combined with extensive quantity determinations. Quantity flow is determined by area and velocity measurements. The more common instruments for velocity measurements include vane anemometer, pitot tube, and smoke tube. Instruments selected will depend on velocity of the airstream.

The quantity survey is useful in locating areas of excessive intake to return leakage, but more important, combined with pressure data, the actual resistance factor for each segment of the system can be calculated. These factors are necessary for accurate analysis of the effect of any major system changes or projections. Results may be determined mathematically or by analog computer.

For the economic solution of problems involving mine ventilation improvement, the U.S. Bureau of Mines employs almost exclusively an electric analog computer designed especially for analyzing mine-ventilation distribution problems. The heart of the analyzer is a nonlinear resistor known as a Fluistor which closely approximates the square-law resistance characteristics of mine airflow; that is, the voltage drop varies approximately as the square of the current.

In developing problem solution the ventilation circuit is divided into segments and the resistance factor for each segment is calculated. The system is reproduced electrically on the analog, using Fluistors of the proper coefficient to simulate airway resistance. Once the analog network is balanced to mine conditions represented by field data, any changes can be readily made and analyzed. The master meter can be connected directly to any segment in the circuit to measure voltage or current. The meter is a digital voltmeter which reads out pressure and airflow in units of inches of water and cubic feet per minute, respectively.

[E. J. HARRIS]

MINE ILLUMINATION

The lighting of mines is accomplished by use of both movable and stationary lamps.

Mobile illumination. Cap lamps, flashlights, portable hand lamps, trip lamps for haulage cars,

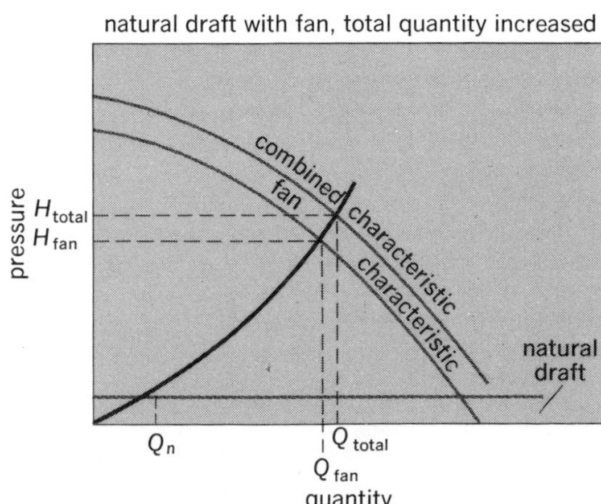

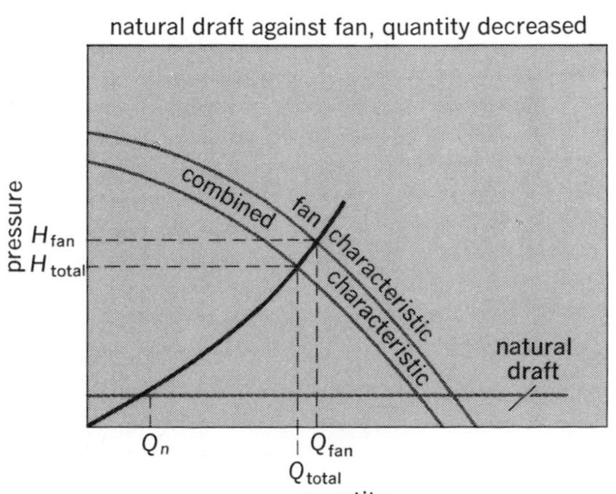

Fig. 5. Graphs of influence of natural draft to mine ventilation.

and lights in mobile machinery provide mobile lighting. The carbide lamp has been outmoded by the safer and more efficient electric cap lamp operated by a 4- or 5-volt battery. Such a lamp, using a polished reflector, can produce a beam candlepower of 1000 candles in a small spot. Less light with greater spread is obtained with a matte reflector.

Permissible continuous mining machines, shuttle cars, and other mining production equipment are fitted with explosion-proof 150-watt, 120-volt lights, front and rear. All permissible lights for use in gassy or dusty mines in the United States are approved by the U.S. Bureau of Mines.

Stationary or area illumination. This type of lighting is provided by incandescent lamps or dual 15-in. fluorescent lights at intervals of 20–90 ft. Incandescent types, 50–150 watts, dc-operated, may be connected singly or in groups in ventilated haulageways but not in gassy areas. Fluorescent lighting, first approved in 1957 for gassy or dusty mines, employs 14-watt lamps on ac power (Fig. 6).

Fluorescent lights have proved very effective in gassy areas, along beltways and shuttle-car roadways, and have been credited with reduction in accidents, better morale among the men, and better production records. Three-wire conventional grounding or two-wire isolated ungrounded systems provide shock protection. Special connectors facilitate the disconnection of lights without arcing. The U.S. Bureau of Mines approved a flashlamp for underground photographic use in 1957. [ROBERT S. JAMES]

MINE DRAINAGE

It is necessary to prevent water infiltration into mines from the surface or from sources underground. Control, transportation, and ultimate disposal of such waters constitute mine drainage.

Mine drainage varies in facilities and importance, at different localities, according to conditions of the mine water. There must be a careful study of its occurrence, corrosive and erosive character, volume to be handled, and the handling and disposal facilities that can economically provide efficient drainage. Major facilities consist of flumes, dikes, and storage ponds to prevent or minimize seepage of surface waters, underground ditches and sumps to collect, store, and control mine water, and pumping plants and drainage tunnels for the disposal of the mine water.

Fig. 6. An installation of permissible fluorescent lights in a coal mine. (*U.S. Bureau of Mines*)

Control of stream pollution by mine drainage is important for both esthetic and economic reasons.

Sources and types of mine waters. Physical characteristics of the individual mining areas create distinctive sources and types of mine waters. Water enters directly into open-pit or underground workings, seeps through pervious strata, drains from normal water channels both on the surface and underground, and infiltrates from impounded waters into mine workings through cracked, crushed, and broken formations.

Mine waters vary widely in character in different mines because of the different geological formations. Waters from mines may be classed as alkaline (pH 7–14), neutral (pH 7), and acidic (pH 0–7). They may carry abrasive matter in suspension and be erosive to a degree dependent on the amount of abrasive substances. Although ground waters are normally alkaline, analyses from the majority of mines show considerable variation because the waters contain varying quantities of dissolved sulfate salts. Acidic water formed by such salts in the majority of mines has resulted in tremendous losses by corrosion, especially to pipes and pumping equipment.

Sumps, drains, and tunnels. The infiltration of water into mines, the volume of water to be handled, and pumping facilities vary with the seasons. The storage of mine water until it can be disposed of satisfactorily is of great importance. Ample sumpage where mine water can be stored is necessary in any drainage system. Sumps should be provided near pumping stations, have ample capacity, and be arranged so as to permit easy cleaning. If practicable, they should be designed to provide gravity feed to the pumps.

Drains, flumes, or ditches. These are used on the surface and underground to divert and convey water and thereby to prevent stream pollution and inflow through crop caves, strippings, cracks and fissures, and stream beds. Large flumes are used for those purposes in subsided areas, particularly where flash floods or quick runoff occur. Because ditches are usually made in mine haulage roads, they should be designed to carry the maximum drainage.

Drainage tunnels between different working levels are preferred means of handling and keeping mine water from reaching lower mine workings. Drainage tunnels from a mine or a group of mines to some disposal point situated favorably with respect to surface disposal areas are a means of draining mines economically and saving reserves of minerals that would be lost otherwise. They are advantageous over a long economic life of a mine for handling large quantities of water that sometimes could not be pumped to the surface. Although the initial cost is high, their upkeep is comparatively low.

Pumping. As long as mines are operated, mine drainage will most likely be done in whole or in part by pumps. Because mine water is usually acidic, suitable metals or alloys should be used for mine pumping equipment. The designing engineer should consult with qualified technicians before attempting to solve corrosion problems. Centrifugal pumps are both horizontal and vertical and known as standard, deepwell, and shaft. Displacement pumps, piston and plunger, are being discarded. Mine tailing pumping is sometimes done by means of centrifugal sand pumps. Deep-well pumps in a shaft or bore hole have proved successful for emergency use and for unwatering abandoned mines. Pumping systems and controls must be designed to handle the maximum inflow of water in mine workings.

[SIMON H. ASH]

MINING SAFETY

The prevention of mine worker injury by precautionary practices in mining operations. This article discusses records of mine safety, methods of fire prevention, government regulations, training programs for mine workers, and mine environment problems causing disease.

Safety records. Vigilance is the price of safety in a mine, as it is in any other activity. Fortunately, the presentation of mining hazards in motion pictures and television bears little relation to fact. The Mine Safety Appliances Co. in the late 1930s originated the John T. Ryan trophies, to be awarded annually to the Canadian mines with the best safety records each year. These awards include two national trophies, one for metalliferous mining and one for coal mining, and six regional trophies, two for coal mining and four for metalliferous mining. The awards are based on the number of lost-time accidents during 1,000,000 worker-hours of exposure for metalliferous mines, and for 120,000 worker-hours of exposure in coal mines.

A lost-time accident is defined as one that results in a compensation payment for a total or partial disability that lasts more than 3 days. In Canada, before Aug. 1, 1968, a 3-day waiting period was required in some provinces before an injury became compensable. The records of the workmen's compensation boards are used to compile safety records, and all locations of employment are included, such as mine, mill, offices, and shops, but smelters and open-pit or strip-mine operations are not included.

The Canada John T. Ryan trophy for metalliferous mines is usually won by a mine with no lost-time accidents as defined. A mine with four or five accidents is likely to be disqualified from the countrywide competition. A fatal accident eliminates a mine. The coal mining record is not as good as that of the metalliferous mines; for one thing, there are fewer coal mines. The low incidence of accidents is not attained without effort. Supervisors, like traffic police, spend a lot of their time trying to keep people from acting foolishly. In North America governments have regulations regarding the safety of workers in mines and inspectors to see that they are observed.

Fire hazards. Government regulations for mine operation and the safety of workers in mines were established in North America during the 19th century, patterned on those of Europe and Great Britain, where there was a longer history of coal mining than in North America. The regulations were first directed to coal mining because it presents the hazard of combustible gas from bituminous coal and flammable dust. Coal is made up of volatile material and fixed carbon. If the ratio

$$\frac{\text{volatile}}{\text{volatile} + \text{fixed carbon}}$$

exceeds 0.12, the dust can burn explosively when mixed with air in dangerous proportions, and the range of dangerous proportions is quite wide. All bituminous coals have a ratio in excess of 0.12. Combustible gas is especially dangerous because it is mobile and if it is ignited the ensuing fire or explosion may affect men remote from the source of trouble. Other accidents, a fall of rock, for instance, do not affect individuals working outside the location of the accident.

Methane gas, the major component of the combustible gas, is dangerous in two ways. It dilutes the mine air and so reduces the amount of life-sustaining oxygen. Coal mines are classified as gassy, slightly gassy, and nongassy. A nongassy mine has less than 0.05% methane in the air. If methane is in a concentration of 5–13.9%, it may be ignited and will burn explosively. The most violent explosive mixture is 9.4% methane. At some point, as the location of the emission is approached, the explosive ratio will be reached; a single spark can set off an explosion. The resultant turbulence in the air will stir up the coal dust and the fire can spread with explosive violence.

A mine fire has side effects. Workers may be caught in the actual fire, but the burning produces carbon monoxide and carbon dioxide which will spread far beyond the fire area. If the roof is sulfurous shale, it may be brought down by the heat.

Prevention of ignition. The propagation of a fire in the dust stirred up by the air turbulence accompanying an explosion may be inhibited by dusting. An incombustible rock dust is spread throughout the workings. When stirred up, it dilutes the mine dust and absorbs heat, so combustion cannot be maintained. The initiation of combustion must be prevented.

Equipment. In the United States only equipment certified by the U.S.B.M. (United States Bureau of Mines) as permissible may be used where gas or combustible dust may be present. The restriction applies to electrical equipment, including cap lamps and machinery. The certification is based on the no-sparking qualities of switches and adequate current-carrying capacities for electrical equipment; and on ensurance that machinery has no operating parts that can heat to the ignition temperature for gas or dust. In Canada a British certification is also accepted.

Explosives. Only permitted explosives may be used. All explosives involve combustion and high temperatures. The U.S.B.M. has certified certain explosives as permissible for use in coal mines. They are compounded to produce a lower temperature and volume of flame than ordinary explosives, and certain salts are included in the formulas to quench the flame rapidly. Permitted explosives must be used as prescribed for prescribed conditions. Special attention is required for stemming to prevent blowouts.

Safety lamp. The biggest single advance in coal mine safety was probably the invention of the Davy safety lamp in 1815. A mantle of wire gauze around the flame dissipated the heat from the flame to below the ignition temperature of methane, which is about 650°C. It provided light and the miner no longer had to work with an open flame. The height and color of the flame gave a measure of the amount of methane in the air. Improved models are still in use, though they are being superseded by no-flame instruments that give a prompt reading or can monitor the methane content continually. The miner is no longer dependent on a flame for light since the electric cap lamp became available.

Ventilation. Coal mine operators have developed ventilation techniques, which until recently, surpassed those used in metal mines. Great care is taken to sweep working places with enough air to prevent dangerous accumulations of gas or of dust. The advent of mechanical coal-cutters and other mechanical equipment has made ventilation more difficult, but the problem has been overcome.

Other mine fires. Methane is not confined to coal mines. It is occasionally released from pockets in metalliferous mines. It is seldom troublesome unless it has accumulated in an unused unventilated working or in a sump at the bottom of a shaft.

Sulfide ores containing certain sulfide minerals may oxidize and generate sulfur dioxide. This may generate enough heat to ignite wood. Pyrrhotite should always be suspected as such an oxidizing sulfide.

Dry, timbered mines in which there is careless smoking are obvious fire hazards. Most metalliferous mines are damp and vigorous fires are not expected, but smoldering fires may be even more hazardous. They may burn for a long time without being detected and the incomplete combustion generates carbon monoxide which is heavier than air, odorless, and lethal. The fire is usually in rubbish accumulated in abandoned places. Good housekeeping is required. Many fires are started in old power cables when the insulation has broken down, and a surprising number start during locomotive battery charging. Sparks from acetylene torch burning are another common source of fire.

Mine rescue teams. The U.S.B.M. officials have accumulated a vast store of fire-fighting knowledge. Most mines have teams trained in mine rescue and in the use of U.S.B.M.-approved equipment in accordance with manuals prepared by the U.S.B.M. officials. They work under their own supervisors but U.S.B.M. officials are available for consultation.

Government regulations. Government regulations for safety in mine operations must be enforced. The higher the government authority and the less localized the enforcement agent, the more effective the enforcement will be. The basic enforcement unit in the United States is the state authority, and even obviously good regulations are often hard to enforce. In addition ot the human tendency to resist change, pressure may be brought to bear on the legislatures of the mining states. Some government inspectors may be appointed for political expediency; in the past the company safety inspector was too often an employee given a sinecure in lieu of a pension.

The U.S.B.M. has been assigned responsibility for the health and safety of United States miners. The period of 1907–1913 was formative. States are jealous of their prerogatives, and so it was 1941 before the U.S.B.M. officers had the right to enter a mine. They were chiefly involved by invitation after a disaster. They attained the right to order a coal mine closed for dangerous practices only in 1952. Legislation in 1969 resulted in a safety code to apply to all mines, coal and metalliferous. The initial responsibility continues to be with the state authority, and the U.S.B.M. will interfere only when the state regulations and enforcement do not meet the requirements of the Federal code.

Regulations in Canada are prepared and enforced by the provinces. The Northwest Territories, Yukon, and the Arctic Islands are not provinces and are under the authority of the federal Canadian government. The Ontario metal mining code is the most comprehensive and has been used as a model for the preparation of the codes in other provinces and countries. There is no coal mining in Ontario. The coal mining provinces based their regulations on British codes originally, and these regulations have since been modified in the light of experience in the United States.

Occurrence of accidents. The following discussion is based on data from the *Ontario Department of Mines Inspection Branch Annual Report for 1967*, which gives accident experience for an average of 17,461 individuals who put in 32,391,000 worker-hours in underground mines and open pits of all sizes and degrees of organization. There were 1887 lost-time accidents. The experience did not vary much from that of the preceding 4- or 5-year period.

The time distribution for lost-time accidents was 1 per 17,100 worker-hours worked or, using 1850 worker-hours per worker-year, 1 per 9.4 worker-years. Whether or not an individual has had an accident during any one year has no bearing on whether the person could have another in the same year or any other year, so this is a Poisson distribution. A person would have a 0.33 probability of having no lost-time accidents in 9.4 years.

Actually, the mathematical probability is misleading because an individual is more apt to have an accident in earlier years of employment before becoming mine-wise, or if employed in a developing mine before a safety program is well organized.

The 16 fatal accidents had an incidence of 1 per 2,020,000 worker-hours worked, or 1 per 1090 worker-years. Mathematically a group of 100 individuals working 10 years would have about 0.37 probability of having no fatal accident. That is subject to the same reservations as those set out above for the lost-time accidents.

Training. The necessity for training people to work safely cannot be overstated. Of the 1887 nonfatal accidents in Ontario in 1967 about 34% (641) were personal accidents—fall of persons, or strains while moving or lifting or handling material other than rock or ore. There is reason to suspect laxity in the use of safety equipment, such as gloves and safety belts and boots; ineffective instruction on how to lift; and insufficient emphasis on the importance of good housekeeping in working places. It must be emphasized that the data are from all sizes of mines and from mines that range from those operated under rather primitive conditions to those using modern equipment and management. Some of them would be close runners-up in the Canada John T. Ryan trophy competition, and one of them won the Regional trophy. Others were barely passing the government inspections. One runner-up for the Canada trophy had 2 accidents and the next had 3 accidents in 1967. The winner

Fig. 7. Rock bolting in a Michigan copper mine. *(Photograph by H. R. Rice)*

of the Ontario Regional trophy had 7 accidents per 1,000,000 worker-hours.

The most lethal class of accident is the fall of rock or ore. About 10% (185) of all accidents, including fatal accidents, in Ontario in 1967, were falls of rock or ore, but 31% (5) of the fatal accidents were from that cause. The real cause was probably the failure to recognize an unsafe condition due to neglect or lack of experience. Either cause requires better training of the supervisors and the workers. Nearly half of the accidents occurred during drilling or scaling of loose rock, clearly showing a lack of skill or the neglect of an unsafe condition.

The advent of mechanized mining in large deposits requires greater areas of exposure to provide room to maneuver equipment. Fortunately, it has been accompanied by the extended use of rock bolts as a means for rock reinforcement, and the incidence of falls of rock has been reduced. Figure 7 shows an experimental stope in a mine where U.S.B.M. engineers tested the effectiveness of rock bolting.

The ultimate in mine safety cannot be achieved without the complete support of top management, but the key person in the achievement is the workman, and in the chain of responsibility it is the supervisor at the lowest level of contact with the worker. Though having the best of intentions, the supervisor is handicapped because workers perform in scattered places, out of the supervisor's sight except for short intervals each day. That condition has improved in mines that work larger deposits with mechanical equipment. There is a concentration of working places and the crews are under less intermittent surveillance.

Most mines have a staff safety engineer responsible only to top management. The safety department usually has no line authority. The most important duty that the safety engineer has is the education of supervisors and workers by lectures and training sessions.

It must be recognized that any accident could be fatal, even a neglected scratch from a nail. Furthermore, an accident need not involve injury to a person. Any undesirable happening at any time or in any place that has not been foreseen is an accident. It may be only by chance that there is no personal injury.

Environmental problems. Instantly recognizable injuries to persons attract more attention, but for ages miners have been subject to pneumoconiosis. This includes all lung diseases, fibrotic or nonfibrotic, caused by breathing in a dusty atmosphere. It takes time to develop, and began to be recognized as something to be eliminated early in the 20th century. Most varieties are not fibrotic and will clear up when the person is out of the dust-laden atmosphere; silicosis and the effects of breathing radioactive dust or gas, however, will not.

Silicosis. South African mine doctors led in the diagnosis of silicosis. About 1920 mine operators in North America became aware of the magnitude of the problem and moved to do something about it. It is caused by inhaling silica dust, and possibly some other mineral dusts. It is not confined to mining. Any industry that produces a silica-dusty atmosphere is dangerous. Furthermore, a clear-looking atmosphere may still be dangerous because the particles that are harmful are less than 5 μm in size. At that size they settle slowly and do not show in a beam of light.

The small particles pass the filtering hairs and mucus in the nose and throat. They reach the lungs and by some action, mechanical or chemical, create fibrosis. The useful volume of the lung is reduced. The disease is seldom lethal in itself but the victim is susceptible to tuberculosis and pneumonia.

As soon as the cause was recognized, the operators and the government authorities increased ventilation requirements and insisted on water sprays, particularly after blasting, to knock down the dust. Wet drilling was well established at that time but there have been some improvements in the drills. The dust content of the mine air at working places and elsewhere was measured, especially in the fine sizes because the presence of coarse dust is easily detected. What are thought to be safe, or acceptable, working levels have been established.

All employees were examined by x-ray for fibrosis in the lungs, and for a proneness to develop silicosis. Some lung shapes are more likely to develop it than others are. Employees with developing fibrosis were given work in dust-free locations, or treatments and pensions. Employees are now examined at least yearly to detect the disease in the primary stage so that prompt action may be taken, and the incidence of silicosis is lessening. Aluminum dust sprayed into the air in mine change houses and underground has been found to have a prophylactic, and possibly therapeutic, effect. Ventilation is the only completely effective remedy.

Radioactivity. It was observed that in some mine areas silicosis seemed to be much more virulent than in others. This was thought to be due to some additive to the silica. It was found that coal dust speeded up the development of the disease.

When uranium mining got under way, it was found that the rocks emitted radon gas which broke down in the lungs and produced lung cancer. When the instruments developed for radioactivity research became available, it was learned that many rocks that had a low level of radioactivity emitted small amounts of radon gas, which in some cases were carried into the mine workings in the mine water and released under the reduced pressure. That provided one reason for the virulence of silicosis in some mining areas.

The government authorities moved quickly and established safe levels of radioactivity and safe exposure time in those levels. The regulations are enforced. The acceptable levels are attained by ventilation.

Other considerations. When diesel engines were introduced underground for motive power, they had to be certified for an acceptable level of carbon monoxide and oxides of nitrogen in the fumes. The required levels were attained by engine design and by scrubbers in the exhaust system. An engine must have a U.S.B.M. certificate of approval, and then it is only approved to travel certain routes for which the government inspectors consider the ventilation to be adequate.

[A. V. CORLETT]

Bibliography: *Eng. Mining J.*, Centennial Issue, June, 1966; D. Frasche, *Mineral Resources,* NAS-NRC Publ. no. 1000, pt. C, 1962; I. A. Given et al., *SME Mining Engineering Handbook*, 1973; M. K. Hubbert, *Energy Resources*, NAS-NRC Publ. no. 1000, pt. D, 1962; R. B. Lewis and G. B. Clark, *Elements of Mining*, 3d ed., 1964; T. A. Rickard, *Man and Metals*, 1932, reprint 1974; G. Robson, *Economics of Mineral Engineering*, 1977; Society of Mining Engineers, *Surface Mining*, 1968; U.S. Bureau of Mines, *Mineral Facts and Problems*, Bull. no. 667, 1975; W. A. Vogely et al., *Economics of the Mineral Industries*, 3d ed., 1976.

Mining excavation

In mining for coal, metallic, and nonmetallic minerals, the process of removing minerals from the Earth. Excavation consists of fragmentation (or in special cases solution) of minerals from their solid state, loading them, and transporting them to the surface. Fragmentation is accomplished by the use of explosives or mechanical means, the former being most commonly applied. Excavation is also involved in establishing mine entries and other development workings in waste rock for access to the minerals. Mechanical fragmentation by means of boring machines (Figs. 1 – 3) is being introduced at some mines for this purpose.

Hard rock is generally broken with explosives to attain fragmentation. Some moderately soft deposits, such as coal, potash, and borax, are fragmented mechanically by machines without the use of explosives. When fragmentation is by machine, the loading device is commonly an integral part of the machine. In special cases (sulfur, salt, and potash) the mineral may be excavated by solution. When solution mining is used, the mineral-bearing solution is pumped to the surface; thus loading and transport become integral parts of the process.

Fig. 1. Raise boring machine (drills pilot holes between levels) in Idaho silver-lead mine. (*Hecla Mining Co.*)

Fig. 2. Raise boring machine set to ream pilot hole to finished dimension of raise. (*Homestake Mining Co.*)

Explosives. Two general classes of explosives are used for mining, black powder and high explosives or dynamites. Black powder is not used underground, where commercial high explosives commonly known as dynamites are preferred. A special class of explosive, designated "permissible" after being tested and passed by the United States Bureau of Mines, is required in gaseous and dusty coal mines. Permissible explosives are especially designed to produce a flame of small volume, short duration, and low temperature. Dynamites are typed according to properties and further subdivided into grades based on strength. The principal properties are strength, density, sensitiveness, velocity, water resistance, freezing resistance, and fume products of detonation. Strength refers to the energy content of an explosive, and is based on the percent by weight of nitroglycerin or equivalent energy when other strength-imparting ingredients are substituted for nitroglycerin.

High explosives are detonated with blasting caps, small metal tubes closed at one end and charged with a highly heat-sensitive explosive.

Blasting agents not technically classified as explosives are increasingly being used for blasting, both in surface mines and underground. They are non-cap-sensitive materials or mixtures in which none of the ingredients are classed as explosives but which can be detonated by a high-explosive primer for blasting purposes. A commonly used blasting agent of this type is a mixture of fertilizer-grade ammonium nitrate and fuel oil (ANFO) that has found wide acceptance because of its safety features and low cost.

Fragmentation. Fragmentation is the process of breaking ground with explosives and machines.

Explosives fragmentation. The object of explosives fragmentation is to break the minerals and produce fragments of a size best suited for handling. It is cheaper and generally desirable to break waste rock in coarse sizes. Customarily, coal has been produced as lumps, but now the trend is toward using fine sizes for more economical handling and combustion. Fine sizes are desirable for most metallic and some nonmetallic ores that are processed after mining. Some nonmetallic minerals can be marketed more profitably in medium coarse sizes (1/4 to 2 in. or 3 to 50 mm), so fine material is wasted and must be minimized. The size of mineral fragments can be controlled partly by the amount and strength of the explosive used in blasting and somewhat by the spacing between shot holes and by their depth.

An assemblage of shot holes drilled into the face of a stope, drift, crosscut, shaft, raise, winze, adit, or tunnel and blasted at one time is a "round" (Fig. 4). The pattern of the round contributes to the effectiveness of a blast. A wide variety of patterns is used, depending upon the character of the material to be broken, the size and shape of the desired opening, and the desired size of fragments. These rounds have been given names such as pyramid, triangle, V, and fan, suggestive of the pattern of the drill holes that are blasted to give the initial cut in the center of the face. Others, such as Norwegian and Michigan, have been named according to place of origin. Michigan is also called burned cut because the holes which are drilled close together in a group near the center of the face, when blasted, pulverize the rock and discharge it at a high velocity.

To break a round, each hole is charged with explosives. The explosives are in the form of cylindrical cartridges, or in the case of blasting agents, may be in bulk. Cartridges are tamped into place in the drill hole and explosives in bulk form are blown into the hole pneumatically. A detonating cap with attached safety fuse or electric wire is inserted into a cartridge and placed in each hole to form a primer. The round is then blasted by igniting the fuse or sending an electric current through the wiring. The sequence in which holes of a round are blasted is important. Usually a group of holes near the center or along one side is blasted first. Then successive groups of holes are blasted in se-

Fig. 3. Tunnel boring machine used in mining. (*James S. Robbins and Associates, Inc.*)

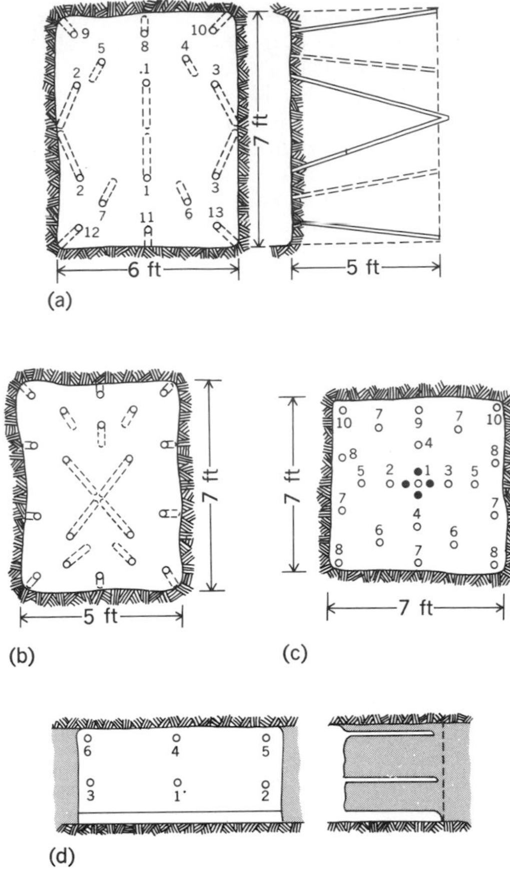

Fig. 4. Typical rounds for explosives fragmentation. (*a*) Horizontal V cut. (*b*) Four-hole pyramid cut. (*c*) Five-hole burn cut. (*d*) Undercut coal face. 1 ft = 0.3 m.

ries. Crude timing between successive blasts is attained by trimming fuses to different lengths. Precision timing is possible with electric blasting caps constructed to vary in detonation by a few thousandths of a second.

Mechanical fragmentation. This method is used primarily in coal mining. The principal machines are the continuous miner, auger, coal plow, and coal cutter. The continuous miner is manufactured in several sizes by both domestic and foreign companies. There are many variations in detail to meet varying conditions in coal seams.

Fig. 5. A continuous miner, finishing cut as it rips overlapping paths in coal seam 18 in. (0.46 m) deep by 42 in. (1.07 m) wide by 90 in. (2.74 m) high. Coal is discharged into waiting shuttle. (20 ft = 6.1 m.) (*U.S. Bureau of Mines, and Joy Manufacturing*)

One representative model is shown in Fig. 5. The coal is fragmented by a front-end cutting head comprising a number of continuously revolving chains upon which are mounted hard-metal-tipped cutters or picks. The chains are mounted on a bar which can be rotated horizontally to cut an entry 12–20 ft (3.7–6.1 m) wide and vertically to cut from 6 in. (15 cm) below the bottom of the machine to 7 or 8 ft (2.1–2.4 m) above. The cutter-bar frame is mounted on a carriage which in turn is mounted on caterpillar treads. The broken coal is collected by a conveyor which transports it from the cutting head to the transportation system located at the rear of the machine.

Augers up to 5 ft (1.5 m) in diameter (Fig. 6) are used under high banks of abandoned strip mines to recover coal to a boring depth of about 200 ft (60 m). The auger is made in sections, the first one having a hard-metal-faced cutting tool that does the cutting as the auger is rotated. The force providing thrust and rotation to the auger is supplied by a diesel engine at the surface. Some machines are self-propelling and handle and store the auger sections mechanically.

The coal plow or planer was developed in Germany to mine coal by the longwall system. The plow is drawn back and forth along the face and cuts or slabs off the coal in a series of slices. The planer unit comprises (1) an armored double-chain (Panzer) conveyor which rests on the mine floor and extends the entire length of the longwall face, (2) an electric driving unit at each end of the conveyor, (3) the planer assembly, and (4) pneumatic conveyor shifters. The electric motors move the plow by means of a rotating drum and chain attached to the planer. Broken coal falls to the conveyor, which moves it to entries adjacent to the mining areas. The pneumatic shifters move the conveyor and the plow toward the face as mining advances.

Where coal is fragmented by explosives, it is first undercut by a coal-cutting machine along the bottom of the seam. It either falls by gravity (undercutting and mass falling), or is broken by permissible explosives or other substitutes, shot holes are placed to take advantage of such natural features as the presence of hard and soft bands and the direction of cleavage planes in the coal.

Three classes of machines are used for cutting coal: disk, bar, and chain, classified by the method employed for holding the cutting tools or cutter picks.

The chain coal cutter (Fig. 7) is in most common use and is manufactured in various sizes. These electrically powered machines are designed to cut in an arc of 180°. The smaller units are maneuvered by power-actuated drums and cable attached to stationary anchors. The larger self-propelled units are mounted on track or use pneumatic tires. The cutting jib can be rotated while cutting without moving the machine. It can also be rotated to cut either vertically or horizontally.

Loading. Since most loading is mechanized, hand shoveling is limited. Underground loaders include the continuous loader, scraper loader, and revolving and overcast shovels. Clamshell loaders are used in sinking shafts. The continuous loader (Fig. 8) is used extensively in underground coal

Fig. 6. A view from the spoil bank of a strip mine showing the operation of auger mining and coal being loaded into a truck in the Pittsburgh coal bed, West Virginia. (*U.S. Bureau of Mines*)

mining and less extensively in other mines. Its essential parts comprise a gathering headframe, or scoop, on the front end that is crowded into the broken material by the machine, and a set of gathering arms to rake the material onto a bar chain conveyor that transfers it to the haulage system. The scraper loader consists of a hoe-type scraper, double-drum electric hoist, and ramp. Fragmented material is scraped up the ramp and discharged into the transportation system. The scraper loader is used in many places in mining to scrape ore and waste short distances up to about 200 ft (60 m). Revolving power shovels are used for excavation in large underground openings. The overcast shovel loader, which runs on rail or caterpillar tracks, is best adapted for use in confined areas. Loading is accomplished by crowding the unit into the broken rock; when the shovel is full, it casts back over the unit into cars or trucks.

Transportation. Moving or hauling of men, supplies, and broken minerals is one of the most complicated operations in mining. Transportation includes (1) transfer of broken material through mine openings by gravity and scraping, (2) rail haulage on surface and underground, (3) trackless, wheeled haulage on surface and underground, (4) hoisting and cable haulage from open pits and underground mines, (5) movement of broken material by numerous types of conveyors, and (6) pumping broken ore through pipelines.

Gravity and scraping methods are used to gather small quantities of material into larger volume for transportation by some other method. Material may fall directly into raises and ore passes where it flows by gravity to a chute or ore pocket.

Underground rail haulage has been adapted from railroad practice, but on a much smaller scale. It is used mostly for collecting broken material from chutes (Fig. 9) and various loaders, and transporting it to the hoist or directly to the surface. Track gage ranges from 18 to 48 in. (0.46 to 1.22 m). Trains are made up of cars ranging from 1 ton to about 20 tons capacity. Cars are equipped for manual dumping or with some dumping device. Small units are drawn by storage battery locomotives and larger units by electric trolley locomotives.

Trackless, wheeled haulage underground usually is by electric cable reel and diesel-electric shuttle cars with self-contained discharging conveyors. They are mounted on pneumatic tires and have capacities ranging to 20 tons (18 metric tons). Dump trucks are used underground to a limited extent. When so used they are equipped with diesel engines for safety. Most trackless underground haulage equipment negotiates grades up to 10%.

Electric hoisting is employed in deep underground mines. Either vertical or inclined shafts are divided into three to six parallel shaftways—one for a manway, pipe, or power lines, and a minimum of two for hoisting. In large deep shafts, two compartments are commonly used for hoisting broken

Fig. 7. Electrically powered chain coal cutter making shear cut in coal face of West Virginia mine. (*U.S. Bureau of Mines, and Joy Manufacturing*)

Fig. 8. Caterpillar-mounted electric machine loading broken limestone into a diesel-powered haulage truck in a deep limestone mine in Ohio. (*U.S. Bureau of Mines*)

Fig. 9. Loading 60-ft^3 (1.70-m^3) capacity, Granby-type mine car at uranium ore-pass station on haulage level, San Juan County, Utah. Underswing arc gate is operated by a compressed-air cylinder controlled by motorman helper on right. (*U.S. Bureau of Mines*)

material, and two others for handling supplies, equipment, and personnel.

The hoisting layout comprises a headframe, usually of timber, concrete, or steel, erected over the collar of the shaft. A sheave wheel is mounted at the top of the headframe for each hoisting compartment. In addition, the headframe contains bins into which hoisted material is discharged.

Most hoists are electrically driven and have two winding drums that can be operated in balance or separately. When four compartments are employed for hoisting, a larger, double-drum unit hoists rock and a smaller one hoists workers and supplies. Each drum is wound with a steel wire cable long enough to reach the sump (bottom) of the shaft. Cables range up to about 1½ in. (38 mm) in diameter. Rock is hoisted from various levels of the mine in skips (buskets, baskets, or open cars) ranging up to 10 tons capacity. Workers and supplies and, at some smaller mines, loaded cars are hoisted in cages with one to three decks.

An important safety feature of the skip and the cage is a device that stops them if the hoisting cable fails. When failure occurs, a spring releases a set of safety catches that engage the wooden guides along each side of the skip or cage, thereby stopping it.

Several different types of conveyors are used in transporting coal in underground mines. The most important of these are belt conveyors. In some mines virtually all the coal is transported by a system of belt conveyors from the mining face through the various entries to surface. *See* UNDERGROUND MINING. [JAMES E. HILL]

Mining machinery

Apparatus used in removing and transporting valuable solid minerals from their place of natural origin to a more accessible location for further processing or transportation. Many of the machines are identical to, or minor adaptations of, those used for excavating in the construction industry. In a wider sense mining machinery could also include all equipment used in finding (exploring and prospecting), removing (mining: developing and exploiting), and improving (processing: ore dressing, milling, concentrating or beneficiating, and refining) valuable minerals; it could even include metallurgical (smelting) and chemical processing equipment used in extracting or purifying the final product for industry. The term is also applied to special equipment for recovery of minerals from beneath the sea. In usual context the term does not include apparatus used principally in the petroleum industry. Perhaps those machines most often considered as uniquely mining machinery are drills, mechanical miners, and specially adapted materials-handling equipment for use in mining underground or on surface (where a large proportion of mines are located). In addition, some unique auxiliary equipment and processing equipment are used in the mining industry.

Design and construction. All components of mining machinery—including primary mechanism, controls, means of powering, and frame—require the following features to a much greater extent than do other machines (with the possible exception of some military, construction, oil well, and marine units).

Ruggedness. Equipment is handled roughly, frequently receiving severe and sudden shock from dropping, striking, and blasting vibration; overloading is common, and long life is demanded by the economics of mining.

Weather resistance. Operations extend over a wide range of climate and altitude.

Abrasion resistance. Minerals include some of the hardest substances known. Dust and fine particles are always present.

Water and corrosion resistance. Moisture and water, often acidic, are common in mining operations.

Infrequent and simple maintenance. Equipment is often widely scattered and in locations with restricted access. Trained mechanics and repair parts are generally limited in availability because of the remoteness of operations.

Easy disassembly and reassembly. Access to the machinery at the site of operation is frequently limited. Also, the working space near it and mechanical aids to moving or lifting it may be limited or nonexistent.

Safety. Mining has had a poor safety record. As a result, most governments have testing bureaus, inspection agencies, and enforcement laws for the approval of mining equipment. Simplicity of operation, low initial cost, and low operating costs are also desirable features of mining machinery.

Exploration machinery (vehicles, drills, and accessory equipment) is subjected to the same operating conditions as other mining equipment. Mineral-beneficiating equipment must have, above all, abrasion resistance. Smelting equipment has the requirement of heat resistance. Chemical refining process equipment must be highly resistant to corrosion. Reliability of all processing equipment is critical because slurries are commonly handled, and they can cause considerable difficulty in restarting after shutdowns.

Underground requirements. Machinery operated underground must meet special design requirements.

Low-ventilation demand. Quantity and geometry of passageways for air are rather rigidly fixed, so that high air consumption is a problem and noxious gases cannot be readily dispersed. Heat removal is a problem in deep mines.

Compactness. Space is at a premium, especially height, particularly in bituminous coal mines.

Easy visibility. Most operating areas are lighted only by individual cap-mounted or hand-held lamps.

Hand portability. Units or components must frequently be hand-carried into an operating area.

Absence of spark and flame. Equipment is often used in or near explosives, timber supports, and natural or human-produced combustible gas or dust. In the presence of hydrocarbons, as well as of certain metal ores such as some sulfides, complete absence of open sparks or flames is a major requirement.

Power source. Mining machinery is very commonly powered by compressed air, but electricity is also widely used and is often the basic source. Compressed air has the advantages of simplicity of transmission and safety under wet conditions. It is especially advantageous underground as an aid to ventilation. Machines powered by compressed air can be easily designed to accommodate overloads or jamming, which is desirable on surface as well as underground. Large central compressors and extensive pipeline distribution systems are common, especially at underground mines.

Electric power, purchased from public sources or locally generated at a large central station, is common in open-pit and strip mines and dredging operations. Underground coal and saline-mineral mines often use electric-powered production machinery, but in other underground mines electricity is normally used only for pumps and transportation systems in relatively dry or permanent locations. Direct-current devices are dominant because of simplicity of speed and power control, but alternating-current apparatus is becoming common. Mobile equipment is often either battery of cable-reel (having a spring-loaded reel of extension power cable mounted on the machine) type. Processing machinery units are almost exclusively powered by individual electric motors.

Diesel engines are popular for generating small quantities of electric power in remote areas and for transportation units. Underground, abundant ventilation is essential, as well as wet scrubbers, chemical oxidizers, and other accessories to aid the removal of noxious and irritant exhaust gases. Hydraulic (oil) control and driving mechanisms are widely used. Transfer of power by wire rope is common, especially for main vertical transportation.

Drills. Drills make openings, of relatively small cross section and long length, which are used to obtain samples of minerals during exploration, to emplace blasting explosives, and to extract natural or artificial solutions or melts of minerals. Exploration holes are generally vertical or inclined steeply downward, less than 6 in. (15 cm) in diameter and up to 10,000 ft (3 km) long. Blastholes range from 3/4 to 12 in. (2 to 30 cm) in diameter and usually are under 50 ft (15 m) long in any direction, with the larger usually downward. Solution wells normally are vertical and 6 to 12 in. (15 to 30 cm) in diameter, sometimes reach depths of several thousand feet, and are equipped with several concentric strings of pipe.

Rock drills. Percussion, rotary, or a combination action of a steel rod or pipe, tipped with a harder metal chisel or rolling gearlike bit, chips out holes up to 12 in. (30 cm) or more in diameter by 125 ft (38 m) or more in length from the surface, and 1–3 in. (2.5–7.6 cm) by 5–200 ft (1.5–60 m) from underground. Crawler or wheeled carriers are used, and the smaller drills are often attached to hydraulically maneuvered booms. Air or liquids flush out the chips.

Diamond drills. Rotation of a pipe tipped with a diamond-studded bit is used in exploration to penetrate the hardest rocks. Large units make holes up to 3 in. (7.6 cm) in diameter and 5000 ft (1.5 km) or more deep; at the other extreme are units so small that they can be pack-carried. A cylindrical core is usually recovered.

Water-jet drills. For drilling in loose or weakly bonded materials, a water jet washes out a hole as a wall-supporting pipe is inserted.

Jet flame drills. For economical surface blast holes in hard abrasive quartzitic rock, a high-velocity flame is used to spall out a hole.

Mechanical miners. There are many machines designed to excavate the valuable mineral or the access openings by relatively continuous dislodgement of material without resorting to the more common practice of intermittent blasting in drill holes. These units also frequently transport the mineral a short distance, and when designed for weakly bonded minerals, they often become primarily materials-handling equipment.

Continuous miners. For horizontal openings in coal and saline deposits, toothlike lugs on moving chains, or rotating drums or disks rip material from the face of the opening as the assembly crawls ahead.

Plows or planers. In coal and other mineral deposits of medium hardness, bladelike devices continuously break off a 6-in. (15-cm) layer as they are pulled by various mechanisms along a wall several hundred feet long and 3–5 ft (0.9–1.5 m) high.

Augers. Coal and soft sediments are mechanically mined by augers up to 5 ft (1.5 m) in diameter and 100 ft (30 m) long, usually used horizontally.

Shaft and raise drills and borers. For vertical and inclined openings up to 8 ft (2.4 m) and even larger in diameter, various rotary coring or full-face boring equipment is used, both in an upward direction (raising) for several hundred feet or downward (sinking) for several thousand feet. These units usually use many rigid teeth or rolling gearlike bits to chip out the mineral.

Tunneling machines. There are similar rotary boring units for horizontal, or nearly so, openings of any length and up to 35 ft (10.7 m) in diameter (in soft rock).

Rock saws. To remove large blocks of material, narrow slots or channels are cut by the action of a moving steel band or blade and a slurry of abrasive particles (sometimes diamonds) rather than teeth. Small flame jets are also used.

Hydraulic monitors. Water jets of medium to high pressure (some to 5000 psi or 34 MPa) are used to excavate weakly cemented surface material and brittle hydrocarbons both on the surface and underground.

Special materials-handling equipment. Loose material (muck) is picked up (mucked or loaded) and transported (hauled or hoisted) by a wide variety of equipment.

Excavator loaders. For confined places underground there are various unique grab-bucket shaft muckers and overcasting shovel tunnel muckers, and also gathering-head loading-conveyor units having eccentric arms, lugged chains, and screws or oscillating pans for handling muck in horizontal openings.

Dragline scrapers. Scrapers (slushers) with a flat plowlike blade or partially open bucket pulled by a wire rope are commonly used to move muck up to a couple hundred feet, especially in underground mines.

Dipper shovels and dragline cranes. In surface mining single-bucket loads can handle up to 100 yd^3 (76 m^3) of material. Many of the intermediate size (20–40 yd^3 or 15–31 m^3) units move on unique walking shoes.

Bucketline and bucketwheel excavators. These are for surface use and can dig up to 5000 yd^3/hr (3800 m^3/hr), using a series of buckets on a moving chain or a rotating wheel supported on crawlers, railcars, or floating hulls (dredges).

Suction dredges and pipelines. On surface, up to 3700 yd^3/hr (2800 m^3/hr) of moderately loose mineral up to several inches in diameter can be picked up and moved as a slurry (mix of water and solid material) by pumps usually mounted on floating hulls.

Trucks. Diesel and electric shuttle cars (short-haul trucks) of unusual design, often having very low profile and conveyor bottoms, are used underground. At surface mines there are diesel-electric- and electric-trolley-type dump trucks of over 100-ton (90-metric-ton) capacity.

Railroads. Underground locomotives range from 1/2 to 80 tons (0.45 to 73 metric tons) in weight, with electric (storage battery, cable-reel, or trolley), diesel, and sometimes compressed-air power units. Cars are usually of special design.

Conveyors. Unique movable, self-propelled, sectional and extensible conveyors are used in underground mining.

Wire rope hoists. Hoists or winders of up to 6000 hp (4.5 kW) are used in shafts for vertical or steeply inclined transportation in single lifts of as much as 6000 ft (1.8 km). Of various particular designs, there are two basic types: drum, simply a powered reel of rope; and friction, in which the rope is draped over a powered wheel and a counterweight is attached to one end and a conveyance to the other.

Auxiliary equipment. Drainage pumps handling hundreds of gallons of water per minute at heads of 1000 ft (300 m) or more are used underground. For underground roof support, primarily in coal, there are mechanically moved jacks of 100-ton (90-metric-ton) capacity. Ventilation fans are capable of moving several hundred thousand cubic feet of air per minute. Crushers can handle pieces of hard rock several feet across in two dimensions. In processing, minerals are sorted by size or density, or both, by a variety of screens, classifiers, and special concentrators, using vibration, fluid flow, centrifugal force, and other principles. Froth flotation and magnetic and electrostatic equipment take advantage of other special properties of minerals. [LLOYD E. ANTONIDES]

Bibliography: Engineering and Mining Journal, *Mining Guidebook*, annual; Engineering and Mining Journal, *Operating Handbook of Mineral Underground Mining*, vol. 3, 1979; R. S. Lewis and G. B. Clark, *Elements of Mining*, 3d ed., 1964; R. Peele (ed.), *Mining Engineers' Handbook*, 2 vols., 3d ed., 1941; Society of Mining Engineers, *SME Mining Engineering Handbook*, 1973.

Monitoring of ionizing radiation

The use of meters and special techniques to determine the absorbed dose of ionizing radiation received by individuals; also, the use of meters and other devices to determine the type of radiation, its energy spectrum and direction, and the absorbed dose in the various areas and inside the human body. Radiation monitoring is accompanied by many health-physics services and functions; for example, if a health physicist is assigned to a radiation survey or monitoring operation, he will not only measure the dose or dose rate, but also will put into effect measures to minimize the exposure. He may specify required shielding, indicate necessary decontamination procedures, or advise the use of appropriate remote-control equipment, protective clothing, respirators, or glove boxes. Monitoring frequently is divided into three categories; personnel monitoring, building surveys, and area monitoring.

Personnel monitoring. This involves those operations directly associated with the measurement and recording of the absorbed dose received by the individual and all health-physics services and functions designed to evaluate, record, and minimize exposure to the individual. Personnel monitoring includes the issuing of dosimeters; the reading, maintenance, and calibration of these devices; the keeping of exposure records; and personal contacts with individuals to determine the causes of exposure and to recommend procedures to limit the recurrence of exposures. Personnel monitoring includes the use of such things as nose swabs to check for inhaled radioactive dust, hand and foot counters to measure hand and foot contamination, friskers to measure clothing contamination, probe counters to measure wound contamination, ring meters to measure β-dose to the fingers, and pocket alarm meters which give a warning when one enters an intense field of radiation. Many personnel monitoring programs require the use of decontamination laundry facilities to decontaminate clothing, the use of total body counters to estimate the buildup of radioactive contamination inside the body, and the operation of a chemical analysis laboratory to measure the level of radioactive material in the blood, urine, and feces. *See* HEALTH PHYSICS.

Building surveys. These are made with many types of survey or monitoring instruments which, for the most part, can be grouped into three classes: Geiger-Müller (GM) counters, scintillation counters, and ionization chambers. There are many specialized instruments in use such as the proportional-counter air sniffer used to monitor airborne ^3H, which emits low-energy β-radiation. This equipment is used to measure the dose rate and accumulated dose in various work areas and to estimate the surface contamination on floors, walls, furniture, and equipment. Many refinements have been made in instruments and techniques to measure separately the absorbed dose and energy spectrum of x-rays, γ-rays, α-particles, β-particles, fast and thermal neutrons, and other types of ionizing radiation. It is not sufficient to measure just the total dose of ionizing radiation for each component and energy of the mixture may have a different quality factor (Q). For measurements where a high gamma sensitivity is required, pressure ion chambers, large banks of GM counters, or liquid scintillators are used, but where high-energy resolution is needed, Ge(Li) spectrometers (some with an energy resolution of 0.5 kV) are put to use. One of the hazards of greatest concern in many types of work with radioactive materials is the inhalation of airborne dusts, fumes, and gases. As a consequence, various types of equipment, such as air filters, precipitators, impingers, and charcoal collectors, have been developed to make collections of airborne contamination for radioisotope analysis. Some monitors are operated continuously in various areas, recording the dose rate and the air contamination, or both, on a graphic recorder. An alarm is sounded if the radiation level or the air contamination exceeds prescribed safe values.

Area monitoring. This is concerned with the measurement of the buildup and spread of radioactive contamination in the air, water, and soil outside the work areas. Many of the instruments used for building surveys are used in area monitoring. In addition, fallout trays are used to collect the dust that settles to the ground, rain samplers are used to collect and measure the radioactive contamination in the rain, and various types of probes are used to measure the level of groundwater contamination in sampling wells and to measure the contamination in river waters and sediment. One of the most useful area monitors for identifying radionuclides in the environment is the Ge(Li) detector. Area monitoring and ecological studies begin in an area before an operation starts producing contamination. This is done in order that the operation's contribution to the buildup of the area background radiation and the accumulation of radioactive materials in the rivers and soil and in plants and animals may be determined and continuously monitored. [KARL Z. MORGAN]

Bibliography: M. Eisenbud, *Environmental Radioactivity*, 2d ed., 1973; G. S. Hurst and J. E. Turner, *Elementary Radiation Physics*, 1970; A. Martin and S. A. Harbison, *An Introduction to Radiation Protection*, 1972; K. Z. Morgan and J. E. Turner, *Principles of Radiation Protection*, 1967; National Council on Radiation Protection and Measurement, *Instrumentation and Monitoring Methods for Radiation Protection*, NCRP Rep. no. 57, 1978; J. Shapiro, *Radiation Protection*, 1972.

Motor

An electric rotating machine which converts electric energy into mechanical energy. Because of its many advantages, the electric motor has largely replaced other motive power in industry, transportation, mines, business, farms, and homes. Electric motors are convenient, economical to operate, inexpensive to purchase, safe, free from smoke and odor, and comparatively quiet. They can meet a wide range of service requirements—starting, accelerating, running, braking, holding, and stopping a load. They are available in sizes from a small fraction of a horsepower to many thousands of horsepower, and in a wide range of speeds. The speed may be fixed (or synchronous), constant for given load conditions, adjustable, or variable. Many are self-starting and reversible. For uniformity and interchangeability, motors are standardized in sizes, types, and speeds. *See* ELECTRIC ROTATING MACHINERY.

Electric motors may be alternating-current (ac) or direct-current (dc). There are many types of each. Although ac motors are more common, dc motors are unexcelled for applications requiring simple, inexpensive speed control or sustained high torque under low-voltage conditions.

Motor classification. Motors are classified in many ways. The following classifications show some of the many available variations in types of motors.

1. Size: flea, fractional, or integral horsepower.
2. Application: general purpose, definite purpose, special purpose, or part-winding start. May be further classified as crane, elevator, pump, and so forth.
3. Electrical type: alternating-current induction, synchronous, or series; direct-current series, permanent magnet, shunt, or compound.
4. Mechanical protection and cooling: (*a*) open: dripproof, splashproof, semiguarded, fully guarded, externally ventilated, pipe ventilated, weather protected; (*b*) totally enclosed: nonventilated, fan cooled, explosionproof, dustproof, ignitionproof, waterproof, water cooled, water-air cooled, air-to-air cooled, pipe ventilated, fan cooled guarded.
5. Speed variability: constant speed, varying speed, adjustable speed, adjustable varying speed, multispeed.
6. Mounting: floor, wall, ceiling, face, flange, vertical shaft.

Characteristics. Each electrical type of motor has its own individual characteristics. Each motor is selected to meet the requirements of the job it must perform. For individual motor characteristics *see* DIRECT-CURRENT MOTOR; INDUCTION MOTOR; REPULSION MOTOR; SYNCHRONOUS MOTOR. For comparison of all ac motors *see* ALTERNATING-CURRENT MOTOR.

Principles of operation. When a conductor located in a magnetic field carries current, a mechanical force is exerted upon it (see illustration). This force has the value shown in Eq. (1), where i

$$F = Bil \qquad \text{newtons} \qquad (1)$$

is the current in amperes, B is the magnetic density in webers per square meter, and l is the conductor length in meters. The illustration shows the relative directions of current, field, and force. The force reverses with either current or field reversal,

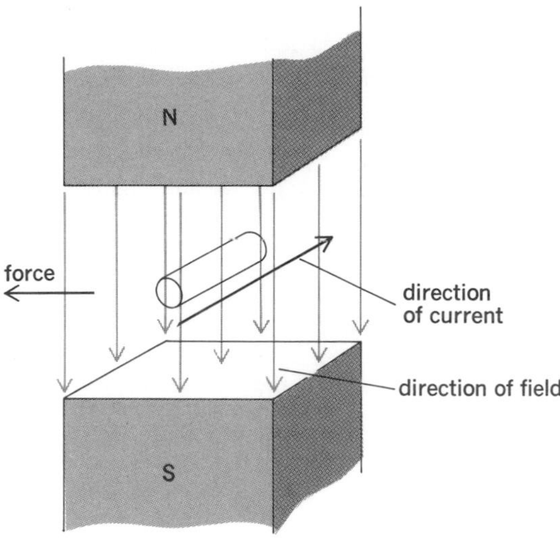

Relative directions of field flux, current, and force.

but not when both are reversed. The torque T is the product of this force and the rotor radius.

If the conductor moves in the direction of F, an emf e is generated which opposes the current (motor action). If the conductor is moved against F, this emf will assist the current (generator action). Its value is shown in Eq. (2), where $v=$ velocity of conductor across the flux in meters per second.

$$e = vBl \qquad \text{volts} \tag{2}$$

The product ei represents the power converted, in watts, shown in Eqs. (3) and (4), which are the

$$\text{Motor output} = ei - \text{rotative loss} \tag{3}$$

$$\text{Generator shaft input} = ei + \text{rotative loss} \tag{4}$$

bases for the emf and output formulas of dynamo machinery. Many machines will operate as either a motor or a generator, but they should be designed for the particular service.

Current may be fed into the field and armature by conduction, as in dc machines and ac series motors, or into the stator by conduction and the rotor by induction, as in ac induction and repulsion motors.

[ALBERT F. PUCHSTEIN]

Bibliography: A. E. Fitzgerald and C. Kingsley, *Electric Machinery*, 3d ed., 1971; V. Gourishankar and D. H. Kelly, *Electromechanical Energy Conversion*, 2d ed., 1973; I. L. Kosow, *Electric Machinery and Transformers*, 1972; L. Matsch, *Electromagnetic and Electromechanical Machines*, 2d ed., 1977.

Motor-generator set

A motor and one or more generators, with their shafts mechanically coupled, used to convert an available power source to another desired frequency or voltage. The motor of the set is selected to operate from the available power supply; the generators are designed to provide the desired output. Motor-generator sets are also employed to provide special control features for the output voltage.

The principal advantage of a motor-generator set over other conversion systems is the flexibility offered by the use of separate machines for each function. Assemblies of standard machines may often be employed with a minimum of engineering required. Since a double energy conversion is involved, electrical to mechanical and back to electrical, the efficiency is lower than in most other conversion methods. In a two-unit set the efficiency is the product of the efficiencies of the motor and of the generator.

Motor-generator sets are used for a variety of purposes, such as providing a precisely regulated dc current for a welding application, a high-frequency ac power for an induction-heating application, or a continuously and rapidly adjustable dc voltage to the armature of a dc motor employed in a position control system. *See* GENERATOR; MOTOR.

[ARTHUR R. ECKELS]

Natural gas

A combustible gas that occurs in porous rock of the Earth's crust and is found with or near accumulations of crude oil. Being in gaseous form, it may occur alone in separate reservoirs. More commonly it forms a gas cap, or mass of gas, entrapped between liquid petroleum and impervious capping rock layer in a petroleum reservoir. Under conditions of greater pressure it is intimately mixed with, or dissolved in, crude oil.

COMPOSITION AND OCCURRENCE

Typical natural gas consists of hydrocarbons having a very low boiling point. Methane, CH_4, the first member of the paraffin series, and with a boiling point of −254°F (−159°C), makes up approximately 85% of the typical gas. Ethane, C_2H_6, with a boiling point of −128°F (−89°C), may be present in amounts up to 10%; and propane, C_3H_8, with a boiling point of −44°F (−42°C), up to 3%. Butane, C_4H_{10}; Pentane, C_5H_{12}; hexane; heptane; and octane may also be present. Structural formulas of four compounds are:

```
     H             H   H
     |             |   |
  H—C—H        H—C—C—H
     |             |   |
     H             H   H
  Methane        Ethane

   H  H  H  H  H           H  H  H  H
   |  |  |  |  |           |  |  |  |
H—C—C—C—C—C—H        H—C—C—C—C—H
   |  |  |  |  |           |  |  |  |
   H  H  H  H  H           H  H  |  H
                                 |
                               H—C—H
                                 |
                                 H
  Normal pentane              iso-Pentane
```

Whereas normal hydrocarbons having 5–10 carbon atoms are liquids at ordinary temperatures, they have a definite vapor pressure and therefore may be present in the vapor form in natural gas. Carbon dioxide, nitrogen, helium, and hydrogen sulfide may also be present.

Types of natural gas vary according to composition and can be dry or lean (mostly methane) gas, wet gas (considerable amounts of so-called higher hydrocarbons), sour gas (much hydrogen sulfide), sweet gas (little hydrogen sulfide), residue gas (higher paraffins having been extracted), and casinghead gas (derived from an oil well by extraction at the surface). Natural gas has no distinct odor. Its main use is for fuel, but it is also used to make carbon black, natural gasoline, certain chemicals, and liquefied petroleum gas. Propane and butane are obtained in processing natural gas. *See* PETROLEUM PRODUCTS.

Distribution and reserves. Gas occurs on every continent (see Table 1). Wherever oil has been found, a certain amount of natural gas is also present. The United States, since 1967, has dropped from first to third in natural gas reserves (behind the Soviet Union and Iran); however, it has retained its position as first in gas production. Six states account for more than 73% of the known United States reserves (Texas, Louisiana, New Mexico, Kansas, Oklahoma, and California). Among these Texas has 27.2% of the total with 54.6×10^{12} ft^3 (1.55×10^{12} m^3), and Louisiana ranks second with 24.8% of the United States total with 49.6×10^{12} ft^3 (1.40×10^{12} m^3). The estimated known reserves in the United States at the end of 1978 were 200.3×10^{12} ft^3 (5.67×10^{12} m^3). Consumption in the United States in 1977 was 19.5×10^{12} ft^3 (0.55×10^{12} m^3). From 1962 through 1967 the amount of new gas found annually in the United States was approximately 20×10^{12} ft^3 (0.57×10^{12} m^3); from 1968 through 1977 annual new gas discoveries had declined to an average of 9.81×10^{12} ft^3 (0.278×10^{12} m^3). In 1978 a total of 10.6×10^{12} ft^3 (0.30×10^{12} ft^3) was found. Production of natural gas in the period 1960–1967 gradually rose from 13×10^{12} ft^3 (0.37×10^{12} m^3) to 18.3×10^{12} ft^3 (0.52×10^{12} m^3); between 1968 and 1977 production averaged 20.9×10^{12} ft^3 (0.59×10^{12} m^3). Gas production in 1978 totaled 19.3×10^{12} ft^3 (0.55×10^{12} m^3). These figures show that as annual production remained fairly steady at about 20×10^{12} ft^3 (0.56×10^{12} m^3), during the 1970s, the annual amount of gas found has decreased by approximately 50%, from 20×10^{12} ft^3 (0.56×10^{12} m^3) to approximately 10×10^{12} ft^3 (0.28×10^{12} m^3). Natural gas is being discovered in western Canada at a rapid rate, and at the beginning of 1978 the proved reserves were 59.4×10^{12} ft^3 (1.68×10^{12} m^3).

Long before supplies of natural gas run out or become expensively scarce, it is expected that some process of coal gasification will produce a gas which is completely interchangeable with natural gas and at a competitive price. This is important because coal makes up a majority of the world's known fossil fuel reserves. But since energy consumers have indicated in the marketplace their preference for fluid and gaseous fuels over the solid forms, coal gasification research, already well under way, will be given additional impetus. *See* COAL GASIFICATION.

In estimating gas reserves, the volumetric method is preferred. The volume of the reservoir is determined by means of the thickness, porosity, and permeability of the producing zones. A study of many depleted fields suggests that about 85% of all gas in dry-gas reservoirs is recovered. Some engineers use the production versus pressure-decline method. They calculate future production by plotting past production against the decline in reservoir pressure.

In California 60% of the gas is associated with oil, but in western Texas the percentage is about 40%. The percentage figure for the United States as a whole is only 23%. This means that a large proportion of the reserves is stored in such dry-gas fields as the Hugoton-Panhandle area of Kansas, Oklahoma, and Texas; Monroe, LA; Carthage, northeastern Texas; Bethany-Waskom, TX; Big Sandy, KY; Sligo, LA; Blanco Mesaverde, NM; Red Oak–Norris, OK; Long Lake, TX; Katy, TX; San Salvador, TX; Joaquin-Logansport, TX; Lake Arthur, LA; Big Piney, WY; Chocolate Bayou, TX; and Kenai and Cook Inlet, AK; there are also about 20 other so-called giant fields in the United States. In addition there are at least 12 giant (10^{12} ft^3 or 2.83×10^{10} m^3 or more) gas fields already discovered in offshore Louisiana in the Gulf of Mexico. Furthermore, prospects of offshore activity in other parts of the world, especially Africa, Asia, the North Sea, and South America, are tremendous for the development of future giant gas fields. This is made possible by the almost incredible technological advances in deep-water drilling, which has enabled explorers to drill and produce in waters in excess of 6000 ft (1800 m) deep as compared with some 200 ft (60 m) in depth in the late 1950s. Of interest are the accomplishments of the *Glomar Challenger*, which in scientific drilling experiments in 1968 cored sediments lying below as much as 17,589 ft (5361 m) of oceanic water from a dynamically positioned vessel. In western Canada some of the large gas fields are the Pincher Creek, the Waterton, and the Jumping Pound. The largest dissolved-gas area in the United States lies along the Gulf Coast of Texas and Louisiana and contains about 50% of the total known reserves of associated gas. Offshore drilling in the waters of the Gulf will add considerably to these reserves. In an average year slightly over 91% of the gas produced is marketed, while 7.7% is used for repressuring, and 1.4% is vented or wasted. In earlier years a much larger percentage was piped away from oil fields and burned. *See* OIL AND GAS, OFFSHORE.

Geological associations. Natural gas is present in every system of rocks down to the Cambrian. The first gas deposits found in the United States were those in the eastern states. In New York and Pennsylvania 85% of the gas came from Devonian rocks. In West Virginia, Kentucky, and eastern Ohio, Devonian and Mississippian rocks rank nearly equally, but Silurian rocks are also important. In Indiana and Illinois, Pennsylvanian rocks outrank the Mississippian. The Hugoton-Panhandle field is one of the largest in the world, and the Permian dolomites produce gas from five different levels. The fact that oil is found lower down in Pennsylvanian and older rocks proves the superior migratory capacity of gas. This field has a high percentage of nitrogen (almost 15%).

Oklahoma and the western part of Texas have gas in many stratigraphic zones, from the Permian

Table 1. World gas reserves, Jan. 1, 1978

Region	Volume 10^{12} m^3	Volume 10^{12} ft^3
North America	8.4	298
South and Central America	2.2	78.5
Western Europe	3.9	138
Africa	5.8	207
Eastern Europe, Soviet Union, and China	27	955
Near East	20.3	719.6
Middle and Far East, and Australia	3.47	122.7

down to the Cambrian. Most of it is associated with crude oil, either in solution or in the form of gas-cap accumulations. In the Carthage field, northeastern Texas, 10 different layers in the Trinity division of the Cretaceous system have been found productive. During 1977, gas production in the whole state of Texas was about 7×10^{12} ft^3 (0.20×10^{12} m^3); of this amount 84.3% was gas-well gas. Cretaceous rocks are the principal reservoir rocks in northern Louisiana and in Mississippi. Throughout the Rocky Mountain states, various layers in the Cretaceous system account for most of the gas. There are many dry-gas pools, of which the outstanding are the Blanco, northwestern New Mexico; the Baxter Basin, southwestern Wyoming; and the Cedar Creek, southwestern Montana. In California, gas production is derived from various layers in the Tertiary system. Although about 60% of the gas is associated with oil reservoirs, there are a number of dry-gas fields. *See* PETROLEUM GEOLOGY.

Of much interest is the trend toward low-temperature transportation and storage of liquid petroleum gas and methane. The ability to store frozen gas (as a liquid) underground will enable pipeline companies to meet more efficiently the cyclical demands of the seasons. Storage of gas close to markets will do away with the necessity of having large pipeline capacity, which is needed to take care of peak seasons but which lies more or less idle during periods of lessened demand. The ability to condense 600 volume units of gas into 1 unit of liquid opens great possibilities for the movement of gas across oceans in tankships. Through such transportation remote areas can become consumer areas, fuel-short consumer areas will have access to needed supplies, and producing areas will benefit from new revenues. *See* LIQUEFIED NATURAL GAS (LNG); OIL AND GAS STORAGE.

Helium, which has many industrial uses, is a byproduct of natural gas and is present in some fields. The Rattlesnake field in New Mexico contains 7.5%, the highest percentage of helium to total gas content found up to 1979.

[MICHEL T. HALBOUTY]

PIPELINE DISTRIBUTION

The analysis and design of natural-gas distribution network systems underwent tremendous change in the 1970s. At present the Hardy Cross method continues to form the analytical basis for all steady-state network studies, but rather than the tedious trial-and-error hand calculations, more efficient and sophisticated computer techniques are used. Design procedures that are based upon the classical optimization theories such as linear programming and dynamic programming have been useful in isolated applications. The most exciting aspect in current transmission and distribution simulations is the inclusion of the unsteady behavior of the natural gas in the system. Along with this added degree of sophistication has come the realization that potential benefits exist, not only for system operation but also for improved design strategies that include the interaction of potential capital expansion and existing system operating characteristics.

Steady-state analysis. Given a piping network configuration with specified deliveries and supply points, the normal analysis problem consists of identifying the pressures and pipeline flows throughout the system. A large number of possible groupings of known and unknown pressures and flows are possible. However, in a specific analysis, the known variables are generally apparent so that a solution to the problem is feasible. A satisfactory solution is achieved when Kirchhoff's first law (continuity equation) is satisfied at each pipe junction or node, and for the system as a whole $\Sigma Q = 0$, where Q is the flow rate. Kirchhoff's second law for each loop in the system must also be satisfied, that is, $\Sigma \Delta p = 0$ along each closed path, where Δp is the pressure drop in each pipeline. When the physical laws governing flow through elements that connect nodes are satisfied and either of the Kirchhoff laws is satisfied, then the other is automatically satisfied and the system is balanced. Such an element relationship—for example, the basic pipeline—is described by an equation of the form $\Delta p^2 = RQ^n$, in which p is pressure, n a constant, and R a function of the pipe geometry, gas properties, and the flow rate Q. Either a loop-balancing procedure or a node-balancing scheme may be used for the analysis. For the fundamental pipeline analysis problem, as opposed to the problem that includes other elements such as compressors, R. Epp and A. G. Fowler provided adequate evidence that the loop-oriented scheme offers some advantage.

In networks that involve different source gases such as liquefied natural gas (LNG), substitute natural gas (SNG), and propane-air mixtures as well as pipeline gas, the specific gravity and the heating value of the components vary widely. M. A. Stoner and M. A. Karnitz described the use of a simulator which balances a system on a thermal or Btu basis and also uses the correct gravity for each branch or element in the system. This scheme enables the user to trace the ultimate destination in the network of the different source gases.

Design. Typical design problems may encompass innumerable variables, ranging from the obvious pressure levels, deliveries, and pipe sizing to conduit alignments, location of compressor stations, location of storage, compression horsepower, and even staging component installation to optimize total cost over the design life of the system. In this article the discussion is limited to determinate design criteria in which feasible alignments, pressure constraints, and loads are provided. This restriction is desirable in order that a more specific problem may be addressed. Even within these limitations a direct procedure to produce the optimal natural gas distribution network design was not available in 1979.

In view of the vastly improved capabilities of current analysis programs, the design of complex network systems, and particularly of additions to existing systems, is entirely feasible through parametric studies. A design is assumed, analyzed, altered, and reanalyzed until the design objectives are satisfied. Computer software companies currently provide programs for use at remote computer consoles at which the designer is able to interact with an analysis program to improve the design of the system. As a side benefit to this type of parametric study, an improved understanding of the operating characteristics of the system is gained by the design engineer.

In 1972 Stoner borrowed a technological development in water distribution analysis and design and presented it along with additional theoretical advances as a convenient, practical, efficient tool for natural-gas network design. Included in the study, in addition to the standard pipeline system elements, are compressors, control valves, and storage fields. The mathematical model is constructed by writing the continuity equation at each node in the system. The flow equation for each element connected to the node is substituted to eliminate the element flow. A set of nonlinear simultaneous equations result, which are solved by the n-dimensional Newton-Raphson method. A very sparse coefficient matrix exists in this formulation, and Stoner's procedure takes advantage of this sparsity in both computer core storage savings and execution time efficiency. A further development was also proposed which involves a sensitivity study, that is, a convenient means of observing the interaction of system variables without independent solutions of the entire system for each new set of specified variables. Such questions as the approximate additional required horsepower at a remote station to provide an increased nearby nodal pressure of 10 psi (69 kPa) may be answered very efficiently. This direct network element design procedure and sensitivity capability are necessary forerunners to packaged optimal design methods.

Unsteady flow. In the actual operation of a distribution system, the volume of gas stored in the pipelines is extremely important in meeting peak-flow demands. This line pack may properly be accounted for in a transient analysis of a system, whereas the potential benefits of the stored volume are ignored in the conventional steady-state study. Inasmuch as real piping systems rarely attain steady-state operation due to changes in demand, supply, failures, additions of equipment, and so forth, there is much motivation for the capability to analyze unsteady flows. H. H. Rachford presented a study of increased operating efficiency and fuel conservation by improved operating strategies in the face of time-varying sales. Operation times for accommodating sudden large loads imposed on a system may be accurately predicted with a transient simulator. In the design process, compressor stations may be optimally positioned for maximum utility. In the operational phase, a transient simulator can provide invaluable experience in the training of operators to handle new and different systems. One of the major difficulties in analyzing natural gas piping, either steady or unsteady, is the accurate evaluation of the frictional characteristics of the system. In 1974 Stoner and Karnitz presented an adaptation of an unsteady flow simulator to utilize real time-varying field data to ascertain the pipeline friction parameters.

In any transient simulation, the partial differential equations of motion and mass continuity and an equation of state are treated numerically to describe the unsteady behavior of the fluid in the pipeline. A total system response is available when boundary conditions are introduced at pipeline junctions and variable supply, or use rates are specified by compressor stations, well fields, or valve operations. In 1974 E. B. Wylie and coworkers presented an analysis procedure that utilizes the method of characteristics to numerically model the equations. Other procedures have also been used in the development of simulators of varying degrees of sophistication, as described by Rachford and T. Dupont in 1974.

The industry has been slow to adopt these simulation procedures. In view of current heavy demands on the limited sources of supply of natural gas, the transmission phase of total operation becomes increasingly important. Transient simulators are essential for leak and failure detection, for optimal system operation. They also represent an essential element in the implementation of automatically controlled systems.

[E. BENJAMIN WYLIE]

UNCONVENTIONAL SOURCES

In the United States there are four distinct sources for so-called unconventional natural gas: (1) low-permeability ("tight") gas-bearing sandstones and limestone formations of the Southwest, the Rockies, and the Northern Great Plains; (2) gas-bearing Devonian age shales of the Appalachian, Michigan, and Illinois basins; (3) coalbed methane, occurring in both eastern and western coal regions; and (4) geopressured aquifers along the Gulf Coast.

Table 2. Estimates of unconventional gas resources

Resources	Gas in place, Tcf*	Technically recoverable, Tcf*	Economically recoverable at \$3.50–5.50/Mcf,† 1979 dollars
Tight gas			
Lewin	420	210	100–190
Federal Power Commission	600	100–300	NA‡
Devonian Shale			
Lewin	NA	NA	8–25
Office of Technology Assessment	NA	NA	15–35
Methane from coal seams			
Lewin	170	NA	1–25
Department of Energy	700	300	NA
Geopressured aquifers			
Lewin	860	42	1–5
U.S. Geological Survey	3100	97	NA

*1 Tcf = 2.83×10^{10} m³. †1 Mcf = 2.83×10^{4} m³. ‡NA = not available.

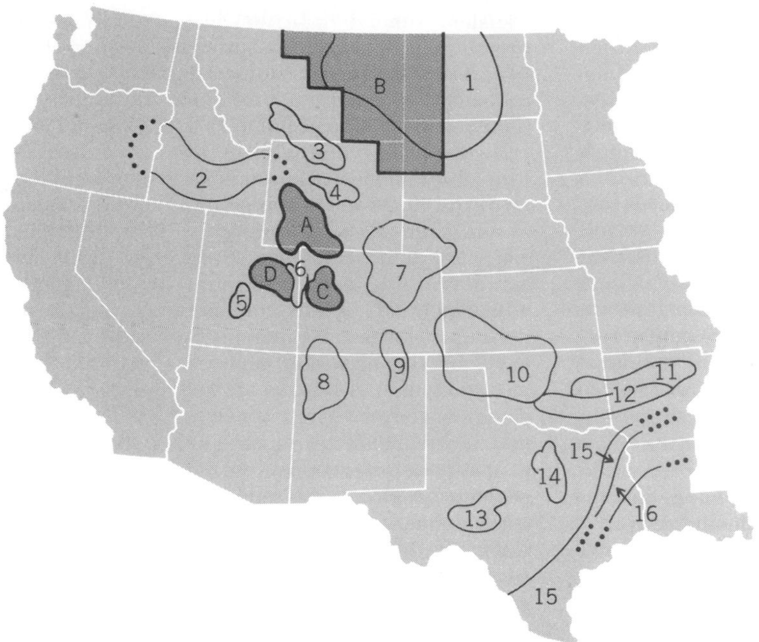

priority basins:

A. Greater Green River Basin
B. Northern Great Plains Province
C. Piceance Basin
D. Uinta Basin

additional low-permeability areas:

1. Williston Basin
2. Snake River Downwarp
3. Big Horn Basin
4. Wind River Basin
5. Wasatch Plateau
6. Douglas Creek Arch
7. Denver Basin
8. San Juan Basin
9. Raton Basin
10. Anadarko Basin
11. Arkoma Basin
12. Ouachita Mountains Province
13. Senora Basin
14. Fort Worth Basin
15. Western Gulf Basin
16. Cotton Valley Trend

Fig. 1. Map of western United States, showing areas of interest for natural gas.

Numerous estimates have been posed for the technical and economical potential of these unconventional resources, as summarized in Table 2. Since domestic natural gas reserves are about 200 trillion cubic feet (Tcf) or 5.7×10^{12} m^3 and yearly gas production is about 19–20 Tcf or $0.54-0.57 \times 10^{12}$ m^3, these unconventional gas sources would make a major contribution to United States energy supplies once their development becomes technically proved and economically feasible.

Tight gas formations. Large quantities of natural gas are trapped in reservoirs too tight to permit recovery by conventional technology. Twenty basins in the United States contain significant amounts of gas in such low-permeability formations (Fig. 1).

In many of these basins (such as Denver and Cotton Valley) the formations are broad, highly continuous "blanket" reservoirs; in others (such as the Green River and Piceance), the reservoirs are highly discontinuous lenses that sharply restrict the potential drainage of an individual well.

Gas has been produced from tight gas formations primarily from the blanket sands of the San Juan, Denver, Cotton Valley, and Sonora basins.

It has been estimated that the volume of gas-in-place in 13 of the more defined basins is over 400 Tcf or 11.3×10^{12} m^3.

Recovery technology. The leading recovery technology for tight gas reservoirs is massive hydraulic fracturing (MHF). The purpose of this technique is to create and prop artificial fractures far into the reservoir and provide a conduit for gas to flow to the well.

Fractures are created by pumping fluid into the formation until the pressure breaks the rock. A propping agent, usually sand, is mixed with the fluid, and is thus carried into the fracture. Fractures are believed to be propagated from 500 to 2000 ft (150 to 600 m) in both directions from the wellbore.

Production. Through 1977 the tight gas formations produced about 0.8 Tcf (23×10^9 m^3) per year. Cumulative production has been about 15 Tcf (0.42×10^{12} m^3), with an equal amount estimated for remaining proved reserves. The bulk of gas from tight formations is from the better "sweet spot" portions of the blanket sands. Considerable technological improvement will be required to economically recover the large gas resource remaining in the lower-permeability blanket and lenticular formations.

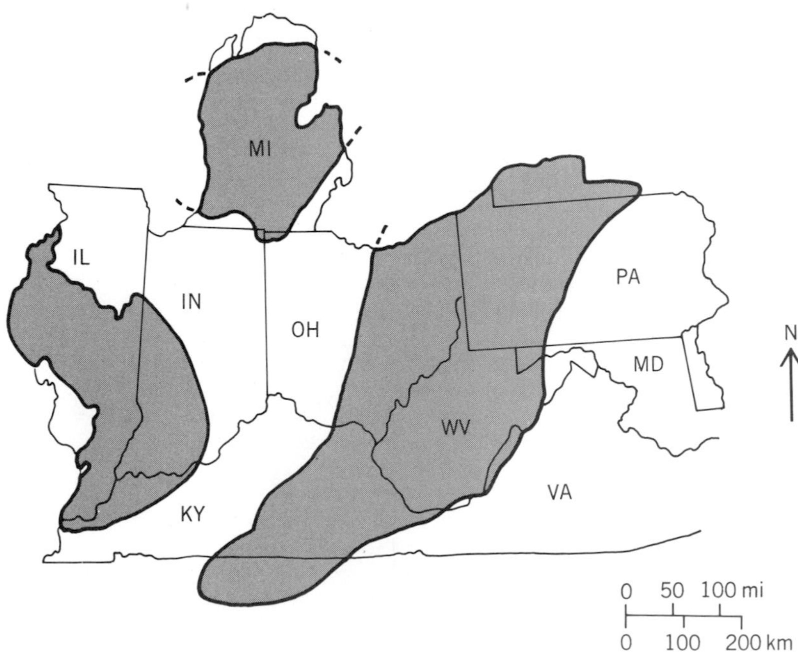

Fig. 2. Location of eastern gas shales basins.

Eastern gas shales. Eastern gas shales are Middle and Late Devonian, and Mississippian formations underlying much of the Appalachian, Michigan, and Illinois basins (Fig. 2). These shale deposits, particularly those in the Appalachian Basin, have produced gas since the 19th century. The attractiveness of this resource is enhanced by its proximity to natural gas markets in the populous East and industrial Midwest.

The largest Devonian Shale accumulation is in the western Appalachian Basin and covers approximately 160,000 mi^2 (410,000 km^2).

Although the gas-in-place has been estimated to be several thousand Tcf absorbed in the impermeable shales, it is the free gas, located in natural fracture systems, that has contributed the production to date. The size of the free-gas accumulation is not well known but has been estimated to range from 100 to 200 Tcf (2.8 to 5.7 × 10^{12} m^3).

Recovery technology. Formerly, Devonian shale wells were stimulated by "shooting" with explosives. This technique creates a rubbled area around the wellbore and helps overcome wellbore damage caused by drilling. However, shooting does not extend fractures a significant distance from the wellbore. Many producers in Devonian Shale areas have turned to small hydraulic water-and-foam fractures. Preliminary data indicate that initial open flows after this fracturing are typically 40–60% greater than in "shot" wells.

Production. A majority of the 9600 producing wells in the Appalachian region have been shot, produce about 0.1 Tcf (2.8 × 10^9 m^3) annually, and in total have produced an estimated 3 Tcf (85 × 10^9 m^3).

Coalbed methane. Approximately 380,000 mi^2 (980,000 km^2) of the 48 contiguous states are underlain by bituminous and subbituminous coal. A large methane resource is contained in those rocks; however, its extent and volume have not been well established. Most of the information on the resource has been acquired from the eastern states, primarily through coal mining; very little is known about the gas content of the large coal resource located in the western states.

Recovery technology. Several techniques for methane removal are being developed: ventilation, gob drainage, and predrainage. In ventilation, the mines are ventilated with large volumes of air to reduce the methane concentration to a safe level (less than 1% by volume). The diluted methane/air mixture is vented to the atmosphere. In gob drainage, methane-rich gases are drained by surface drilling from collapsed areas after coal extraction. This process reduces gas migration into the active mine and reduces the required ventilation. Drainage gas mixtures range between 25 to 90% methane content. In predrainage, methane is removed from virgin coalbeds and adjacent rock strata by drilling and associated stimulation techniques prior to mining.

Production. Virtually all of the gas drained from coalbeds is vented, at estimated rate of 0.1 Tcf (2.8 × 10^9 m^3) per year. Methane venting has usually been for mine safety procedure, and little of produced coal gas has been used for energy to date. With the shortages of energy, several on-site or local applications are being explored for using coalbed methane.

Geopressured aquifers. Geopressured aquifers are subsurface reservoirs containing methane dissolved in water under high pressures and temperatures. Such water-bearing aquifers have been identified in the Gulf and West Coast areas of the United States.

Geopressured aquifers, occurring at depths ranging from 5000 to 20,000 ft (1500 to 6000 m), were formed by rapid deposition of sediments that trapped water and unconsolidated materials under impermeable layers. The aquifers are characterized by pressures exceeding 15,000 psi (100 MPa) and temperatures varying from 200 to 300°F (93 to 149°C).

Recovery technology. Economic production from geopressured aquifers will require production of large amounts of water, which in turn necessitates the drilling of large-diameter wellbores.

Recovery energy from the geopressured brines will require extending existing technology to produce, handle, and dispose of high flow rates of hot, saline water. In addition, the geopressured geothermal zones require special equipment and production schemes that are not conventional practice.

Production. Although several thousand wells have been drilled into and through geopressured aquifers in the search for oil and gas, these geopressured zones have been considered uneconomical. The technological and economical risks of developing a technically unproved and economically marginal resource constrain the development of this natural gas resource. [VELLO A. KUUSKRAA]

Bibliography: *Enhanced Recovery of Unconventional Gas*, Lewin and Associates, Inc., February 1978; R. P. Epp and A. G. Fowler, in *J. Hydraul. Div. ASCE*, vol. 96, no. HY1, 1970; A. M. Leeston, J. A. Chrichton, and J. C. Jacobs, *The Dynamic Natural Gas Industry*, 1963; L. J. P. Muffler (ed.), *Assessment of Geothermal Resources of the United States—1978*, USGS Circ. 790, 1979; E. J. Neuner, *The Natural Gas Industry*, 1960; Office of Technology Assessment, U.S. Congress, *Status Report on the Gas Potential from Devonian Shale of the Appalachian Basin*, November 1977; H. H. Rachford, *Oil Gas J.*, pp. 93–96, July 16, 1973; H. H. Rachford and T. Dupont, *J. Petrol. Eng. AIME*, vol. 14, no. 2, 1974; M. A. Stoner, *J. Petrol. Eng. AIME*, vol. 12, no. 1, 1972; M. A. Stoner and M. A. Karnitz, *Oil Gas J.*, pp. 97–100, Dec. 10, 1973; M. A. Stoner and M. A. Karnitz, *Transport. Eng. J. ASCE*, vol. 100, no. TE3, 1974; U.S. Department of Energy, *Semi-Annual Report for the Unconventional Gas Recovery Program*, March 1979; U.S. Federal Power Commission, *National Gas Survey*, 1973; E. B. Wylie, V. L. Streeter, and M. A. Stoner, *J. Petrol. Eng. AIME*, vol. 14, no. 1, 1974.

Natural gas and sulfur production

Hydrogen sulfide, or sour gas, is a highly toxic gas that can cause failure in certain materials and is associated with the deposition of sulfur in the Earth. Sulfur deposition has been recognized as a serious problem in petroleum reservoirs containing high content of hydrogen sulfide, sulfur, hydrogen, and carbon dioxide. Moreover, depending on pressure and temperature, elemental sulfur may precipitate from the fluid mixtures and cause serious plugging of the formation, tubing, and surface equipment. A practical scheme for producing this type of fluid from a deep reservoir has been proposed, and it may avoid the problem of impairment to the formation by sulfur deposition. This has been accomplished by analysis taking into account the flow of gas mixtures accompanying sulfur deposition in porous media.

Mechanisms of sulfur deposition. A survey of operations of sour gas wells in Canada and Europe was reported by J. B. Hyne. In 9 of 31 cases tabulated, deposition of sulfur in tubing was observed. This survey revealed that sulfur precipitation almost always occurs in wells if the content of pentane-plus in the reservoir fluid is low and that the content of carbon dioxide is not an important factor. High temperatures at bottom hole and wellhead, and low pressure at wellhead, were also found to provide favorable conditions for sulfur deposition in tubing. These results may be accounted for by the solubility behavior of sulfur in a reservoir fluid. For example, J. G. Roof measured the solubility of sulfur in hydrogen sulfide at pressures ranging from 1020 to 4520 psia (1 psia equals 6890 N/m^2) and temperatures varying from 110 to 230°F (43 to 110°C). His results illustrated that the solubility change per unit pressure drop is greater at higher temperature. Therefore, if the temperature in the tubing is high and the pressure at the wellhead is low (large difference between the bottom-hole and wellhead pressures), the solubility reduction between the bottom-hole and wellhead conditions is very significant. Consequently, a large quantity of solid sulfur can precipitate from the solution for a given production rate, and serious plugging in the tubing can occur. This sulfur deposition can be prevented, for instance, by circulating hot fluids under pressure from a specific location in the tubing.

The foregoing evidence confirms that for a given fluid containing dissolved sulfur, the elemental sulfur can precipitate from the solution because of solubility reduction. Furthermore, the solubility of sulfur in the solution is controlled mainly by pressure and temperature. Since the fluid flow in the formation can be considered isothermal under normal operating conditions, pressure remains the most important factor.

Prediction of performance. A mathematical model has been formulated by C. H. Kuo to describe the flow of a fluid accompanied by solid precipitation in porous media. In such a treatment, a mixture of hydrogen sulfide, hydrogen, sulfur, and carbon dioxide can be considered as a homogeneous solution contained in a formation of porosity ϕ_i at its initial condition. As the reservoir, bounded by impermeable cap and base rocks, is depleted, the contained fluid flows radially toward the production well, which is located at the center of the reservoir. Some of the initially dissolved component may precipitate from the solution as the pressure declines. In view of the experimental evidence discussed earlier, it can be assumed that the solid deposits formed are incapable of flowing and accumulate in the voids.

The velocity of the isothermal, horizontal fluid flow through this porous medium is assumed to be governed by the one-dimensional form of Darcy's law, and the vertical velocity component and gravity effects are unimportant. Thus, the equations of continuity for the mixture (solution) and the sulfur component can be written as Eqs. (1) and (2).

$$\frac{1}{r}\frac{\partial}{\partial r}\left(\rho\frac{k}{\mu}r\frac{\partial p}{\partial r}\right)=\frac{\partial}{\partial t}[\rho\phi+\rho^*(\phi_i-\phi)] \quad (1)$$

$$\frac{1}{r}\frac{\partial}{\partial r}\left(R\frac{k}{\mu}r\frac{\partial p}{\partial r}\right)=\frac{\partial}{\partial t}[R\phi+\rho^*(\phi_i-\phi)] \quad (2)$$

In Eqs. (1) and (2) the density of the solution ρ and the solubility of sulfur in the solution R are considered functions of pressure. The density of the deposited sulfur ρ^* is assumed constant, since there is little information in the literature regarding the influence of pressure on this density. The permeability k and the viscosity μ of the fluid are treated as functions of porosity and pressure respectively.

These partial-differential equations can be solved by numerical methods to predict the pressure p and the porosity ϕ as functions of time t and radial distance r. If the fluid is initially undersaturated with the sulfur, it is necessary to solve only Eq. (1) in the early production period to obtain the pressure distribution ($\phi=\phi_i$). Once the reservoir is depleted below the saturation pressure, however, Eqs. (1) and (2) must be solved simultaneously to predict $p(r,t)$ and $\phi(r,t)$, since the pore space is partially filled by deposited sulfur. The fraction of the pore space occupied by the deposited sulfur is obtained, then, as $1-\phi/\phi_i$.

Optimum production scheme. As an example, theoretical performances of a reservoir similar to that discovered in southern Mississippi will be examined. The gas mixture was found to exist in a formation 20,000 ft deep (1 ft is about 0.3 m), with a net pay of 300 ft, at 14,000 psi and 390°F (200°C). A flow test indicated that the gas contains 78% hydrogen sulfide, 20% carbon dioxide, and 2% other compounds. In addition, it was estimated that elemental sulfur is dissolved in the mixture at 7500 lb/MMscf (1 lb is 0.45 kg; MMscf represents million standard cubic feet). Assuming an initial porosity of 0.1, the sour gas initially in place is calculated to be 600 MMscf/acre (1 acre is 4047 m^2), or 380 and 1530 Bscf, billion standard cubic feet, for 640 and 2560 acres respectively. In the study by Kuo using estimated reservoir and fluid properties, it was assumed that the well can be produced at a constant mass rate q varying from 10 to 160 MMscf/day with 640- or 2560-acre well spacing before the pressure at the well face drops

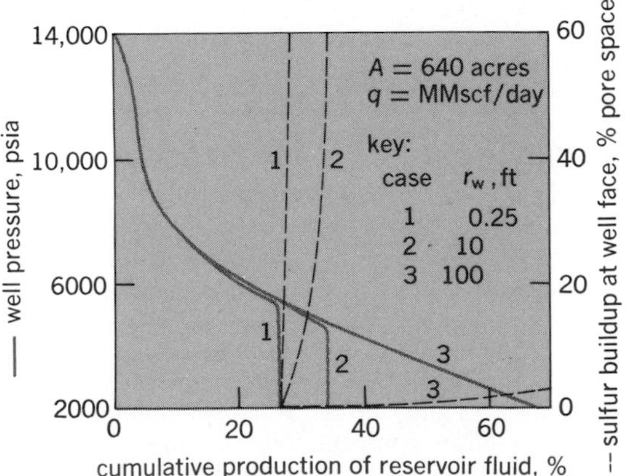

Fig. 1. Effect of well-bore radius. (*From C. H. Kuo, On the production of hydrogen sulfide–sulfur mixture from deep formations, J. Petrol. Technol., 24(9):1142–1146, 1972*)

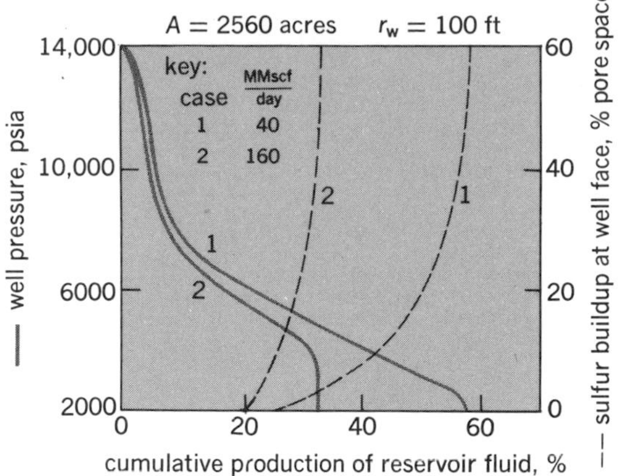

Fig. 2. Effect of production rate. (*From C. H. Kuo, On the production of hydrogen sulfide–sulfur mixture from deep formations, J. Petrol. Technol., 24(9):1142–1146, 1972*)

to a minimum value of 2000 psi. Thereafter, the rate decreases, and the minimum pressure is maintained at the well face. The well is shut in when the rate has declined to less than a certain minimum production rate.

The predicted histories of pressure and of sulfur buildup at the well face are plotted against the cumulative production of the reservoir fluid in Figs. 1 and 2. As illustrated in Fig. 1, assuming an original well-bore radius r_w of 0.25 ft (case 1), 25% of the reservoir fluid is produced before the well pressure declines to the estimated saturation pressure, 5500 psia. The well pressure then drops rapidly, and the pore space in the vicinity of the well bore is filled with deposited sulfur before any significant amount of the reservoir fluid is produced during this period. This implies that the formations must be cleaned frequently to remove deposited sulfur. If the formation can be fractured to create a large, effective well-bore radius, however, sulfur plugging may be alleviated, as indicated by the results for cases 2 and 3.

The effect of well spacing can be investigated by comparing the results predicted for case 3 in Fig. 1 and case 1 in Fig. 2. Although the design for case 3 is preferable to alleviate impairment due to sulfur deposition, the high cost of drilling in deep formations may dictate the choice of case 1, designed with wider spacing. Figures 1 and 2 also demonstrate that the effects of well-bore radius and production rate on the well pressure history are unimportant during early production. Therefore, with an effective well-bore radius equal to that obtained in the drilling of the well, it should be feasible to produce the reservoir fluid at a fairly high flow rate for a certain period. Once the reservoir is depleted to the saturation pressure, the formation should be fractured to create a large, effective well-bore radius. This could be the most desirable scheme because fracturing is relatively inexpensive, and, furthermore, a large well-bore radius makes it possible to produce the reservoir fluid for a long period without seriously impairing the formation. *See* PETROLEUM RESERVOIR ENGINEERING.

[C. H. KUO]

Bibliography: T. W. Hamby and J. R. Smith, *J. Petrol. Technol.*, 24(3):347–356, 1972; J. B. Hyne, *Oil Gas J.*, pp. 107–113, Nov. 25, 1968; C. H. Kuo, *J. Petrol. Technol.*, 24(9):1142–1146, 1972; J. G. Roof, *Soc. Petrol. Eng. J.*, 11(3):272–276, 1971.

Nuclear battery

A battery that converts the energy of particles emitted from atomic nuclei into electric energy. Two basic types have been developed: (1) A high-voltage type, in which a beta-emitting isotope is separated from a collecting electrode by a vacuum or a solid dielectric, provides thousands of volts but the current is measured in picoamperes (pA); (2) a low-voltage type gives about 1 volt with current in microamperes (μA).

High-voltage nuclear battery. In the high-voltage type, a radioactive source is attached to one electrode, emitting charged particles. The source might be strontium-90, krypton-85, or hydrogen-3 (tritium), all of which are pure beta emitters. An adjacent electrode collects the emitted particles. A vacuum or solid dielectric separates the source and the collector electrodes.

One high-voltage model, shown in Fig. 1, employs tritium gas sorbed in a thin layer of zirconium metal as the radioactive source. This source is looped around and spot-welded to the center tube of a glass-insulated terminal. A thin coating of carbon applied to the inside of a nickel enclosure acts as an efficient collector having low secondary emission. The glass-insulated terminal is sealed to the nickel enclosure. The enclosure is evacuated through the center tube, which is then pinched off and sealed.

The Radiation Research Corporation model R-1A is 0.95 cm in diameter and 1.35 cm in height. It weighs 5.7 g and occupies 0.8 cm^3. It delivers about 500 volts at 160 pA. Future batteries are expected to deliver 1 μA at 2000 volts, with a volume of 1048 cm^3.

Earlier models employed strontium-90. This isotope has the highest toxicity in the human body of the three mentioned. Tritium has only one one-thousandth the toxicity of strontium-90. Both strontium-90 and krypton-85 require shielding to reduce external radiation to safe levels. Tritium produces no external radiation through a wall that is thick enough for any structural purpose. Tritium was selected on the basis of these advantages.

The principal use of the high-voltage battery is to maintain the voltage of a charged capacitor. The current output of the radioactive source is sufficient for this purpose.

This type of battery may be considered as a constant-current generator. The voltage is proportional to the load resistance. The current is determined by the number of emissions per second captured by the collector and does not depend on ambient conditions or the load. As the isotope ages, the current declines. For tritium, the inten-

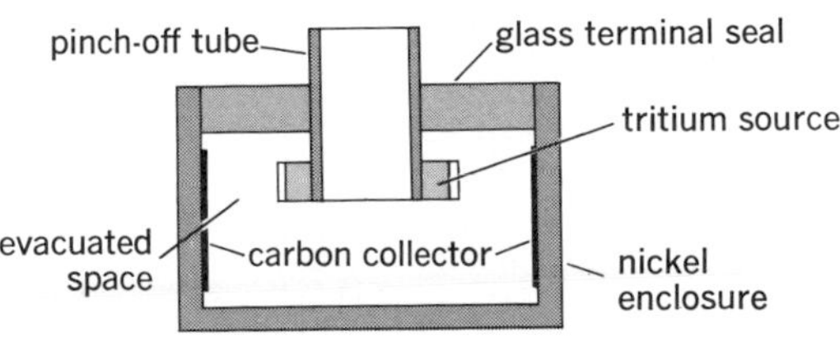

Fig. 1. Tritium battery in cross section.

sity drops 50% in a 12-year interval. For strontium-90, the intensity drops 50% in a 25-year interval.

Low-voltage nuclear battery. Three different concepts have been employed in the low-voltage type of nuclear batteries: (1) a thermopile, (2) the use of an ionized gas between two dissimilar metals, and (3) the two-step conversion of beta energy into light by a phosphor and the conversion of light into electric energy by a photocell.

Thermoelectric-type nuclear battery. This low-voltage type, employing a thermopile, depends on the heat produced by radioactivity (Fig. 2). It has been calculated that a sphere of polonium-210 of 0.25 cm. diameter, which would contain about 350 curies, if suspended in a vacuum, would have an equilibrium surface temperature of 2200°C, assuming an emissivity of 0.25. For use as a heat source, it would have to be hermetically sealed in a strong, dense capsule. Its surface temperature, therefore, would be lower than 2200°C.

To complete the thermoelectric battery, the heat source must be thermally connected to a series of thermocouples which are alternately connected thermally, but not electrically, to the heat source and to the outer surface of the battery. After a short time, a steady-state temperature differential will be set up between the junctions at the heat source and the junctions at the outer surface. This creates a voltage proportional to the temperature drop across the thermocouples. The battery voltage decreases as the age of the heat source increases. With polonium-210 (half-life, 138 days) the voltage drops about 0.5%/day. The drop for strontium-90 is about 0.01%/day (20-year half-life).

A battery containing 57 curies (2.1×10^{12} becquerels) of polonium-210 sealed in a sphere 1 cm in diameter and 7 chromelconstantan thermocouples delivered a maximum power of 1.8 milliwatts. It had an open-circuit voltage of 42 millivolts with a 78°C temperature differential. Over a 138-day period, the total electrical output would be about 1.5×10^4 joules (watt-sec).

Total weight of the battery was 34 g. This makes the energy output per kilogram equal to

$$\frac{1.5 \times 10^4}{3600} \text{ watt-hours (Whr)} \times \frac{1}{34} \times 1000 = 122.5$$

This is the same magnitude as with conventional electric cells using chemical energy. This nuclear energy, however, is being dissipated whether or not the electric energy is being used.

The choice of isotope for a thermoelectric nuclear battery is somewhat restricted. Those with a half-life of less than 100 days would have a short useful life, and those with a half-life of over 100 years would give too little heat to be useful. This leaves 137 possible isotopes. This number is further reduced by the consideration of shielding.

The trend is to use plutonium-238 (unable to support a chain reaction) and strontium-90. The most frequently used thermocouple material is a doped lead-telluride alloy that performs between 200 and 480°C. Another material under investigation is silicon-germanium alloy, which can operate at 800°C (hot junction).

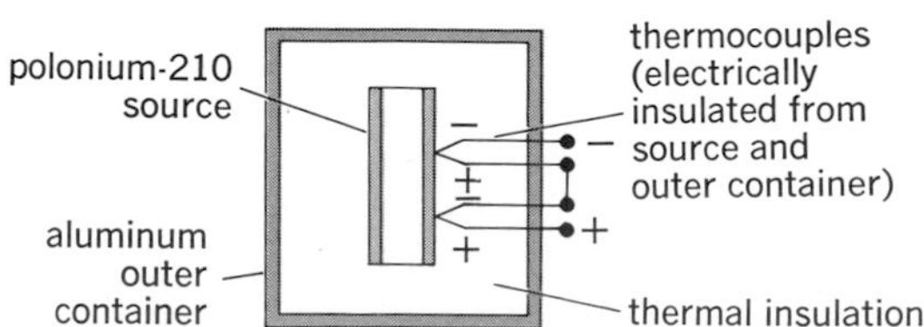

Fig. 2. Thermoelectric nuclear battery.

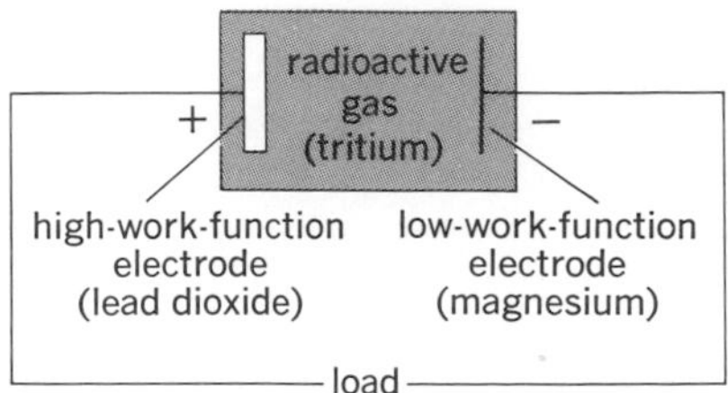

Fig. 3. Gas-ionization nuclear battery (schematic).

The thermoelectric systems developed so far are in the small power range (5–60 watts), and work with a maximum efficiency of 7%.

When shielding is not a critical problem, as in unmanned satellites, they are extremely convenient, being reliable and light. SNAP-3 and SNAP-9A (Systems for Nuclear Auxiliary Power) have weights in the 0.5–1.0 kg/watt range.

The above discussion applies only to a portable power source. Drastic revision would be required if a thermoelectric-type nuclear battery were to be designed for central-station power. *See* THERMOELECTRICITY.

Gas-ionization nuclear battery. In this battery a beta-emitting isotope ionizes a gas situated in an electric field (Fig. 3). Each beta particle produces about 200 current carriers (ions), so that a considerable current multiplication occurs compared with the rate of emission of the source. The electric field is obtained by the contact potential difference of a pair of electrodes, such as lead dioxide (high work function) and magnesium (low work function). The ions produced in the gas move under the influence of the electric field to produce a current.

A cell containing argon gas at 2 atmospheres, electrodes of lead dioxide and magnesium, and a radioactive source consisting of 1.5 millicuries (5.6×10^7 becquerels) of tritium has a volume of 0.16 cm^3 and an active plate area of 1.29 cm^2, and gives a maximum current of 1.6×10^{-9} ampere. The open-circuit voltage per cell depends on the contact potential of the electrode couple. A practical value appears to be about 1.5 volts. Voltage of any value may be achieved by a series assembly of cells.

The ion generation is exploited also in a purely thermal device, the thermionic generator. It consists of an emitting electrode heated to 1200–1800°C and a collecting electrode kept at 500–900°C. The hot electrode is made of tungsten or rhenium, metals which melt at more than 3000°C and have low vapor pressures at the working temperature. The collecting electrode is made out of a metal with a low work function, such as molybdenum, nickel, or niobium. Electrons from the hot electrode traverse a gap of 1–2 mm to the collector and recover the emitter by an external circuit.

The ionization space is filled with cesium vapor which acts in two ways. First, it covers the surface of the two electrodes with adsorbed cesium atoms and thus reduces the work function to the desired level. Second, it creates an ionized atmosphere, thus controlling the electron space charge.

The heating of the emitting electrode can be obtained by any of the known means: concentrated solar energy, radioisotopes, or conventional nuclear reactors. While reaching the high temperature was a problem easy to solve, the cooling of the collecting electrode appeared for a long time to be a critical technical problem. This problem was solved by the development of the "heat pipe."

In principle the heat pipe is an empty cylinder absorbing heat at one end by vaporization of a liquid and releasing heat at the other end by condensation of the vapor; the liquid returns to the heat-absorbing end by capillarity through a capillary structure covering the internal face of the cylinder. The heat transfer of a heat pipe may be 10,000 times or more higher than that of copper or silver.

The prototype developed under the sponsorship of NASA (NASA/RCA Type A1279) represents an advanced design capable of supplying 185 watts at 0.73 volt with an efficiency of 16.5%.

Since no major limitations are foreseen in the development of thermionic fuel cells, it is expected that they will provide the most convenient technique for using nuclear energy in the large-scale production of electricity.

Scintillator-photocell nuclear battery. This type of cell is based on a two-step conversion process (Fig. 4). Beta-particle energy is converted into light energy; then the light energy is converted into electric energy. To accomplish these conversions, the battery has two basic components, a light source and photocells.

The light source consists of a mixture of finely divided phosphor and promethium oxide (Pm_2O_3) sealed in a transparent container of radiation-resistant plastic. The light source is in the form of a thin disk. The photocells are placed on both faces of the light source. These cells are modified solar cells of the diffused-silicon type.

Since the photocells are damaged by beta radiation, the transparent container of the light source must be designed to absorb any beta radiation not captured in the phosphor. Polystyrene makes an excellent light-source container because of its resistance to radiation.

The light source must emit light in the range at which the photocell is most efficient. A suitable phosphor for the silicon photocell is cadmium sulfide or a mixture of cadmium and zinc sulfide.

In a prototype battery, the light source consisted of 50 milligrams (mg) of phosphor and about 5 mg of the isotope promethium-147. This isotope is a pure beta-emitter with a half-life of 2.6 years. It is deposited as a coating of hydroxide on the phosphor particles, which are then dried to give the oxide. For use with the low light level (about 0.001 times sunlight) of the light source, special treatment is necessary to make the equivalent shunt resistance of the cell not less than 100,000 ohms. For a description of the photocell *see* SOLAR CELL.

The prototype battery, when new, delivers 20×10^{-6} ampere at 1 volt. In 2.6 years (half-life) the current drops about 50% but the voltage drops only about 5%.

The power output improves with decreasing temperature, as a result of improved photocell diode characteristics which more than compensate for a decrease in short-circuit current. At −73°C, the power output is 1.7 times as great as at room temperature. At 62°C, the power output is only 0.6 times as great as at room temperature.

The battery requires shielding to reduce the weak gamma radiation to less than 9 milliroentgen per hour (mr/hr), which is the tolerance for continuous exposure of human extremities. The unshielded battery has a radiation level of 90 mr/hr. By enclosing the cell in a case of tungsten alloy, density 16.5, the external radiation becomes less than 9 mr/hr.

The unshielded battery has a volume of 0.23 cm³ and a weight of 0.45 g. Over a 2.5 year period, the total output would be 0.32 Whr (whether or not used). This gives a unit output of 704 Whr/kg, about six times as great as chemical-battery output. But the shielded battery has a volume of 1.15 cm³ and a weight of 1.7 g. This reduced the unit output to 18.7 Whr/kg. The cell can undergo prolonged storage at 93°C. *See* BATTERY.

[JACK DAVIS; L. ROZEANU; KENNETH FRANZESE]

Bibliography: H. C. Carney, Shock support system for nuclear battery, in R. W. Graham (ed.), *Primary Batteries, Recent Advances*, Chem. Technol. Rev. #105, Energy Technol. Rev. #25, pp. 319–326, 1978; B. I. Leefer, Nuclear-thermionic energy converter, *Proceedings of the 20th Annual Power Sources Conferences*, p. 172, 1966; M. Lewis and S. F. Seeman, Performance experience with prototype beta cell nuclear batteries, *Trans. Amer. Nucl. Soc.* 14(2):525, October 1971; F. D. Postula, Miniature nuclear batteries for cardiac pacemakers, *Trans. Amer. Nucl. Soc.* 15(2):704–705, November 1972; *Proceedings of the Ninth Intersociety Energy Conversion Engineering Conference*, pp. 824–825, 1974; L. I. Shure and H. J. Schwartz, Survey of electric power plants for space applications, *Chem. Eng. Progr.*, 63:99, 1967.

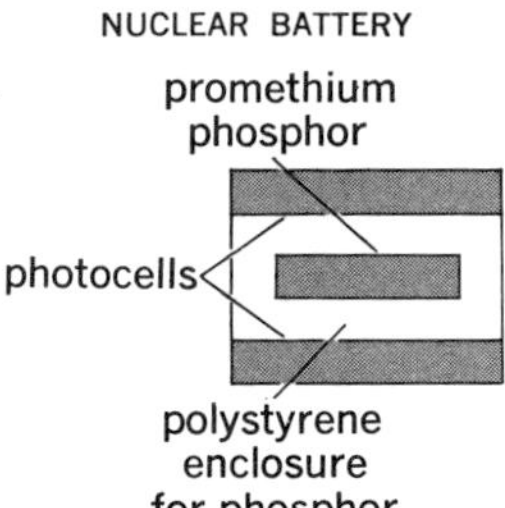

Fig. 4. Scintillator-photocell battery.

Nuclear binding energy

The amount by which the mass of an atom is less than the sum of the masses of its constituent protons, neutrons, and electrons expressed in units of energy. This energy difference accounts for the stability of the atom. In principle, the binding energy is the amount of energy which was released when the several atomic constituents came together to form the atom. Most of the binding energy is associated with the nuclear constituents, or nucleons, and it is customary to regard this quantity as a measure of the stability of the nucleus alone.

A widely used term, the binding energy (BE) per nucleon, is defined by the equation below, where

$$\text{BE/nucleon} = \frac{[ZH + (A-Z)n - {}_ZM^A]c^2}{A}$$

${}_ZM^A$ represents the mass of an atom of mass number A and atomic number Z, H and n are the masses of the hydrogen atom and neutron, respectively, and c is the velocity of light. The binding energies of the orbital electrons, here practically neglected, are not only small, but increase with Z in a gradual manner; thus the BE/nucleon gives an accurate picture of the variations and trends in nuclear stability. The figure shows the BE/nucleon (in million electron volts) plotted against mass number for $A > 40$.

The BE/nucleon curve at certain values of A

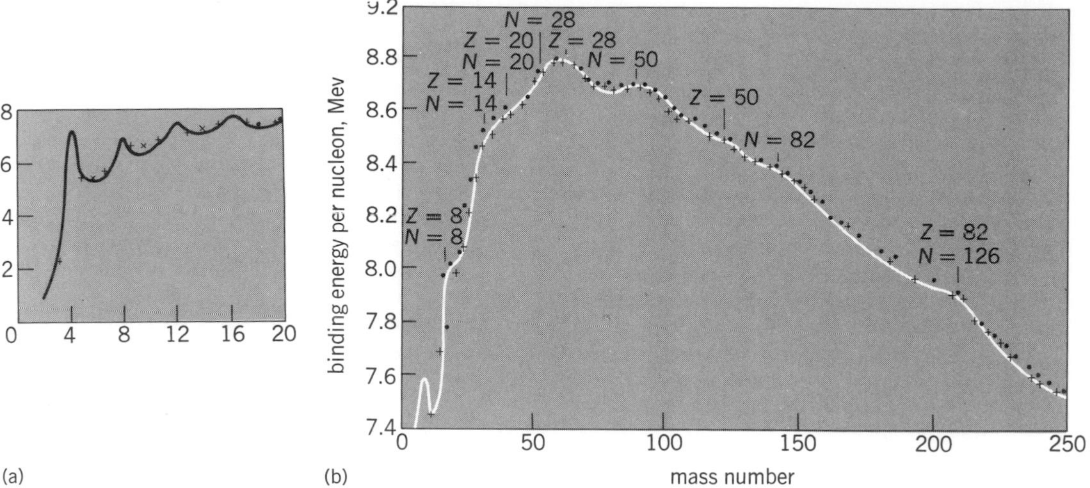

Graph of binding energy per nucleon. (a) Low mass numbers, from 2 to 20. (b) Mass numbers from 12 to 250; N = number of neutrons. (*From A. H. Wapstra, Isotopic measure, part 1, where A is less than 34, Physica, 21:367–384, 1955*)

suddenly changes slope in such a direction as to indicate that the nuclear stability has abruptly deteriorated. These turning points coincide with particularly stable configurations, or nuclear shells, to which additional nucleons are rather loosely bound. Thus there is a sudden turning of the curve over $A=52$ (28 neutrons); the maximum occurs in the nickel region (28 protons, $\sim A=60$); the stability rapidly deteriorates beyond $A=90$ (50 neutrons); there is a slightly greater than normal stability in the tin region (50 protons, $\sim A=118$); the stability deteriorates beyond $A=140$ (82 neutrons) and beyond $A=208$ (82 protons plus 126 neutrons).

The BE/nucleon is remarkably uniform, lying for most atoms in the range 5–9 Mev. This near constancy is evidence that nucleons interact only with near neighbors; that is, nuclear forces are saturated.

The binding energy, when expressed in mass units, is known as the mass defect, a term sometimes incorrectly applied to quantity $M-A$, where M is the mass of the atom. [H. E. DUCKWORTH]

The term binding energy is sometimes also used to describe the energy which must be supplied to a nucleus in order to remove a specified particle to infinity, for example, a neutron, proton, or α-particle. A more appropriate term for this energy is the separation energy. This quantity varies greatly from nucleus to nucleus and from particle to particle. For example, the binding energies for a neutron, a proton, and a deuteron in O^{16} are 15.67, 12.13, and 20.74 Mev, respectively, while the corresponding energies in O^{17} are 4.14, 13.78, and 14.04 Mev, respectively. The usual order of neutron or proton separation energy is 7–9 Mev for most of the periodic table. [D. H. WILKINSON]

Nuclear engineering

That branch of engineering that deals with the production and use of nuclear energy. It is concerned with the development, design, construction, and operation of power plants which convert energy produced by fission or fusion to other useful forms such as heat or electrical energy. Development of these unique sources of energy requires novel solutions to difficult mechanical, electrical, and materials problems. Because many of the components and systems operate in the presence of intense high-energy radiation, special problems that are generated by the interaction of radiation with various materials are encountered. Such problems are unique to nuclear engineering. Training of nuclear engineers places special emphasis on this area. *See* NUCLEAR FISSION; NUCLEAR FUELS; NUCLEAR FUSION; NUCLEAR POWER; NUCLEAR REACTOR.

Radioactive materials are used in a wide variety of industrial processes and equipment, ranging from nondestructive testing of welds to low-temperature sterilization of pharmaceuticals. Handling and storage of the large quantities of radioactive substances used as reactor fuel, produced as by-product material, or generated as waste introduce problems in the protection of personnel, equipment, and the environment. Such protection from high-energy radiation requires the design and construction of a variety of radiation shields and of shipping and handling equipment. *See* DECONTAMINATION OF RADIOACTIVE MATERIALS; NUCLEAR FUELS REPROCESSING; RADIOACTIVE WASTE MANAGEMENT.

Nuclear explosives are being investigated for use in large-scale excavation and for stimulation of the production of natural gas. Nuclear reactors are now used for propulsion of a variety of naval vessels. Serious consideration is being given to the use of nuclear power for propulsion of commercial ships.

More than 60 colleges and universities in the United States offer educational programs in nuclear engineering. Undergraduate curricula emphasize design and analysis of fission reactor power plants, industrial applications of radiation and radioactive isotopes, and radiation protection. Graduate programs typically place emphasis on research in fission reactor fuels management, reactor safety, effects of radiation on materials, generation and control of magnetically confined high-temperature plasmas, laser-generated fusion, design of fusion power plants, radiation measuring devices and systems, and medical applications of radiation. Many nuclear engineering departments operate research or training reactors which are used for laboratory instruction and as intense sources of neutron and gamma radiation for research. [WILLIAM KERR]

Bibliography: T. J. Connolly, *Foundations of Nuclear Engineering*, 1978; J. J. Duderstadt and L. J. Hamilton, *Nuclear Reactor Analysis*, 1976; A. R. Foster and R. L. Wright, *Basic Nuclear Engineering*, 3d ed., 1977; J. R. Lamarsh, *Introduction to Nuclear Engineering*, 1975; R. L. Murray, *Nuclear Energy*, 1975.

Nuclear fission

An extremely complex nuclear reaction representing a cataclysmic division of an atomic nucleus into two nuclei of comparable mass. This rearrangement or division of a heavy nucleus may take place naturally (spontaneous fission) or under bombardment with neutrons, charged particles, gamma rays, or other carriers of energy (induced fission). Although nuclei with mass number A of approximately 100 or greater are energetically unstable against division into two lighter nuclei, the fission process has a small probability of occurring, except with the very heavy elements. Even for these elements, in which the energy release is of the order of 200,000,000 electron volts (eV), the lifetimes against spontaneous fission are reasonably long. *See* NUCLEAR REACTION.

Liquid-drop model. The stability of a nucleus against fission is most readily interpreted when the nucleus is viewed as being analogous to an incompressible and charged liquid drop with a surface tension. Such a droplet is stable against small deformations when the dimensionless fissility parameter X in Eq. (1) is less than unity, where the

$$X = \frac{(\text{charge})^2}{10 \times \text{volume} \times \text{surface tension}} \qquad (1)$$

charge is in esu, the volume is in cm³, and the surface tension is in ergs/cm². The fissility parameter is given approximately, in terms of the charge number Z and mass number A, by the relation $X = Z^2/50\,A$.

Long-range Coulomb forces between the protons act to disrupt the nucleus, whereas short-range nuclear forces, idealized as a surface tension, act to stabilize it. The degree of stability is then the result of a delicate balance between the relatively weak electromagnetic forces and the strong nuclear forces. Although each of these forces results in potentials of several hundred million electron volts, the height of a typical barrier against fission for a heavy nucleus, because they are of opposite sign but do not quite cancel, is only 5,000,000 or 6,000,000 eV. Investigators have used this charged liquid-drop model with great success in describing the general features of nuclear fission and also in reproducing the total nuclear binding energies. *See* NUCLEAR BINDING ENERGY.

Shell corrections. The general dependence of the potential energy on the fission coordinate representing nuclear elongation or deformation for a heavy nucleus such as ^{240}Pu is shown in Fig. 1. The expanded scale used in this figure shows the large decrease in energy of about 200 MeV as the fragments separate to infinity. It is known that ^{240}Pu is deformed in its ground state, which is represented by the lowest minimum of −1813 MeV near zero deformation. This energy represents the total nuclear binding energy when the zero of potential energy is the energy of the individual nucleons at a separation of infinity. The second minimum to the right of zero deformation illustrates structure introduced in the fission barrier by shell corrections, that is, corrections dependent upon microscopic behavior of the individual nucleons, to the liquid-drop mass. Although shell corrections introduce small wiggles in the potential-energy surface as a function of deformation, the gross features of the surface are reproduced by the liquid-drop model. Since the typical fission barrier is only a few million electron volts, the magnitude of the shell correction need only be small for irregularities to be introduced into the barrier. This structure is schematically illustrated for a heavy nucleus by the double-humped fission barrier in Fig. 2, which represents the region to the right of zero deformation in Fig. 1 on an expanded scale. The fission barrier has two maxima and a rather deep minimum in between. For purposes of comparison, the single-humped liquid-drop barrier is also schematically illustrated. The transition in nuclear shape as a function of deformation is schematically represented in the upper part of the figure.

Double-humped barrier. The developments which led to the proposal of a double-humped fission barrier were triggered by the experimental discovery of spontaneously fissionable isomers by S. M. Polikanov and colleagues in the Soviet Union and by V. M. Strutinsky's pioneering theoretical work on the binding energy of nuclei as a function of both nucleon number and nuclear shape. The double-humped character of the nuclear potential energy as a function of deformation arises, within the framework of the Strutinsky shell-correction method, from the superposition of a macroscopic smooth liquid-drop energy and a shell-correction energy obtained from a microscopic single-particle model. Oscillations occurring in this shell correc-

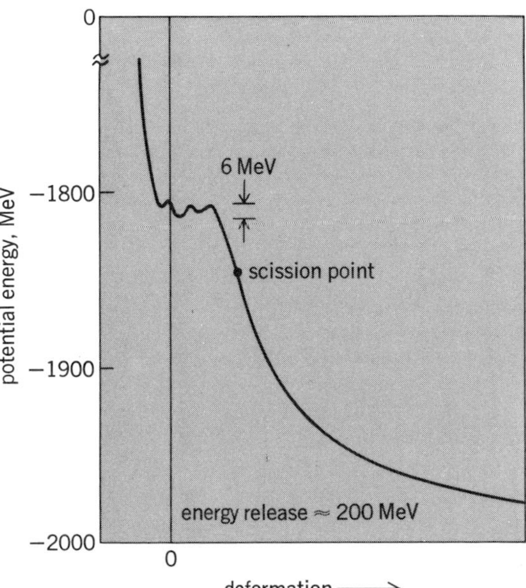

Fig. 1. Plot of the potential energy in MeV as a function of deformation for the nucleus ^{240}Pu. (*From M. Bolsteli et al., New calculations of fission barriers for heavy and superheavy nuclei, Phys. Rev., 5C:1050–1077, 1972*)

tion as a function of deformation lead to two minima in the potential energy, shown in Fig. 2, the normal ground-state minimum at a deformation of β_1 and a second minimum at a deformation of β_2. States in these wells are designated class I and class II, respectively. Spontaneous fission of the ground state and isomeric state arises from the lowest-energy class I and II states, respectively.

The calculation of the potential-energy curve illustrated in Fig. 1 may be summarized as follows. The smooth potential energy obtained from a macroscopic (liquid-drop) model is added to a fluctuating potential energy representing the shell corrections, and to the energy associated with the pairing of like nucleons (pairing energy), derived from a non-self-consistent microscopic model. Calculation of these corrections requires several steps: (1) specification of the geometrical shape of the nucleus, (2) generation of a single-particle potential related to its shape, (3) solution of the Schrödinger equation, and (4) calculation from these single-particle energies of the shell and pairing energies.

The oscillatory character of the shell corrections as a function of deformation is caused by variations in the single-particle level density in the vicinity of the Fermi energy. For example, the single-particle levels of a pure harmonic oscillator potential arrange themselves in bunches of highly degenerate shells at any deformation for which the ratio of the major and minor axes of the spheroidal equipotential surfaces is equal to the ratio of two small integers. Nuclei with a filled shell, that is, with a level density at the Fermi energy that is smaller than the average, will then have an increased binding energy compared to the average, because the nucleons occupy deeper and more bound states; conversely, a large level density is associated with a decreased binding energy. It is precisely this oscillatory behavior in the shell correction that is responsible for spherical or deformed gound states and for the secondary minima in fission barriers, as illutrated in Fig. 2.

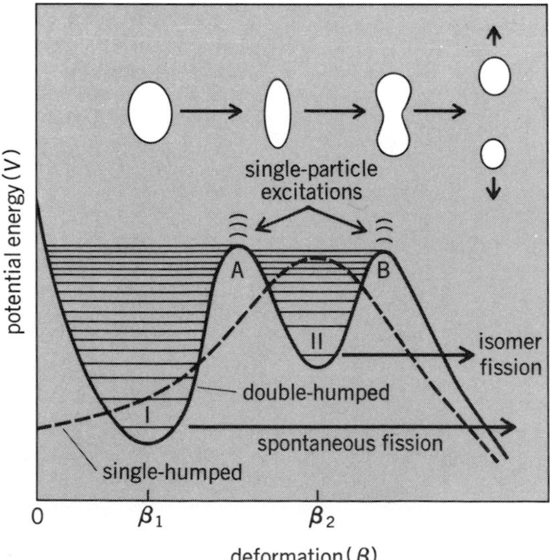

Fig. 2. Schematic plots of single-humped fission barrier of liquid-drop model and double-humped barrier introduced by shell corrections. *(From J. R. Huizenga, Nuclear fission revisited, Science, 168:1405–1413, 1970)*

More detailed theoretical calculations based on this macroscopic-microscopic method have revealed additional features of the fission barrier. In these calculations the potential energy is regarded as a function of several different modes of deformation. The outer barrier B (Fig. 2) is reduced in energy for shapes with pronounced left-right asymmetry (pear shapes), whereas the inner barrier A and deformations in the vicinity of the second minimum are stable against such mass asymmetric degrees of freedom. Similar calculations of potential-energy landscapes reveal the stability of the second minimum against gamma deformations, in which the two small axes of the spheroidal nucleus become unequal, that is, the spheroid becomes an ellipsoid.

Experimental consequences. The observable consequences of the double-humped barrier have been reported in numerous experimental studies. In the actinide region more than 30 spontaneously fissionable isomers have been discovered between uranium and berkelium, with half-lives ranging from 10^{-11} to 10^{-2} s. These decay rates are faster by 20 to 30 orders of magnitude than the fission half-lives of the ground states, because of the increased barrier tunneling probability (see Fig. 2). Several cases in which excited states in the second minimum decay by fission are also known. Normally these states decay within the well by gamma decay; however, if there is a hindrance in gamma decay due to spin, the state (known as a spin isomer) may undergo fission instead.

Qualitatively, the fission isomers are most stable in the vicinity of neutron numbers 146 to 148, a value in good agreement with macroscopic-microscopic theory. For elements above berkelium the half-lives become too short to be observable with available techniques; and for elements below uranium, the prominent decay is through barrier A into the first well, followed by gamma decay. It is difficult to detect this competing gamma decay of the ground state in the second well (called a shape isomeric state), but identification of the gamma branch of the 200-ns ^{238}U shape isomer has been reported.

Direct evidence of the second minimum in the potential-energy surface of the even-even nucleus ^{240}Pu has been obtained through observations of the E2 transitions within the rotational band built on the isomeric 0+ level. The rotational constant (which characterizes the spacing of the levels and is expected to be inversely proportional to the effective moment of inertia of the nucleus) found for this band is less than one-half that for the ground state and confirms that the shape isomers have a deformation β_2 much larger than the equilibrium ground-state deformation β_1. From yields and angular distributions of fission fragments from the isomeric ground state and low-lying excited states some information has been derived on the quantum numbers of specific single-particle states of the deformed nucleus (Nilsson single-particle states) in the region of the second minimum.

At excitation energies in the vicinity of the two barrier tops, measurements of the subthreshold neutron fission cross sections of several nuclei have revealed groups of fissioning resonance states with wide energy intervals between each group where no fission occurs. Such a spectrum is illustrated in Fig. 3*a*, where the subthreshold fis-

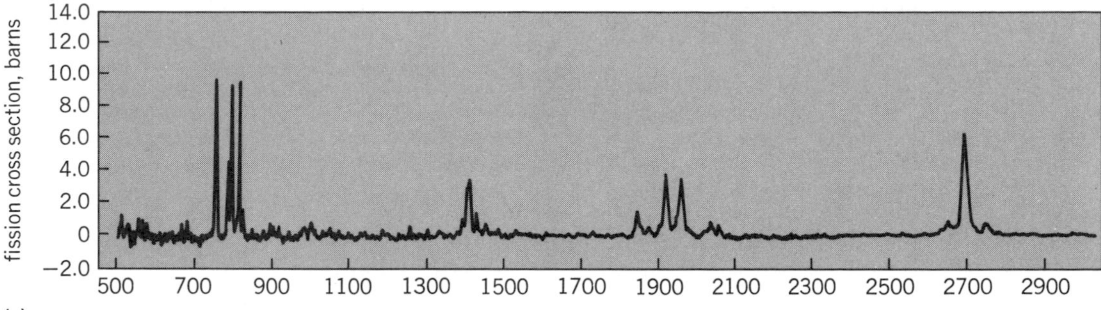

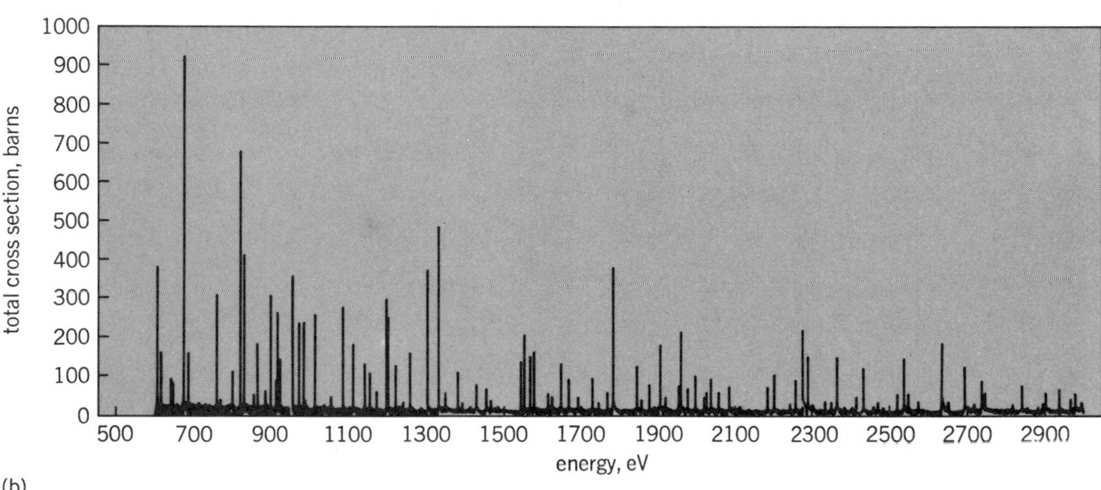

Fig. 3. Grouping of fission resonances demonstrated by (*a*) neutron fission cross section of ^{240}Pu and (*b*) total neutron cross section. 1 barn = 10^{-24} m^2. (*From V. M. Strutinsky and H. C. Pauli, Shell-structure effects in the fissioning nucleus, Proc. of 2d IAEA Symposium on Physics and Chemistry of Fission, Vienna, pp. 155–177, 1969*)

sion cross section of ^{240}Pu is shown for neutron energies between 500 and 3000 eV. As shown in Fig. 3*b*, between the fissioning resonance states there are many other resonance states, known from data on the total neutron cross sections, which have negligible fission cross sections. Such structure is explainable in terms of the double-humped fission barrier and is ascribed to the coupling between the compound states of normal density in the first well to the much less dense states in the second well. This picture requires resonances of only one spin to appear within each intermediate structure group illustrated in Fig. 3*a*. In an experiment using polarized neutrons on a polarized ^{237}Np target, it was found that all nine fine-structure resonances of the 40-eV group have the same spin and parity: $I = 3+$. Evidence has also been obtained for vibrational states in the second well from neutron (n,f) and deuteron stripping (d,pf) reactions at energies below the barrier tops (f indicates fission of the nucleus).

A. Bohr suggested that the angular distributions of the fission fragments are explainable in terms of the transition-state theory, which describes a process in terms of the states present at the barrier deformation. The theory predicts that the cross section will have a steplike behavior for energies near the fission barrier, and that the angular distribution will be determined by the quantum numbers associated with each of the specific fission channels. The theoretical angular distribution of fission fragments is based on two assumptions. First, the two fission fragments are assumed to separate along the direction of the nuclear symmetry axis so that the angle θ between the direction of motion of the fission fragments and the direction of motion of the incident bombarding particle represents the angle between the body-fixed axis (the long axis of the spheroidal nucleus) and the space-fixed axis (some specified direction in the laboratory, in this case the direction of motion of the incident particle). Second, it is assumed that the transition from the saddle point (corresponding to the top of the barrier) to scission (the division of the nucleus into two fragments) is so fast that Coriolis forces do not change the value of K (where K is the projection of the total angular momentum I on the nuclear symmetry axis) established at the saddle point.

In several cases, low-energy photofission and neutron fission experiments have shown evidence of a double-humped barrier. In the case of two barriers, the question arises as to which of the two barriers A or B is responsible for the structure in the angular distributions. For light actinide nuclei like thorium, the indication is that barrier B is the higher one, whereas for the heavier actinide nuclei, the inner barrier A is the higher one. The heights of the two barriers themselves are most reliably determined by investigating the probability of induced fission over a range of several mega-electron volts in the threshold region. Many direct reactions have been used for this purpose, for example, (d,pf), (t,pf), and $(^3\text{He}, df)$. There is reasonably good agreement between the experimental and theoretical barriers. The theoretical barriers are calculated with realistic single-particle potentials and include the shell corrections.

Fission probability. The cross section for particle-induced fission $\sigma(y,f)$ represents the cross section for a projectile y to react with a nucleus and produce fission, as shown by Eq. (2). The quantities $\sigma_R(y)$, Γ_f, and Γ_t are the total reaction cross sections for the incident particle y, the fission width, and the total level width, respectively where $\Gamma_t = \Gamma_f + \Gamma_n + \Gamma_y + \cdots$ is the sum of all partial-level widths. All the quantities in Eq. (2)

$$\sigma(y,f) = \sigma_R(y)\,(\Gamma_f/\Gamma_t) \qquad (2)$$

are energy-dependent. Each of the partial widths for fission, neutron emission, radiation, and so on, is defined in terms of a mean lifetime τ for that particular process, for example, $\Gamma_f = \hbar/\tau_f$. Here $\hbar$, the action quantum, is Planck's constant divided by 2π and is numerically equal to 1.0546×10^{-34} J s $= 0.66 \times 10^{-15}$ eV s. The fission width can also be defined in terms of the energy separation D of successive levels in the compound nucleus and the number of open channels in the fission transition nucleus (paths whereby the nucleus can cross the barrier on the way to fission), as given by expression (3), where I is the angular momentum and i is

$$\Gamma_f(I) = \frac{D(I)}{2\pi} \sum_i N_{fi} \qquad (3)$$

an index labeling the open channels N_{fi}. The contribution of each fission channel to the fission width depends upon the barrier transmission coefficient, which, for a two-humped barrier (see Fig. 2), is strongly energy-dependent. This results in an energy-dependent fission cross section which is very different from the total cross section shown in Fig. 3 for ^{240}Pu.

When the incoming neutron has low energy, the likelihood of reaction is substantial only when the energy of the neutron is such as to form the compound nucleus in one or another of its resonance levels (see Fig. 3*b*). The requisite sharpness of the "tuning" of the energy is specified by the total level width Γ. The nuclei ^{233}U, ^{235}U, and ^{239}Pu have a very large cross section to take up a slow neutron and undergo fission (see table) because both their absorption cross section and their probability for decay by fission are large. The probability for fission decay is high because the binding energy of the incident neutron is sufficient to raise the energy of the compound nucleus above the fission barrier. The very large, slow neutron fission cross sections of these isotopes make them important fissile materials for use in a chain reactor. *See* CHAIN REACTION.

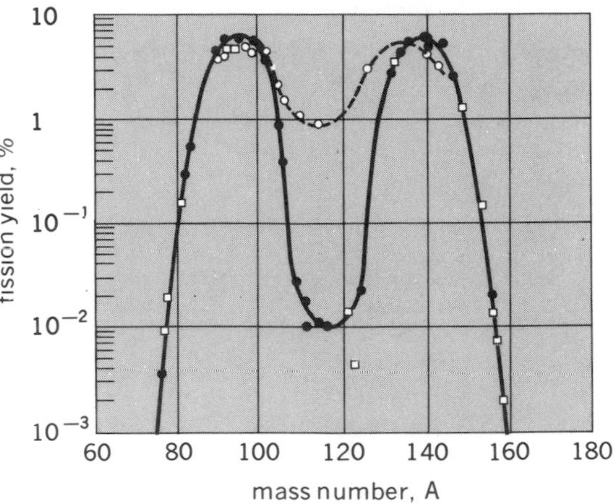

Fig. 4. Mass distribution of fission fragments formed by neutron-induced fission of $^{235}\text{U} + n = {}^{236}\text{U}$ when neutrons have thermal energy, smooth curve (*Plutonium Project Report, Rev. Mod. Phys., 18:539, 1964*), and 14-MeV energy, dashed curve (*based on R. W. Spence, Brookhaven National Laboratory, AEC-BNL* (C-9), *1949*). Quantity plotted is 100 × (number of fission decay chains formed with given mass)/(number of fissions).

Scission. The scission configuration is defined in terms of the properties of the intermediate nucleus just prior to division into two fragments. In heavy nuclei the scission deformation is much larger than the saddle deformation at the barrier, and it is important to consider the dynamics of the descent from saddle to scission. One of the important questions in the passage from saddle to scission is the extent to which this process is adiabatic with respect to the particle degrees of freedom. As the nuclear shape changes, it is of interest to investigators to know the probability for the nucleons to remain in the lowest-energy orbitals. If the collective motion toward scission is very slow, the single-particle degrees of freedom continually readjust to each new deformation as the distortion proceeds. In this case, the adiabatic model is a good approximation, and the decrease in potential energy from saddle to scission appears in collective degrees of freedom at scission, primarily as kinetic energy associated with the relative motion of the nascent fragments.

On the other hand, if the collective motion be-

Cross sections for neutrons of thermal energy to produce fission or undergo capture in the principal nuclear species, and neutron yields from these nuclei*

Nucleus	Cross section for fission, σ_f, 10^{-24} cm^2	σ_f plus cross section for radiative capture, σ_r	Ratio, $1+\alpha$	Number of neutrons released per fission, ν	Number of neutrons released per slow neutron captured, $\eta = \nu/(1+\alpha)$
^{233}U	525 ± 2	573 ± 2	1.093 ± 0.003	2.50 ± 0.01	2.29 ± 0.01
^{235}U	577 ± 1	678 ± 2	1.175 ± 0.002	2.43 ± 0.01	2.08 ± 0.01
^{239}Pu	741 ± 4	1015 ± 4	1.370 ± 0.006	2.89 ± 0.01	2.12 ± 0.01
^{238}U	0	2.73 ± 0.04			0
Natural uranium	4.2	7.6	1.83	2.43 ± 0.01	1.33

*Data from *Brookhaven National Laboratory 325*, 2d ed., suppl. no. 2, vol. 3, 1965. The data presented are the recommended or least-squares values published in this reference for 0.0253-eV neutrons. All cross sections are in units of barns (1 barn $= 10^{-24}$ cm$^2 = 10^{-28}$ m^2).

tween saddle and scission is so rapid that equilibrium is not attained, there will be a transfer of collective energy into nucleonic excitation energy. Such a nonadiabatic model, in which collective energy is transferred to single-particle degrees of freedom during the descent from saddle to scission, is usually referred to as the statistical theory of fission.

The experimental evidence indicates that the saddle to scission time is somewhat intermediate between these two extreme models. The dynamic descent of a heavy nucleus from saddle to scission depends upon the nuclear viscosity. A viscous nucleus is expected to have a smaller translational kinetic energy at scission and a more elongated scission configuration. Experimentally, the final translational kinetic energy of the fragments at infinity, which is related to the scission shape, is measured. Hence, in principle, it is possible to estimate the nuclear viscosity coefficient by comparing the calculated dependence upon viscosity of fission-fragment kinetic energies with experimental values. The viscosity of nuclei is an important nuclear parameter which also plays an important role in collisions of very heavy ions.

The mass distribution from the fission of heavy nuclei is predominantly asymmetric. For example, division into two fragments of equal mass is about 600 times less probable than division into the most probable choice of fragments when ^{235}U is irradiated with thermal neutrons. When the energy of the neutrons is increased, symmetric fission (Fig. 4) becomes more probable. In general, heavy nuclei fission asymmetrically to give a heavy fragment of approximately constant mean mass number 139 and a corresponding variable-mass light fragment (see Fig. 5). These experimental results have been difficult to explain theoretically. Calculations of potential-energy surfaces show that the second barrier (B in Fig. 2) is reduced in energy by up to 2 or 3 MeV, if octuple deformations (pear shapes) are included. Hence, the theoretical calculations show that mass asymmetry is favored at the outer barrier, although direct experimental evidence supporting the asymmetric shape of the second barrier is very limited. It is not known whether the mass asymmetric energy valley extends from the saddle to scission; and the effect of dynamics on mass asymmetry in the descent from saddle to scission has not been determined. Experimentally, as the mass of the fissioning nucleus approaches $A \approx 260$, the mass distribution approaches symmetry. This result is qualitatively in agreement with theory.

A nucleus at the scission configuration is highly elongated and has considerable deformation energy. The influence of nuclear shells on the scission shape introduces structure into the kinetic energy and neutron-emission yield as a function of fragment mass. The experimental kinetic energies for the neutron-induced fission of ^{233}U, ^{235}U, and ^{239}Pu have a pronounced dip as symmetry is approached, as shown in Fig. 6. (This dip is slightly exaggerated in the figure because the data have not been corrected for fission fragment scattering.) The variation in the neutron yield as a function of fragment mass for these same nuclei (Fig. 7) has a "saw-toothed" shape which is asymmetric about the mass of the symmetric fission fragment. Both these phenomena are reasonably well accounted

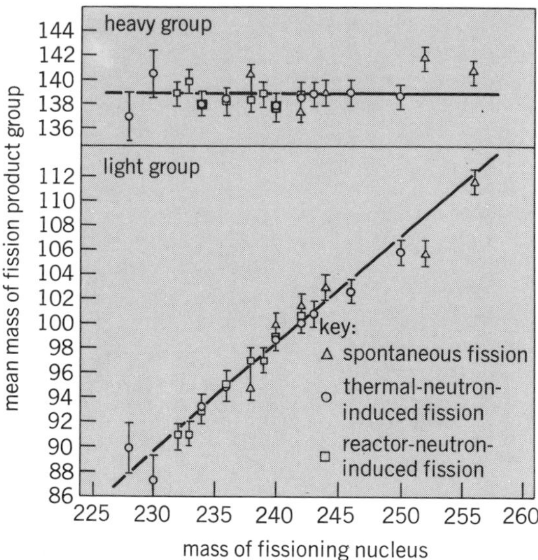

Fig. 5. Average masses of the light- and heavy-fission product groups as a function of the masses of the fissioning nucleus. Energy spectrum of reactor neutrons is that associated with fission. (*From K. F. Flynn et al., Distribution of mass in the spontaneous fission of ^{256}Fm, Phys. Rev., 5C:1725–1729, 1972*)

for by the inclusion of closed-shell structure into the scission configuration.

A number of light charged particles (for example, isotopes of hydrogen, helium, and lithium) have been observed to occur, with low probability, in fission. These particles are believed to be emit-

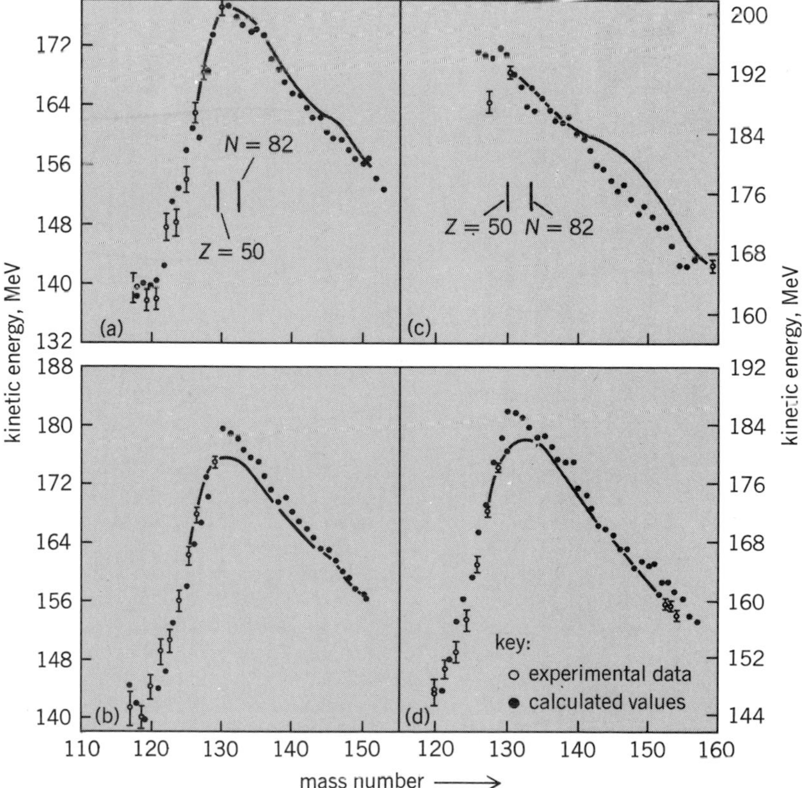

Fig. 6. Average total kinetic energy of fission fragments as a function of heavy fragment mass for fission of (*a*) ^{235}U, (*b*) ^{233}U, (*c*) ^{252}Cf, and (*d*) ^{239}Pu. Curves indicate experimental data. (*From J. C. D. Milton and J. S. Fraser, Time-of-flight fission studies on ^{233}U, ^{235}U and ^{239}Pu, Can. J. Phys., 40:1626–1663, 1962*)

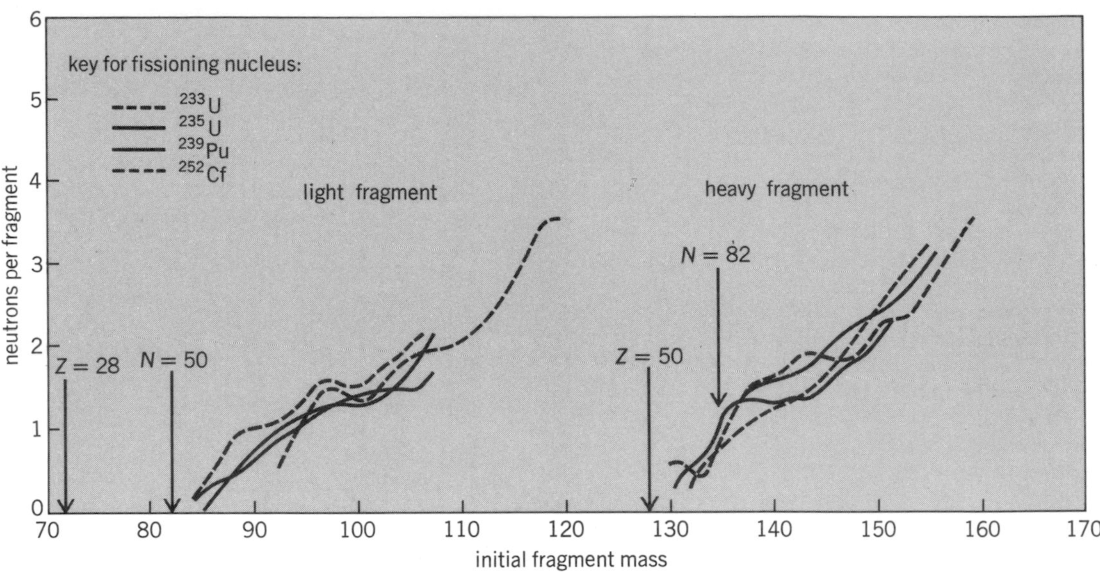

Fig. 7. Neutron yields as a function of fragment mass for four types of fission as determined from mass-yield data. Approximate initial fragment masses corresponding to various neutron and proton "magic numbers" N and Z are indicated. (*From J. Terrell, Neutron yields from individual fission fragments, Phys. Rev., 127:880–904, 1962*)

ted very near the time of scission. Available evidence also indicates that neutrons are emitted at or near scission with considerable frequency.

Postscission phenomena. After the fragments are separated at scission, they are further accelerated as the result of the large Coulomb repulsion. The initially deformed fragments collapse to their equilibrium shapes, and the excited primary fragments lose energy by evaporating neutrons. After neutron emission, the fragments lose the remainder of their energy by gamma radiation, with a lifetime of about 10^{-11} s. The kinetic energy and neutron yield as a function of mass are shown in Figs. 6 and 7. The variation of neutron yield with fragment mass is directly related to the fragment excitation energy. Minimum neutron yields are observed for nuclei near closed shells because of the resistance to deformation of nuclei with closed shells. Maximum neutron yields occur for fragments that are "soft" toward nuclear deformation. Hence, at the scission configuration, the fraction of the deformation energy stored in each fragment depends on the shell structure of the individual fragments. After scission, this deformation energy is converted to excitation energy, and, hence, the neutron yield is directly correlated with the fragment shell structure. This conclusion is further supported by the correlation between the neutron yield and the final kinetic energy. Closed shells result in a larger Coulomb energy at scission for fragments that have a smaller deformation energy and a smaller number of evaporated neutrons.

After the emission of the prompt neutrons and gamma rays, the resulting fission products are unstable against β-decay. For example, in the case of thermal neutron fission of ^{235}U, each fragment undergoes on the average about three β-decays before it settles down to a stable nucleus. For selected fission products (for example, ^{87}Br and ^{137}I) β-decay leaves the daughter nucleus with excitation energy exceeding its neutron binding energy. The resulting delayed neutrons amount, for thermal neutron fission of ^{235}U, to about 0.7% of all the neutrons given off in fission. Though small in number, they are quite important in stabilizing nuclear chain reactions against sudden minor fluctuations in reactivity. [JOHN R. HUIZENGA]

Bibliography: A. Michaudon, Neutrons and fission, *Phys. Today*, 31(1):23–31, 1978; *Proceedings of the 3d IAEA Symposium on Physics and Chemistry of Fission*, Rochester, NY, 1973; R. Vandenbosch and J. R. Huizenga, *Nuclear Fission*, 1973.

Nuclear fuel

The fissionable and fertile elements and isotopes used as the sources of energy in nuclear reactors. Although many heavy elements can be made to fission by bombardment with high-energy alpha particles, protons, deuterons, or neutrons, only neutrons can provide a self-sustaining reaction.

The number of neutrons ν released in the fission process varies from one per many fissions for elements just beyond the fission point (silver) to two or more per fission for the heavier elements, such as thorium and uranium. Even in such elements, neutron capture by the nucleus accompanied by the release of excess energy in the form of a gamma ray occurs in many cases, rather than nuclear fission. This reduces the number of neutrons available for further fission. The ratio of neutron capture to neutron fission varies from nucleus to nucleus and changes with the energy of the bombarding neutrons. Only a few isotopes of the heavy elements have a higher probability of fission than capture. These fissionable isotopes, ^{233}U, ^{235}U, and ^{239}Pu, are the only materials that can sustain the fission reaction and are therefore called nuclear fuels. *See* NUCLEAR REACTOR.

Of these isotopes, only ^{235}U occurs in nature as 1 part in 140 of natural uranium, the remainder being ^{238}U. The other two fissionable isotopes must be produced artificially, ^{233}U by neutron capture in ^{232}TH and ^{239}Pu by neutron capture in ^{238}U. The isotopes ^{232}Th and ^{238}U are called fertile materials.

By using a mixture of fissionable and fertile isotopes in a nuclear reactor, it is possible to reduce the rate of depletion of the nuclear fuel, because capture of excess neutrons by the fertile materials replenishes the fissionable material. Thus, ^{235}U can be burned (fissioned) and the surplus neutrons used to produce plutonium from ^{238}U or ^{233}U from thorium. Such nuclear reactors are called converter reactors.

The efficiency of production of new nuclear fuel depends on the extent of neutron losses due to undesirable neutron absorptions in the reactor or to neutron leakage. In some cases these losses can be kept small enough so that more nuclear fuel is produced than burned. Moreover, the ^{233}U and ^{239}Pu can be subsequently used as fuel in place of the original ^{235}U and by this means a large fraction of fertile material can be gradually converted into fissionable material. Reactors that burn ^{233}U and ^{239}Pu and produce as much fuel as is consumed, or more, are called breeders.

The total energy that can be produced from the fissionable ^{235}U in known resources of high-grade uranium ores corresponds to less than 5% of that from economically recoverable fossil fuels. Thus, nuclear fission will not become an important source of power in the long term unless the breeding and conversion fuel cycles are utilized.

Breeding and conversion. The nuclear reactions governing the consumption and production of nuclear fuel in a reactor are listed in Table 1. Also shown are values for the thermal-neutron (0.004 aJ or 0.025-eV neutron) cross section (probability that the reaction will take place) and the half-life for radioactive decay of the relatively unstable isotopes.

In a mixture of ^{235}U and ^{238}U, three competing reactions take place with thermal neutrons: (1) ^{235}U capture, (2) ^{235}U fission, and (3) ^{238}U capture (numbers in parentheses refer to reactions in Table 1). Reaction (3) leads to the production of ^{239}Pu by successive decay of ^{239}U and ^{239}Np, as shown by reactions (4) and (5). The conversion ratio (relative production and consumption of nuclear fuel) of the system is given by the relative probability that reaction (3) will take place as compared to reactions (1) and (2).

Table 1. Nuclear reactions in a thermal-neutron spectrum

Reaction number	Equation	Cross section, barns*	Half-life
(1)	$^{235}U + n \rightarrow {}^{236}U + \gamma$	107	
(2)	$^{235}U + n \rightarrow \text{Fission} + 2.47\,n$	582	
(3)	$^{238}U + n \rightarrow {}^{239}U + \gamma$	2.74	
(4)	$^{239}U \rightarrow {}^{239}Np + \beta^-$		23.5 m
(5)	$^{239}Np \rightarrow {}^{239}Pu + \beta^-$		2.33 d
(6)	$^{239}Pu + n \rightarrow {}^{240}Pu + \gamma$	277	
(7)	$^{239}Pu + n \rightarrow \text{Fission} + 2.88\,n$	748	
(8)	$^{240}Pu + n \rightarrow {}^{241}Pu + \gamma$	250	
(9)	$^{241}Pu + n \rightarrow {}^{242}Pu + \gamma$	390	
(10)	$^{241}Pu + n \rightarrow \text{Fission} + 3.06\,n$	1025	
(11)	$^{242}Pu + n \rightarrow {}^{243}Pu + \gamma$	19	
(12)	$^{243}Pu \rightarrow {}^{243}Am + \beta^-$		4.98 h
(13)	$^{232}Th + n \rightarrow {}^{233}TH + \gamma$	7.3	
(14)	$^{233}TH \rightarrow {}^{233}Pa + \beta^-$		23.3 m
(15)	$^{233}Pa \rightarrow {}^{233}U + \beta^-$		27.4 d
(16)	$^{233}U + n \rightarrow {}^{234}U + \gamma$	52	
(17)	$^{233}U + n \rightarrow \text{Fission} + 2.51\,n$	527	
(18)	$^{234}U + n \rightarrow {}^{235}U + \gamma$	90	

*Accepted values for monoenergetic thermal neutrons at 2200 m/s (0.00405 aJ or 0.0253 eV); 1 barn = 10^{-28} m^2.

Similarly, in a mixture of ^{239}Pu and ^{238}U the conversion ratio (breeding ratio) is given by the relative probability of reaction (3) as compared to reactions (6) and (7). In this case, however, the higher isotopes of plutonium that are formed have a long half-life and start to absorb neutrons as their concentration builds up by means of reactions (6), (8), and (9). The ^{243}Pu formed by reaction (11) decays rapidly to americium, as shown, to end the chain effectively. Thus, after long exposure to thermal neutrons in a reactor, a mixture of ^{235}U and ^{238}U will contain appreciable concentrations of ^{236}U, ^{239}Pu, ^{240}Pu, ^{241}Pu, and ^{242}Pu, all of which must be taken into consideration in determining the overall conversion ratio.

Reactions (1), (2), and (13)–(18) represent the reactions taking place in a mixture of ^{235}U and thorium. In this fuel cycle secondary isotopes of importance to the conversion ratio are ^{233}U, ^{234}U, ^{236}U, and ^{233}Pa. For most efficient neutron utilization (capture in thorium), it is important to minimize losses due to neutron absorption in ^{233}Pa by keeping the average neutron flux as low as possible.

Maximizing conversion ratio. When it is desired to maximize the neutron-conversion ratio in a reactor, neutron losses are held to a minimum by suitable selection of the materials making up the reactor system, their arrangement in the reactor, and its operating conditions. For example, neutron leakage is reduced if the reactor is made large; fission-product poisons (neutron absorbers) can be lowered by frequent processing of fuel; and nonfission neutron capture by fuel can be minimized by designing the reactor so that the average energy of the neutrons is optimum for causing fission. However, the extent to which these methods of improving neutron utilization can be applied is limited by economic considerations. Thus for any given nuclear power application, there is an optimum reactor size and configuration and an optimum fuel-processing cycle. *See* NUCLEAR FUELS REPROCESSING.

The control of neutron losses due to parasitic capture in fuel by varying the relative amounts of neutron-scattering material (moderator) and fuel to give the proper neutron energy is the most important factor in achieving a high conversion ratio. The effect of neutron energy on α (the ratio of neutrons lost by parasitic capture in fuel to those leading to fission) and on η (the number of neutrons emitted per neutron absorbed in fuel), as seen in the equation below, are shown in Table 2.

$$\eta = \frac{\nu}{1+\alpha}$$

Table 2 indicates that the theoretical maximum conversion ratio (given by $\eta - 1$) is above 1.0 for all three fissionable materials as long as the average energy of the neutrons causing fission is either very high (~ 160 fJ or 1 MeV) or very low (~ 0.004 aJ or 0.025 eV). In a practical reactor design, however, both these neutron energy conditions are difficult to achieve, because for any given mixture of fuel and moderator there will exist neutrons moving at all energies, ranging from those for fission neutrons (fast or high-energy neutrons) down to those moving at approximately the same velocities as the moderator atoms. Even in a highly thermalized reactor (high ratio of moderator to fuel),

Table 2. Capture-to-fission ratio (α) and neutron yield (η) as function of energy

Neutron energy, aJ (eV)	^{233}U		^{235}U		^{239}Pu	
	α	η	α	η	α	η
0.004 (0.025)	0.102	2.28	0.190	2.07	0.380	2.09
0.016 (0.10)	0.08	2.33	0.17	2.11	0.59	1.81
0.048 (0.30)	0.15	2.19	0.25	1.97	0.70	1.70
16 (10^2)			0.52	1.62	0.72	1.67
1.6×10^4 (10^5)			0.18	2.09	0.60	1.80
1.6×10^5 (10^6)	0.03	2.44	0.08	2.28	0.10	2.62

the neutron energy will vary considerably from the mean that is established by the moderator temperature. Because of this, the conversion ratio is affected by the moderator temperature. This is especially true in the case of the ^{235}U, ^{238}U, ^{239}Pu fuel cycle, as shown in Table 3.

In nuclear power reactors that operate with high moderator and coolant temperatures to achieve high thermal efficiencies, it is difficult to get a high conversion ratio because of the effect just described. One solution to this problem is to insulate the moderator thermally from the coolant and to maintain the moderator at a lower temperature. Such a technique cannot be applied in graphite-moderated reactors because of the necessity for keeping the graphite hot to minimize its expansion due to radiation damage and to minimize the build-up of stored energy. *See* RADIATION DAMAGE TO MATERIALS.

Fast reactors. By eliminating the moderator, it is possible to raise the average neutron energy in a reactor to a value close to that of the fission neutrons (320 fJ or 2.0 MeV, average). Coolants, fertile material, and structural material in the core, however, tend to degrade the energy so that the average is normally 96–32 fJ or 0.6–0.2 MeV. Under these conditions, the ratio of parasitic fuel captures to fuel fissions varies from 0.12 to 0.25. Corresponding breeding ratios range from 1.96 to 1.40 for ^{239}Pu-fueled fast (unmoderated) reactors and from 1.34 to 1.08 for ^{235}U-fueled reactors. In all cases the neutron yield from fissions in fuel is increased by fast-neutron fissions in fertile material (^{238}U or ^{232}Th) resulting in a higher breeding ratio than that given simply by $\eta-1$.

It is evident from the foregoing that considerably higher conversion or breeding ratios are possible in a ^{235}U- or ^{239}Pu-fueled fast reactor than in a thermal reactor. Fast reactors, therefore, provide a means of utilizing a far greater proportion of natural uranium than would be otherwise possible.

In the case of thorium utilization by means of the ^{233}U-thorium cycle, breeding is possible with both fast and thermal neutrons. Here, the difference in breeding ratio between thermal and fast reactors is not as great as for the ^{235}U-plutonium cycle, and the choice depends upon other considerations, such as the amount of fissionable material required for criticality in each case.

In addition to achieving a high conversion ratio in a nuclear power reactor, it is also desirable to have a high thermal efficiency and high material economy (heat output per unit weight of fuel and fertile material). Unfortunately, in most cases these three characteristics cannot be maximized simultaneously. For example, in a boiling water reactor, which generates steam inside the reactor core for power production, an increase in the rate of steam generation increases the neutron losses and decreases the neutron economy. Therefore, the optimum design of a boiler reactor and most other reactor types involves a compromise between high power density and high neutron economy.

Fuel requirements and supply. To estimate future fuel requirements for the United States nuclear industry, it is necessary to predict the industry's growth rate, the probable types of nuclear reactors, and the amount of uranium or thorium needed for each type of reactor. Because of uncertainties in these predictions, it is obvious that the resulting estimate of future nuclear fuel requirements must be very approximate. Nevertheless, it is important to make an approximate estimate and to compare it with estimated resources of low-cost uranium to evaluate whether such resources are sufficient to meet long-term needs. Some facts concerning the present nuclear industry help in such an evaluation. It is now clear that up to some time early in the 21st century the industry will consist primarily of light-water reactors, including both pressurized water and boiling water types. In earlier analyses the optimum fueling characteristics of each type of reactor have been calculated on the basis of anticipated fuel cycle economic conditions, including the costs of fresh uranium and recycled plutonium. However, a national policy decision to delay indefinitely the recycle of pluto-

Table 3. Effect of moderator temperature on the nuclear properties of ^{235}U and ^{239}Pu

Average moderator temperature, °C	Average neutron energy (kT), aJ (eV)	Fast neutrons produced per thermal neutron absorbed in:	
		^{235}U	^{239}Pu
75	0.0048 (0.030)	2.083	2.006
200	0.0066 (0.041)	2.094	1.936
350	0.0086 (0.054)	2.102	1.875
600	0.0120 (0.075)	2.103	1.871

Table 4. Status of United States nuclear power plants at end of 1978*

Category	Units	Electrical power, MW
Operating license	72	52,396
Construction permits	92	101,148
Limited work authorization	4	4,112
Reactors on order	30	35,082
Total	198	192,738

From E. Gordon, *Uranium 1978*, Atomic Industrial Forum, Inc., 1979.

nium has established a guideline for the industry to focus on optimization of a once-through slightly enriched uranium fuel cycle. Implicit in this policy is the deferral of fast breeder reactors which operate most efficiently with plutonium as fuel.

Estimates of demand. An estimate of future uranium demand can be made on the basis of the number of nuclear power plants already on the line and those forecast to be on the line in the future. In 1978 there were 72 operating nuclear plants in the United States which accounted for 12.5% of the nation's total electrical output. The status of United States nuclear power plants as of the end of 1978 is shown in Table 4. Since the late 1970s the number of domestic nuclear plant order cancellations have exceeded the number of new orders, and it is not clear when the nation's electric utilities will again initiate orders for new plants.

Nevertheless, the U.S. Department of Energy through its Energy Information Administration routinely prepares forecasts of nuclear power usage. These forecasts are based on projections of annual growth in gross national product, total energy requirements, the share of total energy which is electricity, and the share of electricity which is generated by nuclear power. The forecasts are made in each area for low-, mid-, and high-growth cases and result in a projected range of nuclear plant capacities and concomitant uranium ore requirements. As of the end of 1978, the Department of Energy growth forecasts for domestic nuclear power and uranium ore requirements are those shown in Table 5 for the mid-growth-rate scenario. The U_3O_8 demand shown is based on several assumptions, including a lifetime average plant capacity factor of 66.6%, reactor lifetime of 30 years, no plutonium recycle, and current light-water reactor technology for fuel utilization efficiency. Based on these parameters, a nuclear power plant generating 1000 MW of electric power requires about 5600 metric tons of U_3O_8 over its lifetime.

Estimates of resources. Estimates of United States domestic uranium resources are shown in Table 6. The estimated uranium available is categorized on the basis of the evidence for its existence, ranging from proved "reserves" to "speculative" as indicated in the table. The amount available in each category is shown as a function of cost, which is primarily a reflection of the quality of the ore and the difficulty involved in its recovery. While the numbers presented in Table 6 are subject to change with further uranium exploration, these are the best estimates available and may be used to assess the potential for long-range nuclear power development.

Table 5. Forecast of domestic uranium ore requirements for mid-growth-rate scenario*

Year	Cumulative nuclear power capacity, GW of electric power	Cumulative ore required, 10^3 metric tons of U_3O_8
1979	58	34
1980	66	39
1985	111	139
1990	172	294
1995	250	512
2000	325	795

*From U.S. Department of Energy.

Comparison of resources and demand. From the forecast of United States nuclear installations through the year 2000 (325 GW of electric power), the lifetime uranium commitment to plants existing at that time would be about 1,830,000 metric tons of U_3O_8. From Table 6 it may be seen that "reserves" recoverable at $50 per pound are less than half of the requirement. The balance of the uranium needed to satisfy the commitment to plants forecast for the year 2000, while it is believed to exist as indicated, will require major investment in exploration effort and resource development.

Projection of the availability of uranium resources to sustain expansion of a light-water nuclear reactor industry into the next century is thus highly uncertain. Therefore, the long-range potential for nuclear power appears to depend on either discovery of much larger uranium resources than those now estimated or on much more efficient utilization of known resources than is permitted by current light-water reactor technology.

In the long range, fast breeder reactors, which produce more fissile fuel than they consume, appear to offer a technical solution to the problem posed by limited uranium resources. Fast breeder technology is being developed in the United States and abroad, but commercialization of this technology is not being pursued in the United States. Establishing a fast breeder reactor industry in the United States would require a national policy decision which must consider economics, safety, environmental acceptability, ability to safeguard nuclear-weapons-grade fissile material, and availability of alternative energy sources. In the ab-

Table 6. Estimated United States uranium resources as of Jan. 1, 1979[a]

Cost category, $/lb U_3O_8[b]	10^3 metric tons U_3O_8			
	Reserves[c]	Probable[d]	Possible[e]	Speculative[f]
15	—	378	192	69
15–30	—	536	422	204
30	630	915	614	273
30–50	206	408	395	227
50	836	1368	1064	500

[a]From U.S. Department of Energy.
[b]Does not represent market price. 1 lb = 0.4536 kg.
[c]Proved resources.
[d]Estimated to occur in known productive uranium districts.
[e]Estimated to occur in undiscovered or partly defined deposits in geologic settings productive elsewhere within the same geologic province.
[f]Estimated to occur in undiscovered or partly defined deposits in geologic settings not previously productive within a productive geologic province not previously productive.

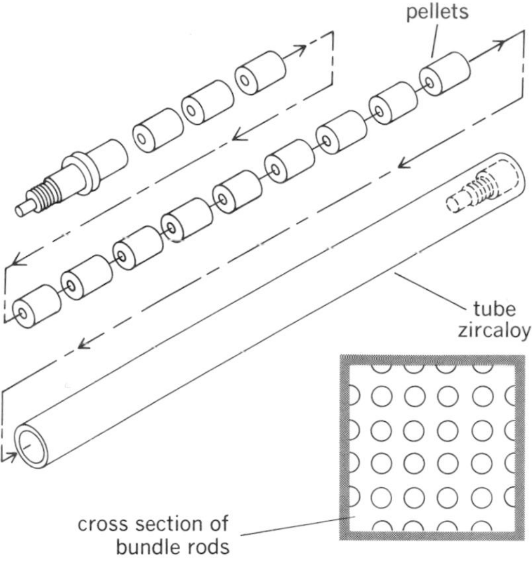

Fig. 1. Cylindrical pellets of UO_2 are pressed to exacting specifications for size and weight. After finishing, pellets are inserted into stainless steel or Zircaloy tubes. Tubes are sealed and welded, then assembled into bundles to form the rod-type element. *(From Nuclear Fuel Elements, General Electric Co.).*

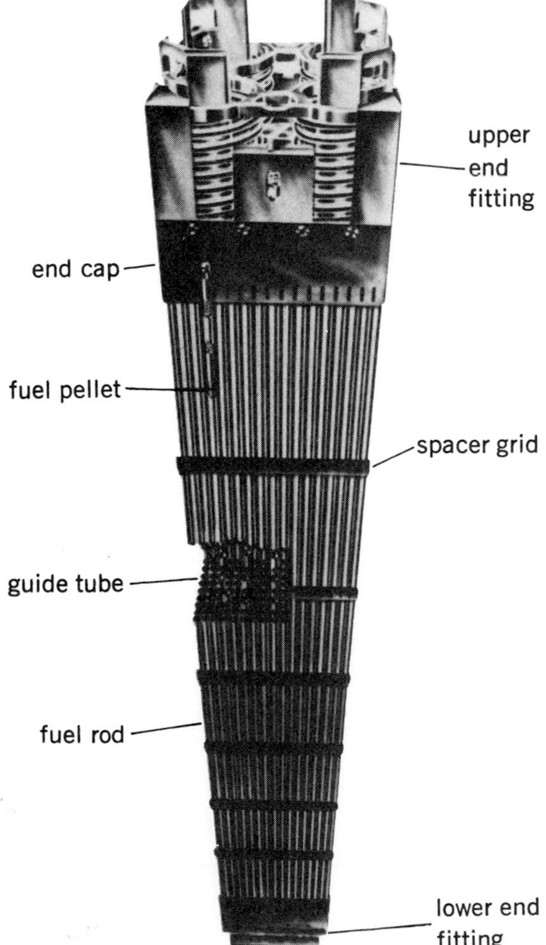

Fig. 2. Pressurized water reactor fuel assembly showing fuel rod and fuel pellet perspective. Total assembly length is about 4 m. *(Babcock & Wilcox Co.)*

sence of commercial fast breeders, the large-scale utilization of nuclear power may be only an interim energy source serving until other renewable energy sources can be developed. *See the feature articles* EXPLORING ENERGY CHOICES; PROTECTING THE ENVIRONMENT; THE RISK OF ENERGY PRODUCTION.

Preparation of uranium fuel. Starting with ore, six major steps are required in the preparation of enriched uranium fuel: (1) recovery of uranium from ore (concentration), (2) purification of crude concentrate, (3) conversion of oxide to UF_6, (4) isotopic enrichment, (5) reduction of enriched UF_6 to UO_2 or metal, and (6) fabrication of the fuel element.

Concentration. Because of the variety of natural sources of uranium, no one concentration method is uniquely suited to all ores. Concentration by gravity methods, for example, is applicable for pitchblende but not for carnotite or autunite, from which uranium is extracted almost exculsively by leaching with acid or alkali carbonate. This is followed by a precipitation process (or by ion exchange or solvent extraction) to recover the uranium from the leach solutions.

Purification. To make uranium suitable for use in a nuclear reactor, it is desirable to reduce the concentration of neutron-absorbing impurities such as boron, cadmium, and the rare earths to levels of 0.1–10 parts per million. This is accomplished either by selective extraction of uranyl nitrate from aqueous solutions by certain oxygenated organic solvents, notably diethyl ether, methyl isobutyl ketone, or tributyl phosphate in kerosine, or by quantitative precipitation or uranium peroxide, $UO_4 \cdot 2H_2O$, from weakly acid solutions of uranyl salts.

Conversion. Conversion of the purified uranyl nitrate or UO_4 to UF_6 is carried out by first calcining the salt to produce UO_3. This is then reduced to UO_2, which is treated with HF to produce green salt, UF_4. The UF_6, which is a gas at temperatures above 56°C, is produced from UF_4 by reaction with fluorine.

Isotope enrichment. Separation of the uranium isotopes, ^{235}U and ^{238}U, depends upon the physical differences arising from the difference in their atomic weights. Gaseous diffusion using UF_6 is the process now employed to enrich the product to its specified ^{235}U content. A centrifuge process is being developed for future additional enrichment capacity because it is much more energy-efficient than gaseous diffusion. *See* ISOTOPE (STABLE) SEPARATION.

UF_6 reduction. The UF_6 product from the diffusion plant must be reduced to uranium oxide or uranium metal for incorporation into fuel elements. To produce UO_2, which is the fuel used in commercial power reactors, the UF_6 is hydrolyzed to uranyl fluoride, UO_2F_2, reacted with ammonia, NH_3, to produce ammonium diuranate, $(NH_4)_2U_2O_7$, and calcined in hydrogen. Uranium metal, which is used in alloy form as fuel only for test reactors and plutonium production reactors, is obtained by reduction with calcium or magnesium metal.

Fuel element fabrication. For power reactor application, fuel elements consist of UO_2 pellets contained in Zircaloy or stainless steel tubes which are grouped into fuel bundles or assemblies (Figs.

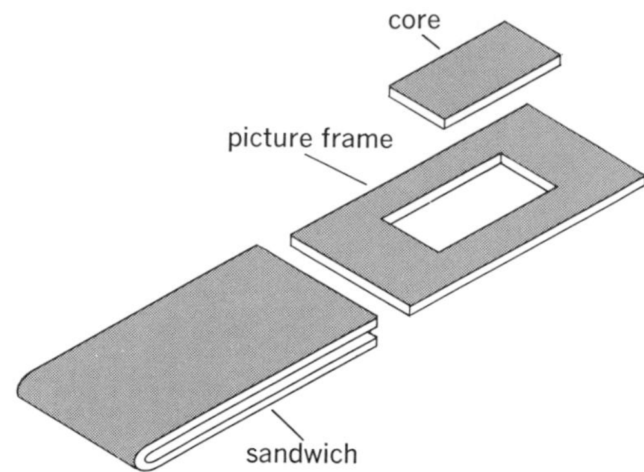

Fig. 3. A fuel plate is assembled with the core, or uranium alloy piece, fitting into a picture frame of aluminum plate. Aluminum plate is placed on either side, and the entire sandwich is hot-rolled to effect bonding. After centering the core by x-ray, the plates are trimmed to size, assembled, and mechanically bonded to the side plates. *(From Nuclear Fuel Elements, General Electric Co.)*

1 and 2). Current light-water reactors use Zircaloy cladding almost exclusively because of its lower neutron absorption relative to stainless steel. Fabrication beginning with the calcined UO_2 described above involves milling and blending, powder granulation, compaction into pellets, sintering to high density, and centerless grinding to obtain the precise pellet diameter needed to maintain close contact with the metal tube for good heat transfer. The UO_2 pellets are inspected, dried, and loaded into the Zircaloy tubes, which are filled with helium and welded shut. The tubes must maintain leak tightness during service to prevent release of radioactive fission products to the reactor coolant circuit. The fuel rods are then inspected and assembled into fuel bundles, which are loaded into the reactor. For test reactor application, fuel elements usually consist of U_3O_8 or uranium alloy metal sandwiched in aluminum plates (Fig. 3). For naval propulsion reactors, the fuel elements are stainless-steel-clad UO_2 dispersed in stainless steel. *See* NUCLEAR FISSION; NUCLEAR FUEL CYCLE; NUCLEAR POWER; PLUTONIUM; THERMONUCLEAR REACTION; THORIUM; URANIUM.

[R. L. BEATTY]

Bibliography: M. Benedict and T. H. Pigford, *Nuclear Chemical Engineering*, 1957; F. R. Bruce, J. M. Fletcher, and H. H. Hyman (eds.), *Process Chemistry*, in *Progress in Nuclear Energy*, ser. 3, vol. 1, 1958; H. Etherington (ed.), *Nuclear Engineering Handbook*, 1958; S. Glasstone, *Sourcebook on Atomic Energy*, 3d ed., 1967; E. Gordon, *Uranium 1978*, Atomic Industrial Forum, Inc., 1979; C. R. Tipton, *Reactor Handbook*, vol. 1: *Materials*, 1960; W. D. Wilkinson and W. F. Murphy, *Nuclear Reactor Metallurgy*, 1958.

Nuclear fuel cycle

The steps by which fissionable (for example ^{233}U, ^{235}U, ^{239}Pu) and fertile (for example, ^{238}U, ^{232}Th) materials are prepared for use in, and recycled or discarded after discharge from, the nuclear reactor. These steps include mining of uranium- or thorium-bearing ore and milling of the ore to form concentrates. The uranium concentrate is converted to the volatile uranium hexafluoride (UF_6) that is used in the separation of isotopes to produce uranium enriched in the fissile ^{235}U. Another part of the fuel cycle is the fabrication of the enriched uranium into fuel assemblies. After the fuel has liberated the desired amount of heat in the reactor, the spent assemblies are reprocessed to separate the remaining fissionable and fertile material (uranium and plutonium) from the nuclear wastes. Other steps in the fuel cycle include the various transportation operations that move materials from one step to another, often connecting plants many hundreds of miles apart. Finally, waste management includes the treatment, storage, and disposal of radioactive wastes from the many other parts of the fuel cycle. The fuel cycle for a thorium-based reactor requires, in addition to the uranium fuel cycle steps outlined above, the mining, milling, and purification of thorium and the reprocessing of thorium-containing fuel into its components, which include unused thorium, uranium in the form of fissionable ^{233}U, and nuclear waste. *See* THORIUM; URANIUM.

Mining and purification. Even though uranium is relatively abundant in the Earth's crust, it is found only at low concentrations in most ores. Concentrations of uranium in typical, high-grade ores in the United States range about 0.2%, and hence the ore, obtained by either open-pit or underground mining, must be treated to concentrate the uranium. For some types of ores, physical processes such as grinding, washing, flotation, and gravity settling will result in separation of the gangue (for example, sandstone, clay, limestone) from uranium-bearing ore. These physical processes are useful only in concentrating the ore to less than 50% uranium content and can be carried out at the mine.

The concentrated ore is treated at a mill to further concentrate and separate the uranium from metallic impurities and the rest of the gangue material. The two chemical methods commonly used are the carbonate leach method and the acid leach method. The choice of method is largely determined by the type of ore to be treated. The product of the purification process is a precipitate, most commonly of ammonium diuranate, that is dried to a powdered oxide containing small amounts of impurities. This material is shipped to conversion facilities, where it may be dissolved and the uranium purified further by solvent extraction methods. The product of this purification is a solution of uranium that is dried and calcined to form U_3O_8.

Conversion to hexafluoride. Conversion of purified uranium to the hexafluoride is done by a multistep process in which uranium oxides are fluorinated with hydrogen fluoride and with fluorine. Purified U_3O_8 is reduced to UO_2 and converted to UF_4 by gaseous HF in a fluidized-bed reaction vessel. Formation of UF_6 requires elemental fluorine and is also carried out in a fluidized bed to simplify removal of heat from the exothermic reaction. Variations exist in the details of the conversion process, including elimination of the initial purification of the uranium concentrate, inclusion of fractional distillation of UF_6 to obtain a pure product, and the use of various types of reaction vessels.

Isotopic enrichment. Natural uranium contains the fissile isotope ^{235}U at a concentration (0.72%) too low to be useful in reactors which use ordinary water as moderator and coolant, that is, light water reactors. The UF_6 is used in the process of separating isotopes by diffusion that increases the ^{235}U content to about 3%, at which concentration it can be conveniently used in reactors. For other purposes such as for fuel used in certain small reactors (for example, research reactors), enrichment of ^{235}U to more than 90% is carried out by the same proccss. *See* ISOTOPE (STABLE) SEPARATION.

Fuel fabrication. Fuel for large nuclear reactors producing electric power is fabricated by conversion of the slightly (that is, about 3%) enriched UF_6 to UO_2. Hydrolysis of UF_6 and precipitation of compounds such as $(NH_4)_2U_2O_7$ is followed by calcination to U_3O_8 and reduction with hydrogen to powdered UO_2. The dioxide, selected because of its chemical stability, is compacted into pellets that are sintered at high temperatures. The fuel assembly is made of an array of sealed tubes of a zirconium alloy (Zircaloy) containing the fuel pellets, of end plates, and of other hardware. Many such assemblies are charged into the core of a reactor. *See* NUCLEAR POWER; NUCLEAR REACTOR.

Reprocessing. Following the discharge of fuel assemblies that no longer contribute efficiently to the generation of heat in the reactor, the spent fuel is allowed to dissipate nuclear radiation and the heat generated by it (that is, to "cool") while submerged in water for some time. The spent fuel is processed to recover the residual ^{235}U and the plutonium that was formed while the fuel was in the reactor, and to separate the radioactive fission product wastes for storage. The processing steps for typical power reactor fuel include disassembly; chopping of the Zircaloy tubes to expose the fuel; chemical dissolutions of the uranium, plutonium, and fission products; and separation and purification of the uranium and plutonium by solvent extraction methods. The waste from this step contains more than 99% of the radioactive fission products and must be handled carefully to ensure confinement and isolation from the biosphere.

The uranium product from the processing plant is converted to UF_6 by procedures similar to those used to make UF_6 from natural uranium, except that the slight enrichment of the spent uranium requires attention to nuclear criticality safety. The product UF_6 is reenriched in the isotope separation plants to about 3% concentration of ^{235}U for recycle as fuel. The plutonium product can be combined with natural uranium to make a "mixed oxide" fuel (that is, PuO_2-UO_2) containing about 5% plutonium that can be used in power reactors otherwise using slightly enriched uranium. Alternate use for plutonium includes fuel for fast breeder reactors. See NUCLEAR FUELS REPROCESSING; PLUTONIUM.

Waste management. Most of the radioactive wastes from reprocessing plants are in the form of aqueous solutions containing high levels of radioactive materials. These solutions can be temporarily stored in cooled tanks. Following interim storage of the solutions, the wastes are solidified by evaporation of water and acid and are converted to stable oxides. In order to decrease the likelihood of leaching of radioactive wastes by water, the solidified wastes can be incorporated into an inert matrix such as glass and can be held in thick metallic containers.

Other wastes include discarded equipment, trash, filters, and miscellaneous materials, all of which may be contaminated to varying degrees. These wastes are treated to reduce their volume and to limit the dispersibility of the radioactive contamination. Waste from the milling operation, called tailings, is treated and stored near the mill. Final disposal of some of the radioactive waste will take advantage of the stability of selected geologic formations to ensure the necessary and prolonged isolation of the waste. *See* NUCLEAR FUEL: RADIOACTIVE WASTE MANAGEMENT.

[MARTIN J. STEINDLER]

Bibliography: D. M. Elliott and L. E. Weaver (eds.), *Education and Research in the Nuclear Fuel Cycle*, 1972; M. Etherington (ed.), *Nuclear Engineering Handbook*, 1958; J. T. Long, *Engineering for Nuclear Fuel Reprocessing*, 1967.

Nuclear fuels reprocessing

The treatment of spent reactor fuel elements to recover fissionable and fertile material. Spent fuel is usually discharged from reactors because of chemical, physical, and nuclear changes that make the fuel no longer efficient for the production of heat, rather than because of the complete depletion of fissionable material. Therefore, discharged fuel usually contains fissionable material in sufficient amounts to make its recovery attractive. In the case of breeder reactors, in order to take advantage of the characteristic of the breeder reactor to produce more fissionable material than is used, reprocessing of fuel must be done to recover the fissile material bred into part of the fuel. If fertile material is also contained in the fuel, it is ordinarily recovered and purified during fuel reprocessing. Purification of the valuable constituents consists of the removal of fission products and extraneous structural material present in the fuel. *See* NUCLEAR FUEL; NUCLEAR REACTOR.

Because of the frequency of fuel discharge and because of the high value of fissionable materials, it is important that the degree of recovery approach 100% as closely as practicable. It is often necessary to reduce the fission product impurity content of discharged fuel by a factor of 10^6 to 10^7 in order to make the recovered material safe to handle during refabrication into new fuel for reuse.

There are several basic steps involved in fuel reprocessing. After fuel has been discharged from a nuclear reactor, it is common practice to store the fuel submerged in 15–20 ft (4.6–6.1 m) of water (for cooling and radiation-shielding purposes) for a period of 50–200 days; this allows the short-lived fission products to decay radioactively. During this period the radioactivity of the fuel decreases rapidly and substantially, so that when reprocessing is commenced, shielding requirements are reduced to practical thicknesses, heat evolution of the spent fuel assemblies is reduced to more easily managed levels, and radiation damage to chemicals or special structural materials in the reprocessing plant can be held to tolerable magnitudes. Following the cooling period, the fuel is mechanically cut or disassembled into convenient sizes. At this point the fuel is ready for chemical treatment to enable recovery and purification of the desired materials.

The specific steps next undertaken depend upon

the particular reprocessing method employed to separate the desired products from each other, from fission products, and from extraneous structural materials. Although many separation methods exist, the one based upon solvent extraction principles is most frequently used for fuel reprocessing. Therefore, the discussion of the sequence of steps for recovery and purification will be based on the use of the solvent extraction method. Further details of this process are given below.

Dissolution of spent fuel. The fuel assembly has been cut into pieces so that the fuel, normally held in metal tubes called cladding, is exposed. Fuel from reactors other than those using UO_2 contained in Zircaloy tubes is disassembled to allow access of chemical reagents to the fuel or fuel assembly components. The cut-up or disassembled fuel is charged, along with an appropriate aqueous dissolution medium, generally nitric acid, into a vessel. Here the solid fuel is dissolved. Except for a few fission products which are volatilized during dissolution, all the constituents initially in the fuel are retained in the dissolver solution as soluble salts or, in the case of minor constituents, as an insoluble sludge. Generally, the resulting solution must be treated by various means to accommodate its use as a feed solution to the solvent extraction process. Important reasons for such pretreatment are the adjustments of oxidation states, removal of sludge, and the adjustment of concentrations of solution constituents for optimum recovery and purification performance in the solvent extraction process.

Solvent extraction. In the solvent extraction steps, purification of the desired constituents (that is, uranium, plutonium and, if present, thorium) is achieved by the selective transfer of these constituents into an organic solution that is immiscible with the original aqueous solution. Fission products and other impurities remain in the aqueous solution and are discarded as waste. Recovery of the desired constituents is achieved by adjusting conditions to retransfer them into a clean aqueous solution. The most commonly used organic solution contains the extractant tributyl phosphate (TBP) dissolved in purified kerosine or similar materials such as long-chain hydrocarbons (for example, *n*-dodecane). The separation process using TBP with nitric acid as the aqueous medium and salting agent is called the Purex process. This is the process in general use for power reactor fuels. The basic requirements for separation during solvent extraction are immiscibility of the organic solvent with the aqueous solution of irradiated fuel, and appreciable differences with which components, initially present in the aqueous fuel solution, distribute or partition themselves between the organic and the aqueous solutions when the organic solution is first thoroughly mixed with the aqueous solution and is later separated from it.

If a quantity of suitably prepared solution of irradiated fuels is well mixed with a similar quantity of TBP-kerosine solution and is allowed to stand, the following results. The TBP-kerosine solution floats, essentially quantitatively, on the aqueous solution because the organic solution is not miscible with, and is less dense than, the aqueous solution. Analyses of the separated liquids show that a large fraction of the uranium and plutonium (and thorium if it is also present) transfers to the organic solution, but only a minute fraction of the fission products and other impurities transfers.

To enhance the transfer of uranium and plutonium into the organic solution without appreciably influencing the transfer of fission products, it is customary to have large concentrations of certain chemicals called salting agents in the aqueous solution. However, if the organic solvent containing the uranium and plutonium is now brought into contact with a clean aqueous solution wherein salting agents are absent, the uranium and plutonium will retransfer almost quantitatively to the new aqueous solution. The solvent can then be reused. Because only a minute fraction of the fission products was initially transferred to the organic mixture, the new aqueous solution contains recovered uranium and plutonium well separated from fission products. It is also possible to separate the uranium and plutonium from each other. The same organic solvent can be used, taking advantage of the fact that under certain conditions (reduced oxidation state of plutonium) plutonium extraction by the solvent is very small. Thus separation of the two heavy elements is achieved in much the same way that the impurities (for example, fission products) were initially removed from uranium and plutonium.

Separation and purification of uranium and plutonium can be done by repeated contact of the fuel solution with quantities of fresh solvent. This is the batch extraction mode and, while useful in the laboratory, it is seldom employed in large-scale practice. Operation of extraction equipment in the continuous countercurrent mode is normally more convenient, efficient, and economic for the purification of large amounts of fuel. The basic principles of separation are the same in the continuous countercurrent mode as in the batch extraction mode. The continuous operation provides repeated mixing and separation of the organic solvent and the aqueous solution of irradiated fuel from which it is desired to remove all of the valuable products freed of impurities. The continuous nature of operation is also applied in the step wherein the purified products are retransferred to a solution free of salting agents. Continuous separation of plutonium and uranium (or thorium and uranium) can also be performed. The principal advantages of continuous overbatch operations are more uniform product quality, greater ease of process control, and economy.

The countercurrent aspects of the operation are derived from having the organic solvent flow in equipment in a direction opposite to that of the aqueous solution. This allows maximum loading of the organic solution with the components to be extracted because fresh solvent encounters initially low concentrations of these components. Progressively higher concentrations occur as the solvent moves toward the point at which the aqueous solution is introduced. The aqueous solution moves counter to the flow of organic extractant and is thereby depleted to a high degree of the valuable products (for example, uranium and plutonium). Similarly, reextraction (stripping) of the products from the organic phase into a clean aqueous phase, when done in a countercurrent mode, reduces product losses to the organic solvent. Thus, a minimum of solvent is needed for maximum recovery of desired materials with solvent extraction equipment of a given efficiency.

With proper process conditions and suitable equipment, the solvent extraction operation yields nearly complete recovery and purification. Products are usually obtained in the form of dilute aqueous solutions. These solutions are subsequently further processed to give the form of plutonium, uranium, or thorium which is suitable for reuse in nuclear reactors. In the case of uranium which has been depleted in its ^{235}U content, the processing may include ^{235}U isotope reenrichment in gaseous diffusion plants. The fission products and other impurities initially present in the fuel are waste and are also obtained in the form of aqueous solutions. The waste is concentrated by evaporation and then may be introduced into underground tanks for interim storage. *See* RADIOACTIVE WASTE MANAGEMENT.

Processing plants. Plants for fuel reprocessing are large and expensive. They can be a few hundred yards in length, are normally built above and below ground level, and may cost hundreds of millions of dollars. Modern reprocessing plants are usually integrated into facilities that also provide other fuel cycle services such as conversion of uranium to UF_6, solidification of solutions of high-level waste, and conversion of plutonium product solutions to a solid suitable for shipment.

There are many factors that contribute to the high cost of these plants. Some of the more important factors are the large amounts of massive shielding (up to 7 ft or 2.1 m of high-density concrete) used to separate the process equipment from normally occupied work areas; the stringent design criteria for resistance of critical parts of the structure and equipment against natural forces such as earthquakes and tornadoes; the high integrity and rigid manufacturing standards of process equipment; the extensive systems to confine particulate radioactivity; and the barriers interposed between stored wastes and the environment.

Operation of the plants, including sampling for process control, is conducted by remote means. In some plants, even repairs and modifications of equipment in high-radiation zones are made by remote techniques. The additional cost of this type of maintenance is large. For those plants in which maintenance is performed by direct methods, the initial capital cost is reduced. This saving may be offset to some extent by increased operating costs when decontamination is difficult and permissible working time of maintenance personnel is limited. Because of the difficulty and cost of maintenance by either remote or direct methods, more spare equipment and higher standards of design, construction, and installation are necessary in fuel-reprocessing plants than in conventional chemical plants. Special precautions, which also contribute to increased capital and operating costs, must be taken in fuel-reprocessing plants to avoid nuclear criticality accidents from inadvertent accumulation of fissionable materials. This is particularly important when highly enriched fuels are reprocessed.

Thus the gross capital and operating costs are high for a fuel-reprocessing plant. The unit cost of recovered products is also very large because the output of moderately large plants is relatively small. They process only a few tons of uranium per day, for uranium containing 3% or less of the ^{235}U isotope, or as little as 10–20 lb/day (4.5–9 kg/day) when uranium containing greater than 90% of the ^{235}U isotope (highly enriched uranium) is processed. Unit cost can be substantially reduced, however, by increased capacity, because total capital and operating costs do not increase proportionally.

The operating experience to date with fuel-reprocessing plants has shown them to be relatively safe in spite of hazards from radiation and nuclear criticality, as well as other hazards of a more conventional nature. *See* HEALTH PHYSICS.

[MARTIN J. STEINDLER]

Bibliography: M. Benedict and T. H. Pigford, *Nuclear Chemical Engineering*, 1957; F. R. Bruce, J. M. Fletcher, and H. H. Hyman, *Process Chemistry*, ser. 3, vol. 3, 1961; H. Etherington (ed.), *Nuclear Engineering Handbook*, 1958; J. T. Long, *Engineering for Nuclear Fuel Reprocessing*, 1967; C. E. Stevenson, A. T. Gresky, and E. A. Mason (eds.), *Process Chemistry*, ser. 3, vol. 4, 1970; S. M. Stoller and R. B. Richards (eds.), *Reactor Handbook*, vol. 2: *Fuel Reprocessing*, 2d ed., 1961.

Nuclear fusion

One of the primary nuclear reactions, the name usually designating an energy-releasing rearrangement collision which can occur between various isotopes of low atomic number. *See* NUCLEAR REACTION.

Interest in the nuclear fusion reaction arises from the expectation that it may someday be used to produce useful power, from its role in energy generation in stars, and from its use in the fusion bomb. Since a primary fusion fuel, deuterium, occurs naturally and is therefore obtainable in virtually inexhaustible supply (by separation of heavy hydrogen from water, 1 atom of deuterium occurring per 6000 atoms of hydrogen), solution of the fusion power problem would permanently solve the problem of the present rapid depletion of chemically valuable fossil fuels. As a power source, the lack of radioactive waste products from the fusion reaction is another argument in its favor as opposed to the fission of uranium.

In a nuclear fusion reaction the close collision of two energy-rich nuclei results in a mutual rearrangement of their nucleons (protons and neutrons) to produce two or more reaction products, together with a release of energy. The energy usually appears in the form of kinetic energy of the reaction products, although when energetically allowed, part may be taken up as energy of an excited state of a product nucleus. In contrast to neutron-produced nuclear reactions, colliding nuclei, because they are positively charged, require a substantial initial relative kinetic energy to overcome their mutual electrostatic repulsion so that reaction can occur. This required relative energy increases with the nuclear charge Z, so that reactions between low-Z nuclei are the easiest to produce. The best known of these are the reactions between the heavy isotopes of hydrogen, deuterium and tritium.

Fusion reactions were discovered in the 1920s when low-Z elements were used as targets and bombarded by beams of energetic protons or deuterons. But the nuclear energy released in such bombardments is always microscopic compared with the energy of the impinging beam. This is because most of the energy of the beam particle is dissipated uselessly by ionization and single-particle collisions in the target; only a small fraction of the impinging particles actually produce reactions.

Nuclear fusion reactions can be self-sustaining, however, if they are carried out at a very high temperature. That is to say, if the fusion fuel exists in the form of a very hot ionized gas of stripped nuclei and free electrons termed a plasma, the agitation energy of the nuclei can overcome their mutual repulsion, causing reactions to occur. This is the mechanism of energy generation in the stars and in the fusion bomb. It is also the method envisaged for the controlled generation of fusion energy.

PROPERTIES OF FUSION REACTIONS

The cross sections (effective collisional areas) for many of the simple nuclear fusion reactions have been measured with high precision. It is found that the cross sections generally show broad maxima as a function of energy and have peak values in the general range of 0.01 barn (1 barn = 10^{-24} cm^2) to a maximum value of 5 barns, for the deuterium-tritium (D-T) reaction. The energy releases of these reactions can be readily calculated from the mass differences between the initial and final nuclei or determined by direct measurement.

Simple reactions. Some of the important simple fusion reactions, their reaction products, and their energy releases in millions of electron volts (MeV) are given by Eqs. (1).

$$\begin{aligned}
&\mathrm{D} + \mathrm{D} \rightarrow \mathrm{He}^3 + n + 3.25\ \mathrm{MeV}\\
&\mathrm{D} + \mathrm{D} \rightarrow \mathrm{T} + p + 4.0\ \mathrm{MeV}\\
&\mathrm{T} + \mathrm{D} \rightarrow \mathrm{He}^4 + n + 17.6\ \mathrm{MeV}\\
&\mathrm{He}^3 + \mathrm{D} \rightarrow \mathrm{He}^4 + p + 18.3\ \mathrm{MeV}\\
&\mathrm{Li}^6 + \mathrm{D} \rightarrow 2\mathrm{He}^4 + 22.4\ \mathrm{MeV}\\
&\mathrm{Li}^7 + p \rightarrow 2\mathrm{He}^4 + 17.3\ \mathrm{MeV}
\end{aligned} \qquad (1)$$

If it is remembered that the energy release in the chemical reaction in which hydrogen and oxygen combine to produce a water molecule is about 1 eV per reaction, it will be seen that, gram for gram, fusion fuel releases more than 1,000,000 times as much energy as typical chemical fuels.

The two alternative D-D reactions listed occur with about equal probability for the same relative particle energies. Note that the heavy reaction products, tritium and helium-3, may also react, with the release of a large amount of energy. Thus it is possible to visualize a reaction chain in which six deuterons are converted to two helium-4 nuclei, two protons, and two neutrons, with an overall energy release of 43 MeV—about 10^5 kilowatt-hours (kWh) of energy per gram of deuterium. This energy release is several times that released per gram in the fission of uranium, and several million times that released per gram by the combustion of gasoline.

Cross sections. Figure 1 shows the measured values of cross sections as a function of bombarding energy up to 100 keV for the total D-D reaction (both D-D,n and D-D,p), the D-T reaction, and the D-He^3 reaction. The most striking feature of these curves is their extremely rapid falloff with energy as bombarding energies drop to a few kilovolts. This effect arises from the mutual electrostatic repulsion of the nuclei, which prevents them from approaching closely if their relative energy is small.

The fact that reactions can occur at all at these energies is attributable to the finite range of nuclear interaction forces. In effect, the boundary of the nucleus is not precisely defined by its classical diameter. The role of quantum mechanical effects in nuclear fusion reactions has been treated by G. Gamow and others. It is predicted that the cross sections should obey an exponential law at low energies. This is well borne out in energy regions reasonably far removed from resonances (for example, below about 30 keV for the D-T reaction). Over a wide energy range at low energies, the data for the D-D reaction can be accurately fitted by a Gamow curve, the result for the cross section being given by Eq. (2), where the bombarding energy W is in kilo-electron-volts.

$$\sigma_{\text{D-D}} = \frac{288}{W} e^{-45.8W^{-1/2}} \times 10^{-24}\ \mathrm{cm}^2 \qquad (2)$$

The extreme energy dependence of this expression can be appreciated by the fact that, between 1 and 10 keV, the predicted cross section varies by about 13 powers of 10, that is, from 3×10^{-42} to 1.5×10^{-29} cm^2.

Energy division. The kinematics of the fusion reaction stipulates that the reaction can occur only if two or more reaction products result. This is because both mass energy and momentum balance must be preserved. When there are only two reaction products (which is the case in all of the important reactions), the division of energy between the reaction products is uniquely determined, the lion's share always going to the lighter particle. The energy division (disregarding the initial bom-

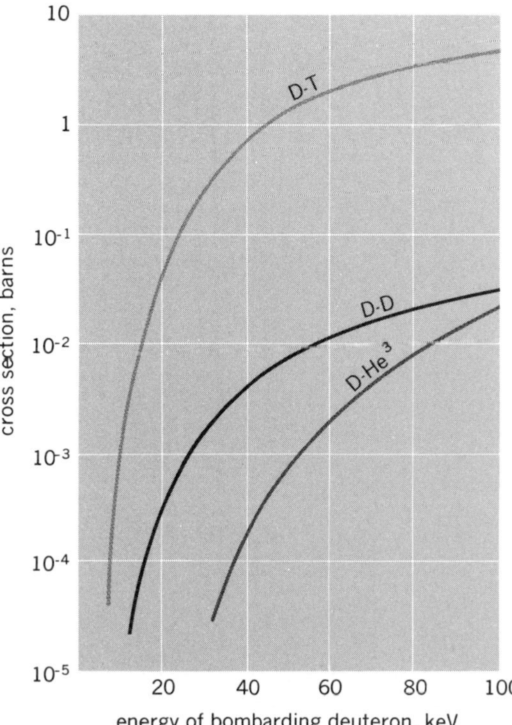

Fig. 1. Cross sections versus bombarding energy for three simple fusion reactions. (*From R. F. Post, Fusion power, Sci. Amer., 197(6):73–84, 1957*)

barding energy) is as in Eq. (3). If Eq. (3) holds,

$$A_1 + A_2 \rightarrow A'_1 + A'_2 + Q \qquad (3)$$

with the As representing the atomic masses of the particles and Q the total energy released, then Eqs. (4) are valid, where $W(A'_1)$ and $W(A'_2)$ are the kinetic energies of the reaction products.

$$W(A'_1) + W(A'_2) = Q$$
$$W(A'_1) = Q\left(\frac{A'_2}{A'_1 + A'_2}\right) \qquad (4)$$
$$W(A'_2) = Q\left(\frac{A'_1}{A'_1 + A'_2}\right)$$

Thus in the D-T reaction, for example, A'_1, the mass of the α-particle, is four times A'_2, the mass of the neutron, so that the neutron carries off four-fifths of the reaction energy, or 14 MeV.

Reaction rates. When nuclear fusion reactions occur in a high-temperature plasma, the reaction rate per unit volume depends on the particle density n of the reacting fuel particles and on an average of their mutual reaction cross sections σ and relative velocity v over the particle velocity distributions. *See* THERMONUCLEAR REACTION.

For dissimilar reacting nuclei (such as D and T), the reaction rate is given by Eq. (5).

$$R_{12} = n_1 n_2 \langle \sigma v \rangle_{12} \quad \text{reactions/(cm}^3\text{)(s)} \qquad (5)$$

For similar reacting nuclei (for example, D and D), the reaction rate is given by Eq. (6).

$$R_{11} = \tfrac{1}{2} n^2 \langle \sigma v \rangle \qquad (6)$$

Note that both expressions vary as the square of the total particle density (for a given fuel composition).

If the particle velocity distributions are known, $\langle \sigma v \rangle$ can be determined as a function of energy by numerical integration, using the known reaction cross sections. It is customary to assume a maxwellian particle velocity distribution, toward which all others tend in equilibrium. The values of $\langle \sigma v \rangle$ for the D-D and D-T reactions are shown in Fig. 2. In this plot the kinetic temperature is given in units of kilo-electron-volts; 1 keV kinetic temperature $= 1.16 \times 10^7$ K. Just as in the case of the cross sections themselves, the most striking feature of these curves is their extremely rapid falloff with temperature at low temperatures. For example, although at 100 keV for all reactions $\langle \sigma v \rangle$ is only weakly dependent on temperature, at 1 keV it varies as $T^{6.3}$ and at 0.1 keV as T^{133}! Also, at the lowest temperatures it can be shown that only the particles in the "tail" of the distribution, which have energies large compared with the average, will make appreciable contributions to the reaction rate, the energy dependence of σ being so extreme.

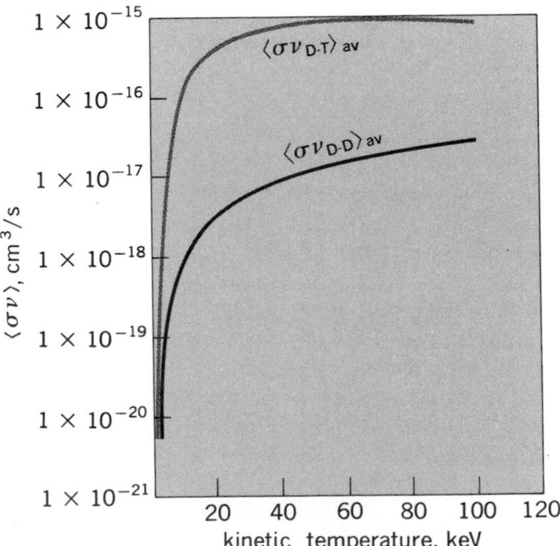

Fig. 2. Plot of the values of $\langle \sigma v \rangle$ versus kinetic temperature for the D-D and D-T reactions.

Critical temperatures. The nuclear fusion reaction can obviously be self-sustaining only if the rate of loss of energy from the reacting fuel is not greater than the rate of energy generation by fusion reactions. The simplest consequence of this fact is that there will exist critical or ideal ignition temperatures below which a reaction could not sustain itself, even under idealized conditions. In a fusion reactor, ideal or minimum critical temperatures are determined by the unavoidable escape of radiation from the plasma. A minimum value for the radiation emitted from any plasma is that emitted by a pure hydrogenic plasma in the form of x-rays or bremsstrahlung. Thus plasmas composed only of isotopes of hydrogen and their one-for-one accompanying electrons might be expected to possess the lowest ideal ignition temperatures. This is indeed the case: It can be shown by comparison of the nuclear energy release rates with the radiation losses that the critical temperature for the D-T reaction is about 4×10^7 K. For the D-D reaction it is about 10 times higher. Since both radiation rate and nuclear power vary with the square of the particle density, these critical temperatures are independent of density over the density ranges of interest. The concept of the critical temperature is a highly idealized one, however, since in any real cases additional losses must be expected to occur which will modify the situation, increasing the required temperature.

FUSION REACTOR

Intense interest in nuclear fusion arises from its promise as a safe and inexhaustible source of energy for the future. Fusion reactors do not yet exist, but studies of the physics and technology that will be needed to construct such reactors have been underway since the 1950s.

The two key problems in achieving net power from a fusion reactor are, first, to heat the fusion fuel charge to its required high temperature, and second, to confine the heated fuel for a long enough time for the fusion energy released to exceed the energy required to heat the fuel to its conbustion temperature, including all relevant losses. *See* LAWSON CRITERION.

The problem of achieving fusion power is in fact dominated by the quantitative requirements associated with the fusion process. The plasma heating technique employed must be capable of raising the fusion fuel charge to kinetic temperatures of order 100,000,000° or higher. The confinement system must be capable of satisfying stringent requirements on confinement time (which could be as long as seconds in some cases). At the same time it must be capable of sustaining the strong outward gas pressure exerted by the fuel charge. Furthermore, since the rate at which fusion power is generated varies as the square of the fuel density, for any continuously operating fusion reactor engineering limits on heat transfer must be taken into account, thus limiting the fuel density for such systems to a small fraction of the particle density of atmospheric air. At higher density it is only possible to conceive of pulsed operation—basically microexplosions. The quantitative requirements of fusion therefore strongly limit the possible approaches to fusion.

Two generically different approaches have emerged as constituting the most promising avenues to the eventual achievement of net fusion power, namely, magnetic confinement and pellet fusion.

Magnetic confinement relies on the fact that at fusion temperatures the fusion fuel charge will be completely ionized, that is, it will exist solely in the plasma state. Charged particles can be held trapped by a properly shaped magnetic field, and are thereby isolated from the reactor chamber walls, physical contact with which would instantly cool the plasma.

Pellet fusion aims at the same objective, but by an entirely different route. Here the idea to rapidly heat and compress a tiny fuel pellet, carrying out the entire operation so quickly that fusion can take place before the pellet flies apart—that is, the confinement is properly called "inertial." The major technical effort on pellet fusion is centered on the use of high-powered lasers to accomplish the heating and compression; substantial activity is also being devoted to pellet fusion induced by bombardment of the pellet with very-high-intensity electron beams; and use of heavy ions as the ignition probe is now receiving serious study.

Magnetic confinement. Magnetic confinement of a fusion plasma depends on the nature of the plasma state. The plasma may be viewed as an electrically conducting gas that exerts an outward pressure, or as a collection of free positive and negative charges. The pressure exerted by the plasma can be resisted by the electromagnetic stresses associated with a strong magnetic field; the individual charged particles can at the same time be guided by a properly shaped magnetic field that forces these particles to execute orbits that remain within the vacuum chamber surrounding the plasma without contact with the walls.

Adequate stability of the confined plasma is a prime requirement for effective magnetic confinement; otherwise, particles can escape prematurely, before having a sufficient probability to fuse. Thus, finding means for suppressing the inherent tendency for confined plasma to become unstable has been one of the central goals of nuclear fusion research since its inception.

There have been many types of magnetic confinement systems proposed since the inception of fusion research. Three generic types appear to have the most promise: the tokamak, the mirror and tandem mirror systems, and field-reversed systems.

Tokamak. The tokamak (Fig. 3) is a closed or toroidal (doughnut-shaped) confinement system. It uses confining fields that represent a combination of a strong toroidal field (that is, field lines directed the long way around the toroid), with a weaker poloidal field (field lines circling the short way around the torus). The toroidal field is generated by external coils that encircle the chamber; the poloidal field is generated by a strong toroidal electric current induced to flow in the plasma by transformer action. The field line pattern that results is helical. In the tokamak, the circulating current not only provides the main confining force through the generation of the poloidal magnetic field but also performs the important function of initial heating (ohmic heating) of the plasma.

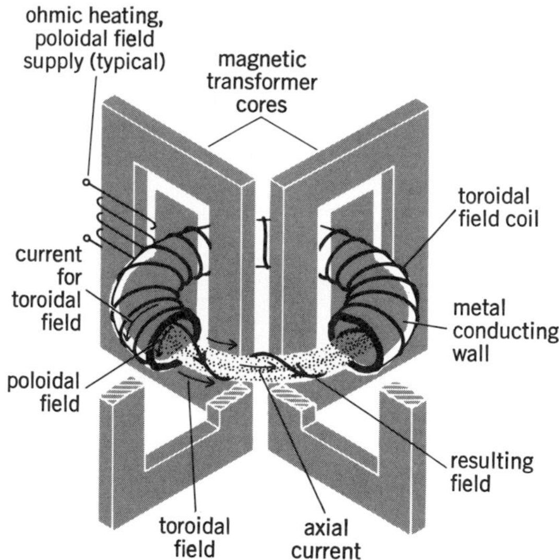

Fig. 3. Schematic illustration of a tokamak device. (*From R. F. Post and F. L. Ribe, Fusion reactors as future energy sources, Science, 186:397–407, 1974*)

The strong toroidal field has the main function of stabilizing the plasma against "kinking" magnetohydrodynamic instability modes. The necessity of having a strong toroidal external field has the disadvantage of limiting the beta value of the tokamak to a few percent. Beta, in magnetic fusion, is the ratio of the energy density of the plasma to that of the applied field. High beta is desirable from an economic standpoint to maximize the utilization of the externally generated field, thus minimizing the capital cost of the magnet system relative to the fusion power output.

As a closed device the tokamak has the important advantage that its plasma confinement time increases as the square of the radius, a, of the plasma column (actually, empirically, as na^2, where n is the plasma density). This property is a consequence of the fact that as long as gross stability is maintained, plasma particles can be lost only by diffusion across the field, and such losses proceed on a time scale which increases with the square of the characteristic distances involved. Thus, adequate confinement times can always be achieved by scaling up the dimensions. Though large (meters), the plasma diameters for tokamak plasmas as projected from present experiments appear to be acceptable for practical fusion power plants.

Mirror machine and tandem mirror. The mirror machine (Fig. 4) is an open-ended system in which a hot plasma is held trapped by repeatedly reflecting its particles between magnetic mirrors (regions of intensified magnetic field at each end of the confinement chamber).

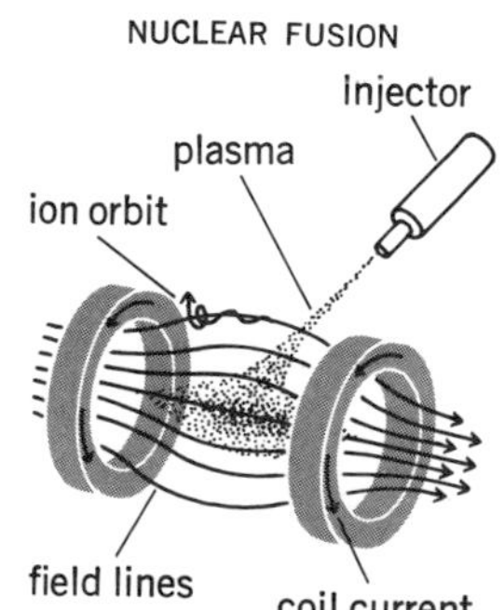

Fig. 4. Schematic illustration of mirror machine using simple mirrors. (*From R. F. Post and F. L. Ribe, Fusion reactors as future energy sources, Science, 186: 397–407, 1974*)

Fig. 5. Fusion power plant based on tandem mirror concept. (*Lawrence Livermore Laboratory*)

An important property of magnetic mirror systems is that their fields can be shaped so as to create a magnetic well, that is, a confining field that has a nonzero minimum surrounded by closed contours of increasing magnetic intensity. One type of magnetic well system is the baseball coil. When confined in a magnetic well, a plasma is restrained from exhibiting any form of gross instability, up to plasma pressures comparable to those of the confining field, that is, up to beta values of order unity.

Mirror systems are in contrast with the tokamak or other closed systems. In the latter there is no need to confine particles as far as their motion along the field lines is concerned. In a mirror machine the required longitudinal confinement is provided by the repelling force exerted on charged particles as they spiral along field lines that are converging, that is, as they move toward regions of increasing field strength. Particles which spiral with sufficiently steep helical pitch angles will be repelled strongly enough to be reflected, that is, trapped, by the mirrors. It follows that this type of mirror system cannot confine an isotropic plasma; only particles whose pitch angles are sufficiently large to lie between the loss cones defined by the strength of the mirror fields relative to the field intensity between the mirrors will be contained.

These systems must therefore rely on the injection of intense beams of energetic neutral atoms to maintain the plasma temperature and density in competition with particle leakage through the mirrors.

An important disadvantage of conventional mirror systems for fusion purposes is that the leakage of particles through the mirrors, arising as it does from collisions that deflect the ions into pitch angles lying within the mirror loss cone, occurs at a rate comparable to the rate of fusion reactions. In this circumstance the energy gain factor Q – the ratio of fusion power to plasma heating power – is at best not much larger than 1, implying an economically unacceptably large fraction of recirculated power needed to maintain the plasma. This deficiency in the mirror concept has stimulated the development of the tandem mirror and the field-reversed mirror concepts, discussed below.

An important aspect of mirror confinement is the positive ambipolar potential of the plasma that arises naturally from its operation. Since the collision frequency for electrons is higher than that for ions (because of the electrons' higher velocities), other factors being equal, they would tend to diffuse into the loss cone and be lost much more rapidly than would the ions. But any such differential loss rate would result in the buildup of a net positive charge in the confined plasma, thus driving it to a positive potential with respect to its surroundings, a potential sufficient to bring the electron loss rate to equality with that of the ions.

In the tandem mirror (Fig. 5) the ambipolar potentials are used to confine a fusion plasma, resulting in greatly improved confinement relative to that in a single mirror cell. A large-volume central chamber, in which the fusion plasma is to be confined, is stoppered at each end by two small-volume mirror cells having a high-ion-temperature plasma. The ambipolar potential of each of these plugs, being relatively more positive than that of the central cell, serves to confine (axially) the ions of the central cell plasma; its electrons are confined by the overall positive potential of the system.

Field-reversal systems. Magnetically confined plasmas exhibit the property of diamagnetism. That is, their confinement necessarily involves the existence of internal electric currents that act to reduce the strength of the confining magnetic field within the plasma relative to the vacuum value it would have in the absence of the plasma. It is in fact these internal diamagnetic currents that, by interacting with the externally generated field, produce a body force balancing the outward-acting pressure of the plasma. If these diamagnetic currents could be employed to provide the major part of the confining force, the confining effect of the external field would be greatly enhanced, with consequent economic and other practical benefits.

The earliest attempted embodiment of this concept was the toroidal pinch, the progenitor of the tokamak, which was unsuccessful because its self-constricted plasma column was subject to rapidly growing kinking instabilities of hydromagnetic origin. However, with increased understanding of magnetic confinement, ways to circumvent such problems in pinch-confined plasmas have been proposed. These ideas, generically describable as field-reversed systems, have taken four distinguishable forms: field-reversing particle rings; the field-reversed mirror; the field-reversed pinch; and self-field tokamaks. *See* PINCH EFFECT.

1. Field-reversing particle rings. These forms employ a ring of high-energy charged particles. The particles circle in large orbits in an external mirror field and generate a diamagnetic current that is sufficiently strong to reverse the direction of the field within the ring, thereby producing a poloidal field with closed field lines that traps the fusion plasma.

2. Field-reversed mirror. This is a mirror system where tangentially injected neutral beams maintain ion diamagnetic currents sufficient to create a field-reversed state, thus inhibiting the rate of escape of plasma ions, which must now diffuse across the field-reversed region before they reach the open field lines of the mirror field

3. Field-reversed pinch. This is a toroidal pinch in which the pinch (poloidal) field has superposed on it a toroidal field, the direction of which is reversed in the deep interior of the plasma, relative to its direction in the exterior parts.

4. Self-field tokamaks. These are tokamaklike configurations in which the main confining fields, both poloidal and toroidal, are the result of diamagnetic currents flowing within the plasma. An example is the spheromak configuration, in which the plasma and the conducting chamber around it (needed to preserve stability) are roughly spherical.

All of the field-reversed magnetic confinement systems described above have as their objective the more efficient use of magnetic field for fusion confinement purposes. Particularly as exemplified by the field-reversed mirror, the plasma beta value, as determined relative to the externally generated magnetic field, is very high. This could result in compact fusion power systems with improved economics, that is, lowered capital costs for the magnet and its support structures. For example, calculations for the field-reversed mirror indicate a fusion power density with the D-T reaction of order 500 times that for conventional tokamak. Alternatively, the increased magnetic efficiency could in principle allow the use of so-called advanced fusion fuel cycles, involving the D-^{3}He or other reactions, having special advantages in terms of reduced neutron fluxes or suitability for direct energy conversion. The basic physics understanding is, however, at present too limited to assure the practical success of any of the above systems, although there are indications that all of them can exhibit stable confinement.

Fig. 6. Portion of the Shiva laser pellet fusion facility at Lawrence Livermore Laboratory, showing 6 of the 20 laser amplifier trains, looking at the output end. System is designed to deliver more than 30 TW of optical power in less than 10^{-9} s. (*Lawrence Livermore Laboratory*)

Progress. In theory, magnetic confinement has been long perceived as an almost ideal solution to the fusion problem. In practice, over the earlier years of fusion research it was found very difficult to achieve confinement that approached the optimistic theoretical predictions, owing to the ubiquitous problem of plasma instabilities. However, refinement of the theory, coupled with experiments performed with technologically sophisticated, scaled-up apparatus, has led to major advances toward the quantitative goals of fusion power. Plasma temperatures and density–confinement time products (discussed below) have both been brought within a factor of 10 or less of breakeven values, and the next generation of tokamaks will probably come close to, or actually achieve, an energy breakeven situation (fusion energy yield equaling heat energy content of the confined plasma).

Increasingly, the emphasis in fusion research is on problems of practical or economic origin. Scientifically, this means a search for magnetic configurations that will result in smaller, more practical, or more efficient confinement geometries. Technologically, this means that more emphasis is being placed on the problems of high-field superconducting magnet coils, high-power neutral beams, vacuum technique, and materials problems associated with the inner wall of the confinement chamber, which must withstand high fluxes of energetic neutrons as well as relatively high-energy neutral atoms and charged parti-

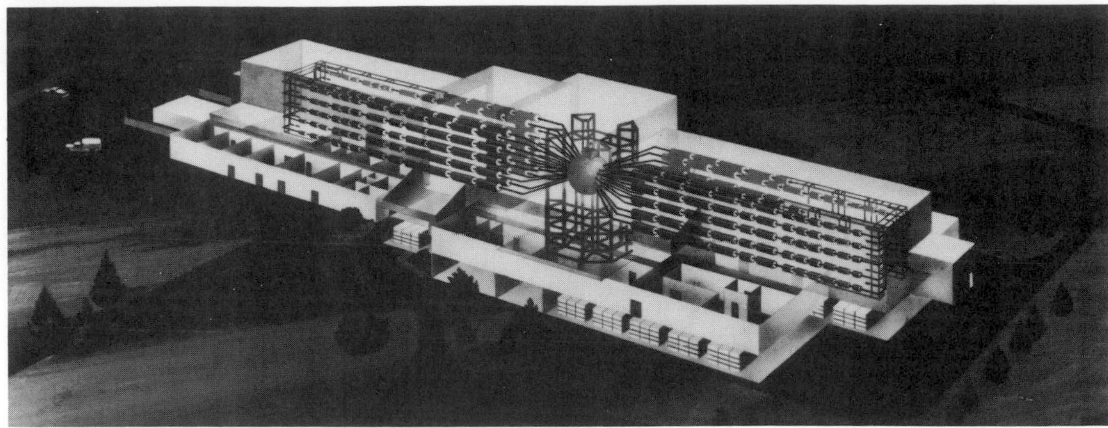

Fig. 7. Design of Shiva Nova laser pellet fusion experimental facility at Lawrence Livermore Laboratory. System is designed to provide enough power (200–300 TW) and energy (200 kJ) to obtain significant gain from laser fusion targets. (*Lawrence Livermore Laboratory*)

cles (hydrogen and helium ions) escaping from the plasma.

A rough (and sometimes misleading) index of progress toward fusion requirements, mentioned above, is the $n\tau$ confinement factor used in the Lawson criterion, where n is the particle density and τ is the average confinement time of the plasma. This parameter and the plasma ion kinetic temperature T_i at which the confinement is achieved, together provide a useful index of progress toward the goal of fusion power. Table 1 compares values $n\tau$ and T_i achieved in experiments in magnetic confinement with nominal values estimated to be required for a fusion power plant.

Pellet fusion. The basic idea behind pellet fusion is the rapid implosion of a high-density fusion fuel pellet to produce a heated core that will fuse before it can fly apart. As usually conceived, the implosion would result from the rapid heating and subsequent ablation of the surface of the pellet, giving rise to an inward-acting reaction force that compresses the core. But for this process to yield a net energy, that is, for it to achieve the required $n\tau$ value, very large compression factors, of order 10,000, are required. That this should be the case can be seen from simple considerations: As matter is compressed spherically, its density (n in the $n\tau$ product) increases as the cube of the radial compression factor. But the confinement time τ, here measured by the time of flight of an average particle out of the compressed core, decreased only as the first power of the radius. Thus $n\tau$ increases with the square of the radial compression factor. However, it is necessary to use tiny pellets to make this approach to fusion technically accessible from the standpoint of engineering limits on the amount of energy deliverable from the pellet drivers (lasers or particle accelerators), and on the amount of fusion energy released from the pellet that can be absorbed in surrounding structures. These limitations, taken together with the Lawson requirement, dictate the need for very large density compression factors, in turn leading to a requirement for very high compression forces [of order 10^{12} atmospheres (10^{17} Pa), ten times greater than those existing at the center of the Sun].

The problems thereby posed are twofold. First, very high pulsed powers of hundreds of terawatts must be focused down to millimeter dimensions and delivered in times of nanoseconds or less. The implied peak power densities are thus of order 10^{16} W/cm^2, made necessary by the requirement that

Table 1. Plasma parameters attained by magnetic confinement experiments compared with values believed needed for a fusion power plant

Approach	Device	Location	Plasma diameter, d (cm)	Particle density, n (cm^{-3})	Ion temperature, T_i (keV)	Confinement time, τ (ms)	Lawson product, $n\tau$ (s/cm^3)	β_{max}
Tokamak	T-4 (1970)	Kurchatov Institute, Moscow	30	3×10^{13}	0.4	10	6×10^{11}	.0006
	TFR (1974)	Fontenay-Aux-rose, Paris	40	4×10^{13}	0.8	15	6×10^{11}	.0016
	Alcator A (1975)	MIT	18	6×10^{14}	0.8	20	1.2×10^{13}	.006
	PLT (1976)	Princeton University	90	6×10^{13}	1.0	50	3×10^{12}	.005
	PLT (neutral beams, 1978)	Princeton University	90	4×10^{13}	5.0	25	$\sim10^{12}$	.013
Conventional mirror	PR-6 (1971)	Kurchatov Institute, Moscow	10	2×10^{12}	0.5	0.15	3×10^{8}	.0016
	Baseball II (1972)	Lawrence	20	2×10^{9}	2.0	1000	2×10^{9}	–
	2XIIB (1977)	Livermore	15	$\sim10^{14}$	13.0	1.	$\sim10^{11}$	>1.0
Tandem mirror	TMX (1979)	Laboratory	50	2×10^{13}	0.2	~5.	$\sim10^{11}$	~.2

Power plant requirements

	Ion temperature, T_i (keV)	Lawson product, $n\tau$ (s/cm^3)	β	Remarks
Tokamak	15	$>10^{14}$	0.1	~1000 MWe, ignited mode
Conv. mirror	150	$>10^{13}$	0.8	High recirculated power required
Tandem mirror	30	$>10^{14}$	0.5	~500 MWe, ignited mode

Table 2. Examples of laser fusion facilities

Name	Location	Energy pulse, kJ	Peak power, TW	Compression achieved (× liquid density)
Zeta	University of Rochester	1.2	3–4	7–20
Chrome I	KMS Fusion, Inc., Ann Arbor, MI	1.0	2	7–35
Helios	Los Alamos Scientific Laboratory	5–10	10–20	8–30
Shiva	Lawrence Livermore Laboratory	10	26	30–160

an almost instantaneous ablation of the outer surface of the pellet should occur before heat can flow into the interior of the pellet; premature heat flow would prevent reaching the required compression factors. This problem can in principle be solved by using a sufficiently large array of high-power lasers, focusing their beams so as to uniformly illuminate the surface of the pellet (Figs. 6 and 7). Alternatively, converging beams of electrons or ions from particle accelerators would be employed. *See* LASER FUSION; PARTICLE-BEAM FUSION.

Second, if the compression process is not carried out with high uniformity, it will go askew; small errors in uniformity can lead to a major reduction in the achievable compression. Sophisticated aiming and timing techniques must be used in the driver, and any instabilities must not be allowed to spoil the symmetry of the compression or lead to undue mixing or preheating.

High temperatures and substantial $n\tau$ values, comparable to those achieved in magnetic fusion, have been attained in pellet fusion experiments. Table 2 lists some of the parameters achieved in laser pellet experiments. Continued progress can be expected, as new and larger facilities come on line, but increase in performance by some orders of magnitude is needed before breakeven can be attained.

POWER PLANTS

Preliminary studies have been made of the forms that fusion power plants might take, following some of the approaches outlined above. These studies cannot of course be definitive, but they have helped to indicate the sizes, capital costs, and special engineering problems that are likely to characterize fusion power plants, insofar as they can now be visualized.

The types of power plants that have been studied encompass both pulsed and steady-state fusion systems, operating in either a driven mode or one in which plasma ignition would be achieved. A driven fusion power system is one in which the plasma temperature is maintained primarily by a continuous input of energy—for example, by the injection of high-intensity beams of energetic neutral fuel atoms—thereby maintaining the required kinetic temperature and density of the plasma. Electrical power needed to produce the beams would be obtained by recirculating a portion of the electrical output of the plant. A positive power balance, that is, net output power ($Q > 1$), would be possible here only when the recirculated power is less than the electrical output as recovered from the plasma, including not only the energy content of the reaction products but that of the unreacted part (electrons and heated fuel ions); economical power would likely be possible only if the recirculated power fraction were to be much smaller than the power produced (that is, Q much greater than 1). By contrast, a fusion system operated in an ignition mode is one where the energy deposited directly within the plasma by charged reaction products (3.5-MeV alpha particles in the case of the D-T reaction, that is, only 20% of the total fusion energy release) is sufficient to maintain the plasma temperature, including the requirement for heating up new cold fuel particles introduced to maintain the fuel plasma density as fusion "combustion" proceeds. The ignition mode is therefore more demanding with respect to confinement time than is a driven mode, but in principle could be technically simpler, since it does not impose as strict requirements on the efficiencies of the energy recovery and plasma heating systems.

By exploiting the increase in confinement time associated with an increase in plasma radius and extrapolating from the performance of present devices, it appears that large tokamaks could operate in an ignition mode. Conventional and tandem mirror systems, requiring as they do the maintenance of plasma in the mirror cells, would need to be operated in a driven mode. However, in the case of the tandem, it seems that it will be possible to achieve ignition in the central cell plasma, with attendant simplifications and other advantages. Laser pellet systems, although the pellet itself would be expected to ignite, necessarily require recirculated power to initiate the burn. To achieve high net power relative to recirculated power would seem to imply the need for pellet energy gains *(Q)* of 100 or more.

Considering the magnetic confinement approach, systems studies have led to some important conclusions: Fusion power plants based on the tokamak principle in its conventional form will be relatively large both in size and in power output; electrical power outputs of 500 to some thousands of megawatts are likely to be typical. Power plants utilizing the tandem idea might be somewhat smaller in physical size than conventional tokamaks, and possibly also capable of somewhat lower plant electrical outputs than tokamaks, still satisfying economic requirements. Power plants based on field reversal, such as the field-reversed mirror, might be the smallest of all, exhibiting the highest fusion power density while being both compact in size and permitting (at least in the demonstration phase) electrical power outputs as low as tens of megawatts.

Another general result of the design studies is to show the importance of choice of materials and heat transfer characteristics of the inner wall of the containment chamber. The flux of 14-MeV neutrons through this wall coming from D-T reactions in the plasma will cause localized heating, radiation damage, and induced radioactivity. Thus the design of this portion of any D-T fusion power plant can be expected to be of critical importance. The materials chosen need to be picked not only for their resistance to radiation damage but also for minimum activation (that is, minimal yield and short half-life for the neutron-induced radioactivity). It does appear that first-wall materials having the desired characteristics can be developed. Another critical factor is that of the generation of the confining magnetic field. Here the criterion is to achieve the required field (which may be very high for some approaches) at the least capital cost and for the least expenditure of energy. Fortunately, the development of practical high-current-density, high-field superconductors appears to provide an almost ideal solution to this problem. *See* RADIATION DAMAGE TO MATERIALS; SUPERCONDUCTIVITY. [RICHARD F. POST]

Bibliography: F. Chen, *Introduction to Plasma Physics*, 1974; N. Krall and A. Trivelpiece, *Principles of Plasma Physics*, 1973; R. F. Post, Controlled fusion research and high temperature plasmas, *Annu. Rev. Nucl. Sci.*, 20:509–558, 1970; G. Schmidt, *Physics of High Temperature Plasmas*, 2d ed., 1979; L. Spitzer, *Physics of Fully Ionized Gases*, 1962.

Nuclear materials safeguards

The term nuclear materials safeguards is used in two contexts, national and international. In the national sense, safeguards refer to government-enforced programs to prevent the theft or seizure of nuclear materials by subnational adversaries or the sabotage of nuclear facilities which might endanger the public. International safeguards refer to the procedures employed by the International Atomic Energy Agency (IAEA) to provide assurance that nuclear materials or facilities subject to the Agency's review are not being used for any military purpose.

National safeguards. A national safeguards system should deter, prevent, or respond to any attempt by a subnational group to steal nuclear materials or to sabotage facilities or shipments containing large amounts of radioactive materials. In the United States, the Nuclear Regulatory Commission is responsible for the safeguards measures applied to privately owned nuclear materials and facilities; the Department of Energy for safeguarding nuclear production and research facilities owned by the Federal government; and the Department of Defense for the protection of its nuclear weapons. Other Federal agencies would also be involved in investigating suspected theft of nuclear materials or in responding to a credible nuclear threat.

Nuclear targets, on the basis of experience, do not appear to be attractive to domestic subversive or terrorist groups. On the other hand, the costs to society of the detonation of even an inefficient nuclear explosive or of the sabotage of a nuclear power reactor could be very great. Consequently, a national safeguards system employs multiple barriers to prevent diversion, theft, or acts of sabotage.

The Nuclear Regulatory Commission, for example, publishes requirements for safeguards in the *Federal Register*. A license from the Commission is required to possess or to process any significant amount of nuclear material. In order to obtain a license, it is necessary to define and to implement safeguards procedures which the Commission considers meet the published requirements. Government personnel inspect the nuclear facilities and enforce the regulations.

Nuclear facility activities. Safeguards at a nuclear facility consist of three complementary activities: physical protection, material control, and material accounting. For protection against surreptitious or forced entry, a nuclear facility must be of substantial construction and surrounded by a protected area and chain link fence. Access to the protected area is limited to persons or vehicles authorized for admittance. Access to the building and to the critical areas within the building is further restricted. Personnel or packages authorized for admittance or exit are subject to search. Armed guards control the entry and exit points, monitor internal and external activities, and are prepared to respond to threats. Arrangements exist to summon the local police, should that be necessary.

Because of their high value and safeguards significance, material control procedures define the responsibilities for management of nuclear materials, for operations involving nuclear materials, and for surveillance by safeguards personnel. Shipments received must be inspected and remeasured by the receiver. Materials are stored in secure vaults. Custodians are responsible for materials in storage and materials being processed. Each internal transfer at a processing plant is recorded and witnessed. The physical protection and material control activities are intended to detect attempts to steal nuclear materials or to sabotage a facility in time to interrupt the activity.

Material accounting is similar to that applied to other valuable materials. Bulk materials are weighed, sampled, and analyzed upon receipt, when transferred from one process stage to another, and when put into a container or fabricated into a fuel element. At periodic intervals a nuclear facility is shut down, the equipment is cleaned out, and a physical inventory is performed. On the basis of the starting inventory, receipts, shipments, and the final inventory, a determination is made as to material gained or lost during the period. In order to determine if this gain or loss is significant, it is necessary to determine the probable errors involved in all of the measurements and to combine them statistically to determine the uncertainty in the material balance due to measurement errors. When the material balance shows a loss that is comparable to or larger than the uncertainty due to measurement errors, it is necessary to find out by investigation whether material may have been diverted or, if not, to determine and rectify the cause of the discrepancy.

At power reactors the nuclear material is in the form of discrete, sealed, fuel assemblies. Accounting for these items is straightforward. Primary emphasis is placed on physical protection and internal controls to prevent sabotage. *See* NUCLEAR REACTOR.

Level of safeguards. National safeguards measures are graded to match the safeguards significance of the materials and facilities involved. Since natural and lowly enriched uranium are neither very radioactive nor useful to a subnational adversary for a nuclear explosive, primary emphasis is placed on material accounting and annual or semiannual material balances. On the other hand, high levels of physical protection and material control are required for facilities that have highly enriched uranium or plutonium. Shipments of highly enriched uranium or plutonium require physical protection by armed guards in the transport vehicle and one or more escort vehicles, and provision for frequent radio or telephone reports.

Prospects. Studies will continue on the nature of the threats which domestic safeguards should counter, and on methods to make the safeguards measures more cost-effective. A most promising area is the extensive application of nondestructive measurement techniques, based on measuring the natural or induced radioactive emissions of uranium and plutonium, in order to supplement the chemical sampling measurement methods and to provide nearly real-time information on the location and flows of nuclear materials throughout the nuclear fuel cycle, taking advantage of advances in computers and automation.

International safeguards. International safeguards are applied by the International Atomic Energy Agency, which was established to assist nations in applications of nuclear energy for peaceful purposes and to provide assurance that nuclear materials are being used only for the declared peaceful purposes. Safeguards are applied according to the terms of an agreement between a nation and the IAEA which states that the nation will not use the declared nuclear facility or the nuclear materials for any military purpose (or in case of agreements under the Nuclear Nonproliferation Treaty, for any nuclear explosive) and specifies the safeguards procedures which the IAEA will employ.

Defining parameters. In order to define the safeguards procedures to be applied at a given nuclear facility, the nation is required to provide the IAEA with information on the location, design, capacity, and processes of the facility, and on the national system of material control and accounting which will be employed. The IAEA may verify this "design information" by inspection. The IAEA and the nation then agree on the records to be maintained by the facility operator, on the reports which the operator will make to the IAEA, and on how IAEA inspectors will conduct their operations.

In many cases, all of the nuclear materials and facilities within a nation are subject to IAEA safeguards. In other cases, some nuclear facilities are subject to IAEA safeguards while others are not. In the former situation, the IAEA can monitor the location and flows of all of the nuclear materials within the nation and transfers into or out of the national system, checking the whole system for consistency. In the latter situation, assurance can be provided only for those facilities which are offered for verification.

IAEA safeguards are designed for the timely detection of diversion of a significant amount of nuclear material for whatever reason. The Agency does not have to determine that the purpose is to fabricate a nuclear explosive. IAEA inspectors are observers. They do not have the powers of national safeguards inspectors to control people or to interfere with plant operations.

Procedures. IAEA safeguards procedures employ material accounting, containment, and surveillance measures.

Material accounting is based on auditing of facility records and reports and on the independent verification of some, or perhaps all, measurements of nuclear materials. By verifying measurements and reports, the IAEA should be able to determine unintentional biases and to detect (and deter) attempts to falsify reports. National safeguards regulations will usually require a processing plant to shut down and be cleaned out for a physical inventory at stated periods. The IAEA will observe and verify the measurements and item counting for the physical inventory and determine a material balance for the period to compare with that reported by the facility operator. Nations are required to submit monthly reports to the IAEA on the location and status of nuclear materials in all safeguarded facilities.

Containment and surveillance measures complement those for accounting. Containment refers to features inherent to facility and equipment construction which constrain access to nuclear materials. Surveillance refers to human or instrumental observation of storage and processing areas to ensure that materials pass through the key points designated for measurements, and that the materials remain within the designated storage, use, or processing areas.

As the number and size of nuclear facilities have continued to increase, many nations have assisted the IAEA to improve instruments and techniques for IAEA safeguards and to demonstrate their effectiveness. *See* NUCLEAR FUEL CYCLE; NUCLEAR FUELS REPROCESSING; NUCLEAR POWER.

[WILLIAM A. HIGINBOTHAM]

Bibliography: International Atomic Energy Agency, *IAEA Safeguards Technical Manual*, IAEA-174, 1976; International Atomic Energy Agency, *Proceedings of the Symposium on Nuclear Safeguards Technology, Vienna, October 1978*, 1979; A Leff and J. V. Roos, *Safeguards against the Theft of Nuclear Materials and the Sabotage of Facilities in the Nuclear Power Industry*, 1977; Office of Technology Assessment, U.S. Congress, *Nuclear Proliferation and Safeguards*, 1977; B. Sanders, *Safeguards against Nuclear Proliferation*, 1975; M. Willrich and T. B. Taylor, *Nuclear Theft: Risks and Safeguards*, 1974.

Nuclear physics

The discipline involving the structure of atomic nuclei and their interactions with each other, with their constituent particles, and with the whole spectrum of elementary particles that is provided by very large accelerators. The nuclear domain occupies a central position between the atomic range of forces and sizes and elementary-particle physics, characteristically within the nucleons themselves. As the only system in which all the known natural forces can be studied simultaneously, it provides a natural laboratory for the testing and extending of many of the fundamental symmetries and laws of nature.

Containing a reasonably large, yet manageable number of strongly interacting components, the nucleus also occupies a central position in the universal many-body problem of physics, falling between the few-body problems, characteristic of elementary-particle interactions, and the extreme many-body situations of plasma physics and condensed matter, in which statistical approaches dominate; it provides the scientist with a rich range of phenomena to investigate—with the hope of understanding these phenomena at a microscopic level.

Activity in the field centers on three broad and interdependent subareas. The first is referred to as classical nuclear physics, wherein the structural and dynamic aspects of nuclear behavior are probed in numerous laboratories, and in many nuclear systems, with the use of a broad range of experimental and theoretical techniques. Second is higher-energy nuclear physics (referred to as medium-energy physics in the United States), which emphasizes the nuclear interior and nuclear interactions with mesonic probes. Third is heavy-ion physics, internationally the most rapidly growing subfield, wherein the accelerated beams of nuclei spanning the periodic table are used to study nuclear phenomena which were previously inaccessible.

Nuclear physics is unique in the extent to which it merges the most fundamental and the most applied topics. Its instrumentation has found broad applicability throughout science, technology, and medicine; nuclear engineering and nuclear medicine are two very important areas of applied specialization. *See* NUCLEAR ENGINEERING.

Nuclear chemistry, certain aspects of condensed matter and materials science, and nuclear physics together constitute the broad field of nuclear science; outside the United States and Canada elementary particle physics is frequently included in this more general classification. *See* NUCLEAR FISSION; NUCLEAR FUSION; NUCLEAR REACTION; NUCLEAR REACTOR.

[D. ALLAN BROMLEY]

Nuclear power

Power derived from fission or fusion nuclear reactions. More conventionally, nuclear power is interpreted as the utilization of the fission reactions in a nuclear power reactor to produce steam for electrical power production, for ship propulsion, or for process heat. Fission reactions involve the breakup of the nucleus of heavy-weight atoms and yield energy release which is more than a millionfold greater than that obtained from chemical reactions involving the burning of a fuel. Successful control of the nuclear fission reactions provides for the utilization of this intensive source of energy, and with the availability of ample resources of uranium deposits, significantly cheaper fuel costs for electrical power generation are attainable. Safe, clean, economic nuclear power has been the objective both of the Federal government and of industry's programs for research, development, and demonstration. Critics of nuclear power seek a complete ban or at least a moratorium on new commercial plants.

Considerations. Fission reactions provide intensive sources of energy. For example, the fissioning of an atom of uranium yields about 200 MeV, whereas the oxidation of an atom of carbon releases only 4 eV. On a weight basis, the 50,000,000 energy ratio becomes about 2,500,000. Only 0.7% of the uranium found in nature is uranium-235, which is the fissile fuel used. Even with these considerations, including the need to enrich the fuel to several percent uranium-235, the fission reactions are attractive energy sources when coupled with abundant and relatively cheap uranium ores. Although resources of low-cost uranium ores are extensive (see Tables 1 and 2), more explorations in the United States are required to better establish the reserves. Most of the uranium resources in the United States are in New Mexico, Wyoming, Colorado, and Utah. Major foreign sources of uranium are Australia, Canada, South Africa, and southwestern Africa; smaller contributions come from France, Niger, and Gabon; other sources include Sweden, Spain, Argentina, Brazil, Denmark, Finland, India, Italy, Japan, Mexico, Portugal, Turkey, Yugoslavia, and Zaire.

Government administration and regulation. The development and promotion of the peaceful uses of nuclear power in the United States was under the direction of the Atomic Energy Commission (AEC), which was created by the Atomic Energy Act of 1946 and functioned through 1974. The Atomic Energy Act of 1954, as amended, provided direction and support for the development of commercial nuclear power. Congressional hearings established the need to assure that the public would have the availability of funds to satisfy liability claims in the unlikely event of a serious nuclear accident, and that the emerging nuclear industry should be protected from the threat of unlimited liability claims. In 1957 the Price-Anderson Act was passed to provide a combination of private insurance and governmental indemnity to a maximum of $560,000,000 for public liability claims. The act was extended in 1965 and again in 1975, each time following congressional hearings which probed the need for such protection and the merits of having nuclear power. The Federal Energy Reorganization Act of 1974 separated the promotional and regulatory functions of the AEC, with

Table 1. United States uranium resources, in tons of U_3O_8 as of Jan. 1, 1979*

Production cost, $/lb U_3O_8†	Proved reserves	Potential reserves			Total reserves
		Probable	Possible	Speculative	
15	290,000	415,000	210,000	75,000	990,000
30	690,000	1,005,000	675,000	300,000	2,670,000
50	920,000	1,505,000	1,170,000	550,000	4,145,000

*1 short ton = 0.907 metric ton.
†Each cost category includes all lower-cost resources. $1/lb = $2.20/kg.
SOURCE: *Statistical Data of the Uranium Industry*, U.S. Department of Energy, 1979.

the creation of a separate Nuclear Regulatory Commission (NRC) and the formation of the Energy Research and Development Administration (ERDA). In 1977 ERDA was absorbed into the newly formed Department of Energy.

Safety measures. The AEC, overseen by the Joint Committee on Atomic Energy (JCAE), a statutory committee of United States senators and representatives, had sought to encourage the development and use of nuclear power while still maintaining the strong regulatory powers to ensure that the public health and safety were protected. The inherent dangers associated with nuclear power which involves unprecedented quantities of radioactive materials, including possible widescale use of plutonium, were recognized, and extensive programs for safety, ecological, and biomedical studies, research, and testing have been integral with the advancement of the engineering of nuclear power. Safety policies and implementation reflect the premises that any radiation may be harmful and that exposures should be reduced to "as low as reasonably achievable" (ALARA); that neither humans nor their creations are perfect and that suitable allowances should be made for failures of components and systems and human error; and that human knowledge is incomplete and thus designs and operations should be conservatively carried out. The AEC regulations, inspections, and enforcements sought to develop criteria, guides, and improved codes and standards which would enhance safety, starting with design, specification, construction, operation, and maintenance of the nuclear power operations; would separate control and safety functions; would provide redundant and diverse systems for prevention of accidents; and would provide to a reasonable extent for engineered safety features to mitigate the consequences of postulated accidents.

Criteria for siting a nuclear power station involve thorough investigation of the region's geology, seismology, hydrology, meteorology, demography, and nearby industrial, transportation, and military facilities. Also included are emergency plans to cope with fires or explosions, and radiation accidents arising from operational malfunctions, natural disasters, and civil disturbances. The AEC Directorates of Licensing, Regulatory Operations, and Regulatory Standards thus functioned to achieve an extraordinary program of safety to be commensurate with the extraordinary risks involved with nuclear power.

Starting about 1970, regulatory safety measures have been significantly augmented in response to the introduction of nuclear power reactors with larger powers and higher specific power ratings; to improved technology, experiences, and more sophisticated analytical methods; to the National Environmental Policy Act of 1969 (NEPA) and the interpretations of its implementations; and to public participation and criticism. A variety of special assessments, studies, and hearings have been undertaken by such parties as congressional committees other than JCAE, the U.S. General Accounting Office (GAO), the American Physical Society, the National Research Council representing the National Academy of Sciences and the National Academy of Engineering, and by organizations which represent public interests in the environmental impacts of the continued use of nuclear power.

Table 2. Foreign resources of uranium, in tons of U_3O_8 as of Jan. 1, 1979*

Production cost, \$/lb U_3O_8†	Reasonably assured	Estimated additional	Total resources
30	1,460,000	870,000	2,330,000
50	2,010,000	1,350,000	3,360,000

*Excluding People's Republic of China, Soviet Union, and associated countries. 1 short ton = 0.907 metric ton.

†Each cost category includes all lower-cost resources. \$1/lb = \$2.20/kg.

Nuclear Regulatory Commission. The independent NRC is charged solely with the regulation of nuclear activities to protect the public health and safety and the environmental quality, to safeguard nuclear materials and facilities, and to ensure conformity with antitrust laws. The scope of the activities include, in addition to the regulation of the nuclear power plant, most of the steps in the nuclear fuel cycle; milling of source materials; conversion, fabrication, use, reprocessing, and transportation of fuel; and transportation and management of wastes. Not included are uranium mining and operation of the government enrichment facilities.

Public issues. Public issues of nuclear power have covered many facets and have undergone some changes in response to changes being effected. Key issues include possible theft of plutonium, with threatening consequences; management of radioactive wastes; whether, under present escalating costs, nuclear power is economic and reliable; and protection of the nuclear industry from unlimited indemnity for catastrophic nuclear accidents.

Special nuclear materials. Guidance for improving industrial security and safeguarding special nuclear materials has been initiated. Scenarios studied include possible action by terrorist groups, and evaluations have been undertaken of effective methods for preventing or deterring thefts and for recovering stolen materials. Loss of plutonium by theft and diversionary tactics could pose serious dangers through threats to disperse toxic plutonium oxide particles in populated areas or to make and use nuclear bombs.

In the commercial nuclear fuel cycle which had been envisioned previous to 1977, the more critical segments in the safeguard program would involve the chemical reprocessing plants where the high-level radioactive wastes would be separated from the uranium and the plutonium, the shipment of the plutonium oxide to the fuel manufacturing plant, and the fuel manufacturing plant where plutonium oxide would be incorporated in the uranium oxide fuel. Plutonium is produced in the normal operation of a nuclear power reactor through the conversion of uranium-238. For each gram of uranium-235 fissioned, about 0.5–0.6 g of plutonium-239 is formed, and about half of this amount is fissioned to contribute to the operation of the power reactor. Reactor operations require refueling at yearly intervals, with about one-fourth to one-third of the irradiated fuel being replaced by new fuel. In the chemical reprocessing, most of the uranium-238 initially present in the fuel would be recovered, and about one-fourth of the uranium-235 initially charged would remain to be recovered along with an almost equal amount of plutonium-239.

The only commercial chemical reprocessing plant, the Nuclear Fuel Services, Inc., facility in West Valley, NY, opened in 1966, recovered uranium and plutonium as nitrates, and stored the high-level wastes in large, underground tanks. The facility was closed in 1972, and in 1976 Nuclear Fuel Services withdrew from nuclear fuels reprocessing because of changing regulatory requirements. The Midwest Fuel Recovery Plant at Morris, IL, was to have begun operations in 1974, but functional pretests revealed that major modifications would have to be undertaken before initiating commercial operations. Maximum use of the facility has been made to accommodate storage of irradiated fuel. A third plant, the Allied General Nuclear Services Barnwell Nuclear Fuel Plant at Barnwell, SC, whose construction was begun in 1971, has not been licensed. Thus, no commercial reprocessing plant is in operation in the United States, and there is only very limited use of test fuel assemblies containing mixed oxides of plutonium and uranium.

In April 1977 the Carter administration decided to defer indefinitely the reprocessing of spent nuclear fuels. This decision reflected a policy of seeking alternative approaches to plutonium recycling and the plutonium breeder reactor for the generation of nuclear power. This policy was motivated by a concern that the use of plutonium in other parts of the world might encourage the use of nuclear weapons.

Prior to the administration's decisions, the NRC had developed a system of reviews, including public participation through hearings and through comments received on draft regulations, to determine whether recycling of plutonium was to be licensed in a generic manner. Consideration has also been given to the possible collocation of chemical reprocessing and fuel manufacturing plants, and whether there are net gains achieved through the concentration of nuclear power reactors and fuel facilities in energy parks.

Some foreign countries have opposed the anti-plutonium policies of the United States, and proceeded with the development of reprocessing plants and plutonium breeder reactors. A small commercial reprocessing plant for oxide fuel began operation in 1976 at La Hague, France, and construction of a much larger facility has been undertaken at this site. Design has begun for an oxide fuel reprocessing plant at Windscale in the United Kingdom to serve overseas markets as well as domestic markets.

Public participation in licensing. The procedures for licensing a nuclear facility for construction and for operation provide opportunities for meaningful public participation. Unique procedures have evolved from the Atomic Energy Act of 1954, as amended, which are responsive to public and congressional inquiry. The applicant is required to submit to the NRC a set of documents called the Preliminary Safety Analysis Report (PSAR), which must conform to a prescribed and detailed format. In addition, an environmental report is prepared. The docketed materials are available to the public, and with the Freedom of Information Act and the Federal Advisory Committee Act, even more public access to information is available. The NRC carries out an intensive review of the PSAR, extending over a period of about a year, involving meetings with the applicant and its contractors and consultants. Early in the review process, the NRC attempts to identify problems to be resolved, including concerns from citizens in the region involved with the siting of the plant. Formal questions are submitted to the applicant, and the replies are included as amendments to the PSAR.

Environmental Impact Statement. Major Federal actions that significantly affect the quality of human environment require the preparation of an Environmental Impact Statement (EIS) in accordance with the provisions of NEPA. The EIS presents (1) the environmental impact of the proposed action; (2) any adverse environmental effects which cannot be avoided should the proposal be implemented; (3) alternatives to the proposed action; (4) relationships between short-term uses of the environment and the maintenance and enhancement of long-term productivity; and (5) any irreversible and irretrievable commitments of resources which would be involved in the proposed action should it be implemented. To better achieve these objectives, the NRC staff supplements the applicant's submittal with its own investigations and analyses and issues a draft EIS so as to gain the benefit of comments from Federal, state, and local governmental agencies, and from all interested parties. A final EIS is prepared which reflects consideration of all comments.

Safety Evaluation Report. A second major report issued by the NRC staff is the Safety Evaluation Report (SER), which contains the staff's conclusions on the many detailed safety items, including discussions on site characteristics; design criteria for structures, systems, and components; design of the reactor, fuel, and coolant system; engineered safety features; instrumentation and control; both off-site and on-site power systems; auxiliary systems, including fuel storage and handling, water systems, and fire protection system; radioactive waste management; radiation protection for employees; qualifications of applicant and contractors; training programs; review and audit; industrial security; emergency planning; accident analyses; and quality assurance.

Independent review. Two additional steps are required before the decision is made regarding the construction license. An independent review on the radiological safety items is made and reported by the Advisory Committee on Reactor Safeguards (ACRS), and a public hearing is held by the Atomic Safety and Licensing Board (ASLB). The ACRS is a statutory committee consisting of a maximum of 15 members, covering a variety of disciplines and expertise. Appointments are made for this part-time activity by the NRC. Members are selected from universities, national laboratories and institutes, and industry, including experienced engineers and scientists who have retired, and, in each case, any possible conflicts-of-interest are carefully evaluated. The ACRS has a full-time staff and has access to more than 90 consultants. The ACRS conducts an independent review on nuclear safety issues and prepares a letter to the chairman of the NRC. Both subcommittee and full committee meetings are held to review the documents available and to discuss the applicant's and NRC staff's views on specific and generic issues.

Public hearing and appeal. Public participation is a major objective in the public hearing conducted by the ASLB. The ASLB is a three-member board, chaired by a lawyer, with usually two technical experts. For each application, a board is chosen from among the members of the Atomic Safety and Licensing Panel. Most members of the panel are part-time, and all members are appointed by the NRC. Prehearing conferences are held by the ASLB to identify parties who may wish to qualify and participate in the public hearing. Attempts are made to improve the understanding of the contentions, to see which contentions can be settled before the hearing, and to agree on the issues to be contested. The hearing may probe the need for additional electrical power, the suitability of the particular site chosen over possible alternative sites, the justification of the choice of nuclear power over alternate energy sources, and special issues regarding environmental impact and safety. The ASLB makes a decision on the construction application and may prescribe conditions to be followed. The decision is reviewed by and may be appealed to the Atomic Safety and Licensing Appeal Board. The appeal board is chosen from a panel completely separate from the ASLB panel. The NRC retains the authority to accept, reject, or modify the decisions rendered. Parties not satisfied by the review process and the decisions rendered can take their case to the courts. In several cases, resort to the U.S. Supreme Court has been utilized.

Authorization. A construction permit license is not issued until the NRC, ACRS, and ASLB reviews have been completed and the application has been approved, including conditions to be met during the construction review phase. Depending upon the justification of need, a Limited Work Authorization may be granted for limited construction activities following satisfactory review of the EIS, but prior to completion of the public hearings. Construction of a nuclear power plant may take 5 to 6 years or more, during which time the NRC Office of Inspection and Enforcement is involved in monitoring and inspection programs. Several years prior to the completion of construction, a Final Safety Analysis Report (FSAR) is submitted by the applicant, and again an intensive review is undertaken by the NRC staff, and later by the ACRS. A detailed Safety Evaluation Report is prepared by the NRC, and a letter is prepared by the ACRS. If all items have been satisfactorily resolved and the construction and preoperational tests have been completed, a license for operation up to the full power is granted by the NRC. Technical specifications accompany the operating license and provide for detailed limits on how the plant may be operated. In some situations where additional information is sought, less than full power is authorized. A public hearing at the operation license stage is not mandatory and would be held only if an intervenor justifies sufficient cause.

Hearings on general matters. Public hearings have also been held on generic matters to establish rules for operation. The two rulemaking hearings conducted by the AEC that have attracted much attention were the Emergency Core Cooling Systems (ECCS) and the "As Low As Practicable" (ALAP) hearings. The hearing on the criteria and conditions for evaluating the effectiveness of the ECCS for a postulated loss of coolant accident lasted from January 1972 to July 1973, and provided more than 22,000 pages of transcript, with probably twice as much additional material in supporting exhibits. EISs on major activities, such as the liquid metal fast breeder reactor (LMFBR) program and the management of commercial high-level and transuranium-contaminated radioactive waste, have provided a process for public interactions and influence.

Reactor Safety Study. A detailed, quantitative assessment of accident risks and consequences in United States commercial nuclear power plants has been carried out (WASH-1400, October 1975). The final report has had the benefit of comments and criticisms on a previous draft from governmental agencies, environmental groups, industry, professional societies, and a broad spectrum of other interested parties. Although the study was initiated by the AEC and continued by the NRC, the ad hoc group directed by N. C. Rasmussen of the Massachusetts Institute of Technology carried out an independent assessment. Aside from the very significant technical advancements made in

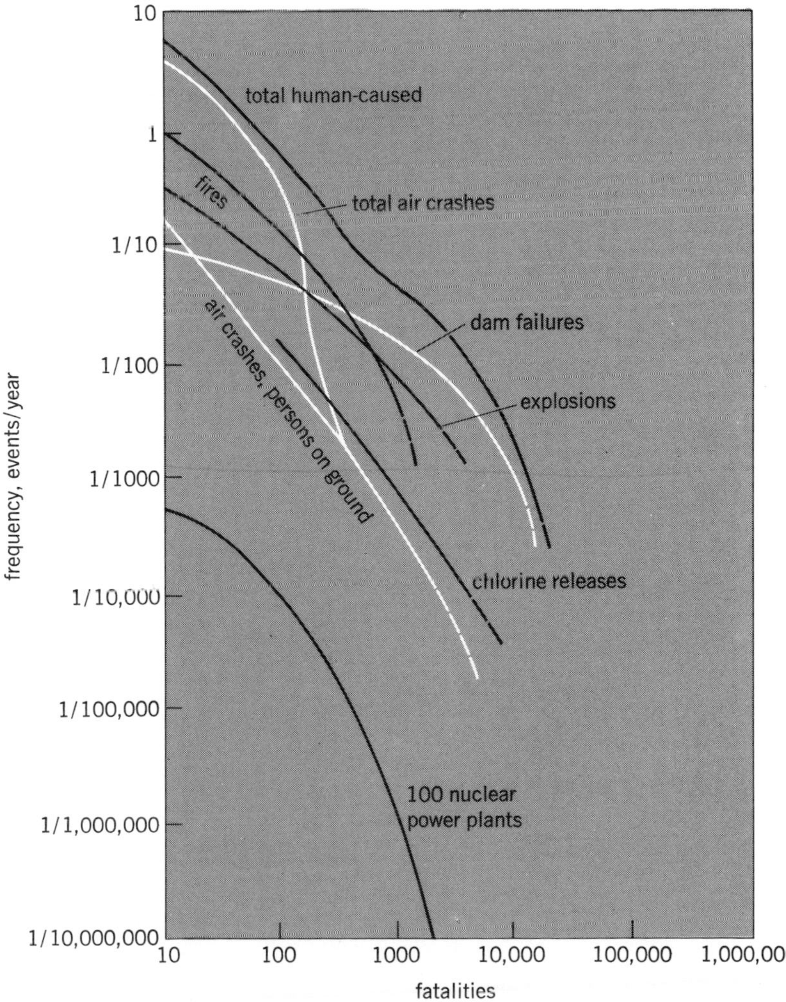

Fig. 1. Estimated frequency of fatalities due to human-caused events.

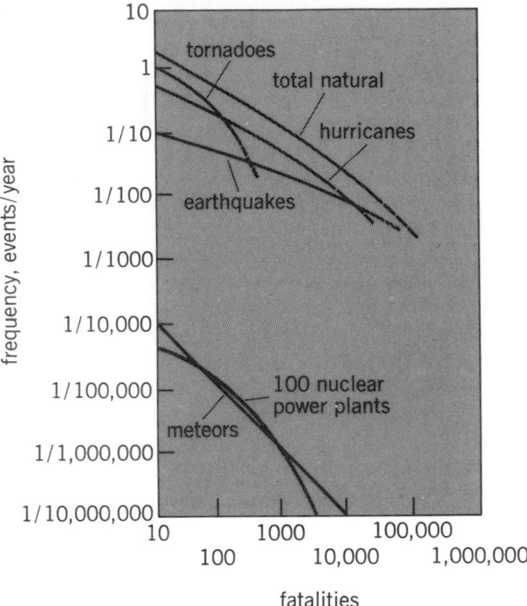

Fig. 2. Estimated frequency of fatalities due to natural events.

the risk assessment methodology, the "Reactor Safety Study" represents an approach to deal with one phase of the controversial impact of technology upon society. The study presents estimated risks from accidents with nuclear power reactors and compares them with risks that society faces from both natural events and nonnuclear accidents caused by people. The judgment as to what level of risk may be acceptable for the nuclear risks still remains to be made.

Figures 1 and 2 illustrate that the frequency of human-caused nonnuclear accidents and natural events is about 10,000 times more likely to produce large numbers of fatalities than accidents at nuclear plants. The study examined two representative types of nuclear power reactors from the 50 operating reactors, and has considered that extrapolation to a base for 100 reactors is a reasonable representation. Improvements in design, construction, operation, and maintenance would be expected to reduce the risks for later expansion in nuclear reactor operations. The fatalities shown in Figs. 1 and 2 do not include potential injuries and longer-term health effects from either the nonnuclear or nuclear accidents. For the nuclear accidents, early illness would be about 10 times the fatalities in comparison to about 8,000,000 injuries caused annually by other accidents. The long-term health effects such as cancer and genetic effects are predicted to be smaller than the normal incidence rates, with increases in incidence difficult to detect even for large accidents. Thyroid illnesses, which rarely lead to serious consequences, would begin to approach the normal incidence rates only for large accidents.

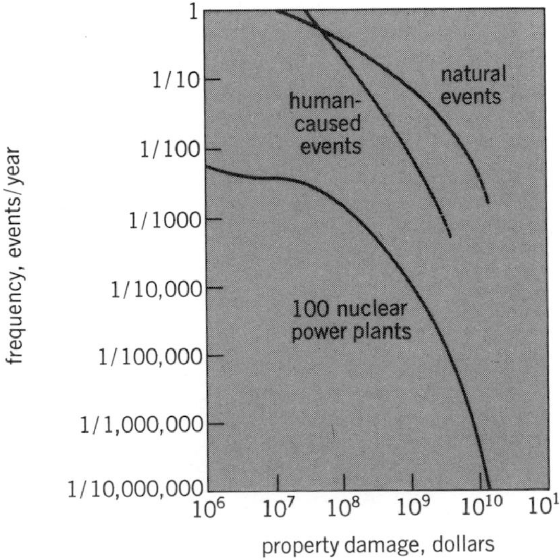

Fig. 3. Estimated frequency of property damage due to natural and human-caused events.

The likelihood and dollar value of property damage arising from nuclear and nonnuclear accidents are illustrated in Fig. 3. Both natural events (tornadoes, hurricanes, earthquakes) and human-caused events (air crashes, fires, dam failures, explosions, hazardous chemicals) might result in property damages in billions of dollars at frequencies up to 1000 times greater than that for accidents arising from the operation of 100 nuclear power plants.

Figures 1, 2, and 3 represent overall risk information. Risk to individuals being fatally injured through various causes is summarized in Table 3. The results of the study indicate that the predicted nuclear accidents are very small compared to other possible causes of fatal injuries.

The probability of an accident leading to the melting of the fuel core was estimated to be one chance per 20,000 reactor-years of operation, or for 100 operating reactors, one chance in 200 per year. The consequences of a core melt depend upon a number of subsequent factors, including additional failures leading to release of radioactivity, type of weather conditions, and population distribution at the particular site. The factors would have to occur in their worst conditions to produce severe consequences. Table 4 illustrates the progression of consequences and the likelihood of occurrence.

There has been considerable controversy concerning the risk estimates given in the Reactor Safety Study as a result of an NRC-sponsored review of them by a panel chaired by H. W. Lewis in 1977–1978. The Lewis panel argued that the Reactor Safety Study had, by its quantitative estimates, suggested a higher estimating precision than was justified by the data and procedures used. The Lewis group suggested that, rather than providing a single numerical estimate for a given risk, a numerical range should be quoted to more correctly reflect the uncertainties involved. However, Lewis also suggested that the Reactor Safety Study estimates were probably overconservative in that the most conservative choice had been made at each branch of the fault tree analyses whereas in an actual incident it would be anticipated that at least some of the branches would involve positive rather than negative choices.

Enrichment facilities. Only 0.7% of the uranium that is found in nature is the isotope uranium-235, which is used for the fuel in the nuclear power reactors. An enrichment of several percent is needed, and the Federal government (ERDA) owns the enrichment facilities. Expansion of the enrichment facilities is needed to meet expected demands.

Breeder reactor program. In a breeder reactor, more fissile fuel is generated than is consumed. For example, in the fissioning of uranium-235, the neutrons released by fission are used both to continue the neutron chain reaction which produces the fission and to react with uranium-238 to produce uranium-239. The uranium-239 in turn decays to neptunium-239 and then to plutonium-239. Uranium-238 is called a fertile fuel, and uranium-235, as well as plutonium-239, is a fissile fuel and can be used in nuclear power reactors. The reactions noted can be used to convert most of the uranium-238 to plutonium-239 and thus provide about a 60-fold extension in the available uranium energy source. Breeder power reactors can be used to generate electrical power and to produce more fissile fuel. Breeding is also possible using the fertile fuel thorium-233, which in turn is converted to the fissile fuel uranium-233. An almost inexhaustible energy source becomes possible with breeder reactors. The breeder reactors would decrease the long-range needs for enrichment and for mining more uranium.

Development. The development of the breeder nuclear power reactors had been initiated by the AEC, and the experimental breeder reactor I (EBR-I) was the first reactor to demonstrate production of electrical energy from nuclear energy (Dec. 20, 1951) and to prove the feasibility of breeding utilizing a fast reactor and a liquid metal coolant. In a special test in 1955, an operator error led to a substantial melting of the fuel, but with no off-site effects. The construction on EBR-II was begun in 1958, and it has been operating since 1963. The EBR-II installation is an integrated nuclear reactor power, breeding, and fuel cycle facility. The first commercial breeder licensed for operation, in 1965, was the Enrico Fermi Fast Breeder Reactor. In 1966, while the reactor was operating at a low power, partial coolant blockage occurred, leading to some melting of fuel and release of some radioactivity within the containment building. Although the operation of the reactor was resumed in 1970, the project was discontinued in 1972, primarily for economic considerations.

Environmental impact studies. The AEC had established that the priority breeder program would utilize the LMFBR concept. In response to a 1973 Court of Appeals ruling requiring an environment impact statement for the total LMFBR research and development program, the AEC issued a seven-volume *Proposed Final Environmental Statement* in December 1974 (WASH-1535), covering environmental, economic, social, and other impacts; alternative technology options; mitigation of adverse environmental impacts; unavoidable adverse environmental impacts; short- and long-term losses; irreversible and irretrievable commitments of resources; cost-benefit analysis; and responses to the many critical comments received during the development of the environmental statements in previous drafts. With due regard to the inherent hazards and the need to carefully manage plutonium, including a vigorous program to strengthen and improve safeguards, the conclusion reached was: "The LMFBR can be developed as a safe, clean, reliable and economic electric power generation system and the advantages of developing the LMFBR as an alternative energy option for the Nation's use far outweigh the attendant disadvantages."

Table 3. Average risk of fatality by various causes

Accident type	Total number	Individual chance per year
Motor vehicle	55,791	1 in 4,000
Falls	17,827	1 in 10,000
Fires and hot substances	7,451	1 in 25,000
Drowning	6,181	1 in 30,000
Firearms	2,309	1 in 100,000
Air travel	1,778	1 in 100,000
Falling objects	1,271	1 in 160,000
Electrocution	1,148	1 in 160,000
Lightning	160	1 in 2,000,000
Tornadoes	91	1 in 2,500,000
Hurricanes	93	1 in 2,500,000
All accidents	111,992	1 in 1,600
Nuclear reactor accidents (100 plants)	–	1 in 5,000,000,000

A reexamination of the LMFBR program was undertaken by ERDA with its issuance of the 10-volume *Final Environmental Statement* (ERDA-1535) to cover additional and supporting information for ERDA's findings and responses to the comment letters on WASH-1535. The possible impact of a new technology upon society has never been so thoroughly questioned. The *Final Environmental Statement* addressed major uncertainties in nuclear reactor safety, fuel cycle performance, safeguards, waste management, health effects, and uranium resource availability. The environmental acceptability, technical feasibility, and economic advantages of the LMFBR cannot be ascertained until additional research, development, and demonstration programs are undertaken. The ERDA decision was to proceed with a plan to continue the research and supporting programs so as to provide sufficient data by 1986 to make the decision on commercialization of the LMFBR technology. The plan contemplated the licensing, construction, and operation of the Clinch River Breeder Reactor; the design, procurement, component fabrication and testing, and the licensing for construction of the prototype large breeder reactor; and the planning of a commercial breeder reactor (CBR-I). The heat trans-

Table 4. Approximate values of early illness and latent effects for 100 reactors

	Consequences			
Chance per year	Early illness	Latent cancer fatalities* per year	Thyroid illness* per year	Genetic effects† per year
1 in 200‡	< 1.0	< 1.0	4	< 1.0
1 in 10,000	300	170	1,400	25
1 in 100,000	3,000	460	3,500	60
1 in 1,000,000	14,000	860	6,000	110
1 in 10,000,000	45,000	1500	8,000	170
Normal incidence per year	4×10^5	17,000	8,000	8,000

SOURCE: Nuclear Regulatory Commission, *Reactor Safety Study*, WASH-1400, 1975.

*This rate would occur approximately in the 10–40-yr period after a potential accident.

†This rate would apply to the first generation born after the accident. Subsequent generations would experience effects at decreasing rates.

‡This is the predicted chance per year of core melt for 100 reactors.

port and power generation system for the Clinch River Breeder Reactor is illustrated in Fig. 4.

United States policies. In April 1977 the Carter administration proposed to cancel construction of the Clinch River Breeder Reactor, reflecting the policies on plutonium recycling and breeding discussed above. However, Congress decided to continue funding for this project in fiscal years 1978 and 1979.

Foreign development. Prototype breeder reactors have been constructed in France, England, and the Soviet Union. Construction of a full-scale (1250 megawatts electric power, or MWe) commercial breeder, the Super Phénix, has been undertaken in France.

Radioactive waste management. In 1975 the responsibility for the development of a proposed Federal repository for high-level radioactive wastes was transferred from the AEC to ERDA (absorbed by the Department of Energy in 1977). The AEC had established regulations that the high-level radioactive wastes from the chemical reprocessing of irradiated fuel must be converted to a stable solid within 5 years of its generation at the reprocessing plant, and that the solids be sealed in high-integrity steel canisters and delivered to the AEC for subsequent management within 10 years of its generation at the reprocessing plant. Initially the AEC had sought to develop a salt mine near Lyons, KS, as the repository, but effective public intervention disclosed deficiencies for the site selected and the proposal was withdrawn. Subsequently, a more extensive study was undertaken to review permanent, safe, geologic formations and to locate possible sites. Stable, deep-lying formations could serve to isolate the wastes and dissipate their heat generation without need for maintenance or monitoring. During the period of time that might be used for the demonstration of a repository on a pilot plant scale, a retrievable surface storage facility (RSSF) would be placed in operation. The RSSF would require maintenance and monitoring, and could be engineered using known technology.

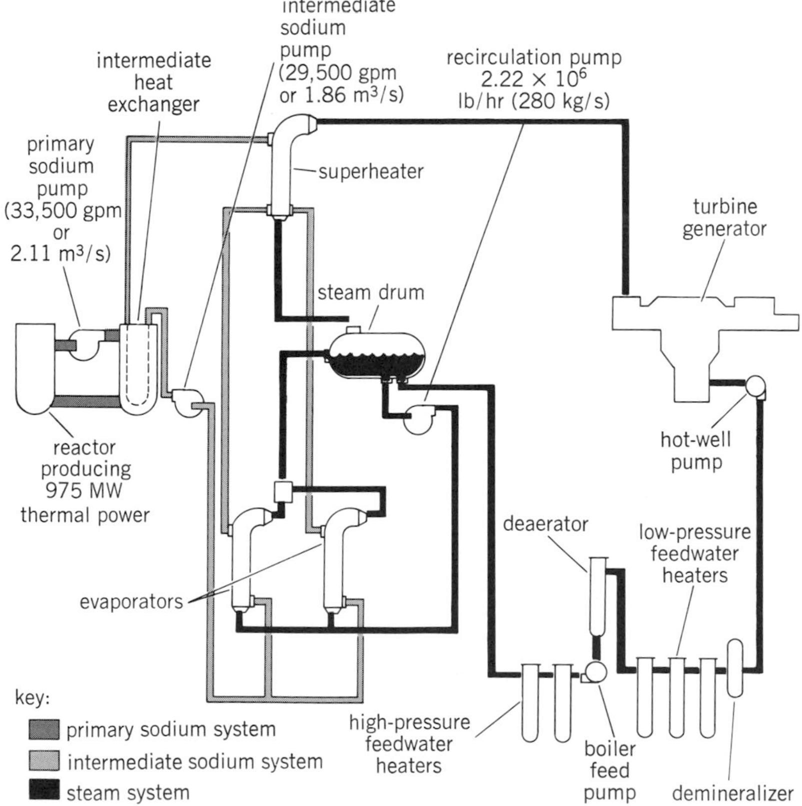

Fig. 4. Heat transport and power generation system for the Clinch River Breeder Reactor.

Critics of nuclear power consider the radioactive wastes generated by the nuclear industry to be too great a burden for society to bear. Since the high-level wastes will contain highly toxic materials with long half-lives, such as about a few tenths of one percent of plutonium that was in the irradiated fuel, the safekeeping of these materials must be assured for time periods longer than social orders have existed in the past. Nuclear proponents answer that the proposed use of bedded salts, for example, found in geologic formations that have prevented access of water and have been undisturbed for millions of years provide assurance that safe storage can be engineered. A relatively small area of several square miles would be needed for disposal of projected wastes.

Research and development has been underway since the 1950s on methods for solidifying the wastes; however, neither ERDA nor the NRC has provided the detailed criteria and guides needed by industry to carry forth the design of the waste solidification portions for the chemical reprocessing plants. Until such time as the chemical reprocessing plants begin operation, irradiated fuel will be retained at the reactor sites and at separate facilities away from reactors in appropriate water storage pools. Spent fuels can be stored in such pools for a period of decades if necessary. The water provides shielding and also carries away residual heat.

The decision by the Carter administration in 1977 to defer indefinitely the reprocessing of spent nuclear reactor fuels forced a reconsideration of the plans and schedules that had been under development for handling spent fuel from reactors in the United States. Utilities have expanded their on-site storage capabilities for spent fuel amid uncertainty about if and when reprocessing would be permitted. With indefinite deferral of reprocessing, some people are calling for geologic disposal of spent fuel assemblies. These assemblies are about the same size as the glass rods of processed waste and could also be placed in permanent geologic storage. However, this concept has drawbacks on a resource basis because it would mean throwing away the remaining fuel value of the uranium and plutonium.

Transportation. Transportation of nuclear wastes has received special attention. With increasing truck and train shipments and increased probabilities for accidents, the protection of the public from radioactive hazards is achieved through regulations and implementation which seek to provide transport packages with multiple barriers to withstand major accidents. For example, the cask used to transport irradiated fuel is designed to withstand severe drop, puncture, fire, and immersion tests.

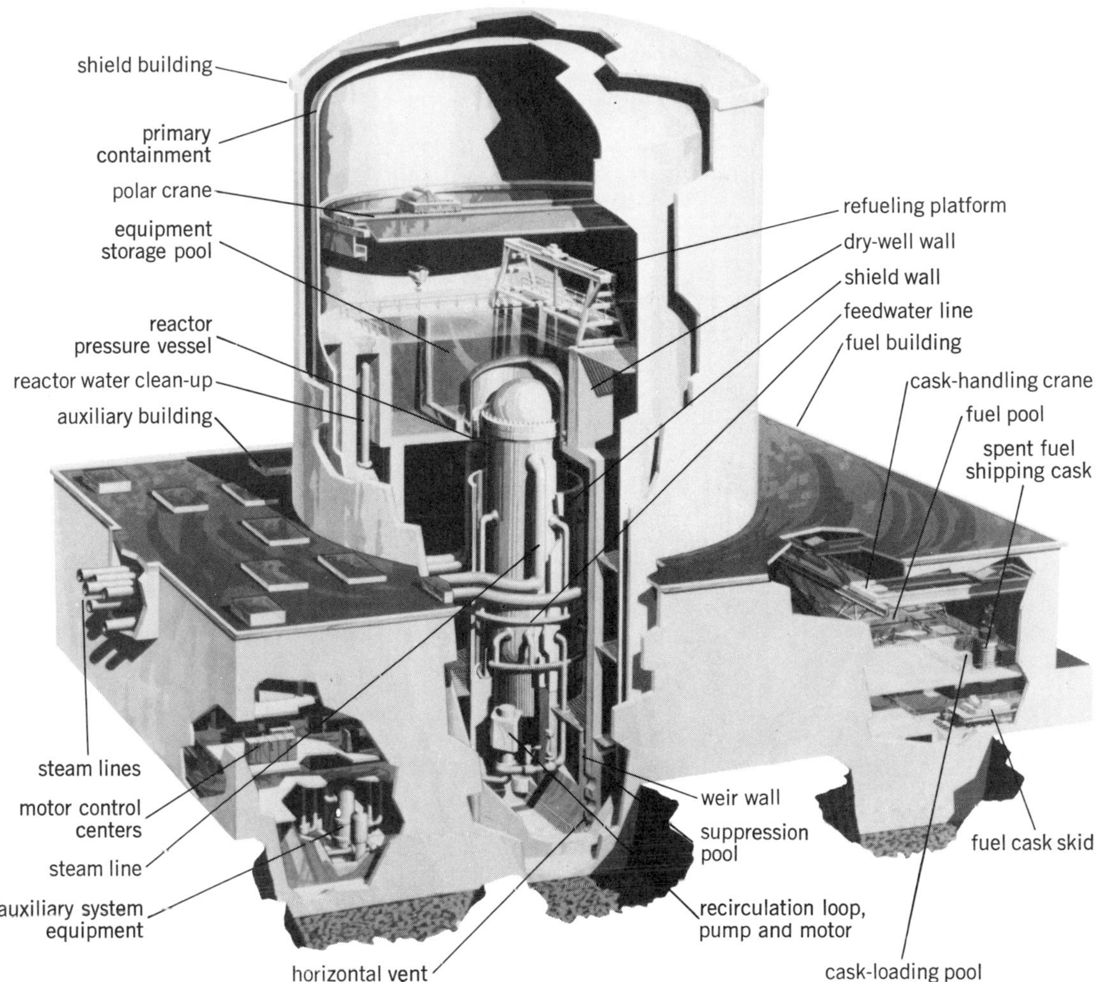

Fig. 5. Mark III containment of a boiling-water reactor, which illustrates safety features designed to minimize the consequences of reactor accidents. (*General Electric Co.*)

Low-level wastes. Management of low-level wastes generated by the nuclear energy industry requires use of burial sites for isolation of the wastes and decay to innocuous levels. Operation of the commercial burial sites is subject to regulations by Federal and state agencies.

Routine operations of nuclear power stations result in very small releases of radioactivity in the gaseous and water effluents. The NRC has adopted the principle that all releases should conform to the ALARA standard. Extension of ALARA guidance to other portions of the nuclear fuel cycle has been undertaken. *See* RADIOACTIVE WASTE MANAGEMENT.

Phase-out of governmental indemnity. In 1954 the Atomic Energy Act of 1946 was revised, making possible the possession and use of fissionable materials for industrial uses. As noted above, the Price-Anderson Act of 1957, extended in 1965 and again in 1975, protects the nuclear industry against unlimited liability in the unlikely event of a catastrophic nuclear accident. The total amount was set at $560,000,000, with private liability insurance starting at $60,000,000 in 1957 and increasing to $140,000,000 by 1977, and the government indemnity commensurately decreasing to $420,000,000. About 67% of the premiums paid to the private sector are placed in a reserve fund, and after a 10-year period, approximately 97 – 98% of this reserve fund has been returned. The smaller premiums paid to the Federal government are not returned. A 1966 amendment to the Price-Anderson Act provided features for no-fault liability and provisions for accelerated payment of claims. The 1975 extension of the act to 1987 provides for phasing out government indemnification, permits the $560,000,000 limit to float upward, and extends coverage to certain nuclear incidents that may occur outside the territorial limits of the United States. Each licensee is assessed a deferred payment, with a maximum level to be set per reactor. For example, in 1978 the licensees of the 66 operating reactors were each liable for a retrospective premium assessment of $5,000,000, making $330,000,000 available in addition to the base level of $140,000,000, and leaving $90,000,000 for the government indemnity. As the number of operating reactors increases, the government indemnity would phase out and would permit increases of total indemnity to exceed $560,000,000.

In addition to the indemnity insurance, private pools have provided property damage up to $175,000,000. New nuclear power stations would have capital costs in excess of $1,000,000,000.

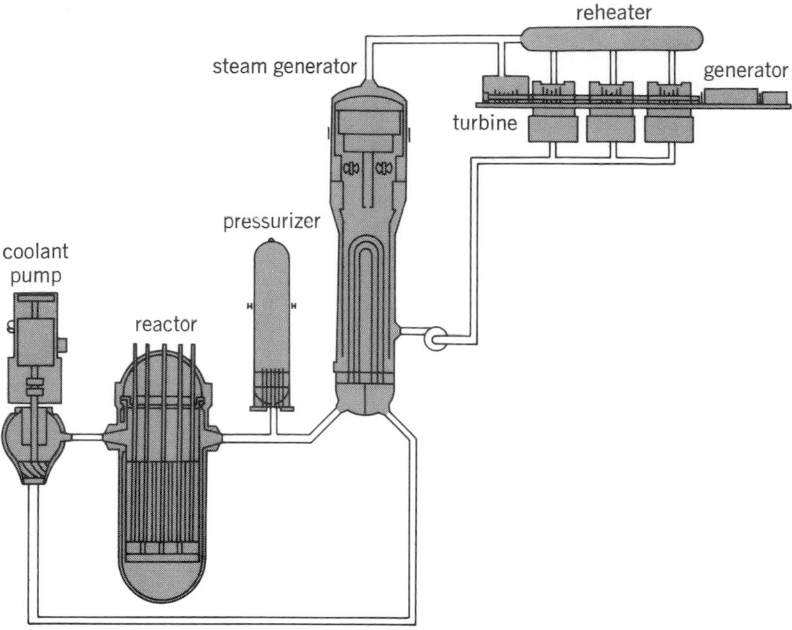

Fig. 6. Closed-cycle PWR. *(Westinghouse Electric Corp.)*

Types of power reactors. There are five commercial nuclear power reactor suppliers in the United States: three for pressurized-water reactors, Combustion Engineering, Inc., Babcock and Wilcox, and Westinghouse Electric Corporation; one for boiling-water reactors, General Electric Company; and one for high-temperature gas-cooled reactors, General Atomic Company. Approximately two-thirds of the orders for nuclear power reactors are shared by General Electric and Westinghouse, and the remaining one-third by Combustion and Babcock and Wilcox. One high-temperature gas-cooled reactor (HTGR), Fort St. Vrain Nuclear, with 330 MW net electrical power, has been licensed for operation, but orders placed for the higher-power units (with up to 1160 MW net electrical power) have been canceled or deferred indefinitely.

Fig. 7. Oconee Nuclear Power Station containment structures.

Boiling-water reactor (BWR). In a General Electric BWR designed to produce about 3580 MW thermal and 1220 MW net electrical power, the reactor vessel is 238 in. (6.05 m) in inside diameter, 5.7 in. (14.5 cm) thick, and about 71 ft (21.6 m) in height. The active height of the core containing the fuel assemblies is 148 in. (3.76 m). Each fuel assembly contains 63 fuel rods, and 732 fuel assemblies are used. The diameter of the fuel rod is 0.493 in. (12.5 mm). The reactor is controlled by the cruciform-shape control rods moving up from the bottom of the reactor in spaces between the fuel assemblies (177 control rods are provided). The water coolant is circulated up through the fuel assemblies by 20 jet pumps at about 70 atm (7 MPa), and boiling occurs within the core. The steam is fed through four 26-in. diameter (66 cm) steam lines to the turbine. About one-third of the energy released by fission is converted into electrical energy, and the remaining heat is removed in the condenser. The condenser operations are typical of both fossil and nuclear power plants, with heat being removed by the condenser having to be dissipated to the environment. Some limited use of the low-temperature heat source from the condenser is possible. The steam produced in the nuclear system will be at lower temperatures and pressures than that from fossil plants, and thus the efficiency of the nuclear plant in producing electrical power is less, leading to proportionately greater heat rejection to the environment.

Shielding is provided to reduce radiation levels, and pools of water are used for fuel storage and when access to the core is necessary for fuel transfers. Among the engineered safety features to minimize the consequences of reactor accidents is the containment. The function of the containment is to cope with the energy released by depressurization of the coolant should a failure occur in the primary piping, and to provide a secure enclosure to minimize leakage of radioactive material to the surroundings. The BWR utilizes a pool of water to condense the steam produced by the depressurization of the primary water coolant. Various arrangements have been used for the suppression pool. Other engineered safety features include the emergency core-cooling systems.

Pressurized-water reactor (PWR). Whereas in the BWR a direct cycle is used in which steam from the reactor is fed to the turbine, the PWR employs a closed system, as shown in Fig. 6. The water coolant in the primary system is pumped through the reactor vessel, transports the heat to a steam generator, and is recirculated in a closed primary system. A separate secondary water system is used on the shell side of the steam generator to generate steam, which is fed to the turbine, condensed, and recycled back to the steam generator. A pressurizer is used in the primary system to maintain about 150 atm (15 MPa) pressure to suppress boiling in the primary coolant. One loop is shown in Fig. 6, but up to four loops have been used.

The reactor pressure vessel is about 44 ft (13.4 m) in height, about 14.5 ft (4.4 m) in inside diame-

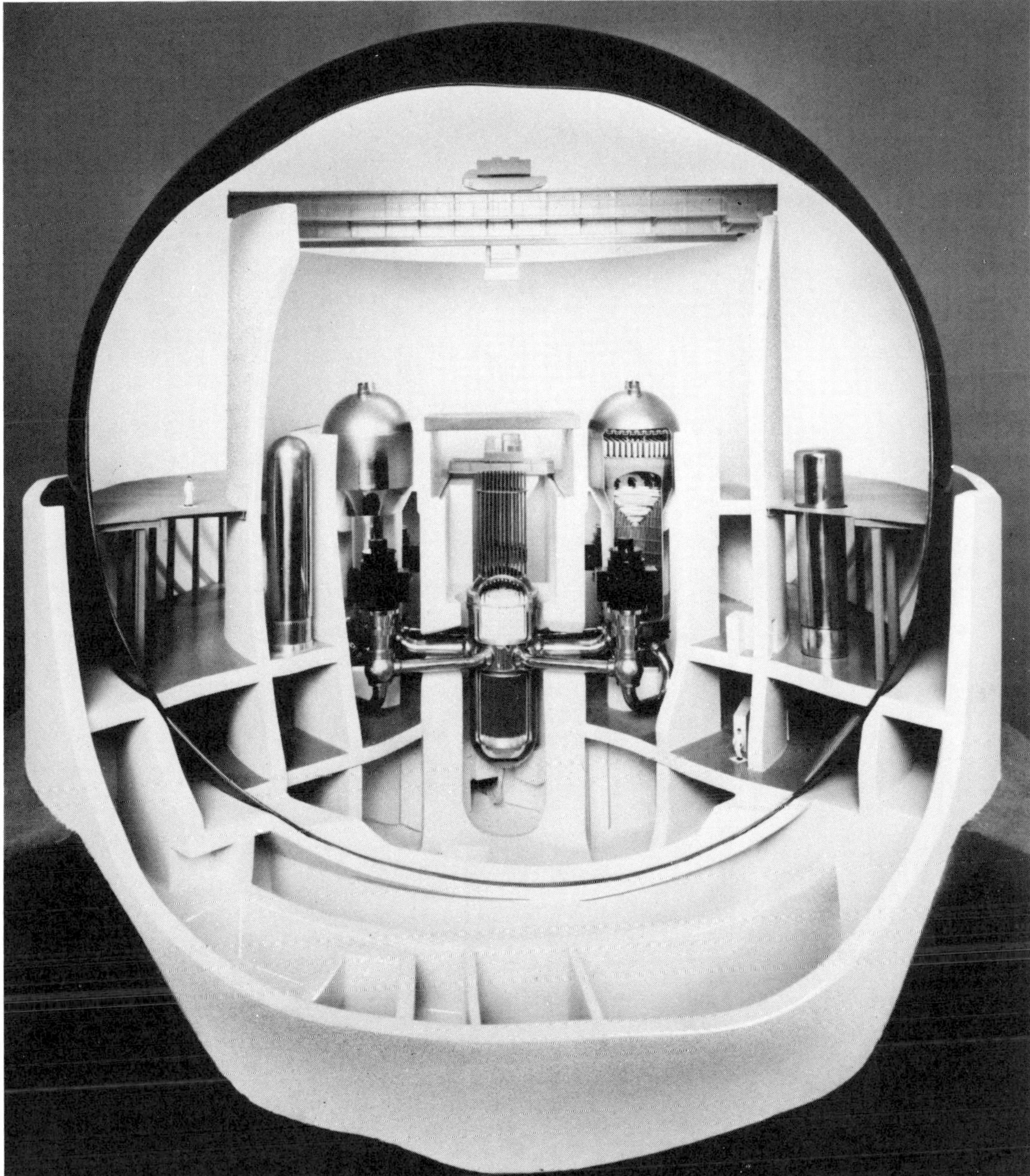

Fig. 8. Spherical containment design for a PWR. *(Combustion Engineering, Inc.)*

ter, and has wall thickness in the core region at least 8.5 in. (22 cm). The active length of the fuel assemblies may range from 12 to 14 ft (3.7 to 4.3 m), and different configurations are used by the manufacturers. For example, one type of fuel assembly contains 264 fuel rods, and 193 fuel assemblies are used for the 3411-MW-thermal, four-loop plant. The outside diameter of the fuel rods is 0.374 in. (9.5 mm). For this arrangement, the control rods are grouped in clusters of 24 rods per cluster, with 61 clusters provided. In the PWR the control rods enter from the top of the core. Control of the reactor operations is carried out by using both the control rods and a system to increase or decrease the boric acid content of the primary coolant.

Figure 7 is an external view of the Oconee Nuclear Power Station with three reactors, each housed in a separate containment building. The prestressed concrete containment buildings are designed for a 4-atm (400 kPa) rise in pressure and have inside dimensions of about 116 ft (35.4 m) in diameter and 208.5 ft (63.6 m) in height. The walls are 45 in. (1.14 m) thick. The containments have cooling and radioactive absorption systems as part of the engineered safety features. Figure 8 is a view of a design for a 3800-MW-thermal nuclear power reactor utilizing a two-loop system placed in a spherical steel containment (about 200, or 60 m, in diameter), surrounded by a reinforced-concrete shield building. A cutaway view of the containment building is shown in Fig. 9.

The instrumentation and control for a nuclear power station involves separation of control and

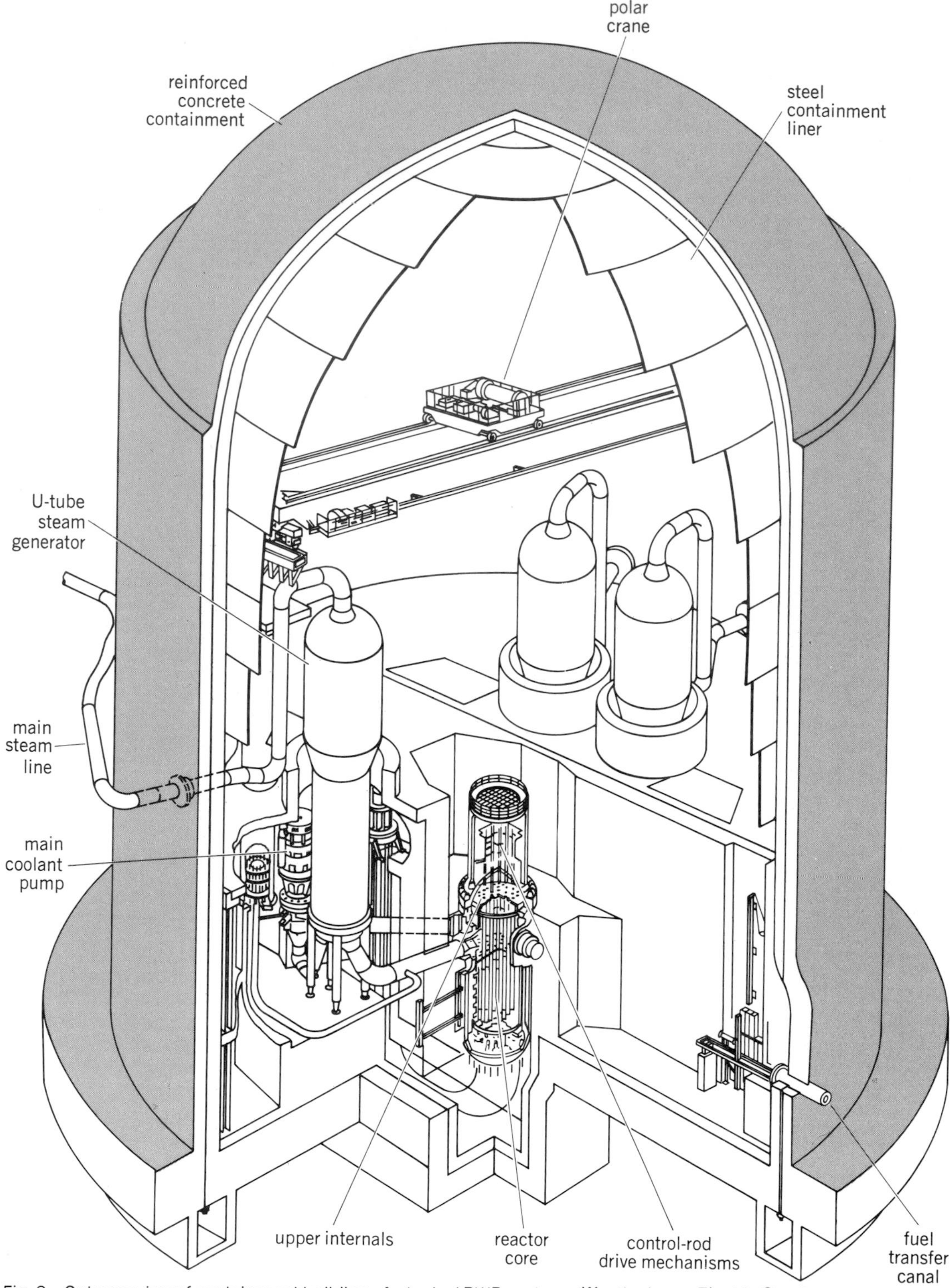

Fig. 9. Cutaway view of containment building of a typical PWR system. (*Westinghouse Electric Corp.*)

protection systems, redundant and diverse features to enhance the safety of the operations, and ex-core and in-core monitoring systems to ensure safe, reliable, and efficient operations. *See* ELECTRIC POWER GENERATION; ENERGY SOURCES.

[H. S. ISBIN]

Bibliography: Atomic Energy Commission, *Proposed Final Environmental Statement*: *Liquid Metal Fast Breeder Reactor Program*, WASH-1535, 1974; Energy Research and Development Administration, *Final Environmental Statement*, ERDA-1535 (including WASH-1535), 1975; International Atomic Energy Agency, *Nuclear Power and Its Fuel Cycle*, vols. 1–7, 1977–1978; H. S. Isbin (coordinator), *Public Issues of Nuclear Power*, 1975; Nuclear Regulatory Commission, *Operating Units Status Reports, such as Licensed Operating Reactors*, NUREG-75/020-12, December 1975; Nuclear Regulatory Commission, *Reactor Safety Study*, WASH-1400, 1975; Subcommittee on Energy and the Environment of the Committee on Interior and Insular Affairs, House of Representatives, 94th Congress, first session, pt. 1: *Overview of the Major Issues*, pt. 2: *Nuclear Breeder Development Program*, 1975; U.S. Congress, Office of Technology Assessment, *Nuclear Proliferation and Safeguards*, 1977.

Nuclear radiation

All particles and radiations emanating from an atomic nucleus due to radioactive decay and nuclear reactions. Thus the criterion for nuclear radiations is that a nuclear process is involved in their production. The term was originally used to denote the ionizing radiations observed from naturally occurring radioactive materials. These radiations were alpha rays (energetic helium nuclei), beta rays (negative electrons), and gamma rays (electromagnetic radiation with wavelength much shorter than visible light).

Nuclear radiations have traditionally been considered to be of three types based on the manner in which they interact with matter as they pass through it. These are the charged heavy particles with masses comparable to that of the nuclear mass (for example, protons, alpha particles, and heavier nuclei), electrons (both negatively and positively charged), and electromagnetic radiation. For all of these, the interactions with matter are considered to be primarily electromagnetic. (The neutron, which is also a nuclear radiation, behaves quite differently.) The behavior of mesons and other particles is intermediate between that of the electron and heavy charged particles.

A striking difference in the absorption of the three types of radiations is that only heavy charged particles have a range. That is, a monoenergetic beam of heavy charged particles, in passing through a certain amount of matter, will lose energy without changing the number of particles in the beam. Ultimately, they will be stopped after crossing practically the same thickness of absorber. The minimum amount of absorber that stops a particle is its range. The greatest part of the energy loss results from collisions with atomic electrons, causing the electrons to be excited or freed. The energy loss per unit path length is the specific energy loss, and its average value is the stopping power of the absorbing substance.

For electromagnetic radiation (gamma rays) and neutrons, on the other hand, the absorption is exponential; that is, the intensity decreases in such a way that the equation below is valid, where I is

$$-\frac{dI}{I} = \mu\, dx$$

the intensity of the primary radiation, μ is the absorption coefficient, and dx is the thickness traversed. The difference in behavior reflects the fact that charged particles are not removed from the beam by individual interactions, whereas gamma radiation photons (and neutrons) are. Three main types of phenomena involved in the interaction of electromagnetic radiation with matter (namely, photoelectric absorption, Compton scattering, and electron-positron production) are responsible for this behavior.

Electrons exhibit a more complex behavior. They radiate electromagnetic energy easily because they have a large charge-to-mass ratio and hence are subject to violent acceleration under the action of the electric forces. Moreover, they undergo scattering to such an extent that they follow irregular paths.

Whereas in the case of the heavy charged particles, electrons, or gamma rays the energy loss is mostly due to electromagnetic effects, neutrons are slowed down by nuclear collisions. These may be inelastic collisions, in which a nucleus is left in an excited state, or elastic collisions, in which the colliding nucleus acquires part of the energy of the nucleus as kinetic energy. In the first instance, the neutron must have enough kinetic energy (of the order of 1 MeV) to excite the collision partner. With less kinetic energy, only elastic scattering can slow down the neutron, a process which is effective down to thermal energies (about 1/40 keV). At this stage the collision, on the average, has no further effect on the neutron's energy.

As noted previously, the other nuclear radiations such as mesons have behaviors which are intermediate between that of heavy charged particles and electrons. Another radioactive decay product is the neutrino; because of its small interaction with matter, it is not ordinarily considered to be a nuclear radiation. *See* NUCLEAR REACTION.

[DENNIS G. KOVAR]

Nuclear reaction

A reaction which is produced as a result of interactions between atomic nuclei when the interacting particles approach each other to within distances of the order of nuclear dimensions (10^{-12} cm). In the usual experimental situation, one of the interacting particles, the target nucleus, is essentially at rest, and the reaction is initiated by bombarding it with nuclear projectiles of some type.

Means of producing reactions. Because of the intense electrostatic field produced by the nuclear charge, positively charged bombarding particles must have a large kinetic energy in order to overcome the electrostatic (Coulomb) repulsion and reach the target nucleus. Whereas for the lightest target nuclei, protons with kinetic energies of a few hundred thousand electron volts are sufficient to cause certain reactions, energies of many hundreds of millions of electron volts (MeV) are required to initiate reactions between the heavier nuclei. Beams of such energetic charged particles are provided by particle accelerators of various types (Van de Graaff generators, cyclotrons, linear accelerators, and so forth).

Since neutrons are uncharged, they are not repelled by the electrostatic field of the target nucleus, and neutron energies of only a fraction of an electron volt are sufficient to initiate nuclear reactions. Neutrons for reaction studies may be obtained from nuclear reactors or from various nuclear reactions. The interaction of electromagnetic radiation with nuclei may also lead to nuclear reactions. So-called photodisintegration may take place if the radiation has sufficient energy to cause the target nucleus to break up into two or more fragments. In a similar manner, high-energy electrons may also cause nuclear disintegrations. Electromagnetic radiation and electrons, however, interact strongly with the atomic electrons surrounding the target nucleus and are relatively less effective in causing nuclear reactions than are nuclear particles such as protons and neutrons.

Typical reactions. The most common and most extensively studied reactions are those which result in two products, one of which, the residual nucleus, is of nearly the same mass number and charge as the target nucleus, while the other product, the emitted particle, is either a single nucleon (a proton or a neutron) or a small assembly of nucleons, such as an α-particle. Examples are the reactions initiated when a layer of carbon of atomic mass 12 is bombarded with deuterons of a few MeV of kinetic energy. Deuterons, protons, neutrons, and α-particles are emitted, reactions (1)–(4) being responsible.

$$^{12}_{6}C + ^{2}_{1}H \rightarrow ^{2}_{1}H + ^{12}_{6}C \quad (1)$$

$$^{12}_{6}C + ^{2}_{1}H \rightarrow ^{1}_{1}H + ^{13}_{6}C \quad (2)$$

$$^{12}_{6}C + ^{2}_{1}H \rightarrow ^{1}_{0}n + ^{13}_{7}N \quad (3)$$

$$^{12}_{6}C + ^{2}_{1}H \rightarrow ^{4}_{2}He + ^{10}_{5}B \quad (4)$$

In these equations the subscript is the atomic number (nuclear charge) of the nucleus indicated by the usual chemical symbol, while the superscript is the mass number of the particular isotope involved. These reactions are conventionally written as $^{12}C(dd)^{12}C$, $^{12}C(dp)^{13}C$, $^{12}C(dn)^{13}N$, and $^{12}C(dd)^{10}B$, respectively. In each case, the interaction of the incident particle with the target nucleus results in the formation of a residual nucleus and an emitted particle. In the (*dd*) reaction, in which the residual nucleus is the same as the target nucleus, the process is referred to as scattering, either elastic or inelastic, depending upon whether the residual nucleus is left in its ground state or in one of its various excited states. The other three reactions lead to the production of a residual nucleus different from the target nucleus and are examples of nuclear disintegrations or transmutations. In these cases, also, the residual nucleus may be formed in its ground state or in one of its excited states. If the latter situation occurs, the residual nucleus will subsequently emit the excitation energy in the form of γ-radiation or, occasionally, electrons. The residual nucleus may also be a radioactive species, as in the case of ^{13}N formed in the $^{12}C(dn)$ reaction. In this case, the residual nucleus undergoes further transformations in accordance with its characteristic radioactive decay scheme.

Q value. In a nuclear reaction, the sum of the kinetic energies of the products may be greater than, equal to, or less than the sum of the kinetic energies before the reaction. The difference between these sums is the Q value for the particular reaction. Through Eq. (5), it is also equal to the

$$m_1c^2 + T_1 + m_2c^2 + T_2 = m_3c^2 + T_3 + m_4c^2 + T_4 \quad (5)$$

difference between the rest (proper) energies of the products and the rest energies of the initial nuclei. Reactions with a positive Q value are called exoergic or exothermic, while those with a negative Q are endoergic or endothermic. In reactions (1)–(4), for those cases in which the residual nucleus is formed in its ground state, the Q values are: $^{12}C(dd)^{12}C$, $Q=0$; $^{12}C(dp)^{13}C$, $Q=2.72$ MeV; $^{12}C(dn)^{13}N$, $Q=-0.28$ MeV; $^{12}C(d\alpha)^{10}B$, $Q=-1.39$ MeV. For reactions with a negative Q, a definite minimum energy, or threshold energy, is necessary for the reaction to take place. While there is no threshold energy for positive Q reactions, the yields of those reactions involving charged incident particles are quite low unless the bombarding energy is high enough to enable the incident particles to overcome the repulsive electric field from the charge on the target nucleus. A nuclear reaction and its inverse are reversible in the sense that their Q values are numerically equal but have opposite signs. Thus, the Q for the $^{10}B(\alpha d)^{12}C$ reaction is +1.39 MeV.

Conservation laws. The probability that a particular reaction will take place when an individual target nucleus interacts with an incident particle is a function of the bombarding energy, and the factors which determine it are not completely understood. However, it has been found experimentally that certain physical quantities are conserved in all nuclear reactions, and these conservation laws restrict the reactions which may take place. Those quantities which are conserved are described in the following paragraphs.

Charge. The total electric charge is always conserved. Except for high-energy reactions involving meson production, the total number of protons is also conserved. In the $^{12}C(d\alpha)^{10}B$ reaction, for example, there are seven protons involved in both the initial components and the final products.

Mass number. The total number of nucleons is always the same both before and after the reaction. For each of the four reactions listed, 14 nucleons are involved. Since, except for reactions which result in meson production, the number of protons is conserved, the number of neutrons is also constant at each stage in nuclear reactions.

Energy. The total energy is conserved in all nuclear reactions. The energy of each particle involved is $mc^2 + T$, where m is its rest mass, c the velocity of light, and T its kinetic energy. For two-particle reactions, such as those of Eqs. (1)–(4), the conservation of total energy is expressed in Eq. (5). In this equation the subscripts 1, 2, 3, and 4 refer to the incident particle, the target nucleus, the residual nucleus, and the emitted particle, respectively. In the common experimental situation, T_2 is so small as to be negligible. In this equation, the kinetic energies and the rest masses are usually expressed in units of millions of electron volts, the conversion factor between the two being 1 atomic mass unit (amu) = 931.502 MeV.

Linear momentum. The total linear momentum is the same before and after any nuclear reaction. A consequence of this conservation law is that the threshold energy necessary to initiate an endoergic reaction is not numerically equal to the negative Q value but is higher by the amount required to enable the final products to have a combined linear momentum equal to that brought into the reaction by the incident particle. The threshold energy of the $^{12}C(dn)^{13}N$ reaction, for example, is 0.33 MeV.

Angular momentum. The total angular momentum in nuclear reactions is the sum of the angular momentum associated with the relative motion of the reaction components and their intrinsic angular momentum, or spin. This total is always conserved.

Parity. Experimental evidence shows that, in most nuclear reactions, the total parity is the same before and after the interaction. Since the parity associated with the wave function describing the motion of a particle is determined by the angular momentum quantum number l (the parity is even if l is even, and odd if l is odd), and since every nucleus in any one of its allowed states has either even or odd parity, this conservation law, together with that for angular momentum, acts to restrict excited states of the residual nucleus which can be formed by an incident particle of given angular momentum.

Statistics. Since the total number of nucleons is conserved during a nuclear reaction, the statistics which govern the system are the same before, during, and after the interaction; Fermi-Dirac statistics are obeyed if the total number of nucleons is odd, and Bose-Einstein if the total number is even.

Reaction mechanisms. A number of mechanisms have been proposed to account for the observed features of nuclear reactions. Although none has been completely successful, they provide means for correlating and at least partially understanding many of the experimental facts. The most generally used models for nuclear reactions involve either compound nucleus formation or a direct interaction.

Compound nucleus formation. According to this point of view, originally proposed by N. Bohr, a nuclear reaction is visualized as proceeding in two distinct steps. The incident nucleus and the target nucleus are assumed to combine to form a compound nucleus, which exists for a time (of the order of 10^{-16} sec) which is much longer than the approximately 10^{-22} sec that would be required for the incident particle to pass through the target nucleus. The compound nucleus is always in a highly excited, unstable state and can subsequently decay into a number of different products, or through a number of so-called exit channels. In the four examples cited earlier, $^{14}_{7}N$ is the compound nucleus formed by the amalgamation of a deutron and $^{12}_{6}C$, and four possible decay modes or exit channels are indicated. Two essential features of this hypothesis are that, during its relatively long lifetime, the compound nucleus "forgets" the particular way in which it was formed, and that the energy brought in by the incident particle is shared by all the nuclear constituents. The probability that a particular reaction will occur is, then, the product of the probability of forming the compound nucleus and the probability that it will decay through a particular exit channel. Experiments indicate that, for a given energy of excitation in the compound nucleus, this latter factor (probability of decay through a particular exit channel) is independent of the manner in which the compound nucleus is formed. In the case of ^{14}N, it can be formed by $^{13}C+p$ or $^{10}B+\alpha$, as well as by $^{12}C+d$. While certain features of various types of interactions cannot be completely explained on the compound nucleus hypothesis, it appears that this mechanism plays some role in nearly all nuclear reactions.

Direct interactions. Some reactions have probabilities or other properties which conflict with the predictions of the compound nucleus hypothesis, and many are better explained on the assumption that the incident nucleus does not combine with the target nucleus as a whole, but rather that it, or some component, interacts only with the surface or with some individual constituent. The entire process is completed in the time required for the bombarding particle to traverse the diameter of the target nucleus.

Many direct reactions can be classified as "pick-up" reactions or as "stripping" reactions. A pick-up reaction is visualized as involving the acquisition by the bombarding nucleus of some component, such as a proton, neutron, or alpha particle, from the target nucleus. In a stripping reaction, some component from the bombarding nucleus is "stripped off" and acquired by the target nucleus. The (dp) reaction is a simple example of a stripping reaction, whereas the (dt) process is a reaction which involves pickup.

Coulomb excitation. It is observed that, in the bombardment of nuclei with charged particles, gamma rays characteristic of transitions between excited states of the target nucleus are produced at bombarding energies so low that the probability of either direct interaction or compound nucleus formation is negligible. This process is well explained by the assumption that the nuclei interact with the rapidly changing electric field caused by the passage of the charged bombarding particle.

Elastic scattering. This process leaves the quantum state of the scatterer unchanged. For charged bombarding particles with low energies, the elastic nuclear scattering is accurately described in terms of the inverse-square force law between electric charges. In this case, the process is known as Rutherford scattering. For higher bombarding energies, where the particle can come within the range of the various nuclear forces (approximately 10^{-13} cm), the scattering deviates from predictions based on the inverse-square law. In the case of neutrons, the elastic scattering is entirely due to the nuclear forces.

Nuclear cross sections. The cross section for a nuclear reaction is a measure of its probability. Consider a reaction initiated by a beam of particles bombarding a region which contains N atoms per unit area (uniformly distributed) and where I particles per second striking the area result in R reactions of a particular type per second. This result can be expressed in terms of the fraction of the bombarded region which is effective in producing reaction products R/I. If this is divided by the number of nuclei per unit area, the effective area or cross section per target nucleus is obtained. The cross section $\sigma=R/IN$. This is referred to as the total cross section, since it involves all the disintegration products of the reaction. The dimensions are those of an area, and total cross sections are expressed in either square centimeters or in barns (1 barn$=10^{-24}$ cm^2). The differential cross section refers to the probability that a particular reaction product will be observed at a given angle with respect to the beam direction. Its dimensions are those of an area per unit solid angle.

Types of reactions. Aside from elastic and inelastic scattering, the most common interactions initiated by the usual bombarding particles are discussed in the following paragraphs.

Proton-induced reactions. Capture reactions, in which the proton combines with the target nucleus to form a compound nucleus in an excited state, occur over a wide range of proton energies. If the compound nucleus decays to its ground state by the emission of a γ-ray, the process is known as a (p,γ) reaction. With higher proton energies, a (p,n) reaction is possible. This always has a negative Q value and leads to a radioactive residual nucleus. For many target nuclei, the (p,α) reaction has a high positive Q, but the yields are low, except at high proton energies, because of the difficulty of the doubly charged α-particle in penetrating the nuclear barrier.

Deuteron-induced reactions. The (d,p), (d,n), and (d,α) reactions usually have positive Q values. Except for light nuclei, in which the nuclear potential barrier is low, the (d,α) reactions have low probabilities. The (d,n) reactions of deuterium, tritium, and beryllium are important as sources of neutrons. Both the (d,p) and (d,n) reactions often lead to radioactive residual nuclei that are useful in various fields of investigation. Deuteron-induced reactions among the very light nuclei, for example, $^{3}H(dn)^{4}He$ ($Q=17.6$ MeV), are important thermonuclear processes. *See* NUCLEAR FUSION; THERMONUCLEAR REACTION.

Neutron-induced reactions. Neutron capture leading to an (n,γ) reaction is important for all stable nuclei and occurs even with very low-energy neutrons. With a given target nucleus, it yields the same final product as the (d,p) reaction. The capture γ-rays usually have maximum energies of about 8 Mev. This reaction is the source of many of the radioactive isotopes produced by nuclear reactors. For high-energy neutrons, the (n,p) and (n,α) reactions are also observed. In very heavy nuclei, neutron capture may lead to disintegration of the compound nucleus into two massive fragments, with the release of large amounts of kinetic energy and several additional neutrons. For a discussion of this phenomenon *See* NUCLEAR FISSION.

Alpha-particle-induced reactions. The (α,p) reactions of various light nuclei using the α-particles from naturally occurring radioactive substances were the first examples of artificially produced nuclear disintegrations. High α-particle energies are required for other than light nuclei because of the Coulomb barrier of the nucleus. At sufficiently high energies (about 30 MeV), (α,p) and (α,n) reactions are observed, even in the bombardment of heavy nuclei.

Heavy-ion reactions. Heavy-ion reactions are those produced by nuclei having masses greater than those of the helium isotopes. High-energy beams of nuclei such as ^{12}C, ^{32}S, and ^{40}Ca can produce reactions of much greater variety and can involve much higher angular momentum transfers than those induced by light ions, and are of great value in nuclear spectroscopy. There is a possibility that the interaction of such heavy ions with other heavy nuclei may lead to the production of new elements with atomic numbers even higher than the known transuranic atoms.

[WILLIAM W. BUECHNER]

Bibliography: J. Cerny (ed.), *Nuclear Spectroscopy and Reactions*, 1974; J. de Boer and H. J. Mang (eds.), *Proceedings of the International Conference on Nuclear Physics*, Munich, 1973; P. E. Hodgson, *Nuclear Reactions and Nuclear Spectroscopy*, 1971; M. Jean and R. A. Ricci (eds.), *Nuclear Structure and Nuclear Reactions*, 1969; R. L. Robinson et al. (eds.), *Proceedings of the International Conference on Reactions between Complex Nuclei*, Gatlinburg, 1974.

Nuclear reactor

A system utilizing nuclear fission in a controlled and self-sustaining manner. Neutrons are used to fission the nuclear fuel, and the fission reaction produces not only energy and radiation, but also additional neutrons. Thus a neutron chain reaction ensues. A nuclear reactor provides the assembly of materials to sustain and control the neutron chain reaction, to appropriately transport the heat produced from the fission reactions, and to provide the necessary safety features to cope with the radiation and radioactive materials produced by the operation of the nuclear reactor. *See* CHAIN REACTION; NUCLEAR FISSION.

Nuclear reactors are used in a variety of ways as sources for energy, for nuclear radiations, and for special tests and feasibility demonstrations. Since the first demonstration of a nuclear reactor, made beneath the West Stands of Stagg Field at the University of Chicago on Dec. 2, 1942, more than 500 nuclear reactors have been built and operated in the United States. Extreme diversification is possible with the materials available, and reactor dimensions may vary from football size to house size. The rates of energy release for controlled operations may vary from a fraction of a watt to thousands of megawatts. The critical size of a nuclear reactor is governed by the factors affecting the control of the neutron chain reaction, and the thermal output of the reactor is determined by the factors affecting the effectiveness of the coolant in removing the fission energy released.

The generation of electric energy by a nuclear power plant requires the use of heat to produce steam or to heat gases to drive turbogenerators. Direct conversion of the fission energy into useful work is possible, but an efficient process has not yet been realized to accomplish this. Thus, in its operation the nuclear power plant is similar to the conventional coal-fired plant, except that the nuclear reactor is substituted for the conventional boiler.

The rating of a reactor is usually given in kilowatts (kW) or megawatts (MW) thermal, representing the heat generation rate. The net output of electricity of a nuclear plant is about one-third of the thermal output. Significant economical gains have been achieved by building improved nuclear reactors with outputs of about 3000 MW thermal and about 1000 MW electrical. *See* ELECTRIC POWER GENERATION; NUCLEAR POWER.

FUEL AND MODERATOR

The fission neutrons are released at very high energies and are called fast neutrons. The average kinetic energy is 2 MeV, with a corresponding neutron speed of 1/15 the speed of light. Neutrons slow down through collisions with nuclei of the surrounding material. This slowing-down process is made more effective by the introduction of light-weight materials, called moderators, such as heavy water (deuterium oxide), ordinary (light) water, graphite, beryllium, beryllium oxide, hydrides, and organic materials (hydrocarbons). Neutrons that have slowed down to an energy state in equilibrium with the surrounding material are called thermal neutrons. The probability that a neutron will cause the fuel material to fission is greatly enhanced at thermal energies, and thus most reactors utilize a moderator for the conversion of fast neutrons to thermal neutrons.

With suitable concentrations of the fuel material, neutron chain reactions also can be sustained at higher neutron energy levels. The energy range between fast and thermal is designated as intermediate. Fast reactors do not have moderators and are relatively small.

Reactors have been built in all three categories. The first fast reactor was the Los Alamos assembly called Clementine, which operated from 1946 to 1953. The fuel core consisted of nickel-coated rods of pure plutonium metal, contained in a 6-in.-diameter (15 cm) mild (low-carbon) steel pot. Coolants for fast reactors may be steam, gas, or liquid metals. Current fast reactors utilize liquid sodium as the coolant and are being developed for breeding and power. An example of an intermediate reactor was the first propulsion reactor for the submarine USS *Seawolf*. The fuel core consisted of enriched uranium with beryllium as a moderator; the original coolant was sodium, and the reactor operated from 1956 to 1959. Examples of thermal reactors are given later.

Fuel composition. Only three isotopes—uranium-235, uranium-233, and plutonium-239—are feasible as fission fuels, but a wide selection of materials incorporating these isotopes is available.

Uranium-235. Naturally occurring uranium contains only 0.7% of the fissionable isotope uranium-235, the balance being essentially uranium-238. Uranium with higher concentrations of uranium-235 is called enriched uranium.

Uranium metal is susceptible to irradiation damage, which limits its operating life in a reactor. The life expectancy can be improved somewhat by heat treatment, and considerably more by alloying with elements such as zirconium or molybdenum. Uranium oxide exhibits better irradiation damage resistance and, in addition, is corrosion-resistant in oxidizing media. Ceramics such as uranium oxide have a very low thermal conductivity and lower density than metals, which are disadvantageous in certain applications.

Uranium metal can be fabricated by relatively well-established techniques, provided proper care is taken to prevent oxidation. The metal is melted in vacuum furnaces and can be cast by gravity or injection. Ingots can be rolled or extruded, and relatively complicated shapes can be fabricated. Most commonly, fuel elements are in shape of rods or plates and are fabricated by casting, rolling, or extrusion.

Current light-water-cooled nuclear power reactors utilize uranium oxide as a fuel, with an enrichment of several percent uranium-235. Cylindrical rods are the most common fuel-element configuration. They can be fabricated by compacting and sintering cylindrical pellets which are then assembled into metal tubes which are sealed.

Developmental programs for attaining long-lived solid-fuel elements include studies with uranium oxide, uranium carbide, and other refractory uranium compounds. *See* URANIUM.

Plutonium-239. Plutonium-239 is produced by neutron capture in uranium-238. It is a by-product in power reactors and is becoming increasingly available as nuclear power production increases. However, plutonium as a fuel is at a relatively early stage of development and the commercial recycle of plutonium from processed spent fuel was deferred indefinitely in the United States by the Carter administration in April 1977.

Plutonium is extremely hazardous to handle because of its biological toxicity and must be fabricated in glove boxes to ensure isolation from operating personnel. It can be alloyed with other metals and fabricated into various ceramic compounds. It is normally used in conjunction with uranium-238; alloys of uranium-plutonium, and mixtures of uranium-plutonium oxides and carbides, are of most interest. Except for the additional requirements imposed by plutonium toxicity, much of the uranium technology is applicable to plutonium. For the light-water nuclear power reactors, the oxide fuel pellets are contained in a zirconium alloy tube. Stainless steel tubes are used for containing the oxide fuel for the fast breeder reactors. *See* PLUTONIUM.

Uranium-233. Uranium-233, like plutonium, does not occur naturally, but is produced by neutron absorption in thorium-232, a process similar to that by which plutonium is produced from uranium-238. Interest in uranium-233 arises from its favorable nuclear properties and the abundance of thorium. However, studies of this fuel cycle are at a relatively early stage.

Uranium-233 also imposes special handling problems because of biological toxicity, but it does not introduce new metallurgical problems. Thorium is metallurgically different, but it has very favorable properties both as a metal and as a ceramic. *See* NUCLEAR FUEL.

Fuel distribution. Fuel-moderator assemblies may be homogeneous or heterogeneous. Homogeneous assemblies include the aqueous-solution-type water boilers and molten-salt-solution dispersions, slurries, and suspensions. The few homogeneous reactors built have been used for limited research and for demonstration of the principles and design features. In the heterogeneous assemblies the fuel and moderator form separate solid or liquid phases, such as solid-fuel elements spaced either in a graphite matrix or in a water phase. Most power reactors utilize an arrangement of closely spaced, solid-fuel rods, about 1/2 in. (13 mm) in diameter and 12 ft (3.7 m) long, in water In the arrangement shown in Fig. 1, fuel rods are arranged in a grid pattern to form a fuel assembly, and over 200 fuel assemblies are in turn arranged in a grid pattern in the reactor core.

The first homogeneous reactor was the Los Alamos Water Boiler, which commenced operations in 1944 at 1/20 watt. Various modifications were carried out to upgrade the thermal output. The aqueous solutions of uranium sulfate and later, uranium nitrate, with enrichments of about 17% were contained in a 1-ft-diameter (0.3 m) sphere. Homogeneous reactors which were used to demonstrate the feasibility of producing electrical power include the Homogeneous Reactor Experiment No. 1 (HRE-1) and HRE-2. HRE-1 operated from 1952 to 1954, generated 140 kW net electrical, contained an aqueous homogeneous solution of $UO_2 SO_4$ with an enrichment in excess of 90%, and was self-stabilizing because of its large negative coefficient. HRE-2 operated from 1957 to 1961 and generated 300 kW net electrical.

HEAT REMOVAL

The major portion of the energy released by the fissioning of the fuel is in the form of kinetic ener-

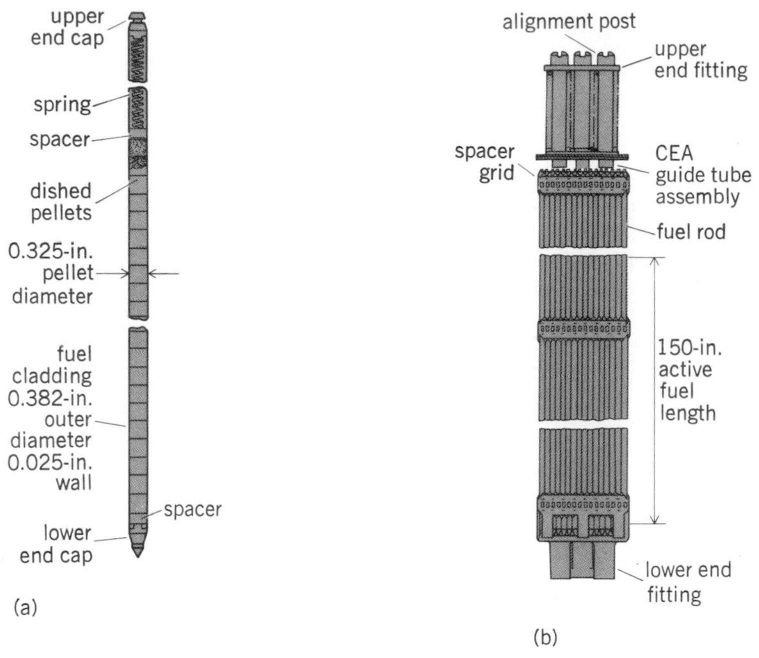

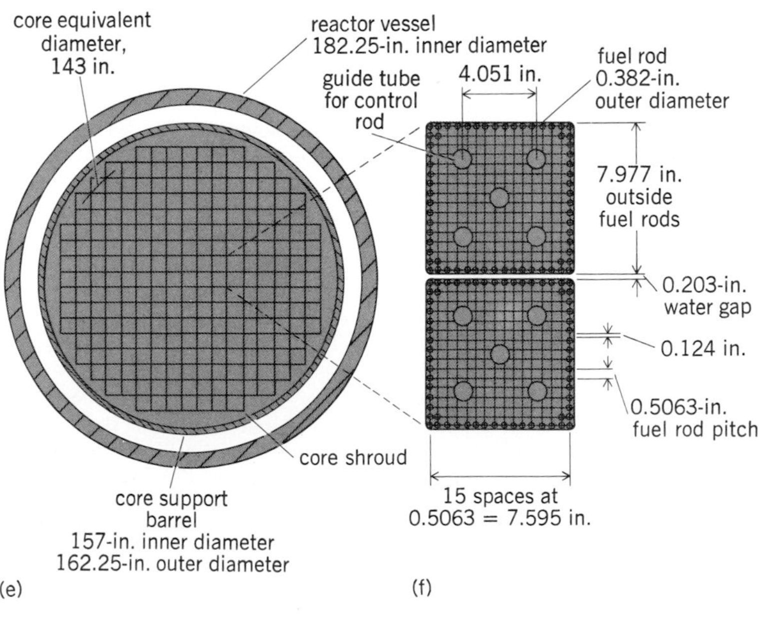

Fig. 1. Arrangement of fuel in the core of a pressurized-water reactor, a typical heterogeneous reactor. (*a*) Fuel rod; (*b*) side view (CEA = control element assembly), (*c*) top view, and (*d*) bottom view of fuel assembly; (*e*) cross section of reactor core showing arrangement of fuel assemblies; (*f*) cross section of two adjacent fuel assemblies, showing arrangement of fuel rods. 1 in. = 25.4 mm. (*Combustion Engineering, Inc.*)

gy of the fission fragments, which in turn is converted into heat through the slowing down and stopping of the fragments. For the heterogeneous reactors this heating occurs within the fuel elements. Heating also arises through the release and absorption of the radiations from the fission process and from the radioactive materials formed. The heat generated in a reactor is removed by a primary coolant flowing through the reactor.

Heat is not generated uniformly in a reactor. The heat flux decreases axially and radially from a peak at the center of the reactor, or near the center if the reactor is not symmetrical in configuration. In addition, local perturbations in heat generation can occur because of inhomogeneities in the reactor structure. These variations impose special considerations in the design of reactor cooling systems, including the need for establishing variations in coolant flow rate through the reactor to achieve uniform temperature rise in the coolant; avoiding local hot-spot conditions; and avoiding local thermal stresses and distortions in the structural members of the reactor.

Nuclear reactors have the unique thermal characteristic that heat generation continues after shutdown because of fission and radioactive decay of fission products. Significant fission heat generation occurs for only a few seconds after shutdown. Radioactive-decay heating varies with the decay characteristics of the fission products.

Accurate analysis of fission heat generation as a function of time immediately after reactor shutdown requires detailed knowledge of the speed and reactivity worth of the control rods. The longer-term fission-product-decay heating depends upon prior reactor operation. Typical values of the total heat generation after shutdown (as percent of operating power) are 10–20% after 1 sec, 5–10% after 10 sec, approximately 2% after 10 min, 1.5% after 1 hr, and 0.7% after 1 day.

Reactor coolants. Coolants are selected for specific applications on the basis of their heat-transfer capability, physical properties, and nuclear properties.

Water. Water has many desirable characteristics. It was employed as the coolant in the first production reactors and most power reactors still utilize water as the coolant. In a boiling-water reactor (BWR; Fig. 2) the water is allowed to boil and form steam that is piped to the turbine. In a pressurized-water reactor (PWR; Fig. 3) the coolant water is kept under increased pressure to prevent boiling, and transfers heat to a separate stream of water in a steam generator, changing that water to steam. Figure 4 shows the relation of the core and heat removal systems to the condenser, electrical power system, and waste management system in the Prairie Island Nuclear Plant, which is typical of plants using pressurized-water reactors. Cool intake water is pumped through hundreds of 1-in.-diameter (25 mm) tubes in the condenser, and the warm water from the condenser is then pumped over cooling towers and returned to the plant. *See* COOLING TOWER; RADIOACTIVE WASTE MANAGEMENT.

For both boiling-water and pressurized-water reactors, the water serves as the moderator as well as the coolant. Both light and heavy water are excellent neutron moderators, although heavy water (deuterium oxide) has a neutron-absorption cross section approximately 1/500 that for light water.

There is no serious neutron-activation problem with pure water; ^{16}N, formed by the (n,p) reaction with ^{16}O (absorption of a neutron followed by emission of a proton), is the major source of activity, but its 7.5-sec half-life minimizes this problem. The most serious limitation of water as a coolant for power reactors is its high vapor pressure. A coolant temperature of 550°F (288°C) requires a system pressure of approximately 1500 psi (10 MPa). This temperature is far below modern power station practice, for which steam temperatures in excess of 1000°F (538°C) have become common. Lower thermal efficiencies result from lower temperatures. Boiling-water reactors operate at about 70 atm (7 MPa), and pressurized-water reactors at 150 atm (15 MPa). The high pressure

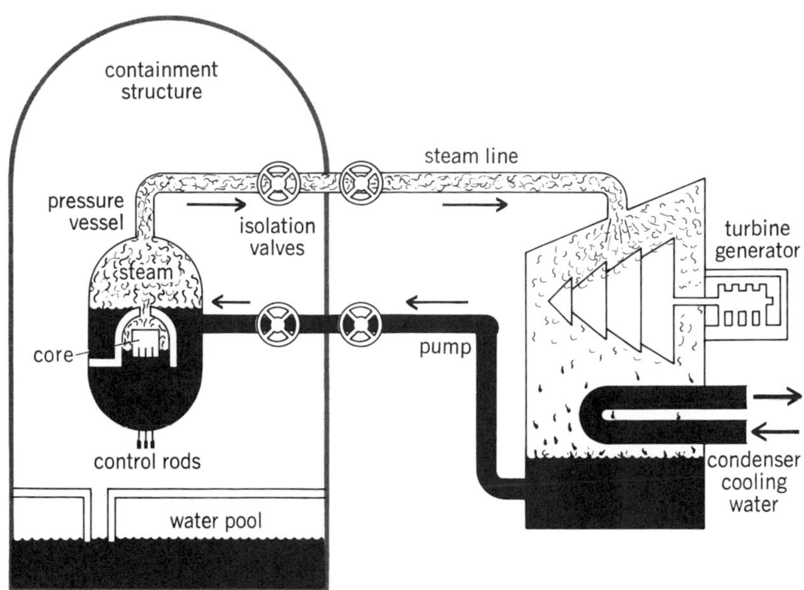

Fig. 2. Boiling-water reactor (BWR). (*Atomic Industrial Forum, Inc.*)

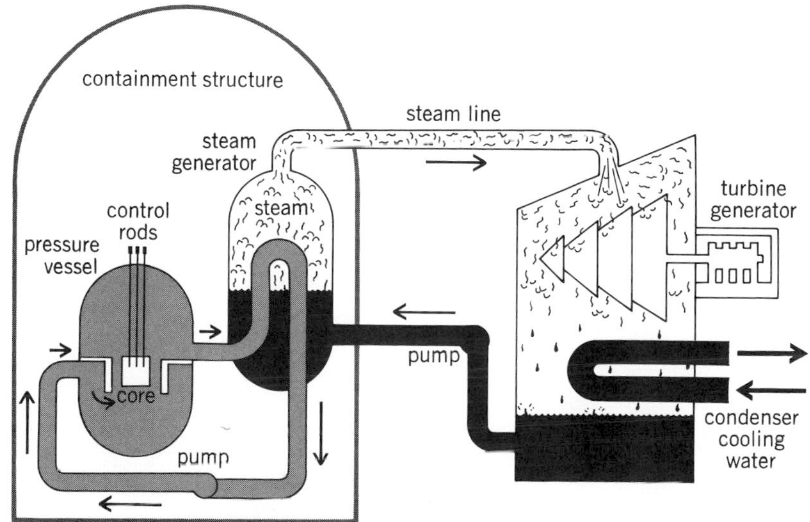

Fig. 3. Pressurized-water reactor (PWR). (*Atomic Industrial Forum, Inc.*)

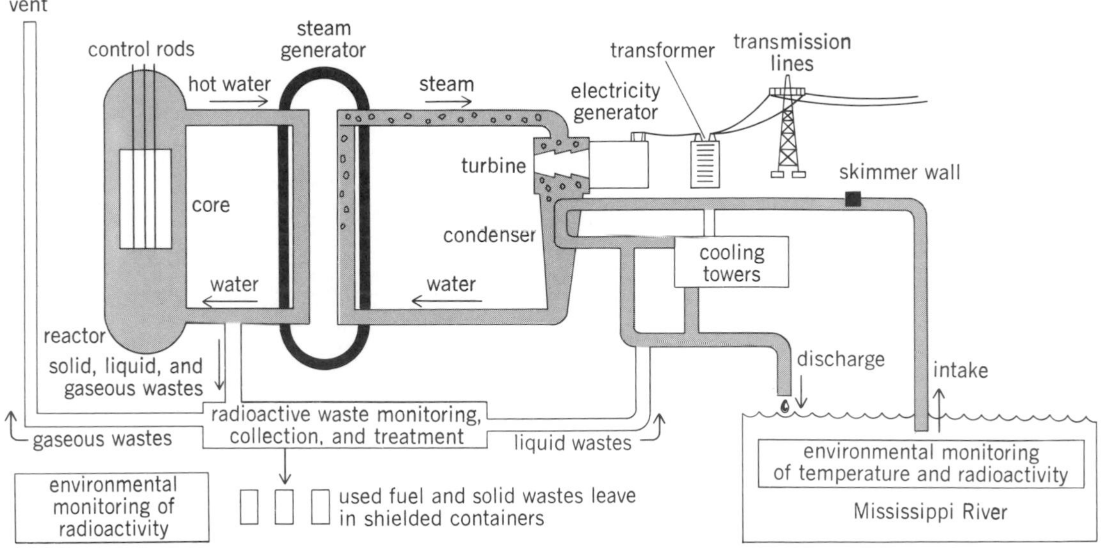

Fig. 4. Prairie Island Nuclear Plant, using pressurized-water reactors. *(Northern States Power Company)*

necessary for water-cooled power reactors imposes severe design problems, which will be discussed later. *See* NUCLEAR REACTION.

Gases. Gases are inherently poor heat-transfer fluids as compared with liquids because of their low density. This situation can be improved by increasing the gas pressure; however, this introduces other problems. Helium is the most attractive gas (it is chemically inert and has good thermodynamic and nuclear properties and has been selected as the coolant for the development of high-temperature gas-cooled reactor (HTGR) systems (Fig. 5), in which the gas transfers heat from the reactor core to a steam generator. Gases are capable of operation at extremely high temperature, and they are being considered for special process applications and direct-cycle gas-turbine applications. Hydrogen was used as the coolant for the reactors developed in the Nuclear Engine Rocket Vehicle Application (NERVA) Program, now terminated. Heated gas discharging through the nozzle developed the propulsive thrust.

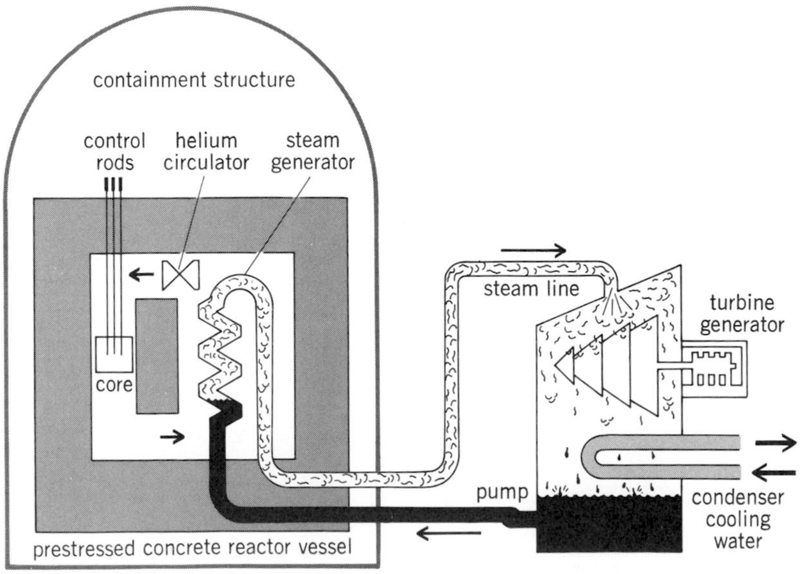

Organic coolants. Diphenyl and terphenyl possess good neutron-moderating properties and have lower vapor pressures than water. Organic coolants are noncorrosive and relatively inexpensive. Their major disadvantage is dissociation or decomposition under irradiation.

Liquid metals. The alkali metals, in particular, have excellent heat-transfer properties and extremely low vapor pressures at temperatures of interest for power generation. Sodium is the most attractive because of its relatively low melting point (208°F; 98°C) and high heat-transfer coefficient. It is also abundant, commercially available in acceptable purity, and relatively inexpensive. It is not particularly corrosive, provided low oxygen concentration is maintained. Its nuclear properties are excellent for fast reactors. In the liquid metal fast breeder reactor (LMFBR; Fig. 6) sodium in the primary loop collects the heat generated in the core and transfers it to a secondary sodium loop in the heat exchanger, from which it is carried to the steam generator.

Sodium presents an activation problem because ^{24}Na is formed by the absorption of a neutron and is an energetic gamma emitter with a 15-hr half-life. The containing system requires extensive biological shielding, and approximately 2 weeks is required for decay of ^{24}Na activity prior to access to the system for repair or maintenance. Sodium does not decompose, and no makeup is required. Sodium reacts violently with water, imposing severe problems in the design of sodium-to-water steam boilers. The poor lubricating properties of sodium and its reaction with air further complicate the mechanical design of sodium-cooled reactors. The other alkali metals exhibit similar characteristics and appear to be less attractive than sodium. The eutectic alloy of sodium with potassium (NaK), however, has the advantage that it remains liquid at room temperature.

Heavy metals have been considered for use as reactor coolants. Uranium is sufficiently soluble in

containment structure
control rods
helium circulator
steam generator
steam line
turbine generator
core
pump
condenser cooling water
prestressed concrete reactor vessel

Fig. 5. High-temperature gas-cooled reactor (HTGR). *(Atomic Industrial Forum, Inc.)*

bismuth at high temperatures to permit a liquid-fuel system. Bismuth also has an extremely small thermal-neutron-absorption cross section. It is a relatively poor heat-transfer fluid and, in addition, the formation of biologically toxic polonium by neutron capture imposes severe leakage restrictions. The high melting point (520°F; 267°C) of bismuth is also a disadvantage. Essentially the same considerations apply to lead-bismuth alloy, except for its more favorable melting point (257°F; 125°C).

Although mercury has seen some application as a heat-transfer fluid, it is not a particularly attractive reactor coolant. As a coolant, mercury has relatively poor heat-transfer and nuclear characteristics and also is toxic and expensive.

Molten salts. Molten salts have been used as reactor coolants because they have favorable high-temperature properties and because mixtures of salts containing fuel permit fluid fuel-coolant systems. In one small experimental power reactor of this type, a molten mixture of the fluorides of beryllium, lithium, zirconium, and uranium was pumped through channels in a graphite moderator within a reactor vessel in which the fuel salt forms a critical mass and generates energy by fissioning the uranium. Design studies have been made of a larger reactor of this type. The fuel salt would be heated to 1300°F (704°C) and would deliver its heat to sodium fluoroborate in an intermediate loop that would isolate the steam boilers from the radioactive fuel circuit. Steam would be generated in the boilers at a temperature of 1000°F (538°C) and a pressure of 3500 psi (24 MPa) to achieve a thermal efficiency of about 45%. In addition to producing high thermal efficiency, this method has the potential of achieving high neutron efficiency because fission products can be removed continuously from the fluid fuel, thus reducing the fraction of neutrons lost nonproductively by capture in fission products. However, the pumping and processing of the intensely radioactive liquid fuel impose special requirements in design, fabrication, and operation.

Fluid flow and hydrodynamics. Because heat removal must be accomplished as efficiently as possible, considerable attention must be given to the fluid-flow and hydrodynamic characteristics of the system.

The heat capacity and thermal conductivity of the fluid at the temperature of operation have a fundamental effect upon the design of the reactor system. The heat capacity determines the mass flow of the coolant required. The fluid properties (thermal conductivity, viscosity, density, and specific heat) are important in determining the surface area required for the fuel to permit transfer of the heat generated at reasonable temperature differences. This, in turn, affects the design of the fuel—in particular, the amount and arrangement of the fuel elements. These factors combine to establish the pumping characteristics of the system because pressure drop and coolant-temperature rise are directly related.

Secondary considerations include other physical properties of the coolant, particularly its vapor pressure. If the vapor pressure is high at the operating temperature, local or bulk boiling of the fluid may occur. This in turn must be considered in establishing the heat-transfer coefficient for the fluid.

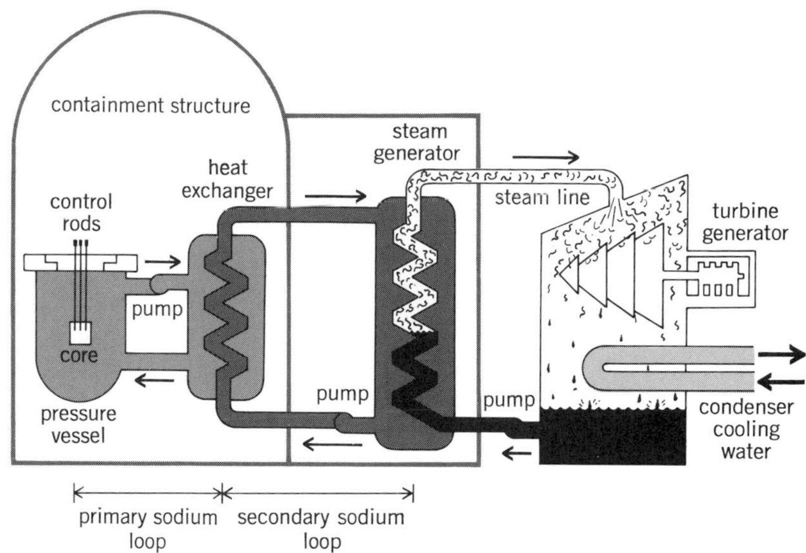

Fig. 6. Liquid metal fast breeder reactor (LMFBR). *(Atomic Industrial Forum, Inc.)*

Because the coolant absorbs and scatters neutrons, variations in coolant density also affect reactor performance. This is particularly significant in reactors in which the coolant exists in two phases, for example, the liquid and vapor phases in boiling systems. Gases, of course, do not undergo phase change, nor do liquids operating at temperatures well below their boiling point; however, the fluid density does change with temperature and may have an important effect upon the reactor.

Power generation and, therefore, the heat-removal rate are not uniform throughout the reactor. If the mass flow rate of the coolant is uniform throughout the reactor, then unequal temperature rise of the coolant results. This becomes particularly significant in power reactors in which it is desired to achieve the highest possible coolant outlet temperature to attain maximum thermal efficiency of the power cycle. The performance limit of the coolant is set by the temperature in the hottest region or channel of the reactor. Unless the coolant flow rate is adjusted in the other regions of the reactor, the coolant will leave these regions at a lower temperature and thus will reduce the average coolant outlet temperature. In high-performance power reactors, this effect is reduced by orificing the flow in each region of the reactor commensurate with its heat generation. This involves very careful design and analysis of the system. In the boiling-type reactor, this effect upon coolant temperature does not occur because the exit temperature of the coolant is at the saturation temperature for the system. However, the variation in power generation in the reactor is reflected by a difference in the amount of steam generated in the various zones, and orificing is still required to achieve most effective use of coolant flow.

In very-high-performance reactors, the flow rate and consequent pressure drop of the coolant are sufficient to create mechanical problems in the system. It is not uncommon for the pressure drop through the fuel assemblies to exceed the weight of the fuel elements in the reactor, with a resulting hydraulic lifting force on the fuel elements. Often this requires a design arrangement to hold the fuel elements down. Although this problem can be overcome by employing downward flow through the system, it is often undesirable to do so because of shutdown-cooling considerations. It is desirable in most systems to accomplish shutdown cooling by natural-convection circulation of the coolant. If downflow is employed for forced circulation, then shutdown cooling by natural-convection circulation requires a flow reversal, which can introduce new problems.

Thermal stress considerations. The temperature of the reactor coolant increases as it circulates through the reactor. This increase in temperature is constant at steady-state conditions. Fluctuations in power level or in coolant flow rate result in variations in the temperature rise. These are reflected as temperature changes in the coolant exit temperature, which in turn result in temperature changes in the coolant system.

A reactor is capable of very rapid changes in power level, particularly, reduction in power level. Reactors are equipped with mechanisms (reactor scram systems) to ensure rapid shutdown of the system in the event of an operational abnormality.

Therefore, reactor-coolant systems must be designed to accommodate the temperature transients that may occur because of rapid power changes. In addition, they must be designed to accommodate temperature transients that might occur as a result of a coolant-system malfunction, such as pump stoppage. The consequent temperature stresses induced in the various parts of the system are superimposed upon the thermal stresses that exist under normal steady-state operations.

In very-high-performance systems, it is not uncommon for the thermal stresses alone to approach the allowable stresses in the materials of construction. In these cases, careful attention must be given to the transient stresses, and thermal shielding is commonly employed in critical sections of the system. Normally, this consists of a thermal barrier, which, by virtue of its heat capacity and resistance to heat transfer, delays the transfer of heat, thereby reducing the rate of change of temperature and protecting critical system components from thermal stresses.

Thermal stresses are also important in the design of reactor fuel elements. Metals that possess dissimilar thermal-expansion coefficients are frequently required. Heating of such systems gives rise to distortions, which in turn can result in flow restrictions in coolant passages. Careful analysis and experimental verification are often required to avoid such circumstances.

Coolant-system components. The development of reactor systems has necessitated concurrent development of special components for reactor coolant systems. These have been required even for systems employing conventional coolants, such as water or air.

Because of the hazard of radioactivity, leak-tight systems and components are a prerequisite to safe, reliable operation and maintenance. Special problems are introduced by many of the fluids employed as reactor coolants.

More extensive component developments have been required for the alkali metals (sodium, NaK, and potassium) which are chemically very active and are extremely poor lubricants. Centrifugal pumps employing unique bearings and seals must be specially designed. Liquid metals are excellent electrical conductors, and, in some special cases, electromagnetic-type pumps have been developed. These pumps are completely sealed, contain no moving parts, and derive their pumping action from electromagnetic forces imposed directly on the fluid.

In addition to the variety of special pumps developed for reactor coolant systems, there is a variety of piping-system components and heat-exchange components. As in all flow systems, flow-regulating devices such as valves are required, as well as flow instrumentation to measure and thereby control the systems. Here again, leak-tightness has necessitated the development of special valves with metallic bellows around the valve stem to ensure system integrity. Measurement of flow and pressure has also required the development of sensing instrumentation that is reliable and leak-tight.

Many of these developments have borrowed from other technologies because toxic or flammable fluids are frequently pumped in other applications. In many cases, however, special equipment has been developed specifically to meet the requirements of the reactor systems. An example of this type of development involves the measurement of flow in liquid-metal piping systems. The simple principle of a moving conductor in a magnetic field is employed by placing a magnet around the pipe and measuring the voltage generated by the moving conductor (coolant) in terms of flow rate. Temperature compensation is required, and calibration is critical.

Although the development of nuclear power reactors has introduced many new technologies, no method has yet displaced the conventional steam cycle for converting thermal energy to mechanical energy. Steam is generated either directly in the reactor (direct-cycle boiling reactor) or in auxiliary steam-generation equipment in which steam is generated by transfer of heat to water from the reactor coolant. These steam generators require very special design, particularly when dissimilar fluids are involved. Typical of these problems are the sodium-to-water steam generators in which absolute integrity is essential because of the violent chemical reaction between sodium and water.

CORE DESIGN AND MATERIALS

A typical reactor core for a power reactor consists of the fuel element rods supported by a grid-type structure inside a vessel (Fig. 1).

The primary function of the vessel is to contain the coolant. Its design and materials are determined by such factors as the nature of the coolant (corrosive properties), operating conditions (temperature and pressure), and quantity and configuration of fuel. To complicate vessel design even further, the vessel is pierced by devices which are used for controlling reactor operation, for loading and unloading the fuel, and for coolant entrance and exit.

Design must also take account of thermal stresses caused by temperature differences in the system. Another problem is radioactivity induced in core materials because of neutron absorption during reactor operation. This precludes normal maintenance of the equipment and, in some areas, makes repairs virtually impossible. For this reason, an exceptionally high degree of integrity is demanded of this equipment. Reactors have been designed to permit removal of the internals from the vessel; this is difficult, however, and tends to complicate the design of the system.

Structural materials. Structural materials employed in reactor systems must possess suitable nuclear and physical properties and must be compatible with the reactor coolant under the conditions of operation. Some requirements are especially severe because of secondary effects; for example, corrosion limits may be established by the rate of deposition of coolant-entrained corrosion products on critical surfaces rather than by the rate of corrosion of the base material.

The most common structural materials employed in reactor systems are aluminum, stainless steel, and zirconium alloys. Aluminum and zirconium alloys have favorable nuclear properties, whereas stainless steel has favorable physical properties. Aluminum is widely used in low-temperature reactors; zirconium and stainless steel are used in high-temperature reactors. Zirconium is relatively expensive, and its use if therefore confined to applications where neutron absorption is critical.

The 18–8 series stainless steels have been used for structural members in both water-cooled reactors and sodium-cooled reactors because of their corrosion resistance and favorable physical properties at high temperatures. Type 304 and type 347 stainless steel have been used most extensively because of their weldability, machinability, and physical properties. To reduce cost, heavy-walled pressure vessels are normally fabricated from carbon steels and clad on the internal surfaces with a thin layer of stainless steel to provide the necessary corrosion resistance.

Although pressure vessels have been constructed for other industries to meet even more severe service requirements, the complex requirements for reactors have introduced new design and fabrication problems. Of particular importance is the dimensional precision required and the special nozzles and other appurtenances required.

As the size of power reactors has increased, it has become necessary, in some instances, to field-fabricate reactor vessels. This involves field-welding of heavy wall sections and subsequent stress relieving. Prestressed concrete vessels for gas-cooled reaction are also field-fabricated and have the potential capability of being fabricated in much larger sizes than steel vessels.

Research reactors operating at low temperatures and pressures introduce special experimental considerations. The primary objective is to provide the maximum volume of unperturbed neutron flux for experimentation. It is desirable, therefore, to extend the experimental irradiation facilities beyond the vessel wall. This has introduced the need for vessels constructed of materials having a low cross section for neutron capture. Relatively large aluminum reactor vessels with wall sections as thin as practicable have been manufactured for research reactors. Special problems with respect to dimensional stability have necessitated unique supporting structures. The vessel design is complicated further by the variety of openings that must be provided to accommodate experimental apparatus. It is highly desirable to provide access to the reactor proper for experiments and, in many cases, to have apparatus installed in so-called through holes that penetrate the vessel from side to side.

In some instances, stainless steel vessels have been employed for research and test reactors at the sacrifice of some experimental flexibility. The experimental irradiations are performed within the reactor vessel and limited in use if made of the space external to the reactor vessel.

A special problem is introduced by research reactors employing heavy water as a moderator and light water as a coolant. A calandria-type design has been employed, consisting of an all-aluminum multitube container for the heavy water, with additional aluminum tubes connected to separate coolant headers for circulation of the light-water coolant. This arrangement introduces the special problems associated with the multitudinous welds to contain a system within a system, each being tight with respect to leakage to the atmosphere and to the other system.

Fuel cladding. Heterogeneous reactors maintain a separation of fuel and coolant by cladding the fuel. The cladding material must be compatible with both the fuel and the coolant.

The cladding materials must also have favorable nuclear properties. The neutron-capture cross section is most significant because the parasitic absorption of neutrons by these materials reduces the efficiency of the nuclear fission process. Aluminum is a very desirable material in this respect; however, its physical strength and corrosion resistance in water decrease very rapidly above about 300°F (149°C).

Zirconium has very favorable neutron properties, and in addition can be made reasonably corrosion-resistant in high temperature water. It has found extensive use for water-cooled power reactors. The technology of zirconium and zirconium-base alloys, Zircaloy, has advanced tremendously under the impetus of the various reactor development programs.

Stainless steel is used for the fuel cladding in fast reactors.

CONTROL AND INSTRUMENTATION

The control of reactors requires the measurement and adjustment of the critical condition. A reactor is critical when the rate of production of neutrons equals the rate of consumption in the system. The neutrons are produced by the fission process and are consumed in a variety of ways, including absorption to cause fission, nonfission capture in fissionable materials, capture in fertile materials, capture in structure or coolant, and leakage from the reactor. A reactor is subcritical (power level decreasing) if the number of neutrons produced is less than the number consumed. The reactor is supercritical (power level increasing) if the number of neutrons produced exceeds the number consumed.

Reactors are controlled by adjusting the balance between neutron production and neutron consumption. Normally, neutron consumption is controlled by varying the absorption or leakage of neutrons; however, the neutron-generation rate can be controlled by varying the amount of fissionable material in the system.

It is essential to orderly control and management of a reactor that the neutron density be sufficiently high to permit reliable measurement. During reactor startup, a source of neutrons is essential, therefore, to the control and instrumentation of reactor systems. Neutrons are obtained from the photo-neutron effect in materials such as beryllium. Neutron sources consist of a photon (γ-ray) source and beryllium, such as antimony-beryllium. Antimony sources are particularly convenient for use in reactors because the antimony is activated by the reactor neutrons each time the reactor operates.

Control drives and systems. The reactor control system requires the movement of neutron-absorbing rods (control rods) in the reactor under very exacting conditions. They must be arranged to increase reactivity (increase neutron population) slowly and under absolute control. They must be capable of reducing reactivity, both rapidly and slowly.

Normal operation of the control drives can be accomplished manually by the reactor operator or by automatic control systems. Reactor scram (very rapid reactor shutdown) can be initiated automatically by one or more system scram-safety signals, or it can be started manually by depressing a scram button convenient to the operator in the control room.

Control drives are normally electromechanical devices that impart linear or swinging motion to the control rods. They are usually equipped with a relatively slow-speed reversible drive system for normal operational control. Scram is usually effected by a high-speed overriding drive accompanied by unlatching or disconnecting the main drive system. To enhance reliability of the scram system, its operation is usually initiated by deenergizing appropriate electrical circuits. This also automatically produces reactor scram in the event of a system power failure. Hydraulic or pneumatic drive systems, as well as a variety of electromechanical systems, have also been developed.

In addition to the actuating motions required, control-rod-drive systems must also provide accurate indication of the rod positions at all times. Various types of selsyn drive, as well as arrangements of switches and lighting systems, are employed as position indicators. It is possible to provide control-rod-position indication accurate to a few thousandths of an inch.

Reactor instrumentation. Reactor control requires measurement of the reactor condition. Neutron-sensitive ion chambers are used to measure neutron flux. These neutron detectors may be located outside the reactor core, and the flux measurements from the detectors are combined to measure an average flux that is proportional to the average neutron density in the reactor. The chamber current is calibrated against a thermal power measurement and then applied over a wide range of reactor power level. The neutron-sensitive detector system must respond to the lowest neutron flux in the system produced by the neutron source.

Normally, many channels of instrumentation are required to cover the entire operating range. Several channels are required for low-level operation, beginning at the source level, whereas others are required for the intermediate and high-power-level ranges. Ten channels of detectors are not uncommon in reactor systems, and some systems contain a larger number. The total range to be covered is in the range of 7 – 10 decades of power level.

The chamber current can be employed as a signal, suitably amplified, to operate automatic control-system devices as well as to actuate reactor scram. In addition to absolute power level, rate of change of power level is also an important measurement which is recorded and employed to actuate various alarm and trip circuits. The normal range for the current ion chambers is approximately 10^{-14} to 10^{-4} A. This current is suitably amplified in logarithmic and period amplifiers, and can be measured directly with a galvanometer.

APPLICATIONS

Reactor applications include production of fissionable fuels (plutonium and uranium-233); mobile, stationary, and packaged power plants; research, testing, teaching-demonstration, and experimental facilities; space and process heat; dual-purpose designs; and special applications. The potential use of reactor radiation for sterilization of food and other products, for chemical processes, and for high-temperature applications has been recognized.

Production reactors. Reactor installations at Hanford, WA, and Savannah River, SC, were designed to produce plutonium-239 from uranium-238. Natural uranium is used as the fuel material. The moderator for the reactors at Hanford is graphite, and heavy water is used as the moderator at Savannah River. Water is used as a coolant in the United States production reactors, whereas in the United Kingdom, gas cooling has been the basis for most designs. The thermal, heterogeneous, natural-uranium, graphite-moderated reactors are representative of the largest reactors. The eight graphite-moderated production reactors at Hanford have been shut down, and the remaining operating production reactor, the N Reactor, is a dual-purpose unit producing special nuclear materials as well as steam for a gross power output of 860 MW electrical.

Breeder reactors. The term "converter" is applied to a reactor that converts a fertile material (for example, uranium-238) to a fissionable material (for example, plutonium). A breeder reactor, strictly speaking, produces the same fissionable material that it consumes (for example, it consumes plutonium fuel and at the same time breeds plutonium). The fuel cycle, of course, could be based on fissionable uranium-233 and fertile thorium-232 rather than uranium-238 and plutonium. In popular usage, however, any reactor that has a conversion ratio of over 100% (that is, produces more fuel than it consumes) is called a breeder, even if the fuels that are consumed and produced are different. The Experimental Breeder Reactor I (EBR-I) operated from 1951 to 1964 and was the first reactor to produce electrical power and to demonstrate the feasibility of breeding. The only operating breeder in the United States in 1976 was EBR-II, which demonstrated the use of integral facility for central station power and a closed remote reprocessing and fabrication system for the fuel cycle. EBR-II is used for fast neutron testing of fuels and materials. Prototype nuclear power breeder reactors are in operation in the United

Kingdom, France, and the Soviet Union. Liquid sodium is used for the coolant (Fig. 6). Studies of fast breeder reactors utilizing gas cooling have been undertaken. Breeding with thermal reactors is also possible, and the molten salt breeder reactor (MSBR) concept, for example, has received some consideration. *See* NUCLEAR FUELS REPROCESSING.

Power reactors. Nuclear power reactors are used extensively by the U.S. Navy for propulsion of submarines and surface vessels, and by the nuclear industry for the generation of electrical power. A variety of organizations and public-interest groups have sought to slow or halt the use of the commercial nuclear power reactors.

As of 1975, more than 130 reactors have been operated by the Navy. The total United States military program has involved more than 200 reactors, operable, being built, planned, or shut down. The prototype of the first reactor used for propulsion operated in 1953, and the first reactor-powered submarine, the USS *Nautilus*, was placed in operation in 1955. Water is used as coolant and moderator and is maintained at 2000 psi (14 MPa) to suppress boiling. Two submarines, the USS *Thresher* and the USS *Scorpion*, were lost in the Atlantic in 1963 and 1968, respectively. Pressurized-water reactors are in use and under further development for submarines, cruisers, aircraft carriers, merchant ships, and (in the Soviet Union) icebreakers. The first civilian maritime reactor application (1961) was the nuclear ship *Savannah*, which utilized a pressurized-water reactor rated at 22,000 shaft horsepower (16.4 MW).

The first reactors for central-station power plant prototypes include the pressurized-water reactors—Shippingport Atomic Power Station (Pennsylvania, 231 MW thermal; 60 MW electrical, 1957) and the Atomic Power Station (Obninsk, Soviet Union, 30 MW thermal; 5 MW electrical, 1954); and the gas-cooled reactors—Calder Hall Station (Sellafield, England, originally 180 MW thermal, increased to 210 MW; 35 MW electrical with four reactors, 1956). The Dresden Nuclear Power Station (Morris, IL) is a boiling-water reactor with an output of 700 MW thermal and 208 MW electrical started in 1959. The 175-MW-electrical Yankee plant (Rowe, MA) is a pressurized-water reactor, started in 1960.

As of December 1979, the electrical generating capacity for the commercial nuclear power reactors was 9% of the total United States generating capacity, with nuclear power in some areas furnishing almost 50% of the energy source for the electric power generation. In the United States, as of April 1980, 72 nuclear power plants had operating licenses with a net electric generating capacity of 53,241 MW; 89 with construction permits, 97,762 MW; and 21 plants on order, 24,532 MW. High capital costs and slow downs experienced in use of electrical power, as well as other factors, have contributed to some cancellations and a number of deferrments for 1 to 4 years in nuclear power as well as in fossil power plant constructions. As of December 1979, the status of nuclear power plants outside the United States was 166 plants operable with a net electric generating capacity of 70,200 MW; 156 plants under construction, 125,364 MW; 33 plants on order, 27,472 MW; and 233 plants planned, 224,003 MW. Nuclear power was used in 22 countries with planned operations extending to 28 additional countries.

Research and test reactors. The research-and-development aspects of a nuclear reactor may be considered from two points of view. One is that the reactor provides experimental irradiation facilities, and the other is that the reactor itself may represent a test of a given design.

Research with reactors covers such activities as measurements of the probabilities of nuclear reactions, shielding measurements, studies of the behavior of materials under neutron and γ-irradiation, and other studies in nuclear physics, solid-state physics, and the life sciences. The irradiation facilities are used extensively for production of isotopes. High-neutron-flux reactors, designed specifically for experimental exposures of materials, are called materials-testing reactors. Reactors built to test design features are called experiments or experimental reactor facilities. Several different types of low-cost reactors, which are called teaching-demonstration reactors, have been promoted to accentuate the teaching aspects. *See* NUCLEAR ENGINEERING.

The four major varieties of research reactors are (1) uranium-fueled, graphite-moderated, air-cooled reactors; (2) uranium-fueled, heavy-water-moderated reactors; (3) enriched-fuel, aqueous-solution-type reactors; and (4) water-moderated, enriched-fuel, pool-type, and tank-type reactors. All the reactors are thermal and, with the exception of the third type, heterogeneous. Both natural and enriched uranium are used in the first two types.

The bulk shielding reactor, or BSR (Oak Ridge, TN, 1950), was the first reactor with the core submerged in an open pool of water—hence the term "swimming-pool reactor." The water is the moderator, coolant, and shield. With forced circulation of water, reactor levels of 1000 kW of heat are possible. Some reactor designs involve the use of a tank instead of a pool. Features of other pool- and tank-type reactors include variability of fuel-element design and configuration, fixed and movable cores, and a lightly pressurized (for tank-type), forced-convection water-cooling system.

The materials-testing reactor, or MTR (1952–1970), was a high-flux irradiation facility designed for studying the behavior of materials for use in power reactors. The maximum neutron fluxes available at 40 MW (thermal) were 5.5×10^{14} thermal neutrons/(cm^2)(s) and 3×10^{14} fast neutrons/(cm^2)(s). Nearly 100 experimental and instrument holes or exposure ports were provided. Other test reactors have been built to accommodate the specialized materials development programs necessary for the continued advancement of the nuclear reactor industry. Included are the engineering test reactor (ETR), 175 MW thermal, in operation since 1957, and the advanced test reactor (ATR), 250 MW thermal, completed in 1967. The ATR provides a flux up to 2.5×10^{15} neutrons/(cm^2)(s).

Test facilities for the fast-breeder-reactor physics program included the zero power reactors (ZPR), zero power plutonium reactors (ZPPR), and the Southwest Experimental Fast Oxide Reactor (SEFOR) reactor. The fast-flux test facility (FFTR) is designed to provide fast neutron environments for testing fuel and materials for fast reactors.

Among the many thermal research reactors is the high-flux isotope reactor (HFIR) at the Oak Ridge National Laboratory. A principal use of this reactor is the production of transplutonium elements such as berkelium, californium, einsteinium, and fermium.

Experimental reactors. A variety of reactors have been built to test the feasibility of given reactor designs. Reactors already noted include the experimental breeder reactors and the homogeneous reactor experiments. Several types of reactors have been designed and operated under severe power excursions to study reactor stability. Five boiling-water reactor experiments (Borax-1 to -5) have been carried out to study the behavior of such reactors at atmospheric and at elevated pressures and with different kinds of fuel elements, including nonmetallic fuels. Power-excursion experiments have been performed with the homogeneous aqueous-solution-type reactors. For example, kinetic experiment water boiler (KEWB, Canoga Park, CA) has successfully handled a power excursion of 0–530 MW in less than 1 s.

The use of boiling water as a coolant for power-producing reactors was established by the experimental boiling water reactor (EBWR, Argonne National Laboratory, Lemont, IL, 1956) and the Vallecitos boiling water reactor (VBWR, Vallecitos, CA, 1957).

The use of sodium as a high-temperature coolant for power reactors was demonstrated by the sodium-graphite reactor experiment (SRE, 1957–1964).

The feasibility of organics as coolants or coolant-moderators for reactors was studied in the organic moderated reactor experiment (OMRE, 1957–1963. The organic was a polyphenyl compound.

Test reactors for the nuclear engine for the NERVA Program included the Phoebus, NRX, and Kiwi reactors, ranging up to 4000 MW thermal. The adaptation and further testing of the reactors for space vehicles was completed successfully with ground experimental engines (XE).

Among the many other reactor experiments, two additional ones are noted here. The feasibility of the molten-salt-reactor concept has been successfully demonstrated by the molten-salt-reactor experiment (MSRE) (1965–1969). The ultra-high-temperature reactor experiment (UHTREX) (1968–1970) employed helium as a coolant and was designed to operate at 2400°F (1316°C).

Thermoelectric power. In early 1959 the AEC Los Alamos Laboratory announced the first successful production of electricity directly from a reactor core without the use of a heat-transfer medium or conventional generating equipment. The experimental unit operated by means of a thermoelectric process. The thermoelectric medium was cesium vapor, and the heat source was enriched uranium. *See* THERMOELECTRICITY.

Specialized nuclear power units. Nuclear power units are being developed for small electrical outputs, but with special purpose for land, sea, and space applications. A 500-W reactor, SNAP-10A, was orbited in 1965 and operated successfully for 43 days. SNAP reactors are used to supply power for lunar surface experiments left behind by Apollo astronauts. Other systems for nuclear auxiliary power (SNAP) include SNAP-8, a 600-kW, thermal unit, and a series of odd-numbered units employing radioisotopes, such as plutonium-238, curium-242, polonium-210, and promethium-147, for the energy source. Other isotopes being considered are cobalt-60, strontium-90, and thulium-171. *See* NUCLEAR BATTERY. [HERBERT S. ISBIN]

Bibliography: J. M. Harrer and J. B. Beckerley, *Nuclear Power Reactor Instrumentation Systems Handbook*, National Technical Information Service, TID-25952-P1 and -P2, vol. 1, 1973, vol.2, 1974; *Nuclear Reactors Built, Being Built, or Planned in the United States*, National Technical Information Service, TID-8200, printed twice yearly as of June 30 and December 31; A. Sesonske, *Nuclear Power Plant Design Analysis*, National Technical Information Service, TID-26241, 1973; U.S. Atomic Energy Commission, *The Safety of Nuclear Power Reactors (Light Water-Cooled) and Related Facilities*, WASH-1250, 1973; J. Weisman, *Elements of Nuclear Reactor Design*, 1977.

Nucleonics

The technology based on phenomena of the atomic nucleus. These phenomena include radioactivity, fission, and fusion. Thus, nucleonics embraces such devices and fields as nuclear reactors, radioisotope applications, radiation-producing machines (such as cyclotrons and Van de Graaff accelerators), the application of radiation for biological sterilization and for the induction of chemical reactions, and radiation-detection devices. Nucleonics makes use of and serves virtually all other technologies and scientific disciplines. *See* NUCLEAR ENGINEERING; NUCLEAR PHYSICS.

That part of the industry concerned with nuclear reactors involves a cross section of the entire industrial complex. The chemical industry is concerned with uranium ore refining, fuel and moderator preparation, and fuel reprocessing; the light and heavy metals industry, with fuel fabrication, special component fabrication to withstand environmental conditions including radiation, and containment materials; the machinery industry with control rods, fuel charge and discharge devices, and manipulators; and the instrument industry with control systems. The many applications of nuclear reactors and isotopes also bring the industries making use of them into the field, so that electrical generation, marine propulsion, process heat, special industrial devices, and agriculture, to name a few, are industries participating to some degree in nucleonics.

A number of service activities such as reactor-design consultation, film-badge reading, special shipping and disposal of radioactive nuclear materials and wastes, and analytical services by such techniques as low-level counting and activation are included in the nucleonics industry. The unique radiation hazards and benefits associated with nuclear technology have also engendered special legal, political, and mercantile aspects.

[BERNARD I. SPINRAD]

Ocean thermal energy conversion

The conversion of energy to a usable form from the temperature difference between warm surface water and cold, deep ocean currents. This process was first proposed by the French physicist Arsène d' Arsonval in the late 1800s. Since a temperature difference of about 27°F (15°C) is necessary to operate an ocean thermal energy conversion (OTEC) system, there are a limited number of sites, primarily between 35°N and S latitudes, where these plants can operate without excessively long cold-water intake pipelines. Both coasts of Africa, the southeastern coast of the United States, the coast of Brazil, and many Caribbean and Pacific islands are located where the sea water decreases in temperature from 75 to 86°F (24 to 30°C) at the surface to 39 to 45°F (4 to 7°C) at a depth of about 2600 ft (800 m).

An open-cycle OTEC pilot plant, in which the working fluid was low-pressure steam evaporated from the warm-water flow, was built and operated in Cuba by Georges Claude in the late 1920s. The plant delivered only 22 kWe (kilowatts of electric power) and ceased to operate when the cold-water pipe was destroyed. Claude later experimented with an 800-kWe floating plant, but was successful only in generally proving the overall OTEC concept.

As a result of these early attempts to generate power from the ocean's thermal gradients, the French formed a partly government-owned company called Energie des Mers, a subsidiary of Electricité de France, and planned the construction of an open-cycle plant for a site near Abidjan, Ivory Coast, during the 1940s and 1950s. This plant was never completed, although several of the subsystems were built, tested, and installed. One of the French land-based plant designs is shown in Fig. 1. The gross power output of the turbine was to be 14.5 MWe (10 MWe net) at a rotational speed of 332 revolutions per minute. The diameter of the power plant building was 122 ft (37.2 m), and the outer diameter of the turbine blades was 47 ft (14.3 m). The basic difficulties encountered by Claude have been overcome, and more advanced concepts for the development of the open-cycle OTEC process have been proposed.

Power can also be generated by using the vapor of a secondary working fluid in a closed cycle instead of the open-cycle process employed by Claude. This approach actually was suggested by d'Arsonval during the late 1800s, and a 100-MWe closed-cycle floating plant with propane as the working fluid was proposed in 1966. A secondary working fluid with a relatively high vapor pressure results in a reduction in the size of the plant's power turbines from the large low-pressure units of the open cycle, but enormous heat-transfer surface areas are required in the evaporators and condensers of the system. *See* HEAT EXCHANGER; HEAT TRANSFER; TURBINE; VAPOR CONDENSER.

The promise of a renewable source of base-load electric power from ocean temperature gradients led to United States government funding of research and development for the OTEC process, beginning in 1972. Other OTEC development projects have been undertaken in Europe and Japan.

The primary unresolved engineering problems in the OTEC concept include: the heat exchangers (evaporators and condensers), including the working fluid, construction materials, and biofouling and corrosion; the ocean platform and cold-water pipe; open-cycle versus closed-cycle processes; underwater electric power transmission lines; and the constructability and long-term reliability of an entire plant.

OTEC process. Theoretically, work can be done by a heat engine when a temperature difference exists between two heat reservoirs, in this case the warm and cold sea water. This temperature difference is hundreds of degrees in conventional and nuclear power plants and gas turbines, but only about 36°F (20°C) in the sea. The maximum theoretical (or Carnot) efficiency of an OTEC power plant is about 3–6%. When the typical operating losses in an actual system are taken into account, the overall efficiency is likely to be only about 2% or less. However, this low thermal efficiency is not necessarily a fatal drawback for such a device, because the boiler of an OTEC plant operates at relatively low pressure and temperature differentials and does not need the complex containment vessel required for operation at both high pressures and temperatures. *See* CARNOT CYCLE; THERMODYNAMIC CYCLE.

A closed-cycle OTEC process is shown in Fig. 2, which illustrates a proposal to combine an OTEC power plant with a floating plant to produce ammonia. This particular process is a Rankine power cycle consisting of a turbine, pump, condenser, and evaporator (or boiler). The working fluid is transformed to a vapor in the evaporator, and power is generated by expansion of the vapor through the turbine. Upon exiting the turbine, the low-pressure vapor is condensed in another heat exchanger by using water from the ocean depths as the cold sink fluid. The working fluid is then pumped back to the boiler pressure and recycled through the system. *See* RANKINE CYCLE.

Typical operating temperatures for an ammonia cycle are shown on the diagram. The temperature drops in the evaporator and condenser, reducing the available temperature difference between the warm- and cold-water inlets from 43°F (24°C) to a temperature difference of about 20°F (11°C) for the closed ammonia loop. For the conditions shown, the maximum thermal efficiency of the ammonia cycle is about 3%; the actual efficiency will be reduced further by the plant's parasitic loads.

There are some inherent advantages in the

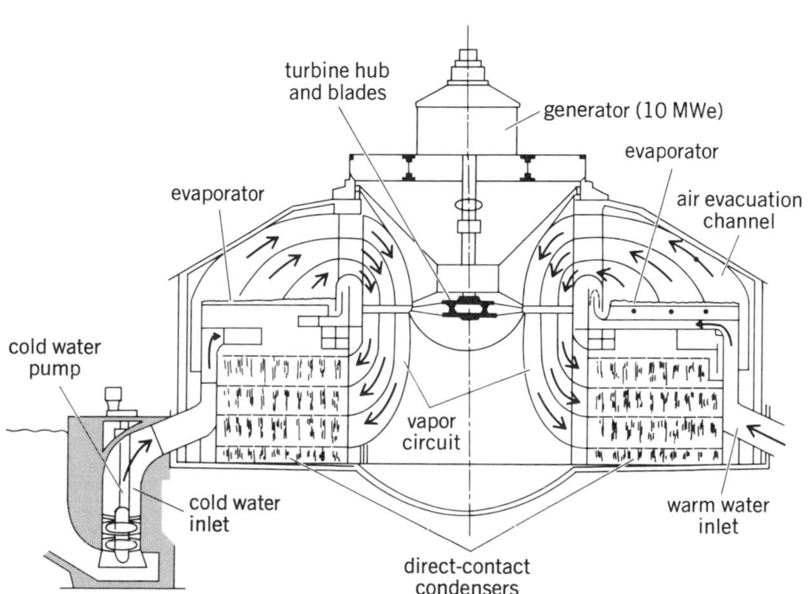

Fig. 1. An open-cycle OTEC power plant designed by the French company Energie des Mers.

closed cycle. As mentioned previously, the power turbines are greatly reduced in size due to the higher operating pressures (typically several atmospheres) and working fluid densities. It is also not necessary to remove the dissolved gases in the warm sea water as must be done in the open cycle. Some disadvantages in using a secondary working fluid are the temperature losses just mentioned and the presence of biofouling and corrosion due to the passage of sea water through the heat exchangers. Other disadvantages are that working fluids such as propane and ammonia require enormous heat-exchanger surface areas and that these fluids pose potential handling and materials compatibility problems in the presence of sea water.

OTEC plant designs. Both open- and closed-cycle OTEC plant designs have evolved since 1950, but the most attention by far has been given to the closed-cycle process.

Open-cycle plants similar in overall layout to the French design shown in Fig. 1 have been proposed for various land-based applications that include an OTEC power plant in combination with desalination and mariculture facilities. Apparently the major faults of Claude's original open-cycle design have been eliminated by utilizing the controlled flash evaporation (CFE) of steam for the production of power and fresh water. In this process the warm sea water enters a vacuum chamber and, as the pressure decreases, the water is evaporated. The low-pressure steam next passes through to an expansion turbine to generate power, and then to a surface condenser for condensation by cold, deep sea water. The CFE process has minimized the problem of removing dissolved gases from the water and has reduced the corrosion problems. It also reduces brine entrainment and temperature losses between the sea water and the vapor. Most open-cycle plant designs have been limited in size to about 10 MWe.

In 1966 an ocean-based, closed-cycle OTEC system was proposed. The plant was to have a net power output of 100 MWe and included submerged turbines, evaporators, and condensers in a floating platform anchored to the ocean bottom. A long vertical pipe was used to pump the cold ocean water from a depth of about 1970 ft (600 m). The working fluid for the closed cycle was propane (later changed to Freon), and compact plate heat exchangers with large surface areas were required to operate the system. Based upon this and other studies, several overall features of the closed-cycle OTEC system have emerged: a plant that is floating or submerged in deep water and anchored to the ocean bottom; modular evaporator, condenser, and turbogenerator systems; ammonia or propane as the most desirable working fluid; and a single, large-diameter, long cold-water pipe. Several industrial and laboratory design teams have undertaken overall systems studies, and three basic plant configurations have emerged. These are described below.

Plant-ship. A proposal was made to construct OTEC plant-ships for the production of ammonia or other energy-intensive products at sea. The OTEC plant-ship is designed to cruise slowly at a nominal speed of about 1 knot (0.5 m/s) in tropical waters, and both 40- and 100-MWe demonstration systems and follow-on 500-MWe production plant-ships have been proposed.

The basic hull and cold-water pipe are to be constructed of reinforced concrete, with the water side of the evaporator and condenser modules making up an integral part of the hull (Fig. 2).

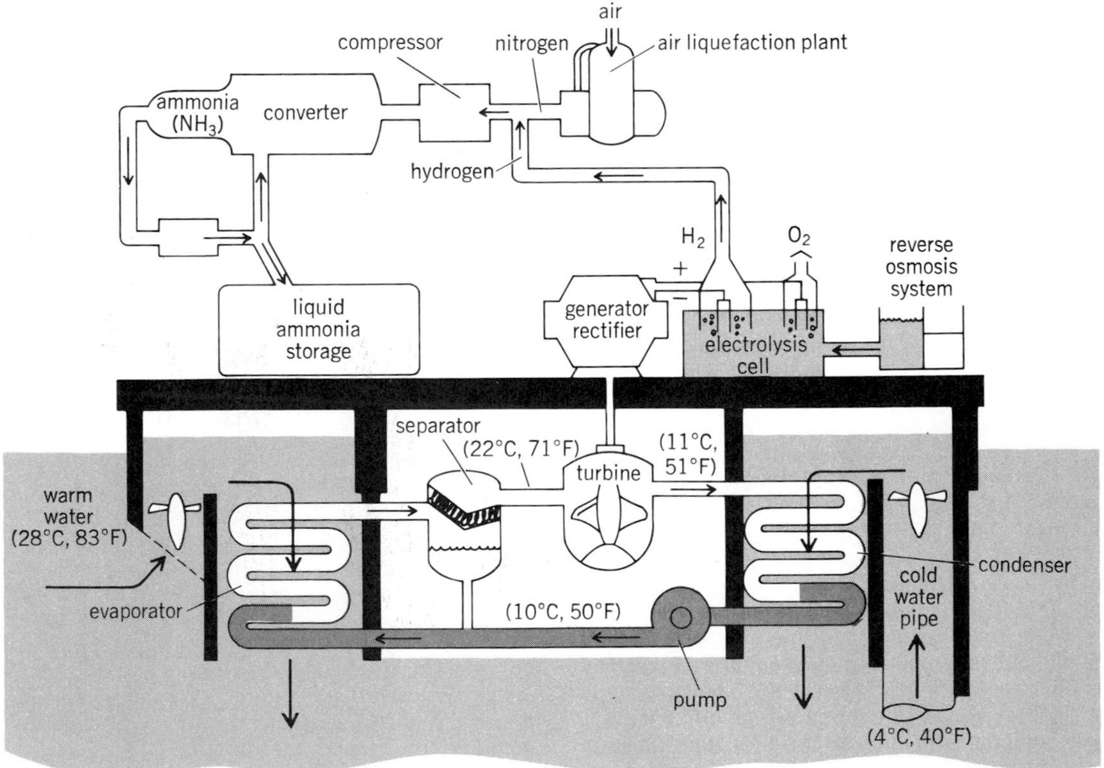

Fig. 2. Sectional view of an OTEC plant-ship for producing both electric power and ammonia. (*From W. H. Avery et al., Maritime and construction aspects of ocean thermal energy conversion plant ships, Johns Hopkins Univ./Applied Phys. Lab. Rep. SR 76-1A, April 1976*)

Ammonia was chosen as the working fluid for the closed-cycle OTEC power modules, and aluminum heat-exchanger tubes were chosen for the evaporators and condensers in order to minimize costs and weight.

The plant-ship has several other interesting features. The absence of a cable to transmit power to shore is an advantage since both the undersea power cable and its riser to the plant pose complex marine and electrical engineering problems. No mooring system is required when the plant is operated in a cruising mode, and this aspect of the design reduces the overall cost and eliminates problems in deep-water mooring design. These advantages are balanced somewhat by the problems and costs associated with production, transportation, and marketing of a chemical product (ammonia) from an ocean-based platform.

Work has been undertaken to ready a converted Navy tanker, renamed *OTEC-I*, for the at-sea testing of OTEC evaporators, condensers, and other plant components on board the ship at an ocean site near the island of Hawaii.

Spar platform. The spar configuration consists of a submerged main platform with access through a slender cylindrical column to the surface. The vertical cold-water pipe is attached to the main platform. The power modules, consisting of evaporators, condensers, and turbine-generators, are mounted externally to the submerged main platform. In one original spar concept, there are four power modules mounted externally to a semisubmersible platform that provides the primary buoyancy. The plant layout includes a 2500-ft-long (776-m) cold-water pipe and a mooring system capable of deployment to water depths of 20,000 ft (6100 m). The power modules are integral units that combine to produce a net plant output of 260 MWe. Ammonia was chosen as the working fluid because of its superior thermodynamic characteristics and relatively wide history of industrial use. Shell-and-tube heat exchangers with titanium tubes were selected for this particular design of the evaporators and condensers.

The platform and the telescoping cold-water pipe are of reinforced concrete construction, and the single-point mooring line is made up of cylindrical steel sections varying in diameter from 6 ft (1.8 m) at the platform to 1 ft (0.3 m) at the anchor. A number of design trade-offs were made before the final plant configuration was agreed upon; these included dynamic positioning versus fixed moorings, and aluminum or stainless steel versus titanium for the heat-exchanger tubes. Other spar-type OTEC plants in the 10–40-MWe size range have been the subject of subsequent design studies.

Floating plant. An early design study of a floating plant includes a floating cylindrical hull with four 25-MWe power modules inside the interior of the hull. Reinforced concrete was chosen for the platform construction, while the cold-water pipe, 4000 ft (1220 m) long and 50 ft (15 m) in diameter, was proposed to be constructed in 18-m sections of stiffened fiber-reinforced plastic. Ammonia was chosen over propane and Freon as the working fluid for the power turbines because of its superior thermodynamic properties and relatively wide history of use. Conventional titanium shell-and-tube heat exchangers also are included in the design since titanium is compatible with ammonia and is corrosion-resistant over long periods of time.

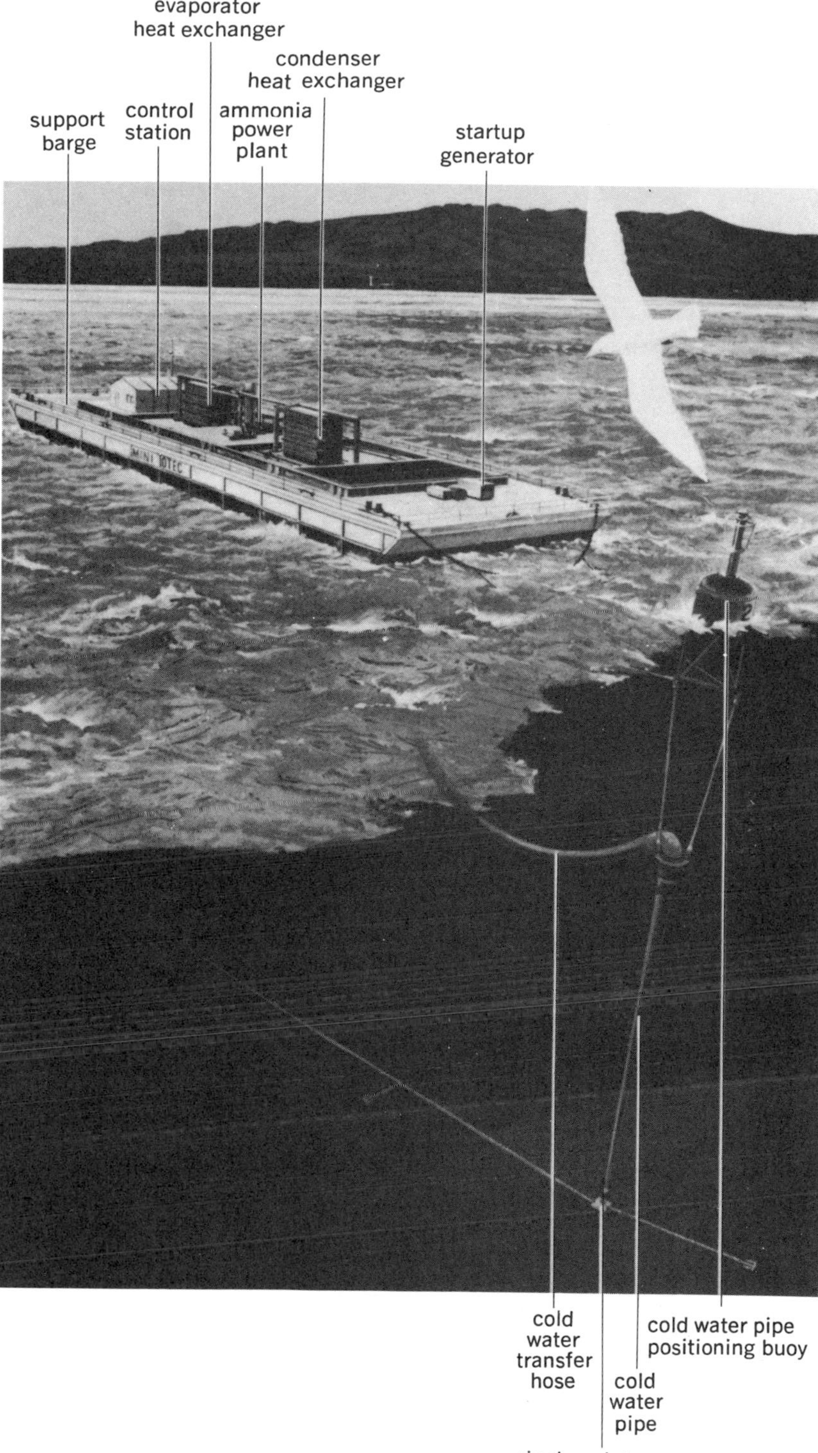

Fig. 3. Mini-OTEC power plant. (*Lockheed Missiles and Space Co., Inc.*)

Some perception of the size of the OTEC heat exchangers can be obtained from the design parameters of this plan. About 1,000,000 ft² (93,000 m²) of heat-transfer surface area are required for each 25-MWe power module. Approximately 65,000 titanium tubes, each 1.5 in. (38 mm) in diameter and 43 ft (13.1 m) long, are necessary to meet this requirement. The total flow rate of sea water through the evaporators and condensers of a 100-MWe OTEC plant is estimated to be 2½ times the flow

Mini-OTEC operating parameters and specifications*

Component	Parameter and specification
Barge	Length: 120 ft (37 m) Beam: 34 ft (10 m) Displacement: 268 long tons (272.29 metric tons)
Cold-water pipe	Length: 2170 ft (663 m) Diameter: 24 in. (0.6 m) Water flow: 2700–3300 gal/min (10,260–12,540 liters/min) Water temperature: 40–45°F (5–8°C)
Warm water	Flow: 2700–3300 gal/min (10,260–12,540 liters/min) Temperature: 75–80°F (24–27°C)
Ammonia	Storage capacity: 879 gal/115 ft³ (3320 liters/3.2 m³) Flow: 5.1 lb/s; 60 gal/min, for liquid (2.5 kg/s; 227 liters/min, for liquid)
Electricity (continuous operation)	Generate: 50 kW Net: 15 kW

*From Lockheed Missiles and Space Co., Inc.

of the Potomac River at Washington, DC.

A dynamic positioning system was chosen for the plant, since wire or chain deepwater mooring systems were considered by the plant's designers to be unproved for water depths greater than 3280 ft (1000 m). The water jet positioning system was designed to make use of the kinetic energy of the plant's enormous water throughput for station-keeping in the open ocean.

Mini-OTEC. An at-sea test of a small, barge-mounted OTEC power plant was begun during 1979. The experiment, called Mini-OTEC, consists of an ammonia-based closed-cycle power plant that is mounted on a barge anchored 1.5 mi (2.4 km) off Keahole Point on the west coast of the island of Hawaii (Fig. 3). The converted Navy barge carries all of the power plant apparatus except for the 2170 ft (662 m) long cold-water pipe, which is suspended from a floating buoy adjacent to the barge. The Mini-OTEC project generates approximately 50 kWe gross, with a net power output of 15 kWe at the design water temperature difference $\Delta T = 36°F$ (20°C). The remaining 35 kWe powers the pumps, instruments, and other auxiliary plant machinery. The table lists the physical and operating characteristics of the Mini-OTEC plant.

Two principal concerns of the plant's designers were biofouling in the plate-type titanium heat exchangers and the structural integrity of the long, slender cold-water pipe, but no problems were encountered during the initial period of testing. An interesting feature of the Mini-OTEC plant is the combination of a single-point mooring system and cold-water pipe. The 24-in.-diameter (0.6-m) high-density polyethylene cold-water pipe is an integral part of the mooring system, with the upper end of the pipe connected to a 20-in.-diameter (0.51-m) flexible hose that transfers cold water to the barge and with the lower end connected to the anchor line cable. As with the proposed open-cycle plants mentioned above, the nutrient-rich deep ocean water pumped to the surface by Mini-OTEC plants is expected to benefit the Hawaiian mariculture industry.

Prospects. Although a number of technical problems have been solved and considerable progress has been made, the low overall thermal efficiency of the OTEC process is still an inherent drawback in the overall plant design. Many of the economic and environmental questions related to OTEC must await answers until problems are further resolved by at-sea testing and reliable operation of pilot- and prototype-scale plants over a long period of time. [OWEN M. GRIFFIN]

Bibliography: O. M. Griffin, Power from the oceans' thermal gradients, in *Ocean Energy Resources*, ASME, vol. 4, pp. 1–21, 1977; Office of Technology Assessment, U.S. Congress, *Renewable Ocean Energy Sources*, pt. 1: *Ocean Thermal Energy Conversion*, May 1978; *Proceedings of the 6th OTEC Conference, June 1979*, John Hopkins University/Applied Physics Laboratory, 1979.

Octane number

A standard laboratory measure of a fuel's ability to resist knock during combustion in a spark-ignition engine. A single-cylinder four-stroke engine of standardized design is used to determine the knock resistance of a given fuel by comparing it with that of primary reference fuels composed of varying proportions of two pure hydrocarbons, one very high in knock resistance and the other very low. A highly knock-resistant isooctane (2,2,4-trimethylpentane, C_8H_{18}) is assigned a rating of 100 on the octane scale, and normal heptane (C_7H_{16}), with very poor knock resistance, represents zero on the scale. Octane number is defined as the percentage of isooctane required in a blend with normal heptane to match the knocking behavior of the gasoline being tested.

The CFR (Cooperative Fuel Research) knock-test engine used to determine octane number has a compression ratio that can be varied at will and a knockmeter to register knock intensity. In the classic method of knock rating, the engine is run on the fuel to be tested, and its compression ratio is adjusted to give a standard level of knock intensity. Without changing the compression ratio, this knock level is then bracketed by running the engine on two primary reference fuel blends, one of which knocks a little more than the test fuel, and the other a little less. The octane number of the fuel being rated is then determined by interpolation from the knockmeter readings of the bracketing reference fuels.

Alternatively, the engine's compression ratio can be adjusted to close to the limit for the fuel being tested, and then, while the engine is run on each of two closely bracketing test fuels, the ratio can be readjusted to a standard intensity reading on the knockmeter. Finally, the engine is again run on the test fuel, and its compression ratio is adjusted to give the same knockmeter reading. The octane number of the test fuel is then interpolated from the compression ratio settings.

Knock tests are performed under one of two sets of engine operating conditions. Results of tests using the so-called Motor (M) method correlated well with the fuels and automobile engines of the 1930s, when the method was developed. The Research (R) method was developed later when improved refining processes and engines gave gasolines better road performance than their M ratings indicated. Today, the arithmetic average of a gasoline's R and M ratings usually is a good indicator of its performance in a typical car on the road. Therefore, that average, (R + M)/2, is posted at service stations to show the antiknock quality of a particular fuel.

For fuels with a rating higher than 100 octane, the rating is usually obtained by determining the amount of tetraethyllead compound that needs to be added to pure isooctane to match the knock resistance of the test fuel. For example, if the amount is 1.3 ml, the fuel's rating is expressed as 100 + 1.3 or extrapolated above 100 (in this case, about 110 octane) by means of a correlation curve.

[JOHN C. LANE]

Bibliography: American Society for Testing and Materials, *Annual Book of ASTM Standards*, pt. 47, published annually.

Oil analysis

Analysis of petroleum, or crude oil, to determine its value in modern refinery operations. The analysis, or assay, must provide the refinery planner with the data needed to predict yields, qualities, and costs for a bewildering variety of refinery operating conditions and product demands.

In a refinery, crude oil is distilled and separated into products according to the boiling points of the crude oil components (see table). *See* PETROLEUM PRODUCTS.

Procedure. A crude assay follows much the same procedure. The oil is distilled and separated into up to 40 narrow-boiling-range cuts. Cuts with normal boiling points higher than about 190°C are distilled at greatly reduced pressure. This allows the oil to vaporize at temperatures below 190°C and avoids thermal reactions which would alter the chemical properties of the oil. Each cut is then subjected to a variety of tests sufficient to characterize it for the products the cut could be included in. The refinery planners may then calculate yield and qualities of any product by blending the yields and qualities of the cuts that are included in the product.

Typical products derived from crude oil

Product	Carbon atom range	Boiling point range, °C
Gas	C_1 and C_2	
Liquefied petroleum gas	C_3 and C_4	
Gasolines	C_4 to C_{10}	15 to 190
Kerosines	C_9 to C_{15}	150 to 280
Middle distillates	C_{12} to C_{20}	200 to 340
Gas oils	C_{20} to C_{45}	340 to 560
Bottoms	Unvaporized residua	

Petroleum consists primarily of compounds of carbon and hydrogen containing from 1 to about 60 carbon atoms. Carbon atoms in natural petroleum occur in straight and branched chains (paraffins), in single or multiple saturated rings (cycloparaffins or naphthenes), and in cyclic structures of the aromatic type such as benzene, naphthalene, and phenanthrene. Cyclic structures may have attached to them side chains of paraffinic carbons. In lubricating oil it is usual to have naphthene rings built onto the aromatic rings and side chains attached. In products produced by cracking in the refinery, olefins or compounds with carbon-carbon double bonds not in aromatic rings are also found. The high-boiling fractions of petroleum contain increasing amounts of oxygen, nitrogen, and sulfur compounds, as well as traces of organic compounds of metals such as vanadium, nickel, and iron.

Three types of tests are used to characterize a crude oil and its narrow boiling range cuts: (1) tests for physical properties such as specific gravity, refractive index, freeze temperature, vapor pressure, octane number, and viscosity; (2) tests for specific chemical species such as sulfur, nitrogen, metals, and total paraffins, naphthenes, and aromatics; (3) test for determining actual chemical composition.

Physical properties. The physical properties are used to predict how petroleum products will perform. Octane number and vapor pressure are two important qualities of gasolines. Jet fuels must be formulated so that they remain fluid, and therefore pumpable, at the low temperatures encountered by high-flying airplanes. Lubricating oils are characterized by their viscosity level together with the change of viscosity with temperature. The many different tests used for physical properties are relatively simple. Methods have been standardized and published by the American Society for Testing and Materials (ASTM).

Chemical species. The amounts of specific chemical species are also important in evaluating petroleum products. In these days of environmental concern and emission controls, the sulfur content often determines whether or not a fuel can be used. Aromatic components are desirable in gasoline and undesirable in diesel fuels. Fractions with vanadium, nickel, and iron compounds should not be used in catalytic processes since these metals permanently deactivate most catalysts. Tests are available to measure all the chemical species of interest. Many of the tests involve controlled burning of an oil sample, followed by analysis of the combustion products.

Chemical composition. The analysis of petroleum in terms of chemical composition can be done with varying degrees of thoroughness. Anything approaching the complete analysis of a crude oil is so time-consuming and expensive that in the whole world only one sample of crude oil is being analyzed thoroughly. This project, known as Project No. 6 of the American Petroleum Institute, was carried on from 1928 to 1968, when the remaining work was transferred to another project. Obviously, such a detailed analysis will never be done on a routine basis.

Since 1965, gas chromatography (GC), mass spectroscopy (MS), and nuclear magnetic resonance (NMR) methods have been developed that are capable of economically determining the detailed chemical composition of petroleum fractions through the C_9 or C_{10} boiling range plus the distribution of hydrocarbon types up to about C_{40}. The hydrocarbon-type analysis includes distinguishing between one-, two-, three-, and four-ring naphthenes and aromatics.

In a GC analysis a small sample is introduced into a column packed with an adsorbent material such as silica gel. The sample is swept along the column by an inert carrier gas such as nitrogen. The carrier gas forces all components in the sample to move along the column, but each component moves at a unique rate depending on how strongly it is adsorbed by the silica gel. Each component leaves the column as a discrete packet, or peak, of noncarrier gas. The time of each peak identifies the component, and amount in the peak gives the composition. The operation of the GC column and the calculation of results have been automated so that very little human effort is required. However, setting up a GC column is a tedious business since each column seems to behave differently and so must be calibrated for all components it will be used to measure. Also, the method has trouble distinguishing between similar isomers that give overlapping peaks.

In an MS analysis the molecules of the sample being tested are bombarded with an electron beam. This breaks the molecules into ionized fragments whose weights are determined by measuring how much the ionized fragments bend as they move through a magnetic field. The fragment pattern obtained is then correlated against the known fragment patterns of pure compounds to find the composition of the sample. It is difficult to apply the method to compounds for which calibration patterns are not available. It is, however, not feasible to obtain calibration patterns for the many thousands of compounds which appear in petroleum. Nevertheless, even in the higher-boiling fractions where these uncertainties exist, the MS analysis is the best available method for hydrocarbon type analysis.

An NMR analyses in its simplest form determines the type of hydrogen present in a sample, that is, whether the hydrogen is attached to an aliphatic, naphthenic, or aromatic type carbon atom. This is determined from a measurement of the magnetic resonance spectrum of the diluted sample under the influence of a very strong magnetic field. Information of this type, coupled with other analytical information, makes possible the development of a rather detailed picture of molecular types and structures.

Oil analysis, then, depends on a wide variety of tests which have been developed over the years and which continue to be extended and improved. Each refiner selects those tests which best characterize the oil for the refinery operations contemplated. A good analysis is critical when evaluating a crude oil. *See* PETROLEUM. [JACK REES]

Bibliography: K. H. Altgelt and T. H. Gouw (eds.), *Chromatography in Petroleum Analysis*, 1979; American Society for Testing and Materials, *ASTM Standards*, annual; ASTM, *Manual on Hydrocarbon Analysis*, 1977; ASTM, *Significance of ASTM Tests for Petroleum Products*, Spec. Tech. Publ. no. 7C, 1977; R. A. Hofstader et al. (eds.), *Analysis of Petroleum for Trace Metals*, 1976.

Oil and gas, offshore

Oil and gas prospecting and exploitation on the continental shelves and slopes. Since Mobil Oil Co. drilled what is considered the first offshore well off the coast of Louisiana in 1945, exploration for petroleum and natural gas on the more than 8,000,000 mi² (20.7 × 10⁶ km²) of the world's continental shelves and slopes, lying between the shore and 1000-ft (300-m) water depth, has expanded rapidly to include exploration or drilling or both off the coasts of more than 75 nations. More than 100 national and private petroleum companies have joined in the worldwide offshore search for oil and gas. The construction costs of equipping mobile offshore drilling fleets has risen tremendously due to inflation and the added sophistication of the equipment. In 1977, with a total fleet of 432 units operating, the drilling costs were estimated to be about $7,000,000,000 per year. For this expenditure, the industry was able to drill between 1200 and 1300 holes per year.

Until early in 1975, most of the offshore rigs were employed in drilling on the United States Outer Continental Shelf. After that time, this changed dramatically, with more than 75% of the offshore drilling being conducted outside the United States waters.

Paralleling the development of drilling vessels able to withstand the rigors of operation in the open sea have been remarkable technological developments in equipment and methods. More than 125 companies in the United States devote a large share of their efforts to the development and manufacture of material and devices in support of offshore oil and gas.

For many years petroleum companies stopped at the water's edge or sought and developed oil and gas accumulations only in the shallow seas bordering onshore producing areas. These activities were usually confined to water depths in which drilling and producing operations could be conducted from platforms, piers, or causeways built upon pilings driven into the sea floor. Major accumulations along the Gulf Coast of the United States, in Lake Maracaibo of Venezuela, in the Persian Gulf, and in the Baku fields of the Caspian Sea were developed from such fixed structures.

Exploration deeper under the sea did not begin in earnest until the world's burgeoning appetite for energy sources, coupled with a lessening return from land drilling, provided the incentives for the huge investments required for drilling in the open sea. There has been a steady annual increase in the number of wells drilled in deep water beyond the continental shelves, although the costs of developing the actual producing systems are enormous.

Geology and the sea. There is a sound geologic basis for the petroleum industry turning to the continental shelves and slopes in search of needed reserves. Favorable sediments and structures exist beneath the present seas of the world, in geologic settings that have proven highly productive onshore. In fact, the subsea geologic similarity—or in some cases superiority—to geologic conditions on land has been a vital factor in the rapid expansion of the free world's investment in offshore exploration and production. More than $12,000,000,000 had been invested in the free world offshore effort by 1969. World offshore expenditures for the year 1972 were approximately $4,000,000,000.

Subsea geologic basins, having sediments considered favorable for petroleum deposits, total approximately 6,000,000 mi² (15.5×10^6 km²) out to a water depth of 1000 ft (300 m), or about 57% of the world's total continental shelf. This 6,000,000-mi² area is equivalent to one-third of the 18,000,000 mi² (46.6×10^6 km²) of geologic basins on land.

Present offshore reserves are currently estimated at 21% of the world's proved and produced reserves, while estimates of ultimate resources of petroleum offshore range from 30 to 35%, indicating that 40–50% of all future resources will come from offshore. While the exploration potential of the continental shelf is fairly well established, the potential of deep-water areas is relatively unknown, has had little petroleum test drilling, and involves a wide range of estimates (3–25% of future resources). At present, deep-water areas have not been included in most estimates of future resources. In addition, petroleum data from offshore areas exhibit a change in the dominance of basin types from those onshore, changes in trap types and age and depth of reservoirs, and a reduction in supergiants and average size of giants—particularly from fields partially onshore to fields that are totally offshore. Presently, many fields with low productivity over an extensive area which are commercial onshore might be uneconomic offshore, where a thick productive column and high productivity over a concentrated area are required.

A high percentage of future petroleum resources is expected to come from offshore, where fields smaller than those on land seem to be prevalent. In addition, productive basins on land seem to have smaller fields in the final stages of development. Therefore it is probable that future resources will involve a greater number of reserves from smaller fields than in the past.

Virtually all of the world's continental shelf has received some geological study, with active seismic or drilling exploration of some sort either planned or in effect. Offshore exploration planned or underway includes action in the North Sea; in the English Channel off the coast of France; in the Red Sea bordering Egypt; in the seas to the north, west, and south of Australia; off Sumatra; along the Gulf Coast and east, west, and northwest coast of the continental United States; in the Cook Inlet and off the west coast of Alaska; in the Persian Gulf; off Mexico; off the east and west coasts of Central America and South America; and off Nigeria, North Africa, and West Africa. In addition, offshore exploration has been undertaken in the Caspian and Baltic seas off the coasts of East Germany, Poland, Latvia, and Lithuania; the China seas; Gulf of Thailand; Irish Sea; Arctic Ocean; and Arctic islands. *See* PETROLEUM GEOLOGY.

Successes at sea. The price of success in the offshore drilling is staggering in the amount of capital required, the risks involved, and the time required to achieve a break-even point.

Offshore Louisiana, long the center of much of the world's offshore drilling, has yielded a large number of oil and gas fields. However, offshore discoveries there have been made at a diminishing rate because the shelf in the area of the northern Gulf of Mexico has been thoroughly explored. Areas under exploration off the United States include the Baltimore Canyon, the Southern Georgia Embayment, the Georges Bank (southeast of Cape Cod, MA), and the eastern Gulf of Mexico (off the coast of Florida). There also has been offshore exploration for hydrocarbon deposits off the coasts of California and Alaska.

Major oil and gas strikes have been made in the North Sea, and oil and gas production has been established in the Bass Straits off Australia. These have the potential for changing historical energy sources and the economies of the countries involved. Discoveries made since 1964 in the Cook Inlet of Alaska are making a major impact on the West Coast market of the United States; although they involve between 1.5×10^9 to 2×10^9 bbl (2.4×10^8 to 3.2×10^8 m³) of oil, the extremely high exploration and development costs will delay the break-even point. The most active offshore areas outside the United States are eastern Canada, the Labrador Sea, the North Sea, Southeast Asian seas, Nigeria, and the Bay of Campeche.

The Okan field of Nigeria reached 45,000 barrels per day (bpd) production in 1966. While offshore drilling in Nigeria has declined, exploration in Africa continues in other areas, extending out into the Atlantic to greater depths. Major activity has centered off West Africa and also in the Gulf of Suez.

The Bay of Campeche off southeastern Mexico has yielded evidence of giant oil and gas reserves, while a large discovery of oil was made in the Norwegian North Sea in 1978.

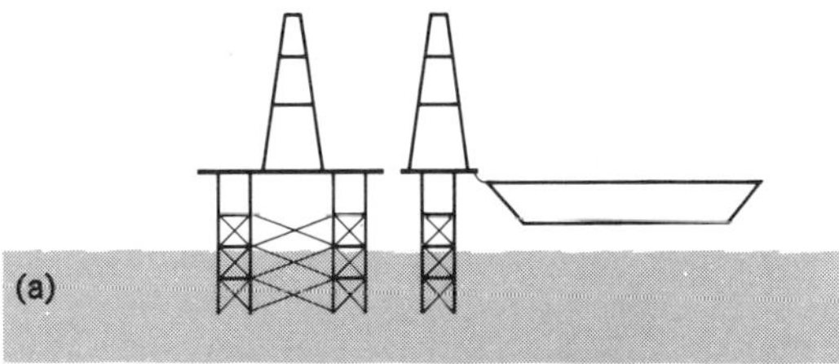

Fig. 1. Offshore fixed drilling platform. (*a*) Underwater design (*World Petroleum*). (*b*) Rig on a drilling site (*Marathon Oil Co., Findlay, Ohio*).

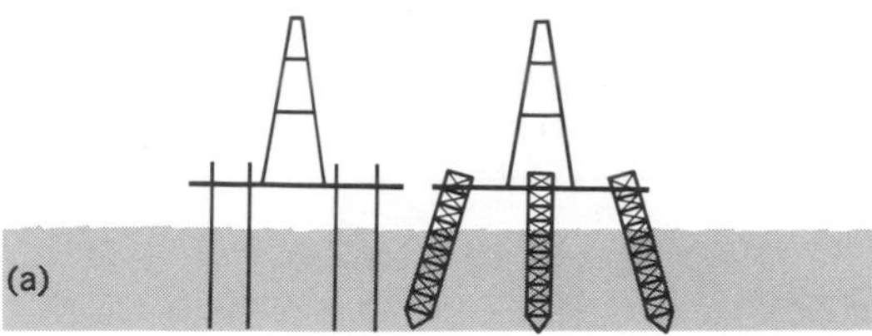

Fig. 2. Offshore self-elevating drilling platform. (*a*) Underwater design (*World Petroleum*). (*b*) Self-elevating drilling platform (*Marathon Oil Co., Findlay, Ohio*).

Mobile drilling platforms. The underwater search has been made possible only by vast improvements in offshore technology. Drillers first took to the sea with land rigs mounted on barges towed to location and anchored or with fixed platforms accompanied by a tender ship (Fig. 1). A wide variety of rig platforms has since evolved, some designed to cope with specific hazards of the sea and others for more general work. All new types stress characteristics of mobility and the capability for work in even deeper water.

The world's mobile platform fleet can be divided into four main groupings: self-elevating platforms, submersibles, semisubmersibles, and floating drill ships.

The most widely used mobile platform is the self-elevating, or jack-up, unit (Fig. 2). It is towed to location, where the legs are lowered to the sea floor, and the platform is jacked up above wave height. These self-contained platforms are especially suited to wildcat and delineation drilling. They are best in firmer sea bottoms with a depth limit out to 300 ft (90 m) of water.

The submersible platforms have been developed from earlier submersible barges which were used in shallow inlet drilling along the United States Gulf Coast. The platforms are towed to location and then submerged to the sea bottom. They are very stable and can operate in areas with soft sea floors. Difficulty in towing is a disadvantage, but this is partially offset by the rapidity with which they can be raised or lowered, once on location.

Semisubmersibles (Fig. 3) are a version of submersibles. They can work as bottom-supported units or in deep water as floaters. Their key virtue is the wide range of water depths in which they can operate, plus the fact that, when working as floaters, their primary buoyancy lies below the action of the waves, thus providing great stability. The "semis" are the most recent of the rig-type platforms.

Floating drill ships (Fig. 4) are capable of drilling in 60-ft (18-m) to abyssal depths. They are built as self-propelled ships or with a ship configuration that requires towing. Several twin-hulled versions have been constructed to give a stable catamaran design. Floating drill ships use anchoring or ingenious dynamic positioning systems to stablize their position, the latter being necessary in deeper waters. Floaters cannot be used in waters much shallower than 70 ft because of the special equipment required for drilling from the vessel subject to vertical movement from waves and tidal changes, as well as minor horizontal shifts due to stretch and play in anchor lines. Exploration in deeper waters necessitates building more semisubmersible and floating drill ships. A conventional exploratory hole has been drilled 50 mi (80 km) off the coast of Gabon in 2150 ft (655 m) of water by Shell Deep Water Drilling Company using the Sedco 445 drill ship. The *Glomar Challenger* has drilled stratigraphic holes in the sea floor to a depth of 3334 ft (1016 m) in water depths of 20,000 ft (6 km).

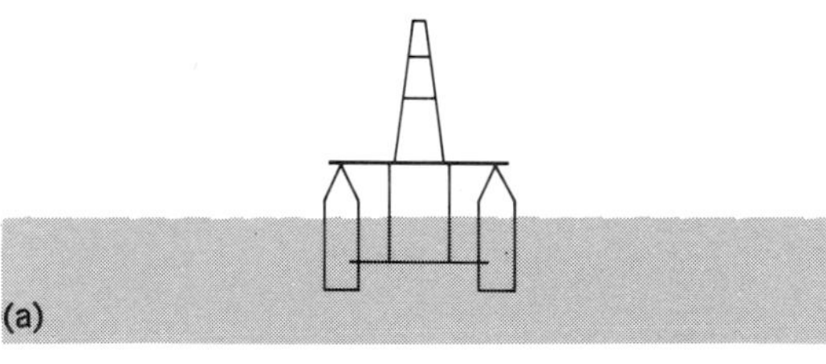

Fig. 3. Offshore semisubmersible drilling platform. (*a*) Underwater design (*World Petroleum*). (*b*) Santa Fe Marine's Blue Water no. 3 drilling rig on a drilling site (*Marathon Oil Co., Findlay, Ohio*).

Production and well completion technology. The move of exploration into the open hostile sea has required not only the development of drilling vessels but a host of auxiliary equipment and techniques. A whole new industrial complex has developed to serve the offshore industry.

Of particular interest is the development of diving techniques and submersible equipment to aid in exploration and the completion of wells. A platform was constructed for installation in 850 ft (260 m) of water in the Santa Barbara Channel. Tentative plans for platforms in 1000 ft (300 m) of water have been announced for the Gulf of Mexico. Economics will soon force sea-bottom completions which require men or robots or both to make the

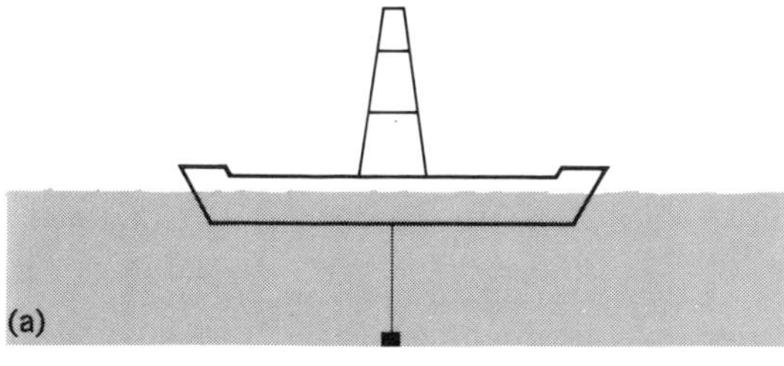

Fig. 4. Floating drill ship. Such ships can drill in depths from 60 to 1000 ft (18 to 300 m) or more. (*a*) Underwater design (*World Petroleum*). (*b*) Floating drill ship on a drilling site (*Marathon Oil Co., Findlay, Ohio*).

necessary pipe and well connections. Such work is necessary even in the water depths now being developed.

A robot device has been developed which operates from the surface and uses sonar and television for viewing; it can excavate ditches for pipelines and make simple pipe connections and well hookups in water depths to 2500 ft (760 m). Limitations of robot devices are such that the more complex needs of well completions and service require the actions of men. To fill this need, diving specialists have been used to sandbag platform bases, recover conductor pipe, survey and remove wreckage, and make pipeline connections and well hookups. Diving depths have been increased to the point where useful work has been performed at depths in excess of 600 ft (180 m). Pressure chambers to take divers to the bottom and to return them to the surface to be decompressed are operational. One has been operating routinely in 425 ft (130 m) of water for Esso Exploration, Norway. This deep diving is made possible by the development of saturation diving, which uses a mixture of oxygen, helium, nitrogen, and argon. This technique has allowed divers to remain below 200 ft (60 m) for 6 days while doing salvage work on a platform. It has also allowed prolonged submergence at 600 ft (180 m) in preparation for actual work on wells at this depth.

Miniature submarines have taken their place in exploration and completion work allowing the viewing of conditions, the gathering of samples, and simple mechanical tasks. Their depth range is for all practical purposes unlimited.

Technical groups are experimenting with the design of drilling and production units that would be totally enclosed and be set on the sea bottom. Living in and working from these units, personnel would be able to carry out all the necessary oil field operations. In effect, such units would resemble a miniature city on the sea floor, from which a man would need to return to the surface only when his tour of work was completed. *See* OIL AND GAS FIELD EXPLOITATION.

Concomitant with the progress of the petroleum industry in its venture into the open sea has been a vast increase in the knowledge of the sea and its contained wealth. Mining of the sea floor using some of petroleum's technology has started in several areas of the world, and actual farming or ranching of the life in the sea is being planned.

Hazards at sea. As the petroleum industry pushed farther into the hostile environment of the sea, it sustained a series of disasters, reflected by the doubling of offshore insurance rates in April 1966. Between 1955 and 1968, for example, 23 offshore units were destroyed by blowout and 6 by hurricane and breakup and collapse at sea. The United States Gulf Coast, where a large percentage of the world's offshore drilling has taken place, was severely hit in 1964 and 1965 by hurricanes, which claimed over $7,500,000 in tow and service vessels, $21,000,000 in fixed platforms, and $28,000,000 in mobile platforms. In 1980 an offshore oil platform in the North Sea was overturned and severely damaged by near-hurricane winds, with a loss of more than half of the personnel aboard.

Expenses sustained from loss of wells, removal of wrecked equipment, and loss of production are huge. Such liabilities have raised insurance rates on platforms valued at millions of dollars to as much as 10% or more per year, depending on the platform type and location. Much of the current design work is aimed at engineering better safety features for the benefit of both the crews and structures.

Despite the hazards and monumental cost involved in extracting oil and gas from beneath the sea, the world's population explosion and its ever increasing demand for petroleum energy will force the search for new reserves into even deeper waters and more remote corners of the world. In truth, the search is only just beginning.

[G. R. SCHOONMAKER]

Bibliography: F. W. Mansvelt-Beck and K. M. Wiig, *The Economics of Offshore Oil and Gas Supplies*, 1977; Ocean engineering, *Petrol. Eng.*, vol. 39, no. 11, 1967; *OCS Statistics*, United States Geological Survey, 1973; Offshore news, *Offshore*, vol. 34, no. 10 and 13, 1974; Offshore 1973, *World Oil*, vol. 177, no. 1, 1973; Offshore report and offshore rigs, *World Petrol.*, vol. 36, no. 5, 1965; Rig insurance, *Oil Gas J.*, vol. 64, no. 19, 1966; Unmanned subsea device, *World Oil*, vol. 165, no. 7, 1968; L. G. Weeks, Petroleum resources potential of continental margins, in C. A. Burk and C. L. Drake (eds.), *The Geology of Continental Margins*, 1974.

Oil and gas field exploitation

In the petroleum industry, a field is an area underlain without substantial interruption by one or more reservoirs of commercially valuable oil or gas, or both. A single reservoir (or group of reservoirs which cannot be separately produced) is a pool. Several pools separated from one another by barren, impermeable rock may be superimposed one above another within the same field. Pools have variable areal extent. Any sufficiently deep well located within the field should produce from one or more pools. However, each well cannot produce from every pool, because different pools have different areal limits.

DEVELOPMENT

Development of a field includes the location, drilling, completion, and equipment of wells necessary to produce the commercially recoverable oil and gas in the field.

Related oil field conditions. Petroleum is a generic term which, in its broadest meaning, includes all naturally occurring hydrocarbons, whether gaseous, liquid, or solid. By variation of the temperature or pressure, or both, of any hydrocarbon, it becomes gaseous, liquid, or solid. Temperatures in producing horizons vary from approximately 60°F (16°C) to more than 300°F (149°C), depending chiefly upon the depth of the horizon. A rough approximation is that temperature in the reservoir sand, or pay, equals 60°F (16°C), plus 0.017°F/ft (0.031°C/m) of depth below surface. Pressure on the hydrocarbons varies from atmospheric to more than 11,000 psi (76 MPa). Normal pressure is considered as 0.465 psi/ft (10.5 kPa/m) of depth. Temperatures and pressure vary widely from these average figures. Hydrocarbons, because of wide variations in pressure and temperature and because of mutual solubility in one another, do not necessarily exist underground in the same phases in which they appear at the surface.

Petroleum occurs underground in porous rocks of wide variety. The pore spaces range from microscopic size to rare holes 1 in. or more in diameter. The containing rock is commonly called the sand or the pay, regardless of whether the pay is actually sandstone, limestone, dolomite, unconsolidated sand, or fracture openings in relatively impermeable rock.

Development of field. After discovery of a field containing oil or gas, or both, in commercial quantities, the field must be explored to determine its vertical and horizontal limits and the mechanisms under which the field will produce. Development and exploitation of the field proceed simultaneously. Usually the original development program is repeatedly modified by geologic knowledge acquired during the early stages of development and exploitation of the field.

Ideally, tests should be drilled to the lowest possible producing horizon in order to determine the number of pools existing in the field. Testing and geologic analysis of the first wells sometimes indicates the producing mechanisms, and thus the best development program. Very early in the history of the field, step-out wells will be drilled to determine the areal extent of the pool or pools. Step-out wells give further information regarding the volumes of oil and gas available, the producing mechanisms, and the desirable spacing of wells.

The operator of an oil and gas field endeavors to select a development program which will produce the largest volume of oil and gas at a profit. The program adopted is always a compromise between conflicting objectives. The operator desires (1) to drill the fewest wells which will efficiently produce the recoverable oil and gas; (2) to drill, complete, and equip the wells at the lowest possible cost; (3) to complete production in the shortest practical time to reduce both capital and operating charges; (4) to operate the wells at the lowest possible cost; and (5) to recover the largest possible volume of oil and gas.

Selecting the number of wells. Oil pools are produced by four mechanisms: dissolved gas expansion, gas-cap drive, water drive, and gravity drainage. Commonly, two or more mechanisms operate in a single pool. The type of producing mechanism in each pool influences the decision as to the number of wells to be drilled. Theoretically, a single, perfectly located well in a water-drive pool is capable of producing all of the commercially recoverable oil and gas from that pool. Practically, more than one well is necessary if a pool of more than 80 acres (32 hectares) is to be depleted in a reasonable time. If a pool produces under either gas expansion or gas-cap drive, oil production from the pool will be independent of the number of wells up to a spacing of at least 80 acres per well (1866 ft or 569 m between wells). Gas wells often are spaced a mile or more apart. The operator accordingly selects the widest spacing permitted by field conditions and legal requirements.

Major components of cost. Costs of drilling, completing, and equipping the wells influence development plans. Having determined the number and depths of producing horizons and the producing mechanisms in each horizon, the operator must decide whether he will drill a well at each location to each horizon or whether a single well can produce from two or more horizons at the same location. Clearly, the cost of drilling the field can be sharply reduced if a well can drain two, three, or more horizons. The cost of drilling a well will be higher if several horizons are simultaneously produced, because the dual or triple completion of a well usually requires larger casing. Further, completion and operating costs are higher. However, the increased cost of drilling a well of larger diameter and completing the well in two or more horizons is 20–40% less than the cost of drilling and completing two wells to produce separately from two horizons.

In some cases, the operator may reduce the number of wells by drilling a well to the lowest producible horizon and taking production from that level until the horizon there is commercially exhausted. The well is then plugged back to produce from a higher horizon. Selection of the plan for producing the various horizons obviously affects the cost of drilling and completing individual wells, as well as the number of wells which the operator will drill. If two wells are drilled at approximately the same location, they are referred to as twins, three wells at the same location are triplets, and so on.

Costs and duration of production. The operator wishes to produce as rapidly as possible because the net income from sale of hydrocarbons is obviously reduced as the life of the well is extended. The successful operator must recover from his productive wells the costs of drilling and operating those wells, and in addition he must recover all costs involved in geological and geophysical exploration, leasing, scouting, and drilling of dry holes, and occasionally other operations. If profits from production are not sufficient to recover all exploration and production costs and yield a profit in excess of the rate of interest which the operator could secure from a different type of investment, he is discouraged from further exploration.

Most wells cannot operate at full capacity because unlimited production results in physical waste and sharp reduction in ultimate recovery. In many areas, conservation restrictions are enforced to make certain that the operator does not produce in excess of the maximum efficient rate. For example, if an oil well produces at its highest possible rate, a zone promptly develops around the well where production is occurring under gas-expansion drive, the most inefficient producing mechanism. Slower production may permit the petroleum to be produced under gas-cap drive or water drive, in which case ultimate production of oil will be two to four times as great as it would be under gas-expansion drive. Accordingly, the most rapid rate of production generally is not the most efficient rate.

Similarly, the initial exploration of the field may indicate that one or more gas-condensate pools exist, and recycling of gas may be necessary to secure maximim recovery of both condensate and of gas. The decision to recycle will affect the number of wells, the locations of the wells, and the completion methods adopted in the development program.

Further, as soon as the operator determines that secondary oil-recovery methods are desired and expects to inject water, gas, steam, or, rarely, air to provide additional energy to flush or displace oil from the pay, the number and location of wells may be modified to permit the most effective secondary recovery procedures.

Legal and practical restrictions. The preceding discussion has assumed control of an entire field under single ownership by a single operator. In the United States, a single operator rarely controls a large field, and this field is almost never under a single lease. Usually, the field is covered by separate leases owned and operated by different producers. The development program must then be modified in consideration of the lease boundaries and the practices of the other operators who are in the field.

Oil and gas know no lease boundaries. They move freely underground from areas of high pressure toward lower-pressure situations. The operator of a lease is obligated to locate his wells in such a way as to prevent drainage of his lease by wells on adjoining leases, even though he may own the adjoining leases. In the absence of conservation restrictions, an operator must produce petroleum from his wells as rapidly as it is produced from wells on adjoining leases. Slow production on one lease results in migration of oil and gas to nearby leases which are more rapidly produced.

The operator's development program must provide for offset wells located as close to the boundary of his lease as are wells on adjoining leases. Further, the operator must equip his wells to produce as rapidly as the offset produces and must produce from the same horizons which are being produced in offset wells. The lessor who sold the lease to the operator is entitled to his share of the recoverable petroleum underlying his land. Negligence by the operator in permitting drainage of a lease makes the operator liable to suit for damages or cancellation of the lease.

A development program acceptable to all operators in the field permits simultaneous development of leases, prevents drainage, and results in maximum ultimate production from the field. Difficulties may arise in agreement upon the best development program for a field. Most states have enacted statutes and have appointed regulatory bodies under which judicial determination can be made of the permissible spacing of the wells, the rates of production, and the application of secondary recovery methods.

Drilling unit. Commonly, small leases or portions of two or more leases are combined to form a drilling unit in whose center a well will be drilled. Unitization may be voluntary, by agreement between the operator or operators and the interested royalty owners, with provision for sharing production from the well between the parties in proportion to their acreage interests. In many states the regulatory body has authority to require unitization of drilling units, which eliminates unnecessary offset wells and protects the interests of a landowner whose acreage holding may be too small to justify the drilling of a single well on his property alone.

Pool unitization. When recycling or some types of secondary recovery are planned, further unitization is adopted. Since oil and gas move freely across lease boundaries, it would be wasteful for an operator to repressure, recycle, or water-drive a lease if the adjoining leases were not similarly operated. Usually an entire pool must be unitized for efficient recycling, or secondary recovery operations. Pool unitization may be accomplished by agreement between operators and royalty owners. In many cases, difference of opinion or ignorance on the part of some parties prevents voluntary pool unitization. Many states authorize the regulatory body to unitize a pool compulsorily on application by a specified percentage of interests of operators and royalty owners. Such compulsory unitization is planned to provide each operator and each royalty owner his fair share of the petroleum products produced from the field regardless of the location of the well or wells through which these products actually reach the surface.

EXPLOITATION—GENERAL CONSIDERATIONS

Oil and gas production necessarily are intimately related, since approximately one-third of the gross gas production in the United States is produced from wells that are classified as oil wells. However, the naturally occurring hydrocarbons of petroleum are not only liquid and gaseous but may even be found in a solid state, such as asphaltite and some asphalts.

Where gas is produced without oil, the production problems are simplified because the product flows naturally throughout the life of the well and does not have to be lifted to the surface. However, there are sometimes problems of water accumulations in gas wells, and it is necessary to pump the water from the wells to maintain maximum, or economical, gas production. The line of demarcation between oil wells and gas wells is not definitely established since oil wells may have gas-oil ratios ranging from a few cubic feet (1 cubic foot $= 2.8 \times 10^{-2}$ m^3) per barrel to many thousand cubic feet of gas per barrel of oil. Most gas wells produce quantities of condensable vapors, such as propane and butane, that may be liquefied and marketed for fuel, and the more stable liquids produced with gas can be utilized as natural gasoline.

Factors of method selection. The method selected for recovering oil from a producing formation depends on many factors, including well depth, well-casing size, oil viscosity, density, water production, gas-oil ratio, porosity and permeability of the producing formation, formation pressure, water content of producing formation, and whether the force driving the oil into the well from the formation is primarily gas pressure, water pressure, or a combination of the two. Other factors, such as paraffin content and difficulty expected from paraffin deposits, sand production, and corrosivity of the well fluids, also have a decided influence on the most economical method of production.

Special techniques utilized to increase productivity of oil and gas wells include acidizing, hydraulic fracturing of the formation, the setting of screens, and gravel packing or sand packing to increase permeability around the well bore.

Aspects of production rate. Productive rates per well may vary from a fraction of a barrel (1 barrel = 0.1590 m^3) per day to several thousand barrels per day, and it may be necessary to produce a large percentage of water along with the oil.

Field and reservoir conditions. In some cases reservoir conditions are such that some of the wells flow naturally throughout the entire economical life of the oil field. However, in the great majority of cases it is necessary to resort to artificial lifting methods at some time during the life of the field, and often it is necessary to apply artificial lifting means immediately after the well is drilled.

Market and regulatory factors. In some oil-producing states of the United States there are state bodies authorized to regulate oil production from the various oil fields. The allowable production per well is based on various factors, including the market for the particular type of oil available, but very often the allowable production is based on an engineering study of the reservoir to determine the optimum rate of production.

Crude oil production in the United States in 1978, as reported by the *Oil and Gas Journal*, averaged 8,680,000 barrels per day (bpd) or 1,380,000 m^3/day. This represents an increase of 6.1% over 1977 production. Imports amounted to 8,658,000 bpd (1,376,500 m^3/day) in 1977. World production in 1978 was estimated at 60,335,000 bpd (9,592,500 m^3/day). Net natural gas production during 1977 was estimated at 53,393.5 $\times 10^6$ ft^3 per day (1,511.936 m^3/day).

Useful terminology. A few definitions of terms used in petroleum production technology are listed below to assist in an understanding of some of the problems involved.

Porosity. The percentage porosity is defined as the percentage volume of voids per unit total volume. This, of course, represents the total possible volume available for accumulation of fluids in a formation, but only a fraction of this volume may be effective for practical purposes because of possible discontinuities between the individual pores. The smallest pores generally contain water held by capillary forces.

Permeability. Permeability is a measure of the resistance to flow through a porous medium under the influence of a pressure gradient. The unit of permeability commonly employed in petroleum production technology is the darcy. A porous structure has a permeability of 1 darcy if, for a fluid of 1 centipoise (cp) [10^{-3} Pa·s] viscosity, the volume flow is 1 $cm^3/(sec)(cm^2)$ [10^{-2} $m^3/(sec)(m^2)$] under a pressure gradient of 1 atm/cm (1.01325 $\times$ 10^7 Pa/m).

Productivity index. The productivity index is a measure of the capacity of the reservoir to deliver oil to the well bore through the productive formation and any other obstacles that may exist around the well bore. In petroleum production technology, the productivity index is defined as production in barrels per day (1 barrel per day $\cong$ 0.1590 m^3/day) per pound per square inch (psi = 6.895 kPa) drop in bottom-hole pressure. For example, if a well is closed in at the casinghead, the bottom-hole pressure will equal the formation pressure when equilibrium conditions are established. However, if fluid is removed from the well, either by flowing or pumping, the bottom-hole pressure will drop as a result of the resistance to flow of fluid into the well from the formation to replace the fluid removed from the well. If the closed-in bottom-hole pressure should be 1000 psi, for example, and if this pressure should drop to 900 psi when producing at a rate of 100 bbl/day (a drop of 100 psi), the well in question would have a productivity index of one.

Barrel. The standard barrel used in the petroleum industry is 42 U.S. gal (approximately 0.1590 m^3).

API gravity. The American Petroleum Institute (API) scale that is in common use for indicating specific gravity, or a rough indication of quality of crude petroleum oils, differs slightly from the Baume scale commonly used for other liquids lighter than water. The table shows the relationship between degrees API and specific gravity referred to water at 60°F (15.6°C) for specific gravities ranging from 0.60 to 1.0.

Viscosity range. Viscosity of crude oils currently produced varies from approximately 1 cp (10^{-3} Pa·s) to values above 1000 cp (1 Pa·s) at temperatures existing at the bottom of the well. In some areas it is necessary to supply heat artificially down the wells or circulate lighter oils to mix with the produced fluid for maintenance of a relatively low viscosity throughout the temperature range to which the product is subjected.

Degrees API corresponding to specific gravities of crude oil at 60°/60°F*

Specific gravity, in tenths	Specific gravity, in hundredths									
	.00	.01	.02	.03	.04	.05	.06	.07	.08	.09
0.60	104.33	100.47	96.73	93.10	89.59	86.19	82.89	79.69	76.59	73.57
0.70	70.64	67.80	65.03	62.34	59.72	57.17	54.68	52.27	49.91	47.61
0.80	45.38	43.19	44.06	38.98	36.95	34.97	33.03	31.14	29.30	27.49
0.90	25.72	23.99	22.30	20.65	19.03	17.45	15.90	14.38	12.89	11.43
1.00	10.00									

*60°F = 15.6°C.

In addition to wells that are classified as gas wells or oil wells, the term gas-condensate well has come into general use to designate a well that produces large volumes of gas with appreciable quantities of light, volatile hydrocarbon fluids. Some of these fluids are liquid at atmospheric pressure and temperature; others, such as propane and butane, are readily condensed under relatively low pressures in gas separators for use as liquid petroleum gas (LPG) fuels or for other uses. The liquid components of the production from gas-condensate wells generally arrive at the surface in the form of small droplets entrained in the high-velocity gas stream and are separated from the gas in a high-pressure gas separator.

PRODUCTION METHODS IN PRODUCING WELLS

The common methods of producing oil wells are (1) natural flow; (2) pumping with sucker rods; (3) gas lift; (4) hydraulic subsurface pumps; (5) electrically driven centrifugal well pumps; and (6) swabbing.

Numerous other methods, including jet pumps and sonic pumps, have been tried and are used to slight extent. The sonic pump is a development in which the tubing is vibrated longitudinally by a mechanism at the surface and acts as a high-speed pump with an extremely short stroke.

The total number of producing oil wells in the United States at the end of 1977 was reported to be 508,340, while the total number of producing gas wells was 145,453.

A total of 49,931 wells were drilled in the United States during 1978. Of this number, 17,755 were productive oil wells, 11,169 were classified as gas wells, 15,437 were nonproductive (dry holes), and 1366 were service wells. Service wells are utilized for various purposes, such as water injection for water flooding operations, salt-water disposal, and gas recycling.

A discussion of production methods, in approximate order of relative importance, follows.

Natural flow. Natural flow is the most economical method of production and generally is utilized as long as the desired production rate can be maintained by this method. It utilizes the formation energy, which may consist of gas in solution in the oil in the formation; free gas under pressure acting against the liquid and gas-liquid phase to force it toward the well bore; water pressure acting against the oil; or a combination of these three energy sources. In some areas the casinghead pressure may be of the order of 10,000 psi, so it is necessary to provide fittings adequate to withstand such pressures. Adjustable throttle values, or chokes, are utilized to regulate the flow rate to a desired and safe value. With such a high-pressure drop across a throttle valve the life of the valve is likely to be very short. Several such valves are arranged in parallel in the tubing head "Christmas tree" with positive shutoff valves between the chokes and the tubing head so that the wearing parts of the throttle valve, or the entire valve, can be replaced while flow continues through another similar valve.

An additional safeguard that is often used in connection with high-pressure flowing wells is a bottom-hole choke or a bottom-hole flow control valve that limits the rate of flow to a reasonable value, or stops it completely, in case of failure of surface controls. Figure 1 shows a schematic outline of a simple flowing well hookup. The packer is not essential but is often used to reduce the free gas volume in the casing.

Flow rates for United States wells seldom exceed a few hundred barrels per day because of enforced or voluntary restrictions to regulate production rates and to obtain most efficient and economical ultimate recovery. However, in some countries, especially in the Middle East, it is not uncommon for natural flow rates to exceed 10,000 bpd/well [1590 m^3/(day)(well)].

Lifting. Most wells are not self-flowing. The common types of lifting are outlined here.

Pumping with sucker rods. Approximately 90% of the wells made to produce by some artificial lift method in the United States are equipped with sucker-rod–type pumps. In these the pump is installed at the lower end of the tubing string and is actuated by a string of sucker rods extending from the surface to the subsurface pump. The sucker rods are attached to a polished rod at the surface. The polished rod extends through a stuffing box and is attached to the pumping unit, which produces the necessary reciprocating motion to actu-

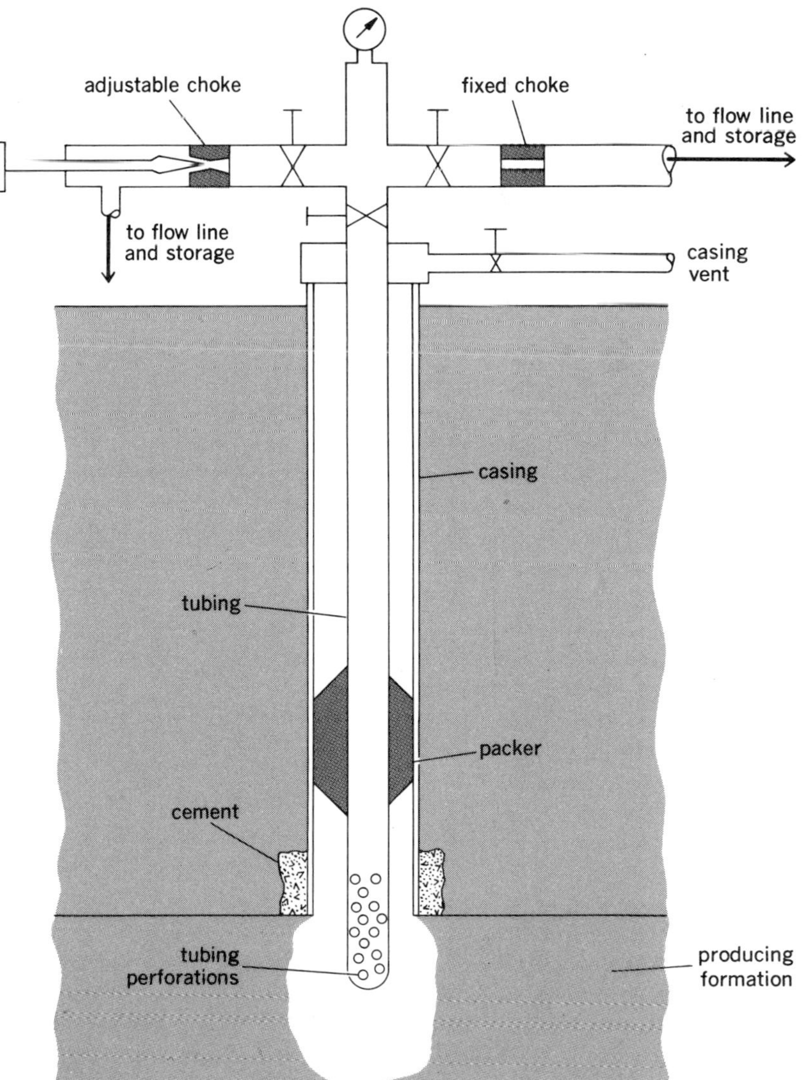

Fig. 1. Schematic view of well equipped for producing by natural flow.

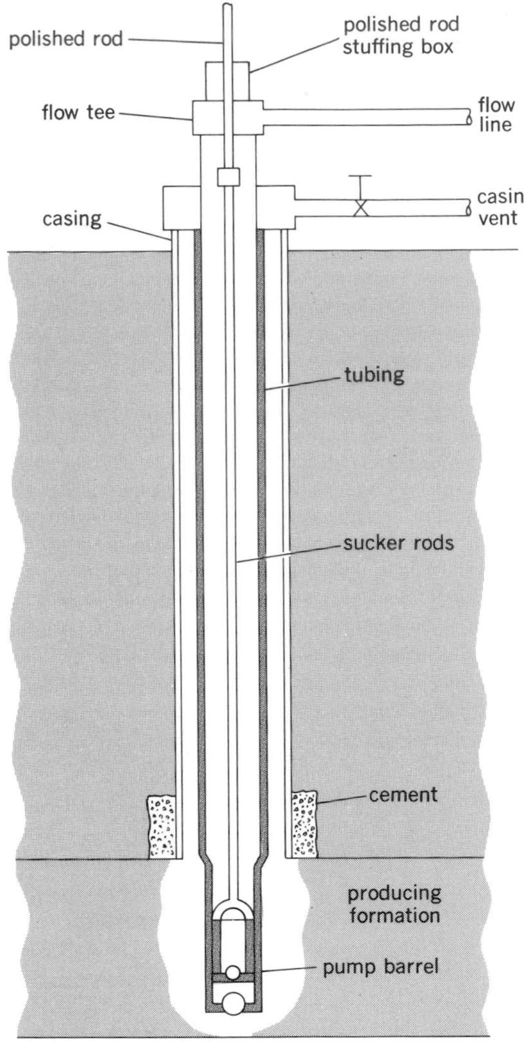

Fig. 2. A schematic view of a well which is equipped for pumping with sucker rods.

ate the sucker rods and the subsurface pump. Figure 2 shows a simplified schematic section through a pumping well. The two common variations are mechanical and hydraulic long-stroke pumping.

1. Mechanical pumping. The great majority of pumping units are of the mechanical type, consisting of a suitable reduction gear, and crank and pitman arrangement to drive a walking beam to produce the necessary reciprocating motion. A counterbalance is provided to equalize the load on the upstroke and downstroke. Mechanical pumping units of this type vary in load-carrying capacity from about 2000 to about 43,000 lb (900 to 19,500 kg), and the torque rating of the low-speed gear which drives the crank ranges from 6400 in.-lb (720 N·m) in the smallest API standard unit to about 1,500,000 in.-lb (170,000 N·m) for the largest units now in use. Stroke length varies from about 18 to 192 in. (46 to 488 cm). Usual operating speeds are from about 6 to 20 strokes/min. However, both lower and higher rates of speed are sometimes used. Figure 3 shows a modern pumping unit in operation.

Production rates with sucker-rod–type pumps vary from a fraction of 1 bpd in some areas, with part-time pumping, to approximately 3000 bpd (480 m^3/day) for the largest installations in relatively shallow wells.

2. Hydraulic long-stroke pumping. For this the units consist of a hydraulic lifting cylinder mounted directly over the well head and are designed to produce stroke lengths of as much as 30 ft (9 m). Such long-stroke hydraulic units are usually equipped with a pneumatic counterbalance arrangement which equalizes the power requirement on the upstroke and downstroke.

Hydraulic pumping units also are made without any provision for counterbalance. However, these units are generally limited to relatively small wells, and they are relatively inefficient.

Gas lift. Gas lift in its simplest form consists of initiating or stimulating well flow by injecting gas at some point below the fluid level in the well. With large-volume gas-lift operations the well may be produced through either the casing or the tubing. In the former case, gas is conducted through the tubing to the point of injection; in the latter, gas may be conducted to the point of injection through the casing or through an auxiliary string of tubing. When gas is injected into the oil column, the weight of the column above the point of injection is reduced as a result of the space occupied by the relatively low-density gas. This lightening of the fluid column is sufficient to permit the formation pressure to initiate flow up the tubing to the surface. Gas injection is often utilized to increase the flow from wells that will flow naturally but will not produce the desired amount by natural flow.

There are many factors determining the advisability of adopting gas lift as a means of production. One of the more important factors is the availability of an adequate supply of gas at suitable pressure and reasonable cost. In a majority of cases gas lift cannot be used economically to produce a reservoir to depletion because the well may be relatively productive with a low back pressure maintained on the formation but will produce very little, if anything, with the back pressure required for gas-lift operation. Therefore, it generally is necessary to resort to some mechanical means of pumping before the well is abandoned, and it may be more economical to adopt the mechanical means

Fig. 3. Pumping unit with adjustable rotary counterbalance. (*Oil Well Supply Division, U.S. Steel Corp.*)

initially than to install the gas-lift system while conditions are favorable and later replace it.

This discussion of gas lift has dealt primarily with the simple injection of gas, which may be continuous or intermittent. There are numerous modifications of gas-lift installations, including various designs for flow valves which may be installed in the tubing string to open and admit gas to the tubing from the casing at a predetermined pressure differential between the tubing and casing. When the valve opens, gas is injected into the tubing to initiate and maintain flow until the tubing pressure drops to a predetermined value; and the valve closes before the input gas-oil ratio becomes excessive. This represents an intermittent-flow–type valve. Other types are designed to maintain continuous flow, proper pressure differential, and proper gas injection rate for efficient operation. In some cases several such flow valves are spaced up the tubing string to permit flow to be initiated from various levels as required.

Other modifications of gas lift involve the utilization of displacement chambers. These are installed on the lower end of the well tubing where oil may accumulate, and the oil is displaced up the tubing with gas injection controlled by automatic or mechanical valves.

Hydraulic subsurface pumps. The hydraulic subsurface pump has come into fairly prominent use. The subsurface pump is operated by means of a hydraulic reciprocating motor attached to the pump and installed in the well as a single unit. The hydraulic motor is driven by a supply of hydraulic fluid under pressure that is circulated down a string of tubing and through the motor. Generally the hydraulic fluid consists of crude oil which is discharged into the return line and returns to the surface along with the produced crude oil.

Hydraulically operated subsurface pumps are also arranged for separating the hydraulic power fluid from the produced well fluid. This arrangement is especially desirable where the fluid being produced is corrosive or is contaminated with considerable quantities of sand or other solids that are difficult to separate to condition the fluid for use as satisfactory power oil. This method permits use of water or other nonflammable liquids as hydraulic power fluid to minimize fire hazard in case of a failure of the hydraulic power line at the surface.

Centrifugal well pumps. Electrically driven centrifugal pumps have been used to some extent, especially in large-volume wells of shallow or moderate depths. Both the pump and the motor are restricted in diameter to run down the well casing, leaving sufficient clearance for the flow of fluid around the pump housing. With the restricted diameter of the impellers the discharge head necessary for pumping a relatively deep well can be obtained only by using a large number of stages and operating at a relatively high speed. The usual rotating speed for such units is 3600 rpm, and it is not uncommon for such units to have 50 or more pump stages. The direct-connected electric motor must be provided with a suitable seal to prevent well fluid from entering the motor housing, and electrical leads must be run down the well casing to supply power to the motor.

Swabs. Swabs have been used for lifting oil almost since the beginning of the petroleum industry. They usually consist of a steel tubular body equipped with a check valve which permits oil to flow through the tube as it is lowered down the well with a wire line. The exterior of the steel body is generally fitted with flexible cup-type soft packing that will fall freely but will expand and form a seal with the tubing when pulled upward with a head of fluid above the swab. Swabs are run into the well on a wire line to a point considerably below the fluid level and then lifted back to the surface to deliver the volume of oil above the swab. They are often used for determining the productivity of a well that will not flow naturally and for assisting in cleaning paraffin from well tubing. In some cases swabs are used to stimulate wells to flow by lifting, from the upper portion of the tubing, the relatively dead oil from which most of the gas has separated.

Bailers. Bailers are used to remove fluids from wells and for cleaning out solid material. They are run into the wells on wire lines as in swabbing, but differ from swabs in that they generally are run only in the casing when there is no tubing in the well. The capacity of the bailer itself represents the volume of fluid lifted each time since the bailer does not form a seal with the casing. The bailer is simply a tubular vessel with a check valve in the bottom. This check valve generally is arranged so that it is forced open when the bailer touches bottom in order to assist in picking up solid material for cleaning out a well.

Jet pumps. A jet pump for use in oil wells operates on exactly the same principle as a water-well jet pump. Advantage is taken of the Bernoulli effect to reduce pressure by means of a high-velocity fluid jet. Thus oil is entrained from the well with this high-velocity jet in a venturi tube to accelerate the fluid and assist in lifting it to the surface, along with any assistance from the formation pressure. The application of jet pumps to oil wells has been insignificant.

Sonic pumps. Sonic pumps essentially consist of a string of tubing equipped with a check valve at each joint and mechanical means on the surface to vibrate the tubing string longitudinally. This creates a harmonic condition that will result in several hundred strokes per minute, with the strokes being a small fraction of 1 in. in length. Some of these pumps are in use in relatively shallow wells.

Lease tanks and gas separators. Figure 4 shows a typical lease tank battery consisting of four 1000-bbl tanks and two gas separators. Such equipment is used for handling production from wells produced by natural flow, gas lift, or pumping. In some pumping wells the gas content may be too low to justify the cost of separators for saving the gas.

Fig. 4. Lease tank battery with four tanks and two gas separators. *(Gulf Oil Corp.)*

Natural gasoline production. An important phase of oil and gas production in many areas is the production of natural gasoline from gas taken from the casinghead of oil wells or separated from the oil and conducted to the natural gasoline plant. The plant consists of facilities for compressing and extracting the liquid components from the gas. The natural gasoline generally is collected by cooling and condensing the vapors after compression or by absorbing in organic liquids having high boiling points from which the volatile liquids are distilled. Many natural gasoline plants utilize a combination of condensing and absorbing techniques. Figure 5 shows an overall view of a natural gasoline plant operating in western Texas.

PRODUCTION PROBLEMS AND INSTRUMENTS

To maintain production, various problems must be overcome. Numerous instruments have been developed to monitor production and to control production problems.

Corrosion. In many areas the corrosion of production equipment is a major factor in the cost of petroleum production. The following comments on the oil field corrosion problem are taken largely from *Corrosion of Oil- and Gas-Well Equipment* and reproduced by permission of NACE-API.

For practical consideration, corrosion in oil and gas-well production can be classified into four main types.

1. Sweet corrosion occurs as a result of the presence of carbon dioxide and fatty acids. Oxygen and hydrogen sulfide are not present. This type of corrosion occurs in both gas-condensate and oil wells. It is most frequently encountered in the United States in southern Louisiana and Texas, and other scattered areas. At least 20% of all sweet oil production and 45% of condensate production are considered corrosive.

2. Sour corrosion is designated as corrosion in oil and gas wells producing even trace quantities of hydrogen sulfide. These wells may also contain oxygen, carbon dioxide, or organic acids. Sour corrosion occurs in the United States primarily throughout Arbuckle production in Kansas and in the Permian basin of western Texas and New Mexico. About 12% of all sour production is considered corrosive.

3. Oxygen corrosion occurs wherever equipment is exposed to atmospheric oxygen. It occurs most frequently in offshore installations, brine-handling and injection systems, and in shallow producing wells where air is allowed to enter the casing.

4. Electrochemical corrosion is designated as that which occurs when corrosion currents can be readily measured or when corrosion can be mitigated by the application of current, as in soil corrosion.

Corrosion inhibitors are used extensively in both oil and gas wells to reduce corrosion damage to subsurface equipment. Most of the inhibitors used in the oil field are of the so-called polar organic type. All of the major inhibitor suppliers can furnish effective inhibitors for the prevention of sweet corrosion as encountered in most fields. These can be purchased in oil-soluble, water-dispersible, or water-soluble form.

Paraffin deposits. In many crude-oil–producing areas paraffin deposits in tubing and flow lines and on sucker rods are a source of considerable trouble and expense. Such deposits build up until the tubing or flow line is partially or completely plugged. It is necessary to remove these deposits to maintain production rates. A variety of methods are used to remove paraffin from the tubing, including the application of heated oil through tubular sucker rods to mix with and transfer heat to the oil being produced and raise the temperature to a point at which the deposited paraffin will be dissolved or melted. Paraffin solvents may also be applied in this manner without the necessity of applying heat.

Mechanical means often are used in which a scraping tool is run on a wire line and paraffin is scraped from the tubing wall as the tool is pulled back to the surface. Mechanical scrapers that attach to sucker rods also are in use. Various types of automatic scrapers have been used in connection with flowing wells. These consist of a form of piston that will drop freely to the bottom when flow is stopped but will rise back to the surface when flow is resumed. Electrical heating methods have been used rather extensively in some areas. The tubing is insulated from the casing and from the flow line, and electric current is transmitted through the tubing for the time necessary to heat the tubing sufficiently to cause the paraffin deposits to melt or go into solution in the oil in the tubing. Plastic coatings have been utilized inside tubing and flow lines to minimize or prevent paraffin deposits. Paraffin does not deposit readily on certain plastic coatings.

A common method for removing paraffin from flow lines is to disconnect the line at the well head and at the tank battery and force live steam through the line to melt the paraffin deposits and flow them out. Various designs of flow-line scrapers have also been used rather extensively and fairly successfully. Paraffin deposits in flow lines are minimized by insulating the lines or by burying the lines to maintain a higher average temperature.

Emulsions. A large percentage of oil wells produce various quantities of salt water along with the

Fig. 5. Modern natural gasoline plant in western Texas. (*Gulf Oil Corp.*)

oil, and numerous wells are being pumped in which the salt-water production is 90% or more of the total fluid lifted. Turbulence resulting from production methods results in the formation of emulsions of water in oil or oil in water; the commoner type is oil in water. Emulsions are treated with a variety of demulsifying chemicals, with the application of heat, and with a combination of these two treatments. Another method for breaking emulsions is the electrostatic or electrical precipitator type of emulsion treatment. In this method the emulsion to be broken is circulated between electrodes subjected to a high potential difference. The resulting concentrated electric field tends to rupture the oil-water interface and thus breaks the emulsion and permits the water to settle out. Figure 6 shows two pumping wells with a tank battery in the background. This tank battery is equipped with a wash tank, or gun barrel, and a gas-fired heater for emulsion treating and water separation before the oil is admitted to the lease tanks.

Gas conservation. If the quantity of gas produced with crude oil is appreciably greater than that which can be efficiently utilized or marketed, it is necessary to provide facilities for returning the excess gas to the producing formation. Formerly, large quantities of excess gas were disposed of by burning or simply by venting to the atmosphere. This practice is now unlawful. Returning excess gas to the formation not only conserves the gas for future use but also results in greater ultimate recovery of oil from the formation.

Salt-water disposal. The large volumes of salt water produced with the oil in some areas present serious disposal problems. The salt water is generally pumped back to the formation through wells drilled for this purpose. Such salt-water disposal wells are located in areas where the formation already contains water. Thus this practice helps to maintain the formation pressure as well as the productivity of the producing wells.

Offshore production. Offshore wells present additional production problems since the wells must be serviced from barges or boats. Wells of reasonable depth on land locations are seldom equipped with derricks for servicing because it is more economical to set up a portable mast for pulling and installing rods, tubing, and other equipment. However, the use of portable masts is not practical on offshore locations, and a derrick is generally left standing over such wells throughout their productive life to facilitate servicing. There are a considerable number of offshore wells along the Gulf Coast and the Pacific Coast of the United States, but by far the greatest number of offshore wells in a particular region is in Lake Maracaibo in Venezuela. Figure 7 shows a considerable number of derricks in Lake Maracaibo with pumping wells in the foreground. These wells are pumped by electric power through cables laid on the lake bottom to conduct electricity from power-generating stations onshore. An overwater tank battery is visible at the extreme right. All offshore installations, such as tank batteries, pump stations, and the derricks and pumping equipment, are supported on pilings in water up to 100 ft (30 m) or more in depth. There are approximately 2300 oil derricks in Lake Maracaibo. A growing number of semipermanent platform rigs and even bottom storage facilities are being used in Gulf of Mexico waters at depths of more than 100 ft (30 m).

Fig. 6. Two pumping wells with tank battery. (*Oil Well Supply Division, U.S. Steel Corp.*)

Instruments. The commoner and more important instruments required in petroleum production operations are included in the following discussion.

1. Gas meters, which are generally of the orifice type, are designed to record the differential pressure across the orifice, and the static pressure.

2. Recording subsurface pressure gages small enough to run down 2-in. ID (inside diameter) tubing are used extensively for measuring pressure gradients down the tubing of flowing wells, recording pressure buildup when the well is closed in, and measuring equilibrium bottom-hole pressures.

3. Subsurface samplers designed to sample well fluids at various levels in the tubing are used to determine physical properties, such as viscosity, gas content, free gas, and dissolved gas at various levels. These instruments may also include a recording thermometer or a maximum reading thermometer, depending upon the information required.

4. Oil meters of various types are utilized to meter crude oil flowing to or from storage.

5. Dynamometers are used to measure polished-rod loads. These instruments are sometimes known as well weighers since they are used to record the polished-rod load throughout a pumping cycle of a sucker-rod–type pump. They are used to determine maximum load on polished rods as well as load variations, to permit accurate counterbalancing of pumping wells, and to assure that pumping units or sucker-rod strings are not seriously overloaded.

6. Liquid-level gages and controllers are used. They are similar to those used in other industries, but with special designs for closed lease tanks.

Fig. 7. Numerous offshore wells located in Lake Maracaibo, Venezuela. (*Creole Petroleum Corp.*)

A wide variety of scientific instruments find application in petroleum production problems. The above outline gives an indication of a few specialized instruments used in this branch of the industry, and there are many more. Special instruments developed by service companies are valued for a wide variety of purposes and include calipers to detect and measure corrosion pits inside tubing and casing and magnetic instruments to detect microscopic cracks in sucker rods.

[ROY L. CHENAULT]

Bibliography: American Petroleum Institute, *History of Petroleum Engineering*, 1961; K. E. Brown, *The Technology of Artificial Lift Methods*, 1977; ETA Offshore Seminars, Inc., *The Technology of Offshore Drilling and Production*, 1976; L. L. Farkas, *Management of Technical Field Operations*, 1970; T. C. Frick (ed.), *Petroleum Production Handbook*, vol. 1: *Mathematics and Production Equipment*, 1962; L. M. Harris, *An Introduction to Deep Water Floating Drilling Operations*, 1972; M. Muskat, *Physical Principles of Oil Production*, 1949; T. E. W. Nind, *Principles of Oil Well Production*, 1964; L. T. Stanley, *Practical Statistics for Petroleum Engineers*, 1973; L. C. Uren, *Petroleum Production Engineering: Oil Field Development*, 4th ed., 1956; L. C. Uren, *Petroleum Production Engineering: Oil Field Exploitation*, 3d ed., 1953; World trends, *World Oil*, vol. 189, no. 3, Aug. 15, 1979; J. Zaba and W. T. Doherty, *Practical Petroleum Engineering*, 5th ed., 1970.

Oil and gas storage

Crude oil and natural gas, after being produced from their natural reservoirs, are stored in great quantities. Large amounts of refined products are stored as well. Storage is necessary to meet seasonal and other fluctuations in demand and for efficient operation of producing equipment, pipelines, tankers, and refineries. Storage also provides ready reserves for emergency use. According to the U.S. Bureau of Mines, 265×10^6 bbl (1 bbl= 0.1590 m^3) of crude oil were in storage in the United States at the end of 1974. In addition, 808×10^6 bbl in the form of refined products, natural gasoline, plant condensate, and unfinished oils were in storage. The American Gas Association reported 4.788×10^{12} ft^3 (1 $ft^3 = 2.832 \times 10^{-2}$ m^3) of natural gas stored in underground reservoirs in the United States at the end of 1978.

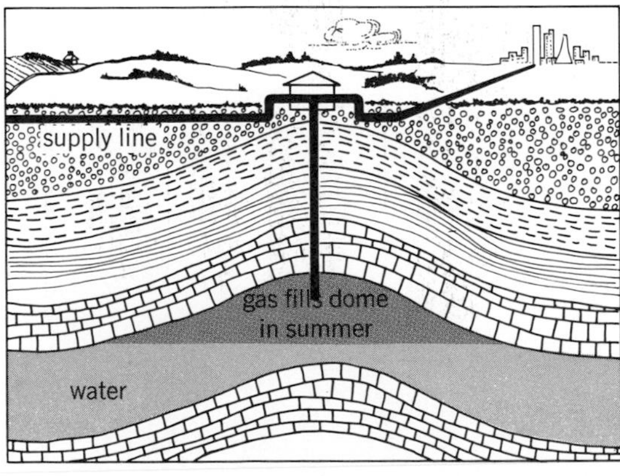

Schematic cross section of typical aquifer gas storage field showing injected gas displacing water. (*Natural Gas Pipeline Company of America*)

Crude oil and refined products. Oil from producing wells is first collected in welded-steel, bolted-steel, or wooden tanks of 100 bbl or greater capacity located on individual leases. These tanks, upright cylinders with low-pitched conical roofs, provide temporary storage while the oil is awaiting shipment. Several tanks grouped together are called a tank battery. Assemblages of large steel tanks, known as tank farms, are used for more permanent storage at pipeline pump stations, points where tankers are loaded and unloaded, and refineries.

With the trend toward giant tankers, accelerated by the closing of the Suez Canal in 1967, large storage facilities are needed at both the loading and unloading ends of the tanker runs. Some tanks with capacities of 1×10^6 bbl are now in use. Large-capacity excavated reservoirs with concrete linings have been used for many years in California to store both crude and fuel oil. One such reservoir with a fixed roof and elliptical in form, is 780 ft (1 ft = 0.3048 m) long, 467 ft wide, and 23 ft deep. It covers $9\frac{1}{4}$ acres and provides storage for more than 1×10^6 bbl. Another reservoir has a capacity of 4×10^6 bbl and covers 16 acres.

Offshore storage. For offshore producing fields a number of unique storage systems have been designed. In several instances old tankers have been adapted for storage, and barges have been constructed especially for offshore storage use. One underwater installation consists essentially of three giant inverted steel funnels. Each unit is 270 ft in diameter and 205 ft high, weighs 28×10^6 lb, and has a capacity of 0.5×10^6 bbl. The bottom is open, and the unit is anchored to the sea floor by 95-ft pilings. A reinforced concrete installation features a nine-module storage unit with 1×10^6 bbl capacity surrounded by a perforated wall 302 ft in diameter that serves as a breakwater. The outer wall is about 270 ft high and extends about 40 ft above the water surface. A submerged floating storage tank 96 ft in diamter and 305 ft high is held in place by six anchor lines and has a capacity of 300,000 bbl. One relatively small unit consists of a platform with four vertical legs, each holding 4700 bbl, and four horizontal tanks at the bottom holding 1850 bbl each, for a total capacity of 26,200 bbl. A second small unit, utilizing bottom tanks as an anchor, holds 2400 bbl underwater and 600 bbl in a spherical tank above the surface. A third small unit consists of a sea-floor base, connected by a universal joint to a large-diameter vertical cylinder, about 350 ft high, which extends above the surface of the water. One proposed design includes an excavated cavern beneath the sea floor, and nuclear cavities have also been suggested.

Volatility problems. To minimize vaporization losses, lease tanks are sometimes equipped to hold several ounces pressure. At large-capacity storage sites, special tanks are generally used. Tanks with lifter or floating roofs are used to store crude oil, motor gasoline, and less volatile natural gasoline. Motor and natural gasolines are also stored in spheroid containers. Spherical containers are used for more volatile liquids, such as butane. Horizontal cylindrical containers are used for propane and butane storage. Refrigerated insulated tank systems enabling propane to be stored at a lower pressure are also in use. One tank has a capacity of 900,000 bbl.

Underground storage. Large quantities of volatile liquid-petroleum products, including propane and butane, are stored in underground caverns dissolved in salt formations and in mined caverns, gas reservoirs, and water sands. In 1973 the underground storage capacity for liquid-petroleum products in the United States was 255.231×10^6 bbl. Underground storage capacity in Canada was 14.996×10^6 bbl. Liquid-petroleum products are also being stored underground in Belgium, France, Germany, Italy, and the United Kingdom. Caverns are also used for storing crude oil in Sweden, Germany, and France. In Pennsylvania an abandoned quarry with a capacity of 2×10^6 bbl has been equipped with a floating roof for storing fuel oil. Refrigerated propane is also being stored in excavations in frozen earth and in underground concrete tanks. To provide security in the event of another oil embargo, the National Petroleum Council has proposed developing salt cavern storage for 500×10^6 bbl of crude oil.

Natural gas. Natural gas is stored in low-pressure surface holders, buried high-pressure pipe batteries and bottles, depleted or partially depleted oil and gas reservoirs, water sands, and several types of containers at extremely low temperature (−258°F, or −161°C) after liquefaction.

Low-pressure holders, which store relatively small volumes of gas, basically use either a water or a dry seal, and variations of each type exist. With the displacement of manufactured gas by natural gas in the United States, the need for surface holders has greatly diminished and they have disappeared almost entirely.

Underground storage. In the United States gas pipeline and utility companies store large quantities of natural gas in underground reservoirs. In most cases these reservoirs are located near market areas and are used to supplement pipeline supplies during the winter months when the gas demand for residential heating is very high. Since gas can be stored in the summer when the gas demand is low, underground storage permits greatly increased pipeline utilization, resulting in lower transportation costs and reduced gas cost to the consumer. Underground storage is the only economical method of storing large enough quantities of gas to meet the seasonal fluctuations in pipeline loads, and has enabled gas companies to meet market requirements which otherwise could not be satisfied.

Gas was first stored underground in 1915 in a partially depleted gas field in Ontario, Canada. The following year gas was injected into a depleted gas field near Buffalo, NY. The table, prepared by the American Gas Association, shows the growth in underground storage capacity in the United States. At the end of 1978 gas was being stored in 311 reservoirs, which utilized depleted gas and oil fields, water sands, salt caverns, and an abandoned coal mine. These reservoirs are located in 26 states and are operated by 86 different companies. They have a total capacity of 7.330×10^{12} ft^3 and, at the end of 1978, held 4.788×10^{12} ft^3 of stored gas plus 1.054×10^{12} ft^3 of negative gas. The maximum volume of gas in storage during 1978 was 5.301×10^{12} ft^3, excluding native gas. During 1978 a total of 2.150×10^{12} ft^3 was withdrawn from storage. The total maximum daily output from all of these reservoirs was 28.3×10^9 ft^3. Canada has 17 storage reservoirs, 14 in gas and oil reservoirs and 3 in salt caverns, with a total capacity of 325×10^9 ft^3. At the end of 1978 these reservoirs held 250×10^9 ft^3 of natural gas. Gas is also being stored in underground reservoirs in France, East and West Germany, Austria, Italy, Poland, Rumania, Czechoslovakia, and the Soviet Union. In the

Growth of underground gas storage in United States, 1916–1974*

Year	Number of reservoirs	Number of states	Total reservoir capacity, $\times 10^9$ ft^3
1916	1	1	
1919	2	2	5
1920	3	3	5
1925	4	4	7
1927	5	5	7
1928	7	5	15
1929	8	6	18
1930	9	6	18
1931	11	6	25
1933	12	7	27
1934	15	7	39
1935	15	7	38
1936	22	8	66
1937	31	8	146
1938	35	8	198
1939	39	9	205
1940	44	9	232
1941	51	10	283
1942	59	12	289
1943	66	12	301
1944	67	13	251
1945	75	14	416
1946	78	14	424
1947	91	15	560
1948	102	15	584
1949	118	15	625
1950	125	15	770
1951	142	15	911
1952	151	16	1285
1953	167	17	1726
1954	172	17	1849
1955	178	18	2084
1956	187	19	2389
1957	199	19	2589
1958	205	19	2703
1959	209	20	2507
1960	217	20	2854
1961	229	21	3201
1962	258	21	3485
1963	278	23	3674
1964	286	24	3942
1965	293	24	4086
1966	303	25	4421
1967	308	25	4520
1968	315	26	4783
1969	320	26	4927
1970	325	26	5178
1971	333	26	5575
1972	348	26	6040
1973	360	26	6279
1974	367	26	6360
1975	376	26	6644
1976	387	26	6927
1977	385	26	7223
1978	388	26	7330

*American Gas Association, *Underground Storage of Gas in the United States and Canada.*

United Kingdom a salt cavern is being used for gas storage.

Gas storage in water sands was first undertaken in 1952, and this method of storage has steadily increased, especially in areas where no gas or oil fields are available. In the United States at the end of 1974, aquifer-type storages numbered 51 and these reservoirs were operated by 25 companies in 10 states. They had a total capacity of 1.408×10^{12} ft^3 and, at the end of 1974, held 849×10^9 ft^3. During 1974 the maximum volume in storage was 899×10^9 ft^3. The total maximum daily output from these reservoirs was 4.7×10^9 ft^3. A cross section of a typical aquifer storage field is shown in the illustration. A geologic trap having adequate structural closure and a suitable caprock is needed. The storage sand must be porous and thick enough and under sufficient hydrostatic pressure to hold large quantities of gas. The sand must also be sufficiently permeable and continuous over a wide enough area so that water can be pushed back readily to make room for the stored gas. In some cases, water removal wells are also utilized.

Reservoir pressure. In operating storage reservoirs only a portion of the stored gas, called working gas, is normally withdrawn. The remaining gas, called cushion gas, stays in the reservoir to provide the necessary pressure to produce the storage wells at desired rates. In aquifer storages some water returns to help maintain the reservoir pressure. The percentage of cushion gas varies considerably among reservoirs. Based on American Gas Association figures, cushion gas amounted to 59% of the maximum gas in storage, including native gas, in the United States in 1974. In some instances, the original reservoir pressure of the oil and gas field is exceeded in storage operations. This has resulted in storage volumes greater than the original content and has substantially increased well deliverabilities. In aquifer storages the original hydrostatic pressure must be exceeded in order to push the water back.

In the Soviet Union aquifer gas storage is being undertaken in one area where no appreciable structure. In an inconclusive field test, air was injected into a center well with control of the lateral spread of the air bubble attempted by injecting water into surrounding wells. Storage of gas in cavities created by nuclear explosions has been proposed and seriously considered.

Liquefied gas. Storage of liquefied natural gas has rapidly increased throughout the world. Storage is in connection with shipment of liquefied natural gas by tanker, and is located at the loading and unloading ends of the tanker runs as well as at peak sharing facilities operated by gas pipeline and local utility companies. In 1974 the United States and Canada had more than 100 liquefied natural-gas storage installations either operational or under construction. These had a total storage capacity of 22.8×10^6 bbl, or 78.5×10^9 ft^3. England, France, the Netherlands, West Germany, Italy, Spain, Algeria, Libya, and Japan also have installations. Storage is in insulated metal tanks, buried concrete tanks, or frozen earth excavations. In two projects using frozen earth excavations, excessive boil-off of the liquefied gas has led to replacement with insulated metal tanks. *See* LIQUEFIED NATURAL GAS (LNG); OIL AND GAS FIELD EXPLOITATION; PETROLEUM PROCESSING; PIPELINE.

[PETER G. BURNETT]

Bibliography: American Gas Association, *The Underground Storage of Gas in the United States and Canada*, Annual Report on Statistics; D. C. Bond, *Underground Storage of Natural Gas*, Illinois State Geological Survey, 1975; G. D. Hobson and W. Pohl, *Modern Petroleum Engineering*, D. L. Katz et al., *Handbook of Natural Gas Engineering*, 1959; D. L. Katz and P. A. Witherspoon, *Underground and Other Storage of Oil Products and Gas*, Proceedings of the 8th World Petroleum Congress, 1971; Stone and Webster Engineering, *Gas Storage at the Point of Use*, Amer. Gas Ass. Proj. PL-56, 1965.

Oil and gas well completion

The operations that prepare for production a well drilled to an oil or gas reservoir. Various problems of well casing during and at the end of drilling are related to modes of completing connection between the proper reservoirs and the surface. Tubing inside the casing and valves and a pumping unit at the surface must deliver reservoir products to the surface at a controlled rate. Variations in reservoir and overlying formations may require special techniques to keep out water or sand or to increase the production rate. *See* OIL AND GAS FIELD EXPLOITATION; OIL AND GAS WELL DRILLING.

Casing. Oil and gas wells are walled with steel tubing which is cemented in place.

Tubing. Steel tubing is manufactured in various diameters, wall thicknesses, lengths, and steel al-

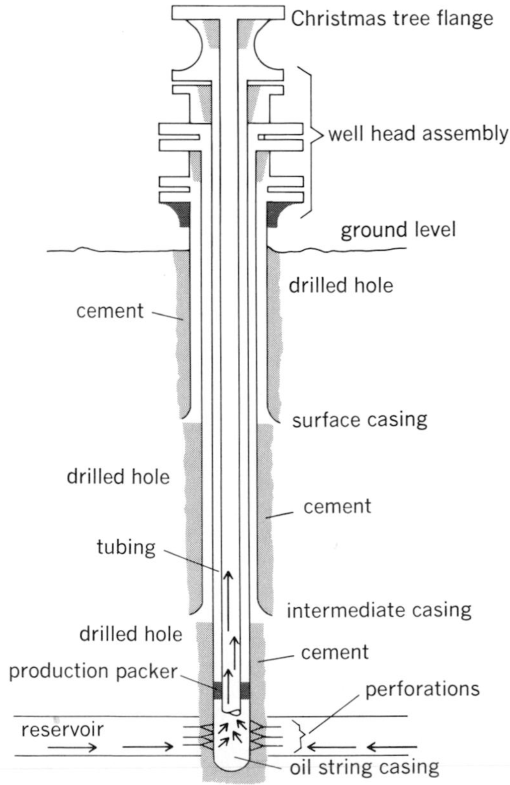

Fig. 1. Casing detail; casing strings in an oil well.

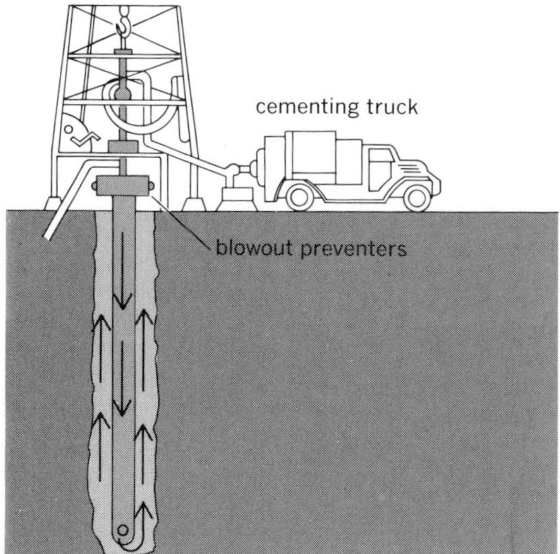

Fig. 2. Diagram of cementing process, showing truck, equipment, and well job.

loys, selected to satisfy specific needs. These lengths, called joints, are threaded and coupled so that they may be joined together in a continuous string in the well bore. Properly placed and cemented in the hole, casing protects fresh-water reservoirs from contamination, supports unconsolidated rock formations, maintains natural separation of formations, aids in the prevention of blowouts and waste of reservoir energy, and acts as a conduit for receiving pipe of smaller diameters through which the well effluent may be brought to the surface under controlled conditions.

It may be necessary to set many strings of casing in one hole before reaching the objective (Fig. 1). The determining factors are many, such as depth of hole, loss of circulation, high-pressure formations, hole sloughing, and wearing out a string of casing while rotating drill pipe through it over a long period of time.

Cementing casing. Basically, ordinary portland cement is used in cementing casing. In order to obtain the protection and fulfil the purposes, it is imperative that each string of casing be securely sealed to the walls of the hole for at least some distance up from the bottom of the casing string. After casing is in place, the cement is pumped down the inside and up the outside to a predetermined height to occupy the space between the casing and the walls of the hole, thereby effecting the desired seal. In the pumping and measuring process, plugs are used to separate the cement from other fluids to eliminate contamination; also, the cement inside the casing is displaced with fluid. Oil or gas well cementing is not performed with the drilling equipment; an outside service company equipped with mobile, high-pressure mixing and pumping equipment and accessories operated by trained personnel is employed (Fig. 2).

Well hole–reservoir connection. Reservoir conditions, known to exist or later defined, determine the type of completion technique to be followed: (1) barefoot completion, (2) preperforated liner, or (3) casing set through and perforated.

Barefoot completion. This type of completion is frequently used when the character of the producing rock is such that it does not require supplemental support or screening, for example, in formations such as limestone, dolomite, or hard sandstone. With this method, the production casing is seated above the producing section in the conventional manner. The casing is cleaned out by insertion of fluid-separating plugs and drilled out through the casing and into the producing formation below. The formation contents enter the bore hole from the bare or unlined producing stratum or strata, hence the term barefoot (Fig. 3).

Liner-type completion. This type of completion is similar to a barefoot completion except that the open portion of the hole is cased with a preperforated section of casing called a liner (Fig. 4). This liner is smaller in diameter than the casing previously set in the hole and is usually suspended from the upper casing near the bottom from a liner hanger. The hanger is attached to the top of the liner, and when it is set, it effects a seal between the liner and the casing. The purpose of the liner is to permit gas and liquids to enter the hole and screen out formation particles.

Gun perforating. Gun perforating is a method of forming holes through the casing and into a formation from within a well bore. The two more popular methods are bullet perforating, as with a rifle, and jet perforating, as with a torch.

The gun is fitted around the outside with barrels containing the perforating medium. Each barrel is wired to fire by remote control from the surface. The gun is run into the hole on a wire line from a service company's shooting truck. The wire line serves to lower and raise the gun in and out of the hole and, when the gun is in position to be fired, the operator sends an electric impulse down the line to trigger it. The hole is thus formed through the steel casing, the cement sheath, and some inches into the reservoir rock, creating an entry for the reservoir content into the well bore. Guns of the bullet type are retrieved and reloaded (Fig. 5). Jet-type guns are expendable and disintegrate (Fig. 6).

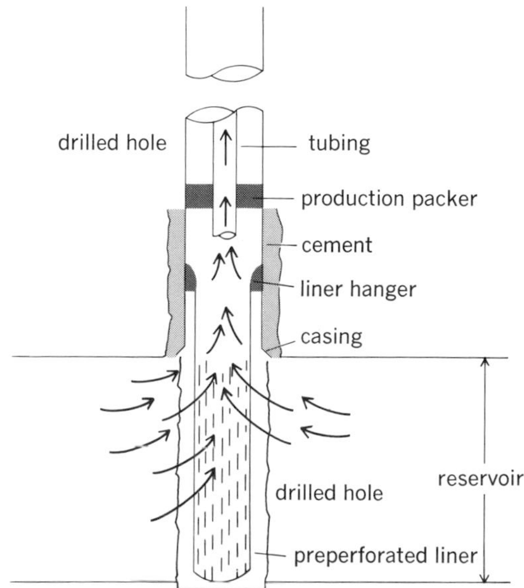

Fig. 4. Liner-type completion; preperforated liner.

OIL AND GAS WELL COMPLETION

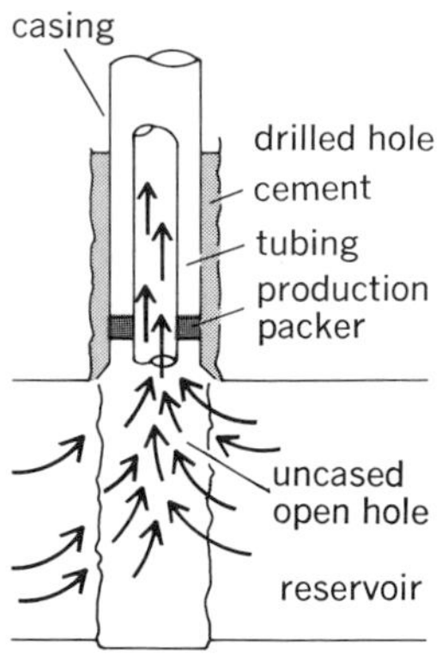

Fig. 3. Diagram of barefoot completion.

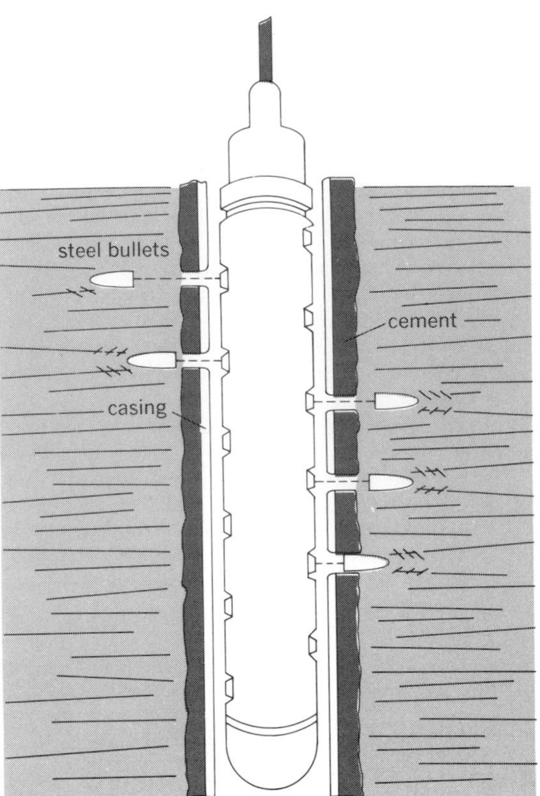

Fig. 5. Gun perforator bullets from bullet-type gun.

Production-flow control. A steel tube, the same as casing except that it is smaller in diameter, serves as a production flow line within the well. The tubing is run inside the casing and is either suspended in the hole or set on a production packer at or near the producing interval. The top of the tubing string terminates at the surface in a sealing element in the wellhead assembly to which the so-called Christmas tree is attached. A manifold constructed of steel valves and fittings, placed on top of the casings protruding above the surface, is called a Christmas tree (Fig. 7). Its purpose is to maintain the well under proper control, to receive the formation products under pressure, and to control the rate of daily production from the reservoir and direct it into a pipeline, generally at reduced pressures, to the oil-gathering station.

Pumping unit. Relatively few oil wells flow from natural pressures. Most require secondary means of removing the reservoir product. The most common of several methods is the pumping unit. A walking beam is operated like a seesaw, raising and lowering a plunger-type pump set near the bottom of the hole. The rods between the walking bean and pump are called sucker rods (Fig. 8).

Multiple completion. An oil or gas well from which several separate horizons are individually, separately, and simultaneously produced is called a multiple completion. Such a completion is accomplished by the use of multiple-zone packers and separate tubing strings. The producing zones are separated one from another by proper placement of packers in the well bore. An individual string of tubing is attached to each packer and extended to the surface where each is interconnected with the Christmas tree or flow assembly (Fig. 9).

OIL AND GAS WELL COMPLETION

jets

(a)

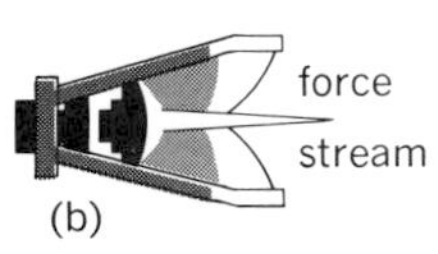

(b)

Fig. 6. Jet-type perforator. (*a*) Charges in firing position. (*b*) Side view of a shaped charge.

The several advantages of such a completion from a single well bore are increase in daily production, more efficient and economic utilization of a well bore with multiple reservoirs, increase in ultimate recovery, accurate measurement of product withdrawal from each reservoir, and elimination of mixing of products of different gravities and basic sediment and water content.

Water problems. The production of water in quantity from an oil or gas well renders it uneconomical; means are provided for water exclusion.

Water-exclusion methods. Water exclusion may be effected by the application of cements or various types of plastic. If it is determined that water is entering from the lower portion of a producing sand in a relatively shallow, low-pressure well, a cement plug may be placed in the bottom of the hole so that it will cover the oil-water interface of the reservoir. This technique is called laying in a plug and may be accomplished by placing the cement with a dump-bottom bailer on a wire line or by pumping cement down the drill pipe or tubing. For deeper, higher-pressure, or more troublesome wells, a squeeze method is used. Squeeze cementing is the process of applying hydraulic pressure to force a cementing material into permeable space of an exposed formation or through openings in the casing or liner. In many conditions cement, plastic, or diesel-oil cement may be squeezed into water-, oil-, or gas-bearing portions of a producing zone to eliminate excessive water without sealing off the gas or oil. A few of the applications are: repair of casing leaks; isolation of producing zones prior to perforating for production; remedial or secondary cementing to correct a defective condition, such as channeling or insufficient cement on a primary cement job; sealing off a low-pressure formation that engulfs oil and gas or drilling fluids; and abandonment of depleted producing zones to prevent migration of formation effluent and to reduce possibilities of contaminating other zones or wells.

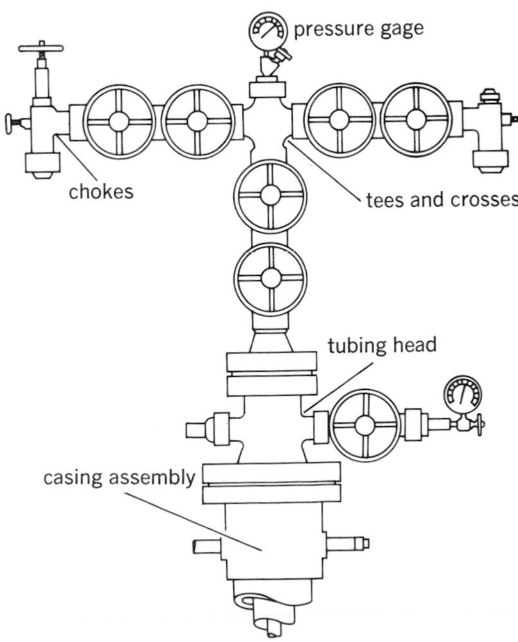

Fig. 7. Typical layout of a Christmas tree.

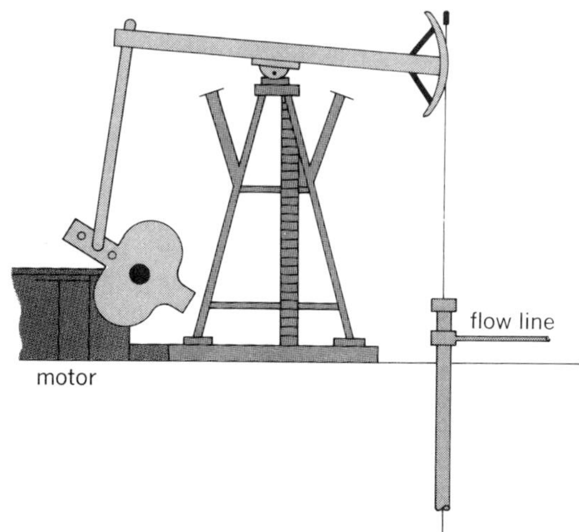

Fig. 8. Pumping unit diagram. These are most common if natural pressure is lacking for well flow.

The squeeze-method tool is a packer-type device designed to isolate the point of entry between or below packing elements. The tool is run into the hole on drill pipe or tubing, and the cementing material is squeezed out between or below these confining elements into the problem area. The well is then recompleted. It may be necessary to drill the cement out of the hole and reperforate, depending upon the outcome of the job performed in the squeeze process.

Water-exclusion plug back. Simple water shutoff jobs in shallow, deep, or high-pressure wells may also be performed in multizone wells in which the lower producing interval is depleted or the remaining recoverable reserves do not justify rehabilitation.

Here, water may be excluded by placing a packer-type plug (cork) above the interval, then producing formations that are already open or perforating additional intervals that may be present higher up the hole (Fig. 10).

Production-stimulation techniques. The initial testing or production history often indicates subnormal production rates, signifying the necessity for remedial action. Any method designed to increase the production rate from a reservoir is defined as production stimulation. Three of the methods used are acidizing, fracturing, and employing explosives.

Acidizing. Varied volumes of hydrochloric acid are used in limestone and dolomite or other acid-soluble formations to dissolve the existing flow-channel walls and enlarge them (Fig. 11). High-pressure equipment, pumps, and wellheads are necessary for satisfactory performance. Fast pumping speeds and acid inhibitors are used to alleviate corrosion of the well equipment.

Fracturing. Formation fracturing is a hydraulic process aimed at the parting of a desired section of formation. Selected grades of sand or particles of other materials are added to the fracturing fluid in varied quantities. These particles pack and fill the fracture, acting as a propping agent to hold it open when the applied pressure is released (Fig. 12). Such fractures increase the flow channels in size and number, improving the fluid-flow characteristics of the reservoir rock. The particle-carrying agent (fluid) is of considerable importance and is varied to fit particular demands. Some of the fluids which are used in this process are crude oil (sand oil fracturing), special refined oils (sand oil fracturing), water (river fracturing), acid (acid fracturing), and oil, water, and chemical emulsion (emulsifracturing).

Explosives. The idea of stimulating production by use of explosives was first used in a well in Pennsylvania on Jan. 21, 1865. The first torpedo consisted of 8 lb (3.6 kg) of gunpowder, contained

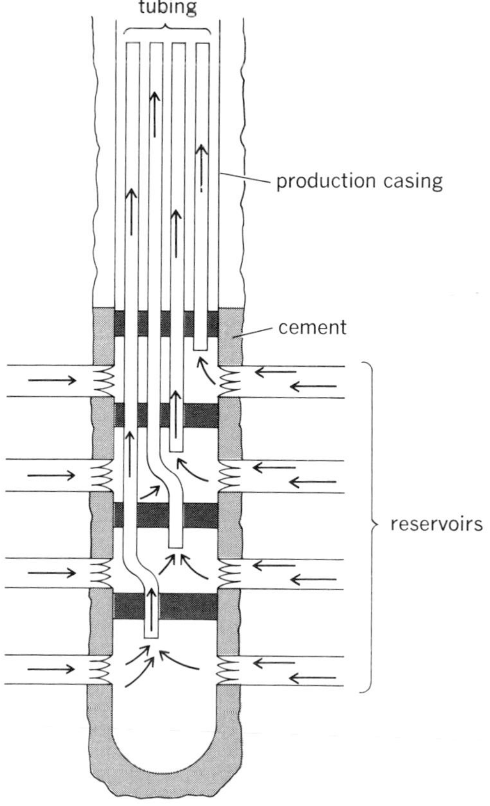

Fig. 9. A representative multiple completion diagram.

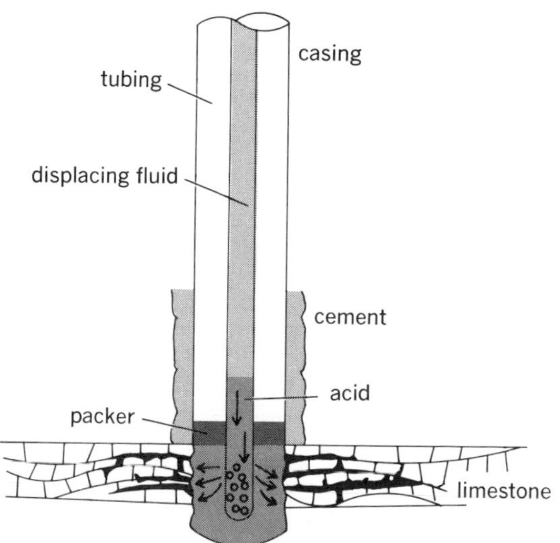

Fig. 11. Outline of the acidizing process.

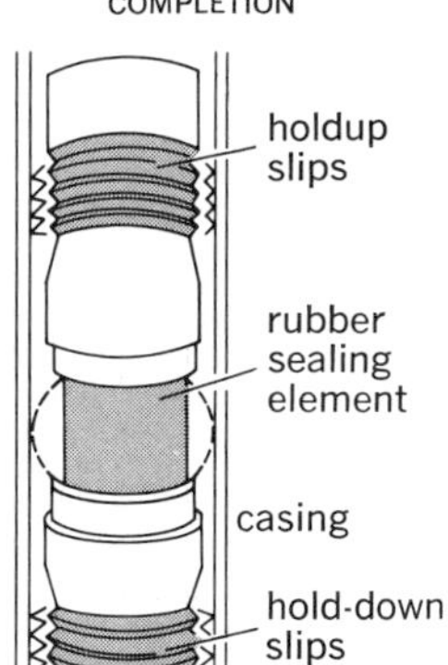

Fig. 10. Bridge plug.

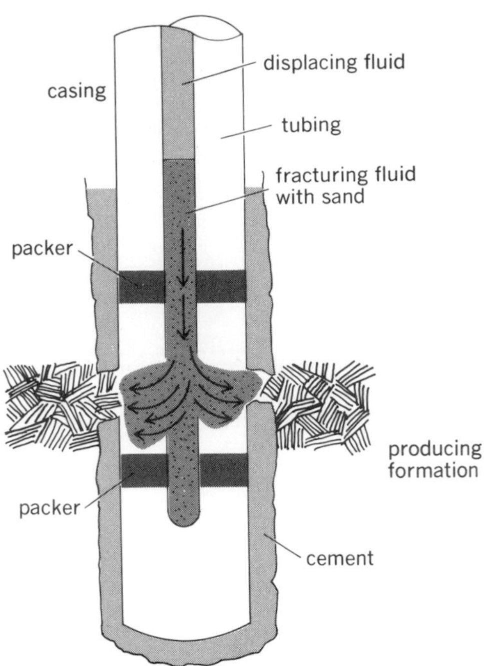

Fig. 12. Outline of the fracturing process.

in a metal tube, which was lowered into the well and detonated. Its more important function, in this shallow well, was to clear away paraffin, which was accomplished, but a decided increase in production was also accomplished. The method has since been improved with new explosives, firing mechanisms, and procedures, but the basic idea is the same, that is, to remove reservoir-blocking material from the reservoir face and to create fractures in the rock to increase production.

Sand exclusion. Some reservoir rock is of an unconsolidated nature similar to beach sands; after having been penetrated with a bore hole, it will slough, if unsupported. This rock also has a tendency to flow with its formation fluids, resulting in plugging of the well bore and restriction or elimination of the entry of formation fluids. Several measures can be used to combat this condition, but no single measure can be used universally.

Screen liner. This type of liner is a segment of preslotted pipe wrapped with wire screen designed to screen out or retain outside the bore hole all except the fine particles that may be brought to the surface with the reservoir fluid. The original design is such that a calculated snug fit is obtained between the liner and bore hole. The coarse screened-out particles form a secondary gravel pack between the liner and the hole, additionally supporting the formation and reducing sand incursion.

Sand consolidation. Sand consolidation is the result of successfully placing a binding material in the producing sand. In effect this glues the sand grains together without completely destroying the porosity and permeability of the sand. The binding material, generally a form of plastic, is forced into the sand through perforations in the casing. The purpose is to consolidate the sand around the well bore to eliminate sloughing and sand incursion.

OIL AND GAS WELL COMPLETION

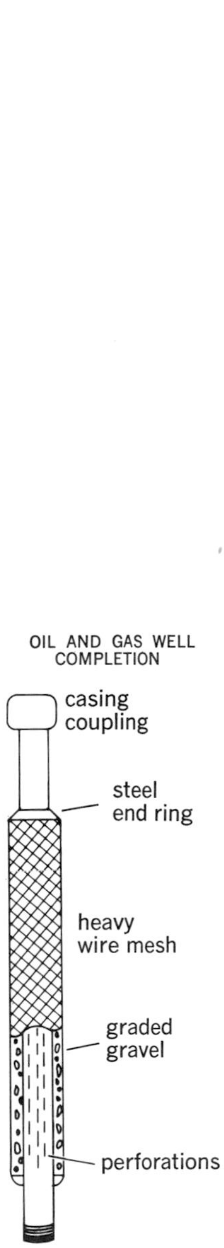

casing coupling

steel end ring

heavy wire mesh

graded gravel

perforations

Fig. 13. Prepacked gravel liner.

Prepacked gravel liners. This type of liner (Fig. 13) is made by using a perforated section of steel pipe, over which has been fitted a tubular sleeve formed of an inner and outer screen of heavy wire mesh or perforated sheet steel. The pipe section is held in concentric relationship by spacers with gravel of proper size packed between the screens and sealed at both ends to retain the gravel. Set in the hole, it serves as a screen liner. *See* PETROLEUM RESERVOIR ENGINEERING.

[HARRY S. BRIGHAM]

Oil and gas well drilling

The drilling of holes for exploration and extraction of crude oil and natural gas. Deep holes and high pressures are characteristics of petroleum drilling not commonly associated with other types of drilling. In general, it becomes more difficult to control the direction of the drilled hole as the depth increases, and additionally, the cost per foot of hole drilled increases rapidly with the depth of the hole. Drilling-fluid pressure must be sufficiently high to prevent blowouts but not high enough to cause fracturing of the bore hole. Formation-fluid pressures are commonly controlled by the use of a high-density clay-water slurry, called drilling mud. The chemicals used in drilling mud can be expensive, but the primary disadvantage in the use of drilling muds is the relatively low drilling rate which normally accompanies high bottom-hole pressure. Drilling rates can often be increased by using water to circulate the cuttings from the hole; when feasible, the use of gas as a drilling fluid can lead to drilling rates as much as 10 times those attained with mud. Drilling research has the objectives of improving the utilization of current drilling technology and the development of improved drilling techniques and tools.

Hole direction. The hole direction must be controlled within permissible limits in order to reach a desired target at depths as great as 25,000 ft (7.6 km). Inclined layers of rocks with different hardnesses tend to cause the direction of drilling to deviate; consequently, deep holes are rarely truly straight and vertical. The drilling rate generally increases as additional drill-collar weight is applied to the bit by adjusting the pipe tension at the surface. However, crooked-hole tendency also increases with higher weight-on-bit. In recent years a so-called packed-hole technique has been used to reduce the tendency to hole deviation. One version of this technique makes use of square drill collars which nearly fill the hole on the diagonals but permit fluid and cuttings to circulate around the sides. This procedure reduces the rate at which the hole direction can change.

Cost factors. Drilling costs depend on the costs of such items as the drilling rig, the bits, and the drilling fluid, as well as on the drilling rate, the time required to replace a worn bit, and bit life. At depths below 15,000 ft (4.6 km), for example, the cost per foot of hole drilled can exceed \$100. Operating costs of a large land rig required for deep holes are about \$2000 per day, whereas comparable costs for an offshore drilling platform or floating drilling vessel can vary from \$500 to \$30,000 per day. Although weak rock at shallow depth can be drilled at rates exceeding 100 ft/hr (30 m/hr), drilling rates often average about 5 ft/hr (1.5 m/hr) in deep holes. Conventional rotary bits have an operating life of 10–20 hr, and 10–20 hr are required for removing the drill pipe, replacing the worn bit, and lowering the pipe back into the hole. Diamond bits may drill for as long as 50–200 hr, but the drilling rate is relatively low and the bits are expensive. For more economical drilling it is desirable to increase both bit life and drilling rate simultaneously.

Drilling fluids. The increased formation or pore-fluid pressures existing at great depths in the Earth's crust adversely affects drilling. Gushers, blowouts, or other uncontrolled pressure conditions are no longer tolerated. High-density drilling fluids maintain control of well pressures. A normal fluid gradient for salt water is about 0.5 psi/ft (11 kPa/m) of depth, and the total stress due to the weight of the overburden increases approximately 1 psi/ft (22 kPa/m). Under most drilling conditions in permeable formations, the well-bore pressure must be kept between these two limiting values. If the mud pressure is too low, the formation fluid can force the mud from the hole, resulting in a blowout; whereas if the mud pressure becomes too high, the rock adjacent to the well may be fractured, resulting in lost circulation. In this latter case the mud and cuttings are lost into the fractured formation.

High drilling-fluid pressure at the bottom of a bore hole impedes the drilling action of the bit. Rock failure strength increases, and the failure becomes more ductile as the pressure acting on the rock is increased. Ideally, cuttings are cleaned from beneath the bit by the drilling-fluid stream; however, relatively low mud pressure tends to hold cuttings in place. In this case mechanical action of the bit is often necessary to dislodge the chips. Regrinding of fractured rock greatly decreases drilling efficiency by lowering the drilling rate and increasing bit wear.

Drilling efficiency can be increased under circumstances where mud can be replaced by water as the drilling fluid. This might be permissible, for example, in a well in which no high-pressure gas zones are present. Hole cleaning is improved with water drilling fluid because the downhole pressure is lower and no clay filter cake is formed on permeable rock surfaces. So-called fast-drilling fluids provide a time delay for filter cake buildup. This delay permits rapid drilling with no filter cake at the bottom of the hole and, at the same time, prevents excessive loss of fluid into permeable zones above.

In portions of wells where no water zones occur, it is frequently possible to drill using air or natural gas to remove the cuttings. Drilling rates with a gas drilling fluid are often 10 times those obtained with mud under similar conditions. Sometimes a detergent foam is used to remove water in order to permit gas drilling in the presence of limited water inflow. In other instances porous formations can be plugged with plastic to permit continued gas drilling. However, in many cases it is necessary to revert to either water or mud drilling when a water zone is encountered. Another possibility is the cementing of steel casing through the zone containing water and then proceeding with gas drilling.

Research. Drilling research includes the study of drilling fluids, the evaluation of rock properties, laboratory simulation of field drilling conditions, and the development of new drilling techniques and tools. Fast-drilling fluids have been developed by selecting drilling-fluid additives which plug the pore spaces very slowly, thereby providing the desired time delay for filter cake buildup. Water-shutoff chemicals have been formulated which can be injected into a porous water-bearing formation in liquid form and then, within a few hours, set to become solid plastics. Well-logging techniques can warn of high-pressure permeable zones so that the change from gas or water drilling fluid to mud can be made before the drill enters the high-pressure zone. Downhole instrumentation can lead to improved drilling operations by providing information for improved bit design and also by feeding information to a computer for optimum control of bit weight and rotary speed. Computer programs can also utilize information from nearby wells to determine the best program for optimum safety and economy in new wells.

Since rock is a very hard, strong, abrasive material, there is a challenge to provide drills which can penetrate rock more efficiently. A better understanding of rock failure can lead to improved use of present equipment and to the development of better tools. Measurements of physical properties of rocks are beginning to be correlated with methods for the theoretical analysis of rock failure by a drill bit. A small 1 -in.-diameter (32-mm) bit, called a microbit, has been used in scale-model drilling experiments in which independent control is provided for bore-hole, formation-fluid, and overburden pressures. These tests permit separation of the effects of the various pressures on drilling rates.

Novel drilling methods which are being explored include studies of rock failure by mechanical, thermal, fusion and vaporization, and chemical means. Some of the new techniques currently have limited practical application while others are still in the experimental stage. For example, jet piercing is widely used for drilling very hard, spallable rocks, such as taconite. Other methods include the use of electric arc, laser, plasma, spark, and ultrasonic drills. Better materials and improved tools can be expected in the future for drilling oil and gas wells to greater depths more economically and with greater safety. *See* OIL AND GAS WELL COMPLETION; PETROLEUM GEOLOGY.

[J. B. CHEATHAM, JR.]

Bibliography: L. W. Ledgerwood, Efforts to develop improved oil-well drilling methods, *J. Petrol. Tech.*, 219:61–74, 1960; W. C. Maurer, *Novel Drilling Techniques*, 1968; A. W. McCray and F. W. Cole, *Oil Well Drilling Technology*, 1976.

Oil burner

A device for converting fuel oil from a liquid state into a combustible mixture. A number of different types of oil burners are in use for domestic heating. These include sleeve burners, natural-draft pot burners, forced-draft pot burners, rotary wall flame burners, and air-atomizing and pressure-atomizing gun burners. The most common and modern type that handles 80% of the burners used to heat United States homes is the pressure-atomizing–type burner shown in Fig. 1.

Characteristics. The sleeve burner, commonly known as a range burner because of its use in kitchen ranges, is the simplest form of vaporizing

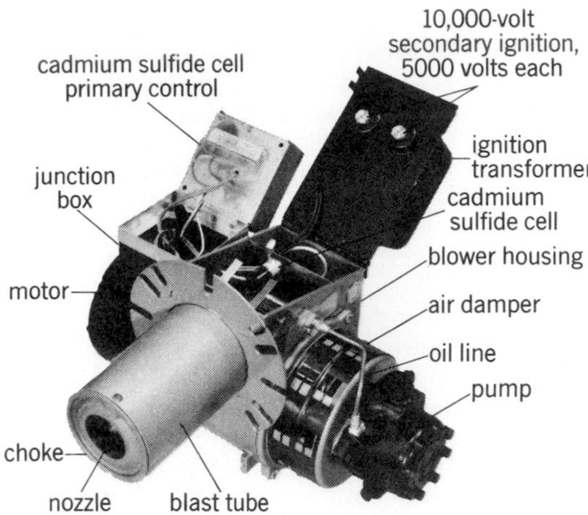

Fig. 1. An oil burner of the pressure-atomizing type. (*Automatic Burner Corp.*)

burner. The natural-draft pot burner relies on the draft developed by the chimney to support combustion. The forced-draft pot burner is a modification of the natural-draft pot burner, since the only significant difference between the two types is the means of supplying combustion air. The forced-draft pot burner supplies its own air for combustion and does not rely totally on the chimney. The rotary wall flame burners have mechanically assisted vaporization. The gun-type burner uses a nozzle to atomize the fuel so that it becomes a vapor, and burns easily when mixed with air.

The most important feature of a high-pressure atomizing gun burner is the method of delivering the air. The most efficient burner is the one which completely burns the oil with the smallest quantity of air. The function of the oil burner is to properly proportion and mix the atomized oil and air required for combustion (Fig. 2).

Efficiency. If a large quantity of excess air is used to attempt to burn the oil, there is a direct loss of usable heat. This air absorbs heat in the heating unit which is then carried away through the stack with the combustion gases. This preheated air causes high stack temperatures which lower the efficiency of the combustion. The higher the CO_2 (carbon dioxide), the less excess air. Overall efficiencies or stack loss can be estimated by the use of the stack loss chart (Fig. 3).

Increased efficiency can be obtained through the use of devices located near the flame end of the burner. CO_2 is not the ultimate factor in efficiency. Burners producing high CO_2 can also produce high smoke readings. Accumulations of 1/8-in. (3-mm) soot layers on the heating unit surface can increase fuel consumption as much as 8%. An oil burner is always adjusting to start smoothly with the highest CO_2 and not more than a number two smoke on a Bacharach smoke scale. For designs in high-efficiency burners commonly known as flame-retention–type burners see Fig. 2.

Nozzle. The nozzle is made up of two essential parts: the inner body, called the distributor, and the outer body, which contains the orifice that the oil sprays through. Under this high pump pressure

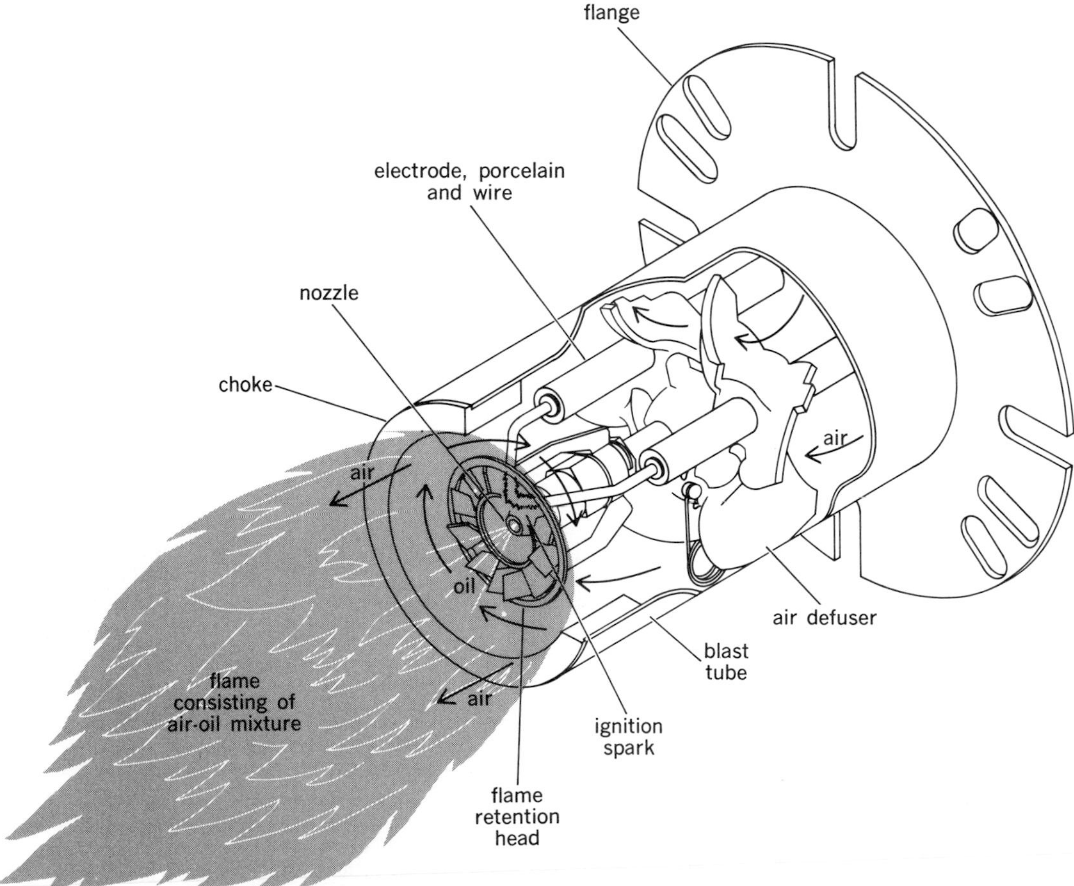

Fig. 2. Flame-retention–type burner, an example of a high-efficiency burner.

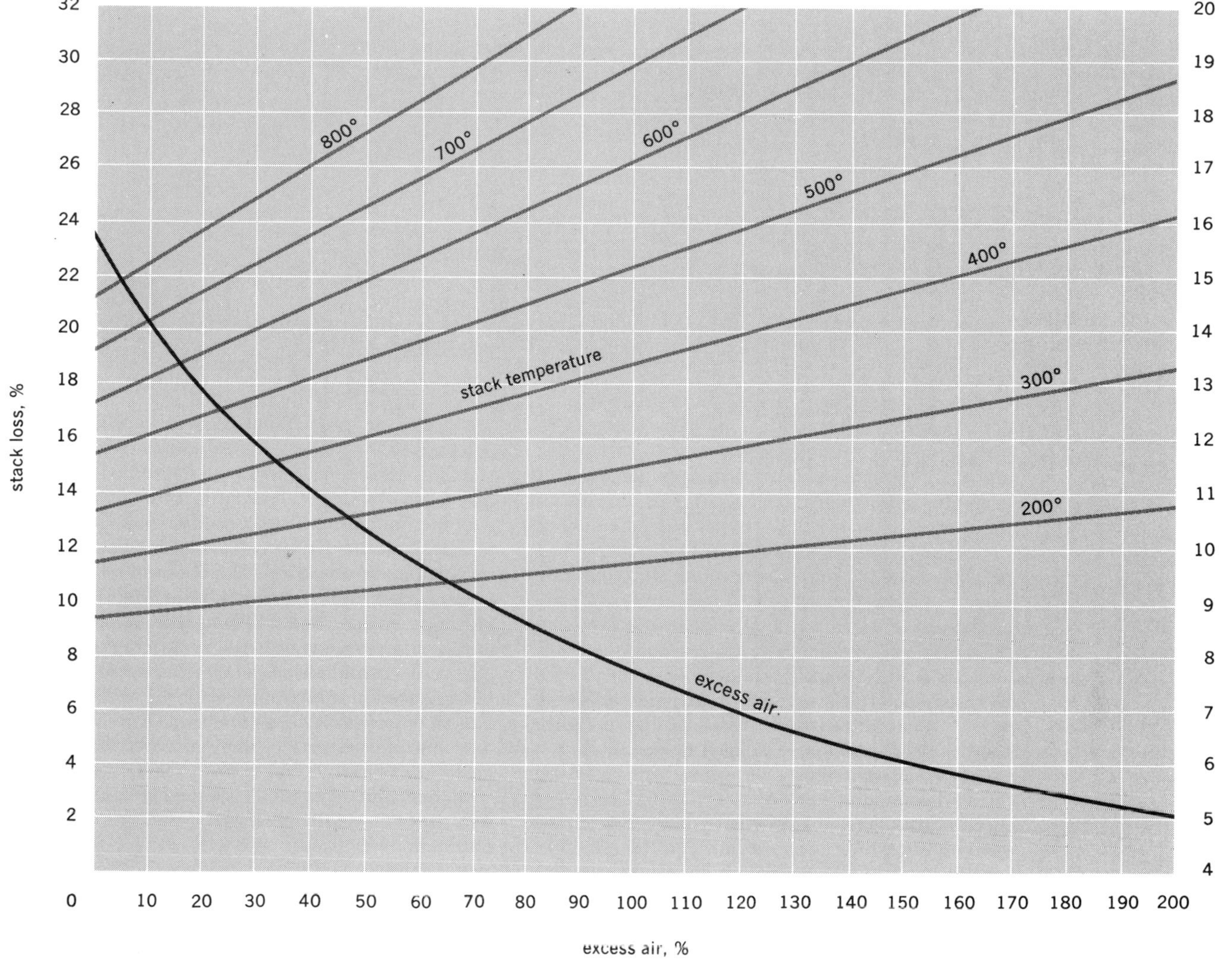

Fig. 3. Typical stack loss chart. To determine stack loss and efficiency, start with correct CO_2 and follow horizontal line to excess air curve, then vertical line to stack temperature, and finally horizontal line to stack loss. Overall efficiency percent= 100− stack loss.

of 100–300 psi (0.7–2.1 MPa) the oil is swirled through the distributor and discharged from the orifice as a spray. The spray is ignited by the spark and combustion is self-sustaining, provided the proper amount of air is supplied by the squirrel cage blower. Air is delivered through the blast tube and moved in such a manner that it mixes well with the oil spray (Fig. 2). This air is controlled by a damper either located on the intake side of the fan or the discharge side of the fan (Fig. 1).

The essential parts of the pressure gun burner are the electric motor, squirrel-cage–type blower, housing, fuel pump, electric ignition, atomizing nozzle, and a primary control (Fig. 1).

The fuel oil is pumped from the tank through the pump gears, and oil pressure is regulated by an internal valve that develops 100–300 psi (0.7–2.1 MPa) at the burner nozzle. The oil is then atomized, ignited, and burned.

The fine droplets of oil that are discharged from the nozzle are electrically ignited by a transformer which raises the voltage from 115 to 10,000 volts. The recently developed electric ignition system called the capacitor discharge system incorporates a capacitor that discharges voltage into a booster transformer developing as much as 14,000 volts. The electric spark is developed at the electrode gap located near the nozzle spray (Fig. 2).

The high velocity of air produced by the squirrel cage fan helps develop the ignition spark to a point where it will reach out and ignite the oil without the electrode tips actually being in the oil spray. Figure 2 shows a typical gun-type burner inner assembly.

Installations. Fuel oils for domestic burners are distilled from the crude oil after the lighter products have been taken off. Consequently, the oil is nonexplosive at ordinary storage temperatures. Domestic fuel oils are divided into two grades, number one and number two, according to the ASTM specification. There are several different methods of installing an oil tank system with an oil burner (Fig. 4). *See* FUEL OIL.

One of the most important safety items of an automatic oil burner is the primary control. This device will stop the operation when any part of the burner or the heating equipment does not function properly. This control protects against such occurrences as incorrect primary air adjustments, dirt in the atomizing nozzle, inadequate oil supply, and

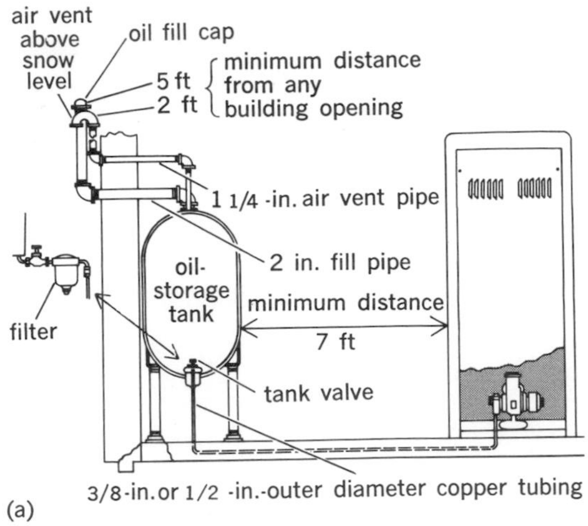

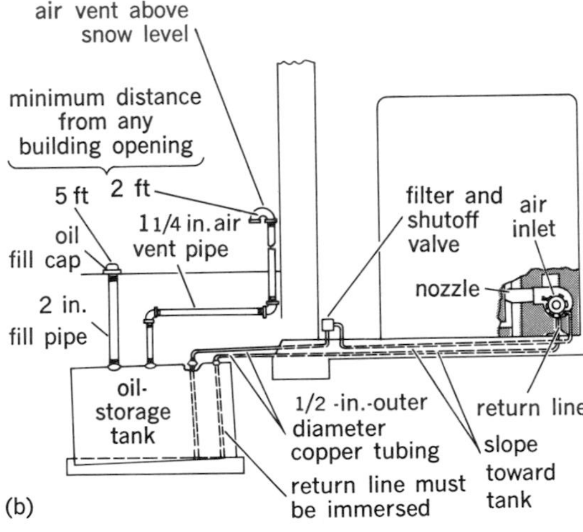

Fig. 4. Two typical installation methods for oil storage tank and oil burner. (*a*) Inside basement. (*b*) Outside underground storage tank. 1 ft = 0.3 m; 1 cm = 25 mm.

improper combustion. The most modern type of primary control is called the cadmium cell control. The cadmium sulfide flame detection cell is located in a position where it directly views the flame. If any of the above functional problems should occur, the electrical resistance across the face of the cell would increase, causing the primary control to shut the burner off in 70 sec or less.

Draft caused by a chimney or technical means is a very important factor in the operation of domestic oil burners. The majority of burners are designed to fire in a heating unit that has a minus 0.02 in. (0.5 mm) water column draft over the fire. Because the burner develops such low static pressures in the draft tube, over fire drafts are an important factor in satisfactory operation.

If the heating unit is designed to create pressure over the fire, a burner developing high static pressures in the blast tube must be used. These burners develop high static by means of higher rpm on the motor and fan. Burners of this nature are available and are being produced. Flame control is very important. Firing heads similar to Fig. 2 are used on applications of this nature.

The oil burner is also used for a wide assortment of heating, air conditioning, and processing applications. Oil burners heat commercial buildings such as hospitals, schools, and factories. Air conditioners using the absorbtion refrigeration system have been developed and fired with oil burners. Oil burners are used to produce CO_2 in greenhouses to accelerate plant growth. They produce hot water for many commercial and industrial applications. *See* AIR COOLING; COMFORT HEATING; HOT-WATER HEATING SYSTEM; OIL FURNACE.

[ROBERT A. KAPLAN]

Bibliography: American Society of Heating, Refrigerating, and Air-conditioning Engineers (ASHRAE), *ASHRAE Handbook and Product Directory*: *Fundamentals*, 1977, and *Equipment*, 1979; C. H. Burkhardt, *Domestic and Commercial Oil Burners*, 3d ed., 1969; E. M. Field, *Oil Burners*, 3d ed., 1977.

Oil field model

A small-scale and commonly simplified replica of subsurface conditions of interest and value in petroleum prospecting and oil-field development. The term model has been applied by geologists to a simplified diagram and by mathematical physicists to a formal analysis with special boundary conditions and related attributes. These and somewhat more complex physical models have value in transmitting concepts and relationships to the nonspecialist, but they are generally designed by geologists and petroleum engineers to aid investigation of a particular problem. To do so, the model is either scaled to size, shape, and like attributes, or according to forces upon it, relative to its archetype.

Size-scaled models. These are commonly models of ore and mineral bodies and of those blocks on which surface topography and surface geology are shown on top and geologic cross sections on the sides, also to scale. Peg models are constructed for oil fields with wells shown as pegs or wires. Cutout block diagrams are also used. Models which are scaled to size are found in many geological museums and in mining companies' offices.

Forces and movement models. This type contains mobile material to simulate by its movement a movement that has taken place or will take place in the prototype. Transparent solid plastics have been incorporated in mobile parts of models for enhanced internal visibility and to obtain photoelastic data. Such models are also used to make graphic representation of changes recorded from observations with geophysical instruments, for example, from microinstruments used to survey physical response of a small mass in an artificial field. Force and movement models have come into use for study of movements of earth materials. By this, past changes resulting in present features can be made as dimensionally credible experiments in periods tremendously short as compared with geologic time. Models with mobility are of three principal types.

Fluid or fluids moving in porous medium. The models for study of fluid movement through a geometrically stable medium have important applications to hydrologic and petroleum engineering. The hydrologic case is mostly of an interface underground between fresh water and salt water caused by extraction or by injection of fresh water through wells. Where the interface owes its position to fresh-water–salt-water patterns found in many coastal zones and along many shorelines, the setting is difficult to describe without a model. Such demonstrations have added to the understanding of water conditions in California's Santa Clara Valley and other critical water areas.

The petroleum engineering application is to movement of fluid in respect to another kind of interface underground. This is between petroleum and natural gas, or both, and salt water, and especially when petroleum is being extracted through wells and salt water simultaneously injected through other wells into the same porous stratum, so that the important flow is radial, that is, two-dimensional.

The model used involves scaling down from the actual distance between wells in the oil field under study, and also the device of an analog for the fluids, whereby the pressure distribution in steady-state porous flow is exactly the same as the potential distribution in an electrical conducting medium.

Uniform high-viscosity materials. The second geologic realm where models are used extensively is that of the movement of large segments of the earth made of materials of essentially uniform high viscosity, comprising nearly all rock species in the Earth's crust. This has little direct bearing on oil-field models.

Adjacent materials of diverse viscosities. Among the movements of the earth which are of interest to geologists are those involving what may be called soft layers between relatively hard ones. This is the realm where diapir structures are inferred to develop and also their subform, the salt dome. Because of the economic importance of salt domes and their associated oil traps, investigations with models have become widespread.

L. L. Nettleton produced a model for salt-dome formation using two viscous fluids of different density to emphasize the role of the lower density of salt to that of adjacent sediments in the process of salt-dome movements. M. B. Dobrin made a parallel mathematical analysis. T. J. Parker and A. N. MacDowell extended salt-dome studies by using materials with higher yield points than Nettleson's first model, and their model showed fractures above and around the "salt" analog in their model.

Also in both the prototype and the model, boundary conditions modify vectors nearby. This seems particularly hard to handle in cases of hydrologic model study, such as at the University of California.

Models versus mathematical analysis. There is a possible alternative to the use of physical models — mathematical analysis, which is more attractive now that computing machines are available. Dobrin used this method for the case of Nettleton's model investigation of salt dome. However, mathematical analysis is as yet a subject of controversy among physicists in many simple geologic settings; physical models are presently available and their value is acknowledged.

Physical models are probably more helpful in educating the nonspecialist in geology; they present the earth features in a readily visible fashion easier to grasp than mathematical analysis. For a discussion of mathematical models *see* PETROLEUM RESERVOIR MODELS. [PAUL WEAVER]

Bibliography: J. W. Amyx, D. M. Bass, and R. L. Whiting, *Petroleum Reservoir Engineering*, vol. 1: *Physical Properties*, 1960; M. B. Dobrin, Some quantitative experiments on a fluid salt-dome model and their geological implications, *Trans. Amer. Geophys. Union*, 22:528–542, 1941; L. L. Nettleton, Recent experimental and geophysical evidence of mechanics of salt-dome formation, *Bull. Amer. Ass. Petrol. Geol.*, 27(1):51–63, 1943; A. E. Scheidegger, *Principles of Geodynamics*, 2d ed., 1963; A. E. Scheidegger, *Theoretical Geomorphology*, 2d ed., 1970; D. K. Todd, *Ground Water Hydrology*, 1959.

Oil field waters

Waters of varying mineral content which are found associated with petroleum and natural gas or have been encountered in the search for oil and gas. They are also called oil field brines, or brines. They include a variety of underground waters, usually deeply buried, and have a relatively high content of dissolved mineral matter. These waters may be (1) present in the pore space of the reservoir rock with the oil or gas, (2) separated by gravity from the oil or gas and thus lying below it, (3) at the edge of the oil or gas accumulation, or (4) in rock formations which are barren of oil and gas. Brines are commonly defined as water containing high concentrations of dissolved salts. Potable or fresh waters usually are not considered oil field waters but may be encountered, generally at shallow depths, in areas where oil and gas are produced.

Oil field waters or oil field brines differ widely in composition and concentration. They may differ from one geologic province to another, from one formation to another within a given geologic province, or from one part of a specific geologic horizon to another. They range from slightly salty water with 1000–3000 parts of dissolved substances in 1,000,000 parts of solution to very nearly saturated brines with dissolved mineral content of more than 270,000 parts per million (ppm).

The most common and abundant mineral found in oil field waters is sodium chloride, or common table salt. Calcium chloride is next in order of abundance. Carbonates, bicarbonates, sulfates, and the chlorides of magnesium and potassium are present in lesser quantities. In addition to the above mentioned salts, salts of bromine and iodine are also found. Traces of strontium, boron, copper, manganese, silver, tin, vanadium, and iron have been reported. Barium has been reported in many of the Paleozoic brines of the Appalachian region. The commercial value of a brine depends upon the concentration of salts, purity of the products to be recovered, and value and practicability of by-product recovery. Concentrations less than 200,000 ppm are seldom of commercial interest.

Slightly salty waters, while not suitable for human consumption, may be used in some industrial processes or may be amenable to beneficiation for municipal supplies in areas lacking fresh waters.

Classified genetically, oil field waters are generally considered connate; that is, they are sea waters which (presumably) originally filled the pore spaces of the rock in which they are now confined. However, few analyses of these waters correspond to present-day sea water, thus indicating some mixing and modification since confinement. Dilute solutions suggest that rainwater has percolated into the rocks along bedding planes, fractures, faults, and other permeable zones. Presence of carbonates, bicarbonates, and sulfates in an oil field water further suggests that at least some of the water had its origin at the surface. Concentrations of dissolved solids greater than that of modern sea water suggest partial evaporation of the water or addition of soluble salts from the adjacent or enclosing rocks.

Waters in most sedimentary rocks increase in mineral concentration with depth. This increase may be due to the fact that, since salt water is heavier than fresh water, the more dense solution will eventually find a position as low as possible in the aquifer. An additional factor would be the longer exposure of the deeper waters to the mineral-bearing rocks. Exceptions have been noted and probably are due to the presence of larger quantities of soluble salts in some geological formations than in others.

Probably the most important geological use of oil field water analyses is their application to the quantitative interpretation of electrical and neutron well logs, particularly micrologs. In order to compute the connate water saturation of a formation in a quantitative manner from electrical data, it is necessary to know with accuracy the connate water resistivity.

Naturally mineralized waters are frequently the only waters available for water-flooding operations. Water analyses are useful in predicting the effect of the water on minerals in the reservoir rock and on the mechanical equipment employed on the project. Waters which exert a corrosive action on the lines and pumps or which tend to plug up the pay zone are not suitable for water-flooding operations.

Oil field water composition may be an important factor in the determination of the source of water in oil wells which have leaky casings or improper completions with resulting communication between wells, and in identifying and correlating reservoirs in multipay oil pools, particularly in those containing lenticular sand bodies.

Industrial wastes, including mineralized water produced with oil, may be disposed of in underground reservoirs. Between the zone of potable water and the horizon of commercial brines, there commonly are rock formations, the waters of which contain chemicals in amounts sufficient to make the waters unsuitable for domestic, municipal, industrial, and livestock consumption, but not in sufficient quantity to be considered as a source for recovery of chemicals. Provided there is sufficient porosity and permeability, these rock formations could receive industrial wastes which would otherwise contaminate surface streams and shallow, fresh groundwater horizons into which they might be discharged. *See* PETROLEUM GEOLOGY. [PRESTON MC GRAIN]

Oil furnace

A combustion chamber in which oil is the heat-producing fuel. Fuel oils, having from 18,000 to 20,000 Btu/lb (42–47 MJ/kg), which is equivalent to 140,000 to 155,000 Btu/gal (39–43 MJ/liter), are supplied commercially. The lower flash-point grades are used primarily in domestic and other furnaces without preheating. Grades having higher flash points are fired in burners equipped with preheaters.

The ease with which oil is transported, stored, handled, and fired gives it a special advantage in small installations. The fuel burns almost completely so that, especially in a large furnace, combustible losses are negligible. *See* FUEL OIL; OIL BURNER.

Domestic oil furnaces with automatic thermostat control usually operate intermittently, being either off or operating at maximum capacity. The heat-absorbing surfaces, especially the convection surface, should therefore be based more on maximum capacity than on average capacity if furnace efficiency is to be high. The combustion chamber should provide at least 1 ft^3 for each 1.5–2 lb (1m^3 for each 24–32 kg) of fuel burned per hour. Gas velocity should be below 40 ft/sec (12m/sec). The shape of the chamber should follow the outline of the flame. [FRANK H. ROCKETT]

Oil mining

The surface or subsurface excavation of petroleum-bearing sediments for subsequent removal of the oil by washing, flotation, or retorting treatments. Oil mining also includes recovery of oil by drainage from reservoir beds to mine shafts or other openings driven into the oil rock, or by drainage from the reservoir rock into mine openings driven outside the oil sand but connected with it by bore holes or mine wells.

Surface mining consists of strip or open-pit mining. It has been used primarily for the removal of oil shale or bituminous sands lying at or near the surface. Strip mining of shale is practiced in Sweden, Manchuria, and South Africa. Strip mining of bituminous sand is conducted in Canada.

Subsurface mining is used for the removal of oil sediments, oil shale, and Gilsonite. It is practiced in several European countries and in the United States. Some authorities consider this the best method to recover oil when oil sediments are involved, because virtually all of the oil is recovered.

Fig. 1. Close-up view of Union Oil Co. of California shale-oil retort near Grand Valley, Colo. Right part of structure is portion of the system for removing oil vapors that would otherwise escape in the gas stream.

Fig. 2. Mine locomotive and cars removing shale from U.S. Bureau of Mines Shale Mine at Rifle, Colo. Just visible left of center is the Colorado River, nearly 3000 ft (900 m) below. (*After R. Fleming, U.S. Bureau of Mines*)

European experience. Subsurface oil mining was used in the Pechelbronn oil field in Alsace, France, as early as 1735. This early mining involved the sinking of shafts to the reservoir rock, only 100–200 ft (30–60 m) below the surface, and the excavation of the oil sand in short drifts driven from the shafts. These oil sands were hoisted to the surface and washed with boiling water to release the oil. The drifts were extended as far as natural ventilation permitted. When these limits were reached, the pillars were removed and the openings filled with waste.

This type of mining continued at Pechelbronn until 1866, when it was found that oil could be recovered from deeper, more prolific sands by letting it drain in place through mine openings, without removing the sands to the surface for treatment.

Subsurface mining of oil shale also goes back to the mid-19th century in Scotland and France. It is not so widely practiced now because of its high cost as compared with that of usual oil production, particularly in the prolific fields of the Middle East.

United States oil shale mining. The U.S. Bureau of Mines carried out an experimental mining and processing program at Rifle, Colo., between 1944 and 1955 in an effort to find economically feasible methods of producing oil shale.

One of the more important phases of this experimental program was a large-scale mine dug into what is known as the Mahagony Ledge, a rich oil shale stratum that is flat and strong, making it favorable for mining. This stratum lies under an average of about 1000 ft (300 m) of overburden and is 70–90 ft (21–27 m) thick.

The Bureau of Mines adopted the room-and-pillar system of mining, advancing into the 70 ft ledge face in two benches. The mine roof was supported by 60-ft (18-m) pillars staggered at 60-ft intervals and supplemented by iron roof bolts 6 ft (1.8 m) long.

Multiple rotary drills mounted on trucks made holes in which dynamite was placed to shatter the shale; the shale was then removed from the mine by electric locomotive and cars (Fig. 2).

The experimental mining program ended in February, 1955, when a roof fall occurred. Despite this occurrence, however, the Bureau is convinced that the room-and-pillar method used in coal, salt, and limestone mines is feasible for shale oil mining in Colorado.

Since 1955, several companies have conducted experimental efforts to produce shale oil. Those continuing to move toward commercialization include companies which plan to use traditional mining techniques combined with some form of surface retorting. One company is testing an onsite process in which both mining and retorting are done underground, and another firm is well along in development work using a method involving a gas-combustion retort similar to that used by the Bureau of Mines in various pilot plants operated during the 1950s and 1960s.

Another possibility is a modified open-cast surface method, which proponents claim has a 95% recovery rate of the minable reserves, compared with a maximum 40% rate by room-and-pillar underground mining and 20% by on-site mining and retorting.

Oil shale does not contain oil, as such. Draining methods, therefore, are not applicable. It does, however, contain an organic substance known as kerogen. This substance decomposes and gives off a heavy, oily vapor when it is heated above 700°F (370°C) in retorts. When condensed, this vapor becomes a viscous black liquid called shale oil, which resembles ordinary crude but has several significant differences.

Colorado's Mahagony Ledge yields an average of about 30 gal of oil per ton (125 liters per metric ton). This means that large amounts of oil shale must be mined, transported, retorted, and discarded for production of commercial quantities of oil. Various types of retort also have been tested in Colorado, but none is in commercial use (Fig. 1).

Gilsonite. Gilsonite is a trade name, registered by the American Gilsonite Co., for a solid hydrocarbon found in the Uinta Basin of eastern Utah and western Colorado. The American Gilsonite Co. uses a subsurface wet-mining technique to extract about 700 tons (635 metric tons) of Gilsonite daily from its mine at Bonanza, Utah.

Conventional mining methods were found unsuitable for mass output of Gilsonite because it is friable and produces fine dust when so mined. This dust can be highly explosive. In the system now being used, tunnels are driven from the main shaft by means of water jetted through a 1/4-in. (6-mm) nozzle under pressure of 2000 psi (14 MPa). The stream of water penetrates tiny fissures and the ore falls to the bottom of the drift. The drifts are cut on a rising grade of about 2.5°. The ore is washed down to the main shaft where it is screened. Particles of sizes smaller than 3/4 in. (19 mm) are pumped to the surface in a water stream; larger pieces are hoisted in buckets. A long rotary drill with carbide-tipped teeth is used to remove ore that cannot be broken with water jets.

Gilsonite is moved through a pipeline in slurry form to a refinery, where it is dried and melted and then heated to about 450°F (232°C). The melted oil is fed to a coker and other processing units to make gasoline and other petroleum products.

[ADE L. PONIKVAR]

Fig. 3. Large draglines dig tar sand at the Syncrude project. Note the relative size of the vehicles in the foreground.

North American tar sands. The world's only tar sand mining operations, the 45,000-barrel-per-day (bpd, 7150 m³ per day) synthetic crude complex of Suncor, Inc. (formerly Great Canadian Oil Sands Limited), which is located 34 km north of Fort McMurray, Alberta, and started production in 1967, and the 129,000-bpd (20,500 m³ per day) project of Syncrude Canada Ltd., which is located approximately 8 km away and started production in 1978, are situated in the minable area of the gigantic Athabasca Tar Sands deposit, the world's largest oil reservoir. The minable area is that portion with an overburden thickness of less than 45 m, embracing about 10% of the estimated 6.24×10^{11} bbl (9.92×10^{10} m³) of total in-place bitumen reserves, may ultimately support five or more such surface mining plants operating at the same time. In 1979 Suncor commenced an expansion program to increase its plant capacity to 58,000 bpd (9200 m³ per day) by 1982. Also in 1979, Alsands, a consortium of nine major oil companies, received preliminary regulatory approval to construct, by 1986, a 140,000-bpd (22,300 m³ per day) project, fashioned after the Syncrude model. *See* OIL SAND.

In both the Suncor and the Syncrude projects, about half of the terrain is covered with muskeg, an organic soil resembling peat moss, which ranges from a few centimeters to 7 m in depth. The major part of the overburden, however, consists of Pliestocene glacial drift and Clearwater Formation sands and shales. The total overburden varies from 7 to 40 m in thickness. Underlying is the McMurray Formation (Lower Cretaceous), in which the oil-impregnated sands reside. The sand strata in the region of the Suncor and Syncrude operations, also variable in thickness, average about 45 m, although typically 5–10 m must be discarded because of bitumen content below the economic cutoff grade of 6 wt %.

Composition. The bitumen content of tar sand can range from 0 to 20 wt %, but feed-grade material normally runs between 10 and 12 wt %. The balance of the tar sand is composed, on the average, of 5 wt % water and 84 wt % sand and clay. The bitumen, which has a gravity on the American Petroleum Institute (API) scale of 8 degrees, is heavier than water and very viscous. Tar sand is a competent material, but it can be readily dug in the summer months; during the winter months, however, when temperatures plunge to −45°C, tar sand assumes the consistency of concrete. To maintain an acceptable digging rate in the winter, mining must proceed faster than the rate of frost penetration; if not, supplemental measures, such as blasting, are required.

Overburden removal. For muskeg removal, a series of ditches are dug 2 or 3 years in advance of stripping to permit as much of the water as possible to drain away. Despite this preparation, the spongy nature of the muskeg persists and removal is best accomplished after freeze-up.

Suncor has tried, and rejected, several overburden removal methods including shovels and trucks, scrapers, and front-end loaders and trucks.

Fig. 4. Bucketwheel excavators get into position to load windrows of tar sand onto the conveyor belts in order to feed the Syncrude plant which is visible in the background.

In 1976 a bucketwheel excavator was purchased to replace seven 14-m³ front-end loaders and five D9G bulldozers. Through twin chutes, the excavator loads a fleet of 21 140-metric-ton-capacity WABCO trucks. Additional equipment is used for maintaining the haul roads and for spreading and compacting the spoiled material.

Bucketwheel excavators. Mining of tar sand is performed at Suncor mainly by two large bucketwheel excavators of German manufacture, each operating on a separate bench. These units, weighing 1600 metric tons, have a 10-m-diameter digging wheel on the end of a long boom. Each wheel has a theoretical capacity of 8600 tons/hr, but the average output from digging is about 4500 tons/hr. Because the availability of these machines is normally 55–60%, the extraction plant has been designed to accept a widely fluctuating feed rate. To facilitate digging of the highly abrasive tar sand and to achieve a reasonable bucket and tooth life, Suncor routinely preblasts the tar sand on a year-round basis. At the rate of 127,000 tons/day, tar sand is transferred from the mine to the plant by a system of 152-cm wide conveyor belts and 183-cm trunk conveyors, operating at 333 m/min. Following extraction by a process using hot water, the bitumen is upgraded by coking and hydrogenation to a high-quality synthetic crude.

Dragline scheme. Syncrude opted for an even more capital-intensive mining scheme. Four large draglines, each equipped with a 70-m³ bucket at the end of a 111-m boom, are employed to dig both a portion of the overburden, which is free-cast into the mining pit, and the tar sand, which is piled in windrows behind the machines (Fig. 3). Four bucketwheel reclaimers, similar to the Suncor excavators but larger to handle the additional capacity, load the tar sand from the windrows onto conveyor belts which transfer it to the plant (Fig. 4). With a peak tar sand mining rate of 300,000 tons/day, the Syncrude project is the largest mining operation in the world.

The advantages of the dragline scheme lie in its ability to open a new mine faster, to handle certain types of overburden at a lower cost, and to reject with greater selectivity lenses of low-grade tar sand and barren material. The disadvantages include the necessity of rehandling the tar sand and the production of an increased percentage of lumps in the plant feed, which can damage conveyor belting. Certain Clearwater overburden strata, occurring mainly on the west side of Syncrude's mine, are too weak to support the draglines and must be prestripped by other methods.

Several years of comparative operation will be required to firmly establish whether one of the schemes enjoys an economic advantage.

United States tar sands. No United States reservoirs have sufficient resources reachable by surface mining to justify the larger-scale operations used in Canada.

Economic exploitation of surface accumulations of tar sands in the United States appears remote because of the high costs associated with developing such small deposits.

Mine-assisted in-situ production. Underground mining may yet play a significant role in tar sand exploitation. For almost a century, production of heavy oil from the Wietze reservoir in Germany has been increased beyond that available from conventional wells, by digging vertical shafts into the formation and collecting the seepage; more recently, horizontal tunnels have attained a higher seepage rate. At the Yarega reservoir in Siberia, the Soviets have taken the underground recovery method a step further by combining tunnels under the formation with 34,000 closely spaced vertical wells rising from the tunnels. Steam is injected into the Yarega formation from above to raise the temperature of the bitumen and enhance its flow characteristics. Approximately 1.8×10^7 bbl (2.9×10^6 m^3) had been produced in this manner by 1979. A major pilot test is under way to investigate the applicability of the mine-assisted in-situ production (MAISP) method to Alberta's Athabasca deposit. The MAISP method may offer the best hope for recovering reserves too deep for surface mining (more than 80 m deep) but too shallow for in-place techniques (less than 300 m deep). *See* PETROLEUM GEOLOGY. [G. RONALD GRAY]

Bibliography: W. L. Oliver and G. R. Gray, Technology and economics of oil sands operations, *10th World Petroleum Congress*, Bucharest, Romania, 1979; G. S. Smith and R. M. Butler, Studies on the use of tunnels and horizontal wells for the recovery of heavy crudes, *United Nations Institute of Training and Research (UNITAR) 1st International Conference on the Future of Heavy Crude and Tar Sands*, Edmonton, Alberta, 1979.

Oil sand

A loose to consolidated sandstone or a porous carbonate rock, impregnated with a heavy asphaltic crude oil, too viscous to be produced by conventional methods; also known as tar sand or bituminous sand.

Oil sands are distributed throughout the world but the largest proven accumulation occurs in Alberta, Canada. A large accumulation appears to be present in the Orinoco Basin in Venezuela, and far smaller deposits occur in the Soviet Union, the United States, Madagascar, Albania, Trinidad, and Romania. Estimates of the world reserves of the contained heavy hydrocarbons range from 2.5×10^{12} to 6×10^{12} barrels (bbl; 4 to 9.5×10^{11} m^3). Estimates of the amount that will ultimately prove to be recoverable—generally of the order of 5–25%—are highly speculative and depend on the development of successful technologies at competitive costs. Because of the fact that only billion-barrel accumulations at or near the surface can be successfully exploited, estimates of recoverable reserves from the total resources must be treated with great caution.

In Venezuela recent drilling has increased the estimated heavy oil resource significantly, and some believe that the total may ultimately reach 0.5×10^{12} bbl (8×10^{10} m^3). Located for the most part in a 60-km-wide band stretching along the northern bank of the Orinoco River for a distance of 700 km, the deposits collectively were known for many years as the Orinoco Tar Belt. Some of the crude has a gravity on the American Petroleum Institute (API) scale of between 15 and 18 degrees, and a fair amount ranges between 10 and 15 degrees; but most of it is between 7 and 10 degrees. Even the heaviest oils have some mobility because of below-normal asphaltene content, the temperature of the reservoir, and the presence of dissolved gases. For this reason, the Venezuelan government is encouraging the use of the term Orinoco Heavy Oil Belt instead of Tar Belt to describe these deposits. Since, however, the vast majority of the reservoirs cannot be produced economically by conventional methods, they should continue to be classified as oil sands. About 80,000 barrels per day (bpd) (12,700 m^3 per day) from fields containing the lightest of the heavy crudes, were being produced in 1979 by cyclic steam injection into the formation.

Canada's Alberta Oil Sands, comprising four enormous deposits—including the famous Athabasca Tar Sands, location of the only existing commercial oil sand mining operations—and a number of lesser deposits, contain the largest and best known of the oil sand reservoirs. Established resources are of the order of 1×10^{12} bbl (1.59×10^{11} m^3).

Hundreds of bitumen occurrences have been charted in the Soviet Union; about 70%, however, are found in carbonate rock. Yarega, in Siberia's Timan-Pechora Province, is the only place in the world where underground mining for commercial production of heavy oil is carried out. Of considerable interest is the Olenek anticline in northeastern Siberia, where, it is speculated, reserves might total 6×10^{11} bbl (9.5×10^{10} m^3).

In the United States 550 tar sand occurrences are known to exist in 22 states. Five states (California, Kentucky, New Mexico, Texas, and Utah) have deposits containing at least 10^6 bbl (1.6×10^5 m^3) in place. Utah is by far the most important tar sand state, with between 2.3 and 2.9×10^{10} bbl (3.7 to 4.6×10^9 m^3) of oil credited to 19 deposits.

The Bemolanga deposit in western Madagascar is worthy of note. Covering approximately 400 km^2, it contains reserves estimated at 175×10^9 bbl (2.8×10^8 m^3).

Alberta deposits. Although the energy crisis is focusing attention on all oil sand deposits, much current experimental and developmental activity is directed toward the Alberta Oil Sands. Twenty-four oil sand research projects were under way in 1979. The deposits range across the northern part of the province (see illustration).

The Athabasca deposit is the largest known petroleum accumulation in the world, with a total area of about 34,000 km^2. The McMurray Formation, in which the oil-impregnated sands reside, belongs to the lower Cretaceous. The formation outcrops along the Athabasca River, north of Fort McMurray. Elsewhere, the deposit is buried under a variable layer of overburden which reaches up to 525 m in thickness. Fortunately, some of the richest parts of the deposit are covered by the thinnest overburden.

In this area Suncor, Inc., a subsidiary of Sun Oil

Company, was producing 45,000 bpd (7150 m^3 per day) of synthetic crude from mined tar sands in 1979. On an adjacent lease, the immense Syncrude Canada Ltd. mining project, constructed at a cost of \$2.5 billion, started production in mid-1978. Plant output, increasing gradually, had reached 100,000 bpd (16,000 m^3 per day) by July 1979, and was scheduled to attain the permit level of 129,000 bpd (20,500 m^3 per day) in 1984. Alsands, a nine-company consortium, has plans to complete a \$5 billion, 140,000-bpd (22,300 m^3/per day) project, 40 km away, by 1986.

Surface mining. Both the Suncor and Syncrude projects employ open-pit mining methods to remove the overburden and underlying tar sand, hot-water extraction units to recover 90% or more of the tarry bitumen, and upgrading units to convert the bitumen to high-quality synthetic crude. Differences in the two projects, aside from size (the Syncrude project, with a peak mining rate of 300,000 metric tons of tar sand per day, is the largest mine in the world), lie mainly in the mining and upgrading methods employed. Suncor uses large bucket-wheel excavators to mine the abrasive tar sand, whereas Syncrude uses large draglines for this purpose; Suncor converts the bitumen to lighter products in a delayed coking unit, whereas Syncrude utilizes two large fluid coking units. In each case the high-sulfur-content coker distillates are upgraded by high-pressure hydrogenation to sweet 32–36 degrees API gravity synthetic crudes. The economically recoverable, or proved, minable reserves in the Athabasca deposit have been calculated at 38×10^{10} bbl (6×10^9 m^3) of bitumen, corresponding to 2.65×10^{10} bbl (4.2×10^9 m^3) of synthetic crude. *See* OIL MINING.

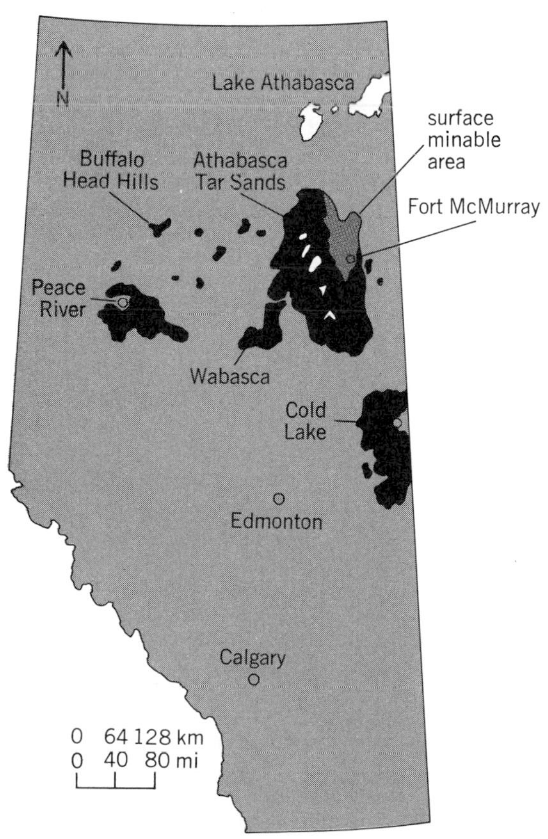

Tar sand deposits in Alberta.

Recovery of deep deposits. Many experimental programs have been undertaken in an attempt to devise a commercially feasible method of recovering the 90% of the bitumen buried too deeply for surface mining. Amoco Canada refers to its process as COFCAW (combination of forward combustion and waterflood). Most of the other companies have utilized some form of steam injection. Esso Resources was producing 8000 bpd (1270 m^3 per day) at its Cold Lake pilot project in 1979. The heavy oil at Cold Lake is 2–3 degrees API lighter then the 8-degree Athabasca oil, and this has a significantly beneficial effect on ease of recovery. Based on its successful pilot plant experience, in 1979 Esso applied for and received preliminary approval to construct, by 1986, a \$4.7 billion, 140,000-bpd in-place project.

United States deposits. Any large-scale tar sand developments in the United States will have to take place in Utah, where 95% of known reserves are located. Over 50 deposits have been identified, but 96% of the oil in place occurs in six giant deposits, four of which are in the Uinta Basin of northeastern Utah. Three of these (P.R. Spring, Hill Creek, and Sunnyside) occur in the Green River Formation (Eocene), but Asphalt Ridge, probably the best-known tar sand deposit in the country, occurs in older strata.

The Tar Sand Triangle, with reserves of approximately 1.4×10^{10} bbl (2.2×10^9 m^3), the largest deposit in the United States, lies in a remote and very rugged area of southeastern Utah, in the White Rim Sandstone Formation. Most of the deposit underlies Federal Recreation and Wilderness areas. Federal agencies also control major sections of the P.R. Spring, Hill Creek, and Sunnyside deposits. The other deposit, Circle Cliffs, is also located in southeastern Utah.

The thickness of the Utah reservoirs varies between 0 and 100 m, and the depth ranges from 0 to 670 m. Some interesting characteristics distinguish the reservoirs: the bitumen occurs in consolidated sandstone; the connate water at 0.6% and the sulfur content of the hydrocarbon at 0.5% are about one-tenth of the corresponding values in the Canadian and Venezuelan tar sands. However, the porosity of the formation and the continuity and magnitude of the bitumen saturation are far less. Economic exploitation is therefore still in question.

Far behind Utah, in second place in the United States, is California, which contains an estimated 2×10^8 bbl (3.2×10^7 m^3) of oil sand reserves. The Edna deposit, located midway between Los Angeles and San Francisco, is the largest. It is a rarity among oil sand deposits in that it is considered a marine facies. Virtually all of the other major oil sand deposits in the world occur in fresh-water fluviatile and deltaic environments. The Sisquoc deposit is amenable to the hot-water extraction process used for the large Canadian projects. The fossiliferous McKittrick deposit is more properly considered a diatomaceous oil shale.

[G. RONALD GRAY]

Bibliography: C. J. Borregales, Production characteristics and oil recovery in the Orinoco Oil Belt, *United Nations Institute of Training and Research (UNITAR) 1st International Conference on the Future of Heavy Crude and Tar Sands*, Edmonton, Alberta, 1979; Energy Resources Conservation Board, *Alberta's Reserves of Crude Oil, Gas, Natural Gas Liquids and Sulphur*, ERCB Rep. no. 78–18, Calgary, Alberta; E. M. Khalimov, R. K. Muslinov, and G. T. Yudin, Bitumen deposits of the U.S.S.R. and the ways of their development, *UNITAR 1st International Conference*, 1979; V. A. Kouskraa, S. Chalton, and T. M. Doscher, *The Economic Potential of Domestic (U.S.) Tar Sands*, U.S. DOE, January, 1978.

Oil shale

A sedimentary rock containing solid, combustible organic matter in a mineral matrix. The organic matter, often called kerogen, is largely insoluble in petroleum solvents, but decomposes to yield oil when heated. Although "oil shale" is used as a lithologic term, it is actually an economic term referring to the rock's ability to yield oil. No real minimum oil yield or content of organic matter can be established to distinguish oil shale from other sedimentary rocks. Additional names given to oil shales include black shale, bituminous shale, carbonaceous shale, coaly shale, cannel shale, cannel coal, lignitic shale, torbanite, tasmanite, gas shale, organic shale, kerosine shale, coorongite, maharahu, kukersite, kerogen shale, and algal shale.

Origin and mineral composition. Oil shale is lithified from lacustrine or marine sediments relatively rich in organic matter. Most sedimentary rocks contain small amounts of organic matter, but oil shales usually contain substantially more. Specific geochemical conditions are required to accumulate and preserve organic matter, and these were present in the lakes and oceans whose sediments became oil shale. R. M. Garrels and C. L. Christ define these conditions in terms of oxidation-reduction potential (Eh) and acid-base condition (pH) of the water in and around the sediment. Organic matter accumulates under the strongly reducing conditions and neutral or basic pH present in euxinic marine environments and organic-rich saline waters. The organic-rich sediments which became oil shale accumulated slowly in water isolated from the atmosphere, a condition relatively rare in natural waters. This isolation was achieved by stagnation or stratification of the water body and the accompanying protection of its sediments.

Quartz, illite, and pyrite (sometimes with marcasite and pyrrhotite) occur in virtually every oil shale. Other clays (particularly montmorillonite) and feldspars, are found in many oil shales. Most oil shale deposits contain small amounts of carbonate minerals, but some, notably the Green River Formation in Colorado, Utah, and Wyoming, contain large amounts of dolomite and calcite. The oil shale minerals were probably formed in the sediment by chemical processes related, at least in part, to the presence of organic matter.

Some oil shales, particularly those called black shales because of the coallike color of their organic matter, have tended to become enriched in trace metals. The reducing conditions necessary to preserve organic matter were conducive to precipitating available trace metals, frequently as sulfides. The Kupferschiefer of Mansfield, Germany, contains an unusually high content of copper, and the Swedish Alum Shale has been exploited for its uranium content. The Devonian Chattanooga Shale of Tennessee and neighboring states contains an average of about 0.006 wt % uranium and has been extensively studied as a potential low-grade source of this element. Vanadium in potentially commercial amounts occurs in the Permian Phosphoria Formation of Wyoming and Idaho. Enrichment of As, Sb, Mo, Ni, Cd, Ag, Au, Se, and Zn has also been noted in black oil shales.

Physical properties. Oil shales are fine-grained rocks generally with low porosity and permeability. Many are thinly laminated and fissile. On outcrop, some oil shales weather to form stacks of thin organic-rich layers which are called paper shale. The colors of oil shales range from black to light tan and are produced or altered by organic matter.

The physical properties of oil shale are strongly influenced by the proportion of organic matter in the rock. The rock's decrease in density with increasing organic content illustrates this most graphically. The mineral components have densities of about 2.6–2.8 g/cm³ for silicates and carbonates and 5 g/cm³ for pyrite, but the density of organic matter is near 1 g/cm³. Larger fractions of organic matter produce rocks with appreciably lower density. The equation below quantifies this

$$D_T = \frac{D_A D_B}{A(D_B - D_A) + D_A}$$

relationship. Here, D_T= density of the rock, A= weight fraction of organic matter, D_A= average density of organic matter, and D_B= average density of mineral matter.

The volume of organic matter in oil shale rock strongly influences its strength properties. The marked increase in volume fraction of organic matter in oil shale with increasing organic weight fraction can be demonstrated by using the above equation. For example, in Green River Formation oil shale the average density of the organic matter (D_A) is about 1.07 g/cm³, and the average density of the mineral matter (D_B) is about 2.72 g/cm³. Thus the density of an oil shale containing 4 wt % organic matter (a lean shale yielding about 2.3 wt % oil or about 6 gal of oil per ton of shale) is calculated to be 2.56 g/cm³. The organic matter occupies about 10 vol % of this rock. An oil shale containing 15 % organic matter (a shale yielding about 10 wt % oil or about 26 gal per ton) has a density of 2.21 g/cm³, and the organic matter occupies 31 vol % of this rock. An oil shale containing 39 wt % organic matter (a relatively rich shale yielding about 20 wt % oil or about 52 gal per ton) has a density of 1.86 g/cm³, and the organic matter makes up about 52 vol %, or more than half the rock's volume. In the Green River Formation, in oil shales containing 15 wt % or more organic matter, the organic material is the largest component by volume, and the physical properties of the organic matter predominate in determining the physical properties of the rock. The organic matter makes this rock tough and resilient and difficult to crush. The richer Green River Formation oil shales tend to deform plastically under load.

The foregoing equation can yield a relationship between oil yield and oil shale density by incorporating a factor for conversion of the organic matter to oil. The resulting relationship is useful in calculating resources and reserves in an oil shale deposit.

Organic composition and oil production. The organic matter in oil shales and other sedimentary rocks has been extensively studied by organic geochemists, but a specific description of it has not been produced. Although some oil shales contain recognizable organic fragments like spores or algae, most do not, because the basic reducing conditions associated with oil shale development digested and homogenized the organic debris. The resulting organic matter (kerogen) is best described as a high-molecular-weight organic mineraloid of indefinite composition. This composition varies from deposit to deposit and is influenced by the depositional conditions and the nature of the organic debris. Variations in the hydrogen content of this organic matter are significant, because the fraction of organic matter converted to oil on heating increases as the amount of hydrogen available in the organic matter increases. To illustrate this relationship, Table 1 compares the proportion of organic carbon converted and recovered as oil during Fischer assay with the weight ratio of organic carbon to organic hydrogen in several oil shales. For petroleum, the carbon-hydrogen values range from 6.2 to about 7.5; for coal, they range upward from 13. The carbon-hydrogen values for organic matter in the world's oil shales range from near petroleum to near coal.

Analytical determination of the elemental composition of the organic matter has been difficult because of the heterogeneous nature of oil shales. Carbon, hydrogen, sulfur, oxygen, and nitrogen are the major elements of the organic matter; but (except for nitrogen) they also occur in the mineral matter of oil shales. The organic matter and the mineral matter in oil shales are difficult to separate from each other either physically or chemically. Analytical techniques designed to distinguish between organic and mineral forms of elements, specialized organic matter enrichment techniques, and other specialized evaluation techniques are still being developed to aid in the study of oil shales.

The Fischer assay is the best known of the specialized analytical procedures. It was developed by the U.S. Bureau of Mines for oil shale evaluation and is now a standard method. This method, employing a modified Fischer retort, determines quantities of liquid oil and other products recoverable from an oil shale sample heated under prescribed conditions. Although the procedure does not measure the total amount of organic matter in the sample, it approximates the oil available by commercial operations. This simple procedure has proved suitable for oil shale evaluation purposes. Resource information for United States oil shales is based on the Fischer assay oil-yield data accumulated by the Laramie Energy Technology Center of the U.S. Department of Energy.

Table 1. Relationship between the organic carbon–to–organic hydrogen ratio and the conversion of oil shale organic matter to oil by heating

Deposit sampled	Carbon-hydrogen ratio	Organic carbon converted, wt %
Pictou County, Nova Scotia, Canada	12.8	13
Top Seam, Glen Davis, Australia	11.5	26
New Albany Shale, Kentucky, United States	11.1	33
Ermelo, Transvaal, South Africa	9.8	53
Cannel Seam, Glen Davis, Australia	8.4	60
Garfield County, Colorado, United States	7.8	69

Oil shale units. The United States expresses oil yield determined by Fischer assay in the volume unit of gallons per ton. This unit is converted from weight percent oil yield by multiplying by a conversion factor, 2.4, and dividing the result by absolute oil specific gravity, determined at 15.6°C. The metric volume unit, liters per metric ton (tonne), corresponds to gallons per ton. One gallon per ton is equivalent to 4.172 liters/tonne. Oil shale resource values are expressed in barrels of oil in the United States. A barrel of oil, the 42-gal volume unit used in Western petroleum commerce, has no direct equivalent in metric countries. A barrel of oil represents 0.159 kiloliter. Specification of oil density, a variable, is necessary to convert the volume unit, barrels, into the metric weight trade unit, tonne. The 1975 World Energy Conference agreed to define a barrel of oil as 0.145 tonne, and 1 gal/ton × 0.29 as 1 kg/tonne, approximations ignoring density variations in oil.

World oil shale resources. The world's organic-rich shale deposits represent a vast store of fossil energy. They occur on every continent in sediments ranging in age from Cambrian to Tertiary. D. C. Duncan and V. E. Swanson estimated the shale oil represented by the world's oil shale deposits. Their evaluations are summarized in Table 2. The values given for known resources refer only to evaluated resources. To these values Duncan and Swanson added possible extensions of known resources and geologically based estimates of undiscovered and unappraised resources to obtain their estimate of order-of-magnitude values for the total in-place oil resource in the world's oil shale deposits.

The size of the potential shale-oil resource is staggering. The richest part of the world's evaluated oil shale resource (9.1×10^{11} bbl or 1.3×10^{11} tonnes; see Table 2) alone is equivalent to the world's crude oil reserves in 1975 (7×10^{11} bbl or 10^{11} tonnes). These petroleum reserves represent only 4% of the projected total resource of rich oil shale.

The resource estimates in Table 2 are separated into three grades according to oil yield, recognizing that the richest deposits are more amenable to economic development. Because many factors besides richness affect the economics of development, the grade designations in Table 2 have limited significance.

World oil shale developments. Although the oil potential of the world's oil shales is great, commercial production of this oil has generally been considered uneconomic. Oil shales are lean ores, producing only limited amounts of oil which historically has been low in price. Mining and heating 1 ton of 25 gal/ton (104 liters/tonne) oil shale produces only 0.6 bbl (0.087 tonne) of oil.

In special situations when other fuels were in short or uncertain supply, or when energy transportation was difficult, energy development from oil shales has been carried out commercially. The

Table 2. Shale-oil resources of the world

Continents	Known resources*			Order of magnitude of total resources*		
Range in grade (oil yield in gal/ton):	25–100	10–25	5–10	25–100	10–25	5–10
Africa	100	Small	Small	4,000	80,000	450,000
Asia	90	14	†	5,500	110,000	590,000
Australia and New Zealand	Small	1	†	1,000	20,000	100,000
Europe	70	6	†	1,400	26,000	140,000
North America	600	1,600	2,200	3,000	50,000	260,000
South America	50	750	†	2,000	40,000	210,000
Totals	910	2,400	2,200	17,000	325,000	1,750,000

*In 10^9 bbl. †Not estimated.

1694 English patent granted for a process "to distill oyle from a kind of stone" is the earliest such record, although medicinal oils were apparently produced from oil shales earlier. A French operation initiated in 1838 is probably the earliest energy development recorded; and Scotland, Canada, and Australia produced shale oil commercially before 1870. From 1875 to 1960, energy equivalent to about 250,000,000 bbl (3.1×10^7 tonnes) of oil was produced from Europe's oil shale deposits. Most of this production was derived from deposits in Estonia, S.S.R., and Scotland. Low-priced oil from the Near East and improving oil transport systems stopped most oil shale developments.

World War II caused sharp increases in petroleum demand and disrupted both petroleum production and petroleum distribution, reactivating interest in oil shale development. Oil shale production operations during and since World War II have been conducted in Germany, France, Spain, Manchuria (China), Estonia and other areas of the Soviet Union, Sweden, Scotland, South Africa, Australia, and Brazil.

Two modern developments, the Manchurian and the Estonian, are relatively large. The Manchurian shale development is near the city of Fushun. The Oligocene oil shale deposit averages about 450 ft (150 m) of shale, yielding approximately 15 gal/ton (63 liters/tonne). The deposit overlies a thick coal seam. Removal of the oil shale deposit to enable the coal to be mined by open-pit methods has resulted in the development of the world's largest oil shale industry. Production information has been difficult to obtain, but a daily output of 40,000 bbl of oil (5800 tonnes) has been reported. Successful development at Fushun appears to have generated other oil shale developments in the area.

Broad Soviet areas in Estonia and the adjacent Leningrad region are underlain by Ordovician oil shale (kukersite) beds at shallow depths. These shales, reaching 10 aggregate feet (3.3 m) of 50 gal/ton (210 liters/tonne) shale, are being used to generate electricity and large quantities of low-heating-value gas for domestic and industrial purposes in Leningrad and Tallin. Production exceeded 30,000,000 tons (2.7×10^7 tonnes) of shale in 1973. Most of this was burned directly to generate electricity.

Several smaller-scale oil shale operations have been conducted during and since World War II. Australia operated an oil shale plant at Glen Davis, New South Wales, during World War II. Problems associated with mining thin seams caused this plant to close about 1950. Brazil, always short of domestic petroleum, has intensively investigated two major deposits for shale-oil production. Petrobras, a corporation partly sponsored by the Brazilian government, collaborated with a United States firm to develop and apply the Petrosix retort to the Permian Irati shale. By 1979 Petrobras had been operating for several years a pilot plant processing about 2200 tons (2000 tonnes) of oil shale per day. Major expansion, under discussion, would include this plant as an operating module. France pioneered destructive distillation of oil shale at Autun, and operated plants on three other deposits after World War II. All had ceased to operate by 1957. Germany operated oil shale plants on the Jurassic deposits in Württemberg during World War II, but these developments did not survive postwar economics. Scotland, an oil shale pioneer, continued to produce shale oil by mining and processing the Carboniferous Lothians oil shale deposits until 1963. In South Africa, the South African Torbanite Mining and Refining Company, Ltd., began operations on a deposit near Ermelo, Transvaal, in 1935. This grew into a large-scale operation which exhausted the 20,000,000–30,000,000-ton oil shale deposit. An integrated company operating at Puertollano in Spain's Ciudad Real Province produced gasoline, diesel and fuel oils, lubricants, and other by-products on a small scale starting about 1922. An enlarged company created in 1942 by the National Institute of Spain built a new installation which incorporated a low-temperature hydrogenation plant to upgrade shale oil. Production was discontinued about 1960. The Swedish government built a large plant at Kvarntorp in Narke Province during World War II to produce oil from the Alum black shale. This plant was in full production by 1947 but closed in the 1950s. In conjunction with this plant, Sweden tested underground gasification of oil shale by electrical heating. In this procedure, known as the Ljungstrom method, hydroelectric power available during times of low demand was used. Its proponents claimed that the calorific value of the oil and gas vaporized was about three times that of the energy used to produce them. Developments being investigated in other areas have been continually reported. Like the more extensive efforts, these came to an end because of the economic pressures caused by the abundant and inexpensive petroleum supplies from the Near East. Political and economic pressures have encouraged investigation of energy production from oil shales in Thailand, Morocco, Yugoslavia, and Israel.

Two major laboratories working primarily on oil shales exist, both government-sponsored. The Oil Shale Institute at Kohtla-Järve, Estonia, S.S.R., was founded in 1950 to investigate Estonian and other oil shales. The Laramie Energy Technology Center in Laramie, WY, began work on oil shale in 1944 under the U.S. Department of the Interior's Bureau of Mines. The center's primary interests are the Green River Formation and other United States oil shales. In 1977 the center became part of the U.S. Department of Energy.

United States oil shale resources. Organic-rich sedimentary deposits underlie about 20% of the United States land area. They range in age from Cambrian to late Tertiary, and most have not been evaluated or have shown only limited shale oil potential. The Cretaceous Niobrara Formation in Wyoming, Colorado, Nebraska, and South Dakota; the Tertiary Humboldt Formation in Nevada; and several Alaskan occurrences are examples of unevaluated organic-rich deposits.

In the United States, the largest deposit in terms of area is the Devonian-Mississippian black shale composite, which extends from Texas to New York and from Alabama to the Canadian border. This vast area, estimated at 250,000 mi² (65,000,000 hectares), is underlain by a time-transgressive continuum of black shale marine sediments occurring in formations locally referred to by names such as Chattanooga, New Albany Shale, Antrim, Ohio, Sunbury, Marcellus, Middlesex, Rhinestreet, Genessee, and Woodford. *U.S. Geological Survey Bulletin no. 523* indicates that the combined deposits offer about 10^{12} bbl (1.45×10^{11} tonnes) of oil from shale yielding 10 gal/ton (42 liters/tonne). Approximately 20% of this total resource is classed as known.

Organic matter in these Devonian-Mississippian black shales tends to be low in hydrogen, yielding only a small fraction (from zero to about one-third) of its weight as oil on heating (see Table 1). The heating value of the organic matter ranges from 8500 to 8900 cal/g. The average elemental composition of organic matter in the Devonian New Albany Shale of Kentucky in weight percent of the organic matter is C, 82.0; H, 7.4; N, 2.3; S, 2.0; O, 6.3. Although the black shales underlie a large area and the organic matter disseminated in them represents a huge amount of fossil energy, the deposits are low-grade and the resource in any area is relatively small. In Kentucky, where the deposit seems richest, the resource reaches 5×10^7 bbl of oil per square mile (28,000 tonnes oil/hectare) in a 100-ft-thick (30-m) section, with an average oil yield of 10 gal/ton (42 liters/tonne). Organic matter in the New Albany Shale in this area is 10 wt %, representing 2.2×10^7 tons of organic matter per square mile (32,000 tonnes/hectare).

The world's largest oil shale resource is the Eocene Green River Formation in Colorado, Utah, and Wyoming (Fig. 1). The oil potential of oil shales in this 16,500-mi² (1 mi² = 259 hectares) deposit exceeds 2×10^{12} bbl (2.9×10^{11} tonnes). Of

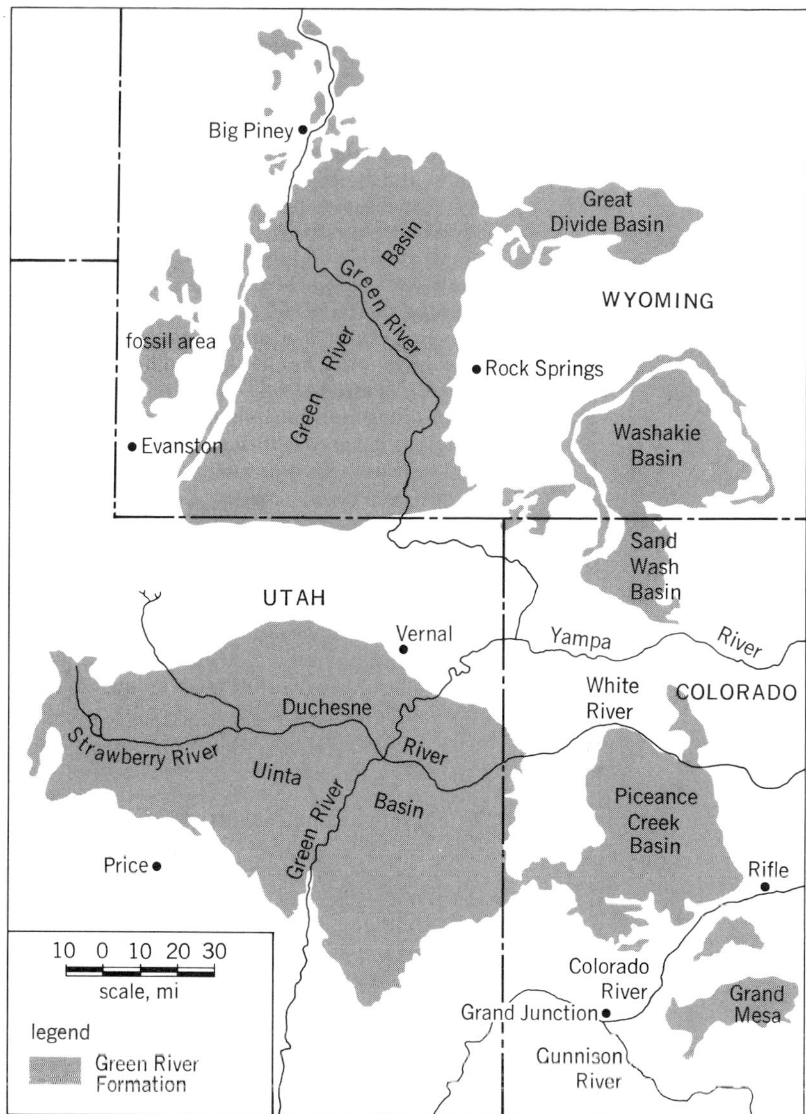

Fig. 1. Location of Green River Formation oil shale deposits. mi = 1.6 km.

this, 6×10^{11} bbl (8.7×10^{10} tonnes) occurs in deposits yielding 25 gal or more of oil per ton of shale (104 liters/tonne) in continuous sections of oil shale at least 10 ft (3 m) thick. In the Piceance Creek Basin of Colorado, the oil shale beds reach a thickness of 2100 ft (645 m) and contain about 5×10^8 tons of organic matter per square mile (700,000 tonnes/hectare). In Utah and Wyoming, the shale beds are not as thick and are sometimes separated by lean or barren rock.

Minerals ubiquitous in the Green River Formation oil shales are dolomite, quartz, sodium and potassium feldspars, illite, and pyrite. In some locations, the Green River Formation also contains large amounts of sodium carbonate minerals, including trona ($Na_2CO_3 \cdot NaHCO_3 \cdot 2H_2O$), nahcolite ($NaHCO_3$), dawsonite [$NaAl(OH)_2CO_3$], and shortite ($Na_2CO_3 \cdot 2CaCO_3$). More than 60% of the soda ash supply in the United States was produced from Green River Formation trona in 1975. Nahcolite and dawsonite occur in some rich Colorado oil shales. Nahcolite can yield soda ash and is an efficient and economic absorbent for removing SO_2 from stack gas. Dawsonite yields readily soluble alumina under appropriate thermal treatment to produce shale oil. Consequently, production of soda ash and alumina together with shale oil is being investigated. The mineral shortite, unique to the Green River Formation, has no known commercial value but occurs in huge amounts in Wyoming and Utah oil shales.

The entire Green River Formation is characterized by a remarkable lateral homogeneity, showing only very gradual geographic changes in its organic and mineral composition. Vertically, however, the formation is extremely variable, most notably in its organic content. Although the oil shale may be vertically continuous, its oil yields range from a few gallons per ton to nearly 100 gal/ton (417 liters/tonne).

The Mahogany Zone, a particularly organic-rich bed in the Green River Formation of Colorado and Utah, has been investigated intensively as a source of fossil energy. Organic matter in the Mahogany Zone has the following average elemental composition, in weight percent: C, 80.5; H, 10.3; N, 2.4; S, 1.0; and O, 5.8, with a gross heating value of 9500–9600 cal/g. The high hydrogen content of this organic matter correlates with the large fraction of oil recovered when the shale is heated (Table 1). Hydrogen-rich organic matter is characteristic of all the Green River Formation oil shales.

Shale oil. Shale oil is produced from the organic matter in oil shale when the rock is heated in the absence of oxygen (destructive distillation). This heating process is called retorting, and the equipment that is used to do the heating is called a retort. The rate at which the oil is produced depends upon the temperature at which the shale is retorted. For example, if Mahogany Zone oil shale is heated rapidly to 500°C and held at that temperature, it will take about 10 min for the reaction to reach completion. However, if the shale is heated rapidly to only 340°C, it will take more than 100 hr; when the shale is heated rapidly to a temperature of 660°C, the reaction takes only seconds. Most references report retorting temperatures as being about 500°C.

Retorting temperature affects the nature of the shale oil produced. Low retorting temperatures produce oils in which the paraffin content is greater than the olefin contents; intermediate temperatures produce oils that are more olefinic; and high temperatures produce oils that are nearly completely aromatic, with little olefin or saturate content.

In those retorting systems not capable of rapidly heating the oil shale to a constant temperature, the nature of the oil is determined by the rate at which the oil shale is heated. Thus at heating rates of about 1°C/min, the reaction is essentially completed by the time the temperature reaches 425°C, and the oil is principally paraffinic. At heating rates of about 100°C/min, the reaction is not complete until the temperature reaches more than 600°C, and the oil is more olefinic. As heating rates are increased above 100°C/min, both the paraffin and olefin contents of the oil decrease, and its aromatic content increases.

In general, shale oils can be refined to marketable products in modern petroleum refineries. There is no really typical shale oil produced from Green River oil shale, but the oils do have many properties in common. They usually have high pour points, 20–32°C; high nitrogen contents, 1.6–2.2 wt %; and moderate sulfur contents, about 0.5 wt %. High pour points make necessary some processing before the oils are amenable to pipeline transportation. The high nitrogen contents make hydrogenation necessary to reduce the nitrogen contents so that the oils can be processed into fuels. Hydrogenation also reduces the sulfur content. *See* DESTRUCTIVE DISTILLATION.

United States technology. The two general approaches to recovering shale oil from Green River Formation oil shales are (1) mining, crushing, and aboveground retorting, called conventional processing; and (2) in-place processing. The basic problems facing conventional processing are handling and heating huge amounts of low-grade ore and disposing of huge volumes of spent shale, the residue remaining after oil production. The in-place approach largely avoids the problems of handling and disposal but faces a different basic problem—the impermeability of the oil shale beds. Progress toward solving the basic problems of both approaches has been made.

With the conventional approach, oil shale mining by the room-and-pillar technique developed by the U.S. Bureau of Mines appears capable of producing the huge amounts of ore necessary to operate a large production plant. The procedure has also been tested by industry in Mahogany Zone shales. Outputs on the order of 2500 tons of oil shale per worker-shift have been reported from the highly mechanized operation. Crushing technology is well demonstrated.

Retorting must be done continuously in order to reach the throughput necessary for economic production of shale oil. Two general systems for heating a continuous stream of oil shale are outlined in Fig. 2. In the internally heated system the oil shale furnishes its own heat because part of its organic matter is burned inside the retort. The Bureau of Mines gas combustion retort, one form of the Paraho retort of Development Engineering, Inc., one form of the Union Oil Company's rock-pump retort, and Superior Oil Company's moving-grate retort are examples of the internally fired retort system. The Oil Shale Corporation's TOS-

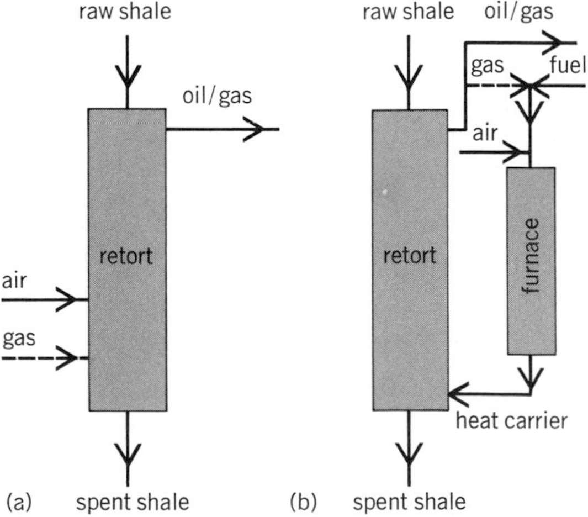

Fig. 2. Oil shale retorting systems. (*a*) Internally heated. (*b*) Externally heated.

CO II retort, in which preheated ceramic balls heat the oil shale; the SGR rock-pump retort of Union Oil; one form of the Paraho retort, in which preheated gas heats the oil shale; and the Lurgi retort, in which preheated sand or spent shale heats the oil shale, are all examples of the externally fired retort system. Several retorting systems, including both internally and externally heated designs, have been tested on pilot or semiworks scales, but a full-scale retort capable of operating in a commercial oil shale development had not been built by 1979.

Spent shale disposal has been studied intensively. More than 80% of the mined oil shale remains as residue after oil production. An oil shale plant producing 50,000 bbl (7250 tonnes) of oil per day from Mahogany Zone shale might mine 75,000 tons (68,000 tonnes) of rock and dispose of 60,000 tons (54,000 tonnes) of spent shale daily. In the vast and largely unpopulated areas of the Green River Formation, dumping these volumes of spent shale is not as large an environmental problem as it appears to be. The spent shale resulting from most surface retorts is largely insoluble, and contoured dumping to control water flow will minimize leaching, already low in an arid region. Native vegetation will establish itself on spent shale dumps, and revegetation procedures can accelerate this process. Returning spent shale to the mine to furnish roof support may permit recovery of additional ore, but this approach has not been tested.

Research and development efforts toward in-place processing have concentrated on creating permeability in the impermeable oil shale. In-place processing may be accomplished by two means: (1) a borehole technique in which oil shale is first fractured undergound and heat is applied, and (2) a process in which some rock is first removed by mining, then the remaining oil shale is fragmented into the voids created by mining, and finally heat is applied. These two methods are referred to as in-place (no mining) and modified in-place (some mining) processing. Several investigators have field-tested in-place methods. Occidental Oil Shale Corporation and Rio Blanco Oil Shale Company (a joint project of Amoco and Gulf) are conducting major experiments on the vertical burn type of modified in-place production. Geokinetics is producing and marketing small amounts of oil by a horizontal burn technique in shallow oil shales of Utah. In-place production experiments have been conducted by the Laramie Energy Technology Center, by Talley Energy Systems in Wyoming oil shale, and by Dow Chemical Company in the Devonian Antrim Shale of Michigan. Equity Oil Company has tested application of heat carried by hot natural gas or steam into a zone of natural permeability in Colorado, and Shell Oil Company has investigated leaching out water-soluble minerals to provide permeable access to Colorado oil shale. Both of the heating methods outlined for retorts (Fig. 2) have also been applied to in-place processing. Proposals for in-place processing include solvent extraction and radio-frequency electrical heating.

Less than one-third of the total Green River Formation oil shale resource is considered minable for conventional processing. In-place processing may make much of the remainder of the resource available to production. Conventional processing offers process control advantages, including ready adaptation of many existing industrial procedures, but it is capital-intensive, requiring huge investments before production begins. In-place processing is less easily controlled and evaluated but requires less capital outlay before production begins.

Prospects for United States oil shale development. By 1979, there was still no commercial production of shale oil. Oil shale is a major fossil fuel resource. The balance between energy supply and energy demand controls the position of oil shale in the United States energy market. Shale oil can compete in the petroleum market when the price of oil exceeds the cost of producing shale oil. The continual petroleum price increases by OPEC members intensified interest in oil shale as a domestic supply of fossil energy. Consumption of domestic petroleum reserves and the ever-increasing domestic demand for petroleum products are additional forces acting toward bringing oil shale into the energy market.

Federal land ownership is one of several factors adversely affecting oil shale development. More than 80% of the Green River Formation oil shale reserves, including the richest and thickest deposits, are federally owned. In 1930 President Herbert Hoover "temporarily" withdrew Green River Formation oil shale from leasing pending evaluation of the resource. Development of the major alumina and nahcolite resources in Colorado is also severely impeded by this withdrawal. The National Petroleum Council believes that oil shale development would be unlikely without leasing of Federal lands. Four tracts for commercial shale oil production were leased in 1974 by bonus bidding. The current Federal attitude is that no more oil shale land will be leased until these prototype tracts are in production. Of the privately held land, two-thirds is held by five major oil companies. Although the total resource is large enough to support any reasonable level of shale oil production, its legal availability may be a major deterrent to development.

Environmental and socioeconomic impacts of oil shale operations also complicate oil shale development. It has been estimated that at least 7 years is required between the time a lease is offered and the time mature production is reached. Much of this time is spent preparing detailed environmental impact statements, obtaining required Federal, state, and local licenses and permits, and preparing to meet housing and service requirements.

Water availability has been frequently cited as a problem in oil shale development. The Federal Energy Administration's 1974 Project Independence oil shale report concludes that water is available for normal development, and that the water requirement of a sharply accelerated development could be met by building additional storage facilities. Similar findings by the U.S. Water Resources Council and the Colorado Department of Natural Resources were reported in 1979.

Many forms of government support for developing synthetic fuel industries, including an oil shale industry, have been proposed to augment the energy supply available in the United States. Implementation of these programs is continually being evaluated, and such support seems perennially imminent. However, continued demand for petroleum in spite of rapidly increasing prices is the largest spur toward development of oil shale. *See* SYNTHETIC FUEL.

[JOHN WARD SMITH; HOWARD B. JENSEN]

Bibliography: Bibliography of Publications Dealing with Oil Shale and Shale Oil from U.S. Bureau of Mines, 1917–1974, and the ERDA Laramie Energy Research Center, 1975–1976, LERC/R1 77–7, 1977; D. C. Duncan and V. E. Swanson, *U.S. Geological Survey Circular no. 523,* 1965; R. M. Garrels and C. L. Christ, *Solutions, Minerals, and Equilibria,* 1965; T. A. Hendrickson, *Synthetic Fuels Data Handbook,* 1975; D. K. Murray, *Energy Resources of the Piceance Creek Basin,* Rocky Mountain Association of Geology, 1974; *Oil Shale: Prospects and Constraints,* Federal Energy Administration Project Independence Blueprint, GPO stock no. 4118–00016, 1974; *Relationship between Rock Density and Volume of Organic Matter in Oil Shales,* LERC/R1-76/6, 1976; J. W. Smith, *Theoretical Relationship between Density and Oil Yield for Oil Shales,* U.S. Bur. Mines Rep. Inv. 7179, 1969.

Open-pit mining

The extraction of ores of metals and minerals by surface excavations. This method of mining is applicable for near-surface deposits which do not have a ratio of overburden (waste material that must be removed) to ore that would make the operation uneconomic. Where these criteria are met, large-scale earth-moving equipment can be used to give low unit mining costs. For a discussion of other methods of surface mining *see* COAL MINING; PLACER MINING; STRIP MINING.

Most open-pit mines are developed in the form of an inverted cone, with the base of the cone on the surface. Exceptions are open-pit mines developed in hills or mountains (Fig. 1). The walls of most open pits are terraced with benches to permit shovels, front-end loaders, or bucket-wheel excavators to excavate the rock and to provide access for trucks, trains, or belts to transport the rock out of the pit. The final depth of the pits ranges from less than 51 feet to nearly 3000 ft (15.5 to 914 m) and is dependent on the depth and value of the ore and the cost of mining. The cost of mining usually increases with the depth of the pit because of the increased distance that the ore must be hauled to the surface, the waste hauled to dumps, and the increased amount of overburden that must be removed. Numerous benches or terraces in the pit permit a number of areas to be worked at one time and are necessary for giving slope, typically less than a 2-to-1 slope, to the sides of the pit and for ore blending. Using multiple benches also permits a balanced operation in which much of the overburden is removed from upper benches at the same time that ore is being mined from lower benches. It is undesirable to remove all of the overburden from an ore deposit before mining the ore because the initial cost of developing the ore is then extremely high.

The principal operations in mining are drilling, blasting, loading, and hauling. These operations are usually required for both ore and overburden. Occasionally the ore or the overburden is soft enough that drilling and blasting are not required.

Rock drilling. Drilling and blasting are interrelated operations. The primary purpose of drilling is to provide an opening in the rock for the placing of explosives. If the ore or overburden is weak enough to be easily excavated without blasting, it is not necessary for it to be drilled. The cuttings from the hole are often used to fill it after the explosive charge has been placed at the bottom. Blast holes drilled in ore serve a secondary function in that the cuttings can be sampled and assayed to determine the mineral content of the ore. Sampling of drill holes is frequently done for ores not readily identified by visual means.

The basic methods of drilling rock are rotary, percussion, and jet piercing. The drill holes are usually vertical and range in size from 11/2 to 15 in. (3.8 to 38 cm) in diameter and are drilled to varying depths and spacings as required for the particular type of rock being mined. The drill hole depth ranges from 20 to 50 ft (6 to 15 m), depending on the bench height used in the mine. The drill hole spacing (distance between holes) is governed by the depth of the holes and the hardness of the rock and generally ranges from 12 to 30 ft (3.7 to 9m). For short holes or hard rock, the hole spacing must be close; and for long holes and soft rock, the spacing may be increased.

Rotary drilling. Drilling is accomplished by rotating a bit under pressure. Compressed air is forced down the hollow drill shank or steel, and is allowed to escape through small holes in the bit for cooling the bit and blowing the rock cuttings out of the hole. Rotary drill holes range from a minimum diameter of 4 in. to a maximum diameter of 15 in. (10 to 38 cm) and are drilled by machines mounted on trucks or a crawler frame for mobility. The weight of these machines ranges from 30,000 to 300,000 lb (13,600 to 136,000 kg) and, in general, the heavier machines are required for the larger-diameter holes. The rotary method is the most common type of drilling used in open-pit mines because it is the cheapest method of drilling; however, it is restricted to soft and medium-hard rock.

Percussion drilling. Percussion drilling is accomplished with a star-shaped bit which is rotated

Fig. 1. Roads and rounds that have developed open cuts on Cerro Bolivar in Venezuela. (*U.S. Steel Corp.*)

Fig. 2. Aerial view of rotary drill. *(Kennecott Copper Corp.)*

while being struck with an air hammer operated by high-pressure air. Air is also forced down the drill steel to cool the bit and to blow the cuttings out of the hole. Small amounts of water are frequently added with the air to reduce the dust. The hole diameters range from 11/2 to 9 in. (3.8 to 23 cm), depending upon the type of drill. The drilling machine used for the smaller-diameter holes (1 1/2 to 41/2 in. or 3.8 to 11.4 cm) is small and usually mounted on a lightweight crawler or rubber-tired frame weighing a few thousand pounds (1 lb = 0.45 kg). The air hammer and rotating device are mounted on a boom on the machine, and the hammer blows are transferred by the drill steel to the drill bit. For the larger-diameter and deeper holes, the air hammer is attached directly to the bit, and the drill unit is lowered down the vertical hole because the impact of the hammer blows is dissipated in the larger and longer columns of drill steel. Percussion drilling is most applicable to brittle rock in the medium-hard to hard range. The depth of the smaller-diameter holes is limited to about 15 to 30 ft (4.6 to 9 m) but the larger holes can be drilled to 40 to 50 ft (12 to 15 m) without significant loss of efficiency.

Jet piercing. The jet piercing method was developed to drill the very hard iron ores (taconite). In this method, a hole is drilled by applying a high-temperature flame produced by fuel oil and oxygen to the rock. The holes range in size from 6 to 18 in. (15 to 46 cm), tend to be irregular in diameter, and require careful control of the flame to prevent over-enlargement. The drilling machines are integrated units which control the fuel oil and oxygen mixture and lower the jet piercing bit down the hole. The jet piercing method is limited to very hard rock for which other types of drilling are more costly.

Blasting. The type and quantity of explosive are governed by the resistance of the rock to breaking. The primary blasting agents are dynamite and ammonium nitrate, which are detonated by either electric caps or a fuselike detonator called Primacord.

Dynamite is available in varying strengths for use with varying rock conditions. It is commonly used either in cartridge form or as a pulverzied, free-flowing material packed in bags. It can be used for almost any blasting application, including the detonation of ammonium nitrate, which cannot be detonated by the conventional blasting cap.

Commercial, or fertilizer-grade, ammonium ni-

Fig. 3. A 27-yd^3 (20.5-m^3) shovel loading a 170-ton (153-metric ton) haulage truck. *(Kennecott Copper Corp.)*

trate has become a popular blasting medium because of its low cost and because it is safer to handle, store, and transport than most explosives. Granular or prill-size ammonium nitrate is commonly obtained in bulk, stored at the mine site in bins, and transported to the blasting site in hopper trucks that mix the agent and fuel oil as the explosive is discharged into the hole. It may also be obtained packed in paper, textile, or polyethylene bags. The carbon necessary for the proper detonation of ammonium nitrate is usually provided by the addition of the fuel oil. Granular or prilled ammonium nitrate is highly soluble and becomes insensitive to detonation if placed in water, and therefore its use is restricted to water-free holes. A portable blast hole dewatering pump is available for use with wet holes.

During the early 1960s ammonium nitrate slurries were developed by mixing calculated amounts of ammonium nitrate, water, and other ingredients such as TNT and aluminum. These slurries have several advantages when compared with dry ammonium nitrate: They are more powerful because of higher densities and added ingredients, and they can be used in wet holes. Further, the slurries are generally safer and less costly than dynamite, and are used extensively in mines which are wet or have rock difficult to blast with the less powerful ammonium nitrate. Both ammonium nitrate and slurries are detonated by a small charge of dynamite.

Mechanical loading. Ore and waste loading equipment in common use includes power shovels for medium to large pits and tractor-type front-end loaders for the smaller pits, or a combination of these. The use of scrapers is becoming more common for overburden removal and in some cases for ore mining. The loading unit must be selected to fit the transportation system, but because the rock will usually be broken to the largest size that can be handled by the crushing plant, the size and weight of the broken material will also have a significant influence on the type of loading machine. Of equal importance in determining the type of loading equipment is the required production or loading rate, the available working room, and the required operational mobility of the loading equipment.

Power shovels. Shovels in open-pit mines range from small machines equipped with 2-yd^3 (1.5-m^3) buckets to large machines with 36-yd^3 (27.5-m^3) buckets. A 6-yd^3 (4.6-m^3) shovel will, under average conditions, load about 6000 tons (5400 metric tons) per shift, while a 12-yd^3 (9.2-m^3) shovel will load about 12,000 tons (10,900 metric tons) per shift. However, the nature of the material loaded will have a significant effect on productivity. If the material is soft or finely broken, the shovel productivity will be high; but if the material is hard or poorly broken with a high percentage of large boulders, the shovel production will be adversely affected. Power may be derived from diesel or gasoline engines or diesel electric or electric motors. The use of diesel or gasoline engines for power shovels is usually limited to shovels of up to 4-yd^3 (3.1-m^3) capacity, and diesel electric drives are not common in shovels of more than 6-yd^3 (4.6-m^3) capacity. Electric drives are used in shovels rang-

Fig. 4. An 8-yd³ (6.2-m³) shovel loading an ore train. *(Kennecott Copper Corp.)*

ing up from 4-yd³ (3.1-m³) capacity and are the most widely used power sources in the open-pit mining operations (Figs. 3 and 4).

Draglines. A dragline is similar to a power shovel but uses a much longer boom. A bucket is suspended by a steel cable over a sheave at the end of the boom. The bucket is cast out toward the end of the boom and is pulled back by a hoist to gather a load of material which is deposited in an ore haulage unit or on a waste pile. Draglines range in size from machines with buckets of a few cubic yards' capacity to machines with buckets of 150-yd³ (115-m³) capacity. Draglines are extensively used in the phosphate fields of Florida and North Carolina. The overburden and the ore (called matrix in phosphate operations) are quite soft, and no blasting is required. The phosphate ore is deposited in slurrying pits, where it is mixed with water and transported hydraulically to the concentrating plant.

Front-end loaders. Front-end loaders are tractors, both rubber-tired and track-type, equipped with a bucket for excavating and loading material (Fig. 5). The buckets are usually operated hydraulically and range in capacity from 1 to 36 yd³ (0.76 to 27.5 m³). Front-end loaders are usually powered by diesel engines and are much more mobile and less costly than power shovels of equal capacity. On the other hand, they are not generally as durable as a power shovel, nor can they efficiently excavate hard or poorly broken material. A front-end loader has about one-half the productivity of a power shovel of equal bucket capacity; that is, a 10-yd³ (7.6-m³) front-end loader will load about the same tonnage per shift as a 5-yd³ (3.8-m³) shovel. However, the loader is gaining in popularity where mobility is desirable and where the digging is relatively easy.

Mechanical haulage. The common modes of transporting ore and waste from open-pit mines

Fig. 5. A 3-yd³ (2.3-m³) front-end loader loading a haulage truck in a small open-pit mine. *(Eaton Corp.)*

are trucks, railroads, and belt conveyors. The application of these methods or combination of them depends upon the size and depth of the pit, the production rate, the length of haul to the crusher or dumping place, the maximum size of the material, and the type of loading equipment used.

Truck haulage. Truck haulage is the most common means of transporting ore and waste from open-pit mines because trucks provide a more versatile haulage system at a lower capital cost than rail or conveyor systems in most pits (Fig. 6). Further, trucks are sometimes used in conjunction with rail and conveyor systems where the haulage distance is greater than 2 or 3 mi. (3 or 5 km). In these cases, trucks haul from the pit to a permanent loading point within the pit or on the surface, where the material is transferred to one or the other system. Trucks range in size from 20- to 350-ton (18- to 318-metric ton) capacity and are powered by diesel engines. Mechanical drives are used almost exclusively in trucks up to 75-ton (68-metric ton) capacity. Between 25- and 100-ton (23- and 91-metric ton) capacity, the drive can be either mechanical or electrical. In the larger trucks, the electric drive, in which a diesel engine drives a generator or alternator to provide power for electric motors mounted in the hubs of the wheels, is most common. Locomotive traction motors are used in some trucks of over 200-ton (181-metric ton) capacity. The diesel engines in the conventional drives range from 175 to 1200 hp (130 to 900 kW), while the engines in the diesel electric units range from 700 to 2400 hp (520 to 1800 kW). Most haulage trucks can ascend road grades of 8 to 12% fully loaded and are equipped with various braking devices, including dynamic electrical braking, to permit safe descent on steep roads. Tire cost is a major item in the operating cost of large trucks, and the roads must be well designed and maintained to enable the trucks to operate efficiently at high speed.

The size of trucks used is primarily dependent upon the size of the loading equipment, but the required production rate and the length of haul are also factors of consideration. Usually 20- to 40-ton-capacity (18- to 36-metric ton) trucks are used with 2- to 4-yd^3 (1.5- to 3.1-m^3) shovels, while the 70- to 100-ton (64- to 91-metric ton) trucks are used with 8- to 10-yd^3 (6.1- to 7.6-m^3) shovels.

Fig. 6. Copper mine waste is dumped from a 170-ton (153-metric ton) haulage truck. *(Kennecott Copper Corp.)*

In the early 1960s major advances were made in truck design which put the truck haulage system in a favorable competitive position compared with the other methods of transportation. Trucks are used almost exclusively in small and medium-size pits and the majority of the large pits, but in some cases are used in conjunction with rail or conveyor haulage. Trucks have the advantage of versatility, mobility, and low cost when used on short-haul distances.

Rail haulage. When mine rock must be transported more than 3 or 4 mi (5 or 6 km), rail haulage is generally employed. Since rail haulage requires a larger capital outlay for equipment than other systems, only a large ore reserve justifies the investment. As a rough rule of thumb, the reserve should be large enough to support for 25 years a production rate of 30,000 tons (27,000 metric tons) of ore per day and an equivalent or greater tonnage of stripping. Adverse grades should be limited to a maximum of 3% on the main lines and 4% for short distances on switchbacks. Good track maintenance requires the use of auxiliary equipment such as mechanical tie tampers and track shifters. The latter are required for relocating track on the pit benches as mining progresses and on the waste dumps as the disposed material builds up adjacent to the track. Ground movement in the pit resulting from disturbance of the Earth's crust or settling of the waste dumps makes track maintenance a large part of mining cost.

Locomotives in use range from 50 to 140 tons (45 to 127 metric tons) in weight, with the largest sizes coming into increased use for steeper grades and larger loads. Most mines operate either all electric or diesel electric models. The use of all-electric locomotives creates the problem of electrical distribution in the pit and on the waste dumps, and requires the installation of trolley lines adjacent to all tracks (Fig. 7).

Mine cars range in capacity from 50 to 100 tons (45 to 91 metric tons) of ore, and to 50 yd^3 (39 m^3) of waste. Ore is transported in various types of cars: solid-bottom, side-dump, or bottom-dump. The solid-bottom car is cheapest to maintain but requires emptying by a rotary dumper. Waste is mostly handled by the side-dump car. Truck haulage has replaced much of the rail haulage in recent years because of advances in truck design and reduction in truck haulage costs.

Conveyors. Rubber belt conveyors may be used to transport crushed material from the pit at slope angles up to 20°. Conveyors are especially useful for transporting large tonnages over rugged terrain and out of pits where ground conditions preclude building of good haulage roads, and where long haul distances are required. Improved belt design is permitting greater loading rates, higher speeds, and the substitution of single-flight for multiple-flight installations. The chief disadvantage of this transport system is that, to protect the belt from damage by large lumps, waste as well as ore must be crushed in the pit before loading on the belt.

Fig. 7. Aerial view of Kennecott's Bingham Mine. *(Kennecott Copper Corp.)*

Waste disposal problems. To keep costs at a minimum, the dump site must be located as near the pit as possible. However, care must be taken to prevent location of waste dumps above possible future ore reserves. In the case of copper mines, where the waste contains quantities of the metal which can be recovered by leaching, the ground on which such waste is deposited must be impervious to leach water. Where the creation of dumps is necessary, problems of possible stream pollution and the effect on farms and on real estate and land values must all be considered.

Slope stability and bench patterns. In open-cut and open-pit mining, the material ranges from unconsolidated surface debris to competent rock. The slope angle, that is, the angle at which the benches progress from bottom to top, is limited by the strength and characteristics of the material. Faults, joints, bedding planes, and especially groundwater behind the slopes are known to decrease the effective strength of the material and contribute to slides. In practice, slope angles vary from 22 to 60° and under normal conditions are about 45°. Steeper slopes have a greater tendency to fail, but may be economically desirable because they lower the quantity of waste material that must be removed to provide access to the ore.

Technology has been developed that permits an engineering approach to the design of slopes in keeping with measured rock and water conditions, making quantitative estimates of the factor of safety of a given design possible. Precise instrumentation can detect the boundaries of moving rock masses and the rate of their movement. Instrumentation to give warning of impending failure is being used in some pits.

Communication. Efficiency of mining operations, especially loading and hauling, is being improved by the use of communication equipment. Two-way, high-frequency radiophones are proving useful for communicating with haulage and repair crews and shovel operators.

[CARL D. BROADBENT]

Otto cycle

The basic thermodynamic cycle for the prevalent automotive type of internal combustion engine. The engine uses a volatile liquid fuel (gasoline) or a gaseous fuel to carry out the theoretic cycle illustrated in Fig. 1. The cycle consists of two isentrop-

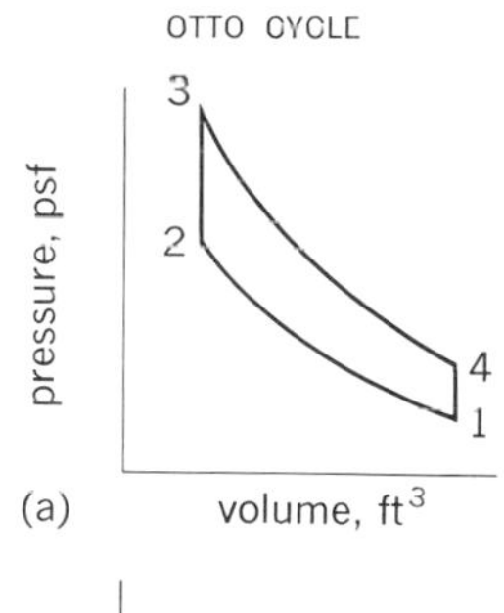

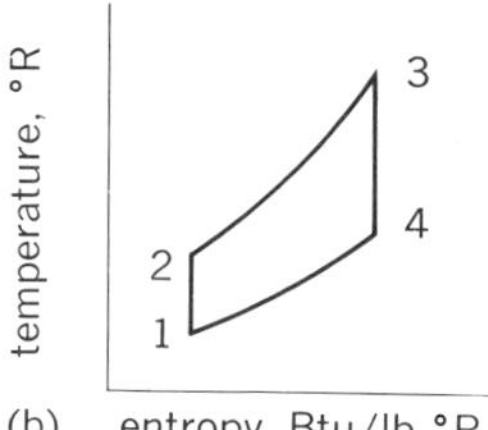

Fig. 1. Diagrams for Otto cycle: *(a)* pressure-volume and *(b)* temperature-entropy. Phase 1–2, isentropic compression; phase 2–3, constant-volume heat addition; phase 3–4, isentropic expansion; phase 4–1, constant volume heat rejection.

ic (reversible adiabatic) phases interspersed between two constant-volume phases. The theoretic cycle should not be confused with the actual engine built for such service as automobiles, motor boats, aircraft, lawn mowers, and other small (generally <300± hp) self-contained power plants.

The thermodynamic working fluid in the cycle is subjected to isentropic compression, phase 1–2; constant-volume heat addition, phase 2–3; isentropic expansion, phase 3–4; and constant-volume heat rejection (cooling), phase 4–1. The ideal performance of this cycle, predicated on the use of a perfect gas, Eqs. (1), is summarized by Eqs. (2) and (3) for thermal efficiency and power output.

$$V_3/V_2 = V_4/V_1 \qquad T_3/T_2 = T_4/T_1$$

$$\frac{T_2}{T_1} = \frac{T_3}{T_4} = \left(\frac{V_1}{V_2}\right)^{k-1} = \left(\frac{V_4}{V_3}\right)^{k-1} = \left(\frac{P_2}{P_1}\right)^{\frac{k-1}{k}} \qquad (1)$$

$$\text{Thermal eff} = \frac{\text{net work of cycle}}{\text{heat added}}$$
$$= [1-(T_1/T_2)] = \left[1-\left(\frac{1}{r^{k-1}}\right)\right] \qquad (2)$$

$$\begin{aligned}\text{Net work of cycle} &= \text{heat added} - \text{heat rejected}\\ &= \text{heat added} \times \text{thermal eff}\\ &= \text{heat added } [1-(T_1/T_2)]\\ &= \text{heat added } [1-(1/r^{k-1})]\end{aligned} \qquad (3)$$

In Eqs. (2) and (3) V is the volume in cubic feet; P is the pressure in pounds per square inch; T is the absolute temperature in degrees Rankine; k is the ratio of specific heats at constant pressure and constant volume, C_p/C_v; and r is the ratio of compression, V_1/V_2.

The most convenient application of Eq. (3) to the positive displacement type of reciprocating engine uses the mean effective pressure and the horsepower equation, Eq. (4), where hp is horsepower;

$$\text{hp} = \frac{\text{mep } Lan}{33{,}000} \qquad (4)$$

mep is mean effective pressure in pounds per square foot; L is stroke in feet; a is piston area in square inches; and n is the number of cycles completed per minute. The mep is derived from Eq. (3) by Eq. (5), where 778 is the mechanical equivalent

$$\text{mep} = \frac{\text{net work of cycle} \times 778}{144\ (V_1 - V_2)} \qquad (5)$$

of heat in foot-pounds per Btu; 144 is the number of square inches in 1 ft²; and $(V_1 - V_2)$ is the volume swept out (displacement) by the piston per stroke in cubic feet.

Air standard. In the evaluation of theoretical and actual performance of internal combustion engines, it is customary to apply the above equations to the idealized conditions of the air-standard cycle. The working fluid is considered to be a perfect gas with such properties of air as volume at 14.7 psia and 492°R equal to 12.4 ft³/lb, and the ratio of specific heats k as equal to 1:4. Figure 2 shows the thermal efficiency for this air-standard cycle as a function of the ratio of compression r, and the mep for a heat addition of 1000 Btu/lb of working gases. These curves demonstrate the intrinsic worth of high compression in this thermodynamic cycle.

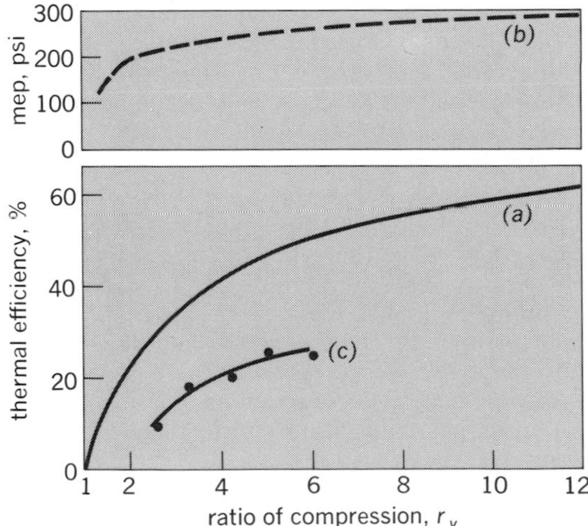

Fig. 2. Effect of compression ratio on thermal efficiency and mean effective pressure of Otto cycle. Curve *a* shows thermal efficiency, air-standard cycle; curve *b* mean effective pressure, air-standard cycle; and curve c thermal efficiency of an actual engine.

Thermal efficiency, mean effective pressure, and peak pressure of air-standard gas-power cycles*

Cycle	Efficiency	Mep	Peak pressures
Otto	60	290	2100
Diesel	42	200	370
Brayton	60	61	370
Carnot	60	Impossibly small	Impossibly high

*Ratio of compression = 10; heat added = 1000 Btu per pound working gases.

The table gives a comparison of the important gas-power cycles on the ideal air-standard base for the case of compression ratio = 10 and 1000 Btu added per pound of working gases. The Otto, Brayton, and Carnot cycles show the same thermal efficiency of 60%. The mean effective pressures, however, show that the physical dimensions of the engines will be a minimum with the Otto cycle but hopelessly large with the Carnot cycle. The Brayton cycle ideal mep is only about one-fifth that of the Otto cycle, and it is accordingly at a distinct disadvantage when applied to a positive displacement mechanism. This disadvantage can be offset by use of a free-expansion, gas-turbine mechanism for the Brayton cycle. The Diesel cycle offers a lower themal efficiency than the Otto cycle for the same conditions, for example, 42 versus 60%, and some sacrifice of mep, 200 versus 290 psi. As opposed to the Otto engine, the diesel can utilize a much higher compression ratio without preignition troubles and without excessive peak pressures in the cycle and mechanism. Efficiency and engine weight are thus nicely compromised in the Otto and diesel cycles. *See* GAS TURBINE.

Actual engine process. This reasoning demonstrates some of the valid conclusions that can be drawn from analyses utilizing the ideal air-standard cycles. Those ideals, however, require the modifications of reality for the best design of internal combustion engines. The actual processes of an internal combustion engine depart widely from the air-standard cycle. The actual Otto cycle uses a mixture of air and a complex chemical fuel which is either a volatile liquid or a gas. The rate of the combustion process and the intermediate steps through which it proceeds must be established. The combustion process shifts the analysis of the working gases from one set of chemicals, constituting the incoming explosive mixture, to a new set representing the burned products of combustion. Determination of temperatures and pressures at each point of the periodic sequence of phases (Fig. 1) requires information on such factors as variable specific heats, dissociation, chemical equilibrium, and heat transfer to and from the engine parts.

N. A. Otto (1832–1891) built a highly successful engine that used the sequence of engine operations proposed by Beau de Rochas in 1862. Today the Otto cycle is represented in many millions of engines utilizing either the four-stroke principle or the two-stroke principle. *See* INTERNAL COMBUSTION ENGINE.

The actual Otto engine performance is substantially poorer than the values determined by the theoretic air-standard cycle. An actual engine performance curve *c* is added in Fig. 2, in which the trends are similar and show improved efficiency with higher compression ratios. There is, however, a case of diminishing return if the compression ratio is carried too far. Evidence indicates that actual Otto engines offer peak efficiencies (25±%) at compression ratios of 15±. Above this ratio, efficiency falls. The most probable explanation is that the extreme pressures associated with high compression cause increasing amounts of dissociation of the combustion products. This dissociation, near the beginning of the expansion stroke, exerts a more deleterious effect on efficiency than the corresponding gain from increasing compression ratio. *See* BRAYTON CYCLE; CARNOT CYCLE; DIESEL CYCLE; THERMODYNAMIC CYCLE.

[THEODORE BAUMEISTER]

Bibliography: T. Baumeister (ed.), *Standard Handbook for Mechanical Engineers*, 8th ed., 1978; J. B. Jones and G. A. Hawkins, *Engineering Thermodynamics*, 1960; J. H. Keenan, *Thermodynamics*, 1970; L. C. Lichty, *Combustion Engine Processes*, 7th ed., 1967; E. F. Obert, *Internal Combustion Engines*, 3d ed., 1973.

Panel heating and cooling

A system in which the heat-emitting and heat-absorbing means is the surface of the ceiling, floor, or wall panels of the space which is to be environmentally conditioned. The heating or cooling medium may be air, water, or other fluid circulated in air spaces, conduits, or pipes within or attached to the panel structure. For heating only, electric current may flow through resistors in or on the panels. *See* ELECTRIC HEATING.

Warm or cold water is circulated in pipes embedded in concrete floors or ceilings or plaster ceilings or attached to metal ceiling panels. The coefficient of linear expansion of concrete is 0.000095; for steel it is 0.000081, or 15% less than for concrete. For copper it is 0.000112, or 20% more than for concrete, and for aluminum it is 0.000154, or 60% more than for concrete. Since the warmest or coolest water is carried on the inside of the pipes and the heat is transmitted to the concrete, only steel pipe should be used for panel heating and cooling systems, except when metal panels are used. Cracks are bound to develop in the concrete or plaster, breaking the bonds between the pipes and the concrete or plaster. The pipes move freely, causing scraping noises. An insulating layer of air is formed between the concrete or plaster and this markedly reduces the coefficient of conductivity between the liquid heating medium and the active radiant surfaces.

Heat transfer. Heat energy is transmitted from a warmer to a cooler mass by conduction, convection, and radiation. Radiant heat rays are emitted from all bodies at temperatures above absolute zero. These rays pass through air without appreciably warming it, but are absorbed by liquid or solid masses and increase their sensible temperature and heat content. *See* HEAT TRANSFER.

The output from heating surfaces comprises both radiation and convection components in varying proportions. In panel heating systems, especially the ceiling type, the radiation component predominates. Heat interchange follows the Stefan-Boltzmann laws of radiation; that is, heat transfer by radiation between surfaces visible to each other varies as the difference between the fourth power of the absolute temperatures of the two surfaces, and is transferred from the surface with the higher temperature to the surface with the lower temperature.

The skin surface temperature of the human body under normal conditions varies from 87 to 95°F and is modified by clothing and rate of metabolism. The presence of radiating surfaces above these temperatures heats the body, whereas those below produce a cooling effect. *See* RADIANT HEATING.

Cooling. When a panel system is used for cooling, the dew-point temperature of the ambient air must remain below the surface temperature of the heat-absorbing panels to avoid condensation of moisture on the panels. In regions where the maximum dew point temperature does not exceed 60°F, or possibly 65°F, as in the Pacific Northwest and the semiarid areas between the Cascade and Rocky mountains, ordinary city water provides radiant comfort cooling. Where higher dew points prevail, it is necessary to dehumidify the ambient air. Panel cooling effectively prevents the disagreeable feeling of cold air blown against the body and minimizes the occurrence of summer colds.

Fuel consumption records show that panel heating systems save 30–50% of the fuel costs of ordinary heating systems. Lower ambient air temperatures produce comfort and air temperatures within the room are practically uniform and not considerably higher at the ceiling, as in radiator- and convector-heated interiors. *See* COMFORT HEATING; HOT-WATER HEATING SYSTEMS.

[ERWIN L. WEBER/RICHARD KORAL]

Bibliography: American Society of Heating, Refrigerating, and Air Conditioning Engineers, *ASHRAE Handbook and Product Directory: Systems*, 1976.

Particle-beam fusion

A method of producing nuclear fusion energy through microexplosions of small fuel pellets driven by intense beams of charged particles. The energy released can be converted to heat and electric power in a reactor. Although this method has not yet been demonstrated experimentally, the existing technology and physics bases provide high confidence that commercial fusion power can be produced.

PRINCIPLES OF INERTIAL FUSION

The fusion reaction, in which light nuclei (isotopes of hydrogen) fuse together to form a heavier nucleus (helium) with the release of excess energy, is accomplished in stars and in thermonuclear weapons. Research has been underway since the early 1950s to achieve in the laboratory the necessary conditions of confinement of the reacting nuclei, as well as ignition temperature, to turn this reaction into the basis for an inexhaustible supply of energy. The first path followed, and in fact that still attracting the greatest emphasis, is the use of large, complex magnetic field geometries to hold the high-temperature thermonuclear fuel together long enough for the reaction to occur.

The other approach being studied is to heat the fuel by compressing it to a density thousands of times that of a solid, inside a rapidly imploded, tiny spherical container. The compressed fuel and the surrounding metal shell serve as an inertial container during the fraction of a billionth of a second that is required for the reaction to take place. The implosion velocity needed to reach ignition in this way is 200 km/s, and this velocity is achieved as a reaction to the explosivelike ablation of the outer surface of the pellet. This energetic rocketlike ablation requires that a highly concentrated beam with an intensity of $10^{13}-10^{14}$ W/cm^2 is efficiently absorbed by the outer layer of the pellet. In this inertial approach, the reactor chamber can be simpler and less costly than in the older approach because no magnets are needed; but instead, one is faced with the problems of producing and repetitively injecting low-cost fuel pellets into the chamber and focusing extremely energetic beams onto the pellet.

Calculational methods using complex hydrodynamic models have been developed to predict the energy input required to obtain various levels of energy gain from the reacting thermonuclear fuel in a single pellet. These methods predict that ignition of thermonuclear fuel pellets with reasonably high burning efficiency and sufficiently high output yield for a power reactor requires deposition of

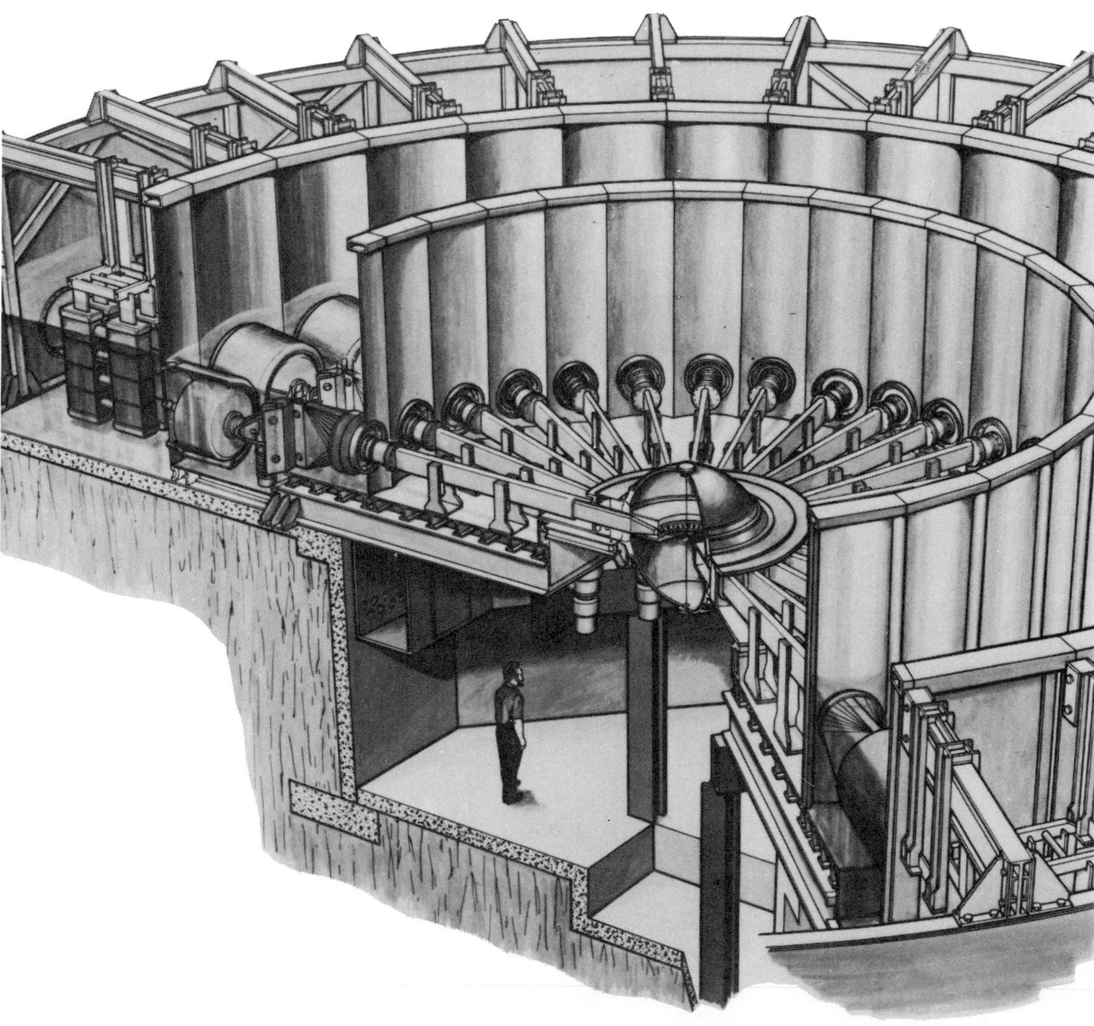

Fig. 1. Particle Beam Fusion Accelerator at Sandia Laboratories.

several megajoules of energy in a time of 10^{-8} s on a target surface area of order 1.0 cm². One calculational prediction indicates that a beam energy of 1–10 MJ is needed to achieve an energy gain G in the range of 10–100. Since the objective of a practical energy demonstration is to produce enough energy with each pellet to compensate for the inefficiencies in the beam generator (or driver), the product of the gain G and the driver efficiency η must have a minimum value ($\eta G > 10$). Thus if $\eta =$ 5%, the gain must be 200, requiring a beam energy of at least 10 MJ; and if $\eta = 40\%$, the gain must be 25, and a beam energy of only 1–2 MJ is needed. The implication of this argument is that the 5% efficient device must receive and handle 200 MJ each time a pellet is ignited, but the 40% efficient device has to provide only 5 MJ.

There are two types of directed energy beams being investigated for pellet implosion and ignition—lasers and charged particle beams. Lasers for fusion produce tightly focused pulsed beams of light. The principal constraint on the high-power laser approach is the problem of ensuring efficient deposition and transport of the light-beam energy into the solid target material. The light beams encounter first a thin, hot outer blow-off corona which interferes with efficient inward energy transport, and the beam energy must be precisely shaped in time within each pulse. However, a beam of charged particles can penetrate into a solid target without difficulty. Another potential disadvantage of laser drivers is their low efficiency of conversion of electric power into beam power compared to particle-beam drivers. Even though solid-state and gas lasers are highly developed, it is thought unlikely that suitable lasers with efficiencies much greater than 5% will be developed. Particle-beam drivers, however, have a demonstrated potential for 40% efficiency, and thus require much less energy, as discussed above. This resulted in 1973 in the initiation of a sizable activity which led to the present-day effort in particle-beam fusion. *See* LASER FUSION.

ELECTRON BEAMS

Programs to investigate the use of intense electron and light-ion beams have been undertaken principally at Sandia Laboratories in the United States and at the Kurchatov Institute in the Soviet Union. At the Kurchatov, development of a 5-MJ 100-TW device, Angara V, has been undertaken; and at Sandia, a 1-MJ 30-TW device, the Particle Beam Fusion Accelerator (PBFA-I), is to be completed in 1980 (Fig. 1), and planning of a 3.5-MJ 100-TW upgrade has begun.

Intense electron and light-ion beams are generated by relatively simple, and thus inexpensive, pulsed megavolt megampere generators. Such devices were initially developed in the mid-1960s to provide powerful x-ray bursts for radiography and for studies of pulsed radiation effects in military electronics. The pulsed generator used is typically a high-voltage source (consisting of capacitors charged at low voltage in parallel and discharged in series, producing high voltage) which transfers its energy through switches into a magnetically insulated pulse-forming transmission line and then into a vacuum diode. The Particle Beam Fusion Accelerator in Fig. 1 consists of 36 such modules, each producing a 2-MeV 400-kA beam.

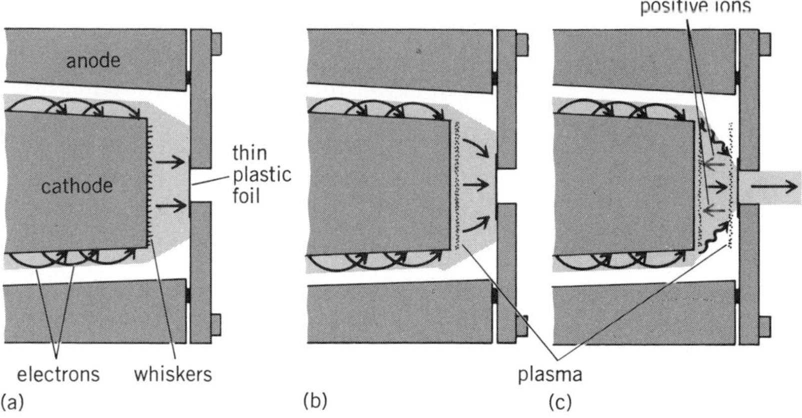

Fig. 2. Focusing and acceleration of electrons in a vacuum diode. (*a*) Initial, relatively unfocused electron flow. (*b*) Intensified electron flow, following formation of plasma layer. (*c*) Sharply focused electron flow, accompanying reverse flow of positive ions. (*From G. Yonas, Fusion power with particle beams, Sci. Amer., 239(5):50–61, November 1978*)

In the diode, electrons are accelerated from the metallic cathode of the magnetically insulated transmission line to an anode as shown in Fig. 2. The self-focusing effect enables the electrons flowing along the cathode of the transmission line to focus onto the anode at the axis where the magnetic field drops to zero. The effect is strongly influenced by the formation of a plasma layer on the face of both the cathode and the anode. Initially, the electrons are emitted from microscopic surface protrusions on the cathode and are accelerated to the anode in a relatively unfocused beam (Fig. 2*a*). The explosion of the heated whiskers forms a plasma on the surface of the cathode, intensifying the flow of electrons across the gap between the cathode and the anode (Fig. 2*b*). The electron flow heats the surface of the anode, forming another plasma layer there. The reverse flow of positively charged ions from the anode plasma to the cathode plasma helps to neutralize the self-repulsive force of the electrons, thereby sharpening the focus of the electron beam as it passes into the reaction chamber through a small hole in the face of the anode (Fig. 2*c*).

The use of electron beams at megavolt energies for pellet implosion requires that the electron depth of penetration into the target be limited by some process other than classical single-particle phenomena. Fortunately, collective beam-stopping effects associated with the enormous self-magnetic field of the focused electron beam can enhance the energy density in irradiated materials. Enhancement factors of 3 to 5 have been reported by Melvin Widner and colleagues at Sandia, confirming earlier observations of the same phenomenon by the Kurchatov group. In spite of this beneficial effect, it is still necessary to scale this phenomenon to higher beam currents on more powerful modules than now exist before its utility can be proved for pellet ignition.

LIGHT-ION BEAMS

Although it is still necessary to demonstrate scaling of enhanced electron deposition, this is not the case for light ions such as those from hydrogen or carbon atoms. For instance, a 1-MeV proton stops classically in matter in a depth similar to that for a

100-keV electron. In addition, such effects as energy backscattering or production of potentially harmful bremsstrahlung (x-rays) do not occur. The bremsstrahlung from decelerating electrons can penetrate and preheat the inner layer of the pellet shell and hinder the subsequent implosive compression of the fuel. For these reasons, interest in light ions has increased since 1975, and the light-ion approach is now favored by many over that using electrons.

In the ion-beam approach, the generator polarity is typically reversed so that the ions flow through the cathode toward the pellet. If no additional steps were taken to suppress the electron flow, such a diode would be extremely inefficient. Self-magnetic or external-magnetic fields are therefore utilized to increase the path length of electrons traveling from the cathode to the anode. In this way, far fewer electrons need to be injected into the diode to achieve an equilibrium flow of charge, and the generator power can be primarily partitioned into the ion beam.

An example of a particularly successful ion-beam diode developed by David J. Johnson at Sandia is shown in Fig. 3. In this diode, more than 80% of the generator output power was delivered from the diode as an ion beam, and this beam was focused onto a thin aluminum conical target 0.8 cm in diameter. The target implosion velocity and other measurements were used to infer the beam current density (approximately 400 kA/cm^2) and power density (approximately 0.5×10^{12} W/cm^2).

Beam transport. In one of the ion-beam diode concepts under consideration for the Particle Beam Fusion Accelerator, each of the 36 modules will produce a separate focused beam. These beams will then be transported to the target in plasma discharge channels which are formed in a dense background gas. Each channel is like a miniature lightning bolt driven by a small capacitor bank that heats, ionizes, and thus produces a narrow tube of low-density magnetized plasma which acts as an efficient conduit for the beam. Because one can confine the focused beam as it propagates over a sizable distance, it is also possible to concentrate the beam in time and space by bunching (varying the beam-accelerating voltage such that the tail end of the beam catches the head) and overlapping many beams at the target. The combination of beam overlap and bunching should afford at least a factor of 10 in beam concentration.

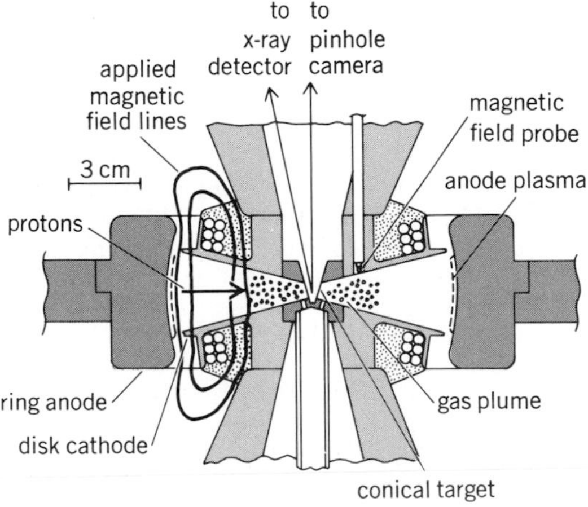

Fig. 3. Ion-beam diode.

Reactor design. These concepts have been considered for application in conceptual reactor design studies. The background gas in the reactor would act as a buffer to shield the wall of the chamber from potentially damaging effects of pellet explosion debris and soft x-rays. In this way, a relatively small-radius chamber can be designed to contain the repetitive explosions. In a sense, the explosion chamber in this future inertial fusion reactor would be similar to that in an internal combustion engine, with the particle accelerator acting as the replacement for the conventional spark plug. The energy released in the form of 14-MeV neutrons could then be utilized to heat a lithium blanket coupled to a conventional thermal-electric converter, or more likely at an early stage be used in a hybrid fusion/fission application. *See* NUCLEAR FUSION. [GEROLD YONAS]

HEAVY-ION BEAMS

Programs to study the production of fusion energy in pellet implosions produced by beams of heavy ions from high-energy accelerators have been undertaken at accelerator laboratories in the United States. In general, particle-beam fusion requires that the kinetic energy of each particle must be low enough to ensure heating of only a small volume of target material. Ions with atomic number greater than about 50 (heavy ions) allow the use of kinetic energies in the gigaelectronvolt range, produced by high-energy particle accelerator systems of relatively conventional design, rather than the pulsed power technology used with light particles.

An advantage of heavy-ion beams, compared to lighter particles, is the relative weakness of collective (plasma) effects expected in the beam propagation to the target. The reason for this is that multigigaelectronvolt ion beams need carry only a few kiloamperes of current each, rather than megamperes for lighter ions of lower kinetic energy. Minimization of plasma effects is expected to facilitate reliable reactor design, as the effects of plasma instabilities are usually difficult to predict precisely.

Adaptation of accelerator technology. In 1975 it was recognized that existing highly mature accelerator technology developed for high-energy and nuclear physics research could be adapted to design a multimegajoule heavy-ion inertial-confinement fusion driver, without any requirements for technological breakthroughs.

The principal driver-design adaptation required was the use of high-vacuum storage rings for circulating currents of heavy ions with low charge-to-mass ratio. High-vacuum proton storage rings have been in operation for about 10 years at the European Center for Nuclear Research at Geneva, Switzerland. Two of these rings exist; each stores 20–40 A of circulating 30-GeV protons for periods of up to 24 h, with a vacuum of order 10^{-10} Pa. The beam energy is 2 MJ per ring. Such a vacuum is sufficient to reduce the rate of charge-changing collisions of the circulating ions with residual background gases to allow a storage lifetime of order 1 s.

Besides the background gas collisions, another source of beam loss is the inter-ion charge-chang-

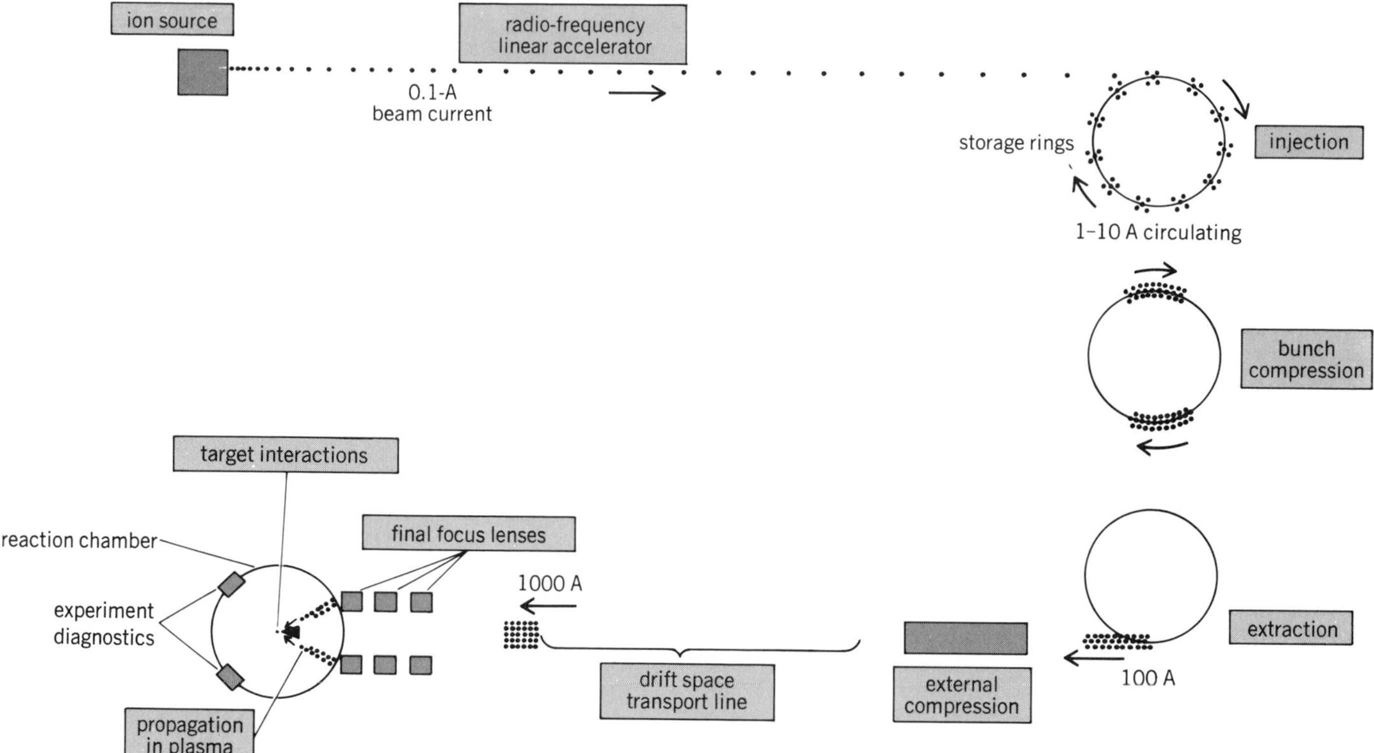

Fig. 4. Radio-frequency linear accelerator driver for heavy-ion-driven inertial confinement fusion.

ing collision process between ions in the circulating beam, which has a small spread of transverse and longitudinal velocities. The loss rate from this source is expected to allow a storage time at least 10–100 ms, depending on the ion species. Such beam lifetimes are adequate to allow filling the storage rings to their full current limit with radio-frequency linear accelerators of relatively conventional design, producing ion currents of order 100 mA.

RFL system. A complete heavy-ion driver system, of conventional radio-frequency linear accelerator (RFL) type, comprises one or more ion sources; a linear accelerator to give the ions several gigaelectronvolts of kinetic energy; several storage rings to accumulate ion currents to the desired total energy; a pulsed extraction and compression system to longitudinally contract the pulses of ions to around 10^{-8} s; and a set of magnetic lenses to transport the pulses to the reactor chamber and focus them on the small target. The logical sequence of events is shown schematically in Fig. 4.

Heavy-ion driver technology development programs are devoted to demonstrating the required ion-beam currents and preserving their quality during acceleration and storage. The principal effort for the RFL approach has been undertaken at Argonne National Laboratory, where a well-focused xenon beam of 50 mA at 1 MeV was produced in 1979, adequate for the acceleration and beam storage studies required before a driver is constructed.

LIA system. Another approach to heavy-ion driver technology, the linear induction accelerator (LIA), has been explored at Lawrence Berkeley Laboratory. Instead of accumulating several megajoules of circulating current by multiturn injection into high-vacuum storage rings, the LIA accelerates a single high-current pulse of ions to an energy adequate to drive the target implosion. The LIA requires low-velocity acceleration of an ion current of several amperes. A cesium source adequate to study this option was operating at Berkeley in 1979.

The technology of the LIA requires somewhat more development than do the RFL systems. LIAs have never been used to accelerate particles other than relativistic electrons. Heavy ions will require more sophisticated pulse control techniques to preserve a beam quality adequate for focusing on a small target. However, the LIA system has an apparent simplicity which may be an advantage if the technology can be adequately demonstrated.

Economics. Since conventional accelerator technology is highly developed, with a great deal of operating experience in existing facilities, it is possible to make fairly reliable estimates of the cost of a heavy-ion driver (at least of the RFL type). The cost scaling with beam energy has been examined. A driver of 1 MJ is estimated to contribute about one-third to one-half of the total price of a fusion power plant with output of 1000 MW of electric power, priced at around $1,500,000,000. Reactor economics improve with total beam energy E: heavy-ion driver costs increase only at $E^{0.4}$ with increasing E above 1 MJ, and target yield is expected to increase faster than linearly with E.

Thus, one may reasonably expect a solid economic foundation for heavy-ion driver power plants with output capacities of 1000 MW or more of electric power. [RICHARD C. ARNOLD]

Bibliography: R. C. Arnold, Heavy-ion beam inertial-confinement fusion, *Nature*, 276:19–23, 1978; T. M. Godlove, Recent progress and plans for heavy-ion fusion, *IEEE Trans. Nucl. Sci.*, NS-26(3):2997–3001, 1979; C. M. Stickley, Laser-driven inertial fusion, *Phys. Today*, 31(5):50–58, May 1978; G. Yonas, Fusion power with particle beams, *Sci. Amer.*, 239(5):50–61, November 1978.

Petroleum

A naturally occurring, oily, flammable liquid composed principally of hydrocarbons, and occasionally found in springs or pools but usually obtained from beneath the Earth's surface by drilling wells. Formerly called rock oil, unrefined petroleum is now usually termed crude oil.

Petroleum is separated by distillation into fractions designed as (1) straight-run gasoline, boiling at up to about 200°C; (2) middle distillate, boiling at about 185–345°C, from which are obtained kerosine, heating oils, and diesel, jet, rocket, and gas turbine fuels; (3) wide-cut gas oil, which boils at about 345–540°C, and from which are obtained waxes, lubricating oils, and feed stock for catalytic cracking to gasoline; and (4) residual oil, which may be asphaltic.

The physical properties and chemical composition of petroleum vary markedly, depending on its source. As it comes from the earth, it ranges from an occasional nearly colorless liquid consisting chiefly of gasoline to a heavy black tarry material high in asphalt content. Although most crudes are black, many are amber, red, or brown by transmitted light and show a greenish fluorescence by reflected light. Their specific gravity is usually in the range about 0.82–0.95.

Hydrocarbons constitute 50–98% of petroleum, and the remainder is composed chiefly of organic compounds containing oxygen, nitrogen, or sulfur, and trace amounts of organometallic compounds. Pennsylvania crude oils contain 97–98% hydrocarbons; some California oils contain only 50%.

Hydrocarbon types. The hydrocarbon types found in petroleum are paraffins (alkanes), cycloparaffins (naphthenes or cycloalkanes), and aromatics. Olefins (alkenes) and other unsaturated hydrocarbons are usually absent.

Paraffins. The paraffins range from methane (found together with ethane, propane, and the butanes in the natural gas which accompanies petroleum) to *n*-hexacontane ($C_{60}H_{122}$, a microcrystalline wax) and compounds of even higher molecular weight. Both straight-chain and branched-chain paraffins are present. The former usually predominate, particularly in the higher-boiling fractions. Commercial paraffin wax ordinarily consists chiefly of straight-chain paraffins of from about 22–30 carbon atoms isolated from the wide-cut, gas-oil fraction.

Cycloparaffins. The cycloparaffins are chiefly those having five or six carbon atoms in the ring. These include not only the monocyclic compounds (cyclopentane, cyclohexane, alkylcyclopentanes, and alkylcyclohexanes) but also polycyclic hydrocarbons, such as the bicycloparaffins (*trans*-decahydronaphthalene and *cis*-bicyclo[3·3·0]octane) as well as tri- and higher cycloparaffins.

Aromatics. Aromatic hydrocarbons are usually present in smaller amounts than the paraffins and cycloparaffins. The aromatic compounds boiling in the gasoline range are chiefly alkylbenzenes (such as toluene, the xylenes, and *p*-cymene). Higher-boiling fractions contain polynuclear aromatics of both fused-ring (alkylnaphthalenes) and linked-ring (biphenyl) types. The fused-ring polycyclics usually predominate. Mono- and polynuclear aromatic rings fused to one or more cycloparaffin rings (as in indan and 1,2,3,4-tetrahydronaphthalene) are also present.

Petroleum fractions. The number of carbon atoms in hydrocarbons of a given boiling range depends on the hydrocarbon type. In general, gasoline will include hydrocarbons having 4–12 carbon atoms; kerosine, 10–14; middle distillate, 12–20; and wide-cut gas oil, 20–36.

A study of the gasoline fractions from representative petroleums from seven different areas in the United States has permitted some interesting conclusions. Five main classes of compounds are present in the gasoline fraction: straight-chain paraffins, branched-chain paraffins, alkylcyclopentanes, alkylcyclohexanes, and alkylbenzenes. Although the relative amounts of the classes vary from petroleum to petroleum, the relative amounts of the individual compounds within a given class are of the same magnitude for the different petroleums. Hence, the gasoline fraction of different crudes is characterized by specifying the relative amounts of the five main classes of compounds in the fraction.

Petroleums may be classified in accordance with their composition. Thus Pennsylvania and Michigan crude oils are largely paraffinic and contain little or no asphalt. Some Texas and California oils are rich in naphthenes, whereas others are unusually high in aromatics; most contain much asphalt.

Asphalt is a dark-brown to black solid or semisolid consisting of carbon, hydrogen, oxygen, sulfur, and sometimes nitrogen. It is made up of three components: (1) asphaltene, a hard, friable, infusible powder; (2) resin, a semisolid to solid ductile and adhesive material; and (3) oil, which is structurally similar to the lubricating oil fraction from which it is derived. The asphalts are almost completely soluble in carbon disulfide, carbon tetrachloride, and pyridine, but are only partly soluble in low-boiling paraffins, which dissolve the oils and resins and precipitate the asphaltenes.

Components other than hydrocarbons. The total oxygen content of crude oils is generally low but may be as high as 2%. The oxygen-containing compounds consist principally of phenols and carboxylic acids. The phenols comprise cresols and higher-boiling alkylphenols. The acids include straight-chain and branched-chain acids, such as hexanoic acid and 3-methylpentanoic acid, and cyclopentane and cyclohexane derivatives, such as cyclopentaneacetic acid and cyclohexanecarboxylic acid. There is also some indication of the presence of acids containing aromatic rings (mono- and dinuclear). Hence the name naphthenic acids, which has been applied to the carboxylic acids derived from petroleum, is a misnomer; petroleum acids is a preferable term.

The nitrogen content of crude oils ranges from less than 0.05 to about 0.8%. Up to about one-half is in the form of basic pyridine and quinoline compounds, the latter predominating. The nonbasic nitrogen compounds or complexes include pyrroles, indoles, and carbazoles.

The sulfur content varies over a wide range, from traces to more than 5%. Pennsylvania and midcontinent crudes usually contain less than 0.25% by weight of sulfur, whereas some California and Texas stocks contain over 2%. Part of the sulfur may be in the form of elemental sulfur and hydrogen sulfide. Most is present as mercaptans (thiols), aliphatic sulfides, and cyclic sulfides. The mercaptans and sulfides exist as both straight-chain compounds, such as *n*-propyl mercaptan and methyl ethyl sulfide, and branched-chain compounds, such as *tert*-butyl mercaptan and methyl isopropyl sulfide. The cyclic sulfides consist of five- and six-membered ring compounds, such as thiacyclopentanes and thiacyclohexanes.

A number of metals have been identified in the ash (about 0.01–0.05% by weight) obtained by burning crude petroleum. These include sodium, magnesium, calcium, strontium, copper, silver, gold, aluminum, tin, lead, vanadium, chromium, manganese, iron, cobalt, nickel, platinum, and uranium. Boron, silicon, and phosphorus have also been detected. It is quite probable that the sodium and strontium are present chiefly in the form of aqueous solutions of salts that are finely dispersed in the oil.

Most of the other metals are present as either oil-soluble salts or organometallic compounds. For example, nickel and vanadium, which are the most abundant of these, occurring in 5–40 ppm in many crude oils of the United States, are probably present as porphyrin complexes. *See* PETROLEUM ENGINEERING; PETROLEUM GEOLOGY; PETROLEUM PROCESSING; PETROLEUM PRODUCTS.

[LOUIS SCHMERLING]

Bibliography: R. F. Goldstein and A. L. Waddams, *The Petroleum Chemicals Industry*, 3d ed., 1967; W. A. Gruse and D. R. Stevens, *Chemical Technology of Petroleum*, 3d ed., 1960; J. M. Hunt, *Petroleum Geochemistry and Geology*, 1979; A. L. Waddams, *Chemicals from Petroleum: An Introductory Survey*, 4th ed., 1979.

Petroleum engineering

An eclectic discipline comprising the technologies used for the exploitation of crude oil and natural gas reservoirs. It is usually subdivided into the branches of petrophysical, geological, reservoir drilling, production, and construction engineering. After an oil or gas accumulation is discovered, technical supervision of the reservoir is transferred to the petroleum engineering group, although in the exploration phase the drilling and petrophysical engineers have played a role in the completion and evaluation of the discovery.

Petrophysical engineering. The petrophysical engineer is perhaps the first of the petroleum engineering group to become involved in the exploitation of the new discovery. By the use of down-hole logging tools and of laboratory analysis of cores made during the drilling operation, the petrophysicist estimates the porosity, permeability, and oil content of the reservoir rock that has been sampled at the drill site. *See* WELL LOGGING.

Geological engineering. The geological engineer, using the petrophysical data, the seismic surveys conducted during the exploration operations, and an analysis of the regional and environmental geology, develops inferences concerning the lateral continuity and extent of the reservoir. However, this assessment usually cannot be verified until additional wells are drilled and the geological and petrophysical analyses are combined to produce a firm diagnostic concept of the size of the reservoir, the distribution of fluids therein, and the nature of the natural producing mechanism. As the understanding of the reservoir develops with continued drilling and production, the geological engineer, working with the reservoir engineer, selects additional drill sites to further develop and optimize the economic production of oil and gas. *See* PETROLEUM GEOLOGY; SEISMIC EXPLORATION FOR OIL AND GAS.

Reservoir engineering. The reservoir engineer, using the initial studies of the petrophysicist and geological engineers together with the early performance of the wells drilled into the reservoir, attempts to assess the producing rates (barrels of oil or millions of cubic feet of gas per day) that individual wells and the entire reservoir are capable of sustaining. One of the major assignments of the reservoir engineer is to estimate the ultimate production that can be anticipated from both primary and enhanced recovery from the reservoir. The ultimate production is the total amount of oil and gas that can be secured from the reservoir until the economic limit is reached. The economic limit represents that production rate which is just capable of generating sufficient revenue to offset the cost of operating the reservoir. The proved reserves of a reservoir are calculated by subtracting from the ultimate recovery of the reservoir (which can be anticipated using available technology and current economics) the amount of oil or gas that has already been produced. *See* PETROLEUM RESERVES.

Primary recovery operations are those which produce oil and gas without the use of external energy except for that required to drill and complete the wells and lift the fluids to the surface (pumping). Enhanced oil recovery, or supplemental recovery, is the amount of oil that can be recovered over and above that producible by primary operation by the implementation of schemes that require the input of significant quantities of energy. In modern times waterflooding has been almost exclusively the supplementary method used to recover additional quantities of crude oil. However, with the realization that the discovery of new petroleum resources will become an increasingly difficult achievement in the future, the reservoir engineer has been concerned with other enhanced oil recovery processes that promise to increase the recovery efficiency above the average 33% experienced to the United States (which is somewhat above that achieved in the rest of the world). The restrictive factor on such processes is the economic cost of their implementation. *See* PETROLEUM ENHANCED RECOVERY.

In the past the reservoir engineer was confined to making predictions of ultimate recovery by using analytical equations for fluid flow in the framework of an overall definition of the geological and lithological description of the reservoir. Extensive references to reservoirs that are considered to have analogous features, and history matching, such as curve fitting and extrapolation of the declining production, are important tools of the reservoir engineer. The analytical techniques are limited in that reservoir heterogeneity and the competition between various producing mechanisms (solution gas drive, water influx, gravity drainage) cannot be accounted for, nor can one reservoir be matched exactly by another. The reservoir engineer has also been able to use mathematical modeling to simulate the performance of the reservoir. The reservoir is divided mathematically into segments (grid blocks), and the appropriate flow equations and material balances are repeatedly applied to contiguous blocks. Extensive history matching is still required to achieve a reliable predictive model, but this can usually be achieved more quickly and reliably than with the use of analytical solutions and analogy. The cost is high, however, and cannot always be justified. *See* PETROLEUM RESERVOIR ENGINEERING; PETROLEUM RESERVOIR MODELS.

Drilling engineering. The drilling engineer has the responsibility for the efficient penetration of the earth by a well bore, and for cementing of the steel casing from the surface to a depth usually just above the target reservoir. The drilling engineer or another specialist, the mud engineer, is in charge of the fluid that is continuously circulated through the drill pipe and back up to surface in the annulus between the drill pipe and the bore hole. This mud must be formulated so that it can do the following: carry the drill cuttings to the surface, where they are separated on vibrating screens; gel and hold cuttings in suspension if circulation stops; form a filter cake over porous low-pressure intervals of the earth, thus preventing undue fluid loss; and exert sufficient pressure on any gas- or oil-bearing formation so that the fluids do not flow into the well bore prematurely, blowing out at the surface. As drilling has gone deeper and deeper into the earth in the search for additional supplies of oil and gas, higher and higher pressure formations have been encountered. This has required the use of positive-acting blow-out preventers that can firmly and quickly shut off uncontrolled flow due to inadvertent unbalances in the mud system. *See* OIL AND GAS WELL DRILLING; ROTARY TOOL DRILL.

Production engineering. The production engineer, upon consultation with the petrophysical and reservoir engineers, plans the completion procedure for the well. This involves a choice of setting a liner across the formation or perforating a casing that has been extended and cemented across the reservoir, selecting appropriate pumping techniques, and choosing the surface collection, dehydration, and storage facilities. The production engineer also compares the productivity index of the well (barrels per day per pounds per square inch of drawdown around the well bore) with that anticipated from the measured and inferred values of permeability, porosity, and reservoir pressure to determine whether the well has been damaged by the completion procedure. Such comparisons can be supplemented by a knowledge of the rate at which the pressure builds up at the well bore when the well is abruptly shut-in. Using the principles of unsteady state flow, the reservoir engineer can evaluate such a buildup to assess quantitatively the nature and extent of well bore damage. Damaged wells, like wells of low innate productivity, can be stimulated by acidization, hydraulic fracturing, additional performation, or washing with selective solvents and aqueous fluids. *See* OIL AND GAS WELL COMPLETION.

Construction engineering. Major construction projects, such as the design and erection of offshore platforms, require the addition of civil engineers to the staff of petroleum engineering departments, and the design and implementation of natural gasoline and gas processing plants require the addition of chemical engineers.

Summary. The relative importance and numbers of petroleum engineers employed in the industry have increased in recent years because of the increasing value of crude oil and natural gas and the need for more economical recovery of these fluids. The technology has become increasingly sophisticated and demanding with the implementation of new recovery techniques and the expanding frontiers of the industry into the hostile territories of arctic regions and deep oceans. *See* OIL AND GAS, OFFSHORE; PETROLEUM.

[TODD M. DOSCHER]

Petroleum enhanced recovery

Novel technology to enhance the fraction of the original oil in place in a reservoir has been under study since the early 1960s. Heightened interest in developing enhanced recovery technology has developed as it has become more certain that over two-thirds of the oil discovered in the United States, and a still greater percentage in the rest of the world, will remain unrecovered through the application of conventional primary and secondary (waterflood) operations. This amounts to some 300×10^9 bbl (4.8×10^{10} m^3) that will be left in the ground of the United States, and it provides a strong incentive for the development and implementation of advanced technology to recover some of this oil. Even a 10% recovery of the "unrecoverable" quantity would more than double the current reserves of crude oil in the United States. *See* PETROLEUM RESERVES.

The problems encountered in developing such technology are very great because of the nature of fluid flow within a subsurface reservoir, and because of the inability of engineers to exercise any intimate degree of control over the flow and distribution of fluids within the reservoir. The only points of contact with the reservoir are at the surface of the producing and injection wells that lead down to the reservoir sands.

Reservoir quality. A subsurface reservoir comprises the interconnected pores of a sandstone whose genesis was the compaction and cementation of sand-rich sediments. Oil is also found in carbonate rocks in which the porosity may have resulted from solution, diagenesis, and fracturing. The flow of fluids through these reservoir rocks is along very tortuous and nonuniform microscopic channels, frequently interrupted by inclusions of shales and clays. Further, a single reservoir is usually composed of several successive beds or layers of rock which have significantly different permeabilities. As a result, there is a great tendency for oil being displaced by encroaching water or an expanding gas to be bypassed by the displacing phase. Since the path of least resistance is always followed, and because of the dendritic nature of the channels that can be delineated through the porous network, the opportunities for bypassing are numerous. The tendency to bypass is exacerbated by the differences in wettability of the reservoir minerals by water and oil. If the encroaching water succeeds in bypassing oil, then pressure gradients, above an attainable level, will be required to effect displacement of the trapped ganglia. *See* PETROLEUM GEOLOGY.

Thus, the occurrence of unrecoverable oil following waterflooding is not unanticipated. The overall recovery of 33% from reservoirs in the United States does not represent the mean of a continuous distribution, but is the volumetric average of discrete values which are quantitatively different from each other. The low-permeability Spraberry reservoir in West Texas, with an original oil in place of 8×10^9 bbl (1.3×10^9 m^3), is anticipated to yield a recovery of less than 10%, whereas many reservoirs in south Louisiana and east Texas will experience recoveries of well over 60% of the original oil in place. The difference between the two is due to the far better permeability and porosity development and the presence of a strong water drive in the latter reservoirs. The target for enhanced oil recovery would superficially appear to be very large in the Spraberry and less in the other reservoirs; however, the nature of the Spraberry mitigates against a high sweep and contact of the remaining oil by any enhanced recovery scheme, and therefore the realistic prospects for additional recovery are small.

Thermal technology. One exception to this relationship between unrecoverable oil and reservoir quality is the heavy-oil reservoirs of California. Numerous shallow reservoirs in that state are at a low temperature and a relatively low pressure, and have a high porosity that is saturated with medium- to high-viscosity oils. The viscosity is so high and the pressure so low that production rates are virtually uneconomic. Because of the much lower mobility of the viscous oils, injected water will readily finger through the reservoir without displacing a significant quantity of oil. During the late 1950s and early 1960s, following the Suez crisis, the industry developed technology by which the viscosity of the crude within the reservoir could be lowered and a suitable pressure gradient applied to achieve significant recovery of crude from such reservoirs. The methods are the steam soak, or cyclic stimulation, in which oil is produced from the same well into which steam had earlier been injected; the steam drive, which is much like a waterflood but with the substitution of steam for injected water; and wet, or quenched, in-situ combustion in which a steam drive is generated in situ by the combustion of a limited fraction of the oil in place.

Steam soak. Cyclic steam injection is very effective in reservoirs which have some modest degree of natural energy either due to gas in solution or due to a substantial thickness to support gravity drainage. It is an extremely profitable operation because of the rather short deferment of costs prior to the production of oil. However, it is usually limited to the recovery of only 5–20% of the oil in place because of the inability of the reservoir energy to maintain suitable rates of influx of oil into the region around the borehole where effective heating occurs. If wells are drilled closely, which is economically feasible in some very shallow reservoirs, the recovery in thick reservoirs can reach values as high as 50%.

Steam drive. The steam drive is more effective in displacing oil; recoveries in excess of 50% have been achieved, although at a lower thermal efficiency than in the case of the steam soak. The major drawback to the process is its intensive use of energy. About a third of the produced oil must be used as fuel for generating the steam. The extension of the process to the recovery of still more viscous oils (tar sands and bitumens), to reservoirs less than 40 ft (12 m) or so in thickness, and to reservoirs containing a low waterflood residual, less than 1000 bbl per acre-foot (13% by volume), will require the consumption of a still greater fraction of the produced oil. This is due to the fact that, for the conditions cited, the amount of heat that is required to raise the temperature of the reservoir and satisfy the heat losses becomes an increasing fraction of the energy contained in the produced crude.

The use of coal as a substitute for produced oil as boiler fuel and the possible use of solar-powered steam generators would increase the salable oil from a steam drive operation. Severe regulatory obstacles are presently in the way of using coal-fired systems, and there are severe economic hurdles that will have to be overcome for the widespread use of solar power.

In-situ combustion. This method, in which air and water are injected simultaneously in the preferred modification, has not fared as well as steam injection in its level of application. Capital and operating costs for air compressors, corrosion and the production of difficult fluid emulsions, and a lower reservoir sweep efficiency have contributed to its being a second choice even though its theoretical efficiency is higher, in that it avoids the heat losses in surface and well bore facilities and generates heat from residual oil.

Production rates. Over 250,000 bbl (40,000 m^3) of oil a day are being produced by steam drive and cyclic steam operations in California, and this rate will probably increase substantially in the years ahead as a result of developing technology, higher needs for liquid fuels, and higher prices. In Canada, where vast accumulations of heavy oil occur in the Athabasca Bituminous Sands and in the Peace River and Cold Lake districts in Alberta, and in eastern Venezuela along the Orinoco River, the application of steam drives will eventually boost the reserves of these nations significantly.

Other technologies. For most of the residual or unrecoverable oil in the United States, the use of thermal technology to increase recovery is out of the question because of the relatively low energy content of the recoverable oil (even if all the residual is recovered) compared to the energy required to heat the reservoir. Instead, technology must be used which focuses on liberating the oil, trapped by capillary forces, by reducing the interfacial tension between the oil and water in the reservoir. There are two basic schemes being pursued: the injection of aqueous solutions of surfactants that can reduce the interfacial tension to very low values, and the injection of solutes which can dissolve in and swell the residual oil and restore mobility to the trapped oil. The latter process ideally would use a displacing phase that, more than merely dissolving in the oil, would be completely miscible with it.

Micellar/polymer fluids. Aqueous solutions of surfactants have been developed that are capable of recovering virtually 100% of the residual oil in a laboratory experiment. These are known as micellar/polymer fluids since they contain surfactants at concentrations above the critical micellar value and polymers (such as polysaccharides and hydrolyzed polyacrylamides) that develop aqueous-phase viscosities that will assure stable displacement (minimum bypassing). The theoretical foundation for the use of such systems is fundamentally sound: a low interfacial tension means a low displacement pressure.

Field tests of the process, however, have been generally disappointing. This performance can be traced back to the fact that, because of the cost of these systems, the solutions of surfactants and polymers can be injected only as slugs, rather than continuously, if economic recovery of crude oil is to be achieved. The slug size is probably limited to less than 5% of the reservoir pore volume, certainly no more than 10%, and the integrity of the slug is weakened by numerous factors: temperature and shear degradation, precipitation by ions occurring in the connate water or released by ion exchange with the reservoir clays, adsorption on mineral surfaces, cross-flow and diffusion into low-permeability layers, and transfer of the active surfactants into the oil phase. Increasing oil prices do not result in a proportionate increase in profit potential, since the surfactants and polymers that are found to be useful are derived from petroleum or use petroleum products in their manufacture, and their price escalates with the price of crude. The micellar/polymer process, however, is potentially the most widely applicable scheme for enhanced oil recovery, and therefore research and development continue. A successful breakthrough in such research would have inestimable value for the United States and the world in increasing the ultimate recovery of crude oil.

Solvents. Again during the 1950s, liquefied petroleum gas (LPG; propane) was sufficiently inexpensive compared to crude oil that it could be considered as a sacrificial solvent for the tertiary recovery of residual oil, or, in some cases, even as a substitute for water in secondary recovery operations. Laboratory studies indicated that the efficiency of the LPG was impaired by its relatively high mobility compared to that of crude oil and water. The LPG tended to finger through the reservoir, recovering only a small fraction of the residual oil, and this at very high and uneconomic ratios of injected LPG to produced crude. A search for cheaper solvents revealed that carbon dioxide under high pressure was very soluble in many crudes and led to significant volumetric expansion and decreased viscosity of the residual oil. With some crudes there was further evidence that a miscible carbon dioxide–rich phase could be generated by repeated contacts of successively enriched phases. These observations gave rise to the expectation that successful recovery of crude could be achieved despite the very high mobility of the carbon dioxide and carbon dioxide–rich phases.

Field pilots have borne out the expectation that carbon dioxide would recover additional crude oil, but in tertiary (after waterflood) pilots the carbon dioxide/produced oil ratios are significantly higher than anticipated. (In reservoirs which have not been waterflooded, the injection of carbon dioxide has resulted in significant improvements in recovery at what appears to be economically viable ratios.) A particularly attractive province for the use of carbon dioxide is in dipping reservoirs on the flanks of salt domes in south Louisiana and offshore in the Gulf of Mexico. The updip injection of carbon dioxide is anticipated to result in a gravity-stabilized displacement of the residual oil.

Research and development activities on the use of carbon dioxide as an enhanced recovery agent are continuing, with several projects being devoted to reducing the mobility of the reagent by injecting it as a foam or taking advantage of intermediate phases created by mixing crude oil with carbon dioxide and additives. Development studies are under way on securing economic sources of carbon dioxide, both from natural reservoirs and from manufacturing operations.

Future productivity. It is difficult to anticipate the amount of reserves that can be added by successful implementation of enhanced oil recovery operations, and similarly difficult to anticipate the production rates that may be achieved. Several studies completed in the past few years have indicated a marked sensitivity of enhanced oil recovery to the real market price of the produced crude oil. The three most respected studies (that of Lewin & Associates for the Department of Energy, that of the National Petroleum Council, and that of the Office of Technological Assessment) yield ranges of additional reserves varying from 3 to 42×10^9 bbl (0.48 to 6.7×10^9 m^3) and production rates ranging from 500,000 to 4,000,000 bbl (80,000 to 640,000 m^3) a day in 1990. These estimates were prepared prior to some of the field results with carbon dioxide, and therefore the higher estimates are considered optimistic. Considering the rapid escalation in the value of crude oil, the chief limiting factor on enhanced recovery will be the development and implementation of technology for amenable reservoirs. A goal of much more than 10^6 bbl (1.59×10^5 m^3) a day by 1990 will be a difficult one to meet. *See* OIL MINING; OIL SAND; OIL SHALE.

[TODD M. DOSCHER]

Petroleum geology

The application of geological concepts to finding and producing petroleum. Petroleum deposits are usually found in sedimentary rocks. The petroleum substances, or hydrocarbons, originate within and are derived from the constituents of sedimentary rocks. With a few notable exceptions, economically important hydrocarbon deposits are found in the subsurface at depths ranging from several meters to several kilometers. In many instances, these buried deposits are totally obscure to direct means of detection, and it is necessary to employ geophysical techniques and finally a drill to locate them.

SUBSURFACE DEPOSITS

Petroleum occurs in porous rocks such as sandstones, limestones, and dolomites. These are reservoir rocks, and they consist of an aggregate of mineral particles such as quartz sand grains, shell fragments from marine organisms, or individual mineral crystals. In the subsurface, porosity, the space between the particles, is usually filled with water, but under certain conditions the fluid is a mixture of water and hydrocarbons, either oil or gas or both.

The conditions necessary for the accumulation of hydrocarbons are a source and a trap to collect them. There are many circumstances which provide the right conditions to form an oil or gas field. A description of one set of circumstances will serve as an example to illustrate how some oil and gas fields might originate. A sequence of events might begin with the deposition of barrier bar and beach sands along the coastline of an ancient land body. The sand extends along the coastline for miles much like the modern beach sands on the seashores today. As the sand is being concentrated along the beach front, mud is being deposited in the swampy areas and bays landward of the beach and also in the deeper water seaward. The mud consists of clay minerals along with considerable amounts of organic material derived from animals that thrive and die in the bay and open sea waters. Following deposition of the sand and mud, the land begins to subside, and the sea inundates the former beach area. A layer of mud is deposited on top of the sand. The beach and barrier sand is now completely surrounded above and on the sides by clay muds. As subsidence continues, the sediments consolidate as they are subjected to higher and higher temperatures and pressures. The loose sands become sandstones and the clay muds turn into shales. When the temperature and pressure reach a critical point, the organic material incorporated in the shale is converted to hydrocarbons. A portion of the hydrocarbons are expelled from the shales, and some of them enter the sandstone and remain there, trapped by the surrounding shales.

In this example, the ancient beach or barrier bar sand is the reservoir rock, and the surrounding shales are the seals that form the trap. There are many other kinds of reservoir rocks and seals; some of the more common ones are:

A. Reservoir rocks
   - Sandstones (beaches, bars, stream channels, sand dunes)
   - Porous limestones (oolitic lime sands, fossil-fragment sands, porous reef deposits)
   - Porous dolomites (intercrystalline porosity)
   - Fractured shales
   - Fractured metamorphic and igneous rocks

B. Seals or barriers
   - Shales
   - Salt beds
   - Gypsum and anhydrite
   - Dense limestones and dolomites

Reservoir traps. There are two basic types of hydrocarbon traps, stratigraphic and structural, but these often occur in combination with each other; that is, many traps are partially stratigraphic and partially structural. Some typical traps are shown in the illustration. The lens (*a*) and the unconformity (*d*) are examples of stratigraphic traps. The anticline (*b*), fault (*c*), and salt dome (*e*) are examples of structural traps.

Reservoir rocks. The quality of a reservoir rock is controlled primarily by porosity and permeability. Porosity is measured as a percentage of total rock volume; in typical hydrocarbon reservoirs it ranges from 5 to 30%. Permeability, the ability of the reservoir rock to transmit or allow movement of fluids, is provided by the interconnection of pores in the rock and allows oil and gas to flow from the reservoir into wells for production. Some rocks contain significant porosity but no permeability because the pores either are not connected or are too small to permit fluid flow. Shales typically have high porosity, but no permeability because the pores are small.

Reservoir fluids. The voids or pore space in all reservoir rocks are filled with fluid, which may be water or various combinations of water, oil, and gas. In traps that contain oil, water, and free gas, the fluids occur in distinct zones: gas, the lightest fluid, occurring at the top of the trap, followed by oil, and then water. Where there is no oil, the gas is immediately above the water.

Interstitial water (absorbed water or wetting water which lines the pore walls or occurs on the surfaces of mineral grains) is present throughout the hydrocarbon column, occupying 10 to 50% of the pore space. Oil field waters are waters associated with oil and gas pools. Most oil field waters are saline. *See* OIL FIELD WATERS.

Oil saturation is the amount of oil contained in a petroleum reservoir. It is measured as a percentage of the pore space. Gas saturation, or natural gas content, of a petroleum reservoir may range from small quantities of gas dissolved in oil up to 100% of the petroleum content. The natural gas found in a reservoir may also occur either as free gas or as gas dissolved in water. Free-gas saturation is measured as a percentage of the pore space. *See* NATURAL GAS.

Terminology. Petroleum geology, like other fields of science, has special terminology. One term frequently used is pool, and is synonymous with the term reservoir. An oil pool is a single reservoir containing oil in a porous rock in the subsurface. It is not a large container full of oil as the term pool might imply. The term field has a broader meaning and is used when referring either to one reservoir or to a group of reservoirs. Oil and gas reservoirs or pools are frequently superimposed in several stratigraphic horizons separated by impermeable strata. In the group, each separate reservoir may be called a pool, and the group may be called a field. A petroleum province is a region in which oil and gas pools or fields occur. The fields within a region often have similar geologic characteristics. An example of a petroleum province is the Salt Dome Province of the Louisiana-Texas Gulf Coast.

SURFACE OCCURRENCES

Petroleum deposits encountered on the surface are often outcrops of former subsurface deposits which have been exposed by erosion; others are leakages of hydrocarbons from subsurface accumulations. In some instances, particles of hydrocarbons (that is, heavy oil, asphalt, or tar) have been mixed with sediments and laid down as a primary part of the deposit. The hydrocarbons in surface deposits are normally heavy tars and asphalts which are residues from normal subsurface oils that have been altered by the escape of volatile fractions and oxidation.

PETROLEUM GEOLOGY

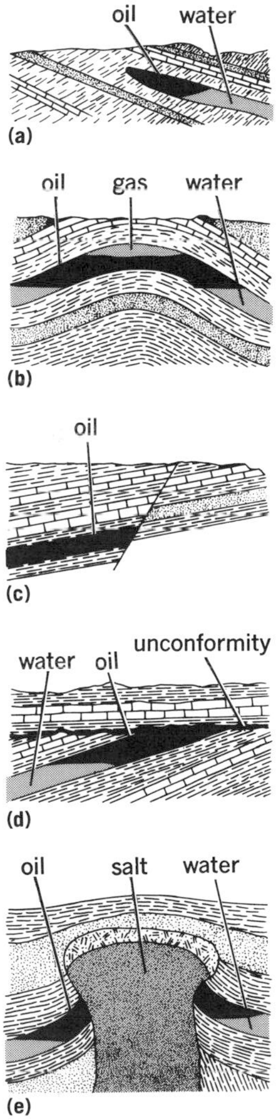

oil water

(a)

oil gas water

(b)

oil

(c)

water oil unconformity

(d)

oil salt water

(e)

Typical petroleum traps. (*a*) Lens. (*b*) Anticline. (*c*) Fault. (*d*) Unconformity. (*e*) Salt dome.

Hydrocarbon seeps. Petroleum may reach the surface along fractures, joints, fault planes, unconformities, or bedding planes. Most seepages (or springs) are formed by the slow escape of petroleum from accumulations that are close to the surface. Many oil fields and producing regions have been discovered by drilling near seepages.

Oil or tar sands. Most oil or tar sand deposits are former subsurface deposits which have been exhumed by erosion. A few are primary deposits in which particles of tar or asphalt have been deposited along with conventional sediments. Only a few surface deposits are economically important. One such deposit, the Athabasca oil sands of Alberta, Canada, contains an enormous reserve of heavy hydrocarbons and covers an area of several thousand square kilometers. The hydrocarbons may be recovered by mining the sands and then extracting the hydrocarbons. The extract, obtained from the sand-oil mixture by steam treatment, is processed to obtain usable petroleum products. *See* OIL SAND.

Oil shale. The world's largest known deposits of oil shale, located in the United States, are the Green River oil shales in Colorado, Utah, and Wyoming. Vast quantities of oil shale are exposed on the surface, with still greater quantities present at only modest depths of burial. The Green River oil shale, curiously, is not a true shale, nor does it contain oil. It is basically a dolomitic marlstone which has high organic content. Under destructive distillation the organic matter, mainly kerogen, will yield oil, gas, and a cokelike substance. Commercial exploitation of oil shale has been held back because the costs of mining and distillation have been very high. As these costs become competitive with costs of recovering conventional oil, oil shale will become an important source of energy. *See* OIL SHALE.

EXPLORATION TECHNIQUES

The methods used by the petroleum geologist fall into two categories, surface and subsurface. While surface geological studies were used almost exclusively in the early years of petroleum geology, most oil and gas exploration is now done with subsurface methods.

Surface methods. It is possible to gather and interpret data at the surface that indicate or suggest the presence of hydrocarbons below the surface. Some surface phenomena, as for example, oil seeps, are direct indications. The most common surface method is to construct maps through aerial photos and physically surveying the terrain. The objective is to find surface indications of subsurface structures which may trap oil or gas. Such indications can be gathered from rock outcrops, stream patterns, and even subtle variations in the character of vegetation. Surface methods are still important in petroleum geology, but they are no longer used as extensively as in the early days of exploration when untested surface features were plentiful.

Subsurface methods. Subsurface geology is the assembly and interpretation of subsurface data. The primary sources of data are the holes drilled to explore for oil and gas. Information from drill holes is carefully recorded by gathering cuttings and core samples, by recording drilling rates, and by wire-line well logs which measure various physical properties of the penetrated rocks. These data together are used to construct maps which depict the structure and distribution of potential reservoir rocks in the subsurface. A good interpretation of the data, along with a fair share of good luck, will provide maps which lead to the discovery of oil and gas. *See* PETROLEUM; PETROLEUM ENGINEERING; PETROLEUM PROSPECTING. [F. E. FOSS]

Bibliography: J. W. Amyx, D. M. Bass, and R. L. Whiting, *Petroleum Reservoir Engineering*, 1960; M. P. Billings, *Structural Geology*, 3d ed., 1972; K. K. Landes, *Petroleum Geology*, 2d ed., 1959; F. J. Pettijohn, *Sedimentary Rocks*, 3d ed., 1975.

Petroleum processing

The recovery and processing of various usable fractions from the complex crude oils. The usable fractions include gasoline, jet fuel, kerosine, fuel oil, asphalt, lubricating oils, and many others.

The petroleum refining industry is one of the largest manufacturing industries. The distribution of refining capacity among regions of the world is shown in Table 1. Almost $9,000,000,000 was spent in 1978 for materials and labor to place new facilities within these refineries. An additional $3,000,000,000 was spent to maintain and modernize existing facilities.

Refineries in the United States produced 7,200,000 barrels per day (bpd) or 1.14×10^9 liters/day of gasoline in 1978. In addition, the following products were produced in the quantities shown: middle distillate (including jet fuel, diesel oil fuels, and others), 4,280,000 bpd (0.68×10^9 liters/day); residual fuel oil (for heating purposes), 1,670,000 bpd (0.27×10^9 liters/day); all others (such as waxes, lubricating oils, asphalt, coke, petrochemi-

Table 1. Refining capacity in the world, Jan. 1, 1979*

Region	Crude oil charge, 10^6 bpcd†
United States	17.44
Other Western Hemisphere countries	11.18
Western Europe	20.96
Soviet Union and Eastern Europe	14.00
Africa	1.73
Middle East	3.52
Asia and Far East	10.40
World total	79.23

*From British Petroleum Company Ltd., *BP Statistical Review of the World Oil Industry*, 1978, except for the United States (see Table2).

†Barrels per calendar day. 1 bpcd = 159 liters/calendar day.

Table 2. United States processing capacity, Jan. 1, 1979*

Process type	Charge capacity, 10^6 bpcd†
Crude distillation	17.44
Catalytic cracking	4.87
Hydrocracking	0.79
Thermal cracking	0.45
Coking	1.00
Catalytic reforming	3.65
Alkylation	0.84‡

*From: U.S. Department of Energy, *Petroleum refineries in the United States and U.S. Territories, January 1, 1979*, June 28, 1979.

†1 bpcd = 159 liters/calendar day.

‡Alkylation given in product capacity.

Table 3. Some chemical compounds found in gasoline*

Name	Formula	Molecular weight	API gravity	Normal boiling point, °F (°C)	Research blending octane number
n-Pentane	C_5H_{12}	72	92.7	97 (36)	62
n-Hexane	C_6H_{14}	86	81.6	158 (70)	19
n-Heptane	C_7H_{16}	100	74.1	209 (98)	0
n-Octane	C_8H_{18}	114	68.7	260 (127)	−18
n-Nonane	C_9H_{20}	128	64.6	310 (154)	−18
n-Decane	$C_{10}H_{22}$	142	61.3	343 (173)	−41
n-Endecane	$C_{11}H_{24}$	156	58.0	387 (197)	−55

*Only the straight-chain paraffin hydrocarbons are shown here to indicate the range. Actually the gasoline contains also branched-chain paraffins, alkenes, naphthenes, aromatics, and other compounds with higher octane numbers.

cal feedstocks), 2,400,000 bpd (0.38×10^9 liters/day).

Crude oil is a mixture of many different hydrocarbon compounds of the paraffin type (wax compounds) and of the naphthene type (asphalt compounds), making the chemistry of petroleum refining extremely complex. The refining processes can be grouped under three main headings: (1) separating the crude oils to isolate the desired products; (2) breaking the remaining large chemical compounds into smaller chemical compounds by cracking; (3) building desired product properties by chemical reactions, such as reforming, alkylation, and isomerization. The capacities for some of these downstream processes are compared to the total crude distillation for United States refining in Table 2.

Refinery products, such as gasoline, kerosine, diesel oil, and others, are not pure chemical compounds but mixtures of chemical compounds. Some of the hydrocarbon compounds contained in gasoline are shown in Table 3, along with the individual specific gravities, molecular weights, and normal boiling points.

A simplified flow sheet of refinery operations is shown in Fig. 1. By means of distillation a typical crude oil may be separated quite easily into many fractions of raw products; see Table 4.

A more complex flow sheet of a refinery for light oils is shown in Fig. 2. Here are included the cracking equipment, reforming equipment, extraction unit, alkylation unit, and other facilities. Figure 3 is a schematic diagram of a refinery for producing lubricating oils.

Separating the crude oil. There are two principal separating procedures: topping of crude oil, and lubricating oil processing. Both of these procedures include combinations of several operations, such as distillation, treating, and blending.

Topping, or distilling, the crude oil. The crude oil is desalted and dehydrated, then passed through heaters where the temperature is raised to

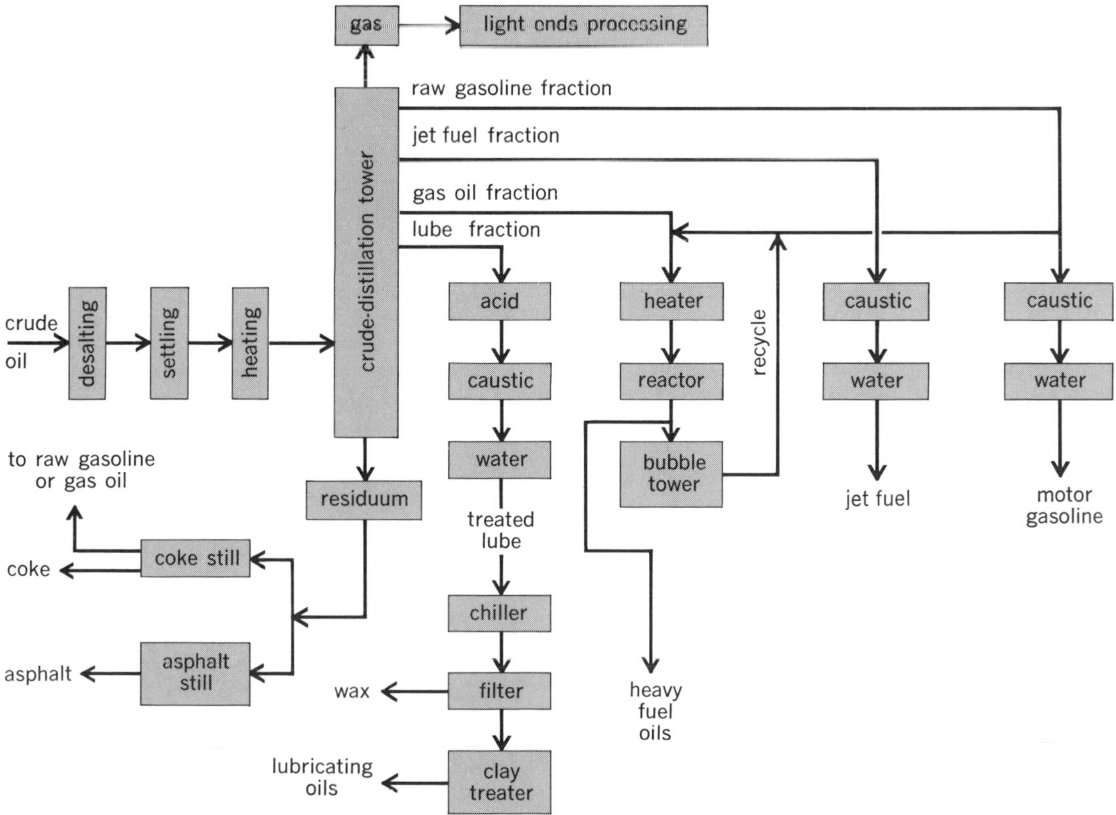

Fig. 1. Simplified flow sheet of crude oil refining.

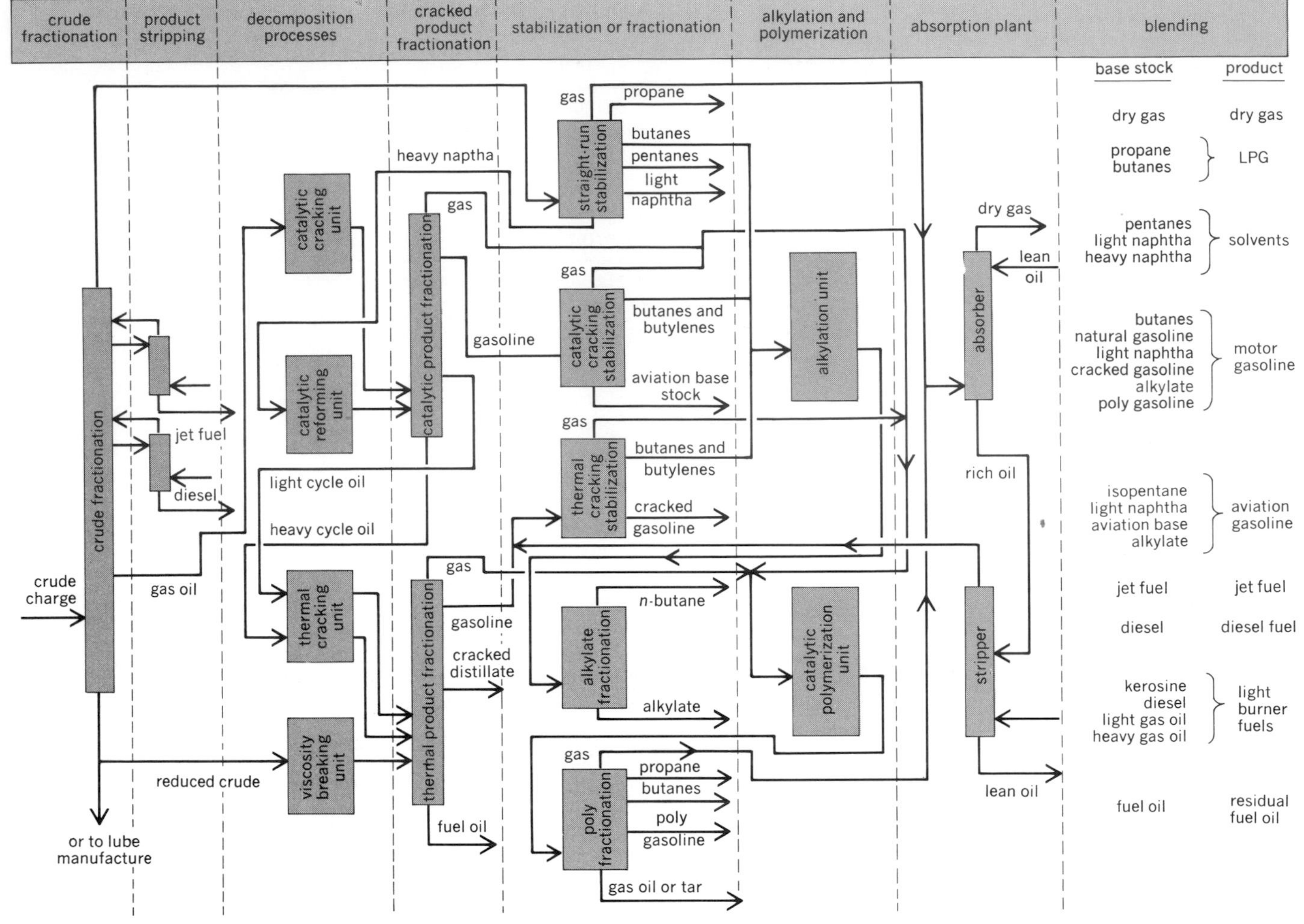

Fig. 2. Refinery for light oils (mainly gasolines, kerosine, and distillates).

about 650°F (343°C), at which temperature all of the gas, gasoline, jet fuel, and light fuel oil fractions are in the vapor phase. This vapor and liquid mixture enters a large distillation tower, about one-third the distance up from the bottom (Figs. 1 and 2). Into the bottom of the tower about 1–2 lb of steam per gallon of crude oil (0.12–0.24 kg/liter) is usually introduced to make the separation easier. From the top of the tower some gases are evolved and sent to units which process light ends. The next higher-boiling fraction is the gasoline, followed successively by the jet fuel, the gas oil, the cracking stock, and the lubricating distillate. Below the feed entrance a fraction called the residuum is removed.

The temperature of the feed to the tower depends considerably upon the ultimate plans for the residual oil. If this residual oil is to be processed further for the manufacture of lubricating oils, the feed is not heated to so high a temperature.

Each of the streams from the distillation unit must be treated further before it can be sold. The gasoline fraction is treated, then blended with other stocks. Finally, chemicals called additives are added to the blend to improve its properties.

Lubricating oil processing. The most important property of lubricating oil is its viscosity. The lube fraction produced in the vacuum distillation column contains some hydrocarbons that give the oil a poor viscosity-temperature characteristic. In addition, the lube oil fraction has poor oxidation resistance and contains wax and other impurities which must be removed. Consequently, the lubricating oil fraction must be treated to remove or to reduce the concentrations of the following: free-carbon–forming material, low viscosity-index materials, wax, unstable compounds which may decompose to form asphaltic substances or coke, and chemicals that affect the color of the lube oil products.

The flow sheet shown in Fig. 3 describes a process for the production of lubricating oils. Not only the lubricating fractions but also a portion of the residuum fraction is used to make the lubricating oils. In this case, the residuum is treated with a solvent to remove the asphaltic material. The deas-

Table 4. Some fractions obtained from crude oil

Fraction	Carbon atoms	Molecular weight	API gravity	Boiling range, °F (°C)
Gas	1–4	16–58		−259 to 31 (−162 to −1)
Gasoline	5–12	72–170	58–62	31–400 (−1 to 204)
Jet fuel	10–16	156–226	40–46	356–525 (180–274)
Gas oil	15–22	212–294	34–38	500–700 (260–371)
Lube oil	19–35	268–492	24–30	640–875 (338–468)
Residuum	36–90	492–1262	8–18	875+ (468+)

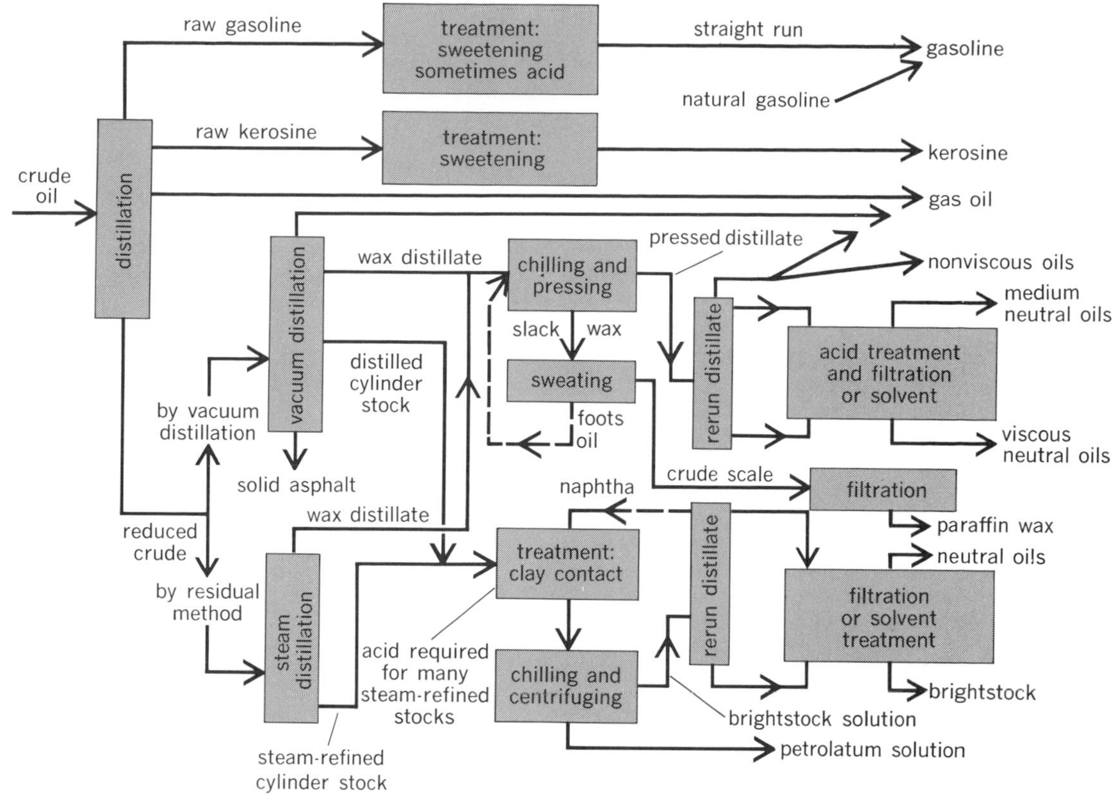

Fig. 3. Refinery for lubricating oils. (*From W. L. Nelson, Petroleum Refinery Engineering, 4th ed., McGraw-Hill, 1958*)

phalted residuum is further extracted along with the other lubricating oil fractions, dewaxed, acid-treated, clay-treated, blended with additives, and then sent to storage.

The solvents used for extraction include furfural, cresylic acid, phenol, sulfur dioxide, chlorex, nitrobenzene, propane, benzene, and many others. Since the solvent must be removed from the oils after the extraction, elaborate distillation equipment is required. The oils are freed from the final traces of solvent by steam stripping or vacuum flashing. Quite often the solvent is more expensive than the oil being treated so that, from the standpoint of economy alone, all of the solvent must be recovered. More important, however, the solvent itself may have properties which are detrimental to the finished oil when they are present in trace amounts.

Distillation. All distillation processes are essentially the same. The factors to be considered for different types of distillation processes include the sensitivity of liquid with respect to heat, the specifications of the product, and the boiling range of the feed.

In topping or skimming procedures, the crude is heated to a certain temperature and fed to a distillation tower where the product fractions are removed at various heights along the column. Figure 1 includes a schematic diagram of such a separation.

Stabilization is the distillation process that removes the lighter hydrocarbons (usually the dissolved gaseous hydrocarbons) from the particular fraction being processed. Here the feed is heated and sent to a fractionation column, where gases are removed overhead and the stabilized product at the bottom. In natural gasoline stabilizers, 40–60 plates are required in the distillation column to remove the dissolved propane and lighter hydrocarbons.

Steam distillation is used to increase the amount of distilled products obtainable at a fixed feed temperature. The feed stock is heated to approximately 550–660°F (288–349°C) in the presence of a large amount of steam. The effect of steam is to reduce the boiling point by partial pressure effects. The boiling point of a material can be reduced either by reducing the total pressure or by adding an inert gas such that the same total pressure will be partially due to the inert gas.

Vacuum distillation is used for additional separation of the crude oil residue, lube stock, and other fractions. Lubricating oil, for example, is thermally sensitive and partially decomposes if exposed to high temperatures. Therefore the distillation is done under a high vacuum to take advantage of the lower temperatures required at the lower operating pressures. Sometimes high vacuum is not sufficient; it is then necessary to combine vacuum distillation with steam distillation in a combination unit. In this case, steam is added to the distillation column operating under the vacuum. The amount of steam required will vary, of course, but may be as high as 1–2 lb/gal (0.12–0.24 kg/liter) of oil processed. The dry vacuum distillation processes have the advantages that smaller towers and smaller condensing equipment are required for a given throughput.

Filtration. This is also an important operation in the refining of petroleum. Regular gravity-type settlers are used wherever possible, but occasionally the solids are too finely divided to settle. There are many types of filters which are used for the removal of finely divided clay from treated stocks in the clay-contact process. These filters are classified as filter presses, leaf filters, rotary filters, and others.

Breaking the large molecules. The major product from the refinery is motor fuel (gasoline). Of course kerosine, diesel oils, jet fuels (mostly kerosine fraction), and others are extremely important also. However, each barrel of oil charged to the distillation tower has a given fraction of gasoline. This varies, but on the average is not over 20% of the total volume of crude. If more gasoline than this 20% obtainable by distillation is desired, and it almost always is, it is necessary to resort to other means than straight distillation to obtain it. This can be done either by recombining the gaseous, or lighter, molecules (alkylation), or by breaking down the heavier molecules (cracking).

Table 3 shows that gasoline molecules seldom contain more than 11–12 atoms of carbon. The crude oil, however, contains many molecules consisting of more than 50–60 atoms of carbon. The heavy naphtha fraction and the jet fuel and gas oil fractions, for example, all contain large molecules compared with the gasoline fraction (Table 4). In order to use these fractions for gasoline production, the long or large molecules must be broken into smaller ones of the gasoline type. This process is called cracking.

The cracking may be done either by thermal means (maintaining the heavy fractions at high temperatures) or by catalytic means. In thermal cracking the charge stocks are usually light and heavy gas oils, residual oils, or any of the topping column fractions heavier than the gasoline fraction. The resulting gasoline yields depend upon the composition of the charge stock but ranges 15–40% by volume of gasoline (100–400°F or 38–204°C boiling range).

In catalytic cracking the fraction to be cracked is contacted with a catalyst under lower pressure conditions than in thermal cracking, although the temperatures are still almost as high. Catalytic cracking gives much better yields of gasoline, lower carbon formation, and a gasoline of much higher octane number. The use of a newer zeolitic-type catalyst gives even better yields of gasoline.

About 80% of the cracking capacity in use in the United States today is of the catalytic type. *See* CRACKING; HYDROCRACKING.

Rebuilding desired chemical compounds. The saturated straight-chain paraffins shown in Table 3 have very low octane numbers. These compounds can be altered, however, by chemical reaction to yield a different kind of molecule with much higher characteristics. In general, the straight octane characteristics. In general, the straight paraffin compounds have the lowest octane rating and the aromatic compounds (benzene family) have the highest. The olefins and the naphthenes have intermediate octane numbers.

Some of the forms of a six-carbon-atom hydrocarbon and their research octane numbers (RON) are shown here. All these forms, and many others, are found in the gasoline fraction. It is possible to convert hexane (RON = 24.8) into benzene, which has an octane number of over 100.

Among the many processes that are used for altering the chemical structure of the molecules are the following:

1. Hydrogenation is used mostly for producing saturated hydrocarbons from unsaturated ones. During World War II, this process was used for making isooctane from isooctene. Later this process was used almost exclusively for desulfurization processes.

2. Dehydrogenation is the removal of hydrogen from a molecule. For example, 1-hexene may be made from *n*-hexane by removing hydrogen. This reaction often results in increased octane number.

3. Aromatization yields aromatic type hydrocarbons from other types, as benzene from hexane or cyclohexane. Aromatization and isomerization predominate in the reforming operation.

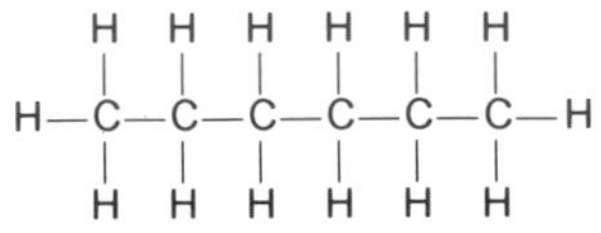

n-Hexane, C_6H_{14} (straight-chain paraffin), RON = 24.8

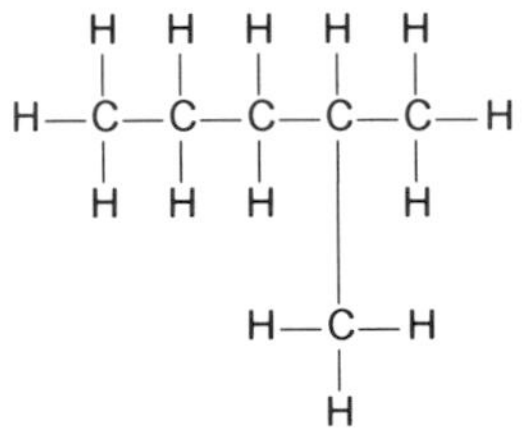

2-Methylpentane or isohexane, C_6H_{14} (branched paraffin or isoparaffin), RON = 73

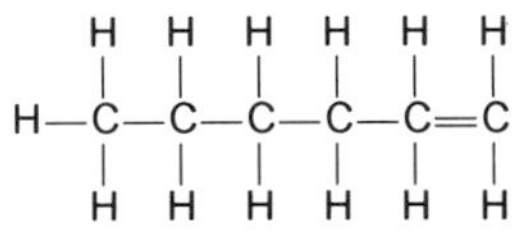

1-Hexene, C_6H_{12} (olefin or alkene), RON = 80

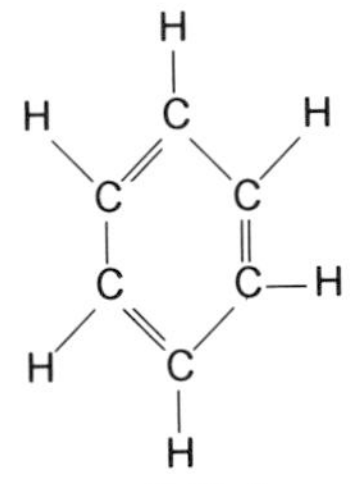

Benzene, C_6H_6 (aromatic), RON = over 100

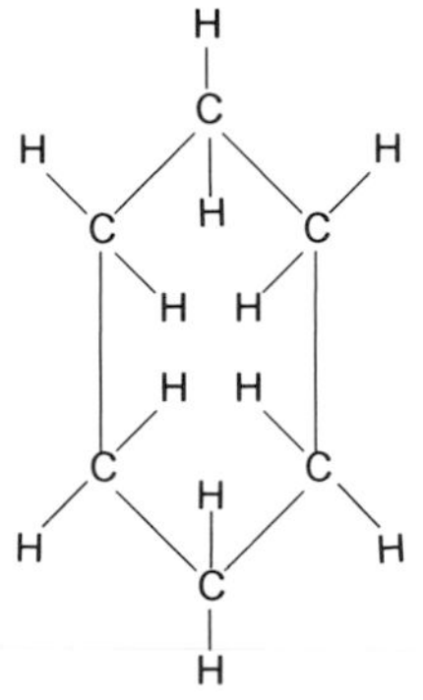

Cyclohexane, C_6H_{12} (naphthene), RON = 83

4. Cyclization is the transformation of a hydrocarbon of the chain type to one of the ring type; for example, making cyclohexane from *n*-hexane.

5. Isomerization is the rearrangement of the atoms in a molecule, such as *n*-hexane, to form an isomer, such as isohexane.

6. Polymerization involves two or more molecules in a building process. For example, as shown by Eq. (1), propene and the butenes, which are

$$\underset{\text{Propene}}{CH_3{=}CHCH_2} + \underset{\text{1-Butene}}{CH_3CH_2CH{=}CH_2} \xrightarrow[\text{catalyst}]{\text{Heat or}} \underset{\text{5-Methyl-1-hexene}}{CH_2{=}CHCH_2CH_2\overset{\overset{\large CH_3}{|}}{C}HCH_3} \quad (1)$$

present in the gases of thermal- or catalytic-cracking operations, are polymerized to form a larger liquid molecule with a high octane number. The catalytic polymerization capacity for making gasoline in the United States had been dropping, but is finding renewed interest.

7. Alkylation also makes use of two or more molecules in the reaction. This process uses an isoparaffin, such as isobutane, and an olefin, such as ethylene, to yield a larger molecule with a high octane number, 2,2-dimethylbutane, as shown by Eq. (2). This reaction uses only one-half as many of

$$\underset{\text{Ethylene}}{CH_2{=}CH_2} + \underset{\text{Isobutane}}{CH_3\underset{\underset{\large CH_3}{|}}{C}HCH_3} \xrightarrow[\text{catalyst}]{\text{Heat or}} \underset{\text{2,2-Dimethylbutane}}{CH_3\overset{\overset{\large CH_3}{|}}{\underset{\underset{\large CH_3}{|}}{C}}CH_2CH_3} \quad (2)$$

the expensive olefin molecules as the polymerization process. By the beginning of 1979 the alkylation capacity in the United States was about 840,000 bpd (0.13×10^9 liters/day). This indicates that alkylation is replacing polymerization as a means of making gasoline from gaseous feedstocks. *See* ALKYLATION (PETROLEUM); OCTANE NUMBER; REFORMING IN PETROLEUM REFINING.

Treating processes. Both the crude oil and the petroleum products must, on occasion, be treated to remove undesirable impurities or to improve the properties of the product. The important treating processes are desalting and dehydrating, sweetening and desulfurization, acid treatment, clay-contact adsorption treatment, vapor-phase treatment, and solvent treatment. Some of these processes are used both on the crude oil and on the products, whereas others are used on only one or the other.

Desalting and dehydration of the crude oil. The salt content of the crude oil which enters the refinery may be as high as 4 or 5%, and the water content may be much higher than the equilibrium amount because water is present as an emulsion.

Because of the high temperatures of the heater tubes, the introduction of the wet crude into the heaters would be dangerous. In addition, the salt would precipitate onto the tube walls, reducing the rate of heat transmission and thereby the efficiency of the heaters.

Many processes are available for the removal of both the salt and the water from the crude oil. These are grouped into four types as shown in Table 5.

The crude oil (containing the salt and the oil) is heated, an emulsion breaker is added, and the resultant mass is settled, or even filtered, to remove the salt and water phase from the oil phase.

Sweetening and desulfurization. Since the original crude oils contain some sulfur compounds, the resulting gasolines and other products also contain sulfur compounds, including hydrogen sulfide, mercaptans, sulfides, disulfides, and thiophenes.

The processes used to sweeten, or desulfurize, the products depend upon the type of sulfur compounds present and the specifications of the finished gasoline or other stocks.

Mercaptans are removed or converted into less undesirable disulfides in the following ways:

1. Mercaptan removal: (*a*) unisol process using alkaline solution of methyl alcohol; (*b*) solutizer processes using sodium hydroxide along with minute amounts of sodium isobutyrate; and (*c*) mercapsol process using an alkaline solution of naphthenic acids and phenols. These are regenerative solution processes.

2. Mercaptan conversion (oxidation to disulfides): (*a*) lead sulfide doctor sweetening; (*b*) copper chloride–oxygen sweetening; (*c*) sodium hypochlorite sweetening; (*d*) oxygen sweetening with chelated cobalt catalyst in either a caustic solution or fixed bed.

3. Hydrogen sulfide removal by regenerative solution processes using aqueous solutions of (*a*) sodium hydroxide, (*b*) calcium hydroxide, (*c*) trisodium phosphate, and (*d*) sodium carbonate.

Hydrotreating is the most widely practiced treating process for all types of petroleum products, whether fuels or lubricants, which is in use in the world today. Total hydrotreating capacity for all products in the United States was in excess of 7,000,000 bpd (1,100,000 m^3/day) in 1978. The process, through the selection of the appropriate catalysts and operating conditions, is used to achieve desulfurization, eliminate other undesirable impurities such as nitrogen and oxygen, decolorize and stabilize products, correct odor problems, and improve many other product deficiencies. Fuel products treated range from naphthas to heavy burner fuels. Specialty products such as lubes and waxes and solvents are also treated to improve various characteristics.

Solvent treatment in petroleum refining. Undesired constituents may also be removed by selective solvent extraction. In this case a liquid that will selectively dissolve the undesired constituents is added to the oil. The solvent processes may be

Table 5. Desalting and dehydrating methods

Method	Temperature, °F (°C)	Type of treatment
Chemical separation	140–210 (60–99)	0.05–4% solution of soap in water 0.5–5% solution of soda ash in water
Electrical separation	150–200 (66–93)	10,000–20,000 volts
Gravity separation	180–200 (82–93)	Up to 40% water added
Centrifugal separation	180–200 (82–93)	Up to 20% water added (sometimes no water added)

divided into two main categories, solvent extraction and solvent dewaxing.

The solvents used in the extraction processes include propane and cresylic acids, 2,2′-dichlorodiethyl ether, phenol, furfural, sulfur dioxide, benzene, and nitrobenzene.

In the dewaxing process, the principal solvents are benzene, methyl ethyl ketone, methyl isobutyl ketone, propane, petroleum naphtha, ethylene dichloride, methylene chloride, and sulfur dioxide.

Before the solvent-extraction processes were developed, only a few types of crudes were considered to be good lubricating oil crudes. By using these solvent processes, the original properties of the crudes can be changed so greatly that almost any crude will make good lubricating oils.

The early developments of solvent processing were concerned with the lubricating oil end of the crude. Solvent-extraction processes are being applied to many useful separations in the purifications of gasoline, kerosine, diesel fuel, and others. In addition, solvent extraction may replace fractionation in many separation processes in the refinery. For example, propane deasphalting has replaced, to some extent, vacuum distillation as a means of removing asphalt from reduced crudes. *See* OIL ANALYSIS; PETROLEUM; PETROLEUM PRODUCTS.

[JOHN J. MC KETTA; HAROLD L. HOFFMAN]

Bibliography: *Hydrocarbon Processing Process Developments*, published September, odd-numbered years; *Hydrocarbon Processing Process Handbook*, published September, even-numbered years; W. L. Nelson, *Petroleum Refinery Engineering*, 4th ed., 1958.

Petroleum products

Crude petroleum is the starting material not only for the fuels used for transportation and energy production but also for petrochemical feedstocks, solvents, lubricants, asphalts, and many other specialities (see table). The more complex the family of petroleum products, the more energy is needed to refine the crude. A simple refinery consumes only 5% of the energy in the crude; to make the quality and types of products to satisfy today's market, however, only about 90% of the input energy emerges in products.

Fuels for transportation. Liquid products such as gasoline, jet fuel, diesel fuel, and marine fuel, which together account for more than half the volume of output, serve the transportation industry. Hydrocarbons from butane to C_{12} appear in gasoline blends, which still constitute the most important product from petroleum, about 40% of crude in the United States. Most specialized energy-consuming process units in refineries, such as catalytic reformers and cracking units, exist to convert crude fractions to gasoline of a quality that meets the needs of the modern automobile engine. Added to the usual engine requirements for antiknock performance and volatility of fuel are environmental demands for emission standards in exhaust. Concern about air pollution has required the introduction of unleaded fuel for cars equipped with catalytic mufflers and, in turn, has made necessary the increased use of aromatic hydrocarbons for antiknock performance. *See* AIR POLLUTION; CRACKING; GASOLINE.

Aviation gasoline, once the leading military and air-transport fuel, is now serving general aviation. Several grades of various antiknock and lead levels are blended from alkylate, pentane, aromatics, and naphthas to the exacting requirements of high-performance aircraft piston engines. Higher-boiling-fraction C_8 to C_{15} hydrocarbons are the blend stocks for jet fuels for the air-transport industry. The U.S. Air Force uses JP-4, a wide-cut naphtha-kerosine blend, and the U.S. Navy uses a kerosine called JP-5. Kerosine used as a jet fuel for the world's airlines accounts for about 6% of crude. *See* JET FUEL.

Diesel fuels are blended from both distilled and cracked fractions—up to C_{22}-type hydrocarbons. They range in volatility from kerosine for lightweight automotive diesel engines to gas oil for large, lower-speed industrial or marine diesel engines, and account for 10% of the crude. Diesel fuels must have a controlled viscosity and must exhibit good combustion performance. Low-temperature properties are important for diesel fuels used in railroads, trucks, and buses that are operated in climatic extremes. *See* DIESEL FUEL.

Marine fuels are made from the highest-boiling crude fraction—the residuum from crude distillation blended with heavy cracked gas oils. In a ship equipped with steam turbines, marine fuel is burned in a boiler to generate steam, just as in an electric power generating station. But the same fuel can be burned directly in a ship's gas turbine engine, provided it meets certain requirements in viscosity and trace metals.

Fuels for energy production. Petroleum fuels heat homes, factories, and offices and also generate electric power. Home-heating oils are similar to diesel fuels in boiling range but must exhibit storage stability and good atomization and combustion performance in small oil burners. Like diesel fuels, they must have a high flash point to ensure safety in handling, and must also display good low-temperature flow characteristics. Their viscosity ranges from that of kerosines, for simple vaporizing-type heating units, to heavy gas oils, for large burners used in the heating plants of large buildings. But the residual oils delivered to the public utilities to generate electricity constitute the largest volume of energy-producing fuels. In this application, petroleum competes with gas, coal, and nuclear power. Heavy fuel oil must be heated to reduce its viscosity and to atomize it successfully in the fireboxes of the boilers that generate steam. Today this heavy fuel must frequently conform to strict limits on sulfur content in order for the power station to meet air-pollution standards on stack gases. The energy-producing fuels account for about 30% of the crude. *See* FUEL OIL.

Nonfuel products. From each of the fractions distilled from crude or its cracked or processed products, valuable and indispensable nonfuel materials are made. Some of the major products are:

Ethylene, propylene, butylene, and other reactive gases for the petrochemical industry's output of polymers, rubbers, chemicals, textiles, and films.

Liquefied petroleum gases for heating, cooking, and drying. *See* LIQUEFIED PETROLEUM GAS (LPG).

Solvents for the paint and dry-cleaning industries and as vehicles for aerosol products such as insecticides.

Lubricants of all types, from light spindle oils to

Crude petroleum and some of its products*

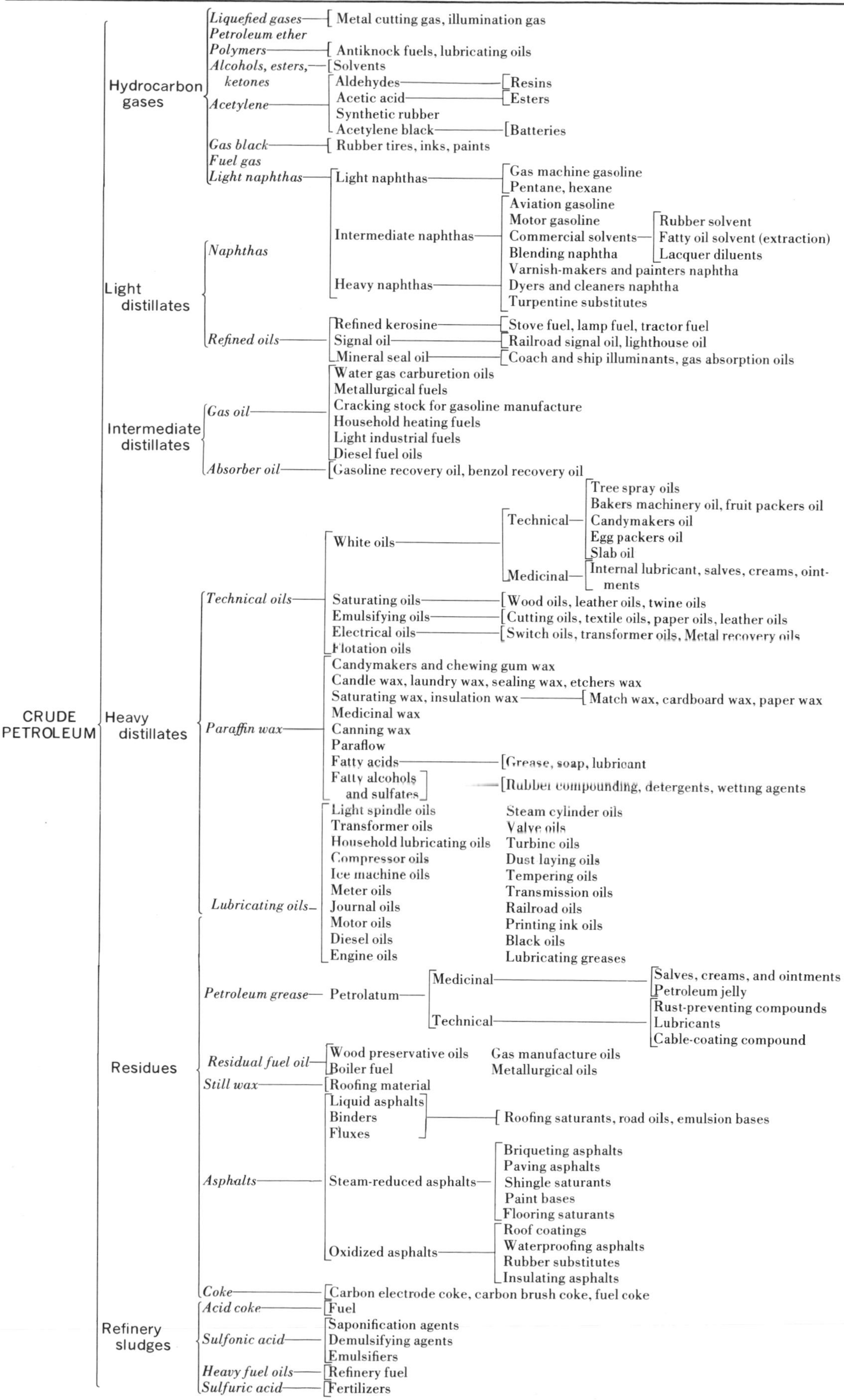

*From P. Albert Washer, Texas A. and M. College Extension Division (First Session).

heavy turbine oils and greases. Engine and machine lubricants are compounded in a wide range of viscosities from both paraffin- and napthene-base stocks.

Specialty oils for hydraulic fluids, transformers, emulsions, cutting fluids, pharmaceuticals, inks, preservatives, and so on.

Wax for a host of applications, from candles, medicines, coatings, and compounding to petrochemical feedstocks.

The residues from processing crude can also yield such familiar products as asphalt, the major road-building material, and coke, burned as fuel or made into electrodes. Useful petroleum products, made directly or indirectly, number in the thousands. No part of the petroleum barrel is wasted, and great efforts are made to recover spills and slops, not only to minimize water pollution but to recover valuable fuel energy.

Crudes differ substantially in their characteristics as sources of valuable nonfuel petroleum products. Some are rich in asphaltenes and are segregated for asphalt manufacture. Others yield excellent lubricating oil feedstocks or high-quality wax. Manufacturing these nonfuel products from the best feedstocks requires an assortment of specialized processing methods such as asphalt oxidation, phenol extraction, hydrogenation, acid treating, dewaxing, and grease blending, and careful blending with additives such as antioxidants, viscosity improvers, rust inhibitors, dispersants, and other specific agents. The distinction between natural and synthetic petroleum products becomes blurred; for example, many synthetic lubricants originate from petrochemical-derived base stocks. Other petroleum products are fortified with chemical agents or compounded with synthetic materials to achieve the proper balance of properties. The amount of petroleum which is not consumed but is turned into valuable and indispensable nonfuel products is about 5% of the supply of crude. *See* PETROLEUM; PETROLEUM PROCESSING.

[W. G. DUKEK]

Petroleum prospecting

The search for commercially valuable accumulations of petroleum. This search at the one extreme may be carried out in a completely haphazard manner with entire dependence on luck for success or, at the other extreme, it may be a highly organized procedure involving the use of complex precision instruments, skilled and experienced personnel, and advanced scientific reasoning. In either case the final and critical step is always the drilling of an exploratory hole. Moreover, in neither case can the successful outcome of the exploratory hole be assured in advance because no infallible means of detecting the presence of a commercial petroleum accumulation ahead of the drill has yet been devised. Much petroleum has been found both by luck and by the application of scientific methods, but statistics demonstrate that at the present time the success ratio of holes located with the benefit of scientific or technical advice is nearly twice as great as that of those located without such advice.

The classic requisities for petroleum accumulations are (1) source or mother rocks from which petroleum can have originated; (2) carrier and reservoir rocks possessing sufficient permeability to provide avenues of migration as well as sufficient porosity to provide storage space; (3) traps adequate to cause commercial concentration of petroleum at local points in the reservoir beds; and (4) proper time and spatial relations in the development of source, reservoir, and trap. A favorable hydrodynamic condition might also be mentioned as a requisite to initial accumulation as well as to later preservation of a petroleum deposit. *See* PETROLEUM RESERVOIR ENGINEERING.

Scientific petroleum prospecting consists of (1) determination of generally favorable regions with respect to source, reservoir, trap, timing, and hydrodynamic conditions; (2) finding local geological features (anticlines, fault traps, and pinch-outs) within these regions, believed to be suited to the trapping of petroleum; (3) location and programming of exploratory holes to test the presence or absence of commercially significant petroleum accumulations on these local features; and (4) after initial discovery, determination of the extent and character of the accumulation.

Prospecting methods are the means employed to gain the information called for in these four steps of petroleum prospecting. Prospecting methods are commonly classified as geological and geophysical, but there is no sharp distinction between the two; all involve geological reasoning and interpretation.

This article outlines two major aspects of petroleum prospecting—geological prospecting and geophysical prospecting.

GEOLOGICAL PETROLEUM PROSPECTING

Nearly all prospecting entails certain preliminary library and cartographic background research. Some mention of base maps is followed in this section by the topics of: surface geology; photogeology; drilling; structure and core drilling; wild-cat wells; subsurface geology; geological laboratory methods; and regional geology.

Base maps. A requisite to petroleum prospecting is accurate base-map control. Horizontal control is necessary for location of property boundaries, physical features, roads, wells, and other cultural features, and for the map location of the points from which geological or geophysical data are obtained. Vertical control is necessary for providing topographic information for operational purposes as well as for adjustment of geological, geophysical, and well data to a common datum. Topography may also be of important geological significance.

Aerial photography and electronic and radio positioning systems are largely replacing the theodolite, alidade, and plane table for mapping and geographic control work both on land and over water, and these methods have advantages both in speed and in accuracy.

Surface geology. The examination and study of outcropping rocks as a clue to the structure and stratigraphy of an area is the oldest of petroleum-prospecting methods and one which is still extremely important. The surface geologist maps the topographic expression and distribution of exposed rock units, determines and plots their structural attitude, measures and describes stratigraphic sections, identifies surface structural anomalies such as anticlines or faults which may reflect deeper structures, and prepares cross sections showing the hypothetical distribution of rocks and structure at depth. Study of the rocks exposed at the surface may yield important information on the presence and position of source and reservoir rocks as well as on structural and stratigraphic accumulation traps. Finally, surface geological examination is the only means of acquiring information on the occurrence and location of petroleum seepages. There is no more encouraging indication of a petroliferous province than the presence of actual oil seepages. In difficult terrain, helicopters commonly attached to surface geological parties aid transportation and communication and facilitate geological observation.

Photogeology. The mapping of surface geologic features is frequently best carried out through study of aerial photographs. In addition to greatly expediting the study of the surface geology of any area, the photographic method provides coverage of regions where access on the ground would be prohibitively difficult. It usually provides more complete detail than is possible by surface-mapping methods and has the additional advantage of greater overall perspective. In regions where outcrops are scarce, photogeology is employed as a means of determining geologic structure and distribution of formations indirectly through interpretation of geomorphologic features, vegetation, fracture patterns, and soil characters. Photogeology does not replace surface study, which is always desirable, but it does constitute an extremely valuable supplement.

Drilling. There is no method of petroleum prospecting so effective as the drilling of a hole to the objective horizon, and if the cost of deep drilling were not so great this method would supplant almost all others. Even so, the great bulk of all money spent on petroleum prospecting goes for drilling of exploratory holes. *See* OIL AND GAS WELL COMPLETION.

Structure drilling and core drilling. These terms are applied to relatively shallow drilling where the purpose is purely that of securing geological information. Highly portable drilling rigs with depth capacities of a few hundred to a few thousand feet are used, and on the basis of cuttings, cores, or electrical logs, information is obtained on near-surface stratigraphy and structure which may guide deeper drilling for petroleum.

Wildcat wells. Exploratory holes drilled with the aim of discovering new petroleum pools are true wildcat wells. Usually these are programmed and equipped for completion as producers if successful, but the so-called stratigraphic test hole, which penetrates potentially productive horizons, is aimed only at providing geological information and is not equipped for production. Even the drill is not always conclusive in prospecting for petroleum. Many potentially productive wells are abandoned as dry each year because of inefficient testing. Under current methods of drilling, the hole is usually kept filled with heavy mud to prevent caving and to hold back excessive fluid pressures; consequently, potentially productive petroleum horizons may frequently be penetrated by the bit with very little indication of their fluid content. To avoid overlooking such horizons, rock samples cut by the bit and brought up in the circulating mud are carefully and concurrently studied for petroleum indications; instrumental equipment is installed on the drilling rig to analyze the mud automatically for traces of petroleum; and electrical and radioactive devices are run down the hole from time to time to record the properties of the rocks penetrated with respect to the probability of their carrying petroleum. Likewise, cores and side-wall samples are taken from intervals suspected of being productive.

Subsurface geology. Regardless of whether production is obtained, an exploratory hole is usually a valuable contribution to prospecting knowledge. The study of the geological and geophysical data made available through drilling is called subsurface geology. The subsurface geologist stationed at the well constantly watches the cores, cuttings, and drilling fluid for direct traces of petroleum and studies the characteristics of electric logs, radioactive logs, and geothermal logs for indirect indications of petroleum. The lithology, paleontology, and mineralogy of the cores and cuttings also yield clues to the stratigraphic position at which the well is drilling and the remaining depth to objective producing horizons. Determination of the attitude of bedding from cores and from dip-meter surveys gives important evidence as to whether the well is off structure with respect to the fold, fault, or other trap structure on which it is being drilled and also gives a factor for correcting the drilled thickness of a formation to its true thickness. *See* PETROLEUM GEOLOGY.

As more wells are drilled in a region, the steadily increasing background of subsurface geological information becomes progressively more effective as a means of locating new structural or stratigraphic traps for testing. Correlation, the identification and tracing of stratigraphic units from one well to another, allows conclusions to be reached on the relative structural positions of wells, on the probable location of new fold and fault structures, on the presence of unconformities, and on lateral changes in thickness and lithology. These correlation data provide the subsurface geologist with the base for cross sections and various kinds of subsurface maps: structure-contour, isopach, paleogeologic, lithofacies, palinspastic, and others.

Geological laboratory methods. Many of the determinations which can usefully be made on rock samples, either from outcrops or from wells, require such specialized knowledge and equipment that the surface or subsurface geologist sends them to specialists in a geological laboratory. Paleontologic and micropaleontologic studies are valuable in determination of the age or stratigraphic position of samples, in correlation, and in determination of past environments of deposition which may bear on source, reservoir, and stratigraphic trap conclusions. Study of the Foraminifera and Ostracoda have been particularly useful in petroleum prospecting, and spore and pollen studies have recently been growing in importance.

Laboratory determination of heavy detrital minerals furnishes useful information for correlation and provenance. Among other laboratory methods which may be useful in identification and correlation are analysis for insoluble residues, size and shape analysis, differential thermal analysis, and calcimetry. Refractive-index determinations made on solvent extracts from rock samples provide useful information on the presence and gravity of even minute traces of oil. Computers and data-processing machines are being employed in some geological laboratories to aid in the sorting and analysis of large batches of data from surface geology and wells.

Regional geology. In petroleum prospecting the various contributions of surface geology, subsurface geology, geophysics, and other methods should all be put together and coordinated to give as complete a regional geologic picture as possible. Given adequate information on the character and attitude of the physical rock framework of a region, its geologic history, and the conditions of movement of its fluids, it should be theoretically possible to predict the location of all its petroleum accumulations. This information is of course never fully forthcoming, but the acquisition of as much of it as can be obtained and the imaginative but intelligent extrapolation of the remainder from experience are essential to long-range success in petroleum prospecting.

[HOLLIS D. HEDBERG]

GEOPHYSICAL PROSPECTING FOR PETROLEUM

Geophysical techniques have contributed decisively to the world supply of oil since about 1930. The seismograph technique has accounted for most of this activity. The seismic method has been the most expensive of those available. For this reason the cheaper gravity and magnetic methods have often been used for reconnaissance purposes, and the more limited anomalous areas thus revealed are then subjected to seismic investigation. Means and procedures have been developed which permit seismic operations in coastal waters at unit costs comparable to those of gravity and magnetic surveys under the same conditions. The rate of expenditure is very high for such operations, but the output rate is also high, so that acceptable unit costs are achieved. The seismic, gravity, magnetic, electrical, and various well-logging methods account for all but a tiny fraction of the geophysical work done in the search for oil.

The geophysical measurements made in oil prospecting are to a large extent related to the configuration and properties of the rocks which enclose oil pools. At best, the results of geophysical surveys indicate the presence, position, and nature of a structure which may or may not contain oil. The discovery of the oil itself is made by drilling a hole into the structure. If oil is found, the well serves the purposes of both discovery and exploitation.

Seismic surveys. Contour maps are generally produced to show the elevation of geological horizons with reference to some datum plane, preferably at the general depth level of formations which are known to produce oil elsewhere or which are thought to be potential reservoirs. The data furnish evidence of possible traps for petroleum accumulation, which must be identified and confirmed by drilling.

This method has progressed considerably in the transition to digital methods of processing the field data. Therefore, large masses of data can be analyzed, and use can be made of new communication theory, velocity filtering, and nondynamite sources. This enables exploration of the remaining more difficult geological provinces, for example, the coastal areas and the deep formations of West Texas and New Mexico, and detection of stratigraphic traps. Seismic surveys furnish the most conclusive evidence available from any geophysical technique. *See* SEISMIC EXPLORATION FOR OIL AND GAS.

Electrical methods. Electrical methods, except for drill hole surveys, have been used very little in prospecting for petroleum in the free world. The Soviet Union, however, is finding them useful, particularly for reconnaissance in unexploited areas.

Remote sensing. Improved air photography, using infrared and microwave, is being tested for reconnaissance work and surface indicators of deeper oil accumulations.

Magnetic surveys. In the search for oil, magnetic surveys are now made almost exclusively from aircraft. Oil is found in deep sedimentary basins, and the magnetic anomalies found in such areas arise from the igneous floor beneath the sediments. The depth below surface to the igneous basement rocks usually is thousands of feet. Aeromagnetic surveys have been made over millions of square miles.

The sedimentary structures which are oil-bearing often lie above uplifts or topographic features of the igneous basement surface. Local magnetic anomalies are associated with such features and therefore are a key to the discovery of basement uplifts. Other anomalies are related to differences in the magnetization of the igneous rocks. If anomalies are present in sufficient numbers and well distributed, the configuration of the sedimentary basin can be predicted and the principal structural features in the basement indicated in advance of any drilling.

In areas where the sedimentary structure arises mainly from thrusting (that is, force applied sidewise on rockbeds), magnetic surveys may be of little help.

Gravity surveys. These have been most successful in discovering and detailing salt domes, a great percentage of which have associated oil accumulation. A newly discovered salt dome therefore represents an oil prospect of high potential. A salt dome usually contains one or more cubic miles of salt, which is ordinarily of lower density than most of the surrounding sediments. A gravity minimum is therefore characteristic of a salt-dome structure.

The gravity manifestation of other structural types is generally more complex. If a structure involves the position of dense beds nearer to the surface, a gravity high will be found. Such anomalies are customarily investigated further by seismic techniques before drilling is undertaken.

A great percentage of the potential oil-producing areas of the United States has been covered by gravity surveys.

Well logging. Geophysical techniques of well logging are now applied to practically every well drilled by the oil industry. As in pregeophysical days, geologists prepare a graphical log of the formations through which a drill hole extends, based on visual examination of drill cuttings brought to the surface by the drilling mud and on core samples. Such logs show lithology and fossil distribution with depth. Structure maps result from correlations between well horizons which can be identified as being the same or substantially equivalent in all of them. Geophysical well logging is merely an extension of this procedure to other physical properties which require physical measurements for their determination.

Various geophysical well-logging methods have been developed, electrical, radioactive, acoustic, and gravity. *See* WELL LOGGING. [G. E. ARCHIE]

Bibliography: M. S. Bishop, *Subsurface Mapping*, 1960; M. B. Dobrin, *Introduction to Geophysical Prospecting*, 2d ed., 1960; J. D. Haun and L. W. LeRoy (eds.), *Subsurface Geology in Petroleum Exploration*, 1958; F. H. Lahee, *Field Geology*, 5th ed., 1952; K. K. Landes, *Petroleum Geology*, 2d ed., 1960; A. I. Levorsen, *Geology of Petroleum*, 2d ed., 1967; G. B. Moody (ed.), *Petroleum Exploration Handbook*, 1961.

Petroleum reserves

Proved reserves are the estimated quantities of crude oil liquids which with reasonable certainty can be recovered in future years from delineated reservoirs under existing economic and operating conditions. Thus, estimates of crude oil reserves do not include synthetic liquids which at some time in the future may be produced by converting coal or oil shale, nor do reserves include fluids which may be recovered following the future implementation of a supplementary or enhanced recovery scheme.

Indicated reserves are those quantities of petroleum which are believed to be recoverable by already implemented but unproved enhanced oil recovery processes or by the application of enhanced recovery processes to reservoirs similar to those in which such recovery processes have been proved to increase recovery.

Thus, crude oil reserves can be called upon in the future with a high degree of certainty, subject of course to the limitations placed on production rate by fluid flow within the reservoir and the capacity of the individual producing wells and surface facilities to handle the produced fluids. It is important to bear in the mind the distinction between resources and reserves. The former term refers to the total amount of oil that has been discovered in the subsurface, whereas the latter refers to the amount of oil that can be economically recovered in the future. The ratio of the ultimate recovery (the sum of currently proved reserves and past production) to the resource or original oil in place is the anticipated recovery efficiency. *See* PETROLEUM ENHANCED RECOVERY.

Levels. In earlier years crude oil reserves were estimated by first defining the volume of the resources from drilling data, the nature of the natural producing mechanism from the performance of the reservoir, particularly the rate of decline in productivity, and then applying a recovery factor based on analogy with similar reservoirs. Although more sophisticated technology is in use today, earlier rule-of-thumb estimates have proved to be surprisingly valid.

Reserves are increased by the discovery of new reservoirs, by additions to already discovered reservoirs by continued drilling, and by revisions due to a better-than-established anticipated performance or implementation of an enhanced recovery project. Reserves are decreased by production, and by negative revisions due to poorer-than-anticipated performance or less-than-projected reservoir volumes. *See* PETROLEUM RESERVOIR MODELS.

In the 15 years between 1954 and 1969, reserves of crude oil in the United States remained at a relatively stable plateau of $31 \pm 0.5 \times 10^9$ barrels ($4.93 \pm 0.08 \times 10^9$ m^3). In 1969 the reserves jumped to a value of 39×10^9 bbl (6.2×10^9 m^3) as a result of the discovery of the gigantic Prudhoe Bay oil field on the North Slope of the Brooks Range in Alaska. In the 1970s the reserves in the United States fell steadily, reaching at the end of 1978 the lowest level since 1952: 27.8×10^9 bbl (4.42×10^9 m^3).

Discovery and production. In retrospect, it can be seen that reserves in the United States were sustained during the 1950s and early 1960s as a result of extensions and revisions, the latter due primarily to the installation of waterfloods, rather than as a result of significant new discoveries.The discovery at Prudhoe Bay indicated the necessity of exploration in new frontiers if new reserves were to be added to the United States total. Subsequent exploration failures in other frontier areas, such as the eastern Gulf of Mexico, the Gulf of Alaska, and the Baltimore Canyon, however, again emphasized the limited occurrence of crude oil in the Earth's crust.

The existence of an accumulation of petroleum fluids in the Earth's crust represents the fortuitous sequence of several natural events: the concentration of a large amount of organic material in ancient sediments, sufficient burial and thermal history to cause conversion to mobile fluids, and the the migration and eventual confinement of the fluids in a subsurface geological trap. Although some oil and gas have been produced in 33 of the 50 states, 4 states account for 85% of the proved reserves: Texas and Alaska account for over 60%, with California and Louisiana ranking third and fourth. Further evidence for the spotty concentration of oil in the crust is gleaned from the fact that although there are some 10,000 producing oil fields in the United States, some 60 of them account for over 40% of the productive capacity.

In this context, it was to be anticipated that oil discovery and production would reach a peak followed by a monotonic decline in reserves and productive capacity. Only the date of peaking might be in doubt, but even this was predicted with a high degree of accuracy at least a decade before it occurred in the late 1960s in the United States. Additional oil discoveries will be made in some of America's remaining new frontiers, such as the deep ocean and in the vicinity of the Arctic Circle; however, most estimates of the amount of

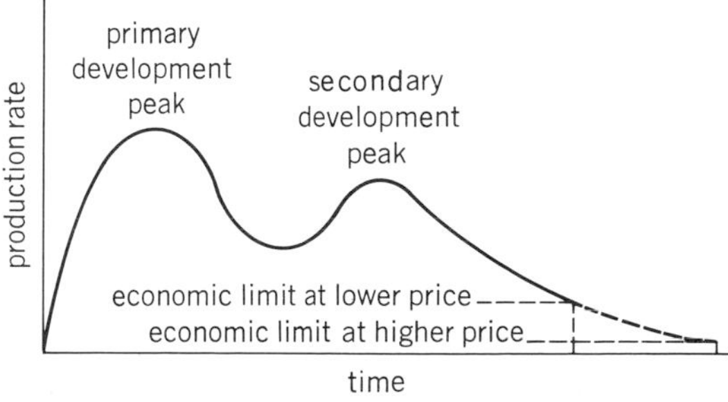

Fig. 1. Oil production rate from a reservoir as a function of time.

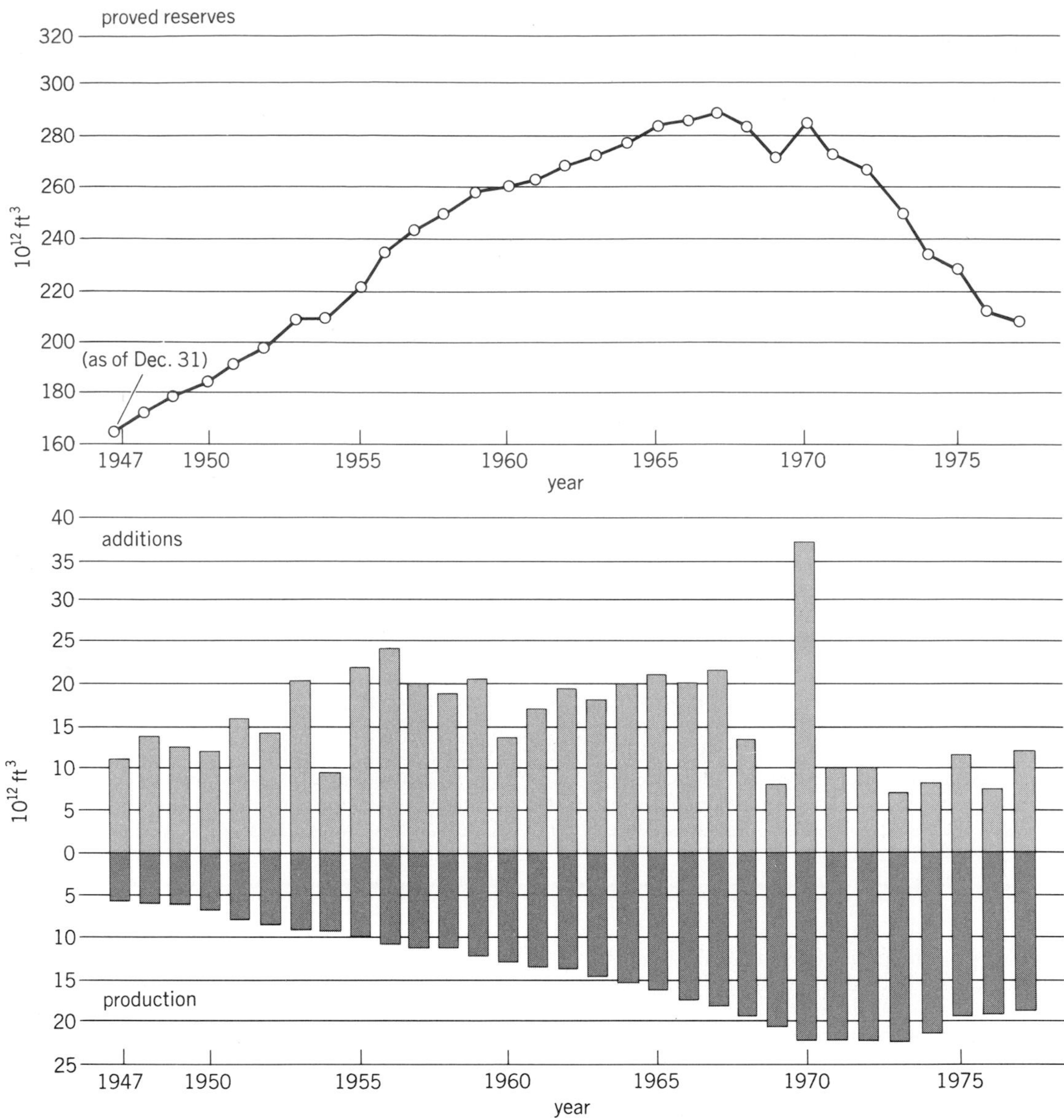

Fig. 2. United States natural gas reserves. 10^{12} ft^3 = 2.83×10^{10} m^3. *(From American Petroleum Institute, Reserves of Crude Oil, Natural Gas Liquids, and Natural Gas in the United States and Canada, May 1979)*

oil remaining to be found have been steadily decreasing, and it is highly unlikely that the decline will be substantially affected by future events.

Economics. The effect of economics on reserves has received much attention because of the fact that recovery of crude oil in the United States is expected to reach an average value of only 33% of the discovered resource, and because of the steadily rising price of crude oil. Despite a more than fivefold increase in the average price of crude oil since 1975, reserve additions have not increased significantly from their prior rate.

The basic reason for this can be deduced from Fig. 1, which shows the oil production rate from a reservoir as a function of time. Production is stopped upon reaching the economic limit, which is the rate of production at which the income from the sale of the produced oil is just sufficient to pay the operating costs, taxes, and overhead encountered in achieving the production. An increase in the sales price (assuming that the associated increase in costs is not completely compensating) will extend the economic limits to a still lower limit. However, the volume of oil produced between any two time points is the area under the curve, and it is obvious that for mature fields the additional production of fluids will be a small percentage of the total producible to the original economic limit. The effect of higher prices for crude oil will provide a stimulant to the exploration for new reserves in hostile, frontier areas and in the development of new technology for enhanced recovery.

World reserves. At the close of 1979, world oil reserves were unchanged from the previous year at 642×10^9 bbl (102×10^9 m^3). However, reserves in the Soviet Union, the United States, and the United Kingdom each dropped 10% or more during 1978–1979. The major addition to reserves in the world occurred in Mexico, where the reserves were increased from 14 to 31×10^9 bbl (2.23 to 4.93×10^9 m^3) during 1978–1979 (see table).

Natural gas. Natural gas reserves are estimated

The 16 nations which held 92% of the estimated world oil reserves on Dec. 31, 1979*

Nation	Oil reserves, 10^9 bbl	Oil reserves, 10^9 m³
Saudi Arabia	163.0	26.1
Soviet Union	67.0	10.7
Kuwait	65.0	10.3
Iran	58.0	9.2
Iraq	31.0	4.9
Mexico	31.0	4.9
Abu Dhabi	28.0	4.5
United States	27.0	4.3
Libya	24.0	3.8
China	20.0	3.2
Venezuela	18.0	2.9
Nigeria	17.0	2.7
United Kingdom	15.0	2.4
Indonesia	10.0	1.6
Algeria	8.4	1.3
Canada	6.8	1.1

*Note that only four nations contained 55%.

in a manner similar to that used for the estimation of crude oil reserves. Some natural gas is found in association with crude oil in the same reservoir, and some is found in reservoirs which do not have a gas-oil contact, that is, nonassociated gas. Gas represents a higher maturation level than that of crude oil, and therefore some geographical areas are more gas-prone than others. The history of reserves of natural gas in the United States parallels that of crude oil, with a peak having been reached in the late 1960s (Fig. 2). The ratio of gas reserves to oil reserves in the United States is approximately twice the ratio for the world. *See* NATURAL GAS; PETROLEUM. [TODD M. DOSCHER]

Petroleum reservoir engineering

The technology concerned with the prediction of the optimum economic recovery of oil or gas from hydrocarbon-bearing reservoirs. It is an eclectic technology requiring coordinated application of many disciplines: physics, chemistry, mathematics, geology, and chemical engineering.

Originally, the role of reservoir engineering was exclusively that of counting oil and natural gas reserves. The reserves—the amount of oil or gas that can be economically recovered from the reservoir—are a measure of the wealth available to the owner and operator. It is also necessary to know the reserves in order to make proper decisions concerning the viability of downstream pipeline, refining, and marketing facilities that will rely on the production as feedstocks.

The scope of reservoir engineering has broadened to include the analysis of optimum ways for recovering oil and natural gas, and the study and implementation of enhanced recovery techniques for increasing the recovery above that which can be expected from the use of conventional technology.

Original oil in place. The amount of oil in a reservoir can be estimated volumetrically or by material balance techniques. A reservoir is sampled only at the points at which wells penetrate it. By using logging techniques and core analysis, the porosity and net feet of pay (oil-saturated interval) and the average oil saturation for the interval can be estimated in the immediate vicinity of the well. The oil-saturated interval observed at one location is not identical to that at another because of the inherent heterogeneity of a sedimentary layer. It is therefore necessary to use statistical averaging techniques in order to define the average oil content of the reservoir (usually expressed in barrels per net acre-foot) and the average net pay. The areal extent of the reservoir is inferred from the completion of dry holes beyond the productive limits of the reservoir. The definition of reservoir boundaries can be heightened by study of seismic surveys and analysis of pressure buildups in wells after they have been brought on production. *See* PETROLEUM GEOLOGY; WELL LOGGING.

If the only mobile fluid in the reservoir is crude oil, and the pressure during the production of the crude remains above the bubble point value (at which point the gas begins to come out of solution), the production of crude oil will be due merely to fluid expansion, and the appropriate equation is Eq. (1). Here c is the compressibility of the fluid, V

$$\text{Production} = N_P = (c \cdot V \cdot \Delta P) B_0 \quad (1)$$

is the volume of crude oil in the reservoir, and ΔP is the decrease in average reservoir pressure associated with the production of N_P barrels of crude oil measured at the surface. The term in parentheses is the expansion measured in reservoir barrels, and B_0 is the formation volume factor (inverse of shrinkage) which corrects subsurface measurements to surface units. Thus, if N_P is plotted as a function of $(c \cdot \Delta P)B_0$, the slope of the resulting straight line will be V, the volume of the original oil is place in the reservoir.

For most reservoirs, the material balance equation is considerably more complicated. Few reservoirs would produce more than 1 or 2% of the oil in place if fluid expansion alone were relied upon. The reasons for the increased complexity of a material balance equation when the pressure falls below the bubble point are as follows: (1) The pore volume occupied by fluids in the reservoir changes with reservoir pressure as does the volume of the immobile (connate) water in the reservoir. (2) As the pressure declines below the bubble point, gas is released from the oil. (3) The volume of gas liberated is greater than the corresponding shrinkage in the oil, and therefore the liberated gas becomes mobile upon establishing a critical gas saturation, and the gas will flow and be produced in parallel with oil production. (4) There may be an initial gas cap above the oil column (above a gas-oil contact), and some of this gas, depending upon the nature of the well completion, may flow immediately upon opening the well to production. (5) Below the oil column, there may be an accumulation of water (below a water-oil contact), and if sufficiently large, the water in this aquifer will expand under the influence of the pressure drop and replace the oil that is produced from the oil column.

Thus the simple material balance equation (1) for fluid expansion must be expanded to take into account all the changes in fluid and spatial volumes and the production of oil and gas. By the suitable manipulation of the various terms in the resulting equation, the material balance can be reduced to the equation of a straight line with the slope and intercept of the line providing information about the magnitude of the original oil in place and the significance of the water encroachment and the magnitude of the gas cap. However, the accuracy of the material balance equation depends strongly on having obtained accurate knowledge about the *PVT* (pressure-volume-temperature) behavior of the reservoir crude, and knowing the average reservoir pressure corresponding to successive levels of reservoir depletion. There are significant limits to acquiring such information.

The material balance equation is not time-dependent; it relates only average reservoir pressure to production. Although it can be extrapolated to provide information on the amount of oil and gas that may be produced in the future when the pressure falls to a given level, it does not provide any information as to when the pressure will fall to such a level. It can be used to estimate the ultimate production from the reservoir if an economically limiting pressure can be specified. This usually can be done since pressure and rate of production are of course explicitly interrelated by the equations for fluid flow, and the economic limit of production is that rate at which the economic income is insufficient to pay for the cost of operation, royalties, taxes, overheads, and other facets of maintaining production.

Fluid flow in crude oil reservoirs. In order to develop some understanding of the future performance of a reservoir on a real-time basis, the reservoir engineer has two options. One is to predict the flow rate history of the reservoir by using the equations for fluid flow, and the second is to extrapolate the already known production history of the reservoir into the future by using empiricism and know-how generated by experience with analogous reservoirs.

The flow of fluids in the reservoir obeys the differential form of the Darcy equation, which for radial flow is Eq. (2), where q is the volumetric flow

$$q=\frac{k\cdot h\cdot 2\pi r(\Delta P/\Delta r)}{\mu} \tag{2}$$

rate measured in subsurface volumes, r is the radial distance from the well bore to any point in the reservoir, $(\Delta P/\Delta r)$ is the pressure gradient at the corresponding value of the radius, μ is the viscosity of the mobile fluid, and k is the permeability of the reservoir to the mobile fluid.

This equation could be integrated directly if $\Delta P/\Delta r$ and the pressure at the outer limits of the reservoir were constant. However, this condition is true only where there exists a strong natural water drive resulting from a contiguous aquifer. Steady-state conditions then prevail and the flow equation is simply Eq. (3), where q is in barrels per day, k in

$$q=\frac{7.08kh(P_e-P_w)}{\mu\ln\left(\frac{r_e}{r_w}\right)} \tag{3}$$

darcies, P_e (pressure at the outer boundary) and P_w (pressure at the well bore) are in pounds per square inch, μ in centipoises, and r in feet.

When steady-state conditions do not exist, the Darcy equation must be combined with the general radial diffusivity equation (4), where ϕ, the porosi-

$$\frac{\partial^2 P}{\partial r^2}+\frac{1}{r}\frac{\partial P}{\partial r}=\frac{\phi\mu c}{k}\frac{\partial P}{\partial t} \tag{4}$$

ty, is the only term not elsewhere defined. Appropriate boundary conditions must be specified in order to obtain an integrated analytical equation for fluid flow in the reservoir. W. Hurst and A. F. Van Everdingen presented solutions for the important two sets of boundary conditions: constant production rate at the well bore, and constant production (well bore) pressure. Solutions are available for both the bounded reservoir in which the pressure declines from its initial value at the physical reservoir boundary, or in the case of multiple wells in a reservoir at the equivalent no-flow boundary between wells, and the infinite reservoir (usually only a transient condition in real reservoirs) in which the pressure has not declined below initial value at the boundary.

However, the use of the solutions to these equations still does not permit prediction of reservoir performance. The solutions are usually presented in terms of dimensionless time and dimensionless pressure or dimensionless cumulative production. The permeability to the mobile fluid and its compressibility enter into these dimensionless parameters, and neither of these functions is constant, nor can they be explicitly calculated. The reasons for this are readily traced.

It has already been noted that as the pressure in the reservoir is reduced, the gas saturation builds up. Accompanying this increase, there is an increase in the permeability to the mobile gas phase, and a decrease in the permeability to oil. The changes are not linear (see illustration). Since the pressure is not uniform throughout the reservoir, neither will the gas saturation be uniform, nor will the corresponding permeabilities to the gas and oil. The compressibility of the reservoir fluids will obviously vary directly as the saturation of the compressible gas. Thus, the integrated forms of the radial diffusivity equation cannot be used with any high degree of accuracy over the entire reservoir. The modern high-speed digital computer has made it possible, however, to represent the entire reservoir by a three-dimensional grid system in which the reservoir is mathematically subdivided, and the fluid flow equations and material balance across each block are iteratively and compatibly solved for adjacent blocks. The use of mathemati-

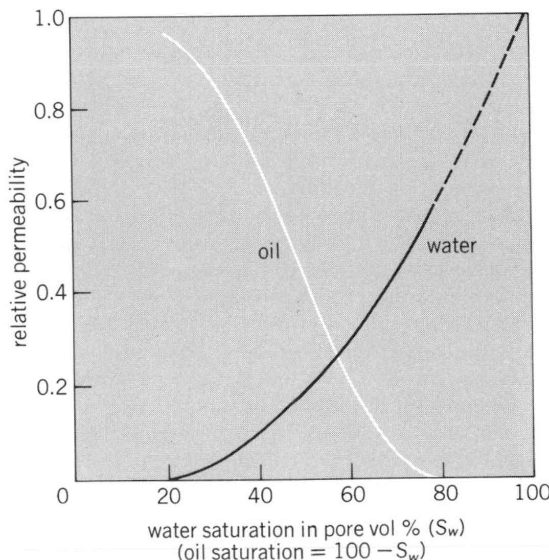

Typical relative permeability relationship for oil and water in a porous medium.

cal simulation techniques requires a detailed specification of the reservoir geometry and lithology, and the relative permeabilities to the mobile fluids under conditions of both diffuse and stratified flow. Again, such knowledge is limited, and as a result it is necessary to calibrate the mathematical simulator by history-matching available production and pressure history for the reservoir. Unique predictions are difficult to achieve and as a result such simulations must be constantly updated. Pressure-transient analyses of individual wells are analyzed by superimposing the flow regimes for the flowing period and the shut-in period to yield data on effective permeability to the flowing fluids and the average reservoir pressure at the time of shutting in the well. Such data are important in calibrating the reservoir simulator for a given reservoir. *See* PETROLEUM RESERVOIR MODELS.

Before the high-speed digital computer was available to the reservoir engineer, and today when the data available for numerical simulation are not available or when the cost of such simulation studies cannot be justified, the reserves may be predicted by decline curve analysis. Depending on the production mechanism, the production rate will decline in keeping with exponential, hyperbolic, or constant-percentage decline rates. The appropriate coefficients and exponents for a particular decline rate are estimated by history matching, and then the decline curves are extrapolated into the future. A variety of nomographs are available for facilitating the history matching and extrapolation.

Reservoir recovery efficiency. The overall recovery of crude oil from a reservoir is a function of the production mechanism, the reservoir and fluid parameters, and the implementation of supplementary recovery techniques. In the United States, recovery shows a geographical pattern because of the differences in subsurface geology in the various producing basins. Many reservoirs in southern Louisiana and eastern Texas, where strong natural water drives, low-viscosity crudes, and highly permeable, uniform reservoirs occur, show a recovery efficiency well over 50% of the original oil in place. Pressure maintenance, the injection of water to maintain the reservoir pressure, is widely used where the natural water influx is not sufficient to maintain the original pressure. In California, on the other hand, strong water drives are virtually absent, the crude oil is generally more viscous than in Louisiana, and the reservoirs, many of which are turbidites, show significant heterogeneity. Both the viscous crude oil and the heterogeneity permit the bypassing of the oil by less-viscous water and gas. Once a volume of oil is bypassed by water in water-wet rocks (the usual wettability), the oil becomes trapped by capillary forces, which then require the imposition of unfeasible pressure gradients or the attainment of very low interfacial tensions for the oil to be released and rendered mobile. As a result, the overall displacement in California reservoirs is relatively poor, and even after the implementation of water flooding, the recovery efficiency in California will probably be only 25–30%.

In general, recovery efficiency is not dependent upon the rate of production except for those reservoirs where gravity segregation is sufficient to permit segregation of the gas, oil, and water. Where gravity drainage is the producing mechanism, which occurs when the oil column in the reservoir is quite thick and the vertical permeability is high and a gas cap is initially present or is developed on producing, the reservoir will also show a significant effect of rate on production efficiency.

The overall recovery efficiency in the United States is anticipated to be only 32% unless new technology can be developed to effectively increase this recovery. Reservoir engineering expertise is being put to making very detailed studies of the production performance of crude oil reservoirs in an effort to delineate the distribution of residual oil and gas in the reservoir, and to develop the necessary technology to enhance the recovery. Significant strides have been made in at least one area in the United States—California, where the introduction of thermal recovery techniques has increased the recovery from many viscous oil reservoirs from values as low as a few percent to well over 50%. *See* PETROLEUM ENHANCED RECOVERY.

[TODD M. DOSCHER]

Petroleum reservoir models

Physical and computational systems whose behavior is designed to resemble that of actual reservoirs. Early in the petroleum industry's history, there was a great deal of effort invested in the development of very complex physical (laboratory) models, while the computational models were quite simple. Since the explosive evolution of computers in the early 1950s, the computational models have become more and more complex, and there has been a diminishing emphasis on the physical models. Today the emphasis is on combining the best features of the physical and computational models to permit evaluation of the complex behavior of enhanced oil recovery processes.

There are three basic requirements for a useful model: (1) a quantitative description of the relevant properties of a reservoir; (2) a means of describing the mechanics of fluid movements in a reservoir; and (3) a means of combining 1 and 2 by constructing analogous physical systems or by constructing mathematical systems which can be solved by computational techniques.

The purpose of a petroleum reservoir model is to estimate field performance (for instance, oil recovery) under a variety of operating strategies. Whereas the reservoir can be produced only once—and at considerable expense—a model can be produced many times. If this production can be accomplished at low expense and over a short period of time, then observation of the model's performance can give great insight into how to optimally produce a reservoir. Petroleum reservoir models can range in complexity from the intuition and judgment of an engineer to complex mathematical or physical systems describing enhanced oil recovery processes. *See* OIL FIELD MODEL.

Simple models. One of the simplest, yet generally used, computational models is the material balance model. This model, in which the flow of a substance into a region minus the flow of that substance out of the region is equated to the change of mass or volume of that substance in the region [Eq. (1)], is the basic building block for the

$$\text{IN} - \text{OUT} = \text{GAIN} \qquad (1)$$

realistic computational models in use today.

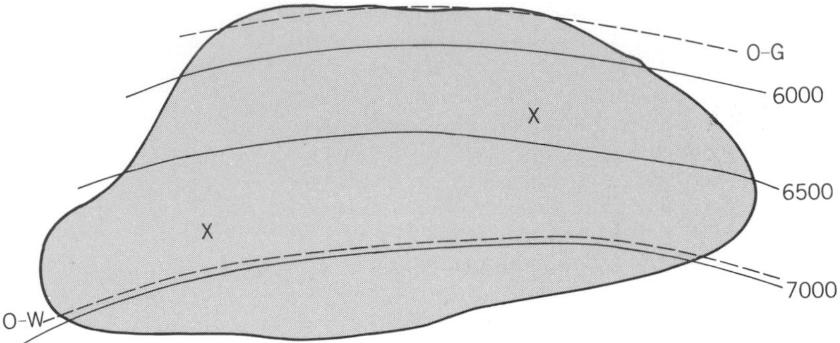

Fig. 1. General material balance. Decrease in oil volume + decrease in gas volume = increase in water volume.

Considering the reservoir depicted in Fig. 1, the material balance equation will permit relating the fluid withdrawals from this reservoir to the water influx, the gas cap expansion, and the change in pressure. This is an excellent tool for evaluating the primary depletion of a reservoir. However, because the solution of the material balance equation gives only average saturations and pressures for the region, one can tell how much of each fluid is present as a function of time, but cannot tell where the fluids are. Since all enhanced oil recovery (EOR) schemes, including waterflooding, are going after the remaining oil in place, it is imperative to know not only how much oil is present but also where it is. For this reason, present computational models break the region of interest into small pieces by means of a grid (Fig. 2) and solve the material balance equation for each small block in the grid system.

Realistic models. If a small three-dimensional block is removed from the reservoir (Fig. 3), the concept of material balance can be generalized as in Eqs. (2)–(4).

$$\text{IN} = [(F_x)_x \Delta y\, \Delta z + (F_y)_y \Delta x\, \Delta z + (F_z)_z \Delta x\, \Delta y]\, \Delta t \quad (2)$$

$$\text{OUT} = q_v\, \Delta x\, \Delta y\, \Delta z\, \Delta t + [(F_x)_{x+\Delta x} \Delta y\, \Delta z + (F_y)_{y+\Delta y} \Delta x\, \Delta z + (F_z)_{z+\Delta z}\, \Delta x \Delta y]\, \Delta t \quad (3)$$

$$\text{GAIN} = [(C)_{t+\Delta t} - (C)_t]\, \Delta x\, \Delta y\, \Delta z \quad (4)$$

where F is a flux, that is, lb of fluid per ft^2 per day
C is a general concentration, lb per ft^3
q_v is a source term, lb per ft^3 per day

Even though the mechanics of fluid flow are complex, it has been found in the laboratory that the flow of a single fluid through the pores of a rock can be described by Darcy's law, Eq. (5).

$$u_x = -\frac{k_x}{\mu}\left[\left(\frac{p_{x+\Delta x} - p_x}{\Delta x}\right) - \gamma\left(\frac{Z_{x+\Delta x} - Z_x}{\Delta x}\right)\right] \quad (5)$$

where u_x = volumetric velocity, ft^3/(ft^2 of area normal to flow)(day)
k_x = rock permeability, millidarcies (md) × 0.00633
μ = fluid viscosity, centipoises (cp)
x = distance, ft
p = fluid pressure, psi
γ = specific weight, psi/ft (g/g_c)
Z = elevation (vertical position) measured positively downward, ft

For multiphase fluid flow, the velocity u_f of each fluid is found from Eq. (5), modified by relative permeability k_r to result in Eq. (6),

$$u_{x,f} = -\frac{k_x k_{r,f}}{\mu_f} \cdot \left[\left(\frac{p_{x+\Delta x,f} - p_{x,f}}{\Delta x}\right) - \gamma_f\left(\frac{Z_{x+\Delta x} - Z_x}{\Delta x}\right)\right] \quad (6)$$

where the subscript f refers to phase f, and $k_{r,f}$ is the relative permeability of phase f and is a function of the relative amounts of the phases present (the saturations S_f). Equations (5) and (6) are empirical relationships which have been developed by laboratory experiments.

For a three-phase black oil system where the phases are oil, water, and gas, the three fluxes are given by Eqs. (7), and the concentrations of the phases are given by Eqs. (8).

$$\left.\begin{aligned} F_{x,w} &= b_w u_{x,w} \\ F_{x,o} &= b_o u_{x,o} \end{aligned}\right\} \quad \frac{\text{stock tank barrel}}{\text{ft}^2\text{-day}} \quad (7)$$

$$F_{x,g} = b_g u_{x,g} + b_o R_s u_{x,o} \qquad 10^3\ \text{ft}^3/\text{ft}^2\text{-day}$$

$$\begin{aligned} C_w &= \phi b_w S_w \\ C_o &= \phi b_o S_o \\ C_g &= \phi(b_o R_s S_o + b_g S_g) \end{aligned} \quad (8)$$

where b_f is the formation volume factor in stock tank barrels (STB) per reservoir barrel for phase f
ϕ is the porosity of the rock
R_x is a solution gas/oil ratio

The fluxes are defined in this manner in order to put them in a consistent set of units, so that all quantities measured are expressed in the standard conditions of 60°F (15.6°C) and atmospheric pressure (101.325 kPa). In all of these equations (7), the material balance model is related to some constant set of conditions. Actually, the conditions in the reservoirs are themselves constantly changing, and the terms are defined in such a way as to represent the amount of oil filling the barrels under standard conditions.

Combining Eqs. (1), (4), (7), and (8) results in the oil material balance equation in two dimensions for the grid shown in Fig. 4. The terms of the oil balance equation are given separately in Eqs. (9), (10), and (11).

$$\text{IN} = [(b_o u_{x,o})_{i\text{-}1/2,\, j,\, n+1/2}\, h_{i\text{-}1/2,\, j} \Delta y_j + (b_o u_{y,\, o})_{i,\, j\text{-}1/2,\, n+1/2}\, h_{i,\, j\text{-}1/2}\, \Delta x_i]\, \Delta t \quad (9)$$

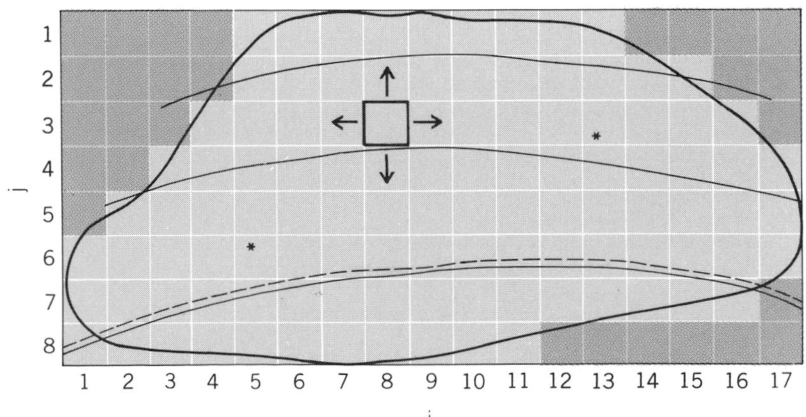

Fig. 2. Grid diagram for a computational model. The material balance equation is solved for each small block in the system.

The above term includes the flow from the blocks labeled i-1, j, and i, j-1 to the block labeled i, j. The variable h refers to vertical thickness.

$$\text{OUT} = [(b_o u_{x,o})_{i+1/2,\, j,\, n+1/2}\, h_{i+1/2,\, j}\, \Delta y_j + (b_o u_{y,o})_{i,\, j+1/2,\, n+1/2}\, h_{i,\, j+1/2}\, \Delta x_i + q_{o,\, i,\, j,\, k,\, n+1/2}]\, \Delta t \quad (10)$$

The above term includes the flow from the block labeled i, j to blocks $i+1, j$ and $i, j+1$. Here q_o is an oil injection or production rate.

$$\text{GAIN} = (\phi b_o S_o)_{i,\, j,\, n+1} - (\phi b_o S_o)_{i,\, j,\, n}\, h_{i,\, j}\, \Delta x_i\, \Delta y_j \quad (11)$$

The above term is the net change of fluid in block i, j. If Eq. (6) is used in Eq. (9), Eq. (12) results, where $t = n, n+1/2, n+1, \ldots$, and $\ell_{i-1/2,\, j} = (\Delta x_i + \Delta x_{i-1})/2$.

$$\left[\frac{b_o k_{ro}}{\mu_o} k_x h\right]_{i-1/2,\, j,\, n+1/2} \Delta y_j \cdot \left[\left(\frac{p_{o,\, i,\, j,\, n+1/2} - p_{o,\, i-1,\, j,\, n+1/2}}{i - 1/2, j}\right) - \gamma_{o,\, i-1/2,\, j,\, n+1/2}\left(\frac{Z_{i,\, j} - Z_{i-1,\, j}}{\ell_{i-1/2,\, j}}\right)\right] \quad (12)$$

If material balance equations for the other two phases are written and definition (13) is included,

$$f_{n+1/2} = \frac{f_n + f_{n+1}}{2} \quad (13)$$

where f may represent any variable, a system of three equations at each grid block (i, j) and each time t for the unknowns $p_o, p_w, p_g, S_o, S_w, S_g$ is developed. Since there are more unknowns than equations, two capillary pressure relationships and one constraint at each grid block and at each point in time are required. These are given by Eqs. (14) and (15), where p_{cow} represents water-oil capillary pressure and p_{cog} represents gas-oil capillary pressure.

$$p_{cow}(S_w) = p_o - p_w \qquad p_{cog}(S_g) = p_o - p_g \quad (14)$$

$$S_o + S_w + S_g = 1 \quad (15)$$

At this stage of the development, there is a deterministic system of equations which can be solved for the average pressures and saturations in each grid block with the aid of large-scale digital computers, provided the required reservoir rock and fluid data are available.

For black oil systems, a great deal of experimental work has been done, and functional relationships (16) are generally accepted. All of these

$$\begin{gathered} k_{ro}(S_w, S_g),\ k_{rw}(S_w),\ k_{rg}(S_g) \\ b_o(p_o),\ b_g(p_g),\ b_w(p_w) \\ \mu_o(p_o),\ \mu_g(p_g),\ \mu_w(p_w) \\ \gamma_o(p_o),\ \gamma_g(p_g),\ \gamma_w(p_w),\ R_s(p_o) \end{gathered} \quad (16)$$

data can be measured by using a fluid sample and some cores from the reservoir. In addition, through the use of geology, petrophysics, well testing, and tracer and pulse tests, as well as history-matching production data using the model itself, the important rock characteristics $h_{i,j}$, $Z_{i,j}$, $k_{x,i,j}$, $k_{y,i,j}$, and $\phi_{i,j}(p_o)$ can be determined.

Once a comprehensive laboratory program is combined with some field tests or production data, a computational model can be used to adequately predict the future performance of the reservoir under various operating strategies. This technology is complete and is currently being used throughout the oil and gas industry for dry gas and black oil reservoirs.

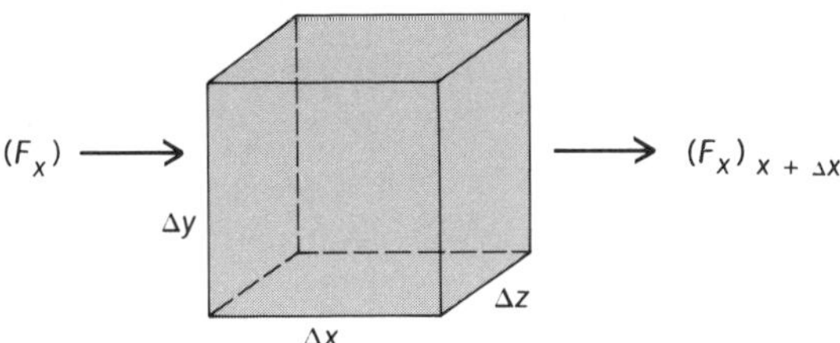

Fig. 3. Material balance in a three-dimensional block from the reservoir.

Enhanced oil recovery. The reservoir models for enhanced oil recovery (EOR) processes are not nearly as advanced as the black oil model. With the possible exception of steamflooding, for which the first model was developed in 1973, even the fluid mechanics of the other EOR processes are uncertain. Today models exist for simulating CO_2 flooding, combustion, and chemical flooding; however, none of these models has been used extensively enough to have reached the level of sophistication of black oil models.

Because the EOR processes are so complex, one of the major uses of the mathematical models is to explain the behavior of the physical models. By using the computational model to simulate laboratory experiments, it can be determined if the model is truly representing the physics of the process. In this manner, insight is gained into the important aspects of the physical processes, and when the model is capable of modeling the laboratory work, it can be used as a tool to scale these results to the field.

Reservoir models require both complex physical models as well as computational models. Each reservoir study requires a comprehensive laboratory program under which actual floods are carried out so that the physics of the EOR process can be properly input to the computational model. In ad-

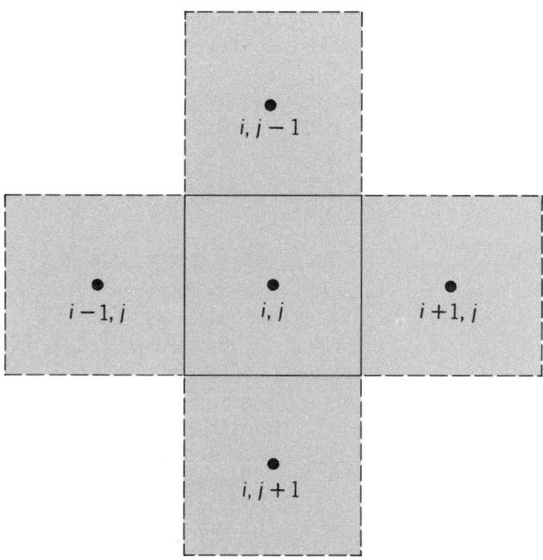

Fig. 4. Oil material balance in two dimensions.

ition to running core floods, a great deal of physical property data regarding the fluids and the reservoir rock must be generated, as is done in the black oil case.

In addition to requiring a comprehensive laboratory program, the EOR models need to be more accurate than the black oil models. This is because small changes in a component's concentration can dramatically affect its ability to mobilize oil, and so the computational model must be able to generate accurate results. Also, because many EOR processes inject slugs of material into the reservoir, the model needs to be capable of computing results on a scale much smaller than is required in black oil models. Finally, all of these complexities lead to requiring a more accurate reservoir characterization than is needed for black oil systems.

In the late 1960s and early 1970s, the need for more accurate computational models was recognized and more accurate approximations were being considered. The most promising of these high-order-accuracy models were the finite-element models; however, at this time (1980) no truly successful finite-element model is being used for EOR. Faster computers (such as the Star and the Cray and conformed grid transformations) have enabled the industry to continue using second-order-correct finite-difference approximations. However, the mathematical models of the complex EOR processes are limited to modeling the behavior of symmetry elements selected from patterns and are not capable of simulating a full field's performance. This, of course, puts a greater burden on the engineers who use these models and requires that the engineers of the future be better trained than their counterparts of a decade ago.

The final chapter on modeling is not being written here. As more and more complex physical processes are modeled, there will be better understanding of which aspects are important and which are not. In that way it will be possible to make reasonable and justifiable assumptions which will simplify the models. At the same time, computers will continue to get faster and less expensive, and numerical methods will continue to improve. While an ever-expanding collection of modeling techniques will become available, it is almost certain that the needs will continue to outnumber the means. Much work remains to make the reservoir models respond more and more like the real reservoirs. However, since more and more complex processes will be used to recover oil, the development of reservoir models will not end until oil is replaced with another form of energy. *See* OIL AND GAS FIELD EXPLOITATION; PETROLEUM ENHANCED RECOVERY; PETROLEUM RESERVOIR ENGINEERING.

[HARVEY S. PRICE]

Bibliography: J. C. Cavendish, H. S. Price, and R. S. Varga, Galerkin methods for the numerical solution of boundary value problems, *Soc. Petrol. Eng. J.*, 246:204–220, June 1969; K. H. Coats, Insitu combustion model, *SPE-AIME 54th Annual Meeting*, Las Vegas, SPE Pap. 8394, Sept. 23–26, 1979; K. H. Coats et al., Three-dimensional simulation of steamflooding, *SPE-AIME 48th Annual Meeting*, Las Vegas, SPE Pap. 4500, Sept. 30–Oct. 3, 1973; H. J. Morel-Seytoux, Analytical-numerical method in waterflooding predictions, *Soc. Petrol. Eng. J.*, 9:247–258, 1965; H. J. Morel-Seytoux, Unit mobility ratio displacement calculations for pattern floods in homogeneous medium, *Soc. Petrol. Eng. J.*, 9:217–227, 1966; P. M. O'-Dell, Use of numerical simulation to improve thermal recovery performance in the Mount Paso Field, California, *SPE-AIME Symposium on Improved Oil Recovery*, Tulsa, SPE Pap. 7078, Apr. 16–19, 1978; A. Settari, H. S. Price, and T. Dupont, Development and application of variational methods for simulation of miscible displacements in porous media, *Soc. Petrol. Eng. J.*, pp. 228–246, June 1976; A. Spivak, H. S. Price, and A. Settari, Solution of the equations for multidimensional, two phase, immiscible flow by variational methods, *SPE-AIME 4th Symposium on Numerical Simulation of Reservoir Performance*, Los Angeles, SPE Pap. 5723, Feb. 19–20, 1976; M. R. Todd and C. A. Chase, A numerical simulator for predicting chemical flood performance, *SPE-AIME 5th Symposium on Reservoir Simulation*, Denver, SPE Pap. 7689, Feb. 1–2, 1979; M. R. Todd and W. J. Longstaff, The development, testing, and application of a numerical simulator for predicting miscible flood performance, *J. Petrol. Technol.*, p. 874, July 1973.

Photoelectrolysis

The process of using optical energy to assist or effect electrolytic processes that ordinarily require the use of electrical energy. In recent years much attention has been directed toward using optical energy from the Sun to effect the photoelectrolysis of water, decomposing liquid water to yield gaseous oxygen and hydrogen. This process is widely regarded as an attractive way to convert solar energy to a useful fuel in the form of hydrogen (H_2), which is an energy-rich material that can burn in the presence of oxygen (O_2) from the air to release heat and regenerate the water. Electricity could also be generated from hydrogen and oxygen by means of devices called fuel cells, which have been used in the space program. Thus, efficient solar-driven photoelectrolysis of water to hydrogen and oxygen represents a potentially important energy resource.

Photoelectrochemical cells. These consist of two solid electrodes immersed in an aqueous electrolyte (for example, NaOH) solution and connected by external circuitry. In the most efficient cells, one or both of the electrodes are a solid semiconducting material that responds to characteristic wavelengths of light. Photoelectrochemical cells for the photoelectrolysis of water involve the production of oxygen at the anode and of hydrogen at the cathode according to reactions (1) and (2) respectively. The net reaction is described in (3).

$$4OH^-_{(aq)} \xrightarrow[\text{anode}]{} O_{2(g)} + 2H_2O_{(l)} + 4e^- \quad (1)$$

$$4H^+_{(aq)} + 4e^- \xrightarrow[\text{cathode}]{} 2H_{2(g)} \quad (2)$$

$$2H_2O_{(l)} \rightarrow 2H_{2(g)} + O_{2(g)} \quad (3)$$

A typical device is shown in Fig. 1. Reactions (1) and (2) represent the so-called half-cell reactions, and the electrons (e^-) appearing in the equations pass through the external circuit. Since the electrons flow in the external circuit it is possible, in principle, to extract electrical energy from the device by putting a load in series in the external circuit. However, efficiently producing both fuel and electricity using a photoelectrochemical cell has proved so far to be quite difficult. Indeed, even effecting the photoelectrolysis of water has been a formidable task. The highest efficiency for the conversion of solar energy to hydrogen fuel is in the vicinity of 1–2% using the photoelectrolysis process.

Efficiency for solar photoelectrolysis. The theoretical efficiency for the solar photoelectrolysis of water is much higher than the efficiencies actually achieved. In order to understand what limiting factors are present and what future prospects for the process might be, some of the fundamental principles of photoelectrolysis must be considered. It is widely believed that a useful solar fuel system must have at least 10% solar efficiency. Researchers are thus a factor of 5 or 10 away. However, high efficiency is no guarantee that photoelectrolysis would be a useful energy technology—the cost must be competitive with other energy technologies for producing fuel, especially coal and oil shale. The questions are very fundamental and include basic concerns such as whether high efficiency is attainable at any cost and whether the photoelectrochemical cells would last long enough to produce more energy than is expended in their fabrication.

Semiconducting materials have been shown to be much more effective as photoelectrodes than any other material. Therefore, much effort is being directed toward understanding the behavior of semiconductors immersed in liquid electrolyte solutions. In the late 1830s the illumination of an electrode immersed in an electrolyte solution was actually the first experiment associated with the photovoltaic effect. It is now known that the electrode was a semiconducting material. Semiconductors are used today in solid-state photovoltaic solar cells (for instance, *pn* silicon solar cells) to convert sunlight into electricity. In fact, the theoretical principles underlying the behavior of a semiconductor/liquid interface follow the treatment for a solid-state photovoltaic cell called a Schottky barrier cell, where the essential element is a semiconductor coated with a thin film of some metal such as gold. In the liquid interface situation the liquid electrolyte solution plays the same role as the metal in the Schottky barrier cell. *See* SOLAR CELL.

An essential difference between the Schottky barrier cell and the liquid-interface-based cells arises from the fact that metals are electronic conductors of electricity whereas the solution is an ionic conductor. This is the fact that makes it possible to generate fuel directly by illumination of the semiconductor immersed in the liquid. At the semiconductor/liquid interface, electron flow can occur only when some solution species accepts or donates electrons. Such electron exchange represents a chemical change; accepting electrons corresponds to reduction of the solutions species [reaction (2)], and donating electrons corresponds to oxidation [reaction (1)]. Schottky barrier cells only directly produce electricity, whereas the semiconductor/liquid interface may yield fuel or electricity, depending on the nature of the oxidation and reduction processes involving the solution species.

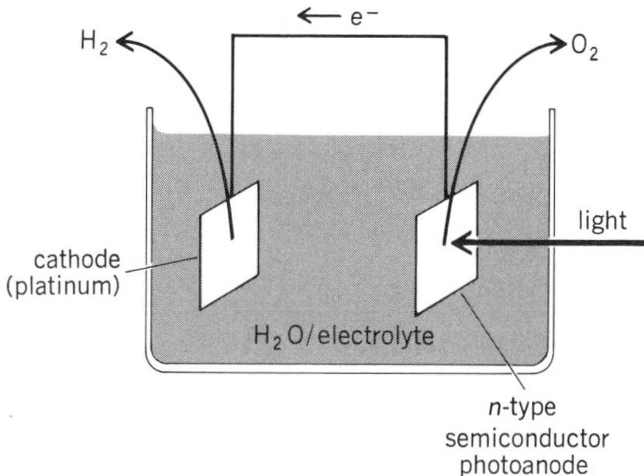

Fig. 1. Photoelectrolysis of water (H_2O) employing an *n*-type-semiconductor–based photoelectrochemical cell.

In both Schottky barrier cells and liquid-interface-based cells, the photosensitive element is the semiconductor. Consequently, the ideal solar efficiency for the devices is the same; it is controlled to a significant extent by the wavelengths of light to which the semiconductor will respond. Semiconductors have a characteristic property called the band gap, E_g, that is the measure of the longest wavelength of light which will be effective in exciting electrons of the semiconductor. Wavelengths of light that are longer than this value are completely transmitted and are not effective. At the same time, wavelengths of light that are shorter (higher-energy) than the wavelength corresponding to E_g are no more effective than light which corresponds exactly to E_g. The Sun itself produces a broad spectrum of light, spanning a range of wavelengths from the ultraviolet to the infrared. A first inclination would be to use a semiconductor with a very low value of E_g such that no light is transmitted, but E_g is unfortunately also the measure of the maximum voltage attainable from the illuminated semiconductor. Thus, a small E_g semiconductor would give an output voltage too low to be useful in making a high-energy fuel. [Electrolysis of water as described by reaction (3) requires a minimum of 1.23 V between the anode and cathode of a cell.] On the other hand, a large-E_g material would give a high output voltage but most of the light would not be used. There is an optimum E_g of ~ 1.4 electron volts (eV) for a photoelectrochemical cell employing a single photoelectrode. The theoretical efficiency for a solar device employing a semiconductor with $E_g = 1.4$ eV is ~ 30%. But producing hydrogen and oxygen may require the use of a semiconductor with $E_g = 1.8$ eV, since conventional electrolytic cells for the generation of hydrogen and oxygen actually operate with a driving force of ~1.5 V. The voltage beyond the minimum of 1.23 V is needed to make the reactions at the anode and cathode proceed at a useful rate; this extra voltage cuts into the efficiency of both conventional electrolyzers and photoelectrochemical cells. However, the efficiency of ~20% for a semiconductor-based cell where $E_g = 1.8$ eV is still well above the minimum practical efficiency and the demonstrated levels of efficiency.

State of the art. It would appear that the best, single-photoelectrode-based photoelectrochemical cell for the photoelectrolysis of water is based on *n*-type semiconducting strontium titanate ($SrTiO_3$). This material gives a solar efficiency of ~1% and yields chemistry according to reaction (2) in a cell like that shown in Fig. 1. The principal contributor

Table 1. Representative photoanode materials for photoelectrolysis of water

Photoanode material	E_g, eV
n-type strontium titanate	3.2
n-type potassium tantalate	3.6
n-type titanium dioxide	3.0
n-type stannic oxide	3.6
n-type tungsten trioxide	2.8

to the subtheoretical efficiency is in fact the value of 3.2 eV for the E_g of $SrTiO_3$. Several other *n*-type semiconducting metal oxide materials can be used as the photoanode (Table 1). Interestingly, the 1% efficiency for $SrTiO_3$ is just about what would be expected for an E_g of 3.2 eV; this provides assurance that the underlying principles are correct, and that once a material of proper E_g is found, the efficiency can be nearly that expected.

Stabilized cells for solar electricity. Research on semiconductors for electronic materials and for use in solid-state photovoltaic solar cells has led to the characterization of a large number of semiconductors, including several which have an E_g nearly equal to the optimum value needed for an efficient solar device. There have been many attempts to use such materials as the photoanode in photoelectrochemical cells to replace $SrTiO_3$, but these studies have revealed a general and fundamental problem associated with illuminated semiconductors immersed in liquid electrolytes: the photoanode itself is susceptible to oxidation processes which lead to its irreversible decomposition. All nonoxide, *n*-type semiconducting photoanode materials have been shown to suffer from this problem. The point is that the anode half-cell reaction for such materials is not that represented by reaction (1), but rather the destruction of the semiconductor. For example, *n*-type cadmium selenide (CdSc), which has $E_g = 1.7$ eV, does not give oxygen when illuminated in a cell. Instead, reaction (4)

$$CdSe_{(s)} \xrightarrow{\text{sunlight}} Cd^{2+}_{(aq)} + Se_{(s)} + 2e^- \qquad (4)$$

occurs, corresponding to the photoanodic destruction of the CdSe. Thus all nonoxide materials studied to date are useless as photoanodes, despite the fact that their E_g may be nearly optimum.

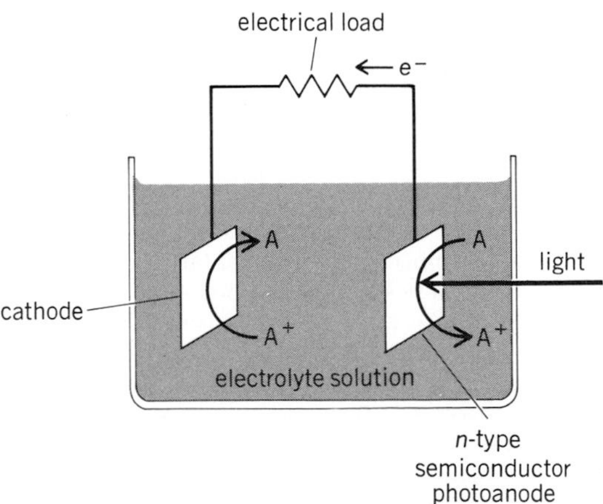

Fig. 2. Photoelectrochemical cell for conversion of light to electricity. No net chemical change occurs in the solution.

The problem of electrode stability has led to some important results relating to the conversion of sunlight to electricity by means of semiconductor-based photoelectrochemical cells. It has been found that many photoanodes can be stabilized by judicious choice of solution species. Stabilized photoanodes then become useful in sustaining the conversion of optical energy. For example, *n*-type semiconducting gallium arsenide (GaAs), which has $E_g = 1.4$ eV, can be a useful photoanode in aqueous solutions in which K_2Se and K_2Se_2 are dissolved. Under these conditions the half-cell reaction occurring at the GaAs photoanode is not the destruction of the photoanode but rather the oxidation of Se^{2-} according to reaction (5). The cathode half-cell reaction (6) is just the reverse of

$$2Se^{2-}_{(aq)} \xrightarrow[\textit{n}\text{-GaAs photoanode}]{\text{sunlight}} Se^{2-}_{2(aq)} + 2e^- \qquad (5)$$

$$Se^{2-}_{2(aq)} + 2e^- \xrightarrow[\text{cathode}]{} 2Se^{2-}_{(aq)} \qquad (6)$$

this process, and there is no net chemical change. However, the electrons do pass through the external circuit and a load can be introduced in series in order to extract useful electrical energy. The efficiency for converting solar energy to electricity is in the vicinity of 12% when the surface of GaAs is properly treated. This efficiency represents a significant milestone in the study of semiconductor/liquid junctions, since it exceeds the 10% regarded as the minimum necessary for practical exploitation.

The high efficiency of the GaAs-based cell is associated with the nearly optimum value of E_g, 1.4 eV, and the fact that Se_2^{2-}/Se^{2-} solution species serve to stabilize GaAs. Figure 2 depicts a general situation where A^+/A represents the crucial solution species and $A \rightarrow A^+ + e^-$ and $A^+ + e^- \rightarrow A$ represent the anode and cathode half-cell reactions, respectively. A large number of photoelectrochemical cells employing such active solution species, including several having photoanodes with E_g near the optimum, have been shown to sustain the conversion of optical energy to electricity (Table 2).

Mediated hydrogen evolution. Photoelectrochemical cells employing a *p*-type semiconductor as a photocathode have also been studied. Fortunately, *p*-type semiconductors do not suffer from the severe stability problems associated with *n*-type semiconductors. Curiously, though, *p*-type semiconducting photocathodes do not yield efficient photoelectrolysis of water. It appears that yet another hurdle is associated with semiconductor/liquid interfaces even when E_g is proper and stability is acceptable. This other problem concerns the fact that the evolution of gases such as oxygen and hydrogen from surfaces is slow, and consequently more driving force (voltage) than is theoretically required is needed to effect useful rates. The study of mechanisms for speeding up half-cell reactions at electrode surfaces constitutes the field of electrocatalysis, which may make an important contribution to improving the efficiency of photoelectrolysis.

Studies of *p*-type silicon (Si: $E_g = 1.1$ eV) and *p*-type GaAs as photocathodes have yielded some results which are promising for the efficient photocathodic generation of hydrogen. Neither *p*-type Si

Table 2. Representative photoelectrode materials for electricity generation

Photoelectrode material (E_g, eV)	Crucial solution species*
Cadmium selenide (1.7)	Se_2^{2-}/Se^{2-}
Gallium arsenide (1.4)	Se_2^{2-}/Se^{2-}
Molybdenum disulfide (1.7)	I_3^-/I^-
Molybdenum diselenide (1.4)	I_3^-/I^-
Indium phosphide (1.3)	Te_2^{2-}/Te^{2-}

*See Fig. 2 and text.

nor p-type GaAs produces hydrogen from water efficiently according to reaction (2). However, it has been discovered that either of these photocathodes can efficiently effect the reduction of the paraquat dication (PQ^{2+}), N,N'-dimethyl-4,4'-bipyridinium, to $PQ^{\dot{+}}$, according to the half-cell reaction (7). The second important point is that $PQ^{\dot{+}}$ can in

$$PQ^{2+}_{(aq)} + e^- \xrightarrow[\text{GaAs or Si photocathode}]{\text{sunlight}} PQ^{\dot{+}}_{(aq)} \qquad (7)$$

turn react with protons (H^+) to liberate hydrogen, regenerating the PQ^{2+}, as shown by reaction (8).

$$2PQ^{\dot{+}}_{(aq)} + 2H^+_{(aq)} \xrightarrow[\text{catalyst}]{} 2PQ^{2+}_{(aq)} + H_{2(g)} \qquad (8)$$

Thus, the evolution of hydrogen by illumination of the photocathode is an indirect process, involving the reduction of PQ^{2+} to produce a species capable of yielding the hydrogen in a second step.

The mechanism for photoelectrolysis involving the $PQ^{2+}/PQ^{\dot{+}}$ solution system is said to involve mediated electron transfer, where the $PQ^{2+}/PQ^{\dot{+}}$ serves as the mediator system. Mediated transfer is necessary since otherwise the evolution of hydrogen is too slow to be efficient. Even the mediated electron-transfer mechanism leads to an efficiency of only about 1% for photoelectrolysis using p-type Si photocathodes. However, a photoelectrochemical cell employing two photosensitive electrodes, for instance p-type Si and n-type $SrTiO_3$, could effect photoelectrolysis with an efficiency exceeding that of a single-photoelectrode device.

Chemical derivatization of surfaces. Stabilizing n-type semiconducting photoanodes and mediating hydrogen evolution at p-type semiconducting photocathodes employ solution species that are crucial. An understanding of both the semiconductor and the solution species is essential to effecting efficient solar photoelectrolysis. These two diverse areas of investigation have been brought together in studies of semiconductor electrodes whose surfaces have been chemically derivatized with reagents that can both stabilize photoelectrodes and mediate electron-transfer processes. Surfaces of semiconductors such as Si, GaAs, and germanium (Ge) invariably have surface functionality such as hydroxyl (—OH) groups that allow covalent anchoring of molecules. The chemistry represented by reaction (9) is representative. The ferrocene-centered reagent is one that has been used, but most of the work has concerned the use of a related substance where the methyl ($-CH_3$) groups are replaced with chlorine (Cl) atoms.

The use of surface-derivatized semiconductor photoelectrodes can be illustrated by considering research on n-type Si photoanodes. In aqueous

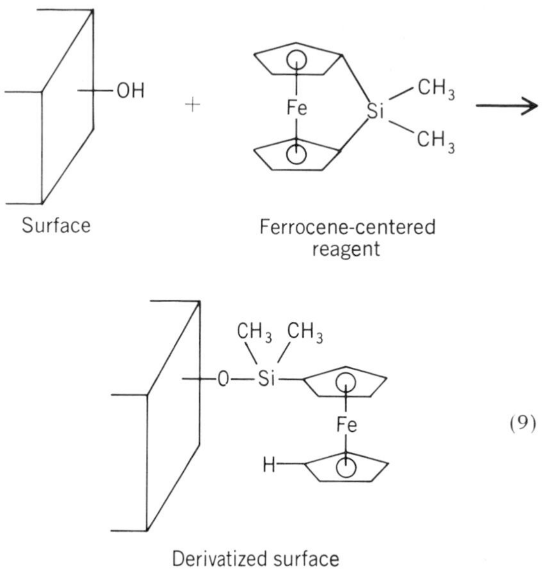

Surface —OH + Ferrocene-centered reagent (Fe, Si, CH_3, CH_3) →

Derivatized surface (—O—Si(CH_3)(CH_3)—ferrocene, Fe, H) (9)

electrolyte solution, illumination of n-type Si results in irreversible oxidation of the Si to form a thick, insoluble layer of silicon oxide on the surface. The photocurrent typically lasts less than 5 min in sunlight, since the silicon oxide is an insulator and prevents the passage of any current. But when the surface of the n-type Si is derivatized with a ferrocene-centered reagent, the situation is markedly different. The result of illuminating the derivatized n-type Si is to oxidize the surface-confined ferrocene, as represented by reaction (10),

$$\text{Surface-ferrocene} \xrightarrow{\text{sunlight}} \text{Surface-ferricenium} + e^- \qquad (10)$$

to form a surface-confined oxidant, ferricenium species. The photogenerated surface-confined ferricenium can then effect the oxidation of some species, B, dissolved in the aqueous electrolyte solution, regenerating the surface-confined ferrocene as shown in reaction (11). The key result is that the

$$\text{Surface-ferricenium} + B \rightarrow \text{Surface ferrocene} + B^+ \qquad (11)$$

ability to effect the photoinduced oxidation of B depends on the interaction with the ferricenium entity and not with the Si surface itself.

In principle, any species that can be oxidized by the ferricenium species can be photooxidized by illuminating the derivatized n-type Si electrode. In this instance the ferricenium/ferrocene system effects the mediated oxidation of B. In the mediated evolution of hydrogen using the $PQ^{2+}/PQ^{\dot{+}}$ system at illuminated p-type Si electrodes, the mediator system is dissolved in the solution. For the n-type Si, the mediator system is confined to the electrode, since the electrode must be protected from the photodecomposition. So far, the mediated evolution of oxygen has not yet been demonstrated, but continued progress in this area can be expected. *See* PHOTOVOLTAIC CELL; PHOTOVOLTAIC EFFECT. [MARK S. WRIGHTON]

Bibliography: D. C. Bookbinder et al., Photoelectrochemical reduction of N,N'-dimethyl-4,4'-bipyridinium in aqueous media at p-type silicon: Sustained photogeneration of a species capable of evolving hydrogen, *J. Amer. Chem. Soc.*, vol. 101, pp. 7721–7723, 1979; F.-R. F. Fan, B. Reichman, and A. J. Bard, Semiconductor electrodes 27: The *p*- and *n*-GaAs-*N,N'*-dimethyl-4,4'-dipyridinium system: Enhancement of hydrogen evolution on *p*-GaAs and stabilization of *n*-GaAs electrodes, *J. Amer. Chem. Soc.*, vol. 102, pp. 1488–1492, 1980; M. S. Wrighton, Photoelectrochemical conversion of optical energy to electricity and fuels, *Acc. Chem. Res.*, 12:303–310, 1979; M. S. Wrighton, Photochemistry, *Chem. Eng. News*, 57:29–47, 1979.

Photovoltaic cell

A device that detects or measures electromagnetic radiation by generating a current or a voltage, or both, upon absorption of radiant energy. Specially designed photovoltaic cells are used in solar batteries, photographic exposure meters, and sensitive detectors of infrared radiation. An important advantage of the photovoltaic cell in these particular applications is that no separate bias supply is needed—the device generates a signal (voltage or current) simply by the absorption of radiation.

Most photovoltaic cells consist of a semiconductor *p-n* junction or Schottky barrier in which electron-hole pairs produced by absorbed radiation are separated by the internal electric field in the junction to generate a current, a voltage, or both at the device terminals. The influence of the incident radiation on the current-voltage characteristics of a photovoltaic cell is to shift the current-voltage characteristic downward by the magnitude of the photogenerated current as shown in the illustration. Under open-circuit conditions (current $I = 0$) the terminal voltage increases with increasing light intensity (points A), and under short-circuit conditions (voltage $V = 0$) the magnitude of the current increases with increasing light intensity (points B). When the current is negative and the voltage is positive (point C, for example), the photovoltaic cell delivers power to the external circuit. In this case, if the source of radiation is the Sun, the photovoltaic cell is referred to as a solar battery. When a photovoltaic cell is used as a photographic exposure meter, it produces a current proportional to the light intensity (points B), which is indicated by a low-impedance galvanometer or microammeter. For use as sensitive detectors of infrared radiation, specially designed photovoltaic cells can be operated with either low-impedance (current) or high-impedance (voltage) amplifiers, although the lowest noise and highest sensitivity are achieved in the current or short-circuit mode. *See* JUNCTION DIODE; PHOTODIODE; PHOTOELECTRIC DEVICES; PHOTOVOLTAIC EFFECT; SEMICONDUCTOR; SOLAR CELL.

[GREGORY E. STILLMAN]

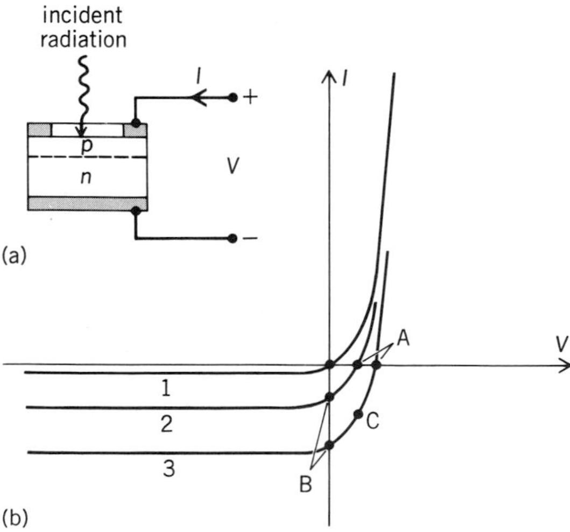

The *p-n* junction photovoltaic cell. (*a*) Cross section. (*b*) Current-voltage characteristics. Curve 1 is for no incident radiation, and curves 2 and 3 for increasing incident radiation.

Bibliography: H. J. Hovel, *Solar Cells*, vol. 11 of R. K. Willardson and A. C. Beer (eds.), *Semiconductors and Semimetals*, 1975; D. Long, Photovoltaic and photoconductive infrared detectors, in R. J. Keyes (ed.), *Optical and Infrared Detectors*, pp. 101–145, 1977; W. F. Wolfe and G. J. Zissis (eds.), *The Infrared Handbook*, 1978.

Photovoltaic effect

A term most commonly used to mean the production of a voltage in a nonhomogenous semiconductor, such as silicon, by the absorption of light or other electromagnetic radiation. In its simplest form, the photovoltaic effect occurs in the common photovoltaic cell, used, for example, in solar batteries and exposure meters. The photovoltaic cell consists of an *np* junction between two different semiconductors, an *n*-type material in which conduction is due to electrons, and a *p*-type material in which conduction is due to positive holes. When light is absorbed near such a junction, new mobile electrons and holes are released, as in photoconduction. An additional feature of a photovoltaic cell, however, is that there is an electric field in the junction region between the two semiconductor types. The released charge moves in this field. This current flows in an external circuit without the need for a battery as required in photoconduction. If the external circuit is broken, an "open-circuit photovoltage" appears at the break.

In certain rather complex electrolytic systems, illumination of the electrodes may give rise to a voltage classed as photovoltaic. *See* PHOTOVOLTAIC CELL; SOLAR BATTERY.

[L. APKER]

Bibliography: L. Azaroff and J. J. Brophy, *Electronic Processes in Materials*, 1963; A. Van der Ziel, *Solid State Physical Electronics*, 3d ed., 1976.

Pinch effect

A name given to manifestations of the magnetic self-attraction of parallel electric currents having the same direction. Since 1952 the pinch effect in a gas discharge has become the subject of intensive study in laboratories throughout the world, since it presents a possible way of achieving the magnetic confinement of a hot plasma (a highly ionized gas) necessary for the successful functioning of a thermonuclear or fusion reactor.

Ampère's law. The law of attraction which describes the interaction between parallel electric currents was discovered by André Marie Ampère in 1820 and can be stated as follows: The force of attraction in dynes per centimeter length between two thin straight wires r cm apart carrying currents of I_1 and I_2 amperes (A), respectively, is $I_1I_2/100r$. The law applies equally to the attraction between the individual components of a current in a single wire, in which case, for a cylindrical wire of radius r cm carrying a total current of I amperes,

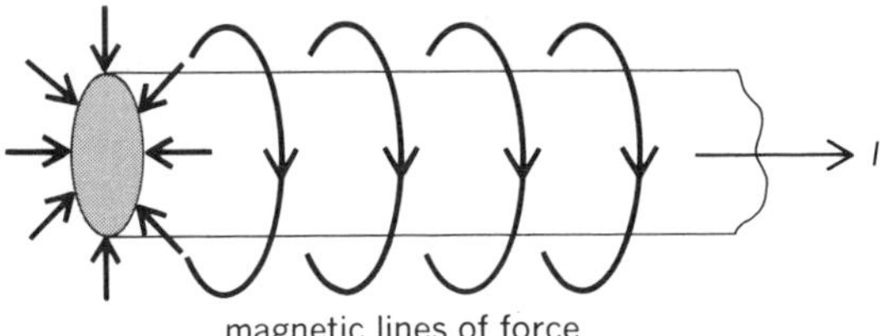

Fig. 1. Pinch pressure on a current-carrying conductor. Arrows at left show direction of pinch pressure.

it manifests itself as a compression force on the material of the wire (Fig. 1), given by $I^2/200\pi r^2$ dynes/cm². For a uniformly distributed current in the wire, the pressure reaches this value on the axis.

For the electric currents of normal experience, this force is small and passes unnoticed, but it is significant that the pressure increases with I^2. At 100,000 A, the pressure amounts to about 1 atm for a wire of 1-cm radius, but at 10^8 A, the pressure is about 10^6 atm, which is considerably greater than the pressure produced by the detonation of trinitrotoluene (TNT).

Manifestations. The pinch effect first showed up practically in certain early types of induction electric furnaces in which large low-frequency alternating currents of the order of 100,000 A were induced at low voltage into a horizontal ring-shaped fused-metal load (Fig. 2). At these currents, the pinch pressure can be larger than the hydrostatic pressure exerted by the fused metal, and as the formula above shows, the pinch pressure increases as the radius of the conductor decreases. Consequently, once the process starts, the pressure at a narrow neck in the ring of fused metal can squeeze out the fluid metal until the neck pinches off completely, cutting off the current. This led to very uneven heating of the charge. The term pinch effect was given to this process by C. Hering in 1907. The technical difficulty was eventually overcome by making the plane of the ring vertical and submerging it deeply below the free surface of the fused metal. The force of the pinch effect has also been known to manifest itself by a crushing of tubular conductors exposed to large impulsive currents such as occur in lightning strokes or high-power short circuits.

Thermonuclear applications. One of the conditions for the attainment of a profitable balance between energy expended in heating and energy released in fusion from a thermonuclear reaction in a plasma composed of deuterium and tritium (DT, the most favorable case) is that the temperature shall be not less than about 10 keV (1.16 times 10^8 K). This is an enormous temperature and can be attained and maintained only if the hot plasma is effectively isolated from the material walls of the container by vacuum. The isolation has the double function of preventing cooling by contact with matter at normal temperatures and of preserving the purity of the plasma from foreign atoms which could upset the energy balance. For the plasma to remain confined under these conditions, its outward pressure must be balanced by inward pressure of nonmaterial origin, that is, a magnetic field. A profitable energy balance also depends on the density n of the confined plasma and on τ, the time it is confined. The product of $n\tau$ must exceed a certain minimum, which is 10^{14} ion cm³ sec for DT. *See* NUCLEAR FUSION; THERMONUCLEAR REACTION.

There are only a limited number of ways in which a magnetic field can be arranged around the plasma to hold it together, and one of these methods is the pinch effect. A fusion reactor using this type of confinement would ideally be a toroidal tube in which the confined plasma would float, the plasma carrying a large electric current induced in it by magnetic induction from a transformer core passing through the axis of the torus. The fundamental equation for the pinch effect in a gas, derived theoretically by W. Bennett in 1934, gives the current I required for the inward pinch pressure to balance the outward gas pressure, as shown in the equation below, where I is the total current in

$$I^2/200 = Nk(T_e + T_i)$$

amperes, N is the number of electrons (also the number of ions) per centimeter length of the pinch (independent of the radius), $k = 1.4 \times 10^{+16}$ erg/K (Boltzmann's constant), and T_i and T_e are the temperatures in kelvins of the ions and electrons, respectively.

Experimental studies. In general, two types of apparatus have been used in studies of the pinch effect: (1) straight discharge tubes composed of quartz or porcelain with a metal electrode at each end, intended for short-duration studies, in which the cooling of the plasma by the relatively cold electrodes is slight during the time of the experiment, and (2) toroidal discharge tubes, also composed of quartz or porcelain, in which the pinch is endless and consequently is more effectively

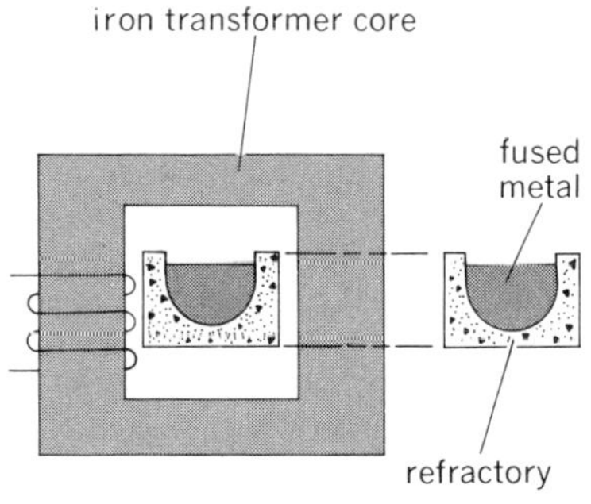

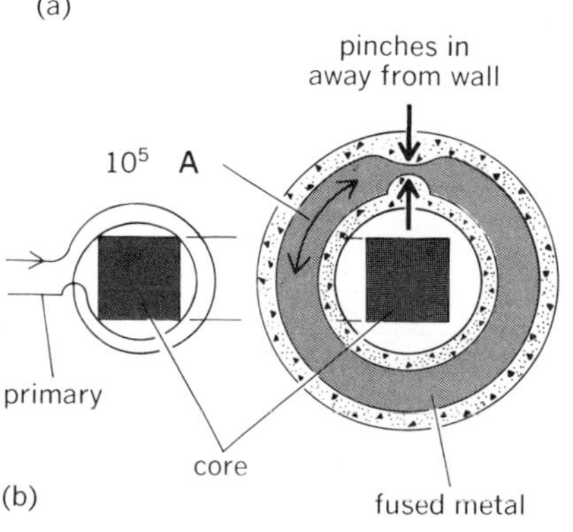

Fig. 2. Early type of ring induction electric furnace. (*a*) Side view. (*b*) Plan view.

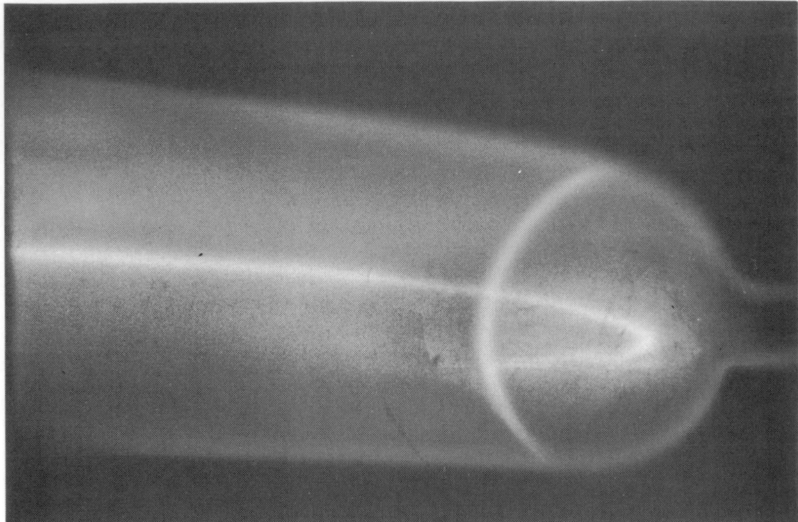

Fig. 3. Xenon pinched discharge in Perhapsatron torus.

confined than in the first type of apparatus, and the current is induced into the discharge by magnetic coupling to a primary winding. In both cases, currents of 50,000–500,000 A are obtained with gradients of 10–100 V/cm along the pinch. The primary power source is a charged condenser with a capacity of 4–50 μF, charged to 10–100 kV. The current in the discharge rises rapidly, reaches a peak in a few microseconds, and decays to zero in a damped oscillation.

Instability. Characteristically, as can be seen by high-speed photography, the discharge forms at the inner surface of the discharge tube wall and contracts inwardly, forming an intense line on the axis (Fig. 3); the wave usually rebounds slightly; the contracted discharge rapidly develops necks and kinks; and in a few microseconds, all structure is lost in an apparently turbulent glowing gas which fills the tube. Thus, the pinch turns out to be unstable, and the plasma confinement is soon lost by contact with the wall. The cause of the instability is easily seen qualitatively; the pinch confinement can well be described as being caused by the magnetic lines encircling the pinch behaving as slippery rubber bands which are stretched longitudinally but which are in compression transversely (Fig. 4). For a uniform cylindrical pinch, the magnetic pinch pressure is everywhere equal to the outward plasma pressure, but at a neck or on the inward side of a kink, the magnetic lines crowd together, creating a higher pressure than the outward gas pressure. Consequently, the neck contracts down further, the kink cuts in on the concave side and bulges out on the convex side, and both perturbations grow.

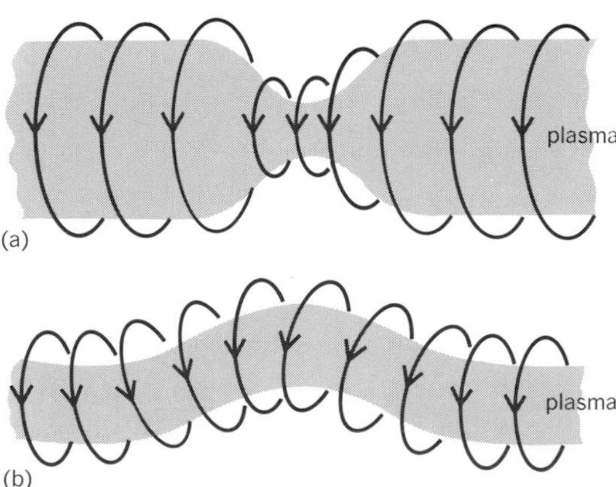

Fig. 4. Instability. (*a*) Sausage type. (*b*) Kink type.

The instability has a disastrous effect on τ, limiting it to 10^{-6} s or less in light atom plasmas such as DT. This not only makes it difficult to reach thermonuclear temperature from room temperature in the short time available for heating, but through its effect on $n\tau$ also forces the n required for a positive energy balance to extremely high values.

It must be noted that neutrons have been produced by deuterium pinches in large numbers. For a time (1952–1953), they were believed to be evidence of thermonuclear reaction, but it has since been shown that they are emitted preferentially in certain directions and are associated with the instability of the pinch and the violent accelerations it produces. Such neutrons (phony neutrons) are not a product of thermal collision and are not thermonuclear.

Great efforts have been devoted to overcoming the basic instability of the simple pinch. One such measure was to add an axial magnetic field by means of an external winding round the pinch tube. This might be expected to resist the sausage and kink deformations by stiffening the discharge. Also, the walls of the tube can be made highly conducting; this has the effect of trapping the magnetic field between the pinch and the wall, cushioning and reflecting the moving pinch back to the center.

Levitrons. A much more powerful measure for stabilizing the pinch consists of adding (1) a stiff current-carrying conductor down the axis inside the pinch and (2) a strong longitudinal magnetic field from an exterior winding outside it. The plasma is, in effect, sandwiched and pinched into a tubular region, between magnetic fields having directions differing by 90°. This so-called hard-core pinch shows greatly diminished instability, so much so that, for short time scales ($\sim 10^{-4}$ s) and straight tubes, it is stable.

Hard-core and levitron pinches have been studied at Lawrence Radiation Laboratory, Livermore, Calif.; Culham, England; and Fontenay-aux-Roses, France, and are associated with S. A. Colgate, H. P. Furth, and P. Rebut, respectively. These geometries are not entirely stable; a more slowly growing tearing mode instability remains. This seems to exhaust the possibility of stabilization of the pinch by static methods, but there remains the possibility of dynamic stabilization by rapidly changing fields. The simple pinch continues to be studied, however. The reason for this is that the reciprocal functional relation between n and τ for power production means that there is always a possibility of achieving a net power output, no matter how much τ is cut down by instabilities, by resorting to very high n. Such plasmas may require heroic extremes of pulse electric power to heat and confine them, but the only real limitation seems to be that the resultant output bursts of thermonuclear power may destroy the machine. For such pulsed systems, the pinch must always

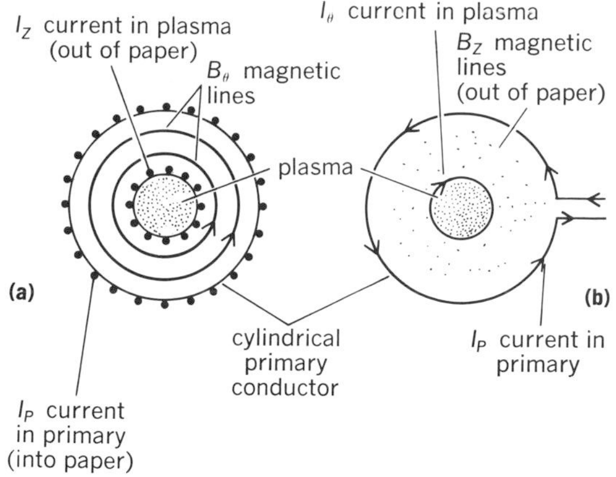

Fig. 5. Two geometries for plasma confinement systems. (*a*) Z pinch, and (*b*) θ pinch.

rank highly—it is uniquely the most efficient magnetic confinement system expressed in terms of magnetic energy expended per unit of plasma energy confined, and furthermore the pressure of the plasma it confines can exceed the strength of any known material.

The term θ pinch had come into wide usage to denote an important plasma confinement system which relies on the repulsion of oppositely directed currents and which is thus not in accord with the original definition of the pinch effect (self-attraction of currents of the same direction). Plasma confinement systems based on the original pinch effect are known as *z* pinches. Figure 5 defines the two geometries.

Tokamak. The toroidal B_z stabilized pinch became the subject of intense interest when L. Artsimovich, director of the Kurchatov Institute, announced in mid-1969 the achievement of a 20-ms confinement and an ion temperature of 0.5 keV in the T3 Tokamak. Tokamak is essentially a low-density, slow *z* pinch in a torus with a very strong longitudinal field, such that the pinch current I_z is below the Kruskal-Shafranov limit. This means that the helical magnetic lines, resultant from the externally applied B_z, and the internal B_0 of the pinch do not complete one revolution (2π) of the minor axis in going round the major axis of the machine once. This is known theoretically to prevent the growth of certain helical distortions of the plasma. The achievements are listed in the table, together with the target values for achieving a positive power balance.

The temperatures achieved are higher than the 0.5 keV that could be reached by resistive heating of the plasma by the pinch current, such as was used in the early 1970s. Auxiliary heating is necessary, and the most successful has been the injection of energetic neutral beams.

The performance of Tokamak experiments has raised the possibility of achieving a net power balance in the mid-1980s. Construction of several large Tokamak installations has been undertaken, including the TFTR (Princeton, United States), JET (European Community at Culham, United Kingdom), JT-60 (Tokyo, Japan), and T-15 (Moscow, Soviet Union).

Tokamak performance

Parameter	Tokamak	Target values
T_i (keV)	6	10
$n\tau$ (ions cm^{-3} s)	3×10^{13}	10^{14}

[JAMES L. TUCK; JAMES D. PHILLIPS]

Bibliography: International Fusion Research Council, Status report on controlled thermonuclear fusion, *Nucl. Fusion*, 18:137–149, 1978; C. L. Longmire, J. L. Tuck, and W. B. Thompson (eds.), Plasma physics and thermonuclear research, *Prog. Nucl. Energ., Ser. 11*, 1963; Theoretical and experimental aspects of controlled nuclear fusion, in *Proc. 2d Int. Conf. Peaceful Uses At. Energ.*, vols. 31–32, 1958; J. L. Tuck, Artsimovich talks about controlled-fusion research, *Phys. Today*, 22(6):54–57, 1969.

Pipeline

A line of piping and the associated pumps, valves, and equipment necessary for the transportation of a fluid. Major uses of pipelines are for the transportation of petroleum, water (including sewage), chemicals, foodstuffs, pulverized coal, and gases such as natural gas, steam, and compressed air. Pipelines must be leakproof and must permit the application of whatever pressure is required to force conveyed substances through the lines. Pipe is made of a variety of materials and in diameters from a fraction of an inch (25 mm) up to 30 ft (9 m). Principal materials are steel, wrought and cast iron, concrete, clay products, aluminum, copper, brass, cement and asbestos (called cement-asbestos), plastics, and wood.

Pipe is described as pressure and nonpressure pipe. In many pressure lines, such as long oil and gas lines, pumps force substances through the pipelines at required velocities. Pressure may be developed also by gravity head, as for example in city water mains fed from elevated tanks or reservoirs.

Nonpressure pipe is used for gravity flow where the gradient is nominal and without major irregularities, as in sewer lines, culverts, and certain types of irrigation distribution systems.

Design of pipelines considers such factors as required capacity, internal and external pressures, water- or airtightness, expansion characteristics of the pipe material, chemical activity of the liquid or gas being conveyed, and corrosion.

Most pipe is jointed, although some concrete pipe is monolithically cast in place. The length of the individual sections of pipe and the method of joining them depend upon the pipe material, diameter, weight, and requirements of use. Steel pipe sections are usually joined by welding, couplings, or riveting. Cast-iron pipe may be joined by couplings or, in the case of bell-and-spigot pipe, by filling the space between the bell and the spigot with calked or melted metal such as lead. Flexible-type joints with rubber gaskets are also used for joining cast iron pipe. The rubber gasket is contained in grooves and is ordinarily the sole element making the joint watertight.

Cement-mortar-filled or lead-filled rigid-type bell-and-spigot joints are usually used for joining concrete or vitrified clay sewer pipe. Tongue-and-groove rigid-type mortar-filled joints are often used for concrete pipe in low-pressure installations. The flexible-type joints are most frequently used for asbestos-cement pipe and concrete pipe under higher pressures. [LESLIE N. MCCLELLAN]

Plutonium

A chemical element, Pu, atomic number 94; the name is derived from the planet Pluto. Plutonium is a reactive, silvery metal in the actinide series of elements. The first isotope to be identified was produced from the irradiation of uranium by G. T. Seaborg, E. M. McMillan, A. C. Wahl, and J. Kennedy. The principal isotope of chemical interest is ^{239}Pu. It is formed in nuclear reactors by the

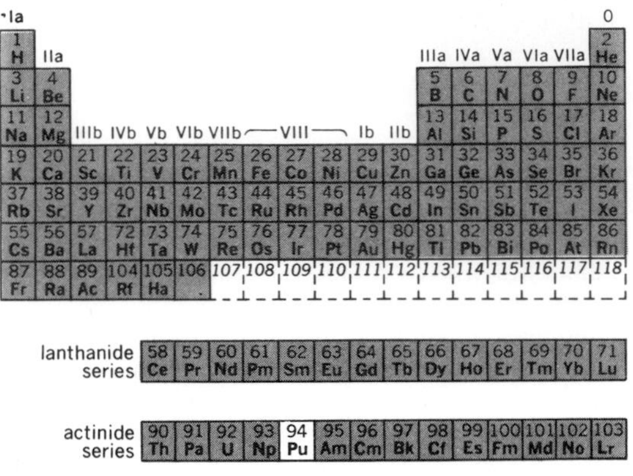

process shown in Eq. (1). Plutonium-239 is fissionable, but may also capture neutrons to form higher

$$^{238}U + n \longrightarrow {}^{239}U \xrightarrow[23.5\ min]{\beta^-} {}^{239}Np \xrightarrow[2.33\ days]{\beta^-} {}^{239}Pu \quad (1)$$

plutonium isotopes ^{240}Pu, ^{241}Pu, ^{242}Pu, ^{243}Pu, and ^{244}Pu. The chemically important isotope ^{239}Pu has a half-life of 24,131 years. The longest-lived species is ^{244}Pu with a half-life of 8.05×10^7 years. Minute quantities of ^{239}Pu are formed in pitchblende and monazite ore by reaction (1), the uranium and plutonium ratio being 10^{11}:1. Primordial ^{244}Pu has been found in the rare-earth mineral bastnasite.

Uses. Plutonium-238, with a half-life of 87.7 years, is utilized in heat sources for space application, and has been used for heart pacemakers. Plutonium-239 is used as a nuclear fuel, in the production of radioactive isotopes for research, and as the fissile agent in nuclear weapons. *See* NUCLEAR FUELS; NUCLEAR REACTOR.

Properties. Plutonium exhibits a variety of valence states in solution and in the solid state. Plutonium metal is highly electropositive. The oxidation states III, IV, V, and VI are known in acid solutions. The ions of the IV, V, and VI states are moderately strong oxidizing agents. The III, IV, and VI states can coexist in 1 *M* perchloric acid. In alkaline solutions, certain compounds of Pu(VI), and an unstable Pu(VII) oxidation state, may be formed under strongly oxidizing conditions (ozone). Solid compounds of all of the III, IV, V, VI, and VII valence states have been prepared.

Preparation. Methods for the isolation and purification of plutonium make use of the fact that the element can exist in a multiplicity of oxidation states, each having different chemical properties. The first isolation of plutonium, which was also the first isolation of a synthetic chemical element, was carried out by B. B. Cunningham in 1942. The first sample was weighed on a microbalance and found to weigh 2.77 μg. Special microchemical techniques were used in the early work to characterize the compounds of plutonium.

Carrier precipitation process. For isolation and purification of plutonium on the industrial scale, processes have been developed using carrier precipitation, solvent extraction, and ion exchange. The earliest plant-scale process was a carrier precipitation process developed by S. C. Thompson, which used bismuth phosphate and lanthanum fluoride as the carrier. In more recently developed processes, solvent extraction techniques are being used. These have the advantage that not only plutonium but uranium as well may be readily recovered and decontaminated from fission products. Some of the more important solvents are listed in Table 1. Control of the extraction behavior is obtained by the use of diluents for the solvent, addition of salting agents to the aqueous layer, and control of the solution pH and temperature. In Table 1, diluents and salting agents commonly employed for some of the more important valence states with two of these solvents are listed, and the relative distribution coefficients, defined as the concentration of the metal in the organic phase divided by the concentration in the aqueous phase) are shown in Table 2. The actual values of the distribution coefficients will change with the conditions, but the (approximate) relative values will be maintained.

The industrial processes are carried out in heavily shielded rows of concrete cells (canyons) in remotely controlled equipment. Charges up to several tons of irradiated fuel may be processed at one time.

Redox process. The redox solvent extraction process employing hexone has superseded the earlier bismuth phosphate separation process. In this procedure the uranium fuel is dissolved in nitric acid. The resulting spent fuel solution is oxidized, and Pu(VI) and U(VI) are separated by extraction from the fission products (codecontamination cycle). Next, the hexone layer is scrubbed to remove impurities, then the solvent is contacted with an aluminum nitrate solution containing a reducing agent (Fe(II) ion, hydroxylamine). The plutonium is back-extracted by this treatment into the aqueous layer as Pu(III), while the uranium remains in the solvent as U(VI) (partition cycle). The aqueous layer may be reoxidized and again extracted as before. By successive cycles of this process the plutonium is purified from fission products to the desired degree. Because the aqueous waste of the process is loaded with aluminum ni-

Table 1. Solvents used in the separation of plutonium and uranium from fission products

Solvent (trivial name)	Diluent	Salting agent
Methyl isobutyl ketone (Hexone)	None	$Al(NO_3)_3$
Tri-*n*-butyl phosphate (TBP)	Kerosine	HNO_3, $Al(NO_3)_3$
Dibutyl ether of ethylene glycol (Carbitol)	None	HNO_3
Dibutyl ether of tetra-ethylene glycol (Pentether)	None or butyl ether	HNO_3
Triglycol dichloride (Trigly)	None	$Al(NO_3)_3$
Thenoyl trifluoroacetone (TTA)	Benzene or toluene	None

Table 2. Distribution coefficients of uranium, plutonium, and fission products from nitrate solutions of 100-day cooled reactor fuel

Solvent	U(VI)	Pu(VI)	Pu(IV)	Pu(III)	Fission products
Hexone	1.5	7.6	1.6×10^{-2}	4.5×10^{-4}	6×10^{-4}
TBP	8.0	0.6	1.5	2×10^{-2}	2×10^{-3}

trate, and hexone is rather flammable, the redox process was discontinued, and was superseded in turn by a process employing a solution of tributyl phosphate in an aliphatic hydrocarbon solvent.

Purex process. The industrial process employing tri-n-butylphosphate (TBP) as the solvent extractant is the process used in most reprocessing plants. It is called the purex process (purex = *p*lutonium *u*ranium *r*eduction *ex*traction). The general operating procedure is similar to that of the redox process. After dissolution of the fuel, the oxidation state is adjusted so that the plutonium is present as Pu(IV), the uranium as U(VI). The nitric acid concentration is adjusted, and Pu(IV) and U(VI), are extracted into a solution of 30% TBP in an aliphatic hydrocarbon solvent (codecontamination cycle). The TBP phase containing Pu(IV) and U(VI) is scrubbed with nitric acid to remove impurities, and the Pu is removed as Pu(III) by back-extracting the solvent with nitric acid containing a reducing agent [NH_2OH, Fe(II) plus sulfamate, U(IV)], or after electrolytic reduction (partition cycle). Both the uranium and the plutonium are further purified by one or two additional purex extractions (second and third decontamination cycles). The major advantage of the purex system is that the waste generated in the process consists of a nitric acid solution of fission products containing very little solids, from which the nitric acid may be recovered by distillation and recycled.

In both the redox and purex processes, the final plutonium product is a nitric acid solution of Pu(IV). From this solution, plutonium may be isolated in a very pure state by precipitation as the peroxide or as the oxalate, either of which may be ignited to the oxide. Crude plutonium solutions containing other heavy metals may be purified from all such metals (except from thorium) by adjusting the acid concentration to 7–8 *M* HNO_3 and passing the solution through a Dowex 1 anion exchange column. Plutonium is retained as the $Pu(NO_3)_6^{2-}$ anion, which is stripped from the column with 3 *M* hydroxylamine nitrate.

Conversion to metal. To prepare the metal, the nitrate solution of Pu(IV) is precipitated as the peroxide or oxalate, which is ignited to PuO_2. The PuO_2 is hydrofluorinated to PuF_3 or PuF_4 by treating the oxide either with an H_2-HF mixture or with O_2-HF mixture. The fluoride is mixed with calcium metal, and charged into a steel bomb fitted with a ceramic liner. The bomb is closed with a flanged

Table 3. Properties of plutonium metal

Phase	Symmetry space group	Lattice Constants at t°C; a,b,c in angstroms β in degrees	Atoms per unit cell	Density (20°C) g cm^{-3}	Transition temperature, °C	Linear expansion coefficient* (20°C), $\alpha\times10^6$	Resistivity (20°C) $\times10^6$, ohm-cm	Temperature coefficient of resistivity†
α	Monoclinic, $P2_1/m$	a = 6.183(1) b = 4.822(1) c = 10.963(1) β = 101.79(1) t = 21°C	16	19.86		54	145	−21
$\alpha\to\beta$					122 (4)			
β	Monoclinic body-centered, I2/m	a = 9.284(3) b = 10.463(4) c = 7.859(3) β = 92.13(3) t = 190°C	34	17.70		42	110.5	−6
$\beta\to\gamma$					207 (5)			
γ	Face-centered orthorhombic, Fddd	a = 3.159(1) b = 5.768(1) c = 10.162(2) t = 235°C	8	17.14		34.6	110	−5
$\gamma\to\delta$					315 (3)			
δ	Face-centered cubic, Fm3m	a = 4.6371(4) t = 320°C	4	15.92		−8.6	103	+7
$\delta\to\delta'$					457 (2)			
δ'	Body-centered tetragonal, I4/mmm	a = 3.34(1) c = 4.44(4) t = 465°C	2	16.00		−65.6	105	+45
$\delta'\to\epsilon$					479 (4)			
ϵ	Body-centered cubic	a = 3.6361(4) t = 490°C	2	16.51		36.5	114	−7
$\epsilon\to$liq					640°C (mp)			
liquid				16.62		93		
ζ	?	?	?		>330 >0.6 kbar			

$^*\alpha = \frac{1}{L}\cdot\frac{L}{T}$ $\quad ^\dagger\frac{1}{\rho}\cdot\frac{\rho}{t}\times10^5$

SOURCE: O. J. Wick, *Plutonium Handbook*, vol. 1, Gordon & Breach, 1969; ζ-plutonium data from J. R. Morgan, in W. N. Miner, Plutonium 1970 and Other Actinides, *Nucl. Metallurgy*, vol. 17, Metallurgical Society of AIME.

Table 4. Plutonium halides and oxyhalides

Compound	Color	Melting point, °C	Density at 20°C	Crystal structure
PuF_3	Purple	1425	9.32	Hexagonal
PuF_4	Pale brown	1037	7.0	Monoclinic
PuF_6	Reddish-brown	50.75		Orthorhombic
$PuCl_3$	Green	760	5.70	Hexagonal
$PuBr_3$	Green	681	6.69	Orthorhombic
PuI_3	Green	777	6.92	Orthorhombic
PuOF	Metallic	1635	9.76	Tetragonal
PuOCl	Blue-green		8.81	Tetragonal
PuOBr	Green		9.07	Tetragonal
PuOI	Green		8.46	Tetragonal

lid and ignited by induction heating. After reduction, the metal is isolated by breaking the liner.

Plutonium metal has unique properties. It exists in six allotropic forms below the melting point (640°C) at ordinary pressure. A seventh allotrope is obtained when the metal is compressed at pressures greater than 0.6 kbar. Particularly puzzling, however, are the contractions which the δ- and δ′-modifications undergo with increasing temperature (see illustration). Noteworthy is the fact that in no phase do both the coefficient of thermal expansion and the temperature coefficient of resistivity have the conventional algebraic signs. If the phase expands on heating, the resistance decreases. The peculiar thermal expansion behavior of the metal prevents the use of unalloyed plutonium metal as a reactor fuel. However, the δ-phase may be stabilized over a wide temperature range by addition of gallium or aluminum (so-called δ-retainers) and may thus be utilized in reactors. A liquid alloy, consisting of plutonium, cobalt, and cerium with a melting point of 415°C, has been used as fuel in the Los Alamos Molten Plutonium Reactor Experiment (LAMPRE). Some of the physical properties of the normal pressure modifications of plutonium are given in Table 3.

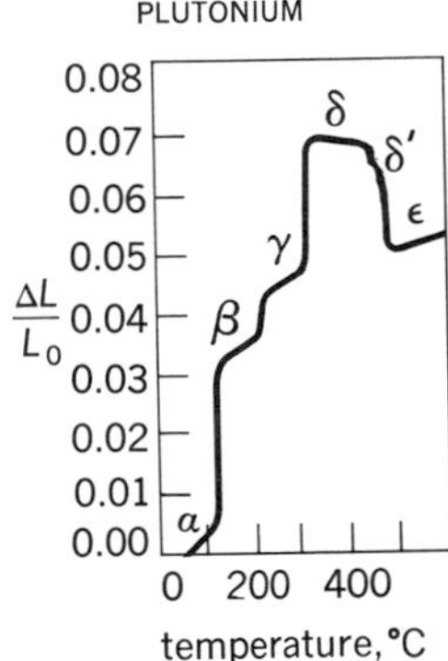

Expansion of high-purity plutonium under conditions of self-heating, L_0 = 0.5 in. (13 mm). (*After E. R. Jette*)

Numerous alloys of plutonium have been prepared, and a large number of intermetallic compounds have been characterized.

Principal compounds. Reaction of the metal with hydrogen yields two hydrides: a nonstoichiometric $PuH_{2.0...2.7}$ and stoichiometric PuH_3. The hydrides are formed at temperatures as low as 150°C. Their decomposition at temperatures above 750°C may be used to prepare reactive plutonium powder.

The most common oxide is PuO_2, which is formed by ignition of hydroxides, oxalates, peroxides, and nitrates of any oxidation state in air of 870–1200°C. PuO_2 crystallizes in a face-centered cubic structure, the exact lattice constant depending on the stoichiometry. Stoichiometric PuO_2 is amber. Deviations from stoichiometry are associated with colors from dark olive green to reddish brown. In the Pu-O system, a number of oxides have been reported which range in composition from $PuO_{1.5}$ to PuO_2, including Pu_2O_3, which is dark blue, hexagonal, and isomorphous with the A-type rare-earth oxides.

A very important class of plutonium compounds are the halides and oxyhalides (Table 4). Plutonium hexafluoride is the most volatile plutonium compound known. It resembles UF_6 and NpF_6 in its properties. PuF_6 is a strong fluorinating agent. Conditions for the preparation of the fluorides are given in Eqs. (2), (3), and (4). The other halides may

$$PuO_2 + \tfrac{1}{2}H_2 + 3HF \xrightarrow{600^\circ C} PuF_3 + 2H_2O \quad (2)$$

$$PuO_2 + O_2 + 4HF \xrightarrow{550^\circ C} PuF_4 + 2H_2O + O_2 \quad (3)$$

$$PuF_4 + F_2 \xrightarrow{750^\circ C} PuF_6 \quad (4)$$

be prepared by a number of methods. Treatment of PuO_2 with powerful halogenating agents such as CCl_4, PCl_5, SCl_2, and hexachloropropene yields $PuCl_3$. $PuBr_3$ may be obtained by dehydrating $PuBr_3 \cdot 6H_2O$ in a HBr atmosphere, and PuI_3 can be prepared by reacting plutonium metal with either HI, I_2, or HgI_2.

A number of other binary compounds are known (Table 5). Among these are the carbides, silicides, sulfides, and selenides, which are of particular interest because of their refractory nature. These compounds are usually prepared by direct combination of the elements.

Plutonium in any of its valency states forms a large number of fluoroplutonates and oxoplutonates with alkali metal and ammonium fluorides, or alkali metal oxides (Table 6). In addition, numerous oxoplutonates and fluoroplutonates are known of similar composition formed with alkaline earths, lanthanides, and a few transition metals.

A large number of plutonium compounds may be prepared from aqueous solution, either by evaporation or by precipitation. In Table 7 a selection of such compounds is given. Furthermore, numerous compounds of plutonium with organic acids are known, and π-complex compounds such as $Pu(C_5H_5)_3$ have also been prepared.

Properties of the ions in aqueous solution. In aqueous solution, plutonium tends to exist in the IV oxidation state. In strongly oxidizing conditions, Pu(VI) is formed; under strongly reducing conditions, Pu(III) is produced. The redox potentials are so close in their values that pure solutions of intermediate oxidation states undergo self-oxidation and reduction reactions (disproportionation). A typical disproportionation reaction is the one in-

Table 5. Binary compounds of plutonium with elements of groups III, IV, V, and VI

III	IV	V	VI		
PuB	PuC_{1-x}	PuN	$PuS_{0.95}$	PuSe	PuTe
PuB_2	Pu_2C_3	PuP	PuS	Pu_2Se_{3-x}	γ-Pu_2Te_3
PuB_4	PuC_2	PuAs	Pu_3S_4	γ-Pu_2Se_3	η-Pu_2Te_3
PuB_6			Pu_5S_7	η-Pu_2Se_3	$PuTe_{2-x}$
PuB_{12}	Pu_5Si_3		α-Pu_2S_3	$PuSe_{1.8}$	$PuTe_3$
"PuB_{100}"	Pu_3Si_2		$PuS_{1.9}$	$PuSe_{1.9}$	
	PuSi		PuS_2		
	Pu_3Si_5				
	$PuSi_2$				

Table 6. Fluoroplutonates and oxoplutonates of alkali metals and ammonium

Valency	Compound	Type	Color
III	$MPuF_4$	M = Na, K	Blue
	MPu_2F_7	M = K	Blue
IV	MPu_3F_{13}	$M = NH_4$	Pink to brownish
	MPu_2F_9	M = K	Pink to brownish
	$MPuF_5$	$M = Li, NH_4$	Pink to brownish
	$M_7Pu_6F_{31}$	$M = Na, K, Rb, NH_4$	Pink to brownish
	M_2PuF_6	$M = Na, K, Rb, Cs, NH_4$	Pink to brownish
	M_3PuF_7	M = Na	Pink to brownish
	M_4PuF_8	$M = Li, NH_4$	Pink to brownish
V	$MPuF_6$	M = Cs	Green
	M_2PuF_7	M = Rb	Green
IV	M_8PuO_6	M = Li	Brown
V	M_3PuO_4	M = Li	Brown
VI	M_2PuO_4	M = Rb, Cs	Brown or black
	M_4PuO_5	M = Li, Na	Brown or black
	M_6PuO_6	M = Li, Na	Dark green
VII	M_5PuO_6	M = Li	Dark green
	M_3PuO_5	M = Rb, Cs	Black

Table 7. Selected plutonium compounds obtained from aqueous solutions

III	IV	V	VI
$PuPO_4 \cdot 0.5H_2O$	$Pu(OH)_4 \cdot xH_2O$	$KPuO_2CO_3$	$PuO_2(NO_3)_2 \cdot 6H_2O$*
$Pu_2(C_2O_4)_3 \cdot 10H_2O$	$Pu(JO_3)_4$	$RbPuO_2CO_3$	$NaPuO_2(CH_3COO)_3$
$PuF_3 \cdot xH_2O$	$PuO_4 \cdot 2H_2O$		$HPuO_2PO_4 \cdot 3H_2O$
$Pu_2(SO_4)_3 \cdot 7H_2O$	$Pu(HPO_4)_2 \cdot xH_2O$		$KPuO_2PO_4 \cdot 3H_2O$
	$Pu(C_2O_4)_2 \cdot 6H_2O$		$NH_4PuO_2PO_4 \cdot 3H_2O$
	$Pu(SO_4)_2$*		$PuO_2C_2O_4$
	$K_4Pu(SO_4)_4 \cdot 2H_2O$		
	$Pu(NO_3)_4 \cdot 5H_2O$*		

*Obtained by evaporation of its aqueous solution.

volving Pu(IV), which can be written as Eq. (5), for which the equilibrium constant is expressed as Eq. (6). K_1, as calculated from the redox potentials, is 0.0089 for Pu(IV) in 1 *M* acid at 25°C.

$$3Pu^{4+} + 2H_2O \rightleftharpoons PuO_2^{2+} + 2Pu^{3+} + 4H^+ \quad (5)$$

$$K_1 = \frac{[PuO_2^{2+}]\,[Pu^{3+}]\,[H^+]^4}{[Pu^{4+}]^3} \quad (6)$$

In solution of acids, such as nitric or hydrochloric acids, whose anions form only weak complexes with plutonium ions, the relative stabilities of the different states are little changed. Qualitatively, it is known that univalent anions, with the exception of fluoride, form relatively weak complexes with plutonium ions in all oxidation states. Higher-valent anions, however, form relatively strong complexes. In general, the relative stabilities of complexes with a given anion decrease in the order

$$Pu^{4+} > PuO_2^{2+} > Pu^{3+} > PuO_2^{+}.$$

The ions of the different oxidation states have characteristic colors. Pu^{3+} is bright blue or blue-violet; Pu^{4+} is reddish brown, brownish red, or, as a nitrate complex, brownish green to dark green; PuO^+, pale purple; and PuO_2^{2+}, pink to orange-red. Like the rare earths, they have characteristic absorption spectra with sharp absorption bands. These have been widely used in the analysis of plutonium solutions to determine the relative amounts of each oxidation state present.

Safety precautions. Because of its radiotoxicity, plutonium and its compounds require special handling techniques to prevent ingestion or inhalation. Therefore, all work with plutonium and its compounds must be carried out inside glove boxes. For work with plutonium and its alloys, which are attacked by moisture and by atmospheric gases, these boxes may be filled with helium or argon. *See* URANIUM. [FRITZ WEIGEL]

Bibliography: J. R. Cleveland, *The Chemistry of Plutonium*, 2d ed., 1979; *Gmelin Handbuch der Anorganischen Chemie*, 8th ed., 10 vols., supplement: Transurane, 1973–1979; G. T. Seaborg (ed.), *Transuranium Elements: Products of Modern Alchemy*, Benchmark Papers in Physical Chemistry and Chemical Physics, 1978; O. J. Wick (ed.), *Plutonium Handbook*, 2 vols., 1967.

Power

The time rate of doing work. Like work, power is a scalar quantity, that is, a quantity which has magnitude but no direction. Some units often used for the measurement of power are the watt (1 joule of work per second) and the horsepower (550 ft-lb of work per second). *See* HORSEPOWER; WORK.

Usefulness of the concept. Power is a concept which can be used to describe the operation of any system or device in which a flow of energy occurs. In many problems of apparatus design, the power, rather than the total work to be done, determines the size of the component used. Any device can do a large amount of work by performing for a long time at a low rate of power, that is, by doing work slowly. However, if a large amount of work must be done rapidly, a high-power device is needed. High-power machines are usually larger, more complicated, and more expensive than equipment which need operate only at low power. A motor which must lift a certain weight will have to be larger and more powerful if it lifts the weight rapidly than if it raises it slowly. An electrical resistor must be large in size if it is to convert electrical energy into heat at a high rate without being damaged.

Electrical power. The power P developed in a direct-current electric circuit is $P=VI$, where V is the applied potential difference and I is the current. The power is given in watts if V is in volts and I in amperes. In an alternating-current circuit, $P=VI\cos\phi$, where V and I are the effective values of the voltage and current and ϕ is the phase angle between the current and the voltage. *See* ALTERNATING CURRENT.

Power in mechanics. Consider a force F which does work W on a particle. Let the motion be restricted to one dimension, with the displacement in this dimension given by x. Then by definition the power at time t will be given by Eq. (1). In this

$$P=dW/dt \tag{1}$$

equation W can be considered as a function of either t or x. Treating W as a function of x gives Eq. (2). Now dx/dt represents the velocity v of the

$$P=\frac{dW}{dt}=\frac{dW}{dx}\frac{dx}{dt} \tag{2}$$

particle, and dW/dx is equal to the force F, according to the definition of work. Thus Eq. (3) holds.

$$P=Fv \tag{3}$$

This often convenient expression for power can be generalized to three-dimensional motion. In this case, if ϕ is the angle between the force $\mathbf{F}$ and the velocity $\mathbf{v}$, which have magnitudes F and v, respectively, Eq. (4) expresses quantitatively that if a

$$P=\mathbf{F}\cdot\mathbf{v}=Fv\cos\phi \tag{4}$$

machine is to be powerful, it must run fast, exert a large force, or do both. [PAUL W. SCHMIDT]

Power plant

A means for converting stored energy into work. Stationary power plants such as electric generating stations are located near sources of stored energy, such as coal fields or river dams, or are located near the places where the work is to be performed, as in cities or industrial sites. Mobile power plants for transportation service are located in vehicles, as the gasoline engines in automobiles and diesel locomotives for railroads. Power plants range in capacity from a fraction of a horsepower (hp) to over 10^6 kilowatts (kW) in a single unit (Table 1). Large power plants are assembled, erected, and constructed on location from equipment and systems made by different manufacturers. Smaller units are produced in manufacturing facilities.

Most power plants convert part of the stored raw energy of fossil fuels into kinetic energy of a spinning shaft. Some power plants harness nuclear energy. Elevated water supply or run-of-the-river energy is used in hydroelectric power plants. For transportation, the plant may produce a propulsive jet, as in some aircraft, instead of the rotary motion of a shaft. As of 1979, other sources of energy, such as winds, tides, waves, geothermal sources, ocean thermal, nuclear fusion, and solar radiation, were of negligible commercial significance in the generation of power despite their magnitudes.

Table 2 shows the scope of United States power-plant capacity. About a third of the world's electric energy (in kilowatt-hours, kWh) is generated by the United States public utility systems (Fig. 1), and the installed generating capacity (in kW) of the United States (including Alaska and Hawaii) is about the same as the total of the next four countries (Fig 2). Figure 3 shows the declining importance of hydroelectric power in the United States, where the kWh output from hydro has dropped from 30 to 9% of the total electric generation in 28 years. These data, coupled with the data of Table 2, reflect the dominant position of thermal power both for stationary service and for the propulsion of land-, water-, and air-borne vehicles. *See* ENERGY SOURCES.

Rudimentary flow- or heat-balance diagrams for important types of practical power plants are shown in Fig. 4. Figure 5 is a diagram for a by-product-type industrial steam plant (also known as cogeneration) which has the double purpose of generating electric power and simultaneously delivering heating steam by extraction or exhaust from the prime mover. *See* HEAT BALANCE.

Table 1. Representative design and performance data on power plants

Type	Unit size range, kW	Fuel*	Plant weight, lb/kW†	Plant volume, ft³/kW‡	Heat rate, Btu/kWh§
Central station					
Hydro	10,000–700,000				
Steam (fossil-fuel–fired)	10,000–1,300,000	CO		20–50	8,500–15,000
Steam (nuclear)	500,000–1,200,000	N			10,000–12,000
Diesel	1,000–5,000	DG			10,000–15,000
Combustion turbine	5,000–10,000	D′G			11,000–15,000
Industrial (by-product) steam	1,000–25,000	COGW		50–75	4,500–6,000
Diesel locomotive	1,000–5,000	D	100–200	2–3	10,000–15,000
Automobile	25–300	G′	5–10	0.1	15,000–20,000
Outboard motor	1–50	G′	2–5	0.1–0.5	15,000–20,000
Truck	50–500	D	10–20		12,000–18,000
Merchant ship, diesel	5,000–20,000	D	300–500		10,000–12,000
Naval vessel, steam	25,000–100,000	DON	25–50		12,000–18,000
Airplane, reciprocating engine	1,000–3,000	G′	1–3	0.05–0.10	12,000–15,000
Airplane, turbojet	3,000–10,000	D′	0.2–1		13,000–18,000

*C, coal; D, diesel fuel; D′, distillate; G, gas; G′, gasoline; N, nuclear; O, fuel oil (residuum); W, waste.
†1 lb/kW = 0.45 kg/kW. ‡1 ft³/kW = 2.83×10^{-2} m³/kW. †1 Btu/kWh = 1.055 kJ/kWh = 2.93×10^{-4} J(heat)/J(output).

Table 2. Approximate 1978 installed capacity of United States power plants

Plant type	Capability, 10^6 kW
Electric central stations	504
Industrial	40
Agricultural	60
Railroad	70
Marine, civilian	40
Aircraft, civilian	70
Military establishment	2,000±
Automotive	10,000±
Total	13,000±

Plant load. There is no practical way of storing the mechanical or electrical output of a power plant in the magnitudes encountered in power plant applications, although several small-scale concepts are in the research and development stage. As of now, however, the output must be generated at the instant of its use. This results in wide variations in the loads imposed upon a plant. The capacity, measured in kW or hp, must be available when the load is imposed. Much of the capacity may be idle during extended periods when there is no demand for output. Hence much of the potential output, measured as kWh or hp-h, cannot be generated because there is no demand for output. This greatly complicates the design and confuses the economics of power plants. Kilowatts cannot be traded for kilowatt-hours, and vice versa. *See* ENERGY STORAGE.

The ratios of average load to rated capacity or to peak load are expressed as the capacity factor and the load factor, Eqs. (1) and (2), respectively. The range of capacity factors experienced for various types of power plants is given in Table 3.

$$\text{Capacity factor} = \frac{\text{average load for the period}}{\text{rated or installed capacity}} \qquad (1)$$

$$\text{Load factor} = \frac{\text{average load for the period}}{\text{peak load in the period}} \qquad (2)$$

Variations in loads can be conveniently shown on graphical bases as in Figs. 6 and 7 for public utilities and in Fig. 8 for air and marine propulsion. Rigorous definition of load factor is not possible with vehicles like tractors or automobiles because of variations in the character and condition of the running surface. In propulsion applications, power output may be of secondary import: performance may be based on tractive effort, drawbar pull, thrust, climb, and acceleration.

Plant efficiency. The efficiency of energy conversion is vital in most power plant installations.

Table 3. Range of capacity factors for selected power plants

Power plants	Factor, percent
Public-utility systems, in general	50–70
Chemical or metallurgical plant, three-shift operation	80–90
Seagoing ships, long voyages	70–80
Seagoing ships, short voyages	30–40
Airplanes, commercial	20–30
Private passenger cars	1–3
Main-line locomotives	30–40
Interurban buses and trucks	5–10

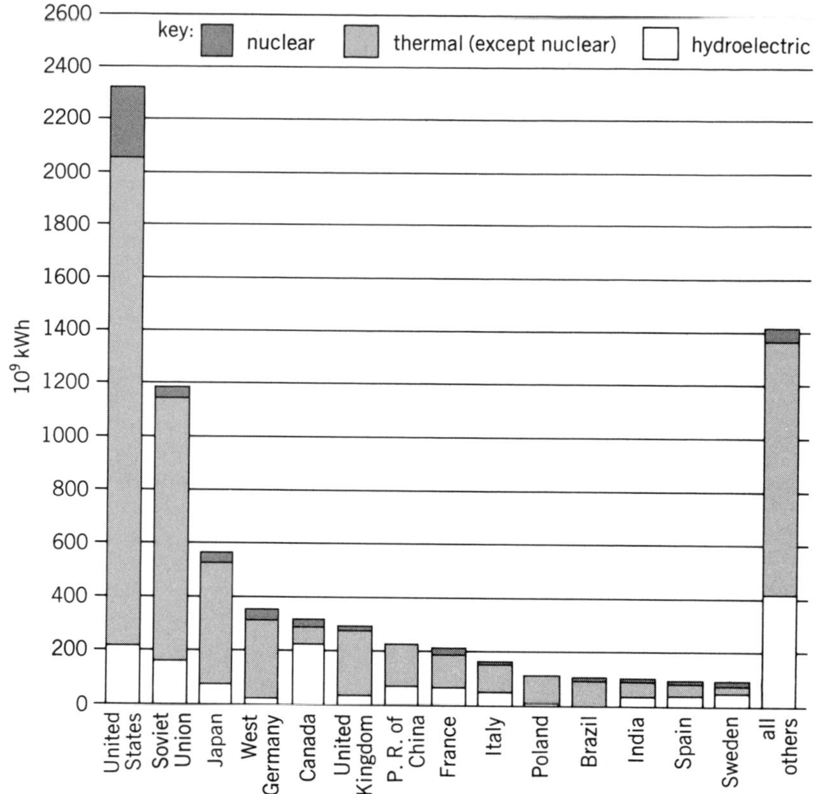

Fig. 1. Electric energy production by type in the 14 highest-producing countries, 1978. Nearly 30% of world electrical energy is generated in United States plants. (*United Nations*)

The theoretical power of a hydro plant in kW is $QH/11.8$, where Q is the flow in cubic feet per second and H is the head (height of water intake above discharge level) at the site in feet. In metric

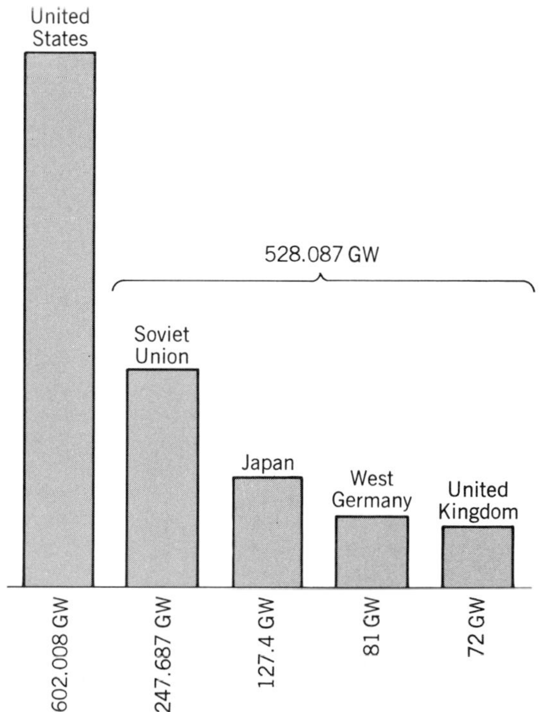

Fig. 2. The five countries with the highest capacities to generate electric power, 1978. (*United Nations*)

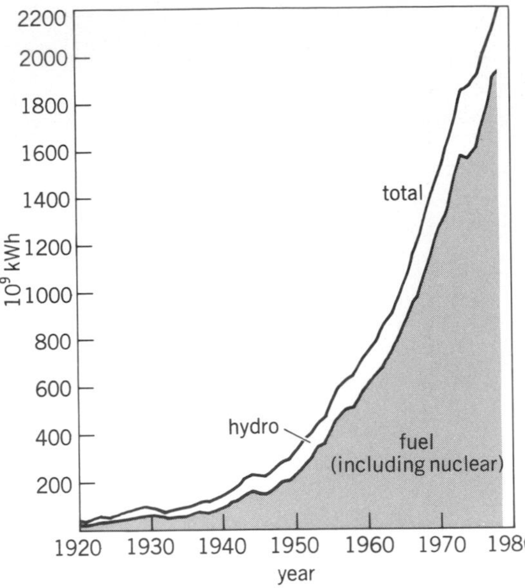

Fig. 3. Total electrical energy generation in the United States utility industry (including Alaska and Hawaii since 1963), by type of prime mover. (*Edison Electric Institute*)

units, the theoretical power of the plant in kW is 9.8 $Q'H'$, where Q' is the flow in cubic meters per second and H' is the head in meters. Losses in headworks, penstocks, turbines, draft tubes, tailrace, bearings, generators, and auxiliaries will reduce the salable output 15–20% below the theoretical in modern installations. The selection of a particular type of waterwheel depends on experience with wheels at the planned speed and on the lowest water pressure in the water path. Runners of the reaction type (high specific speed) are suited to low heads (below 500 ft or 150 m) and the impulse type (low specific speed) to high head service (about 1000 ft or 300 m). The lowest heads (below 100 ft or 300 m) are best accommodated by reaction runners of the propeller or the adjustable blade types. Mixed-pressure runners are favored for the intermediate heads (50–500 ft or 15–150 m). Draft tubes, which permit the unit to be placed safely above flood water and without sacrifice of site head, are essential parts of reaction unit installations. *See* HYDRAULIC TURBINE; WATERPOWER.

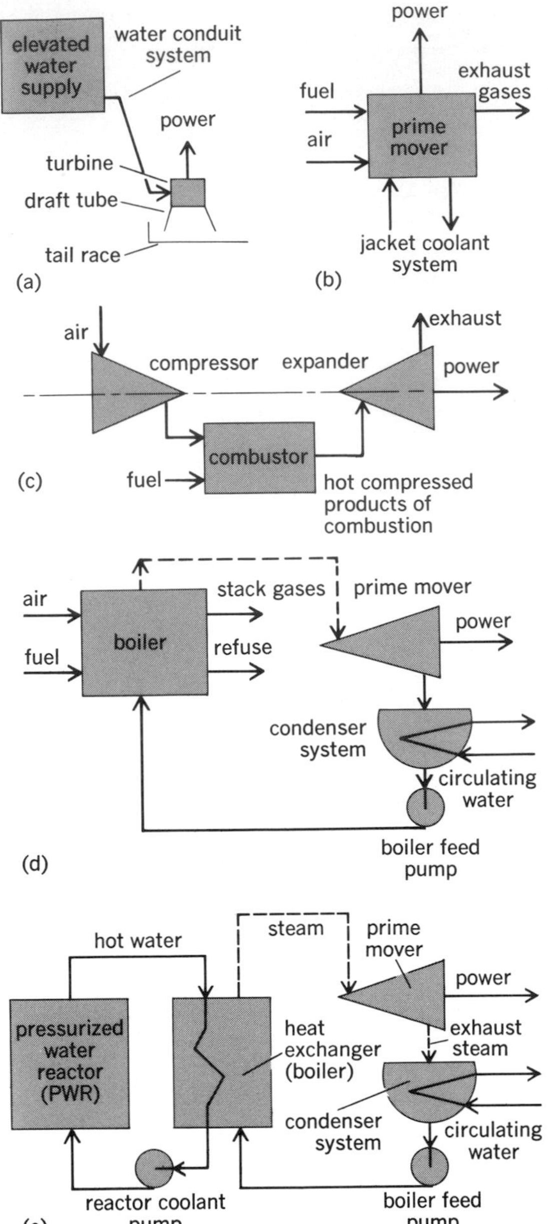

Fig. 4. Rudimentary flow- or heat-balance diagrams for power plants. (*a*) Hydro. (*b*) Internal combustion. (*c*) Gas-turbine. (*d*) Fossil-fuel-fired. (*e*) Nuclear steam (pressurized water reactor, PWR).

Table 4. Costs of representative power plants in 1978

Cost factors	Steam, central station, fossil fuel: Large (2×10^6 kW)*	Steam, central station, fossil fuel: Small (10^5 kW)	Steam, central station, nuclear fuel (10^6 kW)
Investment dollars per kW	354	647	642
Fuel cost, cents per 10^6 Btu (cents per GJ)	140 (131)	193 (183)	43 (41)
Cost of power, mills per kWh:			
Total cost	28.4	42.9	27.5
Carrying cost on investment	12.4	21.4	18.6
Production cost total	16.0	21.5	8.9
Fuel	13.7	16.1	5.9
Labor, maintenance, supplies, and supervision	2.3	5.4	3.0

*Composite of several multiunit stations.

With thermal power plants the basic limitations of thermodynamics fix the efficiency of converting heat into work. The cyclic standards of Carnot, Rankine, Otto, Diesel, and Brayton are the usual criteria on which heat-power operations are variously judged. Performance of an assembled power plant, from fuel to net salable or usable output, may be expressed as thermal efficiency (%); fuel consumption (lb, pt, or gal per hp-h or per kWh); or heat rate (Btu supplied in fuel per hp-h or per kWh). American practice uses high or gross calorific value of the fuel for measuring heat rate or thermal efficiency and differs in this respect from European practice, which prefers the low or net calorific value.

Tables 1 and 3 give performances for selected operations. Figures 9 and 12 reflect the improvement in fuel utilization of the United States electric power industry since 1900, although there have been some minor decreases since 1970. Figure 12 is especially significant, as it shows graphically the impact of technological improvements on the cost of producing electrical energy despite the harassing increases in the costs of fuel during the same period. Such technological improvements, however, were unable to compensate for the steep rise in fuel prices that followed the late-1973 Arab oil embargo and the mid-1979 shortages. These conditions boosted 1974–1975 fuel prices completely out of line with those of earlier years, and led to further large increases in 1979. Figure 10 illustrates the variation in thermal performance as a function of load for an assortment of stationary and marine propulsion power plants. *See* BRAYTON CYCLE; CARNOT CYCLE; DIESEL CYCLE; OTTO CYCLE; RANKINE CYCLE; THERMODYNAMIC CYCLE.

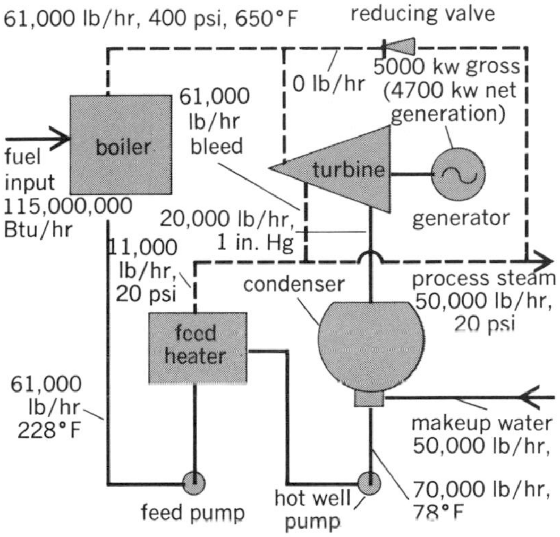

Fig. 5. Heat balance for a by-product industrial power plant delivering both electrical energy and process steam. 1 lb/hr = 1.26×10^{-4} kg/sec. 1 Btu/hr = 0.293 W. 1 psi = 6.895 kPa. 1 in. Hg = 3.386 kPa.

In scrutinizing data on thermal performance, it should be recalled that the mechanical equivalent of heat (100% thermal efficiency) is 2545 Btu/hp-h and 3413 Btu/kWh (3.6 MJ/kWh). Modern steam plants in large sizes (75,000–1,300,000 kW units) and internal combustion plants in modest sizes (1000–5000 kW) have little difficulty in delivering a kWh for less than 10,000 Btu (10.55 MJ) in fuel (34% thermal efficiency). Lowest fuel consumptions per unit output (8500–9000 Btu/kWh or 9.0–9.5 MJ/kWh) are obtained in condensing steam plants with the best vacua, regenerative-reheat cycles using eight stages of extraction feed heating, two stages of resuperheat, primary pressures of 3500 psi or 24 MPa (supercritical) and temperatures of 1150°F (620°C). An industrial plant generating electric power as a by-product of the process steam load is capable of having a thermal efficiency of 5000 Btu/kWh (5.3 MJ/kWh).

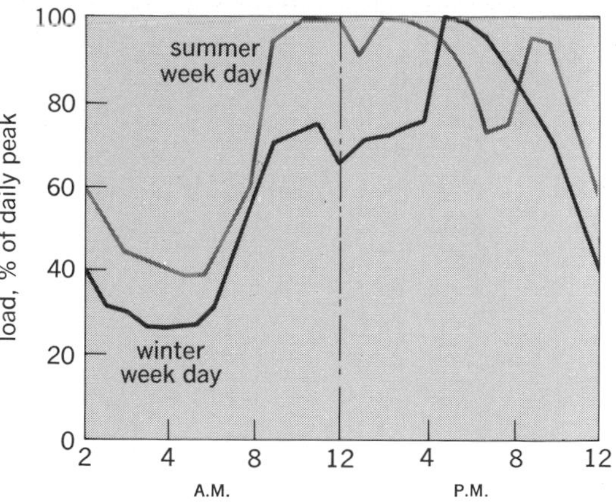

Fig. 6. Daily-load curves for urban utility plant.

The nuclear power plant substitutes the heat of fission for the heat of combustion, and the consequent plant differs only in the method of preparing the thermodynamic fluid. It is otherwise similar to the usual thermal power plant. Low reactor temperatures lead to the overwhelming preference for steam-turbine rather than gas-turbine cycles. When fluid temperatures can be had above 1200°F (650°C), the gas-power cycle will receive more favorable consideration. Otherwise the nuclear power plant is essentially a low-pressure, low-temperature steam operation (less than 1000 psi or 6.9 MPa and 600°F or 320°C). *See* NUCLEAR REACTOR.

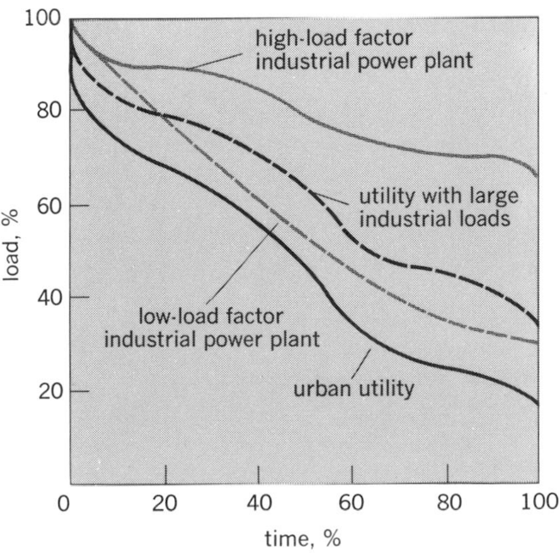

Fig. 7. Annual load-duration curves for selected stationary public utility power plants.

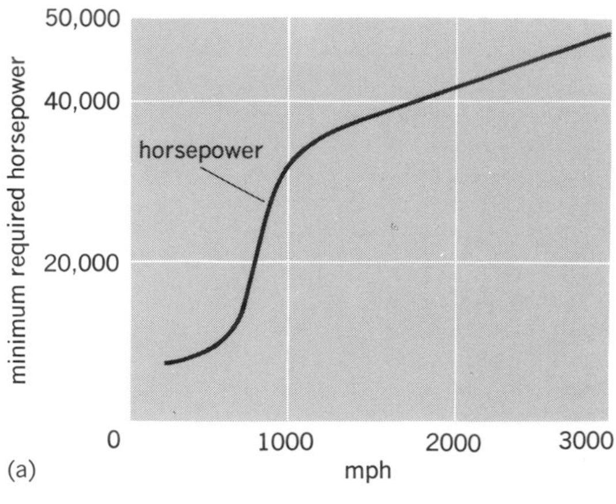

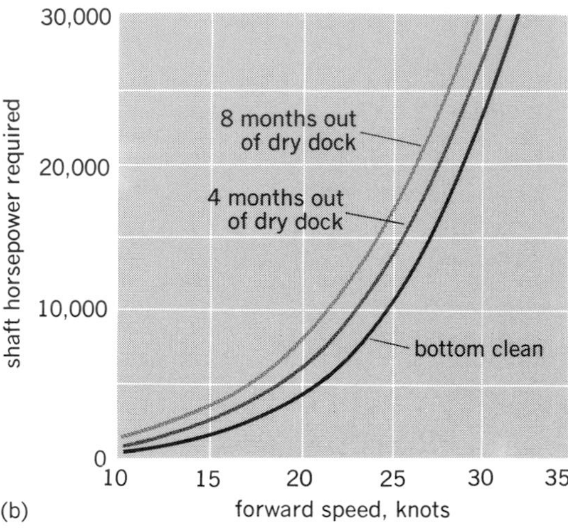

Fig. 8. Air and marine power. *(a)* Minimum power required to drive a 50-ton (45-metric-ton) well-designed airplane in straight level flight at 30,000 ft (9.1 km) altitude. *(b)* Power required to drive a ship, showing effect of fouling. 1 hp = 746 W; 1 mph = 0.447 m/sec; 1 knot = 0.514 m/sec.

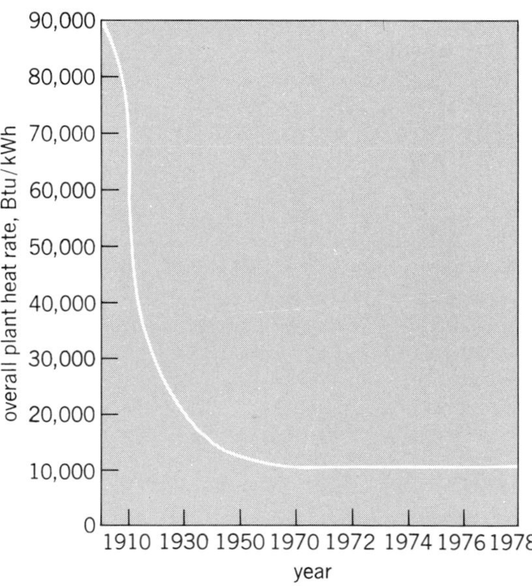

Fig. 9. Thermal performance of fuel-burning electric utility power plants in the United States. 1 Btu/kWh = 1.055 kJ/kWh = 2.93 × 10^{-4} J(heat)/J(output).

Power economy. Costs are a significant, and often controlling, factor in any commercial power plant application. Average costs have little significance because of the many variables, especially load factor. Some plants are short-lived and others long-lived. For example, in most automobiles, which have short-lived power plants, 100,000 mi (160,000 km) and 3000–4000 hr constitute the approximate operating life; diesel locomotives, which run 20,000 mi (32,000 km) a month with complete overhauls every few years, and large seagoing ships, which register 1,000,000 mi (1,600,000 km) of travel and still give excellent service after 20 years of operation, have long-lived plants; electric central stations of the hydro type can remain in service 50 years or longer; and steam plants run round the clock and upward of 8000 hr a year with complete reliability even when 25 years old. Such figures greatly influence costs. Furthermore, costs are open to wide differences of interpretation.

In the effort to minimize cost of electric power to the consumer it is essential to recognize the difference between investment and operating costs, and the difference between average and incremental costs. Plants with high investment (fixed) costs per kW should run at high load factors to spread the burden. Plants with high operating costs (such as fuel) should be run only for the shortest periods to

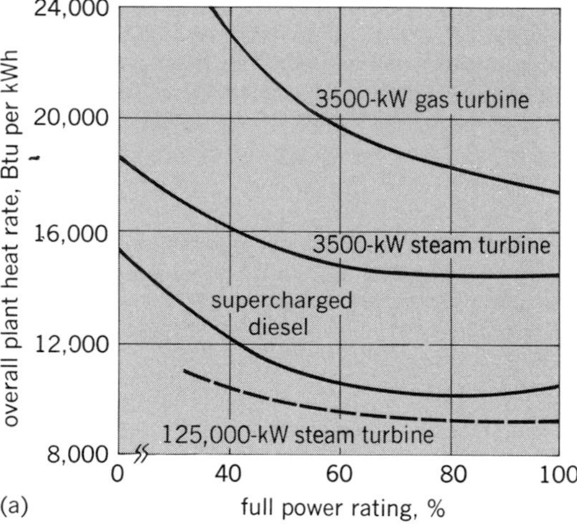

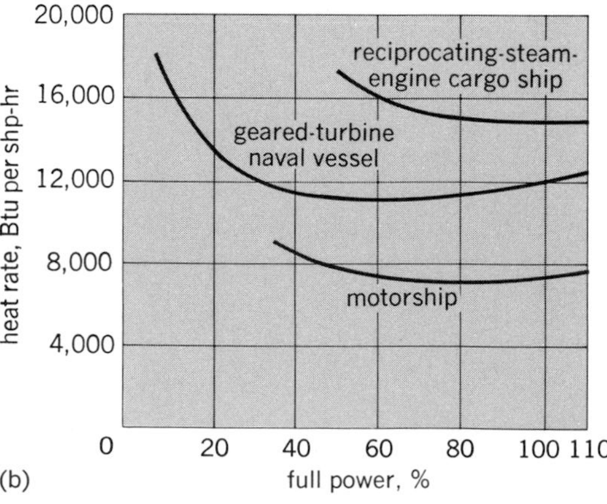

Fig. 10. Comparison of heat rates. *(a)* Stationary power plants. *(b)* Marine propulsion plants. 1 Btu/kWh = 1.055 kJ/kWh = 2.93 × 10^{-4} J(heat)/J(output). 1 Btu per hp-h = 1.415 kJ/kWh = 3.93 × 10^{-4} J(heat)/J(output).

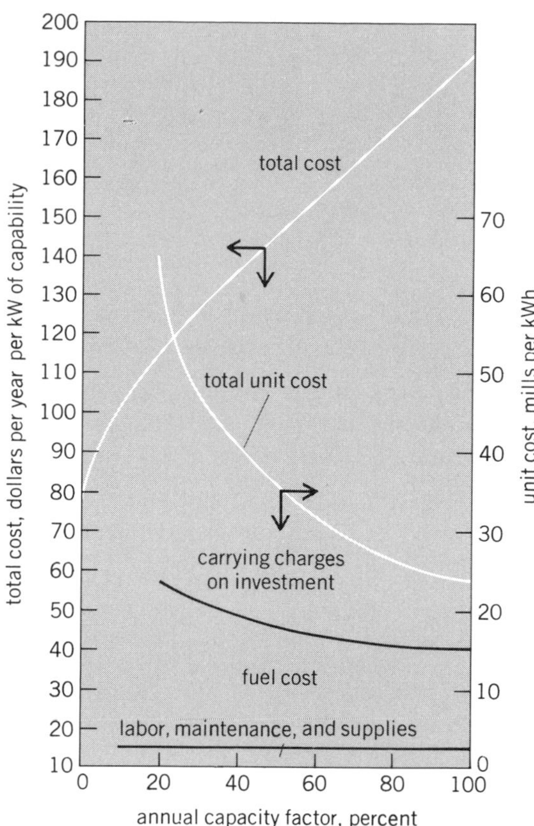

Fig. 11. Representative costs of power from a large steam-electric generating station as of 1978.

meet peak loads or emergencies. To meet these short operating periods various types of peaking plants have been built. Combustion (gas) turbines and pumped-storage plants serve this requirement. In the latter a hydro installation is operated off-peak to pump water from a lower reservoir to an elevated reservoir. On-peak the operation is reversed with water flowing downhill through the prime movers and returning electrical energy to the transmission system. High head sites (for example, 1000 ft or 300 m), proximity to transmission lines, and low incremental cost producers (such as nuclear or efficient fossil-fuel-fired plants) are necessary. If 2 kWh can thus be returned on-peak to the system, for an imput of 3 kWh off-peak, a pumped-storage installation is generally justifiable. *See* GAS TURBINE; PUMPED STORAGE.

In any consideration of such power plant installations and operations it is imperative to recognize (1) the requirements of reliability of service and (2) the difference between average and incremental costs. Reliability entails the selection and operation of the proper number and capacity of redundant systems and components and of their location on the system network. Emergencies, breakdowns, and tripouts are bound to occur on the best systems. The demand for maximum continuity of electrical service in modern civilization dictates the clear recognition of the need to provide reserve capacity in all components making up the power system.

Within that framework the minimum cost to the consumer will be met by the incremental loading of equipment. Incremental loading dictates, typically, that any increase in load should be met by supplying that load with the unit then in service, which will give the minimum increase in out-of-pocket operating cost. Conversely, for any decrease in load, the unit with the highest incremental production cost should drop that decrease in load. This is a complex technical, economic, and management problem calling for the highest degree of professional competence for its proper solution.

Costs are reflected in the curves of Fig. 11 for a steam-electric, investor-owned, central station in the eastern United States. Fixed costs are based on \$354/kW investment with 21% annual carrying charges. If such a plant were government-financed, the annual charges on an investment of \$354/kW might be reduced to 10%. Fuel cost is at 140 cents per 10^6 Btu (133 cents per GJ). Total cost of power can be conveniently expressed as in the following computation.

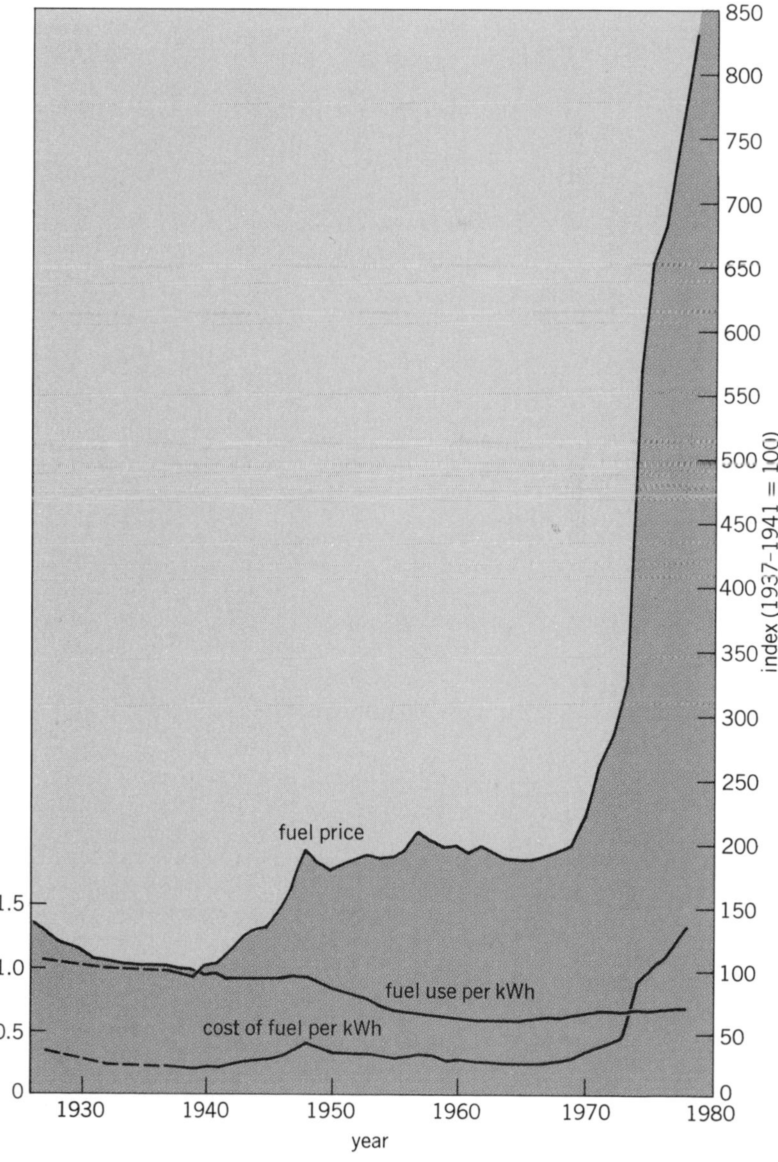

Fig. 12. Effect of fuel price and efficiency of use upon cost of fuel per kilowatt-hour generated in the United States utility industry (including Alaska and Hawaii since 1963). *(Edison Electric Institute)*

\$ per kW per year

$$= K_1 + K_2 + 8760\, K_3 \times \text{capacity factor}$$

where K_1 = capacity charge, \$ per kW per year
K_2 = peak prepared-for charge, \$ per kW per year
K_3 = energy cost, \$ per kWh

Costs of representative power plants are summarized in Table 4. Hydroelectric plants generally range from 20,000 to 200,000 kW and have costs of power about 65% of those for fossil fuel plants. Industrial plants generally range from 1000 to 10,000 kW with power costs 100 to 122% of those for central station units. *See* ELECTRIC POWER GENERATION. [THEODORE BAUMEISTER; LEONARD M. OLMSTED; KENNETH A. ROE]

Bibliography: Babcock and Wilcox Co., *Steam: Its Generation and Use*, rev. 39th ed., 1978; T. Baumeister (ed.), *Standard Handbook for Mechanical Engineers*, 8th ed., 1978; Combustion Engineering, Inc., *Combustion Engineering*, rev. ed., 1966; W. T. Creager and J. D. Justin, *Hydroelectric Handbook*, 1950; Diesel Engine Manufacturing Association, *Standard Practices*, 1972; D. G. Fink and H. W. Beaty, *Standard Handbook for Electrical Engineers*, 11th ed., 1978; G. D. Friedlander, 21st Steam Station Cost Survey, *Elec. World*, 192(10):55–70, Nov. 15, 1979; R. L. Harrington, *Marine Engineering*, 1971; L. C. Lichty, *Combustion Engine Processes*, 7th ed., 1967.

Primary battery

An electric battery designed to deliver only one continuous or intermittent discharge. It cannot be recharged efficiently. Primary batteries are designed to deliver limited amounts of electric energy, determined by the materials used and the size of the cell. When the available energy drops to zero, the battery is usually discarded. Primary batteries may be classified by the type of electrolyte used.

Aqueous-electrolyte batteries. These batteries use solutions of acids, bases, or salts in water as the electrolyte. These solutions have ionic conductivities of the order of 1 mho/cm and practically no electronic conductivity. Practical cells, such as the common Leclanche dry cell and the alkaline-manganese-zinc cell use aqueous electrolytes. Disadvantages of such cells include corrosion of the electrode materials by the electrolyte, a relatively high evaporation rate of water vapor which can cause cell failure, and the difficulties of preventing leakage. For examples of cells with aqueous electrolytes *see* BATTERY; DRY CELL; MERCURY BATTERY; RESERVE BATTERY.

Solid-electrolyte batteries. These use electrolytes of solid crystalline salts which have predominantly ionic conductivity. The conductivity is small compared with aqueous electrolytes, and the current output is of the order of 10^{-8} A/cm^2.

Solid-electrolyte batteries may be classified in two broad categories: (1) cells with solid crystalline salt, such as silver iodide, as the electrolyte; (2) cells with ion-exchange membrane as the electrolyte. In either category, the conductivity must be nearly 100% ionic. Any electronic conductivity causes a continuous discharge of the cell and will limit the stand or shelf life.

A typical cell with solid crystalline salt electrolyte is the lead–lead chloride–silver chloride cell in Fig. 1. Here lead is the anode, lead chloride is the electrolyte, and silver chloride is the cathode. This cell has a potential of 0.49 V. During discharge, lead is oxidized to lead ion and silver chloride is reduced to silver.

PRIMARY BATTERY

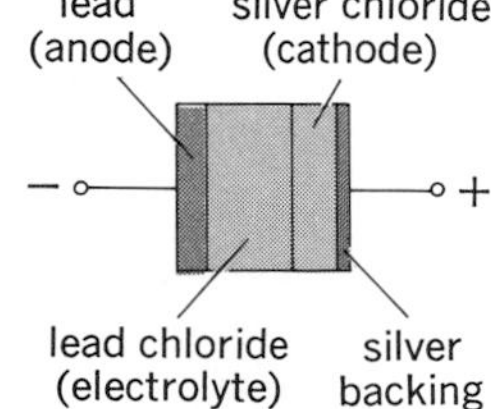

Fig. 1. Typical solid-electrolyte cell with solid crystalline salt electrolyte.

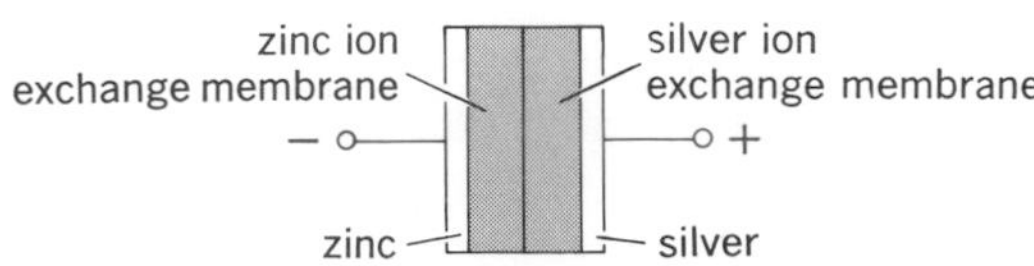

Fig. 2. Ion-exchange solid-electrolyte bimetallic cell.

Cells with solid salt electrolyte have been developed into miniature batteries. One type delivers 90–100 V at 10^{-11} A, and has a capacity of 1 ampere-second. This is over 10^6 days at 10^{-11} A. The practical life of the cell is much less but may be as much as 10 years at room temperature. It can be stored at 71°C for at least 30 days and will operate over the range −54 to +74C. The battery is 0.95 cm in diameter and 2.54 cm in length. With the increasing use of electronic devices and consequent miniaturization, solid-state batteries delivering low currents are finding newer applications. In addition, the use of electrolytes such as Ag_3SI or MAg_4I_5, (where M is K, Rb, NH_4, or Cs), which have better ionic conductivity than the lead or silver halides previously employed, gives cells with flash currents in the low milliampere range.

An example of a cell with ion-exchange membrane as electrolyte is the zinc–zinc ion exchange membrane, silver ion exchange membrane–silver cell shown in Fig. 2. Physically, the metal electrodes are in contact with the solid membrane which contains two regions. The region adjacent to the zinc is in the zinc ion state. The region adjacent to the silver is in the silver ion state. The discharge reaction increases the zinc ion quantity and decreases the silver ion quantity, in proportion to the amount of charge transferred. This cell has a potential of about 1.5 V.

The zinc-silver cell described has serious shortcomings. The shelf life is poor, indicating internal self-discharge, and the capacity is limited by the available supply of silver ions. In strongly ionized types of ion-exchange material, the volume density of ionizing sites is about 1 equiv/liter, or 0.24 ampere-hour/cm^3. This is very low compared with metal oxide cathodes.

A cell with higher capacity can be made by replacing the silver ion exchange material and silver by manganese dioxide plated on an inert metal,

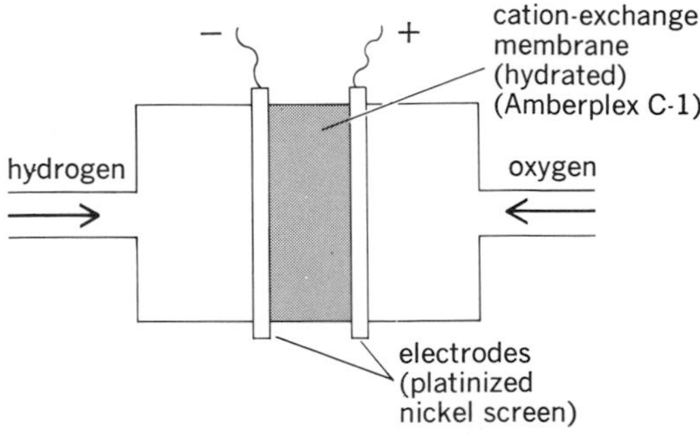

Fig. 3. Solid-electrolyte cell using an ion-exchange membrane as the electrolyte.

such as tantalum. This gives a capacity of about 100 times as much, for equal volume.

Ion-exchange electrolytes are also used with hydrogen and oxygen gas electrodes (Fig. 3). The electrodes consist of platinized metal screens. The electrolyte is a hydrogen ion exchange material. The room temperature emf of this cell is 0.96 V.

Waxy-electrolyte batteries. These use waxy materials, such as polyethylene glycol, in which a small amount of a salt is dissolved in the molten wax. At room temperatures these materials are solid. The conductivity is small and the current output is limited to about 10^{-7} A/cm^2.

Figure 4 shows a battery stack of cells using a waxy electrolyte. The electrodes are sheet zinc and manganese dioxide. The electrolyte is made of polyethylene glycol in which is dissolved a small amount of zinc chloride. This electrolyte is melted and painted on a paper sheet to form the separator.

A 25-cell stack, built as shown in Fig. 4 and measuring 0.86 cm in length and 0.64 cm in diameter, weighed 1.5 g. A 1.27-cm-diameter stack weighed 6.0 g. The initial open-circuit voltage was 37.5 V (1.5 V per cell).

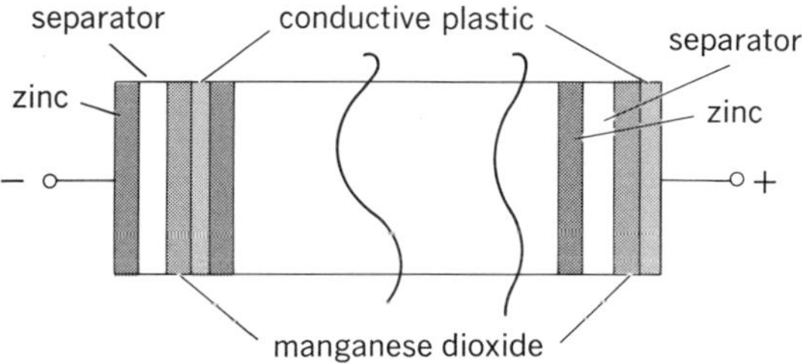

Fig. 4. Waxy-electrolyte battery stack.

The internal resistance of this cell is high, and it increases as temperature decreases. This high internal resistance limits the usefulness of the cell, but it may be suitable for long-life potential sources of miniature size.

Fused-electrolyte batteries. These use crystalline salts or bases which are solid at room temperature. In use, the cell is heated and maintained at a temperature above the melting point of the electrolyte. [JACK DAVIS; KENNETH FRANZESE]

Bibliography: C. R. Argue, B. Owens, and I. J. Grace, *Proceedings of the Power Sources Conference*, vol. 22, 1968; W. J. van der Grinten, *J. Electrochem. Soc.*, 103:210C, 1956; R. Jasinski, *High-Energy Batteries*, 1967; K. Lehovec and J. Broder, *J. Electrochem. Soc.*, 101:208, 1954; A. Sator, *Compt. Rend.*, 234:2283, 1952; S. W. Shapiro, *Proceedings of the Power Sources Conference*, vol. 11, 1957.

Prime mover

The component of a power plant that transforms energy from the thermal or the pressure form to the mechanical form. Mechanical energy may be in the form of a rotating or a reciprocating shaft, or a jet for thrust or propulsion. The prime mover is frequently called an engine or turbine and is represented by such machines as waterwheels, hydraulic turbines, steam engines, steam turbines, windmills, gas turbines, internal combustion engines, and jet engines. These prime movers operate by

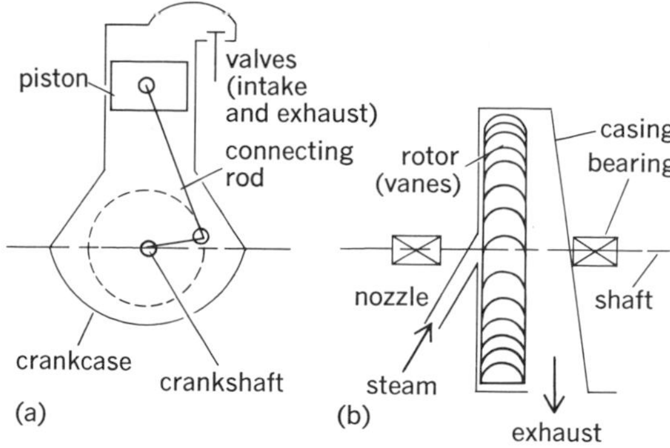

Fig. 1. Representative prime movers. (*a*) Single-acting four-cycle, automotive-type internal combustion engine. (*b*) Single-stage, impulse-type steam turbine.

either of two principles (Fig. 1): (1) balanced expansion, positive displacement, intermittent flow of a working fluid into and out of a piston and cylinder mechanism so that by pressure difference on the opposite sides of the piston, or its equivalent, there is relative motion of the machine parts; or (2) free continuous flow through a nozzle where fluid acceleration in a jet (and vane) mechanism gives relative motion to the machine parts by impulse, reaction, or both. *See* IMPULSE TURBINE; INTERNAL COMBUSTION ENGINE; REACTION TURBINE; STEAM ENGINE.

Displacement prime mover. Power output of a fluid-displacement prime mover is conveniently determined by pressure-volume measurement recorded on an indicator card (Fig. 2). The area of the indicator card divided by its length is the mean effective pressure (mep) in pounds per square inch, and horsepower of the prime mover is given by Eq. (1), where L is stroke in feet, a is piston area

$$\text{Horsepower} = \frac{\text{mep} \times Lan}{33{,}000} \qquad (1)$$

in square inches, and n is number of cycles completed per minute. Actual mep is smaller than the theoretical mep and may be related to the theoretical value by diagram factor or engine efficiency (Table 1).

Acceleration prime mover. Performance of fluid acceleration (hydraulic) prime movers is given by Eq. (2), where Q is water flow rate in cubic

$$\text{Horsepower} = \frac{QH}{8.8} \times \text{efficiency} \qquad (2)$$

feet per second, and H is head in feet. For heat-power prime movers of the fluid acceleration type, actual properties of the thermodynamic fluid, as

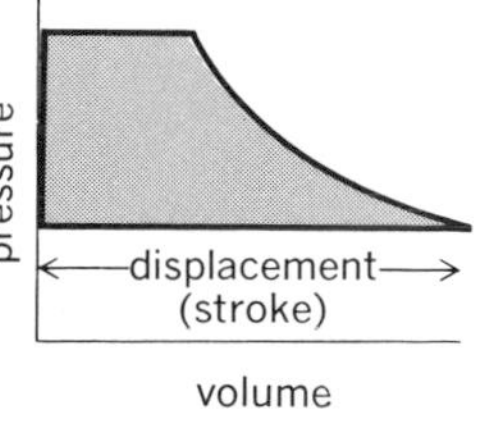

Fig. 2. Pressure-volume diagram (indicator card) for ideal, no-clearance, fluid-displacement types of prime mover.

Table 1. Dimensional and performance criteria of some selected fluid displacement-type prime movers

Type	Size, hp[a]	rpm	Stroke, in.[b]	Bore: stroke ratio	Piston speed, ft/min[c]	Brake mep, psi[d]	Diagram factor, or engine efficiency
Steam engine	25–500	100–300	6–24	0.8–1.2	400–600	50–100	0.6–0.8[e]
Automobile engine	10–300	2000–4000	3–5	0.9–1.1	1000–2000	50–100	0.4–0.6[f]
Aircraft engine	100–3000	2500–3500	4–7	0.8–1.1	1500–3000	100–230	0.4–0.6[f]
Diesel, low-speed	100–5000	100–300	10–24	0.8–1.0	500–1000	40–80	0.4–0.7[f]
Diesel, high-speed	25–1000	1500–2000	3–6	0.8–1.0	800–1500	50–100	0.4–0.6[f]

[a]1 hp = 0.75 kW. [b]1 in. = 25 mm. [c]1 ft/min = 5.1 mm. [d]1 psi = 6.895 kPa. [e]Logarithmic standard. [f]Air-card standard.

Table 2. Dimensional and performance criteria of some selected fluid acceleration-type prime movers

Type	Rating, kW	Number of stages	Head, ft; or pressure, psi[a]	Temperature, °F(°C)	Exhaust pressure in. Hg abs[b]	rpm	Tip speed, ft/sec[c]	Efficiency
Pelton water wheel	1000–200,000	1	500–5000 ft	Ambient	atm	100–1200	100–250	0.75–0.85
Francis hydraulic turbine	1000–200,000	1	50–1000 ft	Ambient	atm[d]	72–360	50–200	0.8–0.9
Propeller (and Kaplan) hydraulic turbine	5000–200,000	1	20–100 ft	Ambient	atm[d]	72–180	70–150	0.8–0.9
Small condensing steam turbine	100–5000	1–12	100–400 psi	400–700 (200–370)	1–5	1800–10,000	200–800	0.5–0.8
Large condensing steam turbine	100,000–1,000,000	20–50	1400–4000 psi	900–1100 (480–690)	1–3	1800–3600	500–1500	0.8–0.9
Gas turbine	500–20,000	10–20	70–100 psi	1200–1500 (650–820)	atm	3600–10,000	500–1500	0.8–0.9

[a]1 ft = 0.3 m; 1 psi = 6.895 kPa. [b]1 in. Hg = 3.4 kPa. [c]1 ft/sec = 0.3 m/sec. [d]Draft tube gives negative pressure on discharge side of runner.

given in tables and graphs, especially the Mollier chart, permit the rapid evaluation of the work or power output from the general energy equation which resolves to the form of Eq. (3), where h is the

$$\Delta W, \text{Btu/lb of fluid} = h_{\text{inlet}} - h_{\text{exhaust}} \qquad (3)$$

enthalpy in Btu/lb, and the inlet and exhaust conditions can be connected by an isentropic expansion for ideal conditions, or modified for irreversibility to a lesser difference by engine efficiency (Fig. 3 and Table 2). Fluid consumption follows from Eq. (4) or Eq. (5).

$$\text{Fluid consumption, lb per hphr} = 2545/\Delta W \qquad (4)$$

$$\text{Pounds per kWh} = 3413/\Delta W \qquad (5)$$

In the fluid acceleration type of prime mover, jet velocities experienced in the nozzles can be found in feet per second, for nonexpansive fluids, by Eq. (6). For expansive fluids, they may be found by Eq. (7), where H and ΔW are as given above and C is

$$\text{Jet velocity} = C\sqrt{2gH} = 8.02\,C\sqrt{H} \qquad (6)$$

$$\text{Jet velocity} = C\sqrt{2g\Delta W} = 223.7\,C\sqrt{\Delta W} \qquad (7)$$

the velocity coefficient, seldom less than 0.95 and usually from 0.98 to 0.99.

Selected representative performance values of some prime movers are presented in Tables 1 and 2. *See* GAS TURBINE; HYDRAULIC TURBINE; STEAM; STEAM TURBINE; TURBINE.

[THEODORE BAUMEISTER]

Bibliography: T. Baumeister (ed.), *Standard Handbook for Mechanical Engineers*, 8th ed., 1978; L. C. Lichty, *Combustion Engine Processes*, 1967.

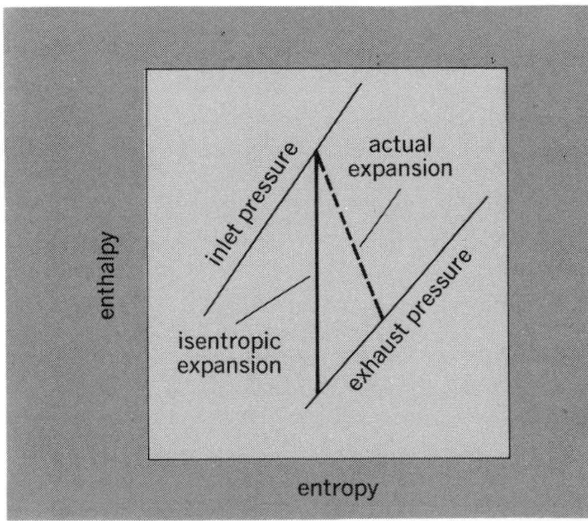

Fig. 3. Enthalpy-entropy (Mollier) chart of performance of steam- or gas-turbine type of prime mover.

Propane

A member of the alkane or paraffin series of hydrocarbons, formula $CH_3CH_2CH_3$. It makes up 3–18% of natural gas. It is readily liquefied (melting point, −187.7°C; boiling point, −42.1°C), and mixtures with liquefied butane are sold as liquefied petroleum gas (LPG) in cylinders under moderate pressure for domestic fuel.

At temperatures above about 650°C, propane undergoes cracking to ethylene and methane. This reaction is the basis of an important commercial source of ethylene and is accompanied by some dehydrogenation to propylene. The yield of propylene is increased in the presence of catalyst.

In the petroleum industry, propane is used as a combined solvent and refrigerant for the refining of lubricants and other products. *See* CRACKING; PETROLEUM PROCESSING. [LOUIS SCHMERLING]

Propellant

Usually, a combustible substance that produces heat and supplies ejection particles, as in a rocket engine. A propellant is both a source of energy and a working substance; a fuel is chiefly a source of energy, and a working substance is chiefly a means for expending energy. Because the distinction is more decisive in rocket engines, the term propellant is used primarily to describe chemicals carried by rockets for propulsive purposes. *See* AIRCRAFT FUEL; FUEL; THERMODYNAMIC CYCLE.

Propellants are classified as liquid or as solid. Even if a propellant is burned as a gas, it may be carried under pressure as a cryogenic liquid to save space. For example, liquid oxygen and liquid hydrogen are important high-energy liquid bipropellants. Liquid propellants are carried in compartments separate from the combustion chamber; solid propellants are carried in the combustion chamber. The two types of propellants lead to significant differences in engine structure and thrust control. For comparison, the effectiveness of either type of propellant is stated in terms of specific impulse.

LIQUID PROPELLANTS

A liquid propellant releases energy by chemical action to supply motive power for jet propulsion. The three principal types of propellants are monopropellant, bipropellant, and hybrid propellant. Monopropellants are single liquids, either compounds or solutions. Bipropellants consist of fuel and oxidizer carried separately in the vehicle and brought together in the engine. Air-breathing engines carry only fuel and use atmospheric oxygen for combustion. Hybrid propellants use a combination of liquid and solid materials to provide propulsion energy and working substance. Typical liquid propellants are listed in the table. Physical properties at temperatures from storage to combustion are important. These properties include melting point, boiling point, density, and viscosity. *See* METAL-BASE FUEL.

The availability of large quantities and their high performance led to selection of liquefied gases such as oxygen for early liquid-propellant rocket vehicles. Liquids of higher density with low vapor pressure (see table) are advantageous for the practical requirements of rocket operation under ordinary handling conditions. Such liquids can be retained in rockets for long periods ready for use and are convenient for vehicles that are to be used several times. The high impulse of the cryogenic systems is desirable for rocket flights demanding maximum capabilities, however, such as the exploration of space or the transportation of great weights for long distances.

Performance. Jet propulsion by a reaction engine, using the momentum of the propellant combustion products ejected from the engine, is not limited to atmospheric operation if the fuel reacts with an oxidizer carried with the engine. Performance of the propellant in such an engine depends upon both the heat liberated and the propellant reaction products. Combustion with air of effective fuels for air-breathing engines gives approximately 18,000 Btu/lb (42 MJ/kg), whereas fuels which are more effective in rocket engines may give only 15,000 Btu/lb (35 MJ/kg). A high heat of reaction is

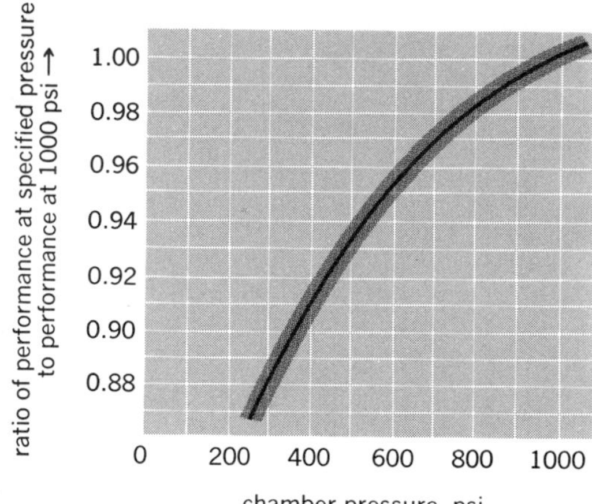

Fig. 1. Graph illustrating approximate effect of chamber pressure on specific impulse. 1 psi = 6.895 kPa.

most effective with gaseous products which are of low molecular weight.

Performance is rated in terms of specific impulse (occasionally specific thrust), the thrust obtained per pound of propellant used in 1 sec. An alternate measure of performance is the characteristic exhaust velocity. The theoretical characteristic exhaust velocity is determined by the thermodynamic properties of the propellant reaction and its products. Unlike the specific impulse, the characteristic exhaust velocity is independent of pressure, except for second-order effects, such as reactions modifying the heat capacity ratio of the combustion gases.

The relationship between these parameters is given by Eq. (1), in which I_s is specific impulse in

$$I_s = \frac{F}{\dot{w}} = \frac{c^* C_F}{g} \tag{1}$$

seconds, F is thrust, and $\dot{w}$ is flow rate of propellant. The characteristic exhaust velocity c^* is given in feet per second. C_F is the thrust coefficient, and g is the gravitational constant.

The actual exhaust velocity of the combustion gases is given by the product of the characteristic

Physical properties of liquid propellants

Propellant	Boiling point, °F (°C)	Freezing point °F (°C)	Density, g/ml	Specific impulse,* sec
Monopropellants				
Acetylene	−119 (−84)	−115 (−82)	0.62	265
Hydrazine	236 (108)	35 (2)	1.01	194
Ethylene oxide	52 (11)	−168 (−111)	0.88	192
Hydrogen peroxide	288 (140)	13 (−11)	1.39	170
Bipropellants				
Hydrogen	−423 (−253)	−436 (−260)	0.07	
Hydrogen-fluorine	−306 (−188)	−360 (−218)	1.54	410
Hydrogen-oxygen	−297 (−183)	−362 (−219)	1.14	390
Nitrogen tetroxide	70 (21)	12 (−11)	1.49	
Nitrogen tetroxide-hydrazine	236 (108)	35 (2)	1.01	290
Red nitric acid	104 (40)	−80 (−62)	1.58	
Red fuming nitric acid – *uns*-dimethyl hydrazine	146 (63)	−71 (−57)	0.78	275

*Maximum theoretical specific impulse at 1000 psi (6.895 MPa) chamber pressure expanded to atmospheric pressure.

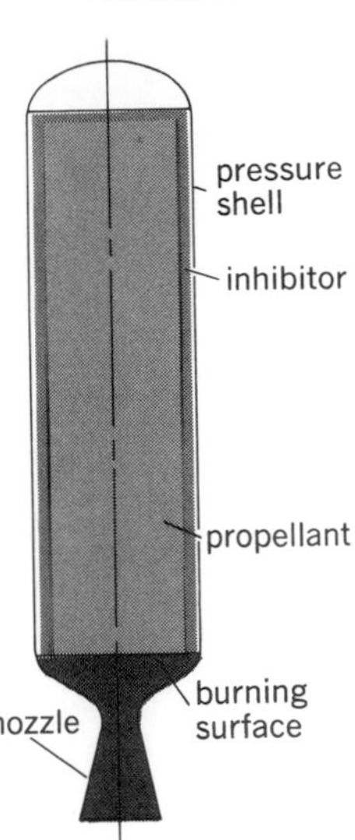

Fig. 2. End-burning grain loaded in rocket.

exhaust velocity and the thrust coefficient, c^*C_F. The thrust coefficient is a function of the heat capacity ratio of the combustion gases (the ratio of the heat capacity at constant pressure to the heat capacity at constant volume) and of the ratio of the chamber pressure to the exhaust pressure. The heat-capacity ratio of common propulsion gases varies from 1.1 to 1.4.

Increase in the combustion-chamber pressure increases the specific impulse (Fig. 1). Variation in the stoichiometry of the propellant reaction (the oxidizer-fuel ratio) also affects performance. A slightly fuel-rich reaction gives higher performance with common liquid propellants despite the lower heat of reaction because of more favorable working-gas composition. Increase in chamber pressure usually moves the optimum performance point toward the stoichiometric reaction ratio. A nonstoichiometric ratio may be used to give low combustion temperatures if required by the structural materials.

A properly designed engine can give 95–100% of the theoretical performance shown in the table.

Combustion. The energy of liquid propellants is released in combustion reactions which also produce the working fluid for reaction propulsion. The liquids in a bipropellant system may ignite spontaneously on contact, or they may require an ignition device to raise them to ignition temperature. In the first case they are called hypergolic liquids; in the second case, anergolic liquids. Combustion can be initiated with a spark, a hot wire, or an auxiliary hypergolic liquid. Monopropellant combustion, or more properly decomposition, can also be ignited by catalysis with an active surface or by a chemical compound in solution. Ignition of common hypergolic bipropellants occurs in a period of 1–100 msec following initial contact of the liquids. Catalytic quantities of detergents or of certain compounds of metals with several oxidation states, such as vanadium pentoxide, decrease the ignition delay period of specific bipropellants.

The combustion chamber in operation contains a turbulent, heterogeneous, high-temperature-reaction mixture. The liquids burn with droplets of various sizes in close proximity and traveling at high velocity. Larger masses of liquid may be present, particularly at the chamber walls. Very high rates of heat release, of the order of 10^5–10^6 Btu/(min)(ft^3)(atm) [$6 \times 10^2 - 6 \times 10^3$ J/(s)(m^3)(Pa)] are encountered.

Oscillations with frequencies of 25–10,000 Hz or more may accompany combustion of liquids in jet-propulsion engines. Low-frequency instability (chugging) can result from oscillations coupling the liquid flowing into the combustion chamber with pressure pulses in the chamber. Higher frequencies (screaming) can result from gas oscillations of the acoustic type in the chamber itself.

Engine performance, in contrast to theoretical propellant performance, depends upon effective combustor design. Mixing and atomization are essential factors in injection of the propellants into the combustion chamber. Injector and chamber design influence the flow pattern of both liquid and gases in the chamber. The characteristic chamber length L^* is given by Eq. (2), in which V_c is the

$$L^* = V_c / A_T \tag{2}$$

chamber volume and A_T is the area of the nozzle throat. In general, monopropellants require larger L^* than bipropellants to provide an equal fraction of theoretical performance in a rocket engine, as expected from the slower combustion exhibited by monopropellants.

[STANLEY SINGER]

SOLID PROPELLANTS

A solid propellant is a mixture of oxidizing and reducing materials that can coexist in the solid state at ordinary temperatures. When ignited, a propellant burns and generates hot gas. Although gun powders are sometimes called propellants, the term solid propellant ordinarily refers to materials used to furnish energy for rocket propulsion.

Composition. A solid propellant normally contains three essential components: oxidizer, fuel, and additives. Oxidizers commonly used in solid propellants are ammonium and potassium perchlorates, ammonium and potassium nitrates, and various organic nitrates, such as glyceryl trinitrate (nitroglycerin). Common fuels are hydrocarbons or hydrocarbon derivatives, such as synthetic rubbers, synthetic resins, and cellulose or cellulose derivatives. The additives, usually present in small amounts, are chosen from a wide variety of materials and serve a variety of purposes. Catalysts or suppressors are used to increase or decrease the rate of burning; ballistic modifiers may be used for a variety of reasons, as to provide less change in burning rate with pressure (platinizing agent); stabilizers may be used to slow down undesirable changes that may occur in long-term storage.

Solid propellants are classified as composite or double base. The composite types consist of an oxidizer of inorganic salt in a matrix of organic fuels, such as ammonium perchlorate suspended in a synthetic rubber. The double-base types are usually high-strength, high-modulus gels of cellulose nitrate (guncotton) in glyceryl trinitrate or a similar solvent.

Propellants are processed by extrusion or casting techniques into what are often intricate shapes that are commonly called grains, even though they may weigh many tons. The double-base types and

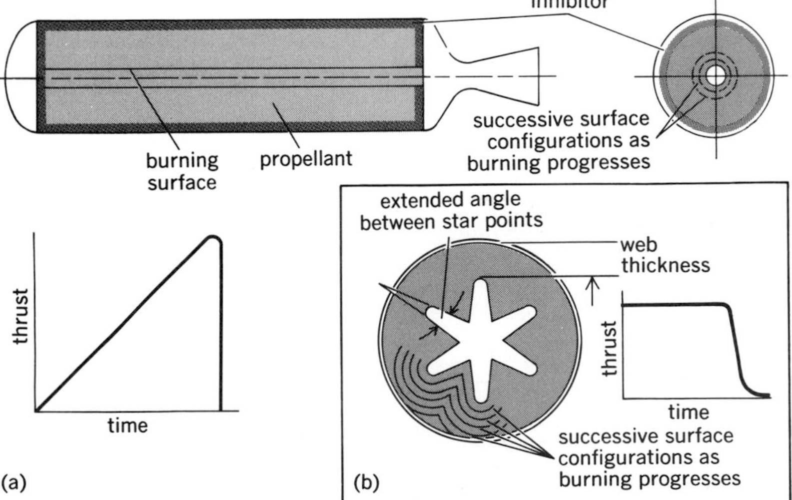

Fig. 3. Internal-burning solid-propellant charge configurations with typical thrust-time characteristics. (*a*) Cylindrical cavity. (*b*) Star-shaped cavity for level, or neutral, thrust-time characteristic.

certain high-modulus composites are processed into grains by casting or extrusion, and are then loaded by insertion of the cartridgelike grain or grains into the combustion chamber of the rocket. This technique requires some type of mechanical support to hold the propellant in place in the chamber. Certain types of composite propellants, bonded by elastomeric fuels, can be cast directly into the chamber, where the binder cures to a rubber and the grain is then supported by adhesion to the walls. Most high-performance rockets are made by this case-bonding technique, for more efficient use of weight and combustion-chamber volume.

Burning rate. The thrust-time characteristic of a solid-propellant rocket is controlled by the geometric shape of the grain. Often it is desired that burning not take place on certain portions of the grain. Such surfaces are then covered with an inert material called an inhibitor or restrictor. Neutral-burning grains maintain a constant surface during burning and produce a constant thrust. Progressive burning grains increase in surface and give an increasing thrust with time. Degressive or regressive grains burn with decreasing surface and give a decreasing thrust.

An end-burning grain is shown in Fig. 2. This type of configuration is neutral, because the surface stays constant while the grain burns forward. For most applications, radial-burning charges which burn outward from the inside perforation are superior because most of the wall area of the chamber can be protected from hot gas generated by combustion. Such protection is a built-in feature of the case-bonded grain; with the cartridge-loaded, inhibited charge, protection is provided by the addition of obturators to prevent gas flow around the outside of the grain.

Figure 3 shows a progressive design called an internal-burning cylinder. Various star-shaped perforations can be used to give neutral or degressive characteristics. By ingenious use of geometry, the thrust-time characteristic can be designed to meet almost any need. Another important neutral-grain design, the uninhibited, internal-external-burning cylinder, is used widely in short-duration applications such as the bazooka rocket weapon.

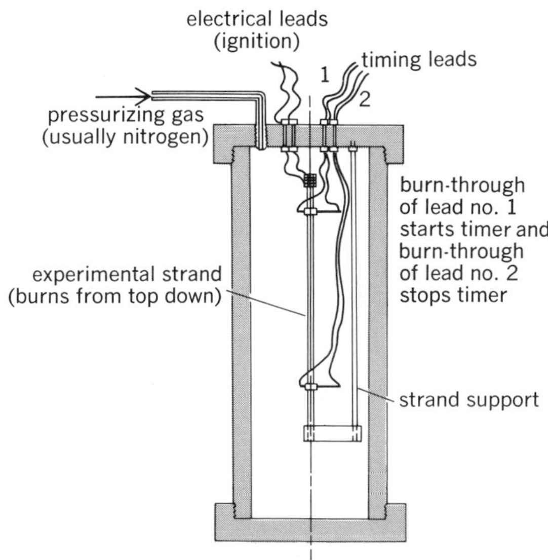

Fig. 4. Strand burner apparatus.

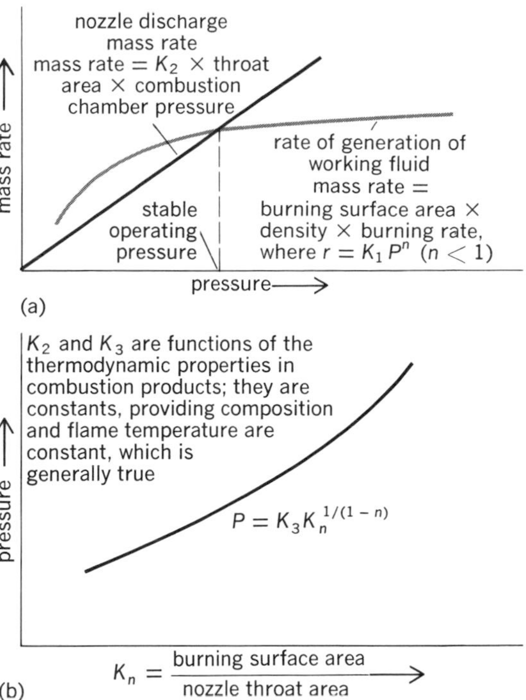

Fig. 5. Solid propellant. (*a*) Stable burning condition. (*b*) Typical pressure characteristic.

which contains many such grains. Obturation is not necessary because the propellant is burned so rapidly that the walls of the chamber do not rise in temperature sufficiently to cause loss of strength.

Propellant burns at a rate r proportional to a fractional power n of the pressure P as expressed by Eq. (3) in which K_1 is the coefficient of proportionality. This rate may be determined at various pressures by measurements of the burning rate of

$$r = K_1 P^n \tag{3}$$

propellant strands in a strand burner (Fig. 4). If the propellant is to operate properly, the exponent n in the burning-rate equation must be less than 1. As illustrated by Fig. 5*a*, if $n < 1$, there is a stable operating pressure at which the lines of fluid generation and fluid discharge intersect. Pressure cannot rise above this value because gas would then be discharged at a faster rate than it is generated by burning of propellant. When the propellant meets the requirement $n < 1$, a relationship exists between operating chamber pressure and a design parameter known as K_n, which is the ratio of the propellant burning area to the nozzle throat area. This ratio is illustrated in Fig. 5*b*.

Specific impulse of solid propellants is normally rated with the rocket operating at chamber pressure of 1000 lb/in.^{2} (6.9 MPa) and exhausting through an optimum nozzle into sea-level atmosphere. Under these conditions, solid propellants in use today can give an impulse of about 250, which is near the ceiling imposed by compositions based on ammonium perchlorate and hydrocarbons, and is 5–10% lower than impulses obtainable from liquid oxygen and gasoline.

The lower specific impulse of solid propellants is partly overcome by their densities, which are higher than those of most liquid propellants. In addition, solid-propellant rockets are easy to launch, are instantly ready, and have demonstrated a high degree of reliability. Because they can be produced by a process much like the casting of concrete, there seems to be no practical limit to the size of a solid-propellant rocket.

[H. W. RITCHEY]

Bibliography: E. M. Goodger, *Principles of Spaceflight Propulsion*, 1970; H. E. Malone, *Analysis of Rocket Propellants*, 1977; P. G. Shepard, *Aerospace Propulsion*, 1972; G. A. Sutton and D. M. Ross, *Rocket Propulsion Elements, An Introduction to the Engineering of Rockets*, 4th ed., 1976.

Propulsion

The process of causing a body to move by exerting a force against it. Propulsion is based on the reaction principle, stated qualitatively in Newton's third law, that for every action there is an equal and opposite reaction. A quantitative description of the propulsive force exerted on a body is given by Newton's second law, which states that the force applied to any body is equal to the rate of change of momentum of that body, and the force is exerted in the same direction as the momentum change.

In the case of a vehicle moving in a fluid medium, such as an airplane or a ship, the required change in momentum is generally produced by changing the velocity of the fluid (air or water) passing through the propulsive device or engine. In other cases, such as that of a rocket-propelled vehicle, the propulsion system must be capable of operating without the presence of a fluid medium; that is, it must be able to operate in the vacuum of space. The required momentum change is then produced by using up some of the propulsive device's own mass, called the propellant. *See* PROPELLANT.

Propulsion principle. Any change in momentum, according to Newton's second law, must be exactly equal to the propulsive force exerted on the body. For example, one may examine the simple propulsive means consisting of a hollow sphere with a small hole of area A (Fig. 1). If the sphere is filled with gas whose pressure is higher than that of the surrounding medium, the gas will exert pressure on the inside surface of the hollow sphere, with the exception of the hole, through which gas will rush out into the surroundings. Directly opposite the hole is an area A of the sphere on which the compressed gas does exert a force. In all other radial directions inside the sphere, pressure forces

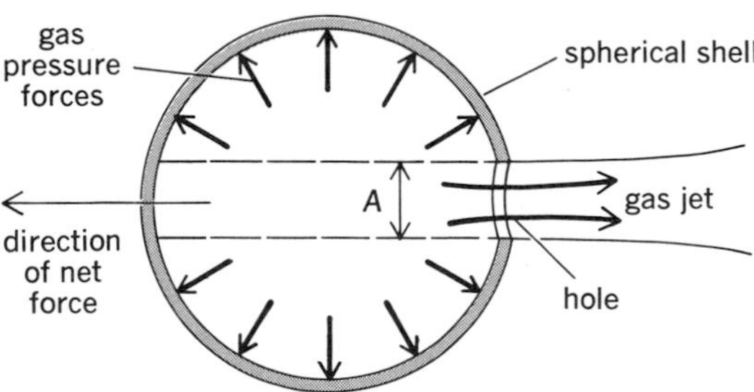

Fig. 1. A shell containing an aperture and saturated with gas at higher pressure than any outside fluid illustrates the principle of propulsion.

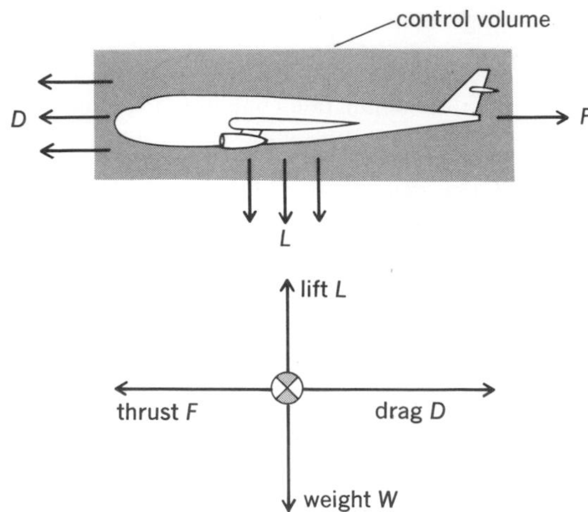

Fig. 2. All forces, including momentum forces, within a control volume about a vehicle are in equilibrium.

balance each other. Consequently, the net force on the sphere is in the direction shown. To compute the size of the unbalanced force, one can either measure and add or subtract all the pressures at the shell surface (including the gas pressure in the hole), or, more simply, in accordance with Newton's second law, calculate the momentum of the escaping gas jet.

This simple propulsive device shows that a rocket gives more thrust in space than in the atmosphere, because atmospheric pressure tends to increase the pressure in the hole, reducing the net force in the forward direction and, of course, reducing by an equal amount the effluent gas momentum. Also, it shows that although the amount of net thrust can be calculated by using the momentum change of the fluid medium, the thrust itself is produced, as it must be, by something pushing against the body—in this case, the pressure exerted by a gas.

Most propulsion engines are more complicated than the simple spherical rocket discussed here. Consider, for example, an airplane in level flight at constant speed (Fig. 2). Within a rectangular region (the control volume) around the airplane, net forces must be in equilibrium. The sum of all momentum changes occurring across the surfaces of the control volume is balanced by the total of all the forces on the airplane, in accordance with Newton's law.

Some air passing over the wings is given downward momentum, resulting in upward reaction forces, called lift, on the airplane. This air is also given some forward momentum as it is accelerated downward, resulting in a rearward force on the airplane called induced drag. Viscous drag is also produced (in the rearward direction) because the airplane carries some air forward with it through viscous adherence, giving the air forward momentum. Finally, the airplane propulsive means (such as an engine-driven propeller or a turbojet engine) takes in quiescent air, which is thereby accelerated to the same speed as the airplane, providing a rearward force called ram drag, and then the propulsive means discharges the air at a much higher rearward velocity (with high rearward momentum), resulting in a forward force called gross thrust. If all these forces are balanced, the airplane flies at constant speed in a straight line; if not, the airplane is accelerated in whichever direction the forces are unbalanced, thereby providing another momentum change to again bring net forces within the control volume into equilibrium.

Thrust. Propulsion capability is measured in terms of the thrust delivered to the vehicle. In general, the net thrust delivered by any propulsive means, neglecting the effect of ambient pressure, is given by Eq. (1).

$$F = (W_a + W_f)\, V_e - W_a V \tag{1}$$

F = net thrust in lb
V_e = velocity of gas leaving the propulsive means relative to the vehicle, in ft/sec
V = flight velocity of the vehicle through still air in ft/sec
W_a = mass flow rate of the ambient medium (air or water) through the propulsive means, in (lb-sec²/sec)/ft
W_f = mass flow rate of fuel (or on-board propellant) through the propulsive means in (lb-sec²/sec)/ft

The first term is the gross thrust; the second is the ram drag. In a rocket engine, no ambient fluid (air or water) is taken into the propulsive means; thus there is no ram drag. The net thrust of a rocket in vacuum is thus given by Eq. (2), where W_f

$$F = W_f V_e \tag{2}$$

represents the total mass flow rate of on-board propellant.

Efficiency. Propulsion efficiency was formerly defined as the useful thrust power divided by the propulsion power developed. Unfortunately, because useful thrust power can only be determined as the thrust multiplied by the flight velocity (FV), and because flight velocity is basically a relative rather than an absolute term, propulsion efficiency has little real meaning as a measure of propulsion effectiveness.

The two terms most generally used to describe propulsion efficiency are thrust specific fuel consumption (SFC) for engines using the ambient fluid (air or water), and specific impulse (I_{sp}) for engines which carry all propulsive media on board. Thrust specific fuel consumption is given by Eq. (3),

$$\text{SFC} = \frac{W_f g_0}{F} = g_0 \frac{W_f}{W_a} \frac{1}{\left[\left(1+\frac{W_f}{W_a}\right) V_e - V\right]} \approx g_0 \frac{W_f}{W_a} \frac{1}{V_e - V} \tag{3}$$

where SFC is expressed in pounds of fuel per second per pound of thrust, g_0 is acceleration due to gravity in ft/sec², and, usually, $W_f/W_a \ll 1$.

Specific impulse is determined similarly (although inversely) from Eq. (4). Obviously,

$$I_{sp} = \frac{F}{W_f g_0} = \frac{V_e}{g_0} \tag{4}$$

effective propulsion performance is indicated by low values of SFC or high values of I_{sp}.

The energy source for most propulsion devices is the heat generated by the combustion of exothermic chemical mixtures composed of a fuel and an oxidizer. An airbreathing chemical propulsion system generally uses a hydrocarbon such as coal, oil, gasoline, or kerosine as the fuel, and atmospheric air as the oxidizer. A non-air-breathing engine, such as a rocket, almost always utilizes propellants that also provide the energy source by their own combustion. Here the choice is wider: Fuels may be hydrocarbons again, but may also be more efficient low-molecular-weight chemicals such as hydrazine (N_2H_4), ammonia (NH_3), or even liquid hydrogen itself. Oxidizers are usually liquid oxygen, nitrogen tetroxide (N_2O_5), or liquid fluorine.

Where nuclear energy is the source of propulsive power, the heat developed by nuclear fission in a reactor is transferred to a working fluid, which either passes through a turbine to drive the propulsive element such as a propeller, or serves as the propellant itself. Nuclear-powered ships and submarines are accepted forms of transportation. Nuclear-powered airplanes and rockets are still in their early developmental stages. In the case of nuclear-powered vehicles, the concept of specific fuel consumption is no longer valid, because the loss in weight of the nuclear fuel is virtually zero. However, the specific impulse is still a meaningful measure of performance, being based on thrust and propellant flow rate, not on mass consumption of fuel. *See* TURBINE PROPULSION.

[JERRY GREY]

Bibliography: E. M. Goodger, *Principles of Spaceflight Propulsion*, 1970; O. E. Lancaster (ed.), *Jet Propulsion Engines*, 1959; P. G. Shepard, *Aerospace Propulsion*, 1972.

Pulse jet

A type of jet engine characterized by periodic surges of thrust. The pulse jet engine was widely known for its use during World War II on the German V-1 missile (Fig. 1). The basic engine cycle was invented in 1908. The inlet end of the engine is provided with a grid to which are attached flap valves. These valves are normally held by spring tension against the grid face and block the flow of air back out of the front of the engine. They can be sucked inward by a negative differential pressure to allow air to flow into the engine. Downstream from the flap valves is the combustion chamber. A fuel injection system is located at the entrance to the combustion chamber. The chamber is also fitted with a spark plug. Following the combustion chamber is a long exhaust duct which provides an inertial gas column.

When the combustion chamber is filled with a mixture of fresh air and fuel, a spark is discharged; it ignites the fuel-air mixture, producing a pressure surge that advances upstream to slam shut the in-

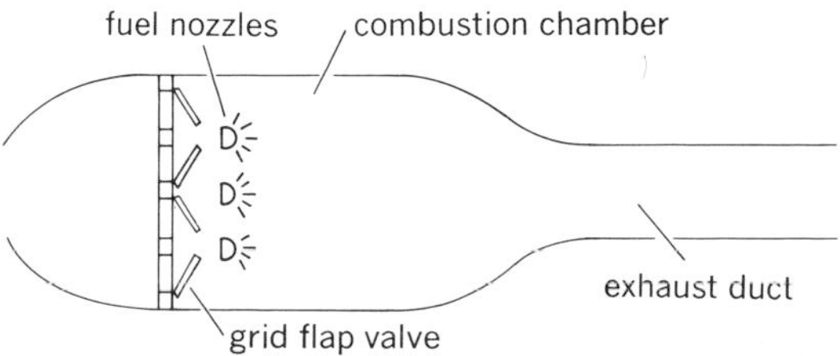

Fig. 1. Diagram of a pulse jet.

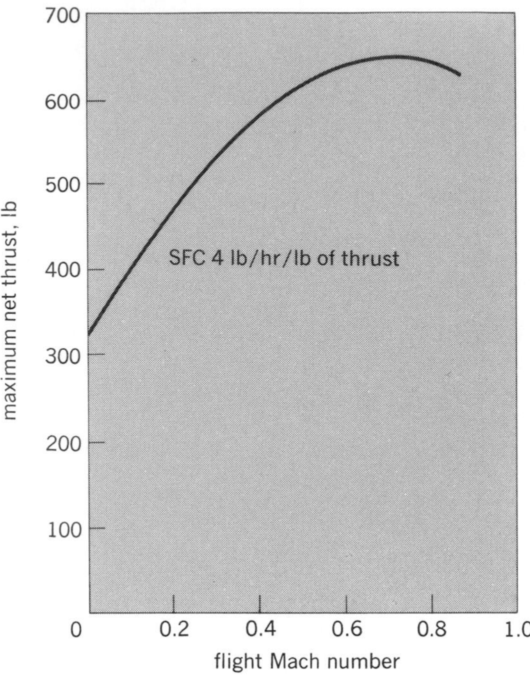

Fig. 2. Effect of flight Mach number on thrust of a German pulse jet; length 137.2 in. (3.48 m), diameter 21.6 in. (0.55 m). 1 lbf = 4.45 N. 4 lb/hr/lbf = 1.13×10^{-4} kg/sec/N.

let valves and to block off the entrance. Simultaneously, a pressure pulse goes downstream to produce a surge of combustion products out the exhaust duct. Thrust results from the rearward discharge of this gas at high velocity. With the discharge of gas from the combustion chamber, its pressure tends to drop. Inertia causes the column of gas in the exhaust duct to continue to flow rearward even after the explosion pressure in the combustion chamber has been dissipated, and this drops the combustion chamber pressure below atmospheric. As a result, the flap valves open and a fresh charge of air enters the combustion chamber. As this air flows past the fuel nozzles, it receives an injection of fuel and the mixture is then ignited by contact with the hot gas residue from the previous cycle. This causes the mixture to explode and the cycle repeats. Thrust increases with engine speed up to a maximum dependent on design (Fig. 2).

Unlike the ramjet, the pulse jet has an appreciable thrust at zero flight speed. However, as the flight speed is increased, the resistance to the flow of air imposed by the flap valves eventually causes substantial loss in performance and the pulse jet becomes less efficient than the ramjet. *See* RAMJET.

Failure of flap valves and valve seats by fatigue was found to be a problem. Research has been conducted on valve systems other than that shown in Fig. 1 and on valveless pulse jets.

In addition to their use on the German V-1 buzz-bomb, pulse jets have been used to propel radio-controlled target drones and experimental helicopters. In the latter case, they were mounted on the blade tips for directly driving the rotor. The high fuel consumption, noise, and vibrations generated by the pulse jet limit its scope of applications. *See* PROPULSION. [BENJAMIN PINKEL]

Pumped storage

A process, also known as hydroelectric storage, for converting large quantities of electrical energy to potential energy by pumping water to a higher elevation, where it can be stored indefinitely and then released to pass through hydraulic turbines and generate electrical energy. An indirect process is necessary because electrical energy cannot be stored effectively in large quantities. Storage is desirable, as the consumption of electricity is highly variable between day and night, between weekday and weekend, as well as among seasons. Consequently, much of the generating equipment needed to meet the greatest daytime load is unused or lightly loaded at night or on weekends. During those times the excess capability can be used to generate energy for pumping, hence the necessity for storage. Normally, pumping energy can be obtained from economical sources, and its value will be upgraded when used for peak loads.

Operation. In a typical operation, night or weekend electrical energy is used to pump water from a lower to a higher elevation, where it is stored as potential energy in the upper reservoir. The water can be retained indefinitely without deterioration or significant loss. During the daylight hours when the loads are greatest, stored water is released to flow from the higher to the lower reservoir through hydraulic-turbine-driven-generators and converted to electrical energy. No water is consumed in either the pumping or generating phase. To provide storage or generation merely requires the transfer of water from one reservoir to the other. Pumped storage installations have attained an overall operating efficiency of about 70%. Projected improvements in equipment design promise an efficiency of 75% or more. Postulating one cycle each of generation and pumping per day plus an allowance to change from one mode to the other, the maximum annual generation attainable is 3500 hr.

Description. A typical pumped-storage development is composed of two reservoirs of essentially equal volume situated to maximize the difference in their levels. These reservoirs are connected by a system of waterways along which a pumping-generating station is located (Fig. 1). Under favorable geological conditions, the station will be located underground, otherwise it will be situated on the lower reservoir. The principal equipment of the station is the pumping-generating unit. In United States practice, the machinery is reversible and is used for both pumping and generating; it is designed to function as a motor and pump in one direction of rotation and as a turbine and generator in opposite rotation. Transformers, a substation, switchyard, and transmission line are required to transfer the electrical power to and from the station. *See* ELECTRIC POWER GENERATION.

The lower reservoir may be formed by impounding a stream or by using an existing body of water. Similarly, an upper reservoir may be created by an embankment across a valley or by a circumferential dike. Frequently, these features produce the most significant environmental impact, which is largely land use and visual. The reservoirs are comparatively small, thus affording some latitude in location to minimize unavoidable effects, such as displacement of developed areas, existing roads, and railways. Problems of emission of particulate matter and gases and of water-temperature rise associated with other generating stations do not exist. On the other hand, the reservoirs and surrounding area afford the opportunity to develop recreational facilities such as camp grounds, marinas, boat ramps, picnic areas, and wildlife preserves. In the United States, recreation commands a high priority, and many existing developments have included recreational facilities.

Economics. The economics of pumped storage are inextricably linked to the system in which it operates. The development must be economically competitive with other types of generation available, namely, nuclear, coal- or oil-fired, gas turbines, hydro, and other storage systems such as compressed air. It has been a generally accepted practice to evaluate the economy of any plant on the basis of its annual cost to the system, but life-cycle costs are now also considered. The annual cost is defined as the sum of fixed costs, operation, maintenance, and fuel. When construction is complete, the fixed costs are established and continue at that level irrespective of the extent of operation. Variations in fuel prices have a major effect. During the late 1970s the cost of coal averaged about twice as much as nuclear fuel and oil about 2½ times as much as coal. Analyses based on current construction and fuel cost or on similar costs projected through 1995 show that between 10 and 20% of the generating capacity should be pumped storage. Pumping energy should be supplied either by coal or nuclear fueled plants; oil has not been considered as a fuel because of the forecast incipient shortage and instability of price. The current emphasis on conservation of oil makes pumped storage a prime candidate to displace gas-turbine peaking capacity in systems having adequate coal and nuclear base-load capacity.

European plants. In European practice, not only system economies receive attention, but also operating requirements and advantages; the latter have materially influenced the design and selection of equipment. Historically, European pumped-storage plants have been used not only to provide peak-load power as in the United States, but also to ensure system stability or frequency control. To accomplish this, the units must have very short response times which will enable them to change mode of operation quickly, to follow changes in

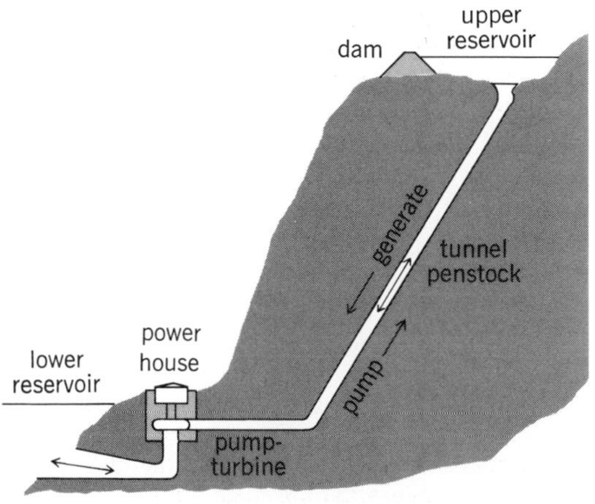

Fig. 1. Schematic of a conventional pumped-storage development.

load, or to provide emergency generation or load. Such operation is the dominant reason for European preference for three-machine sets, that is, separate pump and separate turbine coupled to a motor-generator, as opposed to the two-machine reversible pump-turbine and motor-generator used in the United States. The extent to which Europeans practice incorporates system stability and frequency regulation is shown by the number of pumping and generating starts per day. In one plant, the daily average starts per unit for pumping was 2.5 and for generating 3.5, compared to 1 start for pumping and 1 or possibly 2 per day for generation in typical United States operation.

Extent of use. At the beginning of 1980 the installed pumped storage capacity in the United States was about 13,000 MW or slightly less than 3% of the total capacity, whereas in Western European countries the amount of pumped storage varied from 5 to 10% of the capacity of the connected system. Expansion of pumped storage has been slow in the United States because of slower growth in electric peakloads and environmental opposition. Meanwhile, European construction has kept pace with system expansion. Additionally, a compressed-air storage system has been completed in West Germany. This storage scheme is in direct competition with pumped storage as a means of supplying peaking capacity. Nonetheless, worldwide interest remains high not only in Western areas but also in Taiwan, South Korea, and South Africa, where the first pumped storage plants have been commissioned or are under construction. In the United States the Bath County project, scheduled for completion in 1983, is to become the world's largest at a capacity of 2,100,000 kW.

Prospects. Given favorable economics and improved environmental acceptance, pumped storage should increase as a percentage of overall capacity. One limitation of conventional pumped storage is the need for favorable site conditions accessible electrically to the load centers. Two factors which could further increase interest are conversion of existing hydro developments to pumped storage and the development of the deep underground plant. A number of existing hydro plants could be partially converted to pumped storage at a reasonable cost and minimal environmental impact since the reservoirs already exist.

The deep-underground concept (Fig. 2) is attracting increased attention because of rising fuel costs as well as construction costs for alternate generation, and the flexibility it affords in selecting a site near load centers. Much research has been directed toward the deep-underground scheme owing to its reduced environmental effect and freedom from topographic restraints. Studies have been undertaken in the United States to determine the feasibility of two such plants. The deep-underground concept is similar to conventional pumped storage, having all of the essential features and utilizing similar equipment. There are two notable variations: the upper reservoir is at ground level, while the lower reservoir is a deep underground cavern. Two stages are developed in series to minimize the size and, hence, the construction cost of the lower reservoir. Inherent advantages of the concept are: material reduction in environmental impact since only the upper reservoir is visible, development of heads which utilize the maximum

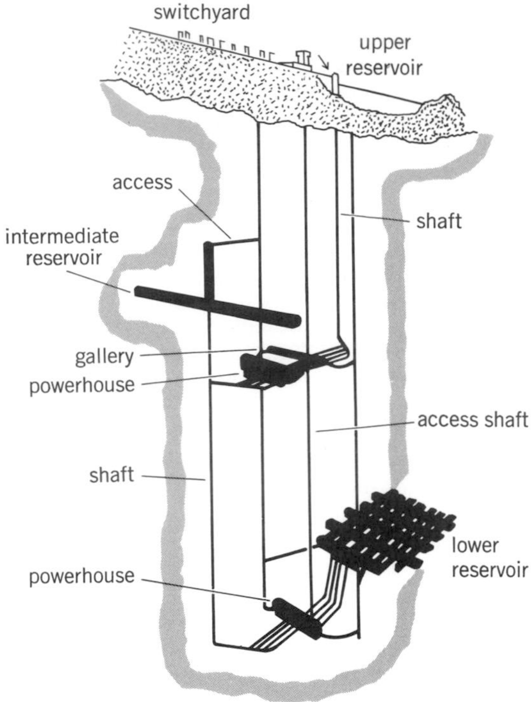

Fig. 2. Schematic of two-stage underground pumped-storage development.

capability of the machines, and elimination of the need for favorable topography. The disadvantages are: the need for large areas to dispose of the excavated material, uncertainties of construction and mining at 3000 ft (900 m) or more below the surface, and substantially increased construction time and costs compared with conventional pumped storage. *See the feature article* EXPLORING ENERGY CHOICES; *see also* ENERGY STORAGE; HYDROELECTRIC GENERATOR; HYDROELECTRIC POWER; TURBINE.

[DWIGHT L. GLASSCOCK]

Bibliography: Chas. T. Main, Inc., *Underground Hydroelectric Pumped Storage: An Evaluation of the Concept*, Department of the Interior, Bureau of Reclamation, 1978; E. Comninellis, Ludington Pumped Storage Project, *ASCE J. Power Div.*, 99 (POI):69–88, May 1973; Engineering Foundation Conference, *Converting Existing Hydroelectric Dams and Reservoirs into Pump Storage Facilities*, American Society of Civil Engineers, 1975; A. Ferreira, Multipurpose aspects of the Northfield Mountain Pumped Storage Project, paper presented at the International Conference on Pumped Storage Development and Its Environmental Effects, University of Wisconsin, September 1972; J. M. Frohnholzer, Operation of pumped-storage stations in the Federal Republic of Germany, *Water Power*, 25(10):374–384, October 1973; Harza Engineering Co., *Underground Pumped Hydro Storage and Compressed Air Storage*, Argonne National Laboratory, 1977; G. M. Karadi, *Pumped Storage Development and Its Environmental Effects: Final Report to the National Science Foundation*, University of Wisconsin, 1974; G. E., Pfafflin, Future trends in hydro pumped-storage equipment, *Proc. Amer. Power Conf.*, 36:390–402, 1974; J. G. Warnock, and D. C. Willet, Underground reservoirs for high-head pumped-storage stations, *Can. Electr. Ass. Eng. Oper. Div. Trans.*, Toronto, vol. 12, pt. 2, Pap. 73-H-107, Mar. 26–30, 1973 (also in *Water Power*, 25(3):81–87, March 1973); A Zagars and J. M. Hagood, Jr., Bath County, a 2100 MW development in the U.S.A., *Int. Water Power Dam Constr.*, 29(10):25–33, 1977.

Radiant heating

Any system of space heating in which the heat-producing means is a surface that emits heat to the surroundings by radiation rather than by conduction or convection. The surfaces may be radiators such as baseboard radiators or convectors, or they may be the panel surfaces of the space to be heated. *See* PANEL HEATING AND COOLING.

The heat derived from the Sun is radiant energy. Radiant rays pass through gases without warming them appreciably, but they increase the sensible temperature of liquid or solid objects upon which they impinge. The same principle applies to all forms of radiant-heating systems, except that convection currents are established in enclosed spaces and a portion of the space heating is produced by convection. The radiation component of convectors can be increased by providing a reflective surface on the wall side of the convector and painting the inside of the enclosure a dead black to absorb heat and transmit it through the enclosure, thus increasing the temperature of that side of the convector exposed to the space to be heated.

Any radiant-heating system using a fluid heat conveyor may be employed as a cooling system by substituting cold water or other cold fluid. This cannot be done with electric radiant-heating systems because, at their present stage of commercial development, they are not reversible; however, experiments on the reversibility of thermocouples may make such a development possible in the future. [ERWIN L. WEBER/RICHARD KORAL]

Bibliography: American Society of Heating, Refrigerating, and Air Conditioning Engineers, *ASHRAE Handbook and Product Directory: Systems*, 1976.

Radiation damage to materials

Harmful changes in the properties of liquids, gases, and solids, caused by interaction with nuclear radiations. The interaction of radiation with materials often leads to changes in the properties of the irradiated material. These changes are usually considered harmful. For example, a ductile metal may become brittle. However, sometimes the interaction may result in beneficial effects. For example, cross-linking may be induced in polymers by electron irradiation leading to a higher temperature stability than could be obtained otherwise.

Radiation damage is usually associated with materials of construction that must function in an environment of intense high-energy radiation from a nuclear reactor. Materials that are an integral part of the fuel element or cladding and nearby structural components are subject to such intense nuclear radiation that a decrease in the useful lifetime of these components can result.

Radiation damage will also be a factor in thermonuclear reactors. The deuterium-tritium (D-T) fusion in thermonuclear reactors will lead to the production of intense fluxes of 14-MeV neutrons that will cause damage per neutron of magnitude two to four times greater than damage done by 1–2 MeV neutrons in operating reactors. Charged particles from the plasma will be prevented from reaching the containment vessel by magnetic fields, but uncharged particles and neutrons will bombard the containment wall, leading to damage as well as sputtering of the container material surface which not only will cause degradation of the wall but can contaminate the plasma with consequent quenching.

Superconductors are also sensitive to neutron irradiation, hence the magnetic confinement of the plasma may be affected adversely. Damage to electrical insulators will be serious. Electronic components are extremely sensitive to even moderate radiation fields. Transistors malfunction because of defect trapping of charge carriers. Ferroelectrics such as $BaTiO_3$ fail because of induced isotrophy; quartz oscillators change frequency and ultimately become amorphous. High-permeability magnetic materials deteriorate because of hardening; thermocouples lose calibration because of transmutation effects. In this latter case, innovations in Johnson noise thermometry promise freedom from radiation damage in the area of temperature measurement. Plastics used for electrical insulation rapidly deteriorate. Radiation damage is thus a challenge to reactor designers, materials engineers, and scientists to find the means to alleviate radiation damage or to develop more radiation-resistant materials.

Damage mechanisms. There are several mechanisms that function on an atomic and nuclear scale to produce radiation damage in a material if the radiation is sufficiently energetic, whether it be electrons, protons, neutrons, x-rays, fission fragments, or other charged particles.

Electronic excitation and ionization. This type of damage is most severe in liquids and organic compounds and appears in a variety of forms such as gassing, decomposition, viscosity changes, and polymerization in liquids. Rapid deterioration of the mechanical properties of plastics takes place either by softening or by embrittlement, while rubber suffers severe elasticity changes at low fluxes. Cross-linking, scission, free-radical formation, and polymerization are the most important reactions.

The alkali halides are also subject to this type of damage since ionization plays a role in causing displated atoms and darkening of transparent crystals due to the formation of color centers.

Transmutation. In an environment of neutrons, transmutation effects may be important. An extreme case is illustrated by reaction (1). The ^{6}Li

$$^6\mathrm{Li} + n \rightarrow {}^4\mathrm{He} + {}^3\mathrm{H} + 4.8\ \mathrm{MeV} \qquad (1)$$

isotope is approximately 7.5% abundant in natural lithium and has a thermal neutron cross section of 950 barns (1 barn $= 1 \times 10^{-24}$ cm^2). Hence, copious quantities of tritium and helium will be formed. (In addition, the kinetic energy of the reaction products creates many defects.). Lithium alloys or compounds are consequently subject to severe radiation damage. On the other hand, reaction (1) is crucial to success of thermonuclear reactors utilizing the D-T reaction since it regenerates the tritium consumed. The lithium or lithium-containing compounds might best be used in the liquid state.

Even materials that have a low cross section such as aluminum can show an appreciable accumulation of impurity atoms from transmutations. The capture cross section of ^{27}Al (100% abundant) is only 0.25×10^{-24} cm^2. Still the reaction

$$^{27}\mathrm{Al} + n \rightarrow {}^{28}\mathrm{Al}^{\beta-} \xrightarrow{2.3\ \mathrm{min}} {}^{28}\mathrm{Si}$$

will yield several percent of silicon after neutron exposures at fluences of 10^{23} n/cm^2.

The elements boron and europium have very large cross sections and are used in control rods. Damage to the rods is severe in boron-containing materials because of the ^{10}B(n,α) reaction. Europium decay products do not yield any gaseous elements. At high thermal fluences the reaction $^{58}\mathrm{Ni} + n \rightarrow {}^{59}\mathrm{Ni} + n \rightarrow {}^{56}\mathrm{Fe} + \alpha$ is most important in nickel-containing materials. The reaction $(n,n') \rightarrow \alpha$ at 14 MeV takes place in most materials under consideration for structural use. Thus, in many instances transmutation effects can be a problem of great importance.

Displaced atoms. This mechanism is the most important source of radiation damage in nuclear reactors outside the fuel element. It is a consequence of the ability of the energetic neutrons born in the fission process to knock atoms from their equilibrium position in their crystal lattice, displacing them many atomic distances away into interstitial positions and leaving behind vacant lattice sites. The interaction is between the neutron and the nucleus of the atom only, since the neutron carries no charge. The maximum kinetic energy ΔE that can be acquired by a displaced atom is given by Eq. (2), where M is mass of the primary

$$\Delta E = \frac{4Mm}{(M+m)^2} \cdot E_N \qquad (2)$$

knocked-on atom (PKA), m is the mass of the neutron, and E_N is the energy of the neutron.

The energy acquired by each PKA is often high enough to displace additional atoms from their equilibrium position; thus a cascade of vacancies and interstitial atoms is created in the wake of the PKA transit through the matrix material. Collision of the PKA and a neighbor atom takes place within a few atomic spacings or less because the charge on the PKA results in screened coulombic-type repulsive interactions. The original neutron, on the other hand, may travel centimeters between collisions. Thus regions of high disorder are dispersed along the path of the neutron. These regions are created in the order of 10^{-12} s. The energy deposition is so intense in these regions that it may be visualized as a temporary thermal spike.

Not all of the energy transferred is available for displacing atoms. Inelastic energy losses (electronic excitation in metals and alloys and excitation plus ionization in nonmetals) drain an appreciable fraction of the energy of the knocked-on atom even at low energies, particularly at the beginning of its flight through the matrix material. The greater the initial energy of the PKA, the greater is the inelastic energy loss; however, near the end of its range most of the interactions result in displacements. Figure 1 is a schematic representation of the various mechanisms of radiation damage that take place in a solid.

A minimum energy is required to displace an atom from its equilibrium position. This energy ranges from 25 to 40 eV for a typical metal such as

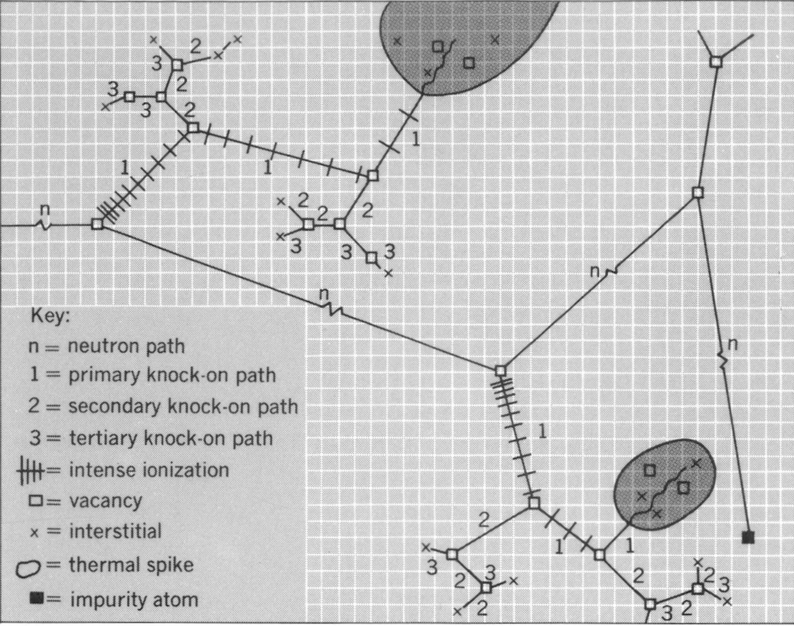

Fig. 1. The five principal mechanisms of radiation damage are ionization, vacancies, interstitials, impurity atoms, and thermal spikes. Diagram shows how a neutron might give rise to each in copper. Grid-line intersections are equilibrium positions for atoms. (*After D. S. Billington, Nucleonics, 14:54–57, 1956*)

iron; the mass of the atom and its orientation in the crystal influence this value. When appropriate calculations are made to compensate for the excitation energy loss of the PKA and factor in the minimum energy for displacement, it is found that approximately 500 stable vacancy-interstitial pairs are formed, on the average, for a PKA in iron resulting from a 1-MeV neutron collision. By multiplying this value by the flux of neutrons [$10^{14\text{-}15}$ n/(cm^2)(s)] times the exposure time [3×10^7 s/yr] one can easily calculate that in a few years each atom in the iron will have been displaced several times.

In the regions of high damage created by the PKA, most of the vacancies and interstitials will recombine. However, many of the interstitials, being more mobile than the vacancies, will escape and then may eventually be trapped at grain boundaries, impurity atom sites, or dislocations. Sometimes they will agglomerate to form platelets or interstitial dislocation loops. The vacancies left behind may also be trapped in a similar fashion, or they may agglomerate into clusters called voids.

Effect of fission fragments. The fission reaction in uranium or plutonium yielding the energetic neutrons that subsequently act as a source of radiation damage also creates two fission fragments that carry most of the energy released in the fission process. This energy, approximately 160 MeV, is shared by the two highly charged fragments. In the space of a few micrometers all of this energy is deposited, mostly in the form of heat, but a significant fraction goes into radiation damage of the surrounding fuel. The damage takes the form of swelling and distortion of the fuel. These effects may be so severe that the fuel element must be removed for reprocessing in advance of burn-up expectation, thus affecting the economy of reactor operation. However, fuel elements are meant to be ultimately replaced, so that in many respects the damage is not as serious a problem as damage to structural components of the permanent structure whose replacement would force an extended shut down or even reconstruction of the reactor.

Damage in cladding. Swelling of the fuel cladding is a potentially severe problem in breeder reactor design. The spacing between fuel elements is minimized to obtain maximum heat transfer and optimum neutron efficiency, so that diminishing the space for heat transfer by swelling would lead to overheating of the fuel element, while increasing the spacing to allow for the swelling would result in lower efficiencies. A possible solution appears to be in the development of low-swelling alloys.

Damage in engineering materials. Most of the engineering properties of materials of interest for reactor design and construction are sensitive to defects in their crystal lattice. The properties of structural materials that are of most significance are yield strength and tensile strength, ductility, creep, hardness, dimensional stability, impact resistance, and thermal conductivity. Metals and alloys are chosen for their fabricability, ductility, reasonable strength at high temperatures, and ability to tolerate static and dynamic stress loads. Refractory oxides are chosen for high-temperature stability and for use as insulators. Figure 2 shows the relative sensitivity of various types of materials to radiation damage. Several factors that enter into susceptibility to radiation damage will be discussed.

Temperature of irradiation. Nuclear irradiations performed at low temperatures (4 K) result in the maximum retention of radiation-produced defects. As the temperature of irradiation is raised, many of the defects are mobile and some annihilation may take place at 0.3 to 0.55 of the absolute melting point T_m. The increased mobility, particularly of vacancies and vacancy agglomerates, may lead to acceleration of solid-state reactions, such as precipitation, short- and long-range ordering, and phase changes. These reactions may lead to undesirable property changes. In the absence of irradiation many alloys are metastable, but the diffusion rates are so low at this temperature that no significant reaction is observed. The excess vacancies above the equilibrium value of vacancies at a given temperature allow the reaction to proceed as though the temperature were higher. In a narrow temperature region vacancy-controlled diffusion reactions become temperature-independent. When the temperature of irradiation is above 0.55 T_m, most of the defects anneal quickly and the temperature-dependent vacancy concentration becomes overwhelmingly larger than the radiation-induced vacancy concentration. However, in this higher temperature region serious problems may arise from transmutation-produced helium. This gas tends to migrate to grain boundaries and leads to enhanced intergranular fracture, thereby limiting the use of many conventional alloys.

Nuclear properties. Materials of construction with high nuclear-capture cross sections are to be avoided because each neutron that is captured in the structural components is lost for purposes of causing additional fissioning and breeding. The exception is in control rods as discussed earlier. Moderator materials, in particular, need to have low capture cross sections but high scattering cross sections. Low atomic weight is an important feature since moderation of fast neutrons to thermal energies is best done by those elements that maximize the slowing-down process. [See Eq. (2).] Beryllium and graphite are excellent moderators and have been used extensively in elemental form. Both elements suffer radiation damage, and their use under high-stress conditions is to be avoided.

Fluence. The total integrated exposure to radiation (flux × time) is called fluence. It is most important in determining radiation damage. Rate effects (flux) do not appear to be significant. The threshold fluence for a specific property change induced by radiation is a function of the composition and microstructure. One of the most important examples is the appearance of voids in metals and alloys. This defect does not show up in the microstructure of irradiated metals or alloys until a fluence of 10^{19} n/cm^2 or greater has been achieved. Consequently, there was no way to anticipate its appearance and the pronounced effect in causing swelling in structural components of a reactor. This and other examples point to the importance of lifetime studies in order to establish the appearance or absence of any unexpected phenomenon during this time.

Lifetime studies in reactors are time-consuming and are virtually impossible if anticipated fluences far exceed the anticipated lifetime of operating test reactors. A technique to overcome this impasse is to use charged-particle accelerators to simulate reactor irradiation conditions. For example, nickel ions can be used to bombard nickel samples. The bombarding ions at 5 to 10 MeV then simulate primary knocked-on atoms directly and create high-density damage in the thickness of a few micrometers. Accelerators are capable of produced beam currents of several $\mu A/cm^2$; hence in time periods of a few hours to a few days ion bombardment is equivalent to years of neutron bombardment. Correlation experiments have established that the type of damage is similar to neutron damage. Moreover, helium can be injected to approximate n,α damage when these reactions do not occur in accelerator bombardments. However, careful experimentation is required to obtain correlation between results obtained on thin samples and thicker, more massive samples used in neutron studies.

Pretreatment and microstructure. Dislocations play a key role in determining the plastic flow properties of metals and alloys such as ductility, elongation, and creep. The yield, ultimate and impact strength properties, and hardness are also expressions of dislocation behavior. If a radiation-produced defect impedes the motion of a dislocation, strengthening and reduced ductility may result. On the other hand, during irradiation point defects may enhance mobility by promoting dislocation climb over barriers by creating jogs in the dislocation so that it is free to move in a barrier-free area. Moreover, dislocations may act as trapping sites for interstitials and gas atoms, as well as nucleation sites for precipitate formation. Thus the number and disposition of dislocations in the metal alloy may strongly influence its behavior upon irradiation.

Heat treatment prior to irradiation determines the retention of both major alloying components and impurities in solid solution in metastable alloys. It also affects the number and disposition of dislocations. Thus it is an important variable in determining subsequent radiation behavior.

Impurities and minor alloying elements. The presence of small amounts of impurities may profoundly affect the behavior of engineering alloys in a radiation field. It has been observed that helium concentrations as low as 10^{-9} seriously reduce the high-temperature ductility of a stainless steel. Concentrations of helium greater than 10^{-3} may conceivably be introduced by the n,α reaction in the nickel component of the stainless steel or by boron contamination introduced inadvertently during alloy preparation. The boron also reacts with neutrons via the n,α reaction to produce helium. The addition of a small amount of Ti (0.2%) raises the temperature at which intergranular fracture takes place so that ductility is maintained at operating temperatures.

Small amounts of copper, phosphorus, and nitrogen have a strong influence on the increase in the ductile-brittle transition temperature of pressure vessel steels under irradiation. Normally these carbon steels exhibit brittle failure below room temperature. Under irradiation, with copper content above 0.08% the temperature at which the material fails in a brittle fashion increases. Therefore it is necessary to control the copper content as well as the phosphorus and nitrogen during the manufacture and heat treatment of these steels to

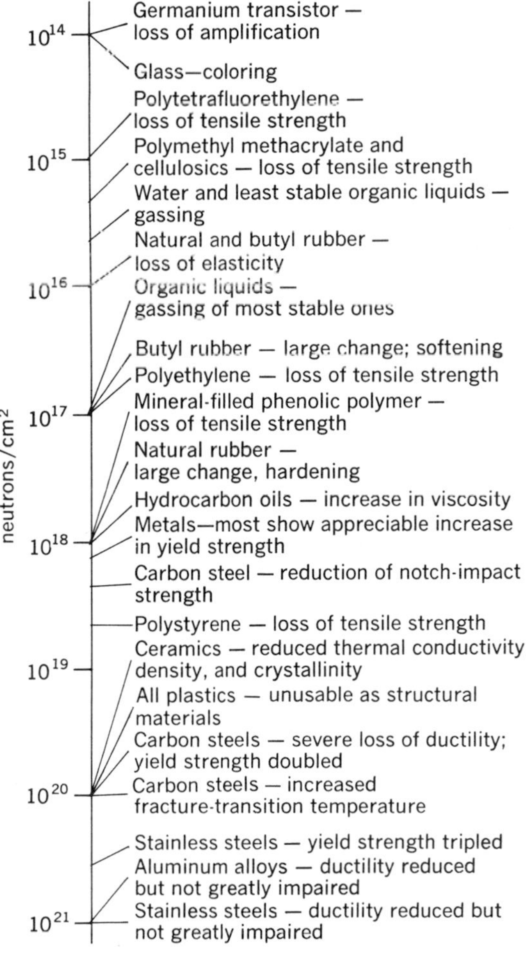

Fig. 2. Sensitivity of engineering materials to radiation. Levels are approximate and subject to variation. Changes are in most cases at least 10%. *(After O. Sisman and J. C. Wilson, Nucleonics, 14:58–65, 1956)*

keep the transition temperature at a suitably low level. A development of a similar nature has been observed in the swelling of type 316 stainless steel. It has been learned that carefully controlling the concentration of silicon and titanium in these alloys drastically reduces the void swelling. This is an important technical and economic contribution to the fast breeder reactor program.

Beneficial effects. Radiation, under carefully controlled conditions, can be used to alter the course of solid-state reactions that take place in a wide variety of solids. For example, it may be used to promote enhanced diffusion and nucleation, it can speed up both short- and long-range order-disorder reactions, initiate phase changes, stabilize high-temperature phases, induce magnetic property changes, retard diffusionless phase changes, cause re-solution of precipitate particles in some systems while speeding precipitation in other systems, cause lattice parameter changes, and speed up thermal decomposition of chemical compounds. The effect of radiation on these reactions and the other property changes caused by radiation are of great interest and value to research in solid-state physics and metallurgy.

Radiation damage is usually viewed as an unfortunate variable that adds a new dimension to the problem of reactor designers since it places severe restraints on the choice of materials that can be employed in design and construction. In addition, it places restraints on the ease of observation and manipulation because of the radioactivity involved. However, radiation damage is also a valuable research technique that permits materials scientists and engineers to introduce impurities and defects into a solid in a well-controlled fashion. [DOUGLAS S. BILLINGTON]

Bibliography: S. Amelinckx et al., *The Interaction of Radiation with Solids*, 1964; D. S. Billington and J. H. Crawford, Jr., *Radiation Damage in Solids*, 1961; E. E. Bloom et al., Austenitic stainless steels with improved resistance to radiation-induced swelling, *Scripta Met.*, 10:303, 1976; E. E. Bloom and J. R. Weir, *Tech. Publ. 457*, American Society for Testing and Materials, 1969; R. O. Bolt and J. G. Carroll (eds.), *Radiation Effects on Organic Materials*, 1963; C. J. Borokowski and T. V. Blalock, A new method of Johnson noise thermometry, *Rev. Sci. Inst.*, 45:151–162, 1974; J. W. Corbett and L. C. Ianniello (eds.), *Radiation Induced Voids in Metals*, CONF-71060 ERDA, 1972; G. J. Dienes (ed.), *Studies in Radiation Effects in Solids*, vol. 2, pp. 1–297, 1967; International Atomic Energy Agency, Vienna, *Interaction of Radiation with Condensed Matter*, 1978; International Atomic Energy Agency, Vienna, *Radiation Damage in Reactor Materials*, vol. 1, 1969; J. F. Kircher and R. E. Bowman (eds.), *Effects of Radiation on Materials and Components*, 1964; C. Lehmann, *Interaction of Radiation with Solids and Elementary Defect Production*, 1977; M. T. Robinson and F. W. Young (eds.), *Fundamental Aspects of Radiation Damage in Metals*, CONF-751006-P1 ERDA, 1976; F. Seitz and D. Turnball (eds.), *Solid State Physics*, vol. 2, 1956; L. E. Steele, *Neutron Embrittlement of Reactor Pressure Vessels*, Tech. Publ. 163, International Atomic Energy Agency, Vienna, 1975; Surface effects in controlled fusion, *J. Nucl. Mater.*, 53:1–357, 1974; F. W. Wiffen and J. S. Watson (eds.), *Radiation Effects and Tritium Technology for Fusion Reactors*, CONF-750989 ERDA, 1976.

Radiation shielding

Physical barriers designed to provide protection from the effects of ionizing radiation; also, the technology of providing such protection. Major sources of radiation are nuclear reactors and associated facilities, medical and industrial x-ray and radioisotope facilities, charged-particle accelerators, and cosmic rays. Types of radiation are directly ionizing (charged particles) and indirectly ionizing (neutrons, gamma rays, and x-rays). In most instances, protection of human life is the goal of radiation shielding. In other instances, protection may be required for structural materials which would otherwise be exposed to high-intensity radiation, or for radiation-sensitive materials such as photographic film and certain electronic components. *See* HEALTH PHYSICS; NUCLEAR RADIATION; RADIATION DAMAGE TO MATERIALS.

Radiation sources. Nuclear-electric generating stations present a variety of shielding requirements. In the nuclear fission process, ionizing radiation is released not only at the instant of fission but also through radioactive decay of fission products and products of neutron absorption. Of the energy released in the fission process, about 2.5% is carried by prompt fission neutrons and 3% by delayed gamma rays. About 2.5 neutrons are released per fission with average energy 2 MeV. About 90% of the prompt neutrons have energies less than 4 MeV, but because of their greater penetrating ability the remaining 10% are of greater concern in radiation shielding. Neutrons are distributed in energy approximately as shown in Fig.

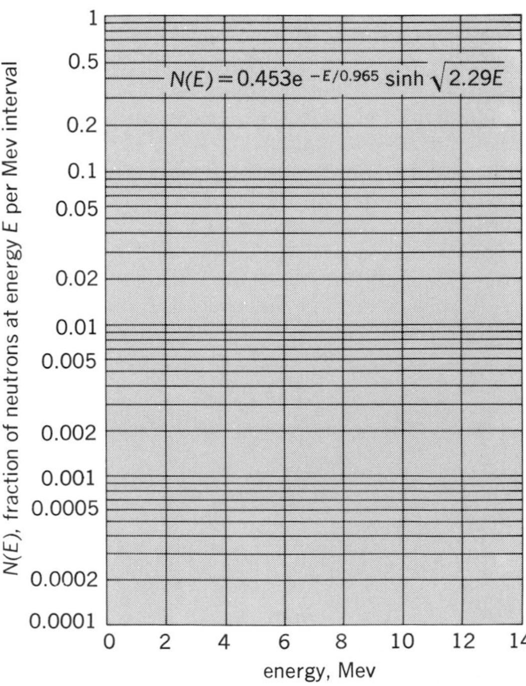

Fig. 1. Neutron spectrum as produced from the fission of U^{235} by thermal neutrons. (*From L. Cranberg et al., Phys. Rev. 103:662–670, 1956*)

1, which gives the distribution for fission of uranium-235. The distribution for other fissionable isotopes is similar. *See* NUCLEAR FISSION; NUCLEAR REACTOR; REACTOR PHYSICS.

Fission of one atom releases about 7 MeV of energy in the form of prompt gamma rays. This energy is distributed over some 10 gamma-ray photons with the energy spectrum as shown in Fig. 2. Evaluation of the intensity of delayed gamma-ray sources within a nuclear reactor requires knowledge of the operating history of the reactor. Absorption of neutrons in structural or shielding materials results in the emission of capture gamma rays. Likewise, gamma rays result from inelastic scattering of neutrons. Because these gamma rays (especially capture gamma rays) are generally of higher energy than prompt fission gamma rays and because they may be released deep within a radiation shield, they require careful consideration in shielding design for nuclear reactors. *See* NUCLEAR REACTION.

Controlled thermonuclear reactors, deriving energy from the nuclear fusion of deuterium and tritium, present radiation shielding requirements similar in kind to those of fission reactors. Highly penetrating (14-MeV) neutrons are released in the fusion process along with charged particles and photons. Capture gamma rays again require careful consideration. *See* NUCLEAR FUSION.

X-ray generators vary widely in characteristics. Typical units release x-rays with maximum energies to 250 KeV, but high-voltage units are in use with x-ray energies as high as tens of MeV. The dominant nature of x-ray energy spectra is that of bremsstrahlung. X-ray generators are but one of many types of charged-particle accelerators and, in most cases, the governing shielding requirement is protection from x-rays. For very-high-energy accelerators, protection from neutrons and mesons produced in beam targets may govern the shield design.

Although many different radionuclides find use in medical diagnosis and therapy as well as in research laboratories and industry, radiation shielding requirements are of special importance for gamma-ray and neutron sources. Alpha and beta particles from radionuclide sources are not highly penetrating, and shielding requirements are minimal. Space vehicles are subjected to bombardment by radiation, chiefly very-high-energy charged particles. In design of the space vehicle and in planning of missions, due consideration must be given to radiation shielding for protection of crew and equipment.

Attenuation processes. Charged particles lose energy and are thus attenuated and stopped primarily as a result of coulombic interactions with electrons of the stopping medium. For heavy charged particles (protons, alpha particles, and such), paths are nearly straight and ranges well defined. Electrons may suffer appreciable angular deflections on collision and may lose substantial energy radiatively. Very-high-energy charged particles may lose energy through nuclear interactions, resulting in fragmentation of the target nuclei and production of a wide variety of secondary radiations.

Gamma-ray and x-ray photons lose energy principally by three types of interactions: photoemission, Compton scattering, and pair production. In photoemission, or the photoelectric effect, the photon transfers all its energy to an atom, and an electron is emitted with kinetic energy equal to the original energy of the photon less the binding energy of the electron in the atom. In Compton scattering, the photon is deflected from its original course by, and transfers a portion of its energy to, an electron. In pair production, the gamma ray is converted to a positron-electron pair. At least 1.02 MeV of gamma-ray energy is required for the rest mass of the pair, and any excess appears as kinetic energy. Ultimately, the positron and an electron recombine and, in annihilation, release two 0.505-MeV gamma rays. Photoemission is especially important for low-energy photons and for stopping media of high atomic number. Compton scattering usually dominates at intermediate photon energies, and pair production at high photon energies.

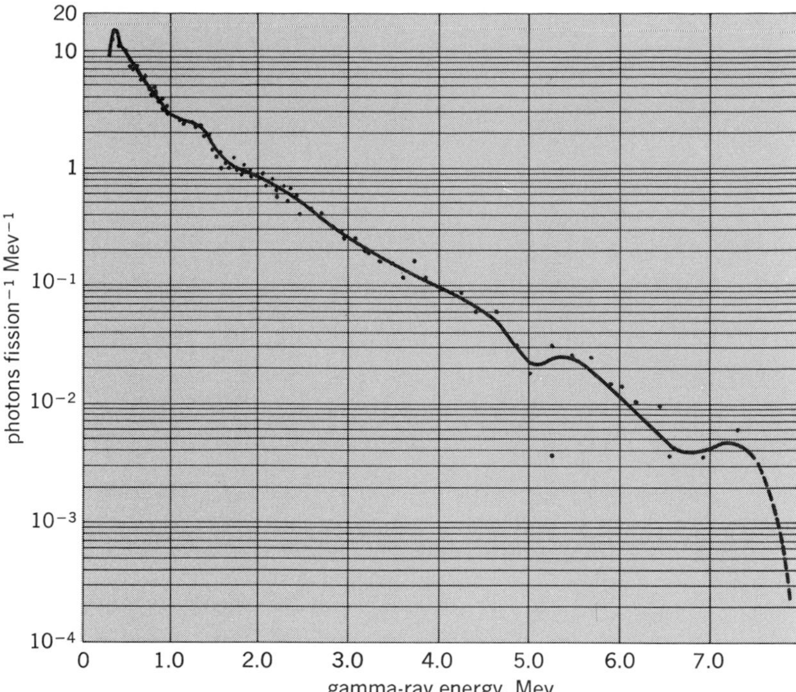

Fig. 2. Energy spectrum of gamma-rays observed within 10^{-7} sec after fission.

Neutrons lose energy in shields by elastic or inelastic scattering. Elastic scattering is more effective with shield materials of low atomic mass, notably hydrogenous materials, but both processes are important, and an efficient neutron shield is made of materials of both high and low atomic mass. The fate of the neutron, after slowing down as a result of scattering interactions, is absorption frequently accompanied by emission of capture gamma rays. Suppression of capture gamma rays may be effected by incorporating elements such as boron or lithium in the shield material. The isotopes ^{10}B and ^{6}Li have large cross sections for neutron capture without gamma-ray emission.

Shielding concepts. The cross section of an atom or electron for interaction with radiation is the effective "target" area presented for the interaction. It is usually given the symbol σ and the units cm^2 or barns (1 barn = 10^{-24} cm^2). The cross section depends on the type of interaction and is a

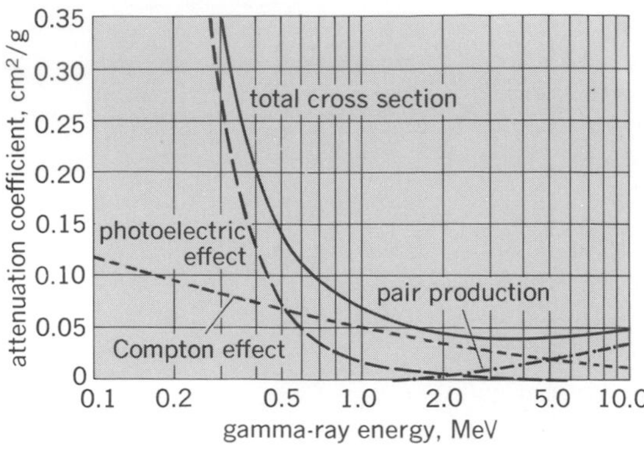

Fig. 3. Gamma-ray attenuation coefficients for lead. (*After G. W. Grodstein*)

function of the energy of the radiation. When the cross section is multiplied by the number of atoms or electrons per unit volume, the product, identified as the linear attenuation coefficient μ or macroscopic cross section Σ, has the units of reciprocal length and may be interpreted as the probability per unit distance of travel that the radiation experiences in an interaction of a given type. The total attenuation coefficient for radiation of a given energy is the sum of attenuation coefficients for all types of interactions in the shielding medium. The quotient of the linear attenuation coefficient and the density, μ/ρ, is called the mass attenuation coefficient. If μ/ρ for gamma rays is weight-averaged by the fraction of the gama-ray energy locally dissipated subsequent to an interaction, the average μ_a/ρ, is called the mass energy absorption coefficient. The total and component parts of the mass attenuation coefficients for lead are illustrated in Fig. 3. In Fig. 4 are shown total mass attenuation coefficients for all elements at several photon energies.

The flux density, or fluence rate, characterizes the intensity of radiation. It may be thought of as the path length traveled by radiation per unit volume per unit time. It is usually given the symbol ϕ and has units $cm^{-2}\ s^{-1}$.

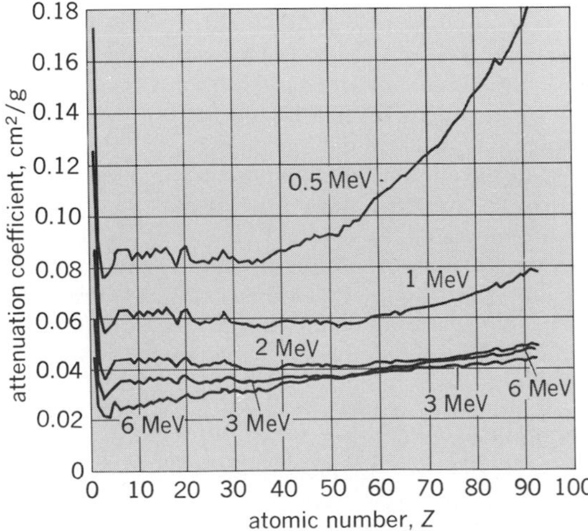

Fig. 4. Gamma-ray attenuation coefficients versus atomic number of absorbing material for various energies.

To illustrate these concepts, consider the flux density at a distance r from a point source isotropically emitting S monoenergetic gamma rays of energy E per second in a uniform medium with total attenuation coefficient μ. If μ were zero, the flux density would be determined just by the inverse square of the distance from the source. If μ were not zero, attenuation of the gamma rays would also be exponential with distance. Thus the equation shown below would hold. This is the flux

$$\phi = \frac{S}{4\pi r^2} e^{-\mu r}$$

density of gamma rays which have traveled distance r without having experienced any interactions in the stopping medium. The energy locally dissipated per unit mass per unit time due to these gamma rays as they experience their first interactions is the product $E(\mu_a/\rho)\phi$. However, some gamma rays reach distance r already having experienced scattering interactions. To account for these secondary gamma rays, a buildup factor B is employed. B is a function of the energy of the source gamma rays and the product μr and depends on the attenuating medium. The total energy locally dissipated per unit mass per unit time is thus the product $BE(\mu_a/\rho)\phi$. Similar concepts apply to neutron shielding; however, the treatment of scattered neutrons is considerably more complicated, and the buildup-factor concept is not well established.

In the final phases of shielding design, digital computers are usually employed to carry out the required calculations. It is also common practice to base shielding design, at least in part, on measurements made on a prototype.

Shielding materials. The most common criteria for selecting shielding materials are radiation attenuation, ease of heat removal, resistance to radiation damage, economy, and structural strength.

For neutron attenuation, the lightest shields are usually hydrogenous, and the thinnest shields contain a high proportion of iron or other dense material. For gamma-ray attenuation, the high-atomic-number elements are generally the best. For heat removal, particularly from the inner layers of a shield, there may be a requirement for external cooling with the attendant requirement for shielding the coolant to provide protection from induced radioactivity.

Metals are resistant to radiation damage, although there is some change in their mechanical properties. Concretes, frequently used because of their relatively low cost, hold up well; however, if heated they lose water of crystallization, becoming somewhat weaker and less effective in neutron attenuation.

If shielding cost is important, cost of materials must be balanced against the effect of shield size on other parts of the facility, for example, building size and support structure. If conditions warrant, concrete can be loaded with locally available material such as natural minerals (magnetite or barytes), scrap steel, water, or even earth.

Typical shields. Radiation shields vary with application. The overall thickness of material is chosen to reduce radiation intensities outside the shield to levels well within prescribed limits for occupational exposure or for exposure of the gen-

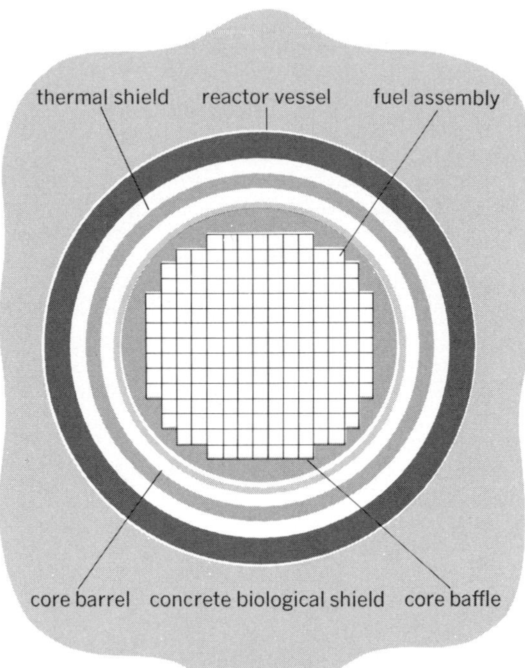

Fig. 5. Typical power reactor shield configuration.

eral public. The reactor shield is usually considered to consist of two regions, the biological shield and the thermal shield. The thermal shield, located next to the reactor core, is designed to absorb most of the energy of the escaping radiation and thus to protect the steel reactor vessel from radiation damage. It is often made of steel and is cooled by the primary coolant. The biological shield is added outside to reduce the external dose rate to a tolerable level (Fig. 5).

[RICHARD E. FAW]

Bibliography: E. P. Blizard, Nuclear radiation shielding, in H. Etherington (ed.), *Nuclear Engineering Handbook*, 1958; E. P. Blizard and L. S. Abbot (eds.), *Reactor Handbook*, vol. 3, pt. B: *Shielding*, 1962; U. Fano, L. V. Spencer and M. J. Berger, *Penetration and Diffusion of X Rays*, vol. 38/2 *of Handbuch der Physik*, 1959; H. Goldstein, *Fundamental Aspects of Reactor Shielding*, 1959; R. G. Jaeger et al. (eds.), *Engineering Compendium on Radiation Shielding*, vol. 1, 1968, and vol. 3, 1970; B. T. Price, C. Horton, and K. Spinney, *Radiation Shielding*, 1957; N. M. Schaeffer (ed.), *Reactor Shielding for Nuclear Engineers*, 1973.

Radioactive waste management

The treatment and containment of radioactive wastes. Radioactive waste management is required to some degree in all operations associated with the use of nuclear energy for national defense or peaceful purposes. Liquid, solid, and gaseous radioactive wastes are produced in the mining of ore, production of reactor fuel materials, reactor operation, processing of irradiated reactor fuels, and numerous related operations. Wastes also result from the use of radioactive materials, for example, in research laboratories, industrial operations, and medical treatment. The magnitude of waste management operations will increase as the nuclear energy program further extends and diversifies and as a large, widespread nuclear power industry develops. The particular problems associated with the decommissioning of a nuclear facility are discussed in the second part of this article.

ASPECTS OF MANAGEMENT

The chief aim in the safe handling and containment of radioactive wastes is the prevention of ra-

Table 1. Typical materials in high-level liquid waste

Material[b]	Grams per metric ton from various reactor types[a]		
	Light water reactor[c]	High-temperature gas-cooled reactor[d]	Liquid metal fast breeder reactor[e]
Reprocessing chemicals			
Hydrogen	400	3800	1300
Iron	1100	1500	26,200
Nickel	100	400	3300
Chromium	200	300	6900
Silicon	–	200	–
Lithium	–	200	–
Boron	–	1000	–
Molybdenum	–	40	–
Aluminum	–	6400	–
Copper	–	40	–
Borate	–	–	98,000
Nitrate	65,800	435,000	244,000
Phosphate	900	–	–
Sulfate	–	1100	–
Fluoride	–	1900	–
SUBTOTAL	68,500	452,000	380,000
Fuel product losses[f,g]			
Uranium	4800	250	4300
Thorium	–	4200	–
Plutonium	40	1000	500
SUBTOTAL	4840	5450	4800
Transuranic elements[g]			
Neptunium	480	1400	260
Americium	140	30	1250
Curium	40	10	50
SUBTOTAL	660	1440	1560
Other actinides[g]	<0.001	20	<0.001
Total fission products[h]	28,800	79,400	33,000
TOTAL	103,000	538,000	419,000

SOURCE: From K. J. Schneider and A. M. Platt (eds.), *Advanced Waste Management Studies: High-Level Radioactive Waste Disposal Alternatives*, USAEC Rep. BNWL-1900, May 1974.

[a]Water content is not shown; all quantities are rounded.

[b]Most constituents are present in soluble, ionic form.

[c]U-235 enriched pressurized water reactor (PWR), using 378 liters of aqueous waste per metric ton, 33,000 MWd/MT exposure. (Integrated reactor power is expressed in megawatt-days [MWd] per unit of fuel in metric tons [MT].)

[d]Combined waste from separate reprocessing of "fresh" fuel and fertile particles, using 3785 liters of aqueous waste per metric ton, 94,200 MWd/MT exposure.

[e]Mixed core and blanket, with boron as soluble poison, 10% of cladding dissolved, 1249 liters per metric ton, 37,100 MWd/MT average exposure.

[f]0.5% product loss to waste.

[g]At time of reprocessing.

[h]Volatile fission products (tritium, noble gases, iodine, and bromine) excluded.

diation damage to humans and the environment by controlling the dispersion of radioactive materials. Harm to humans may result from irradiation by external sources or from the intake (by ingestion, by inhalation, or through the skin) of radioactive materials, their passage through the respiratory and gastrointestinal tract, and their partial incorporation into the body. Radioactive waste contaminants in air, water, food, and other elements of the human environment must be kept below specified concentrations, which differ according to the particular radionuclide or mixture of radionuclides which is present. Liquid or solid waste products containing significant quantities of the more toxic radioactive materials require isolation and permanent containment in media from which any potential escape into the human environment would be at tolerable levels. The radioactive materials of major concern are those that may be readily incorporated into the body and those that have relatively long half-lives, ranging from a few years to thousands of years.

Waste management is focused on those radioisotopes which originate in nuclear reactors. Here the fission products (chemical elements formed by nuclear fragmentation of actinide elements such as uranium or plutonium) accumulate in the nuclear fuel, along with plutonium and other transuranic nuclides. (Transuranic elements, also called actinide elements, are those higher than uranium on the periodic table of chemical elements.) The concentrations of plutonium are substantially higher than those found in nature; they range from 10 to 20 kg per metric ton (1000 kg) of uranium, compared to a high of 17 g per metric ton of uranium in minerals from fumarole areas. *See* PLUTONIUM; URANIUM.

Reprocessing. If fuel discharged from the nuclear reactor is reprocessed, uranium and plutonium are recovered after chemical dissolution. During recovery, the favored treatment processes produce high-level waste in the form of an acidic aqueous stream. Other processes are being considered that would produce high-level waste in different forms. The high-level waste contains most of the reactor-produced fission products and actinides, as well as slight residues of uranium and plutonium (see Table 1). These waste products emit large amounts of potentially hazardous ionizing radiation and generate sufficient heat to require substantial cooling. Because the reprocessing step normally does not dissolve much of the nuclear fuel cladding, high-level waste normally contains only a small amount of the radionuclides formed as activation products within the cladding. The cladding hull waste is managed as a separate solid waste stream, as are several other auxiliary waste streams from reprocessing plants.

The nuclear industry can reuse the recovered uranium and plutonium by reconstituting it into nuclear fuels in which plutonium, instead of uranium-235, is the fissile material. Since these fuels contain both uranium and plutonium mixed oxides (MOX), their fabrication generates additional plutonium-containing wastes.

Policy and treatment. The policy of the United States is to assume custody of all commercial high-level radioactive wastes and to provide containment and isolation of them in perpetuity. Regulations require that the high-level wastes from nuclear fuel reprocessing plants be solidified within 5 years after reprocessing and shipped to a Federal repository within 10 years after reprocessing.

Because of the expected increase in the quantities of waste-containing materials or those contaminated with transuranic elements, and because of the long half-life and specific radiotoxicity of these elements, it has also been proposed that all transuranic wastes be solidified and transferred to the United States government as soon as practicable, but at most within 5 years after generation.

Both of these policies require that high-level, cladding, and other transuranic wastes be converted to solid form. A variety of technologies exist for this conversion, including calcination, vitrification, oxidation, and metallurgical smelting, depending on the primary waste.

A typical solidification process, chiefly for high-level waste, is spray calcination-vitrification (Fig. 1). In this process, atomized droplets of waste fall through a heated chamber where flash evaporation results in solid oxide particles. Glassmaking solid frit or phosphoric acid can be added to provide for melting and glass formation, either in a continuous melter or in the vessel that will serve as the waste canister. The molten glass or ceramic is cooled and solidified. At the present time, however, there are no commercial reprocessing plants operating in the United States, and spent fuel is simply being accumulated in water basins.

Quantities of waste. Civilian nuclear electrical generating capacity in the United States is projected to increase to about 175,000 to 200,000 megawatts (MW) by the year 2000. Assuming the latter figure, a cumulative total of 50,000 metric tons (or megagrams) of heavy metal (MTHM) will have been discharged from reactors as a result of the

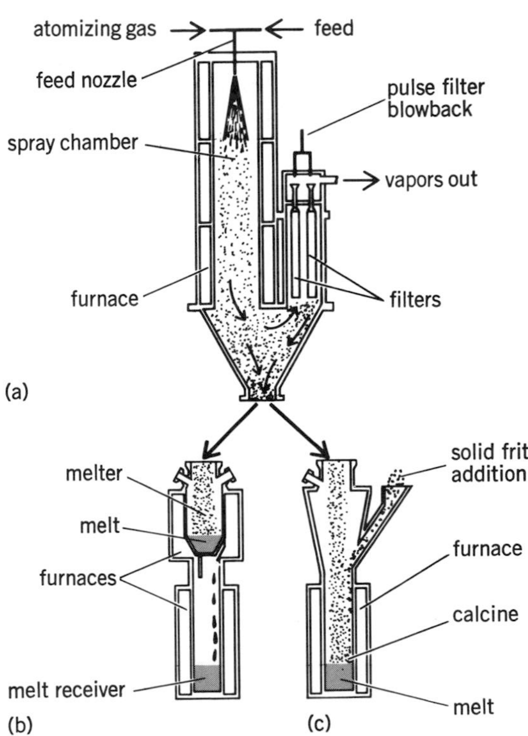

Fig. 1. Spray calcination-vitrification process. (*a*) Spray calciner, producing calcine that drops into either (*b*) continuous silicate glass melter, or (*c*) directly into waste canister vessel for for in-pot melting.

Table 2. Light-water reactor nuclear wastes from 1 GWe-yr of operation, after 10-year cooling

Unreprocessed spent fuel	
Volume	25 m^3
U	36.4 Mg
Pu	0.31 Mg
Radioactivity	13.0 MCi*
Conditioning waste	
Volume	9 m^3
Radioactivity	0.2 kCi
Reprocessing wastes	
Vitrified high-level waste	
Volume	2.8 m^3
Contained Pu	3.4 kg
Radioactivity	10.4 MCi
Intermediate-level waste	
Volume (in concrete)	52 m^3
Contained Pu	1.7 kg
Radioactivity	0.07 MCi
Hulls and spacers	
Volume (compacted)	7.4 m^3
Contained Pu	0.24 kg
Radioactivity	0.03 MCi
Mixed oxide fuel waste	
Volume (in concrete)	18 m^3
Contained Pu	0.65 kg

*1 Ci = 3.7×10^{10} disintegrations per second = 3.7×10^{10} becquerels.

production of some 2500 gigawatt-years (or 21.9×10^{12} kilowatt-hours) of electricity.

The Department of Energy has calculated material balances and the amount of waste expected from operation of light-water reactor fuel cycles, either with spent fuel as the primary waste or with reprocessing and plutonium recycle. The characteristics of wastes requiring disposition in a repository are shown in Table 2. The key point is that reprocessing doubles the volume of waste but reduces the actinide content by some fiftyfold.

If spent fuel were the primary form of waste, the anticipated packaged waste through the year 2000 would be 62,000 m^3. If this were stacked as a solid cube, it would measure nearly 40 m on a side. About 38,000 megacuries of radioactivity (1.4×10^{21} disintegrations per second) and 175 MW of heat would be associated with this quantity of waste.

In addition to the wastes from peaceful uses of nuclear energy, defense activities in the United States have generated substantial nuclear waste. The 1978 quantities of high-level defense wastes are shown in Table 3, as is the volume which would be occupied by the waste existing in 1990 if it was to be solidified as a glass. The location and the quantities of existing defense-activities transuranic wastes as of 1978 are estimated in Table 4.

Before 1970, the Atomic Energy Commission allowed direct burial of transuranic wastes at selected sites. Since that date, policy has directed that the wastes be packaged and stored in such a way that they can be retrieved easily.

Spent fuel packaging. Probably the most comprehensive study on packaging spent fuel has been conducted by the Swedish project Kaernbraenslesaekerhet (KBS) set up in early 1977. The KBS plan calls for 40 years of water pool storage of spent fuel in a granite cavern some 30 m underground.

After the 40-year storage to let the heat dissipate, groups of 500 fuel rods (1.5 MTHM) would be loaded in copper canisters (Fig. 2). The canisters would then be filled with lead and copper covers welded onto the tops. Each canister would weigh about 20 MT.

The canisters would then be transferred to a final storage/disposal location in granite some 500 m underground. Here they would be placed in holes some 7.7 m deep and 1.5 m in diameter, lined (sides, top, and bottom) with 40 cm of isostatically compressed bentonite.

Early investigation by the KBS of three geologic study areas in granite showed the possibility of water flows of 0.1 to 0.2 liter/m^2/yr. This, in part, was the motivation for the sophisticated packaging system.

In the United States, several options are being investigated for encapsulating the spent fuel. The baseline option is placement of spent fuel in canisters with only an inert-gas fill. Other, more advanced methods are under study as technical alternatives. These include a metal matrix fill, sand fill, other glassy or ceramic materials, and multiple-barrier encapsulation of the spent fuel and canister at the time it is declared a waste for disposal. Experimental packaging and dry storage of spent fuel using facilities previously associated with the nuclear rocket program in Nevada have been demonstrated at the Engine Maintenance and Disassembly (EMAD) facility.

High-level waste packaging. If the United States reprocessed spent fuel, the high-level waste would be solidified as a glass, ceramic, or metal matrix material, encapsulated in a metallic container, and possibly surrounded in the repository by additional barriers. Table 5 gives some characteristics of a typical canister. After 1000 years the canister would generate only 20 W, and the

Table 3. United States' high-level defense wastes

Source site	Present form	Radioactivity, 10^6 Ci*	Volume, 10^3 m^3†	Solidified volume, 10^3 m^3‡
Hanford Plants	Alkaline salt/sludge/ liquor	190	173	13
	Separated $^{90}Sr/Y$, $^{137}Cs/Ba$	210		–
Idaho Chemical Processing Plant	Acid calcine	70 (both)	1.5	8 (both)
	Acid liquid		10	
Savannah River Plant	Alkaline salt/sludge	560	82	6

*1 Ci = 3.7×10^{10} disintegrations per second = 3.7×10^{10} becquerels.
†As of 1978.
‡Volume if waste existing in 1990 were solidified as a glass.

Table 4. Existing United States' defense transuranic waste, in 10^3 m^3*

Storage site	Buried	Retrievably stored	Total
Hanford, WA	153	8	161
Idaho National Engineering Laboratory, ID	65	36	101
Los Alamos Scientific Laboratory, NM	116	2	118
Oak Ridge National Laboratory, TN	5	1	7
Savannah River Laboratory, SC	28	2	30
Nevada Test Site	<1	<1	<1
Total	369	49	418
Transuranic content, kg	700	374	1100

*As of 1978.

gamma dose rate would be only 1.6 roentgens/hr (1.147×10^{-7} coulomb/kg·s) at the centerline of the cannister 30 cm from the surface.

Repositories and disposal. The major worldwide thrust has narrowed from the many waste management concepts previously described to the investigation of the sub-seabed and mined cavities in geologic formations as the principal alternatives. Both continental and sub-seabed geologic formations exist that have been physically and chemically stable for millions of years.

The basic requirement for acceptable final storage or disposal of radioactive waste is the capability to contain and isolate the waste safely until decay has reduced the radioactivity to nonhazardous levels—or at least to levels found in nature.

Exactly how long near-total containment and isolation of nuclear waste must be maintained is much debated. However, two characteristics of waste change radically in a few hundred years. First, the heat generation rate decreases an order of magnitude in the period of 10 to 100 years and another order of magnitude in the period of 100 to 1000 years.

Second, as shown in Fig. 3, the toxicity (expressed as the quantity of water required to dilute to drinking-water tolerances) of the high-level waste needed to produce 1 GW-yr of electric power decreases about three orders of magnitude in the first 300 to 400 years due to decay of short-lived fission products. It is then at levels comparable with those of an equal volume of average ores of common toxic elements. After this time, toxicity diminishes slowly, a million years being required for another two orders of magnitude. Thus it is clear that the first 300 to 400 years of the disposal period are the most critical.

Mined cavity disposal. In this disposal option, radioactive wastes would be emplaced in the walls or floor of deep (600-m) tunnels created by conventional mining techniques. Although there are numerous geologic media that could be considered, only salt, granite, basalt, and shale have been studied extensively. Since the repository should provide both containment and isolation, site selection will involve consideration of the properties, dimensions, and characteristics of the host rock, the hydrologic properties of the site, its tectonic stability, its resource potential, and the capability of the site geohydrology to provide natural barriers to the movement of waste.

These natural barriers will be further augmented, as previously described, by a solid waste form, by canisters, and possibly by engineered barriers such as absorption backfill and overpack materials.

Another prime design consideration is the initial and long-term heating of the repository site by the radioactive decay process. Preliminary designs are for initial heating from 100 to 350 kW per hectare. Thus a repository area of 1000 to 3500 hectares would be required to accommodate the nuclear wastes which are expected to be generated through the year 2000.

Sub-seabed geologic disposal. Analysis of the ocean regimes has shown that the most appropriate areas for sub-seabed disposal are abyssal hill regions in the centers of sub-ocean tectonic plates underlying large ocean-surface currents known as gyres. These abyssal hill regions are vast, are remote from human activities, have few known natural resources, are biologically unproductive, have weak and variable bottom currents, and are covered with red clays to a depth of 50 to 100 m.

These clay sediments are currently the prime sub-seabed geologic media under consideration for radioactive waste containment. They are soft and

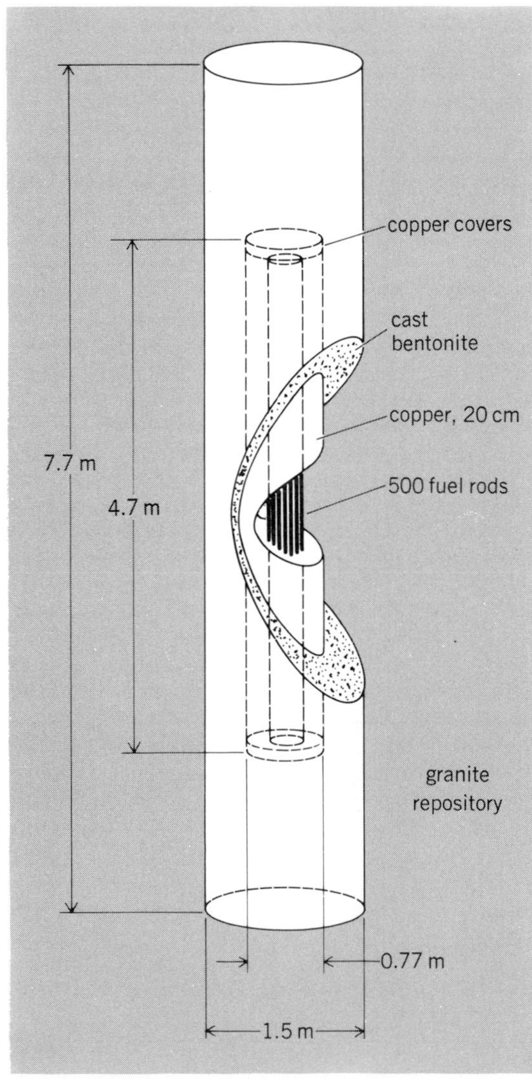

Fig. 2. KBS spent fuel concept.

Table 5. Typical high-level waste canister

Decay time, years	Heat generation rate, kW	Cumulative dose, alpha particles per gram	Dose rate 30 cm from surface, roentgens/hr*
1	22	1.0×10^{17}	1.1×10^{6}
10	3.1	2.5×10^{17}	6.2×10^{4}
100	0.36	7.1×10^{17}	5.8×10^{3}
1000	0.02	1.5×10^{18}	1.6
10,000	0.006	3.0×10^{18}	1.3
100,000	0.003	6.1×10^{18}	0.6

Length: 3 m
Diameter: 30 cm
Can material: 304L stainless steel
Volume: 0.21 m^3
Contents: 2.5 metric tons uranium equivalent

*1 roentgen/hr = 2.58×10^{-4} C/kg · hr = 0.717×10^{-7} C/kg · s.

pliable near the sediment-water interface and become increasingly rigid and impermeable with depth. Tests have shown that these sediments have high sorption coefficients (radionuclide retention) and low natural pore-water movement. They are found by surface acoustic profiling to be uniformly distributed over large areas (tens of thousands of square kilometers) of the ocean floor. Core analysis has shown that deposition in these areas has been continuous and undisturbed for millions of years, so that they can confidently be predicted to remain stable long enough for radionuclides to decay to innocuous levels.

The multiple barriers in this option consist of the waste form and canister for short-term containment and the sediments for long-term radionuclide retention. Predisposal waste treatments (for example, dilution and longer surface storage) will decrease the thermal-related problems during the early disposal period. The main technical problem yet to be addressed is the response of the geologic medium (red clay) to the heat given off by reprocessed high-level waste or spent fuel in the first 500 years.

One emplacement method being studied is the encasement of the wastes or spent fuel in a needle-shaped projectile which, propelled from a ship, will penetrate into a predetermined location in the sediments. It can thereafter be monitored for location, attitude, condition, and temperature through self-contained instrumentation. Laboratory tests have led to the conclusion that the red clay, thus dynamically penetrated, will reseal itself above a buried canister in a comparatively short time. Only about 0.006% of the available abyssal hill area in the central North Pacific area, for instance, would be needed to dispose of all high-level wastes generated in the United States through the year 2040.

[ALLISON M. PLATT]

DECOMMISSIONING OF A NUCLEAR FACILITY

Because of the buildup of radioactive contamination in a nuclear facility such as a nuclear reactor or reprocessing plant, detailed plans must be made, beginning with the design and construction stages of the facility, to provide adequate protection from radiation exposure during the final decommissioning stage. Some of the radionuclides constituting this contamination have half-lives of many thousands of years, and the levels of contamination in the tanks, equipment, plumbing, instruments, hot cells, and the building itself may reach levels of millions of curies (1 curie = 3.7×10^{10} disintegrations per second = 3.7×10^{10} becquerels). Thus in many cases it is necessary to use special techniques, remote-control equipment, and large shields to remove or isolate this contamination in a safe manner from the human environment for hundreds of thousands of years. First, all high-level radioactive materials, such as spent reactor fuel, equipment for reprocessing fuel, and underground tanks for radioactive liquid waste, are removed to an approved site for disposal of such waste. Then surfaces are scoured with remotely operated equipment using water, steam, jets of sand, and various chemicals. Sometimes the equipment and the facility are completely demolished by using demolition balls and explosives, but in that case extreme care must be exercised to prevent spread

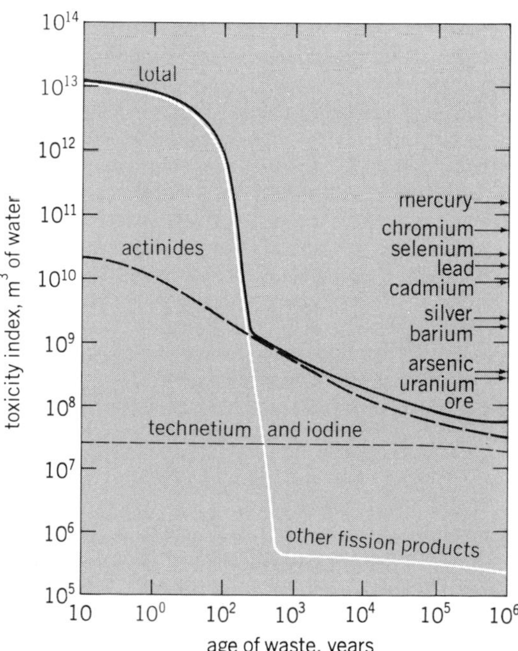

Fig. 3. Toxicity index of high-level waste resulting from production of a gigawatt-year of electric power, and of volumes of average ores of common toxic elements equal to repository volume required by the waste.

of the radioactive contamination to the environment.

There are four principal methods of decommissioning in use: layaway, mothballing, dismantlement, and entombment. Layaway and mothballing are temporary measures to postpone final action until some of the radioactivity has decayed. They involve removing the easily removable high-level radioactive objects. With layaway the facility might be put back into use at some future date, but this is unlikely. Both layaway and mothballing require security measures, such as locked or welded doors and guards on constant duty, to prevent the entry of unauthorized persons who might be exposed to high-level radiation. The barriers to reentry of contaminated areas are more secure and rigid with mothballing than with layaway, so that in the former case fewer guards are required and some uncontaminated areas of the plant may be released to public use. Both of these methods require dismantlement or entombment at a later date unless the levels and half-lives of the residual contamination are very low. Entombment, as the name implies, is making the radioactive contamination inaccessible by the use of demolition techniques and covering the residue with reinforced concrete. The choice of decommissioning method depends on the levels and half-lives of the residual contamination, and on cost, environmental factors (rain, wind, hydrology), population density, and many other factors. In any case, the objective should be to take those measures that will result in minimum radiation exposure and minimum consequent radiation damage to persons now living and those to be born in the future. *See* HEALTH PHYSICS; NUCLEAR FUEL CYCLE; NUCLEAR FUELS REPROCESSING; NUCLEAR REACTOR. [KARL Z. MORGAN]

Bibliography: Department of Energy, *Draft Environmental Impact Statement: Management of Commercially Generated Radioactive Waste*, DOE/EIS0046-D, April 1979; H. W. Dickson, *Standards and Guidelines Pertinent to the Development of Decommissioning Criteria for Sites Contaminated with Radioactive Material*, Oak Ridge National Laboratory, ORNL/DEPA-4, August 1978; Nuclear Regulatory Commission, *Technology, Safety and Costs of Decommissioning a Reference Nuclear Fuel Reprocessing Plant*, NUREG-0278, October 1977; A. M. Platt and J. L. McElroy, *Management of High-Level Nuclear Wastes*, Pacific Northwest Laboratory, PNL-SA-7072, September 1979; K. J. Schneider and A. M. Platt (eds.), *Advanced Waste Management Studies: High-Level Radioactive Waste Disposal Alernatives*, USAEC Rep. BNWL-1900, May 1974; R. I. Smith, G. J. Konzek, and W. E. Kennedy, Jr., *Technology, Safety and Costs of Decommissioning a Reference Pressurized Water Reactor Power Station*, vols. 1 and 2, Nuclear Regulatory Commission, NURE G/eR-0130, 1978.

Ramjet

The simplest of the air-breathing propulsion engines (Fig. 1). In flight, air enters the front of the diffuser at high velocity. The diffuser is shaped to reduce the airspeed and hence its kinetic energy as it passes through. With an efficient diffuser, the reduction in kinetic energy results in a nearly equal increase in potential energy, in the form of an increase in air pressure. This higher-pressure

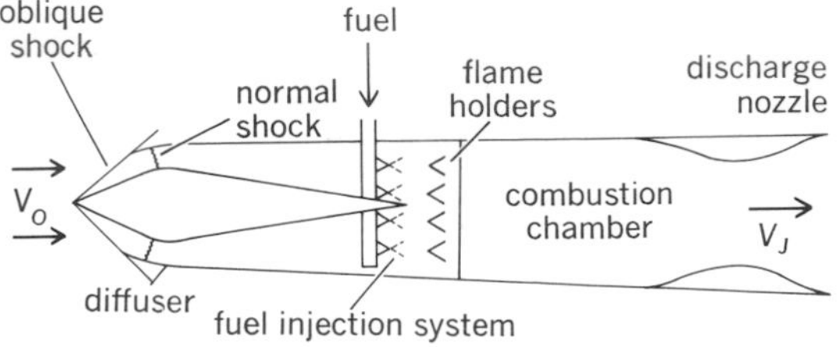

Fig. 1. Diagram of a ramjet engine.

air enters the combustion chamber, where fuel is continuously injected and burned. The hot gas is then ejected rearwardly through the discharge nozzle at velocity V_J greater than flight speed V_0. To a first approximation, thrust F is given in the equation shown below, where M is defined as the

$$F = M(V_J - V_0)$$

mass of air per second which is flowing through the engine.

Characteristics. The objective of the ramjet cycle is to provide a jet velocity V_J that is considerably greater than the initial velocity V_0. This increase in air velocity represents an increase in kinetic energy. The efficiency of a ramjet in converting the chemical energy in the fuel into kinetic energy of the airstream depends upon the ratio of the pressure in the combustion chamber to the ambient air pressure. This pressure ratio in turn depends upon the flight speed or, more exactly, upon flight Mach number (Fig. 2).

At zero flight Mach number, there is of course no increase in pressure through the diffuser, and the efficiency of the ramjet is zero. Thus the ramjet has no thrust at takeoff. As the flight speed increases, the pressure ratio and hence efficiency increase, and an increase in thrust and a reduction in specific fuel consumption result (Fig. 3).

At any given flight Mach number, maximum thrust is developed when sufficient fuel is injected into the combustion chamber to consume substantially all of the oxygen in the air passing through.

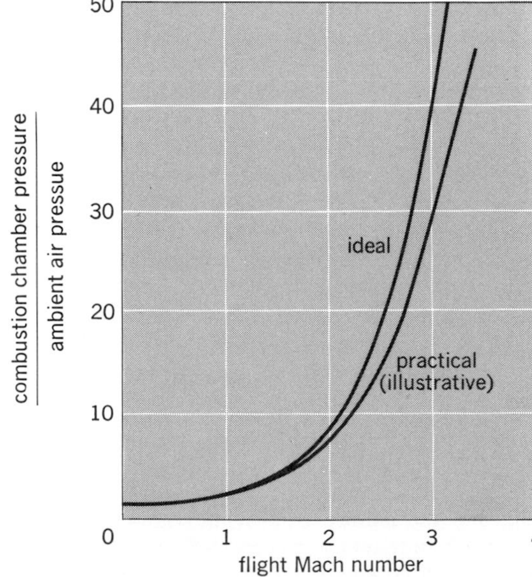

Fig. 2. Effect of flight speed or flight Mach number on pressure ratio across the inlet diffuser.

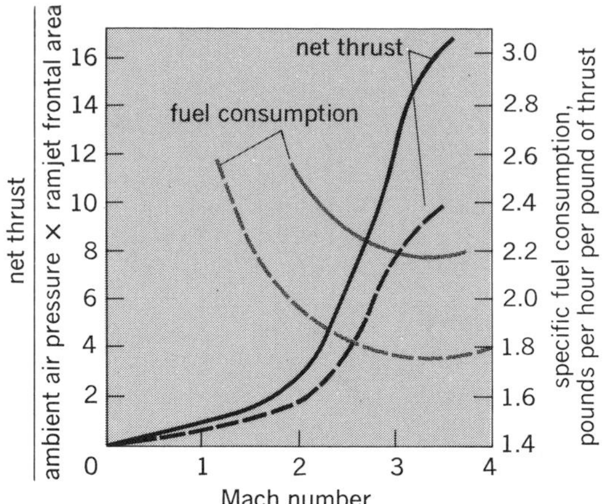

Fig. 3. Effect of flight Mach number on the thrust and the specific fuel consumption of a ramjet engine. Solid lines represent the fuel-air ratio adjusted for maximum thrust; broken lines represent the fuel-air ratio adjusted for maximum efficiency.

This represents the largest amount of heat that can be introduced into the air. Greater efficiency of utilization of the fuel, however, is obtained when less than the maximum burnable amount is injected. The efficiency of utilization of the fuel is represented by the specific fuel consumption (pounds of fuel consumed per hour per pound of thrust). The higher the efficiency, the lower is the specific fuel consumption. Curves in Fig. 3 are for the two operating conditions of maximum power and maximum efficiency.

Combustion. The fuels used in ramjet engines are hydrocarbons obtained from petroleum. These fuels burn in only a narrow range of fuel-air ratios. They have flame speeds which are considerably lower than the speed at which the air must pass through the combustion chamber to obtain the high thrust per unit of cross-sectional area required for practical applications. Flameholders are therefore located in the combustion chamber in the wake of which the airspeeds are reduced locally to accommodate the low flame speeds. The determination of the configuration and location of these flameholders to provide adequate combustion efficiency without imposing excessive drag on the air is one of the crucial development problems of the ramjet. The location and design of the fuel injection nozzles and the design of the control equipment to provide a suitable fuel-air ratio in the vicinity of the flameholder for efficient combustion over the range of flight speeds and altitudes desired in the flight program of a given ramjet vehicle are essential to efficient operation. New high-energy fuels greatly increase the attainable altitude of the ramjet and appreciably reduce the engine length. The absence in this propulsion cycle of moving parts after the burner enables the ramjet to burn metal-based fuels for greater performance.

Takeoff. Because the ramjet has low thrust at low flight speeds, another type of engine is required for takeoff boost. In missiles such as the Bomarc, a rocket is used to take off and accelerate the vehicle to a speed at which the ramjet can take over. In aircraft where successive takeoffs and landings are desired, a turbojet engine can be used for this purpose.

In general, a supersonic ramjet vehicle must be boosted to supersonic flight speeds before the ramjet engines can provide sufficient thrust for propelling the vehicle. By providing diffusers and nozzles in which the configuration and area can be varied, the ramjet can operate efficiently over a wider range of flight speeds and can take over at a lower flight speed. This can result in an appreciable saving in the size of the booster rocket.

The ramjet depends entirely on pressure recovery due to its forward speed. The oblique and normal shock waves at the inlet constitute the first two stages of a compressor deceleration. Control of inlet flow by changing the inlet area through axial translation of the nozzle cone or by varying the back pressure through adjustment of the fuel flow can maintain the optimum positions of the shock waves and thereby minimize aerodynamic drag due to spillage or inadequate capture of air.

Flight speed. Ramjet engines are usually considered for applications in the range of flight Mach numbers between 2.5 and 8, although 8 is not a theoretical upper limit. As the flight Mach number increases above 4, heating of the vehicle by the high air friction becomes progressively a more serious problem, and methods of cooling the structure must be incorporated. At flight Mach numbers of 5 and higher, because of the high gas temperatures in the combustion chamber, dissociation of the gases occurs; that is, the combustion does not go to completion and only part of the heat is released. The remainder of the heat can theoretically be released if combustion continues as the gas expands through the discharge nozzle. However, the occurrence of dissociation, the extent of which is influenced by nozzle design, may be a basic practical limitation on the flight speeds attainable by ramjets.

The working fluid (air) can also be heated by a nuclear reactor, thereby freeing the design from limitations which would be imposed by using the working fluid as oxidizer. The air is decelerated, passed directly through a nuclear reactor that occupies the usual combustion chamber space, and then discharged from a conventional convergent-divergent nozzle.

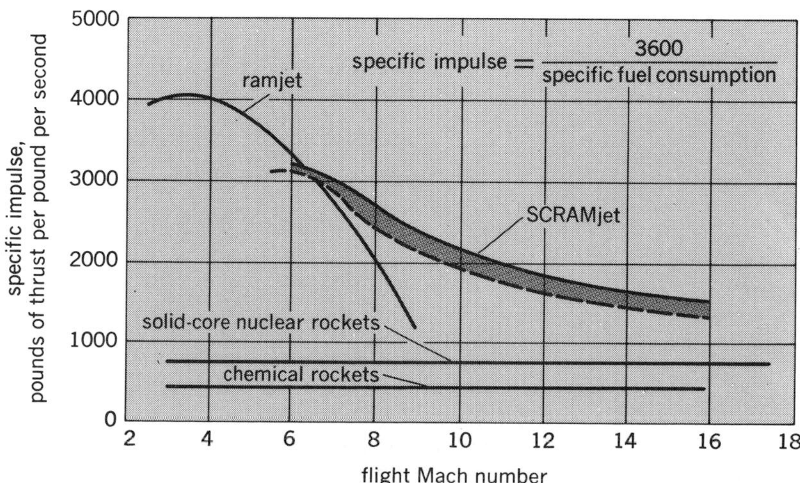

Fig. 4. Comparison of specific impulse of scramjet and ramjet, both burning hydrogen fuel. 1 lbF/(lb · sec) = 9.8 N/(kg · sec). (*From Air Force Mag., May. 1965*)

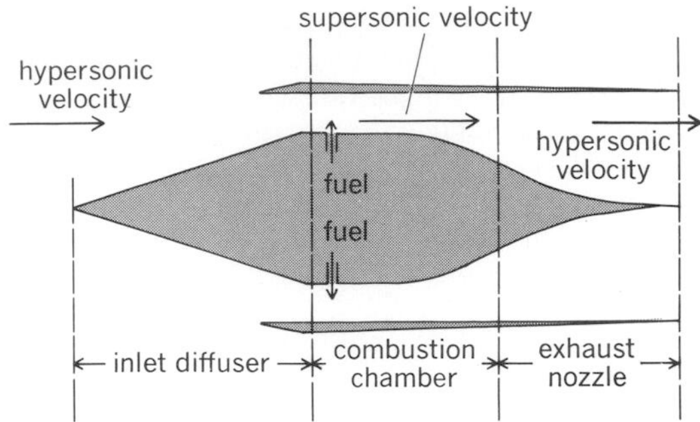

Fig. 5. Diagram of scramjet engine.

Scramjet. The scramjet (supersonic combustion ramjet) is essentially a ramjet engine intended for flight at hypersonic speeds (that is, above Mach 6), with the gases flowing through the combustion chamber and burning at supersonic speeds. (In the ramjet the velocity through the combustion chamber is subsonic.) By designing the engine inlet diffuser to provide supersonic speeds in the combustion chamber for flight at hypersonic speeds, the problem of extreme air temperature and dissociation in the combustion chamber, which limits the practical flight speed of ramjet engines, is avoided in the scramjet. Thus while ramjets cease to be practical at Mach numbers above 8, scramjets, on the other hand, are believed to be feasible up to Mach 25.

The projected specific impulses of scramjets and ramjets with hydrogen fuel are compared in Fig. 4. The specific impulse, which is the pounds of thrust obtained per pound per second of fuel consumed, is a measure of the efficiency of the engine. The scramjet has a much higher specific impulse than the ramjets at Mach numbers above about 6. The specific impulse of the scramjet is much higher than that of the best chemical rocket and the nuclear rocket. The scramjet is considered the logical engine for aircraft and long-range missiles that cruise in the atmosphere at hypersonic speeds.

Various fuels, ranging from hydrocarbons to hydrogen, are being studied for this engine. The specific impulse of the scramjet engine is considerably less with hydrocarbons than with hydrogen, and its flight speeds may be limited to below Mach 15; with hydrogen, flight up to nearly orbital velocity (Mach 26) is believed to be possible.

Because the velocity in the scramjet engine never decreases to subsonic speeds, the diffusers and nozzles have much simpler shapes than their ramjet counterparts; they do not require the converging-diverging contours that characterize ramjet engines. Compare the diagram of a scramjet engine (Fig. 5) with that of the ramjet (Fig. 1). The optimum speed in the combustion chamber is about one-third the vehicle flight speed: that is, for a flight speed of Mach 12 the speed in the combustion chamber is Mach 4.

At hypersonic flight speeds the temperature of the air in the combustion chamber is above the ignition temperature of the fuel. Thus the rate of burning is limited mainly by the rate at which the fuel and air can be mixed. Supersonic combustion of good efficiency has been demonstrated experimentally with hydrocarbons and with hydrogen as fuels. The integration of the diffuser, combustor, and discharge nozzle in an efficient, practical system has yet to be obtained. [BENJAMIN PINKEL]

Bibliography: *Air Force Mag.*, May, 1965; J. V. Casamassa and R. D. Bent, *Jet Aircraft Power Systems*, 3d ed., 1965; A. Ferri, Review of SCRAMjet propulsion technology, *J. Aircraft* (AIAA), vol. 5, no. 1, 1968.

Rankine cycle

A thermodynamic cycle used as an ideal standard for the comparative performance of heat-engine and heat-pump installations operating with a condensable vapor as the working fluid. Applied typically to a steam power plant, as shown in the illustration, the cycle has four phases: (1) heat addition *bcde* in a boiler at constant pressure p_1 changing water at *b* to superheated steam at *e*, (2) isentropic expansion *ef* in a prime mover from initial pressure p_1 to back pressure p_2, (3) heat rejection *fa* in a condenser at constant pressure p_2 with wet steam at *f* converted to saturated liquid at *a*, and (4) isentropic compression *ab* of water in a feed pump from pressure p_2 to pressure p_1.

This cycle more closely approximates the operations in a real steam power plant than does the Carnot cycle. Between given temperature limits it

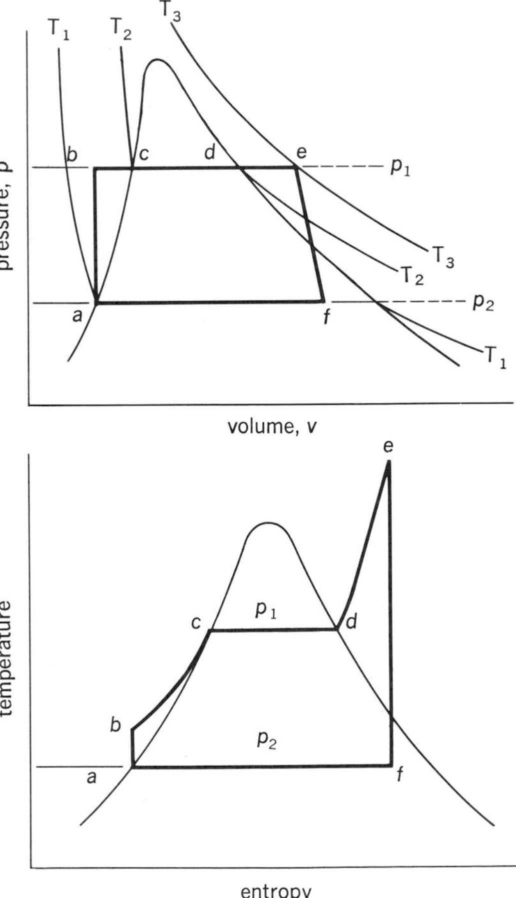

Rankine-cycle diagrams (pressure-volume and temperature-entropy) for a steam power plant using superheated steam. Typically, cycle has four phases.

offers a lower ideal thermal efficiency for the conversion of heat into work than does the Carnot standard. Losses from irreversibility, in turn, make the conversion efficiency in an actual plant less than the Rankine cycle standard. *See* CARNOT CYCLE; REFRIGERATION CYCLE; THERMODYNAMIC CYCLE.

[THEODORE BAUMEISTER]

Bibliography: T. Baumeister (ed.), *Standard Handbook for Mechanical Engineers*, 8th ed., 1978; J. B. Jones and G. A. Howkins, *Thermodynamics*, 1960; J. H. Keenan, *Thermodynamics*, 1970.

Reaction turbine

A power-generation prime mover utilizing the steady-flow principle of fluid acceleration, where nozzles are mounted on the moving element. The rotor is turned by the reaction of the issuing fluid jet and is utilized in varying degrees in steam, gas, and hydraulic turbines. All turbines contain nozzles; the distinction between the impulse and reaction principles rests in the fact that impulse turbines use only stationary nozzles, while reaction turbines must incorporate moving nozzles. A nozzle is defined as a fluid dynamic device containing a throat where the pressure of the fluid drops and potential energy is converted to the kinetic form with consequent acceleration of the fluid. For details of the two basic principles of impulse and reaction as applied to turbine design *see* IMPULSE TURBINE.

[THEODORE BAUMEISTER]

Reactor physics

The science of the interaction of the elementary particles and radiations characteristic of nuclear reactors with matter in bulk. These particles and radiations include neutrons, beta (β) rays, and gamma (γ) rays of energies between zero and about 10^7 electronvolts (eV).

The study of the interaction of β- and γ-radiations with matter is, within the field of reactor physics, undertaken primarily to understand the absorption and penetration of energy through reactor structures and shields. For a discussion of problems of reactor shielding *see* RADIATION SHIELDING.

With this exception, reactor physics is the study of those processes pertinent to the chain reaction involving neutron-induced nuclear fission with consequent neutron generation. Reactor physics is differentiated from nuclear physics, which is concerned primarily with nuclear structure. Reactor physics makes direct use of the phenomenology of nuclear reactions. Neutron physics is concerned primarily with interactions between neutrons and individual nuclei or with the use of neutron beams as analytical devices, whereas reactor physics considers neutrons primarily as fission-producing agents. In the hierarchy of professional classification, neutron physics and reactor physics are both ranked as subfields of the more generalized area of nuclear physics. *See* NUCLEAR PHYSICS.

Reactor physics borrows most of its basic concepts from other fields. From nuclear physics comes the concept of the nuclear cross section for neutron interaction, defined as the effective target area of a nucleus for interaction with a neutron beam. The total interaction is the sum of interactions by a number of potential processes, and the probability of each of them multiplied by the total cross section is designated as a partial cross section. Thus, a given nucleus is characterized by cross sections for capture, fission, elastic and inelastic scattering, and such reactions as (n,p), (n,α) and $(n,2n)$. An outgrowth of this is the definition of macroscopic cross section, which is the product of cross section (termed microscopic, for specificity) with atomic density of the nuclear species involved. The symbols $N\sigma_i^{\,j}$ or $\Sigma_i^{\,j}$ are used for macroscopic cross section, the subscript referring to the nuclear reaction involved, and the superscript to the isotope. The dimensions of Σ are cm^{-1}, and Σ_t (total cross section) is usually of the order of magnitude of unity.

Cross sections vary with energy according to the laws of nuclear structure. In reactor physics this variation is accepted as input data to be assimilated into a description of neutron behavior. Common aspects of cross section dependence, such as variation of absorption cross section inversely as the square root of neutron energy, or the approximate regularity of resonance structure, form the basis of most simplified descriptions of reactor processes in terms of mathematical or logical models.

The concept of neutron flux is related to that of macroscopic cross section. This may be defined as the product of neutron density and neutron speed, or as the rate at which neutrons will traverse the outer surface of a sphere embedded in the medium, per unit of spherical cross-sectional area. The units of flux are neutrons/(cm^2)(sec). The product of flux and macroscopic cross section yields the reaction rate per unit volume and time. The use of these variables is conventional in reactor physics.

The chain reaction, a concept derived from chemical kinetics, is the basis of a physical description of the reactor process. The source of energy in a nuclear reactor is the fission of certain isotopes of heavy elements (thorium, uranium, and plutonium, in particular) when they absorb neutrons. Fission splits the elements into two highly radioactive fragments, which carry away most of the energy liberated by the fission process (about 160 MeV out of about 200 MeV per fission), the bulk of which is rapidly transformed into heat. Neutrons of high energy are also liberated, so that a chain of events alternating between neutron production in fission and neutron absorption causing fission may be initiated. Because more than one neutron is liberated per fission, this chain reaction may rapidly branch out to produce an increasing reaction rate, or divergent reaction; or if the arrangement of materials is such that only a small fraction of the neutrons will ultimately produce fission, the chain will be broken in a convergent reaction. *See* NUCLEAR FISSION.

Criticality. The critical condition is what occurs when the arrangement of materials in a reactor allows, on the average, exactly one neutron of those liberated in one nuclear fission to cause one additional nuclear fission. If a reactor is critical, it will have fissions occurring in it at a steady rate. This desirable condition is achieved by balancing the probability of occurrence of three competing events: fission, neutron capture which does not cause fission, and leakage of neutrons from the system. If ν is the average number of neutrons liberated per fission, then criticality is the condition under which the probability of a neutron causing fission is $1/\nu$. Generally, the degree of approach to criticality is evaluated by computing k_{eff}, the ratio of fissions in successive links of the chain, as a product of probabilities of successive processes.

Reactor constants are the parameters used in determining the probabilities of the various processes which together define k_{eff}. They comprise two sets, those used to characterize nuclear events, and those used to characterize leakage. In the former group are fast effect ε, resonance escape probability p, thermal utilization factor f, and neutrons emitted per fuel absorption η. In the latter group are neutron age or slowing-down area τ or L_s^2, migration area M^2, thermal diffusion area L^2, diffusion coefficient D, and buckling B^2.

Fast effect. The fast effect occurs in thermal reactors containing significant quantities of U^{238} or Th^{232}. These reactors comprise the bulk of plutonium production and civilian power reactors. The isotopes mentioned can undergo fission only when struck by very energetic neutrons; only a little more than one-half of the neutrons born in fission can cause fission in them. Moreover, these fast neutrons (energy greater than about 1.4 MeV) are subject to energy degradation by the competing reaction of inelastic scattering and by collision with moderator. In consequence, only a small number of fast fissions will occur.

The fast effect is characterized by a ratio R of fast to nonfast fissions. From this ratio, a quantity ε, the fast effect, is derived, which determines the neutrons available to the chain reaction after the convergent chain of fast fissions has been completed, per neutron born of nonfast fission.

The common magnitude of ε varies between 1.02 and 1.05. R varies between 0 and 0.15 commonly. When $R < 0.01$, the fast effect is usually omitted from consideration.

Resonance escape probability. The resonance escape probability p is a significant parameter for the same group of reactors that have a significant fast effect and is defined as the probability that a neutron, in the course of being moderated, will escape capture by U^{238} (or Th^{232}) at any of the many energies at which the capture cross section is unusually high (resonance energies). The existence of this resonance absorption prevents most reactors homogeneously fueled with natural uranium from going critical, because resonance absorption, coupled with other losses, causes neutron depletion below the requirement for criticality. Therefore, low-enrichment reactors are heterogeneous, the fuel being disposed in lumps. The lumping of fuel, originally proposed by E. Fermi and E. P. Wigner, increases the probability of resonance escape. Spatial isolation of resonance absorber from moderator increases the number of neutrons which are slowed down without making any collisions with the absorber. Also, because absorption is very probable at resonance energies, the neutrons can travel only very short distances into the fuel lump before being absorbed, so that the interior of the lump is hardly exposed to resonant neutrons. Another effect of lumping is the removal of excess neutron scattering near the absorber so that there is a lesser probability of absorption following multiple collision.

With slight enrichment, homogeneous assemblies can be made critical. The homogeneous problem also provides the formalism by which p can be calculated. Use is made of the resonance integral RI, defined as the absorption probability per absorbing nucleon per neutron slowed down from infinite source energy in a moderator of unit slowing-down power. In a highly dilute system the resonance interval is given by Eq. (1), where σ_a

$$RI = \int_{E_0}^{\infty} \sigma_a \, dE/E \tag{1}$$

is the neutron absorption cross section of the absorber as a function of energy, and E_0 is an energy taken as the lower limit of the resonance region. The slowing-down power of a moderator nucleus is $\xi\sigma_s$, where ξ is the mean increase in lethargy of the neutron per scattering, and σ_s is the moderator scattering cross section. Thus the resonance absorption probability of a system would appear to be given by Eq. (2), where N_r and N_m are, respectively,

$$1 - p = N_r \,(RI)/N_m \xi \sigma_s \tag{2}$$

atomic concentrations of absorber and moderator in the reactor volume. However, because the probabilities of escaping capture by successive nuclei must be multiplied, Eq. (3) is better.

$$p = \exp\{-N_r (RI)/N_m \xi \sigma_s\} \tag{3}$$

The value of E_0 used in the definition of RI follows one of two conventions: it is either taken as some defined thermal cut-off energy between about 0.3 and 1 eV, or as an energy just below the lowest resonance, about 6 eV for U and 20 eV for Th. In the latter case, RI is spoken of as $1/v$ corrected.

Thermal utilization factor. Thermal utilization factor f is the fraction of neutrons which, once thermalized, are absorbed in fuel. In a homogeneous array, f may be calculated from the atomic densities (that is, the number of atoms per cubic centimeter) and thermal-absorption cross sections of the various constituents of the reactor. The problem becomes more complex in a highly absorbing system and in the presence of nuclei (such as Pu or Cd) whose absorption cross sections do not vary with energy in the usual $1/v$ fashion. In this case, it becomes necessary to evaluate the neutron spectrum and average the absorption cross sections over this spectrum in order to obtain reaction ratios.

For low absorptions, the neutron spectrum is given by Eq. (4), the Maxwellian expression, where

$$N(E)\,dE = [2\pi/(\pi kT)^{3/2}]E^{1/2} \exp - \{E/kT\}\, dE \tag{4}$$

k is the Boltzmann constant and T the absolute temperature of the medium. Deviations from this shape caused by absorption were first formulated by Wigner and J. E. Wilkins. This problem has become very significant with the advent of highly absorbing systems operating with considerable quantities of Pu.

In heterogeneous systems the problem is further complicated by the spatial nonuniformity of the neutron flux. It is therefore necessary to calculate reaction rates in various regions of the lattice by multiplying local absorption cross sections and atomic densities by local neutron fluxes; or, in effect the same thing, by introducing flux weights into the cross sections.

The calculation of neutron flux may be performed by diffusion theory or by more exact and elaborate methods for solving the neutron-transport problem. The problem may be further complicated by spatial effects on spectrum.

The term disadvantage factor, applied in simple systems originally to describe the ratio of mean moderator to mean fuel flux, has fallen into disfavor because of vague and local definitions. The symbol F, which is called the fuel disadvantage factor, is still in use to describe the ratio of surface to mean volume flux in a fuel lump with isoperimetric flux.

Fission neutrons per fuel absorption. This constant, η, is a characteristic of the fuel and of the neutron spectrum, but not of the spatial configuration of the system. Pure fissionable materials do not always undergo fission when they absorb neutrons; sometimes they lose their energy of excitation by emission of a gamma ray. The number of neutrons per fission ν varies only slightly with incident neutron energy (a few percent per million electronvolts), but η can fluctuate considerably even within a fraction of an electronvolt. Consequently, the specification of η is dependent upon a good evaluation of neutron spectrum.

For unirradiated, low-enrichment assemblies, the custom of defining fuel as all uranium, U^{235} and U^{238}, persists. Thus, f considers total uranium captures, and η is lowered from the value for pure U^{235} by the fractional absorption rate of U^{235} in uranium. This custom leads to excessive complexity as plutonium builds into the fuel and is therefore declining in use.

Infinite multiplication constant. The infinite multiplication constant, k_∞, is the ratio of neutrons in successive generations of the chain in the absence of leakage. In the formalism just described, the chain is taken from thermal neutron through fission and back to thermal neutron and from the definition of terms, $k_\infty = \eta\varepsilon pf$. This formula is known as the four-factor equation.

For reactors other than weakly absorbing thermal systems, the simplified description breaks down. Thus in a fast reactor significant fission and capture occur at all neutron energies, and no moderator is present; in a very strongly absorbing thermal system, an appreciable fraction of neutrons react at energies between the thermal region and the lowest U^{238} resonance. For such systems, the definition of the neutron chain is usually made in terms of a total time-dependent fission-rate expression, and the parametric representation of k_∞ becomes appreciably more complex. Generally, the spectrum is broken up into energy groups, and k_∞ is defined as the sum of the fission neutron production rate over all groups divided by the sum of absorption rates over all groups.

Neutron age. The neutron age τ is a reactor parameter defined in various ways, all related to the probability of leakage of a fast neutron from a reactor system. The basic definition is that 6τ measures the mean square distance of travel between injection of a neutron at one energy or energy spectrum and its absorption at some other energy in an infinite system. The term age is used because in weakly moderating systems the equation describing neutron slowing down in space and energy has the same form as the time-dependent heat-conduction equation, with τ substituted for time. The British and Canadian usage is L_s^2 for slowing-down length (squared), which more accurately describes the physical parameter.

The common injection spectrum is a fission spectrum, and the common points of measurement or application are absorption at the indium resonance energy, 1.4 ev, or at some arbitrarily defined thermalization energy. The ages of these energies are denoted as τ_{In} or τ_{Th}. When a source other than the fission spectrum is considered, other subscripts are used.

A. M. Weinberg has pointed out that the shape of the spatial distribution in an infinite system for which τ is the second moment can be closely correlated with the fast leakage of a bare finite system. If the finite system has a source and sink distribution which is a solution of Laplace's equation, $\nabla^2\phi + B^2\phi = 0$, then the Fourier transform of the τ distribution for given B will yield a quantity $P_\infty(B^2)$, which is almost exactly the nonleakage probability during slowing down. Two particularly significant cases are those for which the τ distribution is a Gaussian or an exponential curve. In the former case, $P_\infty(B^2) = e^{-\tau B^2}$; in the latter, $P_\infty(B^2) = 1/(1 + \tau B^2)$. The Gaussian distribution is experimentally and theoretically verified for moderators as heavy as Be or heavier. The exponential distribution is crudely applicable to H_2O-moderated systems, and D_2O has a definitely mixed distribution.

In some cases, τ is used in a synthetic way to describe the leakage under some simple approximation to the slowing-down distribution. Thus, τ_{2G} is a number which is used in a two-energy group neutron model to give correct fast nonleakage probability as $P(B^2) = 1/(1 + \tau_{2G}B^2)$. When this model is not a good approximation, τ_{2G} is not the second moment of the τ distribution.

For a heavy moderator, the age between two energies is given by the simple approximation of Eq. (5), where D is the diffusion coefficient, N is

$$\tau_{E_1 \to E_2} = \int_{E_2}^{E_1} D(E)/3(N\xi\sigma_s)\, dE/E \qquad (5)$$

atomic density, σ_s is scattering cross section, and ξ is mean lethargy (logarithmic energy) gain/collision.

Thermal diffusion area. The thermal diffusion area L^2 is one-sixth the mean square distance of travel between thermalization and absorption. It is thus the analog of τ for thermal neutrons. Because thermal migration is usually well represented by a diffusion equation, which has an exponential absorption distribution, thermal nonleakage probability is given by $1/(1 + L^2B^2)$. L^2 is defined by $L^2 = D_{\text{Th}}/N\sigma_{a,\text{Th}}$, where D and σ_a are spectrum-averaged diffusion coefficient and absorption cross section, respectively.

Migration area. The migration area M^2 is one-sixth the mean square distance of travel from birth to death of a neutron. For large, small-leakage systems, $1/(1 + M^2B^2)$ is an excellent approximation to the total nonleakage probability. When only thermal absorption exists, $M^2 = \tau + L^2$.

Diffusion coefficient. The diffusion coefficient D is essentially a scaling factor applied to validate

Fick's law, a relation between neutron flux and current (the latter being defined as a vector describing net rate of flow of neutron density). Fick's law is expressed by Eq. (6), where **j** is current, $\nabla\phi$

$$\mathbf{j} = -D\nabla\phi \qquad (6)$$

flux gradient, and D diffusion coefficient. For generalized systems, D is a tensor, but in regions where $|\nabla\phi|/\phi$ is small, D is approximated by a scalar of magnitude $1/(3\Sigma)$.

Buckling, B^2, is mathematically defined as $\nabla^2\phi/\phi$ in any region of a reactor where this quantity is constant over an appreciable volume. The name is derived from the relationship between force and deflection in a mechanical system with constrained boundaries. In a bare reactor, the buckling is constant except within one neutron mean free path ($\lambda \equiv 1/\Sigma$) of the boundary. For a slab of width t, $B^2 = (\pi/t)^2$; for an infinitely high cylinder of radius a, $B^2 = (2.404/a)^2$; for a sphere of radius a, $B^2 = (\pi/a)^2$; and for other geometries, it is again a geometrical constant. Because B^2 is a definite eigenvalue of Laplace's equation, it is the appropriate number to be used in the formulations of $P(B^2)$ previously described.

Effective multiplication k_{eff} is the ratio of neutron production in successive generations of the chain reaction. It is given by the product of k_∞ and $P(B^2)$, where $P(B^2)$ is itself the product of fast and thermal (or equivalent) nonleakage probabilities.

Reactivity. Reactivity is a measure of the deviation of a reactor from the critical state at any frozen instant of time. The term reactivity is qualitative, because three sets of units are in current use to describe it.

Percent k and millikay are absolute units describing the imbalance of the system from criticality per fission generation. Because $k_{\text{eff}} = 1$ describes a critical system, one says that it is 1% super- or subcritical, respectively, if each generation produces 1.01 or 0.99 times as many neutrons as the preceding one. Millikay are units of 0.1% k, and are given plus sign for supercriticality and minus sign for subcriticality. In both cases, k_{eff} is the base, as in Eqs. (7). These units are used primarily in design and analysis of control rods.

$$\begin{aligned} \%k &= 100|k_{\text{eff}} - 1|/k_{\text{eff}} \\ \text{Millikay} &= 1000\ (k_{\text{eff}} - 1)/k_{\text{eff}} \end{aligned} \qquad (7)$$

Dollars describe reactivity relative to the mean fraction of delayed neutrons per fission. Because the delayed neutrons are the primary agents for permitting control of the reaction, supercriticalities of less than 1 dollar are considered manageable in most cases. Thus, there is 1 dollar to "spend" in maneuvering power level. The dollar is subdivided into 100 cents. Because the delayed neutron fraction is a function of both neutron energy and fissionable material, the conversion rate between dollars and %k varies among reactors.

Inhours are reactivity units based on rate of change of power level in low-power reactors. If a low-power reactor is given enough reactivity so that its level would steadily increase by a factor of e per hour (which is also known as a 1-hr period), that much reactivity is 1 inhour (from inverse hour). It is only for very small reactivities, however, that reactivity in inhours may be obtained from reciprocal periods.

Reactivity is measured in inhours primarily by operators of steady-state reactors, in which only small reactivities are normally encountered.

Reflectors. Reflectors are bodies of material placed beyond the chain-reacting zone of a reactor, whose function is to return to the active zone (or core) neutrons which might otherwise leak. Reflector worth can be crudely measured in terms of the albedo, or probability that a neutron passing from core to reflector will return again to the core.

Good reflectors are materials with high scattering cross sections and low absorption cross sections. The first requirement ensures that neutrons will not easily diffuse through the reflector, and the second, that they will not easily be captured in diffusing back to the core.

Beryllium is the outstanding reflector material in terms of neutronic performance. Water, graphite, D_2O, iron, lead, and U^{238} are also good reflectors. The use of Be, H_2O, C, and D_2O as reflectors permits conversion of neutrons leaking at high energy into thermal neutrons diffusing back to the core. Because the reverse flow of neutrons is always accompanied by a neutron-flux gradient, these reflectors show characteristic thermal flux peaks outside the core. They are, therefore, desirable materials for research reactors, in which these flux peaks are useful for experimental purposes.

The usual measurement of reflector worth is in terms of reflector savings, defined as the difference in the reflected dimension between the actual core and one which would be critical without reflector. Reflector savings are close to reflector dimensions for thin reflectors and approach an asymptotic value dependent upon core size and reflector constitution as thickness increases.

Reactor dynamics. Reactor dynamics is concerned with the temporal sequence of events when neutron flux, power, or reactivity varies. The inclusive term takes into account sequential events, not necessarily concerned with nuclear processes, which may affect these parameters. There are basically three ways in which a reactor may be affected so as to change reactivity. A control element, absorbing rod, or piece of fuel may be externally actuated to start up, shut down, or change reactivity or power level; depletion of fuel and poison, buildup of neutron-absorbing fission fragments, and production of new fissionable material from the fertile isotopes Th^{232}, U^{234}, U^{238}, and Pu^{240} make reactivity depend upon the irradiation history of the system; and changes in power level may produce temperature changes in the system, leading to thermal expansion, changes in neutron cross sections, and mechanical changes with consequent change of reactivity.

Reactor control physics. Reactor control physics is the study of the effect of control devices on reactivity and power level. As such, it includes a number of problems in reactor statics, because the primary question is to determine the absorption of the control elements in competition with the other neutronic processes. It is, however, a problem in dynamics, given the above information, to determine what motions of the control devices will lead to stable changes in reactor output.

Particular problems occurring in the statics of reactor control stem from the particular nature of control devices. Many control rods are so heavily absorbing for thermal neutrons that elegant refinements of neutron-transport theory are needed to estimate their absorption. Other types of control rods include isotopes with heavy resonance absorption, and the interaction of such absorbers with U^{238} resonances must be examined. The motion of control rods changes the material balance of reactor regions, and with water-moderated reactors, peaks in the fission rate occur near empty rod channels. By virtue of high absorption of their constituent isotopes, some absorbing materials (for example, cadmium and boron) burn out in the reactor, and a rod made of these materials loses absorbing strength with time. As a final example, the motion of a control rod may change the shape of the power pattern in the reactor so as to bring secondary pseudostatic effects into play. *See* NUCLEAR REACTOR.

Reactivity changes. Long-term reactivity changes may represent a limiting factor in the burning of nuclear fuel without costly reprocessing and refabrication. As the chain reaction proceeds, the original fissionable material is depleted, and the system would become subcritical if some form of slow addition of reactivity were not available. This is the function of shim rods in a typical reactor. The reactor is originally loaded with enough fuel to be critical with the rods completely inserted. As the fuel burns out, the rods are withdrawn to compensate.

In order to decrease requirements on the shim system, many devices to overcome reactivity loss may be used. A burnable poison may be incorporated in the system. This is an absorbing isotope which will burn out at a rate comparable to or greater than the fuel. Burnable poisons are therefore limited to isotopes with very high effective neutron-absorption cross sections. Combinations of poisons, and the use of self-shielding of poisons can, in principle, make the close compensation of considerable reactivity possible without major control rod motions, but the technological problems in their use are formidable.

A more popular method for compensating reactivity losses is the incorporation of fertile isotopes into the fuel. This is desirable because the neutrons captured in the fertile material are not wasted, but used to manufacture new fissionable material; and also because (as with U^{238} in U^{235} reactors or Pu^{240} in Pu^{239} systems) the fertile material is normally found mixed with the fissionable, and isotopic separation may be circumvented or minimized. Depending on the conversion ratio (new fissionable atoms formed per old fissionable atom burned) and the fission parameters of the materials, a reactor so fueled loses reactivity relatively slowly, and in some cases, may show a temporary reactivity increase. The various isotopes produced by successive neutron capture in uranium and plutonium must be considered in this problem, the higher isotopes becoming prominent at very long exposures.

A final consideration of long-term reactivity is the extra parasitic absorption of the fission products as formed. At long exposure, this absorption becomes significant because of the relatively high absorption of many of the fission products. At shorter exposures, isotopes of very high absorption, mainly Sm^{149} and Xe^{135}, are more prominent. These materials have such high cross sections that they reach a steady-state concentration relatively quickly, burning out by neutron capture as rapidly as they are formed in fission.

Xe^{135} is particularly interesting because its cross section is abnormally large, its fission yield is high, it is preceded by an isotope of low cross section and approximately 7-hr half-life (I^{135}), and it undergoes β-decay to the low absorption Cs^{135} with a half-life of about 10 hr. This combination of properties gives several interesting effects. Chief of these is that, at high flux, a reservoir of I^{135} is formed which continues to decay to Xe^{135} even after the reactor is shut down. Because Xe^{135} is maintained at steady state during operation by a balance of buildup against burnout, the shutdown also removes the chief mode of Xe^{135} destruction. Hence, the Xe^{135} concentration increases immediately after shutdown. The reservoir of I^{135} is so large that reactivity is rapidly lost as Xe^{135} builds up; and in some cases, it may be impossible to restart the reactor after a short shutdown. The operator must then wait almost 2 days for the Xe^{135} to disappear by radioactive decay. At very high fluxes, this effect becomes so severe that even a small temporary reduction in power may lead to ultimate subcriticality of the reactor.

Another effect caused by Xe^{135} in high-flux reactors is that, if the reactor is large enough, the Xe^{135} may force the power pattern into oscillations of 1- or 2-day periods. Although this is not a serious dynamic problem, it does emphasize the necessity of monitoring not only the total power, but also the power pattern, so that appropriate countermeasures may be taken. *See* NUCLEAR FUEL; NUCLEAR FUELS REPROCESSING.

Reactor kinetics. This is the study of the short-term aspects of reactor dynamics with respect to stability, safety against power excursion, and design of the control system. Control is possible because increases in reactor power often reduce reactivity to zero (the critical value) and also because there is a time lapse between successive fissions in a chain resulting from the finite velocity of the neutrons and the number of scattering and moderating events intervening, and because a fraction of the neutrons is delayed.

Prompt-neutron lifetime. This is the mean time between successive fissions in a chain, and it is the basic quality which determines the time scale within which controlling effects must be operable if a reactivity excursion is touched off. In thermal reactors the controlling feature is the time between thermalization and capture, because fast neutrons spend less time between collisions, and the total time for moderation of a neutron is at most a few microseconds (μsec). Prompt-neutron lifetimes vary from 10 to a few hundred microseconds for light-water reactors, the shorter times being found in poorly reflected, highly absorbing systems, and the longer in well-reflected systems, in which time spent in the reflector is the dominating factor. Other thermal reactors have longer lifetimes, with some heavy-water reactors having lifetimes as long as a few milliseconds. Fast reactors have lifetimes of the order of 0.01 – 0.1 μsec, the controlling factor being the amount of scattering material used as diluent.

Delayed neutrons. Delayed neutrons are important because a complete fission generation is not achieved until these neutrons have been emitted by their precursors. A slightly supercritical reactor must wait until the delayed neutrons appear, and this delay allows time for the system to be brought under control. The fraction of delayed neutrons β ranges from about 1/3% for thermal fission of Pu^{239} and U^{233}, to about 3/4% for thermal fission of U^{235}, to several percent for some fast fission events. In these latter cases, however, the extra delayed neutrons have such short lifetimes that they are of only slight extra utility.

In any case, the influence of delayed neutrons is felt only to the extent that they are needed to maintain criticality. When the system is supercritical enough that the delayed neutrons are not needed to complete the critical chain, it is known as prompt critical. Prompt criticality represents in a qualitative sense the threshold between externally controllable and uncontrollable excursions, and it is for this reason that the dollar unit is popular in excursion analysis (a prompt critical system has a reactivity of 1 dollar).

Although it has now been established that a larger number of fission products emit delayed neutrons, the distribution in time after fission of the delayed neutron emission rate is accurately represented for all purposes by a sum of six negative exponentials. For many purposes, however, a three, two, or one group approximation is adequate.

Reactors moderated by D_2O and Be have additional delayed neutrons contributed by photoneutron reactions between the moderator and fission product γ-rays. Although not a large fraction, this effect makes such reactors unresponsive to small reactivity fluctuations and gives them unusual operational smoothness.

Reactor period. This is the asymptotic time required for a reactor at constant reactivity to increase its power by a factor e. When a critical reactor is given extra reactivity, its power will rise. At first, the power production rate has a complex shape on a time plot, but ultimately the power will rise exponentially. The period is the measure of this exponential rate.

The relation between the reactor period and the reactivity is known as the inhour equation. If l is the reactor lifetime in seconds, β_i the fraction of delayed neutrons in group i, λ_i the delay constant of group i delayed neutrons in $\sec^{-1}$, S reactor period in seconds and ρ reactivity in thousands of millikay, then Eq. (8) is the inhour equation. The

$$\rho=\frac{l/S+\sum_i [\beta_i/(1+\lambda_i S)]}{1-\sum_i [\beta_i/(1+\lambda_i S)]} \tag{8}$$

equation has $I+1$ solutions of S for a given ρ, I being the number of groups; and there is always a real value of S with a higher value than the real part of any other solution. This highest S is the period. For very small values of ρ, that is, for very large periods, this value is approximately that given by notation (9). For very large ρ, and therefore small

$$\left(l+\sum_i \frac{\beta_i}{\lambda_i}\right)\Big/\rho \tag{9}$$

S, the period is approximately $l/(\rho-\beta)$. This same result would be found if the delayed neutrons were thrown away completely.

Reactivity coefficients. There are several functions relating changes in reactivity to changes in the physical state of the reactor. The power coefficient is the change in reactivity per unit change in reactor power; the temperature coefficient relates reactivity to temperature change, and is often broken down into fuel, moderator, and coolant coefficients; for low-power graphite reactors there exists a barometric coefficient; one may define also coolant circulation rate coefficients and void coefficients.

Because the reactivity is commonly a complicated function of all the pertinent variables, the reactivity coefficient generally is the coefficient of the first term in a series expansion of the reactivity about the operating point. This in turn describes a linear theory of reactor dynamics. The theory may be extended to reactivity effects of arbitrary type by considering reactivity coefficients as functionals.

The basic problem of reactor dynamics is the specification of the power coefficient of reactivity. The chain, power affects reactivity which affects power, is thereby analyzable, using the power coefficient functional together with the reactor kinetic equations. The power coefficient is, however, predictable only in terms of changes in temperature and flow resulting from power changes in the system. Thus the analysis of power coefficient implies exhaustive knowledge of system behavior.

Reactivity coefficients may be prompt or delayed, and most delayed effects can be characterized as either of decay or transport type. An example of the decay type of coefficient is the contribution of coolant temperature change to power coefficient. Here, a power pulse gives a thermal effect on the coolant which is instantaneously observable, and which decreases exponentially with a time constant imposed by the heat-transfer equations. An example of a transport type of delay is the delay attributable to coolant circuit times. Here, a finite time lapse exists between the cause and the observable response. Effects due to fuel heating are examples of prompt effects.

Some types of power coefficient yield dangerous or unstable situations. Thus a power coefficient may contain a prompt positive (autocatalytic) term and a larger delayed negative term. Even though such a system may be stable against slow power-level increases, it will undergo a violent excursion whenever power is raised rapidly enough to outstrip delayed effects. Again, a system with prominent delayed effects of the transport type is always unstable beyond some critical power, even if the effect opposes the power shift; here, there is a possibility of phase instability.

The dynamic behavior of a reactor is usually analyzed by techniques common to all feedback systems.

[BERNARD I. SPINRAD]

Bibliography: G. I. Bell and S. Glasstone, *Nuclear Reactor Theory*, 1970; A. R. Foster and R. L. Wright, Jr., *Basic Nuclear Engineering*, 2d ed., 1973; A. D. Galanin, *Thermal Reactor Theory*, 1960; S. Glasstone and A. Sesonske, *Nuclear Reactor Engineering*, 1963; D. L. Hetrick, *Dynamics of Nuclear Reactors*, 1971; J. R. Lamarsh, *Introduc-*

tion to Nuclear Engineering, 1975; J. R. Lamarsh, *Nuclear Reactor Theory*, 1966; R. V. Meghreblian and D. K. Holmes, *Reactor Analysis*, 1960; *Nuclear Energy Glossary*, International Organization for Standardization, ISO-921, 1972; *Reactor Physics Constants*, Argonne National Laboratory, ANL-5800, 2d ed., 1963; A. M. Weinberg and E. P. Wigner, *The Physical Theory of Neutron Chain Reactions*, 1958; P. F. Zweifel, *Reactor Physics*, 1975.

Reciprocating aircraft engine

A fuel-burning piston internal combustion engine specially designed and built for light weight in proportion to developed shaft horsepower. The reciprocating engine is currently the principal aircraft engine used for such low-flying, low-speed aircraft as small private and crop-spraying airplanes and helicopters. The reciprocating engine drives a propeller (or rotor) that, in turn, accelerates the surrounding air rearward (or downward), thereby imparting forward momentum (or lift) to the airplane (or helicopter).

Predominantly, aircraft reciprocating engines operate on a four-stroke Otto cycle. These spark-ignition engines burn hydrocarbon fuels and develop shaft power through connecting rods and a crankshaft. Major parts are the crankcase, crankshaft, connecting rods, pistons, cylinders with intake and exhaust valves, camshafts, and such operating auxiliaries as ignition, carburetor, and fuel and oil pumps.

Two-stroke spark ignition and two- or four-stroke compression ignition (diesel) reciprocating engines have been developed and successfully flown, but they have not been widely applied nor have they significantly improved aircraft performance.

The reciprocating engine powered all aircraft for the first 40 years of heavier-than-air flight, including all military, commercial, and private types. The advent of the turbojet engine near the end of World War II started a rapid conversion to turbine power. This conversion has been primarily directed toward military needs and is limited chiefly by availability of government financing and the time required to develop suitable types and sizes of turbine engines and applicable aircraft. The cold war and Korean War accelerated the changeover appreciably. As would be expected under these conditions, engines for combat-type aircraft received primary emphasis, so that funds for further development of reciprocating engines of new types or major model changes were minimized after the end of World War II. Some commercial aircraft were supplied with turbojet and turboprop power in the mid-1950s; however, the major swing to turbines occurred with advanced equipment in late 1958. With availability of small jet and turboprop engines, private aircraft are beginning to employ them, particularly in planes for business purposes for which speed is needed to compete with commercial transportation.

The reciprocating engine has been built for aircraft with numerous arrangements and numbers of cylinders, with a variety of fuels and antiknock ratings, with various cooling systems, with various means and amounts of supercharging, with several types of fuel systems, and with and without reduction gears. The military premium on increased power and performance, the high cost of development, and a relatively rapid rate of obsolescence have combined to discourage purely commercial sponsorship of new types or models of aircraft engines except for low-power applications.

Variety of engines. Through World War I, aircraft had relatively low operating speeds with resultant low velocity head available for cooling. A limited knowledge of air cooling of cylinders also tended to limit use of air-cooled fixed cylinders. Major combat aircraft were therefore mostly powered by water-cooled 6-cylinder in-line, 8-cylinder V, and 12-cylinder V types. There was, however, considerable use by the French of air-cooled rotary radial engines for fighter or pursuit aircraft and some use by the British of air-cooled V engines.

World War I demonstrated the great importance of aircraft as a military weapon. This resulted in the financing and exploration of a great variety of engine types and forms, particularly those promising higher power and improved aircraft performance, during this war and the following decade. Water-cooled engines built and tested included 12-, 18-, and 24-cylinder W, 12- and 16-cylinder parallel-vertical, 16-cylinder X, V, and fan engines. Fixed-radial engines, both air and water cooled, with 3–20 cylinders and one or two rows, were explored. Air-cooled 9-, 11-, and 18-cylinder rotary and 2-cylinder opposed engines were also built and run. As early as 1925, some 31 cylinder combinations in 36 types and nearly 300 models in horsepowers from 30 to 1000 had been or were being developed.

Before 1930 a few outstanding types of engines had emerged from this continued exploration of types, from new invention and design, and from better materials and processes both in the engines and the installations. These were the single-row 9-cylinder and two-row 14-cylinder air-cooled fixed-radial types in medium and high horsepowers, the liquid-cooled 12-cylinder V for high-speed aircraft, and the small 4 and 6 in-line and opposed air-cooled engines in low horsepower. Reciprocating engine development was concentrated during the first half of the 1930s on refinement and performance improvement of the above types. During the last half of the 1930s larger and much higher powered two-row engines of both 14 and 18 cylinders of the radial air-cooled type were developed. Design also emphasized higher outputs and use of higher-temperature cooling on liquid-cooled 12-cylinder V engines. A 24-cylinder double-V of 3420-in.^{3} displacement was the largest liquid-cooled engine developed prior to World War II. Flown in experimental bomber and fighter aircraft, it was scheduled for large production in a fighter airplane in 1943 but was discontinued before production quantities were made, in favor of using the facilities for turbojet engine development and production in late 1944.

The last type of reciprocating engine of medium to high horsepower initiated was a four-row 28-cylinder air-cooled fixed-radial type of 4360-in.^{3} displacement, started in 1940. This engine reached substantial postwar military production and was the only such new engine type to do so.

Fixed-radial air-cooled engines. The 9-cylinder single-row fixed air-cooled engine combined the inherently good features of a short and therefore light crankcase and crankshaft with the maximum number of cylinders arranged peripherally. It also provided uniform airflow at every cylinder and space at the cylinder heads for widely canted valves and the increased head thickness and deep finning needed with aluminum heads. However, considerable research, novel engineering design and development, as well as process improvement and installation-cooling knowledge were needed to utilize fully these basic advantages.

Some of these features, which have been applied to all aircraft engines, included forged and cast aluminum parts, cooled valves, supercharger, and low-drag cowl.

Forged aluminum pistons reduced the reciprocating mass. A one-piece master rod reduced rotating mass; it also decreased distortion of, and improved loading uniformity on, a one-piece master rod bearing. This required a two-piece crankshaft. Forged aluminum crankcases with separate oil sump permitted compact, lighter structure.

Cast aluminum cylinder heads were screwed and shrunk to steel barrels with machined integral fins. The two-valve head with highly canted valves had steel or bronze valve-seat inserts shrunk in. This, in combination with a major improvement in casting technique, permitted more uniform valve, spark plug, and head cooling. Use of integral rocker boxes decreased weight, improved cooling, mechanical strength, and reliability, and simplified forced lubrication without leakage.

Internally cooled exhaust valves permitted higher outputs from a given fuel and improved reliability and durability.

A supercharger impeller, either crankshaft or (preferably) gear driven, improved mixture uniformity and distribution to the various cylinders.

Use of Townend ring or NACA cowl reduced drag in the aircraft.

The above features in combination with many detail refinements in design, such as improved cam design, temperature-compensated valve gears, and avoidance of torsional resonance in the normal operating range, resulted in the late 1920s in specific engine weights for single-row 9-cylinder radials of less than 1.5 lb/hp (0.9 kg/kW) with a new order of reliability and durability. Reduction gearing when added increased specific weights but improved aircraft performance.

Fig. 1. Cross section of 1710-in.³ (28.0 liters) liquid-cooled 12-cylinder engine used in fighters during 1937–1947. (*Detroit Diesel Allison Div., General Motors Corp.*)

Two-row 14-cylinder air-cooled radial engines with features similar to those above were 0.1–0.2 lb/hp (0.06–0.12 kg/kW) heavier than the comparable single-row engine of equal output. Their smaller diameter for the same displacement or power rating as the single-row type and the smaller cylinder size due to the large number of cylinders tended to compensate for the higher specific weight and initial cost except in the smaller engine sizes. However, some structural problems and particularly difficulties with torsional vibration slowed two-row engine development in the larger cylinder sizes for several years, the 1830-in.³ (30.0-liter) being the largest cylinder undertaken in the United States until 1935.

Liquid-cooled 12-cylinder V engine. The development of liquid-cooled engines was sponsored throughout the 1930s by the military because of marked advantages for single-seat fighter or interceptor applications. Schneider trophy contests demonstrated the high specific power outputs obtainable for short periods with liquid cooling and the use of high engine rotational speed, the low drag associated with the low frontal area (form drag), and the low cooling drag obtained with high-temperature coolants and efficient radiators in strategic locations with the 12-cylinder V engine (Fig. 1). The result was that this type of engine predominated in fighters of the British, German, and U.S. air forces until late in World War II, when some large radial air-cooled fighters were introduced. However, the U.S. Navy and the Japanese, both relying primarily on carrier-based fighters, used radial air-cooled engines throughout, the lighter weight and short length being considered of greater carrier utility than high speed.

Lightweight simple reduction gears of the offset spur type centered the propeller shaft in the V engine for low drag and good pilot visibility, and were necessary for the high engine speed inherently available with the adequate bearings and natural balance of the 12-cylinder V engine. High-temperature liquid cooling using either ethylene glycol or ethylene glycol–water mixtures at temperatures of 250–275°F (121–135°C) reduced radiator requirements and drag and provided antifreeze protection. Because of the need for high burst performance, as well as high normal performance in fighter aircraft, many of the detailed features in connection with radial air-cooled engines were first developed or applied on the liquid-cooled V engines. In general, a radial air-cooled engine of 1.5–2.0 times the engine displacement is required to give overall fighter aircraft performance comparable with a given liquid-cooled engine if equal skill is applied in the installation of each.

With the end of World War II, a number of additional liquid-cooled engines were manufactured for military use until suitable jet aircraft could be developed. A relatively small number of foreign liquid-cooled V engines were also converted for commercial use in modified aircraft. However, no liquid-cooled engines were in production and available after the war at comparable cost in the sizes required to compete with the large air-cooled engines, so the liquid-cooled type had been dropped in favor of developing turbojet or turboprop engines with apparent advantages over either the air- or liquid-cooled types of reciprocating engines.

Two-row radial air-cooled engines. In the period 1935–1937 manufacturers in the United States introduced large two-row radial air-cooled engines in sizes of 2600-, 2800-, and 3350-in.^{3} displacement. The first of these had 14 cylinders and the latter two had 18 cylinders. The 2600-in.^{3} engine was used in sizable quantities during World War II, but was not used in new military or commercial postwar applications. The two large 18-cylinder engines were used extensively during World War II and powered almost all advanced postwar reciprocating-engine commercial and military transports in the United States (Fig. 2). The principal exception was the large 28-cylinder radial air-cooled engine of 4360 in.^{3} (71.4 liters) initiated in 1940. Production of this engine was started at the end of World War II, and it was used after the war in both military bomber and cargo aircraft.

The 2800- and 3350-in.^{3} (45.9 and 54.9 liter) 18-cylinder engines represent the ultimate for radial air-cooled types both in variety of features or models and in detail design improvement for high specific outputs. These engines, initiated at power outputs of less than 0.5 hp/in.^{3} (23 kW/liter) displacement on 87 octane fuel, approached 1.0 hp/in.^{3} (46 kW/liter) in the late models on 115–145 octane fuel without use of exhause turbines. In the case of the 3350-in.^{3} (54.9-liter) engine with three blow-down turbines feeding power back to the engine, specific outputs of over 1.1 hp/in.^{3} (50 kW/liter) were attained for takeoff along with specific fuel consumptions under 0.4 lb/hp (0.24 kg/kW) at cruise, while providing a major reduction in exhaust noise. The approximate doubling of specific output was partly a result of major improvement in the octane or antiknock rating of the fuel available and partly a result of major improvement in detail design of the engine, installation, and propeller.

High-output engine features. Some of the features provided in various models of piston engines during World War II for optimum characteristics in various aircraft were (1) various reduction gear ratios as engine speeds were increased and propeller activity factor or disk loadings were changed, and (2) various centrifugal compressors and drives for optimum supercharging, such as single stage, two speed; two stage, two speed; two stage, two speed with intercooler; and turbosupercharging with or without intercooling.

Following are some of the major design, process, and material details and component improvements that have been fed into these engines to permit higher speeds and higher cylinder temperatures and pressures, and to reduce local hot spots for higher overall engine performance on a given fuel: (1) higher-strength and temperature-resistant materials resulting from alloy improvement; (2) use of induction hardening, carburizing, or nitriding steels for higher strength and long wear characteristics, particularly where rubbing occurs, such as on cylinder barrel walls, piston rings, crankshafts, piston pins, link pins, and

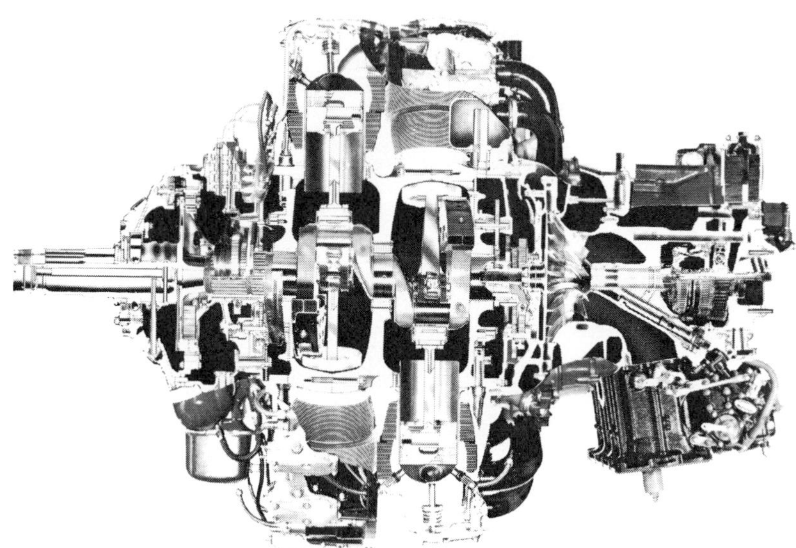

Fig. 2. Cutaway of 18-cylinder twin-row radial air-cooled engine of 2804-in.^{3} (45.9-liter) total piston displacement. (*Pratt and Whitney Aircraft*)

gears; (3) use of shot peening or surface rolling on highly stressed parts to eliminate residual stresses and add light compressive stresses for maximum uniformity, particularly at stress concentrations such as valve springs, connecting rods, rocker arms, and welded areas; (4) surface coating for improved functioning, which included such items as steel-backed, silver-plated master rod bearings with lead-indium coating; flame plating of high-temperature alloys on valve seat faces and insert seats; and silver, copper, or other coating, plating, or surface treatment to eliminate fretting, fretting erosion, or other action leading to fatigue cracks; (5) closer spacing and thinner fins on aluminum cylinder heads and steel barrels, with improved baffling for more uniform and better cylinder cooling; (6) tapered piston rings for higher-temperature piston operation without ring sticking; (7) availability of smaller-diameter spark plugs with ceramic insulation in place of mica; and (8) use of pressure type or floatless carburetors or fuel-injection systems for fuel metering to improve distribution, minimize icing troubles, reduce hazards of backfires (eliminated in the case of fuel injection), and avoid engine cutout, such as occurred with float-type carburetors, in negative-*g* maneuvering.

The result of these features and of other detail refinements in combination with an increase in engine speed and improved supercharger efficiency was a gain of 0.25–0.3 hp/in.^{3} (11–14 kW/liter) exclusive of the gain from improved fuel. Similar improvement by detail refinement and reduction of drag in the engine installation in the nacelle itself further contributed to the satisfactory history of the reciprocating engine in both military and commercial aircraft.

All of these high-output features with the exception of fuel injection and blowdown exhaust turbines were in production on both air- and liquid-cooled engines during World War II in the United States, and many German engines used fuel injection.

Opposed air-cooled engines. The 4-, 6-, and 8-cylinder opposed air-cooled engines in both hori-

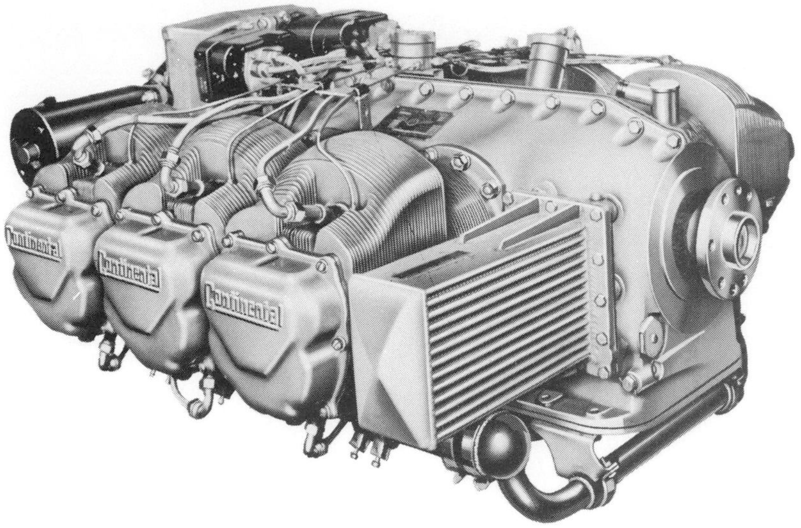

Fig. 3. A 6-cylinder opposed engine of 471-in.^{3} (7.72-liter) displacement rated for 260 hp (194 kW) at 2625 rpm at sea level; dry weight is 426 lb (193 kg). (*Continental Motors Corp.*)

zontal and vertical arrangements took over the aircraft power field up to about 400 hp shortly after World War II, largely eliminating in-line and radial engines in this rating both on commercial and military applications. Developed for commercial single- and twin-engine personal and business aircraft, these engines are also used for military training, reconnaissance, and drone and helicopter applications requiring high reliability and low cost (Fig. 3). They have achieved an excellent balance between automotive cost-saving features and lightweight aircraft construction. An illustration of this is the availability of manifold fuel-injection systems providing good distribution and major reduction in icing and backfiring while avoiding the large expense of cylinder-head injection and while eliminating float-type carburetors.

As a result of improved design and rotor-casting techniques, small turbochargers have become available at reasonable cost and have been applied to several models of opposed piston engines to give much improved altitude performance.

However, seeking to take advantage of the very low specific weights, simplified maintenance, and the capability of using lower-grade fuels, the military has sponsored and applied a variety of low-powered fan jets and turboprop-and-shaft power turbines which are directly competitive with available piston engines for helicopters, training craft, and personal aircraft. As production of the turbines increases, costs come down and the future of small reciprocating engines becomes less secure. *See* AIRCRAFT FUEL; INTERNAL COMBUSTION ENGINE; OTTO CYCLE.

[RONALD HAZEN]

Bibliography: K. Munson, *Military Aircraft*, 1966; Piston engines, *American Aviation Annual World Aviation Encyclopedia*, latest edition; R. Schlaifer and S. D. Heron, *Development of Aircraft Engines and Fuels*, 1970; J. W. R. Taylor (ed.), *Jane's All The World's Aircraft*, revised periodically; P. H. Wilkenson, *Aircraft Engines of the World*, 1970, also 1966–1967, 1964–1965, 1944–1962, and 1941.

Refrigeration cycle

A sequence of thermodynamic processes whereby heat is withdrawn from a cold body and expelled to a hot body. Theoretical thermodynamic cycles consist of nondissipative and frictionless processes. For this reason, a thermodynamic cycle can be operated in the forward direction to produce mechanical power from heat energy, or it can be operated in the reverse direction to produce heat energy from mechanical power. The reversed cycle is used primarily for the cooling effect that it produces during a portion of the cycle and so is called a refrigeration cycle. It may also be used for the heating effect, as in the comfort warming of space during the cold season of the year. *See* HEAT PUMP; THERMODYNAMIC PROCESSES.

In the refrigeration cycle a substance, called the refrigerant, is compressed, cooled, and then expanded. In expanding, the refrigerant absorbs heat from its surroundings to provide refrigeration. After the refrigerant absorbs heat from such a source, the cycle is repeated. Compression raises the temperature of the refrigerant above that of its natural surroundings so that it can give up its heat in a heat exchanger to a heat sink such as air or water. Expansion lowers the refrigerant temperature below the temperature that is to be produced inside the cold compartment or refrigerator. The sequence of processes performed by the refrigerant constitutes the refrigeration cycle. When the refrigerant is compressed mechanically, the refrigerative action is called mechanical refrigeration.

There are many methods by which cooling can be produced. The methods include the noncyclic melting of ice, or the evaporation of volatile liquids, as in local anesthetics; the Joule-Thomson effect, which is used to liquefy gases; the reverse Peltier effect, which produces heat flow from the cold to the hot junction of a bimetallic thermocouple when an external emf is imposed; and the paramagnetic effect, which is used to reach extremely low temperatures. However, large-scale refrigeration or cooling, in general, calls for mechanical refrigeration acting in a closed system.

Reverse Carnot cycle. The purpose of a refrigerator is to extract as much heat from the cold body as possible with the expenditure of as little work as possible. The yardstick in measuring the

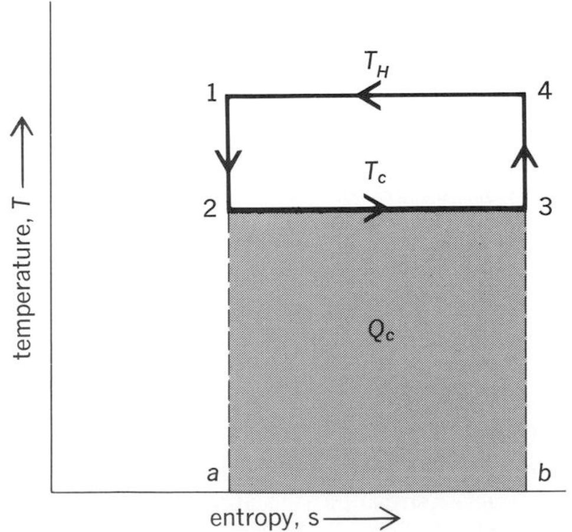

Fig. 1. Reverse Carnot cycle.

performance of a refrigeration cycle is the coefficient of performance, defined as the ratio of the heat removed to the work expended. The coefficient of performance of the reverse Carnot cycle is the maximum obtainable for stated temperatures of source and sink. Figure 1 depicts the reverse Carnot cycle on the T-s plane. *See* CARNOT CYCLE.

The appearance of the cycle in Fig. 1 is the same as that of the power cycle, but the order of the cyclic processes is reversed. Starting from state 1 of the figure, with the fluid at the temperature T_H of the hot body, the order of cyclic events is as follows:

1. Isentropic expansion, 1–2, of the refrigerant fluid to the temperature T_c of the cold body.
2. Isothermal expansion, 2–3, at the temperature T_c of the cold body during which the cold body gives up heat to the refrigerant fluid in the amount Q_c, represented by the area 2–3–b–a.
3. Isentropic compression, 3–4, of the fluid to the temperature T_H of the hot body.
4. Isothermal compression, 4–1, at the temperature T_H of the hot body. During this process, the hot body receives heat from the refrigerant fluid in the amount Q_H represented by the area 1–4–b–a. The difference $Q_H - Q_c$ represented by area 1–2–3–4 is the net work which must be supplied to the cycle by the external system.

Figure 1 indicates that Q_c and the net work rectangles each have areas in proportion to their vertical heights. Thus the coefficient of performance, defined as the ratio of Q_c to net work, is $T_c/(T_H - T_c)$.

The reverse Carnot cycle does not lend itself to practical adaptation because it requires both an expanding engine and a compressor in order to function. Nevertheless, its performance is a limiting ideal to which actual refrigeration equipment can be compared.

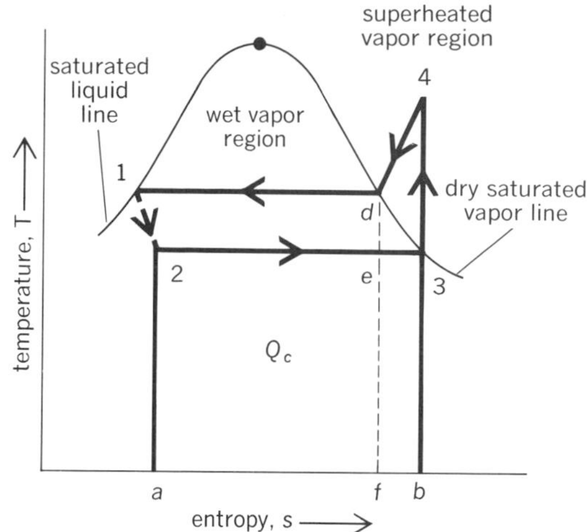

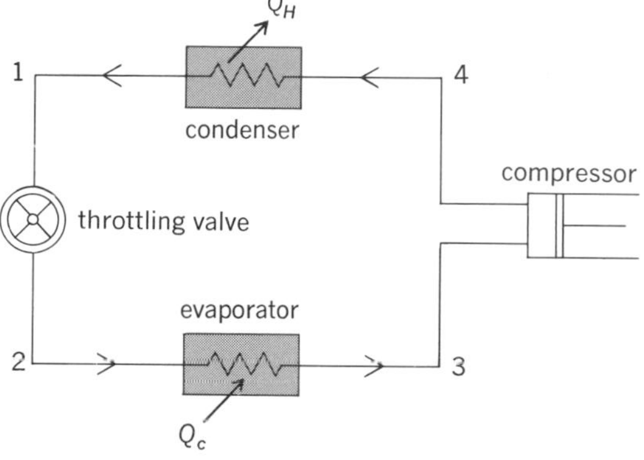

Fig. 2. Vapor-compression refrigeration cycle substitutes valve for expansion engine.

Modifications to reverse Carnot cycle. One change from the Carnot cycle which is always made in real vapor-compression plants is the substitution of an expansion valve for the expansion engine. Even if isentropic expansion were possible, the work delivered by the expansion engine would be very small and the irreversibilities present in any real operations would further reduce the work delivered by the expanding engine. The substitution of an expansion valve, or throttling orifice, with constant enthalpy expansion, changes the theoretical performance but little, and greatly simplifies the apparatus. A typical vapor-compression refrigeration cycle is shown in Fig. 2; it is essentially a reverse Rankine cycle. The irreversible adiabatic expansion 1–2 differs only slightly from the vertical isentropic expansion.

Another practical change from the ideal Carnot cycle substitutes dry compression 3–4 for wet compression e-d in Fig. 2, placing state 4 in the

Ideal refrigeration cycle performance[a]

Refrigerant	Saturation pressure, psia[b]		Refrigerating effect, Btu/lb[c]	Refrigerant required per ton refrigeration[d]		Compressor horsepower per ton refrigeration[e]	Coefficient of performance
	Evaporator	Condenser		lb/min	cfm		
Carnot—any fluid							6.6
Freon, F12	47	151	49	4.1	3.6	0.89	5.31
Ammonia	66	247	455	0.44	1.9	0.85	5.55
Sulfur dioxide	24	99	135	1.5	4.8	0.83	5.67
Free air	15	75	32	6.2	77	2.7	1.75
Dense air	50	250	32	6.2	23	2.7	1.75
Steam	0.1	1.275	1000	0.2	590	0.91	5.18

[a]Performance is based on evaporator temperature = 35°F (2°C), condenser temperature = 110°F (43°C). In vapor-compression cycles, the pressures and temperatures are for saturation conditions; vapor enters the compressor dry and saturated; no subcooling in the condenser. [b]1 psi = 6.895 kPa, absolute pressure. [c]1 Btu/lb = 2326 J/kg. [d]1 lb/min per ton of refrigeration = 2.15×10^{-6} kg/sec per W of refrigeration; 1 cfm per ton of refrigeration = 1.35×10^{-7} m³/sec per W of refrigeration. [e]1 hp per ton of refrigeration = 0.212 W per W of refrigeration.

superheat region above ambient temperature; the process is called dry compression in contrast to the wet compression of the Carnot cycle. Dry compression introduces a second irreversibility by exceeding the ambient temperature, thus reducing the coefficient of performance. Dry compression is usually preferred, however, because it simplifies the operation and control of a real machine. Vapor gives no readily observable signal as it approaches and passes point *e* in the course of its evaporation, but it would undergo a temperature rise if it accepted heat beyond point 3. This cycle, using dry compression, is the one which has won overwhelming acceptance for refrigeration work.

Reverse Brayton cycle. The reverse Brayton cycle constitutes another possible refrigeration cycle; it was one of the first cycles used for mechanical refrigeration. Before Freon and other condensable fluids were developed for the vapor-compression cycle, refrigerators operated on the Brayton cycle, using air as their working substance. Figure 3 presents the schematic arrangement of this cycle. Air undergoes isentropic compression, followed by reversible constant-pressure cooling. The high-pressure air next expands reversibly in the engine and exhausts at low temperature. The cooled air passes through the cold storage chamber, picks up heat at constant pressure, and finally returns to the suction side of the compressor. *See* BRAYTON CYCLE.

The temperature-entropy diagram, Fig. 3, points up the disadvantage of the dense-air cycle. If the temperature at *c* represents the ambient, then the only way that air can reject a significant quantity of heat along the line *b-c* is for *b* to be considerably higher than *c*. Correspondingly, if the cold body service temperature is *a*, the air must be at a much lower temperature in order to accept heat along path *d-a*. If a reverse Carnot cycle were used with a working substance undergoing changes in state, the fluid would traverse path *a-f-c-e* instead of path *a-b-c-d*. The reverse Carnot cycle would accept more heat along path *e-a* than the reverse Brayton cycle removes from the cold body along path *d-a*. Also, since the work area required by the reverse Carnot cycle is much smaller than the corresponding area for the reverse Brayton cycle, the vapor-compression cycle is preferred in refrigeration practice. *See* THERMODYNAMIC CYCLE.

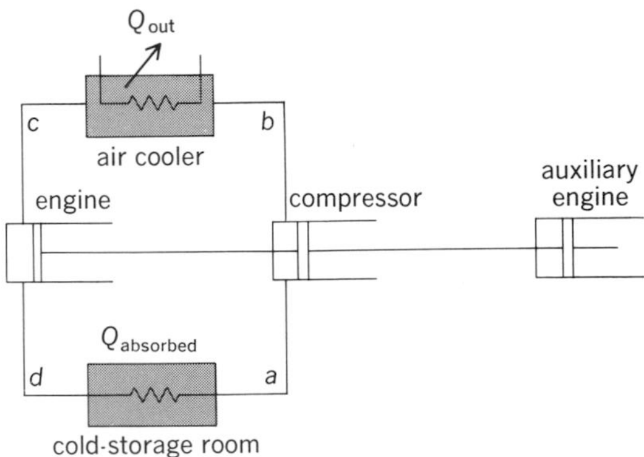

c

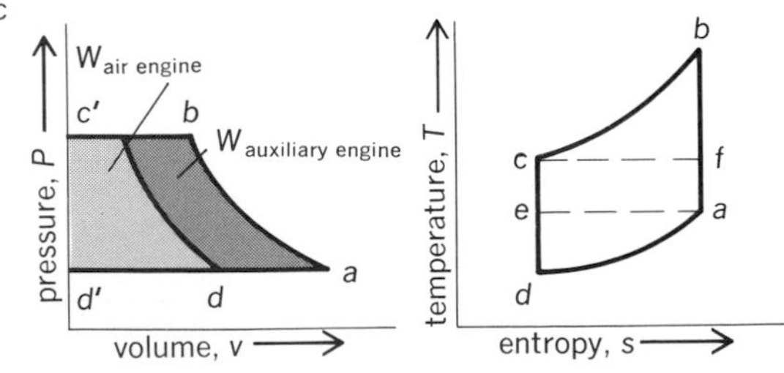

Fig. 3. Schematic arrangement of reverse Brayton, or dense-air, refrigeration cycle.

Comparative performance of refrigerants. The table gives the significant theoretical performance data for a selected group of refrigerants when used in the ideal cycles as outlined above. These data include not only the requisite pressures in the evaporator and condenser for saturation temperatures of 35°F (2°C) and 110°F(43°C), respectively, but also the refrigerating effect, the weight and volume of refrigerant to be circulated, the horsepower required, and the coefficient of performance. The vapor compression cycles presuppose that the refrigerant leaves the evaporator dry and saturated (point 3 in Fig. 2) and leaves the condenser without subcooling (point 1 in Fig. 2). The heat absorbed, Q_c, in the evaporator is called the refrigerating effect and is measured in Btu/lb of refrigerant. The removal of heat in the evaporator at the rate of 200 Btu/min (3517 W) is defined as a ton refrigeration capacity. The weight and volume of refrigerant, measured at the compressor suction, and the theoretical horsepower required to drive the compressor per ton refrigeration capacity are also given in the table. These data show a wide diversity of numerical values. Each refrigerant has its advantages and disadvantages. The selection of the most acceptable refrigerant for a specific application is consequently a practical compromise among such divergent data.

[THEODORE BAUMEISTER]

Bibliography: ASHRAE, *Handbook of Fundamentals*, 1977; T. Baumeister (ed.), *Standard Handbook for Mechanical Engineers*, 8th ed., 1978.

Relay

An electromechanical or solid-state device operated by variations in the input which, in turn, operate or control other devices connected to the output. They are used in a wide variety of applications throughout industry, such as in telephone exchanges, digital computers, motor and sequencing controls, and automation systems. Highly sophisticated relays are utilized to protect electric power systems against trouble and power blackouts as well as to regulate and control the generation and distribution of power. In the home, relays are used in refrigerators, automatic washers and dishwashers, and heat and air-conditioning controls. Although relays are generally associated with electrical circuitry, there are many other types, such as pneumatic and hydraulic. Input may be electrical and output directly mechanical, or vice-versa.

Relays using discrete solid-state components, operational amplifiers, or microprocessors can provide more sophisticated designs. Their use is increasing, particularly in applications where the relay and associated equipment are packaged together. The basic operation may be complex, but frequently is similar or equivalent to the units described.

Basic classifications. A basic, simplified block diagram of a relay is shown in Fig. 1. The actual

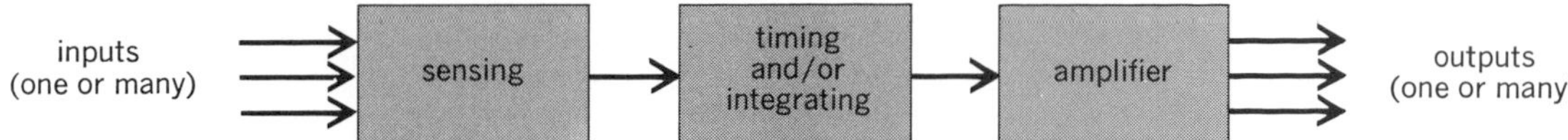

Fig. 1. Basic block diagram of a relay.

physical embodiment of the blocks varies widely, and one or more blocks may not exist or may be combined in practice. Classifying relays by function and somewhat in the order of increasing complexity there are (1) auxiliary, (2) monitoring, (3) regulating, (4) programming, and (5) protective relays.

An auxiliary relay (telephone type) is shown in Fig. 2. The sensing unit is the electric coil; when the applied current or voltage exceeds a threshold value the coil activates the armature, which operates either to close the open contacts or to open the closed contacts. In Fig. 1 the contacts belong to the amplifier block. Therefore relays of this class generally are used as contact multipliers. Other applications are to isolate circuits or to use a low-power input to control higher-power input. The timing-integrating block for these relays is in the armature and springs. Some time delay is inherent, and a number of techniques are used to add more delay or to reduce time of operation to a minimum.

The balanced polar unit of Fig. 3*a* is used as a monitoring relay. The unbalanced type is shown in Fig. 3*b*. A permanent magnet normally polarizes the pole faces on either side of the armature, and direct current is applied to the coil that is around the armature. At nominal values of current, the armature can be adjusted to float in the air gap between the pole pieces by the magnetic shunts, but when this current decreases, or increases significantly, the armature is magnetized and moves to one of the pole pieces to close the contacts. Thus this relay monitors the current level in an electric circuit. The sensing, integrating, and amplifying functions indicated in the block diagram of Fig. 1 are combined in the single mechanical element and its contacts for this polar unit.

The relay unit of Fig. 4, known as the induction-disk type, is widely used in regulating and protec-

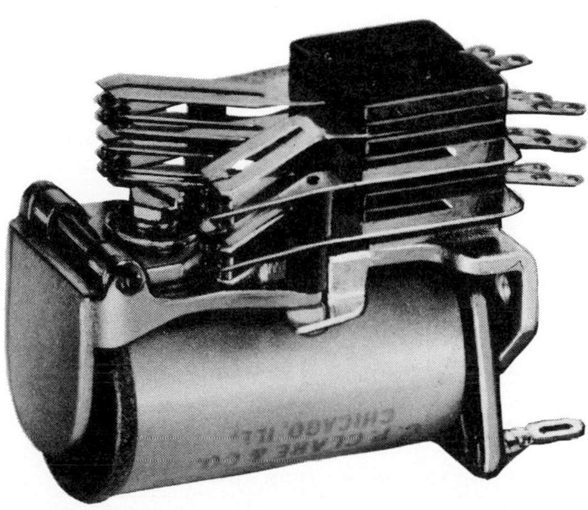

Fig. 2. A multicontrol telephone-type auxiliary relay. (*C.P. Clare and Co.*)

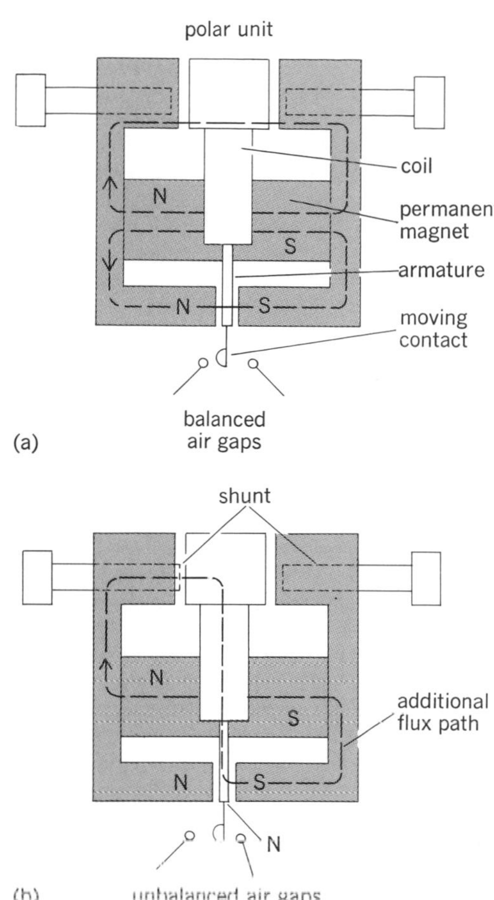

Fig. 3. Polar units. (*a*) Balanced. (*b*) Unbalanced.

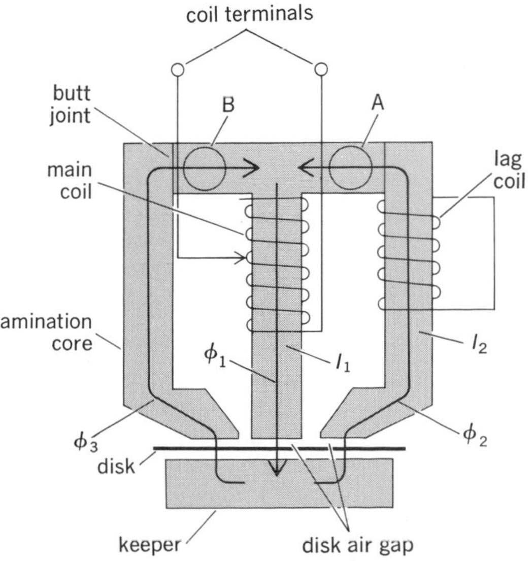

Fig. 4. Schematic of the induction-disk type relay. I_1 = current in main coil; I_2 = current in lag coil; ϕ_1 = flux in main coil; ϕ_2 = flux in lag coil; ϕ_3 = differential induction flux; A = flux field about ϕ_3; B = flux field about ϕ_2.

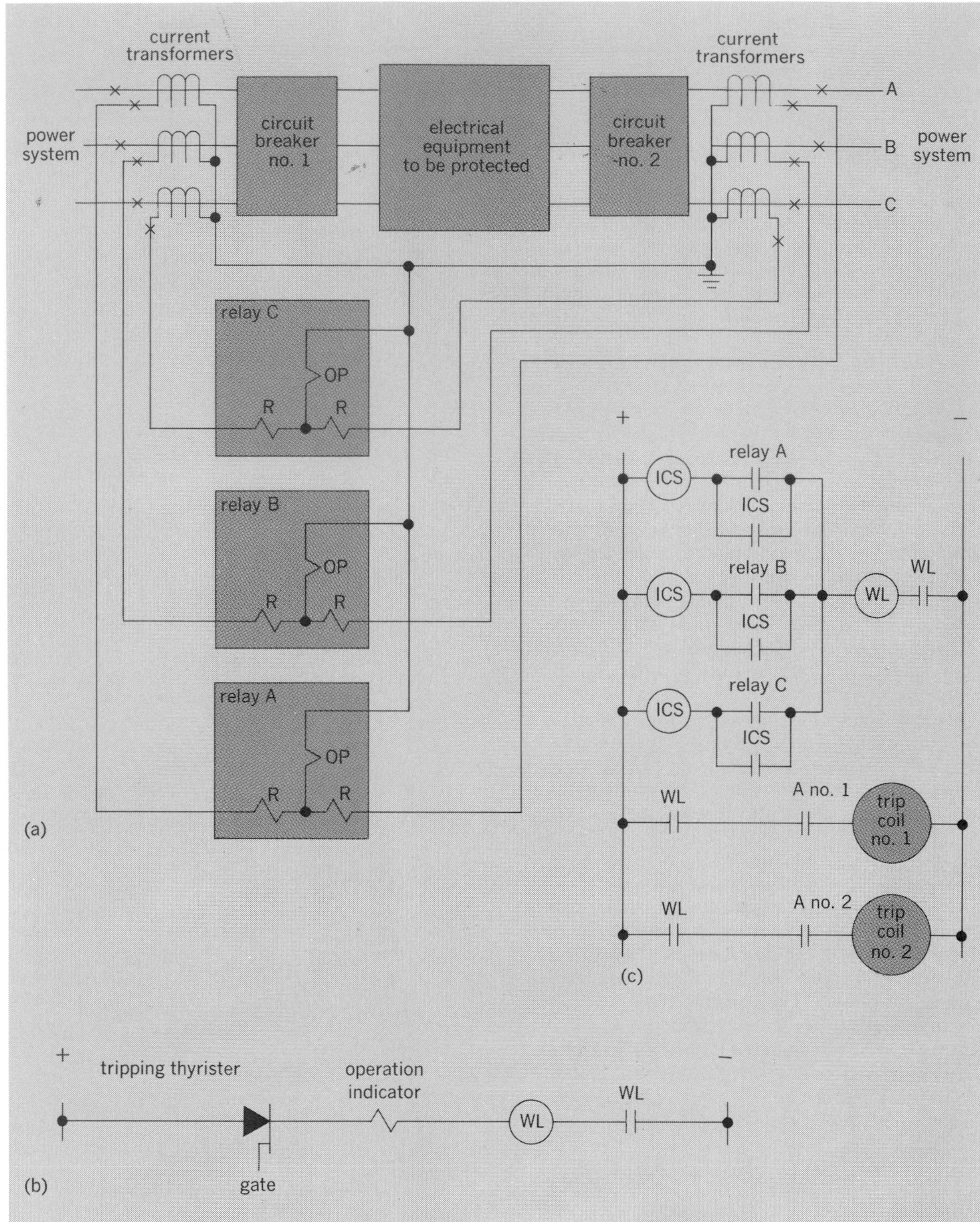

Fig. 5. (*a*) Typical differential relay protection for electrical equipment. Relays A, B, and C are differential relays. (*b*) Simplified alternate trip circuit (one phase) for solid-state relay. Gate is from differential relay integrating circuitry. (*c*) Trip circuit. A = Auxiliary out-off breaker; ICS = operation indicator and circuit seal-in auxiliary relay; OP = operating coil; R = restraining coil; WL = auxiliary hand-reset lock-out relay.

tive relays. Alternating current or voltage applied to the coil produces mechanical torque to rotate the disk, which is of the order of 3 in. (8 cm) in diameter. Contacts are attached to the disk shaft with a spiral spring to reset the contacts to a backstop or normally closed contacts. A permanent magnet provides damping. By design of the coils, damping magnet, and spring, a variety of inverse time overcurrent and under- or overvoltage characteristics are produced. With reference to Fig. 1, the sensing unit is approximated by the coil, by the timing unit, and the contacts by the amplifier.

For regulating, the output is arranged to adjust or regulate the input within prescribed limits; for monitoring, the output is used to alarm or to shut down the input equipment.

Programming relays provide an automatic sequence of operations. Good examples of these are the timer control relays on automatic washers and dishwashers, which provide the necessary program in the various washing operations.

Protective relays. Protective relays are compact analog networks, connected to various points of an electrical system, to detect abnormal conditions occurring within their assigned areas. They initiate disconnection of the trouble area by circuit breakers. These relays range from the simple overload unit on house circuit breakers to complex systems used to protect extra-high-voltage power transmission lines. Heavy-duty protective relay systems detect all intolerable system conditions, such as faults caused by lightning and equipment insulation failure, and initiate tripping of power circuit breakers within 6–10 milliseconds. *See* CIRCUIT BREAKER; LIGHTNING AND SURGE PROTECTION.

Fault detection. Fault detection is accomplished by a number of techniques. Some of the common methods are the detection of changes in electric current or voltage levels, power direction, ratio of voltage to current, temperature, and comparison of the electrical quantities flowing into a protected area with the quantities flowing out. The last-mentioned is known as differential protection and is illustrated by Fig. 5. Differential relays are applied to protect a piece of electrical apparatus. The inputs to the relays are currents from a current transformer. Current through the relay (nominally 5 amperes) is proportional to the high-power current of the main circuit. For load through the equipment or for faults either to the right or left of the current transformers, the secondary current flows through the relay restraining coils, and little or no current flows through the operating coils so that operation is prevented. For a fault between the current transformers, secondary current flows through the restraining windings (with reversed phase in one of the coils) and in the operating coils to operate the relay, trip the two circuit breakers, and isolate the fault or damaged equipment from the rest of the power system.

For transmission lines, in which there are considerable miles between the current transformers, the same principle is used, but a set of relays is used at each end. The intelligence to compare the direction of power flow or phase angle of the currents between the terminals is transmitted over a radio-frequency channel superimposed on the power line, by a telephone pair, or by microwave between the terminals. *See* ELECTROMAGNETIC INDUCTION; TRANSMISSION LINES.

[J. L. BLACKBURN]

Bibliography: D. Beeman, *Industrial Power System Handbook*, 1955; C. R. Mason, *The Art and Science of Protective Relaying*, 1956; National Association of Relay Manufacturers, *Engineers Relay Handbook*, 1966; R. L. Peek, Jr., and H. N. Wagar, *Switching Relay Design*, 1955; A. R. van C. Warrington, *Protective Relays*, 1962; Westinghouse Electric Corp., *Applied Protective Relaying*, 1979.

Reluctance

A property of a magnetic circuit analogous to resistance in an electric circuit.

Every line of magnetic flux is a closed path. Whenever the flux is largely confined to a well-defined closed path, there exists a magnetic circuit. That part of the flux that departs from the path is known as flux leakage. *See* MAGNETIC CIRCUITS.

For any closed path of length l in a magnetic field H, the line integral of $H \cos \alpha \, dl$ around the path is the magnetomotive force (mmf) of the path, as in Eq. (1), where α is the angle between H and

$$\text{mmf} = \oint H \cos \alpha \, dl \qquad (1)$$

the path. If the path encloses N conductors, each with current I, Eq. (2) holds. *See* MAGNETOMOTIVE FORCE.

$$\text{mmf} = \oint H \cos \alpha \, dl = NI \qquad (2)$$

Consider the closely wound toroid shown in the figure. For this arrangement of currents, the magnetic field is almost entirely within the toroidal coil, and there the flux density or magnetic induction B is given by Eq. (3), where l is the mean cir-

$$B = \mu \frac{NI}{l} \qquad (3)$$

cumference of the toroid and μ is the permeability. The flux Φ within the toroid of cross-sectional area A is given by either form of Eqs. (4), which is simi-

$$\Phi = BA = \frac{\mu A}{l} NI$$
$$\Phi = \frac{NI}{l/\mu A} = \frac{\text{mmf}}{l/\mu A} = \frac{\text{mmf}}{\mathcal{R}} \qquad (4)$$

lar in form to the equation for the electric circuit, although nothing actually flows in the magnetic circuit. The factor $l/\mu A$ is called the reluctance $\mathcal{R}$ of the magnetic circuit. The reluctance is not constant because the permeability μ varies with changing flux density. From the defining equation for reluctance, it is seen that when the mmf is in ampere-turns and the flux is in webers, the unit of reluctance is the ampere-turn/weber. *See* MAGNETIC INDUCTION.

Reluctances in series. For the simple toroid, all parts of the magnetic circuit have the same μ and the same A. More complicated circuits may include parts that differ in permeability, in cross section, or in both. Suppose a small gap were cut in the core of the toroid. The flux would fringe out at the gap, but as an approximation, the area of the gap may be considered the same as that of the core.

The magnetic path then has two parts, the core of length l_1 and reluctance $l_1/\mu_1 A$, and the air gap of length l_2 and reluctance $l_2/\mu_2 A$. Since the same flux is in both core and gap, this is considered a series circuit and Eq. (5) holds. Since the

$$\mathcal{R} = \mathcal{R}_1 + \mathcal{R}_2 = \frac{l_1}{\mu_1 A} + \frac{l_2}{\mu_2 A} \qquad (5)$$

relative permeability of the ferromagnetic core is several hundred or even several thousand times that of air, the reluctance of the short gap may be much greater than that of the much longer core. For any combination of paths in series, $\mathcal{R} = \Sigma l/\mu A$. Then Eq. (6) holds.

$$\Phi = \frac{\text{mmf}}{\Sigma \mathcal{R}} = \frac{\text{mmf}}{\Sigma l/\mu A} \qquad (6)$$

Reluctances in parallel. If the flux divides in part of the circuit, there is a parallel magnetic circuit and the reluctance of the circuit has the same relation to the reluctances of the parts as has the analogous electric resistance. Equation (7) is valid for the parallel circuit.

$$\frac{1}{\mathcal{R}} = \frac{1}{\mathcal{R}_1} + \frac{1}{\mathcal{R}_2} + \cdots \qquad (7)$$

See RELUCTANCE MOTOR.

[KENNETH V. MANNING]

RELUCTANCE

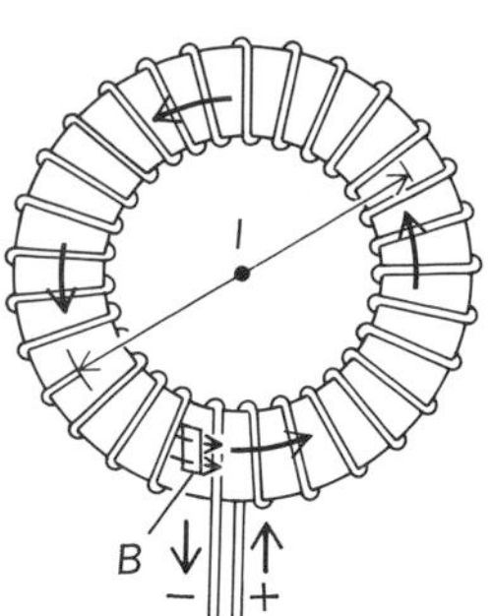

A toroidal coil.

Reluctance motor

A synchronous motor which starts as an induction motor and, upon nearing full speed, locks into step with the rotating field and runs at synchronous speed. The stator and rotor windings are similar to those of an induction motor. The rotor is of squirrel-cage construction, to allow induction-motor starting, and has salient-pole projections which provide synchronous operation at full speed. The reluctance motor is built only in small sizes and for situations in which low cost and simplicity are mandatory and efficiency is of little concern. It can be polyphase, but is usually a single-phase motor with a split-phase or capacitor winding for starting. *See* INDUCTION MOTOR; SYNCHRONOUS MOTOR.

[LOYAL V. BEWLEY]

Repulsion motor

An alternating-current (ac) commutator motor designed for single-phase operation. The chief distinction between the repulsion motor and the single-phase series motors is the way in which the armature receives its power. In the series motor the armature power is supplied by conduction from the line power supply. In the repulsion motor, however, armature power is supplied by induction (transformer action) from the field of the stator winding. *See* ALTERNATING-CURRENT MOTOR.

The repulsion motor primary or stationary field winding is connected to the power supply. The secondary or armature winding is mounted on the motor shaft and rotates with it. The terminals of the armature winding are short-circuited through a commutator and brushes. There is no electrical contact between the stationary field and rotating armature (Fig. 1).

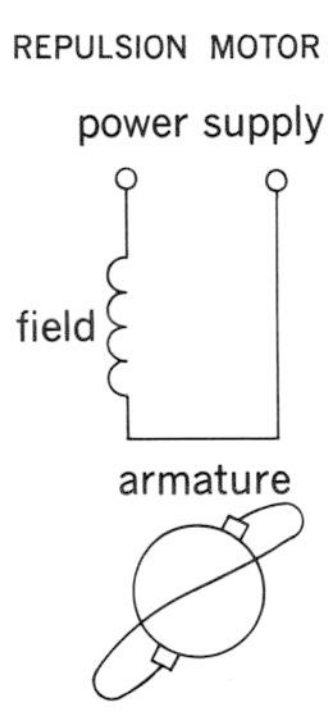

Fig. 1. Schematic of a repulsion motor.

If the motor is at rest and the field coils are energized from an outside ac source, a current is induced in the armature, just as in a static transformer. If the brushes are in line with the neutral axis of the magnetic field, there is no torque, or tendency to rotate. However, if they are set at a proper angle (generally 15–25° from the neutral plane), the motor will rotate.

Repulsion motors may be started with external resistance in series with the motor field, as is done with dc series motors. A more common method is to start the motor with reduced field voltage and increase the voltage as the motor increases speed. This can be done conveniently with a transformer having an adjustable tapped secondary or a variable autotransformer.

It is also possible to doubly excite the motor; that is, the armature may receive its power not only by induction from the stator winding but also by conduction from a transformer with adjustable taps, as shown in Fig. 2.

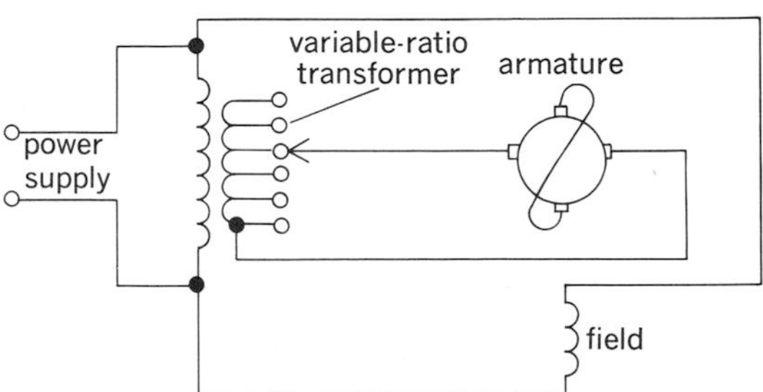

Fig. 2. Doubly excited repulsion motor.

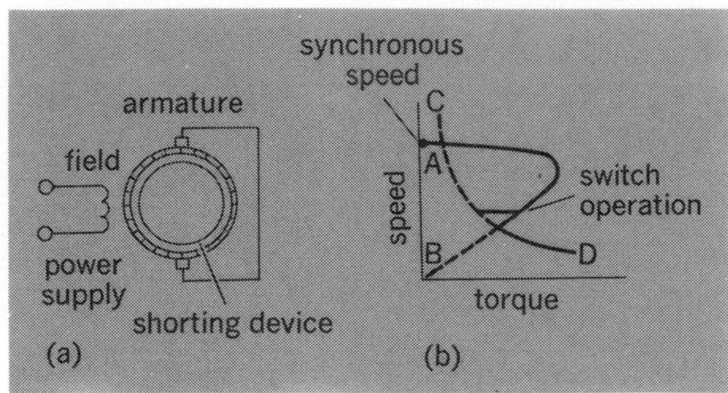

Fig. 3. Repulsion-start, induction-run motor. (*a*) Schematic diagram. (*b*) Speed-torque characteristic.

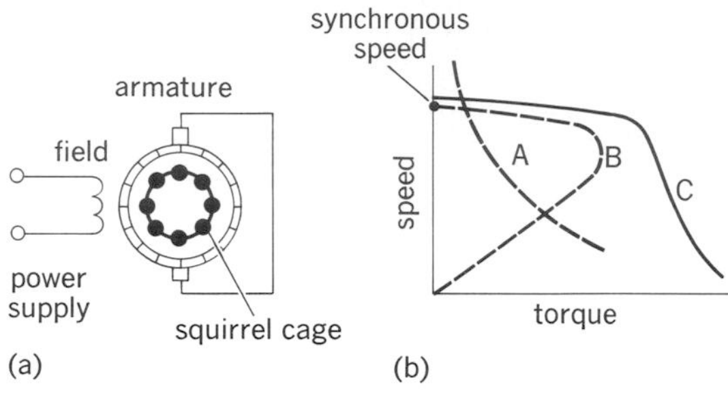

Fig. 4. Repulsion-induction motor. (*a*) Schematic diagram. (*b*) Speed-torque characteristic.

Repulsion-start, induction-run motor. This motor (Fig. 3*a*) possesses the characteristics of the repulsion motor at low speeds and those of the induction motor at high speed. It starts as a repulsion motor. At a predetermined speed (generally at about two-thirds of synchronous speed) a centrifugal device lifts the brushes from the commutator and short-circuits the armature coils, producing a squirrel-cage rotor. The motor then runs as an induction motor. In Fig. 3*b* the curve *AB* represents the characteristics of an induction motor and curve *CD* a repulsion motor. The solid curve *AD* is the combined characteristic of a repulsion-start, induction-run motor. *See* INDUCTION MOTOR.

Repulsion-induction motor. This motor (Fig. 4*a*) is very similar to the standard repulsion motor in construction, except for the addition of a second separate high-resistance squirrel cage on the rotor. Both rotor windings are torque-producing, and the total torque produced is the sum of the individual torques developed in these two windings. In Fig. 4*b*, curve *A* is the characteristic of the repulsion-motor torque developed in this motor. Curve *B* represents the induction-motor torque. Curve *C* is the combined total torque of the motor. The advantages of this machine are its high starting torque and good speed regulation. Its disadvantages are its poor commutation and high initial cost.

[IRVING L. KOSOW]

Bibliography: I. L. Kosow, *Electric Machinery and Control*, 1964; A. F. Puchstein, T. C. Lloyd, and A. G. Conrad, *Alternating-Current Machines*, 3d ed., 1954.

Reserve battery

A battery which is inert until an operation is performed which brings all the cell components into the proper state and location to become active.

High-energy primary batteries are often limited

in their application by poor retention characteristics which are caused by self-discharge between the electrolyte and the active electrode materials. One method of overcoming this problem, particularly when the battery is required to operate at high current levels for relatively short periods of time (minutes or hours), is the use of a reserve battery.

Several types have been developed. In water-activated or electrolyte-activated batteries the water or electrolyte component is not present during storage. It is added just before the cell is put into use. In thermal batteries the electrolyte is a solid at room temperature and has very low conductivity. If the temperature is raised above the melting point, the conductivity of the electrolyte becomes excellent and the cell is capable of delivering significant power.

Water-activated batteries. Practical battery systems have been developed using magnesium anodes against silver chloride or cuprous chloride cathodes (Fig. 1). Cuprous chloride cathodes are less expensive than silver chloride cathodes, but they are also bulkier and less stable, particularly in a humid atmosphere. *Meta*-dinitrobenzene is also finding use as cathode material on account of its high ampere-hour capacity.

The batteries are assembled dry. The active elements may be separated by porous paper or another inert media. Water may be poured into a container holding the elements or may flow continuously through the element. Either fresh or salt water may be used.

The most important design factor for all reserve batteries is to ensure that the electrolyte is delivered as quickly as possible at the time of activation, at the same time avoiding chemical short-circuiting of the cells.

Cells with absorbent separators can be activated by immersion. Subsequent operation may either be in air, using only the water retained in the cells, or while immersed. Performance of a two-cell battery immersed in sea water showed an output of 45 watt-hours/kilogram when fully discharged in 6 min.

The dry elements stored in a sealed container are capable of indefinite storage life.

Electrolyte-activated batteries. Any cell can be made as a reserve-electrolyte cell. If the electrodes are in place, it is necessary only to add the electrolyte to make a complete cell. In practice, however, the separation of the electrolyte is done only when excessive deterioration would occur during wet storage prior to use. Great ingenuity has been shown in designing complete battery packages in which an aqueous electrolyte is stored in a separate chamber. The package contains a mechanism, which may be operated from a remote location, which drives the electrolyte out of the reservoir and into all the cells of the battery. In general, these packaged batteries have been used only in military applications. The following couples have been used in reserve cells containing electrolytes: Zn/Cu, Pb/PbO_2, Zn/AgO, Mg/*meta*-dinitrobenzene, Zn/PbO_2, Zn/MnO_2, Cd/PbO_2. However, not all these couples are amenable to heavy rate discharge.

Gas-activated batteries. The liquid-activated batteries previously described have disadvantages which are difficult to overcome; for example, automatic electrolyte-charging equipment may cause intercell shorting. Unless the design is quite complex, it is difficult to avoid flooding or uneven filling. An alternative approach is to introduce a gas which reacts with the spacer material to form a conducting electrolyte.

Boron trifluoride gas reacts with dry, hydrated barium hydroxide to form a highly acid solution containing barium salts, borates, and fluoborates. Ammonia gas reacts with ammonium salts to form a solution having good conductivity (Fig. 2). Suitable electrode couples are Zn/MnO_2 and Pb/PbO_2. These gas-activated batteries are reported to operate well over a wide temperature range.

Thermal batteries. These are also known as heat-activated or fused-electrolyte batteries. Some compounds, such as sodium chloride and potassium hydroxide, show very low conductivity in the solid state at room temperature but very good conductivity in the molten state. For example, a mixture of sodium hydroxide and potassium hydroxide becomes an excellent ionic conductor when heated above 170°C. If a zinc anode and a silver oxide cathode are combined with solid pads of the eutectic mixture, all the elements of a cell are present. The electrolyte has an appreciable amount of entrained moisture, which plays a role in the discharge, but the cell will also work with carefully dried materials. Such cells are capable of high power output for a few minutes, when heated to 200°C or higher. At .155 A/cm^2 of positive plate, the cell voltage is 1.16 at 200°C, 1.23 at 250°C, 1.30 at 300°C.

Thermal batteries are capable of operation at very low ambient temperatures, provided that a suitable heat source is available to melt the electrolyte. For ordinary temperatures, they are not advantageous as compared with reserve aqueous-electrolyte types. Because the electrolyte in thermal batteries is inert and nonconductive at normal temperatures, they can be stored indefinitely, acting as primary batteries when the temperature of the electrolyte is raised to that at which they become ionically conductive.

A magnesium and manganese dioxide cell with sodium hydroxide electrolyte can operate for longer discharge times than the zinc–silver oxide cell

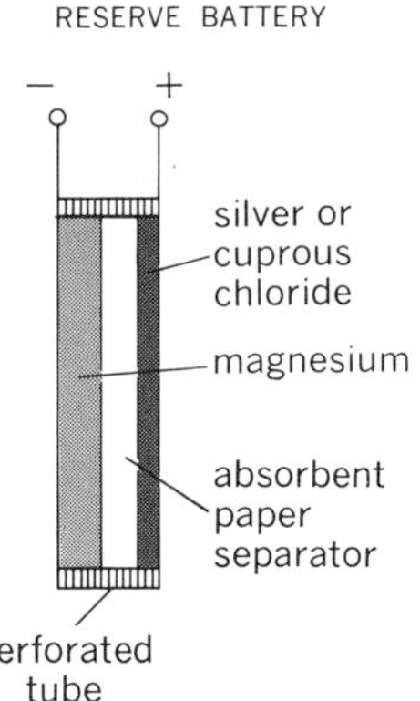

Fig. 1. Schematic of water-activated cell.

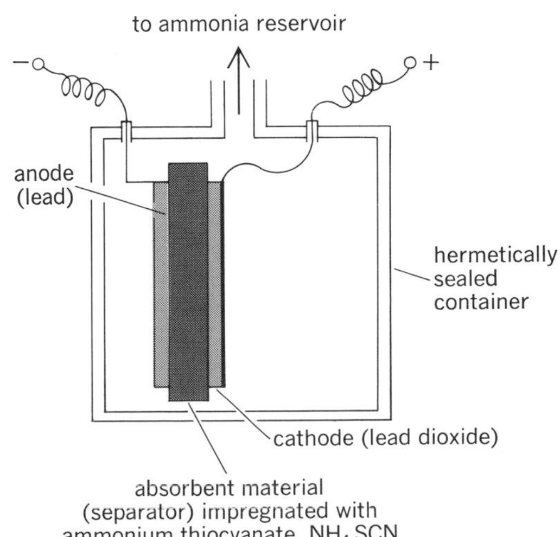

Fig. 2. Schematic of ammonia-vapor-activated reserve-type primary cell.

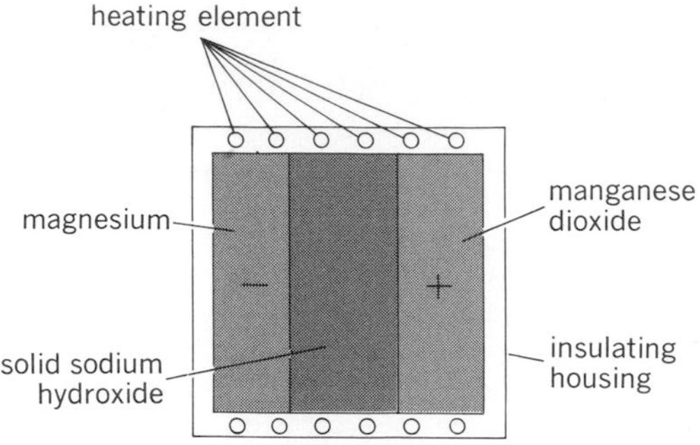

Fig. 3. Schematic of thermal cell.

mentioned previously because of the greater stability of the reactants at high temperatures (Fig. 3).

In addition to the couples already mentioned, successful systems have been employed using Mg, Ca, or Li alloys as anodes and K_2CrO_4, WO_3, MO_3, or $PbCrO_4$ as cathodes.

Thermal batteries are essentially of low energy efficiency because the heat absorbed in melting the electrolyte is not available in the electrical output. Consequently, although their use is restricted to small cell sizes, there is likely to be an increasing demand for thermal cells because of their high energy output, long storage retention over wide temperature ranges, ruggedness, and ability to be used in any position.

[JACK DAVIS; KENNETH FRANZESE]

Bibliography: Burgess Battery Co., *Engineering Manual*, 1964; W. J. Hamer and J. P. Schrodt, Investigations of galvanic cells with solid and molten electrolytes, *J. Amer. Chem. Soc.*, 71:2347, 1949; R. Jasinski, *High Energy Batteries*, 1967; J. P. Mullen and P. L. Howard, Characteristics of the silver chloride-magnesium water activated battery, *Trans. Electrochem. Soc.*, 90:529, 1946; C. B. Root and R. A. Sutula, *Proceedings of 22d Power Sources Conference*, May 14–16, 1968; J. P. Schrodt et al., A lead dioxide cell containing various electrolytes, *Trans. Electrochem. Soc.*, 90:405, 1946; R. Schult and W. Stafford, Electrochemical energy sources: Silver oxide/zinc batteries, *Electro-Technol. (NY)*, 67:84, June, 1961; G. W. Vinal, *Primary Batteries*, 1950; U.S. Army Signal Laboratories, *Proceedings of the 11th Annual Battery Research and Development Conference*, 1957; J. C. White, R. T. Pierce, and T. P. Dirkse, Characteristics of the silver oxide-zinc-alkali primary cells, *Trans. Electrochem. Soc.*, 90:467, 1946.

Resistance heating

The generation of heat by electric conductors carrying current. The degree of heating for a given current is proportional to the electrical resistance of the conductor. If the resistance is high, a large amount of heat is generated, and the material is used as a resistor rather than as a conductor. *See* ELECTRICAL RESISTANCE.

Resistor materials. In addition to having high resistivity, heating elements must be able to withstand high temperatures without deteriorating or sagging. Other desirable characteristics are low temperature coefficient of resistance, low cost, formability, and availability of materials. Most commercial resistance alloys contain chromium or aluminum or both, since a protective coating of chrome oxide or aluminum oxide forms on the surface upon heating and inhibits or retards further oxidation. Some commercial resistor materials are listed in Table 1.

Heating element forms. Since heat is transmitted by radiation, convection, or conduction or combinations of these, the form of element is designed for the major mode of transmission. The simplest form is the helix, using a round wire resistor, with the pitch of the helix approximately three wire diameters. This form is adapted to radiation and convection and is generally used for room or air heating. It is also used in industrial furnaces, utilizing forced convection up to about 1200°F (650°C). Such helixes are stretched over grooved high-alumina refractory insulators and are otherwise open and unrestricted. These helixes are suitable for mounting in air ducts or enclosed chambers, where there is no danger of human contact.

For such applications as water heating, electric range units, and die heating, where complete electrical isolation is necessary, the helix is embedded in magnesium oxide inside a metal tube, after which the tube is swaged to a smaller diameter to compact the oxide and increase its thermal conductivity. Such units can then be formed and flattened to desired shapes. The metal tubing is usually copper for water heaters and stainless steel for radiant elements, such as range units. In some cases the tubes may be cast into finned aluminum housings, or fins may be brazed directly to the tubing to increase surface area for convection heating.

Modification of the helix for high-temperature furnaces involves supporting each turn in a grooved refractory insulator, the insulators being strung on stainless alloy rods. Wire sizes for such elements are 3/16 in. (4.76 mm) in diameter or larger, or they may be edge-wound strap. Such elements may be used up to 1800°F (980°C) furnace temperature.

Another form of furnace heating element is the sinuous grid element, made of heavy wire or strap or casting and suspended from refractory or stainless supports built into the furnace walls, floor, and roof. Some of these forms are shown in Fig. 1.

Silicon carbide elements are in rod form, with low-resistance integral terminals extending through the furnace walls, as shown in Fig. 2.

Direct heating. When heating metal strip or wire continuously, the supporting rolls can be used as electrodes, and the strip or wire can be used as the resistor.

In Fig. 3 the electric current passes from roll A through the strip or wire to roll B. Heating by this method can be very rapid. Disadvantages are that the electric currents are large, and uniform contact between the strip and the rolls is difficult to maintain, since both surfaces must be clean and free of oxides. For these reasons, direct heating is not used extensively.

Molten salts. The electrical resistance of molten salts between immersed electrodes can be used to generate heat. Limiting temperatures are dependent on decomposition or evaporization temperatures of the salt. Parts to be heated are immersed

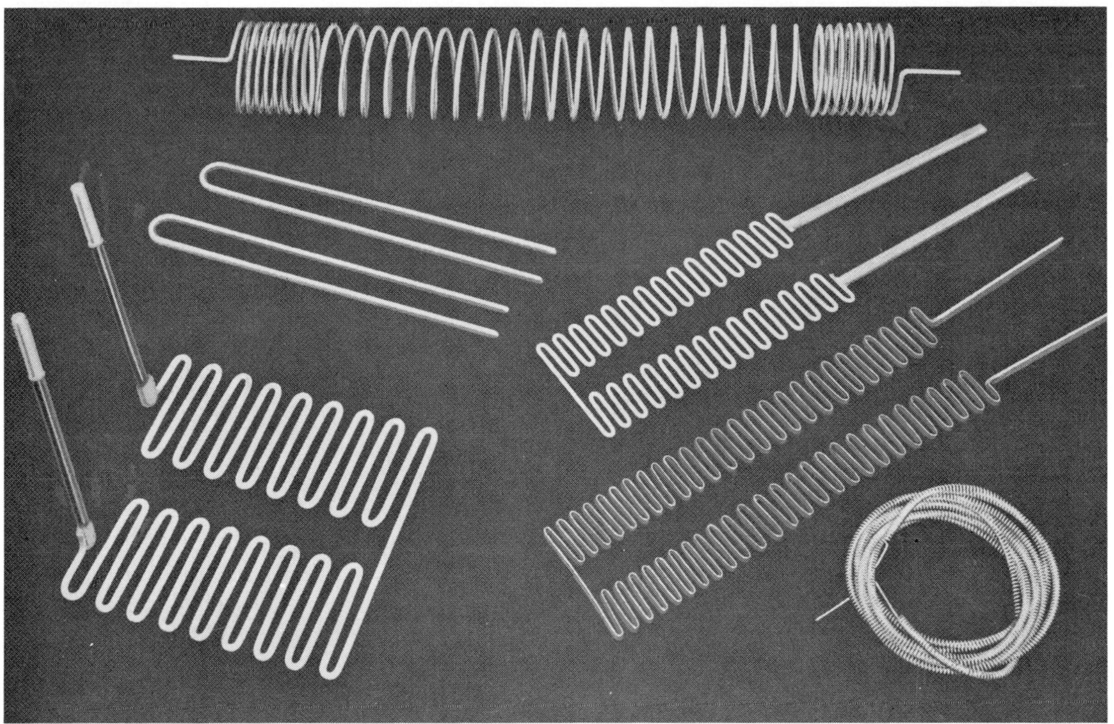

Fig. 1. Metallic resistor forms.

in the salt. Heating is rapid and, since there is no exposure to air, oxidation is largely prevented. Disadvantages are the personnel hazards and discomfort of working close to molten salts.

Major applications. A major application of resistance heating is in electric home appliances, including electric ranges, clothes dryers, water heaters, coffee percolators, portable radiant heaters, and hair dryers. Resistance heating is also finding increasing application in home or space heating; new homes are being designed with suitable thermal insulation to make electric heating practicable. A general rule for such insulation is that heat loss be restricted to 10 watts/ft^2 (108 W/m^2) of floor living area at 75°F (22°C) temperature difference, inside to outside. This applies to areas having 5000–7500 degree days annually. The degree day is the difference in °F (1°F = $^5/_9$°C) between the 24-hr average outdoor temperature and 65°F (19.3°C). For instance, if the 24-hr average were 40°F, there would be 25 degree days for that day. This includes also a three-quarter air change per hour, except in the basement, where one-quarter air change per hour is calculated. The values in Table 2 are applied to determine overall heat loss.

Heating system capacity. For rooms normally kept warm, the installed capacity should be 20% higher than the calculated losses. For rooms intermittently heated or for entries with frequent door openings, the installed capacity should be 50% more than the calculated losses.

A variety of electric heating units may be used for home heating. The simplest is a grid of helical resistance coils mounted in the central heating unit, with a blower to circulate the warm air through the rooms, and adjustable louvers in each room to control the temperature. Another type

Table 1. Electric furnace resistor materials and temperature ranges

Material, major elements	Maximum resistor temperature, °F (°C) In air	In reducing or neutral atmosphere
*35% nickel, 20% chromium	1900 (1040)	2100 (1150)
*60% nickel, 16% chromium	1800 (980)	2000 (1090)
*68% nickel, 20% chromium, 1% cobalt	2250 (1230)	2250 (1230)
78% nickel, 20% chromium	2250 (1230)	2250 (1230)
*15% chromium, 4.6% aluminum	2050 (1120)	Not used
*22.5% chromium, 4.6% aluminum	2150 (1180)	Not used
*22.5% chromium, 5.5% aluminum	2450 (1340)	Not used
Silicon carbide	2800 (1540)	2500 (1370)
Platinum	2900 (1590)	2900 (1590)
†Molybdenum	Not used	3400 (1870)
†Tungsten	Not used	3700 (2040)
†Graphite	Not used	5000 (2760)

*Balance is largely iron, with 0.5–1.5% silicon.

†Usable only in pure hydrogen, nitrogen, helium, or oxygen or in vacuum because of inability to form protective oxide.

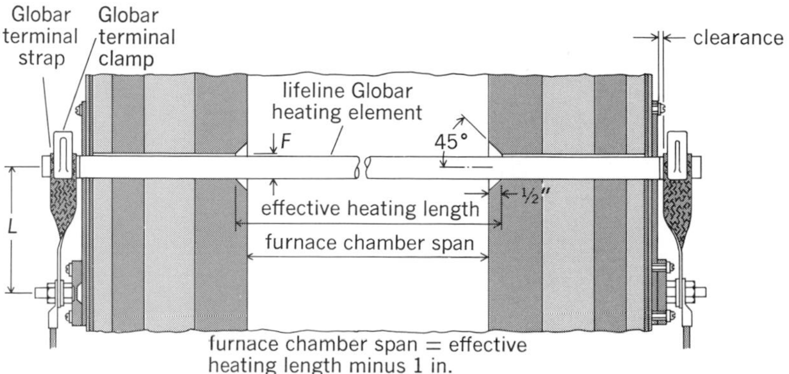

Fig. 2. Silicon carbide heating element. 1 in. = 2.54 cm.

Table 2. Maximum recommended heat loss

Structure	Heat loss, watt/ft²/°F (watt/m²/°C)
Ceiling	0.015 (2.9)
Wall	0.021 (4.1)
Floor	0.021 (4.1)
Floor slab perimeter	12.0/lineal foot (70.9/lineal m)
Basement walls	
Basement walls, 25% of wall aboveground	0.045 (8.7)
Requisites:	0.026 (5.0)
Windows – Double-glazed or storm	
Doors – Storm doors required	
Attic – Ventilating area of 1 ft²/150 ft² (1 m²/150 m²) ceiling area	

is the baseboard heater, consisting of finned sheathed helical resistance coils, with individual room control (thermostat in each room). Still another room unit is the glass or ceramic panel in which resistance wires are embedded. Such panels are designed to radiate infrared wavelengths and are particularly suitable for glassed-in areas or bathrooms. Mounted usually in the ceiling, such panels neutralize the chilling effect of large glass exposures, thus preventing drafts. *See* THERMOSTAT.

For isolated rooms or work areas, integrated units which combine resistors, circulating fans, and magnetic contactors operated by thermostats open and close the power circuits to the heating elements to maintain desired temperature. Where central heating units involving several kilowatts of energy are used, the contactors are arranged in multiple circuits and close in sequence to minimize the transient voltage effects of sudden large energy demands.

Ovens and furnaces. If the resistor is located in a thermally insulated chamber, most of the heat generated is conserved and can be applied to a wide variety of heating processes. Such insulated chambers are called ovens or furnaces, depending on the temperature range and use.

The term oven is generally applied to units which operate up to approximately 800°F (430°C). Ovens use rock wool or glass wool between the inner and outer steel casings for thermal insulation. Typical uses are for baking or roasting foods, drying paints and organic enamels, baking foundry cores, and low-temperature treatments of metals.

The term furnace generally applies to units operating above 1200°F (650°C). In these the thermal insulation is made up of an inner wall of fireclay, kaolin, or high-alumina or zirconia brick, depending on the temperature, with secondary insulating blocks made from such base materials as rock wool, asbestos fiber, and diatomaceous earth. Typical uses of furnaces are for heat treatment or melting of metals, for vitrification and glazing of ceramic wares, for annealing of glass, and for roasting and calcining of ores. *See* ELECTRIC HEATING.

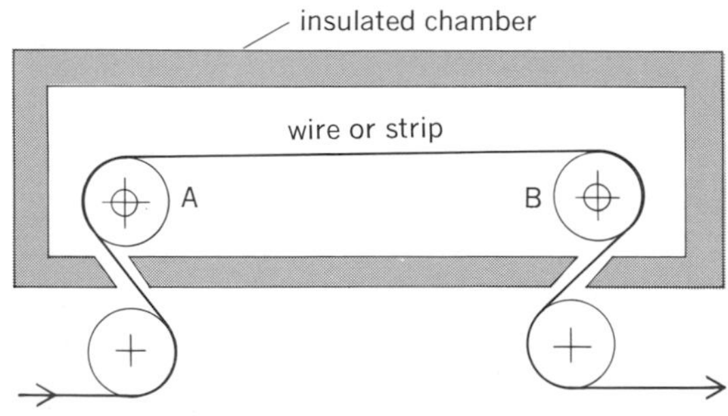

Fig. 3. Direct heating of metal strip or wire.

Actually, ovens and furnaces overlap in temperature range, with ovens being used at temperatures as high as 1000°F (540°C), and furnaces as low as 250°F (120°C). Electrically heated ovens and furnaces have advantages over fuel-fired units. These advantages often compensate for the generally higher cost of electric energy. The main advantages are (1) ease of distributing resistors, or heating elements, to obtain a uniform temperature in the product being heated; (2) ease of operation, since adjustments by operators are usually unnecessary; (3) cleanliness; (4) comfort, since heat losses are low and there are no waste fuel products; (5) adaptability to the use of controlled furnace atmospheres or vacuum, and (6) high temperatures beyond the range attainable with commercial fuels.

Heat transfer in ovens and furnaces. Heating elements mounted in ovens and furnaces may be located to radiate directly to the parts being heated or may be located behind baffles or walls so that direct radiation cannot take place. The heat then is transferred by circulating the furnace air or gas. Determination of which method or combination of methods to use is based on temperature uniformity, speed of heating, and high-temperature strength limitations of fans or blowers. At temperatures below 1200°F (650°C) radiation is slow, and virtually all ovens and furnaces in this temperature range use forced convection or circulation. From 1200 to 1500°F (650 to 820°C) radiation is increasingly effective, while reduced gas density and lower fan or blower speeds (because of reduced strength at high temperatures) make forced convection less effective. Therefore in this temperature range, combinations of radiation and forced convection are used. Above 1500°F (820°C) forced convection is employed only when direct radiation cannot reach all parts of the loads being heated. An example of this is a container filled with bolts or a rack filled with gears.

Obviously, in vacuum furnaces, heat transfer can be only by radiation and conduction.

Ovens. Figure 4 shows a typical oven, for a maximum temperature of 800°F (430°C); heating is by forced convection. The heated air enters through the supply duct, passing downward into the plenum chamber and distributing louvers into and across the load chamber and returning through the right-side plenum to the recirculating duct. The external heater and blower of the circuit are not shown.

Low-temperature furnace. Figure 5 shows a pit-type furnace used for tempering steel. It operates at temperatures of 250–1400°F (120–760°C) and uses forced convection.

The steel parts to be tempered are placed in the load chamber at the right. The electrical heating elements are in the heating chamber at the upper left. The fan, at the lower left, circulates air upward over the heating elements and into the load chamber. The hot air is forced down through the load and back to the fan. The heating chamber is

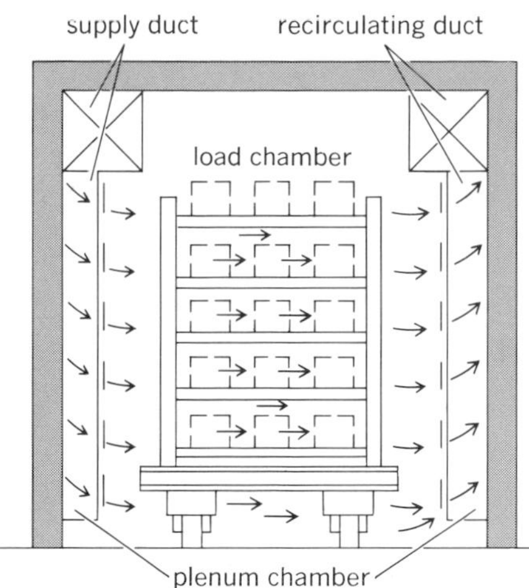

Fig. 4. Typical oven for maximum temperature of 800°F (430°C); heating is by forced convection. (*Carl Mayer Corp.*)

thermally insulated from the load chamber so there is no direct radiation. Such furnaces are designed commercially to hold the load temperature within 10° of the control temperature, and can be designed for closer control if desired.

High-temperature furnace. Figure 6 shows the interior of a pit-type carburizing furnace, which uses radiant-heating elements in the form of corrugated metal bands mounted on the inside of the brick walls. The work basket (not shown) rests on the load support at the bottom. These elements are designed to operate at low voltage (approximately 30 volts) so that soot deposits from the carburizing gases will not cause short circuits.

Electrical input. Electric ovens and furnaces are rated in kilowatts (kW). The electrical input is determined from the energy absorption Q of the load and the thermal losses.

The average rate of energy flow into load during the heating period is Q/t, where Q is in kilowatt-hours and t is the heating time in hours.

However, the rate of energy flow into the load is high when the load is cold and decreases as the load temperature approaches furnace temperature. Therefore the energy input must be high enough to take care of the high initial heating rate. This leads to the following approximate formulas for the input, in which L is the thermal losses in kilowatts at the operating temperature:

For batch furnaces, input in kW $= L + 1.5\ Q/t$
In continuous furnaces, input in kW $= L + 1.25\ Q/t$

Operating voltage for heating elements. Heating elements are usually designed to operate at standard service voltages of 115, 230, or 460 volts, if two conditions can be satisfied. These are, first, that the heating element is sufficiently heavy in cross section to avoid sagging or deformation in service, and second, that there is no appreciable electrical leakage through the furnace refractories tending to short-circuit the heating elements. The latter consideration generally limits voltages to 260 volts at 2100°F (1150°C), and to approximately 50 volts at 3100°F (1700°C), because the refractory walls of the furnace become increasingly better conductors at higher temperatures.

In vacuum furnaces, voltages are limited to 230 volts by the tendency to break down into glow discharge at low pressure.

Temperature control. Almost all commercial electric ovens and furnaces have automatic temperature control. The simplest control uses magnetic contactors which open and close the circuit to the heating elements in response to temperature signals from control thermostats or thermoelectric pyrometers. A refinement of this ON-and-OFF control is to modulate it with a timer, with the ON period becoming a progressively smaller percentage as the control temperature is approached. This prevents the overshooting which results from thermal lag in the control thermocouple and the furnace.

True proportioning control is achieved through the use of saturable reactors in series with the heating elements. Figure 7 shows schematically the arrangement used.

The pyrometer, through the amplifier, controls the direct current to the control winding of the reactor. When the direct current is maximum, the reactor offers virtually no impedance to the alternating current flowing through the heating element, and the normal amount of heat is generated. As control temperature is approached, the direct current is decreased, increasing the impedance of the reactor and reducing the current to the heating elements until equilibrium is reached.

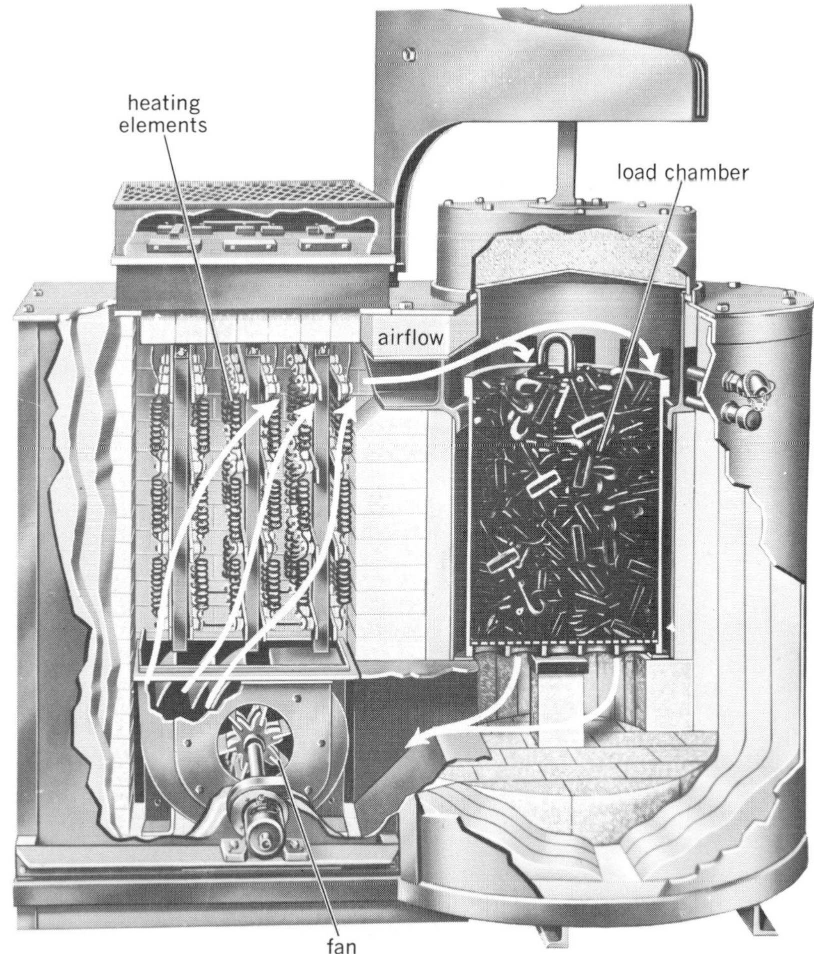

Fig. 5. Low-temperature pit-type furnace, using forced convection and operating at temperatures of 250–1400°F (120–760°C). (*Lindberg Industrial Corp.*)

Fig. 6. Interior of high-temperature pit-type carburizing furnace, using radiant heating. (*Lindberg Industrial Corp.*)

A device which is beginning to be used as a heating contactor is the solid-state silicon-controlled rectifier. While more expensive than the magnetic contactor, it has no moving parts, lower maintenance costs, and silent, vibration-free operation. Moreover, it is directly adaptable to proportioning control, obviating the need for saturable reactors. In this latter case it is cost-competitive.

Snow and ice melting. Low-temperature resistors with moisture-proof insulation are embedded in concrete sidewalks or strung in roof gutters to melt snow and ice as needed. Such cables also are used to prevent freezing of exposed water pipes.

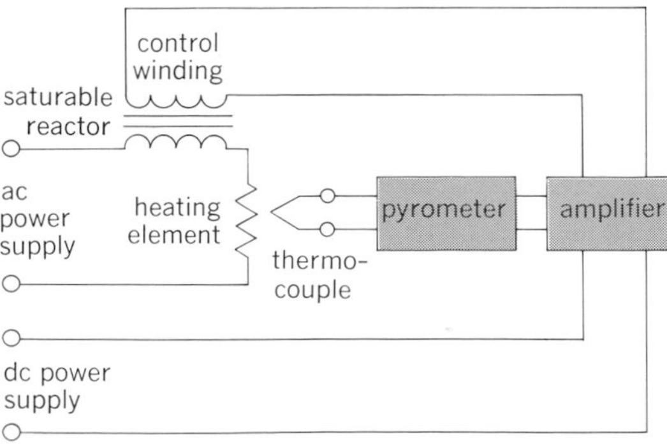

Fig. 7. Temperature control using saturable reactor.

Analogously, very thin resistance wires may be cast in automobile window glass to melt snow and ice.

Metal-sheathed units are used at railroad switches and dam locks to keep the moving parts free in freezing temperatures.

Die and platen heating. Metal-sheathed elements are embedded in grooves or holes in dies or platens to maintain these parts at desired temperatures for hot processing or molding of plastics.

Cryogenics. Metal-sheathed immersion-type units are used to supply heat of evaporation for liquefied gases such as argon, helium, nitrogen, hydrogen, and oxygen. Metal sheathing is usually of stainless steel or aluminum to withstand low temperatures. Gaskets are of Teflon for the same reason.

[WILLARD ROTH]

Bibliography: American Society of Heating, Refrigeration, and Air Conditioning Engineers, *Guide and Data Book*, 1965; Chemical Rubber Publishing Co., *Handbook of Chemistry and Physics*, 1970; D. G. Fink and J. M. Carroll, *Standard Handbook for Electrical Engineers*, 10th ed., 1968; W. H. McAdams, *Heat Transmission*, 1954; W. Trinks, *Industrial Furnaces*, vol. 1, 5th ed., 1961, and vol. 2, 3d ed., 1955.

Rotary engine

Internal combustion engine that duplicates in some fashion the intermittent cycle of the piston engine, consisting of the intake-compression-power-exhaust cycle, wherein the form of the power output is directly rotational.

Four general categories of rotary engines can be considered: (1) cat-and-mouse (or scissor) engines, which are analogs of the reciprocating piston engine, except that the pistons travel in a circular path; (2) eccentric-rotor engines, wherein motion is imparted to a shaft by a principal rotating part, or rotor, that is eccentric to the shaft; (3) multiple-rotor engines, which are based on simple rotary motion of two or more rotors; and (4) revolving-block engines, which combine reciprocating piston and rotary motion. Some of the more interesting engines of each type are discussed in this article.

Cat-and-mouse engines. Typical of this class is the engine developed by T. Tschudi, the initial design of which goes back to 1927. The pistons, which are sections of a torus, travel around a toroidal cylinder. The operation of the engine can be visualized with the aid of Fig. 1, where piston A operates with piston C, and B with D. In chamber 1 a fresh fuel-air mixture is initially injected while pistons C and D are closest together. During the intake stroke the rotor attached to pistons B and D rotates, thereby increasing the volume of chamber 1. During this time the A-C rotor is stationary. When piston D reaches its topmost position, the B-D rotor becomes stationary, and A and C rotate so that the volume of chamber 1 decreases and the fuel-air mixture is compressed.

When the volume of chamber 1 is again minimal, both rotors move to locate chamber 1 under the spark plug, which is fired. The power stroke finds piston D moving away from piston C, with the A-C rotor again locked during most of the power stroke. Finally, when piston D has reached bottom, the B-D rotor locks, the exhaust port has been ex-

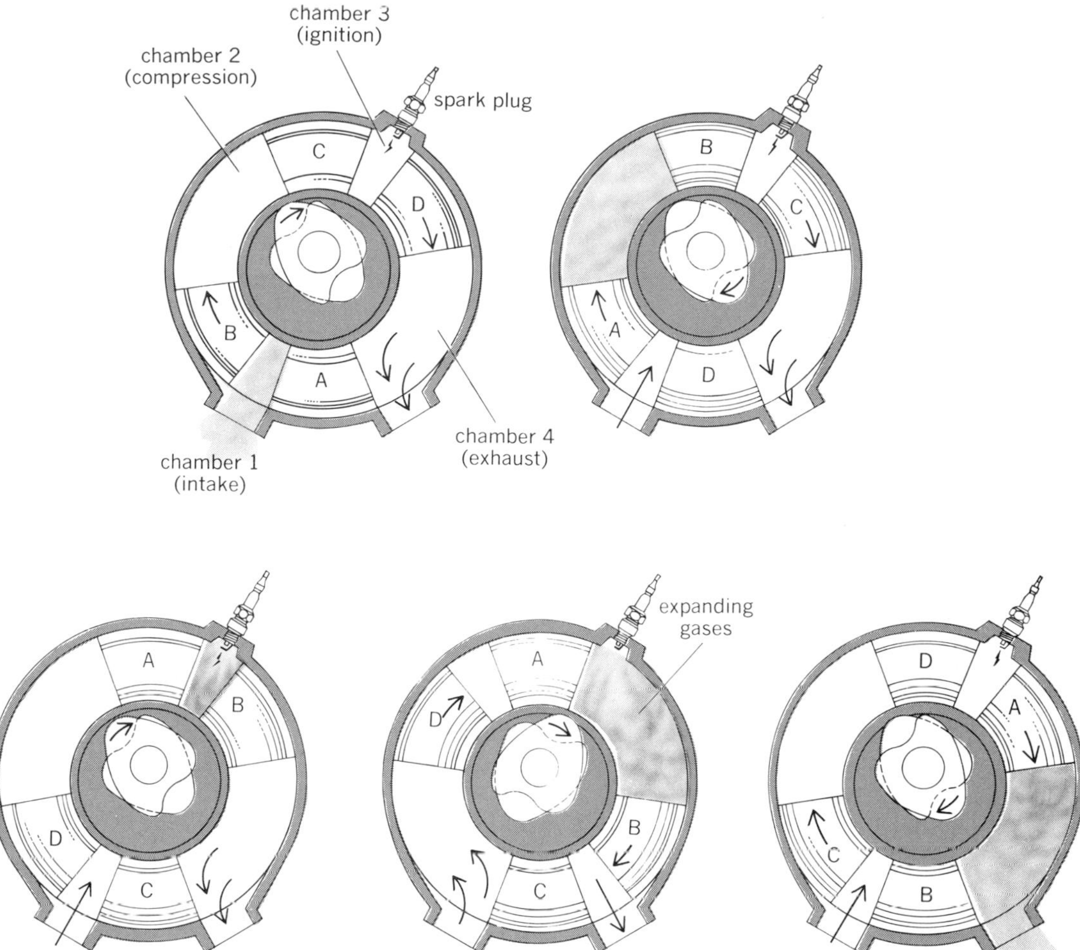

Fig. 1. The Tschudi engine.

posed, and the movement of piston C forces out the combustion product gas. Note that four chambers exist at any time, so that at each instant all the processes making up the four-stroke cycle (intake-compression-power-exhaust) are occurring.

The motion of the rotors, and hence the pistons, is controlled by two cams which bear against rollers attached to the rotors. The cams and rollers associated with one of the rotors disengage when it is desired to stop the motion of that rotor. The shock loads associated with starting and stopping the rotors at high speeds may be a problem with this engine, and lubrication and sealing problems are characteristic of virtually all the engines discussed herein. However, the problem of fabricating toroidal pistons does not appear to be as formidable as was once believed.

An engine similar to the Tschudi in operation is that developed by E. Kauertz. In this case, however, the pistons are vanes which are sections of a right circular cylinder. Another difference is that while one set of pistons is attached to one rotor so that these two pistons rotate with a constant angular velocity, the motion of the second set of pistons is controlled by a complex gear-and-crank arrangement so that the angular velocity of this second set varies. In this manner, the chambers between the pistons can be made to vary in volume in a prescribed manner. Hence, the standard piston-engine cycle can be duplicated. Kauertz tested a prototype which was found to run smoothly and to deliver 213 hp (159 kW) at 4000 rpm. Here again, however, the varying angular velocity of the second set of pistons must produce inertia effects that will be absorbed by the gear-and-crank system. At high speeds over extended periods, problems with this system are likely to be encountered.

An advanced version of the cat-and-mouse concept called the SODRIC engine has been developed by K. Chahrouri. Unlike the Tschudi engine, in which the four processes of intake-compression-power-exhaust are distributed over 360° of arc, the SODRIC engine performs these same four processes in 60° of arc. Hence, six power strokes per revolution are achieved, resulting in very substantial improvements in engine performance parameters. For example, Chahrouri has estimated that 225 hp (168 kW) can be achieved at only 1000 rpm using an engine having a toroid radius of 8 in. (20 cm) and 1-in. (2.5-cm) radius pistons. Chahrouri has also improved upon the method by which alternate acceleration-deceleration of the pistons is achieved, and power is transmitted to the output shaft by using noncircular gears.

The "cat-and-mouse" and "scissors" characterizations of these engines should be clear once the picture of pistons alternately running away from, and catching up to, each other is firmly in mind.

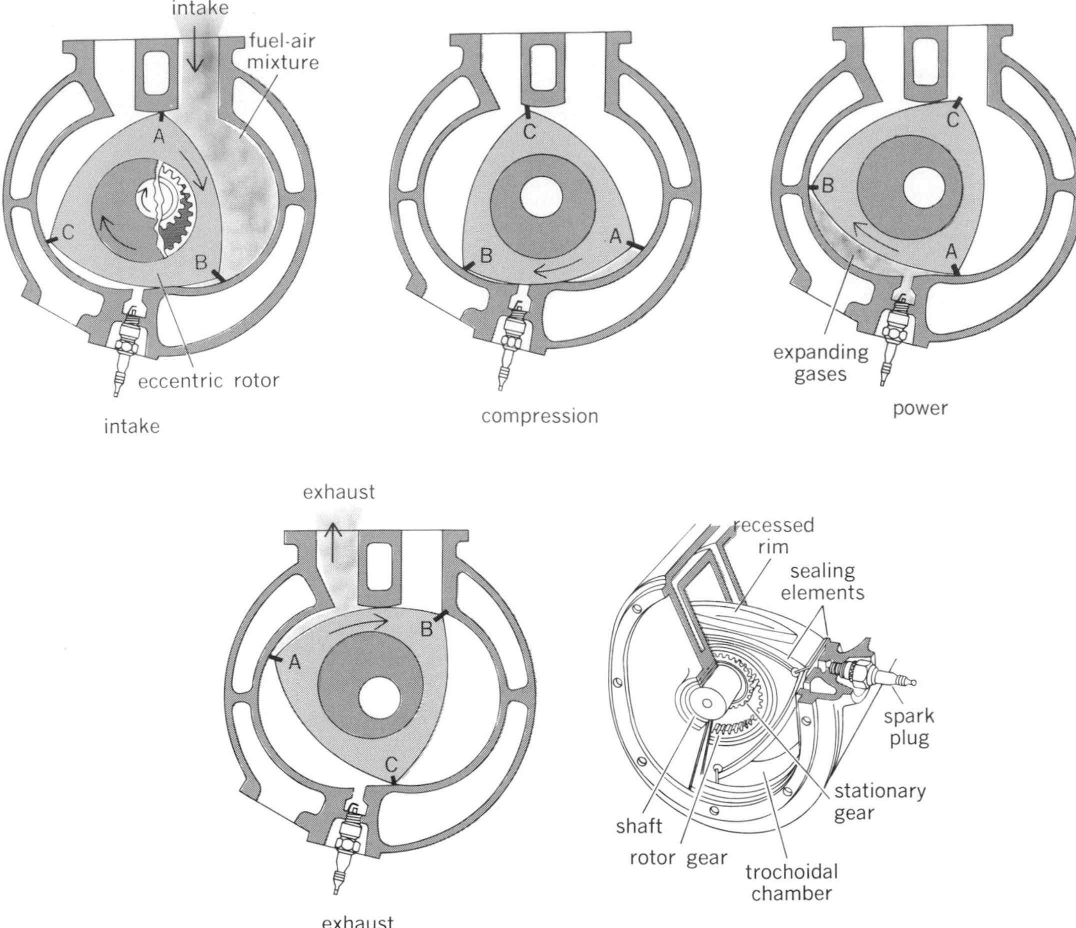

Fig. 2. The Wankel engine.

Other engines of this type, including the Maier, Rayment, and Virmel designs, differ principally in the system used to achieve the cat-and-mouse effect.

It should be noted that since the length of the power stroke is readily controlled in these engines, good combustion efficiencies (close to complete combustion) should be attainable.

Eccentric-rotor engines. The rotary engine which has received by far the greatest development to date is the Wankel engine, an eccentric-rotor type. The basic engine components are pictured in Fig. 2. Only two primary moving parts are present: the rotor and the eccentric shaft. The rotor moves in one direction around the trochoidal chamber, which contains peripheral intake and exhaust ports.

The operation of this engine can be visualized with the aid of Fig. 2. The rotor divides the inner volume into three chambers, with each chamber the analog of the cylinder in the standard piston engine. Initially, chamber AB is terminating the intake phase and commencing its compression phase, while chamber BC is terminating its compression phase and chamber CA is commencing its exhaust phase. As the rotor moves clockwise, the volume of chamber AB approaches a minimum. When the volume of chamber AB is minimal, the spark plug fires, initiating the combustion phase in that chamber. As combustion continues, the point is reached where the exhaust port is exposed, and the products of combustion are expelled from chamber AB.

To increase the chamber volumes, each segment of the rotor rim is recessed (Fig. 2). During the combustion-expansion phase unburned gas tends to flow at high velocity away from the combustion zone toward the opposite corner. As a result, this engine has a tendency to leave a portion of the charge unburned, similar to the problem encountered in ordinary piston engines. In addition to reducing the engine performance, this unburned gas is a source of air pollution. Efforts have been directed toward increasing the turbulence in each chamber, thereby improving the mixing between the burned and unburned gases, leading to better combustion efficiency.

On the other hand, Wankel engines have demonstrated a number of impressive advantages when contrasted with standard engines. Some of these advantages are listed below.

1. It has superior power-to-weight ratio; that is, the Wankel generally produces more or at least comparable horsepower per pound of engine weight when compared with conventional piston engines.

2. To increase power output, additional rotor-

trochoidal chamber assemblies can readily be added, which occupy relatively little space and add little weight. In piston engines, cylinder volumes must be increased, leading to substantial increases in weight and installation space.

3. The rotor and eccentric shaft assembly can be completely balanced; since they usually rotate at constant velocity in one direction, vibration is almost completely eliminated and noise levels are markedly reduced.

4. As with the cat-and-mouse engines, the intake and exhaust ports always remain open, that is, gas flow into and from the engine is never stopped, so that surging phenomena and problems associated with valves which open and close are eliminated.

5. Tests indicate that Wankel engines can run on a wide variety of fuels, including ordinary gasoline and cheaper fuels as well.

6. After considerable development, reasonably effective sealing between the chambers has been achieved, and springs maintain a light pressure against the trochoidal surface.

7. The Wankel has so few parts, relative to a piston engine, that in the long run it will probably be cheaper to manufacture.

The initial application of the Wankel engine as an automotive power plant occurred in the NSU Spider. In the early 1970s, however, the Japanese automobile manufacturer Mazda began to use Wankel engines exclusively. In addition, several American automobile manufacturers have experimented with Wankel-powered prototypes, but as of 1975 no production vehicles have emerged.

The Wankel engine is being used as a marine engine and in engine–electric generator installations, where its overall weight and fuel consumption have proved to be superior to those of a diesel engine or gas turbine generating equivalent power. Other projected applications include lawnmower and chainsaw engines. This wide range of applications is made possible by the fact that almost any size of Wankel engine is feasible.

An engine conceptually equivalent to the Wankel was developed jointly by Renault, Inc., and the American Motors Corp. It is sometimes called the Renault-Rambler engine. In this case, however, the rotor consists of a four-lobe arrangement, operating in a five-lobe chamber (Fig. 3). When a lobe moves into a cavity, which is analogous to the upward motion of the piston in the cylinder, the gas volume decreases, resulting in a compression process. The operation of this engine is detailed in Fig. 3. The fact that each cavity has two valves (intake and exhaust) represents a significant drawback. However, sealing between chambers may be simpler than in the Wankel; since each cavity acts as a combustion chamber, heat is evenly distributed around the housing, resulting in little thermal distortion.

It can be concluded that engines of the eccentric-rotor type are well on their way toward becoming an integral part of the internal combustion engine scene. Their inherent simplicity, coupled with their advanced state of development, should make them attractive alternatives to the piston engine for a wide variety of applications.

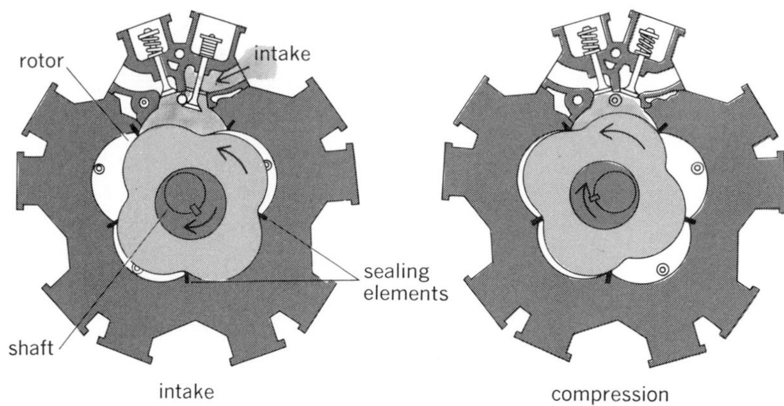

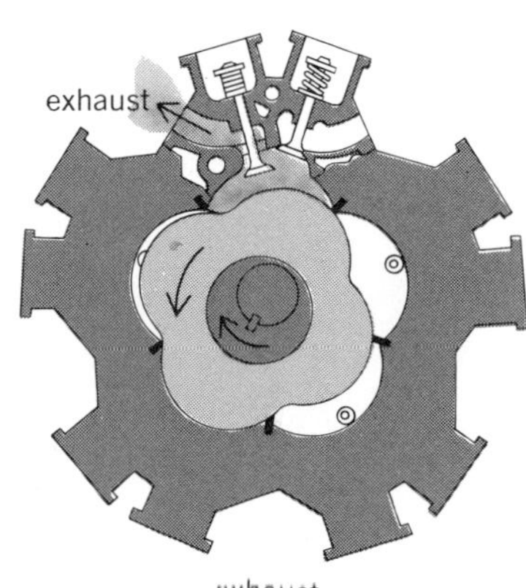

Fig. 3. The Renault-Rambler engine.

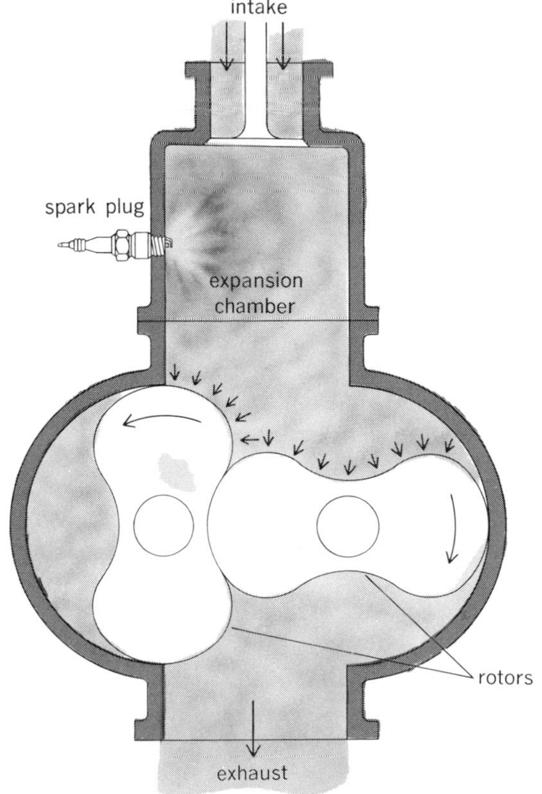

Fig. 4. A simple multirotor engine.

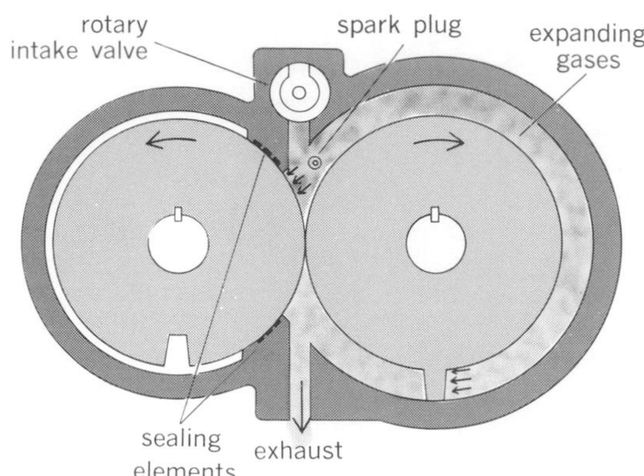

Fig. 5. The Unsin engine.

Multirotor engines. These engines operate on some form of simple rotary motion. A typical design, shown in Fig. 4, operates as follows. A fuel-air mixture enters the combustion chamber through some type of valve. No compression takes place; rather, a spark plug ignites the mixture which burns in the combustion chamber, with a consequent increase in temperature and pressure. The hot gas expands by pushing against the two trochoidal rotors. The eccentric force on the left-hand rotor forces the rotors to rotate in the direction shown. Eventually, the combustion gases find their way out the exhaust.

The problems associated with all engines of this type are principally twofold: The absence of a compression phase leads to low engine efficiency, and sealing between the rotors is an enormously difficult problem. One theoretical estimate of the amount of work produced per unit of heat energy put into the engine (by the combustion process), called the thermal efficiency, is only 4%.

The Unsin engine (Fig. 5) replaces the trochoidal rotors with two circular rotors, one of which has a single gear tooth upon which the gas pressure acts. The second rotor has a slot which accepts the gear tooth. The two rotors are in constant frictional contact, and in a small prototype engine sealing apparently was adequate. The recommendation of its inventor was that some compression of the intake charge be provided externally for larger engines.

The Walley and Scheffel engines employ the principle of the engine in Fig. 4, except that in the former, four approximately elliptical rotors are used, while in the latter, nine are used. In both cases the rotors turn in the same clockwise sense, which leads to excessively high rubbing velocities. (The rotors are in contact to prevent leakage.) The Walter engine uses two different-sized elliptical rotors.

Revolving-block engines. These engines combine reciprocating piston motion with rotational motion of the entire engine block. One engine of this type is the Mercer (Fig. 6). In this case two opposing pistons operate in a single cylinder. Attached to each piston are two rollers which run on a track that consists of two circular arcs. When the pistons are closest together, the intake ports to the chambers behind the pistons are uncovered, admitting a fresh charge. At this moment a charge contained between the pistons has achieved maximum compression, and the spark plug fires. The pistons separate as combustion takes place between them, which results in a compression of the gases behind the pistons. However, the pistons moving apart force the rollers to move outward as well. This latter motion can only occur if the rollers run on their circular track, which consequently forces the entire engine block to rotate. When the pistons are farthest apart, the exhaust ports are uncovered and the combustion gases purged. At this same time the compressed fresh charge behind the pistons is transferred to the region between the pistons to prepare for its recompression and combustion, which must occur because of the continuing rotation of the block.

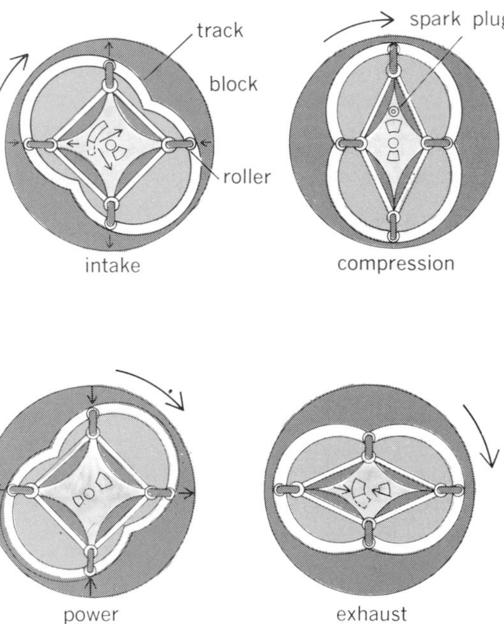

Fig. 7. The Rajakaruna engine.

No doubt some of the fresh charge is lost to the exhaust during the transfer process. In addition, stresses on the roller assembly and cylinder walls are likely to be quite high, which poses some design problems. Cooling is a further problem, since cooling of the pistons is difficult to achieve in this arrangement.

On the other hand, the reciprocating piston motion is converted directly to rotary motion, in contrast with the connecting rod-crank arrangement

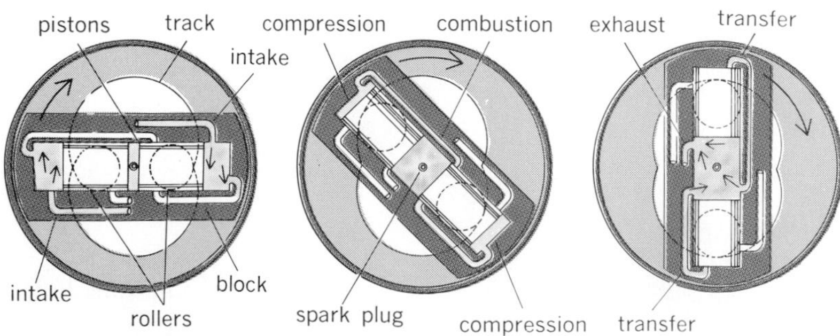

Fig. 6. The Mercer engine.

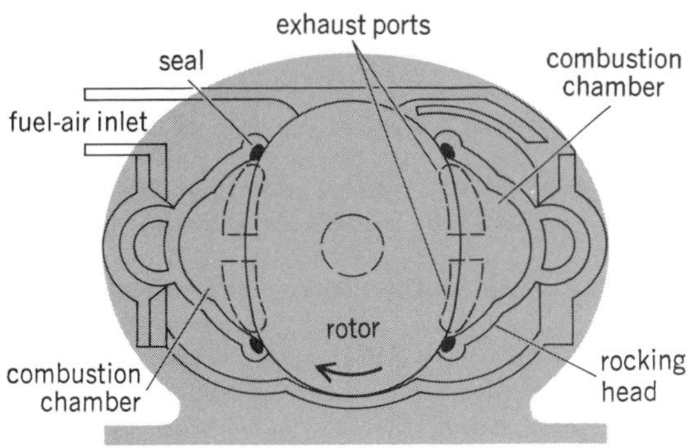

Fig. 8. The Walker engine.

in the conventional piston engine. Also, no flywheel should be necessary since the entire rotating block acts to sustain the rotary inertia. Vibration will also be minimal.

The Selwood engine is similar in operation, except that two curved pistons 180° opposed run in toroidal tracks. This design recalls the Tschudi cat-and-mouse engine, except that the pistons only travel through 30° of toroidal track. This motion forces the entire block to rotate. The Leath engine has a square rotor with four pistons, each 90° apart, with a roller connected to each. As in the Mercer engine, the reciprocating motion of the pistons forces the rollers to run around a trochoidal track which causes the entire block to rotate. The Porshe engine uses a four-cylinder cruciform block. Again, rollers are attached to each of the four pistons. In this arrangement power is achieved on the inward strokes of the piston. Finally, the Rajakaruna engine (Fig. 7) uses a combustion chamber whose sides are pin-jointed together at their ends. Volume changes result from distortion of the four-sided chamber as the surrounding housing, which contains a trochoidal track, rotates. The huge pins are forced against the track. As usual, cooling and lubrication problems will be encountered with this engine, as will excessive wear of the hinge pins and track.

The Ma-Ho engine, invented by G. Hofmann, uses four cylinders welded concentrically around a central shaft. As the pistons oscillate in the cylinders as a result of the intake-compression-power-exhaust processes undergone by the fuel-air mixture, the pistons rotate a barrel cam to which they are connected by cam followers. Hence, the entire block, including the central shaft, is forced to rotate. This operating arrangement recalls the Mercer engine.

Engines of other types. Although the vast majority of rotary engines fall into one of the categories discussed above, several ingenious designs which do not are worthy of mention.

The Walker engine (Fig. 8) involves an elliptical rotor which rotates inside a casing containing two C-shaped rocking heads. The fuel-air mixture is drawn into combustion chambers on each side of the rotor; mixture cut-off occurs when the rotor is in the vertical position. As the rotor turns, the mixture undergoes compression, and combustion is initiated by spark plugs. Rotor momentum coupled with the expansion of the combustion gases forces the rotor to continue turning. Compensation for seal wear is made by adjusting the rocking heads closer to the rotor.

The Heydrich engine (Fig. 9) is a vane-type rotary engine which utilizes a small hole to "store" a quantity of high-temperature combustion gases from a previous firing. This gas is then used to ignite the subsequent fuel-air charge. The first charge is ignited by a glow plug. Floating seals make contact with the chamber wall.

The Leclerc-Edmon-Benstead engine (Fig. 10) has three combustion chambers defined by stationary vanes, a cylindrical stator, and end flanges. An output shaft passing through the chamber is geared to a circular, eccentrically mounted piston. As the chambers are fired sequentially, the movement of the piston forces the shaft to rotate. Slots in the flange control porting of the intake and exhaust gases. *See* DIESEL CYCLE; DIESEL ENGINE; GAS TURBINE; INTERNAL COMBUSTION ENGINE; OTTO CYCLE.

[WALLACE CHINITZ]

Bibliography: R. F. Ansdale, Rotary combustion engines, *Auto. Eng.*, vol. 53, no. 13, 1963, and vol. 54, nos. 1 and 2, 1965; W. Chinitz, Rotary engines, *Sci. Amer.*, vol. 220, no. 2, 1969; W. Froede, *The Rotary Engine and the NSU Spider*, Soc. Automot. Eng. Pap. no. 650722, October, 1965; R. Wakefield, Revolutionary engines, *Road Track*, vol. 18, no. 3, 1966; F. Wankel, in R. F. Ansdale (ed.), *Rotary Piston Machines: Classification of Design Principles for Engines, Pumps and Compressors*, 1965.

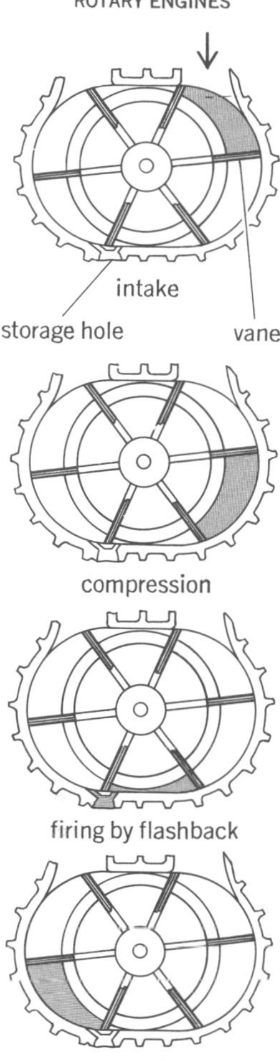

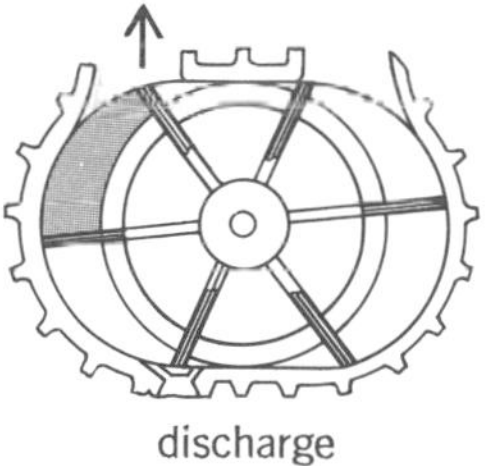

Fig. 9. The Heydrich engine.

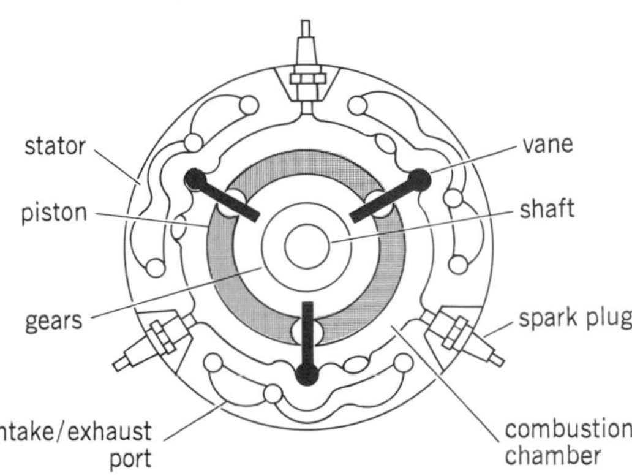

Fig. 10. The Leclerc-Edmon-Benstead engine.

Rotary tool drill

A bit and shaft used for drilling wells. A turntable on the derrick floor rotates a string of hollow steel drill pipe at the bottom of which is a steel bit. The bit grinds the rock. A drilling fluid is pumped down through the drill pipe; the fluid flushes out the rock cuttings and returns up the space between drill string and hole side.

The drilling fluid may be air, water, or, most commonly, mud (a mixture of various clays and chemicals, each having a special function). The mud cools and lubricates the bit, removes cuttings from the hole, and cakes the wall of the hole to prevent caving before steel casing is set. The hydrostatic pressure exerted by the column of mud in the hole prevents blowouts which may result when the bit penetrates a high-pressure oil or gas zone. When the mud reaches the surface, it passes over a vibrating screen to filter out large cuttings. The mud then passes on to a settling tank where smaller particles settle out. The cuttings are examined to determine the type of formation being drilled and for possibilities of oil or gas production. The mud mixture is sucked up from the pit and recirculated by a high-pressure pump. The viscosity, weight, and filtration properties of the mud are altered as drilling proceeds by changing the proportion of its constituents.

Power is transmitted from an engine to a draw works—a winch which drives the rotary table on the derrick floor and also applies power for hoisting or lowering the drill string as shown in the illustration. The string of drill pipe is topped at the surface by a square-sided length of heavy pipe called the kelly. The square shape permits the rotary table to grip and rotate the kelly, and hence the entire drill string, and yet have sufficient freedom so that it can slip vertically through the table as drilling goes deeper. Rotation speeds range from 40 rpm to 500 rpm or more, depending primarily upon the character of the formation being drilled. The drill string usually consists of 30-ft (9-m) lengths of drill pipe coupled together. On the lower end are heavier-walled lengths of pipe, called drill collars, which help regulate weight on the bit.

The drill string is attached to a swivel suspended from a hook which is connected to a traveling block, or pulley, encased in a frame. The drilling cable runs from the draw works over a crown block at the top of the derrick and down to the traveling block. The mud is pumped through a hose attached to the swivel. An opening in the center of the swivel permits the mud to pass down through the attached drill string.

When the bit has penetrated the distance of a pipe section, drilling is stopped, the string is pulled up to expose the top joint, the kelly is disconnected, a new section added, the kelly attached, the string lowered, and drilling resumed. This process continues until the bit becomes worn out, at which time the entire drill string must be pulled. Pipe is usually disconnected in thribbles, or 90-ft (27-m) sections of pipe, and stacked in the derrick. The height of the derrick determines whether doubles, thribbles, or fourbles can be stacked. The process continues until the bit reaches the surface. A new bit is attached, and the drilling string reassembled and lowered into the hole. Such round trips may take up to two-thirds of total rig-operating time, depending upon the depth of the hole. In hoisting or lowering the drill string, the swivel is disengaged from the hook. Elevators, or clamps, which grip the pipe securely, are attached. The elevators are also used when the hole is lined with steel casing. In lowering drill pipe or casing, each new section of pipe is lifted from the derrick floor and suspended on the elevator until it is screwed to the preceding joint, just above the hole opening; the entire column is then lowered into the hole. While new sections of pipe or casing are being attached to the elevators, the pipe in the hole is supported in the rotary table by slips, or gripping devices.

Derricks can be skid-, truck-, or trailer-mounted, but larger units used in very deep drilling are assembled on the site. Derricks usually range in height from 66 ft (20 m) to nearly 200 ft (60 m). The derrick floor is set 7–20 ft (2–6 m) or more above the ground to provide a basement for control devices, such as blowout preventers, below the rotary table. *See* OIL AND GAS WELL DRILLING.

[ADE L. PONIKVAR]

Rural electrification

The generation, distribution, and utilization of electricity in nonurban areas, beyond the confines of incorporated cities, villages, and towns. Electric service is now provided to more than 99% of the farms in the United States. This near-universal availability of electricity has contributed greatly to the magnitude and efficiency of farming operations and has been a contributing factor in the location of nonfarm residential, commercial, and industrial developments in rural areas.

Development. The first central station service was provided by the Pearl Street Station in New York City in 1882. Street lighting and service to some commercial establishments constituted the primary loads of the fledgling industry. However, as early as 1898, electricity was first used on a farm—a 5-hp (3.7-kW) electric motor and irrigation pump was installed by a fruit farmer in northern California.

Rural electrification grew slowly. The problems of low population density and undeveloped uses for electricity on the farm presented many economic and technical problems. Some of these problems were eventually solved, and about 10% of the farms in the United States were electrified by 1930 (Fig. 1). The Depression slowed this effort, and it was not until low-interest loans became available through the Rural Electrification Administration (REA)—an agency created by the Federal government in 1935—that the process of rural electrification again picked up speed.

The technical and economic problems of providing service to rural areas have been solved, and farmers use electricity for many of their operations. In addition to these traditional rural loads, an increasing number of residences and commercial and industrial establishments are found in rural areas.

Uses. The homes found in rural areas, both farm and nonfarm, are as modern and up to date as those found in towns and cities. In some areas, a large percentage of rural homes are heated by

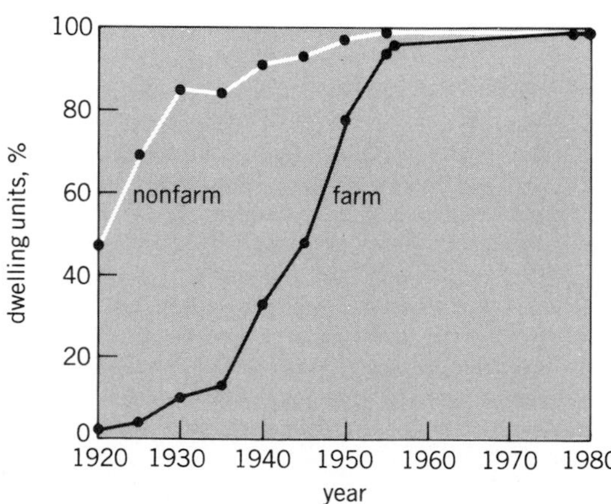

Fig. 1. Percent of United States dwelling units with electric service.

electricity due to the unavailability of lower-cost natural gas. This combination of urban comfort in a rural setting has led to more and more nonfarm residential developments. Such developments constitute a major portion of rural loads, with about 21% of the total nonfarm residences in the United States located in rural areas according to the 1970 census.

Modern farming operations, in terms of electrical load, may be equivalent to a small factory. Large livestock feeding and confinement systems, broiler and egg-laying operations, dairying, and other similar livestock operations require large quantities of reliable electrical energy. Modern harvesting equipment and farming methods depend on drying of crops by artificial means. Loads in some rural areas actually peak during the harvesting season due to the many electric motors required to drive the fans associated with grain-drying equipment. In some areas, irrigation constitutes a major portion of the rural load. Irrigation pumps ranging in size from 30 to 250 hp (22 to 186 kW) dot the countryside, requiring high-capacity distribution lines and much electrical energy (Fig. 2). The relative uses of electrical energy on the farm by category are shown in Fig. 3. In 1978 direct use of electrical energy in agriculture totaled 31.7×10^9 kWh, or about 1.6% of total electric utility industry sales.

The processing of agricultural products and other related agribusiness industries have developed in rural areas. These include: grain handling and storage facilities, livestock feed processing plants, canning and freezing plants, fertilizer and farm chemical complexes, farm machinery manufacturing and maintenance, fuel storage and handling facilities, and transportation depots—all related to the agriculture of the area.

Not all loads in rural areas are related to agriculture. Many other types of industries find it desirable to locate in rural areas because of the availability of land, transportation, labor, raw materials, or other incentives. Recreational developments have also become more numerous.

Many rural areas have a broad, prosperous, economic base—much of which is highly dependent on the availability of electrical energy.

Rural utility systems. Rural areas are served by electric utility systems which may be rural electric cooperatives or extensions of investor-owned or public utility systems serving both urban and rural areas. Generally speaking, rural electric cooperatives serve rural areas only, and statistics relating to them are readily available. Other electric utilities, serving a combination of urban and rural areas, usually do not report separate statistics for urban and rural loads on a regular basis. Rural electric cooperative statistics are believed to be representative of several aspects of rural loads and are used here for illustration.

The rural electric cooperatives and the public power utilities may or may not generate their own electrical energy. They may buy from other utilities, be a joint owner in a generation and transmission utility, or buy from some Federal power agency (Fig. 4).

Load density. Regardless of ownership, the primary difference between a rural electric system and an urban electric system is load density; typical area load densities are given in the table. This is also illustrated by a comparison of consumers served per mile of line. Rural electric cooperatives average about 4 consumers per mile compared to an estimated average of 30–40 consumers per mile for the total electric utility industry, or 8–10 times as many consumers per mile as for the rural electric cooperatives. This lower load density is the major factor contributing to the design differences between rural and urban systems.

Fig. 2. A center pivot irrigation system irrigating corn. A single unit can water as much as 500 acres (200 hectares) or more. *(Eastern Iowa Light and Power Cooperative)*

Distribution system. Rural systems are integrated with the other electric utility systems in the region through transmission system interconnections and reliability coordinating councils. Supply to rural areas is usually provided by a subtransmission system whose voltage is in the range of 33 to 230 kV. This subtransmission system delivers power to many distribution substations, some as small as 1000 kVA, others as large as 15,000 to 25,000 kVA. A typical distribution substation consists of a high-voltage fuse and switch connecting the stepdown transformer to the subtransmission system. The low-voltage side of the transformer is usually connected to voltage regulators—one three-phase or three single-phase—and serves two to six distribution circuits protected by circuit breakers or oil circuit reclosers.

Rural distribution lines may extend up to 50 mi (80 km) or more from a substation, and voltages of 15 or 25 kV are commonly required to provide adequate service. Conductor sizes will range from as small as No. 6 copper equivalent to as large as 350,000 circular mils (1.7735 cm^2) copper equivalent. The voltage drop along a line is roughly proportional to the product of distance from substation times load current. This voltage drop is usually the primary factor in determining the capacity of a rural distribution circuit, whereas conductor current-carrying capacity usually determines the capacity of distribution circuits in higher-load-density urban areas.

Most rural distribution lines are overhead. Circuits are three-phase leaving the substations, but

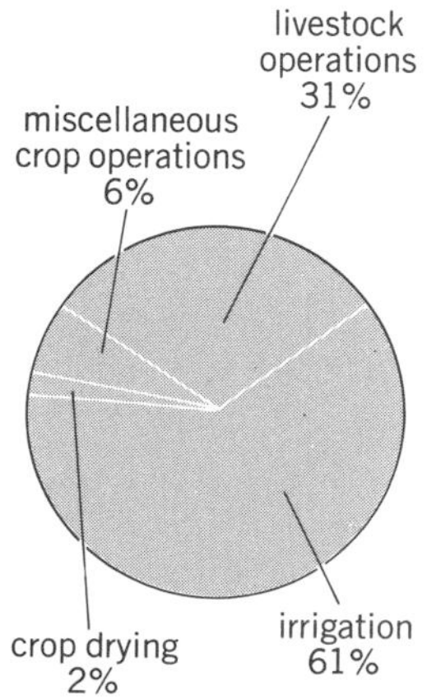

Fig. 3. Major categories of direct electric energy use in United States agriculture, 1978.

Fig. 4. A 55,000-kW-capacity coal-fired steam-electric generating plant which is owned by a rural electric utility in the Midwest. (*Eastern Iowa Light and Power Cooperative*)

may be either three-phase, two-phase, or single-phase at other locations. Single-phase lines provide satisfactory service in many rural areas due to the relatively low load density and small loads.

In some areas, overhead lines are very susceptible to storm damage, particularly from ice storms. A major ice storm can cause havoc in any area, but rural areas are especially susceptible due to the many miles of line required to serve a relatively few customers. Demands for higher service reliability and the high cost of replacing storm-damaged overhead lines have led some utilities to place distribution lines underground. This is especially practical in relatively open areas where underground circuits can be installed by cable plows (Fig. 5).

Rural distribution circuits are usually radial. Because of the relatively low load density, it is seldom practical to provide two-way feed in many rural areas. Most urban distribution systems tend to be "looped," providing two-way feed to most loads. Rural distribution circuits may employ voltage regulators, sectionalizing equipment such as oil circuit reclosers, cutouts, and fuses, and power-factor correction capacitors in order to provide more efficient use of the distribution circuits.

The construction of rural overhead lines differs somewhat from that of urban lines in several ways. Rural lines utilize longer spans and fewer accessories to minimize costs (Fig. 6). The use of common poles by different utilities, such as telephone, cable television, and electric, is not as common in

Range of load densities

Type of area	Load density, kVA/mi² (kVA/km²)	Remarks
Residential, low-density (rural area)	10–300 (4–120)	Number of farms or residences, and demand per farm, can cover a broad range of values, for example, 1 farm at 10 kVA to 150 farms at 2 kVA average each, does not include impact of irrigation load
Residential, medium-density (rural and suburban areas)	300–1,200 (120–460)	Based on home saturation on 70 by 100 ft (21 by 30 m) plots, 20% of total area with average diversified kVA per house from 0.5 to 2
Residential, high-density (urban area)	1,200–4,800 (460–1,900)	Based on home saturation on 70 by 100 ft (21 by 30 m) plots, 80% of total area with average diversified kVA per house from 0.5 to 2
Residential, extra-high-density (all-electric area)	15,000–20,000 (6,000–8,000)	An estimated upper limit for total electric homes with heating and air conditioning, also high saturation of homes in total area
Commercial	10,000–300,000 (4,000–120,000)	Type of area and load includes such variations so that more definitive breakdown is not practical; this range of values covers range of small shopping centers and commercial areas to downtown commercial areas of large cities

Fig. 5. Installation of underground rural distribution line. A crawler-type tractor pulls a cable plow and cable reel. The plow opens a furrow and lays the cable in one operation. *(Eastern Iowa Light and Power Cooperative)*

rural areas. In rural areas a small distribution transformer (5–15 kVA) typically serves a single consumer, whereas in urban areas a large transformer (25–50 kVA) may serve several small businesses or residences.

Rural lines are built according to safety standards similar to those for urban lines, with the National Electrical Safety Code usually providing the governing design criteria. In some instances, the clearance of rural lines must be increased because of irrigation and other farm equipment. The REA provides design standards and guides for use by rural electric cooperatives. *See* ELECTRIC DISTRIBUTION SYSTEMS; TRANSMISSION LINES.

Decentralization of services. Due to the large geographic area typically covered by a rural utility, district offices and service crews are often established to provide faster and more efficient service to the rural consumers. An urban utility may have a single centralized headquarters facility in each metropolitan area, compared to the rural utility which may have several district or division offices.

Impact on United States. The availability of reliable electrical energy to virtually all areas of the United States has had a major impact on the economy of rural America and the life-style there. Farm output per worker-hour increased by over two-thirds between 1960 and 1980. Automation, mechanization, chemical fertilizers, better management practices, and electrification are credited with this change. Many former urbanites have discovered that living in rural areas is as comfortable as in urban areas. Both agriculture-related and nonagricultural industry have found rural areas desirable places to locate.

The unqualified success of rural electrification programs in the United States has led to the exportation of the concept to developing countries around the world. For example, rural electric cooperatives have been formed in several countries in Latin America and the Far East.

Future. Inflation, energy shortages, and environmental issues affect all electric utilities. Rural areas are particularly vulnerable because a rural

Fig. 6. A typical rural single-phase distribution line. The absence of crossarms and longer spans makes lower-cost rural lines possible.

utility tends to be more capital-intensive than an urban utility. However, electricity still is a very desirable form of energy for use by farms, rural industries, and agribusinesses, and rural areas will continue to attract city dwellers to the quieter life of rural-residential areas.

Nonfarm developments in rural areas have caused the rural loads in the United States to grow at a much faster pace than for the average utility industry. In many areas, small farms are consolidated into larger operations, resulting in the loss of farmsteads and farm homes. This is offset to some extent by the new, larger farming operation requiring even more energy than the total of the small ones it replaced. The many uses of energy on the farm have also contributed to the increased electrical loads in rural areas.

The increasing cost of energy and efforts to conserve energy are also evident in rural areas. Efforts are being made to use electricity more efficiently in farming operations. However, many farm homes have typically been heated by fuel oil, and diesel engines have been used to drive irrigation pumps. Electricity may be replacing petroleum fuels in many instances, possibly offsetting the impact of conservation efforts.

The electric utility industry has essentially completed the task of making electric service universally available, regardless of the location. The rural customer—who once was desperate to receive electric service of any kind—has become more and more dependent on electric service for the home, farm, or other business. The task now is to continue to provide electrical energy in a safe, economical, reliable, environmentally acceptable manner to the rural consumer. *See* ELECTRIC POWER SYSTEMS.

[C. MAXWELL STANLEY; RONALD D. BROWN]

Bibliography: Edison Electric Institute, *Statistics of the Electric Utility Industry*; National Rural Electric Cooperative Association, *Facts about America's Rural Electric Systems*, 1979; U.S. Department of Agriculture, *Agricultural Statistics 1978*; USDA and Federal Energy Administration, *Energy and Agriculture, 1974 Data Base*; USDA, Economics, statistics and cooperatives service, *Structure Issues of American Agriculture*, Agric. Econ. Rep. 438, 1979; USDA, Rural Electrification Administration, *Annual Statistical Report: Rural Electric Borrowers*.

Seismic exploration for oil and gas

Exploration seismology is a geophysical method of determining subsurface geologic structure with man-made elastic waves. Energy, usually in the form of an impulse from an explosion, is introduced into the ground at or near the surface. Spreading out from the source, the energy encounters discontinuities in the physical properties of the rocks and is partially reflected back to the surface, detected, and recorded. Discontinuities of exploration interest are interfaces between different types of rock. The time required for the reflected energy to return indicates the depth of a reflector. Plotting this time for each detected signal as he moves along the surface, a geophysicist assembles a picture of the rock layers below him from shallow depths to the deepest interface from which returning energy is measurable. Getting such a picture is the goal of the exploration geophysicist. He has no way of detecting oil or gas directly for, with negligible exceptions, the ways that oil-, gas-, and water-saturated rocks respond to elastic waves are indistinguishable. Even with this limitation reflection seismology is the primary tool of petroleum geophysics. It is the only method which combines penetration many thousand feet into the earth with sufficient resolution to delineate changes in the rock layering.

Reflection seismology. Figure 1 illustrates the basis of the reflection seismic method. Another technique, refraction seismology, preceded the reflection method but is rarely used today. A common seismic source for reflection seismology is dynamite at the bottom of a water-filled hole drilled through the weathered layer. The weathered layer, soil and rock that are poorly consolidated, absorbs the energy of explosions and is avoided whenever possible. An explosion generates in solids both compressional waves in which the motion is in the direction of travel, and shear waves in which the motion is perpendicular to the wave travel direction. In fluids, only compressional waves (that is, sound) travel. Exploration seismology uses compressional waves alone. These are reflected by interfaces separating rocks with different impedances. (Impedance is the product of rock density and compressional wave velocity.) The amplitude of a reflected wave is proportional to the difference of the impedances across an interface.

Reflection coefficients of 0.01 to 0.1 are common for rocks but higher and lower values occur, the unity coefficient of the air-water interface being particularly important in seismic exploration of water-covered areas.

Seismic signals. Detectors for land exploration, called geophones, measure the vertical component of particle velocity, that is, the time derivative of the vertical displacement of the surface. For marine exploration, hydrophones replace geophones. Hydrophones respond to the excess pressure generated by sound waves in the water. In land or marine work, signals from the detectors are in the range of a few microvolts to several millivolts and must be amplified by circuitry that has gains up to several million. Frequencies of interest for petroleum exploration are 10–100 Hz, and the usual signals are pulses that include a fair portion of the frequency range. Twenty-four detector stations are standard but more are used.

In Fig. 1*b* the recorded signals appear as pulses with positive and negative excursions corresponding to positive and negative excursions of the surface from its undisturbed position. The farther a geophone is from the source, the longer the time required for a reflection from an interface to reach it. For reflectors that are plane interfaces, a plot of arrival time against distance along the surface from the source is hyperbolic. If the reflector is a horizontal plane, the time-distance equation takes the simple form shown below, where t is the arrival

$$t^2 = \left(\frac{2D}{v}\right)^2 + \left(\frac{X}{v}\right)^2$$

time of the reflection, X is the source to detector separation, v is the average velocity with which the wave travels to and from the reflector, and D is the corresponding depth. This is the fundamental equation of exploration seismology. It is only an approximation, for there is no consideration of the bending of rays at each interface in accordance with Snell's law of refraction, or of the fact that a reflector is sometimes tilted. Nonetheless, examination of the equation can expose much of important mathematics of the method.

Geophone and hydrophone arrays. Seismic surveys are conducted along lines called profiles. In terms of Fig. 1*a*, another hole would be drilled beyond the most distant geophone, and the geophones would be moved so that the picture would look identical but moved along the profile line. The process would be repeated again and again until the entire line had been surveyed. Source and geophone configurations may be either symmetric (split-spread) or, as shown here, nonsymmetric (single-ended spread). Whatever the arrangement, reflection points along an interface are discrete but without major gaps. This is referred to as continuous subsurface coverage. Nearest and most distant reflection points are separated by half the distance between corresponding geophones, just as a mirror required for full-length viewing needs to be only half the viewer's height. To get an areal picture of each reflector, a geophysicist uses data from a grid of profile lines.

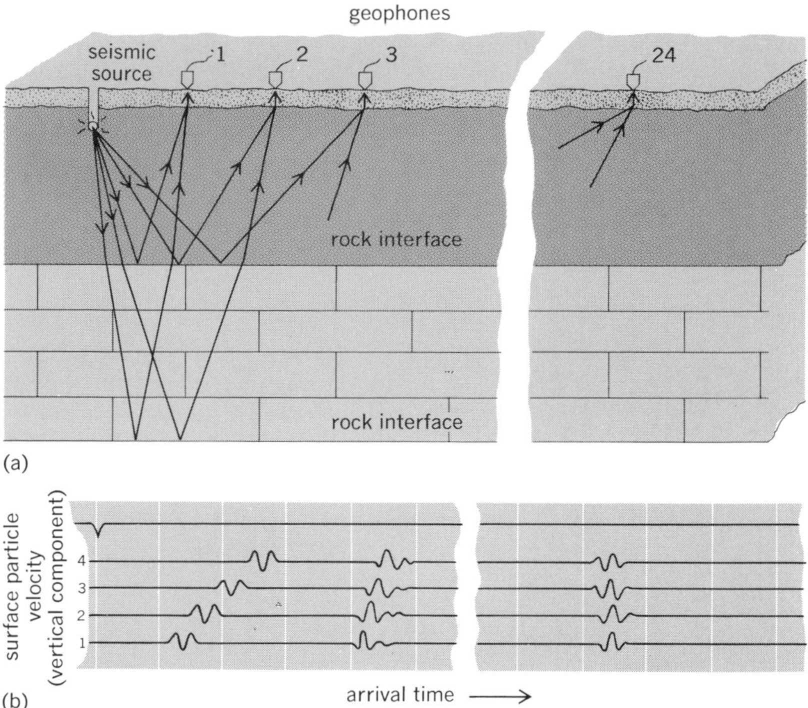

Fig. 1. The basic steps in the reflection seismic method. (*a*) Seismic waves from the source reflect off the rock interfaces and return to the detectors. (*b*) The signals are recorded for each detector.

Figure 1 represents World War II technology. Advances have included surface sources of impactor or vibrator types for land exploration and of gases exploded in elastic sleeves or air bubbles released into water for marine work, the replacement of single geophones by linear or areal arrays of many geophones, and particularly the recording of seismic data on magnetic tape. Surface sources have become popular on land. They permit rapid, relatively inexpensive seismic surveying, although generally with reduced resolution. Replacement of high explosives by other sources in offshore surveys has resulted in operational economies and reduces hazards to marine life and the men involved.

A single ship, moving at 4–6 knots (2–3 m/s), carries sources activated at 10–30-sec intervals and pulls a cable containing 24 hydrophone stations spaced out to 10,000 ft (3 km) from the ship. Shore-based electronic equipment locates the ship and defines the profile lines to be followed, but the location methods fail far from shore. Combinations of several methods, including satellite fixes, are expected soon to locate geophysical ships anywhere in the world with an accuracy of, perhaps, 600 ft (180 m). A typical marine seismic crew averages 50 mi (80 km) of a line a day.

Each of the 24 stations in a marine cable has as its output the summed signals of 10–40 hydrophones arranged as a linear array. Arrays discriminate against energy traveling along the detector cable and favor reflections that reach the cable from below. On land, arrays are formed by placing geophones along a line or spreading them out in an areal pattern. The purpose is the same: to emphasize the reflections which arrive very nearly vertically from below. Arrays of up to 150 geophones have been used, but arrays of 12–36 geophones are more common. They are designed to reduce noise generated by the source and traveling mostly in the weathered layer. The strongest noise of this type is called ground roll by the geophysicist. In addition to limiting ground roll, geophone arrays also reduce random noise by a process of statistical addition of the noise.

Recording and processing data. Recording seismic data in analog form on magnetic tape revolutionized exploration seismology in the 1950s; the advent of recording in digital form has produced a second revolution. Array outputs are sampled every 2 or 4 milliseconds and the samples are recorded as digital numbers for a period of 6–10 sec. One complete record is called a seismogram; a seismogram consists of 24 traces, one from each array output. Processing of the magnetic tapes occurs in centers that include one or more large digital computers and the associated programmers, inputters, and data interpreters.

The first processing step is to compensate for the drop in signal amplitude with record time. Spreading of the wave from the source and attenuation within the earth result in a rapid decrease of signal strength that must be reversed during processing. Next is compensation for the varying thickness of weathered material underlying different geophone stations and for variations in source depths. Then a normal moveout (NMO) correction is applied. This correction flattens the hyperbolic time-distance plots by subtracting from the arrival time the quadratic X term of the equation. Since the velocity v changes from reflector to reflector, usually increasing as depth increases, the NMO correction changes with time along the seismogram. After application of a NMO correction, each seismic trace is the one that would have been produced by a coincident source and geophone. The effect of detectors at different distances from the source has been eliminated. Accompanying the NMO process are procedures for determining v from the curvature of the uncorrected time-distance hyperbolas.

Concurrently with NMO correction or preceding it, the data are filtered to emphasize reflections and discriminate against noise. The usual filters pass that band of frequencies over which reflections are strongest and eliminate frequencies for which noise predominates. Other types of filters find use also. These include filters that remove from marine seismograms signals generated by energy bouncing between the surface and bottom of the water. Another important filtering process is common depth point stacking (Fig. 2). Three to twenty-four signals from the same reflection points but with different source to detector separations are, after NMO correction, summed into a single signal. Common depth point stacking both reduces noise and tends to cancel multiple reflections. These reflections, energy that has been reflected more than once from an interface, interfere with the reflections used for subsurface mapping.

The product of processing is a record section. Each trace is displayed on photographic paper as a constant-width strip with the signal amplitude shown in variable intensity, variable area, wiggly trace form, or some combination of these. Laying the traces side by side without a gap between the strips generates a two-dimensional representation of a three-dimensional surface. The abscissa is distance along the profile line from an arbitrary starting point; the ordinate, record time. Amplitude is the dependent variable. To replace record time by depth below the surface requires that depth be computed from a knowledge of v and reflection times. Fully processed record sections present to the geophysicist the configuration of rock layers down to all depths that drills can reach, albeit with resolution no better than 150 or 200 ft (45 or 60 m). Lines many miles long may be viewed.

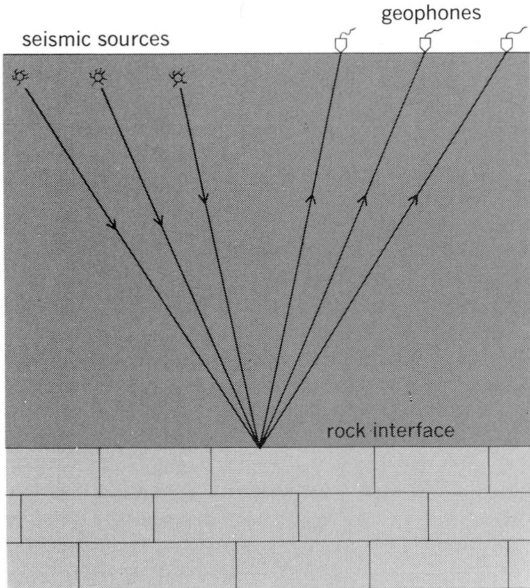

Fig. 2. Seismic signal filtering by the method of common depth point stacking.

Although a record section resembles the subsurface, it is really a picture drawn by elastic waves and may differ from the subsurface in detail. For example, rock layers that end abruptly often seem to continue, as diffracted energy extends the reflections. Reflections from interfaces that are not horizontal appear at positions on a record section that do not correspond to their locations in the earth and must be returned to their true position by a process called migration. When all these difficulties and others not listed here have been admitted, the position of reflection seismology as the most valuable tool of petroleum geophysicists is indisputable. No other method approaches the combination of penetration and resolution that is required for a choice of drilling sites.

[FRANKLIN K. LEVIN]

Bibliography: M. B. Dobrin, *Introduction to Geophysical Prospecting*, 1960; F. S. Grant and G. F. West, *Interpretation Theory in Applied Geophysics*, 1965; D. S. Parasnis, *Principles of Applied Geophysics*, 1962; J. E. White, *Seismic Waves: Radiation, Transmission, and Attenuation*, 1965; P. C. Wuenschel et al., Geophysical research and progress in exploration, 1965–1968, *Geophysics*, April, 1969.

Ship propulsion reactor

A nuclear reactor used to supply motive power to a ship. Nuclear reactors for shipboard propulsion can, in theory, be of any type used for the production of useful power. For basic information applicable to all types of fission reactors *see* NUCLEAR REACTOR; REACTOR PHYSICS.

In all the shipboard nuclear power plants that have been built, energy conversion is based on the steam-turbine cycle, and that portion of the plant is more or less conventional. Gas-turbine applications are also possible, and these have been studied.

Shipboard problems. There are four radical differences between shipboard reactors and similar installations ashore, involving (1) problems of weight and space limitations, (2) problems of plant reliability and on-board maintenance, (3) problems in plant safety, and (4) problems inherent in location on a moving platform.

Weight and space. The actual size of the shipboard reactor itself does not present a problem, but the size and configuration of the shielded volume around the entire cooling system are of considerable significance. Figure 1 shows a simplified schematic diagram of a typical marine nuclear propulsion plant. Both the primary shield around the pressure vessel containing the nuclear reactor itself and the secondary shield located around all radioactive components of the cooling system are shown. For a discussion of the theory and application of reactor shielding *see* RADIATION SHIELDING.

Weight considerations usually result in a compact layout to reduce the size and hence the weight of the secondary shield. However, the conflicting requirements for access and maintenance result in a balance being required such that the permeability (ratio of free volume of space to total volume of space) of the portion of a nuclear power plant within the secondary shield usually runs from 65 to 70% as opposed to approximately 80% permeability for normal shipboard machinery spaces exclu-

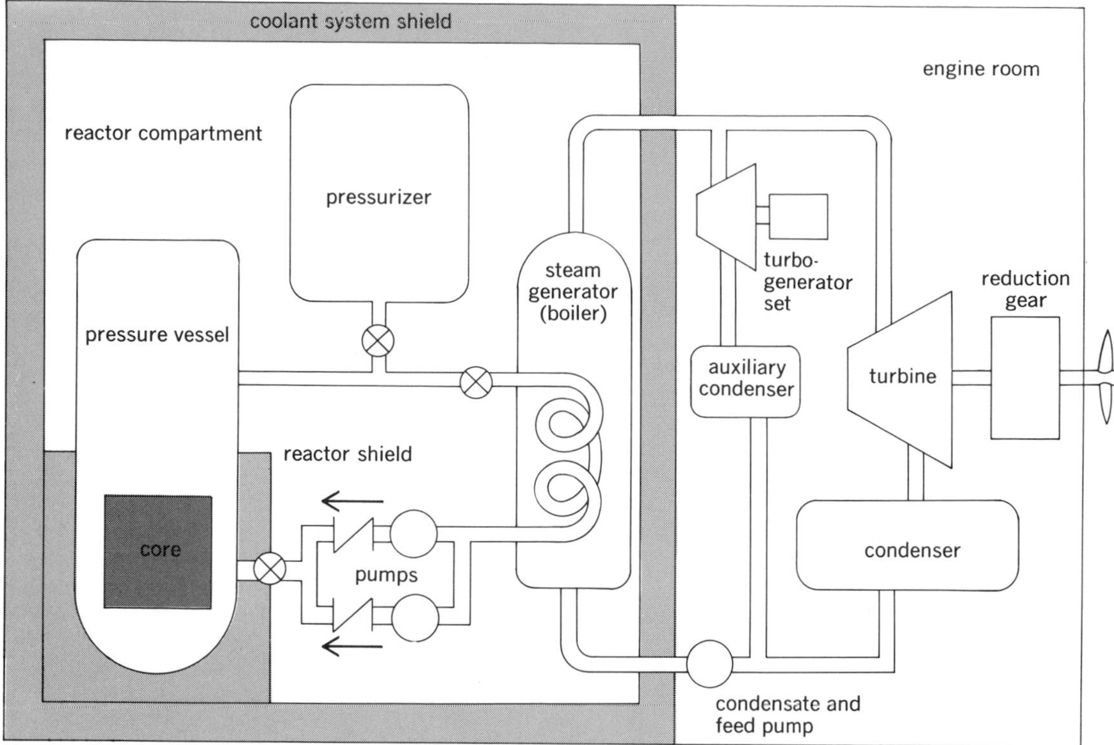

Fig. 1. Schematic diagram of a typical pressurized nuclear propulsion plant (relative sizes of the various components are not indicated). (*Society of Naval Architects and Marine Engineers*)

sive of uptakes, air vents, or tonnage openings.

It will be difficult to reduce the weight of present-day shields for pressurized water plants because the principal γ-ray attenuation results from the use of heavy materials.

Reliability and maintenance. Nuclear power confers on any vessel to which it is applied an exceptionally long time interval between refuelings. Therefore, the importance of plant reliability and the ability to perform reasonable maintenance functions on board ship assume increased importance as compared to conventional shipboard power plants. In particular, planning for access to plant components and planning for necessary maintenance functions to be performed on these plant components is essential. Many of these components will transfer radioactive fluids, and therefore special provisions must be made either in the form of installed standby equipment or by the provisions of cleaning and servicing equipment.

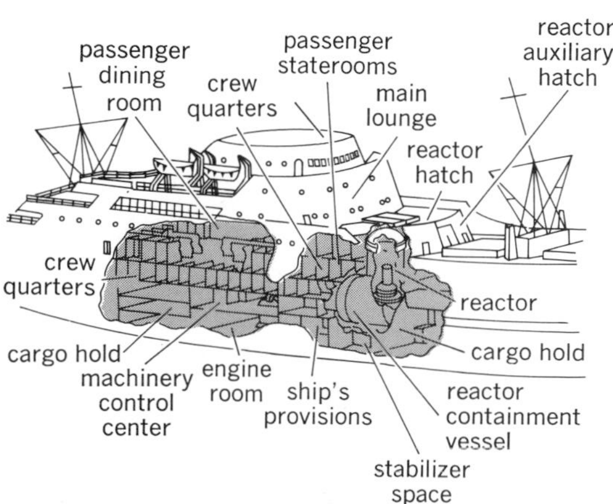

Fig. 2. NS *Savannah,* showing relationship of reactor to ship's spaces. (*From Nucleonics, vol. 16, no. 9, 1958*)

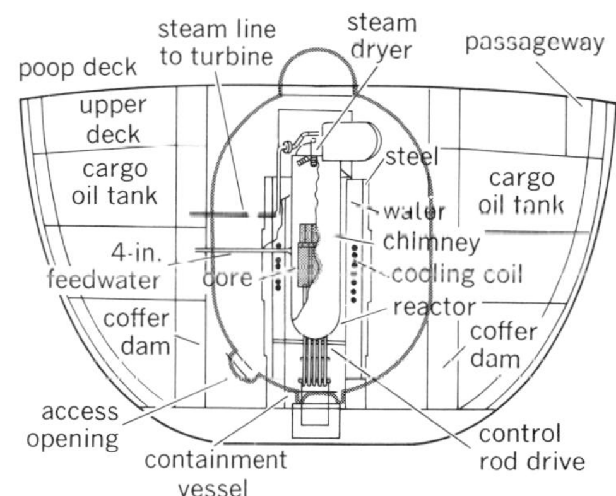

Fig. 3. Typical arrangement of boiling-water reactor machinery in containment vessel. 4 in. = 10 cm. (*Society of Naval Architects and Marine Engineers*)

Plant safety. A nuclear reactor on board ship is considerably more subject to external hazards, with resulting possibility of nuclear accident, than is a similar plant located ashore. Special precautions in the form of collision damage protection, secondary containment, and special safety devices are necessary. Figures 2 and 3 show how the vital parts are protected from collision damage to the greatest extent possible by careful location. The principal hazard from a nuclear accident results from the spreading of radioactivity. Therefore, any reactor system which has the fission-product activ-

Fig. 4. Two views of the USS *Patrick Henry*, a nuclear-powered Polaris-missile submarine. This vessel, launched Sept. 18, 1959, is equipped with a pressurized, water reactor. (*Official U.S. Navy photographs*)

ity held in a reasonably permanent form inside solid fuel elements and which has a coolant of low long-term activity possesses an advantage from a safety standpoint.

Another important problem affecting safety is the characteristic of a nuclear reactor that, even when shut down, it continues to evolve heat at a low rate from residual radioactivity. Adequate means must be provided for disposing of this radiation during periods of inoperation and even in cases where the vessel is sunk.

Seakeeping. Any nuclear power reactor for shipboard propulsion and its associated equipment must be designed in accordance with all the accepted basic principles of marine engineering. If large free surfaces are present, special precautions must be taken to protect them. If fluids that are rare and difficult to obtain are used in the reactor system (either as coolants or as auxiliary fluids), provision must be made either for their generation on board or for adequate storage of a reserve supply. If high-melting-point materials are involved, preheating systems with adequate energy sources must be provided. Exceptionally reliable auxiliary power sources are required because of the safety and control problems involved.

Reactor types. Only two types of reactors—the pressurized-water reactor and the sodium-cooled reactor—have actually been applied to operating vessels. The pressurized-water plant has been the favorite because it can be more easily shielded and maintained in working order. Other nuclear power systems use boiling water, organic coolants, and gas coolants. These systems, however, all have drawbacks for shipboard application and were not used on the earlier atomic-powered vessels (1955–1959). Even the sodium-cooled reactor originally installed on the submarine USS *Seawolf* has been replaced by a pressurized-water reactor (Fig. 4).

All the principal maritime nations are studying the application of nuclear power to naval and commercial ships. The United States, Great Britain, and the Soviet Union were the first nations actually to construct nuclear vessels.

The U.S. Navy has in operation and under construction a large number of nuclear-powered submarines and surface vessels.

The first nonnaval marine installations of nuclear power were on the Soviet icebreaker *Lenin* (Fig. 5) and on the United States demonstration merchant ship NS *Savannah*, both of which used pressurized-water reactors. The *Lenin* was launched on

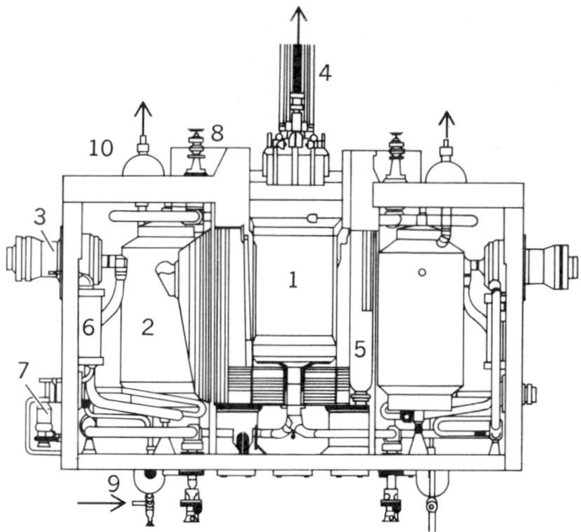

Fig. 5. General layout of a reactor unit in *Lenin*: 1, reactor; 2, steam generator; 3, main circulating pump; 4, control-rod mechanism; 5, filter; 6, cooler; 7, secondary circuit pump; 8, primary steam valve; 9, feedwater inlet; 10, steam outlet. (*From Lenin, The Russian icebreaker, Nucl. Eng., 3(31):432–433, 1958*)

Dec. 5, 1957. The *Savannah* was launched on July 21, 1959 and was retired in the early 1970s. The *Otto Hahn*, a nuclear-powered demonstration ore carrier built in West Germany, first underwent sea trials in 1969; its reactor is a modified pressurized-water reactor that operates at saturation pressure with a free water surface as in a boiling water reactor. The *Mutsu*, a small pressurized-water-reactor-powered demonstration cargo training ship built in Japan, was completed in 1974. Due to minor technical problems and major political problems, the ship has never become operational. The *Arktika*, a pressurized-water-reactor-powered icebreaker, was completed in the Soviet Union in 1975. Studies of other merchant nuclear plants have continued throughout the world, but none has reached the construction stage. The Babcock and Wilcox Consolidated Nuclear Steam Generator (CNSG) design has undergone considerable development in the United States with Maritime Administration support, but the economic climate has not been favorable for the construction of any further nuclear merchant vessels.

[L. H. RODDIS, JR.; MICHAEL G. PARSONS]

Bibliography: J. W. Landis, The power plant for the first nuclear merchant ship, NS *Savannah*, *J. Amer. Nav. Eng.*, 70(4):629–641, 1958; L. H. Roddis, Jr., and J. W. Simpson, The nuclear propulsion plant of the USS *Nautilus*, SSN-571, *Trans. Soc. Nav. Architects Mar. Eng.*, 62:491–521, 1954; G. L. West, Jr., and E. J. Roland, Some aspects of nuclear safety and ship design, *Trans. Soc. Nav. Architects Mar. Eng.*, 75: 385–419, 1967.

Solar cell

A semiconductor electrical junction device which absorbs and converts the radiant energy of sunlight directly and efficiently into electrical energy. Solar cells may be used individually as light detectors, for example in cameras, or connected in series and parallel to obtain the required values of current and voltage for electric power generation.

Most solar cells are made from single-crystal silicon and have been very expensive for generating electricity, but have found application in space satellites and remote areas where low-cost conventional power sources have been unavailable. Research has emphasized lowering solar cell cost by improving performance and by reducing materials and manufacturing costs. One approach is to use optical concentrators such as mirrors or Fresnel lenses to focus the sunlight onto smaller-area solar cells. Other approaches replace the high-cost single-crystal silicon with thin films of amorphous or polycrystalline silicon, gallium arsenide, cadmium sulfide, or other compounds.

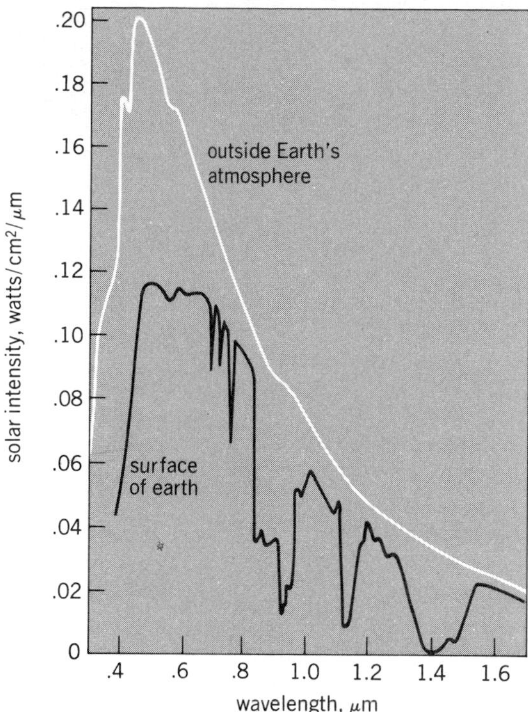

Fig. 1. Variation of solar intensity with wavelength of photons for sunlight outside the Earth's atmosphere and for a typical spectrum on the surface of the Earth.

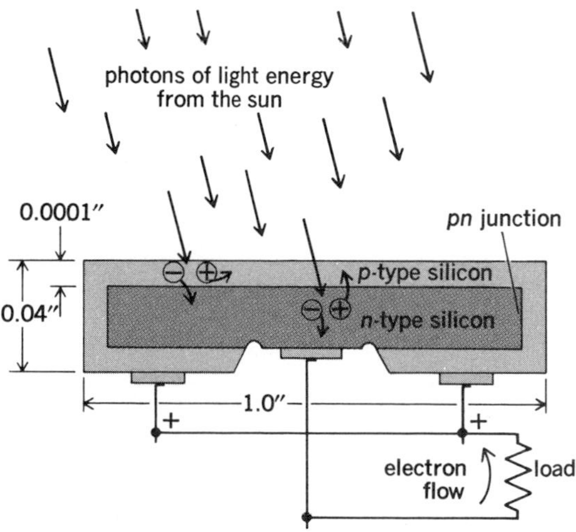

Fig. 2. Cross-sectional view of a silicon *pn* junction solar cell, illustrating the creation of electron pairs by photons of light energy from the Sun. 0.0001″ = 2.5 μm; 0.04″ = 1 mm; 1.0″ = 25 mm.

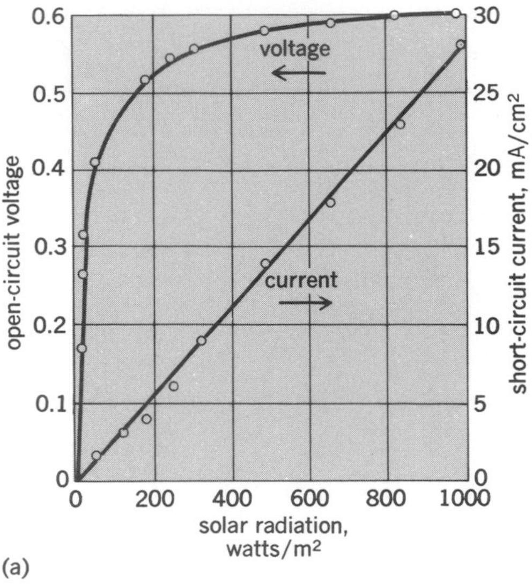

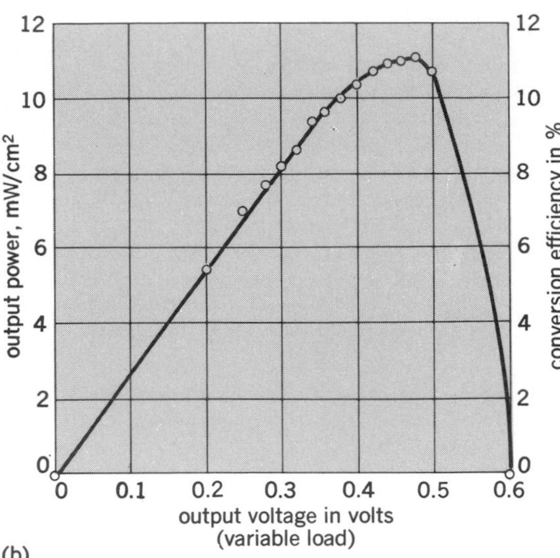

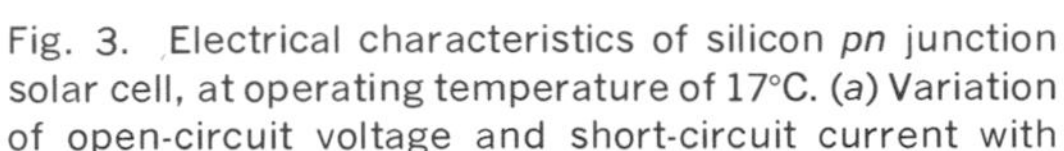

Fig. 3. Electrical characteristics of silicon *pn* junction solar cell, at operating temperature of 17°C. (*a*) Variation of open-circuit voltage and short-circuit current with light intensity. (*b*) Variation in power output as load is varied from short to open circuit.

Solar radiation. The intensity and quality of sunlight are dramatically different outside the Earth's atmosphere from that on the surface of the Earth, as shown in Fig. 1. The number of photons at each energy is reduced upon entering the Earth's atmosphere due to reflection, to scattering, or to absorption by water vapor and other gases. Thus, while the solar energy at normal incidence outside the Earth's atmosphere is 1.36 kW/m² (the solar constant), on the surface of the Earth at noontime on a clear day the intensity is about 1 kW/m².

On clear days the direct radiation is about 10 times greater than the diffuse radiation, but on overcast days the sunshine is entirely diffuse. The mean annual solar energy falling on the Earth's surface varies greatly from one location to another. The sunniest regions of the globe receive about 2500 kWh/m² per year of total sunshine on a horizontal surface. The Earth receives about 10^{18} kWh of solar energy each year. The worldwide annual energy consumption is about 80×10^{12} kWh, so that from a purely technical viewpoint, the world

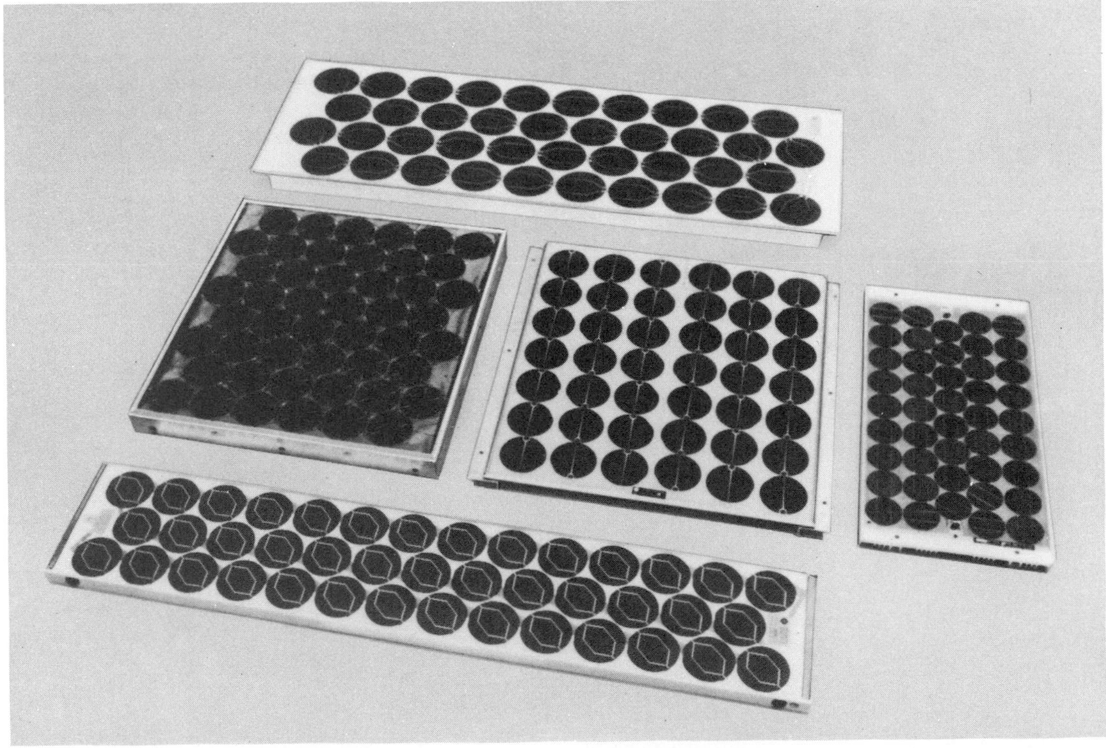

Fig. 4. Silicon solar cells assembled into panels and arrays to obtain higher voltage and power output. Peak power of modules is at 60°C and 15.8 V. (*U.S. Department of Energy*)

Fig. 5. A 3-kW solar cell–powered communication link in southern California. (*Spectrolab, Inc., Sylmar, CA*)

energy consumption corresponds to the sunlight received on about 0.008% of the surface of the Earth. *See* SOLAR RADIATION.

Principles of operation. The conversion of sunlight into electrical energy in a solar cell involves three major processes: absorption of the sunlight in the semiconductor material; generation and separation of free positive and negative charges to different regions of the solar cell, creating a voltage in the solar cell; and transfer of these separated charges through electrical terminals to the outside application in the form of electric current.

In the first step the absorption of sunlight by a solar cell depends on the intensity and quality of the sunlight, the amount of light reflected from the front surface of the solar cell, the semiconductor bandgap energy which is the minimum light (photon) energy the material absorbs, and the layer thickness. Some materials such as silicon require tens of micrometers' thickness to absorb most of the sunlight, while others such as gallium arsenide, cadmium telluride, and copper sulfide require only a few micrometers.

When light is absorbed in the semiconductor, a negatively charged electron and positively charged hole are created. The heart of the solar cell is the electrical junction which separates these electrons and holes from one another after they are created by the light. An electrical junction may be formed by the contact of: a metal to a semiconductor (this junction is called a Schottky barrier); a liquid to a semiconductor to form a photoelectrochemical cell; or two semiconductor regions (called a *pn* junction).

The fundamental principles of the electrical junction can be illustrated with the silicon *pn* junction. Pure silicon to which a trace amount of a fifth-column element such as phosphorus has been added is an *n*-type semiconductor, where electric current is carried by free electrons. Each phosphorus atom contributes one free electron, leaving behind the phosphorus atom bound to the crystal structure with a unit positive charge. Similarly, pure silicon to which a trace amount of a column-three element such as boron has been added is a *p*-type semiconductor, where the electric current is carried by free holes. Each boron atom contributes one hole, leaving behind the boron atom with a unit negative charge. The interface between the *p*- and *n*-type silicon is called the *pn* junction. The fixed charges at the interface due to the bound boron and phosphorus atoms create a permanent dipole charge layer with a high electric field. When photons of light energy from the Sun produce electron-hole pairs near the junction, the build-in electric field forces the holes to the *p* side and the electrons to the *n* side (Fig. 2). This displacement of free charges results in a voltage difference between the two regions of the crystal, the *p* region being plus and the *n* region minus. When a load is connected at the terminals, an electron current flows in the direction shown by the arrow, and useful electrical power is available at the load.

Characteristics. The electrical characteristics of a typical silicon *pn*-junction solar cell are shown in Fig. 3. Figure 3*a* shows open-circuit voltage and short-circuit current as a function of light intensity from total darkness to full sunlight (1000 W/m^2). The short-circuit current is directly proportional to light intensity and amounts to 28 mA/cm^2 at full sunlight. The open-circuit voltage rises sharply under weak light and saturates at about 0.6 V for radiation between 200 and 1000 W/m^2. The variation in power output from the solar cell irradiated by full sunlight as its load is varied from short circuit to open circuit is shown in Fig 3*b*. The maximum power output is about 11 mW/cm^2 at an output voltage of 0.45 V.

Under these operating conditions the overall conversion efficiency from solar to electrical energy is 11%. The output power as well as the output current is of course proportional to the irradiated surface area, whereas the output voltage can be increased by connecting cells in series just as in an

Fig. 6. A 25-kW photovoltaic array which is operating in an agricultural irrigation experiment located at Mead, NE. (*U.S. Department of Energy and MIT Lincoln Laboratory*)

ordinary chemical storage battery. Experimental samples of silicon solar cells have been produced which operate at efficiencies up to 18%, but commercial cell efficiency is around 10–12% under normal operating conditions.

Using optical concentration to intensify the light incident on the solar cell, efficiencies above 20% have been achieved with silicon cells and above 25% with gallium arsenide cells. The concept of splitting the solar spectrum and illuminating two optimized solar cells of different bandgaps has been used to achieve efficiencies above 28%, with expected efficiencies of 35%. Thin-film solar cells have achieved between 4 and 9% efficiency and are expected in low-cost arrays to be above 10%.

[KIM W. MITCHELL]

Arrays. Individual silicon solar cells are limited in size to about 40 cm² of surface area. At a 15% conversion efficiency, such a cell can deliver about 0.6 W at 0.5 V when in full sunlight. To obtain higher power and higher voltage, a number of cells must be assembled in panels or arrays (Fig. 4). Cells may be connected in series to multiply their output voltage and in parallel to multiply their output current. Cells operated in series must be closely matched in short-circuit current since the overall performance of a solar cell array is limited by the cells having the lowest current.

Applications. Although the photovoltaic effect was discovered by A.C. Becquerel in 1839, practical solar cells made of silicon crystals were not developed until 1955. Beginning with *Vanguard 1*, launched in 1958, silicon solar cell arrays have become the almost exclusive power source for satellites. *Skylab*, launched in 1973, had a 20-kW solar cell array, the most powerful to be used in space so far (early 1980).

Terrestrial applications of solar cells have increased rapidly since 1970. Solar cell arrays have been used primarily to power small remote electrical loads that would otherwise be impractical or uneconomical to power by conventional means such as storage batteries or motor-generator sets. In 1979 a total of approximately 1 MW of solar cell arrays were sold worldwide to power such equipment as remote radio repeaters, navigational aids, consumer products, railroad signals, cathodic protection devices, and water pumps. Figure 5 shows a remote solar cell–powered communication link installed in 1976. Since most of the aforementioned uses require power to the load at times even when the Sun is not shining, electrical storage batteries are typically used in conjunction with solar cell arrays to provide reliable, continuous power availability.

Although extended terrestrial uses await cheaper solar arrays, a number of experiments have been undertaken to explore the use of solar cell arrays in larger agricultural, residential, commercial, and industrial applications (Fig. 6). When powering loads which require ac voltage, a static inverter is used to convert the dc voltage from the solar cell array into usable ac power.

Future prospects. The growing worldwide demand for energy, its increasing cost, and the depletion of nonrenewable energy reserves make solar cell power systems an attractive alternative for supplying electricity for a wide range of uses. Despite significant progress in this technology, the cost of solar cells must be significantly reduced before they can economically supply a substantial amount of electricity. Research has been undertaken to develop new approaches lowering the cost of solar cell materials and manufacturing processes. *See* SOLAR ENERGY.

[DONALD G. SCHUELER]

Bibliography: J. A. Merrigan, *Sunlight to Electricity: Prospects for Solar Energy Conversion by Photovoltaics*, 1975; W. Palz, *Solar Electricity: An Economic Approach to Solar Energy*, 1978; D. L. Pulfrey, *Photovoltaic Power Generation*, 1978.

Solar energy

The energy transmitted from the Sun. This energy is in the form of electromagnetic radiation. The Earth receives about one-half of one-billionth of the total solar energy output. In 1971, based on radiation measurements in space, the National

Aeronautics and Space Administration proposed a new space solar constant of 1353 watts per square meter (W/m^2), and a standard spectral irradiance in W/m^2 over a small range of wavelengths (bandwidth) centered at the wavelength (in millionths of meters, or μm) shown in Fig. 1. Accordingly, the solar radiation energy in the ultraviolet is 105.8 W/m^2 (7.82% of the solar constant), in the visible 640.4 W/m^2 (47.33%), and in the infrared 606.8 W/m^2 (44.85%). The solar radiation energy output is essentially constant. However, because of the ellipticity of the Earth's orbit, the solar constant varies between 1398 W/m^2 at the winter solstice and 1308 W/m^2 at the summer solstice, or 3% about the mean value. Based on its cross-sectional area, the rotating Earth receives therefore 751.10^{15} kWh annually.

The following are the most frequently used metric units for the radiative input area: langley ($1\ l = 1\ cal/cm^2 = 0.001163\ Wh/cm^2 = 4.186\ J/cm^2$); calorie ($1\ cal/cm^2\ min = 0.0697\ W/cm^2 = 1\ l/min = 0.00418\ J/cm^2\ min$); kilowatt-hour ($1\ kWh/m^2 = 860{,}000\ cal/m^2 = 86.2\ l = 3.6 \cdot 10^6\ J/m^2$); joule ($1J = 0.239\ cal = 2.78 \cdot 10^{-4}\ Wh = 0.239\ l\text{-}cm^2$); and terawatt-year ($1\ TW\text{-}yr = 10^9\ kW\text{-}yr = 8.766 \times 10^{12}\ kWh$).

Actually, passage through the atmosphere splits the radiation reaching the surface into a direct and a diffuse component, and reduces the total energy through selective absorption by dry air molecules, dust, water molecules, and thin cloud layers, while heavy cloud coverage eliminates all but the diffuse radiation. Figure 2 specifies the conditions for a surface perpendicular under a clear sky at mid- and low latitudes within 4 h on either side of high noon. If these conditions were to prevail for 12 h each day of the year (4383 h), the energy received would lie between some 4200 and 5200 kWh/m^2 yr. Actually, the number of sunshine hours even in high-insolation areas ranges from 78 to 89% of the possible, resulting in a reduction in radiative energy ("solar crude") received to the values shown in Fig. 2. Since atmospheric absorption and scattering increase strongly at low solar elevation, the average solar crude received in the most favorable areas by a horizontal surface is about 2550 kWh/m^2 yr or 2.55 terawatt-hours per square kilometer per year (TWh/km^2 yr). Figure 3 shows the global distribution of the average solar radiation energy incident on a horizontal surface. By following the Sun's diurnal and seasonal motion, thus facing it from near-sunrise to sunset, an instrument such as a heliostat can attain values between 3 and 3.5 TWh/km^2 yr.

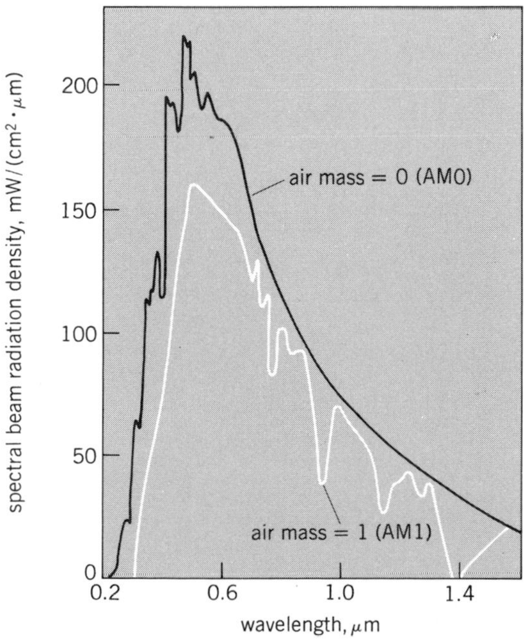

Fig. 1. Spectral distribution of sunlight in space (AMO) and on surface of Earth with the Sun at the zenith (AM1).

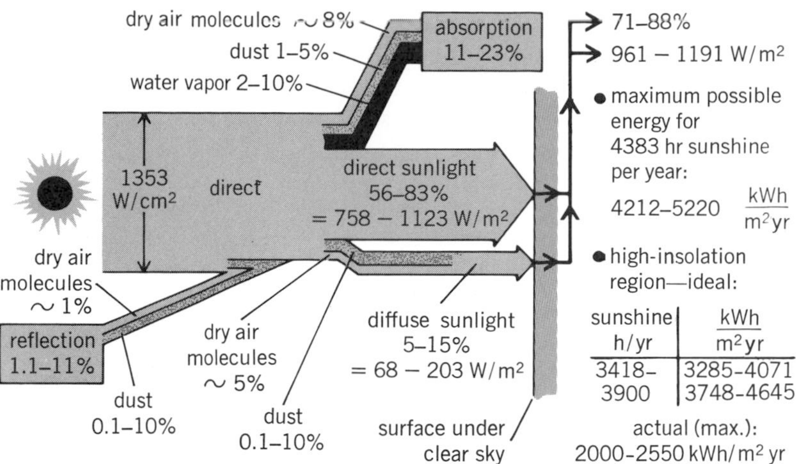

Fig. 2. Sunlight penetration of atmosphere (clear sky).

On a global basis, about 50% of the total incident radiation of 751.10^{15} kWh/yr is reflected back into space by clouds, 15% by the surface, and about 5.3% is absorbed by bare soil. Of the remaining 29.7%, only about 1.7% ($3.79 \cdot 10^{15}$ kWh/yr) is absorbed by marine vegetation and 0.2% ($4.46 \cdot 10^{14}$ kWh/yr) by land vegetation. By far the largest portion is used to evaporate water and lift it into the atmosphere. The evaporation energy is radiated into space by vapor condensation to clouds. The solar energy spent to lift the water can be partly recovered in the form of waterpower (hydraulic energy). Solar energy can be utilized in the form of heat, organic chemical energy through photosynthesis, and wind power, and also in the form of photovoltaic power (generating electricity by means of solar cells). The two greatest problems in utilizing solar energy are its low concentration and its irregular availability due to the diurnal cycle and to seasonal and climatic variations. Improved technology and investment capital are also needed.

In 1974, the U.S. Congress established the Energy Research and Development Agency (ERDA) and charged it with the development of energy conservation techniques and new technologies for extracting less readily accessible fossil energy reserves (such as shale oil) and for broadening the use of coal (coal gasification and liquefaction); with the advancement of nuclear (fission) technology; and with spearheading new energy technologies (geothermal, fusion, and all forms of solar energy). The basic energy systems associated with solar power generation are surveyed in Fig. 4. This energy is used essentially in three forms: light and power, high-temperature process heat, and low-temperature heat. The distribution of users in each category is different (Fig. 5).

SOLAR HEATING AND COOLING

The simplest way of utilizing solar energy is in the form of low-temperature heat. Therefore, the first large-scale solar-technical application is to residential and commerical buildings by means of passive systems. *See* SOLAR HEATING AND COOLING.

Passive solar-thermal systems. In the order of increasing technological complexity, a passive solar-thermal system (PSTS) can be used to heat water, to heat a building, and to heat and cool a building.

Water heating. The simplest form of water heater is shown in Fig. 6*a*. The water is heated in a solar collector and circulated by natural convection through a storage tank which, in this case, must be positioned above the collector, to force the water through the heater. The forced circulation water heater (Fig. 6*b*) is more flexible, but requires a pump, a flow check valve (controls), and a temperature sensor system (T) to operate the pump based on temperature difference between tank bottom and collector outlet. A forced circulation system can also use closed-cycle operation (Fig. 6*c*), with antifreeze as working fluid in climates where freezing occurs.

Interior space heating. The interiors of buildings can be heated by air (Fig. 7*a*) or by water (Fig. 7*b*). Although the basic technology for solar heating exists, standardized designs and quantity production, to reduce costs, are not yet available and may never be. The reasons include differences in climate, in maintenance problems, and the fact that cities are not conceived as "solar cities" (hence, problems with shadowing by buildings or trees and associated novel legal problems). Area-intensive city cores will probably use community solar heating systems (for example, servicing a shopping center or a block of apartment houses), but these are quite different from residential house systems and probably are even less readily standardized.

Cooling. Solar cooling requires the most complex arrangement (Fig. 8) and therefore is still less developed for the general market than house or water heating. The technique is the same as in refrigerators: high-grade energy (electricity, fuel) is used to remove low-grade energy from a relatively low-temperature space, thereby cooling it down further. Solar energy can replace conventional high-grade energy sources. Of the two main applications—refrigeration of food and cooling of buildings—the former will initially be of greater significance, because of the importance of food preservation in developing countries, many of which are better supplied with solar than with other high-grade energy sources.

The first solar cold (food) storage houses have become operational in the Middle East. A solar-heated fluid replaces the fuel used in gas-driven refrigerators which employ so-called absorption cooling, as distinct from electric compression cooling. The need for relatively large-scale electric energy generation is thus avoided. The heat extracted from the storage space, along with the solar heat, must be removed. While air cooling is possible, water cooling is more efficient. Thus, conditions are near-optimum for floating cold storage systems in sun-drenched tropical harbors (or offshore anchoring) for refrigerating fish or imported foods prior to distribution.

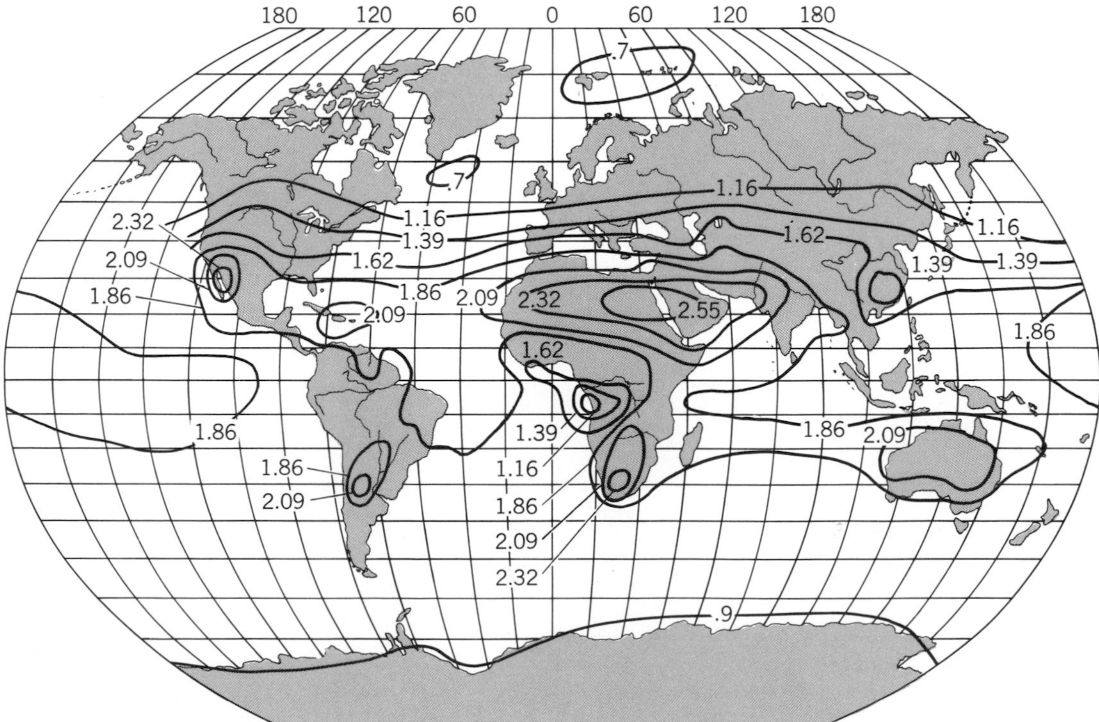

Fig. 3. Global distribution of the average annual solar radiation energy incident on a horizontal surface at the ground. The units are terawatt-hours per square kilometer year (TWh/km² yr).

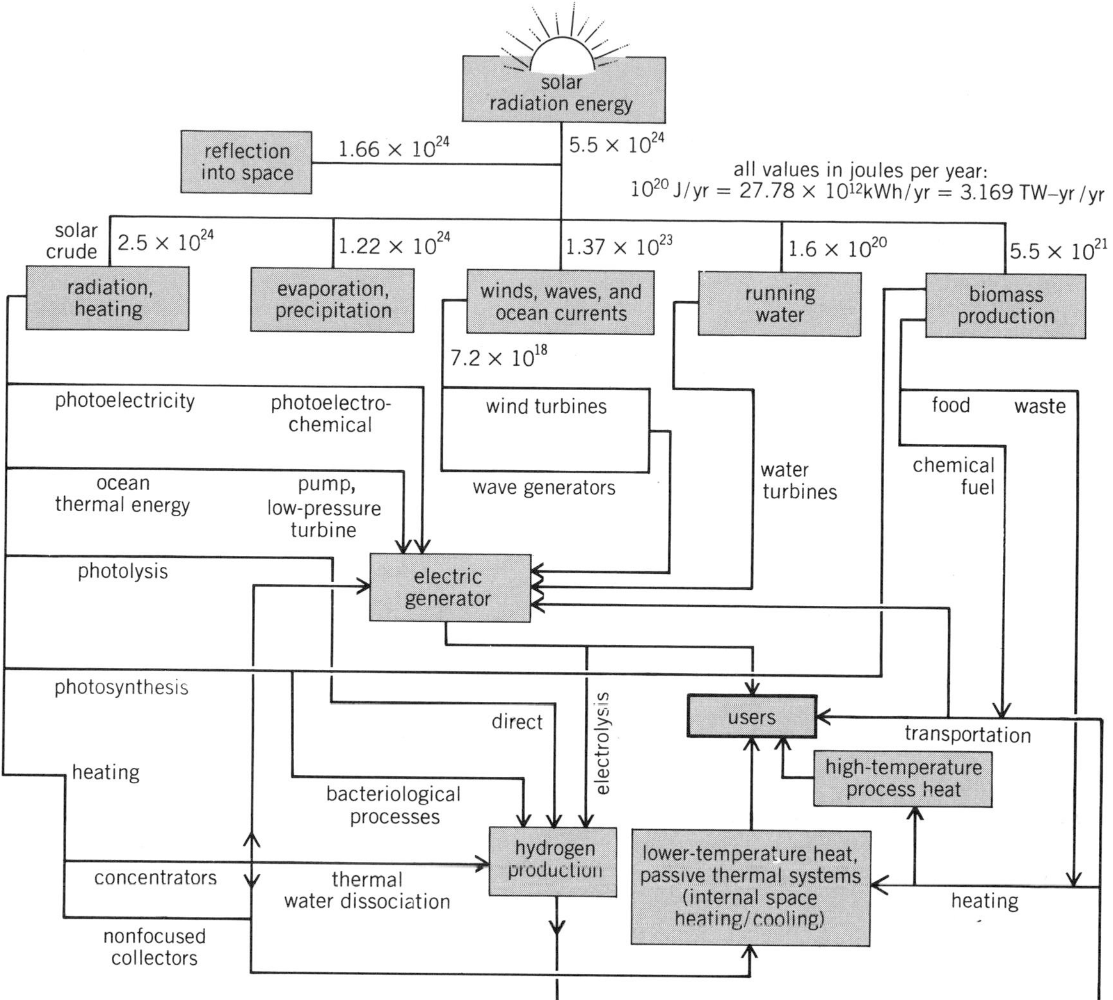

Fig. 4. Solar energy distribution and solar power generation systems.

Drying and distillation. In Japan, the United States, the southern Soviet Union, and other countries, solar energy is utilized for drying of fruits and vegetables by means of solar-heated air. Water evaporation in solar stills is an attractive method only where fresh water is extremely expensive, as in isolated arid regions with ready access to brackish or sea water, or under special conditions such as emergency provisions for downed flyers or astronauts, or for shipwrecked sailors. Chile, Greece, Australia, and Israel are operating and developing solar stills. Outputs may run as high as 0.15 gal of fresh water per square foot per day (6 liters per square meter per day). Generally, it is of the order of 0.1 to 0.12 gal per square foot per day (4 to 5 liters per square meter per day).

Nonfocusing collectors. The most suitable collector for the above applications is the simple flat plate collector (Fig. 9*a*), because it can be integrated most readily into the house structure, forming the roof, which improves the overall economics of the solar house system. The collector has the advantage of absorbing direct and diffuse sunlight. The covers minimize long-wave radiation losses from the heated black absorbing surface without themselves reflecting or absorbing too much of the incoming solar radiation. The insulation prevents conductive losses. Therewith a maximum of the incoming radiation energy is available to heat the fluid in the tubes.

Dust and chemical pollutants can be a major enemy of flat plate collectors. Chemical pollution can affect the choice of cover materials and mitigate against less expensive clear plastics as compared to the (usually) more expensive glass covers. An innovative approach is the construction of flexible heat exchanger/absorber mats (Fig. 9*b*) constructed of an elastomer ethylene-propylene-diene-monomer. The black absorber mat consists of parallel and continuous-circulation tubes, 0.73 in. (9 mm) apart, separated by strips of webbing that act as heat collection plates.

It is possible to utilize other collector configurations; for example, one in which the flat absorbing surface is replaced by fins between which the incoming solar rays are reflected repeatedly to virtually blackbody absorption, while the reradiation surface is minimized. This collector type is particularly useful for air heating, as air is forced between the closely spaced, intensely absorption-heated fins.

Focusing collectors. These collectors consist of concentrator, receiver, and absorber (usually part of the receiver). The purpose of the concentrators is to raise the operating temperature from between 90 to 150°C for nonfocused collectors to between

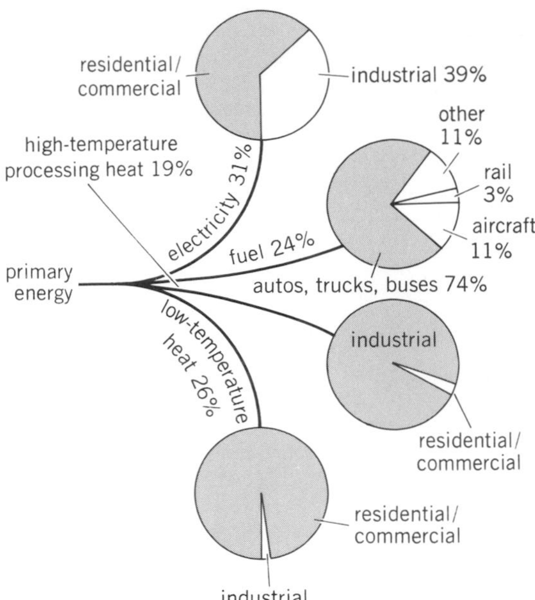

Fig. 5. Distribution of main users of the four principal energy categories (1980 estimate). Conversion losses for electricity and other energy processing are included in each energy category.

1000 and 1500°C (and also to intermediate temperatures). The upper temperatures are in the range needed for industrial process heat or thermoelectric power generation. Fig. 10 shows the basic concentrator-receiver configurations, where each cross section may be applied to a cylindrical configuration with tubular receivers or to surfaces of revolution with near-spherical receivers with which the highest concentration ratios and temperatures can be attained. The Fresnel concentrators (Fig. 10*e* and 10*f*) are two-dimensional configurations with a three-dimensional effect (cylindrical or surface of revolution). They are less expensive to fabricate, but must be large, due to space lost between the segments. Generally, the concentration ratio of Fresnel reflectors is inferior to that of a continuously curved reflector, but not necessarily very much. Thus the Fresnel reflector offers the possibility of trading off facet size, volumetric requirements, optical quality, and cost, to obtain an optimum compromise for the function intended.

The intense focal energy influx causes high receiver temperatures, hence, intense thermal radiation, which can result in high reradiation losses. One means to reduce these losses is to provide a receiver coating material with high absorptivity α to solar radiation and low infrared emissivity ϵ at receiver temperatures of 950–1200°C. The higher the ratio α/ϵ, the higher the collector's efficiency. No single material is satisfactory, since the receiver surface should be optically black for wavelengths shorter than 1.3 μm and highly reflective for wavelengths between 1.8 and 10 μm. One approach is to provide a sandwich coating of different materials with the proper spectral characteristics. However, no economically, or even technologically, satisfactory solution to achieving the desired values of $\alpha/\epsilon = 10$ at 1000–1200°C over periods of decades has been found so far.

Efficiency is also increased by choosing a collector geometry that makes the absorbing (irradiated) area large and the emitting area small, as in the case of a finned receiver configuration (Fig. 11). Thus, by the right combination of receiver surface materials, receiver geometry, and heat transfer efficiency, it is possible to operate at high concentration ratios and still extract a high fraction (60–80%) of the solar "crude" focused into the system by the concentrator.

SOLAR ENERGY STORAGE

Since solar energy is more time-variant than its use, it is necessary to supply either nonsolar make-up energy or to store solar energy. Solar energy storage systems may be divided into two groups: controlled energy storage (CES) and environmental energy storage (EES). Controlled energy storage systems store the energy in as concentrated form as possible in a variety of forms—as thermal, chemical, mechanical, or potential (gravitational) energy; environmental energy storage systems utilize the thermal energy stored by solar irradiation in the ground, air, and water. In contrast to thermal controlled energy storage, environmental energy storage generally is thermal energy storage at temperatures below the user level.

Controlled energy storage. Thermal controlled energy storage can be accomplished in the form of latent or of sensible heat. In the first case, the state of the material is changed, primarily from liquid to solid and vice versa, with the release or absorption of heat of fusion. Sodium hydroxide and sodium nitrite display high performance as fusion materials but are difficult to handle. As storage material for sensible heat, the weight-specific capacity of water is excellent, but its volumetric capacity is poor. High volumetric capacity is desirable particularly for smaller (nonindustrial) storage systems, because of their less favorable volume-to-surface area ratio. For this reason, water can be less suitable than concrete or steel. For solar interior heating systems, concrete is particularly suitable, because it can serve as storage as well as

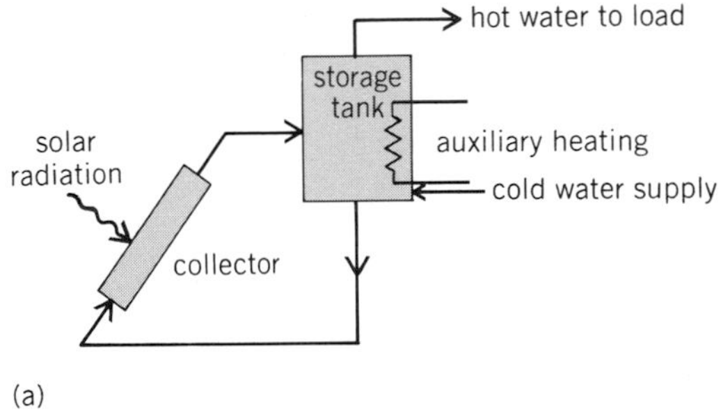

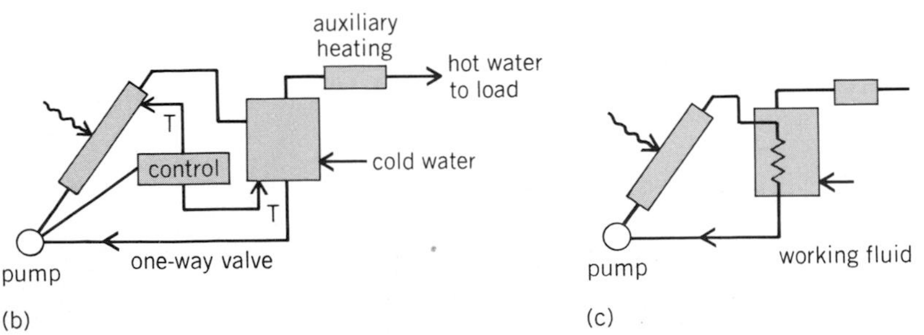

Fig. 6. Solar water heating. *(a)* Natural circulation water heater. *(b)* Forced circulation water heater with open cycle, and *(c)* with closed cycle.

structural (floor) material, providing even and very effective (from the ground up) room heating.

In chemical storage, excess electricity is used to produce a fuel. Solar energy or its derivative (for example, wind) may be abundant in places where the need for electricity is far below the energy offered. In this case, a fuel may be distributed more effectively over greater distances than electricity. Among possible fuels for this purpose, hydrogen is least difficult to produce. Its source (water) is inexpensive, it burns cleanly, and it is nearly ideal as fuel for heat engines. In nonliquid form, it can be effectively stored as a hydride.

An alternative form of chemical storage is the use of batteries. They are important means of energy storage for electric cars and as a source of electricity from households to utilities, but their power and energy densities remain essentially limited to 0.1 kW/kg and 0.1 kWh/kg. Only fuel cells, whose development is still in an initial stage, offer significantly higher performance. A photoelectrochemical cell operating in the fuel storage mode with sunlight can be reversed to function as fuel cell when there is no sunlight.

Finally, it is possible to store electrical energy in the form of compressed air, recoverable by driving gas turbines with generators; as mechanical energy in the form of flywheels; or as potential energy by pumping water into an elevated reservoir and recovering the energy by running the water back out through the pumps then serving as turbines driving generators. *See* ENERGY STORAGE.

Environmental energy storage. Environmental energy storage simply is solar energy stored as thermal energy in the ground, underground waters, rivers, and lakes, and in the air. The energy is extracted by means of the refrigerating process in reverse—the heat pump. Any refrigerator is a heat pump. Its backside is hot, because the heat it extracts out of one space has to be accumulated in another. The heat pump may be viewed as an environmental refrigerator whose hot "backside" is a house heating system. The volume of any house is negligibly small compared to the volume of the outside. The cooling of the environment, therefore, is negligible, while the house can be heated comfortably. By means of the heat pump, solar energy continues to be used, albeit in another form, reducing the requirement for a separate energy source to a greater or lesser degree, depending on climatic conditions. Along with passive solar heating and cooling systems, heat pumps will be among the fastest-growing solar energy systems. Together, they can reduce significantly the amount of fossil energy consumed for low-temperature heat (Fig. 5). *See* HEAT PUMP.

The utilization of low-temperature thermal energy in the oceans by means of the heat pump process for large-scale generation of electricity is described below.

WATERPOWER

Of the world power consumption, only about 2% is derived from waterpower, while contributions by wind power are negligible. In Europe, about 23% of available waterpower is utilized, In North America about 22%. This contrasts sharply with the lower utilization level of waterpower in other parts of the globe.

Waterpower is continuously resupplied by the Sun. Its use does not diminish a given reservoir as in the case of fossil fuels (oil, coal, gas). Waterpower is concentrated solar power and is more regularly available. It can be readily regulated and stored in the form of reservoirs. Today hydroelectric conversion efficiency from waterpower to electric power approaches 80%, compared to about 33% conversion efficiency from coal or oil. Utilization of waterpower avoids air pollution.

These are compelling reasons for increasing utilization of the world's waterpower. It is entirely possible to raise the hydroelectric power supply to some 6,400,000,000 kWh, by the year 2000, benefiting primarily the developing areas. The main obstacle is the availability of investment capital to build the large installations required and, in some cases, the transmission facilities to distant load centers.

The development of waterpower is being advanced particularly in the Soviet Union and the People's Republic of China. Large unused waterpower resources exist still in New Guinea, Africa, South America, and Greenland.

Another form of waterpower generation is through utilization of the thermal gradient in oceans, which serve as the storage system for a vast amount of solar energy. The temperature difference between surface and bottom waters is a potentially very large source of electric power. Tropical regions, particularly between 10°N and 10°S latitude, are especially suitable, because relatively high surface temperatures provide a larger temperature gradient. Temperature differences of at least 20–23°C are available between surface and depth throughout the year, wind speeds are moderate (25 knots or 13 m/s or less; no hurricanes), and currents are below 1 knot or 0.5 m/s at all depths. One method (Fig. 12) to extract energy is to heat a suitable working fluid (for example, propane or ammonia) which is evaporated in the warm surface water and ducted to an underwater turbine, where it is allowed to expand, driving a turbogenerator system, and to condense in cool depth-waters. From there it is returned to the surface, and the process is repeated (closed Rankine cycle). *See* OCEAN THERMAL ENERGY CONVERSION.

WIND POWER

Wind is the next largest solar-derivative power, after solar crude itself and the energy contained in the oceans. Its utilization is environmentally even more benign than fresh waterpower, because no dams and land floodings are involved.

The bulk of wind power, which lies in the upper troposphere and the lower stratosphere, is not accessible to present-day technological potential.

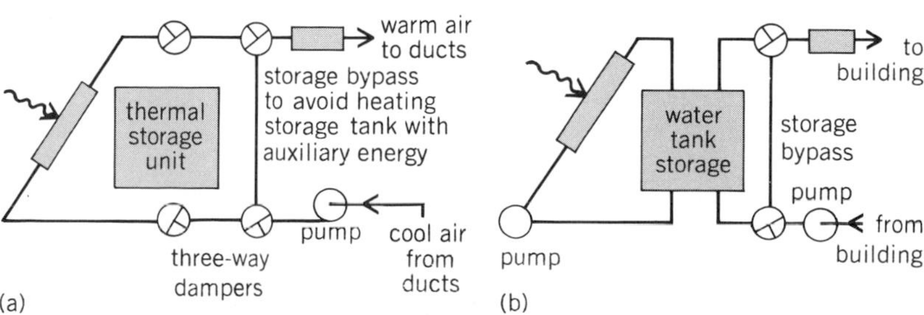

Fig. 7. Solar interior space heating. *(a)* Air heating system. *(b)* Water heating system.

Table 1. Comparison of energy sources in terms of electric energy output at the bus-bar

Energy source	at electric conversion efficiency of	yields the following electric energy in kilowatt-years*
1000 metric tons of oil	35%	490 = 1.0
1000 metric tons of coal†	35%	312 = 0.637
1000 metric tons of enriched uranium in light-water reactor	31.7%	28,500,000 = 58,160
1000 metric tons of plutonium in liquid-metal fast-breeder reactor	40.4%	36,000,000 = 73,470
1000 km² solar absorber area‡ (solar crude: 2.2×10^9 kWh/km² yr)	20%	50,300,000§ = 102,700

*The numbers at the right are indices, based on the yield of 1000 metric tons of oil as unity (1.0). For example, 1000 km² solar absorber area under specified conditions yields 102,700 times as much electric energy as 1000 tons of oil, and 1000 tons of coal yields 63.7% of the energy from 1000 metric tons of oil.

†Based on mean heating value of 26×10^6 Btu/metric ton $= 27 \times 10^9$ J/ton.

‡Since 1000 km² solar crude yields the equivalent of 102.7×10^6 metric tons of oil and since a metric ton of oil corresponds to 7.14 barrels (1.135 m³), the yield of the 1000 km² corresponds to $102.7 \times 10^6 \times 7.14 = 734 \times 10^6$ bbl $= 116.6 \times 10^6$ m³. Coal yield equivalent per 1000 km² solar follows from 102,700/0.637 = 161,000 per 1000 metric tons of coal, or the yield of 1000 km² corresponds to the yield of 161×10^6 metric tons of coal.

§United States electricity sales in 1980 ≅ 271×10^6 kW-yr.

However, there are large areas with moderate to strong surface winds in the United States, particularly along the Aleutian chain, through the Great Plains, and along portions of the East and West Coasts. This is shown in Fig. 13 for the contiguous 48 states.

A study that was conducted at Oklahoma State University showed that the average wind energy in the Oklahoma City area is about 0.2 kW/m² (18.5 W/ft²) of area perpendicular to the wind direction. This is roughly equivalent to the solar energy received by the same area, averaging the sunlight over 24 h per day, all seasons, and all weather conditions. However, in contrast to solar energy, the wind energy could be converted at an efficiency of some 40%. *See* WIND POWER.

AGRICULTURAL UTILIZATION

The basis of the biological utilization of solar energy is the process called photosynthesis, in which solar energy provides the power within plants to convert carbon dioxide (CO_2) and water (H_2O) into sugars (carbohydrates) and oxygen. The prime conversion mechanism is the chlorophyll molecule. Organic-chemical solar energy conversion operates at very low efficiency of 0.1–0.2%; that is, for every light quantum used, 1000–500 quanta are reflected by the vegetation. However, research on algae, especially the alga *Chlorella*, has shown that higher efficiencies can be achieved. On the basis of extensive experimentation, the practically achievable yield of "chlorella farms," using sunlight as the energy source, has been estimated to be at least of the order of 35 tons of dry algae per acre (78 metric tons per hectare). This corresponds to about 0.6% efficiency and compares very favorably even with the highest agricultural yields (10–15 tons per acre or 22–24 metric tons per hectare), let alone the much lower yields in less developed countries (2–2.5 tons per acre or 4.5–5.6 metric tons per hectare). By building large algae farms on nonarable ground, a growing portion of the solar energy presently absorbed by bare soil, that is, about 5.3% or 400×10^{14} kWh per annum, could be utilized for the production of organic matter for food and for conversion into synthetic liquid fuels as a complement to the world's oil supply. Again, capital and local or regional requirements determine the feasibility and worthwhileness of such endeavors.

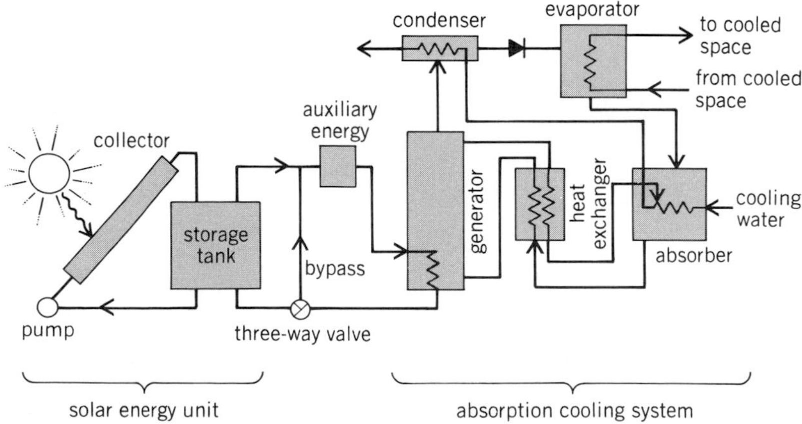

Fig. 8. Solar-powered absorption air cooling system.

SOLAR-ELECTRIC POWER GENERATION

For solar-electric power plants, the annual sunshine hours should be as large as possible, and the humidity, which causes absorption and scattering, should be low. With its "sun bowl" (Fig. 14), the United States is the only major industrial country with high-insolation (2–2.5 TWh/km² yr) territory within its borders; the highest values lie around 2.2 TWh/km² yr (Fig. 3). Table 1 shows that if in such high-insolation areas the incident solar radiation over 1000 km² is converted to electric energy at only 20% efficiency, the output is equivalent to the annual consumption of 734×10^6 bbl (117×10^6 m³) of crude oil or 161×10^6 metric tons of coal.

The energy budget of a solar station is compared with that of average desert ground in Fig. 15. Of the solar irradiation, the desert surface reflects about 40% and retains 60%, whereas the absorber area

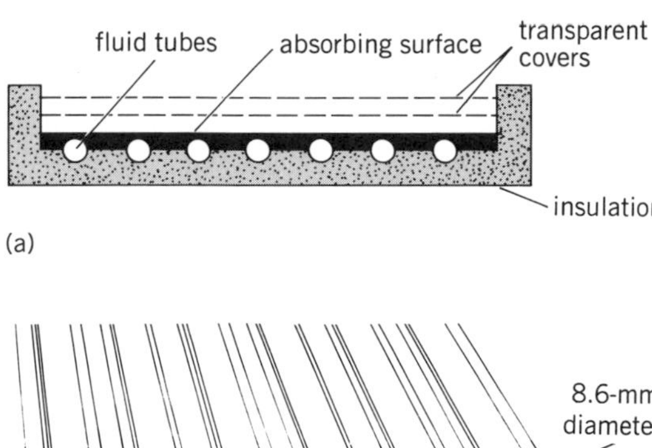

Fig. 9. Flat plate collectors. (*a*) Conventional collector. (*b*) Plastic collector mat.

retains about 90%. Thus, 30% more of the incident solar energy is trapped. Of this, about two-thirds appears as electric energy and one-third as true, that is, extrinsic, heat production at 20% overall conversion. At 30% overall conversion from solar crude to electricity, no extrinsic heat would be generated. At still higher conversion efficiency, the area occupied by the power station would be cooled, compared to the natural environment, rather than slightly heated as with the 20% system.

The overall unused thermal energy is concentrated in the power station's thermal storage system and ultimately in its electric conversion system. If the electric conversion system uses cooling water rather than air cooling, advantage can be taken of the large amount of thermal energy concentrated in the water. For this, two alternatives are available: desalination, and power generation by temperature-gradient utilization.

Four methods can be used to generate solar-electric power: (1) the solar-thermal distributed receiver system (STDS); (2) the solar-thermal central receiver system (STCRS); and (3) the photovoltaic system (PVS); and (4) photoelectrochemical conversion (PECC). In each case, the overall system may serve as backup, that is, operating only when the sun shines, equipped only with a minor energy storage capacity (for example, for 1 h full output) to bridge temporary cloud coverage; alternatively, the overall system may include a conventional fuel system to replace solar energy at night or during cloudy days; and finally, the independent solar-electric system includes a storage system to ensure continuous power-generating capacity based on solar energy only.

Solar thermal systems. In the STDS, solar radiation is absorbed over a large area covered with flat-plate (nonconcentrating) collectors or parabolic-trough concentrators, which focus the sunlight on a heat pipe carrying the working fluid (Fig. 16). The heat pipes are covered with selective coating characterized by high absorptivity to solar radiation and low emissivity at the temperature of the heat pipe. The flat-plate collector operates at turbine inlet temperatures of 250–500°F (121–260°C). With the parabolic trough, temperatures between 550–1000°F (288–538°C) at turbine inlet are attainable. The higher the temperature, the higher the efficiency and the smaller the land area needed for a given power level, but the more expensive is the system, especially the piping and the coating.

In the STCRS, sunlight is concentrated on a receiver by a large number of mirrors designed to follow the Sun (heliostats). The receiver is a heater located atop a tower (Fig. 17) which is served by a certain collector area. The power output of such a module depends on the collector array area which, in turn, determines the tower height. The horizontal distance of the farthest mirror from the foot of the tower is about twice the tower height, which may range from 250 m to 500 m. A given solar power plant may consist of an arbitrary number of these modules.

Both the STDS and STCRS require large areas, due to the nature of the energy source. It is possible, of course, to subdivide the STDS into small modules, each with a small standardized electric power station, and to collect the electric current in an overall power-conditioning station. In the STCRS, the working fluid can be heated to higher temperatures, since the collector area acts as a giant parabolic mirror consisting of individual facets (heliostats). The higher temperatures attainable with the STCRS yield higher overall efficiencies (25–35%), or 150,000–200,000 kWe per square kilometer of *collector* area (not total area covered), compared to about 1.5 km² for the same

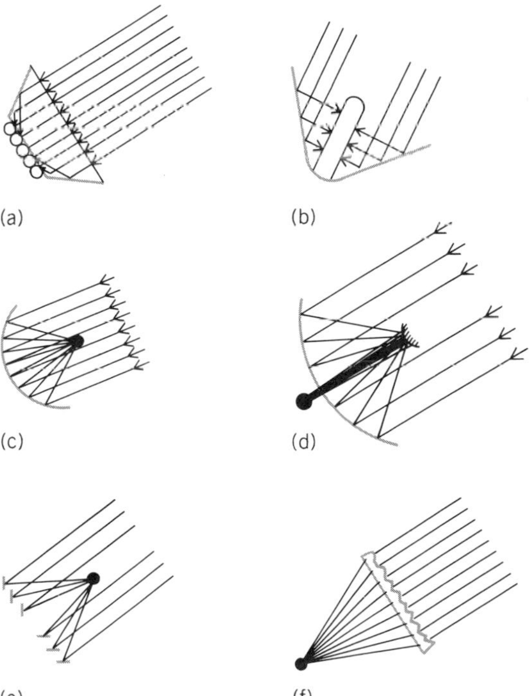

Fig. 10. Focusing collector configurations. (*a*) Planar concentrator and receiver. (*b*) Conical concentrator and cylindrical or flat plate receiver. (*c*) Paraboloidal concentrator with tubular or spherical receiver. (*d*) Dual reflection concentrator with paraboloidal primary reflector and tubular or spherical receiver. (*e*) Fresnel reflecting concentrator with tubular or spherical receiver. (*f*) Fresnel lens concentrator with tubular or spherical receiver.

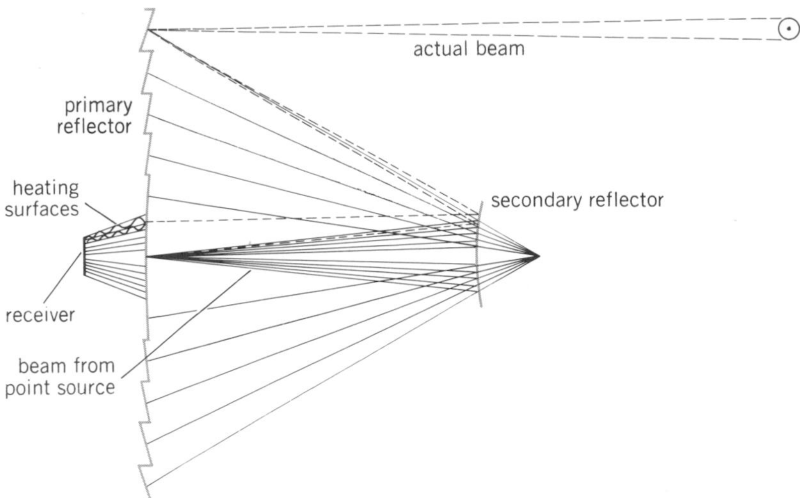

Fig. 11. Dual reflection concentrator with finned receiver.

power output in an STDS. The working fluid can be water (steam), sodium, or helium (in the order of increasing system temperature). In the simplest case, superheated steam is generated in a heater atop the tower. The steam is ducted to the ground and used in a high-pressure and low-pressure turbogenerator system.

To provide power during off-radiation hours, part of the energy generated during sunshine hours must be stored. This means that part of a solar power module's (SPM) power output, or the output of entire modules, is not available during sunshine hours. It may be possible to design the central receiver towers so that wind generators can be attached, causing the towers to generate power in the absence of sunshine as well.

Photovoltaic systems. In the photovoltaic systems, electricity is produced by means of solar cells. The photovoltaic system offers the advantage that its performance and, to a large extent, its economy are independent of size over a wide range from watts to gigawatts. Thus, relatively small photovoltaic systems for communities, buildings, and homes which greatly reduce or eliminate altogether the distribution cost component are feasible. Since electricity for residential and commercial consumption makes up a substantial part of the overall demand (Fig. 5), a large potential for saving fossil fuel is indicated by the introduction of photovoltaic electricity just for daytime demands in the residential and commercial sector. The difference between daytime peaks and nocturnal valleys in the electrical load of central power plants would be reduced, and regional peak requirements could be met more readily with fewer power plants.

Even if photovoltaic power generation is used in a large central power plant, the fact that the output is in direct current (dc) reduces the cost of distribution, since long-distance transmission losses are lower than for alternating current (ac). For most uses, dc must be converted to ac; but if extensive transmission is involved, as in countries with large areas, dc-to-ac conversion equipment can be installed at terminal points rather than at the plant.

The silicon solar cell was developed in the United States in 1954. Besides silicon, there are other semiconductor materials suitable for photovoltaic conversion, but silicon is, and will be in the foreseeable future, the most frequently used semiconductor. For a discussion of the operating principles of solar cells *see* SOLAR CELL.

Efficiency. A solar cell is activated by all light energy contained in that part of the spectrum that is defined by wavelengths equal to and shorter than the wavelength corresponding to the bandgap photon energy for the particular semiconductor, the energy required for the production of electron-hole (e^-/h^+) pairs. Photons with energies greater than the bandgap energy can create electron-hole pairs, but the excess energy shows up in the form of heat. The lower the bandgap, the larger the part of the spectrum capable of generating electricity, but the more excess heat is created; this heat tends to reduce the electron-hole flow and must be removed to prevent degradation of the cell. Thus, the efficiency of the solar cell, defined as the ratio of electrical energy extracted to solar radiation energy offered, falls off if the wavelength associated with the bandgap energy is either too short or too long. The maximum theoretical efficiency of a semiconductor is limited to about 25%. Because of losses such as reflection of light from the cell surface (at least 5%), covering of the cell surface (about 10%) by current collectors (thin wiring), resistivity losses, and recombination of electron-hole pairs within the crystal, the actual efficiency remains below that of an ideal photocell.

The efficiency is not increased by focusing light on a cell. Besides the greater complexity of requiring a driving mechanism to keep the concentrator focused on the solar cells, light concentrators tend to heat up the cell. However, concentration reduces the costly solar panel area for a given electrical output and replaces it by usually far less expensive reflector area. Light concentration also has a major advantage that thereby the light intensity in the different spectral ranges is increased. From the wavelength-bandgap correlation, it follows that a nonhomogeneous solar cell—one consisting of several semiconductors with selective spectral absorption—can exceed the limit of a single-semiconductor efficiency and reach values of 30–45%. However, at one solar constant, a beam splitter could assign to each cell type only the fraction of intensity contained in the solar spectral distribution. Around 1.1-μm wavelength at which silicon cells respond, the solar radiation intensity is only one-third of that around 0.6 μm. Concentration raises the intensity level in each spectral range. At the same time, cooling requirements are reduced when photons with excess energy for one semiconductor are split off and directed to another semiconductor whose bandgap is larger.

Manufacture and cost. The cost of photovoltaic electricity must be reduced substantially before it can become a large-scale energy source. The high cost of high-quality solar cells is due to their demanding manufacturing process. For silicon, the most commonly used material, this process consists of the following steps: (1) preparation of almost completely pure silicon; (2) growth of almost perfect cylindrical silicon crystals by the Czochralski method; (3) slicing of the crystal into thin wafers (about 0.75 mm thick), each of which will eventually be an individual *p*-type cell; (4) doping of a thin surface layer of the wafer into an *n*-type crystal by

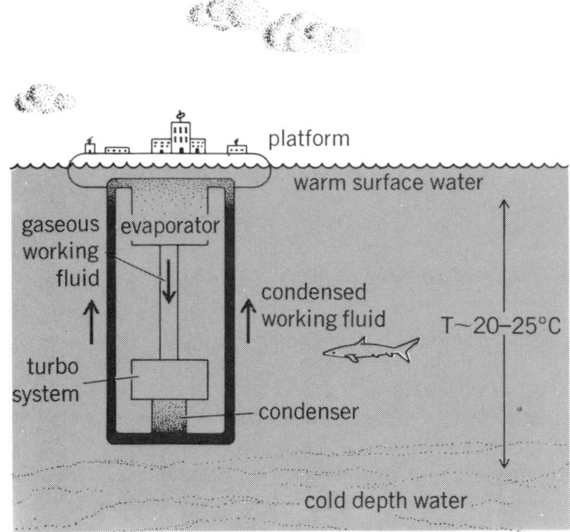

Fig. 12. Closed Rankine cycle, ocean thermal energy conversion system.

exposure to hot phosphorus so that its atoms can penetrate the wafer to a depth of about 0.5 μm; (5) attachment of electrical contacts to front and back of the wafer (cell wiring); (6) application of antireflection coating (for example, titanium oxide); and (7) encapsulation of the entire cell in a protective skin.

This high-precision manufacturing process is difficult to mechanize, and is time-consuming and skill-intensive. For these reasons, and because of losses of up to 35% in slicing and cutting the crystals into wafers with nonround shapes, the cost of solar cells is still very high, and will have to be drastically reduced before their large-scale use as generators for buildings or central power stations becomes economically feasible. Promising approaches toward cost reduction can be made in three directions: use of other materials than silicon; extensive automation of the manufacturing process, or changes in the manufacturing process that permit more extensive automation; and acceptance of reduced efficiency.

Thin films of substances that can be vapor-deposited onto a substrate may eventually be much less costly than silicon wafers. However, the cost of silicon wafers can be lowered by more automated and less wasteful manufacturing both on Earth and in orbiting processing facilities, especially if the quality requirements may be reduced. If the efficiency of silicon cells is reduced from 18 to 10–12%, the cost could be reduced by a factor of 10 to 20, perhaps more at very large mass production. One manufacturing process for producing cheaper cells, known as edge-defined film-fed growth, applies capillary action to cause molten silicon to rise through a slot in a graphite die. Typically, a silicon-ribbon, 15 mm thick and 10 cm wide, can be grown at rates of at least 150 cm per hour. The ribbon is then cut into wafers at minimum waste.

Research was undertaken in the early 1970s to investigate replacing the hand-soldering technique for cell wiring by a microwelding technique. The cell wiring process has been fully automated with attendant cost reduction, and microwelding machines have contributed essentially to the drastic reduction in the cost of solar panels for use in space since the late 1950s. In order to achieve a significant cost reduction for terrestrial mass production, the welding rate needs to be increased by integrating welding heads for simultaneous welding of a larger number of connections. The associated tooling costs, however, pay off only if other preconditions for mass production are met.

Basic silicon single-crystal material contributes about 30% to the cost of presently available solar cells. Therefore, numerous efforts have been undertaken to cut base material cost by using silicon of lesser purity and higher degree of crystallographic imperfection than in conventional semiconductor-grade silicon. The use of polycrystalline silicon is an important part of these efforts. Problems of low efficiency normally associated with polycrystalline silicon can be improved substantially by growing silicon whose individual grains are of controlled size and structure, to minimize losses caused by recombination of electron-hole pairs at grain boundaries. Controlled polycrystalline base material, together with optimized processing, offers a potential for significant cost reduction, because they lend themselves to automated fabrication techniques, hence, to mass production.

Prospective utilization. The principal advantage of the photovoltaic system over the thermal systems is the absence of moving parts and fluids at high temperatures. Its major disadvantages are its comparatively low efficiency and the fact that thermal storage cannot be used. Considerable development is required before the system can be economically competitive with the thermal systems.

The potential global market for solar cells is very large, because of the many small- and medium-sized applications. Integration of solar panels for electricity during sunshine hours with solar house heating and cooling, along with heat pumps and wind generators (where suitable), yields maximum utilization of local solar energy (80–90%) on a widely distributed basis, saving distribution costs. Supplemental electricity and thermal energy will be required in most cases. But the energy assurance of homes and small communities would

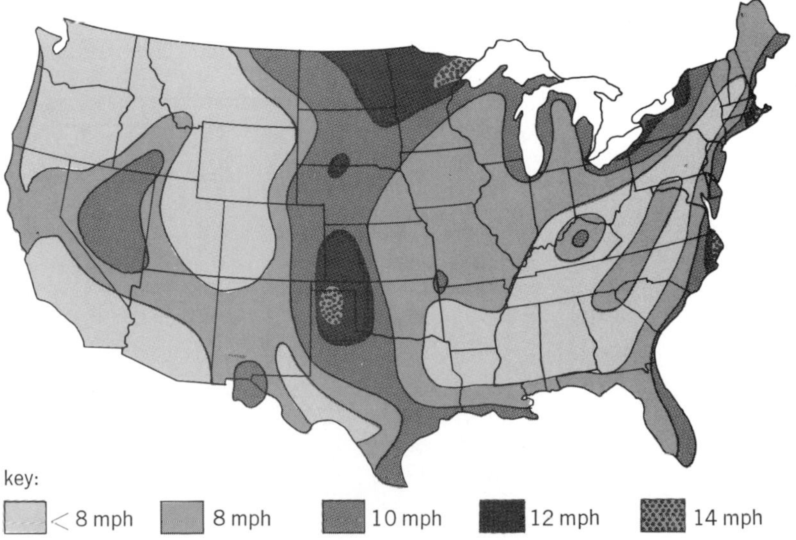

Fig. 13. Average surface wind velocities in high-wind regions of the 48 contiguous states. 1 mph = 0.447 m/sec.

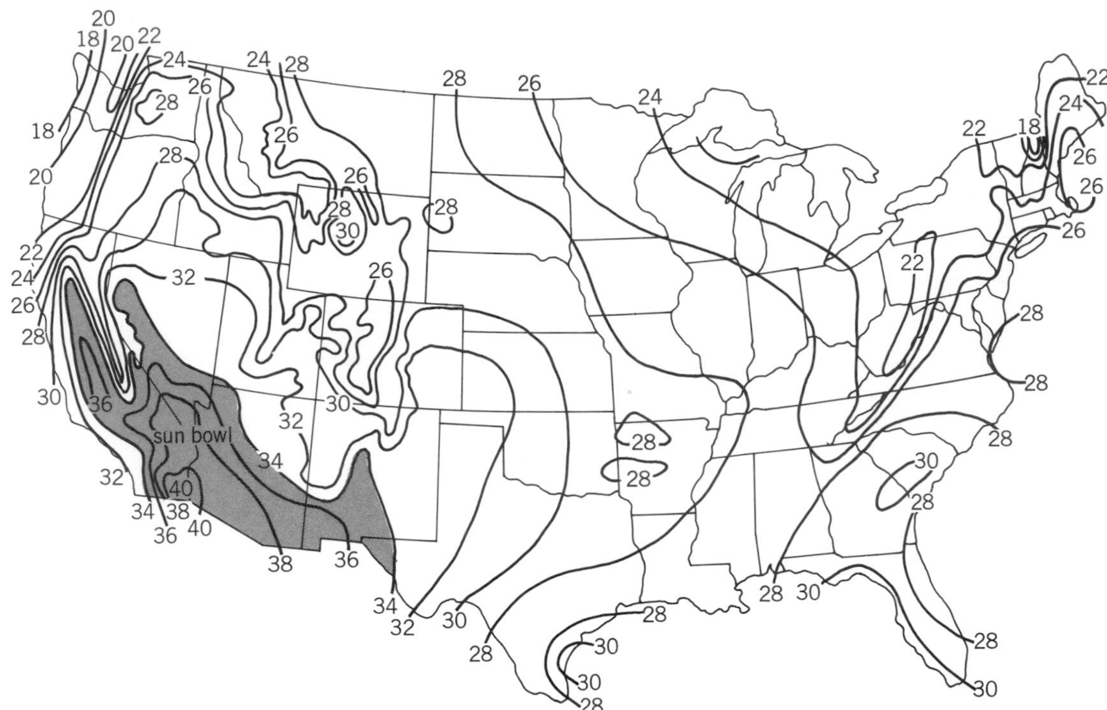

Fig. 14. National climatic center annual sunshine hours with "sun bowl" accented; values in hundreds.

be maximized on a basis of minimized dependence on central electric power plants and outside supplies of fuel.

However, only high-insolation countries that also have relatively large areas available can hope to utilize large solar central power plants. By 1990, a total of roughly 200 GW-yr of electric energy must be furnished by central power plants in the United States. If only 50% of this was to be generated by photovoltaic power plants even at 15% conversion efficiency, some 2000 km² of collecting area would be required in the deserts of the United States "sun bowl." This would meet with many environmental objections, even if otherwise practical. Thus it is unlikely that solar energy alone can meet the electrical energy requirements even in the United States with its (for an industrial nation) optimal solar assets.

Photoelectrochemical conversion. By immersing semiconductors in liquid electrolytes, solar energy can be changed by the same converter either to electricity or to chemical energy by generat-

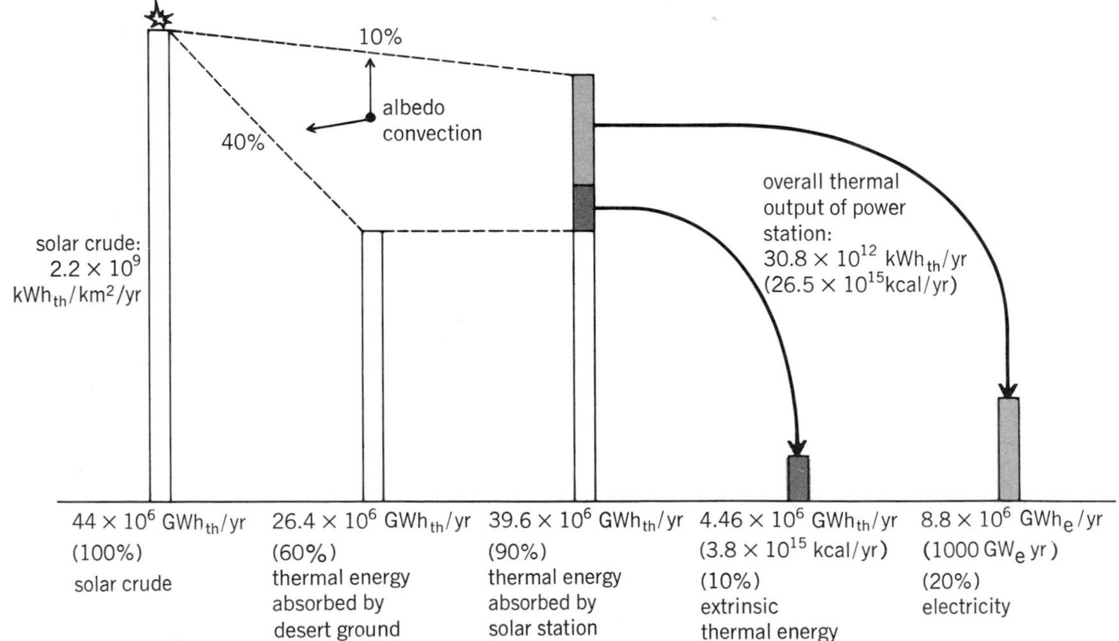

Fig. 15. Typical energy budget of a 1000 GWe-yr solar power station.

Fig. 16. Solar thermal distributed power station system.

ing chemical fuels, especially hydrogen. In a photoelectrochemical converter (Fig. 18), light replaces a conventional power source in driving the electrolytical decomposition of water. The solar radiation energy is converted to electricity by a semiconductor anode. In this arrangement, the electrolyte plays the same role as the *p*-type layer of a solar cell. The holes move into the electrolyte, while the electrons stay in the outer layers of the semiconductor. However, back reaction, that is, charge recombination between photoanode and electrolyte, would prevent the generation of an electric current. This back reaction is prevented automatically by a phenomenon called band-bending. At first, after the photovoltaic process is started, the semiconductor must come into electric equilibrium with the electrolyte, so that eventually there is no longer a driving force for net charge transfer from semiconductor to electrolyte or vice versa. In the course of this process, electrons are transferred from semiconductor to electrolyte. After equilibrium is established, it takes energy to transport an electron from the bulk of the semiconductor to the surface (that is, to the conduction band), because the surface is in touch with the now negatively charged electrolyte which tends to repel electrons. As a result, the conduction band is "bent" from the surface to some depth within the photoanode. By this band-bending, an electron-depleted zone is created between conduction band and electrolyte. This depletion region prevents the electric current in the conduction band from being short-circuited by the electrolyte. The electrons, being unable to overcome the charge barrier across the depletion zone, are forced to flow through the circuit and enter the electrolyte through the cathode. Thus a current flows that can handle the load of electrolytic decomposition. The returning electrons recombine with holes, preventing the electrolyte from becoming positively charged by the holes from the photoreaction in the semiconductor and thereby destroying the depletion zone.

The total energy output of the photoelectrochemical converter (that is, the sum of stored chemical energy plus electrical energy) is proportional to the amount of band-bending, that is, the potential across the electron-depleted zone. This potential equals the net voltage drop experienced by the electron as it passes through the external circuit. The energy conversion efficiency of the cell cannot exceed the ratio of band-bending energy to bandgap energy. The band-bending energy is about 0.63 eV. For a cadmium telluride (CdTe) semiconductor, the bandgap is 1.4 eV, so that the maximum conversion efficiency of the cell in this case is 0.45.

Such efficiencies have not yet been reached, but they show that the potential of the photoelectrochemical converter is at least as good as that of a photovoltaic panel. In contrast to the latter, the photoelectrochemical converter has the advantage of flexibility. The cell stores light energy in the form of chemical fuels and also converts light to electricity, so that it will eventually be possible to vary the cell output from exclusively producing hydrogen to exclusively producing electricity and any ratio in between. For the cell, fabrication problems are reduced, compared to the "dry" photovoltaic panel. Band-bending is introduced automatically upon immersion into the electrolyte, which also provides antireflection coating and otherwise pro-

Table 2. Lunetta systems

Application	Purpose	Main beneficiaries	Required brightness level (clear sky, full moon equivalent)	Nominal brightness at 75% average broken cloud cover (moons)
Rural area illumination	Reduction of crop losses due to inclement weather	Developing countries	5-10	25-50
	Better utilization of expensive agricultural machines by co-operatives	Developing countries		
	Increased number of crops per year (multicropping) where other conditions are met	Developing countries in tropical belt		
Urban illumination	Brighter and more even lighting Increased public safety Back-up at blackouts and brownouts	Industrialized and developing countries	10-40	50-200
Illumination of centers of activity in polar regions	Urban and rural areas Industrial operations (oil, gas, minerals)	Northern countries	10-20	50-150
Illumination of disaster areas	Lighting for aid and rescue operations	Earthquake and typhoon regions	5-10	25-50

Table 3. Soletta systems

System	General function	Special applications	Purpose; benefit	Nominal radiation intensity, suns	Typical orbit	Typical reflector size	System area
Agrisoletta	Local weather modification	Temperature stabilization (avoidance of night frosts)	Better harvests	0.3–0.7	Sun-synchronous, 3500–4200 km	3–10 km^2	2700–6300 km^2
		Modulation of local precipitation	Better harvests, hydropower		6-hr orbit, 10,400 km, 20–30° inclination	4–10 km^2	3000–7000 km^2
		Wind generation	Electric power				
		Water desalination	Fresh water				
		Crop drying	Reduction of crop losses				
Metsoletta	Weather modification on mesoscale	Manipulation of meteorological highs and lows	Prevention of prolonged dry, rainy, or cold periods	0.8–1.0	Geosynchronous	50–100 km^2	80,000–100,000 km^2
		Transport of water over larger distances in form of clouds	Irrigation of dry areas; limited local cloud control				
Powersoletta	Serving as night sun for power generation	Irradiation of photovoltaic and photoelectrochemical ground power stations	Generation of electricity	1.0 at 70° elevation	Sun-synchronous, 3500–4200 km	10–15 km^2	11,500 km^2
			Production of hydrogen				
		Supply of solar heat at night	Heating of solar ponds as source of heat and power				
Biosoletta	Bioproduction enhancement	Food production	Increased cereal production	0.4–0.8	6-hr orbit (as above)	4–10 km^2	3600–7200 km^2
		Growth of biomass	Increased seafood production		Elliptical 24-hr orbit, perigee/apogee 6/7.2 earth radii; 63.4° inclination	50–100 km^2	100,000 km^2
			Fuel production				

tects the semiconductor from environmental influences. Problems of temperature sensitivity encountered with the solid-solid junction of a solar cell do not exist in the solid-liquid junction of the photoelectrochemical converter.

On Earth, liquids can be used with ease, and the production of hydrogen as chemical fuel without the consumption of separately generated electricity is of great importance. Therefore, the photoelectrochemical converter appears to have a tremendous potential as a device for energy generation and storage. *See* PHOTOELECTROLYSIS.

SOLAR ENERGY AND SPACE

Extensive use of solar energy can be anticipated in spacecraft and by extraterrestrial communities in orbit or on the Moon. Solar cells are the primary power supply for uncrewed spacecraft. They are an attractive source of electric power for crewed space stations, especially in near-Earth orbits where the use of large nuclear installations might be less desirable, and in orbits where solar exposure is practically continuous so that little storage capacity is required. On the Moon, undiminished solar energy can be soaked up by solar cells

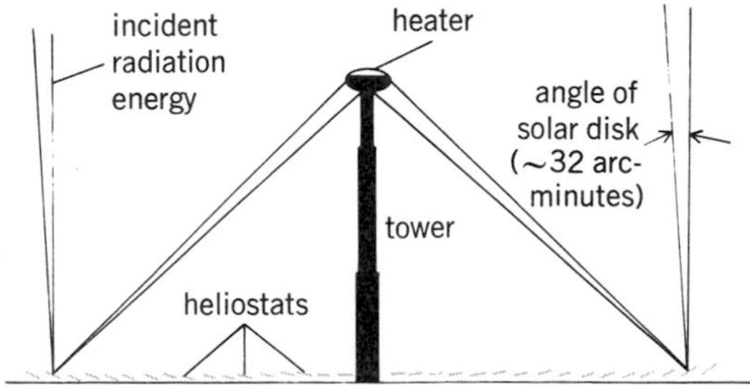

Fig. 17. Solar-thermal central-receiver system. 32 ft = 9.75 m.

and solar concentrators during 14 days for immediate use.

The use of sunlight-spectrally-tailored semiconductor crystals of corresponding bandgaps together with high concentration ratios appears particularly promising for space applications, hence also for use on the Moon, because there are no atmospheric absorptive effects to interfere with the more effective utilization of the energy spectrum offered. High concentration ratios (more electric power per unit solar panel area) become feasible at comparatively lower cooling requirements, since fewer photons of higher-than-bandgap energy are used on a particular semiconductor to promote electrons into the conduction band. However, because of large storage requirements for the lunar night, solar power on the Moon is inferior to nuclear power on a kilowatt per unit mass basis for power outputs above 50 kW.

Another important relation between space and solar energy is provided by weather satellites. With improving long-term weather prediction, the practical aspects of utilizing solar energy on Earth will improve also. Advancements in the industrial utilization of space beyond the application satellites will make it possible to intercept large amounts of solar radiation in space for use on Earth or as processing heat in extraterrestrial production facilities. For terrestrial applications, sunlight can be transmitted optically by reflectors (space light), or it can be converted to electric energy in space and transmitted to Earth via microwave beam (solar power satellites, SPS).

Space light. The general objective of space light is to transmit sunlight to selected areas of Earth. Orbiting reflectors beam to Earth measured amounts of solar light energy. The light is beamed by a number of reflectors trained at a given service area. Reflector size and number are tailored to their functional requirements, which fall into two

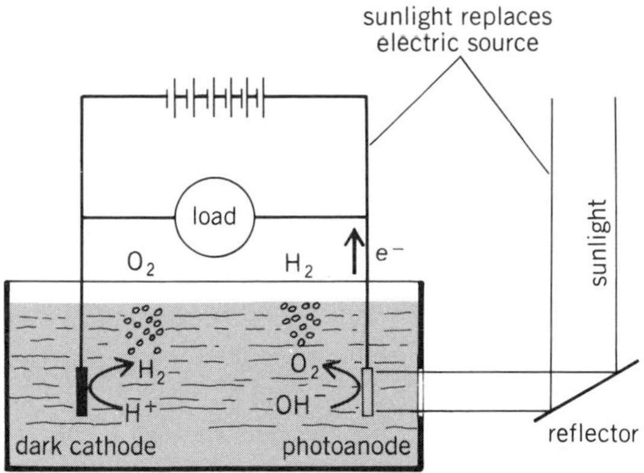

Fig. 18. Photoelectrochemical cell.

main categories: night illumination (Lunetta systems) and solar-type irradiation intensity (Soletta systems). The most suitable orbits are either highly inclined near-Earth orbits or more nearly equatorial distant Earth orbits.

Lunetta and Soletta systems are characterized by a high degree of functional flexibility. The Lunetta functions are listed in Table 2 in the order of declining mass benefit. Rural night illumination can provide tens of millions of added labor hours in the agricultural industry and could economically assist in generating additional food annually for 25 persons or more per hectare (0.01 km^2 or 2.47 acres) of reflecting area. The reflectors preferably circle in Sun-synchronous orbits where, for given service areas, illumination patterns can be established which do not vary throughout the year.

Table 3 surveys the Soletta systems, designated according to their functions. The system size overlaps for several functions. This means not only that the systems can be similar, but that a system can serve different functions at different times, unless the system is "dedicated" to a single function, such as Powersoletta and Biosoletta. Individual reflectors can even be engaged as Lunettas temporily, for special purposes.

Powersoletta reflectors illuminate the ground station from several and varying directions in the sky. The ground station, therefore, cannot be an STCRS type (Fig. 17), but should be of the photovolatic, photoelectrochemical, or solar pond variety. At least three ground stations can be serviced by one Powersoletta system in Sun-synchronous orbit, located in a belt between 30 and 45° latitude. The Powersoletta reflectors are not subject to seasonal variations as is solar irradiation, and can therefore contribute more energy than the Sun on an annual cumulative basis. Powersoletta also removes to some extent geographic constraints on solar central power station siting.

Solar power satellite. As distinct from the Powersoletta reflector system, the solar power satellite (SPS) absorbs the solar energy in geosynchronous orbit. In the photovoltaic solar power system, solar cells convert the light energy to electricity. The electricity is converted to microwave energy which, in turn, is transmitted to Earth as a phase-coherent (laserlike) beam. An antenna-rectifier (rectenna) system (the microwave equivalent of a photovoltaic receiver system in the shorter-wave spectrum) absorbs the energy and converts it back to dc electricity. Phase coherence prevents the microwave beam from spreading beyond a controlled amount, thereby keeping the rectenna area within limits determined by safety considerations. The photovoltaic receiver in space and rectenna on Earth are sizable, because only large systems delivering 5 GW or more (ground receiver output) promise to be economically competitive at the expected state of the art for the late 1990s.

Most configurations studied use single-crystal silicon cells. The fact that the waste heat is released in space rather than on Earth is a favorable characteristic of the solar power satellite. In lieu of photovoltaic conversion, thermal cycles can be considered. Large reflectors concentrate sunlight at heaters cooled by a working fluid which, in turn, drives a turbo-generator system.

Since the voltage per solar cell is small, the high voltages required for conversion to microwave energy must be produced by wiring many cells in series. The dc current is collected via superconducting cables and converted to radio frequency (rf) power by dc-rf conversion tubes and fed to two large phased-array transmitter antennas, consisting of subarrays of about 100 m^2 each. Their radiating surfaces are slotted waveguides.

The atmosphere is highly transparent at wavelengths around 10 cm. For safety reasons, the beam generated by the antenna creates a tapered power density profile such that power density is highest at beam center (boresight) and falls off toward the periphery. At the antenna surface in space, the power density is very high (subject to heating limitations); but the beam is designed to spread moderately, causing the power density to decline with growing distance from the antenna. When passing through the ionosphere, the density at beam center is down to a maximum value of 23 mW/cm^2, to prevent nonlinear heating which could interfere with radio communication and cause losses in the power beam.

Since most of the power losses occur in space, the thermal load of the ground station on the environment is very small. However, the land use is considerable, which makes the use of solar power satellites questionable for highly industrialized small-area countries. In the United States, by the year 2010, at least 500 GW-yr will have to be furnished by central power stations. Fifty solar power satellites would be required to meet this demand, each with two 5-GW delivery beams, each beam requiring a rectenna area of 135 km^2 if the United States standards for exposure to microwave radiation at the rectenna boundary are to be satisfied, or 700 km^2 if the more stringent standards used in the Soviet Union are to be realized. This adds up to 13,500 and 70,000 km^2, respectively. The upper value amounts to slightly less than half of the size of Michigan. [KRAFFT A. EHRICKE]

Bibliography: I. Beckey and W. Blocker, High efficiency low cost solar cell power, *Astronaut. Aeronaut.*, 16:32–38, 1978; W. Blocker, High-efficiency solar energy conversion through flux concentration and spectrum splitting, *Proc. IEEE*, 66(1):104–105, 1978; K. A. Ehricke, Space light: Space industrial enhancement of the solar option, *Astronaut. Acta*, 6(8/9):1515–1633, 1979; Federal Energy Administration, *National Energy Plan*,

1977; H. Fischer and W. Pschunder, Low-cost solar cells based on large-area unconventional silicon, *IEEE Trans. Electr. Devices*, ED-24(4): 438–442, 1977; P. E. Glaser et al., *Feasibility of a Satellite Solar Power Station*, NASA CR-2357, 1974; F. R. Kalhammer, Energy-storage systems, *Sci. Amer.*, 241(6):56–65, December 1979; H. Kelly, Photovolatic power systems: A tour through the alternatives, *Science*, 199:634–643, 1978; A. B. Meinel and M. P. Meinel, Physics looks at solar energy, *Phys. Today*, 25(2):44–49, February 1972; National Academy of Sciences, Space Science Board, Ad Hoc Panel on Solar Cell Efficiency, *Solar Cells: Outlook for Improved Efficiency*, 1972; Office of Technology Assessment, U.S. Congress, *Application of Solar Energy to Today's Energy Needs*, 1978; *Satellite Power System Engineering and Economic Analysis: Summary*, NASA TM X-73344, 1976.

Solar heating and cooling

The use of solar energy to produce heating or cooling for technological purposes. When the Sun's short-wave radiation impinges upon a blackened surface, much of the incoming radiant energy can be absorbed and converted into heat. The temperature that results is determined by: the intensity of the solar irradiance; the ability of the surface to absorb the incident radiation; and the rate at which the resulting heat is removed. By covering the absorbing surface with a material such as glass, which is highly transparent to the Sun's short-wave radiation but is opaque to the long-wave radiation emitted by the Sun-warmed surface, the effectiveness of the collection process can be greatly enhanced. The energy which is collected can be put to beneficial use at many different temperature levels to accomplish: distillation of sea water to produce salt or potable water; heating of swimming pools; space heating; heating of water for domestic, commercial, and industrial purposes; cooling by absorption or compression refrigeration; cooking; and power generation by thermal or photovoltaic means. *See* SOLAR ENERGY.

DISTILLATION

The oldest of these applications dates back to a 50,000-ft² (4600-m²) installation built in Chile in 1872 to distill saline water and make it potable. Figure 1 shows in cross section the type of glass-roofed solar still used more than a century ago in Chile and used again in modern times on the Greek islands in the eastern Mediterranean and in central Australia. When the Sun shines through the glass cover into the salt or brackish water contained within the concrete channel, the water is warmed and some evaporates to be condensed on the underside of the glazing. The condensate runs down into the scuppers and then into a suitable container. More than 6000 gal (23,000 liters) of pure drinking water were produced on each sunny day by the Chilean still, and the larger version now operating at Coober Pedy, Australia, produces nearly twice as much.

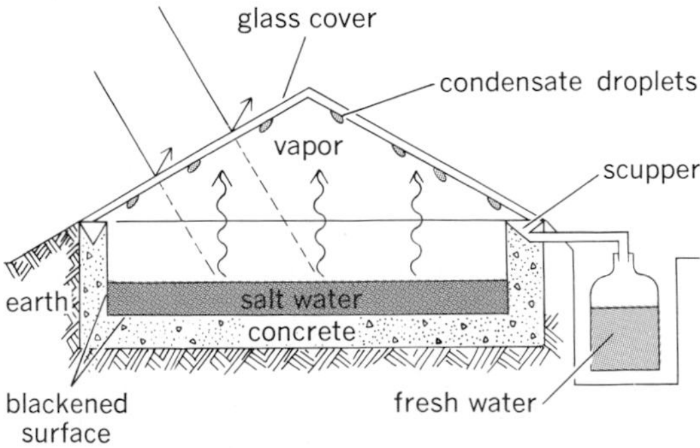

Fig. 1. Roof-type solar still.

Production of salt from the sea has been accomplished for hundreds of years by trapping ocean water in shallow ponds at high tide and simply allowing the water to evaporate under the influence of the Sun. The residue contains all of the compounds that were present in the sea water, and it is sufficiently pure for use in many industrial applications.

SWIMMING POOL HEATING

Swimming pool heating is a moderate-temperature application which, under suitable weather conditions, can be accomplished with a simple unglazed and uninsulated collector similar to the black polyethylene extrusion shown in Fig. 2. When the water to be heated is at almost the same temperature as the surrounding air, little or no heat will be lost from the absorber and so there is no need for either glazing or insulation. Such collectors are usually designed to empty themselves when the circulating pump is shut off so they can avoid the danger of freezing at night. For applications where a significant temperature difference exists between the fluid within the collector passages and the ambient air, both glazing and insulation are essential.

SPACE HEATING

Space heating can be carried out by active systems which use separate collection, distribution, and storage subsystems (Figs. 3 and 4), or by passive designs which use components of a building to admit, store, and distribute the heat resulting from absorbing the incoming solar radiation within the building itself.

Passive systems. Passive systems can be classified as direct-gain when they admit solar radiant energy directly into the structure through large south-facing windows (Fig. 5), or as indirect-gain when a wall (Fig. 6) or a roof (Fig. 7) absorbs the solar radiation, stores the resulting heat, and then transfers it into the building. If the absorber and the storage components are not a part of the building fabric but are separate subsystems which operate by the natural circulation of warmed or cooled air, the term isolated-gain is generally used (Fig. 8).

Passive systems are generally effective where the number of hours of sunshine during the winter months is relatively high, where moderate indoor temperature fluctuations can be tolerated, and where the need for summer cooling and dehumidification is moderate or nonexistent.

Direct-gain systems. Most passive systems make use of the fact that, in winter, whenever the Sun is above the horizon it is in the southern part of the sky, and its altitude above the horizontal plane at noon is relatively low compared to the much higher position which it will attain in summer. This means that in winter when heat is needed, vertical south-facing windows admit solar radiation freely as long as the Sun is shining. The use of double glazing reduces the transmission of the incoming solar radiation by about 16%, but it can halve the thermal loss due to the indoor-outdoor temperature difference. The use of movable insulation which can be placed over the windows at night and removed during the sunlit hours of the day can also greatly improve the performance of direct-gain systems. Window area on surfaces other than those facing south should be kept to the minimum permissible under local building codes. The building components (walls and floors) on which the solar radiation falls should possess as much mass as possible so that the excess solar heat gained during the day can be stored for use at night.

Indirect-gain systems. The indirect-gain concepts shown by Figs. 6 and 7 interpose a thermal mass between the incoming solar radiation and the space to be heated. The system shown in Fig. 6 uses a south-facing glazed wall of concrete or masonry with an air space between the wall's outer surface and the single or double glazing. This is known as a Trombe wall. Vents with dampers are provided near the floor and at the ceiling level so that cool air can be drawn by chimney action from the room into the space between the glass and the wall. The air is heated by contact with the wall, and rises to reenter the room at the ceiling level. At night, the chimney action stops and the dampers are closed to prevent the downflow of cooled air back into the room.

As the Sun's rays warm the outer surface of the concrete, a wave of heat begins to move slowly (at about 5 cm or 2 in. per hour) through the wall. The thickness of the wall is chosen to delay the arrival of the wave of warmth at the indoor surface until after sunset, when the long-wave radiation emitted by the wall will be welcome. Heat also flows outward from the wall to the glazing and thence to the outdoor environment, but this can be minimized by the use of movable insulation within the air space. Windows may be introduced into the Trombe wall, and the building need not be restricted to one story.

Another indirect-gain system, shown in Fig. 7, uses enclosed bags of water, called thermoponds, which are supported by a heat-conducting roof-ceiling and covered by horizontally movable insulating panels. This system collects solar energy during winter days by rolling the insulation away to a storage area, thus admitting Sun's rays into the thermoponds. At night, the insulating panels are rolled back to provide the insulation needed to retain the collected heat.

In summer, the operation is reversed, and the insulating panels are rolled back at night to expose the thermoponds to the sky and thus to enable them to dissipate heat that has been absorbed from the building during the day. Convection to cool night air, when it is available, and radiation to the sky on clear nights are two of the natural processes by which heat can be rejected. The third process, and the most potent, is evaporation, which can occur when the ponds are provided with exposed water surfaces by flooding or spraying them. *See* HEAT TRANSFER.

Another indirect-gain system makes use of a greenhouse attached to the south side of a building to gather heat during the day. The wall between the glazed space and the building is warmed by the Sun's rays and by contact with the warm air in the greenhouse. Additional heat storage can be transferred into the house by opening windows or vents.

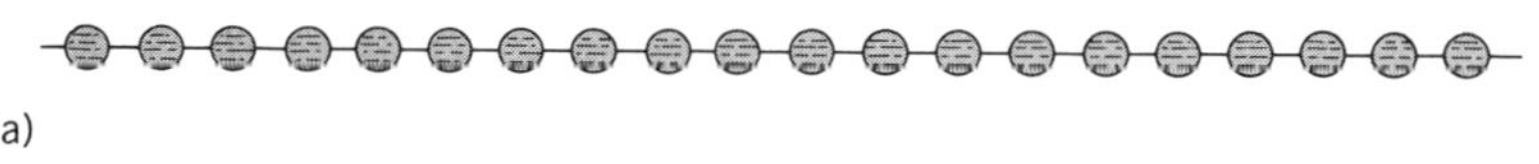

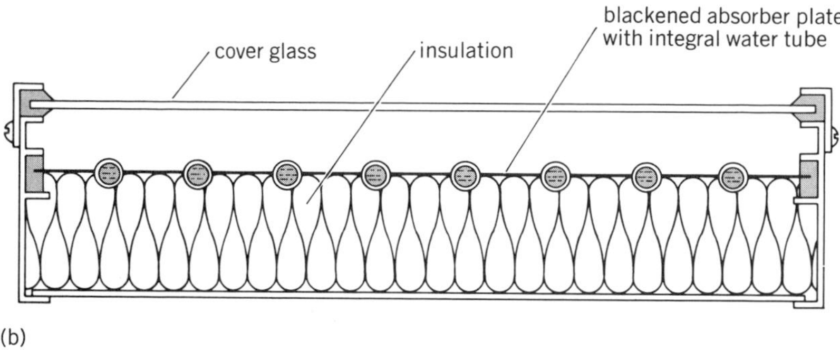

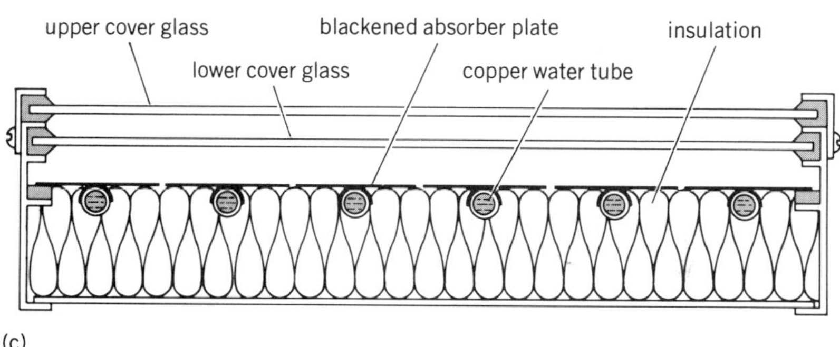

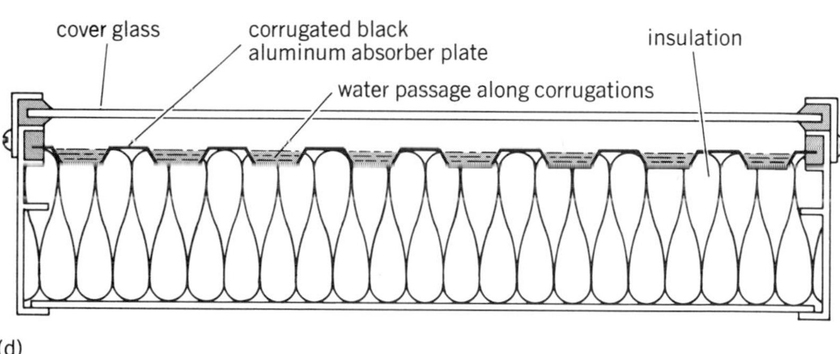

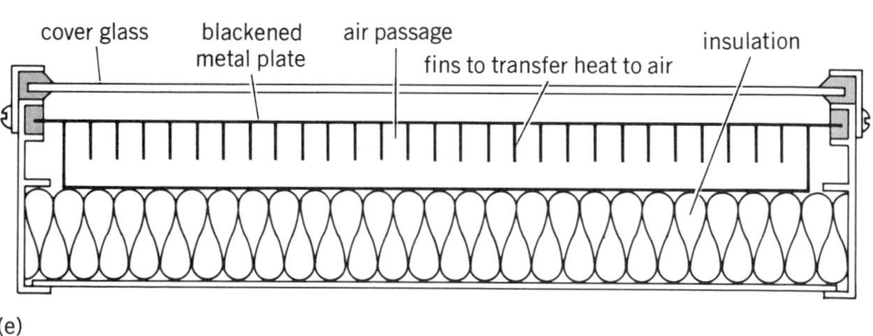

Fig. 2. Flat-plate collector types. (*a*) Extruded black plastic swimming pool heater. (*b*) Single-glazed collector with tube-in-sheet. (*c*) Double-glazed collector with extruded aluminum absorber element and copper water tube. (*d*) Open-flow collector used in the Thomason "Solaris" system. (*e*) Air heater using finned aluminum heat absorber.

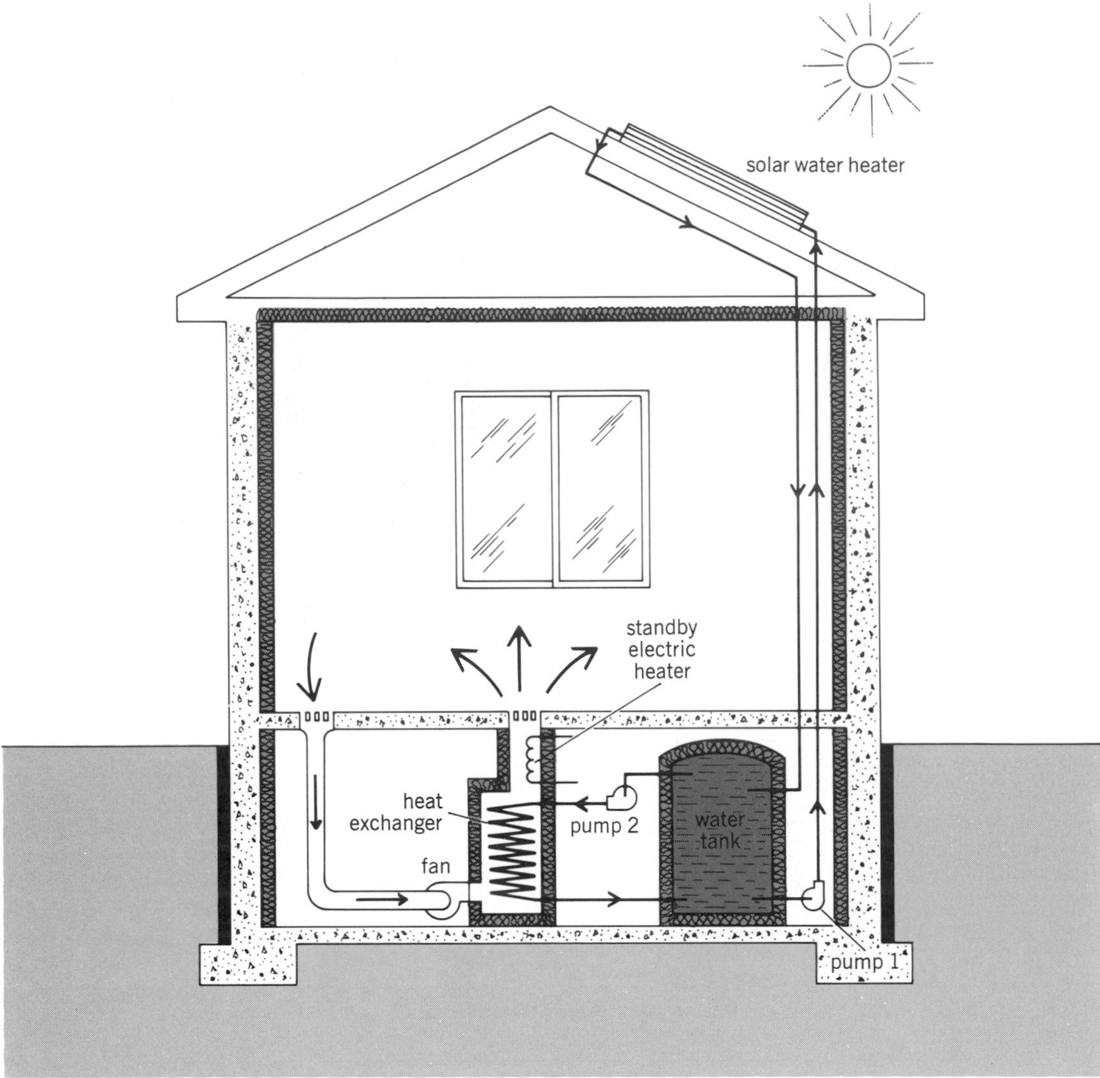

Fig. 3. Active system using roof-mounted solar water heater with storage tank in basement.

The warmth which is gathered even during cold winter days can thus be used to aid in heating the residence as well as to provide an environment in which plants can grow.

Isolated-gain systems. The isolated-gain system shown in Fig. 8 uses a thermosyphon heater to warm air, which rises as it becomes hot, until it encounters the entrance to the rock-bed thermal storage component. The rocks abstract heat from the air which moves downward as it becomes cooler. The air eventually descends to the bottom of the rock bed and thence to the inlet of the heater. The cycle continues as long as the Sun shines on the collector, and heat is stored throughout most of the day. At night, when the occupants of the house sense the need for heat, the damper settings are changed and warm air from the top of the rock bed is admitted to the living space. At the same time, cool air from the north wall of the house is allowed to flow downward through ducts to enter the base of the rock bed. There the air is warmed, and the heat which has been stored during the day is used to warm the house at night.

Active systems. Active systems may use either water (Fig. 3) or air (Fig. 4) to transport heat from roof-mounted south-facing collectors to storage in rock beds or water tanks. The stored heat may be withdrawn and used directly when air is the transfer fluid. When the heat is collected and stored as hot water, fan-coil units are generally used to transfer the heat to air which is then circulated through the warmed space. Each system has both advantages and disadvantages.

Air cannot freeze or cause corrosion, and leakage is not a serious problem. Water requires relatively small pipes compared with the ducts needed to transport the same amount of heat in the form of warm air. Water tanks can store more than three times as much heat as rocks in a given volume per degree of temperature change, but rock beds are considerably lower in cost than water tanks, and rocks can tolerate virtually any temperature while water will boil at 212°F (100°C) unless its pressure is raised above atmospheric. The choice between water and air systems must be made carefully, taking into account the many features of each.

Heat storage. Heat storage can be accomplished with specific-heat materials such as water or rocks, which can store and discharge heat by simply undergoing a change in temperature. A differ-

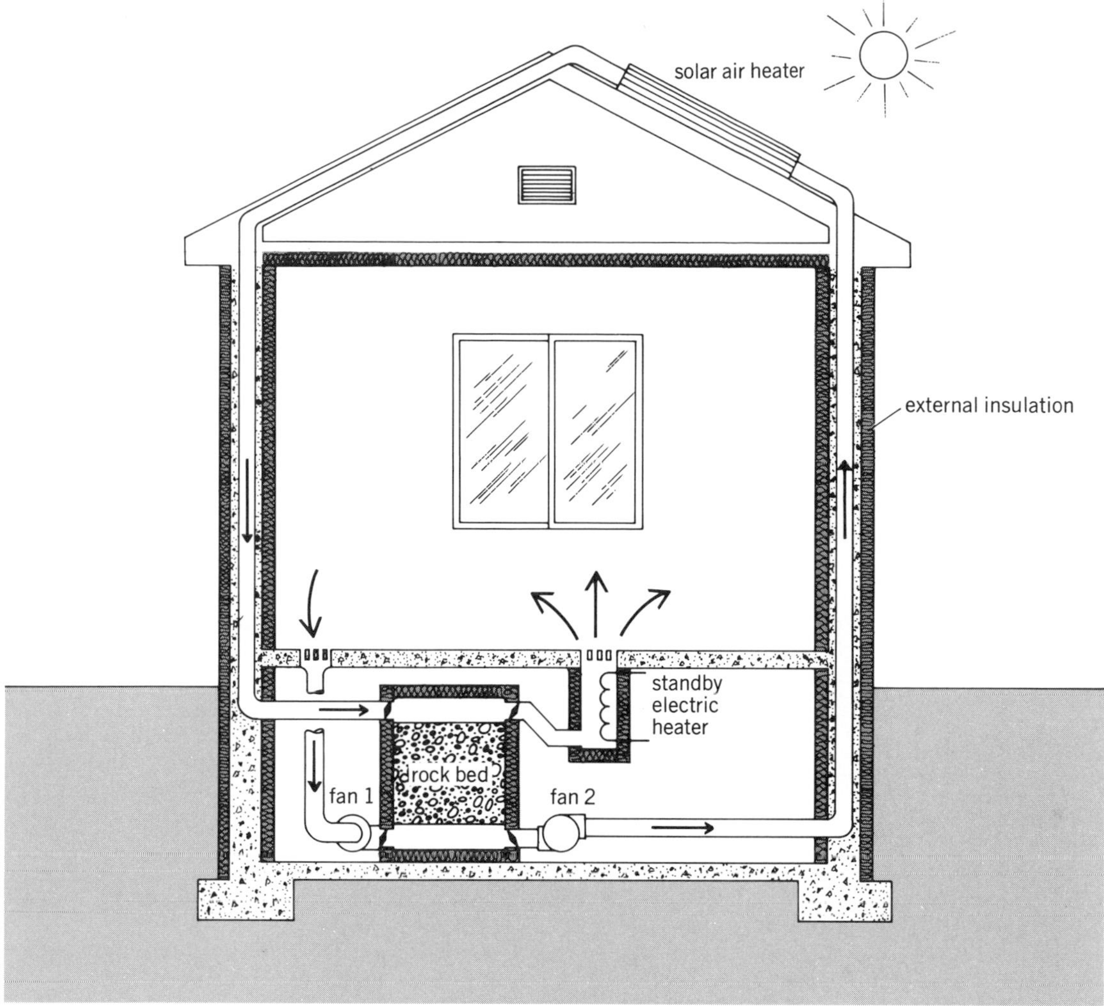

Fig. 4. Active system using air heater with rock-bed heat storage in basement.

ent process is involved in the heat-of-fusion materials, which, like water at 32°F (0°C), can freeze when heat is removed and melt again when the same amount of heat is returned to them. Ice has the unique property of increasing in volume by 7% compared with the water from which it is formed, but virtually all other substances become denser when they solidify and so their solid components sink instead of floating. A number of materials are available which, like the well-known Glauber's salt, change their state from liquid to solid and back again at temperatures which are useful for either heating or cooling. None has yet been found, however, which possesses all of the attributes required for successful application to solar heating and cooling systems.

Standby electrical sources. Standby electrical sources are shown in Figs. 3 and 4, since some method of providing warmth must be included for use when the Sun's radiant energy is inadequate for long periods of time. The standby heater may be something as simple as a wood-burning stove or fireplace, or as complex as an electrically powered heat pump. Simple electrical resistance heaters are frequently chosen for this service because of their low first cost, but their operating cost can become excessive in applications where they must be used for long periods of time. The rapidly escalating cost of electricity must be borne in mind when the standby energy source is selected. *See* COMFORT HEATING; HEAT PUMP.

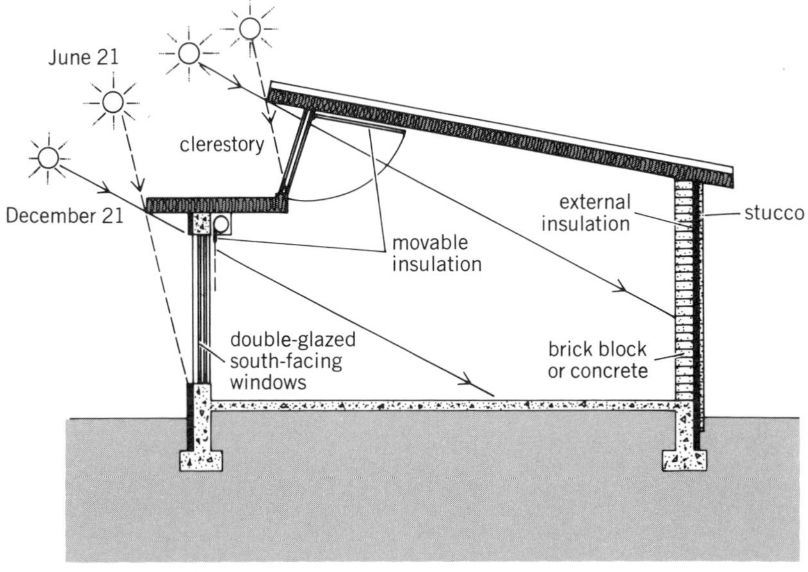

Fig. 5. Direct-gain passive system using large south-facing windows.

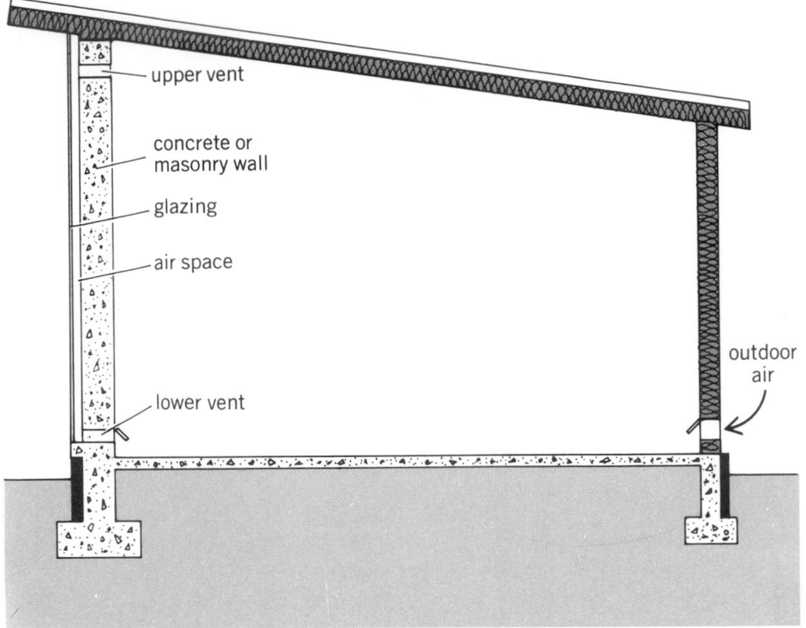

Fig. 6. Indirect-gain passive system using Trombe wall.

SERVICE WATER HEATING

Solar water heating for domestic, commercial, or industrial purposes is one of the oldest and most successful applications of solar-thermal technology. The most widely used water heater, and one that is suitable for use in relatively warm climates where freezing is a minor problem, is the thermosyphon type (Fig. 9). A flat-plate collector of one of the types shown in Fig. 2 is generally used with a storage tank which is mounted above the collector. A source of water is connected near the bottom of the tank, and the hot water outlet is connected to its top. A downcomer pipe leads from the bottom of the tank to the inlet of the collector, and an insulated return line runs from the top of the collector to the upper part of the storage tank which is also insulated.

The system is filled with water, and when the Sun shines on the collector, the water in the tubes is heated. It then becomes less dense than the water in the downcomer, and the heated water rises by thermosyphon action into the storage tank. It is replaced by cool water from the bottom of the tank, and this action continues as long as the Sun shines on the collector with adequate intensity. When the Sun moves away, the thermosyphon action stops, but the glazing on the collector minimizes heat loss from this component and the insulation around the storage tank enables it to retain the collected heat until needed. The system must be prevented from freezing, and the components must be strong enough to withstand operating water pressure.

For applications where the elevated storage tank is undesirable or where very large quantities of hot water are needed, the tank is placed at ground level. A small pump circulates the water in response to a signal from a controller which senses the temperatures of the collector and the water near the bottom of the tank. Heat exchangers may also be used with water at operating pressure within the tubes of the exchanger and the collector water outside to eliminate the necessity of using high-pressure collectors. Antifreeze substances may be added to the collector water to eliminate the freezing problem, but care must be taken to prevent any possibility of cross-flow between the collector fluid and the potable water. Many plumbing codes require special heat exchangers which interpose two metallic walls between the two fluid circuits. *See* HOT-WATER HEATING SYSTEM.

COOLING

Cooling can be provided by both active and passive systems.

Active cooling systems. The two feasible types of active cooling systems are Rankine cycle and absorption. The Rankine cycle system uses solar collectors to produce a vapor (steam or one of the fluorocarbons generally known as Freon) to drive an engine or turbine. A condenser must be used to condense the spent vapor so it can be pumped back through the vaporizer. The engine or turbine drives a conventional refrigeration compressor which produces cooling in the usual manner. *See* RANKINE CYCLE.

The absorption system uses heat at relatively high temperature (180 to 200°F or 82 to 93°C) and employs a hygroscopic solution of lithium bromide to absorb water vapor at a very low pressure. The evaporating water is cooled to a temperature low enough (about 45°F or 7°C) to provide the chilled water needed to produce air conditioning. The heat introduced into this cycle by the high-temperature activating water and the low-temperature chilling water is removed by a third stream of water which has been cooled to below the outdoor air temperature by passing through an evaporative cooling tower. The cycle, simple in principle but much more complex in practice, is in wide use in large buildings where the necessary hot activating water is available either from solar collectors or as waste heat from other processes. Use of either Rankine cycle or absorption solar cooling systems for residences is restricted to a few experimental installations, but both hold promise for future development.

Passive cooling systems. Passive cooling systems make use of three natural processes: convec-

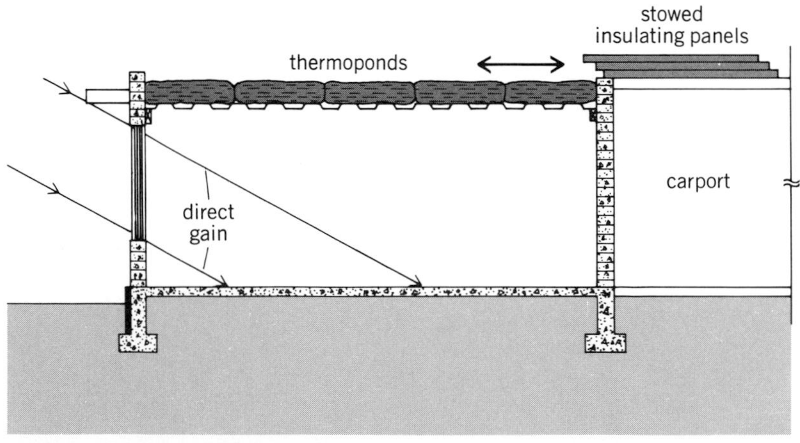

Fig. 7. Indirect-gain "Skytherm" system with movable insulation.

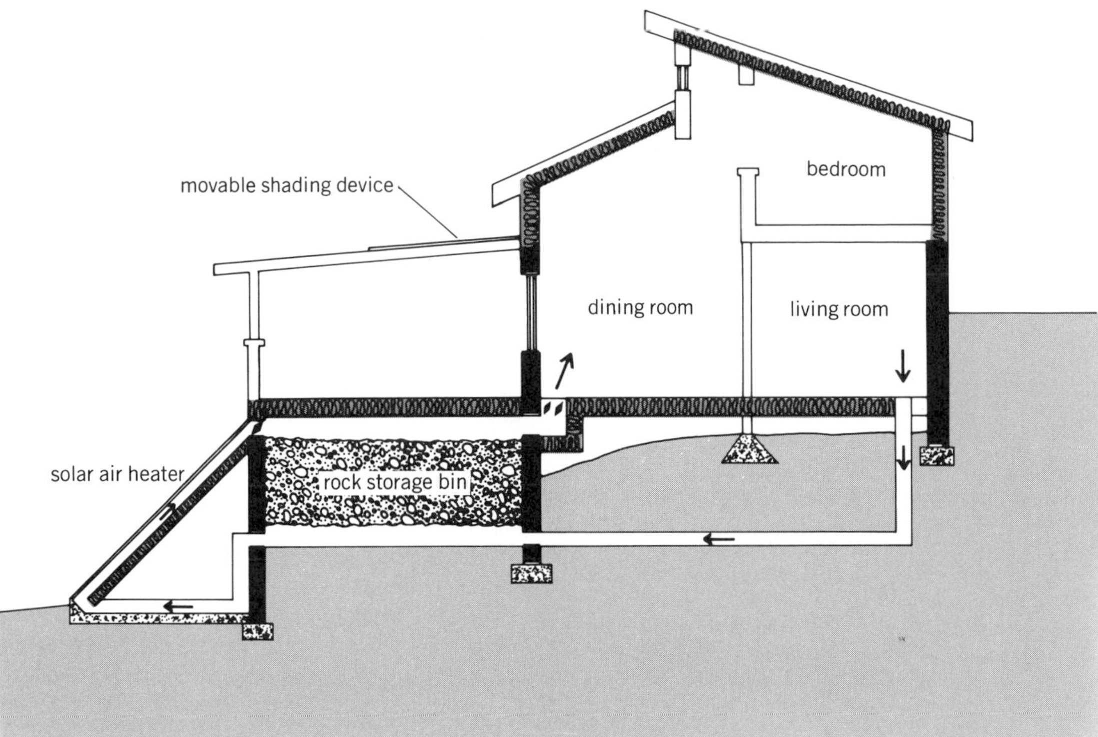

Fig. 8. Isolated-gain system with thermosyphon air heater and rock bed.

tion cooling with night air; radiative cooling by heat rejection to the sky on clear nights; and evaporative cooling from water surfaces exposed to the atmosphere. The effectiveness of each of these processes depends upon local climatic conditions. In the hot and dry deserts of the southwestern United States, radiation and evaporation are the two most useful processes. In hot and humid regions, other processes must be employed which involve more complex equipment that can first dehumidify and then cool the air.

POWER GENERATION

Electric power can be generated from solar energy by two processes, the first of which uses the Rankine cycle described previously. In order to attain vapor temperatures high enough to make the cycle operate efficiently, concentrating collectors must be used. For installations of moderate size, parabolic troughs (Fig. 10*a*) are adequate, but for very large utility-type plants the "power tower" concept (Fig. 10*b*) is preferred. Concentrators can use only the radiation which comes directly from the solar disk, so some means of tracking must be provided and energy must be stored during the daylight hours to ensure continuity of operation during intermittent cloudy periods. Storage systems for night-long operation are not yet available. Concentrating systems cannot make effective use of the diffuse radiation that comes from the sky, and so their use at the present stage of development of solar technology is limited to areas where bright sunshine is generally available.

Electric power can also be produced from solar radiation, without need for a thermal cycle, by photovoltaic cells which convert radiant energy directly into electricity. Originally used primarily in the United States and Soviet space programs, silicon solar batteries can now be produced at a cost low enough to justify their use in special circumstances for terrestrial applications. One development uses concentrators similar to those in Fig. 10*a* to enable a single solar cell to do the work of ten; the heat which must be removed to make the cells function efficiently can be used for many purposes requiring hot water or hot air. Solar cells produce direct current which can be stored in ordinary automobile-type batteries; the direct current can be converted to alternating current at the standard frequency, 60 Hz in the United States, and at any necessary voltage. One of the goals of the Department of Energy is the reduction of the cost of photovoltaic cells to the point where residential rooftop units become economically feasible. *See* SOLAR CELL.

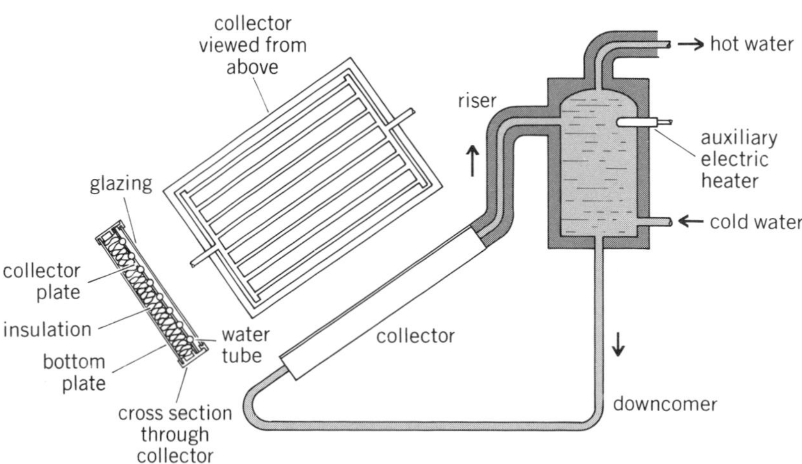

Fig. 9. Passive domestic water heater using thermosyphon system.

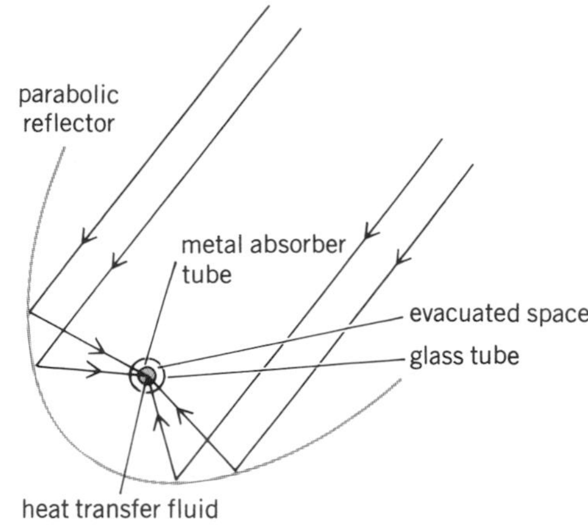

(a)

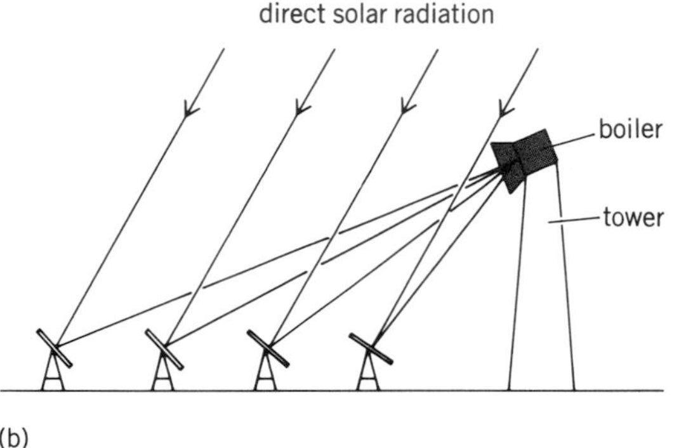

(b)

Fig. 10. Concentrating collectors. (*a*) Parabolic trough, for use in installations of moderate size. (*b*) Power tower, for large utility-type plants.

CONCLUSION

Heating of space and service hot water is currently practical and economical. Passive heating and cooling are coming into wide use as the cost of alternative fuels continues to rise. Active cooling and power generation are too expensive for general use, but it is expected that the cost of these systems will diminish in the future. The major unsolved problem is how to store large quantities of energy for long periods of time. *See* ENERGY STORAGE.

[JOHN I. YELLOTT]

Bibliography: AIA Research Corporation, *A Survey of Passive Systems*, 1979; American Society of Heating, Refrigerating and Air Conditioning Engineers, *ASHRAE Handbook of Fundamentals*, chap. 26, 1977; Copper Development Association, Inc., *Solar Energy Systems*, 1979; F. Daniels, *Direct Use of the Sun's Energy*, 1964; B. Liu and R. C. Jordan (eds.), *Solar Energy Utilization for Heating and Cooling of Buildings*, ASHRAE Pub. GRP 170, 1977; E. Mazria, *The Passive Solar Energy Book*, 1979; A. A. M. Sayigh (ed.), *Solar Energy Engineering*, 1977; D. Watson (ed.), *Energy Conservation through Building Design*, 1979; J. I. Yellott, in Portola Institute, *Energy Primer*, chap. 1, 1975; J. I. Yellott, Solar energy utilization for heating and cooling, in *ASHRAE Handbook of Applications*, chap. 59, 1978.

Solar radiation

The electromagnetic radiation and particles (electrons, protons, and rarer heavy atomic nuclei) emitted by the Sun. Electromagnetic energy has been observed over the whole spectrum with wavelengths varying from 0.1 A (10^{-9}cm) to 30 km. The bulk of the energy is in the spectrum of visible light (4000–8000 A). The solar spectrum doubtless extends far beyond the observed limits in both directions. The total power is 3.86×10^{33} ergs/sec.

The Sun also emits a continuous stream of electrons with shorter bursts of electron and proton showers sufficiently intense to affect the ionization of the upper terrestrial atmosphere. These sporadic particles have energies from a few thousand to a few billion electron volts. The lower-energy particles are much more abundant, but those of high energy are sufficient to occasionally damage the solid-state circuitry of spacecraft. The physical mechanism of high-energy particle emission is not understood, but is closely associated with the more energetic forms of solar activity. *See* SOLAR ENERGY.

[JOHN W. EVANS]

Solid-state battery

A battery in which both the electrodes and the electrolyte are solid-state materials. Solid-state batteries have been developed as a logical application for a new class of materials, the solid electrolytes, otherwise known as superionic conductors. These materials are very good conductors for ions, but they maintain a very high resistance toward electronic conduction. These properties are necessary for any electrolyte; the high ionic conductivity minimizes the internal resistance of the battery, while the high electronic resistance minimizes its self-discharge rate. Typical materials are β-Al_2O_3 for Na^+ conduction, Ag_4RbI_5 for Ag^+ conduction, and Li-β-Al_2O_3 for Li^+ conduction. The ionic conductivities of these materials are high. At room temperature the ionic conductivity of a single crystal of β-Al_2O_3 is 0.035 $(\Omega\text{-cm})^{-1}$, similar to the conductivity of a 0.1 *M* HCl solution. The polycrystalline electrolytes typically employed in batteries have ionic conductivities about one-fifth that of the corresponding single crystals, the actual value being a function of the grain size and the packing density.

Use. By far the greatest use of solid electroytes in the battery field is in the high-energy-density Na/β-Al_2O_3/S battery. However, since this battery operates at about 350°C, where Na and S are in the liquid phase, this is not a solid-state battery.

Solid-state batteries generally fall into the low-energy-density category. This limitation arises from the polarization, at high current densities, at the solid-solid interfaces. However, in certain applications the inherent advantages of solid-state batteries are more important than the net energy content or the current output. From a manufacturing point of view, they are easier to miniaturize, and there is no problem with the electrolyte leaking out of the container. They have long shelf life and are resistant to shock and vibration. They have a wider operating temperature range, and there is no abrupt decrease in performance such as that which occurs when a liquid electrolyte freezes or boils.

Configuration. The configuration of a solid-state battery is shown in the illustration. The polycrystalline, pressed electrolyte is interspaced between a metallic anode and the solid cathode material. The electrodes are applied to the electrolyte by mechanically pressing the materials together. It is also possible to deposit a thin film of the electrode materials on the electrolyte by a vacuum deposition technique. A carbon current collector is generally employed on the cathode side, with this carbon sometimes being intermixed with the cathode material. These batteries are generally employed in cardiac pacemakers, cameras, and electronic watches. Batteries are classified according to the ion transported across the solid electrolyte.

Silver battery. Silver batteries are based on the Ag^+-ion-conducting Ag_4RbI_5 electrolyte. This material has an ionic conductivity of 0.26 $(\Omega\text{-cm})^{-1}$ and an electronic conductivity of less than 10^{-11} $(\Omega\text{-cm})^{-1}$. An available battery is $Ag/Ag_4RbI_5/RBI_3$, C. This battery has an open-circuit voltage of 0.66 V. Depending on the application, different numbers of cells can be stacked inside the battery. For low-current-drain applications a battery with five cells supplies 25 μA at 2.5–3.0 V for 144 hr at 25°C. High current drain can be achieved by stacking 13 cells in a case 2 mm high and 24 mm in diameter. This provides currents of 1.2–1.5 A for 8 ms with voltages in the range of 2.9–3.7 V. At low current densities, less than 10 mA/cm², more than 85% of the theoretical battery capacity can be utilized. Although the electrolyte is thermodynamically unstable below 27°C, the battery can be operated down to −60°C in a dry environment. Such a low operating temperature is unique to a solid-state battery.

Lithium battery. The lithium-iodine battery employed for use with cardiac pacemakers is constructed with a lithium anode and an iodine–polyvinyl pyridine (PVP) cathode. This cathode has a tarlike consistency in the fresh battery, then solidifies gradually as the battery is discharged. A unique aspect of the battery is that the electrolyte is formed in place; when the above two components are assembled, the Li and I_2 react to form a layer of LiI between them. The LiI is a solid electrolyte conductive toward Li^+ ions. More LiI is continually formed during the discharge of the battery; hence the resistance of the electrolyte continually increases, with a subsequent decrease in the output voltage. The battery has an initial potential of 2.8 V.

The most important characteristics of this battery, for this application, are its high reliability and very low rate of self-discharge (6% over a period of years). A typical implantable battery has a volume of 12 cm³, weighs 33 g, and has a 3 A-hr capacity. Since pacemakers typically draw only 20–30 μA, the lifetime of such a battery is 8–10 years. The battery is considered to be discharged when the output voltage decreases to 1.8 V. Clinically, this is detected by monitoring the pulse rate of the patient; the battery is exchanged when the pulse reaches some predefined minimum value. *See* BATTERY; LITHIUM PRIMARY CELL.

[GABRIEL G. BARNA]

Bibliography: B. B. Scholtens and W. Van Gool, *Solid Electrolytes*, 1978; R. M. Dell, *Electrode Processes in Solid State Ionics*, 1976.

Steam condenser

A heat-transfer device used for condensing steam to water by removal of the latent heat of steam and its subsequent absorption in a heat-receiving fluid, usually water, but on occasion air or a process fluid. Steam condensers may be classified as contact or surface condensers.

In the contact condenser, the condensing takes place in a chamber in which the steam and cooling water mix. The direct contact surface is provided by sprays, baffles, or void-effecting fill such as Rashig rings. In the surface condenser, the condensing takes place separated from the cooling water or other heat-receiving fluid (or heat sink). A metal wall, or walls, provides the means for separation and forms the condensing surface.

Both contact and surface condensers are used for process systems and for power generation serving engines and turbines. Modern practice has confined the use of contact condensers almost entirely to such process systems as those involving vacuum pans, evaporators, or dryers, and to condensing and dehumidification processes inherent in vacuum-producing equipment such as steam jet ejectors and vacuum pumps. The steam surface condenser is used chiefly in power generation but is also used in process systems, especially in those in which condensate recovery is important. Air-cooled surface condensers are used increasingly in process systems and in power generation when the availability of cooling water is limited.

Water-cooled surface condensers used for power generation are of the high-vacuum type. Their main purpose is to effect a back pressure at the turbine exhaust. To achieve an economical station heat rate and fixed cost, these condensers must be designed for high heat-transfer rates, minimum steam-side-pressure loss, and effective air removal. The usual power plant steam condenser incorporates a single steam-condensing chamber into which the steam turbine exhausts. Multipressure condensers, having two or more steam-condensing chambers operating at different pressures, have been found to be more economical than single-pressure condensers for large installations, which generally use cooling towers as the final heat sink. Most multipressure condensers are compartmented internally, on the steam-condensing side, with the cooling water arranged to flow through the tubes in one direction from inlet to outlet. The mean cooling water temperature progressively increases as the water flows from compartment to compartment, with back pressure on the turbine increasing proportionately. Economical performance is effected with multiple-exhaust pressure turbines when each exhaust pressure section of the turbine is designed for the back pressure produced in its respective pressure section in the condenser. A rise in water temperature in the order of 20–30°F (11–17°C) is necessary to effect required pressure differences in the turbine.

Air-cooled surface condensers used for power generation are designed to operate at higher back pressures than water-cooled condensers. This is consistent with their lower overall heat-transfer rate, which is of the order of 10–12 Btu/(°F)(ft²)(hr) [57–68 W/(°C)(m²)] in contrast with the normal 600–700 Btu/(°F)(ft²)(hr) [3.4–4.0 kW/(°C)(m²)]

SOLID STATE BATTERY

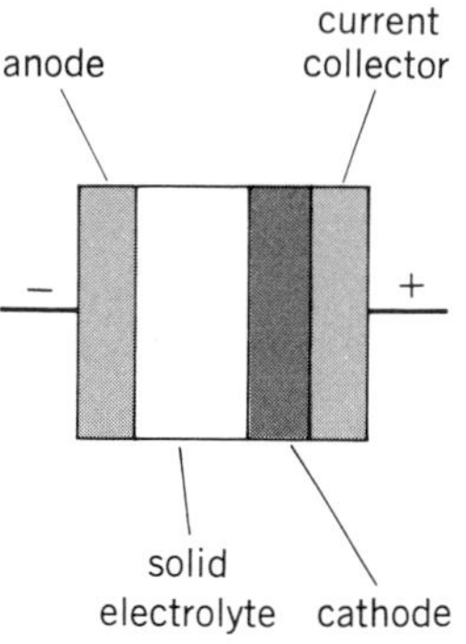

Schematic diagram of solid-state battery.

characteristic of water-cooled steam condensers. To compensate for these large differences in heat transfer, greater temperature differences are required between the condensing steam and the cooling air than those needed with conventional water cooling. Steam is condensed on the inside of the heat-transfer tubes, and airflow is usually produced by fans across their outside surface. The tubes are of the extended-surface type with fins on the outside. The normal ratio of outside to inside surface is in the order of 10 or 12 to 1. The use of air as the cooling medium for steam condensers used in smaller industrial steam power plants is increasing rapidly. Air-cooled steam condensers for the production of electric power have been confined to smaller plants; however, with the decreasing availability of water for cooling, the application of air-cooled steam condensers to the larger steam electric plants may be anticipated.

Condenser sizes have increased with the increase in size of turbine generators. The largest, in a single shell, is approximately 485,000 ft² (45,000 m²); the steam-condensing space occupies a volume of 80,000 ft³ (2,265 m³). Large units have been built with the surface divided equally among two or three shells. Power plant condensers require 35–100 lb of cooling water to condense 1 lb (35–100 kg for 2 kg) of steam, or, if air-cooled, 2500–5000 ft³ of air for 1 lb (150–300 m³ for 1 kg) of steam. Normally, about 0.5 ft² (0.05 m²) of surface is required for water-cooled condensers, and about 10 ft² (0.9 m²) of inside tube surface for air-cooled condensers is required for each kilowatt of generating capacity. *See* STEAM TURBINE; VAPOR CONDENSER. [JOSEPH F. SEBALD]

Bibliography: *See* VAPOR CONDENSER.

Steam engine

A machine for converting the heat energy in steam to mechanical energy of a moving mechanism, for example, a shaft. The steam engine dominated the industrial revolution and made available a practical source of power for application to stationary or transportation services. The steam power plant could be placed almost anywhere, whereas other means of power generation were more restricted, experiencing such site limitations as an elevated water supply, wind, animal labor, and so on. The steam engine can utilize any source of heat in the form of steam from a boiler. It was developed in sizes which ranged from that of children's toys to 25,000 hp (18.6 MW), and it was adaptable to pressures up to 200 psi (1.4 MPa). It reached its zenith in the 19th century in stationary services such as drives for pumping plants; drives for air compressor and refrigeration units; power supply for factory operations with shafting for machine shops, rolling mills, and sawmills; and drives for electric generators as electrical supply systems were perfected. Its adaptability to portable and transportation services rested largely on its development of full-rated torque at any speed from rest to full throttle; its speed variability at the will of the operator; and its reversibility, flexibility, and dependability under the realities of stringent service requirements. These same features favored its use for many stationary services such as rolling mills and mine hoists, but the steam engine's great contribution was in the propulsion of small and large ships, both naval and merchant. Also, in the form of the steam locomotive, the engine made the railroad the pratical way of land transport. Most machine elements known today had their origin in the steam engine: cylinders, pistons, piston rings, valves and valve gear crossheads, wrist pins, connecting rods, crankshafts, governors, and reversing gears.

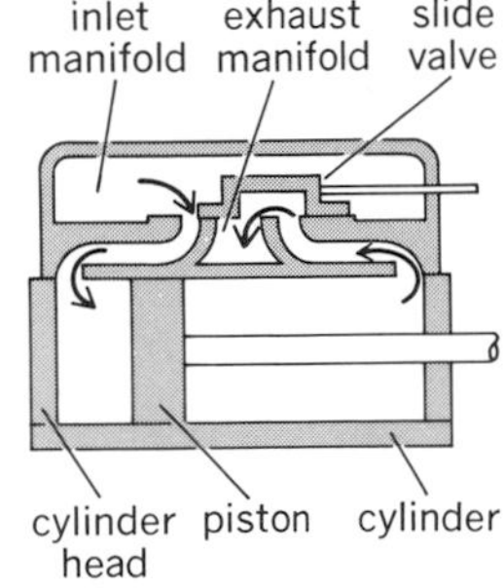

Fig. 2. Single-ported slide valve on counterflow double-acting cylinder.

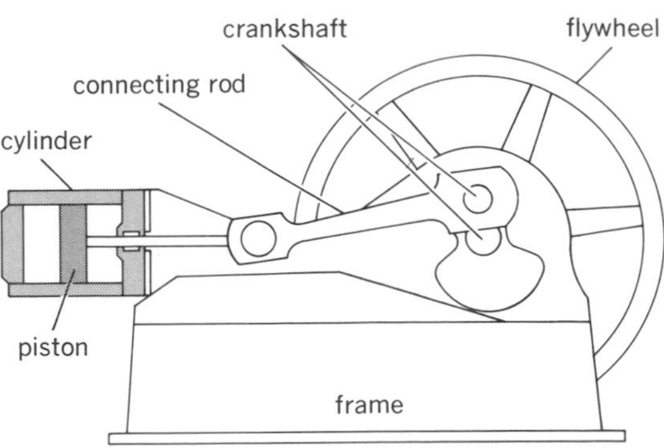

Fig. 1. Principal parts of horizontal steam engine.

The 20th century saw the practical end of the steam engine. The steam turbine with (1) its high speed (for example, 3600± rpm); (2) its utilization of maximum steam pressures (2000–5000 psi or 14–34 MPa), maximum steam temperatures (1100±°F or 600±°C), and highest vacuum (29± in. Hg or 98± kPa); and (3) its large size (1,000,000± kW) led to such favorable weight, bulk, efficiency, and cost features that it replaced the steam engine as the major prime mover for electric generating stations. The internal combustion en-

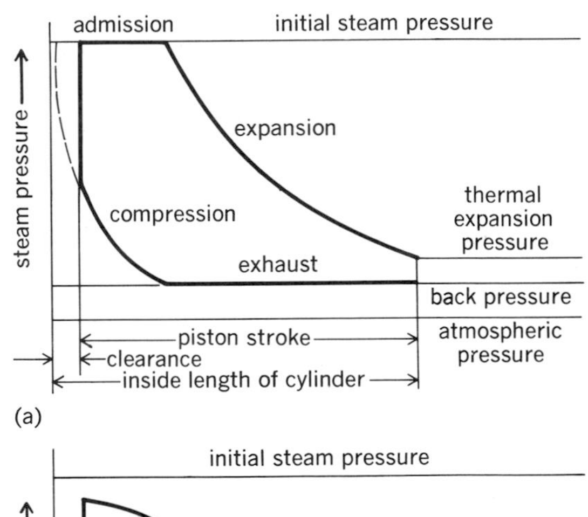

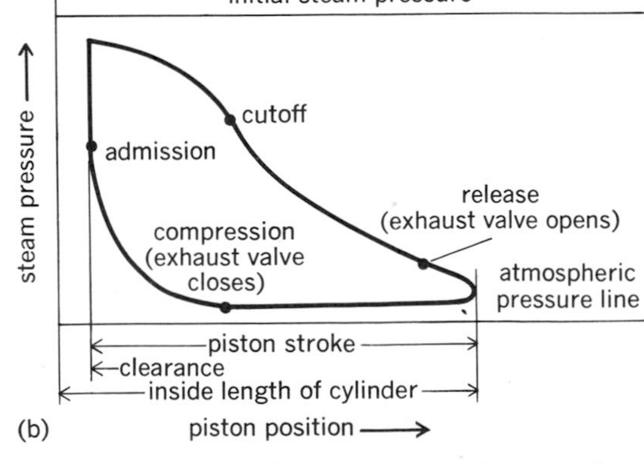

Fig. 3. Events during one cycle of a piston operation. (*a*) In ideal engine. (*b*) As depicted on the indicator card of a noncondensing steam engine.

gine, especially the high-speed automotive types which burn volatile (gasoline) or nonvolatile (diesel) liquid fuel, offers a self-contained, flexible, low-weight, low-bulk power plant with high thermal efficiency that has completely displaced the steam locomotive with the diesel locomotive and marine steam engines with the motorship and motorboat. It is the heart of the automotive industry, which produces in a year 10,000,000± vehicles that are powered by engines smaller than 1000 hp (750 kW). Because of the steam engine's weight and speed limitations, it was excluded from the aviation field, which has become the exclusive preserve of the internal combustion piston engine or the gas turbine. *See* DIESEL ENGINE; GAS TURBINE; INTERNAL COMBUSTION ENGINE; STEAM TURBINE; TURBINE.

Cylinder action. A typical steam reciprocating engine consists of a cylinder fitted with a piston (Fig. 1). A connecting rod and crankshaft convert the piston's to-and-fro motion into rotary motion. A flywheel tends to maintain a constant-output angular velocity in the presence of the cyclically changing steam pressure on the piston face. A D slide valve admits high-pressure steam to the cylinder and allows the spent steam to escape (Fig. 2). The power developed by the engine depends upon the pressure and quantity of steam admitted per unit time to the cylinder.

Indicator card. The combined action of valves and piston is most conveniently studied by means of a pressure-volume diagram or indicator card (Fig. 3). The pressure-volume diagram is a thermodynamic analytical method which traces the sequence of phases in the cycle. It may be an idealized operation (Fig. 3*a*), or it may be an actual picture of the phenomena within the cylinder (Fig. 3*b*) as obtained with an instrument commonly known as a steam engine indicator. This instrument, in effect, gives a graphic picture of the pressure and volume for all phases of steam admission, cutoff, expansion, release, exhaust, and compression. It is obtained as the engine is running and shows the conditions which prevail at any instant within the cylinder. The indicator is a useful instrument not only for studying thermodynamic performance, but for the equally important operating knowledge of inlet and exhaust valve leakage and losses, piston ring tightness, and timing correctness. *See* THERMODYNAMIC PROCESSES.

The net area of the indicator card shows, thermodynamically, the work done in the engine cylinder. By introducing the proper dimensional quantities, power output can be measured. Thus if the net area within the card is divided by the length, the consequent equivalent mean height is the average pressure difference on the piston during the cycle and is generally called the mean effective pressure p, usually expressed in lb/in.^{2} With a cylinder dimension of piston area a (in in.^{2}), length of piston stroke l (in ft), and n equal to the number of cycles completed per minute, the equation given below holds. It has often been referred to as

$$\text{Indicated horsepower} = \frac{plan}{33{,}000}$$

the most important equation in mechanical engineering.

Engine types. Engines are classified as single- or double-acting, and as horizontal (Fig. 1) or vertical

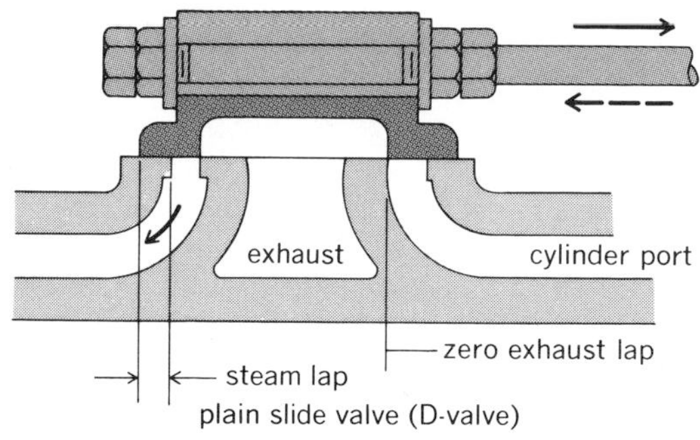

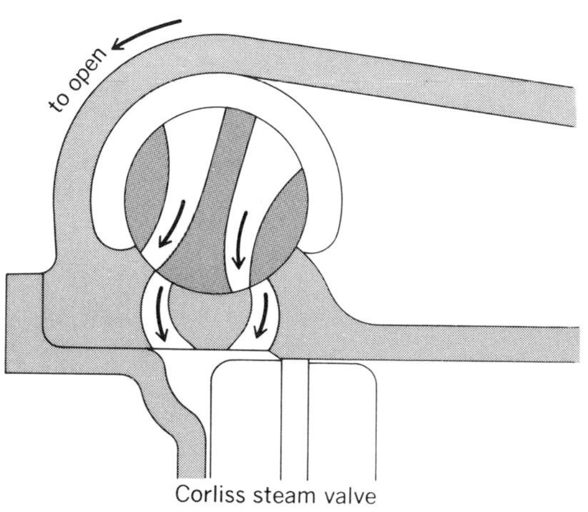

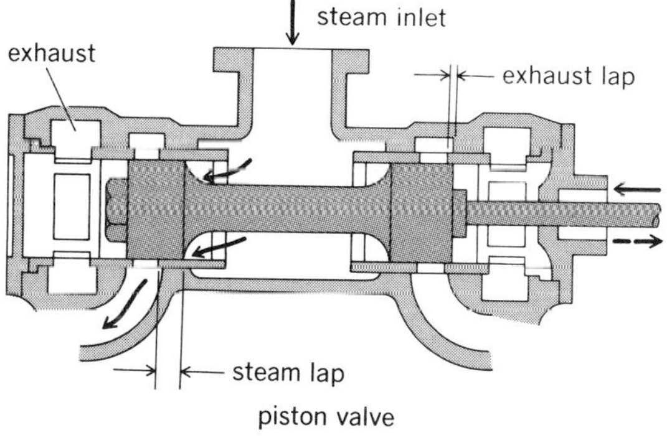

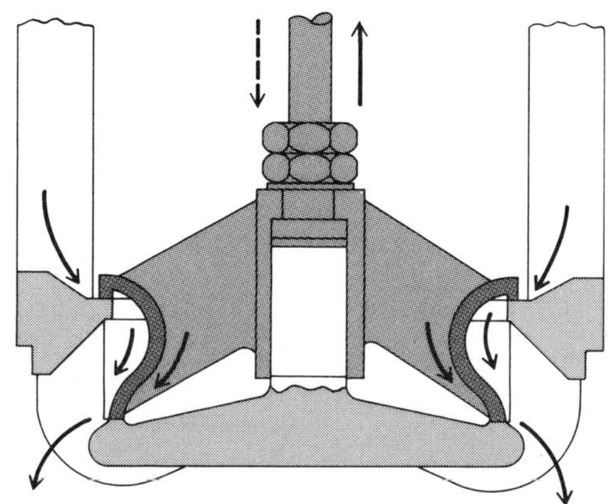

Fig. 4. Steam engine valves in closed positions. The arrows indicate the path steam will travel when the valves are open.

depending on the direction of piston motion. If the steam does not fully expand in one cylinder, it can be exhausted into a second, larger cylinder to expand further and give up a greater part of its initial energy. Thus, an engine can be compounded for double or triple expansion. In counterflow engines, steam enters and leaves at the same end of the cylinder; in uniflow engines, steam enters at the end of the cylinder and exhausts at the middle.

Steam engines can also be classed by functions, and are built to optimize the characteristics most desired in each application. Stationary engines drive electric generators, in which constant speed is important, or pumps and compressors, in which constant torque is important. Governors acting through the valves hold the desired characteristic constant. Marine engines require a high order of safety and dependability.

Valves. The extent to which an actual steam piston engine approaches the performance of an ideal engine depends largely on the effectiveness of its valves. The valves alternately admit steam to the cylinder, seal the cylinder while the steam expands against the piston, and exhaust steam from the cylinder. The many forms of valves can be grouped as sliding valves and lifting valves (Fig. 4). *See* CARNOT CYCLE; THERMODYNAMIC CYCLE.

D valves (Fig. 2) are typical sliding valves where admission and exhaust are combined. A common sliding valve is the rocking Corliss valve; it is driven from an eccentric on the main shaft like other valves but has separate rods for each valve on the engine. After a Corliss valve is opened, a latch automatically disengages the rod and a separate dashpot abruptly closes the valve. Exhaust valves are closed by rods, as are other sliding valves.

Lifting valves are more suitable for use with high-temperature steam. They, too, are of numerous forms. The poppet valve is representative.

Valves are driven through a crank or eccentric on the main crankshaft. The crank angle is set to open the steam port near dead center, when the piston is at its extreme position in the cylinder. The angle between valve crank and connecting-rod crank is slightly greater than 90°, the excess being the angle of advance. So that the valves will open and close quickly, they are driven at high velocity and thus travel further than is necessary to open and close the ports. The additional travel of a sliding valve is the steam lap and the exhaust lap. The greater the lap, the greater the angle of advance to obtain the proper timing of the valve action.

Engine power is usually controlled by varying the period during which steam is admitted. A shifting eccentric accomplishes this function or, in releasing Corliss and in poppet valves, the eccentric is fixed and cutoff is controlled through a governor to the kickoff cams or to the latch that allows the valves to be closed by their dashpots.

For high engine efficiency, the ratio of cylinder volume after expansion to volume before expansion should be high. The volume before expansion into which the steam is admitted is the volumetric clearance. It may be determined by valve design and other structural features. For this reason, valves and ports are located so as not to necessitate excessive volumetric clearance.

[THEODORE BAUMEISTER]

Bibliography: T. Baumeister (ed.), *Standard Handbook for Mechanical Engineers*, 8th ed., 1978.

Steam-generating furnace

An enclosed space provided for the combustion of fuel to generate steam. The closure confines the products of combustion and is capable of withstanding the high temperatures developed and the pressures used. Its dimensions and geometry are adapted to the rate of heat release, to the type of fuel, and to the method of firing so as to promote complete burning of the combustible and suitable disposal of ash. In water-cooled furnaces the heat absorbed materially affects the temperature of gases at the furnace outlet and contributes directly to the generation of steam.

Furnace walls. Prior to 1925, most furnaces were constructed of firebrick. As the capacity and the physical size of boiler units increased and the suspension burning of pulverized coal was developed, limitations were imposed by the height of the refractory walls that could be made self-supporting at high temperatures and by the inability of refractories to resist the fluxing action of molten fuel ash. Thus refractory is used primarily for special-purpose furnaces, for the burner walls of small boilers, or for the arches of stoker-fired boilers to promote the ignition of the fuel.

The limitations of refractory constructions can be extended somewhat by cooling the brickwork with air flowing through channels in the structure, or by sectionalizing the wall into panels and transferring the structural load to external air-cooled steel or cast-iron supporting members. The heat absorbed can be recovered by using the cooling air for combustion, and the low rate of heat transfer through the refractory helps to maintain high furnace temperatures, thus accelerating ignition and the burning of the fuel.

The low tensile strength of refractories restricts the geometrical shapes that can be built, making it difficult to provide overhanging contours or roof closures. Thus sprung arches or shaped tiles suspended from steel must be used. Many refractory mixtures having air-setting or hydraulic-setting properties are available, and they can be used to form monolithic structures by ramming, guniting, or pouring in forms.

Water-cooling of all or most of the furnace walls is almost universally used in boiler units of all capacities and for all types of fuel and methods of firing. The water-cooling of the furnace walls reduces the transfer of heat to the structural members and, consequently, their temperatures are within the limits that provide satisfactory strength and resistance to oxidation. Water-cooled tube constructions facilitate large furnace dimensions and optimum arrangements of the roof, hopper, and arch, as well as the mountings for the burners and the provision for screens, platens, or division walls to increase the amount of heat-absorbing surface exposed in the combustion zone. External heat losses are small and are further reduced by the use of insulation.

Furnace wall tubes are spaced on close centers to obtain maximum heat absorption and to facilitate ease of ash removal (see illustration). The so-called tangent tube construction, wherein adjacent tubes are almost touching with only a small clearance provided for erection purposes, has been frequently used. However, most new boiler units use membrane-tube walls, in which a steel bar or

STEAM-GENERATING FURNACE

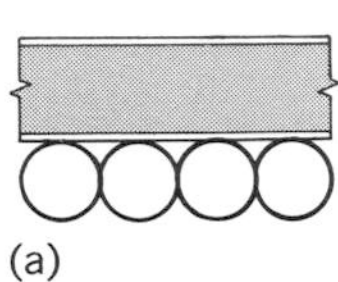

(a)

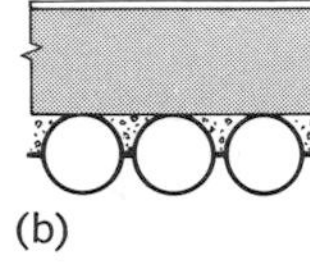

(b)

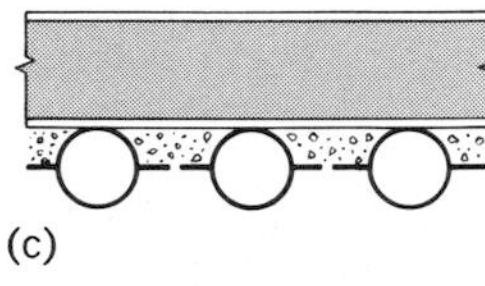

(c)

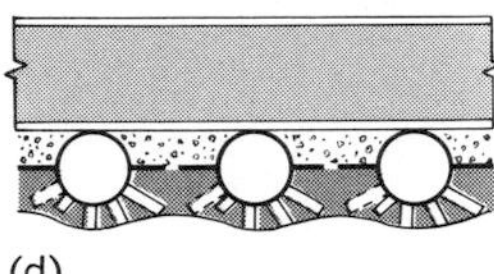

(d)

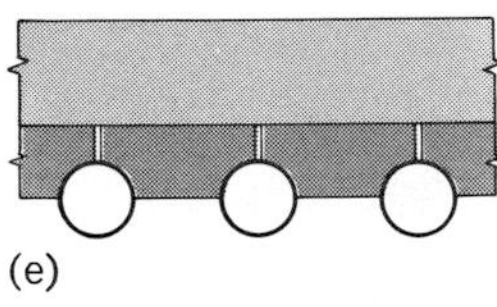

(e)

(f)

Water-cooled furnace-wall constructions. (*a*) Tangent tube. (*b*) Welded membrane and tube. (*c*) Flat stud and tube. (*d*) Full stud and refractory-covered tube. (*e*) Tube and tile. (*f*) Tubes backed by refractory. (*From T. Baumeister, ed., Marks' Standard Handbook for Mechanical Engineers, 7th ed., McGraw-Hill, 1967*)

membrane is welded between adjacent tubes. This construction facilitates the fabrication of water-cooled walls in large shop-assembled tube panels. Less effective cooling is obtained when the tubes are wider spaced and extended metal surfaces, in the form of flat studs, are welded to the tubes; if even less cooling is desired, the tube spacing can be increased and refractory installed between or behind the tubes to form the wall enclosure.

Furnace walls must be adequately supported, with provision for thermal expansion and with reinforcing buckstays to withstand the lateral forces resulting from the difference between the furnace pressure and the surrounding atmosphere. The furnace enclosure walls must prevent air infiltration when the furnace is operated under suction, and they must prevent gas leakage when the furnace is operated at pressures above atmospheric.

Additional furnace cooling, in the form of tubular platens, division walls, or wide-spaced screens, often is used; in high heat input zones the tubes may be protected by refractory coverings anchored to the tubes by studs.

Heat transfer. Heat-absorbing surfaces in the furnace receive heat from the products of combustion and, consequently, lower the furnace gas temperature. The principal mechanisms of heat transfer take place simultaneously and include intersolid radiation from the fuel bed or fuel particles, nonluminous radiation from the products of combustion, convection from the furnace gases, and conduction through any deposits and the tube metal. The absorption effectiveness of the surface is greatly influenced by the deposits of ash or slag.

Analytical solutions to the transfer of heat in the furnaces of steam-generating units are extremely complex, and it is most difficult, if not impossible, to calculate furnace-outlet gas temperatures by theoretical methods. Nevertheless, the furnace-outlet gas temperature must be predicted accurately because this temperature determines the design of the remainder of the unit, particularly that of the superheater. Calculations therefore are based upon test results supplemented by data accumulated from operating experience and judgment predicated upon knowledge of the principles of heat transfer and the characteristics of fuels and slags. *See* STEAM-GENERATING UNIT.

[GEORGE W. KESSLER]

Bibliography: Babcock and Wilcox Co., *Steam: Its Generation and Use*, 1955; T. Baumeister (ed.), *Standard Handbook for Mechanical Engineers*, 8th ed., 1978; M. W. Thring, *The Science of Flames and Furnaces*, 1962.

Steam-generating unit

The wide diversity of parts, appurtenances, and functions needed to release and utilize a source of heat for the practical production of steam at pres-

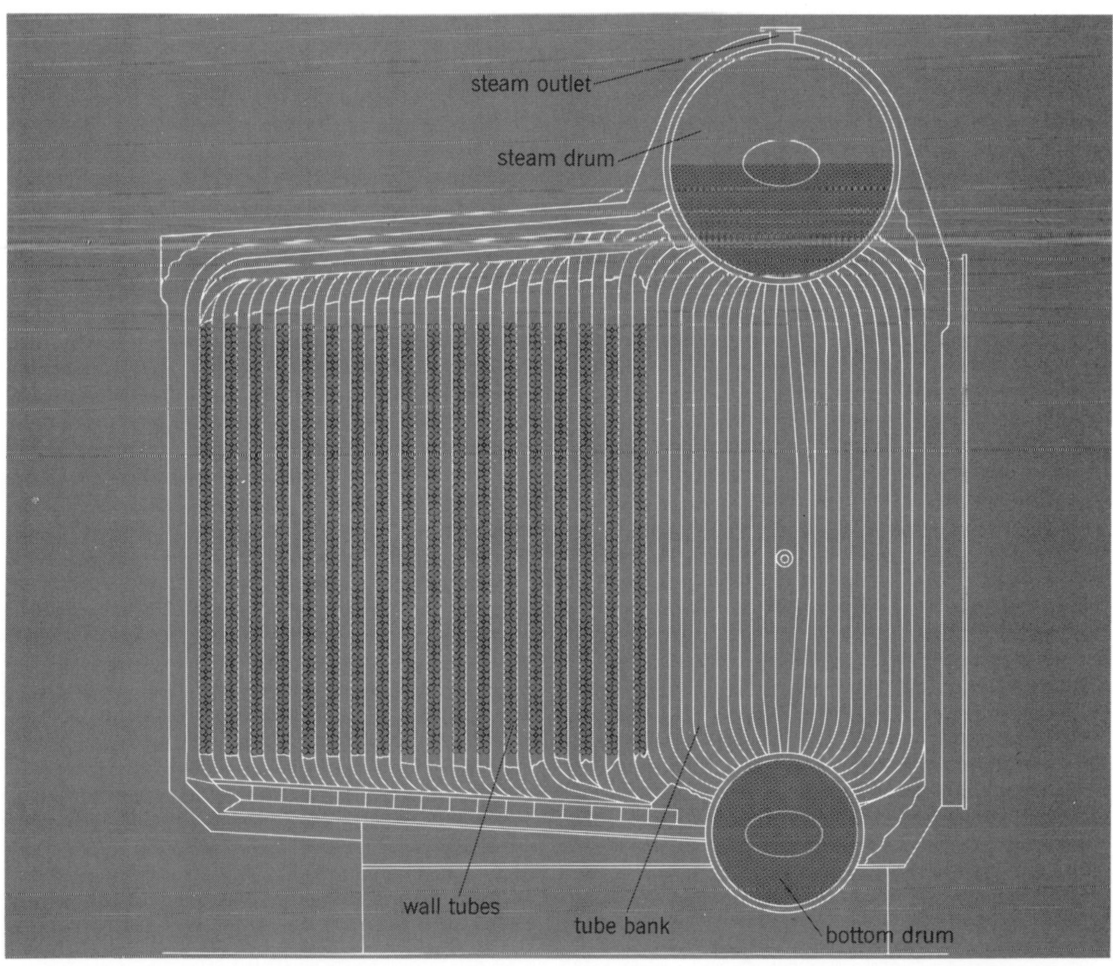

Fig. 1. Package-type water-tube boiler, single gas pass, oil- or gas-fired, 50,000 lb/hr (6.3 kg/sec) steam output for industrial service. (*From T. Baumeister, ed., Standard Handbook for Mechanical Engineers, 7th ed., 1967*)

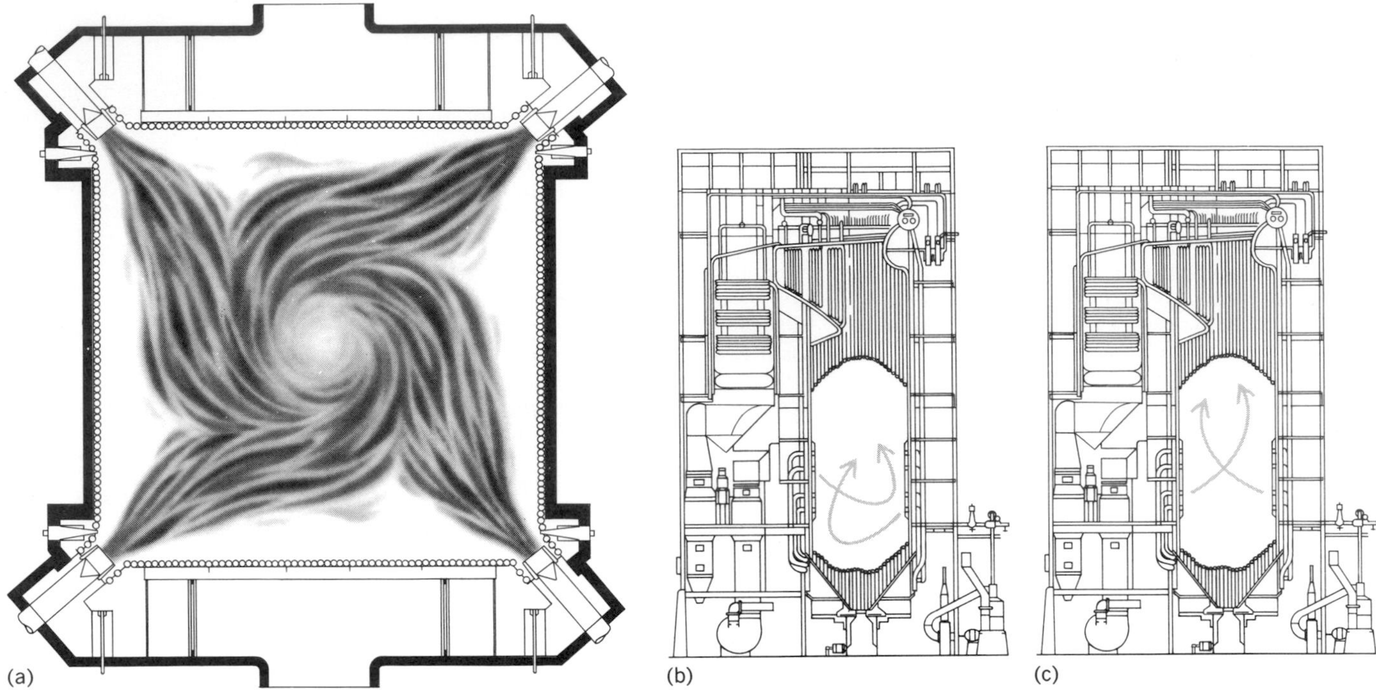

Fig. 2. Large, central-station steam-generating unit, with pump-assisted circulation, corner-fired with tangential, tilting burners for pulverized coal, oil, or gas. (a) Plan; (b) burners tilted down; (c) burners tilted up. Output 2,000,000 lb steam/hr (250 kg/sec) at 2400 psi (16.5 MPa), 1000°F (540°C), and 1000°F reheat. (*Combustion Engineering, Inc.*)

sures to 5000 psi (34 MPa) and temperatures to 1100°F (600°C), often referred to as a "steam boiler for brevity.

The essential steps of the steam-generating process include (1) a furnace for the combustion of fuel, or a nuclear reactor for the release of heat by fission, or a waste heat system; (2) a pressure vessel in which feedwater is raised to the boiling temperature evaporated into steam, and generally superheated beyond the saturation temperature;

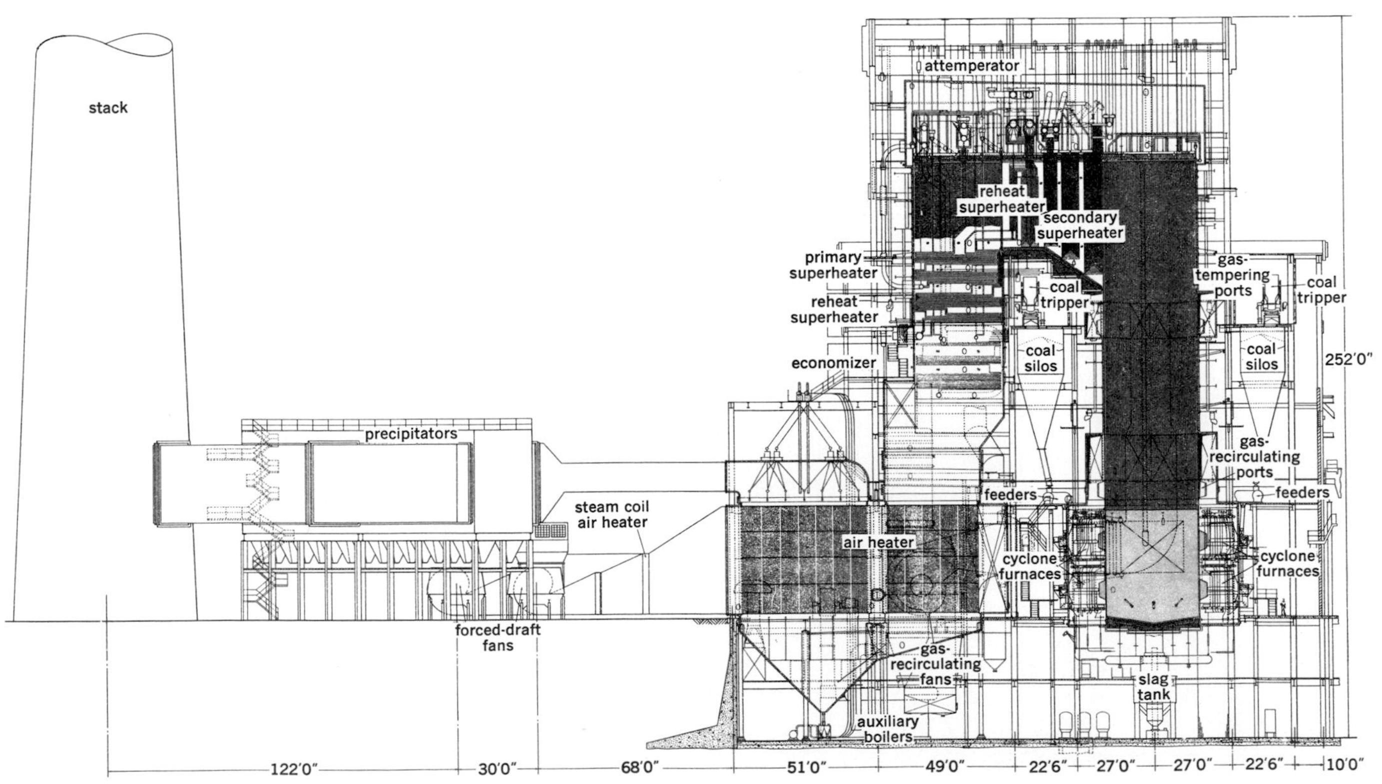

Fig. 3. Supercritical central-station, steam-generating unit with pressurized cyclone-furnace coal firing. Output 5,000,000 lb steam/hr (630 kg/sec) at 3600 psi (24.8 MPa), 1000°F (540°C), and 1000°F reheat. 1 ft = 0.3048 m. (*Babcock and Wilcox Co.*)

and (3) in many modern central station units, a reheat section or sections for resuperheating steam after it has been partially expanded in a turbine. This aggregation of functions requires a wide assortment of components, which may be variously employed in the interests, primarily, of capacity and efficiency in the steam-production process. Their selection, design, operation, and maintenance constitute a complex process which, in the limit, calls for the highest technical skill.

Service requirements. The wide diversity of commercial steam-generating equipment stems from attempts to accommodate design to the dictates of the imposed service conditions.

Safety. The pressure parts are a potential bursting hazard and must be designed, maintained, and operated under the most stringent codes, such as the ASME Boiler and Pressure Vessel Code. The codes and regulations are backed by the police power of the state.

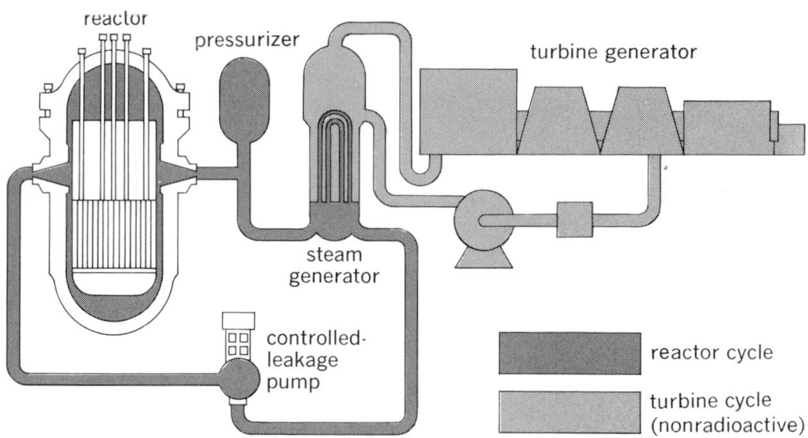

Fig. 4. Pressurized water-cooled reactor (PWR) and steam generator for a large central-station electric-generating plant illustrating the essential components. Typical performance 6,000,000 lb steam/hr (750 kg/sec) at 700 psi (4.8 MPa), saturated steam. (*Westinghouse Electric Corp.*)

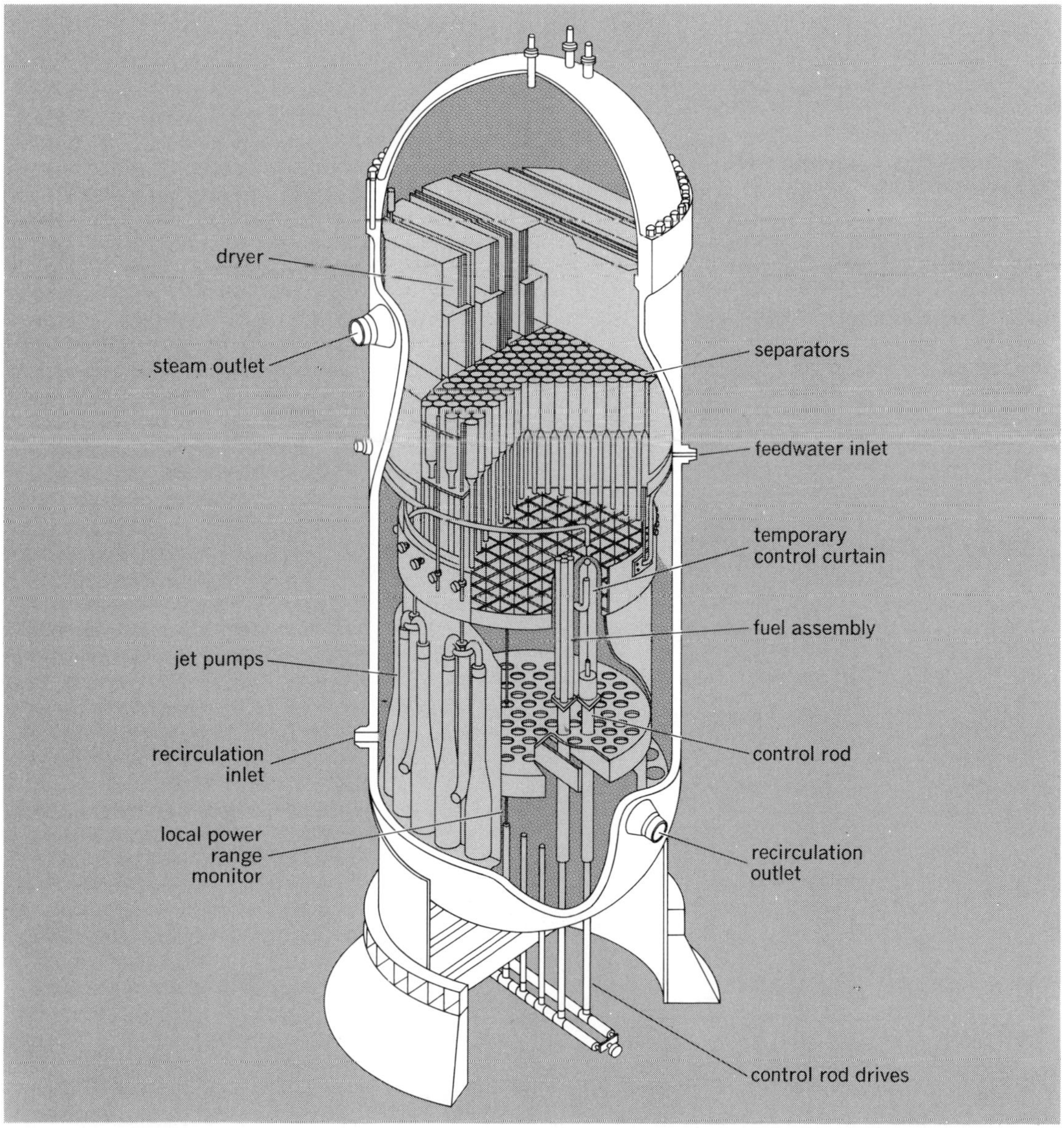

Fig. 5. Boiling-water reactor (BWR) for a large central-station electric-generating plant showing essential components. Typical performance 6,000,000 lb/hr (750 kg/sec) at 1000 psi (6.9 MPa), saturated steam. (*General Electric Co.*)

Performance characteristics of some steam-generating units

Boiler type	Fuel (potential)	Steam output, lb/hr (kg/sec)	Steam pressure, psi (MPa)	Steam temperature, °F (°C)	Boiler efficiency, %
Horizontal return tubular (HRT), fire-tube	Coal, oil, gas	10,000 (1.3)	150 (1.0)	Dry and saturated	70
Large industrial, natural-circulation, water-tube	Coal, oil, gas	300,000 (38)	1400 (9.7)	900 (480)	88
Large central station, natural-circulation	Coal, oil, gas	2,000,000 (250)	2700 (18.6)	1050/1050 (570/570)	90
Large central station, supercritical, once-through	Coal, oil, gas	5,000,000 (630)	3600 (24.8)	1000/1000/1000 (540/540/540)	90
Boiling-water reactor (BWR)	Uranium	6,000,000 (750)	1000 (6.9)	Dry and saturated (550±°F or 290±°C)	98±
Pressurized-water reactor (PWR), hot water, 4 loops	Uranium	6,000,000 (750)	700 (4.8)	Dry and saturated (500±°F or 260±°C)	98±
Electric boiler, small	200± volts	2,000 (0.25)	150 (1.0)	Dry and saturated	98±

Shape. Physical shape must fit the limits imposed. The differences between marine and stationary applications are an example.

Bulk. Space occupied has various degrees of value, and the equipment must best comply with these dictates.

Weight. Frequently, weight is the lowest common denominator in determining portability and economic suitability.

Setting. Confinement of heat source operations and the ensuing transport function will vary the details of the setting, for example, the use of metal or of refractory enclosure.

Character of labor. Highly skilled labor, as in atomic power plants, can utilize design details that would be prohibitive in areas where the labor is essentially primitive.

Cleanability. Impurities on the heat-transfer and heat-release surfaces must not impair safe or efficient operation; surfaces must be cleanable to give acceptable reliability and performance.

Life. Short or long life and short or long operating periods between overhauls are vital to the selection of a preferred design.

Efficiency. Inherent high efficiency can be built into a design, but economics will dictate the efficiency chosen; the full realization of built-in high efficiency requires the services of skilled operators and firemen.

Cost. The overall cost of steam in cents per 1000 lb or 1000 kg is the ultimate criterion, reflecting investment, operating, and maintenance expenses in commercial operations.

Adaptation to use. These service requirements differ in each installation so that a wide variety of designs is available in the commercial markets. There is no one best design for universal application, but propriety rests in the proper selection. A sampling of the diversity of acceptable steam-generating units is represented in Figs. 1–5. These designs illustrate the various degrees of applicability of component parts such as steam drums and tubes; fuel-burning equipment and furnaces; draft systems for air supply and flue gas removal; blowers, fans, and stacks; heat traps such as economizers and air preheaters; structural steel for the support of parts with ample provision for expansion; a casing or setting with possible utilization of water walls, refractory, and insulation; pumps for boiler feed and for water circulation; fuel- and refuse-handling systems; feedwater purification systems; blowdown systems and soot-blowing equipment; and a wide assortment of accessories and instruments such as pressure gages, safety valves, water-level indicators, and sophisticated automatic control with its elaborate interlocks for foolproof, efficient operation and complete computerization.

The maintenance of a safe working temperature for metal parts requires ample circulation of water over the steam-generating parts and ample circulation of steam over the superheating and reheating parts. Water circulation in the generating sections may be by natural convection processes where ample physical height gives the circulatory pressure difference between a column of water in the downcomer and a column of steam mixed with water in the riser parts. With high operating steam pressures, such as 1400 psi (9.7 MPa), pump-assisted circulation is often selected. With supercritical operation (above 3200 psi or 22.1 MPa) there is no difference in density between liquid and vapor so that forced-flow, once-through principles supersede water recirculation practice (Fig. 3). Such supercritical steam generators have no steam or separating drums and, like nuclear steam generators (Figs. 4 and 5), require the highest purity of feed and boiler water (typically 10^{-9} part of impurities) to avoid deposits in heated circuits or the transport of solids to the turbine.

The illustrations give an indication of the diversity of problems met by a selected group of commercial steam generators. The table gives some performance data for an assorted group of representative units. *See* BOILER ECONOMIZER; FIRE-TUBE BOILER; WATER-TUBE BOILER.

[THEODORE BAUMEISTER]

Bibliography: American Society of Mechanical Engineers, *Boiler and Pressure Vessel Code*, 1968; Babcock and Wilcox Co., *Steam: Its Generation and Use*, 1955; T. Baumeister (ed.), *Standard Handbook for Mechanical Engineers*, 7th ed., 1967; Combustion Engineering, Inc., *Combustion Engineering*, 1966.

Steam heating

A heating system that uses steam generated from a boiler. The steam heating system conveys steam through pipes to heat exchangers, such as radiators, convectors, baseboard units, radiant panels, or fan-driven heaters, and returns the resulting condensed water to the boiler. Such systems normally operate at pressure not exceeding 15 pounds per square inch gage (psig), and in many designs the condensed steam returns to the boiler by gravity because of the static head of water in the return piping. With utilization of available operating and safety control devices, these systems can be designed to operate automatically and safely with minimum maintenance and attention.

One-pipe system. In a one-pipe steam heating system, a single main serves the dual purpose of supplying steam to the heat exchanger and conveying condensate from it. Ordinarily, there is but one connection to the radiator or heat exchanger, and this connection serves as both the supply and return; separate supply and return connections are sometimes used. Because steam cannot flow through the piping or into the heat exchanger until all the air is expelled, it is important to provide automatic air-venting valves on all exchangers and at the ends of all mains. These valves may be of a type which closes whenever steam or water comes in contact with the operating element but which also permits air to flow back into the system as the pressure drops. A vacuum valve closes against subatmospheric pressure in order to prevent the return of air.

Two-pipe system. A two-pipe system is provided with two connections from each heat exchanger, and in this system steam and condensate flow in separate mains and branches. A vapor two-pipe system operates at a few ounces above atmospheric pressure, and in this system a thermostatic trap is located at the discharge connection from the heat exchanger which prevents steam passage, but permits air and condensation to flow into the return piping.

When the steam condensate cannot be returned by gravity to the boiler in a two-pipe system, an alternating return lifting trap, condensate return pump, or vacuum return pump must be used to force the condensate back into the boiler. In a condensate return-pump arrangement, the return piping is arranged for the water to flow by gravity into a collecting receiver or tank, which may be located below the steam-boiler waterline. A motor-driven pump controlled from the boiler water level then forces the condensate back to the boiler.

In large buildings extending over a considerable area, it is difficult to locate all heat exchangers above the boiler water level or return piping. For these systems a vacuum pump is used that maintains a suction below atmosphere up to 25± in. (max) of mercury in the return piping, thus creating a positive return flow of air and condensate back to the pumping unit. Subatmospheric systems are similar to vacuum systems, but in contrast provide a means of partial vacuum control on both the supply and return piping so that the steam temperature can be regulated to vary the heat emission from the heat exchanger in direct proportion to the heat loss from the structure.

Figure 1 depicts a two-pipe vacuum heating system which uses a condensation pump as a mechanical lift for systems where a part of the heating system is below the boiler room. Note that the low section of the system is maintained under the same vacuum conditions as the remainder of the system. This is accomplished by connecting the vent from

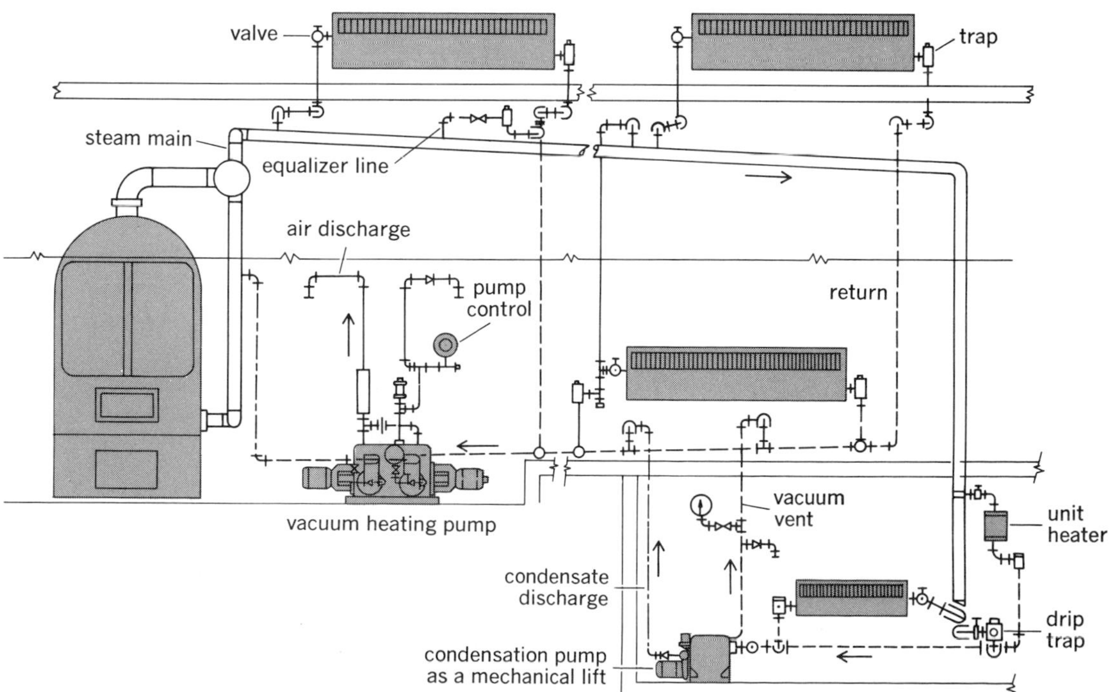

Fig. 1. Layout of a vacuum heating system with condensation and vacuum pumps.

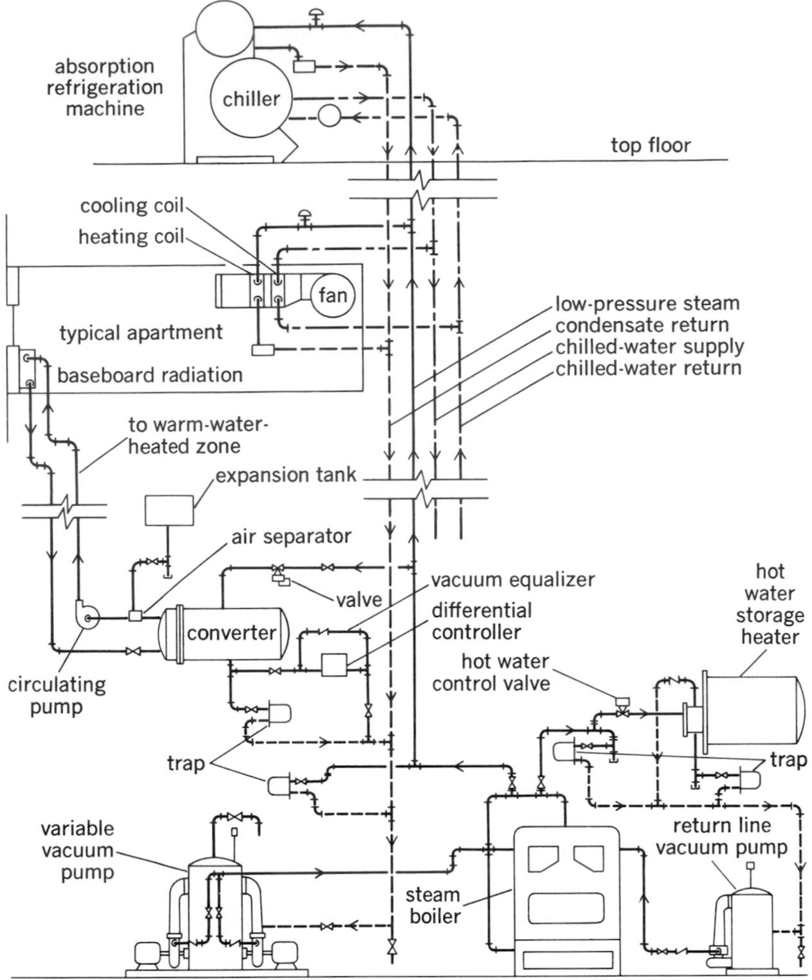

Fig. 2. Diagram of a heating and cooling system for an apartment building employing a low-pressure steam boiler.

the receiver and pump discharge to the return pipe located above the vacuum heating pump.

With the wide acceptance of all-year air conditioning, low-pressure steam boilers have been used to produce cooling from absorption refrigeration equipment. With this system the boiler may be used for primary or supplementary steam heating as diagrammed in Fig. 2.

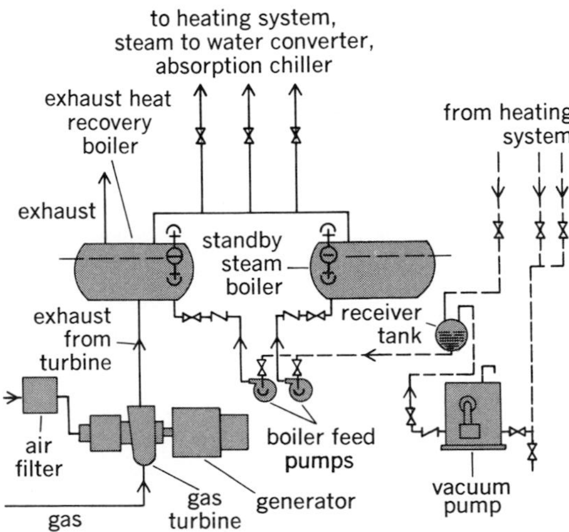

Fig. 3. Steam system using waste heat.

Exhaust from gas-driven or oil-driven turbines or engines may be used in waste heat boilers or separators, along with a standby boiler to produce steam for a heating system as is illustrated in Fig. 3.

Another source of steam for heating is from a high-temperature water source (350–450°F or 180–230°C) through the use of a high-pressure water to low-pressure steam heat exchanger. *See* COMFORT HEATING; OIL BURNER.

[JOHN W. JAMES]

Bibliography: R. H. Emerick, *Heating Handbook*, 1964; E. B. Woodruff and H. B. Lammers, *Steam-Plant Operation*, 4th ed., 1976.

Steam turbine

A machine for generating mechanical power in rotary motion from the energy of steam at temperature and pressure above that of an available sink. By far the most widely used and most powerful turbines are those driven by steam. In the United States well over 85% of the electrical energy consumed is produced by steam-turbine-driven generators. By the mid-1970s, over 25,000 MW (1 MW = 1341 hp) of steam turbine capacity for electrical power generation was shipped in the United States in a single typical year. Individual turbine ratings historically have tended to follow the increasing capacity trend but are now reaching limits imposed by material and machine design considerations. The largest unit shipped during the 1950s was rated 500 MW. Units rated about 1100 MW were in service by the close of the 1960s, and ratings up to 1300 MW were seeing frequent application in the 1970s. Units of all sizes, from a few horsepower to the largest, have their applications. Manufacturers of steam turbines are located in every industrial country.

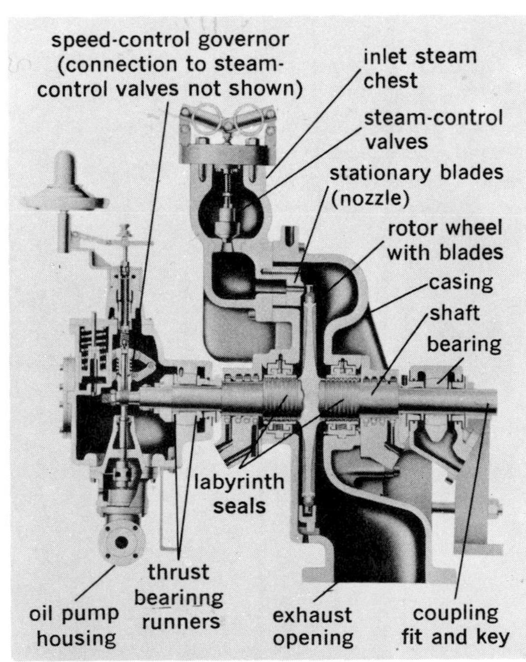

Fig. 1. Cutaway of small, single-stage steam turbine (*General Electric Co.*)

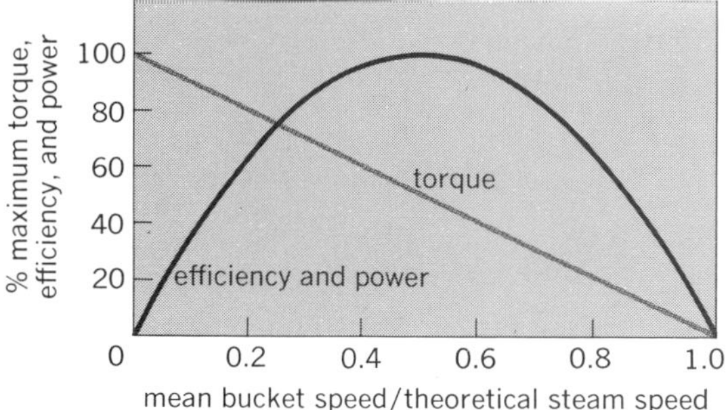

Fig. 2. Illustrative stage performance versus speed.

Until the 1960s essentially all steam used in turbine cycles was raised in boilers burning fossil fuels (coal, oil, and gas) or, in minor quantities, certain waste products. The 1960s marked the beginning of the introduction of commercial nuclear power. About 50% of the steam turbine capacity ordered from 1965 to 1975 was designed for steam from nuclear reactor steam supplies. Approximately 10% of the power generated in 1975 was from nuclear steam plants, about 75% from fossil fuel-fired steam plants, and the balance from other sources.

Turbine parts. Figure 1 shows a small, simple mechanical-drive turbine of a few horsepower. It illustrates the essential parts for all steam turbines regardless of rating or complexity: (1) a casing, or shell, usually divided at the horizontal center line, with the halves bolted together for ease of assembly and disassembly; it contains the stationary blade system; (2) a rotor carrying the moving buckets (blades or vanes) either on wheels or drums, with bearing journals on the ends of the rotor; (3) a set of bearings attached to the casing to support the shaft; (4) a governor and valve system for regulating the speed and power of the turbine by controlling the steam flow, and an oil system for lubrication of the bearings and, on all but the smallest machines, for operating the control valves by a relay system connected with the governor; (5) a coupling to connect with the driven machine; and (6) pipe connections to the steam supply at the inlet and to an exhaust system at the outlet of the casing or shell.

Applications. Steam turbines are ideal prime movers for driving machines requiring rotational mechanical input power. They can deliver constant or variable speed and are capable of close speed control. Drive applications include centrifugal pumps, compressors, ship propellers, and, most important, electric generators.

The turbine shown in Fig. 1 is a small mechanical-drive unit. Units of this general type provide 10–1000 hp with steam at 100–600 pounds per square inch gage (psig) inlet pressure and temperatures to 800°F. [See keys to Figs. 6 and 7 for conversion factors from United States customary to metric (SI) units.] These and larger multistage machines drive small electric generators, pumps, blowers, air and gas compressors, and paper machines. A useful feature is that the turbine can be equipped with an adjustable-speed governor and thus be made capable of producing power over a wide range of rotational speeds. In such applications efficiency varies with speed (Fig. 2), being 0 when the rotor stalls at maximum torque and also 0 at the runaway speed at which the output torque is 0. Maximum efficiency and power occur where the product of speed and torque is the maximum.

Many industries need steam at one or more pressures (and consequently temperatures) for heating and process work. Frequently it is more economical to raise steam at high pressure, expand it partially through a turbine, and then extract it for process, than it would be to use a separate

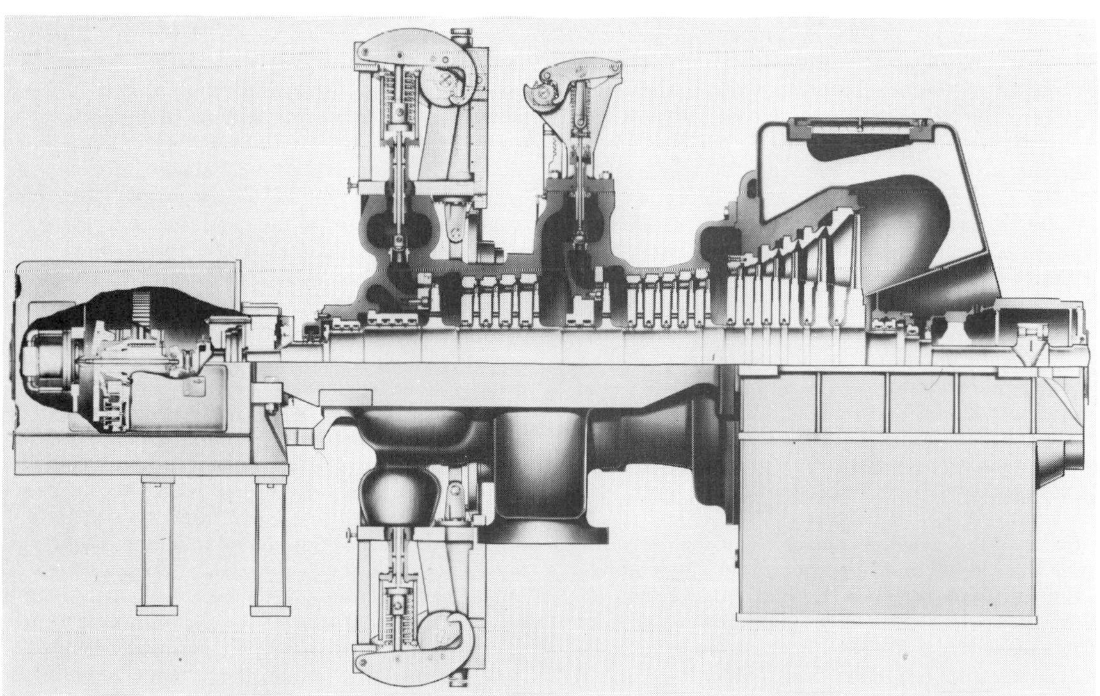

Fig. 3. Cross-section view of single-automatic-extraction condensing steam turbine. (*General Electric Co.*)

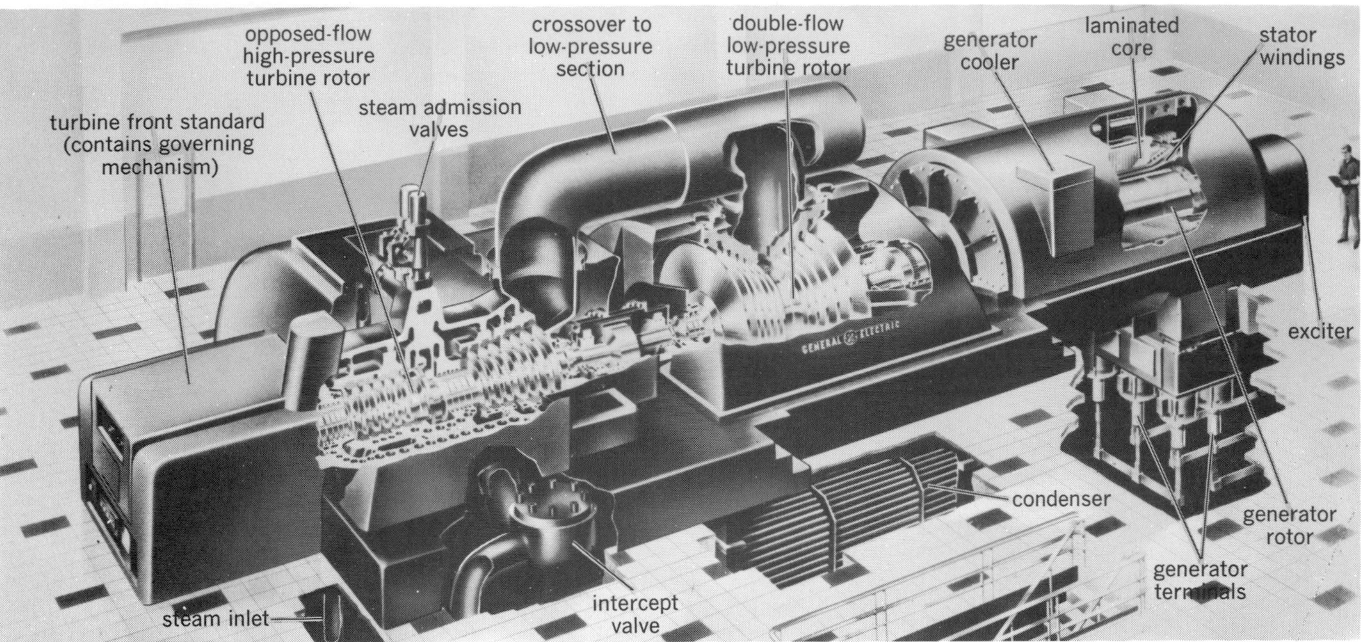

Fig. 4. Partial cutaway view of 3600-rpm fossil-fuel turbine generator. (*General Electric Co.*)

boiler at the process steam pressure. Figure 3 is a cross section through an industrial automatic extraction turbine. The left set of valves admits steam from the boiler at the flow rate to provide the desired electrical load. The steam flows through five stages to the controlled extraction point. The second set of valves acts to maintain the desired extraction pressure by varying the flow through the remaining 12 stages. Opening these internal valves increases the flow to the condenser and lowers the controlled extraction pressure.

Industrial turbines are custom-built in a wide variety of ratings for steam pressures to 2000-psig, for temperatures to 1000°F, and in various combinations of nonextracting, single and double automatic extraction, noncondensing and condensing. Turbines exhausting at or above atmospheric pressure are classed as noncondensing regardless of what is done with the steam after it leaves the turbine. If the pressure at the exhaust flange is less than atmospheric, the turbine is classed as condensing.

Turbines in sizes to about 75,000 hp are used for ship propulsion. The drive is always through reduction gearing (either mechanical or electrical) because the turbine speed is in the range of 4000–10,000 rpm, while 50–200 rpm is desirable for ship propellers. Modern propulsion plants are designed for steam conditions to 1450 psig and 950°F with resuperheating to 950°F. Fuel consumption rates as low as 0.4 lb of oil per shaft-horsepower-hour are achieved.

Central station generation of electric power provides the largest and most important single application of steam turbines. Ratings smaller than 50 MW are seldom employed today; newer units are rated as large as 1300 MW. Large turbines for electric power production are designed for the efficient use of steam in a heat cycle that involves extraction of steam for feedwater heating, resuperheating of the main steam flow (in fossil-fuel cycles), and exhausting at the lowest possible pressure economically consistent with the temperature of the available condenser cooling water.

In fossil-fuel-fired cycles, steam pressures are usually in the range of 1800–3500 psig and tend to increase with rating. Temperatures of 950–1050°F are used, with 1000°F the most common. Single resuperheat of the steam to 950–1050°F is almost universal. A second resuperheating is occasionally employed. Figure 4 shows a typical unit designed for fossil-fuel steam conditions. Tandem-compound double-flow machines of this general arrangement are applied over the rating range of 100–400 MW. Initial steam flows through the steam admission valves and passes to the left through the high-pressure portion of the opposed flow rotor. After resuperheating in the boiler it is readmitted through the intercept valves and flows to the right through the reheat stages, then crosses over to the double-flow low-pressure rotor, and exhausts downward to the condenser.

The water-cooled nuclear reactor systems common in the United States provide steam at pressures of about 1000 psig, with little or no initial superheat. Temperatures higher than about 600°F are not available. Further, reactor containment and heat-exchanger considerations preclude the practical use of resuperheating at the reactor. The boiling-water reactor, for example, provides steam to the turbine cycle at 950-psig pressure and the saturation temperature of 540°F. Such low steam conditions mean that each unit of steam flow through the turbine produces less power than in a fossil-fuel cycle. Fewer stages in series are needed but more total flow must be accommodated for a given output. Nuclear steam conditions often produce a turbine expansion with water of condensation present throughout the entire steam path. Provisions must be made to control the adverse effects of water: erosion, corrosion, and efficiency loss. In consequence of the differences in steam conditions, the design for a nuclear turbine differs considerably from that of a turbine for fossil fuel application. The former tend to be larger, heavier, and more costly. For example, at 800 MW, a typical fossil-fuel turbine can be built to run at 3600 rpm, is about 90 ft long, and weighs about 1000 tons. A comparable unit for a water-cooled nuclear reactor requires 1800 rpm, is about 125 ft long, and weighs about 2500 tons. *See* NUCLEAR REACTOR.

Figure 5 represents a large nuclear turbine generator suitable for ratings of 1000 to 1300 MW. Steam from the reactor is admitted to a double-flow high-pressure section, at the left, through four parallel pairs of stop and control valves, not shown. The stop valves are normally fully open and are tripped shut to prevent dangerous overspeed of the unit in the event of loss of electrical load combined with control malfunction. The control valves regulate output by varying the steam flow rate. The steam exhausting from the high-pressure section is at 150 to 200 psia and contains about 13% liquid water by weight. The horizontal cylinder alongside the foundation is one of a pair of symmetrically disposed vessels performing three functions. A moisture separator removes most of the water in the entering steam. Two steam-to-steam reheaters follow. Each is a U-tube bundle which condenses heating steam within the tubes and superheats the main steam flow on the shell side. The first stage uses heating steam extracted from the high-pressure turbine. The final stage employs reactor steam which permits reheating to near initial temperature. Alternate cycles employ reheat with initial steam only or moisture separation alone with no reheat. Reheat enhances cycle efficiency at the expense of increased investment and complexity. The final choice is economic, with two-stage steam reheat, as shown, selected most frequently.

Reheated steam is admitted to the three double-flow low-pressure turbine sections through six combined stop-and-intercept intermediate valves. The intermediate valves are normally wide open but provide two lines of overspeed defense in the event of load loss. Exhaust steam from the low-pressure sections passes downward to the condenser, not shown in Fig. 5.

Turbine cycles and performance. Figures 6 and 7 are representative fossil-fuel and nuclear turbine thermodynamic cycle diagrams, frequently called heat balances. Heat balance calculations establish turbine performance guarantees, provide data for sizing the steam supply and other cycle components, and are the basis for designing the turbine generator.

The fossil-fuel cycle (Fig. 6) assumes a unit rated 500 MW, employing the standard steam conditions of 2400 psig (2415 psia) and 1000°F, with resuperheat to 1000°F. As can be seen in the upper left corner, the inlet conditions correspond to a total heat content, or enthalpy, of 1461 Btu/lb of steam flow. A flow rate of 3,390,000 lb/hr is needed for the desired output of 500 MW (500,000 kW). For efficiency considerations the regenerative feedwater heating cycle is used. Eight heaters in series are employed so that water is returned to the boiler at 475°F and 459 Btu/lb enthalpy, rather than at the condenser temperature of 121°F. Because of the higher feedwater temperature, the boiler adds heat to the cycle at a higher average temperature, more closely approaching the ideal Carnot cycle, in which all heat is added at the highest cycle temperature. The high-pressure turbine section exhausts to the resuperheater at 530 psia pressure and 1306 Btu/lb enthalpy. The reheat flow of 3,031,000 lb/hr returns to the reheat or intermediate turbine section at 490 psia pressure and 1520 Btu/lb enthalpy. These data are sufficient to calculate the turbine heat rate, or unit heat charged against the turbine cycle. The units are Btu of heat added in the boiler per hour per kilowatt of generator output. Considering both the initial and reheat steam, the heat rate is given by Eq. (1).

$$\text{Turbine heat rate} = \frac{3{,}390{,}000\,(1461-459)+3{,}031{,}000\,(1520-1306)}{500{,}000} = 8090\ \text{Btu/kWh} \qquad (1)$$

The typical power plant net heat rate is poorer than the turbine heat rate because of auxiliary power required throughout the plant and because of boiler losses. Assuming 3% auxiliary power (beyond the boiler-feed pump power given in the turbine cycle in Fig. 6) and 90% boiler efficiency, the net plant heat rate is given by Eq. (2).

$$\text{Net plant heat rate} = \frac{3{,}390{,}000\,(1461-459)+3{,}031{,}000\,(1520-1306)}{500{,}000\,((100-3)/100)(90/100)} = 9270\ \text{Btu/kWh} \qquad (2)$$

The heat rates of modern fossil-fuel plants fall in the range of 8600–10,000 Btu/kWh. Considering that the heat-energy equivalent of 1 kWh is 3412

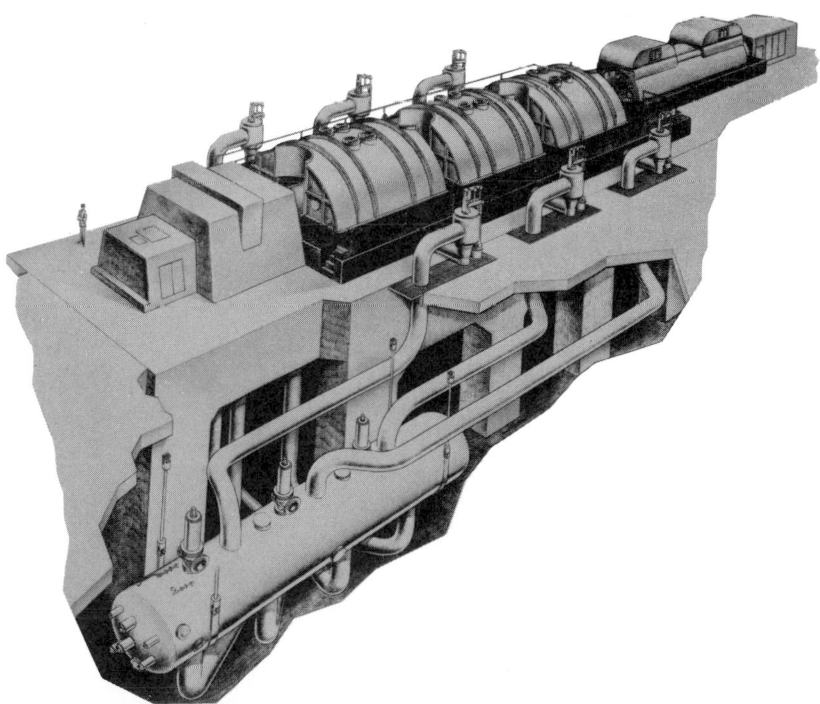

Fig. 5. An 1800-rpm nuclear turbine generator with combined moisture separator and two-stage steam reheater. (*General Electric Co.*)

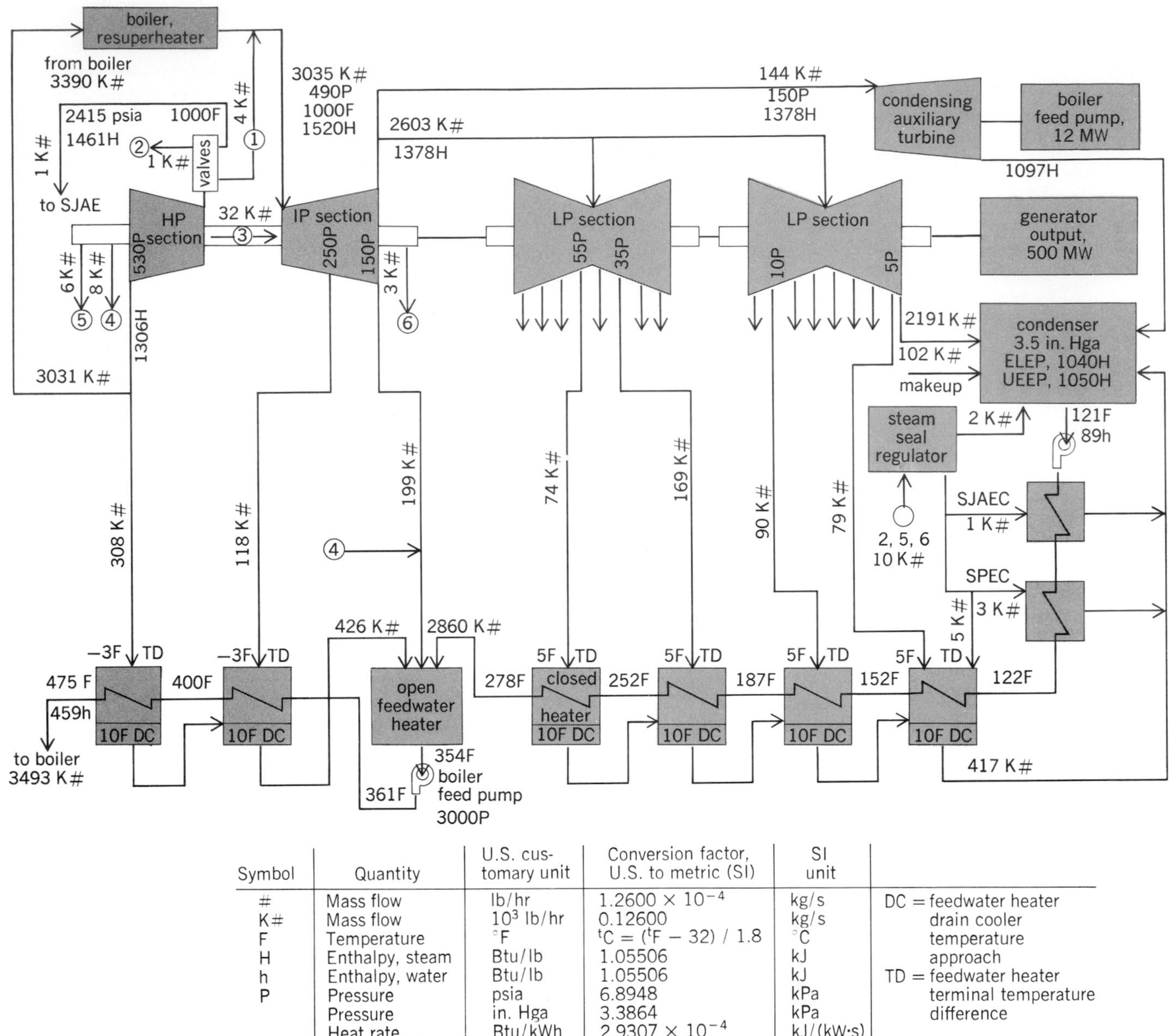

Symbol	Quantity	U.S. customary unit	Conversion factor, U.S. to metric (SI)	SI unit
#	Mass flow	lb/hr	1.2600×10^{-4}	kg/s
K#	Mass flow	10^3 lb/hr	0.12600	kg/s
F	Temperature	°F	$^tC = (^tF - 32) / 1.8$	°C
H	Enthalpy, steam	Btu/lb	1.05506	kJ
h	Enthalpy, water	Btu/lb	1.05506	kJ
P	Pressure	psia	6.8948	kPa
	Pressure	in. Hga	3.3864	kPa
	Heat rate	Btu/kWh	2.9307×10^{-4}	kJ/(kW·s)

DC = feedwater heater drain cooler temperature approach
TD = feedwater heater terminal temperature difference

Fig. 6. Typical fossil-fuel steam turbine cycle.

Btu, the thermal efficiency of the example is given by Eq. (3).

$$\eta_t = (3412/9270)\,100 = 37\% \qquad (3)$$

Figure 6 shows 2,191,000 lb/hr of steam exhausting from the main unit to the condenser at an exhaust pressure of 3.5 in. of mercury absolute. The theoretical exhaust enthalpy (ELEP), without considerations of velocity energy loss and friction loss between the last turbine stage and the condenser, is 1040 Btu/lb. The actual used energy end point (UEEP) is 1050 Btu/lb. The exhaust heat at the condenser pressure is thermodynamically unavailable and is rejected as waste heat to the plant's surroundings. The exhaust steam is condensed at a constant 121°F and leaves the condenser as water at 89 Btu/lb enthalpy.

On a heat-rate basis, this cycle rejects heat to the condenser at the approximate rate given by Eq. (4).

Net station condenser heat rejection rate

$$= \frac{2{,}191{,}000\,(1050-89) + 144{,}000\,(1097-89) + 417{,}000\,(100-89)}{500{,}000\,(0.97)}$$

$$= 4650 \text{ Btu/kWh} \qquad (4)$$

If evaporating cooling towers are used, each pound of water provides about 1040 Btu cooling capacity, which is equivalent to a required minimum cooling-water flow rate of 4.5 lb/kWh. The cooling-water needs of a large thermal plant are a most important consideration in plant site selection.

The nuclear cycle (Fig. 7) assumes a unit rated 1210 MW and the steam conditions of the boiling-

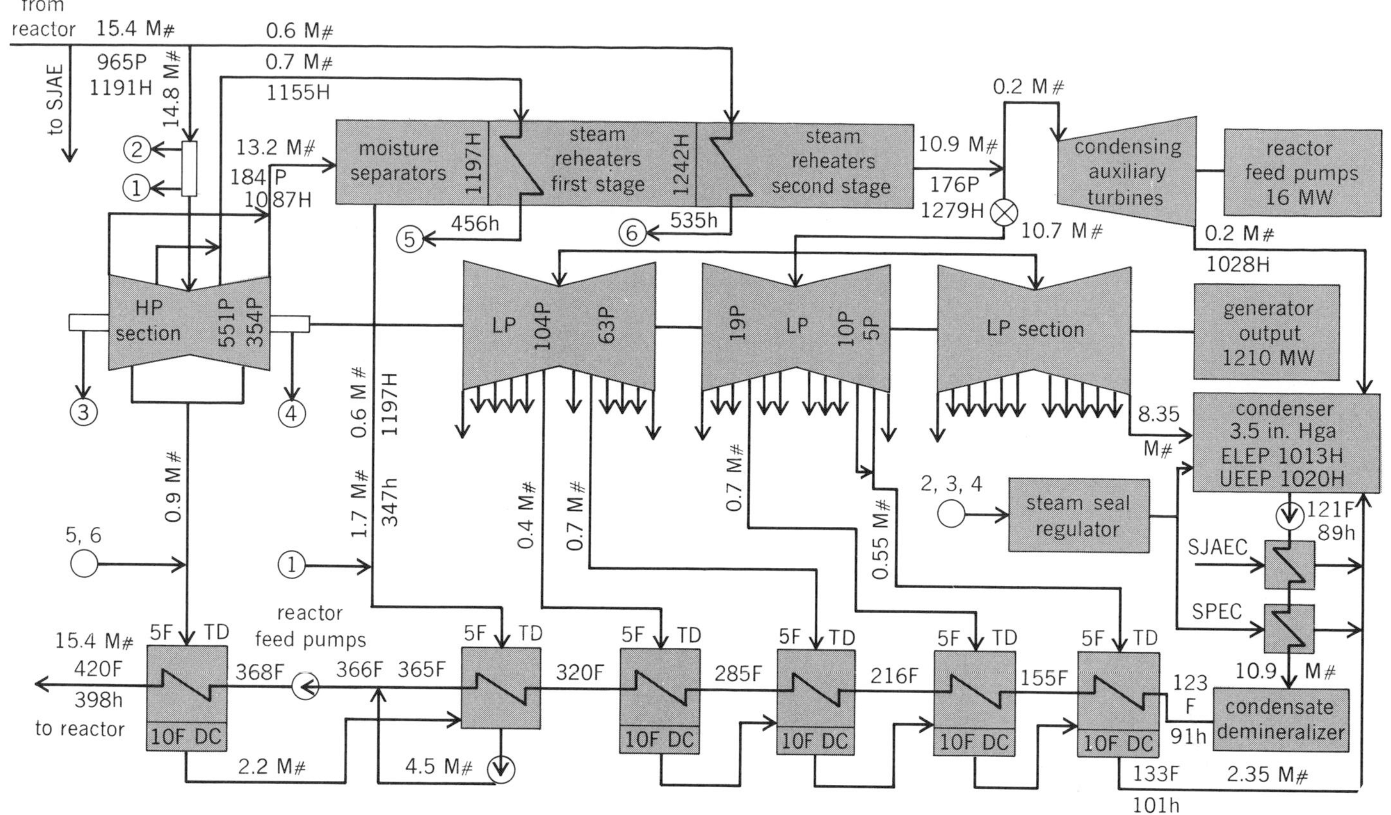

Symbol	Quantity	U.S. customary unit	Conversion factor, U.S. to metric (SI)	SI unit	
#	Mass flow	lb/hr	1.2600×10^{-4}	kg/s	DC = feedwater heater
M#	Mass flow	10^6 lb/hr	0.12600	kg/s	drain cooler
F	Temperature	°F	$^tC = (^tF - 32) / 1.8$	°C	temperature
H	Enthalpy, steam	Btu/lb	1.05506	kJ	approach
h	Enthalpy, water	Btu/lb	1.05506	kJ	TD = feedwater heater
P	Pressure	psia	6.8948	kPa	terminal temperature
	Pressure	in. Hga	3.3864	kPa	difference
	Heat rate	Btu/kWh	2.9307×10^{-4}	kJ/(kW·s)	

Fig. 7. Typical nuclear steam turbine cycle.

water reactor. Many similarities can be seen to Fig. 6. The major differences include moisture separation and steam reheating and the lack of need for an intermediate pressure element. The low steam conditions are apparent. The consequent turbine heat rate is given in Eq. (5).

Turbine heat rate

$$= \frac{15{,}400{,}000\ (1191 - 398)}{1{,}210{,}000}$$

$$= 10{,}090 \text{ Btu/kWh} \qquad (5)$$

A typical nuclear plant also requires about 3% auxiliary power beyond the reactor feed-pump power already included in Fig. 7. The equivalent boiler efficiency approaches 100% however, and leads to the equivalent net plant heat rate given by Eq. (6).

Net plant heat rate

$$= \frac{15{,}400{,}000\ (1191 - 398)}{1{,}210{,}000\ [(100 - 3)/100]}$$

$$= 10{,}400 \text{ Btu/kWh} \qquad (6)$$

The corresponding thermal efficiency is given by Eq. (7).

$$\eta_t = (3412/10{,}400)\ 100 = 33\% \qquad (7)$$

Heat is rejected at the condenser at a rate given approximately by Eq. (8).

Net station condenser heat rejection rate

$$= \frac{8{,}350{,}000\ (1020 - 89) + 200{,}000\ (1028 - 89) + 2{,}350{,}000\ (101 - 89)}{1{,}210{,}000\ (0.97)}$$

$$= 6810 \text{ Btu/kWh} \qquad (8)$$

Comparison of the heat rates shows that the nuclear cycle requires about 12% more input heat than does the fossil-fuel cycle and rejects about 46% more heat to the condenser, thus requiring a correspondingly larger supply of cooling water. It consumes heat priced at about half that from coal or one quarter that from oil, and is essentially free of rejection to the atmosphere of heat and combustion products from the steam supply.

Turbine classification. Steam turbines are classified (1) by mechanical arrangement, as single-casing, cross-compound (more than one shaft side by side), or tandem-compound (more than one casing with a single shaft); (2) by steam flow direction (axial for most, but radial for a few); (3) by steam cycle, whether condensing, noncondensing, automatic extraction, reheat, fossil fuel, or nuclear; and (4) by number of exhaust flows of a condensing unit, as single, double, triple flow, and so on. Units with as many as eight exhaust flows are in use.

Often a machine will be described by a combination of several of these terms.

The least demanding applications are satisfied by the simple single-stage turbine of Fig. 1. For large power output and for the high inlet pressures and temperatures and low exhaust pressures which are required for high thermal efficiency, a single stage is not adequate. Steam under such conditions has high available energy, and for its efficient utilization the turbine must have many stages in series, where each takes its share of the total energy and contributes its share of the total output. Also, under these conditions the exhaust volume flow becomes large, and it is necessary to have more than one exhaust stage to avoid a high velocity upon leaving and consequent high kinetic energy loss. Figure 5 is an example of a large nuclear turbine generator which has six exhaust stages in parallel.

Machine considerations. Steam turbines are high-speed machines whose rotating parts must be designed for high centrifugal stress. Difficult stress problems are found in long last-stage blading, hot inlet blading, wheels, and rotor bodies.

Casing or shell stresses. The casings or shells at the high-pressure inlet end must be high-strength pressure vessels to contain the internal steam pressure. The design is made more difficult by the need for a casing split at the horizontal center line for assembly. The horizontal flange and bolt design must be leakproof. Shell design problems lead to the use of small-diameter, high-speed turbines at high pressure, and the use of double shell construction (Fig. 4).

Rotor buckets or blades. Turbine buckets must be strong enough to withstand high centrifugal, steam bending, and vibration forces. Buckets must be designed so that their resonant natural frequencies avoid the vibration stimulus frequencies of the steam forces, or are strong enough to withstand the vibrations.

Sealing against leakage. It is necessary to minimize to the greatest possible extent the wasteful leakage of steam along the shaft both at the ends and between stages. The high peripheral velocities between the shaft and stationary members preclude the use of direct-contact seals. Seals in the form of labyrinths with thin, sharp teeth on at least one of the members are utilized. In normal operation these seals do not touch one another, but run at close clearance. In the case of accidental contact, the sharp teeth can rub away without distorting the shaft.

Vibration and alignment. Shaft and bearings should be free of critical speeds in the turbine operating range. The shaft must be stable and remain in balance.

Governing. Turbines usually have two governors, one to control speed and a second, emergency governor to limit possible destructive overspeed. The speed signal is usually mechanical or electrical. A power relay control, usually hydraulic, converts speed signals to steam valve position. Great reliability is required.

Lubrication. The turbine shaft runs at high surface speed; consequently its bearings must be continuously supplied with oil. At least two oil pumps, a main pump and a standby driven by a separate power source, are usually provided on all but the smallest machines. A common supply of oil is often shared between the governing hydraulic system and the lubrication system.

Aerodynamic design. The vane design for highest efficiency, especially for the larger sizes of turbines, draws upon modern aerodynamic theory. Classic forms of impulse and reaction buckets merge in the three-dimensional design required by modern concepts of loss-free fluid flow. To meet the theoretical steam flow requirements, vane sections change in shape along the bucket. To minimize centrifugal forces on the vanes and their attachments, long turbine buckets are tapered toward their tips. *See* CARNOT CYCLE; HEAT BALANCE; TURBINE.

[FREDERICK G. BAILY]

Bibliography: F. G. Baily, Steam turbines, in T. Baumeister (ed.), *Standard Handbook for Mechanical Engineers*, sec. 9, pp. 38–54, 8th ed., 1978; F. G. Baily, K. C. Cotton, and R. C. Spencer, Predicting the performance of large steam turbine-generators operating with saturated and low superheat steam conditions, *Combustion*, 3(3):8–13, 1967; R. L. Bartlett, *Steam Turbine Performance and Economics*, 1958; J. K. Salisbury, *Steam Turbines and Their Cycles*, 1950; B. G. A. Strotzki, Steam turbines, *Power*, 106(6):S1–S40, 1962.

Stirling engine

An engine in which work is performed by the expansion of a gas at high temperature to which heat is supplied through a wall. Like the internal combustion engine, a Stirling engine provides work by means of a cycle in which a piston compresses gas at a low temperature and allows it to expand at a high temperature. In the former case the heat is provided by the internal combustion of fuel in the cylinder, but in the Stirling engine the heat, which is obtained from externally burning fuel, is supplied to the gas through the wall of the cylinder (Fig. 1). *See* INTERNAL COMBUSTION ENGINE.

The rapid changes desired in the gas temperature are achieved by means of a second piston in the cylinder, called a displacer, which in moving up and down transfers the gas back and forth between two spaces, one at a fixed high temperature and the other at a fixed low temperature. When the displacer in Fig. 2 is raised, the gas will flow from the hot space via the heater and cooler tubes into the cold space. When it is moved downward, the gas will return to the hot space along the same path. During the first transfer stroke the gas has to yield up a large amount of heat to the cooler; an equal quantity of heat has to be taken up from the heater during the second stroke.

The regenerator shown in Figs. 1 and 2 has been inserted between the heater tube and cooler tube in order to prevent unnecessary waste of this heat.

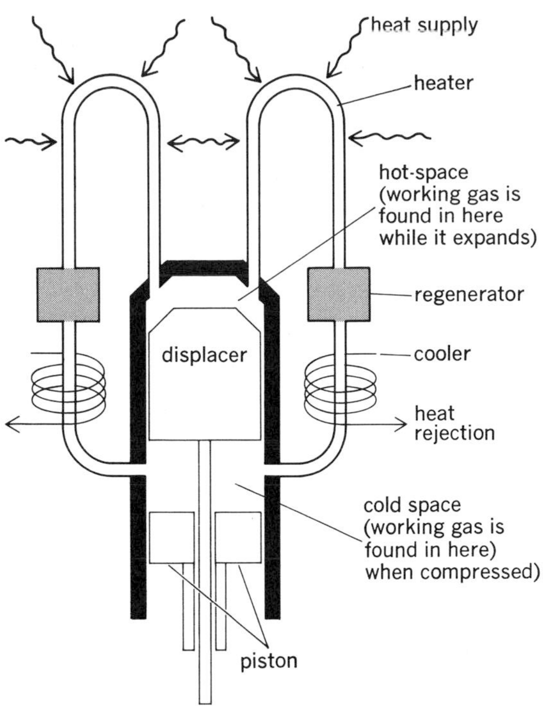

Fig. 1. Principle of Stirling engine, displacer type.

It is a space filled with porous material to which the hot gas gives up its heat before entering the cooler; the cooled gas recovers the stored heat on its way back to the heater.

The displacer system serves to heat and cool the gas periodically; associated with it is a piston which compresses the gas while it is in the cold space of the cylinder and allows it to expand while in the hot space. Since expansion takes place at a higher temperature than compression, it produces a surplus of work over that required for the compression.

Stirling cycle. Any practical version of the engine will embody some kind of crank and connecting rod mechanism, in consequence of which there will be no sharp transitions between the successive phases indicated in Fig. 2; but this will not alter the principle of the cycle (nor detract from its efficiency). If, for the sake of simplicity, the piston and displacer are assumed to move discontinuously, the cycle can be divided into the following four phases. (1) The piston is in its lowest position, the displacer in its highest. All the gas is in the cold space. (2) The displacer is still in its highest position; the piston has compressed the gas at low temperature. (3) The piston is still in its highest position; the displacer has moved down and transferred the compressed gas from the cold to the hot space. (4) The hot gas has expanded, pushing the pistons, followed by the displacer, to their lowest positions. The displacer is about to rise and return the gas to the cold space, the piston remaining where it is, to give phase 1 again.

The actual piston and displacer movements might be as indicated in Fig. 3, which shows that

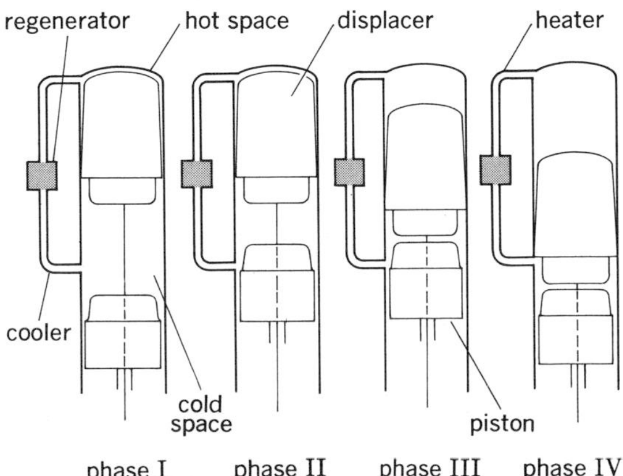

Fig. 2. Stylized Stirling process.

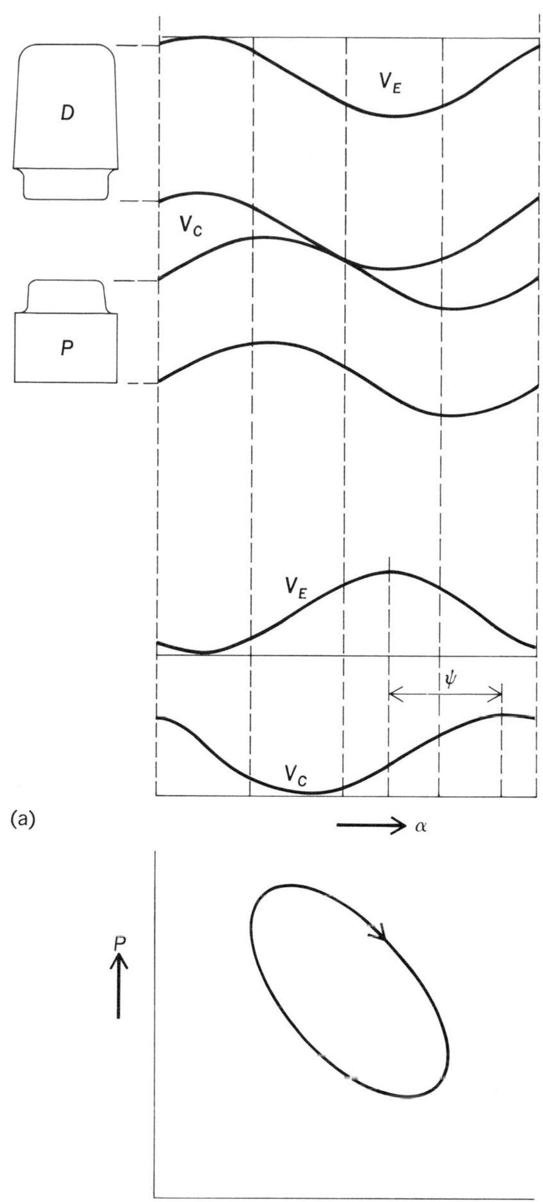

Fig. 3. Piston and displacer movement for a Stirling engine. (*a*) The continuous movement of the piston and displacer shown as a function of crank angle α. There is no clear-cut division between the various phases of the cycle. The variations in volume of the hot space V_E and the cold space V_c are plotted separately in the lower part of the diagram. (*b*) The $p-V$ diagram of the cycle.

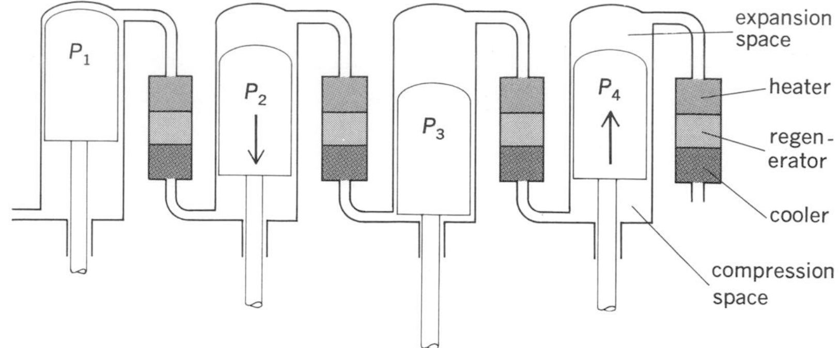

Fig. 4. Principle of the double-acting engine.

the only essential condition for obtaining a surplus of work is that the maximum volume of the hot space occur before that of the cold space. This condition shows that more configurations with pistons and cylinders are possible than just the type with displacer in order to get a Stirling cycle. One of the most compact systems is shown in Fig. 4, which is the system known as the double-acting engine. In the double-acting engine there is a hot space (expansion space) at the top and a cold one (compression space) at the bottom of each of the four cylinders shown. The hot space of a cylinder is connected to the cold one through a heater, a regenerator, and a cooler. The pistons P_n of the cylinders move with a suitable phase shift between them. In the case of four cylinders, as shown in Fig. 4, this shift is 90°.

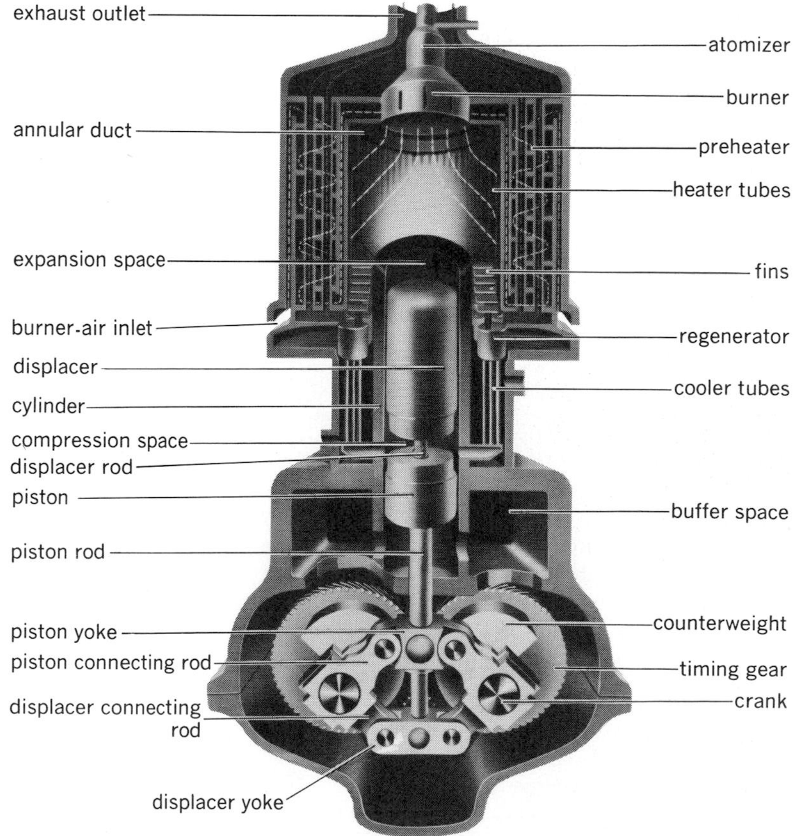

Fig. 5. Philips Stirling engine in cross section. (*Philips Gloeilampenfabrieken*)

The theory of an actual engine is very complicated. In order to provide an understanding of the quantities that play a role, the formulas of the power and efficiency of an engine of highly idealized form will be given. On the assumption that the volumes of the hot (expansion) space V_E and that of the cold (compression) space V_C (Fig. 3) vary with the crank angle in a purely sinusoidal way, that both expansion and compression are isothermal at respectively T_E and T_C, and that all kinds of losses caused by flow resistances in the tubes and regenerator, heat losses in the regenerator, and so on are ignored, the power P can be expressed as in Eq. (1). Here $\tau = T_C/T_E$, n is the speed in rpm, p_m is

$$P = (1-\tau)\frac{\pi n}{60}V_{E_{max}} \cdot p_m \frac{\delta}{1+\sqrt{1-\delta^2}}\sin\Theta \quad (1)$$

the mean pressure of the pressure variation, and δ and Θ are functions of the temperature ratio, swept volume ratio, dead spaces, and phase angle between the two swept volumes. Under these conditions the thermal efficiency η is, of course, that of the Carnot cycle, and is expressed in Eq. (2). *See* CARNOT CYCLE; THERMODYNAMIC CYCLE.

$$\eta = \frac{T_E - T_C}{T_E} = 1 - \tau \quad (2)$$

Engine design. Actual engines have been built in the Philips Laboratories at Eindhoven, Netherlands, as prototypes in the range of 10–500 hp per cylinder. After 30 years of research on the Stirling engine, actual thermal efficiencies of 30–45% (depending on the specific output and temperature ratio) and a specific power of 115 hp per liter swept volume of the piston have been obtained with the displacer type of engine equipped with rhombic drive, as shown in Fig. 5. In the figure the piston and displacer drive concentric rods, which are coupled to the rhombic drive turning twin timing gears. The cooler, regenerator, and heater are arranged as annular systems around the cylindrical working space. The heater tubes surround the combustion chamber. In the preheater the gas at 800°C from the heater is cooled to 150–200°C while heating the combusion air to about 650°C.

The rhombic drive mechanism allows complete balancing even of a single-cylinder engine and of a separate buffer space, thus avoiding heavy forces acting on the drive. The results of measurements on the first engine of this type, with hydrogen as the working fluid, are shown in Fig. 6. An approximate heat balance is output, 40%; exhaust and radiation, 10%; and heat rejection by cooling water, 50%. Control of engine output is by regulation of the pressure of the working fluid in the engine, while the temperature of the heater is being kept constant by a thermostat; hence the efficiency shows little dependence on the load.

The closed system of the Stirling engine endows this engine with many advantages and also some shortcomings. The continuous external heating of the closed system makes it possible to burn various kinds of liquid fuels and gases, without any modification whatsoever. This multifuel facility can be demonstrated with a 10-hp or 7.5-kW (at 3000 rpm) generator set (Fig. 7). The engine can operate on alcohol, various lead-containing gasolines, diesel fuel, lubricating oil, olive oil, salad oil, crude oil, propane, butane, and natural gas. Furthermore,

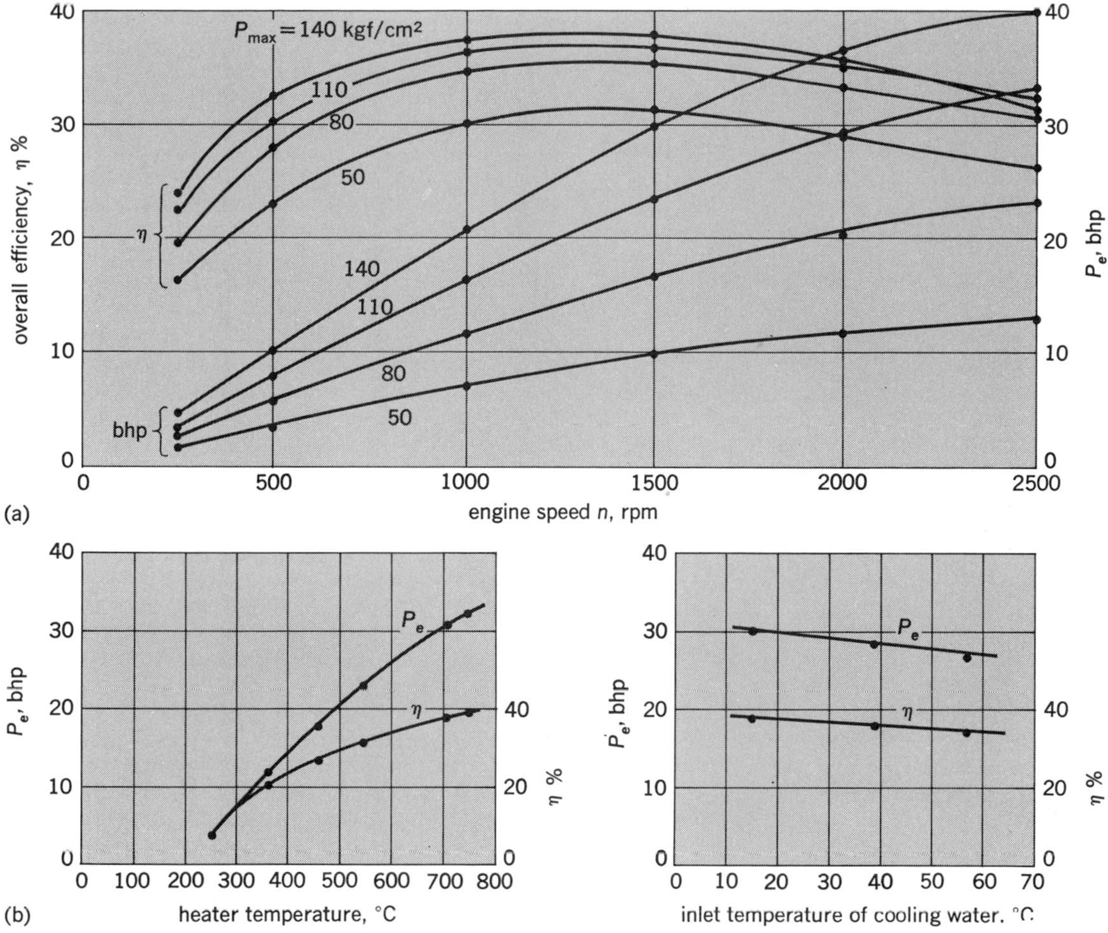

Fig. 6. Test measurements for the Stirling engine. (*a*) Measured shaft power and efficiency of a 40-hp (30-kW) single-cylinder test engine with rhombic drive, plotted as a function of the engine speed *n*, at different values of the maximum pressure of the working fluid in the engine (p_{max}). 1 hp = 0.75 kW. (*b*) Power efficiency of the 40-hp test engine, given as a function of the heater temperature and as a function of the inlet temperature of the cooling water. The curves apply to n = 1500 rpm and p_{max} = 140 kgf/cm² (13.7 MPa).

Fig. 7. Philips Stirling engine with generator to demonstrate its multifuel capacity. (*Philips Gloeilampenfabrieken*)

Fig. 8. Three-kilowatt Stirling engine generator set. (*General Motors Corp.*)

it allows combustion to take place in such a way that air pollution is some orders of magnitude less than that due to internal combustion engines. Through the intermediary of a suitable heat transport system (for example, heat pipes) any heat source at a sufficiently high temperature can be used for this engine—radioisotopes, a nuclear reactor, heat storage, solar heat, or even the burning of coal or wood.

The almost sinusoidal cylinder pressure variation and continuous heating make the Stirling engine very quiet in operation. An engine having four or more cylinders gives a virtually constant torque per revolution, as well as a constant dynamometer torque over a wide speed range, which is particularly valuable for traction purposes. The present configuration makes complete balancing possible, thus eliminating vibrations. There is no oil consumption and virtually no contamination because a new type of seal for the reciprocating rods shuts off the cycle hermetically from the drive mechanism. Figure 8 shows an engine of this configuration.

Where direct or indirect air cooling is required, the closed cycle has the drawback that more heat has to be removed from the cooler than in comparable engines with open systems, where a greater quantity of heat inevitably escapes through the exhaust.

If it is envisaged as someday taking the place of existing engines, the Stirling might be ideal as a propulsion engine in yachts and passenger ships, and in road vehicles, such as city buses, where a large radiator is acceptable. The system of continuous external heating is also able to open fields of application inaccessible to internal combustion engines. [ROELOF J. MEIJER]

Bibliography: T. Finkelstein, Air engines, *Engineer*, pp. 492–497, 522–527, 568–571, 720–723, March–May, 1959; J. H. Lienesch and W. R. Wade, Stirling engine progress report: Smoke, odor, noise and exhaust emissions, *SAE Trans.*, 77: 292–307, 1968; J. N. Mattavi, F. E. Heffner, and A. A. Miklos, *The Stirling Engine for Underwater Vehicle Applications*, Pap. no. 690731, SAE Meeting, Cleveland, Ohio, Oct. 27–29, 1969; R. J. Meijer, The Philips Stirling engine, *De Ingenieur*, pp. W69–W79, W81–W93, May, 1969; R. J. Meijer, Rebirth of the Stirling engine, *Sci. J.*, 5A(2):31–37, 1969.

Storage battery

An assembly of identical voltaic cells in which the electrochemical action is reversible so that the battery may be recharged by passing a current through the cells in the opposite direction to that of discharge. While many nonstorage batteries have

Comparison of the principal types of storage battery

Battery type	Volts per cell	Energy, Wh/lb	Density, Wh/in.^{3}
Lead-acid	2.0	10–15	0.6–1.3
Nickel-iron	1.2	10–14	0.6–1.0
Nickel-cadmium	1.2	8–11	0.4–0.8
Nickel-cadmium sintered	1.2	10–13	1.0
Silver-zinc	1.5	20–100	3
Silver-cadmium	1.1	15–50	2.5

a reversible process, only those that are economically rechargeable are classified as storage batteries. *See* BATTERY; PRIMARY BATTERY.

Storage batteries, sometimes known as electric accumulators or secondary batteries, have two general classifications: lead-acid and alkaline. Active materials and electrolytes for both classes of batteries will be explained later. The table gives an approximate comparison of the several principal types of storage battery couples in terms of output per unit weight and unit volume.

Some of the important uses of storage batteries are to start gasoline and diesel engines; to operate communications circuits; switch tripping and closing in power-generating and -handling systems; emergency lighting; emergency power both with and without conversion to alternating current; railway car lighting and air conditioning; rapid transit car controls; marine power systems; power for underwater exploratory vehicles and submarines; to activate photographic and portable sound systems as well as portable TV and radio; and various military applications.

LEAD-ACID STORAGE BATTERY

The lead-acid type of storage battery is so classified because the electrolyte is an acid and the plates are largely lead. The positive active material is lead peroxide and the negative active material is lead sponge. The active materials are supported by grids made of lead alloys.

The lead-acid battery maintains a preeminent place among all commercial types of storage batteries in volume of manufacture.

Principles of operation. A great many types of lead-acid cells are produced, but all have certain features in common. One is the open-circuit cell electromotive force (emf), which exists between a positive lead peroxide (PbO_2) electrode and a negative sponge lead (Pb) electrode when the two are immersed in sulfuric acid electrolyte ($H_2SO_4 + H_2O$). This value is independent of the quantities of lead peroxide, lead, or electrolyte present but does vary with temperature and sulfuric acid (H_2SO_4) concentration. At 25°C the emf varies from 2.050 volts with acid at 1.200 sp gr to 2.148 volts with acid at 1.300 sp gr. The relatively small variation with temperature is given in millivolts/°C over a range 0–40°C, as 0.30 for 1.200 sp gr electrolyte, 0.22 for 1.250 sp gr, 0.19 for 1.280 sp gr, and 0.18 for 1.300 sp gr.

Equation (1) represents the cell reactions insofar

$$PbO_2 + Pb + 2H_2SO_4 \underset{\text{Charge}}{\overset{\text{Discharge}}{\rightleftharpoons}} 2PbSO_4 + 2H_2O \quad (1)$$

as beginning and end materials are concerned. It is known as the double-sulfate theory, since lead sulfate ($PbSO_4$) is formed at both electrodes.

Equation (1) can be split into Eqs. (2) and (3), indicating the reactions at the two electrodes.

$$PbO_2 + 2H^+ + H_2SO_4 + 2e^- \underset{\text{Charge}}{\overset{\text{Discharge}}{\rightleftharpoons}} PbSO_4 + 2H_2O \quad \text{(at positive)} \quad (2)$$

$$Pb + SO_4^{--} \underset{\text{Charge}}{\overset{\text{Discharge}}{\rightleftharpoons}} PbSO_4 + 2e^- \quad \text{(at negative)} \quad (3)$$

On discharge the overall effect is a reduction of PbO_2 at the positive electrode and an oxidation of Pb at the negative electrode, accompanied by sulfation in both cases. In charging, a counter voltage is imposed on the cell terminals, and current is forced through the cell in a direction opposite to that in which the cell discharges. This reverses the ionic movements in relation to the electrodes and, in effect, reverses the cell reactions. On discharge the electrolyte specific gravity decreases, and on charge it increases. Specific gravity serves as a measure of the sulfuric acid concentration and thus as an index of state of charge.

Reactants. For a given quantity of electricity, such as ampere-hours, the three reactants PbO_2, Pb, and H_2SO_4 take part in the reaction in amounts governed by Faraday's law. Thus, for a 1 ampere-hr discharge, 3.866 g of sponge lead are converted to $PbSO_4$, 4.463 g PbO_2 are converted to $PbSO_4$, and 3.660 g of H_2SO_4 are consumed.

A cell constructed to contain exactly the amounts of reactants given above, however, would not yield 1 ampere-hr of capacity even under optimum practical conditions. Action at each electrode is slowed drastically when the concentration of H_2SO_4 in the electrolyte approaches a low figure because it is required in the electrode reaction. But even if ample H_2SO_4 were present, 1 ampere-hr

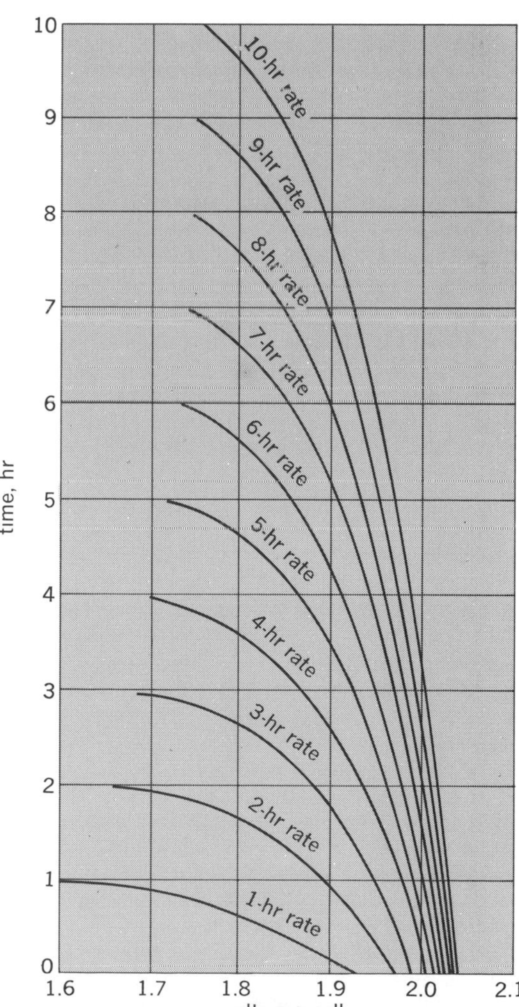

Fig. 1. Typical volt-time curves for a lead-acid cell for various discharge rates. (*ESB Inc.*)

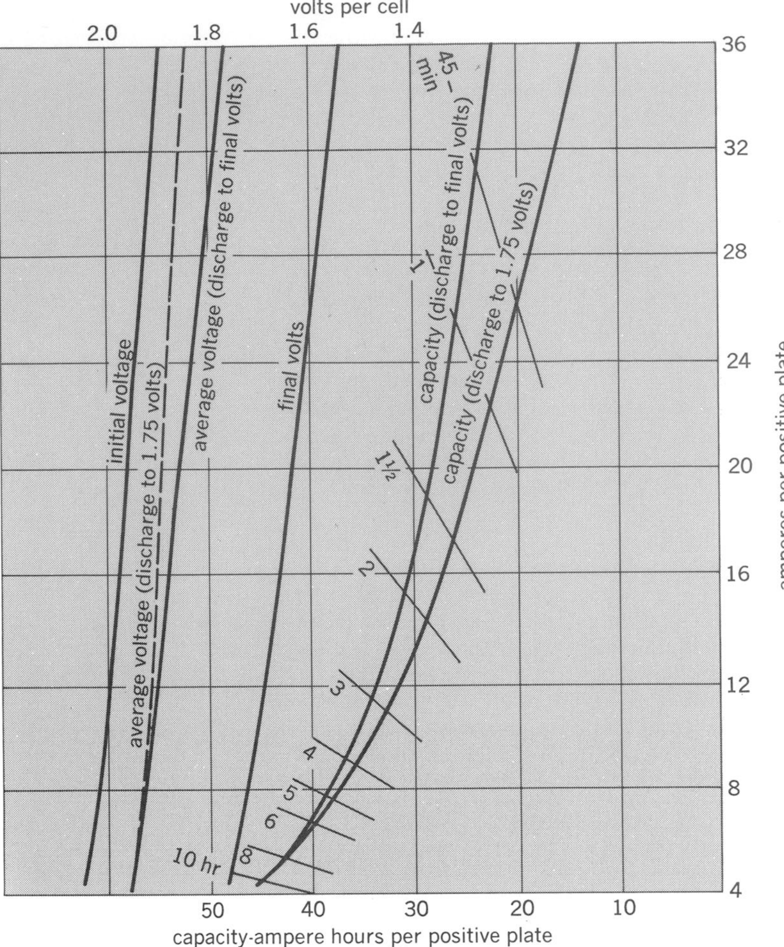

Fig. 2. Rated-discharge curves for a lead-acid cell. (*ESB Inc.*)

of capacity still would not be attained, since there will always remain an appreciable amount of PbO_2 or Pb, or both, in the solid electrodes, which cannot be reached by the electrolyte. The capacity attained in practice divided by what should be obtained in principle from the amount of reactants present is known as the utilization coefficient. This coefficient varies with types of cells, rate of discharge, and temperature. Unfortunately, it is a low value even under the best of conditions.

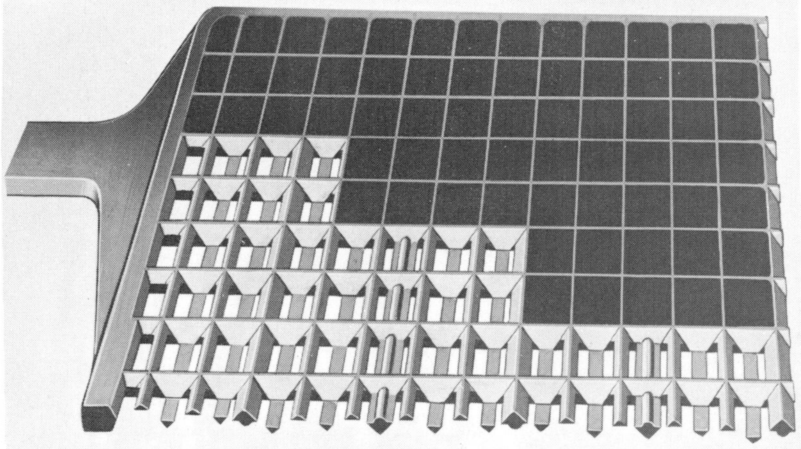

Fig. 3. Typical ladder-type grid showing a portion of it pasted. (*ESB Inc.*)

This problem is aggravated by the coating of nonconducting sulfate that forms on the active materials. Another is the diminishing conductivity of the electrolyte as the H_2SO_4 content decreases.

The utilization coefficient is decreased by the use of high-current rates. At higher current densities, the electrode reaction is concentrated at the surface of the plates. As a result, the pore openings at the plate surfaces become blocked with sulfate, restricting conduction and diffusion to the interiors of plates. Plates destined for high-discharge current densities are therefore made relatively thin. By substituting many thin plates for a few thick plates containing the same amounts of active materials, the utilization coefficient at high rates, and hence the capacity attainable, will be increased.

Figure 1 shows a typical set of volt-time curves for for different discharge rates of a lead-acid cell, illustrating the variations from a 1-hr rate at high discharge current to a 10-hr rate at low discharge current. Figure 2 illustrates the decrease in ampere-hour capacity with increase in discharge current. If a short time is allowed for diffusion after a high-rate discharge, more of the unused possible capacity of the cell becomes available.

Cell temperature has an appreciable effect on capacity, largely because the viscosity of the electrolyte changes. Thus the diffusion of H_2SO_4 is retarded at low temperatures and the capacity is lowered.

Also, the capacity is decreased if the acid concentration becomes too low. On the other hand experience has shown that negative plates do not function well if the full-charge specific gravity is over 1.300, although positive plates operate more efficiently in high specific gravity. The usual range of full-charge specific gravity is 1.200–1.280, the choice depending on the application of the cell, the ambient operating temperature, and susceptibility of the cell to self-discharge. Specific gravity is usually determined by a hydrometer. It can also be measured by chemical methods.

Cell construction. Aside from cost, first consideration must be given to the kind of service for which a cell is destined, and second consideration to design features that reduce operational troubles. A compromise, for example, between life and weight or between life and cost, is usually required.

Pasted plates. In the most familiar type of plate construction, the basic structural number is a die-cast grid, such as the ladder type shown in Fig. 3. This grid is made of a lead alloy containing, for example, (3–11%) antimony or (.01–0.1%) calcium. The grid is pasted with a slurry of lead oxide, sulfuric acid, and water. This is followed by a processing which finally converts the active material to PbO_2 for positive plates (chocolate-brown color) and to sponge lead for negative plates (gray color).

The negative plates are always sponge lead whether used with pasted positive plates or with positive plates of other types.

The life of a lead-acid battery plate is closely related to the thickness of the metal bars used in the positive plate. In applications such as engine cranking where light weight and high-rate performance are of more importance than life, the battery plates are usually made as thin as possible. Thicker plates are used where long life and reliability are more important than first cost, space, and weight. The thinnest plates in use are about 0.05 in. (1.3 mm) thick; the thickest plates range up to 0.75 in (19 mm).

Manchester plate. These plates consist of a heavy alloy grid with circular openings into which pure lead "buttons" are pressed. These buttons are made from lead tape by crimping and rolling to develop a large surface area (Fig. 4). A forming agent in dilute sulfuric acid electrochemically forms a layer of PbO_2 on the surface of the button. Manchester plates are usually mounted in a cell with pasted negatives and in a relatively large quantity of low-gravity acid.

The cells are heavy and bulky. They are used in stationary installations, as for telephones, switch operation, and large emergency lighting, where they are "floated" on a line of constant voltage or trickle-charged with a constant current and are only occasionally discharged. Under such conditions of service, Manchester plates give exceptionally long life.

Gould spun plates. This type of positive plate, shown in Fig. 5, is manufactured from heavy sheet lead by passing the plate between disks which cause the lead to flow in between them to form leaves and spaces. After PbO_2 is formed on this developed surface, the plates are assembled with pasted negative plates and used in substantially the same types of service as Manchester plates. An advantage of this plate is elimination of antimony, hence local action, from the cell construction. This advantage is usually gained at some sacrifice of life.

Tubular-type plate. In this positive plate the active material is held in a porous-walled tube with a central alloy spine as conductor. The tube is made of felted or woven chemically inert fibers. This plate has many applications but is particularly successful where the service calls for repeated or routine deep-discharge cycles, such as in industrial trucks and mine locomotives.

Freezing of electrolyte. The freezing points of the usual range of sulfuric acid electrolytes at full-charge specific gravities (from −52°C for 1.250 at 15°C to −70°C for 1.300 at 15°C) are well below most arctic temperatures, but the end-of-discharge specific gravities can result in freezing points above arctic temperatures unless precautions are taken. With the proper choice of separator, a high-gravity acid can be used in severe arctic conditions without detriment to the negative electrodes and yet can have a relatively high end-of-discharge gravity.

Charging. For fast, yet efficient and noninjurious, charging, the modified constant-potential method is recommended. A high current rate is used until a voltage, such as 2.38, is obtained. This voltage is then maintained with a decreasing current until the finishing rate recommended by the manufacturer is reached. The finishing rate is continued to the end of the charge. Several methods are used for setting the end of the charge, the best known being arrival at a constant potential, arrival at a constant specific gravity, or charging a certain percent of ampere-hours in excess of the ampere-hours that have been taken out. A lengthy but efficient charge can be made using the finishing rate from the start. A two-step charge can be made using first a high current and then the finishing rate, the change being made automatically by a voltage relay or ampere hour meter.

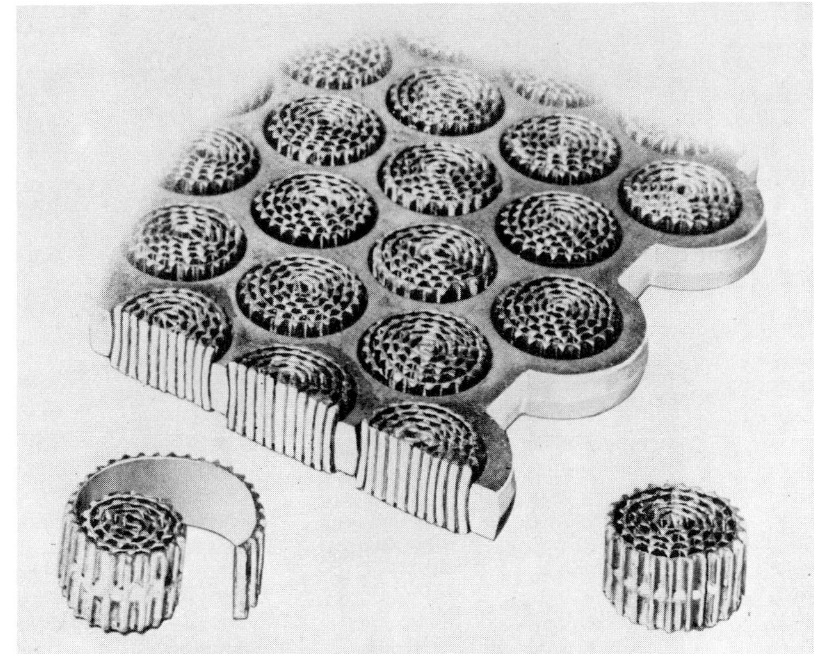

Fig. 4. Section of Manchester plate with detail of lead button. (*ESB Inc.*)

Cell containers are currently made of hard rubber or plastic. For automobile, railway, and motive power batteries the cell containers are made of highly shock-resistant materials such as semihard rubber, modified polystyrene, or polyporpylene. Stationary batteries use jars made of clear modified polystyrene. For extreme shock resistance such as submarine service, rubber-lined polyester fiber-glass jars are used.

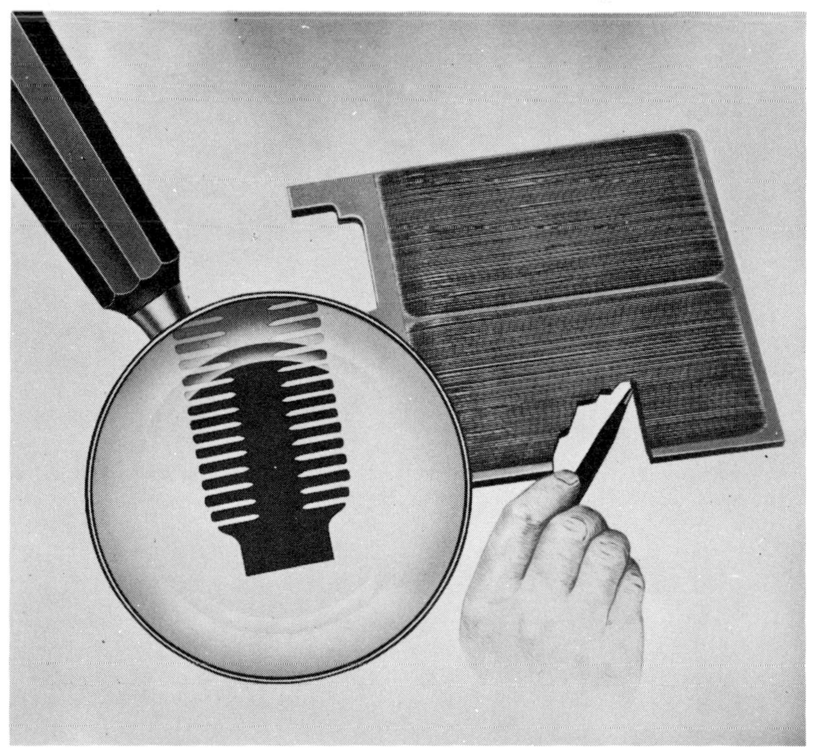

Fig. 5. Gould spun Plante positive plate. (*Gould-National Batteries, Inc.*)

Fig. 6. Plate construction of nickel-iron cell. (*a*) Positive plate. (*b*) Negative plate. (*ESB Inc.*)

Battery troubles and remedies. Some of the important battery troubles are corrosion of the grid, shedding of active materials, and self-discharge.

Corrosion. Gradual wearing away of the electrode grid containing the active PbO_2 results in subsequent disintegration of the plate. Cells subjected to repeated deep discharges and overcharges are particularly prone to this trouble. The antimony used in the grid alloy aids in resisting corrosion under certain conditions. The addition of small percentages of arsenic and silver increases this resistance to a notable extent.

Shedding of active materials. This usually pertains to cells that are subjected to overcharging, whereby gas formation strains and loosens particles of material near the surface of the plate. Unless retainers, such as slotted rubber or plastic or mats of glass fiber, are used against the positive plates, active material may drop to the bottom as sediment and cause short circuits between plates. The material may also be transported by gas streams to the top of the cell to pile up and short-circuit there.

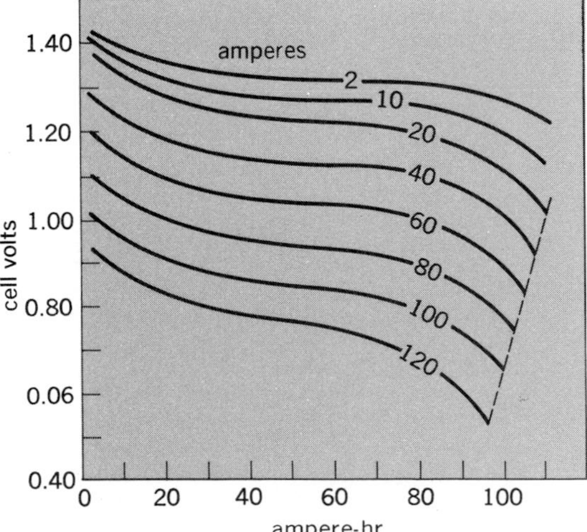

Fig. 7. Typical volt-time curves of nickel-iron alkaline cells for various discharge rates. (*From D. G. Fink and J. M. Carroll, eds., Standard Handbook for Electrical Engineers, 10th ed., McGraw-Hill, 1968*)

Self-discharge or local action. Self-discharge of negative plates is caused by deposition of certain metals on the plate to form a voltaic couple with the sponge lead. Since these couples are actually local short circuits, which have an emf in excess of the hydrogen overvoltage on the metal deposit, an evolution of hydrogen ensues, and the adjacent lead is sulfated. Metals most frequently producing local action include antimony, copper, silver, and, less frequently, tin, arsenic, bismuth, platinum, and nickel.

Other metals, such as iron and manganese, whose salts readily exist in solution in two stages of oxidation, can reduce the positive plate by diffusion or convection and oxidize the negative plate with $PbSO_4$ formation at both electrodes.

If self-discharge of the negative electrodes becomes rapid or is allowed to act over a long period, or if a cell stands for some time in a discharged condition, the sulfate crystals become large, hard, and difficult to reduce to lead.

Undercharging. This causes buckling, or warping, of plates and sulfation. This condition is usually due to unequal work on the two sides of the positive plates, resulting from unequal electrolytic attack or unequal expansion of active materials.

Overcharging. This causes corrosion, buckling, washing, and overheating. It is caused by high charging rates, which should be tapered off as the battery becomes charged.

Densification of negative material. A deficiency of certain organic and inorganic compounds that are used as additives to the active material permits coalescence of lead particles with a consequent loss of porosity. Compounds added to the negative plate material to prevent this are commonly called expanders.

Separator shorts. The separators between positive and negative plates may be oxidized through contact with PbO_2, permitting lead bridges to form between positive and negative plates and thus short-circuiting the cell.

Sealed lead-acid cells. Ways have been found to operate smaller lead-acid cells in a completely sealed container. These batteries are built with capacities up to about 10 ampere-hr. They are used for portable TV, electric hand tools, and so on. In order to keep these batteries operable, it is necessary to use a very carefully controlled charge.

ALKALINE-TYPE STORAGE BATTERY

The alkaline-type storage battery is so classified because the electric energy is obtained from chemical action of an alkaline solution. One type of battery has positive plates of some nickel compound and negative plates of iron. Another type uses a nickel compound and cadmium. A third uses silver oxide and zinc.

Nickel-iron alkaline cell. This battery is composed of cells having a hydrated nickel oxide and iron in an alkaline solution. It was invented by Thomas Edison early in the 20th century. The positive active material in this cell is a higher oxide or hydroxide of nickel. The negative material is fine

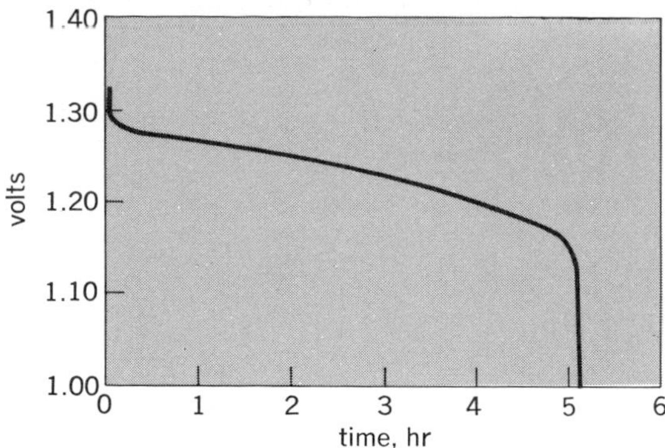

Fig. 8. Typical discharge curve for sintered-plate nickel-cadmium cell. (*ESB Inc.*)

iron powder. The electrolyte is 1.200 sp gr (at 15°C) potassium hydroxide, to which a little lithium hydroxide is sometimes added.

The chemical behavior of the nickel-iron cell is shown in Eq. (4).

$$2NiOOH \cdot H_2O + Fe \underset{Charge}{\overset{Discharge}{\rightleftharpoons}} 2Ni(OH)_2 + Fe(OH)_2 \quad (4)$$

The nickel hydrate formed by charging the battery is not an exact chemical compound. Directly after charging, it contains some excess dissolved oxygen. The dissolved oxygen is not tightly held and is released in the 10–24-hr period following charge. It is electrically active, and a battery discharged immediately after charge will have greater output than if it stands until the oxygen is lost.

The KOH electrolyte supplies ions for conductivity, but unlike the lead-acid battery, the concentration of electrolyte in alkaline cells (nickel-iron, nickel-cadmium, silver-zinc, and silver-cadmium) does not undergo any net change in the chemical action of the cell. As a consequence, the specific gravity of the electrolyte does not change and cannot be used to indicate the state of charge of an alkaline battery, as in the case of the lead acid battery. However, it also means that the gravity stays up at all times, and the battery is much less susceptible to accidental damage from freezing than lead acid.

In the tubular nickel-iron cell a perforated steel tube is tightly packed with alternate layers of nickel hydrate and thin nickel flake to provide electrical conductivity. The layers are thin and there are about 300 layers in a 4-in. tube.

The negative material is packed into long pockets of perforated sheet steel. The pockets are laced together and pressed to form a single structural member. The top rail and bottom rail are welded in place to complete the plate. Positive and negative plates are shown in Fig. 6. Figure 7 shows typical discharge curves of a 100-ampere-hr nickel-iron cell.

It has been found desirable to charge the tubular nickel-iron cell at comparatively high rates (10–20 amperes for a 100-ampere-hr battery) in order to reduce the time required for a full charge. However, great numbers of tubular-iron cells are floated at rates of 0.002–0.004 milliampere per ampere-hour of capacity with excellent results.

The open-circuit potential of the negative iron electrode is very close to the hydrogen potential. This makes the electrode susceptible to rather high local action or self-discharge. The battery when on open circuit or on charge continually gives off hydrogen. Therefore nickel-iron cells must be well ventilated. They require higher float currents and more frequent watering than most other types of cell.

Manufacturers warn against operating nickel-iron cells above 46°C. Also, depending somewhat on discharge rate, capacities drop rapidly below a critical temperature of about 2.2°C.

Nickel-cadmium alkaline cells. Equation (5), the chemical equation for the nickel-cadmium cell, is exactly the same as the one for the previous cell

$$2NiOOH \cdot H_2O + Cd \underset{Charge}{\overset{Discharge}{\rightleftharpoons}} 2Ni(OH)_2 + Cd(OH)_2 \quad (5)$$

type except for the use of cadmium instead of iron. The earlier remarks about positive electrode and electrolyte apply equally to this cell. The cadmium negative electrode differs from the iron negative in that its potential is below the hydrogen potential. Therefore, the cadmium electrode is completely inert to the electrolyte. It requires almost no float current to keep charged, and consequently the water consumption and float charge currents are extremely low.

The original nickel-cadmium cell, now known as the "pocket" type, was invented by Waldmar Jungner at about the same time that Edison invented the tubular cell. In this cell the positive and negative plates are of the same construction as that described for the iron electrode used in the tubular-iron cell. The positive electrode uses graphite as conductor instead of the nickel flake used in the tubular cell.

The pocket nickel-cadmium battery is widely used for emergency power use. The cadmium electrode has extremely low stand loss, and it can be kept at a state of full charge with very little maintenance. In most forms it is not well suited for cycle service. It complements the tubular-iron cell. Nickel-cadmium cells may be floated at voltages of 1.40–1.45 per cell.

After an emergency discharge it is desirable to fully charge the battery before it is shifted to the float circuit.

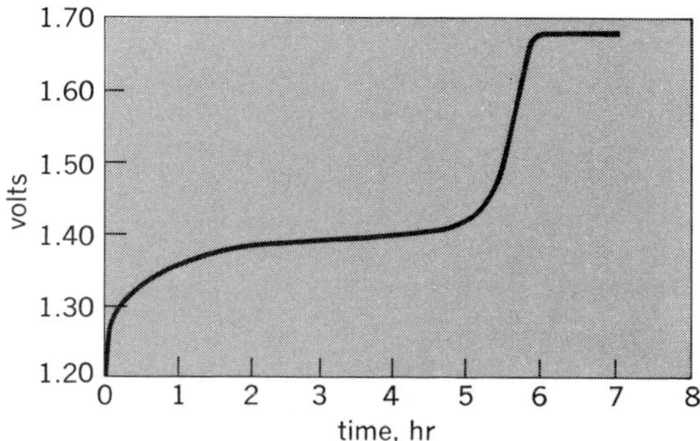

Fig. 9. Typical charge curve for sintered-plate nickel-cadmium cell. (*ESB Inc.*)

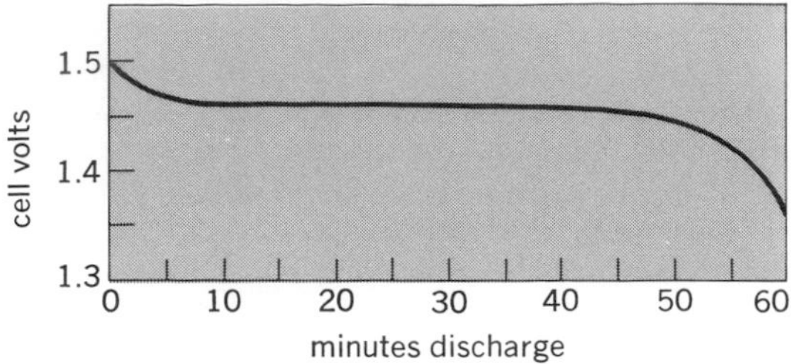

Fig. 10. Typical discharge curve exhibited by a silver oxide–zinc cell. (*ESB Inc.*)

Sintered plate cells. During World War II the Germans developed a sintered-plate type of nickel-cadmium cell. Extremely fine nickel powder, obtained from decomposition of nickel carbonyl, is sintered in a mold around a nickel or nickel-plated screen. For positive plates these plaques are impregnated with a nickel salt (usually nitrate) and processed to produce nickel hydrate in the pores. Plaques for the negative electrodes are impregnated with a cadmium salt (nitrate or chloride) and processed in a manner like that for the positive.

The electrolyte is a solution of KOH made with specific gravities ranging 1.240–1.300.

Sintered-plate cells are displacing the original types, being superior in several respects. They have less internal resistance and a higher utilization coefficient, and they perform better at both higher and lower temperatures.

They are especially suited to extremely-high-rate discharges, low-temperature operation, and other severe applications. They are used for aircraft and diesel starting and for many military services.

Sealed cells. It has been found that the smaller sizes of nickel-cadmium cells can be operated in the fully sealed state. Sealed cells are made from very small hearing-aid sizes up to the larger flashlight sizes. In order to work in the sealed state, the cell must have a very limited amount of electrolyte, and the ratio and relative states of charge of positive and negative plates must be carefully controlled. Containers are made of nickel-plated steel or plastic. In the flashlight types the plates and separators are often rolled up in a spiral coil. They have many of the features of sintered-plate cells.

Charging is not critical. It can be done rapidly and efficiently by constant-current, constant-potential, and modified constant-potential methods; gassing begins around 1.47 volts, and when using normal charge rate (5-hr), the end voltage will be 1.75.

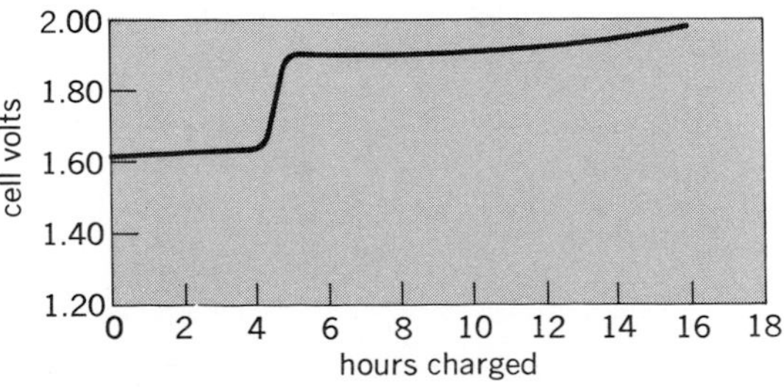

Fig. 11. Typical charge curve exhibited by a silver oxide–zinc cell. (*ESB Inc.*)

Typical discharge and charge curves are shown in Figs. 8 and 9.

Silver oxide–zinc alkaline cell. Silver oxide positive plates and sponge-zinc negative plates came into use during the late 1940s. They have high ampere-hour and watt-hour capacities per unit of volume or weight. A high-specific-gravity KOH solution, up to 1.450, has been found advantageous in minimizing local action. The cell reaction can be expressed as Eq. (6).

$$AgO + Zn + H_2O \underset{\text{Charge}}{\overset{\text{Discharge}}{\rightleftharpoons}} Ag + Zn(OH)_2 \qquad (6)$$

Charging can be accomplished by a constant-current or modified constant-potential charge, as long as the cell voltage does not exceed 2.1 volts at any time. Typical discharge and charge curves are shown in Figs. 10 and 11.

Silver oxide–zinc cells are used both as primary and secondary cells for military use and for nonmilitary applications where battery power with minimum weight is an essential consideration.

Freezing of alkaline electrolyte. The use of high-gravity KOH electrolyte for nickel-cadmium and silver oxide–zinc cells eliminates freezing under severe arctic conditions. High-specific-gravity electrolyte cannot be used with nickel-iron cells.

Venting of storage cells. Venting must be provided for all storage cells to permit escape of local-action gas or gas generated in the charging process. The only exceptions are the special sealed cells, in which gassing is held to a minimum and any hydrogen or oxygen generated is recombined through catalysis.

The provision for escape of gas has led to numerous devices to prevent spillage of electrolyte from cells in aircraft and other applications.

[W. W. SMITH]

Bibliography: H. Bode, *Lead Acid Batteries*, 1977; J. T. Crennell and F. M. Lea, *Alkaline Accumulators*, 1928; S. U. Falk and A. J. Salkind, *Alkaline Storage Batteries*, 1969; D. G. Fink and H. W. Beaty (eds.), *Standard Handbook for Electrical Engineers*, 11th ed., 1978; G. W. Vinal, *Storage Batteries*, 4th ed., 1955.

Strip mining

A surface method of mining by removing the material overlying the bed and loading the uncovered mineral, usually coal. It is safer than underground mining because neither the workers nor the equipment is subjected to such hazards as roof falls and explosions caused by gas or dust ignitions. Coal near the outcrop or at shallow depth can be stripped not only more cheaply but more completely than by deep mining, and the need for leaving pillars of coal to support the mine roof is eliminated. The roof over coal at shallow depth is weak and difficult to support in underground workings by conventional methods, yet this same weakness, of cover and of coal seam, facilitates stripping.

Stripping techniques. Power shovels, draglines, bulldozers, and other types of earth-moving equipment slice a cut through the overburden down to the coal. The cut ranges from 40 to 150 ft (12 to 46 m) wide, depending on the type and size of equipment used. The stripped overburden (spoil) is stacked in a long ridge (spoil bank) parallel with

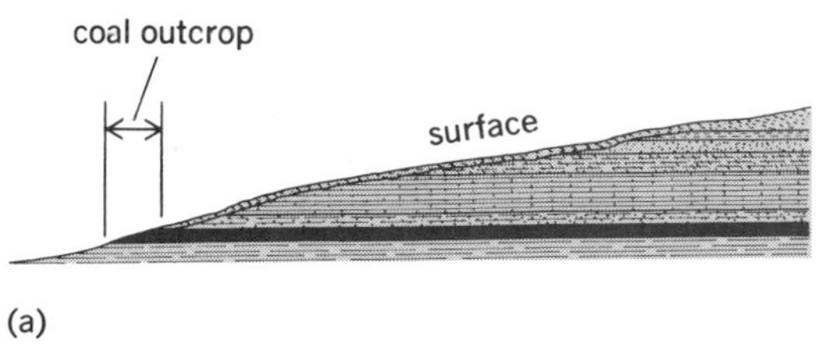

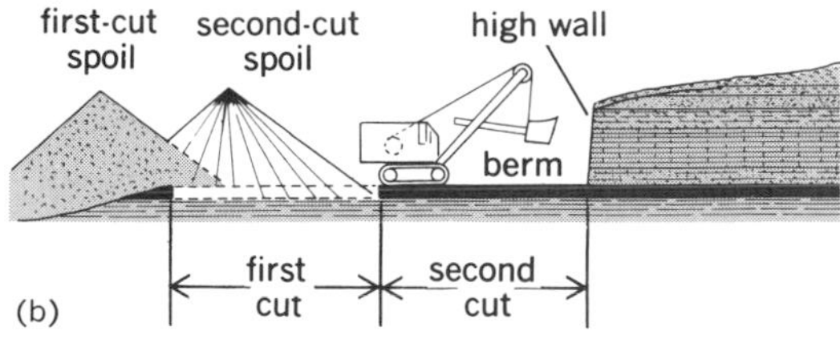

Fig. 1. Representative cross-section profile diagrams of contour strip mining of coal. (*a*) Section before stripping. (*b*) Section after second cut.

the cut and as far as possible from undisturbed overburden (high wall). The slope of a spoil bank is approximately 1.4:1 and that of a high wall under average conditions is 0.3:1. The uncovered coal (berm) is then fragmented, loaded, and transported from the pit. Spoil from each succeeding cut is stacked overlapping and parallel with the previous ridge and also fills the space left by the coal removed (Fig. 1).

Techniques of stripping methods are similar, but the size of equipment used depends on whether the mine is in prairie or hill country. In prairie areas the thickness of overburden is nearly uniform, the coal bed is extensive, and equipment can be used for years at one mine without dismantling and moving to another location. Large-capacity shovels costing $1,000,000 or more and requiring many months to erect on the site are used at prairie mines. A unit of this type is the 60-yd^3 (46-m^3) rig shown in Fig. 2.

In 1958 the world's largest power shovel was the 70-yd^3 (54-m^3) unit at Peabody Coal Co.'s River King mine near Freeburg, Ill. By 1968 the 180-yd^3 (138-m^3) shovel at Southwestern Illinois Coal Corp.'s Captain mine near Percy, Ill., was the largest.

Most coal underlying hills is mined by underground methods, but where the working approaches the outcrop and the overburden is thin, the roof becomes difficult and expensive to support. The coal between the actual or potential underground workings and the outcrop is then more suitable for stripping. Usually, only two or three cuts 40–50 ft (12–15 m) wide can be made on the contour of the coal bed, after which the shovel has to be moved to another site. Thus, in contour stripping, mobility of shovels up to 5-yd^3 (3.8-m^3) capacity is more important than those of larger capacity.

Large draglines are used instead of shovels to strip pitching beds of anthracite to depths surpassing 400 ft; however, this use of large draglines could more properly be classed as open pit. *See* OPEN-PIT MINING.

Removal of unconsolidated overburden by hydraulic monitoring is a technique used especially in Alaska. Water under a high-pressure head is directed through a nozzle against the overburden to wash it into deep valleys where swift streams carry it away.

Although the character of the overburden determines the thickness of overburden that can be stripped, the maximum for shovels up to 5-yd^3 (3.8-m^3) capacity is about 50 ft (15 m) and for the largest equipment about 110 ft (33.5 m). To reach these goals it frequently is necessary to use a dragline, carryall, or bulldozer on the high wall or a dragline in tandem with the shovel on the berm to strip the upper few feet of the overburden.

Digging equipment known as the wheel can be used ahead of a large power shovel to remove the upper 20–40 ft (6–12 m) of unconsolidated soil, clay, or weathered strata. This spoil is discharged onto a belt conveyor, then onto a stacker, and finally deposited several hundred feet from the high wall. Overburden thus removed improves the shovel productivity rate materially. Also, coal reserves can be mined that would have been too deep for the power shovel alone to handle.

Rocks overlying coal beds present some diversity of conditions for removal. Materials generally comprise shale, sandstone, and limestone with shale predominating. Proper fragmentation before stripping may be necessary to produce sizes that are smaller than the shovel dipper. Probably more research has been done on overburden drilling and blasting than on any other phase of stripping. The diameter, depth, and spacing of drill holes, the type of drill (whether vertical or horizontal), and the amount and type of explosive for each blast hole are the variables that must be determined for optimum production. Truck-mounted rotary drills have replaced churn drills for drilling vertical holes, and for horizontal drilling, auger drills are used. Package explosives, Airdox, Cardox, and

Fig. 2. Large electric shovel of the type used in prairie regions, high wall (right), and spoil bank (left) at Hanna Coal Co.'s Georgetown mine, eastern Ohio. (*U.S. Bureau of Mines and Marion Power Shovel Co.*)

commercial ammonium nitrate mixed with diesel fuel are used for blasting. The ammonium nitrate–diesel fuel mixture is one of the cheaper explosives and has gained favor rapidly. Equipment for mixing this explosive and automatically injecting it through a plastic tube into horizontal drill holes is being tested. If perfected, it will mechanize the only manual operation remaining in the stripping cycle.

Before coal is loaded, spoil remaining on the berm is removed by bulldozer, grader and rotary brooms, or at small mines by hand brooms. Small-diameter vertical holes are augered into the bed and blasted with small charges of explosive to crack the coal. A ripper pulled by a bulldozer is effective in replacing coal drilling and blasting. Broken coal is loaded by $\frac{1}{2}$–5-yd^3 (0.38–3.8-m^3) capacity shovels into trucks of 5–80-ton (4.5–73-metric ton) capacity and transported from the stripping.

Land reclamation. Surface mining for coal can create many problems. The drastic disturbance of the overburden severely changes the chemical and physical properties of the resulting spoils. These altered properties often create a hostile environment for seed germination and subsequent plant growth. Unless vegetative cover is established almost immediately, the denuded areas are subject to both wind and water erosion that pollute surrounding streams with sediment.

Topsoil. Stripmining removes the developed soils and vegetative cover and creates a heterogeneous mass of topsoil, subsoil, and substrata rock fragments. The 1977 United States stripmine law requires that topsoil be removed and reapplied on the spoil surface during regrading and reclamation. Even when topsoil is reapplied over the spoil surface, an organic mulch is generally required for good seed germination and development. These mulching materials alter the surface microclimate and help to conserve soil moisture during the critical seedling establishment period.

The removal of topsoil before mining and its replacement on the spoil surface after final grading have aided materially in the reclamation process. Many of the chemical and physical limitations previously associated with mine spoils have been alleviated or eliminated. Surface grading techniques and seedbed preparation are very important in obtaining good vegetative cover for erosion and sedimentation control. Research has indicated that the surface should be left rough—preferably with small contour furrrows perpendicular to the slope. These furrows catch seed and fertilizer carried in runoff water, thus increasing germination and seedling development and decreasing soil and fertilizer loss from erosion.

Plant species adaptation. Many plant species of economic importance can be used to produce hay, pasture, various horticultural crops, and major row crops. Commercial varieties of grasses that have shown promise on United States eastern stripmine spoils with moderate pH (5.0–6.0) include orchard grass, tall fescue, bromegrass, ryegrass, and timothy. Other grasses that have shown promise on low-pH (4.5 or less) stripmine spoils are weeping lovegrass, bermuda grass, switch grass, bent grass, deer's-tongue, and redtop. Legumes that have shown promise on eastern acid stripmine spoil areas include alfalfa, white clover, crimson clover, bird's-foot trefoil, lespedeza, red clover, crown vetch, hairy vetch, flat pea, kura clover, zigzag clover, and white and yellow sweet clovers.

In the western United States a number of species have been tested. In general, several of the wheatgrass varieties, green needle grasses, side oats grama grass, smooth brome, wheat, and wild rye species have been used with varying degrees of success. Legumes species tested include bird's-foot trefoil, sweet clover, alfalfa, flat pea, and crown vetch. Most spoils in the western United States are returned to perennial grasses for eventual use by grazing livestock and wildlife.

Several woody species can be used on stripmine areas where rainfall is adequate. Tree species should be planted only with or after herbaceous species, such as grasses and legumes, that have stabilized the soil. Almost any species of trees can be grown on stripmine areas as long as their nutrient and environmental needs are met.

Many of the high-organic waste materials such as sewage sludge and composted sewage and garbage materials, are highly beneficial for establishing plant material on stripmine areas, especially with a low pH. These materials can contribute significantly to the plant nutrition on stripmine spoils for the production of vegetative cover.

[ORUS L. BENNETT]

Bibliography: W. S. Doyle, *Strip Mining of Coal: Environmental Solutions*, 1976; R. F. Munn, *Strip Mining: An Annotated Bibliography*, 1973; R. Peele, *Mining Engineers' Handbook*, 3d ed., 1948; U.S. Department of the Interior, *Surface Mining and Our Environment*, 1967; R. A. Wright (ed.), *The Reclamation of Disturbed Arid Lands*, 1978.

Substitute natural gas (SNG)

Any synthesized high-Btu-value gas—usually manufactured by the chemical conversion of a hydrocarbon fossil fuel, or by biological conversion of renewable materials such as seaweed (biomass)—that is chemically and physically interchangeable with the natural gas sold today by utilities. *See* NATURAL GAS.

The principal feedstocks used or anticipated for producing substitute natural gas (SNG) are coal, oil shale, peat, petroleum, and petroleum-related products; however, SNG might also be produced by using vegetation or biological waste as feedstocks with either thermal or biological methods of conversion.

Specifications. SNG is distinguished from other manufactured gases by its high methane content—usually above 85%, and by its high heating value—above 900 Btu/standard cubic foot (SCF), or 33.5 megajoules/m^3. Minimum specifications for SNG were set by the Office of Coal Research (OCR), formerly in the Department of the Interior, now incorporated into the Department of Energy (DOE). The specifications include a pipeline delivery pressure of 1000 psig (6.895 MPa gage), minimum heating value of 900 Btu/SCF, 0.1% maximum carbon monoxide (CO) content, 0.3% maximum carbon dioxide (CO_2) content, 0.25 grains/100 SCF (0.057225 grams/m^3) hydrogen sulfide (H_2S), 10 grains/100 SCF (2.289 grams/m^3) total sulfur, 7 lb/10^6 ft^3 (3.18 kg/28,317 m^3) water, specific gravity (air standard) of 0.59–0.62, 5% maximum total inert constituents, and hydrocarbon dew point of

Summary of characteristics for conceptual and emerging coal-to-SNG processes

Process	Type of process*	Feedstock and technique	Gasifier stages	Reactor parameter: Temp., °F(°C)	Reactor parameter: Press psig (MPa, gage)	Gas-solids contacting scheme	Nitrogen barrier	Fraction of total methane produced directly	Process status
Koppers-Totzek	G	Pulvarized coal in entrained flow	1	2700–3300 (1480–1830)	Atmos+	Dilute phase co-current slagging	Oxygen plant	Minor	Commercial gasifier (for synthesis gas)
Wellman-Galusha	G	3/16 to 5/16 in. (5 to 8 mm) coal lockhopper	1	1000–1200 (540–650)	Atmos+	Moving-bed countercurrent	Oxygen plant	Minor	Commercial gasifier (air-blown, for low-Btu gas)
Winkler	G	Crushed 0 to 3/8 in. (0 to 10 mm) coal screw feeder	1	1500–1850 (820–1010)	Atmos+	Fluidized-bed	Oxygen plant	Minor	Commercial gasifier (for synthesis gas)
Texaco	G	Pulverized coal in slurry	1	2700–3300 (1480–1830)	800 (5.5)	Dilute phase cocurrent	Oxygen plant	Minor	Coal plant piloted; commercial plant in design stage
British Gas Corporation Slagging Gasifier (Lurgi)	G/HD	1 to 1½ in. (25 to 38 mm) × 0 through lockhopper	1	2700 (1480) (slag) to 1100 (590) (outlet gas)	350–400 (2.4–2.8)	Descending-bed, countercurrent, slagging	Oxygen plant	26%	Pilot plant was operated at Westfield, Scotland; in design stage
COGAS	G/HD	Direct-fed char from pyrolysis stage	1 + regenerator	1600–1700 (870–930)	Atmos+	Fluidized-bed	Heat carrier	Minor	A portion of process piloted at Leatherhead, England; demonstration plant in design stage
Lurgi	G/HD	¼ to 1¼ in. (6 to 32 mm) coal lockhopper	1	1150–1600 (620–870)	350–400 (2.4–2.8)	Fixed-bed countercurrent	Oxygen plant	40%+	Commercial gasifier (for synthesis gas)
IGT HYGAS–EG	G/H			1200–1400 (650–760) (1st stage)		Dilute phase cocurrent (1st stage)	Electrical process	64%	Pilot plant shut down
IGT HYGAS–SQ	G/H	−8 mesh coal in slurry	2 + hydrogen source	1600–1800 (870–980) (2d stage)	1000+ (6.9+)		Oxygen plant	58%	Pilot plant operating; demonstration design completed
IGT HYGAS–SI	G/H					Fluidized-bed (2d stage)	Iron oxide solids	78%	Pilot plant shut down
Exxon Catalytic Coal Gasifier	G (Catalytic)	−8 mesh	1	1300 (700)	500 (3.4)	Catalysis in fluidized bed	Exothermic heat from catalytic methanation	100%	Pilot plant operations in development
BI–GAS	G/HD	70%, −200 mesh pulverized coal; lockhopper or slurry	2	3000 (1650) (1st stage) 1700 (930) (2d stage)	1000 (6.9) 1000 (6.9)	Dilute phase cocurrent (1st stage) Cyclone slagging gasifier (2d stage)	Oxygen plant	50%†	Pilot plant shut down
Synthane	G/H	−60 mesh in lockhopper or slurry	1	1800 (980)	1000 (6.9)	Fluidized bed	Oxygen plant	65%†	Pilot plant shut down

*H = hydrogasification, G = gasification, HD = hydrodevolatilization.
†Estimated or calculated from limited available data.

−40°F (−40°C) at 1000 psig. *See* HEATING VALUE.

There was no recognized standard for SNG in the United States at the end of 1979. However, the American Society for Testing and Materials (ASTM) has undertaken development of a new standard for natural gas and synthetic equivalents for pipeline use that presumably would apply to SNG. Specifications are expected to resemble those of OCR noted above, but to differ from them in several areas.

In addition, an interchangeability index, to determine the compatibility of SNG with appliances that burn natural gas, has been compiled by the American Gas Association (AGA) on the basis of SNG calorific value, specific gravity, flame velocity, and flame stability. The index is used to predict such undesirable flame features as lifting, flashback, and yellow tipping. Additional research on interchangeability has been undertaken by the International Gas Union (IGU) to assess the various requirements of different nations.

Commercial production. As a result of early research supported by the AGA and OCR, a number of advanced American processes are emerging that are expected to provide the basis for commercial production of SNG from both coal and oil shale beginning before 1985; through 1979, however, there was no such commercial SNG production in the world. Gas manufactured from coal actually preceded the worldwide use of natural gas, and was the basis for the United States gas industry until pipeline technology developments in the 1940s permitted nationwide conversion to less expensive natural gas. Historically, gas, manufactured from coal in unaerated retorts, was used for lighting as early as the 17th century. After the gas industry conversion to processed wellhead gas in the 1940s, manufactured gases in the United States served for the most part to supply utility peakload needs. In this instance, an SNG was produced by pyrolysis of oil on hot checker-brick. *See* COAL GASIFICATION.

The resurgence of interest in synthetic gas by the United States gas industry in the 1970s resulted from predictions and experience of diminishing domestic natural gas producibility and gas reserves, and rising fuel consumption rates, plus the reexamination of the commodity and political values of energy after the 1973 Organization of Petroleum Exporting Countries (OPEC) oil embargo and the 1979 crisis in Iran. Commercial SNG production was advocated by the AGA in 1979. Other interim developments are widely advocated—including both fissionable and other fossil sources—to make the nation independent of foreign fuel sources until fission, fusion, solar energy, and other resources can be advanced to large-scale commercial application. *See* ENERGY SOURCES.

Feedstocks. Alternative liquid and solid feedstocks have been proposed for SNG production. Preference is based on the technological status of a conversion process, feedstock availability and cost, and the economics of conversion. Preferably, the feedstock should have a low sulfur content, and a low carbon-hydrogen weight ratio approaching that of the methane end product. A feedstock of this type is desirable because: (1) high-carbon liquid feedstocks tend to result in coke deposition that can plug the gasifier apparatus and can interfere with catalyst performance (such feedstocks require greater quantities of expensive hydrogen to produce methane); and (2) sulfur deactivates most gasification process catalysts and is a potential pollutant. On these bases, the preferred feedstocks are light (low-molecular-weight) petroleum fractions, which include propane, butane, ethane, naphtha, and natural gasolines. These feedstocks, however, being required in increasing amounts which must be largely imported, represent a solution that is not a rational approach to solving the energy dilemma. In terms of domestic availability, abundant coal and oil shale are the most practical feedstocks for long-term domestic SNG production that ameliorates supply-demand pressures.

Coal is found in 30 of the 50 states in beds that often traverse natural-gas pipelines. Proved and currently recoverable coal reserves of the United States (all types) total about 4.8×10^{18} Btu (5.1×10^{21} J) and the estimated total remaining recoverable coal may total as much as 20.7 to 35.8×10^{18} Btu (21.8 to 37.8×10^{21} J). The total of United States oil shale resources is estimated at 4.4×10^{12} barrels (7.0×10^{11} m^3) of shale oil (about 25.52×10^{18} Btu or 26.93×10^{21} J). This includes 1.8×10^{12} bbl (2.9×10^{11} m^3) of shale oil (about 8.44×10^{18} Btu or 8.90×10^{21} J) in the Green River Formation (Eocene) of the West, including 700×10^9 bbl or 1.1×10^{11} m^3 (about 4.06×10^{18} Btu or 4.28×10^{21} J) of 25 gal per ton (104 liters per metric ton) richness or greater; and 2.6×10^{12} bbl or 4.1×10^{11} m^3 (about 14.30×10^{18} Btu or 15.09×10^{21} J) in the Devonian shales of the East, including up to 10^{12} bbl or 1.6×10^{11} m^3 (about 5.5×10^{18} Btu or 5.8×10^{21} J) recoverable. United States peat resources have been estimated to total 1.4×10^{18} Btu (1.5×10^{21} J). The first United States commercial SNG plants have been designed to reform naphtha and liquid petroleum gas feedstocks while emerging coal, oil shale, and peat gasification processes are perfected. *See* COAL; LIQUEFIED PETROLEUM GAS (LPG); OIL SHALE.

Reforming processes. The three principal light-hydrocarbon reforming processes of current (1980) interest are the Catalytic Rich Gas (CRG, developed by the British Gas Council), Gasynthan (Lurgi GmbH), and Methane Rich Gas (MRG, Japan Gasoline and Universal Oil Products) processes. Each makes methane by reacting light hydrocarbons with low-temperature steam (800–1000°F, or 425–540°C), over a fixed bed of nickle-based catalyst to produce methane and carbon dioxide (which is later removed).

Despite the simplicity and economy of light-hydrocarbon reforming, the shortage of such feedstocks and their cost make coal by far the greatest potential domestic source of SNG through new processes. Whether coal is converted to a gas or a liquid in a process depends on the nature of process conditions.

Coal conversion. Four principal conversion routes for coal are possible to produce a gaseous or liquid fuel convertible to the principal SNG constituent—methane:

1. Thermal cracking (pyrolysis) thermally breaks chemical bonds of the molecules in coal, generating a broad spectrum of lower-molecular-weight products. *See* CRACKING.

2. Solvation dissolves the coal in a solvent that transfers hydrogen to the molecules in coal, which break up, resulting in lower-molecular-weight sta-

ble hydrocarbons. Based on the process, this reaction is carried out with or without a catalyst, and with or without free hydrogen being present. Methane is produced during this processing, and product oils can subsequently be reformed to SNG. *See* COAL LIQUEFACTION.

3. Gasification reacts the coal with steam, producing a synthesis gas that, after suitable processing, can be converted to methane over a nickel catalyst. Oxygen can be added with the steam to react with a portion of the coal to supply the necessary heat of reaction, or heat can be introduced by other indirect means.

4. Hydrogasification reacts the coal with hydrogen or hydrogen-steam mixtures, usually at high pressures (30–100 atm or 3–10 MPa), and produces methane directly.

The amount of methane produced directly in these various routes depends on the process conditions, mainly pressure (high, giving increased methane content) and temperature (low, giving high methane content); on the method of contacting the coal with the reacting gases; and on the feed composition.

On leaving the reactor, the hot product gas, in addition to methane, will have steam, hydrogen, carbon monoxide, and carbon dioxide as principal constituents, combined with impurities, which may include ammonia, nitrogen, hydrogen cyanide, hydrogen sulfide, oils, and tars. After the impurities are removed, the gas is upgraded to SNG by reacting the carbon oxides and hydrogen over a catalyst to produce additional methane.

Of the four routes outlines for fuel conversion to SNG, principal interest focuses on the conversion of coal to SNG by direct hydrogasification, or by gasification followed by methanation, or by both.

New processes. Emerging processes now being developed in the United States vary principally on the following points: (1) the technique of introducing heat while excluding nitrogen; (2) the technique of gas-solids contacting, by which gases are placed in contact with solids; and (3) the technique of high-carbon utilization. These differences account for different reactor concepts. The principal candidate processes for commercial coal-to-SNG conversion are compared in the table.

Each of these processes employs a series of basic steps, which are: (1) coal receiving and storage; (2) coal processing into the size required for gasification, and often into a form, such as coal oil or coal-water slurry, for convenient introduction into the gasification reactor; (3) conversion to a gas; (4) gas cleaning and cooling; (5) shift conversion of carbon monoxide in the gas to carbon dioxide, with the generation of more hydrogen (usually a 3:1 hydrogen-to-carbon monoxide ratio is sought prior to catalytic methanation); (6) acid-gas removal and purification, taking out carbon dioxide and gaseous sulfur compounds, for the most part as hydrogen sulfide; (7) methanation, to maximize the heating value of the gas by reacting hydrogen and carbon monoxide to form additional methane; and (8) drying and compressing the gas (if required) to meet pipeline requirements mentioned above. Peat can be processed in a manner similar to coal; however, because of the ease with which peat is converted to gases, less severe conditions are required.

Domestic coal. The various emerging American processes that are listed along with existing technology in the table are all designed specifically to produce SNG and to utilize domestic coal as a feedstock. American coal generally differs from European coal, for which the older gasification processes were designed. Eastern United States coals have somewhat higher sulfur contents and may tend to agglomerate; therefore, these new processes must take advantage of the most modern mechanical and chemical engineering concepts to overcome these objectionable features.

HYGAS process. The only United States process to have reached an advanced operational pilot plant stage through 1979 is the Institute of Gas Technology's HYGAS process, sponsored jointly through 1979 by the Gas Research Institute and DOE. In its first 8 years of pilot plant operation, the process achieved many milestones, including the first production of SNG on a large scale in continuous plant operation using coal as a feedstock. The HYGAS process is basically a two-stage high-pressure (over 1000 psi or 7 MPa) process that feeds any rank of coal, utilizing dilute-phase gas-solids contacting in the first stage at 1200–1400°F (650–760°C), and fluidized-bed gas-solids contacting in the second stage at 1600–1800°F (870–980°C), as well as in the steam-oxygen char gasification section of the reactor, where the necessary hydrogen is produced.

Underground processes. In-place processes are being considered again for the production of gas from coal. By this technique the coal is fractured in place, and the gasification reactions are conducted underground. Because of difficulties in controlling these underground reactions and associated gas flow, this concept—although used for low-Btu gas production in Russia with varying degrees of success—does not at this time offer as promising an approach as do aboveground techniques. Work on in-place processes is continuing, however, in an attempt to achieve a practical solution.

Oil shale conversion. Oil shale is the second most likely feedstock for commercial SNG production; however, the concept of producing SNG from oil shale is not as advanced in development as the coal-based processes. Two approaches can be taken. In one concept, the shale is first retorted to convert the kerogen in the shale to oil, which is subsequently gasified by techniques already discussed. The second approach, and the one that holds the most promise, is the direct hydrogasification of the kerogen in the shale to directly form a high-methane content gas. Because of the differences between kerogen and coal, considerable oil is coproduced when producing gas by this route. *See the feature article* OUTLOOK FOR FUEL RESERVES. [FRANK C. SCHORA]

Bibliography: W. Bodle and J. Huebler, Coal gasification, in R. D. Meyers (ed.), *Coal Handbook*, publication pending; J. R. Donnell, Western United States oil shale resources and geology, *IGT Symposium: Synthetic Fuels from Oil Shale*, Atlanta, Dec. 3–6, 1979; J. C. Janka and J. M. Dennison, Devonian oil shale: A major American energy resource, *IGT Symposium*, 1979; G. Long, *Amer. Gas Ass. Mon.*, 54(6):31–33, 1972; Office of Coal Research (formerly in the U.S. Department of the Interior, now in the Department of Energy), *Standard for Acceptable Quality Pipeline Gas from Coal*, Apr. 23, 1965; J. D. Parent, *A Survey of Unit-*

ed States and Total World Production, Proved Reserves, and Remaining Recoverable Resources of Fossil Fuels and Uranium, IGT, March 1979; Peat resources data from A. M. Rader, Testimony before the House of Representatives Environment, Energy and Natural Resource Subcommittee of the Committee on Government Operations, Sept. 29, 1977; F. C. Schora et al., Oil shale: Present technology and the IGT/AGA process, *Proceedings of the 8th Synthetic Pipeline Gas Symposium, AGA/ERDA/IGU*, Chicago, Oct. 18–20, 1976; F. C. Schora, B. S. Lee, and J. Huebler, The HYGAS process, *12th World Gas Conference*, Nice, France, 1973.

Superconductivity

A phenomenon occurring at low temperatures in many electrical conductors, in which their electrical resistance vanishes and they display other remarkable properties. In 1911 H. Kammerlingh Onnes discovered the phenomenon when he observed that the electrical resistance of a sample of mercury dropped to an immeasurably small value when it was cooled below a critical temperature of 269°C (or about 4 K). Subsequent, more accurate measurements confirmed the complete absence of ohmic losses in the superconducting state. It was obvious from the outset that the absence of ohmic losses should have important practical applications. In fact, Kammerlingh Onnes suggested the possibility of obtaining very high critical fields with superconducting coils, since the absence of ohmic losses permitted the use of very high current densities.

Superconductivity found its first large-scale application in the construction of large-size magnets needed for high-energy physics experiments. The superconducting coil (Fig. 1) of the bubble chamber at the European Commission for Nuclear Research (CERN) produces a field of 3.5 teslas in a volume of the order of 50 m^3. The stored energy is of the order of 1 gigajoule. The magnet consumes less than 1 MW (for refrigeration), while a similar one with conventional coils would require 70 MW.

The fundamental characteristic of superconductors—their zero electrical resistance—should allow significant energy savings in the areas of electricity generation and transport. Superconducting alternators and power cables are being developed. Energy storage in very large superconducting coils may also be an economic proposition: utility companies are considering this solution for peak shaving. More importantly, superconductivity is a key element for the development of two radically new

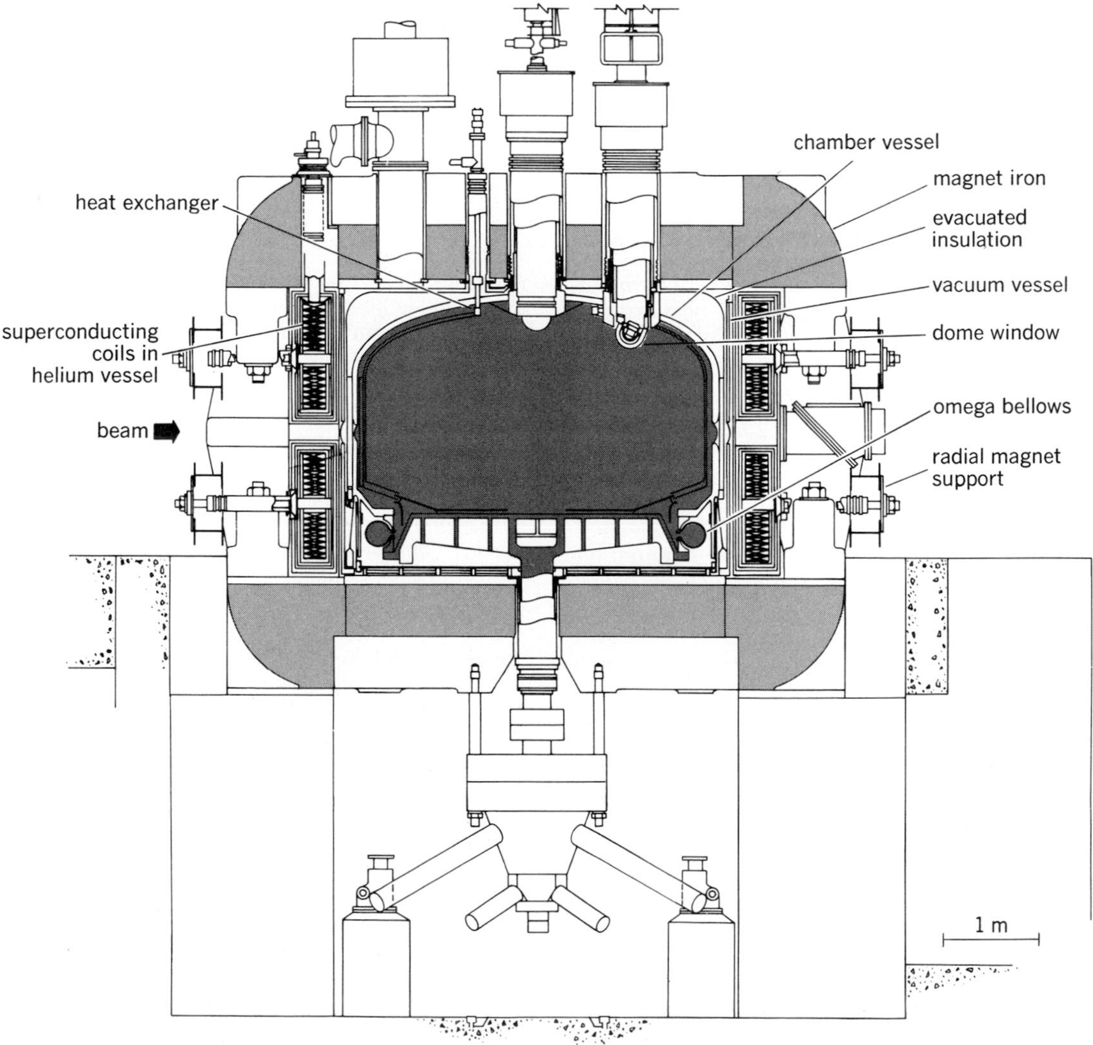

Fig. 1. Hydrogen bubble chamber at CERN. *(Proceedings of the 1967 CERN Conference on Bubble Chambers)*

types of power plants that may become major energy suppliers by the 21st century. These are magnetohydrodynamics and controlled-fusion power plants. Both use fuels that are in large supply (coal and hydrogen, respectively), but their operation requires large magnetic fields in large volumes. Such fields can be produced economically (at a low energy cost) only by superconducting magnets: with conventional copper windings, the energy output of these power plants would not be sufficient to produce the required magnetic field. *See the feature article* EXPLORING ENERGY CHOICES.

SCIENTIFIC BACKGROUND

The original attempts of Kammerlingh Onnes to produce high magnetic fields failed because mercury was not an appropriate superconductor for this purpose. A low magnetic field of a few hundred gauss (1 G = 0.0001 T) was sufficient to quench superconductivity in this material. Many more superconducting elements were subsequently discovered—in fact, most metals are superconducting at low enough temperatures with the exception of the alkalies, the monovalent metals (copper, silver, and gold), and the magnetic metals. However, they all were found to have the limitations of a low critical field and relatively low critical temperatures. The strongest superconducting element is niobium, which has a critical temperature of 9.2 K and a critical field of 0.2 T.

Type II superconductivity. The application of superconductivity to the construction of high-field coils was made possible by the discovery of type II superconductivity: superconducting alloys composed of a superconducting element with impurities in solution (such as bismuth in lead) have a much higher critical field than the pure superconductor. In the early 1960s, practical superconducting wires became commercially available with critical fields of 5–10 T; only then did projects for large-scale applications of superconductivity receive serious attention. The problem of manufacturing superconducting wire with high enough critical fields and current-carrying capability for most uses has been essentially solved.

Higher critical temperatures. Progress in finding alloys with higher critical temperatures has been much slower, and superconductivity remains more or less a low-temperature phenomenon. The highest critical temperature known, that of the niobium-germanium alloy (Nb_3Ge), is equal to 23 K, and commercial wires have to be operated below 10 K, thus requiring the use of expensive refrigerators. The question remains whether this is a fundamental limitation of superconductivity, or whether a superconducting alloy that can be operated at liquid hydrogen temperature (21 K) or liquid nitrogen temperature (77 K) will be discovered eventually. The conservative opinion is that superconductors will remain essentially in their present form, and that researchers should focus on building cheaper, more reliable refrigerators, rather than hope for discovery of a better superconductor. But one cannot rule out the discovery of an alloy that will superconduct at, say, 30 K and thus could be used at liquid hydrogen temperatures.

PRACTICAL SUPERCONDUCTING MATERIALS

Normal conductors such as copper or aluminium have a current-carrying capability limited to 2×10^7 A/cm^2 even with forced water cooling. Although it is possible to manufacture coils that produce very high fields in small volumes (up to 25 T in cores a few centimeters in diameter), they are ruled out for large-scale applications because of their large size and huge power consumption.

On the contrary, superconducting coils can be used to produce high fields in large volumes, the only energy cost being that required to refrigerate the coil. The additional capital outlay involved in the refrigeration equipment must also be considered. For many applications, refrigeration at 5 K is not a prohibitive proposition. Liquid nitrogen superconductors would be preferable, but liquid helium superconductors are already, in many instances, a more economic solution than the conventional (room-temperature) conductors.

Critical current density. The field intensity that can be reached is determined by the current-carrying capability of the superconducting wire in the presence of the field generated by the coil. This current-carrying capability is expressed in terms of a critical current density, the maximum current density that can be reached before a finite resistance appears. Critical current density is a decreasing function of temperature and magnetic field. Typical current densities used in superconducting coils are 10^9 A/cm^2 or more, about two orders of magnitude higher than values used in normal conductors.

Vortex lines. In strong magnetic fields, type II superconductors are permeated by thin normal regions, called vortex lines, that run parallel to the applied field. In the presence of a current, these vortex lines are subjected to a Lorentz force. The zero resistance state will persist only as long as the Lorentz force, which is proportional to the current, is not strong enough to start the vortex lines moving. Fortunately, vortex lines can be efficiently pinned down by metallurgical defects such as

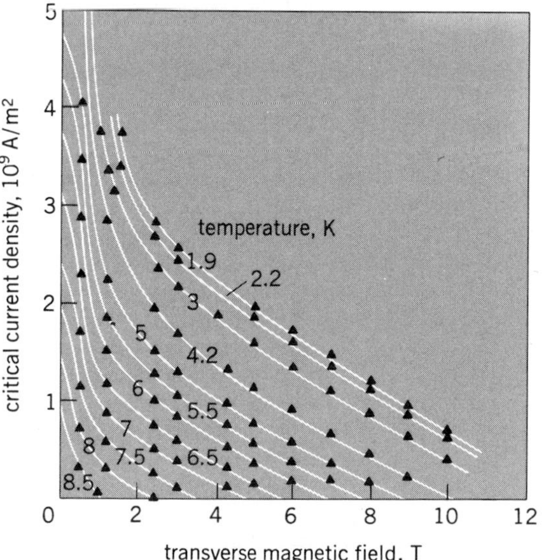

Fig. 2. Critical current density of NbTi-alloy–copper wire as function of transverse magnetic field in the temperature range of 1.9–8.5 K. (*From G. Bogner, Large-scale applications of superconductivity, in B. B. Schwartz and S. Foner, eds., Superconductor Applications: SQUIDS and Machines, chap. 20, pp. 547–719, 1977*)

Comparison of superconducting and conventional 1200-MVA generators*

Characteristic	Superconducting	Conventional
Phase to phase voltage, kV	26–500	26
Line current, kA	26.6–1.4	26.6
Active length, m	2.5–3.5	6–7
Total length, m	10–12	17–20
Stator outer diameter, m	2.6	2.7
Rotor diameter, m	1	1
Rotor length, m	4	8–10
Synchronous reactance, pu	0.2–0.5	1.7–1.9
Transient reactance, pu	0.15–0.3	0.3–0.4
Subtransient reactance, pu	0.1–0.2	0.3
Field exciter power, kW, continuous	6	5000
Generator weight, tons†	160–300	600–700
Total losses, MW	5–7	10–15

*From M. Rabinowitz, Cryogenic power generation, *Cryogenics*, 17(6):319–330, 1977.

†1 short ton = 0.9 metric ton.

small normal precipitates. These pinning centers allow the practical operation of type II superconductors in high fields and high currents, as long as the fields and currents are constant (direct-current mode). The situation is much less favorable in the alternating-current mode because in that mode vortex lines can "vibrate" between the pinning points, resulting in finite losses. In practice, ac operation is limited to the low-field (vortex-free) regime, that is, it is ruled out for coils but possible for power cables and perhaps for transformers.

Niobium-titanium alloy. The most commonly used superconducting alloy is the niobium-titanium alloy (NbTi). This consists of titanium impurities in solution in niobium, the effect of the titanium impurities being to transform the pure niobium into a strong type II superconductor. Commercial niobium-titanium-alloy wires are composed of thin niobium-titanium-alloy filaments, typically 25 μm in diameter, embedded in a copper matrix. The purpose of the copper matrix is to stabilize the superconducting wire: it provides a parallel low-resistance conducting path for the case where a normal spot would appear accidentally along one of the niobium-titanium-alloy filaments. Current is then diverted momentarily into the copper, giving time for the hot spot to cool and return to the superconducting state, thus avoiding an avalanche effect that could quench the whole superconducting coil. The superconducting characteristics of a typical niobium-titanium-alloy–copper wire are given in Fig. 2.

Niobium-tin alloy. Another superconducting alloy that has found commercial application is the niobium-tin alloy (Nb_3Sn). It has a much higher critical temperature (18 K) and critical field than that of niobium-titanium alloy, and belongs in fact to a completely different family of alloys, called A15 compounds after their crystallographic classification. They have the highest known critical temperatures and critical fields (the niobium-germanium alloy, Nb_3Ge, belongs to that category). Unfortunately, A15 compounds are very brittle, which makes wire manufacturing difficult and expensive. Filamentary niobium-tin alloy in a copper-tin matrix is used for very-high-field application, and niobium-tin-alloy tape on a copper substrate has been developed mostly for power cable application, because it has a higher operating temperature and lower ac losses than niobium-titanium alloy.

POWER GENERATION, TRANSPORT, AND STORAGE

Electric power generation, transport, and storage have emerged as some of the most promising areas of application of superconductivity.

Superconducting ac generators. Superconducting ac generators may eventually become established as the standard generators of the future. However, their reliability will be established only after large-scale machines have been built, tested, and operated for a few years.

Conventional ac generators are composed of a rotating multipole electromagnet (the rotor) and a fixed armature (the stator). Power is produced in the stator windings as the magnetic flux lines produced by the rotor cut them periodically. As the machine rotates at a fixed speed, the power output is determined by the amplitude of the magnetic field produced by the rotor.

Superconducting ac generator models already built or currently developed have a superconducting rotor and a normal armature. The reason for this combination is the poor behavior of superconductors in strong ac fields, the losses becoming quickly prohibitive.

The advantages of the superconducting machine over conventional generators (see table) derive from the higher magnetic field intensity produced by the superconducting coils of the rotor and from the absence of electrical losses in the rotor. The higher field results in a smaller volume and weight, while the absence of losses results in a higher conversion efficiency. Also, the weight and size are reduced by a factor of about 2, while the efficiency is increased by approximately 0.5%.

In the early 1970s, the interest in a reduced size and weight generator came from the perceived need for larger generators to accommodate expanding energy demand. Present-day reasoning is quite different. Energy production is not expanding as fast as in the past, and the justification for the construction of ever larger power plants is being seriously questioned. On the other hand, there is increased emphasis on conservation. Because the superconducting generator is lighter and smaller, it is generally estimated that it will be cheaper; in addition, the increase in efficiency of 0.5%, although it seems a modest improvement, results in savings that will pay for the generator in less than 25 years—the normal lifetime of this type of machine.

A 5-MVA machine has been built and tested in the United States, testing of a 20-MVA machine has been undertaken, and construction of a 300-MVA generator has been planned. The last machine will have a 5-T rotor field produced by niobium-titanium-alloy windings, as compared to the maximum 2-T field of a conventional machine. The 300-MVA target was selected so that design and construction of a commercial-size 1000-MVA machine would be the next stage in development. *See* ALTERNATING-CURRENT GENERATOR; GENERATOR.

Superconducting power cables. Practically all developed countries have established research programs on superconducting generators along similar lines, and interest has been accentuated by the energy crisis. On the contrary, programs on superconducting power cables have been adversely affected in every country except the United States. This reflects economic studies which have shown that up to a rating of the order of 5 GVA the conventional forced cooled solution is less expensive than the superconducting one. Only in the United States will there be a need for such high-power cables before the end of the 20th century.

The construction of a 138-kV 4-kA three-phase cable (circuit capacity 1000 MVA) is shown in Fig. 3*a*. Niobium-tin-alloy tape is wound as a double helix to produce a flexible cable, while the cryogenic envelope is rigid (Fig. 3*b*) and includes a permanent, sealed, vacuum-insulating space.

Superconducting ac cables will be used for underground transport of large power blocks in urban areas where overhead lines are environmentally unacceptable. In the more distant future, dc cables may be used for the transport of large power blocks over long distances—for instance, for the transport of solar electricity from the Sun Belt to the northern states in the United States. *See* DIRECT-CURRENT TRANSMISSION; ELECTRIC POWER TRANSMISSION; TRANSMISSION LINES.

Of all the large-scale applications of superconductivity, power cables are probably the one whose economic value is most sensitive to the critical temperature of the superconductor used. The use of the niobium-germanium alloy (Nb_3Ge), which has a critical temperature of 23 K and can be operated between 15 and 20 K with subcooled liquid hydrogen, would radically change present prospects, since the hydrogen coolant could be used as a fuel after its coolant role.

Energy storage. In present utility systems, the installed generating capacity is about twice that required to meet the average load. There is therefore a need for large energy storage units for power leveling. Large superconducting plants for energy storage are attractive for the following reasons: unlike hydroelectric storage, they are not limited by geographical constraints and can be placed near the load centers; they have a higher efficiency (90% versus 70%); and they have a very short response time (milliseconds), a very important feature to improve network stability.

Construction of a 100-MJ storage system has been undertaken at Los Alamos, NM. Design of a 10-GWh unit, about the size of a football field, to be placed 500 m underground to avoid stray fields at the surface has also begun. This is about the size required for economic power leveling. *See* ENERGY STORAGE.

SUPERCONDUCTING MAGNETS FOR MHD AND CONTROLLED FUSION

The base load power plants of the next century will have to meet two main requirements: they must burn fuels that are in abundance, and be environmentally acceptable. Magnetohydrodynamics (MHD) and controlled fusion are two of the few options available.

MHD power plants. In an MHD power plant, an electrically conducting expanding fluid (combustion gases seeded with ions) moves across magnetic flux lines, and electricity is produced in the form of a dc voltage that appears in a direction perpendicular to that of the fluid velocity and of the magnetic field (Fig. 4), similar to a Hall voltage. This principle has been known for over a century, but progress toward commercialization has been very slow for two reasons: the corrosion problem of the electrodes and insulating material in the hot expansion chamber; and the high intensity of the magnetic field (about 5 T) required in the large-volume (approaching 100 m^3) of the expansion chamber necessary to make the MHD plant economical. Renewed interest in MHD plants has derived from the realization that they can burn coal (even with high sulfur content) with very high efficiency (about 55%, as compared with a maximum of 40% in a conventional coal-fired plant). It now appears that MHD plants could be one of the most efficient and ecologically acceptable ways to make use of the large coal reserves in the United States and elsewhere (especially in the Soviet Union). Many of the corrosion problems have been solved, and the large-scale magnets necessary for economic operation can be built.

A large-scale cooperative program between the United States and the Soviet Union underlines the worldwide significance of MHD generators. The first 25-MW MHD plant was built in Moscow, and testing of a superconducting magnet built by the Argonne National Laboratory has been undertak-

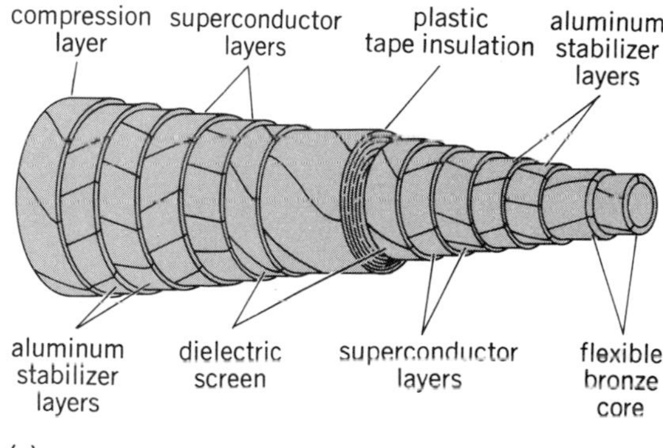

Fig. 3. Superconducting cable. (*a*) Diagram of cable construction (*from M. Garber, 10m Nb_3Sn cable for 60 Hz power transmission, IEEE Trans. Magnet., MAG-15(1): 155–158, 1979*). (*b*) Prototype cryogenic envelope under construction; technician is applying aluminum-coated Mylar sheets which prevent heat input to the cold inner pipe by radiation (*from E. B. Forsyth, Cryogenic engineering for the Brookhaven power transmission project, Cryogenics, 17(1):3–7, 1977*).

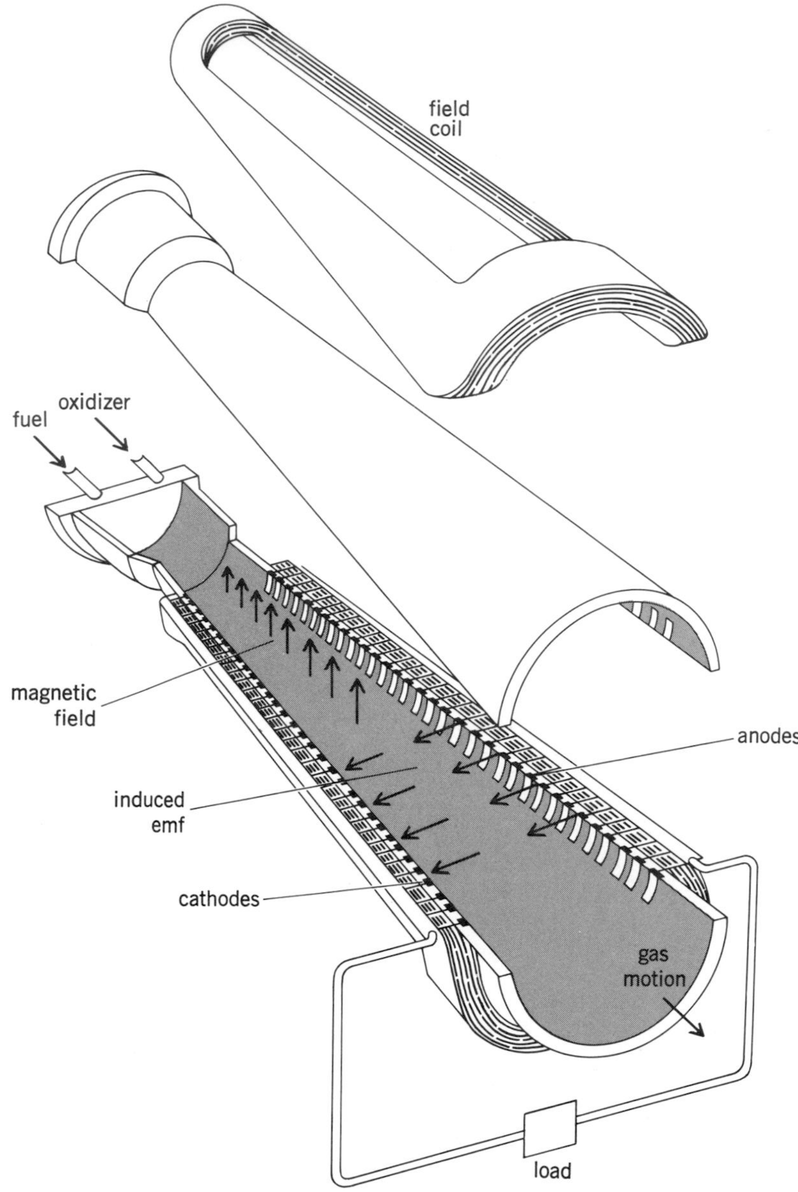

Fig. 4. Exploded view of an MHD expansion chamber and magnet. (*From G. Bogner, Large-scale applications of superconductivity, in B. B. Schwartz and S. Foner, eds., Superconductor Applications: SQUIDS and Machines, chap. 20, pp. 547–719, 1977*)

en in the Soviet plant. The next step in MHD development is the construction of a 500-MW commercial-size plant, probably at the end of the 1980s. *See* MAGNETOHYDRODYNAMIC POWER GENERATOR.

Controlled fusion. A controlled thermonuclear reaction can be in principle achieved in two ways: magnetic confinement of the plasma and laser ignition. In magnetic confinement, large-scale coils are needed to produce the field; in laser ignition, large energies have to be stored to fire the lasers. In both cases, superconductivity provides the most economical (and probably the only practical) way to achieve the very strong fields needed for these applications. Magnetic confinement will require fields in excess of 12 T in very large volumes for economic plant operation. Such fields are much beyond anything that has been attained, and will require considerable development work, for the dc as well as for the pulsed field coils. A tokamak magnetic confinement device with superconducting coils has been constructed in the Soviet Union. *See* NUCLEAR FUSION.

[GUY DEUTSCHER]

Bibliography: S. L. Ackerman et al., Practical aspects of designing and manufacturing MHD superconducting base load magnets in 1988 time frame, *IEEE Mag.*, 15:310–313, January 1979; J. S. Edmonds, Superconducting generator technology: An overview, *IEEE Mag.*, 15:673–679, January 1979; W. A. Fietz, NB_3Sn in 1978: The state of the art, *IEEE Mag.*, 15:67–75, January 1979; S. Foner and B. B. Schwartz (ed.), *Superconducting Machines and Devices: Large Systems Applications*, 1974.

Superport

An oil export terminal where many oil tankers can be loaded simultaneously. A striking example is located in the Persian Gulf off the coast of Kharg Island, Iran. Designed to store and ship oil produced in fields on the Iranian mainland, it consists of large-diameter storage tanks, pipelines, and tanker mooring and loading facilities. At the Kharg Island Superport it is possible to load up to 12 tankers simultaneously, ranging in size up to 500,000 deadweight tons (1 DWT = 1017 kg).

Design and construction. From its initial design in 1958, the terminal has developed and been constructed in stages (Fig. 1). The original installation, commissioned in 1960, consisted of a 2,700,000-barrel (1 bbl = 0.159 m³) tank farm, a 1500-ft (1 ft = 0.3 m) causeway and small-craft harbor, and a 2500-ft pile-supported steel trestle, connected to a deep-water pier 2016 ft long for mooring and loading tankers. An expansion program completed late in 1966 raised the tank farm capacity to approximately 8,000,000 bbl, widened the trestle, and lengthened the pier to 6000 ft overall. Further expansion added more tank farm storage and a sea island structure, providing two berths, one for 500,000-DWT tankers.

In the mid-1950s, the Iranian government reached an agreement with a consortium consisting of the world's major oil companies to explore, produce, and develop the oil fields in an agreed area of Iran. During 1957, preliminary planning for a suitable terminal progressed. At first, locations along the coast of the mainland were explored, principally in the northeastern corner of the Persian Gulf, because of the proximity to the fields. Eventually, the natural attributes of a terminal site off Kharg Island became apparent. It was relatively close to the mainland (27 mi; 1 mi = 1.6 km) and could be supplied by pipeline from the fields. It had deep water close to the island for pier facilities, and there was adequate land elevation on the island for storage to allow the tankers to be loaded by gravity flow. The east side of the island was chosen for the tanker marine terminal facilities because protection is provided by the island against the prevailing northwesterly winds and waves.

The facilities engineered for the original terminal included a continuous pier for berthing four tankers, two on each side, located in 65 ft of water. A trestle containing a roadway and oil pipelines connects the pier to shore through a causeway constructed where water depth is shallower. A breakwater arm at the deep end of the causeway provides a harbor for smaller vessels, such as tugs and launches used in mooring tankers. Onshore, 12 tanks were erected for crude oil storage. Pipelines, 36 in. (1 in. = 2.5 cm) in diameter, connect the tank farm with the loading positions on the pier (Fig. 1). *See* PIPELINE.

Considerable study resulted in selecting the 100,000-DWT tanker as the design ship for the terminal. Sufficient capacity in the design allowed future expansion to handle 150,000-DWT tankers. This was a major decision, as the largest vessel afloat at the time was approximately 80,000 DWT and drew 45 ft of water fully loaded. The parameters selected added 1500 ft to the distance from shore for the location of the pier and placed it in a 65-ft water depth to accommodate the loaded draft of 150,000-DWT tankers. The original design provided a continuous berthing face for tankers to approach the pier, as berthing was conceived to be without the use of tugs. The energy of impact on berthing could, therefore, be applied anywhere along the length of the pier for the contact length between tanker and fender system. The resulting design was a continuous steel superstructure of transverse trusses at regular intervals connected in the horizontal plane with bracing. The trusses sit on pairs of steel batter piles which provide the reaction for the forces of impact. A fender system of rubber blocks acting in shear and placed along the pier face absorbs the energy of impact. Horizontal and vertical steel members with timber cladding and supported by vertical piles are the contact surface with the tanker, and distribute the berthing impacts to fenders and piles. The structure is all steel with timber decking. No concrete is used except for foundation structures onshore.

Terminal expansion. The original terminal was expanded as additional capacity was required to meet growing world oil demand. The existing pier was lengthened at both ends to allow 10 ships to berth simultaneously. Eleven 500,000-bbl storage tanks were added to the tank farm. Additional pipelines, 48 in. in diameter, were added to the delivery system. The expansion provided the terminal with a capacity to deliver 337,500 barrels per hour (bph) to the ships berthed at the pier, and a maximum of 75,000 bph to a tanker. A 100,000-DWT tanker could be loaded in 10 hr, and with about 4 hr for berthing, deballasting, documentation, and deberthing, it could be turned around in a total of 14 hr. Tankers of up to 200,000 DWT could be accommodated at the outside berths under controlled conditions of berthing and with suitable weather and tide levels. The export capacity of the terminal with this expansion was rated at 3,250,000 barrels per day (bpd).

To keep pace with increasing oil demands, additional upgrading was required and resulted in improvements to the energy-absorbing capacity of the fender system and localized dredging. These improvements allowed 250,000-DWT tankers to be loaded at selected berths and raised the capacity of the terminal to approximately 3,750,000 bpd.

Crude oil supply areas, especially in the Middle East, were under continuing pressure to increase export throughout the late 1960s and into the 1970s. The rapid advancement in the size of tankers was making obsolete many existing facilities. The original pier facility of Kharg Island for loading tankers had reached the limit of expansion possibilities. No extensions to the pier were practical, and a new generation of tankers had to be ac-

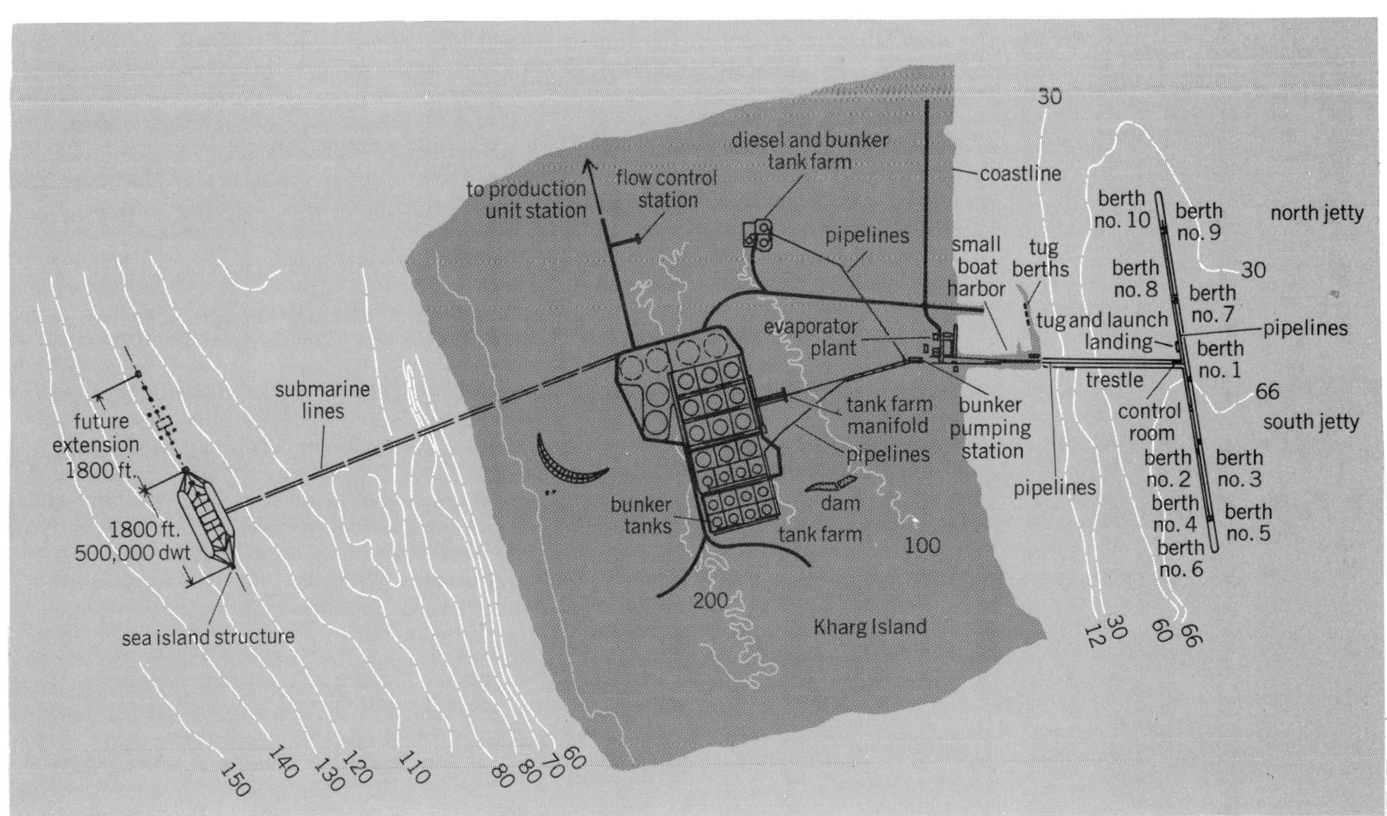

Fig. 1. Plot plan of Kharg Export Terminal. Water depths and land elevation are in feet (1 ft = 0.3 m).

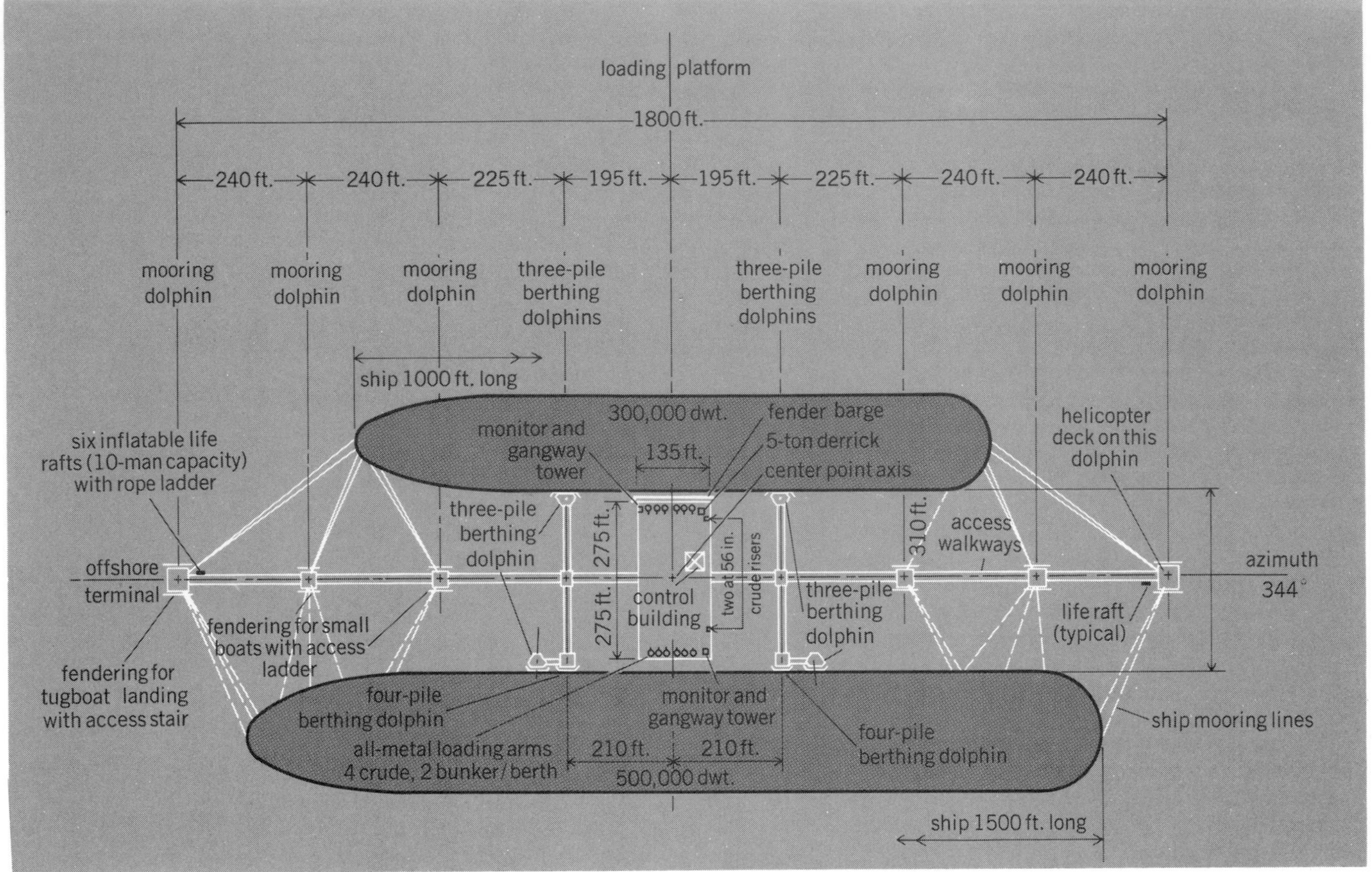

Fig. 2. Plan of the arrangement of structures for the sea island portion of Kharg Export Terminal. 1 ft = 0.3 m; 1 in. = 2.5 cm; 1 DWT = 1017 kg.

commodated – the 500,000-DWT class. Also, the loading rate had to be substantially increased to keep the turn-around time in port of these tankers within reasonable limits to avoid demurrage costs.

New facilities. Three sites were considered for new facilities. All had 100 ft of water depth available so that they could berth vessels with drafts of 90 ft, the best estimate of the draft for the 500,000-DWT tanker. One site was east of the existing 10-berth pier on the east side of Kharg Island. Another was east of the nearby island of Khargu, which afforded some protection from the prevailing northwesterly swells. A third site, west of Kharg Island but exposed to the prevailing winds, was much closer to shore and the existing tank farm than the other sites, and was selected.

Single-point mooring (SPM) systems were considered when studying alternate tanker berths. However, two berths were necessary, and as SPM needs more sea area per berth to allow for a 360° mooring swing from critical water depths and obstructions, the submarine lines became long and the system uneconomic. A fixed-structure island design was chosen. Two berths were immediately available in a small area closest to shore. Also, it best suited the very high loading rates (30,000 tons per hour; 200,000 bph) for which the new facilities were to be designed. Although the site was exposed to the prevailing northwesterly winds and it was estimated there could be as many as 60 days a year in which the berths were unavailable due to unfavorable weather, it proved economic to adopt this site and the sea island concept.

Sea island. The completed sea island structure incorporates the anticipated extension to four berths. It consists of two modules, side by side, of the structure arrangement shown in Fig. 2, with an end mooring dolphin being common to each module. Construction was performed in two stages, the first module completed in 1972, the second in 1976. Component structures have been kept to the minimum necessary for function, either berthing, mooring, or trolling tankers. Walkways are provided for access to the essential elements.

The sea island structure is located in approximately 105 ft of water depth and has an overall length of 3600 ft. Four tankers can be loaded simultaneously, two each at the outer and inner berths. Tankers ranging from 500,000 to 250,000 DWT will use the other berths, while the inner berths will cater to tankers from 300,000 to 150,000 DWT. Vessels smaller than these can be loaded at the 10-berth jetty on the eastern side of the island. Two central platforms, each 275 ft by 135 ft, are the loading areas for each module and control the flow to the tankers at the adjacent berths. Berthing dolphins are arranged to protect these platforms from the impacts of arriving tankers. These dolphins contain sets of 72-in.-diameter vertical piles which deflect on contact and absorb berthing energies. The minimum number of dolphins and their strategic placement reflects the tanker berthing operation. Tankers are stopped alongside the sea island and are pushed sideways into the structure. In the latest module, only one cluster of six piles was used in the breasting dolphin on each side of the central platform for the outer berth, instead of the two clusters shown on Fig. 2, with no reduction in energy-absorbing capacity. In spacing the breasting dolphins, the distance was made as large as possible consistent with the length of the parallel sides of the tankers. This increases the distance from the center of mass of the tanker to the point of contact with the structure, thus decreasing the impact. The spacing between the berthing faces of outer and inner berth was calculated to limit the maximum vertical angle of a breasting line from the tanker to no more than 30° through all stages of tide.

The design of the berthing dolphins was controlled by the energy of impact of the tankers. The design speed perpendicular to the berth was taken as 6 in. per second, increased by the tidal current set onto the outside berthing face, to a total of 8.5 in. per second. This energy results in extreme berthings of the 500,000-DWT vessel, and the berthing dolphin is designed to the level of yield stresses in the material. Normal berthings are expected to be no greater than 50% of this energy, with resultant lower working stresses.

A scale model of the tanker was tested for random sea conditions, winds, and current to determine loads from mooring lines. The tanker was ballasted for deadweight tonnages varying from 20 to 100%. Tests showed that a tanker at the inner berth produced the maximum mooring dolphin loads of 300 tons (1 ton force = 8896 N) when remaining alongside in seas having a 6-ft significant wave height and westerly winds of up to 40 knots (20 m/s).

Six storage tanks, each 1,000,000 bbl capacity, were added to the tank farm. Oil is delivered to the sea island through two parallel 78-in.-diameter lines running aboveground to the west shoreline. Here they reduce to two pairs of 56-in. diameter, one pair going to each central platform of the sea island as buried submarine pipelines.

Piping on the central platform is manifolded to deliver crude to each berth from the incoming submarine lines through turbine meters. Four 24-in. all-metal loading arms load oil to tankers at rates to 30,000 tons per hour (200,000 bph) at the outer berths and four 16-in. arms allow loading at a rate of 15,000 tons (100,000 bph) at the inner berths. The Kharg Island terminal, with all its facilities, can now export in excess of 7,500,000 bpd.

The terminal at Kharg Island is a fine example of a high-capacity export facility engineered to the highest state of the art in each development phase. With continuing dependence on crude oil as a source of world energy, and the ability of Iran to supply this oil, Kharg will remain a major terminal with future expansions as necessary. *See* PIPELINE. [JOSEPH A. FERENZ]

Bibliography: C. L. Bretschneider, A theory of waves of finite height, *7th Conference of Coastal Engineering*, 1961; P. Bruun, *Port Engineering*, 1973; J. R. Morrison et al., The force exerted by surface waves on piles, *Petrol. Trans.*, 1950; A. D. Quinn, *Design and Construction of Ports and Marine Structures*, 2d ed., 1972.

Surface condenser

A heat-transfer device used to condense a vapor, usually steam, by absorbing its latent heat in a cooling fluid, ordinarily water. Most surface condensers consist of a chamber containing a large number of 0.5- to 1-in. (13- to 25-mm) diameter corrosion-resisting alloy tubes through which cooling water flows. The vapor contacts the outside surface of the tubes and is condensed on them. The tubes are arranged so that the cooling water passes through the vapor space one or more times. About 90% of the surface is used for condensing vapor and the remaining 10% for cooling noncondensable gases. Air coolers are normally an integral part of the condenser but may be separate and external to it. The condensate is removed by a condensate pump and the noncondensables by a vacuum pump. *See* STEAM CONDENSER; VAPOR CONDENSER.

[JOSEPH F. SEBALD]

Synchronous motor

An alternating-current (ac) motor which operates at a fixed synchronous speed proportional to the frequency of the applied ac power. A synchronous machine may operate as a generator, motor, or capacitor depending only on its applied shaft torque (whether positive, negative, or zero) and its excitation. There is no fundamental difference in the theory, design, or construction of a machine intended for any of these roles, although certain design features are stressed for each of them. In use, the machine may change its role from instant to instant. For these reasons it is preferable not to set up separate theories for synchronous generators, motors, and capacitors. It is better to establish a general theory which is applicable to all three and in which the distinction between them is merely a difference in the direction of the currents and the sign of the torque angles. *See* ALTERNATING-CURRENT GENERATOR; ALTERNATING-CURRENT MOTOR.

SYNCHRONOUS MOTOR

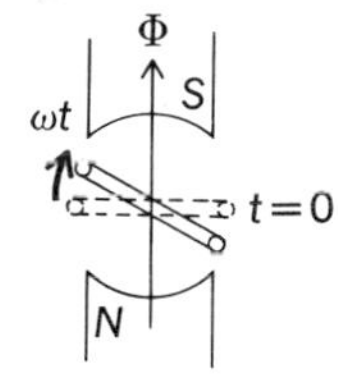

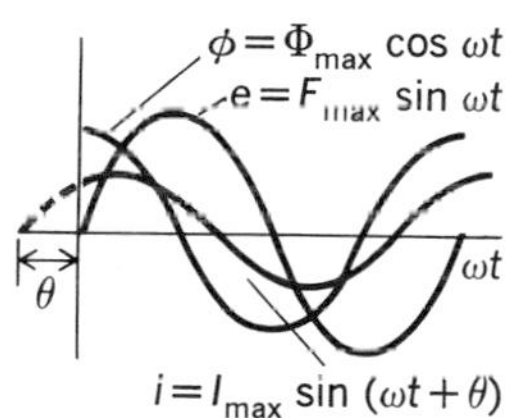

Fig. 1. Single-phase, two-pole synchronous machine.

Basic theory. A single-phase, two-pole synchronous machine is shown in Fig. 1. The coil is on the pole axis at time $t=0$, and the sinusoidally distributed flux ϕ linked with the coil at any instant is given by Eq. (1), where ωt is the angular displacement of the coil and Φ_{max} is the maximum value of the flux. This flux will induce in a coil of N turns an instantaneous voltage e, given by Eq. (2). The effective (rms) value E of this voltage is given by Eq. (3).

$$\phi = \Phi_{max} \cos \omega t \qquad (1)$$

$$e = -N\frac{d\phi}{dt} = \omega N \Phi_{max} \sin \omega t = E_{max} \sin \omega t \qquad (2)$$

$$E = \frac{E_{max}}{\sqrt{2}} = \sqrt{2}\,\pi f N \Phi_{max} = 4.44\, f N \Phi_{max} \qquad (3)$$

If the impedance of the coil and its external circuit of resistance R_t and reactance X is given by Eq. (4), there will flow a current, with a value given by Eq. (5), in which the phase angle θ is taken positive for a leading current. This current will develop a sinusoidal space distribution of armature

$$Z = R_t \pm jX = Z\underline{/\pm\theta} \qquad (4)$$

$$\mathbf{I} = \frac{\mathbf{E}}{\mathbf{Z}} = \frac{E}{Z}\underline{/\mp\theta} \qquad (5)$$

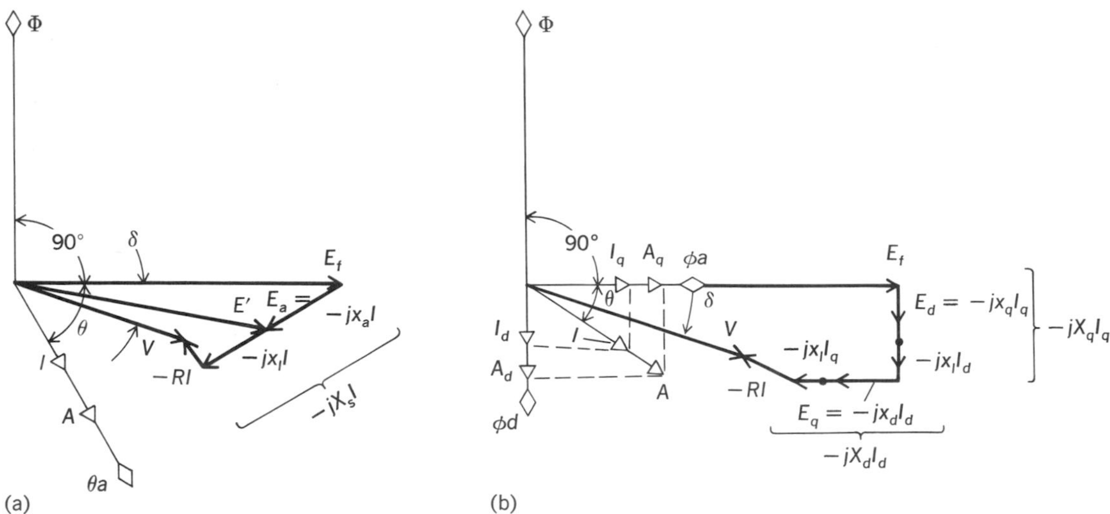

Fig. 2. Vector diagrams of synchronous generators. (*a*) Smooth-rotor machine. (*b*) Salient-pole machine.

reaction as in Eq. (6). If this single-phase mmf is

$$A = 0.8NI_{\max} \sin(\omega t + \theta) \tag{6}$$

expressed as a space vector and resolved into direct (in line with the pole axis) A_d and quadrature A_q components, it is given by Eq. (7).

$$\begin{aligned}\mathbf{A} &= A_d + jA_q \\ &= 0.4\,NI_{\max}\{[\sin\theta + \sin(2\omega t + \theta)] \\ &\qquad + j[\cos\theta - \cos(2\omega t + \theta)]\}\end{aligned} \tag{7}$$

In a three-phase machine with balanced currents, the phase currents are given by Eqs. (8).

$$\begin{aligned} i_a &= I_{\max}\sin(\omega t + \theta) \\ i_b &= I_{\max}\sin(\omega t + \theta - 120°) \\ i_c &= I_{\max}\sin(\omega t + \theta - 240°)\end{aligned} \tag{8}$$

Upon writing Eq. (7) for ωt, $\omega t - 120°$, and $\omega t - 240°$, respectively, and adding, Eq. (9) results for the polyphase armature reaction.

$$\mathbf{A} = A_d + jA_q = 1.2NI_{\max}(\sin\theta + j\cos\theta) \tag{9}$$

Equation (10) gives the three-phase power of the

$$P = 3EI\cos\theta \tag{10}$$

machine, and Eq. (11) gives the developed torque.

$$T = \frac{P}{\omega} = \frac{3}{\omega}EI\cos\theta \tag{11}$$

The above equations constitute the essential description of the synchronous generator. The same equations apply for a motor if the currents are reversed, that is, by changing the sign of the current I. They may also be interpreted in the form of phasor diagrams, and show the two cases of a smooth-rotor and a salient-pole machine.

Smooth-rotor synchronous machine. In the smooth-rotor machine, the reluctance of the magnetic path is essentially the same in either the direct or quadrature axes. In Fig. 2*a* let the flux Φ be selected as reference phasor and drawn vertically. Then comparing Eqs. (1) and (2) it is seen that the induced voltage E_f lags the flux by 90°. By Eq. (5) the current I lags the voltage by an angle θ for an inductive circuit, and by Eq. (9) causes a constant mmf of armature reaction A in phase with the current. This armature reaction causes a flux ϕ_a, stationary in space with respect to the field poles, which in turn induces a voltage E_a lagging it by 90°. The two induced voltages E_f (due to the field flux Φ) and E_a (due to the armature reaction flux ϕ_a) combine vectorially to give the resultant voltage E'. But the terminal voltage V is less than E' by the resistance and reactance drops, $R\mathbf{I}$ and $jx_l\mathbf{I}$ in the winding, and Eq. (12) applies.

$$\mathbf{V} = \mathbf{E}' - (R + jx_l)\mathbf{I} \tag{12}$$

The leakage reactance drop $jx_l\mathbf{I}$ lags the current by 90° as does the armature reaction voltage E_a. If a fictitious reactance of armature reaction x_a is introduced to account for E_a, it is obvious that Eq. (12) may be rewritten to give Eqs. (13), in which

$$\begin{aligned}\mathbf{V} &= \mathbf{E}_f - jx_a\mathbf{I} - (R + jx_l)\mathbf{I} \\ &= \mathbf{E}_f - R\mathbf{I} - j(x_a + x_l)\mathbf{I} \\ &= \mathbf{E}_f - (R + jX_s)\mathbf{I}\end{aligned} \tag{13}$$

$X_s = x_a + x_l$ is called the synchronous reactance of the machine.

Salient-pole synchronous machine. In a similar fashion the phasor diagram for a salient-pole machine, Fig. 2*b*, may be set up. Here the effects of saliency result in proportionately different armature reaction fluxes in the direct and quadrature axes, thereby necessitating corresponding direct, X_d, and quadrature, X_q, components of the synchronous reactance. The angle δ in Fig. 2 is called the torque angle. It is the angle between the field-induced voltage E_f and the terminal voltage V and is positive when E_f is ahead of V.

The foregoing equations and phasor diagrams were established for a generator. A motor may be

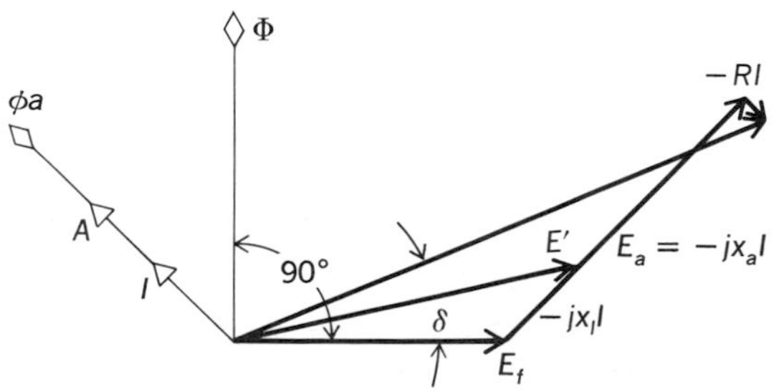

Fig. 3. Phasor diagram of synchronous motor.

regarded as a generator in which the power component of the current is reversed 180°, that is, becomes an input instead of an output current. The motor vector diagram is shown in Fig. 3. Here the torque angle δ is reversed, since V is ahead of E_f in a motor (it was behind in the generator). Therefore a motor differs from a generator in two essential respects: (1) The currents are reversed, and (2) the torque angle has changed sign. As a result the power input, Eq. (10), for a motor is negative, or has become a power output, and the torque is reversed in sign.

When the current **I** is 90° out of phase with the terminal voltage V the torque angle δ is nearly zero, being just sufficient to account for the power lost in the resistance.

Therefore, a synchronous machine is a generator, motor, or capacitor, depending on whether its torque angle δ is positive, negative, or zero. For these conditions the output current is, respectively, at an angle in the first or fourth quadrant, second or third quadrant, or essentially ±90° with respect to the terminal voltage at zero degrees. For any given power input or output, the machine can be made to operate at either leading or lagging power factor by changing the magnitude of the field current producing the flux Φ.

Synchronous capacitor. A synchronous capacitor can be made to draw a leading current and to behave like a capacitance by overexciting its field. Or, it will draw a lagging current on underexcitation. This characteristic thus presents the possibility of power-factor correction of a power system by adjusting the field excitation. A machine so employed at the end of a transmission line permits a wide range of voltage regulation for the line. One used in a factory permits the power factor of the load to be corrected. Of course a synchronous motor can also be used for power-factor correction, but since it must also carry the load current, its power-factor correction capabilities are more limited than for the synchronous condenser.

Power equations. The power output P_o and reactance power output Q_o of a round-rotor synchronous machine are given by Eq. (14), in which $\tan \alpha = R/X_s$ and $\mathbf{Z}_s = R + jX_s = Z_s < 90° - \alpha$

$$P_o + jQ_o = \left(\frac{VE_f}{Z_s}\sin(\delta+\alpha) - \frac{RV^2}{Z_s^2}\right) + j\left(\frac{VE_f}{Z_s}\cos(\delta+\alpha) + \frac{X_sV^2}{Z_s^2}\right) \quad (14)$$

For a round-rotor motor, both the torque angle δ and the electrical power output are negative.

For a salient-pole machine, neglecting resistance, Z_s is equal to the direct-axis reactance X_d, α is zero, and P_o is given by Eq. (15). Thus the

$$P_o = \frac{VE_f}{X_d}\sin\delta + V^2\frac{X_d - X_q}{2X_dX_q}\sin 2\delta \quad (15)$$

power or torque depends essentially on the product of the terminal and induced voltages and sine of the torque angle δ; but in the case of the salient-pole machine there is also a second harmonic term which is independent of the excitation voltage E_f. This term, the so-called reluctance power, vanishes for nonsaliency when $X_d = X_q$. The small synchronous motors used in some electric clocks and other low-torque applications depend solely on this reluctance torque. *See* RELUCTANCE MOTOR.

Excitation characteristics. The so-called V curves of a synchronous motor are curves of armature current plotted against field current with power output as parameter. Usually a second set of curves with input power factor (pf) as parameter is superimposed on the same plot. Such curves (Fig. 4), where armature current is plotted against generated voltage, can be determined from design calculations or from test; they yield a considerable amount of data on the performance of the motor. Thus, given any two of the four variables E_f, I, pf, mechanical power P_m, the remaining two may be easily determined, as well as the conditions of maximum power, constant pf, minimum excitation, stability limit, and so forth.

Circle diagrams. Voltage equation (13) and current equation (5) can be combined in such a fashion as to yield Eq. (16), which is the equation of

$$I^2 = \frac{V^2}{Z_s^2} + \frac{E_f^2}{Z_s^2} - 2\frac{V}{Z_s}\frac{E_f}{Z_s}\cos\delta \quad (16)$$

a set of circles with offset center and with different radii (E_f/Z_s). The locus of these circles is the current.

A companion set of circles can be developed giving the locus of I as a function of its pf angle for

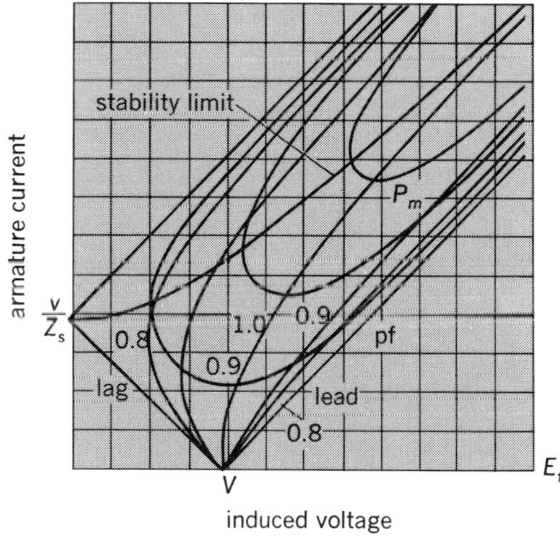

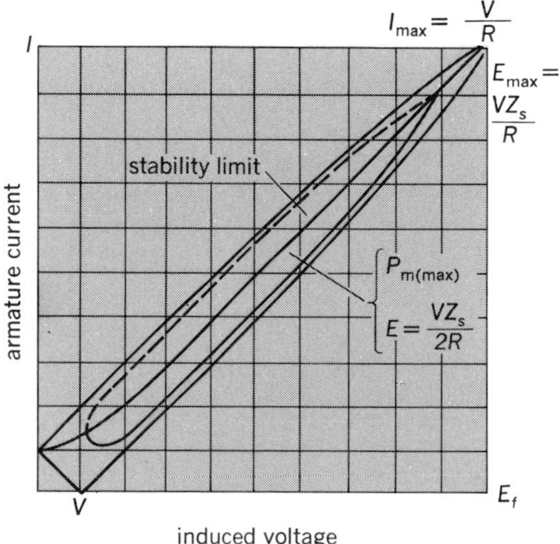

Fig. 4. V curves (armature current versus induced voltage) of synchronous motor.

different values of constant developed mechanical power.

These two sets of circles are shown in Fig. 5. Such circle diagrams relate the mechanical power, pf angle, armature current, torque angle, and excitation.

Losses and efficiency. The losses in a synchronous motor comprise the copper losses in the field, armature, and amortisseur windings; the exciter and rheostat losses of the excitation system; the core loss due to hysteresis and eddy currents in the armature core and teeth and in the pole face; the stray loss due to skin effect in conductors; and the mechanical losses due to windage and friction. The efficiency of the motor is then given by Eq. (17).

$$\text{Eff} = \frac{\text{output}}{\text{input}} = \frac{\text{output}}{\text{output} + \text{losses}} \tag{17}$$

Mechanical oscillations. A synchronous motor subjected to sudden changes of load, or when driving a load having a variable torque (for example, a reciprocating compressor), may oscillate about its mean synchronous speed. Under these conditions the torque angle δ does not remain fixed, but varies. As a result the four separate torques expressed in Eq. (18) act on the machine rotor. The

$$\begin{pmatrix}\text{Synchronous}\\ \text{motor torque}\\ \text{Eq. (15)}\end{pmatrix} + \begin{pmatrix}\text{induction motor}\\ \text{torque of}\\ \text{amortisseur}\end{pmatrix} = \begin{pmatrix}\text{torque to}\\ \text{overcome}\\ \text{inertia}\end{pmatrix} + \begin{pmatrix}\text{torque}\\ \text{required}\\ \text{by the load}\end{pmatrix} \tag{18}$$

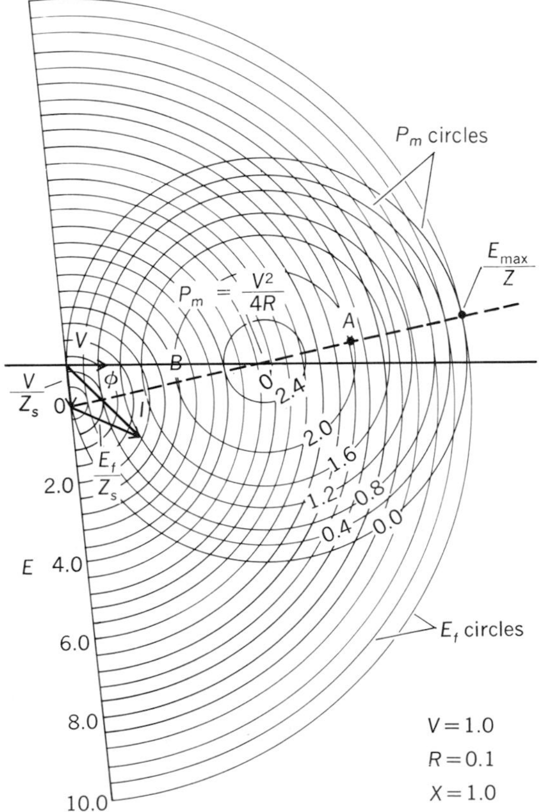

Fig. 5. Circle diagram of synchronous motor.

possibility exists that cumulative oscillations will build up and cause the motor to fall out of step.

Starting of synchronous motors. Synchronous motors are provided with an amortisseur (squirrel-cage) winding embedded in the face of the field poles. This winding serves the double purpose of starting the motor and limiting the oscillations or hunting. During starting, the field winding is either closed through a resistance, short-circuited, or opened at several points to avoid dangerous induced voltages. The amortisseur winding acts exactly as the squirrel-cage winding in an induction motor and accelerates the motor to nearly synchronous speed. When near synchronous speed, the field is excited, and the synchronous torque pulls the motor into synchronism. During starting, Eq. (18) applies, since all four types of torque may be present. Of course, up to the instant when the field is excited the portion of the synchronous motor torque depending on E_f does not exist, although the reluctance torque will be active.

Other methods of starting have been used. If the exciter is direct-connected and a dc source of power is available, it may be used to start the synchronous motor. In the so-called supersynchronous motor the stator is able to rotate in bearings of its own, and is provided with a brake band. For stator allowed to come up to nearly synchronous speed by virtue of the amortisseur windings; the field is then excited and the stator brought to synchronous speed, the rotor remaining stationary. Then as the brake band is tightened, the torque on the rotor causes it to accelerate while the speed of the stator correspondingly slackens; finally the stator comes to rest and is locked by the brake band. In this way maximum synchronous motor torque is made available for acceleration of the load. For other types of synchronous motors *see* HYSTERESIS MOTOR; RELUCTANCE MOTOR.

[RICHARD T. SMITH]

Bibliography: D. G. Fink and H. W. Beaty (eds.), *Standard Handbook for Electrical Engineers*, 11th ed., 1978; A. E. Fitzgerald, C. Kingsley, Jr., and A. Kusko, *Electric Machinery*, 3d ed., 1971; L. W. Matsch, *Electromagnetic and Electromechanical Machines*, 2d ed., 1977.

Synthetic fuel

A fuel which must be synthesized, manufactured, or artificially formulated. A synthetic material is a substance that does not exist in nature, or is a manufactured or artificially formulated substitute for a material that does occur naturally but in such limited quantities that its recovery becomes costly. In popular usage, the term synthetic fuels signifies fuels derived from other forms of fossil fuels which are less convenient or environmentally less acceptable for direct use than gaseous or liquid fuels. (Potential fuels such as products of biomass conversion or chemicals such as ammonia or hydrogen are not usually referred to as synthetic fuels, even though the term is equally applicable.) Thus, substitute natural gas (SNG), manufactured from coal, shale, and crude oil fractions, and synthetic solid and liquid fuels from coal, shale, and tar sands are known as synthetic fuels. *See* FOSSIL FUEL.

Fossil fuels. Coal is one of the most abundant and most common of all fossil fuels. However, gaseous and liquid fuels such as natural gas and petroleum crude, the so-called fluid fuels, are the most desirable forms of fossil fuels because they are more conveniently stored, transported, and controlled in use. They are also very useful as raw materials for the production of chemicals and plastics. *See* COAL; LIQUID FUEL; NATURAL GAS; PETROLEUM.

Since the end of World War II, the use of natural gas and petroleum crude has increased as the supply of these fuels, at low cost, has increased. As a result, the technologically advanced industrial nations have experienced significant growth both in industry and in the general standard of living. However, demand for such fuels in many instances has far exceeded the domestic supply. The resulting shortfall has been met by importing oil and liquefied natural gas from other countries.

The shortage of gaseous and liquid fuels, coupled with price escalation, has provided incentive for development of substitute or synthetic fuels. Environmental concerns about pollutants given off when fossil fuels are burned have also been factors in the development of synthetic fuels. They can be manufactured under controlled conditions with a minimum emission of pollutants.

Although coal, oil shale, and tar sands are all used to produce a variety of synthetic fuels, coal has been the raw material most often selected. Bituminous coal was the source of kerosine and illuminating gas in the United States prior to the 1940s. It was the major source of synthetic liquid fuels used by Germany in World War II. Coal is also the raw material for the only existing commercial liquefaction plant located in South Africa.

Energy program. The impact of the fuel shortage on social, economic, and political life in the United States has been such that a national energy plan with the goal of achieving a greater degree of energy self-sufficiency is being developed. A Department of Energy was created and charged with the development and implementation of such a plan. Unlike the earlier Project Independence program, a sophisticated and ambitious national energy plan is evolving which includes the promotion of energy conservation, new sources of energy, and synthetic fuel technologies. Activities in the new energy program include the development and commercialization of methods for converting coal into liquid fuels and SNG, as well as parallel efforts for developing oil shale conversion technologies, in collaboration with private industry.

Economic factors. Synthetic fuels are more expensive than their natural counterparts because of the high costs of synthesis processes. In addition, some energy is inevitably consumed during the process. As crude oil prices have sharply increased, however, the cost differences have decreased and made the economics of synthetic fuel manufacture more favorable. According to present plans, synthetic fuel production may reach 397,000 m^3 (2,500,000 bbl) per day by the year 1990. This will provide a hedge against future supply interruptions, since an assured supply of fuels for transportation is necessary for the stability of American society.

Production. The national energy plan calls for building several pioneering synthetic-fuel-producing facilities to demonstrate technologies for coal liquefaction, coal gasification, and recovery of oil from shale. These facilities will be built with commercial-size equipment. The facilities for successfully demonstrating technology will ultimately be expanded to larger and more economical plants by adding more demonstration-size units.

Coal conversion. Synthetic fuels may be produced from coal with either direct or indirect methods. The three major processes for direct production are the Solvent Refined Coal (SRC-I), H-Coal, and Exxon Donor Solvent processes. SRC-I produces primarily a solid fuel low in sulfur and ash. In a variation of SRC-I, called SRC-II, the process is modified to produce a variety of liquid fuels and fuel gases. In all direct liquefaction processes, pulverized coal is suspended in a solvent obtained from coal. The resultant slurry is then heated to about 425°C (800°F) and variously exposed to hydrogen gas under a pressure of approximately 140 bars (14 MPa or 2000 psig). The coal structure is cracked open and hydrogen attaches to smaller molecules, forming different liquids and gases. The liquids are separated from the ash and distilled to obtain fuel gas and fuel oils. In the SRC-I process, the heavy liquid residue is chilled to give a solid product which is utilized in coal-burning boilers.

In the indirect coal liquefaction processes, coal is first thermally decomposed to generate a mixture of combustible gases in a one- or two-stage gasifier under pressure of up to 70 bars (7 MPa or 1000 psig). Air or oxygen is used to burn part of the coal to provide heat. A large number of commercial and developmental processes exist to convert coal into fuel gases of different heating values. At the Sasol plant in Sasolburg, Republic of South Africa, coal is first converted to a low-heating-value gas. The gas is next processed to separate the combustible components—carbon monoxide, hydrogen, and methane—which are then synthesized to yield gasoline, alcohols, and waxes. *See* COAL GASIFICATION; COAL LIQUEFACTION.

Methanol. Commercial processes are now available to convert carbon monoxide and hydrogen generated from natural gas or petroleum oils to methanol. These processes can be applied to gases from coal. Mobil Oil Corporation has developed a process to convert methanol to gasoline. A new route, therefore, has opened up to obtain gasoline from coal. The U.S. Department of Energy is considering a proposal to build a pioneering plant to make 1200 tons per day of methanol, or 480 TPD of gasoline, from coal. Methanol is useful as a motor fuel by itself, mixed with a large volume of gasoline, as "gasohol." *See* GASOHOL.

Substitute natural gas. SNG of low or medium heating value is obtained from coal via purification and methanation steps. SNG consists essentially of methane, the main ingredient of natural gas; its heating value is about 37 MJ/m^3 (1000 Btu/SCF). *See* SUBSTITUTE NATURAL GAS (SNG).

Oil shale. Oil shale, which contains kerogen and bitumen, may be thermally decomposed in a kiln or retort to yield oil, gas, and residual carbon. The oil is refined to yield a synthetic crude oil, or syncrude, which resembles petroleum crude. Oil yields can be as low as 20 liters/ton and as high as 150 liters/ton of shale. Techniques for in-place processing of shale are under study. The shale deposit is first fractured by underground explosion, and then steam or air is injected to propagate a high-temperature wave which heats the shale and vaporizes the oil for recovery. Plans for several pioneering surface and in-place shale-processing plants using less than commercial-size equipment are under way. Commercial production of oil may reach only about 8000 m^3 (50,000 bbl) per day by the year 1985. *See* OIL SHALE.

Tar sands. Tar sands found in Alberta, Canada, consist of a mixture of sand grains, water, and bitumen. The latter is a viscous petroleum liquid. Tar sands are at present strip-mined and treated with steam, hot water, and caustic soda in surface extraction units. The mixture is then transferred to settling tanks, where naphtha is added as a diluent. Bitumen and the diluent are skimmed off as froth, and sand is removed from the bottom. The froth containing the bitumen-naphtha solution is skimmed to separate the latter from the water and sand. Bitumen and naphtha are then separated by distillation, and the bitumen is thermally cracked to yield a synthetic crude oil and a residual coke or pitch. Some fuel gas is produced, and the synthetic crude may be further processed by hydrogen treatment and conventional processing schemes. There are no significant accumulations of tar sands in the United States as there are in Canada. *See* COKE: OIL SAND.

Future prospects. Whether the synthetic fuels are obtained by direct liquefaction of coal, by coal gasification, or from oil shale, one of the objectives of the United States' energy plan is to develop enough synthetic fuel capacity to increase domestic liquid fuel supplies. There has been some concern about the possible increase in pollution resulting from the coal conversion plants and additional coal required to feed them, although changes in pollution levels are not at all certain to occur. The U.S. Environmental Protection Agency has been continuously updating its regulations in these areas. A pact has been made between the Department of Energy and the Environmental Protection Agency so that only those regulations would apply to the plants which are current when the letter of intent to build a commercial synthetic fuel plant has been filed with the Department. This pact helps remove any uncertainty about future regulations that may hamper the commercialization of a synthetic fuel process. Private industries, sponsoring various synthetic fuel processes, are competing for public funds and favorable economic incentives to get into the marketplace. As the oil shortages persist and the price and demand for imported oil continue to rise, efforts toward energy independence through synthetic fuels continue to accelerate. *See the feature article* EXPLORING ENERGY CHOICES. [LALIT H. UDANI]

Bibliography: American Chemical Society, *Preprints*, vol. 22, no. 7 (Division of Comparative Economics of Coal Conversion Processes), 1977; American Chemical Society, *Preprints*, vol. 23, no. 3 (Division of Fuel Chemistry), no. 4 (Division of Petroleum Chemistry), 1978; D. M. Considine (ed.), *Energy Technology Handbook*, 1978; W. Goldstein, Politics of energy policy, *Energy Policy*, vol. 6, no. 3, September 1978; R. F. Naill and G. T. Backus, Evaluating National Energy Plan, *Techol. Rev.*, 79(8):51–56, July–August 1977.

Thermionic power generator

A device in which heat energy is directly converted to electric energy, frequently called a thermionic converter. The free electrons of good electric conductors flow around suitably arranged conducting paths to create the infinity of useful applications of electricity. At normal temperatures the escape of these electrons from the conducting material can hardly be detected, but at higher temperatures (from 1000 to 2500 K) large numbers of electrons do escape from a heated conductor. This is called thermionic emission of electrons.

Two metallic elements, an emitter and a collector, are the minimum needed for a thermionic converter. The thermionic electron emitter must be capable of yielding electrons to the space that separates the emitter from the electron collector. The collector must be operated at a significantly lower temperature than the emitter so that the collector does not also emit electrons. The general term for such a thermionic device is thermionic diode.

Classification. If the space between the two elements of the diode is evacuated sufficiently so that the residual gas has no significant influence on the flow of electrons from the emitter to the collector, the device is known as a vacuum thermionic converter. Electrons are negatively charged particles and thus repel each other. The presence of electrons in transit between the emitter and the collector can, therefore, interfere seriously with the free flow of additional charges and thus set up a space-charge limitation on the current density and the efficiency attainable.

Two methods are used to minimize the space-charge limitation. One depends on a diode construction with fantastically close spacing—of the order of 5 μm or approximately 0.0002 in. The second method is to introduce an ionized gas. The number density of the ions of the gas must be equal to or locally greater than the electron density in order to neutralize the negative space charge otherwise present. Since the ions used are positively charged, the net charge can be zero even though a high density of electrons is present to provide the means of conduction from the emitter to the collector. The term plasma is applied to a medium in which the net electric charge is zero. A thermionic converter that depends on the presence of an ionized gas to give good conduction in the space between the emitter and the collector is known as a plasma thermionic converter.

Emitter and collector properties. The maximum possible current density in any diode depends on the temperature of the emitter and on the ease of electron removal. The work function of a substance is a direct measure of the energy per electron required for its removal. The emitter work function is defined as the energy difference between the Fermi level within the conductor and the potential energy of an electron at rest just outside the conductor. If current flows through any conductor, the value of the current is generally directly proportional to a measured voltage difference at the ends of the conductor. This voltage difference is equal to the voltage displacement of the Fermi levels at each end of the conductor. In a thermionic converter under actual operating conditions, the Fermi level of the collector must therefore be negative with respect to the Fermi level of the emitter for electric power to be delivered to an external circuit. The work function of the collector must be as small as possible in order to make the voltage available as large as possible.

These points are illustrated in the figure, known

as a motive diagram, by which energy relations may be shown. The difference in potential between the Fermi level of the emitter and its surface potential is represented by the vertical arrow designated ϕ_1 and is equal to the emitter work function. For illustrative purposes, the surface potential of the collector is set at the same energy value as that of the emitter, and the Fermi level is positive with respect to this point by the amount of ϕ_2, which is the work function of the collector. Thus, if ϕ_2 is smaller than ϕ_1 and the surface potentials under operating conditions are practically equal, an output voltage designated by V_0 will appear at the terminals of the converter. This voltage can be used to drive current through the external load and, under the circumstances illustrated, this voltage is equal to the difference between the emitter work function and the collector work function.

Two conditions of operation are illustrated, that of the vacuum-type converter and that of the preferred plasma converter. In the plasma converter the current transported across the space between the emitter and the collector could be nearly equal to the maximum possible current density J available at the emitter, which is given by Eq. (1), where

$$J = 120T_1^2\, e^{-(\phi_1 q/kT_1)} \quad \text{amp/cm}^2 \qquad (1)$$

T_1 is the temperature of the emitter in degrees Kelvin, ϕ_1 is the work function, q is the charge on the electron, and k is Boltzmann's constant. If space charge is present, as in the vacuum diode, then the energy difference represented by ϕ_m must be used in place of ϕ_1 in Eq. (1) to determine the maximum current that will be available as the output current of the converter. In the plasma converter there is no inhibiting action, and a current may approach the full emitter current available. This is important because the product of the current and output voltage is the power delivered to the external circuit by the converter.

Inspection of Eq. (1) shows that the ratio (ϕ_1/T_1) must be as small as is practical in order to achieve a high current density. In order not to sacrifice output voltage, this desirable result can best be obtained by having as high a temperature as is possible. Refractory materials such as tungsten, molybdenum, and tantalum can be operated at high temperatures. All these metals have relatively high work-function values unless the emitter surface is partially covered by an electropositive metal, such as cesium. The cesium plasma converter has great promise of being an efficient device.

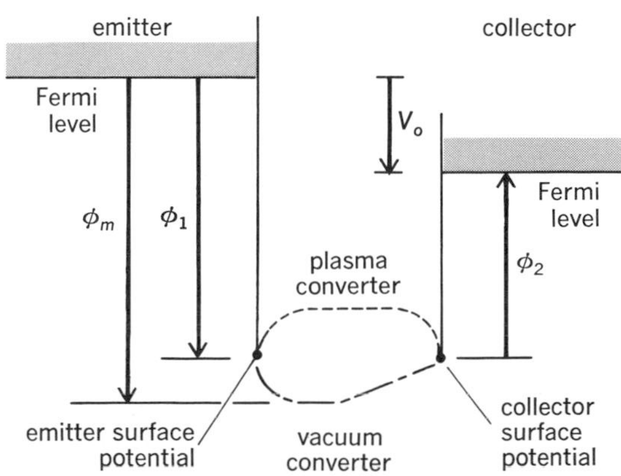

Motive diagram showing energy relations in vacuum and plasma thermionic converters.

In the temperature range of 550–650°K, the vapor pressure of cesium changes from 1 to 10 torr. Associated with a thermionic converter, a cesium reservoir maintained within this range of temperature can supply enough cesium to an emitter surface so that even the refractory materials will have work functions as low as 3 ev. Cesium ions are produced at the heated emitter surface in sufficient quantity to neutralize the electron space charge and give a motive function qualitatively represented by that for the plasma converter of the illustration. The adsorption of cesium on the colder collector surface serves to lower its work function to a value of about 1.8 ev, or even less under well-controlled conditions.

Since a low-work-function collector is necessary for an efficient thermionic converter, a correspondingly low temperature must be maintained at the collector to stop the back emission of electrons, which would produce a reduction in current. A satisfactory estimate of the collector temperature T_2 needed to limit back current to less than 2% of the forward current is given by Eq. (2), where ϕ_2 is the work function of the collector.

$$T_2 = T_1 \frac{\phi_2}{\phi_1 + 2.6 \times 10^{-4} T_1} \qquad (2)$$

State of development. Although many engineering details remain to be worked out, it is anticipated that a high-temperature plasma converter will be capable of delivering to an external circuit power corresponding to a density at the emitter of not less than 10 watts/cm² and probably not greater than 40 watts/cm². This operation will be done at an efficiency of approximately 20%, measured in terms of the heat actually delivered to the emitter structure. This heat may be obtained in space vehicles from solar radiation received on a reflector and a suitably designed concentrator. Since applications of this type are so important, research and development related to the direct conversion of heat to electricity by thermionic converters is well warranted.

[W. B. NOTTINGHAM/T. F. STRATTON]

Bibliography: S. W. Angrist, *Direct Energy Conversion*, 3d ed., 1976; G. N. Hatsopoulos and E. N. Giftopoulis, *Thermionic Energy Conversion*, vol. 1: *Processes and Devices*, 1974, vol. 2: *Theory, Technology and Application*, 1979; R. O. Jenkins and W. C. Trodden, *Electron and Ion Emission from Solids*, 1965; N. W. Snyder, Energy conversion for space power, in M. Summerfield (ed.), *Progress in Astronautics and Rocketry*, vol. 3, 1961.

Thermocouple

A device that uses the voltage developed by the junction of two dissimilar metals to measure temperature difference. Two wires of dissimilar metals welded together at the ends make up the basic thermocouple (Fig. 1). One junction, called the sensing or measuring junction, is placed at the point where temperature is to be measured. The other junction, called the reference or cold junction, is maintained at a known reference temperature. The voltage developed between the two junctions is approximately proportional to the difference between the temperatures of the two

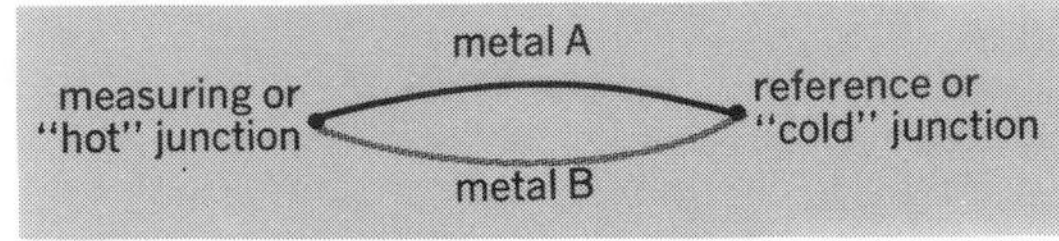

Fig. 1. Elements of a thermocouple.

junctions, and may be measured by including a suitable voltmeter in the circuit, as in Fig. 2. *See* THERMOELECTRICITY.

Although all dissimilar metals exhibit the thermoelectric effect, only a few are in wide use. The major characteristics which make certain metals or combinations of metals outstanding for this purpose are (1) stability or reproducibility, the emf does not change rapidly with time; (2) constant or controllable composition, small impurities or changes in composition of wire from end to end or lot to lot can result in varying or nonidentical emf curves; (3) corrosion resistance, wires do not deteriorate or change properties in oxidizing or reducing atmospheres; (4) sensitivity, the emf generated per degree temperature change is large; (5) range, the couple can be used through a broad range of temperatures; (6) ruggedness, tough but easily worked metals give good service; and (7) cost.

The temperature-emf relation of a homogeneous thermocouple is a definite physical property and does not depend upon the details of the apparatus. In a homogeneous thermocouple each element is homogeneous in both chemical composition and physical condition throughout its length. The temperature-emf curves for the six most commonly used pairs of wire materials are shown in Fig. 3, and the normal temperature ranges and corrosion characteristics are listed in the table.

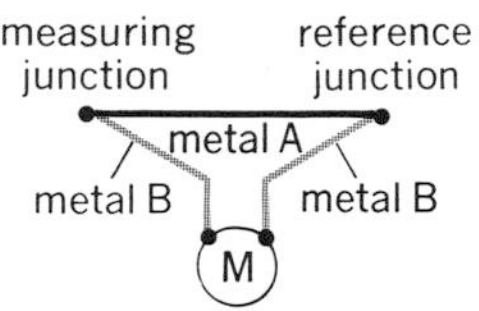

Fig. 2. Thermocouple circuit for measuring metal temperatures.

As implied in the preceding paragraph, the emf generated by a thermocouple is independent of the size of wire, and thus small fine wires may be used so that the mass and thermal lag of the sensing junction can be very low. However, small fine wires have limited mechanical strength, and generally such junctions are encased in protective tubing. Thus the fast-response feature of the fine-wire junction can be masked by the protective tubing. In most cases a balance has to be found between a fast response, a long service life, and the protection needed at the place of use.

Usually, extension leads are used to connect the measuring junction with the reference junction and voltmeter when the lengths required are over about 10 ft. The extension wire may be alloyed to match the thermoelectric characteristics of the thermocouple wires or copper wire used (Fig. 4*a*). When the reference junction can be located close to the measuring junction, copper-wire leads to the meter can be used as required (Fig. 4*b* and *c*).

A common practice is to maintain the reference junction at 0°C, or 32°F, by an ice and water bath. Alternatively, the reference junction may be maintained at a carefully controlled temperature somewhat above atmospheric temperature, and the instrument may be zeroed by adding the corresponding voltage. Other instruments allow the reference-junction temperature to vary with ambient conditions within the voltage-measuring instrument and to provide a calibrated electrical or me-

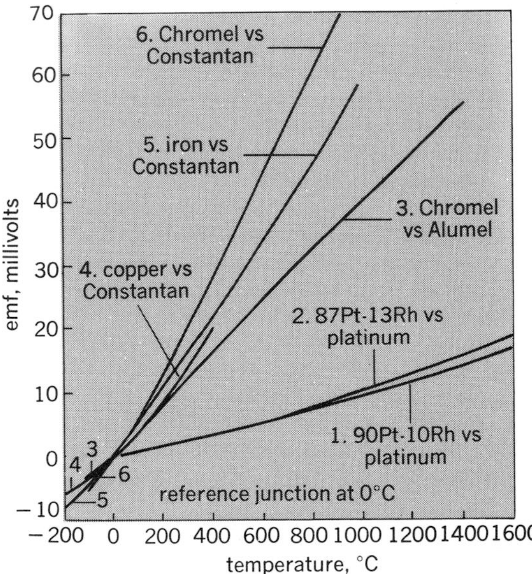

Fig. 3. Temperature-emf curves for thermocouples.

Temperature limits and corrosion characteristics of thermocouples*

Positive element	Negative element	Temperature range, °C	Influence of temperature and gas atmospheres
90% Pt, 10% Rh; 87% Pt, 13% Rh	Platinum	0 to 1450	Resistance to oxidizing atmosphere very good; resistance to reducing atmosphere poor; platinum corrodes easily above 1000°C and should be used in gastight ceramic protecting tube
Chromel-P	Alumel	−200 to 1100	Resistance to oxidizing atmosphere good to very good; resistance to reducing atmosphere poor; affected by sulfur, reducing or sulfurous gas, SO_2, and H_2S
Iron	Constantan	−200 to 750	Oxidizing and reducing atmospheres have little effect on accuracy, best used in dry atmospheres; resistance to oxidation good up to 400°C, poor above 700°C; resistance to reducing atmosphere good up to 400°C; protection from oxygen, moisture, and sulfur required
Copper	Constantan	−200 to 350	Subject to oxidation and alteration above 400°C due to copper, above 600°C due to Constantan wire; contamination of copper affects calibration greatly; resistance to oxidizing atmosphere good; resistance to reducing atmosphere good; requires protection from acid fumes
Chromel-P	Constantan	−100 to 1000	Chromel attacked by sulfurous atmosphere; resistance to oxidation good; resistance to reducing atmosphere poor

*From D. M. Considine (ed.), *Process Instruments and Controls Handbook*, McGraw-Hill, 1957.

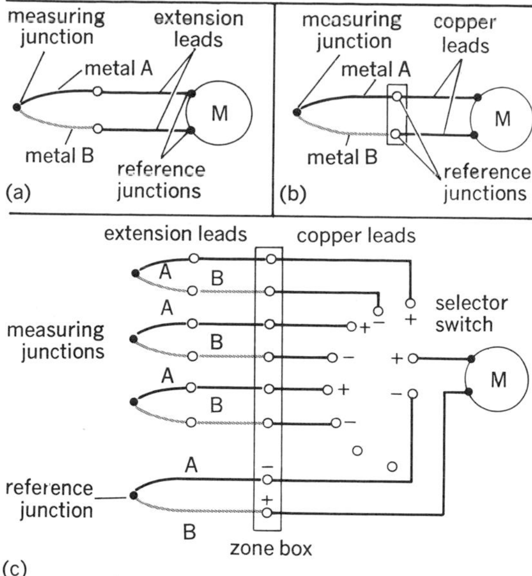

Fig. 4. Thermocouple circuits. (*a*) Single thermocouple with reference junction(s) at a distance from measuring junction. (*b*) Single thermocouple circuit with reference junction close to measuring junction. (*c*) Multiple thermocouple system with reference junction at a distance from both measuring junction and meter.

chanical manual adjustment so that with proper adjustment the instrument reading is correct. Most of the newer instruments that are used with only one type of thermocouple have automatic reference-junction compensation, and their scale or chart is calibrated directly in temperature units.

Any method of measuring small voltages accurately may be used for thermocouple voltages. The millivoltmeter and the potentiometer in various forms are in wide use. Precision measurements with thermocouples (say to 0.1°F or 0.06°C) are possible in certain ranges with specially calibrated units. On industrial applications, errors of 5°F (3°C) and larger occur, but the actual magnitude depends upon the thermocouple materials, the temperature level, the lead wires, the compensation, the voltage-measuring system, and the installation. Although thermocouple systems have many limitations, they are particularly advantageous for applications requiring remote indication or recording, for those on which the measuring junction must be replaced relatively frequently, and for those requiring measurements in the temperature range between 800 and 2400°F (430 and 1320°C). In the laboratory and in experimental work, the thermocouple is frequently a convenient substitute for the more accurate, but less rugged, liquid-in-glass thermometer.

The thermopile is a number of thermocouples connected in series or parallel. The series circuit provides a higher sensitivity (greater emf per degree) than one thermocouple alone and is often selected for this reason. Either the series or parallel circuit may be used for obtaining an indication which approaches the average of the several temperatures. [HOWARD S. BEAN]

Bibliography: American Society for Testing and Materials, *Manual of the Use of Thermocouples in Temperature Measurement*, 2d ed., 1974; American Society for Testing and Materials, *Thermocouples*, 1977; D. M. Considine and S. D. Ross (ed.), *Process Instruments and Controls Handbook*, 1957; P. A. Kinzie, *Thermocouple Temperature Measurement*, 1973.

Thermodynamic cycle

A procedure or arrangement in which one form of energy, such as heat at an elevated temperature from combustion of a fuel, is in part converted to another form, such as mechanical energy on a shaft, and the remainder is rejected to a lower temperature sink as low-grade heat.

Common features of cycles. A thermodynamic cycle requires, in addition to the supply of incoming energy, (1) a working substance, usually a gas or vapor; (2) a mechanism in which the processes or phases can be carried through sequentially; and (3) a thermodynamic sink to which the residual heat can be rejected. The cycle itself is a repetitive series of operations.

There is a basic pattern of processes common to power-producing cycles. There is a compression process wherein the working substance undergoes an increase in pressure and therefore density. There is an addition of thermal energy from a source such as a fossil fuel, a fissile fuel, or solar radiation. There is an expansion process during which work is done by the system on the surroundings. There is a rejection process where thermal energy is transferred to the surroundings. The algebraic sum of the energy additions and abstractions is such that some thermal energy is converted into mechanical work. *See* HEAT.

A steam cycle that embraces a boiler, a prime mover, a condenser, and a feed pump is typical of the cyclic arrangement in which the thermodynamic fluid, steam, is used over and over again. An alternative procedure, after the net work flows from the system, is to employ a change of mass within the system boundaries, the spent working substance being replaced by a fresh charge ready to repeat the cyclic events. The automotive engine and the gas turbine illustrate this arrangement of the cyclic processes, called an open cycle because new mass enters the system boundaries and the spent exhaust leaves it.

The basic processes of the cycle, either open or closed, are heat addition, heat rejection, expansion, and compression. These processes are always present in a cycle even though there may be differences in working substance, the individual processes, pressure ranges, temperature ranges, mechanisms, and heat transfer arrangements.

Air-standard cycle. It is convenient to study the various power cycles by using an ideal system such as the air-standard cycle as illustrated. This is an ideal, frictionless mechanism enveloping the system, with a permanent unit charge of air behaving in accordance with the perfect gas relationships.

The unit air charge is assumed to have an initial state at the start of the cycle to be analyzed. Each process is assumed to be perfectly reversible, and all effects between the system and the surroundings are described as either a heat transfer or a mechanical work term. At the end of a series of processes, the state of the system is the same as it was initially. Because no chemical changes take place within the system, the same unit air charge is conceivably capable of going through the cyclic processes repeatedly.

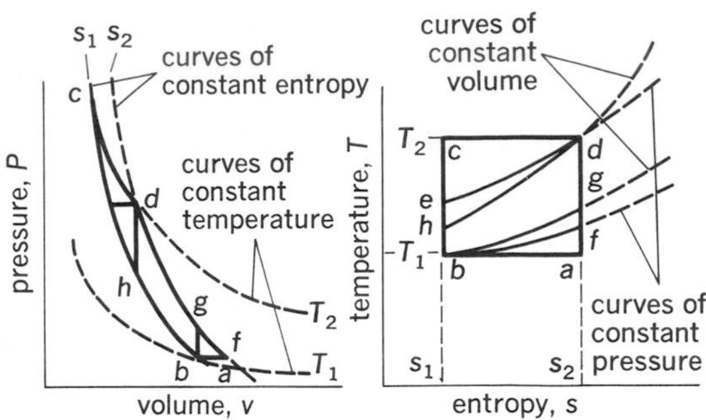

cycles are, in the order of decreasing efficiency:
Carnot cycle *(a-b-c-d-a)*
Brayton cycle *(b-e-d-f-b)*
Diesel cycle *(b-e-d-g-b)*
Otto cycle *(b-h-d-g-b)*

Comparison of principal thermodynamic cycles.

Whereas this air-standard cycle is an idealization of an actual cycle, it provides an amenable method for the introductory evaluation of any power cycle. Its analysis defines the upper limits of performance toward which the actual cycle performance may approach. It defines trends, if not absolute values, for both ideal and actual cycles. The air-standard cycle can be used to examine such cycles as the Carnot and those applicable to the automobile engine, the diesel engine, the gas turbine, and the jet engine.

Cyclic standards. Many cyclic arrangements, using various combinations of phases but all seeking to convert heat into work, have been proposed by many investigators whose names are attached to their proposals, for example, the Diesel, Otto, Rankine, Brayton, Stirling, Ericsson, and Atkinson cycles. All proposals are not equally efficient in the conversion of heat into work. However, they may offer other advantages which have led to their practical development for various applications. Nevertheless, there is one overriding limitation on efficiency. It is set by the dictates of the Carnot cycle, which states that no thermodynamic cycle can be projected whose thermal efficiency exceeds that of the Carnot cycle between specified temperature levels for the heat source and the heat sink. Many cycles may approach and even equal this limit, but none can exceed it. This is the uniqueness of the Carnot principle and is basic to the second law of thermodynamics on the conversion of heat into work. *See* BRAYTON CYCLE; CARNOT CYCLE; DIESEL CYCLE; OTTO CYCLE; THERMODYNAMIC PROCESSES.

[THEODORE BAUMEISTER]

Bibliography: T. Baumeister (ed.), *Standard Handbook for Mechanical Engineers*, 8th ed., 1978; J. B. Jones and G. A. Hawkins, *Engineering Thermodynamics*, 1960; J. H. Keenan, *Thermodynamics*, 1970; M. W. Zemansky et al., *Basic Engineering Thermodynamics*, 2d ed., 1975.

Thermodynamic principles

Laws governing the conversion of energy from one form to another. Among the many consequences of these laws are relationships between the properties of matter and the effects of changes in pressure, temperature, electric field, magnetic field and composition. The great practicality of the science arises from the foundations of the subject. Thermodynamics is based upon observations of common experience that have been formulated into the thermodynamic laws. From these few laws all of the remaining laws of the science are deducible by purely logical reasoning. There is a choice as to which few are considered independent laws, from which the remainder may be derived. A modern tendency is to choose basic laws or postulates that are different from those first discovered. Some of these choices are most useful in that the derivation of the remainder may be accomplished very efficiently. However, those laws that arose from the historical development will be discussed here since they are less abstract and lend themselves to a clearer physical interpretation.

One may say that the whole development of thermodynamic principles was completed when three state functions, the absolute temperature T, the internal energy U, and the entropy S, were defined. The zeroth law formalizes the concept of temperature, the first law defines the internal energy, and the second law brings in the concept of entropy as well as the absolute scale of temperature. Finally, the third law describes the behavior of entropy and internal energy as the absolute temperature approaches zero.

For exposition, it is necessary to define a few terms. A *system* is that part of the physical world under consideration. The rest of the world is the *surroundings*. An *open system* may exchange mass, heat, and work with the surroundings. A *closed system* may exchange heat and work but not mass with the surroundings. An *isolated system* has no exchange with the surroundings. A closed or isolated system is sometimes referred to as a body. Those parts of a system spatially uniform and homogeneous are called *phases*. For example, a liquid together with its vapor may be considered a two-phase system. Systems may be made quite elaborate when required, but since focus is on the thermal properties, single-phase isotropic systems not acted upon by electric or magnetic fields are considered so that the only force allowed is that due to a uniform normal pressure. This restriction is not a basic limitation on the generality of thermodynamics but is simply a pedagogical device.

Specification of equilibrium state. The material properties of concern to thermodynamics are the macroscopic properties such as temperature, pressure, volume, concentration, surface tension, and viscosity. Molecular properties such as interatomic distances are not used. The state of a system is specified by all of the macroscopic properties together with their spatial variation. It is a fact of experience, however, that an isolated system approaches a particularly simple terminal state such that the properties are constant and spatially uniform. This simple state is called an equilibrium state. If one confines attention to a given quantity of a single-phase system, the equilibrum state is completely specified by $r+1$ of its properties, where r is the number of components. For a single-component, single-phase system not subject to magnetic or electric fields, one may fix two properties such as pressure and volume; all the remaining properties such as viscosity, surface tension, and so forth then assume fixed values. In other words, any macroscopic property of the system may be expressed as a function of the pressure and volume.

Temperature. It is within the scope of thermodynamics to refine the primitive notion of hotness and coldness into an operational and precise concept of temperature. The equilibrium states of a single-component, single-phase fluid provide a starting point. For such a fluid the equilibrium state is defined by fixing two of its properties. For example, one could construct a mercury-in-glass thermometer that has its pressure held constant; then only one other property could be varied independently. If the volume (height of mercury) is observed at any equilibrium state, there is a 1:1 correspondence between it and any other property excepting the pressure. The degree of hotness is one of these properties.

Also note that thermal equilibrium between different systems exists. For example, if the mercury thermometer is placed in contact with a body of quiet water, the mercury will either expand or contract. The volume change of the mercury will eventually stop, and the properties of the mercury will be constant, indicating an equilibrium state; moreover, the water will also have the constant properties of an equilibrium state. Bodies in thermal equilibrium are said to have the same temperature. Thus, one arrives at a method of measuring the temperature of a body of water.

Now suppose there is a large body of water in thermal contact through a wall with a large body of another fluid such as alcohol, both at equilibrium. It is an experimental fact that the mercury-in-glass thermometer will register the same volume when placed in either of the fluids. This fact is most important if one is to attach meaningful numbers to temperature. This fact of experience is designated as the zeroth law of thermodynamics: If two bodies A and B are separately in thermal equilibrium with C, then A and B are in equilibrium with each other. Thus, a useful empirical temperature measurement based upon the volume of mercury under constant pressure is established.

But the empirical temperature scale is unfortunately unique to the choice of the fluid. The mercury-in-glass thermometer is calibrated by bringing it into equilibrium with an ice-water mixture and then with boiling water, all under a pressure of 1 atm. The two mercury levels are marked 0° and 100°, respectively, and linear interpolation is used to assign numbers between the two fixed points. If one constructs and calibrates another thermometer but uses another fluid instead of mercury, one would find that the numerical values between the two fixed points do not agree. Indeed, water cannot be used as a working fluid at all since it has a minimum value of volume at 4°C and yields anomalous readings in this region. It is necessary to choose some fluid as a standard and calibrate all other thermometers by comparison with the standard. The second law of thermodynamics removes this dependence upon a particular material by defining an absolute temperature scale that is independent of the working fluid. Meanwhile the empirical temperature serves a useful operational purpose.

One could have used a low-pressure gas as the thermometric fluid. The volume could have conveniently been held constant, and the pressure of the fluid would have had a 1:1 correspondence with temperature. Here one would have found that all gases at low pressures yield the same temperature scale. This ideal gas temperature scale proves to be identical with the thermodynamic scale of the second law.

In summary, the relationship between the properties of the equilibrium state, the notion of thermal equilibrium, and the zeroth law have been used to establish the property of temperature. It may be noted in passing that temperature is a state property, and for a given mass of a single-phase, single-component fluid the temperature is a function of pressure and volume as in Eq. (1). This can be inverted as in Eqs. (2).

$$t=t(p,V) \qquad (1)$$

$$\begin{aligned} p&=p(t,V) \\ V&=V(p,t) \end{aligned} \qquad (2)$$

Internal energy. Thermodynamics does not define the concepts of energy or work but adopts them from the other macroscopic sciences of mechanics and electromagnetism. Also, the conservation of energy is taken as axiomatic. Therefore, if an isolated system is formed from any part of the world, a definite amount of energy will be trapped in the system. The energy resides in the kinetic and potential energy of the trapped molecules. The trapped energy is of a definite quantity because the isolated system cannot gain or lose energy to the surroundings and remains constant because of the conservation principle. This trapped energy is called the internal energy U.

Because of the conservation of energy, the internal energy of a closed system can be altered only by an exchange of energy with the surroundings. There are only three modes by which the exchange can occur: by mass transfer, heat transfer, or work exchange. So for a closed (no mass transfer), adiabatic (no heat transfer) system the change in internal energy ΔU is equal to the work done by the surroundings on the system, as defined by Eq. (3).

$$\Delta U=W_{\mathrm{AD}} \qquad (3)$$

Here a convention has been adopted that work done on the system is positive. There is a great mass of experimental information where work has been done on a closed system enclosed within adiabatic walls. Among these are experiments performed by J. Joule more than a century ago. He caused work to be done on an adiabatically enclosed mass of water in several different ways. A measured amount of work was used to drive an agitator in the water, to create an electric current which was then passed through a coil in the water, to compress a gas in a cylinder immersed in the water, and to rub metal blocks together in the water. In Joule's experiments the same temperature increase was always obtained with the same expenditure of work. It may be concluded from Joule's experiments that the expenditure of a given quantity of work always causes the same change of state regardless of how the work is carried out. Both W_{AD} and ΔU are independent of the path. It is concluded that U is a state function. So for a single-phase, single-component fluid Eq. (4) is

$$W_{\mathrm{AD}}=U_2-U_1=\Delta U \qquad (4)$$

written, where U_2 and U_1 depend only on the final and initial state, respectively. Also one may write Eqs. (5).

$$U = U(p,V) \qquad U = (t,p) \tag{5}$$

It is known from experience that the same change in state of a system can be effected by either supplying work to the system in an adiabatic enclosure or by contacting the system through a conducting wall with a higher temperature system. The latter method is a different means of transferring energy than work and is termed heat and given the symbol Q. Measuring the amount of work required to cause the same change in state as an amount of heat enables one to express heat quantities in terms of work quantities. For example, 1 calorie is taken to be the amount of heat necessary to raise 1 g of water 1°C at 15°C and 1 atm. The same change in state can be effected by 4.186×10^7 ergs of work. Therefore, Eq. (6) holds.

$$1 \text{ calorie} = 4.186 \times 10^7 \text{ ergs} \tag{6}$$

The first law of thermodynamics may now be derived by the useful device of a composite system. Imagine a very large system of water that transfers neither heat nor work to the surroundings. Within this large system there is a small system of a cylinder of gas in thermal contact with the water and having a piston connected to the outside. Work can be done on the small system, and it can in turn interchange heat with the large system called a reservoir. Using subscripts s for the small system and r for the reservoir, the process of doing work can be described as in Eq. (7). But $\Delta U_r =$

$$W_s = \Delta U_s + \Delta U_r \tag{7}$$

$Q_r = -Q_s$. Therefore, for the small system that is exchanging both heat and work with its surroundings, one writes Eq. (8), omitting the subscript.

$$\Delta U = Q + W \tag{8}$$

This is the first law of thermodynamics and states that the algebraic sum of heat and work during a process is equal to the change in the state function U. The term $(Q + W)$ is therefore independent of the path taken between the two states. One could, for example, cause 1 g of water to undergo the change in state as in notation (9) by

$$15°\text{C}, 1 \text{ atm} \rightarrow 16°\text{C}, 1 \text{ atm} \tag{9}$$

supplying 1 calorie of heat and no work or by doing 4.186×10^7 ergs of work alone, or one could do a great deal of work and abstract all of this energy in the form of heat excepting 1 calorie. Thus, although $(Q + W)$ is independent of the path, neither Q nor W by itself is independent of the path.

It is important to realize that U is a state function and a property of the system whereas W and Q are not. The work, as well as the heat, simply represents energy in transit. Once the energy is in the system, it is not possible to determine whether it came from heat transfer or work transfer; it is simply internal energy.

The differential form of the first law is given by Eq. (10), where q and w represent small quantities.

$$dU = q + w \tag{10}$$

In general, one may not treat q or w as well-behaved differential coefficients dQ and dW. However, if the change is such that either q or w depends only on the initial and final states and not on the path, Eq. (11) may be correctly written. If two

$$dU = dQ + dW \tag{11}$$

terms are independent of the path, the third must be also. Obviously, if either q or w is zero, one may properly write Eqs. (12). There is a third case

$$\begin{aligned} dU &= dW \\ dU &= dQ \end{aligned} \tag{12}$$

where neither q nor $w = 0$, but nevertheless $dU = dQ + dW$ is still proper. Before treating this interesting case one needs to develop the notion of a reversible process.

Reversible and irreversible processes. Any process that occurs in nature is in agreement with the first law, but many processes permissible by the first law never occur. It has already been noted that systems approach an equilibrium state if left to themselves. There is an overwhelming preference for processes to proceed in one direction. Consider Joule's experiment. A falling weight caused a paddle to do work on an adiabatically enclosed body of water. The total effect of the experiment was to increase the internal energy of the water and to lower the weight. The surroundings remained unchanged. The water temperature increased and the volume increased slightly. There is no way one can reverse this process, that is, restore the water to its original state and raise the weight to its original height without also making some additional change in the surroundings. The process is irreversible.

Consider some other processes occurring within an adiabatic enclosure by examining only the initial and final states. Two blocks of copper are initially at different temperatures and finally at the same temperature which is intermediate between the two initial temperatures. A gas is initially filling just half of a container and finally the whole of the container. Again these processes are irreversible; that is, they cannot be reversed without causing some permanent change in the surroundings. The reverses of these processes do not violate the first law; therefore, there must be some condition other than the conservation of energy which is obeyed by those processes which actually take place.

If one were presented with a description of only the initial and final states of these irreversible processes, as was done in the last two examples, one could unerringly decide from the description which state was initial and which was final. The direction of the process is entirely determined by the nature of the states. It may be expected, therefore, that there is some state function that shows which state precedes the other. The function which tells whether a process is possible or not is the entropy S and will be derived from the information that some adiabatic processes are impossible.

Thermodynamics makes use of an idealization, called a reversible process, that is a limiting case of the natural or irreversible process. The reversible process may be defined as one which can be completely reversed without leaving more than a vanishingly small change in the surroundings. It is a consequence of the definition that a reversible process proceeds through a succession of equilibrium states and may be reversed by an infinitesimal change in the external conditions. Imagine having a cylinder of gas fitted with a frictionless piston. If the piston is moved so slowly that pressure gradients are absent, the gas will be in an equilibrium state at all times. The difference between the gas pressure and the external pressure needs only to be infinitesimal in order to move the frictionless piston. Under the rather restrictive conditions of a reversible process, Eq. (13) may be

$$dW = -pdV \tag{13}$$

quite properly written, where p is the gas pressure and V is the gas volume. The first law now may be written as Eq. (14).

$$dQ = dU + pdV \tag{14}$$

Entropy. The discussion on irreversible processes has led to the second law of thermodynamics, which is just a general statement of the idea that there is a preferred direction for a given process. There are many physical statements of the second law, all being equivalent and leading to the same mathematical statement. The statement of R. Clausius is: "It is *not* possible that, at the end of a cycle of changes, heat has been transferred from a colder to a hotter body without producing some other effect." Lord Kelvin's statement is: "It is *not* possible that, at the end of a cycle of changes, heat has been extracted from a reservoir and an equal amount of work has been produced without producing some other effect."

A specific example of Kelvin's statement may be useful. Work can be converted continuously and completely into heat. For example, work could be expended on rubbing blocks in a large mass of water. The blocks would become infinitesimally hotter than the water and transfer energy to the water by heat flow. The process could be continued indefinitely with the only effect being a complete conversion of work into heat. If, however, heat is converted from the large water reservoir completely into work, some other effect occurs. For example, a gas within a cylinder can be expanded reversibly causing a transfer of heat from the bath to the gas. All of the heat extracted from the bath is converted into work. However, the gas, in this process, has changed its state since its volume is larger. The gas cannot be returned to its original state without undoing the conversion of heat into work already accomplished.

The most efficient way of developing the mathematical consequences of the second law is to proceed from Caratheodory's principle, which can be either taken as another physical expression of the second law or derived from the Clausius or Kelvin statement. Caratheodory's principle is: "In the neighborhood of any equilibrium state of a system there are states which are not accessible by an adiabatic process."

Caratheodory used this principle together with a mathematical theorem that he developed to infer the existence of a state function S and an integrating factor $1/T$, where T is the thermodynamic temperature such that Eq. (15) holds for a reversible

$$dQ_{\mathrm{REV}} = TdS \tag{15}$$

change. The state function S is called the entropy. It can also be shown that the entropy in an adiabatic system increases for an irreversible change and remains constant for a reversible change as in Eq. (16). The implication is that entropy increases for a

$$\Delta S_{\mathrm{AD}} \geq 0 \tag{16}$$

natural change until equilibrium is reached, and then it remains constant at its maximum value.

The first part of the mathematical statement of the second law allows one to write one of the most important thermodynamic equations, Eq. (17). Al-

$$dU = TdS - pdV \tag{17}$$

though this equation was derived for reversible changes, it is valid for all changes. All the quantities are functions of state. Therefore, for a change between two states the integral of the equation will be valid even if the path is not reversible. In other words, for a change from a state characterized by (p_1, T_1) to a state characterized by (p_2, T_2), the values of ΔU, ΔV, and ΔS will all have definite values dependent only upon the two states and independent of how the change came about. From this equation are obtained some of the most fruitful applications of thermodynamics to physical problems.

The second part of the mathematical statement is a concise summary of physical statements on the direction of processes. As a simple example, consider pure heat transfer to a body. The heat transfer causes a definite change of state such that $dQ = dU$. The definite change of entropy is then given by Eq. (18). If heat dQ is transferred from a

$$dS = \frac{dQ}{T} \tag{18}$$

body at temperature T_2 to a body at temperature T_1, the change in entropy is given by Eq. (19).

$$\begin{aligned} dS &= dS_1 + dS_2 \\ &= \frac{dQ}{T_1} - \frac{dQ}{T_2} \\ &= \frac{dQ\,(T_2 - T_1)}{T_1 T_2} \end{aligned} \tag{19}$$

Since dS must be positive or zero, $T_2 > T_1$. Therefore heat flows from the hotter body to the colder body.

Also contained in the second part of the entropy statement is the key idea of equilibrium. The equilibrium state of an adiabatic or isolated system is characterized by entropy being at its maximum value consistent with the physical constraints. Therefore, equilibrium states can be determined by setting $dS = 0$. Also, for a maximum in entropy, $d^2S < 0$. This latter condition leads to the notion of stability that is important in the study of phase equilibrium.

When a system is in thermal contact with its surroundings, the entropy of the system may decrease. For example, a gas being compressed isothermally decreases its entropy, but a greater increase of entropy occurs in the surroundings. The total entropy change is always positive. Clausius stated the first and second laws of thermodynamics as: "The energy of the world is constant. The entropy of the world tends toward a maximum."

By way of completeness the third law of thermodynamics needs comments. In the main body of thermodynamics one is mostly interested in changes of entropy and internal energy between states. However, the third law defines an absolute scale for entropy: The entropy of all perfect crystalline solids is zero at absolute zero temperature. The third law is used primarily in classical thermodynamics for the calculation of absolute entropies which combined with thermochemical data permits the calculation of chemical equilibrium. The foundations of the third law of thermodynamics, however, are to be found in molecular theory and therefore require a statistical mechanical treatment.

Summary. By way of summary, for a closed system all of the fundamentals of thermodynamics are contained in notation (20).

$$\begin{aligned} dU &= q + w \\ dS &= \frac{dQ}{T} && \text{reversible change} \\ dS &\geq 0 && \text{for an isolated system} \\ dU &= TdS - pdV \end{aligned} \tag{20}$$

The equations are applicable when work is restricted to volume changes only. But the generalization to include changes of polarization, magnetization, surface area, and so forth is quite straightforward. Also, the equations are not applicable to systems that involve irreversible chemical changes, but here too the extension to include these situations presents no difficulty. In actual application it is convenient to define other state functions in terms of those already introduced, but no additional basic principles are needed.

[WILLIAM F. JAEP]

Bibliography: H. B. Callen, *Thermodynamics*, 1960; K. Denbigh, *The Principles of Chemical Equilibrium*, 1964; E. A. Guggenheim, *Thermodynamics*, 1967; J. G. Kirkwood and I. Oppenheim, *Chemical Thermodynamics*, 1961; A. B. Pippard, *Classical Thermodynamics*, 1964; W. C. Reynolds, *Thermodynamics*, 1965; M. W. Zemansky, *Heat and Thermodynamics*, 1957.

Thermodynamic processes

Changes of any property of an aggregation of matter and energy, accompanied by thermal effects. The participants in a process are first identified as a system to be studied; the boundaries of the system are established; the initial state of the system is determined; the path of the changing states is laid out; and, finally, supplementary data are stated to establish the thermodynamic process. These steps will be explained in the following paragraphs. At all times it must be remembered that the only processes which are allowed are those compatible with the first and second laws of thermodynamics: Energy is neither created nor destroyed and the entropy of the system plus its surroundings always increases.

A system and its boundaries. To evaluate the results of a process, it is necessary to know the participants that undergo the process, and their mass and energy. A region, or a system, is selected for study, and its contents determined. This region may have both mass and energy entering or leaving during a particular change of conditions, and these mass and energy transfers may result in changes both within the system and within the surroundings which envelop the system.

As the system undergoes a particular change of condition, such as a balloon collapsing due to the escape of gas or a liquid solution brought to a boil in a nuclear reactor, the transfers of mass and energy which occur can be evaluated at the boundaries of the arbitrarily defined system under analysis.

A question that immediately arises is whether a system such as a tank of compressed air should have boundaries which include or exclude the metal walls of the tank. The answer depends upon the aim of the analysis. If its aim is to establish a relationship among the physical properties of the gas, such as to determine how the pressure of the gas varies with the gas temperature at constant volume, then only the behavior of the gas is involved; the metal walls do not belong within the system. However, if the problem is to determine how much externally applied heat would be required to raise the temperature of the enclosed gas a given amount, then the specific heat of the metal walls, as well as that of the gas, must be considered, and the system boundaries should include the walls through which the heat flows to reach the gaseous contents. In the laboratory, regardless of where the system boundaries are taken, the walls will always play a role and must be reckoned with.

State of a system. To establish the exact path of a process, the initial state of the system must be determined, specifying the values of variables such as temperature, pressure, volume, and quantity of material. If a number of chemicals are present in the system, the number of variables needed is usually equal to the number of independently variable substances present plus two such as temperature and pressure; exceptions to this rule occur in variable electric or magnetic fields and in some other well-defined cases. Thus, the number of properties required to specify the state of a system depends upon the complexity of the system. Whenever a system changes from one state to another, a process occurs.

Whenever an unbalance occurs in an intensive property such as temperature, pressure, or density, either within the system or between the system and its surroundings, the force of the unbalance can initiate a process that causes a change of state. Examples are the unequal molecular concentration of different gases within a single rigid enclosure, a difference of temperature across the system boundary, a difference of pressure normal to a nonrigid system boundary, or a difference of electrical potential across an electrically conducting system boundary. The direction of the change of state caused by the unbalanced force is such as to reduce the unbalanced driving potential. Rates of changes of state tend to decelerate as this driving potential is decreased.

Equilibrium. The decelerating rate of change implies that all states move toward new conditions of equilibrium. When there are no longer any balanced forces acting within the boundaries of a system or between the system and its surroundings, then no mechanical changes can take place, and the system is said to be in mechanical equilibrium. A system in mechanical equilibrium, such as a mixture of hydrogen and oxygen, under certain conditions might undergo a chemical change.

However, if there is no net change in the chemical constituents, then the mixture is said to be in chemical as well as in mechanical equilibrium.

If all parts of a system in chemical and mechanical equilibrium attain a uniform temperature and if, in addition, the system and its surroundings either are at the same temperature or are separated by a thermally nonconducting boundary, then the system has also reached a condition of thermal equilibrium.

Whenever a system is in mechanical, chemical, and thermal equilibrium, so that no mechanical, chemical, or thermal changes can occur, the system is in thermodynamic equilibrium. The state of equilibrium is at a point where the tendency of the system to minimize its energy is balanced by the tendency toward a condition of maximum randomness. In thermodynamics, the state of a system can be defined only when it is in equilibrium. The static state on a macroscopic level is nevertheless underlaid by rapid molecular changes; thermodynamic equilibrium is a condition where the forward and reverse rates of the various changes are all equal to one another. In general, those systems considered in thermodynamics can include not only mixtures of material substances but also mixtures of matter and all forms of energy. For example, one could consider the equilibrium between a gas of charged particles and electromagnetic radiation contained in an oven.

Process path. If under the influence of an unbalanced intensive factor the state of a system is altered, then the change of state of the system is described in terms of the end states or difference between the initial and final properties.

The path of a change of state is the locus of the whole series of states through which the system passes when going from an initial to a final state. For example, suppose a gas expands to twice its volume and that its initial and final temperatures are the same. Various paths connect these initial and final states: isothermal expansion, with temperature held constant at all times, or adiabatic expansion which results in cooling followed by heating back to the initial temperature while holding volume fixed.

Each of these paths can be altered by making the gas do varying amounts of work by pushing out a piston during the expansion, so that an extremely large number of paths can be followed even for such a simple example. The detailed path must be specified if the heat or work is to be a known quantity; however, changes in the thermodynamic properties depend only on the initial and final states and not upon the path.

There are several corollaries from the above descriptions of systems, boundaries, states, and processes. First, all thermodynamic properties are identical for identical states. Second, the change in a property between initial and final states is independent of path or processes. The third corollary is that a quantity whose change is fixed by the end states and is independent of the path is a point function or a property. However, it must be remembered that by the second law of thermodynamics not all states are available (possible final states) from a given initial state and not all conceivable paths are possible in going toward an available state.

Pressure-volume-temperature diagram. Whereas the state of a system is a point function, the change of state of a system, or a process, is a path function. Various processes or methods of change of a system from one state to another may be depicted graphically as a path on a plot using thermodynamic properties as coordinates.

The variable properties most frequently and conveniently measured are pressure, volume, and temperature. If any two of these are held fixed (independent variables), the third is determined (dependent variable). To depict the relationship among these physical properties of the particular working substance, these three variables may be used as the coordinates of a three-dimensional space. The resulting surface is a graphic presentation of the equation of state for this working substance, and all possible equilibrium states of the substance lie on this *P-V-T* surface. The *P-V-T* surface may be extensive enough to include all three phases of the working substance: solid, liquid, and vapor.

Because a *P-V-T* surface represents all equilibrium conditions of the working substance, any line on the surface represents a possible reversible process, or a succession of equilibrium states.

The portion of the *P-V-T* surface shown in Fig. 1 typifies most real substances; it is characterized by contraction of the substance on freezing. Going from the liquid surface to the liquid-solid surface onto the solid surface involves a decrease in both temperature and volume. Water is one of the few exceptions to this condition; it expands upon freezing, and its resultant *P-V-T* surface is somewhat modified where the solid and liquid phases abut.

Gibbs' phase rule is defined in Eq. (1). Here f is

$$f = c - p + 2 \tag{1}$$

the degree of freedom; this integer states the

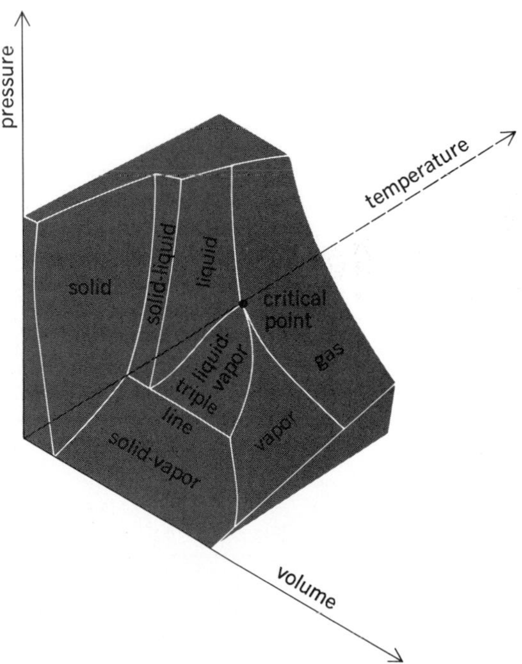

Fig. 1. Portion of pressure-volume-temperature (*P-V-T*) surface for a typical substance.

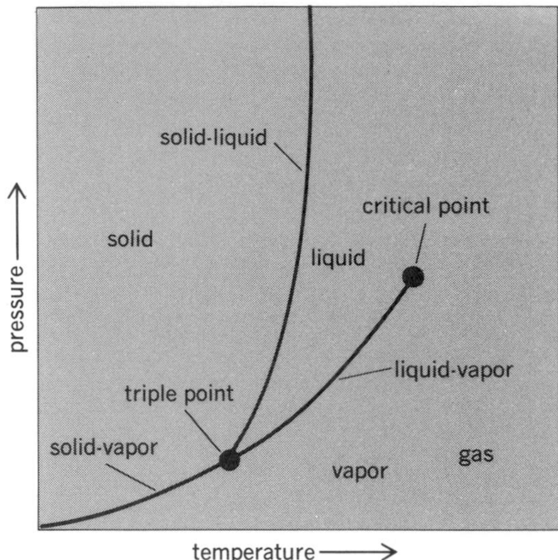

Fig. 2. Portion of equilibrium surface projected on pressure-temperature (*P-T*) plane.

number of intensive properties (such as temperature, pressure, and mole fractions or chemical potentials of the components) which can be varied independently of each other and thereby fix the particular equilibrium state of the system (see discussion under Temperature-entropy diagram, below). Also, p indicates the number of phases (gas, liquid, or solid) and c the number of component substances in the system. Consider a one-component system (a pure substance) which is either in the liquid, gaseous, or solid phase. In equilibrium the system has two degrees of freedom; that is, two independent thermodynamic properties must be chosen to specify the state. Among the thermodynamic properties of a substance which can be quantitatively evaluated are the pressure, temperature, specific volume, internal energy, enthalpy, and entropy. From among these properties, any two may be selected. If these two prove to be independent of each other, when the values of these two properties are fixed, the state is determined and the values of all the other properties are also fixed. A one-component system with two phases in equilibrium (such as liquid in equilibrium with its vapor in a closed vessel) has $f=1$; that is, only one intensive property can be independently specified. Also, a one-component system with three phases in equilibrium has no degree of freedom. Examination of Fig. 1 shows that the three surfaces (solid-liquid, solid-vapor, and liquid-vapor) are generated by lines parallel to the volume axis. Moving the system along such lines (constant pressure and temperature) involves a heat exchange and a change in the relative proportion of the two phases. Note that there is an entropy increment associated with this change.

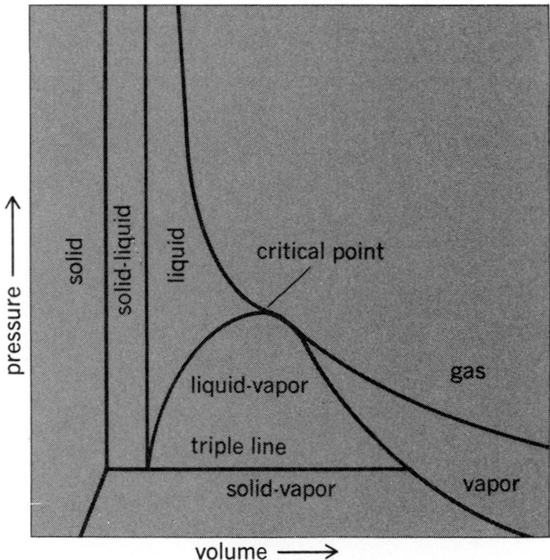

Fig. 3. Portion of equilibrium surface projected on pressure-volume (*P-V*) plane.

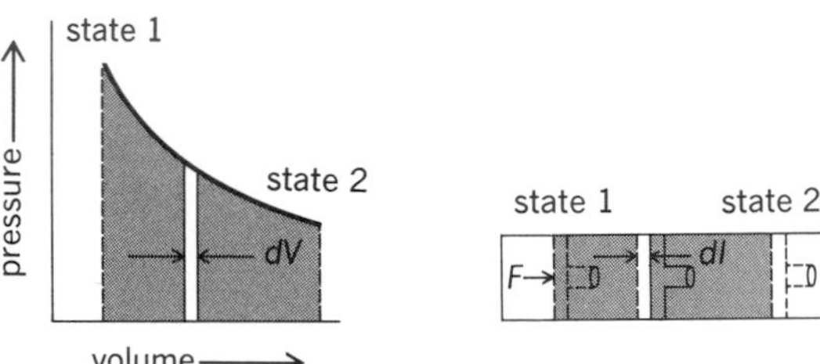

Fig. 4. Area under path in *P-V* plane is work done by expanding gas against piston.

One can project the three-dimensional surface onto the *P-T* plane as in Fig. 2. The triple point is the point where the three phases are in equilibrium. When the temperature exceeds the critical temperature (at the critical point), only the gaseous phase is possible. The gas is called a vapor when it can coexist with another phase (at temperatures below the critical point). The *P-T* diagram for water would have the solid-liquid curve going upward from the triple point to the left (contrary to the ordinary substance pictured in Fig. 2). Then the property so well known to ice skaters would be evident. As the solid-liquid line is crossed from the low-pressure side to the high-pressure side, the water changes from solid to liquid: Ice melts upon application of pressure.

Work of a process. The three-dimensional surface can also be projected onto the *P-V* plane to get Fig. 3. This plot has a special significance: The area under any reversible path on this plane represents the work done during the process. The fact that this *P-V* area represents useful work can be demonstrated by the following example.

Let a gas undergo an infinitesimal expansion in a cylinder equipped with a frictionless piston, and let this expansion perform useful work on the surroundings. The work done during this infinitesimal expansion is the force multiplied by the distance through which it acts, as in Eq. (2), wherein dW is

$$dW = F\,dl \tag{2}$$

an infinitesimally small work quantity, F is the force, and dl is the infinitesimal distance through which F acts.

But force F is equal to the pressure P of the fluid times the area A of the piston, or PA. However, the product of the area of the piston times the infinitesimal displacement is really the infinitesimal volume swept by the piston, or $A\ dl = dV$, with dV equal to an infinitesimal volume. Thus Eq. (3) is valid. The work term is found by integration, as in Eq. (4).

$$dW = P\,A\,dl = P\,dV \qquad (3)$$

$${}_1W_2 = \int_1^2 P\,dV \qquad (4)$$

Figure 4 shows that the integral represents the area under the path described by the expansion from state 1 to state 2 on the P-V plane. Thus, the area on the P-V plane represents work done during this expansion process.

Temperature-entropy diagram. Energy quantities may be depicted as the product of two factors: an intensive property and an extensive one. Examples of intensive properties are pressure, temperature, and magnetic field; extensive ones are volume, magnetization, and mass. Thus, in differential form, work has been presented as the product of a pressure exerted against an area which sweeps through an infinitesimal volume, as in Eq. (5). Note that as a gas expands, it is doing work on

$$dW = P\,dV \qquad (5)$$

its environment. However, a number of different kinds of work are known. For example, one could have work of polarization of a dielectric, of magnetization, of stretching a wire, or of making new surface area. In all cases, the infinitesimal work is given by Eq. (6), where X is a generalized applied

$$dW = Xdx \qquad (6)$$

force which is an intensive quantity such as voltage, magnetic field, or surface tension; and dx is a generalized displacement of the system and is thus extensive. Examples of dx include changes in electric polarization, magnetization, length of a stretched wire, or surface area.

By extending this approach, one can depict transferred heat as the product of an intensive property, temperature, and a distributed or extensive property defined as entropy, for which the symbol is S.

If an infinitesimal quantity of heat dQ is transferred during a reversible process, this process may be expressed mathematically as in Eq. (7),

$$dQ = T\,dS \qquad (7)$$

with T being the absolute temperature and dS the infinitesimal entropy quantity.

Furthermore, a plot of the change of state of the system undergoing this reversible heat transfer can be drawn on a plane in which the coordinates are absolute temperature and entropy (Fig. 5). The total heat transferred during this process equals the area between this plotted line and the horizontal axis.

Reversible processes. Not all energy contained in or associated with a mass can be converted into useful work. Under ideal conditions only a fraction of the total energy present can be converted into work. The ideal conversions which retain the maximum available useful energy are reversible processes.

Characteristics of a reversible process are that the working substance is always in thermodynamic equilibrium and the process involves no dissipative effects such as viscosity, friction, inelasticity, electrical resistance, or magnetic hysteresis. Thus, reversible processes proceed quasistatically so that the system passes through a series of states of thermodynamic equilibrium, both internally and with its surroundings. This series of states may be traversed just as well in one direction as in the other.

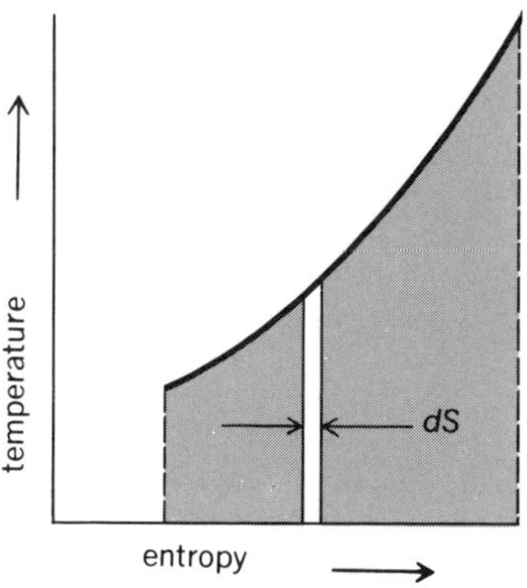

Fig. 5. Heat transferred during a reversible process is area under path in temperature-entropy (T-S) plane.

If there are no dissipative effects, all useful work done by the system during a process in one direction can be returned to the system during the reverse process. When such a process is reversed so that the system returns to its starting state, it must leave an effect on the surroundings since, by the second law of thermodynamics, in energy conversion processes the form of energy is always degraded. Part of the energy of the system (including heat source) is transferred as heat from a higher temperature to a lower temperature. The energy rejected to a lower-temperature heat sink cannot be recovered. To return the system (including heat source and sink) to its original state, then, requires more energy than the useful work done by the system during a process in one direction. Of course, if the process were purely a mechanical one with no thermal effects, then both the surroundings and system could be returned to their initial states. *See* CARNOT CYCLE; THERMODYNAMIC CYCLE.

It is impossible to satisfy the conditions of a quasistatic process with no dissipative effects; a reversible process is an ideal abstraction which is not realizable in practice but is useful for theoretical calculations. An ideal reversible engine operating between hotter and cooler bodies at the temperatures T_1 and T_2, respectively, can put out $(T_1 - T_2)/T_1$ of the transferred heat energy as useful work.

There are four reversible processes wherein one of the common thermodynamic parameters is kept constant. The general reversible process for a closed or nonflow system is described as a polytropic process.

Irreversible processes. Actual changes of a system deviate from the idealized situation of a quasistatic process devoid of dissipative effects. The extent of the deviation from ideality is correspondingly the extent of the irreversibility of the process.

Real expansions take place in finite time, not

infinitely slowly, and these expansions occur with friction of rubbing parts, turbulence of the fluid, pressure waves sweeping across and rebounding through the cylinder, and finite temperature gradients driving the transferred heat. These dissipative effects, the kind of effects that make a pendulum or yo-yo slow down and stop, also make the work output of actual irreversible expansions less than the maximum ideal work of a corresponding reversible process. For a reversible process, as stated earlier, the entropy change is given by $dS = dQ/T$. For an irreversible process even more entropy is produced (turbulence and loss of information) and there is the inequality $dS > dQ/T$.

[PHILIP E. BLOOMFIELD, WILLIAM A. STEELE]

Bibliography: H. A. Bent, *The Second Law*, 1965; M. Mott-Smith, *The Concept of Energy Simply Explained*, 1934; F. W. Sears and G. L. Salinger, *Thermodynamics, the Kinetic Theory of Gases and Statistical Mechanics*, 3d ed., 1975; K. Wark, *Thermodynamics*, 3d ed., 1977.

Thermoelectric power generator

A device that converts heat energy directly into electric energy by using the Seebeck effect. A thermoelectric generator is composed of at least two dissimilar materials, one junction of which is in contact with a heat source and the other junction of which is in contact with a heat sink.

The power converted from heat to electricity is dependent upon the materials used, the temperatures of the heat source and sink, the electrical and thermal design of the thermocouple, and the load of the thermocouple.

Theory and operation. All thermoelectric effects are related to the transport properties of electrons in materials. The most important transport parameter for analysis of thermoelectric phenomena is called the Seebeck coefficient, which is the open-circuit voltage per unit temperature difference between the hot and cold junctions. The Seebeck coefficient is also called the thermoelectric power. For definition and typical values *see* THERMOELECTRICITY.

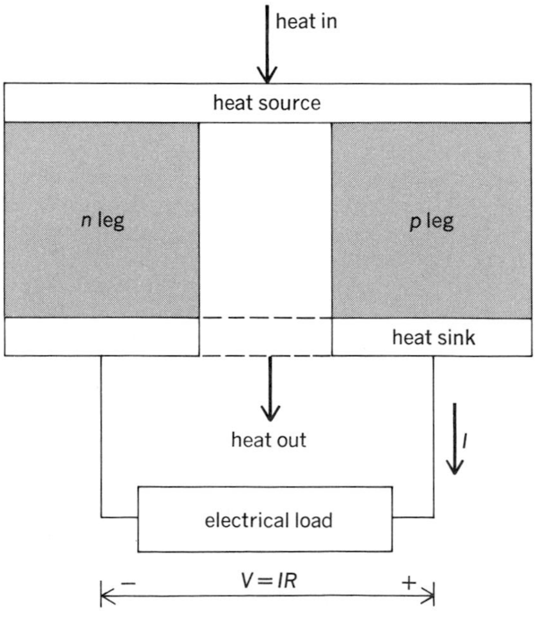

A single two-element generator is shown in Fig. 1. One leg is n-type semiconductor material; the other is p-type material. The effective composite parameters of this simple generator are the total Seebeck coefficient of the junction S, the total internal resistance r, and the total thermal conductivity K. If the Seebeck coefficients of each leg S_n and S_p, the electrical resistivities of each leg ρ_n and ρ_p, and the thermal conductivities of each leg k_n and k_p are all assumed to be independent of temperature, the composite parameters can be defined as

$$S = S_p - S_n = |S_p| + |S_n|$$

$$r = \frac{\rho_n l_n}{A_n} + \frac{\rho_p l_p}{A_p}$$

$$K = \frac{k_n A_n}{l_n} + \frac{k_p A_p}{l_p}$$

where l_n, l_p and A_n, A_p refer to the length and area of the n-type material.

The thermocouple of Fig. 1 operating as a generator has a heat source of constant temperature T_h, a heat sink at temperature T_c, and an electrical load of resistance R. The efficiency η of the generator is the power out I^2R divided by the heat in Q_{in}, which consists of the Peltier heat ST_hI plus the conduction heat $K(T_h - T_c)$ less one-half of the Joule heat I^2r liberated in the thermocouple legs:

$$\eta = \frac{P_{out}}{Q_{in}} = \frac{I^2R}{ST_hI + K(T_h - T_c) - \frac{1}{2}I^2r}$$

This efficiency does not consider losses in maintaining the temperature T_h and is therefore not a total efficiency including heat-source losses.

The ratio of load resistance R to internal resistance r is defined as $m = R/r$; the open-circuit voltage is $V_o = S(T_h - T_c)$; and the current is $I = V_o/(R + r)$. Using these quantities the efficiency expression becomes

$$\eta = \frac{T_h - T_c}{T_h}\left[\frac{\dfrac{m}{m+1}}{1 + \left(\dfrac{m+1}{ZT_h}\right) - \dfrac{1}{2}\dfrac{T_h - T_c}{T_h}\left(\dfrac{1}{m+1}\right)}\right]$$

where Z, the figure of merit, is

$$Z = \frac{S^2}{Kr}$$

If efficiency is optimized by selecting m to give optimum loading, the efficiency expression becomes

$$\eta = \frac{T_h - T_c}{T_h}\left[\frac{M - 1}{M + \dfrac{T_c}{T_h}}\right]$$

where $M = m\Big|_{\frac{d\eta}{dm} = 0} = \sqrt{\dfrac{Z(T_h - T_c)}{2}}$

This efficiency for an optimum load consists of a Carnot efficiency

Fig. 1. Simple thermoelectric generator.

$$\eta_C=\frac{T_h-T_c}{T_h}$$

and a device efficiency

$$\eta_d=\left[\frac{M-1}{M+\dfrac{T_c}{T_h}}\right]$$

The device efficiency η_d will be a maximum value for the largest value of M. For a fixed T_h and T_c this requires a maximum value of Z. Z depends upon the material parameters S_p, k_p, ρ_p, S_n, k_n, ρ_n and the dimensions of the two legs A_p, l_p, A_n, l_n. Maximizing Z with respect to X (the area-to-length ratio of the legs) gives

$$Z\Bigg|_{\frac{dZ}{dX}=0}=\frac{(S_p-S_n)^2}{[(k_p\rho_p)^{1/2}+(k_n\rho_n)^{1/2}]^2}$$

when $\left[\dfrac{k_n\rho_n}{k_p\rho_n}\right]^{1/2}=\dfrac{A_pl_n}{A_nl_p}=X$

For the optimum area-to-length ratio of the legs, the figure of merit Z depends only upon the specific properties of the thermoelectric materials. The effect of Z in determining the efficiency is shown in Fig. 2, where η is plotted versus Z and T_h for a fixed cold temperature $T_c=300$ K.

In general, the parameters S, k, and ρ are not independent of temperature, and in fact the temperature dependence of the n and p legs may differ radically. The simple figure of merit shown above does not apply to the temperature-dependent case. The general solution of the thermocouple with temperature-dependent parameters is difficult even if parameters and temperature dependence are known, and requires numerical methods.

Several approximate methods of treating temperature-dependent parameters have been proposed. A. H. Boerdijk has shown that there is no gain in performance by varying the cross section of the elements along their length. A. F. Ioffe has proposed that replacing (S_p-S_n), $k_p\rho_p$, and $k_n\rho_n$ by their average over the temperature range in the figure of merit equation of the temperature-independent thermocouple is a reasonable approximation. C. Zener has proposed that the thermocouple with temperature-dependent parameters be treated by considering an infinite series of differential thermocouples with each material matched to its optimum load.

To check the Ioffe and Zener approximations against several exact calculations, B. Sherman, R. R. Heikes, and R. W. Ure, Jr., assumed three hypothetical thermocouples and calculated the efficiency by all three methods, using numerical calculations on a digital computer. These calculations show that both approximate methods—average parameters (Ioffe) and infinite staging (Zener)—lead to reasonably correct results, yielding accuracy of the order of ±10%.

State of the art. The present state of the art in materials development indicates that existing thermoelectric p and n materials operate from 300 to 1300°K and yield an overall theoretical thermal efficiency of 18%. The most widely used generator material is lead telluride, which has a maximum figure of merit Z of approximately 1.5×10^{-3} reciprocal kelvins. It can be doped to produce both p- and n-type material and has a useful temperature range of about 300–700 K. In segmented couples at the low-temperature end, bismuth telluride and its alloys are sometimes used. Couples of bismuth telluride and its alloys can be obtained both as raw materials and as finished couples. Lead telluride is similarly available commercially.

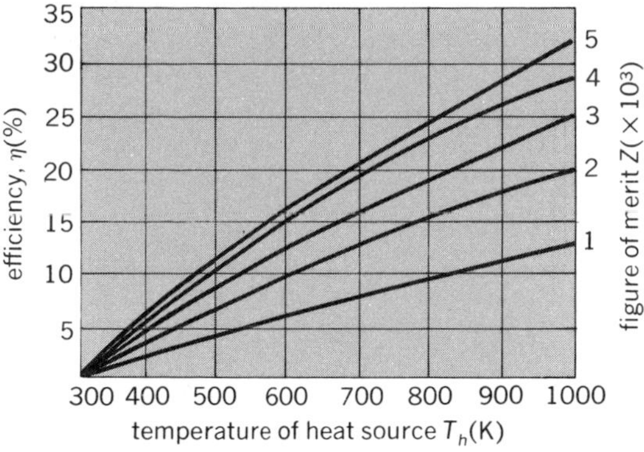

Fig. 2. Relation of efficiency η to the temperature of the heat source T_h for a constant temperature T_c of the heat sink of 300 K. Different curves are for different values of the figure of merit Z.

Thermoelectric generators have been built in sizes up to 5 kW. Primary energy sources are hydrocarbon fuels, radioisotopes, and solar energy.

The maximum theoretical thermal efficiency for materials over a temperature range of 300–1300 K is approximately 18%. The best actual thermal efficiency of a physically constructed device is between 6 and 10% and is obtained by operation between 300 and 950 K. The overall conversion efficiency of hydrocarbon-fueled generators is 2–3%; and overall efficiency of radioisotope generators is approximately 5%. Specific powers of 12 watts/lb have been obtained, but advances in electrical contacts indicate that higher values are obtainable.

Major problems still exist in the development of materials with higher figures of merit that are capable of operation at higher temperature. Even with present materials, there are severe engineering problems involving such items as low electrical and thermal contact resistances, mechanical strength to withstand thermal and mechanical shocks, minimum heat losses due to packaging of thermoelements, efficient fuel combustion, heat transfer from heat source to thermoelements to heat sink, packaging to minimize weight, and long-term contamination of thermoelements by diffusion.

The state of the art can probably best be summarized by stating that thermoelectric generators have been built and are under construction for many special applications.

[DAVID C. WHITE]

Bibliography: A. H. Boerdijk, Contribution to a general theory of thermocouples, *J. Appl. Phys.*, 30(7):1080–1083, 1959; I. B. Cadoff and E. Miller (eds.), *Thermoelectric Materials and Devices*, 1960; S. F. DeGroot, *Thermodynamics of Irreversible Processes*, 1952; C. A. Domenicali, Irreversible thermodynamics of thermoelectricity, *Rev. Mod.*

Phys., 26(2):237–275, 1954; P. H. Egli (ed.), *Thermoelectricity*, 1960; A. F. Ioffe, *Semiconductor Thermoelements and Thermoelectric Cooling*, 1957; B. Sherman, R. R. Heikes, and R. W. Ure, Jr., Calculation of efficiency of thermoelectric devices, *J. Appl. Phys.*, 31(1):1–16, 1960.

Thermoelectricity

The direct conversion of heat into electrical energy, or the reverse, in solid or liquid conductors by means of three interrelated phenomena—the Seebeck effect, the Peltier effect, and the Thomson effect—including the influence of magnetic fields upon each. The Seebeck effect concerns the electromotive force (emf) generated in a circuit composed of two different conductors whose junctions are maintained at different temperatures. The Peltier effect refers to the reversible heat generated at the junction between two different conductors when a current passes through the junction. The Thomson effect involves the reversible generation of heat in a single current-carrying conductor along which a temperature gradient is maintained. Specifically excluded from the definition of thermoelectricity are the phenomena of Joule heating and thermionic emission.

The three thermoelectric effects are described in terms of three coefficients: the absolute thermoelectric power (or thermopower) S, the Peltier coefficient Π, and the Thomson coefficient μ, each of which is defined for a homogeneous conductor at constant temperature. These coefficients are connected by the Kelvin relations, which convert complete information about one into complete information about all three. It is therefore necessary to measure only one of the three coefficients; usually the thermopower S is chosen. The combination of electrical resistivity, thermal conductivity, and thermopower is sufficient to provide a complete description of the electronic transport properties of conductors for which the electric current and heat current are linear functions of both the applied electric field and the temperature gradient.

Thermoelectric effects have significant applications in both science and technology and show promise of more importance in the future. Studies of thermoelectricity in metals and semiconductors yield information about electronic structure and about the interactions between electrons and both lattice vibrations and impurities. Practical applications include the measurement of temperature, generation of power, cooling, and heating. Thermocouples are widely used for temperature measurement, providing both accuracy and sensitivity. Research has been undertaken concerning the direct thermoelectric generation of electricity using the heat produced by nuclear reactors. Cooling units using the Peltier effect have been constructed in sizes up to those of home refrigerators. Development of thermoelectric heating has also been undertaken.

THERMOELECTRICITY

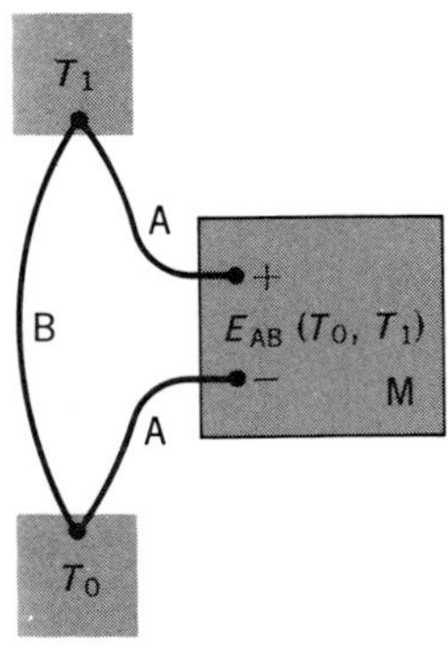

Fig. 1. Diagram of apparatus usually used for measuring thermoelectric (Seebeck) emf $E_{AB}(T_0,T_1)$. M is an instrument for measuring potential.

SEEBECK EFFECT

In 1821 T. J. Seebeck discovered that when two different conductors are joined into a loop, and a temperature difference is maintained between the two junctions, an emf will be generated. Such a loop is called a thermocouple, and the emf generated is called a thermoelectric (or Seebeck) emf.

Measurements. The magnitude of the emf generated by a thermocouple is standardly measured by using the system shown in Fig. 1. Here the contact points between conductors A and B are called junctions. Each junction is maintained at a well-controlled temperature (either T_0 or T_1) by immersion in a bath or connection to a heat reservoir. This bath or reservoir is indicated by the dashed rectangles. From each junction, conductor A is brought to a measuring device, usually a potentiometer. When the potentimeter is balanced, no current flows, thereby allowing direct measurement of the open-circuit emf, undiminished by resistive losses and unperturbed by spurious effects arising from Joule heating or from Peltier heating and cooling at the junctions. This open-circuit emf is the thermoelectric emf.

Equations. According to the experimentally established law of Magnus, for homogeneous conductors A and B the thermoelectric emf depends only upon the temperatures of the two junctions and not upon either the shapes of the samples or the detailed forms of the temperature distributions along them. This emf can thus be described by the symbol $E_{AB}(T_0,T_1)$. According to both theory and experiment, if one of the conductors, say B, is a superconductor in its superconducting state, it makes no contribution to E_{AB}. That is, when B is superconducting, $E_{AB}(T_0,T_1)$ is determined solely by conductor A and can be written as $E_A(T_0,T_1)$. It is convenient to express this emf in terms of a property which depends upon only a single temperature. Such a property is the absolute thermoelectric power (or, simply, thermopower) $S_A(T)$, defined so that Eq. (1) is valid. If $E_A(T,T+\Delta T)$ is known—for example, from measurements involving a superconductor—then $S_A(T)$ can be determined from Eq. (2). If Eq. (1) is valid for any homogeneous

$$E_A(T_0,T_1) = \int_{T_0}^{T_1} S_A(T)\,dT \qquad (1)$$

$$S_A(T) = \lim_{\Delta T \to 0} \frac{E_A(T,T+\Delta T)}{\Delta T} \qquad (2)$$

conductor, then it ought to apply to both sides of the thermocouple shown in Fig. 1. Indeed, it

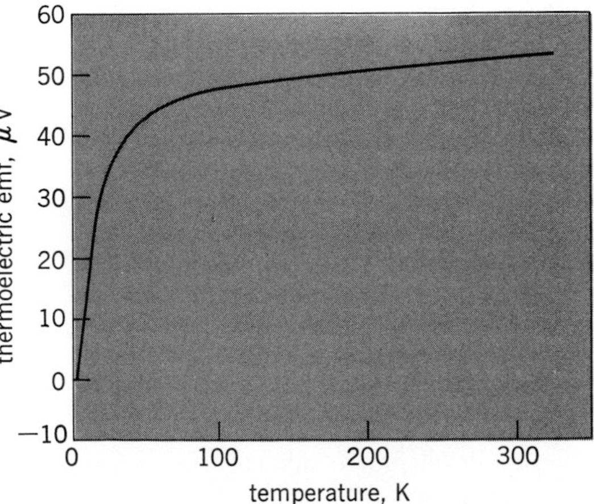

Fig. 2. The thermoelectric emf of a thermocouple formed from pure annealed and pure cold-worked copper. The cold junction reference temperature is 4.2 K. (*From R. H. Kropschot and F. J. Blatt, Thermoelectric power of cold-rolled pure copper, Phys. Rev., 116:617–620, 1959*)

has been verified experimentally that the emf $E_{AB}(T_0,T_1)$ produced by a thermocouple is just the difference between the emfs, calculated using Eq. (1), produced by its two arms. This result can be derived as follows. Employing the usual sign convention, to calculate $E_{AB}(T_0,T_1)$, begin at the cooler bath, labeled T_0, integrate $S_A(T)dT$ along conductor A up to the warmer bath labeled T_1; and then return to T_0 along conductor B by integrating $S_B(T)dT$. This circular excursion produces $E_{AB}(T_0,T_1)$, given by Eq. (3). Inverting the last integral in Eq. (3) gives Eq. (4), which, from Eq. (1), can be rewritten as Eq. (5). Alternatively, combining the two integrals in Eq. (4) gives Eq. (6). Defining S_{AB} according to Eq. (7) then yields Eq. (8).

$$E_{AB}(T_0,T_1) = \int_{T_0}^{T_1} S_A(T)\,dT + \int_{T_1}^{T_0} S_B(T)\,dT \quad (3)$$

$$E_{AB}(T_0,T_1) = \int_{T_0}^{T_1} S_A(T)\,dT - \int_{T_0}^{T_1} S_B(T)\,dT \quad (4)$$

$$E_{AB}(T_0,T_1) = E_A(T_0,T_1) - E_B(T_0,T_1) \quad (5)$$

$$E_{AB}(T_0,T_1) = \int_{T_0}^{T_1} [S_A(T) - S_B(T)]\,dT \quad (6)$$

$$S_{AB}(T) = S_A(T) - S_B(T) \quad (7)$$

$$E_{AB}(T_0,T_1) = \int_{T_0}^{T_1} S_{AB}(T)\,dT \quad (8)$$

Equation (6) shows that $E_{AB}(T_0,T_1)$ can be calculated for a given thermocouple whenever the thermopowers $S_A(T)$ and $S_B(T)$ are known for its two constituents over temperature range T_0 to T_1. By convention, the signs of $S_A(T)$ and $S_B(T)$ are chosen so that, if the temperature difference $T_1 - T_0$ is taken small enough for $S_A(T)$ and $S_B(T)$ to be presumed constant, $S_A(T) > S_B(T)$ when the emf $E_{AB}(T_0,T_1)$ has the polarity shown in Fig. 1.

Results of equations. These equations lead directly to the following experimentally and theoretically verified results.

Uniform temperature. In a circuit kept at a uniform temperature throughout, $E=0$ even though the circuit may consist of a number of different conductors. This follows directly from Eq. (8), since $dT=0$ everywhere throughout the circuit. It follows also from thermodynamic reasoning. If E did not equal 0, the circuit could drive an electric motor and make it do work. But the only source of energy would be heat from the surroundings, which, by assumption, are at the same uniform temperature as the circuit. Thus, a contradiction with the second law of thermodynamics would result.

Homogeneous conductor. A circuit composed of a single, homogeneous conductor cannot produce a thermoelectric emf. This follows from Eq. (6) when $S_B(T)$ is set equal to $S_A(T)$. It is important to emphasize that, in this context, homogeneous means perfectly uniform throughout. A sample made of an isotropic material can be inhomogeneous either because of small variations in chemical composition or because of strain. Figure 2 shows the thermoelectric emf generated by a thermocouple in which one arm is a cold-rolled copper (Cu) sample, and the other arm is the same material after annealing to remove the effects of the strain introduced by the cold-rolling. Figure 3 shows how the addition of impurities can change the thermopower of a pure metal. An additional effect can occur in a noncubic material. As illustrated in Fig. 4, two samples cut in different directions from a noncubic single crystal may be thermoelectrically different even if each sample is highly homogeneous. A thermocouple formed from these two samples will generate a thermoelectric emf.

If material B is superconducting, so that $S_B=0$, Eq. (5) reduces to $E_{AB}(T_0,T_1)=E_A(T_0,T_1)$, as assumed above.

Source of emf. Finally, Eq. (6) makes it clear that the source of the thermoelectric emf in a thermocouple lies in the bodies of the two materials of which it is composed, rather than at the junctions. This serves to emphasize that thermoelectric emfs are not related to the contact potential or Volta effect, which is a potential difference across the junction between two different metals arising from the difference between their Fermi energies. The contact potential is present even in the absence of temperature gradients or electric currents.

PELTIER EFFECT

In 1834 J. C. A. Peltier discovered that, when an electric current passes through two different conductors connected in a loop, one of the two junctions between the conductors cools, and the other warms. If the direction of the current is reversed, the effect also reverses: the first junction warms, and the second cools. In 1853 Quintus Icilius showed that the rate of heat output or intake at each junction is directly proportional to the current i. The Peltier coefficient Π_{AB} is defined as the heat generated per second per unit current flow through the junction between materials A and B. By convention, Π_{AB} is taken to be positive when cooling occurs at the junction through which current flows from conductor A to conductor B. Quintus Icilius's result guarantees that the Peltier coefficient is independent of the magnitude of the current i. Additional experiments have shown that it is also independent of the shapes of the conductors. It therefore depends only upon the two materials and the temperature of the junction, and can be written as $\Pi_{AB}(T)$ or, alternatively, $\Pi_A(T) - \Pi_B(T)$, where Π_A and Π_B are the Peltier coefficients

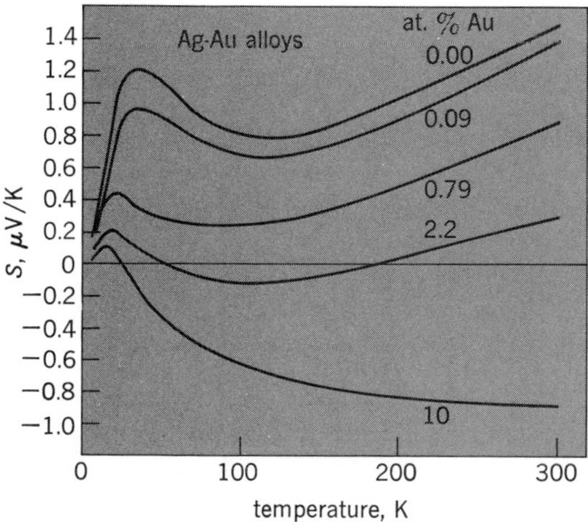

Fig. 3. The thermopower S from 0 to 300 K for pure silver (Ag) and a series of dilute silver-gold (Au) alloys. (*From R. S. Crisp and J. Rungis, Thermoelectric power and thermal conductivity in the silver-gold alloy system from 3–300°K, Phil. Mag., 22:217–236, 1970*)

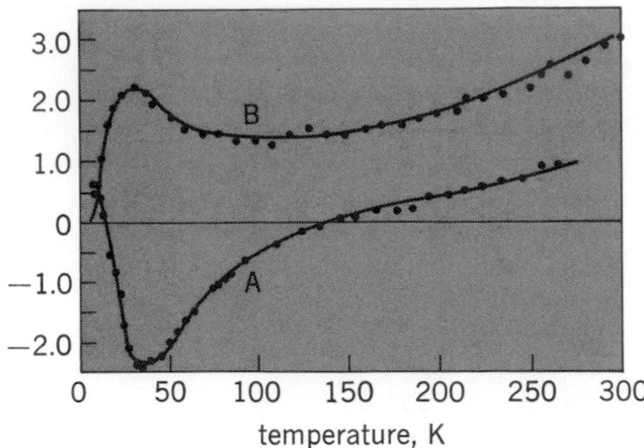

Fig. 4. The thermopower S of zinc (Zn) parallel (A) and perpendicular (B) to the hexagonal axis. (*From V. A. Rowe and P. A. Schroeder, Thermopower of Mg, Cd and Zn between 1.2° and 300°K, J. Phys. Chem. Sol., 31:1–8, 1970*)

for materials A and B respectively. The second form emphasizes that the Peltier coefficient is a bulk property definable for a single conductor.

Because of the small amount of heat transfer associated with the Peltier effect, as well as complications resulting from the simultaneous presence of Joule heating and the Thomson effect, $\Pi_{AB}(T)$ is usually difficult to measure accurately, and has therefore rarely been carefully studied. Rather, its value is usually determined from the Kelvin relations, using experimental values for S_{AB}.

THOMSON EFFECT AND KELVIN RELATIONS

When an electric current passes through a conductor which is maintained at a constant temperature, heat is generated at a rate proportional to the square of the current. This is called Joule heat, and its magnitude for any given material is determined by the electrical resistivity of the material. In 1854 William Thomson (Lord Kelvin), in an attempt to explain discrepancies between experimental results and a relationship between Π_{AB} and S_{AB} which he had derived from thermodynamic analysis of a thermocouple, postulated the existence of an additional reversible generation of heat when a temperature gradient is applied to a current-carrying conductor. This heat, called Thomson heat, is proportional to the product of the current and the temperature gradient. It is reversible, in the sense that the conductor changes from a generator of Thomson heat to an absorber of Thomson heat when the direction of either the current or the temperature gradient (but not both at once) is reversed. By contrast, Joule heating is irreversible, in that heat is generated for both directions of current flow.

The magnitude of Thomson heat generated (or absorbed) is determined by the Thomson coefficient μ. Using reasoning based upon equilibrium thermodynamics, Thomson derived results equivalent to Eqs. (9) and (10), called the Kelvin (or Kelvin-Onsager) relations.

$$\frac{\Pi_A}{T} = S_A \tag{9}$$

$$\frac{\mu_A}{T} = \frac{dS_A}{dT} \tag{10}$$

Here, μ_A is the Thomson coefficient, defined as the heat generated per second per unit current flow per unit temperature gradient when current flows through conductor A in the presence of a temperature gradient. Equation (10) can be integrated to give Eq. (11), in which the third law of

$$S_A(T) = \int_0^T \frac{\mu_A(T')}{T'}\,dT' \tag{11}$$

thermodynamics has been invoked to set $S_A(0) = 0$. By using Eq. (11), $S_A(T)$ can be determined from measurements on a single conductor. In practice, however, accurate measurements of μ_A are very difficult to make; therefore, they have been carried out for only a few metals—most notably lead (Pb)—which then serve as standards for determining $S_B(T)$ by using measurements of $S_{AB}(T)$ in conjunction with Eq. (7).

Long after the Thomson heat was observed and the Kelvin relations were verified experimentally, debate raged over the validity of the derivation employed by Thomson. But the theory of irreversible processes, developed by L. Onsager in 1931, and by others, yields the same equations and thus provides them with a relatively firm foundation.

THERMOPOWERS OF METALS AND SEMICONDUCTORS

Since the Kelvin relations provide recipes for calculating any two of the thermoelectric coefficients, S, Π, and μ, from the third, only one of the three coefficients need be measured to determine the thermoelectric properties of any given material. Although there are some circumstances under which one of the other two coefficients may be preferred, because of ease and accuracy of measurement it is almost always the thermopower S which is measured.

Reference materials. Because S must be measured by using a thermocouple, the quantity determined experimentally is $S_A - S_B$, the difference between the thermopowers of the two conductors

Table 1. The absolute thermoelectric power S of pure lead between 0 and 300 K*

T (K)	$S(\mu V/K)$†	T (K)	$S(\mu V/K)$†
0	0	60	-0.77_9
5	0	70	-0.78_4
7.5	-0.22_1	80	-0.79_4
8	-0.25_7	90	-0.82_4
8.5	-0.29_7	100	-0.86_5
9	-0.34_3	113.2	−0.91
10	-0.43_4	133.2	−0.96
11	-0.51_6	153.2	−1.02
12	-0.59_3	173.2	−1.06
14	-0.70_6	193.2	-1.10_5
16	-0.77_1	213.2	−1.15
18	-0.78_{45}	233.2	−1.18
20	-0.78_4	253.2	−1.21
30	-0.77_4	273.2	−1.25
40	-0.76_4	293.2	-1.27_5
50	-0.77_4		

*From J. W. Christian et al., Thermoelectricity at low temperatures: VI. A redetermination of the absolute scale of thermo-electric power of lead, *Proc. Roy. Soc.*, A245:213–221, 1958.

†Subscripts indicate figures which are uncertain.

constituting the couple. Only when one of the arms of the thermocouple is superconducting, and therefore has zero thermopower, can the absolute thermopower of the other arm be directly measured. At temperatures up to about 18 K it is possible to use the superconducting alloy Nb_3Sn for conductor B and thereby determine S_A. For higher temperatures no superconducting wires are generally available. It is thus necessary to have a standard thermoelectric material to use above 18 K. For historical reasons, the reference material for temperatures up to 293 K has been chosen to be Pb. Based upon Thomson coefficient measurements made in the early 1930s, the thermopower of Pb has been calculated from Eq. (11) to have the values given in Table 1. All accepted values for S in this temperature range are ultimately traceable to this table. A redetermination of the Thomson coefficient of Pb has been undertaken. For temperatures above 293 K, no standard exists for which S is as well known as for Pb, but platinum (Pt) is often used because of its high melting temperature, resistance to chemical attack, and availability in high purity.

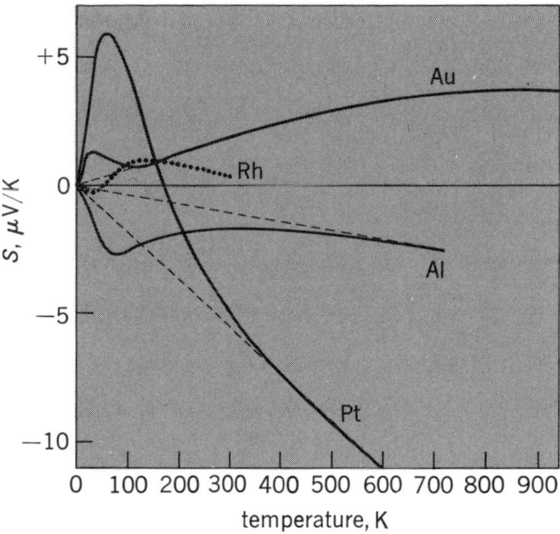

Fig. 5. The thermopower S of the metals gold (Au), aluminum (Al), platinum (Pt), and rhodium (Rh) as a function of temperature. The differences between the solid curves for Pt, Al, and Au and the broken lines indicate the magnitude of the phonon-drag component S_g.

Temperature variation. Figure 5 shows the variation with temperature of the thermopowers of four different pure metals. The data for the three metals gold (Au), aluminum (Al), and platinum (Pt) are typical of those for most simple metals and for some transition metals as well. The thermopower S consists of a slowly varying portion which increases approximately linearly with absolute temperature, upon which a "hump" is superimposed at lower temperatures. In analyzing these results, S is written as the sum of two terms, as in Eq. (12), where S_d, called the electron-diffusion

$$S = S_d + S_g \tag{12}$$

component, is the slowly varying portion, and S_g, called the phonon-drag component, is the hump. For some transition metals, on the other hand, the behavior of S is more complicated, as illustrated by the data for rhodium (Rh) in Fig. 5. Figure 6 shows comparable data for a simple p-type semiconductor, illustrating that the separation of S into S_d and S_g is still valid.

Theory. When a small temperature difference ΔT is established across a conductor, heat is carried from its hot end to its cold end by the flow of both electrons and phonons (quantized lattice vibrations). If the electron current is constrained to be zero, for example, by the insertion of a high resistance measuring device in series with the conductor, the electrons will redistribute themselves in space so as to produce an emf along the conductor. This is the thermoelectric emf. If the phonon flow could somehow be turned off, this emf would be just $S_d\Delta T$. However, the phonon flow cannot be turned off, and as the phonons move down the sample, they interact with the electrons and "drag" them along. This produces an additional contribution to the emf, $S_g\Delta T$.

Source of S_d. The conduction electrons in a metal are those having energies near the Fermi energy η. Only these electrons are important for thermoelectricity. As illustrated in Fig. 7, the energy distribution of these electrons varies with the temperature of the metal. When it is at high temperatures, a metal has more high-energy electrons, and less low-energy electrons, than when it is at low temperatures. This means that if a temperature gradient is established along a metal sample, the total number of electrons will remain constant, but the hot end will have more high-energy electrons than the cold end, and the cold end will have more low-energy electrons. The high-energy electrons will then diffuse toward the cold end, and the low-energy electrons will diffuse toward the hot end. However, in general, the diffusion rate is a function of electron energy, and thus a net electron current will result. This current will cause

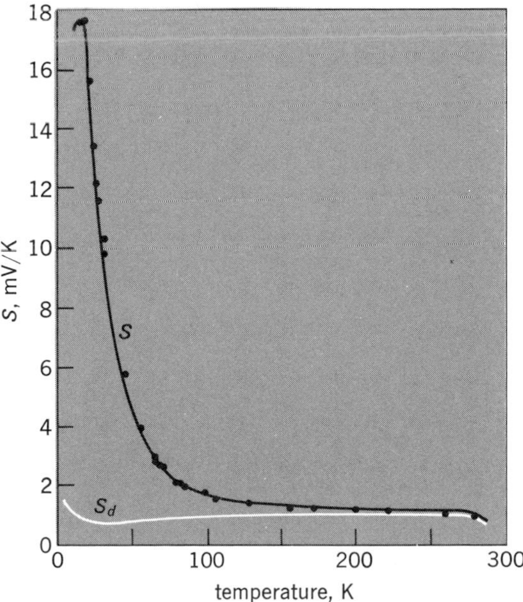

Fig. 6. The thermopower S of p-type germanium (Ge) (1.5×10^{14} acceptors per cubic centimeter) and calculated value for the electron-diffusion thermopower S_d. (*From C. Herring, The role of low-frequency phonons in thermoelectricity and thermal conductivity, Proc. Int. Coll. 1956, Garmisch-Partenkirchen, Vieweg. Braunschweig, p. 184, 1958*)

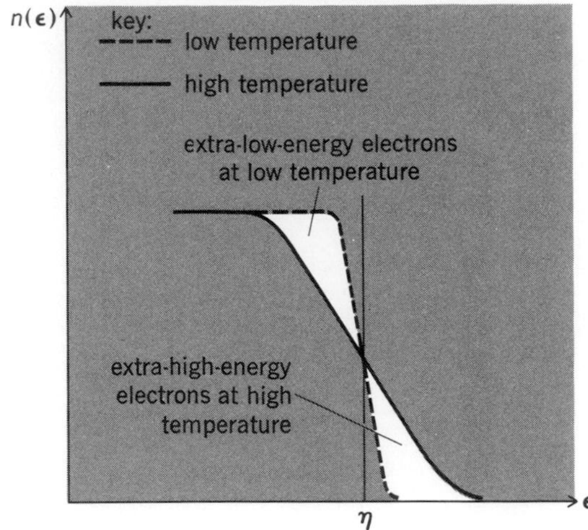

Fig. 7. The variation with energy ϵ of the number of conduction electrons $n(\epsilon)$ in a metal in the vicinity of the Fermi energy η for two different temperatures. A small variation of η with temperature has been neglected.

electrons to pile up at one end of the metal (usually the cold end) and thereby produce an emf which opposes the further flow of electrons. When the emf becomes large enough, the current will be reduced to zero. This is the thermoelectric emf arising from electron diffusion. Essentially the same argument is applicable also to semiconductors, except that in this case the conduction electrons (or holes) are those electrons just above (or below) the band gap.

S_d *for a metal.* For a completely free-electron metal, S_d would be given by Eq. (13), where k is

$$S_d = \frac{\pi^2}{2}\frac{k}{e}\left(\frac{kT}{\eta}\right) \tag{13}$$

Boltzmann's constant, e is the charge on an electron, T is the absolute (Kelvin) temperature, and η is the Fermi energy of the metal. According to Eq. (13), S_d should be negative—since e is a negative quantity—and should increase linearly with T. In Table 2 the predictions of Eq. (13) are compared with experiment for a number of the most free-electron-like metals. Equation (13) is not very satisfactory, in several cases even predicting the wrong sign. To understand the thermopowers of real metals, it is necessary to use a more sophisticated model which takes into account interactions between the electrons in the metal and the crystal lattice, as well as scattering of the electrons by impurities and phonons. The proper generalization of Eq. (13) is Eq. (14), where $\sigma(\epsilon)$ is a generalized

$$S_d = \frac{\pi^2 k^2 T}{3e}\left[\frac{\partial \ln \sigma(\epsilon)}{\partial \epsilon}\right]_\eta \tag{14}$$

conductivity defined so that $\sigma(\eta)$ is the experimental electrical conductivity of the metal, and the logarithmic derivative with respect to the energy ϵ is to be evaluated at $\epsilon = \eta$. Equation (14) is able to account, at least in principle, for all the deviations of experiment from Eq. (13). If the logarithmic derivative is negative, S_d will be positive; S_d will differ in magnitude from Eq. (13) if the logarithmic derivative does not have the value $(3/2)\eta^{-1}$; and $S_d(T)$ will deviate from a linear dependence on T if the logarithmic derivative is temperature-dependent. Research interest in S_d in metals centers upon understanding changes in S_d resulting from alloying with both magnetic and nonmagnetic impurities, strain, application of pressure, and application of magnetic fields. In some cases the changes can be dramatic. Figure 8 shows that the addition of very small amounts of the magnetic impurity iron (Fe) can produce enormous changes in S_d for copper (Cu). Sample 1 (in which the deviation of the thermopower from zero is too small to be seen with the chosen scales) is most representative of pure copper because the iron is present as an oxide and is thus not in "magnetic form." Figure 9 shows that, at low temperatures, application of a magnetic field H to Al can cause S_d to change sign. (To obtain a temperature-independent quantity, S_d has been divided by the absolute temperature T. In order to remove the effects of varying impurity concentrations, H has been divided by $\rho(4.2)nec$, where $\rho(4.2)$ is the sample resistivity at 4.2 K, n is the number of electrons per unit volume in the sample, and c is the speed of light.) Figure 10 illustrates the significant changes which occur in S when a metal melts. Substantial effort has been devoted to the study of thermoelectricity in liquid metals and liquid metal alloys.

S_d *for a semiconductor.* Equation (13) is appropriate for a free-electron gas which obeys Fermi-Dirac statistics. The conduction electrons in a metal obey these statistics. However, there are so few conduction electrons in a semiconductor that, to a good approximation, they can be treated as though they obey a different statistics—Maxwell-Boltzmann statistics. For electrons obeying these statistics, S_d is given by Eq. (15), which predicts

$$S_d = \frac{3}{2}\frac{k}{e} \tag{15}$$

that S_d will be temperature-independent and will have the value $S_d = -130 \times 10^{-6}$ V/K. For a p-type extrinsic semiconductor, in which the carriers are approximated as free holes, S_d would be just the negative of this value. An examination of the data of Fig. 6 reveals that S_d is very nearly independent of temperature, but is considerably larger than predicted by Eq. (15). Again, a complete understanding of the thermopowers of semiconductors requires the generalization of Eq. (15). The appropriate generalizations are different for single-band and multiband semiconductors, the latter being considerably more complicated. For a single-band (extrinsic) semiconductor, the generalization is relatively straightforward and yields predictions for S_d which, in agreement with experiment, vary slowly with temperature and are several times

Table 2. Comparison between theoretical values for S and experimental data

	Thermopower S (μV/K)	
Metal	Theoretical values at 0°C according to Eq. (13)	Experimental data at approximately 0°C
Lithium (Li)	−2	+11
Sodium (Na)	−3	− 6
Potassium (K)	−5	−12
Copper (Cu)	−1.5	+ 1.4
Gold (Au)	−2	+ 1.7
Aluminum (Al)	−0.7	− 1.7

larger than the prediction of Eq. (15). (The curve for S_d in Fig. 6 is calculated from this generalization.). Experimental interest in the thermopower of semiconductors concerns topics similar to those for metals. In addition, the large magnitudes of the thermopowers of semiconductors continue to spur efforts to develop materials better suited for electrical power generation and thermoelectric cooling.

Source of S_g. Unlike the behavior of S_d, which is determined in both metals and semiconductors primarily by the properties of the charge carriers, the behavior of S_g is determined in both cases primarily by the properties of the phonons. At low temperatures, phonons scatter mainly from electrons or impurities rather than from other phonons. The increase in S_g with increasing temperature shown in Figs. 5 and 6 results from an increasing number of phonons being available to drag the electrons along. However, at higher temperatures the phonons begin to scatter more frequently from each other. At sufficiently high temperatures, phonon-phonon scattering becomes dominant, the electrons are no longer dragged along, and S_g falls off in magnitude. Interest in phonon drag is based on such questions as whether it is the sole source of the humps shown in Figs. 5 and 6, how it changes upon the addition of impurities, and how it varies in the presence of a magnetic field.

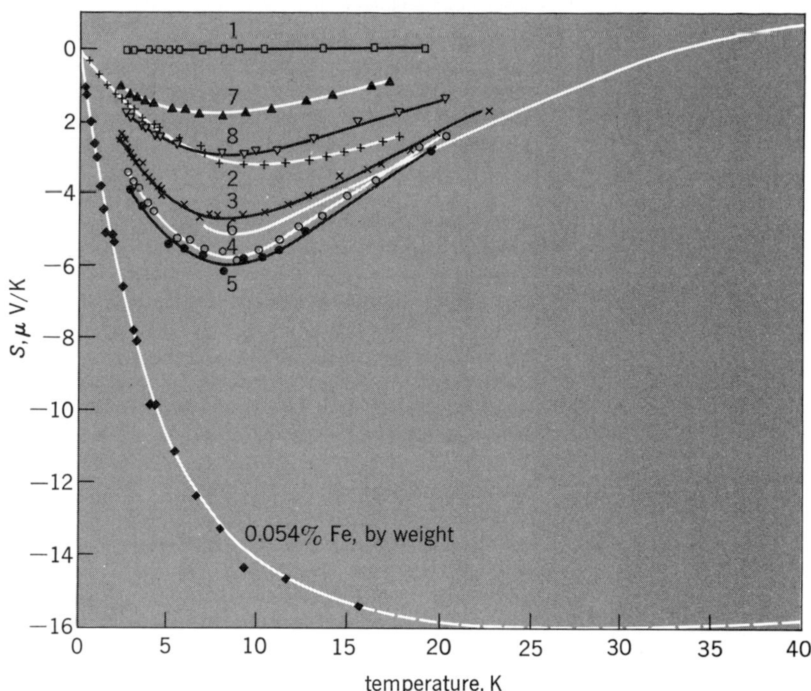

Fig. 8. The low-temperature thermopowers of various samples of copper (Cu) containing very small concentrations of iron (Fe). Specific compositions of samples 1–8 are unknown. (*From A. V. Gold et al., The thermoelectric power of pure copper, Phil. Mag., 5:765–786, 1960*)

APPLICATIONS

The most important practical application of thermoelectric phenomena is in the accurate measurement of temperature. The phenomenon involved is the Seebeck effect. Of lesser importance are the direct generation of electric power by application of heat (also involving the Seebeck effect) and thermoelectric cooling and heating (involving the Peltier effect).

A basic system suitable for all four applications is illustrated schematically in Fig. 11. Several thermocouples are connected in series to form a thermopile, a device with increased output (for power generation or cooling and heating) or sensitivity (for temperature measurement) relative to a single thermocouple. The junctions forming one end of the thermopile are all at the same low temperature T_L, and the junctions forming the other end are at the high temperature T_H. The thermopile is connected to a device D which is different for each application. For temperature measurement, the temperature T_L is fixed, for example, by means of a bath; the temperature T_H becomes the running temperature T, which is to be measured; and the device is a potentiometer for measuring the thermoelectric emf generated by the thermopile. For power generation, the temperature T_L is fixed by connection to a heat sink; the temperature T_H is fixed at a value determined by the output of the heat source and the thermal conductivity of the thermopile; and the device is whatever is to be run by the electricity which is generated. For heating or cooling, the device is a current generator which passes current through the thermopile. If the current flows in the proper direction, the junctions at T_H will heat up, and those at T_L will cool down. If T_H is fixed by connection to a heat sink, thermoelectric cooling will be provided at T_L. Alternatively, if T_L is fixed, thermoelectric heating will be provided at T_H. Such a system has the advantage that at any given location it can be converted from a cooler to a heater merely by reversing the direction of the current.

Temperature measurement. In principle, any material property which varies with temperature can serve as the basis for a thermometer. In practice, the two properties most often used for preci-

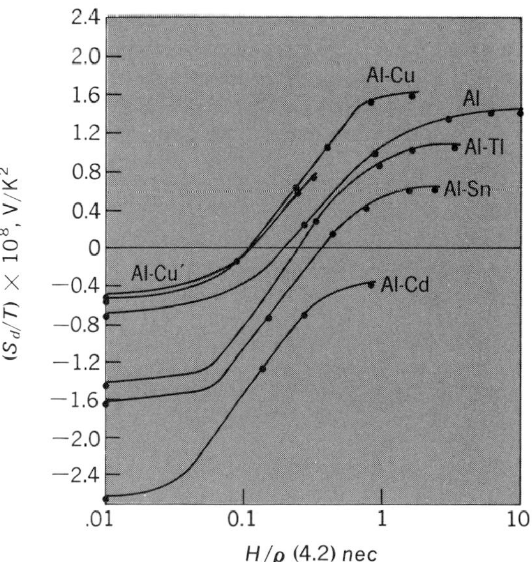

Fig. 9. The variation with magnetic field H of the low-temperature electron-diffusion thermopower S_d of aluminum (Al) and various dilute aluminum-based alloys. Sample labeled Al-Cu′ is a second sample of Al-Cu. (*From R. S. Averback C. H. Stephan, and J. Bass, Magnetic field dependence of the thermopower of dilute aluminum alloys, J. Low Temp. Phys., 12:319–346, 1973*)

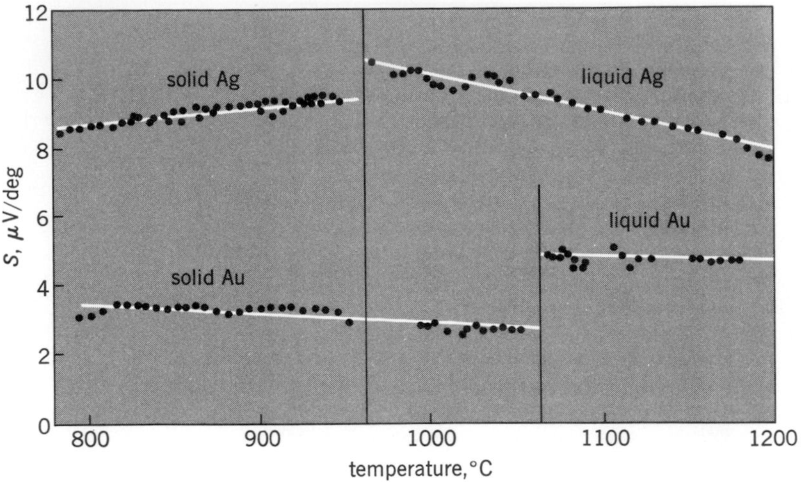

Fig. 10. The changes in the thermopowers of gold (Au) and silver (Ag) upon melting. (*From R. A. Howe and J. E. Enderby, The thermoelectric power of liquid Ag-Au, Phil. Mag., 16:467–476, 1967*)

sion thermometry are electrical resistance and thermoelectric emf. Thermocouples are widely employed to measure temperature in both scientific research and industrial processes. In the United States alone, several hundred tons of thermocouple materials are produced annually.

Construction of instruments. In spite of their smaller thermopowers, metals are usually preferred to semiconductors for precision temperature measurement because they are cheaper, are easier to fabricate into convenient forms such as thin wires, and have more reproducible thermoelectric properties. With modern potentiometric systems, standard metallic thermocouples provide temperature sensitivities adequate for most needs; small fractions of a degree Celsius are routinely obtained. If greater sensitivity is required, several thermocouples can be connected in series to form a thermopile (Fig. 11). A 10-element thermopile provides a temperature sensitivity 10 times as great as that of each of its constituent thermocouples. However, the effects of any inhomogeneities are also enhanced 10 times.

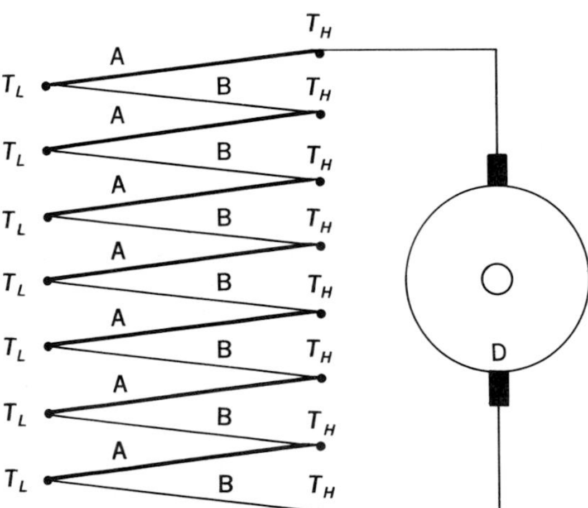

Fig. 11. Thermopile, a battery of thermocouples connected in series; *D* is a device appropriate to the particular application.

The thermocouple system standardly used to measure temperature is shown in Fig. 12. It consists of wires of three metals A, B, and C, where C is usually the metal copper. The junction between the wires of metals A and B is located at the temperature to be measured T. Each of these two wires is joined to a wire of metal C at the reference temperature T_0. The other ends of the two wires of metal C are connected to the potentiometer at room temperature T_r. Integrating the appropriate thermopowers around the circuit of Fig. 12 yields the total thermoelectric emf E in terms of the separate emfs generated by each of the four pieces of wire, as given in Eq. (16).

$$E = E_A(T_0,T_1) - E_B(T_0,T_1) + E_{C1}(T_0,T_r) - E_{C2}(T_0,T_r) \quad (16)$$

If the two wires C_1 and C_2 have identical thermoelectric characteristics, the last two terms in this expression cancel, and, with the use of Eq. (5), Eq.

$$E = E_{AB}(T_0,T_1) \quad (17)$$

(17) results. That is, two matched pieces of metal C produce no contribution to the thermoelectric emf of the circuit shown in Fig. 12, provided their ends are maintained at exactly the same two temperatures. This means that it is not necessary to use either of the sometimes expensive metals making up the thermocouple to go from the reference-temperature bath to the potentiometer. This portion of the circuit can be constructed of any uniform, homogeneous metal. Copper is often used because it is inexpensive, is available in adequate purity to ensure uniform, homogeneous samples when handled with care, can be obtained in a wide variety of wire diameters, and can be either spot-welded or soldered to the ends of the thermocouple wires. Special low-thermal emf alloys are available for making solder connections in thermocouple circuits.

Choice of materials. Characteristics which make a thermocouple suitable as a general-purpose thermometer include adequate sensitivity over a wide temperature range, stability against physical and chemical change under various conditions of use and over extended periods of time, availability in a selection of wire diameters, and moderate cost. No single pair of thermocouple materials satisfies all needs. Platinum versus platinum–10% rhodium can be used up to 1700°C. A combination of the two alloys chromel versus alumel gives greater sensitivity and an emf which is very closely linear with temperature, but this thermocouple cannot be used to so high a temperature. A combination of copper versus the alloy constantan also has high sensitivity above room temperature, and maintains adequate sensitivity down to as low as 15 K. For temperatures of 4 K or lower, special gold-cobalt alloys versus copper or gold-iron alloys versus chromel are used.

Thermocouple tables. To use a thermocouple composed of metals A and B as a thermometer, it is necessary to know how $E_{AB}(T_0,T)$ varies with temperature T for some reference temperature T_0. According to Eq. (6), $E_{AB}(T_0,T_1)$ can be determined for any two temperatures T_0 and T_1 if both

$S_A(T)$ and $S_B(T)$ are known for all temperatures between T_0 and T_1. Once $S_A(T)$ and $S_B(T)$ are known, it is possible to construct a table of values for $E_{AB}(T_0,T)$ using any arbitrary reference temperature T_0. Such tables are available for the thermocouples mentioned above, and for some others as well, usually with a reference temperature of 0°C. A table of $E_{AB}(T_0,T)$ for one reference temperature T_0 can be converted into a table for any other reference temperature T_1 merely by subtracting a constant value $E_{AB}(T_0,T_1)$ from each entry in the table to give Eq. (18). Here $E_{AB}(T_0,T_1)$ is a positive

$$E_{AB}(T_1,T) = E_{AB}(T_0,T) = E_{AB}(T_0,T_1) \quad (18)$$

quantity when T_1 is greater than T_0 and when $S_{AB}(T)$ is positive between T_0 and T_1.

Other uses. Thermoelectric systems made from even the best available materials have the disadvantages of relatively low efficiencies and concomitant high cost per unit of output. Their use in power generation, heating, and cooling has therefore been largely restricted to situations in which these disadvantages are outweighed by such advantages as small size, low maintenance due to lack of moving parts, quiet performance, light weight, and long life.

Figure of merit. A measure of the utility of a given thermoelectric material for power generation, cooling, or heating at a given temperature T is provided by a dimensionless parameter zT, where z is called the figure of merit. The dimensionless parameter zT is given by Eq. (19), where S is the

$$zT = \frac{S^2\sigma T}{\kappa} \quad (19)$$

thermopower of the material, σ is its electrical conductivity, and κ is its thermal conductivity. The largest values for zT are attained in semimetals and highly doped semiconductors, which are therefore the materials normally used in practical thermoelectric devices. As illustrated in Fig. 13, for most materials z varies substantially with temperature. The best available thermoelectric materials, such as lead-telluride (Pb-Te) and bismuth-telluride (Bi-Te), have values of z as large as 2 to $4 \times 10^{-3}K^{-1}$ at their best temperatures. Unfortunately, however, z falls off substantially at both higher and lower temperatures. Combining these materials into thermocouples therefore results in values of z which average less than $1 \times 10^{-3}K^{-1}$ over a temperature range sufficiently wide to be useful. Such values of z yield conversion efficiencies of only a few percent. A commercially competitive efficiency of about 30% would require a constant value of $z = 5 \times 10^{-3}K^{-1}$ over the temperature range 300–1000 K. *See* THERMOELECTRIC POWER GENERATOR.

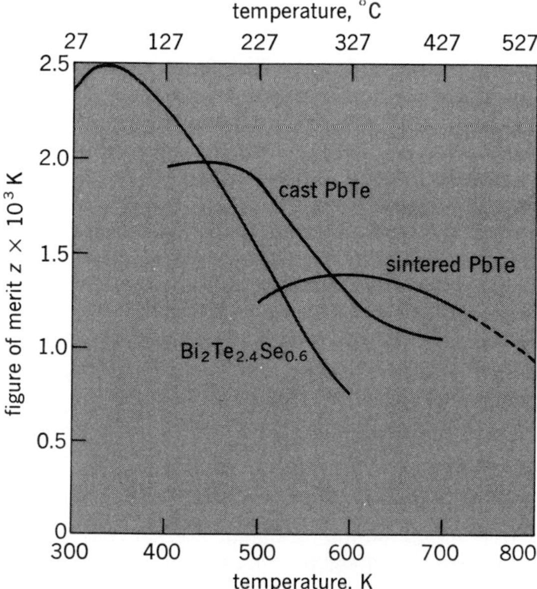

Fig. 13. Temperature variation of the figure of merit for some *n*-type semiconductors. (*From R. R. Heikes and R. W. Ure, Thermoelectricity: Science and Engineering, p. 538, Interscience Publishers, 1961*)

Just as the figures of merit for single materials vary with temperature, so do the figures of merit for thermocouples formed from two such materials. This means that one thermocouple can be better than another at one temperature but less effective at a second temperature. To take maximum advantage of the different properties of different couples, thermocouples are often cascaded as shown in Fig. 14. Cascading produces power generation in stages, the higher temperature of each stage being determined by the heat rejected from the stage above. Thus, in Fig. 14 the highest and lowest temperatures T_4 and T_1 are fixed by connection to external reservoirs, whereas the middle temperatures T_3 and T_2 are determined by the properties of the materials. By cascading, a series of thermocouples can be used simultaneously in the temperature ranges where their figures of merit are highest. Cascaded thermocouple systems have achieved conversion efficiencies as high as 10–15%.

The quantity of importance in power generation is the figure of merit of a thermocouple rather than the separate figures of merit of its constituents. Although at least one constituent should have a high figure of merit, two constituents with high figures of merit do not necessarily guarantee that the figure of merit of the thermocouple will be high. For example, if the thermopowers of the two constituents are the same, the figure of merit of the couple will be zero.

Thermoelectric generators. A thermoelectric generator requires a heat source and a thermocouple. Kerosine lamps and firewood have been used as heat sources in producing a few watts of electricity in locations where electricity was otherwise unavailable. In the future, sunlight may also be used. Radioactive sources, especially strontium-90, have provided the heat to activate small, rugged thermoelectric batteries for use in lighthouses, in navigation buoys, at isolated weather stations or oil platforms, in spaceships, and in heart pacemakers. Small nuclear batteries have been operating pacemakers implanted in humans since 1970. One such battery, powered by Pu^{238} and using a bismuth telluride thermopile module, supplies a few tenths of a volt over a design lifetime of more than 10 years. Nuclear-powered batteries for medical use must be designed to remain intact following the maximum credible accident. Capabilities such as retention of integrity after crushing by 1 ton, or impact at 50 m/s, or saltwater corrosion for centuries, or cremation at temperatures up to 1300°C for half an

THERMOELECTRICITY

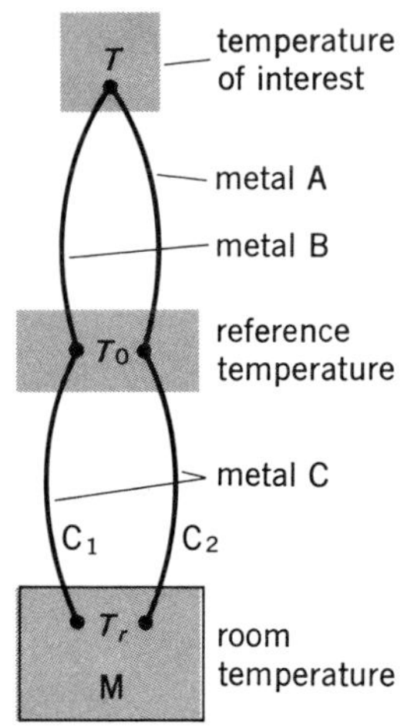

Fig. 12. The thermocouple system standardly used to measure temperature; M is a measuring device, usually a potentiometer, which is at room temperature.

THERMOELECTRICITY

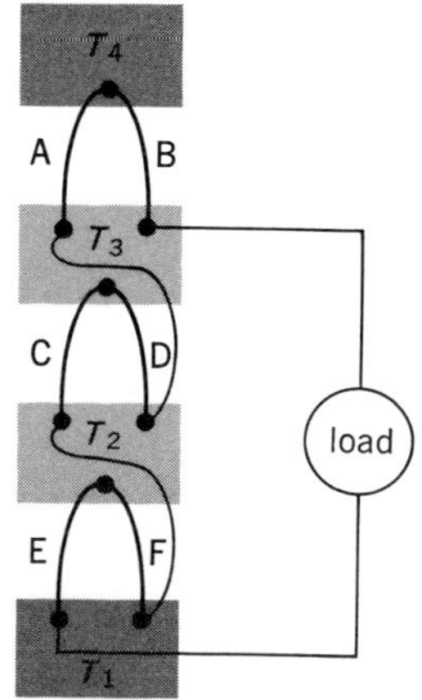

Fig. 14. Three-level cascade consisting of three different thermocouples (A versus B, C versus D, and E versus F) at four Temperatures (*T*).

hour are required. Investigation of the feasibility of thermoelectric generation using the copious heat generated by nuclear reactors has also been undertaken. Here, one major problem lies in the development of efficient thermoelectric materials capable of operating for a long time at the high temperatures which are encountered. *See* NUCLEAR BATTERY; NUCLEAR POWER.

Peltier cooling. With available materials, thermoelectric refrigerators suitable for use in homes are more expensive and less efficient than standard vapor-compression-cycle refrigerators. Their use is thus largely restricted to situations in which lower maintenance, increased life, or quiet performance are essential, or in situations (such as in space vehicles or artificial satellites) in which the compressor type of refrigerator is impractical. A number are in use in hotels and other large facilities. A typical unit having about 50 liters' capacity requires a dc power input of 40 W, has a refrigerative capacity of 20 kcal/hr (23 W), and a cooling time of 4–5 hr.

For lower temperatures, the proper choice of thermoelectric materials and the use of cascading can result in a reduction in temperature at the coldest junctions of as much as 150°C. Temperature drops of 100°C have been obtained in single crystals of the semimetal bismuth through use of the thermomagnetic Ettingshausen effect.

Small cooling units with capacities of 10 W or less have been developed for miscellaneous applications such as cold traps for vacuum systems, cooling controls for thermocouple reference junctions, cooling devices for scientific equipment such as infrared detectors, and cold stages on microscopes or on microtomes used for sectioning cooled tissues. However, the real commercial success of thermoelectric refrigeration appears to await development of thermocouple materials with higher figures of merit.

Thermoelectric heating. As noted earlier, a thermoelectric heater is nothing more than a thermoelectric refrigerator with the current reversed. No large heaters have been marketed. However, various small household convenience devices have been developed, such as a baby-bottle cooler-warmer which cools the bottle until just before feeding time and then automatically switches to a heating cycle to warm it, and a thermoelectric hostess cart. *See* ELECTRICITY. [JACK BASS]

Bibliography: American Institute of Physics, *Temperature, Its Measurement and Control in Science and Industry*, vol. 1, 1941, and vol. 2, 1955; R. D. Barnard, *Thermoelectricity in Metals and Alloys*, 1972; F. J. Blatt et al., *Thermoelectric Power of Metals*, 1976; F. J. Blatt and P. A. Schroeder (eds.), *Thermoelectricity in Metallic Conductors*, 1978; T. C. Harman and H. M. Honig, *Thermoelectric and Thermomagnetic Effects and Applications*, 1967; R. R. Heikes and R. W. Ure, Jr., *Thermoelectricity: Science and Engineering*, 1961; D. K. C. MacDonald, *Thermoelectricity: An Introduction to the Principles*, 1962; A. C. Smith, J. F. Janak, and R. B. Adler, *Electronic Conduction in Solids*, 1967.

Thermonuclear reaction

A nuclear fusion reaction which occurs between various nuclei of the light elements when they are constituents of a gas at very high temperatures. Thermonuclear reactions, which are the source of energy generation in the Sun and the stable stars, are utilized in the fusion bomb. *See* NUCLEAR FUSION.

Thermonuclear reactions occur most readily between isotopes of hydrogen (deuterium and tritium) and less readily among a few other nuclei of higher atomic number. At the temperatures and densities required to produce an appreciable rate of thermonuclear reactions, all matter is completely ionized; that is, it exists only in the plasma state. Thermonuclear fusion reactions may then occur within such an ionized gas when the agitation energy of the stripped nuclei is sufficient to overcome their mutual electrostatic repulsions, allowing the colliding nuclei to approach each other closely enough to react. For this reason, reactions tend to occur much more readily between energy-rich nuclei of low atomic number (small charge) and particularly between those nuclei of the hot gas which have the greatest relative kinetic energy. This latter fact leads to the result that, at the lower fringe of temperatures where thermonuclear reactions may take place, the rate of reactions varies exceedingly rapidly with temperature.

The reaction rate may be calculated as follows: Consider a hot gas composed of a mixture of two energy-rich nuclei, for example, tritons and deuterons. The rate of reactions will be proportional to the rate of mutual collisions between the nuclei. This will in turn be proportional to the product of their individual particle densities. It will also be proportional to their mutual reaction cross section σ and relative velocity v. Thus Eq. (1) gives the

$$R_{12} = n_1 n_2 \langle \sigma v \rangle_{12} \text{ reactions/(cm}^3\text{)(sec)} \quad (1)$$

rate of reaction. The quantity $\langle \sigma v \rangle_{12}$ indicates an average value of σ and v obtained by integration of these quantities over the velocity distribution of the nuclei (usually assumed to be maxwellian). Since the total density $n = n_1 + n_2$, then if the relative proportions of n_1 and n_2 are maintained, R_{12} varies as the square of the total nuclear particle density.

The thermonuclear energy release per unit volume is proportional to the reaction rate and the energy release per reaction, as in Eq. (2).

$$P_{12} = R_{12} W_{12} \text{ ergs/ (cm}^3\text{) (sec)} \quad (2)$$

If this energy release, on the average, exceeds the energy losses from the system, the reaction can become self-perpetuating. *See* NUCLEAR REACTION; PINCH EFFECT.

[RICHARD F. POST]

Bibliography: *See* NUCLEAR FUSION.

Thermostat

An instrument which directly or indirectly controls one or more sources of heating and cooling to maintain a desired temperature. To perform this function a thermostat must have a sensing element and a transducer. The sensing element measures changes in the temperature and produces a desired effect on the transducer. The transducer converts the effect produced by the sensing element into a suitable control of the device or devices which affect the temperature.

The most commonly used principles for sensing changes in temperature are (1) unequal rate of expansion of two dissimilar metals bonded together (bimetals), (2) unequal expansion of two dissimilar metals (rod and tube), (3) liquid expansion (sealed diaphragm and remote bulb or sealed bellows with or without a remote bulb), (4) saturation pressure of a liquid-vapor system (bellows), and (5) temperature-sensitive resistance element.

The most commonly used transducers are (1) switches that make or break an electric circuit, (2) potentiometer with a wiper that is moved by the sensing element, (3) electronic amplifier, and (4) pneumatic actuator.

The most common thermostat application is for room temperature control. Figure 1 shows a typical on-off heating-cooling room thermostat. In a typical application the thermostat controls a gas valve, oil burner control, electric heat control, cooling compressor control, or damper actuator.

To reduce room temperature swings, high-performance on-off thermostats commonly include a means for heat anticipation. The temperature swing becomes excessive if thermostats without heat anticipation are used because of the switch differential (the temperature change required to go from the break to the make of the switch), the time lag of the sensing element (due to the mass of the thermostat) in sensing a change in room temperature, and the inability of the heating system to respond immediately to a signal from the thermostat.

To reduce this swing, a heater element (heat anticipator) is energized during the on period. This causes the thermostat to break prematurely. Figure 2 shows a comparison of the room temperature variations when a thermostat with and without heat anticipation is used.

The same anticipation action can be obtained on cooling thermostats by energizing a heater (cool anticipator) during the off period of the thermostat. Room thermostats may be used to provide a variety of control functions, such as heat only; heat-cool; day-night, in which the night temperature is controlled at a lower level; and multistage, in which there may be one or more stages of heating, or one or more stages of cooling, or a combination of heating and cooling stages.

Thermostats are also used extensively in safety and limit application. Thermostats are generally of

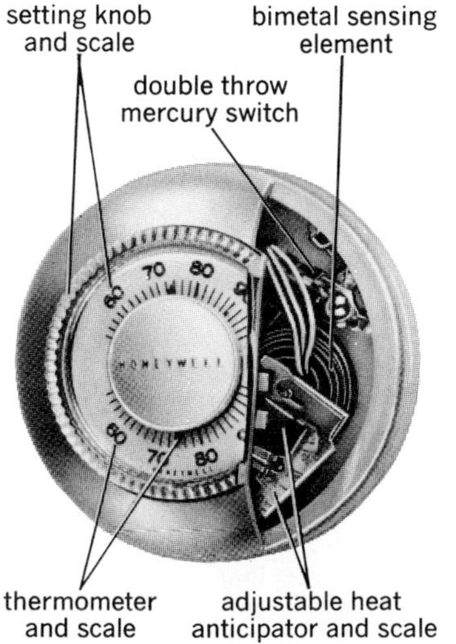

Fig. 1. Typical heat-cool thermostat. (*Honeywell Inc.*)

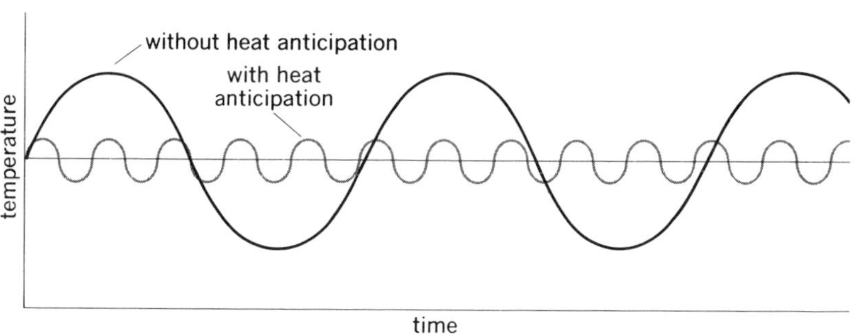

Fig. 2. Comparison of temperature variations using a timed on-off thermostat with and without heat anticipation.

the following types: insertion types that are mounted on ducts with the sensing element extending into a duct, immersion types that control a liquid in a pipe or tank with the sensing element extending into the liquid, and surface types in which the sensing element is mounted on a pipe or similar surface. *See* COMFORT HEATING; OIL BURNER.

[NATHANIAL ROBBINS, JR.]

Bibliography: American Society of Heating, Refrigeration, and Air-Conditioning Engineers, *ASHRAE Handbook and Product Directory: Systems*, 1976; J. E. Haines, *Automatic Control of Heating and Air Conditioning*, 2d ed., 1961; V. C. Miles, *Thermostatic Control*, 2d ed., 1974.

Thorium

A chemical element, Th, atomic number 90. It was discovered by J. J. Berzelius in 1828. However, little use was found for thorium before the development of the incandescent gas mantle by C. A. von Welsbach in 1885. Several thousand

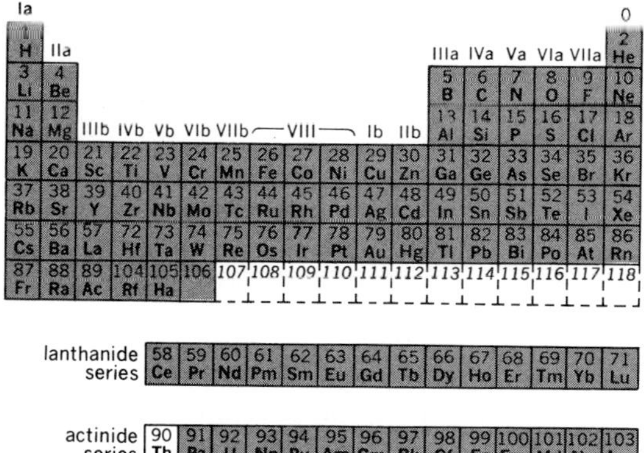

pounds of thorium oxide still go into the annual production of these mantles. For many years thorium oxide has been incorporated in tungsten metal, which is used for electric light filaments; recently, small amounts of the oxide have been found to be useful in other metals and alloys. The oxide is employed in catalysts for the promotion of certain organic chemical reactions. Thorium oxide has special uses as a high-temperature ceramic material. The metal or its oxide is employed in some electronic tubes, photocells, and special welding electrodes. The metal can serve as a getter in vacuum systems and in gas purification, and it is also used as a scavenger in some metals.

Because of its high density, chemical reactivity, mediocre mechanical properties, and relatively high cost, thorium metal has no market value as a structural material. However, many alloys containing thorium metal have been studied in some detail and thorium does have important applications as an alloying agent in some structural metals. Perhaps the major use for thorium metal, outside the nuclear field, is in magnesium technology. Approximately 3% thorium, added as an alloying ingredient, imparts to magnesium metal high-strength properties and creep resistance at elevated temperatures. The magnesium alloys containing thorium, because of their light weight and desirable strength properties, are being used in aircraft engines and in airframe construction.

Thorium can be converted in a nuclear reactor to uranium-233, an atomic fuel. The system of thorium and uranium-233 gives promise of complete utilization of all thorium in the production of atomic power. The energy available from the world's supply of thorium has been estimated as greater than the energy available from all of the world's uranium, coal, and oil combined.

Natural occurrence. Monazite, the most common and commercially most important thorium-bearing mineral, is widely distributed in nature. Important deposits occur along the shores of India, Brazil, and Ceylon. Other extensive deposits of monazite are found in South Africa, the Soviet Union, Scandinavia, and Australia. Sources in the United States include deposits in Florida, Idaho, and the Carolinas. Monazite is chiefly obtained as a sand, which is separated from other sands by physical or mechanical means, following dredging operations. The monazite sand concentrate is essentially an orthophosphate of rare-earth elements, and generally contains 3–10% ThO_2. Other thorium-bearing minerals of lesser importance include thorite, thorianite, and uranothorite.

Metallurgical extraction. Processes for thorium recovery generally start by digestion of the monazite sand with either hot concentrated sulfuric acid or hot concentrated caustic. Subsequent chemical treatments, varying greatly even with the same initial treatment, yield a concentrate of impure thorium. This impure concentrate may be further treated by a liquid-liquid extraction process to yield high-purity thorium. For a system consisting of water, tributyl phosphate, nitric acid, thorium, and the associated impurities, an extractor can be set up to remove the thorium with the water-immiscible tributyl phosphate phase, while the impurities are carried away in the aqueous phase. Generally, the purified thorium is back extracted to an aqueous solution and either crystallized from solution as the nitrate or precipitated as the oxalate. From these pure salts the oxide or other compounds of thorium can be prepared.

Because thorium is quite reactive, some difficulty is experienced in preparing thorium metal. Only by electrolysis or by treatment with elements high in the electromotive force series (the alkali and alkaline-earth metals), has good-quality thorium metal been satisfactorily prepared directly from its compounds.

The calcium reduction of ThO_2 has been widely used for many years to prepare thorium metal. In this process, granular calcium metal is mixed with thorium oxide and charged into a lined iron crucible which is then filled with an inert gas and heated to almost 1000°C to form thorium metal powder and calcium oxide. After cooling to room temperature, the thorium powder is recovered by leaching and then drying. Powder metallurgy techniques are employed to obtain massive metal.

The electrodeposition of thorium from a bath, consisting of thorium chlorides or fluorides dissolved in fused alkali halides, yields granular thorium which may be pressed and sintered to give massive pieces of ductile metal.

Large-scale production of thorium metal has been carried out by a bomb process. The charge, consisting of a mixture of thorium tetrafluoride, granular calcium metal, and zinc chloride, is placed in a refractory-lined vessel that is closed by a lid. The charged bomb is placed in a furnace held at about 650°C, where, after several minutes, the charge ignites spontaneously and the resulting reaction yields a slag of calcium fluoride and calcium chloride and an alloy of thorium and zinc. The temperature reached by the reaction in the charge is sufficient to melt the products, and the thorium-rich alloy collects as a molten pool under the liquid slag. The bomb is allowed to cool, and then the solid piece of thorium alloy is removed and cleaned of adhering slag. Next, the zinc is removed by heating the alloy in a vacuum at a temperature of 1100°C, leaving the thorium metal as a sponge. Solid ingots of thorium metal are prepared by vacuum-induction-melting the sponge in a crucible or by shaping the sponge in the form of bars and melting these by consumable electrode arc melting. Good-quality thorium metal can be readily worked to shape by standard methods of fabrication.

Properties. Thorium has an atomic weight of 232. The metal has a density of 11.7 g/cm^3. Good-quality thorium metal is relatively soft and ductile. It can be shaped readily by any of the ordinary metal-forming operations. It must be protected, however, to prevent oxidation in treatments involving high temperatures. The massive metal is silvery in color, but it tarnishes on long exposure to the atmosphere; finely divided thorium has a tendency to be pyrophoric in air.

The atoms of thorium in the metal are arranged in a face-centered cubic system at all temperatures below 1400°C. On heating, the atoms rearrange at this temperature into a body-centered cubic pattern which is stable up to the melting temperature. However, the temperature at which pure thorium melts is not known with certainty; it is thought to be not far from 1750°C.

Thorium is a member of the actinide series of elements, which includes protactinium, uranium, and the synthetic transuranic elements. It is radioactive with a half-life of about 14×10^9 years. It is the first member of the radioactive decay series which in a chain of 10 successive disintegrations (α and β combined) finally terminates as lead-208.

All of the nonmetallic elements, except the rare gases, form binary compounds with thorium. Binary intermetallic compounds have been reported for thorium with beryllium, magnesium, boron, aluminum, and silicon, and with all of the metallic elements in the three long periods of the periodic chart in groups positioned to the right of group VIb. A number of the intermetallic compounds of thorium, especially those with copper, silver, and gold, are quite pyrophoric. A study of the binary alloy systems formed by thorium metal and metals of the IIIb, IVb, Vb, and VIb groups, including the rare earths, shows no evidence of intermetallic compound formation.

Principal compounds. Thorium does not impart any visible spectrum colors to its inorganic compounds or their solutions. With minor exceptions, thorium exhibits a valence of 4+ in all of its salts. Chemically, it has some resemblance to zirconium and hafnium. The most common soluble compound of thorium is the nitrate which, as generally prepared, appears to have the formula $Th(NO_3)_4 \cdot 4H_2O$.

The common oxide of thorium is ThO_2, thoria, which can be obtained by thermal decomposition of the nitrate, hydroxide, oxalate, or other compounds of thorium. A peroxide of thorium, Th_2O_7 with water of hydration, and the hydroxide, $Th(OH)_4$, can be precipitated from solutions of thorium salts.

The halogens form a variety of salts with thorium. Thorium tetrahalides of the general formula ThX_4 (X = halogen), anhydrous and with varying degrees of hydration, are known. $ThOX_2$ and $Th(OH)_2X_2$ with and without water of hydration are known. The halides of thorium also tend to form double salts with other halides, such as those of the alkali metals.

Thorium sulfate can be obtained in the anhydrous form or with 2, 4, 6, 8, or 9 molecules of water of crystallization. A somewhat insoluble basic sulfate forms when a dilute water solution of thorium sulfate is boiled. Double sulfates of thorium with alkali metals or with ammonium are known. The hydrosulfate and the thiosulfate of thorium are water-insoluble compounds.

Thorium carbonates, phosphates, iodates, chlorates, chromates, molybdates, and other inorganic salts of thorium are well known. Thorium also forms salts with many organic acids, of which the water-insoluble oxalate, $Th(C_2O_4)_2 \cdot 6H_2O$, is important in preparing pure compounds of thorium.

Analytical methods. Thorium in small quantities in rocks and other natural sources can be estimated by a study of the radioactivity of the sample. The chemical analysis of materials for thorium, however, generally involves getting the thorium into solution with sulfuric acid. The thorium must then be isolated from other interfering ions, and this may involve a separation by ion exchange or solvent extraction or a precipitation as iodide or pyrophosphate. The final determination of thorium is made by gravimetric, titrimetric, or colorimetric means.

The gravimetric method generally depends on subsequent formation of the oxalate, which is calcined to ThO_2 and weighed. A titrimetric determination can be made on a thorium solution by titration with EDTA, an organic chelating agent. The end point of the titration may be detected visually by using an indicator such as xylenol orange. In one colorimetric method, a complex organic compound referred to as Thorin gives, with thorium ion, a color that can be measured to indicate the quantity of thorium present.

[HARLEY A. WILHELM]

Bibliography: O. N. Carlson and E. R. Stevens, *A Compilation of Thorium Binary Phase Diagrams*, USAEC Rep. no. IS-1752, 1968; C. J. Rodden (ed.), *Analysis of Essential Nuclear Reactor Materials*, 1964; G. T. Seaborg and J. J. Katz (eds.), *The Actinide Elements*, NNES, Div. IV, vol. 14A, 1954; G. T. Seaborg and L. I. Katzin (eds.), *Production and Separation of U-233*, USAEC Rep. no. TID-5222, 1951; H. A. Wilhelm (ed.), *The Metal Thorium*, 1958.

Tidal power

Tidal-electric power is obtained by utilizing the recurring rise and fall of coastal waters in response to the gravitational forces of the Sun and the Moon. Marginal marine basins are enclosed with dams, making it possible to create differences in the water level between the ocean and the basins. The oscillatory flow of water filling or emptying the basins is used to drive hydraulic turbines which propel electric generators.

Large amounts of electric power could be developed in the world's coastal regions having tides of sufficient range, although even if fully developed this would amount to only a small percentage of the world's potential water (hydroelectric) power. The estimated annual energy production at known tidal sites in the world amounts to about 1.2×10^{12} kWh. This amount is equivalent to approximately two-thirds of the annual production of electrical energy in the United States. Nevertheless, tidal-electric power may become locally important, particularly because it produces no air or thermal pollution, consumes no exhaustible resource, and produces relatively minor impacts on the environment.

The use of ocean tides for power purposes dates back to the tidal mills in Europe during the Middle Ages and to those in America during colonial times. Two tidal developments producing electric power are now in operation. The Rance development in northwestern France was completed in 1967. It has an installed capacity of 240,000 kW in 24 units and is capable of producing about 500×10^6 kWh annually. The experimental 400-kW Kislayaguba development, located north of Murmansk in the Soviet Union, was completed in 1969.

Tidal range requirements. Tidal range is measured as the difference in level between the successive high and low waters. Although there are variations at certain locations in the intervals between successive high tides, at most places the tides reach the highest levels at intervals of about 12 hr 25 min. The tidal ranges vary from day to day. The highest tides, known as spring tides, occur twice monthly near the time of the new moon and the full moon when the Sun and Moon are in line with the Earth. The lowest tides, known as neap tides, occur midway between the spring tides when the Sun and Moon are at right angles with the Earth. The highest spring tides occur near the time of the equinoxes in the spring and fall of the year. Except for variations caused by meteorological changes, the tides are predictable and follow similar patterns from year to year.

Large tidal ranges occur when the oscillation of the ocean tides is amplified by relatively shallow bays, inlets, or estuaries. There is a limited number of locations where the tidal ranges are sufficiently large to be considered favorable for power development. The largest tidal ranges in the world, reaching a maximum of over 50 ft (1 ft = 0.3 m), are

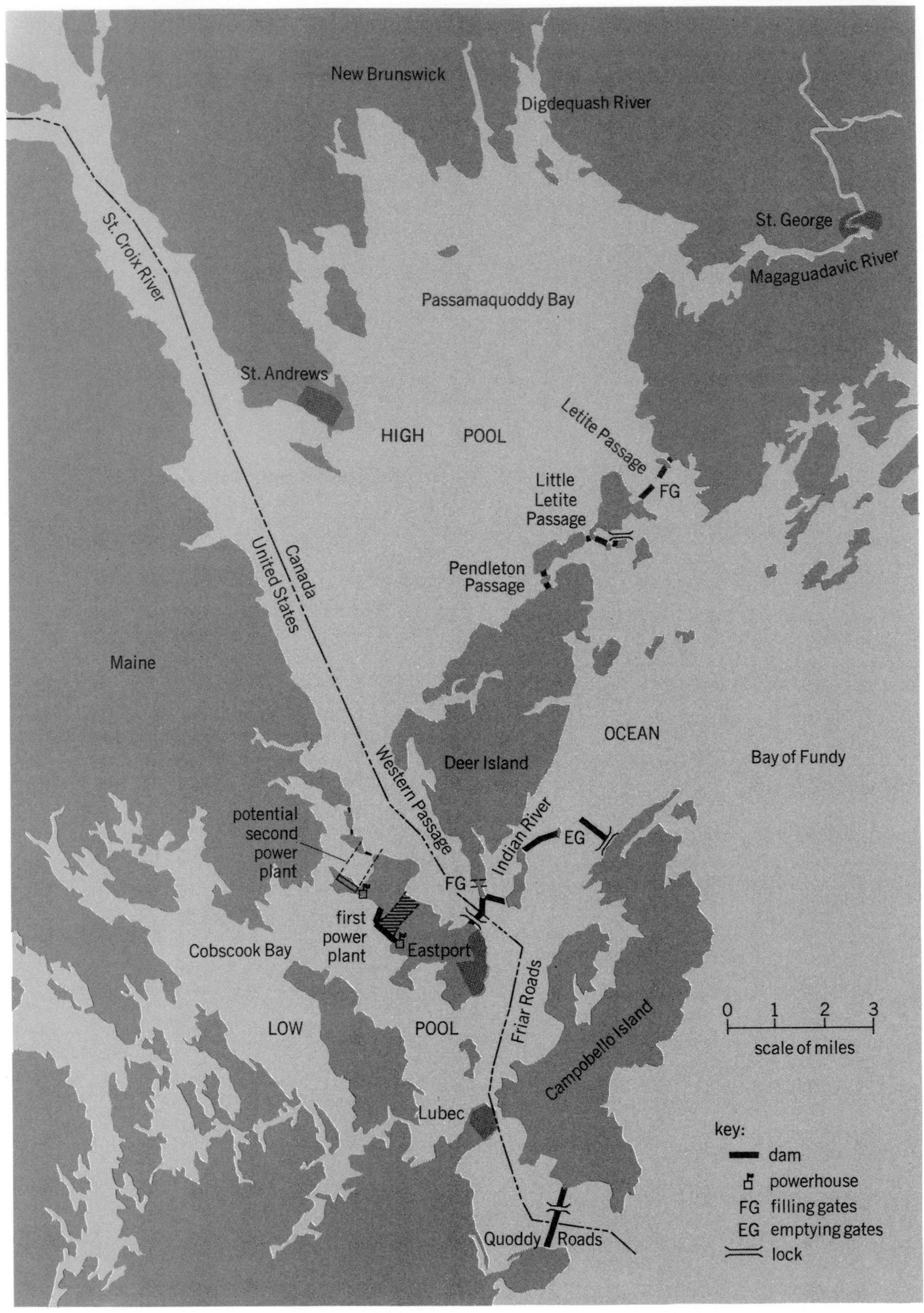

Fig. 1. Tidal power project, selected plan, general arrangement. 1 mi = 1.6 km. (*From S. L. Udall, The International Passamaquoddy Tidal Power Project and Upper Saint John River Hydroelectric Power Development: A report to President John F. Kennedy, July 1963*)

said to occur in the Bay of Fundy in Canada. Other locations with large maximum tidal ranges are the Severn Estuary in Great Britain, 45 ft; the Rance Estuary in France, 40 ft; Cook Inlet in Alaska, 33 ft; and the Gulf of California in Mexico, 30 ft. Large tidal ranges also occur at locations in Argentina, India, Korea, Australia, and on the northern coast of the Soviet Union.

Choosing the site. In order to be favorable for power development, a site not only must have a large tidal range, but it must also be capable of storing large amounts of water for energy production with minimum dam and dike construction. The site should also be in reasonably close proximity to population centers for the purpose of minimizing transmission requirements. The arrangement of facilities depends on the site conditions, and a number of types of development have been proposed. *See* DAM.

Single-pool scheme. The simplest type of development would be a single-pool scheme, consisting of dams and dikes to separate the pool from the ocean, sluiceways to fill the pool, and a power plant containing hydroelectric generating units. Water would be stored in the pool during high tide and subsequently released through the power plant during low tide. Additional energy could be produced if generation took place during both the filling and emptying of the pool. Since the power would be produced in accordance with the lunar cycle of the tides, much of the output would occur during periods of low power demands. Also, there would be no generation during portions of the operating cycle. Thus, there woud be no firm power production.

A variation of the single-pool scheme, adopted for the existing Rance development, involves the installation of units capable of generating or pumping in either direction. In addition to generating during both the filling and emptying phases, the plant could use energy generated elsewhere to increase the available head by pumping into or out of the pool. Although power could not be produced continuously, a certain amount of dependable power could be provided by pumping into the pool and holding the water for release during peak-load periods. However, such operation would reduce the total amount of energy generated.

Two-pool scheme. A two-pool development would include a high pool which would be filled through sluices or filling gates during high tides, a low pool which would discharge through other sluices or gates during low tides, and a power plant utilizing the head between the two pools to generate power. Some power could be produced continuously. This is the type of development proposed for the Passamaquoddy site on the United States–Canadian border (Fig. 1). Additional power could be provided for peak-load use, at the expense of some loss in energy production, by holding the high pool near the high-tide level and the low pool near the low-tide level until the power is required. The installation of reversible pumping-generating units would permit pumping from the low pool to the high pool during off-peak hours to provide greater head and power production during peak-load periods.

Generating units. Three types of units are considered suitable for tidal power installations. The Rance development contains bulb-type units (Fig. 2*a*), each of which consists of a horizontal-shaft turbine connected to a generator, with both housed in a metal bulb-shaped casing. Each unit is supported on struts in the horizontal water passage. These units can be operated for pumping and generation in both directions. The tube-type unit (Fig. 2*b*), considered for installation at the proposed Passamaquoddy development, consists of an axial-flow turbine connected by an inclined shaft to a generator located outside of the water passage. By means of a gear arrangement, the speed of the generator is increased and its size is reduced. The units can be used for both generating and pumping. Designs also have been prepared for a straight-flow turbine unit (Fig. 2*c*) in which the rotor of the generator would be mounted as a rim on the tips of the turbine blades and would

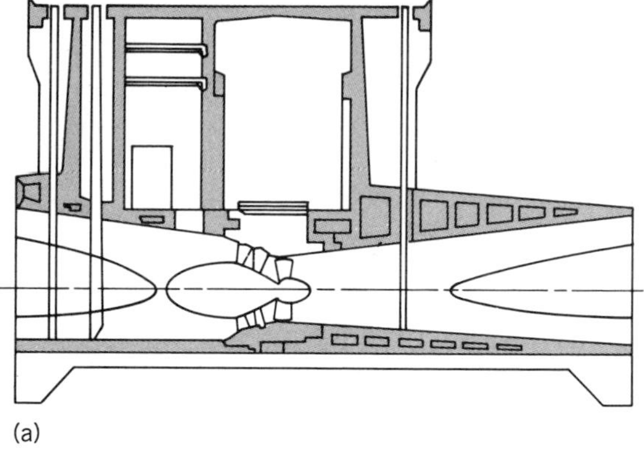

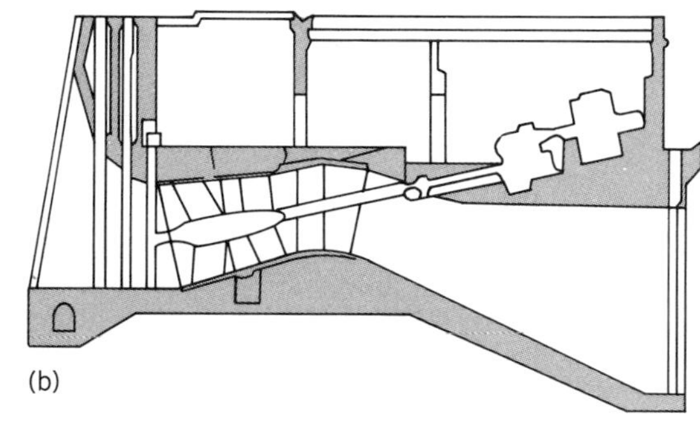

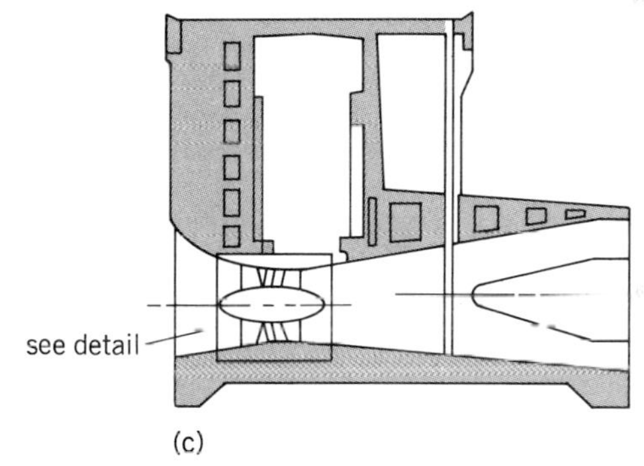

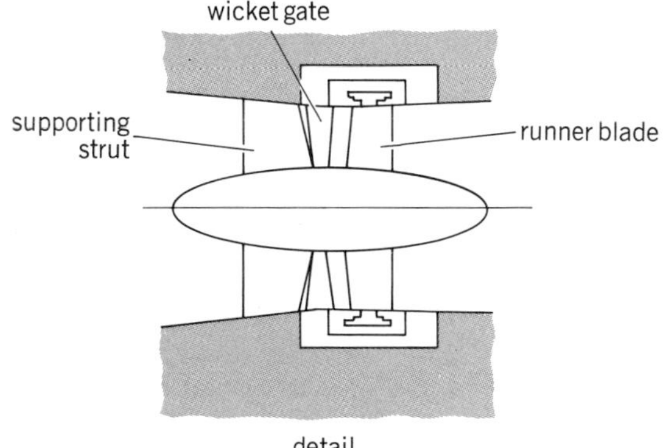

Fig. 2. General arrangements for generating units. (*a*) Bulb turbine. (*b*) Tube turbine. (*c*) Straight-flow (rim-type) turbine. *(From R. H. Clark, Fundy Tidal Power, Energy Int., 9(11):21–26, November 1972)*

turn in a sealed recess in the water conduit. The stator would be located in the dry area surrounding the rotor recess.

Major elements of cost in developing tidal power include the construction of dams and powerhouses. Difficult and costly construction may be involved in building dams in deep water with swift currents produced by the tides. An innovative scheme was used in building the powerhouse for the Kislayaguba development. The plant was constructed in the dry, floated to the site, and sunk onto a prepared foundation.

Future developments. Although only two tidal power plants are now in operation, studies and plans have been made for developing other sites. In the Soviet Union, plans have been announced for building a 6×10^6 kW plant at Mezenskaya on the White Sea. A number of plans have been prepared for developing power in the Severn Estuary. Detailed studies have been made for several tidal power developments near the head of the Bay of Fundy. The most favorable site would have an installed capacity of over 2×10^6 kW.

The Passamaquoddy site on the Bay of Fundy, where the maximum tidal range is 26 ft, has been studied at various times over half a century. Detailed plans completed in 1961 by the International Joint Commission, United States and Canada, as subsequently modified by the U.S. Department of the Interior, provide for an installation of 500,000 kW, capable of an annual generation of approximately 1.9×10^9 kWh. The studies to date have not found the development of this site to be economically justified. *See* ELECTRIC POWER GENERATION; HYDROELECTRIC POWER.

[GEORGE G. ADKINS]

Bibliography: R. H. Clark, *Energy Int.*, 9(11): 21–26, November 1972; J. Cotillon *Water Power*, 26(10):314–322, October 1974; T. K. Gray and O. J. Gashus (eds.), *Tidal Power*, 1972; E. Jeffs, *Energy Int.*, 11(12):19–21, December 1974; U.S. Department of Commerce, National Oceanic and Atmospheric Administration, National Ocean Survey, *Tide Tables, East Coast, North and South America (including Greenland)*, and *Tide Tables, West Coast of North and South America (including the Hawaiian Islands)*; E. M. Wilson, *Underwater J.*, 5(4):175–186, August 1973.

Transformer

An electrical component used to transfer electric energy from one alternating-current (ac) circuit to another by magnetic coupling. Essentially, it consists of two or more multiturn coils of wire placed in close proximity to cause the magnetic field of one to link the other. In general, the transformer accomplishes one or more of the following between two circuits: (1) a difference in voltage magnitude, (2) a difference in current magnitude, (3) a difference in phase angle, (4) a difference in impedance level, and (5) a difference in voltage insulation level, either between the two circuits or to ground.

Transformers are used to meet a wide range of requirements. Pole-type distribution transformers supply relatively small amounts of power to residences. Power transformers are used at generating stations to step up the generated voltage to high levels for transmission. The transmission voltages are then stepped down by transformers at the substations for local distribution. Instrument transformers are used to measure voltage and currents accurately. Audio- and video-frequency transformers must function over a broad band of frequencies. Radio-frequency transformers transfer energy in narrow frequency bands.

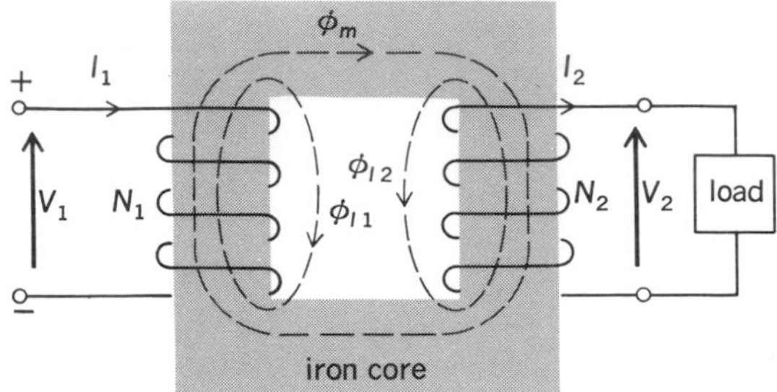

Fig. 1. Basic transformer.

A power transformer consists of two or more multiturn coils wound on a laminated iron core. At least one of these coils serves as the primary winding.

Principle of operation. When the primary of a power transformer is connected to an alternating voltage, it produces an alternating flux in the core. The flux generates a primary electromotive force, which is essentially equal and opposite to the voltage supplied to it. It also generates a voltage in the other coil or coils, one of which is called a secondary. This voltage generated in the secondary will supply alternating current to a circuit connected to the terminals of the secondary winding. A current in the secondary winding requires an additional current in the primary. The primary current is essentially self-regulated to meet the power (or volt-ampere) demand of the load connected to the secondary terminals. Thus in normal operation, energy (or volt-amperes) can be transferred from the primary to the secondary electromagnetically.

Figure 1 shows a transformer with a primary of N_1 turns and a secondary of N_2 turns. A primary voltage V_1 causes a current I_1 to flow through the coil. Since all quantities shown are alternating, the arrows indicate only instantaneous polarities.

The magnetic flux ϕ set up by the primary consists of two components. One part passes completely around the magnetic circuit defined by the iron core, thus linking the secondary coil. This is the mutual flux ϕ_m. The second part is a smaller component of flux that links only the primary coil. This is the primary leakage flux ϕ_{l1}. If the secondary circuit is completed through a load, a secondary current I_2 flows and in turn creates a secondary leakage flux ϕ_{l2}. These leakage fluxes contribute to the impedance of the transformer. If the leakage flux is small, the coupling between primary and secondary is said to be close. The use of an iron core decreases the leakage flux by providing a low-reluctance path for the flux. *See* MAGNETIC CIRCUITS.

In a power transformer the voltage drops due to winding resistance and leakage are small; therefore V_1 and V_2 are essentially in phase (or 180° out of phase, depending on the choice of polarity). Since the no-load current is small, I_1 and I_2 are essentially in phase (or 180° out of phase). Therefore, Eq. (1) applies, and the voltage ratio is expressed

$$V_1 I_1 \simeq V_2 I_2 \quad (1)$$

by Eq. (2), in which a is the transformation ratio.

$$\frac{V_1}{V_2} \simeq a \quad (2)$$

Substituting Eq. (2) into Eq. (1) demonstrates that the current ratio is inversely proportional to the transformation ratio, as in Eq. (3). A transform-

$$\frac{I_1}{I_2} \simeq \frac{1}{a} \quad (3)$$

er therefore may be used to step up or down a voltage from a level V_1 to a level V_2 according to the transformation ratio a. Simultaneously the current will be transformed inversely proportional to a.

Equation (1) may be rewritten in the form of Eq. (4).

$$I_1^2 \frac{V_1}{I_1} \simeq I_2^2 \frac{V_2}{I_2} \quad (4)$$

Since V_2/I_2 is the impedance Z_2 of the load on the secondary and V_1/I_1 is the impedance Z_1 of the load as measured on the primary, Eq. (5) applies.

$$I_1^2 Z_1 \simeq I_2^2 Z_2 \quad (5)$$

Equation (5) may be rewritten in the form of Eq. (6).

$$\frac{Z_1}{Z_2} \simeq \left(\frac{I_2}{I_1}\right)^2 \simeq a^2 \quad (6)$$

The transformer is thus capable of transforming circuit impedance levels according to the square of the transformation ratio; this property is used in telephone, radio, television, and audio systems.

The transmission of power from primary coil to secondary coil is via the magnetic flux. The flux is proportional to the ampere turns in either coil. Since the power in each coil is nearly the same, Eqs. (7) and (8) are obtained. The transformation ratio is therefore approximately equal to the turns ratio.

$$N_1 I_1 \simeq N_2 I_2 \quad (7)$$

$$\frac{N_1}{N_2} \simeq \frac{I_2}{I_1} \simeq a \quad (8)$$

Construction. Transformer cores are made of special alloy steels rolled to approximately 0.014 in. thick. These thin sheets, or laminations, are stacked to form the transformer core, each sheet being insulated from the others to reduce unwanted eddy-current loss. The steel is heat-treated to obtain low hysteresis loss, low exciting current, and low sound level.

Copper conductors are used almost universally. Conductor wires are round in smaller transformers, and rectangular in larger ones.

The conductors are insulated with special paper or cotton covering, with enamel, or with a combination of both. Large outdoor transformers are immersed in oils to obtain good electrical insulation within small spacings and to provide a cooling medium. When lightweight or nonflammable materials are important, transformers may be made with compressed gases as the insulating and cooling medium. Increasing the pressure raises the dielectric strength. The gas is pumped through the transformer and through a gas-to-air heat exchanger for cooling.

The low-voltage (LV) winding is usually in the form of a cylinder next to the core. The high-voltage (HV) winding, also cylindrical, surrounds the LV windings as in Fig. 2*a*. These windings are often described as concentric windings. The number of turns N may be obtained from Eq. (9), where

$$E = \frac{fBAN}{22{,}500} \quad (9)$$

E is rms voltage, f frequency in cps, B maximum flux density in kilolines/in.^{2}, and A cross-sectional area of the iron core in square inches.

Some manufacturers use a winding arrangement having coils adjacent to each other along the core leg as in Fig. 2*b*. The coils are wound in the form of a disk, with a group of disks for the LV winding stacked alternately with a group of disks for the HV windings. This construction is referred to as interleaved windings.

The core sheets are stacked sheet by sheet to form the desired cross-sectional area. The closed magnetic circuit typically has joints between adjacent sheets, but cores of moderate cross section may be made with a long continuous sheet which has been coiled up to give the required cross section. Passages may be provided between groups of sheets for circulation of the cooling oil.

For single-phase transformers (Fig. 3), the HV and LV coils may be on one leg of a core, with the return path in one, two, or more other legs. The total area of the return legs is equal to that of the main leg. An alternative construction has two legs, each with half of the primary windings and half of the secondary windings.

Figure 4 shows a typical three-phase transformer core with coils. A typical three-phase core has three legs, with the HV and LV windings for one phase on each leg. The yokes of the core connect between the two outer legs and the middle leg on top and bottom. This core-type construction is shown in Fig. 5*a*. The iron in another construction that is sometimes used (shell type) is as shown in Fig. 5*b*. Either concentric windings or interleaved windings may be used with either core.

The core and coils are placed in a steel tank with openings for the electrical connections to the windings, and for the cooling equipment.

Cooling. Small transformers are self-cooled. Radiation, conduction, and convection from the tank or from radiating surfaces remove the heat generated by the power losses of the transformer. On larger units, fans are sometimes added to the radiating surfaces. A transformer may have one rating with a basic method of cooling and a higher rating with supplemental cooling. Pumps may be

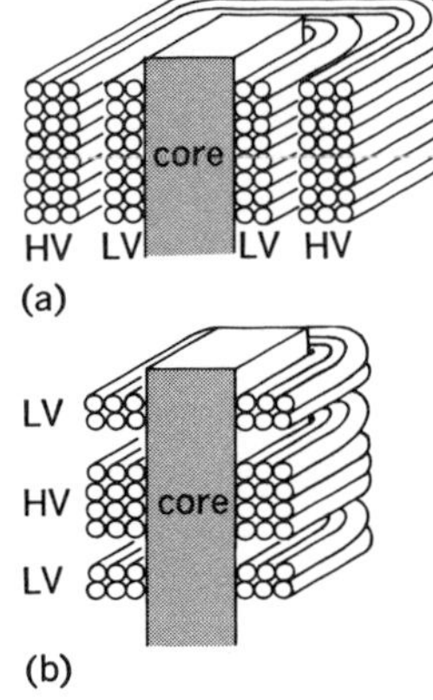

Fig. 2. Winding arrangements. (*a*) Concentric. (*b*) Interleaved.

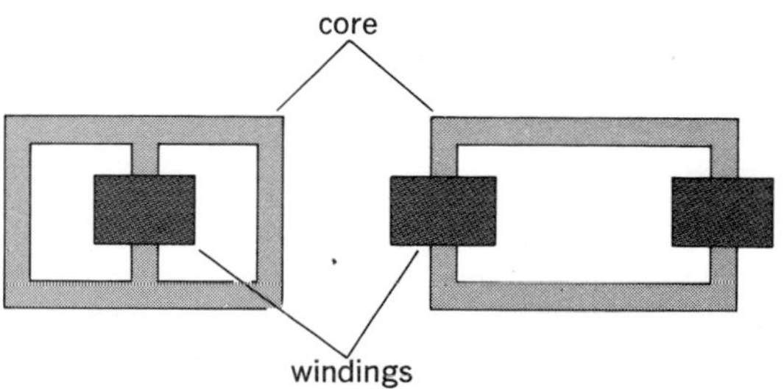

Fig. 3. Location of windings in single-phase cores.

Fig. 4. Three-phase core and coils, rated at 50,000 kVA, 115,000 V.

added to give further cooling. An oil-to-air heat exchanger with finned tubes is used on the very large units. This equipment has a pump for circulating oil and fans for forcing the air against the heat exchanger. Water cooling may be used with cooling coils or with an oil-to-water heat exchanger having an oil pump.

Characteristics. The service conditions for a particular transformer are considered by the designer in choosing materials and the arrangement of parts.

The final design then may be measured by test with respect to a number of characteristics.

No-load loss. The sum of the hysteresis and eddy loss in the iron core is the no-load loss.

Exciting current. The exciting current is that supplied to the transformer at no load when operating at rated voltage. This current energizes the core and supplies the no-load loss. Owing to the characteristic shape of the *B-H* curve of iron, the current is not a true sine wave, but has higher frequency harmonics. In a typical power transformer the exciting current is so small (usually less than 1%) that I_2 is approximately $(N_1/N_2)I_1$. In this sense the ampere-turns in the two windings are said to balance.

Load loss. This is the sum of the copper loss, due to the resistance of the windings (I^2R loss), plus the eddy-current loss in the winding, plus the stray loss (loss due to flux in metallic parts of the transformer adjacent to the windings, the flux resulting from current in the windings).

Total loss and efficiency. The total loss in a transformer is the sum of the no-load and full-load losses. Representative values for a 20,000-kVA, three-phase, 115-kV power transformer are no-load loss, 42 kW; load loss, 85 kW; and total loss, 127 kW. Equation (10) expresses the efficiency of a

$$\text{Efficiency} = \frac{\text{output in kw}}{\text{input in kw}} = \frac{\text{output}}{\text{output} + \text{losses}} \qquad (10)$$

transformer. For this transformer the efficiency is 20,000/20,127, or 99.37%.

Voltage ratio. This is the ratio of voltage on one winding to the voltage on another winding at no load. It is the same as the turns ratio.

Impedance. Consider a transformer having equal turns in the primary and secondary windings. If one side is connected to a generator and the other side to a typical power system load, the voltage measured on the load side will be less than that on the generator side, by the amount of the impedance drop through the transformer.

Impedance is measured by connecting the secondary terminals together (short-circuited) and applying sufficient voltage to the primary terminals to cause rated current to flow in the primary winding. The transformer impedance in ohms equals the primary voltage divided by the primary current. Impedance is usually referred to the transformer kva and kv base and given as percent impedance, as in Eq. (11). Percent reactance is usually close in value to percent impedance, since the percent resistance, given by Eq. (12), is small.

$$\%\ \text{impedance} = \frac{1}{10} \frac{\text{kVA}}{(\text{kV})^2} \times \text{ohms} \qquad (11)$$

$$\%\ \text{resistance} = \frac{\text{load loss in kVA}}{\text{kVA rating}} \times 100 \qquad (12)$$

Typical values for a 20,000-kVA, three-phase, 115-kV self-cooled power transformer are resistance, 0.4%, and impedance, 7.5%.

Regulation. Regulation is the change in output (secondary) voltage that occurs when the load is reduced from rated value to zero, with the primary impressed terminal voltage maintained constant. This is usually expressed as a percent of rated output voltage at full load (E_{FL}), as in Eq. (13),

$$\%\ \text{regulation} = \frac{E_{NL} - E_{FL}}{E_{FL}} \times 100 \qquad (13)$$

where E_{NL} is the output voltage at no load. When a transformer supplies a capacitive load, the power factor may cause a higher full-load voltage than no-load voltage.

Cooling. Temperature tests (heat run tests) are made by operating the transformer with total losses until the temperatures are constant. In the United States the standard winding rise is 55°C over a 30°C air ambient.

Insulation. Sufficient insulation strength must be built into a transformer so that it can withstand normal operation at its rated voltage and system voltage transients due to lightning and switching surges.

Audio sound. The iron core lengthens and shortens because of magnetostriction during each voltage cycle, giving rise to a hum having a frequency twice that of the voltage. This and other frequencies may cause mechanical vibrations in different parts of the transformer due to resonance.

TRANSFORMER

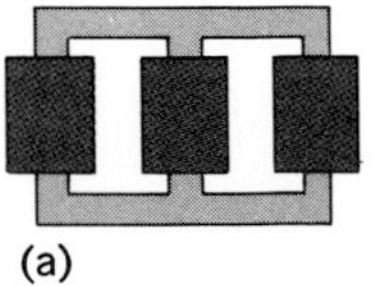

(a)

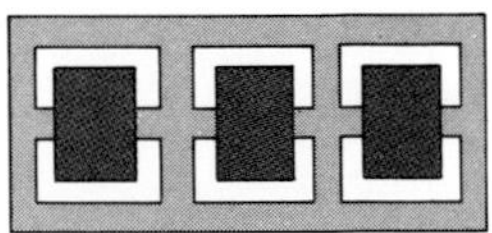

(b)

Fig. 5. Typical three-phase cores showing location of windings. (*a*) Core type. (*b*) Shell type.

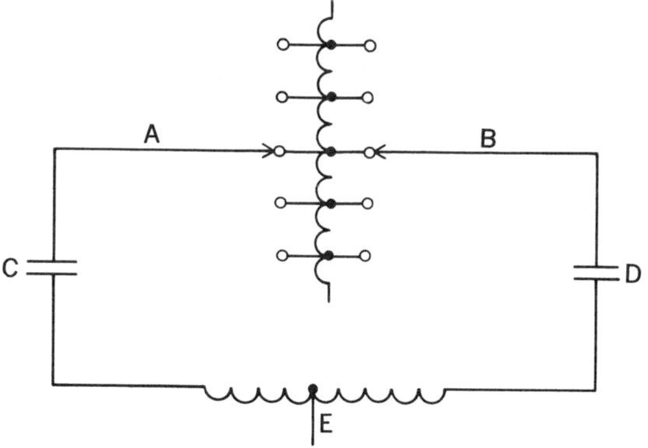

Fig. 6. Typical circuit for tap changing under load.

Taps. The application of a transformer to a power system involves a correct choice of turns ratio for average operating conditions, and the selection of proper taps to obtain improved voltage levels when average conditions do not prevail.

Tap changers are frequently used in the HV winding to give plus or minus two 2½% taps (5% above and 5% below rated voltage). These taps may be changed only when the transformer is deenergized, that is, when the service is interrupted.

Tap changing under load. A special motor-driven tap changer is used to permit tap changing when the transformer is energized and carrying full load. One of its simpler forms is shown in Fig. 6. The transformer taps are brought to a tap changer having two sets of fingers A and B. Initially these are on the same tap. When a change is required, a contactor C opens, and A moves to the next lower tap. C now closes. Next D opens and B moves down to the same tap as A. The current, which initially divided half-and-half through A and B, has changed, first to be all in B, then partly in A and partly in B, then all in A, and finally half-and-half in A and B. E is a center-tapped reactor, which limits the current when A and B are not on a common tap.

This equipment is essential where a constant voltage is required under changing loads. It is frequently applied with a tap range of plus or minus 10% of rated voltage. It may be made to operate automatically, maintaining a specified voltage at a predetermined point remote from the transformer.

Tap changing under load equipment is used on power transformers supplying residential loads, where variations in voltage would adversely affect the use of lights and appliances. It is also used for chemical and industrial processes, such as on pot lines for the manufacture of aluminum.

Parallel operation. Two transformers may be operated in parallel (primaries connected to the same source and secondaries connected to the same load) if their turns ratios and per unit impedances are essentially equal. A slight difference in turns ratio would cause a relatively large out-of-phase circulating current between the two units and result in power losses and possible overheating.

Phase transformation. Polyphase power may be changed from 3-phase to 6-phase, 3-phase to 12-phase, and so forth, by means of transformers. This is of value in the power supply to rectifiers, where the greater number of phases results in a smoother dc voltage wave.

Overloads. Transformers have a capacity for loading above their rating. Such factors as low ambient temperature and type of load carried may be used to increase the continuous load possible on a given transformer. In emergencies it is possible to increase the load further for short times with a calculable loss of transformer life. Such a load would permit, for instance, a 50% overload for 2 hr following full load. [J. R. SUTHERLAND]

Transmission lines

A system of conductors suitable for conducting electric power or signals between two or more termini. For example, commercial-frequency electric power transmission lines connect electric generating plants, substations, and their loads. Telephone transmission lines interconnect telephone subscribers and telephone exchanges. Radio-frequency transmission lines transmit high-frequency electric signals between antennas and transmitters or receivers. In this article the theory of transmission lines is considered first, followed by its application to power transmission lines.

Although only a short cord is needed to connect an electric lamp to a wall outlet, the cord is, properly speaking, a transmission line. However, in the electrical industry the term transmission line is applied only when both voltage and current at one line terminus may differ appreciably from those at another terminus as a result of the electrical properties of the line. Transmission lines are described either as electrically short if the difference between terminal conditions is attributable simply to the effects of conductor series resistance and inductance, or to the effects of a shunt leakage resistance and capacitance, or to both; or as electrically long when the properties of the line result from traveling-wave phenomena.

TRANSMISSION-LINE THEORY

Depending on the configuration and number of conductors and the electric and magnetic fields about the conductors, transmission lines are described as open-wire transmission lines, coaxial transmission lines, cables, or waveguide transmission lines.

Open-wire transmission lines. Open-wire lines may comprise a single wire with an earth (ground) return or two or more conductors. The conductors are supported at more or less evenly spaced points along the line by insulators, with the spacing between conductors maintained as nearly uniform as feasible, except in special-purpose tapered transmission lines, discussed later in this section.

Open-wire construction is used for communication or power transmission whenever practical and permitted, as in open country and where not prohibited by ordinances.

Open-wire lines are economical to construct and maintain and have relatively low losses at low and medium frequencies. Difficulties arise from electromagnetic radiation losses at very high frequencies and from inductive interference, or crosstalk, resulting from the electric and magnetic field coupling between adjacent lines accompanying the characteristic field configuration (Fig. 1).

Coaxial transmission lines. A coaxial transmission line comprises a conducting cylindrical shell, solid tape, or braided conductor surrounding an isolated, concentric, inner conductor which is solid, stranded, or (in certain video cables and delay cables) helically wound on a plastic or ferrite core. The inner conductor is supported by ceramic or plastic beads or washers in air- or gas-dielectric lines, or by a solid polyethylene or polystyrene dielectric.

The purpose of this construction is to have the shell prevent radiation losses and interference from external sources. The electric and magnetic fields shown in Fig. 1*b* are nominally confined to the space inside the outer conductor. Some external fields exist, but may be reduced by a second outer sheath.

Coaxial lines are widely used in radio, radar, television, and similar applications.

Sheathed cables. Also termed shielded cables, these comprise two or more conductors surrounded by a conducting cylindrical sheath, commonly supported by a continuous solid dielectric. The sheath provides both shielding and mechanical protection.

Coaxial lines, sheathed cables, or shielded cables are often termed simply cables.

Traveling waves. When electric power is applied at a terminus of a transmission line, electromagnetic waves are launched and guided along the line. The steady-state and transient electrical properties of transmission lines result from the superposition of such waves, termed direct waves, and the reflected waves which may appear at line discontinuities or at load terminals.

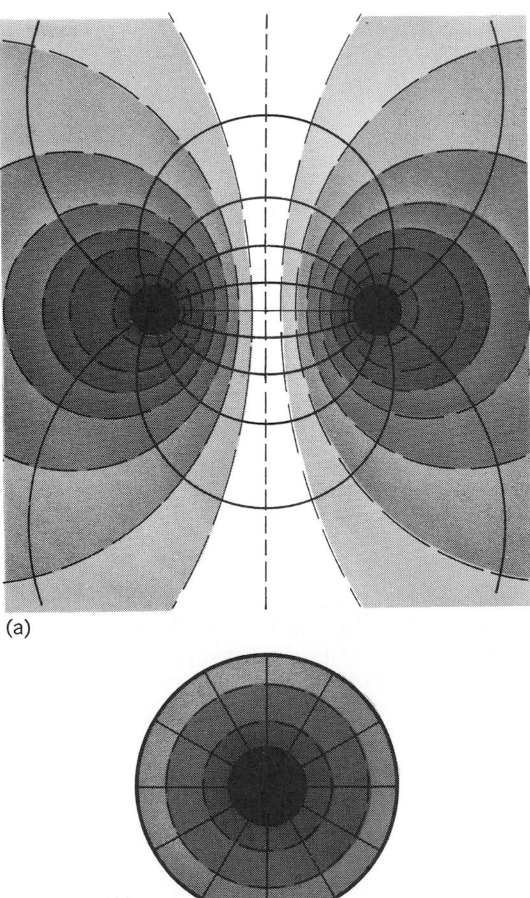

Fig. 1. Electric (solid lines) and magnetic (dashed lines) fields about two-conductor (*a*) open-wire and (*b*) coaxial transmission lines in a plane normal to the conductors, for continuous and low-frequency currents.

Principal mode. When the electric and magnetic field vectors are perpendicular to one another and transverse to the direction of the transmission line, this condition is called the principal mode or the transverse electromagnetic (TEM) mode. The principal-mode electric- and magnetic-field configurations about the conductors are essentially those of Fig. 1. Modes other than the principal mode may exist at any frequency for which conductor spacing exceeds one-half of the wavelength of an electromagnetic wave in the medium separating the conductors. Such high-frequency modes are called waveguide transmission modes.

In a uniform (nontapered) transmission line, the voltage or current applied at a sending terminal determines the shape of the initial voltage or current wave. In a line with negligible losses the transmitted shape remains unchanged. When losses are present, the shape, unless sinusoidal, is altered, because the phase velocity and attenuation vary with frequency.

If a wave shape is sinusoidal, the voltage and current decay exponentially as a wave progresses. The voltage or current, at a distance x from the sending end, is decreased in magnitude by a factor of $\epsilon^{-\alpha x}$, where ϵ is the Napierian base (2.718), and α is called the attenuation constant. The voltage or current at that point lags behind the voltage or current at the sending end by the phase angle βx, where β is called the phase constant.

The attenuation constant α and the phase constant β depend on the distributed parameters of the transmission line, which are (1) resistance per unit length r, the series resistance of a unit length of both going and returning conductors; (2) conductance per unit length g, the leakage conductance of the insulators, conductance due to dielectric losses, or both; (3) inductance per unit length l, determined as flux linkages per unit length of a line of infinite extent carrying a constant direct current; and (4) capacitance per unit length c, determined from charge per unit length of a line of infinite extent with constant voltage applied.

The values of α and β may be found from complex equation (1), where j is the notation for the

$$\alpha+j\beta=\sqrt{(r+j2\pi fl)(g+j2\pi fc)} \tag{1}$$

imaginary number $\sqrt{-1}$, and f is the frequency of the alternating voltage and current. The complex quantity $\alpha+j\beta$ is often called the propagation constant γ. Since $r+2\pi fl$ is the impedance z per unit length of line, and $g+2\pi fc$ is the admittance y per unit length of line, the equation for the propagation constant is often written in the form of Eq. (2). The velocity at which a point of constant phase

$$\gamma=\sqrt{zy} \tag{2}$$

is propagated is called the phase velocity v, and is equal to $2\pi f/\beta$. For negligible losses in the line (when r and g are approximately zero) the phase velocity is $1/\sqrt{lc}$, which is also the velocity of electromagnetic waves in the medium surrounding the transmission-line conductors.

The distributed inductance and resistance of the lines may be modified from their dc values because of skin effect in the conductors. This effect, which increases with frequency and conductor size, is usually, but not always, negligible at power frequencies.

Characteristic impedance. The ratio of the voltage to the current in either the forward or the reflected wave is the complex quantity Z_0, called the characteristic impedance.

When line losses are relatively low, that is, when relationships (3) apply, the characteristic imped-

$$\begin{aligned} r &\ll 2\pi fl \\ g &\ll 2\pi fc \end{aligned} \qquad (3)$$

ance is given by Eq. (4), and is a quantity nearly

$$Z_0 = \sqrt{l/c} \qquad (4)$$

independent of frequency (but not exactly so since both l and c may be somewhat frequency-dependent). The magnitude of Z_0 is used widely, at high frequencies, to identify a type of transmission line such as 50-ohm line, 200-ohm line, and the popular 300-ohm antenna lead-in line used with television antennas.

Distortionless line. Transmission lines used for communications purposes should be as free as possible of signal waveshape distortion. Two types of distortion occur. One is a form of amplitude distortion due to line attenuation, which varies with the signal frequency. The other, delay distortion, occurs when the component frequencies of a signal arrive at the receiving end at different instants of time. This occurs because the velocity of propagation along the line is a function of the frequency.

Theoretically, a distortionless line can be devised if the line parameters are adjusted so that $r/g = l/c$. In practice this is approached by employing loading circuits. Under these conditions the propagation constant is given by Eq. (5).

$$\gamma = \alpha + j\beta = \sqrt{r/g}\,(g + j2\pi fc) \qquad (5)$$

The attenuation constant α is $\sqrt{rg}$, which is independent of frequency f. Therefore, there will be no frequency distortion.

The phase constant β is $2\pi f\sqrt{lc}$ which depends upon frequency. The velocity of propagation along any transmission line is $2\pi f/\beta$, and for the distortionless line this becomes $1/\sqrt{lc}$. Thus the velocity of propagation is independent of frequency, and there will be no delay distortion.

Transmission-line equations. The principal-mode properties of the transmission-line equations are described by Eqs. (6) and (7), in which e and i

$$\frac{\partial e}{\partial x} = -\left(ri + l\frac{\partial i}{\partial t}\right) \qquad (6)$$

$$\frac{\partial i}{\partial x} = -\left(ge + c\frac{\partial e}{\partial t}\right) \qquad (7)$$

are instantaneous values of voltage and current, respectively, x is distance from the sending terminals, and t is time.

For steady-state sinusoidal conditions, the solutions of these equations are given by Eqs. (8) and (9) for voltage E and current I at a distance x from

$$E = E_s \cosh \gamma x - I_s Z_0 \sinh \gamma x \qquad (8)$$

$$I = I_s \cosh \gamma x - \frac{E_s}{Z_0} \sinh \gamma x \qquad (9)$$

the sending end in terms of voltage E_s and current I_s at the sending end. In Eqs. (8) and (9) $Z_0 = \sqrt{(r + j2\pi fc)/(g + j2\pi fc)}$. All values of current and voltage in these and the following equations are complex.

In terms of receiving-end voltage E_r and current I_r, these solutions are given by Eqs. (10) and (11), where x is now the distance from the receiving end.

$$E = E_r \cosh \gamma x + I_r Z_0 \sinh \gamma x \qquad (10)$$

$$I = I_r \cosh \gamma x + \frac{E_r}{Z_0} \sinh \gamma x \qquad (11)$$

Reflection coefficient. If the load at the receiving end has an impedance Z_r, the ratio of reflected voltage to direct voltage, known as the reflection coefficient ρ, is given by Eq. (12).

$$\rho = \frac{Z_r - Z_0}{Z_r + Z_0} \qquad (12)$$

When the load impedance is equal to Z_0, the reflection coefficient is zero. Under this condition the line is said to be matched.

Pulse transients. The transient solutions of Eqs. (6) and (7) are dependent on the particular problem involved. Typical physical phenomena with pulse transients are shown in Fig. 2. The characteristic

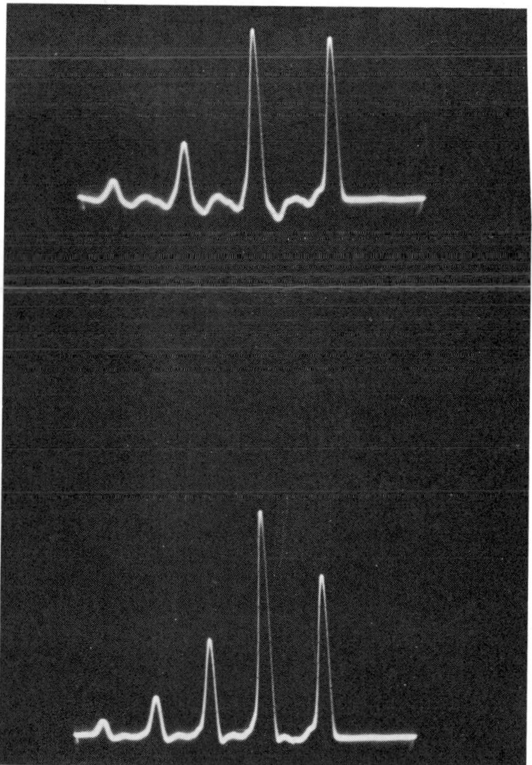

Fig. 2. Typical transient phenomena in a transmission line. These are oscillographic recordings of voltage as a function of time at the sending end of a 300-m transmission line with the receiving end open-circuited. Time increases from right to left; the first (right-hand) pulse is delivered by a generator, equivalent to an open circuit, so that a new forward wave results from each reflected wave arriving at the sending end. At the end of each 2-μs interval, an echo arrives from the receiving end. In the upper trace, minor discontinuities in the line at intermediate points result in intermediate echos. Intermediate discontinuities are minimized in lower trace.

time delay in transmission is often advantageously employed in radar systems and other pulse-signal systems.

Standing waves. The superposition of direct and reflected waves under sinusoidal conditions in an unmatched line results in standing waves (Fig. 3).

Voltage standing-wave ratio. When losses are negligible, successive maxima are approximately equal; under this condition a quantity, the voltage standing-wave ratio, abbreviated VSWR, is defined by Eq. (13).

$$\text{VSWR} = \frac{V_{\text{max}}}{V_{\text{min}}} \tag{13}$$

Power standing-wave ratio. This quantity, abbreviated PSWR, is equal to $(\text{VSWR})^2$. Measurements of voltage magnitude and distribution on a line of known characteristic impedance Z_0 can be used to determine the magnitude and phase angle of an unknown impedance connected at its receiving end. Lines adapted for such impedance measurements, known as standing-wave lines, are widely used.

Transmission-line circuit elements. The impedance Z_s at the sending end of a loss-free section of transmission line that has a length d, in terms of its receiving-end impedance Z_r, is given by Eq. (14).

$$Z_s = \frac{Z_r \cos \beta d + jZ_0 \sin \beta d}{\cos \beta d + j(Z_r/Z_0) \sin \beta d} \tag{14}$$

This equation describes the property of a length of line which transforms an impedance Z_r to a new impedance Z_s. In the simple cases, in which Z_r is a short circuit or open circuit, Z_s is a reactance. Various lengths of line may be used to replace more conventional capacitors or inductors. These properties are widely applied at high frequencies, where suitable values of βx require only physically short lengths of line.

Tapered transmission lines. Transmission lines with progressively increasing or decreasing spacing are used as impedance transformers at very high frequencies and as pulse transformers for pulses of millimicrosecond duration. Although tapers designed to produce exponential-varying parameters, as in the exponential line, are most common, a number of other tapers are useful.

[EVERARD M. WILLIAMS]

POWER TRANSMISSION LINES

In an electric power system the facility used to transfer large amounts of power from one location to a distant location is termed a power transmission line. Techniques of power transmission are presented in this section.

Power transmission lines are distinguished from subtransmission and distribution lines by their higher voltages, greater power capabilities, and greater lengths. With the exception of a few high-voltage dc lines for satisfying special requirements, power transmission lines employ three-phase alternating currents. Such lines require three conductors. The standard frequency in the United States is 60 hertz (Hz). In Europe it is 50 Hz, while in the rest of the world both of these frequencies are used. For transmitting large amounts of power over long distances, high voltages are necessary. Standard transmission voltages in the United States are 69, 115, 138, 161, 230, 345, 500, and 765 kilovolts (kV). These figures refer to the nominal effective voltages between any two of the three conductors. The line conductors are usually placed overhead, supported by poles or towers; however, they may form part of an underground or underwater cable. *See* ELECTRIC DISTRIBUTION SYSTEMS.

Requirements of transmission. Power transmission systems must be reliable, have good voltage regulation and adequate power capability, and be capable of economical operation.

Reliability. This requirement is met by sturdy construction, by protection against overvoltages, by rapid automatic disconnection of accidentally short-circuited lines, by suitable transmission layouts, and by automatic rapid reconnection of lines experiencing only transitory faults.

Good voltage regulation. When the load voltage does not vary appreciably as the load increases from no load to full load, the regulation is said to be good. The inherent voltage regulation depends mainly on the inductive reactance of the line and the power factor of the load. If the inherent regulation is unsatisfactory, the voltage can be controlled by switched shunt capacitors or synchronous condensers connected at the load.

Power capability. The maximum power that can be transmitted, with due regard to limitations imposed by losses, temperature of the conductors, voltage regulation, and system stability, is the power capability of the line. It varies approximately as the square of the voltage.

Economy. Fulfillment of this requirement depends on a balance between low first cost and low operating cost, including cost of power loss. The principal loss is the I^2R loss in the conductors.

Constants. From a knowledge of the size and type of conductors and the spacing between them, one can obtain the values of series resistance r and inductive reactance x per phase per unit length of line and of shunt capacitive susceptance b and leakage conductance g per phase per unit length of line. All of these values are multiplied by the length of the line, giving constants R, X, B, and G, respectively. These are then combined to give the complex impedance, admittance, hyperbolic angle, and characteristic impedance, Eqs. (15)–(18), respectively.

$$Z = R + jX \tag{15}$$

$$Y = G + jB \tag{16}$$

$$\theta = \sqrt{ZY} \tag{17}$$

$$Z_0 = \sqrt{\frac{Z}{Y}} \tag{18}$$

The approximate value of characteristic impedance for a line with low losses, given by $Z_0 = \sqrt{l/c}$, is often called the surge impedance and is real. The power carried by a transmission line is often expressed in terms of its natural power or surge-impedance loading (SIL), which is defined by Eq. (19), where E_n is the nominal voltage, and Z_0 is the

$$P_n = \left|\frac{E_n^{\,2}}{Z_0}\right| \tag{19}$$

surge impedance. Units of kilovolts for E_n, ohms for Z_0, and megawatts for P_n are convenient for power lines. If E_n is the voltage between conductors, P_n is the three-phase power.

If a line is operating at its surge-impedance loading, there are no standing waves (Fig. 3), but the graph of voltage magnitude versus distance along the line is flat. In addition, the reactive power E^2B produced by the shunt capacitance is balanced by the reactive power I^2X consumed by the series inductance, and at every point of the line the current is in phase with the voltage.

Equivalent circuit. This circuit indicates lumped values which represent values distributed along the line. A short line can be represented adequately by its nominal π circuit shown in Fig. 4*a*. Here the shunt admittance Y, actually distributed uniformly along the line, is assumed to be lumped and divided into two equal parts, one at each end of the line. For a long line the theoretically exact equivalent π circuit should be used. Each of its branch impedances is calculated by multiplying the corresponding branch impedance of the nominal π by a correction factor which is given in Fig. 4*b*.

By use of equivalent circuits, the steady-state electrical performance of a line can be calculated by the ordinary theory of ac circuits with lumped constants. These circuits can be combined with circuits representing series capacitors, shunt reactors, transformers, and loads. If a complicated network is to be studied, it can be represented by a low-power model in a network analyzer wherein each line is represented by its π circuit or can be solved on a digital computer by use of a suitable power-flow program.

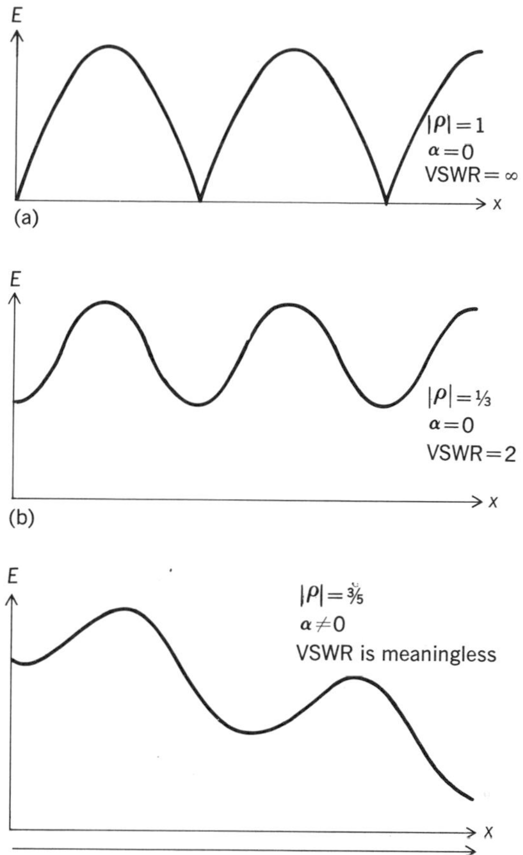

Fig. 3. Voltage distribution under sinusoidal steady-state conditions on a section of transmission line, illustrating three standing-wave conditions: (*a*) line with negligible losses, reflection coefficient of unity, (*b*) line with negligible losses, reflection coefficient of one-third, and (*c*) line with finite losses, reflection coefficient of three fifths. Position of the voltage wave in each case is dependent on angle of phasor value of reflection coefficient. In each case a current maximum (not shown) appears at a voltage minimum in the wavelength.

Alternating-current overhead lines. An overhead transmission line consists of a set of conductors, usually bare, which are supported at a specified distance apart and with specified clearances from the ground and from the supporting structures.

Routes. Lower-voltage transmission lines are usually built along highways, whereas higher-voltage lines are put on a special right of way, cleared of trees and brush. Such routes are often chosen from results of aerial surveys.

Supporting structures. Lower-voltage overhead lines are usually supported by wooden poles and higher-voltage lines by wooden H frames or steel towers. Rigid steel towers give the greatest strength and reliability. The higher the voltage, the greater must be the spacing between conductors and the clearance from conductor to ground. The farther apart the towers are placed, the greater is the sag of the conductors and the taller and stronger the towers must be. Figure 5 shows some typical structures.

The towers shown with vertical strings of suspension insulators are tangent towers or suspension towers. Dead-end towers, used at the ends of a line, and angle towers, used at large angles in the line, have almost horizontal strings of insulators. The center conductor of Fig. 5*f* is supported by V strings, which prevent the conductor from swinging sideways in a cross wind, keeping it from the grounded tower.

Insulators. Conductor supports, or insulators (Fig. 6), are generally made of glazed porcelain or of glass. On lower-voltage lines, they are usually of the pin or post type. On higher-voltage lines, they are of the suspension type, consisting of several units connected by swivel joints. The number of units per string depends on the desired impulse flashover voltage, but is not proportional to it, because the voltage does not divide equally between the several units.

Insulators exposed to industrial dust deposits or to salt spray will, when moist from fog, carry leakage currents which may lead to flashovers at normal operating voltage. Remedies are the use of special fog-type insulators with deeper corrugations on their lower sides for increasing the length of leakage paths, occasional washing of insulators by a stream of water drops from a nozzle, and coating of insulators with silicone grease.

Conductors. For overhead lines, conductors are usually bare, multilayered, concentrically stranded aluminum cables. Adjacent layers of strands are spiraled in opposite directions. Additional tensile strength is provided usually by a core of steel strands (Fig. 7*a*) or sometimes by inclusion of strands of a strong aluminum alloy. These two types of conductor are known by the abbreviations ACSR (aluminum cable, steel reinforced) and ACAR (aluminum cable, alloy reinforced), respectively. If the conductor diameter required for low

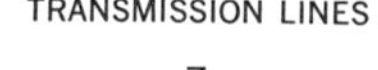

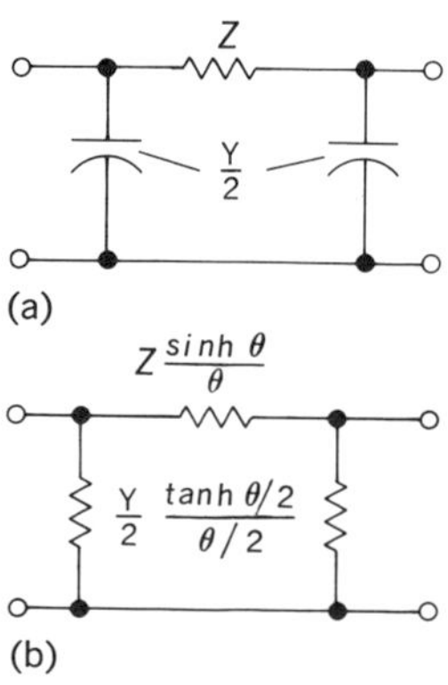

Fig. 4. Lumped-constant representations of a single-phase transmission line or of one phase of a three-phase line. (*a*) Nominal π. (*b*) Equivalent π.

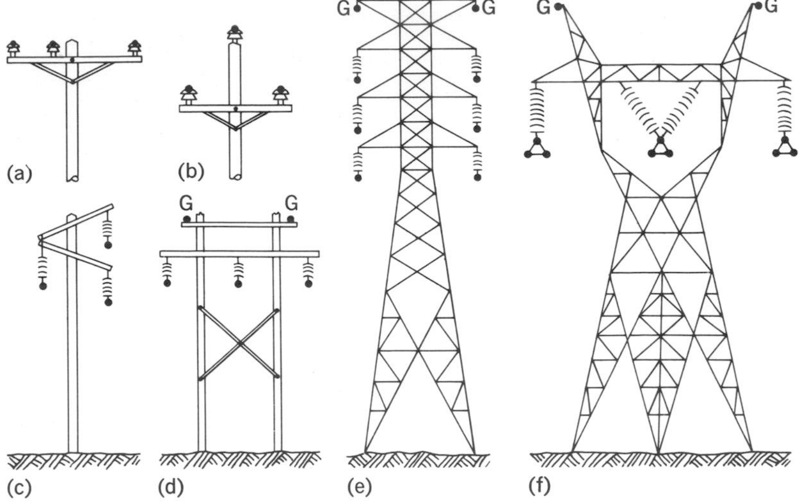

Fig. 5. Supporting structures for electric power transmission lines: (*a-d*) wood poles; (*e, f*) steel towers. Structures in *a–d* and *f* are for single-circuit lines; *e* is for double-circuit lines. In *a* and *b*, pin-type insulators are used; in *c–f*, suspension insulators are used. Structures in *d–f* have ground wires (G) above the line conductors.

corona loss is greater than that of an ordinary stranded conductor having the cross-sectional area required for the desired resistance, special "expanded" conductors are used. Some of these have a paper filler between the steel core and the outer layers of aluminum strands; others have two filler layers each of four large aluminum strands (Fig. 7*b*).

Many very high voltage lines (400 kV and above) use multiple, or bundle, conductors, each consisting of two, three, or four stranded subconductors connected one to another through metal spacers and hung from the same insulators (Fig. 5*f*). Bundle conductors have the advantages of lower inductive reactance; higher natural power; lower corona loss, radio interference, and audible noise; and better cooling than single conductors of the same total cross-sectional area.

Splices in large conductors are usually made with metal sleeves squeezed over the butted ends of the conductor by hydraulic jacks.

Sag and tension. Conductors between adjacent supports hang in a curve called a catenary (Fig. 8). For a given length of span, the greater the tension in the conductor, the smaller is the sag. High mechanical tension is desirable to reduce sag and thus to permit use of longer spans or shorter towers, while maintaining adequate ground clearance.

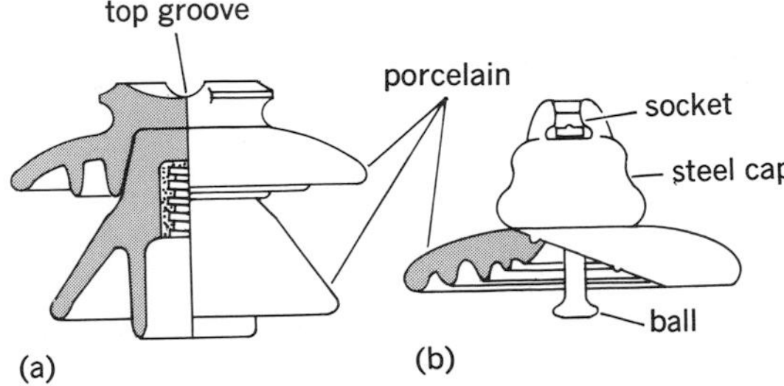

Fig. 6. Insulators used on transmission lines. (*a*) Large two-piece pin-type insulator. (*b*) One unit of a suspension insulator. (*From H. Pender and W. A. Del Mar, Electrical Engineers' Handbook, vol. 1, 4th ed., copyright © 1949 by John Wiley and Sons; used with permission*)

However, the tension must not exceed the tensile yield strength of the conductor under the worst condition, which occurs under a combination of low temperature (causing shortening) simultaneously with the thickest coating of ice on the conductor and the strongest wind.

Vibration. At times the wind causes the conductors to vibrate with low amplitude and audible frequency. This vibration bends the conductor where it is clamped to the insulators and eventually may produce fatigue breakage. A device with duplex weights is used on some lines to reduce conductor vibration. One or more of these dampers are fastened to the conductors several feet from the insulator clamp (Fig. 9).

Sleet. An ice-covered conductor acts as an air foil and is lifted by the wind, so that "dancing" occurs, of such amplitude that one conductor may strike another, producing a short circuit. Conductors should be located so that contact will not be made. Formation of ice is prevented if the current heats the conductor sufficiently. Some power companies make a practice of periodically taking endangered lines out of service and sending high currents through them.

Corona. When the voltage gradient, or electric field strength, at the surface of the conductor exceeds the breakdown gradient of air, the air near the conductor surface becomes ionized. This condition, called corona, is evidenced by a visible glow at night and by a buzzing noise.

Corona results in a loss of power, interference with radio reception, and audible noise, all of which increase rapidly with voltage. Transmission lines are normally operated at a voltage near that at which corona becomes appreciable. The larger the conductor diameter and the greater the number of subconductors, the higher the operating voltage may be.

Inductive coordination. If a telephone line runs near and parallel to a power line for some distance, the high currents and voltages in the power line may induce currents and voltages in the telephone line. These signals may be comparable in strength to the telephone signals and thus produce objectionable noise in telephone receivers. The worst noise is produced by magnetic coupling from harmonic currents having a ground-return path. The coupling between the two lines can be reduced by greater physical separation, by transposition of the telephone wires, and by shielding, such as that provided by grounded cable sheaths.

Inspection and fault location. Transmission lines should have both periodic general inspections and special immediate inspections of points where short circuits have occurred, to detect damage, such as broken insulators, which might impair the reliability of the line. Faults can be located approximately by electrical measurements made from the ends of the line and then exactly by visual patrol of the vicinity. *See* CIRCUIT.

Lightning protection. Lightning is the most detrimental factor affecting the reliability of electric power service, but its damaging effects have been greatly reduced by proper design. Lightning striking a transmission line momentarily impresses a very high voltage on the line, causing spark-over to ground, usually at an insulator. Power current then follows the spark path, producing an arc, which constitutes a short circuit and which can be extin-

Natural and economic loading of single-circuit, three-phase, 60-Hz lines

Voltage, kV	Surge impedance, ohms	Natural loading, MW	Economic loading, MW
69	400	12	15
138	400	47	60
230	400	130	170
345	350	340	430
500	300	830	1000
700	300	1600	2000
1000	250	4000	5000

guished only by disconnecting the faulted line from the rest of the power network. Lines built where severe thunderstorms are prevalent are equipped with overhead ground wires (Fig. 5*d*–*f*) for intercepting the lightning stroke and leading it to ground at the nearest tower.

Switching surges. Another source of overvoltage, which has become important enough on extra-high-voltage lines (500 kV and above) to determine their insulation levels, is switching. The transient overvoltages caused by reenergization of a line which still has trapped charges left from a recent deenergization may be as high as 3.5 times normal line-to-ground crest voltage. By use of circuit breakers which energize the line through one step of series resistance before making direct connection from the power source to the line, the overvoltage can be reduced to about twice normal. By use of two or three steps of decreasing resistance, the overvoltage can be limited to a still lower value, say 1.5. The values cited are representative, but actual overvoltages vary with the length of line section, the time during which the resistors are inserted, the time span from closure of the first pole of the breaker to the last pole, and so on.

Another source of transient overvoltage similar to a switching surge is a short circuit from one line conductor to ground. The overvoltage, having a crest value up to 2.0 times normal crest voltage, appears on the conductor whose voltage phase leads that of the faulted conductor.

Overhead power line constants. Overhead 60-Hz power transmission lines with single conductors per phase have series inductive reactance of about 0.8 ohm/mi (0.5 ohm/km), shunt capacitive susceptance of about 5 micromhos/mi (3 micromhos/km), and surge impedance of about 400 ohms. For lines with two-conductor bundles per phase these become, respectively, 0.6 ohm/mi (0.37 ohm/km), 6.7 micromhos/mi (4.2 micromhos/km), and 300 ohms. For any 60-Hz open-wire line, the phase constant β is about 0.0020 radian/mi (0.0012 radian/km).

Power capability. The economic loading of an overhead power line is usually in the range of 1.0 to 1.5 times its natural power (SIL loading), depending on conductor size. At much higher loadings, the losses become excessive. The table gives typical values of natural power and economic loading of single-circuit, three-phase, 60-Hz lines.

Cost of transmission. The cost of building a transmission line is very nearly proportional to the voltage and to the length of the line. Its power capability is almost proportional to the square of the voltage. Consequently, the cost per unit of power varies directly as the distance and inversely as the square root of the power. If the amount of power to be transmitted is quadrupled, it can be transmitted twice as far for the same unit cost. This explains why it is not economical to transmit for a long distance unless a large quantity of power is involved.

Reactive compensation. There are two kinds of reactive compensation: shunt and series. In shunt compensation the distributed shunt capacitive susceptance B_C of the transmission line is partially or entirely compensated by the addition of lumped inductive susceptance B_L in the form of shunt reactors. The net shunt susceptance is $B=B_C-B_L$. The degree of shunt compensation is the ratio B_L/B_C. The net charging current is reduced in proportion of 1 minus this ratio. Shunt compensation is needed principally on extra-high-voltage overhead lines (500 kV and above) when they are lightly loaded and on long cables.

In series compensation the distributed series inductive reactance X_L of the transmission line itself is partially compensated by the addition of lumped series capacitive reactance X_C in the form of series capacitors. The net series reactance is $X=X_L-X_C$. The degree of series compensation is the ratio X_C/X_L.

Series compensation is used for several different purposes. The first is to reduce the voltage fluctuation (flicker) due to rapidly changing loads, as in electric furnaces and the starting of large motors. The second purpose is to obtain proper division of current between transmission lines connected in parallel. If two such lines have conductors of different cross section, the total losses can be reduced by connecting a capacitor in series with the line having the larger conductors. The capacitor should have such capacitive reactance that the ratio of net reactance to resistance is the same for both lines.

The third purpose of series compensation is to increase the distance that a given amount of power can be transmitted, to increase the power that can be transmitted over a given distance, or, more generally, to increase the possible product of power and distance. Analysis of the equivalent circuits of Fig. 4 for a line with negligible losses shows that the power P transmitted for terminal voltages of fixed magnitudes E_s and E_r but of variable phase difference δ varies according to Eq. (20), where X

$$P=\frac{E_sE_r}{X}\sin\delta \tag{20}$$

is the series inductive reactance of the line (Z in Fig. 4*a*) or, more exactly, for a long line, of the horizontal branch of the equivalent π of Fig. 4*b*. This

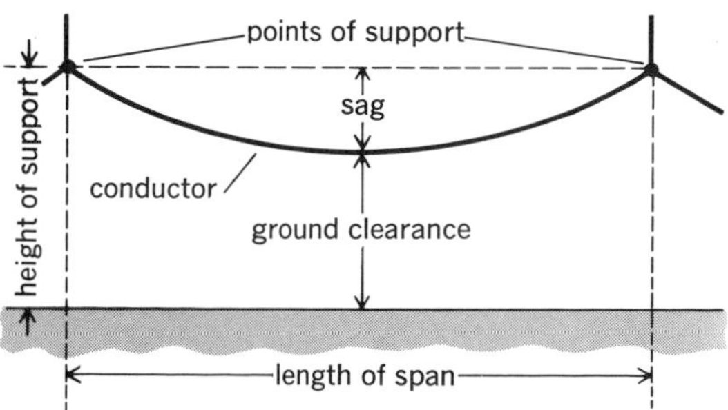

Fig. 8. Catenary curve assumed by one span of a flexible conductor with supports at equal elevations.

TRANSMISSION LINES

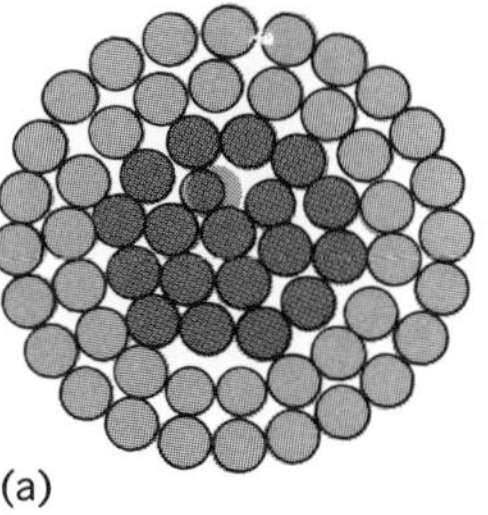
(a)

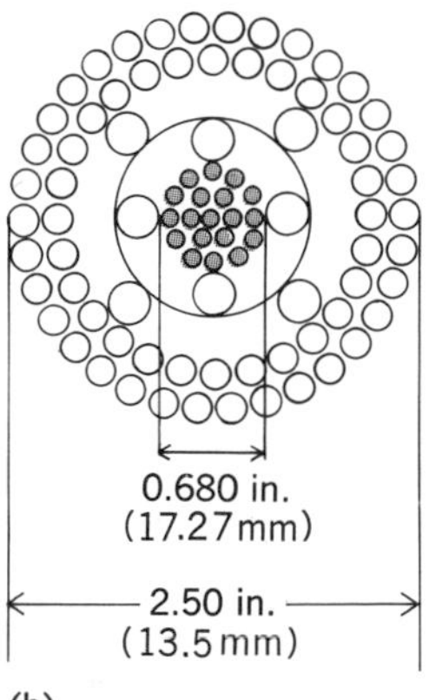

(b)

Fig. 7. Cross sections of typical overhead conductors. (*a*) Steel-reinforced aluminum cable (ACSR) with 19 steel and 42 aluminum strands. (*b*) Expanded ACSR. (*Aluminum Company of America*)

Fig. 9. Stockbridge dampers on transmission-line conductor. (*From H. Pender and W. A. Del Mar, Electrical Engineer's Handbook, vol. 1, 4th ed., copyright © 1949 by John Wiley and Sons; used with permission*)

equation shows that maximum power is obtained at $\delta=90°$, for which $\sin\delta=1$. Considerations of stability require that the line be operated with δ less than 90°, and experience indicates that it can be operated prudently at $\delta=30°$, $\sin\delta=0.5$. With any fixed values of the terminal voltages and fixed maximum value of $\sin\delta$, there is a maximum value of PX, the product of power and series reactance. Since the frequency is fixed in practice, the series reactance of an overhead line of given frequency is directly proportional to the length of line d, and hence there is a limit to the product Pd (in megawatt-miles) of an uncompensated line. Assuming the economic power to be 1.25 P_n, $f=60$ Hz, and $\sin\delta=0.5$, the limiting distance for that power is about 200 mi or 320 km.

In the past, this limitation was not seriously felt because transmission for distances much greater than 200 mi (320 km) between points of maintained voltage was generally not economical in comparison with supplying power from generating stations nearer to the load centers. However, the use of electric power has grown to a point where transmission of large amounts of power over greater distances (perhaps up to 1000 mi or 1600 km) is economical in some circumstances. Transmission over such distances becomes feasible through series compensation.

Underground ac lines. Insulated cables are used in congested areas where the cost of right of way for overhead lines would be excessive, in city streets where overhead lines would be too unsightly or hazardous, in and around power stations, and for crossing wide bodies of water. About 1% of the total transmission mileage of the United States is underground, located mostly in congested urban areas. Because the cost of underground transmission is so much higher than that of overhead transmission, it is used only where necessary. The cost of constructing an underground line ranges from 8 times (at 69 kV) to 20 times (at 500 kV) that of an overhead line of the same length and power capability. Cables are now made for alternating voltages up to 500 kV, and cables for 700 kV are being developed. The conductors of most cables are of stranded copper, insulated by wrapping with layers of paper tape saturated with mineral oil. The thickness of the insulation depends on the voltage, and varies from 0.285 in. in 69-kV cables to 1.34 in. for 500-kV cables. The dielectric constant of this insulation is from 3.5 to 3.7.

Solid and oil-filled cables. Low-voltage cables are of the solid type, in which the only oil is that put in the paper during manufacture. The disadvantage of this type of cable for high voltages is that small voids may form between layers of paper and that corona in such voids causes insulation to deteriorate, leading to puncture. Thus, high-voltage cables are provided with oil under pressure.

In the low-pressure oil-filled cable, there are oil channels in the center of the conductor of single-conductor cables or between the insulated conductors of three-conductor cables. Oil reservoirs, connected to the cable at intervals, keep the cable full of oil at a pressure of from 3 to 20 pounds per square inch (psi) or 21 to 138 kPa in spite of contraction and expansion caused by changes in cable temperature.

The solid and low-pressure oil-filled types of cables have a lead sheath surrounding the insulation to keep moisture out of the cable and the oil in. Such cables are pulled into concrete or fiber ducts which give mechanical protection. At intervals of 600 ft (180 m) or less, these ducts terminate in underground chambers called manholes, where sections of cables are spliced together.

Pipe-type cables. In these, three single-conductor, paper-insulated cables are pulled into a buried stell pipe, which is later filled with oil at high pressure (200 psi or 1.4 MPa). Manholes may be spaced as far apart as 1/2 mi (0.8 km). Oil reservoirs and pumps are required. Power capability ranges from about 225 MVA at 138 kV to 650 MVA at 550 kV.

Compressed-gas-insulated cables. Each of the three isolated phases consists of two coaxial aluminum tubes separated mechanically by epoxy insulating spacers and electrically by these spacers and the compressed gas. The inner tube is the main conductor, while the outer tube, which is grounded, normally carries little current but serves as a sheath, contains the gas, and gives some mechanical protection. Expansion joints incorporated in the central tube allow differential thermal expansion. The gas cable is shipped in factory-assembled sections about 40 ft (12 m) long. These are then welded together in the field. The outer surface of the sheath is provided with a protective coating to prevent corrosion. Additional protection against corrosion may be given by a cathodic protection such as that used on long pipes for oil or gas. The compressed gas currently used is sulfur hexafluoride (SF_6) at pressures of 45 to 60 psi (300 to 400 kPa). It combines excellent chemical stability with high electric breakdown strength, good heat-transfer characteristics, and low dielectric constant compared to paper and oil. Cables employing SF_6 can be designed to withstand rated voltage even with the gas pressure reduced to normal atmospheric pressure. Research is under way to find other gases (or mixtures) as good as SF_6 but less expensive. Power ratings range from 600 MVA at 230 kV to 11,000 MVA at 765 kV.

Low-temperature (cryogenic) cables. Cryogenic cables are being investigated. The resistance of aluminum or copper decreases as the temperature decreases. If a cable is cooled by circulating liquid nitrogen at temperature 77 K through a hollow conductor, for example, the power loss (I^2R) in the conductor is greatly reduced. However, the power required for refrigeration of the coolant partly offsets the saving in I^2R loss. Good thermal insulation of such a cable is obviously necessary.

Certain special alloys become superconductive at very low temperatures (below 18.2 K for niobium-tin), and their resistance to direct current becomes zero, leaving only the power for refrigeration as the loss of the cable. With alternating current, there is also some loss in the conductors but much less than that at or above room temperature.

If the current or the temperature should exceed its critical value, the conduction would become resistive. For this reason, the superconductive alloy is either plated onto the surface of an ordinary metal or is made in fine filaments embedded in the ordinary metal. Thus, if the alloy should lose its superconductivity, the ordinary metal could carry the current for a long enough time to permit reduction of the current before melting of the conductors could occur.

Terminations. Connections of cables to overhead lines or to substations are made through potheads which provide oil seals and longitudinal as well as radial insulation. High-voltage potheads are encased in porcelain with a corrugated outer surface and are similar in appearance to the insulating bushings used on transformers.

Splices in high-voltage cables must be made with great care and require many hours of skilled labor. The distance between splices depends largely upon the length of cable that can be wound on one shipping reel.

Losses and power capability. The power capability of a cable is limited by the rise of temperature in the conductor and adjacent insulation, because too high a temperature will char the insulation and cause its breakdown. The temperature rise depends on the power losses in the cable and on the rate of conduction of heat from the cable into the surrounding soil. Whereas the loss in the conductor is the only important loss in an overhead line, the cable has also a considerable dielectric loss, which increases with voltage because of the increasing thickness of the insulation. Also, whereas the overhead conductor is bare and directly exposed to the cooling air, the heat produced by losses in a cable must pass through the insulation, the duct walls (if ducts are used) and a considerable thickness of earth before reaching cool earth or air. Thus the permissible current in a cable is less than that in an overhead conductor of equal resistance, and the permissible power for a given voltage is correspondingly less. This is the apparent power, $S = \sqrt{P^2 + Q^2}$, where P is the active power and Q is the reactive power.

Improved cooling. Since the power capability of cables is limited by their temperature rise, capability can be increased by improved heat removal. Among the means that have been used—but so far only to limited extent—are (1) backfilling the trench in which cable or cable ducts are buried with material of better heat conductivity than that of the original soil, (2) burying cooling pipes in the earth near the cables and circulating water or other fluid through these pipes, and (3) circulating the oil of pipe-type cables through heat exchangers.

Charging current and critical length. Because of the closer spacing between conductors and the higher dielectric constant of the oil-paper insulation, cables have a much higher shunt capacitance per unit length than do overhead lines. Hence for a given voltage the charging current and the charging reactive power of cables are correspondingly higher. Compressed-gas-insulated cables have an advantage in this respect.

The critical length of a cable is that length for which the charging current equals the rated current. A cable longer than the critical length would be overloaded at the sending end even if nothing were connected to the receiving end.

Shunt compensation. The limitation of length or active-power capability due to charging current can be raised by connecting inductive reactors in shunt with the cable at its terminals and at intermediate points. The total shunt current taken by the reactors should be approximately equal in magnitude to the charging current of the whole length of the cable but in phase opposition to it. The economic spacing of shunt reactors on a 60-Hz cable would be between 5 and 10 mi (8 and 16 km).

Shunt compensation of power cables is seldom, if ever, used. Most underground cables are too short to require it.

Submarine ac cables. These are used in crossing rivers, bays, or straits too wide for overhead spans and to transmit power to offshore islands. They are mostly of the solid type. Water pressure prevents formation of voids and the natural cooling is good. Length of uncompensated cables is limited by charging current to about 25 mi (40 km), and shunt compensation is deemed impractical because of the additional complications in laying the cables and in retrieving them when repairs are needed. The lead sheath is protected by an armor of steel wires, sometimes covered with jute. Submarine cables are liable to damage by trawling and by dragging of ship's anchors. Shore sections, used in shallow water and across beaches, usually differ in diameter and amount of armor from the deep-water sections.

Direct-current lines. Although most electric power transmission is by alternating current, there is an increasing number of direct-current transmission lines. These require converter stations at both ends to connect the line to an ac system.

Overhead lines. Bipolar overhead lines are similar in construction to overhead three-phase lines except that they have only two conductors instead of three. For the same conductor size and insulation level, a dc line can carry the same power on two conductors that a three-phase line can carry on three conductors; and the cost of the dc line is about two-thirds the cost of the corresponding ac line. In some places, two monopolar lines spaced well apart (about 1½ mi or 2.4 km) are used instead of one bipolar line for the sake of improved reliability.

Since internal overvoltages are somewhat less on dc lines than on ac lines, lower insulation levels are used for the same crest voltage to ground. Under these conditions leakage currents at normal operating voltage become more important, especially when insulators are dirty from industrial wastes and moist from fog. For this reason, insulators for dc lines are usually of special design, having a higher ratio of length of leakage path to flashover distance.

Cable lines. Most of the dc lines built before 1969 where wholly or partly submarine cables. Since that date the trend has been toward overhead lines, although several cases have been considered for underground cables to bring power to metropolitan areas. Direct-current cables have no charging current and therefore are not subject to the limitation on their length that applies to ac cables. In addition, a dc cable has no dielectric loss and can safely withstand a higher direct voltage than root-mean-square alternating voltage. As a result, a dc cable can carry about six times as much power as the rated apparent power when the same cable is used for alternating current. Single-conductor solid cables are used.

Ground return. The resistance of the ground to direct current is very much lower than to alternating current, being essentially only that in the vicinity of the ground electrodes. The use of ground or sea return for a monopolar line saves most of the cost of one conductor and of its power loss. A bipolar dc line can operate with one pole and ground return while there is a fault on the other pole of the line or of the terminal equipment.

[EDWARD W. KIMBARK]

Bibliography: American Radio Relay League, *ARRL Antenna Book*, revised periodically; L. N. Dworsky, *Modern Transmission Line Theory and Applications*, 1979; Federal Power Commission, *National Power Survey*, 1970; Federal Power Commission, *Underground Power Transmission*, April, 1966; D. G. Fink and H. W. Beaty, *Standard Handbook for Electrical Engineers*, 11th ed., 1978; P. Graneau, *Underground Power Transmission*, 1979; E. W. Kimbark, *Direct Current Transmission*, vol. 1, 1971; H. Pender and W. A. Del Mar (eds.), *Electrical Engineers' Handbook*, vol. 1, 4th ed., 1949; E. R. Schatz and E. M. Williams, Pulse transients in exponential transmission lines, *Proc. IRE*, vol. 38, pt. 2, 1950; W. D. Stevenson, Jr., *Elements of Power System Analysis*, 3d ed., 1975; B. M. Weedy, *Electric Power Systems*, 2d ed., 1972.

Turbine

A machine for generating rotary mechanical power from the energy in a stream of fluid. The energy, originally in the form of head or pressure energy, is converted to velocity energy by passing through a system of stationary and moving blades in the turbine. Changes in the magnitude and direction of the fluid velocity are made to cause tangential forces on the rotating blades, producing mechanical power via the turning rotor.

The fluids most commonly used in turbines are steam, hot air or combustion products, and water. Steam raised in fossil fuel-fired boilers or nuclear reactor systems is widely used in turbines for electrical power generation, ship propulsion, and mechanical drives. The combustion gas turbine has these applications in addition to important uses in aircraft propulsion. Water turbines are used for electrical power generation. Collectively, turbines drive over 95% of the generating capacity in the world. *See* GAS TURBINE; HYDRAULIC TURBINE; STEAM TURBINE; TURBOJET.

Turbines effect the conversion of fluid to mechanical energy through the principles of impulse, reaction, or a mixture of the two. Illustration *a* shows the impulse principle. High-pressure fluid at low velocity in the boiler is expanded through the stationary nozzle to low pressure and high velocity. The blades of the turning rotor reduce the velocity of the fluid jet at constant pressure, converting kinetic energy (velocity) to mechanical energy. *See* IMPULSE TURBINE.

The reaction principle is shown in illustration *b*. The nozzles are attached to the moving rotor. The acceleration of the fluid with respect to the nozzle causes a reaction force of opposite direction to be applied to the rotor. The combination of force and velocity in the rotor produces mechanical power. *See* REACTION TURBINE. [FREDERICK G. BAILY]

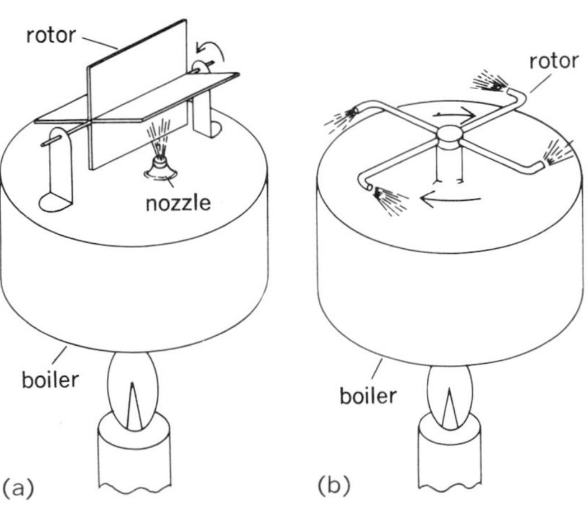

Turbine principles. (*a*) Impulse. (*b*) Reaction.

Turbine propulsion

Propulsion of a vehicle by means of a gas turbine. Widely used to propel aircraft, the gas turbine is being developed for many ground and marine applications as well. *See* GAS TURBINE; PROPULSION.

Basic engine. The gas turbine consists essentially of an air compressor, a combustion chamber, and a turbine wheel (Fig. 1). The compressor, which can be axial flow, centrifugal flow, or a combination of the two, produces the highly compressed air needed for efficient combustion. In the widely used axial-flow compressor, this is accomplished as follows: Air from the atmosphere is sucked or rammed into the compressor inlet ports and compressed progressively through each blade stage. As the air pressure increases, air volume decreases until maximum compression is reached at the last blade stage. The highly compressed (and thus high-temperature) air is then discharged into the duct leading to the combustion chamber, which has one or more fuel nozzles through which fuel is sprayed to mix with the moving air. To start combustion, the fuel spray is ignited by electric resistance ignition plugs. Once ignited, the fuel-air mixture burns continuously, so that the ignition can be switched off.

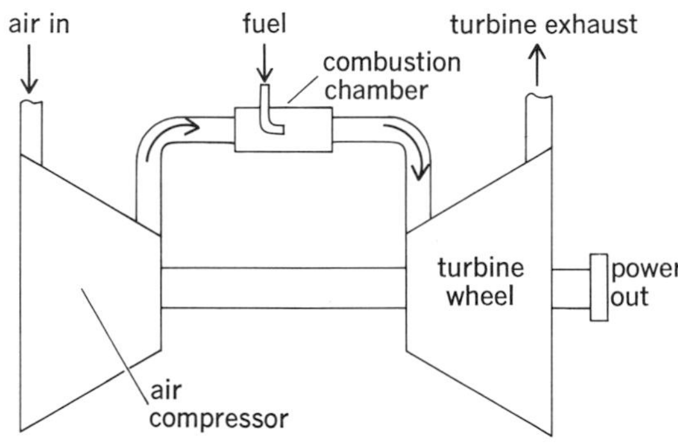

Fig. 1. Schematic diagram of simple gas turbine.

Instead of a single (annular or ring-shaped) combustion chamber, which is employed in many axial-flow compressor gas turbines, a number of separate chambers, known as can-type or cannular chambers, may be used. Thus, in a typical aircraft gas turbine, 8 to 10 can-type combustors, each with its own fuel nozzle, are employed.

The products of combustion leave the combustor through a duct, passing along fixed guides or nozzles to strike the cupped blades of the turbine wheel at high speed. The gases flow axially between the fixed guides and the rotating blades of the turbine, falling in both temperature and pressure as they deliver most of their energy to the turbine wheel. This wheel drives the compressor, since both are on the same shaft. *See* BRAYTON CYCLE.

Aircraft applications. The turbine engine has already replaced the piston engine on military aircraft, except light planes. All new long- and medium-range commercial transports are turbine-powered, as are most new short-haul transports, cargo aircraft, helicopters, business aircraft, and utility aircraft. Only light private and business planes, trainers, and utility aircraft requiring less than 300–400 hp are powered by piston engines. Small turbine engines of low enough cost for even light aircraft may soon be developed. *See* AIRCRAFT ENGINE; AIRCRAFT PROPULSION.

The main reasons for using the turbine engine are its small frontal cross section and its inherent simplicity. The turbine-compressor unit, with its relatively simple shaft and bearings—usually of the ball or roller type—replaces the complicated piston-engine mechanism for converting the reciprocating motion of the piston to a rotary one. For example, a 12-cylinder V-type reciprocating engine has at least 90 pairs of sliding or bearing members, each a source of friction loss, as compared with four main shaft bearings of the gas turbine.

Other advantages of the turbine over the piston engine include: (1) higher mechanical efficiency, (2) inherently better balance, (3) no upper limit to size or output, (4) lower weight and smaller frontal area for a given output, (5) ability to operate on less expensive fuels, (6) simpler lubrication, (7) negligible warmup period, (8) greater reliability, and (9) longer life.

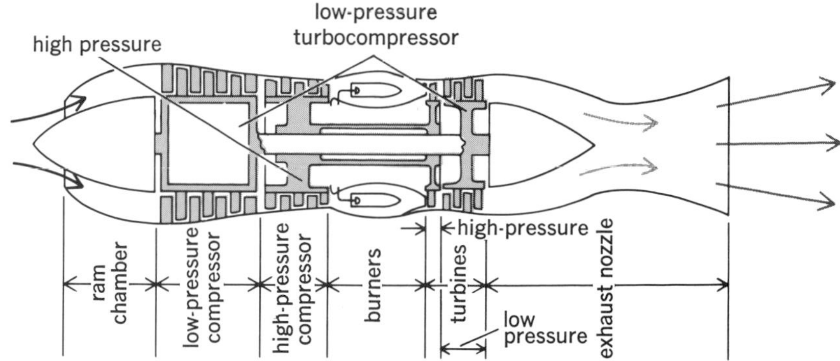

Fig. 2. Dual-rotor turbojet engine has a two-part compressor. The low-pressure compressor is driven by a single-stage turbine, and operates at a lower speed than does the high-pressure compressor, which is driven by a two-stage turbine.

The simple single-shaft aircraft turbine engine does have inherent disadvantages, such as: (1) high specific fuel consumption, (2) low rate of change in acceleration or power response, and (3) sensitivity to air inlet temperature. In actual engines, however, these shortcomings are minimized or avoided by a variety of devices and techniques, the most important of which will be discussed later in this article.

The two major types of aircraft turbine engines are the turbojet and the turboprop.

Turbojet engine. The thrust or push of a turbojet engine (Fig. 2) comes from reactions set up inside the engine as the gases are accelerated through it and out the exhaust nozzle. *See* TURBOJET; TURBORAMJET.

The propulsive ability of a turbojet engine is measured in pounds of thrust rather than in horsepower (which is used for piston and turboprop en-

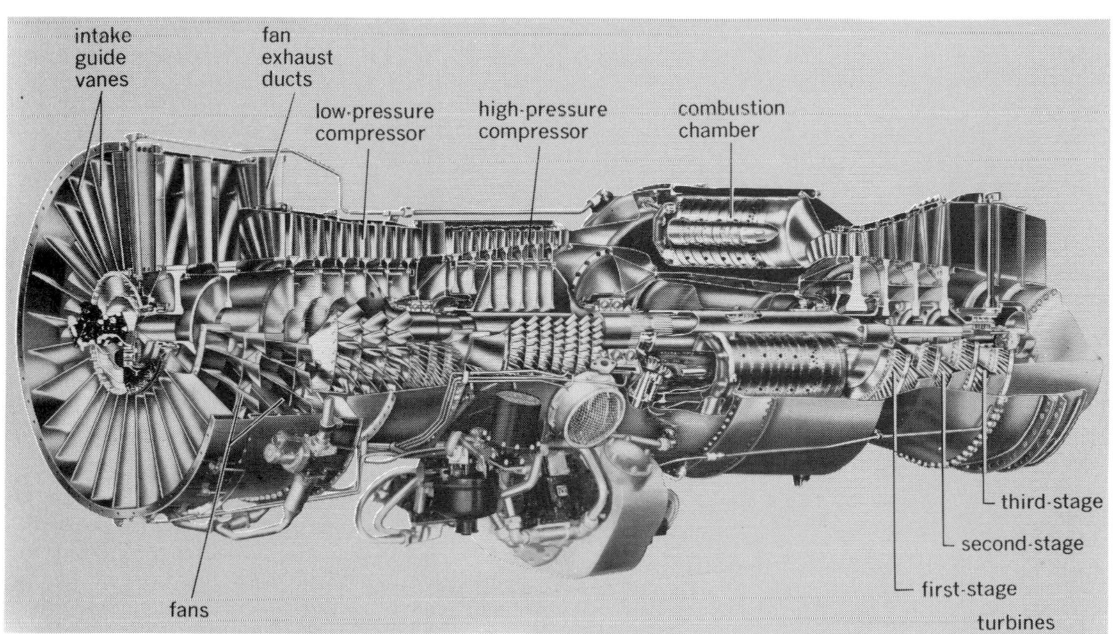

Fig. 3. Pratt and Whitney JT9D turbofan engine produces 43,500 lb (193.5 kN) thrust, is 10.8 ft (3.3 m) long, and weighs 8430 lb (3824 kg). Engine has one fan plus 13 compressor stages and 6 turbine stages. Turbine section is air-cooled.

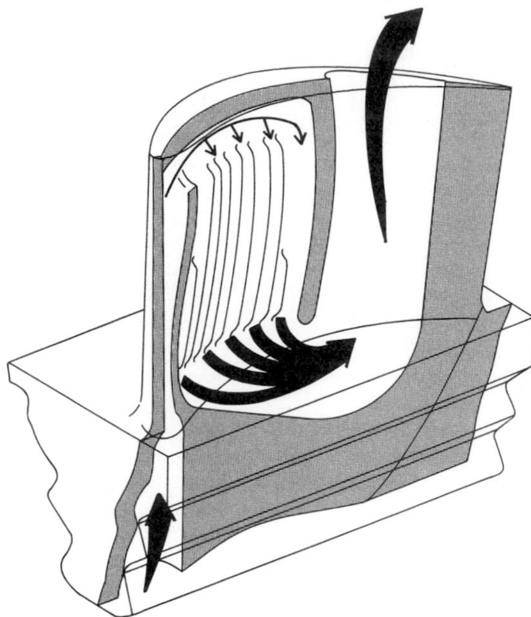

Fig. 4. Internally finned, convection-cooled turbine blade can operate at 2200°F (1200°C) turbine inlet temperature.

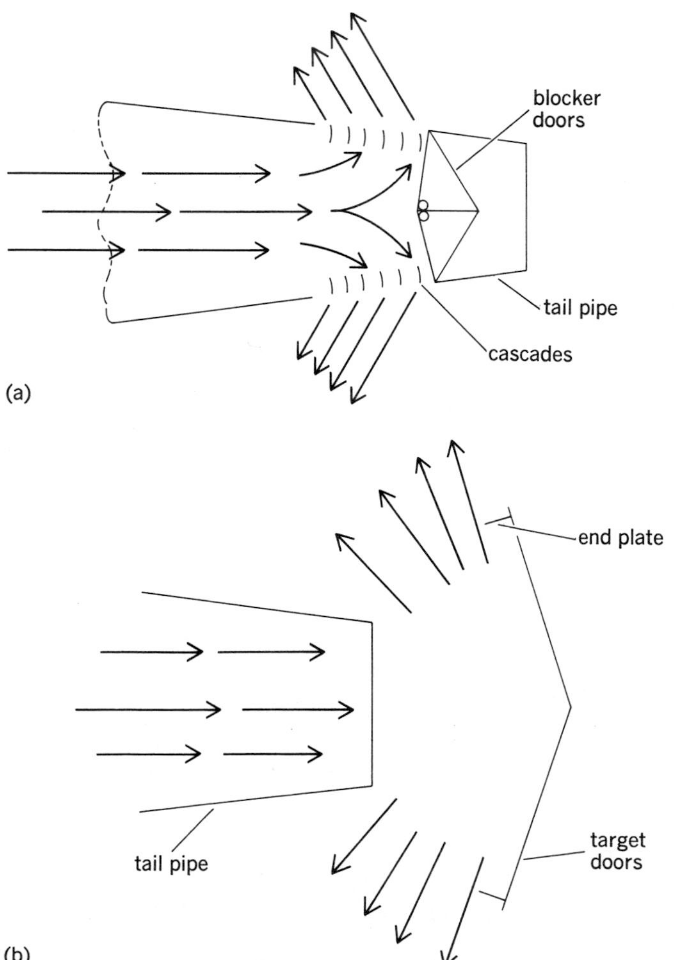

Fig. 5. Thrust reversers, such as (*a*) cascade type and (*b*) target type, reverse the direction of the jet exhaust gases to produce a braking force.

gines) because no actual engine motion is involved when the aircraft is standing still. Once the airplane starts to move, however, a comparison between thrust and horsepower can be made with the equation $T_{HP}=TS/375$, where T_{HP} is thrust horsepower, T is thrust in lb, and S is speed in mph. Thus, at 375 mph, one pound of thrust equals one thrust horsepower. In metric units, $T_p=TS$, where T_p is thrust power in watts, T is thrust in newtons, and S is speed in m/sec. Aircraft turbojet engines range from small units delivering a few hundred pounds of thrust to the huge 67,000-lb-thrust (298-kN) General Electric engine being developed for the Boeing supersonic transport aircraft.

The nine-stage axial compressor of this engine is connected by a shaft to the two-stage turbine that drives it. The engine has an annular combustor, and uses an afterburner to improve performance.

Afterburning, a technique for increasing the maximum thrust output of a turbojet engine for short periods, may increase output by as much as 50% at take off and several times this amount at high flight speeds. Normally, only about 25% of the atmospheric oxygen passing through the main engine is used for combustion. In the afterburner some of this excess oxygen is burned with a large amount of fuel to increase the energy of the gases, and thereby the exit jet velocity and thrust.

Fanjet. Turbojet engine performance can also be improved by adding a fan which operates in an annular duct surrounding the engine. The resulting turbofan (fanjet) engine is becoming increasingly popular for subsonic transports, military aircraft, and business aircraft, because it operates so much more efficiently than the simple turbojet engine, thereby using less fuel. The fan may be located behind the compressor turbine, where it is driven by exhaust gases, but in actual practice it is usually ahead of the compressor, where it serves somewhat like a propeller, except that it is inside the engine (Fig. 3). Considerably more air is drawn through the fan than is needed for the compressor. The excess bypasses the main engine; that is, before it reaches the compressor, it is directed through ducts to the outside, producing additional thrust. In an engine of high bypass ratio (ratio of air ducted to the outside to air passing through the basic engine) as much as 85% of the thrust may be developed by the fan. *See* TURBOFAN.

Another way to increase jet engine performance is by raising turbine inlet temperature. However, even with the best materials available this generally means that turbine stators and blades must be air-cooled (Fig. 4).

Large turbojet and turbofan engines have thrust reversers that accomplish the same purpose as blade-pitch reversal on propeller-driven engines. They reverse the direction of the jet exhaust gases (or bypassed air from the turbofan engine) to provide a braking force directly opposed to the normal forward thrust force. They are used for braking during landing and for taxi control, and may eventually be used in flight. Two common types are the target and the cascade thrust reversers (Fig. 5).

Turboprop engine. The turboprop engine (propjet) consists of the basic turbine engine plus a reduction gear through which the turbine drives a propeller to produce thrust. Additional thrust is also obtained from the jet exhaust. (For this reason, turboprop engines are rated in equivalent shaft horsepower—horsepower produced at the output shaft plus a correction to take into account the exhaust thrust.) When designed with shafting suitable for operating helicopters and other shaft-driven vehicles, this engine is termed a turboshaft engine. *See* TURBOPROP.

The prime advantage of the turboprop arrangement is that the engine can be designed to deliver the best possible shaft power, while the propeller can be designed for best possible handling of large masses of air. This combination of turbine and propeller is highly efficient up to medium-high subsonic speeds. At takeoff and in climb the turboprop produces more thrust than the same size of turbofan or turbojet engine, but thrust falls off sharply at higher speeds and higher altitudes.

Turboprop engine performance may be improved by (1) regeneration, the use of the exhaust gas to preheat the compressed air before it enters the combustors; (2) reheating, the injection and burning of fuel between stages of the turbine after the first stage; and (3) afterburning, also used with turbojet engines, as already discussed.

Another important modification of the turboprop engine uses a free turbine. The propeller is driven by this separate turbine unit (also termed the power turbine) and is supplied with exhaust gases from the compressor-driving front turbine unit (termed the compressor gas generator turbine). This arrangement (Fig. 6) gives greater control flexibility, easier starting, better takeoff performance, and lower propeller speed for cruising than the fixed-shaft system when applied to lightweight fixed-wing business and commercial aircraft.

The extremely small size of the turboprop engine shown in Fig. 6 is achieved by some unusual design features. The air, brought directly to the inlet of the compressor at the front of the engine, is compressed by the axial-centrifugal compressor and then delivered through a vaned diffuser to a collector scroll. The compressed air then flows through two external tubes (located on either side of the engine at centerline height), which carry the air to the single combustion chamber. This, in turn, is located at the rear of the engine coaxial with the compressor turbine shaft. Combustion gases leaving the combustor move forward through the turbine into an exhaust collector in the middle of the engine, where they are collected and exhausted downward.

Turbine trends. The immediate future of the turbine engine appears to lie in improvements to the present basic types rather than in the development of hybrids, such as the already discarded (for aircraft) turbocompound engine (piston engine combined with turbine) or the turboramjet. (The latter may eventually power hypersonic transports at 3000–7200 mph or 1.3–3.2 km/sec.)

The present 4:1 to 6:1 thrust-to-weight ratios will reach 10:1 or 12:1 during the 1970s. Engine life will equal airframe life, with hot parts being replaced as necessary.

To help make these advancements a reality, compressor ratios will rise to 25:1 and, if fans are used, the overall cycle pressure ratio may be above 35:1. This increase will improve thermal efficiency, lowering specific fuel consumption. For the same pressure ratio, the number of compressor stages might be reduced by 30%.

Turbine inlet temperatures may eventually reach 3700°F or 2000°C (the maximum possible with hydrocarbon fuel). Experimental turbine engines have been run at over 5000°F (2800°C) using high-energy fuel. These extreme temperatures

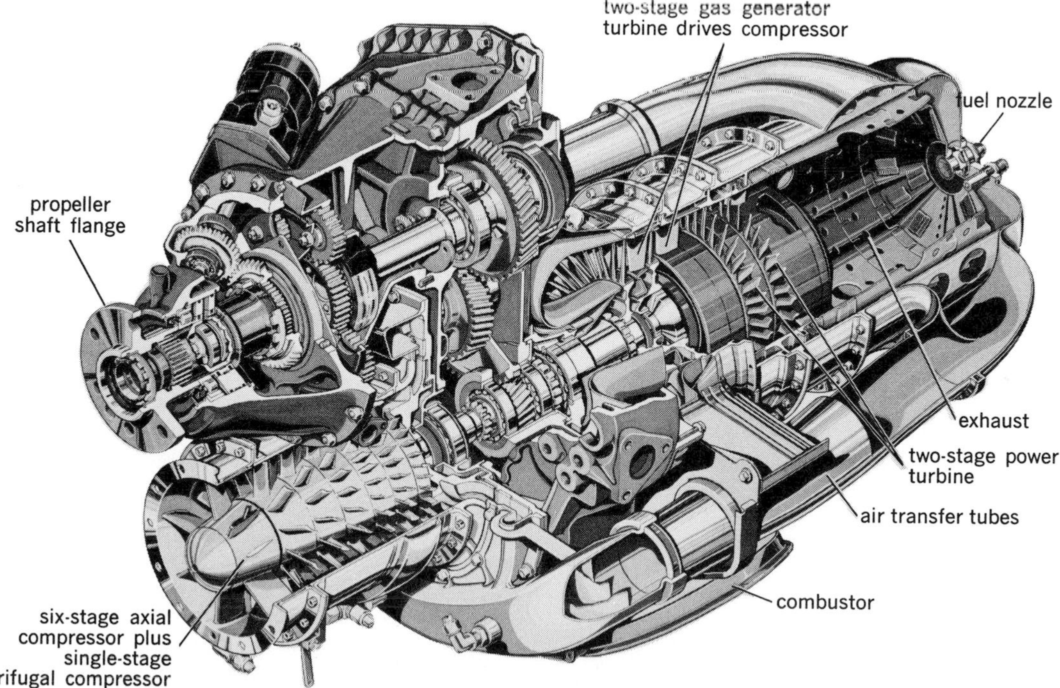

Fig. 6. Allison 250–B15 turboprop engine produces 330 equivalent shaft horsepower (246 kW), weighs 164 lb (74 kg). It has a six-stage axial compressor, centrifugal compressor, two-stage gas generator turbine, and two-stage power turbine. (*Allison Division, General Motors Corp.*)

will require further refinements in turbine blade cooling techniques, such as film and convection cooling.

Combustors will become more compact. For example, an experimental 20 in. (51-cm) combustor has been designed that uses scoops to get higher air pressure into the mixing zone than is possible with conventional four-hole combustors.

Materials that can maintain strength and resist oxidation and sulfidation at high temperatures must also be developed. Such materials may include new nickel- and titanium-base alloys, special coatings able to resist high temperatures, and plastic- and metal-matrix composites which provide stiffness and strength at reduced weight. The last may consist of continuous filaments of boron or of silicon carbide embedded in a matrix of epoxy, aluminum, or titanium.

Automotive applications. The turbine engine, either separately or in combination with a piston engine, is beginning to find application in some types of surface vehicles because of its light weight, compactness, inherent reliability, starting ease, and ability to burn a wide range of fuels.

Two divergent trends are evident. A small but increasing number of boats and ships, trains, and racing cars requiring large amounts of power are being powered by suitably modified aircraft-type turbine engines.

For cars, trucks, buses, tractors, and other vehicles requiring smaller, more compact, and generally less powerful engines, research continues. Companies are road-testing experimental units, but no commerically marketable turbine-powered vehicle of this type has yet been developed.

The many advantages of the turbine engine make it a potentially serious threat to the piston engine, especially in heavy-duty vehicles, but its use is limited by high fuel consumption and other economic factors. Although fuel consumption has been reduced nearly to that of the diesel engine, manufacturing and materials costs, as well as operating costs, are higher than those of the diesel. Mass production will help bring production costs down, but the starting costs of retooling for an entirely different kind of engine remain. It will also be necessary to provide production and maintenance facilities manned with trained and experienced personnel. *See* DIESEL ENGINE; INTERNAL COMBUSTION ENGINE.

The following brief descriptions of a few of the many engines being developed show the great interest in making the turbine engine more commercially feasible.

Turbine test car. Among the advanced railroad cars being built and tested is a turbine-driven unit powered by two AiResearch single-shaft 535-shaft-horsepower turbine engines mounted underneath the floor and geared directly to the wheels at either end (Fig. 7). The train has attained a top speed of 100 mph (45 m/sec).

Turbine test tractor. A heat-exchanger turbine engine is being road-tested in a 170,000-lb (77,000-kg) tractor-trailer. The 600-hp (450-kW) engine weighs 1475 lb (670 kg), about one-third that of a comparable diesel engine. It has two centrifugal compressors, three turbines, and two combustors (Fig. 8).

Turbine engine truck. The General Motors GT-309 280-hp (208-kW) turbine engine is designed for vehicles weighing 78,000 lb (35,000 kg) and has been tested in trucks and a bus. Its fuel consumption is close to that of a diesel, and its installed weight is about half that of a comparable piston engine.

Turbine engine car. The Chrysler 130-hp (97-kW) A831, which weighs 410 lb (186 kg), was installed in 50 cars for assessment by selected customers. Extensive field tests showed that the engine had poorer fuel consumption and acceleration than expected, and was likely to have electrical trouble. Testing was terminated, but from the findings Chrysler is continuing turbine engine development.

Turbine engine racer. The first turbine-driven automobile was demonstrated in England in 1950 by Rover, a company since in the forefront in the

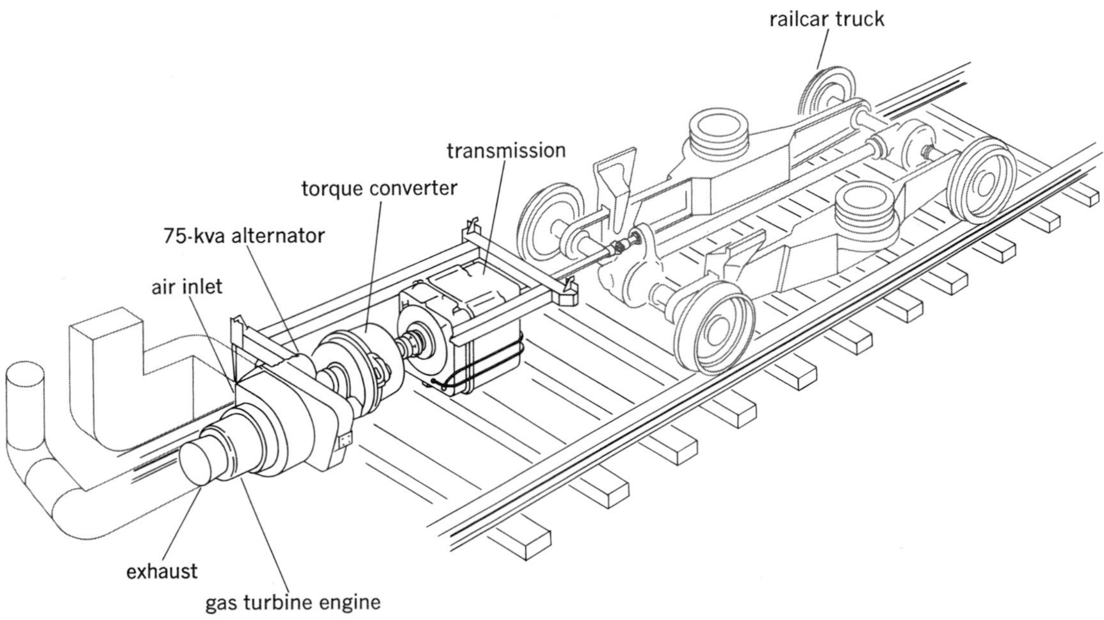

Fig. 7. Propulsion system for Budd turbine-powered rail car.

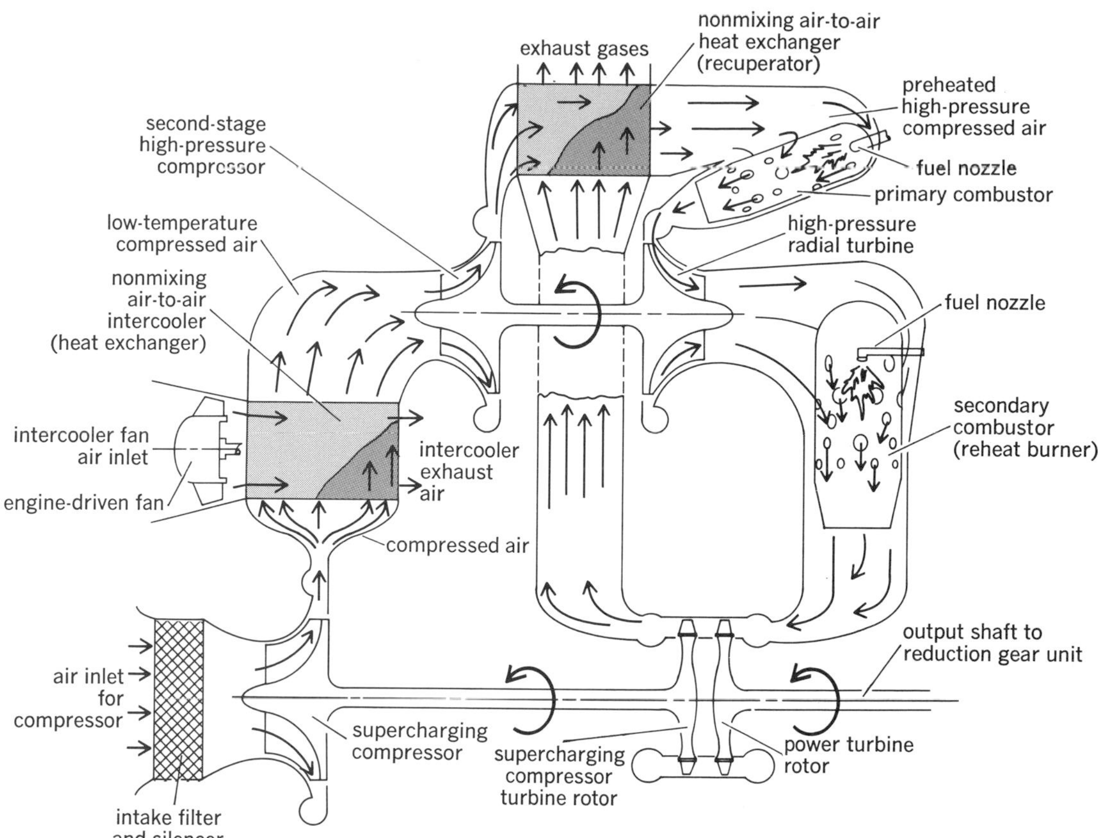

Fig. 8. Ford 705 turbine engine is supercharged to maintain fuel economy at low power conditions. Supercharging is accomplished by adding a supercharging compressor, turbine and burner, and an intercooler to the basic turbine engine.

development of turbine engines for cars. The first turbine-powered racing car, which ran at Le Mans in 1963, used a non-heat-exchanger 2S/150 Rover turbine engine. Rover's latest model, the 2S/150R is a twin-shaft recuperative engine developing 145 hp (108 kW). Its single compressor is driven by a turbine that also drives accessories, including twin-disc rotary heat exchangers mounted on either side of the engine. The annular combustor is between these heat exchangers.

Free-piston engine. The free-piston engine combines a free-piston gasifier, which is an air compressor run by a diesel engine, with a power turbine run by the gasifier exhaust. After considerable work with this type of engine, Ford and General Motors suspended studies. There are many difficult problems to be solved before it can be commercially useful. In Great Britain, France, the Soviet Union, and in Japan, research is continuing on this type of power plant, which has already been used as a stationary unit and has been applied to locomotives and ships. Its advantages include useful power-weight ratio, fuel economy, ability to run on a wide range of fuels, and the fact that the gasifier and power turbine can be located far apart, connected only by hot gas ducts.

[ELEANOR ALLEN]

Bibliography: T. Baumeister (ed.), *Standard Handbook for Mechanical Engineers*, 7th ed., 1967; D. G. Shepherd, *Introduction to the Gas Turbine*, 2d ed., 1960; F. G. Swanborough, *Turbine-Engined Airliners of the World*, 1963.

Turbofan

An air-breathing jet engine with operational characteristics between those of the turbojet and the turboprop. Like the turboprop, the turbofan consists of a compressor-combustor-turbine unit, which is called a gas generator and a power turbine, which in this case drives a low-pressure-ratio compressor called a fan.

Operating principle. The gas generator produces useful energy in the form of hot gas under pressure. Part of this energy is converted through the turbine and the fan that it drives into increased total pressure of the fan airflow, which is expanded

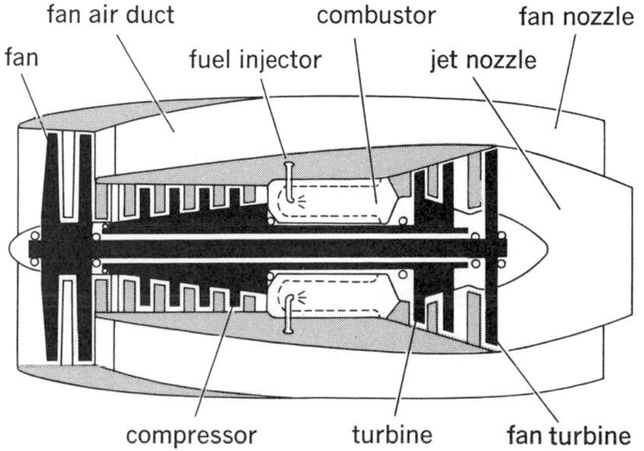

Fig. 1. Aft-fan configuration.

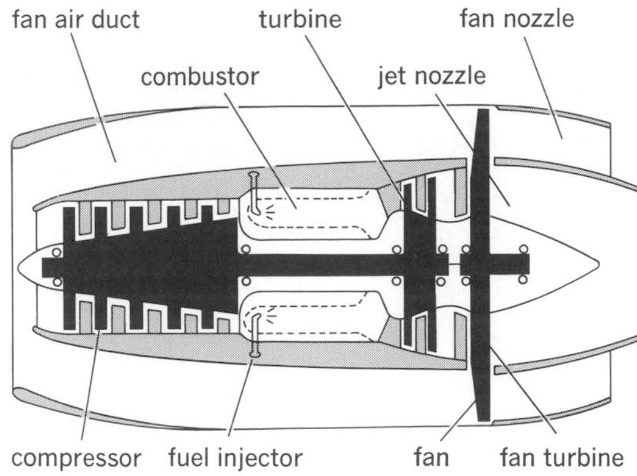

Fig. 2. Dual-rotor front-fan configuration.

in a fan air nozzle and is thereby converted into kinetic energy. The rest of the gas generator energy is converted into kinetic energy through expansion in a jet nozzle. In this manner the turbofan produces useful thrust through two separate streams, the gas generator flow and the fan flow.

There is an optimum energy split between both streams which is a function of component efficiencies, temperature ratios, and flight speed. Both streams may be separately discharged through concentric jet nozzles or mixed before expansion in a common jet nozzle. Mixing the mass flows improves the efficiency slightly and reduces the noise level.

Design characteristics. The ratio of fan mass flow to gas generator mass flow is called the bypass ratio. Bypass ratios may range from 0.5 to 8, depending upon the desired operational characteristics.

The turbofan cycle is inherently superior in propulsive efficiency at subsonic flight speeds, and provides increased thrust levels for takeoff and climb compared to the turbojet cycle. It produces the required thrust by accelerating a larger mass of air by a smaller velocity increment which requires less energy. Compared to the turboprop, it offers higher speed and therefore higher productivity for transport aircraft, as well as less mechanical complexity. For these reasons turbofan engines are the preferred power plants for subsonic jet-type aircraft. *See* TURBOJET; TURBOPROP.

Higher turbine inlet temperatures in connection with high cycle-pressure ratios, permit increased bypass ratios and therefore improve the overall efficiency of the turbofan. The recent trend toward increased bypass ratios is a direct result of progress toward higher turbine inlet temperatures made possible through cooled turbines.

Configurations. A variety of turbofan configurations is possible. In most cases the fan rotor is kept mechanically separate from the gas generator rotor because proper matching of the tip speeds of fan, compressor, and turbines requires independently variable rotational speeds in each rotor.

The aft-fan configuration (Fig. 1) is attractive if an existing turbojet engine is to be converted into a turbofan with minimum modifications. The dual-rotor front-fan configuration (Fig. 2) utilizes the fan to supercharge the gas generator. It is generally used for advanced technology turbofans suitable for subsonic or, with reheat, for supersonic applications.

High-bypass turbofans (Fig. 3) are excellent power plants for efficient subsonic passenger or cargo transports of all sizes. Their use may ultimately also penetrate the field of small business jets and private aircraft. Fans with high bypass ratios are also suitable to produce vertical lift in VTOL aircraft, designed to take off and land vertically without extensive runways.

[PETER G. KAPPUS]

Bibliography: N. E. Borden, Jr., *Jet Engine Fundamentals*, 1967; J. V. Casmassa, *Jet Aircraft Power Systems*, 3d ed., 1965; N. W. Sawyer, *Gas Turbine Engineering Handbook*, 1966; D. G. Shepherd, *Introduction to the Gas Turbine*, 2d ed., 1960; M. J. Zucrow, *Aircraft and Missile Propulsion*, vol. 2, 1958.

Turbojet

A propulsion engine used in many high-speed military and commercial aircraft.

Operating principle. In a turbojet, as illustrated, air approaches the inlet diffuser at a relative velocity equal to the flight speed. In passing through the diffuser the velocity of the air is decreased and its pressure is increased. The air pressure is increased further as it passes through the compressor. In the combustion chamber a steady stream of fuel is injected into the air and combustion takes place continuously. The high-pressure hot gas passes through the turbine nozzles, which direct it at high velocity against the buckets on the turbine wheel, thereby causing the wheel to rotate. The turbine wheel drives the compressor to which it is connected through a shaft. This is the sole function of the turbine.

After the hot gas leaves the turbine, it is still at a high temperature and at a pressure considerably above atmospheric. The hot gas is discharged rearward through the exhaust nozzle of the engine at a high velocity. *See* BRAYTON CYCLE.

The thrust obtained is equal to the overall increase in momentum of the gas as a result of its passage through the engine. This thrust is given by the equation below, where M is the mass flow of

$$F = M(V_j - V_o)$$

gas per second through the engine, V_j is the exhaust jet velocity, and V_o is the airplane velocity.

Compressors. Two types of compressors are used on turbojet engines—centrifugal-flow and

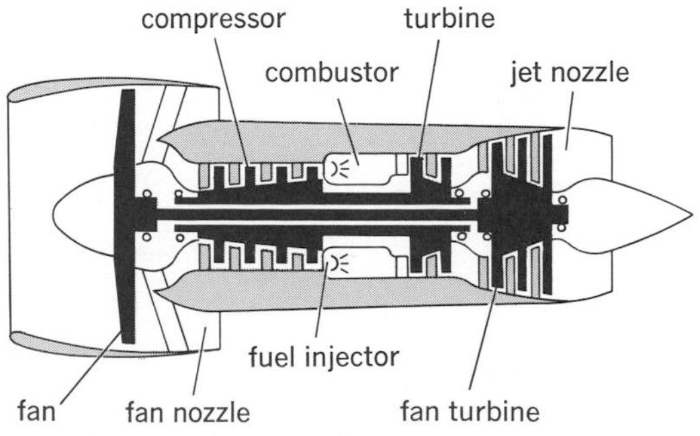

Fig. 3. High-bypass turbofan.

axial-flow compressors. The centrifugal compressor is the simpler of the two and was used on the early versions of the engine. For example, it was used on the first British engine designed by Frank Whittle, which was the forerunner of a series of early American engines, namely, the I-A, I-16, and I-40 (which became the J-33).

The trend has been consistently toward the application of the axial-flow compressor because of its greater efficiency and greater air-handling capacity per unit frontal area, in spite of its complexity and fragility. Engines such as the J-57, the J-75, and the J-79 are of this type.

Compressor stall. One of the problems associated with the axial-flow compressor is that of designing the stator vanes to direct the air into the rotor vanes at the proper angle at all rotational speeds. If the air is directed at a vane at too sharp an angle, it will not follow the vane but will break away. Loss in pressure and efficiency occurs and even possibly vibration of the blade.

Axial-flow compressors are designed for the full-power condition. At engine rotative speeds below 70% of maximum, the compressors on some engines develop a rotating stall. One or more stall cells or areas of stall form around the compressor; in these cells the flow strikes the blades at an improper angle of attack. The stalled areas move around the compressor, and each time one passes a given blade, the blade is subject to an impact. If these impacts come in resonance with a natural frequency of the blade, severe and sometimes destructive vibration may result.

The problems of low efficiency and stall at low rotative speeds are particularly severe in compressors with many stages. One solution is to separate the compressor into two sections, each mounted on a separate shaft coaxial with the other and driven by separate shaft coaxial with the other and driven by separate turbines. An engine of this type, called a two-spool engine, is the J-57. Another solution to these problems is to provide a mechanism for adjusting the angular setting of the stator vanes with change in rotative speed. The J-79 engine uses this method. Interstage bleedoff of air at low rotative speeds is also used to assist in combating stall on some engines, as in the J-57. Other schemes have also been investigated.

Foreign object damage. Another serious problem has been the ingestion of foreign objects by the engine. A foreign object passing through an axial compressor may cause blade failure or may produce a nick in a blade, which can become the nucleus of a fatigue failure. The failure of one blade can set off an avalanche of failures as it passes through the compressor. Compressor failures have been caused by objects picked up from the external environment, by objects shaken loose from the engine, or by objects left by mechanics. These later two sources can be eliminated by careful design and maintenance.

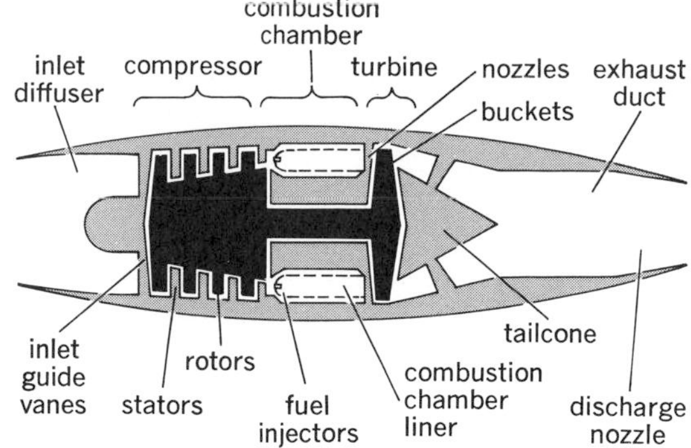

Diagram of an axial-flow subsonic turbojet engine.

Foreign objects have been sucked up from runways during takeoff and landing with the assistance of vortexes generated by the engine. Forward-directed air jets, which operate during takeoff to break up these vortexes, have been suggested. Screens are also used to stop the larger particles. However, particles smaller than the screen mesh can get through. Screens have the disadvantages that their resistance to the airflow imposes a loss in performance; furthermore, they provide a surface to which ice can adhere in an icing atmosphere. For this reason, they are designed to be retracted once the airplane is aloft. By keeping runways clean and by application of these various aids, the possibility of foreign object damage can be greatly reduced.

Control. The following are some conditions encountered in turbojet operations which are associated with engine control:

1. Overtemperature of the combustion gas during start-up can damage turbine buckets.
2. Overspeed of the turbine and overtemperature of the combustion gas at maximum thrust can overstress turbine wheels and buckets.
3. Compressor surge during acceleration of the engine may cause damaging vibration of compressor blades.
4. Flameout (cessation of combustion) can occur when engine speed is reduced at high altitude.
5. Flameout and also compressor surge can occur when an engine is jockeyed in the course of landing the airplane.

Automatic engine controls are designed to relieve the pilot of the task of avoiding these undesirable operating conditions. The earliest automatic

Maximum continuous rating of Pratt and Whitney J-57 engine*

Flight speed, knots	Sea level: Thrust, lbf	Sea level: Specific fuel consumption, lb/(hr)(lbf)	25,000 ft (7.62 km): Thrust, lbf	25,000 ft (7.62 km): Specific fuel consumption, lb/(hr)(lbf)	45,000 ft (15.24 km): Thrust, lbf	45,000 ft (15.24 km): Specific fuel consumption, lb/(hr)(lbf)
0	9500	0.76				
200	8000	0.87	4700	0.82	2005	0.85
400	7650	0.99	4750	0.92	2250	0.93
600	7600	1.07	5000	0.98	2500	0.98

*1 lbf = 4.45 N. 1 lb/(hr)(lbf) = 2.83×10^{-5} kg/(sec)(N).

controls incorporated only a maximum rotative speed governor and an ambient-pressure-sensing element that adjusted fuel flow for change in altitude.

The trend has been toward more sophisticated automatic controls that handle a greater number of the control requirements at the expense of greater complexity.

Typical performance. For the purpose of illustrating the performance of turbojet engines, the maximum continuous performance of an engine is displayed in the table. *See* PROPULSION.

[BENJAMIN PINKEL]

Turboprop

A gas turbine power plant producing shaft power for aircraft using a propeller. The turboprop engine has basic components similar to those of a turbojet: compressor, combustor, and turbine. In addition, it has a power turbine. This power turbine extracts usable shaft power from the engine mass flow and drives a conventional propeller through a reduction gear. Figure 1 illustrates the simplest possible turboprop configuration. Here the power turbine is rigidly connected to the main engine rotor; hence this type is commonly called a single-shaft engine. This has certain disadvantages in regard to starting, partial-load fuel consumption, and windmilling drag. *See* GAS TURBINE; TURBINE.

Figure 2 shows a more advanced turboprop type. Here the power turbine has no mechanical connection with the main engine rotor and consequently operates at a different rotational speed, this type being called a free-turbine engine. This arrangement improves the operating characteristics and, especially, the cruise efficiency under partial-load conditions. This form of engine is also used to drive helicopter rotors and is then called a gas turbine.

Performance characteristics. Compared to the turbojet and, to a lesser degree, to the turbofan, the turboprop offers lower fuel consumption and a higher takeoff thrust. It has low engine noise level. Its propellers can be reversed to shorten the landing run. For these reasons the turboprop is an excellent power plant for aircraft in which these qualities are important. Its disadvantages are heavier weight and increased complexity and maintenance cost. Because of propeller characteristics, the turboprop usually reaches peak operating efficiency at lower cruise speeds than the turbofan and is, therefore, better suited for transports in the speed range below 450 mph (200 m/sec), although it is basically possible to reach high subsonic and even supersonic fight speeds with a turboprop. At high altitudes the turboprop achieves lower cruise fuel consumption levels than those of the best reciprocating engines (approximately 0.34 lb of fuel per equivalent shaft horespower-hour or 0.21 kg of fuel per kWh).

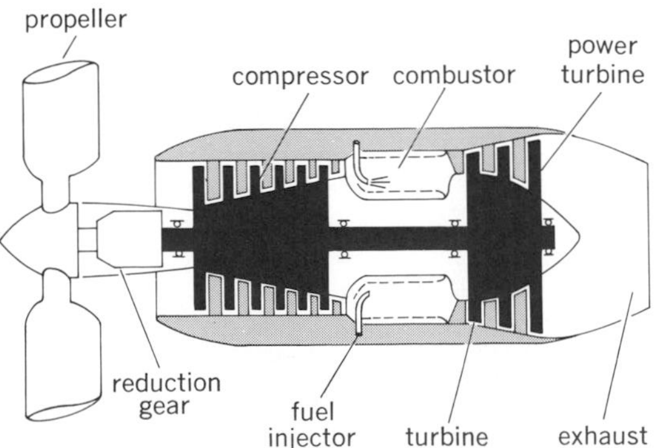

Fig. 1. Turboprop engine with connected power turbine.

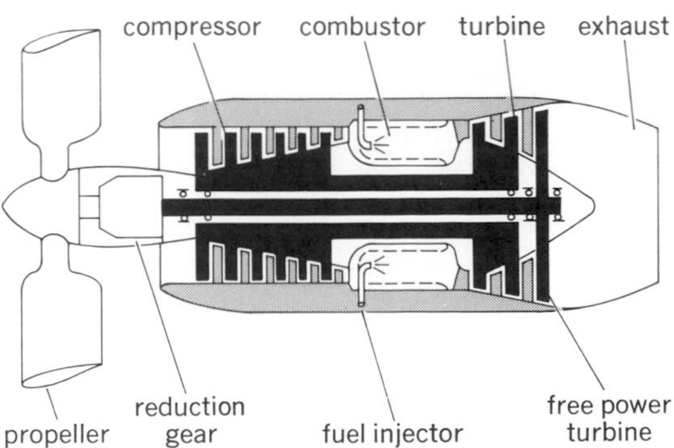

Fig. 2. Turboprop engine with free power turbine.

Design considerations. For maximum efficiency the turboprop engine should have a high pressure ratio (between 8 and 14) and high turbine-inlet temperatures (2000–2400°F or 1100–1300°C). Its specific fuel consumption, in contrast to that of the turbojet cycle, decreases with increasing turbine-inlet temperatures.

The design limitations of the turboprop are similar to those of the turbojet, and it is subject to similar problems. *See* TURBOJET.

Its high pressure ratio leads to compressor stall under adverse operating conditions and requires appropriate measures for control; interstage bleed, variable stators, or a twin spool rotor are examples.

Control problems. The turboprop has the same control variables as the turbojet, as well as a number of additional ones. These include fuel flow, affecting rotor speed, and variable stator position or bleed valve position, depending upon the system selected. In addition, there is the propeller pitch control which, with fuel flow, affects the speed of the power turbine. It is generally desirable to maintain gas generator speed at a high level and to select the desired horsepower level through proper adjustment of the fuel flow and propeller pitch. The free power turbine configuration allows propeller rotational speed to be an additional variable.

Compensation for altitude and ram air density is required as in the turbojet engine.

Regeneration. The efficiency of a turboprop engine can be increased through use of a regenerator, especially on engines with only a moderate pressure ratio. Early designs attempted to incorporate this feature. Advances in compressor technology made it possible to obtain almost equivalent fuel consumption levels through high pressure ratios and high turbine-inlet temperatures at lower overall complexity and weight, so that regenerators are not widely used in modern turboprop engines. *See* BRAYTON CYCLE. [PETER G. KAPPUS]

Bibliography: J. V. Casamassa and R. D. Bent, *Jet Aircraft Power Systems*, 3d ed., 1965.

Turboramjet

An aircraft engine in which intake air is compressed at high flight speeds by the ram effect of the intake moving into the air ahead of the engine and at low flight speeds by the compressor rotating behind the intake, and in which propulsion is produced by a high-speed exhaust. For operation from standstill to high flight Mach numbers, the turboramjet engine provides useful performance characteristics. The compressor is driven by a turbine for engine operation at low flight speeds. At high supersonic flight speeds, the ram entry of air into the engine makes the compressor and its turbine superfluous. Under this operating condition, the fuel is burned to greater advantage in a combustion chamber downstream of the turbine instead of between compressor and turbine as in the turbojet mode. Also, this second combustion chamber in back of the turbine can provide additional thrust at low speed when the engine is operated as a turbojet. *See* RAMJET; TURBOJET.

Turbojet with afterburning. The turbojet with afterburning is the simplest of the class of engines called turboramjets. The chamber for afterburning is provided with fuel injection nozzles and flameholders (Fig. 1). The temperature of the gas from the primary combustion chamber is limited to a value below stoichiometric to avoid weakening the blades of the turbine. This limits the fuel burned in the primary burner to an amount that consumes about one-third of the available oxygen. The remaining oxygen is used for combustion of the fuel in the secondary combustion chamber (afterburner). A typical temperature of the gas from the turbine in subsonic flight is in the order of 1300°F (1760°R or 980 K). This temperature can be raised to about 3000°F (3460°R or 1920 K) in the afterburner. If the pressure drop in the secondary burner is negligible, then, to a first approximation, exhaust jet velocity V_j is proportional to the square root of the temperature of the gas approaching the exhaust nozzle. Hence Eq. (1) is obtained. The

$$\frac{V_{jab}(\text{with afterburning})}{V_j(\text{without afterburning})} = \sqrt{\frac{3460}{1760}} = 1.40 \quad (1)$$

thrust F (lb) of the engine to a first approximation is given by Eq. (2), in which V_0 is the flight speed

$$F = M(V_j - V_0) \quad (2)$$

(ft/sec) and M is the mass flow rate of gas passing through the engine (slugs/sec). (One slug equals 32.2 lb. In metric units, F is in newtons, M in kg/sec, and V in m/sec.) The mass flow of gas through a given engine with and without afterburning is the same. Hence, the thrust, with and without afterburning, is given by Eq. (3). At take-

$$\frac{F(\text{with afterburning})}{F(\text{without afterburning})} = \frac{V_{jab} - V_0}{V_j - V_0} \quad (3)$$

off, V_0 is close to zero; thus afterburning increases the thrust for takeoff by a factor of 1.4.

Effect of afterburning. The thrust augmentation by afterburning increases substantially with increase in flight speed. For example, at a flight Mach number of 2 above 36,000 ft (11 km) altitude, the value of V_0 is about 2000 ft/sec (600 m/sec), the turbine outlet temperature is about 1500°R (830 K), and jet velocity V_j without afterburning is about 300 ft/sec (900 m/sec). Under these conditions, afterburning to an exhaust temperature of 3460°R (1920 K) causes an increase in jet velocity given by Eq. (4). The jet velocity V_{jab}

$$\frac{V_{jab}}{V_j} = \sqrt{\frac{3460}{1500}} = 1.5 \quad (4)$$

is then 1.5 × 300 = 4500 ft/sec (2.7 km/sec). The thrust augmentation from Eq. (3) is given by Eq. (5).

$$\frac{F(\text{with afterburner})}{F(\text{without afterburner})} = \frac{4500 - 2000}{3000 - 2000} = 2.5 \quad (5)$$

For this case with afterburning, the thrust is 2.5 times as great as that without it. As a consequence, specific fuel consumption is 3/2.5 = 1.2 times as great. In spite of this disadvantage in higher specific fuel consumption, the afterburning engine has the desirable operational flexibility of augmented thrust for takeoff and for maneuvers. At a Mach number above 2 it is useful because of its lower specific engine weight and specific frontal area, although at Mach 2 the range of an airplane is about the same with or without afterburning. As flight Mach number increases, the disadvantage in specific fuel consumption of the afterburning engine tends to reduce. At Mach 3–4 and higher the afterburner engine is a more efficient type than the simple turbojet. *See* AIRCRAFT PROPULSION.

The afterburning engine must be provided with a variable-area discharge nozzle (Fig. 1) because of the large difference in discharge area required with and without afterburning and because of the area adjustment required with large variation in flight Mach number. For most efficient operation, an engine required to operate at supersonic flight speed should also have an adjustable inlet diffuser.

Engine controls for the secondary fuel flow, the variable-area nozzle, and the inlet diffuser are integrated with the primary engine control in the automatic control system to avoid serious instability in the operation of the engine. This integration of control functions is one of the difficult design problems associated with this type of engine. Another problem that usually imposes considerable development testing is the design of the flameholder and fuel injection system. This system should provide efficient combustion over the required range of afterburner fuel flow rates and flight altitudes, without introducing excessive pressure drop in the flow through the afterburner.

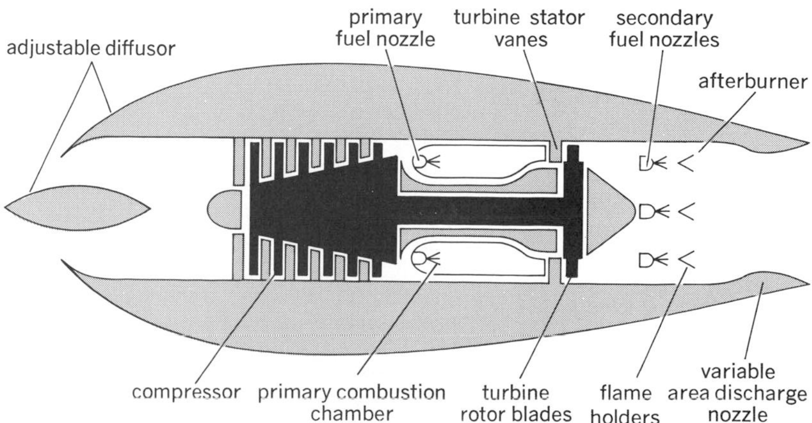

Fig. 1. Turbojet engine with afterburner.

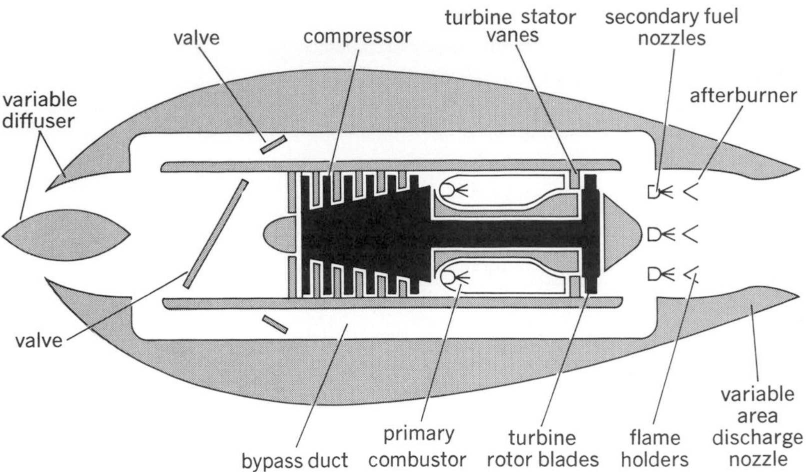

Fig. 2. Schematic diagram of dual-cycle engine.

Dual-cycle engines. Overall engine pressure ratio is the ratio of the pressure in the combustion chamber to the ambient air pressure. This overall pressure ratio is a major factor in determining the efficiency of such engines as the turbojet, ramjet, and turboramjet. A high pressure ratio has associated with it a large increase in compression temperature. In the turbojet engine, the combustion discharge temperature must be limited to a value that the turbine can withstand. The compressor discharge temperature must be considerably lower than this value to allow an appreciable amount of fuel to be burned in the primary combustor. These considerations place an upper limit on the overall pressure ratio desired in a turbojet engine.

As flight Mach number is increased, the pressure ratio across the inlet diffuser of the turbojet engine increases, and the pressure ratio provided by the engine compressor must therefore be decreased. In addition, ram compression raises the temperature of the air entering the compressor. The pressure ratios provided by a typical inlet diffuser at flight Mach number of 3 and 4 are 27 and 73, respectively. As flight Mach number increases through this range, it is desirable to eliminate the compressor and turbine from the path of the propulsion gas and thus convert the engine to a ramjet. Unfortunately, elimination of the turbojet engine from the system involves complications in ducting and valving. One method of accomplishing this is to provide ducts from the diffusor to the afterburner that bypass the turbojet engine (Fig. 2). Valves are provided in the bypass flow system and the turbojet engine system to divert the airflow from one system to the other.

Because of the importance for supersonic aircraft of drag minimization, an engine with a low frontal area is essential. This poses the problem in the system shown in Fig. 2 of designing a compact installation without introducing excessive loss in engine efficiency because of sharp bends and restrictions in the bypass ducts. The attainment of high reliability in the valves, valve activators, and controls is another requirement.

Simply to stop the flow of fuel to the primary burners when ramjet operation is desired and allow the turbine rotor to windmill is inefficient.

[BENJAMIN PINKEL]

Underground mining

An underground mine is a system of underground workings for the removal of ore from its place of occurrence to the surface, and involves the deployment of men and services.

There are several basic physical elements in an underground mining system. The passageways (openings) in a mine are called drifts if they are parallel to the geological structure, and cross-cuts if they cut across it. They range in size about 60–200 ft^2 in cross section, depending on their functions. The workings on a level (horizontal plane) are joined with those on another level by passageways of similiar cross section, called raises if they are driven upward and winzes if driven downward.

The passageways give access to, and provide transportation routes from, the stopes, which are the excavations where the ore is mined. The stopes are between levels. There may be rooms on the level, such as pump rooms, service shops, and lunchrooms. This article discusses exploration of a mine site, methods of removing ore material, and design of underground openings for ore removal and mine facilities. For other aspects *see* COAL MINING; MINING EXCAVATION; MINING MACHINERY.

EXPLORATION

A mine is designed and the mine openings specified after the exploration phase of a mine's history. Exploration in this context is not to be confused with prospecting. Exploration is the process of finding the characteristics of the mineralized rocks and the environmental rocks that make up the mine site. These attributes are absolute and unchanging, but they can only be predicted from sampling so there is always the risk that the predictions may be wrong. Some of the attributes of a mineral deposit and their limits of variation that are found by sampling and measurement are: shape, tabular or curvilinear; attitude, flat or vertical; dimensions, thick or thin, uniform or variable, long or short, or shallow or deep; physical character, hard or soft, strong or weak, laminar, jointed, or massive; mineralization, massive, globular, or disseminated, intense or sparse, or chemically stable or unstable; and surface and overlying formations, expendable or not expendable.

During the exploration there is a feedback from predictions of the revenue and expense that would result from operation. The end result of the exploration phase is a forecast of the grade (amount of valuable mineral per ton) and tonnage that can be mined at a specified rate. The ore grade acceptable could be different at another mining rate.

Parts of the risk involved are (1) a change in the mineralization or the environmental rocks, or both, as mining progresses, (2) drastic changes in the exchange value of the production relative to wages, equipment, and supplies, (3) the availability of new or better equipment, and (4) a change in the governmental attitude, such as in taxation. Any of these will affect the grade and the tonnage for the mine site.

MINE OPERATION

A mine is designed, developed, and worked in blocks of levels and stopes. The size of the blocks may be determined by the amount of ore that has been sufficiently explored, by geological boundaries, or by the need for effective supervision. The design must meet ventilation requirements and the openings must be maintained as long as they are needed. When mining in a block is completed it may, if expedient, be cut off from the ventilating system and the workings may be allowed to collapse. This can be done only if the failure will not disrupt other operations, for example, by causing rock bursts.

There are two basic plans of attack. The choice of plan depends on whether or not the surface, or the rocks overlying the ore, may be disturbed, and on stress redistribution problems.

Longwall. The principle of longwall mining is to advance in line all the stopes and pillars being mined in a block. No remnants are left either to support the back (overlying rocks) or to constitute stress concentrators. The line of attack may advance toward the shaft or other entrance or retreat from it. In the latter case, or if there are blocks beyond the current mining to be mined later, passageways must be maintained. Longwall mining is the method generally used to mine flat-lying deposits such as coal. Rooms with uniform dimensions separated by pillars with uniform dimensions are mined. The pillars support the backs. If the preservation of the backs or the surface is not a factor, the pillars may be systematically mined (robbed) in a longwall retreat or advance.

The result of mining with rooms and pillars is a cellular pattern. Most mining methods available to the mine designer involve the creation of a cellular pattern made up of stopes and pillars. Generally the ore from the stopes is won with less expense than is involved in recovering the pillars. Often the pillars are not mined because they are worth more as pillars than they would be as ore. They are stronger than any material that could economically be used to replace them, and they fit better.

Fill. There are few situations in which it is possible to recover a worthwhile amount of pillar ore unless the adjacent stopes or rooms have been filled. No filling will support the overlying rocks as well as the ore or pillars can, but if there is some settlement onto the fill, it will be limited because the rock is dilated and occupies more volume than the solid rock does. The swell will finally support the back.

The fill may be any incombustible available material—waste rock, sand or gravel, or mill tailings. The tailings from the mill are treated to meet the mine specifications for settling and percolation rates by removing some of the fine sizes. They are then transferred from the mixing plant to the stopes as a slurry by a pipeline and distributed in the stopes through hoses. Fine-grained natural sand may be placed that way as well. The water must be taken out by decantation or by percolation, or both. Some operators add portland cement to form a weak concrete. A mixture of smelter slag and sulfide-bearing mill tailings has been used to form a weak rock by chemical action.

When fill is placed in a stope alongside a pillar that is to be recovered, a partition, usually of light timber, is installed between the fill and the pillar. Some operators use an enriched mixture of cement at the interface to form a concrete.

MINE DESIGN

Fundamentally, mining is materials handling, and a mine is designed accordingly. Three functions are involved: breaking the ore from the face, delivering it to the surface, and delivering supplies to where they are needed. Ancillary functions are getting men and services, air, power, and water to where they are needed. Power may be electrical or compressed air.

Ore breaking and transporting together constitute 30–55% of the total cost of mining. If the ore has to be broken by explosives, the drilling and blasting cost is 10–20% of the total mine cost and the transportation portion is 20–35%.

Primary breaking cost varies inversely, and the transportation cost varies directly, with the size of the broken ore. There is an optimum size for the product of blasting. Each stage of the transportation phase has a limiting size that can be accommodated, so expense saved in the breaking phase may be exceeded by that of secondary breaking between stages in the transportation system.

A mine is designed around the method of primary breaking that is chosen, and the choice is governed by the forecasts from exploration. Stopes may be open or filled if the ore has to be broken by explosives. Caving may be used if conditions are favorable, and the ore will be broken by natural forces that make up the stress field in the ore and in the environmental rocks. Mining methods will be described under major headings, but since each ore body is unique and operators are ingenious, there are variations and hybrid methods.

Open stopes. There is a further qualification to this type of stope—with or without delayed filling. Using open stopes without delayed filling may be expedient, but if there are other extensive workings, the open stopes may redistribute the ground stresses in a manner which will interfere with subsequent work.

The mining method is chosen according to the thickness and inclination of the deposit. The breaking point between thick and thin in a tabular deposit is about 16 ft (4.9 m). That is about the maximum length of a stull (round timber) that can be handled conveniently in an open stope to give casual support where it is thought that loose rock might develop, or to provide a working platform for the miners. The breaking point between steep and flat is about 40° from horizontal, the limit at which rock will move by gravity on a rough surface.

The stope may be worked either overhand or underhand. If the ore will move by gravity to the drawpoints, the overhand attack is by advancing a breast (face) parallel to the level and so breaking out a slice. Because a platform to work off must be constructed with stulls and lagging or plank for each drilling site, the method is limited to thin deposits. The underhand attack is started from the top of the stope, usually from the access raise, and a block is broken out by drilling and blasting down holes. The bench must be cleaned off after each blast before drilling is resumed. This method has an advantage in gold mines, where coarse gold may lodge on the footwall and have to be swept out. Once cleaned, an area will not have to be cleaned again.

If the deposit is steep enough to deliver the broken ore to the drawpoints by gravity and thick enough to prevent the broken ore from arching over the opening, it probably should not be mined off platforms in an open stope.

Shrinkage stoping. An open stope requires successive platforms for the miners to work off. If the broken ore is drawn off just enough to give working room for the miners and they can work off the broken ore, the stope is called a shrinkage stope. The length of the stope is established by two raises, one at each end, which serve as manways and service entrances for air and water lines. By cribbing up with timber, they are maintained through the broken ore. The draw is from one-third to one-half of the break. The rest of the ore is retained in the stope until the stope is complete, and then the stope may be drawn empty through chutes which are installed during stope preparation.

The broken ore, which moves when drawn, has little ground-support capability. Shrinkage stoping may be used only up to the limit where the back is not self-supporting, or the walls will slab off and give unacceptable dilution. The limit may be extended by using casual timber support, or rock bolts for the back, and by rock-bolting potential slabs to the walls. Once a series of shrinkage stopes has been established and some of the stopes completed, the storage available provides flexibility in production beyond that of any other method.

If the ore body is too wide for the span of the back to be self-supporting or to be cheaply supported when mined parallel to the long dimension of the ore, it may be worked with transverse shrinkage stopes and intervening pillars. This involves delayed filling to permit mining the pillars.

Sublevel stoping. A deposit that is wide enough (about 40 ft) and has walls and ore sufficiently strong to permit shrinkage stoping may be worked by sublevel stopes (Fig. 1).

A vertical face is maintained. At vertical intervals, spaced to accommodate the drilling method to be used for breaking, sublevels are driven in the long direction of the stope from the entry raise. Two procedures are available after the initial slot has been made, which is generally done by widening a raise to the width of the stope. Holes may be drilled radially from the sublevels to give an acceptable distribution of explosive behind the face or slice to be broken off, or a cut may be slashed out across the face from each sublevel and drilled with vertical holes, quarry fashion, to give a better explosive distribution. The choice depends in part on the equipment available and in part on the control wanted at the sides. When the stope is wide, or for better control of the sides (walls), the radial drilling may be done from each of two sublevels at the same level which have been driven on either side of the stope.

Fig. 1. Sublevel stoping.

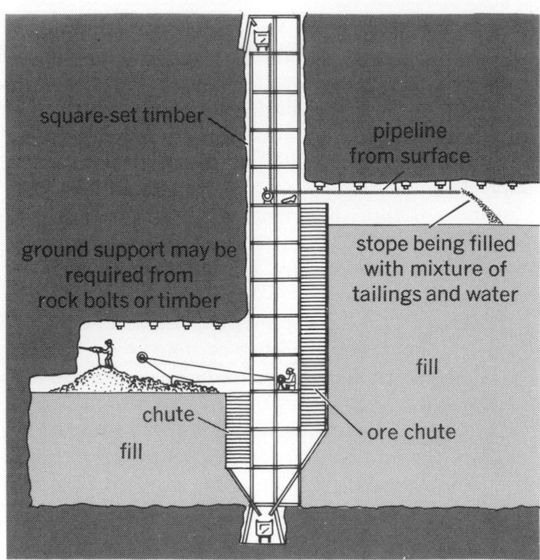

Fig. 2. Hydraulic-fill stope.

The stope may be longitudinal with respect to the long dimension of the ore body, or it may be transverse and separated from the adjacent stope by a pillar. It follows that it will be a delayed-fill stope if the pillar is to be recovered. Even delayed-fill stopes are not satisfactory if the pillar material is sufficiently valuable to require complete and clean recovery.

Filled stopes. When the wall rock or ore is not sufficiently strong to permit the use of one of the open-stope methods, or if clean pillar recovery is important, methods have been devised to mine with only a small area of unsupported rock or ore exposed.

Cut-and-fill stoping. The breaking phase of this method is not different from that which has been described for the overhand shrinkage stope, but the broken ore is removed after each breast (slice) is completed and a layer of fill is placed. The cycle is: breaking, removing the broken ore, picking up the floor, raising the fill level, and replacing the floor (Figs. 2 and 3). Before the new fill is placed, the manways and the ore chute are raised with cribbing. The cribbing is covered with burlap if hydraulic fill is used. The manway at one end of the stope is built with two compartments. One compartment is used as the ore chute, called a mill hole in filled stopes.

The back must be strong enough to be self-supporting over the span, but the method is sufficiently lower in cost than the alternative so that the back may be supported with casual timber or rock bolts, which are broken out with the new breast.

Square-set stopes. When the backs, and perhaps the walls, are not strong enough to permit cut-and-fill stoping, temporary support may be provided by carefully framed and placed timbers called sets. The sets are filled when no longer needed, at a rate that depends on the rate at which they take weight and might collapse (Fig. 4).

A set is made up of posts 8 or 9 ft (2.4–2.7 m) long, and caps and girts about 6 ft (1.8 m) long, cut to exact lengths and framed to give a good fit at the corners. The timber is usually about 8 in. (20 cm) square, though round timber may be used.

A set is installed in an opening just large enough to accommodate it, and then blocked against the surrounding unstable rock or ore. Little blasting is needed because of the characteristics of the rock that make square-setting the best choice. The problem is to hold back the broken rock until the set can be placed. This may be accomplished by extending boom timbers out over the caps.

Some of the ore may be moved manually. Generally, however, by retaining open sets in the fill and fitting them with inclined slides, provision is made for gathering the ore for scraping to the mill hole.

The sets may be installed either overhand or underhand, depending on the problem, or, if the choice is not critical, on the skill of the miners. The overhand technique is more common but in some camps the miners work better underhand.

Square-setting is the usual method for removing pillars if clean, complete extraction is needed. It is a flexible method and can be used to recover ore in offsets from the main deposit.

Caving. When the surface is expendable and other characteristics are favorable, one of the caving methods may be used.

Top slicing. In some respects this is like square-setting but it is less expensive after it is underway. It is used when the surface is expendable and the ore is too weak to stay in place over a useful span.

The mining block is developed by driving a two-compartment raise through the ore to serve as an access manway and a mill hole, and by driving a longitudinal drift from the raise, at the top of the ore, to the extremity of the ore or the end of the proposed stope.

The initial unit of mining is a timbered crosscut driven each way from the drift to the edge of the block to be mined. Subsequent units are crosscuts driven adjacent to each preceding crosscut to take out a slice of ore. As the face of the slice is retreated toward the mill hole, the timbers in abandoned workings (several sets back) are permitted to fail, or forced to fail by blasting. The routine is continued until the slice is completed to the raise. The overlying formations collapse onto the broken timbers. As the routine is continued by taking successive slices, the broken timbers form a feltlike mat that has some tensile strength, and little timber support is needed in the crosscuts. Several slices are mined concurrently, step fashion. The overlying caving formations must follow the mat. No caverns can be left which could collapse and create an air blast.

Sublevel caving. If the ore is sufficiently strong, and after a timber mat has been developed, one or more slices may be omitted. The cantilever shelf formed when the next slice is taken will collapse under the load of caved material and its own weight. The broken ore is moved to the mill hole as in top slicing. Several slices are advanced simultaneously as in top slicing.

An adaption has been used in which the slices are taken out as sublevels in open stoping are taken, and the over lying formations are caved against the face. The sublevel slices are advanced in steps as in top slicing (Fig. 5). There is no mat and some ore is lost into the cave material. However, it is low-grade and the overall low cost of mining makes up for the loss.

Block-caving. If the ore texture (blockiness) and strength are suitable, and if the primitive stress field is favorable, an entire block 150–250 ft on a side and several hundred feet high may be induced to cave after it is undercut. The broken ore is drawn off through bell-shaped drawpoints (Fig. 6).

Fig. 3. A cut-and-fill stope. (*a*) Ready for hydraulically placed sand fill, except for the burlap lining. (*b*) Placing the floor over the fill.

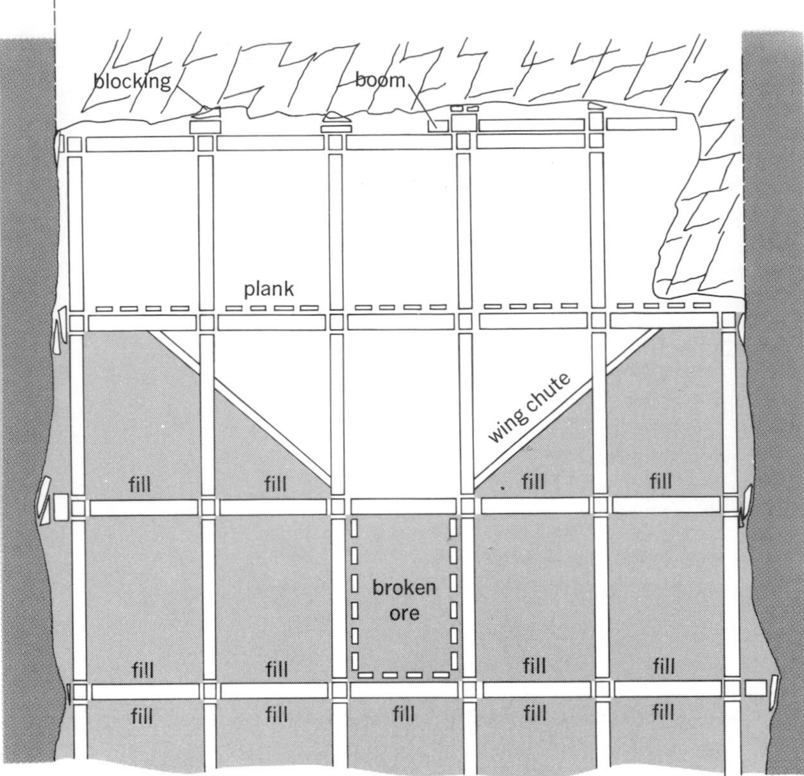

Fig. 4. Square-set stoping.

The drawing cycle is critical. It must keep the undersurface of the block unstable and continuing to fail. The lateral dimensions of the block are controlled by weakening the perimeter with raises and lateral workings, or even short shrinkage stopes. No large cavities are permitted to develop. In the final phase, when caving has reached the overlying formation, care must be taken to avoid drawing it

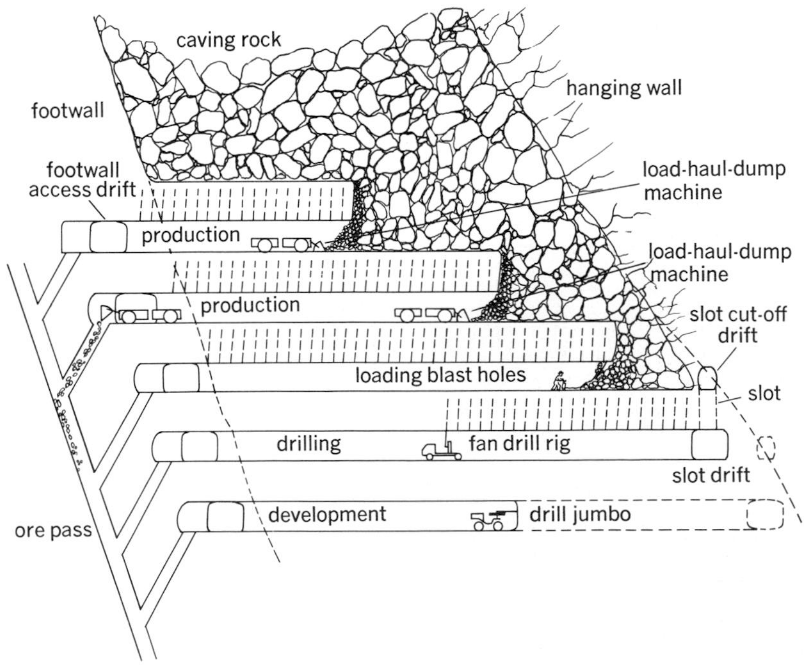

Fig. 5. An adaption of sublevel caving. Cutaway view shows progress of caving.

with the ore. There is no primary breaking expense but the cost of secondary breaking for transportation may be high.

Ground control. Mine openings must be kept open as long as they are needed. Mining engineers recognize that rock is not necessarily solid or inert. The study of the behavior of rocks when subjected to force is called rock mechanics. It is a comparatively new field, although knowledge of the phenomena under study has been utilized for years without formal analyses of what was going on.

The observations that rocks around a mine opening do not always behave in a manner that would be predictable by classical mechanics imply that there are other than gravitational forces involved, and that there is strong lateral component of strain energy. The source and reservoir of the strain energy have been less obscure since geologists have measured the rate of spreading of the North Atlantic Ocean floor and associated continental land masses (average 6 cm per year since Carboniferous times). The resultant force vector from the combination of gravitational force and tectonic force is referred to as primitive stress in this description of underground mining. The rock mass is in equilibrium until a mine opening is made; that is, it is in equilibrium for a relatively short time involved in the mine operation. The mine opening accepts no force and the force is diverted to around the opening.

Ground support. The ideal support is a pillar of appropriate size, but the use of pillars is not always feasible.

Timber. Traditionally timber has been the usual support for the perimeters of mine openings. It is usually supplied as stulls or as lagging, depending on the slenderness ratio, diameter to length, and to some extent on the use. If it is slender, it is lagging. If a log (stull) is placed vertically, it is called a post. If it is placed nearly horizontally, it is usually called a stull, whether it is acting as a beam or as a column. Both posts and stulls are installed with lower ends in hitches in the rock and upper ends loosely fitted to the back or wall, depending on the location. The final fit is achieved by driving wooden wedges between the end of the timber and the rock. The hitch may be chiseled into the rock, but in hard rock a natural recess is generally used.

When a lateral working requires timbered support, the stulls are usually framed to give a neat fit at the corners, and flatted on two sides to save space in the working. Sawn square timber is often used. The unit is two posts and a cap (stull), usually with a sill on the floor. Whether or not the sill is used depends on the expected loading. The posts and the cap are wedged tightly to the walls and the back at the corners. Lagging or plank is laid over the caps to provide overhead protection and placed behind the posts if a loose wall is expected. Raises and winzes are similarly protected unless they are to be used for hoisting and more precise timbering is needed. Steel is frequently used in the same manner as timber.

Concrete. Openings are frequently lined with concrete if permanence or added strength is needed, if the ventilation friction factor must be reduced, or if the operator does not trust timber because of the fire hazard. Generally the opening is made round or ovaloid so the concrete will be in compression. Forms and poured concrete are commonly used, generally with reinforcing bars. Circumferential steel reinforcement is not effective if the concrete is loaded in circumferential compression.

Concrete may be blown onto the rock surface with a cement gun (guniting). Sand and small-sized aggregate are mixed dry and blown through a hose to the face to be coated. Water is added as the mixture passes the nozzle. The low-moisture mixture hits the face and a portion of the aggregate falls out. A tight bond is formed at the concrete-rock interface. It is thought that the peining action of the aggregate helps to make the bond and to produce a dense concrete. In treacherous rock quite large rooms, such as underground hoist rooms 30 ft or more across, have been successfully secured in this way. The angle of impingement for the application is critical. The thickness of the coating is not more than a few inches (say 3 or 4) over the depressions in the rock surface and thinner over the bumps (Fig. 7).

Rock bolts. The systematic use of rock bolts for rock reinforcement has increased rapidly. These are steel bolts about 3/4 in. (19 mm) in diameter and generally 3–5 ft (1.5–2.1 m) long, anchored at the bottoms of holes drilled at a right angle to the rock surface, and tensioned by a nut over a small plate at the rock surface. The anchorage is generally a split shell forced against the wall of the bore hole by a wedge as tension is applied at the bolt end. Some suppliers offer a method to anchor the bolt in an epoxy resin, and some others supply a bolt that is to be embedded in concrete or cement for its entire length.

There is no consensus as to the reason that rock bolts are effective, but it is agreed that they should be installed as soon as possible after an opening is made, and should be under high tension (Fig. 8). The mechanism offered most frequently for the effectiveness of bolts in a bedded formation is that a compound beam is built up by binding several laminar beds together to act as a single thick beam. For massive rocks that have no bedding it is commonly accepted that the bolts must extend into the compression arch that is postulated to be formed above the opening, and for that reason long bolts are often specified.

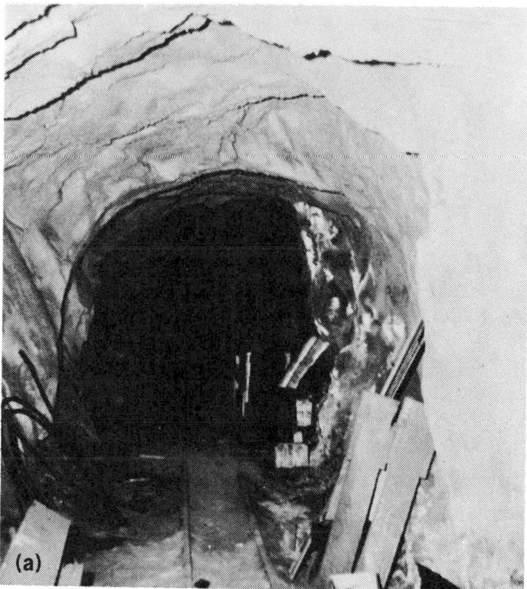

Fig. 7. The reinforcement of a mine opening with concrete. (*a*) A drift reinforced with gunite. (*b*) The same drift beyond the gunite.

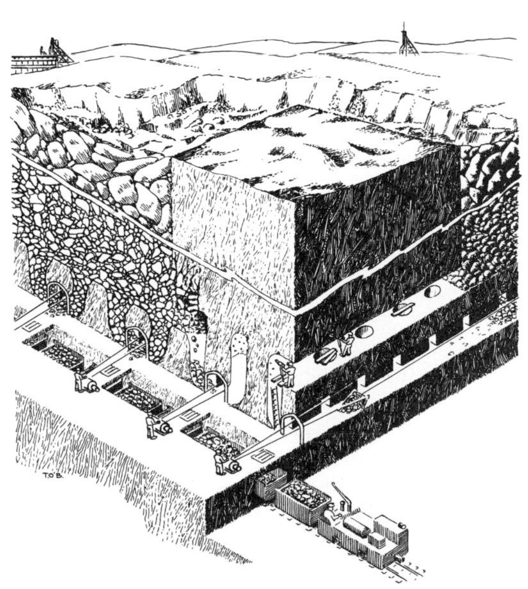

Fig. 6. Block-caving.

Actually, the abutments of the arch are restrained from moving outward, and the tendency is to move inward, especially if there is a high primitive stress lateral component. In narrow openings there is compression close to the skin of the opening and the function of the bolt in tension is to reduce the tendency for failure in oblique shear by preventing the thickening of the rock. At some width, as an opening is enlarged from narrow to wide, the compressive primitive stress is neutralized and the back goes into tension. Rock has little tensile strength because of discontinuities. If the rock is blocky, the blocks may be held together by bolts and form a flat, or nearly flat, voussoir arch.

Transportation. Gravity is used wherever it can be effective in the movement of ore toward the surface. Ore from open stopes that are steep enough for gravity flow is loaded through chutes directly

Fig. 8. Testing the tension on a rock bolt by using a simple instrument to measure torque.

into the level transportation units. When a stope is wider than about 25 ft, bell-shaped openings (drawpoints) are driven into the floor of the stope. If more than one row of them is needed, these drawpoints are driven on about 25-ft centers on a regular pattern so one crosscut can serve the outlets of several drawpoints. When the ore is loaded into the haulage equipment, it is taken to an ore pass and moved by gravity either to a loading pocket at the shaft or to a crushing plant, and thence to the shaftpocket. An inclined ore pass will give considerable lateral movement. A mine will also have a system of waste passes.

A drawpoint may discharge through a chute into a haulage vehicle or into a short branch off the haulage line and be loaded into the main-line vehicle mechanically. If the stope is wide, the drawpoints may discharge onto the floor of a scraper drift at a higher elevation than the back of the haulage level. The ore is then delivered to the main-line vehicle by scraping (Fig. 6).

When a lode is too flat to permit the use of gravity, ore is moved to a central gathering point by a scraper, either in one stage or two. If the lode is flat enough to permit the use of wheeled or crawler-tracked vehicles, the broken ore may be loaded into a gathering vehicle and taken either to the ore pass or to the main-line transportation unit. An alternative is to use a load-haul-dump vehicle. Smaller versions of this type of machine are being introduced into large stopes. They are displacing the scraper, which in turn had displaced the small railcar and the wheelbarrow (Fig. 5).

Equipment is designed to do a specific job and its value in use beyond that job decreases rapidly. The primary gathering equipment is designed for a short haul.

Entry from surface. When the topography of the area has low relief, the entry will be by a shaft or a ramp. Sometimes both means are employed, the ramp being used for moving heavy, large equipment within the mine. If the relief is high, an adit (tunnel) may be used.

A shaft is usually located in the footwall far enough from the mine workings to avoid ground movement. It is designed for specific functions which determine the area (cross section) and shape, if the shape is not modified to accommodate ground stresses and sinking problems. It may be vertical or inclined, though the vertical shaft is the more common. Functionally either kind should be rectangular to accommodate the equipment used in it.

Many vertical shafts are circular or elliptical, but a rectangular framework is fitted in them to guide the shaft vehicles. There is an exception, not common in North America, when rope (steel-cable) guides are hung in the shaft to guide the vehicles. On the other hand, a round or elliptical shaft is better for ventilation because it may be smooth-lined and offers less air resistance.

A shaft is designed after its functions have been decided and the rock conditions have been forecast. It may be multipurpose and the cross section (plan) must include space for each of the functions, as well as a ladderway for an emergency exit. A shaft may be specialized, that is, designed exclusively for ore hoisting, for services, or for ventilation. A mine must have two shafts to provide alternate routes to the surface in case one shaft is out of commission in an emergency. [A. V. CORLETT]

Bibliography: R. S. Lewis and G. B. Clark, *Elements of Mining*, 3d ed., 1964; R. Peele (ed.), *Mining Engineers' Handbook*, vol. 1. 3d ed., 1941; Society of Mining Engineers, *SME Mining Engineering Handbook*, 1973.

Universal motor

A series motor built to operate on either alternating current (ac) or direct current (dc). It is normally designed for capacities less than 1 hp (0.75 kW). It is usually operated at high speed, 3500 rpm loaded and 8000 to 10,000 rpm unloaded. For lower speeds, reduction gears are often employed, as in the case of electric hand drills or food mixers. As in all series motors, the rotor speed increases as the load decreases and the no-load speed is limited only by friction and windage. To obtain more constant speed with variations in load a centrifugal governor may be used to switch in or out a small resistor in series with the armature (Fig. 1).

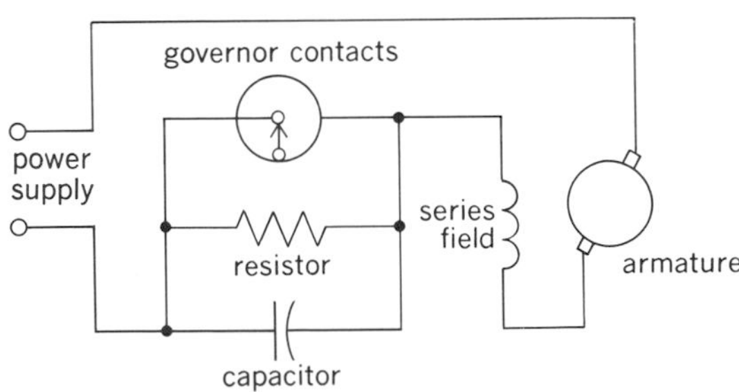

Fig. 1. Universal motor diagram.

If an alternating current is applied to any dc series motor, the motor would still rotate. Since the current is reversed simultaneously in the armature and the field, the torque would pulsate but would not reverse direction. However, a universal motor designed to operate on ac should have certain modifications: laminated cores to avoid excessive eddy currents, fewer turns in the field coils than in a dc motor, and more poles and usually more commutator segments. *See* DIRECT-CURRENT MOTOR.

UNIVERSAL MOTOR

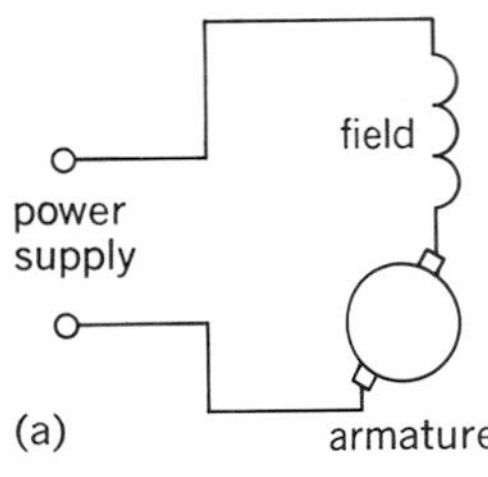

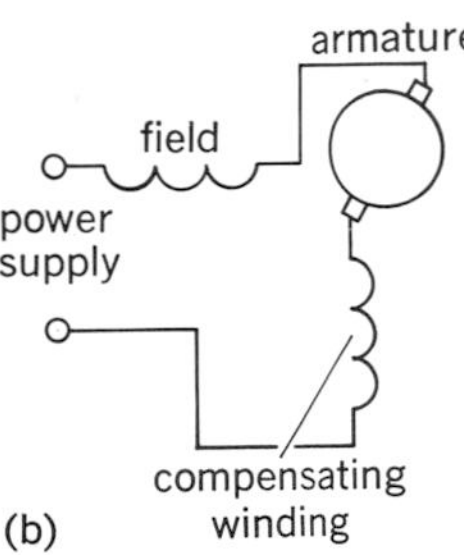

Fig. 2. Series motor diagram. (*a*) Without compensating winding. (*b*) With compensating winding.

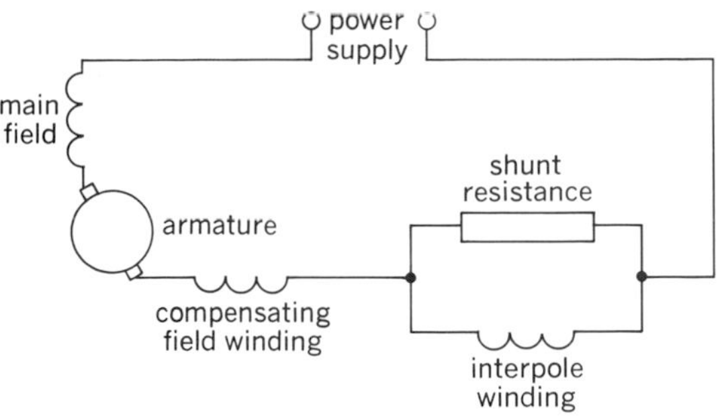

Fig. 3. Single-phase traction motor circuit diagram. (*General Electric Co.*)

The series ac motor is an alternating-current commutator motor which has great flexibility of performance. It can be operated over a wide range of speeds and is readily controllable. The series ac commutator motor is in many respects similar to the dc series motor and the universal motor.

The ac series motor, like the dc series motor and the repulsion motor, consists fundamentally of these windings or their equivalent: (1) rotating armature winding, (2) stationary field winding, and (3) compensating winding (Fig. 2).

A major problem in larger (up to 1000 hp or 750 kW) ac series motors is in commutation. Because of the transformer action between the field and armature coils, voltage is produced in the armature coils which are short-circuited by the brushes as the commutator bars pass under them. The coils which are short-circuited act like a short-circuited secondary of a static transformer. The resulting large currents are interrupted as the bars pass the brushes, causing bad sparking. In addition, these induced currents reduce the magnetic flux of the field and reduce the torque of the motor. Interpoles shunted with noninductive resistance are required on ac series motors as in Fig. 3.

The single-phase commutator motor usually has a large number of turns in the armature winding, more commutator segments, and a small number of turns in the field winding, as compared with the dc motor, which is designed for relatively strong field and weaker armature. The ac motor usually has more poles and operates at a lower voltage than its dc counterpart.

[IRVING L. KOSOW]

Bibliography: A. E. Knowlton (ed.), *Standard Handbook for Electrical Engineers*, 9th ed., 1957; I. L. Kosow, *Electric Machinery and Control*, 1964; R. R. Lawrence and H. E. Richards, *Principles of Alternating-Current Machinery*, 4th ed., 1953.

Uranium

A chemical element, symbol U, atomic number 92, atomic weight 238.03; one of the actinide series, in which the 5*f* shell is being filled. The name is derived from the planet Uranus. The valence electron configuration is $5f^36d7s^2$. Uranium was isolated in 1789 by Martin Heinrich Klaproth in a sample of pitchblende from Saxony. In 1841 Eugène-Melchior Péligot showed that the "semimetallic" element obtained by Klaproth was actually the dioxide. Péligot succeeded in preparing the metal by reduction of uranium tetrachloride with potassium. In 1896 Antoine-Henri Becquerel discovered that uranium undergoes radioactive decay. Discovery of the nuclear fission phenomenon by Otto Hahn and Fritz Strassmann in 1939 vaulted uranium from a position of relative obscurity to a role of major importance.

Uranium in nature is a mixture of three isotopes: ^{234}U (0.00054%), ^{235}U (0.72 ± 0.030%), and ^{238}U (99.275%). These abundance values may vary somewhat, depending on the origin or on the degree of depletion of the sample. Half-lives of the three isotopes are $(2.446 \pm 0.007) \times 10^5$ years (^{234}U), $(7.038 \pm 0.005) \times 10^7$ years (^{235}U), and $(4.4683 \pm 0.0024) \times 10^9$ years (^{238}U). Uranium-235, which was discovered by A. J. Dempster in 1936, undergoes fission with slow neutrons to release large amounts of energy. Uranium-238 absorbs slow neutrons to form uranium-239, which in turn decays to fissile plutonium-239 by the emission of two β-particles.

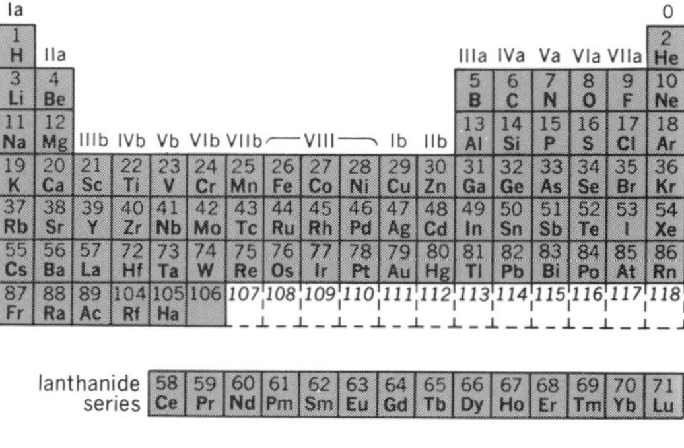

Other isotopes of uranium ranging in mass from uranium-226 to uranium-240 have been prepared by radioactive processes. Among these, fissile uranium-233 is obtained by the irradiation of natural thorium with neutrons. Thorium 232, the major component in natural thorium, absorbs slow neutrons to form thorium-233, which decays to fissile uranium-233 by the emission of two β-particles.

Natural occurrence. Uranium is believed to be concentrated largely in the Earth's crust, where the average concentration is 4 parts per million (ppm). For comparison, the crust contains 0.1 ppm silver and 0.5 ppm mercury. Basic rocks (basalts) contain less than 1 ppm uranium, whereas acidic rocks (granites) may have 8 ppm or more. Estimates for sedimentary rocks are 2 ppm, and for ocean water 0.0033 ppm. The total uranium content of the Earth's crust to a depth of 25 km is calculated to be 10^{17} kg; the oceans may contain 10^{13} kg of uranium.

Several hundred uranium-containing minerals have been identified, but only a few are of commercial interest. Table 1 summarizes data on some of the more important minerals. All uranium minerals contain lead, which results from radioactive decay of uranium, and thus contain an excess of the isotope ^{206}Pb. Uraninite, as found in pegmatites, usually occurs in rather small amounts which are of little economic significance. The euxenite-polycrase series, brannerite, and davidite are complex pegmatitic minerals. Pitchblende, a variety of uraninite found in hydrothermal veins, is the most important mineral of uranium. It is usually poorly crystalline, contains very little thorium or rare earths, and is frequently found associated with sulfide minerals. Coffinite, first identified in

Table 1. Uranium minerals

Mineral	Chemical composition	Color	Specific gravity, 20°C	Typical occurrence
Uraninite	UO_2 (contains Th, rare earths)	Black	8–10.6	Arendal, Norway
Pitchblende (var.)	UO_{2+x}	Black	6–8	Shinkolobwe, Zaire
Euxenite-polycrase	$(Y,Ca,Ce,U,Th)(Nb,Ta,Ti)_2O_6$	Dark brown	4–6	Nipissing, Ontario
Samarskite	$(Y,Ca,Fe,U,Th)(Nb,Ta)_2O_6$	Black	5–6	Mitchell Co., N.C.
Brannerite	$(Y,Ca,Fe,U,Th)_3(Ti,Si)_5O_{16}$	Black	4–5	Blind River, Ontario
Davidite	$(Fe,Ce,U)(Ti,Fe,V,Cr)_3(O,OH)_7$	Black	4–5	Rum Jungle, Australia
Coffinite	$USiO_4$	Black	5–6	Colorado Plateau
Carnotite	$K_2(UO_2)_2(VO_4)_2 \cdot xH_2O$	Yellow	3–5	Colorado Plateau
Tyuyamunite	$Ca(UO_2)_2(VO_4)_2 \cdot xH_2O$	Yellow	3–4	Ferghana, Turkestan
Autunite	$Ca(UO_2)_2(PO_4)_2 \cdot xH_2O$	Greenish-yellow	3–4	Autun, France
Torbernite	$Cu(UO_2)_2(PO_4)_2 \cdot xH_2O$	Green	3–4	Erzgebirge, Saxony
Uranophane	$Ca(UO_2)_2Si_2O_7 \cdot xH_2O$	Greenish-yellow	3–4	Congo Republic

1951, has become recognized as an important mineral on the Colorado Plateau. The remaining compounds are secondary minerals. Near-surface uranium mineralization invariably consists of oxidized ore.

Prior to 1942, uranium was obtained principally as a by-product of radium mining operations. With the discovery of nuclear fission and the potential of atomic power, the possession of uranium reserves became vitally important. Uranium reserves containing more than 1g U_3O_8/kg of ore, for that part of the world for which statistics are available, are estimated at about 2.2×10^9 kg U_3O_8, and those of the United States are about 10^9 kg U_3O_8. Deposits containing as little as 0.1% uranium are being mined. Some of the largest occurrences are the sandstone-impregnated Colorado Plateau deposits, the Blind River conglomerates (Ontario, Canada), and the reefs of the Witwatersrand (South Africa), from which uranium is produced as a by-product of the gold industry. The vein deposits at Great Bear Lake (Northwest Territories, Canada) and Lake Athabasca (west-central Canada) are also important sources of uranium, but the Shinkolobwe, Zaire, deposits are virtually exhausted. An interesting deposit is the one at Oklo, Gabon, where in primordial times a spontaneous fission chain reaction occurred which caused a shift in the isotopic composition of the uranium in the deposit. In addition to the occurrence mentioned, extensive reserves of low-grade ore (0.005 to 0.02% uranium) exist in phosphate deposits (Florida, Brazil, Soviet Union, and North Africa), in bituminous shales (Soviet Union, Sweden, and Tennessee), and in lignites (the Dakotas).

The Earth's oceans also are a potential source of uranium. It is estimated that in the total volume of the oceans, 1.37×10^9 km^3, a uranium quantity of 4.5×10^9 tons (4.1×10^9 metric tons) is dissolved, probably as carbonate complex. Also, it is estimated that 27,000 tons (24,500 metric tons) of uranium per year are carried to the ocean. There are experiments in progress to isolate uranium from sea water, but these experiments are still in an early development state.

Separation of uranium isotopes. Because of the great importance of the fissile isotope ^{235}U, rather sophisticated industrial methods for its separation from the natural isotope mixture have been devised. Some of these methods date as far back as the Manhattan Project. In the Calutron process, uranium tetrachloride was charged to the ion source of an electromagnetic 180° mass separator (Calutron) and ionized. The beams of the ^{235}U and ^{238}U ions, which were separated in the magnetic field, were collected separately in individual pockets of a collector made of graphite. The pockets were mechanically separated and ignited to burn the graphite, and the separated isotopes were isolated from the ignition residue. Because of its complexity the Calutron process is no longer in use for uranium separation, but it has been utilized to separate stable isotopes and long-lived radioactive isotopes of almost any element in the periodic table. In the gaseous diffusion process, the major industrial process used at the present time, gaseous uranium hexafluoride kept at elevated temperatures is passed through porous barrier tubes contained in so-called converters; $^{235}UF_6$ passes through the barriers slightly faster than $^{238}UF_6$. Thousands of individual successive stages (a cascade) are required to separate pure ^{235}U from the natural mixture. The gaseous diffusion process, which in the United States is operated in three large plants—at Oak Ridge, TN, Paducah, KY, and Portsmouth, OH—has been the established industrial process. Other processes applied to the separation of uranium include the centrifuge process, in which gaseous uranium hexafluoride is separated in centrifuge cascades, the liquid thermal diffusion process, the separation nozzle, and laser excitation. However, with the exception of the gas centrifuge method, none of these methods has progressed beyond the pilot plant stage. *See* ISOTOPE (STABLE) SEPARATION.

Uranium metal. Uranium is a very dense, strongly electropositive, reactive metal; it is ductile and malleable, but a poor conductor of electricity. It is most conveniently prepared by the reduction of a halide (UF_4) with calcium or magnesium in a sealed bomb at 1200–1400°C. The steps involved in preparation of the metal from uranyl nitrate are summarized by reactions (1)–(4).

$$UO_2(NO_3)_2 \cdot 6H_2O \xrightarrow{500^\circ C} UO_3 + 2NO_2 + \tfrac{1}{2}O_2 + 6H_2O \quad (1)$$

$$UO_3 + H_2 \xrightarrow{700^\circ C} UO_2 + H_2O \quad (2)$$

$$UO_2 + 4HF \xrightarrow{550^\circ C} UF_4 + 2H_2O \quad (3)$$

$$UF_4 + \begin{matrix} 2Ca \\ 2Mg \end{matrix} \longrightarrow U + \begin{matrix} 2CaF_2 \\ 2MgF_2 \end{matrix} \quad (4)$$

Selected physical and thermal properties of uranium are listed in Table 2. Uranium metal exists in three crystalline modifications. α-Uranium

Table 2. Physical and thermal properties of uranium

Property	Value
Melting point	1132.4 ± 0.8°C
Boiling point	3818°C
Vapor pressure, 1720–2340 K	$\log p\ (\text{atm}) = -\dfrac{26210 \pm 270}{T} + (5.920 \pm 0.135)$
Heat of fusion	19.7 kJ/g-atom
Heat of vaporization	446.4 kJ/g-atom
Heat of sublimation (0 K)	487.9 kJ/g-atom
Heat of transition $\alpha \rightarrow \beta$	2.791 kJ/g-atom
Heat of transition $\beta \rightarrow \gamma$	4.757 kJ/g-atom
Enthalpy at 25°C	6.3655 kJ/g-atom
Heat capacity at 25°C	27.664 J/K·g-atom
Entropy at 25°C	50.170 ± 0.008 J/K·g-atom
Thermal conductivity (70°C)	0.29 J/cm·s (K)
Electrical conductivity	$2-4 \times 10^4\ (\Omega\cdot\text{cm})^{-1}$

(25–668°C) is orthorhombic ($a = 0.2854$, $b = 0.5896$, $c = 0.4956$ nm), with four atoms per unit cell, and a density of 19.04 g/cm³. Its structure is interpreted as a distorted hexagonal lattice containing corrugated sheets of uranium atoms. The β-phase (668–775°C) is a complex tetragonal structure ($a = 1.0754$ nm, $c = 0.5623$ nm), with 30 atoms per cell, and a density of 18.13 at 720°C. γ-Uranium (775–1132°C) is body-centered cubic ($a = 0.3525$ nm), with two atoms per cell, and a density of 18.06 g/cm³ at 805°C. The β-form can be stabilized at room temperature by addition of small amounts of chromium, the γ-form with molybdenum.

The unique nature of the room-temperature γ-structure curtails solid solution of uranium with many metals. Extensive solid solution without compound formation has been found only with molybdenum and niobium. Aluminum, beryllium, bismuth, cadmium, cobalt, gallium, germanium, gold, indium, iron, lead, manganese, mercury, nickel, tin, titanium, zinc, and zirconium all form one or more intermetallic compounds with uranium. Chromium, magnesium, silver, tantalum, thorium, tungsten, and vanadium, as well as calcium, sodium, and some of the rare-earth metals, form neither compounds nor extensive solid solutions. Many uranium alloys are of great interest in nuclear technology because the pure metal is chemically active and anisotropic and has poor mechanical properties. However, cylindrical rods of pure uranium coated with silicon and canned in aluminum tubes (slugs) are used in production reactors. Uranium alloys can also be useful in diluting enriched uranium for reactors and in providing liquid fuels. Uranium depleted of the fissile isotope ^{235}U has been used in shielded containers for storage and transport of radioactive materials.

Uranium reacts with nearly all nonmetallic elements and their binary compounds. Table 3 lists a number of its reactions. Uranium dissolves in hydrochloric acid to leave a black residue of uranium hydroxy hydride. Addition of fluosilicate prevents formation of this residue. Nitric acid dissolves the metal, but nonoxidizing acids, such as sulfuric, phosphoric, or hydrofluoric acid, react very slowly. Usually a trace of mercuric nitrate tends to catalyze the dissolution. Uranium metal is inert to alkalies, but addition of peroxide causes formation of water-soluble peruranates. *See* URANIUM METALLURGY.

Uranium ions in solution. Four oxidation states of uranium exist in solution, but only two of these, U^{4+} and U^{6+} (or rather UO_2^{2+}), are stable. Aqueous solutions of U^{3+} decompose with hydrogen evolution, even at temperatures as low as 0°C. Solutions of U^{3+} are obtained by electrolytic reduction of U^{4+} or UO_2^{2+} solutions, preferably in sulfuric acid. They have a green color in daylight or fluorescent light, but are deep red in incandescent light.

Green solutions of tetravalent uranium are readily obtained by electrolytic reduction or reduction with strong chemical reductants. They are readily oxidized in air on standing. Although the ion U^{4+} is extensively hydrolyzed, its existence in acid solutions has been proved. The first hydrolysis step is believed to give the monomeric ion $U(OH)^{3+}$. Numerous salts and complexes of uranium(IV) may be prepared from a U(IV) solution.

Pentavalent uranium exists in aqueous solution as the ion UO_2^{+}. It is unstable in solution and tends to disproportionate to U^{4+} and UO_2^{2+}; its lowest rate of decomposition is in the pH range 2–4.

Hexavalent uranium, in the form of the uranyl ion (UO_2^{2+}), is the most stable oxidation state. The lemon-yellow, fluorescent uranyl solutions can be reduced by moderately strong reductants, such as Sn^{2+}, Ti^{3+}, hydrogen in the presence of a catalyst, or sodium dithionite. Hydrolysis of UO_2^{2+} leads to formation of UO_2OH^{+}; further hydrolysis gives polymeric ions of the type $UO_2\,[(OH)_2UO_2]_n^{2+}$, in-

Table 3. Chemical reactions of uranium metal

Reactant (reaction temperature, °C)†	Products	Heat of formation of underlined product, kJ/mole at 25°C
H_2(250)	$\alpha, \underline{\beta\ UH_3}$	−127
O_2(100–350)	UO_2, $\underline{U_3O_8}$	−3572
F_2(250)	$\underline{UF_6}$	−2186
Cl_2(500)	$\underline{UCl_4}$, UCl_5, UCl_6	−1051
Br_2(650)	$\underline{UBr_4}$	−826
I_2(350)	UI_3, $\underline{UI_4}$	−529
B(1650)	$\underline{UB_2}$, UB_4, UB_{12}	−148
C(1500)	UC, U_2C_3, $\underline{UC_2}$	−82
Si(1700)	U_3Si, U_3Si_2, $\underline{USi}$, USi_2, USi_3	−80
N_2(500)	UN, $\underline{U_2N_3}$, UN_2	−709
P(400–1100)*	$\underline{UP}$, U_3P_4, UP_2	−316
S(400)	$\underline{US}$, US_2	−306
H_2O(100)	$\underline{UO_2}$	−1084
H_2S(500)*	US, $\underline{US_2}$	−502
HF(350)*	$\underline{UF_4}$	−1882
HCl(300)*	$\underline{UCl_3}$	−893
NH_3(700)	$\underline{UN}$, UN_2	−303
CH_4(650)*	$\underline{UC}$	−97
CO(750)	UO_2, UC	
CO_2(600)	UO_2, UC	
NO(400)	U_3O_8	
N_2O_4(25)	$UO_2(NO_3)\cdot 2NO_2$	

*Powdered metal. †Values are given for massive metal.

volving sheetlike complexes with double hydroxyl bridges.

Extensive studies have been made of complex formation of UO_2^{2+} and, to a lesser extent, of U^{4+}, with many anions (for example, fluoride, chloride, bromide, thiocyanate, nitrate, bisulfate, sulfate, acetate, phosphate, oxalate, citrate, carbonate, and acetyl acetonate). In general, the strength of such a complex is inversely proportional to the strength of the acid from which the complexing anion is derived.

The oxidation potentials of uranium in 1 *M* acid solution are as in notation (5); those in 1 *M* basic solution are as in notation (6).

$$U \xrightarrow{+1.80\ V} U^{3+} \xrightarrow{+0.63\ V} U^{4+} \xrightarrow{-0.58\ V} UO_2^{+} \xrightarrow{-0.06\ V} UO_2^{2+} \quad (U^{4+} \xrightarrow{-0.32\ V} UO_2^{2+}) \tag{5}$$

$$U \xrightarrow{2.17\ V} U(OH)_3 \xrightarrow{2.14\ V} U(OH)_4 \xrightarrow{0.62\ V} UO_2(OH)_2 \tag{6}$$

Hydride. Uranium reacts reversibly with hydrogen to form UH_3 at 250°C. Correspondingly, the hydrogen isotopes form uranium deuteride, UD_3, and uranium tritide, UT_3. Because of the low reaction temperatures, and because UH_3, UD_3, and UT_3 are easily decomposed above 430°C to form pyrophoric, finely dispersed uranium powder and the corresponding hydrogen isotope, the reaction is useful both to prepare powdered uranium and to store D_2 or T_2 as the deuteride or tritide in an easily retrievable form.

Oxides. The uranium-oxygen system is characterized by an extremely complicated phase diagram. Table 4 gives a general survey of known uranium oxides. Uranium monoxide, UO, is a gaseous species which is not stable below 1800°C. In the range UO_2 to UO_3, a large number of phases exist. The complexity of the uranium-oxygen system is best understood if one considers that the addition of oxygen to UO_2 produces continually increasing distortions of the original fluorite lattice. The added oxygen (or oxygen deficiency) can be distributed at random to produce a single phase of constant space group and variable stoichiometry, such as $UO_{2\pm x}$, or it can be distributed in an ordered fashion, forming cubic, tetragonal, or monoclinic superlattices. The stoichiometry range of a phase at a particular temperature is a measure of its capacity to add randomly distributed oxygen without change in long-range ordering. When this limit is reached, at least part of the added oxygen atoms become ordered in a superlattice structure, producing a new phase possibly also of variable composition. Such behavior occurs up to $UO_{2.4}$. Above $UO_{2.4}$, an abrupt change to lower-density structures occurs. The phases in the range $UO_{2.4}$ to UO_3 appear to contain uranyl-type bonding for at least part of the uranium atoms present. In this bonding there are two short collinear primary uranium-oxygen bonds with four to six weaker bonds more or less in a plane perpendicular to these bonds.

This general behavior leads to rather strange stoichiometries, and because of the subtle changes from one phase to the next the earlier literature contains many conflicting results. Another peculiarity of the uranium-oxygen system is the fact that UO_3 exists in one amorphous and at least six crystalline modifications, the most stable being γ-UO_3.

The range of each oxide phase is determined by the temperature and by the oxygen partial pressure above the solid. Ignition of any uranium oxide to 750°C in air leads to the formation of U_3O_8.

The dioxide, UO_2, which is obtained by reduction of UO_3 or U_3O_8 with hydrogen or CO at 400–600°C in a fluidized bed, is an important ceramic fuel for nuclear reactors. For this application it may be mixed with plutonium dioxide, and it is compacted into pellets which are then loaded into the fuel tubes.

Uranium trioxide reacts with water to form either $UO_3 \cdot 2H_2O$, α-, β-, or γ-$UO_2(OH)_2$, α-$UO_3 \cdot 0.8H_2O$, or $U_3O_8(OH)_2$, depending on the conditions. All these compounds are yellow or orange. These uranyl hydroxides may also be considered as acids, H_2UO_4 or $H_2U_3O_{10}$. Uranium peroxide,

Table 4. Survey of defined uranium-oxygen phases

Compound	Symmetry	Color	Density (gcm^{-3})
UO	(Gaseous)	–	–
UO_2	Cubic	Cinnamon brown	10.950
U_4O_{9-y}*	Cubic	Black	(11.299)
$U_{16}O_{37}$	Tetragonal	Black	(11.366)
U_8O_{19}	Monoclinic	Black	11.34
α-U_2O_5	Monoclinic	Black-purple	10.5
β-U_2O_5	Hexagonal	Black-purple	10.76
γ-U_2O_5	Monoclinic	Black-purple	10.36
$U_{13}O_{34}$	Orthorhombic		(8.40)
U_8O_{21}	Orthorhombic		(8.341)
$U_{11}O_{29}$	Orthorhombic		(8.40)
α-U_3O_8	Orthorhombic	Greenish-black	(8.395)
β-U_3O_8	Orthorhombic	Greenish-black	(8.326)
$U_{12}O_{35}$	Orthorhombic		7.72
UO_3 (A)	Amorphous	Orange	6.80
α-UO_3	Hexagonal	Beige	7.3
β-UO_3	Monoclinic	Orange red	8.25
γ-UO_3	Orthorhombic	Yellow	7.80
δ-UO_3	Cubic	Deep red	6.69
ϵ-UO_3	Triclinic	Red	(8.67)
ζ-UO_3	Orthorhombic	Brown	(8.86)

*U_4O_{9-y} is a nonstoichiometric compound, defined by its x-ray structure.

$UO_4 \cdot xH_2O$, where $x = 1$ or 2, precipitates from uranyl salt solutions on addition of H_2O_2. It cannot be dehydrated without decomposition.

Ternary oxide systems of hexa-, penta-, and tetravalent uranium with alkali, alkaline earth, rare earths, and group IV elements have been investigated. Extensive regions of solid solutions are generally encountered. Hexa-, penta-, and tetravalent uranium form ternary oxides (uranates) exhibiting a wide range of composition with other metals. These compounds are insoluble in water but are readily soluble in acids. A survey of typical compounds of this kind is given in Table 5.

For these compounds, solid-state reactions of the type shown in reactions (7) and (8), involving

$$6Na_2CO_3 + 2\,U_3O_8 + O_2 \rightarrow 6Na_2UO_4 + 6CO_2 \quad (7)$$

$$CaO + UO_3 \rightarrow CaUO_4 \quad (8)$$

stoichiometric quantities of the reactants, yield single-phase products, while precipitation from aqueous solutions generally leads to nonstoichiometric mixtures. Addition of hydrogen peroxide and alkali to uranyl solutions leads to the formation of soluble peroxyuranates, for example $Na_4UO_8 \cdot xH_2O$.

Halides. The uranium halides constitute an important group of compounds. Uranium tetrafluoride is an intermediate in the preparation of the metal and the hexafluoride. Uranium hexafluoride, which is the most volatile uranium compound, is used in the isotope separation of ^{235}U and ^{238}U. The halide volatilities increase in the order $UX_3 < UX_4 < UX_5 < UX_6$. Reactions for the preparation of uranium halides are summarized in Table 6. Uranium hexafluoride boils at 56.54°C and melts at 64°C (1140 mmHg or 152 kilopascals pressure). It is a reactive substance and a strong fluorinating agent. Equipment for containing the compound may be constructed of copper, nickel, aluminum, Monel, or fluorine-containing polymers (Teflon, Kel-F). The hexafluoride reacts with water to form UO_2F_2. Uranium tetrachloride, UCl_4, has been used as charge material in the electromagnetic isotope separation process. Like all the other chlorides, bromides, and iodides, it is hygroscopic and soluble in water. The penta- and hexachlorides are also soluble in some nonpolar solvents (CCl_4, CS_2). Besides the binary halides, a number of ternary and quaternary halides of U^{3+} and U^{4+} are known, such as $UBrCl_2$, UCl_2Br_2, and many others. The halides react with oxygen at elevated temperatures to form uranyl compounds and ultimately U_3O_8.

Halogeno complexes. Many halogeno complexes of uranium are known. Of particular interest are those based on UF_4 and UF_3, because they may be used as nuclear fuels in molten salt reactors. In particular, $LiF-BeF_2-UF_4-$ and $LiF-ZrF_4-UF_4-$ eutectics are used as fuel materials. Table 7 summarizes the various types of fluoro complexes observed. Many phase diagrams of these systems are known. In the chloro system, complexes of the type UCl_6^- with U^{5+}, and UCl_6^{2-} with U^{4+} are known; in the bromo and iodo system, only a few compounds with the UBr_6^{2-} and UI_6^{2-} ions have been prepared.

Uranyl salts. These salts, which have the cation UO_2^{2+}, are the most common uranium salts. They generally have a yellow to greenish-yellow color and show bright green fluorescence in ultraviolet light. Uranyl nitrate is obtained as the hexahydrate $UO_2(NO_3)_2 \cdot 6H_2O$ (so-called UNH) from dilute nitric acid, as the trihydrate from concentrated nitric acid, and as the dihydrate from fuming nitric acid. The lower hydrates may also be obtained by careful dehydration of the hexahydrate. Anhydrous uranyl nitrate may be prepared by reacting N_2O_5 with UO_3 under strictly anhydrous conditions. Uranyl nitrate is probably the most frequently encountered compound in uranium chemistry.

Uranyl sulfate, also very soluble in water, crystallizes from aqueous solution as the trihydrate, $UO_2SO_4 \cdot 3H_2O$. A monohydrate is formed by careful dehydration or by equilibration in water at 180°C. The anhydrous sulfate is formed by heating the hydrate to 300°C. With alkali sulfates, double sulfates such as $K_2UO_2(SO_4)_2 \cdot 2H_2O$ are formed. Phase diagrams of uranyl nitrate, uranyl sulfate, uranyl fluoride, uranyl carbonate, and uranyl phosphate have been constructed. Acid uranyl phosphate, $HUO_2PO_4 \cdot 4H_2O$, and the corresponding arsenate, $HUO_2AsO_4 \cdot 4H_2O$, form numerous alkali and alkaline-earth salts, many of which occur in nature as minerals. Some of the important uranyl salts of organic acids are the formate $UO_2(HCOO)_2 \cdot H_2O$, the acetate $UO_2(CH_3COO)_2 \cdot 2H_2O$, the sodium double acetate $NaUO_2(CH_3COO)_3$, and the oxalate $UO_2C_2O_4 \cdot 3H_2O$.

Uranium(IV) salts. By electrolytic reduction or the use of suitably strong reducing agents, the yellow UO_2^{2+} solutions may be reduced to green U^{4+} solutions. From these latter solutions, green uranium(IV) salts may be prepared. Of importance is the dark green sulfate $U(SO_4)_2 \cdot 8H_2O$. From the octahydrate, lower hydrates with 6, 5, 4, and 2 H_2O, as well as the anhydrous salt, may be prepared by thermal dehydration in inert atmospheres. Ura-

Table 5. Ternary oxides of uranium(V) and uranium(VI) with alkali oxides

Oxidation state	Compound type	Cation
+6	$M_8U_{16}O_{52}$	M = K
+6	$M_2U_7O_{22}$	M = K, Rb
+6	$M_2U_6O_{19}$	M = Li
+6	$M_2U_4O_{13}$	M = Rb
+6	$M_2U_3O_{10}$	M = Li, K, Tl
+6	$M_2U_2O_7$	M = Li, Na, K, Rb, Tl
+6	M_2UO_4	M = Li, Na, K, Rb, Cs
+6	M_4UO_5	M = Li, Na, K, Rb
+6	M_6UO_6	M = Li
+5	MUO_3	M = Li, Na, K, Rb, Tl
+5	M_3UO_4	M = Li, Na
+5	M_7UO_6	M = Li

Table 6. Preparation of uranium halides

Compound	Reaction
UF_3	$UF_4 + Al \rightarrow UF_3 + AlF\uparrow$ (900°C)
UF_4	$UO_2 + HF \rightarrow UF_4$ (550°C)
U_4F_{17}, U_2F_9, UF_5	$xUF_4 + yUF_6 \rightarrow U_4F_{17}, U_2F_9, UF_5$ (250°C)
UF_6	$UF_4 + F_2 \rightarrow UF_6$ (350°C)
UCl_3, UBr_3, UI_3	$UH_3 + HX \rightarrow UX_3$ (350°C)
UCl_4	$UO_3 + Cl_2C{=}CClCCl_3 \rightarrow UCl_4$ (210°C)
UCl_5	$UCl_4 + Cl_2 \rightarrow UCl_5$ (500°C)
UCl_6	$UCl_5 \rightarrow UCl_6 + UCl_4$ (125°C in vac.)
UBr_4, UI_4	$U + X_2 \rightarrow UX_4$ (550°C)

Table 7. Ternary fluorides of uranium(VI), (V), (IV), and (III)

Oxidation state	Compound type	Cation
+6	M_3UF_9	M = Na
+6	M_2UF_8	M = Na, K
+6	MUF_7	M = Na, K, NH_4^+, $N_2H_5^+$, NO^+, NO_2^+
+5	M_3UF_8	M = Na, K, Rb, Cs, Tl, Ag
+5	M_2UF_7	M = K, Rb, Cs, NH_4^+
+5	MUF_6	M = Li, Na, K, Rb, Cs, Ag, NH_4^+, H^+
+4	MU_6F_{25}	M = K, Rb
+4	MU_4F_{17}	M = Li
+4	MU_3F_{13}	M = K, Rb
+4	MU_2F_9	M = Na,.K
+4	MUF_5	M = Li, Rb, Cs
+4	$M_7U_6F_{31}$	M = Na, K, Rb, Cs, NH_4^+
+4	M_2UF_6	M = Na, K, Rb, Cs, NH_4^+
+4	M_3UF_7	M = Li, Na, K, Rb, Cs
+3	MUF_4	M = Na
+3	M_3UF_6	M = K

nium(IV) nitrate is not stable, and it may be obtained only as an *NN*-dimethyl-acetamide complex, $U(NO_3)_4 \cdot 2.5CH_3CON(CH_3)_2$. However, a few unstable salts with the anion $U(NO_3)_6^{2-}$ have been prepared. The oxalate, $U(C_2O_4)_2 \cdot 6H_2O$, is relatively stable.

Uranium(III) salts. From uranium(III) solutions in sulfuric acid obtained by strong electrolytic reduction, uranium(III) sulfate, $U_2(SO_4)_3 \cdot 8H_2O$, precipitates on addition of ethanol as olive-green crystals which are brick red in incandescent light. On heating, the octahydrate may be dehydrated to the orange trihydrate. With alkali sulfates, complexes of the type $MU(SO_4)_2$ (where M = NH_4, K, Rb) or $M_5U(SO_4)_4$ (with M = K, Tl) may be precipitated. All U(III) salts are extremely sensitive to air.

Behavior in nonaqueous solvents. The solubility of uranium salts, especially uranyl nitrate, in certain organic solvents can be used to separate uranium from other metal ions by solvent extraction. Numerous types of liquid (aqueous)–liquid (organic) separation techniques are employed. Distribution of uranium between the two phases may be controlled by "salting" the aqueous phase with mineral acids or their salts, changing pH in the aqueous phase, or using different types of organic extractants. The three types of organic phase are a neutral extractant, such as diethyl ether, methyl isobutyl ketone (hexone), or tributyl phosphate (TBP); an acidic extractant, such as octanoic acid or thenoyl trifluoro acetone (TTA); and a basic extractant, such as tri-octylamine or trilaurylamine (TLA). In some instances, the organic phase may be diluted with a "carrier" diluent such as kerosine. The most frequently used extraction procedure is the purification of uranium with TBP dissolved in kerosine or hexane. The extraction is based on the formation of a neutral nonionized complex according to reaction (9).

$$UO_2^{2+}(aq) + 2NO_3^-(aq) + 2TBP\,(org) \rightleftharpoons UO_2(NO_3)_2 \cdot 2TBP\,(org) \quad (9)$$

Quantitative determination. Numerous procedures are available for the quantitative determination of uranium. Macro quantities of uranium may be analyzed by gravimetric or volumetric methods. Gravimetric procedures usually utilize U_3O_8 ignited in air or 8-hydroxy-quinolinate. Volumetric methods are based on reduction of uranium to U(IV) with lead or zinc, followed by titration with an oxidizing agent such as potassium dichromate, ceric sulfate, potassium bromate, or potassium permanganate. Small amounts of uranium may be determined by coulometric, polarographic, colorimetric, fluorescence, or spectroscopic methods. The isotopic composition may be determined by mass spectroscopy or, in the case of ^{235}U, by fission counting. *See* NUCLEAR FISSION; NUCLEAR FUEL; NUCLEAR POWER; NUCLEAR REACTION.

[FRITZ WEIGEL]

Bibliography: I. I. Chernyaev, *Complex Compounds of Uranium*, 1966; E. H. P. Cordfunke, *The Chemistry of Uranium*, 1969; *Gmelin Handbuch der Anorganischen Chemie*, System no. 55: *Uran*, including supplementary volumes, 1979; J. J. Katz and E. Rabinowitch, *The Chemistry of Uranium*, NNES VIII 5 (*Survey*), 1951, and TID-5290, vols. 1 and 2 (*Collected Papers*), 1958; National Lead Company of Ohio, *Analytical Chemistry Manual of the Feed Materials Production Center*, TID-7022, vol. 1, books 1–8, and vol. 2, 1964.

Uranium metallurgy

The processing treatments for the production of uranium concentrates and the recovery of pure uranium compounds, as well as the conversion chemistry to produce uranium metal and the processes employed for preparing uranium alloys.

Ore processing. The procedures to recover uranium from its ores are numerous, because of the great variety in the nature of uranium minerals and associated materials and the wide range of concentration in the naturally occurring ores. Recovery of uranium requires chemical processing; however, preliminary treatment of the ore may involve a roasting operation, a physical or chemical concentration step, or a combination of these.

Roasting can bring about significant chemical

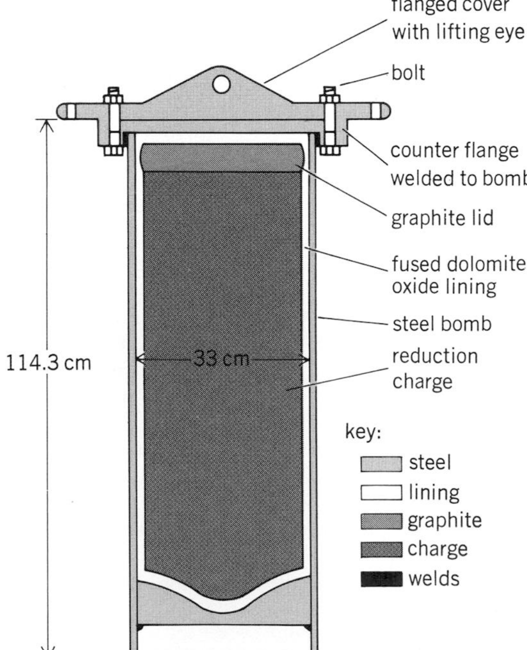

Fig. 1. Standard Derby bomb for reduction of UF_4 with magnesium by the Ames process. Capacity is 144.2 kg of uranium.

change on its own, but moreover it can improve the filtering and settling characteristics of an ore for subsequent processing. Only a limited number of deposits are sufficiently rich in uranium-bearing minerals and have appropriate physical characteristics to make a preliminary physical concentration of the uranium a feasible operation by present techniques. Chemical concentration is most common as a preliminary treatment.

In general, one of two leaching treatments—acid leaching and carbonate leaching—is used as the initial step in chemical concentration. The choice depends on the nature of the ore, which largely determines the efficiency and the cost of the process employed.

Acid leaching is carried out universally with sulfuric acid. The finely ground ore is treated with the dilute acid, and usually an oxidant is added to make sure that all the uranium goes into solution as the uranyl ion. When an ore is carbonate-leached, the action of sodium carbonate on the uranium forms water-soluble $[UO_2(CO_3)_3]^{4-}$ ion. This solution is stabilized by adding bicarbonate and by keeping the uranium in the oxidized state. From either the acid or carbonate leaching process, the uranium can be recovered as ammonium uranate or sodium uranate (yellow cake) by precipitation with ammonia or with sodium hydroxide.

Recovery of pure uranium. The concentrate, whether obtained by chemical or physical means, is treated chemically to give a uranyl nitrate solution that can be further purified by solvent extraction. The impurities remain in the aqueous phase, while the uranium is extracted into the organic phase. Formerly, ether or hexone was used as an extractant; presently, tributyl phosphate (TBP) is favored. From the organic phase, the uranium is stripped by means of water, and the high-purity uranium can be recovered as nitrate crystals or by precipitation from the solution. The nitrate serves as the starting material for other compounds, such as the oxides. In large-scale processing, the nitrate is decomposed thermally to give uranium trioxide, UO_3, which is subsequently reduced with hydrogen to form uranium dioxide, UO_2. Uranium tetrafluoride, UF_4 (green salt), is prepared by treating UO_2 with hydrogen fluoride (HF) gas.

Metal reduction. Uranium metal can be obtained from its halides by fused-salt electrolysis or by reduction with more reactive metals. The reaction of UO_2 with calcium yields metal of fair quality.

The electrolysis of UF_4 in a fused-salt bath of sodium chloride and calcium chloride has been used to prepare tons of high-purity metal. The metal is deposited as small granules on a molybdenum cathode. In the calcium reduction of UO_2, the ingredients of the charge are mixed and placed in an inert atmosphere and are heated to about 1000°C to form finely divided uranium and calcium oxide. Calcium hydride can be employed instead of calcium in this process. The metal from either electrolysis or the calcium-UO_2 process is leached, pressed, and sintered or melted to give solid uranium.

The largest tonnages of good-quality uranium have been produced by metallothermic reduction of finely divided UF_4 with calcium or magnesium in steel bombs lined with fused dolomitic oxide (Ames process). The charge, consisting of an intimate mixture of UF_4 with the reductant metal in granular form together with a suitable booster (usually calcium plus iodine), is placed into the reduction bomb, the lid is bolted down, and the bomb is heated to ignition temperature. In the case of calcium, the reaction heat is such that the charge can be ignited at room temperature, and the products are sufficiently molten to allow separation of metal and slag in the reduction vessel. For magnesium, the whole bomb has to be preheated in a heat-soaking pit. The heat generated in the reduction reaction is sufficient to melt both metal and slag, and the molten metal collects by gravity as a pool under the molten slag, where it remains during solidification. The shape of the metal ingot (also referred to as a biscuit) depends on the shape of the reduction bomb. A flat, pancake-shaped ingot, which has to be remelted for reshaping, is obtained in the standard bomb (Fig. 1), while reduction in a Dingot bomb (Fig. 2). yields a so-called direct-reduction ingot (dingot), which is shaped in such a way that it may undergo further metallurgical treatment without remelting. The Dingot process is particularly suited for large-scale reductions (50 kg to 1.5 metric tons). The metal obtained in the metallothermic reduction process is quite pure. Metal obtained by reduction in a standard bomb has to be recast by vacuum melting in a graphite crucible and casting into graphite molds. The metal can be fabricated by conventional means with due consideration for its chemical reactivity and allotropic transformations. The low-ductility modification of uranium is usually avoided in fabrication processes. In the metallurgy of ^{233}U and ^{235}U, strict weight and configuration limitations must be imposed to avoid criticality.

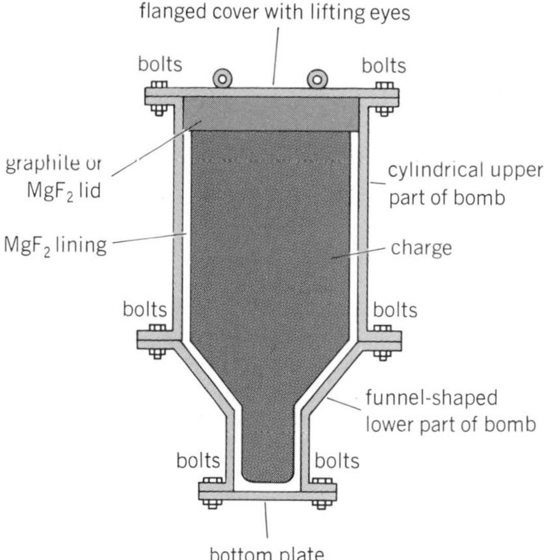

Fig. 2. Dingot bomb for large charges.

Alloys. Uranium alloys are prepared by fusing the components together. All procedures following conventional metallurgical techniques, however, may have to be carried out inside inert-gas glove boxes because many alloys are attacked by oxygen or moisture. *See* NUCLEAR FUEL; NUCLEAR FUELS REPROCESSING; URANIUM. [F. WEIGEL]

Bibliography: G. A. Akin et al., *Chemical Processing Equipment: Electromagnetic Separation*

Process, NNES-I-12 (TID-5232) Oak Ridge, 1951; C. D. Harrington and A. E. Ruehle, *Uranium Production Technology*, 1959; F. S. Patton, J. M. Googin, and W. L. Griffith, *Enriched Uranium Processing*, 1963; W. D. Wilkinson, *Uranium Metallurgy*, vols. 1 and 2, 1962.

Vapor condenser

A heat-transfer device that reduces a thermodynamic fluid from its vapor phase to its liquid phase. The vapor condenser extracts the latent heat of vaporization from the vapor, as a higher-temperature heat source, by absorption in a heat-receiving fluid of lower temperature. The vapor to be condensed may be wet, saturated, or superheated. The heat receiver is usually water but may be a fluid such as air, a process liquid, or a gas. When the condensing of vapor is primarily used to add heat to the heat-receiving fluid, the condensing device is called a heater and is not within the normal classification of a condenser.

Classification by use. Condensers may be divided into two major classes according to use: those used as part of a processing system, and those used for serving engines or turbines in a steam power plant cycle.

Process condensers. In a processing system condensers selectively recover liquid from a mixed vapor, recover pure liquid from the vapor of an impure liquid, recover noncondensables from a mixture of gas and vapor, or extract heat from heat-pump, refrigeration, and cryogenic cycles. The vapor condensed may be steam or the vapor from any nonaqueous liquid. Condensers used for any purpose except as vapor condensers in the power plant cycle, even though they provide the heat sink for a process cycle, are classified as process condensers.

One of the more important uses of process condensers is that of condensate recovery. Condensers are frequently used with fractionating and distillation columns in the production of hydrocarbon liquids and in processes for liquefying gases. Similarly they are used in the production of distilled water as distiller condensers in single- and multiple-effect evaporating systems. The multistage flash evaporator with integral vapor condensers has been established as one of the more economical means for the production of potable water from sea water. This apparatus contains a multiplicity of condensers in series, each condenser producing a successively lower pressure in which flashed vapor from sea-water brine is produced, condensed, and collected. Both distillate and the flashing brine streams reduce in temperature from stage to stage as they flow through the apparatus (Fig. 1).

Power cycle condensers. Used as part of the power plant cycle, condensers serve as the heat sink in the cycle and reduce the back pressure on the turbine or engine so that a maximum of heat energy becomes available as useful work. Steam (water) is the most common substance used for the power generation cycle. Other thermodynamic fluids can be used but seldom are. Another important function of a condenser in the steam power plant cycle is the recovery of condensate as the major source of pure boiler feedwater for the boilers. *See* THERMODYNAMIC CYCLE.

Classification of operation. Condensers may be further classified according to mode of operation as surface condensers or as contact condensers.

Surface condensers. In surface condensers the condensing vapor and the cooling fluid remain separated from each other by a dividing wall or walls, which form the heat-transfer surface. This heat-transfer surface is most often in the form of tubes but may also be plates or partitions of various geometries. Condensate, cooling fluid, and noncondensable gases are usually removed separately, although in some designs the condensate and noncondensable gases are withdrawn from the condenser as a mixture. *See* SURFACE CONDENSER.

Contact condensers. In the contact condenser the vapor and cooling liquid come into direct contact with each other and are mixed in the condensing process. The condensed vapor and cooling liquid combine and are withdrawn from the apparatus together. The noncondensable gases are usually withdrawn separately, although in some contact condensers they are entrained in the mixed condensate and cooling liquid and removed through a common outlet. *See* CONTACT CONDENSER.

Removal of noncondensables. Condensers are required, almost without exception, to condense impure vapors, that is, vapors containing air or other noncondensable gases. Because most condensers operate at subatmospheric pressures, air leaking into the apparatus or system becomes a common cause for vapor contamination. Many process condensers are used to condense vapors

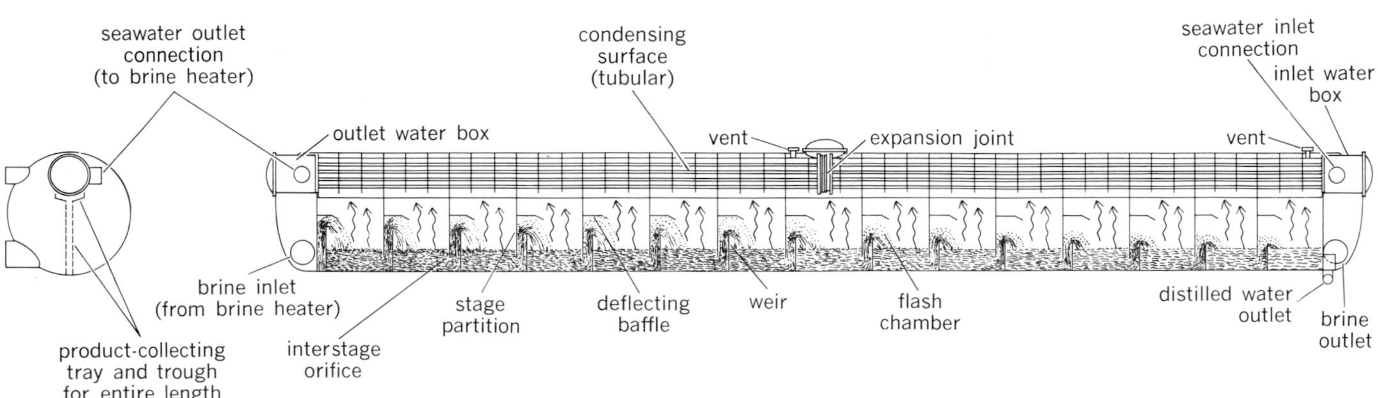

Fig. 1. Simplified views of multistage flash evaporator for production of potable water from sea water.

whose noncondensable impurities are relatively independent of air that may leak into a vacuum system. Because the vapor supplied to the condenser is continuously reduced to liquid, the noncondensable gases in the gas-vapor mixture collect and concentrate. Their accumulation seriously affects heat transfer, and means must be provided to direct them to a suitable outlet. Most surface and contact condensers are arranged with a separate zone of heat-transfer surface within the condenser and located at the outlet end of the vapor flow path for efficient removal of the noncondensable gases through dehumidification. A vapor flow path of the displacement type, free from zones of stagnation and short circuiting, is essential for achieving maximum condensing heat-transfer rates (Figs. 2 and 3).

Separate external vapor condensers arranged in series with the vapor flow path of the main condenser are used when the ratio of noncondensable gases to condensing vapor is high, or when the vapor content of the noncondensable gases must be reduced to low values.

Removal of noncondensables from condensers operating at subatmospheric pressure requires vacuum pumps. The noncondensables are removed from condensers operating above atmospheric pressure by venting to atmosphere, or venting to aftercondensers to reduce further the moisture content of the removed gases before discharging them to atmosphere or to a recovery process.

Heat-receiving fluid. Water is the most commonly used liquid for absorbing heat from condensing vapors. Liquid hydrocarbons and other chemical compounds in liquid state are used primarily as heat receivers for condensers, used as a part of a petroleum-refining or chemical-manufacturing process. Because of natural evaporation, surface waters are normally at a lower temperature than ambient air when air temperatures exceed 32°F (0°C), and thus provide the lowest-temperature cooling medium readily available. High specific heat, ease in pumping, and rapid heat transfer characterize water as an excellent medium for use as a heat receiver in a condensing system. The tendency of water to corrode metal surfaces is not usually a serious disadvantage. Thin-walled heat-transfer surfaces of the less costly corrosion-resistant alloys are generally satisfactory. Ferritic materials used for the water-containing parts of condensers can be made suitable for the corrosion environment by increasing thickness as a corrosion allowance or by the use of corrosion-resistant linings or by cathodic protection.

Air is the gas most commonly used for absorbing heat from a condensing process. However, its low heat capacity, low density, and relatively low heat transfer preclude it from being an ideal fluid for this purpose. Its availability and the fact that air does not readily corrode or foul heat-transfer surfaces offer some advantages in its use. Shortages of water for cooling or for use as a heat sink and the danger of thermal pollution of surface water have markedly increased the use of air. Application of extended-surface tubes to process condensers, to process coolers, and to steam condensers serving turbines has been shown to be reasonably economical and practical where water supplies are critical.

The cooling of heat-transfer surfaces by evaporation involves the use of both water and air. The heat-transfer surfaces are continuously wetted by water. Air, blown over the wetted surfaces, absorbs the water vapor as it is released from the evaporating cooling water, thereby removing the heat released in condensers.

Condensate cooling. In those process systems where recovery of condensate is of primary importance, it is also desirable to cool the condensate below the saturation temperature of its condensing vapor. Some condensate cooling results as condensate falls from tube to tube in the process of being collected and accumulated. If additional cooling is needed, condensate cooling sections in the condenser can be provided by flooding a selected amount of surface within the condenser with condensate. Flooding is done either by installing

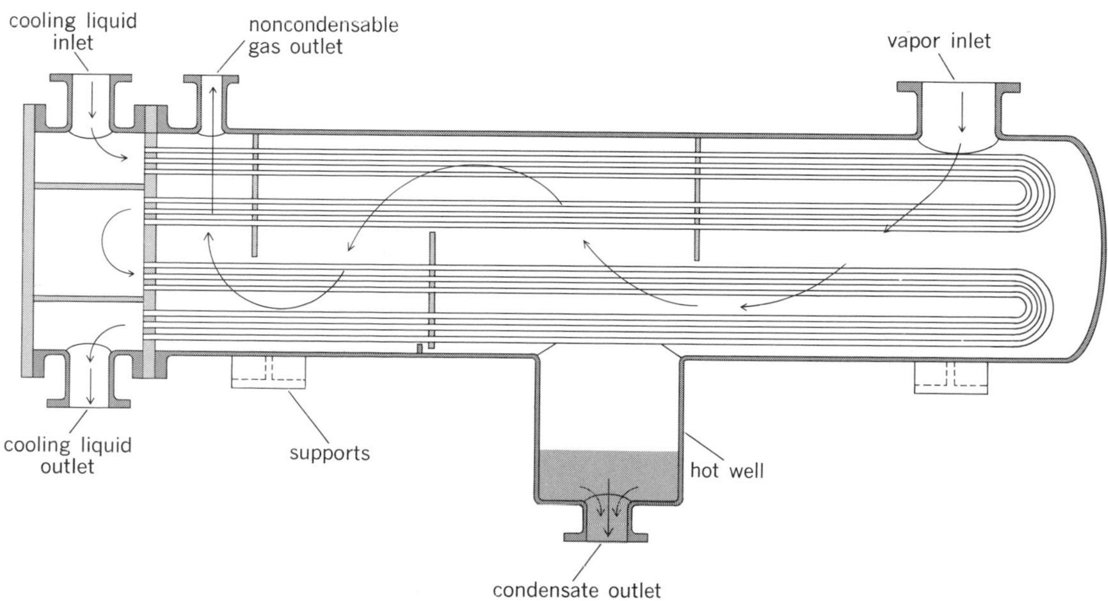

Fig. 2. Process condenser with one baffled-shell pass of condensing vapor and four tube passes of cooling liquid.

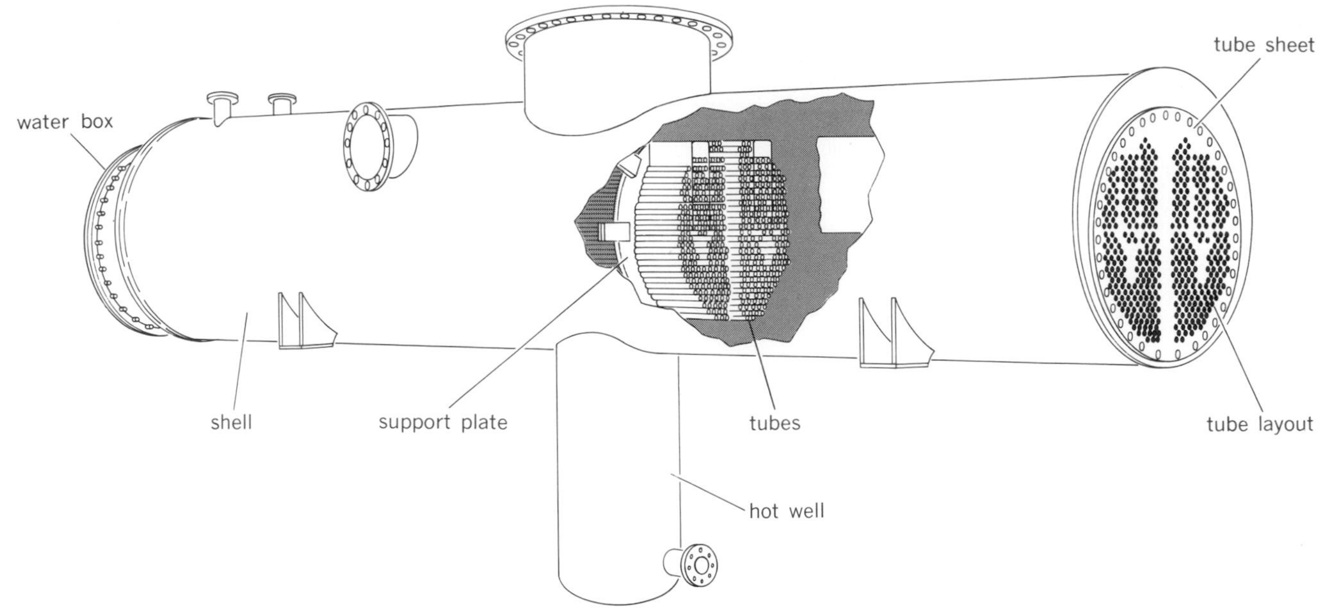

Fig. 3. Typical sections of small steam surface condenser.

baffles in the condensate drain zone to flood the tubes in these sections or by effecting a similar result with a loop in the condensate drain piping. Baffles are most effective in process condensers of the horizontal type, and loops are more adaptable to process condensers of the vertical type.

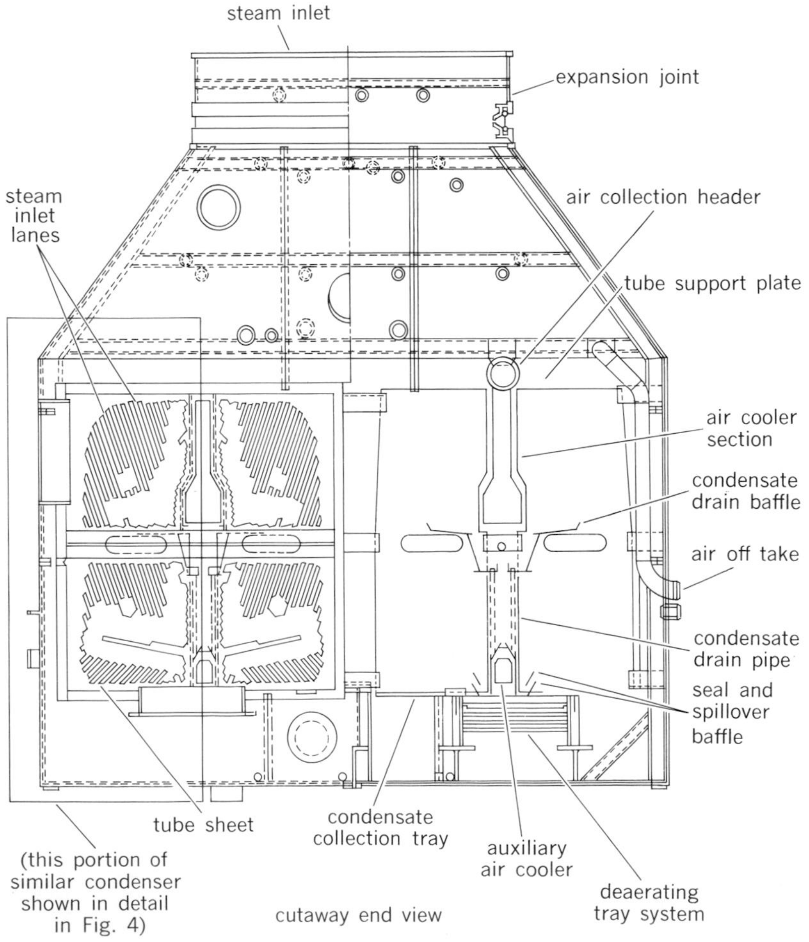

Fig. 4. Typical section of steam surface condenser. (*Worthington Corp.*)

Condensate reheating. Modern steam surface condensers for turbines are provided with means for condensate reheating (Fig. 4). Condensate falling from tube to tube becomes cooled below the temperature corresponding to condenser operating pressure; unless provision is made for reheating this condensate before it is removed from the condenser, a measurable amount of heat energy is lost from the cycle, and cycle efficiency is reduced. In addition, subcooled condensate absorbs air from the condensing vapor and, if allowed to remain in the boiler feedwater, may cause serious corrosion of the feed system and boilers.

The condensate is reheated by a portion of the incoming steam. Baffles direct this steam to the condenser hot well in such a manner that its velocity energy is converted to pressure (Fig. 5). As a result, the local static pressure and corresponding temperature become equal to, or greater than, the static pressure and corresponding temperature at the condenser stream inlet. Condensate falling from the tube bundle to the hot well is reheated, the degree of reheat depending on the height of fall and the flow rate per unit area. If vertical height is limited, effective reheating and the associated deaeration can be achieved by redirecting the condensate falling from the tubes over collecting baffles and a series of trays or plates. This higher zone of pressure and temperature is vented along with the noncondensables removed from the condensate into the condenser tube bank and finally to the air cooler joining the mainstream of noncondensables being cooled and expelled. Under favorable conditions the condensate may reach temperatures as much as 5°F (3°C) above the temperature corresponding to static pressure at the condenser steam inlet, and the dissolved oxygen content of the deaerated condensate may be consistently less than 0.01 cm^3 per liter.

Condenser capacity. The quantity of cooling fluid required to condense a given quantity of vapor may be obtained by applying the general heat-balance equation. For a condensing vapor which may be superheated, saturated, or wet, with its condensate subcooled or reheated and with the noncondensable gases cooled, the relation is given by Eq. (1). (The standard engineering practice is to express quantities in lb/hr and enthalpies in Btu/lb.)

$$W_v(h_v - h_c) + W_n(h_{na} - h_{nb}) = W_f(h_{fb} - h_{fa}) \qquad (1)$$

W_v = quantity of vapor
h_v = enthalpy of entering vapor
h_c = enthalpy of leaving condensate
W_n = quantity of noncondensable gases
h_{na} = enthalpy of entering noncondensables
h_{nb} = enthalpy of leaving noncondensables
W_f = quantity of cooling fluid
h_{fa} = enthalpy of entering cooling fluid
h_{fb} = enthalpy of leaving cooling fluid

The temperature of the leaving cooling fluid is usually selected to be 5–10°F (3–6°C) less than the condensing temperature of the saturated vapor. Ordinarily, low temperature differences are associated with contact condensers; larger temperature differences are characteristic of economic design for surface condensers. Small temperature differences require enormous amounts of condensing surface and are seldom a practical design criterion.

For both contact and surface condensers the conditions for determining the hourly heat transferred can be expressed by Eq. (2) for undirection-

$$Q = UA\,\Delta t_m \qquad (2)$$

Q = hourly heat, Btu/hr
U = overall heat-transfer coefficient, Btu/(hr)(ft²)(°F)
A = area of heat-transfer surface, ft²
Δt_m = logarithmic mean-temperature difference, °F

al steady-state heat flow. Logarithmic mean-temperature difference Δt_m is based on the assumption of constant specific heat and substantially constant condensing temperature with the cooling-fluid temperature increasing from inlet to outlet in its flow path. Overall heat-transfer coefficient U and surface A are generally applicable to surface condensers only.

For direct-contact condensers where surface is not readily determined, it is usual to consider heat transferred per unit volume of condensing space rather than per unit area. In this case Eq. (3) de-

$$Q = KV\,\Delta t_m \qquad (3)$$

scribes the heat flow, where K = volumetric heat-transfer coefficient, Btu/(hr)(ft³)(°F); V = volume of direct contact condensing space, ft³; and the other symbols are as previously defined.

The magnitude of K depends on the physical properties of the condensing vapor and heat-receiving fluid, the type of direct-contact surface (spray, tray, or packing), and the operating temperature. It is ordinarily determined experimentally for each type of contact condenser.

The magnitude of U, for use in Eq. (2), can be obtained from the summation of Eq. (4) for the individual resistances in the heat-transfer system, resistance being expressed in (hr)(ft²)(°F)/Btu.

$$U = 1/(r_v + r_l + r_f + r_w) \qquad (4)$$

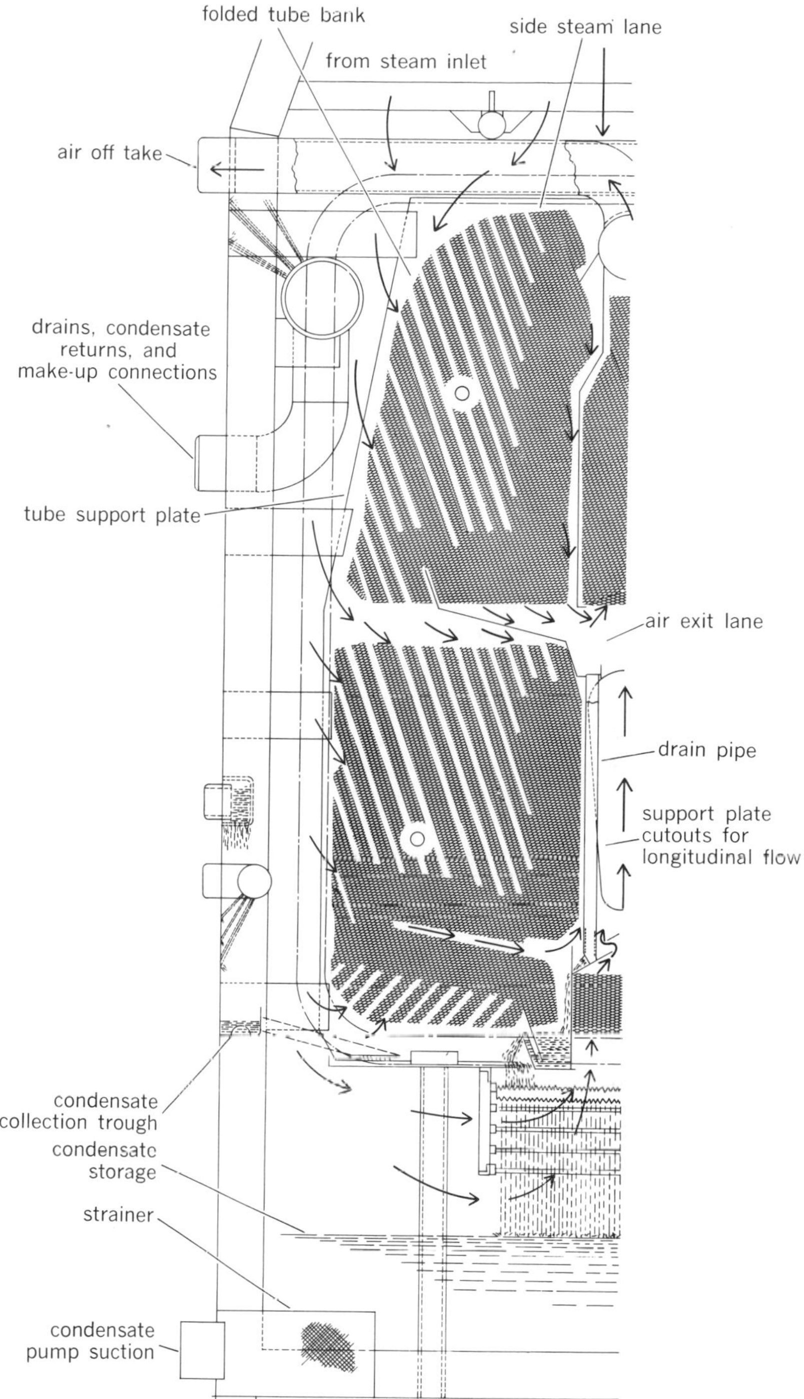

Fig. 5. Detail of portion of condenser similar to that of Fig. 4. (*Worthington Corp.*)

$1/r$ = conductance across any boundary, Btu/(hr)(ft²)(°F)
r_v = condensing boundary resistance
r_l = cooling-fluid boundary resistance
r_f = fouling, dirt, or scale resistance
r_w = separating wall, resistance

Conductance of the condensing boundary $1/r_v$ may be determined analytically for pure vapors. For steam on the outside of horizontal tubes, $1/r_v$ may be 2000–2500 Btu/(hr)(ft²)(°F)[11,400–14,200 J/(s)(m²)(°C)] for well-designed multitube condensers. Conductance of the cooling-fluid boundary $1/r_l$, for fluids flowing inside of tubes, may be determined from the relation of the Nusselt

Conductance values for some thermodynamic fluids

Fluid	Condensing vapor conductance*	Cooling fluid conductance*
Isopropyl alcohol	400	360
Benzene	600	520
Water	2500	1400
Ammonia	3200	2300

*Btu/(hr)(ft²)(°F) = 5.8 J/(s)(m²)(°C).

number to the product of the Reynolds and Prandtl numbers with the use of empirical constants such as those suggested by W. H. McAdams. Excellent agreement between computed values and test values results from the use of these equations. For water inside of tubes $1/r_t$ may be in the order of 1000–1400 Btu/(hr)(ft²)(°F) [5700–8000 J/(s)(m²)(°C)] for the normal range of water velocities.

Values for conductance of the condensing boundary and the cooling-fluid boundary for some common thermodynamic fluids are listed in the table for cooling-fluid velocities in the range of 6–7 ft per second (1.8–2.1 m/s), condensing temperatures in the range of 100°F (38°C), and condensing rates of approximately 10 lb/(ft²)(hr) [1.4×10^{-2} kg/(m²)(s)].

Conductance for fouling or dirt films $1/r_f$ depends on the characteristics of the cooling fluid and the condensing vapor in relation to the accumulation (of dirt and corrosion products on the separating wall or tube surfaces) in service. In addition, the oxide film on new clean tubes or other types of metal heat-transfer surfaces contributes to overall fouling resistance r_f. For water-cooled shell and tube steam condensers a conductance of 2000 Btu/(hr)(ft²)(°F) [11,400 J/(s)(m²)(°C)] is reasonable for copper-base alloy tubes after mechanical cleaning. Conductance of the separating wall $1/r_w$ may be computed from the thickness of the wall and the thermal conductivity of the material from which it is made. The overall heat-transfer coefficient varies with the physical properties, the flow rates of the fluids, and the geometry of the condenser and condensing surfaces. For steam condensers U may be in the order of 350–800 Btu/(hr)(ft²)(°F) [2000–4500 J/(s)(m²)(°C)]; with few exceptions, heat transfer is lower with condensing vapors other than steam.

Condenser components. The components of a representative contact condenser are shown in Fig. 6. The cooling-liquid distribution system shown consists of baffles and impingement surfaces to distribute the liquid uniformly and to allow it to cascade in counterflow relationship with the condensing vapor. In the illustrated condenser the cooling section for noncondensable gases coincides with the coolant distribution section where vapor is condensed and the noncondensable gases dehumidified and concentrated. Other constructions for distributing the coolant include rings or slats made of metal, plastic, or ceramic. Most process condensers of the contact type are of counterflow design. In jet condensers noncondensables are entrained in the cooling liquid and thereby removed. The tail pipe provides the barometric leg and discharges into a sealing well, thus eliminating the need for an extraction or tail pump. Low-level contact condensers, those without barometric legs, usually require pumps to extract the cooling water.

Components of typical surface condensers are shown in Figs. 2, 3, 4, 5, and 7. The condensing surface consists of tubes of 0.5–1.5-in. (13–38-mm) outside diameter made of copper base alloys, less frequently of aluminum, nickle alloys, chromium steel, chromium-nickel steels, and titanium. The cooling surface for noncondensables is ordinarily the same material as the main condensing surface, although special corrosion-resistant alloys may be used when the noncondensables are

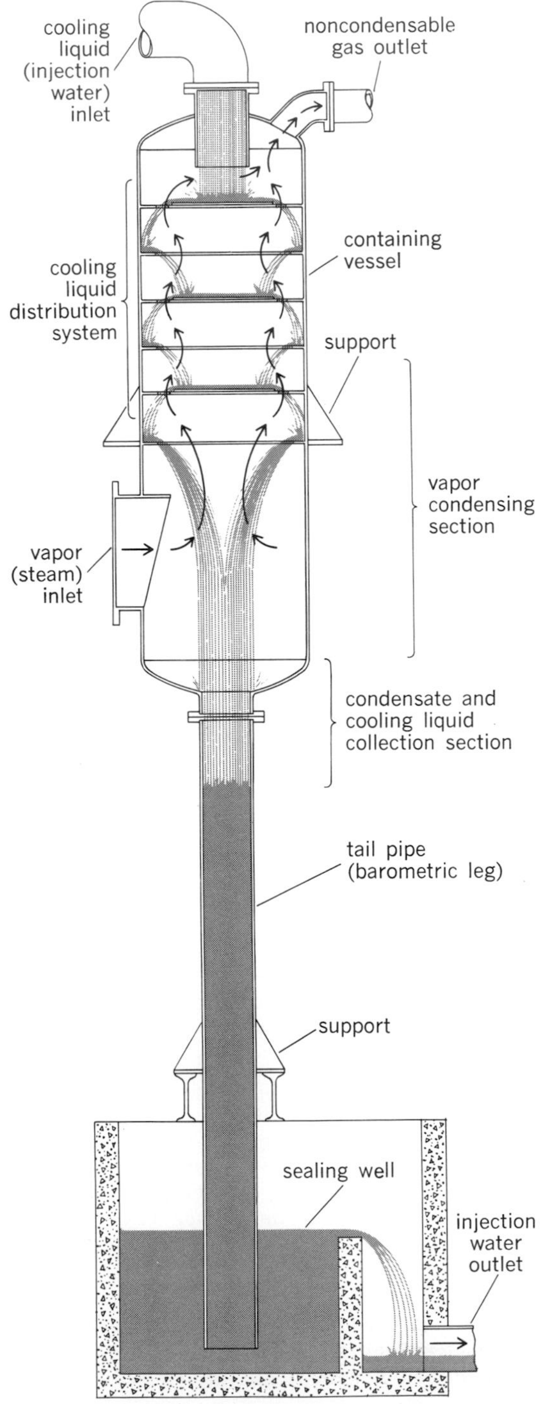

Fig. 6. Counterflow barometric-type contact condenser.

especially corrosive. Usual practice is to condense the vapors on the outside of tubes (Figs. 2, 3, 4, and 5), but when the vapors are especially corrosive they may be condensed on the inside of tubes (Fig. 7). Tubes are usually plain, but for condensing vapors from low-conductance fluids, finned tubes may be used. For this purpose low fins, approximately equal in height to basic tube wall thickness, are used on the condensing side spaced so that the condensate formed does not bridge the fins and act like insulation. Gas- or air-cooled condensers use high fins on the cooling side with the extended surface generally at least 10 times the equivalent plain tube area. Bimetal tubes are most frequently used in process condensers when the corrosion environment is severe and different on one side of the tube than on the other.

Tube vibration of magnitudes sufficient to cause tube failure is controlled by the spacing of tube support plates or baffles. Tube vibration seldom results from transmitted mechanical energy but rather from vapor velocity, causing tubes to deflect and vibrate at their natural frequency. With low-density vapors at high velocity, generally in the sonic range or greater, tube vibration is caused by the lift and drag effects of the vapor flowing around the tubes. Severe vibration is usually independent of the von Karman vortex street with low-density vapors, but may be associated with both the lift and drag effects of velocity and the von Karman vortex street at high-vapor densities even at relatively low velocities.

The tube sheets (into which tubes are welded, expanded, or packed) are generally of Muntz metal, naval brass, copper-silicon alloy, or other corrosion-resistant alloys. Carbon-steel tube sheets may be used with carbon or chromium-nickel steel tubes. In nuclear power plants steam condenser tube sheets may be double, that is, with an air gap between to detect leakage and to prevent cooling water from leaking into the steam space. The inner tube sheets are usually made of carbon steel and the outer sheets of copper alloy.

The cooling fluid system for surface condensers, when liquid is used, usually consists of chambers attached to the tube sheets and arranged with inlet and outlet connections for the circulation of the cooling medium. When this fluid is liquid, the chambers are designed to distribute the liquid over the face of the tube sheets and into the tubes with little or no cavitation.

Condensers may be designed with one or more liquid passes, dependent on the thermal design conditions, the quantity of cooling liquid available or desired, and the space conditions for installation. Steam surface condensers in large steam power plants are of single-pass design except where the cooling water supply is limited, in which case two-pass condensers are used. Large installations (those that require cooling towers, spray, or evaporation ponds as a means for controlling cooling-water temperature) ordinarily use two-pass condensers. Condensers with three or more passes seldom prove economical for serving engines or turbines. Process condensers are frequently designed with more than two cooling liquid passes and seldom with less.

Auxiliary equipment. Operation of condensers requires pumps (1) for injecting or circulating the cooling fluid, (2) for removing condensate or mixed condensate and injection water, and (3) for removing noncondensables.

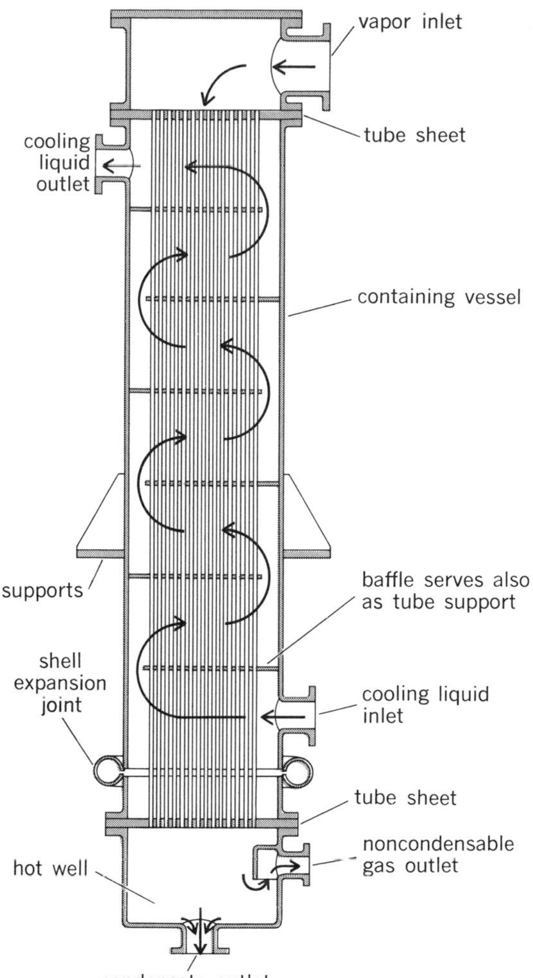

Fig. 7. Vertical-process surface condenser with condensing vapor inside the tubes and the cooling liquid outside the tubes, within the shell.

Centrifugal pumps are usually used for injecting or circulating cooling water. Conventional volute pumps are used for circulating cooling water for small surface condensers and as injection water pumps for contact condensers. Mixed-flow volute pumps and axial-flow pumps of vertical design are best suited to large surface condensers.

Centrifugal pumps of the horizontal volute type, equipped with pressure-sealed stuffing boxes, are well suited for removing condensate. Vertical, multistage condensate pumps are more suited to large steam power plant installations, where they effect significant installation economies. Positive displacement pumps are used for withdrawing condensate from small condensers. Tail pumps of the conventional double-suction centrifugal type are used for pumping mixed condensate and injection water from low-level jet (contact) condensers.

Pumps for removing noncondensable gases are classified as displacement and ejector types. The displacement machine is built either as a reciprocating vacuum pump, similar to a piston-type air compressor, or as a rotary machine, similar to a gear-pump or sliding-vane rotary compressor. The displacement-type vacuum pump is widely used on condensers of all sizes. It is not suitable for use at high vacuum with high noncondensable gas loads because it must be disproportionately large for such conditions. It is suitable for use with extremely large surface condensers serving turbines, because of the low noncondensable gas loads characteristic of these installations.

Steam jet ejectors are widely used as vacuum pumps. Having no moving parts, they require little or no operating attention and are simple to install. They have excellent capacity characteristics at high vacuum and are especially suited to applications where the noncondensable gas load is high. When used with surface condensers serving turbines, steam jet ejectors are equipped with surface inter- and after-condensers. The exhaust steam is condensed and returned to the feed system, and the heat from the exhaust is also recovered in the feedwater. Thus they become highly efficient machines for removing noncondensable gases from condensers used in the generation of power. *See* COOLING TOWER.

[JOSEPH F. SEBALD]

Bibliography: *Feedwater Heater Workshop Proceedings*, Palo Alto, CA, sponsored by Electric Power Research Institute and Joseph Oat Corp., EPRI WS-78-133, July 1979; M. Jakob and G. A. Hawkins, *Elements of Heat Transfer*, 3d ed., 1957; W. H. McAdams, *Heat Transmission*, 3d ed., 1954; C. C. Peake, G. F. Gerstenkorn, and T. R. Arnold, Some reliability considerations for large surface condensers, *Proceedings of the American Power Conference*, Chicago, vol. 37, pp. 562-574, 1975; B. W. Pendrick, *The Surface Condenser*, 1935; *The Performance of Condensers in Nuclear and Fossil Power Plants*, pts. 1 and 2, Seminar (Columbus, OH) sponsored by Electric Power Research Institute, American Society of Mechanical Engineers, and Ohio State University, June 2-4, 1975; J. F. Sebald, Main and auxiliary condensers, *Marine Engineering*, chap. 13, Society of Naval Architects and Marine Engineers, 1971; J. F. Sebald and W. D. Nobles, Control of tube vibration in steam surface condensers, *Proceedings of the American Power Conference*, Chicago, vol. 24, pp. 630-643, 1962.

Vapor cycle

A thermodynamic cycle, operating as a heat engine or a heat pump, during which the working substance is in, or passes through, the vapor state. A vapor is a substance at or near its condensation point. It may be wet, dry, or slightly superheated. One hundred percent dryness is an exactly definable condition which is only transiently encountered in practice. Vapor behavior deviates so widely from the ideal gas laws that calculation requires the use of tables and graphs that give the experimentally determined properties of the fluid.

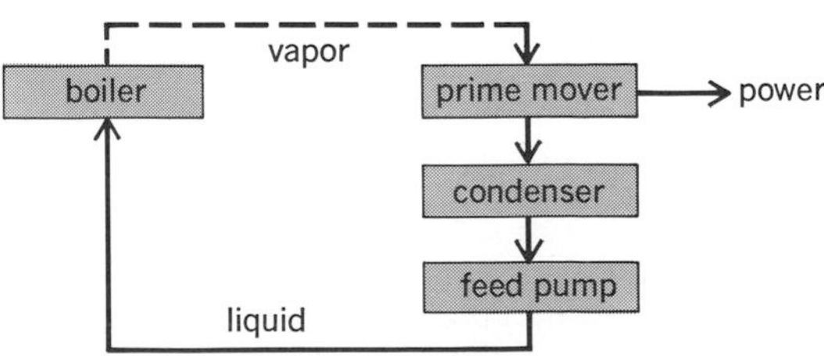

Fig. 1. Rudimentary steam power plant flow diagram.

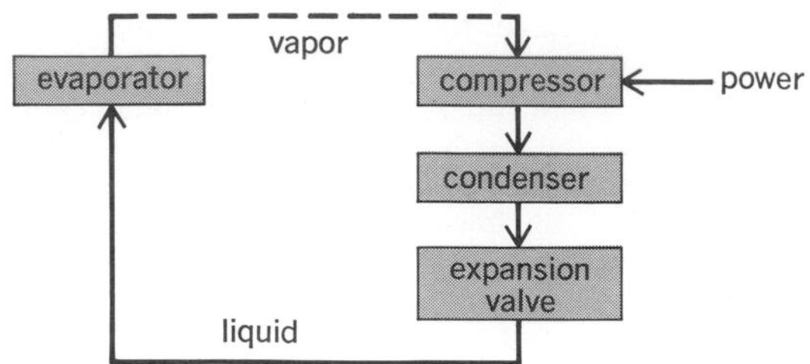

Fig. 2. Rudimentary vapor-compression refrigeration plant flow diagram.

Power and refrigeration plants. A steam power plant operates on a vapor cycle where steam is generated by boiling water at high pressure, expanding it in a prime mover, exhausting it to a condenser, where it is reduced to the liquid state at low pressure, and then returning the water by a pump to the boiler (Fig. 1).

In the customary vapor-compression refrigeration plant, the process is essentially reversed with the refrigerant evaporating at low temperature and pressure, being compressed to high pressure, condensed at elevated temperature, and returned as liquid refrigerant through an expansion valve to the evaporating coil (Fig. 2).

The Carnot cycle, between any two temperatures, gives the limit for the efficiency of the conversion of heat into work (Fig. 3). This efficiency is independent of the properties of the working fluid. Although the thermal efficiency is independent of the properties of the fluid, the mean effective pressure (and consequent physical dimensions of the engine) will be vitally influenced by choice of fluid (compare Fig. 3*b* with 3*c*). The Carnot cycle is not realistic for the evaluation of steam power plant performance because the cycle precludes the use of superheat and calls for the isentropic compression of vapor. It is useful, however, for specifying the limiting efficiency that a real cycle might approach. It is so used in judging performance of vapor-cycle heat engines and heat pumps. *See* CARNOT CYCLE.

The Rankine cycle is more realistic in describing the ideal performance of steam power plants and vapor-compression refrigeration systems.

Vapor steam plant. In the case of the steam power plant (Fig. 1), the Rankine cycle (Fig. 4) has two constant pressure phases joined by a reversible adiabatic (isentropic) phase 1-2. From the properties of the fluid, the work of the prime mover, ΔW_{PM}, is most conveniently evaluated as in Eq. (1), where h is the enthalpy, Btu/lb. The feed pump

$$\Delta W_{PM} = h_1 - h_2 \qquad (1)$$

uses some of this work, ΔW_{FP}, to return the water from the condenser to the boiler so that the net output of the cycle is $\Delta W_{PM} - \Delta W_{FP}$. This net output can be related to the heat that must be added to produce steam by consideration of the *T-s* diagram (Fig. 4*b*). The area under line *bcd*-1 is the heat supplied in the boiler (heat source); phase 1-2 is the isentropic expansion in the prime mover; the area under line 2-*a* is the heat rejected to the condenser (heat sink); phase *a-b* is the isentropic

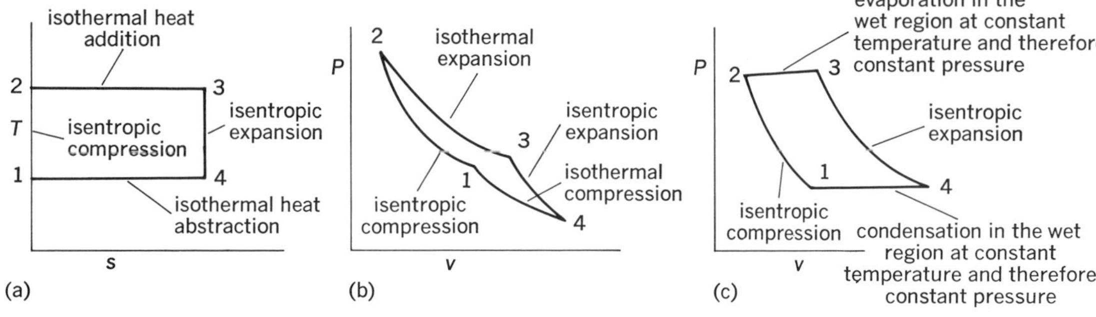

Fig. 3. Carnot cycle. (*a*) Temperature-entropy. (*b*) Pressure-volume for fixed gas. (*c*) Pressure-volume for vapor.

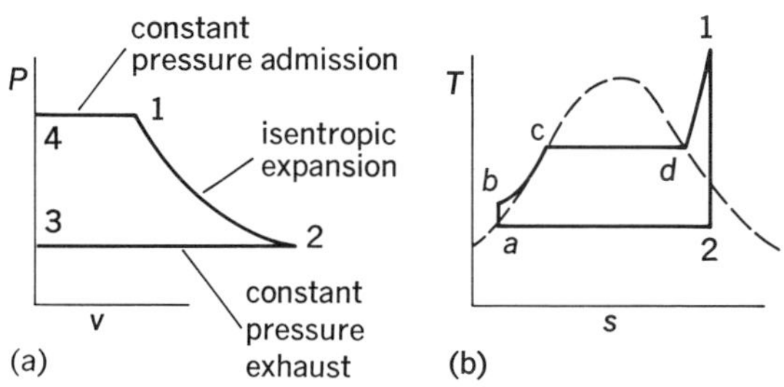

Fig. 4. Rankine cycle for heat-engine plant. (*a*) Pressure-volume diagram of prime mover. (*b*) Temperature-entropy diagram for the power plant.

compression of the liquid in the feed pump. Thus, the thermal efficiency is given by Eq. (2).

$$\frac{\text{work done}}{\text{heat added}}=\frac{\Delta W_{PM}-\Delta W_{FP}}{h_1-h_a-\Delta W_{FP}} \qquad (2)$$

There are many variables which influence the performance of the Rankine cycle. For steam the thermal efficiency is a function of pressure, temperature, and vacuum (Fig. 5). High pressure, high superheat, and high vacuum lead to high efficiency. *See* RANKINE CYCLE.

Vapor refrigeration plant. The Rankine cycle can be used to evaluate the performance of the vapor-compression system of refrigeration (Fig. 2). A counterclockwise path is followed on the *P-v* and *T-s* cycle diagrams (Fig. 6). The refrigerant enters the compressor as low-temperature, low-pressure vapor, (4–1); isentropic compression follows (1–2), and then high-pressure delivery (2–3). The work to drive the compressor is h_2-h_1. The machine cooling coefficient of performance *cp*, which is essentially the reciprocal of thermal efficiency, is given by Eq. (3); for a warming machine, it is given by Eq. (4).

$$cp=\frac{\text{refrigeration}}{\text{work done}}=\frac{h_1-h_d}{h_2-h_1} \qquad (3)$$

$$cp=\frac{\text{heat delivered}}{\text{work done}}=\frac{h_2-h_c}{h_2-h_1} \qquad (4)$$

The difference h_1-h_d is heat removed in the refrigerating coils, the area under phase $d-1$ on the *T-s* diagram. The difference h_2-h_c is the heat delivered to the condensing coils, the area under phase (2*abc*). Because the flow is throttled through the expansion valve, the enthalpy is constant, $h_c=h_d$. Some ideal performance values of a vapor-compression refrigeration system are plotted in Fig. 7. For further details on heat-pump vapor cycles *see* HEAT PUMP.

The remainder of this article is concerned with the vapor cycle as it is applied to power-generation purposes.

Regenerative heat cycle. The regenerative cycle is a modification of the simple Rankine cycle. Feed water is heated by extracted steam (Fig. 8). As a result, less heat must be added in the boiler to evaporate a pound of steam and in turn, to deliver a kilowatt hour of work output from the associated steam engine.

A cycle with a single stage of regenerative heating can be viewed as two Rankine cycles superimposed on one another (Fig. 8*b*). In one the exhaust pressure and consequent temperature are substantially higher than in the other (point *f* versus point 2). The heat of condensation represented by the area under *c-f* can be used to raise the feed temperature from T_b to T_c. The area under phase *b-c* is

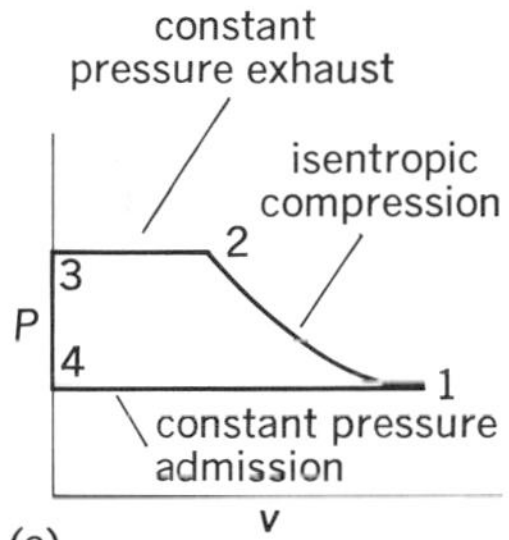

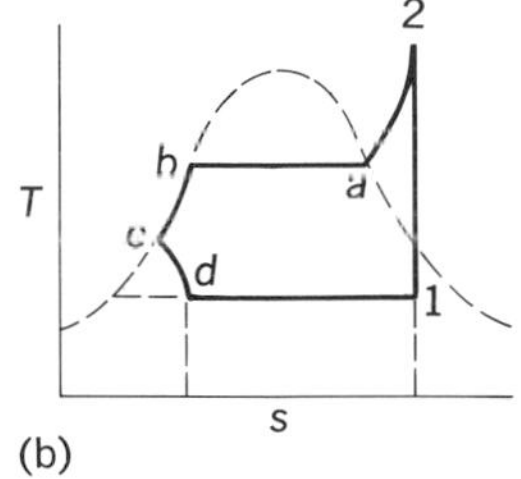

Fig. 6. Rankine cycle for heat-pump plant. (*a*) Pressure-volume diagram of compressor. (*b*) Temperature-entropy diagram for a vapor-compression refrigeration plant.

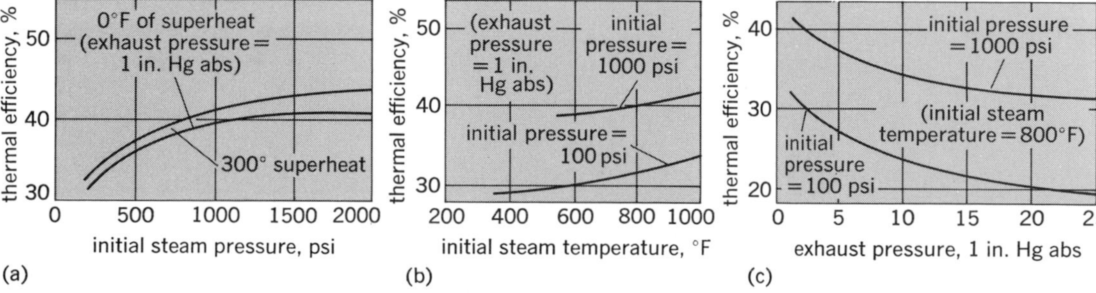

Fig. 5. Thermal efficiency of ideal Rankine steam cycle. (*a*) Steam pressure. (*b*) Steam temperature. (*c*) Vacuum. 1 psi = 6.895 kPa. 1 in. Hg = 3.386 kPa. Temperature interval of 1°F = 5/9°C. Temperature (°C) = (5/9) [temperature (°F) − 32].

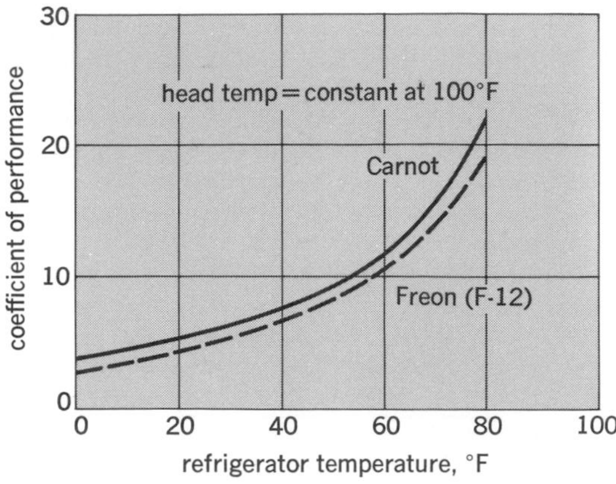

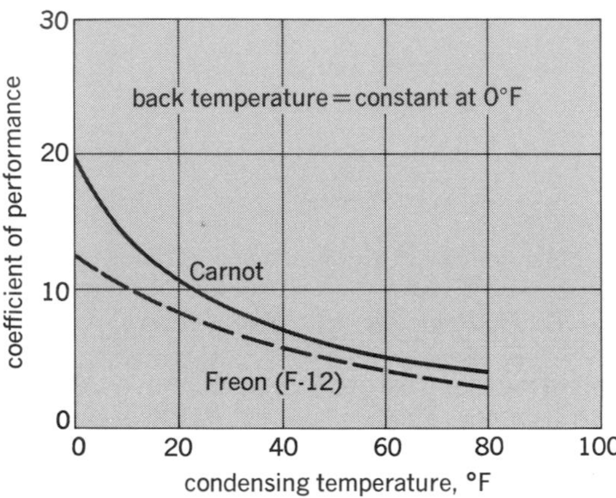

Fig. 7. Coefficient of performance of ideal Rankine and Carnot cycles as influenced by condensing and refrigerating temperatures. Temperature (°C) = (5/9)[temperature (°F) − 32].

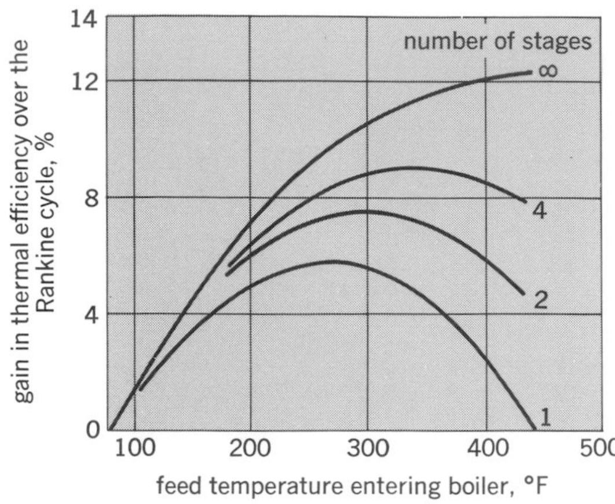

Fig. 9. Gain in thermal efficiency by use of regenerative instead of Rankine cycle; steam conditions, 400 psi (2.76 MPa) and 700°F (370°C); exhaust pressure, 1 in. Hg abs (3.39 kPa). Temperature (°C) = (5/9)[temperature (°F) − 32].

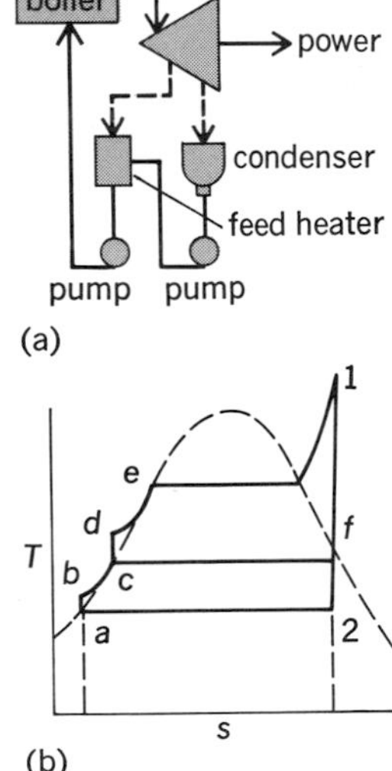

Fig. 8. Regenerative cycle. (*a*) Rudimentary steam power plant flow diagram with single-stage feed heating. (*b*) Temperature-entropy diagram.

smaller than under phase *c-f* so a fraction of a pound of steam is needed to raise 1 lb of water to the common temperature level T_c.

The principle of regeneration can be extended to multiple-stage heating with different final feed temperatures and in the limit reaching the boiler saturation temperature T_e with an infinite number of heating stages. Some consequences of the process are reflected in the data of Fig. 9. The gain in thermal efficiency is the consequence of reducing the quantity of heat rejected to sink 2-*a* in Fig. 8. The weight of steam flow to the prime mover for the production of a kilowatt-hour is larger than with the simple Rankine cycle (Fig. 4). But the heat required to make a pound of steam is so much less that there is an overall thermodynamic gain per kilowatt-hour. The weight flow of steam, from the prime mover to the condenser, is less than with the simple Rankine cycle. It is this reduction in heat rejected to the thermodynamic sink that raises the overall thermal efficiency of the regenerative above the nonregenerative cycle. Modern steampower practice uses the steam turbine for up to ten stages of regenerative heating.

Reheat cycle. The resuperheat or reheat cycle is another improvement in vapor cycles favored in current central station practice. Steam expanding isentropically (Fig. 4*b*) grows wetter with consequent increased erosion of machinery parts and loss in mechanism efficiency: Superheat tends to correct these weaknesses but metallurgical limitations fix the maximum allowable steam temperature. Reheating the steam after a partial expansion (Fig. 10*a*) gives a practical correction. The reheating can be carried out at various pressure and temperature levels and in multiple stages. Thermal efficiency is improved over the simple Rankine cycle (Fig. 10*b*). Current practice uses a single reheat stage with temperatures approximately equal to primary steam temperature. Supercritical pressure plants favor two reheating stages.

Binary vapor cycle. Comparison of the Rankine (Fig. 4*b*), the reheat (Fig. 10*a*), and Carnot (Fig. 3*a*) cycles shows that there are considerable thermodynamic losses in the first two by failure to approach the rectangular *T-s* configuration of the Carnot cycle. The binary vapor cycle uses two fluids with totally different vapor pressures, such as mercury and water (Fig. 11). If a Rankine cycle using mercury is superimposed on that using steam, it is possible to operate the mercury condenser as the steam boiler, transferring the heat

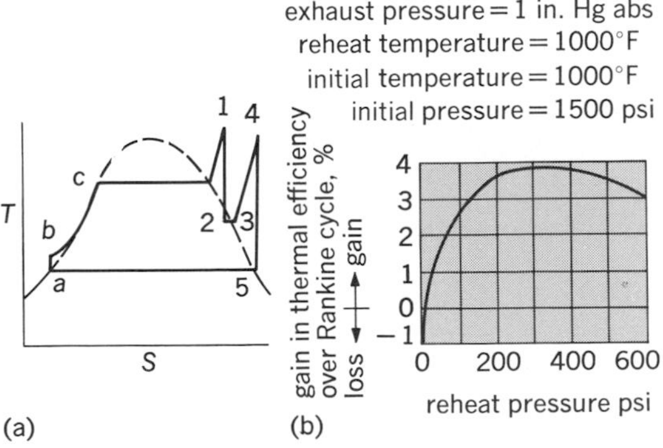

Fig. 10. Resuperheat or reheat cycle. (*a*) Temperature-entropy diagram. (*b*) Gain in thermal efficiency as function of reheat pressure; primary pressure, primary temperature, and reheat temperature constant at 1500 psi (10.3 MPa), 1000°F (540°C), and 1000°F, respectively.

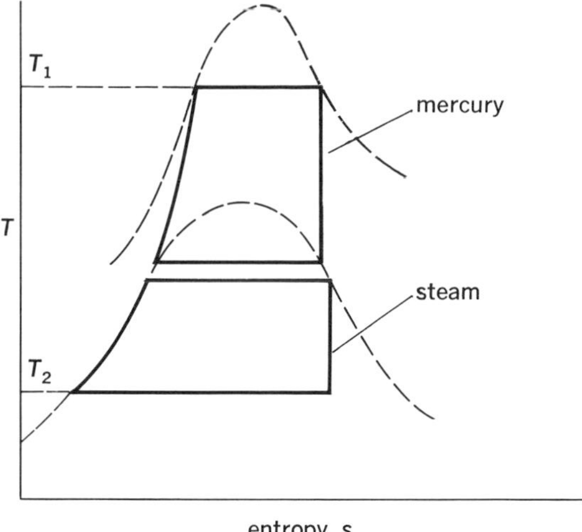

Fig. 11. Binary-vapor cycle using mercury and water, temperature-entropy diagram.

from the one fluid to the other. Because of the differences in the latent heats and specific heats of the two fluids, several pounds of mercury must be circulated to make 1 lb of steam. However, the combined *T-s* diagram (Fig. 11) approximates the rectangular specification with appreciable gain in thermal efficiency. For many years the binary vapor cycle was thought to have practical as well as theoretical advantages, but its economic difficulties have led to its eclipse by, and abandonment in favor of, the high-pressure, regenerative, reheat, steam cycles. *See* THERMODYNAMIC CYCLE.

[THEODORE BAUMEISTER]

Bibliography: T. Baumeister (ed.), *Standard Handbook for Mechanical Engineers*, 7th ed., 1967; V. M. Faires, *Thermodynamics*, 4th ed., 1962; G. A. Hawkins, *Thermodynamics*, 2d ed., 1951; J. H. Keenan, *Thermodynamics*, 1941.

Warm-air heating system

In a general sense, a heating system which circulates warm air. Under this definition both a parlor stove and a steam blast coil circulate warm air. Strictly speaking, however, a warm-air system is one containing a direct-fired furnace surrounded by a bonnet through which air circulates to be heated (see illustration).

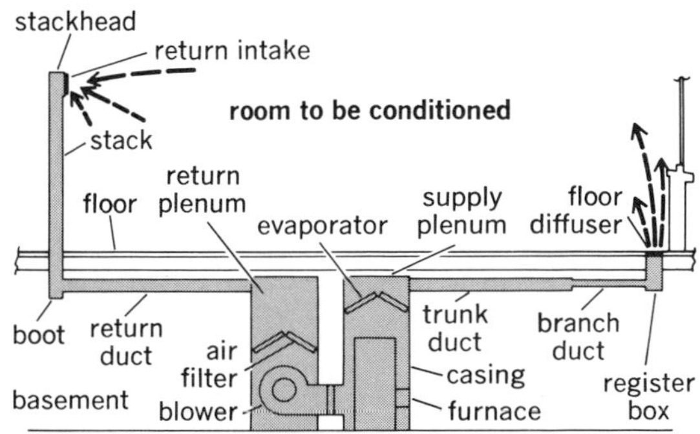

Air passage in a warm-air duct system. (*From S. Konzo, J. R. Carroll, and H. D. Bareither, Winter Air Conditioning, Industrial Press, 1958*)

When air circulation is obtained by natural gravity action, the system is referred to as a gravity warm-air system. If positive air circulation is provided by means of a centrifugal fan (referred to in the industry as a blower), the system is referred to as a forced-air heating system.

Direct-fired furnaces are available for burning of solid, liquid, or gaseous fuels, although in recent years oil and gas fuels have been most commonly used. Furnaces have also been designed which have air circulating over electrical resistance heaters. A completely equipped furnace-blower package consists of furnace, burner, bonnet, blower, filter, and accessories. The furnace shell is usually of welded steel. The burner supplies a positively metered rate of fuel and a proportionate amount of air for combustion. A casing, or jacket, encloses the furnace and provides a passage for the air to be circulated over the heated furnace shell. The casing is insulated and contains openings to which return-air and warm-air ducts can be attached. The blower circulates air against static pressures, usually less than 1.0-in. water gage (250 Pa gage pressure). The air filter removes dust particles from the circulating air. The most common type is composed of 1- to 2-in. (25- to 50-mm) thick fibrous matting, although electrostatic precipitators are sometimes used. *See* OIL BURNER.

Accessories to assure effective operation include automatic electrical controls for operation of burner and blower and safety control devices for protection against (1) faulty ignition of burner and (2) excessive air temperatures.

Ratings of warm-air furnaces are established from tests made in laboratories under industry-specified conditions. The tests commonly include heat-input rate, bonnet capacity, and register delivery. Heat-input rate is the heat released inside the furnace by the combustion of fuel, in Btu/hr. Bonnet capacity refers to the heat transferred to the circulating air, in Btu/hr. Register delivery is the estimated heat available at the registers in the room after allowance for heat loss from the ducts has been made, in Btu/hr.

The recommended method for selection of a furnace is to estimate the total heat loss from the structure under design weather conditions, including the losses through the floor and from the basement, and to choose a furnace whose bonnet capacity rating is equal to, or greater than, the total design heat loss.

The complete forced-air heating system consists of the furnace-blower package unit; the return-air intake, or grille, together with return-air ducts leading from the grille to the return-air plenum chamber at the furnace; and the supply trunk duct and branch ducts leading to the registers located in the different spaces to be heated.

The forced-air system is no longer confined to residential installations. The extreme flexibility of the system, as well as the diversity of furnace types, has resulted in widespread use of the forced-air furnace installations in the following types of installations, both domestic and commercial: residences with basement, crawl space, or with concrete floor slab; apartment buildings with individual furnaces for each apartment; churches with several furnaces for different zones of the building; commercial buildings with summer-winter arrangements; and industrial buildings with individual furnace-duct systems in each zone. *See* COMFORT HEATING.

[SEICHI KONZO]

Bibliography: Air Conditioning Contractors of America, *Load Calculation for Residential Winter and Summer Air Conditioning*, Manual J, 1975; American Society of Heating, Refrigerating, and Air-Conditioning Engineers, *ASHRAE Handbook and Product Directory*: *Systems*, 1976; S. Konzo, J. R. Carroll, and H. D. Bareither, *Winter Air Conditioning*, 1958.

Waste heat management

The systematic process of minimizing the release of energy to the environment by placing priority emphasis on technical sophistication in the use of energy-related machines and products over short-term economic return.

For more than a century, physical growth and financial expansion have been generally regarded as synonymous with progress. Many of the notions of modern resources management continue to reflect this idea. Certainly, there is no single segment of Western culture that has contributed more to the advancement of the human race than energy technology, yet all the achievements will be of little value if the management of energy is unsuccessful.

In order to evaluate the potential effects of the present course of action in energy management, it is important to understand that energy demand and energy release to the environment are synonymous. All of the energy converted from nuclear or fossil sources appears in the environment as heat to be added to the Earth's thermal budget. Hence, in a realistic discussion of total energy demand, it is necessary to consider the subject in terms of the thermal energy released to satisfy the demand. In this context, because of prime-mover cycle limits, the projected growth of electrical power assumes a positon of paramount importance.

ENERGY DEMAND FORECAST

Energy demand projections for the contiguous United States vary widely, depending on the assumptions of the investigator, ranging from 90 to 140×10^{15} Btu (95 to 148×10^{18} J) per year. Persons who place a high degree of faith in economic elasticity tend to forecast on the low side, with the implicit hope that increasing technological sophistication will in some undefined way reduce the energy demand. Persons who represent energy interests tend to forecast higher figures. Table 1 represents the author's best view of the many forecasts available. *See the feature article* ENERGY CONSUMPTION.

Current study. Using these forecasts and other prominent forecasts, this author conducted a study of the breakdown of energy requirements by major expected use. This study is the basis for the interpretive comments which follow. Figure 1 summarizes the results of the study. The projections of energy allocations on a per capita basis or by gross national product are derived from this summary.

Figure 2, which is based on a total United States population of 250×10^6 by the year 2000, places the expected overall energy usage per capita in perspective. Energy usage per capita is expected to increase by some 180% over 1960 estimates, and electrical energy (kilowatts at the bus-bar) is expected to increase by 350% in the same period.

Table 1. United States energy demand by major sources in 1973 and estimated demand in 1985 and 2000

Resources	1973	1985	2000
Petroleum (includes natural gas liquids and synthetic oil)			
10^6 bbl[a]	5,982	6,500	7,150[f]
10^6 bpd[a]	16.4	17.8	19.6
10^{12} Btu[b]	34,700	37,700	41,500
Percent of gross energy inputs	45.9	39.2	37.7
Natural gas (includes gaseous fuels)			
10^9 ft[3c]	22,868	21,700	18,500
10^{12} Btu[b]	23,600	22,400	19,100
Percent of gross energy inputs	31.2	23.3	17.4
Coal (bituminous, anthracite, lignite)			
10^3 short tons[d]	515,500	737,000	910,000
10^{12} Btu[b]	13,500	19,300	23,600
Percent of gross energy inputs	17.9	20.1	21.4
Hydropower, utility			
10^9 kWh[e]	286	385	650
10^{12} Btu[b]	2,900	3,600	6,000
Percent of gross energy inputs	3.8	3.7	5.5
Nuclear power (all types)			
10^9 kWh[e]	83.5	1,250	1,850
10^{12} Btu[b]	900	13,200	19,800
Percent of gross energy inputs	1.2	13.7	18.0
Total gross energy inputs, 10^{12} Btu (solar excluded)	75,600	96,200	110,000

[a] 10^6 bbl $= 1.590 \times 10^5$ m^3.
[b] 10^{12} Btu $= 1.055 \times 10^{15}$ J.
[c] 10^{12} ft$^3 = 2.83 \times 10^{10}$ m^3.
[d] 10^3 short tons $= 9.07 \times 10^2$ metric tons.
[e] 10^9 kWh $= 3.6 \times 10^{15}$ J.
[f] Including synthetic oil estimated at 5.5% of total coal usage.

The growth shown in Fig. 1 will occur primarily in the use of electricity to replace traditional heat sources in automotive propulsion and in the conversion industries. Solar power, expected to provide some 5% of the nation's energy, is not counted in the aggregate. Because of the high cost of energy for air conditioning and lighting, better design and restrictive building standards are expected to reduce the per capita consumption in large buildings. This will be accomplished largely by the elimination of incandescent lighting and low-efficiency fluorescent lighting, and by the employment of regenerative systems to reduce heating and cooling loads.

People-related energy requirements present some interesting contrasts, as shown by Table 2. The largest energy by far is released by the private motor car. Even total conversion to electric automobiles would cut the total energy release to the environment by only a factor of 2 because a 50% thermal efficiency appears to be barely achievable under current energy policies by the year 2000.

Of particular interest to the chemical and conversion industry is the projection of excellent progress in reducing of unit heat (energy) requirements for the production of steel by the basic oxygen process and other related new reduction processes. However, total energy use by industry is expected to increase about twice as fast as people-related use (Fig. 3). Also of interest is the projection that the use of exothermic energy in the production of petroleum will show a decrease of about 7% to about 590,000 Btu/barrel (3.9×10^9 J/m^3) by the year 2000.

Legislation. The electrical energy requirements are expected to be the predominant growth segment in the remainder of the century, rising from 21.5% of all energy requirements in 1965 to 43% of all requirements by 2000. It is also in this segment that legislation has delayed research and development in the multipurpose use of related materials in the production of electrical energy to the greatest extent. The Holding Company Act of 1935 limited the effectiveness of mergers in the utility field as a means of integrating the transmission system of the industry. The act has also served to concentrate the electrical power industry's research and development on conversion processes which have a single purpose and are limited to the direct interest of the industry as constituted under the act of 1935. This is in contrast to other industries, where aggressive development of related business opportunities has permitted a wide cross-fertilization of technology. *See* ELECTRICAL UTILITY INDUSTRY.

The modern pulverized coal furnace-boiler, for example, has followed a uniquely single-purpose path. This type of furnace is commonly used to convert the chemical energy in coal to heat energy in steam for electrical generation. Ever since its adoption in the early 1920s as a means of creating an ash so fine as to be readily dispersed to the atmosphere, to avoid the sooty particles from incomplete carbon combustion prevalent from stoker systems of that era, designers of this combustion system have thought of it as an incinerator instead of as a chemical reactor, which it really should more properly emulate.

The chemical industry is not hampered in this way. Combustion processes are common in the oil refinery business, and a majority of the fuels and major petroleum products are involved in high-temperature transformations using flame technology. As a result, the chemical industry is in a unique position to retain valuable side streams of otherwise useless materials and recycle them for other in-plant processes or sell them.

However, even chemical processes are cheaper to operate with relatively pure input materials, and the availability of relatively cheap sulfur from underground deposits has not produced a high incentive for aggressive recycling of the various sulfur by-products. The low cost of sulfur is attrib-

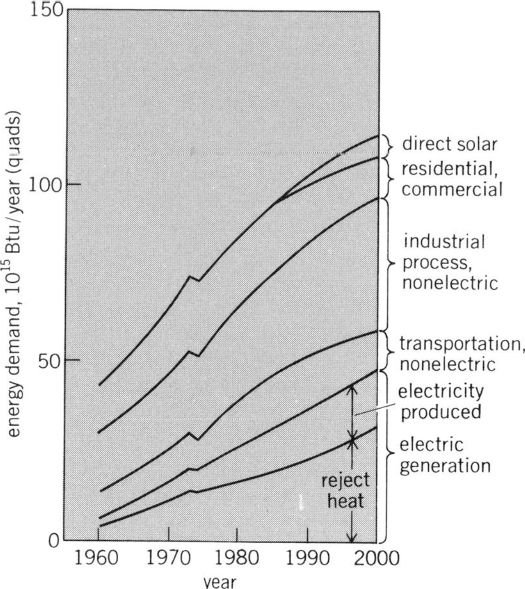

Fig. 1. Projected total energy demand by use. Top three bands do not include electrical energy consumption. 10^{15} Btu = 1.055×10^{18} J.

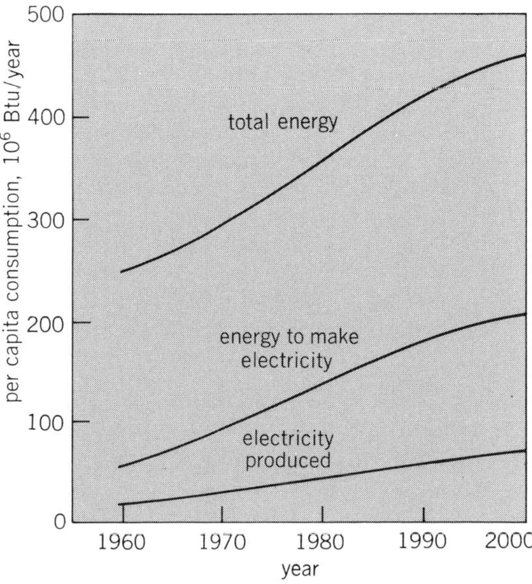

Fig. 2. Relative growth of energy use by decades through 2000.

Table 2. People-related heat release rates

Parameter	Rate, Btu-hr*
One human being at rest	450
Electric lighting for a typical office	
Incandescent	2.500
Fluorescent	900
Theoretical (100% efficient)	82
Air conditioning for typical office	600
Heating to maintain 25°F (13.9°C) ΔT for:	
Typical office	
Electrical resistance*	1,000
Heat pump†	350
Furnace, oil or gas (clean)	1,600
4000-lb (1800-kg) automobile at 70 mph (31 m/s)	
Internal combustion (piston)	750,000
Electric†	150,000
Electric (including losses)	450,000

*1 Btu/hr = 1055 J/hr = 0.293 W.

†Does not include generation or transmission losses. Total energy and resulting heat are about three times these values at efficiency of conversion predicted through the year 2000.

utable not only to large supplies of pure materials but to national policy on depletion allowances and related incentives to the mining and conversion industry. In general, such policies do not encourage the use of scrap or reclaimed materials. During the 1930s and continuing through World War II, the simple pragmatism of more and more production at lower and lower prices held full sway. In that period, annual sulfur production (estimated at 10×10^6 tons or 9×10^6 metric tons) was closely matched by the discharge of wasted sulfur as SO_2 totaling about 7×10^6 tons (6.4×10^6 metric tons) by the utility industry. Even so, with new emission standards resulting from the Clean Air Act of 1977, there is no market for recovered sulfur, although a small market for recovered calcium sulfate is developing. The recovery of valuable elements such as vanadium from combustion residuals has not been done to any extent.

While these two industries—utilities on the one hand and the chemical industry on the other—have been perfecting their individual processes, the availability of cheap, relatively pure input fuels and ores has declined. Until 1979, because of the availability of foreign oil and also natural gas in liquid form, research in the use of native resources such as coal and oil shale for the production of liquid and gaseous fuel had not yeilded commercially attractive processes. *See* COAL; LIQUEFIED NATURAL GAS (LNG); OIL SHALE.

Similarly, policies which depress fuel prices in relation to other indexes of productivity have given little incentive for advanced research and development on alternative methods of energy conversion. However, an impending global shortage of oil coupled with ever-increasing prices has given new impetus to the early resolution of the price-supply question for that fuel. *See* ENERGY CONVERSION.

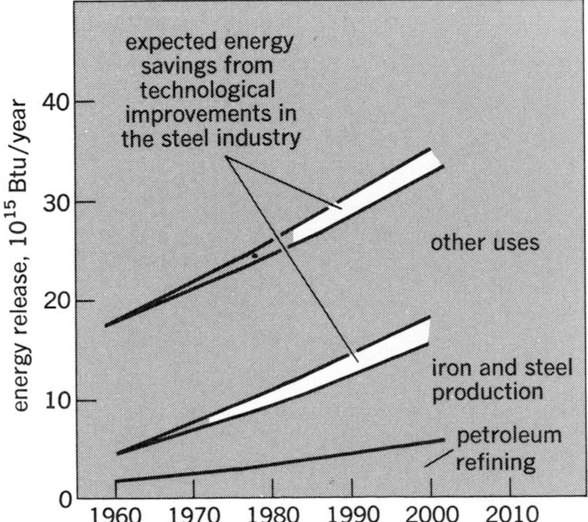

Fig. 3. Nonelectrical energy release by industry through 2000.

WASTE HEAT AND THE ATMOSPHERE

Many persons have argued that the energy release per square foot in the United States is so low that there is no cause for any long-range concern over the potential effects of thermal release to the environment. On the basis of a toal United States land area of 84×10^{12} ft^2 (7.8×10^{12} m^2), it can be computed that the total release of electricity as heat to the environment in 1978 amounted to about 0.006 W/ft^2 (0.065 W/m^2), or only about 0.018% of the Sun's averaged incident energy on the surface of the land and water of the country. However, this number fails to take into account the approximately 2 W of heat that is released for each watt of electrical power generated. Moreover, no matter what the efficiency, eventually all 3 W of heat is released as waste heat to the environment. In addition, the argument fails to include the large contribution of thermal releases from transportation, industry, and residential heating. In 1970 such contributions were about 10 times the amount of energy used in electrical systems.

Forecasted to the year 2000, the figures for energy release per square foot become quite significant. For example, in the energy forecast of Fig. 1, the total energy release per square foot increases from 0.22 W/ft^2 (2.4 W/m^2) in 1970 to 0.53 W/ft^2 (5.7 W/m^2) in 2000. If the energy release in key metropolitan areas is related to local areas and climatic conditions, this rises to as much as 15 W/ft^2 (161 W/m^2). Thus, enough energy would be concentrated to cause a mean air temperature increase of 13°F (7°C) in Los Angeles and 5°F (3°C) in the New York–Washington corridor (Fig 4). Currently, the highest thermal island effects occur around Japanese cities—some as high as 9–10°F (5–6°C) on a seasonal average.

The effects are more pronounced in the city than in the country, largely because of the increased roughness of the surface resulting from the erection of structures and the changes in albedo resulting from the modifications of the surface by humans (Table 3). A number of investigators, such as Helmut E. Landsberg and Steven R. Hanna, have developed empirical models of the effects of increased roughness of the city landscape. However, much more research on this subject is needed to arrive at a clear and unequivocal concept which can be used to generate constructive action.

City and country. As poorly conducting and evaporating vegetable surfaces (rural areas) are replaced by well-conducting, essentially dry surfaces of high heat capacity and low albedo (urban areas), a higher thermal balance is favored. As thermal releases increase from energy decay, and as human sources (including metabolic sources), industry, and transportation activity become more concentrated, and as the surface roughness of the urban landscape increases, a relatively stable low-pressure area develops which prevents normal nocturnal circulation. The lack of circulation concentrates the effluents of the urban aggregation. The energy budget of the metropolitan area is thus greatly affected by the prodigious production of pollutants. As a result, ultraviolet radiation is much reduced, and net incoming radiation is concentrated in upper air masses instead of being distributed. These redistributed fractions of the energy spectrum then react with the trapped chemical ejecta of this artificial "volcano" to produce a fantastic array of photochemical products known to be injurious to all higher plants and animals. *See* AIR POLLUTION.

Development of policy. Despite a lack of clear agreement on the numerical modeling of such "heat islands" and their microclimates, the following observations appear to be a sound basis for the development of a policy for waste heat management:

1. From a global perspective, the foreseeable effects of the release of energy and the by-products of energy conversion do not provoke the need for crisis thinking in the institutional management of resources. The single exception is particle emission in the upper atmosphere and stratosphere, trophosphere, and ionosphere.

2. Particle emission and changes in the opacity of the atmosphere resulting from artificial cloud seeding and from heat islands are not likely to cause serious dislocations in the weather patterns, rainfall, and general surface energy budget of the Earth.

Table 3. Climatic changes produced by cities

Parameter	Comparison with rural environs
Radiation	
Total on horizontal	15–20% less
Ultraviolet, winter	30% less
Ultraviolet, summer	5% less
Cloudiness	
General cloud cover	5–15% more
Fog, winter	100% more
Fog, summer	30% more
Precipitation	
Total amounts	5–10% more
Heavy rains over 0.2 in./day (5 mm/day)	10% more
Temperature	
Annual mean	1–1.5°F (0.6–0.9°C) more
Winter minimum	2–3°F (1.1–1.7°C) more
Relative humidity	
Annual mean	4–6% less
Winter	2% less
Summer	8% less
Wind speed	
Annual mean	20–30% less
Extreme gusts	12–20% less
Calms	5–20% less
Contaminants	
Dust and particulate	10 times more
Sulfur dioxide	5 or more times
Carbon dioxide	10 times more
Carbon monoxide	25 times more

SOURCE: H. E. Landsberg, *City Air: Better or Worse?*, 1961.

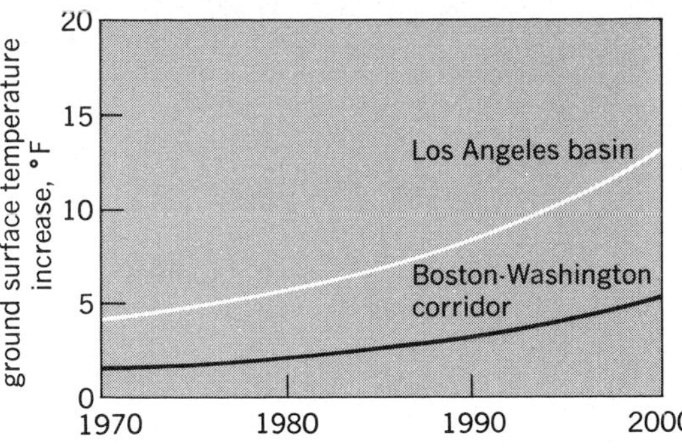

Fig. 4. Regional temperature elevations computed from energy forecasts, based on all energy and waste heat sources. 1°F = 5/9°C ≅ 0.56°C.

3. On a regional scale of 360,000 mi² (10⁶ km²), distinctly observable patterns are related to the release of products of combustion, moisture, and thermal energy (the end state of all useful energy of interest to this discussion). These patterns are increasing in size and intensity. One can predict with certainty that unless accelerating urbanization is controlled, climatic changes will increase from the present purely local scale to a regional scale and will appreciably affect the existing ecological balance. Known climatic changes include increased cloudiness, reduction of winds, concentration of pollutants (smog), drastic changes in runoff of rainfall, and profound, but as yet quantitatively unmodeled, effects of the immense quantities of gaseous and suspended materials ejected to the metropolitan atmosphere.

The role of the urban heat island in stabilizing and concentrating chemical pollutants and particulates released by various combustion processes is becoming increasingly apparent. Ulrich H. Czapski has suggested that the climatic effects in and near cities may in fact be more a result of the added heat than of the condensation and freezing nuclei to which they are usually attributed. Czapski predicts severe consequences of large latent and sensible heat emisions: cumulus clouds prevail most of the time downwind from a large power plant or aggregation of power plants; rainfall is increased downwind for a considerable distance; severe thunderstorms, and even tornadoes, can be caused in very unstable weather by dry and clear heat.

A BASIS FOR CONCERN

The microclimatic effects described here make it clear that waste heat management may be one of the seriously underestimated problems of this century. There is not only a lack of appreciation for the size of the problem but also a lack of understanding on the part of many people, including those with technical training, of the pervasive nature of heat itself.

It is time to begin to stress the fact that all energy produced and consumed modifies the environment and pollutes it. This occurs regardless of how efficient the conversion is or how many uses of "waste heat" are cascaded. Norman H. Brooks has put it this way: "Heat as a pollutant has a unique characteristic. Because of the basic laws of thermodynamics there is no treatment as such; any efforts to concentrate it (by heat pumps) simply require more mechanical energy, which means more waste heat is generated at the power plants. Other types of environmental pollution can be alleviated by various processes, which usually consume power and ultimately produce waste heat. Thus, heat is an ultimate residual of society's activities." *See* HEAT PUMP.

Currently, water-quality criteria for streams are stated in terms which do not relate to thermal modification of the atmosphere. But as the thermal islands related to metropolitan areas expand in size, streams will be affected. Therefore any policy regarding the increase of thermal releases to the metropolitan atmosphere is in fact involved in the modification of regional stream systems. An outstanding example of the influence of metropolitan air temperature on streams occurs in the Chicago area, where a 30-year record of Illinois waterway temperatures, even when corrected for thermal releases of the Commonwealth Edison Company and other industry, still shows a 3°F (1.7°C) increase. This increase is a result of a general increase in the mean air temperature of the Chicago metropolitan area.

WASTE HEAT AND ENERGY IN WATER

If waters are warmed, life activity in them is increased. The rate at which organic materials decompose and the rate at which microorganisms die and are replaced by succeeding generations are therefore generally increased. The rate at which oxygen is consumed and the rate at which carbon dioxide is evolved go up with the increasing temperature. Conversely, the solubility of gases declines. In addition, the overall process of self-purification of streams tends to be augmented, although this is an extremely general statement. *See* WATER POLLUTION.

Microorganisms. Observations of plankton and microorganisms passing through industrial processes involving sudden temperature increases usually show a rather dramatic and usually fatal effect. The death of plankton in turn may produce a regional rise in biochemical oxygen demand as a result of the accelerated formation of decaying materials. It is also true that dissolved oxygen concentration caused by natural reaeration of streams increases geometrically at about 1.5%/°C. However, even as reaeration occurs, the dissolved oxygen concentration of the water at saturation declines, and the rate at which organic matter consumes oxygen also rises. The net result is that, during colder seasons, oxygen deficits appear earlier in the downstream flow of a warmer stream than in its thermally polluted equivalent.

Although small increases in water temperature tend to provide a more congenial condition for the multiplication of microorganisms in a high-nutrient media, natural water usually provides a rather lean diet. Many tests have indicated that increased stream temperatures improve the rate at which disease-causing microorganisms are eliminated in lakes and streams. This relation between microorganism count and temperature is currently receiving considerable attention, which should lead to additional sophistication in separation of the variables affecting reactions in water. It may also lead to rejuvenation of microorganism counting as a measure of stream pollution.

Drinking water. Along these lines, it is interesting to examine the effects of changing temperature on the effectiveness of chlorine as a disinfectant for potable water. These effects have been studied for many years by many industrial and public authorities. It is generally true that greatly increased quantities of chlorine are required as temperature decreases in the range of 25 to 4°C.

However, both the pH of the water and the presence of contaminants play a part in determining the overall activity of disinfectants. Furthermore, there are few cases on record, if any, that give much insight into the effects, other than seasonal, on a standardized pathogen with only temperature as the variable parameter. Since the formation of free chlorine residual is also temperature-dependent, the question of chemistry versus biology remains unresolved.

Of considerable concern to the public consumer of water is the matter of taste. Generally, increased water temperature increases the taste problems that result from the contaminants remaining after treatment. Algae are the culprits, but not all species are equally offensive. The grass-green algae are most abundant in water supplies during the summer but are seldom found in winter. The seasonal distribution of the blue-greens is similar but their maximum growth occurs later in the summer, and they often show a great increase after a period of sustained warm weather. Since the blue-greens seem to be the biggest problem in regard to both taste and oxygen depletion in high-nutrient waters, improved understanding of the role of temperature in the acclimatization of these species is essential.

Fish. Perhaps the most highly publicized effect of temperature on life forms is that on fish. A substantial amount of research has gone into the relationship of temperature to the propagation of desirable and undesirable species. Undoubtedly, much more will continue to be done as thermal problems

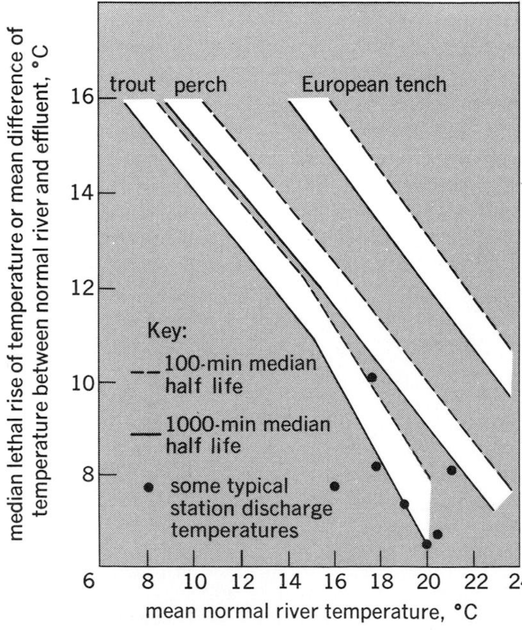

Fig. 5. Effect of temperature rise on lifetime of three fish species.

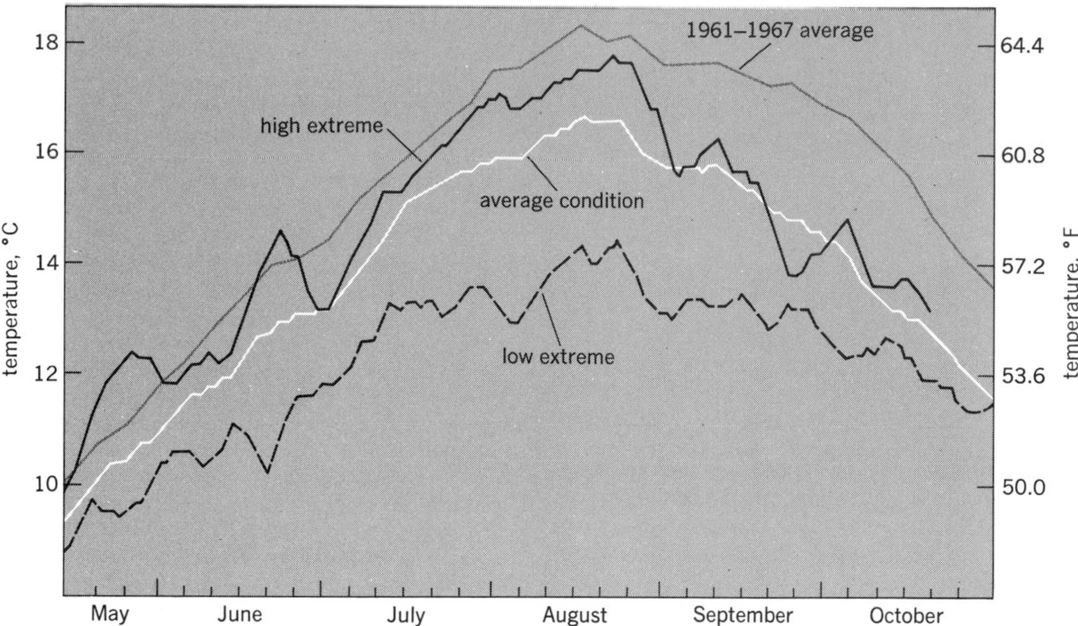

Fig. 6. Expected temperature variations of the Upper Columbia River after completion of reservoirs in 1975.

are better defined. Figure 5 shows the results of a statistical study conducted in the United Kingdom. The shaded area between the broken lines for each species represents the range in which a temperature rise (vertical axis) above the normal water temperature (horizontal axis) will affect the lifetime of the fish. Figure 5 shows conclusively the better adaptability of the perch-type fish forms over the salmonoid forms, such as trout, and the high adaptability of the carplike species, of which the European tench is an example. The points at lower right indicate the discharge temperatures of some major power-generating stations and their relationship to the life expectancy of these three statistical groups.

Cooling water passed through power station condensers is frequently warmed as much as 10°C (18°F). Poorly mixed effluents of this type disturb normal river conditions, particularly in streams with restricted flows. In addition, these heated effluents, by virtue of high temperatures—and temperature fluctuations alone—are lethal to trapped trout and even to coarse fish acclimatized to normal river temperature. However, small free-living fish are affected to a lesser extent, and larger fish seem to be able to swim away from warmer regions without difficulty. Sudden temperature changes, nevertheless, are problems to all species.

The effect of temperature on fish increases dramatically when dissolved oxygen is low and CO_2 is high. This is another case where proper definition of temperature as a parameter is necessary to draw meaningful and objective conclusions.

HYDRO POWER AND THERMAL EFFECTS

The impoundment of streams for hydroelectric purposes has a number of significant effects on the water quality. Reservoirs which are relatively deep tend to produce cooling of summer extremes and warming of winter lows with relatively small changes to the average temperature. However, the timing of thermal maximums and minimums is changed from the normal seasonal cycle. The larger the system, the more severe the modification of timing. Associated with this change in timing are variations in the total annual available water flow.

Reactions which tend to favor reducing conditions and oxygen depletion are enhanced in the lower depths of larger reservoirs, even those with relatively low pollution burdens, because of the high degree of organic loading which occurs. On the other hand, in relatively shallow reservoirs on river main stems such as the Columbia and the St. Lawrence, such effects are minimal. The reservoir regime is relatively homogeneous, although other problems, such as gas supersaturation, can result from the entraining effect of spillway discharges.

As an example of the extent of thermal modification in a hydroelectric system, Fig. 6 shows the

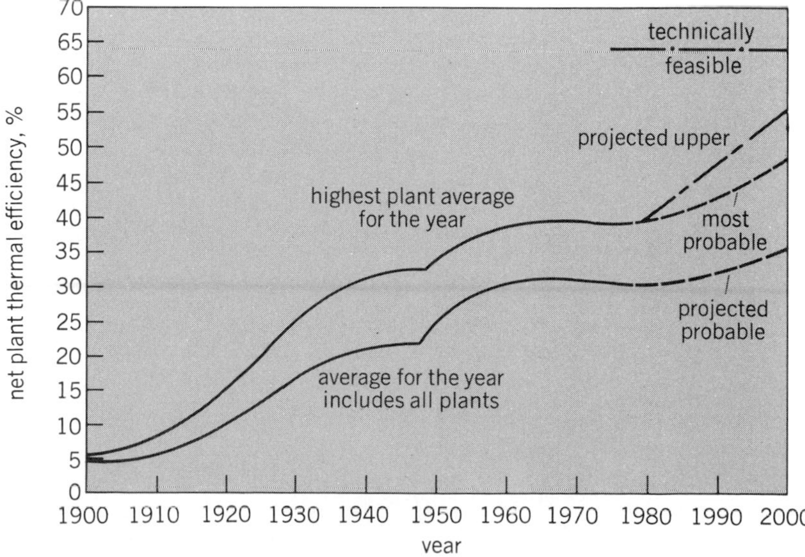

Fig. 7. Trends in the efficiency of steam electric plants since 1890.

changes in Upper Columbia River temperatures expected from the Columbia Treaty projects. The projects will impound some 20×10^6 acre-ft (25×10^9 m^3) of water and increase the monthly mean flow of the normally low-flow months by some 20,000 cfs (566 m^3/s). Temperature reductions forecast for the system in the summer months are the equivalent of some 25,000 MW of thermal energy. Figure 6 shows that August temperatures will average 2°C (3.6°F) lower under normal water conditions, and range downward to 4°C (7.2°F) lower in years with excess water. In critical years, temperatures will approach historical maximums and the timing of the peaks will move forward to the middle of June, in contrast to present conditions where peaks are expected in the middle of September. *See* HYDROELECTRIC POWER.

The data from this and a large number of other carefully studied cases show that the management of streams for the purpose of relating thermal history to power needs is feasible and, in fact, is of vital necessity.

EQUIPMENT DEVELOPMENT TRENDS

In the face of the staggering increase in total energy requirements forecast for the next 30 years, it seems logical that equipment, principally electrical generation prime movers, would reflect increasing emphasis on extending available fuel resources. The facts show the contrary to be the case. *See the feature article* OUTLOOK FOR FUEL RESERVES.

Historical trends. Figure 7 shows the historical trends of efficiency in the utilization of energy in the power industry. Commencing in the late 1800s, the efficiency of thermal plants rose steadily, paced by fuel cost/supply balances that favored the growth of technical efficiency. This progress was seriously affected by the lack of research and development during the 1929–1939 depression and came to a complete halt during World War II. After the war, until 1960, plant efficiency rose again as the repressed innovations of the Depression and the wartime were tried. This technical innovation proved both disappointing and expensive. After a long period of little or no substantial investment in new plants, an explosive period of growth occurred, spurred on by the Vietnam War. Efficiency came in second to the lure of cheap fuels, both fossil and nuclear, and improvement of the average efficiency of plants leveled off and even decreased. This trend still persisted in 1979. Even though fuels are expensive, it will take another 15 years to return the average efficiency to the 1964 level. This illustrates the enormous leverage exerted in sunk capital for low-efficiency plants and the need for national efficiency goals independent of short-term financial considerations.

Causes. There are several reasons for the lack of technical progress. First of all, reliance in the recent past on the economy of scale has tended to distort traditional measurements of technical efficiency. This is because insufficient experience is available to fully weigh the economic consequences of unit size over a statistically significant operating period. Economic policies which have led to relatively cheap fuel prices with respect to plant investment and interest costs have had a large impact on these trends.

Misguided policies. Policies which tend to arbitrarily require off-stream cooling cause further deterioration of technical efficiency. Recent practices which favor the production of high-speed single-shaft prime movers, instead of cross-compound systems, produce machines which are optimized for purposes other than total energy conservation. Similarly, the relatively low-efficiency light-water reactors place additional emphasis on capital optimization at the cost of large increases in energy consumption and release. Such practices have culminated in "high back-end loading" of single-shaft machines to reduce capital cost, meanwhile incurring higher energy consumption. (High back-end loaded machines are generally used on peaking plants or plants served by cooling towers in order to reduce the sensitivity of the system to water temperature changes. However, once such a machine is built and the operating parameters set, the net losses cannot be reduced by the later use of cooler water because of the inherent losses in the terminal blading of such machines.)

Price policies. Figure 8 reveals that policies which preserve low energy costs in contradiction to the overall inflationary trend cause energy waste and exacerbate the potential energy shortage and the environmental impacts of energy production and waste heat. The data of Table 4 suggest strongly that raw fuel prices are seriously undervalued in terms of gross national product and that increased costs to the consumer, though they may be burdensome, are a correct direction for energy prices to move in order to correct the situation. Thus, as a first principle in the management of waste heat, prices must be allowed to increase enough to make it worthwhile to conserve energy and promote incentives for research and development on improved efficiency in prime movers and energy-consuming equipment.

The cost-push effect of the economy of scale is being fully felt in the nuclear area with extensive cancellations and deferments. Much higher fuel prices have arrested the decay of efficiency, but a

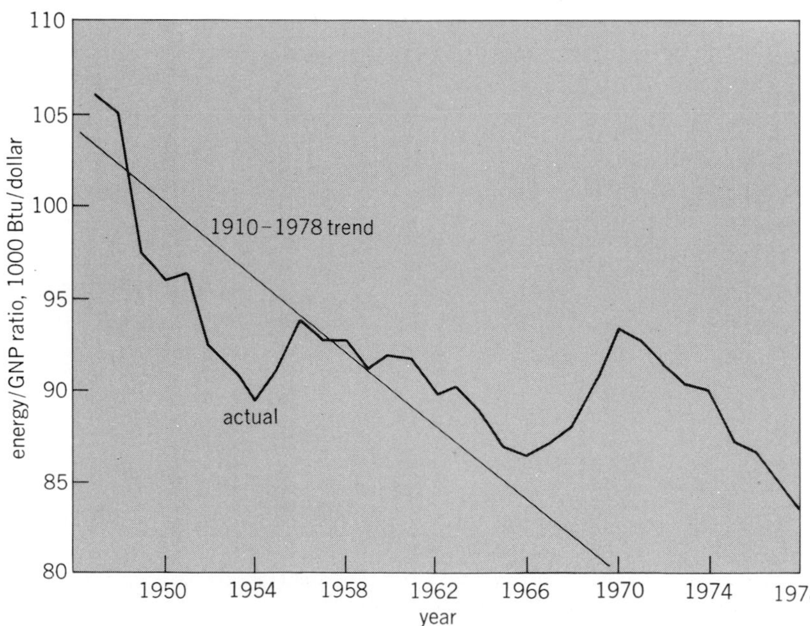

Fig. 8. Energy/GNP ratio for 1947–1974.

Table 4. Energy consumption, gross national product, and the energy/GNP ratio for 1947–1978

	Aggregate energy consumption		GNP		Energy/GNP ratio	
Year	Amount, Btu × 10^{12}*	Change from preceding year, %	Amount 1958 dollars, × 10^9	Change from preceding year, %	Amount, 1000 Btu/dollar*	Change from preceding year, %
1947	32,870	—	309.9	—	106.1	—
1948	33,994	+2.5	323.7	+4.5	105.0	−1.0
1949	31,604	−7.0	324.1	+0.1	97.5	−7.1
1950	34,153	+8.1	355.3	+9.6	96.1	−1.4
1951	36,913	+8.1	383.4	+7.9	96.3	+0.2
1952	36,576	−0.9	395.1	+3.1	92.6	−3.8
1953	37,697	+3.1	412.8	+4.5	91.3	−1.4
1954	36,360	−3.5	407.0	−1.4	89.3	−2.2
1955	39,956	+9.9	438.0	+7.6	91.2	+2.1
1956	42,007	+5.1	446.1	+1.8	94.2	+3.3
1957	41,920	−0.2	525.5	+1.4	92.6	−1.7
1958	41,493	−1.0	447.3	−1.1	92.8	+0.2
1959	43,411	+4.6	475.9	+6.4	91.2	−1.7
1960	44,960	+3.6	487.7	+2.5	92.2	+1.0
1961	45,573	+1.4	497.2	+1.9	91.7	−0.5
1962	47,620	+4.5	529.8	+6.6	89.9	−2.0
1963	49,649	+4.3	551.0	+4.0	90.1	+0.2
1964	51,515	+3.8	581.1	+5.5	88.7	−1.6
1965	53,785	+4.4	617.8	+6.3	87.1	−1.8
1966	56,948	+5.9	658.1	+6.5	86.5	−0.7
1967	58,868	+3.4	675.2	+2.6	87.2	+0.8
1968	62,448	+6.1	707.2	+4.7	88.3	+1.3
1969	65,832	+5.4	727.1	+2.8	90.5	+2.5
1970	67,099	+2.0	722.7	−0.7	92.8	+2.5
1971	68,700	+2.4	745.4	+3.1	92.2	−0.6
1972	72,100	+4.9	790.7	+6.1	91.2	−1.1
1973	75,561	+4.8	837.6	+5.9	90.2	−1.1
1974	72,260	−4.4	803.7	−4.0	89.9	−0.3
1975	70,860	−1.9	800.2	−0.4	87.5	−2.7
1976	74,311	+4.9	858.0	+6.7	86.6	−1.0
1977	76,000	+2.7	892.0	+4.0	85.2	−1.6
1978	78,000	+2.6	934.6	+4.8	83.5	−2.0

*1 Btu = 1055 J.

long time will be needed to undo the damage of the Korean and Vietnam wars in overstimulating investment in poorly performing equipment. Figure 8 shows that the energy/GNP ratio is still some 18,000 Btu/dollar (19×10^6 J/dollar) of gross national product above the historical trend line, a cost to the public which has not been included in conventional evaluations.

CAUSE FOR OPTIMISM

In the face of all these adverse trends, it is important to avoid defeatist attitudes. Quite to the contrary, there is reason for optimism about the future. However, the amount of wishful thinking and false notions about energy and energy policy are so pervasive that it is imperative that a realistic examination of open alternatives be conducted. In the following paragraphs, the most promising alternatives will be examined, all based on the philosophy that economic elasticity will balance usage and provide a test of reality.

Whenever possible, environmental enhancement on a broad scale should be a principal consideration in all future utility planning. Interim measures adopted as expedients without a long-range goal should be rejected as soon as such measures are revealed to be incorrect or misleading.

Simulation modeling should be used for the advance planning of alternatives to improve coordination of water resources and energy technology. This will provide more systematic use of water for low flow augmentation, recreational development, enhancement of fisheries, irrigation, and new development for cities, including interbasin diversions.

Both short- and long-run solutions to the waste heat problem should fully recognize the capacity of water in dispersing heat whenever these uses can be shown to have long-range public benefits.

Fuel cell concept. New systems deserve more attention than they are currently receiving. An example is the electrically interconnected neighborhood fuel cell concept, with full recovery of thermal energy for water heating and other domestic uses. In conjunction with better insulation, improved lighting efficiency, and more efficient air-conditioning equipment employing air, such systems could provide improved sophistication in home living with total energy use reduced by as much as 75% over present uncoordinated systems. The fuel gas supplies needed for the fuel cell concept are much easier to provide, and gas service corridors require an order of magnitude less space than electrical transmission lines for equivalent energy-transmission rates. It is entirely possible that the average residential load could be held to as little as 3500 kWh per year using a fully coordinated optimized residential energy system. This is only one-quarter of the present load in the Pacific Northwest as reported by the Bonneville Power Administration. Electrical systems of the present high-voltage, alternating-current type are ideal for high industrial and commercial loads and would be retained for service to the transportation and conversion industries as the total energy source. In this way, transmission lines would avoid residential areas. The development of improved utilization technology could make surprising reductions in predicted requirements for energy. *See* FUEL CELL; FUEL GAS.

New equipment. The efficiency and flexibility of equipment available to the utilities need order-of-magnitude improvements. Presently, such equipment is far too limited by convention, standards, and lack of fundamental research. Equipment offered by manufacturers should reflect less attention to quick response, near-term considerations, and regulatory requirements, and should focus more attention on long-term, national objectives. Equipment designs that minimize energy consumption and that decrease environmental impact should be encouraged by suitably appropriate economic incentives.

Basic research need. Also needed is a closer identification of the specific changes associated with the impact of energy on the environment. Almost all of the existing study programs place so little emphasis on the specific identification of temperature as an independent variable that it is obscured in search for an oversimplified solution. Separation of the variables, with emphasis on temperature, would properly recognize the interdisciplinary nature of heat and provide background data of real usefulness.

Finally, the question of energy in the environment must be faced from an integrated, sociological viewpoint. The impact of increased industrialization, along with the predicted need for recreational facilities, provides the technical basis for a rational synthesis. What are needed are appropriate guiding factors, a constructive attitude, and a sincere appreciation of the total broad needs of the community. *See the feature article* PROTECTING THE ENVIRONMENT.

[ROBERT T. JASKE]

Bibliography: David Freeman et al., *Exploring Energy Choices*, Energy Policy Project, Ford Foundation, 1974; G. Gude et al., *Project Interdependence: U.S. and World Energy Outlook through 1990*, Congressional Research Service, GPO Pub. no. 95-31, June 1977; G. H. Hutchinson et al., *The Biosphere*, 1970; R. T. Jaske, An evaluation of energy growth and use trends as a potential upper limit in metropolitan development, *Sci. Total Environ.*, 2:45–60, 1973; P. A. Krenkel and F. L. Parker, *Biological Aspects of Thermal Pollution*, 1969; National Academy of Engineering, Committee on Power Plant Siting, *Engineering for Resolution of the Energy Environment Dilemma*, 1972; S. H. Schurr et al., *Energy in America's Future: The Choices before Us*, Resources for the Future, June 1979; C. Starr et al., *Energy and Power*, 1971.

Water pollution

Any change in natural waters which may impair their further use, caused by the introduction of organic or inorganic substances, or a change in temperature of the water. The growth of population and the concomitant expansion in industrial and agricultural activities have rapidly increased the importance of the field of water-pollution control. In the attack on environmental pollution, higher standards for water cleanliness are being adopted by state and Federal governments, as well as by interstate organizations.

Historical developments. Ancient humans joined into groups for protection. Later, they formed communities on watercourses or the seashore. The waterway provided a convenient means of transportation, and fresh waters provided a water supply. The watercourses then became receivers of wastewater along with contaminants. As industries developed, they added their discharges to those of the community. When the concentration of added substances became dangerous to humans or so degraded the water that it was unfit for further use, water-pollution control began. With development of wide areas, pollution of surface water supplies became more critical because wastewater of an upstream community became part of the water supply of the downstream community.

Serious epidemics of waterborne diseases such as cholera, dysentery, and typhoid fever were caused by underground seepage from privy vaults into town wells. Such direct bacterial infections through water systems can be traced back to the late 18th century, even though the germ or bacterium as the cause of disease was not proved for nearly another century. The well-documented epidemic of the Broad Street Pump in London during 1854 resulted from direct leakage from privies into the hand-pumped well which provided the neighborhood water supply. There were 616 deaths from cholera among the users of the well within 40 days.

Eventually, abandoning wells in such populated locations and providing piped water to buildings improved public health. Further, sewers for drainage of wastewater were constructed, but then infections between communities rather than between the residents of a single community became apparent. Modern public health protection is provided by highly refined and well-controlled plants both for the purification of the community water supply and treatment of the wastewater.

Relation to water supply. Water-pollution control is closely allied with the water supplies of communities and industries because both generally share the same water resources. There is great similarity in the pipe systems that bring water to each home or business property, and the systems of sewers or drains that subsequently collect the wastewater and conduct it to a treatment facility. Treatment should prepare the flow for return to the environment so that the receiving watercourse will be suitable for beneficial uses such as general recreation, and safe for subsequent use by downstream communities or industries.

The volume of wastewater, the used water that must be disposed of or treated, is a factor to be considered. Depending on the amount of water used for irrigation, the amount lost in pipe leakage, and the extent of water metering, the volume of wastewater may be 70–130% of the water drawn from the supply. In United States cities, wastewater quantities are usually 75–200 gal (284–757 liters) per capita daily. The higher figure applies to large cities with old systems, limited metering, and comparatively cheap water; the lower figure to smaller communities with little leakage and good metering. Probably the average in the United States for areas served by sewers is 125–150 gal (473–568 liters) of wastewater per person per day. Of course, industrial consumption in larger cities increases per capita quantities.

Related scientific disciplines. The field of water-pollution control encompasses a part of the broader field of sanitary or environmental engineering. It includes some aspects of chemistry, hydrology, biology, and bacteriology, in addition to public administration and management. These scientific disciplines evaluate problems and give the civil and sanitary engineer basic data for the designing of structures to solve the problems. The solutions usually require the collection of domestic and industrial wastewaters and treatment before discharge into receiving waters.

Self-purification of natural waters. Any natural watercourse contains dissolved gases normally found in air in equilibrium with the atmosphere. In this way fish and other aquatic life obtain oxygen for their respiration. The amount of oxygen which the water holds at saturation depends on temperature and follows the law of decreased solubility of gases with a temperature increase. Because water temperature is high in the summer, oxygen dissolved in the water is then at a low point for the year.

Degradable or oxidizable substances in wastewaters deplete oxygen through the action of bacteria and related organisms which feed on organic waste materials, using available dissolved oxygen for their respiration. If this activity proceeds at a rate fast enough to depress seriously the oxygen level, the natural fauna of a stream is affected; if the oxygen is entirely used up, a condition of oxygen exhaustion occurs which suffocates aerobic organisms in the stream. Under such conditions the stream is said to be septic and is likely to become offensive to the sight and smell.

Domestic wastewaters. Domestic wastewaters result from the use of water in dwellings of all types, and include both water after use and the various waste materials added: body wastes, kitchen wastes, household cleaning agents, and laundry soaps and detergents. The solid content of such wastewater is numerically low and amounts to less than 1 lb per 1000 lb of domestic wastewater. Still, the character of these waste materials is such that they cause significant degradation of receiving

Table 1. General nature of industrial wastewaters

Industry	Processes or waste	Effect
Brewery and distillery	Malt and fermented liquors	Organic load
Chemical	General	Stable organics, phenols, inks
Dairy	Milk processing, bottling, butter and cheese making	Acid
Dyeing	Spent dye, sizings, bleach	Color, acid or alkaline
Food processing	Canning and freezing	Organic load
Laundry	Washing	Alkaline
Leather tanning	Leather cleaning and tanning	Organic load, acid and alkaline
Meat packing	Slaughter, preparation	Organic load
Paper	Pulp and paper manufacturing	Organic load, waste wood fibers
Steel	Pickling, plating, and so on	Acid
Textile manufacture	Wool scouring, dyeing	Organic load, alkaline

Table 2. Dilution ratios for waterways

Type	Stream flow, $ft^3/(sec)$ (1000 population)*
Sluggish streams	7–10
Average streams	4–7
Swift turbulent streams	2–4

*1 $ft^3/sec = 28.3 \times 10^{-3}\ m^3$.

waters, and they may be a major factor in spreading waterborne diseases, notably typhoid and dysentery.

Characteristics of domestic wastewater vary from one community to another and in the same community at different times. Physically, community wastewater usually has the grayish colloidal appearance of dishwater, with floating trash apparent. Chemically, it contains the numerous and complex nitrogen compounds in body wastes, as well as soaps and detergents and the chemicals normally present in the water supply. Biologically, bacteria and other microscopic life abound. Wastewaters from industrial activities may affect all of these characteristics materially.

Industrial wastewaters. In contrast to the general uniformity of substances found in domestic wastewaters, industrial wastewaters show increasing variation as the complexity of industrial processes rises. Table 1 lists major industrial categories along with the undesirable characteristics of their wastewaters.

Because biological treatment processes are ordinarily employed in water-pollution control plants, large quantities of industrial wastewaters can interfere with the processes as well as the total load of a treatment plant. The organic matter present in many industrial effluents often equals or exceeds the amount from a community. Accommodations for such an increase in the load of a plant should be provided for in its design.

Discharge directly to watercourses. The industrial revolution in England and Germany and the subsequent similar development in the United States increased problems of water-pollution control enormously. The establishment of industries caused great migrations to the cities, the immediate result being a great increase in wastes from both population and industrial activity. For some years discharges were made directly to watercourses, the natural assimilative power of the receiving water being used to a level consistent with the required cleanliness of the watercourse. Early dilution ratios required for this method are shown in Table 2. Because of the more rapid absorption of oxygen from the air by a turbulent stream, it has a high rate of reaeration and a low dilution ratio; the converse is true of slow-flowing streams.

Development of treatment methods. With the passage of time, the waste loads imposed on streams exceeded the ability of the receiving water to assimilate them. The first attempts at wastewater treatment were made by artificially providing means for the purification of wastewaters as observed in nature. These forces included sedimentation and exposure to sunlight and atmospheric oxygen, either by agitated contact or by filling the interstices of large stone beds intermittently as a means of oxidation. However, practice soon outstripped theory because bacteriology was only then being born and there were many unknowns about the processes.

In later years testing stations were set up by municipalities and states for experimental work. Notable among these were the Chicago testing station and one established at Lawrence by the state of Massachusetts, a pioneer in the public health movement. From the results of these direct investigations, practices evolved which were gradually explained through the mechanisms of chemistry and biology in the 20th century.

Thermal pollution. An increasing amount of attention has been given to thermal pollution, the raising of the temperature of a waterway by heat discharged from the cooling system or effluent wastes of an industrial installation. This rise in temperature may sufficiently upset the ecological balance of the waterway to pose a threat to the native life-forms. This problem has been especially noted in the vicinity of nuclear power plants. Thermal pollution may be combated by allowing the wastewater to cool before it empties into the waterway. This is often accomplished in large cooling towers.

Current status. Modern water-pollution engineers or chemists have a wealth of published information, both theoretical and practical, to assist them. While research necessarily will continue, they can draw on established practices for the solution to almost any problem. A challenging problem has been the handling of radioactive wastes. Reduction in volume, containment, and storage constitute the principal attack on this problem. Because of the fundamental characteristics of radioactive wastes, the development of other methods seems unlikely.

Public desire for complete water pollution control continues, but there is an increasing realization that solution to the problem is costly. While cities have had little concern about the initial construction cost because of the large Federal share, the expenditures for operation fall entirely on the local community. During the life of a project, operating costs may exceed the initial construction outlay, and with present rates of increase, they may become a major financial burden.

The control of 100% of the organic pollution reaching the watercourses is the goal of many people, but such an ideal cannot even be approached, since about one-third of the total is from nonpoint sources. Essentially, this third is from vegetable matter carried by surface drains and direct runoff. To achieve the public health protection which is the primary purpose of collection and treatment of wastewater, proper measures are essential, and should be the principal focus of a program of water pollution control. Interestingly, such was the original purpose of sewers and drains employed by ancient civilizations, as manifested by the Romans, who gave Venus the title of Goddess of the Sewers, in addition to her other titles associated with health and beauty. The pediment of her "lost" statue in the Roman Forum identifies her as *Venus Cloacinae.*

However, there are strong manifestations of improved quality in the waters throughout the United States. This is apparent not only in chemical and biological measurements, but in more readily observed effects such as better appearance, eliminated smells, and the return of fish life to watercourses which had become "biological deserts" because of the effects of pollution from municipal and industrial wastewaters. The overall results are a living tribute to the cooperative efforts of local citizens and local, state, and Federal agencies working together to improve the quality of the nation's waters. A measure of the activity in the field is indicated by the employment of nearly 90,000 in the local wastewater collection and treatment works of the United States.

Table 3. Federal funds for wastewater treatment plant construction, 1973–1979, in $\$10^9$

Fiscal year	Authority	Appropriated	Obligated	Outlays
1973	5	2	1.531	0
1974	6	3	1.444	.159
1975	7	4	3.616	.874
1976*	0	9	4.814	2.563
1977†	1.48	1.48	6.664	2.710
1978	4.5	4.5	2.301	2.960
1979‡	5	4.2	.953	1.984

*Includes transition quarter, July-September 1976.

†Includes $480,000,000 under Public Works Employment Act.

‡Obligated and Outlays as of Apr. 30, 1979.

Federal aid. Because of public demands and the actions of state legislatures and the Congress of the United States, there has been a surge of interest in, and a demand for, firm solutions to water-pollution problems. Although the Federal government granted aid for construction of municipal treatment plants as an employment relief measure in the 1930s, no comprehensive Federal legislation was enacted until 1948. This was supplemented by a major change in 1956, when the United States government again offered grants to municipalities to assist in the construction of water-pollution control facilities. These grants were further extended to small communities for the construction of both water and sewer systems.

Since 1965, Federal activity in water-pollution control has advanced from a minor activity in the Public Health Service, through the Water Pollution Control Administration in the Department of the Interior, to a major activity in the Environmental Protection Agency. In the 1972 act (P. L. 92-500) Congress authorized a massive attack on municipal pollution problems by a grant-in-aid program eclipsing any previous effort. Federal funds for 1973–1979 are given in Table 3.

State and Federal regulations are increasing constantly in severity. This tendency is expected to continue until the problem of water pollution is brought under complete control. Even then, water quality will be monitored to make certain that actual control is achieved on a day-to-day or even an hour-to-hour basis.

[RALPH E. FUHRMAN]

OIL SPILL

The problem of oil spillage came to the public's attention following the grounding of the tanker *Torrey Canyon* in March 1967 at the southwest coast of England near the entrance to the English Channel. Subsequent major oil spills such as the Santa Barbara channel California oil spill in January 1969 have further raised the level of concern until today the terminology "oil spill" has become a household word. Few problems have had greater impact on the petroleum industry than those associated with oil spills. This industry in the United States is faced with the problem of supplying the ever-increasing demand for oil and petroleum products to customers who are demanding that the oil be supplied without a risk of oil spills. This large disparity between the production and use of oil throughout the world results in the requirement that enormous quantities of oil be transported large distances, primarily by tanker.

To counter the threat of environmental damage as a result of oil spills, extensive research is being performed in the United States by private industry as well as by the Environmental Protection Agency and the Coast Guard. This research is primarily directed at developing methods to combat oil spills which minimize the damage to the environment. Treating the spilled oil with dispersants was the primary method used to fight oil spills at the time of the *Torrey Canyon.* Dispersants cause oil to spread farther and disperse in a manner similar to the way soap removes oil from one's hands, allowing the oil to be emulsified and washed away with the water. The dispersants used during the *Torrey Canyon* cleanup effort were not developed specifically for use in waters containing marine life and contained aromatic solvents which are toxic. Since that time specific dispersants less toxic to marine life and biota have been developed. Today it is generally accepted that the most extensive damage to marine life resulting from the *Torrey Canyon* incident was caused by the excessive use of dispersants in the coastal zone. In fact, the areas of the shore where dispersants were not used, but which were heavily polluted with oil alone, showed very minimum damage according to J. E. Smith, director of the Plymouth Laboratory, who has studied the biological effects of the *Torrey Canyon* oil spill. At present in the United States regulations severely limit the use of dispersants, and research efforts place emphasis on containment and recovery of oil by mechanical means.

Effects of oil pollution. When oil is spilled on water, it spreads rapidly over the surface. The forces which cause the oil to spread include the force of gravity, which results in the lighter oil seeking constant level by spreading horizontally on the heavier water. A second force is the surface-tension force, which acts at the edge of the oil slick as shown in Fig. 1. It is the surface-tension force which can result in the oil spreading to a thickness approaching a monomolecular layer. This limiting thickness is almost never achieved in large oil spills, however, because the oil interfacial surface tensions change, and the net surface tension becomes negative. The interfacial surface-tension forces change as a result of the natural processes that affect oil. One of the most important natural processes is the evaporation of the oil. Evaporation occurs rapidly, the rate depending upon the nature of the oil, the rate of thinning of the slick, wave intensity, strength of the wind, temperature, and so forth. Crude oil is a mixture of a very large number of components, each with its own properties. The most volatile components evaporate first, but with all crude oils there will undoubtedly be a residue left which is virtually involatile. In addition to evaporation, some of the oil goes into solution with the water, some is oxidized, and some is utilized by microorganisms. The most important of these processes, and the one receiving the most extensive research, is the process of microbial degradation. Many microorganisms present in seas, fresh-water lakes, and rivers have a great capacity to utilize hydrocarbons. The hydrocarbons are used as an energy source and are incorporated into new cell mass. Seeding oil slicks with special bacterial cultures has been suggested to accelerate the rate of microbial decay. However, the rate of microbial degradation of oil is limited not only by the quantity of organisms but by the availability of the oxygen and nutrients needed to support the metabolic process. Acceleration of the natural process by adding nutrients such as phosphorus or nitrogen compounds, particularly in open seas where the nutrients are not naturally available, is presently being considered. Unfortunately the rate of bacterial degradation (accelerated or natural) of floating oil is slow and is therefore not effective if the oil is threatening a coastline.

As with most types of problems, the short-term deleterious effects associated with an oil spill are better understood than the long-term effects. Marine birds, especially diving birds, appear to be the most vulnerable of the living resources to the effects of oil spillage. Harm to birds from contact with oil is reported to be a result of breakdown of the natural insulating oils and waxes which shield the birds from water, as well as due to plumage damage and ingestion of oil. Efforts to cleanse or rehabilitate birds have been generally unsuccessful because of the excessive stress that the bird experiences. If treatment is prolonged for any reason, most if not all of the birds will die. Shellfish are another segment of marine life directly affected by oil spillage in the coastal zone. Many shellfish have a relatively high tolerance to oil, but their flesh can become tainted for a period subsequent to heavy pollution. Shellfish are particularly vulnerable to most chemical dispersants. Fish are not generally affected by an oil spill because of their mobility which allows them to avoid heavily contaminated areas. The effects of oil on the marine food chain which consists of plants, bacteria, and small organisms is not well understood because of its complexity and because of the wide fluctuations that occur naturally and are independent of the effects of oil. In contrast to the ecological damage, the damage caused by an oil spill which is associat-

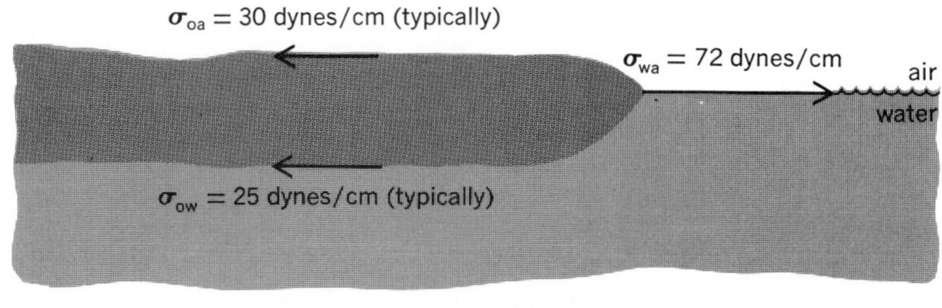

Fig. 1. Cross section of oil-water-air interface of a spreading oil slick showing the relevant surface-tension forces; 1 dyne is equivalent to 10^{-5} N.

ed with recreational beach areas, coastline areas used for water sports, and areas where personal property such as docks or boats are located is well understood. A large expense is associated with loss of use of these areas for even a few days.

Cleanup procedures. Experience in attempting to clean up an oil spill has shown that no perfect method exists for all situations. Cleanup methods must be evaluated and chosen on a case-by-case basis. Spills can be more easily dealt with if they are confined to a small area on the water surface. At present, however, confinement devices have not been developed which are successful in all situations. The methods being studied and which are being used to dispose of oil floating on the surface of the sea include mechanical removal of the floating oil, the use of absorbents to facilitate the removal of the oil, sinking the oil, dispersion of the oil, and burning the oil.

Mechanical removal of oil. The primary ingredient of many mechanical oil spill cleanup systems is the use of mechanical booms or barriers. Containment of the oil spill at its source is the most important single action which can be taken when the oil spill is first detected. The use of booms has only had limited success to date because at moderate currents as low as 1 knot many booms have failed by allowing oil to pass below the boom. Figure 2 shows the types of failure which booms can experience in the presence of currents. At low currents, approximately 1.5 knots or less, oil will pile down against the barrier until the boom reaches its capacity. Additional oil will cause the boom to fail as shown in the figure. At higher currents the situation is less manageable since the boom will fail

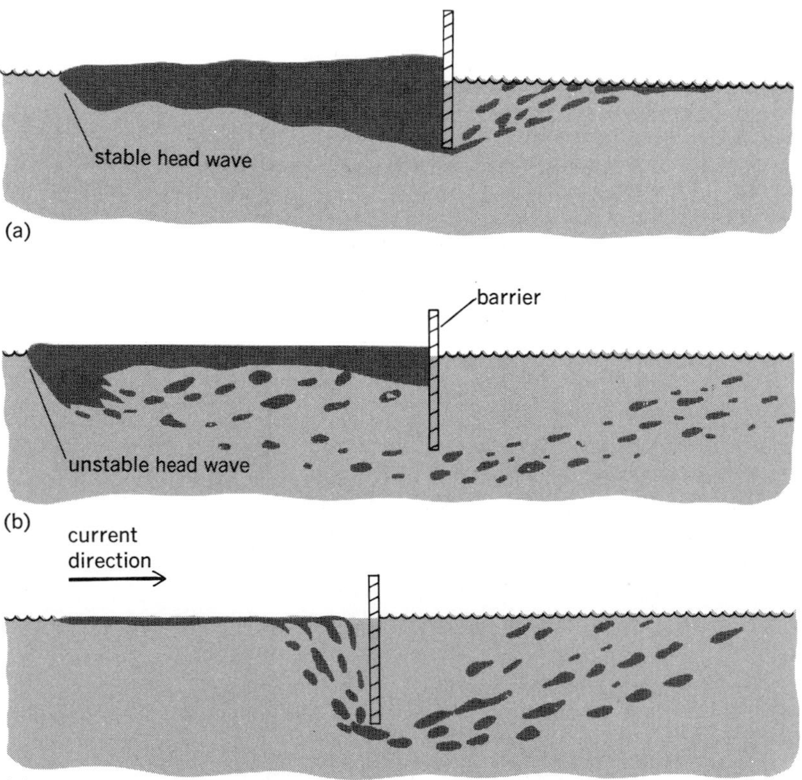

Fig. 2. Types of mechanical boom failure. (*a*) Low current. (*b*) Moderate current. (*c*) High current.

before a large quantity of oil has been collected within the barrier. For the low-current situation booms will perform satisfactorily if the oil is continually skimmed from the region in front of the barrier. Booms also are satisfactory for directing or sweeping oil, provided the angle between the boom and the current or drift direction of the oil is small so that oil does not accumulate along the boom. In addition, the relative velocity of the water at right angles to the boom must be less than the critical velocity for boom failure.

The performance of skimmers depends upon the thickness of the oil. When the film thickness is below 1/4 in. (6 mm), many techniques require pumping large amounts of water and very small amounts of oil. However, once the oil-and-water mixture is removed from the water surface, the separation of the oil from the water is easily accomplished by gravity when the oil and water are allowed to settle in a tank. Skimmers now available generally fall into one of two categories. The first, mechanical surface skimming, removes the top layers of the water and oil from the surface. These devices suffer particularly in wave action, where they gulp large amounts of water unless some provision is provided to allow the weir or suction port to follow the water surface. A second type of skimmer operates on the principle of selective wetting of a surface by oil rather than by water. Rotating metal disks or conveyor belts dip into the water surface through the oil slick. When the moving surface is drawn from the water, a surface layer of oil is removed.

Absorbents. Absorbents are used to facilitate the cleaning up of oil spills. When they are applied to the slick, they absorb the oil and prevent it from spreading, and when the absorbent material is removed from the water, the oil is removed. A class of absorbents which is commonly used consists of natural materials such as peat moss, straw, sawdust, pine bark, talc, and perlite. A second class of absorbents is derived from synthetics or plastics such as high-molecular-weight polyethylene and polystyrene, polypropylene, and polyurethane. Of all synthetic absorbents, polyurethane is generally accepted to be the most promising. The natural absorbents are generally less expensive and are attractive whenever there is a chance of losing the absorbent material. The natural absorbents are either inert or biodegrade more quickly than the synthetic materials. Alternatively the synthetic material has a greater buoyancy and a higher affinity for oil. One of the problems with absorbents is distributing them in large enough quantities on the slick. Unless the absorbents can be applied and recovered from the shore, special equipment must be available. One of the most recent concepts including absorbents involves recycling the absorbent material by wringing the oil from the absorbent and returning the absorbent material to the slick. Synthetic absorbents are most suitable for this application, and a feasibility study of such a system for operation in offshore conditions is presently being conducted by the Environmental Protection Agency.

Sinking the oil. A sinking agent which consisted of 3000 tons (2700 metric tons) of calcium carbonate with about 1% of sodium stearate was applied to an oil slick which originated from the *Torrey Can-*

yon and reportedly resulted in the sinking of about 20,000 tons (18,000 metric tons) of oil. The oil was sunk in the Bay of Biscay off the coast of France in 60 to 70 fathoms (110–128 m) of water. The sinking of the oil prevented the French coast from being contaminated, and after a period of 14 months no sign of the oil was found. Other materials such as specially treated sand, fly ash, and similar synthetic material have also been used to sink oil. Opinion is still divided as to the possible environmental effects of treating the oil with a dense material and sinking it. Opposition to sinking centers around the fact that sinking the oil reduces the contact surfaces between the oil and air and between the oil and water by preventing natural diffusion of the oil. Hardening of the oil subsequent to sinking would lead to a more persistent and concentrated pollution of the sea bed compared with a lower level of more dispersed pollution on the surface. In any case, utilization of this technique would be most advantageous in deeper waters outside the heavy fishing zones and where there will be a minimum of adverse effects to productive biological life in the coastal zones.

Dispersants. A dispersant is a substance which, when applied to an oil slick, causes the oil to spread farther and disperse. A dispersant contains a surfactant, a solvent, and a stabilizer. The solvent usually comprises the bulk of the dispersant and enables the surface-active agent or surfactant to mix with, and penetrate into, the oil slick and thus form an emulsion. The stabilizer fixes the emulsion and prevents it from coalescing once it is formed. The process of dispersion of the slick is shown in Fig. 3. Dispersion is similar to sinking in that it simply displaces the oil from the water surface rather than removing it from the water altogether. Dispersion has one advantage over sinking in that it increases the slick surface area and allows a rapid increase in the rate of microbial decomposition. Dispersants are not useful in coastal regions because the process of dispersion leads to an increased extent of contamination. In addition, the most effective dispersants use solvents which are toxic to marine life. To reduce this toxic effect, reduced effectiveness must be accepted. Primarily due to the question of toxicity, the use of dispersants in the open sea appears doubtful, although their use there has potential pending additional study and field data.

Burning. Burning oil slicks on the open sea has generally met with little success because the more volatile light ends evaporate quickly from the oil slick. Also, the water generally can remove heat faster than it can be created to support the combustion. Various attempts have been made to treat oil slicks to facilitate burning. The most promising of these techniques involves spreading a wicking material over the oil slick which acts to physically separate the flame from the water. Wicking agents also aid in confining the fire to a particular location, but air pollution must be expected when oil is burned in this fashion.

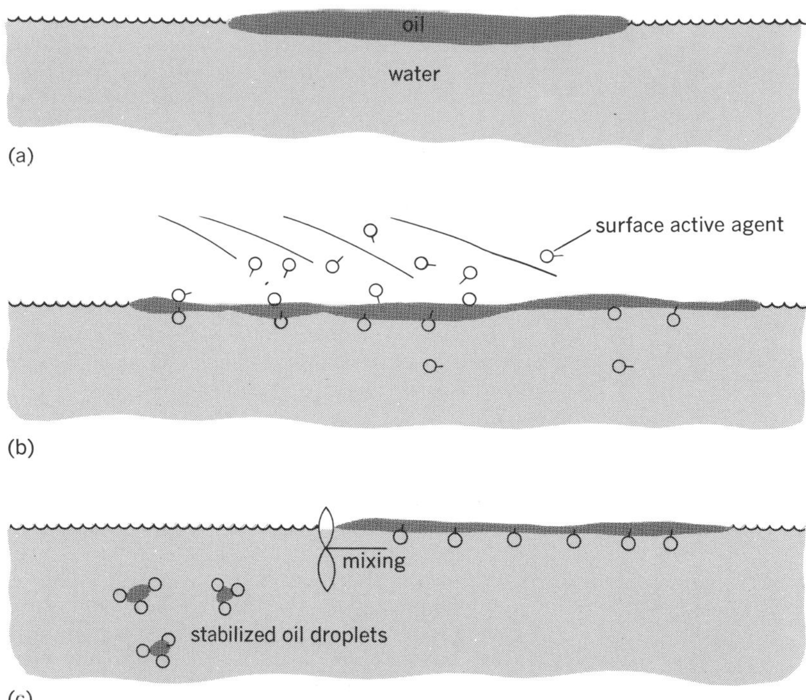

Fig. 3. Mechanism of oil slick dispersion. (*a*) Initial slick. (*b*) Application of chemical dispersant. (*c*) Mixing forms droplets which become stable emusion.

Prevention of oil spills. Although oil spills can probably not be eliminated entirely, steps are being taken to reduce the probability that they will occur. The United States Environmental Protection Agency and others have sponsored studies that apply reliability engineering principles to the problem and that will result in recommending procedures to be adopted to reduce the oil-spill threat. Steps are also being taken to quickly discover oil spills and to develop methods to continually monitor the water where spills are likely so that, in the event of a spill, it can be discovered quickly and proper action can be taken. The United States Coast Guard is sponsoring considerable research toward the development of remote sensing techniques. These studies have shown that, using radar and passive microwave techniques, large areas can be surveyed on a 24-hr basis even in adverse weather conditions. For the purpose of cleaning up oil spills, 67 private cooperatives are now in operation throughout the United States. These cooperatives will become more numerous in the future. Cooperatives operate in coordination with the Coast Guard, the Army Corps of Engineers, and the Environmental Protection Agency. Assessing the ability of these cooperatives, or of any group for that matter, to clean up a large spill is difficult, but it appears that near the shore mechanical techniques can be used effectively to remove oil from the water. In the offshore situation, oil recovery is more complicated and will depend upon the given situation and other environmental conditions encountered. Most current research efforts are directed toward developing systems which will operate effectively in offshore conditions.

[ROBERT A. COCHRAN]

Bibliography: G. M. Fair et al., *Elements of Water Supply and Wastewater Disposal*, 1971; Federal Water Quality Administration, *Santa Barbara Oil Pollution, 1969: A Study of the Biological Effects of the Oil Spill Which Occurred at*

Santa Barbara, California, in 1969, University of California, Santa Barbara, October 1970; R. E. McKinney, *Microbiology for Sanitary Engineers*, 1962; N. L. Nemerow, *Liquid Waste of Industry*, 1971; *Proceedings of the 1975 Conference on Prevention and Control of Oil Pollution*, San Francisco, Mar. 25–27, 1975, American Petroleum Institute, 1975; C. N. Sawyer and P. L. McCarty, *Chemistry for Sanitary Engineers*, 1962; J. E. Smith (ed.) *"Torrey Canyon" Pollution and Marine Life: A Report by the Plymouth Laboratory of the Marine Biological Association of the United Kingdom*, 1968; J. Snow, *Mode of Communication of Cholera*, 1855; Water Pollution Control Federation, *Careers in Water Pollution Control*.

Water-tube boiler

A steam boiler in which water circulates within tubes and heat is applied from outside the tubes. The outstanding feature of the water-tube boiler is the use of small tubes (usually 1–3 in. or 25–75 mm outside diameter) exposed to the products of combustion and connected to steam and water drums which are shielded from these high-temperature gases. Thus, possible failure of boiler parts exposed to direct heat transfer is restricted to the small-diameter tubes and, in the event of failure, the energy released is reduced and explosion hazards are minimized.

The water-tube construction facilitates greater boiler capacity by increasing the length and the number of tubes and by using higher pressure, since the relatively small-diameter tubes do not require an abnormal increase in thickness as the internal pressure is increased. In addition, water-tube boilers offer great versatility in arrangement, and this permits efficient use of the furnace, superheater, and other heat-recovery components.

There are many types of water-tube boilers but, in general, they can be grouped into two categories: the straight-tube and the bent-tube types. In essence, both types consist of banks of parallel tubes which are connected to, or by, headers or drums. The early water-tube boilers were fitted with refractory furnaces. However, most modern water-tube boilers utilize a water-cooled surface in the furnace, and this surface is an integral part of the boiler's circulatory system. Further, modern water-tube boilers generally incorporate the use of superheaters, economizers, or air heaters to utilize more efficiently the heat from the fuel and to provide steam at a high potential for useful work in an engine or turbine.

Straight-tube boiler. The straight-tube boiler (Fig. 1), often called the header-type boiler, has the advantage of direct accessibility for internal inspection and cleaning through handholes, located opposite each tube end, in the headers. The steam-generating sections are joined to one or more steam-and-water drums located above and parallel or transverse to the boiler tube bank. The circulation of water in the downcomer headers and of the water-steam mixture in the boiler tubes and riser headers is the result of the differential density between the water in the downcomers and the steam and water mixture in the heated boiler tubes and riser portion of the circuit. In many designs the tube bank can be baffled to increase the rate of flow of the products of combustion and, thus, improve heat transfer and the resultant absorption efficiency.

The straight-tube boiler is not applicable to high-pressure designs because of header limitations, and capacity is restricted by space requirements. Thus, bent-tube instead of header-type boilers are used in most modern industrial installations.

Bent-tube boiler. In bent-tube boilers (Fig. 2), commonly referred to as drum-type boilers, the boiler tubes terminate in upper and lower steam and water drums which have few access openings. Although internal inspection is restricted, the tubes can be mechanically or chemically cleaned, and developments in water treatment and cleaning methods have overcome the early objections to the use of bent tubes. Circulation in drum-type boilers, as in header-type boilers, is due to the difference between the density of the water in the downcomer tubes and the water-steam mixtures in the riser tubes.

Because of the greater slope of tubes, or even the use of vertical tubes, in drum-type boilers, there is less possibility of steam pockets forming in the tubes and, thus, the rate of heat absorption can be increased appreciably. Consequently, drum-type boilers can be used for both low and high pressures and for capacities ranging from a few thousand to more than a million pounds of steam per hour. In addition, they can be designed to meet almost any space allotment and can readily accommodate heat-absorbing components such as superheaters, reheaters, economizers, and air heaters. An evolution of the bent-tube boiler is the so-called radiant-type boiler in which the boiler tubes (steam-generating surface) form the boundary of the furnace and the containment for the superheater, reheater, and economizer surface (Fig. 3). Such boilers can deliver several million pounds of steam per hour at high steam pressures and temperatures. Further, by eliminating the steam drum and using a once-through flow of water and steam, they can operate at supercritical steam pressures.

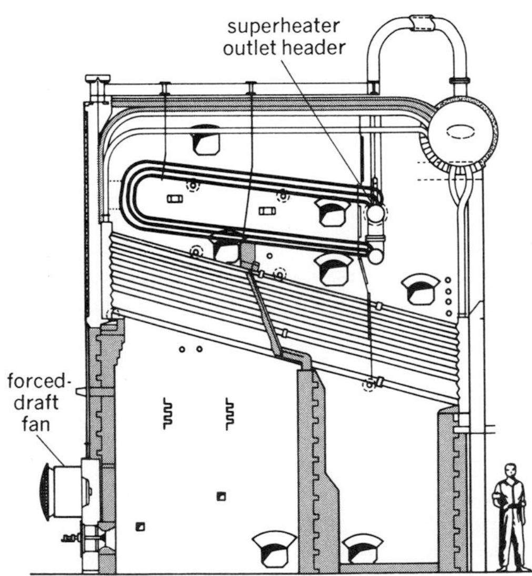

Fig. 1. Diagram of a straight-tube-type boiler.

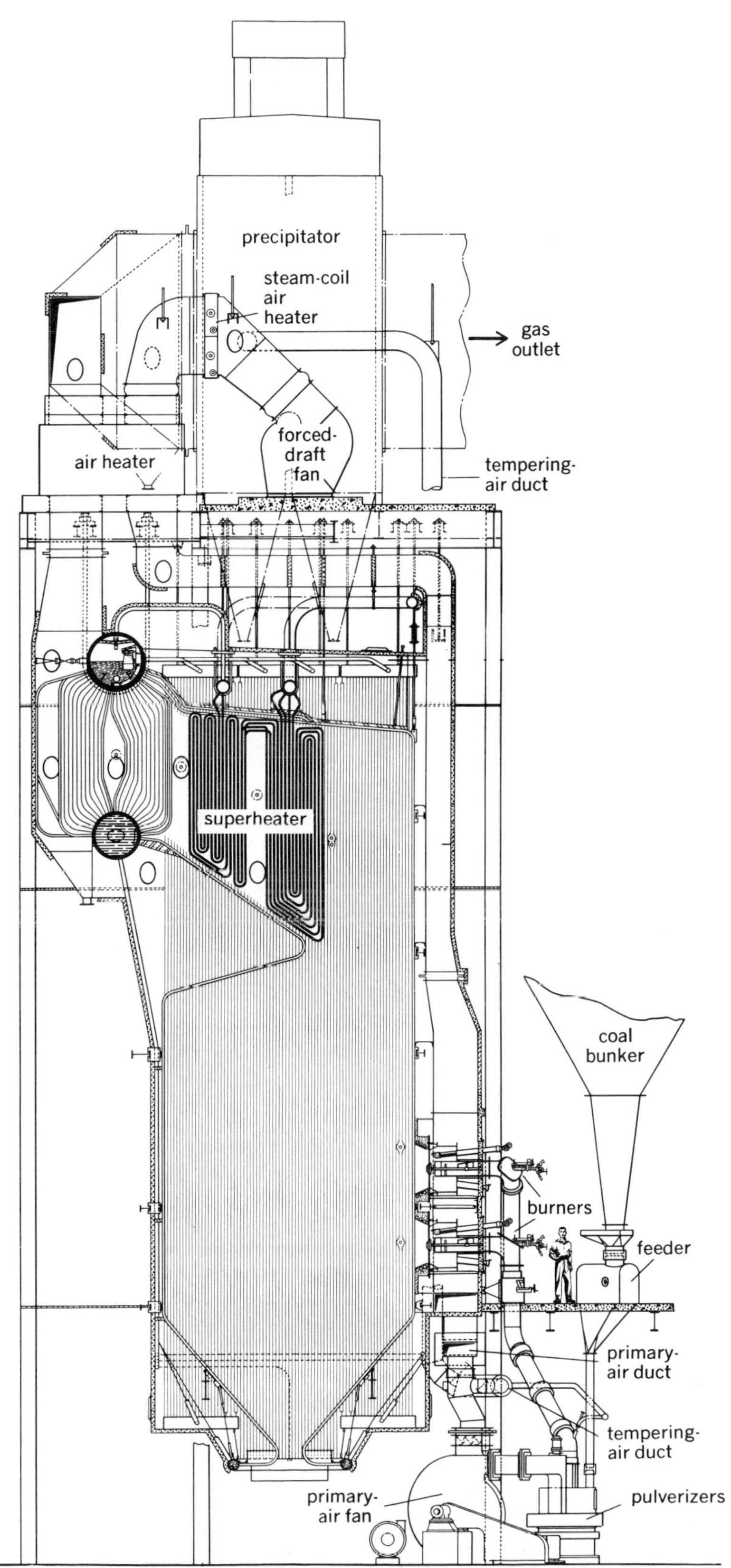

Fig. 2. Diagram of a bent-tube-type boiler.

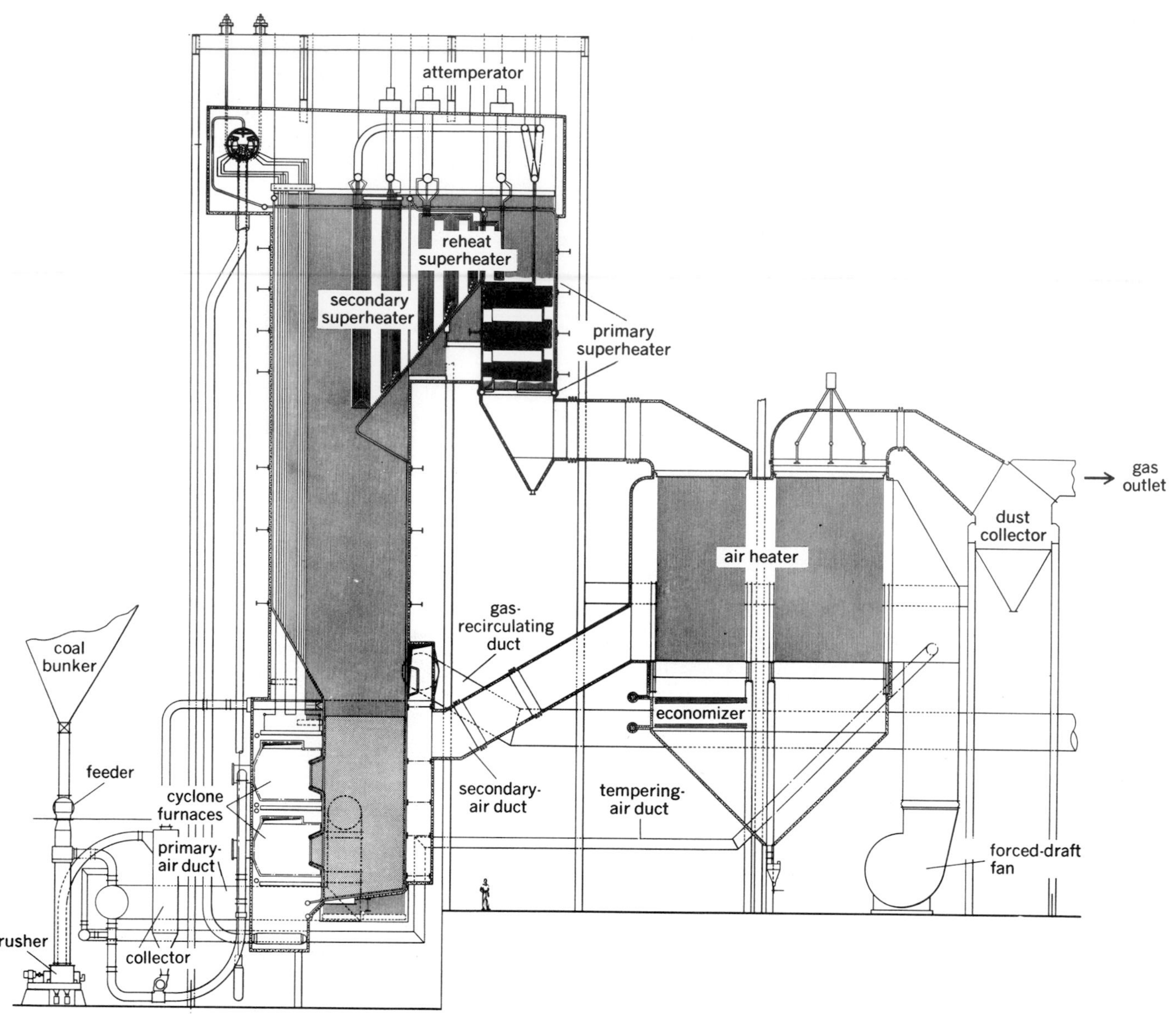

Fig. 3. Diagram of a radiant-type boiler.

See STEAM-GENERATING FURNACE; STEAM-GENERATING UNIT.

[GEORGE W. KESSLER]

Bibliography: Babcock and Wilcox Co., *Steam: Its Generation and Use*, 1955: T. Baumeister (ed.), *Standard Handbook for Mechanical Engineers*, 7th ed., 1967; G. R. Fryling (ed.), *Combustion Engineering*, 1966.

Waterpower

Power developed from movement of masses of water. Such movement is of two kinds: (1) the falling of streams through the force of gravity, and (2) the rising and falling of tides through lunar (and solar) gravitation.

While that part of solar energy expended to lift water vapor against Earth gravity is a minute fraction of the total, the absolute amount of energy that is theoretically recoverable from resulting streams is an enormous but unknown quantity. Of this, but a tiny portion is actually suitable for harnessing.

WATER RESOURCES

The capacity of world waterpower plants in use at the end of 1978 was about 408×10^6 kW, which produced in that year about 1558×10^9 kWh. This is about 21% of the total world electric power generated. Of this, the United States accounted for about one-fifth of the total. As of January 1980, the total conventional hydropower developed in the contiguous United States was 64,686 MW, provided by about 1400 installations. Almost one-half of this hydropower was installed in Washington, Oregon, and California. Under construction or planned were installations of some 24,000 MW, of which 90% was in the western United States.

In 1978 the waterpower plants of Europe totaled 138×10^6 kW; North America, 111×10^6; Asia, 62×10^6; Soviet Union, 41×10^6; leaving 56×10^6 for the remainder of the world. Thus the principal waterpower development of the future can be expected in Africa, Asia, South America, and island countries, whose potential has been little investigated.

The contribution of waterpower installations to the nation's electric power supply at the beginning of World War II was about 30%. While the output from hydroelectric plants has grown (to 274×10^9 kWh in 1978), their contribution to the total electric power has dropped to about 13%, because steam-electric plants have grown at a much more rapid rate.

As of March 1980, the Energy Information Agency of the Department of Energy estimated that by the end of 1993 the total developed capacity of waterpower installations could be 99,346 MW. However, that figure does not include a number of potential sites that could at some future time be considered. In theory, the Energy Information Agency estimates an eventual maximum potential of 187,000 MW, most of the undeveloped sites being in the Pacific Northwest and Alaska. Only a fraction of that will be developed for a variety of reasons. The most attractive sites have already been utilized. Hydro plants, with their initial high cost and generally long distances from major load centers, must compete with the large, efficient, fuel-fired stations, and the burgeoning, economical, large nuclear plants. Large dam sites usually must be justified not alone on the value of the power developed, but also on the benefits from flood control, irrigation, and recreation. Problems of migrating fish, conservation, and preservation of esthetic values are also factors. On the other hand, waterpower developments add greatly to power-system flexibility in meeting peak and emergency loads. Modern excavation and tunneling techniques are lowering construction costs. The economies of lowhead sites are improved by the new, efficient, axial-flow turbines of the tubular type. *See* ENERGY SOURCES.

Silting. The capacity of hydro plants cannot be counted on for perpetuity because of gradual filling of reservoirs with sediment. This effect is serious for irrigation, flood control, and navigation. Even when a lake behind a power dam becomes filled completely with silt, electric power can be generated on the run-of-the-river flow, although output would vary with stream flow.

The rate of silting varies widely with drainage basins. Because the Columbia River carries comparatively little silt, the reservoirs at Grand Coulee and Bonneville dams should have lives of many hundreds of years. The Colorado River, on the other hand, is muddy. In the first 13.7 years after Hoover Dam went into operation in 1935, 1,424,000 acre-feet (175,600 ha-m) of silt was dumped into Lake Mead. That is equivalent to a layer 1 ft deep over 2225 mi^2 (or 1 m deep over 1756 km^2). This inflow of silt has been diminished by about 22% by the construction of other dams upstream, for example, the Glen Canyon Dam. It is now expected that Lake Mead will have a useful life of more than 500 years.

Pumped storage. The most significant waterpower development of the 1960s was the rapid growth of pumped-storage hydroelectric systems. In these schemes, water is pumped from a stream or lake to a reservoir at a higher elevation. Pumping up to a storage reservoir is most commonly done by reversing the hydraulic turbine and generator. The generator becomes a motor driving the turbine as a pump. Power is drawn from the power system at night or on weekends when demand is low. It is not practical to shut down large, high-temperature steam stations or nuclear units for a few hours at night or even over a weekend. Because they must run anyway, the cost of pumping power is low, whereas the power generated from pumped storage at peak periods is valuable. Also, the pumped-storage system provides a means of supplying power quickly in an emergency situation, for example, during the failure of a large steam or nuclear unit. A pumped-storage system can be changed over from pumping to generation in 2 to 5 min.

Pumped-storage systems are not new. They date from 1928, but by 1972 only 13 pumped-storage plants of a capacity greater than 50 MW had been installed in the United States. These had a combined capacity of 3718.5 MW. By December 1979, pumped storage capacity had risen to 10.6 GW, with 10.6 GW additional planned or in construction. *See* PUMPED STORAGE.

Tidal power. A portion of the kinetic energy of the rotation of the Earth appears as ocean tides. The mean tide of all the oceans has been calculated as 2.1 ft (0.64 m), and the mean power as $54{,}000 \times 10^6$ hp (40 TW) or, on a yearly basis, the equivalent of 36×10^{12} kWh. Unfortunately, only a minute amount of this is likely to be harnessed for use. For tidal sites to be of sufficient engineering interest, the fall would have to be at least 15 ft (4.5 m). There are few such falls, and some of these are in remote areas. The only tide-power sites that have received serious attention are on the Severn River in England, the Rance River and Mont St. Michel in northern France, the San José and Deseado rivers of Argentina, the Petitcodiac and Memramcook estuaries in the Bay of Fundy, Canada, the Passamaquoddy River where Maine joins New Brunswick, Canada, and, lately, the Cambridge Gulf of Western Australia.

The Passamaquoddy site, with a potential of 1800 MW (peak), is the only important tidal-power prospect in the United States. However, as late as 1980 engineers did not consider its electrical output to be economically competitive with power produced by other means.

A second major handicap to tidal power is that, with a simple, single-basin installation, power is available only when there is a several-feet difference between levels in the sea and the basin. Thus, firm power is not available. Also, periods of generation occur in consonance with the tides—not necessarily when power is needed.

The only major tidal power plant in operation is the one near the mouth of the Rance River in Normandy, France. This plant operates on 40-ft (12-m) tides. It began operation in November, 1966. It consists of twenty-four 10-MW bulb-type turbine-generator units of novel design. The system embodies a reservoir into which sea water is pumped during off-peak hours. Turbines are then run as pumps, power being drawn from the French electrical grid. The plant produces 500×10^6 kWh annually, including a significant amount of firm power.

Tidal power is an appealing and dramatic technique, and some other large plants may be constructed. However, the total contribution of the tides to the world's energy supply will be miniscule. *See* TIDAL POWER.

[CHARLES A. SCARLOTT]

APPLICATION

The basic relation for power P in kilowatts from a hydrosite is $P=QH/11.1$, where Q is water flow in ft^3/sec, and H is head in feet. Actual power will be less as occasioned by inefficiencies such as (1) hydraulic losses in conduit and turbines; (2) mechanical losses in bearings; and (3) electrical losses in generators, station use, and transmission. Overall efficiency is always high, usually in excess of 80% to the station bus bars.

Choice of site. The competitive position of a hydro project must be judged by the cost and reliability of the output at the point of use or market. In most hydro developments, the bulk of the investment is in structures for the collection, control, regulation, and disposal of the water. Electrical transmission frequently adds a substantial financial burden because of remoteness of the hydrosite from the market. The incremental cost for waterwheels, generators, switches, yard, transformers, and water conduit is often a smaller fraction of the total investment than is the cost for the basic structures, real estate, and transmission facilities. Long life is characteristic of hydroelectric installations, and the annual carrying charges of 6–12% on the investment are a minimum for the power field. Operating and maintenance costs are lower than for other types of generating stations.

The fundamental elements of potential power, as given in the equation above, are runoff Q and head H. Despite the apparent basic simplicities of the relation, the technical and economic development of a hydrosite is a complex problem. No two sites are alike, so that the opportunity for standardization of structures and equipment is nearly nonexistent. The head would appear to be a simple surveying problem based largely on topography. However, geologic conditions, as revealed by core drillings, can eliminate an otherwise economically desirable site. Runoff is complicated, especially when records of flow are inadequate. Hydrology is basic to an understanding of water flow and its variations. Runoff must be related to precipitation and to the disposal of precipitation. It is vitally influenced by climatic conditions, seasonal changes, temperature and humidity of the atmosphere, meteorological phenomena, character of the watershed, infiltration, seepage, evaporation, percolation, and transpiration. Hydrographic data are essential to show the variations of runoff over a period of many years. Reservoirs, by providing storage, reduce the extremes of flow variation, which are often as high as 100 to 1 or occasionally 1000 to 1.

Economic factors. The economic factors affecting the capacity to be installed, which must be evaluated on any project, include load requirements, runoff, head, development cost, operating cost, value of output, alternative methods of generation, flood control, navigation, rights of other industries on the stream (such as fishing and lumbering), and national defense. Some of these factors are components of multipurpose developments with their attendant problems in the proper allocation of costs to the several purposes. The prevalence of government construction, ownership, and operation, with its subsidized financial formulas which are so different from those for investor-owned projects, further complicates economic evaluation. Many people and groups are parties of interest in the harnessing of hydrosites, and stringent government regulations prevail, including those of the U.S. Corps of Engineers, Federal Power Commission, Bureau of Reclamation, Geological Survey, and Securities and Exchange Commission.

Capacity. Prime capacity is that which is continuously available. Firm capacity is much larger and is dependent upon interconnection with other power plants and the extent to which load curves permit variable-capacity operation. The incremental cost for additional turbine-generator capacity is small, so that many alternatives for economic development of a site must be considered. The alternatives include a wide variety of base load, peak load, run-of-river, and pumped-storage plants. All are concerned with fitting installed capacity, runoff, and storage to the load curve of the power system and to give minimum cost over the life of the installation. In this evaluation it is essential clearly to distinguish capacity (kW) from energy (kWh) as they are not interchangeable. In any practical evaluation of water power in this electrical era it should be recognized that the most favorable economics will be found with an interconnected electric system where the different methods of generating power are complementary as well as competitive.

As noted above, there is an increasing tendency in many areas to allocate hydro capacity to peaking service and to foster pumped-water storage for the same objective. Pumped storage, to be practical, requires the use of two reservoirs for the storage of water—one at considerably higher elevation, say, 500 to 1000 ft (150 to 300 m). A reversible pump-turbine operates alternatively (1) to raise water from the lower to the upper reservoir during off-peak periods, and (2) to generate power during peak-load periods by letting the water flow in the opposite direction through the turbine. Proximity of favorable sites on an interconnected electrical transmission system reduces the investment burden. Under such circumstances the return of 2 kWh on-peak for 3 kWh pumping off-peak has been demonstrated to be an attractive method of economically utilizing interconnected fossil-fuel, nuclear-fuel, and hydro power plants. *See* ELECTRIC POWER GENERATION; HYDRAULIC TURBINE; POWER PLANT; NUCLEAR REACTOR.

[THEORDORE BAUMEISTER]

Bibliography: Annual Statistical Report for 1980, *Elec. World*, 193(6):49–80, Mar. 15, 1980; T. Baumeister (ed.), *Standard Handbook for Mechanical Engineers*, 8th ed., 1978; Department of Agriculture, *Summary of Reservoir Sediment Deposition Surveys Made in the U.S. through 1960*, USDA Misc. Publ. no. 964, 1964; Department of the Interior, *Rate of Sediment Accumulation Drops at Lake Mead*, 1967; Edison Electric Institute, *Statistical Year Book of the Electric Utility Industry*, published annually; Federal Power Commission, *Development of Pumped Storage Facilities in the United States*, 1972, *Hydroelectric Power Resources of the United States*, Jan. 1, 1972, *The Role of Hydroelectric Developments in the Nation's Power Supply*, May 1974, *World Power Data*, 1966; D. G. Fink and H. W. Beaty,

Standard Handbook for Electrical Engineers, 11th ed., 1978; French stem the tides, *Elec. World*, 166:17, Nov. 7, 1966; Pumped storage power, at last, comes into its own, *Eng. News Rec.*, 182:22–25, Jan. 2, 1969; W. H. Hunt, Pumped storage: A major hydro power resource, *Civil Eng.*, 38:48–53, March, 1968; Rance tidal power station, *Smokeless Air*, no. 145, pp. 174–176, spring, 1968: W. A. Schoales, Prospects of tidal power in western Australia, *Elec. World*, 167:61–63, Feb. 20, 1967: Worldwide pumped storage projects, *Power Eng.*, p. 58, Oct., 1968.

Watt-hour meter

An electricity meter which measures and registers the integral, with respect to time, of the power in the circuit in which it is connected. In effect, it is an electric motor, the torque of which is proportional to the electric power in the circuit. The speed of the rotor is proportional to the torque, making each revolution of the rotor a measurement in watt-hours. Summation of the watt-hours is accomplished by gearing a counter, or register, to the rotor. *See* MOTOR.

The basic elements of a watt-hour meter are the stator, rotor, retarding magnet or magnets, register, and meter housing.

Principle of operation. Watt-hour meters may be classified into three types, according to fundamental differences in principle of operation.

Mercury type. This type is used for measuring energy on a dc circuit. It differs from other types in that the driven portion of the rotor consists of a radially slotted copper disk immersed in mercury. The load current flows diametrically through the disk, interacting with a flux produced by the line-voltage electromagnet to cause the disk to rotate.

Mercury-type meters are readily applied to high-current loads, since they are used with shunts and since only a portion of the load current passes through the current circuit of the meter.

Commutator type. This is also used for measuring energy on a dc circuit. It may also be used on ac circuits if all windings are of air-core construction. This meter is a shunt-type motor. The field coils, part of the stator, produce a field proportional to the load current. The armature is mounted on the rotor and is energized by the line voltage through a commutator and brushes, producing a rotor torque proportional to the power in the circuit.

Induction type. This is the common meter found in homes. It is used for measuring energy on an ac circuit. Figure 1 is a schematic sketch showing the basic elements of an induction meter. Figure 2 shows the stator in more detail.

The potential-circuit winding of the stator is made highly inductive to obtain a quadrature-time relationship between the potential-circuit and the current-circuit working fluxes in the disk air gap. These fluxes, displaced in time and space, produce a rotor-disk torque proportional to the circuit power. Retarding magnets control the rotor speed, making it proportional to the power. The register, which is geared to the rotor, records the watt-hours.

This type of meter has been developed to a high degree of accuracy under extreme environmental conditions and over great ranges of load and voltage. Although it has the appearance of simplicity and low cost, its magnetic circuitry is extremely complex. Its calibration is stable, and there are almost no maintenance requirements.

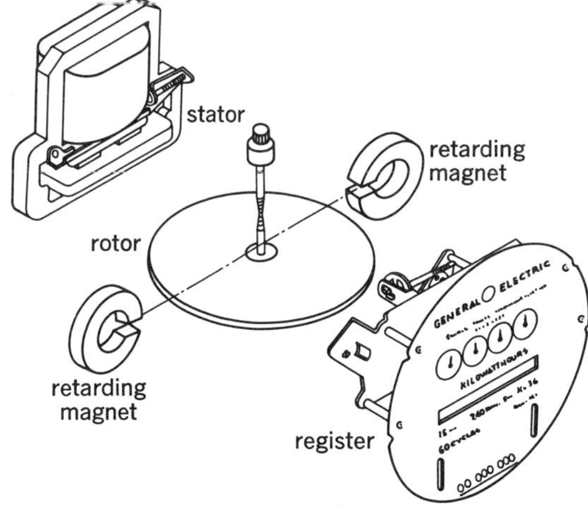

Fig. 1. The basic elements of the induction-type watt-hour meter. (*General Electric Co.*)

Multistator, or polyphase, watt-hour meters employ the same principle of operation as do single-stator meters. Magnetic interference between stators and the need for an adjustment to balance the torques of all stators are added complications of multistator meters.

Special watt-hour meters. Many types of watt-hour meter are available for special needs and applications.

The switchboard-type meter is used for industrial or central-station applications. It differs from the basic types only in housing construction.

The totalizing meter records in one meter the energy used in two or more circuits. This meter may have four or more stators acting on a single rotor.

The portable watt-hour meter standard is a specially developed, high-accuracy watt-hour meter having a multiplicity of current and voltage cir-

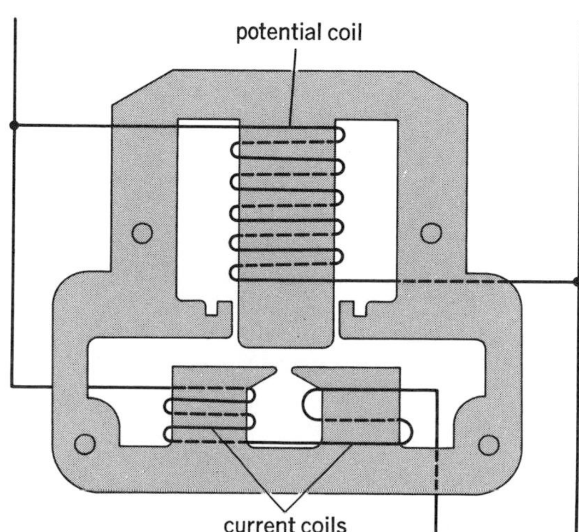

Fig. 2. Stator of induction-type watt-hour meter, showing the coils. (*General Electric Co.*)

cuits. It is generally used in the meter shops of utilities and in the field for testing watt-hour meters.

A combination watt-hour meter and time switch, consisting of a standard single-phase meter and a time switch combined in one housing, is used on water heaters. The time switch opens the main heater circuit during predetermined peak-load conditions.

A combination watt-hour and demand meter is used to indicate the maximum demand in addition to recording watt-hours. This device is made in two different constructions. The thermal type combines in a single housing a thermal-demand indicator with a single or multistator watt-hour meter. The mechanical, or integrated-demand, type consists of a single or multistator watt-hour meter equipped with a demand register instead of a conventional watt-hour register. The demand register has, in addition to the watt-hour register parts, a timing means and a mechanism for integrating the energy consumed over the demand interval.

A watt-hour meter with a contact device is used for measurement of demand, particularly when large blocks of energy are involved. A demand meter, located externally to the watt-hour meter, is actuated through a contact device contained in the watt-hour meter and geared to the rotor shaft. *See* ELECTRIC POWER MEASUREMENT.

[GEORGE R. STURTEVANT; J. ANDERSON]

Bibliography: American National Standard, *Code for Electricity Metering*, C-12, 1975; Edison Electric Institute, *Electrical Metermen's Handbook*, 7th ed., 1965; F. K. Harris, *Electrical Measurements*, 1952; F. W. Kirk and N. R. Rimboi, *Instrumentation*, 3d ed., 1974; National Electrical Manufacturers Association, *Standards for Watthour Meters*, Publ. no. EI 20-1975, 1975.

Wattmeter

An instrument that measures electric power. For a complete discussion of the power in various types of electric circuits *see* ELECTRIC POWER MEASUREMENT.

A variety of wattmeters is available to measure the power in ac circuits. They are generally classified by names descriptive of their operating principles. Determination of power in dc circuits is almost always done by separate measurements of voltage and current. However, some of the instruments described will also function in dc circuits.

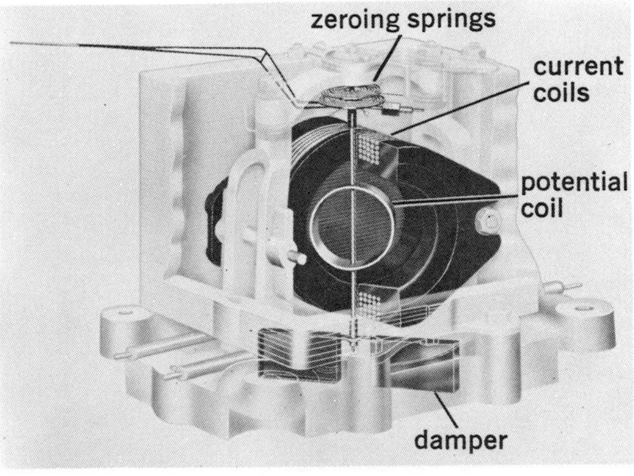

Fig. 1. Single-element electrodynamic wattmeter. (*Weston Instruments, Inc.*)

Electrodynamic wattmeter. Probably the most useful instrument in the measurement of ac power at commercial frequencies is the indicating (deflecting) electrodynamic wattmeter. It is similar in principle to the double-coil dc ammeter or voltmeter in that it depends on the interaction of the fields of two sets of coils, one fixed and the other movable. The moving coil is suspended, or pivoted, so that it is free to rotate through a limited angle about an axis perpendicular to that of the fixed coils. As a single-phase wattmeter, the moving (potential) coil, usually constructed of fine wire, carries a current proportional to the voltage applied to the measured circuit, and the fixed (current) coil carries the load current. This arrangement of coils is due to the practical necessity of designing current coils of relatively heavy conductors to carry large values of current. The potential coils can be lighter because the operating current is limited to low values.

If i_1 is the instantaneous current in the potential coil and i_2 is the instantaneous current in the current coil, and there is no iron or disturbing magnetic field due to current in neighboring conductors, then Eq. (1) holds. Since i_1 is proportional to the

$$\text{Instantaneous torque} = k(i_1 i_2) \qquad (1)$$

instantaneous voltage e across the circuit, Eq. (2) is valid.

$$\text{Instantaneous torque} = k(ei_2) \qquad (2)$$

The moving system, however, is designed with sufficient inertia that it is unable to follow the rapid alternations of the alternating current, but it will rotate, opposed by light zeroing springs, to a position corresponding to the average torque, which is proportional to the average power being supplied. A modern wattmeter of the electrodynamic type is shown in Fig. 1.

To avoid a loss of power in the instrument due to the current flowing in the potential coil, a fine-wire coil is wound with the current coil and connected in series with the potential coil. Its effect is to cancel the magnetic effect of the potential coil current from the field of the current coils. If this compensation is correct, with the load circuit open, the instrument will read zero.

The presence of inductance in the potential circuit would normally introduce phase displacement between voltage and current, which is theoretically inadmissible. Simple tuning with capacity would introduce frequency error, but capacity in series with the moving coil and shunted by noninductive resistance is found to reduce the net reactance to a tolerably low value.

Other disturbing effects are those of ambient temperature, eddy currents in metal structures close to the moving coils, transformer effect due to mutual induction between current and potential coils, and skin effect in which a change in distribution of current in the current coil causes frequency error. In precision instruments these effects are taken care of as much as possible by proper choice of materials and by shielding.

This type of wattmeter is capable of great accuracy within its range and is commonly used as a laboratory standard. It has the great advantage of

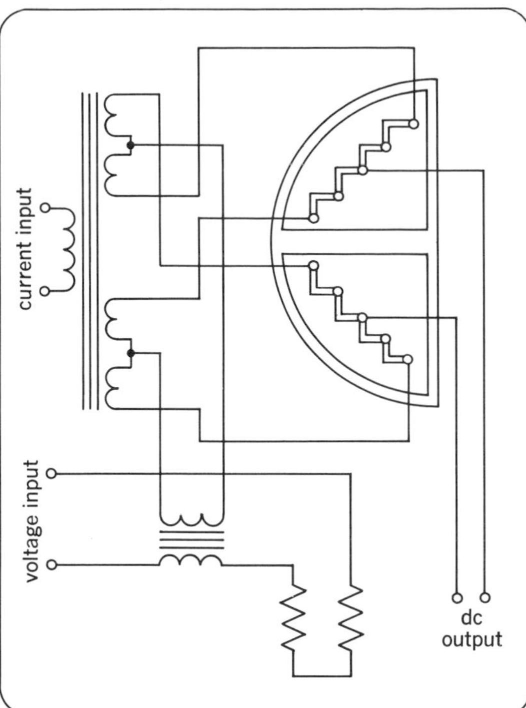

Fig. 2. Thermal wattmeter circuit.

being usable as a transfer instrument; that is, it can be calibrated with direct current and used for ac measurements. It is useful chiefly at power frequencies. In modern practice, the ranges of voltage are limited to about 150 volts, though external multipliers to extend the voltage range may be used. Maximum current is usually about 20 amperes. Shunting of current coils is not recommended. The preferred method of extending the range of both current and voltage is through current and potential transformers, by which the voltages and current in the load circuit are reduced to nominal values by definite transformation ratios without introducing appreciable phase errors, or, if such errors are introduced, their values are known and may be accounted for.

Thermal wattmeter. The thermal wattmeter is a versatile means of power measurement, since its operation is based on the heating effect of current, the I^2R relationship. It is applicable to both direct current and alternating current and is usable at frequencies up into the radio range without regard to waveform. Various electronic wattmeters for audio and radio power measurements are largely based on this principle. Thermocouples are commonly used as transducers, and the derived voltage is available for the common types of dc indicators and for potentiometers of either the manual or automatic self-balancing type. A limitation is that the thermal wattmeter must be calibrated against another wattmeter. Also, the thermal wattmeter is relatively slow in response and may be affected by temperature.

A typical system used in commercial power measurements is shown in Fig. 2. The transducer elements are placed in the semicylindrical structure shown, which is divided into two compartments designed to limit cooling by convection. Each compartment contains a resistance-thermocouple assembly, which is made up of a series of thermocouples having the cold junctions *a* mounted on short metal posts and the hot junctions *b* suspended in air. The mounting posts are electrically insulated from, but in close thermal association with, the base of the structure. The materials of the thermocouples are so selected and proportioned to obtain not only a suitable value of thermoelectric effect but also an ohmic resistance, which may be heated by the load current. Thus, in each of these identical elements, a thermoelectric potential is developed with a polarity dependent on the arrangement of the thermoelectric pairs and a magnitude proportional to the square of the current that is flowing. The load current is applied to the primary of a current transformer with two identical secondary windings I_1 and I_2, which are differentially connected to the transducer elements. The line voltage is connected in series with a resistance to a potential transformer E whose secondary voltage is tapped into the midpoints of secondaries of the current transformer. If the circuit elements are symmetrical, the temperature rise of the free ends of the thermocouples is proportional to the power dissipated in each element or to the squares of the corresponding currents. These values are different, since they depend on sum and difference terms, and thus the value and polarities of the thermoelectric elements are such that the dc potential existing between the midpoints of the two elements is proportional to the respective temperatures and, therefore, to the squares of the respective currents and, in turn, to the power in the ac circuit.

As with other methods, the range of voltage and current is extended through the use of additional current and potential transformers in the circuit external to the equipment shown in Fig. 2.

Electrostatic wattmeter. The elementary quadrant electrometer has been adapted for power measurements. A schematic diagram of this instrument is shown in Fig. 3. The mechanism consists of two quadrants a and b charged by the voltage drop across a noninductive shunt resistance R_1 through which the load current passes. The

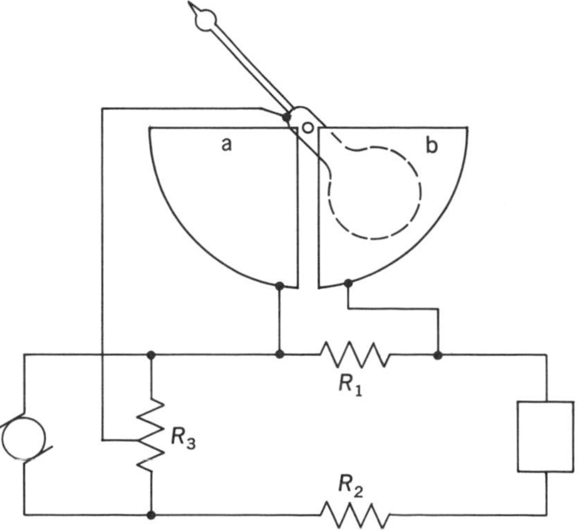

Fig. 3. Electrostatic wattmeter circuit.

line voltage is applied between the moving vane and one of the quadrants. In the circuit shown, the voltage to the moving vane is taken from the voltage divider R_3. This, in combination with the series resistance R_2, is a common means for providing compensation for power losses in the shunt. All resistances are noninductive. The deflection θ of the indicator is proportional to the average power, as defined by Eq. (3), where T is the period of the alternating wave.

$$K\theta=\frac{1}{T}\int_0^T ei\,dt \qquad (3)$$

The method is unique among others in use in that it is a voltage method rather than one of the more usual current methods. It has the advantages of (1) the possibility of wide ranges of measured current and voltage, (2) readings essentially unaffected by ambient conditions, and (3) measurements generally free of errors due to frequency or waveform.

Its disadvantages are that it is relatively insensitive, being limited by the voltage drop permitted across the shunt, and that it requires careful screening to eliminate all possible effects of charged bodies in the vicinity. It also must be calibrated at the voltages at which it is used. Moreover, the movement is generally low in torque and weight and hence relatively delicate mechanically.

This instrument has found its chief use in the laboratory for standardizing purposes and in capacity testing where small values of power, low power factors, and high voltages are involved.

Polyphase wattmeter. The instruments thus far considered are designed for single-phase power measurement. In polyphase circuits, the total power is the algebraic sum of the power in each phase. This summation is assisted by simple modifications of single-phase instruments.

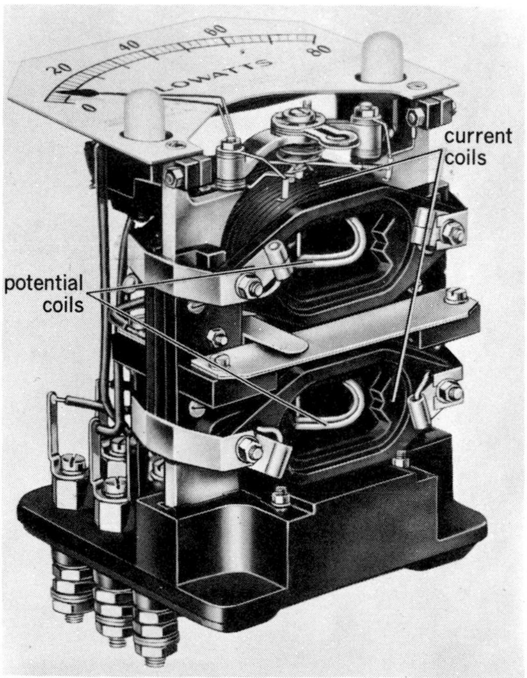

Fig. 4. Two-element, polyphase electrodynamic wattmeter. (*Weston Instruments, Inc.*)

For example, an electrodynamic wattmeter may contain a second coil system similar to the coils of a single-phase meter, with the second potential coil on the same shaft as the potential coil in the first system. The two systems are mounted in the same case and are designed to have matched characteristics, but care is taken that there is no magnetic interaction between them (Fig. 4). The deflection is proportional to the sum of the torques of the two elements; thus, total power is read from the instrument scale. Electrical connections are the same as for two single-phase wattmeters.

The thermal wattmeter is also commonly made in a two-element system, the second element being identical with the first and the two transducer networks being connected so that the combined dc output is the algebraic sum of the ac power in the two systems.

[DONALD B. SINCLAIR]

Bibliography: W. D. Cooper, *Electric Instrumentation and Measurement Techniques*, 2d ed., 1978; E. Frank, *Electrical Measurement Analysis*, 1959; F. K. Harris, *Electrical Measurements*, 1952.

Well logging

The technique of making measurements in drill holes with probes designed to measure the physical and chemical properties of rocks and their contained fluids. Over a million holes are drilled annually around the world to probe for oil, gas, solid minerals, and fresh water, and to better understand the subsurface geology. Much information can be obtained from samples of rock brought to the surface in cores or bit cuttings, or from other clues while drilling, such as penetration rate; however, the greatest amount of information comes from well logs.

Well logs result from a probe lowered into the borehole at the end of an insulated cable. The resulting measurements are recorded graphically or digitally as a function of depth. These records are known as geophysical well logs, petrophysical logs, or more commonly well logs, or simply logs.

An enormous number of logging devices have been developed since the first experiments with borehole resistivity measurements, in 1927. From a few dozen basic logging tools several hundred different types and variations of curves can now be recorded; however, in any one hole only about a dozen or fewer curves are commonly run. An outline of logging methods and some of their main applications are shown in Table 1.

In addition to these logging methods, there are a number of closely related electromechanical tools that can be run on the same wire line used for logging. While these tools perform important functions in the borehole, they are not actually well logs. Some of these tools and their main purposes are listed in Table 2.

Although the most common uses of logs are for correlation of geological strata and location of hydrocarbon zones, there are many other important subsurface parameters that need to be detected or measured. Also, different borehole and formation conditions can require different tools to measure the same basic property. In petroleum engineer-

Table 1. Logging methods

Logs and curves	Property investigated	Primary uses
Electrical		
Self-potential	Natural currents	Define sand/shale sequences, bed thickness; determine correlation water salinity
Normal resistivity	Resistivity	Shallow investigation of invaded zone to aid in quantifying hydrocarbon saturations; aid correlation hydrocarbon
Lateral resistivity	Resistivity	Moderate to deep investigation of formation; used on older logging devices
Induction	Resistivity	Moderate to deep investigation of formations in high-resistivity muds
Focused	Resistivity	Shallow to deep investigation of formations in relatively low-resistivity muds
Microresistivity	Resistivity	Very shallow investigation to detect permeable rocks and aid in measurement of porosity
Acoustical		
Velocity	Interval transit time of compressional wave	Determine porosity, lithology, abnormal pressures; construct synthetic seismograms; calibrate seismic records
Wave train	Attenuation of acoustic waves	Detect fractures, gas zones; determine cement bonding
Nuclear		
Gamma ray	Natural radioactivity	Determine shale content, correlation, bed thickness; locate radioactive tracers
Spectral gamma ray	Energy spectrum of natural radioactivity	Identify radioactive elements
Density	Bulk density	Determine porosity; help identify gas zones
Neutron	Hydrogen content	Determine porosity; help identify gas zones
Pulsed neutron capture	Capture cross section of elements	Distinguish between salt water and hydrocarbon behind casing
Carbon/oxygen	Ratio of carbon and oxygen atoms	Identify oil behind casing
Nuclear magnetism	Hydrogen atoms in larger pore spaces	Identify productive zones; distinguish heavy oil from fresh water
Production		
Flowmeter	Fluid flow	Determine location and rate of fluid entry
Fluid density	Density of fluid in borehole	Determine type of fluid
Temperature	Temperature	Locate fluid entry, especially gas
Radioactive tracer	Fluid flow	Trace flow path of radioactive material released in borehole
Noise	Fluid entry and flow	Detect fluid movement inside or behind pipe and located entry
Capacitance	Type fluid in borehole	Diagnose fluid interfaces
Casing inspection	Electromagnetic properties of casing	Evaluate corrosion or damage of casing
Other		
Caliper	Borehole diameter	Estimate cement volumes behind casing; supplement log analysis
Dipmeter	Azimuth and inclination of borehole and correlation of beds in borehole	Dip and strike of bedding; geometry of hole
Borehole gravity meter	Bulk density	Determination of density further from borehole; calibrate surface gravity surveys
Dielectric constant	Dielectric properties	Distinguish fresh water from oil

Table 2. Electromechanical devices run on logging cables

Tool	Purpose
Sidewall sampling	To retrieve small core samples of formations from the side of an open borehole at selected intervals
Wire-line tester	To test formation pressures and fluids at selected intervals from cased hole or open hole
Perforators	To perforate casing for completion with bullets or shaped charge jets
Pipe cutters	To sever drill pipe or casing with mechanical, explosive, or chemical devices
Directional survey tool	To determine azimuth and inclination of borehole at selected intervals
Wire-line packer	To set a plug in casing or tubing
Free-point indicator	To determine where pipe is stuck
Wireline shot	To loosen pipe so it may be backed off threaded joint above where it is stuck
Fluid sampler	To catch a sample of fluid at a selected depth in borehole
Television or photographic cameras	To inspect casing or other borehole conditions
Impression block	To determine nature of top of "fish" or lost tools in hole

ing, logs are used to: identify potential reservoir rock; determine bed thickness; determine porosity; estimate permeability; locate hydrocarbons; estimate water salinity; quantify amount of hydrocarbons; estimate type and rate of fluid production; estimate formation pressure; identify fracture zones; measure borehole inclination and azimuth; measure hole diameter; aid in setting casing; evaluate quality of cement bonding; locate entry, rate, and type of fluid into borehole; and trace material injected into formations (such as artificial fractures). In geology and geophysics, they are used to: correlate between wells; locate faults; determine dip and strike of beds; identify lithology; deduce environmental deposition of sediments; determine thermal and pressure gradients; create synthetic seismograms; calibrate seismic amplitude anomalies to help identify hydrocarbons from surface geophysics; calibrate seismic with velocity surveys; and calibrate gravity surveys with borehole gravity meter. Other applications include: locating freshwater aquifers; locating solid minerals; and studying soil and rock conditions for foundations of large structures.

Electrical devices. These employ instruments that generate data based on electrical measurements.

Spontaneous potential. The spontaneous potential, usually recorded along with resistivity curves, is a simple but valuable aid to help geologists correlate from one well to another and to assist them in inferrring the depositional environment of the sediments. It defines permeable zones from surrounding nonpermeable shales and can be used to estimate the salinity of formation water. In the early 1930s, observations were made in boreholes of naturally occurring currents. These currents, now referred to as spontaneous potential, self-potential, or simply SP, are recorded from a simple electrode on the probe with a reference electrode usually grounded at the surface. Spontaneous potential values opposite shales are relatively constant; however, when the electrode passes a permeable bed containing water more saline than the borehole fluid, the spontaneous potential will deflect to the left or in the negative direction (Fig. 1). The greater the salinity contrast between formation water and borehole water base mud, the greater the spontaneous potential deflection. It is this electrochemical property that allows estimates of formation water salinity to be made.

Borehole size and bed thickness along with the density, chemical composition, and invasion of mud into the formation will also affect the magnitude of the spontaneous potential. The presence of hydrocarbons in the permeable zone may suppress the spontaneous potential a slight amount, but there is still formation water trapped in the smaller pores which will contribute to the spontaneous potential deflection.

Clays in permeable sand will also tend to suppress the spontaneous potential. Because the amount of clays deposited with sands can give valuable clues to the environment of deposition, the spontaneous potential shape is used by geologists to help piece together the subsurface geology. An example of the spontaneous potential curve, along with several other curves, is shown in Fig. 1. Note how the shaly sands tend to suppress the spontaneous potential toward the bottom of the lower sand interval. This sand could have been deposited as part of an ancient marine beach.

Resistivity. While resistivity was the first logging measurement to be made, it is still one of the most important of physical properties to record. It is easy to visualize that a sand filled with salt water will conduct electricity much easier (and therefore have a low resistivity) than if most of the salt water is replaced with a nonconducting oil or gas. The basic measurement of resistivity, the ohm-meter, is the electrical resistance of a cube 1 m on a side; it is the reciprocal of conductivity. Like most log measurements, the resisitivity log response depends not only upon water saturation but also upon secondary properties of the rock/fluid system. Quantification of the water saturation S_w depends on knowledge of the rock pore space or porosity ϕ, cementation exponent m, saturation exponent n, and water resistivity R_w. In addition, the response characteristics of the particular tool for the surrounding physical environment, including mud and formation geometry, should be understood. In 1942 G. E. Archie published an empirical relation between water saturation, rock and fluid properties, and true resistivity R_t. This classic equation is still widely used today [Eq. (1)].

$$S_w = \left[\frac{R_w \phi^{-m}}{R_t}\right]^{1/n} \qquad (1)$$

Where significant amounts of conductive minerals such as clay are also in the formation, allowance must be made for their influence on the total resistivity. Many improvements have been made over the years in tools to record resistivity, but the primary purpose—to determine hydrocarbon saturation $(1-S_w)$ by calculation of the true formation resistivity—remains the same.

Some resisitivity devices (listed below) are designed to investigate deep into the formation away from the borehole, while others respond to the formation near the borehole. In general, the resistivity tools with shallow investigation can define thin beds better, but are more influenced by the

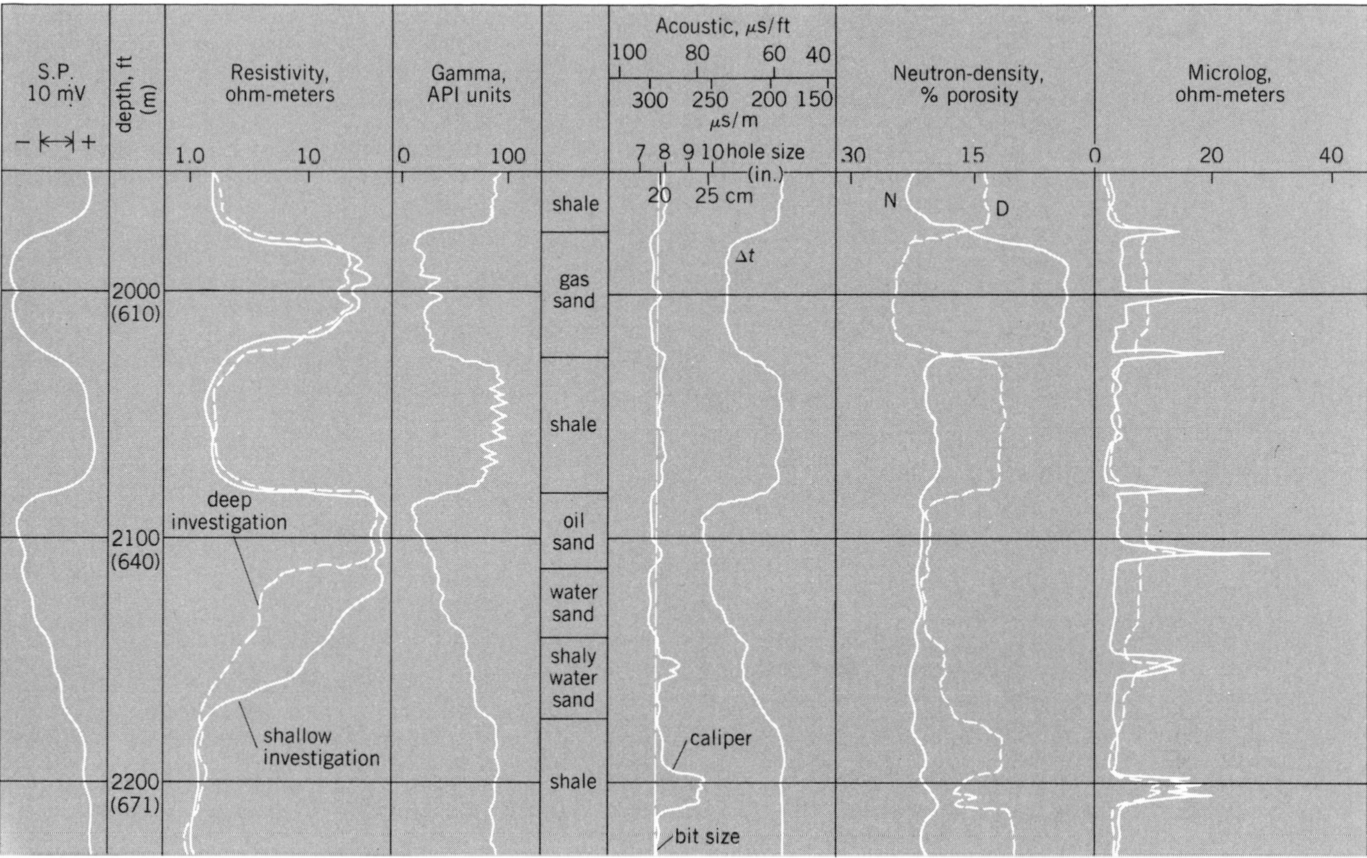

Fig. 1. Example of log response in a sand/shale sequence.

annular zone of mud filtrate invasion. The resistivity tools with deep investigation approach true resistivity, but still may need correction due to influence of filtrate invasion, borehole geometry, and bed thickness.

1. Normal and lateral. The short normal, long normal, and lateral curves were common on older electrical logs and had shallow, intermediate, and deep investigations, respectively. The short normal, which has 16–18 in. (41–46 cm) spacing, is still used on the induction electrical log. In spite of its shallow investigation, it has been one of the basic geological correlation tools.

2. Induction log. The induction log is probably the most commonly run logging tool today; it is designed for medium and deep investigation. As shown in Fig. 2, this tool consists of one or more coils that generate an alternating electromagnetic field. The induced currents which are dependent upon the conductivity of the formation are then detected by receiver coils. Multiple coils are used to focus the measurement deep into the formation; this minimizes effects of the borehole, invaded zone, and nearby zones. The induction log works best in relatively resistive borehole fluids such as freshwater or oil base muds.

3. Focused log. Where the borehole mud is more saline than the formation water, a focused log such as the Laterolog is recommended. In this tool a sheet of current from an electrode is forced into the formation by bucking currents from surrounding electrodes. This bucking current keeps the measuring current thin and allows excellent resolution of thin beds.

4. Microresistivity. There are several microresistivity tools where relatively shallow investigations of resistivity are made from pads pressed against the borehole wall. These auxiliary logs aid in the interpretation of true resistivity and porosity. One of these tools, called the microlog (also known as minilog or contact log), is of particular interest. It is the oldest of all pad devices. This tool combines two resistivity measurements, one measuring about 1 1/2 in. (4 cm) away from the pad and the other about 4 in. (10 cm). When the pad is pressed against mud cake built up over a permeable zone, a separation appears between the two curves (see Fig. 1). Shales or other nonpermeable zones do not have such separation; therefore, under favorable conditions the microlog can give excellent vertical resolution of permeable zones.

Acoustical devices. Quantitative interpretation of hydrocarbon saturation requires a knowledge of porosity. One of the most widely used porosity logs is the acoustic log, also known as a sonic log, acoustilog, or velocity log. This log is used for many other purposes, including identification of lithology, prediction of pressures, and assisting with geophysical interpretation. In addition, information from the amplitude of the acoustic wave aids in detection of fractures and gas zones, determination of mechanical properties of rocks, and analysis of cement bond behind pipe.

The acoustical log measures the shortest time for sound waves to travel through 1 ft (or 1 m) of formation. This interval travel time (Δt), measured in microseconds per foot (or per meter), is the inverse of velocity. A sketch of a modern borehole compensated acoustic tool is shown in Fig. 3. The dual receivers with each transmitter are arranged so that travel time through mud and hole rugosity will not detract from measuring formation velocity. The formation travel time is dependent upon the porosity ϕ, the travel time of the fluids in the pore space (Δt_f), and the travel time of the rock matrix (Δt_{ma}). An empirical formula for clean sands known as the Wyllie time average equation relates porosity ϕ_{sv} to the sonic velocity interval travel time Δt of a water-filled rock as shown in Eq. (2).

$$\phi_{sv} = \frac{\Delta t - \Delta t_{ma}}{\Delta t_f - \Delta t_{ma}} \qquad (2)$$

While Δt depends only upon the time of the first arrival of the compressional wave, other information may be extracted from the amplitude of the compressional wave and later arrivals.

In open hole logging, the amplitude of the acoustical wave train may be attenuated by natural fractures or faults. In cased hole logging, the cement bond log is an important aid to determine cement bonding in the annular space between the casing and the formation. This tool measures the amplitude of an acoustical signal imposed upon the casing. A good cement bond against the pipe will prevent it from "ringing" and therefore will

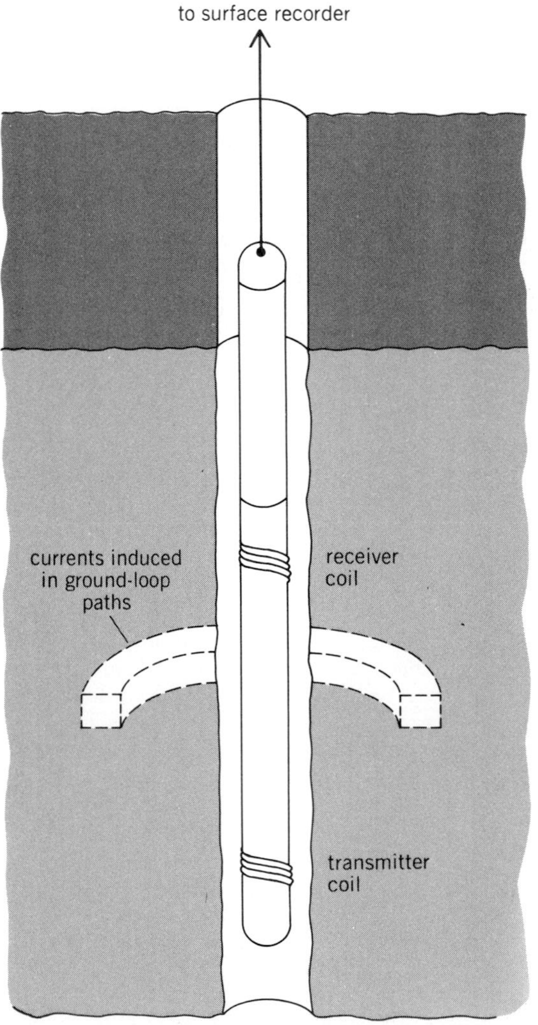

Fig. 2. Schematic of basic two-coil induction system.

greatly attenuate the amplitude of the compressional wave before it reaches the receiver. Supplemental information, including a display of acoustical wave trains, gives further insights into the degree of bonding.

Nuclear devices. These are instruments which generate data based on measurements of nuclear particles.

Gamma ray. In most cases the gamma-ray log may be considered as a shale log; that is, clays or shaly rocks will give a higher radioactive count than clean sands or carbonates. Like the spontaneous potential, this provides a good correlation curve, defines bed thicknesses, and aids interpretation of environmental deposition. It is usually a more diagnostic indicator of beds in a carbonate sequence than the spontaneous potential, and unlike spontaneous potential, can detect formations behind casing. The gamma log is relatively unaffected by borehole conditions. Often it is used to trace material injected into a formation, such as a radioactively tagged sand used as a proppant in a hydraulic fracture. Measurement is made in terms of arbitrary units, API (American Petroleum Institute) units.

Spectral gamma ray. Almost all the naturally occurring radioactivity that is detected by the gamma-ray log comes from three elements: potassium, thorium, and uranium. Although it is usually assumed that high radioactive values on the gamma log come from potassium-rich shales, the spectral analysis log allows one to verify the contribution from each element. Experience with this log has shown that a great number of "shales" are, in reality, uranium-rich stringers in sands, dolomites, and limestones. Some of these formations originally assumed to be impermeable shales have turned out to be hydrocarbon productive zones. The spectral breakdown also allows for more sophisticated geological environmental interpretation.

Density. Another important porosity tool is the density log; this tool emits a beam of gamma rays into the rock from a source such as cesium-137. These gamma rays interact with electrons in the formation through Compton scattering. The resulting lower-energy gamma rays are sensed by two detectors above the source. Although the tool responds to electron density, for most formation rocks the apparent bulk density is practically identical to the actual bulk density. For a few substances such as sylvite and rock salt (and to a lesser extent gypsum, anhydrite, and coal), small corrections are needed to arrive at true bulk density values. With knowledge of the bulk density ρ_B and reasonable assumptions of rock matrix density ρ_{ma} and fluid density ρ_f, the porosity derived from a density log (ϕ_D) can be calculated from Eq. (3).

$$\phi_D = \frac{\rho_{ma} - \rho_b}{\rho_{ma} - \rho_f} \qquad (3)$$

Neutron. Another important device for determination of porosity is the neutron log. These logs respond primarily to hydrogen atoms. Therefore, in clean or shale-free formations the neutron log reflects the amount of liquid-filled porosity. Because the neutron log does not recognize gas-filled porosity, it can be compared with another porosity device such as the density to detect gas-filled zones. It can also be used in combination with other porosity logs for interpretation of lithology, including shaly sands. In neutron logging, high-energy neutrons are emitted from a source such as plutonium and beryllium. When these neutrons collide with hydrogen atoms which have a nucleus of nearly the same mass, the greatest amount of energy is lost. When the neutrons have slowed down to thermal velocities after successive collisions, they are subject to capture by nuclei of other atoms; this causes an emission of gamma rays. Either these gamma rays or the low-energy neutrons are counted by the detector in the sands.

Pulsed neutron capture. Besides resistivity, another physical characteristic that distinguishes salt water from hydrocarbons is that the chlorine atom in salt water has a significantly higher capture cross section than the hydrogen or carbon atoms found in oil or gas. Since resistivity measurements of the formation cannot be made in cased holes, a new tool sensitive to capture cross section was developed to distinguish hydrocarbon from salt water behind casing. The pulsed neutron capture tool employs a neutron generator that repeatedly emits pulses of high-energy neutrons. This is done using a miniaccelerator to bombard a tritium target with deuterium atoms. After these neutrons are slowed down to the thermal state, they are captured by nuclei of the various atoms surrounding the tool. With each capture a corresponding emission of gamma rays occurs; these can be detected a short distance from the source. The total or bulk capture cross section (Σ_t) of the formation is the sum of the component cross sections of the rock matrix (Σ_{ma}) and the fluids such as water (Σ_w) and hydrocarbons (Σ_h) within the rock pores. This may be expressed as Eq. (4). Therefore, water saturation S_w can be estimated where there is sufficient contrast in capture cross section between the water and hydrocarbons, and reasonable estimates can be made of the other variables. This tool has proved to be a valuable aid in identifying bypassed oil stringers in cased holes. Also it is used to monitor changes in the saltwater level within a producing horizon.

$$\Sigma_t = \Sigma_{ma}(1-\phi) + \Sigma_w S_w \phi + \Sigma_h (1-S_w)\phi \quad (4)$$

Carbon/oxygen. The carbon/oxygen log is designed to directly measure hydrocarbons in any salinity environment in open hole or through casing. It does this by measuring the relative amounts of certain elements such as carbon and oxygen. Since oil is composed largely of carbon, a measure of this element provides a direct evaluation of oil in the formation. A primary use of this tool has been to detect oil in the presence of fresh or unknown salinity water behind casing. This tool is similar to the pulsed neutron capture tool in that it utilizes a pulsed 14-MeV neutron source and a gamma-ray detector. However, it is optimized for production of desired prompt gamma rays from inelastic scattering of neutrons by carbon in the formation. Energies characteristic of other elements are also measured. Because changes in lithology can also affect the carbon/oxygen ratio, a ratio of calcium/silicon is recorded to help distinguish a change in carbonate content from hydrocarbon saturation.

Nuclear magnetism log. Except for limited periods of field experimentation, this tool has essentially been in a state of development since 1960, which indicates the difficulty of making such measurements in a borehole. Basically, measurements are made by applying a strong, magnetic field to the formation for a limited time (about 2 s). After the imposed field is relaxed, measurements are made of the free precession of protons (hydrogen atom nuclei) in pore space immediately adjacent to the borehole. Two independent parameters are derived from these measurements: free-fluid index I_{Ff} and thermal relaxation time T_1. Water molecules in very fine pores such as those found in clays will not respond. The magnitude of I_{Ff} is a value which approaches total porosity in clean formations with relatively large pores. The value of T_1 is dependent upon fluid-solid interactions and pore size. This tool is of great importance because of its ability to estimate permeability (at least within an order of magnitude). By killing signals from water treated with paramagnetic ions that have been injected into an oil zone, the nuclear magnetism log can measure the amount of residual oil with great accuracy.

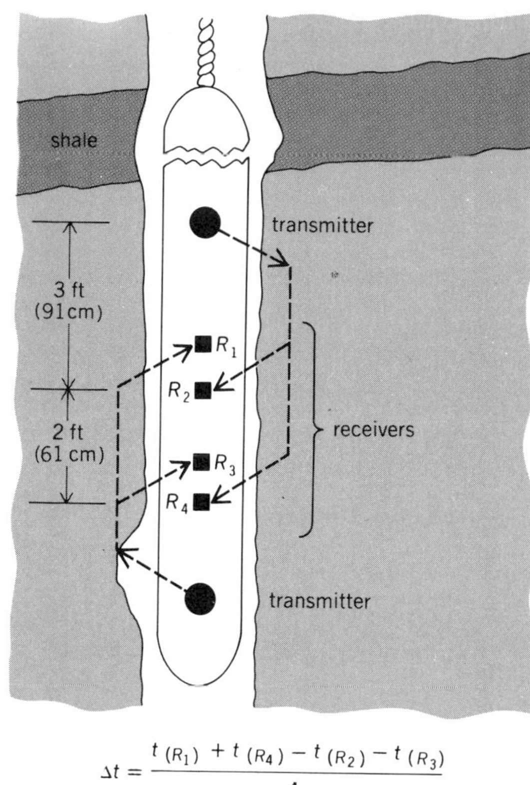

$$\Delta t = \frac{t(R_1) + t(R_4) - t(R_2) - t(R_3)}{4}$$

Fig. 3. Diagram of a borehole compensated acoustic tool.

Production logging. A very important suite of tools has been developed for production logging. These tools, usually run after casing has been set, are designed to locate fluid movement behind the casing and into the well bore, to detect the type of fluid, and to determine the flow rate. Production logging tools include various types of flow meters, fluid density devices, sensitive thermometers, radioactive tracer devices, noise loggers, and capacitance logs. In addition, these tools usually record some correlation curve such as gamma-ray or collar locator to tie the production curves to the for-

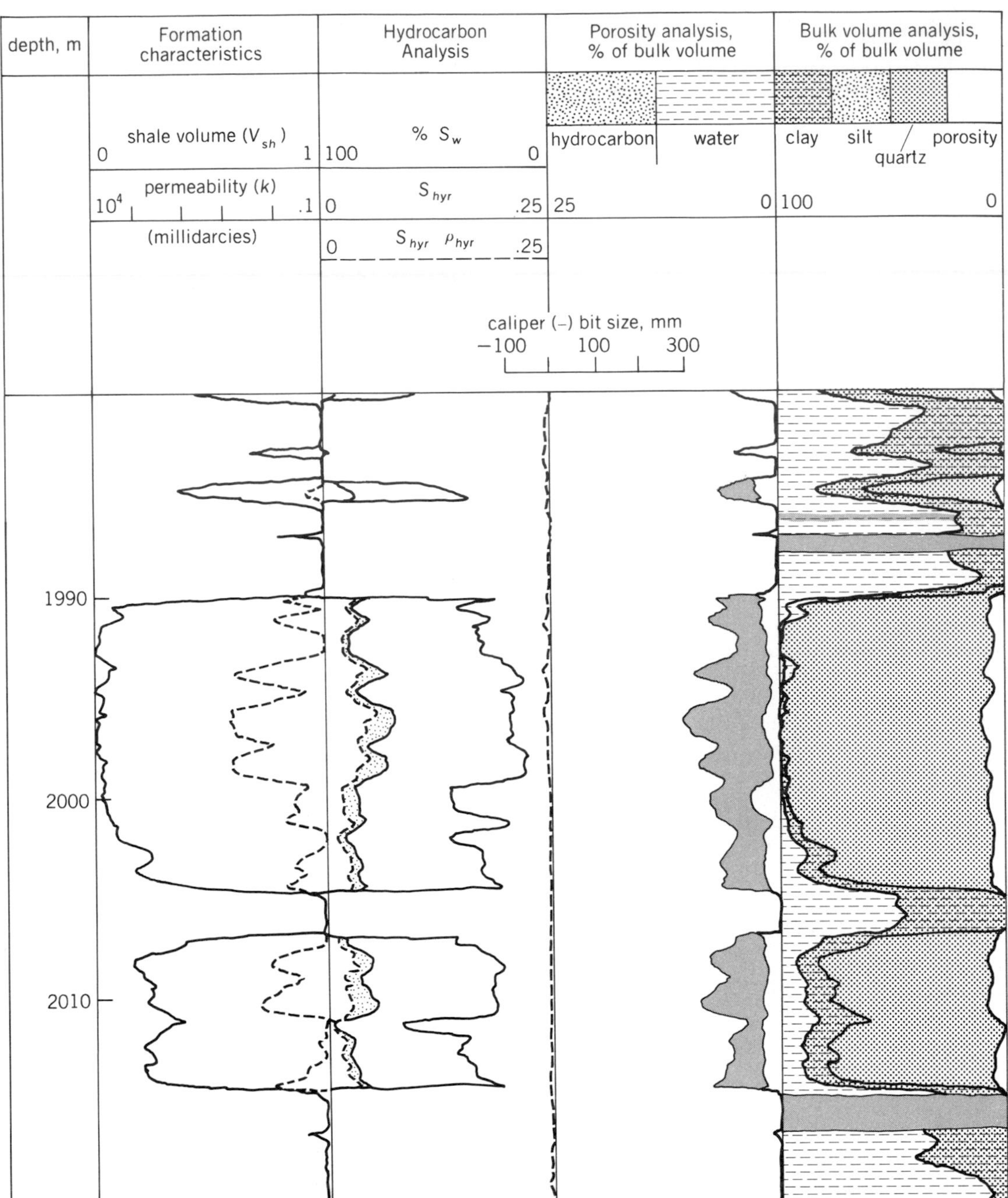

Fig. 4. Example of a computed log.

mation and casing. Closely related to production tools are logging devices that either magnetically or mechanically inspect the condition of the casing in the borehole.

Other devices. There are a number of miscellaneous devices that give additional information.

Caliper. The caliper is an important auxiliary curve that records the diameter of the hole with depth. This is essential for estimating the amount of cement needed in the annular space between the casing and formation. It also aids the interpretation of other curves where hole diameter or hole rugosity affects the readings. When a mud cake builds up over a permeable zone, the caliper can often detect such permeable streaks with excellent vertical resolution. The caliper is commonly run with each of the porosity tools.

Dipmeter. The dipmeter has three or four pads pressed against different sides of the hole. In the most common type of tool, these pads record microresistivity curves which can be correlated to determine depth displacement of formations from one side of the borehole to another. This information, combined with borehole size, deviation, and azimuth, is used to determine formation dip and azimuth. Such information is useful in correlating from well to well and in structural and stratigraphic interpretations. In addition, information necessary to plot the course of the borehole is provided by this tool.

Borehole gravity meter. Precise measurements of relative gravity can be made at intervals in the borehole to give estimates of formation density. These measurements are determined by large volumes of rocks that extend from tens to hundreds of feet away from the borehole. They are unaffected by borehole conditions such as mud type, hole rugosity, invaded zone, or casing and cement. Primary applications of the borehole gravity surveys include detection of significant porous zones or ore bodies away from the borehole, detection of oil and gas zones behind casing, and vertical density profiling to help interpret surface gravity mapping.

Dielectric constant. The dielectric constant log reflects the relative change in the dielectric constant of the rock fluids being surveyed. Since the relative dielectric constant of fresh water is significantly greater than oil, this tool can be used to distinguish oil from fresh water. The tools developed for open hole logging measure phase and amplitude of a 16- and 30-MHz electromagnetic wave and at much higher frequencies such as 1.1×10^9 Hz. At these high frequencies, conductivity becomes much less important than dielectric permittivity, unlike the much lower frequencies used in the induction log. Because of the very shallow depth of investigation, this tool works best in smooth boreholes.

Computer analysis of logs. Logging analysis has taken full advantage of the emergence of digital computing and the ability to take small but rugged and powerful computers to the field. It is now commonplace to record logs on magnetic tapes as they are run and to mathematically manipulate them in the logging trucks at the well site. Detailed computations may be quickly made on a foot by foot basis to give information on fluid and rock properties. Examples of information that are commonly computed and recorded with depth are shown in Fig. 4; these computed curves include shale index, permeability index, water saturation, hydrocarbon volume and weight, porosity, and lithology. In addition to analysis for open hole logging, special programs are designed for cased hole logging, solid mineral analysis, dip and structure of formations, synthetic seismic traces, and many others. In fact, the digital information from practically every curve can be incorporated into some type of analysis or interpretation program.

[RICHARD E. WYMAN]

Bibliography: L. A. Allaud and M. H. Martin, *Schlumberger: The History of a Technique*, 1977; H. Guyod and L. E. Shane, *Geophysical Well Logging*, vol. 1, 1969; E. J. Lynch, *Formation Evaluation*, 1974; S. J. Pirson, *Geologic Well Log Analysis*, 2d ed., 1977; S. J. Pirson, *Handbook of Well Log Analysis: For Oil and Gas Formation Evaluation*, 1963; Schlumberger Well Services, *Log Interpretation*, vol. 1: *Principles*, 1972, vol. 2: *Applications*, 1974; M. R. Wyllie, *Fundamentals of Well Log Interpretation*, 3d ed., 1963.

Wet cell

A primary cell in which there is a substantial amount of free electrolyte in liquid form. Important examples are the Lalande or caustic soda cell, the air-depolarized alkaline cell, the Weston standard cell, and the organic electrolyte cell.

Lalande cell. This cell (Fig. 1) uses a zinc anode, a cupric oxide cathode, and an electrolyte of sodium hydroxide in aqueous solution (caustic soda).

The amalgamated zinc electrodes are cast as flat plates or as hollow cylinders with thin sections, which corrode through as the copper oxide electrode approaches exhaustion. The bright copper color of the cathode can be seen through these openings to warn the user of approaching exhaustion of the cell.

The cathode is made by molding cupric oxide into flat plates or hollow cylinders. The oxide is mixed with a binder, pressed, and roasted. As it is used, the cupric oxide is partially reduced to metallic copper, which greatly increases the conductivity of the cathode.

The electrolyte is a solution of sodium hydroxide in water of good purity. The specific gravity is about 1.21. Normally, the surface of the electrolyte is covered with a layer of oil which retards evaporation of water and absorption of carbon dioxide from the atmosphere.

The anode reaction in the Lalande cell is the oxidation of zinc to form zinc oxide, which dissolves in the electrolyte to form sodium zincate. With sufficient electrolyte, no solid phase forms until the cell is nearly exhausted. Then precipitation occurs. Because of this reaction, it is necessary to provide about 8 ml of 1.21 sp gr solution per rated ampere-hour. The zinc must be mounted near the top of the electrolyte to prevent premature cutting off of the discharge by anodic polarization.

The cathodic reaction is the reduction of cupric oxide to metallic copper. For this reason, the plate always has the appearance of metallic copper.

The cell potential is 0.95–1.0 V, but the closed-circuit voltage on normal discharge starts at about 0.65 and decreases slowly to a cutoff voltage of about 0.50. Commercial cells are intended for relatively long continuous service. A 500-ampere-hour

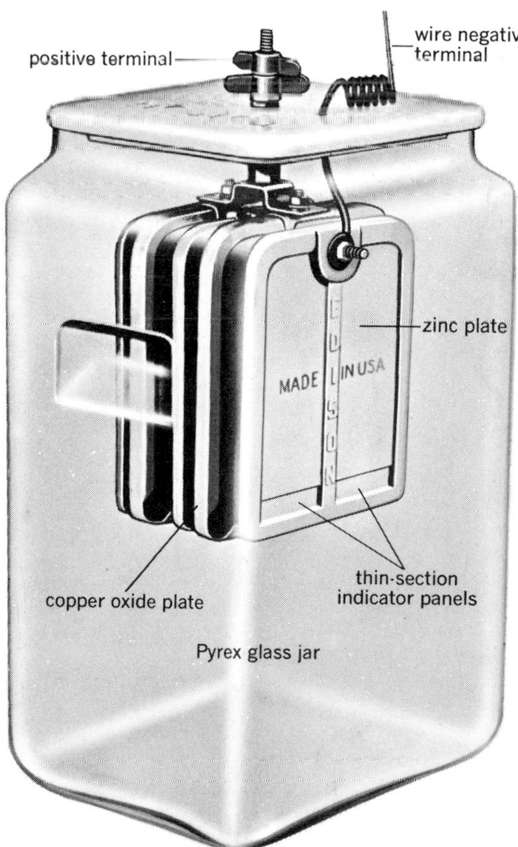

Fig. 1. Lalande-Edison copper oxide primary battery.

(A-h) light duty cell has an output rating of 1.75 A at 21°C. A heavy-duty cell of the same ampere-hour rating is capable of continuous output at 6.5–12.0 A. Unit energy output is about 38.8 kWh/m³ of cell.

At temperatures below 21°C, full output (ampere-hours) can be obtained at a reduced current. Current is reduced by 40% at 4°C, by 67% at −7°C, and by 83% at −18°C.

Air-depolarized alkaline cell. This cell (Fig. 2) uses a zinc anode, a porous carbon cathode exposed to air on one face, and an alkaline electrolyte. The carbon cathode utilizes atmospheric oxygen. Its zinc anode and alkaline electrolyte are like those of the Lalande cell, but it has twice the operating voltage and twice the watt-hour output of an equal-size Lalande cell.

The cathodic reactions have been shown to be Eqs. (1)–(3). The oxygen reacts to form hydrogen

$$2e + O_2 + H_2O \rightarrow O_2H^- + OH^- \quad (1)$$

$$O_2H^- + OH^- + 2H^+ \rightarrow H_2O_2 + H_2O \quad (2)$$

$$H_2O_2 \rightarrow H_2O + {}^1\!/_2 O_2 \quad (3)$$

peroxide, which decomposes readily to water and oxygen. The net amount of oxygen, then, is 0.3 g/A-h. At standard temperature and pressure, 210 ml of oxygen is consumed per ampere-hour.

The porous carbon has been reported to have an apparent density of only 0.65 with a porosity of 60%. To function properly, the inner surfaces should be dry. To resist penetration of electrolyte, the pore size must be small and the surfaces partially waterproofed by impregnation with paraffin. The cathode acts as a pump, drawing oxygen from the air. If too great a current is drawn, the pressure in the pore may drop sufficiently to allow electrolyte penetration. This reduces the activity of the carbon and may cause cell failure. Hence it is important that the cell should not be overloaded. For a railway cell, rated at 500 A-h, the recommended continuous drain is 2.0 A at temperatures above 7°C. This is a current density of 3.3 mA/cm². Much higher current densities can be obtained with porous carbon electrodes by special design.

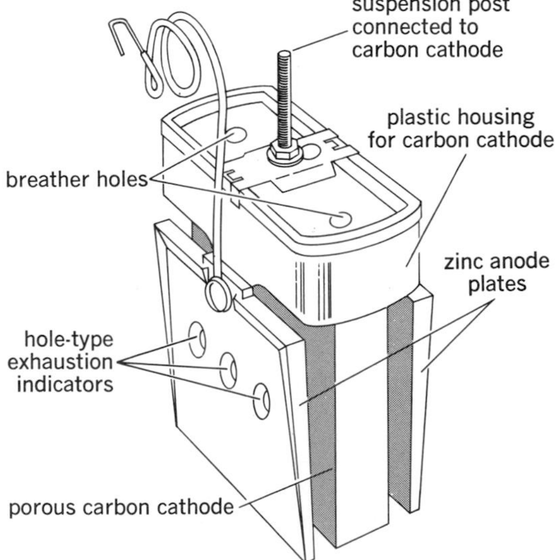

Fig. 2. Construction of air-depolarized alkaline cell.

The cell open-circuit voltage is 1.46. The railway cell at room temperature (24°C), rated at 500 A-h, will deliver 2 A at an average of 1.13 V to a cutoff of 1.05 V. On 3-A intermittent signal test, the cell delivers rated capacity of 1.09 V, average at 24°C, 0.98 V average at 0°C.

Air-depolarized cells are now available in sizes up to 2500 A-h. The 2500 A-h cell delivers 199 kWh/m³ and 0.163 kWh/kg when discharged at low rates.

Weston standard cell. The Weston cell of 1893 has become the accepted standard of electromotive force (emf). The Weston normal or saturated cadmium cell has an emf of 1.01864 absolute volts at 20°C. When purified materials are used, cells having the same emf to within a few microvolts may be made. These cells maintain their emf very well. Reference standards, in daily use for many years, are remarkably constant. These standard cells are made with materials of spectroscopic purity and are maintained under diffuse light in a thermostatically controlled oil bath in a room maintained at 25°C and 60% relative humidity.

The cell uses a two-phase amalgam of cadmium as the anode. For a 10% amalgam, one part of cadmium by weight and nine parts of mercury are required. These materials may be combined either by heating them together or by electrolytic deposition of cadmium into mercury. At ordinary temperatures a liquid phase is in equilibrium with a solid phase. This gives a very stable potential which depends only on the temperature.

The cathode is mercurous sulfate, Hg_2SO_4, in contact with mercury.

The electrolyte is a saturated solution of cadmium sulfate in equilibrium with the solid phase $CdSO_4 \cdot \frac{8}{3}H_2O$. In some cells, sulfuric acid is added in order to prevent the hydrolysis of mercurous sulfate.

The cell is usually made of glass in the form of an H, as in Fig. 3. Platinum-wire leads are sealed in the base of each arm. Mercury, carefully purified, is placed at the bottom of one arm and a 10% cadmium amalgam is placed, while warm and in a single phase, in the other arm. When the amalgam has cooled and separated into two phases, crystals of mercurous sulfate are placed above the mercury and crystals of cadmium sulfate are placed above the amalgam. A saturated solution of cadmium sulfate is then added to about 2–3 mm above the crossbar, and the cell is hermetically sealed.

The best-saturated cells of this type may be measured to the ten-millionth part of a volt at specified temperatures which must be known to within 0.01°C.

The saturated Weston cell has a relatively large temperature coefficient of emf. For portable use, it is general practice to use a cell with an unsaturated electrolyte. This has a temperature coefficient which is only one-fourth as great as that of the saturated cell.

Standard cells are not intended as power sources. They should be used only for comparison of voltages. Ordinary voltmeters put too heavy a drain on the cell for any reliable voltage measurements.

Organic electrolyte cell. A different class of cells is that based on the use of particularly reac-

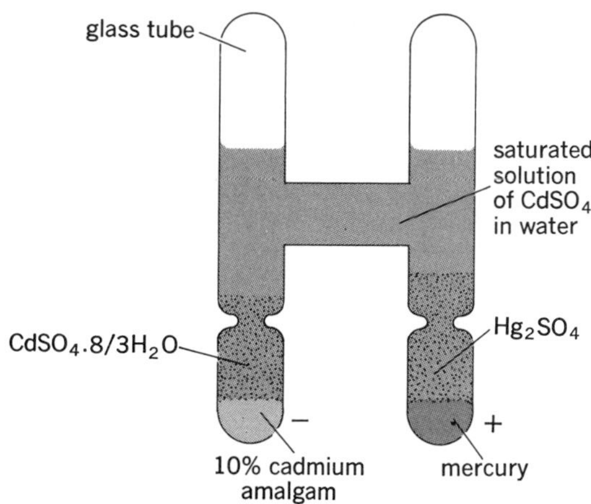

Fig. 3. Schematic of Weston saturated standard cell.

tive metals (Li, Ca, Mg) in conjunction with organic electrolytes. The best-known type in this class is the lithium-cupric fluoride cell, theoretically capable of delivering over 0.15 kWh/kg, more than three times the capacity of the highly ranked Zn-AgO cell.

The lithium anode is usually made in sheet form, but variants using lithium powder trapped in appropriate grids are also known.

The cathode is a mixture of CuF_2 and various conductive materials, most often graphite and carbon black either together or separately, in order to obtain the electronic conductivity that CuF_2 does not possess.

The electrolyte considered most compatible with these electrodes is lithium perchlorate (1 *M*) in propylene carbonate or butyrolactone.

The cell works with 80% cathodic current efficiency, delivering 25% of its energy to the 2-V end point, the initial voltage being 3.2 V. This represents 0.176 kWh/kg, that is, 0.704 kWh/kg available energy. On the other hand the current density is only 0.5–3 mA/cm², too little except for special applications. Nonetheless, owing to their considerable thermodynamic energy density, the organic electrolyte cells remain the great hope for the future.

For the present, the most favored anode material is lithium which, besides providing fully reversible anodic reaction, possesses the exclusive properties of a low specific gravity (0.534) and a high electrode potential (3.045 V). In fact after beryllium, lithium has the highest energy availability per atom and is by far the lightest metal known.

The cathode problem is still under investigation, and at least two other salts, CoF_3 and $NiCl_2$, are claimed to be superior, working with cathodic current efficiencies close to 100%.

The critical problem remains the electrolyte because of the difficulty of combining high specific conductivity, low viscosity, hydrophobicity, chemical inertness toward electrodes, and adequate electrolytic properties. The conditions of high specific conductivity (2.85×10^{-2} ohm^{-1} cm^{-1}) and low viscosity (0.35 centipoise) are fulfilled by acetonitrile. Unfortunately acetonitrile reacts powerfully with lithium. The compromise solution at this stage is propylene carbonate, 8 times more viscous and 30 times less conductive than acetonitrile but chemically less aggressive.

Another problem in which progress is expected is the construction of the electrodes. The techniques presently used, sintering in a grid or dispersing the cathode powder in a porous structure under the protection of filter paper or some ion-exchange membrane, seem to be only a step in an evolutionary process. Finally, the short shelf-life, due to uncontrolled ionic circulation and dendrite formation, is a problem that must be improved before organic electrolyte cells enter the industrial production stage.

[KENNETH FRANZESE; JACK DAVIS]

Bibliography: D. P. Boden, Electrolytes for nonaqueous batteries, *Proceedings of the 20th Annual Power Sources Conferences*, p. 63, 1966; K. H. M. Braeuer, Organic electrolyte, high energy density batteries, *Proceedings of the 20th Annual Power Sources Conferences*, p. 57, 1966; W. E. Elliot et al., Active metal anode-electrolyte systems, *Proceedings of the 20th Annual Power Sources Conferences*, p. 67, 1966; G. W. Heise, E. A. Schumacher, and C. R. Fisher, The air-depolarized primary cell with caustic alkali electrolyte II, *Trans. Electrochem. Soc.*, 92:173, 1947; G. W. Vinal, *Primary Batteries*, 1950.

Wind power

Kinetic energy in the Earth's atmosphere used to perform useful work. Total atmospheric wind power is of the order of 10^{14} kW. Annual kinetic energy is of the order of 10^{17} kWh (1 kWh = 3.60×10^6J). Practical land-based wind generators could extract as much as 10^{14} kWh of energy per year worldwide. Energy, and thus productivity of winds, varies markedly with geographic location. Annual energy available to a conversion machine at a site is very reproducible (±15% variability). Annual average power per square meter of vertical area at a 10-m height across and over water near the United States is estimated in Fig. 1. Southern Wyoming shows the greatest chance for productive wind power systems on land (>400 W/m²), and the edge of the shelf off New England shows >800 W/m².

Wind power has lifted water for centuries, but there is new interest in wind-powered irrigation in the United States. Electricity was generated by wind power in 1880. Thousands of small wind generators and water pumps worked in the United States as late as 1940. Subsidized rural electrification, low fossil-fuel prices, and desire for more powerful farm machines caused near extinction of wind power systems in this country. In other nations, simple wind power systems are still the key to material sufficiency. The exponential energy appetite of the industrialized world consuming finite resources and the exponential growth in pollution associated with energy production has renewed interest in wind power.

Water lifting. Water spilled on agricultural soil by a random wind is still the best example of a storage subsystem buffering between the wind and the energy consumer. Most large-scale irrigation pumping in the United States today consumes natural gas, but wind power pumping could come back. Wind-powered pump-back could expand the capacity of hundreds of smaller hydroelectric installations.

Heating. At least 20% of United States energy is consumed in heating buildings. In colder climates,

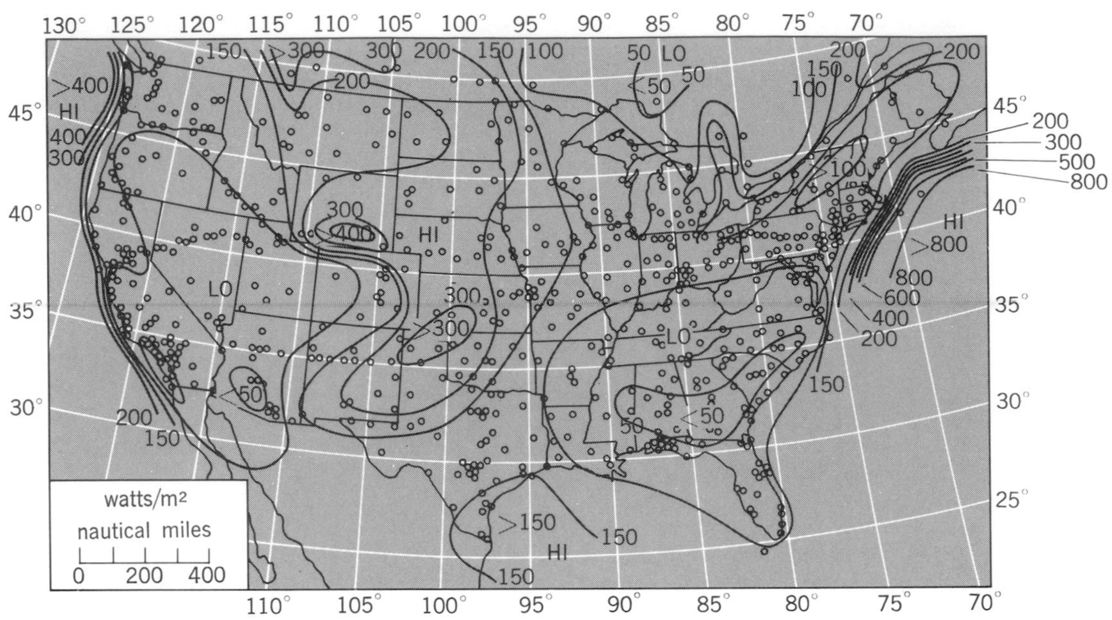

Fig. 1. Available wind power—annual average.

heating demand frequently matches high winds. Wind-generated electricity feeding thermal storage units, in some instances combined with solar thermal collectors feeding adjacent storage units, offers excellent potential for reducing fuel oil consumption. Large-scale wind-generator systems located off the Atlantic coast could take over a large fraction of building heating load in the largest urban areas.

Wind-generated electricity. Complete sets of hardware combining wind generators and electric storage batteries large enough to supply 500–1000 kWh/month electricity on demand can be purchased. Where the generator can be placed at moderate height into a strong and persistent wind regime, delivered electricity is economically competitive. In milder wind regimes, electricity is still expensive compared with most utility prices. Larger centralized wind electricity systems sharing larger storage subsystems have a greater chance of being economic. When winds produce more electricity than the market demands, the energy must be stored, then recalled and used when demand exceeds windpower. Wind generators in large numbers located in productive winds and equipped with hydrogen generation, storage, and reconversion devices have been proposed. It has been estimated that winds available to the United States could generate as much as 2×10^{12} kWh of firm power on demand, equivalent to the total 1975 United States electricity consumption. *See* ELECTRIC POWER GENERATION; ELECTRICAL UTILITY INDUSTRY; STORAGE BATTERY.

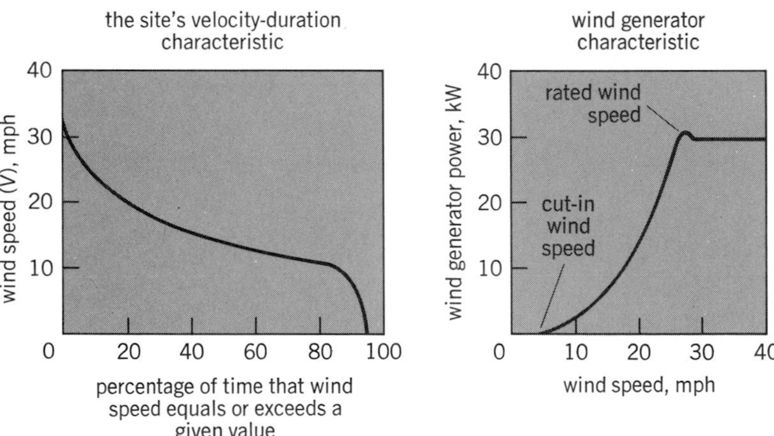

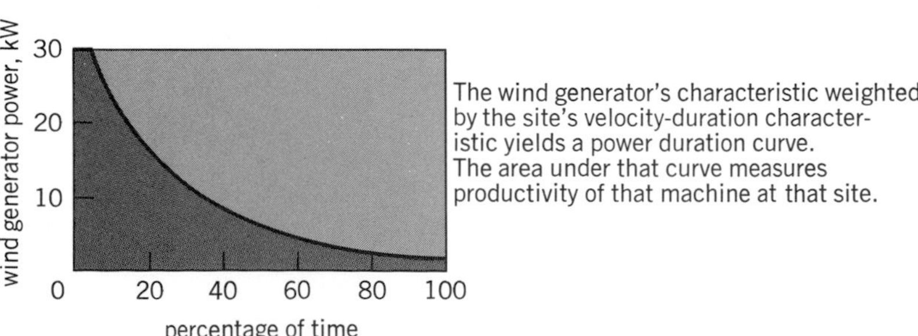

Fig. 2. Productivity of a wind generator is a function of wind regime, height, the machine's aero-mechanical-electrical characteristics, and the delivered product.

Mechanics. A windmill is a rotating machine capable of interchanging (extracting) momentum with particles of air mass that flow through its swept area. Power available in the wind in a swept area varies with the size of that area, the density of the air, and the square of the velocity. Energy extractable from an oncoming wind stream over a period of time varies as the size of swept area, density, and cube of the velocity, as described in Eq. (1), where K.E. is the kinetic energy available,

$$\text{K.E.} = kDV^3 \quad (1)$$

k is the density of air, D is the sweeping blades' diameter, and V is the average wind speed. At sea level, K.E. $= 0.000935\ D^2V^3$ lb-ft/s, corresponding to $1.7 \times 10^{-6}\ D^2V^3$ horsepower, for D in feet, and V in feet per second. Adolph Betz showed that no more than 59% of the energy in an oncoming stream tube of wind could be extracted without bypassing the momentum exchanger. The ability of any windmill to approach that 59% maximum extraction capability is thus an excellent indicator of its performance, and has been named the coefficient of performance, C_p. Thus, the realizable power P_R from a windmill is as shown in Eq. (2).

$$P_R = C_p \times 1.7 \times 10^{-6} D^2 V^3 \text{ horsepower} \quad (2)$$

A C_p as high as 0.48 has been observed for a modern high-speed propeller-type wind machine, whereas the sheet-metal-bladed American Fan Mill for water pumping seldom achieves a C_p in excess of 0.30. Coefficient of performance is related to the aerodynamic features of a machine, particularly to the tip speed ratio, defined by Eq. (3).

$$\text{Tip speed ratio} = \frac{\text{tangential speed of blade tip}}{\text{speed of oncoming wind}} \quad (3)$$

Different types of wind machines work best at their own optimum tip speed ratio. The American Fan Mill wants a tip speed ratio near 1; the very-high-speed twisted and tapered two- or three-bladed propeller type wants a ratio between 7 and 12. Vertical-axis machines of the S-rotor (Savonius) type, cross-flow vertical-bladed machines, and the Darrieus twirling rope (troposkien) rotor operate best at a ratio between 1 and 2. It has been suggested that any configuration of horizontal-axis machine can be given characteristics that will permit it to have a high C_p, but economics and other practical considerations seem to favor the three-bladed twisted and tapered propeller type operating at a tip speed ratio of about 8. Advocates of the vertical-axis Darrieus type hope to prove that type most economic, however.

System and economic considerations. The utility and competitiveness of a wind power system depend upon the wind regime and height at which windmills are placed, size and characteristics of the machine, nature of delivered product, and productivity of that product. Figure 2 shows how a wind regime can be characterized by a velocity duration curve (for any specific time period) and the overall power out versus wind speed of the wind machine (in this case a generator of electricity). If the desired product is simply raw electrical energy, the product of delivered voltage times delivered amperage suffices. Figure 2 shows how a power duration curve measures that productivity. Wind systems delivering other products would be assessed differently, but with the same philosophy. Wind systems can thus be compared with one another and with other systems capable of delivering the same product. *See* ENERGY SOURCES.

[WILLIAM HERONEMUS]

Bibliography: A. Betz, *Windmills in the Light of Modern Research*, Nat. Advis. Comm. Aeronaut. Nat. Mem. no. 474 (available from NTIS), 1928; Department of Energy, *Federal Wind Energy Program: Program Summary*, 1979; F. R. Eldridge (ed.), *Proceedings of the 2d Workshop on Wind Energy Conversion Systems*, Washington, DC, 1975; H. Glauert, Windmills and fans, in W. F. Durand (ed.), *Aerodynamic Theory*, vol. 4, 1935; E. W. Golding, *Generation of Electricity by Windpower*, 2d ed., 1976; W. E. Heronemus, Pollution free energy from offshore winds, in *Preprints of the 8th Annual Conference and Exposition, Marine Technology Society*, Washington, DC, September, 1972; D. R. Inglis, *Wind Power and Other Energy Options*, 1978; P. C. Putnam, *Power from the Wind*, 1948.

Work

In physics, the term work refers to the transference of energy that occurs when a force is applied to a body that is moving in such a way that the force has a component in the direction of the body's motion. Thus work is done on a weight that is being lifted, or on a spring that is being stretched or compressed, or on a gas that is undergoing compression in a cylinder.

When the force acting on a moving body is constant in magnitude and direction, the amount of work done is defined as the product of just two factors: the component of the force in the direction of motion, and the distance moved by the point of application of the force. Thus the defining equation for work W is Eq. (1), where f and s are the

$$W = f \cos \phi \cdot s \quad (1)$$

magnitudes of the force and displacement, respectively, and ϕ is the angle between these two vector quantities (Fig. 1). Because $f \cos \phi \cdot s = f \cdot s \cos \phi$, work may be defined alternatively as the product of the force and the component of the displacement in the direction of the force. In Fig. 2 the work of the constant force f when the application point moves along the curved path from P to P', and therefore undergoes the displacement $\overline{PP'}$, is $f \cdot \overline{PP'} \cos \phi$, or $f' \overline{PE}$.

Work is a scalar quantity. Consequently, to find the total work done on a moving body by several different forces, the work of each may be computed separately and the ordinary algebraic sum taken.

Examples and sign conventions. Suppose that a car slowly rolls forward a distance of 30 ft along a straight driveway while a man pushes on it with a constant magnitude of 50 pounds of force (50 lbf) and let Eq. (1) be used to compute the work W done under each of the following circumstances: (1) If the man pushes straight forward, in the direction of the car's displacement, then $\phi = 0°$, $\cos \phi = 1$, and $W = 50 \text{ lbf} \times 1 \times 30 \text{ ft} = 1500$ foot-pounds of force (ft-lbf); (2) if he pushes in a sideways direction making an angle ϕ of 60° with the displacement, then $\cos 60° = 0.50$ and $W = 750$ ft-lbf; (3) if he pushes against the side of the car and therefore at right angles to the displacement, $\phi = 90°$, $\cos \phi = 0$, and $W = 0$; (4) if he pushes or pulls backward, in the direction opposite to the car's displacement, $\phi = 180°$, $\cos \phi = -1$, and $W = -1500$ ft-lbf.

Notice that the work done is positive in sign whenever the force or any component of it is in the same direction as the displacement; one then says that work is being done *by* the agent exerting the

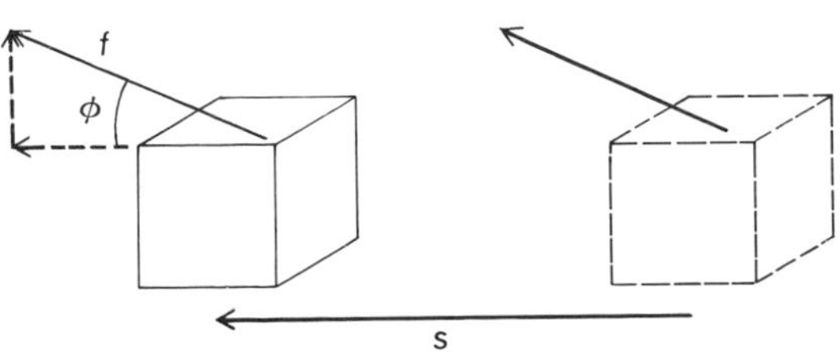

Fig. 1. Work of constant force f is $fs \cos \phi$.

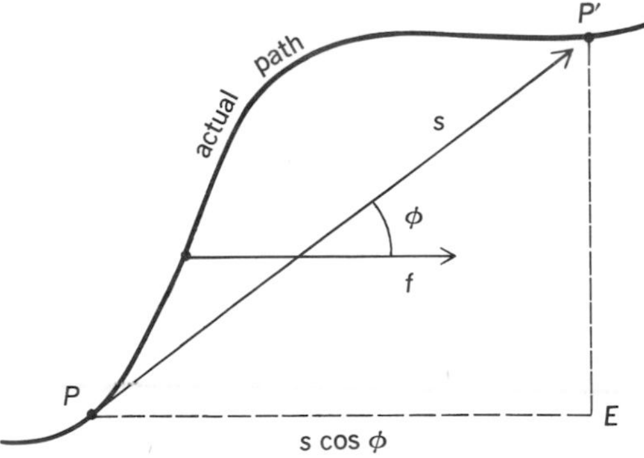

Fig. 2. The work done in traversing any path connecting points P and P' is $f \cdot \overline{PE}$, assuming the force f to be constant in magnitude and direction.

force (in the example, the man) and *on* the moving body (the car). The work is said to be negative whenever the direction of the force or force component is opposite to that of the displacement; then work is said to be done *on* the agent (the man) and *by* the moving body (the car). From the point of view of energy, an agent doing positive work is losing energy to the body on which the work is done, and one doing negative work is gaining energy from that body.

Units of work and energy. These consist of the product of any force unit and any distance unit. Units in common use are the foot-pound, the foot-poundal, the erg, and the joule. The product of any power unit and any time unit is also a unit of work or energy. Thus the horsepower-hour (hp-hr) is equivalent, in view of the definition of the horsepower, to 550 ft-lbf/sec × 3600 sec, or 1,980,000 ft-lbf. Similarly, the watt-hour is 1 joule/sec × 3600 sec, or 3600 joule; and the kilowatt-hour is 3,600,000 joule. *See* HORSEPOWER.

Work of a torque. When a body which is mounted on a fixed axis is acted upon by a constant torque of magnitude τ and turns through an angle θ (radians), the work done by the torque is $\tau\theta$.

Work principle. This principle, which is a generalization from experiments on many types of machines, asserts that, during any given time, the work of the forces applied to the machine is equal to the work of the forces resisting the motion of the machine, whether these resisting forces arise from gravity, friction, molecular interactions, or inertia. When the resisting force is gravity, the work of this force is mgh, where mg is the weight of the body and h is the vertical distance through which the body's center of gravity is raised. Note that if a body is moving in a horizontal direction, h is zero and no work is done by or against the gravitational force of the Earth. If a person holds an object or carries it across level ground, he does no net work against gravity; yet he becomes fatigued because his tensed muscles continually contract and relax in minute motions, and in walking he alternately raises and lowers the object and himself.

The resisting force may be due to molecular forces, as when a coiled elastic spring is being compressed or stretched. From Hooke's law, the average resisting force in the spring is $-\frac{1}{2}ks$, where k is the force constant of the spring and s is the displacement of the end of the spring from its normal position; hence the work of this elastic force is $-\frac{1}{2}ks^2$.

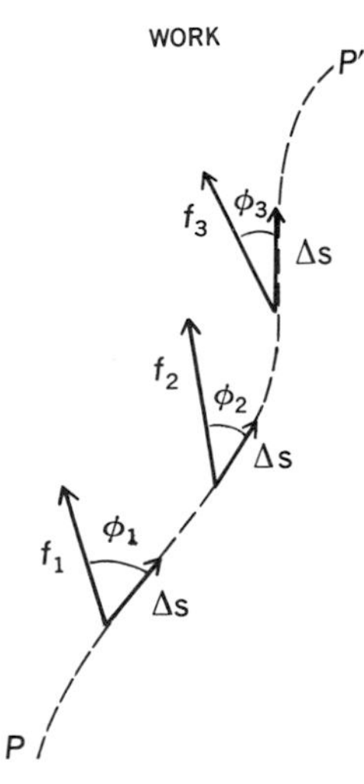

Fig. 3. Work done by a variable force.

If a machine has any part of mass m that is undergoing an acceleration of magnitude a, the resisting force $-ma$ which the part offers because of its inertia involves work that must be taken into account; the same principle applies to the resisting torque $-I\alpha$ if any rotating part of moment of inertia I undergoes an angular acceleration α.

When the resisting force arises from friction between solid surfaces, the work of the frictional force is $-\mu f_n s$, where μ is the coefficient of friction for the pair of surfaces, f_n is the normal force pressing the two surfaces together, and s is the displacement of the one surface relative to the other during the time under consideration. The frictional force μf_n and the displacement s giving rise to it are always opposite in direction ($\phi = 180°$).

The work done by any conservative force, such as a gravitational, elastic, or electrostatic force, during a displacement of a body from one point to another has the important property of being path-independent: Its value depends only on the initial and final positions of the body, not upon the path traversed between these two positions. On the other hand, the work done by any nonconservative force, such as friction due to air, depends on the path followed and not alone on the initial and final positions, for the direction of such a force varies with the path, being at every point of the path tangential to it.

Since work is a measure of energy transfer, it can be calculated from gains and losses of energy. It is useful, however, to define work in terms of forces and distances or torques and angles because these quantities are often easier to measure than energy changes, especially if energy changes are produced by nonconservative forces.

Work of a variable force. If the force varies in magnitude and direction along the path $\overline{PP'}$ of its point of application, one must first divide the whole path into parts of length Δs, each so short that the force component $f \cos \phi$ may be regarded as constant while the point of application traverses it (Fig. 3). Equation (1) can then be applied to each small part and the resulting increments of work added to find the total work done. Various devices are available for measuring the force component as a function of position along the path. Then a work diagram can be plotted (Fig. 4). The total work done between positions s_1 and s_2 is represent-

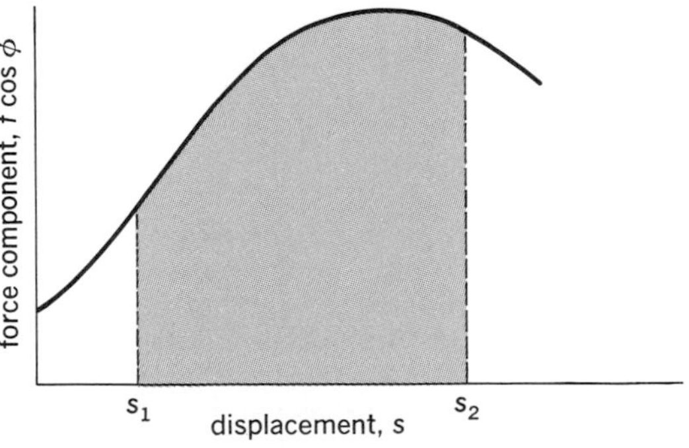

Fig. 4. A work diagram.

ed by the area under the resulting curve between s_1 and s_2 and can be computed by measuring this area, due allowance being made for the scale in which the diagram is drawn.

For an infinitely small displacement ds of the point of application of the force, the increment of work dW is given by Eq. (2), a differential expres-

$$dW = f \cos \phi \, ds \qquad (2)$$

sion that provides the most general definition of the concept of work. In the language of vector analysis, dW is the scalar product of the vector quantities $\mathbf{f}$ and ds; Eq. (2) then takes the form $dW = \mathbf{f} \cdot ds$. If the force is a known continuous function of the displacement, the total work done in a finite displacement from point P to point P' of the path is obtained by evaluating the line integral in Eq. (3).

$$W = \int_P^{P'} f \cos \phi \, ds = \int_P^{P'} \mathbf{f} \cdot ds \qquad (3)$$

When a variable torque of magnitude τ acts on a body mounted on a fixed axis, the work done is given by $W = \int_{\theta_1}^{\theta_2} \tau \, d\theta$, where $\theta_2 - \theta_1$ is the total angular displacement expressed in radians. *See* ENERGY.

[LEO NEDELSKY]

Bibliography: R. Resnick and D. Halliday, *Physics*, 3d ed., 1977; E. M. Rogers, *Physics for the Inquiring Mind*, 1960.

List of Contributors

List of Contributors

A

Abrahams, Elihu. *Department of Physics, Rutgers University.* MAGNETISM (coauthored).

Adkins, George G. *Chief, Department of River Basins, Bureau of Power, Federal Power Commission.* TIDAL POWER.

Allen, Eleanor. *Society of Automotive Engineers, Inc., New York.* TURBINE PROPULSION.

Altman, Dr. David. *Division Vice President and Technical Director, United Technology Center, Sunnyvale, CA.* METAL-BASE FUEL.

Anderson, J. *Manager, Meter and Instrument Business Department, General Electric Company, Somersworth, NH.* WATT-HOUR METER (coauthored).

Antonides, Lloyd E. *Development Engineer, Fenix and Scisson, Inc., Tulsa.* MINING MACHINERY.

Apker, Dr. L. *General Electric Research Laboratory, Schenectady.* PHOTOVOLTAIC EFFECT.

Applegate, Charles E. *Consulting Engineer, Weston, MA.* ELECTRICAL RESISTANCE.

Archie, G. E. *Manager (retired), Petroleum Engineering Research, Shell Development Company, Houston.* PETROLEUM PROSPECTING (in part).

Arnold, Dr. Richard C. *High Energy Physics Division, Argonnne National Laboratory.* PARTICLE-BEAM FUSION (in part).

Aron, Dr. Walter. *Science Applications, Inc., Palo Alto, CA.* ELECTRICITY.

Ash, Col. Simon H. *Consulting Civil and Mining Engineer, Santa Rosa, CA.* MINING (in part).

B

Baily, Frederick G. *Large Steam Turbine-Generator Department, General Electric Company, Schenectady.* STEAM TURBINE; TURBINE.

Barna, Dr. Gabriel G. *Texas Instruments Inc., Central Research Laboratories, Dallas.* SOLID-STATE BATTERY.

Bass, Prof. Jack. *Department of Physics, Michigan State University.* THERMOELECTRICITY.

Batchelder, Howard R. *Consulting Chemical Engineer, Battelle Memorial Institute, Columbus, OH.* FOSSIL FUEL.

Baumeister, Theodore. *Deceased; formerly, Consulting Editor, and Stevens Professor Emeritus of Mechanical Engineering, Columbia University.* BOILER; ENERGY CONVERSION; HEAT PUMP (in part); POWER PLANT (coauthored); PRIME MOVER; STEAM-GENERATING UNIT; WATERPOWER (in part). Articles on thermodynamic cycles, engines, and turbines.

Bean, Howard S. *Consultant on Fluid Metering, Liquids and Gases, Sedona, AZ.* THERMOCOUPLE.

Beatty, R. L. *Oak Ridge National Laboratory.* NUCLEAR FUELS.

Beaty, H. Wayne. *"Electrical World," McGraw-Hill Publishing Company, New York.* CONDUCTOR (ELECTRICITY).

Beggs, H. L. *Consultant, Fluid Processing Division, Selas Corporation of America, Bresher, PA.* Validator for FURNACE CONSTRUCTION.

Bennett, Dr. Orus L. *Supervisory Soil Scientist, Science and Education Administration, U.S. Department of Agriculture, and Department of Plant Science, West Virginia University.* STRIP MINING.

Bewley, Dr. Loyal V. *Dean (retired), College of Engineering, Lehigh University.* HYSTERESIS MOTOR; RELUCTANCE MOTOR; SYNCHRONOUS MOTOR.

Billington, Dr. Douglas S. *Senior Staff Advisor for Materials Science, Metals and Ceramics Division, Oak Ridge National Laboratory.* RADIATION DAMAGE TO MATERIALS.

Binning, Dr. Robert C. *Associate Director, Dayton Laboratory, Monsanto Research Corporation.* HEATING VALUE (coauthored).

Blackburn, J. L. *Consulting Engineer, Bothell, WA.* RELAY.

Blanton, J. W. *General Manager, Advanced Component Technology Department, General Electric Company, Cincinnati.* JET PROPULSION.

Bloomfield, Prof. Philip E. *Department of Physics, University of Pennsylvania, and City College, City University of New York.* THERMODYNAMIC PROCESSES (coauthored).

Bolt, Prof. Jay A. *Department of Mechanical Engineering, University of Michigan.* Validator of GAS TURBINE.

Breuer, Glenn D. *Manager, HVDC Transmission Engineering, General Electric Company, Schenectady.* LIGHTNING AND SURGE PROTECTION.

Brewer, Dr. Charles P. *Shell Development Corporation, Emeryville, CA.* HYDROCRACKING.

Brigham, Harry S. *Executive Vice President, Dixilyn Corporation, New Orleans.* OIL AND GAS WELL COMPLETION.

Broadbent, Carl D. *Manager, Mining Engineering, Kennecott Copper Corporation, Salt Lake City.* OPEN-PIT MINING.

Bromley, Dr. D. Allan. *Professor, Department of Physics, and Director, Wright Nuclear Structure Laboratory, Yale University.* NUCLEAR PHYSICS.

Brown, Ronald D. *Stanley Consultants, Muscatine, IA.* RURAL ELECTRIFICATION (coauthored).

Brzustowski, T. A. *Department of Mechanical Engineering, University of Waterloo, Ontario.* COMBUSTION OF LIGHT METALS.

Burnett, Peter G. *Petroleum Engineer, Chicago.* OIL AND GAS STORAGE.

C

Campbell, Harold E. *Senior Engineer, Power Distribution Systems Engineering, General Electric Company, Schenectady.* ELECTRIC DISTRIBUTION SYSTEMS.

Casten, Thomas R. *General Manager, Cummins Cogeneration, Cummins Corporation, Columbus, IN.* COGENERATION.

Cheatham, Dr. J. B., Jr. *Professor of Mechanical Engineering and Chairman of Department of Mechanical and Aerospace Engineering and Materials Science, Rice University.* OIL AND GAS WELL DRILLING.

Chenault, Roy L. *Chief Research Engineer (retired), Oilwell Division, U.S. Steel Corporation.* OIL AND GAS FIELD EXPLOITATION.

Childs, Dr. Orlo E. *Texas Technical University.* LIQUID FUEL.

Chinitz, Dr. Wallace. *Cooper Union for the Advancement of Science and Art, School of Engineering, New York.* CHEMICAL FUEL; ROTARY ENGINE.

Clayton, James M. *Advisory Engineer, Advanced Systems Technology, Westinghouse Electric Corporation, Pittsburgh.* ELECTRIC PROTECTIVE DEVICES.

Cleveland, Prof. Laurence Fuller. *Department of Electri-*

cal Engineering, Northeastern University. DIRECT-CURRENT MOTOR.

Cochran, Robert A. *Shell Development Company, Bellaire Research Center, Houston.* WATER POLLUTION (in part).

Conrad, Prof. Albert G. *Dean Emeritus and Professor of Electrical Engineering, College of Engineering, University of California, Santa Barbara.* INDUCTION MOTOR.

Cooper, Franklin D. *Bureau of Mines, U.S. Department of the Interior.* COAL LIQUEFACTION.

Corlett, A. V. *Mining Engineer, Kingston, Ontario.* MINING (in part); UNDERGROUND MINING.

Coroniti, S. C. *Climatic Impact Assessment Program, Office of the Secretary of the Treasury.* AIR POLLUTION (coauthored).

Crittenden, Dr. Charles V. *Geographer, Economic Development Administration, U.S. Department of Commerce.* GAS FIELD AND GAS WELL.

D

Davis, Dr. Jack. *Vice President, Research and Development, Bright Star Industries, Inc., Clifton, NJ.* DRY CELL; FUEL CELL; LITHIUM PRIMARY CELL; WET CELL (all coauthored). Articles on types of batteries.

Dean, H. Clark. *Harza Engineering Company, Consulting Engineers, Chicago.* HYDROELECTRIC POWER (coauthored).

Dempsey, Dr. Jerry E. *President, Borg-Warner Corporation, Chicago.* HEAT PUMP (in part).

Deutscher, Prof. Guy. *Department of Physics, Tel Aviv University.* SUPERCONDUCTIVITY.

Doscher, Dr. Todd. *Department of Petroleum Engineering, University of Southern California, Los Angeles.* EXPLORING ENERGY CHOICES (feature); GASOHOL; PETROLEUM ENGINEERING; PETROLEUM ENHANCED RECOVERY; PETROLEUM RESERVES; PETROLEUM RESERVOIR ENGINEERING.

Duckworth, Dr. Henry E. *Department of Physics, University of Manitoba.* ATOMIC NUCLEUS; NUCLEAR BINDING ENERGY (in part).

Dukek, W. G. *Exxon Research and Engineering Company, Linden, NJ.* JET FUEL; PETROLEUM PRODUCTS.

E

Eckels, Dr. Arthur R. *Department of Electrical Engineering, North Carolina State University.* DYNAMO; MOTOR-GENERATOR SET.

Ehricke, Dr. Krafft A. *President, Space Global, La Jolla, CA.* SOLAR ENERGY.

Ellis, Dr. A. J. *Director, Chemistry Division, Department of Scientific and Industrial Research, Petone, New Zealand.* GEOTHERMAL POWER (in part).

Evans, Dr. John W. *Director, Sacramento Peak Observatory, Air Force Cambridge Research Laboratories, Sunspot, NM.* SOLAR RADIATION.

Everetts, Dr. John, Jr. *Professor of Architectural Engineering, Pennsylvania State University.* DEHUMIDIFIER.

F

Faw, Prof. Richard E. *Head, Department of Nuclear Engineering, Kansas State University.* RADIATION SHIELDING.

Ferenz, Joseph A. *Senior Vice President, Frederic R. Harris, Inc., New York.* SUPERPORT.

Field, Joseph H. *Benfield Corporation, Pittsburgh.* FISCHER-TROPSCH PROCESS.

Fisher, Dr. John C. *Research and Development Center, General Electric Company, Schenectady.* ENERGY CONSUMPTION (feature); ENERGY FLOW.

Foss, F. E. *Evans Energy Company, Denver.* PETROLEUM GEOLOGY.

Foster, Prof. Mark G. *Department of Electrical Engineering, University of Virginia.* MAGNETIC CIRCUITS.

Francis, Arthur W. *Linde Division, Union Carbide Corporation, New York.* LIQUEFIED NATURAL GAS (LNG).

Franzese, Dr. Kenneth. *Research and Development, Bright Star Industries, Inc., Clifton, NJ.* DRY CELL; LITHIUM PRIMARY CELL (both coauthored).

Fremed, Raymond F. *Burson-Marsteller Associates, New York.* HEAT EXCHANGER.

Fuhrman, Dr. Ralph E. *Black and Veatch, Consulting Engineers, Washington, DC.* WATER POLLUTION (in part).

G

Gambs, Gerard C. *Vice President, Ford, Bacon, and Davis, Inc., New York.* ENERGY SOURCES.

Glasscock, Dwight L. *Vice President, Harza Engineering Company, Chicago.* PUMPED STORAGE.

Goodheart, Prof. Clarence F. *Chairman, Department of Electrical Engineering, Union College, Schenectady.* CIRCUIT.

Gray, G. Ronald. *Director, Syncrude Canada Ltd., Edmonton, Alberta.* OIL MINING (in part); OIL SAND.

Greensfelder, Dr. Bernard S. *Deceased; formerly, Director of Oil Research, Shell Development Company, Emeryville, CA.* CRACKING.

Gregory, Dr. G. Robinson. *Department of Forestry, School of Natural Resources, University of Michigan.* FOREST RESOURCES.

Grey, Jerry. *President, Greyad Corporation, Princeton.* PROPULSION.

Griffin, Dr. Owen M. *Ocean Technology Division, Naval Research Laboratory, Washington, DC.* OCEAN THERMAL ENERGY CONVERSION.

Grobecker, Dr. A. J. *Project Manager, Climatic Impact Assessment Program, Office of the Secretary of Transportation.* AIR POLLUTION (coauthored).

H

Haensel, Dr. Vladimir. *Vice President and Director of Research, Universal Oil Products Company, Des Plaines, IL.* CATALYTIC REFORMING (coauthored).

Halbouty, Dr. Michel T. *Consulting Geologist and Petroleum Engineer, Houston.* NATURAL GAS (in part).

Harris, E. J. *Assistant Chief, Health and Safety Technical Support Center, U.S. Bureau of Mines, Pittsburgh.* MINING (in part).

Hartwig, Dr. William H. *Department of Electrical Engineering, University of Texas, Austin.* ELECTRIC ENERGY MEASUREMENT; ELECTRIC POWER MEASUREMENT.

Harza, Richard D. *President, Harza Engineering Company, Chicago.* HYDROELECTRIC POWER (coauthored).

Hayes, William C. *Editor in Chief, "Electrical World," McGraw-Hill Publications Company, New York.* ELECTRIC POWER SYSTEMS; ELECTRICAL UTILITY INDUSTRY.

Hazen, Ronald. *Consultant, Indianapolis.* AIRCRAFT ENGINE; RECIPROCATING AIRCRAFT ENGINE.

Hedberg, Dr. Hollis D. *Department of Geology, Princeton University.* PETROLEUM PROSPECTING (in part).

Heronemus, Prof. William. *Department of Engineering, University of Massachusetts.* WIND POWER.

Hewson, Dr. E. Wendell. *Chairman, Department of Atmospheric Sciences, Oregon State University.* AIR POLLUTION (coauthored).

Higinbotham, Dr. William A. *Deceased; formerly, Brookhaven National Laboratory, Upton, NY.* NUCLEAR MATERIALS SAFEGUARDS.

Hill, James E. *Assistant Director (retired), Mining Research, U.S. Bureau of Mines.* MINING EXCAVATION.

Hingorani, Dr. Narain G. *Electric Power Research Insti-*

tute, Palo Alto, CA. ELECTRIC POWER SUBSTATION.

Hirst, Eric. *Energy Division, Oak Ridge National Laboratory.* ENERGY CONSERVATION (feature).

Hockett, Richard S. *Research Physicist, Dayton Laboratory, Monsanto Research Corporation.* HEATING VALUE (coauthored).

Hoffman, Dr. Harold L. *Gulf Publishing Company, Houston.* PETROLEUM PROCESSING (coauthored).

Holm, Prof. Jens T. *Webb Institute of Naval Architecture, Glen Cove, NY.* MARINE ENGINE.

Hubbert, Dr. M. King. *Research Geophysicist, U.S. Department of the Interior Geological Survey, Reston, VA.* OUTLOOK FOR FUEL RESERVES (coauthored feature).

Huebler, Dr. Jack. *Senior Vice President, Institute of Gas Technology, IIT Center, Chicago.* FUEL GAS.

Huizenga, Dr. John R. *Nuclear Structure Research Laboratory, University of Rochester.* NUCLEAR FISSION.

Hynes, Lee P. *Consulting Engineer, Electrical and Mechanical Engineering, Haddonfield, NJ.* ELECTRIC HEATING.

I

Ingram, William T. *Consulting Engineer, Whitestone, NY.* AIR-POLLUTION CONTROL.

Inhaber, Dr. Herbert. *Atomic Energy Control Board, Ottawa.* RISK OF ENERGY PRODUCTION (feature).

Isbin, Prof. Herbert S. *Department of Chemical Engineering and Materials Science, University of Minnesota.* NUCLEAR POWER; NUCLEAR REACTOR.

J

Jackson, Dr. William D. *U.S. Energy Research and Development Administration.* MAGNETOHYDRODYNAMIC POWER GENERATOR.

Jaep, William F. *Central Research Department, Experimental Station, E. I. du Pont de Nemours and Company, Wilmington.* HEAT BALANCE; THERMODYNAMIC PRINCIPLES.

James, John W. *Vice President, Research, McDonnell and Miller, Inc., Chicago.* STEAM HEATING.

James, Robert S. *Pennsylvania State University.* MINING (in part).

Jaske, Robert T. *Technical Adviser to the Director, Office of State Programs, U.S. Nuclear Regulatory Commission.* WASTE HEAT MANAGEMENT.

Jensen, Dr. Howard B. *Research Supervisor, Laramie Energy Research Center, ERDA.* OIL SHALE (coauthored).

Jones, Prof. Lawrence W. *Harrison M. Randall Laboratory of Physics, University of Michigan.* HYDROGEN-FUELED TECHNOLOGY.

Just, Prof. Evan. *Department of Mining and Geology, Stanford University.* MINING (in part).

K

Kalhammer, Dr. Fritz. *Electric Power Research Institute, Palo Alto, CA.* ENERGY STORAGE (coauthored).

Kaplan, Robert A. *Vice President in Charge of Engineering, Automatic Burner Corporation, Chicago.* OIL BURNER.

Kappus, Peter G. *Flight Propulsion Laboratory, General Electric Company, Cincinnati.* TURBOFAN; TURBOPROP.

Kaufmann, R. H. *Kaufmann Engineering, Schenectady.* GROUNDING.

Keffer, Prof. Frederic. *Professor of Physics, University of Pittsburgh.* MAGNETISM (coauthored).

Kerr, Dr. William. *Chairman, Department of Nuclear Engineering, University of Michigan.* NUCLEAR ENGINEERING.

Kessler, George W. *Vice President, Engineering and Technology, Power Generation Division, Babcock and Wilcox Company, Barberton, OH.* BOILER AIR HEATER; BOILER ECONOMIZER; CYCLONE FURNACE; FIRE-TUBE BOILER; STEAM-GENERATING FURNACE; WATER-TUBE BOILER.

Kimbark, Dr. Edward W. *Consulting Engineer, Portland, OR.* DIRECT-CURRENT TRANSMISSION; TRANSMISSION LINES (in part).

Kingery, Donald S. *Director, Health and Safety, Research and Testing Center, U.S. Bureau of Mines.* MINING (in part).

Kinnard, Dr. Isaac F. *Deceased; formerly, Manager of Engineering, Instrument Department, General Electric Company, Lynn, MA.* ELECTRICAL MEASUREMENTS (in part).

Kirchmayer, Dr. L. K. *Advanced System Technology & Planning, Electric Utility Systems Engineering Department, General Electric Company, Schenectady.* ELECTRIC POWER SYSTEMS ENGINEERING.

Konzo, Prof. Seichi. *Department of Mechanical Engineering, University of Illinois.* WARM-AIR HEATING SYSTEM.

Koral, Richard L. *"Building Systems Design," New York.* AIR CONDITIONING; AIR COOLING; DISTRICT HEATING. Validator for PANEL HEATING AND COOLING; RADIANT HEATING.

Kosow, Dr. Irving L. *Staten Island Community College.* ALTERNATING-CURRENT MOTOR (in part); REPULSION MOTOR; UNIVERSAL MOTOR.

Kovar, Dr. Dennis G. *Argonne National Laboratories.* NUCLEAR RADIATION.

Kuo, C. H. *Department of Chemical Engineering, Mississippi State University.* NATURAL GAS AND SULFUR PRODUCTION.

Kuuskraa, Dr. Vello A. *Lewin & Associates, Washington, DC.* NATURAL GAS (in part).

L

Lane, James A. *Oak Ridge National Laboratory.* ATOMIC ENERGY.

Lane, Dr. John C. *Ethyl Corporation, Ferndale, MI.* OCTANE NUMBER.

Lee, Dr. Thomas H. *Strategic Planning Operation, General Electric Company, Fairfield, CT.* BLOWOUT COIL; BUS-BAR; CIRCUIT BREAKER.

Levin, Dr. Franklin K. *Esso Production Research Company, Houston.* SEISMIC EXPLORATION FOR OIL AND GAS.

Lewis, Dr. Bernard. *Combustion and Explosives Research, Inc., Pittsburgh.* COMBUSTION.

Luckenbach, Edward C. *Exxon Research and Engineering Company, Exxon Engineering—Petroleum Department, Florham Park, NJ.* Validator for CRACKING.

Luebbers, Dr. Ralph H. *Professor of Chemical Engineering, University of Missouri.* HEAT TRANSFER.

M

McClellan, Leslie N. *Consulting Editor, Engineering Consultants, Inc., Denver.* PIPELINE.

McClelland, Richard H. *Manager, SNG Project Planning, CNG Energy Company, Pittsburgh.* COAL GASIFICATION.

McCormack, Congressman Mike. *U.S. House of Representatives.* PROTECTING THE ENVIRONMENT (feature).

MacCoull, Neil. *Lecturer in Mechanical Engineering, Columbia University.* INTERNAL COMBUSTION ENGINE.

McGrain, Preston. *Assistant State Geologist, Kentucky Geological Survey, University of Kentucky.* OIL FIELD WATERS.

McKetta, Dr. John J. *Professor of Chemical Engineering, University of Texas.* PETROLEUM PROCESSING (coauthored).

McNish, Alvin G. *Chief, Meteorology Division, National Bureau of Standards, Chevy Chase, MD.* ELECTRICAL UNITS AND STANDARDS.

Manning, Dr. Kenneth V. *Professor Emeritus, Pennsylvania State University.* ELECTROMAGNETIC INDUCTION; ELECTROMAGNETISM; MAGNETIC FIELD; MAGNETIC INDUCTION; MAGNETOMOTIVE FORCE; RELUCTANCE.

Meijer, Dr. Roelof J. *Assistant Director of Research, Philips Research Laboratories, Eindhoven, Netherlands.* STIRLING ENGINE.

Mitchell, Dr. Kim W. *Solar Energy Research Institute, Golden, CO.* SOLAR CELL (in part).

Morgan, Dr. Karl Z. *Neely Professor of Nuclear Engineering, Georgia Institute of Technology, Atlanta.* HEALTH PHYSICS; MONITORING OF IONIZING RADIATION; RADIOACTIVE WASTE MANAGEMENT (in part).

Moser, J. F., Jr. *Formerly, Esso Research Laboratories, Humble Oil Refining Company.* COKING (PETROLEUM).

Muffler, Dr. L. J. Patrick. *Geologist, Branch of Field Geochemistry and Petrology, U.S. Geological Survey, Department of the Interior, Menlo Park, CA.* GEOTHERMAL POWER (in part).

N

Nedelsky, Prof. Leo. *Professor of Physical Science, University of Chicago.* Validator for CONSERVATION OF ENERGY; ENERGY; WORK.

Newsham, Dr. Robert S. *Petroleum Economist, Stanford Research Institute, Menlo Park, CA.* ALKYLATION (PETROLEUM).

Nottingham, Prof. W. B. *Deceased; formerly, Professor Emeritus of Physics, Massachusetts Institute of Technology.* THERMIONIC POWER GENERATOR.

O

Olmsted, Leonard M. *Technical Consultant, South Orange, NJ.* POWER PLANT (coauthored).

Ott, Paul A., Jr. *Automotive Engineering Consultant, Mount Clemens, MI.* AUTOMOTIVE ENGINE.

P

Pangborn, Dr. Jon B. *Director, Alternative Energy Systems Research, Institute of Gas Technology, Chicago.* AIRCRAFT FUEL.

Perch, Michael. *Koppers Company, Inc., Monroeville, PA.* COKE.

Perry, Stephen F. *Exxon Research, Florham Park, NJ.* DEWAXING OF PETROLEUM.

Phelan, Prof. Richard M. *Department of Mechanical Systems and Design, Cornell University.* MACHINE.

Pinkel, Benjamin. *Engineering Sciences Department, Rand Corporation, Santa Monica.* AIRCRAFT PROPULSION; PULSE JET; RAMJET; TURBOJET; TURBO-RAMJET.

Platt, Allison M. *Manager, Nuclear Waste Technology Department, Battelle Pacific Northwest Laboratories, Richland, WA.* RADIOACTIVE WASTE MANAGEMENT (in part).

Pollitzer, Dr. Ernest L. *Associate Director of Research, Corporate Research Center, Universal Oil Products Company, Des Plaines, IL.* CATALYTIC REFORMING (coauthored).

Ponikvar, Ade L. *Formerly, "Modern Plastics," McGraw-Hill Publications Company, New York.* OIL MINING (in part); ROTARY TOOL DRILL.

Pope, Michael. *Pope, Evans and Robbins, Consulting Engineers, New York.* FLUIDIZED-BED COMBUSTION.

Post, Dr. Richard F. *Lawrence Livermore Laboratory, University of California, Livermore.* LAWSON CRITERION; NUCLEAR FUSION; THERMONUCLEAR REACTION.

Price, Dr. Harvey S. *Vice President, Intercomp, Houston.* PETROLEUM RESERVOIR MODELS.

Priester, Gayle B. *Consulting Engineer, Baltimore.* COMFORT HEATING.

Pryke, John K. *Slocum and Fuller, New York.* CENTRAL HEATING AND COOLING.

Puchstein, Albert F. *Consulting Engineer, Columbus, OH.* ALTERNATING-CURRENT MOTOR (in part); MOTOR.

Putz, Dr. Thomas J. *Deceased; formerly, Westinghouse Electric Corporation, Philadelphia.* GAS TURBINE.

R

Rasmussen, Dr. Norman C. *Department of Nuclear Engineering, Massachusetts Institute of Technology.* CHAIN REACTION.

Rees, Jack. *Corporate Services Staff, Exxon Research and Engineering Company, Florham Park, NJ.* OIL ANALYSIS.

Reilly, Dr. James D. *Vice President, Consolidation Coal Company, Pittsburgh.* COAL MINING.

Remde, Dr. Harry F. *Chief, Basic Physics Research Section, Johns-Manville Research and Engineering Center, Manville, NJ.* HEAT INSULATION.

Riggs, Harold C. *Manager (retired), Marketing New Product Development, Electric Storage Battery Company, Philadelphia.* BATTERY.

Ritchey, Dr. H. W. *President, Thiokol Chemical Corporation, Ogden, UT.* PROPELLANT (in part).

Robb, Dr. D. D. *Consulting Engineer, Salina, KS.* DIRECT CURRENT.

Robbins, Nathanial, Jr. *Director of Engineering, Residential Division, Honeywell Inc., Minneapolis.* THERMOSTAT.

Rockett, Frank H. *Engineering Consultant, Charlottesville, VA.* ENGINE; OIL FURNACE.

Roddis, L. H., Jr. *President, Consolidated Edison Company of New York, Inc.* SHIP PROPULSION REACTOR.

Roe, Kenneth A. *President, Burns and Roe, Oradell, NJ.* POWER PLANT (coauthored).

Rogowski, Prof. Augustus R. *Department of Mechanical Engineering, Massachusetts Institute of Technology.* CETANE NUMBER.

Roller, Duane E. *Deceased; formerly, Harvey Mudd College.* CONSERVATION OF ENERGY; ENERGY.

Root, Dr. David H. *Office of Resource Analysis, U.S. Department of the Interior Geological Survey, Reston, VA.* OUTLOOK FOR FUEL RESERVES (coauthored feature).

Rosenberg, Leon T. *Senior Consultant, Generator Design, Power Generation, Installation and Service Department, Allis-Chalmers Manufacturing Company, Milwaukee.* ALTERNATING-CURRENT GENERATOR; ELECTRIC ROTATING MACHINERY; GENERATOR; HYDROELECTRIC GENERATOR.

Roth, Willard. *Engineer, Sunbeam Equipment Corporation, Meadville, PA.* RESISTANCE HEATING.

Rozeanu, Prof. L. *Department of Material Science, Technion, Israel Institute of Technology.* FUEL CELL; NUCLEAR BATTERY; WET CELL (all coauthored).

S

Scarlott, Charles A. *Manager of Publications Department (retired), Stanford Research Institute.* WATERPOWER (in part).

Schmerling, Dr. Louis. *Research Associate, Universal Oil*

Products Company, Des Plaines, IL. METHANE; PETROLEUM; PROPANE.

Schmidt, Dr. Paul W. *Department of Physics, University of Missouri.* HORSEPOWER; POWER.

Schmitt, Walter R. *Scripps Institution of Oceanography, La Jolla.* MARINE RESOURCES.

Schneider, Dr. Thomas R. *Electric Power Research Institute, Palo Alto, CA.* ENERGY STORAGE (coauthored).

Schoonmaker, G. R. *Vice President, Production-Exploration, Marathon Oil Company, Findlay, OH.* OIL AND GAS, OFFSHORE.

Schora, Frank C. *Vice President, Process Research Institute of Gas Technology, IIT Center, Chicago.* SUBSTITUTE NATURAL GAS (SNG).

Schueler, Dr. Donald G. *Sandia Laboratories, Albuquerque.* SOLAR CELL (in part).

Schutt, H. C. *Deceased; formerly, Consulting Engineer.* FURNACE CONSTRUCTION.

Schwieger, Robert C. *Associate Editor, "Power," McGraw-Hill Publications Company, New York.* COMBUSTION TURBINE.

Sebald, Joseph F. *Consulting Engineer, and President, Heat Power Products Corporation, Bloomfield, NJ.* CONTACT CONDENSER; COOLING TOWER; STEAM CONDENSER; SURFACE CONDENSER; VAPOR CONDENSER.

Sell, Dr. Heinz G. *Metals Development Section, Westinghouse Lamp Divisions, Bloomfield, NJ.* GRAYBODY (coauthored).

Shannon, Dr. Hugh F. *Products Research Division, Exxon Research and Engineering Company, Linden, NJ.* GASOLINE.

Siegmund, C. W. *Exxon Research and Engineering Company, Linden, NJ.* FUEL OIL.

Silfvast, Dr. William T. *Bell Labs, Holmdel, NJ.* LASER FUSION.

Simon, Dr. Jack A. *Illinois State Geological Survey, Urbana.* COAL; LIGNITE.

Sinclair, Dr. Donald B. *President (retired), General Radio Company, Concord, MA.* ELECTRIC POWER MEASUREMENT; ELECTRICAL MEASUREMENTS (in part); WATTMETER.

Singer, Prof. Stanley. *Director, ATHENEX Research Associates.* PROPELLANT (in part).

Skilling, Prof. H. H. *Department of Electrical Engineering, Stanford University.* ALTERNATING CURRENT.

Smith, Dr. John Ward. *Research Supervisor, Laramie Energy Research Center, ERDA.* OIL SHALE (coauthored).

Smith, W. W. *Technical Adviser, Legal Department, ESB Inc., Philadelphia.* STORAGE BATTERY.

Souders, Dr. Mott. *Deceased; formerly, Director, Oil Development, Shell Oil Company, Emeryville, CA.* DISTILLATE FUEL; LIQUEFIED PETROLEUM GAS (LPG).

Spinrad, Dr. Bernard I. *Senior Physicist, Applied Physics Division, Argonne National Laboratory.* NUCLEONICS; REACTOR PHYSICS.

Stanley, C. Maxwell. *Chairman of the Board, Stanley Consultants, Muscatine, IA.* RURAL ELECTRIFICATION (coauthored).

Starr, Dr. Eugene C. *U.S. Department of the Interior, Bonneville Power Administration, Portland.* ELECTRIC POWER GENERATION.

Steele, Dr. William A. *Department of Chemistry, Pennsylvania State University.* THERMODYNAMIC PROCESSES (coauthored).

Steindler, Martin J. *Chemical Engineering Division, Argonne National Laboratory.* NUCLEAR FUEL CYCLE; NUCLEAR FUELS PROCESSING.

Stephenson, Prof. Richard M. *Department of Nuclear Engineering, University of Connecticut.* ISOTOPE (STABLE) SEPARATION.

Stevenson, Dr. Edward C. *Department of Electrical Engineering, School of Engineering and Applied Sciences, University of Virginia.* Validator for ELECTRICAL MEASUREMENTS (in part).

Stewart, Dr. John W. *Department of Physics, University of Virginia.* ELECTRIC CURRENT; ELECTRICAL CONDUCTION; ELECTRODYNAMICS.

Stillman, Dr. Gregory E. *Department of Electrical Engineering, University of Illinois.* PHOTOVOLTAIC CELL.

Storch, Dr. H. H. *Deceased; formerly, Assistant Professor of Chemistry, New York University.* DESTRUCTIVE DISTILLATION.

Stratton, Dr. T. F. *Los Alamos Scientific Laboratory.* Validator for THERMIONIC POWER GENERATOR.

Sturtevant, George R. *Manager (retired), Engineering Meter Department, General Electric Company.* WATT-HOUR METER (coauthored).

Suryanarayana, Prof. N. V. *Department of Mechanical Engineering and Engineering Mechanics, College of Engineering, Michigan Technological University, Houghton.* HEAT PUMP (in part).

Sutherland, J. R. *Power Transformer Department, General Electric Company, Pittsfield, MA.* TRANSFORMER.

T

Teasley, Robert E., Jr. *Cummins Engine Company, Columbus, IN.* DIESEL FUEL.

Thompson, Jack R. *U.S. Army Corps of Engineers, Office of the Secretary of the Army, Department of Army.* DAM.

Tuck, Dr. James L. *Associate Division Leader, Los Alamos Scientific Laboratory.* PINCH EFFECT.

U

Udani, Dr. Lalit H. *Catalytic, Inc., Philadelphia.* SYNTHETIC FUEL.

W

Waddington, Prof. Thomas C. *Department of Chemistry, University of Durham.* CHEMICAL ENERGY.

Wainwright, H. W. *Coal Research Center, U.S. Bureau of Mines, Morgantown, WV.* Validator for DESTRUCTIVE DISTILLATION.

Walsh, Dr. Peter J. *Department of Physics, Fairleigh Dickinson University.* GRAYBODY (coauthored).

Watson, Prof. W. W. *Professor Emeritus of Physics, Yale University.* MASS DEFECT.

Weaver, Dr. Paul. *(Retired) Texas A&M College.* OIL FIELD MODEL.

Weber, Erwin L. *Deceased; formerly, Trust Department, National Bank of Commerce, Seattle.* HOT-WATER HEATING SYSTEM; PANEL HEATING AND COOLING; RADIANT HEATING.

Weber, Prof. Harold C. *Department of Chemical Engineering, Massachusetts Institute of Technology.* BRITISH THERMAL UNIT (BTU); HEAT; HEAT CAPACITY.

Weigel, Dr. Fritz. *Institut fur Anorganische Chemie, Universitat Munchen.* PLUTONIUM; URANIUM; URANIUM METALLURGY.

Weil, Robert T., Jr. *Dean, School of Engineering, Manhattan College.* DIRECT-CURRENT GENERATOR.

Wheeler, Prof. John A. *Department of Physics, Joseph Henry Laboratories, Princeton University.* CRITICAL MASS.

White, Prof. David C. *Department of Electrical Engineering, Massachusetts Institute of Technology.* THERMOELECTRIC POWER GENERATOR.

Wilhelm, Dr. Harley A. *Associate Director, Institute for Atomic Research and Ames Laboratory, Ames, IA.* THORIUM.

Wilkinson, Prof. D. H. *Department of Nuclear Physics, Oxford University.* NUCLEAR BINDING ENERGY (in part).

Williams, Prof. Everard M. *Department of Electrical Engineering, Carnegie-Mellon University.* TRANSMISSION LINES (in part).

Winch, Prof. Ralph P. *Department of Physics, Williams*

College. Electric charge; Electric field; Electromotive force (emf).

Wrighton, Dr. Mark S. *Department of Chemistry, Massachusetts Institute of Technology*. Photoelectrolysis.

Wylie, Prof. E. Benjamin. *Department of Civil Engineering, University of Michigan*. Natural gas (in part).

Wyman, Dr. Richard. *Director of Research, Canadian Hunter Exploration Ltd., Calgary, Alberta*. Well logging.

Y

Yellott, John I. *Professor Emeritus, College of Architecture, Arizona State University*. Solar heating and cooling.

Yonas, Dr. Gerold. *Director, Pulsed Energy Programs, Sandia Laboratories, Albuquerque*. Particle-beam fusion (in part).

Appendix

Appendix

The Appendix covers the following: (I) Measurement systems — U.S. Customary, metric, and International — are discussed, and conversion tables are provided. Usage of the Fahrenheit, Celsius, and Kelvin temperature scales is also described. (II) Important Federal energy legislation of the 1970s is outlined, and U.S. government organizations responsible in energy matters are detailed. (III) Energy-related publications are described, including addresses for ordering.

I. MEASUREMENT SYSTEMS

U.S. Customary System and the metric system

Scientists and engineers have been using two major systems of units in measurement. These are commonly called the U.S. Customary System (inherited from the British Imperial System) and the metric system.

In the U.S. Customary System the units yard and pound with their divisions, such as the inch, and multiples, such as the ton, are basic. The metric system was evolved during the 18th century and has been adopted for general use by most countries. Nearly everywhere it is used for precise measurements in science. The meter and kilogram with their multiples, such as the kilometer, and fractions, such as the gram, are basic to the metric system.

In the U.S. Customary System, units of the same kind are related almost at random. For example, there are the units of length, the inch, yard, and mile. In the metric system the relationships between units of the same kind are strictly decimal (millimeter, meter, and kilometer).

However, to complicate matters in scientific writing, there is no uniformity within each of these two systems as to the choice of units for the same quantities. For example, the hour or the second, the foot or the inch, and the centimeter or the millimeter could be chosen by a scientist as the unit of measurement for the quantities time and length.

Introduction of the International System, or SI

To simplify matters and to make communication more understandable, an internationally accepted system of units is coming into use. This is termed the International System of Units, which is abbreviated SI in all languages.

Fundamentally the system is metric with the base units derived from scientific formulas or natural constants. For example, the meter in the SI is defined as the length equal to 1 650 763.73 wavelengths in vacuum of the radiation corresponding to the transition between the electronic energy levels $2p_{10}$ and $5d_5$ of the krypton-86 atom. Previously, in the metric system, the meter was

Introduction of the International System, or SI (cont.)

defined as the distance between two marks on a specific metal bar.

In a similar way the second in the SI is defined as the duration of 9 192 631 770 periods of the radiation corresponding to the transition between two hyperfine levels of the ground state of the cesium-133 atom.

Interestingly, the kilogram, the SI unit of mass, is still the mass of the kilogram kept at Sèvres, France. However, it is possible that eventually the unit will be redefined in terms of atomic mass.

Although the SI is increasing in usage by scientists and engineers, there are some units in everyday use which will probably remain, for example, minute, hour, day, degree (angle), and liter. The point should be made, however, that these terms will not be employed in a scientific context if the SI is fully adopted.

Because of their extremely common use among scientists, several units are still permitted in conjunction with SI units, for example, the electron volt, gauss, barn, and curie. In time their usage might be phased out.

One further point is that in October, 1967, the Thirteenth General Conference of Weights and Measures decided to name the SI unit of thermodynamic temperature "kelvin" (symbol K) instead of "degree Kelvin" (symbol °K). For example, the notation is 273 K and not 273°K.

The base units and derived units of the SI are shown in **Table 1** and **Table 2.**

In the SI the prefixes differ from a unit in steps of 10^3. A list of prefix terms, symbols, and their factors is given in **Table 3**. Some examples of the use of these prefixes follow:

$$1000\text{ m} = 1\text{ kilometer} = 1\text{ km}$$

$$1000\text{ V} = 1\text{ kilovolt} = 1\text{ kV}$$

$$1\,000\,000\,\Omega = 1\text{ megohm} = 1\text{ M}\Omega$$

$$0.000\,000\,001\text{ s} = 1\text{ nanosecond} = 1\text{ ns}$$

Only one prefix is to be employed for a unit. For example:

$$1000\text{ kg} = 1\text{ Mg} \qquad \text{not } 1\text{ kkg}$$

$$10^{-9}\text{ s} = 1\text{ ns} \qquad \text{not } 1\text{ m}\mu\text{s}$$

$$1\,000\,000\text{ m} = 1\text{ Mm} \qquad \text{not } 1\text{ kkm}$$

Also, when a unit is raised to a power, the power applies to the whole unit including the prefix. For example:

$$\text{km}^2 = (\text{km})^2 = (1000\text{ m})^2 = 10^6\text{ m}^2 \qquad \text{not } 1000\text{ m}^2$$

Table 1. Base units of the International System

Quantity	*Name of unit*	*Unit symbol*
length	meter	m
mass	kilogram	kg
time	second	s
electric current	ampere	A
temperature	kelvin	K
luminous intensity	candela	cd
amount of substance	mole	mol

Table 2. Derived units of the International System

Quantity	*Name of unit*	*Unit symbol or abbreviation, where differing from basic form*	*Unit expressed in terms of base or supplementary units**
area	square meter		m^2
volume	cubic meter		m^3
frequency	hertz	Hz	s^{-1}
density	kilogram per cubic meter		kg/m^3
velocity	meter per second		m/s
angular velocity	radian per second		rad/s
acceleration	meter per second squared		m/s^2
angular acceleration	radian per second squared		rad/s^2
volumetric flow rate	cubic meter per second		m^3/s
force	newton	N	$kg \cdot m/s^2$
surface tension	newton per meter, joule per square meter	N/m, J/m^2	kg/s^2
pressure	newton per square meter, pascal	N/m^2, Pa	$kg/m \cdot s^2$
viscosity, dynamic	newton-second per square meter, pascal-second	$N \cdot s/m^2$, $Pa \cdot s$	$kg/m \cdot s$
viscosity, kinematic	meter squared per second		m^2/s
work, torque, energy, quantity of heat	joule, newton-meter, watt-second	J, $N \cdot m$, $W \cdot s$	$kg \cdot m^2/s^2$
power, heat flux	watt, joule per second	W, J/s	$kg \cdot m^2/s^3$
heat flux density	watt per square meter	W/m^2	kg/s^3
volumetric heat release rate	watt per cubic meter	W/m^3	$kg/m \cdot s^3$
heat transfer coefficient	watt per square meter kelvin	$W/m^2 \cdot K$	$kg/s^3 \cdot K$
heat capacity (specific)	joule per kilogram kelvin	$J/kg \cdot K$	$m^2/s^2 \cdot K$
capacity rate	watt per kelvin	W/K	$kg \cdot m^2/s^3 \cdot K$
thermal conductivity	watt per meter kelvin	$W/m \cdot K, \dfrac{J \cdot m}{s \cdot m^2 \cdot K}$	$kg \cdot m/s^3 \cdot K$
quantity of electricity	coulomb	C	$A \cdot s$
electromotive force	volt	V, W/A	$kg \cdot m^2/A \cdot s^3$
electric field strength	volt per meter		V/m
electric resistance	ohm	Ω, V/A	$kg \cdot m^2/A^2 \cdot s^3$
electric conductivity	ampere per volt meter	$A/V \cdot m$	$A^2 \cdot s^3/kg \cdot m^3$
electric capacitance	farad	F, $A \cdot s/V$	$A^3 \cdot s^4/kg \cdot m^2$
magnetic flux	weber	Wb, $V \cdot s$	$kg \cdot m^2/A \cdot s^2$
inductance	henry	H, $V \cdot s/A$	$kg \cdot m^2/A^2 \cdot s^2$
magnetic permeability	henry per meter	H/m	$kg \cdot m/A^2 \cdot s^2$
magnetic flux density	tesla, weber per square meter	T, Wb/m^2	$kg/A \cdot s^2$
magnetic field strength	ampere per meter		A/m
magnetomotive force	ampere		A
luminous flux	lumen	lm	$cd \cdot sr$
luminance	candela per square meter		cd/m^2
illumination	lux, lumen per square meter	lx, lm/m^2	$cd \cdot sr/m^2$
activity (of radionuclides)	becquerel	Bq	s^{-1}
absorbed dose	gray	Gy	J/kg
dose equivalent	sievert	Sv	J/kg

*Supplementary units are: plane angle, radian (rad); solid angle, steradian (sr).

Table 3. Prefixes for units in the International System

Prefix	*Symbol*	*Power*	*Example*	*Prefix*	*Symbol*	*Power*	*Example*
exa	E	10^{18}		deci	d	10^{-1}	
peta	P	10^{15}		centi	c	10^{-2}	
tera	T	10^{12}		milli	m	10^{-3}	milligram (mg)
giga	G	10^{9}		micro	μ	10^{-6}	microgram (μg)
mega	M	10^{6}	megahertz (MHz)	nano	n	10^{-9}	nanosecond (ns)
kilo	k	10^{3}	kilometer (km)	pico	p	10^{-12}	picofarad (pf)
hecto	h	10^{2}		femto	f	10^{-15}	
deka	da	10^{1}		atto	a	10^{-18}	

Introduction of the International System, or SI (cont.)

Some common units defined in terms of SI units are given in **Table 4** (the definitions in the fourth column are exact).

Table 4. Some common units defined in terms of SI units

Quantity	*Name of unit*	*Unit symbol*	*Definition of unit*
length	inch	in.	2.54×10^{-2} m
mass	pound (avoirdupois)	lb	0.45359237 kg
force	kilogram-force	kgf	9.80665 N
pressure	atmosphere	atm	$101325\ N \cdot m^{-2}$
pressure	torr	Torr	$(101325/760)\ N \cdot m^{-2}$
pressure	conventional millimeter of mercury*	mmHg	$13.5951 \times 980.665 \times 10^{-2}\ N \cdot m^{-2}$
energy	kilowatt-hour	kWh	3.6×10^{6} J
energy	thermochemical calorie	cal	4.184 J
energy	international steam table calorie	cal_{IT}	4.1868 J
thermodynamic temperature (T)	degree Rankine	°R	(5/9) K
customary temperature (t)	degree Celsius	°C	$t(°C) = T(K) - 273.15$
customary temperature (t)	degree Fahrenheit	°F	$t(°F) = T(°R) - 459.67$
radioactivity	curie	Ci	$3.7 \times 10^{10}\ s^{-1}$
energy†	electron volt	eV	$eV \approx 1.60219 \times 10^{-19}$ J
mass†	unified atomic mass unit	u	$u \approx 1.66057 \times 10^{-27}$ kg

*The conventional millimeter of mercury, symbol mmHg (not mm Hg), is the pressure exerted by a column exactly 1 mm high of a fluid of density exactly 13.5951 $g \cdot cm^{-3}$ in a place where the gravitational acceleration is exactly $980.665\ cm \cdot s^{-2}$. The mmHg differs from the torr by less than 2×10^{-7} torr.
†These units defined in terms of the best available experimental values of certain physical constants may be converted to SI units. The factors for conversion of these units are subject to change in the light of a new experimental measurements of the constants involved.

Conversion factors for the measurement systems

Because it will take some years for all scientists and engineers to convert to the SI, the Encyclopedia has retained the U.S. Customary and metric systems, but has incorporated SI units when preparation of the text permitted. Conversion factors between the three measurement systems are given in **Table 5** for some prevalent units; in each of the subtables the user proceeds as follows:

To convert a quantity expressed in a unit in the left-hand column to the equivalent in a unit in the top row of a subtable, multiply the quantity by the factor common to both units.

The factors have been carried out to seven significant figures, as derived from the fundamental constants and the definitions of the units. However, this does not mean that the factors are always known to that accuracy. Numbers followed by ellipses are to be continued indefinitely with repetition of the same pattern of digits. Factors written with fewer than seven significant digits are exact values. Numbers followed by an asterisk are definitions of the relation between the two units.

Table 5. Conversion factors for the U.S. Customary System, metric system, and International System

A. UNITS OF LENGTH						
Units	cm	m	in.	ft	yd	mile
1 cm	= 1	0.01*	0.3937008	0.03280840	0.01093613	6.213712×10^{-6}
1 m	= 100.	1	39.37008	3.280840	1.093613	6.213712×10^{-4}
1 in.	= 2.54*	0.0254	1	0.08333333...	0.02777777...	1.578283×10^{-5}
1 ft	= 30.48	0.3048	12.*	1	0.3333333...	$1.893939... \times 10^{-4}$
1 yd	= 91.44	0.9144	36.	3.*	1	$5.681818... \times 10^{-4}$
1 mile	$= 1.609344 \times 10^{5}$	1.609344×10^{3}	6.336×10^{4}	5280.*	1760.	1
B. UNITS OF AREA						
Units	cm^2	m^2	$in.^2$	ft^2	yd^2	$mile^2$
1 cm^2	= 1	10^{-4}*	0.1550003	1.076391×10^{-3}	1.195990×10^{-4}	3.861022×10^{-11}
1 m^2	$= 10^{4}$	1	1550.003	10.76391	1.195990	3.861022×10^{-7}
1 $in.^2$	= 6.4516*	6.4516×10^{-4}	1	6.944444×10^{-3}...	7.716049×10^{-4}	2.490977×10^{-10}
1 ft^2	= 929.0304	0.09290304	144.*	1	0.1111111...	3.587007×10^{-8}
1 yd^2	= 8361.273	0.8361273	1296.	9.*	1	3.228306×10^{-7}
1 $mile^2$	$= 2.589988 \times 10^{10}$	2.589988×10^{6}	4.014490×10^{9}	2.78784×10^{7}*	3.0976×10^{6}	1

continued

Conversion factors for the measurement systems (cont.)

Table 5. Conversion factors for the U.S. Customary System, metric system, and International System (cont.)

C. UNITS OF VOLUME

Units	m^3	cm^3	liter	$in.^3$	ft^3	qt	gal
1 m^3	= 1	10^6	10^3	6.102374×10^4	35.31467	1.056688×10^3	264.1721
1 cm^3	= 10^{-6}	1	10^{-3}	0.06102374	3.531467×10^{-5}	1.056688×10^{-3}	2.641721×10^{-4}
1 liter	= 10^{-3}	1000.*	1	61.02374	0.03531467	1.056688	0.2641721
1 $in.^3$	= 1.638706×10^{-5}	16.38706*	0.01638706	1	5.787037×10^{-4}	0.01731602	4.329004×10^{-3}
1 ft^3	= 2.831685×10^{-2}	28316.85	28.31685	1728.*	1	2.992208	7.480520
1 qt	= 9.46353×10^{-4}	946.353	0.946353	57.75	0.0342014	1	0.25
1 gal (U.S.)	= 3.785412×10^{-3}	3785.412	3.785412	231.*	0.1336806	4.*	1

D. UNITS OF MASS

Units	g	kg	oz	lb	metric ton	ton
1 g	= 1	10^{-3}	0.03527396	2.204623×10^{-3}	10^{-6}	1.102311×10^{-6}
1 kg	= 1000.	1	35.27396	2.204623	10^{-3}	1.102311×10^{-3}
1 oz (avdp)	= 28.34952	0.02834952	1	0.0625	2.834952×10^{-5}	3.125×10^{-5}
1 lb (avdp)	= 453.5924	0.4535924	16.*	1	4.535924×10^{-4}	0.0005
1 metric ton	= 10^6	1000.*	35273.96	2204.623	1	1.102311
1 ton	= 907184.7	907.1847	32000.	2000.*	0.9071847	1

E. UNITS OF DENSITY

Units	$g \cdot cm^{-3}$	$g \cdot l.^{-1}$, $kg \cdot m^{-3}$	$oz \cdot in.^{-3}$	$lb \cdot in.^{-3}$	$lb \cdot ft^{-3}$	$lb \cdot gal^{-1}$
1 $g \cdot cm^{-3}$	= 1	1000.	0.5780365	0.03612728	62.42795	8.345403
1 $g \cdot l.^{-1}$, $kg \cdot m^{-3}$	= 10^{-3}	1	5.780365×10^{-4}	3.612728×10^{-5}	0.06242795	8.345403×10^{-3}
1 $oz \cdot in.^{-3}$	= 1.729994	1729.994	1	0.0625	108.	14.4375
1 $lb \cdot in.^{-3}$	= 27.67991	27679.91	16.	1	1728.	231.
1 $lb \cdot ft^{-3}$	= 0.01601847	16.01847	9.259259×10^{-3}	5.7870370×10^{-4}	1	0.1336806
1 $lb \cdot gal^{-1}$	= 0.1198264	119.8264	4.749536×10^{-3}	4.3290043×10^{-3}	7.480519	1

Table 5. Conversion factors for the U.S. Customary System, metric system, and International System (cont.)

F. UNITS OF PRESSURE

Units	Pa, N · m^{-2}	dyn · cm^{-2}	bar	atm	kg(wt) · cm^{-2}	mmHg (torr)	in. Hg	lb (wt) · in.$^{-2}$
1 Pa, 1 N · m^{-2}	= 1	10	10^{-5}	9.869233 $\times 10^{-6}$	1.019716 $\times 10^{-5}$	7.500617 $\times 10^{-3}$	2.952999 $\times 10^{-4}$	1.450377 $\times 10^{-4}$
1 dyn · cm^{-2}	= 0.1	1	10^{-6}	9.869233 $\times 10^{-7}$	1.019716 $\times 10^{-6}$	7.500617 $\times 10^{-4}$	2.952999 $\times 10^{-5}$	1.450377 $\times 10^{-5}$
1 bar	= 10^5*	10^6	1	0.9869233	1.019716	750.0617	29.52999	14.50377
1 atm	= 101325.0*	1013250.	1.013250	1	1.033227	760.	29.92126	14.69595
1 kg (wt) · cm^{-2}	= 98066.5	980665.	0.980665	0.9678411	1	735.5592	28.95903	14.22334
1 mmHg (torr)	= 133.3224	1333.224	1.333224 $\times 10^{-3}$	1.3157895 $\times 10^{-3}$	1.3595099 $\times 10^{-3}$	1	0.03937008	0.01933678
1 in. Hg	= 3386.388	33863.88	0.03386388	0.03342105	0.03453155	25.4	1	0.4911541
1 lb (wt) · in.$^{-2}$	= 6894.757	68947.57	0.06894757	0.06804596	0.07030696	51.71493	2.036021	1

G. UNITS OF ENERGY†

Units	g mass (energy equiv)	J	int J	cal	cal$_{IT}$	Btu$_{IT}$	kWh	hp hr	ft-lb (wt)	cu ft-lb (wt) in.$^{-2}$	l.-atm
1 g mass (energy equiv)	= 1	8.987552 $\times 10^{13}$	8.986069 $\times 10^{13}$	2.148076 $\times 10^{13}$	2.146640 $\times 10^{13}$	8.518555 $\times 10^{10}$	2.496542 $\times 10^{7}$	3.347918 $\times 10^{7}$	6.628878 $\times 10^{13}$	4.603388 $\times 10^{11}$	8.870024 $\times 10^{11}$
1 J	= 1.112650 $\times 10^{-14}$	1	0.999835	0.2390057	0.2388459	9.478172 $\times 10^{-4}$	2.777777... $\times 10^{-7}$	3.725062	0.7375622	5.121960 $\times 10^{-3}$	9.869233 $\times 10^{-3}$
1 int J	= 1.112834 $\times 10^{-14}$	1.000165	1	0.2390452	0.2388853	9.479735 $\times 10^{-4}$	2.778236 $\times 10^{-7}$	3.725676 $\times 10^{-7}$	0.7376839	5.122805 $\times 10^{-3}$	9.870862 $\times 10^{-3}$
1 cal	= 4.655328 $\times 10^{-14}$	4.184*	4.183310	1	0.9993312	3.965667 $\times 10^{-3}$	1.1622222... $\times 10^{-6}$	1.558562 $\times 10^{-6}$	3.085960	2.143028 $\times 10^{-2}$	0.04129287
1 cal$_{IT}$	= 4.658443 $\times 10^{-14}$	4.1868*	4.186109	1.000669	1	3.968321 $\times 10^{-3}$	1.163000 $\times 10^{-6}$	1.559609 $\times 10^{-6}$	3.088025	2.144462 $\times 10^{-2}$	0.04132050
1 Btu$_{IT}$	= 1.173908 $\times 10^{-11}$	1055.056	1054.882	252.1644	251.9958*	1	2.930711 $\times 10^{-4}$	3.930148 $\times 10^{-4}$	778.1693	5.403953	10.41259
1 kWh	= 4.005540 $\times 10^{-8}$	3600000.*	3599406.	860420.7	859845.2	3412.142	1	1.341022	2655224.	18439.06	35529.24
1 hp hr	= 2.986931 $\times 10^{-8}$	2684519.	2684077.	641615.6	641186.5	2544.33	0.7456998	1	1980000.*	13750.	26494.15
1 ft-lb (wt)	= 1.508551 $\times 10^{-14}$	1.355818	1.355594	0.3240483	0.3238315	1.285067 $\times 10^{-3}$	3.766161 $\times 10^{-7}$	5.050505... $\times 10^{-7}$	1	6.944444... $\times 10^{-3}$	0.01338088
1 cu ft-lb (wt) in.$^{-2}$	= 2.172313 $\times 10^{-12}$	195.2378	195.2056	46.66295	46.63174	0.1850497	5.423272 $\times 10^{-5}$	7.272727... $\times 10^{-5}$	144.*	1	1.926847
1 l.-atm	= 1.127393 $\times 10^{-12}$	101.3250	101.3083	24.21726	24.20106	0.09603757	2.814583 $\times 10^{-5}$	3.774419 $\times 10^{-5}$	74.73349	0.5189825	1

†The electrical units are those in terms of which certification of standard cells, standard resistances, and so forth, is made by the National Bureau of Standards. Unless otherwise indicated, all electrical units are absolute.

Units of temperature in measurement systems

Temperature is a basic physical quantity. It is a measure of the thermal energy of random motion of particles in a system. As such it has been chosen as one of the basic quantities in the SI. It is to be treated as are the units length, mass, time, electric current, and luminous intensity. In the SI the unit of length is the meter, the unit of time the second, and so on. The question arises as to the choice of the unit of temperature in the SI.

In the past it was customary to refer to scales of temperature, for example, the Celsius and Fahrenheit scales. On the Celsius scale, 0 designates the freezing point (ice point) and 100 the boiling point (steam point) of water. Corresponding numbers on the Fahrenheit scale are 32 and 212. There are 100 units between the ice point and steam point on the Celsius scale, and 180 units between these points in the Fahrenheit system.

By measuring the volume changes of a gas within the 100-unit interval of the ice point and steam point of water on the Celsius scale, it was found that a numerical value could be assigned for a basic unit of temperature. Careful measurement of this ice-steam interval in a gas thermometer determined that the ice point of water should be assigned the value of 273.15 kelvins. The unit of temperature was thus called the kelvin with the symbol K. Further experiments led to the decision to define the kelvin in the SI along the same lines but in terms of the triple point of water. This is the temperature and pressure at which ice, liquid water, and water vapor coexist at equilibrium. The triple point was chosen because it was a more reproducible value than the ice point.

This change led to the SI definition of temperature in terms of the triple point of water, which contains exactly 273.16 kelvins.

It follows that the Celsius temperature (°C) is an intermediate scale. It is useful in defining Kelvin temperature in the SI. Celsius temperature (t) is related to Kelvin temperature (K) as follows:

$$t_{\text{ice point}} = 0°\text{C}$$

$$t_{\text{steam point}} = 100°\text{C}$$

$$0\ \text{K} = -273.15°\text{C}$$

A summary of the conventions in the SI as proposed in the Thirteenth General Conference of Weights and Measures pertaining to temperature units is given below.

1. The unit of SI temperature is the kelvin, symbol K.
2. The word "scale" is not to be used except in terms of measurement of temperature between certain fixed points on the Celsius scale.
3. The terms "thermodynamic scale" or "absolute scale" are not to be used to describe temperature. The degree sign is to be eliminated with the symbol K.
4. When Celsius temperatures are used (°C), it is understood that the temperature unit is the kelvin.

Not all scientists and engineers have adopted the SI of temperature terminology. For this reason the contributors to the Encyclopedia have retained the term "scale" in relation to thermodynamic temperature. Furthermore, many engineers in the United States still use the Fahrenheit system in discussing practical engineering systems.

In converting Fahrenheit (°F) to Celsius (°C) the following formula applies.

$$°\text{C} = \frac{°\text{F} - 32°}{1.8}$$

Units of temperature in measurement systems (cont.)

In converting Celsius to Fahrenheit the following formula can be used.

$$°F = (°C \times 1.8) + 32°$$

In changing from Celsius terminology (t) to kelvin units (K) the following formula can be used.

$$K = t + 273.15$$

II. FEDERAL ENERGY LEGISLATION AND ORGANIZATIONS

A. Highlights of energy legislation

1. Emergency Petroleum Allocation Act. Signed by President Nov. 27, 1973. Designed to assure that shortages were shared equally by empowering the President to issue regulations for the allocation of oil and oil products. Also established "passthrough," which permitted retailers to pass on to their customers the increases in the wholesale prices of oil and oil products.

2. Federal Energy Administration Act. Signed by President May 7, 1974. Created the Federal Energy Administration to take over the various allocation and fuel policy functions scattered throughout the executive branch. (The FEA was supposed to be a temporary agency, but was subsequently extended.)

3. Energy Supply and Environmental Coordination Act. Signed by President June 22, 1974. Directed the Federal Energy Administration to prohibit the use of oil or natural gas in electric power plants that were able to burn coal, allowed the FEA administrator to prohibit the use of oil or gas in other facilities if they could use coal, and provided for temporary suspension of certain air-pollution requirements.

4. Energy Reorganization Act. Signed by President Oct. 11, 1974. Established the Nuclear Regulatory Commission to discharge the licensing and related functions formerly assigned to the Atomic Energy Commission and the Energy Research and Development Administration and to coordinate and direct all Federal activities relating to research and development of various sources of energy, including the general basic research, military, and production activities of the AEC.

5. Energy Policy and Conservation Act. Signed by President Dec. 22, 1975. Expanded the Federal Energy Administration's authority to order major power plants to use coal in place of oil or gas, and assigned various broad powers in the energy field. Established a national energy policy and energy conservation measures designed to: (1) maximize the domestic production of energy and provide for the strategic storage of reserves of oil and petroleum products; (2) minimize the consequences of disruptions in energy supplies by providing for emergency standby measures; (3) provide a level of oil prices which would both encourage production and not impede economic recovery; (4) reduce domestic energy consumption through voluntary and mandatory energy conservation programs.

6. Energy Conservation and Production Act. Signed by President Aug. 14, 1976. Further expanded the control of natural gas and petroleum in terms of the prohibition of use in power plants.

7. Department of Energy Organization Act. Signed by President Aug. 4, 1977. Consolidated the major Federal energy functions into one Cabinet-level department, transferring to the Department of Energy all the responsibilities of the Energy Research and Development Administration, Federal Energy Administration, Federal Power Commission, the Alaska, Bonneville, Southeastern, and Southwestern power administrations—and also the marketing functions of the Department of the Interior's Bureau of Reclamation.

A. *Highlights of energy legislation (cont.)*

8. "National Energy Acts." This group of energy bills was signed by the President Nov. 9, 1978:

Public Utility Regulatory Policies Act. Reformed electrical utility rates; required state utility commissions and other regulatory agencies to price electricity lower in off-peak hours and discontinue discounts for large-volume users.

Energy Tax Act. Provided tax credits for homeowners and businesses for installing energy-saving devices in their buildings.

National Energy Conservation Policy Act. Required utilities to give customers information about energy conservation devices. Allowed utilities to arrange for the installations (but not sell them or actually do the installation) and allowed customers to pay for the improvements through utility bills.

Power Plant and Industrial Fuel Use Act. Provided authority to force utility conversion from oil and gas to coal. Required new industrial and utility plants to be built to use coal or a fuel other than oil or gas.

Natural Gas Policy Act. Permitted price of natural gas to increase gradually until 1985, when price controls would be eliminated.

B. *Federal organizations relating to energy*

1. U.S. Department of Energy (DOE)

(a) Economic Regulatory Administration (ERA). Created by Department of Energy Organization Act, August 1977. Major responsibilities: (1) Administration of pricing and allocation programs. (2) Administration of oil import programs. (3) Administration of gas import and export programs. (4) Administration of electricity import programs. (5) Supervision of mandatory conversion from oil and gas to coal and other abundant or renewable fuel resources. (6) Development of programs to promote energy conservation by requiring utility companies to offer insulation services and to revise electrical utility rate structures to promote efficient use of facilities. (7) Right of intervention in proceedings before the Federal Energy Regulatory Commission as well as Federal and state regulatory bodies. (8) Supervision of electrical industry programs for voluntary coordination, interconnection, planning for continued reliable supplies, and assessments of the need for power. (9) Development of plans for emergency allocation of fuels during periods of shortage; also responsible for overseeing administration of those plans, including contingency gas rationing plans.

(b) Federal Energy Regulatory Commission (FERC). Created by the Department of Energy Organization Act (August 1977) to replace the Federal Power Commission. Major responsibilities: (1) Regulating the transportation and sale (wholesale prices) of natural gas in interstate commerce. (2) Regulating the construction and operation of interstate natural gas pipeline facilities. (3) Establishing and enforcing curtailment plans proposed by gas companies to reduce service to certain areas. (4) Regulating the transmission and wholesale of electricity in interstate commerce. (5) Authorizing interconnections of electric utilities. (6) Licensing non-Federal hydroelectric projects. (7) Regulating security (stock) issues and mergers of electrical utilities and approving interlocking directorships. (8) Regulating rates charged for interstate transportation of oil by pipeline. (9) Establishing pipeline valuations.

B. Federal organizations relating to energy (cont.)

(c) Energy Information Administration (EIA). Created by Department of Energy Organization Act, August 1977. Major responsibilities: (1) Collection, processing, and publication of accurate and timely data on energy reserves, the financial status of energy-producing companies, production, demand and consumption. (2) Analysis of data to assist government and nongovernment users.

2. National Bureau of Standards (NBS)

Part of the Department of Commerce. Major responsibilities: (1) Research, testing product performance and safety, informing public on product standards. (2) Overseeing and coordinating energy conservation programs for appliances, solar heating and cooling of residential and commercial buildings, community energy systems, and industry.

3. Nuclear Regulatory Commission (NRC)

An independent agency which in 1975 took over the functions of the Atomic Energy Commission in nuclear regulating and licensing. Major responsibilities: (1) Nuclear reactor regulations. (2) Nuclear material safety and safeguards. (3) Nuclear regulatory research. (4) Standards development. (5) Inspection and enforcement.

III. ENERGY-RELATED PUBLICATIONS

This listing is largely from the *Federal Regulatory Directory 1979–80*, published by Congressional Quarterly, Inc., Washington, DC.

Publication	Description	Available from
Bitting and Harrington's Oil Industry Gazette	Monthly newsletter on the oil industry.	Bitting and Co. Inc., 127–111 S. Beniston Ave., Clayton, MS 63105
Capital Energy Letter	Weekly coverage of legislation related to energy.	Capital Energy Letter Inc., 1061 National Press Bldg., Washington, DC 20045
Chilton's Oil and Gas Energy	Monthly newsletter on the gas and oil industries.	Chilton Co., Chilton Way, Radnor, PA 19089
Crude Petroleum, Petroleum Products and Natural Gas Liquids	Annual summary of crude oil supply and demand, refined petroleum products, and natural gas liquids. Compiled by the Division of Fossil Fuels, Bureau of Mines, Department of the Interior.	U.S. Government Printing Office, Washington, DC 20402
Economic Regulatory Administration Enforcement Manual	Guide to agency policies and procedures, updated periodically. Originally published as the Federal Energy Administration Compliance Manual.	Commerce Clearing House, 4025 W. Peterson Ave., Chicago, IL 60646
Energy Atlas	Guide to Federal, regional, and state agencies and personnel dealing with conservation, energy management, fossil energy resources, nuclear energy, and research and development activities; lists energy publications. Updated every 6 months.	Fraser/Rudd & Finn, 1701 K St. N.W., Washington, DC 20006
Energy Controls	Looseleaf service updated weekly. Covers all new energy control developments, laws, regulations, and related material.	Prentice-Hall Inc., Englewood Cliffs, NJ 07632
The Energy Daily	Daily coverage of developments in the energy field.	The Energy Daily, 1239 National Press Bldg., Washington, DC 20045

III. ENERGY-RELATED PUBLICATIONS (cont.)

Publication	Description	Available from
Energy Directory Update Service: A Comprehensive Guide to the Nation's Energy Organizations, Decision Makers and Information Sources	Data on energy organizations, their programs, projects, publications, research, and personnel; updated bimonthly.	Environment Information Center Inc., 124 E. 39th St., New York, NY 10016
Energy Information	Weekly newsletter on domestic energy industry and developments in Washington.	Petroleum Information Corp., P.O. Box 2612, Denver, CO 80201
Energy Index: A Selected Guide to Energy Documents, Laws and Statistics	Annual guide to statistics on energy production and consumption; Federal research and development spending and priorities; analysis of energy trends and developments; status and summary of energy legislation.	Environment Information Center Inc., 124 E. 39th St., New York, NY 10016
Energy Information Abstracts	Bimonthly guide to energy sources; subject, Standard Industrial Classification code, and author indexes to reports, documents, journals, and hearings; lists new energy books; cumulated annually.	Environment Information Center Inc., 124 E. 39th St., New York, NY 10016
Energy Intelligence and Analysis for Energy Consumers	Weekly report on consumer issues related to energy use and development.	Newsletter Services, 1120 19th St. N.W., Washington DC 20036
Energy Legislative Service	Semimonthly newsletter on legislative proposals related to energy.	McGraw-Hill Publishing Co., 437 National Press Bldg., Washington, DC 20045
Energy Management	Weekly report on Federal, state, and local energy programs and regulations; documents; texts of fuel allocation programs; legislative activities; new developments in energy.	Commerce Clearing House, 4025 W. Peterson Ave., Chicago, IL 60646
Energy News	Semimonthly coverage of energy developments.	Petroleum Engineer Publishing Co., P.O. Box 1589, 800 Davis Bldg., Dallas, TX 75201

III. ENERGY-RELATED PUBLICATIONS (cont.)

Publication	Description	Available from
Energy Regulation Update Service	Covers Federal regulatory activities in energy areas; texts of major laws, regulations, rulings, and guidelines; updated monthly.	Environmental Information Center Inc., 124 E. 39th St., New York, NY 10016
Energy Research Reports	Monthly information on energy and energy-related topics.	Advanced Technology Publications Inc., 385 Elliot St., Newton, MA 02164
Energy Resources Report	Weekly coverage of energy supplies.	Business Publishers Inc., 818 Roeder Rd., Silver Spring, MD 20910
Energy Today	Semimonthly reporting on developments in energy.	Trends Publishing Inc., National Press Bldg., Washington, DC 20045
Energy Users Report	Weekly coverage of energy policy, supply, and technology; Federal controls, regulations, decisions, and rulings.	Bureau of National Affairs, 1231 25th St. N.W., Washington, DC 20037
Energy Week	Semimonthly newsletter on energy and related fields.	Petroleum Engineer Publishing Co., P.O. Box 1589, 800 Davis Bldg., Dallas, TX 75201
Federal Energy Guidelines	Texts of decisions, orders, appeals, rulings, regulations, and final rules issued by the Department of Energy, including the Economic Regulatory Administration.	Commerce Clearing House, 4025 W. Peterson Ave., Chicago, IL 60646
Federal Regulatory Directory	Annual listing of all U.S. government regulatory agencies, both independent and within executive departments, and extensive related information.	Congressional Quarterly, Inc., 1414 22 Street N.W., Washington, DC 20037
Fossil Energy Program Report	Programs and projects relating to the availability, consumption, and use of coal, petroleum, natural gas, oil shale, and tar sands resources. Issued by the Department of Energy, 1977.	U.S. Government Printing Office, Washington, DC 20402

III. ENERGY-RELATED PUBLICATIONS (cont.)

Publication	Description	Available from
Inside DOE	Weekly newsletter covering developments within the Department of Energy, including policy decisions.	Inside DOE, McGraw-Hill Inc., 1221 Avenue of the Americas, New York, NY 10020
International Oil News	Weekly newsletter on the international oil industry.	William F. Bland Co., Box 1421, Stamford, CT 06904
Nucleonics Week	Weekly newsletter on nuclear power industry.	McGraw-Hill Publishing Co., 1221 Avenue of the Americas, New York, NY 10020.
Ocean Oil Weekly Report	Newsletter on topics related to deep-water oil exploration and drilling.	Petroleum Publishing Co., 1200 S. Post Oak Rd., Houston, TX 77056
Oil and Gas Federal Income Taxation (by Kenneth G. Miller)	Annual compilation of Federal income tax laws, rulings, and regulations affecting the oil and gas industry; depletion deduction, taxable income, drilling and development costs; court cases.	Commerce Clearing House, 4025 W. Peterson Ave., Chicago, IL 60646
Oil Daily	Daily newspaper on the energy industry.	Oil Daily, Circulation Dept., 850 Third Ave., New York, NY 10022
Oil Express	Weekly newsletter on petroleum marketing trends.	United Publishing Co., 128 C St. N.W., Washington, DC 20001
Oil and Gas Journal	Weekly magazine on the oil and gas industries.	Oil and Gas Journal, P.O. Box 1260, Tulsa, OK 74101
Petroleum Intelligence Weekly	Weekly newsletter on international energy matters.	Petroleum and Energy Intelligence Weekly Inc., 49 West 45th St., New York, NY 10036
Platt's Oilgram News Service	Daily reporting on oil news.	McGraw-Hill Publishing Co., 1221 Avenue of the Americas, New York, NY 10020
Platt's Oilgram Price Service	Daily quotations of prices.	McGraw-Hill Publishing Co., 1221 Avenue of the Americas, New York, NY 10020

III. ENERGY-RELATED PUBLICATIONS (cont.)

Publication	Description	Available from
Platt's Regulatory Insight	Biweekly reports on developments related to energy regulation.	McGraw-Hill Publishing Co., 759 National Press Bldg., Washington, DC 20045
Reserves of Crude Oil, Natural Gas Liquids, Natural Gas in the United States and Canada and United States Productivity Capacity	Annual report published jointly by the American Gas Association, Canadian Petroleum Association, and American Petroleum Institute; reserve data for crude oil, natural gas liquids, and natural gas in the United States and Canada.	Division of Statistics, Publication Section, American Petroleum Institute, 2101 L St. N.W., Washington, DC 20037
Resources Tomorrow	Bimonthly newsletter on status of natural resources, related topics.	World Future Society, 4916 St. Elmo Ave., Washington, DC 20014
The Touhey-Oil Express	Weekly oil industry newsletter.	The Touhey-Oil Express, 1052 Finegrove Ave., Hacienda Heights, CA 91745
United States Government Manual	Annual official handbook of the Federal government providing comprehensive information on the agencies of the legislative, judicial, and executive branches.	U.S. Government Printing Office, Washington, DC 20402
U.S. Oil Week	Weekly coverage of topics related to oil and the oil industry.	U.S. Oil Week, 1054 31st St. N.W., Washington, DC 20007
Utilities Law Reports	Weekly coverage of court and commission decisions on public utility regulations; Federal laws and regulations.	Commerce Clearing House, 4025 W. Peterson Ave., Chicago, IL 60646
Washington Information Directory	Annual listing of information concerning agencies of the executive branch of the United States government, the Congress, and Washington-based private (nongovernmental) organizations.	Congressional Quarterly, Inc., 1414 22 Street N.W., Washington, DC 20037

III. ENERGY-RELATED PUBLICATIONS (cont.)

Publication	Description	Available from
Washington Monitor on Energy	Weekly newsletter on energy developments in the national capital.	Washington News Service Inc., 9908 Hillridge Dr., Kensington, MD 20795

Index

Index

A

B

C

F

G

H

M

S

T

U

V

W

Z